16th Conference on Applied Mathematics (APLIMAT 2017)

Bratislava, Slovak Republic
31 January – 2 February 2017

Volume 1 of 2

ISBN: 978-1-5108-3698-3

Printed from e-media with permission by:

Curran Associates, Inc.
57 Morehouse Lane
Red Hook, NY 12571

proceedings.com

Some format issues inherent in the e-media version may also appear in this print version.

Printed by Curran Associates, Inc. (2017)

Institute of Mathematics and Physics, Faculty of Mechanical Engineering
Slovak University of Technology in Bratislava
Namestie Slobody 17
812 31 Bratislava 1
Slovak Republic

Phone: +421 2 572 96 111
Fax: +421 2 529 25 749

www.stuba.sk

Additional copies of this publication are available from:

Curran Associates, Inc.
57 Morehouse Lane
Red Hook, NY 12571 USA
Phone: 845-758-0400
Fax: 845-758-2633
Email: curran@proceedings.com
Web: www.proceedings.com

STU
SjF

SLOVAK UNIVERSITY OF TECHNOLOGY IN BRATISLAVA
FACULTY OF MECHANICAL ENGINEERING

16th CONFERENCE ON APPLIED MATHEMATICS

APLIMAT 2017 PROCEEDINGS

January 31 - February 2, 2017

Bratislava, Slovak Republic

Slovenská asociácia strojných inžinierov SASI

Vydavatel'stvo Spektrum STU
Bratislava

Proceedings, 16th Conference on Applied Mathematics Aplimat 2017
First edition

Editors: Dagmar Szarková, Peter Letavaj, Daniela Richtáriková, Monika Prašílová

ISBN 978-80-227-4650-2

The purpose of Aplimat Conference is to promote discussions and interactions between researchers and practitioners focused on disciplinary, interdisciplinary and transdisciplinary issues, ideas, concepts, theories, methodologies and applications. We are particularly interested in fostering the exchange of concepts, research ideas and other results which could contribute to the academic area and also benefit business and the industrial community.

Conference main topics

- Algebra and geometry and their applications
- Differential equations, dynamic systems and their applications
- Mathematics and art, section dedicated to the memory of Mauro Francaviglia
- Modeling and simulation in engineering and scientific computations
- New trends in mathematical education
- Statistical methods in technical and economic sciences and in practise
- Mathematical Aspects of Network Science and Applications
- Mathematics and Physics in Contemporary Science and Technique

Conference Chair

VELICHOVÁ Daniela	–	Slovak Republic

Conference Scientific Board

BAŠTINEC Jaromír	–	Czech Republic
CARKOVS Jevgenijs	–	Latvia
DOLEŽALOVÁ Jarmila	–	Czech Republic
FERREIRA Manuel A. M.	–	Portugal
JANIGA Ivan	–	Slovak Republic
LORENZI Marcella G.	–	Italy
NEHÉZ Martin	–	Slovak Republic
TEDESCHINI-LALLI Laura	–	Italy

List of reviewers

BAŠTINEC Jaromír, Czech Republic
BERÁNEK Jaroslav, Czech Republic
BOHÁČ Zdeněk, CZech republic
BUIKIS Māris, Latvia
CAPANNA Alessandra, Italy
CARKOVS Jevgenijs, Latvia
CARRANTI Andrea, Italy
CHALMOVIANSKÝ Pavel, Slovakia
CHVALINA Jan, Czech Republic
COCCHIARELLA Luigi, Italy
ČERMÁK Jan, Czech Republic
DOLEŽALOVÁ Jarmila, Czech Republic
FALCOLINI Corrado, Italy
FERREIRA Manuel A. M., Portugal
FILPA Andrea, Italy
FRONC Marek, Slovakia
Ferdinand HAVELKA, Slovakia
HOŠKOVÁ-MAYEROVÁ Šárka, Czech Republic
JANIGA Ivan, Slovakia
JUKL Marek, Czech Republic
KOLÁŘOVÁ Edita, Czech Republic
KOLYSHKIN Andrej, Latvia
KREML Pavel, Czech republic
LETAVAJ Peter, Slovakia
LINNERTOVÁ Dagmar, Czech Republic
LORENZI Marcella G., Italy
MAGDOLEN Ľuboš, Slovakia
MAGRONE Paola, Italy
MARIN Marin, Romaia
MIKEŠ Josef, Czech Republic
MOUČKA Jiří, Czech Republic
NEHÉZ Martin, Slovakia
NUCCI Lucia, Italy
PĂUN Marius, Romania
RICHTÁRIKOVÁ Daniela, Slovakia
RŮŽIČKOVÁ Miroslava, Slovakia
SIVÝ Július, Slovakia
SPADAFORA Giovanna, Italy
SVOBODA Zdeněk, Czech Republic
ŠMARDA Zdeněk, Czech Republic
TEDESCHINI-LALLI Laura, Italy
ÚRADNÍČEK Juraj, Slovakia
VALA Jiří, Czech Republic
VELICHOVÁ Daniela, Slovakia
VOLNÁ Jana, Czech Republic
VOLNÝ Petr, Czech Republic

CONTENTS

First Edition

Bratislava 2017

Slovak University of Technology in Bratislava
Faculty of Mechanical Engineering
16th Conference on Applied Mathematics APLIMAT 2017
Proceedings

PSEUDO REGULAR RINGS DERIVED FROM MULTIPLICATIVE HYPERRINGS

AMERI Reza (IR), HOSKOVA-MAYEROVA Sarka (CZ), KORDI Ali (IR)

Abstract. In this paper we introduce the notation of pseudo regular rings and obtain some properties of such rings. Further, we will proceed by considering a multiplicative hyperrings R, we obtain the smallest equivalence relation η^*, such that the quotient R/η^*, the set of all equivalence classes, is a pseudo regular ring. Finally, we characterize pseudo regular and strongly regular relations on R and obtain several results on the topic.

Keywords: Multiplicative hyperring, hyperideal, regular element, pseudo regular

Mathematics subject classification: Primary 20N20; Secondary 16Y99

1 Introduction

Various classes of hyperstructures have been investigated in many countries of the world since 1934. One motivating example of such generalizations can be found among other in the paper [25] by Marty. Let G be a group and H any subgroup of G. Then $G/H = \{xH : x \in G\}$ becomes a hypergroup where the hyperoperation is defined in usual manner: $\bar{x} \odot \bar{y} = \{\bar{z} : z \in \bar{x} \cdot \bar{y}\}$, where $\bar{x} = xH$. The motivation for the generalization of the notion of group resulted naturally from various problems in non-commutative algebra, another motivation for such an investigation came from geometry.

The theory of hyperstructures has been widely reviewed Over the following decades, new and interesting results appeared as the theory of hyperstructures has many applications in both pure and applied sciences; e.g. semi-hypergroups are the simplest algebraic hyperstructures that possess the properties of closure and associativity. The relation β^* was studied in depth, the relations γ^* and ε^* were defined. These fundamental relations are defined on H_v-groups, H_v-rings and H_v-vector spaces, respectively, as the smallest equivalences so that the quotient would be a group, ring and vector spaces, respectively (see [36]). The fundamental relations are one of the most important and interesting concepts in algebraic hyperstructures that ordinary algebraic structures are derived from algebraic hyperstructures by them.

The fundamental relation β^* on hypergroups was defined by Koskas [23], mainly studied by Corsini [11], Freni [19], Vougiouklis [37] and many others. [2, 10, 13, 15, 27, 30, 35, 37].

Moreover, special classes of hyperstructures as canonical hypergroups, feebly canonical hypergroups, cogroups, very thin H_v, P-hypergroups, hyper-lattices, hyperrings, hyperfields, and hyper-vector spaces were studied.[11, 12, 15, 25, 36].

In [12] Corsini and Leoreanu-Fotea have presented numerous applications of algebraic hyperstructures to the following subjects: algebra, geometry, convexity, combinatorial problems of coloring, lattice theory, Boolean algebras, logic, hypergraphs, binary relations, lattices, fuzzy sets and rough sets, automata theory, cryptography, codes, median algebras, relation algebras, artificial intelligence, and probabilities. [22, 21] The hyperrings were introduced and studied by Krasner [24], Massouros [26], by Davvaz and Leoreanu-Fotea [16], Ameri and Norouzi [2]. The study on hyperrings in [36], ends with an outline of applications in chemistry and physics, analyzing several special kinds of hyperstructures: e-hyperstructures and transposition hypergroups. [9, 20]. The theory of suitable modified hyperstructures can serve as a mathematical background in the field of quantum communication systems. A well-known type of a hyperring is called the Krasner hyperring [24]. Krasner hyperrings are essential hyperrings, with modified axioms in which addition is a hyperoperation, while the multiplication is an operation. This concept has been studied by a variety of authors. Some principal notions of hyperring theory can be found also in [3, 4, 15, 16, 31, 35, 37].

In 1982 Rota introduced a hyperring where the multiplication is a hyperoperation, while the addition is an operation, and it is called it a *multiplicative hyperring.* [28, 32, 33, 34]. De Salvo [18] introduced hyperrings in which the additions and the multiplications are hyperoperations. Moreover, we have to mentioned another types of hyperrings that both the addition and multiplication are hyperoperations and instead the associativity, commutativity and distributivity satisfy weak associativity, weak commutativity and weak distributivity. They are called H_v-hyperrings. [36, 37]. Furthermore, there are other types of hyperrings which were completely studied in [17] and also by Rahnamai Barghi in [31]. Asokkumar and Velrajan [6, 7] have studied Von Neumann regularity in Krasner hyperrings.

Recall that a hyperoperation "." on nonempty set H is a mapping of $H \times H$ into the family of all nonempty subsets of H. Let "." be a hyperoperation on H. Then, $(H, .)$ is called a *hypergroupoid.* We can extend the hyperoperation on H to subsets of H as follows: For $A, B \subseteq H$ and $h \in H$, then $AB = \cup_{a \in A, b \in B} ab$, $Ah = A\{h\}$, $hB = \{h\}B$.

A *semihypergroup* is a hypergroupoid $(H, .)$, which is associative, that is $(a.b).c = a.(b.c)$ for all $a, b, c \in H$.

A *hypergroup* is a semihypergroup $(H, .)$, that satisfies the reproduction axioms, that is $a.H = H = H.a$ for all $a \in H$.

A non-empty set R with two hyperoperations $+$ and . is said to be a *hyperring* if $(R, +)$ is a *canonical hypergroup,* $(R, .)$ is a semihypergroup with $r.0 = 0.r = 0$ for all $r \in R$ (0 as a bilaterally absorbing element) and the hyperoperation . is distributive with respect to $+$, i.e., for every $a, b, c \in R$; $a(b + c) = ab + ac$ and $(a + b)c = ac + bc$.

A *multiplicative hyperring* is an additive commutative group $(R, +)$ endowed with a hyperoperation . which satisfies the following conditions:
$(1.)$ $\forall a, b, c \in R : a(bc) = (ab)c$;

(2.) $\forall a, b, c \in R : (a + b)c \subseteq ac + bc, a(b + c) \subseteq ab + ac$;
(3.) $\forall a, b \in R : (-a)b = a(-b) = -(ab)$.

If in (2) we have equalities instead of inclusions, then we say that the multiplicative hyperring is *strongly distributive*.

In this paper, we introduce the notation of pseudo regular ring and we show some properties of it. Then we introduce the smallest equivalence relation η^* on a multiplicative hyperring R such that the quotient R/η^* is a pseudo regular ring. The characterization of pseudo regular via strongly regular relations is investigated and several results on the topic are presented.

2 Pseudo Regular Rings

Let R be a ring. An element $a \in R$ is *regular* if there exists $x \in R$, such that $a = axa$. R is a *regular ring* if every elements of R is regular. The set of all regular elements in R is denoted by $V(R)$. In this section we introduce the notation of pseudo regular rings. In 1950, Brown and McCoy [8], defined the set of elements of a ring such that generated ideal of that elements is regular and they denoted this set by $\mathcal{M}(R)$. They proved that $\mathcal{M}(R)$ is an ideal and clearly $\mathcal{M}(R) \subseteq V(R)$.

Definition 2.1. *Let $(R, +, \cdot)$ be a ring. We define*
(1) $T_0(R) = R$
(2) $T_{k+1}(R) = \{x - xrx | x \in T_k(R), r \in R\}$, *for* $k \geq 0$.

Definition 2.2. *Let R be a ring. An ideal series of R is a finite chain of ideals of R such that*

$$\{0\} = R_0 \lhd R_1 \lhd \cdots \lhd R_k = R$$

such that $1 \leq i \leq k$, $R_{i-1} \lhd R_i$. Then k is said to be the length *of series and denoted by $\ell(R)$.*

Definition 2.3. *Let R be a ring. An ideal series $\{0\} = R_0 \lhd R_1 \lhd \cdots \lhd R_k = R$ is called* regular series, *if for all $1 \leq i \leq k$,*

$$\frac{R_i}{R_{i-1}} \lhd \mathcal{M}\Big(\frac{R}{R_{i-1}}\Big),$$

where $\mathcal{M}(R) = \{x \in R| < x >\}$ is a regular ideal.

Proposition 2.1. [8] *Let R be a ring. Then $\mathcal{M}(\frac{R}{\mathcal{M}(R)}) = \{0\}$.*

Remark 2.1. *Let R be a ring. We denote $\mathcal{M}_0(R) = \{0\}$.*

Definition 2.4. *Let R be a ring. A lower ideal series is an ideal series*

$$R = R^0 \rhd R^1 \rhd R^2 \rhd \cdots ,$$

where $R^i =< T_i(R) >$, for all $1 \leq i \leq k$.

Definition 2.5. *A ring R is said to be pseudo regular if it has a regular series. The smallest length of a regular series of R is called* regularity class *of R.*

Example 2.1. *Let R be a nontrivial pseudo regular ring. Then $\mathcal{M}(R) \neq \{0\}$, because on the otherwise R will be trivial.*

Example 2.2. *Let $R = \mathbb{Z}_p[i] = \{a + ib | a, b \in \mathbb{Z}_p\}$, $i = \sqrt{-1}$ be the Gaussian integer modulo p, for some odd prime p. Then by Corollary* 3.11 *of* [29], $\mathcal{M}(R) \neq \{0\}$, *and hence R is pseudo regular with length ≥ 1, where $\mathbb{Z}_p[i]$.*

If $R = \mathbb{Z}_{p^k}[i]$ for some odd prime p and $k \neq 1$, then $\mathcal{M}(R) = \{0\}$ and in this case R is not pseudo regular.

Example 2.3. [29] *Let $R = \mathbb{Z}_{2^k}[i]$ for all k. Then $\mathcal{M}(R) = \{0\}$. Therefore, R is not a pseudo regular ring.*

Theorem 2.1. *Let R be a ring and $n \geq 1$. Then the following statements are equivalent:*
(i) $R^n = \{0\}$;
(ii) R is pseudo regular.

Proof. $(i) \Rightarrow (ii)$ Since $R^n = \{0\}$, then

$$R = R^0 \rhd R^1 \rhd R^2 \rhd \cdots \rhd R^n = \{0\},$$

is a regular series with finite length for R. Hence R is pseudo regular.
$(ii) \Rightarrow (i)$ Since R is pseudo regular, there exists a regular series

$$\{0\} = R_0 \lhd R_1 \lhd \cdots \lhd R_n = R,$$

such that $\frac{R_i}{R_{i-1}} \lhd \mathcal{M}(\frac{R}{R_{i-1}})$, for $1 \leq i \leq n$. First, by induction on i we prove that $R^i \lhd R_{n-i}$. If $i = 0$ then $R_{n-0} = R_n = R = R^0$ and we have $R^0 \lhd R_n$. Now, assume that $R^i \lhd R_{n-i}$ for $i > 0$. We need to show that $R^{i+1} \lhd R_{n-i-1}$. For all $a \in R^i$ and $r \in R$, we have $a \in R_{n-i}$ and since $\frac{R_{n-i}}{R_{n-i-1}} \lhd \mathcal{M}(\frac{R}{R_{n-i-1}})$, so $a + R_{n-i-1}$ is in $\mathcal{M}(\frac{R}{R_{n-i-1}})$, therefore

$$(a + R_{n-i-1}) - (a + R_{n-i-1})(r + R_{n-i-1})(a + R_{n-i-1}) = R_{n-i-1},$$

so $a - ara$ is in R_{n-i-1}, hence $R^{i+1} \lhd R_{n-i-1}$. For $i = n$, we have

$$R^n \lhd R_0 = \{0\}.$$

Therefore, $R^n = \{0\}$. □

Corollary 2.1. *Let R be a ring and $n \geq 1$. If $\mathcal{M}(R) = R$, then R is pseudo regular ring..*

Proof. Since $\mathcal{M}(R) = R$, then it is clear that there exists a regular series with finite length for R. Hence R is pseudo regular. □

Proposition 2.2. [8] *If r is an element of R such that $a - ara$ is regular, then a is regular.*

Theorem 2.2. *Let R be a ring. If $\mathcal{M}(R) = V(R)$ and R is pseudo regular, then $\mathcal{M}(R) = R$.*

Proof. Since R is pseudo regular, then by Theorem 2.1, $R^n = \{0\}$. By induction on i we prove that $R^{n-i} \lhd \mathcal{M}(R), i \geq 0$. If $i = 0$, then by our hypothesis $R^n = \{0\}$. Now, let $R^{n-i} \lhd \mathcal{M}(R)$ for $i \geq 0$. For all $a \in R^{n-i-1}$ and $r \in R$ we have

$$a - ara \in R^{n-i} \lhd \mathcal{M}(R) = V(R),$$

now by Proposition 2.2, a is in $V(R)$, and so $a \in \mathcal{M}(R)$. Hence $R^{n-i-1} \lhd \mathcal{M}(R)$. Now, we have $R^{n-n} \lhd \mathcal{M}(R)$, so $R^0 \lhd \mathcal{M}(R)$, i.e., $\mathcal{M}(R) = R$. □

Theorem 2.3. *Let R be a non trivial pseudo regular ring with an unitary element and I be an non zero ideal of R. Then $\mathcal{M}(I) \neq \{0\}$.*

Proof. Since R is pseudo regular ring, then there exists $n \geq 1$ such that $R^n = \{0\}$. Let m be the smallest positive integer such that $I \bigcap R^m \neq \{0\}$, but $I \bigcap R^{m+1} = \{0\}$. Thus there exists $0 \neq a \in I \bigcap R^m$ and for some $r \in R$, $a - ara \in R^{m+1}$, but $a - ara \in I$ so $a - ara \in I \bigcap R^{m+1} = \{0\}$. Therefore $a = ara$ for some $r \in R$. Hence a is in $I \bigcap R^m$, so $I \bigcap \mathcal{M}(R) \neq \{0\}$, then in virtue of Theorem 3 of [29], $\mathcal{M}(I) \neq \{0\}$. □

Theorem 2.4. *Let I be an ideal of a pseudo regular ring R. Then I and $\frac{R}{I}$ are pseudo regular.*

Proof. Assume that R is pseudo regular, then by 2.1, there exists $n \in \mathbb{N}$ such that $R^n = \{0\}$. Now, by induction we prove that for all $k \geq 0$, $I^k \lhd R^k$. If $k = 0$, then $I^0 = I \lhd R = R^0$. Suppose that $I^k \lhd R^k, k \geq 0$. Thus $I^{k+1} =< L_{k+1}(I) > \lhd < L_{k+1}(R) >= R^{k+1}$. Now, by getting $k = n$, we have $I^n \lhd R^n = \{0\}$, so $I^n = \{0\}$, i.e., I is pseudo regular.

Suppose that I is an ideal of pseudo regular ring R, so there is $n \in \mathbb{N}$ such that $R^n = \{0\}$. By induction on k, we will prove that $(\frac{R}{I})^k \lhd \frac{R^k I}{I}, k \geq 0$. If $k = 0$, then $(\frac{R}{I})^0 = \frac{R}{I} = \frac{R^0 I}{I}$. Let for $k > 0$, $(\frac{R}{I})^k \lhd \frac{R^k I}{I}$. Thus

$(\frac{R}{I})^{k+1} =< L_{k+1}(\frac{R}{I}) >=< \{a - ara | a \in L_k(\frac{R}{I}), r \in \frac{R}{I}\} >$
$\lhd < \{a - ara | a \in \frac{L_k(R)I}{I}, r \in \frac{R}{I}\} >:= S.$

Now, for all $a \in R^k$ and $r \in R$ we have

$$(a - ara) + I \in \frac{R^{k+1} I}{I},$$

therefore $S \lhd \frac{R^{k+1}I}{I}$. Hence $(\frac{R}{I})^{k+1} \lhd \frac{R^{k+1}I}{I}$. Letting $k = n$. Then

$$(\frac{R}{I})^n \lhd \frac{R^n I}{I} = \frac{I}{I} = I = 0_{\frac{R}{I}},$$

i.e., $(\frac{R}{I})^n = \{0_{\frac{R}{I}}\}$. Hence $\frac{R}{I}$ is pseudo regular. □

Theorem 2.5. *If $R_1, R_2, \cdots, R_r$ are pseudo regular rings, then $R = \prod_{i=1}^{r} R_i$ is pseudo regular ring.*

Proof. Straightforward. □

Corollary 2.2. *Let R be a ring and I and J be two ideals of R. If $\frac{R}{I}$ and $\frac{R}{J}$ are pseudo regular, then $\frac{R}{I \bigcap J}$ is also pseudo regular.*

Proof. Define $\phi : R \longrightarrow \frac{R}{I} \times \frac{R}{J}$ by $\phi(x) = (x + I, x + J)$. Then ϕ is a homomorphism with kernel $I \bigcap J$. Thus, by first isomorphism theorem $\frac{R}{I \bigcap J}$ is an isomorphic with $\phi(R)$. Since $\frac{R}{I}$ and $\frac{R}{J}$ are pseudo regular, then by Theorem 2.5, $\frac{R}{I} \times \frac{R}{J}$ is pseudo regular. Hence by Theorem 2.4, $\frac{R}{I \bigcap J}$ is pseudo regular. □

3 Strongly regular relation η^*

Let $(H, \cdot)$ be a hypergroup and $\rho \subseteq H \times H$ be an equivalence relation. For a non-empty subsets A and B of H, define

$$A\bar{\bar{\rho}}B \Leftrightarrow a\rho b, \forall a \in A, \forall b \in B.$$

Recall that a relation ρ is called strongly regular on the left (on the right) on H if $x\rho y \Rightarrow a \cdot x\bar{\bar{\rho}}a \cdot y$ ($x\rho y \Rightarrow x \cdot a\bar{\bar{\rho}}y \cdot a$, respectively), for all $(x, y, a) \in H^3$. Moreover, ρ is called strongly regular if it is strongly regular on the right and on the left.

Proposition 3.1. [11] *If $(H, \cdot)$ is a semihypergroup(resp. hypergroup) and ρ is a strongly regular relation on H, then the quotient H/ρ is a semigroup(resp. group) under the operation:*

$$\rho(x) \otimes \rho(y) = \rho(z), \quad \forall z \in x \cdot y.$$

We denote $\rho(x)$ by $\bar{x}$ and instead of $\bar{x} \otimes \bar{y}$ we write $\bar{x}\bar{y}$.

For all $n > 1$, we define the relation β_n on a semihypergroup H, as follows:

$$a\beta_n b \Leftrightarrow \exists (x_1, \cdots, x_n) \in H^n : \{a, b\} \subseteq \prod_{i=1}^{n} x_i, \quad and \quad \beta = \bigcup_{n \geq 1} \beta_n$$

where $\beta_1 = \{(x, x); x \in H\}$, is the diagonal relation on H. This relation was introduced by Koskas [23] and studied by many researchers in the theory of algebraic hyperstructures (for more see [1, 2, 11, 12, 13, 16, 36, 37]). Consider β^* as the transitive closure of β. It is proved that the relation β^* is a strongly regular relation, and it is called the *fundamental relation* of H [11].

Let $(R, +, .)$ be a hyperring. Define the relation γ as follows:

$$x\gamma_n y \Leftrightarrow \exists (x_1, \cdots, x_n) \in H^n, \exists \tau \in \mathbb{S}_n : x \in \prod_{i=1}^{n} x_i, y \in \prod_{i=1}^{n} x_{\tau(i)},$$

and $\gamma = \bigcup_{n \geq 1} \gamma_n$. We denote the transitive closure of γ by γ^*. The relation γ^* is the smallest equivalence relation on a hyperring $(R, +, .)$ such that the quotient R/γ^*, the set of all equivalence classes, is a ring. The relation γ^* is called fundamental relation on R, and R/γ^* is called the fundamental ring. Suppose that $\gamma^*(a)$ is the equivalence class containing $a \in R$. Then both the sum $\oplus$ and the product $\odot$ in R/γ^* are defined as follows:

$\gamma^*(a) \oplus \gamma^*(b) = \gamma^*(c)$ for all $c \in \gamma^*(a) + \gamma^*(b)$ and $\gamma^*(a) \odot \gamma^*(b) = \gamma^*(d)$ for all $d \in \gamma^*(a).\gamma^*(b)$. Then R/γ^* is a ring, which is called fundamental ring of R (for more details see also [33]).

Definition 3.1. *Let R be a multiplicative hyperring. We say that $a \in R$ is regular if there exists $x \in R$ such that $a \in axa$, and R is a regular multiplicative hyperring, if all the elements of R are regular. The set of all regular elements in R is denoted by $V(R)$.*

Definition 3.2. *Let $(R, +, \cdot)$ be a multiplicative hyperring. Define*

(1) $L_0(R) = R$;
(2) $L_{k+1}(R) = \{h | h \in (x - xrx) \bigcap (xr - rx), x \in L_k(R), r \in R\}$; *for* $k \geq 0$.

Suppose that $n \in \mathbb{N}$ and

$$\eta_n = \bigcup_{m>1} \eta_{m,n}$$

, *where $\eta_{1,n}$ is the diagonal relation and for every integer $m \geq 1$, the relation $\eta_{m,n}$ is defined as follows:*

$$x\eta_{m,n}y \Leftrightarrow \exists(z_1, \cdots, z_m) \in R^m, \exists\sigma \in \mathbb{S}_m : \sigma(i) = i \ if \ z_i \notin L_n(R) \text{ such that } x \in \prod_{i=1}^{m} z_i \ and \ y \in \prod_{i=1}^{m} z_{\sigma(i)}.$$

Obviously, for every $n \geq 1$, the relation η_n is reflexive and symmetric. Now suppose that η_n^* is the transitive closure of η_n.

Corollary 3.1. *For every $n \in \mathbb{N}$, we have $\beta^* \subseteq \eta_n^* \subseteq \gamma^*$.*

Proof. It is an immediate consequences of definitions of these three relations. □

Theorem 3.1. *For every $n \in \mathbb{N}$, the relation η_n^* is a strongly regular relation.*

Proof. Suppose that $n \in \mathbb{N}$. Clearly, η_n^* is an equivalence relation. For strongly regular, first we have to prove that:

$$x\eta_n y \Rightarrow x \cdot z \overline{\overline{\eta}}_n^* y \cdot z, z \cdot x \overline{\overline{\eta}}_n^* z \cdot y$$

for every $z \in R$. Suppose that $x\eta_n y$. Then, there exists $m \in \mathbb{N}$ such that $x\eta_{m,n}y$. Hence, there exist $(z_1, z_2, \cdots, z_m) \in R^m, \sigma \in \mathbb{S}$ with $\sigma(i) = i$ if $z_i \notin L_n(R)$, such that $x \in \prod_{i=1}^{m} z_i$ and $y \in \prod_{i=1}^{m} z_{\sigma(i)}$. Suppose that $z \in H$. We have

$$x \cdot z \subseteq (\prod_{i=1}^{m} z_i) \cdot z, \ y \cdot z \subseteq (\prod_{i=1}^{m} z_{\sigma(i)}) \cdot z$$

and $\sigma(i) = i$ if $z_i \notin L_n(R)$.

Now, suppose that $z_{m+1} = z$ and we define the permutation $\sigma' \in \mathbb{S}_{m+1}$ as follows:

$$\sigma'(i) = \sigma(i), \quad for \ all \ 1 \leq i \leq m \ and \ \sigma'(m+1) = m+1.$$

Thus, $x \cdot z \subseteq \prod_{i=1}^{m+1} z_i$ and $y \cdot z \subseteq \prod_{i=1}^{m+1} z_{\sigma'(i)}$ such that $\sigma'(i) = i$ if $z_i \notin L_n(R)$.

Therefore, $x \cdot z \overline{\overline{\eta}}_n^* y \cdot z$. Similarly, we have $z \cdot x \overline{\overline{\eta}}_n^* z \cdot y$. Now, if $x\eta_n^* y$ then there exists $k \in \mathbb{N}$ and $(x = u_0, u_1, \cdot, u_k = y) \in H^{k+1}$ such that $x = u_0\eta_n u_1\eta_n \cdots \eta_n u_{k-1}\eta_n u_k = y$.

Hence, by the above results, we obtain

$$x \cdot z = u_0 \cdot z\overline{\overline{\eta}}_n^* u_1 \cdot z\overline{\overline{\eta}}_n^* u_2 \cdot z\overline{\overline{\eta}}_n^* \cdots \overline{\overline{\eta}}_n^* u_{k-1} \cdot z\overline{\overline{\eta}}_n^* u_k \cdot z = y \cdot z$$

and so $x \cdot z \overline{\overline{\eta}}_n^* y \cdot z$.

Similarly, we can prove that $z \cdot x \overline{\overline{\eta}}_n^* z \cdot y$.

Therefore, $\overline{\overline{\eta}}_n^*$ is a strongly regular relation on R. □

Theorem 3.2. *Let R be a multiplicative hyperring. Let ρ be a strongly regular relation on R. Then $L_{k+1}(R/\rho) = \{\bar{h}|\bar{h} = \bar{x} - \bar{x}\bar{r}\bar{x} = \bar{x}\bar{r} - \bar{r}\bar{x}, x \in L_k(R), r \in R\}$, where $\bar{r}$ is the class of r with respect to ρ.*

Proof. Assume that $\bar{R} = R/\rho$ and for all $r \in R, \bar{r} = \rho(r)$. By induction, suppose that $k = 0$, then $L_1(\bar{R}) = \{\bar{h}|\bar{h} = \bar{x} - \bar{x}\bar{r}\bar{x} = \bar{x}\bar{r} - \bar{r}\bar{x}, x \in L_0(R), r \in R\}$.

Let $\bar{a} = \bar{x} - \bar{x}\bar{r}\bar{x} = \bar{x}\bar{r} - \bar{r}\bar{x}$, where $x \in L_{k+1}(R), s \in R$.

Thus there exists $z \in L_k(R)$, such that $x \in (z - zrz)\bigcap(zr - rz)$. Therefore, $\bar{z} - \bar{z}\bar{r}\bar{z} = \bar{x} = \bar{x}\bar{r} - \bar{r}\bar{x}$.

Now, by induction hypothesis we have $\bar{x} \in L_{k+1}(\bar{R})$, and so $\bar{a} \in L_{k+2}(\bar{R})$.

Conversely, let $\bar{a} \in L_{k+2}(\bar{R})$.

Then $\bar{a} = \bar{x} - \bar{x}\bar{r}\bar{x} = \bar{x}\bar{r} - \bar{r}\bar{x}$, where $\bar{x} \in L_{k+1}(\bar{R})$.

So, we have $\bar{x} = \bar{\ell} - \bar{\ell}\bar{r}\bar{\ell} = \bar{\ell}\bar{r} - \bar{r}\bar{\ell}$, where $\ell \in L_k(R)$ and $r \in R$.

Thus, there exists $t \in R$ such that $t \in (\ell - \ell r\ell)\bigcap(\ell r - r\ell)$.

Hence $\bar{t} = \bar{\ell} - \bar{\ell}\bar{r}\bar{\ell} = \bar{\ell}\bar{r} - \bar{r}\bar{\ell} = \bar{x}$ and $t \in L_{k+1}(R)$.

Thus $\bar{a} = \bar{x} - \bar{x}\bar{r}\bar{x} = \bar{t} - \bar{t}\bar{r}\bar{t}$, and $\bar{a} = \bar{x}\bar{r} - \bar{r}\bar{x} = \bar{t}\bar{r} - \bar{r}\bar{t}$.

Therefore, $\bar{a} \in \{\bar{h}|\bar{h} = \bar{t} - \bar{t}\bar{r}\bar{t} = \bar{t}\bar{r} - \bar{r}\bar{t}, t \in L_{k+1}(R)\}$. □

Theorem 3.3. *R/η_n^* is a pseudo regular ring of the class at most $n + 1$.*

Proof. Let $R' = R/\eta_n^*$. We show that for all $r \in R$, $L_{n+1-i}(R') \subseteq \mathcal{M}(R')$. Let $i = 0$, then $L_{n+1}(R') \subseteq \mathcal{M}(R')$. Now let $a \in L_{n-i}(R')$.

Since $a \in L_{n-i}(R')$, so $a - a\bar{r}a, a\bar{r} - \bar{r}a \in L_{n+1-i}(R')$. Since $L_{n+1-i}(R') \subseteq \mathcal{M}(R')$, thus $a + \mathcal{M}(R') = a\bar{r}a + \mathcal{M}(R'), a\bar{r} + \mathcal{M}(R') = \bar{r}a + \mathcal{M}(R')$. So $a \in \mathcal{M}(R')$.

Now suppose that $i = n + 1$, so $L_{n+1-(n+1)}(R') = \mathcal{M}(R')$, and hence we have $L_0(R') = \mathcal{M}(R')$. Thus $R' = \mathcal{M}(R')$. Therefore, R' is pseudo regular ring, as desired. □

Definition 3.3. *Let R be a multiplicative hyperring and $r \in R$ be a fix element. Then we define the relation η^* on R by $\eta^* = \bigcap_{n\geq 1} \eta_n^*$.*

Corollary 3.2. *The relation η^* is a strongly regular relation on a multiplicative hyperring R, such that R/η^* is a pseudo regular ring.*

Theorem 3.4. *The relation η^* is the smallest strongly regular relation on a multiplicative hyperring such that R/η^* is a pseudo regular ring.*

Proof. Suppose that ρ is a strongly regular relation on R such that $R' = \frac{R}{\rho}$ is a pseudo regular of class c. Suppose that $x\eta y$. Then, $x\eta_n y$, for some $n \in \mathbb{N}$ and so there exists $m \in \mathbb{N}$ such that
$x\eta_{m,n}y \Leftrightarrow \exists(z_1, \cdots, z_m) \in R^m, \exists\sigma \in \mathbb{S}_m : \sigma(i) = i \; if \; z_i \notin L_n(R)$ such that
$x \in \prod_{i=1}^m z_i \; and \; y \in \prod_{i=1}^m z_{\sigma(i)}$.
Thus

$$\rho(x) = \prod_{i=1}^{n} \rho(z_i) \quad and \quad \rho(y) = \prod_{i=1}^{n} \rho(z_{\sigma(i)}).$$

Now, since R' is pseudo regular of class $c+1$, we have:

$$L_{c+1}(R') = \{\bar{h} | \bar{h} = \rho(t) - \rho(t)\rho(s)\rho(t) \\ = \rho(t)\rho(s) - \rho(s)\rho(t), t \in L_c(R), s \in R\} \subseteq R_0 = \{0\},$$

so $\rho(t) = \rho(t)\rho(s)\rho(t), \rho(t)\rho(s) = \rho(s)\rho(t)$ for every $t \in L_c(R), s \in R$.

Therefore $\rho(x) = \rho(y)$, i.e., $x\rho y$. □

4 Conclusion

The pseudo regular rings were defined and some its properties were shown. Moreover, the smallest equivalence relation η^* on a multiplicative hyperring R such that the quotient R/η^* is a pseudo regular ring was described. The characterization of pseudo regular via strongly regular relations was investigated and several results on the topic were presented. For the future the authors are going to study other properties, mainly the transitivity.

Acknowledgement

The first author has been partially supported by "Algebraic Hyperstructure Excellence, Tarbiat Modares University, Tehran, Iran" and "Research Center in Algebraic Hyperstructures and Fuzzy Mathematics, University of Mazandaran, Babolsar, Iran".

The second author was supported within the project for "Development of basic and applied research developed in the long term by the departments of theoretical and applied bases FMT (Project code:"VÝZKUMFVT (DZRO K-217)") supported by the Ministry of Defence the Czech Republic.

References

[1] R. Ameri, E. Mohamadzadeh, Derived Engel Groups from Hypergroups, European J. Combin., 44 (2015) 191–197.

[2] R. Ameri, M. Norouzi, New fundamental relation of hyperrings, European J. Combin., 34 (2013) 884–891.

[3] R. Ameri, S. Hoskova-Mayerova, M. Amiri-Bideshki, A. Borumand Saeid, Prime filters of hyperlattices. An. Stiint. Univ. "Ovidius" Constanta Ser. Mat., 24(2) (2016) 15–26.

[4] R. Ameri, A. Kordi and S. Hoskova-Mayerova, Multiplicative hyperring of fractions and coprime hyperideals, An. Stiint. Univ. "Ovidius" Constanta Ser. Mat., 25(1) (2017).

[5] H. Aghabozorgi, B. Davvaza, M. Jafarpour, Nilpotent groups derived from hypergroups, J. Algebra, 382 (2013) 177-184.

[6] A. Asokkumar, M. Velrajan, Characterizations of regular hyperrings, Ital. J. Pure Appl. Math., 22 (2007) 115-124.

[7] A. Asokkumar and M. Velrajan, Von Neumann regularity on Krasner hyperring, Algebra, Graph Theory and their Applications, Editors: T. Tamizh Chelvam, S. Somasundaram, R. Kala, Narosa Publishing House, New Delhi, India, (2010) 9-19.

[8] B. Brown, N. H. McCoy, The maximal regular ideal of a ring, Proc. Amer. Math. Soc., 1 (1950), 165-171.

[9] J. Chvalina, J., Š. Křehlík, M. Novák, Cartesian composition and the problem of generalizing the MAC condition to quasi- multiautomata, An. Stiint. Univ. "Ovidius" Constanta Ser. Mat., 24(3) (2016) 79–100.

[10] I. Cristea, S. Jancic-Rasovic, Compositions Hyperrings, An. Stiint. Univ. "Ovidius" Constanta Ser. Mat., 21(2) (2013) 81–94.

[11] P. Corsini, Prolegomena of hypergroup theory, Second ed., Aviani Editore, 1993.

[12] P. Corsini, V. Leoreanu, Applications of hyperstructures theory, Adv. Math., Kluwer Academic Publishers, 2003.

[13] B. Davvaz, S. Mirvakili, On α-relation and transitive condition of α, Commun. Algebra, 36 (5) (2008) 1695-1703.

[14] B. Davvaz, M. Karimian, On the γ^*-complete hypergroups, European J. Combin. 28 (2007) 86-93.

[15] B. Davvaz, T. Vougiouklis, Commutative rings obtained from hyperrings (H_v-rings) with α^*-relations, Comm. Algebra 35 (2007) 3307-3320.

[16] B. Davvaz, Applications of the γ^*-relation to polygroups, Comm. Algebra 35 (2007) 2698-2706.

[17] B. Davvaz, V. Leoreanu-Fotea, Hyperring theory and applications, International Academic Press, USA, 2007.

[18] M. De Salvo, G. Lo Faro, On the n^*-complete hypergroups, Discrete Math. 208/209 (1990) 177-188.

[19] D. Freni, A new characterization of the derived hypergroup via strongly regular equivalences, Comm. Algebra, 30 (8) (2002) 3977-3989.

[20] S. Hoskova, J., Chvalina, Discrete transformation hypergroups and transformation hypergroups with phase tolerance space, Discret. Math. 308(18) (2008), 4133–4143.

[21] K. Hasilova, Iterative Method for Bandwidth Selection in Kernel Discriminant Analysis. In: 32nd International Conference Mathematical Methods in Economics MME2014. Olomouc: Palacky University, Olomouc, (2014), 263-268.

[22] I. Horova, J. Kolacek, K. Hasilova. Full bandwidth matrix selectors for gradient kernel density estimate. Computational Statistics & Data Analysis, 57(1), (2013), 364-376.

[23] M. Koskas, Groupoids, demi-groupes et hypergroupes, J. Math. Pures Appl., 49 (1970) 155-192.

[24] M. Krasner, A class of hyperrings and hyperfields, Int. J. Math. Math. Sci. 2 (1983) 307-312.

[25] F. Marty, Sur une generalization de la notion de groupe, In: $8iem$ Congres des Mathematiciens Scandinaves, Stockholm, (1934) 45-49.

[26] CH. Massouros, Homomorphic Relations on Hyperringoids and Join Hyperrings, Ratio Mathematica 13, (1999), 61-70.

[27] M. Novák, n-ary hyperstructures constructed from binary quasi-orderer semigroups, An. Stiint. Univ. "Ovidius" Constanta Ser. Mat., 22(3) (2014) 147–168.

[28] D. M. Olson, V.K. Ward, A note on multiplicative hyperrings, Italian J. Pure Appl. Math., 1 (1997) 77-84.

[29] E. A. Osba, M. Henriksen, O. Alkam, Combining local and Von Neumann regular rings, Comm. Algebra , 32 (7), 2639-2653, 2004.

[30] C. Pelea, Hyperrings and α^*-relations. A general approach, J. Algebra, 383, 1 (2013) 104-128.

[31] A. Rahnamai Barghi, A class of hyperrings, J. Discrete Math. Sci. Cryptogr., 6 (2003) 227-233.

[32] R. Rota, Strongly distributive multiplicative hyperrings, J. Geom., 39 (1990) 130-138.

[33] R. Rota, Sugli iperanelli moltiplicativi, Rend. Di Mat., Series VII (4), 2 (1982) 711-724.

[34] R. Rota, Congruenze sugli iperanelli moltiplicativi, Rend. Di Mat., Series VII (1), 3 (1983) 17-31.

[35] S. Spartalis, T. Vougiouklis, The fundamental relations on H_v-rings, Math. Pure Appl., 13 (1994) 7-20.

[36] T. Vougiouklis, Hyperstructures and Their Representations, Hadronic Press, Inc., Palm Harber, USA, 1994.

[37] T. Vougiouklis, The fundamental relation in hyperrings, The general hyperfield, in: Proc. Fourth Int. Congress on Algebraic Hyperstructures and Applications, AHA, 1990, World Scientific, 1991, 203-211.

Current address

Ameri Reza, Ph.D. Full Professor
School of Mathematics, Statistics and Computer Science,
College of Sciences, University of Tehran
P.O. Box 14155-6455, Teheran, Iran
E-mail: rameri@ut.ac.ir

Hoskova-Mayerova Sarka, Ph.D. Associative Professor
Department of Mathematice and Physics, University of Defence
Kounicova 65, 662 10 Brno, Czech Republic
Tel.: 00420973442225, e-mail: Sarka.mayerova@unob.cz

Kordi Ali, Ph.D. lecturer
School of Mathematics, Statistics and Computer Science,
College of Sciences, University of Tehran
P.O. Box 14155-6455, Teheran, Iran
E-mail: ali.kordi@gmail.com

Slovak University of Technology in Bratislava
Faculty of Mechanical Engineering
16th Conference on Applied Mathematics **APLIMAT 2017**
Proceedings

PROBLEM OF SEARCHING THE MST

ANTOŠ Karel (CZ)

Abstract. This article describes searching for The Minimum Spanning Tree of the graph using the Weighted Adjacency Matrix. It describes the Weighted Adjacency Matrix as a new element, and shows how it could be used for searching for the Minimum Spanning Tree of the graph. This creates a new procedure for searching for the Minimum Spanning Tree of the graph which completes previously known algorithms of searching for the Minimum Spannig Tree. The proposed Weighted Adjacency Matrix could be used in similar issues in the field of the graph theory, where graphs with weighted edges are given. The procedure is shown on the attached example.

Keywords: graph theory, minimum spanning tree, weighted adjacency matrix, Prim´s algorithm

Mathematics Subject Classification: 05C05

1 Introduction

All graphs in this article are finite, simple and connected. The graph theory can help us with solutions of various problems which can by displayed by a graph. Example of problem where we can use weighted graph is the problem of electricity network of any area. In the graph vertices represent villages, edges represent connections between these villages and weights of edges represent costs required for the construction of electrical networks. The task is to connect all villages through electricity and to make connection as cheap as possible To model this situation we create connected graph $G = (V, E)$ with weighted edges. The spanning tree of a connected graph G is a subgraph G' which connects all vertices and which does not contain any cycles [3], thus the spanning tree is a tree. The spanning tree satisfies the condition that all the villages should be joined in only way and the minimum spanning tree satisfies the condition that the sum of weights of all edges of the spanning tree is minimal.

The spanning tree we denote $T = (V', E')$, where $V' = V$ and E' is the set of $n - 1$ edges of the minimum spanning tree, and it applies that $E' \subseteq E$. In weighted graph the edges have the evaluations which in our example are relevant to costs of construction power lines between the two villages. When the sum of the weights of edges of spanning tree is minimal from all possible spanning trees, we denote it as The Minimal Spanning Tree. In the subsequent text we use the abbreviation MST (short for the Minimum Spanning Tree) [6].

For searching for the minimum spanning tree there are several obviously known algorithms which search for the MST in different ways. For example The Kruskal's algorithm, Prim's algorithm or Borůvka's algorithm are the generally known. In the article we use some principles of Prim´s algorithm for searching the MST [4]. But this article shows a new procedure for searching for the MST, which is Weighted Adjacency Matrix.

The idea of this proposed algorithm is that we create the MST by selecting edges starting with the edge of the smallest weight and then we gradually add following edges according to increasing sequence of weights of edges. Those edges that would make cycles are not added. This process is similar to the Prim´s algorithm but we begin with the element of the smallest weight and the Prim´s algorithm begins with any arbitrary vertex [2]. Similarity is also in the fact that the MST is gradually and continuously growing. This process works only in the proposed Weighted Adjacency Matrix. Selected matrix elements represent the selected edges belonging to the MST. Nonselected elements should be discarded and denoted by a cross. So this algorithm combines features from assigning and deleting algorithm.

But it is important to say that knowledge of previously mentioned algorithms is not necessary and author did not used any elements of them but only the idea of so called „greedy“ algorithm.

2 Starting conditions

Let $G = (V, E)$ be a connected, finite and non-oriented graph with positively weighted edges, where V is a set of n vertices and E is the set of m edges. The set of vertices V we denote $V = \{v_1, v_2, \dots, v_n\}$ and the set of edges we denote E, where e_{ij} denotes the edge between vertices v_i and v_j, then it is $e_{ij} = \{v_i, v_j\} \in E$. $W(e_{ij})$ denotes the weight of the edge connecting vertices v_i and v_j, where $e_{ij} = \{v_i, v_j\} \in E$.

The spanning tree of a connected graph G is a subgraph G' which connects all vertices and which does not contain any cycles [5]. For this subgraph G' it holds that $G' = (V', E')$, where $V' = V$ and $E' \subseteq E$. Note that the set E'contains $n - 1$ edges [2].

For subgraph $G' = (V', E')$ of the graph G we put $w(G') = \sum_{e \in E'} w(e)$. Because of the spanning tree is a tree we denote it T.

The spanning tree $T_1 = (V, E_1)$ of the graph G we call the minimum spanning tree if for each spanning tree $T_2 = (V, E_2)$ of the graph G it holds that

$$w(T_1) \leq w(T_2)$$

The minimum spanning tree we denote $T = (V, E')$, where $V' = V$ and E' is the set of $n - 1$ edges of the minimum spanning tree, and it applies that $E' \subseteq E$. In the subsequent text we use the abbreviation MST (short for the Minimum Spanning Tree).

If we define the function $w: E \rightarrow R$ (ie. evaluation of edges), then the minimum spanning tree is such a spanning tree for which it holds that the sum of the weights of edges of MST is minimal, i.e.

$$w(E') = \sum_{e \in E'} w(e) \text{ is minimal.}$$

In the following capture there is displayed a new algorithm which uses some new elements for searching the MST and adopts them to one of the previously mentioned, to the Prim's algorithm.

3 Weighted Adjacency Matrix

At first in the proposed algorithm we create a modified adjacency matrix, which we call " Weighted Adjacency Matrix". This matrix is similar to the Adjacency Matrix where in positions of elements of the matrix are either 1 or 0 if there is an edge between vertices v_i and v_j or not. In this modified Weighted Adjacency Matrix the positive number w_{ij} on the position of the element v_i and v_j indicates the weight of the edge connecting vertices v_i and v_j, if the edge between vertices v_i and v_j exists. A value of 0 indicates that there is no edge between vertices v_i and v_j [6].

Weighted Adjacency Matrix (Fig. 1) is thus a square matrix $W = n \, x \, n$, where n denotes the number of vertices and the value of the element at the position w_{ij} corresponds to the weight of the edge between vertices v_i and v_j.

$$w_{ij} = \begin{cases} w(e_{ij}) & \text{if } e_{ij} \in E \\ 0 & \text{otherwise} \end{cases}$$

	v_1	...	v_i	...	v_j	...	v_n
v_1	0	...	w_{1i}	...	w_{1j}	...	w_{1n}
...			...	...	...	...	...
v_i			0	...	w_{ij}	...	w_{in}
...					...	...	...
v_j					0	...	w_{jn}
...							...
v_n							0

Fig. 1. Weighted Adjacency Matrix

Weighted Adjacency Matrix is symmetric with respect to the main diagonal, the diagonal elements have a value of 0, the algorithm will only use the elements of the triangle above the main diagonal.

4 Searching for the MST in the matrix

The algorithm of searching for the MST works in the Weighted Adjacency Matrix and works with elements in the triangle above the main diagonal .

Description of the algorithm

a) Let and $|V| = n$.
b) Go through the matrix and find the smallest nonzero element $a = \min_{i \neq j}(w_{i,j})$, where indices i and j denote number of the row and column in the position of the smallest element. Then set $E_1 = \{e_{ij}\}, V_1 = \{v_i, v_j\}$ and $I_1 = \{i, j\}$.
c) Go through the matrix and find another smallest nonzero element w_{ij} where $i \in I$ and $j \notin I$ or element w_{ji} where $i \in I$ and $j \notin I$. Then set $E_i = E_{i-1} \cup \{e_{ij}\}$ or $E_i = E_{i-1} \cup \{e_{ji}\}$, $V_i = V_{i-1} \cup \{v_j\}$, $I_i = I_{i-1} \cup \{j\}$.
d) Continue until $V_i = V$.

Algorithm procedure

Search through the elements of the matrix and find the one with the smallest positive value w_{ij}. Denote chosen matrix element in bold and underlined, then mark the rows v_i, v_j and columns v_i, v_j (Denote the columns and rows with arrows at the top of the table). If there is more than one element with the same smallest positive value, it is possible to choose arbitrary one of these. Then more than one MST exists.

	v_1	…	↓ v_i	↓ v_j	…	v_k	…	v_n
v_1	0	…	w_{1i}	w_{1j}	…	w_{1k}	…	w_{1n}
…		0	…	…	…	…	…	…
→ v_i			0	**$\underline{w_{ij}}$**	…	w_{ik}	…	w_{in}
→ v_j				0	…	w_{jk}	…	w_{jn}
…					0	…	…	…
v_k						0	…	w_{kn}
…							0	…
v_n								0

Fig. 2

Search again through the elements of the matrix and find another smallest positive element, search between elements in the marked rows and columns (Fig. 2). Chosen element denote in the matrix in bold and underlined. Let the new element be w_{jk}. According to the index position of the element mark the row v_k and column v_k. Rows and columns marked in the previous steps remain marked.

This step ensures the connection of the generated MST because the connecting edge has one of the indexes the same as the previous selected element, so this element connects to any of previously connected vertices .

Furthermore delete (ie. replace by the cross) all the elements in positions where newly marked row and column intersect with rows and columns previously marked. Here delete the element w_{ik}. This step prevents creating cycles.

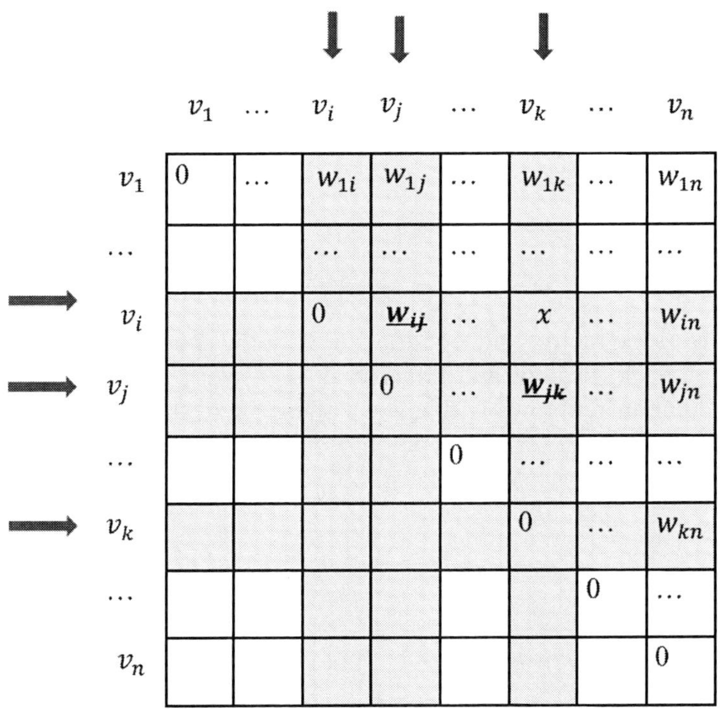

Fig. 3

Search again through the elements of the matrix and find another smallest positive element, search between elements in the marked rows and columns. Chosen element denote in the matrix in bold and underlined (Fig. 3). Let the new element be w_{1i}. According to the index position of the element mark the row v_1 and column v_1. Rows and columns marked in the previous steps remain marked. Furthermore delete all the elements in positions where newly marked row and column intersect with rows and columns previously marked. Here delete the elements w_{1j} and w_{1k} (Fig. 4).

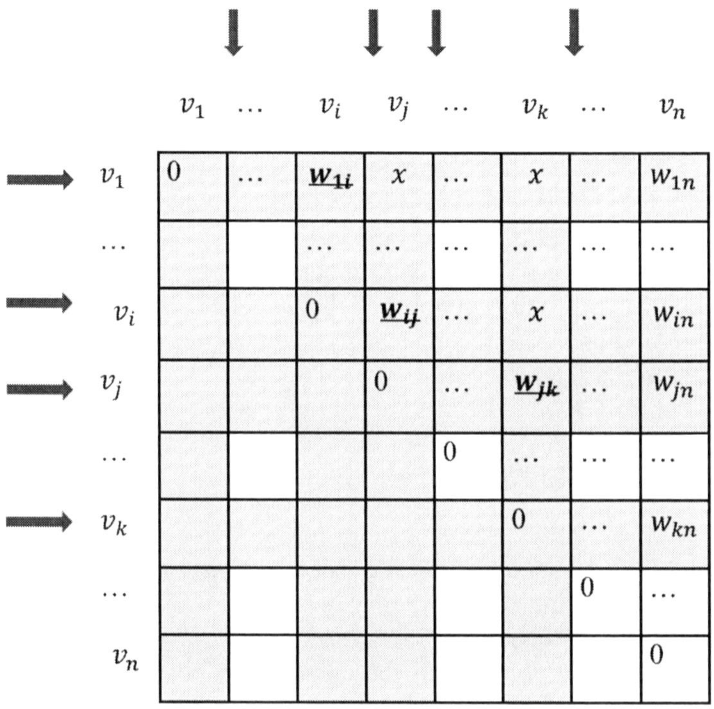

Fig. 4

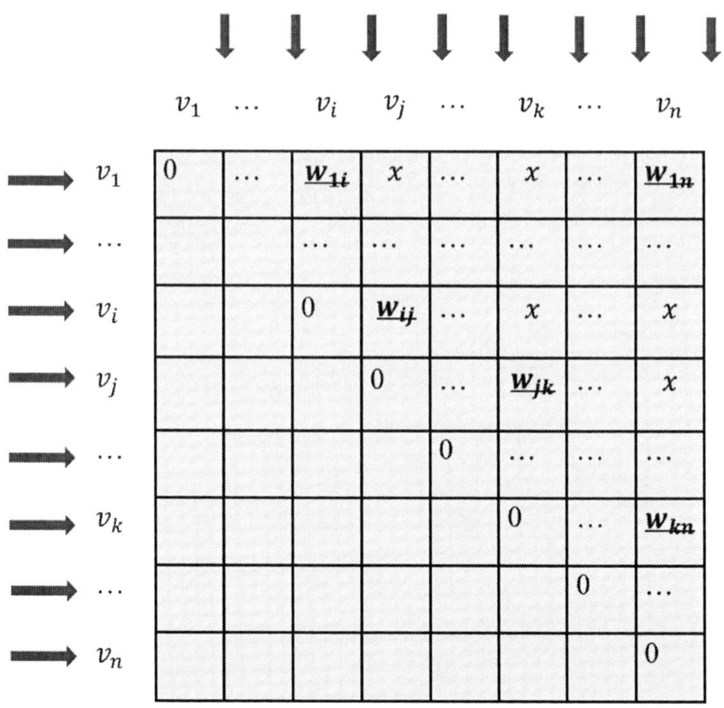

Fig. 5. Final Matrix

Suppose that our algorithm made k steps.

- If $k = n - 1$, algorithm stops, we have made all steps
- If $k < n - 1$, we make $(k + 1)$-th step analogously

After we make the $(n - 1)$-th step in our Weighted Adjacency Matrix $(n - 1)$ chosen elements are labeled (in bold and underlined), the other elements (which were not chosen) are replaced by a cross. At the same time all rows and columns in our matrix are labeled (Fig. 5). Elements denoted in the matrix in bold and underlined are values of weights of edges of the MST. Labeling the rows and columns of the selected element indicates the vertices v_i and v_j that the edge on this position connects. The sum of the values of all chosen elements give the total weight of the MST.

5 Verification of the algorithm (Idea of the proof)

1. Continuity of generated MST is guaranteed by the fact that newly connected edge has one of the indices the same as the indices of previously selected elements [4]. Therefore it connects to one of the previously connected vertices.
2. Avoiding the cycles is ensured by deleting all the elements in positions where newly marked row and column intersect with rows and columns previously marked [5].
3. The algorithm is a variant of the Prim´s algorithm, with the difference that in the first step we do not begin by selecting the arbitrary initial vertex, but in our Weighted Adjacency Matrix we begin by selecting the edge with the smallest weight. From the second step our algorithm works analogously as in the Prim´s algorithm (which has been proven, see [1]). This guarantees selection of the minimum spanning tree.

6 Demonstration of solved Example

Let the weighted graph be given by (Fig. 6):

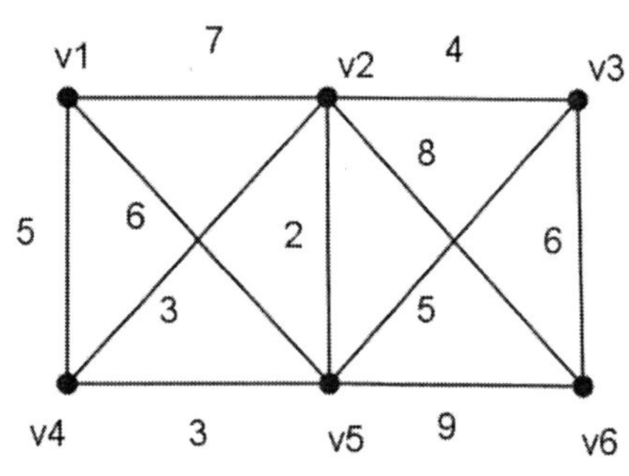

Fig. 6

Corresponding Weighted Adjacency Matrix is (Fig. 7):

	v_1	v_2	v_3	v_4	v_5	v_6
v_1	0	7	0	5	6	0
v_2		0	4	3	2	8
v_3			0	0	5	6
v_4				0	3	0
v_5					0	9
v_6						0

Fig. 7. Given Matrix

Steps of algorithm:

1. Search through the elements of the matrix and find the one with the smallest positive value $w_{25} = 2$. Denote chosen matrix element in bold and underlined, then mark the rows v_2, v_5 and columns v_2, v_5 (Fig. 8).

	v_1	v_2	v_3	v_4	v_5	v_6
v_1	0	7	0	5	6	0
v_2		0	4	3	**2**	8
v_3			0	0	5	6
v_4				0	3	0
v_5					0	9
v_6						0

Fig. 8

Search again through the elements of the matrix and find another smallest positive element $w_{24} = 3$, search between elements in the marked rows and columns. Chosen element denote in the matrix in bold and underlined. According to the index position of the element mark the row v_4 and column v_4. Rows and columns marked in the previous steps remain marked (Fig. 9).

Furthermore we delete (ie. replace by the cross) all the elements in positions where newly marked row and column intersect with rows and columns previously marked. Here we delete the element w_{45}.

	v_1	v_2	v_3	v_4	v_5	v_6
v_1	0	7	0	5	6	0
v_2		0	4	**3**	**2**	8
v_3			0	0	5	6
v_4				0	x	0
v_5					0	9
v_6						0

Fig. 9

Search again through the elements of the matrix and find another smallest positive element w_{23} = 4, search between elements in the marked rows and columns. Chosen element denote in the matrix in bold and underlined. According to the index position of the element mark the row v_3 and column v_3. Rows and columns marked in the previous steps remain marked. Furthermore delete all the elements in positions where newly marked row and column intersect with rows and columns previously marked. Here delete the elements, w_{34} = x, w_{35} = x (Fig. 10).

	v_1	v_2	v_3	v_4	v_5	v_6
v_1	0	7	0	5	6	0
v_2		0	**4**	**3**	**2**	8
v_3			0	x	x	6
v_4				0	x	0
v_5					0	9
v_6						0

Fig. 10

Search again through the elements of the matrix and find another smallest positive element $w_{14} = 5$, search between elements in the marked rows and columns. Chosen element denote in the matrix in bold and underlined. According to the index position of the element mark the row v_1 and column v_1. Rows and columns marked in the previous steps remain marked. Furthermore delete all the elements in positions where newly marked row and column intersect with rows and columns previously marked. Here delete the elements w_{12} = x, w_{13} = x, w_{15} = x (Fig. 11).

	v_1	v_2	v_3	v_4	v_5	v_6
v_1	0	x	x	**5**	x	0
v_2		0	**4**	**3**	**2**	8
v_3			0	x	x	6
v_4				0	x	0
v_5					0	9
v_6						0

Fig. 11

	v_1	v_2	v_3	v_4	v_5	v_6
v_1	0	x	x	**5**	x	x
v_2		0	**4**	**3**	**2**	x
v_3			0	x	x	**6**
v_4				0	x	x
v_5					0	x
v_6						0

Fig. 12

Search again through the elements of the matrix and find another smallest positive element $w_{36} = 6$, search between elements in the marked rows and columns. Chosen element denote in the matrix in bold and underlined. According to the index position of the element mark the row v_6 and column v_6. Rows and columns marked in the previous steps remain marked. Furthermore delete all the elements in positions where newly marked row and column intersect with rows and columns previously marked. Here delete the elements w_{16} = x, w_{26} = x, w_{46} = x, w_{56} = x (Fig. 12).

Termination of the algorithm

When graph G has n vertices then the MST has $n - 1$ edges [3]. At each step of the algorithm we add to the gradually rising MST one edge, then algorithm makes n - 1 steps. In our example, the graph has 6 vertices, then the MST has 5 edges. That is the reason why algorithm makes 5 steps.

After the final step, all the elements in the Weighted Adjacency Matrix went through processing, ie. the edges chosen for the MST are denoted in the agreed way, i.e. here in bold and underlined, deleted edges are marked by symbol x.

At the same time, after the last step all the rows and columns of the matrix are marked with the arrows next to the rows and above columns.

Final graph of the MST is here (Fig. 13):

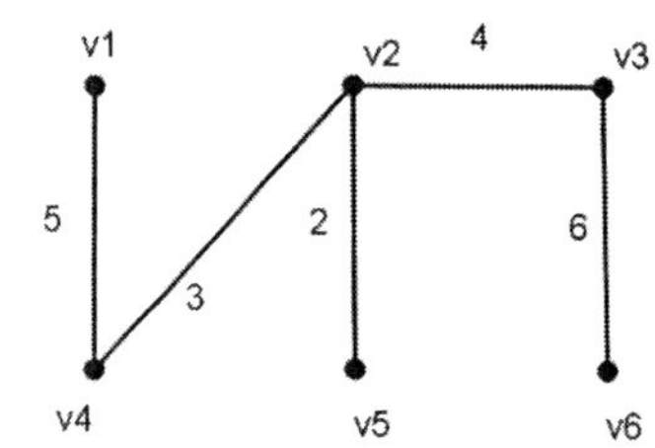

Fig. 13. Final Graph of the MST

Value of the final MST is:

$$\sum_{e \in E'} w(e) = (2 + 3 + 4 + 5 + 6) = 20$$

7 Conclusion

This article describes the design of the algorithm for searching for the minimum spanning tree. The proposed algorithm is similar to Prim's algorithm [1], which creates the minimum spanning tree as gradually growing set of edges of the MST. In this regard there is a compliance with Prim's algorithm. The Prim's algorithm starts with the arbitrary vertice. Here, however, the first element the algorithm starts with is the edge with the lowest weight [7].

The new tool here is the "Weighted Adjacency Matrix (abr. WAM)". It follows the principle of adjacency matrix known known in the graph theory, but in the positions of matrix elements there are values of weights of edges connecting the vertices. The vertices denote the rows and columns of the matrix.

The whole process of searching MST begins with choosing the smallest element of the matrix, representing the edge with the lowest weight. Gradually we add elements so that another new element has one index same as some of the elements that have been chosen in previous steps. This step guarantees the continuity of MST.

Elements which in denoted rows and columns are not chosen, must be removed because these edges would create cycles. The entire process takes place in WAM, the original graph is not needed.

Benefits of the proposed algorithm is that the searching the MST by using WAM is efficient and fast. The time complexity of the presented algorithm is O(m.n). So the time complexity seems to be O(m.n). According to my knowledge the searching for the MST by using WAM is a new tool and it can be assumed that the WAM could be used for solving other similar problems in the graphs, where wighted edges are given. Working of the algorithm is shown on solved example.

References

[1] R. C. Prim: Shortest connection networks and some generalisations. In: Bell System Technical Journal, 36 (1957)

[2] Kučera L., *Kombinatorické algoritmy*, SNTL, 1989

[3] Harris J. M., Hirst J. L., Hossinghofer M. J., Combinatorics and Graph Theory, Springer, New York, (2000)

[4] J. Matoušek , J. Nešetřil, Kapitoly z diskrétní matematiky, vydání čtvrté (4th edition), Prague, Charles University in Prague, (2009)

[5] J. Kleinberg, E. Tardos, Algorithm Design, New York: Pearson Education, Inc.. (2006)

[6] T. S. Jackson, and N. Read, Theory of minimum spanning trees, Phys Rev E 81 (2010) 021130

[7] Kučera L., *Kombinatorické algoritmy*, SNTL, 1989

Current address

Antoš Karel, Ing. Bc.
Department of Natural Sciences, VŠTECB in České Budějovice
Okružní 517/10, 370 01 České Budějovice, Czech Republic
Tel. number: +420 387 842 111, e-mail: kaant@seznam.cz

Slovak University of Technology in Bratislava
Faculty of Mechanical Engineering
16th Conference on Applied Mathematics **APLIMAT 2017**
Proceedings

ARE BUSINESS CYCLES PHASE TRANSITIONS?

BARIĆ PISAROVIĆ Gordana (CRO), GREBLIČKI Marijana (CRO), RAGUŽ Andrija (CRO)

Abstract. In this note we interpret the well-known one-dimensional Cahn-Hilliard functional as the model for achieving the dynamic equilibrium over large periods of time in given national economy with known initial aggregate economic parameters. The economy is viewed as the dynamical system, without strict separation to exogenous and endogenous variables, but with assumed existence of business cycles (viewed as "phases" or "states"), which bring both balance and imbalance to the state of the economy. Our conclusions are based on a number of results obtained by A. Raguž outside the economic context, which are related to standard analysis of Cahn-Hilliard functionals in one dimension. Mathematical apparatus used is Gamma convergence of functionals defined on the space of appropriately chosen functions, as well as the relaxation method in the variational calculus over the space of Young measures. We prove that under relatively mild assumption on double well potential function and the optimal profile function we can indeed deduce both economic interpretation of the mentioned model and the desired equilibrium state of the national economy in terms of balancing the usual economic parameters.

Keywords: Convergence, dynamical system, equilibrium, minimization, Cahn-Hilliard functional

Mathematics subject classification: Primary 34E15, 34C25, 35M10; Secondary 49J45

1 Introduction

In the last several decades numerous attempts have been made to bridge the gap between natural sciences like physics, applied mathematics and corresponding classical sub-fields of study (already saturated with rigorous theoretical results) on the one hand, and modern theoretical economy (which is viewed as a borderline discipline which includes both social science and natural science methods and terminology) on the other. Herein we provide a contribution to this effort by proposing to interpret already independently developed model from the physical context in the economic context. As such, our approach must be classified as an econophysical approach.

2 Preliminaries

In this section we present the summary of our consideration. While a more complex interplay between various economic parameters is presumed to exist in real-life situations, we focus on simplified model where aggregated quantities provide sufficiently good approximation for market behavior. More than that, we extend the usual assumption that minimizing the imbalance leads to balance, i.e., to equilibrium state of the market. The chief notion in that respect is the notion of finite-energy sequence (FE sequence for short) instead of the notion of minimizing sequence, representing unsuccessful attempts of the market to balance itself, whereby attempts are such that state of the market is still rather close to the actual equilibrium (with some small error measured in terms of units of trend). We do not consider extremely bad states of the market which are very far off. Our model, which can be best described as variational formulation of the problem of finding equilibrium of the economy, assumes that economy is demand-driven. We propose our model as the model which explains appearance of the phase-cycling (state-switching). If discrepancy between say, demand and supply, defines the entropy function function $t \mapsto u(t)$, its slope -1 ($+1$, resp.) represents decline (growth, resp.) of the economy, but can be interpreted in a number of different ways: for instance, in microeconomic (macroeconomic, resp.) context, slope $+1$ can be seen as (aggregate, resp.) surplus of supply, and slope -1 as (aggregate, resp.) surplus of demand. As is customary in the physical context, we assume that for large values of t the entropy of the system is small. Since simplified case of slope belonging to a two-point set $\{-1, 1\}$ is not feasible in real-life situations, in the end of our paper we set up more realistic time-paths of entropy function $t \mapsto u(t)$.

In the sequel we present mathematical background necessary for our consideration. To this end, we introduce the notation and precisely describe the objects which will be used throughout the paper. We conform to the notation of [1] whenever possible. We work on the unit interval $(0, 1)$ endowed with Lebesgue measure, which is denoted by λ. The letter s denotes the slow variable in $(0, 1)$, t the fast variable in $\mathbf{R}$, and u, v are always real functions on $\mathbf{R}$. The letter x stands for (Borel) measurable function of the fast variable t which is defined almost everywhere. The letter C denotes a generic constant. We also consider a compact metric space (K, d) (the space of patterns), which is defined as follows: K is the set of all measurable mappings $x : \mathbf{R} \longrightarrow [-\infty, +\infty]$, endowed with the metric d defined by

$$d(x_1, x_2) := \sum_{k=1}^{\infty} \frac{1}{2^k \alpha_k} \Big| \int_{\mathbf{R}} y_k \Big(\frac{2}{\pi} \arctan x_1 - \frac{2}{\pi} \arctan x_2 \Big) d\lambda \Big| \,,$$

where (y_k) is a sequence of bounded functions which are dense in $\mathrm{L}^1(\mathbf{R})$, such that the support of y_k is a subset of $(-k, k)$, with $\alpha_k := \|y_k\|_{\mathrm{L}^1} + \|y_k\|_{\mathrm{L}^\infty}$. As shown in [1], $\mathrm{L}^p_{loc}(\mathbf{R})$ embeds continuously in K (cf. [1], Proposition 5.1, Remark 5.2). $Meas((0, 1); K)$ is a class of all measurable maps from Ω to K, the Banach space $\mathrm{C}(K)$ is the space of all continuous real functions on K, whose dual is, as usual, identified with the space of all finite real Borel measures on K, denoted by $\mathcal{M}(K)$, by the duality pairing

$$\langle \mu, g \rangle := \int_K g d\mu \,, \quad g \in \mathrm{C}(K) \,, \quad \mu \in \mathcal{M}(K) \,.$$

$\mathcal{M}(K)$ is always endowed with the corresponding weak-star topology. Weak-star topology on $\mathcal{M}(K)$ is induced by norm ϕ defined by

$$\phi(\nu) := \sum_i \frac{1}{2^i \alpha_i} \langle \nu, g_i \rangle \ , \tag{1}$$

where the countable family (g_i) is dense in $\mathrm{C}(K)$, and $\alpha_i := \|g_i\|_{\mathrm{L}^\infty}$. The set of all probability Borel measures acting on space K is denoted by $\mathcal{P}(K)$. To introduce the notion of the Young measure in the next subsection, we recall that the Banach space $\mathrm{L}^1((0,1);\mathrm{C}(K))$ is the set of all measurable mappings $\varphi : (0,1) \longrightarrow C(K)$ which satisfy $\int_0^1 |\varphi(s)| ds < +\infty$. Dual of such a space is isometrically isomorphic to the Banach space of all weakly-star measurable mappings $\boldsymbol{\nu} : (0,1) \longrightarrow \mathcal{M}(K)$ which are essentially bounded with respect to a one-dimensional Lebesgue measure, equipped with the obvious norm. The later space will be denoted by $\mathrm{L}^\infty_{w*}((0,1);\mathcal{M}(K))$. The duality pairing used in this case is

$$\langle \boldsymbol{\nu}, \varphi \rangle := \int_0^1 \langle \nu_s, \varphi(s) \rangle ds \ , \quad \varphi(s) \in \mathrm{C}(K) \ .$$

It is a well-known fact that every bounded set of $\mathrm{L}^\infty_{w*}((0,1);\mathcal{M}(K))$, if endowed with the weak-star topology, is pre-compact and metrizable. Since the tensor products $h \otimes g$, $h \in \mathrm{L}^1(0,1)$, $g \in \mathrm{C}(K)$, span the dense subspace in $\mathrm{L}^1((0,1);\mathrm{C}(K))$, it turns out that the weak-star topology on bounded sets in $\mathrm{L}^\infty_{w*}((0,1);\mathcal{M}(K))$ is induced by the following norm:

$$\Phi(\boldsymbol{\nu}) := \sum_{i,j} \frac{1}{2^{i+j} \alpha_{i,j}} \int_0^1 \langle \nu_s, g_i \rangle h_j(s) ds \ , \tag{2}$$

where the countable family (g_i) is dense in $\mathrm{C}(K)$, and (h_j) are bounded and dense in $\mathrm{L}^1(0,1)$, with $\alpha_{i,j} := \|g_i\|_{\mathrm{L}^\infty} \|h_j\|_{\mathrm{L}^\infty}$. For every $\tau \in \mathbf{R}$ $T_\tau : K \longrightarrow K$ is the functional translation defined by $T_\tau x(t) := x(t - \tau)$, for $x \in K$ and $t \in \mathbf{R}$. It is not difficult to see that the group G of functional translations acts continuously on K (see [1], Proposition 5.3). Similarly, it can be easily checked that an associated group $G^\# := \{T^\#_\tau : \tau \in \mathbf{R}\}$ acts continuously on $\mathcal{M}(K)$, where a notation $T^\#_\tau \mu$ is reserved for a push-forward measure of the measure μ according to the map T_τ, and is defined by the relation

$$\langle T^\#_\tau \mu, g \rangle := \langle \mu, g \circ T_\tau \rangle \ , \quad g \in \mathrm{C}(K) \ .$$

We say that a measure μ is *invariant* if it is invariant with respect to action of $G^\#$, that is to say, if there holds $T^\#_\tau \mu = \mu$ for every $\tau \in \mathbf{R}$. $\mathcal{I}(K)$ denotes the class of all invariant probability measures on K. The *orbit* of $x \in K$, denoted by $\mathcal{O}(x)$ is the set of all translations of x. If x is periodic, $\mathcal{O}(x)$ is compact. Every h-periodic function x generates the measure ϵ_x given by

$$\langle \epsilon_x, g \rangle := \fint_0^h g(T_\tau x) d\tau \ , \quad g \in \mathrm{C}(K) \ .$$

Such a measure is the unique invariant probability measure supported on $\mathcal{O}(x)$. We refer to ϵ_x as to an *elementary invariant measure* associated to x. By $\mathrm{H}^2_{per}(0,1)$ we denote a set of all real functions on $(0,1)$, extended to $\mathbf{R}$ by periodicity, which belong to Sobolev space $\mathrm{H}^2_{loc}(\mathbf{R})$. If $g : (0,1) \longrightarrow \mathbf{R}$ is a Lipschitz-continuous function satisfying $\mathrm{Lip}(g) < 1$. $\mathcal{S}^s(I)$ stands for a class of all continuous piecewise affine functions (sawtooth functions) on bounded open interval I with slope $\{-1+g'(s), 1+g'(s)\}$ only. Finally, $\mathcal{S}^s_{per}(I)$ denotes a set of all real functions on I extended to $\mathbf{R}$ by periodicity which are of the class $\mathcal{S}^s(J)$ for every open interval J.

We are primarily interested in proving a Γ-convergence result for certain functionals defined over the space of Young measures, and in this subsection we outline general results for Young Measures, relaxation, and continuity properties of mappings extended from $Meas((0,1);K)$ to $YM((0,1);K)$ (cf. Definitions below). We point out that proofs, as well as a more detailed list of references, can be found in [1] and [32].

Definition. (Young Measures) A K-valued Young measure on $(0,1)$ (or Young measure on micropatterns) is a map $\boldsymbol{\nu} \in \mathrm{L}^\infty_{w*}((0,1);\mathcal{M}(K))$ such that ν_s is a probability measure for almost every $s \in (0,1)$. The set of all Young measures is denoted by $YM((0,1);K)$ and it is always endowed with the topology of $\mathrm{L}^\infty_{w*}((0,1);\mathcal{M}(K)$. The *elementary Young measure* associated to a measurable map $u : (0,1) \longrightarrow K$ is the map $\underline{\delta}_u : (0,1) \longrightarrow \mathcal{M}(K)$ given by $\underline{\delta}_u(s) := \delta_{u(s)}$, $s \in (0,1)$. We say that a sequence of measurable maps $u^k : (0,1) \longrightarrow K$ *generates the Young measure* $\boldsymbol{\nu}$, if the corresponding elementary Young measures $\underline{\delta}_{u^k}$ converges to $\boldsymbol{\nu}$ in the weak-star topology of $\mathrm{L}^\infty_{w*}((0,1);\mathcal{M}(K))$.

Definition. (Relaxation) Let X be a metric space and let $F : X \longrightarrow [0,+\infty]$. The *relaxation* $\overline{F}$ of F on X is the lower-semicontinuous envelope of F, that is, the supremum of all lower-semicontinuous functions below F. Alternatively, $\overline{F}$ is characterized by the following formula:

$$\overline{F}(x) := \inf\{\liminf_{k\longrightarrow+\infty} F(x^k) : x^k \longrightarrow x\}\,.$$

Definition. (Γ – convergence and continuous convergence) Let X be a metric space. A sequence of functions $F^\varepsilon : X \longrightarrow [0,+\infty]$ *Γ-converges* to F on X, and we write $F^\varepsilon \xrightarrow{\Gamma} F$, if the following is fulfilled:

(i) Lower-bound inequality: For every $x \in X$ and a sequence (x^ε) in X such that $x^\varepsilon \longrightarrow x$ it holds $\liminf_\varepsilon F^\varepsilon(x^\varepsilon) \geq F(x)$, and

(ii) Upper-bound inequality: For every y in X there exists a sequence (y^ε) in X such that $y^\varepsilon \longrightarrow y$ and $\limsup_\varepsilon F^\varepsilon(y^\varepsilon) \leq F(y)$.

Functions F^ε *continuously converge* to F on X if $F^\varepsilon(x^\varepsilon) \longrightarrow F(x)$ whenever $x^\varepsilon \longrightarrow x$, which is written as $F^\varepsilon \xrightarrow{C} F$.

The following property shows that the proper setting for studying asymptotic minimization problems is precisely Γ-convergence setting.

Theorem. (Stability of minima) For every sequence (F^ε) such that $F^\varepsilon \xrightarrow{\Gamma} F$ as $\varepsilon \longrightarrow 0$ there holds $\lim_{\varepsilon\longrightarrow 0} \min F^\varepsilon = \min F$.

Theorem. (Fundamental theorem of Young measures) For every sequence of measurable maps $u^k : (0,1) \longrightarrow K$ there exists a subsequence (not relabelled) which generates a Young measure $\boldsymbol{\nu}$, i.e., $\underline{\delta}_{u^k} \xrightarrow{*} \boldsymbol{\nu}$. Moreover, $\boldsymbol{\nu}$ has the following properties:

(i) If $f : (0,1) \times \mathbf{R} \longrightarrow \mathbf{R}$ is a Carathéodory function satisfying $|f(s,x)| \leq h(s)$ for some $h \in \mathrm{L}^1((0,1))$, then

$$\lim_{k \longrightarrow +\infty} \int_0^1 f(s, u^k(s))ds = \int_0^1 \langle \nu_s, f_s \rangle ds \ .$$

(ii) The maps u^k converge in measure to some $u : (0,1) \longrightarrow K$ if and only if $\boldsymbol{\nu}$ is an elementary Young measure associated to u.

3 Formulation of the problem

We include a brief review of traditional rudimentary macroeconomic equilibrium models. The common feature of these models is an a priori assumption that equilibrium is achieved at times when certain quantities obtain corresponding optimal values. First, we mention the so-called static linear market equilibrium model. As such, this model assumes that $D = D(p)$ ($S = S(p)$, resp.), a given aggregate demand (supply, resp.), is affine function in independent variable p, where p is price of a single (or aggregate) product. Then equilibrium occurs when $S = D$, which results in single optimal value of the price. Dynamic variant of this model include more realistic assumption that price itself depends on time variable t, so that $p = p(t)$, and $D = D(t, p(t))$, $S = S(t, p(t))$. Classical dynamic equilibrium models include balancing of higher-order derivatives of $p(t)$, so that the equilibrium equation can be stated in the form of lower-order ordinary differential equation, endowed with suitable initial conditions. According to coefficients in such equation, the market can, with respect to time-path of aggregated price, behave as oscillatory stable (cf. Figure 1), oscillatory unstable (with steady or increasing amplitude of oscillations (cf. Figure 2 and Figure 3)), decreasingly stable (cf. Figure 4) or explosively increasing unstable market (cf. Figure 5). Such model is known as the model of the market equilibrium with price expectation. In its simplest form, this model assumes that $t \mapsto D(t,\xi)$ ($t \mapsto S(t,\xi)$, resp.) is constant for every $\xi \in \mathbf{R}$, and that $Q_D(t) := D(p(t)) = \alpha - \beta p(t) + mp'(t) + np''(t)$ ($Q_S(t) := S(p(t)) = -\gamma + \delta p(t) + \eta p'(t) + \theta p''(t)$, resp.), where m, n, η, θ are constants (the so-called speculative elements) and where $\alpha, \beta, \gamma, \delta$ are positive constants. Then the equilibrium equation $S - D = 0$ reads $(\theta - n)p''(t) + (\eta - m)p'(t) + (\delta + \beta)p(t) = \gamma + \alpha$. Usually it is written in the form $p''(t) + a_1 p'(t) + a_2 p(t) = b$, where $a_1 := \frac{\eta - m}{\theta - n}$, $a_2 = \frac{\delta + \beta}{\theta - n}$ and $\frac{\gamma + \alpha}{\theta - n}$. We choose w and n in such a way that $\theta \approx 0$, $n \approx 0$, $\varepsilon := \theta - n > 0$, $\varepsilon \approx 0$. That is to say, trend coefficients θ and n are small, and overall impact of trend $p''(t)$ on equilibrium is rather small in comparison to impact of inflation $p'(t)$ and price time-path $t \mapsto p(t)$. This is reasonable, since trend affects equilibrium in a significant way only over small intervals on time.

In classical economic literature, the noise picked up by trends is filtered out by applying suitable computational algorithms, such as Kalman's filtering algorithm etc. This is the reason why we assume that $\varepsilon \approx 0$: we are only interested in behavior of the optimal price $p(t)$ "in the mean" over some finite time interval, but not so much at particular times t. Thus, we smooth out excessive time-paths by letting ε tend to zero.

Obviously the models above assume absence of probability aspects in economic dynamics, and also perfect availability of data for sampling. We do not consider situations where these two assumptions fail. Therefore, we take the view that economic dynamics is as predictable as any mechanical dynamical system, which places our method of modeling diametrally op-

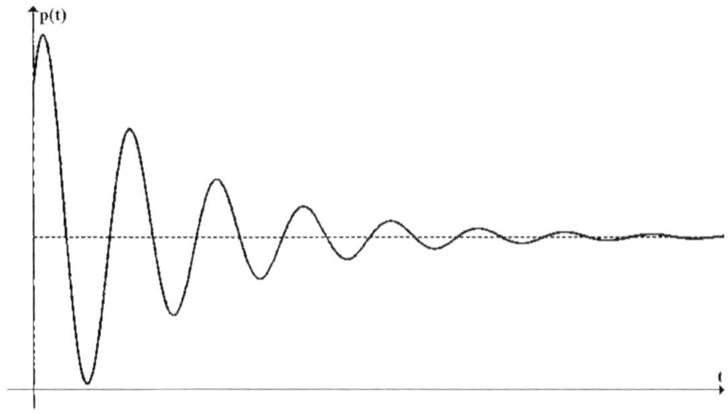

Fig. 1.

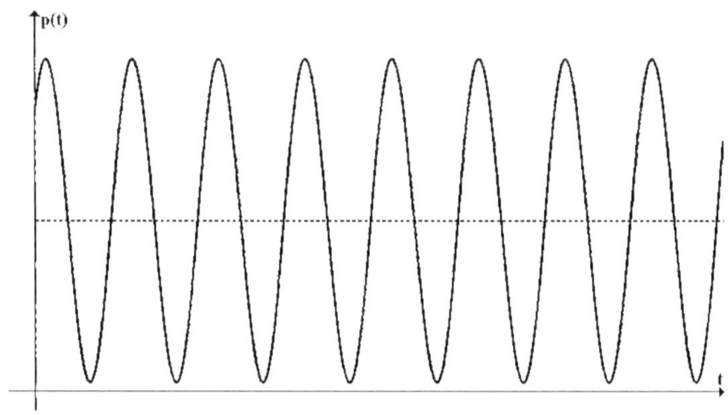

Fig. 2.

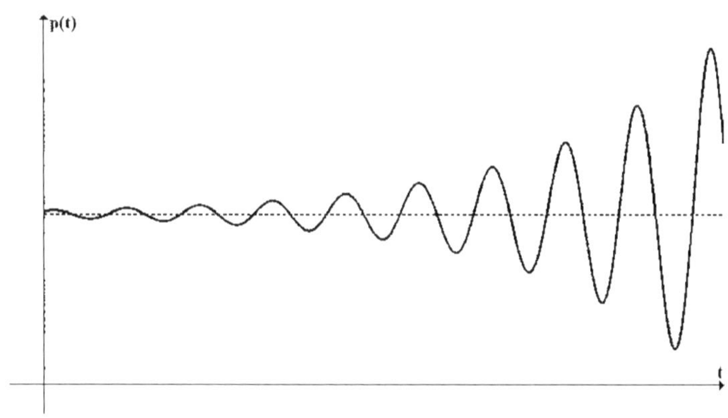

Fig. 3.

posite to behavioral economy. We assume that consumers are perfectly rational, and that aggregate demand (supply, resp.) is perfectly elastic with respect to price p and time t. This, from certain viewpoint, makes our economy an ideal economy, where no outside disturbances (like, for instance, government intervention, external shocks...) affect behavior of "players" (consumers, suppliers...) in economic exchange.

Even in the case of oscillatory behavior, the classical model does not take into account rectifiability of the graph of solution u as a criterion for quality of state of the market. In papers [25]

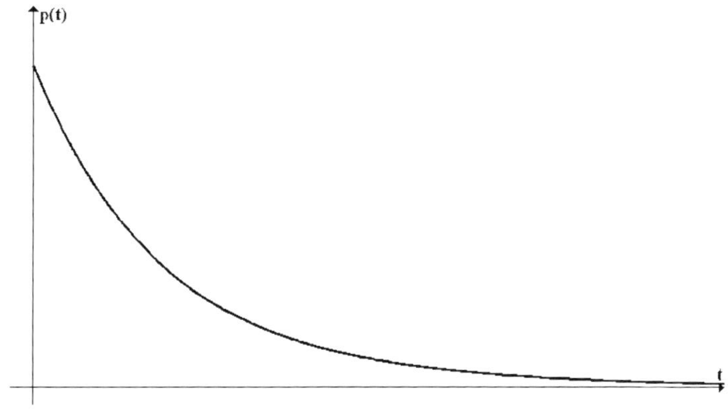

Fig. 4.

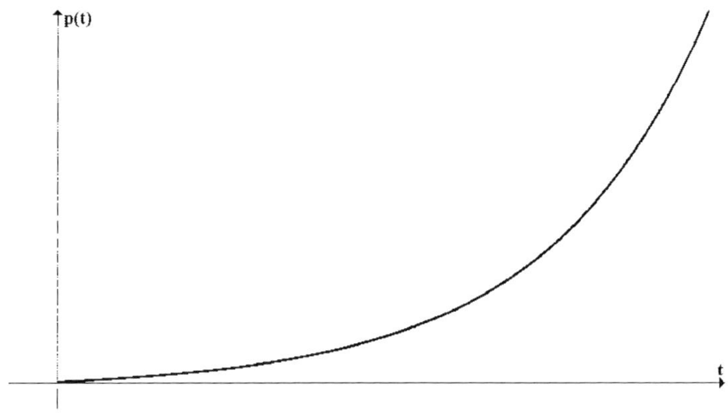

Fig. 5.

and [26] (and references therein), the authors incorporated such criterion in order to describe different oscillatory behaviors which in our context can provide a finer insight into market dynamics.

Our goal is to provide an explanation of appearance of business cycles as an inherent property of the closed national economy, starting from generic rudimentary model from physics, known as Cahn-Hilliard model, which we eventually upgrade while attempting to describe more precisely the behavior of economic balance/imbalance over some finite time-interval. In other words, we attempt to model business cycles as phase transitions. Rather than assuming that equilibrium is necessarily achieved (and possibly preserved as a favored state of the market), we start from the assumption that equilibrium conditions occur relatively rare (if ever), which results in increasing unsuccessful attempts of the market to balance itself, which, in turn, provides economic dynamics.

We mention that our choice of model prescribes a very different role to double-well function W in comparison with the usual models which aim to predict recessive state of the economy. Then latter models usually rely on a study time-path of demand (supply, resp.) of a given product, whereby graphic depiction of tends and long-term behaviors is considered to be telling enough for making appropriate conclusions. Typical types of market behavior with recession classified according to shape of time-paths are: V-shaped (cf. Figure 6), U-shaped (cf. Figure 7), W-shaped (cf. Figure 8), and L-shaped recession (cf. Figure 9), where without loss of generality we can assume that the lowest quantity of demand equals zero.

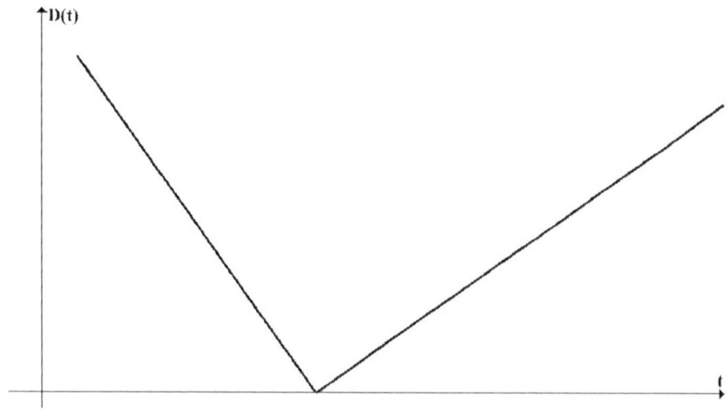

Fig. 6.

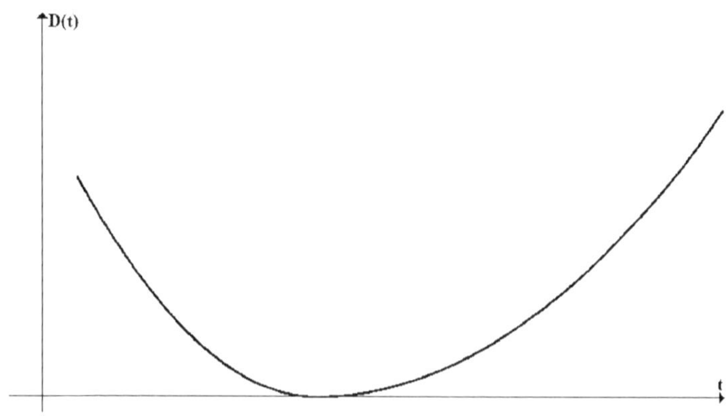

Fig. 7.

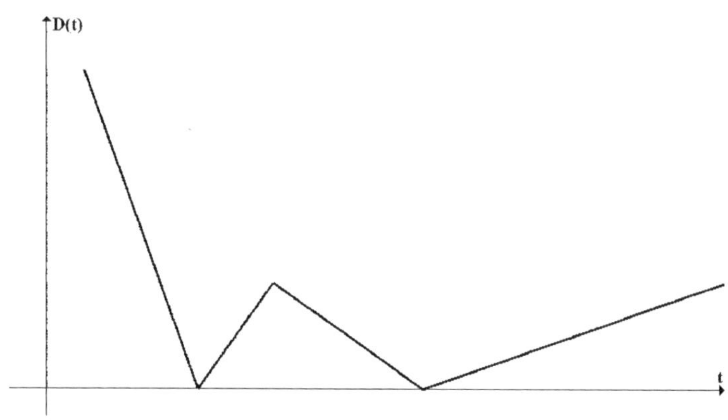

Fig. 8.

In our model the role of function W is quite different, and it does not mean W-shaped recession. For us the function W is a parameter which is built into the model, and it describes the shape of wells (or states) of the economy. The situation when economy is in decline is represented by -1 well, and when it is in ascension is represented by $+1$ well. The fact that acceleration is very near zero "in the mean" regardless of initial conditions can be

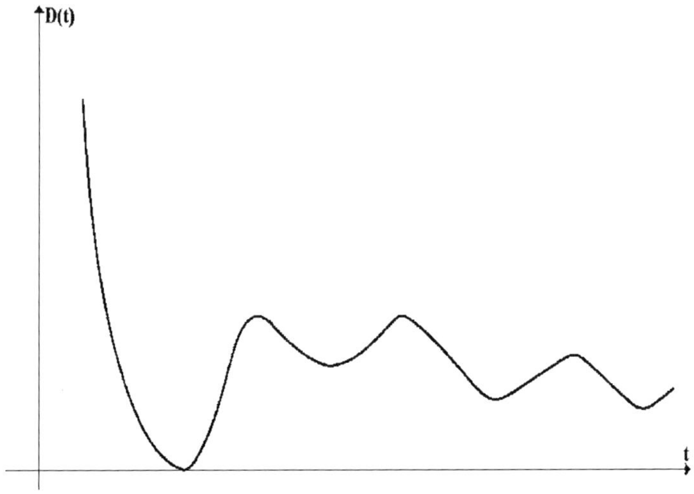

Fig. 9.

best understood as flatness property of economic dynamics: while economic dynamics is assumed to be elastic, aggregate coefficient of elasticity is non-elastic at most points in time t. We believe that this matches both intuition and observation: it takes time to switch from one phase to another, and compelling factor in switching regime is acceleration, where the impact of acceleration is measured in units of small parameter ε (the mean field parameter in physics, which here is interpreted as the weight in which of acceleration affects the balancing of the market). Then it makes sense to consider only small values of $\varepsilon > 0$, which is mathematically expresses as limiting behavior of the market as $\varepsilon \longrightarrow 0$. The case $\varepsilon = 0$ does not allow for transition layers to develop and therefore does not makes much sense: it assumes that switching is immediate and that phases can be ideally flat since in such a case we do not take into consideration the effect of the trend. Trend, however, is crucial in determining the dynamics of the market, in the absence of the trend the market will be to steady. This is expressly the reason why we can desired information only after taking the limit as $\varepsilon \longrightarrow 0$, whereas we do not consider the model if ε is equal to zero.

4 Assumptions

W is a non-negative empirical function, $W(\xi) = 0 \iff \in Z$, where Z is a simple set of roots (it can be finite or infinite, but without cluster points). For simplicity we assume that $Z = \{-1, 1\}$. Leoni's results in [17] shows that Z can be chosen in a more general way as well. We introduce a function of entropy/imbalance u, $u(t) = |Q_S(t) - Q_D(t)|$, where $Q_S(t)$ ($Q_D(t)$) is the quantity of supply (demand) of certain goods/workforce (e.g. with the perfectly elastic supply and the perfectly elastic demand) at the moment t. The result will be the long-term market equilibrium (the constitutive assumption of the market efficiency): $\lim_{t\to\infty} u(t) = 0$. We strengthen assumptions on time-path of u by imposing $\int_0^\infty |u(t)|dt < +\infty$. Speed constraint u' (we refer to u' as to phase function or state function of u) is $\int_0^\infty W(u'(t))dt \to 0$, which means that the stock accretion/ unemployment is 'dictated' by factors such as demographic conditions, economic structure, consumer preferences, etc. Intuitively, $u'(t) \simeq Z$, which results in flatness of accretion. In that state, the accretion is able to oscillate rather significantly, but not uncontrollably, and still changes relatively slowly (using mathematical term: u is a function of bounded variation, meaning that it cannot function as the Brownian motion or the Wiener process). This particular premise is easier to uphold in macroeconomic, rather than microeconomic context. The counterexample of its validation in the macroeconomic context may be the state of economy during the change of political regimes.

Roughly speaking, we distinguish three cases in terms of of behavior of W at infinity, each treated separately:

- $\lim_{\xi \longrightarrow \pm\infty} W(\xi) = +\infty$,
- $+\infty > \limsup_{\xi \longrightarrow \pm\infty} W(\xi) \geq \liminf_{\xi \longrightarrow \pm\infty} W(\xi) > 0$,
- $\lim_{\xi \longrightarrow \pm\infty} W(\xi) = 0$.

Typical assumption on W which is widely used in the literature is that W is continuous function with superlinear growth at infinity, i.e., the Tartar-Fonseca condition holds (cf. [12] and [18]): there exists $c_1 > 0$ such that $W(\xi) \geq c_1|\xi|$ for every $\xi \in \mathbf{R}$ such that $|\xi| >> 1$. The publication of Leoni's results in [17] provided the framework in which general W can be considered, and therefore applied to economic context. We conjecture that theory can be extended to the case of W with discontinuities.

In Figure 10-Figure 17 below it is seen how distinguishable shape of "wells" around points $(\pm 1, 0)$ affect the amount of "energy" needed to switch from one phase to another. Variability of internal consumption dictates the pace of switching regime.

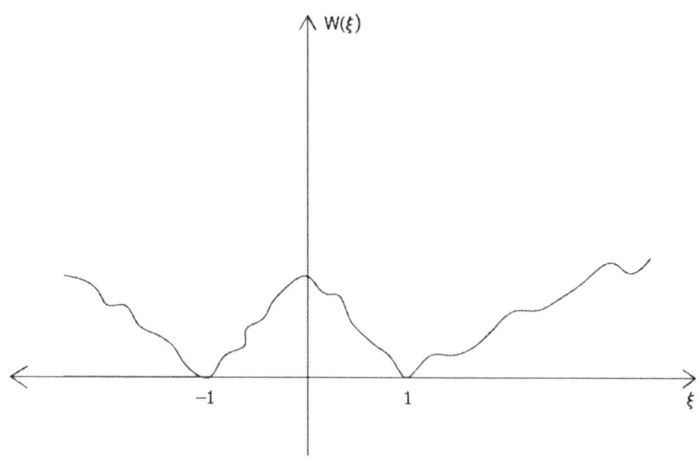

Fig. 10. The case of asymmetric W with linear growth at $\pm\infty$

In Figure 15-Figure 17 below we depict the case when we allow W to have unremovable discontinuity at one point, say point zero.

In Figure 18-Figure 24 we depict how bad FE sequences of $\varepsilon^{-1} I_0^{\varepsilon}$ can get if we abandon Tartar-Fonseca assumption on W. The economic interpretation of such behavior would be that, apart from the preferred normal ± 1 states of the economy, the approximate equilibrium can be attained with rapid oscillations of the entropy function with increasing slope as $\varepsilon \longrightarrow 0$. In Figure 21-Figure 24 we depict the shape of v_ε for FE sequence (v_ε) of $\varepsilon^{-1} I_0^{\varepsilon}$ in the case $W(\xi) \approx \frac{1}{|\xi|^q}$ for large values of $|\xi|$, with $q > 2$. Layers denoted by the letter "B", marked in

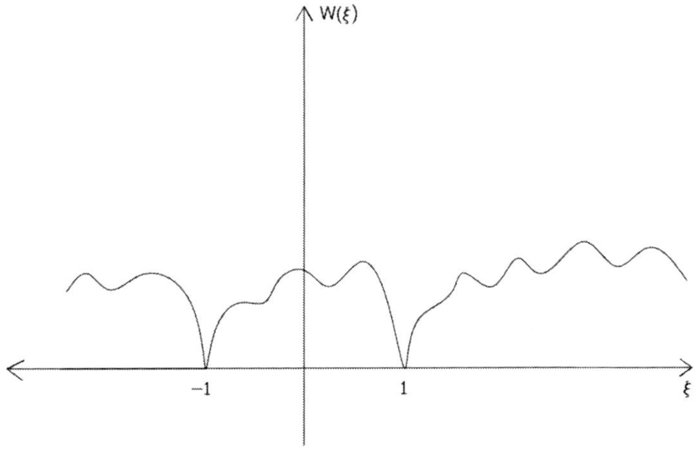

Fig. 11. The case of asymmetric W with sub-linear growth at $\pm\infty$

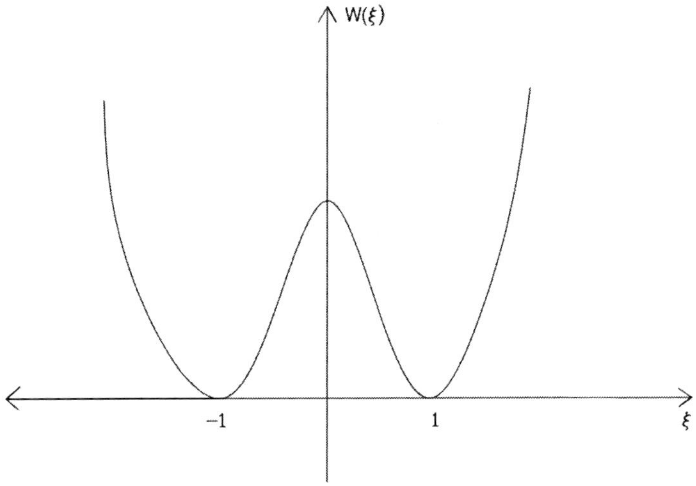

Fig. 12. The case of symmetric W with polynomial growth at $\pm\infty$, typical example being $W(\xi) = (\xi^2 - 1)^2$

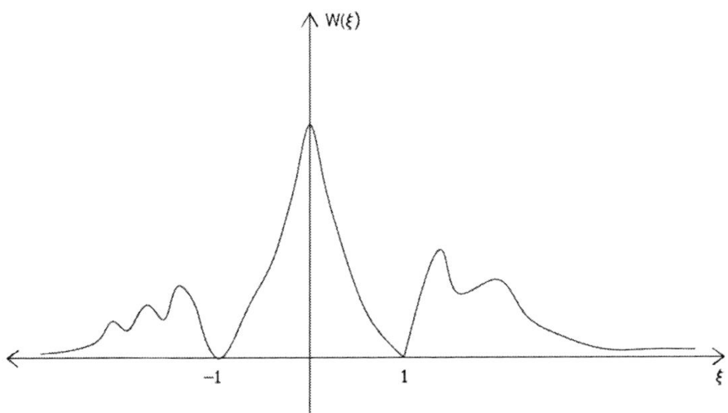

Fig. 13. The case of asymmetric W which vanishes at $\pm\infty$, typical example being $W(\xi) \approx \frac{1}{|\xi|^q}$ for large values of $|\xi|$, where $q > 0$

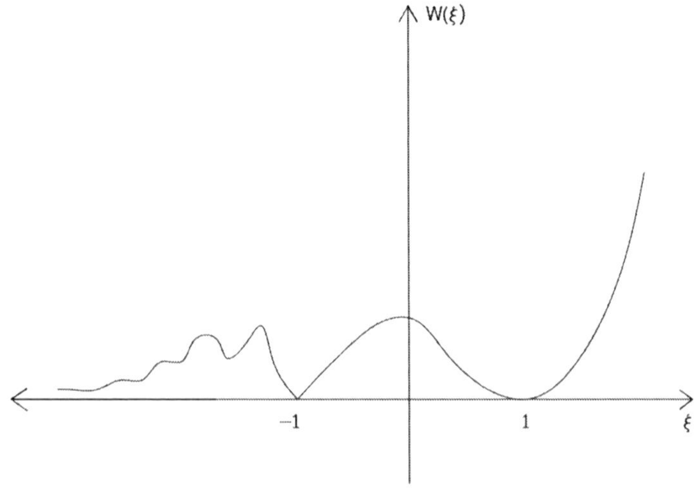

Fig. 14. The case of asymmetric W which vanishes at $-\infty$, but which grows at least quadratically at $+\infty$

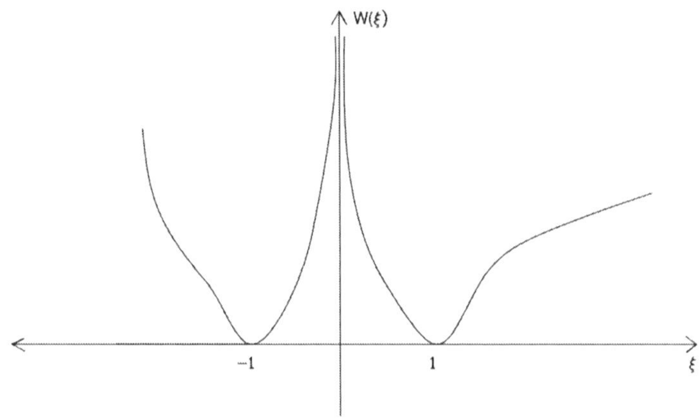

Fig. 15.

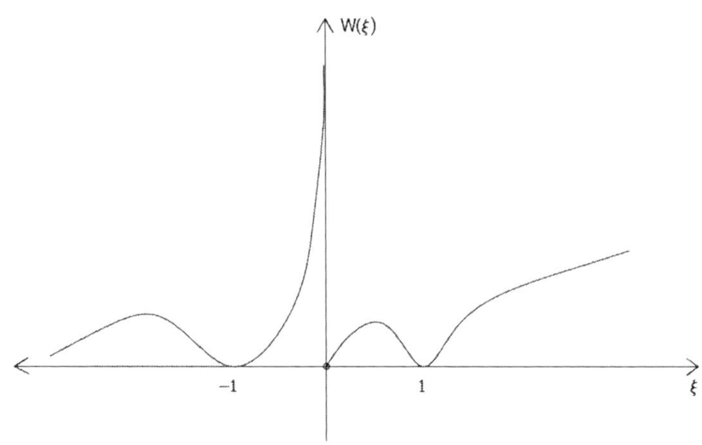

Fig. 16.

gray, represent slopes which tend to $\pm\infty$ as $\varepsilon \longrightarrow 0$. Subintervals denoted by the letter "A" represent slopes approximately equal to ± 1 as $\varepsilon \longrightarrow 0$. In such situations the economy is

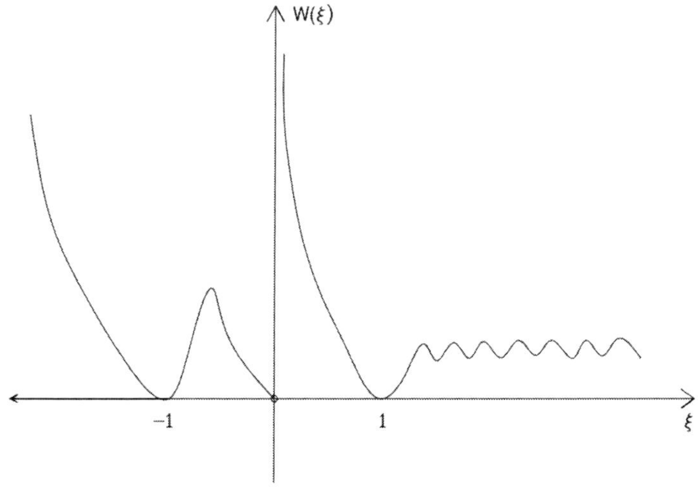

Fig. 17.

unbalanced, even though some kind of equilibrium can be reached.

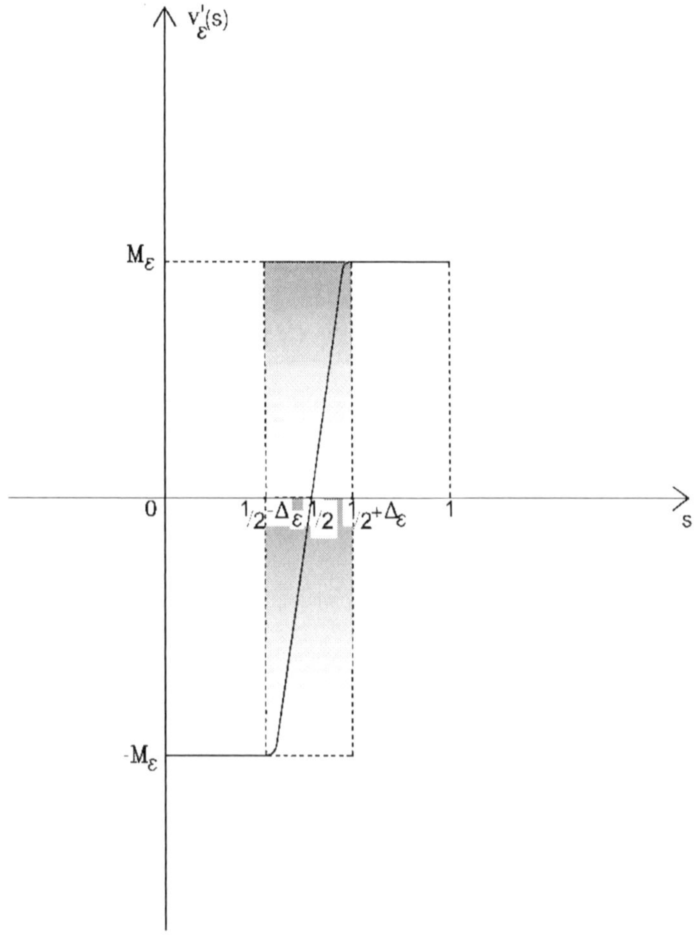

Fig. 18. The shape of v_ε' for FE sequence (v_ε) of $\varepsilon^{-1}I_0^\varepsilon$ in the case $W(\xi) \approx \frac{1}{|\xi|^q}$ for large values of $|\xi|$, with $q > 2$. Here there holds $\lim_{\varepsilon \longrightarrow 0} M_\varepsilon = +\infty$, $\lim_{\varepsilon \longrightarrow 0} \Delta_\varepsilon = 0$

5 Model

We define the functional of the market imbalance

$$I_0(u) := \int_0^\infty \Big(W(u'(t)) + |u(t)| \Big) dt,$$

and set up the optimization problem:

$$I_0(u) \to \min$$

trying to determine an optimal time path $t \to u(t)$. The time-path in such a case, however, does not tell much: minimizing sequences are zig-zag functions (or sawtooth functions) with slope (near to) ± 1, with very few other underlying geometric properties. By introducing the second derivative u'' (the trend of imbalance), we can obtain the additional state description. In the first variant of the model we assume that trend has a significant impact on equilibrium: $I_1^1(u) = \int_0^\infty \Big(u''(t)^2 + W(u'(t)) + |u(t)|^2 \Big) dt$ The problem $I_1^1(u) \to \min$ is also called 'the infinite horizon problem' and is a well-known, not easily solved, problem in economic

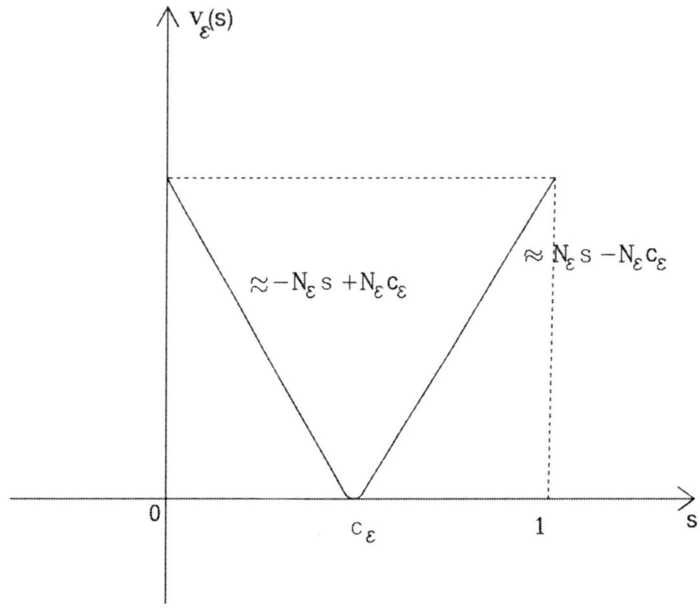

Fig. 19. The shape of v_ε for FE sequence (v_ε) of $\varepsilon^{-1} I_0^\varepsilon$ in the case $W(\xi) \approx \frac{1}{|\xi|^q}$ for large values of $|\xi|$, with $q > 2$. Here there holds $\lim_{\varepsilon \longrightarrow 0} N_\varepsilon = +\infty$, $\lim_{\varepsilon \longrightarrow 0} c_\varepsilon = 1/2$

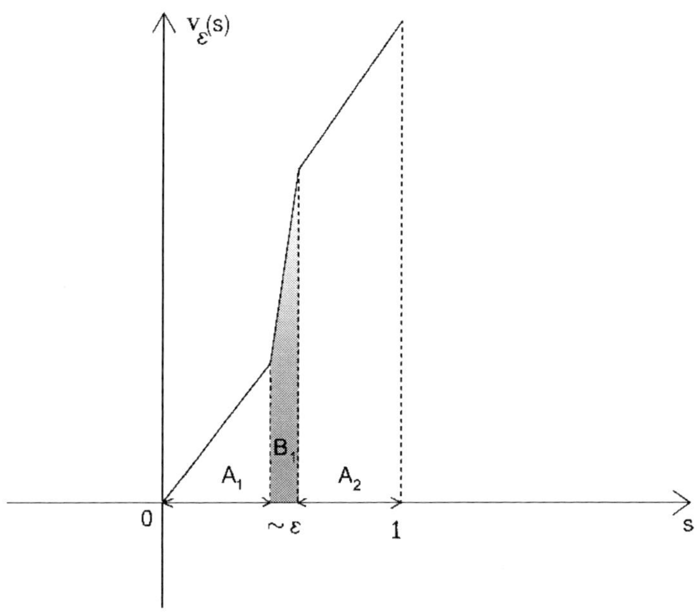

Fig. 20. The shape of v_ε for FE sequence (v_ε) of $\varepsilon^{-1} I_0^\varepsilon$ in the case $W(\xi) \approx \frac{1}{|\xi|^q}$ for negative ξ and for large values of $|\xi|$, with $q > 2$. Here there holds $\lim_{\xi \longrightarrow +\infty} W(\xi) = O(1)$

modeling and variational calculus (cf. for instance the work of A. Leizarowitz and V. J. Mizel in 1989 presented in paper [19]) In the second variant of the model we assume that trend affects the equilibrium to a certain but small degree: $I_a^\varepsilon(u) = \int_0^\infty \Big(\epsilon^2 u''(t)^2 + W(u'(t)) +$

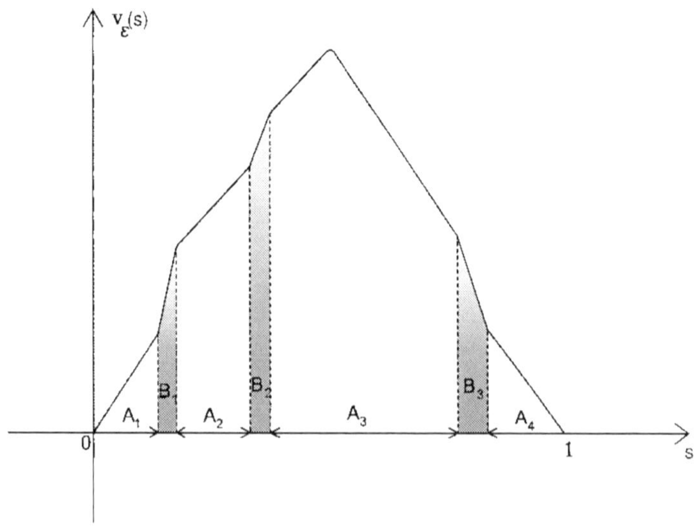

Fig. 21.

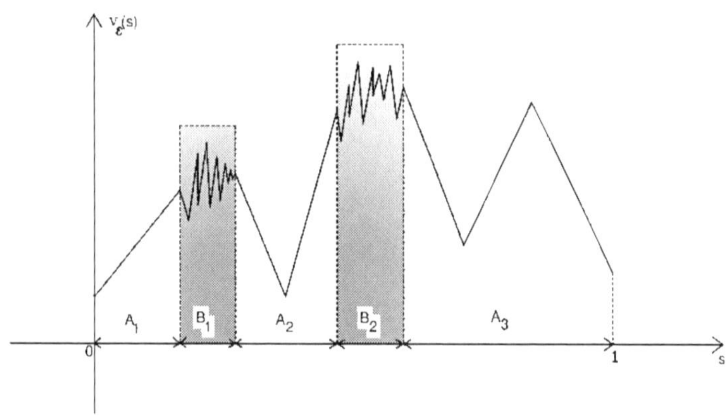

Fig. 22.

$a|u(t)|^2\Big)dt$, where $a > 0$ and $\epsilon \approx 0$. The problem $I_a^\epsilon(u) \to \min$ when $\epsilon \to 0$ is also called 'singular perturbation problem' for a non-convex integrand W, and is connected to the so-called problem of non-attainment of extrema in the calculus of variation. It is considered a 'relaxation' of the problem of the first type. Key observation in both cases is that the non-convexity of W function generates oscillations in minimizer u. In the sequel we state the main conclusions of mathematical Ginzburg-Landau theory, associated to the functional I_1^ε. Leoni observed in [17] that, even if Z is countable (with no cluster points), in the case of functional I_0^ε, without loss of generality we can assume that Z is a finite set, the fact corresponding to macroeconomic intuition: in the case of closed economy there will be a finite set of optimal states/phases. First results regarding asymptotic behavior of functionals I_0^ε as $\varepsilon \longrightarrow 0$ (see below) in terms of Γ-convergence were obtained by L. Modica and S. Mortola in the 1970's (cf. [20]): since we are interested primarily in minimization as $\varepsilon \longrightarrow 0$, the proper notion of convergence of such functionals is exactly Γ-convergence, a notion introduced by De Giorgi in 1970's. In the following, for the sake of simplicity, we

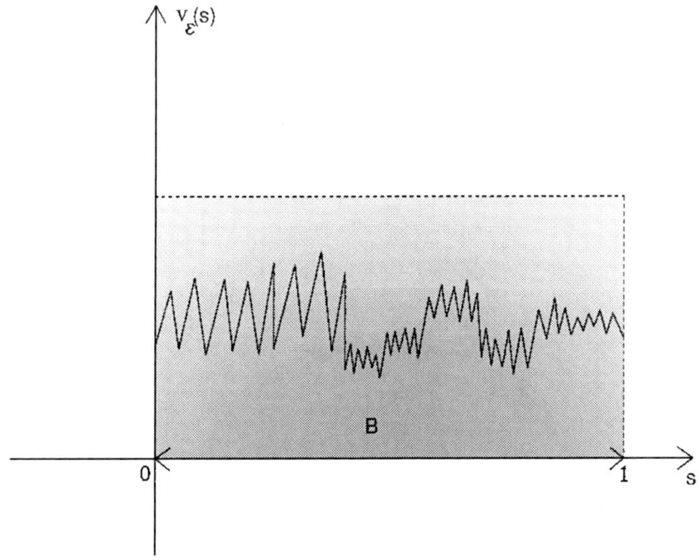

Fig. 23.

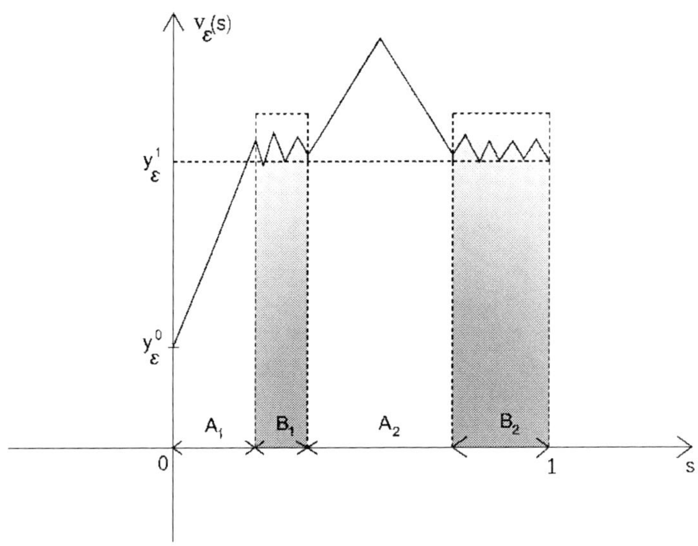

Fig. 24.

consider integration over finite time-interval, such as the unit interval, whereby

$$I_a^\varepsilon(u) = \int_0^1 \Big(\epsilon^2 u''(s)^2 + W(u'(s)) + a|u(s)|^2\Big)ds \ ,$$

where $a > 0$ is strictly positive constant (or, more generally, strictly positive integrable function). This is not loss of generality, since, for instance, in the macroeconomic context, we can assume that time is measured in years, decades or centuries. The first results for the functional I_a^ε were presented by G. Alberti and S. Müller in 2001 in [1]. It can be shown that there holds:

- $\lim_{\varepsilon \longrightarrow 0} \min_v I_a^\varepsilon(v) = 0$, $0 < \lim_{\varepsilon \longrightarrow 0} \min_v \varepsilon^{-2/3} I_a^\varepsilon(v) < +\infty$,

- $\lim_{\varepsilon\longrightarrow 0}\min_v I_0^\varepsilon(v) = 0$, $0 < \lim_{\varepsilon\longrightarrow 0}\min_v \varepsilon^{-1} I_0^\varepsilon(v) < +\infty$,
- $\inf_v I_a^0(v) = 0$, where infimum is not attained.

Such asymptotic behavior suggests that it is reasonable to introduce the generalization of the notion of minimizer, namely, the notion of FE sequence: a sequence (v_ε) is said to be FE sequence of $(\varepsilon^{-2/3} I_a^\varepsilon)$ ($\varepsilon^{-1} I_0^\varepsilon$, resp.) if there holds $\limsup_{\varepsilon\longrightarrow 0}\varepsilon^{-2/3} I_a^\varepsilon(v_\varepsilon) < +\infty$ ($\limsup_{\varepsilon\longrightarrow 0}\varepsilon^{-1} I_0^\varepsilon(v_\varepsilon) < +\infty$, resp.).

Typical FE sequence of $(\varepsilon^{-2/3} I_a^\varepsilon)$ under Tartar-Fonseca assumption on W behaves as depicted in Figure 25: derivative takes consecutive values $\approx \pm 1$ on intervals of order $\varepsilon^{1/3}$, with the exception of transition layers of order ε at the end of each such interval.

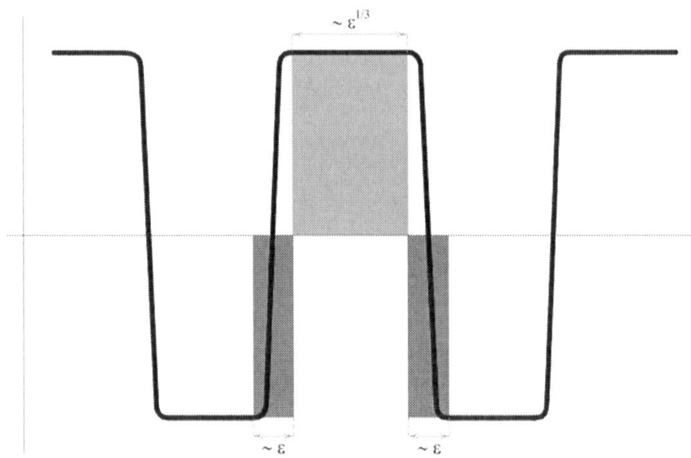

Fig. 25. Two-scale structure of $s \mapsto v_\varepsilon'(s)$ for $\varepsilon \approx 0$

Main steps in the approach of Alberti and Müller are the following:

Step 1. Characterize the class of all Young measures $\boldsymbol{\nu} \in YM((0,1);K)$ which are generated by sequences of ε-blowups of functions v^ε.

Step 2. Rewrite the rescaled functionals $\varepsilon^{-2/3} I_a^\varepsilon(v)$ as $\int_0^1 f_s^\varepsilon(R_s^\varepsilon v)ds$ for a suitable choice of $R_s^\varepsilon v$ and f_s^ε.

Step 3. Identify the Γ-limit f_s of f_s^ε in the topology of K for almost every $s \in (0,1)$.

Step 4. Identify the Γ-limit of naturally defined relaxed functionals on the space $YM((0,1);K)$.

Step 5. Determine the minimizer for the relaxed functional in the limit and prove its uniqueness.

6 Advanced model variants

One possible generalization of the model above is to consider the transition from efficient to inefficient market, where the source of the market dynamics is precisely in its imbalance. Consequently, the assumption $\int_0^\infty |Q_S(t) - Q_D(t)| \to 0$ is replaced by the assumption

$\int_0^\infty |u(t) - g(t)| \to 0$, where $t \to g(t)$ is an exogenous function, such as the external shock function, which can, for instance, describe the seasonal variation as is the case with Croatian national economy (heavily dependent on tourism), the oscillations of currency, etc.

The following result (also valid in the case $g = 0$) was proved by A. Raguž in [32].

Theorem. (Main Γ-convergence result) Let us suppose that $\Omega := (0, 1)$ and that the functions $g : \Omega \longrightarrow \mathbf{R}$ and $W : \mathbf{R} \longrightarrow [0, +\infty)$ satisfy

$$g \in C^1(\Omega), |g'(s)| < 1 \quad (\text{a.e. } s \in \Omega) \quad (\text{or } g \in \mathrm{W}^{1,\infty}(\Omega), \mathrm{Lip}(g) < 1)\,, \tag{3}$$

$$W \in \mathrm{C}(\mathbf{R})\,, \tag{4}$$

$$W(\xi) = 0 \quad \text{if and only if} \quad \xi \in \{\pm 1\}\,, \tag{5}$$

$$W(\xi) \geq C|\xi| \quad \text{if } |\xi| > R \text{ for some } R > 0 \text{ and } C > 0\,. \tag{6}$$

Consider the functional $I^\varepsilon : \mathrm{H}^2_{per}(\Omega) \longrightarrow \mathbf{R}$

$$I^\varepsilon(v) := \int_\Omega \Big(\varepsilon^2 v''^2(s) + W(v'(s)) + a(v(s) + g(s))^2\Big) ds\,.$$

Set ε-blowups as:

$$R^\varepsilon_s v(t) := \varepsilon^{-1/3}\Big(v(s + \varepsilon^{1/3}t) + g(s) + g'(s)\varepsilon^{1/3}t\Big)\,.$$

Let us define the functionals $F^\varepsilon_g, F_g : YM(\Omega; K) \longrightarrow [0, +\infty]$ by

$$F^\varepsilon_g(\boldsymbol{\nu}) := \begin{cases} \int_\Omega \langle \nu_s, f^\varepsilon_s \rangle ds, & \text{if } \boldsymbol{\nu} = \underline{\delta}_{R^\varepsilon v} \text{ for some } v \in \mathrm{H}^2_{per}(\Omega) \\ +\infty & \text{otherwise} \end{cases} \tag{7}$$

$$F_g(\boldsymbol{\nu}) := \begin{cases} \int_\Omega \langle \nu_s, f_s \rangle ds, & \text{if } \nu_s \in \mathcal{I}(K) \text{ for a.e. } s \in \Omega \\ +\infty & \text{otherwise.} \end{cases} \tag{8}$$

Then there holds $F^\varepsilon_g \xrightarrow{\Gamma} F_g$. Moreover, ε-blowups of minimizers (v_ε), $R^\varepsilon v_\varepsilon$, generate, as $\varepsilon \longrightarrow 0$, Young measure $\overline{\boldsymbol{\nu}} \in YM(\Omega; K)$, $\overline{\nu}_s = \epsilon_{y^*_s}$, $s \in \Omega$, which is the unique minimizer for the functional F_g, where for any admissible $h > 0$ and $b := g'(s) \in (-1, 1)$ $y_{h,b}$ is a function in the class $\mathcal{S}^s_{per}(0, h)$ (depicted in Figure 18) defined by

$$y_{h,b}(t) := \begin{cases} (-1 + b)t - \frac{h}{4}(1 - b^2), & \text{if } t \in [-\frac{h}{2}(1 + b), 0] \\ (1 + b)t - \frac{h}{4}(1 - b^2), & \text{if } t \in [0, \frac{h}{2}(1 - b)). \end{cases}$$

and where $y^*_s = y_{h_*(s), g'(s)}$ for some unique $h_*(s) > 0$ (which depends on a, W and $g'(s)$).

All previously discussed variants are so-called local models (the interaction between integrands is local in time). In reality, one should also include 'non-local' integrand members, that is, those depending on their own average value both globally and in the past, such as GDP. One of those examples is the subject of the paper [42], where $a = a(s)$ is replaced by $A = A(s, v, v')$. The problem of minimization of the functional endowed with local lower-order term $a = a(s, v(s), v'(s))$ is solved in [43].

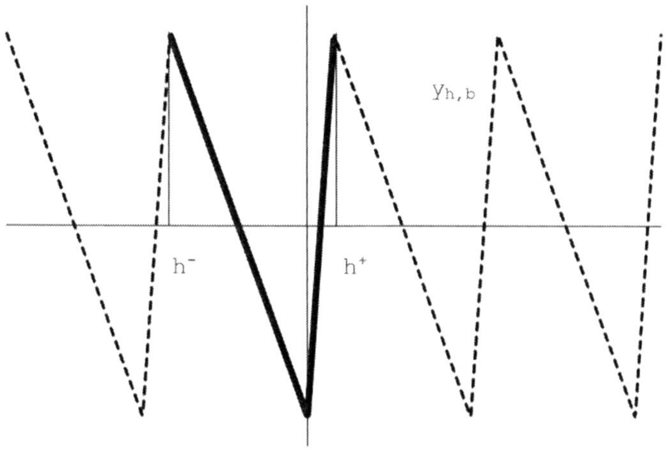

Fig. 26. The asymmetric sawtooth function $t \mapsto y_{h,b}(t)$ for $b = 2/3$ and $h = 1$; $h^- = -\frac{h}{2}(1+b)$, $h^+ = \frac{h}{2}(1-b)$

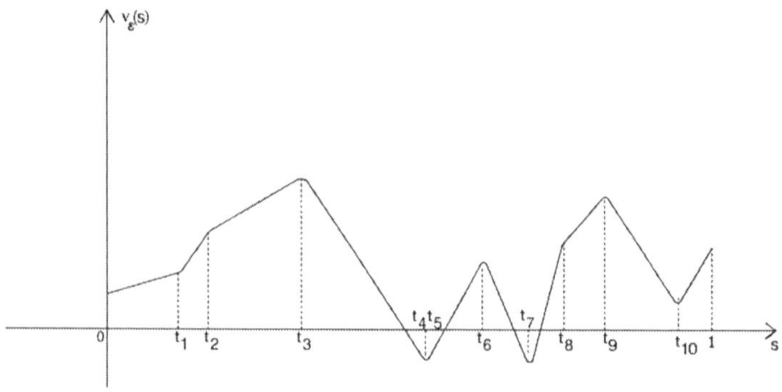

Fig. 27. Approximate zig-zag pattern of the minimizer (v_ε) with several different slopes in the case of $W = W(t, \xi)$

If we allow the optimal states Z to be variable in time, that is if $Z = Z(t)$ and $W(\xi)$ is replaced by $W(t, \xi)$, we can expect non-local effect in the process of minimization even with the initial local formulation of the equilibrium model. This work is in progress. Figure 27 shows approximate structure of the minimizers in this case, where we assume that $t \mapsto W(t, \xi)$ is piecewise constant function. The central flaw of the model: this is a deterministic/mechanical model which lacks the stochastic features of empirical data and presumes data to be 'perfectly available for sampling'. The attempts to involve probabilistic components in the integrand function were made by N. Dirr and E. Orlandi in 2010 (cf. [23]), using the application of martingale central limit theorem (it is a complex theory still in its early developmental stage). Even more realistic would be to introduce a recurrence relation between times $t \in [0, s)$ and $t \in (s, 1]$ (for a given $s \in (0, 1)$) and/or some additional probabilistic term in terms of variable t, but analysis of such situations is beyond the scope of this paper.

References

[1] G. Alberti, S. Müller: A new approach to variational problems with multiple scales, Comm. Pure Appl. Math. **54**, 761–825 (2001)

[2] N. Antonić, A. Raguž, M. Vrdoljak: Homogenisation of nonlinear elliptic systems /

Applied mathematics and computation / Rogina, Mladen ; Hari, Vjeran ; Limić, Nedžad ; Tutek, Zvonimir (eds.). pp. 81-90, Zagreb, PMF-Matematički odjel 2001

[3] J. M. Ball, A version of the fundamental theorem for Young measures, in *PDE's and Continuum Models of Phase Transitions*, eds. M. Rascle *et al.*, Lecture Notes in Physics, Vol. 344, pp. 207–215, Springer-Verlag, New York (1989)

[4] J. M. Ball, R. D. James: Fine phase mixtures as minimizers of energy, Arch. Rational Mech. Anal. **100**, 13–52 (1987)

[5] A. Braides: Gamma-convergence for beginners, Oxford Lecture Series in Mathematics and its Applications, 22, Oxford, Oxford University Press, 2002

[6] R. Choksi: Scaling laws in microphase separation of diblock copolymers, *J. Nonlinear Sci.* **11**, 223–236 (2001)

[7] G. DalMaso: *An Introduction to Γ-convergence*, Progress in Nonlinear Differential Equations, Birkhäuser, Boston, 1993

[8] A. DeSimone, R. V. Kohn, S. Müller, F. Otto: A compactness result in the theory of phase transitions, The Royal Society of Edinburgh Proceedings **A-131**, 833–844 (2001)

[9] A. DeSimone, R. V. Kohn, S. Müller, F. Otto: Magnetic microstructures - a paradigm of multiscale problems, ICIAM 99 J. M. Ball and J. C. R. Hunt, eds., Oxford Univ. Press, 175–190 (2000)

[10] I. Fonseca and C. Mantegazza: Second order singular perturbation models for phase transitions, SIAM J. Math. Anal. (5) **31**, 1121-1143 (2002)

[11] I. Fonseca and G. Leoni: Modern Methods in Calculus of Variations: L^p Spaces, Springer, 2007.

[12] I. Fonseca, L. Tartar: The gradient theory of phase transitions for systems with two potential wells, Proc. Roy. Soc. Edinburgh Sect. A 111, 89–102 (1989)

[13] A. Khachaturyan: Theory of structural transformations in solids, New York, John Wiley and Sons 1983.

[14] A. Khachaturyan: Some questions concerning the theory of phase transformations in solids, Sov. Phys. - Solid State **8**(9), 2163–2168 (1967)

[15] A. Khachaturyan, G. Shatalov: Theory of macroscopic periodicity for a phase transition in a solid state, Sov. Phys. JETP **29**(3), 557–561 (1969)

[16] R. V. Kohn, S. Müller: Branching of twins near an austensite-twinned-martensite interface, Philosophical Magazine A **66**, 697–715 (1992)

[17] G. Leoni: A Remark on Compactness for the Cahn-Hilliard Functional, ESAIM COCV **20**(2), 517–523 (2014)

[18] G. Leoni: Gamma Convergence and Applications to Phase Transitions, CNA Lecture Notes, CMU, in preparation

[19] A. Leizarowitz, V. J. Mizel: On dimensional infinite-horizon variational problems arising in continuum mechanics, Archive for Rational Mechanics and Analysis **106**, 161-194 (1989)

[20] L. Modica, S. Mortola: Un esempio di Γ-convergenca, Bull. Un. Mat. Ital. (5), **14-B**, 285–299 (1977)

[21] S. Müller: Singular perturbations as a selection criterion for periodic minimizing sequences, Calc. Var. Partial Differential Equations **1**(2), 169–204 (1993)

[22] S. Müller: Variational models for microstructure and phase transitions, Lecture notes, Max Planck Institut für Mathematik in den Naturwissenschaften (1998).

[23] N. Dirr, E. Orlandi: Unique minimizer for a Random functional with double-well potential in dimension 1 and 2, Communications in Mathematical Sciences, vol. 9, 331–351 (2011)

[24] T. Ohta and K. Kawasaki: Equilibrium morphology of block copolymer melts, *Macromolecules* **19**, 2621–2632 (1986)

[25] M. Pašić, A. Raguž: Rectifiable oscillations and singular behaviour of solutions of second-order linear differential equations, International Journal of Mathematical Anal. **Vol. 2 no. 10**, 477-490 (2008)

[26] M. Pašić, A. Raguž: Singular behavior of bounded radially symmetric solutions of p-Laplace nonlinear equation, International Journal of Mathematical Anal. **Vol. 3 no. 36**, 1775-1788 (2009)

[27] P. Pedregal: Parametrized measures and variational principless, Progress in Nonlinear Equations, Vol. 30, Birkhäuser, Boston, 1997.

[28] A. Raguž: Compensated compactness for higher-order differential relations, Z. Angew. Math. Mech. **81**, Suppl. 3, 609-610 (2001)

[29] A. Raguž: Remarks on Γ-convergence of penalized functionals of Ginzburg-Landau type in one dimension, *Proceedings of the Conference on Multiscale Problems in Science and Technology, Dubrovnik, Croatia, September 2000, N. Antonić and al., eds.),* pp. 271–283, Springer, 2002

[30] A. Raguž: Relaxation of some energy functionals related to formation of microstructure // Applied Mathematics and Scientific Computing / Drmač Zlatko, Hari Vjeran, Sopta Luka, Tutek Zvonimir, Veselić Krešimir (eds.), pp. 251-254, New York, Kluwer Academic Press, 2001.

[31] A. Raguž: On some questions related to relaxation and minimization of a class of functionals of Ginzburg-Landau type in one dimension, PhD thesis (in croatian), University of Zagreb (2004)

[32] A. Raguž: Relaxation of Ginzburg-Landau functional with 1-Lipschitz penalizing term in one dimension by Young measures on micro-patterns, Asymptotic Anal. **41**(3,4), 331–361 (2005)

[33] A. Raguž: A note on calculation of asymptotic energy for Ginzburg-Landau functional with ε-dependent 1-Lipschitz penalizing term in one dimension, Glas. Mat. Ser. III **41(61)**, 89–99 (2006)

[34] A. Raguž: A note on calculation of asymptotic energy for Ginzburg-Landau functional with externally imposed lower-order oscillatory term in one dimension, Boll. Un. Mat. Ital. **(8)**10-B, 1125–1142 (2007)

[35] A. Raguž: A result in asymptotic analysis for the functional of Ginzburg-Landau type with externally imposed multiple small scales in one dimension, Glas. Mat. Ser. III **44(64)**, 401–421 (2009)

[36] A. Raguž: On stability of asymptotic energy for a functional of the Ginzburg-Landau type with ε-dependent weakly-star convergent 1-Lipschitz penalizing term, Proc. Appl. Math. Mech. 9, 535–536 (2009)/DOI 10.1002/pamm.200910240

[37] A. Raguž: Γ-convergence for one-dimensional Ginzburg-Landau functional with generalized Lipschitz penalizing term, Proc. Appl. Math. Mech. 10, 523–524 (2010)/DOI 10.1002/pamm.201010254

[38] A. Raguž: Selection of minimizers for one-dimensional Ginzburg-Landau functional with 3-well potential by singular lower-order term, Proc. Appl. Math. Mech. 11, 689–690 (2011)/DOI 10.1002/pamm.201110334

[39] A. Raguž: On asymptotic behavior for 1-dimensional functional of Ginzburg-Landau type with internally-externally created oscillations of minimizers, Proc. Appl. Math. Mech. 12, 595–596 (2012)/DOI 10.1002/pamm.201210286

[40] A. Raguž: Gamma-convergence for Ginzburg-Landau functional with degenerate 3-well potential in one dimension, Proc. Appl. Math. Mech. 13, 371–372 (2013)/DOI 10.1002/pamm.201310181

[41] A. Raguž: Some properties of asymptotic energy of a class of functionals of Ginzburg-Landau type endowed with generalized lower-order oscillatory term in one dimension, Proc. Appl. Math. Mech. 14, 753–754 (2014)/DOI 10.1002/pamm.201410359

[42] A. Raguž: Relaxation of Ginzburg-Landau functional perturbed by continuous nonlinear lower-order term in one dimension, Anal. Appl. **13**(1), 101–123 (2015)

[43] A. Raguž: Minimization of asymptotic energy of Müller's functional endowed with local nonlinear lower-order term, Proc. Appl. Math. Mech. 16, 661–662 (2016)/DOI 10.1002/pamm.201610319

[44] L. Tartar: Beyond Young measures, Meccanica **30**, 505–526 (1995)

[45] W. P. Ziemer: Lectures on the calculus of variations and optimal control theory, Chelsea, 1980

[46] L. C. Young: Lectures on the calculus of variations and optimal control theory, Saunders, Philadelphia-London-Toronto, 1969. Reprinted with corrections by the American Mathematical Society, Providence, R.I., 2000.

Current address

Barić Pisarović Gordana
Zagreb School of Economics and Management
Jordanovac 110, 10000 Zagreb, Croatia
E-mail: gbaric@zsem.hr

Greblički Marijana
University of Zagreb, Faculty of Transport and Traffic Sciences
Vukelićeva 4, 10000 Zagreb, Croatia
E-mail: mgreblicki@fpz.hr

Raguž Andrija
Zagreb School of Economics and Management
Jordanovac 110, 10000 Zagreb, Croatia
(Luxembourg School of Business, 19 Rue Eugene Ruppert L-2453, Luxembourg)
E-mail: araguz@zsem.hr

Slovak University of Technology in Bratislava
Faculty of Mechanical Engineering
16th Conference on Applied Mathematics **APLIMAT 2017**
Proceedings

ASYMPTOTIC BEHAVIOR OF POSITIVE SOLUTIONS OF A DISCRETE DELAYED EQUATION

BAŠTINEC Jaromír (CZ), DIBLÍK Josef (CZ)

Abstract. In the paper the scalar discrete equation with delay (1) is considered. We prove that there exists a positive solution of the equation (1) and give its upper estimation. The proof of the main results is based on the asymptotic decomposition and discrete version of Wazevski's topological method. Main results gives sufficient conditions for the existence of a positive solution and it cannot be further improved.

Keywords: Discrete delayed equation, Positive solution, Critical case

Mathematics subject classification: Primary 39A10.; Secondary 39A11

1 Introduction

Denote $\mathbb{Z}_s^q := \{s, s+1, \dots, q\}$ where s and q are integers such that $s \leq q$. Similarly a set $\mathbb{Z}_s^\infty$ is defined. In the paper the scalar discrete equation with delay

$$\Delta x(n) = -\left(\frac{k}{k+1}\right)^k \frac{1}{k+1}\left[1+\omega(n)\right] x(n-k) \tag{1}$$

is considered where function $\omega\colon \mathbb{Z}_a^\infty \to \mathbb{R}$ will be more precisely defined later, $k \geq 1$ is fixed integer, $n \in \mathbb{Z}_a^\infty$, and a is a whole number. We prove that there exists a positive solution $x = x(n)$ of the equation (1) and give its upper estimation.

The paper is organized as follows. In Part 2 necessary auxiliary notions and results are collected. Main result is given in Part 3.

2 Preliminary

Let $b, c\colon \mathbb{Z}_{a-k}^\infty \to \mathbb{R}$ be given functions such that $b(n) < c(n)$, $n \in \mathbb{Z}_{a-k}^\infty$ and

$$\Omega(n) := \{n \in \mathbb{Z}_{a-k}^\infty : b(n) < x < c(n)\}.$$

Let us consider the scalar discrete equation

$$\Delta u(n) = f(n, u(n), u(n-1), \ldots, u(n-k)), \tag{2}$$

with $f\colon \mathbb{Z}_a^\infty \times \mathbb{R}^{k+1} \to \mathbb{R}$. Below we utilize a nonlinear result concerning the existence of a solution of (2) with the graph remaining in a prescribed set (see [1, 2]).

Lemma 2.1 *Let $f\colon \mathbb{Z}_a^\infty \times \mathbb{R}^{k+1} \to \mathbb{R}$ be continuous with respect to last $(k+1)$ coordinates. If, moreover, the inequalities*

$$f(n, b(n), u_1, \ldots, u_k) - b(n+1) + b(n) < 0, \tag{3}$$

$$f(n, c(n), u_1, \ldots, u_k) - c(n+1) + c(n) > 0 \tag{4}$$

hold for every $n \in \mathbb{Z}_a^\infty$ and every

$$u_1 \in \Omega(n-1), \ldots, u_k \in \Omega(n-k),$$

then there exists an initial problem

$$u^*(a-m) = u_m^* \in \Omega(a-m),\ \ m = 0, 1, \ldots, k$$

such that the corresponding solution $u = u^*(n)$ of equation (2) satisfies for every $n \in \mathbb{Z}_{a-k}^\infty$ the inequalities

$$b(n) < u^*(n) < c(n).$$

In the following lemma the symbol O means the Landau order symbol.

Lemma 2.2 *Let $\sigma \in \mathbb{R}$ and $d \in \mathbb{R}$ be fixed. Then the asymptotic decomposition*

$$\left(1 + \frac{d}{n}\right)^\sigma = 1 + \frac{\sigma d}{n} + \frac{\sigma(\sigma-1)d^2}{2n^2} + O\left(\frac{1}{n^3}\right) \tag{5}$$

holds for $n \to \infty$.

Due to the form of the right-hand side of the linear discrete equation (1) the Lemma 2.1 implies the following statement.

Lemma 2.3 *If inequalities*

$$-\left(\frac{k}{k+1}\right)^k \frac{1}{k+1}\left(1 + \omega(n)\right) b(n-k) - b(n+1) + b(n) < 0, \tag{6}$$

$$-\left(\frac{k}{k+1}\right)^k \frac{1}{k+1}\left(1 + \omega(n)\right) c(n-k) - c(n+1) + c(n) > 0 \tag{7}$$

hold for every $n \in \mathbb{Z}_a^\infty$, then there exists an initial problem

$$u^*(a-m) = u_m^* \in \Omega(a-m),\ \ m = 0, 1, \ldots, k$$

such that the corresponding solution $u = u^*(n)$ of equation (1) satisfies for every $n \in \mathbb{Z}_{a-k}^\infty$ the inequalities

$$b(n) < u^*(n) < c(n).$$

3 Main result

Let $p \in (0,1)$ be fixed and δ be a positive constant. Define

$$b(n) := (n - \delta n^p)\left(\frac{k}{k+1}\right)^n, \quad n \in \mathbb{Z}_{a-k}^{\infty}, \tag{8}$$

$$c(n) := (n + \delta n^p)\left(\frac{k}{k+1}\right)^n, \quad n \in \mathbb{Z}_{a-k}^{\infty}. \tag{9}$$

Theorem 3.1 *Let $p \in (0,1)$ and $\delta > 0$. If inequalities (6), (7) with functions b and c, defined by formulas (8), (9) hold for every fixed $n \in \mathbb{Z}_a^{\infty}$ and $\omega(n) = O(1/n^3)$, then there is an initial problem*

$$x^*(a_1 - m) = x^*_{-m} \in \Omega(a_1 - m), \; m = 0, 1, \ldots, k$$

where $a_1 \geq a$ exists, such that the corresponding solution $x = x^*(n)$ of equation (1) satisfies for every $n \in \mathbb{Z}_{a_1-k}^{\infty}$ the inequalities

$$(n - \delta n^p)\left(\frac{k}{k+1}\right)^n < x^*(n) < (n + \delta n^p)\left(\frac{k}{k+1}\right)^n.$$

Proof. Without loss of generality assume that the number $a_1 \geq a$ is sufficiently large and that for $n \geq a_1 - k$

$$b(n) < c(n).$$

We prove inequality (7). Let

$$H := -\left(\frac{k}{k+1}\right)^k \frac{1}{k+1}(1 + \omega(n))\, c(n-k) - c(n+1) + c(n) =$$

$$= -\left(\frac{k}{k+1}\right)^k \frac{1}{k+1}(1 + \omega(n))\left((n-k) + \delta(n-k)^p\right)\left(\frac{k}{k+1}\right)^{n-k}$$

$$-\left((n+1) + \delta(n+1)^p\right)\left(\frac{k}{k+1}\right)^{n+1} + (n + \delta n^p)\left(\frac{k}{k+1}\right)^n =$$

$$= \left(\frac{k}{k+1}\right)^n \left\{-\frac{1}{k+1}(1 + \omega(n))\left((n-k) + \delta(n-k)^p\right)\right.$$

$$\left. -\left((n+1) + \delta(n+1)^p\right)\left(\frac{k}{k+1}\right) + (n + \delta n^p)\right\} =$$

$$= \left(\frac{k}{k+1}\right)^n \left\{\left[-\frac{1}{k+1}(n-k) - (n+1)\frac{k}{k+1} + n\right]\right.$$

$$\left. -\frac{1}{k+1}\left(\delta(n+k)^p + \omega(n)\left((n-k) + \delta(n-k)^p\right)\right) - \frac{k}{k+1}\delta(n+1)^p + \delta n^p\right\} = (*).$$

The term in square brackets equals

$$\left[-\frac{1}{k+1}(n-k) - (n+1)\frac{k}{k+1} + n\right] = \frac{-(n-k) - k(n+1) + n(k+1)}{k+1} = 0,$$

so we have

$$(*) = \left(\frac{k}{k+1}\right)^n \left\{ -\frac{1}{k+1}\left(\delta(n-k)^p + \omega(n)\left((n-k)+\delta(n-k)^p\right)\right) \right.$$

$$\left. -\frac{k}{k+1}\delta(n+1)^p + \delta n^p \right\} =$$

$$n^p \left(\frac{k}{k+1}\right)^n \left\{ \left[-\frac{\delta}{k+1}\frac{(n-k)^p}{n^p} - \frac{\delta k}{k+1}\frac{(n+1)^p}{n^p} + \delta \right] \right.$$

$$\left. -\frac{\omega(n)}{n^p(k+1)}\left((n-k)+\delta(n-k)^p\right) \right\} =$$

$$n^p \left(\frac{k}{k+1}\right)^n \left\{ \left[-\frac{\delta}{k+1}\left(1-\frac{k}{n}\right)^p - \frac{\delta k}{k+1}\left(1+\frac{1}{n}\right)^p + \delta \right] \right.$$

$$\left. -\frac{\omega(n)}{n^p(k+1)}\left((n-k)+\delta(n-k)^p\right) \right\} =$$

$$n^p \left(\frac{k}{k+1}\right)^n \frac{1}{k+1} \left\{ \left[-\delta\left(1-\frac{k}{n}\right)^p - \delta k\left(1+\frac{1}{n}\right)^p + \delta(k+1) \right] \right.$$

$$\left. -\frac{\omega(n)}{(k+1)n^p}\left((n-k)+\delta(n-k)^p\right) \right\} = (**).$$

The last term we denote as Θ, i.e.

$$\Theta := \frac{-\omega(n)}{(k+1)n^p}\left((n-k)+\delta(n-k)^p\right).$$

For $n \to +\infty$

$$\Theta = O\left(\frac{1}{n^{2+p}}\right).$$

Now we use asymptotical decomposition (see Lemma 2.2), for $d = -k$ and $\sigma = p$:

$$\left(1-\frac{k}{n}\right)^p = 1 - \frac{pk}{n} + \frac{p(p-1)k^2}{2n^2} + O\left(\frac{1}{n^3}\right),$$

and for $d = 1$ and $\sigma = p$:

$$\left(1+\frac{1}{n}\right)^p = 1 + \frac{p}{n} + \frac{p(p-1)}{2n^2} + O\left(\frac{1}{n^3}\right).$$

After substitution we have

$$(**) = n^p \left(\frac{k}{k+1}\right)^n \frac{1}{k+1} \times$$

$$\times \left[-\delta\left(1 - \frac{pk}{n} + \frac{p(p-1)k^2}{2n^2} + O\left(\frac{1}{n^3}\right)\right) \right.$$

$$\left. -\delta k\left(1 + \frac{p}{n} + \frac{p(p-1)}{2n^2} + O\left(\frac{1}{n^3}\right)\right) + \delta(k+1) + \Theta \right].$$

The term in square brackets is

$$-\delta\left(1-\frac{pk}{n}+\frac{p(p-1)k^2}{2n^2}+O\left(\frac{1}{n^3}\right)\right)-\delta k\left(1+\frac{p}{n}+\frac{p(p-1)}{2n^2}+O\left(\frac{1}{n^3}\right)\right)$$
$$+\delta(k+1)+\Theta=$$
$$-\delta+\frac{\delta pk}{n}-\frac{\delta p(p-1)k^2}{2n^2}-\delta O\left(\frac{1}{n^3}\right)-\delta k-\frac{\delta kp}{n}-\frac{\delta kp(p-1)}{2n^2}-\delta kO\left(\frac{1}{n^3}\right)$$
$$+\delta(k+1)+\Theta=$$
$$-\frac{1}{2n^2}\left[\delta p(p-1)k^2+\delta p(p-1)k\right]+O\left(\frac{1}{n^3}\right)+O\left(\frac{1}{n^{2+p}}\right)=$$
$$\delta k\left[-\frac{p(p-1)}{2n^2}(k+1)\right]+O\left(\frac{1}{n^{2+p}}\right)=(***).$$

Because $p\in(0;1)$, then $(p-1)$ is negative and $(***)$ is positive. So H is positive too. Similarly we can prove that inequality (8) holds for sufficiently large n. □

Remark 3.2 *It is well-known (we refer, e.g., to [7]) that the equation with constant and positive coefficient*

$$\Delta x(n)=-px(n-k)$$

has a positive solution if

$$p\le\left(\frac{k}{k+1}\right)^k\frac{1}{k+1}$$

and every solution is oscillating for $n\to\infty$ *if*

$$p>\left(\frac{k}{k+1}\right)^k\frac{1}{k+1}.$$

Therefore, Theorem 3.1 gives sufficient conditions for the existence of a positive solution when $\lim_{n\to\infty}p(n)$ *exists and equals to a critical constant, i.e., when*

$$\lim_{n\to\infty}p(n)=\lim_{n\to\infty}\left(\frac{k}{k+1}\right)^k\frac{1}{k+1}[1+\omega(n)]=\left(\frac{k}{k+1}\right)^k\frac{1}{k+1}.$$

Acknowledgement

The work was supported by the project FEKT-S-14-2200 of Brno University of Technology.

References

[1] Baštinec, J., Diblík, J., Zhang, B.: *Existence of bounded solutions of discrete delayed equations,* Proceedings of the Sixth International Conference on Difference Equations, CRC, Boca Raton, FL, 359–366, 2004.

[2] Diblík, J.: *Asymptotic behavior of solutions of discrete equations,* Functional Differential Equations, **11** (2004), 37–48.

[3] Diblík, J., Hlavičková, I.: *Asymptotic upper and lower estimates of a class of positive solutions of a discrete linear equation with a single delay.* Abstract and Applied Analysis, 2012, ArticleID 764351, 1–12. ISSN: 1085– 3375.

[4] Diblík, J., Hlavičková, I.: *Asymptotic behavior of solutions of delayed difference equations.* Abstract and Applied Analysis, 2011, Article ID 67196, 1–24. ISSN: 1085–3375.

[5] Diblík, J., Hlavičková, I.: *Combination of Liapunov and retract methods in the investigation of the asymptotic behavior of solutions of systems of discrete equations.* Dynamic systems and applications, 2009, 18, 3– 4, 507-538. ISSN: 1056–2176.

[6] Diblík, J., Růžičková, M.: *Asymptotic behavior of solutions and positive solutions of differential delayed equations,* Functional Differential Equations **14**, 1, 2007, 83–105. (ISSN 0793–1786)

[7] Györi, I., Ladas, G.: *Oscillation Theory of Delay Differential Equations,* Clarendon Press (1991).

Current address

Baštinec Jaromír, doc. RNDr. CSc.
Department of Mathematics, Faculty of Electrical Engineering and Communication
Brno University of Technology
Technická 10, 616 00 Brno, Czech Republic
Tel. number: +420 541 146 081, e-mail: bastinec@feec.vutbr.cz

Diblík Josef, prof. RNDr. DrSc.
Department of Mathematics, Faculty of Electrical Engineering and Communication
Brno University of Technology
Technická 10, 616 00 Brno, Czech Republic
Tel. number: +420 541 146 064, e-mail: diblik@feec.vutbr.cz

Slovak University of Technology in Bratislava
Faculty of Mechanical Engineering
16th Conference on Applied Mathematics **APLIMAT 2017**
Proceedings

SOLUTION OF SPECIAL TYPE STOCHASTIC DIFFERENTIAL EQUATION

BAŠTINEC Jaromír (CZ), KLIMEŠOVÁ Marie (CZ)

Abstract. The natural world is influenced by randomness therefore stochastic models are used to test various situations because only the stochastic model can approximate the real model. This article studies the solution of stochastic differential linear equations and systems of stochastic differential equations with Brownian motion. There is used a Itô formula for proofs of main results. Results are illustrated by examples.

Keywords: Brownian motion, stochastic differential equation, Itô formula

Mathematics subject classification: 60G05

1 Introduction

Stochastic modeling has come to play an important role in many branches of science and industry where more and more people have encountered stochastic differential equations. Stochastic model can be used to solve problem which evinces by accident, noise, etc. Definition of probability spaces, stochastic process, stochastic differential equation and an existence and uniqueness of solution of these equations, were mentioned in [5], [6], [8].

We use definitions of the theory of the stochastic system defined by B. Øksendal [13], E. Kolářová [9], B. Maslowski [11], S. Ditlevsen [2], M. Navara [12], R. Z. Khasminskii [4], X.Mao [10] and J. Staněk [14] and others. In this paper we focus on the description of the solution stochastic equation and stochastic matrix system. We derived sufficient conditions for finding solution of the stochastic differential equation using Itô formula.

Definition 1 *[13] If Ω is a given set, then a σ-algebra $\mathcal{F}$ on Ω is a family $\mathcal{F}$ of subsets of Ω with the following properties:*

(i) $\emptyset \in \mathcal{F}$

(ii) $F \in \mathcal{F} \Rightarrow F^C \in \mathcal{F}$, *where* $F^C = \Omega \setminus \mathcal{F}$ *is the complement of* $\mathcal{F}$ *in* Ω

(iii) $A_1, A_2, \cdots \in \mathcal{F} \Rightarrow A := \bigcup_{i=1}^{\infty} A_i \in \mathcal{F}$.

The pair $(\Omega, \mathcal{F})$ *is called a measurable space.*

Definition 2 *[13] A probability measure* P *on a measurable space* $(\Omega, \mathcal{F})$ *is a function* $P : \mathcal{F} \longrightarrow [0, 1]$ *such that*

(a) $P(\emptyset) = 0, P(\Omega) = 1.$

(b) if $A_1, A_2, \cdots \in \mathcal{F}$ *and* $\{A_i\}_{i=1}^{\infty}$ *is disjoint (i.e.* $A_i \cap A_j = \emptyset$ *if* $i \neq j$*) then*

$$P\left(\bigcup_{i=1}^{\infty} A_i\right) = \sum_{i=1}^{\infty} P(A_i).$$

The triple $(\Omega, \mathcal{F}, P)$ *is called a probability space. It is called a complete probability space if* $\mathcal{F}$ *contains all subsets* G *of* Ω *with* P*-outer measure zero, i.e. with*

$$P^*(G) := \inf\{P(F); F \in \mathcal{F}, G \subset F\} = 0.$$

Definition 3 *The stochastic process* B_t *is called Brownian motion or Wiener process if the process has some basic properties:*

(i) $B_0 = 0$

(ii) $B_t - B_s$ *has the distribution* $N(0, t-s)$ *for* $t \geq s \geq 0$

(iii) B_t *has independent increments, i.e.*

$$B_{t_1},\ B_{t_2} - B_{t_1},\ \ldots,\ B_{t_k} - B_{t_{k-1}}$$

are independent for all $0 \leq t_1 < t_2 \cdots < t_k$.

Note. The unconditional probability density function at a fixed time t

$$f_{B_t}(X_t) = \frac{1}{\sqrt{2\pi t}} \exp\left(-\frac{X_t^2}{2t}\right).$$

The expectation is zero; $E[B_t] = 0$ for $t > 0$. The variance is t; $E[B_t^2] = t$.

Theorem 1 *Let* B_t *be Brownian motion. Then*

$$E[B_t B_s] = min\{t, s\} \;\; for\; t \geq 0,\; s \geq 0.$$

Proof: [9], pp. 14.

Definition 4 *Let $B_i(t), t = 1, 2, \ldots, m$, be a Brownian motion. Then $B(t) = (B_1(t), ..., B_m(t))$ denote m-dimensional Brownian motion.*

Definition 5 *[13] Let $(\Omega, \mathcal{F}, P)$ be a probability space. Let $B_t = (B_1(t), ..., B_m(t))$ be m-dimensional Brownian motion and $b : [0, T] \times R^n \to R^n$, $\sigma : [0, T] \times R^n \to R^{n \times m}$ be measurable functions. Then the process $X_t = (X_1(t), ..., X_m(t)), X_t \equiv X(t)$, $t \in [0, T]$ is the solution of the stochastic differential equation*

$$\mathrm{d}X_t = b(t, X_t)\mathrm{d}t + \sigma(t, X_t)\mathrm{d}B_t, \tag{1}$$

where $b(t, X_t) \in R \times R^n$, $\sigma(t, X_t)B_t \in R \times R^n$.
After the integration of equation (1) we give the different form of the solution of the SDE

$$X_t = X_0 + \int_0^t b(s, X_s)\mathrm{d}s + \int_0^t \sigma(s, X_s)\mathrm{d}B_s.$$

Assume that for every initial value $X(t_0) = X_0 \in R^n$ equation (1) has a unique global solution that is denoted by $X(t; t_0, X_0)$. We know that the solution has continuous sample paths and its every moment is finite.

2 Main results

2.1 The One-Dimensional Itô formula

Theorem 2 *[13] Let X_t be an Itô process given by*

$$\mathrm{d}X_t = u\mathrm{d}t + v\mathrm{d}B_t.$$

Let $g(t, x) \in C^2([0, \infty) \times R)$(i.e. f is twice continuosly differentiable on $[0, \infty) \times R$). Then

$$Y_t = g(t, X_t)$$

is again an Itô process and

$$\mathrm{d}Y_t = \frac{\partial g}{\partial t}(t, X_t)\mathrm{d}t + \frac{\partial g}{\partial x}(t, X_t)\mathrm{d}X_t + \frac{1}{2}\frac{\partial^2 g}{\partial x^2}(t, X_t)(\mathrm{d}X_t)^2,$$

where $(\mathrm{d}X_t)^2 = (\mathrm{d}X_t) \cdot (\mathrm{d}X_t)$ is computed according to the rules

$$\begin{aligned} \mathrm{d}t \cdot \mathrm{d}t &= \mathrm{d}t \cdot \mathrm{d}B_t = \mathrm{d}B_t \cdot \mathrm{d}t = 0, \\ \mathrm{d}B_t \cdot \mathrm{d}B_t &= \mathrm{d}t. \end{aligned}$$

Theorem 3 *Let X_t be an Itô process given by*

$$\mathrm{d}X_t = AX_t\mathrm{d}t + GX_t\mathrm{d}B_t. \tag{2}$$

Let $f(t,x) \in C^2([0,\infty) \times R)$(i.e. f is twice continuosly differentiable on $[0,\infty) \times R$). Then

$$Y_t = f(t, X_t)$$

is again an Itô process and

$$\mathrm{d}Y_t = \frac{\partial f}{\partial t}(t,X_t)\mathrm{d}t + \frac{\partial f}{\partial x}(t,X_t)(AX_t\mathrm{d}t + GX_t\mathrm{d}B_t) + \frac{1}{2}\frac{\partial^2 f}{\partial x^2}(t,X_t)(G^2X_t^2\mathrm{d}t), \tag{3}$$

where $(\mathrm{d}X_t)^2 = (\mathrm{d}X_t)\cdot(\mathrm{d}X_t)$ is computed according to the rules

$$\mathrm{d}t\cdot\mathrm{d}t = \mathrm{d}t\cdot\mathrm{d}B_t = \mathrm{d}B_t\cdot\mathrm{d}t = 0, \mathrm{d}B_t\cdot\mathrm{d}B_t = \mathrm{d}t. \tag{4}$$

Proof
First observe that if we substitute

$$\mathrm{d}X_t = AX_t\mathrm{d}t + GX_t\mathrm{d}B_t$$

in (3) and use (4) we get the equivalent expression

$$\begin{aligned} f(t,X_t) &= f(0,X_0) + \int_0^t \left(\frac{\partial f}{\partial s}(s,X_s) + AX_s\frac{\partial f}{\partial x}(s,X_s) + G^2X_s^2\frac{\partial^2 f}{\partial x^2}(s,X_s)\right)\mathrm{d}s \\ &+ \int_0^t GX_s\frac{\partial f}{\partial x}(s,X_s)\mathrm{d}B_s \end{aligned}$$

Assume that AX_t and GX_t are elementary functions. Using Taylor's theorem we get

$$\begin{aligned} f(t,X_t) &= f(0,X_0) + \sum_j \Delta f(t_j,X_j) = f(0,X_0) + \sum_j \frac{\partial f}{\partial t}\triangle t_j + \sum_j \frac{\partial f}{\partial x}\triangle X_j \\ &+ \frac{1}{2}\sum_j \frac{\partial^2 f}{\partial t^2}(\Delta t_j)^2 + \sum_j \frac{\partial^2 f}{\partial t\partial x}(\Delta t_j)(\Delta X_j) \\ &+ \frac{1}{2}\sum_j \frac{\partial^2 f}{\partial x^2}(\Delta X_j)^2 + \sum_j R_j, \end{aligned}$$

where $\frac{\partial f}{\partial t}, \frac{\partial f}{\partial x}$, etc. are evaluated at the points (t_j, X_j), $\Delta t_j = t_{j+1} - t_j$,
$\triangle X_j = X_{t_{j+1}} - X_{t_j}, \Delta f(t_j,X_j) = f(t_{j+1}, X_{t_{j+1}}) + f(t_j,X_j)$ and
$R_j = o(|\triangle t_j|^2 + |\triangle X_j|^2)$ for all j.
If $\Delta t_j \to 0$ then

$$\begin{aligned} \sum_j \frac{\partial f}{\partial t}\triangle t_j &= \sum_j \frac{\partial f}{\partial t}(t_j,X_j)\triangle t_j \to \int_0^t \frac{\partial f}{\partial s}(s,X_s)\mathrm{d}s, \\ \sum_j \frac{\partial f}{\partial x}\triangle X_j &= \sum_j \frac{\partial f}{\partial x}(t_j,X_j)\triangle X_j \to \int_0^t \frac{\partial f}{\partial x}(s,X_s)\mathrm{d}X_s. \end{aligned}$$

Moreover, since AX_t and GX_t are elementary we get

$$\begin{aligned}\sum_j \frac{\partial^2 f}{\partial x^2}(\triangle X_j)^2 &= \sum_j \frac{\partial^2 f}{\partial x^2}(A_j X_{t_j})^2(\triangle t_j)^2 + 2\sum_j \frac{\partial^2 f}{\partial x^2} A_j X_{t_j} G_j X_{t_j}(\triangle t_j)(\triangle B_j)\\ &+ \sum_j \frac{\partial^2 f}{\partial x^2}(G_j X_{t_j})^2(\triangle B_j)^2.\end{aligned}$$

The first two terms here tend to 0 as $\triangle t_j \to 0$. For example,

$$E\left[\left(\sum_j \frac{\partial^2 f}{\partial x^2} A_j X_{t_j} G_j X_{t_j}(\triangle t_j)(\triangle B_j)\right)^2\right] = \sum_j E\left[\left(\frac{\partial^2 f}{\partial x^2} A_j X_{t_j} G_j X_{t_j}\right)^2\right](\triangle t_j)^3 \to 0.$$

We claim that the last term tends to

$$\int_0^t G^2 X_s^2 \frac{\partial^2 f}{\partial x^2}(s, X_s)\mathrm{d}s \to 0.$$

That completes the proof of the formula (3).

Example 1 *Let us consider the following stochastic differential equation* (2) *with constant coefficients* A *and* G, $X_t \neq 0, X_0 = \eta, B_0 = 0$. *We can rewrite this equation as follows*

$$\frac{\mathrm{d}X_t}{X_t} = A\mathrm{d}t + G\mathrm{d}B_t$$

and integrate both sides over $[0, t]$ *and get the expression*

$$\int_0^t \frac{\mathrm{d}X_s}{X_s} = A\int_0^t \mathrm{d}s + G\int_0^t \mathrm{d}B_s.$$

First of all we evaluate the part $I = \int_0^t \mathrm{d}B_s$ *using Itô's formula* (3). *Choose* $X_t = B_t$ *and* $f(t, X_t) = X_t$. *Then*

$$Y_t = f(t, B_t) = B_t.$$

Then

$$\begin{aligned}\mathrm{d}Y_t &= 0 + 1\cdot \mathrm{d}B_t + 0\cdot \mathrm{d}t\\ \mathrm{d}B_t &= \mathrm{d}B_t,\end{aligned}$$

hence

$$\int_0^t \mathrm{d}B_s = B_t.$$

We get the expression

$$\int_0^t \frac{\mathrm{d}X_s}{X_s} = At + GB_t.$$

Now we solve the left-hand side.
Replacing $AX_t\mathrm{d}t + GX_t\mathrm{d}B_t$ for $\mathrm{d}X_t$ in the equation (3) we give

$$\begin{aligned}\mathrm{d}Y_t &= \frac{\partial f}{\partial t}(t,X_t)\mathrm{d}t + \frac{\partial f}{\partial x}(t,X_t)(AX_t\mathrm{d}t + GX_t\mathrm{d}B_t) + \frac{1}{2}\frac{\partial^2 f}{\partial x^2}(t,X_t)(AX_t\mathrm{d}t + GX_t\mathrm{d}B_t)^2\\ &= \frac{\partial f}{\partial t}(t,X_t)\mathrm{d}t + \frac{\partial f}{\partial x}(t,X_t)(AX_t\mathrm{d}t) + \frac{\partial f}{\partial x}(t,X_t)(GX_t\mathrm{d}B_t)\\ &+ \frac{1}{2}\frac{\partial^2 f}{\partial x^2}(t,X_t)(G^2X_t^2\mathrm{d}t).\end{aligned}$$

We use the Itô's formula of the function $f(t,X_t) = \ln X_t$. Then

$$Y_t = f(t,X_t) = \ln X_t.$$

$$\begin{aligned}\mathrm{d}Y_t &= \left(0 + \frac{1}{X_t}AX_t + \frac{1}{2}\cdot\frac{-1}{X_t^2}G^2X_t^2\right)\mathrm{d}t + \frac{1}{X_t}GX_t\mathrm{d}B_t\\ &= \frac{-1}{2}G^2\mathrm{d}t + A\mathrm{d}t + G\mathrm{d}B_t.\end{aligned}$$

We get

$$\mathrm{d}(\ln X_t) = \frac{-1}{2}G^2\mathrm{d}t + \frac{\mathrm{d}X_t}{X_t} \Rightarrow \frac{\mathrm{d}X_t}{X_t} = \mathrm{d}(\ln X_t) + \frac{1}{2}G^2\mathrm{d}t,$$

hence

$$\begin{aligned}\int_0^t \mathrm{d}(\ln X_s) + \int_0^t \frac{1}{2}G^2\mathrm{d}s &= At + GB_t,\\ \ln X_t - \ln X_0 &= At + GB_t - \frac{1}{2}G^2t,\\ X_t &= X_0e^{At+GB_t-\frac{1}{2}G^2t}.\end{aligned}$$

For $X_0 = \eta$ we obtain $X_t = \eta e^{At+GB_t-\frac{1}{2}G^2t}$.

Example 2 *Let X_t be an Itô process given by*

$$\mathrm{d}X_t = AX_t\mathrm{d}t + G\mathrm{d}B_t$$

with constant coefficients A and $G, X_t \neq 0, X_0 = \eta, B_0 = 0$. We can rewrite the stochastic equation as follows

$$\frac{\mathrm{d}X_t}{X_t} = A\mathrm{d}t + \frac{G}{X_t}\mathrm{d}B_t$$

and integrate both sides over $[0,t]$

$$\int_0^t \frac{\mathrm{d}X_s}{X_s} = A\int_0^t \mathrm{d}s + G\int_0^t \frac{1}{X_s}\mathrm{d}B_s$$

and get the expression

$$\int_0^t \frac{\mathrm{d}X_s}{X_s} = At + G\int_0^t \frac{1}{X_t}\mathrm{d}B_s.$$

Let's solve the left-hand side. Replacing $AX_t\mathrm{d}t + G\mathrm{d}B_t$ for $\mathrm{d}X_t$ in the Itô's formula (3) we give

$$\begin{aligned}\mathrm{d}Y_t &= \frac{\partial f}{\partial t}(t, X_t)\mathrm{d}t + \frac{\partial f}{\partial x}(t, X_t)(AX_t\mathrm{d}t + G\mathrm{d}B_t) + \frac{1}{2}\frac{\partial^2 f}{\partial x^2}(t, X_t)(AX_t\mathrm{d}t + G\mathrm{d}B_t)^2 \\ &= \frac{\partial f}{\partial t}(t, X_t)\mathrm{d}t + \frac{\partial f}{\partial x}(t, X_t)(AX_t\mathrm{d}t) + \frac{\partial f}{\partial x}(t, X_t)(G\mathrm{d}B_t) + \frac{1}{2}\frac{\partial^2 f}{\partial x^2}(t, X_t)(G^2\mathrm{d}t).\end{aligned}$$

We use the Itô's formula of the function $f(t, X_t) = \ln X_t$. Then

$$Y_t = f(t, X_t) = \ln X_t.$$

$$\mathrm{d}Y_t = 0 + \frac{1}{X_t}AX_t\mathrm{d}t + \frac{1}{2}\cdot\frac{-1}{X_t^2}G^2\mathrm{d}t + \frac{1}{X_t}G\mathrm{d}B_t$$

We get

$$\mathrm{d}(\ln X_t) = -\frac{G^2}{2X_t^2}\mathrm{d}t + \frac{\mathrm{d}X_t}{X_t} \Rightarrow \frac{\mathrm{d}X_t}{X_t} = \mathrm{d}(\ln X_t) + \frac{G^2}{2X_t^2}\mathrm{d}t,$$

hence

$$\begin{aligned}\int_0^t \mathrm{d}(\ln X_s) + \int_0^t \frac{G^2}{2X_s^2}\mathrm{d}s &= At + G\int_0^t \frac{1}{X_s}\mathrm{d}B_s, \\ \ln X_t - \ln X_0 &= At + G\int_0^t \frac{1}{X_s}\mathrm{d}B_s - \frac{G^2}{2}\int_0^t \frac{1}{X_s^2}\mathrm{d}s\end{aligned}$$

Now we solve the part $\int_0^t \frac{1}{X_s}\mathrm{d}B_s$ replacing $X_t \equiv B_t$

$$\int_0^t \frac{1}{X_s}\mathrm{d}X_s = \ln X_t - \ln X_0.$$

We get the expression

$$\begin{aligned}\ln X_t - \ln X_0 &= At + G(\ln X_t - \ln X_0) - \frac{G^2}{2}\int_0^t \frac{1}{X_s^2}\mathrm{d}s \\ (\ln X_t - \ln X_0)(1 - G) &= At - \frac{G^2}{2}\int_0^t \frac{1}{X_s^2}\mathrm{d}s\end{aligned}$$

Finally result is given

$$\begin{aligned}\ln X_t - \ln X_0 &= \frac{At}{1-G} - \frac{G^2}{2(1-G)}\int_0^t \frac{1}{X_s^2}\mathrm{d}s \\ X_t &= X_0 e^{\frac{At}{1-G} - \frac{G^2}{2(1-G)}\int_0^t \frac{1}{X_s^2}\mathrm{d}s}\end{aligned}$$

For $X_0 = \eta$ we obtain $X_t = \eta e^{\frac{At}{1-G} - \frac{G^2}{2(1-G)}\int_0^t \frac{1}{X_s^2}\mathrm{d}s}$.

2.2 The Multi-dimensional Itô formula

There is a very useful analogue of Itô formula (3) in many dimensions. Let $B_t = (B_1(t), \ldots, B_m(t))$ denote m-dimensional Brownian motion.

Theorem 4 *[13] Let*

$$\mathrm{d}X_t = u\mathrm{d}t + v\mathrm{d}B_t$$

be an n-dimensional Itô process, where

$$X_t = \begin{pmatrix} X_1(t) \\ \vdots \\ X_n(t) \end{pmatrix}, u = \begin{pmatrix} u_1 \\ \vdots \\ u_n \end{pmatrix}, v = \begin{pmatrix} v_{11} & \cdots & v_{1m} \\ \vdots & & \vdots \\ v_{n1} & \cdots & v_{nm} \end{pmatrix}, B_t = \begin{pmatrix} B_1(t) \\ \vdots \\ B_m(t) \end{pmatrix}.$$

Let $g(t,x) = (g_1(t,x), ..., g_p(t,x))$ be a twice continuously differentiable function from R^n into R^p. Then the process

$$Y(t) = g(t, X(t))$$

is again an Itô process, whose component number k, Y_k, is given by

$$\mathrm{d}Y_k = \frac{\partial g_k}{\partial t}(t, X_t)\mathrm{d}t + \sum_i \frac{\partial g_k}{\partial x_i}(t, X_t)\mathrm{d}X_{i,t} + \frac{1}{2}\sum_{i,j} \frac{\partial^2 g_k}{\partial x_i \partial x_j}(t, X_t)\mathrm{d}X_{i,t}\mathrm{d}X_{j,t},$$

where $\mathrm{d}X_{i,t}\mathrm{d}X_{j,t}$ is computed according to the rules

$$\begin{aligned} (\mathrm{d}B_i)^2 &= \mathrm{d}t, \\ \mathrm{d}B_i\mathrm{d}t &= \mathrm{d}t\mathrm{d}B_i = \mathrm{d}t\mathrm{d}t = \mathrm{d}B_i\mathrm{d}B_j = 0. \end{aligned}$$

Theorem 5 *Let*

$$\mathrm{d}X_t = AX_t\mathrm{d}t + G\mathrm{d}B_t \tag{5}$$

be an n-dimensional Itô process, where

$$X_t = \begin{pmatrix} X_1(t) \\ \vdots \\ X_n(t) \end{pmatrix}, A = \begin{pmatrix} a_{11} & \cdots & a_{1n} \\ \vdots & & \vdots \\ a_{m1} & \cdots & a_{mn} \end{pmatrix}, G = \begin{pmatrix} g_{11} & \cdots & g_{1m} \\ \vdots & & \vdots \\ g_{n1} & \cdots & g_{nm} \end{pmatrix}, B_t = \begin{pmatrix} B_1(t) \\ \vdots \\ B_m(t) \end{pmatrix}.$$

Let $f(t,x) = (f_1(t,x), ..., f_p(t,x))$ be a twice continuously differentiable function from R^n into R^p. Then the process

$$Y(t) = f(t, X(t))$$

is again an Itô process, whose component number k, Y_k, is given by

$$\mathrm{d}Y_k = \frac{\partial f_k}{\partial t}(t, X_t)\mathrm{d}t + \sum_i \frac{\partial f_k}{\partial x_i}(t, X_t)(AX_{i,t}\mathrm{d}t + G\mathrm{d}B_{i,t}) \tag{6}$$

$$+\frac{1}{2}\sum_{i,j} \frac{\partial^2 f_k}{\partial x_i \partial x_j}(t, X_t)(G^2\mathrm{d}B_{i,t}^2),$$

where $\mathrm{d}X_{i,t}\mathrm{d}X_{j,t}$ is computed according to the rules

$$(\mathrm{d}B_i)^2 = \mathrm{d}t, \mathrm{d}B_i\mathrm{d}t = \mathrm{d}t\mathrm{d}B_i = \mathrm{d}t\mathrm{d}t = \mathrm{d}B_i\mathrm{d}B_j = 0. \tag{7}$$

Proof

The proof is similar to the 1-dimensional version. First observe that if we substitute

$$\mathrm{d}X_t = AX_t\mathrm{d}t + G\mathrm{d}B_t$$

in (6) and use (7) we get the equivalent expression

$$\begin{aligned} f(t,X_t) &= f(0,X_0) + \int_0^t \left(\frac{\partial f}{\partial s}(s,X_s) + AX_s\frac{\partial f}{\partial x}(s,X_s) + G^2\frac{\partial^2 f}{\partial x^2}(s,X_s) \right) \mathrm{d}s \\ &+ \int_0^t G\frac{\partial f}{\partial x}(s,X_s)\mathrm{d}B_s \end{aligned}$$

Assume that AX_t and G are elementary functions. Using Taylor's theorem we get

$$\begin{aligned} f(t,X_t) &= f(0,X_0) + \sum_j \Delta f(t_j,X_j) = f(0,X_0) + \sum_j \frac{\partial f}{\partial t}\triangle t_j + \sum_j \frac{\partial f}{\partial x}\triangle X_j \\ &+ \frac{1}{2}\sum_j \frac{\partial^2 f}{\partial t^2}(\Delta t_j)^2 + \sum_j \frac{\partial^2 f}{\partial t \partial x}(\Delta t_j)(\Delta X_j) \\ &+ \frac{1}{2}\sum_j \frac{\partial^2 f}{\partial x^2}(\Delta X_j)^2 + \sum_j R_j, \end{aligned}$$

where $\frac{\partial f}{\partial t}, \frac{\partial f}{\partial x}$, etc. are evaluated at the points (t_j, X_j), $\Delta t_j = t_{j+1} - t_j$,
$\triangle X_j = X_{t_{j+1}} - X_{t_j}$, $\Delta f(t_j, X_j) = f(t_{j+1}, X_{t_{j+1}}) + f(t_j, X_j)$ and
$R_j = o(|\triangle t_j|^2 + |\triangle X_j|^2)$ for all j.
If $\Delta t_j \to 0$ then

$$\begin{aligned} \sum_j \frac{\partial f}{\partial t}\triangle t_j &= \sum_j \frac{\partial f}{\partial t}(t_j,X_j)\triangle t_j \to \int_0^t \frac{\partial f}{\partial s}(s,X_s)\mathrm{d}s, \\ \sum_j \frac{\partial f}{\partial x}\triangle X_j &= \sum_j \frac{\partial f}{\partial x}(t_j,X_j)\triangle X_j \to \int_0^t \frac{\partial f}{\partial x}(s,X_s)\mathrm{d}X_s. \end{aligned}$$

Moreover, since AX_t and G are elementary we get

$$\begin{aligned} \sum_j \frac{\partial^2 f}{\partial x^2}(\Delta X_j)^2 &= \sum_j \frac{\partial^2 f}{\partial x^2}(A_jX_{t_j})^2(\Delta t_j)^2 + 2\sum_j \frac{\partial^2 f}{\partial x^2}A_jX_{t_j}G_j(\Delta t_j)(\Delta B_j) \\ &+ \sum_j \frac{\partial^2 f}{\partial x^2}(G_j)^2(\Delta B_j)^2. \end{aligned}$$

The first two terms here tend to 0 as $\Delta t_j \to 0$. For example,

$$E\left[\left(\sum_j \frac{\partial^2 f}{\partial x^2}A_jX_{t_j}G_j(\Delta t_j)(\Delta B_j)\right)^2\right] = \sum_j E\left[\left(\frac{\partial^2 f}{\partial x^2}A_jX_{t_j}G_j\right)^2\right](\Delta t_j)^3 \to 0.$$

We claim that the last term tends to

$$\int_0^t G^2\frac{\partial^2 f}{\partial x^2}(s,X_s)\mathrm{d}s \to 0.$$

That completes the proof of the formula (6).

Example 3 *Suppose the stochastic system* (5) *with* $X_t \neq 0, X_0 = \eta, \eta$ *is a constant vector,* $B_0 = 0$,

$$\mathrm{d}X_t = AX_t\mathrm{d}t + G\mathrm{d}B_t.$$

First we compute the deterministic part

$$\begin{aligned} \mathrm{d}X_t &= AX_t\mathrm{d}t, \\ \frac{\mathrm{d}X_t}{X_t} &= A\mathrm{d}t, \\ \int_0^t \frac{1}{X_s}\mathrm{d}X_s &= A\int_0^t \mathrm{d}s, \\ \ln X_t - \ln X_0 &= At, \\ X_t &= e^{At}\eta. \end{aligned}$$

Suppose that $\eta = \phi(t), \phi(t)$ *is a function*

$$\begin{aligned} X_t &= e^{At}\phi(t) \\ \mathrm{d}X_t &= e^{At}\mathrm{d}\phi(t) + e^{At}A\phi(t) \\ e^{At}\mathrm{d}\phi(t) + e^{At}A\phi(t) &= Ae^{At}\phi(t) + G\mathrm{d}B_t \\ e^{At}\mathrm{d}\phi(t) &= G\mathrm{d}B_t \\ G^{-1}e^{At}\mathrm{d}\phi(t) &= \mathrm{d}B_t \end{aligned}$$

At this moment let's solve the right-hand side using Itô formula (6). Choose $X_t = B_t$ *and* $f(t, X_t) = X_t$. *Then*

$$Y_t = f(t, B_t) = B_t.$$

Then by Itô's formula,

$$\begin{aligned} \mathrm{d}Y_t &= 0 + 1 \cdot \mathrm{d}B_t + 0 \cdot \mathrm{d}t \\ \mathrm{d}B_t &= \mathrm{d}B_t = Bt \end{aligned}$$

hence

$$\begin{aligned} G^{-1}e^{At}\mathrm{d}\phi(t) &= \mathrm{d}B_t, \\ G^{-1}e^{At}\int \mathrm{d}\phi(t) &= \int \mathrm{d}B_t, \\ G^{-1}e^{At}\phi(t) &= B_t, \\ \phi(t) &= Ge^{-At}B_t. \end{aligned}$$

Solution of the stochastic system (5) *is*

$$X_t = e^{At}Ge^{-At}B_t.$$

3 Conclusion

There was defined the stochastic differential equation. Conditions for finding solution were established on the model of the stochastic differential equation and system by using Itô theorem. Results were illustrated on trivial examples. Such type of equations can be used also in biomedical engineering, in meteorology, epidemic modeling, predicting economics, etc.

Acknowledgement

The work was supported by the project FEKT-S-14-2200 of Brno University of Technology.

References

[1] BAŠTINEC, J.; KLIMEŠOVÁ, M. *Stability of the Zero Solution of Stochastic Differential System with Three- dimensional Brownian motion.* In Matematika, Informační technologie a aplikované vědy. Brno: UNOB, 2016. s. 1-8. ISBN: 978-80-7231-464- 5.

[2] DITLEVSEN, S., BATZEL, J., BACHAR, M.: *Stochastic biomathematical models*, Heidelberg: Springer, 2013, 206 p.

[3] GILBERT, G. T. *Positive Definite Matrices and Sylvester's Criterion.* The American Mathematical Monthly 98.1 (1991): 44-46. DOI:10.2307/2324036.

[4] KHASMINSKII, R.: *Stochastic stability of differential equations.* New York: Springer Berlin Heidelberg, 2011, 358 s. ISBN 978-3-642-23279-4.

[5] KLIMEŠOVÁ, M.: *Stochastic Differential Equations*, Student EEICT. Brno: LITERA, 2014. s. 150-154. ISBN: 978-80-214-4924-4.

[6] KLIMEŠOVÁ, M.: *Stability of the Stochastic Differential Equations*, Student EEICT. Brno: BUT, 2015. s. 526-530. ISBN: 978-80-214-214-5148-3.

[7] KLIMEŠOVÁ, M.; BAŠTINEC, J. *Application of Stochastic Differential Equations.* MITAV. Brno: UO, 2014. s. 1-6. ISBN: 978-80-7231-961-9.

[8] KLIMEŠOVÁ, M.; BAŠTINEC, J. *Stability of the Zero Solution of Stochastic Differential Systems with Two-dimensional Brownian motion.* Mathematics, Information Technologies, and Applied Science 2015. Brno: UNOB, 2015. s. 8-20. ISBN: 978-80-7231-436- 2.

[9] KOLÁŘOVÁ, E.: *Stochastické diferenciální rovnice v elektrotechnice.* Thesis. Brno: BUT, 2006, 26 s. ISBN 80-214-3330-2.

[10] MAO, X.: *Stochastic differential equations and applications.* Chichester: Horwood Pub., 2008, 422 s. ISBN 978-1-904275-34-3.

[11] MASLOWSKI, B., MILOTA, J.: *Proceedings of Seminar in Differential Equations.* Plzeň: Vydavatelský servis, 2006, 118 s. ISBN 80-86843-14-9.

[12] NAVARA, M.: *Pravděpodobnost a matematická statistika.* Skriptum. Praha: FEL ČBUT, 2007, 240 s. ISBN 978-80-01-03795-9.

[13] ØKSENDAL, B.: Stochastic Differential Equations. *An Introduction with Applications*, Springer-Verlag, 1995.

[14] STANĚK, J.: *Stochastické diferenciální rovnice*, KDM - MFF UK, 2011.

[15] STEPANOV, V. V.: *Course of differential equations (review)*, Uspekhi Mat. Nauk, 1939, no. 6, 288-289.

Current address

Baštinec Jaromír, doc. RNDr. CSc.
Department of Mathematics, Faculty of Electrical Engineering and Communication
Brno University of Technology
Technická 2848/8, Žabovřesky, 61600, Brno, Czech republic
Tel. number: +4205 4114 6081, e-mail: bastinec@feec.vutbr.cz

Klimešová Marie, Ing.
Department of Mathematics, Faculty of Electrical Engineering and Communication
Brno University of Technology
Technická 2848/8, Žabovřesky, 61600, Brno, Czech republic
Tel. number: +4205 4114 6097, e-mail: xklime01@stud.feec.vutbr.cz

Slovak University of Technology in Bratislava
Faculty of Mechanical Engineering
16th Conference on Applied Mathematics **APLIMAT 2017**
Proceedings

THE IMPACT OF IBOOKS ON GEOMETRIC CONCEPTIONS OF STUDENTS ABOUT ISOMETRIES

BAYERL Erik (SK), ŽILKOVÁ Katarína (SK)

Abstract. The aim of this paper is to describe the experience of the implementation of iBook textbooks into geometric education, to analyze the data obtained and to summarize the impact of the use of iBook textbooks on learning outcomes. According to our findings, we can conclude that the learning outcomes through iBook textbooks are not only comparable with the traditional forms of education, but in some aspects, have showed a positive impact, particularly when solving the tasks aimed at higher cognitive processes.

Keywords: interactive textbooks, electronic textbooks, iBooks, mobile learning, math education

Mathematics Subject Classification: 97U20, 97G50, 97U50

1 Introduction

Currently, there is no doubt about the importance of mobile learning and education through new technology platforms. There are a number of reasons and arguments why it is necessary to reflect on developments in the field of technology and to implement them into education. Warschauer (2006) states that „study of mobile learning and literacy suggests that mobile devices have contributed to students gaining broad skills, knowledge and abilities that support learning and literacy development“. In order to create really high-quality and useful electronic sources, it is necessary to continually analyze their impact on learning outcomes. According to Drigas and Pappas (2016) „online and mobile learning applications motivated the students, making mathematics course more enjoyable and interactive than the ordinary teaching practices". By Skillen (2015), the introduction of mobile learning and subsequent technology innovations into schools allows for a significant change and revision of educational strategies. In order to support and further develop mobile learning, a creation of applications for mobile phones and tablets, which are aimed at solving mathematical tasks from geometry, or algebra, has been taking place in Slovakia (Voštinár and Hanzel, 2016; Pokorný, 2013; Híc and Pokorný, 2013). Recently, there has been a growing interest abroad, to the use of tablets, specifically the iPads, as a tool to support learning and teaching, especially in mathematics, because the iPads create an environment, that facilitates the learning process of pupils and provides interactive learning with immediate feedback (Banister, 2010; Enriquez, 2010).

A similar view is held by Bannon, Martin and Nunes-Bufford (2012), who also see a high value in the integration of iPads in education process of mathematics, as a tool to support education of pupils. The team of authors, Broda, Schmidt and Wereley (2011), have investigated the meaningful strategies of the use of iPads in the area of educating teachers, to be able to better integrate Ipads into educational process. This is a new area in which it is necessary to carry out research and identify the potential of mobile phone devices in the teaching of mathematics and its various themes. This is one of the many reasons why we have decided to verify the effectiveness of iBooks textbooks through iPads in the process of teaching mathematics to upper secondary students. At the same time, we examine whether the use of iBooks in mathematical education can have an impact on the quality of the knowledge of students, respectively to develop their geometric thinking. We draw on theoretical knowledge of van Hiele theory of levels of geometric reasoning.

2 Design of research

Before carrying out the research, it was necessary to prepare an iBook textbook on isometrics of the plane. The textbook was created by the iBook Author, which allowed us to include some interactive elements, including dynamic GeoGebra applets. The content of the textbook is the basic theoretical knowledge of the Euclidean plane isometry and its properties, but the core of the textbook consists of a collection of solved geometric tasks, along with the methodology of their solutions.

Besides the common static elements (e.g. texts and images), there is an interactive and dynamic applet implemented into each task that allows students to change positional and metrical properties of the geometric elements. Exactly this attribute was the most essential for us and our research, because we wanted to find whether the active use of interactive geometric constructions may have an impact on cognitive process of upper secondary students.

The research problem can be characterized by means of a causal question: What is the effectiveness of using iBook textbooks in the topic of central and axial symmetry on students′ knowledge?

The research method was an experiment, at which the dependant variable was the use of the interactive textbook and the independent variable was our expectations of better learning outcomes of students about isometries. We assume that if we use iBook textbooks for teaching isometries, these students will gain better results than the students educated by traditional methods (without the use of any interactive and dynamic elements). For the implementation of the experiment, it was necessary to select a control and experimental group. The selection was not random, but it was subjected to the availability and technical capabilities of the school. The school, where the experiment was carried out, was the school, where the author of the textbook works as a teacher, and so he was able to carry out the research personally. The experiment was conducted in a standard manner. After selecting the relevant and comparable groups, we carried out the experiment in one of the groups, while the teaching process in the control group was carried out by using standard educational strategies, through using common drawing aids. The students of the experimental group used interactive boards and devices like iPads during their teaching process, and worked mostly with the interactive textbook, that had been created by us.

After the experimental period that lasted for four weeks, the both groups were again subjected to testing from the area.

The post-test was administrated in a form of a written test, during which, students could use common drawing aids. The test contained 4 tasks aimed at isometries of the plain.

2.1 The sample characteristics and comparability of groups

The research groups (sample groups) were created by the students of the second grade classes of the selected grammar school in Banská Bystrica (Slovakia), where we maintained compactness. In order to gain the most objective results, we decided, before the research, to test all the students on their general knowledge of geometry. This was carried out in four classes. Based on the results of the test, we chose two classes, which were comparable, and could create a control and an experimental group. The control group consisted of 28 students (20 girls and 8 boys) and the experimental group consisted of 29 students (20 girls and 9 boys).

To determine the most suitable statistical method it was needed to assess the data obtained from the pre-test (sample group), whether they correspond to a normal distribution. The alpha level was set on 0.05. As the first, we used Kolmogorov-Smirnov normality test, then Shapiro-Wilk normality test. The results of the both tests in the control group are shown in Tab. 1, wherein can be seen that the p-value (columns Sig.) is higher than the significance level (alpha = 0.05). Therefore, with the control group we do not reject the hypothesis about a normal distribution of data, but we state that the distribution of the data in the control group is normal.

Control group	Kolmogorov-Smirnov			Shapiro-Wilk		
	Statistics	Df	Sig.	Statistics	df	Sig.
	0,080	26	0,200*	0,990	26	0,994

Tab. 1. Normality Tests – the control group

However, the analysis of the results of the experimental group showed that p-values with both tests are quite low (Tab. 2). Kolmogorov-Smirnov test had a p-value at 0,086, which still meets the criteria of a normal distribution, but Shapiro-Wilk test showed a p-value at 0,049, which is under the significance level, therefore, we reject a normal distribution of data.

Experimental group	Kolmogorov-Smirnov			Shapiro-Wilk		
	Statistics	Df	Sig.	Statistics	df	Sig.
	0,157	27	0,086	0,924	27	0,049

Tab. 2. Normality Tests – the experimental group

Since the normal distribution with both groups was not confirmed, we decided to use a nonparametric version of the 2-sample Mann-Whitney test (Tab. 3. and Tab. 4.). We formulated statistical null and alternative hypothesis and examined them at the significance level alpha of 0.05.

- H_0: There are no differences between the pre-test results obtained in the both groups.
- H_A: There are differences between the pre-test results obtained in the both groups.

Groups	No.	Mean Rank	Sum of Ranks
Control	26	25,96	675,00
Experimental	27	28,00	756,00
Total	53		

Tab. 3. Ranks – a comparation of the groups

Mann-Whitney U	324,000
Z	-0,481
Asymp. Sig. (2-tailed)	0,630

Tab. 4. Test Statistics – a comparation of the groups

Since the test statistics Z has the asymptote of the normal distribution, so we compare the calculated value Z = - 0.481 with the critical value of the normal N (0,1) distribution. The value Z is in the range of (-1.96, 1.96), and thus we do not reject the null hypothesis. When comparing the p-values at the level of 0.630, we see that it is higher than the level of the importance, thus we do not reject the null hypothesis. We note that between the control and experimental groups were no differences, we consider them as equal, and therefore we were able to carry out our experimental operation.

2.2 Analysis and interpretation of the results

After the implementation of the experiment, the post-test was administered in the both experimental and control groups. The post-test tasks reflected on lower as well as higher cognitive levels.

The analysis and the interpretation of the results shall be reported for each task, and we (as with verifying the equivalence of groups) used the Mann-Whitney U test. We formulated null and alternative hypothesis:

- H_0: There are no differences in post-tests results achieved in the both experimental and control groups.
- H_A: There are differences in post-tests results achieved in the both experimental and control groups.

We realized the verification at the significance level of 0.05.

Task 1. Display the given broken line in the square grid in the centre of symmetry with the centre S, and then in axial symmetry to the axis o. Draw directly to the square grid.

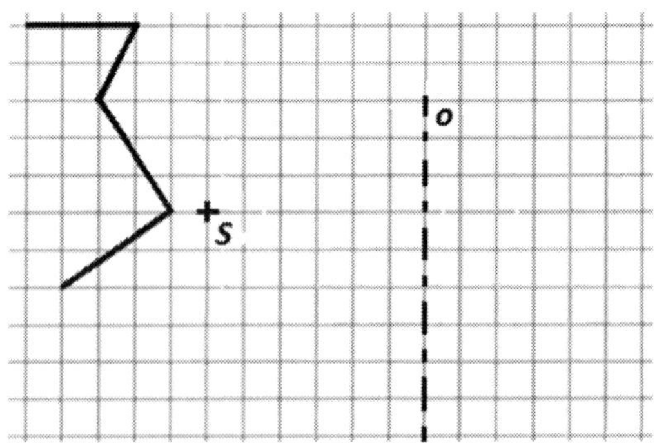

Fig. 1. Task 1 – post-test

According to the revised Bloom´s taxonomy, in terms of the knowledge dimension on the level of procedural knowledge and in terms of dimension of cognitive processes, we classified the task as application task. The student must know the process of the construction of the image points, if he knows the centre of the symmetry or the axis of the symmetry. The solution of the task is simplified by the square grid. The results of the two groups are shown in Tab. 5 and Tab. 6.

Groups	No.	Mean Rank	Sum of Ranks
Control	28	22,41	627,50
Experimental	29	35,36	1025,50
Total	57		

Tab. 5. Ranks – Task 1

Mann-Whitney U	221,500
Z	-3,525
Asymp. Sig. (2-tailed)	0,000

Tab. 6. Test Statistics – Task 1

From the above data (Tab. 5 and Tab. 6) it is evident that the test statistic Z = - 3.525 does not fall within the interval (-1.96, 1.96), and therefore we reject the null hypothesis. The rejection of the null hypothesis is confirmed by the p-value of 0.00, which is lower than the defined level of the significance. We state that there is a statistically significant difference between the results of the both groups. At the same time, we can confirm, that the experimental group has a higher rating respectively (Table. 5, column sum of ranks and mean rank), thus reaching better results in this task than the control group.

Task 2. Determine whether the shape in the picture is centrally or axially symmetrical. Circle the correct answer.

Shape		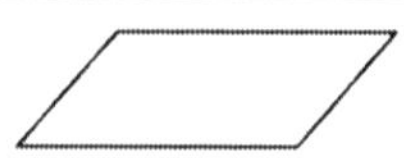	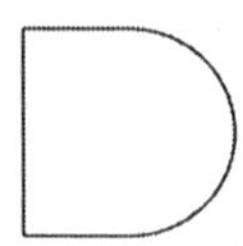	
Axially symmetrical	Yes – No	Yes – No	Yes – No	Yes – No
Centrally symmetrical	Yes – No	Yes – No	Yes – No	Yes – No

Tab. 7. Task 2 post-test

This task is from the cognitive view of a different nature than the previous task. Its solution requires conceptual knowledge and the ability to analyze. The results of the solution of this task are listed in Table 8 and Table 9.

Groups	No.	Mean Rank	Sum of Ranks
Control	28	30,05	841,50
Experimental	29	27,98	811,50
Total	57		

Tab. 8. Ranks – Task 2

Mann-Whitney U	376,500
Z	-0,494
Asymp. Sig. (2-tailed)	0,621

Tab. 9. Test Statistics – Task 2

From the data obtained it appears that the test statistics Z = - 0.494, and it is in the range of (-1.96, 1.96). Therefore, we do not reject the null hypothesis and conclude that there is not a statistically significant difference between the results of the control and the experimental groups. This statement is confirmed by the p-value (0.621) that is higher than the level of significance.

Task 3. In the square ABCD there is a point S, which is the center of side BC. Circle, which of the lines AB, BC, CD, AD, AC, BD are invariant in the central symmetry with the center S.

To solve this problem the student needs to master the fact, what an invariant straight line is, but at the same time, he/she needs to evaluate whether these lines correspond to the required criteria. Therefore, we classify this task on the level of conceptual knowledge and analysis. In terms of difficulty, we consider this task to be comparable to the Task 2 and we also expected analogous results.

Groups	No.	Mean Rank	Sum of Ranks
Control	28	26,89	753,00
Experimental	29	31,03	900,00
Total	57		

Tab. 10. Ranks – Task 3

Mann-Whitney U	347,000
Z	-1,252
Asymp. Sig. (2-tailed)	0,211

Tab. 11. Test Statistics – Task 3

From the data, which were processed on the results of the third post-test task (Table 10 and Table 11), it is evident that the test statistic Z = - 1.252 is within the interval of (-1.96; 1.96), and also the p-value (0.221) exceeds the level of significance of 0.05. Therefore we do not reject the null hypothesis and assume that there was no significant difference of the results of the third task between the control and the experimental groups. This will also confirm our assumption that from the cognitive point of view were tasks 2 and 3 comparable. Although no statistical significance was confirmed, nevertheless, slightly better results were demonstrated in favor of the experimental group (Tab. 10).

Task 4a. Given are two different circles k_1, k_2 and a straight line m. Construct all equilateral triangles ABC, which height v_c is a subset of a line p, point A is on the circle k1 and point B lies on the circle k_2.

Task 4b. Have a discussion with respect to the number of the solutions of the task. How many solutions should be relevant if the centres of the circles k1 and k_2 lie on a straight line p and had the same radius?

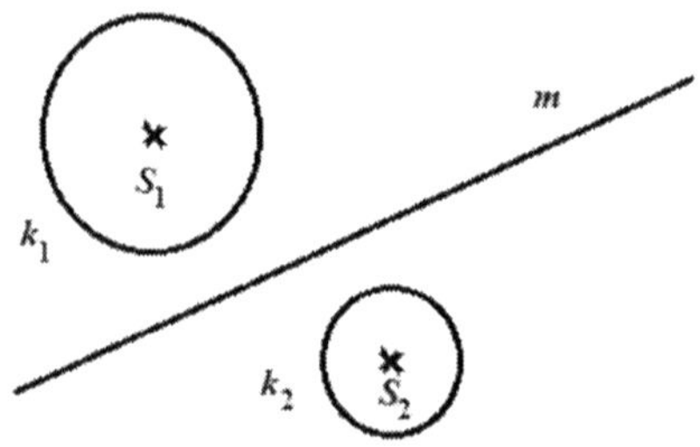

Fig. 2. Task 4 – post-test

Task 4 was divided into two parts:

- Part A) focused on solving a construction problem with all its components (outline, analysis, construction process, construction and the answer on the number of solutions in a given situation). To solve the task requires to analyse and to assess the situation, to identify and apply the correct type of isometries. The student has to know which technique and procedure to use in the tasks of a similar type. Therefore we classify this task as the one developing procedural knowledge at the level of evaluation.
- Part B) was aimed at a wider debate about the number of solutions and analysis of the particular described situation. It requires a creativity for the reorganization of the elements of the task. Therefore, we consider this part of the task as the hardest from cognitive view.

We expected, that in the tasks of developing higher cognitive processes, the results of the students of the experimental group, could be significantly better. The basis for this assumption was the idea, that the interactive and dynamic elements in GeoGebra applets allow to simulate various conditions, various positional and metrical changes. Time experience with the use of such dynamics could have an impact on the development of the imagination and abstraction, mainly in the ability of the student to mentally manipulate with geometric figures and abstract the important elements. We present the results for both parts of the task separately.

Groups	No.	Mean Rank	Sum of Ranks
Control	28	27,23	762,50
Experimental	29	30,71	890,50
Total	57		

Tab. 12. Ranks – Task 4a

Mann-Whitney U	356,500
Z	-0,820
Asymp. Sig. (2-tailed)	0,412

Tab. 13. Test Statistics – Task 4a

It was shown that the test statistic Z = - 0.820 falls within the interval (-1.96, 1.96) and the p-value (0.412) exceeds the level of significance of 0.05. Therefore, we do not reject the null hypothesis with the assumption that the results of the Task 4 part a) is between control and experimental group statistically significant difference. Students in both groups solved the tasks comparable.

Groups	No.	Mean Rank	Sum of Ranks
Control	28	24,64	690,00
Experimental	29	33,21	963,00
Total	57		

Tab. 14. Ranks – Task 4b

Mann-Whitney U	284,000
Z	-2,002
Asymp. Sig. (2-tailed)	0,045

Tab. 15. Test Statistics – Task 4b

The analysis of the data of the job 4b) by means of the Mann-Whitney test (Tab. 14 and Tab. 15) shows that the test statistic Z = - 2.002 minute intervals of (-1.96, 1.96) a p-value of 0.045 at the level of the below the level of significance. Therefore we reject the null hypothesis and claimed that the results of the fourth task B) are between the control and experimental groups were significantly different. At the same time we note that the experimental group had a higher sum order, that experimental group students were solving this task significantly more successful than students of the control group. In this part of the job is confirmed our assumption of a more efficient development of higher thought processes through iBooks textbooks.

2.3 A summary of the results

More comprehensive view is shown by the overall statistical evaluation of the post-test. Therefore, by using the Mann-Whitney U test, we evaluated the total score points from the post-test, in the both control and experimental groups and compared them to each other. The results of the statistical test (Tab. 16 and Tab. 17) show that the test statistic Z has the value of - 3.144, and do not fall within the interval (-1.96, 1.96), and the p-value is 0.02, which is less than the significance level. Therefore, we reject the null hypothesis and conclude that the results achieved in the post-test control and experimental groups, showed statistically significant differences. The experimental group had a higher sum of ranks, and therefore students of the experimental group achieved significantly better results than the students of the control group.

Groups	No.	Mean Rank	Sum of Ranks
Control	28	21,98	615,50
Experimental	29	35,78	1037,50
Total	57		

Tab. 16. Ranks – total point score

Mann-Whitney U	209,500
Z	-3,144
Asymp. Sig. (2-tailed)	0,002

Tab. 17. Test Statistics – Comparison of total point scores

Overall, we can therefore conclude that iBooks integration into the educational process at the particular school allowed the students to gain the knowledge not only comparable with the students educated by traditional methods, but in some tasks, even deeper, more abstract knowledge. The students educated through iBooks aquired qualitatively better knowledge, especially in tasks that required the application of higher cognitive processes, including mental manipulation. Thus, we consider the results of the experiment for more than satisfactory. Nevertheless, both groups will be subjected to repeated testing.

This time it will be in a six-month interval and we will be concerned about the quality of geometric knowledge of the topic in terms of retention.

3 Conclusions

The aim of this paper was to describe the experience with the integration of iBook textbooks into geometric education of the students of a Slovak secondary school, to analyze the data obtained and to summarize the impact of the use of iBook textbooks on learning outcomes. Whereas, there is no available iBook textbook of geometry in Slovakia, it was necessary first to create its interactive and dynamic content. We prepared an iBook textbook, aimed at isometries of the plane. We offered students to work with alternative teaching materials during their learning process, whereby the essential attributes were the interactivity and dynamics of geometrical constructions.

Based on the above results of the experiment, we can conclude that learning outcomes using iBooks textbooks (by interactive boards and tablets) are not only comparable to traditional forms of education, but in some respects even exceeded our expectations. We identified a positive and statistically significant impact of new forms of education, particularly when solving the tasks aimed at higher cognitive processes. This can have a major impact on the development of skills to mentally manipulate with geometric objects, to understand them as abstract concepts in general, which is a necessary condition for the development of geometric reasoning in van Hiele higher levels.

Acknowledgement

The authors gratefully acknowledge the Scientific Grant Agency of the Ministry of Education, Science, Research and Sport of the Slovak Republic for supporting this work under the Grant No. VEGA 1/0440/15: "Geometric conceptions and misconceptions of pre-school and school age children".

References

[1] BANNON, S., MARTIN, G., NUNES-BUFFORD, K. (2012). *Integrating iPads into Maths Education*. In *P. Resta (Ed.)*, Proceedings of Society for Information Technology & Teacher Education Australian Educational Computing Volume 27 Number 3, 2013 84 International Conference 2012 (pp. 3519-3522). Chesapeake, VA: AACE.

[2] BANISTER, S. (2010). *Integrating the iPod touch in K 12 education: Visions and vices*. In *Computers in the schools*, 27(2) 121-131.

[3] BRODA, M., SCHMIDT, A., WERELEY, M. (2011). *Moses was on to something: A reflective analysis of the iPad tablet in field and clinical experiences*. In *M. Koehler & P. Mishra (Eds.)*, Proceedings of Society for Information Technology & Teacher Education International Conference 2011 (pp. 3149-3153). Chesapeake, VA: AACE.

[4] DRIGAS, A. S., PAPPAS, M. A. (2015). *A Review of Mobile Learning Applications for Mathematics*. In *International Journal of Interactive Mobile Technologies* (iJIM) 9(3), https://www.researchgate.net/profile/Athanasios_Drigas/publication/280645657_A_Review_of_Mobile_Learning_Applications_for_Mathematics/links/55c0b4de08ae9289a09b9cd1.pdf

[5] ENRIQUEZ, A. G. (2010). *Enhancing student performance using tablet computers*. In *College teaching*, 58(3), 77-84.

[6] HÍC, P., POKORNÝ, M. (2013). Efficiency of blended learning in basics of statistics teaching. Didactica Slovenica = Pedagoška Obzorja: scientific journal for didactics=znanstvena revija za didaktiko. ISSN 0353-1392. Vol. 28, no. 3-4 (2013), p. 184-192.

[7] POKORNÝ, M. (2013). *Blended Learning as an Efficient Method for Discrete Mathematics Teaching*. Advances in Education Sciences, Vol. 1 (2013), p. 249-252. ISBN 978-981-07-5946-9, ISSN 2339-5141.

[8] SKILLEN, M. A. (2015). *Mobile Learning: Impacts on Mathematics Education*. Proceedings of the 20th Asian Technology Conference in Mathematics, Leshan (China), 2015.

[9] VOŠTINÁR, P., HANZEL, P. (2016). *Mobile application – a tool for teachers, pupils and their families*. In *15th Conference on Applied Mathematics Aplimat 2016*. Bratislava: Nakladateľstvo STU, 2016. ISBN 978-80-227-4531-4.

[10] WARSCHAUER, M. (2006). *Laptops and literacy: Learning in the wireless classroom*. New York, NY: Teachers College Press, 2006.

Current address

Bayerl Erik, Mgr.
Department of Mathematics, Faculty of Natural Sciences, Matej Bel University
Tajovského 40, 974 01 Banská Bystrica, Slovak Republic
E-mail: erik.bayerl@gmail.com

Žilková Katarína, doc. PaedDr. PhD.
Faculty of Education, Catholic University in Ružomberok
Hrabovská cesta 1, 034 01 Ružomberok, Slovak Republic
E-mail: katarina@zilka.sk

Slovak University of Technology in Bratislava
Faculty of Mechanical Engineering
16th Conference on Applied Mathematics **APLIMAT 2017**
Proceedings

STRUCTURAL EQUATION MODELING USING THE AMOS AND REGRESSION OF EFFECTIVE ORGANIZATIONAL COMMITMENT INDICATORS IN LITHUANIAN MILITARY FORCES

BEKESIENE Svajone (LT), HOSKOVA-MAYEROVA Sarka (CZ), DILIUNAS Paulius (LT)

Abstract. The recognition of their importance to the scientific research suggests the need to evaluate and compare different types of techniques so that the scientific research designs can be selected adequately. The article presents a consecutively example, which analyzes the same dataset via two very different statistical techniques. After that it compares two modeling possibilities: modeling using Analysis of Moment Structures (AMOS) and statistical software package SPSS. Finally, the article discusses linear regression models and offers guidelines as to when SEM techniques and when regression techniques should be used.

Keywords: structural equation modeling (SEM), AMOS, regression; research techniques, formative constructs, construct validity, the leading behavioral traits

Mathematics Subject Classification: Primary 00A71; Secondary 97M10

1 Introduction

Structural Equation Modeling techniques such as AMOS are a covariance-based SEM. It was developed by Dr. Arbuckle and promoted by SPSS as a statistically equivalent tool to Linear Structural Relations (LISREL) [11]. AMOS is the second-generation data analysis techniques that can be used for high quality statistical analysis.

Being different to first generation statistical tools such as linear regression, SEM enables researchers to answer a set of interconnected research questions in a single, efficient, and all-inclusive analysis by modeling the relationships between multiple independent and dependent constructs at the same time [3]. SEM analysis differs greatly from the first generation regression models such as linear regression, ANOVA, and MANOVA because they cannot analyze more than one layer of linkages between independent and dependent variables at the same time. The comparative analyses between these techniques are explained in the Tab. 1 [4].

Issue	LISREL /AMOS	Linear Regression
Objective of Overall Analysis	Show that the null hypothesis of the entire proposed model is plausible, while rejecting path-specific null hypotheses of no effect.	Reject a set of path specific null hypotheses of no effect.
Objective of Variance Analysis	Overall model fit, such as insignificant $\chi 2$ or high AGFI.	Variance explanation (high R-square)
Required Theory Base	Requires sound theory base. Supports confirmatory research.	Does not necessarily require sound theory base. Supports both exploratory and confirmatory research.
Assumed Distribution	Multivariate normal, if estimation is through ML. Deviations from multivariate normal are supported with other estimation techniques.	Relatively robust to deviations from a multivariate distribution, with established methods of handling non multivariate distributions.
Requirements for Sample Size	At least 100-150 cases.	Supports smaller sample sizes, although a sample of at least 30 is required.

Tab. 1. Comparative Analysis between Techniques [4]

The capabilities for linear regression and AMOS/ LISREL are shown in the Tab. 2 [4]. As we can see, unlike first generation regression tools, SEM assesses the structural model and at the same time assumed causation among a set of dependent and independent constructs.

Capabilities	LISREL /AMOS	Linear Regression
Maps paths to many dependent (latent or observed) variables in the same research model and analyze all the paths simultaneously rather than one at a time.	Supported	Not supported
Maps specific and error variance of the observed variables into the research model.	Supported	Not supported
Maps reflective observed variables	Supported	Supported
In one analysis can do both measurement and structural analyzes.	Supported	Not supported
Can perform a confirmatory factor analysis	Supported	Not supported
Provides a statistic to compare alternative confirmatory factor analyses models	Supported	Not supported

Tab. 2. Capabilities for first generation regression tools and SEM [4]

In addition, in the same analysis are also evaluated the measurement model loadings of observed items (measurements) on their expected latent variables (constructs). Moreover, the combined analysis of the measurement and the structural model enables measurement errors of the observed variables to be analyzed as an integral part of the model, and factor analysis to be combined in one operation with the hypotheses testing.

The purpose of this study was to examine the influence of structural model measurement capabilities on organizational commitment of Lithuanian servicemen. The main objective of this research was to improve the understanding of the determinants of effective analysis style.

2 Theoretical considerations

As mentioned above, the organizational commitment of Lithuanian servicemen was chosen to show SEM competitive abilities vis-à-vis first generation regression tools.
The servicemen organizational commitment has been conceptualized and measured in the Meyer and Allen's three-component model which has become the dominant model for study of workplace commitment and integrates conceptualizations: the affective component (ACS), the continuance component (CCS) and the normative component (NCS) [1, 8].

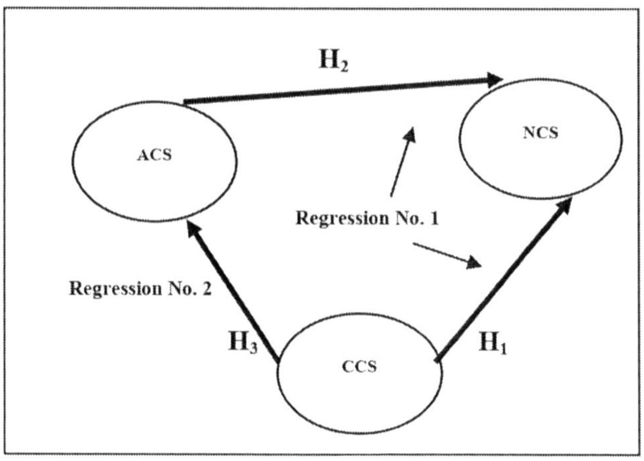

Fig. 1. Basic organizational commitment model for linear regression

The affective component organizational commitment, proposed by the model, refers to employees' emotional attachment to, identification with, and involvement in the organization. The continuance component refers to commitment based on the costs that employees associate with leaving the organization. Finally, the normative component refers to employees' feelings of obligation to remain with the organization. [1], [10].

Previous studies have suggested that, as predicted by the model, the affective and continuance components of Organizational commitment are empirically distinguishable constructs with different correlates. The affective and normative components, although distinguishable, appear to be somewhat related. [7]

Factors	Measurable Items	Label in Model
Continuance Commitment Scale (CCS)	Right now, staying with my organization is a matter of necessity as much as desire	C1
	It would be very hard for me to leave my organization right now, even if I would like to	C2
	Too much in my life would be disrupted if I decided that I would like to leave my organization now	C3
	I feel that I have too few options to consider leaving this organization	C4
	One of the major reasons I continue to work for this organization is that leaving would require considerable personal sacrifice – another organization may not match the overall benefits I have here	C5
	One of the few serious consequences of leaving this organization would be the scarcity of available alternatives	C6
Affective Commitment Scale (ACS)	I would be very happy to spend the rest of my career with this organization	A1
	I really feel as if this organization's problems are my own	A2
	I do not feel a strong sense of belonging to my organization (R)	A3
	I do not feel 'emotionally attached' to this organization (R)	A4
	I do not feel like 'part of the family' at my organization (R)	A5
	This organization has a great deal of personal meaning for me	A6
Normative Commitment Scale (NCS)	I do not believe that a person must always be loyal to his or her organization (R)	N1
	If I got another offer for a better job elsewhere I would not feel it was right to leave my organization	N2
	Jumping from organization to organization seem unethical to me	N3
	One of the major reasons I continue to work for this organization is that I believe that loyalty is important	N4
	Now I have to stay in the organization because I feel committed to the people who work there	N5
	I feel indebted to this organization	N6

Tab. 3. Indicator variables of organizational commitment analysis of Lithuanian servicemen

Thus, our initial literature review suggested that the model presented in Figure 1 above involves three hypotheses and could guide our empirical investigation analysis as the first generation regression model. The hypotheses are with these constructs:

H_1: CCS will affect the *normative* commitment component;
H_2: ACS will affect the *normative* commitment component;
H_3: CCS will affect the *affective* commitment component.

The indicator labels for variables CCS, ACS and NCS are shown in Tab. 3. As we know, the linear regression cannot test all three relationships at the same time, so it is necessary to use two separate regressions to test the model drawn in the Figure 1. Moreover, at the first we have to create the variables by responses to items C1-C2 for CCS, to items A1-A2 for ACS and to items N1-N2 for NCS, which are described in the Tab. 3.

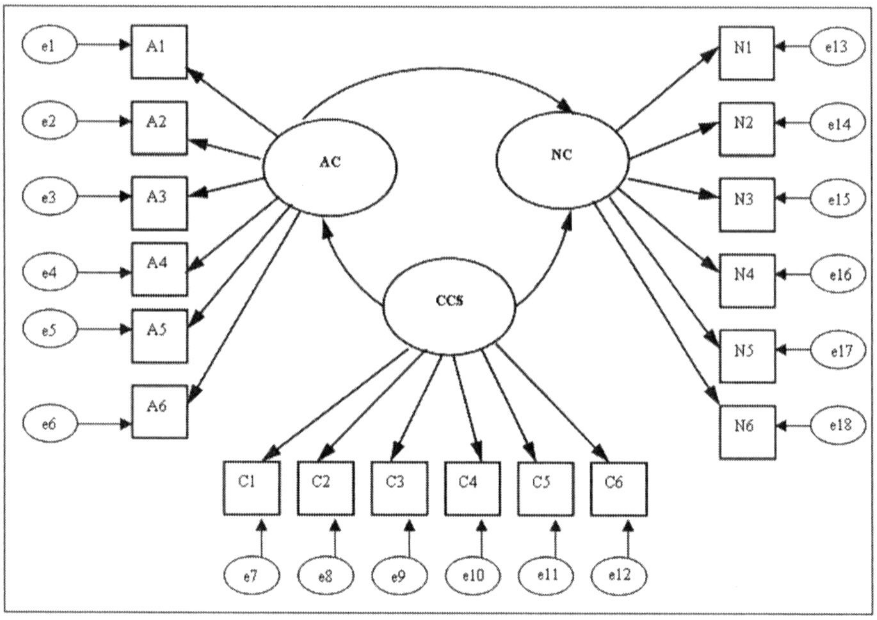

Fig. 2. Basic organizational commitment model for SEM

For the moment, it is important to say that in SEM like AMOS preliminary factor and reliability analyses are not necessary because the instrument measurement is simultaneous with the testing of hypotheses. The organizational commitment model for SEM is presented in the Figure 2.

3 Methodology

3.1 Data Collection

The data for study were collected from March till April in 2016. The participants were given a questionnaire within web page manoapklausa.lt. The link is sent personally or through Facebook. They were asked their verbal consent and confidentiality was guaranteed. The survey involved 204 respondents, the professional soldiers who are serving in the Lithuanian army units. The demographic and social characteristics of the respondents who served as study respondents are summarized in the Tab. 4 below. There we can see that the majority of them 52.9% (N=108) live in the town of Vilnius and the majority of them have the bachelor level of education 56.9% (N=116). [6]

Variables	Frequency	Percentage	Variables	Frequency	Percentage
Gender:			**Level of education:**		
Male	168	82.4	secondary	8	3.9
Female	36	17.6	bachelor	116	56.9
Living place:			Higher Education	4	2.0
Vilnius town	108	52.9	Unfinished higher education	4	2.0
Kaunas town	8	3.9	Master's degree	68	33.3
Alytus district	8	3.9	Higher than Master's degree	4	2.0
Kaunas distr.	32	15.7			
Klaipėda distr.	32	15.7			
Šiauliai distr.	4	2.0			
Tauragė distr.	8	3.9			
Vilnius distr.	4	2.0			

Tab. 4. The study participant's distributions for: Gender, Living place and Level of education

In the Tab. 5 below there is summarized the categorized number of years the study participants have been working in the military forces. The distribution in Tab. 5 shows that a vast majority of the respondents, comprising about 47.1 % (N= 96) have been working in the Lithuania army for a time period of more than twelve years. As Tab. 5 indicates, the 23.5 % (N = 48) have been working in their forces for a time period of more than eighteen years, while 7.8 % (N=16) have been working for a period ranging from one year to five years. The 21.6 % (N = 44) of the respondents spent from one year to five years in the Lithuania army.

Working Years	Frequency (N)	Percentage	Valid Percentage	Cumulative Percentage
1-5 years	44	21.6	21.6	21.6
6-11 years	16	7.8	7.8	29.4
12-17 years	96	47.1	47.1	76.5
18+ years	48	23.5	23.5	100.0
Total	204	100.0	100.0	

Tab. 5. Summary of study participants work years in the Lithuania army

The respondents were asked what the Lithuanian Armed Forces serving. The majority of subjects 86% (N=174) stated that serves Land forces, about one tenth of the subjects 12% (N=24) serves the Navy forces and 2% (N=4) subjects serving in the Air force (Tab. 6).

Military Forces	Frequency (N)	Percent	Valid Percent	Cumulative Percent
Land Forces	172	84.3	86.0	86.0
Air Force	4	2.0	2.0	88.0
Navy	24	11.8	12.0	100.0
Total	200	98.0	100.0	
Missing System	4	2.0		
Total	204	100.0		

Tab. 6. The Lithuanian Armed Forces serving summary for respondents

In this survey was also found out the participants' military degree. More than half of them 68.6% (N=136) were the senior officers, about one fifth of them 21.6% (N=48) were the junior officers and only the 9.8% (N=20) - enlisted grade. The descriptive analysis numbers implies that the study participants are fairly knowledgeable about the managerial and structural characteristics of their specific work environments.

3.2 Data Analysis

The first generation regression tools by using SPSS program and the SEM by using analysis of moment structures (AMOS) program were done. Before regression analysis all items were tested. Their formed unities and measurement properties (construct validity and reliability) were checked. In this case, for the primary constructs the Principal Component Analysis (PCA) was done. As for SEM analysis the testing of measurement properties of the instruments was done simultaneously with the hypotheses testing.

4 Results and Discussion

In this section, the data analysis results using linear regression and AMOS 24 are presented. The first findings was from the reliability analyses, which shows, that ACS has the lowest reliability because the Cronbach's alpha coefficient is 0.689, for CCS Cronbach's alpha shows the – 0.768 and the highest reliability (0.895) was counted for NCS. The Principal Component Analysis of the constructs did not show clean loadings in the factor structure, but we continue with analysis.

4.1 Data Analysis using Linear Regression

The first generation statistical tools such as SPSS 20 software were chosen for linear regression analyses. First there were created variables by responses to items C1-C2 for CCS, to items A1-A2 for ACS and to items N1-N2 for NCS and after that three hypotheses were checked in two analyses. The first analysis was expected that ACS and CCS influence outcome variable such as the normative commitment (NCS). In this regression NCS was the dependent variable when ACS and CCS were independent variables. So, first there were checked the H1 and H2 and then H3, where the CCS was regressed on ACS. The obtained results are presented in the Tab. 7 and it shows that the H1 and H3 are significant and in the posited directions while H2 is not.

Regression	Dependent variable	$F(R^2)$	Independent variable	Coefficient (T-value)
H_1& H_2	NCS	75.47** (0.43)	CCS	0.54 (12.26**)
			ACS	-0.12 (-1.26)
H_3	ACS	5.72* (0.03)	CCS	0.075 (2.39*)

**-Significant at the 0.05 level;*
***- Significant at the 0.01 level.*

Tab. 7. Summary of linear regression analyses

The F statistics, which provides a test for statistical significance of observed differences among the means shows us that first (75.47**) and second (5.72*) regressions are significant, however, the R2 shows that measure of the proportion of the variance of dependent variable mean explained by the independent variables (CCS & ACS) is good enough only for the first model (0.43). The R2 value for the second model is very close to zero (0.03), so it shows that not much variation in the data is explained by the fitted model.

4.2 Data Analysis using AMOS 24 software

At the start of the SEM analysis, the predicted model structure (see Figure 2) outlining the relationships among the organizational commitment model components (i.e. the affective component (ACS), the continuance component (CCS) and the normative component (NCS)) were loaded into the AMOS 24 software. We have one exogenous construct (CCS) and two endogenous constructs (NCS &ACS), all-in-all the 18 predictive indicators and three latent variables. For SEM analysis, we use the same collected data as had been used in the linear regression analysis. The logic of SEM links to start with theory, including labeled constructs and then test for model fit. The covariance – based AMOS 24, which applies Maximum Likelihood functions for minimize parameter estimates, tested all model and the minimum was achieved.

AMOS 24 Notes for model-fitted		**Link**	**Coefficient (T-value)**	**SMC**
X2=762.23		CCS-->NCS	0.57 (3.42**)	0.70
df=122		ACS-->NCS	0.3(2.04*)	
AGFI= 0.61		CCS-->ACS	0.81 (6.56**)	0.65
RMR= 0.12				

**-Significant at the 0.05 level;*
***- Significant at the 0.01 level.*

Tab. 8. Summary of SEM analyses

The standardized model-fits generated by AMOS 24 software are shown in the Figure 3. Some of obtained results (notes for model-fitting, links, coefficients, squared multiple correlations (SMC)), which are discussed below are presented in the Tab. 8.

The adjusted goodness of fit (AGFI) index (0.61) is bound above by one and the root mean square residual (RMR) index (0.12), which is the square root of the average squared amount, by which the sample variances and covariances differ from their estimates obtained under the assumption tell us that the model fits. However, the estimated X2of 762.23 (df =122) is so large that null hypothesis of a good fit is rejected at the 0.05 level ($p<0.000$). So it seems that the overall model fit does not appear quite good. This is because the AMOS 24 software recognized some of predictive indicators as latent variables and thus was not able to provide a model fit. [10]

The standardized regression estimates obtained from AMOS 24 for indicators associated with the arrows from latent variables (CCS, ACS, NCS) to the respective indicator variables by the rule should have loadings of 0.7 or higher on the latent variable to be significant [6]. The figure 3 shows that for latent variable CCS only indicator C3 (0.82) should be left, for other latent variables ACS and NCS more or less all indicators are acceptable.

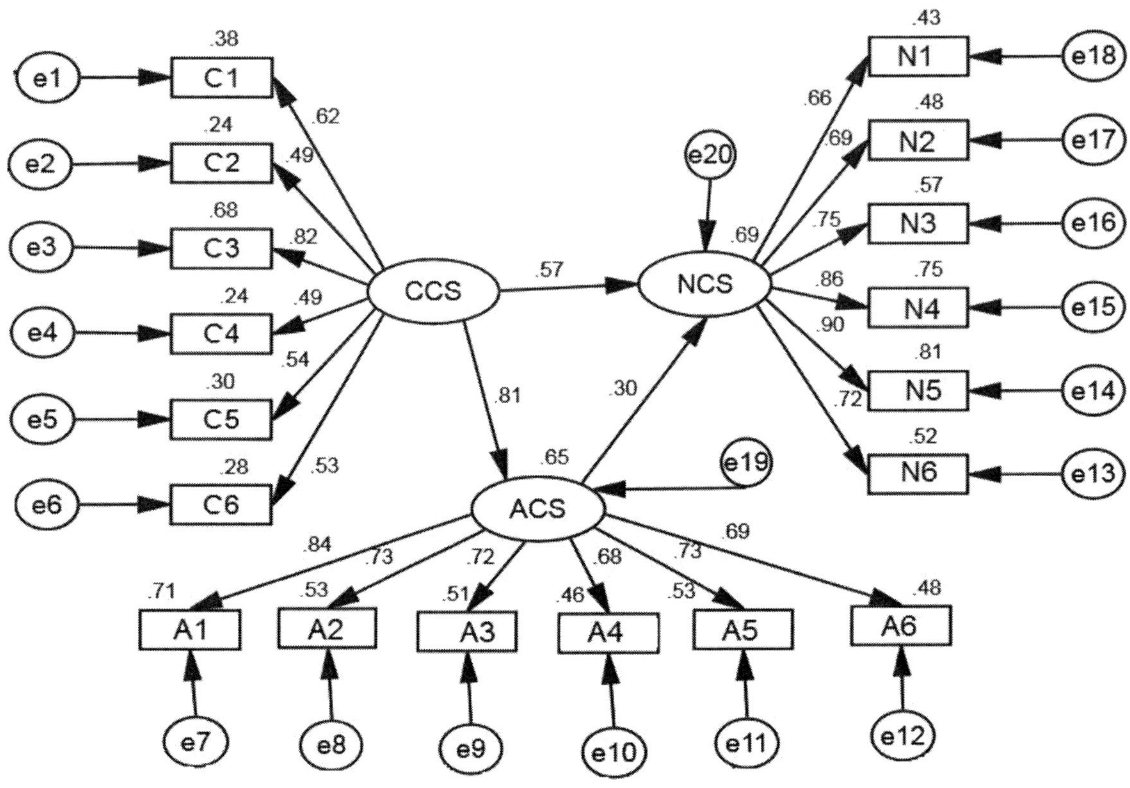

Fig. 3. Standardized indicator loadings in the path diagram generated by AMOS 24 graphics for the organizational commitment model

The AMOS 24 software considered the underlying structural relationships of all the latent variables for the predicted model structure (i.e. figure 2). This allowed us to know more about indicators and the relationships in the organizational commitment model of Lithuanian servicemen. From this analysis were found the indicators that could not provide the basis for the continuance component (CCS) of organizational commitment model: C1- "Right now, staying with my organization is a matter of necessity as much as desire" ($r= 0.62$, $\alpha= 0.38$); C2 - " It would be very hard for me to leave my organization right now, even if I " ($r= 0.49$,$\alpha= 0.24$); C4 - "I feel that I have too few options to consider leaving this organization" ($r= 0.49$, $\alpha= 0.24$); C5 – "One of the major reasons I continue to work for this organization is that leaving would require considerable personal sacrifice – another organization may not match the overall benefits I have here" ($r= 0.54$, $\alpha=0.30$); C6 - "One of the few serious consequences of leaving this organization would be the scarcity of available alternatives"($r= 0.53$, $\alpha=0.28$).

Nevertheless, at the end we can see that the SEM results are very similar as in the linear regression analysis because only the H1 and the H3 are significant at the 0.01 level, and H2 is significant only at the 0.05 level ($p=0.042$) (see in Tab. 8).

4 Conclusion

The findings from this study have shown that the statistical objective of linear regression is to show high R2 and significant t-values, thus rejecting the null hypothesis of no-effect.

The covariance- based SEM provides a set of overall model-fit indices that include a wide set of types of fit. More over structural equation modeling is suited to show that the assumed research model is significant with all its paths, and the suitability of theory being examined is corroborated and not disconfirmed by the data.

At the end we have to say that the linear regression seems to be more amenable to meta-analysis or can be appropriate if it is important to make specific and direct comparisons to other studies where these techniques were used.

Acknowledgements

The work presented in this paper was supported within the project "Development of basic and applied research in the long term developed by the departments of theoretical and applied foundation FMT" (Project code VÝZKUMFVT) supported by the Ministry of Defence the Czech Republic.

References

[1] ALLEN, N. J.,MEYER, J. P. "The Measurement and Antecedents of Affective, Continuance, and Normative Commitment to the Organization", Journal of Occupational Psychology, 63, 1990, 1-18.

[2] ALLEN Natalie J. Organizational Commitment in the Military: A Discussion of Theory and Practice. Military Psychology. 15 (3), 2003, 237-253.

[3] ANDERSON, J.C. and GERBING D.W. Structural Equation Modeling in Practice: A Review and Recommended Two - Step Approach. Psychological Bulletin, 103, 1988, 411-423.

[4] GEFEN, D., D.W. STRAUB, and M. BOUDREAU. Structural Equation Modeling and regression: Guidelines for Research Practice. Communications of the Associations for Information Systems, 4, Article 7, 2000, 1-77.

[5] HAIR. Jr., J. F., BLACK., W. C., BABIN., B. J., ANDERSON., R. E., & L.TATHAM., R. Multivariant Data Analysis. 2006, New Jersey: Pearson International Edition.

[6] HASILOVÁ, K. Iterative Method for Bandwidth Selection in Kernel Discriminant Analysis . In: 32nd International Conference Mathematical Methods in Economics MME2014. Olomouc: Palacký University, Olomouc, 2014, 263-268. ISBN 978-80-244-4209-9.

[7] HOFMANN A., HOSKOVA-MAYEROVA S. Development of applications in the analysis of the natural environment, *In: Aplimat - 15th Conference on Applied Mathematics 2016 Proceedings*. STU - Slovak University of Technology in Bratislava, 467-481.

[8] MEYER J. P., ALLEN, N. J. A Three Component Conceptualization of Organizational Commitment, Human Resource Management Review, 1(1), 1991, 61-89.

[9] POTŮČEK, R.: Construction of the smallest common coarser of two and three set partitions. Analele Stiintifice ale Universitatii "Ovidius" Constanta, Seria Matematica, 2014, vol. XXII, no. 1/2014, p. 237-246. ISSN 1224-1784

[10] TALHOFER, V. & HOSKOVA-MAYEROVA, S., (2016). Verification Tests of Mathematical Models of Passage throug Terrain. Aplimat - 15th Conference on Applied Mathematics 2016 Proceedings. STU - Slovak University of Technology in Bratislava, 1010-1025.

[11] SPSS AMOS. Available at http://www-03.ibm.com /software /products / en/spss-amos

Current address

Bekesiene Svajone, Assoc. Prof. Ph.D.
The General Jonas Žemaitis Military Academy of Lithuania
Šilo Str. 5A, LT-10322 Vilnius, Lithuania
Tel. number: +370 5 212 6923, e-mail: svajone.bekesiene@lka.lt

Hoskova-Mayerova Sarka, Assoc. Prof. Ph.D.
University of Defence
Kounicova 65, 662 10 Brno, Czech Republic
Tel. number: +420973442225, e-mail:sarka.mayerova@unob.cz

Diliunas Paulius
The General Jonas Žemaitis Military Academy of Lithuania
Šilo Str. 5A, LT-10322 Vilnius, Lithuania
Tel. number: +370 5 212 6923, e-mail: paulius.diliunas@lka.lt

Slovak University of Technology in Bratislava
Faculty of Mechanical Engineering
16th Conference on Applied Mathematics **APLIMAT 2017**
Proceedings

THE PARAMETRIC ARCHITECTURE PROJECT TOWARDS COGNITIVE COMPUTING

BELIBANI Rosalba (IT)

Abstract. The work aims to investigate what can be the possible perspectives with the use of algorithms and digital tools in creative architecture, specifically in the process of design and shape control. There is always a relationship between architecture and mathematics, both in conception and transcription. Today, architecture makes use of digital tools that support the design and impact the language used allowing for new spatial experiments. The relationship between the project's design and mathematics arrives today in cognitive computing.

Keywords: parametric architecture, algorithm, cognitive computing, Watson

Mathematics Subject Classification: Primary 68Uxx, Secondary 68U07, 68D18

1 Mathematics and the relationships with other arts

Architecture, as a producer of forms and even minor art until the Middle Ages, has always had a relationship to mathematics. Moreover, shapes are determined by size and, therefore, by numbers that, accordingly, govern all art forms, including architecture.

Mathematics has influenced and interacted with music and painting and all the great masters and geniuses of the past have studied these materials and applied them simultaneously and inseparably. Leon Battista Alberti[1], the father of the Renaissance architecture, practiced all major arts such as painting and music and theorized on the close relationship between these and mathematics and numbers. All his works bear witness to this relationship and his writings[2] deal, in fact, with the operational and experimental canons for the study and application of the arts.

[1] Leon Battista Alberti (1404 - 1472) for the careful modern reworking of the old is considered, along with Brunelleschi, the architect of Renaissance architecture founder. Important figure of the Italian Renaissance, mathematician, musician (and also a writer, linguist and philosopher) published treatises showing his search for rules, theoretical or practical, in the study, experimentation and application of the arts.

[2] Reference treatises on art: The *De Pictura* provides the first definition of the scientific perspective and the *De statua* exposes the human body proportions. It also develops the first encryption system Polyalphabetic, on the cipher disk, described in *De Componendis Cifris* or *De Cifris*, a treaty that represents a turning point in Western cryptographic

The observation of nature has brought man to question the processes and the organization of nature, taking in its mathematical structure for inspiration and to create works with the same harmonic principles. Today there emerges new and deeper links between mathematics and design, in a contradictory scenario that sees the transformation of architecture as a physical element immersed in a flow of information. A process is taking place whereby mathematics, understood as generating forms, appears as a new genetic code for architecture.

Fig. 1. Leon Battista Alberti, *Definitor*

The algorithmic approach to the problems of modeling requires an investigation into the mathematical aspects of the objects since the general form and the details of a model are no longer the result of a manipulation of forms, but become a strengthening of thought in the interaction between man and machine, the result which combine geometry, mathematics and logic.

Fig. 2. Leon Battista Alberti. S. Andrea di Mantova, 1472. Front view (drawing by Marta Anselmi)

technique. Lastly, in *De re aedificatoria* of 1472, he deals with cases related to modern architecture in which he is the protagonist.

2 The parametric architecture

The generic term 'parameter' used in mathematics for the most part has the meaning of an independent variable, while in electronics parametric devices are operations in which depends a parameter that can be varied from the outside. The parameter is a data that changes a situation and a parametric design is based on using these parameters. The concept often refers to the use of specific software rather than to an organic definition of the relationship between the various parameters during the design phase.

The definition for architecture is not exhaustive. An embryonic idea of parametric procedures is given by some projects that pursue the idea of generating the shape through rigorous geometric relationships between quantifiable parameters, derived from the concept and design intentions.

Moretti also argued that the parameters and their interrelationships are the expression, the codes of the new architectural language, the structure in the original rigorous sense of the word. The parameters and their relationships should assist the techniques and instruments of scientific thought; particularly logic-mathematics, operations research and computers, especially those for the opportunity to express self-correcting cyclical series of probable solutions for parameter values and their relationships.

Now it is universally recognized that the use of specific software and the spread of new technologies have encouraged the expression of new architectural languages and the creation of contemporary architecture that before was hardly conceivable. The impressive stylistic performances have allowed for great possibilities of expression and ongoing research. The architect has been able to free himself from predefined shapes and thanks to the manipulation of modeling software has expanded his ability to experiment and give expression to his creative potential.

Already in the nineties some projects, while using manual or analog processes, seemed to possess parametric capabilities such as the paintings by Zaha Hadid, that appeared as digital models. The same can be observed for the deconstructed projects of Peter Eisenman and the additive over layering techniques by Bernard Tschumi, that characterized the project of the Parc de la Villette. The generative process of these examples is structured in such a way as to allow the designer to keep track of each intermediate stage with the final result presented as an body open to subsequent modifications without distorting the overall logic.

Fig. 3. Sergio Musumeci. Ponte sul Basento, 1967

Despite initial criticism and skepticism, the indisputable revolutionary contribution of parametric design in architectural processes must be recognized and not only as a facilitator of forms.

To counter formalism judgements, Fulvio Wirz of Zaha Hadid Architects declares "By its very nature a parametric concept is independent from the formal result that at the initial stage is often undefined in the designer's mind. It is a metaphor, a mental algorithm, the same logical process that we place at the base of a design project that supports, unifies and strengthens it. If this substrate is strong, the formal outcome will be not only coherent but also open to infinite variations within the parametric diagram".

The complex nature of the parametric software allows one to transfer this consistency also during the construction phase, planning and integrating it into the design process through the generation of drawings, and the interaction with the database and the production processes. The process in the case of the parametric design simplifies everything by following a logical structure that goes from the particular to the general, according to a bottom-up process.

The architecture that results from generative design sums up clearly the new transition from product design to the entire production process of the product. The project resources, the need for optimal behavior on the part of the buildings, the connection between printing and prototyping techniques have evolved architecture to a scientific approach. According to the digital generative paradigm, architecture is considered, like the natural world and bio-mimetic characteristics, as a kind of artificial life conditioned by the principles of morphogenesis, the genetic code, replication and selection.

The developers of Rhinoceros, McNeel & Associates, have written a revolutionary plug-in for the Rhinoceros 3D program, software for creating three-dimensional models, based on NURBS[3] curves and surfaces, initially called Explicit History and subsequently renamed Grasshopper. The potential of scripting, realizing a visual type of application based on the most accessible algorithmic logic, thus without the need to enter codes.

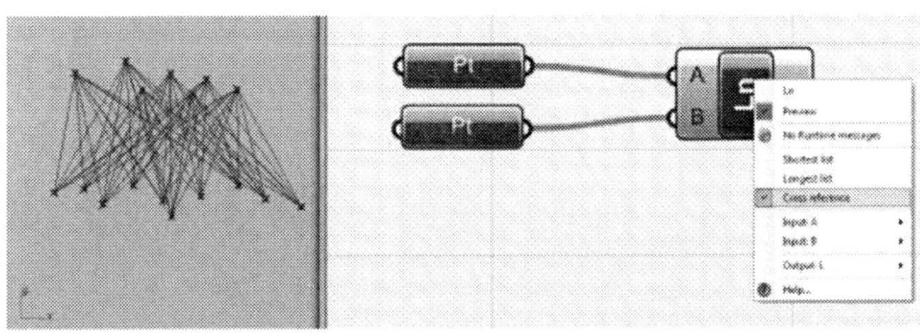

Fig. 4. A combination between some points in space represented by the algorithm of Grasshopper with setting a circuit

Grasshopper is a visual editor of algorithms that, through an intuitive graphical method based on interface nodes, allows one to define a sequence of instructions translated into three-dimensional models in the Rhinoceros window and to draw complex shapes thanks to the design. It promotes, with programming codes, a whole process until the prototyping in 3D. The new perspective in the definition of shapes is modeling through algorithms. The geometries by mathematical functions, the

[3] Non-Uniform Rational Basic Spline, a mathematical representation of 2D and 3D geometries.

mathematical/logical approach and the generation of parametric models enable fast and deep variations of the initial geometries and rapidly obtain forms of great complexities with the repetition of geometric elements.

We speak of Generative Algorithms, or Generative Algorithm of the form, whenever the result of a sequence of algorithms consists of a more or less complex geometry.

In order to satisfy the needs of a very complex project the process uses basic geometry, linked to editing and processing tools in the program. This highlights new links between the project and mathematics, understood as a generator of forms that appears as a new architecture genetic code, reiterating once again the historical need of man to question the processes and the organization of nature in order to grasp itsmathematical structure (see the Golden Section) and be inspired from it to create works based on the same principles. The generative and parametric software allows you to control more complex situations and surmount the scale gap of BIM software.

Arturo Tedeschi, in his book *Architettura Parametrica. Introduzione a Grasshopper* declares that the approach to modeling algorithm requires special attention to the mathematical aspects of objects: "the general form and details of a model are no longer the result of a manipulation - albeit virtual - of forms, but they become the result of instruction that combine geometry, mathematics and logic".

Before the digital revolution, the designer took charge of managing complexity. The digital revolution has offered the designer the ability to focus exclusively on the project, controlled through three-dimensional models, entrusting the task to the program to obtain floor plans, sections and details useful to the production process.

Software permits a virtual and realistic display of an object at each stage of the project, and also to transfer data directly to production systems that turn them into concrete objects. You get closer and closer to the ideal self-manufacturing model, the transition from the idea to the actual object is directed and the control of the generative process is free and rich in expressive possibilities.

3 Parametric trials

Frank Gehry, Zaha Hadid, Patrick Schumacher, Ben Van Berkel and Makoto Sei Watanabe were the first to use parametric programs, and these programs have become a fundamental means (tools) of expression for their projects.

An earlier project, the Sun-God City, is one of the first parametric design experiences. The entire project tries to solve two problems: the excessive density of the fabric of the Japanese metropolis and the exposure to natural light. The parameters in play were therefore, the shape and the solar vector. The use of algorithms allowed for the placement of all units along the axis of the solar vector without having to calculate the optimal exposure for each dwelling, and based on the needs of each resident.

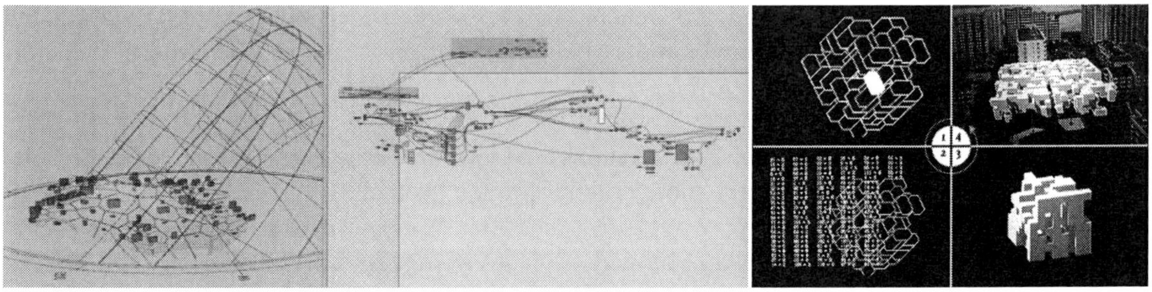

Fig. 5. Makoto Sei Watanabe. Generative algorithms for the Sun-God City, 1993, and their representation in Rhinoceros

Watanabe does not use scripts for his projects, but he uses specific software, developed by his team. The Ildabashi Station is a highly parametric design, a reticular structure that, sinking into the ground, seems to be a natural extension of the Japanese underground network. The performance is the formal aspect, made of contractions and expansions, which seems to accompany the flow of users of the station. The shape is the result of a proper use of generative software. Watanabe uses a logic which is opposed to that which sees the definition of complex shapes at random - as through a completely computerized design - but sets a formal definition of a precise basic idea, transposing it into a tight algorithm and manages to free himself from the cold mathematics that often characterizes these transactions giving his architecture a quality and a depth of feeling connected to the potential feeling of the user.

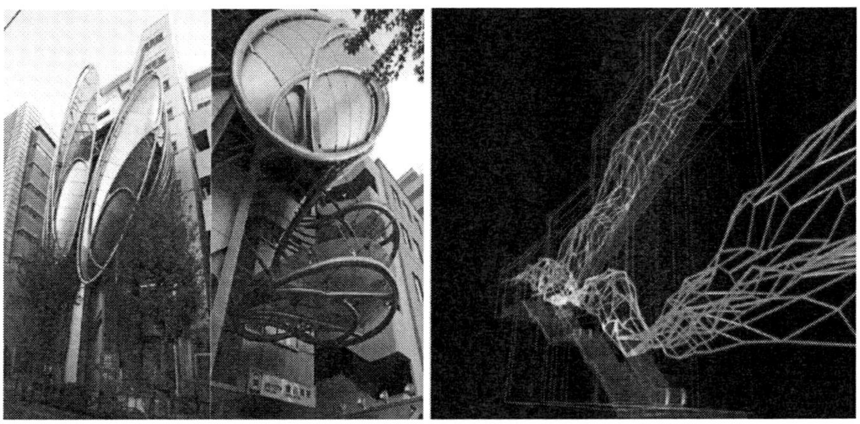

Fig. 6. Makoto Sei Watanabe. Stazione Iidabashi, 2000. Ventilation chimneys and structure of the tubes

The project is composed of three elements: the structure of the platforms and the exit systems, a plant system developed through a mesh of tubes that accompanies the path, and the ventilation tower whose complex structure is the result of a structural optimization program that considers the different materials and the different sections and surfaces that come into play, making the most of the performance of the three-dimensional structure.

Fig. 7. Zaha Hadid. Heydar Aliyev Center, 2013. Render and roofing under construction

Zaha Hadid and Patrik Shumacker have used the parametric design in its most complex meaning, making the building a real sculpture. It would be unthinkable to imagine such a complex structure using simple controls of a design program. The complex and winding structures they produced are unique organisms in perfect consistency from the smallest element to the macrostructure. The generative and parametric software allows the control of more complex situations and eliminates the scale gap of BIM software, which works better with architectural structures without specific complexes and of small dimension.

The Heydar Aliyev Center is a fluid form that emerges from the bending of the natural topography of the landscape that surrounds the individual functions: the structure is the space, an indivisible continuum between form and function.

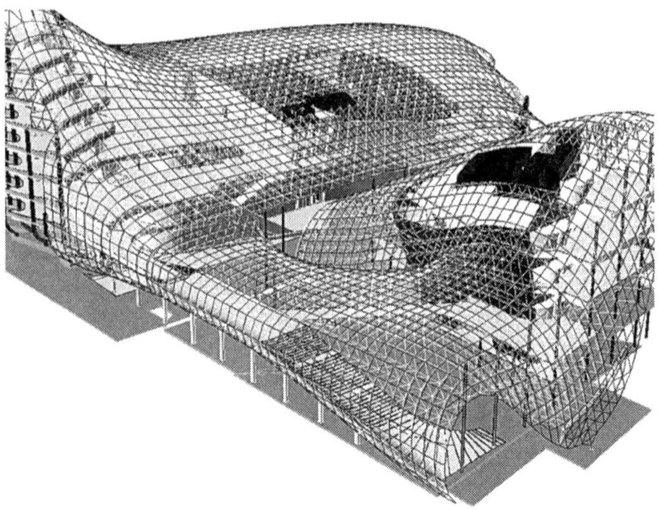

Fig. 8. Zaha Hadid. Heydar Aliyev Center, 2013. Structure realized with the use of generative software

Gaetano De Francesco and Antonio De Pasquale developed for the Urban Green Line project a project that recycles urban chaos into electricity through the use of nanotechnology. The project "Atinnitus" is characterized by a steel frame, result of an hypothetical breakdown of the noise in different sound waves that give formal life to a series of tapes that become shelters, docks, chairs and new facades.

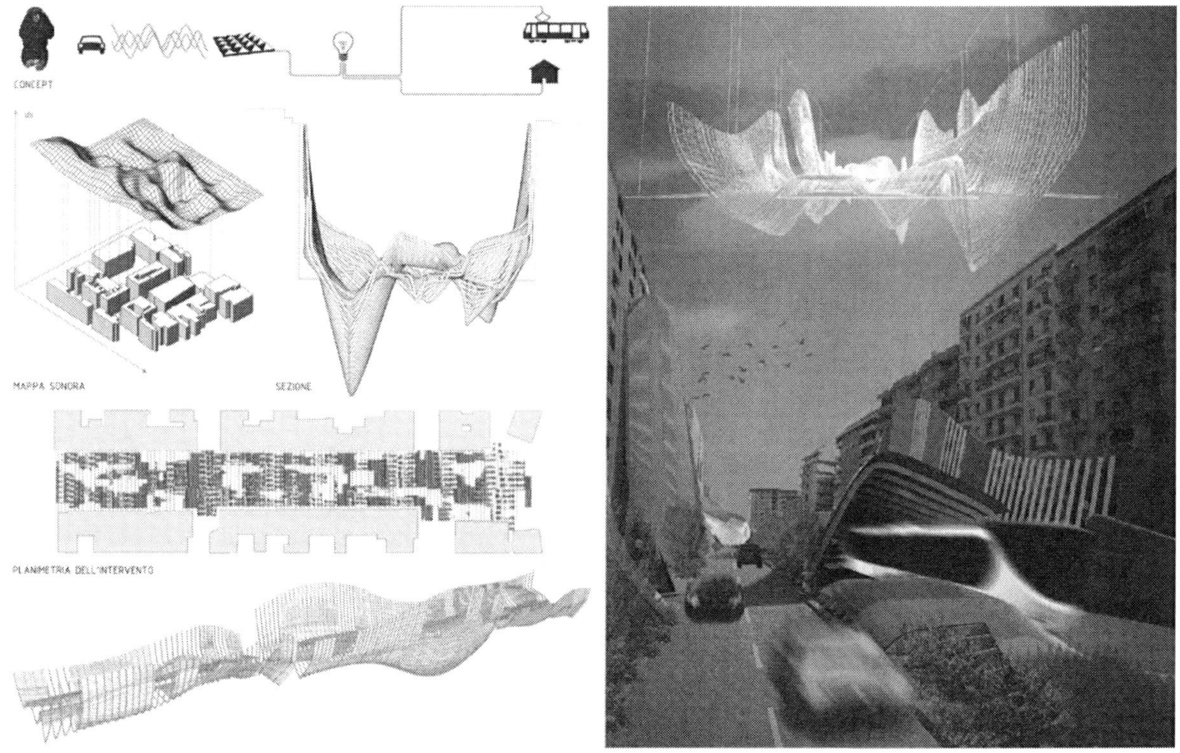

Fig. 9. Gaetano De Francesco, Antonio De Pasquale. Atinnitus: interactive membranes for processing noise into energy

The position of the panels on the structure is the result of the transformation of a sound map into numerical data, through the use of Grasshopper with a 3D model that enabled simulation of the interaction between architecture designed and acoustic oscillations. The end result is an architecture that makes interactive Noise the protagonist and the generating element of the project. A structure that captures the sound, absorbs it, turns it into energy and changes with the movement of the sound waves. "The sound, the protagonist of our research, is for us a shimmering immersive space of which we take the continuous movement of intangible entities that inhabit it. An immaterial space that changes around us according to external sound waves to produce new energy, new realities, new interconnections and thus ecologies. A structure that captures the noise, the absorbs, transforms into energy and varies with the change of the sound waves."

Fig. 10. (Un)Plug, R&Sie. La Defense, Paris, 2001. Office building

Gaetano De Francesco writes: "We were interested in creating mutant architecture that interacts with the hiss and noise of the wired city, with the chaos of the contemporary city, in which the word relationship became the key term. Like many architectures capable of interacting fully with the surrounding environment, in an input/output relationship with climatic parameters, we imagined an architecture that interacts with sound waves while creating clean energy and a dynamic landscape, not made of masses and of measurable spaces in lengths, but consists of an energy environment in which the only measure was the time. So our project, "Interactive panel for urban noise reduction", is a noise barrier that can creep in and wrap between the gorges of the consolidated city, performing multiple destinations, infrastructure, public space, environmental system. The most fascinating aspect was discovering the convergence of architecture and technology, in a process in which technology becomes architecture and the architectural element is technology: piezoelectric cells become a changeable skin which may take different configurations, it is the Nanoarchitecture of the world."

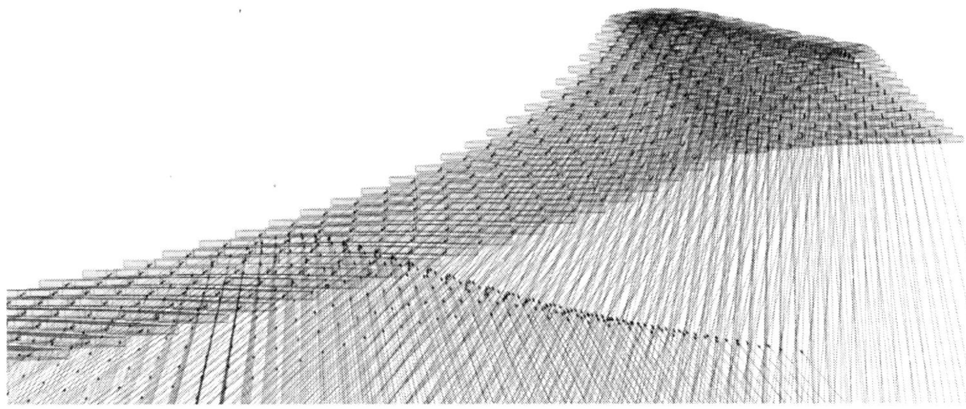

Fig. 11. Gaetano De Francesco, 2011. Soundscape. Structure realized with the use of generative software

4 Cognitive Computing and Watson

The new frontier of design process is cognitive computing, like Watson, a system that acquires information and integrates processes in different ways, through inclusion in the database and through access to the infinite web database. Watson will soon be applied to the design process and is expected to be able to implement the design, construction and management of the entire architectural project.

The novelty is the network connection that allows one to search an immense database and to extract key information from documents and everything that is available and stored in the network and produce certain assumptions.

It is possible to query the system with simple questions. In fact, Watson derives from the jeopardy quiz questions that can be trivial and not directly formulated. For example, the question "When was Einstein born?" is put in another way: "At the beginning of the century the scholar responsible for the theory of relativity was born, when?"

The game is indirect and the winner is he who provides the correct answer in the shortest time.

This artificial, robotics intelligence, will be put at the service as a decision support tool. Will it be possible to describe the space through words or descriptions with visual support of thousands of images/concepts evoked by the same words? Could it be possible to interact plastically with a hologram, as a physical model of a project? And could it be possible to do it in connection with a digital transcriber in the PC which processes the physical and material data?

The logical path made by computers follows the methodology of the thought of man and the model of dialogue, obeying to verbal commands, and suggests many formal solutions. The strength of the system is its ability to analyze a huge database to help you come to a synthesis and the resulting product exhibits an architecture project protean, a semantic generator of energy.

Conclusions

The development of mathematical models of new generative sofwares and the simulation of logical processes, similar to the man dendriniche networks, will allow a more interesting man-machine interaction in the design process and in the formulation of digital architecture. Interaction with parametric models and cognitive computing will become greater facilitators in the creation of forms of architecture, facilitators but not a substitute for a creative and selective process whose spatial synthesis will always be human.

References

[1] DI RAIMO, A., *François Roche. Eresie meccaniche e architetture viventi di New-Territories.com,* Edilstampa, Roma, 2015

[2] SAGGIO, A., *Introduzione alla Rivoluzione Informatica in Architettura,* Carocci, Roma, 2007

[3] SAGGIO, A., *Architettura e Modernità. Dal Bauhaus alla Rivoluzione Informatica,* Carocci, Roma, 2010

[4] WATANABE, M.S., *Induction Design. Un metodo per la progettazione evolutiva,* Testo&Immagine, Roma, 2004

[5] HEIDEGGER, M., *Pensieri-guida sulla nascita della metafisica, della scienza contemporanea e della tecnica moderna,* Bompiani, Milano, 2014

[6] TEDESCHI, A., *Architettura Parametrica. Introduzione a Grasshopper*, Le Penseur, 2010.

[7] MEREDITH, M., *From Control to Design: Parametric. Algorithmic Architecture*, Actar, 2008.

[8] TERZIDIS, K., *Algorithmic Architecture*, Oxford, Architectural Press, 2006.

[9] RUBY, A., DURANDIN, B., R & SIE, et al, *Spoiled Climate: R&Sie_ Arquitects*, Birkhäuser-Publishers for Architecture, Boston, 2004

[10] http://www.grasshopper3d.com

[11] https://www.rhino3d.com/it/

[12] http://data-tribe.net/wework4her/

[13] http://andreagraziano.blogspot.it

[14] http://btlb.blogspot.it

[15] http://madeincalifornia.blogspot.it

[16] http://www.makoto-architect.com/subway/subway_2e.html

[17] www.liftarchtects.com

Current address

Belibani Rosalba, researcher Ph.D.
Dipartimento di Architettura e Progetto - DIAP
Via Flaminia 359, 00196 Roma, Italy
Tel. number: +39 0632191226, e-mail: rosalba.belibani@uniroma1.it

Slovak University of Technology in Bratislava
Faculty of Mechanical Engineering
16th Conference on Applied Mathematics **APLIMAT 2017**
Proceedings

APPLICATION OF HEDGING ON NATURAL GAS PRICES

BENADA Luděk (CZ), KONEČNÁ Kateřina (CZ)

Abstract. The following paper examines the problematic of hedging natural gas in the USA. We apply the method of moving window to provide dynamic hedging. The hedge ratio calculated from the last 60 days is employed on the prices for the next 30 days. To estimate the hedge ratio naive portfolio, OLS, copula and wavelet were applied.

Keywords: Spot, Futures, Risk, Hedging, Hedge ratio, Hedging effectiveness, Copula, Wavelet

Mathematics subject classification: 91G10

1 Introduction

Energy and energy commodities are an integral part of modern society life. Abundance of energy which is processed and used by people guarantees comfort of living. Growing consumption of energy is characteristic for developed countries and even more for developing countries. However, concerns about environment could be observed in the last years. A higher pressure on energy mix in the developed countries could be noted as a result of this circumstance. This fact, as well as a superior availability of natural gas, improvement in transportation and relative easy production push the commodity to the top position between others. Given the low environmental encumbrance and the way of unconventional extraction will further increase its importance in the future [15].

In our paper we examined the opportunity for hedging applied on the spot and futures prices of natural gas in the USA. We worked with the most representative benchmark Henry Hub. We focused on daily closing prices. In this study, different forms for estimating hedge ratio from the point of subsequent ability to risk reduction were compared. Measured risk is expressed as a variance of spot and portfolio composed from spot and futures. In our research, the expression for computation of the hedging effectiveness was used that one proposed by Johnson [11]. Then the percentage change in the variance of portfolio over the variance of spot is compared. We determined the optimal hedge ratio by using an objective function that was proposed by Johnson as portfolio variance [11]. We considered an application of three models to obtain an optimal hedge ratio. The variances and covariance of spot and futures were calculated adopting ordinary least squares regression, copula and wavelet theory. Additionally, we apply a hedging by naive portfolio to compare obtained results. From calculated variances and covariance further calculated, hedge ratio will be subsequently verified on future data. The results of risk reduction were compared. Since we supposed time-varying covariance, a dynamic approach on hedge ratio was used. We worked with rolling window over 30 days. Dynamic view of the relation between spot and futures and its impact on risk reduction together with comparison of four distinct methods promised to our paper brought a piece of innovation in given problematic.

2 Literature review

Papers concerning a protection against unexpected price movement came from the first half of the 20th century. Scientists focused on application of futures to keep a price risk of agricultural product on low level [6]. Likewise, Yamey noted the potential of futures to protect manufacturers and merchants against adverse movements of raw material prices [9]. Since, the price of Cash and futures demonstrated strong dependence a simple hedging in the range 1:1 was deemed as appropriate [7]. Later, this attitude was released and a new methodology based on knowledge from modern portfolio theory was introduced [11]. The author determined the optimal hedge ratio as a ratio of covariance of spot and futures to futures variance. He gave a measurement of hedging effectiveness as well. Afterwards, regression of spot on futures were employed to find a hedge ratio [4]. Nevertheless, some authors pointed out that such methodology could be inconsistent, because of the non-stationary data. Cherubini tried to solve this issue by copula approach [3]. Whereas Lien focused on wavelet transformation for the hedging purposes [12]. Due to liberalization and market development in energy sector, emerged some articles concentrated on hedging within energy commodities, where markets are characterized by high price volatility. Mostly they were crude oil, or its derivatives, natural gas and electricity. In the eighties Chen dealt with the eventuality of futures for hedging purposes in oil prices [1]. Woo et al. applied co-integration on natural gas in California to provide effective hedging. They recommended to use financial futures with underlying of Henry Hub spot [16].

3 Models and data

Our examination was entirely focused on short hedging. Such a concept implies that we were in the position of possessing natural gas in spot prices and we provided a protection against an unintentional price risk via selling certain amount of futures. Besides the short hedging we concerned merely in financial hedging, that means we did not contemplate a physical delivery of the explored commodity. The optimal hedging ratio was derived in accordance with the objective function represented by portfolio variance, the portfolio variance for spot and futures is

$$\sigma_p^2 = w_s^2\sigma_s^2 + w_f^2\sigma_f^2 + 2w_s w_f \sigma_{s,f},$$

where w_s is weight of spot, σ_s^2 is variance of spot, w_f is weight of futures, σ_f^2 is variance of futures and $\sigma_{s,f}$ is covariance between spot and futures. The weights restriction is distinct to classical Markowitz [14]. It is assumed one unit of spot. The object of hedging here is to find an appropriate weight proportion of futures to the spot. Solving the extreme of the objective function will provide following expression

$$\frac{\partial \sigma_p^2}{\partial w_f} = 2w_f\sigma_f^2 + 2\sigma_{s,f}.$$

The optimal hedge ratio h^* represents actually the weights for futures. Assuming that the weight for spot is 1, h^* is derived as

$$h^* = -\frac{\sigma_{s,f}}{\sigma_f^2}.$$

The sign of the of the hedge ratio refers to short position in futures. By expressing the second derivative, we can confirm that the extreme value of the objective function is minimum

$$\frac{\partial^2 \sigma_p^2}{\partial w_f^2} = 2\sigma_f^2.$$

Expecting some (nonzero) variance in prices, positive value of the expression in the right side confirmed a minimum of the examined function.

As soon as the ratio is calculated, the value is applied on real data to calculate variances and covariance. Evaluation of hedging performance is based on percentage reduction in spot variance compared to portfolio variance. The metrics was made in accordance with the methodology of Johnson [11]

$$HE = \frac{\sigma_{Up}^2 - \sigma_{Hp}^2}{\sigma_{Up}^2}.$$

HE stands for hedging effectiveness and σ_{Hp}^2 is variance of hedged portfolio and σ_{Up}^2 is variance of unhedged portfolio. Negative futures weights must be taken into account in the calculation of the portfolio variance. The reverse weights leads to the reduction of the covariance risk, since the both variances are additive

$$\sigma_p^2 = \sigma_s^2 + (h^*)^2\sigma_f^2 - 2h^*\sigma_{s,f}.$$

It is evident that the closer price co-movement exists, the better hedging will be provided. So the hedging efficiency is determined by correlation, because it is obviously true

$$\rho = \frac{\sigma_{s,f}}{\sqrt{\sigma_s^2}\sqrt{\sigma_f^2}}.$$

The effect of risk reduction decreases unless the correlation release. In the case of significant drop in correlation or if the correlation even changes to negative value, the risk of hedged portfolio paradoxically increases.

We applied four models which were afterwards compared. The simplest hedging approach was *naive portfolio*, where weights of spot and futures were equal.

The second model was application of *ordinary least squares* on daily logarithmic return. Since the closing prices were not stationary, we transformed them to percentage change. Then we regressed spot return on futures return

$$r_s = \alpha + \beta r_f + \varepsilon,$$

where $r_s = \ln \frac{P_t}{P_{t-1}}$, $r_f = \ln \frac{P_t}{P_{t-1}}$ and P_t stands for closing prices of spot and futures respectively in time t. The parameter α is intercept in the regression and ε stands for standard errors. It must hold $\varepsilon \approx iid$. Here, the coefficient $\beta = -h^*$.

Then we examine the *copula calculus*, at first let us introduce a formal definition of copula. We consider a 2–dimensional copula which is basically a bivariate cumulative distribution function with uniformly distributed margins in interval $[0,1]$. Consider a pair (X,Y) of continuous random variables with marginal cumulative distribution functions

$$F_1(x) = P(X \leq x) \text{ and } F_2(y) = P(Y \leq y)$$

defined for all $x, y \in \mathbb{R}$ and joint cumulative distribution function

$$H(x,y) = P(X \leq x, Y \leq y).$$

The transformed variables $U = F_1(X)$ and $V = F_2(Y)$ are uniformly distributed on $[0,1]$ and their joint distribution function

$$C(u,v) = P(U \leq u, V \leq v), u, v \in [0,1].$$

is called copula. The two mentioned functions are related through equation

$$H(x,y) = C(F_1(x), F_2(y)) = C(u,v),$$

where the copula provides all the information about the dependence structure of X and Y [5]. In our application, the random variables X a Y stand for the closing prices of spot and futures. At first, we created pseudo-observations from the original closing prices, i.e. we transformed them to the interval $[0,1]$. Because of a few possible choices of the copula function, we had to find the appropriate copula family. The closing prices of both spot and futures where identified to belong to Frank copula defined as

$$C(u,v) = -\frac{1}{\theta} \ln \left(1 + \frac{\left(\mathrm{e}^{-\theta u} - 1\right)\left(\mathrm{e}^{-\theta v} - 1\right)}{\mathrm{e}^{-\theta} - 1}\right), \quad \theta \in \mathbb{R} \setminus \{0\}.$$

The next step included the estimation of the parameters of chosen copula. The set of new observation was generated from the copula function, the example of one such simulation is displayed in Fig. 1. After inverse transformation of simulated values into the original scale, the covariance matrix was computed. This process was repeated one hundred times and the covariance matrix was computed as an average of simulations based on the suggested copula [10].

As the last model, we chose *wavelet transformation*. At first, let us consider discrete wavelet transformation (DWT). The signal x_t can be expressed by

$$x_t \approx \sum_k s_{J,k}\phi_{J,k}(t) + \sum_k d_{J,k}\psi_{J,k}(t) + \sum_k d_{J-1,k}\psi_{J-1,k}(t) + \cdots + \sum_k d_{1,k}\psi_{1,k}(t)$$

with the wavelet transform coefficients $s_{J,k} = \int \phi_{J,k}(t)x_t \, \mathrm{d}t$ and $d_{J,k} = \int \psi_{J,k}(t)x_t \, \mathrm{d}t$, where $j = 1, 2, \ldots, J$, J is the number of scales such that $J = \log_2 n$, n is a number of data points

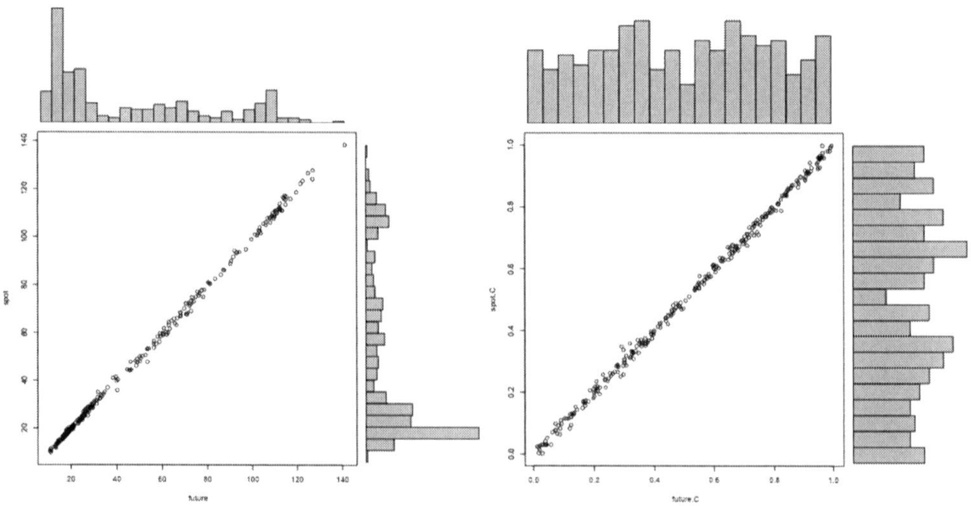

Fig. 1. A plot of original closing prices (x axis: futures, y axis: spot) on the left side and simulations from Frank copula (x axis: futures, y axis: spot) on the right side

and k ranges from 1 to number of coefficients in given component. The $\phi(\cdot)$ and $\psi(\cdot)$ functions stand for the father and mother wavelet, see [8] for more information.

Because of restriction on the sample size to be an integer multiple of 2^J, we considered maximum overlap discrete wavelet transform (MODWT) [13]. In case of sample size divisible by 2^J, the wavelet coefficients $\tilde{d}_{j,k}$, $\tilde{s}_{J,k}$ of MODWT can be expressed by

$$d_{j,k} = 2^{j/2}\tilde{d}_{j,2^j(k+1)-1}, \quad k = 0, \ldots, n/2^j - 1$$

and the scaling coefficient

$$s_{J,k} = 2^{J/2}\tilde{s}_{J,2^J(k+1)-1}, \quad k = 0, \ldots, n/2^J - 1,$$

where $s_J = (s_{J,1}, s_{J,2}, \ldots, s_{J,n/2^J})^T$ and $d_j = (d_{j,1}, d_{j,2}, \ldots, d_{j,n/2^j})^T$ for $j = 1, 2, \ldots, J$. See [13] for detailed explanation.

The signals x_t, y_t are represented by the closing prices of spot and futures. Our goal is to estimate the wavelet variances of X_t and Y_t at scale λ_j and covariance which can be defined as a covariance between the wavelet coefficients of X_t and Y_t at scale λ_j. The variance of the time series x_t, expressed in terms of the MODWT coefficients, is given by

$$\upsilon_X^2(\lambda_j) = \frac{1}{\tilde{N}_j}\sum_{t=L_j}^{N}\left(\tilde{d}_{j,t}^X\right)^2$$

with $\tilde{N}_j = N - L_j + 1$ and $\tilde{L}_j = (2^j - 1)(L - 1) + 1$, where L is the width of the wavelet filter [2]. The covariance between two times series x_t and y_t can be expressed as follows

$$\sigma_{XY,j} = \frac{1}{2^j\tilde{N}_j}\sum_{k=L_j}^{\tilde{N}_j}\tilde{d}_{j,k}^X\tilde{d}_{j,k}^Y.$$

After variances on the data from next 30 days were calculated, we measured and compared hedging effectiveness by each model. For our research we investigated data provided by Energy Information Agency. The examined data were on the Natural Gas index Henry Hub. We used daily closing prices on spot and futures on Henry Hub, traded in New York Mercantile Exchange. The investigated period were January 7, 1997 to July 15, 2015, data are displayed in Fig. 2. We assumed an unstable process in surveyed data as well as unstable affinity. To

Fig. 2. The closing prices of spot and futures

reflect the circumstance we incorporated rolling window methodology - that means we were continuously calculating input data for determining the hedge ratio. Examined data for determining the coefficient was from the last 60 observations. The obtained hedge ratio was afterwards used for the next 30 days. The model was provided over the whole examined period, i.e. hedging was not provided for the first 60 days from January, 1997. Since, the data to March, 1997 was used to establish hedge ratio.

We measured the efficiency of risk reduction by the Johnson's methodology, in other words we measured the decline in variances. We also introduced alternative measurements to evaluate the hedging effectiveness over the whole examined period. Firstly, we compared means and medians of computed hedging effectiveness. Furthermore, we compared the distances of the hedging effectiveness from a reference value that was one. We investigated both, the absolute distances (D_*) of hedging from the reference value of one as well as the squared distances (D_*^2) from the reference value, where

$$D_* = |HE_* - 1|,$$
$$D_*^2 = (HE_* - 1)^2,$$

the subscript $*$ stands for one of the suggested methods (wavelet, naive, copula, ols) and HE_* for the hedging performance corresponding to $*$.
The results could be seen as a residual risk. As the global criterion of the distance from the reference value we computed the absolute sum and the squared sum of these distances.

4 Results

The instability of hedge ratio was confirmed in all of the three employed models. This fact confirmed the variability in correlation. The circumstances are well illustrated in Fig. 3.

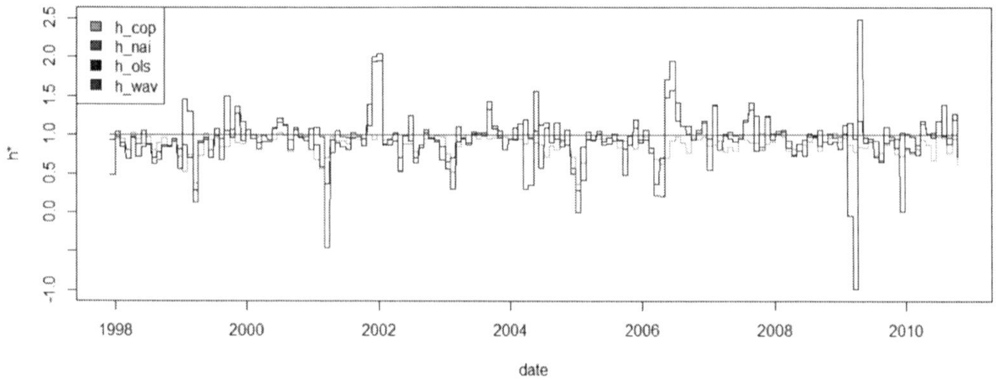

Fig. 3. Instability of hedge ratio

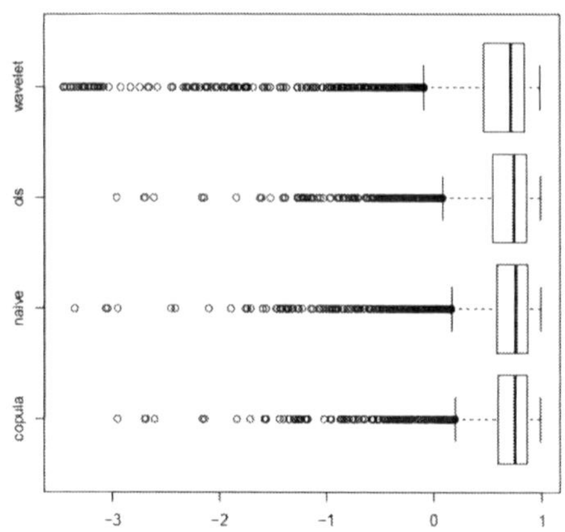

Fig. 4. Boxplots for hedging performance for all methods

As represented in Fig. 3, except for naive portfolio, there is evidence of changeability in all cases over the examined period. As well it is obvious that the performance results are more prevalent below the naive portfolio. The figure also shows hedge ratio received by the copula methodology with the closest oscillation to naive portfolio. On the other hand, results received by the wavelet transformation exhibit the greatest oscillation.

The mean of hedge ratio computed via the wavelet transformation was the highest in comparison to the copula methodology and OLS as well. Simultaneously, wavelet and OLS exhibit often the ratio one. In some cases it could be regarded as overreaction.
The data of hedging effectiveness for all suggested methods are displayed by boxplots in Fig. 4. The performances of means and medians of hedging for all the models over the entire period were calculated and they are shown in Fig. 5.

As we can see in Fig. 6, the hedging performance reflects the hedge ratio, the trend is identical in all models. It is apparent that the ability to reduce risk fluctuates and the risk of hedged

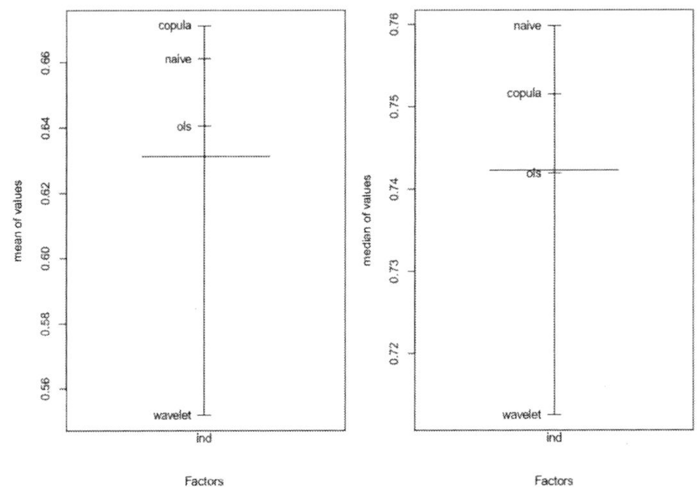

Fig. 5. Means and medians of HE for all the models. The horizontal line describes the mean and median

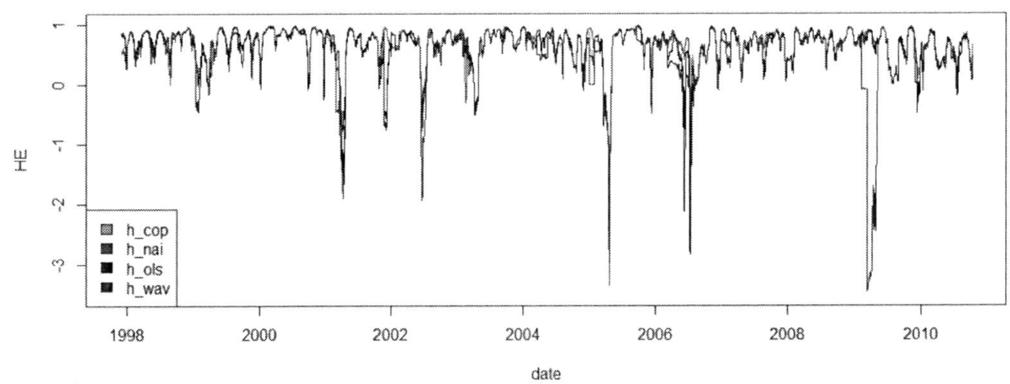

Fig. 6. Hedging effectiveness

portfolio increases dramatically in some intervals. The hedging is altogether counterproductive in the last mentioned cases. As can be seen from the time progress of HE in Fig. 6, wavelet seems to be related with the highest excesses.

As already mentioned, the hedging effectiveness is determined by the state of correlation. Our analysis confirmed this hypothesis. The hedging effectiveness of copula approach with the correlation is displayed in Fig. 7.

The distances of values of HE from the reference value of one were compared for all the models over the entire period. Due to long-time period, we present only the table of global values (see Tab. 1). Furthermore, we compared the hedging performance of each method together. The absolute and squared distances of hedging performance for each method are shown in Tab. 2 and Tab. 3.

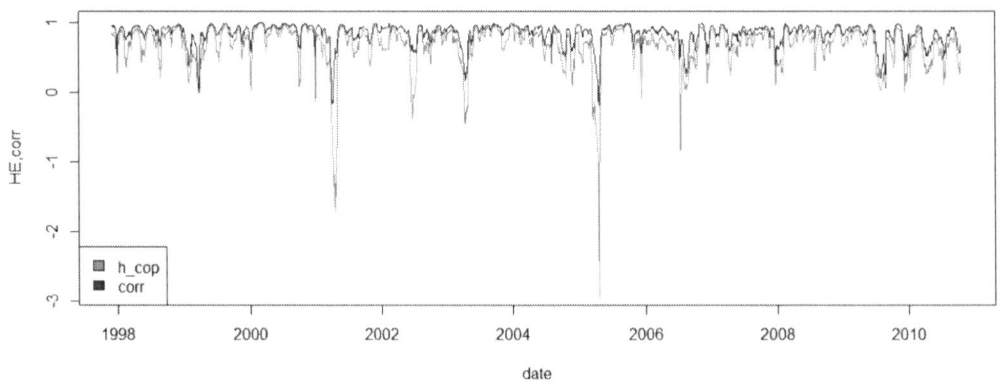

Fig. 7. Correlation and hedging effectiveness

	D_*	D_*^2
wavelet	2105.208	2317.829
naive	1591.539	1114.468
copula	1544.468	977.946
ols	1688.462	1164.357

Tab. 1. Global measure of distance of hedging performance and reference value: absolute and squared distances

	wavelet	naive	copula	ols
wavelet	0	704.5758	705.9027	554.0117
naive		0	176.1576	311.9415
copula			0	252.1855
ols				0

Tab. 2. Absolute distances of hedging performance of each method

	wavelet	naive	copula	ols
wavelet	0	1004.901	1005.5843	825.6868
naive		0	27.8852	108.2101
copula			0	90.1609
ols				0

Tab. 3. Squared distances of hedging performance of each method

5 Discussion

For the future research, it may be interesting to apply more complex models, for instance a dynamic conditional correlation. It would also be useful to determine an appropriate scope of input data for determining hedge ratio and for the application. Further research could be focused on comparison of the performance in other objective functions or if the risk aversion will be append. Further research may utilize a distinct objective functions with a measurement of risk aversion. Regarding the wavelet transformation more research will be convenient with respect to the appropriateness of analysed data and the level of decomposition. Very likely, an application on high frequency data could provide an interesting result.

Conclusion

In our paper we researched a way for reducing price risk on natural gas spot represented from Henry Hub. We examined the daily closing price of spot and futures from the period July 1997 to November 2016. We applied the hedging strategy based on usage of futures in appropriate rate to spot. For hedge ratio we applied four methodologies, especially naive portfolio, OLS, copula and wavelet. Our objective function for optimization was the Markowitz's portfolio variance. It means that we focused on minimum risk from the utility function. To reflect the dynamic in variances and covariance we decided to provide the technique of rolling window. Afterwards, we mutually compared the results. The copula approach gave the best results in our study, naive portfolio also gave satisfactory results. Maybe surprisingly, the ordinary least squares did not over-performed the naive portfolio. The wavelet transformation ended with worst results on the last place. According to our research, we could not recommend to use wavelet to estimate hedge ratio for hedging purposes with futures. Besides this method had the worst overall score, wavelet produced in some periods inconvenient hedge ratio, that was for instance over 2.

Acknowledgement

This research was supported by the project MUNI/A/1039/2016 *Modelling volatility in financial markets and its application in the field of risk management and asset pricing* (Masaryk University) and the project 22738 *Institutional support for development research organization* (Brno University of Technology).

References

[1] Chen, K., Sears, R. S., Tzang, D.-N.: *Oil prices and energy futures*, Journal of Futures Markets, vol. 7, no. 5, (1987), pp. 501–518.

[2] Chen Dar-Hsin, Chuang C.-L., Chen Wei-Ning: *Multiscale hedge ratio between taiwan stock and futures index: an application of wavelet analysis.*, Investment Management and Financial Innovation, vol. 6, no. 1, (2009), pp. 217–230.

[3] Cherubini, U., Luciano, E., Vecchiato, W.: *Copula methods in finance*, John Wiley & Sons, 2004.

[4] Ederington, L. H.: *The hedging performance of the new futures markets*, The Journal of Finance, vol. 34, no. 1, (1979), pp. 157–170.

[5] Genest, C., Segers, J.: *On the covariance of the asymptotic empirical copula process*, Journal of Multivariate Analysis, vol. 101, no. 8, (2010), pp. 1837–1845.

[6] Howell, L. D., Watson, L. J.: *Relation of spot cotton prices to prices of futures contracts and protection afforded by trading in futures*, vol. 601, US Dept. of Agriculture, 1938.

[7] Howell, L. D., et al.: *Analysis of hedging and other operations in grain futures*, US Department of Agriculture, 1948.

[8] In F., Kim S.: *An introduction to wavelet theory in finance: a wavelet multiscale approach*, 9814397849, World Scientific Pub., 2013.

[9] James, T.: *Energy Markets: Price Risk Management and Trading*, Wiley Finance, Wiley, 2012, ISBN 9781118580479, URL: `https://books.google.cz/books?id=iiWrEVkZUj0C`.

[10] Jansons, V., Kozlovskis, K., Lace, N.: *Portfolio modeling using the theory of copula in latvian and american equity market*, Faculty of Engineering Economics, Riga Technical University.

[11] Johnson, L. L.: *The theory of hedging and speculation in commodity futures*, The Review of Economic Studies, vol. 27, no. 3, (1960), pp. 139–151.

[12] Lien, D., Luo, X.: *Estimating the extended mean-gini coefficient for futures hedging*, Journal of Futures Markets, vol. 13, no. 6, (1993), pp. 665–676.

[13] Lien, D., Shrestha K.: *An empirical analysis of the relationship between hedge ratio and hedging horizon using wavelet analysis*, Journal of Futures Market, vol. 27, no. 2, (2006), pp. 127–150.

[14] Markowitz, H.: *Portfolio selection*, The journal of finance, vol. 7, no. 1, (1952), pp. 77–91.

[15] Vivek, C.: *Fundamentals of Natural Gas: An Interntional Perspective*, 1-59370-08-1, Penn Well Corporation, Tulsa, Oklahama USA, 2006.

[16] Woo, C.-K., Olson, A., Horowitz, I.: *Market efficiency, cross hedging and price forecasts: California's natural-gas markets*, Energy, vol. 31, no. 8, (2006), pp. 1290–1304.

Current address

Benada Luděk, Ing.
Department of Finance, Faculty of Economic and Administration,
Masaryk University, Brno
Lipová 41a, 602 00 Brno, Czech republic
Tel. number: +420 549 495 088, e-mail: benada@mail.muni.cz

Konečná Kateřina, Mgr.
Institute of Mathematics and Descriptive Geometry, Faculty of Civil Engineering,
University of Technology, Brno
Žižkova 17, 602 00 Brno, Czech republic
Tel. number: +420 541 147 604, e-mail: konecna.k@fce.vutbr.cz

Slovak University of Technology in Bratislava
Faculty of Mechanical Engineering
16th Conference on Applied Mathematics **APLIMAT 2017**
Proceedings

ALMOST GEODESIC MAPPINGS OF THE FIRST TYPE ONTO SYMMETRIC SPACES

BEREZOVSKI Vladimir (UA), JUKL Marek (CZ), JUKLOVÁ Lenka (CZ)

Abstract. The article is devoted to the theory of almost geodesic mappings of the first type onto symmetric spaces. There are found certain necessary and sufficient conditions when a space with affine connection admits a canonical almost geodesic mapping of the first type onto symmetric space.

Keywords: Almost geodesic mapping, symmetric spaces

Mathematics subject classification: Primary 53B05; Secondary 53B99

1 Introduction

The article deals with further study of theory of almost geodesic mappings of spaces with affine connection. The idea of this theory is based on the article [13] of Levi-Civita. He has formulated and solved (in a special coordinate system) a task to find Riemannian spaces with general geodesics.

Let us remark, that it was related with study of equations of mechanical systems. Subsequently, the theory of geodesic mappings has been developed by Thomas, Weyl, Shirokov, Solodovnikov, Sinyukov, Mikeš and others. Some questions which has appeared during studying of geodesic mappings were developed by Kagan, Vrançeanu, Shapiro and others. Above authors have found special classes of $(n-2)$-projective spaces. See [15, 16, 22].

A.Z. Petrov [19] has introduced a notion of quasi geodesic mappings. In particular, special quasi geodesic mappings are holomorphically projective mappings of Kähler spaces. These mappings were first studied by Otsuki, Tashiro, Prvanovich, Sakaguchi, Mikeš, and others, see [15, 16, 14, 17, 18, 22].

A natural generalization of these classes of mappings is a notion of *almost geodesic mappings* which has been introduced by Sinyukov [22]. He introduced three types of almost geodesic mappings: π_1, π_2 and π_3. See also [2, 3, 18], [15] (pp. 455–462).

We note, that to a theory of almost geodesic mappings π_1 is devoted many papers, for example [1, 4, 5, 6, 7, 9, 8, 10, 18], see also in books [22] and [15] (pp. 463–480).

Fundamental equations of canonical almost geodesic mappings of the first type of spaces with affine connection onto symmetric spaces are described in [22] in the form of closed Cauchy system in covariant derivatives. In this book there is determined a set of basic parameters on which a general solution of such system depends.

Let us consider a simply-connected space the dimension n of which is greater than 2. Suppose that geometric objectives are continuous and sufficiently smooth.

2 Fundamental notions of the theory of almost geodesic mappings of spaces with affine connection

Let us remind basic notions and theorems of theory of almost geodesic mappings of spaces with affine connection as presented by Sinyukov in [22], see [15] (pp. 455–480).

We consider a space A_n with affine connection ∇ without torsion, which is related to a local coordinate system $x^1, x^2, \dots, x^n$ with components $\Gamma^h_{ij}(x)$ of connection ∇. A curve ℓ: $x^h = x^h(t)$ in a space with affine connection A_n, $n > 2$, is called an *almost geodesic* if its tangent vector $\lambda^h = dx^h(t)/dt$ fulfils equalities

$$\lambda^h_2 = a(t)\, \lambda^h + b(t)\, \lambda^h_1 \tag{1}$$

where $\lambda^h_1 \equiv \lambda^h_{,\alpha}\lambda^\alpha$, $\lambda^h_2 \equiv \lambda_{1,\alpha}\lambda^\alpha$, comma denotes covariant derivative respective connection ∇ of space A_n, $a(t)$ and $b(t)$ are some functions of a said argument.

The mapping π of a space A_n with affine connection ∇ onto a space $\bar{A}_n$ with affine connection $\bar{\nabla}$ is called an *almost geodesic mapping* if every geodesic in space A_n is mapped onto an almost geodesic in space $\bar{A}_n$.

Theorem 1. (Sinyukov [22]) *A mapping A_n onto $\bar{A}_n$ is almost geodesic if and only if in a common coordinate system $x^1, x^2, \dots, x^n$ with respect to this mapping a connection deformation tensor*

$$P^h_{ij}(x) = \bar{\Gamma}^h_{ij}(x) - \Gamma^h_{ij}(x) \tag{2}$$

identically fulfils with respect to the $x^1, x^2, \dots, x^n$ and $\lambda^1, \lambda^2, \dots, \lambda^n$ the following condition

$$(P^h_{\alpha\beta,\gamma} + P^h_{\delta\alpha} P^\delta_{\beta\gamma})\lambda^\alpha\lambda^\beta\lambda^\gamma = b\, P^h_{\alpha\beta}\lambda^\alpha\lambda^\beta + a\,\lambda^h \tag{3}$$

where Γ^h_{ij} and $\bar{\Gamma}^h_{ij}$ are components of affine connection on A_n and $\bar{A}_n$, $\lambda^1, \lambda^2, \dots, \lambda^n$ are components of some vector, a and b are functions depending on $x^1, x^2, \dots, x^n$ and $\lambda^1, \lambda^2, \dots, \lambda^n$.

Relating to character of a dependency of functions a and b on $\lambda^1, \lambda^2, \dots, \lambda^n$ Sinyukov has defined three types of almost geodesic mappings π_1, π_2 and π_3.

We have proved (see [2, 3], [15] (p. 459)) that in the case $n > 5$ there does not exists any other type of geodesic mapping.

Especially, a mapping π_1: $A_n \to \bar{A}_n$ is a almost geodesic mapping of the type π_1 if in a common coordinate system with respect to π_1 the following conditions are fulfil

$$P^h_{(ij,k)} + P^\alpha_{(ij} P^h_{k)\alpha} = \delta^h_{(i} a_{jk)} + b_{(i} P^h_{jk)} \tag{4}$$

where a_{ij} is a symmetric tensor, b_i is some covector and brackets denote a symmetrization with respect to denoted indices.

If a covector b_i is identically equal to zero then the mappings is called *canonical almost geodesic mapping of the type* π_1.

It is known (see [22]) that any almost geodesic mapping of the type π_1 may be represented in the form of a composition of canonical geodesic mapping of the type π_1 and a geodesic mapping.

3 Diffeomorphisms of spaces with affine connection onto symmetric spaces

A space with affine connection $\bar{A}_n$ is called *(local) symmetric*, if the Riemann tensor is absolutely parallel (P.A. Shirokov [20, 21], É. Cartan [11], S. Helgason [12]). By this way, symmetric spaces $\bar{A}_n$ are characterized by

$$\bar{R}^h_{ijk|m}(x) \equiv 0 \tag{5}$$

where $\bar{R}^h_{ijk}$ is the Riemannian tensor of $\bar{A}_n$, symbol "$|$" denotes covariant derivative with respect to connection $\bar{\nabla}$ of a space $\bar{A}_n$.

By diffeomorphism of spaces with affine connection we mean a bijective smooth mapping, the inverse of which is also a smooth mapping. Almost geodesic mappings have an important role between diffeomorphisms of spaces with affine connection.

Let us suppose that a space A_n with affine connection ∇ admits a diffeomorphism f onto a space $\bar{A}_n$ with affine connection $\bar{\nabla}$ and these mappings are related to a common coordinate system $(x^1, x^2, \dots, x^n)$ with respect to the mapping f. For covariant derivative of the Riemannian tensor $\bar{R}^h_{ijk}$ of space $\bar{A}_n$ with affine connection $\bar{\nabla}$ we have

$$\bar{R}^h_{ijk|m} = \frac{\partial \bar{R}^h_{ijk}}{\partial x^m} + \bar{\Gamma}^h_{m\alpha}\bar{R}^\alpha_{ijk} - \bar{\Gamma}^\alpha_{mi}\bar{R}^h_{\alpha jk} - \bar{\Gamma}^\alpha_{mj}\bar{R}^h_{i\alpha k} - \bar{\Gamma}^\alpha_{mk}\bar{R}^h_{ij\alpha}. \tag{6}$$

Considering (2) we may formula (6) written in the form:

$$\bar{R}^h_{ijk|m} = \bar{R}^h_{ijk,m} + P^h_{m\alpha}\bar{R}^\alpha_{ijk} - P^\alpha_{mi}\bar{R}^h_{\alpha jk} - P^\alpha_{mj}\bar{R}^h_{i\alpha k} - P^\alpha_{mk}\bar{R}^h_{ij\alpha}. \tag{7}$$

In what follows we suppose that a space $\bar{A}_n$ is symmetric. With respect to the formula (5) we obtain from (7):

$$\bar{R}^h_{ijk,m} = -P^h_{m\alpha}\bar{R}^\alpha_{ijk} + P^\alpha_{mi}\bar{R}^h_{\alpha jk} + P^\alpha_{mj}\bar{R}^h_{i\alpha k} + P^\alpha_{mk}\bar{R}^h_{ij\alpha}. \tag{8}$$

Formulas (8) hold for diffeomorphisms of any nature of spaces with affine connection onto symmetric spaces respective to common coordinate system $(x^1, x^2, \dots, x^n)$.

4 Canonical almost geodesic mappings of the first type of spaces with affine connection onto symmetric spaces

Let us consider a canonical almost geodesic mappings of spaces A_n with affine connection onto symmetric spaces $\bar{A}_n$. It is well known (see [22], [15] (p. 463)) that equations (4) may be expressed in the form

$$3\,(P^h_{ij,k} + P^\alpha_{ij}P^h_{\alpha k}) = R^h_{(ij)k} - \bar{R}^h_{(ij)k} + \delta^h_{(k}a_{ij)} + b_{(i}P^h_{jk)} \tag{9}$$

where R^h_{ijk} is the Riemannian tensor of a space A_n.

Therefore canonical almost geodesic mappings of the first type of spaces with affine connection will be characterized by equations

$$P^h_{ij,k} = -P^\alpha_{ij} P^h_{\alpha k} + \frac{1}{3}\,(R^h_{(ij)k} - \bar{R}^h_{(ij)k} + \delta^h_{(k} a_{ij)}). \tag{10}$$

Covariant differentiate (10) with respect to x^m. Considering formulas (8) and (10) we may derive:

$$\begin{aligned} P^h_{ij,km} = &- \tfrac{1}{3}\,R^\alpha_{(ij)m} P^h_{\alpha k} - \tfrac{1}{3}\,R^h_{(\alpha k)m} P^\alpha_{ij} + \tfrac{1}{3}\,R^h_{(ij)k,m} + P^\beta_{ij} P^\alpha_{\beta m} P^h_{\alpha k} + \\ & P^\beta_{\alpha k} P^h_{\beta m} P^\alpha_{ij} + \tfrac{1}{3}\,(\bar{R}^\alpha_{(ij)m} P^h_{\alpha k} + \bar{R}^h_{(\alpha k)m} P^\alpha_{ij} + P^h_{m\alpha} \bar{R}^\alpha_{(ij)k} - P^\alpha_{mk} \bar{R}^h_{(ij)\alpha} - \\ & P^\alpha_{mi} \bar{R}^h_{(j\alpha)k} - P^\alpha_{mj} \bar{R}^h_{(i\alpha)k} - \delta^\alpha_{(m} a_{ij)} P^h_{\alpha k} - \delta^h_{(\alpha} a_{km)} P^\alpha_{ij} + \delta^h_{(k} a_{ij),m}). \end{aligned} \tag{11}$$

Alternating (11) with respect to k and m and using Ricci identity and properties Riemannian tensor we may obtain:

$$\begin{aligned} \delta^h_{(m} a_{ij),k} - \delta^h_{(k} a_{ij),m} = &-3P^h_{\alpha j} R^\alpha_{ikm} - 3P^h_{i\alpha} R^\alpha_{jkm} - R^\alpha_{(ij)m} P^h_{\alpha k} + R^\alpha_{(ij)k} P^h_{\alpha m} + \\ R^h_{(ij)k,m} - R^h_{(ij)m,k} &+ 3\bar{R}^h_{\alpha km} P^\alpha_{ij} - P^\alpha_{mi} \bar{R}^h_{(j\alpha)k} + P^\alpha_{ki} \bar{R}^h_{(j\alpha)m} - P^\alpha_{mj} \bar{R}^h_{(i\alpha)k} + \\ & P^\alpha_{kj} \bar{R}^h_{(i\alpha)m} - \delta^\alpha_{(m} a_{ij)} P^h_{\alpha k} + \delta^\alpha_{(k} a_{ij)} P^h_{\alpha m}. \end{aligned} \tag{12}$$

Contracting equation (12) with respect to indices h and m we may derive that

$$\begin{aligned} &(n+1)\,a_{ij,k} - a_{ik,j} - a_{jk,i} = \\ &-3P^\beta_{\alpha(j} R^\alpha_{i)k\beta} - P^\beta_{\alpha k} R^\alpha_{(ij)\beta} + P^\beta_{\alpha\beta} R^\alpha_{(ij)k} + R^\beta_{(ij)k,\beta} - R_{(ij),k} + 3P^\alpha_{ij} \bar{R}_{\alpha k} - \\ &P^\alpha_{\beta i} \bar{R}^\beta_{(j\alpha)k} + P^\alpha_{ki} R_{(j\alpha)} - P^\alpha_{\beta j} \bar{R}^\beta_{(i\alpha)k} + P^\alpha_{kj} \bar{R}_{(i\alpha)} - \delta^\alpha_{(\beta} a_{ij)} P^\beta_{\alpha k} + a_{(ij} P^\beta_{k)\beta} \end{aligned} \tag{13}$$

where R_{ij}, $\bar{R}_{ij}$ are Ricci tensors od spaces A_n and $\bar{A}_n$, respectively.

Alternating equations (13) with respect indices j and k we may derive that

$$\begin{aligned} a_{ij,k} - a_{ik,j} = &\tfrac{1}{n+2}\,\big[P^\beta_{\alpha k}(R^\alpha_{ij\beta} + R^\alpha_{\beta ji}) - P^\beta_{\alpha j}(R^\alpha_{ik\beta} + R^\alpha_{\beta ki}) + 3P^\beta_{\alpha\beta} R^\alpha_{ijk} + \\ 3R^\beta_{ijk,\beta} - R_{(ij),k} &+ R_{(ik),j} + 2P^\alpha_{ij} \bar{R}_{\alpha k} - 2P^\alpha_{ik} \bar{R}_{\alpha j} + P^\alpha_{ki} \bar{R}_{j\alpha} - P^\alpha_{ji} \bar{R}_{k\alpha} - \\ & P^\alpha_{\beta j} \bar{R}^\beta_{(i\alpha)k} + P^\alpha_{\beta k} \bar{R}^\beta_{(i\alpha)j} - a_{ij} P^\alpha_{\alpha k} - a_{\alpha j} P^\alpha_{ik} + a_{ik} P^\alpha_{\alpha j} + a_{\alpha k} P^\alpha_{ij}\big]. \end{aligned} \tag{14}$$

Wit respect to equations (14) we may the equation (13) write in the form:

$$\begin{aligned} (n-1)\,a_{ij,k} = &-3P^\beta_{\alpha(j} R^\alpha_{i)k\beta} - P^\beta_{\alpha k} R^\alpha_{(ij)\beta} + P^\beta_{\alpha\beta} R^\alpha_{(ij)k} + R^\beta_{(ij)k,\beta} - R_{(ij)k} + \\ & 3P^\alpha_{ij} \bar{R}_{\alpha k} - P^\beta_{\alpha i} \bar{R}^\alpha_{(j\beta)k} + P^\alpha_{ki} \bar{R}_{(j\alpha)} - P^\beta_{\alpha j} \bar{R}^\alpha_{(i\beta)k} + P^\alpha_{kj} \bar{R}_{(i\alpha)} - \\ & \delta^\alpha_{(\beta} a_{ij)} P^\beta_{\alpha k} + \delta^\alpha_{(k} a_{ij)} P^\beta_{\alpha\beta} - \tfrac{1}{n+2}\,B_{(ij)k} \end{aligned} \tag{15}$$

where

$$B_{ijk} = P^{\beta}_{\alpha k}(R^{\alpha}_{ij\beta} + R^{\alpha}_{\beta ji}) - P^{\beta}_{\alpha j}(R^{\alpha}_{ik\beta} + R^{\alpha}_{\beta ki}) + 3P^{\beta}_{\alpha\beta}R^{\alpha}_{ijk} + 3R^{\beta}_{ijk,\beta} -$$

$$R_{(ij),k} + R_{(ik),j} + 2P^{\alpha}_{ij}\bar{R}_{\alpha k} - 2P^{\alpha}_{ik}\bar{R}_{\alpha j} + P^{\alpha}_{ki}\bar{R}_{j\alpha} - P^{\alpha}_{ji}\bar{R}_{k\alpha} -$$

$$P^{\alpha}_{\beta j}\bar{R}^{\beta}_{(i\alpha)k} + P^{\alpha}_{\beta k}\bar{R}^{\beta}_{(i\alpha)j} - a_{ij}P^{\alpha}_{\alpha k} - a_{\alpha j}P^{\alpha}_{ik} + a_{ik}P^{\alpha}_{\alpha j} + a_{\alpha k}P^{\alpha}_{ij}.$$

Evidently, equations (8), (10) and (15) in a given space A_n present a form of closed Cauchy system with respect to functions $\bar{R}^h_{ijk}(x)$, $P^h_{ij}(x)$ and a_{ij}, which, naturally, must fulfil the following algebraic conditions

$$\bar{R}^h_{i(jk)} = \bar{R}^h_{(ijk)} = 0, \quad P^h_{ij} = P^h_{ji}, \quad a_{ij} = a_{ji}. \tag{16}$$

Theorem 2. *A space A_n with affine connection admits a canonical almost geodesic mapping of the type π_1 onto symmetric space $\bar{A}_n$ if and only if in A_n there exists a solution of mixed closed Cauchy system (8), (10), (15) and (16) with respect to functions $\bar{R}^h_{ijk}(x)$, $P^h_{ij}(x)$ and $a_{ij}(x)$.*

Evidently, general solution of a mixed closed Cauchy system (8), (10), (15) and (16) depends at most on $\frac{1}{6}\,n\,(n+1)\,(n^2+2n+3)$ real parameters.

Theorem 2 above precises results of theory of canonical almost geodesic mappings onto symmetric spaces which have been obtained in [22] and [8, 7].

References

[1] Berezovski, V., Bácsó, S., Mikeš, J.: *Almost geodesic mappings of affinely connected spaces that preserve the Riemannian curvature*, Ann. Math. Inform., vol. 45, (2015), pp. 3–10.

[2] Berezovski, V., Mikeš, J.: *On the classification of almost geodesic mappings of affine-connected spaces*, in *Differential geometry and its applications (Dubrovnik, 1988)*, Univ. Novi Sad, Novi Sad, 1989, pp. 41–48.

[3] Berezovski, V., Mikeš, J.: *On a classification of almost geodesic mappings of affine connection spaces*, Acta Univ. Palack. Olomuc. Math., vol. 35, (1996), pp. 21–24 (1997).

[4] Berezovski, V., Mikeš, J., Vanžurová, A.: *Almost geodesic mappings onto generalized Ricci-symmetric manifolds*, Acta Math. Acad. Paedagog. Nyházi. (N.S.), vol. 26, no. 2, (2010), pp. 221–230.

[5] Berezovski, V. E., Guseva, N. I., Mikeš, J.: *On special first-type almost geodesic mappings of affine connection spaces preserving a certain tensor*, Math. Notes, vol. 98, no. 3, (2015), pp. 515–518.

[6] Berezovski, V. E., Mikeš, J., Vanžurová, A.: *Canonical almost geodesic mappings of type $\tilde{\pi}_1$ onto pseudo-Riemannian manifolds*, in *Differential geometry and its applications*, World Sci. Publ., Hackensack, NJ, 2008, pp. 65–75, URL: `http://dx.doi.org/10.1142/9789812790613_0007`.

[7] Berezovski, V. E., Mikeš, J., Vanžurová, A.: *Fundamental PDE's of the canonical almost geodesic mappings of type $\tilde{\pi}_1$*, Bull. Malays. Math. Sci. Soc. (2), vol. 37, no. 3, (2014), pp. 647–659.

[8] Berezovskii, V. E., Mikeš, J.: *On canonical almost geodesic mappings of the first type of affinely connected spaces*, Russian Math. (Iz. VUZ), vol. 58, no. 2, (2014), pp. 1–5.

[9] Berezovskij, V., Mikeš, J.: *On almost geodesic mappings of the type π_1 of Riemannian spaces preserving a system n-orthogonal hypersurfaces*, Rend. Circ. Mat. Palermo (2) Suppl., vol. 59, (1999), pp. 103–108.

[10] Berezovsky, V., Mikeš, J.: *On special almost geodesic mappings of type π_1 of spaces with affine connection*, Acta Univ. Palack. Olomuc. Fac. Rerum Natur. Math., vol. 43, (2004), pp. 21–26.

[11] Cartan, E.: *Les espaces riemanniens symétriques*, Verhandlungen Kongress Zürich, vol. 1, no. 3, (1932), pp. 152–161.

[12] Helgason, S.: *Geometric analysis on symmetric spaces*, AMS, Providence, 2008.

[13] Levi-Civita, T.: *Les espaces riemanniens symétriques*, Verhandlungen Kongress Zürich, vol. 1, no. 3, (1932), pp. 152–161.

[14] Mikeš, J.: *Holomorphically projective mappings and their generalizations*, J. Math. Sci. (New York), vol. 89, no. 3, (1998), pp. 1334–1353.

[15] Mikeš, J., et al.: *Differential geometry of special mappings*, Palacky Univ. Press, Olomouc, 2015.

[16] Mikeš, J., Vanžurová, A., Hinterleitner, I.: *Geodesic mappings and some generalizations*, Palacky Univ. Press, Olomouc, 2009.

[17] Mikeš, J.: *On geodesic and holomorphic-projective mappings of generalized m-recurrent Riemannian spaces*, Sib. Mat. Zh., vol. 33, no. 5, (1992), p. 215.

[18] Mikeš, J., Berezovski, V., Stepanova, E., Chudá, H.: *Geodesic mappings and their generalizations*, J. Math. Sci., New York, vol. 217, no. 5, (2016), pp. 607–623.

[19] Petrov, A. Z.: *Modeling of the paths of test particles in gravitation theory*, Gravit. and the Theory of Relativity, vol. 4–5, (1998), pp. 7–21.

[20] Shirokov, A. P.: *P. A. Shirokov's work on the geometry of symmetric spaces*, J. Math. Sci., New York, vol. 89, no. 3, (1998), pp. 1253–1260.

[21] Shirokov, P. A.: *Selected investigations on geometry*, Kazan Univ. Press, 1966.

[22] Sinyukov, N. S.: *Geodesic mappings of Riemannian spaces*, Nauka, Moscow, 1979.

Current address

Berezovski Vladimir, doc. CSc.
Department of Mathematics and Physics, Uman National University of Horticulture
Instytutska 1, Uman, Ukraine
Tel. number: +380 662 901 879, e-mail: berez.volod@rambler.ru

Jukl Marek, doc. RNDr. PhD.
Department Algebra and Geometry, Faculty of Natural Science, Palacky University
17. listopadu 12, 77146 Olomouc, Czech republic
Tel. number: +420 777 144 274, e-mail: marek.jukl@upol.cz

Juklová Lenka, RNDr. PhD.
Department Algebra and Geometry, Faculty of Natural Science, Palacky University
17. listopadu 12, 77146 Olomouc, Czech republic
Tel. number: +420 607817 705, e-mail: lenka.juklova@upol.cz

APPLIED MATHEMATICS SEMINAR

BIBA Vladislav (CZ), SMETANOVÁ Dana (CZ), VARGOVÁ Michaela (CZ)

Abstract. The paper describes the initial state of basic university courses of mathematics taught at The Institute of Technology and Business in České Budějovice, introduction of new applied mathematics seminar. The model examples especially from part of the ordinary differential equations are presented.

Keywords: applied mathematics, seminar, ordinary differential equations

Mathematics Subject Classification: Primary 97B10, 97B40, Secondary 97A80

1 Introduction

Mathematics is an integral part of teaching technical subjects for which the mathematical apparatus is one of the basic "working tools". It has extensive applications in all disciplines taught at The Institute of Technology and Business in České Budějovice (hereinafter ITB).

Given our experience and that of our colleagues from other departments who teach technical subjects, we know that students do not link the knowledge of mathematics lessons with specific tasks in technical seminars. From our experience we know that students learn mathematics primarily to cope with the exam. Subsequently, the acquired skills are quickly forgotten. During direct use of mathematical methods in technical subjects they do not perceive it as the same subject matter, they study it as new and they do not see it in the context of prior knowledge. Specifically, we can mention searching for the function extreme values, which students use e.g. in physics subjects and in Construction and Engineering subjects. Matrices are used extensively in the field of Technology of traffic and transportation.

We can illustrate this situation by the following two statements that were written via the information system into an anonymous questionnaire on teaching mathematics by students in combined form of study.

Mathematics (specialization in Business Economics): *"Beautifully explained subject matter that can be well understood - what a pity that I feel that mathematics as such is a subject utterly useless in practice and for life - matrices, vectors, functions and other cannot be used in practice by an office woman or in other business field. Therefore to learn and to forget the substance matter is total waste of time on studying. "*

Mathematics II (Technology of traffic and transport or Civil Engineering fields of study): *"After 25 years of practice I have to learn higher mathematics which is impossible to learn or to use practically even in combined study, and which I have never needed, and all of that only to be forgotten immediately because I do not need it either at work or for further studies."*

Note that the previous two views are not entirely relevant. Students use mathematical apparatus in many technical subjects, especially in further studies.

Currently, teaching mathematics at technical colleges is a huge challenge due society-wide inclination towards humanities. Often, the teachers at engineering colleges meet students in mathematics lessons who do not have a good foundation in mathematics. They must examine the depth of this deficiency [11] and resolve such deficiencies in preparatory courses [14]. Other current issues in mathematics education at university are the contents of these works [3-9, 11, 12, 14].

In previous years we conducted a survey among students of daily and combined form of study about their relationship to the natural sciences - mathematics, physics and chemistry using an anonymous questionnaire. Among other things we were interested in how they perceive the practical usefulness of science in everyday life. We asked the question: "Have you ever used mathematical (or physical or chemical) knowledge in your life? a) yes b) no c) I do not know." If the answer is yes, then write an example of the use of such knowledge (c.f. [1, 2, 10, 13]).

The results were surprising to us, even though most respondents perceived that they use the knowledge of these sciences, a small number were able to give a concrete example of this usage.

The situation observed in the teaching and the results of a questionnaire survey prompted us to attempt to introduce a new mathematical seminar named Applied Mathematics for Technicians which is devoted to applications of mathematical knowledge gained in the basic university courses of mathematics. More details about questionnaire survey can be found in [1, 2, 10, 13].

The following sections describe the initial state of basic university courses of mathematics taught at ITB, new seminar topics, the applied examples choice. We also selected model examples.

2 Initial situation to applied mathematics

Starting with the academic year 2016/17 the situation in teaching the basic course of higher mathematics bachelor programs is following: The students complete three courses in mathematics (Mathematics or Mathematics I, Mathematics II, Mathematics III) and one course of statistics. The subject Mathematics with 4 lessons per week is intended for students of Business Economics. Mathematics I and Mathematics II have two variants of lessons per week and are intended for all engineering students. The options are 2 hours of lectures and 2 hours of seminars per week or 2 hours of lectures and 4 hours of seminars. Mathematics III is meant only for students majoring in construction of buildings in the form of two-hour seminars weekly. Statistics is meant for students of economy and most technical fields in the range of 2 hours lectures and 2 hours of seminar.

The situation with two different numbers of lessons for mathematics was due to the gradual re-accreditation of technical fields. Currently Technology of traffic and transport is re-accredited with higher number of lessons of mathematics. Other fields will have a higher subsidy after re-accreditation in the respective years. This has opened the way to a better explanation of the substance matter.

Syllabi of Mathematics and Mathematics I are fairly extensive, there are basics of linear algebra, differential and integral calculus. In Mathematics II there are presented special types of integration and foundation solutions of ordinary differential equations of the first order and higher order. Mathematics III covers the basics of differential and integral calculus of several variables. The objective of Statistics course is to familiarize the students with the basic procedures of statistical inference, methods of analysis of qualitative and quantitative traits and elements of time series analysis.

The idea of creating a seminar devoted to applied mathematics prompted us to applying for an internal grant competition at ITB in spring 2016. We received an internal grant for introducing a new subject, which supported the creation of study materials and purchase of necessary literature.

We have created a new subject called Applied Mathematics for technicians intended for students in bachelor's degree.

Course syllabus of Applied Mathematics for Engineers:

- applications of matrices calculus (e.g., cryptography) – 1 seminar,
- linear optimization (graphic method, simplex method, introduction to game theory) – 2 seminars,
- local extremas of real functions (volume, costs, path length …) – 2 seminars,
- applications of integral calculus (geometric and physical) – 2 seminars,
- applications of ordinary differential equations of 1st order, separation of variables (decline, growing population, …) – 1 seminar,
- applications of ordinary linear differential equations of 1st order – 1 seminar,
- applications of ordinary differential equations of higher order with constant coefficients – 1 seminar,
- probability (games of chance, conditional probability) – 1 seminar,
- financial mathematics (interest rate, compound interest, redemption of loan and annuity calculus) – 2 seminars.

The syllabus has been created with respect to the following factors: the semester at the ITB has 13 weeks, optional seminars are always in the range of 2 hours per week.

The seminar content followed the subjects that all students in technical fields have, namely Mathematics I, II, and to a lesser extent Statistics. Our intention was to complement the subject matter with appropriate illustrative examples of mathematical knowledge applications. The plan is to start the lessons with a brief revision followed by solving specific problems.

The aim of the course is to provide students with knowledge of the possibilities of mathematical applications into the technical (or economical) practice of linear algebra, differential and integral calculus of functions of one variable, differential equations, probability and financial mathematics.

After completing the course the students will be able to solve applied problems from subject matter individually.

3 Selecting practical problems solving

In order to solve application examples successfully, the students require the knowledge of basic math calculations. Therefore, we have connected the selection of examples to subjects Mathematics I, Mathematics II and Statistics. Substance matter of Mathematics III was not included in the selection of examples, because it is attended the only by students majoring in Building Construction.

To select the examples we are limited by several factors:

- students with only the foundations of higher mathematics knowledge,
- examples, the calculation of which does not take too long given the limited time of lessons,
- aptness of the example,
- possibility to use the solution for more applied problems.

Approach to the choice of examples:

- straightforward - examples illustrating technical problem directly,
- "detour" - examples of "attractive and funny nature," which do not solve a specific technical problem, but their solutions can be used for technical examples.

In the following section we present some selected examples illustrating both approaches from part the ordinary differential equations.

"Detour" choice:

A. Alcohol in blood

Let $y(t)$ be the level of alcohol in the blood during alcohol consumption. Suppose that the average rate of drinking is constant. Further assume that alcohol is detoxified in the liver at a rate proportional to its blood level. Then an equation in the form $y' = a - by$, $y(0) = y_0$ would describe the blood level over the period of drinking. Condition $y(0) = 0$ would signify the absence of alcohol in the body at the beginning of the evening. The constant a would reflect the rate of intake per unit volume of the individual's blood: larger people take longer to "get drunk" for a given amount consumed. The constant b would represent the rate of decay of alcohol per unit time due to degradation by the liver. In the phase of "recovery", after drinking stops, the above differential equation no longer describes the level of alcohol. Instead, the process of recovery is represented by $y' = -by$, $y(0) = y_0$. The level of alcohol then decays exponentially with rate b from its level at the moment that drinking ends.

B. Forensic mathematics

A detective discovers a murder victim in a hotel room at 9:00 am one morning. The temperature of the body is 27° C. One hour later, at 10:00 am, the body has cooled to 24° C. The room is kept at a constant temperature of 21° C. Assume that the victim had a normal temperature of 36° – 37° C at the time of death. We'll use differential equations to find the time the murder took place. Let $u(t)$ be the temperature of the body after t hours. By Newton's Law of Cooling we have the differential equation $u' = k(u - 20)$, where k is a constant (to be determined). We'll solve the differential equation and get a formula for $u(t)$. Note that $u(t) = 21 + A\,e^{kt}$. In time $t = 0$ the formula $u(0) = 27 = 21 + A$ is satisfied, hence $u(t) = 21 + 6\,e^{kt}$. After 1 hour $t = 1$ the equation $u(1) = 24 = 21 + 6\,e^{k}$ reveals $k = \ln\frac{1}{2} = -\ln 2$.

The solution of the differential equation has the following form $u(t) = 21 + 6\,e^{-t\,ln2}$. Now we can easily calculate that the murder time is approximately 7:35 am - 7:40 am.

Straightforward choice:

C. The catenary

If we want to describe the shape of hanging cable, such as telephone wire hanging between telephone poles or an electrical cable suspended between pylons, we need to solve the (second order non-linear) differential equation in the form $y'' = k\sqrt{1+(y')^2}$ with initial conditions $y(0)=0$, $y'(0)=0$, where k is material constant. By solving of that differential equation we can find that the cable takes the shape of a curve with equation $y = \frac{1}{k}\cosh kx$ called the catenary.

D. Sandwich panel

In the construction of building often there is used a so-called sandwich panel. The sandwich panel is made from concrete, where its the inner and outer wall is interposed a layer of polystyrene. Assume the temperature of the inner wall is 20° C and the outer wall has a temperature of 5° C. For simplicity we consider unit area. Determine how much more heat escapes using polystyrene a width of 7 cm instead of polystyrene with a width of 10 cm. Thermal conductivity of polystyrene 0.157 $Wm^{-1}\,°C^{-1}$. The solution of the above example is given in a similar way as example B.

E. Linear optimization

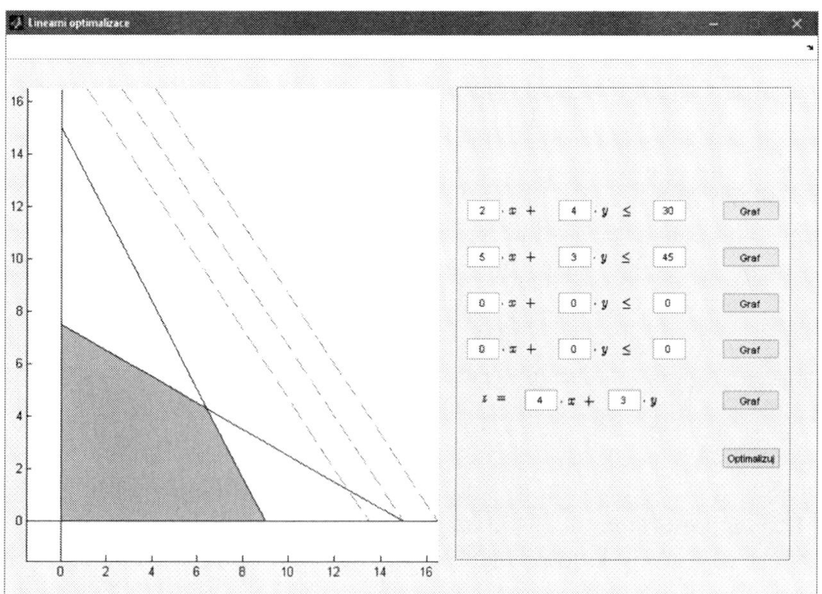

Fig. 1. MatLab applet for linear optimization

Students of Technology of traffic and transport often need to solve an optimization problem. When dealing with the types of problem, it is beneficial to have a geometric view of the situation. Unfortunately, students' ability to use tools of analytic geometry is very poor due to the fact that on High Schools they have limited time for learning specialized parts of mathematics.

Applet on above Fig. 1 shows graphical approach to solving simple two-dimensional linear optimization problem. The applet was created in SW MatLab. It should be useful for acquiring a geometrical notion of this type of problems.

4 Conclusions

As is obvious from the beginning, integrating applied mathematics into teaching is more than desirable. It is necessary to make the students of bachelor's degree realize the future usefulness of the substance matter taught in specialized subjects and in subsequent study and familiarize them with the mathematical description of particular technical problems.

Acknowledgement

The authors gratefully acknowledge the Institute of Technology and Business in České Budějovice for supporting this work under the internal grant The Applied Mathematics for Technicians.

References

[1] BIBA, V., KLEPANCOVÁ, M. Názor študentov na predmet chémia - skúmanie pomocou testu nezávislosti. *Trendy ve vzdělávání*, Olomouc: Univerzita Palackého v Olomouci, 2015, vol. 8, č. 1, s. 22-26. ISSN 1805-8949.

[2] BIBA, V., SMETANOVÁ, D. Zkoumání vztahu k matematice pomocí testu nezávislosti. *Trendy ve vzdělávání*, Olomouc: Univerzita Palackého v Olomouci, 2015, vol. 8, č. 1, s. 27-30. ISSN 1805-8949.

[3] BOHÁČ, Z., DOLEŽALOVÁ, J., KREML, P. Problems of studying in the first year at VSB - Technical university of Ostrava. In *APLIMAT 2014: 13th Conference on Applied Mathematics*, Proceedings, 34 – 43, 2014, ISBN 978-80-227-4140-8.

[4] BOHÁČ, Z., DOLEŽALOVÁ, J., KREML, P. Selected aspects affecting study results in mathematics, In *APLIMAT 2015: 14th Conference on Applied Mathematics*, Proceedings, 80 – 88, 2015 ISBN 978-80-227-4143-3.

[5] ČEJKA, J. Metody pro efektivní metody vzdělávání studentů technicky zaměřených oborů. *Trendy ve vzdělávání*, Olomouc: Univerzita Palackého v Olomouci, 2016, roč. 9, číslo 1, s. 37-42. ISSN 1805-8949.

[6] DEDECIUS, K., KALOVÁ, J., VYSOKÁ, J. Využití svobodného matematického softwaru při výuce statistiky. In *Setkání učitelů matematiky všech typů a stupňů škol*. Plzeň: Vydavatelský servis, Republikánská 28, Plzeň, 2008. 6 s. ISBN 978-80-86843-21-6.

[7] CHLÁDEK, P., NÝDL, V., ŠULISTA, M. Struktura zápočtových testů z matematiky a jejich vyhodnocování. In L. Hráčková. *Sborník Aktuální otázky výuky matematiky na ekonomických oborech*. Brno: Masarykova univerzita, 2011. s. 34-37, 4 s. ISBN 978-80-210-5669-5.

[8] HLAVÁČOVÁ, I. Motivační účinky vícesložkového hodnocení studentů v prezenčním studiu. *Trendy ve vzdělávání*, Olomouc: Univerzita Palackého v Olomouci, 2016, roč. 9, číslo 1, s. 84-89. ISSN 1805-8949.

[9] JELÍNEK, J. Application of the Simulation Model of Social Network in Education. In Ľudovít Balko, Dagmar Szarková, Daniela Richtáriková. In *APLIMAT 2016: 15th Conference on Applied Mathematics*. Bratislava: Slovak University of Technology in Bratislava, 2016. s. 627-635, 9 s. ISBN 978-80-227-4531-4.

[10] KLEPANCOVÁ, M., SMETANOVÁ, D. Vztah studentů k přírodním vědám – chemie. *Media4u Magazine*, Praha: Ing. Jan Chromý, Ph.D., 2015, roč. 12, č. 2, s. 6-9. ISSN 1214-9187.

[11] RICHTÁRIKOVÁ, D., Mathematics preparatory course and its effectivness, *In APLIMAT 2015: 14th Conference on Applied Mathematics*, Proceedings, 672 – 681, 2015 ISBN 978-80-227-4143-3.

[12] RICHTÁRIKOVÁ, D., Supporting Methods in Mathematics. Effectveness of Mathematics Workshop, In *APLIMAT 2016: 15th Conference on Applied Mathematics*, Proceedings, 959 – 966, 2016 ISBN 978-80-227-4531-4.

[13] VYSOKÁ, J., SMETANOVÁ, D. Analysis of Attitude of Students Towards Mathematics and Physics, In *APLIMAT 2016: 15th Conference on Applied Mathematics*, Proceedings, 1126 – 1138, 2016 ISBN 978-80-227-4531-4.

[14] ZÁHONOVÁ, V., Influence of Students' High School Mathematics Knowledge on Their Success in Subject Mathematics I, In *APLIMAT 2015: 14th Conference on Applied Mathematics*, Proceedings, 767 – 773, 2015 ISBN 978-80-227-4143-3.

Current address

Biba Vladislav, Mgr. Ph.D.
The Institute of Technology and Bussiness in České Budějovice
Okružní 517/10, České Budějovice, Czech Republic
Tel. number: +420 387 842 200, e-mail: biba@mail.vstecb.cz

Smetanová Dana, RNDr. Ph.D.
The Institute of Technology and Bussiness in České Budějovice
Okružní 517/10, České Budějovice, Czech Republic
Tel. number: +420 387 842 133, e-mail: smetanova@mail.vstecb.cz

Vargová Michaela, Mgr. PhD.
The Institute of Technology and Bussiness in České Budějovice
Okružní 517/10, České Budějovice, Czech Republic
Tel. number: +420 387 842 186, e-mail: vargova@mail.vstecb.cz

Slovak University of Technology in Bratislava
Faculty of Mechanical Engineering
16th Conference on Applied Mathematics **APLIMAT 2017**
Proceedings

HOW ARE THE WAGE DISTRIBUTIONS OF CZECHS AND SLOVAKS DIFFERENT? COMPARISON: DEVELOPMENT, MODELS AND FORECASTS

BÍLKOVÁ Diana (CZ)

Abstract. The present paper aims at a comparison of wage trends since 2000, focusing on wage level, differentiation and concentration since the adoption of the euro in Slovakia. In this study, theoretical wage models are constructed, using three-parameter lognormal curves, taking into account the minimal wage in both countries. Predictions of wage developments – not just those of the level, variability and concentration of future wages, but also forecasts of the entire wage distribution – employing the methods of time series trend analysis are also a part of the research.

Keywords: Probability density function, three-parameter lognormal curves, maximum likelihood method of parameter estimation, time series trend analysis, interpolation criteria, theoretical wage models, wage level, wage differentiation, wage diversification, Gini coefficient and Lorenz curve, minimal wage, forecasts of wage distribution for the Czech and Slovak Republics

Mathematics Subject Classification: Primary 60E05, 62E20, 62H12; Secondary 65C20

1 Introduction

Wages for work performed constitute the largest part of household income. Over the last twenty years, average wages and salaries have increased more than three times in the Czech Republic and Slovakia. The claim that the proportion of people threatened by poverty can rise with increasing income may sound paradoxical. It is, however, true since the poverty threshold lies at 60 per cent of the national median of equalized income. If income grows, the poverty line also rises, as may the number of people who find themselves below it.

In the former Czechoslovakia, wages were going up by an average of 3.2 per cent a year in the period 1955–1970, the largest increases being recorded in 1968 and 1969 (8.5 per cent and 7.9 per cent, respectively). The average wage for the whole period grew by 60.7 per cent, i.e. from 1,192 CSK[1] in 1955 to 1,915 CSK in 1970. Between 1971 and 1989, the average wage growth was

[1] Czechoslovakia having disintegrated into two successor states – the Czech and Slovak Republics – on 1 January 1993,

steady without any surprising fluctuations. Having ranged from 2.4 per cent to 4.7 per cent in the 1970s, the annual growth of wages slowed in the years 1980–1989 (between 1.5 per cent and 2.4 per cent), the average growth rate being 2.0 per cent then. From 1971 to 1989, the average wage increased by 58.7 per cent – from 1,997 to 3,170 CSK. (At the end of the 1980s, it was the miners who earned the most of all employee groups – over 7,000 CSK gross; the mining industry is currently in decline.)

Significant changes in the labour market took place after 1989. Throughout the 1990s, the Czech and Slovak Republics underwent transformation to the market economy. Companies were privatized, prices liberalized and manufacturing sectors restructured. The above processes also affected the methodology of average wage calculations; due to the growth of small businesses, it was necessary to use a combination of census and sample surveys. The average wage almost quadrupled in this period, the highest annual growth being achieved in 1992 and 1993 (22.5 percent and 25.3 per cent). However, the growth of nominal wages was slower than that of consumer prices which significantly influenced the development of real wages, the largest decline (by 26.3 per cent) having been witnessed after the price liberalization in 1991. Despite the real wage increasing annually since 1992, the 1990 level was not exceeded until 1996.

Employee wages and household income have been receiving considerable academic research attention. The knowledge of their distribution and comparison from socio-economic and time-spatial perspectives allows for assessing the standard of living as well as social security and equity levels. Statistical analysis of the wage and income distribution is supposed to be the basis for decision-making about the government budgetary, tax and social policies. Since wages and income are closely related to the purchasing power, it is important to monitor the level, structure and changes in their distribution when identifying sales opportunities for both durable and non-durable goods. Wage and income distribution outcomes can be used indirectly to improve survey sampling of other correlated variables, such as household expenditure, equipment, time management, purchasing power, etc. They are also reflected in sociological surveys, allowing for the quantitative assessment of the living standards of distinctive population segments, such as particular social groups and ethnic and regional minorities, both within a country and in international comparisons.

When conducting a statistical analysis of the standard of living, we focus only on its measurable wage-and-income-dependent factors. For their correct quantification, it is necessary to comprehensively capture the level and structure of employee wages and household income, finding appropriate probability distribution models for both individual social groups and the entire population. The choice of optimal statistical methods is thus a precondition for a thorough evaluation and comparison of the living standards of the populations of the Czech Republic and Slovakia.

2 Database

The present research deals with employee wages in the Czech and Slovak Republics from 2000 to 2015, focusing on a comparison of the wage distribution development models since the introduction of the euro in Slovakia (2009). The gross monthly wage is a research variable. Since it is the nominal wage, Tab. 1 presents information on inflation rate trends in both countries. As the wage

the federal currency (CSK) was converted into Czech and Slovak crowns (CZK, SK), respectively.

distribution is substantially affected by the minimum wage, Tab. 2 offers an overview of its rates since 2000. The research data were taken from the official websites of the Czech and Slovak Statistical Offices in the form of interval frequency distributions, separate figures not being commonly available. For the research purposes, both Czech and Slovak national currencies (CZK and SK, the latter until 2008) had to be converted in euros using average annual exchange rates of the Czech and Slovak National Banks; see Tab. 3.

The minimum wage in euros depends also on average annual CZK and SK exchange rates. (From 1 January 2007 to 31 July 2013, for example, the minimum wage in the Czech Republic amounted to 8,000 CZK, the indicated fluctuation of the minimum wage in euros being caused by changes in the average annual exchange rate over this period.)

A wage is paid to an employee in the private (business) sector, while a salary in the budget (national, public, non-business) organizations. The present study sticks to common usage employed by the Czech and Slovak Statistical Offices, using the word "wage" as an umbrella term covering earnings in both the business and non-business sectors.

	Year							
Country	2000	2001	2002	2003	2004	2005	2006	2007
CR	3.9	4.7	1.8	0.1	2.8	1.9	2.5	2.8
SR	12.2	2.8	3.5	8.5	7.5	2.8	4.3	1.9
	Year							
Country	2008	2009	2010	2011	2012	2013	2014	2015
CR	6.3	1.0	1.5	1.9	3.3	1.4	0.4	0.3
SR	3.9	0.9	0.7	4.1	3.7	1.5	-0.1	1.0

Source: www.czso.cz; slovak.statistics.sk

Tab. 1. Inflation rate (in %) in the Czech and Slovak Republics in 2000–2015

Czech Republic				Slovak Republic			
Year	Min. wage	Year	Min. wage	Year	Min. wage	Year	Min. wage
2000	112.33[1)] 126.37[2)]	2008	320.74	2000	94.30[7)] 103.73[8)]	2008	267.59
2001	146.70	2009	302.51	2001	104.88[9)] 117.28[10)]	2009	295.50
2002	184.99	2010	316.33	2002	110.75[11)] 125.38[12)]	2010	307.70
2003	194.70	2011	325.39	2003	134.14[13)] 146.42[14)]	2011	317.00
2004	210.01	2012	318.18	2004	152.80[15)] 163.36[16)]	2012	327.20
2005	241.24	2013	308.00[5)] 327.25[6)]	2005	169.85[17)] 180.30[18)]	2013	337.70
2006	267.09[3)] 280.67[4)]	2014	308.72	2006	179.74[19)] 197.98[20)]	2014	352.00
2007	288.16	2015	337.21	2007	226.90[21)] 241.83[22)]	2015	380.00

Source: www.mpsv.cz; www.employment.gov.sk

2) From 1 July 2000 to 31 December 2000
3) From 1 January 2006 to 30 June 2006
4) From 1 July 2006 to 31 December 2006
5) From 1 January 2013 to 31 July 2013
6) From 1 August 2013 to 31 December 2013
7) From 1 January 2000 to 30 September 2000
8) From 1 October 2000 to 31 December 2000
9) From 1 January 2001 to 18 October 2001
10) From 19 October 2001 to 31 December 2001
11) From 1 January 2002 to 30 September 2002
12) From 1 October 2002 to 31 December 2002
13) From 1 January 2003 to 30 September 2003
14) From 1 October 2003 to 31 December 2003
15) From 1 January 2004 to 30 September 2004
16) From 1 October 2004 to 31 December 2004
17) From 1 January 2005 to 30 September 2005
18) From 1 October 2005 to 31 December 2005
19) From 1 January 2006 to 30 September 2006
20) From 1 October 2006 to 31 December 2006
21) From 1 January 2007 to 30 September 2007
22) From 1 October 2007to 31 December 2007

Tab. 2. Development of monthly minimum wage (in €) in the Czech and Slovak Republics since 2000

Czech Republic				Slovak Republic			
Year	Exch. rate	Year	Exch. rate	Year	Exch. rate	Year	Exch. rate
2000	35.610	2008	24.942	2000	42.418	2008	30.270
2001	34.083	2009	26.445	2001	41.951	2009[1)]	–
2002	30.812	2010	25.290	2002	44.426	2010	–
2003	31.844	2011	24.586	2003	41.525	2011	–
2004	31.904	2012	25.143	2004	39.790	2012	–
2005	29.784	2013	25.974	2005	38.270	2013	–
2006	28.343	2014	27.533	2006	38.388	2014	–
2007	27.762	2015	27.283	2007	33.495	2015	–

Source: www.cnb.cz, www.nbs.sk

1) Slovakia joined the Eurozone on 1 January 2009

Tab. 3. Average annual exchange rates (CZK/EUR, SK/EUR) since 2000

3 Brief Review to Literature

The wage and income issue has been constantly examined worldwide. Bachmann, Bechara & Schaffner (2016), for example, investigate the wage inequality and mobility in Europe. Blaise & Puhani (2013), using personnel records of a large privatized public company, conclude that employment contract liberalization leads to a relative reduction of wages of women, permanent residents as well as older, high-tenure, low-skilled and part-time workers. Chang (2015) deals with non-linear effects of financial and income trends on the energy consumption. Ebner & Helbling (2016) explore an impact of the social distance between immigrants and Swiss-born residents on wage inequalities. Malá (2013) uses finite mixtures of lognormal distributions for the modelling of net annual income per capita and an equivalised income of Czech households in 2004–2010. Malá (2015) also analyses their total equivalised net annual income between 2007 and 2010. Marek (2010) identifies the income distribution trend in the Czech Republic over the years 1995–2008. Ozanne (1959) examines the influence of trade unions on wage levels and income distribution. Pencavel (2015) investigates "Keynesian controversies" over wages. Piketty (2003), using the data from income, wage and inheritance tax returns (from 1915/1919–1998 and 1902–1994, respectively), calculates homogeneous yearly estimates of income, wage, and wealth inequality in France. Schafgans (1998) applies parametric and semiparametric estimated wage equations, correcting for sample selection bias, to assess gender and ethnic wage differences in Malaysia. Stanovnik & Verbič (2005) research the dynamics of wage and income inequality in Slovenia from 1993 to 2002, using two data sources – a personal income tax database and publicly available figures on wage distributions.

4 Theory and Methods

4.1 Three-Parametric Lognormal Curves

Random variable *X* has three-parametric lognormal distribution with parameters μ, σ^2 and θ, where $-\infty < \mu < \infty$, $\sigma^2 > 0$, $-\infty < \theta < \infty$, if its probability density function have the form

$$f(x;\,\mu,\,\sigma^2,\,\theta) = \begin{cases} \dfrac{1}{\sigma\cdot(x-\theta)\cdot\sqrt{2\pi}}\cdot\exp\left[-\dfrac{[\ln(x-\theta)-\mu]^2}{2\sigma^2}\right], & x > \theta, \\ = 0, & \text{else.} \end{cases} \tag{1}$$

Lognormal distribution with parameters μ, σ^2 and θ (beginning of distribution, theoretical minimum) is marked LN(μ, σ^2, θ). Probability density function of three-parametric lognormal distribution is asymmetric, positively skewed. Figs. 1 and 2 show the graphs of the probability density function of three-parametric lognormal distribution depending on the values of the parameters of this distribution.

Probability density function of three-parametric lognormal distribution is sometimes presented in the form

$$f(x;\,\gamma,\,\delta,\,\theta) = \begin{cases} \dfrac{\delta}{(x-\theta)\cdot\sqrt{2\pi}}\cdot\exp\left\{-\dfrac{1}{2}[\gamma+\delta\cdot\ln(x-\theta)]^2\right\}, & x > \theta, \\ = 0, & \text{else,} \end{cases} \tag{2}$$

where it is valid $\mu = -\dfrac{\gamma}{\delta}$ and $\sigma = \dfrac{1}{\delta}$ between the expressions for probability density function (1) and (2).

If we substitute $\theta = 0$ (distribution minimum) into expressions for the probability density function of three-parametric lognormal distribution (1) and (2), we obtain formulas for the probability density function of two-parametric lognormal distribution.

Distribution function of three-parametric lognormal distribution has the form

$$F(x) = \Phi\left[\frac{\ln(x-\theta)-\mu}{\sigma}\right], \quad x > \theta. \tag{3}$$

If the random variable *X* has three-parametric lognormal distribution LN(μ, σ^2, θ), then the random variable

$$Y = \ln(X-\theta) \tag{4}$$

has normal distribution N(μ, σ^2) and the random variable

$$U = \frac{\ln(X-\theta)-\mu}{\sigma} = \gamma + \delta \cdot \ln(X-\theta) \quad (5)$$

has standardized normal distribution N(0; 1). Parameter μ is the expected value of random variable (4) and parameter σ^2 is the variance of this random variable. Parameter θ is the beginning of the distribution, i.e. theoretical minimum of the random variable X.

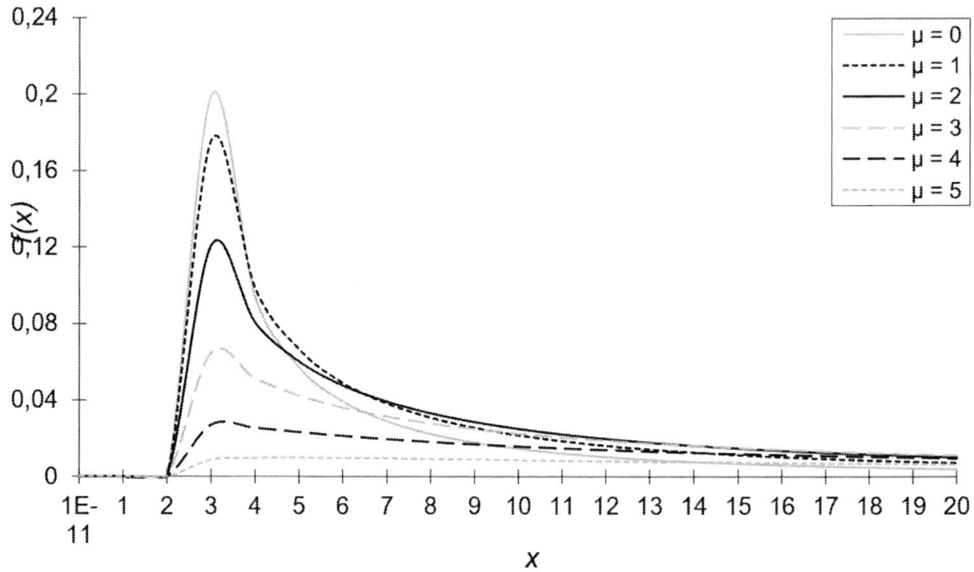

Source: Own research

Fig. 1. Probability density function of three-parametric lognormal distribution for the values of parameters $\sigma = 2$ ($\sigma^2 = 4$); $\theta = 2$

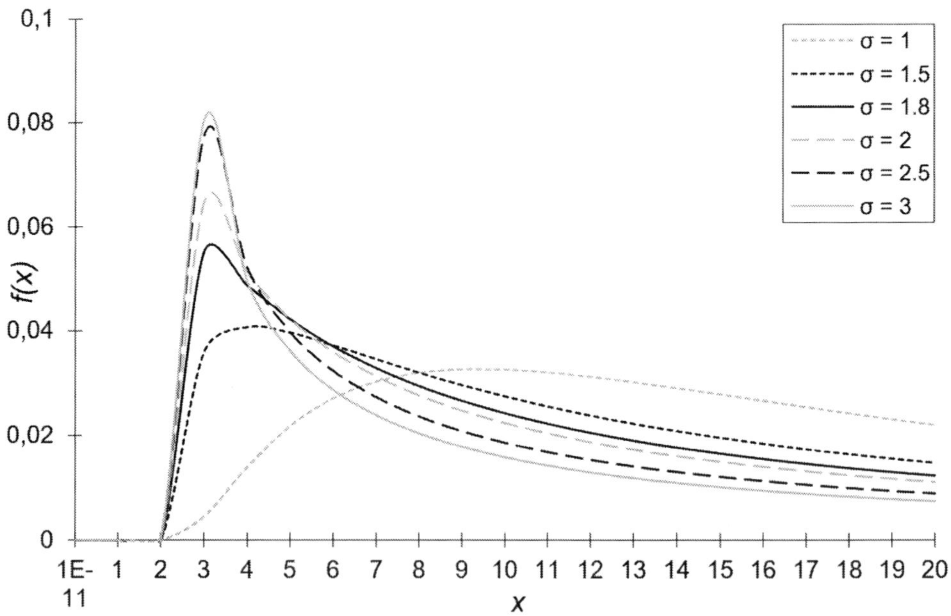

Source: Own research

Fig. 2. Probability density function of three-parametric lognormal distribution for the values of parameters $\mu = 3$; $\theta = 2$

For $\omega = \exp(\sigma^2)$ the r-th common and central moments of three-parametric lognormal distribution have the form

$$\mu_r^{\backslash} = E(X^r) = \theta + \exp\left(r \cdot \mu + \frac{r^2 \sigma^2}{2}\right), \tag{6}$$

$$\mu_r = E[(X - \mu_1^{\backslash})^r] = \omega^{r/2} \cdot \left[\sum_{j=0}^{r} (-1)^j \cdot \binom{r}{j} \cdot \omega^{(r-j)\cdot(r-j-1)/2}\right] \cdot \exp(r \cdot \mu), \tag{7}$$

specifically

$$\mu_3 = \omega^{3/2} \cdot (\omega - 1)^2 \cdot (\omega + 2) \cdot \exp(3 \cdot \mu), \tag{8}$$

$$\mu_4 = \omega^2 \cdot (\omega - 1)^2 \cdot (\omega^4 + 2\omega^3 + 3\omega^2 - 3) \cdot \exp(4 \cdot \mu). \tag{9}$$

We obtain the expressions for the expected value and variance of random variable X having three-parametric lognormal distribution from (6) and (7)

$$E(X) = \theta + \exp\left(\mu + \frac{\sigma^2}{2}\right), \tag{10}$$

$$D(X) = \exp(2\mu + \sigma^2) \cdot [\exp(\sigma^2) - 1] = \exp(2\mu) \cdot \omega \cdot (\omega - 1). \tag{11}$$

The expression for median

$$Median(X) = \theta + \exp(\mu) \tag{12}$$

comes from the expression for $100 \cdot P\%$ quantile of this distribution

$$x_P = \theta + \exp(\mu + \sigma \cdot u_P). \tag{13}$$

Three-parametric lognormal distribution is unimodal with one mode

$$Mode(X) = \theta + \exp(\mu - \sigma^2) = \theta + \frac{\exp(\mu)}{\omega}. \tag{14}$$

The relationship between the expected value, median and mode follows from the equations (10), (12) and (14)

$$E(X) > Median(X) > Mode(X), \tag{15}$$

which is typical just for positively skewed distribution.

The coefficient of variation of three-parametric lognormal distribution is a function of all three parameters μ, σ^2 and θ of this distribution

$$V(X) = \frac{\exp\left(\mu + \frac{\sigma^2}{2}\right)\sqrt{\exp(\sigma^2) - 1}}{\theta + \exp\left(\mu + \frac{\sigma^2}{2}\right)} = \frac{\exp\left(\mu + \frac{\sigma^2}{2}\right)\sqrt{\omega - 1}}{\theta + \exp\left(\mu + \frac{\sigma^2}{2}\right)}. \tag{16}$$

Gini's coefficient of three-parametric lognormal distribution depends on all three parameters μ, σ^2 and θ of this distribution, too

$$G = \frac{\exp\left(\mu + \frac{\sigma^2}{2}\right) \cdot \operatorname{erf}\left(\frac{\sigma}{2}\right)}{\theta + \exp\left(\mu + \frac{\sigma^2}{2}\right)}. \tag{17}$$

Moment measurement of skewness and kurtosis depend on a single parameter σ^2

$$\beta_1 = \sqrt{\exp(\sigma^2) - 1} \cdot [\exp(\sigma^2) + 2] = \sqrt{\omega - 1} \cdot (\omega + 2), \tag{18}$$

$$\beta_2 = [\exp(4\sigma^2) + 2\exp(3\sigma^2) + 3\exp(2\sigma^2) - 3] = (\omega^4 + 2\omega^3 + 3\omega^2 - 3). \tag{19}$$

4.2 Four-parametric lognormal curves

Random variable X has four parametric lognormal distribution with parameters μ, σ^2, θ and τ, where $-\infty < \mu < \infty$, $\sigma^2 > 0$, $-\infty < \theta < \tau < \infty$, if its probability density function has the form

$$f(x; \gamma, \delta, \theta, \tau) = \begin{cases} \dfrac{(\tau - \theta)}{\sigma \cdot (x - \theta) \cdot (\tau - x) \cdot \sqrt{2\pi}} \cdot \exp\left[-\dfrac{\left(\ln\dfrac{x - \theta}{\tau - x} - \mu\right)^2}{2\sigma^2}\right], & \theta < x < \tau, \\ = 0, & \text{else.} \end{cases} \tag{20}$$

Lognormal distribution with parameters μ, σ^2, θ a τ is marked LN(μ, σ^2, θ, τ). The probability density function of four-parametric lognormal distribution can have very different shapes depending on the values of the parameters of the distribution, see Figs. 3–5. Distribution may be also bimodal for $\sigma^2 > 2$ and $|\mu| < \sigma^2 \cdot \sqrt{(1 - 2/\sigma^2)} - 2\tanh^{-1}\sqrt{(1 - 2/\sigma^2)}$. Probability density function of four-parametric lognormal distribution is often presented in the form

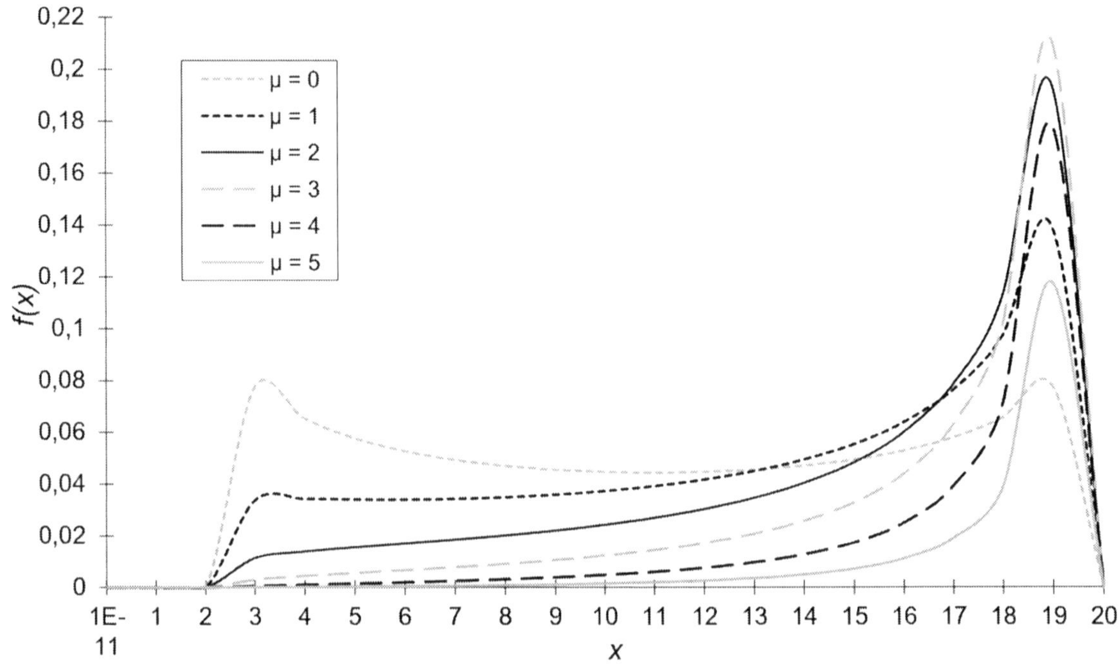

Source: Own research

Fig. 3. Probability density function of four-parametric lognormal distribution for the values of parameters $\sigma = 2$ ($\sigma^2 = 4$); $\theta = 2$; $\tau = 20$

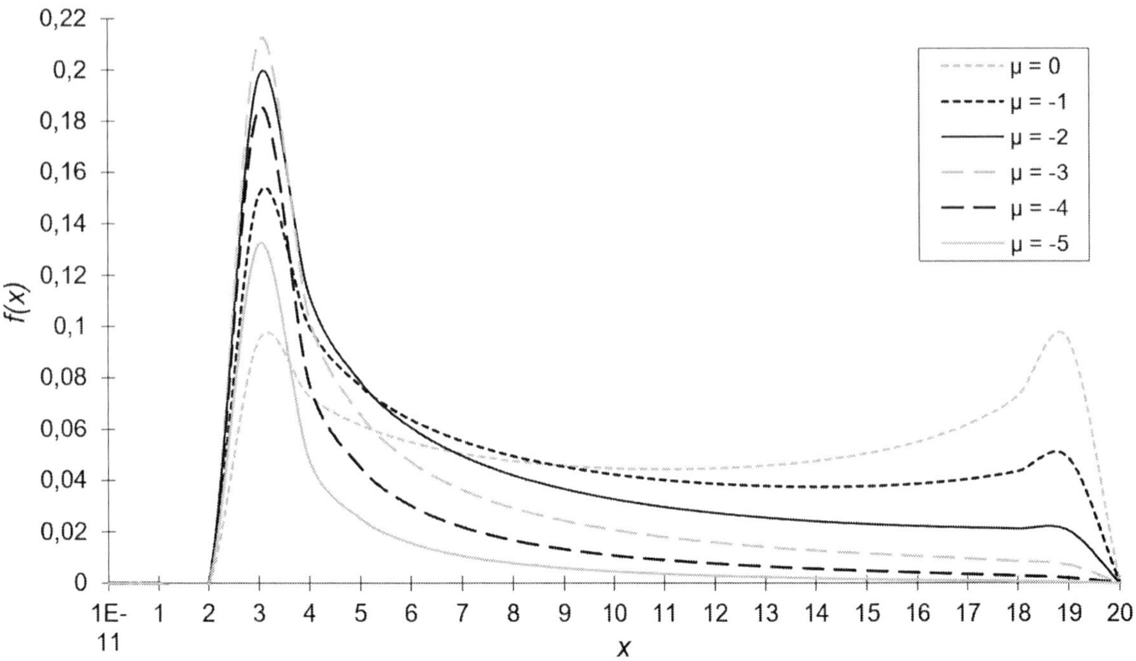

ource: Own research

Fig. 4. Probability density function of four-parametric lognormal distribution for the values of parameters $\sigma = 2$ ($\sigma^2 = 4$); $\theta = 2$; $\tau = 20$

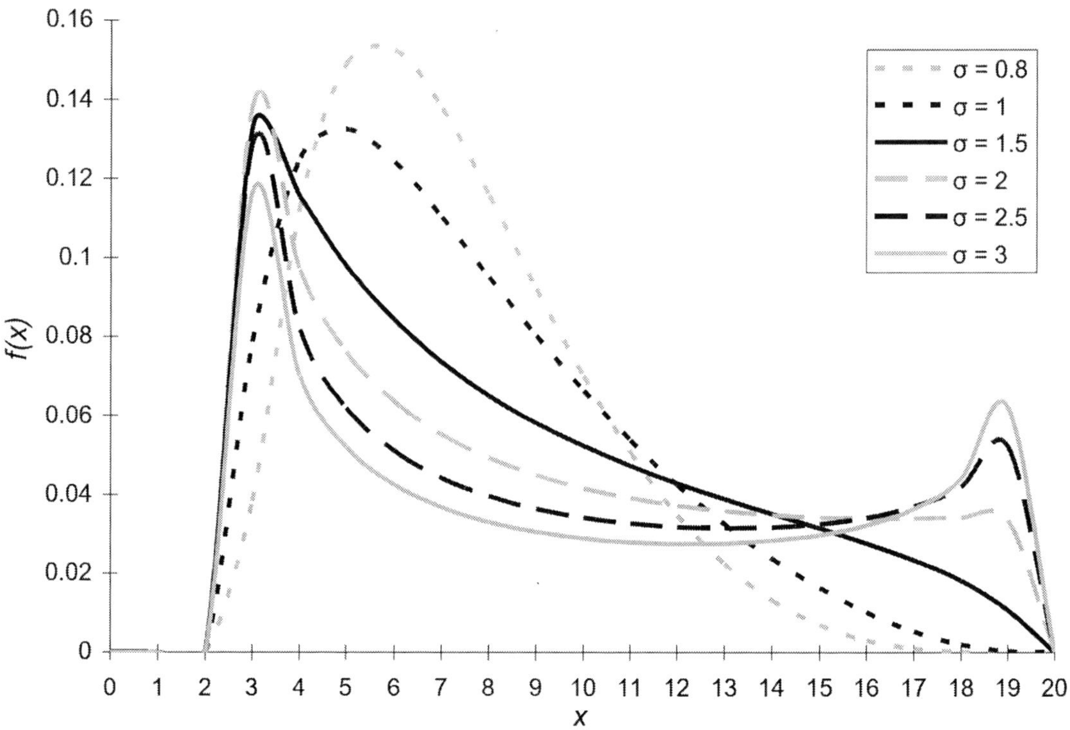

Source: Own research

Fig. 5. Probability density function of four-parametric lognormal distribution for the values of parameters $\mu = -1$; $\theta = 2$; $\tau = 20$

$$f(x;\, \mu, \sigma^2, \theta, \tau) = \frac{\delta \cdot (\tau - \theta)}{(x - \theta) \cdot (\tau - x) \cdot \sqrt{2\pi}} \cdot \exp\left[-\frac{1}{2}\left(\gamma + \delta \cdot \ln\frac{x - \theta}{\tau - x}\right)^2\right], \qquad \theta < x < \tau,$$
$$= 0, \qquad \text{else,} \tag{21}$$

where it is valid between the expressions for probability density function (20) and (21) $\mu = -\frac{\gamma}{\delta}$ and $\sigma = \frac{1}{\delta}$.

If the random variable X has four-parametric lognormal distribution LN(μ, σ^2, θ, τ), then the random variable

$$Y = \ln\frac{X - \theta}{\tau - X} \tag{22}$$

has normal distribution N(μ, σ^2) and the random variable

$$U = \frac{\ln\dfrac{X - \theta}{\tau - X} - \mu}{\sigma} = \gamma + \delta \cdot \ln\frac{X - \theta}{\tau - X} \tag{23}$$

has standardized normal distribution N(0; 1). Parameter μ is therefore the expected value of a random variable (22) and the parameter σ^2 is the variance of this random variable. The parameter θ is the beginning of the distribution (theoretical minimum) of a random variable X and the parameter τ represents the end point of the distribution (theoretical maximum) of the random variable X.

4.3 Maximum Likelihood Method

Let the random sample of the sample size n comes from three-parametric lognormal distribution with probability density function (1) or (2). Then the likelihood function has the form

$$L(\boldsymbol{x};\mu,\sigma^2,\theta)=\prod_{i=1}^{n} f(x_i;\mu,\sigma^2,\theta)=$$

$$=\frac{1}{(\sigma^2)^{n/2}\cdot(2\pi)^{n/2}\cdot\prod\limits_{i=1}^{n}(x_i-\theta)}\cdot\exp\left\{\sum_{i=1}^{n}-\frac{[\ln(x_i-\theta)-\mu]^2}{2\sigma^2}\right\}. \tag{24}$$

We determine the logarithm of the likelihood function

$$\ln L(\boldsymbol{x};\mu,\sigma^2,\theta)=\sum_{i=1}^{n}-\frac{[\ln(x_i-\theta)-\mu]^2}{2\sigma^2}-\frac{n}{2}\cdot\ln\sigma^2-\frac{n}{2}\cdot\ln(2\pi)-\sum_{i=1}^{n}\ln(x_i-\theta). \tag{25}$$

We put in the equality to zero the first partial derivation of the logarithm of the likelihood function according to μ and according to σ^2 by

$$\frac{\partial\ln L(\boldsymbol{x};\mu,\sigma^2,\theta)}{\partial\mu}=\frac{\sum\limits_{i=1}^{n}[\ln(x_i-\theta)-\mu]}{\sigma^2}=0, \tag{26}$$

$$\frac{\partial\ln L(\boldsymbol{x};\mu,\sigma^2,\theta)}{\partial\sigma^2}=\frac{\sum\limits_{i=1}^{n}[\ln(x_i-\theta)-\mu]^2}{2\sigma^4}-\frac{n}{2\sigma^2}=0. \tag{27}$$

We obtain maximum likelihood estimations of the parameters μ and σ^2 for the given parameter θ after treatment

$$\hat{\mu}(\theta)=\frac{\sum\limits_{i=1}^{n}\ln(x_i-\theta)}{n}, \tag{28}$$

$$\hat{\sigma}^2(\theta)=\frac{\sum\limits_{i=1}^{n}[\ln(x_i-\theta)-\hat{\mu}(\theta)]^2}{n}. \tag{29}$$

If the value of the parameter θ in known, we get the maximum likelihood estimations of the remaining two parameters of three-parametric lognormal distribution using the expressions (28) and (29). However, if the value of the parameter θ is unknown, the problem is more complicated. It can be proved that if the parameter θ closes to $\min\{X_1, X_2, \ldots, X_n\}$, then the maximum likelihood approaches to infinity. The maximum likelihood method is also often combined with Cohen's method, where we put the smallest sample value to be equal to the $100 \cdot (n+1)^{-1}$-percentage quantile

$$x_{\min}^{V} = \hat{\theta} + \exp(\hat{\mu} + \hat{\sigma} \cdot u_{(n+1)^{-1}}). \tag{30}$$

Equation (30) is then combined with a szstem of equations (28) and (29).

For solving of maximum likelihood equations (28) and (29) it is also possible to use $\hat{\theta}$ satisfying the equation

$$\sum_{i=1}^{n}(x_i - \hat{\theta}) + \frac{\sum_{i=1}^{n} \frac{z_i^{\backslash}}{(x_i - \hat{\theta})}}{\hat{\sigma}(\hat{\theta})} = 0, \tag{31}$$

where

$$z_i^{\backslash} = \frac{\ln(x_i - \hat{\theta}) - \hat{\mu}(\hat{\theta})}{\hat{\sigma}(\hat{\theta})}, \tag{32}$$

where $\hat{\mu}(\hat{\theta})$ and $\hat{\sigma}(\hat{\theta})$ satisfy equations (28) and (29) with the parameter θ replaced by $\hat{\theta}$. We may also obtain the limits of variances

$$n \cdot D(\hat{\theta}) = \frac{\sigma^2 \cdot \exp(2\mu)}{\omega \cdot \left[\omega \cdot (1 + \sigma^2) - 2\sigma^2 - 1\right]}, \tag{33}$$

$$n \cdot D(\hat{\mu}) = \frac{\sigma^2 \cdot [\omega \cdot (1 + \sigma^2) - 2\sigma^2]}{\omega \cdot (1 + \sigma^2) - 2\sigma^2 - 1}, \tag{34}$$

$$n \cdot D(\hat{\sigma}) = \frac{\sigma^2 \cdot [\omega \cdot (1 + \sigma^2) - 1]}{\omega \cdot (1 + \sigma^2) - 2\sigma^2 - 1}. \tag{35}$$

Especially difficulties related with the use of the equations (28), (29) and (31) lead us to think about other methods.

4.4 Appropriateness of the Model

It is also necessary to assess the suitability of constructed model or choose a model from several alternatives, which is made by some criterion, which can be a sum of absolute deviations of the observed and theoretical frequencies for all intervals

$$S = \sum_{i=1}^{k} \left| n_i - n\,\pi_i \right| \tag{36}$$

or known criterion χ^2

$$\chi^2 = \sum_{i=1}^{k} \frac{(n_i - n\,\pi_i)^2}{n\,\pi_i}, \tag{37}$$

where n_i are the observed frequencies in individual intervals, π_i are the theoretical probabilities of membership of statistical unit into the i-th interval, n is the total sample size of corresponding statistical file, $n \cdot \pi_i$ are the theoretical frequencies in individual intervals, $i = 1, 2, ..., k$, and k is the number of intervals.

The question of the appropriateness of the given curve for model of the distribution of wage is not entirely conventional mathematical-statistical problem in which we test the null hypothesis

H_0: The sample comes from the supposed theoretical distribution

against the alternative hypothesis

H_1: non H_0,

because in goodness of fit tests in the case of wage distribution we meet frequently with the fact that we work with large sample sizes and therefore the tests would almost always lead to the rejection of the null hypothesis. This results not only from the fact that with such large sample sizes the power of the test is so high at the chosen significance level that the test uncovers all the slightest deviations of the actual wage distribution and a model, but it also results from the principle of construction of the test. But practically we are not interested in such small deviations, so only gross agreement of the model with reality is sufficient and we so called "borrow" the model (curve). Test criterion χ^2 can be used in that direction only tentatively. When evaluating the suitability of the model we proceed to a large extent subjective and we rely on experience and logical analysis.

4.5 The Gini Coefficient of Diversification

In economic applications, more illustrative measures of the degree of concentration are preferred to moment kurtosis. The Lorenz curve is a frequently used tool for a graphical representation and measure of the degree of concentration of a numeric variable (see Fig. 6).

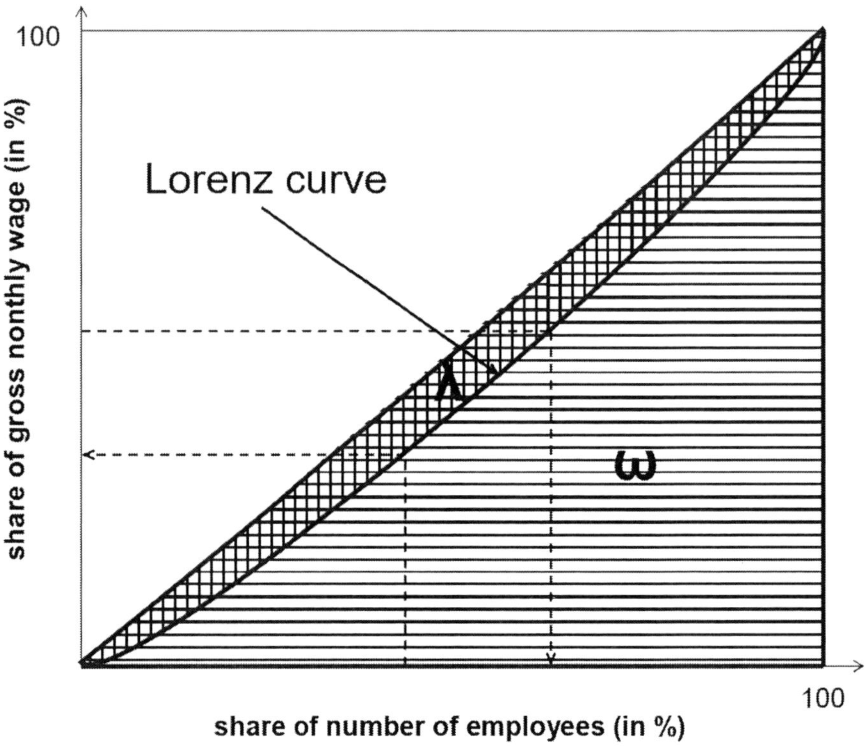

Source: Own research

Fig. 6. Lorenz curve

The Lorenz curve is plotted in a rectangular chart with two scales from 0 to 100 per cent. On the horizontal axis, there are the cumulative relative frequencies of statistical units – the holders of the monitored variable. Cumulative sums of the concentrated variable are on the vertical axis. The coordinates of points on the Lorenz curve are the cumulative relative frequencies of statistical units (holders of the research variable) and corresponding cumulative sums of the concentrated variable. In the case of zero concentration, the Lorenz curve merges into a diagonal, which means that all the monitored variable values are allocated equally to each statistical unit. The more the Lorenz curve bends, the larger is the concentration of the research variable; i.e. the concentration of a relatively large part of the total sum of the monitored variable values in a small number of statistical units. In the case of maximum possible concentration, the Lorenz curve transforms into two mutually orthogonal line segments, merging with the horizontal axis and the right edge of the graph, which means that the total sum of the values of the research variable is concentrated only in one statistical unit. The Gini coefficient of diversification G – the characteristic of concentration – is the ratio of the area between the diagonal and the Lorenz curve and the whole area of the total triangle below the diagonal

$$G = \frac{\lambda}{\lambda + \omega}, \tag{38}$$

(see Fig. 6). At the zero concentration level (extreme equalization when all employees earn the same wage), the Gini coefficient of diversification is zero (0 per cent). At the highest possible concentration level (maximum differentiation when a single employee receives the whole amount of wages paid), the Gini coefficient is one (100 per cent), commonly ranging from 0 to 1 (i.e. from 0 to 100 per cent).

4.6 Other Statistical Methods Used

The present study also applies other methods as well as approaches to time series modelling, whose detailed overviews are provided, e.g., in Triola (1989), Brockwell & Davis (2002) or Cowpertwait & Metcalfe (2009).

5 Results and Discussion

Fig. 7 presents the development of average and middle wages in relation to the minimum wage and Fig. 8 shows the growth rates of average and middle wages in relation to inflation in the Czech and Slovak Republics since 2000. While at the beginning of the century, the annual growth rate of the gross monthly wage median did not fall below four per cent in the Czech Republic and Slovakia, the financial and economic downturn led to the wage growth slowdown and freeze in the former country in particular (see Tab. 4). In terms of the wage level development, the critical year was 2009, when average and middle wages went down by 6.46 per cent and 5.19 per cent, respectively. The decline can be also caused by exchange rate fluctuations (see Tab. 3). Despite having slowed the wage growth, the economic crisis did not result in wage level decline in Slovakia – unlike the Czech Republic where the wage level decreased particularly in 2013–2014, partly perhaps because of exchange rate changes.

Fig. 9 shows that the wage distribution in Slovakia is more skewed than that in the Czech Republic, the ratio of the median wage to the average wage fluctuating from 88 per cent to 95 per cent, while in Slovakia it reaches only 78–79 per cent. For a positively skewed distribution, the median < average relationship is typical, while median = average equality holds for a symmetrical distribution. In other words, the closer the median wage to the average wage (i.e. the higher the median-to-average wage ratio), the less positively skewed the frequency distribution. It means that low wages are more likely in Slovakia than in the Czech Republic. This corresponds to a lower wage level in the former country, very high wages being not so rare in the latter.

Period	Czech Republic		Slovak Republic	
	Average	Median	Average	Median
Ø 2000-08	11,97	11,66	9,07	8,17
Ø 2008-13	0,46	–0,69	3,36	3,38
Ø 2013-15	0,21	0,76	4,56	3,82

Source: Own research

Tab. 4. Average annual growth rate (in %) of average and median wage in the Czech and Slovak Republics since 2000

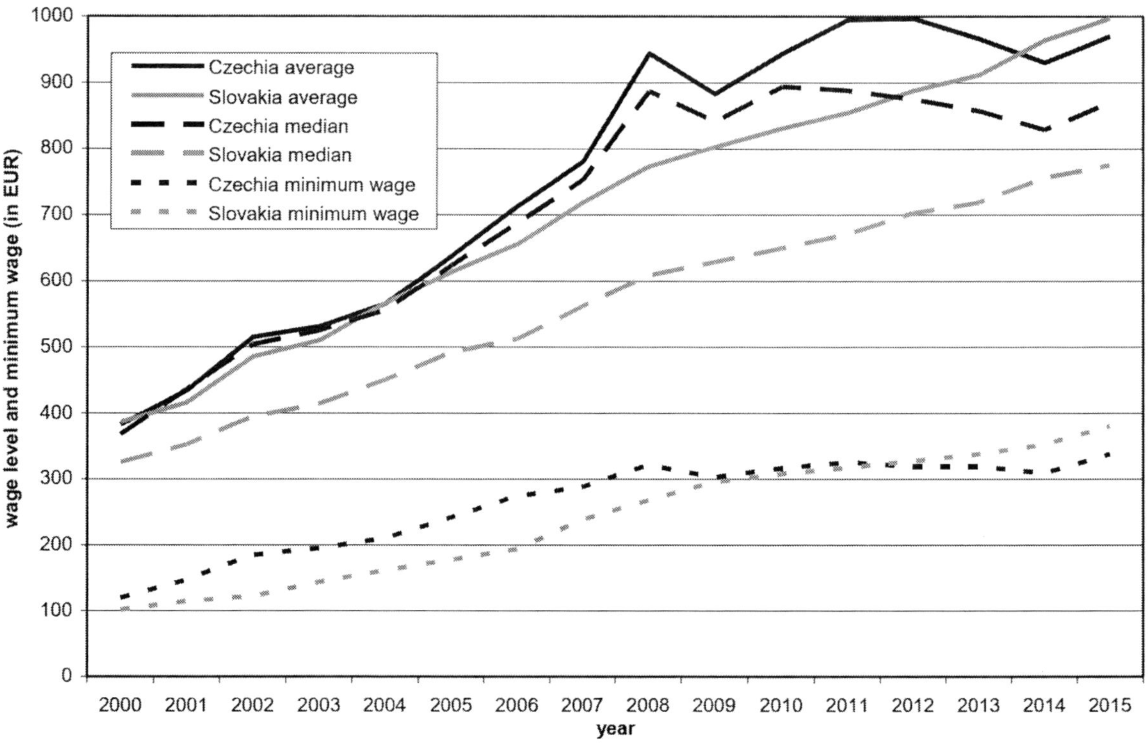

The minimum wage in the Czech and Slovak Republics (the latter until 2008) in EUR depends not only on the minimum wage in CZK and SK, but also on the average exchange rate in a given year, which may result, in some years, in a decrease of the minimum wage in EUR.

Source: Own research

Fig. 7. Wage level development in relation to minimum wage (in €) in the Czech and Slovak Republics since 2000

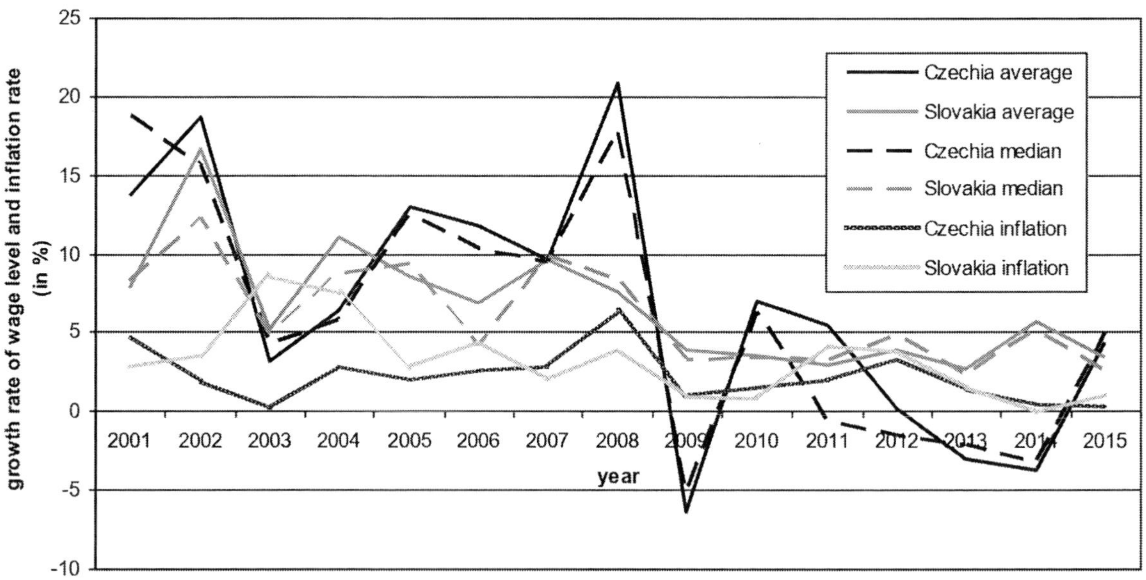

Source: Own research

Fig. 8. Average and median wage growth rates and inflation rate (in %) in the Czech and Slovak Republics since 2000

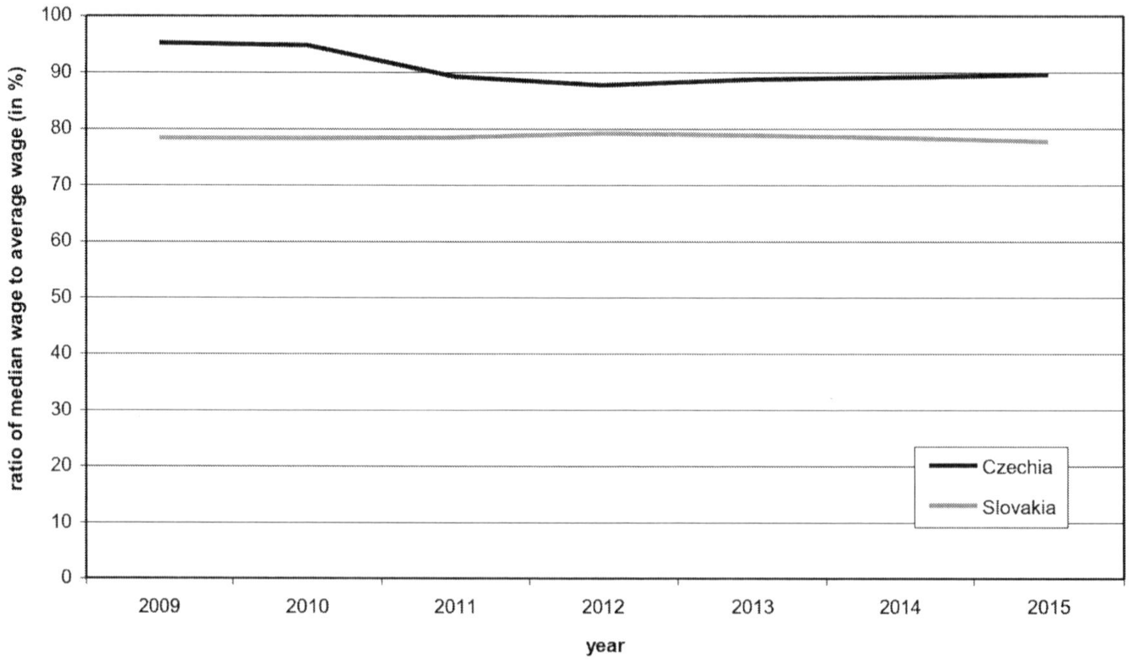

Source: Own research

Fig. 9. Median-to-average wage ratio trend (in %) since euro introduction in Slovakia

Fig. 10 indicates the development of standard deviation (in EUR) and the coefficient of variation (in per cent) since the introduction of the euro in Slovakia. Standard deviation represents a characteristic of absolute variability, defined as the quadratic average of individual wage deviations from the average wage, indicating the average extent to which an individual wage deviates from the average wage.

The coefficient of variation is a characteristic of relative variability because it is defined as a ratio of the standard deviation of wage to the average wage. When multiplied by a hundred, it indicates by how many percentage points the standard deviation of wage contributes to the average wage. Overall, Fig. 10 shows that the variability of wages is higher in the Czech Republic than in Slovakia. However, we can see a rather decreasing trend in the wage variability in the Czech Republic (except a noticeable upswing in 2012), probably caused by the economic downturn. In the Slovak Republic, on the other hand, wages are not particularly sensitive to the world recession, absolute variability having increased, while relative variability decreased since 2009.

Wage diversification is not to be confused with wage variability. Theoretically, if an employee receives the full amount of wages, the Gini coefficient of diversification acquires the value of one (100 per cent), i.e. its maximum possible value, while that of standard deviation depends on the sample size and the actual amount earned by an employee.

In the Czech Republic, the value of the Gini coefficient of diversification rose considerably in 2011, when the effect of the economic crisis got fully manifested in the field of remuneration for work. Between 2011 and 2015, the coefficient's value was going down steadily. In Slovakia, having reached a relatively high value in 2009, the Gini coefficient has dropped from 37.60 per cent to 24.63 per cent since then without any significant oscillation (see Fig. 11).

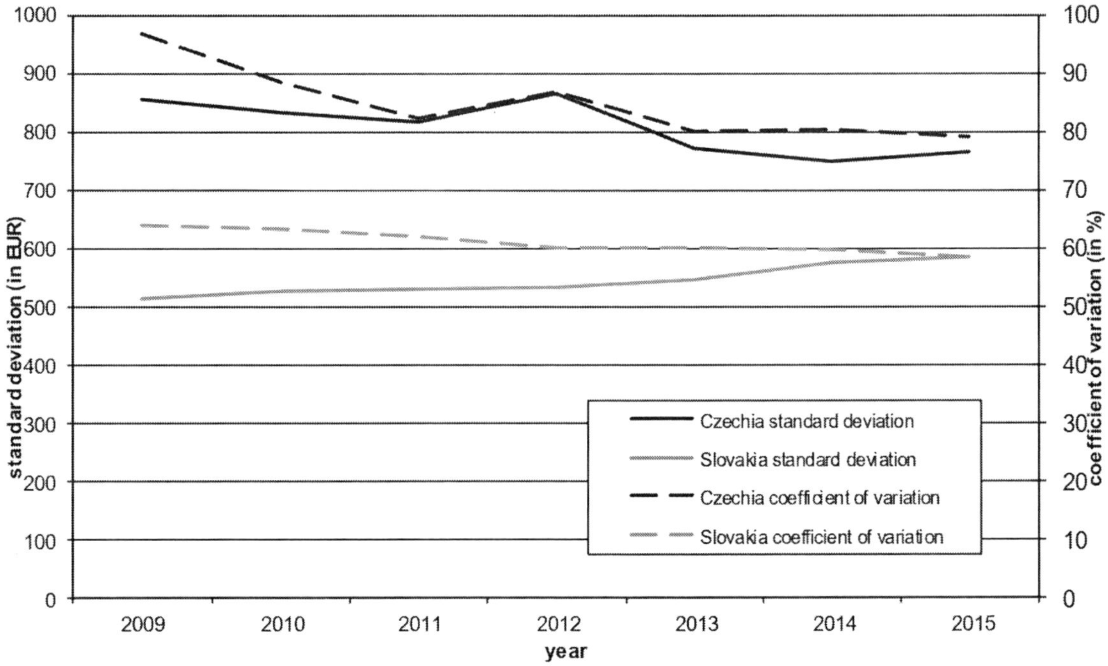

Source: Own research

Fig. 10. Development of variability characteristics in the Czech and Slovak Republics since euro introduction in Slovakia

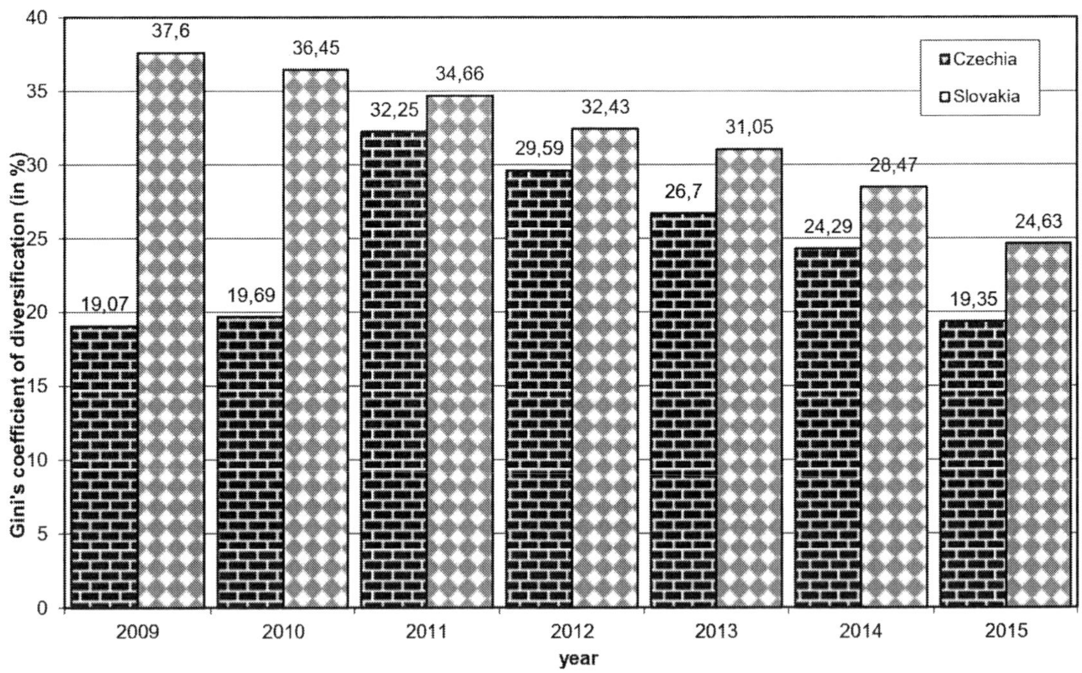

Source: Own research

Fig. 11. Development of Gini coefficient of diversification (in %) in the Czech and Slovak Republics since euro introduction in Slovakia

Figs. 12 and 13 represent a comparison of theoretical model wage distributions in the Czech and Slovak Republics between 2009 and 2015; empirical frequency distribution cannot be drawn because of the former country's different widths of wage distribution intervals. It is clear from these figures that the wage distributions in the Czech Republic have a greater overlap than those in Slovakia. This indicates that the wage distribution in the latter country is more changeable, the level and absolute variability of wages increasing and the skewness and kurtosis of the wage distribution fluctuating in time. (The vertical axis scale is different for better clarity of the two figures.) Figs. 14 and 15 allow a comparison of model wage distributions in the Czech and Slovak Republics in 2013 and 2015. Higher skewness and kurtosis of the wage distribution in Slovakia is evident, a higher concentration of low wages and a rare appearance of very high earnings indicating a lower wage level than in the Czech Republic.

Figs. 16–19 display the trend function of average and median wage time series in the Czech and Slovak Republics since the beginning of the century (see also Tab. 5). Interpolation criteria having been used, appropriate trend functions were chosen. On their basis, the prediction of average and median wage for the period 2016–2020 was made (see Tab. 5). The findings are consistent with the economists' expectations which anticipate further growth in wage levels in the aftermath of every recession. Economic recovery continues in the Czech Republic, GDP growing, the number of vacancies having climbed to its maximum since the onset of the crisis in autumn 2008. Figs. 20–23 show the trend functions and corresponding predictions of the standard deviation and skewness of the wage distribution for 2016 and 2017 in the Czech and Slovak Republics. Tab. 6 displays the estimated parameters of three-parameter lognormal curves for 2016 and 2017, representing the basic theoretical model wage distribution. Based on the forecast of these theoretical distributions, the predictions of entire wage distributions for 2016 and 2017 were constructed.

It is obvious from Figs. 24 and 25 that low-income intervals in Slovakia are more capacious than those in the Czech Republic. As far as 801–1,000 EUR wage interval (with the middle of 900 EUR), in Slovakia are expected higher frequencies than in the Czech Republic in the following two years 2016 and 2017. In the interval between 1,001 and 1,200 EUR (1,100 EUR in the middle), on the other hand, the frequencies in Slovakia are lower than those in the Czech Republic. It allows to state that a higher level of the wage distribution in the Czech Republic in comparison with Slovakia can be expected in the future as well. Figs. 24 and 25 provide more detailed information on the capacity of particular wage intervals in both countries for 2016 and 2017.

In November 2013, the Czech National Bank launched foreign exchange interventions in order to weaken the crown and to further ease the monetary conditions. According to the CNB governor, the above measures against the strong crown were intended for an indefinite period, i.e. until the desired value of the crown and a two-percent inflation target is reached. Using the exchange rate as a monetary policy instrument encounters its limits when it becomes necessary to toughen the monetary policy so as to achieve the target even under strong inflationary pressures. Then, currency interventions will be terminated, interest rates raised (technically) above the zero level. However, easing monetary policy will take a long time. The CNB board declared that the exchange rate would remain in operation as an instrument of monetary policy until 2017 at least. For this reason, the 2014/2015 average exchange rate (27.408 CZK = 1 EUR) was used (corresponding to the CNB intervention period) for the conversion of the proposed 2016 and 2017 minimum wage from CZK to EUR.

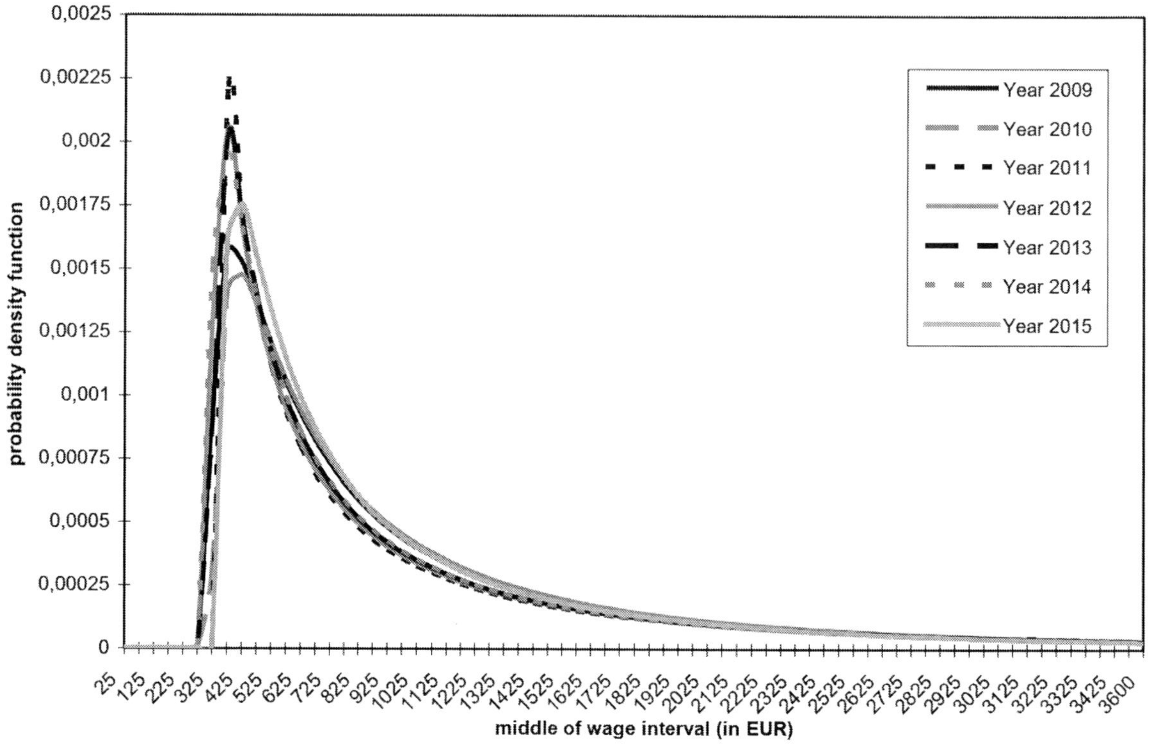

Source: Own research

Fig. 12. Development of model wage distributions in the Czech Republic since euro introduction in Slovakia

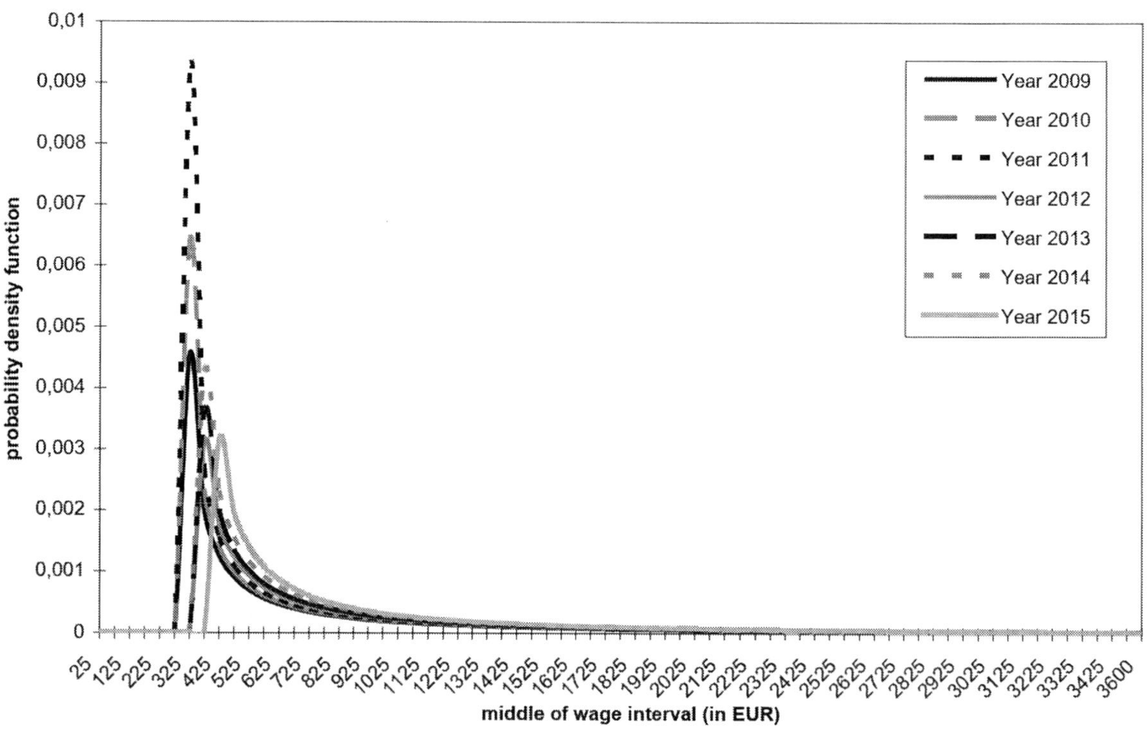

Source: Own research

Fig. 13. Development of model wage distributions in Slovakia since euro introduction

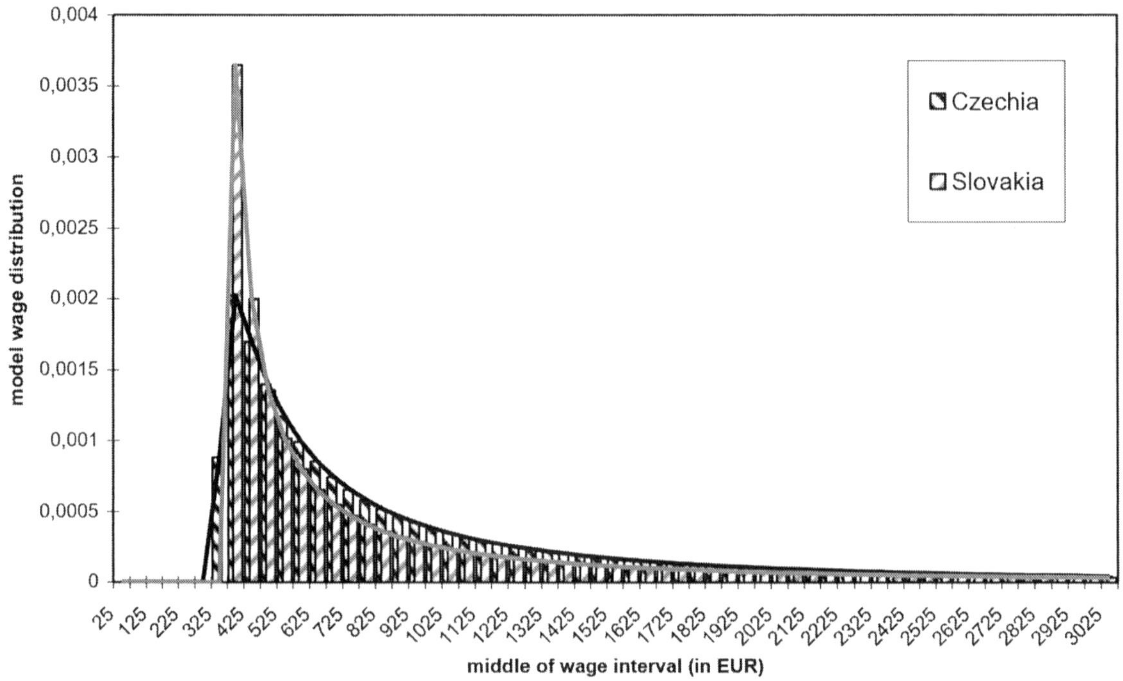

Source: Own research

Fig. 14. Comparison of 2013 model wage distributions in the Czech Republic and Slovakia

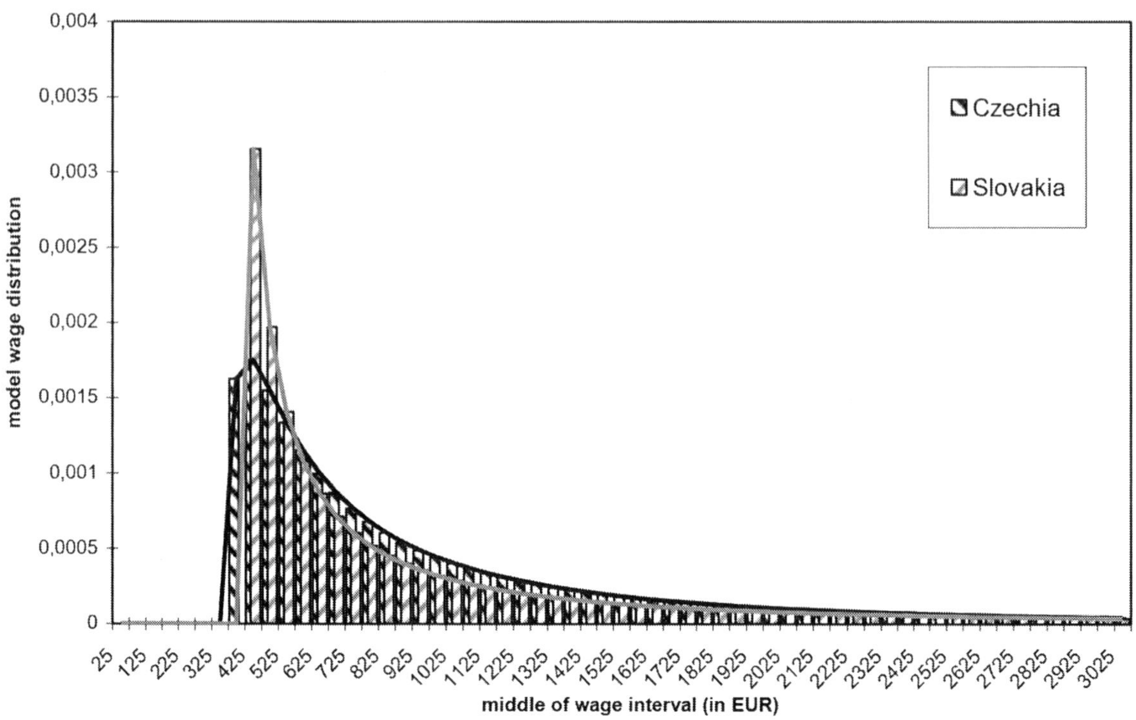

Source: Own research

Fig. 15. Comparison of 2015 model wage distributions in the Czech Republic and Slovakia

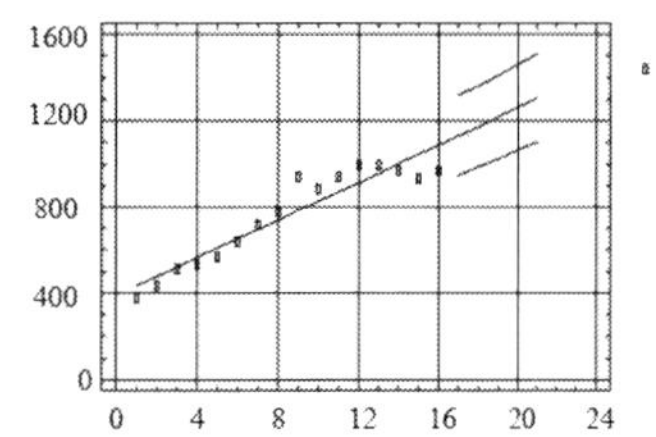

Source: Own research

Fig. 16. Average wage trend function (Czech Republic)

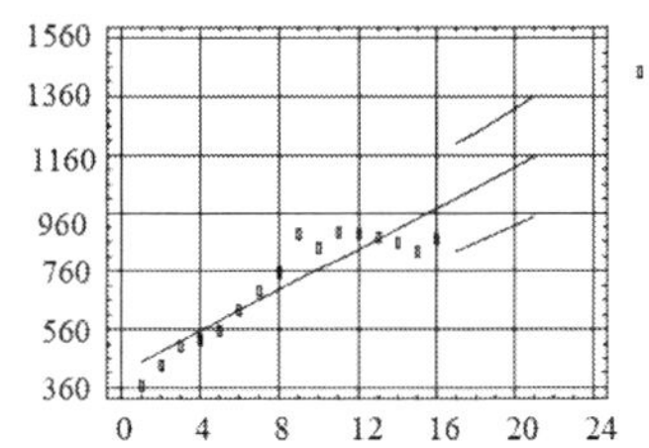

Source: Own research

Fig. 17. Median wage trend function (Czech Republic)

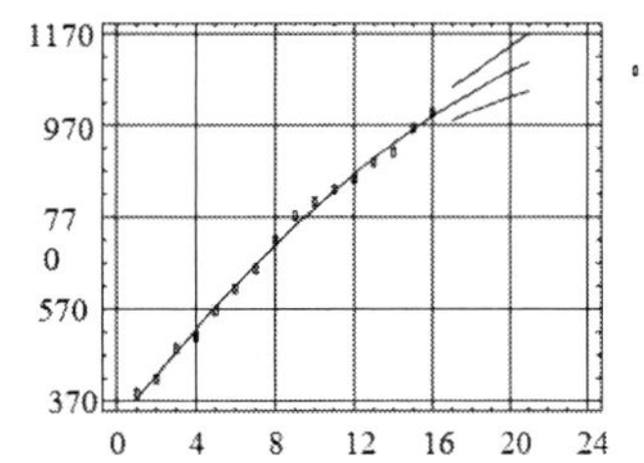

Source: Own research

Fig. 18. Average wage trend function (Slovakia)

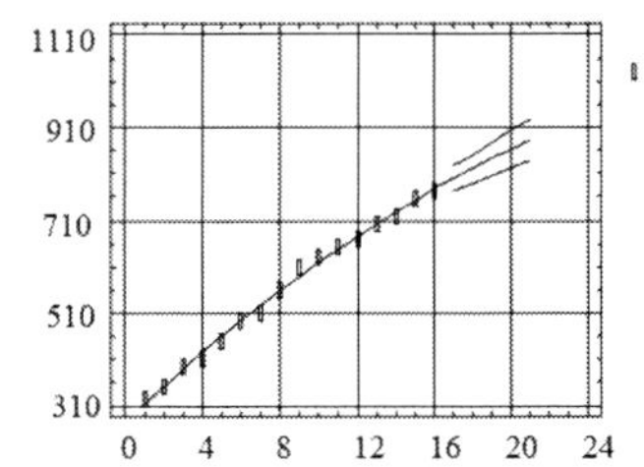

Source: Own research

Fig. 19. Median wage trend function (Slovakia)

Set		Trend function	Set			
			Czech Republic		Slovak Republic	
			Average	Median	Average	Median
CR Average	Linear trend = 391,3 + 43.5824 t	2016	1,132	1,012	1,017	800
CR Median	Linear trend = 412,6 + 35.2368 t	2017	1,176	1,047	1,042	821
SR Average	Quadratic trend = 316.791+ 55.9162 t – 0.8678 t^2	2018	1,219	1,082	1,066	842
SR Median	Quadratic trend = 276.993 + 39.1073 t – 0.4923 t^2	2019	1,263	1,117	1,088	862
		2020	1,307	1,153	1,108	881

Source: Own research

Tab. 5. Trend function of time series of gross monthly wage level and annual predictions of gross monthly wage level (in €) until 2020

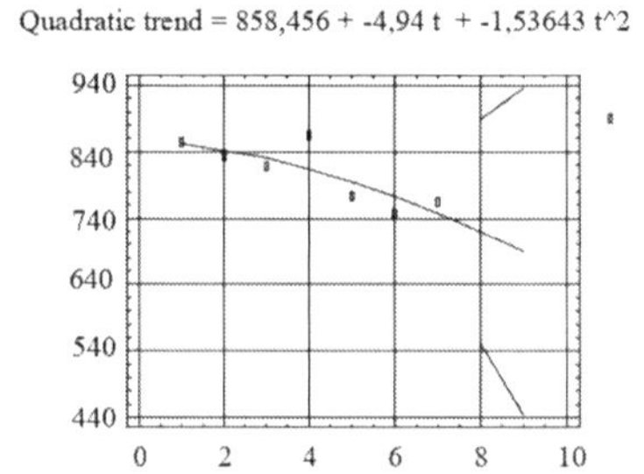

Source: Own research

Fig. 20. Trend analysis of wage standard deviation in the Czech Republic incl. predictions for 2016 and 2017

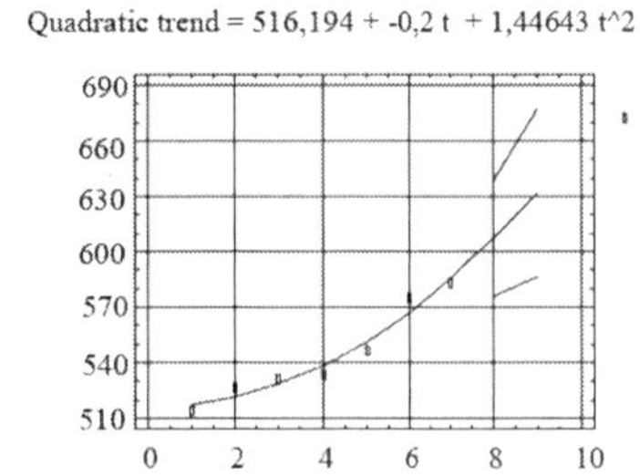

Source: Own research

Fig. 21. Trend analysis of wage standard deviation in Slovakia incl. predictions for 2016 and 2017

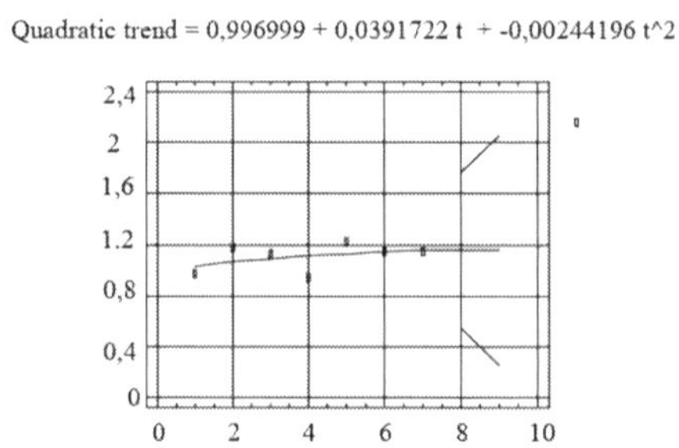

Source: Own research

Fig. 22. Moment skewness trend analysis of wage in the Czech Republic incl. predictions for 2016 and 2017

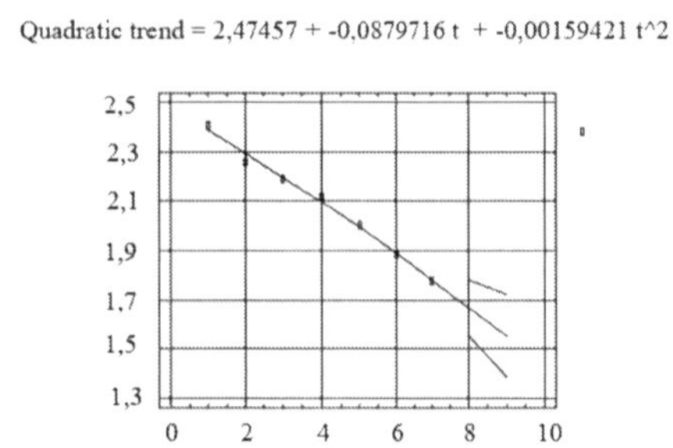

Source: Own research

Fig. 23. Moment skewness trend analysis of wage in Slovakia incl. predictions for 2016 and 2017

In 2016, the minimum wage in the Czech Republic amounted to 9,900 CZK, i.e. 361.27 EUR. The unions' proposal for 2017 increases the minimum wage to 11,500 CZK, which is 419.66 EUR. The minimum wage in Slovakia was 405 EUR in 2016 and the proposed increase is to reach 442 EUR in 2017. The above minimum wage rates represent the beginning of three-parameter lognormal curves used in the construction of the wage distribution prediction for 2016 and 2017. Since the anticipated values of the minimum wage for 2018–2020 are not currently available, predictions of the whole wage distribution for this period cannot be made. 2016 and 2017 forecasts of the average wage (see Tab. 5) as well as those of moment skewness using the trend analysis (Figs. 22 and 23) were applied when constructing the prediction of the entire wage distribution for the two years. For the sake of completeness, trend analysis-based prediction of wage standard deviation was also made (see Figs. 20 and 21).

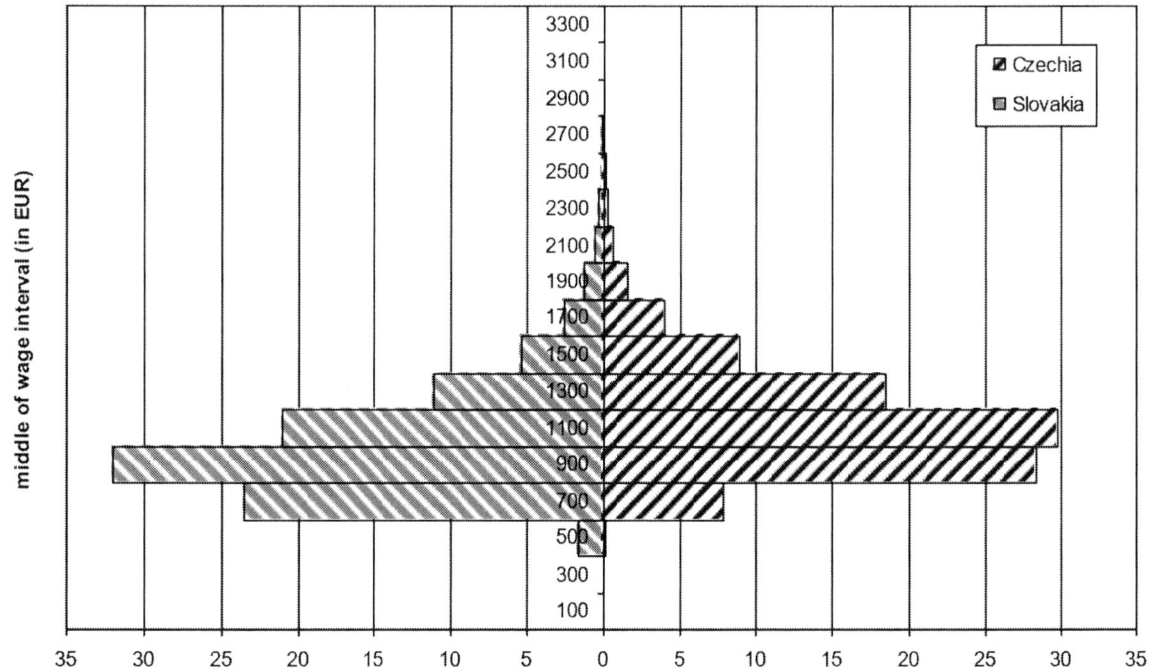

Source: Own research

Fig. 24. 2016 wage distribution prediction (in %) for the Czech and Slovak Republics

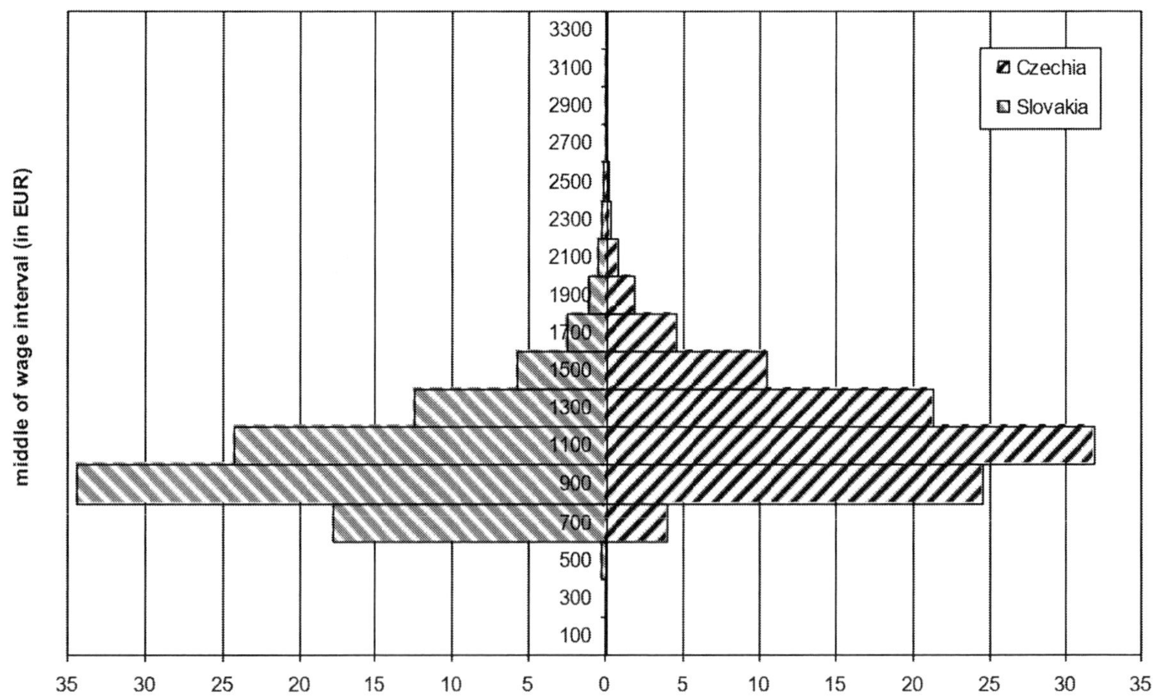

Source: Own research

Fig. 25. 2017 wage distribution prediction (in %) for the Czech and Slovak Republics

	Czech Republic			Slovakia		
Year	θ	μ	σ	θ	μ	σ
2016	361.27	6,584	0,356	405	6,300	0,482
2017	419.66	6,565	0,356	442	6,293	0,456

Source: Own research

Tab. 6. Prediction parameters of three-parameter lognormal curves as theoretical wage distributions for 2016 and 2017

6 Conclusion

In addition to GDP trends, wage and income global developments are dependent on other factors as well. A diverse labour market is one of them, neither employees nor companies being homogeneous entities since different types of labour are in demand and supply. Apart from these labour market aspects, there are other off-market factors affecting the level of wages, such as the limited mobility of workforce, labour legislation, trade union action and the minimum wage. In 2015, the average salary in the non-business sector slightly exceeded the average wage in the larger business sector. This seems to be due to considerably higher educational levels in the former sector, where only every fifth employee does not have at least a secondary school diploma, while 30 per cent received a bachelor's degree or higher. In the latter sector, on the other hand, half of the employees have less than a secondary school diploma and only one in seven is a university graduate. Generally, salaries in the non-business sector are much more equalized than wages in the business sphere, professionals having the highest and unskilled workers the lowest earnings.

Acknowledgement

This work is processed with the contribution of long term institutional support of research activities by the Faculty of Informatics and Statistics, University of Economics, Prague (IP 400040).

References

[1] BACHMANN, R., BECHARA, P., SCHAFFNER, S. Wage Inequality and Wage Mobility in Europe. *The Review of Income and Wealth*, Vol. 62, No. 1, 2016: pp. 181–197, ISSN 1475-4991.

[2] BLAISE, M., PUHANI, P. A. Do Public Ownership and Lack of Competition Matter for Wages and Employment? Evidence from Personnel Records of a Privatized Firm. *Journal of the European Economic Association*, Vol. 11, No. 4, 2013: pp. 918–944, ISSN 1542-4774.

[3] BROCKWELL, P. J., DAVIS, R. A. *Introduction to Time Series and Forecasting*. Springer, 2002, ISBN 978-3319298528.

[4] CHANG, S. C. Effects of Financial Developments and Income on Energy Consumption. *International Review of Economics and Finance*, Vol. 35, No. 1, 2015: pp. 28–44, ISSN 1059-0560.

[5] COWPERTWAIT, P. S. P., METCALFE, A. V. *Introductory Time Series with R*. Springer, 2009, ISBN, 978-0387886978.

[6] EBNER, C., HELBLING, M. Social Distance and Wage Inequalities for Immigrants in Switzerland. *Work, Employment and Society*, Vol. 30, No. 3, 2016: pp. 436–454, ISSN 0950-0170.

[7] MALÁ, I. The Use of Finite Mixtures of Lognormal Distribution for the Modelling of Household Income Distributions in the Czech Republic. *Politická ekonomie*, Vol. 61, No. 3, 2013: pp. 356–372, ISSN 0032-3233.

[8] MALÁ, I. Multivariate Probability Model for Incomes of the Czech Households. *Politická ekonomie*, Vol. 63, No. 7, 2015: pp. 895–908, ISSN 0032-3233.

[9] MAREK, L. The Trend of Wage Distributions in Czech Republic in the Years 1995-2008 Analysis. *Politická ekonomie*, Vol. 58, No. 2, 2010: pp. 186–206, ISSN 0032-3233.

[10] OZANNE, R. Impact of Unions on Wage Levels and Income Distribution. *The Quarterly Journal of Economics*, Vol. 73, No. 2, 1959: pp. 177–196, ISSN 0033-5533.

[11] PENCAVEL, J. Keynesian Controversies on Wages. *The Economic Journal*, Vol. 125, No. 583, 2015: pp. 295–349, ISSN 1468-0297.

[12] PIKETTY, T. Income inequality in France, 1901–1998. *Journal of Political Economy*, Vol. 111, No. 5, 2003: pp. 1004–1042, ISSN 0022-3808.

[13] SCHAFGANS, M. M. A. Ethnic Wage Differences in Malaysia: Parametric and Semiparametric Estimation of the Chinese-Malay Wage Gap. *Journal of Applied Econometric*, Vol. 13, No. 5, 1998: pp. 481–504, ISSN 1099-1255.

[14] STANOVNIK, T., VERBIČ, M. Wage and Income Inequality in Slovenia, 1993–2002. *Post-Communist Economies*, Vol. 17, No. 3, 2005: pp. 381–397, ISSN 1463-1377.

[15] TRIOLA, M. F. *Elementary Statistics*. The Benjamin/Cummings Publishing Company, 1989, ISBN 978-0321836960.

Current address

Bílková Diana, doc. Ing. Dr.
University of Economics, Prague, Faculty of Informatics and Statistics,
Department of Statistics and Probability
Sq. W. Churchill 1938/4, 130 67 Prague 3, Czech Republic
Tel. Number: +420 224 095 484, e-mail: diana.bilkova@vse.cz

Slovak University of Technology in Bratislava
Faculty of Mechanical Engineering
16th Conference on Applied Mathematics **APLIMAT 2017**
Proceedings

A COMPARISON OF NORMALS CALCULATION FOR THE CONSTRUCTION OF INTERPOLANTS ABOVE TRIANGULAR MESHES

BOHDAL Róbert (SK)

Abstract. In this paper, we compare selected methods for calculation of normal vectors, which are necessary for the construction of interpolants above the given triangulation. The normals at individual vertices of the underlying triangulation greatly affect the shape and the smoothness of the resulting interpolation surface. We compare results of selected methods for normals calculation with analytically calculated normals of test functions. We also calculate the difference between the created interpolants and the corresponding test functions. The best results were achieved by the method of calculating normals using the local thin plate interpolation spline. The weighted average method, created by combining Little's and Max's method came as the second in order.

Keywords: normals calculation, thin plate spline, Clough-Tocher, Powell-Sabin

Mathematics subject classification: Primary 65D05; Secondary 65D17

1 Introduction

In many applications, we are often confronted with the problem of scattered points interpolation. In case of a small number of input points, it is usually the best to choose so-called global interpolation methods, which interpolate the input points using only one function. The global interpolation methods yield good overall smoothness and accuracy of the interpolation. Interpolation methods using radial basis functions [4] such as Hardy multiquadrics or thin plate splines, etc. are often used. For a bigger number of input points (of the order ten thousand and more), it is not possible to use global interpolation methods because they need to solve large systems of equations. In such case, local interpolation methods are used with an advantage, which use piecewise functions, each interpolates only a few points. The classical examples are the Clough-Tocher method [1, 8], the Powell-Sabin [1, 10], the natural neighbour interpolation [7], and others. Many of them require normal vectors at the given points. Since these vectors are not known, they must be estimated.

The quality of the created interpolation surfaces (in the sense of visual smoothness, continuity, accuracy, etc.) greatly depends on the accuracy of the calculated normals. The gradient of the surface in a given point has often a greater influence on the shape of the surface than the chosen degree of polynomials or the degree of smoothness.

In this paper we compare some known methods for normals calculation. We also suggest a small modification which includes removing "long and thin" triangles on the boundary of the given triangulation and removing "unsuitable" normals from the calculations. Next, we suggest a combination of the Max's and Little's method into one formula. We then compare the calculated normals with the "true" analytical normals, and we also quantify the accuracy with which the interpolation surfaces using these normals match the test functions.

Jin et al. [5] also compare some methods for normals calculation on the selected test models with known analytical normals, but their comparison technique is based on the cumulative histograms of angular discrepancies. Moreover, they exclude vertices of non-closed test surfaces from the comparison.

2 The methods for normals calculation

There are many methods how to estimate the normals. The most commonly used methods are weighted average methods and methods using local interpolants and approximants. Other global methods usually based on minimization of a given integral on the triangular network are used rarely. Another approach for normal calculation based on the linear regression and on the finite difference method we can find in the paper [6].

Almost all methods for normals calculation use an in-advance constructed triangulation of the input points. The normals are then calculated at the triangulation vertices using some weighted average of the normals of adjacent triangles.

We search such method that is robust enough with respect to input points. It often happens that the considered triangulation includes "long and thin" triangles on the boundary that negatively affect the accuracy of calculating normals using the weighted average methods. Thin triangles have at least one angle much smaller than the other ones. In addition, long triangles have at least two sides much longer than all other triangles in the triangulation. The constructed interpolation surface which uses such normals then can create unwanted shapes on the surfaces boundary, see figure 3(a).

2.1 Normals calculated using weighted average

Weighted average methods use the existing triangulation of the input points. It is good to use Delaunay triangulation that minimizes the number of "long and thin" triangles, because it maximizes the minimal angles of all the triangles of the triangulation.

Let we have the vertices $B_i[x_i, y_i, z_i]$ of the triangulation $\mathcal{T}$ created from the set of given points $(x_i, y_i, z_i) \in \mathbb{E}^3$. According to [3], we can calculate the normals at the vertices by the formula

$$\hat{\boldsymbol{n}}_i = \sum_{\mathcal{N}_i} \omega_{ijk} \boldsymbol{n}_{\triangle ijk}, \qquad (1)$$

where the weight ω_{ijk} is calculated from the relation

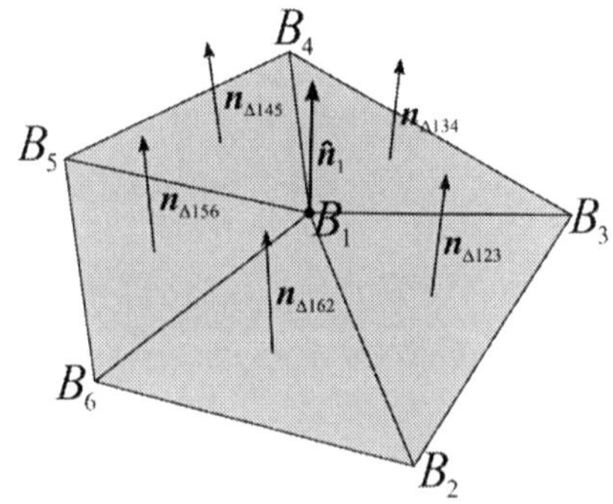

Fig. 1. Example of adjacent triangles for calculation $\hat{n}_1$ with indices belongs to the set $\mathcal{N}_1$

$$\omega_{ijk} = \frac{\sigma_{ijk}}{\sum_{\mathcal{N}_i} \sigma_{ijk}} \tag{2}$$

and where $n_{\triangle ijk}$ denotes the normal of the triangle $B_iB_jB_k$. The sum in equation (1) is enumerated for all triple indices (i, j, k) of the triangulation vertices from the set

$$\mathcal{N}_i = \{(i, j, k) \in \mathbb{N}^3; j \neq k, \\ \text{where } B_j, B_k \text{ meet "selection criterion"}\}.$$

The selection criterion includes all the vertices which either create an edge with the vertex B_i or lie in some neighbourhood of the vertex B_i (see figure 1).

According to [3], the values σ_{ijk} in formula (2) are given by one of the following possibilities (see figure 2):

- The arithmetic average (Gouraud)

  $$\sigma_{ijk} = 1.$$

  Every triangle contributes with the same weight to the calculated normal.

- The inverse value of lengths (Little)

  $$\sigma_{ijk} = \frac{1}{|B_iB_j|^r |B_iB_k|^r},$$

  where $r = 1$, 2 or $1/2$ as a rule. The weight depends inversely on the distance of triangulation vertices adjacent to the vertex B_i. The longer is the triangle, the less it contributes to the resulting normal.

- The angle at the vertex (Thurmer)

  $$\sigma_{ijk} = \alpha_i.$$

  The symbol α_i denotes the angle $\measuredangle B_iB_jB_k$ at the vertex B_i. The triangle that makes the biggest angle between the edges B_jB_i and B_kB_i contributes the most to the resulting normal.

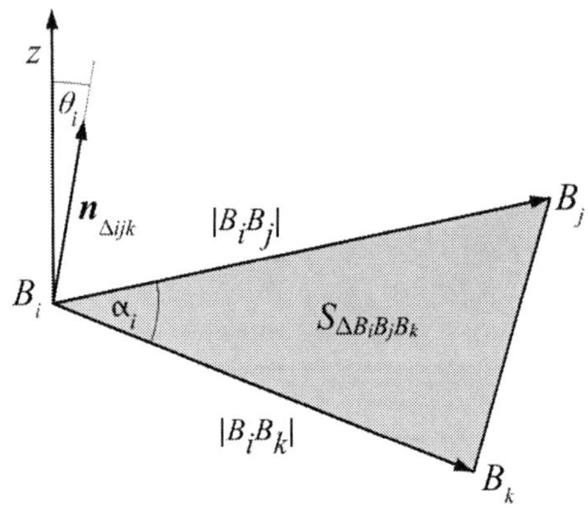

Fig. 2. Elements of a triangle for calculation of normals

- The area of the triangle (Akima)

$$\sigma_{ijk} = S_{\triangle B_i B_j B_k}.$$

The triangle with the greatest surface contributes the most to the resulting normal. The symbol $S_{\triangle B_i B_j B_k}$ denotes the area of the triangle $B_i B_j B_k$.

- The gradient of the surface (Akima2)

$$\sigma_{ijk} = \cos\theta_i S_{\triangle B_i B_j B_k}.$$

The symbol θ_i denotes the angle between the z-axis and $\boldsymbol{n}_{\triangle ijk}$. A triangle whose normal makes a smaller angle with the z-axis, contributes more to the resulting normal than one with the same area. This ensures the equality of the triangles which have the same area in projection to the xy plane.

Max in paper [9] proposes to calculate σ_{ijk} using the formula

$$\sigma_{ijk} = \frac{\sin\alpha_i}{|B_i B_j|^r |B_i B_k|^r}.$$

This method prefers triangles whose angle at the vertex B_i approaches the right angle and whose edges are short.

We can also combine Little's and Max's method into one to achieve a better "average accuracy":

$$\sigma_{ijk} = \frac{1 + \sin\alpha_i}{2|B_i B_j|^r |B_i B_k|^r}.$$

Although weighted average methods have low computational complexity, they give low-grade results. They are not suitable for triangulations with smaller number of vertices, in which the gradient of the triangles often changes its direction.

2.2 Normals calculated using local interpolation or approximation

The next possibility how to estimate a normal at a point B_i is to use a local function $f_i(x, y)$, which interpolates or approximates the set of "close neighbours" of the point B_i, and calculate the normal using partial derivatives:

$$\hat{\boldsymbol{n}}_i = \left(\frac{\partial f_i(x_i, y_i)}{\partial x}, \frac{\partial f_i(x_i, y_i)}{\partial y}, -1 \right).$$

Stead compared several kinds of functions in [13], among them:

- The Shepard's interpolant

$$f(x, y) = \sum_{i=1}^{n} \omega_i(x, y) z_i,$$

  where the weight ω_i is given by $\omega_i(x, y) = \frac{d_i^2(x,y)}{\sum_{j=1}^{n} d_j^2(x,y)}$ and $d_i^2(x, y) = (x - x_i)^2 + (y - y_i)^2$.

- The Hardy's multiquadrics interpolant

$$f(x, y) = \sum_{i=1}^{n} c_i \sqrt{d_i^2(x, y) + R^2},$$

  where the shaping parameter R is given, and the unknowns c_i are calculated from a set of equations following the interpolation conditions $f(x_i, y_i) = z_i$.

- Linear polynomial function given by the least squares method.

- Quadratic polynomial function given by the least squares method.

According to the performed tests, the best results are achieved by Hardy's multiquadrics interpolant.

Instead of Hardy's multiquadrics $\phi(d_i^2(x, y)) = \sqrt{d_i^2(x, y) + R^2}$ we can use thin plate splines $\phi(d_i^2(x, y)) = d_i^2(x, y) \ln(d_i^2(x, y))$ because they do not require estimation of the parameter R.

Renka and Cline [11] described method for finding the normal at the vertex B_i using partial derivatives of the quadratic polynomial function $f(x, y) = z_i + a(x - x_i)^2 + b(x - x_i)(y - y_i) + c(y - y_i)^2 + d(x - x_i) + e(y - y_i)$, which interpolates the vertex B_i and approximates the set of "close neighbours" in the sense of the least squares method. For each vertex B_j from the set of neighbour vertices, a weight ω_j is calculated so that the furthest vertices do not contribute to the resulting normal

$$\omega_j = \frac{(r_i - |B_i B_j|)_+}{r_i |B_i B_j|},$$

where r_i is the radius of neighbourhood influence of the vertex B_i and $(\cdot)_+ = \max\{0, \cdot\}$.

2.3 Normals calculated using global methods

We can get the best results for normals estimation if we use global methods which are based on finding the minimum of appropriate integral functions. An example is the method of Nielson's minimum norm network described in [12].

Global methods for estimating normals give better results than local methods, but we have to solve large systems of equations for their calculation. Moreover, this estimation is only a little more accurate than results obtained by methods based on local interpolants [3].

3 Testing the methods

For testing purposes, we have used data which were created by 9 test functions, so we have been able to calculate all normals using partial derivatives. We have added two own functions (see figure 4) to the seven test functions presented in [2].

The samples contained from 100, 900, 2500, and 4900 randomly selected points lying in the interval $[0, 1] \times [0, 1]$, and they were different for each test function. As a result, we have gathered 36 different input triangulations. We have compared 7 methods of calculating normals using various weighted averages and one method using the local thin plate spline interpolation function.

Some of the weighted average methods (e.g. the Akima method) make unwanted shapes at the border of the constructed interpolant (see figure 3(a)) if the underlying triangulations include very "long and thin" triangles, which often occur at the boundaries of the triangulations. After deleting them[1], the results significantly improved for the less robust methods (see figure 3(b)). Moreover, the problem of calculating the values of the interpolation function outside of the triangulation occurred. The next attempt for normals calculation on the boundary of the triangulation was not to use normals which differ a lot from the "average normal" (see algorithm 1). The combination of both modifications gave noticeable improvement (see figure 3(c)).

Little's and Max's methods have the best results in the category of methods using the weighted average. Little's method was better for some data, and Max's method was better for the other ones (see the plots in figure 5). Consequently, we have decided to combine both methods into one relation (Little-Max). Using this new method we have achieved a better "average accuracy" (see plots in figure 5).

From the methods using local interpolants, we have selected thin plate splines instead of Hardy's multiquadrics because they do not require estimation of the parameter R.

In our test we have calculated:

1. The deviations of the estimated normals $\hat{\boldsymbol{n}}_i$ from the normals of the test function $\nabla f_t(x_i, y_i)$ calculated using partial derivatives at the given point (x_i, y_i)

$$\mathrm{RMSE}_\alpha = \sqrt{\frac{\sum_{i=1}^{n}\left(\cos^{-1}\left(\frac{\hat{\boldsymbol{n}}_i \cdot \nabla f_t(x_i,y_i)}{|\hat{\boldsymbol{n}}_i||\nabla f_t(x_i,y_i)|}\right)\right)^2}{n}}.$$

   Results are visualized in the plots in figure 5.

[1] We have deleted triangles whose inner angle exceed 178 degrees.

Algorithm 1 "Unsuitable" normals removing
input: $num\_normals, trng\_normals[], const = 1.0$
output: $trng\_normals$

if $num\_normals < 5$ **then**
    return
$sum\_norm = 0$
for all $normal$ in $trng\_normals$ **do**
    $sum\_norm \mathrel{+}= normal$
$avg\_normal = sum\_norm/num\_normals$
$sum\_sqd = 0.0$
for $i = 0$ to $num\_normals$ **do**
    $square\_dif[i] = \|trng\_normals[i] - avg\_normal\|^2$
    $sum\_sqd \mathrel{+}= square\_dif[i]$
$average\_sqd = sum\_sqd/num\_normals$
for $i = 0$ to $num\_normals$ **do**
    $delta\_sqd = |square\_dif[i] - average\_sqd|$
    if $delta\_sqd > const * average\_sqd$ **then**
        remove $trng\_normals[i]$ from list

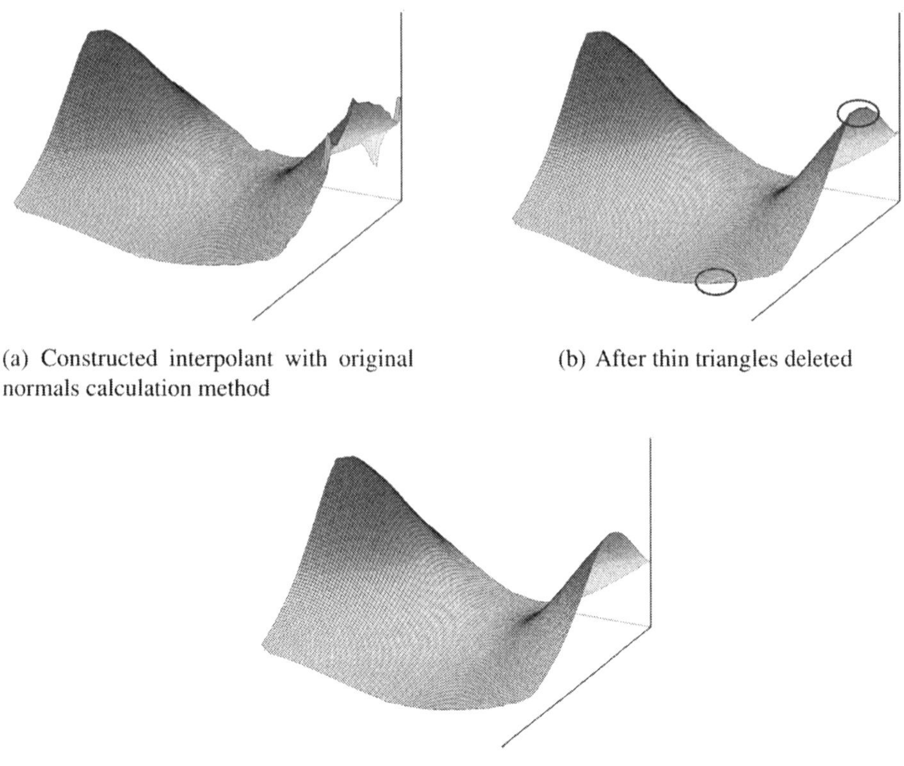

(a) Constructed interpolant with original normals calculation method

(b) After thin triangles deleted

(c) After "unsuitable" normals deleted

Fig. 3. Effect of a modification for calculating the normals on the shape of constructed interpolant. The normals have been calculated by Akima method

2. The accuracy of "matching" the resulting interpolation surfaces calculated by the Clough-Tocher and the Powell-Sabin method with the corresponding test function. To test the accuracy of the interpolation surface, we have calculated the deviation from the square of the difference between the function value of interpolation function $\hat{f}(x, y)$ and the function value of the test function $f_t(x, y)$ for $m = 2500$ randomly selected points from interval $[0, 1] \times [0, 1]$:

$$\text{RMSE} = \sqrt{\frac{\sum_{i=1}^{m}(\hat{f}(x_i, y_i) - f_t(x_i, y_i))^2}{m}}.$$

   The plots in figure 6 present the result. Since the order of accuracy of the tested methods was almost identical for both Clough-Tocher and Powell-Sabin interpolation method, we present only the results for the first of them.

The approximate time in milliseconds of the individual normals calculation methods is presented in the table 3. All methods have been tested on the desktop PC with Intel(R) Core(TM) i5-4670K CPU @3.40GHz processor with 8GB RAM.

4 Conclusion

As expected, the method using the local interpolant (see plots in figures 5 and 6) has come out as the best one from all tested methods. The combination of Little's and Max's method was the second best method in general. Both approaches of calculations were sufficiently robust with respect to the used input data.

For applications not needing great speed of normals calculating, we recommend the method which uses local thin plate spline interpolation. In other cases, we suggest to use the above described combination of Little's and Max's method or one of them.

From table 3 it is clear that the method using local thin plate interpolation splines (TPS) is approximately 5 times more computationally expensive than all the weighted average methods.

Our suggested combination of Little's and Max's methods is sufficiently robust considering the triangulation of the input points and it could be used for the the construction of interpolants with the Clough-Tocher or Powell-Sabin method even without "unsuitable" normals removing.

Methods/number of points	**4900 points**	**8100 points**
All methods except TPS	$\approx$ 90ms	$\approx$ 250ms
TPS method	$\approx$ 480ms	$\approx$ 1310ms

Tab. 1. A comparison of the times of computations

List of the test functions

$$f_1(x,y) = \frac{3e^{-\frac{(9y+1)^2}{10}-\frac{(9x+1)^2}{49}}}{4} + \frac{3e^{-\frac{(9y-2)^2}{4}-\frac{(9x-2)^2}{4}}}{4} + \frac{e^{-\frac{(9y-3)^2}{4}-\frac{(9x-7)^2}{4}}}{2} - \frac{e^{-(9y-7)^2-(9x-4)^2}}{5}$$

$$f_2(x,y) = \frac{1-\tanh(9y-9x)}{9}$$

$$f_3(x,y) = \frac{\cos\left(\frac{27}{5}y\right)+\frac{5}{4}}{6(3x-1)^2+6}$$

$$f_4(x,y) = \frac{e^{-\frac{81\left(\left(y-\frac{1}{2}\right)^2+\left(x-\frac{1}{2}\right)^2\right)}{16}}}{3}$$

$$f_5(x,y) = \cos(10y)+\sin(10(x-y))$$

$$f_6(x,y) = \frac{\sqrt{64-81\left(\left(y-\frac{1}{2}\right)^2+\left(x-\frac{1}{2}\right)^2\right)}}{9} - \frac{1}{2}$$

$$f_7(x,y) = \frac{1}{\sqrt{2e^{-3\left(\sqrt{y^2+x^2}-\frac{67}{10}\right)}+1}}$$

$$f_8(x,y) = 50e^{-200\left(\left(y-\frac{3}{10}\right)^2+\left(x-\frac{3}{10}\right)^2\right)} + e^{-50\left(\left(y-\frac{1}{2}\right)^2+\left(x-\frac{1}{2}\right)^2\right)}$$

$$f_9(x,y) = \sin(\pi x)\sin(2\pi y)$$

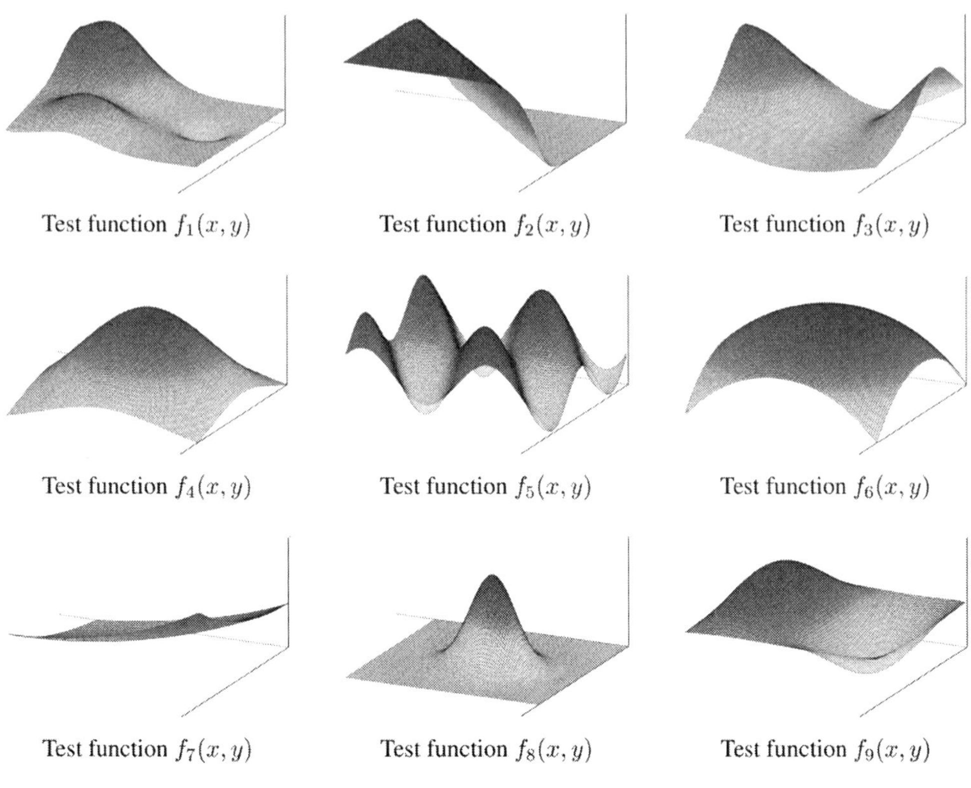

Fig. 4. Pictures of the test functions

Deviations between the calculated and the analytical normals

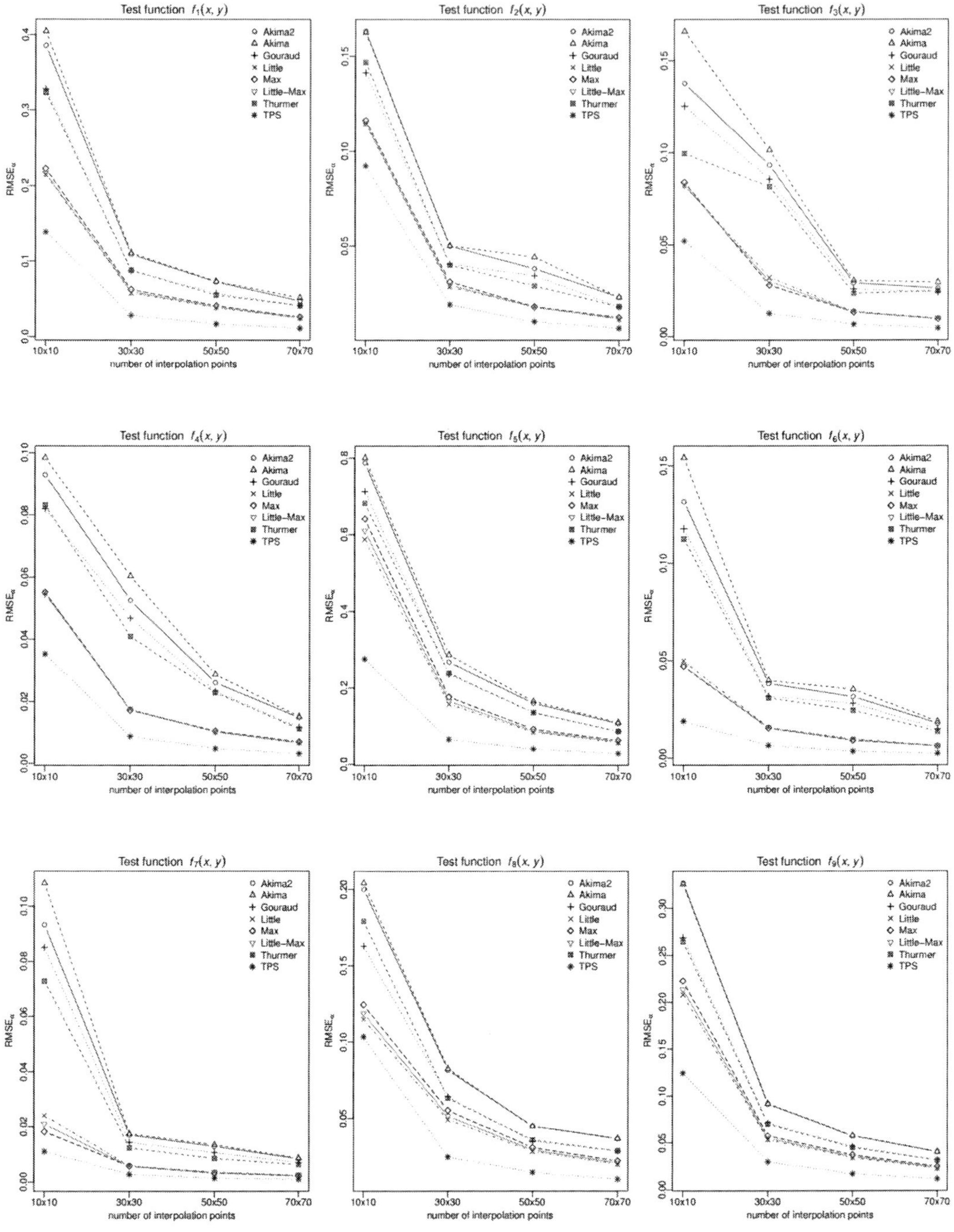

Fig. 5. Plots of deviations between the calculated normal and the analytical normal of the test function

Deviations between the Clough-Tocher interpolant and the test function

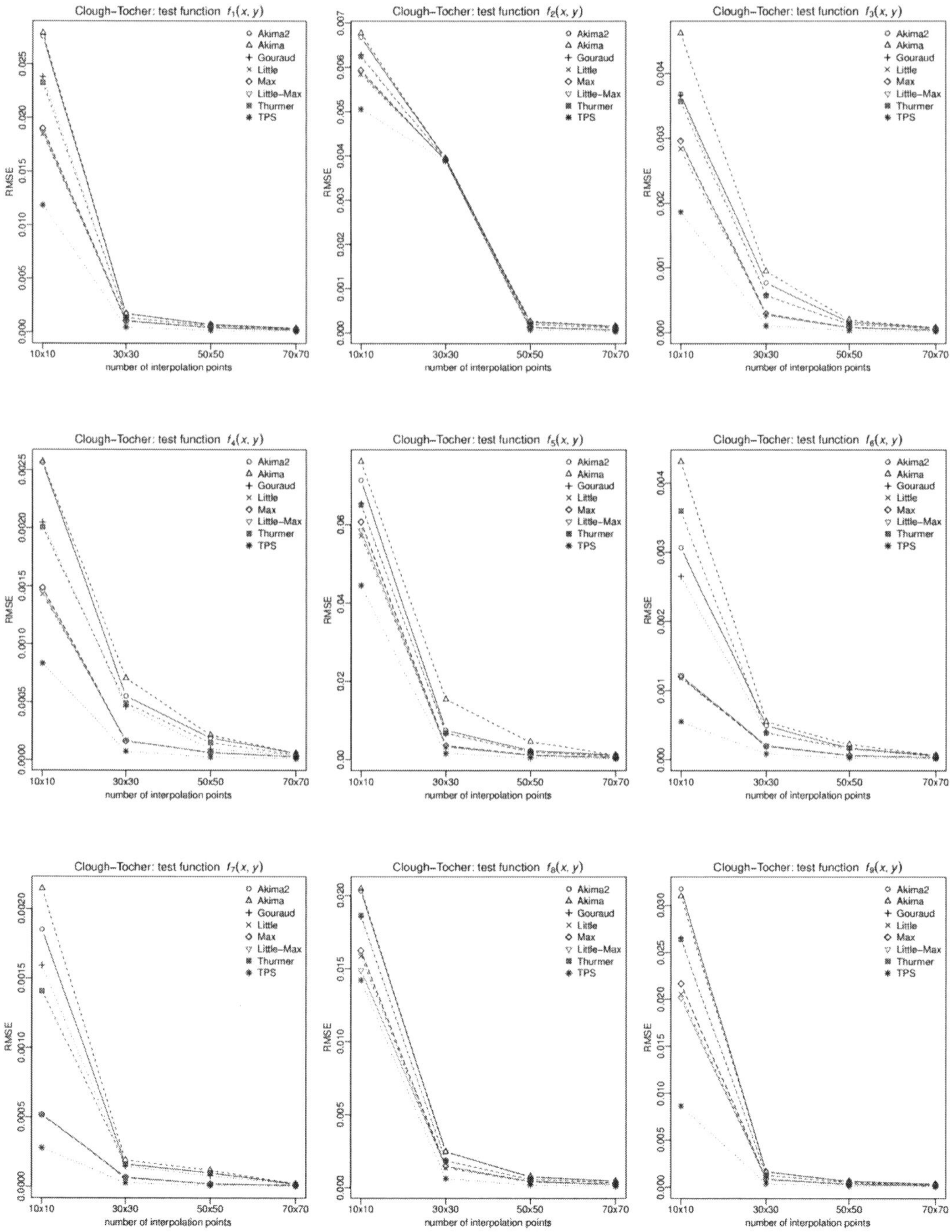

Fig. 6. Plots of deviations between the function value of the interpolation and the test function

References

[1] AMIDROR, I. *Scattered data interpolation methods for electronic imaging systems: a survey.* Journal of electronic imaging 11, 2 (2002), 157–176.

[2] FRANKE, R. *Scattered data interpolation: Tests of some methods.* Mathematics of computation 38, 157 (1982), 181–200.

[3] HOSCHEK, J., LASSER, D. *Fundamentals of Computer Aided Geometric Design.* A. K. Peters, Ltd., Natick, MA, USA, 1993.

[4] ISKE, A. *Radial basis functions: basics, advanced topics and meshfree methods for transport problems.* Rendiconti del Seminario Matematico, Polytechnic University of Turin 61, 3 (2003), 247–285.

[5] JIN, S., LEWIS, R. R., West, D. *A comparison of algorithms for vertex normal computation.* The Visual Computer 21, 1-2 (2005), 71–82.

[6] JIRKA, T., SKALA, V. *Gradient vector estimation and vertex normal computation.* Tech. rep., University of West Bohemia, Pilsen, 2002.

[7] LEDOUX, H., GOLD, C. *An efficient natural neighbour interpolation algorithm for geoscientific modelling.* Developments in Spatial Data Handling. Springer, 2005, pp. 97–108.

[8] MANN, S. *Cubic precision clough-tocher interpolation.* Computer aided geometric design 16, 2 (1999), 85–88.

[9] MAX, N. *Weights for computing vertex normals from facet normals.* Journal of Graphics Tools 4, 2 (1999), 1–6.

[10] POWELL, M., SABIN, M. *Piecewise quadratic approximations on triangles.* ACM Transactions on Mathematical Software (TOMS) 3, 4 (1977), 316–325.

[11] RENKA, R., CLINE, A. *A triangle-based c1 interpolation method.* Rocky Mountain Journal (1984).

[12] SCHEUERMANN, G., TRICOCHE, X., HAGEN, H. *C1-interpolation for vector field topology visualization.* Visualization'99. Proceedings (1999), IEEE, pp. 271–533.

[13] STEAD, S. *Estimation of gradients from scattered data.* Rocky Mountain Journal of Mathematics 14, 1 (1984).

Current address

Bohdal Róbert, RNDr. PhD.
Department of Algebra, Geometry and Didactics of Mathematics
Faculty of Mathematics, Physics and Informatics, Comenius University in Bratislava
Mlynska dolina, 842 48 Bratislava, Slovakia
Tel. number: +421-2-602 95 ext. 185, e-mail: robert.bohdal@fmph.uniba.sk

Slovak University of Technology in Bratislava
Faculty of Mechanical Engineering
16th Conference on Applied Mathematics **APLIMAT 2017**
Proceedings

MATH-MAPS A NEW ROAD THAT OPENS UP A WORLD PART I DYNAMIC MODELS WITHOUT DERIVATIVES

BRANDI Primo (IT), SALVADORI Anna (IT)

Abstract. Math-MAPS is a kind of Google-maps for Mathematical Education, created by Mathematics&Real-Life (M&R), with the aim to offer innovative educational paths to the Schools of any order. M&R (www.matematicaerealta.eu) is a national project which promotes the *education to mathematical modelling* as innovative direction. Modelling education involves a completely new way to approach Mathematics, based on a dynamic interaction between real world and mathematical world. Math-MAPS suggests the *learning highways* to connect different topics. In our contribution we illustrate one of these recommended tracks: the introduction to dynamic models. The fist part deals with the *discrete case*, suitable for High Schools (also of vocational type) and weak math University courses. The second part, which presents the traslation to the *continuous case*, is devoted to Scientific High Schools and University base courses.

Keywords: Mathematical skills, elementary mathematical models, teaching innovation

Mathematics Subject Classification: Primary 97D50; Secondary 97U40, 97U80

1 Introduction to dynamic models

Elementary mathematical models are usually *static models*; in a figurative language they can be compared to a *photograph* of reality. When we study a phenomenon *evolution* in time we must necessarily pass to a sequence of photos, in other words to a *movie*. Precisely we adopt a *dynamic model*, which describes, step by step, the changes that occur in the system over time.

In this part we propose an introduction to elementary dynamic models, suitable also for undergraduate education. The present first part is devoted to *discrete dynamic models*, described by sequences. The main aim is to show how some discrete dynamic models can be proposed also in school contexts in which infinitesimal computational tools are not available. In the second part we propose the transfer to *continuous dynamic models*, which leads to differential equations.

The comparison between the discrete and the continuous version of some elementary dynamic models will be discussed into details.

M&R educational approach is based on the interaction between real world and mathematical world. In this context, the passage from linear phenomena/functions to non-linear ones offers the opportunity to introduce dynamic models. As usual, we start from a real life situation.

Savings Plan. Lucia has inherited from the grandparents a large sum of money. She planes to deposit the sum in a savings account and to use the accrued annual interest for supporting her university studies.

The construction of the model. Denoted by C the sum and by i the annual interest rate, we can adopt the following *iterative* process to describes the investment plane

$$\begin{cases} M_0 = C & start \\ M_{n+1} = M_n + I \end{cases} \quad n \in N$$

where $I = iC$. It is easy to see that the model admits the *closed formula*

$$M_n = C + nI \quad n \in N$$

The discussion of the solution. The closed formula has a remarkable advantage: it allows to determine directly the n-th post, without the calculation of all the previous steps.
From the closed formula we deduce that the sequence $(M_n)_{n\in N}$ is an *arithmetic progression*, thus the *capital growth is directly proportional to the investment life*, as Fig. 1 shows.

Remark. Note that $M_n = (1+ni)C \quad n \in N$, thus the *capital is directly proportional to the invested sum*. For this reason often the financial experts do their considerations on the unitary initial sum.

Revised savings plan. Lucia wonders how the saving would change if she did not withdrew the accrued interests.

The modification of the model. In this case also the accrued annual interests capitalize and the process becomes

$$\begin{cases} M_0 = C & start \\ M_{n+1} = (1+i)M_n & n = 0,1,2,\ldots \end{cases} \Rightarrow \quad M_n = C(1+i)^n \quad n \in N$$

The discussion of the solution. The closed formula shows that the sequence $(M_n)_{n\in N}$ moves into a *geometric progression*. The capital growth is no more linear, but becomes exponential.
Fig. 2 shows the comparison between the two saving planes.

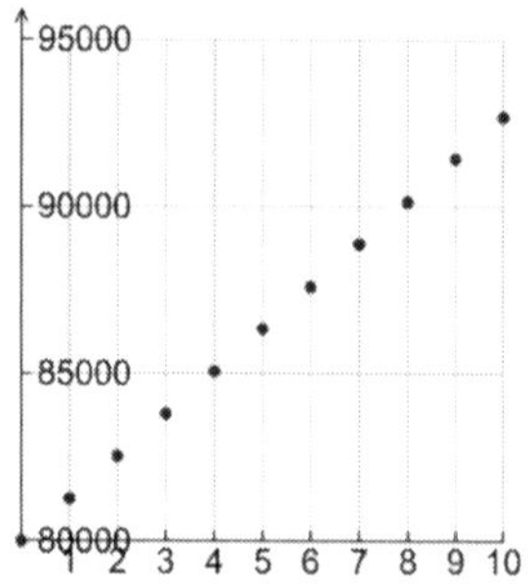

Fig. 1

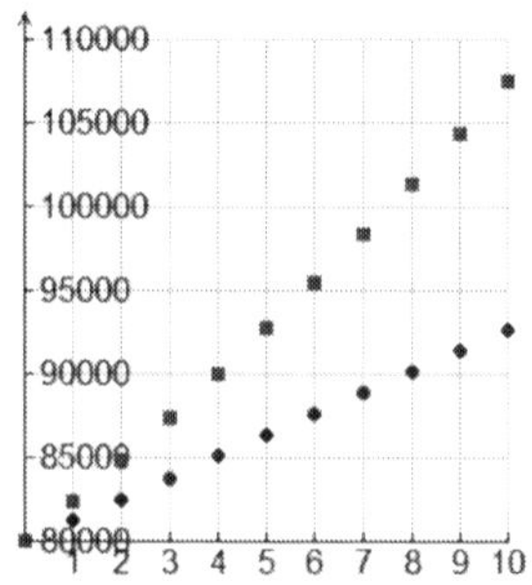

Fig. 2

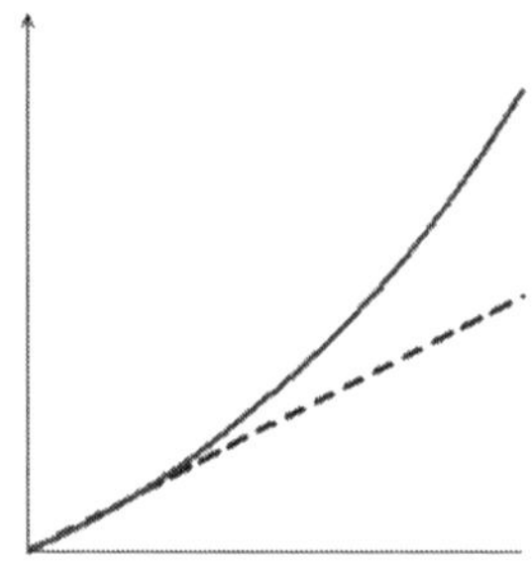

Fig. 3

Moreover, it is worth remembering that while the support function of an arithmetic progression is a linear function, a geometric progression is supported by an exponential function. May be the two support functions allow a better graphic comparison (see Fig. 3) .

Note that the present iterative process is suitable to describe many other phenomenoa as biological growth, radioactive decay, filtering, ... (see for example [1-4]) as we will show in the next section.

2 Some elementary dynamic models

After the introduction of Section 1, we are ready to introduce a some remarkable dynamic model.

2.1 Malthus model in population dynamics

In 1798 the English economist Thomas Malthus proposed a first model to study the evolution of an isolated population. Malthus assumed that

(M) The growth rate is directly proportional to the number of individuals.

In mathematical language the statement becomes

$$x_{n+1} - x_n = ax_n$$

where x_n denotes the population at the n-th stadium.
Malthus principle leads to the following dynamic model

$$\begin{cases} x_0 & start \\ x_{n+1} = x_n + ax_n = (1+a)x_n \end{cases} \Rightarrow \quad x_n = (1+a)^n x_0 \quad n \in N$$

where a is the *growth factor.*

Note that very different evolutions can be described, according to different growth factors. The table below and Fig. 4 summarizes the situations.

growth factor	population evolution
$a < 0$	extinction
$a = 0$	stable process
$a > 0$	explosion

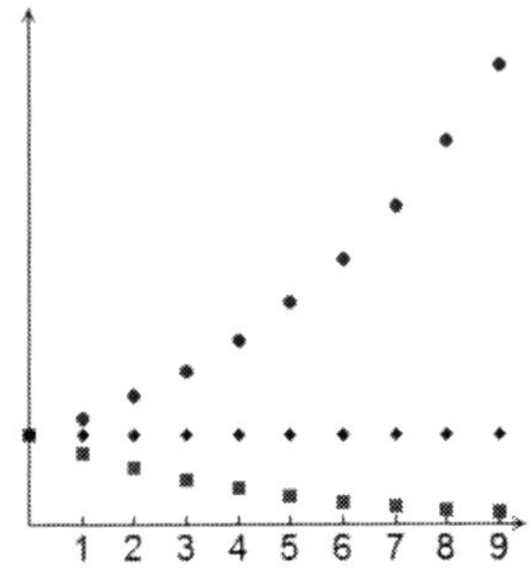

Fig. 4

We propose two current real-life phenomena connected to Malthus model, taken from a newspaper.

Venice's death. "Venice in 2030: an empty city, no people, only tourists." [Font: La Repubblica, 25.8.06] (see also Fig. 5)

The demographic boom in Africa. The population explosion in sub-Saharan Africa is near to trap the world's poorest countries into a spiral of misery and disease, with unpredictable consequences. [Font: La Repubblica, 26.8.06] (see Fig. 6).

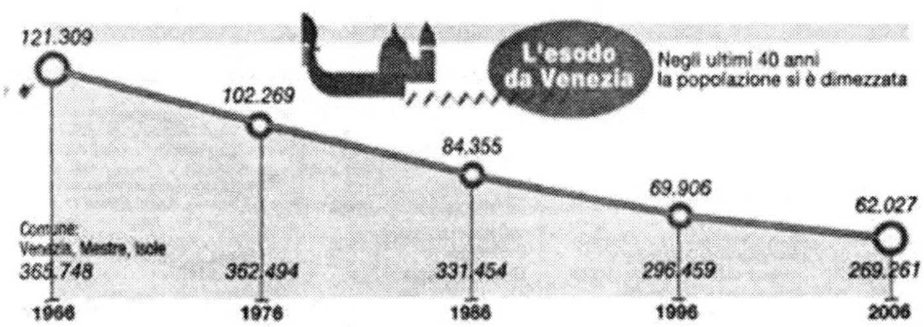

Fig. 5

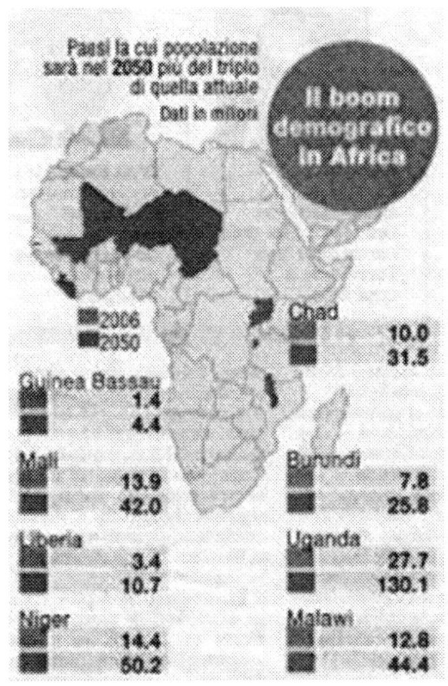

Fir. 6

We refer to [2,4] for the discussion of these phenomena and the related teaching suggestions. In the same references other applications of Malthus model to real life situations can be found.

2.2 Newton's law on heating and cooling

An other remarkable evolution model was introduced by Newton to describe heating/cooling. He assumed that

(N) The temperature variation of a fluid is directly proportional to the difference between the fluid temperature and that of the environment.

To discuss the related model, let us start with a daily life situation.

Coffee cooling.
According to Newton law, let us describe the cooling a coffee poured into a cap.

The construction of the model. We denote by
- T_a the room temperature;
- T_0 the temperature of the coffee just poured ($T_0 > T_a$);
- T_n the coffee temperature at the n-th stadium.

Newton's law in mathematical language leads to the iterative process

$$\begin{cases} T_0 & start \\ T_{n+1} - T_n = k(T_n - T_a) \end{cases} \qquad n \in N$$

where $k < 0$ is a constant.

In order to find a closed formula, we can change variable, assuming the heat drop $S_n = T_n - T_a$ as new variable. It is easy to see that the system becomes

$$\begin{cases} S_0 & start \\ S_{n+1} = (1+k)S_n \end{cases} \quad n \in N \quad \Rightarrow \quad S_n = (1+k)^n S_0 \quad n \in N$$

i.e. it turns into a Malthus process!

The discussion of the solution. The closed formula of the sequence $(S_n)_n$ can be re-written in terms of the old variable (the temperature) as it follows

$$T_n = T_a + (1+k)^n (T_0 - T_a) \quad n \in N$$

In conclusion Newton's law leads to the model

$$\begin{cases} T_0 & start \\ T_{n+1} - T_n = k(T_n - T_a) \end{cases} \quad \Rightarrow \quad T_n = T_a + (1+k)^n (T_0 - T_a) \quad n \in N$$

The coffee temperatures decreases untill it reaches the room temperature (note that $k < 0$). Fig. 5 shows the perfect agreement between the model and the real temperature (measured in a sequences of experiments whose specific data are listed in the table below).

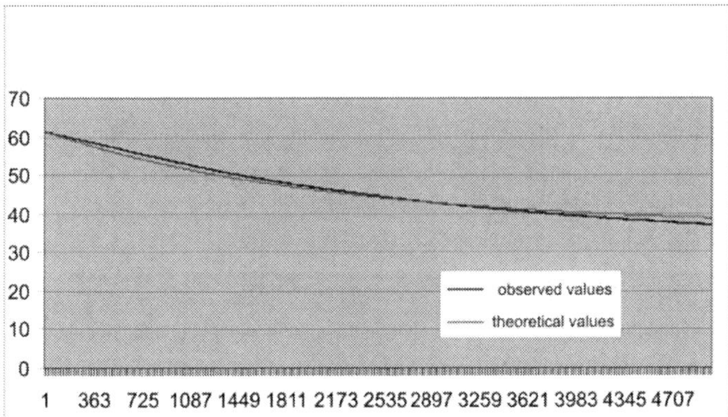

Fig. 5

Some experiment specifics	
Room temperature	$29\ ^{0}C$
Duration of each test	45 minutes
Number of readings for each test	4500
Reading approximation	$\pm 0{,}01\ ^{0}C$
Coffee initial temperature	$70 \div 75\ ^{0}C$
Final measured temperature	$34 \div 38\ ^{0}C$

Note that, in the general case, Newton model

$$\begin{cases} T_0 & start \\ T_{n+1} - T_n = k\left(T_n - T_a\right) \end{cases} \Rightarrow \quad T_n = T_a + \left(1+k\right)^n\left(T_0 - T_a\right) \quad n \in N$$

allows to deal with both cooling ($k < 0$) Fig. 6 or heating ($k > 0$) Fig. 7.

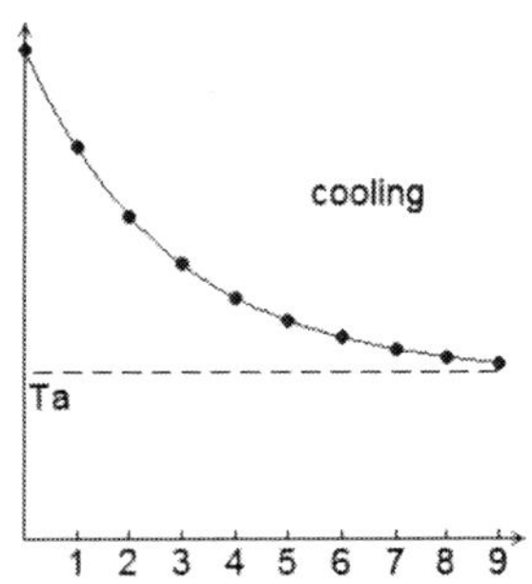

Fig. 6

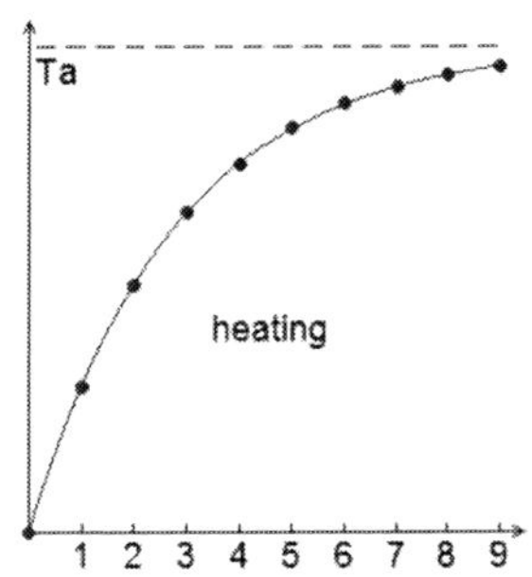

Fig. 7

2.3 General formulation of growth-decay discrete dynamic models

Both Malthusian principle and Newton's law reduce to the following general iterative pattern

$$\begin{cases} x_0 & start \\ x_{n+1} = k \cdot x_n & n \in N \end{cases} \Rightarrow \quad x_n = x_0 \cdot k^n \quad n \in N$$

This elementary iterative system reveals ductile and operative to describe a wide range of different growth-decay situation, as we will show in the next section.

3 Some examples of discrete dynamic models in real life situations

Let us discuss some examples of discrete dynamic models in order to describe real life situations.

Fish length. According to the biologist Ludwig von Bertalanffy, the growth rate of the length in some fish species is proportional to the difference between the maximum length of the species and the present length.
In particular, for the North Sea cod, we can take into account the following estimates (obtained by experimental tests):

the maximum length of the species $53\,cm$,
the length of newborn fish $L_0 = 10\,cm$,
the growth factor $k = -0.2$.

The construction of the model. We can adopt the general Newton model to describe the phenomenon. Precisely, denoted by
- L^* the maximum length of the species;
- L_n the fish length at the n-th stadium;

the fish growth is described by the iterative process

$$\begin{cases} L_0 & start \\ L_{n+1} - L_n = k\left(L^* - L_n\right) \end{cases} \Rightarrow \quad L_n = L^* - (1-k)^n \left(L^* - L_0\right) \quad n \in N$$

where $k > 0$.

The discussion of solution. In the particular case of North Sea cod, taking into account the estimates, we obtain the increasing sequence

$$L_n = 53 - 43(0{,}8)^n \quad n \in N$$

Fig. 8 shows the growth of North sea cod: the fishes length progressively increase approaching the maximim growth (unless they are caught!)

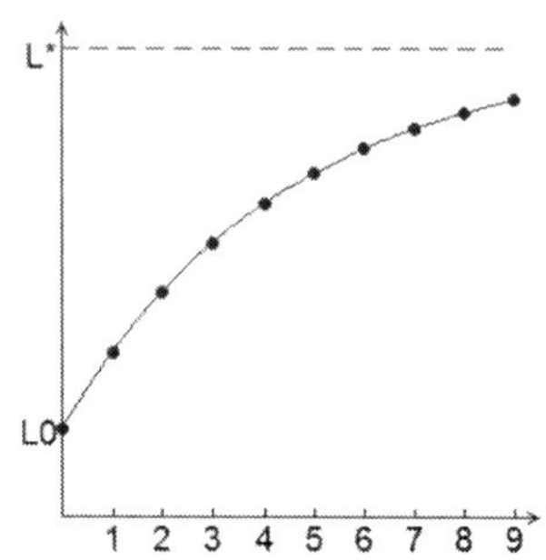

Fig. 8

The jam tart. Taking into account Newton's law, we can describe the cooking of a tart by means of the geometric progression

$$T_n = a(1+b)^n \quad n \in N$$

where T_n is the jam temperature after $n-$minutes and a, b are parameters.

A suitable discussion of the model allows to detect the meaning of the parameters.
This aspect of modeling reveals very instructive from the educational point of view

The famine in Ireland. In 1841 the Irish census assessed a population of 8.175.124 inhabitants, the population growth rate was estimated at 1,6%. The next census, ten years later, registered just 6.522.285 people. In that decade a severe famine, caused by a fungus that attacked potato plants, had produced a strong emigration to Canada, the United States and Wales. It could be interesting to estimate the emigration average rate.

The construction of the model. We can dsecribe the phenomenon by means of a suitable adaptation of Malthusian model

$$\begin{cases} P_0 \quad start \\ P_{n+1} - P_n = \gamma P_n + m P_n \end{cases} \quad n \in N$$

where P_n denotes the population in the n-th year (from the start year 1841), $\gamma > 0$ is the growth rate and $m < 0$ is the emigration rate. The process can be written as it follows

$$\begin{cases} P_0 \quad start \\ P_{n+1} = (1+\gamma+m)P_n \end{cases} \quad n \in N \quad \Rightarrow \quad P_n = (1+\gamma+m)^n P_0$$

The discussion of the solution. Taking into account our informations, we get the equation (in the unknown m)

$$6{,}522 \cdot 10^6 = (1+0{,}016+m)^{10}\ 8{,}175 \cdot 10^6$$

$$\log(1+0{,}016+m) = \frac{\log 6{,}522 - \log 8{,}175}{10} \cong -0{,}02259$$

$$1{,}016 + m = e^{-0{,}02259} \Leftrightarrow m = e^{-0{,}02259} - 1{,}016 = -0{,}03834$$

Thus we can conclude that the emigration average rate was about 3,8%.

The liquid mixing. In a tank containing salt water at concentration α a salt water at a different concentration β is introduced, at a constant rate r. In order to keep the liquid volume constant, water exits from the tank with the same speed. The solution is mixed continuously. Discuss the evolution of salt concentration.

The construction of the model. Denoted by V the liquid volume and by C_n the salt concentration al the n -th minute , we can describe the phenomenon by means of the iterative model

$$\begin{cases} C_0 = \alpha \\ C_{n+1} = C_n \dfrac{V-r}{V} + \beta \dfrac{r}{V} \end{cases} \quad n \in N$$

In order to find the *closed formula* we can denote

$$\frac{V-r}{V} = \lambda_1 \qquad \frac{r}{V} = \lambda \qquad 0 < \lambda < 1 \quad 0 < \lambda_1 < 1 \quad \lambda + \lambda_1 = 1$$

and note that

$$C_1 = C_0\lambda_1 + \beta\lambda = \alpha\lambda_1 + \beta\lambda \qquad C_2 = C_1\lambda_1 + \beta\lambda = (\alpha\lambda_1 + \beta\lambda)\lambda_1 + \beta\lambda = \alpha\lambda_1^2 + \beta\lambda(1+\lambda_1)$$

$$C_3 = \left[\alpha\lambda_1^2 + \beta\lambda(1+\lambda_1)\right]\lambda_1 + \beta\lambda = \alpha\lambda_1^3 + \beta\lambda\left(\lambda_1^2 + \lambda_1 + 1\right) \ldots$$

In conclusion we get the sequence

$$C_n = \alpha\lambda_1^n + \beta\lambda\left(\lambda_1^{n-1} + \lambda_1^{n-2} + \cdots + \lambda_1 + 1\right) = \alpha\lambda_1^n + \beta\lambda\frac{1-\lambda_1^n}{1-\lambda_1} = (\alpha - \beta)\lambda_1^n + \beta \qquad n \in N$$

The discussion of the solution. The following immages show the concentration evolution. In both the cases $\alpha < \beta$ (Fig. 9) or $\alpha > \beta$ (Fig. 10) the concentration is monotone and evolves asymptotically up to the concentration β of the incoming liquid.

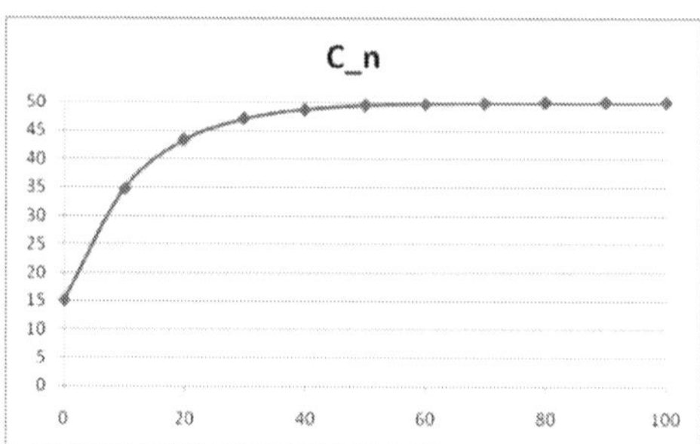

Fig. 9

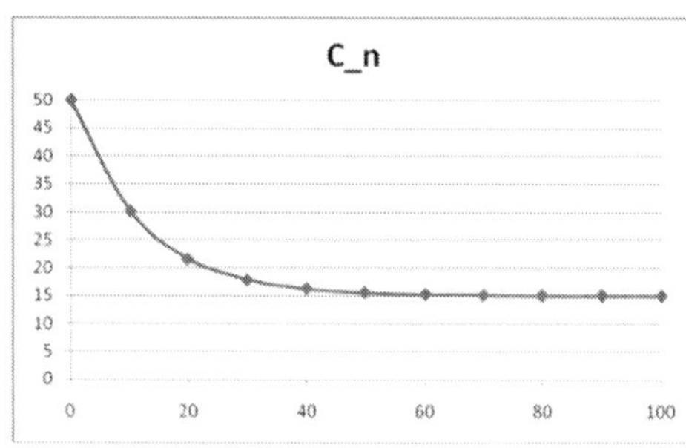

Fig.10

References

[1] BRANDI, P., SALVADORI, A. *Prima di iniziare*. Aguaplano Officina del libro, 2011, ISBN 978-88-905726-5-4.

[2] BRANDI, P., SALVADORI, A. *Nuovi percorsi di Matematica Vol.1* Aguaplano Officina del libro, 2015, ISBN 978-88-97738-65-7.

[3] BRANDI, P., SALVADORI, A. *Math MAPS II Itinerari per le competenze* Alice&Bob 45 EGEA, Milano, 2016, ISBN 978-88-238-6199-2.

[4] BRANDI, P., SALVADORI, A. *Math-MAPS. Introduction to dynamic models*, work in progress.

[5] BRANDI, P., SALVADORI, A. *Math-MAPS a new road thet opens up a world. Part II Elementary dynamic models, Proceedings APLIMAT 2017.*

Current address

Brandi Primo, full professor
Dipartimento di Matematica e Informatica, Università degli Studi di Perugia
via Vanvitelli 1, 06123 Perugia, Italy
E-mail: primo.brandi@unipg.it

Salvadori Anna, assiciated professor
Dipartimento di Matematica e Informatica, Università degli Studi di Perugia
via Vanvitelli 1, 06123 Perugia, Italy
E-mail: anna.salvadori@unipg.it

Slovak University of Technology in Bratislava
Faculty of Mechanical Engineering
16th Conference on Applied Mathematics **APLIMAT 2017**
Proceedings

MATH-MAPS A NEW ROAD THAT OPENS UP A WORLD PART II ELEMENTARY DYNAMIC MODELS

BRANDI Primo (IT), SALVADORI Anna (IT)

Abstract. Math-MAPS is a kind of Google-maps for Mathematical Education created by Mathematics&Real-Life (M&R) with the aim to offer innovative educational paths to the Schools of any order. M&R (www.matematicaerealta.eu) is a national project which promotes the *education to mathematical modelling* as innovative direction. In our contribution we will illustrate one of main recommended tracks: the introduction to dynamic models. This second part is devoted to the *continuous case*, introduced as the traslation of the *discrete case*, discussed in the first part.

Keywords: Mathematical skills, elementary mathematical models, teaching innovation

Mathematics Subject Classification: Primary 97D50; Secondary 97U40, 97U80

1 From discrete to continuous dynamic models

In the first part we have discussed the modelization of evolution phenomena by means of *discrete models*. More precisely the time is divided into steps (i.e. the time variable is an integer) and the model is an iterative process. In this second part we discuss a *continuous version* of these models, where the time flows continuously (i.e. the time is real variable).

The main discrete models of the first part are inspired by Malthus principle and Newton law. Both Malthus and Newton models involve the concept of *variation*. Since they are governed by a sequence $(x_n)_{n\in N}$, for variation we mean the difference between the values corresponding to two consecutive steps $\Delta x = x_{n+1} - x_n$.

Thus the variation coincides with the *rate of change*

$$\Delta x = \frac{x_{n+1} - x_n}{(n+1) - n} = \frac{\textit{variation of the variabile}}{\textit{variation of the stadium}} = \textit{rate of change}$$

When a phenomenon is described by a function of real variable, the *variation* and the *rate of change* are two distinct concepts.

More precisely

- the *variation* of the function f in the interval $[x_0, x_0 + h]$ is

$$\Delta f = f(x_0 + h) - f(x_0)$$

while

- the *rate of change* is the given by

$$\frac{\Delta f}{\Delta x} = \frac{f(x_0 + \mathrm{h}) - f(x_0)}{h}$$

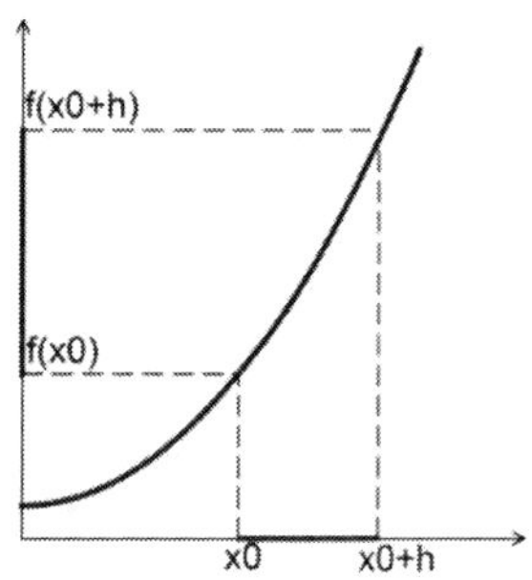

Fig. 1

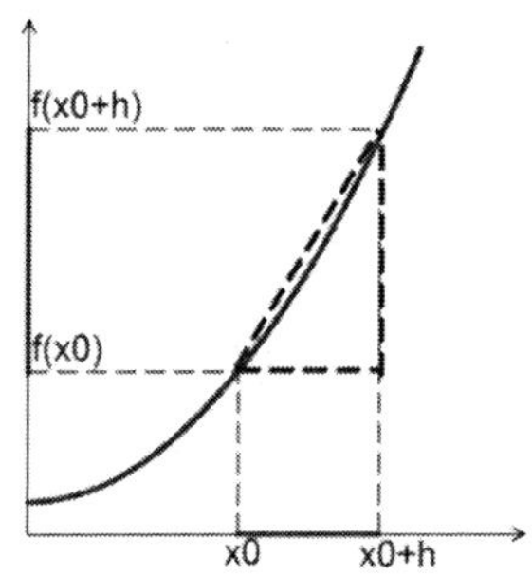

Fig. 2

An example from everyday experience points out the fundamental difference between the two concepts easily.

Example. If a Bank refunds 1.050 € for an investment of 1.000 €, the capital growth

$$\Delta C = 1050 - 1000 = 50 \text{ €}$$

is not so important as it is the capital growth over the time

$$\frac{\Delta C}{\Delta t}.$$

Depending on Δt is one, two or ten years, the investment reveals a very different performance!

Usually the rate of change is not constant (unless the phenomenon is linear) as the discrete models show. Thus, in the continuous case, it could be interesting to discuss the change of $\Delta C / \Delta t$ with respect to Δt, in particular when Δt gradually is shorter and shorter.

As we will see, in some situations we need to move from the *average rate of change* to an *instantaneous rate of change*, called the *derivative*

$$\lim_{x \to x_0} \frac{f(x) - f(x_0)}{x - x_0} = f'(x_0).$$

In force of the derivative, we are able to update the patterns of the first part switching from discrete to continuous dynamic models.

2 Some continuous dynamic models

Let us discuss now the translation to continuous of the dynamic models we have illustrated in the first part.

2.1 Malthus model in population dynamics

We recall that the English economist Thomas Malthus proposed in 1798 a first model to study the evolution of an isolated population, based on the assumption:

(M) the growth rate is directly proportional to the number of individuals.

If we adopt a real function $P: R_0^+ \to R$ to describe the population with respect to time, Malthusian principle is translated into the relation

$$P'(t) = kP(t) \tag{3.1}$$

Note that (3.1) is a *functional equation*, more precisely a *differential equation*, since the unknown function P, appears in the relation together with its derivative P'.
If we take into account the population at the starting time (as in the discrete case), the Malthus model is the Cauchy problem

$$\begin{cases} P'(t) = kP(t) \\ P(0) = P_0 \end{cases}$$

Remark. Note that we can look at Malthus principle from a different point of view. Precisely, assumed $P(t) \neq 0$, the equation (3.1) can be written

$$\frac{P'(t)}{P(t)} = k$$

where the term on the left hand side represents the *relative growth rate*: it describes the population rate of change with respect to the consistency of the population itself. The relative growth rate is also called *growth factor* and Malthus principle can be read

(M) the relative growth factor is constant.

2.2 Newton's law on heating and cooling

According to Newton's law

(N) the rate of change of temperature is proportional to the thermal gradient (the difference between the fluid temperature and the environment temperature)

Denoted by $T : R_0^+ \to R$ the function that describe the temperature with respect to the time, the statement is translated into the Cauchy problem

$$\begin{cases} T'(t) = k[T(t) - T_a] \\ T(0) = T_0 \end{cases}$$

where T_a denotes the environment temperature.

The transition from static situations (the system is immutable in time) to dynamic ones (the system evolution changes in time) turns numerical equations (the unknown is a number) into functional equations (the unknown is a function).

Now a big question arises: how to approach the study of such equations?

3 Introduction to differential equations

We propose here an approach to differential equations that mimics the resolution of a numerical equation. More precisely, we refer to the *revisitation* of the resolution of a numerical equation, based on the *inverse technique* we have proposed in [1].

3.1 The inverse tecnique for a numerical equation

A numerical equation is a relation of the type

$$f(x) = c$$

where the function $f : Dom f \to R$ and the number $c \in R$ are given.
The resolution of the equation consists on determining the set

$$S = \{x \in Dom f : \quad f(x) = c\}.$$

The following equivalence result is fundamental.

Proposition. If f admits the inverse function f^{-1} and $c \in \mathrm{Im} f$, then

$$f(x) = c \quad \Leftrightarrow \quad x = f^{-1}(c)$$

As a consequence the equation admits the unique solution

$$S = \{f^{-1}(c)\}.$$

As we proved in [1], most of the devices adopted to solve numerical equations is inspired by this technique, which plays the role of unifying approach to resolution.

3.2 The Inverse of derivative operator

In order to adopt the inverse technique for differential equations, we must first discuss the invertibility of derivative operator.

The derivative operator D associates to any derivable function $h: I \to R$, its derivative Dh.
The inverse operator, if it exists, acts backward: given a function $g: I \to R$, it finds a derivable function h such that $Dh = g$.

The question, as it is well known, is non easy. However, in the particular case of a functions $g \in C^1$ (derivable with continuous derivative), the problem is solved by the integral function

$$h(x) = G(x) = \int_{x_0}^{x} g(t)dt .$$

More precisely, the derivateve $D: C^1 \to C^1$ admits as inverse operator $D^{-1}: C^1 \to C^1$

$$D^{-1}g = G .$$

3.3 The inverse tecnique for differential equations

We are now able to transport the inverse tecnhique to differential equations.

First case. Let us consider first the differential equations

$$x'(t) = f(t) \tag{3.2}$$

where $f: Dom f \to R$ is a given continuos function. To solve the equation we have to reconstruct a function from its derivative.

Adopting the *inverse technique*, we apply the *integral function* (i.e. the inverse operator of derivative) to both the sides

$$\int_0^t x'(s)ds = x(t) - x_0 = \int_0^t f(s)ds .$$

In this way we obtain the *integral form* of equation (3.2)

$$x(t) = x_0 + \int_0^t f(s)ds .$$

Second case. Let us consider now an equation of the type

$$x'(t) = f(t)g(x(t)) \tag{3.3}$$

where f ang g are given function.

Note that both Malthus and Newton equations belongs to this family .

Assumed $g(x) \neq 0,$ the equation can be written

$$\frac{x'(t)}{g(x(t))} = f(t).$$

Applying the *integral function* to both the sides

$$\int_0^t \frac{x'(s)}{g(x(s))} ds = \int_0^t f(s)\, ds.$$

by virtue of the integration by substitution, we get

$$\int_0^t \frac{x'(s)}{g(x(s))} ds = \left[\int_{x_0}^x \frac{dy}{g(y)} \right]_{x=x(t)}.$$

Finally, if we put $G^*(x) = \int_{x_0}^x \frac{1}{g(y)} dy$, the solution of the equation (in implict form) is

$$G^*(x(t)) = \int_{t_0}^t f(s)ds + c \qquad c \in R.$$

We can now discuss the solution of Malthus and Newton differential equations.

3.4 Malthus model

In the particular case of Malthus equation

$$\begin{cases} P'(t) = k\,P(t) \\ P(0) = P_0 \end{cases}$$

the inverse technique gives an exponential function as a solution (see Fig. 3 and Fig. 4).

$$\int_0^t \frac{P'(s)}{P(s)} ds = \int_0^t k\, ds \quad \Rightarrow \quad \log P(t) = k\,t + \log P_0 \quad \Rightarrow \quad P(t) = P_0\, e^{kt}$$

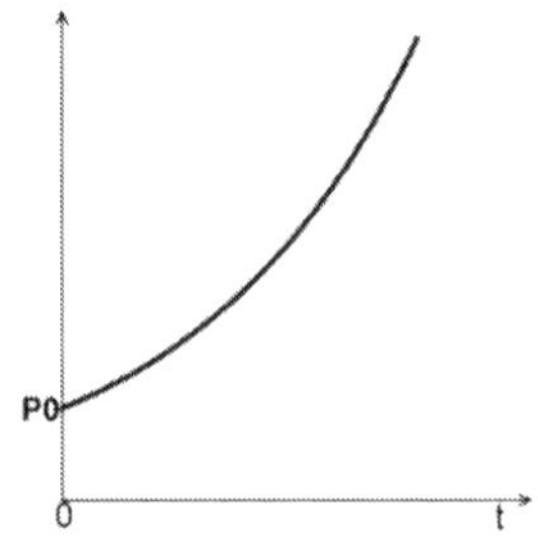

Fig. 3. Case $k > 0$

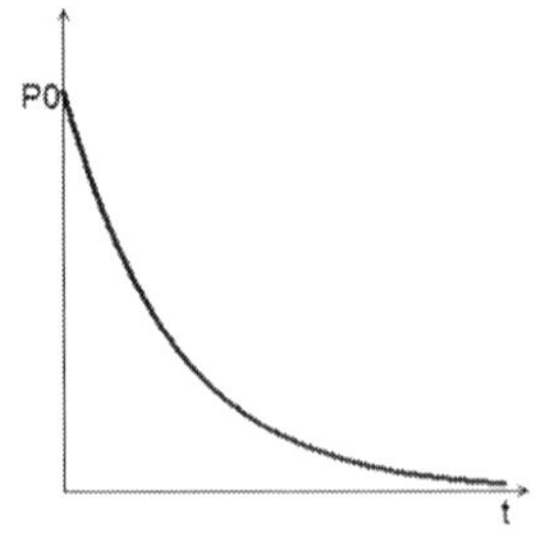

Fig. 4. Case $k < 0$

3.5 Newton model

Let us solve now the general growing/decay equation inspired by Newton's law

$$\begin{cases} T'(t) = k[T(t) - T_a] \\ T(0) = T_0 \end{cases} \quad t \geq 0 \qquad k < 0$$

First note that if $T_0 = T_a$ the solution is a constant function $T(t) = T_a$ (called *equilibrium solution*). If $T(t) \neq T_a$, the equation can be written

$$\frac{T'(t)}{T(t) - T_a} = k$$

and the *inverse technique* leads to the family of solutions

$$T(t) = (T_0 - T_a)e^{k(t-t_0)} + T_a$$

It is easy to see that two possible evolutions occur (see Fig. 5 and 6)

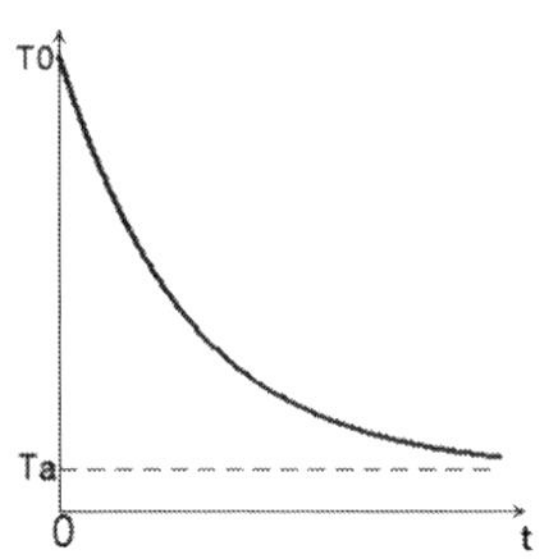

Fig. 5. Case $T_a < T_0$

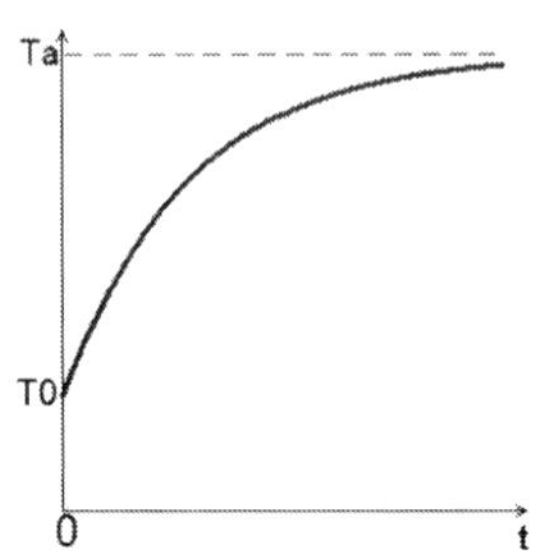

Fig. 6. Case $T_a > T_0$

3.6 Three examples

Let us discuss the continuos dynamic models for three problems that we have solved in the first part adopting a discrete dynamic model.

Coffee cooling.
Let us describe, according to Newton law, the temperature evolution of a coffee poured into a cap.

If $T: R_0^+ \to R$ is the function which describe the coffee temperature, with respect to time, applying the formula in Section 3.4, we obtain (see also Fig. 7)

$$T(t) = (T_0 - T_a)e^{kt} + T_a \quad t \geq 0 \quad \text{where } k < 0$$

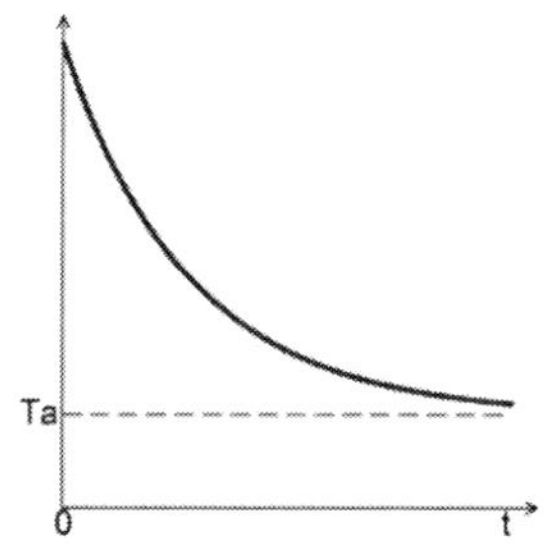

Fig. 7

Where T_a and T_0 denote the room temperature and the temperature of the coffee just poured, respectively. As Fig. 7 shows, the solution of the continuous model do agrees with that obtained with the discrete model in the first part of the paper.

Fish length. According to the biologist Ludwig von Bertalanffy the growth rate of the length of some fish species is proportional to the difference between the maximum length of the species and the present length. In particular, for the North Sea cod, we can take into account the following estimates:
the maximum length of the species $53\,cm$, the length of newborn fish $L_0 = 10\,cm$, the growth factor $k = -0.2$.

The constraction of the model. Denoted by $L: R_0^+ \to R$ the function that describe the length of the fish in time, the assumption of von Bertalanffy is translated into the Cauchy problem

$$\begin{cases} L'(t) = k(L^* - L(t)) \\ L(0) = L_0 \end{cases} \quad t \geq 0 \quad \text{where } k > 0.$$

Note thet the model is very similar to Newton model.

By applying the inverse technique, we obtain the solution

$$L(t) = L^* - (L^* - L_0)e^{-kt}.$$

Thus the growth of North Sea cod is described by the function (see also Fig. 8)

$$L(t) = 53 - 43e^{-0,2t}.$$

The discussion of the solution. In Fig. 8 we compare the present solution L with the sequence that solves the discrete model discussed in the first part

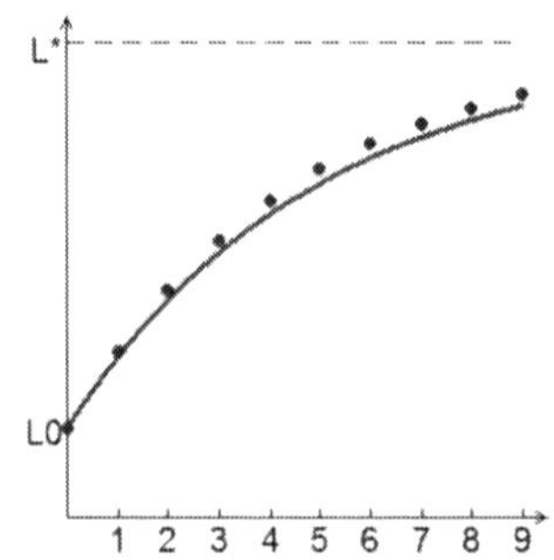

Fig. 8

Both the solutions (of discrete or continuous dynamic model) emphasize an esponential growth. We wish to remark that the continuous solution does not coincides with the supportt function of the discrete sequence. However the two solution are very closed as Fig. 8 shows.

The liquid mixing. In a tank containing salt water at concentration α a salt water at a different concentration β is introduced, at a constant rate (r). In order to keep the liquid volume constant, water exits from the tank with the same speed. The solution is mixed continuously. Discuss the evolution of salt concentration.

The construction of the model. Denoted by $C: R_0^+ \to R$ the function that describes the salt concentration in time, the phenomenon can be described by the model

$$C(t+\Delta t) = C(t)(1-\lambda_1 \Delta t) + \beta\lambda_1 \Delta t \quad \Rightarrow \quad C'(t) = -C(t)\lambda_1 + \beta\lambda_1$$

where V denote the liquid volume and $\lambda_1 = \frac{r}{V}$.

Thus we get the Cauchy problem

$$\begin{cases} C'(t) = -C(t)\lambda_1 + \beta\lambda_1 \\ C_0 = \alpha \end{cases}$$

By applying the inverse technique, we obtain the solution

$$C(t) = \beta + (\alpha - \beta)e^{-\lambda_1 t}$$

The discussion of the solution. It is easy to see that two possible evolutions occur, according to $\alpha < \beta$ (Fig. 9) or $\alpha > \beta$ (Fig. 10).

In both the cases the concentration is monotone and evolves asymptotically up to the concentration β of the liquid incoming.

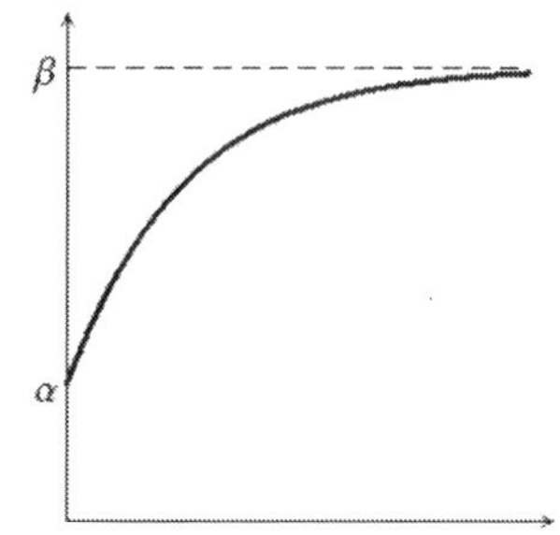

Fig. 9

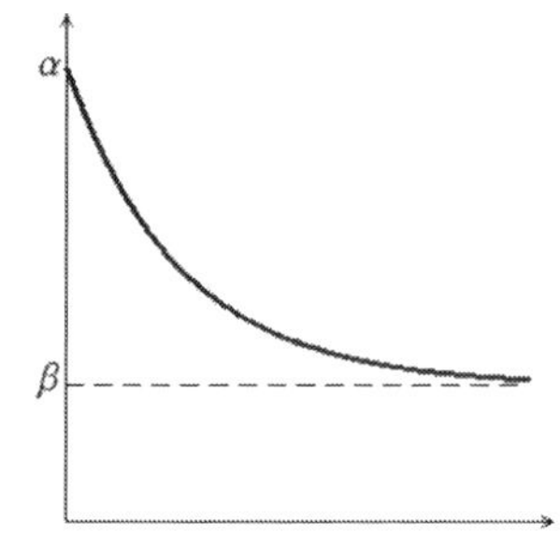

Fig. 10

References

[1] BRANDI, P., SALVADORI, A. *Prima di iniziare.* Aguaplano Officina del libro, 2011, ISBN 978-88-905726-5-4

[2] BRANDI, P., SALVADORI, A. *Nuovi percorsi di Matematica Vol.1* Aguaplano Officina del libro, 2015, ISBN 978-88-97738-65-7

[3] BRANDI, P., SALVADORI, A. *Math MAPS II Itinerari per le competenze* Alice&Bob 45 EGEA, Milano, 2016, ISBN 978-88-238-6199-2

[4] BRANDI, P., SALVADORI, A. *Math-MAPS. Introduction to dynamic models*, work in progress

[5] BRANDI, P., SALVADORI, A. *Math-MAPS a new road thet opens up a world. Part I Dynamic models without derivative, Proceedings APLIMAT 2017*

Current address

Brandi Primo, full professor
Dipartimento di Matematica e Informatica, Università degli Studi di Perugia
via Vanvitelli 1, 06123 Perugia, Italy
E-mail: primo.brandi@unipg.it

Salvadori Anna, assiciated professor
Dipartimento di Matematica e Informatica, Università degli Studi di Perugia
via Vanvitelli 1, 06123 Perugia, Italy
E-mail: anna.salvadori@unipg.it

Slovak University of Technology in Bratislava
Faculty of Mechanical Engineering
16th Conference on Applied Mathematics **APLIMAT 2017**
Proceedings

HEURISTIC STRATEGY SYSTEMATIC EXPERIMENTATION

BŘEHOVSKÝ Jiří (CZ), EISENMANN Petr (CZ), PŘIBYL Jiří (CZ)

Abstract. The paper describes a heuristic strategy for problem solving called Systematic experimentation. The strategy is described and illustrated by several problems. In our conception we count Systematic experimentation strategy among experimental strategies. The fundamental definition of this strategy is based on systematic exhausting of all potential results. The strategy Systematic experimentation is an advisable option for the pupil to solve a problem that is to be solved through a straight-forward way.

Keywords: problem solving, systematic experimentation, heuristic strategy, ICT in mathematics education, brute force

Mathematics Subject Classification: Primary 97D50; Secondary 97C70

1 Introduction

It has been an accepted fact for a longer period of time now that efficient teaching of mathematics is supported where pupils are taught to solve problems (see [1], [2]). We understand problem solving to be a cognitive process which can be carried out in three approaches, depending on the solver's engagement, abilities and skills.

The first approach is referred to as *trial*. It is the most primitive way of dealing with a problem. It only requires a solver's external motivation. The solver does not ask if they solved the problem correctly. Their only goal is to 'solve the problem', usually once without any need to check that the solution is correct.

The second approach is referred to as *straight-forward way* and is based on the application of some acquired knowledge. The solver knows the solving process that is required, is able to recognize how they should use it and applies it.

The third approach is referred to as *using a heuristic strategy* (sometimes only strategy). The solver does not have the needed knowledge or does not know how to use it and therefore cannot solve the problem in a straight-forward way. The solver is motivated to solve the problem. It is heuristic strategy that allows them to solve the problem. ([3], p. 538)

Our definition of heuristic strategies corresponds to the works of Pólya [4] and Schoenfeld [5]. Experimental strategies based on carrying out experiments are irreplaceable among the heuristic strategies. The solver usually executes more experiments as a problem is in progress of being solved, whereas new experiments are modified, based on the results gained in previous experiments.

2 The heuristic strategy Systematic experimentation

Systematic experimentation (SE) belongs to the very family of experimental strategies. The principle behind the strategy Systematic experimentation is the idea that the result may be reached in a finite number of attempts. These attempts are carried out systematically. The solver starts by selecting some value (either the first possible or some closer to the solution) and then works systematically. After each conducted experiment (therefore it is a type of experimental strategy) the solver checks whether the result is the sought solution. Each subsequent attempt will be "slightly" modified with respect to the previous one. The principle is based on the fact that the solver selects an initial value and then gradually gets nearer to the sought solution. This strategy seems to be very efficient in connection with the use of technological devices as these make it possible to conduct the experiments in real time. It is usually also possible to conduct these experiments (often numerical) manually, but it considerably increases the time needed for the solution of the problem. The use of computer within systematic experimentation is very often referred to as so-called brute force use. The brute force solution is defined as a systematic exhausting of all potential results

The following lines offer an outline of some relevant researches on the use of SE strategy in teaching mathematics.

[6] report on a study conducted with 28 Turkish pupils. The teacher tried to teach his pupils to use selected heuristic strategies in the period of three months. The outcome of this research was the finding that this strategy (the authors refer to it as "make a systematic list") can be taught relatively easily and is easy to use in problem solving. However, our research does not confirm the authors' claim that younger pupils are more successful when using this strategy in problem solving.

[7] uses the findings from her research that compares the abilities and skills of mathematically gifted and non-gifted students aged 6-10 years to claim that mathematically gifted students show more ability to solve problems in a systematic way. This group of pupils used Systematic experimentation more often and spontaneously when solving appropriate problems.

The following conclusions based on experiments are presented on the strategy Systematic experimentation by [8].

- By generating the results of a series of calculations for a variety of input values, the computer tool strengthened the students' notion of a function as a calculation procedure that transforms input values into output values.
- By generating output values for series of input values, by generating tables of input and output values, and by enabling the students to move up and down the values in these tables, the computer tool supported the students in developing a dynamic notion of a variable that can move in a space of possible values, and the corresponding idea of co-variation.

Our research [3] has proved that pupils aged 12–18 are able to adopt the SE strategy and use it efficiently, all within a span of six months. It may well be stated that after about three months of using computers to solve problems with the strategy Systematic experimentation, the pupils grew sensitive to the selection of the initial value.

3 Illustrative problems

As stated above, the use of the computer in systematic experimentation is sometimes referred to as the use of "brute" force. The solution by brute force is specified as exhausting all single possibilities from the set of all potential results. The specification allows us to classify the presented illustrative problems into the following categories:

1. Problems solved using brute force (exploring all the possibilities).
2. Problems solved using brute force with justified termination of the experiment (exploring only a selected subset of all possible solutions).
3. A specific group is formed by problems in which systematic experimentation results in formulation or verification of some hypothesis.

Problem 1

Assignment: Some tickets in a theatre were sold for 220 CZK and some at the price of 160 CZK. How many tickets at each price were sold if the total sum of 97 tickets was 19 300 CZK?

Solution: A spreadsheet will be used for experimentation. As mentioned above there are two possibilities (category 1 and 2). The first one starts as follows

Number of tickets at 220 CZK	Price in CZK	Number of tickets at 160 CZK	Price in CZK	Total price in CZK
97	21340	0	0	21340
...	...	...	...	...

Tab. 1

and the second starts by a well-founded estimation (see Tab. 2).

Number of tickets at 220 CZK	Price in CZK	Number of tickets at 160 CZK	Price in CZK	Total price in CZK
65	14300	32	5120	19420
64	14080	33	5280	19360
63	**13860**	**34**	**5440**	**19300**
62	13640	35	5600	19240
...	...	...	...	...
1	220	96	15360	15580
0	0	97	15520	15520

Tab. 2

It is possible to stop experimentation in a qualified way exactly at the place where is stopped being meaningful (see Tab. 3):

Number of tickets at 220 CZK	Price in CZK	Number of tickets at 160 CZK	Price in CZK	Total price in CZK
65	14300	32	5120	19420
64	14080	33	5280	19360
63	**13860**	**34**	**5440**	**19300**
62	13640	35	5600	19240

Tab. 3

Answer: 63 tickets at the price of 220 CZK and 34 tickets the price of 160 CZK were sold in the theatre.

Let us now discuss three problems from category 3 above, i.e. problems where systematic experimentation results in verification or formulation of a hypothesis.

Problem 2

Assignment: The numbers that are read the same from left to right and right to left, e.g. 452 254, are called palindromes. My friend claims that all four digit palindromes are divisible by 11. Is it true?

Solution: Palindromes are e.g. numbers: 127 721, 94 749, 8 338, 565, 44, 8. In the task, we are only interested in four digit numbers such as 6 776, 1 001, 2 992. Let us now take several four digit palindromes at random and divide them by eleven. We will get e.g. the following:

$$4\,554 = 414 \times 11$$
$$1\,001 = 91 \times 11$$
$$8\,338 = 758 \times 11$$

It seems my friend could be right. However, we will not be one hundred percent sure unless we check all the 90 existing four digit palindromes. This can be done efficiently using a spreadsheet (see Tab. 4). Table 4 – in its header – shows a formula for the calculation in relevant cells.

a	*b*	**"=*b2*"**	**"=*a2*"**	**"=1000*A2+100*B2+10*C2+1*D2"**	**"=E2/11"**
1	0	0	1	1 001	91
1	1	1	1	1 111	101
...	...	...	...	...	...
2	9	9	2	2 992	272
...	...	...	...	...	...
9	1	1	9	9 119	829
...	...	...	...	...	...
9	9	9	9	9 999	909

Tab. 4

The last column shows that all quotients are whole numbers.

Answer: My friend was right.

In contrast, the use of spreadsheet can result in rejection of the hypothesis.

Problem 3

Assignment: Decide if the sequence $x_n = \frac{20^n}{n!}$ is monotonous.

Solution: Let us determine the first few members of the sequence x_n.

$$\begin{aligned} x_1 &= 20 \\ x_2 &= 200 \\ x_3 &= 1\,333.3\ldots \\ x_4 &= 6\,666.6\ldots \end{aligned}$$

The natural hypothesis is that the sequence x_n is increasing. Most solvers will come to this conclusion without hesitation.

It is necessary to be aware of the definition of an increasing sequence and to consider whether the conclusion really corresponds to the condition in the definition that inequality $x_n < x_{n+1}$ holds for all natural numbers n. Only very few solvers will be able to realize at this point that it is not the case. However, using a spreadsheet they will see it all at once (see Tab. 5).

N	x_n
1	20.0
2	200.0
3	1 333.3
...	...
17	36 850 332.5
18	40 944 813.9
19	43 099 804.1
20	43 099 804.1
21	41 047 432.5
22	37 315 847.7
23	32 448 563.2
...	...

Tab. 5

Answer: It is obvious that the sequence x_n is not monotonous. The member $x_{19} = x_{20}$ and the following members are decreasing.

The last problem shows a situation in which systematic experimentation results in a formulation of a hypothesis.

Problem 4

Assignment: Decide: Can $n^2 + n + 1$ be a square of a natural number for some natural $n \geq 1$?

Solution: We conduct experimentation using a spreadsheet as shown in table 6.

N	n^2+n+1
1	3
2	7
...	...
10	111
...	...

Tab. 6

Even if we carry on calculation using the spreadsheet, the situation is not likely to change. None of the numbers in the right column will be the square of a natural number. Thus it is only natural to formulate the following hypothesis.

Hypothesis: Number $n^2 + n + 1$ is never a square of a natural number for any natural $n \geq 1$.

One must bear in mind that in this case of non-finite numbers of n, the computer can never exhaust all elements of the given set in proving the hypothesis using systematic experimentation. The hypothesis can be falsified if we find a counterexample. To verify it, we must carry out a proof. For completeness we provide it here.

Proof: Suppose number $n^2 + n + 1$ is a square of some natural number (let it be e.g. x). Then the following holds

$$n^2 + n + 1 = x^2, \qquad (1)$$

or

$$(x + n)(x - n) = n + 1. \qquad (2)$$

i) If $x > 1$, then $n + x > n + 1$. It follows from equation (2) and the preceding condition that $n + 1$ is not divisible by $n + x$ and so there is no natural number n that would meet equation (1).

ii) Due to the condition $n \geq 1$, number $x = 1$ obviously is not a square of number $n^2 + n + 1$.

The hypothesis is now proved and has become a mathematical theorem.

Answer: Number $n^2 + n + 1$, where $n \geq 1$, is not a square of any natural number.

4 Problems solvable through the strategy SE in textbooks of mathematics

The goal of this section is to offer an overview of individual textbook sets and topic-based lessons examined at elementary and secondary schools. To follow our purpose, we have carried out searching activities focused on the occurrence of topics and suitable problems possibly to efficiently solvable through the strategy SE. We have opted for two series of textbooks commonly used at elementary and secondary schools in the Czech Republic. The results have not been significantly different in either of the series.

4.1 Typified problems in elementary school textbooks

Series of textbooks used in our search activities: [9], [10].

We may state that strategy SE is absent from any task solved in the textbook. The tasks open to successful use of the strategy SE are to be find especially in the 6th, 8th and 9th forms. These are chiefly word problems used to exercise arithmetic operations or lead to constructing a linear equation or system of linear equations. Here we show a sample task from an 8th grade textbook:

Assignment: Mrs Hruška's garden is larger by two metres than what its width is; its perimeter is 160 metres. What is the size of the garden?

4.2 Typified problems in secondary school textbooks

Series of textbooks used in our search activities: [11], [12].

The situation with secondary schools is nevertheless slightly different and the pupils are expected to prove different skills. We may state once again that strategy SE is absent from any task solved in the textbook. Despite of the fact there is a number of problems that would easily be solved with strategy SE; these are especially word problems leading to constructing a system of linear equations or problems searching into the properties of sequences. Apart from these, the strategy SE can be also used in solving more challenging problems where it is necessary to hypothesize, verify the hypothesis or reject it with a suitable counterexample. Here we offer the following problem.

Assignment: Decide if all numbers $n^2 + n + 41$, where n is a natural number, are prime numbers. Hypothesize and verify the hypothesis.

5 Conclusion

As shown in section 4, problems efficiently solvable through the strategy SE occur not insignificant in the curriculum of mathematics at Czech elementary and secondary schools. The SE strategy is an advisable option for the pupil who is unable to solve the problem that is to be solved through a straight-forward way – e.g. where they are not able to construct a relevant equation. As seen in researches shown in section 2, pupils are able to learn to use the strategy SE with great effect. The use of spreadsheet with problem solving through the strategy SE is in accordance with stressing the use of ICT in curricular documents in the Czech Republic as well as abroad (e.g., [1] or [2]). Our contribution herewith is an appeal to introduce this heuristic strategy into the group of ways to solve problems in mathematics at elementary and secondary schools.

Acknowledgement

The paper was supported by Czech Science Foundation project P407/12/1939.

References

[1] NTCM. *Principles and Standards for School Mathematics*. Reston: The National Council of Teachers of Mathematics, Inc., 2000.

[2] JEŘÁBEK, J., LISNEROVÁ, R., SMEJKALOVÁ, A., & TUPÝ, J. (Eds.). *Rámcový vzdělávací program pro základní vzdělávání: (verze platná od 1. 9. 2013) úplné znění upraveného RVP ZV*. Praha: MŠMT, 2013.

[3] EISENMANN, P., NOVOTNÁ, J., PŘIBYL, J., & BŘEHOVSKÝ, J.: The development of a culture of problem solving with secondary students through heuristic strategies and modes. *Mathematics Education Research Journal, 27*(4), 2015, pp. 535–562.

[4] PÓLYA, G. *How to Solve It: A New Aspect of Mathematical Method*. Princeton: Princeton University Press, 2004.

[5] SCHOENFELD, A. H. *Mathematical problem solving*. London: Academic Press Inc. (London) Ltd. 1985.

[6] ARSLAN, Ç., & ALTUN, M. Learning to solve non-routine mathematical problems. *Elementary Education Online, 6*(1), 2007, pp. 50–61.

[7] HEINZE, A. Differences in problem solving strategies of mathematically gifted and non-gifted elementary students. *International Education Journal, 6*(2), 2005, pp. 175–183.

[8] DOORMAN, M., DRIJVERS, P., GRAVEMEIJER, K., BOON, P., & REED, H. Design research in mathematics education: The case of an ICT-rich learning arrangement for the concept of function. In T. PLOMP & N. NIEVEEN (Eds.), *Educational design research – Part B: Illustrative cases*. Enschede, the Netherlands: SLO, 2013, pp. 425–446.

[9] PŮLPÁN, Z. a kol. *Matematika 6–9*. Série učebnic pro základní školy. Praha: SPN, 2007–2010.

[10] MOLNÁR, J. a kol. *Matematika 6–9: Učebnice s komentářem pro učitele*. Série učebnic pro základní školy. Olomouc: Prodos, 1998–2001.

[11] *Matematika pro střední odborné školy a studijní obory středních odborných učilišť 1.–6. část*. Série učebnic pro střední odborné školy. Praha: SPN, 1985–1988.

[12] BOČEK, L. a kol. *Matematika pro gymnázia*. Série učebnic pro gymnázia. Praha: Prometheus, 1993–1999.

Current address

Břehovský Jiří, Mgr. Ph.D.
Technical University of Liberec, Faculty of Sciences, Humanities and Education
Studentská 1402/2, 461 17 Liberec 1, Czech Republic
Tel. number: +420 485 352 923, e-mail: jiri.brehovsky@tul.cz

Eisenmann Petr, Doc. PaedDr. CSc.
University of J. E. Purkyně in Ústí nad Labem, Faculty of Science
České mládeže 8, 400 96 Ústí nad Labem, Czech Republic
Tel. number: +420 475 283 401, e-mail: petr.eisenmann@ujep.cz

Přibyl Jiří, Mgr. Ph.D.
University of J. E. Purkyně in Ústí nad Labem, Faculty of Science
České mládeže 8, 400 96 Ústí nad Labem, Czech Republic
Tel. number: +420 475 283 410, e-mail: jiri.pribyl@ujep.cz

Slovak University of Technology in Bratislava
Faculty of Mechanical Engineering
16th Conference on Applied Mathematics **APLIMAT 2017**
Proceedings

COMBINATORIAL PROBLEMS IN TEXTBOOKS OF MATHEMATICS FOR ELEMENTARY SCHOOL

BŘEHOVSKÝ Jiří (CZ), PŘÍHONSKÁ Jana (CZ)

Abstract. This article describes an analysis of textbooks that contain tasks, which develop combinatorial reasoning of pupils in elementary schools, and the categorization of these tasks. It is a quantitative evaluation that will provide a basis for further research.

Keywords: analysis of textbooks, combinatorial problems, combinatorial reasoning, quantitative evaluation

Mathematics Subject Classification: Primary 97-02

1 Introduction

The article is one of the outputs of the research project SGS 21162 „Combinatorics in Elementary School - Encouraging Activities for Developing Logical Combinatorial Thinking“. The primary research orientation is creating a file of the stimulating activities. That will help develop logical thinking and combinatorial reasoning of pupils and their verification in practical teaching.

Combinatorics is one of the oldest branches of discrete mathematics, dating back to the 16th century when games of chance played a key role in society life (Abramovich & Pieper, 1996). Combinatorics is a significant component of the mathematics curriculum, comprising a rich structure of powerful principles that underlie several other areas such as counting, computation, and probability (Borovcnik, Peard 1996). Solving combinatorial problems still makes many students considerable difficulties. Despite its importance in the mathematics curriculum, combinatorics continues to remain neglected, particularly at the elementary school level (Lyn 2005). Therefore we searched for selected mathematics textbooks that are used in elementary schools in the Czech Republic, to detect the representation of combinatorial problems in these textbooks. In the next step we propose a new categorization of the problems.

2 Theoretical background

Combinatorics is an essential component of discrete mathematics and, as such, it has an important role to play in school mathematics. In 1970, Kaput presented the following reasons, which could still be valid, to justify the teaching of elementary Combinatorics at school:

- Since it does not depend on calculus, it has suitable problems for different grades; usually very challenging problems can be discussed with pupils, so that they discover the need for more mathematics to be created;
- It can be used to train pupils in enumeration, making conjectures, generalization and systematic thinking; it can help the development of many concepts, such as equivalence and order relations, function, sample, etc.
- Many applications in different fields can be presented.

That is why our research focuses on the development of elementary school children's combinatorial reasoning. Combinatorics is an essential part of mathematics education. Although combinatorial problems are easy to formulate, their solutions are often very difficult. The importance of children's combinatorial reasoning in analysing sample space has been evident in several studies (e.g., Benson & Jones 1999; Johnson, Jones, Thornton, Langrall, & Rous 1998; Nisbet et al., 2000; Zimmermann & Jones, 2002). The results of these studies show that children have difficulties solving tasks that require combinatorial reasoning.

The question of how to increase children's combinatorial skills is addressed in Lynn (2005). She emphasizes the importance of including combinatorial problems at the elementary school level. Combinatorial problems can help children construct meaningful representations, reason mathematically, and abstract and generalize mathematical concepts (Sriraman & English, 2004). Therefore word tasks in mathematics textbooks have a vital role (Čeretková 2003).

One of the objectives of mathematics education is to develop children's ability to solve simple word and practical problems without the use of classical mathematical approaches and algorithms. Mathematics textbooks contribute significantly to the development of such competency and thus are very important in the mathematics education (Molnár 2007).

3 Combinatorial problems

A combinatorial task/problem is a problem that leads towards the creation of various configurations and diagrams (Fuchs 2000). We can create configurations and diagrams by e.g. selecting elements from a predetermined finite set. In that case, it is important to distinguish whether the order of the selections matters or not. However, it does not necessarily have to be a selection of elements we are talking about. A combinatorial task might also involve a reorganisation of the set, change of its properties, etc. Such combinatorial problems also facilitate the development of enumeration processes, as well as conjectures, generalisations, and systematic thinking (Lyn 2007). The solutions to these problems depend upon the understanding of the following:

Basic Combinatorial Concepts and Models

- Combinatorial operations: These comprise combinations, arrangements, permutations, and the associated concepts, notations, and formulae;

- Combinatorial models: These include the sampling model (population, sample, ordered/non-ordered sampling, replacement), the distribution model (correspondence and application), and the partition model (sets, subsets, union);

and

Combinatorial Procedures

- Logical procedures: Included here are classification, systematic enumeration, inclusion/exclusion principle, and recurrence;
- Graphical procedures: Common procedures here include tree diagrams and graphs;
- Numerical procedures: These include addition, multiplication, and division principles, combinatorial and factorial numbers, Pascal's triangle, and difference equations;
- Tabular procedures: The construction of tables and arrays are most common here;
- Algebraic procedures: These include the generation of functions.

(Batanero, Godino, et al., 1997, p.240).

Each verbal task that must be solved with using application of these concepts or models we consider to be combinatorial.

4 Textbook analysis

The analysis focuses on the percentage representation of combinatorial tasks in mathematics textbooks for elementary school. We analysed five series of mathematics textbooks and three extended studies textbooks by different authors and publishers, as shown in Table 1. A total of 58 textbooks were analysed. We started by detecting all the tasks in each textbook, then we determined the number of the combinatorial tasks and calculated their percentage representation in the textbooks.

Title	Author	Publisher
Matematika pro 1 -5. Ročník 1-3. díl	Potůčková. J., Potůček. V.	Studio 1+1
Matematika pro 1-2. ročník 1-2. díl	Cihlář, Melichar, Zelenka	Fortuna
Barevná matematika pro prvňáčky - páťáky	Fialová	SPN
Matematika pro 1-5. ročník 1-3. díl	Molnár, Mikulenková	Prodos
Matematika pro 3-5. ročník 1-3. díl	Blažková, Vaňurová, Matoušková	ALTER
Matematika a její aplikace pro 1. ročník 1. díl	Mikulenová, Molnár	Prodos
Zajímavá matematika pro druháky	Mikulenová, Molnár	Prodos
Zajímavá matematika (nejen) pro páťáky	Mikulenová, Molnár	Prodos

Tab. 1. Analysed textbooks

Based on this analysis we proposed a basic classification of combinatorial tasks. Different tasks were divided into 9 categories (C1 - C9), according to their representation in the textbooks. The categories are shown in Ta. 2. Each category has the potential for the development of certain areas of combinatorial reasoning in children.

Category group	Category name	Category description, examples
C1	Games with numbers	generating numbers via different arithmetic operations, looking for possible solutions
C2	Quantifying tasks	the most/the least/at least; e.g. Bara ate 7 pieces of candy. Teresa ate less the Bara. How many could Teresa have eaten?
C3	Sudoku and Magic squares	
C4	Graph theory tasks	maze solving, liquid-pouring and weighing puzzles sport tasks
C5	Basic combinatorial principles	rules of sum and product tasks, Pascal's triangle related tasks, combining k-tuples from n elements
C6	Geometry tasks	2D/3D tasks: dividing shapes into pieces, and vice versa, painting the faces of a cube with distinct colors, matchstick puzzles
C7	Decision making tasks	pulling out a number of pieces to guarantee getting a certain amount, e.g. how many socks to guarantee a matching pair
C8	Classification tasks	e.g. paying a certain amount of money using a designated number/type of coins
C9	Other	tasks that cannot be put into any of the categories above

Tab. 2. Classification of combinatorial tasks

5 Results and discussions

Each of the analyzed textbooks were assigned a number from 1 to 58 see Table 3 in the appendix. For the percentage representation of combinatorial tasks in the textbooks, see Graph 1 and 2. Graph 1 shows the representation in 1st and 2nd grade textbooks, graph 2 shows the representation in 3rd, 4th, and 5th grade textbooks.

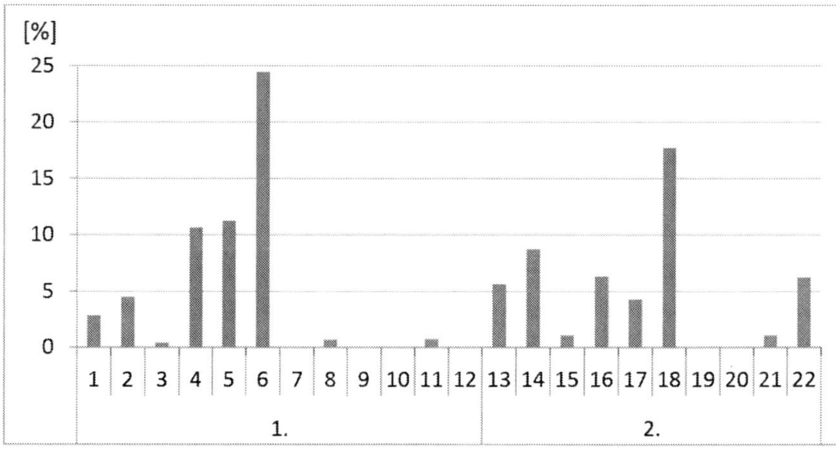

Graph 1. Percentage representation of combinatorial tasks in mathematics textbooks for 1st and 2nd grades

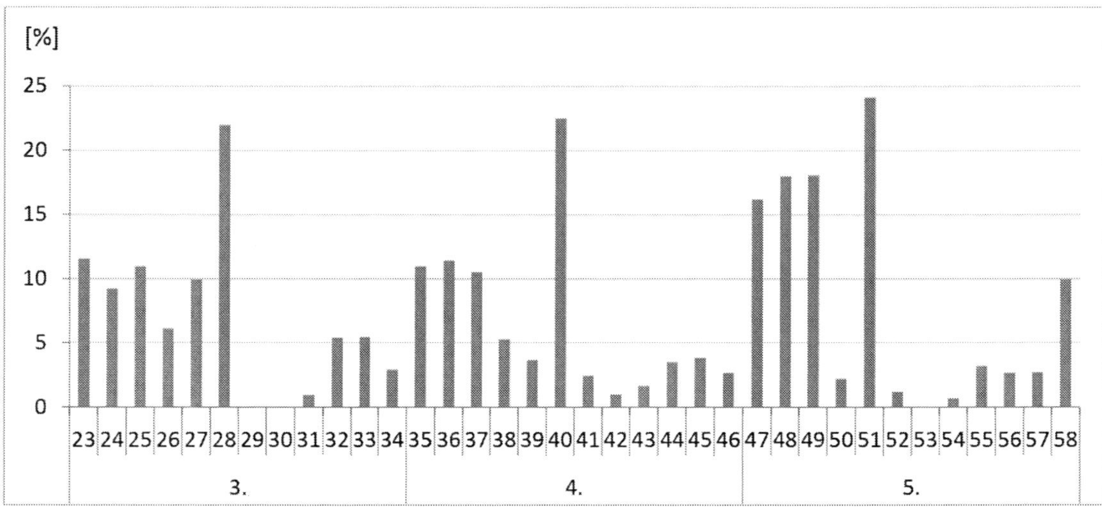

Graph 2. Percentage representation of combinatorial tasks in mathematics textbooks for 3rd, 4th, and 5th grades

It is apparent that the number of combinatorial tasks differs from textbook to textbook, depending on the author. The average percentage representation of combinatorial tasks in the textbooks is 6.3%. The average total number of all tasks in a textbook is 216, meaning the average number of combinatorial tasks in each textbook is 13. However, 24 of the analyzed textbooks contained less than 2.5% of them. That means only 5 tasks that promote combinatorial thinking, per textbook on average. Despite the fact that combinatorial tasks (problems) can help children construct meaningful representations, reason mathematically, and abstract and generalize mathematical concepts (Sriraman & English, 2004). Therefore we think that the average representation of 6.3% is poor. Another thought-provoking aspect is the representation of the different combinatorial task categories (C1 - C9) in the textbooks (see Graph 3).

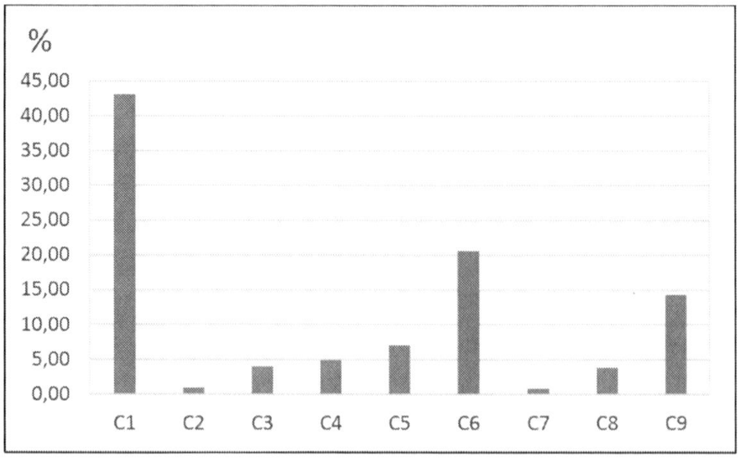

Graph 3. Percentage representation of combinatorial task categories in mathematics textbooks

As you can see, there are two categories of tasks that are the most frequent: C1 - Games with numbers (43%), and C6 - Geometry tasks (20%). That corresponds with the subject matter taught in elementary school. C9 category - Other follows with 14%. The other categories, with the exception of C5 - Basic combinatorial rules, have a representation of less than 5%. This rather poor representation of tasks that promote the development of combinatorial reasoning in children may be the reason for low success rates in the TIMMSS tests carried out in 2011. The results of the tests show that only a small amount of the children tested, were able to attain average scores in reading and mathematics simultaneously (PIRLS & TIMSS, 2011).

6 Conclusion

The research was focused on the analysis of selected mathematics textbooks, to detect the representation of tasks that promote combinatorial reasoning in children in elementary schools. Based on the analysis, we proposed a new classification of combinatorial problems (see Table 2). The tasks in each category have the potential for the development of a specific area of combinatorial reasoning in children. The average percentage representation of such tasks in the textbooks was 6.3%. Sadly, representation of 2.5% and lower was identified in 24 of the 58 textbooks. The problem is not only the small number of the tasks in the textbooks, but also their limited variety. Tasks from categories C1 and C6 are the most frequent. This leaves only a little space for the development of combinatorial skills in class. The analysis of the textbooks allows us to focus on further research in the future. It will help us propose new activities that will stimulate the development of combinatorial reasoning in children, which are not included in the mathematics textbooks yet. Special attention will be paid to word tasks that are the most frequently occurring in both the textbooks and mathematics competitions.

Acknowledgement

The research was supported by SGS project 2016 at FP TUL in Liberec.

Appendix

No.	name	autors	publisher
1	Matematika pro 1. ročník 1. díl	Potůčková, J., Potůček, V.	Studio 1+1
2	Matematika pro 1. ročník 2. díl	Potůčková, J., Potůček, V.	Studio 1+1
3	Matematika pro 1. ročník 3. díl	Potůčková, J., Potůček, V.	Studio 1+1
4	Matematika pro 1. ročník 1. díl	Cihlář, Melichar, Zelenka	Fortuna
5	Matematika pro 1. ročník 2. díl	Cihlář, Melichar, Zelenka	Fortuna
6	Barevná matematika pro prvňáčky	Fialová	SPN
7	Matematika pro 1. ročník 1. díl	Molnár, Mikulenková	Prodos
8	Matematika pro 1. ročník 2. díl	Molnár, Mikulenková	Prodos
9	Matematika pro 1. ročník 3. díl	Molnár, Mikulenková	Prodos
10	Matematika a její aplikace pro 1. ročník 1. díl	Mikulenová, Molnár	Prodos
11	Matematika a její aplikace pro 1. ročník 2. díl	Mikulenová, Molnár	Prodos
12	Matematika a její aplikace pro 1. ročník 3. díl	Mikulenová, Molnár	Prodos
13	Matematika pro 2. ročník 1. díl	Potůčková, J., Potůček, V.	Studio 1+1
14	Matematika pro 2. ročník 2. díl	Potůčková, J., Potůček, V.	Studio 1+1
15	Matematika pro 2. ročník 3. díl	Potůčková, J., Potůček, V.	Studio 1+1
16	Matematika pro 2. ročník 1. díl	Cihlář, Melichar, Zelenka	Fortuna
17	Matematika pro 2. ročník 2. díl	Cihlář, Melichar, Zelenka	Fortuna
18	Barevná matematika pro druháky	Fialová	SPN
19	Matematika pro 2. ročník 1. díl	Molnár, Mikulenková	Prodos
20	Matematika pro 2. ročník 2. díl	Molnár, Mikulenková	Prodos
21	Matematika pro 2. ročník 3. díl	Molnár, Mikulenková	Prodos
22	Zajímavá matematika pro druháky	Mikulenová, Molnár	Prodos
23	Matematika pro 3. ročník 1. díl	Potůčková, J., Potůček, V.	Studio 1+1
24	Matematika pro 3. ročník 2. díl	Potůčková, J., Potůček, V.	Studio 1+1
25	Matematika pro 3. ročník 3. díl	Potůčková, J., Potůček, V.	Studio 1+1
26	Matematika pro 3. ročník 1. díl	Cihlář, Melichar, Zelenka	Fortuna
27	Matematika pro 3. ročník 2. díl	Cihlář, Melichar, Zelenka	Fortuna
28	Barevná matematika pro třeťáky	Fialová	SPN
29	Matematika pro 3. ročník 1. díl	Molnár, Mikulenková	Prodos
30	Matematika pro 3. ročník 2. díl	Molnár, Mikulenková	Prodos
31	Matematika pro 3. ročník 3. díl	Molnár, Mikulenková	Prodos
32	Matematika pro 3. ročník 1. díl	Blažková, Vaňurová, Matoušková	ALTER
33	Matematika pro 3. ročník 2. díl	Blažková, Vaňurová, Matoušková	ALTER
34	Matematika pro 3. ročník 3. díl	Blažková, Vaňurová, Matoušková	ALTER
35	Matematika pro 4. ročník 1. díl	Potůčková, J., Potůček, V.	Studio 1+1
36	Matematika pro 4. ročník 2. díl	Potůčková, J., Potůček, V.	Studio 1+1
37	Matematika pro 4. ročník 3. díl	Potůčková, J., Potůček, V.	Studio 1+1
38	Matematika pro 4. ročník 1. díl	Cihlář, Melichar, Zelenka	Fortuna
39	Matematika pro 4. ročník 2. díl	Cihlář, Melichar, Zelenka	Fortuna
40	Barevná matematika pro čtvrťáky	Fialová	SPN

41	Matematika pro 4. ročník 1. díl	Molnár, Mikulenková	Prodos
42	Matematika pro 4. ročník 2. díl	Molnár, Mikulenková	Prodos
43	Matematika pro 4. ročník 3. díl	Molnár, Mikulenková	Prodos
44	Matematika pro 4. ročník 1. díl	Blažková, Vaňurová, Matoušková	ALTER
45	Matematika pro 4. ročník 2. díl	Blažková, Vaňurová, Matoušková	ALTER
46	Matematika pro 4. ročník 3. díl	Blažková, Vaňurová, Matoušková	ALTER
47	Matematika pro 5. ročník 1. díl	Potůčková, J., Potůček, V.	Studio 1+1
48	Matematika pro 5. ročník 2. díl	Potůčková, J., Potůček, V.	Studio 1+1
49	Matematika pro 5. ročník 3. díl	Potůčková, J., Potůček, V.	Studio 1+1
50	Matematika pro 5. ročník 1. díl	Cihlář, Melichar, Zelenka	Fortuna
51	Barevná matematika pro páťáky	Fialová	SPN
52	Matematika pro 5. ročník 1. díl	Molnár, Mikulenková	Prodos
53	Matematika pro 5. ročník 2. díl	Molnár, Mikulenková	Prodos
54	Matematika pro 5. ročník 3. díl	Molnár, Mikulenková	Prodos
55	Matematika pro 5. ročník 1. díl	Blažková, Vaňurová, Matoušková	ALTER
56	Matematika pro 5. ročník 2. díl	Blažková, Vaňurová, Matoušková	ALTER
57	Matematika pro 5. ročník 3. díl	Blažková, Vaňurová, Matoušková	ALTER
58	Zajímavá matematika (nejen) pro páťáky	Mikulenová, Molnár	Prodos

Tab. 3. The list of selected textbooks of mathematics witch we used for research

References

[1] ABRAMOVICH, S., PIEPER, A.: Fostering recursive thinking in combinatorics through the use of manipulatives and computer technology. In The Mathematics Educator. 1996. Volume 7. Issue 1. pp 4-12.

[2] BATANERO, C., NAVARRO-PELAYO, V., GODINO, J. D.: Effect of the implicit combinatorial model on combinatorial reasoning in secondary school pupils. In Educational Studies in Mathematics. 19897. Volume 32. pp 181-199.

[3] BENSON, C. T., JONES, G. A.: Assessing students' thinking in modelling probability contexts. In The Mathematics Educator. 1999. Volume 4. Issue 2. pp 1-21.

[4] BOROVCNIK, M., PEARD, R.: Propability. In Internacional Hnadbook in Mathematics Education. 1996. Dordrecht, Netherlands. Part 1, pp 239-288. ISBN 9789401071550.

[5] ČERETKOVÁ, S., ŠEDIVÝ, O., MOLNÁR, J., PETR, D.: The Role and Essessment of Textbooks in Mathematics Educetion. In Problems of Education in 21 Century. 2008. Volume 6.

[6] ENGLISH, L. D.: Combinatorics and the Developement of Childrens Combinatorial Reasoning. In Mathematics Education Library, Exploring Probability in School. 2005. Volume 40. pp 121-142. ISSN 0924-4921.

[7] FUCHS, E.: Diskrétní matematika a Teorie množin pro učitele. 2000. [CD ROM]. Brno: Masarykova univerzita.

[8] JOHNSON, T. M., JONES, G. A., THORNTON, C. A., LANGRALL, C. W., ROUS, A.: Students' thinking and writing in the context of probability. In Written Communication, 1998. Volume 15. Issue 2. pp 203-229.

[9] KAPUR, J. N.: Combinatorial Analysis and School Mathematics. In Educational Studies in Mathematics 3. 1970. pp 111-127.

[10] NISBET, S., JONES, G. A., LANGRALL, C. W., & THORNTON, C. A.: A dicey strategy to get your M & Ms. In Australian Primary Mathematics Classroom. 2000. Volume 5. Issue 3. pp 19-22.

[11] MOLNÁR, J.: Učebnice matematiky a klíčové kompetence. 2007.UPOL Olomouc. ISBN 978-80-244-1722-6.

[12] SRIRAMAN, B., ENGLISH, L. D.: Combinatorial mathematics: Research into practice. In The Mathematics Teacher. 2004. Volume 98. Issue 3. pp 182-191.

[13] ZIMMERMANN, G. M., & JONES, G. A.: Probability simulation: What meaning does it have for high school students? In Canadian Journal of Science, Mathematics and Technology Education. 2002. Volume 2. Issue 2. pp 221-236.

[14] PIRLS 2011 & TIMSS 2011 vybraná zjištění., Česká školní inspekce. Praha 2013.

[15] Curriculum Document of Czech Republic, RVP. VUP 2005.

Current address

Břehovský Jiří, Mgr. Ph.D.
Faculty of Sciences, Humanities and Education, Technical University of Liberec
Studentská 1402/2, 461 17 Liberec 1, Czech Republic
Tel. +420 48 535 2923, e-mail: jiri.brehovsky@tul.cz

Příhonská Jana, Doc. RNDr. Ph.D.
Faculty of Sciences, Humanities and Education, Technical University of Liberec
Studentská 1402/2, 461 17 Liberec 1, Czech Republic
Tel. +420 48 535 2870, e-mail: jana.prihonska@tul.cz

Slovak University of Technology in Bratislava
Faculty of Mechanical Engineering
16th Conference on Applied Mathematics **APLIMAT 2017**
Proceedings

THE LIMIT AS A RESOURCE. VISUAL DESIGN AND EXPLORATIONS OF EXTREME SPACE GEOMETRY: FROM THE M.C. ESCHER GRAPHIC RESEARCH TO THE SPHERICAL CAMERAS APPLICATION

BRUNETTI Federico Alberto (IT)

Abstract. The work of the '900 graphic researcher M.C. Escher concerned his intellectual interest to the limits between sight and vision, drawing and geometry, imagination and paradoxes in understanding and representation of space. One of the themes that intrigued the artist were the oppositions that seems to lead the relation between Nature and human existence. Through his graphic images we can easily understand how many hidden significance can be embedded in the more spontaneous visual relational action. An action in the instant of light, but for future memory. Both mathematical researchers, and many other scientist and literature artists or publishers, founded in his work a source for further inspiration. From 1954 he had the opportunity to have appreciation and relation with some mathematicians that identified the conceptual meaning of his visual and figurative topological research, reaching a deeper consideration about the complete process of decoding of visual informations is engaged, since to the limits that some graphic paradoxes can challenge. After quite an half of century of his creative experience, the legacy of these iconography re-appears in some digital devices defined as spherical digital cameras, in which combined optical wideangles lenses obtain surrounding 360° ("stiched") virtual maps, that can be so observed by VR / AR (virtual /augmented reality) visors and glasses. The visual experience and the iconographic heritage awareness that we must have towards authors like C.M. Escher if so evident, and even quite unexplored if considered under the point of view not only of the effective works that he leaved us, but in the light of the creative methodology that he represented. The opportunity, and possible duty for mathematicians, artist and designer is full opened to disclose patiently a new world already around us.

Keywords: non-Euclidean perspective, spherical digital cameras

Mathematics Subject Classification: Primary AMS01A99; Secondary 97M80

1 Images from XX century graphic research as paradigm of new vision

The event of a recent exhibition in Milan (curated by Marco Bussagli and Federico Giudiceandrea),[1] gave us the opportunity to observe closer and study M.C. Escher: his texts, letters and rare conferences -as far as his well notorious masterpieces- gave us a view of a very interesting path through the osmotic borders between Arts and Science.[2]

Fig. 1.a (left): Milano, piazza e Museo del Duomo. The entrance in Palazzo Reale of the C.M. Escher exhibition reflected in a spherical ball; picture taken in outdoor by the author with a wide-angle lens camera

Fig. 1.b (right) The main indoor entrance on the large staircase reaching the C.M. Escher exhibition reflected in a spherical ball, picture taken by the author with a wide-angle lens camera

[1] "Between 24 June 2016 and 22 January 2017, the Palazzo Reale di Milano is hosting an exhibition fully devoted to Maurits Cornelis Escher, the Dutch engraver and graphic artist. The exhibition is designed to highlight the artist's inclination to gaze at nature from an original angle through a perspective which enables him to bring out the gossamer beauty of regular geometries verging at time on magic and playfulness. The Escher exhibition, promoted by the Comune di Milano – Cultura, is produced and organized by the Palazzo Reale di Milano, Arthemisia Group and 24 ORE Cultura – Gruppo 24 ORE, in collaboration with the M. C. Escher Foundation, and is curated by Marco Bussagli and Federico Giudiceandrea. Featuring over 200 works spread across six sections, the exhibition itinerary offers a journey to discover the creative development of the artist: from the very roots of art history it extends to the figurative culture of Art Nouveau, focusing on Escher's love of Rome and Italy, and identifying his visit to the Alhambra and Cordoba as what triggered his interest in geometric patterns. The eyes of the great artist rest on the wonderful views Italy has to offer – as in *Roofs of Siena* (1922) and *Nocturnal Rome: Colosseum* (1934) – but especially on small details forming an extraordinary natural architecture, as illustrated by works such as *Dandelion* (1943), *Beetles* (1935) and *Grasshopper* (1935). Escher's gaze proves capable of grasping the geometric web behind things and of turning it into the cornerstone for the development of what he himself came to describe as "inner images". A key turning point in the exhibition is the stage of Escher's artistic maturity, with the themes of tessellation, reflective surfaces and impossible objects: these are exemplified by masterpieces such as *Hand with Reflecting Sphere* (1935), *Relativity* (or *House of Stairs*) (1953), *Metamorphosis* (1939) and *Belvedere* (1958). No monographic exhibition on the artist would be complete without a section documenting how Escher has influenced culture, publishing and music during the 20th century: his art has crept into the world of comics, advertising, music videos and films, unleashing a real #Eschermania." Source: www.mostraescher.it/la-mostra/

[2] The author of this paper asked the authorization, concerning its declared scientific purposes, to the C.M. Escher Foundation to quote and reproduce some artist's quotations and selected pictures of his works. In actual absence of any answers, the present extended paper results as a simple enquiry about the artist's well famous works (referring anyway to the artist's General Catalogue titles and codes), orientated in comparison with the iconographical legacy inspirating the applications of actual 360° digital VR/AR (virtual reality / augemented reality) devices. As iconographical examples of spherical visual empirical experiences, some pictures realized by the author around the public exteriors of the Exhibition at Palazzo Reale in Milano, or settled in a specific photo worksop, are included.

1.1 Maurits Cornelis Escher (1898-1972) creative itinerary

The artistic research of C.M. Escher begins from a typical north European artist: one of his first case study has been in an Italian tour observing the Mediterranean and Appenninians landscape. The central Italy complex and sweet countryside, characterized by an historical integration between geography, little urban centers, agricultural landscapes and tortuous roads itineraries astonished the artist. This sites were particularly meaningful for him since these landscapes, with multiple relations of both quantitative than qualitative dimensional scales – so different from the Netherland's flat panoramic views- suggested to him a concave and convex morphology that evoked a specific experience for the sight. [9] .
[i.e.: Cat.no. 96. *San Giminiano*, 1923, woodcut; Cat.no. 102. *Vitorchiano nel Cimino*, 1925 woodcut; Cat.no. 120. *Bonifacio, Corsica*, 1928 woodcut; Cat.no. 132. *Castrovalva (Abruzzi)*, 1930 lithograph.] [3]
He managed also a travel on a cruise ship, as a guest of the travel company in change of the gift of a sequence of travel destinations, drawing and carving some typical open horizon views, concerning the sea, skies (by night and by day) and water waves observations.
[i.e.: Cat.no. 231. *Phosphorean sea,* 1931 lithograph; Cat.no. 286. *Freighter,* 1936 woodcut; Cat.no. 97. *Dolphins,* 1923, woodcut].
Starting always from freehand sketches, he used to realize graphic compositions based on geometrical accurancy, and finally manually carving in woodcut - and even later for a most accurate definition also in lithograph - the matrix for limited serial prints. On the base of this rather typical, but anyway original experience, Escher moved and developed a deeper quality of enquiry through the sight experience, combined with sharp considerations about the observation of the graphic visual effect of the same drawing. His diaries and letters are a clear proof this awareness.

1.2 Drawing black & white in woodcut: a binary perceptive communication

In fact, both in his full contrast black and white engravures, or in his following multitonal lithographs, the basic principle of the print technique is founded -and remains- the engraving of the scratch on the surface of the cliché plate, which is as thin or large following the author's expressive intentions. Nowadays we could define that procedure as undoubtedly a binary (black or white, 0 or 1) signal. Far from researching the quality of his prints as a mere mannerist perfectionism, Escher explored the power of engravure binary textures for the visual effect of the perception of tones, as depth impression or simulation, also related to the perspective scaling of the perceived space.

1.3 Drawing and text: a template dichotomy evocate a further quality for the visual meaning

Starting some ex–libris that he realized for several friends, and private customers and other magazine publisher covers, he combines the deep effect of simple b&w visual icons with the framed motto that announce and reveal the meaning of the engraved picture.
In this way the apparently descriptive images gain a metaphoric substance when related to the word: a further significant dimension of the image as symbol is here explored into the close interaction – apparently only graphic, but essentially intellectual- between icon and text.
[i.e.: Cat.no. 91. *Bookplate B.G. Escher [Beer],* 1922 woodcut; Cat.no. 155-187. *Emblemata XXIV,* 1931; 1931 woodcut].

---

[3] The quotations official classifications of titles and datations of C.M. Escher works here adopted derives from the fundamental catalogue edited by J.L Locher (Author, Editor), With: F. H. Bool, J. R. Kist, F. Wierda, M. C. Escher, with an essay by Bruno Ernst (Translations by Tony Langham, Plym Peters), *M.C. Escher.His life and complete graphic work.* N.Y. Harry N. Abrams Inc. Publishers 1992. ISBN 0810981130. This scientific reference method has been also confirmed to us from the Milan exhibition curator Federico Giudiceandrea.

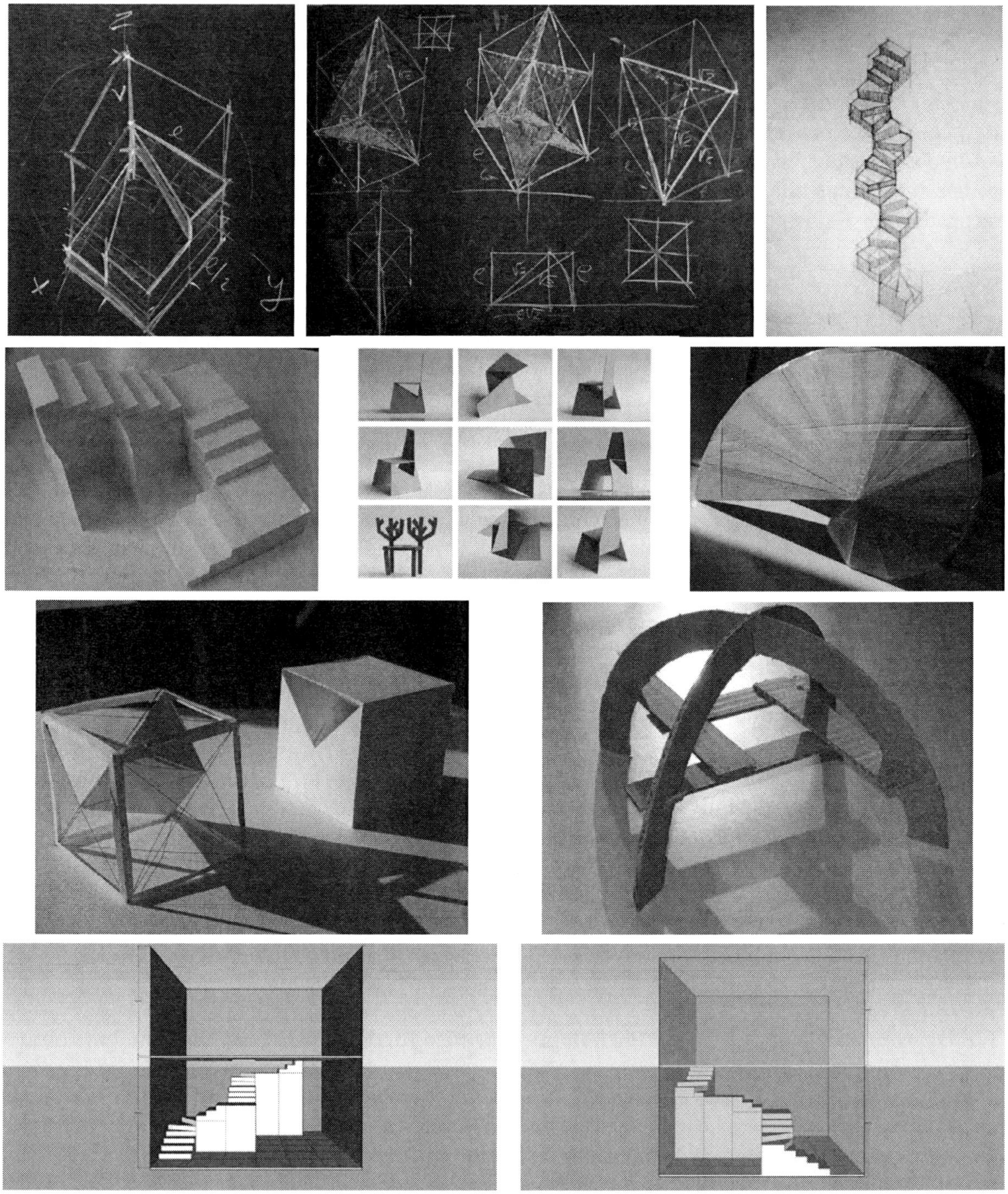

Fig. 2. Some didactic examples of three main tonal perceptive decoding of cartesian volumes and light shading for visual interpretation of spaces, usually adopted in any design drawing, photo-graphic or photo-realistic illustration
Fig. 2. a (up, left) Blackboard (white&black, negative), (up, right) sketches (black&white positive);
Fig. 2. b (middle) photographic pictures of cardboard maquettes;
Fig. 2. c (down) virtual sketchup modelling with different transparenct renedering simulation.

Source: didactic database of images from Design and Architecture Drawing Workshop, by Dott.Arch.PhD Federico Brunetti. Design Discipline Progettuali e Laboratorio (concept & workshop). Coordinatori Alternanza Scuola-Lavoro, Liceo Artistico di Brera, Milano, Italy.

1.4 Representing the three dimensional vision of spaces, between perception and imagination

Already since these first sequence of works a clear intuition appears as the power of a very essential technique- if applied to the visual codes of drawing or enquiring the dynamics of perception- can open incredible windows to mind's understanding of imaginative thought. Not only towards the description or reinterpretation of the figurative expression, but exploring the cognitive connections and multiple purposes suggestions of images, vision and space mental modelling.
[i.e. Cat.no. 401. *Three worlds*, 1955 lithograph; Cat.no. 128. *Selfportrait,* 1929 lithograph.]

In this directions many of his famous works investigated the three dimensional visual paradoxes, already suggested from mathematical and topological researches, about impossible figures, ambivalent or even multivalent representation of "impossible" spaces, or objects.
[Cat.no. 366. *Contrast (Order and Chaos)* 1950; Cat.no. 386. *Cubic Space Division (cubic space filling)* 1952 lithograph; Cat.no. 387. *Concentric rinds (concentrics Space Filling / regular Sphere division),* 1953 wood engraving.]

In a double binded relation -deeply implicitly profiled in what will be successively defined by Karl Popper (see § 5.2) [14] [16] we can recognize two cognitive procedures: the three main tonal perceptive decoding of light and volume in visual interpretation of spaces, usually adopted in any graphic or photorealistic illustration (see fig. 2, 2a, 2b, 2c,) and the imaginative, or conceptual, invention in theoretical mathematical spaces (not necessary real), and new topologies represented even exceeding the conventional flat perspective view.

2 The diligent and sharp excursus between vision, graphic and geometry. Drawing as an algorithm: sublimating the contrasts of experience and facing conceptual antinomy.

It is quite evident that, since these stating elements of the mere black and white reproduction technique used by Escher, one of the themes that intrigued the artist were the oppositions that seems to lead the relations between Nature and human existence.
[.i.e.: Cat.no. 67. *Paradise,* 1921, woodcut; Cat.no. 104 -> 114. *The Creation;* Cat.no. 303. *Day and night* 1938 woodcut; Cat.no. Cat.no. 366. *Contrast (Order and Chaos),* 1950]

We could remember, simply as a biographical detail, that he was a left handedness artist; his interest for the hand (or hands) at work, and the symmetric mirroring of reality and visions, could find in this physiological orientation a subtle suggestion. Anyway surely both the specific, and very original, iconography and the arguments and titles choose for several of his works, indicates clearly this particular attention.
[i.e.: Cat.no. 104 -> 114. *The Creation;* Cat.no. 248. *Still life with mirror,* 1934 lithograph.]

2.1 Subjective / objective

On other hand the main opposition that any human being can refer to is the phenomenological ***juxtaposition*** between the observer, as a ***subject***, and any ***object*** external to the person. The deep capability of astonishment that Escher reveils – and his specific kind of sentiment and knowledge research – that he was able to recognize in the more usual detail of everyday life the same complexity that ruled the whole world.
[i.e.: Cat.no. 128. *Selfportrait*, 1929 lithograph; Cat.no. 248. *Still life with mirror*, 1934 lithograph; Cat.no. 296. *Still life and street*, 1937 woodcut (see drawing: The street in Savona cat. N.° 296)].

2.2 From reflections to selfportait

Without having to spell things out, the artist act in deep awareness about the fact that an effective observation is possible only if a conscious observer is in action. The theme of the reflections is a symbolic argument in which natural phenomena exemplify the multiplicity of relations that happen around us (see case study §.6).

The recursive mirroring of the person who draws, which seems quite unavoidably represented - or simply evoked - in the same picture, is an evidence of this deep understanding. The quite ritual exercise of the self portrait reveals not only a canonic iconography of the author, but - as in the impressive sequence of the enlarged details of the observer eye, where not the face of the observe is represented, but the skull of his future inevitable destiny (if not even the symbolic representation of the time passing during the light / sight travel) - puts in evidence the idea that even the narrow distance of everyday experience contains an inevitable enigma.
[i.e.: Cat.no. 355. *Drawing hands,* 1948 lithograph. (see: preliminary studies cat.no.: 355; note that Escher used his right hand for from life drawing); Cat.no. 344 *Eye,* october 1946 Mezzotint, seventh and final state].

Under this point of view we cannot avoid to feel the same approach of French existentialism of Jean Paul Sartre or, on other side, the quite contemporary considerations of Karl Popper in epistemology of science about the inevitable participation of the observer to the phenomena which is in examination (see also § 5.2) [8] [4].

2.3 Self-portraits in mirroring spheres

We can notice that in Escher's works, regarding the representation of a spherical mirror, quite often but not always, the author's drawing reflected act is explicitly reported.
[i.e.; Cat.no.80. *The sphere,*1921, woodcut; and in: Cat.no.339. *Three spheres II,* 1946, lithograph.].

In fact usually the drawing hand and pencil of the author is in first evidence; but in the former and famous: Cat.no. 268. *Hand with reflecting Sphere (self portrait in spherical mirror),* 1935 lithograph, it is difficult to decipher the act of drawing in the reflected self-portrait. In fact the left hand is the one holding the sphere, and since Escher was a left handedness artist, we could ipothetize that this work has been made indirectly from true, but in well memorized and very skilled way, by an author well expert in such kind of these extreme visuals.

With the aim of empathizing Escher'e experience, I tryed to realize some similar pictures in the exterior of the Milan exibition. Doing so I noted the not secondary difference occurring in this experience in taking pictures to a spherical mirror, if shooting with a middle format digital camera (framing with viewfinder in front of my face) or otherwise using some recent smartglasses with embedded camera (simply looking at the sphere holded in one hand, and taking the picture by the same glasses)

---

[4] For a deeper interest about these enquiry we cannot forget another text very concerned with M.C. Escher: Hofstadter, Douglas R., *Gödel, Escher, Bach: An Eternal Golden Braid,* 1979, Basic Books, ISBN 0-465-02656-7

Fig. 3.a (left): The exterior of the entrance gallery room of the C.M. Escher exhibition in Milan, reflected in a spherical ball; picture taken by the author by using high-res. Wide-angle lens camera

Fig. 3.b (right): The entrance of the C.M. Escher exhibition, reflected in a spherical ball; picture taken by the author by using a smart-glasses mid-res. embedded camera (with free hand; see :§ 7.1)

3 The act of seeing

If literally the *the act of seeing,* if considered as "observation", has the same etymology from the latin verb "obserbo", meaning also "to conserve" - i.e. to maintain into the time - and otherwise the verb *to watch* has the same origin of the idea of "having care of" -i.e. to maintain into the space- [13] [5]; we can easily understand how many hidden significance can be embedded in the more spontaneous visual relational action. An action in the instant of light, but for future memory.

3.1 Observation experiences, Drawing Images: invention and antinomy

In these contexts the works of Escher seem to find the way where the Drawing assume the role as an algorithm able to elaborate a patient regulation in the observation of antinomies (effective, simulated or invented), realizing into the artifact of the image an interactive framework to approach the reality and face its contrasts through pictures. In other words the work of C.M. Escher remembers to us that observation results as an interaction and interrogation (inquiry) of the reality; the space of Drawing could be so as a sort of middle ground between an attentive recording and a proactive invention; Images con be considered the result of a creative elaboration to deal (sublimate) the contradictions.

3.2 Drawing, images and metaphors: an essential knowledge representation process

Drawing and images occupy a relevant place among the languages and tools used to communicate both in arts than in sciences. The iconic component becomes particularly important in this type of path and, although it usually takes only the form of para-textual apparatus, when it is adequately

[5] Alberto Munari, the pedagogist scientist, son of the wellknown italian designer and pedagogue Bruno Munari, remembered this origin of the visual concept in a unforgettable conference at Rotonda della Besana in the MuBa event on 18.1.2016 in Milan. The philosopher Silvano Petrosino underlined the anthropologic value of visual knowledge in several essays, see for instance: Petrosino Silvano, *Piccola metafisica della luce,* Milano 2004.

conceived, it can be critically important and can even shape the thought that generated it. Among such iconic expressions three main elements can be identified: *drawing, image and metaphor* [2].

Drawing. Through the formalization process of drawing, events and concepts previously invisible become visible, in accordance to a compositional order that in itself establishes the analysis of the object of a research and makes it shareable. Like in geometry, in describing physical phenomena the dot is taken as the essential element used to trace a sequence of events; this minimal reference becomes the starting point for analyzing and investigating their dynamic development.

Image. During the heuristic phase of scientific (and artistic) research, the image is recognized as the mental configuration (even before becoming the visual configuration) of the synthetic comprehension of a problem and its data. Then the interpretation can be further analyzed logically and formally.

Metaphor . The classical origin of this rhetorical figure and its use mainly in poetic expressions has kept the scientific epistemology of the early modernity away from this concept, which has recently been re-appraised in the analysis of phenomena that cannot be empirically perceived in quantum physics. Metaphor is a fundamental cognitive and communicative modality that uses a linguistic procedure in order to define, even if figuratively, the form or the concept of the object that is being studied, by introducing and referring to the figure of another object which is better known. This mode of communication, found both in text form and in iconography, uses an hypothetical model, not only to represent the formal structure of the phenomena, but also to enable the comprehension of the real data according to specific explanatory hypotheses or at least according to certain elements that make such comprehension possible through shared experiences.

4 The unexpected fortune of a new iconographical paradigm

So as far as in any creative or scientific skills these three steps, just here defined, are deeply interrelated in a continuous process. The creative intuition of Escher endeavor is well rooted in this epistemological etymologies: this is a clear reason, even if unexpected for him, for the interdisciplinary good fortune of his iconographical paradigm. Both mathematical researchers, and many other scientists and literature artists or publishers, founded in his work a source for further inspiration. Even if the iconography of the curved mirror reflecting images can bet easily related to a specific Netherland's arts heritage [see for instance: Jan van Eyek, *Portrait of Arnolfini couple,* (1434). National Gallery, London; or other well known history of art sources, or discovered during his researches, by Escher (Giovanni B. Piranesi, *Carceri d'Invenzioni,* tav. XIV, 1761; Luca Pacioli, *Solidi Platonici*; travel to Alahambra and geometrical studies about the division of plane, 1936), we must here shortly remember the attention and reciprocal dialogues and suggestions with important his contemporary mathematical researchers.

4.1 Escher and the mathematicians.

In 1954 C.M. Escher met at an Amsterdam Congress of Geometry Prof. Harold Scott MacDonald Coxeter (1907- 2003), who sent him a copy of his paper *"Crystal Symmetry and Its Generalizations".* The geometrical illustrated complex geometric figures inspired Escher to use his figures for his series of "Circle Limit". Many of his most famous works, including his "Ascending and Descending" (1960),

graphic illusion of interlocking staircases, are related to Coxeter's ideas about polytopes. In 1996, Coxeter published a paper proving that Escher's "Circle Limit III" was mathematically perfect.
[I.e.: cat.no: 436. *Circle limit IV;* woodcut in two colours, 1960; cat.no: 446. *Snakes*, woodcut in three colours, 1960] Here he explores the limits of infinity on board or the center of the circle, graphically considered as the geometric isomorphic structure, ideal model for recursive research.

Following these planar enquiry he develop also further enquiry about three dimensional spaces.
[Cat.no. 386. *Cubic Space Division (cubic space filling)* 1952 lithograph. Cat.no. 387. *Concentric rinds (concentric Space Filling / regular Sphere division)* 1953 wood engraving].
We cannot avoid to remember the unforgettable sequence of the film *Interstellar* (Christopher Nolan 2014), evidently inspired by these drawings.

In 1956 Escher met the mathematics teacher Bruno Ernst (pseudonomyn of J. A. F. Hans de Rijk, Rotterdam, 1926) that defined the artist's research following some scientific taxonomies. For example he recognized specific categories of geometrical meanings in his images: 1) regular spatial figures; 2) regular division of the plane; 3) spirals; 4) Mobius strips; 5) perspective; 6) metamorphoses and cycles; 7) approaches to infinity; 8) the conflict between depicting something on a plane and the three-dimensional reality which is depicted; 9) the penetration of more worlds; 10) spatial anomalies (impossible figures); 11) relativities. (other themes and detailed study of B. Ernst has been published in the main Escher catalogue mentioned at the beginning of this essay) [9]

Other admired interests (and interesting actual web resources) came from professor Roger Penrose (1931, Colchester; 1988 Wolf Prize for Physics) about impossible figures and triangle studies [12]
[*Relativity, and Convex and Concave*, 1955, lithograph], (ref. Bool *et al.* 1982, p. 147 [9]), configuring an ambiguous figure that depicts two different staircases at the same time.

Following his first enquiry about the regular division of the plane (1942), in 1957 the De Roos Foundation invited Escher to write an essay on the regular division of the plane (original Dutch edition of 1958). Following his methodological procedure we can find in these visual way of thinking the original source of many analogical fundaments who inspired further mathematical and even digital actual research developments in the treatment of space image.

5 Graphic language as a means of cognitive representation

In this context, looking at the use of the graphic language as a means of cognitive representation, it could be interesting to distinguish different levels of increasing complexity and abstraction in the relationship between the individual thinking and the surrounding world, which correspond to similar level of in-depth analysis.

5.1 Description – interpretation – prefiguration

In contemporary epistemology, within the trinomial: *description - interpretation – pre-figuration* [3], in the practice of identifying the three criteria of increasing involvement between the subject who reproduces the real object (through drawing or other means of representation) and the virtual object (that will be projected) of such representation are commonly accepted.

Description can be defined as the transcription of the most extensive collection of data the "observing" subject is able to capture / extract from the object being observed in the real world, through the expressive characteristics of the languages and the technical instruments at his disposal.

Interpretation must be understood as the critical selection of the collected data, separated and actively directed/applied to the subject according to his working needs.

Prefiguration is the invention of new possible configurations of reality starting from the collected data, which are tampered with, modified or related to each-other according to a specific project, obtained by sublimating the state of things and the pre-existing real world to achieve the project scenery.

Clearly these three *cognitive criteria* must be considered as a *hypothetical procedural scheme*, repleted with possible interferences and retro-actions, since it is often necessary and suitable that these three actions take place in a mutually synergic and communicating style.

In fact, ***description*** cannot be naively neutral, but it carries within the instrumental methodologies and the thematic approaches characterized and conformed by the used disciplines and the languages. Cognitive competences and descriptive abilities are inescapably intertwined.

Similarly, the selective modalities of ***interpretation*** must inevitably conform to the project's goals as they mature, thus creating in the mind of the observer- a sort of project -destination, which makes some elements more interesting than others, so as to stress the need to obtain further information on portions of reality or on different and heterogeneous orders of magnitude. Finally, by continuity or by opposition the acts of ***prefiguration*** - which methodologically consist of the act of overcoming the existing conditions. More precisely, the problem of *invention* finds its place in our discipline as a factor that characterizes the intelligent forms of life, and therefore originates the projectual ability.

5.2 "Three Worlds"

Here it can be needed and very interesting to refer explicitly to the fundamental epistemological work carried out by Karl Popper that we can read in the published accounts of his memorable conversations with John Eccles [15]. Unfortunately we need to drastically synthesize this epistemological contribution in order to include it in this article, but we are well aware to the importance and to the methodological vastness this quotation would deserve in a broader publication.

Karl Popper's three hypothetical "*worlds*" are named as it follows:
- ***World 1***, made up by concrete materiality;
- ***World 2***, sensory perception elaborated inside the human subject;
- ***World 3***, abstract mental construction, conceived inside one's mind and therefore able to theorize and to plan activities relating to the outside environment.

The three areas identified by Popper help to comprehend the different relations between the real / concrete world, the subject who observes it and the representations of reality the observer is able to build.

World 1, which is totally outside the subject, exists notwithstanding the observer, and it is a solid foundation of material elements; it is the object of every research that pre-exists and persists through the cognitive activity.

World 2, stems from the relation between the material data and the subject scrutinizing it: it is the sum of the relations and sensations generated inside the observer, starting from an interactive perception of world 1.

World 3 is the sum of images and elaborations the intelligent subject is able to develop in his inner self, in an autonomous and abstract way, producing theories, imaginations, projects.

In fact World 3 is able to retro-act in the way described in order to draw perceivable representations that can be shared in the sphere of the perception of World 2, so that the ideas can become projects - shared and completed works – that can affect and modify the pre-existing nature found in World 1.

In this theoretical configuration the ***act of drawing*** finds an extraordinary confirmation of its disciplinary value. In fact, if one acknowledges the inevitable otherness and complexity of World 1, one can also see the essential and incoercible value of its subjective formulation in the sphere of World 2 - where the representations belong to those mnemonic apparatuses needed for an in-depth and shared knowledge - prelude to that inventive transformation typical of World 3, that can re-descend [go back] to the preceding worlds to modify their status.

The idea of ***"conjecture"***, as a necessary condition for any cognitive activity, is another essential emphasis introduced by Popper in his epistemological path. In other words, no representation can be offered as an absolute and unquestionable reference, but it must be considered as a hypothetical conclusion, which is structurally falsifiable and liable to criticism from the very act of its definition, explicitly stating the terms of a possible verification or contradiction of itself.

In this context the desire to know and represent is no longer inhibited by the "fear to make a mistake", but is totally focused on the human attempt to understand the world surrounding us and ourselves at work with the reality to which we also belong with our ideas, intuitions, desires and projects. These clearly systematic and interactive working assumptions seem to us extremely befitting/pertinent in describing all the phases – ***hypothetical, conjectural*** *and* ***projectual*** – that characterize the different ways, meanings and valences of drawing.

The objects and the places that have been explored are considered a universe (*world 1*) that must be discovered through a consciously sensitive experience (*World 2*: spatial-temporal, analytic or synthetic, subjectively or collectively oriented sensitivity), by developing artefacts and graphic expressions able to represent the mental and abstract reconstruction brought forth by the subject (*World 3*). Such formulations, transmitted outside the subject, can become iconographic communication, shared knowledge, projectual reinterpretation, all the way up to the material transformation of the concrete datum at the source of the whole process. This procedural scheme, now customary for scientific protocols, provides an interesting paradigm to synthesize the considerations put forward up to this point. We can in fact trace a synoptic comparison between:

description - interpretation – pre-figuration
World 1 – World 2 - World 3

Within their sequence these three conceptual groups show a progressive trend that puts the acting subject at the centre, starting from the original condition of the object under observation, leading the interest towards the projectual capabilities, the conditions of artificialization, the transformative potentials. Such considerations do not necessarily point to univocal solutions regarding the ways to represent the existing reality or the projectual pre-figuration, but they put the possible graphic-visual formulations in the context of a critical and intentional method of drawing as a cognitive act.

6 Case study workshop, following C.M. Escher drawing *"Three spheres II"*

[Cat.no. 339. *Three spheres II*, 1946, lithograph.].
In this visual and graphical enquiry C.M. Escher, observing the multiple relations of simple visual effect through geometrical elements realized in different material, obtains a deep investigation into the light and visual phenomenon. In fact if the mirrored surface reflects proportionally curved the surrounding space (and the image of the observer), also the other two spheres imagines are viewed from their internal lateral sight (otherwise invisible from the observer since covered by their alignments).

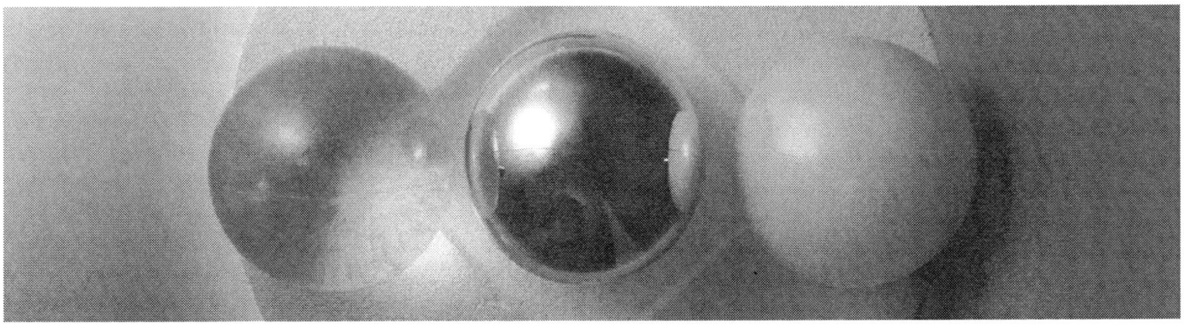

Fig. 4.: Phographic azimuth setting of the light reciprocity effects represented in C.M. Escher drawing "*Three spheres*" [Cat.no. 339. *three spheres II*, 1946, lithograph.]. The central mirrored ball reflects at 360° the environmental shapes and both the two lateral spheres. On the left, the full crystal ball performs as a lens – reflecting partially the light source by its curved glass surface - and project though its transparent material a light image on the floor upon it is based; on the right side the white opaque ball shapes its round surface scaling continuous grey tones and sets its shadow on the floor. Picture taken by Federico Brunetti

6.1 Vision as relation

Besides we can consider how a simple variation of monochromatic tones let us understand and decipher visually both the shape of the objects, than the optical characteristic of the material in which they are built, and therefore the optical game of the light acting in their multiple relations.
Even if we can note that the effective experience of these geometrical – material – brightness phenomena are dramatically more effective in the real view – probably in reason of the coupling stereoscopic deepness in comparison with of the monoscopic observations or reproductions.
It is quite impressive to consider how a mere image can unveil the complex dynamics embedded in the real world, if carefully observed. In other words, throughout these kind of images, C.M. Escher introduce us into a wondering dimension of our capability to see, where not only our ability to observe is limited to a quite passive and direct optic physiological evidence, but in a deeper consideration where the complete process of decoding of visual information is engaged, due to the limits that some graphic paradoxes can challenge.

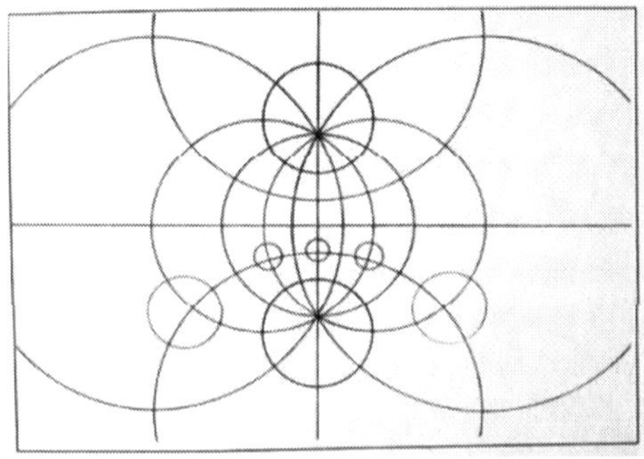

Fig. 5. Theorical diagram of circular perspective, from: Liliana Curcio e Roberto Di Martino *"La prospettiva a A 180° e oltre Il punto di vista di Emilio Frisia,* Quaderni n°4.ed. Accademia Raffaello, Urbino [6]

7 Geometrical algorithms for digital spherical optical capture and visual immersive display devices

After quite an half of century of his creative experience - that has been quite fashionably perceived as an "alternative" research about vision instead of his rigorous true nature of sharp enquiry overpassing the limits of Euclidean flat screen perspective - the legacy of these iconography re-appears in some digital devices defined as spherical digital cameras, in which combined optical wide-angles lenses obtain surrounding 360° ("stitched") virtual maps, that can be so observed by VR / AR (virtual /augmented reality) visors and glasses.

7.1 360° Spherical digital views and interactive immersive or augmented virtual visual experience. [5]

Some particular experience have been recently offered to the geometrical research and applications combining the smart-glasses technologies, with spherical cameras at 360° -built integrating two counterposed fish-eye devices- being suitable to transform in rational framed picture some otherwise a continuous –enclosed and curved- source image. Such spherical optical devices, and data processing software for related interactive experiences, are recently available for visual works and aesthetic quality concerning these 360°image camera which enable immersive shooting on sites. Those visualization results that can be obtained by appropriated head mounted selective vision systems. The image experience realized by these devices, at the end, get closer to the mental map processed about the surrounding environment, more than what we selectively perceive by visual senses.

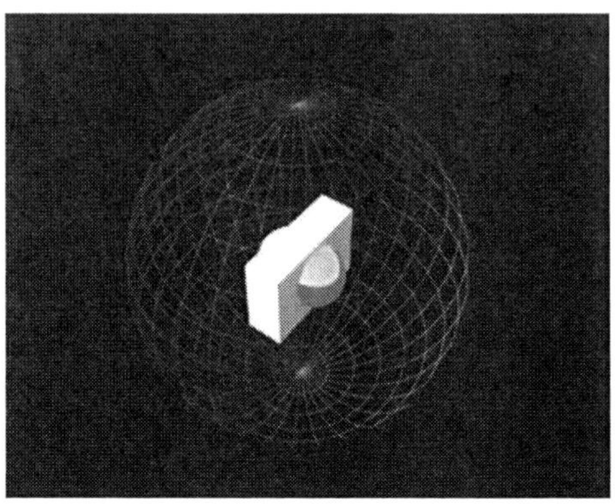

Fig. 6.a. Spherical digital camera diagram concept framework 360° view. Concept layout graphic design: Federico Alberto Brunetti PhD. Politecnico di Milano, School and Dept. of Design, 2016; Digital graphic visualization: Arch. Marco Valentino (I.S.A.D.) Milan

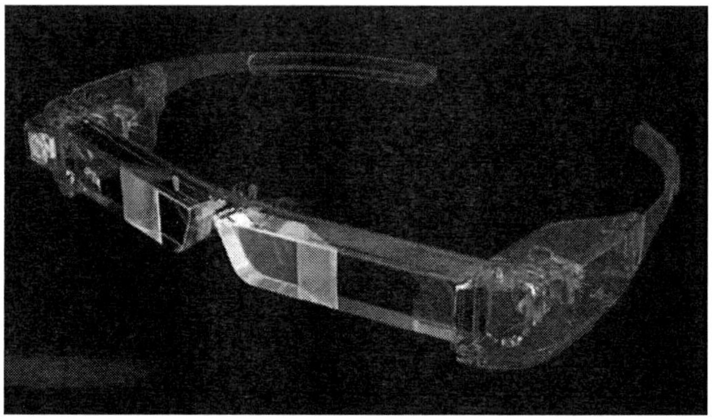

Fig. 6.b. Design structure in transparent simulation of stereo prisms in the see-trough glasses lens (equipped with embedded compass, gyroscope and accelerometer)

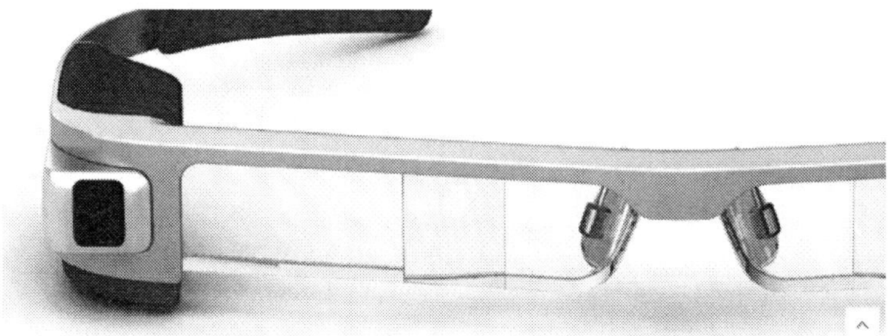

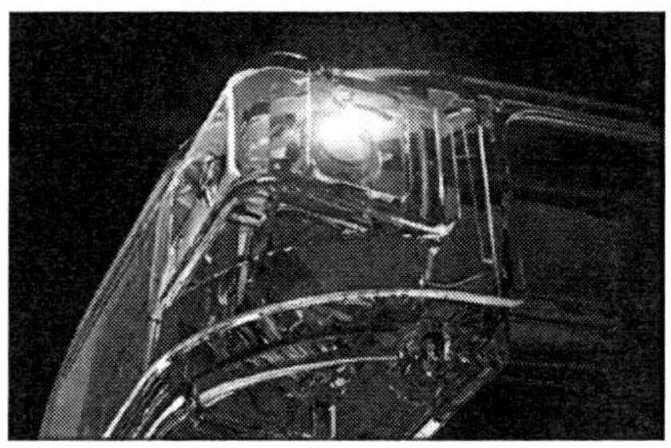

Fig. 7.a (left): Augmented reality vision in see-trough smart glasses, with integrated camera. (front detail)

Fig 7.b) (right): Detail in transparent simulation of the integrated coaxial camera unit [17] [6]

7.2 VR/AR for spherical vision

Usually VR devices are proposed for these 360°camera monitoring, but specially AR smart-glasses are particularly suitable to these applications since they maintain both the direct surrounding vision, and the seeing of an "augmented reality" through transparent visors with dichroic screens. It is interesting to compare the experience between the geometric nature of the visual field optically detectable and needed to the "reduction to the perceivable" effect obtainable by these digital viewers. The issue of algorithmic conversion from a spherical image of its sectors is relatively reducible to the framed flat plane display (the original theme of each geometry of the Earth's mapping). But the viewer interface of geometric conversion algorithms from the sphere to the framed plan needs to be intertwined with the interpretative models of cognitive psychology and symbolic meanings, since the configuration of physiology space perceived by our mind is shaped by our sensory organs: orientated and symmetrical, and mnemonically interactive. We must consider how such a 360° spherical device (and detection software), for its structure is full inclusive, isomorphic and deliberately undifferentiated. This results must be available selectively surveyed to our view by a prospective and rational monitor (literally capable of dividing into parts a unique spherical image) in order to allow the observer a mental understanding of a global scene, in some of its interested parts, i.e. to be able to choose the images to then take operational decisions.

[6] Source : with courtesy of: *https://files.support.epson.com/docid/cpd4/cpd40542.pdf*
www.epson.it/products/see-through-mobile-viewer/moverio-bt-300?mwd_contactregistration=1

7.3 Vision in movement

As further progress of this "state of the art" I propose an illustration [and the following case studies] of same evolved dynamic use [1] of these interactive devices, considering the time-space dimension in which a 360° camera could be applied. In fact even if already a spherical vision could be interesting to explore, it is an evidence that –under a projectional geometry point of view – the evolving sequence of curved perspectives introduced by these linear, circular and orbital (or even spherical spiral) visual survey perspectives could offer an interesting conceptual argument. In fact not only a first still punctual grade of full surrounding environment could be explored, but dynamically also the bilateral view in a trajectory, or the inner and outer horizontal rotational views, and even a complete spherical exploration could be conceivable, starting and around an initial position.

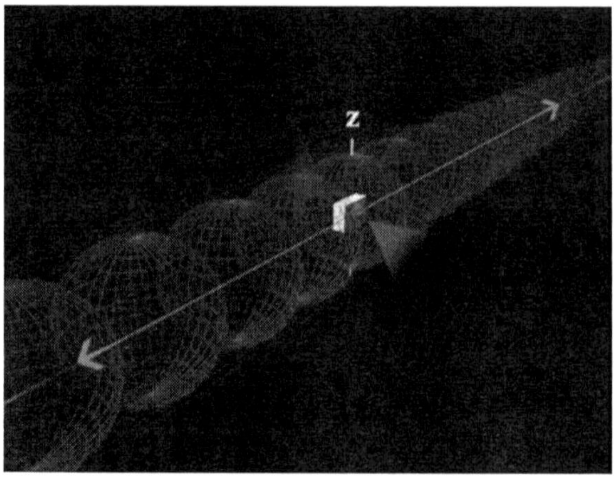

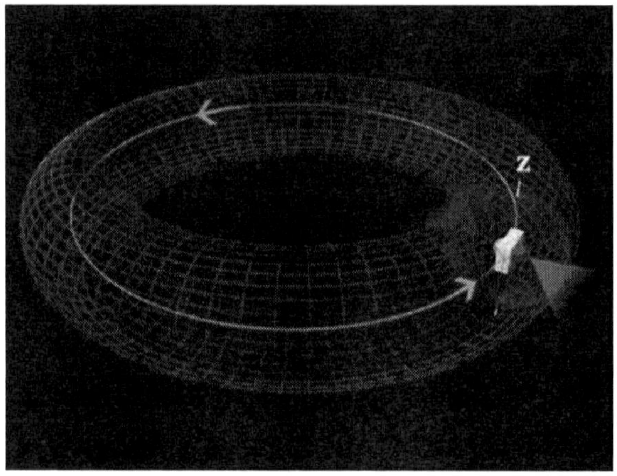

Fig.8.a (left): linear doublesided translation space view   Fig. 8.b (right): torus orbital rotation space view.

Concept layout graphic design: Federico Alberto Brunetti PhD. Politecnico di Milano, School and Dept. of Design, 2016; Digital graphic visualization: Arch. Marco Valentino (I.S.A.D.) Milan

In this applications scenery both mathematics, or artistic and technological visual research are involved: Spherical geometry, Projective geometry, Cognitive Psychology and a specific appreciation of the Symbolic semiology [11]: all can intertwine into this new methods of representation. A new interpretive attitude is requested to the observer, in the ambitious aim of a visual and immersive "empathy" proposed by these new technologies for the images of arts and science.

8 Imagination concept environment. Heuristic spherical immersive case study:

I had the opportunity to test a particularly advanced immersive system, interesting for the simulation of a gestural, dynamic design in a virtual environment. The costumer is first scanned in 3D and made visible to the designer himself through stereo VR eyewear connected to a dynamic detection system of his gestures. A virtual palette is then made available, so every movement of the body can draw in the virtual space user desired shapes in real time. At the center of this VR stereo application is the 360° 3D experience by the interaction of drawing by the body itself.

My first intuition was to build an open window to the virtual landscape; so the first position memorized in the representation of my body concerns the opening of the wings and the first sign traced in the virtual space has been the open square of the window. Afterwards I drew the first wall to support it and therefore all around a sort of observatory cabin where to stand to observe this landscape during the day, with a wide door for visitors and a small table and chair for reading.

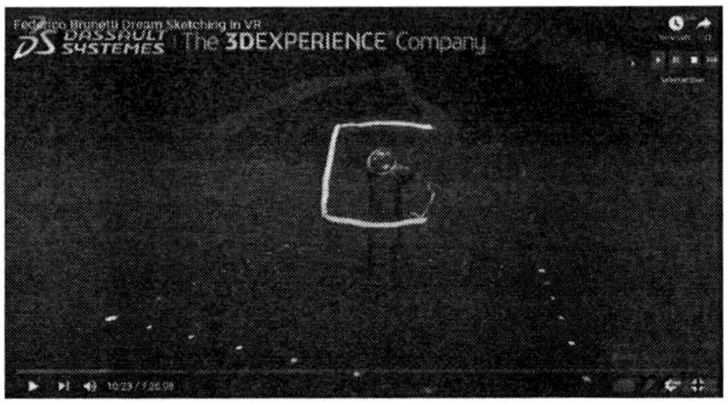

Fig. 9. Observation studio with open window toward virtual landscape

Fig. 9.a. Body scanning phase

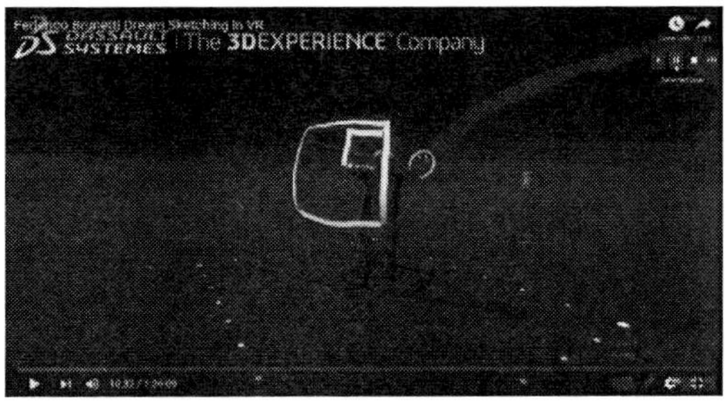

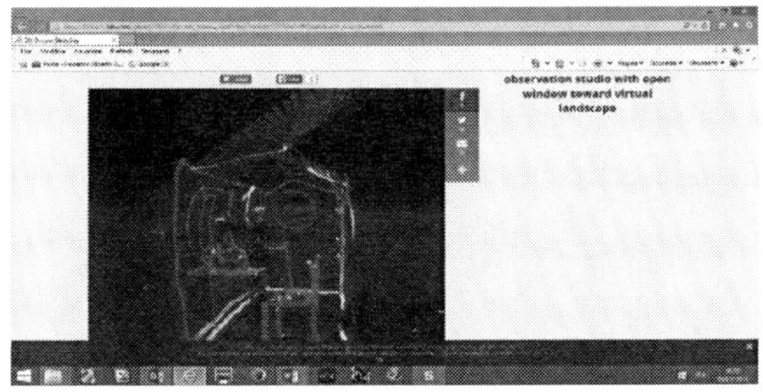

Fig.9.b. 3D interactive gesture virtual immersive drawing Fig. 9.c. 3D design result of virtual immersive drawing

Milano 2016.04.12 "Design in the Age of Experience", Federico Brunetti Dream Sketching in VR.
[HTC Vive + 2 Microsoft Kinects + Some inhouse technology] [18]

The experience of this space with horizon around me -and the transparent, but serene atmosphere and bright night where to draw with signs of colored light- was very joyful. The framing of the window allows me the time to internalize and measure the inner knowledge of this new kind of space to live and to observe it calmly: I thought that this was the first condition in order to be able to live there. I also realized that I had tried to test the limits conditions of these instruments toward the available field that was concerned and limited by the extension of explorable soil and ergonomic measures of gestures. The accuracy and interactivity of the software is impressive; the palette ribbons sympathetically reminds the art of weaving, which is the anthropological basis of the production of handcrafted artifacts, but also resemble strongly to the graphic effect of urban graffiti. This experience of immersive technology could be a booster for the imagination, and the feeling of living.

9 Open conclusions

The actual possible interactions between different technologies and available devices is only an example of the numerous possibilities to consider in the near future, due to the implementation supported by adequate digital platforms. The visual experience and the iconographic heritage awareness that we must have towards authors like C.M. Escher if so evident, and even quite unexplored if considered under the point of view not only of the effective works that he leaved us, but in the light of the creative methodology that he represented. In fact we can appreciate the example of his multiple connection between the intelligent observation of reality, the full technique mastery, and the ability to conceive deep conceptual innovation regarding the relation between the subject and the world, the space and its representation, the idea and the image. We can quite simply suppose that a certain creative part of his work, as in any true artistic or scientific research path, happened during the execution of his pieces, in the same wondering experience the leaded him toward the whole reality.

In a contemporary paradox, nowadays we are surrounded by powerful technologies, but we are sometimes unable to conceive and understand their effective, useful, and finalized customized utility to serve the human needs. In this context the new technological experiences of representation of spaces seems to offer us many opportunity for gaming into - and within- the virtual world, but more rarely to face with ability, wonder and joy the real one.

Actually the geometric assumptions and algorithms can realize a connection among:

- **3D visualization of digital models,**
- **images generated from the optical survey by 360 ° spherical view of the real image,**
- **the visual representation and immersive stereoscopic view in augmented reality.**

These options will enable visual experiences that we are opened totally to invent and design.
In the context of digital cinema, R&D concerning special effects already obtain indistinguishable results between real and fantastic. The mathematics and scientific research, technology and creativity, IT and visionary skills can find here fertile interdisciplinary areas for joining into cross-science research. Some theoretical concepts on multidimensional space-time date back more than a century ago, and at present are reflected in the physics science exploration iconography of the micro and macrocosm [4]. Some of the theoretical concepts sometimes does not appear immediately apparent to be finalized. Anyway sometimes new ideas or inventions are very useful in fields different from the ones initially supporting the ideas. For these reasons, it is a good chance anyway to properly evaluate the potential of any representation concepts, as they could open future application scenarios still unknown and innovative [10].

The spherical perspective -as a representation of space method- in the specific feature as a mediation algorithm between the viewer, the outside world and the perceived image, can be outlined how in phenomenological way has assumed to three main schematic roles:

- ***Survey,*** (perspective and camera based technologies)
- ***Imagination,*** (qualitative imaginary pictures, scenography)
- ***Design*** (quantitative visualization of invention and project, and possible multiple choice problems solving).

The opportunity, and possible duty for mathematicians, artist and designer is full opened to disclose patiently a new world already around us.

At the end of this even basic enquiry, we must remember that the first spherical device which allows us the visual experience is the same eye: inside of these couple of incredible natural creations happens the first optical step of the complex (up-down reversed) further physiological elaboration of the real time image recognition mental process. It is quite evident why the sphere, even for its intrinsic conceptual perfect form, evokes the artists' and scientist' inspiration.

In the meantime and same place of Escher's exhibition, another important Italian artist great presentation took place in the 90° living anniversary of Arnaldo Pomodoro, well known for his dramatic sculptural exploration of the imaginative inner spaces of external geometrical polished surface.

Another re-creational world dimension so appear into the fractures and structures of mathematical equilibrium. But this matter could disclose another, quite endless, and different discussion.

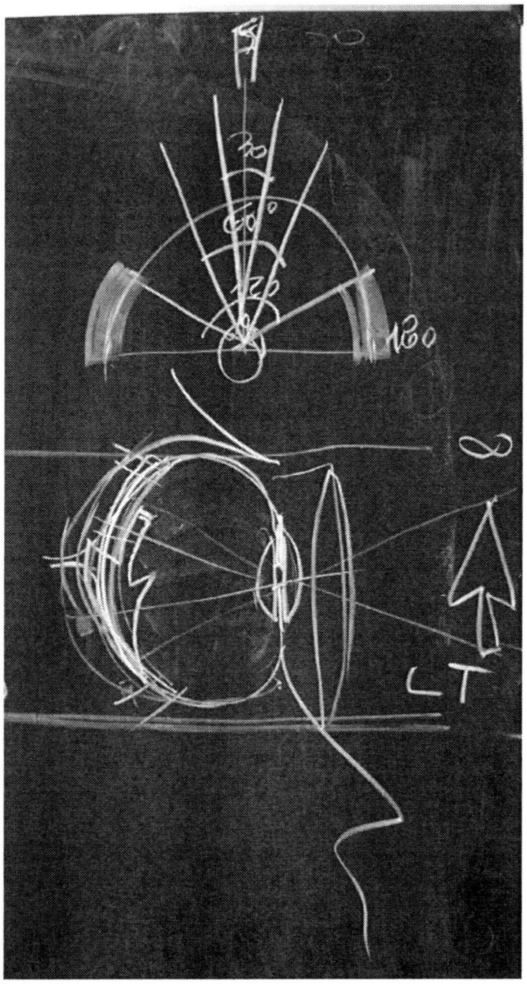

Fig. 10. (left): Sketches on the blackboard about visual fields and optical reversed vision. Politecnico di Milano, Scuola del Design. prof. PhD. Federico Brunetti: workshop: Elementi di Percezione Visiva, Laboratorio di elementi visivi del progetto

Fig. 11. (right): Arnaldo Pomodoro, 90° anniversary of the birth, Milano, palazzo Reale; side event Exhibition about C.M. Escher; self-portait: "*Three spheres in vertical*", by the author of this article. 2016.11

References

[1] Brunetti Federico, *Spazio, tempo ed architettura. Esperienze di luoghi, forme e costruzioni attraverso le percezioni dinamiche della restituzione visiva multimediale: dall'architettura in video al cantiere nell'era digitale.* In: *Studiare con l'architettura,* a cura di Alberto Fratelli e Maria Letizia Gagliardi a atti convegno: *AAA Arti Artefatti Architetture, Video, Architettura, Città.* Udine, 18.02.2005. Università degli Studi, Corso di Studi in Architettura, Consorzio Universitario del Friuli. *Colloqui di Architettura 2004/2005* Udine 2009 Forum Ed. pp. 55-74. ISBN 88-8420-392-9 *http://qui.uniud.it/notizieEventi/cultura/documento.2005-02-17.9471858927*

[2] Brunetti Federico. *"Disegno, Immagini e metafore nella comunicazione scientifica" in: ComunicareFisica.07.* 2007.10.1-6 Trieste *Conferenza-Workshop nazionale sulle tematiche e sulle metodologie della comunicazione della fisica e delle altre scienze. Linguaggi e strumenti della comunicazione scientifica verso il pubblico. Sponsor: I.N.F.N. Istituto Nazionale di Fisica Nucleare; Università di Trieste; ICTC International Centre for theoretical Physics, Sincrotrone Trieste S.C. p.a.; Università degli studi di Udine. http://www.ts.infn.it/eventi/ComunicareFisica/presentazioni/brunetti.pdf*

[3] Brunetti Federico, *Morphology and mental imagery. Geometry and its applications to architectural and design drawing learning, (APLIMAT 2011, Bratislava),* p.15-21, in : Federico Brunetti, (editor and essays) *Attraverso il Disegno. Relazioni tematiche, tracce metodologiche e atti didattici per l'architettura ed il design. Reprint with translations.* pp.436. Milano 2012 ISBN 97888-8132-649-5.

[4] Brunetti Federico Alberto, (Politecnico di Milano – Scuola del Design), *Image as algorithm; visions of science in the digital era.* In: *Scientific and Technological culture 14th International Art and Technology Meeting (#14.ART): Art and Human Enhancement,* University of Aveiro, Portugal, October 7-11, 2015. Joint organization between Universities of: Aveiro (PT), Brasília (BR) and Goiás (BR).; Procedings Bastos P.B.; 2015, University of Aveiro; ISBN 978-972-789-458-1.

[5] Brunetti Federico, *The legacy of perspective in the technological design of visual digital devices.* in: *Proceedings of: ICGG2016* 17th International Conference on Geometry and Graphics ©2016 ISGG 4–8 august 2016, BEIJING; in: *Proc. 17th Intern. Conf. Geometry and Graphics,* Aug. 4-8, 2016, Beijing, PRC, article No. 149.

[6] Curcio Liliana e Di Martino Roberto, *La prospettiva a 180° e oltre,* Centro Internazionale di Studi sulla Prospettiva di Urbino. Quaderni n°4.ed. Accademia Raffaello, Urbino 2009

[8] Hofstadter, Douglas *R. Gödel, Escher, Bach: An Eternal Golden Braid, 1979, Basic Books, ISBN 0-465-02656-7*

[9] Locher J.L (Author, Editor), With: Bool F. H. , Kist J. R. , Wierda F. , Escher M. C. , with an essay by Ernst Bruno (Translations by Langham T. , Plym Peters), *M.C. Escher.His life and complete graphic work.* N.Y. Harry N. Abrams Inc. Publishers 1992. ISBN 0810981130.

[10] Manzini Enzo, *Cultura tecnologica. Oggetti virtuali. L'ambiente simulato come nuova dimensione del reale. L'ibridazione tra materiale e immateriale come nuovo territorio del progetto. Una guida ragionata al mondo virtuale.* In: Ottagono 91, dicembre 1988. Milano.

[11] Panofsky Erwin , *Idea. Contributo alla storia dell'estetica*, Torino, 2006.

[12] Penrose Roger, *The Art of the Impossible: MC Escher and Me - Secret Knowledge,* BBC Scotland, *Art documentaries,* Sep 2015. *www.youtube.com/watch?v=f7kW8xd8p4s; www.youtube.com/watch?v=1CYrGpd8k5w&t=8s*

[13] Petrosino Silvano, *Piccola metafisica della luce,* Milano 2004.

[14] Popper Karl, *Epistemology Without a Knowing Subject,* (1967), published as chapter three in his book *Objective Knowledge: An Evolutionary Approach*, Oxford University Press, 1972;

[15] Popper Karl with Eccles John C., *The Self and Its Brain: An Argument for Interactionism*, Berlin 1977.

[16] Popper Karl, *Three Worlds. The Tanner Lecture on Human Values.* The University of Michigan, April 7, 1978.

[17] Source: *https://files.support.epson.com/docid/cpd4/cpd40542.pdf*

www.epson.it/products/see-through-mobile-viewer/moverio-bt-300?mwd_contactregistration=1

[18] *http://perspectives.3ds.com/virtualreality/all-about-the-3d-dream-sketcher*
http://perspectives.3ds.com/tag/immersive-virtuality/
http://3dream.3ds.com/photo/352834?access_token=cda934fe62b61cfd332258a0c9f83aeb&utm_source=email

Acknowledgement

The authors gratefully acknowledge the Scientific Grant Agency: Politecnico di Milano; Scuola del Design and Design Department. And: *14th International Art and Technology Meeting (# 14.ART): Art and Human Enhancement, 2015 edition*, held by the Universities of Aveiro (PT), Brasília (BR) and Goiás (BR), with the participation of Municipality of Aveiro (PT). and : 17TH *INTERNATIONAL CONFERENCE ON GEOMETRY AND GRAPHICS* ISGG 4–8 AUGUST, 2016, BEIJING, CHINA.

This paper follows the friendly gratitude to Mauro Francaviglia that take care from the beginning, with accurate and eminent supervision and encouragement, to my interest in science communication and visual design of scientific iconography. With his rare and enthusiastic vision of dialogue between science and arts, he has been a rare mentor for me. Now, in an impossible direct colloque, we simply try to look to his original example, exploring the challenges of the present. The expert survey of some collegues and friends, as Marcella Lorenzi, supported me in the concern of this work.

Current Adress

Brunetti Federico Alberto, Dott. Arch. PhD.
Politecnico di Milano Scuola del Design; Dept. of Design
Visiting professor: *Strumenti e tecniche per il Disegno* (Laboratorio del Disegno), *Produzione dell'immagine per il Design* (Laboratorio di Progettazione visiva)
Via Durando 38/a, 20133 Milano, Italy
E-mail: federico.brunetti@polimi.it
www.federico.brunetti.it, www.sciencecommunicationdesign.eu

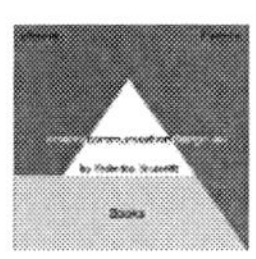

Slovak University of Technology in Bratislava
Faculty of Mechanical Engineering
16th Conference on Applied Mathematics **APLIMAT 2017**
Proceedings

RUBRICS AS ASSESSMENT TOOL OF MATHEMATICAL OPEN-ENDED PROBLEMS

BULKOVÁ Kristína (SK), ČERETKOVÁ Soňa (SK)

Abstract. The aim of educational process in mathematics should be focused on developing the student's mathematical competencies, so they are also useful in the real life. In this paper, a flexible and progressive tool, named rubrics, for the assessment of mathematical problems solutions, is described. The Revised Bloom Taxonomy of Educational Objectives in posing open-ended mathematics problems is reckoning with creating the rubrics for the particular problem solution assessment. The Mathematics B-day contest problems are open-ended problems based on inquiry based learning principles. Rubrics for assessment of selected solutions of the Mathematics B-day contest are used as an example of the rubrics in this paper.

Keywords: mathematical competency, rubrics, inquiry based learning, mathematical open ended problems, assessment

Mathematics Subject Classification: Primary 97A01; Secondary 97G01

1 Introduction

The implementation of Inquiry Based Learning into educational process is an actual topic. Through the mathematical open-ended problems, it is possible to point out to students' connections between mathematics principles and the real world. Problem solving, process of open-ended mathematical problems solution, requires analyse of the situation and the use of complex knowledge. Compared to standard mathematics tasks solutions, there is possibility of wide range of solutions in open-ended problems solving. The natural question during the process of assessment is, how to assess also the way, the process of thinking demonstrated in the solutions, and not only mathematical correctness of result.

Present studies are looking for a universal tool for assessment of mathematical open-ended problems. The goal of this paper is to propose and describe an objective tool of analysing students approach by observing and assessing their mathematical competencies.

2 Mathematical competencies used in open-ended problems solving

In inquiry in science, researchers submit questions, they search for ways to answer them, and operate with obtained results. In the same way teachers can lead lessons in the classroom, where students do not memorize knowledge, but they are looking for answers themselves. Asking questions is the core of Inquiry Based Learning (IBL). IBL is based on posing, exploring and evaluating. In [1] authors mention characteristics of IBL by Yackel and Rasmussen, as developing personally meaningful solutions, explaining students' approaches, ability to listen to each other and attempt to make sense of the thinking of other students.

Mathematical open-ended problems represent a tool for Inquiry Based Learning implementation. Problems contexts are close to real situations, and solutions require multiple-solution strategies. Contrary to standard mathematical tasks, which are well known within the educational process and which solutions demand only using the known algorithm. Zhouf [12] defines mathematical problem as a situation requiring the use of higher mathematical thinking. According to [2] the problem solving involves "understanding the task, formulating an appropriate sequence of actions or strategy, applying the strategy to produce a solution, and then reflecting on the solution to ensure that it produced an appropriate response". The proposal of mathematical open-ended problem depends on the mathematical experience of students. In open-ended problem not only result is assessed, but also approach and the steps of approach used to solve the problem are in the focus of the assessment. This approach has to be handout in written form as a report of results.

Assessing process of mathematical open problems is not uniformed and it presents one of disadvantages. Assessor usually in subjective view corrects each of students' solutions. It means that one assessor can assess student's solution differently than another one [12]. Recent studies are looking for objective tool for assessing of mathematical open-ended problems. The essential base for creating this assessing tool is Revised Bloom Taxonomy of Educational Objectives. Two-dimensional system of assessing by Bloom Taxonomy is represented by a table, which contains six main levels of cognitive processes (Remember, Understand, Apply, Analyse, Evaluate and Create) in horizontal axis and four levels of knowledge (Factual, Conceptual, Procedural and Metacognitive Knowledge) in vertical axis of the table. Cells of the table represent an intersection of the knowledge and cognitive process dimensions. The Revised Bloom Taxonomy of Educational Objectives can be used for assessment of students how deeply the aim of problem was mastered, as well as a tool to classify the level of difficulty of the assignment of the particular problem [4].

Observation of knowledge manifested by cognitive processes in student's approach is useful in assessing of mathematical problem with close end or standard mathematical tasks. Open-ended problem, as well as a problem with the creation of an answer are characterised in [12]. The assessing should question how students use their knowledge, not which knowledge was used. The option for evaluating level of mathematical thinking is observation of mathematical competencies. Likely a definition of mathematical competency correlates with the ability of solving of mathematical open-ended problems.

Mathematical competency is defined in [2] as an ability to develop and apply mathematical thinking in solving different problems in everyday, real life, situations. "Mathematical competencies are components of mathematical literacy, which means ability of students to formulate, employ, and interpret mathematics in a variety of contexts, which includes reasoning mathematically and using mathematical concepts, procedures, facts and tools to describe, explain and predict phenomena" [7].

There are several different sets and structures of mathematical competencies defined. For purposes of this paper, the structure of competencies [5] was used. The following structure of mathematical competencies is dominantly beneficial to students during mathematics lessons:

- Mathematical thinking and reasoning.
- Knowledge of concepts, facts, assertions and approaches in mathematics.
- Application of symbolic, formal and technical operations.
- Description of mathematical objects and situations.
- Defining the problem by asking a question
- Mathematical modelling.
- Mathematical arguments and proofs.
- The use of tools and information processing.

In solving process of mathematical open-ended problems students operate with mathematical competencies, which are manifested in written form of solution. Therefore, suitable assessing tool may be based on analysis of students' mathematical competencies in the hand out written solutions.

3 Mathematics B-day contest

The team of experts from Freudenthal Institute of Utrecht University in the Netherlands issues the assignment of the Mathematics B-day. The Mathematics B-day aims to mathematical problem solving and higher order thinking skills. The pilot year in Slovak Republic happened in 2011, although Mathematics B-day has a long lasting tradition in Netherlands, Belgium and Germany. In Netherlands, Mathematics B-day is based on the educational program of Mathematics B, which is needed for university level, technical studies and studies in science and mathematics [10]. 138 students from 13 high schools in Nitra, Banská Bystrica, Sučany, Bratislava, Košice, Žilina, Poprad and Michalovce participated at the contest in 2015.

The Mathematics B-day contest gives students a unique opportunity to solve mathematical open problem in teamwork. Students solve assignment created with intent to motivate inquiring in mathematics. Teams of three or four students work on written final report by conjecturing and proving in mathematics, demonstrate their mathematical knowledge and competences.

Assignment from 2015 "Around the corner" was inspired by old problem about moving furniture around corners. "*The Sofa Problem*" is two-dimensional problem in combinatorial geometry given by question "*What is the region of largest area which can be moved around a right-angled corner in a corridor of width one*?" [11]. In Mathematical B-day 2015 students were gradually guided to create some mathematical model of this situation. Open-ended problem is usually opened task, where students explore firsts connections of situation with some mathematical knowledge. For example, in The Sofa Problem: if it is possible to move figures with given dimensions around the right-angled corner of one-meter-wide corridor, see Fig. 1.

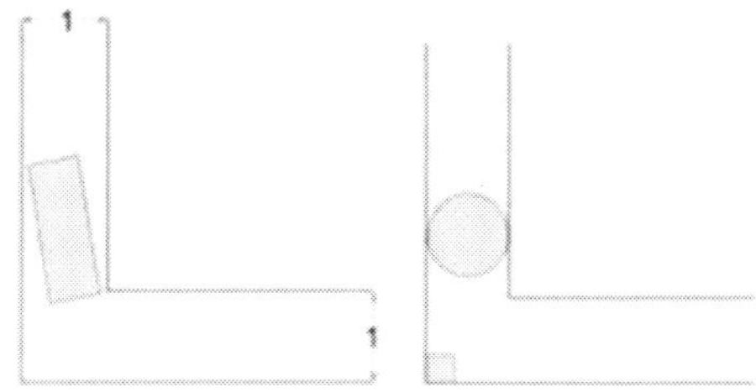

Fig. 1. Examples of explanation tasks from Mathematical B-day 2015

The initial problem is followed by several assignments. In assignments students solve concrete open problems related to main idea of problem. The most common are generalization of previous explorations: moving figures with largest areas (length of longest stick or shape) around the right-angled corner of one-meter-wide corridor. To proof the conjectures and solutions, students are encouraged to use experiments, sketches, dynamic mathematics software and other tools. Main part of Mathematics B-day contest lies in performing own original research of the posed problem. The final research is based on solved several exploration and assignments. After completing first part of the assignment, students are challenged for performing their own research in mathematics: "*Did you know that it is actually an unsolved problem what the largest object is that you can get through the corridor? ... You may choose a shape that has not yet been considered in the assignment so far and find maximum length or area of the object, different strategies to move the object through the corridor or the area that the object covers while being moved.*" [6]. All founded results, assignments and research, are implemented into written understandable final report, and supported by mathematical arguments. Final report is the assessed outcome.

4 Rubric for open-ended problems

Solving process of mathematical open-ended problem is not predictable. Students can use different ways how to come to answers. For this reason, observation of knowledge only would require extensive assessing criteria. Also the quantity of knowledge is not a guarantee that students know how to use them in a proper way. In assessing mathematical open-ended problems solutions, it is necessary to review more than theoretical side of the particular solution. Assessor should be focused on performance and level of mathematical thinking. The suitable tool for assessing of mathematical open problems represents rubrics.

Rubric is a coherent set of criteria for students' work that includes descriptions of levels of performance quality on the criteria. The purpose lies in description of level of performance. In contrast with state standards and curriculum goals, rubrics do not serve as learning outcomes [3]. For creating rubric performance, it is necessary to define two major aspects: coherent sets of criteria in vertical axis of rubric and descriptions of level of performance for these criteria in horizontal axis of rubric. Core of rubrics, created in this paper, come out also from Revised Bloom Taxonomy of Educational Objectives. Dimension of cognitive processes is suitable division to assess of students' performance. Sets of criteria are therefore considered as levels of cognitive processes defined by Bloom Taxonomy, from the lowest level "Remember", until highest level of cognitive processes "Create". Levels are coherent and subsequent.

On first processual phase (level Remember) students operate just with knowledge from long-term memory and use known standard approaches. Students are able to process assignment on basic level and to interpret result in a simple way. Second level of cognitive processes (Understand) is represented by skills determining the meaning of assignment and thinking about problem in mathematical way. Students use basic mathematical terminology, mathematical operators or graphic sketch of situation. But their solutions are missing some systematic characteristics. Using a procedure in a given situation is significant for the third level (Apply), where students look at problem in more complex way: students know how to overwrite assignment to formal mathematical notation, they work with algebraic symbols and they are familiar with spatial imagination. Asking a question represents first step of fourth level criteria. Indicators of higher cognitive processes (level Analyse) in mathematical thinking are represented by breaking situation into constituent parts, detecting how the parts relate to one another and to overall goal. Students have to be able to evaluate their approach. Fifth level includes the proof of correct solution by mathematical argumentation. Students on the highest level (Create) achieve level of IBL [4].

To complete the rubric is needed to identify the criteria of mathematical competences. Based on characteristics of level of cognitive processes in the Revised Bloom Taxonomy of the Educational Objectives, above mentioned mathematical competencies are sorted and assigned.

	MATHEMATICAL COMPETENCIES
1	Using of tools and information processing.
2	Knowledge of concepts, facts, assertions and approaches. Application of symbolic, formal and technical operations.
3	Description of mathematical objects and situations.
4	Define the problem by asking a question. Mathematical thinking and reasoning.
5	Mathematical arguments and proofs.
6	Mathematical modelling.

Tab. 1. Rubrics of mathematical competencies

Rubrics of mathematical competences, (see Tab. 1) represents the proposal of assessing tool for mathematical open-ended problems solutions. Assessor has opportunity to assess written solution of problems by observing and analysing mathematical competences. Rubrics can help also to describe level of students' mathematical thinking, whether students achieve level of mathematical modelling by open-ended problem solving.

The Mathematics B-day assignment is composed by set and/or sequences open-ended problems. The next problem solution is one selected from written reports.

> Assessment: Given is an isosceles triangle ABC, where M is the midpoint of AC. Also, F is a point between B and C on the line BC and the line FM intersects the line AB in point E. Show that line segment EF is longer than line segment AC [6].

Students choose the strategy to prove the veracity of theorem by relation (1) based on Pythagorean theorem:

$$(x-k)^2+(x+k)^2=a^2-2ak+k^2+a^2+2ak+k^2=2a^2+2k^2, \quad (1)$$

Where x are legs of the triangle ABC, a is a hypotenuse of the triangle ABC and k is a length made x shorter or longer, see Fig. 2. Students made the conclusion: "*The hypotenuse of triangle ABC is always shorter as the hypotenuse of others right-angle triangles with sum of lengths of legs of the triangle ABC.*"

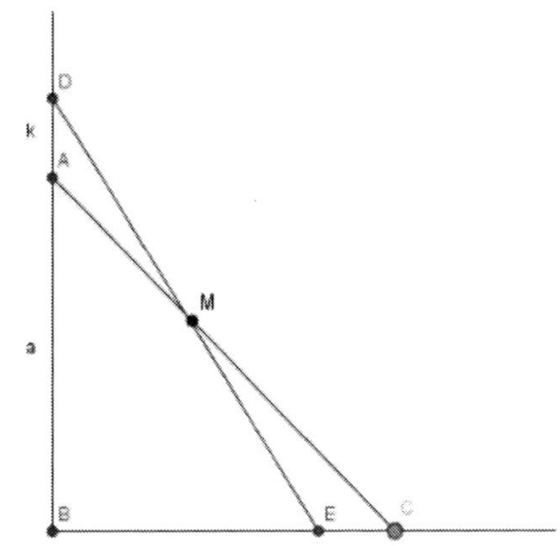

Fig. 2

Students started the proof by describing the mathematical objects. By mathematical thinking and reasoning, they made the premise. Subsequently students used mathematical arguments to prove the theorem. Compared to rubrics of mathematical competences, see Table 1, students' formulation of conclusion signifies the level of mathematical modelling.

5 Rubric for mathematical modelling

Mathematical modelling presents the highest level of mathematical thinking based on IBL structure. By rubrics it is possible to observe development of aspects of IBL like exploring, experimenting, planning strategies, and others. For assessment of mathematical modelling level there is possibility to create another rubric. Coherent set of criteria of new rubric is described by processes of IBL, which are determined by [9] as: simplifying and structuring, observing systematically, measuring, classifying, creating definitions, quantifying, inferring, predicting, hypothesizing, controlling variables, experimenting, visualizing, discovering relationships and connections, communicating. Processes of IBL are coherent and each process is a condition for another one. As well as mathematical competencies, sorted processes were assigned with respect to characteristics of levels of cognitive processes from Revised Bloom Taxonomy of Educational Objectives.

	PROCESSES OF INQUIRY BASED LEARNING
1	Sorting information. Observing systematically and visualising.
2	Measuring and quantifying. Controlling variables.
3	Discovering connections and relationships. Simplifying and structuring.
4	Classifying and creating definitions. Hypothesizing and predicting.
5	Experimenting. Inferring.
6	Mathematical model.

Tab. 2. Rubric of processes of IBL

In contrast with previous rubrics, see Tab. 1, sorted processes of IBL in rubrics indicate level of mathematical modelling. Some mathematical open-ended problems directly require students to create mathematical models as the solution. In this case it is useful to observe processes demonstrated in the particular solution. The next example of using the rubrics (Tab. 2) shows assessment of students' mathematical competences in mathematical modelling.

Example of observing and assessing the processes of IBL was taken from the same report as above-mentioned. Students created model of moving figures through the corridor while considering the other point of view: "*Blue lines on the picture* (see Fig. 3) *are walls of corridor in different time during the rotation of object in it (rotation wall around object). If object fits into the white annulus, it can rotate across the corridor without moving (maximum dimension of object must be one).*"

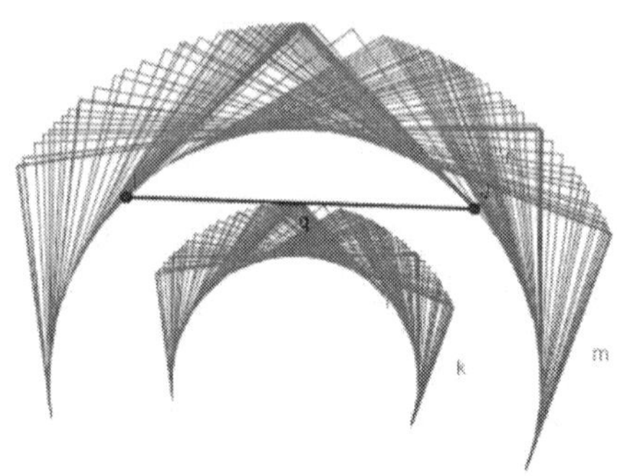

Fig. 3

"We focus how large area can be added to stick, which length is $2\sqrt{2}$, see Fig. 4. The object represents rectangle with dimensions $2\sqrt{2}$ x 1, in which there is cut a circular section (radius is $\sqrt{2}$, height 1.)"

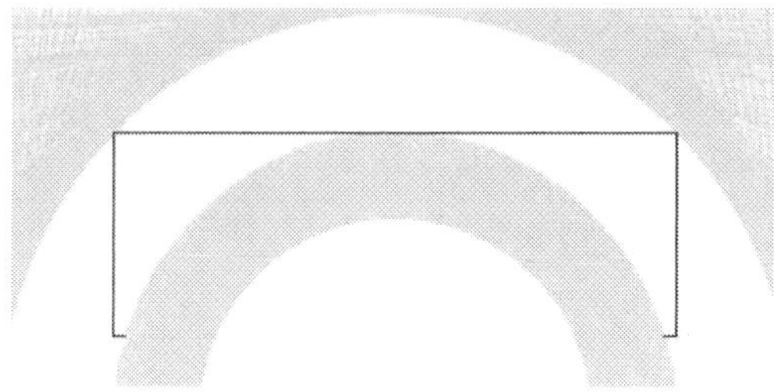

Fig. 4

Students laid the foundation of known solution of "The Sofa Problem". If this team continues in solving with other figures in the same way, students may conclude largest area, which can be moved around a right-angled corner in a corridor. Based on rubric of processes of IBL, see Tab. 2, students achieved the highest level of mathematical modelling. Solution started by discovering connections and relationships (level 3) related to previous assignments. Then students defined rotation space around the object, and by experiment in dynamic mathematical software they created model for solving of problem.

5 Findings

For illustration of rubrics' purpose in the paper, there are three more descriptions of students' solutions from Mathematics B-day 2015. Solutions of Teams A, B, C were distinguished by different ways of approaches.

TEAM A: The first part of Team A final report was composed from defining the concepts and terms, which were operated then in the whole report. Based on created system of definitions and theorems students deduced properties of objects and created hypothesis. Open problem solving process of Team A lies in proving of assignments, supported by mathematical argumentation. Students were on the best way to achieve the highest level of inquiring. Approach was closed just by written conclusion *"There are two strategies for rotation of object around the right-angled corner across the corridor. In both cases it is necessary to "push" the object to the wall. In first case on the outside wall, in second case on inside wall."*

TEAM B: Approach in Team B inquiry was based on formulation of logical consideration about largest area, which can be moved across the corridor around 45° angle: "*That area is given by the smallest possible circle and the largest possible circle. Circles represent the path of the edge of the object, which rotates around the point of reflection. The smallest possible circle has the centre in the point R and its radius represents the distance between the point R and the corner around which we rotate the circle. There are two biggest possible circles, both have the centre R, but the first one has the radius determined by the distance between the point R and one edge of the longest stick. The second one has the radius given by the distance between the R and the other edge of the longest stick.*" Students constructed premise for next steps of inquiring, but they finished their research by the calculation of the largest area.

TEAM C: Students in Team C inquired moving of specific object around right-angled corner of one-meter-wide corridor. Letter T was the chosen object, linked by the possible longest stick ($2\sqrt{2}$) and stick represents "leg of T". Inquiring lied in looking for the largest area depended from length of "leg of letter T". The most suitable way how to move object across corridor is rotation. Trajectory of rotation the longest possible stick is marked out by circuits, see Fig. 5. It means that the difference in the radii of circuits, and trajectories of rotation, represent the longest "leg of T".

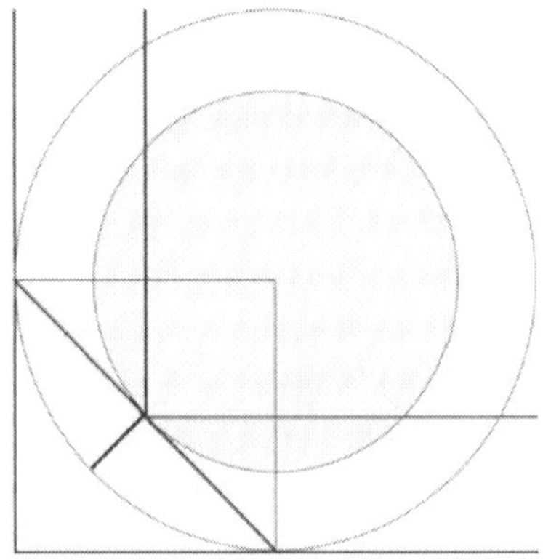

Fig. 5

Selected solutions show impossibility to observe and assess knowledge. It would be necessary to construct assessment tool composed of all possible knowledge, which could be manifested in solutions. Selected aspects for assessing inquiry in mathematics represent processes of IBL. The classification of team solutions, see was marked into the rubrics of processes of IBL, see Tab. 3.

	PROCESSES OF INQUIRY BASED LEARNING	**TEAMS**
1	Sorting information. Observing systematically and visualising.	
2	Measuring and quantifying. Controlling variables.	TEAM B
3	Discovering connections and relationships. Simplifying and structuring.	TEAM C
4	Classifying and creating definitions. Hypothesizing and predicting.	
5	Experimenting. Inferring.	TEAM A
6	Mathematical model.	

Tab. 3

6 Conclusion

The proper and objective analysis of the solutions to mathematical open-ended problems is important yet challenging for every assessor. Observing and analysing of manifested mathematical competencies may be the way, how to minimize subjective view in assessing. The advantage of using rubric lies in clear and accepted description of assessed aspects.

Solutions of Mathematics B-day represent examples, how to observe defined aspects. It is necessary to point out, that teams of pupils composed the solutions. For complex assessment of teamwork on mathematical open-ended problems, to observe social competencies of students (mutual interaction and communication) is also recommended and/or needed.

Acknowledgement

The paper was supported by project APVV-15-0368; *Practice in the centre of the subject field didactics, subject field didactics in the centre of preparation for practice.*

References

[1] ANDERSEN, M. S. et al: *Best experiences with implementing student active methods: results and challenges, ideas and tasks.* In: *Staircase to Even More Interesting Mathematics Teaching.* Nitra: UKF, 2016, ISBN 978-80-558-0973-1, pp. 11-68.

[2] BLAŠKO, M. *Kvalita v systéme modernej výučby.* Košice: Technická univerzita, 2013. ISBN 978-80-553-1281-1, pp. 126-127.

[3] BROOKHART, S. M. *How to Create and Use Rubrics for Formative Assessment and Grading.* Alexandria, Virginia, USA: ASCD, 2013, ISBN 978-1-4166-1507-1, pp.4-8.

[4] KRATHWOHL, D. R. *A Revision of Blooms Taxonomy: An Overview.* In theory into Practice: 41 no4 Aut, 2002, pp. 212-216. http://rt3region7.ncdpi.wikispaces.net/file/view/8+Perspectives+on+RBT.pdf

[5] LUKÁČ, S., SEKERÁK, J. *Interaktívne vzdelávacie activity stimulujúce rozvíjanie kľúčových kompetencií.* In: *Mathematika Informatika Fyzika: didaktický časopis učiteľov matematiky, informatiky, fyziky.* Vol 36 (XIX), Prešov: Metodicko-Pedagogické Centrum. pp. 61 http://www.mcpo.sk/downloads/Publikacie/MIF/MIF36.pdf

[6] Mathematics B-day 2015: http://www.primas.ukf.sk/download/bday/Mathematics%20B-day_2015_EN_for_SK.pdf

[7] PISA 2012: *Assessment and Analytical Framework: Mathematics, Reading, Science, Problem Solving and Financial Literacy*, OECD Publishing, 2013, ISBN 978-92-64-19052-8, pp. 24

[8] POWEL, A. B. (Ed.) *Challenging Tasks and Mathematics Learning. In Challenging Mathematics In and Beyond the Classroom*, New ICMI Study Series 12, 2009, pp. 133. http://www.math.mun.ca/~mkondra/papers/ICMI16_Chapter4.pdf

[9] THE PRIMAS PROJECT: *Promoting inquiry-based learning (IBL) in mathematics and science education across Europe*: http://www.primas-project.eu/servlet/supportBinaryFiles?referenceId=10&supportId=1247

[10] URL: http://www.uu.nl/en/education/mathematics-b-day

[11] Wagner, N. R. *The Sofa Problem.* In *American Mathematical Monthly* Vol 83, No. 3, 1976, pp. 188-189. http://www.cs.utsa.edu/~wagner/pubs/corner/corner_final.pdf

[12] ZHOUF, J. *Tvorba matematických problémů pro talentované žáky.* Praha: Univerzita Karlova, 2010, ISBN 978-80-7290-432-7, pp. 28-33.

Current address

Bulková Kristína, Mgr.
Constantine the Philosopher University in Nitra
Tr. A. Hlinku 1, 979 74 Nitra, Slovak Republic
Tel. number: +421 37 6408 709, e-mail: kristina.bulkova@ukf.sk

Čeretková Soňa, doc. PaedDr. PhD.
Constantine the Philosopher University in Nitra
Tr. A. Hlinku 1, 979 74 Nitra, Slovak Republic
Tel. number: +421 37 6408 692, e-mail: sceretkova@ukf.sk

Slovak University of Technology in Bratislava
Faculty of Mechanical Engineering
16th Conference on Applied Mathematics **APLIMAT 2017**
Proceedings

ART OF GEOMETRY. THE USE OF PERSPECTIVE IN THE WOODEN CHOIR OF SANTA MARIA MAGGIORE IN BERGAMO

BURATTI Giorgio (IT)

Abstract. The choir of Santa Maria Maggiore in Bergamo is a unique furniture in the tradition of wooden choirs because is characterized by a number of innovative solutions resulting from the successful collaboration between the painter Lorenzo Lotto and the inlayer master Giovan Francesco Capoferri. The inlays that decorate the backrests are among the elements of greatest interest. This paper studies the marquetry *Amon violante Tamar*. The geometrical analysis of the perspective will be detailed references to Lombard culture.

Keywords: Perspective, Geometry, Inlay, Wooden choir

Mathematics Subject Classification: 00A05, 01A05, 00A66, 51N05, 51N15

1 Historical development of perspective and inlays

The technique of wood inlay is a consequence of the flourishing of the arts progress manifested in Italy during the Renaissance. As already mentioned[1] success of genre is closely linked to the study of perspective in Florence in early fifteenth century, when representation of three-dimensionality, defined by Filippo Brunelleschi through the exercise of mathematics, promoted the spread of a practice that gradually affected all the arts.

Brunelleschi's studies are the fulfillment of a long journey. Representation's attempts of three-dimensional reality were already present in ancient Greek and Roman painting. During ancient times and the Middle Ages, greatly influenced by the work of Euclid, study of space was sometimes identified with the optics and investigated according to geometric criteria, while at other times it was connected with the physiology of vision.

In Middle Ages the representation criteria were related to the importance of portrayed subjects: the most important character of a picture was often represented with larger size, regardless of any perspective view. Despite this, practical arrangements for the pictorial representation of three-dimensionality can already be identified in the works of many artists, including Duccio di Boninsegna (1255-1319) and Giotto da Bondone (1267-1336). Also Ambrogio Lorenzetti (about 1330) from Siena was very skilled in intuitive drawing of the third dimension: in his works appeared for the first time in the history of painting the convergence to a single point of the straight lines perpendicular to a horizontal straight line.

However all these experiences are empirical, they are not supported by a theoretical apparatus that demonstrates the exemplarity of a proceedings. Brunelleschi was the first to develop an internally consistent system for all the spatial elements of which it is demonstrated the non-arbitrariness.

This finding led artists and mathematicians to break away from empiricism to elaborate precise rules for the representation of the real, codified norms written in systematic treaties that finally surpassed the ancient Euclidean theories.

Despite the chronological priority, the lack of a Brunelleschi's treatise on perspective makes Leon Battista Alberti the first Renaissance writer on this topic. Alberti is the author of *De Pictura* (printed in 1511) in which the ideas of Brunelleschi are codified and extended.

The geometrical theories of Alberti's work appear however still linked to a vague empiricism, which it will be exceeded with the work of Piero della Francesca, one of the greatest painters of the fifteenth century who was also a capable mathematician. The author in 1475 wrote *De Prospectiva Pingendi*, the most important treatise on Renaissance perspective, wherein it is evident the awareness to link the pictorial work to a coherent system of mathematical procedures able to translate the real space in a represented space.

The commitment in searching rules for a correct representation of reality through the application of geometry will develop mathematical research which will bring in the following centuries Desargues, Pascal and Monge to the theoretical arrangement of projective geometry.

The course of the inlay covers the crucial period of the perspective studies. It shall take from time to time specific connotations depending on territorial variations underlying the genesis of different cycles, but succeeding in developping a code of shared values. In spite of this, inlay is an artistic genre often investigated as an episode of lesser crafts parallel to the major arts development, a patient mechanical extension of the painting that has no direct links with Drawing "*padre delle tre arti nostre*"[2] .

The lack of recognition of a formal autonomy is probably due to the heterogeneous character of the work inlaid: developed on a two-dimensional plane, according to chromatic canons of painting, the marquetry is made up of material carved, sculpted, constituting a unique case that does not belong completely to sculpture, or to painting. The creation of a inlaid work takes place by using the combination of different types and different colored wooden tiles. Historically the pieces are following a preparatory drawing, setting out the main lines and the image backgrounds, to which the tiles must perfectly match.

The onset of the genre is attributed to Sienese craftsmen that, at the beginning of the fifteenth century, received the greatest number of commissions to realize choirs' inlays for the main cathedrals and churches of the time.

The subsequent influence of Florence in the mid-fifteenth century, directed the art of inlay to the representation of geometric solids or perspective views describing innovative Renaissance architecture. In this period the marquetry evolves: initially used in ecclesiastical context as a favourite option to decorate dossals of choirs, it was subsequently used in *studioli* rooms, places of reflection for Renaissance humanist, researcher of science and letters[3] .

Choirs and *studioli* often share the same iconographic program: life transience symbols, biblical stories, half-open doors that reveal open or piled books, science tools, allegorical weapons, polyhedra, musical instruments, cages with birds and views of cities to symbolize the universal harmony theorized by neo-Platonism.

The Bergamo cycle analyzed in this work, realized in the twenties of sixteenth century, is the wooden inlay's art apex. Built by Giovanni Francesco Capoferri[4] on preparatory studies made by Lorenzo Lotto[5] the choir of the Basilica of Santa Maria Maggiore in Bergamo denotes a clear mastery of perspective rules and a hardly replicable craftsmanship.

The work is distinguished by the search for more free figurative expressions, far from the topics listed previously and aimed at the representation of a single narrative program, in urban environments, buildings or landscapes, of biblical events.

2 The choir of Santa Maria Maggiore

The choir by Lotto and Capoferri is part of the plan of improvements provided for the main altar of Santa Maria Maggiore in Bergamo, in 1522 . Built at the behest of citizenship in a time of strong consolidation for the free town of Bergamo, following the annexation of the *Serenissima* domains, the church is overlooking the square, which for centuries has been the focus of political, economic and social life of the city. The church's autonomy is obvious since 1449, when the management was entrusted to *Consorzio della Misericordia Maggiore*, an important laical city institution that continues to this day to administer the building.

Fig. 1. Basilica of Santa Maria Maggiore, Bergamo. The northern side of the nave

The implementation of work is headed by twenty five years-old Capoferri thanks to the likely intercession of Lotto. The artist, which previously collaborated with the Consortio[6], appreciated the young master's skill and especially his willingness to experiment with new ways and new languages. Capoferri will turn to be the right man for the creation of innovative narrative cycles focused on the biblical stories imagined by Lotto, capable of translating in marquetry the "well colored in perspective" paintings as required by the demands of the contract.[7]

The demand for painted, colorful works, is atypical and contrasts with the tradition of marquetry. Artist normally employed a drawing. The Bergamo customers fit the specific, following talks with Lotto that led them to consider problems for the inlay master in representing biblical stories very different from traditional models to which Capoferri was accustomed. The indications on lights and tones facilitate the analysis and the composition required to the work of translation into wooden work, contributing to get the results that concur to the choir's uniqueness.

The choir, built in walnut with the exception of coniferous wood structure, brings together in a single plan seating of the clergy, the benches beside the altar where there are celebrants and public

Fig. 2. Lotto's and Capoferri's inlayed choir. The benches for the secular representation, without protective covers

representations of the state and the benches of city delegation, placed in the hemicycle of the apse behind the altar. In the visual prominence of the main chapel, raised four steps above the floor of the choir's presbytery tribune, it reflects the hierarchy of the institutions and representatives of the community. The hierarchical ordering included all participants in the liturgy, symbolically differentiated according to reasons of celebration and the one who officiates. On rare occasions, when the bishop was invited to celebrate, he took place in a dedicated seat, to emphasize the role of civic temple, center of a democratic Municipality opposed to the old ecclesiastic and aristocratic government. Located next to the *sanctum sanctorum* inside the choir, the representation of political and civil power assumes a leading position respect to the resident clergy, to highlight its mediator role between the celebrant, the people and God.

The original project foresees a division of religious choir in two wings of thirteen stalls each, arranged in a single order. Choir stalls differ from the traditional typology because they lack the high *postergale*[8] where the inlays were usually placed. In Bergamo the space reserved to the panels with the biblical stories is therefore the backrest, a solution that exposes the works inlaid to increased wear. To overcome the problem Lotto, principal proponent of this solution, provided a cover as a protection for the precious artifact for each inlaid framework, by take off only in solemn festivities. For this purpose more than seventy studies were prepared, including drawings and sketches, ranked according to a program that started from Creation and continued chronologically to the Old Testament texts. Regarding the choir of the laity, although the banks were visible and equipped with a large seatback, apparently that they were not provided for ornamental inlays. According to the original project, in fact, attention had to be focused on an altarpiece, in silver and gilded copper, which had to shine above the altar. Sadly, although it was started, the altarpiece was never completed. So the Consorzio della Misericordia decided during the construction to allocate to the benches of lay choir inlaid prepared for the choir of the religious, thinking thereby to compensate for the absence of ornaments. Symbolic inlays provided for covers were then used as a substitute in the canons choir's seatback. Lotto's iconographic plan is therefore destroyed: to deprive the covers by reference images makes the allegorical interpretation impossible. Following the changes a large number of inlays are separate although in the original project they had been conceived as sequential and contiguous.

Fig. 3. Amon violante Tamar

3 *Amon violante Tamar* geometrical analysis

For the present detailed analysis it has been selected the inlay called *Amon violante Tamar*. The work illustrates how the eldest son of David, Amon falls in love with the stepsister Tamar. Amon confides his infatuation to his friend Jonabad who suggested to him to pretend to be ill and asked his father to make Tamar call to bring him consolation. Amon consumed violence against Tamar and then repelled her, horrified by his own action.[9].

The story is articulated into four narrative units, distributed in a circular path that allows the painter to make the best use of the whole stage space. In the building, whose rooms upstairs have no walls to permit the vision of what is happening, the story takes place in a counterclockwise direction. In the foreground, near to the vanishing point, the conversation between Amon and Jonabad is focused, while the aggression narrated in the biblical text is shown in the corner room, to the left of the upper floor.

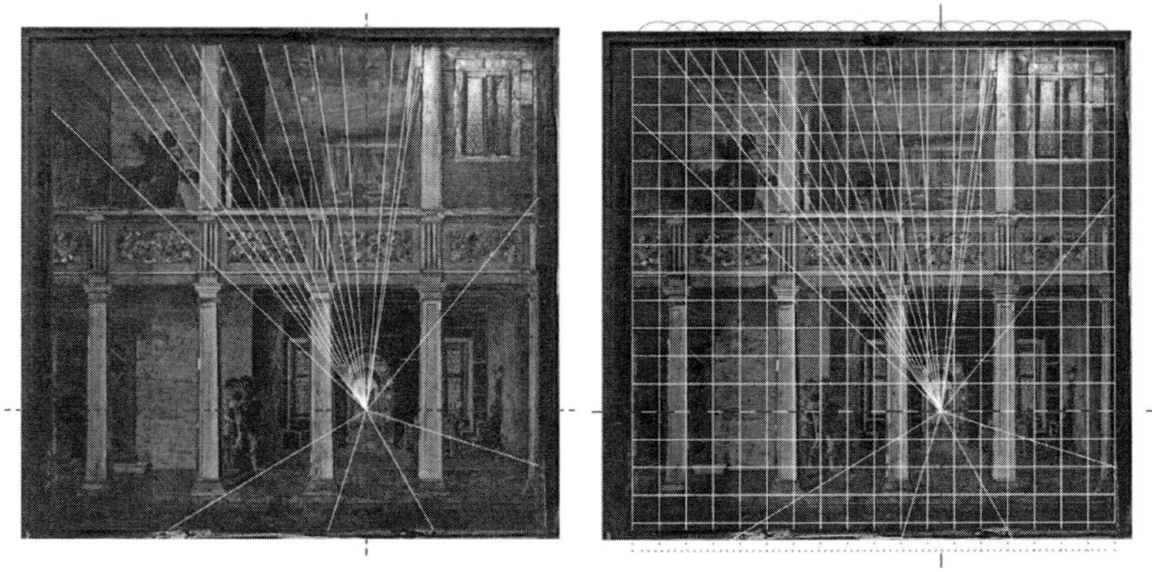

Fig. 4. Identification of the main point and construction grid

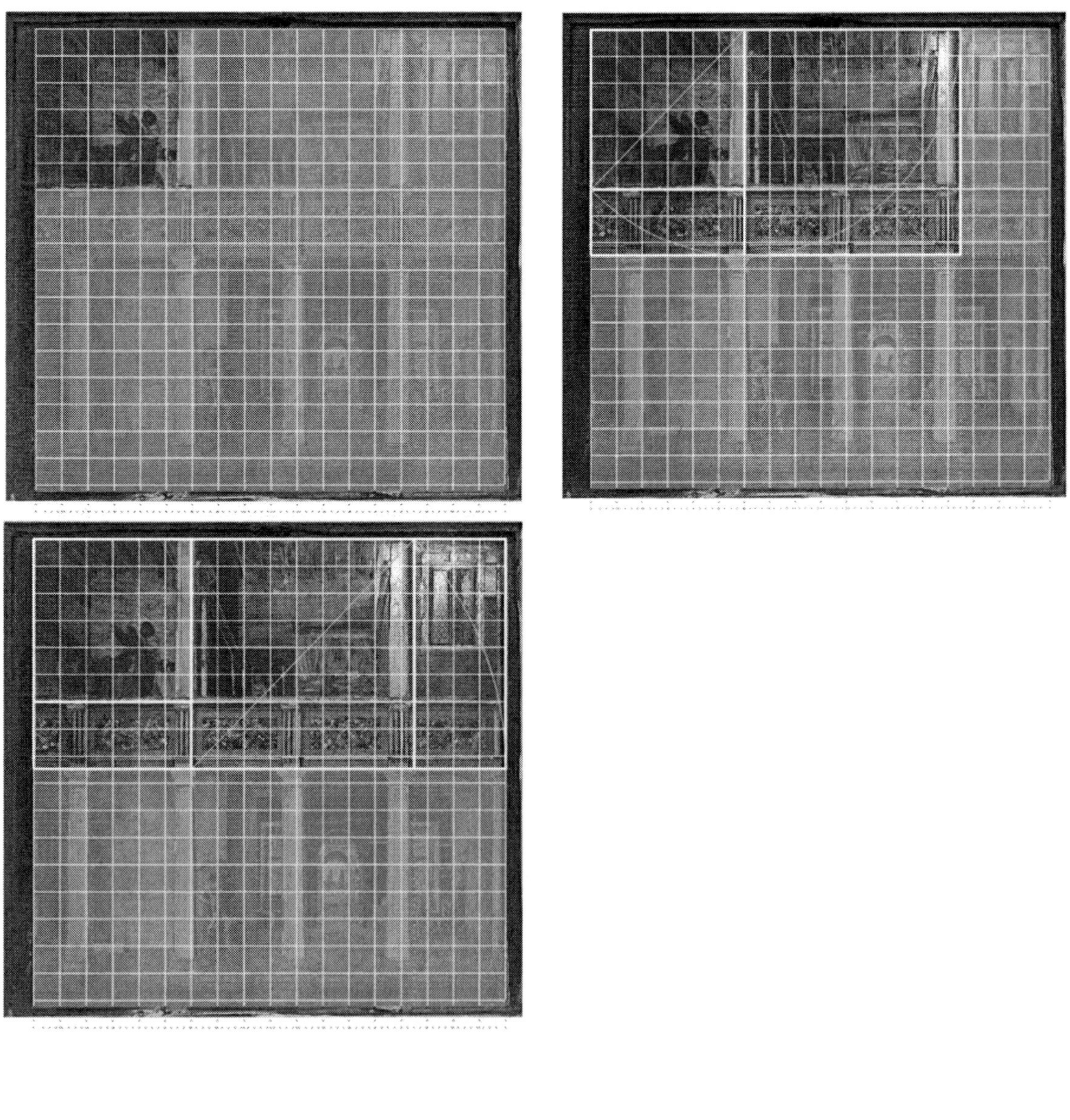

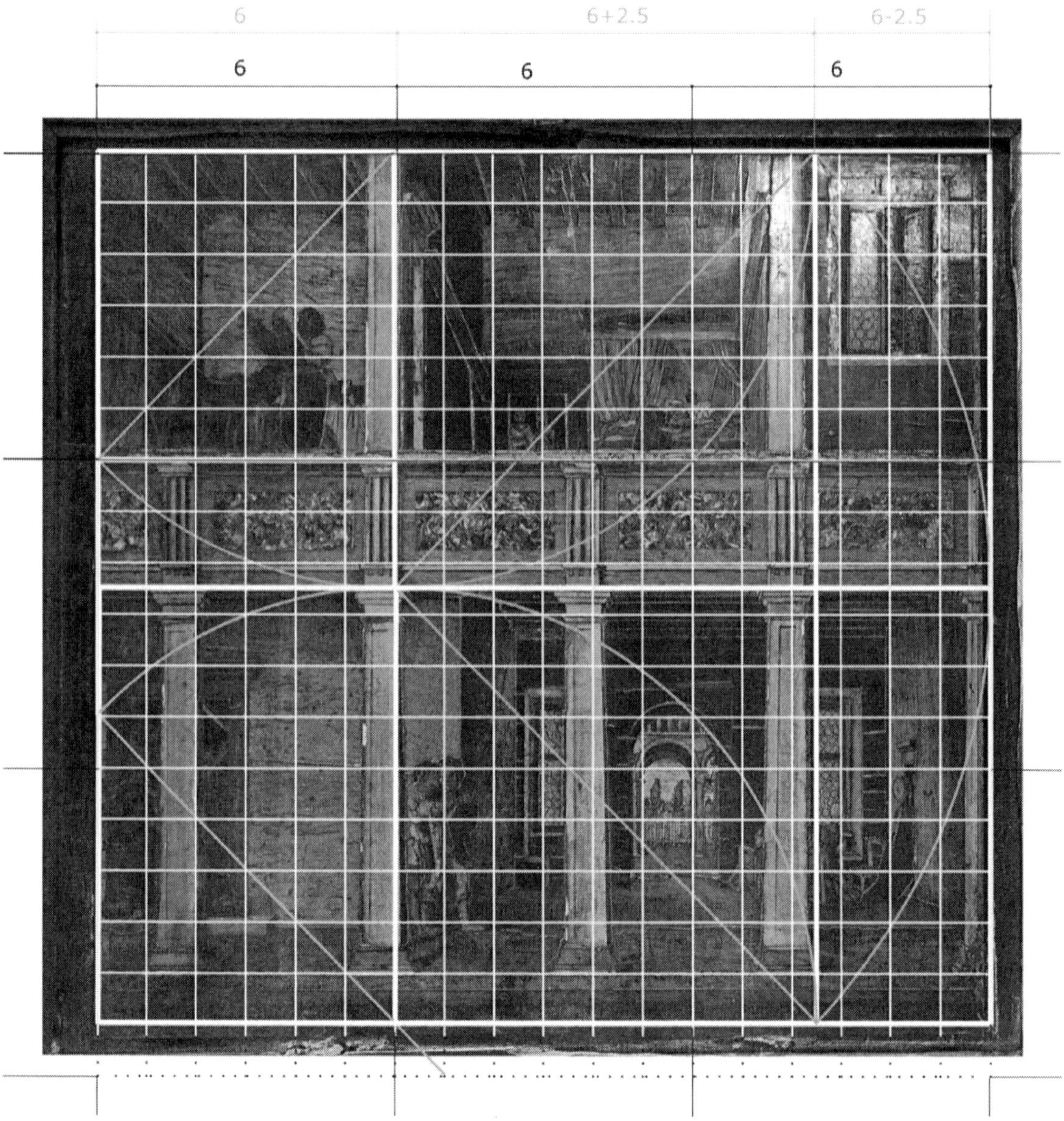

Fig. 6. Dimensional relationships among the different narrative units

According to the usual classification it is a central perspective. The convergence of the upper floor beams, together with that of the walls on the lower level, identify the vanishing point at top of the gate, visible outside of the courtyard with arcade.

The straight lines that coincide with the beams define, with good approximation, a regular module which divides the inlay in a grid of 17 units (about) in the vertically direction and 18 horizontally.

Some imperfections in the lower part of the artifact can be highlighted. The work, in fact, is actually larger than the picture frame, making plausible the hypothesis that the real size was of 18x18 units, later reduced in the installation phase. The inlay is 44 cm wide and 42.5 cm high. If one considers that *piede Bergamasco* unit, used in Bergamo in that period, corresponded to 43.7767 cm, it is reasonable to assume that the inlay has been dimensioned on the measurement of a foot wide by one foot long, while single units are squares of side 2,432 cm.

The grid obtained coincides with prospective plane that Lotto used to build the four narrative units: considering the rape scene corresponding to a 6x6 unit module it's possible to obtain through the diagonal tipping operations the inlay's whole structure.

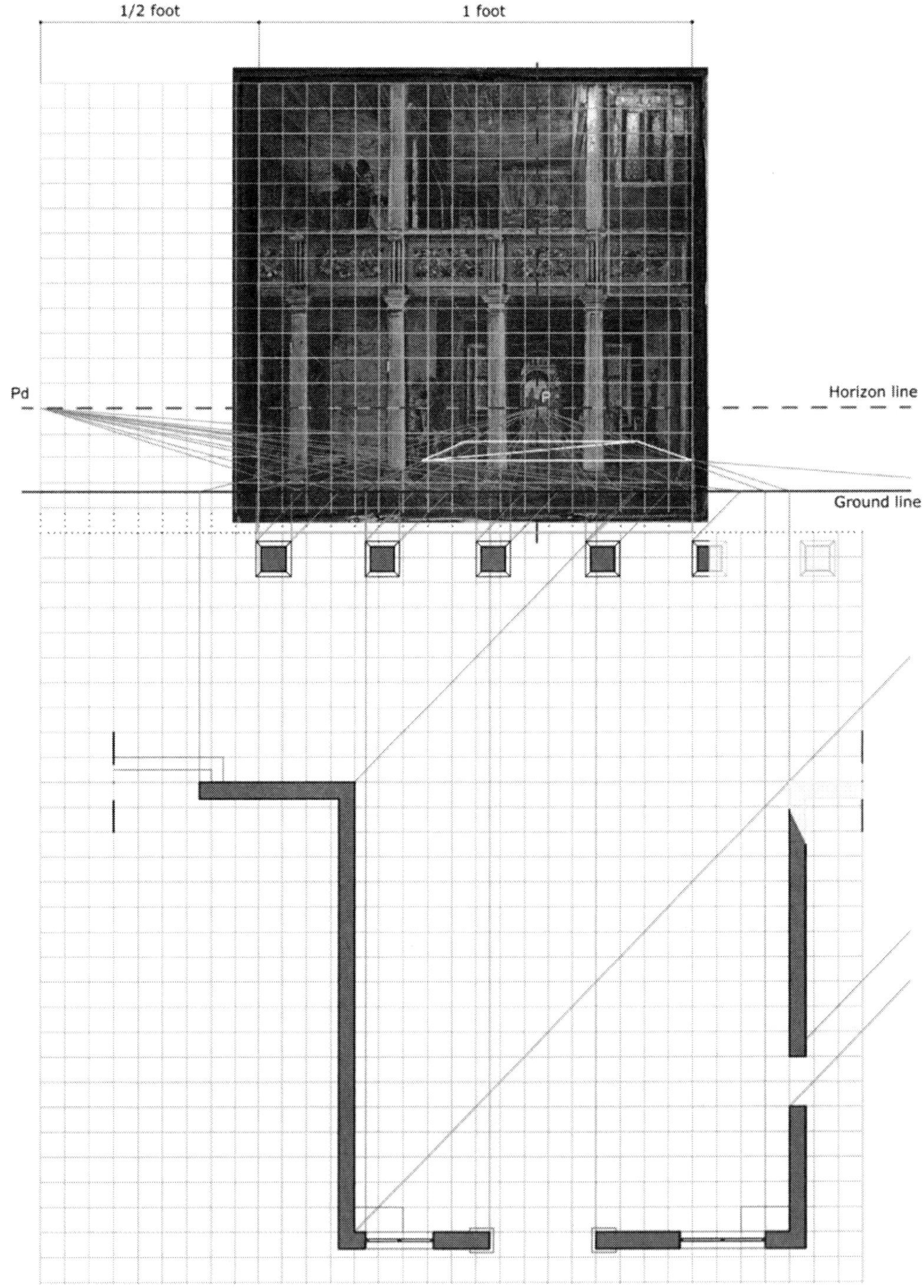

Fig. 7. Lay-out derived from the analysis of architecture represented in the painting

4 Conclusion

Studying the floor plan obtained by reconstruction of the represented space it is possible to identify some elements reminding to models of classicist Lombard culture: the wide porch with architraves, the adjacent courtyard with columns opening onto the garden, the spacious rooms upstairs are characteristic elements of the Milanese palaces of the period. Especially the use of a perspective closed by a garden, the presence of the side doors, the architraved ceiling and the pillar square section with beaded edges make reference to some illustrations of Cesariano[10] .

Nevertheless, the building designed by Lotto is characterized by slender columns to support heavy structures. It is clear that the author does not intend to represent real or possible architecture but symbolic buildings, with balconies and openings on the landscape.

To conclude one can say that in this case perspective is closely connected to the inlay features and allows you to chain and control the size of the shapes in a measured way.

The attention given by the choir makers at Lombard classicism is in line with the intention of representing in the choir inlays a realistic and modern setting, which would break away from the traditional Inlay themes. The narrative language is no longer the Platonic number, proportion and regular bodies, but psychological analysis of characters and of inhabited environment.

Fig. 8. Cavedia illustrated by Cesare Cesariano in a lithograph of the commentary on the De Architectura of Vitruvius (Como 1521)

Notes

[1] See BURATTI, G. *The use of perspective in inlay "Amon violante tamar" in Choir of Santa Maria Maggiore in Bergamo* in The reasons of Drawing: Thought, Shape and Model in the Complexity Management, edited by Bini M., Bertocci S., Gangemi editore, 2016

[2] "Father of our three arts", namely architecture, sculpture and painting. VASARI, G. Le Opere della Pittura cap.XV, Firenze 1832-1838

[3] Famous are inlays made by Maiano brothers for the studio of Federico da Montefeltro at Palazzo Ducale in Urbino executed from 1474 to 1476 and considered among the best creations in laical context. For details, see FERRETTI, M. *I maestri della prospettiva* in Storia dell'arte italiana, vol.11, 1982, pp. 518-523

[4] Giovan Francesco Capoferri (1497-1534) Carpenter master. Pupil of Fra Damiano Zambelli participated in the realization of San Domenico main chapel's benches, where he was probably noticed by Lotto who worked here at Pala Martinengo. Thirty-seven died without seeing fulfilled the choir, the main work of his life.

[5] Lorenzo Lotto (1480-1557). Painter among the principal representatives of the Venetian early sixteenth century Renaissance. He was a prolific artist, working mainly in areas considered peripheral to the great artistic centers, such as Bergamo and cities of Marche. In reality, during the period covered by this paper, Bergamo, the city further west in the Venetian possessions, was a place of confrontation between the Lombardy and Veneto experiences. A stimulating reality that Lotto was able to interpret creating works that are a successful synthesis of the currents, thanks to the local client, cultured and exigent, and relationships with artists and architects such as Correggio and Zenale.

[6] Lotto works with Consorzio della Misericordia since 1516, to build the Martinengo Pala realized for the Dominican church of Santo Stefano. The Pala, together with inlays made by Zambelli, is now preserved in Bergamo in the church of San Bartolomeo.

[7] FERRETTI, M. I maestri della prospettiva in Storia dell'arte italiana vol.11, 1982, pp. 459-480

[8] The postergale is the seatback of choir chairs and is usually characterized by a height greater than that required to support your back.

[9] Bible (2 Kings 13: 1-19)

[10] Cesare di Lorenzo Cesariano (1475–1543) was an Italian painter, architect and architectural theorist. He authored the first Italian-language version of Vitruvius' De architectura. Most of his activity was in Milan where he was reputed a good geometer and architect. For more information on Lot and the relationship with the Milanese culture see CORTESI BOSCO, F. Il coro intarsiato di Lotto e Capoferri, Edizioni Amilcare Pizzi, Bergamo, 1987, pag. 408

References

[1] ANDERSON, M., et al. (eds.) *Educational Perspectives on Mathematics as Semiosis: From Thinking to Interpreting to Knowing*, Legas, Ottawa, pp. 49–79.

[2] ARCANGELI, F. Tarsie, in Quaderni d`arte edited by Emilio Cecchi, Roma: Ed. Tumminelli,

[3] BERENSON, B. *Lorenzo Lotto*, edited by Vertova L., Milano: Abscondita, 2008.

[4] BRUSATIN, M. *La macchina come soggetto d'arte*, in Storia d'Italia. Torino: Ed. Einaudi, 1980.

[5] BURATTI, G. *The use of perspective in inlay "Amon violante tamar" in Choir of Santa Maria Maggiore in Bergamo* in The reasons of Drawing: Thought, Shape and Model in the Complexity Management, edited by Bini M., Bertocci S., Gangemi editore, 2016

[6] CALABRESE, O. *La macchina della pittura. Pratiche teoriche della rappresentazione figurativa fra Rinascimento e Barocco*. Firenze: La Casa Usher, 2012.

[7] CARLEVARO, G. *Materiale per lo studio di Bernardo Zenale*, in Arte Lombarda, n. 63,Milano, 1982.

[8] CHIODI, L. (edited by). *Lettere inedite di Lorenzo Lotto*. Bergamo: Centro culturale Nicolò Rezzara, 1998.

[9] CORTESI BOSCO, F. *Il coro intarsiato di Lotto e Capoferri*. Bergamo: Edizioni Amilcare Pizzi, 1987.

[10] DE MARCHI, A., GIANNATIEMPO LOPEZ, M. (edited by). *Il Quattrocento a Camerino, luce e prospettiva nel cuore della marca*, Milano: Federico Motta, 2002.

[11] FERRETTI, D. *I maestri della prospettiva* in Storia dell'arte italiana vol.11 (Parte terza "situazioni momenti indagini", edited by Federico Zeri, vol. IV) Forme e modelli. Torino: Editore Enaudi, 1982

[12] GATTI, S. *L'attività milanese del Cesariano dal 1512 al 1519* in Arte lombarda, n.16, Milano 1971.

[13] GONZÁLEZ-PALACIOS, A. *Mobili d'arte*. Milano: Fabbri Editore, 197.

[14] PINETTI, A. *Cronistoria artistica di S. Maria Maggiore, V, Il coro ligneo di Gianfrancesco Capoferri*. Bergamo, 1926.

[15] TREVISA, L. (edited by). *Tarsie lignee del rinascimento in Italia*. Vicenza: Sassi editore, 2011.

Current address

Buratti Giorgio, Phd-Adjunct professor
Scuola del Design, Politecnico di Milano
Via Giuseppe Candiani, 72 Milano, Italy
Tel. number: 02 2399 9699, e-mail: giorgio.buratti@polimi.it

Slovak University of Technology in Bratislava
Faculty of Mechanical Engineering
16th Conference on Applied Mathematics **APLIMAT 2017**
Proceedings

CLAUDE FAYETTE BRAGDON, ARCHITECT. THE FOURTH DIMENSION EXPOSED

CAPANNA Alessandra (IT)

Abstract. At the beginning of the 20th century, in the United States of America, some treatises about the diffusion of scientific knowledge, about the notion of space and time and on the fourth dimension, were published. The authors were mathematicians only in some cases.
Claude Fayette Bragdon (1866-1946) [1], a valued architect and stage designer, quoted in the renowned book of American Architectural History "The Architecture of America: A Social and Cultural History" (John Burchard and Albert Bush-Brown, 1961), published a chain of essays, books and articles about the higher space geometries and related visualizations. The present paper investigates the connections and consequences among the issues that concern the understanding and construction of space in four dimensions and its application in Bragdon's architectural projects.

Keywords: Architecture, fourth dimension, magic squares, magic cubes

Mathematics Subject Classification: Primary 00A67, 00A66, 00A65; Secondary 97M80

1 Introduction. Why architectural space should have only three dimensions?

The perception of the physical space in architecture is experiential. It means that, although the sense of sight is able to capture the geometric, figurative and aesthetic characters, ultimately the harmony of the shapes, the quality of the built environment is perceived with all the senses, in a dynamic approach, as a continuous sequence of space-events and at the same time in separate instants, as a cognitive act.

To quote Immanuel Kant, space is "a pure form of sensuous intuition", and performing architectural space is something that can be perceived all at once, at an intellectual level, but always practicing it materially, in an empirical way, as an absolute necessity for our human nature.

A primer to the topic of architecture related to the higher dimensional space has to ponder about matters discussed within the phenomenological tradition, including the nature itself of perception, involving the so-called time-consciousness and self-consciousness, the awareness of the body and the consciousness of "others" and, in a way, could face some transcendental problems. First of all the problem of "pure knowledge", that actually in Kant's *Transcendental Aesthetic* [2], [3], [4][1] is introduced by his analysis of cognition and by the concept itself of space and time, dealing with the way we acquire awareness of the objects.

In fact, at the basis of phenomenology there is the idea that any everyday object can become a knowledge engine. In this context, Architecture, as a constructed space inside which we live, represents the epitome of everyday life, and "lived space" represents the "object" to discover and to understand, either as a formal continuum in terms of geometry and physics, and like a sort of external theater, within which human beings are spectators and actors. In so doing, the center of the interest in modern architecture is no longer representative but in the hunt of atmospheres, no longer in search of an external figurative harmony or even perfection, but in a space-time experience in which memories and actions take place.

> *"In a fragment of a second you can understand: Things you know, things you don't know, things you don't know that you don't know, conscious, unconscious, things which in a fragrant of a second you can react to: we can all imagine why this capacity was given to us as human beings. Architecture to me has the same kind of capacity. It takes longer to capture, but the essence to me is the same. I call this* atmosphere. *When you experience a building and it gets to you, it sticks in your memory and your feelings"*[2]

It is time, therefore, together with memory, to become the main component of the architectural experience, to become a physical measure as well, inseparably tied to the spatial dimension.

One of the most important historians of Modern Architecture, Sigfried Giedion, considers architecture as a set of "phenomena of transition" [8], in which the present time is correlated with the historical development of the past. Therefore, he sets out three spatial conceptions. To summarize: "plastic and corporeal" is the first, from Greek Architecture, "close and compact" the second, characterizing ancient Roman Architecture and the third, as a fusion of the first two, belongs to Modern Architecture.

Starting from *Space, Time, Architecture* (1941) [7], to *The Eternal Present. The Beginning of Art* (1957) [8] and *The Eternal Present. The Beginning of Architecture* (1957) [9][3], up to *Architecture and the Phenomena of Transition* (1971)[4] [10], the philosophical assumption of Giedion's space-time triad introduces and establishes an analysis of the organic space formation-transformation process, not only in a stylistic sense, but also as sort of character of the formative creative process itself. Moreover, the analysis of the way of perception in space and in time leads to deepen questions of phenomenological type and on multiple space and multisensory, which are directly related to the concept of higher dimension itself.

---

[1] I. Kant *Prolegomena* with enclose editorial explanations by Jonathan Bennet (2004) [3].

[2] Interview with Peter Zumthor, Pritzker Prize-winning in 2009, *Seven Personal Observations on Presence In Architecture* [6].

[3] In the field of the Mellon Lectures at the Washingtonian National Gallery of Art, Giedion presented two lectures entitled *The Eternal Present*, the first about art [8] and the second about architecture [9]. The common subtitle "a Contribution on Constancy and Change" well defines the topic of the lectures.

[4] Giedion's last work, published posthumously, examines the fundamentals of modern construction, through the changings of the characteristics of architectural space in the different eves of the past.

The individual is in physical place and at the same time in his personal place of memory; he perceives the place in the instant, in continuity with the instant before, aware of being already in the next one.

This is a typical question intersecting the *general problem of setting and solving* in architecture, emblematic in music as well [11]. We are in an in-between[5] place, we are here and elsewhere.
In order to make some examples: where do we are when we listen to the music? How do pedestrians crowding narrow sidewalks not collide without "taking measurements"?

Those are the questions: whether space really possessed these properties or if it is our mind that owned this intuitive capacity and just for this reason could conceive space-time and many dimensions and at the same time, even if common people are not able to, succeed in rendering (at least formulating) a true visualization of 4d ambiances.

The phenomenological matter would require a long dissertation. Following a renewed interest in this field, Steven Holl[6] in 2000 published a book entitled *Parallax* [12] which presents 15 projects as representative of the consequences in Architecture of the achievements of the scientific knowledge. The Parallax is both the fourth dimension of the space-time and that of the human experience itself; it is also the fourth dimension of the elementary particles of light (let's say photons or quanta of light), whose scientific empathizing through the Holl's watercolor contribution at the design process has been so important; it is, above all, the purely geometric astronomical parallax, which needs an angular measurement to calculate a dynamic distance. In this field too, the issue of visualization and light[7] as a material of construction, provides a fundamental contribution to the understanding of architectural space.

Claude Fayette Bragdon (1866-1946), whose design method derived from the geometrical development of architectural elements conceived in a higher dimensional space, is the subject of the present study, anticipated the topic of visualization of the fourth dimension in architecture, fitting in the cultural debate of the time, that even in mathematicians' academic dissertations, frequently trespassed in dealing with cosmological matters.

2 Context. Charles Hinton, Henry Manning, Eric Neville, Cora Lenore Williams William Granville, Robert Reeves and the diffusion of scientific knowledge in the United States

In 1905, Einstein submitted the theory of Special Relativity and in 1915 that of General Relativity, making the concept of space-time and the fourth dimension in some way "official" although still not accessible to all.

[5] It is a sort of place of no-where, the space of the experience, in which the opposites: natural/artificial, indoor/outdoor, insight/outsight become physical place of the architectural experience.
[6] Steven Holl (1947 -) is an American architect, Alvaar Aalto medal in 2000, whose philosophical interests in Maurice Merleau-Ponty studies on the foundational role of perception in all the arts (from literature to architecture, up to politics, as the ancient art of oratory) leaded to a unique compositional method and distinguishable architectural language, and to the important collaboration with the Finnish architect Juhani Pallasmaa, author of the book "The eye of the Skin", a milestone in the New Phenomenology in Architecture. Holl previous interests in various implementation of topology in the conception of his architectures had a concrete realization, for example, in his "Experiments in porosity" (2005) and Open Architecture Sliced Porosity Block (Chengdu, China 2012).
[7] whose measurement is in terms of luminous flux in which space (distance) and time ("vectorial transportation" from the light source to the illuminated object), in so consisting as a matter in the fourth space-time dimension.

The history of a fascination goes back to the Greek philosophers and to the lost Ptolemy's *On Dimensionality*, but at the beginning of the 1900's, some eminent mathematicians published treatises on the fourth dimension, and on projective-geometry problems.
Some of them went quickly to literature with novels of science fiction that retraced the way open by Lewis Carroll and Edwin Abbott.

A still incomplete list of publications on this subject that has been a source of inspiration for Claude Bragdon, testified by the architect himself in his autobiography [28], is set forth here below:

- Alfred T. Schofield, *Another World; Or, The Fourth Dimension,* London 1888
- Charles Howard Hinton, *A New Era of Thought*, 1888
- Charles Howard Hinton, *The Fourth Dimension*, 1904
- Henry Edward Manning, *The Fourth Dimension simply explained,* 1910
- Henry Edward Manning, *Geometry of Four Dimensions,* 1914.
- Cora Lenore Williams *Creative Involution,* 1914
- Eric Harold Neville, *The Fourth Dimension,* 1921
- William Anthony Granville, *The Fourth Dimension and the Bible* 1922
- Robert Reeves, *Space and Fourth Dimension*, 1922

In particular Claude Bragdon payed explicit tribute to Charles Hinton's and Henry Manning's works [17][18][14][15].

This part of the research concerning the study of the space described in these treaties, is still ongoing.

3 Claude F. Bragdon's *A Primer of Higher Space* and other writings

Bragdon[8], who was an insatiable reader and a writer, just like his father George, while moving the first step in architecture joining the Art Nouveau movement, was also involved in the field of editorial activities. In fact he published a conspicuous number of articles and seventeen books on architectural theory, fourth dimension, esotericism, and even yoga. The reason of the interdisciplinary approach is that fourth dimension, in Bragdon opinion, is a new way to rehabilitate symbolism in architecture and cosmic consciousness achieved as a common language speaking "mathematics": a kind of theory of everything that fully explains the physical aspects of the architectural space as our human microcosms encapsulated in a higher-dimensioned universe.

"Nothing new", he wrote, "mathematics has long been made to serve man's use" [27, p. 72]. What is new in the fall of the new century is that since the advent of non-Euclidean geometry the field of mathematics has been enlarged, and thanks to the theory of relativity, popularized. "Space-time, curved space, the fourth dimension having become catchwords" [27 p. 72].

Shown below the synthesis of the seven Bragdon's books concerning the fourth dimension and the relationships with Architecture.

[8] A thorough biographical profile with detailed reports on Bragdon's buildings and professional and cultural activities can be read in Rochester History, October 1967 [15].

The Beautiful Necessity. Seven Essays on Theosophy and Architecture (1910) [21] is the first in which Bragdon introduced Mathematics as the common denominator of Nature, Music and Architecture. It consists of series of essays on architecture as "frozen music"[9]. Goethe's famous statement associated with the observation of the latent geometry of Nature and with the analysis of the arithmetic of beauty, became the *leitmotif* of his design research. In the late 19th-early 20th centuries Theosophy emerged in America due to the questioning of science and religion incongruity, finding a new point of view in the common search for a deeper meaning and new ways of creation and perception in the field of artistic production were introduced. Some American Historians stated that the development of modernity in America was coincident with an occultism that was paradoxically supported by various technological advances such as electricity, with its implication of invisible force enlightened according to the strong American Spiritualist interpretation of many scientists of that time. At the Columbian Exposition (1893), in fact, the popular use of electricity was presented, and pervasive. Theosophical atmosphere coupled science with religion, together with a sort of reconsideration of mental/spiritual relationships and a redefinition of space/time concepts.

Bragdon was active in the Theosophical Society in America in continuity with his father's interests, whose collection of books on Buddism, Induism, ancient Eastern Religions, as well as literary works by important Transcendentalist authors such as Shakespeare, Carlyle, Ruskin, Thoreau the young Claude attended with curiosity.

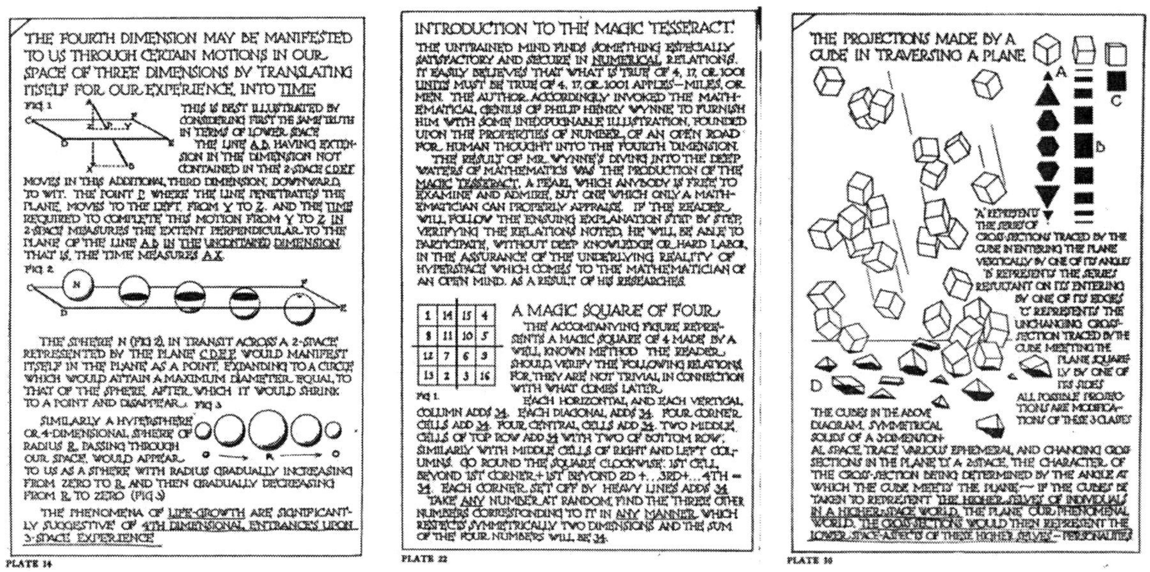

Fig. 1. Illustrated plates from the set of thirty, in Bragdon's *A Primer of Higher Space*

In this contest, Bragdon's studies and writings were born. Geometries, starting from the cube – on one hand – and from the organic metaphor – on the other hand - became a chain of books on Architecture and higher space dimension once developed as the construction of architectural elements, once as a sort of revelation of the human capacities, through which the concept followed a transversal logic, with repetitions and variations.

9 Johann Wolfgang von Goethe, *Conversation with Eckermann* (March 23, 1829).

Man the square: a higher space parable (1912) [22] opens with an abstract from Kepler's Mysterium Cosmographicum introducing the parable of the Man depicted as "a three-dimensional projection of a higher-space unity"

> *Artful nature has given to the most perfect animal the same six limits as the cube has, most perfectly marked. ... Man himself is, as it were, a cube*[10]. Kepler

Like some contemporary treaties described in the previous paragraph, a sort of mystical truth corresponds to the fact that man is the cube. We can note in the illustrated plates in the book, interesting correspondences with Abbott *Flatlandia* graphic representation of vanishing volumes under plane horizons (fig.1, left) and with linguistic dramaturgical settings in an Ideal Heavenly City.

In the first edition of *A Primer of Higher Space* (1913) [23] the "parable" *The Man and the Cube* is re-edited, following 30 pages of graphic plates describing step by step the geometrical construction of the phenomenon of the fourth dimension. Of particular interest Plate 14 (see fig.1) entitled *The Fourth Dimension May Be Manifested To Us Through Certain Motions In Our Space Of Three Dimensions By Translating Itself For Our Experience Into Time* and Plate 22 *Introduction to the magic Tesseract* (fig.1, center), revealing the design method to derivate architectural patterns exposed in his last publication *The Frozen Fountain 1932 Being Essays on Architecture and the Art of Design in Space.* [27]

Bragdon annotates that, before dismissing the idea of fourth dimension as absurd just because we cannot have direct or sensory knowledge of it, we have to become aware of the analogy with astronomy, whose method is to proceed by hypothesis of the real presence of something that no one can see, but only suppose, and on the historical evidence of the developing of human knowledge, once convinced about the flatness of the earth and then forced to accept that human idea of space is partial and evolving, in the sense of the constant improving of knowledge.

The drawing opening the first chapter in *A Primer of Higher Space* represents the Man and the Hypercube, an identical image respect to Salvator Dalì's *Crucifixion.* Dalì utilized his theory of "nuclear mysticism" - a fusion of Catholicism, mathematics, and science, in a similar way to Bragdon's Theosophical approach - to create this interpretation of Christ's crucifixion in a geometric, multidimensional form.

---

[10] We can find the same quote in Hinton's *The Education of the Imagination* (1898).

Fig. 2. C.F. Bragdon, *A Primer of Higher Space*, 1913, drawing opening the first chapter: "The Fourth Dimension". Frontal view of a crucified man on a Tesseract

Fig. 3. Detail of the cover of C.F. Bragdon, *A Primer of Higher Space*, 1913

Fig. 4. Salvator Dalì, *Crucifixion* (Corpus Hypercubus), 1954, oil on canvas 194.3 cm × 123.8 cm

In *Projective ornament* (1915) [26] as well, Bragdon recommends geometry as an appropriately modern, scientific inducement to ornament, and explores the mystical mathematics of magic lines and magic squares, of tesseracts and hyperspheres, demonstrating their beauty and blaming the urban disconnect from the natural world.

In *Architecture and Democracy*[11] (1918) [25], Democracy is presented as condition of consciousness and in the chapter *Ornament from Mathematics. The fourth dimension* [25 pp. 104-120] Bragdon describes step by step the construction of the hypercube, so that each line is the genesis of patterns developed into ornamentations. We have to underline that, in his opinion, "four dimensional geometry yields numberless patterns whose beauty and interest could not possibly be impeached-patterns beyond the compass of the cleverest designer unacquainted with projective geometry" [25 p 117-118].

The book *Four Dimensional vistas* (1916) [24] is actually an essay on fourth dimension, without drawings or images. It begins with reference to the Theory of Relativity, to the overcoming of Euclide's axiom of the parallels and on the nature of curved space. Chapter by chapter the Fourth Dimension is exposed passing through questions of symmetry and order, chemistry, electric current, astronomy, gravitation and transcendental physics, and finally border on mysticism and transcendence.

He presents the fourth dimension as a mystical hyperspace that explained some of the mysteries of life and the after-life. Bragdon based his new ornamental style on a complex system of projecting and manipulating two-dimensional shapes such as squares and triangles into four-dimensional

[11] Collection and re-edition of essays that Bragdon published in important architectural reviews such as "The Architectural Review", "The Architectural Forum" and "The American Architect".

shapes such as tesseracts and pentahedroids[12], deeply explained in his last book, *The Frozen Fountain. Being Essays on Architecture and the Art of Design in Space* (1923) [27].

3.1 Magic Squares, Magic Cubes, Nature's Geometry and Brsgdon's Organic Architecture

Bragdon developed several methods of generating plan and façade settings, as well as ornaments, based on mathematical principles. In general, he derived his patterns either from the magic path of the magic squares and magic cubes, or referring his ideas on the platonic solids output, or even, as main topics of his figurative research on the hyperspace geometry of fourth dimensions. Mathematics is the common ground of these three source for architectural ornament drawings, because according to him "mathematical truth is absolute within its own limits" [27, pag. 73] and, following Keat's dictum "Beauty is Truth; Truth, Beauty", he had only to find some method to translate mathematics as "truth-to-the-mind" into architectural "beauty-to-the-eye".

By means of drawings with handmade texts he described his method of designing, previously exposed in *A Primer of Higher Space* (1913) [23], then in 1915 in the book *Projective Ornament* [26], and finally, in 1932, in one of his very last publications, *The Frozen Fountain* [27], with direct reference to Henry Parker Manning's *Geometry of Fourth Dimensions* [15], W.S. Andrews's *Magic Squares an Cubes* (1908), E. Jouffret's *Traité élémentaire de géométrie à quatre dimensions* (1903), and Matila Ghyka's *Le Nombre d'or. Le Rythmes* (1931)[13].

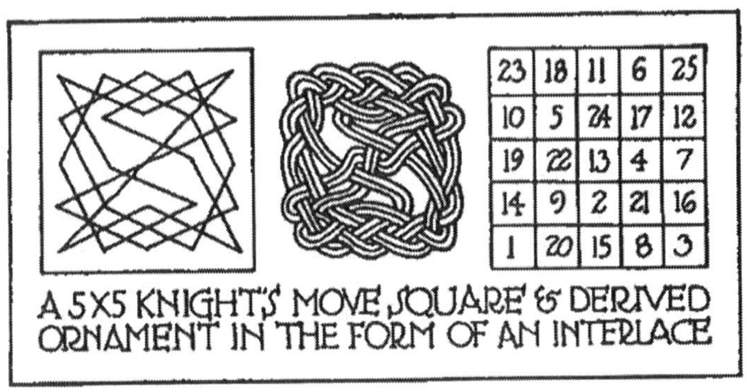

Fig. 5a – 5b. Genesis and final drawing of the ventilation grille in Rochester Chamber of Commerce, in Bragdon *Frozen Fountain* pag 76

Fig. 6. A Celtic interlace derived from a magic square, of 5, in Bragdon *Frozen Fountain* pag 79

The book publishes numerous examples of stained glass, furniture, architectural elements, in which he tested this method, which consists in tracing the magic lines generated by the conjunction of the magic square numbers, in orderly sequence from 1 to X. The starting case study is the 3x3 magic square, both in the example of straight lines connections and in the example of curved lines (see fig.5). Particularly interesting are the variants derived from magic squares 5x5 e 8x8 (fig.6, fig.7.). We can follow the construction of different patterns generated as examples in eight different magic squares of 5x5, following cell to cell the numbers in their natural order, or tracking the lines

[12] The Pentahedroid is the 5-cell as described in H.P. Manning *The fourth dimension simply explained*, [14, pp. 33-34].

[13] *Le Nombre d'or* was published just a year before the *Frozen Fountain*.

connecting odd numbers and even numbers, or with different intervals, as in the example with an interval of four.

> *The three mathematical sources from which I was able to derive ornaments were magic paths in magic squares, the Platonic solids, and the diagrammatic representations of regular hypersolids of four-dimensional space. My first experiments* – he wrote - *were with magic squares because they constitute such a conspicuous instance of the intrinsic harmony of number – of mathematical truth.* [27 p. 74]

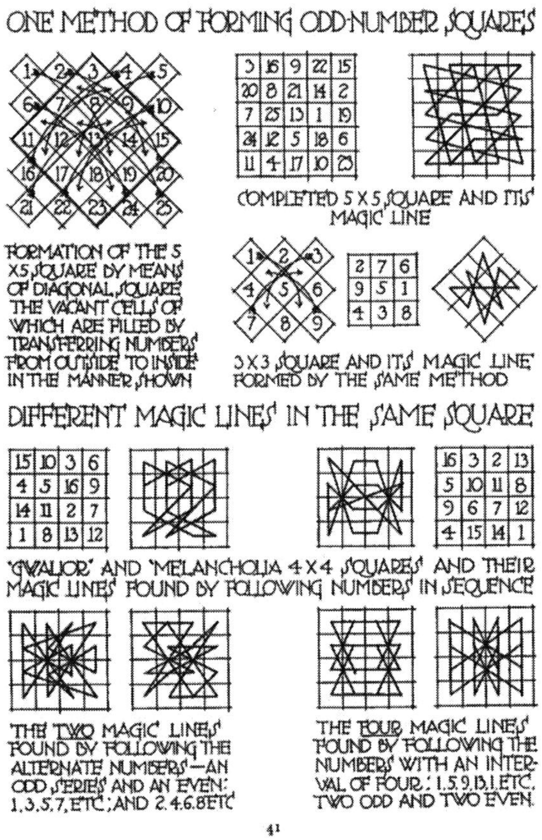

Fig. 7. Magic lines in magic squares in Bragdon *Frozen Fountain* pag 75

Fig. 8. Example of the use of textile patterns derived from 5x5 magic squares, in Bragdon *Frozen Fountain* pag 86

Moreover, Bragdon promoted regular geometry and musical proportion as ways for architects to harmonize buildings in the interior space and with their urban context.

> *The New York Central Station was by far my most important commission. I was anxious--over-anxious-- to make the building an illustrative example of architectural art. This was the cause of a false start: I had in mind to build a monument to the railroad, to the city, and incidentally to myself, (...) On a piece of paper I drew five equal, aligned tangential circles, like the driving-wheels of a locomotive engine. Of these the two end ones defined the height and width of the office divisions at right and left of the waiting-room, and the three remaining circles circumscribed the great round-arched windows which gave it light. The station was built in just that way.* [28 pp. 163-164]

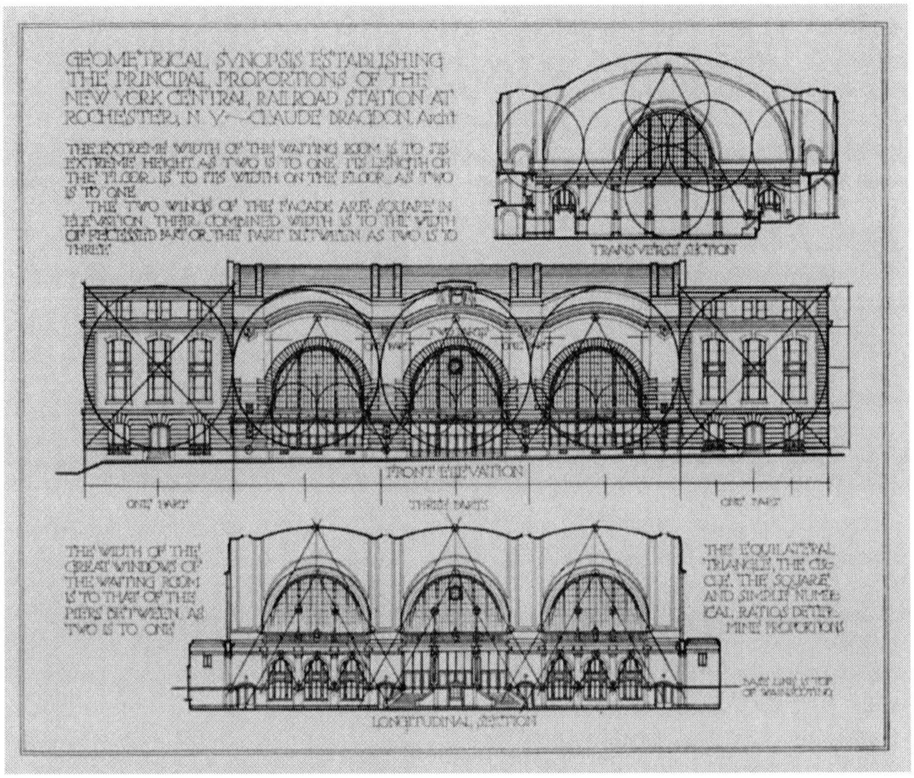

Fig. 9. New York Central Railroad Terminal, Rochester. Diagram of the harmonic proportions
https://www.tumblr.com/search/claude%20fayette%20bragdon
http://ronisolomondds.tumblr.com/post/20310105290/new-york-centrail-railroad-station-rochester-ny

The First Universalist Church (1907), New York Central Railroad Terminal (1909 – 1913) (Fig. 10), and Rochester Chamber of Commerce (1915 – 1917) exemplified his distinctive synthesis of Chicago School organicism[14] with harmonic proportion, number symbolism, and Renaissance architecture.

In the Central Railroad Terminal building and in Rochester Chamber of Commerce as well, Bragdon applied the mathematical common denominator exposed in *The Frozen Fountain,* with particular reference to music. In the Central Station, the waiting room become a "resonant chamber" when a friend of him who actually was an Opera singer sang in the empty station, before the official opening, a diatonic scale, just as a test. Witnesses of the event referred that the entire building seemed "to cry aloud" [20 pag.12].

> *I made use of* (…) *those numerical ratio subsisting between the consonant intervals within the octave – namely: 1:2, the octave; 2:3, the fifth; 3;4, the fourth; 4:5, the major third; and 4:7, the subminor seventh. The waiting room is twice as wide as it is high and*

[14] Some of the distinguishing features of the Chicago School in Architecture are the use of steel-frame structures with masonry cladding (usually terra cotta), allowing large plate-glass window and limiting the amount of exterior decorations. Sometimes light elements of neoclassical architecture were used to reduce the effect of toughness and heaviness of the first tall buildings of the American History of Architecture whose character of regularity was the consequence of squared geometries and window grids which are the essence of American Architecture as we actually know it.

twice as long as it is wide – the interval of the octave; or, ... the proportions of a root-four rectangle of Dynamic Symmetry. [20 pag.11]

Fig 10. Claude Bragdon, Rochester Grand Central Train Station, 1914. Demolished in 1978

Fig. 11. DeWoff Partnership Architects, The study of Bragdon's Rochester Grand Central Train Station submitted to the exhibition *The Arts & Cultural Council for Greater Rochester Architectural Design Showcase* as an example of Bragdon's use of regular geometry and musical proportion, connecting art and mathematics, to harmonize buildings with their urban context

3.2 Festivals of Song and Lights. Sound and light in four-dimensional settings

In 1916 Bragdon began a collaboration with the Community music reformers and the Song and Light Festival of the State of New York organizers, to whom he proposed projective ornaments, exposed in his books of theories, as the main component for outdoor settings of the Festival, as a universal ornamental language applicable to all manner of designs.

Bragdon staged Festivals of Song and Light in four different cities between 1915 and 1918. In 1916 it took place in New York, at Central Park. It was one of eight staged, at the peak of Progressive Era reform activism. These nocturnal community chorus festivals materialized also in Rochester, Buffalo and Syracuse.

A central role for theatrical lighting as an aid to access the dimension "beyond the tyranny of appearances", had ornamental lamps and decorations derived from the axonometric drawing method. Spotlights and screen, a system of color combination, together with music event, performed a four-dimensional cosmic vision where he believed all past and present consciousness could merge, through an empathic involvement that gave to the participants the opportunity to project themselves into the performing space.

Lit by electricity, the space-time matter of the century, Bragdon's projective ornament designs created an otherworldly, weird, ethereal outdoor environment likened to a "cathedral without walls."

Fig. 12. C.F. Bragdon, Setting for sound, light screens, lanterns, and song, at Central Park, New York City, September 1916. Lakefront stage

In the Song and Light Festival at Central Park, adjacent to the Bow Bridge and nearby the Bethesda Terrace, he installed a temporary wooden stage with a podium at its center, bore a circular shield, and tall pylons at either side supported large hexagonal lanterns; similar lanterns hanging from trees. A sixty-five-piece orchestra seated on the platform.

Above, stretched cables carried an array of circular and rectangular shields decorated with unusual geometric patterns: kaleidoscopic visionary appearances in the form of colored lights evoking the Indian Mandala. The patterns originated from the center point of the circle in the square with four gates of the original ancient Mandala, generate interlocking figures – symbolic representation of cosmos – with remarkable similarity to the bi-dimensional graphic image of the 5d regular polytope.

4 Conclusions

The study of Claude F. Bragdon's writings and works is part of a research on fourth dimension in Architecture within which I am analyzing the treaties of the beginning of the twentieth century, that faced for the first time the problem of scientific popularization and the discussion on the visualization and the representation of higher order geometries.

"No one is able to show the fourth dimension, yet it surrounds us – is written on the cover lap of Rudy Rucker's *The Fourth Dimension: A Guided Tour of the Higher Universes* – (…) It is food for thought in many disciplines".

Bragdon's work is aimed to do just that: to make visible the invisible and make it visible in terms of beauty. Just like in the past, when artists – whether they was painters, musicians, architects – searched for a formula for beauty results, exploring the magic of Nature through the discovering of its secret harmony. Bragdon's research on the higher dimensions was deep founded into the exploration of the growth of consciousness and on the magic of geometry which carried him into unexplored path in a long-life journey through the fourth dimension.

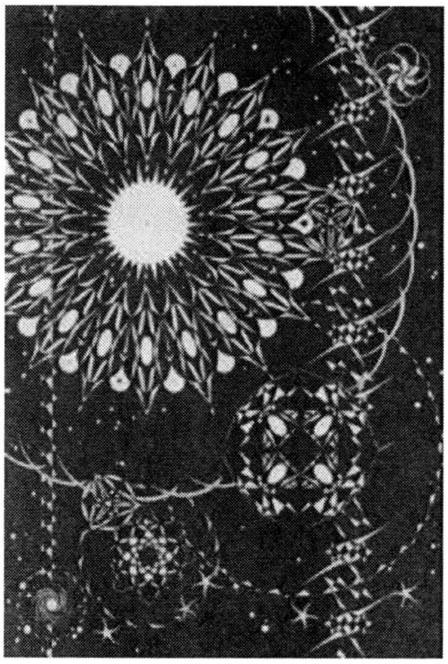

Fig. 13. C.F. Bragdon, harmonic composition in projective ornament. Design for Decorative Lighting, 1920-1929 India Ink on thin board 16 1/2 in. x 11 5/8 in. (41.91 cm x 29.53 cm) part of the Memorial Art Gallery's collection at the Ever Better's University of Rochester. Also in *The Frozen Fountain* frontispiece

Fig. 14. C.F. Bragdon, 'Study incomplementaries'. *MathematicalAbstraction* No.5 (ca. 1940) Watercolor, charcoal, ink and pen with red ink 28 3/8 in. x 20 5/16 in. (72.07 cm x 51.59 cm), sheet

Fig. 15. C.F. Bragdon, 'The sun by day and the moon by night'. *MathematicalAbstraction* No.14 (ca. 1939-1941) Watercolor and graphite on four ply board 36 in. x 24 in. (91.44 cm x 60.96 cm) *Mathematical Abstraction* are a set 15 of watercolor "Episodes from an Unwritten History: Claude Bragdon and Fritz Trautmann", part of the Memorial Art Gallery's collection at the University of Rochester

References

[1] ELLIS, E.V., REITHMAYR, A., *Claude Bragdon and the Beautiful Necessity,* RIT Cary Graphic Arts Press, Rochester 2010.

[2] KANT, I., *Critica della Ragion Pura*, Laterza, Roma-Bari 2000, (G. Gentile, G. Lombardo-Radice first Italian translation Laterza 1944; first Deutsch ed.: *Kritik der reinen Vernunft*, Riga: J. F. Hartknoch 1781).

[3] KANT, I., *Prolegomeni ad ogni metafisica futura che vorrà presentarsi come scienza*, (P. Carabellese Italian translation, rev. R. Assunto, intr. H. Hohenegger), Laterza, Roma-Bari

2007; (first Deutsch ed *Prolegomena zu einer jeden künftigen Metaphysik die als Wissenschaft wird auftreten können,* Riga: J. F. Hartknoch 1783).

[4] BENNETT, J., *Prolegomena [= Preliminaries] to any Future Metaphysic that can Present itself as a Science. Immanuel Kant,* Copyright ©2010–2015. http://www.earlymoderntexts.com/assets/pdfs/kant1783.pdf (First launched: July 2004. Last amended: April 2007) .

[5] KAZIG, R., *Presentation of Hermann Schmitz' paper, "Atmospheric Spaces"*, in *Ambiances* International Journal of sensory environment, architecture and urban space, http://ambiances.revues.org/709.

[6] ZUMTHOR, P., *Seven Personal Observations on Presence In Architecture,* http://www.archdaily.com/452513/peter-zumthor-seven-personal-observations-on-presence-in-architecture.

[7] GIEDION, S., *Space, Time and Architecture: The Growth of a New Tradition*, Harvard University Press 1941.

[8] GIEDION, S., *The Eternal Present. The Beginning of Art*, Bollingen Series xxxv. 6, I, New York, Pantheon Books 1957.

[9] GIEDION, S., *The Eternal Present. The Beginning of Architecture*, Bollingen Series xxxv. 6, II, New York, Pantheon Books 1957.

[10] GIEDION, S., *Architecture and the Phenomena of Transition. The Three Space Conceptions in Architecture*, Harvard University Press1971.

[11] VIZZARDELLI, S., *Spazi dell'ascolto e nuova estetica fenomenologica*, http://www.treccani.it/enciclopedia/spazi-dell-ascolto-e-nuova-estetica-fenomenologica_(XXI-Secolo)/, 2009.

[12] HOLL, S., *Parallax*, Princeton Architectural Press, New York 2000.

[13] NEVILLE, E.H., *The Fourth Dimension,* Cambridge at the University Press, 1921

[14] MANNING, H.P., *The Fourth Dimension simply explained,* Munn & Co, New York 1910

[15] MANNING, H.P., *Geometry of Four Dimensions,* The Macmillan Company, New York 1914.

[16] SCHOFIELD,A. T., *Another World; Or, The Fourth Dimension,* Swan Sonnenschein & Co., London 1888

[17] HINTON, C.H., *A New Era of Thought,* Swan Sonnenschein & Co., London 1888

[18] HINTON, C.H., *The Fourth Dimension,* Swan Sonnenschein & Co., New York 1904

[19] WILLIAMS, C.L., *Creative Involution,* Alfred A. Knopf, New York 1914

[20] COSTA, E., *Claude F. Bragdon, Architect, Stage Designer, and Mystic*, in *Rochester History*, vol. XXIX October 1967, No.4.

[21] BRAGDON, C.F. *The Beautiful Necessity. Seven Essays on Theosophy and Architecture*, Manas Press, Rochester 1910.

[22] BRAGDON, C.F. *Man the square: a higher space parable*, Manas Press, Rochester 1912.

[23] BRAGDON, C.F. *A Primer of Higher Space (The Fourth Dimension),* Manas Press, Rochester 1913.

[24] BRAGDON, C.F. *Four Dimensional vistas,* Alfred A. Knopf, New York 1916.

[25] BRAGDON, C.F. *Architecture and Democracy,* Alfred A. Knopf, New York 1918.

[26] BRAGDON, C.F. *Projective ornament,* Manas Press, Rochester 1915; reprinted Alfred A. Knopf, New York 1927.

[27] BRAGDON, C.F. *The Frozen Fountain. Being Essays on Architecture and the Art of Design in Space,* Alfred A. Knopf, New York 1932.

[28] BRAGDON, C.F. *More Lives Than One*, Alfred A. Knopf, New York 1938.

[29] ELLIS, E.V., *Geomantic (Re)Creation: Magic Squares And Claude Bragdon's Theosophic Architecture*, in Nexus V: Architecture and Mathematics, ed. Kim Williams and Francisco Delgado Cepeda, Fucecchio (Florence): Kim Williams Books, 2004. pp. 79-92 http://www.nexusjournal.com/conferences/N2004-Ellis.html.

[30] MASSEY, J., *Organic Architecture and Direct Democracy: Claude Bragdon's Festivals of Song and Light*, in *Journal of the Society of Architectural Historians*, Vol. 65 No. 4, Dec., 2006, (pp. 578-613) DOI:10.2307/25068329.

Current Address

Capanna Alessandra, Architect Ph.D, Researcher - Assistant Professor
Dipartimento di Architettura e Progetto - University of Rome, "La Sapienza"
Via Flaminia 359, 00196 Roma, Italy
Tel. number: +39 06 32101244, e-mail: alessandra.capanna@uniroma1.it

Slovak University of Technology in Bratislava
Faculty of Mechanical Engineering
16th Conference on Applied Mathematics **APLIMAT 2017**
Proceedings

ON GAUSSIAN APPROXIMATION FOR THE STATIONARY SIS MODEL WITH RANDOM CONTACT RATE

CARKOVS Jevgeņijs (LV), GOLDŠTEINE Jolanta (LV)

Abstract. The paper deals with a mathematical model of the classical dynamic epidemic susceptible-infected-susceptible (SIS) model for the population community given by a scalar differential equation with dependent on an impulse type compound Poisson process of large intensity. Applying the stochastic asymptotic approximation procedure we deduced the approximative stochastic differential equations for the steady stay stationary process and discussed the probabilistic characteristics of the infected population dynamics.

Keywords: Epidemic model, diffusion approximation

Mathematics Subject Classification: Primary 60H10, 34C12; Secondary 37H10, 92D30

1 The Markov SIS model

In mathematical biology the classical dynamic epidemic susceptible-infected-susceptible (SIS) model for the population community of size N is usually expressed [3] by differential equation for the amount of the infectious individuals $I(t)$

$$\frac{d}{dt}I = \frac{\beta}{N}(N-I)I-(b+\gamma)I \tag{1.1}$$

and the amount $S(t)=N-I(t)$ for the susceptible individuals. This model supposes that a susceptible individual with intensity rate $\beta>0$ comes in contact with an infected individual, becomes infected and infectious, and after recovery with rate $\gamma>0$ the individual returns to the susceptible community. The parameter $b\geq 0$ is the birth rate for the susceptible individuals' community. To discuss the population dynamics we can use the time scaling $t=(b+\gamma)^{-1}s$ and turn into well-behaved equation for the normalized infected individuals amount $x(s)=N^{-1}I((b+\gamma)^{-1}s)$:

$$\frac{d}{ds}x = R(1-x)x-x \tag{1.2}$$

with only one coefficient $R=\dfrac{\beta}{b+\gamma}$ called *the basic reproduction number* [3]. It is not so difficult to be sure that the equation (1.2) has two stationary solutions $x_1=0$ and $x_2=1-R^{-1}=\dfrac{\beta-b-\gamma}{\beta}$, which define the asymptotic of the defined by equation (1.1) the infectious individuals *I(t)* amount:

$$\lim_{t\to\infty} I(t)=\begin{cases}0, & \text{if } \beta\le b+\gamma,\\ N\dfrac{\beta-b-\gamma}{\beta}, & \text{if } \beta>b+\gamma\end{cases} \tag{1.3}$$

In reality the parameters of the SIS model vary in random manner and therefore the model (1.1) is in need of stochastic generalization, which has been done by many authors (see the survey [8]). Our proposal stochastic SIS model supposes a random pulsating character of the intensity rate of infection be given in a form of function $\beta=\beta(\xi(t))$, where $\{\xi(t),t\in\mathbf{R}\}$ is the right continuous stationary Poisson process with values in the segment $[0,1]$ with the weak infinitesimal operator [2]:

$$v\in \mathrm{C}(0,1):\quad (Qv)(y)=\lambda\int_0^1[v(z)-v(y)]dz\,. \tag{1.4}$$

This means that we can define such a right continuous filtration [10] $\{\Omega,\mathfrak{F},\mathfrak{F}^t,t\in\mathbf{R}\}$ that the process $\{\xi(t),t\in\mathbf{R}\}$ is $\mathfrak{F}^t$-adopted and defined as follows: after some time moment $t_0\in\mathbf{R}$ up to time moment $t_0+\tau$, where τ is nondependent on $\mathfrak{F}^{t_0}$ exponentially distributed random variable with parameter λ. This process is the $\mathfrak{F}^{t_0}$-measurable random variable: $\forall s\in[t_0,t_0+\tau):\xi(s)=\xi(t_0)$, and at the time moment $t_0+\tau$ this process has the jump from $\lim_{s\uparrow\tau}\xi(t_0+s):=\xi(t_0+\tau-)=\xi(t_0)$ to $\xi(t_0+\tau)=\varsigma$, where ς is nondependent on τ and $\mathfrak{F}^{t_0}$ uniform $\mathrm{R}(0,1)$ distributed random variable. Under the above assumptions the basic reproduction number of this random SIS model is also the pulsating process $R(\xi(t))=\dfrac{\beta(\xi(t))}{b+\gamma}$. Now taking into account the above assumptions we have to rewrite the equation (1.1) in a following form:

$$\frac{d}{dt}I=\frac{\beta(\xi(t))}{N}(N-I)I-(b+\gamma)I \tag{1.5}$$

and after substitution

$$I(t)=N^{-1}X^{-1}((b+\gamma)t) \tag{1.6}$$

write the linear differential equation for the random process $X(t)$:

$$\frac{d}{dt}X(t)=(1-R(\eta(t))X(t)+R(\eta(t)) \tag{1.7}$$

where $\eta(t)=\xi((b+\gamma)^{-1}t)$ is compound Poisson process defined by the weak infinitesimal operator

$$v\in C(0,1):\quad (Qv)(y)=\frac{\lambda}{b+\gamma}\int_0^1[v(z)-v(y)]dz\,. \tag{1.8}$$

The operator Q has two eigenvalues: $\lambda_1=0$ and $\lambda_2=-\frac{b+\gamma}{\lambda}<0$. Therefore [2] the random process $\{\eta(t),t\in\mathbf{R}\}$ possesses an ergodic property and for any continuous function $v\in C(0,1)$ and any initial value $\eta(t_0)=\eta_0$ we can use the formulae:

$$\lim_{t\to\infty}\mathrm{E}\{v(\eta(t)/\eta(t_0)=\eta_0\}=\bar{v}:=\int_0^1 v(x)dx,\ \mathrm{P}\left(\lim_{t\to\infty}\frac{1}{t}\int_{t_0}^t v(\eta(s)ds=\bar{v}/\eta(t_0)=\eta_0\right)=1\,. \tag{1.9}$$

If $\bar{R}=\lim_{t\to\infty}\mathrm{E}\{R(\eta(t)\}=\int_0^1 R(x)dx=(b+\gamma)^{-1}\int_0^1\beta(x)dx>1$ then almost sure there exists a limit $\lim_{t\to\infty}\exp\left\{\int_{t_0}^t(1-R(\eta(s))ds\right\}=\lim_{t\to\infty}\exp\left\{\left[\left[\frac{1}{t}\int_{t_0}^t(1-R(\eta(s))ds\right]t\right]\right\}=\lim_{t\to\infty}\exp\left\{(1-\bar{R})t\right\}=0$ and any solution of the corresponding to (1.7) the linear homogeneous differential equation

$$\frac{d}{dt}\tilde{X}(t)=(1-R(\eta(t))\tilde{X}(t) \tag{1.10}$$

almost sure exponentially tends to zero. Therefore [5] the inequality $\bar{R}>1$ guarantees existence of the stationary solution $\{\hat{X}(t),t\in\mathbf{R}\}$ of the equation (1.7) and for any solution $X(t,t_0,X_0)$ with the initial condition $X(t_0)=X_0$ of equation (1.7) we can be sure that $\mathrm{P}\{\lim_{t\to\infty}(X(t,t_0,X_0)-\hat{X}(t))=0\}=1$. The solution $X(t,t_0,X_0)$ may be presented as a sum of two terms:

$$X(t,t_0,X_0)=X_0\exp\left\{\int_{t_0}^t(1-R(\eta(s))ds\right\}+\int_{t_0}^t\exp\left\{\int_s^t(1-R(\eta(u))du\right\}R(\eta(s))ds\,. \tag{1.11}$$

If $\bar{R}>1$ and $t_0\to-\infty$ then almost sure the first term in (1.11) tends to zero with probability one and therefore the stationary solution of the equation (1.7) may be written in a form of the improper integral

$$\hat{X}(t)=\lim_{t_0\to\infty}\int_{t_0}^t\exp\left\{\int_s^t(1-R(\eta(u))du\right\}R(\eta(s))ds=\int_{-\infty}^t\exp\left\{\int_s^t(1-R(\eta(u))du\right\}R(\eta(s))ds\,. \tag{1.12}$$

Therefore if $\bar{R} > 1$ then the distribution of the amount of the infectious individuals *I(t)* with the course of time stabilizes and may be approximated by the distribution of the random variable

$$\hat{I} = N^{-1}\left(\hat{X}(0)\right)^{-1} = N^{-1}\left(\int_{-\infty}^{0} \exp\left\{\int_{s}^{0}(1 - R(\eta(u))du\right\}R(\eta(s))ds\right)^{-1}. \quad (1.13)$$

Unfortunately we cannot find the formula for the distribution of the above random variable. At the next paragraph we will prove that for sufficiently large intensity λ of switching Markov process one can approximate the stationary distribution (1.13) using the Gaussian approximation. In the third paragraph of this paper applying a simulation technique we will discuss the dependence of the distribution of the random variable (1.13) on the parameters of model.

2 The Gaussian approximation for stationary distribution

In this section we suppose that the susceptible individual intensity rate is a very fast changing random process $\beta(\xi(\varepsilon^{-1}t))$, where ε is a small positive parameter. The supplied with this parameter amount of the infectious individuals satisfies the differential equation:

$$\frac{d}{dt}I_\varepsilon(t) = \frac{\beta(\xi(\varepsilon^{-1}t))}{N}(N - I_\varepsilon(t))I_\varepsilon(t) - (b+\gamma)I_\varepsilon(t). \quad (2.1)$$

By the definition the pair $\{I_\varepsilon(t), \xi(\varepsilon^{-1}t), t \in \mathbf{R}\}$ is the homogeneous Markov process with a weak infinitesimal operator

$$\begin{aligned}\left(\mathcal{L}(\varepsilon)v\right)(x,y) &:= \lim_{t\downarrow 0}\frac{1}{t}\left[\mathrm{E}\{v(I_\varepsilon(t),\xi(\varepsilon^{-1}t))/I_\varepsilon(0)=x,\xi(0)=y\} - v(x,y)\right] = \\ &= \left[\frac{\beta(y)}{N}(N-x)x - (b+\gamma)x\right]\frac{\partial}{\partial x}v(x,y) + \frac{1}{\varepsilon}(Qv)(x,y)\end{aligned} \quad (2.2)$$

We will apply the well known stochastic approximation procedure [1] to analyze the above Markov dynamical system. The first step of this method is a construction of the averaged equation, which may be used for an approximate analysis of the infectious individuals' dynamics. It is easy to prove that the equation $Qu = -g$ in the space $C(0,1)$ has a solution if and only if $\bar{g} := \int_0^1 g(y)dy = 0$. Then we can find such continuous function $h \in C(0,1)$ that $(Qh)(y) = \bar{\beta} - \beta(y)$ and for any function $v \in C^2(\mathbf{R})$ construct the continuous function

$$V_\varepsilon(x,y) = v(x) + \varepsilon v_1(x,y) \quad (2.3)$$

where $v_1(x,y) = \frac{h(y)}{N}(N-x)x\frac{\partial}{\partial x}v(x)$. By definition the function $v_1(x,y)$ is continuous differentiable by *x* and therefore we may apply the operator $\mathcal{L}(\varepsilon)$ to the function V_ε:

$$\left(\mathcal{L}(\varepsilon)V_\varepsilon\right)(x,y)=\left(\mathcal{L}(\varepsilon)v\right)(x)+\varepsilon\left(\mathcal{L}(\varepsilon)v_1\right)(x,y)=$$

$$=\left[\frac{\beta(y)}{N}(N-x)x-(b+\gamma)x\right]\frac{\partial}{\partial x}[v(x)+\varepsilon v_1(x,y)]+(Qv_1)(x,y)= \tag{2.4}$$

$$=\left[\frac{\bar{\beta}}{N}(N-x)x-(b+\gamma)x\right]\frac{\partial}{\partial x}v(x)+\varepsilon\left[\frac{\beta(y)}{N}(N-x)x\right]\frac{\partial}{\partial x}v_1(x,y)$$

and to calculate a limit operator

$$\lim_{\varepsilon\to 0}\left(\mathcal{L}(\varepsilon)V_\varepsilon\right)(x,y)=\left(\bar{\mathcal{L}}v\right)(x)=\left[\frac{\bar{\beta}}{N}(N-x)x-(b+\gamma)x\right]\frac{\partial}{\partial x}v(x). \tag{2.5}$$

The operator $\bar{\mathcal{L}}$ defines the first order ordinary differential equation

$$\frac{d}{dt}\bar{I}(t)=\frac{\bar{\beta}}{N}(N-\bar{I}(t))\bar{I}(t)-(b+\gamma)\bar{I}(t). \tag{2.6}$$

If $\bar{I}(0)=I_\varepsilon(0)$ then [1]

$$\lim_{\varepsilon\to 0}\sup_{0\le t\le T}\mathrm{P}(|\bar{I}(t)-I_\varepsilon(t)|>C)=0 \tag{2.7}$$

for any $C>0$ and $T>0$. The next step of stochastic approximation procedure [1] is an analysis of the normalized deviations

$$z_\varepsilon(t)=\frac{I_\varepsilon(t)-\bar{I}(t)}{\sqrt{\varepsilon}}. \tag{2.8}$$

Taking into account the formulae (2.1) and (2.6) one can derive the differential equation for this random process:

$$\begin{aligned}
&\frac{d}{dt}z_\varepsilon(t)=\frac{1}{\sqrt{\varepsilon}}\left[\frac{d}{dt}I_\varepsilon(t)-\frac{d}{dt}\bar{I}(t)\right]=\\
&=\frac{1}{\sqrt{\varepsilon}}\frac{\beta(\xi(\varepsilon^{-1}t))}{N}(N-\bar{I}(t)-\sqrt{\varepsilon}z_\varepsilon(t))\left(\bar{I}(t)+\sqrt{\varepsilon}z_\varepsilon(t)\right)-\frac{1}{\sqrt{\varepsilon}}(b+\gamma)\left(\bar{I}(t)+\sqrt{\varepsilon}z_\varepsilon(t)\right)-\\
&-\frac{1}{\sqrt{\varepsilon}}\left[\frac{\bar{\beta}}{N}(N-\bar{I}(t))\bar{I}(t)-(b+\gamma)\bar{I}(t)\right]=\\
&=\frac{1}{\sqrt{\varepsilon}}\frac{\beta(\xi(\varepsilon^{-1}t))}{N}(N-\bar{I}(t)-\sqrt{\varepsilon}z_\varepsilon(t))\left(\bar{I}(t)+\sqrt{\varepsilon}z_\varepsilon(t)\right)-(b+\gamma)z_\varepsilon(t)-\\
&-\frac{1}{\sqrt{\varepsilon}}\left[\frac{\bar{\beta}}{N}(N-\bar{I}(t))\bar{I}(t)\right]=\frac{1}{\sqrt{\varepsilon}}\frac{\beta(\xi(\varepsilon^{-1}t))-\bar{\beta}}{N}(N-\bar{I}(t))\bar{I}(t)+\\
&+\left[\frac{\beta(\xi(\varepsilon^{-1}t))}{N}(N-2\bar{I}(t))-(b+\gamma)\right]z_\varepsilon(t)-\sqrt{\varepsilon}z_\varepsilon(t)\bar{I}(t)-\varepsilon\left(z_\varepsilon(t)\right)^2\\
&=\frac{1}{\sqrt{\varepsilon}}F_1(\bar{I}(t),\xi(\varepsilon^{-1}t))+F_2(z_\varepsilon(t),\bar{I}(t),\xi(\varepsilon^{-1}t))-\sqrt{\varepsilon}z_\varepsilon(t)\bar{I}(t)-\varepsilon\left(z_\varepsilon(t)\right)^2
\end{aligned} \tag{2.9}$$

where

$$F_1(r,y)=\frac{\beta(y)-\bar{\beta}}{N}(N-r)r, F_2(z,r,y)=\left(\frac{\beta(y)}{N}(N-2r)-(b+\gamma)\right)z. \qquad (2.10)$$

By construction the triple $\{z_\varepsilon(t),\bar{I}(t),\xi(t/\varepsilon),t\in\mathbf{R}\}$ defines the homogeneous Markov process on a phase space $\mathbf{R}\times\mathbf{R}\times[0,1]$ with a weak infinitesimal operator

$$\begin{aligned}&\left(\mathcal{D}(\varepsilon)u\right)(z,r,y)=\\&=\lim_{s\downarrow 0}\frac{1}{s}\left[\mathrm{E}\{u(z_\varepsilon(s),\bar{I}(s),\xi(\varepsilon^{-1}(s))/z_\varepsilon(0)=z,\bar{I}(0)=r,\xi(0)=y\}-u(z,r,y)\right]=\\&=\left[\frac{1}{\sqrt{\varepsilon}}F_1(r,y)+F_2(z,r,y)-\sqrt{\varepsilon}zr-\varepsilon z^2\right]\frac{\partial}{\partial z}v(z,r,y)+\\&+\left[\frac{\bar{\beta}}{N}(N-r)r-(b+\gamma)r\right]\frac{\partial}{\partial r}v(z,r,y)+\frac{1}{\varepsilon}(Qv)(z,r,y)\end{aligned} \qquad (2.11)$$

According to the stochastic approximation algorithm [1] we have to construct such a function

$$U_\varepsilon(z,r,y)=u(z,r)+\sqrt{\varepsilon}u_1(z,r,y)+\varepsilon u_2(z,r,y) \qquad (2.12)$$

that permits to calculate the limit expression

$$\lim_{\varepsilon\to 0}\left(\mathcal{D}(\varepsilon)U_\varepsilon\right)(z,r,y)=(\mathbf{L}u)(z,r) \qquad (2.13)$$

with diffusion type operator $\mathbf{L}$. So far as after substitution the expression (2.12) in (2.11) we will take the limit (2.13) the resulting formula for $\mathcal{D}(\varepsilon)U_\varepsilon(z,r,y)$ may be written accurate with the terms of the order $\sqrt{\varepsilon}$:

$$\begin{aligned}&(\mathcal{D}(\varepsilon)U_\varepsilon)(z,r,y)=(\mathcal{D}(\varepsilon)u)(z,r)+\sqrt{\varepsilon}\left(\mathcal{D}(\varepsilon)u_1\right)(z,r,y)+\varepsilon\left(\mathcal{D}(\varepsilon)u_2\right)(z,r,y)=\\&=\left[\frac{1}{\sqrt{\varepsilon}}F_1(r,y)+F_2(z,r,y)\right]\frac{\partial}{\partial z}u(z,r)+\left[\frac{\bar{\beta}}{N}(N-r)r-(b+\gamma)r\right]\frac{\partial}{\partial r}u(z,r)+F_1(r,y)\frac{\partial}{\partial z}u_1(z,r,y)+\\&+\frac{1}{\sqrt{\varepsilon}}(Qu_1)(z,r,y)+(Qu_2)(z,r,y)+\mathrm{O}(\sqrt{\varepsilon})=\\&=\frac{1}{\sqrt{\varepsilon}}\left[\frac{\beta(y)-\bar{\beta}}{N}(N-r)r\frac{\partial}{\partial z}u(z,r)+(Qu_1)(z,r,y)\right]+\left[\frac{\bar{\beta}}{N}(N-r)r-(b+\gamma)r\right]\frac{\partial}{\partial r}u(z,r)+\\&+\left(\frac{\beta(y)}{N}(N-2r)-(b+\gamma)\right)z\frac{\partial}{\partial z}u(z,r)+\frac{\beta(y)-\bar{\beta}}{N}(N-r)r\frac{\partial}{\partial z}u_1(z,r,y)+(Qu_2)(z,r,y)+\mathrm{O}(\sqrt{\varepsilon})\end{aligned}$$

By definition (1.8) if $\int_0^1 v(y)dy = 0$ then $(Qv)(y) = -v(y)$. This means that substituting $u_1(z,r,y) = \frac{\beta(y)-\bar{\beta}}{N}(N-r)r\frac{\partial}{\partial z}u(z,r)$ we made equal to zero the first singular term have to look for a limit of the expression

$$(\mathcal{D}(\varepsilon)U_\varepsilon)(z,r,y) = \left[\frac{\bar{\beta}}{N}(N-r)r-(b+\gamma)r\right]\frac{\partial}{\partial r}u(z,r)+\left(\frac{\beta(y)}{N}(N-2r)-(b+\gamma)\right)z\frac{\partial}{\partial z}u(z,r)+$$

$$+\left(\beta(y)-\bar{\beta}\right)^2\left(\frac{(N-r)r}{N}\right)^2\frac{\partial^2}{\partial z^2}u(z,r)+(Qu_2)(z,r,y)+\mathrm{O}(\sqrt{\varepsilon})$$

Now we can choose

$$u_2(z,r,y) = \left(\frac{\beta(y)-\bar{\beta}}{N}(N-2r)-(b+\gamma)\right)z\frac{\partial}{\partial z}u(z,r)+$$

$$+\left[\left(\beta(y)-\bar{\beta}\right)^2-\left(\int_0^1\left(\beta(y)-\bar{\beta}\right)^2dy\right)\right]\left(\frac{(N-r)r}{N}\right)^2\frac{\partial^2}{\partial z^2}u(z,r)$$

and pass to limit in (2.13):

$$\begin{aligned}\lim_{\varepsilon\to 0}(\mathcal{D}(\varepsilon)U_\varepsilon)(z,r,y) &= \left[\frac{\bar{\beta}}{N}(N-r)r-(b+\gamma)r\right]\frac{\partial}{\partial r}u(z,r)+\\ &+\left(\frac{\bar{\beta}}{N}(N-2r)-(b+\gamma)\right)z\frac{\partial}{\partial z}u(z,r)+\\ &+\left(\int_0^1\left(\beta(y)-\bar{\beta}\right)^2dy\right)\left(\frac{(N-r)r}{N}\right)^2\frac{\partial^2}{\partial z^2}u(z,r) := (\mathbf{L}u)(z,r)\end{aligned} \tag{2.14}$$

The operator $\mathbf{L}$ defined by this equality corresponds [2] to the system of two equations: the equation (2.6) and the stochastic differential equation

$$dz(t) = \left(\frac{\bar{\beta}}{N}(N-2\bar{I}(t))-(b+\gamma)\right)zdt+\sigma\frac{(N-\bar{I}(t))\bar{I}(t)}{N}dw(t) \tag{2.15}$$

where $\sigma^2 = \frac{1}{2}\left(\int_0^1\left(\beta(y)-\bar{\beta}\right)^2dy\right)$ and the standard Wiener process. Now for sufficiently small ε we can approximate the distributions of the infected individuals $I_\varepsilon(t)$ by distributions of the stochastic process

$$\forall t\in\mathbf{R}, \forall C\in\mathbf{R}: \quad \mathrm{P}(I_\varepsilon(t)\le C) = \mathrm{P}\left(z(t)\le\frac{C-\bar{I}(t)}{\sqrt{\varepsilon}}\right)+0(\varepsilon) \tag{2.16}$$

Applying the results of the first paragraph to the equation (2.6) we can assert that under assumption $\bar{\beta} > b+\gamma$ there exists unique asymptotically stable solution, that is:

$$\lim_{t\to\infty} \bar{I}(t) = \hat{I} = N\frac{\bar{\beta} - b - \gamma}{\bar{\beta}} = N(1 - \bar{R}^{-1}) \tag{2.17}$$

for any initial condition $\bar{I}(0)$. Substituting $\bar{I}(t) \equiv \hat{I}$ in the equation (2.16) we can analyze the random dynamics of defined by the equation (2.1) amount of the infectious individuals $I_{\varepsilon}(t)$ by an approximate formula:

$$\hat{I} \approx N(1 - \bar{R}^{-1}) + \sqrt{\varepsilon}\hat{z}(t) \tag{2.18}$$

where $\hat{z}(t)$ is the Gaussian process given by the solution of equation

$$d\hat{z}(t) = -\bar{\beta}(1 - \bar{R}^{-1})\hat{z}(t)dt + \sigma N\bar{R}^{-1}(1 - \bar{R}^{-1})dw(t) \tag{2.19}$$

with the initial condition $\hat{z}(0) = 0$.

3 Simulation study for the amount of the infectious individuals

In this paragraph we will analyze the dynamics of the amount of infectious individuals applying a simulation technique. By substitution $x(s) = N^{-1}I((b+\gamma)^{-1}s)$ in the equation (1.5) we may rewrite this equation in a form of a differential equation with only one random parameter

$$\frac{dx(s)}{dt} = (1 - x(s))x(s) - x(s) \tag{3.1}$$

where $R(\eta(s)) = (b+\gamma)^{-1}\beta(\eta(s))$, and $\eta(s) = \xi((b+\gamma)^{-1}s)$ is the Poisson process with infinitesimal operator

$$v \in \mathrm{C}(0,1): \quad (Qv)(y) = \frac{\lambda}{b+\gamma}\int_0^1 [v(z) - v(y)]dz\,. \tag{3.2}$$

At any time-interval $\tau_{k-1} < t < \tau_k$ the coefficient $R(\eta_{k-1})$ in the equation (3.1) is a random constant $R(\eta(t)) \equiv R(\eta_{k-1}), \eta_{k-1} \square R(0,1)$. Therefore by substitution $z(t) = (x(t))^{-1}$ we can move from (3.1) to an ordinary no homogeneous differential equation

$$\frac{d}{dt}z(t) = (R(\eta_{k-1}) - 1)z(t) + R(\eta_{k-1})\,. \tag{3.3}$$

Not so difficult to find the solution of this equation at the above time interval $\tau_{k-1} < t < \tau_k$

$$t \in (\tau_{k-1}, \tau_k): \quad z(t) = \left(z(\tau_{k-1}) - \frac{R(\eta_{k-1})}{R(\eta_{k-1}) - 1} \right) e^{-(t-\tau_{k-1})(R(\eta_{k-1})-1)} + \frac{R(\eta_{k-1})}{R(\eta_{k-1}) - 1}. \tag{3.4}$$

In order to simulate a solution of the equation (3.1) at any finite time interval (0,T) at first we have to find the values for the times of jumps by iterative formula

$$\tau_k = \tau_{k-1} + \delta_k, k = 1, 2, ... \tag{3.5}$$

for $k \le n(T) := \max\{k : \tau_k < T\}$ where $\{\delta_k, k = 1, 2, ...\}$ are independent exponentially distributed random values with parameter λ, to find the random values

$$k = 1, 2, ..., n(T): \quad z(\tau_{k-1}) = \left(z(\tau_{k-1}) - \frac{R(\eta_{k-1})}{R(\eta_{k-1}) - 1} \right) e^{-\delta_k (R(\eta_{k-1})-1)} + \frac{R(\eta_{k-1})}{R(\eta_{k-1}) - 1} \tag{3.6}$$

and finally to calculate the solution value $x(T)$ for the equation (3.1) by formula

$$x(T) = \left[\left(z(\tau_n) - \frac{R(\eta(\tau_n))}{R(\eta(\tau_n)) - 1} \right) \mathrm{e}^{-(R(\eta(\tau_n))-1)(T-\tau_n)} + \frac{R(\eta(\tau_n))}{R(\eta(\tau_n)) - 1} \right]^{-1}. \tag{3.7}$$

The MATLAB M-file performing 1000 simulations of $x(T)$ values and generating the histogram of these values is given in Fig. 1.

```
% SIS
clear all;
T = 10;                          % Final time
lambda = 2;                      % Positive parameter
N = 1000;                        % Repeat number
for j=1:N
    clear n; clear eta; clear delta, clear tau, clear z;
    tau(1)=0;                    % Initial time
    z(1) = 2;                    % Initial z value
    m = 50;
    eta = unifrnd(0,1,[m,1]);    % Unif(0,1)
    delta = exprnd(1/lambda,[m,1]); % Exp(lambda), time intervals
    % Definition of swithing times
    for k=1:m
        tau(k+1)=tau(k)+delta(k);
        if tau(k+1)>T
            n = k;
            break
        end
    end
    for i=1:n
        z(i+1)=z(i)*exp(-eta(i)*delta(i))+((1+eta(i))/eta(i))*(1-exp(-eta(i)*delta(i)));
        if i==n
            xT(j)=z(i+1)*exp(-eta(i)*(T-tau(i)))+((eta(i))^(-1))*(1+eta(i))*(1-exp(-eta(i)*(T-tau(i))));
        end
    end
end
hist(xT)
title('Histogram for x(T)')
```

Fig. 1. M-file for simulations of $x(T)$ values

Here we chose the time interval (0, 10), the initial z value equal to 2 and $\lambda = 2$. Afterwards we increased λ value to demonstrate how it impacts the $x(T)$ values. If λ value increases also m

value in the M-file should be increased in order to provide the possibility to cover all time interval with the switching times. To find the values for the times of jumps the exponential distributed random number generator was used. The obtained histogram in the case when $\lambda = 2$ is given in Fig. 2.

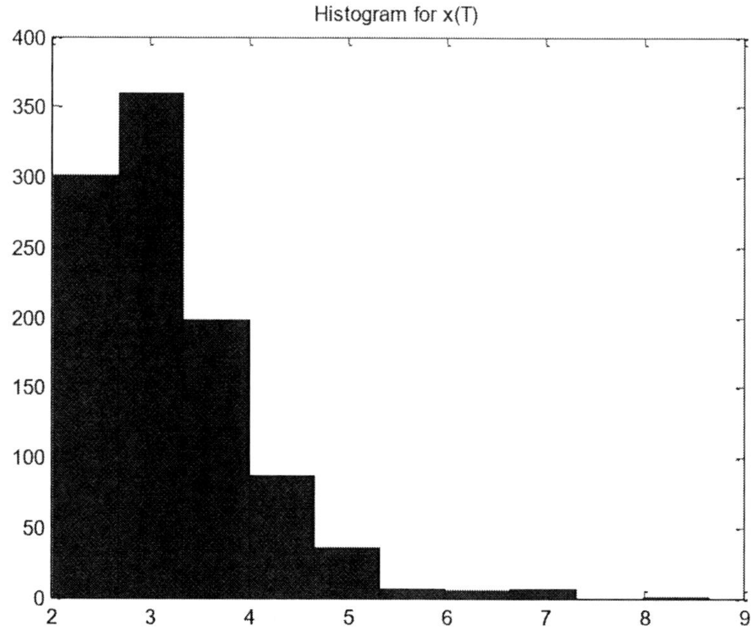

Fig. 2. Histogram of $x(T)$ values if $\lambda = 2$

If λ value is increased the resulting $x(T)$ values shows the different tendency. We analyzed in more details the case when $\lambda = 1000$. For demonstrating the data behavior there was fixed one sample of $x(T)$ values saved in MATLAB MAT-file. The MATLAB M-file used for constructing a histogram and testing for normality is showed in Fig. 3.

```
load('xT_values_lambda_1000.mat')
h = lillietest(xT)
test_cdf = makedist('Normal','mu',mean(xT),'sigma',std(xT))
[h,p] = kstest(xT,'CDF',test_cdf)
histfit(xT,31,'Normal')
title('Histogram for x(T) with a normal distribution fit')
```

Fig. 3. M-file for histogram and normality test ($\lambda = 1000$)

A Lilliefors test was used to test a decision for the null hypothesis that the obtained data $x(T)$ comes from a distribution in the normal family, against the alternative that it does not come from such a distribution. It gave the result $h = 0$, that means we cannot reject the null hypothesis at the 5% significance level. Next we used Kolmogorov-Smirnov test to test the hypothesis that these data comes from a normal distribution with sample mean and standard deviation. The obtained result affirmed that the data is normally distributed with the indicated parameters. p-value in this case was 0.8686, since small values of p cast doubt on the validity of the null hypothesis, we can be quite sure about the data normality. To visualize these results the data histogram with a normal distribution fit line was constructed, see Fig. 4.

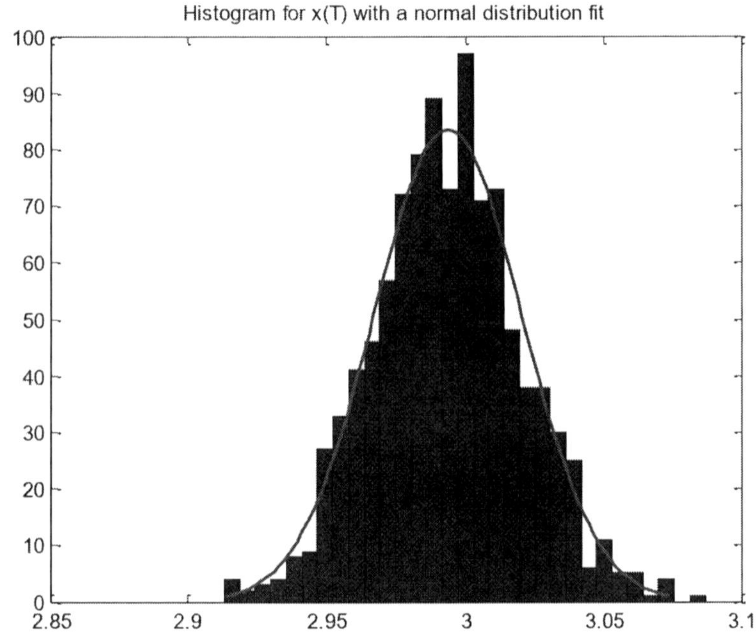

Fig. 4. Histogram of $x(T)$ values if $\lambda = 1000$

As we used 1000 values of $x(T)$, the number of bins was chose as square root of this number, that is, 31. This graph convincing confirms data normality for large λ values.

References

[1] SKOROKHOD, A. V. *Asymptotic methods in the theory of stochastic differential equations.* (Transl. Math. Monographs 78), American Mathematical Society, Providence, RI, 1989.

[2] DYNKIN, Ye. B. *Theory of Markov Processes.* Prentice-Hall, 1961.

[3] ANDERSON, R. M. *Population Dynamics of Infectious Diseases: Theory and Applications.* Chapman and Hall, London-New York, ed. 1982.

[4] ALLEN, LINDA J. S. *An introduction to stochastic epidemic models.* In *Mathematical Epidemiology*. Springer, 2008, s. 81–130.

[5] KHASMINSKY, R. Z. *Stochastic Stability of Differential Equations*. Springer, NY, 2012.

[6] BAILEY, N. T. J. *The Mathematical Theory of Infectious Diseases.* Second Edition, Grin, 1975.

[7] BAILEY, N. T. J. *Some Stochastic Models for Small Epidemics in Large Populations.* Journal of the Royal Statistical Society, Series C (Applied Statistics), 13 (1), 1964, s. 9–19.

[8] BRITTON, T. *Stochastic Epidemic Models: A Survey.* Mathematical Biosciences, 225(1), 2010, s. 24–35.

[9] GRAY, A., GREENHALGH, D., MAO, X., PAN, J. *The SIS Epidemic Model with Markovian Switching.* Journal of Mathematical Analysis and Applications, 394 (2), 2012, s. 496–516.

[10] SHIRYAEV, A. *Probability*. Volume 95 of Graduate Texts in Mathematics, Springer, NY, 2015.

Current address

Carkovs Jevgeņijs, Prof. Dr. hab. math.
Riga Technical University
1 Kalku street, Riga, LV-1658, Latvia
Tel. number: +371 67089517, e-mail: carkovs@latnet.lv

Goldšteine Jolanta, Dr. math.
Riga Technical University
1 Kalku street, Riga, LV-1658, Latvia
Tel. number: +371 67089517, e-mail: jolanta.goldsteine@rtu.lv

Slovak University of Technology in Bratislava
Faculty of Mechanical Engineering
16th Conference on Applied Mathematics **APLIMAT 2017**
Proceedings

ON DIFFUSION APPROXIMATION FOR THE VAN DER POL OSCILLATOR SUBJECT TO RANDOM IMPULSE PERTURBATION

CARKOVS Jevgenijs (LV), MATVEJEVS Andrejs (LV), PAVLENKO Oksana (LV)

Abstract. The paper deals with a mathematical model of the Van der Pol oscillator given by second order differential equation dependent on an impulse type compound Poisson process of large intensity. Applying the stochastic asymptotic approximation procedure we deduce the approximative stochastic differential equations for the phase and amplitude of vibrations. These equations permit us to analyze an existing capability and to describe the probabilistic characteristics of a stationary amplitude, and to discuss the behavior of a vibration phase.

Keywords: The Van der Pol oscillator, random oscillations, diffusion approximation

Mathematics Subject Classification: Primary 60H10, 60H30; Secondary 37H10

1 The model and the Skorokhod limit theorem

The oscillating type nonlinear systems of differential equations with a small parameter are very popular mathematical models for analysis of vibratory systems of the contemporary engineering (see [1] and references there). One of commonly encountered models is the well-known Van der Pol oscillator given by the second order differential equation:

$$\frac{d^2x}{ds^2} + \omega^2 x + \varepsilon f\left(\varepsilon, s, x, \frac{dx}{ds}\right) = 0 \tag{1}$$

where $f\left(\varepsilon, s, x, \frac{dx}{ds}\right)$ is a quasi-periodic function on s, ε is a small positive parameter and $f\left(\varepsilon, s, x, \frac{dx}{dt}\right)$ is analytical function on ε. It has long since been shown by Van der Pol, Lagrange, and other mathematicians that applied the different methods of slowly changing coefficients one can found the solution of the equation (1)-(2) in form of the function $x = a(s,\varepsilon)\, cos(\omega s + \varphi(s,\varepsilon))$ with close to constant amplitude $a(s,\varepsilon)$ an phase $\varphi(s,\varepsilon)$ of vibration. Later the famous Ukrainian mathematicians N.M.Krylov and N.N.Bogolybov proposed the very powerful algorithm, called *the averaging method*, [1] for a decomposition of the amplitude $a(s,\varepsilon)$ and phase $\varphi(s,\varepsilon)$ by the powers of a small parameter ε. Unfortunately if $f\left(\varepsilon, s, x, \frac{dx}{ds}\right)$ as a function on time is a random process this method does not work [3] because the amplitude $a(t,\varepsilon)$ and phase $\varphi(s,\varepsilon)$ are not sufficiently smooth functions of a parameter ε. Applying proposed by the averaging method substitution we can derive only the differential equations for the first approximation of the amplitude $a(t,\varepsilon)$ and phase $\varphi(s,\varepsilon)$. The second approximative term has an order $\sqrt{\varepsilon}$ and we have to look for this term applying the theory of stochastic differential equations [2]. We assume that auto oscillating system given by equation (1) with the nonlinearity

$$f\left(\varepsilon, s, x, \frac{dx}{ds}\right) = -2\delta\varepsilon\frac{dx}{ds}(1-x^2) - h(y(s))x \tag{2}$$

is subject to random impulse perturbations $h(y(s))$, where $y(t)$ is a compound Poisson process [4] with the infinitesimal operator $Qv(y) = \lambda\int_0^1[v(z) - v(y)]dz$. The realizations of this process are [4] piecewise constant functions with the uniform in $(0;1)$ mutually independent jumps values ξ_n at the time moments $\tau_n, n \in N$. The time intervals $\tau_n - \tau_{n-1}$ are also mutually independent and exponentially distributed with the parameter λ. We assume that $E\{h(\xi_n)\} = 0$ and $D\{h(\xi_n)\} = \gamma^2$. Applying the substitution $x(s) = R(s)cos\left(\omega s + \frac{\theta(s)}{2}\right)$ in the equation (1)-(2) we can write the equations

$$\frac{d\theta}{ds} = \varepsilon^2 2\delta sin(2\omega s + \theta)(0.5R^2(cos(2\omega s + \theta) + 1) - 1) + \frac{\varepsilon}{\omega}\big(1 + cos(2\omega s + \theta)\big)h(y(s))$$

$$\frac{dR}{ds} = \varepsilon^2 R\delta\big(1 - cos(2\omega s + \theta)\big)(0.5R^2(cos(2\omega s + \theta) + 1) - 1) + \frac{\varepsilon}{2\omega}Rh(y(s))sin(2\omega s + \theta)$$

for the radius $R(s)$ and for the phase $\theta(s)$ of the oscillations. According to the stochastic averaging procedure [5] we have to substitute $s = \frac{t}{\varepsilon^2}$ and rewrite the above equations in a singular form:

$$\dot{\psi} = \frac{1}{\varepsilon}\left[\frac{1}{\omega}\big(1 + cos(2\omega t + \psi)\big)h(y)\right] + 2\delta\big(0.5r^2\big(1 + cos(2\omega t + \psi)\big) - 1\big)sin(2\omega t + \psi) \tag{3}$$

$$\dot{r} = \frac{1}{\varepsilon}\left[\frac{1}{2\omega}rh(y)sin(2\omega t + \psi)\right] + r\delta\big(1 - cos(2\omega t + \psi)\big)\big(0.5r^2\big(1 + cos(2\omega t + \psi)\big) - 1\big) \tag{4}$$

where $(t) = R(\frac{t}{\varepsilon^2})$, ψ (t)= θ $(\frac{t}{\varepsilon^2})$. Not so difficult to prove that the triple $\{\psi(t), r(t), y(t)\}$ is the Markov process [4] with the weak infinitesimal operator

$$Q(\varepsilon) = \varepsilon^{-2}Q_0 + \varepsilon^{-1}Q_1 + Q_2 \tag{5}$$

where

$$Q_0V^\varepsilon = \left(\frac{\partial}{\partial t} + Q\right)V^\varepsilon,$$

$$Q_1V^\varepsilon = \left[\frac{1}{\omega}\big(1 + cos(2\omega t + \psi)\big)h(y)\right]\frac{\partial V^\varepsilon}{\partial\psi} + \left[\frac{1}{2\omega}rh(y)sin(2\omega t + \psi)\right]\frac{\partial V^\varepsilon}{\partial r}$$

$$Q_2V^\varepsilon = \left[\left[r\delta\big(1-cos(2\omega t+\psi)\big)\big(0.5r^2\big(1+cos(2\omega t+\psi)\big)-1\big)\right]\frac{\partial V^\varepsilon}{\partial r} + \left[2\delta\big(0.5r^2\big(1+cos(2\omega t+\psi)\big)-1\big)sin(2\omega t+\psi)\right]\frac{\partial V^\varepsilon}{\partial \psi}\right]$$

To construct an approximative equations for the defined by the dynamical system (3)-(4) Markov dynamical system we will be governed by the derived in [6] the stochastic averaging procedure. The algorithm of this procedure consists of the following procedure: for any sufficiently smooth on r and ψ function $V(r,\psi)$ we have to choose such functions $V_1(t,r,\psi,y)$ and $V_2(t,r,\psi,y)$ that for function

$$V^\varepsilon(t,r,\psi,y) = V(r,\psi) + \varepsilon V_1(t,r,\psi,y) + \varepsilon^2 V_2(t,r,\psi,y) \tag{6}$$

there exists a limit

$$\lim_{\varepsilon\to 0} Q(\varepsilon)\, V^\varepsilon(t,r,\psi,y) = LV(r,\psi) \tag{7}$$

where L is the diffusion operator:

$$L = a_1(r,\psi)\frac{\partial}{\partial r} + a_2(r,\psi)\frac{\partial}{\partial \psi} + \frac{1}{2}\left(b_{11}(r,\psi)\frac{\partial^2}{\partial r^2} + 2b_{12}(r,\psi)\frac{\partial^2}{\partial r\partial \psi} + b_{22}(r,\psi)\frac{\partial^2}{\partial \psi^2}\right)$$

Now we may approximate the distributions of the solutions of the equations (3)-(4) by the corresponding distributions of the solutions of the system of stochastic differential equations

$$d\tilde{r} = a_1(\tilde{r},\tilde{\psi})dt + \sigma_{11}(\tilde{r},\tilde{\psi})dw_1(t) + \sigma_{12}(\tilde{r},\tilde{\psi})dw_2(t)$$
$$d\tilde{\psi} = a_2(\tilde{r},\tilde{\psi})dt + \sigma_{12}(\tilde{r},\tilde{\psi})dw_1(t) + \sigma_{22}(\tilde{r},\tilde{\psi})dw_2(t)$$

where $w_1(t)$ and $w_2(t)$ are the independent standard Wiener processes and

$$\begin{pmatrix}\sigma_{11} & \sigma_{12}\\ \sigma_{12} & \sigma_{22}\end{pmatrix} = \begin{pmatrix}b_{11} & b_{12}\\ b_{12} & b_{22}\end{pmatrix}^2$$

As it has been proved in [6] we may be sure that

$$\max_{0\le t\le T} P(|\, r(t) - \tilde{r}(t)| + |\psi(t) - \tilde{\psi}(t)| > C) = O(\varepsilon)$$

for any T>0 and C>0 .

$$\left[\frac{1}{\omega}\big(1+cos(2\omega t+\psi)\big)h(y)\right]\frac{\partial V}{\partial \psi} + \left[\frac{1}{2\omega}rh(y)sin(2\omega t+\psi)\right]\frac{\partial V}{\partial r}$$

2 The diffusion approximation

The expression $Q(\varepsilon)\, V^{\varepsilon}(t,r,\psi,y)$ may be decomposed in a following way:

$$\begin{aligned}&\left(\varepsilon^{-2}Q_0+\varepsilon^{-1}Q_1+Q_2\right)\left(V(r,\psi)+\varepsilon V_1(t,y,r,\psi)+\varepsilon^2 V_2(t,y,r,\psi)\right)=\\&=\varepsilon^{-2}Q_0V(r,\psi)+\varepsilon^{-1}\left(Q_0V_1+Q_1V\right)+\left(Q_0V_2+Q_1V_1+Q_2V\right)+0(\varepsilon)\end{aligned} \tag{8}$$

This function of ε has a singularity at the point $\varepsilon=0$ bet we have choose a limit expression (7). Remember that the operator Q_0 is defined on the space of sufficiently smooth functions on $t\in\mathbf{R}$ and $y\in[0,1]$ by the equalities:

$$Q_0g(t,y):=\left(\frac{\partial}{\partial t}+Q\right)g(t,y),\ \ Qg(t,y):=\lambda\int_0^1[g(t,z)-g(t,y)]dz$$

Therefore the first singular term in (8) is equal to zero. To avoid a singularity of the second term in (8) we have to choose such a function $V_1(t,r,\psi,y)$ that $Q_0V_1=-Q_1V$. In conformity with a structure of the function $(Q_1V)(t,r,\psi,y)$ we will look a solution of the above equation in a form of a sum:

$$V_1(t,r,\psi,y)=V_{11}(r,\psi,y)+V_{12}(t,r,\psi,y)$$

where V_{11} and V_{12} satisfy the equations:

$$Q_0V_{11}(r,\psi,y)=-h(y)\frac{1}{\omega}\frac{\partial}{\partial\psi}V(r,\psi) \tag{9}$$

$$Q_0V_{12}(t,r,\psi,y)=-h(y)\frac{1}{\omega}\left(\frac{\partial V(r,\psi)}{\partial\psi}cos(2\omega t+\psi)+\frac{1}{2}r\frac{\partial V(r,\psi)}{\partial r}sin(2\omega t+\psi)\right) \tag{10}$$

Not so difficult to insure that therewith assumption $\overline{h(y)}:=\int_0^{\pi}h(z)dz$ a solution of the equation (9) may be chosen as follows:

$$V_{11}(r,\psi,y)=\frac{h(y)}{\lambda\omega}\frac{\partial}{\partial\psi}V(r,\psi) \tag{11}$$

To solve the equation (10) we may apply an undefined multiplier method and look for solution in a form of a sum:

$$V_{12}(t,r,\psi,y)=h(y)\big(A(r,\psi)cos(2\omega t+\psi)+B(r,\psi)sin(2\omega t+\psi)\big) \tag{12}$$

with unknown expressions $A(r,\psi)$ and $B(r,\psi)$. After a substitution (12) in equation (10) and an equating the coefficients near $sin(2\omega t+\psi)$ and $cos(2\omega t+\psi)$ we can find the above expression in a following form:

$$A(r,\psi) = \left(-\frac{2}{\lambda^2+4\omega^2}\frac{\partial}{\partial\psi} + \frac{1}{\lambda^2+4\omega^2}\frac{\lambda}{2\omega}r\frac{\partial}{\partial r}\right)V(r,\psi)$$
$$B(r,\psi) = \left(\frac{1}{\lambda^2+4\omega^2}\frac{\lambda}{\omega}\frac{\partial}{\partial\psi} + \frac{1}{\lambda^2+4\omega^2}r\frac{\partial}{\partial r}\right)V(r,\psi)$$

Finely we may choose

$$V_1(t,r,\psi,y) = \left\{\frac{h(y)}{\lambda}\frac{1}{\omega}\frac{\partial}{\partial\psi} - \frac{h(y)}{\lambda^2+4\omega^2}\frac{1}{\omega}\big(2\omega sin(2\omega t+\psi) - \lambda cos(2\omega t+\psi)\big)\frac{\partial}{\partial\psi} + \frac{h(y)}{\lambda^2+4\omega^2}\frac{1}{2\omega}r\big(\lambda sin(2\omega t+\psi) + 2\omega cos(2\omega t+\psi)\big)\frac{\partial}{\partial r}\right\}V(r,\psi) \tag{13}$$

and avoid a singularity in (8). Now by selection such a function $V_2(t,r,\psi,y)$ that

$$(Q_0V_2)(t,r,\psi,y) + (Q_1V_1)(t,r,\psi,y) + (Q_2V)(t,r,\psi,y) = F(r,\psi) \tag{14}$$

we can take an advantage of the opportunity for simplification of the expression $(Q_0V_2 + Q_1V_1 + Q_2V)$. Naturally we can do that if and only if the equation

$$(Q_0V_2)(t,r,\psi,y) = -\big((Q_1V_1)(t,r,\psi,y) + (Q_2V)(t,r,\psi,y)\big) + F(r,\psi) \tag{15}$$

has a solution. Now after substitution

$$F(r,\psi) = \overline{(Q_1V_1)(t,r,\psi,y) + (Q_2V)(t,r,\psi,y)} \tag{16}$$

with

$$Q_1V_1(t,r,\psi,y) = \left[\frac{1}{\omega}\big(1+cos(2\omega t+\psi)\big)h(y)\right]\frac{\partial}{\partial\psi}\left\{\left(\frac{h(y)}{\lambda}\frac{1}{\omega} - \frac{h(y)}{\lambda^2+4\omega^2}\frac{1}{\omega}\big(2\omega sin(2\omega t+\psi) - \lambda cos(2\omega t+\psi)\big)\right)\frac{\partial}{\partial\psi} + \frac{h(y)}{\lambda^2+4\omega^2}\frac{1}{2\omega}r\big(\lambda sin(2\omega t+\psi) + 2\omega cos(2\omega t+\psi)\big)\frac{\partial}{\partial r}\right\}V + \\ + \left[\frac{1}{2\omega}rh(y)sin(2\omega t+\psi)\right]\frac{\partial}{\partial r}\left\{\left(\frac{h(y)}{\lambda}\frac{1}{\omega} - \frac{h(y)}{\lambda^2+4\omega^2}\frac{1}{\omega}\big(2\omega sin(2\omega t+\psi) - \lambda cos(2\omega t+\psi)\big)\right)\frac{\partial}{\partial\psi} + \frac{h(y)}{\lambda^2+4\omega^2}\frac{1}{2\omega}r\big(\lambda sin(2\omega t+\psi) + 2\omega cos(2\omega t+\psi)\big)\frac{\partial}{\partial r}\right\}V(r,\psi)$$

and

$$(Q_2V)(t,r,\psi,y) = \left[r\delta\big(1-cos(2\omega t+\psi)\big)\big(0.5r^2(1+cos(2\omega t+\psi)) - 1\big)\right]\frac{\partial}{\partial r}V(r,\psi) + \left[2\delta\big(0.5r^2(1+cos(2\omega t+\psi)) - 1\big)sin(2\omega t+\psi)\right]\frac{\partial}{\partial\psi}V(r,\psi)$$

we can find the final expression for the desired limit (7) in an operator form:

$$(LV)(r,\psi) = F(r,\psi)=\left(\frac{3\lambda}{8\omega^2}\frac{\gamma^2}{\lambda^2+4\omega^2} - \delta + \frac{\delta}{4}r^2\right) r\frac{\partial}{\partial r}V + \frac{\lambda}{8\omega^2}\frac{\gamma^2}{\lambda^2+4\omega^2}r^2\frac{\partial^2 V}{\partial r^2} -$$

$$-\frac{1}{\omega}\frac{\gamma^2}{\lambda^2+4\omega^2}\frac{\partial V}{\partial\psi} + \frac{1}{\omega^2}\frac{\lambda^2+2}{2\lambda}\frac{\gamma^2}{\lambda^2+4\omega^2}\frac{\partial^2 V}{\partial\psi^2} \quad (17)$$

The diffusion operator L defines [4] the Markov process given by a system of the stochastic Ito differential equations:

$$dr=\left\{\frac{3\lambda}{8\omega^2}\frac{\gamma^2}{\lambda^2+4\omega^2} - \delta + \frac{\delta}{4}r^2\right\} rdt + \frac{\gamma\sqrt{\lambda}}{2\omega\sqrt{\lambda^2+4\omega^2}}rdw_1(t)\,, \quad (18)$$

$$d\psi=-\frac{1}{\omega}\frac{\gamma^2}{\lambda^2+4\omega^2}dt + \frac{\gamma\sqrt{\lambda^2+2}}{\omega\sqrt{\lambda(\lambda^2+4\omega^2)}}dw_2(t) \quad (19)$$

where $w_1(t), w_2(t)$ are the independent standard Brownian motion processes. This means that the processes $\{r(t), t \in \mathbf{R}\}$ and $\{\psi(t), t \in \mathbf{R}\}$ are independent and we may analyze the above equations separately.

3 The steady-state dynamics of the oscillator

The equation (19) defines the Gaussian process with the mean

$$m(t) = E\{\psi(t)\} = \psi(t_0) - \frac{1}{\omega}\frac{\gamma^2}{\lambda^2+4\omega^2}(t-t_0)$$

and the correlation function

$$K(t,s) = \frac{\gamma^2(\lambda^2+2)}{\omega^2\lambda(\lambda^2+4\omega^2)}\min\{t,s\}$$

To analyze the defined by the equation (18) dynamics of amplitude $r(t)$ we can make a substitution $z = r^{-2}$ and by Ito formula [2] write the linear no homogeneous equation for $z(t)$:

$$dz= 2\delta zdt - \frac{\gamma\sqrt{\lambda}}{\omega\sqrt{\lambda^2+4\omega^2}}zdw_1(t) - \frac{\delta}{2}dt \quad (20)$$

By the Cauchy formula [7] the solution of this equation with initial condition $z(t_0) = z_0$ may be written as the sum of two terms

$$z(t) = Z(t-t_0)z_0 - \frac{\delta}{2}Z(t-t_0)\int_{t_0}^{t} Z^{-1}(s-t_0)ds$$

where $Z(t)$ is the solution of the correspondent homogeneous equation

$$dZ= 2\delta Zdt - \frac{\gamma\sqrt{\lambda}}{\omega\sqrt{\lambda^2+4\omega^2}}Zdw_1(t)$$

with initial condition $Z(0) = 1$. The solution of this equation has an exponential form [2]

$$Z(t) = \exp\left\{\left(2\delta - \frac{\gamma\sqrt{\lambda}}{2\omega\sqrt{\lambda^2+4\omega^2}}\right)t - \frac{\gamma\sqrt{\lambda}}{\omega\sqrt{\lambda^2+4\omega^2}} w_1(t)\right\} \tag{21}$$

and therefore

$$z(t) = z_0 \exp\left\{\left(2\delta - \frac{\gamma\sqrt{\lambda}}{2\omega\sqrt{\lambda^2+4\omega^2}}\right)(t-t_0) - \frac{\gamma\sqrt{\lambda}}{\omega\sqrt{\lambda^2+4\omega^2}} w_1(t-t_0)\right\} -$$

$$-\frac{\delta}{2}\int_{t_0}^{t} \exp\left\{\left(-2\delta + \frac{\lambda}{2\omega^2}\frac{\gamma^2}{\lambda^2+4\omega^2}\right)(t-s) + \frac{\gamma\sqrt{\lambda}}{\omega\sqrt{\lambda^2+4\omega^2}}\left(w_1(s-t_0) - w_1(t-t_0)\right)\right\} ds=$$

$$=z_0 \exp\left\{\left(2\delta - \frac{\gamma\sqrt{\lambda}}{2\omega\sqrt{\lambda^2+4\omega^2}}\right)(t-t_0) - \frac{\gamma\sqrt{\lambda}}{\omega\sqrt{\lambda^2+4\omega^2}} w_1(s-t_0)\right\} +$$

$$+\frac{\delta}{2}\int_{0}^{t-t_0} \exp\left\{\left(-2\delta + \frac{\lambda}{2\omega^2}\frac{\gamma^2}{\lambda^2+4\omega^2}\right)u + \frac{\gamma\sqrt{\lambda}}{\omega\sqrt{\lambda^2+4\omega^2}}\left(w_1(t-t_0-u) - w_1(t-t_0)\right)\right\} du$$

We may rewrite the formula (21) as follows:

$$Z(t) = \exp\left\{t\left[\left(2\delta - \frac{\gamma\sqrt{\lambda}}{2\omega\sqrt{\lambda^2+4\omega^2}}\right) - \frac{\gamma\sqrt{\lambda}}{\omega\sqrt{\lambda^2+4\omega^2}}\frac{w_1(t)}{t}\right]\right\}$$

By definition of the Wiener process [2] $\mathbf{P}\left(\limsup_{t\to\infty}\left|\frac{w_1(t)}{t}\right| = 0\right) = 1$

This mean that to analyze the behavior of the stochastic exponent (21) for the large t>0 defines a sign of the expression $\left(2\delta - \frac{\gamma\sqrt{\lambda}}{2\omega\sqrt{\lambda^2+4\omega^2}}\right)$. Under assumption

$$\delta > \frac{\gamma^2\lambda}{4\omega^2\lambda^2+4\omega^2} \tag{22}$$

the improper integral

$$\int_0^\infty Z^{-1}(s)ds = \int_0^\infty \exp\left\{\left(-2\delta + \frac{\lambda}{2\omega^2}\frac{\gamma^2}{\lambda^2+4\omega^2}\right)s + \frac{\gamma\sqrt{\lambda}}{\omega\sqrt{\lambda^2+4\omega^2}} w_1(s)\right\} ds$$

there exists with probability one, bet $\mathrm{P}(\lim_{t\to\infty} Z(t) = \infty) = 1$. Therefore under assumption (22) the amplitude $r(t)$ exponential decreases with probability one:

$$\mathbf{P}(\lim_{t\to\infty} r(t) = 0) = \mathbf{P}(\lim_{t\to\infty} z(t) = \infty) = 1$$

If $\delta < \frac{\gamma^2\lambda}{4\omega^2\lambda^2+4\omega^2}$ then

$$P(\lim_{t\to\infty} Z(t) = 0) = 1$$

and there exists the random process

$$\tilde{U}(t) = Z(t)\int_{-\infty}^{t} Z^{-1}(s)ds=$$

$$=\frac{\delta}{2}\int_0^{\infty} \exp\left\{\left(-2\delta + \frac{\lambda}{2\omega^2}\frac{\gamma^2}{\lambda^2+4\omega^2}\right)u + \frac{\gamma\sqrt{\lambda}}{\omega\sqrt{\lambda^2+4\omega^2}}(w_1(t) - w_1(t-u))\right\}du \quad (23)$$

which satisfies the equation (18). Because by definition [2] the difference $\{w_1(t) - w_1(t-u), u \geq 0\}$ is the standard Wiener process with no dependent on *t* distributions we can insure that the process (23) is possessed of stationary properties. If the random process *y(t)* satisfies the no homogeneous stochastic differential equation with initial condition $y(t_0) = y_0$ then $u(t) = y(t) - \tilde{U}(t) = Z(t)(y(t_0) - \tilde{U}(t_0)$ and $P(\lim_{t\to\infty} u(t) = 0) = 1$. Therefore for any solution of the equation (18) and for any C>0:

$$\lim_{t\to\infty} \mathbf{P}(y(t) \leq C) = \mathbf{P}(\tilde{U} \leq C){:=}F(C)$$

where

$$\tilde{U} = \frac{\delta}{2}\int_0^{\infty} \exp\left\{\left(-2\delta + \frac{\lambda}{2\omega^2}\frac{\gamma^2}{\lambda^2+4\omega^2}\right)u + \frac{\gamma\sqrt{\lambda}}{\omega\sqrt{\lambda^2+4\omega^2}}w_1(u)\right\}du \quad (24)$$

This means that there exists the stationary amplitude $\{\tilde{r}(t), t \in \mathbf{R}\}$ with distribution defined by formula

$$\mathbf{P}(\tilde{r}(t) \leq C) = \mathbf{P}(\tilde{U} \geq C^{-0.5}) = 1 - F(C^{-0.5}) \quad (25)$$

Joining the above formulae and applying the results of the book [2] (Part1, §15) we can made the following conclusion:

- the finite dimensional distributions of any solution of the Van der Pole equation (1)-(2) for sufficiently small positive ε may be approximated by corresponding distributions of a periodic function with random phase and random amplitude

$$x(s) = R(s)cos\left(\left(\omega - \frac{1}{2\omega}\frac{\gamma^2}{\lambda^2+4\omega^2}\varepsilon^2\right)s + \frac{\gamma\sqrt{\lambda^2+2}}{2\omega\sqrt{\lambda(\lambda^2+4\omega^2)}}w_2(\varepsilon^2 s)\right) \quad (26)$$

  where $R(s) = r(\varepsilon^2 s)$ and r(t) satisfies the stochastic Ito differential equation (18);

- if $\delta < \frac{\gamma^2\lambda}{4\omega^2\lambda^2+4\omega^2}$ then $\mathbf{P}\left(\limsup_{t\to\infty}|R(t)| = 0\right) = 1$;

- if $\delta > \frac{\gamma^2\lambda}{4\omega^2\lambda^2+4\omega^2}$ then with probability one the amplitude in (26) become established to a stationary process $\tilde{R}(s)$ with distribution (25);

- the steady state amplitude $\tilde{R}(s) = \tilde{r}(\varepsilon^2 s)$ is such a very fast oscillating process that for any *b>a>0*: if $\tilde{R}(0) = \rho \in (a,b)$ the stay time $\tau_\rho(a,b) := \inf\{s: \tilde{R}(s) \in (a,b)\}$ of this process within the interval *(a,b)* has an expectation of order ε^2:

$$\mathbf{E}\{\tau_\rho[a,b]\} = \varepsilon^2 \frac{4\omega^2(\lambda^2+1)}{\gamma^2\lambda}\left\{-\int_a^\rho\int_a^y \frac{2\Phi(y)}{z^2\Phi(z)}dz\,dy + \int_a^b\int_a^y \frac{2\Phi(y)}{z^2\Phi(z)}dz\,dy \frac{\int_a^\rho \Phi(x)dx}{\int_a^b \Phi(x)dx}\right\}$$

where

$$\Phi(x) = \left(\frac{a}{x}\right)^{\frac{(\lambda^2+1)\left[3\gamma^2\lambda - 8\delta\omega^2(\lambda^2+4\omega^2)\right]}{(\lambda^2+4\omega^2)\gamma^2\lambda}} e^{\frac{\delta(\lambda^2+1)\omega^2(a^2-x^2)}{\gamma^2\lambda}}$$

References

[1] Bogoliubov, N.N., Mitropolsky, Yu.A., *Asymptotic methods in the theory of non-linear oscillations*, Gordon and Breach, New York, 1961.

[2] Gikhman, I.I., Skorokhod, A.V. Stochastic differential equations, Springer-Verlag, Jan 1, 1972.

[3] Khasminskii, R. , Stochastic Stability of Differential Equations, Springer Berlin. 2011.

[4] Dynkin, E.B., Markov Processes. Spriger, Berlin, 1965.

[5] Tsarkov, Ye. (J.Carkovs). Asymptotic methods for stability analysis of Markov impulse dynamical systems.// Advances of Stability Theory of the End of XXth Century. Stability and Control: Theory, Methods and Applications.- Gordon and Breach Science Publishers, London, 13 (2000).- p.p. 251-264.

[6] Skorokhod, A. V., Asymptotic Methods in the Theory of Stochastic Differential Equations, AMS, Providance. Rhode Islands, 1989.

[7] A.M.Sadovyak, E.F.Tsar'kov (J.Carkovs). An Analogue of the Cauchy Formula for Stochastic Differential Equations", *Teor. Veroyatnost. i Primenen.*, 18:2 (1973), 415–417.

Current address

Carkovs Jevgenijs, professor
1 Kaļķu Street, Riga, LV-1658
e-mail: carkovs@latnet.lv

Matvejevs Andrejs, professor
1 Kaļķu Street, Riga, LV-1658
e-mail: Andrejs.Matvejevs@rtu.lv

Pavlenko Oksana, doc.
1 Kaļķu Street, Riga, LV-1658
e-mail: Oksana.Pavlenko@rtu.lv

Slovak University of Technology in Bratislava
Faculty of Mechanical Engineering
16th Conference on Applied Mathematics **APLIMAT 2017**
Proceedings

ELLIPSES AND OVALS IN THE PHYSICAL SPACE OF ST. PETER'S SQUARE IN ROME

CARLINI Alessandra (IT), MAGRONE Paola (IT)

Abstract. We discuss the spatial and perceptive implications of the geometry at urban scale of St. Peter's square in Rome. The case study of Bernini's project for the colonnade makes explicit the relation between the geometry of the plan and the spatial development. We will show that the geometry of the circle affects intentionally the perception of the walking visitor. There is an ambiguity of use between the terms ellipse and oval, especially in touristic literature and some school textbooks. The specific mathematical properties, of one curve respect to the other, were well known at the time and were obviously intentionally exploited in the design of the square along with the colonnade. We also report of a "geometric flash mob", we organized in occasion of the 2015 European Researcher's Night.

Keywords: ovals, ellipses, Bernini, spatial development, flash mob

Mathematics Subject Classification: 51M15, 00A67

1 Introduction

In this paper, we aim to investigate the 3d spatial implications, in architecture and at urban scale, of some curved geometries, giving an occasion of active citizenship and teaching. In particular, we will render explicit the relation between the geometry of the plan and the its development in the three dimensions.

The case study of Bernini's project for the colonnade of St Peter's square can be used to show the geometric differences between the ellipse and the ovate through the actual experience in the physical space. For this purpose, we will show some of the many meanings that the word elliptic has in different mathematical frameworks and describe the geometric features of the ovate.

In the architectural culture this difference has often been subject of interest because this kind of curved shapes were used for the design of buildings, construction elements or urban spaces. Some well-known examples are the ancient amphitheaters, the plan of churches, the base curve of domes, the design of low arches [17; 20].

Fig. 1. St Peter's square. Designed by G. L. Bernini (1656)

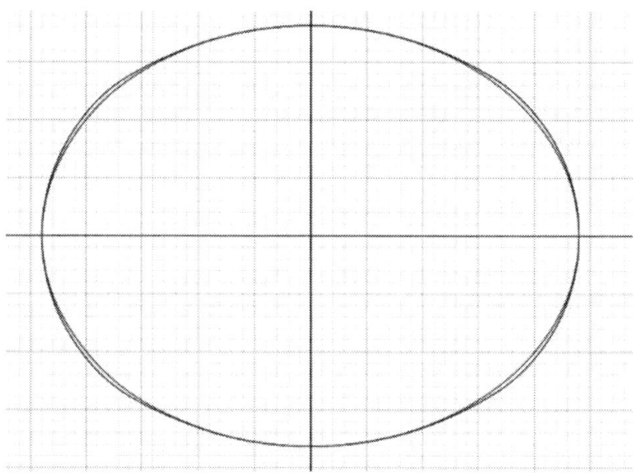

Fig. 2. An ellipse (blue), an ovate (black), inscribed in the same rectangle. The proportions are the same of those of the internal dimensions of the square. The shape of the two plots is very similar

The literature that has been produced followers two main directions: the analytical investigation [22,23] that aims to measure how much ovals constructed with arcs of circles can be similar to ellipses; the comparison between the original drawings and the relief of the state of fact [6,10,12], focusing on the analysis of the two-dimensional plan.

We propose to observe the geometry of St Peter's square from another point of view, taking the plan along with the elevation and moving in the three-dimensional space of the city. We will show that the drawing of the colonnade influence the visitor's observation who moves on foot in the square. Moreover, that the perceptive analysis of the spatial phenomenon makes explicit the geometry of the construction.

This paper is divided as follows. Section 2 is devoted to show some of the many aspects that the word elliptic addresses in mathematical frameworks, and the geometric features of the ovate.

In Section 3 we describe examples of the use, sometimes ambiguous, of the terms oval and elliptic taken from historical treatises, contemporary articles, touristic guides and schoolbooks. In Section 4 we describe the shape of St. peter's square, and Section 5 is devoted to a description of the plan together with the elevation, in order to explain how the geometry of the circle influences the perception of the visitor who moves on foot in the square. Finally, in Section 6 we report of a flash mob organized in St Peter's square.

2 Ellipse, elliptic and oval

The ellipse is a conic section, a geometric locus, draws the orbit of planets. It has a primary role in the background of mathematicians, so that the mathematics community keeps using the name "elliptic" to describe new concepts. In this section, we want to show some of these aspects.

On the other hand, every architect knows exactly what ovate means, this word is very common in architecture, while it is not in mathematical culture.

The ellipse is a conic curve, i.e. can be obtained as a plane section of a cone (a cone is the surface generated by the rotation of a line *r* around another line *l*, being the two lines incident; the line *r* and all its rotated copies will be called generatrixes). When the plane intersects all the generatrixes (fig 1 left), the section is an ellipse.

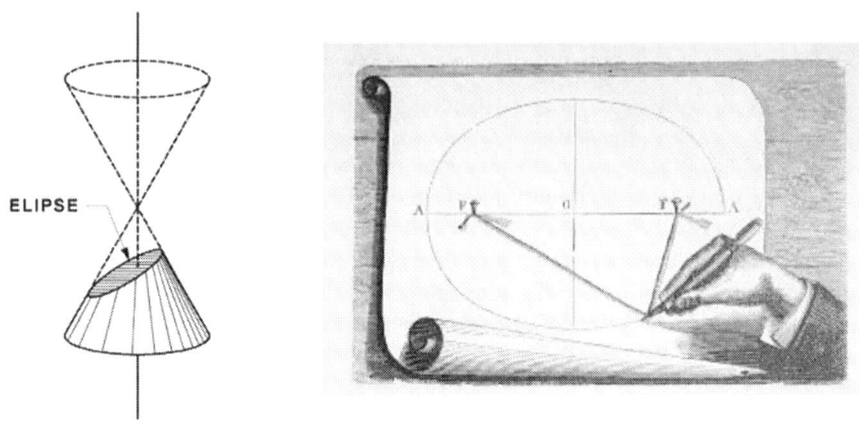

$$\frac{X^2}{a^2} + \frac{Y^2}{b^2} = 1$$

Fig. 3. From left to right: the ellipse as a plane section of the cone, the ellipse as geometric locus and its equation. Central picture: method to track an ellipse, called "the gardener's method". Illustration from da A. Guillemin (1882), *El mundo físico: gravedad, gravitación, luz, calor, electricidad, magnetismo*, etc.

Apollonius (who lived between 262 and 190 B.C.) in his texts about conics sections writes down the equations in this way [18]:

Y^2=Lx for the parabola, Y^2=Lx+ L/$a$$X^2$ for the hyperbola, Y^2= Lx- L/$a$$X^2$ for the ellipse (L is a parameter, a is a transverse axis of the ellipse, see also [42]).

One possible interpretation for the name ellipse comes from the equation, since the word derives from the greek *elleipsis* which means lack or shortage, and in the equation for the ellipse, the term L/aX^2 is subtracted.

A conic section can also be defined as the projection of a circle on a plane. So, the ellipse, which is one of the possible projections, inherits all the property of the circle that are invariant under projective transformations, such for example the cross ratio.

One of the most common definition of ellipse is that of geometric locus, "*the ellipse is a plane curve such that the sum of the distances of each point from two fixed points called foci is constant*". The mathematical machine in figure 3 allows to plot an ellipse with a continuous track and embodies the definition of the curve as geometric locus. The two pins are the foci, the inextensible rope obliges the sum of the distances to remain constant.

What is really striking is that a curve that was known since two thousand years, reveals to be the orbit of planets around the Sun. In his De Rivolutionibus, Copernicus (1473-1543) when describing the eliocentric system speaks about orbits that are not perfectly circular, but does not mention explicitly the ellipse. It was Kepler (1571-1630) who asserted that the orbit of planets around the sun are perfect ellipses, and that the sun is in one of the foci. More in general the conic curves are three different solutions of the differential equation of the two body problem, which involves the motion of two objects, supposed to be point-like, under the action of their gravitational fields.

The measure of the length of an arc of ellipse leads to an integral, that is called elliptic, in particular of the kind

$$L(\theta) = \int_0^{\theta} \sqrt{1 - k^2(\sin t)^2}dt$$

almost impossible to be solved. The so called *elliptic functions* are related to the inversion of another kind of elliptic integral, not related to the computation of the arc length of an ellipse, but anyway called in this way. They were created by Abel, who had to deal with an elliptic function when aimed to divide Bernoulli's Lemniscate in *n* equal parts by ruler and compass [21]. The theory of elliptic functions and that of elliptic curves, the geometric counterpart, is today a lively field of research, and played an important role in the proof of Fermat's last theorem.

The equation of a conic curve, a second degree polynomial, can be associated to the 2x2 matrix, called the quadratic invariant, composed by the coefficents of the second degree terms. The determinant of the quadratic invariant allows to distinguish between the three different conic curves. In other words only the second degree coefficents determine which kind of conic is represented by the equation. When the conic is an ellipse, the determinant is strictly positive.

Transferring the information on the higher degree terms to differential operators, elliptic refers to a particular class among them. A differetial operator is called uniformly elliptic if coefficients of the highest-order derivatives are positive. These operators are a generalization of the Laplace operator (actually of the negative Laplacian) and elliptic equations are the partial differential equations related to elliptic operators.

In the framework of non-Euclidean geometry, the elliptic or Riemannian geometry is the one that generalizes the geometry on the sphere. It was introduced by Riemann in the inaugural speech of 1851 at Gottinghen University. It is a geometry where lines are not infinite, but finite and closed [8], in particular, on a sphere, lines are great circles. In a more general example of Riemannian geometry, there can be points of two different kind. Elliptic points: in the neighborhood of the point the surface is like a sphere, the tangent plane can be shifted towards the surface and cuts the surface in a curve that looks like an ellipse. Hyperbolic points, if the tangent plane is shifted towards the surface, the intersection is similar to a hyperbola.

Finally, if it is possible to calculate the Gaussian curvature of the surface at a point, elliptic points are those where this curvature is strictly positive, which, in terms of differential calculus means that the homogeneous second order derivatives have the same sign.

We will now briefly discuss the definition of the term ovate. In architectural framework, the word oval or ovate refers to a particular polycentric curve, that is a curve formed by arcs of circles. As it will be showed in section 3, polycentric curves, in particular ovals, were created for practical purposes and widely used in architecture.

A rigorous definition of oval can be: "*an oval is a closed convex curve, continuous and differentiable*". So, an ellipse can be an oval, since it agrees with the definition above. A polycentric curve is defined as: "*line formed by several arcs of circumference, mostly of different radii, which, at the connection points, have the same tangent line*" (cit. Treccani Enciclopedia online).

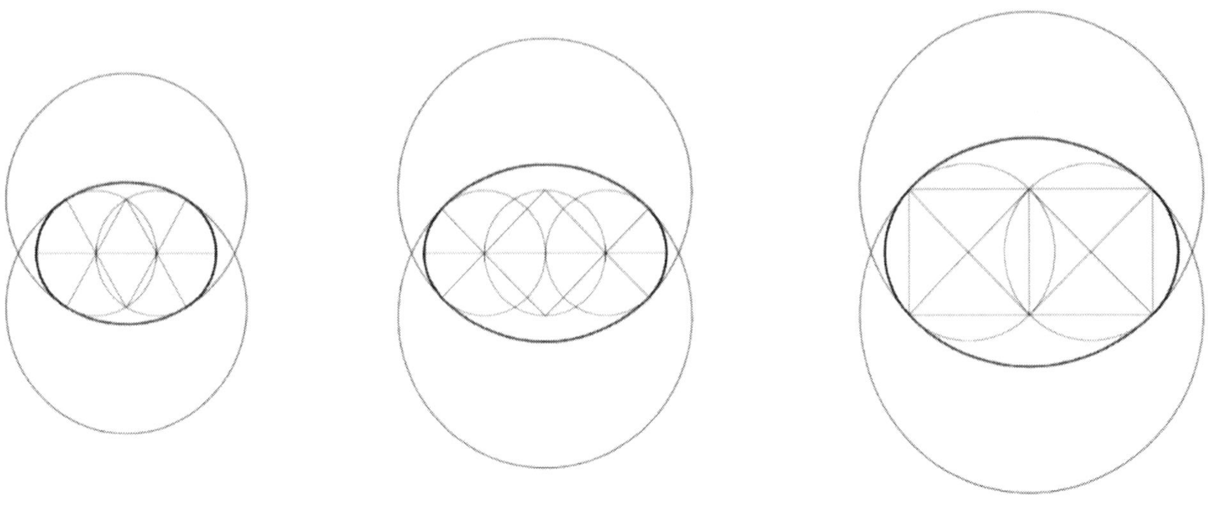

Fig. 4. An oval curve, i.e. a polycentric curve obtained by the junction of arcs of different circles. Some geometric constructions of the oval with ruler and compass. The first two starting by the major axis dimension. The arcs of different rays are highlighted with different colours.

The condition on the tangent line at the connection points implies that the polycentric curves are differentiable.

The property that is exploited in the design of the colonnade of St Peter's square involves the normal lines to the curves. The normal lines to a circle have the direction of the radius, so they all meet at a single point, the center.

On the other hand the geometric locus where the normal to an ellipse meet *is not* a single privileged point. The normal lines do envelope a curve, called the evolved of the ellipse.

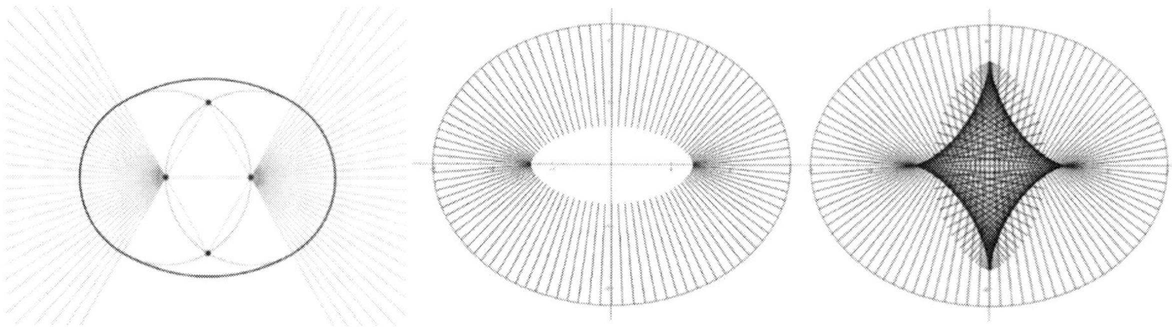

Fig. 5. Left: the oval used for the design of St Peter's square, and the normal lines from the centers of the smaller circles. Center and right: the normal lines to an ellipse do not meet all at one point. The enveloped curve which can be seen on the right: hand side is called evolute of the ellipse. The center and right pictures where realized with the software Mathematica by Corrado Falcolini

The way this property is exploited in the design of St Peter's square is shown briefly in picture 6 and will be discussed in sections 4 and 5. The perceptor positioned on the the center of the circular arc of the colonnade (left picture), where all the normal lines meet, sees one column for each row: the columns are aligned along the normal. Right picture: the shape is elliptic and the perceptor on the focus. The normal lines do not meet in one point, so there is not a point from which the visitor could perceive the same effect.

Fig. 6. The sketch shows the perceptual effect of the visitor looking at the colonnade from the center of the circular arc of the colonnade (left), drawing the plan as the actual ovate. Supposing that the plan was elliptical, the visitor being on the focus (right)

3 Ellipse and ovate in architectural culture

There is an ambiguity of use that often occurs between the terms ellipse and oval. In the historic architectural literature related to the manuals and treatises, the expression "elliptic" is used to address a curved geometry, not circular, in other words any unspecified oval.

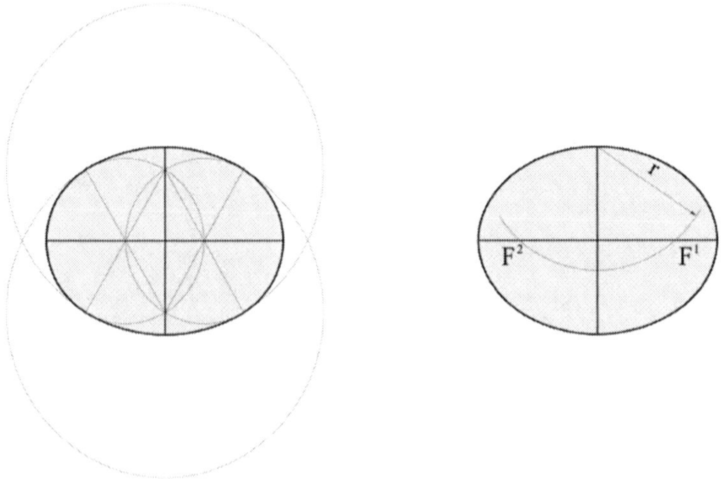

Fig. 7. Left: An oval and an ellipse with the same horizontal and vertical dimensions. The segment *r* has the length of half of the major axis

Migliari [20], in his studies on curved geometries taken from existing historical architecture and construction practices, points out the spread of this ambiguity; he shows examples from treatises, starting from the ones of the Renaissance, such as that of Peter Cataneo 1567 [7 Libro VII, proposizione XIV].

The two curves are similar in their aspect but their tracking is very different and so are the spatial implications of this on the perceptual level. If one traces an ellipse and a round ovate with the same dimensions (length, width) the difference between the two curves appears minimal (fig. 2).

The reasons for the choice of an oval with respect to an ellipse are, in an architectural and spatial framework, therefore of other kind: ease of tracking, convenience in the execution on the construction yard, perceptual and spatial factors.

In any case, the differences between ellipses and ovals for the usual proportions are so small [...]. Even the most precise mensuration does not serve to settle the matter. It is not a matter of mensuration, but of the history of building traditions. [17]

Like many authors uphold [17, 20, 22], the reasons for such a widespread use of polycentric curves in architecture are related to practical problems, as demonstrated by their spread in the Renaissance treatises (from Serlio onwards).

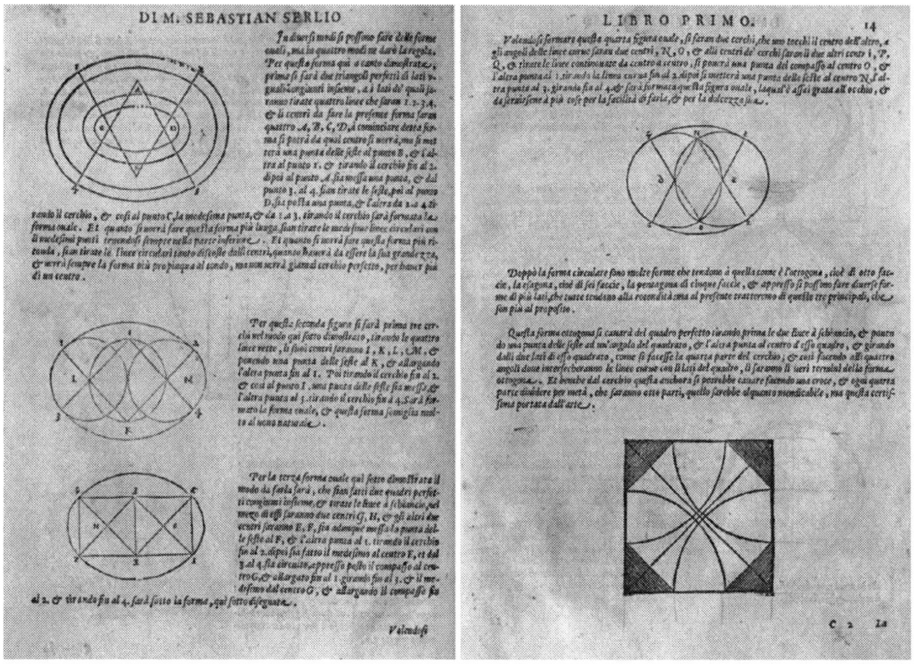

Fig. 8. The four geometric constructions of the oval in the Treatise on Architecture by Sebastiano Serlio (1584) [30]. The Treatise, like other analogue texts, was an asset of operational and stylistic knowledge fot the architects of the next generation

The mathematicians Mancini Proia and Menghini discuss in [18,19] the introduction and the use of the oval and elliptic shapes in baroque architecture, and the meaning which was attributed to these two terms. The great architects of Baroque were influenced by the astronomic new theories of Copernicus (1473-1543) and Kepler (1571-1630), and "*they become infatuated of the elliptical shape*" (cit [19] pag 334). It was Sebastiano Serlio, an architect, artist and geometer who writes about the geometric constructions of ovals which will help the architects in their design. From the citation of Serlio in [18] we understand which was the use of ellipses and ovals in architecture (here we report the transposition of the author of [18] of the archaic text of Serlio, first book pag. 11)

> "*vuole l'architetto fare un ponte o un arco, o anche una volta di minore altezza del mezzo cerchio, cosa che molti muratori hanno certa pratica e col filo fanno grandi volte quali corrispondono all'ellisse e si raccordano con alcune ovali fatte col compasso. Nondimeno se l'architetto vorrà provare teoricamente di essere indirizzato dalla ragione deve tener conto di questa linea.* " [18, pag 65]

> "the architect wants to make a bridge or arch, or a vault of lesser height of the half circle, which many bricklayers have some practice and with wire they make great vaults which correspond to the ellipse. They then connect them with some ovals made with a compass. Nevertheless, if the architect wants to try theoretically to be addressed by the reason it must take in account this method." [18, pag 65]

Serlio does not expose any of the properties of the ellipse, neither does he report the "gardener's method" to track this curve (fig. 3). He describes a method to draw an ellipse by points, and observes that in any case the resulting points should be connected drawing by hand, which is not simple. Therefore, he points out that "in a number of ways you can track an oval, but I will expose only four ways".

Oval shapes were developed in order to approximate the ellipse with portions of circles, so they were born from the very beginning to solve practical issues. Specialized texts included geometric constructions for the tracing by points and instruments for the continuous tracking of ellipses, but these methods were not suitable for drawing at large scale and carried theoretical and practical issues such as measuring portions of an ellipse, or making a large number of different bricks to cover an elliptical arch [20, 22].

We observe also that it was necessary to wait until the nineteenth century to find a veritable repertoire of curves inside a treatise on architecture (see for example Rondelet [31] section devoted to stereotomy).

The debate between oval or elliptical covered many popular historic buildings: the Roman amphitheaters, (see the debate on the forum Nexus journal [40] and [22, 23,10]); some churches, from the late Renaissance onwards [14], and Baroque churches such as San Carlino by Borromini, [6,12] or San Andrea al Quirinale of Bernini [24]; Bernini's colonnade in St. Peter's Square [3, 15,16]; the Ellipse in Washington [15].

An ambiguity of use between the terms ellipse and oval remains in some kind of literature, as touristic literature and some school textbooks. For example, the Coliseum and the plan of St Peter's square are both described as elliptical shaped in [36,37,38,39]. Moreover, in touristic guides (in the four we consulted) St Peter's square is described as elliptical [32,33,34,35].

We found a further ambiguity in the bestseller book "Angels and Demons" by Dan Brown [5].

> *«Two fountains flanked the obelisk in perfect symmetry. Art historians knew the fountains marked the exact geometric focal points of Bernini's elliptical piazza, but it was an architectural oddity Langdon had never really considered until today. It seemed Rome was suddenly filled with ellipses, pyramids, and startling geometry»*
>
> [5 pag 73]

4 The geometry of St Peter's square in Rome.

As it can be seen from an hand-drawn sketch for St. Peter's Square design, Bernini adopts very soon a flattened shape curve.

Fig. 8 (right) reproduces the drawing of the square realized by Fontana in 1694, a few years after the end of the works for the buindilg of the square. In the notes accompanying the figure (top left), still we find the habit to use either the term ellipse or oval. In the notes, in Section A, "*Elipse quale racchiude li portici*" (ellipse which encloses the arcades) and, Section B, "*Elipse di data proportione quale racchiude l'anfiteatro Flavio cioè Colosseo di minor capacità del Vaticano*" (Ellipse of proportion able to enclose the Flavian Amphitheatre, the Coliseum, smaller than the Vatican capacity).

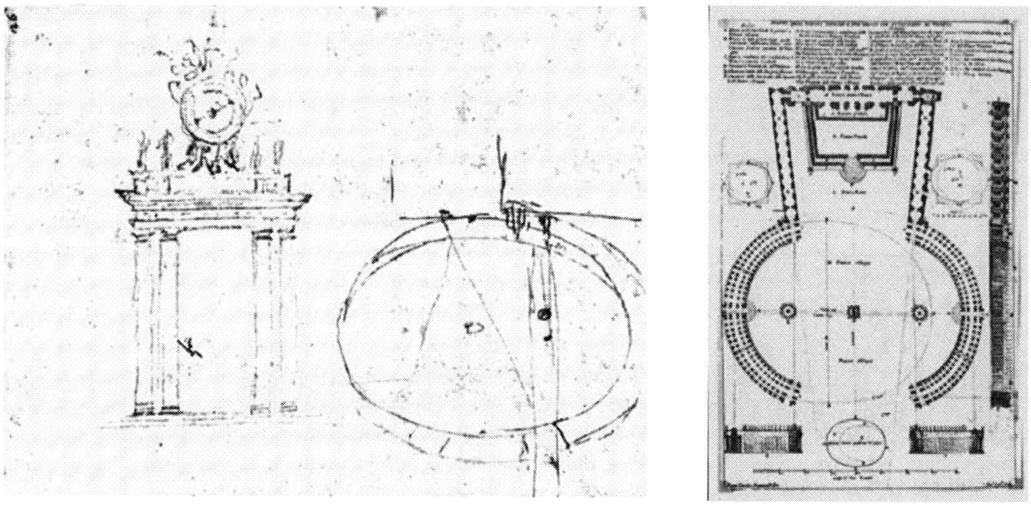

Fig. 9. Left: hand drawing by Bernini (1957, Roma, Bibl. Vaticana, cod. Chigi a I 19, f. 26r) [2,3]. Right: engraving, Carlo Fontana, 1694 [11]

The author in [18] points out that "Bernini to substantiate the elliptical shape, poses the two fountains where more or less should be the foci of the ellipse which has the same axes of the ovate". So in literature we have found two possible geometric interpretations on the design of the square:

A_ the oval geometry of the perimeter of the colonnade, with the smaller circumferences centers positioned where the existing porphyry discs are;

B_ the perimeter of the colonnade suggests an ellipse, with the two foci close to the so-called "twin fountains".

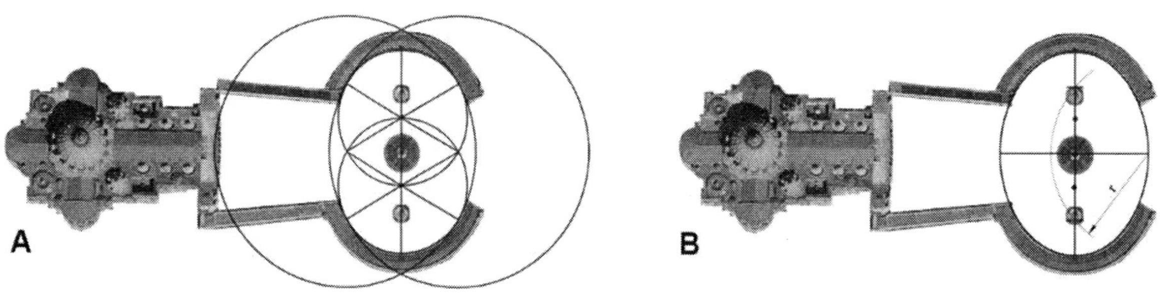

Fig. 10. Reconstructions of geometric assumptions. (A) the perimeter of the oval colonnade, the smaller circumferences centers in correspondence of the existing discs of porphyry. (B) the hypothesis of elliptical colonnade, with foci close to the so-called "twin fountains"

In the historical-critical type literature we found no doubts about the interpretation of the oval design of the square. The fundamental studies of the Roman Baroque, from those of Wittkower [25], to Fagiolo dall'Arco [11], to Brandi [4], speak of ovate and other studies [3,11,16] identify accurately the polycentric construction adopted by Bernini.

5 The relation between the geometry of the plan and the spatial development.

The specific properties of the oval were well known and were deeply exploited in the design of St Peter's colonnade.
From the point of view of the perception of the form [1,4,11] polycentric constructions allow to preserve the typical spatial tensions of the ellipses, but use circular curves which are more practical to treat on the architectural plan. A geometric analysis of the shape of the square can be done through observation on the premises. The perceptual experience of the spatial phenomenon makes explicit the oval geometry of the building.

In St Peter's square the theatricality of the Baroque city can be already appreciated before being swallowed up from the breadth of the square. Actually, when coming from the Borgo district, the colonnade looks dense and almost impenetrable, a forest of backlit columns.

Coming from *Via di Porta Angelica*, and not from *Via della Conciliazione*, through the arches of the *Passetto di Borgo*, it is impossible to imagine the void that opens a little further. Almost as walking in a forest, the visitor passes the shafts of the columns, one after the other. Then he passes through the thickness of the porch; at the end of it he almost falls into the vacuum of the square.

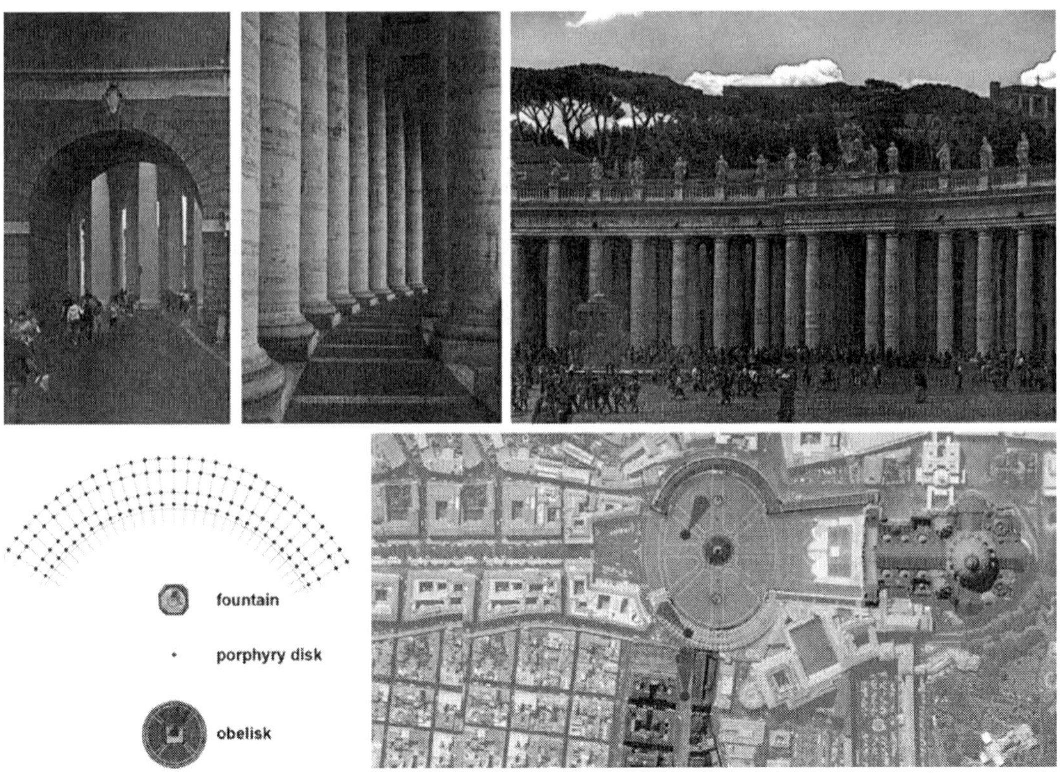

Fig. 11. How to access and how to get around in St. Peter's Square to enjoy the spatial properties coming from his plane geometry. From left: the arrival from Borgo district; the crossing of galleries of the colonnade; the arrival in the square

Crossing the square, that forest of ever changing columns, which are apparently disordered, becomes transparent and ordered only from two observation points, placed on opposite positions with respect to the obelisk in the center of the square. These two notable points are now marked on the ground by two porphyry discs: on the edge of the disc, the writing "center of the colonnade".

Fig. 12. The porphyry disk and one of the sundial nodes which mark the pavement of the square

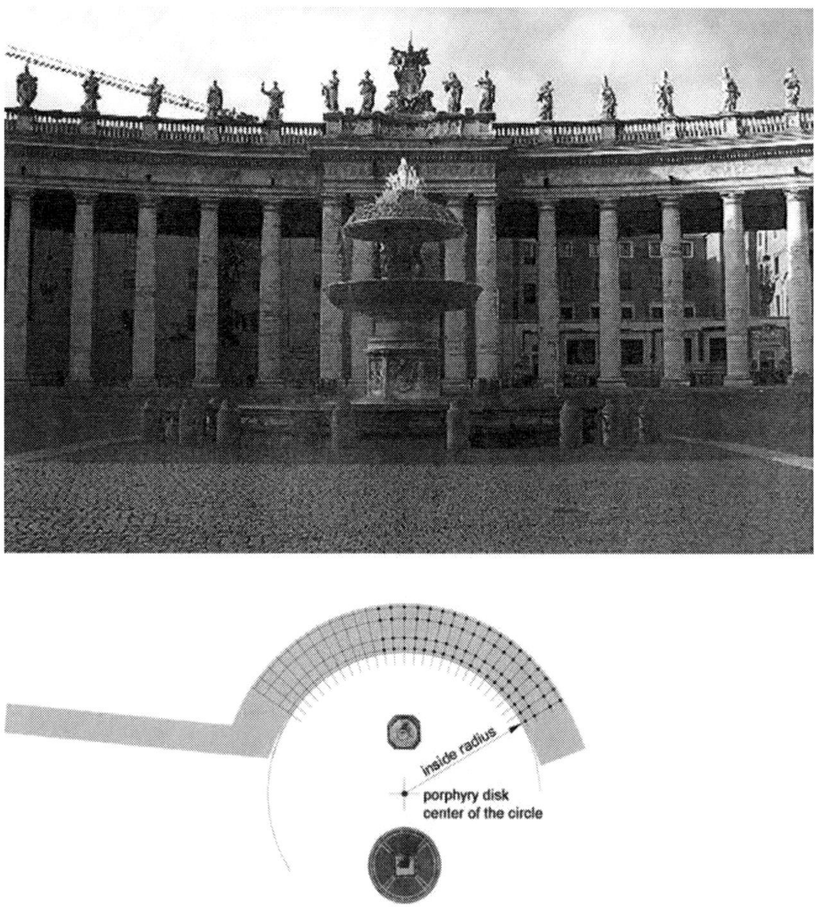

Fig. 13. Image taken by the observer positioned on the porphyry disk.
The development of the plane geometry in the physical space.

On the porphyry disk, the columns are aligned along the normal to the circumference and appear as if they were one. When you start moving into the square, even slightly, the columns in the rows appear again

The design of the colonnade is dictated by the circle geometry: joining any point of a circle with the center, you get the radius, that is a straight line perpendicular to the curve. The columns are aligned along the rays and arranged on four arcs of concentric arcs of circumferences. For this reason, the observer positioned in the center of these circles, marked on the ground by the marble disk, sees only the columns on the innermost curve. Here the multiplicity of three galleries vanishes, the four rows of columns become one. So, the phenomenon that is appreciated in two specific points of the square, where the four rows of columns appear aligned in a single row, allows to acknowledge in the three dimensional space the underlying geometric consistency.

On the other hand, as soon as we rotate our gaze to the opposite side of the square, this unifying view vanishes and again we see a forest of shafts. As soon as we step outside of the porphyry disk, the single rows are broken, the columns become immediately visible as a forest. The disk has a ray of not more than 30 cm. The ray of the circle that defines the colonnade is about 80 meters. A step outside of this small disk means that the perceptor finds himself positioned not more than 50 cm from the geometrical center of the semi-circle, a very small distance if compared to the radius. The perceptive effect is very remarkable.

If the curve was an ellipse, the same perceptual effect could not be achieved, since the perpendicular lines to an ellipse, do not converge all at a single point (see fig. 5-6).

6 The geometric flash mob

Fig. 14. The geometric flash mob in St. Peter's Square

In the occasion for "European Researchers' Night 2015: exploring science, having fun" we had a geometric flash mob in St Peter's square. The activity was inspired by a "Progetto Lauree Scientifiche", a project funded by the Italian Ministry of Education, which dealt with mathematical drawing machines for tracking conic curves, realized by the two authors together with Laura Farroni (Architect, Researcher in the Department of Architecture, Roma Tre University).

Actually, the first flash mob was experimented by the first author and her students. Since it was a success, it was also proposed for the European Researcher's Night together with www.formulas.it, a group of mathematicians and architects from the Department of Architecture of Roma Tre University.

The flash mob was preceded by a "Scientific trip" (in the picture, the two authors speaking about the geometry of the square, in particular about ovals and ellipses).

Fig. 15. "Scientific trip". In the picture, the two authors speaking about the geometry of the square

Before entering the square the visitor do not perceive the enveloping geometry that characterizes it. For this reason, the flash mob started from the neighborhood: so, the surprise effect is more conspicuous and consistent with the theatricality of the Baroque space. Once through the colonnade, the group was guided towards the obelisk and invited to look around, to let them develop their own perceptive relationship with the square and with the spatial scales involved. The colonnade appears at this moment still very consistent, with rows of columns that alternate between the opening and closing of the intercolumns, which are still deep and in the shadow. At the same time, at the top, the colonnade profile, linear and continuous, draws the curved geometry of the plant.

Only after this perceptual experience, gradually, the group was invited to move forward and to conquer the center of the circle marked on the ground by the porphyry disk. Now the relationship between the planar geometry and the physical space is clear and surprising. What seemed obvious on the plan, becomes plastically almost magic. This result that can only be appreciated in this way, walking on foot, moving the eyes, trying to look for a physical contact with the object, reclaiming the public space.

The group of participants was then asked to guess how much could the square be wide. The given answers were quite much under the real measure. Then the group was proposed to measure one of the rays of the "ovato tondo" by forming a line of people, actually to experiment an anthropomorphic sizing. The group was not enough numerous to cover the ray, so other tourists joined the line. It took 58 people, standing with their arms wide. The average arm width is 1.50 mt, so the estimate of the ray was 87 mt. The geometrical composition of the ovato shows that the width of the whole square is three times the ray, so 261 mt. The oval is actually 240 meters wide [11, plate 166], so the anthropomorphic measurement of the amplitude reached a startling precision. Knowing the geometry allows to involve fewer tourists to measure the square.

A final remark: during the flash mob the police officers in the square asked us what we were doing. We explained to them that we were from the School of Architecture, and we were doing a scientific experiment. They let us do and were amused by out intention of forming a human line. We don't actually know if nowadays the flash mob could be done again, thinking of internationals facts that sometimes lead to stiffen the security measures. We hope that it will be possible to perform it again. But the view of the colonnade will always be possible and the wonder of its geometry always be appreciated.

Acknowledgement

This work developed under continuous discussions with Prof. Laura Tedeschini Lalli.
The authors thank Corrado Falcolini for the discussions about ovato tondo and for making the pictures in fig. 6 with the software Wolfram Mathematica.

References

[1] ARNHEIM, R. *Il potere del centro*, Abscondita, Milano 2011.

[2] BENEVOLO, L., *San Pietro e la città di Roma*, Roma-Bari, Laterza, 2004

[3] BIRINDELLI, M. *Piazza San Pietro*, Bari, 1981

[4] BRANDI, C. *La prima architettura barocca*, Bari, 1970

[5] BROWN, D., Angels & Demons, Washington Square Press, 2006

[6] CANCIANI, M. Il disegno della cupola del San Carlino alle Quattro Fontane di Borromini: ovale canonico?, in *Disegnare con*, Vol 8, N° 15, 2015

[7] CATANEO, P., *L'architettura di Pietro Cataneo senese*, 1567

[8] COURANT, R., ROBBINS, H., *Che cos'è la matematica?*, Universale Scientifica Boringhieri, 1988

[9] COXETER, H.,S.,M, *Introduction to geometry*, John Wiley and sons, 1989

[10] DOCCI, M., MIGLIARI, R.M.R. Architettura e Geometria nel Colosseo di Roma, in *Matematica e Architettura – Metodi analitici, metodi geometrici e Rappresentazione in Architettura*. Firenze, 2001

[11] FAGIOLO DELL'ARCO, M. e M. *Bernini*, Roma, 1967

[12] FALCOLINI, C. Algorithms for Geometrical Models in Borromini's San Carlino alle Quattro alle Quattro Fontane, in *Handbook of research on visual computing and emerging geometrical design tools*, 2016

[13] FARRONI L, MAGRONE, P., *A multidisciplinary approach to teaching mathematics and architectural representation: historical drawing machines. Relations between mathematics and drawing, Proceedings of History and Pedagogy of Mathematics*, Monpellier, 2016

[14] FASOLO, V. Sistemi ellittici nell'architettura, in *Guida metodica per lo studio della storia dell'architettura*, Roma, 1954.

[15] KIMBERLING, C. The Shape and History of The Ellipse in *Washington, D.C . BRIDGES Mathematical Connections in Art, Music, and Science*

[16] KITAO, T. K. *Circle and Oval in the Square of Saint Peter's*, New York, 1974

[17] HUERTA, S. Oval Domes: History, Geometry and Mechanics. In *Nexus Network Journal 9*, 2007

[18] MANCINI PROIA, L. Il cinquecento e il seicento: rapporti fra matematica, astronomia, architettura, in *L'Unità della cultura: in memoria di Lucio Lombardo Radice*, ed Dedalo, 1985

[19] MANCINI PROIA, L., MENGHINI M. Dalle ovali policentriche alle ellissi nell'architettura barocca, *Atti d. Acc. d. Scienze di Torino, 118,* 1984

[20] MIGLIARI,R., Ellissi e ovali, epilogo di un conflitto, *Palladio* 16, 1995

[21] PRASLOV, V., SOLOVYEV, Y., Elliptic functions and elliptic integrals, *translations of Mathematical Monographs*, vol. 170, AMS, 1997

[22] ROSIN P.L., On Serlio's constructions of oval, *Mathematical Intelligencer*, vol. 23, no. 1, 2001

[23] ROSIN P. L., E. TRUCCO (2005) The Amphitheatre Construction Problem, *Incontro Internazionale di Studi Rileggere L'Antico*, Rome 13-15 december 2004

[24] SMYTH-PINNEY, J.M. The Geometries of S. Andrea al Quirinale, *Journal of the Society of Architectural Historians*, Vol. 48 No. 1, Mar., 1989

[25] WITTKOWER, R. *Arte e architettura in Italia. 1600-1750*, Torino, 1972

[26] YATES, C. A Handbook on Curves and their Properties,1947

Historical treatises

[27] Bullet, P. L'architecture pratique, qui comprend le détail du toisé, & du devis des ouvrages de massonnerie, charpenterie, menuiserie, serrurerie, plomberie, vitrerie, ardoise, tuille, pavé de grais & impression, *Geometria pratica*, proposizione XII, 1691

[28] Cataneo, P. *L'architettura di Pietro Cataneo senese*, Libro VII, proposizione XIV, 1567

[29] Guillemin, A. *El mundo físico: gravedad, gravitación, luz, calor, electricidad, magnetismo, etc.*, 1882

[30] SERLIO, S. Tutte l'Opere d'Architettura di Sebastiano Serlio Bolognese [...], Primo Libro, pag.13-14, 1584

[31] RONDELET, G.B. *Trattato teorico e pratico dell'arte di edificare. Prima traduzione italiana sulla sesta originale con note e giunte importantissime*, per cura di Basilio Soresana, seconda edizione. Mantova: Fratelli Negretti, 1832

Touristic guides consulted

[32] Guida Edt Roma, 9 edizione, ISBN 9788859225843

[33] National Geographic traveler, Roma, 2014, ISBN 8854025798

[34] Guida Roma, collana Best of, Touring Editore, 2013, ISBN 9788836562022

[35] Roma e città del Vaticano, collana Guide Verdi, 2015, ISBN 9788836567959

Textbooks of art history consulted

[36] CRICCO, G., DI TEODORO, F.P. Il Cricco Di Teodoro. Itinerario nell'arte Terza edizione – Versione verde, Ed Zanichelli, 2010

[37] DORFLES, G. *Civiltà d'arte. Edizione arancio*, Atlas ed., 2015

[38] PESCIO, C. *Dossier arte*, Edizione Giunti-Treccani, 2015

[39] BRANDUARDI, K., MORO W., Arte & emozioni, ed. La Nuova Italia, 2012

Webliography

[40] http://www.nexusjournal.com/readers-queries-online/208-why-are-roman-amphitheatres-elliptical.html, consulted on october 28 2016

[41] http://www.mathcurve.com/courbes2d/descartes/descartes.shtml, consulted on october 28 2016

[42] http://www.macchinematematiche.org/images/fontistoriche/apollonio%20ellisse%2007.pdf

[43] web.mit.edu/8.01t/www/materials/modules/guide17.pdf

Current address

Carlini Alessandra, PhD. architect
Liceo Scientifico Cavour
Via delle Carine 1 I-00146 Rome, Italy
E-mail: alessandracarlini@yahoo.com

Magrone Paola, PhD. researcher
Dipartimento di Architettura
Via Madonna dei Monti 40 I-00146 Rome, Italy
E-mail: magrone@mat.uniroma3.it

Slovak University of Technology in Bratislava
Faculty of Mechanical Engineering
16th Conference on Applied Mathematics **APLIMAT 2017**
Proceedings

IMPERIAL ROMAN COCHLEAR COLUMNS AND THEIR HISTORICAL NARRATIVE FRIEZES

CARLINI Alessandra (IT), TEDESCHINI-LALLI Laura (IT)

Abstract. Cochlear columns in imperial Rome bore the historic visual narration of campaigns, battles, victories and conquers by emperors. This was possible as a long sculpture narration evolved along a helical frieze, permitting representation without aberration or visual distortion, due to the locally Euclidean property of the surface. We also describe the narrative correspondences relating to the vertical periodicity of the helix.

Keywords: locally Euclidean geometries, cochlear columns, helices, Roman art

Mathematics Subject Classification: Primary 00A06, 00A66, Secondary 00A67

1 Introduction

We present here a study of the surface of cochlear columns in imperial Rome. Helices appear to be part of what we have called *persistence of forms*, a repertory of shapes handed down over centuries and geographies [5]. In [7] we approached the three-dimensional study of medieval tortile columns, which are quite interesting for us, because they are complicated, being the compounding of several different three-dimensional curves. Moreover, they are quite torqued, in a mathematical sense. This compound aspect commanded a subtle work of 3d survey methods, and in particular we used point cloud documentation to infer the horizontal sections.

In the present paper we discuss, instead, a seemingly elementary surface: that of a cylinder spanned by a helix of the same radius. Today we regard the surface of a cylinder as a locally Euclidean geometry; in particular, notwithstanding its obviously curved aspect, it allows to report on it flat representations without any aberration. This attribute was exploited in long historic narrations in imperial Rome, who devised for this the cochlear column. The composition is in fact typically Roman. Mathematically, all aspects of a helix on a cylinder are exploited in the historiated frieze. The frieze develops as a time-line narration of events; the rhythm of repetition of some key characters introduces a diachronicity, being modulated by the vertical rhythm of the helix; we also study the extent the possibility of visual representation and perception without aberration, computing the radius of the neighborhoods where the geometry is locally Euclidean, and intersecting it with that of the visual cone of an onlooker standing in some specific place.

This paper is organized as follows. In Section 2 you will find the history of this celebrative monument, its origins, and its later uses. In Section 3 we discuss the mathematical part: a modern view, as locally Euclidean surface highlights possibilities of visual representation; we also report on the origin of the term cochlear column, which has definitely a scientific derivation. In Section 4, finally, we draw together the mathematical periodicities with a reading strategy for the actual onlooker.

2 Roman Cochlear columns: their historical derivation and fortune, an excursus.

The decorated cochlear column is a celebrative architecture of imperial Rome, and in fact, only developed here, as far as we know. From a figurative standpoint, the historiated column transposes and accommodates the "triumphal painting", a typical Roman tradition, and so figurative language becomes an effective mean of Imperial propaganda. It combines two main typologies of the celebrative art in Roman tradition: the honorary column, and the relief sculpture on a continuous freeze [12].

Fig. 1. Two typologies of the celebrative art in Roman tradition
The relief sculpture on a continuous freeze: A) Parthenon, frieze with Panathenaic Procession, Athens; B) Ara Pacis: processional frieze showing members of the Imperial household, Rome.
The honorary column: C) Column of Phocas, Roman Forum, Rome

Continuous sculpted friezes, on "flat" background surfaces, were also already in use in Greece, e.g. the Panathenaic Procession described on the Ionic frieze of the Parthenon (Athens, 5th century BC)[1]. The large frieze, though, is not so much narrative, as descriptive of a spatially long ritual

[1] This beautiful frieze is so long that, alas, the original has been separated into pieces and scattered around the world: The appearance of entire frieze has been reconstructed in cast in the Acropolis Museum (Athens, architect B. Tschumi, 2008). From its parts held at the British Museum in London, those still onsite at Athens, and several fragments from other Museums and institutions.

procession and games. A Roman example of historic narrative frieze is the relief of the Augustus's procession on the Ara Pacis. The Ara Pacis celebrates the *pax augusti* (peace) "donated" by the emperor to Hispania and Gaul (Rome, 9 BC).

In cochlear columns, object of the present study, the continuous frieze is arranged on a ribbon, winding up the column's shaft; this allows for long narrations of events taking place over a long time span.

The idea of a "figurative ribbon" winding up and spiraling around the column's shaft seems to be new and peculiar to Roman art. Several hypothesis have been set forward to its genesis. We report those cited by Settis [12]:

1. The ribbon is inspired by the *rotulo*, early form for a book, or *volumen* (volume) used in the ancient libraries and consisting of a strip of flexible material so that it could be rolled in various ways. It could be considerably long [1].
2. The ribbon is inspired to the habit of winding some fabric around columns and piers on festivities; the fabric was an ornament, and in turn, could be decorated with images.

We report these two because they are also both congruent with our studies. In fact, the first instance ever of such art, appeared in the Trajan column, situated in the inner courtyard of libraries, this the visual art would reflect visually the *rotuli* contained in the library in their narrative development. On the other hand, a ribbon of fabric can be wound around a cylinder without folds and creases, depending on the relation between its height and the slope of the helix.

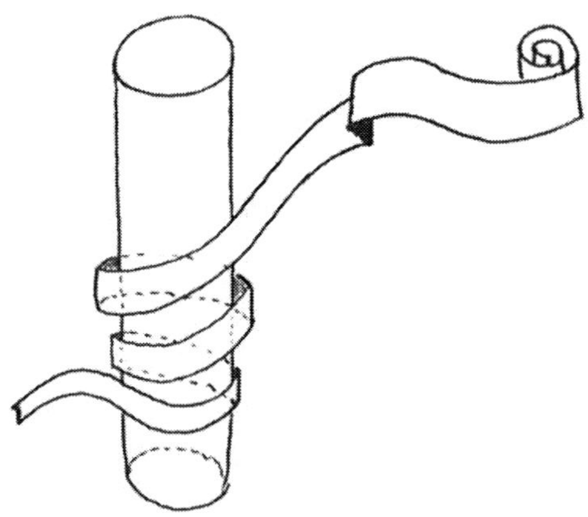

Fig. 2. A historiated frieze on a flat ribbon. The ribbon arranged on the cylinder without loosing its local flatness

The first cochlear historiated column is that of Trajan, built to celebrate the emperor's victory over the Dacians (Rome, Trajan's Forum, 110-113 A.D.). The architect of the forum was probably Apollodorus from Damascus, who also conceived the column. The sculpture are due to one or more sculptors under the guide of a *maestro*, whose name we do not know [2].

The Trajan column was hereafter taken as reference model for other imperial columns:

- In Rome, not much later, was built another historiated column, the Colonna Antonina (180-193 A.D.) or "of Marco Aurelio".

- A couple of centuries later, in Constantinople two more columns were built: the Columns of Theodosius (386 A.D.) and that of Arcadius. Neither one is preserved to our days, but we have some fragments and some historical drawings.

Fig. 3. Trajan's Column, the first cochlear historiated columns (Rome, 110-113 AD) and Column of Marcus Aurelius (Rome, 180-193 AD)

The celebrative cochlear column becomes hereafter a reference model crossing different centuries, styles and narrated historical or political events. Some other examples of celebrative cochlear columns are (Fig. 4):

- Hildesheim, Germany, 1022: the Column of Saint Bernnward, depicting scenes of Christ's life;
- Vienna, Austria, 1715-1737: the two columns flanking the Karlskirche, depicting scenes of the life of Saint Carlo Borromeo;
- Paris, France, 1805-1811: the famous column in Place Vendôme, commemorating Napoleon's victories at Austerlitz;
- Astoria, Oregon Usa 1926, depicting the American epic.

Fig. 4. The mentioned cochlear columns respectively from: Hildesheim; in Vienna; Paris; Oregon

3 Helices, some mathematical aspects.

A helix is the composition of a rigid uniform rotation in a horizontal plane and a uniform translation along an orthogonal axis. Thus defined, a cylindrical helix evolves around a solid cylinder, maintaining a constant slope:

$$X(t) = r\cos(t)$$
$$Y(t) = r\sin(t) \qquad (1)$$
$$Z(t) = p\,t$$

Where r is the radius of the cylinder and p is the *slope* of the helix.

We will discuss only the surface of such cylinder, which has zero Gaussian curvature, and therefore we prefer to regard it as a locally Euclidean geometry [10]. A modern way to read a locally Euclidean 2d geometry is to see it as the quotient space of the plane R^2 with respect to an equivalence relation introduced by an isometry, in this case a translation (Fig. 5).

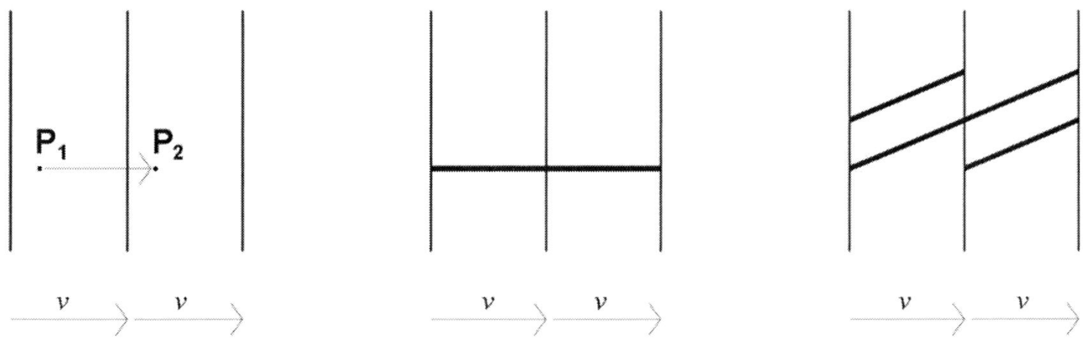

Fig. 5. Surface of cylinder defined as equivalence on the plane: points are equivalent if taken into each other by a translation of vector ***v***

Several sets can represent the quotient space in the plane; when the set is connected it is called the "fundamental domain" of the geometry defined by the equivalence relation. One can then think of the cylinder as the realization in 3d of this fundamental domain, for instance by imagining that we could glue lines that are equivalent under the translation. In such description lines orthogonal to the translation vector, remain as such; lines parallel to the translation vector become finite segments (or circles on the surface). All other straight lines become helices on the final imagined surface (Figs. 5, 6); the two representations (helix on a surface, straight line on a plane) share the same slope. This representation highlights the uniformity of the cylindrical surface, while thinking of a three-dimensional object naturally divides the surface into front and back (not visible), and into more areas depending on perception. Images can locally be represented exactly as on a flat surface, and will be perceived correctly by an onlooker moving around the solid.

To the spectator today, standing at street level, or to our readers, this might seem as a sophisticated geometrical condition, and out of the point: to appreciate the narration, today, we go to printed photographs. On the contrary, we think that a locally Euclidean geometry is implicitly weaved in the construction of the narration. Documentation is available [3, 4] as to the fact that spectators looked at the Trajan column by going around the column on balconies of the two-story Libraries that surrounded it (Fig. 8).

Some go further by thinking that spectators would also climb to next level of the library, but this idea is not necessary to our argument. What is important is to reconstruct what a spectator could see from that exact distance, even without moving, which we will discuss in next section.

There is evidence that besides the linear organization along the helix, that narration was also planned to have episodes read on the same side, in vertical, thus respecting both the parametric order of the helix, and the behavior displayed vertically due to the periodicity of the helix. In system (1) points corresponding to parameter values of $t_1 = t_0+2\pi$ are on the same vertical line.

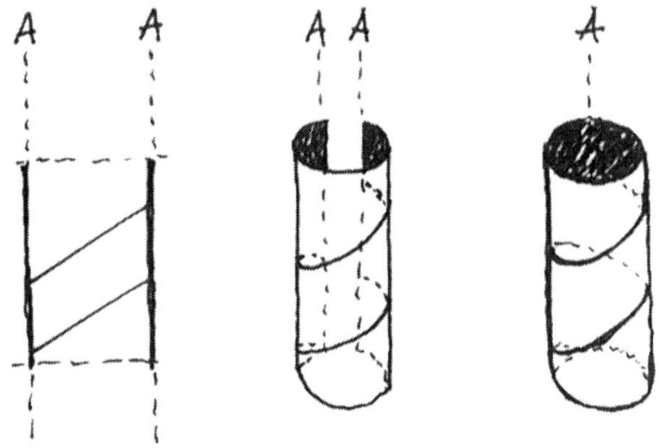

Fig. 6. Recomposition of three dimensional object by identification at the borders

3.1 The "*cochlea*".

A cochlea can denote several objects, all deriving their name from the Greek *koklias*, or the Latin *cochlea*. It was one of the first historical idraulic machines, the Archimedes' screw, devised to lift water and still in use to lift incoherent materials. It has also been used to denote the spiral stair. Mathematically, all the material realizations of *cochleae* have in common the constant inclination with respect to the horizontal plane, which while being functional in the "machine" makes the device geometrically very clear. In fact, representations of drawing machines to plan a crew can already be found in the drawings by Heron of Alexandria, passed on to us via the Persian manuscripts [8]. The early supporter of Galileo Guidobaldo Del Monte, himself a mathematician and astronomer, devoted a whole book to the cochlea [6].

Cochlear columns are called this way both for the helix resulting on their surface, and because internally they carry a spiral staircase (Fig. 7). The stair allowed going up to the summit, to the base of Emperor's statue. Common to all such mathematical and mechanical uses, is a curve or a surface of constant vertical slope. As well known, in the mechanical case the constant slope was functional to lifting materials, much as in the spiral stairs it is to walking up. In our case the constant slope is the mathematical expression of this flat ribbon with parallel borders, i.e. very patently a locally Euclidean surface. In fact, also Giacomo Martines, a historian of architecture, sees in fact the Trajan's columns as an "exercise in Hellenistic mechanics" [9].

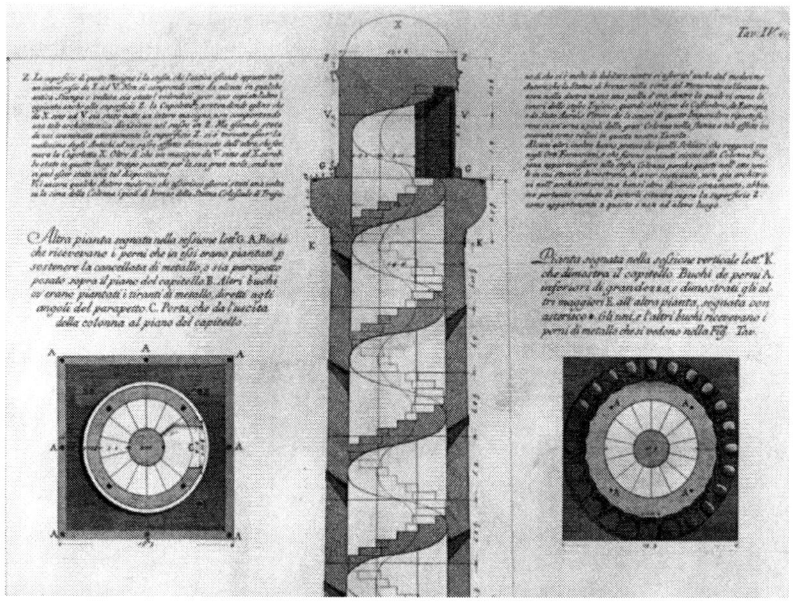

Fig. 7. G.B. Piranesi, Trajan's column (original title: *Trofeo o sia magnifica colonna coclide di marmo [...]*, 1774-79). Vertical section with spiral staircase

4 Mathematical Rhythms, Narration Rhythms.

The present condition of visibiltiy of the column is quite different from the original. Today the onlooker perceives the reliefs either form the level of the present road, or from the level of the ancient Forum. The Column is thus isolated in a wide space, including Piazza Venezia, and can thus be seen from very far away. In ancient times, the Column was in a narrow courtyard (a square of 20mt side), in the trajan's Forum; two libraries were the sides of this courtyard [3, 4]. Fig. 8 gives a reconstruction of this situation. The onlooker had a close-up vision, and some of the reliefs underwent aberrations due to distance, many of which were perceived in foreshortening (Fig. 9).

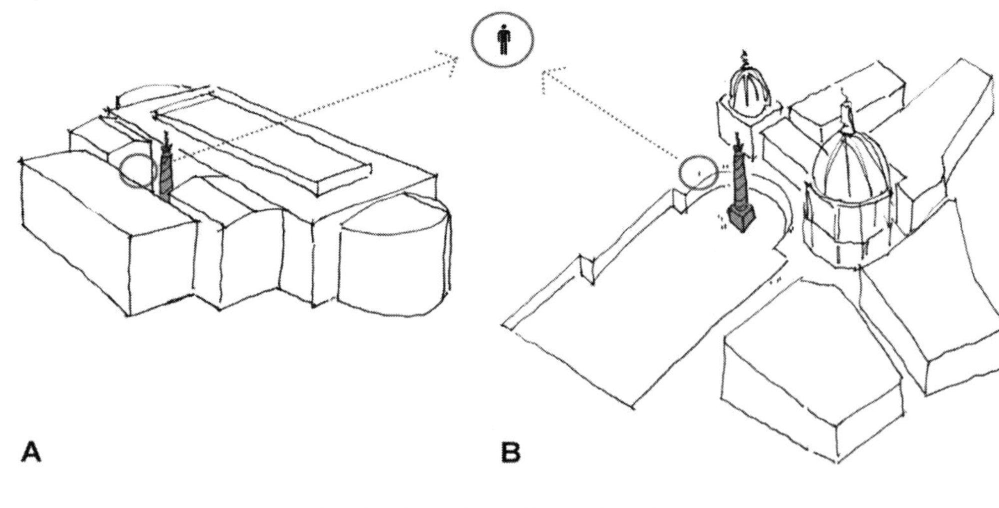

Fig. 8. Onlooker at Trajan's Column:
A) at the windows in the original forum complex. B) at street level in its urban context today.

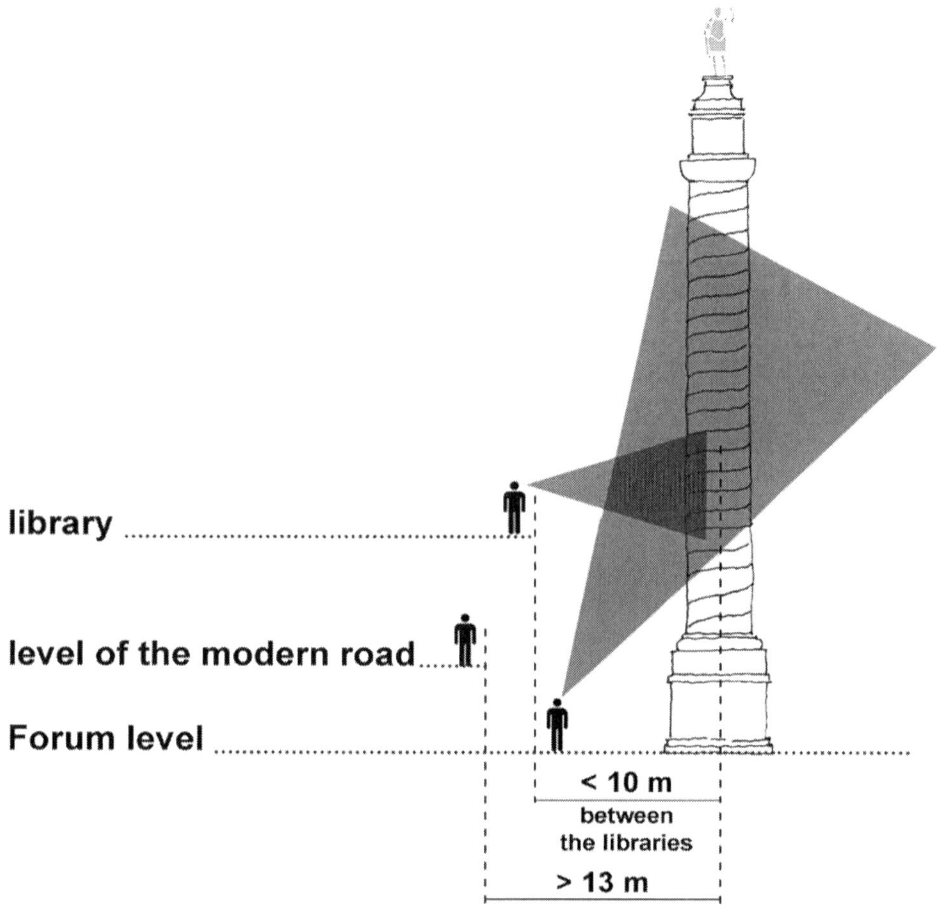

Fig. 9. The two levels at which the ancient onlooker would stand: library level, and Forum level, at the bottom of the Column. The corresponding optical cone is highlighted

A ribbon winding up a column therefore suggest two different paths of reading.

A historical, chronological reading that is continuous and develops along the spiral. The reading follows the narration much as in writing, i.e. from left to right, which, in this case, implies also from bottom to top. One then imagines the onlooker moving around the columns and raising his eyes to follow the narration. His interpretation will be temporal.

A discontinuous order, going from one coil to next, along the vertical. The reading then follows figurative correspondences among successive coils, and gives another interpretation of the events: an a-temporal interpretation, digging into symbols and yielding to propaganda [11].

We will only comment on three examples, but there are more, depending on the side of the column.

The Emperor's image is repeated over and over, in several instances. Some of the scenes are aligned along vertical correspondences. This creates redundancy and a "visual vocabulary" [11].

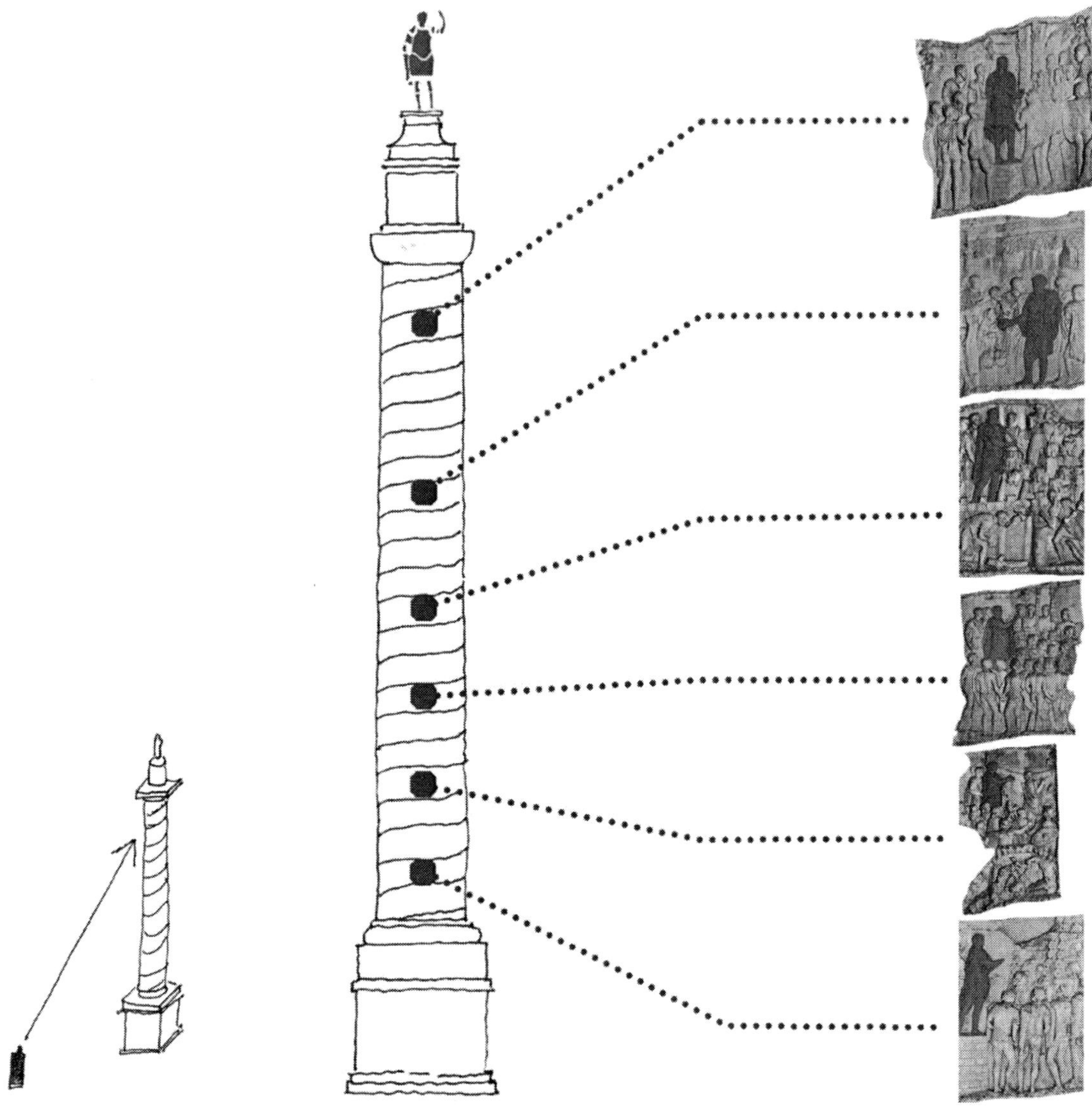

Fig. 10. The Emperor's portraits, aligned in one strip visually available to a standing onlooker

Celebration of Victory. An a-temporal vertical narration depicts the symbolic unraveling and disentangling of victorious battles. All on the same vertical line, we find, starting from the bottom: the premonition of victory, the Winged Victory, and the suicide of the Dacian king, sanctioning the defeat of the Dacians.

A pontoon bridge is laid for crossing by the troops. The repetition of the scene exalts the sophisticate abilities of military engineering. This type of scene appears only three times in the almost 200mt long frieze; alignment along the vertical favors its understanding.

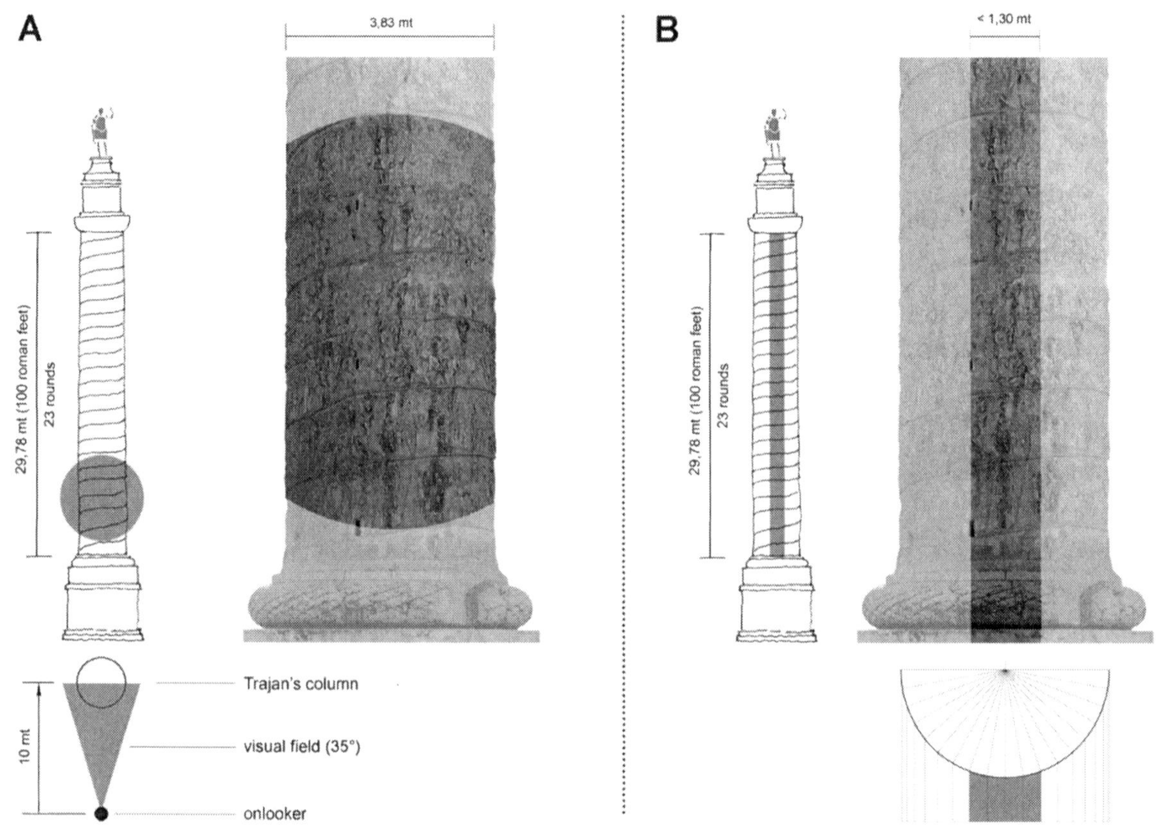

Fig. 11. A) Computed "visual cone" of a perceiver standing in original spot.
B) Frontal area: vision not affected by curvature of the cylinder

Acknowledgements

We thank Elisa Conversano for first pointing out the vertical correspondences studied by historians. We thank Aplimat for providing a forum of discussion for these interdisciplinary points of view.

References

[1] BERTELLI C., entry "Rotulo", *Enciclopedia dell'Arte Antica Treccani* (1965).

[2] BIANCHI BANDINELLI R., Il maestro delle imprese di Traiano, Venezia 2003.

[3] CARANDINI A., *Atlante di Roma antica*, vol.1-2, Milano 2012.

[4] COARELLI, F. *La Colonna Traiana*, Roma 1999.

[5] CONVERSANO, E., FRANCAVIGLIA, M., LORENZI M. G., TEDESCHINI LALLI L. Persistence of Form in Art and Architecture: Catenaries, Helicoids and Sinusoids, APLIMAT J. of Appl. Math., 4 (2011)

[6] DEL MONTE, G., *Sulla Coclea, Libri Quattro [...]* (Venezia 1615) modern edition MAGNINI E., MOLARI P.G. eds, Forni editore, Bologna (2013).

[7] FALCOLINI C., TEDESCHINI LALLI L. Compounds of helical curves: Medieval twisted

columns, proceedings 15th conference on Applied Mathematics APLIMAT 2016 (2016).
[8] FERRIELLO G., GATTO M., GATTO R., *The «Baroulkos» and the «Mechanics» of Heron,* Firenze, 2016.
[9] MARTINES, G. La struttura della colonna Traiana: un'esercitazione di meccanica alessandrina, in *Prospettiva, 32*, 1983, p. 60.
[10] NIKULIN V.V., SHAFAREVICH R. *Geometry and Groups*, Springer (1983).
[11] SETTIS, S. *La Colonna Traiana*, Torino 1988.
[12] SETTIS, S., MAFFEI, S., entry "Colonna", *Enciclopedia dell'Arte Antica Treccani* (1994).
[13] STRANEO, S. L., entry "Coclea", *Enciclopedia italiana Treccani* (1931).

Current Address:

Carlini Alessandra, Ph.D. architect
Liceo Scientifico Cavour
Via delle Carine 1, I-00146 Rome, Italy
E-mail: alessandracarlini@yahoo.com

Tedeschini-Lalli Laura, Professor
Dipartimento di Architettura
Via Madonna dei Monti 40, I-00146 Rome, Italy
E-mail: tedeschi@mat.uniroma3.it

Slovak University of Technology in Bratislava
Faculty of Mechanical Engineering
16th Conference on Applied Mathematics **APLIMAT 2017**
Proceedings

OPTIMISATION OF THE LINE ROUTING USING THE MULTI-CRITERIA EVALUATION OF ALTERNATIVES

ČEJKA Jiří (CZ), DOBIÁŠ Václav (CZ), LUDVÍK Miroslav (CZ)

Abstract. The paper is concerned with the assessment and design of the transport organization in the town of TÁBOR. The examined issue was the urban public transport analysis focused only on the lines routed in parallel towards Planá nad Lužnicí. The solution designed using the TOPSIS method was based on the carrier and the orderer's interest in optimizing the above stretch. Based on the results, the draft transport model of the backbone line No. 13 was created and the inappropriate lines were removed.

Keywords: multi-criteria evaluation of alternatives, TOPSIS, analysis, optimization, line routing

Mathematics Subject Classification: 90B50

1 Introduction

The addressed issue of this paper is to assess and design the transportation organization in the selected town, i.e. the optimization of the parallel line loops using the TOPSIS method.

This method of multi-criteria evaluation of the alternatives is aimed to determine the order of benefits of individual alternatives from the perspective of chosen criteria, in which case the alternative of the best location represents the best compromise variant. The results obtained using the various methods are of a subjective nature and may differ from each other. For designing the solution, the TOPSIS (Technique for Order Preference by Similarity to Ideal Solution) method was chosen.

The TOPSIS method is based on the selection of the variant which is as close as possible to the ideal variant and as far as possible from the basal variant. The maximised nature of all criteria is expected. Unless all criteria are maximised, they need to be converted to the maximised criteria.

The order of benefit of the given decision-making alternatives set by some of the multi-criteria evaluation methods depends especially on the weights of individual criteria and the method used. Examining of the sensitivity of the preferential order of alternatives to determining the importance of individual criteria falls into the area of doing experiments on the models when the multi-criteria evaluation of alternatives is carried out using the changing weights of criteria. Only when the preferential order of the alternatives (or at least the best evaluated alternative) does not change under the changes, this order is not sensitive to inaccuracy of the determined weights and it can be said that the importance of individual criteria was properly evaluated by the decision-maker. If the preferential order is considerably sensitive to the changes in the criteria weights, their reliability needs to be increased.

The dependency of the preferential order of alternatives on the used method of their multi-criteria evaluation is based on the fact that individual methods stem from various, basically, simplifying assumptions and, as a result of a different approach to the "compromise alternative", they use various calculation procedures.

For determining the weights of the criteria, the Füller´s method was chosen. This method compares only two criteria. As a result, the benefits of the criteria being compared can be assessed more easily. One of the options for evaluation of these comparisons is provided by the Füller triangle. Provided that individual criteria are numbered with the order numbers 1, 2, . . . , n, the Füller triangle is formed by the double-lines in which each pair of criteria occurs only once (see Tab. 1). For each pair the decision-maker will circle or otherwise indicate the number of that criterion which he considers more important so that the sum of the numbers $_j$ in the circle represents the number of its preferences for the criterion Kj. [1]

1.1 Analysis of the Current State of the Examined Issue

With respect to the necessary optimization of the transport direction Planá nad Lužnicí via Sezimovo Ústí, Sídliště nad Lužnicí, the bus station and Klokoty in Tábor, the multi-criteria decision-making will be used to choose the line which is best for the passengers from the routing perspective. The urban public transport in Tábor, Planá nad Lužnicí and Sezimovo Ústí is provided by the Urban Public Transport Division of the Comett Plus company. In this area, the company operates a total of 14 lines of the urban public transport, served in the peak hours by 31 buses. The whole transport network is formed by five main directions of the lines. The total length of the transport network is 65 km and of all lines over 270 km (both directions). More than 640 connections are dispatched during a working day. The optimization will concern exclusively one important direction where there are a few lines routed in parallel in more than 90 percent of their routes. In this direction there are the lines which are used most and also the lines which are, on the other hand, used less or inappropriate due to the identical routes with the minor detours. Hence, only the four lines (11, 13, 16, and 17) will be assessed. [2]

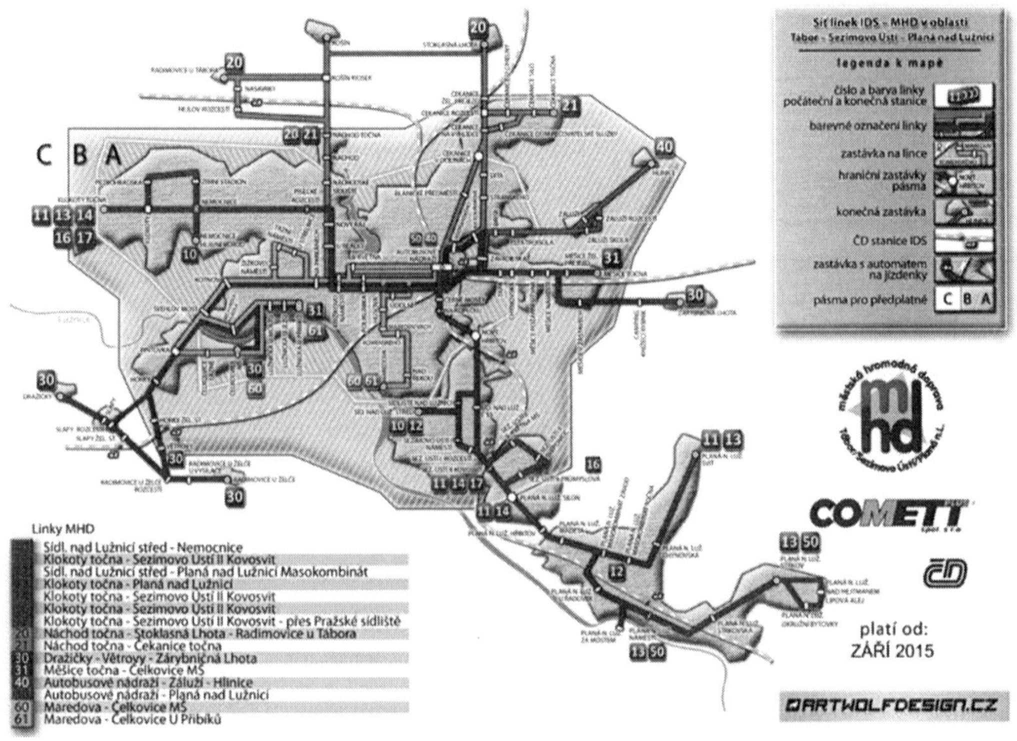

Source: Official website of comettplus.cz, 2015

Fig. 1. Map of the urban public transport in Tábor

1.1.1. Line No. 11

The line No. 11 regularly connects the bus stop next to the Kovosvit company located in Sezimovo Ústí II, the bus station and the Klokotské housing estate. The route of five connections per day will be extended up to the bus stop next to the meat-processing plant (Masokombinát) in Planá nad Lužnicí and three of them continue to the bus stop called Planá nad Lužnicí - Svit. This bus stop is located only a few hundred of meters from the motorway D3. The line No. 11 offers three connections between midnight and 5 a.m. Then, the connection is more intensive as it is used by the school buses. This line has even one request stop (Nový hřbitov, New cemetery) on its route. The route length from the Klokotské housing estate to the bus stop Sezimovo Ústí II – Kovosvit is approximately 12.5 km and is travelled by the bus in approximately 25 minutes. If the bus continues to the bus stop Planá nad Lužnicí - Svit, its total route is extended to nearly 16 km and the bus journey takes approximately 33 minutes.

1.1.2. Line No. 13

The main goal of this line is to ensure the transportation of passengers between the Klokotské housing estate, the bus station and the square in Planá nad Lužnicí. The bus, however, extends its route eight times a day to Strkov, an urban part of Planá nad Lužnicí. This line is used by the meat-processing plant staff (maso Planá) which is listed as the line No. 13´s bus stop. The first morning bus leaves the bus stop Klokoty točna (turntable) at 3.45 a.m. and the last night bus at 11:45 p.m.

This line is very important not only for the employees of the local companies but also the pupils who live in Planá nad Lužnicí and attend the schools in Tábor. The route between the Klokoty turntable and the square is 15 km long and is travelled by the bus in approximately 33 minutes. If the bus goes to Strkov, the entire length is extended to 16.5 km and the total time to 36 minutes.

1.1.3. Line No. 16

The line No. 16 has the same beginning as the lines No. 11 and 13 (i.e. Klokotské housing estate) and continues via the bus station to Sezimovo Ústí II, namely the bus station Průmyslová. This line is not too busy as it is served by 11 connections only from Klokoty and 11 connections from Sezimovo Ústí a working day. This line was supposed to ensure the transportation of the passengers to the bus station at the Tesco store. This store and COMETT PLUS spol. s r. o., however, terminated the contract and this bus stop was cancelled in December 2014. The line, however, remained, except for the above bus stop. There are a total of 31 bus stops on the line No. 16 at present. The route is 12.3 km long and is travelled by the bus in approximately 29 minutes. The first bus leaves the Klokoty turntable bus stop at 6.33 a.m. and the last bus departs from Sezimovo Ústí II at 9:00 p.m.

1.1.4. Line No. 17

This line again starts at the Klokotské housing estate, however, unlike the others it goes along the Petrohradská street, the ice pool to the bus station and then it continues to Sezimovo Ústí II to the Kovosvit bus stop where it ends. The first morning bus leaves the Kovosvit bus stop at 5:20 a.m. and the last night bus leaves the Klokoty turntable bus stop at 9:50 p.m. The route of this line is approximately 13.5 km long and takes 28 minutes. The low-floor buses are predominantly used on this line.

2 Design of Solution using the TOPSIS Method (Multi-Criteria Decision-Making)

To solve the transportation problem in Tábor, the evaluation criteria which usually concern the passenger load factor of the means of transport, the peak hours, etc. were chosen. The capacity of the buses is, thus, enhanced in the peak hours and then reduced.

The problem was designed using the TOPSIS method. Individual weights were calculated using the method of the pair comparison of criteria – the Füller triangle method. Individual alternatives and preferences of individual weights were set by COMMET PLUS.

The basic criteria include:

- Passenger load factor of the connection in the morning peak hours,
- Passenger load factor of the connection in the afternoon peak hours,
- Passenger load factor of the connection in the morning slack period,
- Passenger load factor of the connection in the afternoon slack period,
- No. of vehicles.

After finding the times of the peak hours and the slack periods, the documents on the occupancy provided by COMETT PLUS s.r.o. were intensively analysed to implement the optimization. The documents show the detail occupancy of individual lines throughout the week. The drawback of these electronic documents is the fact that they contain a vast amount of data which could not be processed into a simpler form which can be used as a preview.

Füller triangle method (Tab. 1):
Chosen criteria:

- K1 - Passenger load factor in the peak hours (7:00 a.m. - 7:45 a.m.)
- K2 - Passenger load factor in the peak hours (2:30 p.m. - 3:30 p.m.)
- K3 - Passenger load factor in the slack period (11:00 a.m. - 12:00 p.m.)
- K4 - Passenger load factor in the slack period (7:30 p.m. - 8:30 p.m.)
- K5 - No. of vehicles.

The chosen criteria include twice the passenger load factor of the line in the peak hours and twice the passenger load factor of the line in the slack period as viewed by the carrier. The slack period is the period of time when the passenger transport is not as busy as during the peak hours which represent the increase in the number of passengers. The obtained material showed that the peak hours are between 7:00 a.m. and 7:45 a.m. when most passengers commute to their schools or jobs. The second peak hours are in the afternoon between 2:30 p.m. and 3:30 p.m. when the passengers usually return home. On the other hand, the slack periods are between 11:00 a.m. and 12:00 p.m. and between 07:30 p.m. and 08:30 p.m. where the lower amount of passengers uses the urban public transport. It should be, however, noted that the peak hours and the slack periods were examined on the weekdays. (Tab. 2) [1- 5].

Fuller's method is one of scoring methods. It is used in the case when large number of criteria is applied and the user will have difficulties to score each criterion. The principle of this method is based on the fact that the criteria are submitted to the user in pairs gradually. The user selects only one criterion which is more important for him from each pair. The important criteria will be evaluated by points and this point will be summed. (Tab. 1) [1- 5].

Füller triangle method				Criterion	Preference	Preference + 1	Weight
K1	K1	K1	K1	K1	3	4	0.27
K2	K3	K4	K5				
	K2	K2	K2	K2	4	5	0.33
	K3	K4	K5				
		K3	K3	K3	1	2	0.13
		K4	K5				
			K4	K4	0	1	0.07
			K5				
				K5	2	3	0.2

Source: Author

Tab. 1. Füller triangle

TOPSIS method evaluates variants in terms of their distance from the ideal and basal variant. The algorithm of the process is (1-7):

The minimization criteria are transferred to the maximization criteria (Tab. 2)
Normalized criterial matrix R_{ij} is designed (see (1), (2))
Weighted criterial matrix Z_{ij} is calculated (3)
Ideal and basal variant relative to the values of matrix Z_{ij} are found (4)
Distances between variants and the basal variant are calculated (5), (6)
Relative indicators of distances between variants and the basal variant are calculated (7)

Line	Average passenger load factor peak/slack period				No. of vehicles
	(7:00 a.m. - 7:45 a.m.)	(02:30 p.m. - 03:30 p.m.)	(11:00 a.m. - 12:00 p.m.)	(07:30 p.m. - 08:30 p.m.)	
11	38.47	43.44	19.29	20.24	4
13	45.2	39.45	24.96	18.41	4
16	42.31	29.63	25.39	18.6	2
17	38.57	35.11	24.39	15.32	3
	max	max	max	max	min

Source: Author

Tab. 2. Evaluation from the carrier's viewpoint

Calculation of the standardized criteria matrix:

To simplify the work, we will first calculate the numerator which is a part of the main formula:

$$\sqrt{\sum_{i=1}^{m} y^2{}_{ij}} \tag{1}$$

Then we will add it in the main formula:

$$r_{ij} = \frac{y_{ij}}{\sqrt{\sum\limits_{i=1}^{m} y_{ij}^2}}; \; i = 1, 2, \ldots, m; j = 1, 2, \ldots, n. \tag{2}$$

0.466	0.583	0.408	0.555	0.000
0.548	0.529	0.528	0.505	0.000
0.513	0.398	0.537	0.510	0.894
0.468	0.471	0.516	0.420	0.447

Source: Author

Tab. 3. Standardized table (r_{ij})

Weights: (w_j)

1st column 0.27

2nd column 0.33

3rd column 0.13

4th column 0.07

5th column 0.20

Each column will be gradually multiplied by the weights belonging to the certain column:

$$z_{ij} = w_j r_{ij}. \tag{3}$$

0.126	**0.192**	**0.053**	**0.039**	**0**
0.148	**0.175**	**0.069**	**0.035**	**0**
0.139	**0.131**	**0.070**	**0.036**	**0.179**
0.126	**0.155**	**0.067**	**0.029**	**0.089**

Source: Author

Tab. 4. Multiplying by the weights z_{ij}

Creation of the ideal and basal alternatives (H-ideal, D-basal)

$$h_j = \max_i z_{ij};\ j = 1, 2, \ldots, n$$
$$d_j = \min_i z_{ij};\ j = 1, 2, \ldots, n \quad (4)$$

H ={0.148; 0.192; 0.070; 0.039; 0.179}

D ={0.126; 0.131; 0.053; 0.029; 0}

Calculation of distance from the ideal alternative:

$$d_i^+ = \sqrt{\sum_{j=1}^{n}(z_{ij} - h_j)^2};\ i = 1, 2, \ldots, m \quad (5)$$

$(0{,}126 - 0{,}148)^2 + (0{,}192 - 0{,}192)^2 + (0{,}053 - 0{,}070)^2 + (0{,}039 - 0{,}039)^2 + (0 - 0{,}179)^2 = \sqrt{0{,}032814} = \mathbf{0{,}181}$

$(0{,}148 - 0{,}148)^2 + (0{,}175 - 0{,}192)^2 + (0{,}069 - 0{,}070)^2 + (0{,}035 - 0{,}039)^2 + (0 - 0{,}179)^2 = \sqrt{0{,}032347} = \mathbf{0{,}180}$

$(0{,}139 - 0{,}148)^2 + (0{,}131 - 0{,}192)^2 + (0{,}070 - 0{,}070)^2 + (0{,}036 - 0{,}039)^2 + (0{,}179 - 0{,}179)^2 = \sqrt{0{,}003811} = \mathbf{0{,}062}$

$(0{,}126 - 0{,}148)^2 + (0{,}155 - 0{,}192)^2 + (0{,}067 - 0{,}070)^2 + (0{,}029 - 0{,}039)^2 + (0{,}089 - 0{,}179)^2 = \sqrt{0{,}010062} = \mathbf{0{,}100}$

Calculation of distance from the basal alternative:

$$d_i^- = \sqrt{\sum_{j=1}^{n}(z_{ij} - d_j)^2};\ i = 1, 2, \ldots, m \quad (6)$$

$(0{,}126 - 0{,}126)^2 + (0{,}192 - 0{,}131)^2 + (0{,}053 - 0{,}053)^2 + (0{,}039 - 0{,}029)^2 + (0 - 0)^2 = \sqrt{0{,}003821} = \mathbf{0{,}062}$

$(0{,}148 - 0{,}126)^2 + (0{,}175 - 0{,}131)^2 + (0{,}069 - 0{,}053)^2 + (0{,}035 - 0{,}029)^2 + (0 - 0)^2 = \sqrt{0{,}002712} = \mathbf{0{,}052}$

$$(0{,}139-0{,}126)^2+(0{,}131-0{,}131)^2+(0{,}070-0{,}053)^2+(0{,}036-0{,}029)^2+(0{,}179-0)^2=\sqrt{0{,}032548}=\mathbf{0,180}$$

$$(0{,}126-0{,}126)^2+(0{,}155-0{,}131)^2+(0{,}067-0{,}053)^2+(0{,}029-0{,}029)^2+(0{,}089-0)^2=\sqrt{0{,}008693}=\mathbf{0,093}$$

Calculation of distance of alternatives from the basal alternative:

$$c_i=\frac{d_i^-}{d_i^+ + d_i^-};\ i=1,2,\ldots,m \tag{7}$$

$$\frac{0{,}062}{0{,}181+0{,}062}=\mathbf{0,255}$$

$$\frac{0{,}052}{0{,}180+0{,}052}=\mathbf{0,224}$$

$$\frac{0{,}180}{0{,}062+0{,}180}=\mathbf{0,744}$$

$$\frac{0{,}093}{0{,}100+0{,}093}=\mathbf{0,482}$$

Listing the results in the table:

Line No. 11	3rd place
Line No. 13	4th place (backbone line)
Line No. 16	**1st place (can be combined)**
Line No. 17	2nd place (can be combined)

Source: Author

Tab. 5. Evaluation

3 Results Discussion

The results obtained using the TOPSIS method make it clear that the least used line is the line No. 16 due to the lower number of the vehicles deployed. The weight allocated to the criterion and concerning the number of the vehicles may be distorted due to unfamiliarity with the transport practice. The line can, thus, be cancelled and the remaining buses can be operated on the busier lines.

The line No. 13 is, on the other hand, most loaded and will be used to optimize the less used lines, i.e. the lines determined by the method as less loaded (the numbers of passengers and the vehicles were minimized). [1-3]

4. Proposed Solution for the Transport Model of the Line Routing in Tábor

Although Tábor ranks amongst the smaller towns, it has a wide range of lines serving the town. The basis of the proposed optimization of the number of the existing lines in the chosen direction will be the fact that most lines will be cancelled. Only the line No. 10 and the line No. 13 will remain (the former will remain without any changes).

The line No. 10 is currently the most loaded and most popular line from the passengers´ perspective. An important fact is that it interconnects the biggest housing estate in Tábor and, unlike the other lines in this direction, provides the transport up to the entrance to the Tábor Hospital. This line will, therefore, be kept as this is also required by the municipal office. The resulting proposal will be transferring of the backbone line from the No. 10 to the line No. 13. The lines No. 11, 16 and 17 will be cancelled and the line No. 10 (not included in the evaluation) will become the additional line. The line No. 13 is the longest line and serves an overwhelming majority of the cancelled line bus stops. Practicality of the final solution proves that using of the mathematical methods can have a fixed place in the transport practice. [1- 5]

References

[1] BARTUŠKA, L., ČEJKA, J., CAHA, Z, The application of mathematical methods to the determination of transport flows, Journal Nase More, University of Dubrovnik 2016, ISSN: 04696255.

[2] ČEJKA, J., Bartušková, P., Bartuška, L., Application of mathematical methods in transport and logistic area, APLIMAT 2016 - 15th Conference on Applied Mathematics 2016, Pages 225-235

[3] MIKLÓŠ, R. Surfaces. In Proceedings of 5th International Conference of Mathematics in Technical and Natural Sciences, Kraków, 2000, ISBN 83-86888-02, s. 63–68.

[4] SMETANOVÁ, D., VARGOVÁ, M., BIBA, V., HINTERLEITNER, I. Mercator's Projection – a Breakthrough in Maritime Navigation. Nase More, Dubrovnik: University of Dubrovnik, 2016, roč. 63, č. 3, s. 182-184, ISSN 0469-6255.

[5] VACKA, M., SMETANOVÁ, D. Project Solution in Designing Communications and Routes for Fieldwork. Nase More, Dubrovnik: University of Dubrovnik, 2016, 2016, roč. 63, č. 3, s. 223-226, ISSN 0469-6255.

Current address

Čejka Jiří, Ing. Ph.D.
Institute of Technology and Business České Budějovice
Okružní 10, 370 01 České Bbudějovice, Czech Republic
Tel. number: +420 602 88 66 48, e-mail: cejka@mail.vstecb.cz

Dobiáš Václav, Mgr.
Institute of Technology and Business České Budějovice
Okružní 10, 370 01 České Bbudějovice, Czech Republic
Tel. number: +420 602 43 04 56, e-mail: dobias@mail.vstecb.cz

Ludvík Miroslav
Institute of Technology and Business České Budějovice
Okružní 10, 370 01 České Bbudějovice
Tel. number: +420 777 97 87 77, e-mail: miroslav.ludvik@znalecka-kancelar.com

Slovak University of Technology in Bratislava
Faculty of Mechanical Engineering
16th Conference on Applied Mathematics APLIMAT 2017
Proceedings

MATHEMATICAL MODELS FOR FOOD SANITIZATION PROCESS

CEPPITELLI Rita (IT), LA TONA Alice (IT), MONTANARI Luigi (IT)

Abstract. The expression mild technologies is used to define those food treatments able to achieve safe food by significantly reducing all the possible negative side-effects. In the specific case of thermal pasteurization and sterilization, the concept of mild technologies is linked to treatments with familiar abbreviations for the consumer: UHT (Ultra High Temperature) and HTST (High Temperature Short Time). Examining the technological parameters (time and temperature) to sterilize milk, we constructed a predictive mathematical model to determine in advance couples time-temperature that will ensure the sanitization of the product with minimizing the degradation of milk quality, in accord with the mild technologies treatments. The proposed model could be applied to other foods and to other food process technologies and could be used in future research projects.

Keywords: Bigelow model, Arrhenius model, first-order reaction kinetics, food sanitization process, mild technologies

Mathematics subject classification: Primary 34A34; Secondary 80A30

1 Introduction

The main food production control system has been, and it is still, focused on food safety. In the near past, very often some *hard treatment* were used to reach the maximum safety level, with particular regards to the use of long time temperature treatment applied for minimize the microbial food content. The use of long time temperature treatment to minimize the microbial food content, e.g. in milk and dairy products, implies the reduction of healthy nutrients (e.g. vitamins) and the deterioration of sensorial aspects.

In the last twenty years, one of the most relevant food technology topics is the research targeted to minimize this technological negative impact, maintaining the maximum food safety level.

The expression *mild technologies*([11]) is used to define those food treatments able to achieve safe food by significantly reducing all the possible negative side-effects. In the specific case of thermal pasteurization and sterilization, the concept of *mild technologies* is linked to treatments with familiar abbreviations for the consumer: UHT (Ultra High Temperature) and HTST (High Temperature Short Time) ([10]). As much as possible, the final product is similar to the fresh one, e.g. avoiding vitamins reduction and strong flavour changes.

The optimization of operations based on the heat transport is a consequence of an increased knowledge of the desired positive effects and negative side reactions of heating treatment ([4]).

In this paper we are interested in some kinetics occurring during the process of sterilization of milk.
The optimization of sterilizing effect is achieved by increasing the temperature and reducing time. This is made possible thanks to new technologies with an high efficiency of heat transfer.
Microbial mortality at constant lethal temperatures is traditionally considered a process following a first-order reaction kinetics with the rate constant of decrease (*First Bigelow model*).
During sterilization process several food compounds undergo changes. We deal with three different side-effects: the production of *furosine*, a typical *Maillard reaction* product ([1, 14]), the destruction of *thiamine* ([13]) and the regression of *lipase activity*.
Of course, the rate of each reaction above described increases in the temperature (*Arrhenius model*). We unify every kinetic model with the Arrhenius model and we construct a predictive mathematical model to determine in advance couples time-temperature that will ensure the sanitization of the product with minimizing the degradation of milk quality, in accord with the *mild technologies* treatments.

Of course, the proposed model could be also applied to other foods and to other food process technologies and could be used in future research projects. Its performances should guarantee a fast way of providing a simultaneous evaluation of data useful for individuate the parameters for producing safe and high quality food.

2 Thermal processes at a constant temperature

Microbial mortality at constant lethal temperatures is traditionally considered a process following a first-order reaction kinetics [7]. According to this model the organism's survival can be described by the *First Bigelow model*

$$\frac{dN}{dt} = -kN$$

where $N(t)$ is the microbial population survived after t seconds of thermal process and $k > 0$ is the rate constant of microbial destruction relating a homogeneous microbial population.

The solution, in semi-logarithmic coordinates, $N(t) = \ln N_0 - kt$ represents the *straight line microbial survival.*
The aim of thermal destruction is to reach the commercial sterility of the product.

During heat treatment several food compounds undergo changes. Concerning sterilization of milk, we are interested in three different side-effects: the destruction of *thiamine*, the regression of *lipase activity* and the production of *furosine*, a typical *Maillard reaction* product; these are considered as *quality markers*.

The Maillard reaction ([1, 14]) is strongly negative because leads to the production of compounds that give yellow-brown colour and smell of cooked milk. This is a typical example of negative side effects, not desired by consumers, and synonymous of the decay of product quality.
The formation of furosine follows a linear growth $F(t) = k_F t$ where $F(t)$ is the concentration of furosine after t seconds of heating treatment at a constant temperature and k_F is the constant increase rate.

The thiamine (vitamin B_1) is present in small amounts in milk, approximately 45 μg in 100 g of product; the losses during sterilization process are valued approximately 30-50 %.
The thiamine degradation ([13]) is considered a process following the first order kinetics

$$\frac{dB}{dt} = -k_B B^2$$

where $B(t) = \frac{B_0}{1+K_B B_0 t}$ is the concentration of thiamine after t seconds of heating treatment at a constant temperature, B_0 is the initial concentration and $k_B > 0$ is the rate constant of degradation.

The lipase is an enzyme naturally present in milk that catalyzes the hydrolysis of fats by causing the so-called *hydrolytic rancidity* that gives the food unpleasant odors and tastes. The lipase activity is normally evaluated to measure the acidimetry. The normal value in milk is 0.5, values greater than 1.5 cause the rancid taste. During thermal processes at a constant temperature, the kinetics of regression of lipase activity $\frac{dL}{dt} = -k_L L$ follows an exponential decrease $L(t) = L_0 e^{-k_L t}$, where L_0 is the initial value and $K_L > 0$ is the rate constant of regression.

3 Arrhenius model and non-isothermal heat treatmens

According to the Collision Theory, a reaction occurs only if the molecules that collide possess a minimal kinetic energy (*activation energy*) sufficient to break chemical bonds. Thus the rate constants k of each reaction depends in particular from the activation energy and the number of collisions, i.e. from temperature. Raising the temperature, the rate of reaction increases [2].

The temperature dependence of the kinetic parameters is adequately described by the *Arrhenius model*

$$\frac{dk}{dT} = \frac{E_a}{RT^2} k$$

where R is the universal gas constant and E_a is the activation energy, typical of each reaction. The solution is $k(T) = k_0 e^{-\frac{E_a}{RT}}$ where k_0 is a constant feature of the reaction (*collision frequency*) whose value depends on the frequency of collisions with the correct geometry. Note that $k_0 = \lim_{T \to +\infty} k(T)$ i.e. k_0 is the upper bound of $k(T)$.
In semi-logarithmic coordinates, we have

$$\ln k(T) = \ln k(T_0) + \frac{E_a}{R}\left(\frac{1}{T_0} - \frac{1}{T}\right)$$

which shows that an increase in temperature results in increased in the reaction rate, for equal E_a and concentration.

Arrhenius's model jointly with First Bigelow's model and the above described kinetic models, allows us to consider predictive models of non-isothermal heat treatmens.

We have unified every kinetic model with Arrhenius's model and considered systems of differential equations

$$\begin{cases} \frac{\partial N}{\partial t} &= -k(T)N \\ \frac{dk}{dT} &= \frac{E_a}{RT^2}k \end{cases}$$

$$\begin{cases} \frac{\partial F}{\partial t} &= k(T) \\ \frac{dk}{dT} &= \frac{E_a}{RT^2}k \end{cases} \quad \begin{cases} \frac{\partial B}{\partial t} &= -k(T)C_B^2 \\ \frac{dk}{dT} &= \frac{E_a}{RT^2}k \end{cases} \quad \begin{cases} \frac{\partial L}{\partial t} &= -k(T)L \\ \frac{dk}{dT} &= \frac{E_a}{RT^2}k \end{cases}$$

where $k(T)$ is the temperature dependent rate constant. The solutions of the above systems

$$N(t,T) = N_0 \exp(-k_0 \exp(-\frac{E_a}{RT})t)$$

$$F(t,T) = k_{0_F} \exp(-\frac{E_a}{RT})t$$

$$B(t,T) = \frac{C_0}{1 + C_0 k_{0_B} \exp(-\frac{E_a}{RT})t}$$

$$L(t,T) = L_0 \exp(-k_{0_L} \exp(-\frac{E_a}{RT})t)$$

represent the kinetics considered with respect to the two independent variables t, T where N_0, C_0 and L_0 are the initial conditions. The value of $k_0, k_{0_F}, k_{0_B}, k_{0_L}$ can empirically be calculated. To compute their values we use the data in the following Tab.1 [12] representing the constant rates (at 393 Kelvin temperature) and activation energies of the reactions we are studying.

reaction	$k_{393K}(s^{-1})$	$E_a(J/mol)$
microbial destruction	$1.10 \cdot 10^{-2}$	345357
production of furosine	$62.622 \cdot 10^{-3}$	81637
degradation of thiamine	$7.52 \cdot 10^{-4}$	100800
regression of lipase activity	$1.15 \cdot 10^{-2}$	53037

Tab. 1

The formulation of $N(t,T), F(t,T), B(t,T), L(t,T)$ as functions of time-temperature allows us to study various aspects of the problem by applying simple mathematical tools.

For example the trace $T = const$ give the spore's survival rate or the furosine, thiamine, lipasi concentrations as a function of time treatment at a given temperature. The level curve $N(t,T) = 10^{-4}$ is known as *Commercial sterility curve* (Fig.1) and it is formed by all the pairs time-temperature that assure the reduction of the spore concentration is equal to 10^{-4} $(spore/ml)$ (commercial sterility).

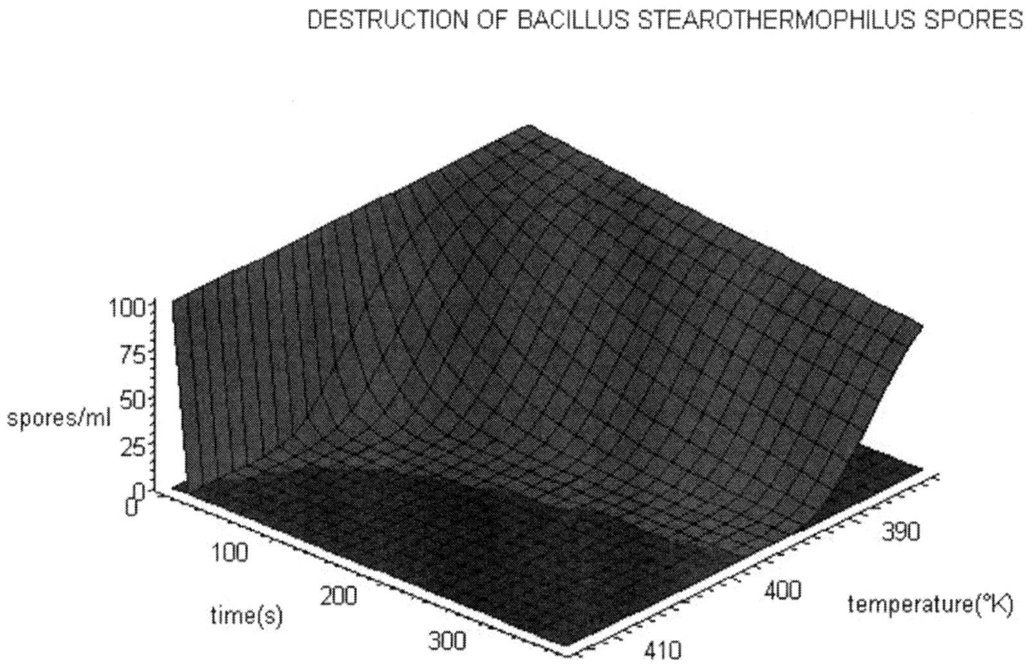

Fig. 1

In particular, we have used the level curves of $N(t,T), F(t,T), B(t,T)$ and $L(t,T)$ to test the validity of the model in a sterilization treatment of milk for the destruction of *Bacillus stearothermophilus* spores. We assumed $N_0 = 100 spores/ml, F_0 = 0\ mg/l, B_0 = 100\%$ of concentration of Thiamine and $L_0 = 100\%$ of lipase activity. The following evaluations prove that the obtained results are in accord with those of the *mild technologies* treatments.

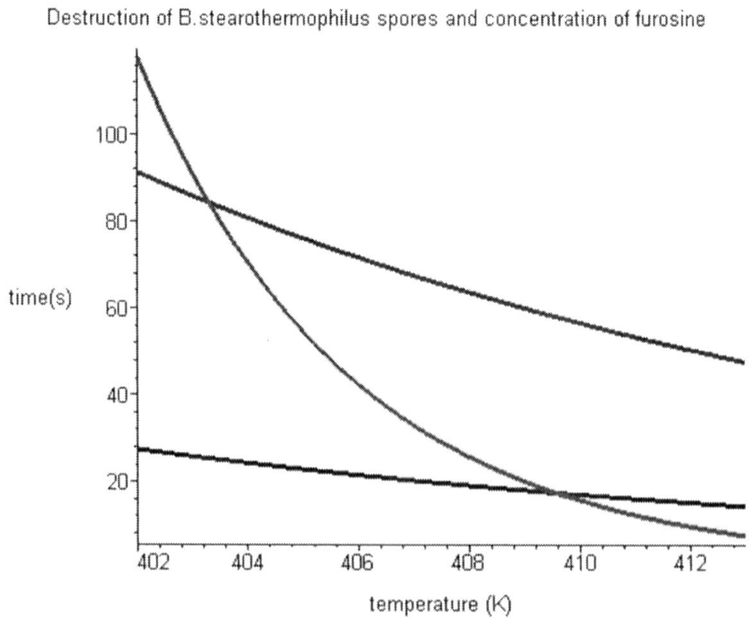

• Commercial sterility curve $N(t,T) = 10^{-4}$
• Level curve $F(t,T) = 10mg/l$
• Level curve $F(t,T) = 3mg/l$

Fig. 2

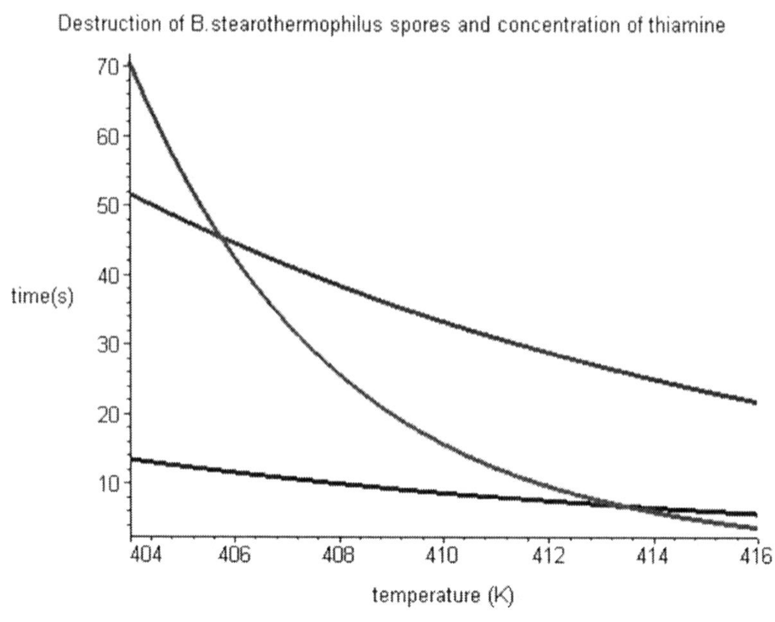

• Commercial sterility curve $N(t,T) = 10^{-4}$
• Level curve $B(t,T) = 10\%$
• Level curve $B(t,T) = 30\%$

Fig. 3

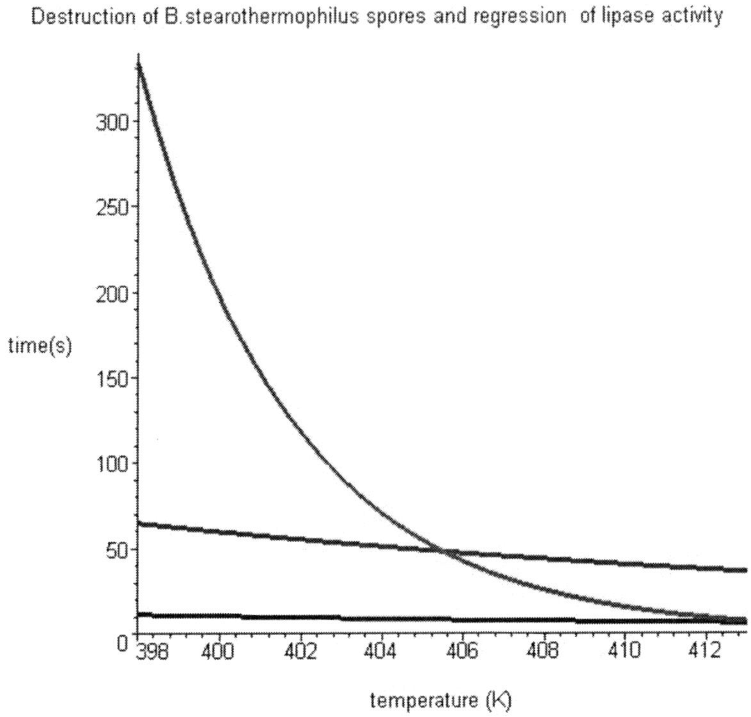

• Commercial sterility curve $N(t,T) = 10^{-4}$
• Level curve $L(t,T) = 60\%$ of lipase activity.
• Level curve $L(t,T) = 15\%$ of lipase activity.

Fig. 4

As may be noted in each figure, our mathematical model allows us to determine couples time-temperature that ensure the sanitization of the product with minimizing the degradation of milk quality, in accord with the mild technologies treatments.

In conclusion, we would like to underline the following consideration regarding this type of predictive mathematical model applied to food technologies.

First of all, this model could be conveniently applied directly to the on-line food process control systems, increasing their capacity to manage and ensuring the production of safe food with the highest quality level.

Moreover it could be used in future research projects. Its performances should guarantee a fast way of providing a simultaneous evaluation of data useful for individuate the parameters for producing safe and high quality food.

Finally, the proposed model could be also applied to other foods and to other food process technologies.

References

[1] Ames J. M.: *Application of Maillard reaction in the food Industry*, Food Chemistry, **62** (1998), 431–439.

[2] Cunha L. M., Oliveira F. A. R.: *Optimal experimental design for estimating the kinetic parameters of processes described by first-order Arrhenius model under linearly increasing temperature profiles*, J. Food Engineerig, **46** (2000), 53–60.

[3] Hills B. P., Mackey B. M.: *Multi-compartment kinetic models for injury, resuscitation, induced lag and growth in bacterial cell populations*, Food Microbiology, **12** (1995), 333–346.

[4] Karel V: *Prediction of nutrient losses and optimization of processing condictions* Nutritional and safety aspects of food processing, Marcel Dekker Inc., (1979).

[5] Körmendy I., Körmendy L., Ferenczya A.: *Thermal inactivation kinetics of mixed microbial population. A hypothesis paper*, J. Food Engineerig, **38** (1998), 439–453.

[6] Leniger H. A., Bruin S.: *The state of the art of food dehydration in food quality and nutrition*, Applied Science Publishers, London, (1997), 265–295.

[7] Peleg M.: *Mathematical interpretation of dose-reponse curves*, Bull. Math. Biology, **59** (1997), 747–761.

[8] Peleg M.: *Microbial survival curves - the reality of flat "shoulders" and absolute thermal death times*, Food Research International, **33** (2000), 531–538.

[9] Peleg M., Penchina C. M., *Estimation of the survival curve of Listeria monocytogenes during non-isothermal heat treatments*, Food Research International, **34** (2001), 383–388.

[10] Peri C.: *Influenza comparativa dei diversi processi di conservazione sui principi nutrizionali degli alimenti*, La Rivista delle Soc. Ital. di Scienza dell'Alimentazione, **12** (1983), 363.

[11] C. Peri, *Miglioramento della qualità attraverso la riduzione del danno tecnologico: mild technologies*, Nuovi orientamenti dei consumi e delle produzioni alimentari. Innovazione e ricerca nel settore delle tecnologie alimentari. C.N.R. (1985), 13–14.

[12] Peri C., Zanoni B.: *Manuale di Tecnologie Alimentari* Milano, (1998).

[13] Oste R., Viberg U.: *Development and evaluation of microscale apparatus for the generation of kinetic data at high temperatures, applied to the degradation of thiamine*, Food Chemistry, **52** (1995), 29–33.

[14] M.A.J.S. Van Boeckel, *Formation of flavour coumpounds in the Maillard reaction*, Biothecnology Advances, **24** (2006), 230–233.

Current address

Ceppitelli Rita, Associate Professor
Dipartimento di Matematica e Informatica, Università di Perugia
via L. Vanvitelli 1, 06123 Perugia, Italy
Tel. number: +390755855032/6046, e-mail: rita.ceppitelli@unipg.it

La Tona Alice, Product Development Manager
Dr. Schär R&D Centre, Area Science Park
Padriciano 99, 34149 Trieste, Italy
Tel. number: +390403755581, e-mail: alice.latona@drschaer.com

Montanari Luigi, Associate Professor
Dipartimento di Agraria, Università di Sassari
via De Nicola 9, 07100 Sassari, Italy
Tel. number: +39079229275, e-mail: lmontanari@uniss.it

Slovak University of Technology in Bratislava
Faculty of Mechanical Engineering
16th Conference on Applied Mathematics **APLIMAT 2017**
Proceedings

COMPLEX SUBMANIFOLDS OF LCK-MANIFOLDS, PSEUDO-VAISMAN AND VAISMAN MANIFOLDS

CHEREVKO Yevhen (UKR), BEREZOVSKI Vladimir (UKR), CHEPURNA Olena (UKR)

Abstract. The article is dedicated to the immersions of submanifolds in LCK-manifolds that a tangent space in all points of the submanifolds to be normal to Lee field and we find conditions under which LCK-manifold admits the immersion of complex submanifolds. Also we explore properties of Lee form of Vaisman and pseudo-Vaisman manifold.

Keywords: Complex manifold, submanifold, LCK-manifold, almost complex structure, Lee form

Mathematics subject classification: Primary 53B20; Secondary 53B35, 32C15

1 Introduction

Differential geometric aspects of submanifolds of manifolds with certain structures are very fruitful fields for Riemannian geometry. Study of complex submanifolds immersed in locally conformal Kähler manifolds (for brevity, LCK-manifolds) was begun by Vaisman in [14], and more attention was paid to so called Generalized Hopf manifolds. Special mappings of Kähler and conformal Kähler manifolds were studied in [7, 8, 10, 11]. Certain questions on conformal mappings and conformal tensor are solved in [5, 6, 9, 12, 13].

We continue to study the immersions of submanifolds that a tangent space in all points of the submanifolds to be normal to Lee field. Also we explore properties of Lee form of Vaisman and pseudo-Vaisman manifolds.

2 Preliminaries

A Hermitian manifold (M^{2m}, J, g) is called a *locally conformal Kähler manifold* (*LCK-manifold*) if there is an open cover $\mathfrak{U} = \{U_\alpha\}_{\alpha\in A}$ of M^{2m} and a family $\{\sigma_\alpha\}_{\alpha\in A}$ of C^∞ functions $\sigma_\alpha : U_\alpha \to \mathbb{R}$ so that each local metric

$$\hat{g}_\alpha = e^{-2\sigma_\alpha} g|_{U_\alpha}$$

is Kählerian. An LCK-manifold is endowed with some form ω, so called a *Lee form* which can be calculated as [2]

$$\omega = \frac{1}{m-1}\delta\Omega \circ J \quad \text{or} \quad \omega_i = -\frac{2}{n-2}J^{\alpha}_{\beta,\alpha}J^{\beta}_{i}, \tag{1}$$

The form should be closed:

$$d\omega = 0.$$

One can compute covariant derivative an almost complex structure with respect of the Levi-Civita connection of (M^{2m}, J, g) using the formulae

$$J^k_{i,j} = \frac{1}{2}\big(\delta^k_j J^{\alpha}_i \omega_{\alpha} - \omega^k J_{ij} - J^k_j \omega_i + J^k_{\alpha}\omega^{\alpha} g_{ij}\big). \tag{2}$$

Let (M^{2m}, J, g) be a complex m-dimensional Hermitian manifold, g is its Hermitian metric, J is its comlex structure. Consider an immersion of an m-dimensional manifold $\bar{M}^k$ in M^{2m}:

$$\Psi : \bar{M}^k \longrightarrow M^{2m}.$$

Let ∇ and $\overline{\nabla}$ be operators of covariant differentiations on M^{2m} and $\overline{M}^k$, respectively. Then the Gauss and Weingarten formulas are given by [1, p. 2] :

$$\nabla_X Y = \bar{\nabla}_X Y + h(X, Y), \tag{3}$$

$$\nabla_X \xi = -A_{\xi}X + \nabla^{\perp}_X \xi, \tag{4}$$

respectively, where X and Y are vector fields tangent to $\overline{M}^k$ and ξ normal to $\overline{M}^k$. $h(X, Y)$ is the *second fundamental form*, $\nabla^{\perp}$ the linear connection induced in the normal bundle $E(\Psi)$, called the *normal connection*, and A_{ξ} the *second fundamental tensor* at ξ.

We call $\overline{M}^k$ *$CR-$ submanifold* of (M^{2m}, J, g) if $\overline{M}^k$ carries a C^{∞} distribution D so that

1. D is *holomorphic* (i.e. $J_x(D_x) = D_x$) for any $x \in \overline{M}^k$,
2. the orthogonal complement $D^{\perp}$ with respect to $\overline{g} = \Psi^* g$ of D in $T(\overline{M}^k)$ is *anti-invariant* (i.e. $J_x(D^{\perp}{}_x) \subseteq E(\Psi)_x$) for any $x \in \overline{M}^k$)[2, p. 153]

Let $(\overline{M}^k, D)$ be a $CR-$submanifold of the Hermitian manifold M_0^{2m}. Set $p = \dim_{\mathbb{C}} D_x$ and $p = \dim_{\mathbb{R}} D^{\perp}_x$; for any $x \in \overline{M}^k$ such that $2p + q = k$. If $q = 0$ then $\overline{M}^k$ is a *complex submanifold*, i.e. it is a complex manifold and Ψ is a holomorphic immersion. If $p = 0$ then $\overline{M}^k$ is an *anti-invariant submanifold* (i.e. $J_x(T_x(\overline{M}^k)) \subseteq E(\Psi)_x$ for any $x \in \overline{M}^k$). A $CR-$submanifold $(\overline{M}^k, D)$ is *proper* if $p \neq 0$ and $q \neq 0$. Also $(\overline{M}^k, D)$ is *generic* if $q = 2m - k$ (i.e. $J_x(T_x(\overline{M}^k)) = E(\Psi)_x$ for any $x \in \overline{M}^k$). A submanifold $\overline{M}^k$ of the complex manifold (M^{2m}, J) is *totally real* if

$$T_x(\overline{M}^k) \cap J_x(T_x(\overline{M}^k)) = \{0\}$$

for any $x \in \overline{M}^k$

3 Complex hypersurfaces of LCK-manifolds

Let submanifold $\overline{M}^k$ is immersed in LCK-manifold M^{2m}

$$\Psi : \bar{M}^k \longrightarrow M^{2m},$$

so that $k = 2p$ and for any $x \in \overline{M}^{2p}$

We are concerned with finding conditions under which LCK-manifold M^{2m} admits immersion of complex submanifolds. Then we obtain the following Theorem.

Theorem 1 *LCK-manifold M^{2m} admits immersion of complex hypersurface $\overline{M}^{2m-2}$ such that the Lee field $B = \omega^{\#}$ and the anti-Lee field $A = -JB = -J\omega^{\#}$ are normal to the hypersurface $\overline{M}^{2m-2}$ if and only if the Lee form of M^{2m} satisfies the condition*

$$\Phi_4(\nabla_X \omega(Y)) = \frac{\|\omega\|^2}{2} g(X,Y).$$

Here Φ_4 is the fourth Obata's projector:

$$\Phi_4(\omega_{i,j}) = \frac{1}{2}(\delta_i^a \delta_j^b + J_i^a J_j^b)\omega_{a,b}.$$

Necessity. Let us consider an LCK-manifold M^{2m}. Let $\theta = \omega \circ J$ and $A = -JB$ be respectively the anti-Lee form and the anti-Lee vector field. Then, we can rewrite (2) as

$$\nabla_X(J)Y = \frac{1}{2}(\theta(Y)X - \omega(Y)JX - g(X,Y)A - \Omega(X,Y)B),$$

and hence we get

$$\nabla_X A = -J\nabla B + \frac{1}{2}(\|\omega\|^2 JX + \omega(X) - \theta(X)B)$$

for any $X \in T(M^{2m})$. Let M^{2m-2} be a complex hypersurface of an M^{2m}. If $B \in E(\Psi)$, then $A \in E(\Psi)$ since the immersion is analytic one. Moreover, if $X, Y \in T(M^{2m-2})$, then $[X,Y] \in T(M^{2m-2})$ according to the classical Frobenius theorem. Hence

$$\begin{aligned} 0 = g([X,Y],A) = g(\nabla_X Y, A) - g(\nabla_Y X, A) \\ = -g(Y, \nabla_X A) + g(X, \nabla_Y A) = \\ = g(Y, J\nabla_X B) - g(X, J\nabla_Y B) + \|\omega\|^2 \Omega(X,Y) \end{aligned} \tag{5}$$

Rewriting (5) in local coordinates, we obtain

$$\omega_{t,j} J_i^t - \omega_{t,i} J_j^t - \|\omega\|^2 J_{ij} = 0. \tag{6}$$

Next, multiply (6) by J_k^j:

$$\omega_{t,j} J_i^t J_k^j + \omega_{k,i} - \|\omega\|^2 g_{ik} = 0. \tag{7}$$

We can rewrite (7) as

$$2\Phi_4(\omega_{i,j}) - \|\omega\|^2 g_{ij} = 0.$$

where Φ_4 is the fourth Obata's projector [4]. For instance, applying the operator to a tensor Q^h_{ij} means

$$\Phi_4(Q^h_{ij}) = \frac{1}{2}(\delta^a_i\delta^b_j + J^a_i J^b_j)Q^h_{ab}.$$

Hence

$$\Phi_4(\omega_{i,j}) = \frac{\|\omega\|^2}{2} g_{ij}. \tag{8}$$

Sufficiency. Tangent bundle $T(M^{2m})$ should satisfy the system since the bundle is normal to both Lee field B and anti-Lee field A

$$\begin{cases} \omega = 0 \\ \theta = 0. \end{cases} \tag{9}$$

According to the Frobenius theorem the system (9) is completely integrable if and only if both Lee-form and anti-Lee form identically satisfy the conditions

$$\begin{array}{ll} 1) & d\omega \wedge \omega \wedge \theta = 0 \\ 2) & d\theta \wedge \omega \wedge \theta = 0. \end{array} \tag{10}$$

Identity (10_1) is satisfied since an M^{2m} is LCK-manifold, hence $d\omega = 0$. We have to explore (10_2). Let us take the exterior differential of the anti-Lee form $\theta = \omega \circ J$ [15, p. 6].

$$\begin{aligned} d\theta &= \frac{1}{2}\big(\nabla_k(\omega_i J^i_j) - \nabla_j(\omega_i J^i_k)\big)dx^k \wedge dx^j \\ &= \frac{1}{2}\big(\omega_{i,k}J^i_j + \omega_i J^i_{j,k} - \omega_{i,j}J^i_k - \omega_i J^i_{k,j}\big)dx^k \wedge dx^j \end{aligned}$$

According to (2) we obtain:

$$\begin{aligned} d\theta &= \big(\omega_{i,k}J^i_j + \omega_i J^i_{j,k}\big)dx^k \wedge dx^j \\ &= \Big(\omega_{i,k}J^i_j + \omega_i\frac{1}{2}\big(\delta^i_k J^t_j\omega_t - \omega^i J_{jk} - J^i_k\omega_j + J^i_t\omega^t g_{jk}\big)\Big)dx^k \wedge dx^j \\ &= \Big(\omega_{i,k}J^i_j + \frac{1}{2}\omega_k J^t_j\omega_t - \frac{1}{2}\|\omega\|^2 J_{jk} - \frac{1}{2}\omega_t J^t_k\omega_j\Big)dx^k \wedge dx^j \\ &= \frac{1}{2}\big(\omega_{i,k}J^i_j - \omega_{i,j}J^i_k - \|\omega\|^2 J_{jk} + \omega_k J^t_j\omega_t - \omega_t J^t_k\omega_j\big)dx^k \wedge dx^j \end{aligned}$$

Then,

$$\begin{aligned} d\theta \wedge \omega \wedge \theta &= \frac{1}{2}\big(\omega_{i,k}J^i_j - \omega_{i,j}J^i_k - \|\omega\|^2 J_{jk} \\ &\quad + \omega_k J^t_j\omega_t - \omega_t J^t_k\omega_j\big)dx^k \wedge dx^j \wedge \omega_l dx^l \wedge \omega_s J^s_h dx^h \\ &= \frac{1}{2}\big(\omega_{i,k}J^i_j - \omega_{i,j}J^i_k - \|\omega\|^2 J_{jk}\big)dx^k \wedge dx^j \wedge \omega_l dx^l \wedge \omega_s J^s_h dx^h, \end{aligned} \tag{11}$$

since the equation

$$\frac{1}{2}\big(\omega_k J^t_j\omega_t - \omega_t J^t_k\omega_j\big)dx^k \wedge dx^j \wedge \omega_l dx^l \wedge \omega_s J^s_h dx^h = 0;$$

is identically satisfied. Hence the equation

$$\omega_{i,k}J^i_j - \omega_{i,j}J^i_k - \|\omega\|^2 J_{jk} = 0,$$

gives us a sufficient condition for the identity $d\theta \wedge \omega \wedge \theta = 0$ to be satisfied. The condition coincides with (6) which is equivalent to (8):

$$\Phi_4(\omega_{i,j}) = \frac{1}{2}\|\omega\|^2 g_{ij}.$$

Hence (10_2) is satisfied too. Sufficiency is proved.

Taking into account that LCK-manifolds with Lee form which satisfies the condition

$$\Phi_4(\nabla\omega(X,Y)) = \frac{\|\omega\|^2}{2} g(X,Y) \tag{12}$$

have very particular properties, we propose call such LCK-manifolds as the *Pseudo-Vaisman manifolds*.

4 Pseudo-Vaisman manifolds

Let us discover the condition (12). An LCK-manifold is a Hermitian since its almost complex structure is integrable. This means the existence of the coordinate system in which J and metric g take correspondingly the forms

$$(J^i_j) = \begin{pmatrix} \sqrt{-1}\delta^\kappa_\lambda & 0 \\ 0 & -\sqrt{-1}\delta^{\hat\kappa}_{\hat\lambda} \end{pmatrix}, \qquad (g_{ij}) = \begin{pmatrix} 0 & g_{\mu\hat\lambda} \\ g_{\hat\mu\lambda} & 0 \end{pmatrix}.$$

In the system of coordinates we can rewrite (12) as

$$\omega_{\alpha,\hat\beta} = \frac{\|\omega\|^2}{2} g_{\alpha\hat\beta}. \tag{13}$$

One knows that the components of the connection with respect to the Hermitian metric g_{ij} by formulas [15, p. 64]

$$\begin{aligned} &1)\ \Gamma^\kappa_{\mu\lambda} = \frac{1}{2}\, g^{\kappa\hat\rho}(\partial_\mu g_{\lambda\hat\rho} + \partial_\lambda g_{\mu\hat\rho}), \qquad \text{conj.} \\ &2)\ \Gamma^\kappa_{\mu\hat\lambda} = \frac{1}{2}\, g^{\kappa\hat\rho}(\partial_{\hat\lambda} g_{\mu\hat\rho} - \partial_{\hat\rho} g_{\mu\hat\lambda}), \qquad \text{conj.} \\ &3)\ \Gamma^\kappa_{\hat\mu\hat\lambda} = 0, \qquad \text{conj.} \end{aligned} \tag{14}$$

where conj. means that there exists a formula which is the complex conjugate of the formula written at the left.

For LCK-manifolds (14_2) takes the form

$$\Gamma^\kappa_{\mu\hat\lambda} = \frac{1}{2} g^{\kappa\hat\rho}(\omega_{\hat\lambda} g_{\mu\hat\rho} - \omega_{\hat\rho} g_{\mu\hat\lambda}) = \frac{1}{2}(\omega_{\hat\lambda}\delta^\kappa_\mu - \omega^\kappa g_{\mu\hat\lambda}), \qquad \text{(conj.)}. \tag{15}$$

Hence taking into account (15) we can rewrite (13) as

$$\omega_{\alpha,\hat\beta} = \partial_{\hat\beta}\omega_\alpha - \omega_\kappa \Gamma^\kappa_{\alpha\hat\beta} - \omega_{\hat\kappa}\Gamma^{\hat\kappa}_{\alpha\hat\beta} = \partial_{\hat\beta}\omega_\alpha - \omega_{\hat\beta}\omega_\alpha + \frac{\|\omega\|^2}{2} g_{\alpha\hat\beta} = \frac{\|\omega\|^2}{2} g_{\alpha\hat\beta}.$$

Then we obtain

$$\partial_{\hat\beta}\omega_\alpha - \omega_{\hat\beta}\omega_\alpha = 0. \tag{16}$$

Let us multiply the both sides (16) by $e^{-\sigma}$ where σ is a function such that $d\sigma = \omega$. Then

$$-e^{-\sigma}\partial_{\hat{\beta}}\omega_\alpha + e^{-\sigma}\omega_{\hat{\beta}}\omega_\alpha = \partial_{\hat{\beta}}(-e^{-\sigma}\omega_\alpha) = 0.$$

Finally, we obtain the equation

$$\frac{\partial^2\psi}{\partial z^{\hat{\beta}}\partial z^{\alpha}} = 0. \tag{17}$$

The general solution of (17) is given by

$$\psi = f(z) + f(\hat{z}),$$

where $f(z)$ is a some analytic function of coordinates $z^1, z^2, \ldots z^m$, and $f(\hat{z})$ is a complex conjugate of the function $f(z)$, since ψ must be a real number which is subject to the condition $Ref(z) > 0$. Hence the function σ determining the conformal mapping is

$$\sigma = \ln \frac{1}{f(z) + f(\hat{z})},$$

and Lee form is

$$\omega_\alpha = -\frac{f'_\alpha(z)}{f(z) + f(\hat{z})}, \quad \omega_{\hat{\alpha}} = -\frac{f'_{\hat{\alpha}}(\hat{z})}{f(z) + f(\hat{z})}.$$

Hence we obtain the theorem.

Theorem 2 *If the Lee form of an LCK M^{2m} satisfies the condition*

$$\Phi_4(\nabla_X\omega(Y)) = \frac{\|\omega\|^2}{2}g(X,Y).$$

then the form is

$$\omega_\alpha = -\frac{f'_\alpha(z)}{f(z) + f(\hat{z})}, \quad \omega_{\hat{\alpha}} = -\frac{f'_{\hat{\alpha}}(\hat{z})}{f(z) + f(\hat{z})},$$

where $f(z)$ is a some analytic function such that $Ref(z) > 0$.

One might conclude that it is easy to construct pseudo-Vaisman manifold because it seems sufficient the latter admits a Kählerian metric. But that is wrong because not always it is possible to find an analytic function such that $Ref(z) > 0$ in every manifold's point. In particular, there exists a theorem.

Theorem 3 *There is no compact pseudo-Vaisman manifold.*

Proof. Let M is an LCK-manifold such that its Lee form satisfies the condition

$$\Phi_4(\omega_{i,j}) = \frac{1}{2}\|\omega\|^2 g_{ij}.$$

Transvecting this with g^{ij}, we find

$$\nabla_i\omega^i = \frac{n}{2}\|\omega\|^2.$$

On the other hand, according to the Theorem of Green [15, p. 21]

$$\int\limits_{M^n} \nabla_i\omega^i d\tau = 0,$$

where $d\tau$ is the volume element

$$d\tau = \sqrt{g} d\xi^1 \wedge d\xi^2 \wedge ... \wedge d\xi^n.$$

In this case we obtain

$$\frac{n}{2} \int\limits_{M^n} \|\omega\|^2 d\tau = 0,$$

that is impossible for an LCK-manifold. The theorem is proved.

5 Vaisman manifods and Kähler-Vaisman potential

Let us concider locally conformal Kähler manifolds with a parallel Lee form.

$$\nabla_j \omega_i = 0.$$

In particular, using so called a holomorphic coordinate system we have

$$\nabla_{\hat{\beta}} \omega_\alpha = 0. \tag{18}$$

Taking into account (15), we can rewrite (18) as

$$\omega_{\alpha,\hat{\beta}} = \partial_{\hat{\beta}} \omega_\alpha - \omega_\kappa \Gamma^\kappa_{\alpha\hat{\beta}} - \omega_{\hat{\kappa}} \Gamma^{\hat{\kappa}}_{\alpha\hat{\beta}} = \partial_{\hat{\beta}} \omega_\alpha - \omega_{\hat{\beta}} \omega_\alpha + \frac{\|\omega\|^2}{2} g_{\alpha\hat{\beta}} = 0.$$

We get

$$\partial_{\hat{\beta}} \omega_\alpha - \omega_{\hat{\beta}} \omega_\alpha = -\frac{\|\omega\|^2}{2} g_{\alpha\hat{\beta}}. \tag{19}$$

Multiplying the both sides (19) by $e^{-\sigma}$ where σ is a function such that $d\sigma = \omega$ we obtain

$$-e^{-\sigma} \partial_{\hat{\beta}} \omega_\alpha + e^{-\sigma} \omega_{\hat{\beta}} \omega_\alpha = \partial_{\hat{\beta}} \big(-e^{-\sigma} \omega_\alpha \big) = \frac{\|\omega\|^2}{2} e^{-\sigma} g_{\alpha\hat{\beta}}.$$

Finally we get

$$\partial_{\hat{\beta}} \partial_\alpha \big(e^{-\sigma} \big) = \frac{\|\omega\|^2}{2} \bar{g}_{\alpha\hat{\beta}},$$

or, since on a Vaisman manifold condition $\|\omega\|^2 = const$ holds,

$$\partial_{\hat{\beta}} \partial_\alpha \Big(\frac{2}{\|\omega\|^2} e^{-\sigma} \Big) = \bar{g}_{\alpha\hat{\beta}}, \tag{20}$$

where $\bar{g}_{\alpha\hat{\beta}} = e^{-\sigma} g_{\alpha\hat{\beta}}$ is a local Kählerian metric. The theorem follows from (20).

Theorem 4 *For any Vaisman manifold M^n there locally exists a function*

$$V(z, \hat{z}) = \frac{2}{\|\omega\|^2} e^{-\sigma},$$

which determines a local Kählerian metric.

The function $V(z, \hat{z}) = \frac{2}{\|\omega\|^2} e^{-\sigma}$ is said to be a *Kähler-Vaisman potential.*

References

[1] CHEN, B. *Geometry of submanifolds and its applications*, Sci. Univ. Tokyo, 1981.

[2] DRAGOMIR, S., ORNEA, L. *Locally conformal Kähler geometry*, Birkhäuser, 1998.

[3] DRAGOMIR, S. *Generalized Hopf manifolds, locally conformal Kaehler structures and real hypersurfaces*, Kodai Math. J., vol. 14, (1991), 366–391.

[4] ISHIHARA, S. *Holomorphically projective changes and their groups in an almost complex manifold*, Tôhoku Math. J., vol. 9, (1957), 273–297.

[5] HINTERLEITNER, I., CHEPURNA, O. *On the mobility degree of (pseudo-) Riemannian spaces with respect to concircular mappings*, Miskolc Math. Notes, vol. 14, no. 2, (2013), 561–568.

[6] MIKEŠ, J.: *Holomorphically projective mappings and their generalizations*, J. Math. Sci., New York, vol. 89, no. 3, (1998), 1334–1353.

[7] MIKEŠ, J., at al. *Differential geometry of special mappings*, Palacky Univ. Press, Olomouc, 2015.

[8] MIKEŠ, J., VANŽUROVÁ, A., HINTERLEITNER, I. *Geodesic mappings and some generalizations*, Palacky Univ. Press, Olomouc, 2009.

[9] MIKEŠ, J., JUKL, M., JUKLOVÁ, L. *Some results on traceless decomposition of tensors*, J. Math. Sci., New York, vol. 174, no. 5, (2011), 627–640.

[10] RADULOVICH, Zh., MIKEŠ, *Geodesic and holomorphically-projective mappings of conformally-Kählerian spaces*, in *DGA Opava 1992*, Math. Publ. Silesian Univ. Opava, vol. 1, (1993), 151–156.

[11] RADULOVICH, Zh., MIKEŠ, J. *Geodesic mappings of conformal Kähler spaces*, Russ. Math., vol. 38, no. 3, (1994), 48–50.

[12] STEPANOV, S.E., JUKL, M., MIKEŠ, J. *On dimensions of vector spaces of conformal Killing forms*, in *Algebra, geometry and mathematical physics*, Springer Proc. Math. Stat., vol. 85, (2014), 495–507.

[13] STEPANOV, S.E., JUKL, M., MIKEŠ, J. *Vanishing theorems of conformal Killing forms and their applications to electrodynamics in the general relativity theory*, Int. J. Geom. Methods Mod. Phys. vol. 11, no. 9 (2014), 1450039, 8 pp.

[14] VAISMAN, I. *Generalized Hopf manifolds*, Geometriae Dedicata, vol. 13, (1982), 231–255.

[15] YANO, K. *Differential geometry on complex and almost complex spaces*, Pergamon Press Book, New York, 1965.

Current address

Cherevko Yevhen
Dept. of Economic Cybernetics, Odessa National Economic University
Odessa 65082, Ukraine
E-mail: cherevko@usa.com

Berezovski Vladimir, Doc. CSc.
Dept. of Mathematics and Physics, Uman National University of Horticulture
Instytutska 1, Uman, Ukraine
Tel. number: +380 662 901 879, e-mail: berez.volod@rambler.ru

Chepurna Olena, Doc. PhD.
Department of MMAE, Odessa State Economic University
Odessa 65082, Ukraine
E-mail: chepurna67@gmail.com

Slovak University of Technology in Bratislava
Faculty of Mechanical Engineering
16th Conference on Applied Mathematics **APLIMAT 2017**
Proceedings

LIMITS OF ELECTRONIC TESTING OF MATHEMATICAL KNOWLEDGE

CHLÁDEK Petr (CZ), SMETANOVÁ Dana (CZ)

Abstract. In the paper we discuss limits of multiple choice testing in Mathematics. This problem is illustrated on disadvantages which appeared during three consecutive years.

Keywords: mathematics, computer-adaptive testing, multiple-choice test

Mathematics Subject Classification: Primary 97B40, Secondary 97R20

1 Introduction

In recent decades and especially the last few years there has been a significant increase in use of computer technology in everyday life. PC user skills are at a very good level for most people entering the college. In the context of increasing quality and availability of computer technology, with increasing user skills of students, with the development of software technology the computer technology is being incorporated into education. Because continuous testing of knowledge and testing students is one of the considerable parts of education, computer technology begins to play an increasingly larger role in it as well.

For this reason we have involved electronic testing in mathematics lessons. In the years 2013-2015 it included testing of all students in combined forms, and in 2015 voluntary testing of candidates from among students of combined and daily form of studying. Further details on the first year of testing can be found in [8].

This paper deals mainly with the disadvantages and limits of electronic testing of mathematical knowledge. How they interfere with the testing and how they can be at least partially restricted for the full advantage of its capabilities. Among the main benefits of electronic testing using multiple-choice tests belongs large time saving, objectivity and impartiality of the evaluation. The more detailed classification of the advantages and disadvantages can be found in [13].

Note that university teachers are facing diverse challenges in the classroom - to get acquainted with some of them we recommend [1-5, 7, 9-12, 14].

2 Computer-adaptive-testing in Mathematics I

In 2013 we began to use electronic testing using multiple-choice problems in Mathematics I for students of combined studies at The Institute of Technology and Business in České Budějovice. The students of combined form of study have significantly smaller number of lessons compared to the students of daily form of study. For this reason it is necessary to lead the education efficiently and with minimum time losses. Electronic knowledge tests are very useful for continuous testing of these students because of significant time saving.

Two approaches can be selected for creating examples for electronic testing:

- calculation is required to gain the response,
- the result can be drawn from theoretical knowledge.

Bearing in mind that the subject is finished with a written exam we have chosen the first option for creating exercises for continuous testing - the need to perform calculations. Final written exam evaluates the students' knowledge better, during exam evaluation the progress in solving can be taken into account, which is not possible in the case of multiple-choice tests.

In three consecutive years (2013 - 2015) 563 students were tested, 175 students in the first year, 193 in the second year and 195 in the third year. Each of the students solved 3 examples. Points from the continuous assessment of knowledge were included in the evaluation of the course. Numbers of students and examples answered correctly are listed in the following Tab. 1.

	0 examples	1 example	2 examples	3 examples
Year 1	8	25	41	101
Year 2	5	8	45	135
Year 3	12	28	50	105

Tab. 1. Evaluation knowledge testing - the numbers of students and examples answered correctly

In 2015 we also enabled voluntary testing of the students of combined and daily forms of study. The students could respond from home at given time. Unlike previous testings, this one was not included into evaluation of the course. Students were asked to fill out answers after starting the test even though it would not be evaluated.

The total of 146 students enrolled for the testing from home. A total of 48 of them generated a test and then did not respond. Numbers of correct answers are presented in Tab. 2.

	0 examples	1 example	2 examples	3 examples
Year 3	28	44	21	5

Tab. 2. Voluntary testing of knowledge - the numbers of students and examples answered correctly

It is obvious at first sight that the students did not have any motivation to get correct solutions. While in the first three compulsory testing correct answers prevailed, they are not dominant in voluntary testing.

3 Disadvantages in practise

Now we will discuss what disadvantages appeared in multiple choice testing in several testing in Mathematics I.

Disadvantages of multiplechoice testing: [13]

- some skills (e. g. to formulate an opinion, to sketch a picture, graph or diagram etc.) cannot be tested with the use of close questions or only to a limited extent;
- it handicaps inattentive or distracted students (they may overlook between alternative answers or they make a mistake when ticking the right answer), or contemplative students who think deeply over the alternative answers which makes them unsure or who try to identify a trick answer;
- there is certain probability that students tip the right answers, which rises a small number of alternatives (2–3) ,
- it is not possible to track the thinking process of the students;
- the creation of a good close question is not easy, the revelation of construction deficiencies is not always clear from students' answers and it requires a deeper analysis;
- students may cheat more easily;
- the community of teachers (at least in the Czech Republic) have a rather sceptical attitude to close questions and they are used to a limited extent in the teaching and learning process.

Each of the disadvantages mentioned above is commented in terms of our experience with the electronic testing. It says how the disadvantage shows or what reasons limit it.

Some types of exercises cannot be tested:
There can be types of exercises in the basic course of higher mathematics, the testing of which via electronic form is problematic, difficult, even impossible. For instance, an exercise requiring function graph drawing, which is one of the basic requirements in the study of mathematical analysis. Closed tasks do not enable this kind of testing. Partly this can be treated with a variant when tested student selects the correct graph of the specified function from offered images, but it is a fundamentally different type of task.

Probability:
The structure of the responses and their number should be chosen so that the probability of a random correct tip is sufficiently small. In our case each question asks the student to select four options a), b), c), d). The student selects one from four options. Additionally, the order of offered answers is chosen randomly. Thus, with the same questions in two different tests the correct answer could be a) in one test, b) in the second.

Furthermore, the probability that a student randomly chooses the correct answer at one particular questions is 0.25. The probability of accidental filling the whole test correctly is 0.015. This value is sufficiently small, and therefore four options in the test are sufficient. For more details about calculation of the probability see [6].

Disadvantaged inattentive students:
Closed questions tested electronically disadvantage inattentive students more than traditional written exams. There is a wider range of options, which enable inattention mistakes. It is of course expected that the student performs the appropriate calculation. But then it is necessary to find the

answer matching the observed outcome from the options provided, which may not always be identical. E.g. fraction vs. real number, or in the case of several partial results in one example there is the possibility of different order of the answers, etc.

The skill-testing using multiple-choice test can also contain such exercises with more correct answers. Such tests can be created for mathematics use. However, the choice of tests with multiple correct answers for testing mathematical skills is not too frequent. Such tests disadvantage inattentive students even more. The students mostly assume that they have to choose only one option and this creates the wrong solution.

Furthermore, there is a risk of failure during saving the answers – a student forgets to save the results, although the example has been properly solved and answered.

No possibility to track the thinking process:
This is a crucial and fundamental drawback of electronic testing. Students may find the correct answer via a completely wrong procedure. But even from the other side, using the proper procedure may result in an incorrect result. This may be due to numerical error in the calculation. Neither the teacher or the student have feedback on the correctness of the method used for solutions.

Examples of bad methods that lead to the right result:
Example 1: Calculate the derivation of the function $f(x) = \frac{x^2}{x+1}$ at the point $x = -\frac{3}{2}$. Wrong solution is $f'(x) = \frac{(x^2)'}{(x+1)'} = \frac{2x}{1} = 2x$, hence $f'\left(-\frac{3}{2}\right) = -3$. Note that $f'\left(-\frac{3}{2}\right) = -3$ is solution of correct way of calculus:

$$f'(x) = \frac{2x.(x+1) - x^2}{(x+1)^2} = \frac{x^2 + 2x}{(x+1)^2}.$$

Example 2: Students should calculate the surface area between the *x*-axis and the parabola

$$y = x^2 + \frac{5}{3}$$

on interval $\langle 0,1 \rangle$. The student forms an integral correctly but instead integrating the appropriate functions the student differentiates it

$$\int_0^1 x^2 + \frac{5}{3}\,dx = [2x]_0^1 = 2.$$

But correct method is

$$\int_0^1 x^2 + \frac{5}{3}\,dx = \left[\frac{x^3}{3} + \frac{5}{3}x\right]_0^1 = \frac{1}{3} + \frac{5}{3} = 2.$$

In recent testing [8] we have also found that some of the students solve certain types of examples via correct procedure which could not be used in the case of a written test. For example when the student does not know the set from which the result must be chosen. This was particularly the case of calculating indefinite integrals. Some students did not calculate integrals. They used their knowledge of inverse functions and differentiated the results. They compared the derivation with the function after integral and thus found the correct result. In fact, we tested them from knowledge of the derivation procedure instead of knowledge of the integration process.

Cheating:
In our experience, cheating in electronic testing is not easier than in writing. We understand that the ease of copying is higher in the case of paper multiple-choice tests (see the disadvantages [13]). And it may be a problem. For electronic tests high prevention of cheating can be set. In our particular case the prevention is as follows:
• generating random tests (each student gets a test randomly created from the tasks from a reservoir of examples)
• In each example the responses are mixed randomly, therefore for one student the right answer may be d), and for other b)
• generating a random sequence of examples in the test.

Even if the case of the system generating test of the same examples for two students, the order would be different. And the correct answers will be under different letters in each test. A brief look at the neighbor´s computer screen does not have to reveal that they both have the same test.

Disorders of computing:
There is a risk that cannot be eliminated by a teacher or a student. It is a technical error, which includes mainly electrical outage, current network failure or collapse of the PC and the need to reboot. These errors usually result in data loss and problems with storing and sending the filled test.

In our case the test system was implemented into the school information system and the system uses the internet. Internet failure occurred during testing in one of the groups. This disabled the testing and we had to choose an alternate date.

Creation of test questions:
Electronic form of testing has its own specifics that need to be taken into account when creating the test. The test questions and formulations of individual tasks should be carefully selected. Not all of the tasks commonly used in traditional written tests are also suitable for closed tasks.

Selecting good test tasks is very challenging. We constantly need to optimize the procedures and possibly even change the task types in following testing, see paragraph *No possibility to track the thinking process*.

4 Conclusion

It is not possible to claim that the benefits of electronic testing significantly prevail over the disadvantages or vice versa. In each test, it is necessary to realize the purpose for which the students are tested. The advantages of testing prevail in cases of testing partial knowledge, large groups of people and in limited time. From the perspective of a thorough and detailed testing of mathematical knowledge electronic testing is disadvantageous, we are not able to determine students' way of thinking and the process of solving problems.

References

[1] BIBA, V., HUSÁR, Š., VARGOVÁ, M. Numerical methods in teaching of mechanical engineering. In Balko L'udovít, Szarková Dagmar, Richtáriková Daniela. *APLIMAT 2016: 15th Conference on Applied Mathematics*. Bratislava: Slovak University of Technology in Bratislava, 2016. s. 66-74, 9 s. ISBN 978-80-227-4531-4.

[2] BOHÁČ, Z., DOLEŽALOVÁ, J., KREML, P. Problems of studying in the first year at VSB - Technical university of Ostrava. In *APLIMAT 2014: 13th Conference on Applied Mathematics*, Proceedings, 34 – 43, 2014, ISBN 978-80-227-4140-8.

[3] BOHÁČ, Z., DOLEŽALOVÁ, J., KREML, P. Selected aspects affecting study results in mathematics, In *APLIMAT 2015: 14th Conference on Applied Mathematics*, Proceedings, 80 – 88, 2015 ISBN 978-80-227-4143-3.

[4] ČEJKA, J. Metody pro efektivní metody vzdělávání studentů technicky zaměřených oborů. *Trendy ve vzdělávání*, Olomouc: Univerzita Palackého v Olomouci, 2016, roč. 9, číslo 1, s. 37-42. ISSN 1805-8949

[5] DEDECIUS, K., KALOVÁ, J., VYSOKÁ, J. Využití svobodného matematického softwaru při výuce statistiky. In *Setkání učitelů matematiky všech typů a stupňů škol*. Plzeň: Vydavatelský servis, Republikánská 28, Plzeň, 2008. 6 s. ISBN 978-80-86843-21-6

[6] DOBBS, S., MILLER, J. *Statistics 1. Adv. Level Mathematics*. Cambridge University Press 2002.

[7] CHLÁDEK, P., NÝDL, V., ŠULISTA, M. Struktura zápočtových testů z matematiky a jejich vyhodnocování. In L. Hráčková. *Sborník Aktuální otázky výuky matematiky na ekonomických oborech*. Brno: Masarykova univerzita, 2011. s. 34-37, 4 s. ISBN 978-80-210-5669-5.

[8] CHLÁDEK, P., SMETANOVÁ, D. Elektronické testování z matematiky. *Media4u Magazine*, Ing. Jan Chromý, Ph.D., 2014, roč. 11, č. 2, s. 42-44. ISSN 1214-9187.

[9] HLAVÁČOVÁ, I. Motivační účinky vícesložkového hodnocení studentů v prezenčním studiu. *Trendy ve vzdělávání*, Olomouc: Univerzita Palackého v Olomouci, 2016, roč. 9, číslo 1, s. 84-89. ISSN 1805-8949.

[10] JELÍNEK, J. Application of the Simulation Model of Social Network in Education. In *APLIMAT 2016: 15th Conference on Applied Mathematics*. Bratislava: Slovak University of Technology in Bratislava, 2016. s. 627-635, 9 s. ISBN 978-80-227-4531-4.

[11] RICHTÁRIKOVÁ, D, Mathematics preparatory course and its effectivness, *In APLIMAT 2015: 14th Conference on Applied Mathematics*, Proceedings, 672 – 681, 2015 ISBN 978-80-227-4143-3.

[12] RICHTÁRIKOVÁ, D. Suporting Methods in Mathematics. Effectivenes of Mathematics Workshop, In *APLIMAT 2016: 15th Conference on Applied Mathematics*, Proceedings, 959 – 966, 2016 ISBN 978-80-227-4531-4.

[13] SCHINDLER, R. et al. *Rukověť autora testových úloh,* Tauris Press 2006.

[14] ZÁHONOVÁ, V. Influence of Students´ Hight School Mathematics Knowledge on Their Success in Subject Mathematics I, In *APLIMAT 2015: 14th Conference on Applied Mathematics*, Proceedings, 767 – 773, 2015 ISBN 978-80-227-4143-3.

Current address

Chládek Petr, Mgr. Ph.D.
Faculty of Economics, University of South Bohemia
Studentská 13, 370 05 České Budějovice, Czech Republic
Tel. number: +420 389 032 464, e-mail: chladek@jcu.cz

Smetanová Dana, RNDr. Ph.D.
The Institute of Technology and Bussiness in České Budějovice
Okružní 517/10, České Budějovice, Czech Republic
Tel. number: +420 387 842 133, e-mail: smetanova@mail.vstecb.cz

Slovak University of Technology in Bratislava
Faculty of Mechanical Engineering
16th Conference on Applied Mathematics **APLIMAT 2017**
Proceedings

COMPATIBLE RELATIONS ON STATE SETS OF QUASI-AUTOMATA

CHVALINA Jan (CZ), SMETANA Bedřich (CZ)

Abstract. Using analogy with relations between hypergoups and automata there is developed access to relations between these subjects. In this paper some concepts of compatible relations on state sets of quasi-automata are investigated.

Keywords: Hypergroup, parametrizability, substitution property, G-homomorphic relation

Mathematics subject classification: Primary 93A10; Secondary 20N35, 20N20, 08A60, 06A06

1 Introduction

Many interesting results concerning the investigation of compatible relations on algebras have been obtained by I. Chajda, J. Niederle, B, Pondělíček,J. Rachůnek, B. Zelinka and others (e.g. [3, 8-12, 22]), In papers of the above mentioned authors the Substitution Property (SP) of binary relations on a given algebra is considered as a compatibility type. This notion is motivated by the concept of a congruence. However, in the general system theory in the sense of [20, 22] a notion of the functional parametrization or the functional decomposition plays the essential role. In the theory of general input-output systems, which are triads S = (X,S,Y), notions such as functional decomposition of general systems and global reaction, which is in a certain sense similar to the previous one, are defined as important concepts. Let us remind the definition from [20]:

Let $\mathcal{S} = (X, S, Y)$ be a general input-output system, i.e. X, Y are sets, $S \subseteq X \times Y$ (the input-output relation of $\mathcal{S}$). Let C be a nonempty set, $F : C \times X \to Y$ be a mapping such that $(x, y) \in S$ iff $F(c, x) = y$ for some $c \in C$. Then C is called the set (or object) of global states of $\mathcal{S}$, elements $c \in C$ are called global states and the mapping F is said to be the global reaction of the system $\mathcal{S}$.

These concepts are analogical to the notion of continuity of binary relations in the sense of Grimeisen. It means that the binary relation $R \subseteq X \times Y$ between topological spaces (these can be different or identical) is continuous in the sense of Grimeisen [7] or briefly *G-continuous* if there exists a system Φ of continuous mappings $f : (X, \tau) \to (Y, \sigma)$ such that for any pair of points $(a, b) \in X \times Y$ there is $f \in \Phi$ with the property $f(a) = b$ and $\{(x, f(x)) \mid x \in X\} \subseteq R$ or (which is equivalent) the relation R is a set-theoretical union of mappings from Φ. This is a direct modification of the so called relational form of the Axiom of Choice. Also a certain variability of compatibility concepts of binary relations for topological spaces can be transferred into hyperstructures.

As another example of the parametrizability of multifunctions by measurable functions allowing to define the concept of an indefinite integral of a set-valued mapping [i.e. a domain-full binary relation) can serve the following definition taken from [6], p, 1:

Let F be a mapping of the compact real interval $\langle t_0, t_1\rangle$ into the class $\mathbb{K}$ of nonempty compact subset of a real Banach space B. Then an indefinite integral of $F : \langle t_0, t_1\rangle \to \mathbb{K}$ is dehned hy taking all Lebesgue measurable functions $f : \langle t_0, t_1\rangle \to B$ that satisfy $f(t) \in F(t)$ almost everywhere on $\langle t_0, t_1\rangle$ and setting

$$\int_{t_0}^{t} F(\tau)\,\mathrm{d}\tau = \left\{ \int_{t_0}^{t} f(\tau)\,\mathrm{d}\tau \right\}, \qquad t \in \langle t_0, t_1\rangle.$$

For other details see [6] and literature devoted to analytical theory of multifunctions, (e.g., [19]).

On the other hand binary relations compatible with algebras are also of standing interest - [9-13] and other papers related to these, In this investigation, the so called Substitution Property (SP) of binary relations on a given algebra plays fundamental role, This notion is motivated by the concept of a congruence. (The present contribution can serve as a certain extension or a supplement of results of the paper [8] devoted tn investigation of hypergroups determined by binary relations [4,16, 17, 23]).

2 Preliminaries

For further development it is necessary to clarify basic concepts:
A hypergrupoid is a pair $(H, \cdot)$, where $\cdot : H \times H \to \mathcal{P}(H)$ (the power set of H) is a hyperoperation,i.e. $\emptyset \neq a \cdot b \subseteq H$ for any pair $a, b \in H$.
A hypergroup is an associative hypergrupoid $(H, \cdot)$, i.e. $a \cdot (b \cdot c) = (a \cdot b) \cdot c$ for any triad $a, b, c \in H$, satisfying the reproduction axiom:
$a \cdot H = H \cdot a = H$ for any element $a \in H$.

Now, let A^* be the free monoid of words over an arbitrary (nonempty) alphabet A; suppose e stands for the empty word.

In accordance with [5], [18] and other publications, by an automaton we mean a triad $\mathbb{A} = (S, A, \delta)$, where S, A are arbitrary sets ($A \neq 0$), which are called – in the given order – a set of states (or a state set), a set of input symbols (or an input alphabet) and $\delta : S \times A \to S$ is a mapping called a transition function, which satisfies these two conditions:

- $\delta(s, e) = s$ for any state $s \in S$, (the identity axiom).

- $\delta(s, ab) = \delta(\delta(s, a), b)$ for any state $s \in S$ and any pair of words $a, b \in A^*$ (the homomorphism axiom). It is to be noted that the transition function $\delta : S \times A^* \to S$ is a usual extension of the next-state function $\delta_{\mathbb{A}}^{(0)} : S \times A \to S$.

An automaton $\mathbb{A}_1 = (S_1, A, \delta_1)$ is said to be a subautomaton of an automaton $\mathbb{A} = (S, A, \delta)$ if $S_1 \subseteq S, \delta(s, a) \in S_1$ for any state $s \in S_1$ and any word $a \in A^*$ and further $\delta_1 = \delta \uparrow (S \times A^*)$, i.e. δ_1 is a restriction of the transition function δ on $S_1 \times A^*$.

For $\emptyset \neq T \subseteq S$ we denote $\delta(T, A^*) = \{\delta(t, a); t \in T, a \in A^*\}$. (In [3], [4] this operator is denoted only by δ, i. e. $\delta(T)$ stands for the above set, in the mentioned papers). If T is a singleton, e.g. $T = \{t\}$, we write $\delta(t, A^*)$ instead of $\delta(\{t\}, A^*)$.

Let $\mathbb{A} = (S, A, \delta)$ be an automaton. We define a binary hyperoperation on the state set S of $\mathbb{A}$ by

$$s \circ t = \delta(s, A^*) \cup \delta(t, A^*).$$

for any pair of states $s, t \in S$. Since A^* contains the empty word e, we have $s, t \in s \circ t$, thus $s \circ t. \neq \emptyset$ and commutativity of this hyperoperation is also evident. Further, $s^2 = \delta(s, A^*)$ thus $s \circ t = s^2 \cup t^2$ and

$$\begin{gathered} s^3 = s \circ s^2 = s \circ \delta(s, A^*) = \bigcup_{t \in \delta(s, A^*)} s \circ t = \delta(s, A^*) \cup \bigcup_{t \in \delta(s, A^*)} \delta(t, A^*) = \\ = s^2 \cup \delta(\delta(s, A^*), A^*) = s^2 \cup \{\delta(\delta(s, a), b); a \in A^*), b \in A^*)\} = \\ = s^2 \cup \{\delta(s, ab); ab \in A^*A^* = A^*\} = s^2 \cup \{\delta(s, c); c \in A^*\} = s^2 \cup \delta(s, A^*) = \\ s^2 \cup s^2 = s^2 \end{gathered}$$

i.e. $s \in s^2 = s^3$ for any state $s \in S$. Consequently, the hyperoperation $\circ$ is associative and we get that $(S, \circ)$ is a quasi-order hypergroup. The same also follows from the fact, that $s \circ t = \{u \in S; (s, u) \in r$ or $(t, u) \in r\} = r(s) \cup r(t)$, where r is the transitive cover of the inertial relation ν on the state set of $\mathbb{A}((s, t) \in \nu$ if $\delta(s, a) = t$ for some input symbol $a \in A)$ which has been used and studied by M. W. Warner in [26]. The just defined quasi-order hypergroup $(s, \circ)$ will be called the state hypergroup of the automaton $\mathbb{A} = (S, A, \delta)$ and denoted also $\mathbb{H}(\mathbb{A})$.

3 Compatibility types of binary relations

Now it is useful describe concepts of relation compatibility on hypergroups $(H, \cdot)$:

Substitution Property We say that a binary relation $R \subseteq H \times H$ has *a substitution property*

- **of the first type(SP-1)**if for any quadruple $a, b, c, d \in H$ such that aRb, cRd we have $(a \cdot c)\ \overline{R}\ (b \cdot d)$, i.e. for each $x \in a \cdot c$ there is a $y \in b \cdot d$ such that xRy and vice versa, thus for any $y \in b \cdot d$ there exists an $x \in a \cdot c$ with xRy,

- **of the second type(SP-2)** if for any quadruple $a, b, c, d \in H$ and aRb, cRd implies $R(a \cdot c) = R(b \cdot d)$ (here, as above, $R(X) = \{y \in H | \exists x \in X$ with $xRy\})$,

- **of the third type(SP-3)** if $a, b, c, d \in H$, aRb, cRd implies $(a \cdot c)\ \overline{\overline{R}}\ (b \cdot d)$, i.e. for any pair $(x, y) \in (a \cdot c) \times (b \cdot d)$ we have xRy.

Parametrizability by endomorphisms. A relation $R \subseteq H \times H$ is said to be parametrizable by a set Φ of mappings of dom R into H if for any pair $(a, b) \in R$ there exists $f \in \Phi$ such that $f(a) = b$ and $\{(x, f(x)) \mid x \in \text{dom } R\} \subseteq R$, i.e. $R = \cup\{grf | f \in \Phi\}$, where the binary relation $grf = \{(a, f(a)|a \in \text{dom } R\})$ is the graph of the mapping $f : \text{dom } R \to H$. For the sake of brevity we write only f instead of grf. A relation $R \subseteq H \times H$ is said to be *parametrizable by a set of endomorphisms (good endomorphisms)* or briefly *parametrizable by endomorphisms (good endomorphisms)* if it is parametrizable by $\Phi \subseteq End(H, \cdot)$ $(\Phi(H, \cdot)$

Preservation of set-systems. Let $\emptyset \neq \mathfrak{G}_H \subseteq \mathcal{P}(H)$. A relation $R \subseteq H \times H$ is called $\mathfrak{G}_H$ - preserving ($\mathfrak{G}_H$ - reflecting or $\mathfrak{G}_H$ - returning) if it preserves (returns) members of $\mathfrak{G}_H$, i.e. $X \in \mathfrak{G}_H$ implies $R(X) \in \mathfrak{H}$ $(R^-(X) \in \mathfrak{G}_H$, i.e. R^- is $\mathfrak{G}_H$ - preserving). Here R^- is the opposite relation to R (cf. [19], which is denoted also by R^{-1}, $R^{\cup}$ elsewhere and called inverse or converse of R).

Theorem 1. *Let $(G, \cdot)$,$(H, \cdot)$(be qoset-hypergroups, $R \subseteq G \times H$ be a domain-full relation parametrizable by hoznomorphisrns of the hypergroup $(G, \cdot)$ into the hypergroups $(H, \cdot)$. Then the following assertions hold:*

1. *The relation R is $\mathfrak{C}Q$-reflecting.*
2. *For any pair of elements $x, y \in G$ we have $R(x.y) \subseteq R(x).R(y)$.*
3. *For arbitrary nonempty subset $X, Y \subseteq G$ we have $R(X.Y) \subseteq R(X).R(Y)$(in particular $R(X^2) \subseteq [R(X)]^2$).*

Theorem 2. *Let $(G, \cdot)$,$(H, \cdot)$ be qoset-hypergroups, $R \subseteq G \times H$ be a domain-full relation parametrizable by good homomorphisms of the hypergroup $(G, \cdot)$ into the hypergroup $(H, \cdot)$. Then R satisfies (1), (2), (3) from Theorem 1 and moreover the following conditions:*

1. *The relation R preserves multiplicatively closed subsets, i.e. it is Q-preserving.*
2. *For any pair $(a, b), (c, d) \in R \times R$ and any pair $(x, y) \in (a.c) \times (b.d)$ we have $R(x) \cap (b.d) \neq \emptyset \neq (a.c) \cap R^-(y)$.*

4 Parametrizations of relations by morphisms

Proposition 1. *Let $(G, *)$, $(H, \cdot)$ be commutative hypergroups, $R \subseteq G \times H$ be a domain-full relation parametrizable by homomorphisms of $(G, .)$ into $(H, .)$. Then $R(\langle X \rangle) \subseteq \langle R(X) \rangle$ for any nonempty subset $X \subseteq G$.*

Proof. Suppose $R = \bigcup\limits_{\gamma \in \Gamma} f_\gamma$ for a suitable system of homomorphisms $f_\gamma : (G, \cdot) \to (H, \cdot)$, $\emptyset \neq X \subseteq G$. Since $f_\gamma \subseteq R$ for each $\gamma \in \Gamma$ we have $f_\gamma(X) \subseteq R(X)$ which implies $\langle f_\gamma(X) \rangle \subseteq \langle R(X) \rangle)$ in virtue of monotonicity of the operator $\langle\ \rangle$ and thus $\bigcup\limits_{\gamma \in \Gamma} \langle f_\gamma(X) \rangle \subseteq \langle (R(X) \rangle$. Then we have

$$R(\langle X \rangle) = \left(\bigcup_{\gamma \in \Gamma} f_\gamma \right) (\langle X \rangle) = \bigcup_{\gamma \in \Gamma} f_\gamma(\langle X \rangle) \subseteq \bigcup_{\gamma \in \Gamma} \langle f_\gamma(X) \rangle \subseteq \langle R(X) \rangle.$$

Now, let us consider a mono-unary algebra (H, f), i.e. a nonempty set H and a mapping $f : H \to H$. Denote $\mathbb{N}_0 = N \cup \{0\} = \{0, 1, 2, ...\}$ and define a binary relation on H by

$$R_f = \{(x, f^m(x)) | x \in H, m \in \mathbb{N}_0 - \{1\}\}.$$

It is easy to see that R_f is a quasi-order on H which is an order if and only if the set-transformation $f : H \to H$ has at most one-element cycles, i.e. loops. A KW-equivalence $\sim_f$ (Kuratowski-Whyburn-type) on a mono-unary algebra (H, f) is defined by: $x \sim_f y \equiv f^n(y)$ for some pair m, n of non-negative integers. Blocks of this equivalence relaton are-called orbits of the set-transformation $f : H \to H$, subalgebras $(K, f \restriction K)$, where $K \in H/\sim_f$ are called components of the mono-unary algebra (H, f).

Example. This simple example shows that the parametrizability by good endo-morphisms of a binary relation on a hypergroup does not imply substitution property. Consider the Peano algebra of all positive integers $(\mathbb{N}, f)$, where $f(n) = n + 1$ for each $n \in \mathbb{N}$. Denote $\mathbb{N}_0 = \mathbb{N} \cup \{0\}$ and define mappings $\varphi, \psi : \mathbb{N} \to \mathbb{N}$ by $\varphi(1) = 1$, $\varphi(n) = n$ - 1 for $n \geq 2$, $\varphi(1) = \varphi(2) = 1$ and $\varphi(n) = n$ - 2 for $n \geq 3$. Now define $C_f X = \bigcup_{k \in \mathbb{N}_0} f^k(X)$, where $f^k$is the k-th iteration of f. Now for any $x, y \in \mathbb{N}$ we define $x \cdot y = C_f\{x, y\}$. Then $(\mathbb{N}, \cdot)$ is a commutative poset-hypergroup and the relation $R = \varphi \cup \psi$ is parametrizable by endomorphisms $\varphi, \psi : (\mathbb{N}, \cdot) \to (\mathbb{N}, \cdot)$. Further $(1, 1) \in R$, however $(f(1), f(1)) = (2, 2) \notin R$ thus R does not possess SP on the mono-unary algebra $(\mathbb{N}, f)$. In a similar way we can define a relation on $\mathbb{N}$, parametrizable by endomorphisms of the hypergroup $(\mathbb{N}, \cdot)$, which does not possess SP-1 on $(\mathbb{N}, \cdot)$ as well.

Proposition 2. *Let H be a nonempty set and $f : H \to H$ be a surjective mapping which does not possess any cycle, i.e. $f(H) = H$ and $f^n(x) \neq x$ for any pair $(x, n) \in H \times \mathbb{N}$. Let $(H, *_f)$ be a hypergroup defined by $x *_f y = R_f(x) \cup R_f(y)$ for any pair $x, y \in H$. Then the following assertions hold:*

(1) A relation $R \subseteq H \times H$ is parametrizable by endomorphisms of the mono-unary algebra (H, f) if and only of it is parametrizable by good endmorphisms of the hypergroup $(H, *_f)$.

(2) Any relation $R \subseteq HxH$ parametrizable by good endomorphisms of the hypergroup $(H, *_f)$ has substitution property (SP) on (H, f) and it has Substitution Property of the first type (SP-1) on $(H, *_f)$.

(1) In virtue of Theorem 7.1 from [17], p. 186 the endomorphism monoid of the mon-unary algebra (H, f) (i.e. the monoid of all selfmaps of H which commute with f) coincides with the monoid of all good endomorphisms of the hypergroup $(H, *_f)$. (In fact monoids of set-transformations which are carrying corresponding morphism are equal.) From this equality assertion (1) follows immediately.

(2) Let $x, y \in H$ be a pair of elements such that xRy. Then for a suitable good endomorphism φ of the hypergroup $((H, *_f)$ we have $\varphi \subseteq R$ and $\varphi(x) = y$. Since by [17], Theorem 7.1 $\varphi \circ f = f \circ \varphi$ then $f(y) = f(\varphi(x)) = \varphi(f(x))$, hence $f(x)Rf(y)$ and we get the relation R has SP with respect to mono-unary algebra (H, f). Now suppose $(a, b), (c, d) \in H \times H$, aRb, cRd. Then there are good endomorphisms $\varphi, \psi : (H, \cdot) \to (H, \cdot)$ such that $\varphi \subseteq R, \psi \subseteq R, \varphi(a) = b, \psi(c) = d$. Suppose $x \in a \cdot c$. Then for some $k \in \mathbb{N}_0 \backslash \{1\}$ either $x = f^k(a)$ or $x = f^k(c)$. Since by Theorem 7.1 [17]. $\varphi, \psi \in End(H, f)$, i.e. $f\varphi = \varphi f$, $f\psi = \psi f$, we have either

$$\varphi(x) = \varphi(f^k(a)) = f^k(\varphi(a)) = f^k(b) \in b \cdot d \text{ or}$$

$$\psi(x) = \psi(f^k(c)) = f^k(\psi(c)) = f^k(d) \in b \cdot d\,,$$

thus there exists $y \in b \cdot d$, $y = \varphi(x)$ or $y = \psi(x)$ such that xRy. If $y \in b \cdot d$ then —similarly as above for some $m \in \mathbb{N}_0 \backslash \{1\}$ either $y = f^m(b)$ or $y = f^m(d)$ and then either

$$y = f^m(\varphi(a)) = \varphi(f^m(a)) \text{ and } f^m(a \in a \cdot c \text{ or}$$
$$y = f^m(\psi(c)) = \psi(f^m(c)) \text{ where } f^m(c) \in a \cdot c,$$

thus there exists $x \in a \cdot c$ such that xRy. Consequently R has Substitution Property of the first type (SP-1) on the hypergroup $(H, *_f)$.

Definition 1. A binary relation R on an algebra (A, F) or a closure space (A, C) respectively is said to be *homomorphic in the sense of Grimeisen*, briefly *G-homomorphic*, if for every pair of elements $a, b \in A$ with $\langle a, b\rangle \in R$ there exists $\varphi \in End(A, F)$ or $\varphi \in End(A, C)$ respectively such that $\varphi \subset R$ and $\varphi(a) = b$.

Remark 1. It is evident that a binary relation R on (A, F) (similarly for a closure space) is G-homomorphic iff there exists a set $\Phi \subset End(A, F)$ such that $R = \bigcup_{\varphi \in \Phi} \varphi$.

Definition 2. A binary relation R on an algebra (A, F) is said to be *G-closure homomorphic*, if for every pair of elements $a, b \in A$ with $\langle a, b\rangle \in R$ there exists $\varphi \in End(A, C_F)$ such that $\varphi \subset R$ and $\varphi(a) = b$, i.e. if it is G-homomorphic on (A, C_F).

Evidently G-homomorphic and G-closure homomorphic relations are domain full. In what follows all considered relations are domain full. In the agreement with the notion of the Substitution Property we shall also use terms: the G-homomorphness and the G-closure homomorphness.

Proposition 3. *Every G-homomorphic relation on any state set of an automaton is G-closure homomorphic.*

Remark 2. (The G-closure homomorphness implies neither (SP) nor the G-homomorphness, (SP) implies neither the G-homomorphness nor the G-closure homomorphness and the G-homomorphness does not imply (SP) in general. Indeed, let us consider a 1-unary algebra $(\mathbf{N}, f)$, where $f(n) = n + 1$ for each $n \in \mathbf{N}$. Define mappings φ, ψ of the set $\mathbf{N}$ into itself by putting $\varphi(1) = 1$, $\varphi(n) = n - 1$ for $n \geq 2$, $\psi(1) = \psi(2) = 1$ and $\psi(n) = n - 2$ for $n \geq 3$. Put $R = \varphi \cup \psi$. Then R is evidently a G-homomorphic relation on $(\mathbf{N}, C_f)$, where $C_f X = \bigcup_{0 \leq k < \infty} f^k(X)$ for every subset $X \subset \mathbf{N}$, i.e. R is G-closure homomorphic on $(\mathbf{N}, f)$. It holds $\langle 1, 1\rangle \in R$ but $\langle f(1), f(1)\rangle = \langle 2, 2\rangle \notin R$, thus R does not possess (SP). Since for any mapping $g : \mathbf{N} \to \mathbf{N}$ with the property $g \subset R$ there is $g(1) = g(2) = 1$, consequently $g \notin End(\mathbf{N}, f)$, the relation R is not G-homomorphic.

For the proof of the second assertion put $G = \{a, b, c\}$ and denote by $\mathfrak{G}$ a commutative groupoid (a semigroup in fact) $(G, \circ)$, where the binary operation $\circ$ is given by this rule: For $x, y \in G$ we put $x \circ y = b$ iff $x = y = a$ and $x \circ y = c$ otherwise. Let P be a reflexive relation on G such that $\langle b, a\rangle \in P$ and $\langle c, b\rangle \in P$. Then P has (SP) with respect to $\mathfrak{G}$, but any mapping $\varphi \subset P$ with $\varphi(b) = a$ is not an endomorphism of $\mathfrak{G}$ for $\varphi(a) = a, \varphi(a \circ a) = a \neq b = \varphi(a) \circ \varphi(a)$. Further (denoting by C the corresponding algebraic closure operation on $\mathfrak{G}$) we have $C\{a\} = G$, $C\{b\} = b, c$, $C\{c\} = \{c\}$, hence the considered mapping φ is not a closure homo-morphism of the space (G, C), as well.

Finally, consider the additive semigroup of positive integers $(\mathbf{N}, +)$ and a reflexive relation

$R=\{\langle m,n\rangle : m,n\in \mathbf{N}, \frac{n}{m}\in\{1,2,3\}\}$. Denoting by φ_1 the identity mapping $id_{\mathbf{N}}$ and by $\varphi_k:\mathbf{N}\to\mathbf{N}, k=2,3$ mappings defined by $\varphi_k(n)=k.n$ for k = 2, 3 and each $n\in\mathbf{N}$, we have $R=\bigcup_{k=1}^{3}\varphi_k$ and simultaneously $\varphi_k, k=1,2,3$ are endomorphisms of $(\mathbf{N},+)$, thus R is G-homomorphic. On the other hand, $\langle 1,2\rangle\in R$, $\langle 1,3\rangle\in R$ e R but $\langle 2,5\rangle\notin R$. It is to be noted that the finiteness of $\mathfrak{G}$ in the first part of the above proof is not essential for the given counterexample. It can be replaced by the groupoid $\mathfrak{G}_1=(G_1,\circ)$ defined as follows: Let $(G_1,\geq)$ be a chain of the type ω^* with the greatest element s and for $a,b\in G_1$ we put $a\circ b=min\{x:max\{a,h\}<x\}$ if $a\neq s\neq b$ and $a\circ b=s$ in the opposite case. Further, $\langle a,b\rangle\in R$ if $a=b$ or a is a predecessor of the element b. If we restrict ourselves to unary algebras (with arbitrary many unary operations) we get:

Proposition 4. *Every G-homomorphic relation on a unary algebra has (SP).*

Proof. Let R be a G-homomorphic relation on a unary algebra (A,F). For a,beA such that $\langle a,b\rangle\in R$ there exists $\varphi\subset R,\varphi\in End(A,F)$ with $\varphi(a)=b$. Then for any operation $f\in F$ we have $\varphi(f(a))=f(\varphi(a))=f(b)$, i.e. $\langle f(a),f(b)\rangle\in R$ for each $f\in F$.

Theorem 3. *Let (A,f) be a 1-unary algebra, $\{(A_i,,f_i);i\in I\}$ the system of all its components, $I_0=\{i\in I;cardA_i>1, F=\{f^2,f^3\}$. Suppose that there is satisfied exactly one of these conditions:*
1° $i\in I$ implies either $f_i^2=f_i or f_i^2=id_{A_i}$,
2° $i\in I_0$ implies $A_i=A_i^{\infty 1}$,
3° $i\in I_0$ implies $A_i=A_i^{\infty 1}\cap A_i^0$ where $(A_i^{\infty 1},f_i)$ is a two-way infinite chain and $A_i^0\neq\emptyset$,
4° $i\in I_0$ implies either (A_i,f_i) is a two-way infinite chain or $A_i=B_i\cap D_i$ where $(D_i,f_i|D_i)$ is a one-way infinite chain with the first element d, $B_i=A_i^0$ and $f(x)=d$ for each $x\in B_i$. Then every G-closure homomorphic relation on the 2-unary algebra (A,F) has (SP) with respect to (A,f) and (A,F).

Proof. Let $a,b\in A$ be a pair of elements with the property $\langle a,b\rangle\in R$. There exists an endomorphism g of (A,C_F) such that $g(a)=b$ and $\langle x,g(x)\rangle\in R$ for every $x\in A$. Since for each $a\in A$ it holds $C_F\{a\}=\{f^k(a):k=0,2,3,...\}$ and $C_FX=\bigcup_{x\in X}C_FF\{x\}$ for $\emptyset\neq X\subset A$, we have by Theorems 3.2,4.1 from [15] $g\in End(A,f)$, thus $g(f(a))=f(g(a))=f(b)$. Hence $\langle f^k(a),f^kb\rangle\in R$ for $k=1,2,3$, q.e.d. From the above theorem we can obtain some corollaries as e.g. the following one:

Corollary. *Let $\mathfrak{A}=(A,f_1,f_2\})$ be a 2-unary algebra such that f_1,f_2 are permutations of the set A and*
1° $f_k^nh\neq hf_k^n$ for each $n\in\mathbf{N}$, any constant selfmap h of A and $k=1,2$,
2° $f_1f_2=f_2f_1$,
3° $f_1^3=f_2^2$ Then every G-closure homomorphic relation on the algebra $\mathfrak{A}$ has (SP).
Proof. Let $a\in A$ be an arbitrary element. There exists $b\in A$ with $f_1(b)=a$. Put $g(a)=f_2(b)$. Evidently g is a selfmap of the set A. For each $x\in A$ there exists $y\in A$ such that $f_1(y)=x$ and there exists $z\in A$ such that $f_1(z)=f_2(y)$. Then $g^2(x)=g(g(x))=g(f_2(y))=f_2(z)$ and $f_1^2(x)=f_1^3(y)=f_2^2(y)=f_2f_1(z)=f_1f_2(z)=f_1(g^2(x))$, i.e. $f_1^2=f_1g^2$. Since the mapping f_1 is one-to-one, we have $f_1==g^2$. Further, $g^3(x)=g^2g(x)=f_1f_2(y)=f_2f_1(y)=f_2f(x)$, i.e. $f_2=g^3$.

5 Conclusion

It has been mentioned above, there is often used the inertial relation ν on the state set of a quasi-automaton $\mathbb{A} = (S, A, \delta)$ (i.e. for $s, t \subset S$ we have $s\nu t$, whenever $\delta(s, a) = t$ for some input symbol $a \in A$). Then the transitive closure of the mentioned relation yields a quasi-ordering (i.e. a reflexive and transitive binary relation) on the state set S. As usually, using a quasi-ordering we can define a topology on the state set S; there is a relationship allowing to us to express structural properties of automata using topological terms. This is derived e.g. in the paper *A topology for automata, A Note* by Wagish SHUKLA and Arun K. SRIVASTAVA in Information and Control 32 (1976), pp. 163-168, cf. References and comments in [17]. In connection with generalized topologies, determined e.g. by Fréchet-Čech closure operations cl : exp $M \to$ exp M, satisfying $\text{cl}\emptyset = \emptyset$, $X \subseteq \text{cl}X$, $X \subseteq Y \subseteq M$ implies $\text{cl}X \subseteq \text{cl}Y$, we can endowed state sets of quasi-automata with closure operations in such a way that we fix a nonempty subset $P \subseteq A$ and then we put $\text{cl}X = \{\delta(s, a); (s, a) \in X \times P\}$.

A generalized filter base in a set M is defined as a nonempty family of nonempty subsets of M and this set-system is also called a stackbase. There can be constructed a functor from the category of Fréchet-Čech closure spaces and continuous mappings into the category of sets endowed with stackbases with morphisms-mappings reflecting members of stackbases. It can be verified that there is preserved the parametrizability of suitable relations between state sets of quasi-automata. Thete can be also established relationship to other special structures - [1]. However the thema in question will be investigated in a separate paper.

In connection with the extreme development of the rough sets theory, in particular devoted to the Pavlak approximations there occure possibility to investigate compatibility of created relations on rough automata with structures obtained for such automata or in fuzzy concept.[2]

Acknowledgement

The work was supported by the project FEKT-S-14-2200 of Brno University of Technology.

References

[1] AMIRI-BIDESHKI, M; SAEID, A. B.; AMERI, R.; HOŠKOVÁ-MAYEROVÁ, Š.*Prime filters of hyperlattices*. Analele Stiintifice ale Universitatii Ovidius Constanta, 2016, vol. 24(2), 15-26.

[2] AMERI, R.; HOŠKOVÁ-MAYEROVÁ, Š. *Fuzzy continuous polygroups*. Aplimat - 15th Conference on Applied Mathematics 2016 Proceedings. Bratislava: Vydavatelstvo STU - Slovak University of Technoology in Bratislava, 2016, 13-19.

[3] BANDELT H. J., *Tolerance relations on lattices*, Bull. Austral. Math. Soc. 23 (1981), 367-381.

[4] CORSINI P., *Graphs and join spaces*, J. of Combinatorics, Information and System Science 4 (1991), 313-318.

[5] DÖRFLER, W. *The cartesian composition of automata*, Math. Systems Theory (1978) 11: 239 - 257.

[6] GOODMANN, G. S., HOFFMANN-JORGENSEN, J., *Remarks on the calculus of set-valued mappings*, Preprint Series 1968/69 No. 10, Mat. Institut Aarhus Univ., 3pp.

[7] GRIMEISEN G., *Continuous relations*, Math. Zeit. 127 (1972), 35-44.

[8] HORT, D., CHVALINA, J., *On certain compatibility types of binary relations on hypergroups determined by quasi-orders*, Contributions to Gen. Algebra (Klagenfurt), 11(1999), 12.

[9] CHAJDA I., *Algebraic Theory of Tolerance Relations*, UP Olomouc, 1991.

[10] CHAJDA I., *Two constructions of compatible relations*, Czech. Math. J. 28 (1978), 439-444.

[11] CHAJDA I., ZELINKA B., *Compatible relations on algebras*, Časopis pěst. mat. 100 (1975), 355-360.

[12] CHAJDA, I., NIEDERLE, J., ZELINKA, B., *On existence conditions for compatible tolerances*, Czech. Math. J. 26(1976), 304-311.

[13] CHVALINA J., *Note on compatible binary relations on algebras*, Arch. Math. (Brno) 4 (1980), 185 - 190.

[14] CHVALINA, J. *Parametrizable relations on hypergroups determined by quasi- orders.* 1998, č. 6, s. 22 .

[15] CHVALINA, J. *Set transformations with centralizers formed by closed deformations of quasi- discrete topological spaces.* In In: Proc. IV. Prague Topological Symp. 1976, Part B, Soc. Czech. Math. Phys. Praha: 1977. s. 83 (s.)

[16] CHVALINA, J., *Commutative hypergroups in the sense of Marty and ordered sets*, Proceedings of the Summer School an General Algebra and Ordered Sets, Olomouc (Czech Republic), 1994, 19-30. .

[17] CHVALINA, J., *Functional Graphs, Quasi-ordered Sets and Commutative Hypergroups*, Vydavatelstvi Masarykovy univerzity, Brno, 1995 (Czech).

[18] CHVALINA, J., CHVALINOVÁ, L.,*State hypergroups of automata.* Acta Mathematica et Informatica Universitatis Ostraviensis 04.1 (1996): 105-119.

[19] KLEIN, E., THOMPSON, A. C., T*heory of Correspondences*, Wiley et Sons, New York-Brisbane-Toronto-Singapore, 1984.

[20] MESAROVIČ M. D., TAKAHARA Y., *General Systems Theory*, New York 1975, Academic Press.

[21] PICHLER, F., *Mathematische Systemtheorie*, Walter de Gruyter, Berlin 1975.

[22] RACHŮNEK J., *Quasi-orders of algebras*, Časopis pěst. mat. 104 (1979), 327-337.

[23] ROSENBERG, I. G., *Hypergroups introduced by paths of a directed graph*, Ital. J. Pure Appl. Math 4 (1998), 133 - 142.

[24] ROSENBERG, I. G., *Hypergroups and join spaces determined by relations*, Ital. J. Pure Appl. Math 4 (1998), 93 - 101.

[25] SMITHSON R. E., *Multifunctions*, Niew Arch. Wisk. 20 (1972), 31-55.

[26] WARNER, M. W., *Semigroup, group quotient and homogeneous automata*, Inform. Control 47 (1980), 59-66.

Current address

Chvalina Jan, Prof. RNDr. DrSc.
Brno University of Technology,
Faculty of Electrical Engineeering and Comunication, Department of Mathematics
Technická 8, 61600 Brno, Czech Republic
Tel. number: +420 54114 6058, e-mail: chvalina@feec.vutbr.cz

Smetana Bedřich, RNDr.
Brno Univerzity of Defence,
Faculty of Military Leadership, Department of Econometrics
Kounicova 65,662 10 Brno, Czech Republic
Tel. number: +420 973443429, e-mail: bedrich.smetana@unob.cz

Slovak University of Technology in Bratislava
Faculty of Mechanical Engineering
16th Conference on Applied Mathematics **APLIMAT 2017**
Proceedings

SEQUENCES OF AUTOMATA FORMED BY GROUPS OF POLYNOMIALS AND BY SEMIGROUPS OF LINEAR DIFFERENTIAL OPERATORS

CHVALINA Jan (CZ), STANĚK David (CZ)

Abstract. The concept of an automaton is a mathematical interpretation of real systems that work in a discrete time-scale. Using linear differential operators there constructing quasi-automata formed by actions of linear differential operator of n-th order on rings of real polynomials and also by actions of polynomials on groups of differential operators. Moreover decreasing and increasing of such quasi-automata are also constructed.

Keywords: Quasi-automaton, ordinary linear differential operator, abelian groups of polynomials of one real variable, sequences of quasi-automaton, sequences of actions of additive groups of polynomials

Mathematics subject classification: Primary 11B85, 37B05; Secondary 08A40, 47E05

1 Introduction

Automata, theory of which belongs to the classical direction of the mathematical structure theory were considered as systems that can be used for the transmission of information of certain type. They are belonging to systems involving the modelling of various processes and simulating of inner connections of time developable processes. With respect to concept of an action of semigroups or groups on a set there are useful such concepts as quasi-automata or semiautomata or simply automata without outputs which are useful generalizations of Mealy-type automata. The concept of an automaton is mathematical interpretation of real systems that work in a discrete time-scale. Using the binary operation of concatenation of chains of input symbols we obtain automata with input alphabets endowed with the structure of a semigroup or a group. Considering mainly the structure given by transition function and neglecting output functions with output sets we are coming to very useful generalization of concept of an automaton called a quasi-automaton, [1, 3, 4, 11, 15, 16] and other papers.

In more detail a quasi-automaton is a system (A, S, δ) which consists of a non-void set A, an arbitrary semigroup S and a mapping $\delta : A \times S \to A$ such that

$$\delta(\delta(a, r, s)) = \delta(a, r, s) \tag{1}$$

for arbitrary $a \in A$ and $r, s \in S$. Notice that the concept of the quasi-automaton has been introduced by S. Ginsberg under the term of quasi-machine as a generalization of the Mealy-type automaton.

In fact theses structures can be considered as discrete dynamical systems or transformation groups (semigroups). It is to be noted that bibliography of the theory of dynamical systems or the dynamical topology up to 1972 was compiled by Walter Hellbig Gottschalk and this bibliography contains 178 pages and lists 2475 membered items. Let us also notice that literature devoted to actions of groups or semigroups treated from the algebraic point of view is also very comprehensive.

Rapid development of algebraic hyperstructures - formerly called multistructures - leads naturally also to investigations of actions of hyperstructures on sets of various objects especially sets of continuous or differentiable functions. Now, hyperstructures formed by various operators (transformation operators on spaces of real or complex functions, differential or integral operators) show that generalizations of above mentioned structures (quasi-automata) with input hyperstructures on centralizers of elements of suitable semigroup(groups) are of a certain interest. More precisely, let A be a nonempty set, $(H, \cdot)$ be a semihypergroup or a hypergroup and $\delta : A \times H \to A$ by a mapping satisfying the condition

$$\delta(\delta(s, a), b) \in \delta(s, a \cdot b) \tag{2}$$

for any triad $(s, a, b) \in A \times H \times H$, where $\delta(s, a \cdot b) = \{\delta(s, x); x \in a \cdot b\}$. Then the triad (A, H, δ) is called multiautomaton with the state set A and the input semihypergroups $(H, \cdot)$. The mapping $\delta : A \times H \to A$ is called the transition function or the next-state function of the multiautomaton (A, H, δ). The above condition is called the *Generalized Mixed Associativity Condition,* shortly GMAC - [7, 8, 9, 10].

Following a number of books and articles concerning actions of semigroups and groups we also say that the *(semi-)hypergroup H acts on the phase set* (also known as *phase space*) S or that the triad (S, H, δ) which satisfies the above mentioned general mixed associativity condition, is the *multi-act* (also known as *multiaction*) of H on S. However the present contribution is devoted mainly to basic non-deterministic quasi-automata which can serve as a base for generalizations in the above mentioned sense.

Investigation of the below used groups of ordinary linear differential operators is motivated by important models formed by a linear homogeneous ordinary differential equations of the form

$$y^{(n)}(x) + \sum_{k=0}^{n-1} p_k(x) y^{(k)}(x) = 0 \tag{3}$$

cf. [17].

By an ordered group we mean (as usually) a triad $(G, ., \leq)$, where $(G, .)$ is a group and $\leq$ is a reflexive, symmetrical and transitive binary relation on the set G such that for any triad $x, y, z \in G$ with the property $x \leq y$ also $x.z \leq y.z$, $z.x \leq z.y$ is satisfied. Further, $[a)_{\leq} = \{x \in G, a \leq x\}$ is the principal end generated by $a \in G$.

The following lemma which is crucial for further constructions is proved in [18] (the Czech version is proved in [5], pp. 146, 147.

Lemma 1.1 *Let $(G, ., \leq)$ be a quasi-ordered group. Define a hyperoperation $\star : G\times G \to P(G)$ by*

$$a \star b = [a.b)_{\leq} (= \{x \in G; a.b \leq x\}) \tag{4}$$

for all pairs of elements $a, b \in G$. Then $(G, \star)$ is a hypergroup which is commutative if and only if the group $(G, .)$ is commutative.

The just presented lemma, now termed as the *Ends Lemma,* was used and treated in many papers cf. e.g. [18, 19, 20, 21].

In physics applications of signal modelling very often, one needs to recover an unknown signal from its garbled and noise influenced form which is a result of passing through a garbling or noisy environment (such as a communication channel). As an example recall a recovery method called *Kalman filtering.* Apart from the scalar Kalman filter the vector filter is often used in applications since the state vector need not contain the delayed values of one state variable only but instead it can be compose of various variables that have various physical meaning in the modelled system. This enables a straightforward creation of well understandable models of physical systems, the internal variables of which (i.e. the original signals) can be Kalman filters estimated using external observation. Denote vectors of delayed values of the original and observed signals by

$$\vec{x}_n = \left(x_n^1, x_n^2, \ldots, x_n^R\right)^T, \quad \vec{y}_n = \left(y_n^1, y_n^2, \ldots, y_n^R\right)^T, \tag{5}$$

where R is the order of the given model. In a similar way denote the vectors of the driving noise and undesired noise $\vec{g}_n$ and ν_n. The state equation of the model describing the signal generation is

$$\vec{x}_{n+1} = \mathrm{A}_{n+1}\vec{x}_n + \vec{g}_{n+1} \tag{6}$$

while the output equation is

$$\vec{y}_{n+1} = \mathrm{C}_{n+1}\vec{x}_n + \vec{\nu}_{n+1} \tag{7}$$

In contrast to the standard form of the state description, we have that it is the vector of undesired noise that is used in the output equation of the system (instead of the input of the system), otherwise there is obviously $\mathrm{B} = \mathrm{d} = \mathrm{I}$, i.e. the output equation of the system is

$$\vec{y}_{n+1} = \mathrm{A}_{n+1}\mathrm{C}_{n+1}\vec{x}_n + \mathrm{C}_{n+1}\vec{g}_{n+1} + \vec{\nu}_{n+1}. \tag{8}$$

In cases when we want to create signals of complicated structure or model complicated systems (provided we choose sufficiently high order R) we need to use vector and matrix mathematical apparatus (e.g. in finite impulse response filters).

Upon composition of the state and output equation we get

$$\vec{y}_{n+1} = \mathrm{A}_{n+1}\mathrm{C}_{n+1}\vec{x}_n + \mathrm{C}_{n+1}\vec{g}_{n+1} + \vec{\nu}_{n+1}. \tag{9}$$

Now we denote by $\mathbb{M}_{m,m}(\mathbb{R})$ the set of square matrices of order m. One can construct a Mealy-type automaton $\mathbb{A} = (\mathbb{M}_{m,m}(\mathbb{R}), X, Y, \delta, \sigma)$, where $\delta : \mathbb{M}_{m,m}(\mathbb{R}) \times X \to X$, $\sigma : \mathbb{M}_{m,m}(\mathbb{R}) \times X \to Y$, $\delta(\mathrm{A}, \vec{x}) = \mathrm{A}\vec{x} + \vec{g}$, $\sigma(\mathrm{C}, \vec{u}) = \mathrm{C}\vec{u} + \vec{\nu}$, where $\vec{g}, \vec{\nu}$ are fixed column numerical vectors. Notice that if $(\mathbb{M}_{m,m}(\mathbb{R}), \cdot)$ is the usual multiplicative semigroup, then the defining condition of (finite) deterministic automata, sometimes -and in this paper - referred to MAC, i.e. mixed associativity condition, does not hold. However, it is easy to construct an automaton satisfying this condition once its input alphabet is a non-commutative semigroup or group

2 Linear differential operators

As usual, $\mathbb{R}$ stands for the set of all reals, $J \in \mathbb{R}$ is an open interval (bounded or unbounded) of real numbers, $\mathbb{C}^k(J)$ is the ring (with respect to usual addition and multiplication of functions) of all real functions with continuous derivatives up to the order $k \geq 0$ including. We write $\mathbb{C}(J)$ instead of $\mathbb{C}^0(J)$. For a positive integer $n \geq 2$ we denote $\mathbb{A}_n$ (according to papers of F. Neuman) the set of all linear homogeneous differential equations of the n-th order with continuous real coefficients on J, i.e.

$$y^{(n)} + p_{n-1}(x)y^{(n-1)} + \cdots + p_0(x)y = 0, \tag{10}$$

(cf. [8]), where $p_k \in \mathbb{C}(J), k = 0, 1, \ldots, n-1$, $p_0(x) > 0$ for any $x \in J$ (this is not essential restriction). Denote $L_0(p_{n-1}, \ldots, p_0) : \mathbb{C}^n(J) \to \mathbb{C}^n(J)$ the above linear operator defined by

$$L_0(p_{n-1}, \ldots, p_0)(y) = y^{(n)} + p_{n-1}(x)y^{(n-1)} + \cdots + p_0(x)y \tag{11}$$

and put

$$\mathbb{LA}_n(J) = \{L(p_{n-1}, \ldots, p_0); p_k \in \mathbb{C}(J), p_0 > 0\}. \tag{12}$$

Further denote $\mathbb{N}_0 = \{0, 1, \ldots, n-1\}$ and δ_{ij} stands for the Kronecker δ, $\overline{\delta_{ij}} = 1 - \delta_{ij}$. For any $m \in \mathbb{N}_0(n)$ we denote by $\mathbb{LA}_n(J)_m$ the set of all linear differential operators of the n-th order $L_0(p_{n-1}, \ldots, p_0) : \mathbb{C}^n(J) \to \mathbb{C}(J)$, where $p_k \in \mathbb{C}(J)$ for any $k \in \mathbb{N}_0(n)$, $p_m \in \mathbb{C}_+(J)$, (i.e. $p_m(x) > 0$ for each $x \in J$). Using the vector notation $\vec{p}(x) = (p_{n-1}(x), \ldots, p_0(x))$, $x \in J$ we can write $L_n(\vec{p})y = y^{(n)} + (\vec{p}(x), (y', y, \ldots, y^{n-1}))$.

We define a binary operation " $\circ_m$ " and a binary relation " $\leq_m$ " on the set $\mathbb{LA}_n(J)_m$ in this way:

For arbitrary pair $L(\vec{p}), L(\vec{q}) \in \mathbb{LA}_n(J)_m$, $\vec{p} = (p_{n-1}, \ldots, p_0)$, $\vec{q} = (q_{n-1}, \ldots, q_0)$ we put $L(\vec{p}) \circ_m L(\vec{q}) = L(\vec{u})$, $\vec{u} = (u_{n-1}, \ldots, u_0)$, where

$$u_k(x) = p_m(x)q_k(x) + (1 - \delta_{km})p_k(x), \ x \in J \tag{13}$$

and $L(\vec{p}) \leq_m L(\vec{q})$ whenever $p_k(x) \leq q_k(x)$, $k \in \mathbb{N}_0(n)$, $p_m(x) = q_m(x)$, $x \in J$.

Evidently, $(\mathbb{LA}_n J_m, \leq_m)$ is an ordered set.

The following lemma is proved in the paper [9].

Lemma 2.1 *The triad* $(\mathbb{LA}_n(J)_m, \circ_m, \leq_m)$ *is an ordered (noncommutative) group.*

A certain construction of an action of the additive group of all integer on the phase space $\mathbb{LA}_n(J)_m$ (or in terminology of O. Boruvka - an algebraic space with operators] with use of the group of operators $(\mathbb{LA}_n(J)_m, \circ_m)$ is presented in paper [8].

Now, using the construction described in Lemma 1.1 we obtain the final join space.

Denote by $P(\mathbb{LA}_n(J)_m)$ the power set of $\mathbb{LA}_n(J)_m$ consisting of all subsets of the last set and define a binary hyperoperation

$$\star_m : \mathbb{LA}_n(J)_m \times \mathbb{LA}_n(J)_m \to P(\mathbb{LA}_n(J)_m) \tag{14}$$

by the rule

$$L(\vec{p}) \star_m L(\vec{q}) = \{L(\vec{u}); L(\vec{p}) \circ_m L(\vec{q}) \leq_m L(\vec{u})\} \tag{15}$$

for all pairs $L(\vec{p}), L(\vec{q}) \in \mathbb{LA}_n(J)_m$. More in detail, if $\vec{u} = (u_{n-1}, \ldots, u_0), \vec{p} = (p_{n-1}, \ldots, p_0)$, $\vec{q} = (q_{n-1}, \ldots, q_0)$, then $p_m(x)q_m(x) = u_m(x)$, $p_m(x)q_k(x) + p_k(x) \leq u_k(x)$ if $k \neq m$, $x \in J$. Then by Lemma 1.1 we have that $(\mathbb{LA}_n(J)_m, \star_m)$ is a (non-commutative) hypergroup. The above defined invariant (or normal) subgroup $\mathbb{L}_1\mathbb{A}_n(J)_m$ of the group $(\mathbb{LA}_n(J)_m, \circ_m)$ is the carrier set of a subhypergroup of $(\mathbb{LA}_n(J)_m, \star_m)$ and it has certain significant properties.

Using certain generalization of methods in [8, 10] there were obtain after investigation of constructed structures this result:

Theorem 2.2 *Let* $J \subset \mathbb{R}$ *be an open interval,* $2 \leq n$ *positive integer and* $0 \leq m \leq n-1$. *Let* $\mathbb{LA}_n(J)_m = \{L(p_{n-1}, \ldots, p_0); p_k \in \mathbb{C}(J), p_m > 0\}$. *Then the hypergroup* $(\mathbb{LA}_n(J)_m, \star_m)$ *is a join space of linear ordinary differential operators of the n-th order such that* $(\mathbb{L}_1\mathbb{A}_n(J)_m, \star_m)$ *is a subhypergroup of* $(\mathbb{LA}_n(J)_m, \star_m)$*, which is normal (i.e.* $L(\vec{p}) \star_m \mathbb{L}_1\mathbb{A}_n(J)_m \star_m L(\vec{p})$ *for any operator* $L(\vec{p}) \in \mathbb{LA}_n(J)_m$*).*

There seems to be a quite natural approach to actions of binary hyperstructures $\mathbb{LA}_n(J)_m$ of differential operators on tyhe space of infinitely smooth function $\mathbb{C}^\infty(J)$ or - in another words - to multiautomata with the state space $\mathbb{C}^\infty(J)$, $J \subset \mathbb{R}$ and with the input alphabet $\mathbb{LA}_n(J)_m$ consisting in definition of the transition function (i.e. the next state function)

$$\delta : \mathbb{LA}_n(J)_m \times \mathbb{C}^\infty(J) \to \mathbb{C}^\infty(J) \tag{16}$$

as an evaluation map, i.e.

$$\delta(D, f) = D(f) = f^{(n)}(x) + \sum_{k=0}^{n-1} p_k(x) f^{(k)}(x), \tag{17}$$

$x \in J$, for any pair $[D, f] \in \mathbb{LA}_n(J)_m \times \mathbb{C}^\infty(J)$. However, the just mentioned approach creates the following objectionable problem: Demanding validity of the generalized mixed associativity condition (GMAC) which has the form $\delta(D_1, \delta(D_2, f)) \in \delta(D_1 \cdot D_2, f)$ for some pairs of operators $D_1, D_2 \in \mathbb{LA}_n(J)_m$ and arbitrary function $f \in \mathbb{C}^\infty(J)_m$, we get immediately a contradiction because $\delta(D_1, \delta(D_2, f)) = \delta(D_1, D_2(f)) = D_1(D_2(f))$ and usual composition of differential operators of n-th order D_1, D_2 yields the differential operator of the 2-*nd* order which does not belong to (semi-)hypergroup $\mathbb{LA}_n(J)_m$. This unacceptable situation can be solved in a certain special case using of a certain idea from the formal language theory.

Actions of semigroups or groups on groups of linear differential operators are presented in some papers as [22]. In this contribution we construct the action of certain groups of linear differential operators on phase spaces of smooth functions, concretely on vector spaces of polynomials.

As above, considered linear differential operators $L(m, p_{n-1}, \ldots, p_0) : \mathbb{C}^\infty(\mathbb{R}) \to \mathbb{C}^\infty(\mathbb{R})$ defined

$$L(m, p_{n-1}, \ldots, p_0)f = m\frac{\mathrm{d}^n f(x)}{\mathrm{d}x^n} + \sum_{k=0}^{n-1} p_k(x)\frac{\mathrm{d}^k f(x)}{\mathrm{d}x^k}. \tag{18}$$

Denote by $\mathbb{L}_{A1}\mathbb{A}_n(\mathbb{R})$ the additive abelian group of differential operators $L(m, p_{n-1}, \ldots, p_0)$, where for $L(m, p_{n-1}, \ldots, p_0), L(k, q_{n-1}, \ldots, q_0) \in \mathbb{L}_{A1}\mathbb{A}_n(\mathbb{R})$ we define

$$L(m, p_{n-1}, \ldots, p_0) + L(k, q_{n-1}, \ldots, q_0) = L(m+k, p_{n-1}+q_{n-1}, \ldots, p_0+q_0), \tag{19}$$

where $L(m+k, p_{n-1}+q_{n-1}, \ldots, p_0+q_0)f = (m+k)\frac{\mathrm{d}^n f(x)}{\mathrm{d}x^n} + \sum_{k=0}^{n-1}(p_k(x)+q_k(x))\frac{\mathrm{d}^k f(x)}{\mathrm{d}x^k}$.
Now, suppose $p_k \in \mathbb{R}_{n-1}[x]$ and define

$$\delta_n : \mathrm{R}_n[x] \times \mathrm{L}_{A1}\mathbb{A}_n(\mathbb{R}) \to \mathbb{R}_n[x] \tag{20}$$

by $\delta_n(f, L(m, p_{n-1}, \ldots, p_0)) = m\frac{\mathrm{d}^n f(x)}{\mathrm{d}x^n} + f(x) + m + \sum_{k=0}^{n-1} p_k(x), f \in \mathbb{R}_n[x]$.
This group is a subgroup of the group $[\mathbb{C}^\infty(\mathbb{R})]^n = \mathbb{C}^\infty(\mathbb{R}) \times \cdots \times \mathbb{C}^\infty(\mathbb{R})$ of the n-dimensional vectors of smooth (infinitely differentiable) functions of one real variable.

We are going to verify the mixed associativity condition (MAC) which should satisfy the above defined action:

Suppose $f \in \mathbb{R}_n[x], f(x) = \sum_{k=0}^{n} a_k x^k, L(m, p_{n-1}, \ldots, p_0), L(k, q_{n-1}, \ldots, q_0) \in \mathbb{L}_{A1}\mathbb{A}_n(\mathbb{R})$.

Then

$$\begin{aligned}
&\delta_n\left(\delta_n\left(f, L\left(m, p_{n-1}, \ldots, p_0\right)\right), L\left(k, q_{n-1}, \ldots, q_0\right)\right)= \\
&=\delta_n\left(m \frac{\mathrm{d}^n f(x)}{\mathrm{d} x^n}+f(x)+m+\sum_{k=0}^{n-1}, L(k, q_{n-1}, \ldots, q_0)\right)= \\
&=\delta_n\left(m \cdot n! \cdot a_n+m+f(x)+\sum_{k=0}^{n-1} p_k(x), L\left(k, q_{n-1}, \ldots, q_0\right)\right)= \\
&=k \frac{\mathrm{d}^n f(x)}{\mathrm{d} x^n}+m \cdot n! \cdot a_n+m+f(x)+\sum_{k=0}^{n-1} p_k(x)+\sum_{k=0}^{n-1} q_k(x)+k= \\
&=(m+k) n! \cdot a_n+(m+k)+f(x)+\sum_{k=0}^{n-1}\left(p_k(x)+q_k(x)\right)= \\
&=(m+k)(n! \cdot a_n+1)+f(x)+\sum_{k=0}^{n-1}\left(p_k(x)+q_k(x)\right)= \\
&=(m+k) \frac{\mathrm{d}^n f(x)}{\mathrm{d} x^n}+f(x)+(m+k)+\sum_{k=0}^{n-1}\left(p_k(x)+q_k(x)\right)= \\
&=\delta_n\left(f, L(m+k, p_{n-1}+q_{n-1}, \ldots, p_0+q_0)\right)= \\
&=\delta_n\left(f, L(m, p_{n-1}, \ldots, p_0)+L(k, q_{n-1}, \ldots, q_0)\right),
\end{aligned} \tag{21}$$

thus the mixed associativity condition is satisfied.

On the other hand we can construct the action of abelian additive groups of continuous functions $\mathbb{C}^k(J)(J \subseteq \mathbb{R})$ on groups $\mathbb{L}_{A1}\mathbb{A}_n(J)$ of linear differential operators $L(m, p_{n-1}, \ldots, p_0)$ defined by $\sigma_n(L(m, p_{n-1}, \ldots, p_0, f) = L(m, p \circ f + p_{n-1}, \ldots, p \circ f + p_1, p_0)$. We show that the mapping (the next state function)

$$\sigma_n : \mathbb{L}_{A1}\mathbb{A}_n(J) \times \mathbb{C}^k(J) \to \mathbb{L}_{A1}\mathbb{A}_n(J) \tag{22}$$

satisfies the MAC: For $L(m, p_{n-1}, \ldots, p_0) \in \mathbb{L}_{A1}\mathbb{A}_n(J)$, $f, g \in \mathbb{C}^k(J)$ we have

$$\begin{aligned}
&\sigma_n\left(\sigma_n\left(L(m, p_{n-1}, \ldots, p_0), f\right), g\right)= \\
&=\sigma_n\left(L(m, p \circ f+p_{n-1}, \ldots, p \circ f+p_1, p_0), g\right)= \\
&=L(m, p \circ g+p \circ f+p_{n-1}, \ldots, p \circ g+p \circ f+p_1, p_0)= \\
&=L(m, p \circ(g+f)+p_{n-1}, \ldots, p \circ(g+f)+p_1, p_0)= \\
&=\sigma_n\left(L(m, p_{n-1}, \ldots, p_0), f+g\right),
\end{aligned} \tag{23}$$

thus the mixed associativity condition is satisfied.

Now, considering vector spaces $\mathbb{R}_n[x]$ of polynomials instead of $\mathbb{C}^k(J)$ (here $J = \mathbb{R}$) we obtain the action of the abelian group $\mathbb{R}_n[x]$ on the group $\mathbb{L}_{A1}\mathbb{A}_n(\mathbb{R})$:

$$\sigma_n : \mathbb{L}_{A1}\mathbb{A}_n(\mathbb{R}) \times \mathbb{R}_n[x] \to \mathbb{L}_{A1}\mathbb{A}_n(\mathbb{R}). \tag{24}$$

Using actions we can construct other composed mappings

$$\phi_n : \mathbb{R}_n[x] \times \mathbb{L}_{A1}\mathbb{A}_n(\mathbb{R}) \times \mathbb{R}_n[x] \to \mathbb{R}_n[x] \tag{25}$$

and

$$\psi_n : \mathbb{L}_{A1}\mathbb{A}_n(\mathbb{R}) \times \mathbb{R}_n[x] \times \mathbb{L}_{A1}\mathbb{A}_n(\mathbb{R}) \to \mathbb{L}_{A1}\mathbb{A}_n(\mathbb{R}), \tag{26}$$

in this way:

Denoting $\mathrm{id}_{\mathbb{R}_n[x]} : \mathbb{R}_n[x] \to \mathbb{R}_n[x]$ the identity mapping and similarly $\mathrm{id}_{\mathbb{L}} : \mathbb{L}_{A1}\mathbb{A}_n(\mathbb{R}) \to \mathbb{L}_{A1}\mathbb{A}_n(\mathbb{R})$ we obtain

$$\mathbb{R}_n[x] \times \mathbb{L}_{A1}\mathbb{A}_n(\mathbb{R})[x] \times \mathbb{R}_n[x] \xrightarrow{\mathrm{id}_{\mathbb{R}_n[x]}\times\sigma_n} \mathbb{R}_n[x] \times \mathbb{L}_{A1}\mathbb{A}_n(\mathbb{R}) \to \delta_n\mathbb{R}_n[x] \tag{27}$$

and

$$\mathbb{L}_{A1}(\mathbb{R}) \times \mathbb{R}_n[x] \times \mathbb{L}_{A1}\mathbb{A}_n(\mathbb{R}) \xrightarrow{\mathrm{id}_{\mathbb{L}}\times\delta_n} \mathbb{L}_{A1}\mathbb{A}_n(\mathbb{R}) \times \mathbb{R}_n[x] \to \sigma_n\mathbb{L}_{A1}\mathbb{A}_n(\mathbb{R}), \tag{28}$$

thus

$$\phi_n = \delta_n \circ \left(\mathrm{id}_{\mathbb{R}[x]} \times \sigma_n\right), \; \psi_n = \sigma_n \circ \left(\mathrm{id}_{\mathbb{L}} \times \delta_n\right). \tag{29}$$

However, investigation of properties of constructed mappings and other mappings derived from the above presented ones should be contained in forthcoming paper.

3 Actions of abelian groups and hypergroups

Using the *Ends Lemma* we are going to construct hypergroups from groups $\mathbb{R}_n[x], \mathbb{L}_{A1}\mathbb{A}_n(\mathbb{R})$ endowed with a naturally defined orderings, i.e. from ordered abelian groups. Let us mention that very useful *Ends Lemma* - introduced in [5] - was used in many papers and also generalized, in particular by Michal Novák [18, 19, 20, 21], for constructions of various multistructures. As usually, for polynomials $f, g \in \mathbb{R}_n[x]$ we put $f \leq g$ whenever $f(x) \leq g(x)$ for each $x \in \mathbb{R}$.

From the obtained actions (automata without outputs) we will construct multi-automata, i.e. actions of hyperstructures - semihypergroups and hypergroups.

For a pair of polynomials $f, g \in \mathbb{R}_n[x]$ we put $f \leq g$, whenever (as usually) $f(x) \leq g(x), z \in \mathbb{R}_n[x]$. Then $(\mathbb{R}_n[x], \leq)$ is an ordered abelian group.

We define a binary hyperoperation

$$\# : \mathbb{R}_n[x] \times \mathbb{R}_n[x] \to \mathcal{P}^{\star}\left(\mathbb{R}_n[x]\right) \tag{30}$$

(here $\mathcal{P}^{\star}(\mathbb{R}_n[x])$ means the collection of all nonempty subsets of $\mathbb{R}_n[x]$) by the rule

$$f\#g = \{h; h \in \mathbb{R}_n[x], f(x) + g(x) \leq h(x), x \in \mathbb{R}\} = [f+g)_{\leq}. \tag{31}$$

By the *Ends Lemma* [18], p. or by [ ], *Theorem 4* we have that the hypergroupoid $(\mathbb{R}_n[x], \#)$ is a commutative hypergroup (by [22], *Theorem 4* moreover a transposition hypergroups, i.e. a join space).

Defining a binary hyperoperation

$$\# : \mathbb{L}_{A1}\mathbb{A}_n(\mathbb{R}) \times \mathbb{L}_{A1}\mathbb{A}_n(\mathbb{R}) \to \mathcal{P}^{\star}(\mathbb{L}_{A1}\mathbb{A}_n(\mathbb{R})) \tag{32}$$

by $L(m,\overrightarrow{p(x)})\#L(k,\overrightarrow{q(x)}) = \left[L(m,\overrightarrow{p(x)}) + L(k,\overrightarrow{q(x)})\right)_{\leq} = \left[L(m+k,\overrightarrow{p(x)}+\overrightarrow{q(x)}\right)_{\leq} = \{L(r,\overrightarrow{u(x)}); m+k \leq r, \overrightarrow{p(x)}+\overrightarrow{q(x)} \leq \overrightarrow{u(x)}\}$, which means $p_j(x)+q_j(x) \leq u_j(x), j = 0,1,\ldots,n-1$, we obtain also by *Ends Lemma* that the hypergroupoid $(\mathbb{L}_{A1}\mathbb{A}_n(\mathbb{R}), \#)$ is a commutative semihypergroup.

We show that the transition function

$$\sigma_n : \mathbb{L}_{A1}\mathbb{A}_n(\mathbb{R}) \times \mathbb{R}_n[x] \to \mathbb{L}_{A1}\mathbb{A}_n(\mathbb{R}) \tag{33}$$

satisfies *GMAC* under the supposition that $(\mathbb{L}_{A1}\mathbb{A}_n(\mathbb{R}), \#)$ is a semihypergroup and $(\mathbb{R}_n[x], \#)$ is a join space (i.e. a commutative transposition hypergroup).

Suppose $L(m,\vec{p}) \in \mathbb{L}_{A1}\mathbb{A}_n(\mathbb{R})$, $f, g \in \mathbb{R}_n[x]$. Then

$$\begin{aligned}
&\sigma_n\left(\sigma_n(L(m,\vec{p}), f), g\right) = \sigma_n(L(m, p\circ f + p_{n-1}, \ldots, p\circ f + p_1, p_0), g) = \\
&= L(m, p\circ g + p\circ f + p_{n-1}, \ldots, p\circ g + p\circ f + p_1, p_0) = \\
&= L(m, p\circ(g+f) + p_{n-1}, \ldots p\circ(g+f) + p_1, p_0) \in \\
&\in \{\sigma_n(L(m, p\circ h + p_{n-1}, \ldots p\circ h + p_1, p_0); f, g, h \in \mathbb{R}_n[x], f+g \leq h\} = \\
&= \sigma_n(L, m, p_{n-1}, \ldots, p_1, p_0), [f+g)_{\leq}) = \sigma_n(L(m,\vec{p}), f\#g),
\end{aligned} \tag{34}$$

hence the *Generalized Mixed Associativity Condition* is satisfied. We Obtained that the structure $((\mathbb{L}_{A1}\mathbb{A}_n(\mathbb{R}), \#), (\mathbb{R}_n[x], \#), \sigma_n)$ is a multiautomaton (also termed a quasi-multiautomaton).

Similarly we can construct a multiautomaton from the action $(\mathbb{R}_n[x], \mathbb{L}_{A1}\mathbb{A}_n(\mathbb{R}, \delta_n)$.

Now we are coming to actions on objects of different dimensions.

4 Homomorphisms of automata and sequences of those

Recall that a homomorphism of the automata (the action) (S, G, δ_S) into the automata (the action) (T, H, δ_T) is a mapping $F = \phi \times \psi : S \times G \to T \times H$ such that $\phi : S \to T$ is a mapping and $\psi : G \to H$ is a homomorphism (of semigroups or groups) such that for any pair $[s, g] \in S \times G$ we have

$$\phi(\delta_S(s,g)) = \delta_T(\phi(s), \psi(g)), \text{ i.e. } \phi\circ\delta_S = \delta_T \circ (\phi\times\psi). \tag{35}$$

Definition of homomorphism of our actions and particularly for construction of the sequence of automata (actions) with decreasing dimension of corresponding objects we need another construction of automata (actions).

If $f \in \mathbb{R}_n[x]$, $f(x) = \sum_{k=0}^{n} a_n x^k$ and $L(m,\vec{p}) \in \mathbb{L}_{A1}\mathbb{A}_n(\mathbb{R}), \overrightarrow{p(x)} = (p_{n-1}(x), p_{n-2}(x), \ldots, p_0(x))$, we define

$$\tau_n(L(m, p_{n-1}, \ldots, p_0), f) = L(m, a_n + p_{n-1}, \ldots, a_1 + p_0 + a_0). \tag{36}$$

Now, if $g \in \mathbb{R}_n[x], g(x) = \sum_{k=0}^{n} b_k x^k$, we have

$$\begin{aligned}
&\tau_n(\tau_n(L(m, p_{n-1}, \ldots, p_0), f), g) = \\
&= \tau_n(L(m, a_n + p_{n-1}, \ldots, a_1 + p_1 + a_0), g) = \\
&= L(m, a_n + b_n + p_{n-1}, \ldots, a_1 + b_1 + p_0 + a_0 + b_0) = \\
&= \tau_n\left(L(m, p_{n-1}, \ldots, p_0), \sum_{k=0}^{n}(a_k + b_k)x\right) = \\
&= \tau_n(L(m, p_{n-1}, \ldots, p_0), f + g).
\end{aligned} \tag{37}$$

Hence $\tau_n : \mathbb{L}_{A1}\mathbb{A}_n(\mathbb{R}) \times \mathbb{R}_n[x] \to \mathbb{L}_{A1}\mathbb{A}_n(\mathbb{R})$ is the transition function (in satisfies *MAC*) of the automaton $\mathcal{A} = (\mathbb{L}_{A1}\mathbb{A}_n(\mathbb{R}), \mathbb{R}_{n-1}[x], \tau_{n-1})$.

Consider two automata - the above one and $\mathcal{A}_{n-1} = (\mathbb{L}_{A1}\mathbb{A}_{n-1}(\mathbb{R}), \mathbb{R}_{n-1}[x], \tau_{n-1})$. Define mappings

$$\phi_n : \mathbb{L}_{A1}\mathbb{A}_n(\mathbb{R}) \to \mathbb{L}_{A1}\mathbb{A}_{n-1}(\mathbb{R}), \quad \psi_n : \mathbb{R}_n[x] \to \mathbb{R}_{n-1}[x] \tag{38}$$

in this way: For $L(m, p_{n-1}, \ldots, p_0) \in \mathbb{L}_{A1}\mathbb{A}_n(\mathbb{R})$ we put $\phi_n(L(m, p_{n-1}, \ldots, p_0)) = L(m, p_{n-2}, \ldots, p_0) \in \mathbb{L}_{A1}\mathbb{A}_{n-1}(\mathbb{R})$ and for $f \in \mathbb{R}_n[x]$, $f(x) = \sum_{k=0}^{n} a_k x^k$ we define $\psi_n(f) = \psi_n\left(\sum_{k=0}^{n} a_k x^k\right) = \sum_{k=0}^{n-1} a_k x^k \in \mathbb{R}_{n-1}[x]$.

Evidently $\psi_n(f + g) = \psi_n(f) + \psi(g)$ for any pair of polynomials $f, g \in \mathbb{R}_n[x]$. Now we define a homomorphism $F_n : \mathcal{A}_n \to \mathcal{A}_{n-1}$ as the map

$$F_n = \phi_n \times \psi_n : \mathbb{L}_{A1}\mathbb{A}_n(\mathbb{R}) \times \mathbb{R}_n[x] \to \mathbb{L}_{A1}\mathbb{A}_{n-1}(\mathbb{R}) \times \mathbb{R}_{n-1}[x]. \tag{39}$$

In order to verify the equality (35) we show that the diagram

$$\begin{array}{ccc}
\mathbb{L}_{A1}\mathbb{A}_n(\mathbb{R}) \times \mathbb{R}_n[x] & \xrightarrow{\tau_n} & \mathbb{L}_{A1}\mathbb{A}_n(\mathbb{R}) \\
\Big\downarrow{\scriptstyle \phi_n \times \psi_n} & & \Big\downarrow{\scriptstyle \phi_n} \\
\mathbb{L}_{A1}\mathbb{A}_{n-1}(\mathbb{R}) \times \mathbb{R}_{n-1}[x] & \xrightarrow{\tau_{n-1}} & \mathbb{L}_{A1}\mathbb{A}_{n-1}(\mathbb{R})
\end{array} \tag{40}$$

is commutative.

Let $[L(m, \vec{p}), f] \in \mathbb{L}_{A1}\mathbb{A}_n(\mathbb{R}) \times \mathbb{R}_n[z], f(x) = \sum_{k=0}^{n} a_k x^k$.

Then

$$\begin{aligned}
&(\phi_n \circ \tau_n)(L(m, \vec{p}), f) = \phi_n\left(\tau_n\left(L(m, p_{n-1}, \ldots, p_0), \sum_{k=0}^{n} a_k x^k\right)\right) = \\
&= \phi_n((m, a_n + p_{n-1}, \ldots, a_1 + p_0 + a_0)) = L(m, a_{n-1} + p_{n-2}, \ldots, a_1 + p_0 + a_0) = \\
&= \tau_{n-1}\left(L(m, p_{n-2}, \ldots, p_0), \sum_{k=0}^{n-1} a_k x^k\right) = \\
&= \tau_{n-1}\left((\phi_n \times \psi_n)\left(L(m, p_{n-1}, \ldots, p_0), \sum_{k=0}^{n} a_k x^k\right)\right) = \\
&= (\tau_{n-1} \circ (\phi_n \times \phi_n)))(L(m, \vec{p}), f),
\end{aligned} \tag{41}$$

thus the diagram (40) is commutative.

Using the above defined homomorphism of automata, we obtain the sequence of automata with connecting homomorphisms $F_k : \mathcal{A}_k \to \mathcal{A}_{n-1},\ k \in \mathbb{N},\ k \geq 2$:

$$\begin{aligned} \ldots \xleftarrow{F_{n-1}} (\mathbb{L}_{A1}\mathbb{A}_{n-1}(\mathbb{R}), \mathbb{R}_{n-1}[x], \tau_{n-1}) \xleftarrow{F_n} (\mathbb{L}_{A1}\mathbb{A}_n(\mathbb{R}), \mathbb{R}_n[x], \tau_n) \\ \ldots \xleftarrow{F_2} (\mathbb{L}_{A1}\mathbb{A}_2(\mathbb{R}), \tau_2) \xleftarrow{F_3} \ldots \xleftarrow{F_{n-2}} (\mathbb{L}_{A1}\mathbb{A}_{n-2}(\mathbb{R}), \mathbb{R}_{n-2}[x], \tau_{n-2}) \xleftarrow{F_{n-1}} \ldots \\ (\mathbb{L}_{A1}\mathbb{A}_1(\mathbb{R}), \tau_1) \xleftarrow{F_2} \ldots \end{aligned} \tag{42}$$

Here, for $L(m, P-1, P-0) \in \mathbb{L}_{A1}\mathbb{A}_2(\mathbb{R})$ we have

$$L(m, p_1, p_0)y(x) = m\frac{\mathrm{d}^2 y(x)}{\mathrm{d}x^2} + p_1(x)\frac{\mathrm{d}y(x)}{\mathrm{d}x} + p_0(x)y(x) \tag{43}$$

for any $y(x) \in \mathbb{C}^2(\mathbb{R})$ and any $L(m, p_0) \in \mathbb{L}_{A1}\mathbb{A}_1(\mathbb{R})$ it holds $L(m, p_0)y(x) = m\frac{\mathrm{d}y(x)}{\mathrm{d}x} + p_0(x)y(x)$.

Now we are going a sequence of automata (actions) with increasing dimension, i.e. in a certain sense "dual" to the previous sequence.

First of all, we need a certain "reduction" member to definition of a transition function

$$\lambda_n : \mathbb{L}_{A1}\mathbb{A}_n(\mathbb{R}) \times \mathbb{R}_n[x] \to \mathbb{L}_{A1}\mathbb{A}_n(\mathbb{R}), \tag{44}$$

namely $\mathrm{red}_{n-1} : \mathbb{R}_r[x] \to \mathbb{R}_{n-1}[x]$ whenever $r > n-1$. In detail, if $f(x) = \sum_{k-0}^{r} a_k x^k \in \mathbb{R}_r[x]$, then

$$\begin{aligned} \mathrm{red}_{n-1}(f) = \mathrm{red}_{n-1}\left(\sum_{k=0}^{r} a_k x^k\right) = \mathrm{red}_{n-1}\left(\sum_{k=n}^{r} a_k x^k + \sum_{k=0}^{n-1} a_k x^{k-1}\right) = \\ = \sum_{k=0}^{n-1} a_k x^{k-1} \in \mathbb{R}_{n-1}[x]. \end{aligned} \tag{45}$$

Further, $L(m, p_{n-2}, \ldots, p_0)_n \in \mathbb{L}_{A1}\mathbb{A}_n(\mathbb{R})$ is acting by $L(m, p_{n-2}, \ldots, p_0)_n y = m\frac{\mathrm{d}^n y(x)}{\mathrm{d}x^n} + \sum_{k=0}^{n-2} p_k(x)\frac{\mathrm{d}^k y(x)}{\mathrm{d}x^k}$ whereas $L(m, p_{n-2}, \ldots, p_0)_{n-1} \in \mathbb{L}_{A1}\mathbb{A}_{n-1}(\mathbb{R})$,
i.e. $L(m, p_{n-2}, \ldots, p_0)_{n-1} y = m\frac{\mathrm{d}^{n-1} y(x)}{\mathrm{d}x^{n-1}} + \sum_{k=0}^{n-2} p_k(x)\frac{\mathrm{d}^k y(x)}{\mathrm{d}x^k}$.

Then for any pair $[L(m, p_{n-1}, \ldots, p_0), f] \in \mathbb{L}_{A1}\mathbb{A}_n(\mathbb{R}) \times \mathrm{R}_n[x]$, where $f(x) = \sum_{k=0}^{n} a_k x^k$ we obtain

$$\lambda_n = \lambda_n^\star \circ (\mathrm{id} \times \mathrm{red}_{n-1}) \tag{46}$$

and

$$\begin{aligned} (\lambda_n^\star \circ (\mathrm{id} \times \mathrm{red}_{n-1}))\,(L(m, p_{n-1}, \ldots, p_0, f) = \lambda_n^\star\left(L(m, p_{n-1}, \ldots, p_0), \sum_{k=0}^{n-1} a_k x^k\right) = \\ = L(m, p_{n-1} + a_{n-1}, \ldots, p_0 + a_0) \in \mathbb{L}_{A1}\mathbb{A}_n(\mathbb{R}), \end{aligned} \tag{47}$$

thus the mapping $\lambda_n : \mathbb{L}_{A1}\mathbb{A}_n(\mathbb{R}) \times \mathbb{R}_n[x] \to \mathbb{L}_{A1}\mathbb{A}_n(\mathbb{R})$ is well defined. We should verify validity of *MAC* and commutativity of the square diagram determining a homomorphism between automata.

Suppose $L(m, p_{n-1}, \dots, p_0) \in \mathbb{L}_{A1}\mathbb{A}_n(\mathbb{R})$, $f, g \in \mathbb{R}_n[x]$, $f(x) = \sum_{k=0}^{n} a_k x^k$, $g(x) = \sum_{k=0}^{n} b_k x^k$. Then

$$\begin{aligned}
&\lambda_n(\lambda_n(L(m, p_{n-1}, \dots, p_0), f), g) = \\
&= \lambda_n(L(m, p_{n-1} + +a_{n-1}, \dots, p_0 + a_0, g) = \\
&= L(m, p_{n-1} + a_{n-1} + b_{n-1}, \dots, p_0 + a_0 + b_0) = \\
&= \lambda_n\left(L(m, p_{n-1}, \dots, p_0), \sum_{k=0}^{n}(a_k + b_k)x^k\right) = \\
&= \lambda_n(L(m, p_{n-1}, \dots, p_0), f + g),
\end{aligned} \tag{48}$$

thus *MAC* is satisfied.

Further, we are going to verify commutativity of this diagram

$$\begin{array}{ccc}
\mathbb{L}_{A1}\mathbb{A}_{n-1}(\mathbb{R}) \times \mathbb{R}_{n-1} & \xrightarrow{\lambda_{n-1}} & \mathbb{L}_{A1}\mathbb{A}_{n-1}(\mathbb{R}) \\
\downarrow{\scriptstyle \xi_{n-1}\times\eta_{n-1}} & & \downarrow{\scriptstyle \eta_{n-1}} \\
\mathbb{L}_{A1}\mathbb{A}_n(\mathbb{R}) \times \mathbb{R}_n[x] & \xrightarrow{\lambda_n} & \mathbb{L}_{A1}\mathbb{A}_n(\mathbb{R})
\end{array} \tag{49}$$

where $\xi_{n-1}(L, m, p_{n-1}, \dots, p_0)_{n-1}) = L(m, p_{n-1}, \dots, p_0)_n$, i.e. $m\frac{\mathrm{d}^{n-1}}{\mathrm{d}x^{n-1}} + p_{n-1}(x)\frac{\mathrm{d}^{n-2}}{\mathrm{d}x^{n-1}} + \dots + p_0(x)\mathrm{id} \to m\frac{\mathrm{d}^n}{\mathrm{d}x^n} + p_{n-2}(x)\frac{\mathrm{d}^{n-2}}{\mathrm{d}x^{n-2}} + \dots + p_0(x)\mathrm{id}$ and $\eta_{n-1}\left(\sum_{k=0}^{n-1} a_k x^k\right) = a_{n-1}x^n + \sum_{k=0}^{n-1} a_k x^k$.

Considering $L(m, p_{n-2}, \dots, p_0)$ and the polynomial $f(x) = \sum_{k=0}^{n-1} a_k x^k$ similarly as above, we have

$$\begin{aligned}
&(\eta_{n-1})(L(m, p_{n-2}, \dots, p_0), f) = \eta_{n-1}(L(m, p_{n-2} + a_{n-2}, \dots, p_0 + a_0)_{n-1} = \\
&= \eta_{n-1}(\lambda_{n-1}(L(m, p_{n-2}, \dots, p_0), f) = (\eta_{n-1} \circ \lambda_{n-1})(L(m, p_{n-2}, \dots, p_0), f),
\end{aligned} \tag{50}$$

thus the above diagram is commutative. Now denoting by T_{n-1} the pair of mappings (ξ_{n-1}, η_{n-1}) we obtain that $T_{n-1} : (\mathbb{L}_{A1}\mathbb{A}_{n-1}(\mathbb{R}), \mathbb{R}_{n-1}[x], \lambda_{n-1}) \to (\mathbb{L}_{A1}\mathbb{A}_n(\mathbb{R}), \mathbb{R}_n[x], \lambda_n)$ is a homomorphism of the given automata. Finally, using T_k as connecting homomorphism, we obtain the sequence

$$\begin{aligned}
&(\mathbb{L}_{A1}\mathbb{A}_1(\mathbb{R}), \mathbb{R}_1[x], \lambda_1) \xrightarrow{T_1} (\mathbb{L}_{A1}\mathbb{A}_2(\mathbb{R}), \mathbb{R}_2[x], \lambda_2) \xrightarrow{T_2} (\mathbb{L}_{A1}\mathbb{A}_3(\mathbb{R}), \mathbb{R}_3[x], \lambda_3) \xrightarrow{T_3} \dots \\
&\dots \xrightarrow{T_{n-2}} (\mathbb{L}_{A1}\mathbb{A}_{n-1}(\mathbb{R}), \mathbb{R}_{n-1}[x], \lambda_{n-1}) \xrightarrow{T_{n-1}} (\mathbb{L}_{A1}\mathbb{A}_n(\mathbb{R}), \mathbb{R}_n[x], \lambda_n) \xrightarrow{T_n} \dots
\end{aligned} \tag{51}$$

5 Conclusion

Construction of endomorphisms of the above automata which are members of obtained sequences yielding commutative squares will be presented in a later paper. Moreover, the lifting of constructed sequences of automata in the category of multi-automata and their morphisms will be investigated in future. Here, the *Ends Lemma* [5] serves as a very suitable tool. Concerning consideration and results in this direction we mention at least publications [18, 19, 20, 21].

Acknowledgement

The work was supported by the project FEKT-S-14-2200 of Brno Unviersity of Technology.

References

[1] Bavel, Z.: *The source as a tool in automata*, Inform. and Control 18 (1971), 140-155.

[2] Borzooei, R. A., Varasteh, H. R., Hasankhani, A.: *$\mathcal{F}-$multiautomata on join spaces induced by differential operators*, Appl. Math. (Irvine) 5 (2014), 1386-1391.

[3] Dörfler, W.: *The cartesian composition of automata*, Math. System Theory 11 (1978), 239-257.

[4] Gécsek, F., Peák, I.: *Algebraic theory of automata*, Budapest, Akademiai Kiado, 1972.

[5] Chvalina, J.: *Functional graphs, quasi-ordered sets and commutative hypergroups*, (MU Brno 1995), in Czech.

[6] Chvalina, J., Chvalinová, L.: *Join spaces of linear ordinary differential operators of the second order*, Colloquium on Differential and Difference Equations, CDDE 2002. In: Folia Fac. Sci. Nat. Univ. Masarykianae Brunensis, mathematica 13, MU Brno, 2003, 77-86.

[7] Chvalina, J., Chvalinová, L.: *Actioin of join spaces of continuous functions on the underlying hypergroups of $2-$dimensional linear spaces of functions*, Aplimat, J. of Applied Math. Vol. II, n. 1(2009), 23-32.

[8] Chvalina, J., Chvalinová, L.: *Modelling of join Spaces by $n-$th order linear ordinary differential operators*, Proc. of the 4th International Conference Aplimat, Dept. math. FME Slovak Univ. of Technology, Bratislava (2005), 279-284.

[9] Chvalina, J., Hošková-Mayerová, Š., Dehghan Nezhad, A.: *General actions of hyperstructures and some applications*,An. St. Univ. Ovidius Constanta, 21. no. 1, (2013) 59-82.

[10] Chvalina, J., Moučka, J.: *Actions of join spaces of continuous functions on hypergroups of second-order linear differential operators*, Proc. 6-th Math. Workshop with International Attendance, FCI Univ. of Tech. Brno 2007, CD-ROM, 10 pp.

[11] Chvalina, J., Moučka, J., Vémolová, R.: *Functorial passage from quasi-automata to multiautomata*, XXIV Internat. Colloqium on the Acquisition Process Management, Proc. of Contributions, Brno, University of Defence, 2006, 8 pp.

[12] Chvalina, J., Račková, P.: *Join spaces of smooth functions and their actionis on transposition hypergroups of second order linear differential operators.* Aplimat, Journal of Applied Math. Vol.1, 1 (2008), 55-63.

[13] Křehlík, Š.: *Quasi-automata formed by continuous functions and by second-order linear differential operators in the Jacobi form*, EEICT Proc. 19th Conf. Vol. 3 (2013), 149-153.

[14] Křehlík, Š., Novak, M.: *From lattices to H_v-matrices*, Analele St. Univ. Ovidius Constanta, 23(3)(2016), to apper.

[15] Massouros, G. G.: *Hypercompositional structures in the theory of languages and Automata*, An. St. Univ. A.I. Cuza Iasi, Sect. Inform, III (1994) 65-73.

[16] Massouros, G. G., Mittas, J.: *Languages, automata and hypercompositional structures*, Proceedings of the 4th International Congress on Algebraic Hyperstructures and Applications. Xanthi 1990, World Scientific, 1991.

[17] Neuman, F.: *Global properties of linear ordinary differential equations*, Mathmatics and its Applications, East European Series 52, Kluwer Academic Publishers (with Academia Praha) Dordrecht-Boston-London 1991.

[18] Novák, M.: *EL-hyperstructures: an overview*, Ration Mathematica 23(2012), 65-80.

[19] Novák, M.: *Some basic properties of EL-hyperstructures*, European J. of Combinatorics 34 (2013), 446-459.

[20] Novák, M.: *n-ary hyperstructures constructed from binary quasi-ordered semigroups*, Analele St. Univ. Ovidius Constanta, Vol. 22(3), 2014, 147-168.

[21] Novák, M.: *On EL-semihypergroups.* European J. of Combinatorics 44(2015), 274-286.

[22] Račková, P.: *Hypergroups of symmetric matrices*, 10th International Congress on Algebraic Hyperstructures and Applications, Proc. of Contributions, Brno, 2008, pp. 267-271.

[23] Staněk, D.: *Rough sets of state spaces of automata.* Student EEICT Proceedings of the 22nd conference, 2016, 753-757.

Current address

Chvalina Jan, Prof. RNDr. DrSc.
Department of Mathematics, Faculty of Electrical Engineering and Communication
Brno University of Technology
Technicka 8, 616 00 Brno, Czech Republic
Tel. number: +420 541 146 058, e-mail: chvalina@feec.vutbr.cz

Staněk David, Mgr.
Department of Mathematics, Faculty of Electrical Engineering and Communication
Brno University of Technology
Technicka 8, 616 00 Brno, Czech Republic
Tel. number: +420 541 146 097, e-mail: xstane41@stud.feec.vutbr.cz

Slovak University of Technology in Bratislava
Faculty of Mechanical Engineering
16th Conference on Applied Mathematics **APLIMAT 2017**
Proceedings

MAXIMUM FLOW IN A NETWORK WITH AN OVERESTIMATED ARC CAPACITY

CIUPALĂ Laura (RO)

Abstract. There are several types of problems that can be modeled as maximum flow problems in networks. Sometimes is necessary to determine a maximum flow in a network differing only by an arc capacity (which is reduced by a units) from another network in which a maximum flow is already established. We will describe an algorithm that determines a maximum flow in the new network starting from a maximum flow in the initial network in $O(am)$ time.

Keywords: network flow, maximum flow, preflow, pseudoflow

Mathematics Subject Classification: Primary 90B10; Secondary 90C90

1 Introduction

The maximum flow problem is a fundamental problem in network flow theory and it has been studied extensively in the last 6 decades because it arises in a wide variety of situations and in several forms. Sometimes the maximum flow problem occurs as a sub-problem when solving more difficult network problems, such as minimum cost flow problems or generalized flow problems, but it also arises in a number of combinatorial applications that on the surface might not appear to be maximum flow problems at all.

The maximum flow problem was first formulated and solved using the well-known augmenting path algorithm by Ford and Fulkerson in 1956. Since then, two types of maximum flow algorithms have been developed: augmenting path algorithms and preflow algorithms.

2 Notation and definitions

Let $G = (N, A, c, s, t)$ be a capacitated network with n nodes, m arcs and with a nonnegative integer capacity $c(i, j)$ associated with each arc $(i, j)\in A$. We distinguish two special nodes in the network G: a source node s and a sink node t.
A *flow* is a function $f : A \rightarrow \mathbf{R}_+$ satisfying the next conditions:

$$f(s, N) - f(N, s) = v \quad (1)$$
$$f(i, N) - f(N, i) = 0,\ i \neq s,t \quad (2)$$
$$f(t, N) - f(N, t) = -v \quad (3)$$
$$0 \leq f(i,j) \leq c(i,j),\ (i,j) \in A \quad (4)$$

for some $v \geq 0$. We refer to v as the *value* of the flow f.
The maximum flow problem is to determine a flow f for which v is maximized.
A *preflow* is a function $f: A \rightarrow \mathbf{R}_+$ satisfying (4) and the next conditions:

$$f(i, N) - f(N, i) \geq 0,\ i \neq s,t \quad (5)$$

Let f be a preflow. We define the *excess* of a node $i \in N$ as $e(i) = f(i, N) - f(N, i)$.
Thus, for any preflow f, we have that: $e(i) \geq 0,\ i \in N \setminus \{s, t\}$.

A preflow f for which $e(i) = 0,\ i \in N \setminus \{s, t\}$ is a flow. Consequently, a flow is a particular case of preflow.

A *pseudoflow* is a function $f: A \rightarrow \mathbf{R}_+$ satisfying the conditions (4).
For any pseudoflow f, we define the *imbalance* of node i as $e(i) = f(N, i) - f(i, N)$, for all $i \in N$. If $e(i) > 0$ for some node i, we refer to $e(i)$ as the *excess* of node i; if $e(i) < 0$, we refer to $-e(i)$ as the *deficit* of node i. If $e(i) = 0$ for some node i, we refer to node i as *balanced*. Consequently, a preflow is a particular case of psedoflow.

The *residual capacity* $r(i,j)$ of any arc $(i,j) \in A$, with respect to a given psedoflow f, is given by

$$r(i,j) = c(i,j) - f(i,j) + f(j, i). \quad (6)$$

By convention, if $(i, j) \in A$ and $(j, i) \notin A$, then we add the arc (j, i) to the set of arcs A and we set $c(j, i) = 0$. The residual capacity $r(i,j)$ of the arc (i,j) represents the maximum amount of additional flow that can be sent from the node i to node j using both of the arcs (i,j) and (j, i).

The network $G(f) = (N, A(f))$ consisting only of those arcs with strictly positive residual capacity is referred to as the *residual network* (with respect to the given pseudoflow f).

A directed path from the source node s to the sink node t in the residual network $G(f) = (N, A(f))$ is called an *augmenting path*.

3 Determining a maximum flow in a network with an overestimated arc capacity

Now we will study the problem of determining a maximum flow in a network after decreasing the capacity of the an arc (k, l) by a units.

Let $G=(N, A, c, s, t)$ be a network in which a maximum flow f is already determined. Let (k, l) be an arbitrary arc of G, whose endpoints are intermediate nodes and whose capacity will be decreased by a units. In this way we obtain the network $G' = (N, A, c', s, t)$, where:

$$c'(i,j) = c(i,j),\ \forall (i\ j) \in A \setminus \{(k,l)\} \quad (7)$$
$$c'(k,l) = c(k,l) - a \quad (8)$$

There are two cases that might appear:

1. $f(k,l) \le c'(k,l)$. In this case f is a maximum flow in the network G' also.

2. $f(k,l) > c'(k,l)$. In this case, we will denote by b the difference $f(k,l)$ - $c'(k,l)$ and we will set $f(k,l) = c'(k,l)$, creating in this way in node k an excess equal to b and a deficit in node l equal also to b. So, now f is a pseudoflow. The algorithm will try to push the excess of node k toward node l along directed paths in the residual network, decreasing in this way both the excess of node k and the deficit of node l. If both of them become 0, then the pseudoflow becomes a flow, which, moreover, is a maximum flow; otherwise the remaining excess of k wil be sent toward the sink node t or, if this isn't possible because there aren't directed paths from k to t, back to the source node s. For eliminating the remaining deficit of node l, the algorithm will augment the flow along directed paths from s to l if there are such paths; otherwise the remaining deficit will be sent to the sink node t. Consequenlty, the value of the maximum flow in the network G' might be smaller than the value of the maximum flow in G.

Overestimated Arc Capacity Algorithm;
Begin
  let f be a maximum flow in the network G;
  determine the network G';
  if $f(k,l) > c'(k,l)$ **then**
  begin
    $f(k,l) = c'(k,l)$;
    determine the residual network $G'(f)$;
    while $e(k) > 0$ **and** there is a directed path from k to l in $G'(f)$ **do begin**
      determine a directed path D from k to l in $G'(f)$;
      $r(D) = \min\{r(i,j) \mid (i,j) \in D\}$;
      send $\min\{e(k), r(D)\}$ units of flow from k to l along the path D;
    end;
    while $e(k) > 0$ **and** there is a directed path from k to t in $G'(f)$ **do begin**
      determine a directed path D from k to t in $G'(f)$;
      $r(D) = \min\{r(i,j) \mid (i,j) \in D\}$;
      send $\min\{e(k), r(D)\}$ units of flow from k to t along the path D;
    end;
    while $e(k) > 0$ **do begin**
      determine a directed path D from k to s in $G'(f)$;
      $r(D) = \min\{r(i,j) \mid (i,j) \in D\}$;
      send $\min\{e(k), r(D)\}$ units of flow from k to s along the path D;
    end;
    while $e(l) < 0$ **and** there is a directed path from s to l in $G'(f)$ **do begin**
      determine a directed path D from s to l in $G'(f)$;
      $r(D) = \min\{r(i,j) \mid (i,j) \in D\}$;
      send $\min\{-e(l), r(D)\}$ units of flow from s to l along the path D;
    end;
    while $e(l) < 0$ **do begin**
      determine a directed path D from t to l in $G'(f)$;

$r(D) = \min\{r(i, j) \mid (i, j) \in D\}$;
send $\min\{-e(l), r(D)\}$ units of flow from t to l along the path D;
end;
end;
end.

Since by each augmentation of flow along a directed path the excess of node k and/or the deficit of node l decreases by at least 1 unit and at the beginning of the algorithm they were both equal to b, which is at most a, it follows that after at most $2a$ flow augmentations along directed paths in $G'(f)$, the algorithm ends with a maximum flow in G'. Consequently, our algorithm, which is an augmenting path algorithm, runs in $O(am)$ time.

Another approach for solving the maximum flow in a network with an overestimated arc capacity is a preflow approach that can be obtained by modifying the incremental algorithm developed by Kumar and Gupta in [3]. Their algorithm determines a maximum flow in the network obtained after deleting the arc (k, l) from a network G, in which a maximum flow f is already established. It is possible that f is no longer a flow, but a pseudoflow. More precisely, if in G the flow on the arc (k, l) was strictly positive, then, in the new network, f will no longer satisfy the mass balance constraints (2) for both nodes k and l. This means that node k will have a strictly positive excess and node l will have a strictly positive deficit. The incremental algorithm performs flow pushes until the pseudoflow becomes a maximum flow and it runs in $O(|AN|^2m)$ time, where AN is the set of affected nodes. These are nodes contained in directed paths from s to k or in directed paths from l to t in the residual network and they can be determined by applying a modified Backward Breadth First Search algorithm from k to s and a modified Breadth First Search algorithm from l to t.

The algorithm developed by Kumar and Gupta in [3] solves a problem which is a particular case of the maximum flow in a network with an overestimated arc capacity. But, in case of a slight decrease of the capacity (if a is at most the square of the number of affected nodes), our algorithm outperforms theirs.

References

[1] AHUJA, R.,MAGNANTI, T., ORLIN, J. Network Flows. Theory, Algorithms and Applications. Printice Hall, 1993.

[2] BANG-JENSEN, J., GUTIN, G. Digraphs: Theory, Algorithms and Applications. Springer Verlag, 2001.

[3] KUMAR, S., GUPTA, P. An incremental algorithm for the maximum flow problem. *Journal of Mathematical Modelling and Algorithms*, 2003, s. 1–16.

Current address

Ciupală Laura, Ph.D. Assoc. Prof.
University Transilvania of Braşo, Faculty of Mathematics and Computer Science
Iuliu Maniu Str., 50, Braşov, Romania
E-mail: laura_ciupala@yahoo.com

Slovak University of Technology in Bratislava
Faculty of Mechanical Engineering
16th Conference on Applied Mathematics **APLIMAT 2017**
Proceedings

BACK TO THE BABEL TOWER, WARNING AND ARCHETYPE

COCCHIARELLA Luigi (IT)

Abstract. As graphic educators, looking at the city as a special case in the field of architectural configurations, we should not avoid considering the links between *forma urbis* and *forma mentis*, and consequently the *vision* behind the built settlements, either plan-based or sprawl-based - whose variety reminds the Babylonian archetype - together with the role of *image* in supporting urban design, so evoking a sort of genetic-iconic materialism, and the role of Geometry and Graphics in creating and supporting our whole *idea of city*.

Keywords: descriptive geometry, drawing, geometry and graphics, urban design, urban patterns, urba-tecture

Mathematics Subject Classification: Primary 00A05, 00A66, 01A05, 51N05; Secondary 97U99

1 City as Architecture

Probably no artificial human systems show a complexity level comparable with that of a city. Almost every disciplinary field and activity takes place in the city, as well as it contributes to form and to transform the city. From the point of view of an architect, whose main focus is on *space*, namely a space where human beings can live, the city is a special kind of spatial system, and consequently, of architecture. That is consistent with one of the most impressive and complete modern definitions of architecture, proposed by William Morris, who stated that it concerns every spatial transformation of the terrestrial crust finalized to provide adequate places for living to human beings. More recently, the idea has been significantly reconfirmed by Jan Lubicz-Nycz with the neologism *urba-tecture*, and it has been also analyzed in the very well-known book - we could say a revolutionary best-seller - titled *The Architecture of the City*, by Aldo Rossi. Given the complexity of the city, no exhaustive analyses are really possible, and in fact this is not our purpose, as well as any attempt to investigate the relationships between Mathematics and the city could defeat even the most brilliant and talented researcher. Instead, given our interest in Geometry and Graphics, we would like to focus more strictly on *space*. Although a city can not be reduced to a pure geometrical configuration, Geometry can be seen as a meta-structure representative of the physical, sensory, and symbolic components of a urban space, working as a special glue making each urban context real, unique and different from any other. We could say that without geometrical identity urban spaces collapse into a *literary* form.

2 Why Babel?

Fig. 1. The Tower of Babel, by Pieter Brueghel the Elder (detail), 1653

Looking at the present urban spaces, and especially at the current *idea* of urban space, we can easily feel confused, but at the same time, city is an extraordinary, and unique, architectural archive of spatial models and experimentations. The rich series of spatial patterns and solutions we find in a city clearly reveals, in fact, the effects of the forces that have contributed to modeling it in the long history. In this sense we can trust on the intrinsic power of Geometry to understand *processes* behind *shapes*. On this purpose, the tangle of geometric patterns and characters displayed in the contemporary cities shows a vivid and interesting example. Walking from downtown to outskirts, as well as floor-by-floor from ground to rooftops, we can quickly through the tangible and intangible vicissitudes of an urban settlement. As Christian Norberg-Schulz taught us, we enter the real dimension of its architectonic space, which is of course measurable in any direction as an isotropic geometric entity, but it is at the same time extremely much dense and directionally oriented, not only owing to gravity force and other geographic specificities, but also, and most of all, due to signs and traces of the human actions and experience there, reason why we use to call architectonic spaces *places*, instead of simply *spaces*. However, the point is that the features and all the abovementioned characteristics of a *place* are strictly reflected by certain specifiable geo-metrical properties of the place itself, in the same way the geometry of the human body is the result of the interaction between genetics and lifestyle. Compared to the past, in the last decades, the hybridization between analogue and digital - and more or less predictably soon, virtual - dimensions inside the cities is making the situation even more confused, with profound changes in the social behaviors, as well as in the image of the city. Disconnections and discontinuities from part to part seem to show many cities at the same time in the same urban context, which is sign of the variety of processes generating those spaces. In part this variety is attributed to the weakness of the policies, in part it reflects the many changes in the *vision* of the city occurring in different moments of the history and in different areas of the settlement. Therefore, assuming the correspondence between *place* and *space*, and assuming Geometry as the *language of space*, we can state that the current cities speak several languages in their administrative areas, as well as the social groups inside them do, what suggested us to look back to *Babel*. One never minds the crucial symbolic role of Babel in the story of the development of the world, until she/he does not realize that the story - or the legend - of Babel is narrated in the book of Genesis, that is, the initial part of the Bible dedicated to the Origins. In that literary story, Babel stopped being built when God confused languages and people could not longer understand each other. But we know that this action coincides at the same time with the origin of languages, which can be considered the last act of The Creation.

In fact, the loss of a unique language and the flowering of a multiplicity of idioms is not the end of the story, instead it can be seen as the real start of the human story, as the multiplicity of languages pushed human beings to form many communities that gradually abandoned Babylon to explore and populate other places of the world, that is, precisely, the real genesis of the human colonization of the Earth. In this wider world human beings could discover new things and do new experiences, becoming responsible for their own knowledge. In this wider world they could finally build new cities, each one different from the others, and this was also the early origin of urban planning.

3 Drawing and the City

Fig. 2. A detail from a wall painting in Catal Höyuk showing a settlement (Anatolia, about 7200-6800 BC)

Since drawing is the main tool for architects, graphic representations are of course relevant documents to them, not less important than the real cities. Graphic representations concerning cities have always shown a specific physiognomy, being distinguishable as a particular iconographic category. This is clear since the most remote examples, like the well-known wall painting found in Catal Höyuk, Anatolia, date back to about 7000 BC. Geometry and graphic codes are astonishingly similar to those normally used in the present urban cartography. Emphasis is given on the public spaces and the courtyards, which are urban topics and therefore the main subject of the painting, and to better achieve the goal architectural blocks are not represented in detail but as abstract units. The whole graphic construction shows the natural connection of the settlement with the landscape, that gradually becomes a system of public areas confined in-between the buildings. A special graphic feature is adopted to represent the volcano located in the upper part, which has also effects on the distribution of the buildings and again reveals a native idea of city as something that can not be thought separately from its natural geographic location. But at the same time the graphic as well as geometric distinctions between *nature* and *artifice* are clearly remarked in the image. The graphic language is suspended between formal analogy and abstract symbolism, anyway, all in all an *urba-tectural* spatial system is there represented, where architecture and the urban pattern are integrated - together with territorial elements - in a way that no changes can be made on each of them without effects on the other. Furthermore, the foundation of the city in the remote past was not only a physical activity, but also a *cosmogonical* ritual, which took place at a precise moment of the year. Since the settlements had to be consistent with cosmic as well as with social sets, Geometry was relevant in several ways, that is, in terms of orientation of the main axes, of hierarchies between sacred and residential areas, and of proportions and dimensions of the land allotments.

Fig. 3. Details from a map of Palmanova (Italy about 1600)

As we know, at least for the Western culture, main models in this vein were that of Hippodamus of Miletus from the Ancient Greece, and the Roman Castrum from the Ancient Rome. These approaches, *rational* and Geometry based, lasted long centuries, and, apart from the cosmogonical aspects, are still valid nowadays. Later on, after the fall of the Roman Empire, in the long era of the Middle Age, more spontaneous settlements grew up, where buildings and urban spaces appear integrated as never before, forming *organic* patterns which are very far from the linearity and the orthogonality of the former settlements inspired by Greek and Roman models. In these cities, the more fluid shapes of the public spaces reflect a different social organization of the communities. Once more, Geometry also works as an indicator of deeper aspects of the built environment. After this period, with the advent of Renaissance, Drawing reaches a leading function, then the re-appropriation of the trust in the human opinion is well symbolized in the absolute dominance of the sight point in perspective, and the trust in the rational approach to knowledge is clearly shown in the well-known diagram of the *homo ad circulum* by Leonardo da Vinci. We know how much important became Geometry in this process of cultural renovation. As a consequence, Drawing became the main design tool with direct effects not only on architecture but also on the city, whose favorite model soon became that of the *ideal city*. In our opinion this was the origin of the urban theory but at the same time of the crisis of the *real city*. To better explain our opinion, we will focus on two exemplary cases. Looking at the map of Palmanova, our first example, we notice an unnatural *architectural mimicry* of the urban structure, where the city appears forcibly configured as a big - too big - architecture, confined into a rigid system of walls which, although spectacular, fix a limit to the natural growth of that city, like in any *utopian* urban model, that is, going in the opposite direction of God's decision, when He confused languages to encourage people to expand their presence on the Earth. In a way, the *ideal city* follows the furrow of Babel Tower. From our point of view, the equilibrium between architectural and urban space is the weak point here, architectural approach is dominant and the urban identity starves to emerge. In terms of Geometry, the structure seems to be too much rigidly Euclidean to be acceptably defined as urban. At the opposite side, we can see examples, like the recent case of the masterplan of Kartal Pendik in Istanbul, proposed by Zaha Hadid. Apart from the indisputable formal innovation and the aesthetic beauty of the proposal, here a *naturalistic mimicry* seems to emerge, where starting from a precise streets pattern, the architectural blocks are distorted and elevated as tectonic plates. Since the city-effect is produced through a direct Topologic similarity between architecture and landscape, again the urban identity is the weak point in this spatial model. At the end, in both the cases the *urban identity* seems to have escaped from the Drawing, with ascertainable real effects in the city of Palmanova, and with foreseeable effects in the urban area of Kartal Pendik.

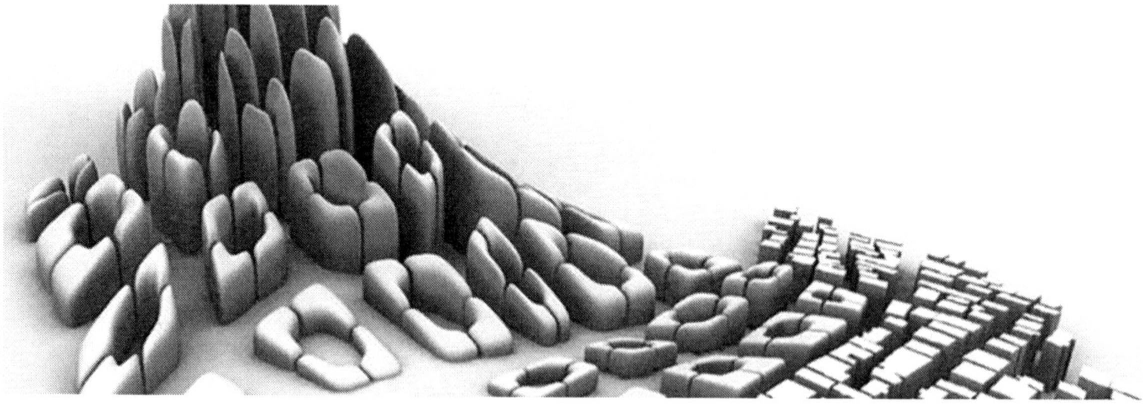

Fig. 4. Detail of the masterplan of Kartal Pendik, Istanbul (Turkey), by Zaha Hadid, 2006

4 After Babel Tower, the case of Ferrara

The Babel Tower can be also regarded as a *eulogy of artificiality*, aiming at daring the physical gravity law and, finally, the same sense of a city. In most of the iconic representations, it appears as a huge volume, almost independent on the surrounding landscape, showing at the same time itself as a symbol of the human conceit. At the end *it is not a city*, and trying to give a possible metaphoric sense to the following mass-exodus, maybe it tells us that a real city requires an active research by the communities, carried and developed on the specific territorial areas of the settlements, and of course, many languages as indicators of a variety of opinions and levels of knowledge. On the purpose, the case of *Ferrara* between XV and XVI centuries, called "the first modern European city", as remarked by Bruno Zevi, is one of the most relevant examples of integration and balance between architectural and urban sensitivity, let us say, of *urba-tecture*. The map shows the adaptive connection between the new northern addition, based on two long crossing axes, and the southern preexisting town. The two urban areas have been integrated thanks to wise extensions of some streets across the two patterns, in order to obtain a *sewing effect* on the whole urban texture. From a geometrical point of view, the Euclidean-like pattern of the new district appears very well combined with the Topology-like structure of the former town, with a very positive effect also in Projective terms, it means, in terms of visual perception of the urban space, which is very different from the one obtainable walking through a pure ideal city. Ferrara has been designed by Biagio Rossetti, an architect very sensitive to the specificities of the urban space and of its relation with architecture. Moreover, all the elements, including fortifications and green areas appear displayed in accordance with the large-scale surrounding territory, as well as with the smaller urban places inside the city. Again referring to Babel Tower, we could affirm that in this case all the languages have been appropriately used, without conflicts and therefore generating a coherent and rich urban system. Unfortunately, in more recent times, and especially during the XIX century, architects abdicated to the large scale in favor of Engineers, who took care of it from a different point of view. As a consequence, for long time landscape, territory, and infrastructures in general have been considered not strictly consistent with the architectural field. The aesthetic qualities of the urban environments of modern cities show us every day the effects of this detachment. As architects went far from the city, they gradually lost the ability to draw and design them, and since what is not drawn does not appear in the project, as well as what does not appear in the project is not built in the real world, for that reason, and for a too long time, the contribution of architects and their spatial sensitivity have not appeared much in the built spaces of modern cities.

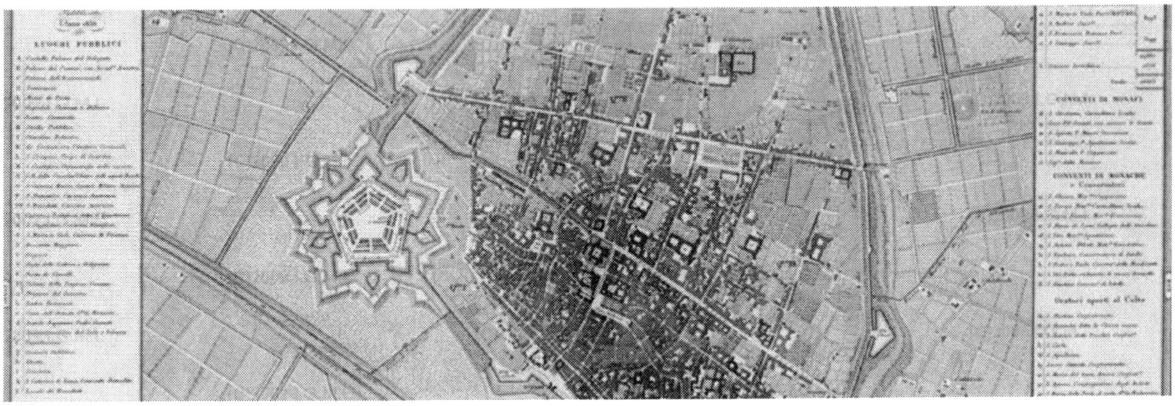

Fig. 4. Map of Ferrara, about three centuries after Biagio Rossetti's plan (Italy), 1836

5 Conclusions

It seems that in our very post-Babel era, the great variety of languages and forms, as well as their reciprocal contaminations, can be precious sources of inspiration for a serious updating of the *idea of city* as well as of *real cities* and their *spatial patterns*, as results of the mentioned interactions. However, this complexity should not be completely operated in the design processes without an appropriate *multi-dimensional* drawing, that is, without a deep appeal to Geometry and Graphics as key languages in designing life spaces. In order to avoid new mass-exoduses from Babel.

References

[1] COCCHIARELLA, L., *La forma oltre il codice: ambiente architettonico, teoria, rappresentazione*. Milano: Academia Universa Press, 2009, ISBN 9788864440057.

[2] HOVESTADT, *L., Jenseits des Raster. Architektur und Informationstechnologie. Anwendungen einer digitalen Architektonik / Beyond the Grid. Architecture and Information Technology. Applications of a Digital Architectonic.* Basel – Boston – Berlin: Birkhauser Verlag, 2010, ISBN 9783034600965.

[3] MUMFORD, L., *The culture of cities*. London: Seker & Warburg, 1938.

[4] ZEVI, B., *Saper vedere l'urbanistica: Ferrara di Biagio Rossetti, la prima città moderna europea*. Torino: Einaudi, 1971, ISBN 8806295616.

[5] URL SkyscraperCity: www.skyscrapercity.com.

Current address

Cocchiarella Luigi, associate professor
Politecnico di Milano – Campus Leonardo,
Department of Architecture and Urban Studies (DASTU)
Via Bonardi 9, Building 14 „La Nave", 20133 Milano, Italy
Tel. number: +39 02 2399 9402, e-mail: luigi.cocchiarella@polimi.it

Slovak University of Technology in Bratislava
Faculty of Mechanical Engineering
16th Conference on Applied Mathematics **APLIMAT 2017**
Proceedings

ON PROPERTIES OF LENGTH-SCALED BETWEENNESS CENTRALITY

CORONIČOVÁ HURAJOVÁ Jana (SK), MADARAS Tomáš (SK)

Abstract. The length-scaled betweenness centrality of a vertex x in a graph G is the sum of reciprocally distance-normalized relative numbers of shortest paths in G which have x as an inner vertex. In this paper, we explore selected basic properties of this local graph invariant along with properties of graphs having the same length-scaled betweenness for each of their vertices.

Keywords: length-scaled betweenness

Mathematics subject classification: Primary 05C12; Secondary 68R10, 91D30

1 Introduction

Throughout this paper we consider connected graphs without loops or multiple edges. We use the standard terminology of graph theory taken from [14]. The distance of vertices u, v (or, from u to v) in a graph G (that is, the length of a shortest $u - v$-path in G) is denoted by $d_G(u, v)$ (we omit the subscript if the graph is known from the context). By symbols $\mathrm{diam}(G)$, $\overline{\ell}(G)$ and $\mathrm{ad}(G)$, we denote the diameter (the largest distance in G), the average distance and the average vertex degree of a graph G. For a positive integer i, we put $N_i(x) = \{y : d(x, y) = i\}$ and $n_i(x) = |N_i(x)|$; we write $N_1(x) = N(x)$.

In social network analysis, one of the most common tasks to be performed concerns the identification of key actors within a network. The importance of these actors is usually expressed in terms of centrality indices. In general, a vertex centrality index is a positive real-valued function defined on vertex set of a graph which is invariant under graph isomorphism. Today, many centrality indices are known (for a survey, see, for example, Chapter 3 of [3] or the interactive visualization at [1]); among them, the important position has the *vertex betweenness centrality*, introduced first in [5] (also, independently, in [2]) as the sum of relative numbers of shortest paths that pass through a selected vertex. More precisely, for a graph $G = (V, E)$, its vertex x and two distinct vertices $u, v \neq x$, let $\sigma_{G,u,v}$ be the number of all $u - v$-shortest paths, and $\sigma_{G,u,v}(x)$ be the number of those shortest $u - v$-paths that pass through x (for fixed G, we write just $\sigma_{u,v}$ and $\sigma_{u,v}(x)$). Then the betweenness centrality of x is equal to

$$B(x) = \sum_{u,v \in V} \frac{\sigma_{G,u,v}(x)}{\sigma_{G,u,v}}$$

(the summation being taken over the unordered pairs of vertices). This graph invariant is widely used in real-world applications, but has been also investigated from purely theoretical point of view, see [6, 7, 9, 8].

Several variants of the betweenness centrality are known. This paper concentrates on exploring graph-theoretical properties of *length-scaled betweenness* defined first in [4] as the sum

$$B_{dist}(G, x) = \sum_{u,v \in V} \frac{1}{d_G(u, v)} \frac{\sigma_{G,u,v}(x)}{\sigma_{G,u,v}}$$

(once again, if G is known from the context, we write just $B_{dist}(x)$); an example of calculation of length-scaled betweennesses is included in Figure 1. The motivation for this variant comes from the premise that, when the information flow is propagated within a network, long paths are not often used, and, moreover, it has less benefit, for actors, to control the communication along such paths; hence, their contribution to the betweenness value of a vertex should be reduced in some way. Although less common than the original vertex betweenness, the length-scaled betweenness recently has found its way in applied graph theory in the connection with the extraction of keywords from documents (see [13]); nevertheless, still it is not explored in deeper details. Hence, the aim of our paper is to initiate the graph-theoretic study of this invariant by pointing out selected basic properties (Section 2) and looking for graphs whose vertices possess the same value of lebgth-scaled betweenness (Section 3).

2 Basic properties of length-scaled betweenness

We start with several observations which are fully analogous to previously known theorems on vertex betweenness. Taking into account that no pair of adjacent vertices u, v contributes with a positive value to the sum for $B_{dist}(x)$ and any positive contribution is not greater than $\frac{1}{2}$, we easily derive

Proposition 2.1 *For a vertex x of a graph G, $B_{dist}(x) = 0$ if and only if the neighbours of x induce a clique in G.*

Proposition 2.2 *For any vertex x of an n-vertex graph G, $B_{dist}(x) \leq \frac{1}{2}\binom{n-1}{2}$; the equality holds if and only if x is the center of the star $K_{1,n-1}$.*

We show also analogous result (see [9] for standard betweenness version) for 2-connected graphs, where, instead of the star, the extremal graph is the fan graph (recall that the fan graph $F_{1,r}$ is obtained from r-vertex path by connecting its vertices with new edges to the single new vertex):

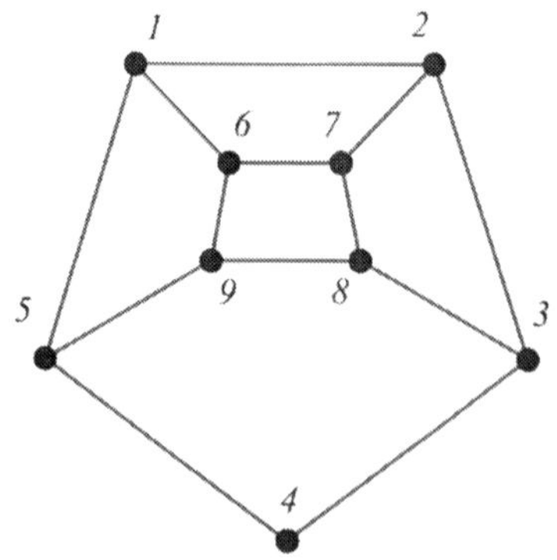

$$B_{dist}(1) = \frac{1}{2}\cdot\frac{1}{1}+\frac{1}{2}\cdot\frac{1}{2}+\frac{1}{3}\cdot\frac{2}{5}+\frac{1}{3}\cdot\frac{1}{4}+\frac{1}{3}\cdot\frac{1}{2}+\frac{1}{3}\cdot\frac{1}{2}+\frac{1}{3}\cdot\frac{2}{4} = \frac{13}{20}$$

$$B_{dist}(3) = \frac{1}{3}\cdot\frac{1}{5}+\frac{1}{2}\cdot\frac{1}{1}+\frac{1}{2}\cdot\frac{1}{2}+\frac{1}{3}\cdot\frac{1}{5}+\frac{1}{3}\cdot\frac{1}{1}+\frac{1}{2}\cdot\frac{1}{1}+ = \frac{103}{60}$$

$$B_{dist}(4) = \frac{1}{2}$$

1,3	1-2-3
1,4	1-5-4
1,7	1-6-7, 1-2-7
1,8	1-6-9-8, 1-6-7-8, 1-5-9-8, 1-2-7-8, 1-2-3-8
1,9	1-6-9, 1-5-9
2,4	2-3-4
2,5	2-1-5
2,6	2-7-6, 2-1-6
2,8	2-7-8, 2-3-8
2,9	2-7-8-9, 2-7-6-9, 2-3-8-9, 2-1-6-9, 2-1-5-9
3,5	3-4-5
3,6	3-8-9-6, 3-8-7-6, 3-2-7-6, 3-2-1-6
3,7	3-8-7, 3-2-7
3,9	3-8-9
4,6	4-5-9-6, 4-5-1-6
4,7	4-3-8-7, 4-3-2-7
4,8	4-3-8
4,9	4-5-9
5,6	5-9-6, 5-1-6
5,7	5-9-6-7, 5-9-8-7, 5-1-6-7, 5-1-2-7
5,8	5-9-8
6,8	6-9-8, 6-7-8
7,9	7-8-9, 7-6-9

Fig. 1. An example of calculation of length-scaled betweenness centralities in a graph (the table contains shortest paths between nonadjacent vertices)

Lemma 2.3 *For any vertex x of a 2-connected n-vertex graph G, $B_{dist}(x) \leq \left(\frac{n-3}{2}\right)^2$; the equality holds if and only if x is the vertex of the maximum degree in the fan graph $F_{1,n-1}$.*

Proof: We will rephrase the proof of similar result for standard vertex betweenness from [9] (Proposition 3.2 on p. 3126) with some additional details. Let x be a vertex of G. Since G is 2-connected, the graph $G - x$ is connected and has at least $m(G-x) \geq n-2$ edges (that is, $m(G-x)$ pairs of vertices which contribute 0 to $B_{dist}(x)$). Let $m_2(G-x)$ be the number of pairs of vertices whose distance in $G-x$ is 2 (for these pairs, either $\sigma_{G,u,v} \geq 2$ and $\sigma_{G,u,v}(x) = 1$, or $\sigma_{G,u,v}(x) = 0$, so they contribute to $B_{dist}(x)$ at most $\frac{1}{2} \cdot \frac{1}{2}$).

We show first that $m(G-x) + \frac{m_2(G-x)}{2} \geq \frac{3n-7}{2}$. This is true when the minimum degree of $G-x$ is at least 3, because then $m(G-x) \geq \frac{3(n-1)}{2} > \frac{3n-7}{2}$. Otherwise, we proceed by induction: if $G-x$ contains a vertex y of degree 1, then $m(G-x-y) + m_2(G-x-y) \geq \frac{3(n-1)-7}{2} = \frac{3n-10}{2}$ and so $m(G-x) + m_2(G-x) \geq \frac{3n-10}{2} + 1 + \frac{1}{2} \cdot 1 = \frac{3n-7}{2}$ (as y contributes to $m(G-x) + m_2(G-x)$ with one incident edge and at least one pair of distance 2). Finally, if $G-x$ has no leaves, but contains a vertex y of degree 2, then (again using induction on at most two components of $G-x-y$ and the fact that y contributes to $m(G-x) + m_2(G-x)$ with two incident edges) $m(G-x) + m_2(G-x) \geq \frac{3n-10}{2} + 2 > \frac{3n-7}{2}$.

Note that any pair of vertices in $G-x$ of distance at least 3 contributes at most $\frac{1}{2} \cdot 1$ to $B_{dist}(x)$ in G (as these vertices may be connected in G with unique path of length 2 passing through x). Hence,

$$B_{dist}(x) = \sum_{u,v \in V} \frac{1}{d_G(u,v)} \frac{\sigma_{G,u,v}(x)}{\sigma_{G,u,v}} \leq \frac{1}{2} \sum_{u,v \in V} \frac{\sigma_{G,u,v}(x)}{\sigma_{G,u,v}} \leq$$

$$\frac{1}{2}\left(\binom{n-1}{2} - \left(m(G-x) + \frac{m_2(G-x)}{2}\right)\right) \leq \frac{1}{2}\left(\binom{n-1}{2} - \frac{3n-7}{2}\right) = \left(\frac{n-3}{2}\right)^2.$$

Here, the equality is attained if and only if

- all distances in G are at most 2,
- the graph $G-x$ has exactly $n-2$ edges, and
- $G-x$ has exactly $n-3$ pairs of vertices in distance 2 (in $G-x$)

The first property yields that x is adjacent to each vertex of G, the second one gives that $G-x$ is a tree. Moreover, it is not hard to see that the number of pairs of $G-x$ in distance 2 is minimal if and only if $G-x$ is a path. Thus, the graph G is isomorphic with a fan graph. □

Next, we show that the average length-scaled betweenness $\overline{B}_{dist}(G)$ is closely related to other graph invariants, namely, the average betweenness $\overline{B}(G)$ and the Harary index (defined independently in [12] and [10]) $H(G) = \sum_{u,v \in V} \frac{1}{d(u,v)} = \frac{1}{2} \sum_{u \in V} c_R(u)$ (where $c_R(u) = \sum_{\substack{v \in V \\ v \neq u}} \frac{1}{d(u,v)}$ is so called harmonic centrality of u, see [11]).

Lemma 2.4 *For any graph G,*

$$\overline{B}_{dist}(G) = \frac{n-1}{2} - \frac{1}{n} H(G). \qquad (✱)$$

Proof: Observe that

$$\overline{B}_{dist}(G) = \frac{1}{n} \sum_{w \in V} B_{dist}(w) = \frac{1}{n} \sum_{w \in V} \sum_{u,v \in V} \frac{1}{d(u,v)} \cdot \frac{\sigma_{u,v}(w)}{\sigma_{u,v}} = \frac{1}{n} \sum_{u,v \in V} \left(\frac{1}{d(u,v)} \sum_{w \in V} \frac{\sigma_{u,v}(w)}{\sigma_{u,v}} \right) =$$

$$\frac{1}{n} \sum_{u,v \in V} \left(\frac{1}{d(u,v)} (d(u,v) - 1) \right) = \frac{n-1}{2} - \frac{1}{n} \sum_{u,v \in V} \frac{1}{d(u,v)}.$$

□

Proposition 2.5 *For any graph G,*

$$\overline{B}_{dist}(G) \leq \frac{n-1}{2} \left(1 - \frac{1}{\overline{\ell}(G)} \right).$$

Proof: Using the harmonic-arithmetic mean inequality in the form $\frac{r}{\sum_{i=1}^{r} a_i} \leq \frac{1}{r} \sum_{i=1}^{r} \frac{1}{a_i}$ on the right side of (✱) (with a_i being the distances in G), we obtain

$$\overline{B}_{dist}(G) \leq \frac{n-1}{2} - \frac{1}{n} \cdot \frac{\binom{n}{2}^2}{\sum_{u,v \in V(G)} d(u,v)} = \frac{n-1}{2} - \frac{n-1}{2\overline{\ell}(G)} = \frac{n-1}{2} \left(1 - \frac{1}{\overline{\ell}(G)} \right)$$

□

The relation of average length-scaled betweenness with the Harary index helps us also to determine the maximum average length-scaled betweenness for 2-connected graphs:

Lemma 2.6 *For any n-vertex 2-connected graph G and its vertex x, $c_R(x) \geq \frac{H(C_n)}{n}$.*

Proof: Let x be a vertex of G with eccentricity t (note that $t \leq \left\lfloor \frac{n}{2} \right\rfloor$). Then $c_R(x) = \sum_{i=1}^{t} \frac{n_i(x)}{i}$. Since G is 2-connected, $n_i(x) \geq 2$ for each $i \in \{1, \ldots, t-1\}$. Consider now the function $f(t, n_1, \ldots, n_t) = \sum_{i=1}^{t} \frac{n_i}{i}$ where $t, n_1, \ldots, n_i$ are integers, and the integer linear program

$$\begin{aligned} f(t, n_1, \ldots, n_t) &\rightarrow \min \\ \sum_{i=1}^{t} n_i &= n-1 \\ n_i &\geq 2 \text{ for each } i = 1, \ldots, t-1 \end{aligned}$$

It is not hard to see that, for fixed t, its minimum is attained when $n_1 = \cdots = n_{t-1} = 2, n_t = n-1-2(t-1)$, and $f^*(t) = f(t, 2, \ldots, 2, n-2t+1)) = 2H_t + \frac{n}{t} - 2$ where $H_t = \sum_{i=1}^{t} \frac{1}{i}$ is t-th harmonic number. Observe that $f^*(t)$ is decreasing (as $f^*(t) - f^*(t+1) = \frac{n-2t}{t(t+1)} \geq 0$); hence, the global minimum of $f(t, n_1, \ldots, n_t)$ is attained for $t = \left\lfloor \frac{n}{2} \right\rfloor$, $n_1 = \cdots = n_{t-1} = 2, n_t \in \{1, 2\}$ which is exactly the value of harmonic centrality of a vertex of C_n.

□

Corollary 2.7 *The cycle C_n has the smallest Harary index among all 2-connected n-vertex graphs.*

Corollary 2.8 *For each 2-connected n-vertex graph, $\overline{B}_{dist}(G) \leq \overline{B}_{dist}(C_n)$.*

Proposition 2.9 *For any graph G on n vertices,*

$$\frac{n-1-\mathrm{ad}(G)}{4} \leq \overline{B}_{dist}(G) \leq \frac{(n-1-\mathrm{ad}(G))}{2}\left(1 - \frac{1}{\mathrm{diam}(G)}\right).$$

Proof: Let A_1 be the set of all (unordered) pairs of vertices which are adjacent in G (with $m = |A_1|$ being the number of edges of G), and let $A_2 = \binom{V(G)}{2} \setminus A_1$. Then

$$\sum_{u,v \in V} \frac{1}{d(u,v)} = \sum_{u,v \in A_1} 1 + \sum_{u,v \in A_2} \frac{1}{d(u,v)} = m + \sum_{u,v \in A_2} \frac{1}{d(u,v)}$$

We first prove the upper bound. Note that

$$m + \sum_{u,v \in A_2} \frac{1}{\mathrm{diam}(G)} = m + \frac{1}{\mathrm{diam}(G)}\left(\binom{n}{2} - m\right)$$

and, since $d(u,v) \leq \mathrm{diam}(G)$ for any two vertices u and v,

$$m+\sum_{u,v\in A_2}\frac{1}{\mathrm{diam}(G)}\le m+\sum_{u,v\in A_2}\frac{1}{d(u,v)}.$$

By series of inequality adjustments, we obtain

$$\begin{aligned}
m+\frac{1}{\mathrm{diam}(G)}\left(\binom{n}{2}-m\right) &\le m+\sum_{u,v\in A_2}\frac{1}{d(u,v)} \\
m+\frac{1}{\mathrm{diam}(G)}\left(\binom{n}{2}-m\right) &\le \sum_{u,v\in V}\frac{1}{d(u,v)} \quad \Big/\cdot\left(-\frac{1}{n}\right) \\
-\frac{m}{n}-\frac{1}{\mathrm{diam}(G)}\left(\frac{n-1}{2}-\frac{m}{n}\right) &\ge -\frac{1}{n}\sum_{u,v\in V}\frac{1}{d(u,v)} \quad \Big/\text{set } \frac{m}{n}=\frac{\mathrm{ad}(G)}{2} \\
-\frac{\mathrm{ad}(G)}{2}-\frac{1}{\mathrm{diam}(G)}\left(\frac{n-1}{2}-\frac{\mathrm{ad}(G)}{2}\right) &\ge -\frac{1}{n}\sum_{u,v\in V}\frac{1}{d(u,v)} \quad \Big/+\frac{n-1}{2} \\
\frac{n-1}{2}-\frac{\mathrm{ad}(G)}{2}-\frac{1}{\mathrm{diam}(G)}\left(\frac{n-1}{2}-\frac{\mathrm{ad}(G)}{2}\right) &\ge \frac{n-1}{2}-\frac{1}{n}\sum_{u,v\in V}\frac{1}{d(u,v)} \\
\frac{n-1-\mathrm{ad}(G)}{2}\left(1-\frac{1}{\mathrm{diam}(G)}\right) &\ge \frac{n-1}{2}-\frac{1}{n}H(G) \\
\frac{n-1-\mathrm{ad}(G)}{2}\left(1-\frac{1}{\mathrm{diam}(G)}\right) &\ge \overline{B}_{dist}(G).
\end{aligned}$$

Concerning the lower bound, we proceed in a similar way: for any pair u,v of vertices of A_2, $d(u,v)\ge 2$ and $\frac{1}{d(u,v)}\le\frac{1}{2}$; thus, we obtain

$$\begin{aligned}
m+\sum_{u,v\in A_2}\frac{1}{d(u,v)} &\le m+\sum_{u,v\in A_2}\frac{1}{2} \\
\sum_{u,v\in V}\frac{1}{d(u,v)} &\le m+\frac{1}{2}\left(\binom{n}{2}-m\right) \quad \Big/\cdot\left(-\frac{1}{n}\right) \\
-\frac{1}{n}\sum_{u,v\in V}\frac{1}{d(u,v)} &\ge -\frac{m}{n}-\frac{1}{2}\left(\frac{n-1}{2}-\frac{m}{n}\right) \Big/\text{set } \frac{m}{n}=\frac{\mathrm{ad}(G)}{2} \\
-\frac{1}{n}\sum_{u,v\in V}\frac{1}{d(u,v)} &\ge -\frac{\mathrm{ad}(G)}{2}-\frac{1}{2}\left(\frac{n-1}{2}-\frac{\mathrm{ad}(G)}{2}\right) \Big/+\frac{n-1}{2} \\
\frac{n-1}{2}-\frac{1}{n}H(G) &\ge \frac{n-1-\mathrm{ad}(G)}{4} \\
\overline{B}_{dist}(G) &\ge \frac{n-1-\mathrm{ad}(G)}{4}
\end{aligned}$$

□

Corollary 2.10 *If G is a graph of diameter two, then* $\overline{B}_{dist}(G)=\dfrac{n-1-\mathrm{ad}(G)}{4}$.

The above theorem implies also the following result from [15] (where the upper bound is equivalently given in the form $\frac{n(n-1)}{4}+\frac{m}{2}$):

Corollary 2.11 *For a graph G, $H(G) \leq \dfrac{n(n-1+\mathrm{ad}(G))}{4}$.*

Proposition 2.12 *For any graph G, $\dfrac{1}{\mathrm{diam}(G)}\overline{B}(G) \leq \overline{B}_{dist}(G) \leq \frac{1}{2}\overline{B}(G)$.*

Proof: Using the notation of Proposition 2.9 and the fact that, for any pair u, v of vertices of A_2, $\dfrac{1}{\mathrm{diam}(G)} \leq \dfrac{1}{d(u,v)} \leq \dfrac{1}{2}$, we obtain

$$\frac{1}{\mathrm{diam}(G)}\overline{B}(G) = \frac{1}{\mathrm{diam}(G)}\left(\frac{1}{n}\sum_{w\in V(G)}\sum_{u,v\in A_2} b_{u,v}(w)\right) \leq$$

$$\frac{1}{n}\sum_{w\in V(G)}\sum_{u,v\in A_2}\frac{1}{d(u,v)}\cdot b_{u,v}(w) = \overline{B}_{dist}(G) \leq \frac{1}{n}\left(\sum_{w\in V(G)}\sum_{u,v\in A_2}\frac{1}{2}\cdot b_{u,v}(w)\right) =$$

$$\frac{1}{2}\left(\sum_{w\in V(G)}\sum_{u,v\in A_2}\frac{1}{n}\cdot b_{u,v}(w)\right) = \frac{1}{2}\overline{B}(G).$$

□

Corollary 2.13 *If G is a graph of diameter two then $\overline{B}_{dist}(G) = \frac{1}{2}\overline{B}(G)$.*

The following theorem is analogue of results from [6].

Theorem 2.14 *Let G be an n-vertex graph and G' be a graph obtained from G by adding the new edge uv for a pair of nonadjacent vertices u, v. Then*

$$\overline{B}_{dist}(G') \leq \overline{B}_{dist}(G) - \frac{1}{n}\left(1 - \frac{1}{d_G(u,v)}\right).$$

Proof: Since, for each pair $[x,y] \neq [u,v]$, $d_{G'}(x,y) \leq d_G(x,y)$, we obtain $H(G') \geq H(G) + 1 - \dfrac{1}{d_G(u,v)}$. Hence

$$\overline{B}_{dist}(G') = \frac{n-1}{2} - \frac{1}{n}H(G') \leq \frac{n-1}{2} - \frac{+}{n}\left(H(G) + 1 - \frac{1}{d_G(u,v)}\right) =$$

$$\overline{B}_{dist}(G) - \frac{1}{n}\left(1 - \frac{1}{d_G(u,v)}\right).$$

□

3 Length-scaled betweenness-uniform graphs

In this section, we explore the graphs whose vertices have the same value of scaled vertex betweenness (called *length-scaled betweenness-uniform graphs*). We start with exhibiting all length-scaled betweenness-uniform graphs up to 10 vertices. The testing was performed using Wolfram Mathematica computer algebra system; since the computation of length-scaled betweenness is – according to our knowledge – only occasionally implemented in available general purpose graph theory software (notable exception is the "sna" package within R environment, however, the computed centrality is in floating-point form), we have implemented also the procedure in Wolfram Language for its exact computation (as a rational number). A bit surprisingly, we have not found length-scaled betweenness-uniform graphs other than those betweenness-uniform graphs listed in [7].

Since the length-scaled betweenness is a centrality index, it follows that each vertex-transitive graph is length-scaled betweenness-uniform. Further, in a graph of diameter two, a positive contribution of a pair u, v to $B(x)$ is just twice the contribution of the same pair to $B_{dist}(x)$; hence each betweenness-uniform graph of diameter two is also length-scaled betweenness-uniform, and vice versa. Thus, for nontransitive examples of diameter two, we can take nontransitive strongly regular graphs (see [8]).

The situation is much more difficult for higher diameters: for example, there are nontransitive betweenness-uniform graphs which are not length-scaled betweenness-uniform, such as Folkman graph (for each its vertex x, $B(x) = 12$ but $B_{dist}(x) \in \{\frac{17}{4}, \frac{37}{8}\}$) or Gray graph (for each its vertex x, $B(x) = 75$ but $B_{dist}(x) \in \{\frac{272}{15}, \frac{539}{30}\}$). On the other hand, Tutte 12-cage is the nontransitive graph which is both betweenness-uniform and length-scaled betweenness-uniform (for each its vertex x, $B(x) = 227$ and $B_{dist}(x) = \frac{1411}{30}$), and the same property has $(16, 4)$-Hadamard graph ($B(x) = \frac{65}{2}$, $B_{dist}(x) = \frac{317}{24}$) which is also not vertex transitive. So far, we do not know a graph which would be length-scaled betweenness-uniform but not betweenness-uniform.

Examples of nontransitive length-scaled betweenness-uniform graphs of higher diameters can be found among distance-regular graphs. Recall that the graph G is called *distance-regular* if, for any two vertices u and v with $d(u, v) = k$ and any i, j, the number $p^k_{i,j}$ of vertices which are at distance i from u and simultaneously at distance j from v is the same (that is, not depending on u and v); in this case, for each $k \in \{1, \ldots, \mathrm{diam}(G)\}$, the numbers $c_k = |N_{k-1}(u) \cap N(v)|$ are positive and not dependent on u, v.

In [7], it was proved that each distance-regular graph is betweenness-uniform. Here, we show

Theorem 3.1 *Each distance-regular graph is length-scaled betweenness-uniform.*

Proof: Let G be distance-regular graph and x be its vertex; we express the length-scaled betweenness of x as $B_{dist}(x) = \sum_{k=2}^{\mathrm{diam}(G)} \sum_{\substack{u,v \in V(G), d(u,v)=k \\ u,v \neq x}} \frac{1}{k} \cdot \frac{\sigma_{u,v}(x)}{\sigma_{u,v}}$, discussing further the total contribution of all pairs u, v in distance k.

As pointed out in [7], having a pair of vertices u, v with $d(u, v) = k$, we have $\sigma_{u,v} = \prod_{i=1}^{k} c_i$.

Further, for a vertex w lying on a shortest $u-v$-path such that $d(u,w)=l$, the ratio $\frac{\sigma_{u,v}(w)}{\sigma_{u,v}}$ is equal to $\frac{\sigma_{u,w}\cdot\sigma_{w,v}}{\sigma_{u,v}} = \frac{\prod_{i=1}^{l} c_i}{\prod_{j=k-l+1}^{k} c_j}$. Now, for each positive integers k,l with $l<k$ and each vertex x of G, the number of pairs u,v with $d(u,v)=k$ and $d(u,x)=l$ is equal to $\sum_{i=1}^{\mathrm{diam}(G)} p_{i,l}^{k}$, thus, it does not depend on x. These facts yield that the total contribution of pairs in distance k to $B_{dist}(x)$ does not depend on x, hence $B_{dist}(x)$ is the same for every $x\in V(G)$. □

In the light of this theorem and the previous notes, it would be interesting to find nontransitive length-scaled betweenness-uniform graphs which would not be distance regular. Also, we think that the following conjecture might hold:

Conjecture 3.2 *Each length-scaled betweenness-uniform graph is 2-connected.*

References

[1] http://schochastics.net/sna/periodic.html

[2] Anthonisse, J.M.: *The rush in a directed graph*, Technical Report BN 9/71, Stichting Mathematisch Centrum, 2e Boerhaavestraat 49, Amsterdam, October 1971

[3] Koschützki, D. et al.: *Centrality indices*, in: Network Analysis (U. Brandes and T. Erlebach, Eds.), Lecture Notes in Computer Science 3418, Springer, 2005

[4] Everett, M.G., Borgatti, S.P.: *A graph-theoretic perspective on centrality*, Social Networks 28 (2006) 466-484

[5] Freeman, L.C.: *A set of measures of centrality based on betweenness*, Sociometry 40 (1977) 35—41

[6] Gago, S., Coroničová Hurajová, J., Madaras, T.: *Notes of the betweenness centrality of a graph*, Mathematica Slovaca 62 No. 1 (2012) 1–12

[7] Gago, S., Coroničová Hurajová, J., Madaras, T.: *On betweenness-uniform graphs*, Czechoslovak Mathematical Journal 63 (138) (2013) 629–642

[8] Gago, S., Coroničová Hurajová, J., Madaras, T.: *Betweenness centrality in graphs*, in: Quantitative graph theory: Mathematical Foundations and Applications (M. Dehmer and F. Emmert-Streib, Eds.), CRC Press, 2014

[9] Klisara, J., Coroničová Hurajová, J., Madaras, T., Škrekovski, R.: *Extremal graphs with respect to vertex betweenness for certain graph families*, Filomat 30:11 (2016) 3123—3130

[10] Ivanciuc, O., Balaban, T.-S., Balaban, A.T.: *Design of topological indices. Part 4. Reciprocal distance matrix, related local vertex invariants and topological indices*, Journal of Mathematical Chemistry 12 (1993) 309–318

[11] Newman, M.E.J.: *The Structure and Function of Complex Networks*, SIAM Review 45 (2) (2003) 167–256

[12] Plavšić, D., Nikolić, S., Trinajstić, N.: *On the Harary index for the characterization of chemical graphs*, Journal of Mathematical Chemistry 12 (1993) 235–250

[13] Schluter, N.: *Centrality measures for non-contextual graph-based unsupervised single document keyword extraction* , in: In Actes de la 21ème conférence sur le traitement automatique des langues naturelles, Volume 2, TALN '14 (2014) 455–-460

[14] West, D.: *Introduction to Graph Theory*, 2nd ed., Prentice Hall, 2001

[15] Zhou, B., Cai, X., Trinajstić, N.: *On Harary index*, Journal of Mathematical Chemistry 44 (2008) 611–618

Current address

Coroničová Hurajová Jana, RNDr. PhD.
Faculty of business economics with seat in Košice, University of Economics
Tajovského 2, 040 01 Košice, Slovak republic
E-mail: jana.coronicova.hurajova@euke.sk

Madaras Tomáš, doc. RNDr. PhD.
Institute of Mathematics, Faculty of Science, P.J. Šafárik University in Košice
Jesenná 5, 040 01 Košice, Slovak republic
E-mail: tomas.madaras@upjs.sk

Slovak University of Technology in Bratislava
Faculty of Mechanical Engineering
16th Conference on Applied Mathematics **APLIMAT 2017**
Proceedings

TEACHING ARITHMETIC THROUGH GEOMETRY: ORIGAMI PYTHAGOREAN TREE, NATURAL NUMBERS' POWERS, SUM AND SERIES

CUMINO Caterina[1] (IT), SPREAFICO Maria Luisa[1] (IT)

Abstract. We show an example of a lesson, tested with teen-agers and children, in which, by developing an idea of Albert E. Bosman, we propose an origami model to depict geometric forms in order to help learning of arithmetical topics such as powers of natural numbers and their sum, and also the principle of induction, with a tactile, visual and motion experience.

Keywords: natural numbers, powers of natural numbers, origami, fractals

Mathematics subject classification: Primary 97C70; Secondary 97F30, 97F10

Origami is a very versatile tool for hand-on made mathematical activities in every school degree. Origami technique has many strengths: it allows to visualize mathematics, it shows maths from another points of view, it is experienced as a game from students and it is cheap. Any fold on the paper produces (segments of) lines and so it is natural to apply it in teaching geometric subjects according to van Hiele model ([5], [12]), at any educational level from kindergarten to university as shown in many projects (see, above all, [8], [9] and [10]).

However it is possible to use origami also for arithmetical topics as fractions ([7]), powers of natural numbers and integers: it is known indeed how geometric intuition helps the learning of arithmetics by bestowing meaning and existence to matters of numbers, as suggested by R. Thom ([11]).

In this paper we would like to show an example of a lesson for students aged 10-13, in which origami is a tool to depict geometric forms, helping the representation of natural numbers, of their powers and sum of powers. In fact, one of the basic arithmetical subject for students in 5th degree primary Italian school and in the first classes of the secondary Italian school are powers of natural numbers and the properties of their operations.

[1]The authors were partially supported by the "National Group for Algebraic and Geometric Structures, and their applications" (GNSAGA-INDAM)

In this lesson/workshop we propose an origami model to visualize the concept of $a^0 = 1$ ($a \neq 0$ a natural number) and to show that, in addition to the rules about product and quotient of powers, one can deduce the formula for the sum of the first n powers of a given natural number. The need of this last formula follows naturally by the design of the origami model and it allows students to research the solution on their behalf. They have the opportunity to create the formula and not only to learn it! To do this, we think that students need to visualize the problem, as suggested also in [6], and origami is a good technique to do it. Folding paper is a not traditional way to explain mathematics and involves all the students, in particular those who have difficulty in understanding mathematics, in a tactile, visual and motion experience.

In this workshop, students fold a simple model, created by the authors referring to the Pythagorean tree introduced by A. Bosman in 1942 (see [4]): this is a plane fractal constructed from squares and it is named after Pythagoras because each triple of touching squares encloses a right triangle, in a configuration traditionally used to depict the Pythagorean theorem; authors' Pythagorean tree is constructed from suitable similar rectangles instead of squares.

In the first part of the activity, students fold the basic module used in the construction and they can see the final tree on the blackboard, in other words they are invited to achieve a 2D representation, by folding a piece of paper in 3D; then they are requested to answer questions about the numbers of modules needed to obtain that tree. This gives the opportunity to visualize 2^0 and to deduce the rule to sum the first $n+1$ powers of base 2. Then they fold the tree, and finally, they can find the rules for the first $n+1$ powers of $3, 4$ and, generalizing, for any natural number $a > 1$.

Due to students age, the rules in the workshop are predicted and verified at a given step. The same workshop is also proposable to older students, introducing and using the principle of induction to prove the above rule (see section 4) and generalizing it for any real number $a \neq 1$. Even for them, in fact, who for training reasons are more accustomed to abstract thinking, it can still be useful and fun to start from a visual and tactile experience. We remark that, even if the focus of this lesson are powers, the same model can be used to introduce fractals (with the computation of perimeter and area using series, for undergraduate students) and also a generalized Pythagoras Theorem (see Section 4).

We tested this workshop in some classes of the primary school *C. Gallazzi* and secondary school *C. Costamagna* in Busto Arsizio (Italy) and we thank them both for their helpfullness. We thank very much Francesco Decio for drawing diagrams and the referee for her/his helpful suggestions.

1 Prerequisites

Students need to know the definition of power of natural numbers and the multiplication rule of two powers with same base.

2 The model

There are many origami models to construct Pythagorean trees.

We propose a new easy model starting from paper size A3, A4, and submultiples. In particular, to fold 4 levels for the tree we suggest to start from an A3 sheet.

In this case 4 A3 sheets are needed, one for each level. As A4 sheets are more readily available, 3 A3 sheets may be substituted with 6 A4. Two sheets A4 must be cut in A5 size and two A4 must be cut in A6 size, as shown in the following picture:

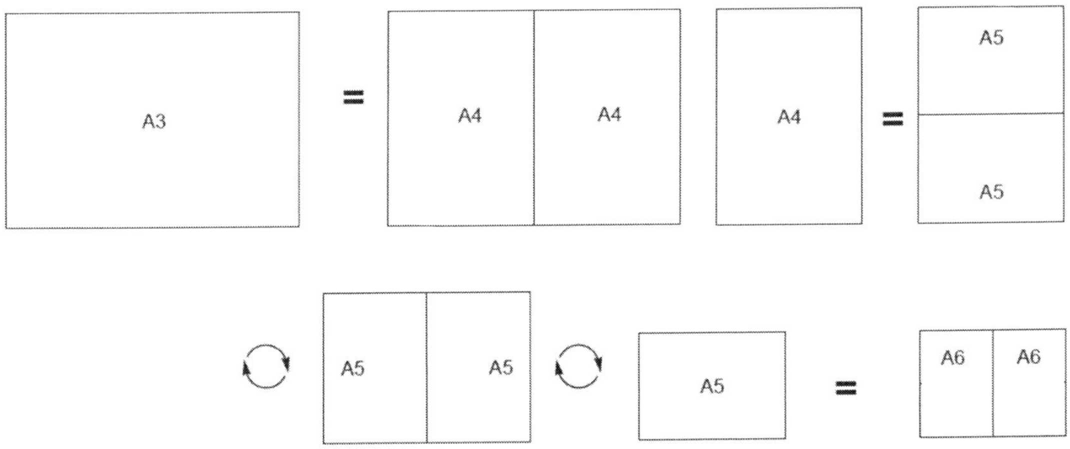

If it is possible, we suggest to use different colors depending on the size of the cut sheet, to put in evidence tree ramifications connected with the different powers.
To add another level, you need another A3 sheet or a couple of A4, cut in 8 pieces of size A6. Remember that in all A format there is always the ratio $\sqrt{2}$ between length and width.

We give also some folding instruction and the resulting folding diagram:

A. Fold the segment connecting the medium points of the shorter edges of the rectangle, overlapping one longer edge over the other. Reopen (picture 1).

B. Overlap the longer edges of the rectangle over the fold realized in point A (picture 2).

C. Considering one of the shorter edges, overlap half of it over the central line (picture 3, bottom edge).

D. Repeat the instruction C for the other short edge (picture 4, top edge).

E. Always referring to one of the shorter edges (for example the top one), push the triangles inside, obtaining two triangular pockets (picture 5). Obtain a house shape.

F. Now fold the model moving the middle point of the house base over the top of the roof (picture 6). Picture 7 shows the magnification of the resulting model.

G. Fold again the two triangles of the double upper paper layer and put them in the pockets (picture 8). We obtain the cottage, drawn in smaller size in picture 9.

During the workshop, students will fold all the different size sheets in the same way. We consider now the bigger cottage (obtained from A3 size sheet), and we put on its roof two small cottages (each of them obtained from A4 size sheets). At the next level we need four smaller cottages, then eight even smaller cottages and so on. This sequence of power of two will be the central point of the lesson.

Remark that this easy model allows students to fold the cottage using only one sheet at a time.

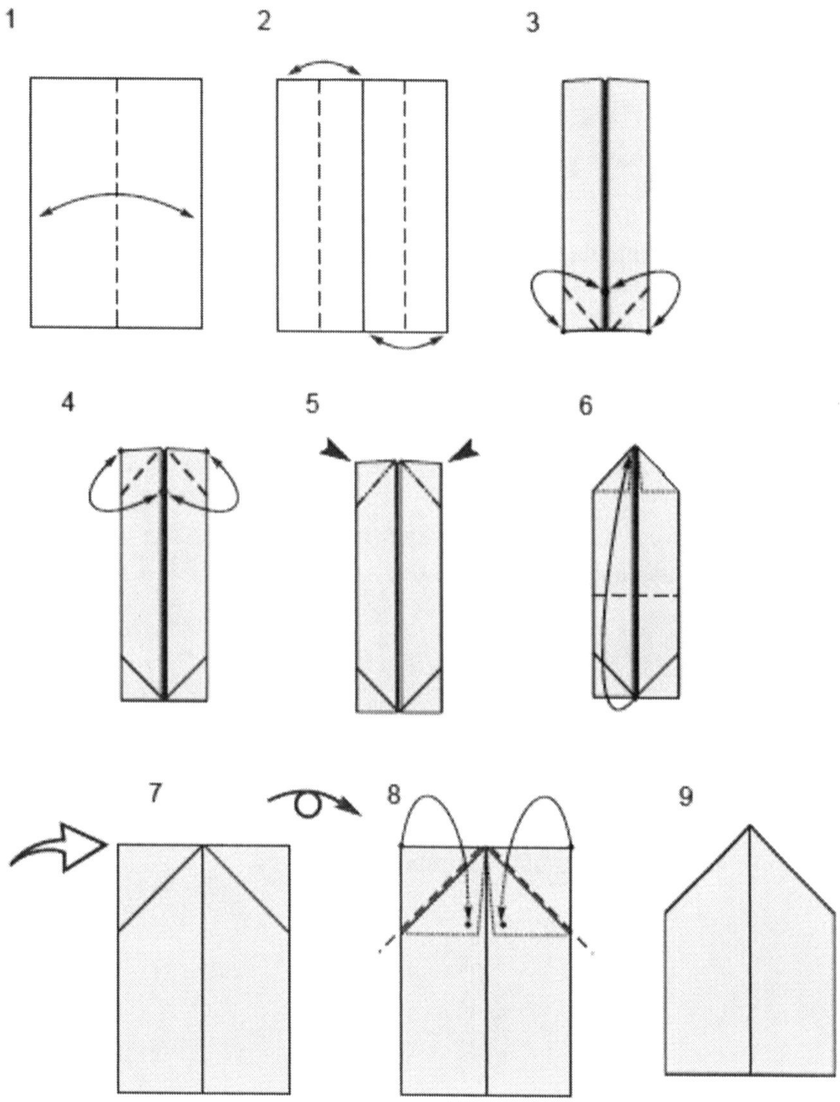

3 The lesson

We propose a lesson that we tested in some classes of a primary school (Step 1 to 5) and in some classes of the secondary school (students aged 12).

Every teacher can modify the lesson and the step's sequence depending on the student group and on their prior knowledge.

We point out a list of key questions that help students to research the solution.

Step 1: Preliminaries.

We remind the definition of power and we invite students to recall properties related to the multiplication and division. In particular:

$a^n \times a^m = a^{n+m}$, for every $a \in \mathbb{N}$.

Step 2: The Project.

We draw the final result on the blackboard to make students curious and to give them a target.

Step 3: Folding the module.

We distribute one size A4 paper each student (paper recycling is good); the students fold a single module, called "cottage" in the following, to learn the fold sequences.

Step 4: Visualization of 2^0.

Following the drawing on the blackboard, we can ask:

◇ *How many cottages have we fold at any level?*

The cottage at the base (ground floor, i.e. $0-$level), is 1. On its roof ($1^{th}-$level) we have to put other 2 cottages. Each of these two cottages needs other two cottages yet more ($2^{nd}-$level), that is $2 \times 2 = 4$ cottages and $2 \times 2 = 2^2$; as pointed out in the Step 1. Finally, each of the $4 = 2^2$ cottages at $2^{nd}-$level hosts other 2 cottages, that is $2 = 2 \times 2^2 = 2^3 = 8$ cottages at $3^{rd}-$level.
If the teacher wants to fold another level, students need other $2 \times 2^4 = 2^5 = 16$ cottages. Summarizing the previous computations, we get:

Levels	Number of cottages
0	1
1	2
2	$2^2 = 4$
3	$2^3 = 8$
4	$2^4 = 16$

Tab. 1. Calculating the number of cottages, as powers of 2

It is easy for students to find a regularity in the formula which gives the number of cottages at each step: the numbers are expressed as growing powers of 2, where the exponent is exactly the level of the cottages. But this is apparently not true for the $0-$level, where we have only 1 cottage.

◇ *Can we solve this situation? In which way?*

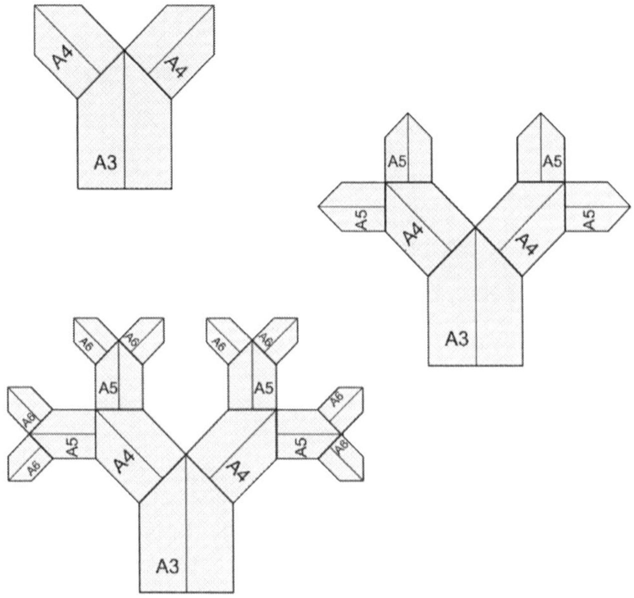

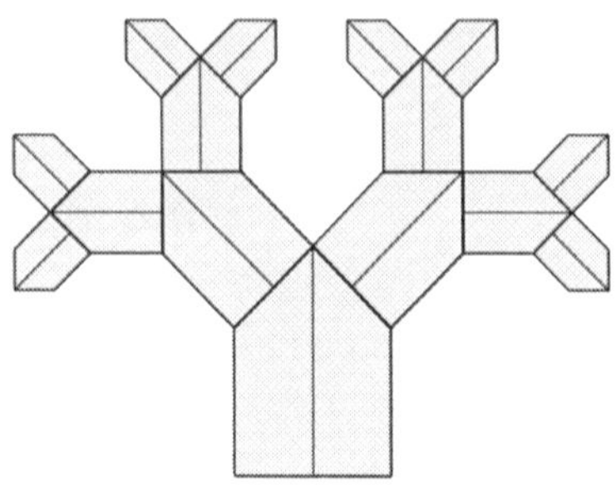

In our experience students propose to set : $1 = 2^0$.

Teachers observed that, after this workshop, students remember better the rule $a^0 = 1$ for every natural number $a \geq 1$.

Step 5: Folding the model.

We have to produce 15 cottages, if we stop at level 3, and 31 cottages, if we stop at level 4.
In the first case, if there are many students in the class, they can fold two trees.
We distribute one sheet to each student and we let them remember the folding sequence that they used to obtain the basic model in Step 3.
The teacher helps students.

Then we collect the models and we glue them on a cardboard, obtaining the three showed in the picture below:

Step 6: The formula for the sum $2^0 + 2^1 + 2^2 + \cdots + 2^n$.

We ask students:

◇ *How many cottages have we folded?*

Before all, we compute the number directly on the cardboard. If we fold until level 3, we get

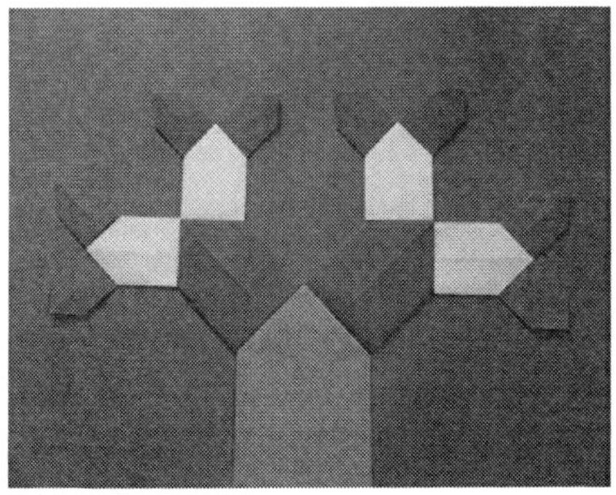

15 cottages. Computing level by level we have: $2^0 + 2^1 + 2^2 + 2^3 = 15$. We observe that 15 is not a power of the number 2, but we can ask:

◇ *Is there the possibility to relate* 15 *to any power of* 2*?*

In our experience, many students propose immediately the equality: $15 = 16 - 1 = 2^4 - 1$.

◇ *Can we give a general rule?*

At level 3 the result of the sum is $2^4 - 1$. It is possible that, at level n, we get $2^{n+1} - 1$.

◇ *Let us verify the rule, if we were at level* 4.

To construct level 4 we need $2^4 = 16$ more cottages. We have:

$$2^0 + 2^1 + 2^2 + 2^3 + 2^4 = 15 + 16 = 31 = 32 - 1 = 2^5 - 1.$$

It seems that the rule works. We cannot ask for a proof, but we let students know that the rule they found is correct and proved by mathematicians. So, for the sum of the first powers of the number 2 there exists a rule:

$$2^0 + 2^1 + 2^2 + \cdots + 2^n = 2^{n+1} - 1$$

To write this formula suggests to students the use of symbols and the abstraction.

Step 7: Generalization of the previous formula.

We can continue our investigation of the sums, changing the base of the considered first $n+1$ powers. It is natural consider the base 3.

◇ *Which formula can we propose to sum* $3^0 + 3^1 + 3^2 + \cdots + 3^n$*?*

To help students, we can visualize the situation on the blackboard, proposing another type of cottage: we can modify the roof making it a three pitched roof. The following picture shows a possible new cottage:

We calculate the 3−cottages at each step, to obtain their sum.

At the end, the number of all cottages will be: $3^0 + 3^1 + 3^2 + 3^3 = 40$

We remark immediately that the previous rule doesn't work! In fact $40 \neq 3^4 - 1 = 80$. But students may observe that $40 = 80 : 2$ and this gives them the idea to propose the following rule:

$$3^0 + 3^1 + 3^2 + \cdots + 3^n = \frac{3^{n+1} - 1}{2}.$$

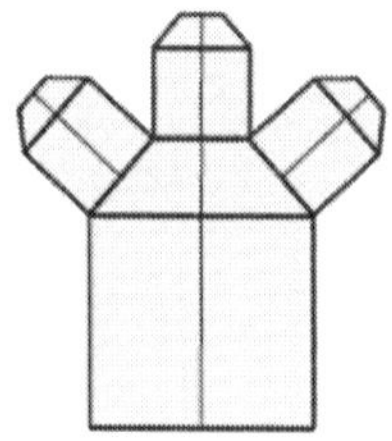

Levels	Number of cottages
0	$1=3^0$
1	$3=3^1$
2	$3 \times 3 = 3^2 = 9$
3	$3 \times 3^2 = 3^3 = 27$

Tab. 2. Calculating the number of cottages, as powers of 3

We can test it at level 4 :

$$3^0 + 3^1 + 3^2 + 3^3 + 3^4 = 40 + 81 = 121 = \frac{3^5 - 1}{2}.$$

As we say before, we confirm to students that the rule has been proved by mathematicians, for every n.

We can also continue the discussion, changing the base of the powers; we can ask:

◇ *Which formula holds for the sum* $4^0 + 4^1 + 4^2 + \cdots + 4^n$*?*

In this case we invite students to abstract the problem and we don't draw anything on the blackboard.

We start computing low powers, we expect to find a formula in which a power of the number 4, appears, minus 1 and the result is divided by another number, such as:

$$4^0 + 4^1 + 4^2 = 21 = \frac{4^3 - 1}{?} = \frac{64 - 1}{?} = \frac{63}{?}$$

We realize immediately that the formula works using the fraction $\frac{63}{3}$. We can verify this idea in the next level:

$$4^0 + 4^1 + 4^2 + 4^3 = 21 + 64 = 85 = \frac{4^4 - 1}{3} = \frac{256 - 1}{3}.$$

The rule would hence be:

$$4^0 + 4^1 + 4^2 + \cdots + 4^n = \frac{4^{n+1} - 1}{3}.$$

In conclusion we can conjecture a generalized formula for any natural base $a \geq 2$.

◇ *Which formula do you propose to calculate* $a^0 + a^1 + a^2 + \cdots + a^n$*, where* a *is a natural number greater or equal to* 2*?*

We collect the previous formulas to deduce the general one:

$$2^0 + 2^1 + 2^2 + \cdots + 2^n = 2^{n+1} - 1$$
$$3^0 + 3^1 + 3^2 + \cdots + 3^n = \frac{3^{n+1} - 1}{2}$$
$$4^0 + 4^1 + 4^2 + \cdots + 4^n = \frac{4^{n+1} - 1}{3}$$

We remark that the behavior of the numerator is clear: if the base is a, the numerator will be $a^{n+1} - 1$.

If students had difficulties to deduce the rule for the denominator, we could write the first formula in the following manner:

$$2^0 + 2^1 + 2^2 + \cdots + 2^n = \frac{2^{n+1} - 1}{2 - 1}.$$

Now it should be evident that the denominator is always $a - 1$. We are now ready to write the formula that we deduced folding the paper: if a is a natural number greater or equal than 2, we have

$$a^0 + a^1 + a^2 + \cdots + a^n = \frac{a^{n+1} - 1}{a - 1}.$$

◇ *Does the formula work even for $a = 1$?*

No it does't. In fact $a = 1$ clears the denominator. In this case, a direct computation, gives

$$1^0 + 1^1 + 1^2 + \cdots + 1^n = 1 + 1 + \cdots + 1 = n + 1.$$

4 Other spin off

This model can be used to develop other topics; we give here two suggestions both for students aged 10-13 and for students aged 14-18.

4.1 Primary school and secondary school of prime degree (10-13 years)

The adjective "pythagorean" suggests a link between this tree and the well known Pythagoras Theorem. In fact each isosceles right triangle, roof of the cottages, supports three shapes which satisfy Pythagoras Theorem generalized to similar rectangles: each small rectangle on the roof has half extension with respect to the basis of the cottage, because it is obtained using half of the paper respect to the bigger one. So the sum of the extensions of the two small rectangles (constructed on the cathetuses) is equal to the extension of the rectangle constructed on the hypotenuse.

A model, which is faithful to the original pythagorean tree, was designed by Francesco Mancini and is currently used by the museum of Mathematics "Il Giardino di Archimede" in Firenze ([1]), to visualize Pythagoras Theorem in the case of squares constructed on cathetuses and hypotenuse.

Moreover, this tree is an easy example of fractal.

For student aged 10-13, another beautiful example of fractal origami (three-dimensional one), which involves the power of number 3, is due to Bahamani, Sharif and Hudson [3]. The authors show many examples of fractals in the arts, in particular Arab art and they design an origami model of great effectiveness.

4.2 Secondary school of second degree (14-18 years)

As one can easily see, during the lesson, the principle of Mathematical induction come out naturally. The use of this principle, however, needs a good level of abstraction, hence it is only suitable for high school students.

To introduce the induction principle during the lesson, we can propose at first the case of powers of 2, then the case of powers of $a \in \mathbb{N}$, with $a \neq 1$. At the end we will point out to students that the found proof (and hence the found formula) holds for each $a \in \mathbb{R}$ with $a \neq 1$.

Apply for example the induction principle to show the formula $\sum_{i=0}^{n} a^i = \frac{a^{n+1}-1}{a-1}$ in the case of $a \in \mathbb{N}$. It is obvious that this formula holds for $n = 0$ (base step); we suppose then that the sum of the first n terms is $\frac{a^{n+1}-1}{a-1}$ (inductive hypothesis) and we deduce that the sum of the first $n+1$ terms $n+1$ is (inductive step):

$$(a^0 + a^1 + \cdots + a^n) + a^{n+1} = \left(\frac{a^{n+1}-1}{a-1}\right) + a^{n+1} = \frac{a^{n+1} - 1 + a^{n+1}(a-1)}{a-1} = \frac{a^{n+2}-1}{a-1}$$

And so the formula holds for the all $n \in \mathbb{N}, a \neq 1$.

Moreover, exploring the idea of fractal using our models, for high school students we can suggest to compute the area of our fractal tree model at any level and its perimeter.

About area, it's very easy to remark that the cottages at level n have half area with respect to any cottage at level $n-1$ because each of them is folded using the half paper with respect the previous one. But, as we use exactly A3 equivalent sheets at each level (cut in parts which are numbered by power of two), the resulting area A_n of the tree at level n is $n+1$ the area A of the cottage at level 0 i.e.

$$A_n = (n+1)A$$

The computation of the tree perimeter is a little bit more challenging: consider at first the lateral walls of the cottages, let L be the size of one of them; starting with a sheet having shorter edge of size a (and hence longer edge of size $a\sqrt{2}$) and following the folding instructions of the cottage, one has $L = a\frac{\sqrt{2}}{2} - \frac{a}{4}$; since at the next level the shorter edge of the sheet has size $a\frac{\sqrt{2}}{2}$, one has that the wall at this level has size $L' = L\frac{\sqrt{2}}{2}$. To complete the perimeter it is necessary to add the ground floor of the 0−level cottage, which is of size $\frac{a}{2}$; from the expression of the wall size L one gets $\frac{a}{2} = \frac{2L}{2\sqrt{2}-1}$.

Then, if the initial cottage has wall size L, the situation is the following (recalling that at level 0 we have also the ground floor):

Students can deduce that the perimeter's formula at level n is:

$$p_n = \frac{2L}{2\sqrt{2}-1} + 2L + 2L(\sqrt{2}) + 2L(\sqrt{2})^2 + \cdots + 2L(\sqrt{2})^n$$

Levels	Perimeters
0	$\frac{2L}{2\sqrt{2}-1}+2L$
1	$4(L\frac{\sqrt{2}}{2})=2L(\sqrt{2})$
2	$8(L\frac{\sqrt{2}}{2}\frac{\sqrt{2}}{2})=8L(\frac{\sqrt{2}}{2})^2=2L(\sqrt{2})^2$
3	$16(L(\frac{\sqrt{2}}{2})^2\sqrt{2})=2L(\sqrt{2})^3$

Tab. 3. Perimeter growing

and so it seems that the perimeter grows quickly. To find an effective formula for calculating the perimeter we can rewrite p_n:

$$p_n=\frac{2L}{2\sqrt{2}-1}+2L+2L(\sqrt{2})+2L(\sqrt{2})^2+\cdots+2L(\sqrt{2})^n=\frac{2L}{2\sqrt{2}-1}+2L(1+\sqrt{2}+\cdots+(\sqrt{2})^n)$$

Now if we put $p'_n=1+\sqrt{2}+\cdots+(\sqrt{2})^n$, we can compute this sum in two different ways:

1. For students who followed the introduction about induction on the sum of powers, we get immediately:

$$p'_n=\frac{(\sqrt{2})^{n+1}-1}{\sqrt{2}-1}$$

   and so

$$p_n=\frac{2L}{2\sqrt{2}-1}+2L(p'_n)=\frac{2L}{2\sqrt{2}-1}+2L\frac{(\sqrt{2})^{n+1}-1}{\sqrt{2}-1}$$

2. Otherwise we can propose the classical computation about p'_n:

$$(\sqrt{2})p'_n-p'_n=(\sqrt{2})^{n+1}-1\Rightarrow p'_n(\sqrt{2}-1)=(\sqrt{2})^{n+1}-1\Rightarrow p'_n=\frac{(\sqrt{2})^{n+1}-1}{\sqrt{2}-1}$$

   which gives the same value of p_n, as above.

If one uses this model for undergraduate students, one can also write the limit of the sum as geometric series of reason $\sqrt{2}$:

$$p=\frac{2L}{2\sqrt{2}-1}+2L\sum_{n=0}^{+\infty}(\sqrt{2})^n$$

It may also be interesting let students see some technological application of Pythagorean tree; for example, in the domain of wireless communication, fractal concepts have emerged as a method for designing wideband planar antennas and a novel printed Pythagorean tree fractal monopole antenna was recently presented by a group of iranian researchers: in addition to its excellent space-filling characteristics, it is also flexible (due to its multi fractal structure) in terms of controlling resonances and bandwidth (see [2]).

References

[1] URL: `http://web.math.unifi.it/archimede`.

[2] URL: `http://ieeexplore.ieee.org/document/5766703/`.

[3] Bahamani, A., Sharif, K., Hudson, A.: *Using Origami to Enrich Mathematical Understanding of Self Similarity and Fractals*, Origami6: II. Technology, Art, Education, Proceeding of the sixth international meeeting on Origami Science, Mathematics and Education, 2015, Ed. AMS, pp. 723-734.

[4] Bosman, A. E. : *Het wondere onderzoekingsveld der vlakke meetkunde*, N.V. Uitgeversmaatschappij Parcival, Breda, 1957.

[5] Crowley, M.L.: *The van Hiele model of the development of geometric thought*, In Learning and Teaching Geometry, K-12, 1987 Yearbook of the National Council of Teachers of Mathematics, Mary Montgomery Lindquist, Reston, Va.: National, 1987, pp. 1-16.

[6] de Finetti, B.: *Ciò che ritengo essenziale raccomandare agli insegnanti*, Annuario dell'insegnante 1976-77, a cura dell'Istituto Nazionale delle assicurazioni, 1977, pp.85-98.

[7] Frigerio, E., Spreafico, M.L.: *Area and Optimization Problem*, Origami6: II. Technology, Art, Education, Proceeding of the sixth international meeeting on Origami Science, Mathematics and Education, 2015, Ed. AMS, pp. 679-689.

[8] Golan, M., Jackson, P. : *A Program to Teach Geometry and to Develop Learning Skills Using the Art of Origami*, Origami4, Proceeding of the fourth international meeeting on Origami Science, Mathematics and Education, 2009, Ed. A. K. Peters Ltd, pp. 459-469.

[9] Hull, T.: *Project Origami*, 2006, Ed. A. K. Peters Ltd.

[10] Origami6: II. Technology, Art, Education, Proceeding of the sixth international meeeting on Origami Science, Mathematics and Education, 2015, Ed. AMS.

[11] Thom, R. : *Modern Mathematics: does it exists?*, in Howson, A.G. ed., Developments in Mathematical Education, Cambridge: CU, 1973, pp. 195-209.

[12] van Hiele, P. : *The Child's Thought and Geometry*, Brooklyn, NY: City University of New York, 1985, pp. 243-252.

Current address

Cumino Caterina, Associate Professor
Department of Mathematical Sciences, Politecnico di Torino
corso Duca degli Abruzzi 24, Torino, Italy
Tel. number: +39 011 090 7500, e-mail: caterina.cumino@polito.it

Spreafico Maria Luisa, Assistant Professor
Department of Mathematical Sciences, Politecnico di Torino
corso Duca degli Abruzzi 24, Torino, Italy
Tel. number: +39 011 090 7500, e-mail: maria.spreafico@polito.it

Slovak University of Technology in Bratislava
Faculty of Mechanical Engineering
16th Conference on Applied Mathematics **APLIMAT 2017**
Proceedings

ORIGAMI MODELING FOR ARCHITECTURAL SHAPES: MATHEMATICAL AND DESIGN INVESTIGATION

CUMINO Caterina[1] (IT), SPREAFICO Maria Luisa[1] (IT), ZICH Ursula (IT)

Abstract. We propose a construction of 3D origami model of mansard roofs, using interactive approaches between Mathematics and Architecture, through the common language of Geometry.

Keywords: folding design, developable surfaces, roofing systems, real modeling

Mathematics subject classification: Primary 00A66; Secondary 00A67

1 Introduction

Origami is the ancient Japanese art of paper folding and in this sense it has been known for a long time. However, it is immediate that each fold creates a segment and that subsequent folds describe symmetries and geometric loci.

In the nineties it was carried out a formal study of mathematics described by this art. In particular, H. Huzita [3] gave an axiomatic formalization of origami folds and B. Scimemi [7] characterized geometric constructions based on Galois theory. Moreover, the majority of scientific and technological applications of origami have come within the past 50 years. Advances in mathematics and computer science have paved the way for new analysis of the folding process and design techniques, which now allow to realize structures with specified properties, in a predictable and precise way, far beyond the art itself. For example, R. Lang and J. Mitani developed software programs (respectively Treemaker [4] and Ori-revo [13]), which allow to generate the so called Crease Pattern, a specification document consisting of fold lines, needed to obtain the desired origami model.

We expose a research experience in this context. We are interested in constructing 3D models of architectural shapes from a sheet of paper or other suitable material, using origami techniques. To achieve this goal, we make use of different approaches based on the interaction

[1]The first and the second author were partially supported by the "National Group for Algebraic and Geometric Structures, and their applications" (GNSAGA-INDAM)

between mathematical and architectural languages, employing Geometry as communication medium. In the present paper we show an example in which this methodology is applied to the analysis of a particularly widespread type of roofing system: the mansard roof.

The paper is organized as follows.

After a brief presentation of the methodology followed in our research, we describe how to get a theoretical model via description of architectural forms: a preliminary phase which concerns recognition and interpretation of the object of interest, to achieve a geometrical description of it.

Secondly, we illustrate how to construct the Crease Pattern with mathematical tools borrowed from Analytic and Differential Geometry, starting from the analysis of the identified geometric object.

We pass then to apply our approach to the specific case of mansard roofs. In the last section, we deal with some general technical aspects of origami folding, that arise in particular when we model these types of roofs, such as: managing the excess paper or possible use of materials different from paper, die-cut or laser cut techniques.

2 Methodology

We give now a brief account of the methodology we have applied in our research. In the subsequent sections we shall go into the details for each item.

Description of architectural forms.
Starting from the knowledge of the real architectural object, one can get a theoretical model, through various levels of accuracy, using survey: the choice of tools and methods in data acquisition (measurement devices, objectives, processing procedures) may strongly affect the resulting origami modeling.

Design of the crease pattern.
Mathematical description of the identified geometric objects and of their developments on the plane, positioning of these developments (suitably translated) onto a single sheet to get the "pattern".

Folding process and management of the excess material.
In this phase, first of all we have to addition to the pattern the fold lines necessary to give volume to the geometric developments and to eliminate those lines which do not have to be folded. Moreover, we have to solve problems of paper managing.

The picture below shows the method applied in a very easy case.

Fig. 1. Origami modeling of a barrel vault

3 Description of architectural forms

Each representation is only a representation and, as such, it differs from reality; therefore an origami model can be representative of a complex shape but it is necessary to assess its level of approximation.

To project an origami model of a roof, we need to have a geometrical description of it. So we need a preliminary phase which concerns recognition and interpretation of the object, in order to design its reproduction. We present two possible approaches to the analysis of an architectural object.

3.0.1 Recognizing the pure form

Analyzing the architectural object, one can look for a decomposition in simple shapes to achieve, with different levels of approximation, a symbolic representation of the pure form.

This mode is specific to the architectural context: one moves from general to particular, starting from the definition of maximum volumes, one proceeds to describe more specific volumes up to a visualization as a sum of simple elements: we could draw a parallel with a raw material block roughed down to get a sculpture.

Similarly, in origami modeling, it is possible, for example, to describe a building as a simple parallelepiped to which we can add or subtract volumes that create openings and we can add inclined planes that create the roofing systems. Therefore it is not necessary for the model to be very detailed in order to be significant: a simple form is expressive if it is evocative. This approach is, for example, typical in the origami modeling of Mitani [5].

3.0.2 Survey and mathematical equations

Conversely, if we are interested in real architecture modeling, we need a precise geometric description of the elements that compose it.

Depending on the purpose, on the modeling scale and on the materials for the realization, the data required for a modeling starting from the development of surfaces, like in our case (architectural shape generated by cylinders), are different.

Minimum requirement is the knowledge of the geometry of translating section. As the information is conjecturable on the basis of a first perceptive survey and/or photographic or graphic acquired documentation, it is possible to formulate a first geometric interpretation, arising from the sequence of graphic construction steps that identify centers of curvature and radii of the different circumferences which gradually approximate the single portions of the section.

This hypothesis can and should be compared with the results of a survey, because the modeling has to be not only symbolic but representative of the real form.

The operating modes to do this are many:

(i) Direct survey: after identifying the most significant planes of section, we proceed with measurements on the plane via trilateration or rectangular coordinates to obtain the directrix of the cylinder and to verify the possible variants. After obtaining sections we proceed to the graphic development of them.

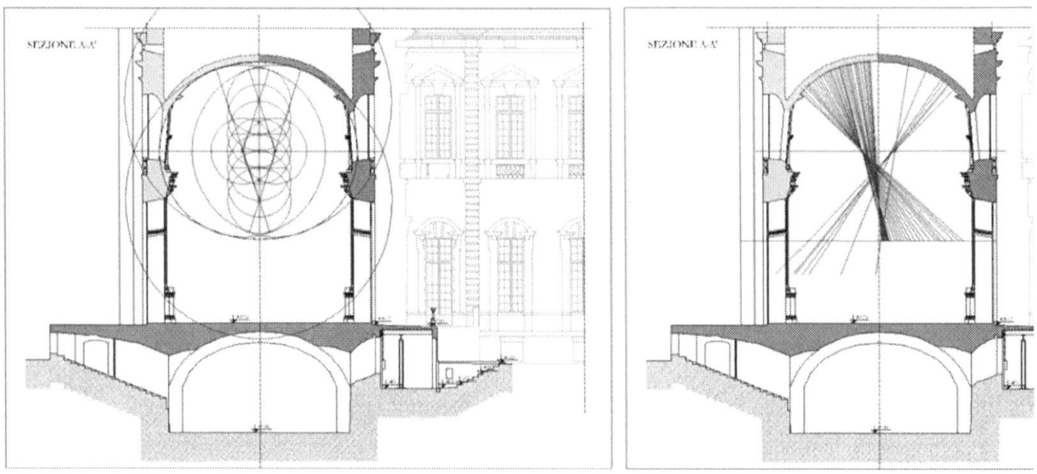

Fig. 2. An example of geometric interpretation and direct survey, see [11]

(ii) Instrumental survey with 3D laser scanning: given the cloud of points we proceed to texturing and then the model may be sectioned to get all sections needed to the surfaces development. After obtaining sections, their graphic development can be achieved.

(iii) Instrumental survey with 3D laser scanning and numerical analysis: from the point cloud, one can deduce the equation of the cylinder approximating the real form, using numerical analysis, as D. Salvaggio does in his thesis [6].
His idea is to create an algorithm to approximate cylinders equations and implement it, using the software Python. Starting from a 3D point cloud, the points are projected onto the Oxy plane and a 2D Delaunay Triangulation of vault points is obtained.
Then, for each point P, an approximating quadric surface is computed together with its curvatures and principal directions. As we expect to find locally a cylinder, we can look for the direction related to the null curvature (corresponding to the null eigenvalue of the quadric matrix; its eigenvectors give the ruling direction).
This method allows to divide points in two regions, depending on the cylinder to whom they belong. Finally, choosing one cylinder at a time, one can project on a vertical plane, using a statistic median of the punctual ruling direction.
The interpolation of the obtained cloud of points gives a directrix of the considered cylinder and it allows to obtain the cylinder equation.

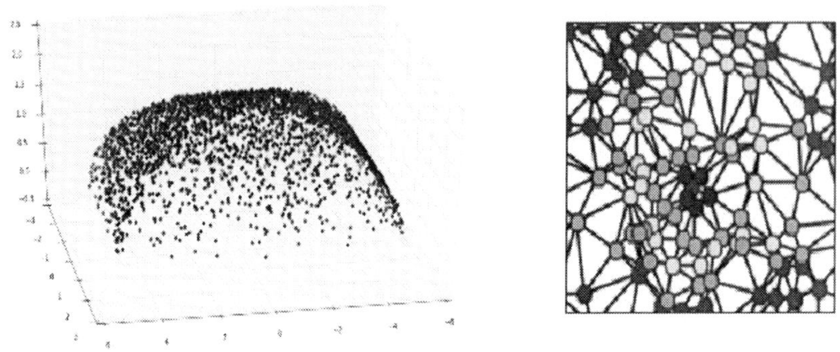

Fig. 3. Point cloud and Delaunay triangulation, [6]

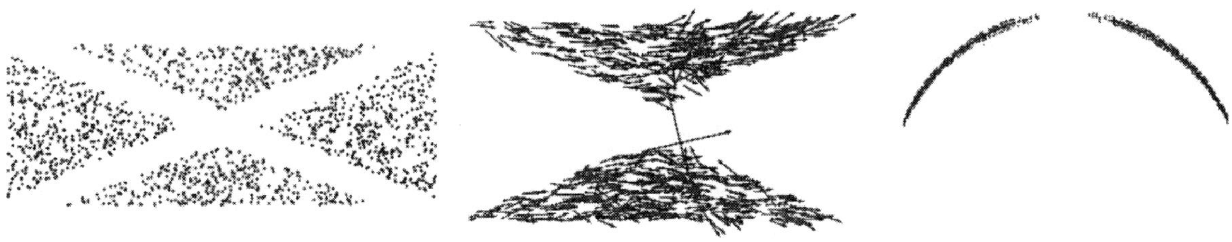

Fig. 4. Data numerical analysis, [6]

4 Design of the crease pattern

We now describe how to construct a crease pattern, starting from the analysis of the pure form.

In a recent paper (see [1]) the authors create origami models of compound vaults: groin vault, barrel vault with lunettes, cloister vault, barrel vault with cloister heads. Mathematical description of these vaults is given by developable surfaces, i. e. surfaces that can be mapped onto the plane, preserving the length of any curve drawn on them (see [2]). In regard to an architectural point of view, they consider two mutually orthogonal cylinders C_1 and C_2, with any cross-section. To obtain foldability, they need some good condition about the intersection curve of the two cylinders, which in fact must be piecewise planar (to solve an overlapping folding problem). Moreover, the developments γ_1 and γ_2 of this curve, viewed respectively as a curve on C_1 and on C_2, in an appropriate plane, must be piecewise mutually symmetric, up to translation.

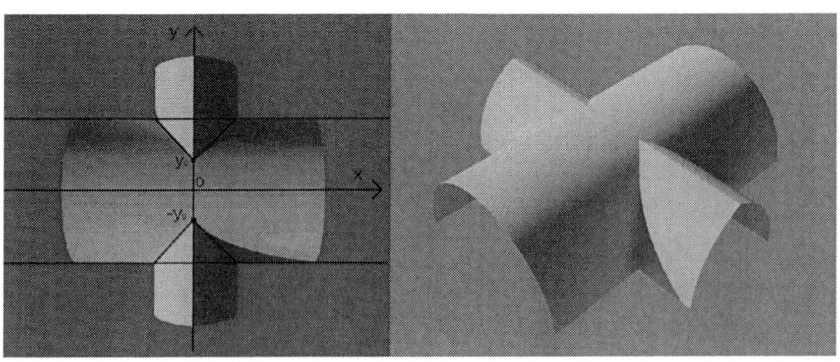

Fig. 5. Two intersecting cylinders, [1]

To be more explicit, we fix some notation and we recall the main result of [1].

Let C_1 be a cylinder of equation: $z = f(y)$, where $f : [-a, a] \to \mathbb{R}$ is a continuous function satisfying the following properties:

$$\begin{array}{l} f(y) \geq 0,\ \forall y \in [-a,a],\ \text{ and }\ f(-a) = f(a) = 0; \\ f(-y) = f(y),\ \forall y \in [-a,a]; \\ f \in C_2((-a,0) \cup (0,a)); \\ f'(y) \leq 0, \forall y \in (0,a)\ \text{ and monotonic.} \end{array} \tag{1}$$

Consider an analogous cylinder C_2 orthogonal to C_1 and symmetric with respect to the plane $x = 0$. Unroll the cylinders C_1 and C_2 onto appropriate planes parallel to $z = 0$ and project their developments onto this plane. Put $\gamma = C_1 \cap C_2$. Because of symmetry, it is sufficient to develop the portion of γ, which represents the intersection of the two cylinders in the first octant. For example, the portion of the curve γ in the first octant, viewed as a curve on C_1, develops into the curve:

$$\gamma_1(v) = (\frac{v - y_0}{m}, \int_0^v \sqrt{1 + f'^2(t)}dt, 0), \text{ with } y_0 \leq v \leq a \tag{2}$$

Analogously we define γ_2.

The following Theorem shows how to obtain the piecewise planar and symmetric conditions mentioned above.

Theorem 4.1 *Using the previous notation,*

1. *the intersection curve $\gamma := C_1 \cap C_2$ is piecewise planar, i.e. it lies in the union of the four planes $y = \pm mx \pm y_0$, where $y_0 \in [0, a), m \in \mathbb{R}$, if and only if the cylinder C_2 has cartesian equation:*
$$z = f(m|x| + y_0).$$
2. *In the plane $z = 0$ the curve γ_1 is symmetric to the curve γ_2, up to translation, if and only if one of the following holds:*
   a) $f(v) = \alpha x + q$;
   b) $m = \pm 1$;

Remark 1 *(i) Claim 1. means that, in each octant, the intersection curve γ is contained in a plane.*

(ii) From Claim 1. we deduce the development of γ viewed as a curve on C_2, in the first octant :

$$\gamma_2(u) = (\int_0^u \sqrt{1 + m^2 f'^2(mt + y_0)}dt, mu + y_0, 0), \text{ with } 0 \leq u \leq \frac{a - y_0}{m}. \tag{3}$$

(iii) From Claim 2. it follows that there are two possibilities: or the cylinders are formed by portions of planes (planar cylinders) or the intersection curves are obtained by cutting the cylinders with a plane forming an angle of 45° with their generatrices and the second cylinder is constituted by portions of the first.

In paper [1], starting from this result the authors analyzed non planar cylinders. In this paper, we study how to obtain the CP in the planar case; we remark that this case is useful to describe the particular roofing system generated by plane surfaces, called mansard roof.

Fig. 6. Example of hystorical and contemporary mansard roofs in Torino (IT), (photos by U. Zich), and in Bratislava (SK), (views taken from www.google.it/maps)

Let us imagine the roof divided in horizontal slices, as in the following picture

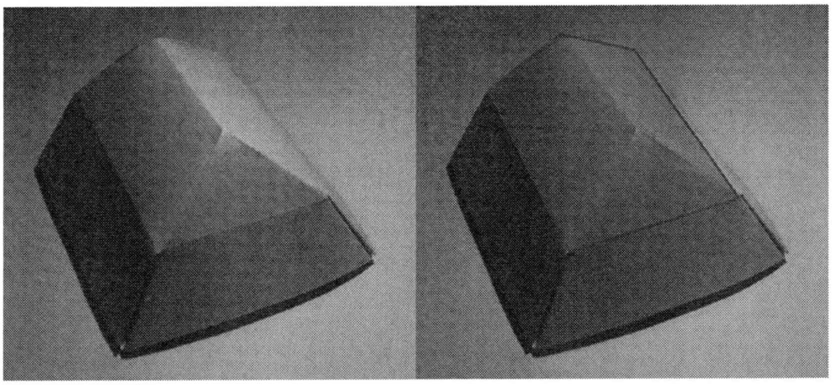

Fig. 7. Mansard roof divided in horizontal slices

If we consider one slice at a time, the cylinders have equation $C_1 : z = my + q, \quad y \in [-a, a], a \in \mathbb{R}$, and $C_2 : z = m'x + q', \quad x \in [-b, b], b \in \mathbb{R}$, respectively.

For these cylinders the Theorem above holds, hence we can get the CP of the desired model. Note that this CP is characterized by the fact that all its creases (obtained either unrolling the intersections of the two cylinders, or the folding lines) are segments.

Now, if we have at least two slices, we can obtain the CP of each of them, and, moreover, we can glue together the two CP's moving opportunely the segments on each CP so that they have common extremes. Note that the way to move unrolled segments and to add folding lines is not unique.

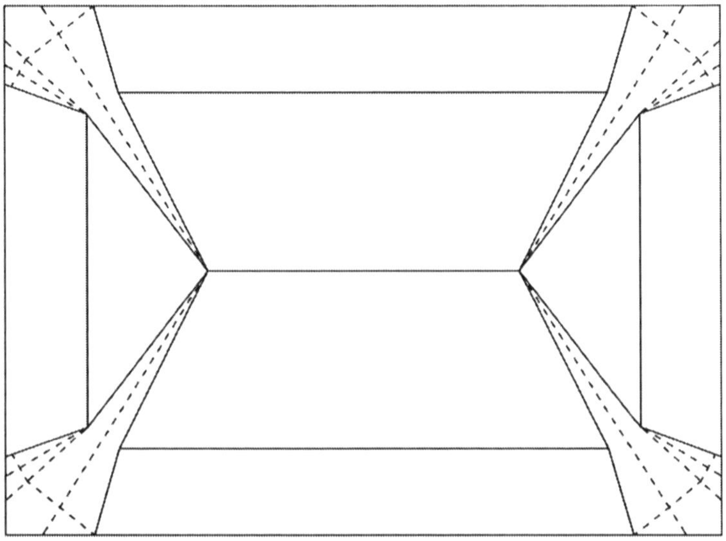

Fig. 8. Example of crease pattern of a mansard roof

Observe that the above procedure also holds in some "anomalous case", see photo below.

Fig. 9. Example of mansard roof (Bratislava) where the pitches intersection does not belong to a vertical plane, (view taken from www.google.it/maps)

5 From idea to realization through design

Origami modeling is a synthesis of creative, scientific and artistic acts. However, for not considering the model as an "art work", each step of its creation needs argued and not random choices, that make the result repeatable and sharable. Mathematical premises, sketched above as a basis for drawing our CP, shall support the theoretical precision but, at the same time, do not ensure it in the practical, because origami requires manipulation of physical medium, and this operation is very less rigorous.

In our experience the same shape may have different CP and/or equal CP with different folding sequences: on many occasions the origamist himself, with his cultural background and his practical attitude, chooses the way more congenial to him.

At first glance the outcome may look identical, but a closer examination points out elements that lead to prefer one them, because it responds better to certain expectations.

For example, our mansard origami model was folded in different ways giving priority now to an origami practical approach, in order to optimize paper handling (emphasizing sometimes the external perception and sometimes the inside one), now to a geometric approach, in order to find a solution through development of planes, according to a concatenation of steps.

Fig. 10. Examples of folding sequences

Actually we have to relate this multiplicity of approaches, all seemingly useful and equally possible to achieve the designed model, to the variety of possible supports.

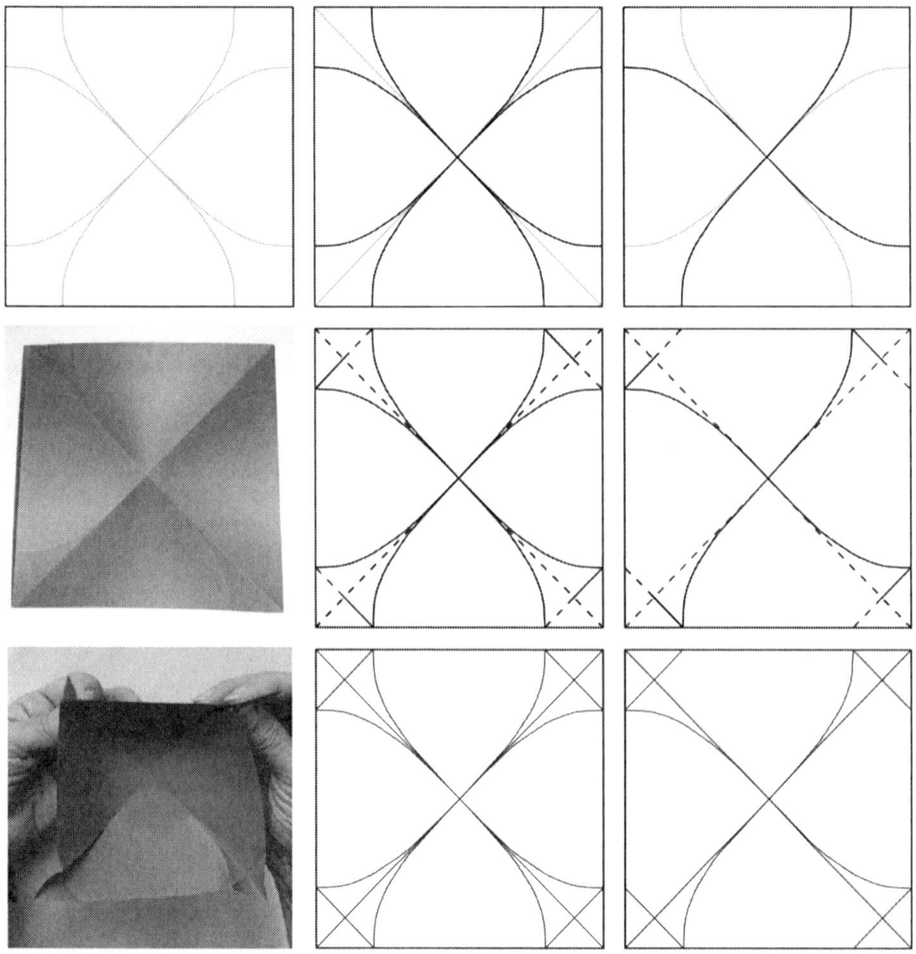

Fig. 11. Study for a die-cut production of a cloister vault model

The experience done to industrialize a model of cloister vault, (i.e. optimization of the folds as outcome of a papermaking process and/or of a laser printer drawing the specific CP), guided us in the project of the mansard model; initially we designed simple mountain/valley folds in order to assess its feasibility even with rigid materials. It should be recalled that modeling with rigid materials would require the presence of hinges and the gradual increase of thickness for the overlapping of layers at certain points (points where the model has to be closed), which would make the management more complex.

Our experimentation with direct folding has shown that, despite the small number of folds and the defined sequence of them, folding dynamics are actually made possible by flexibility and elasticity of the paper and therefore, at the moment, our models are feasible with papers and cards of various weights, folded directly or partially prepared with a laser cut (paper weight no more than $250g/m^2$) or pre-engraved with a die-cut (paper weight no more $200g/m^2$): our red model, (see figure above), created thanks to collaboration of the company Bottega Fagnola, Torino [12] is of weight $140g/m^2$.

References

[1] Cumino, C., Frigerio, E. , Gallina, S., Spreafico, M. L., and Zich, U.: *Modeling Vaults in Origami: a Bridge between Mathematics and Architecture*, in Origami6: II. Technology, Art, Education, Proceeding of the sixth international meeeting on Origami Science, Mathematics and Education, Ed. AMS, 2015.

[2] Do Carmo, M.P.: *Differential geometry of curves and surfaces*, Pearson Education: Canada, 1976.

[3] Huzita, H.: *Understanding Geometry through Origami Axioms* , Proceeding of the first International conference on Origami in Education and Therapy, (COET91), J. Smith ed., British Origami Society, 1992.

[4] Lang, R. J.: *Origami Design Secrets, $2^{n}d$ Edition* , 2011, Ed. A. K. Peters Ltd.

[5] Mitani, J., Igarashi, T. : *Interactive Design of Planar Curved Folding by Reflection*, Pacific Conference on Computer Graphics and Applications, Bing-Yu Chen, Jan Kautz, Tong-Yee Lee and Ming C. Lin (Editors), 2011.

[6] Salvaggio, D.: *Metodi di approssimazione per la ricostruzione di superfici a volta a pertire da dati di laser-scanning 3D*, Thesis Mathematics Engineering, Politecnico di Torino, 2015.

[7] Scimemi, B.: *Geometria Sintetica; Trasformazioni, Triangoli, Coniche*, Ed.Cleup, 2012.

[8] Demaine, E.D., Demaine, M.L. Huffman, D.A., Koschitz, D., Tachi, T.. : *Chracterizations of Curved Creases and Rulings: Design and Analysis of Lens Tesselations*, Origami6: I. Mathematics, Proceeding of the sixth international meeeting on Origami Science, Mathematics and Education, 2015, Ed. AMS.

[9] Tachi, T.: *Geometric considerations for the design of Rigid Origami Structures*, Proceedings of the International Association for Shell and Spatial Structures (IASS) Symposium, Vol. 12, 2010.

[10] Tachi, T.: *Simulation of Rigid Origami*, Proceeding of the fourth international meeeting on Origami Science, 2009, Ed. A. K. Peters Ltd.

[11] Volpiano, M., Zich,U., *Scientific monitoring and documentation of the Venaria Reale restoration sites*, Proceedings of the XX International Symposium: International Cooperation to Save the World's Cultural Heritage, CIPA, 2005.

[12] URL: `http:www.bottegafagnola.it`

[13] URL: `http://mitani.cs.tsukuba.ac.jp/ori-revo/`.

Current address

Cumino Caterina, Associate Professor
Department of Mathematical Sciences, Politecnico di Torino
corso Duca degli Abruzzi 24, Torino, Italy
Tel. number: +39 011 090 7500, e-mail: caterina.cumino@polito.it

Spreafico Maria Luisa, Assistant Professor
Department of Mathematical Sciences, Politecnico di Torino
corso Duca degli Abruzzi 24, Torino, Italy
Tel. number: +39 011 090 7500, e-mail: maria.spreafico@polito.it

Zich Ursula, Assistant Professor
Department of Architecture and Design, Politecnico di Torino
viale Pier Andrea Mattioli 39, Torino, Italy
Tel. number: +39 011 090 4364, e-mail: ursula.zich@polito.it

Slovak University of Technology in Bratislava
Faculty of Mechanical Engineering
16th Conference on Applied Mathematics APLIMAT 2017
Proceedings

"THE KING AND THE ORIGAMI", EDUCATIONAL TOURS AT THE ROYAL RESIDENCE OF VENARIA REALE (ITALY) TO OBSERVE, ANALYZE AND MODEL ARCHITECTURE THROUGH MATHEMATICS

CUMINO Caterina[1] (IT), SPREAFICO Maria Luisa[1] (IT), ZICH Ursula (IT)

Abstract. We present a set of guided tours and educational workshops, where the halls of a magnificent baroque royal residence become classrooms and visitors are led to discover architectural shapes and to understand their geometry with the use of various origami modeling activities, rigorously supported by mathematics (descriptive geometry and basics in analytic and differential geometry).

Keywords: visualization, analysis, abstraction, deduction, rigor, education, modeling, gaming

Mathematics subject classification: Primary 00A67; Secondary 97D40, 97M80

1 Introduction

We present an educational project for schools and general visitors, which is a joint work of a research group of architects and mathematicians, implemented at Venaria Reale (Torino).

Our proposal concerns the interaction between mathematics and architecture; it consists of a set of activities: many of them are performed as guided tours, others as educational workshops.

The Venaria Royal Palace, one of the european royal residences, is a huge baroque architectural complex near Torino (northwestern Italy), dating from the second half of the XVII century onwards, where important architects such as Amadeo di Castellamonte, Filippo Juvarra, Benedetto Alfieri have left their mark [9]; it has undergone a great restoration and

[1]The first and the second author were partially supported by the "National Group for Algebraic and Geometric Structures, and their applications" (GNSAGA-INDAM)

redevelopment since 1999 and, after the restoration, it became, for the most part, a museum of Savoy court life and court rituals.

The choice of the site, where we could implement our project, is due to the fact that one of the author of the present paper (U. Zich) has been monitoring team Survey responsible of Venaria Reale restoration in the years 2000-2007 (see [6]); this has made available a large amount of materials, sources, surveys, photos, drawings, as well as a deep knowledge of the location.

The implementation of the project has been possible thanks to a partnership agreement between Politecnico di Torino and Centro Studi Venaria Reale.

Fig. 1. Venaria Royal Residence (partial view from www.google.it/maps) and details of Portrait Gallery, Great Gallery, Diana's Hall (photo by Centro Studi Venaria Reale), St. Hubert's Church (photo by U. Zich)

The activities included in our project aim to offer the visitors a new perception of the complex: the halls of the royal residence become classrooms where visitors discover architectural shapes and the geometry hidden in them and where origami is used as a tool in different contexts: for modeling, in order to observe and for observing a model, in order to understand geometry. A particular attention is devoted to roofing systems and to the analysis of various types of vaults and roofs.

2 The project

The project described below is the result of an experimentation conducted at the Royal Residence from december 2014 to december 2015. During this period a set of various activities was planned and tested on numerous classes from kindergarten up to high school and also on the general public. The purpose was in fact to tailor the proposed activities to students of all school levels and to a broad audience of visitors and families, taking into account the different levels of knowledge of participants, following a precise theory of teaching geometry, as we explain below.

At the end of each experimentation, we asked teachers and pupils to fill out a questionnaire in order to collect comments and suggestions and we could consequently adapt and reformulate some parts of our work.

Since last year, it is possible for everyone to book this type of tours at the Royal Residence of Venaria Reale (see [10]).

More in detail, the theoretical support (Descriptive Geometry and basics in Analytic and Differential Geometry) allows to propose various origami modeling activities (calibrated on the level of participants) and provides visitors the opportunity of touching with hands the shapes just observed, in the spirit of "learn by doing", in a new teaching language and knowledge sharing; as J. Piaget said (see [4]): *"Intelligence is a system of operations... In order that the child becomes able to arrange operations (both numerical and spatial ones), he need to have manipulated, ... experienced ... on a real material, on physical objects"*.

This way of introducing architectural elements, figures and geometric properties in an intuitive, practical and unconventional manner presents undoubted advantages even for people with disability (for example, visually impaired persons or people with spatial cognition deseases).

Most of the proposed activities are marked by gaming dynamics: from body modeling techniques, for the little children, to 2D/3D treasure hunts, for the older students.

Visits of this type take place in parallel with the normal guided tour which are proposed to all visitors: local guides take care of logistics operations, introduction, hystorical remarks, while visitors are guided by properly trained tutors, which are students of architecture, design and visual communication, mathematics and mathematics for engineering.

Visits are planned into different typologies, in accordance to the age of participants, following the pattern of 5 levels in learning Geometry, postulated by Pierre and Dina van Hiele in their doctoral dissertation in 1957 and still basic, nowadays, to set up any teaching of Geometry ([5], [1]).

Van Hiele theory puts forward a hierarchy of levels (that goes from small children to adults) defined as: visualization, analysis, abstraction, deduction, rigor.

Actually, these levels are a product of experience and instruction, rather than age; however it is known that, for example, most elementary schoolers are at the visualization or analysis level, while some middle-school children are at the deduction level and so on.

3 The tours

In this section we summarize the principal activities that students can do in the tours planned in our project.

All participants are welcomed outside the palace, where the guides explain a bit of history of the complex and the tutors show some shapes that visitors can see from the outside and that they will observe from the inside during the tour, recognizing various relationships.

Fig. 2. Welcome at the entrance of the Royal Residence (photo by U. Comollo)

The main internal spaces used for the visit are: the Portrait Gallery, the royal apartments, the kitchens, the Diana Hall and the Juvarra's Great Gallery (above the Portrait Gallery), which is a magnificent hall, probably the most famous hall of the complex.

Small children start their visit straightaway, while to the oldest ones an educational workshop is offered, in a teaching room, then their visit continues with the same scheme, adding at the end a passage through other halls.

3.1 Students aged 3 to 7 years: Visualization

Aim of the visit is teaching the children to observe and recognize the space where they are and the shapes that are above and around them, in accordance to van Hiele first level (visualization): at this level, in fact, children are able to identify prototypes of basic geometrical figures. The tour starts in the Portrait Gallery. The tutor invites people to look upward, so that they can see different "ceilings"; he tells them that they are called "vaults".

Visitors are in a hall with visible bricks and the tutor remarks that the bricks draw straight lines on the barrel vault and semicircles on the hemispherical vault, introducing the definitions of translational cylinder and rotational surfaces, then he makes the definitions concrete, involving children in some mimic game.

The proposed activities, based on origami modeling, take place in the wonderful Great Gallery.

Children are guided to fold simple 2D figures (explained in a mathematical language) in order to realize a bidimensional treasure hunt by identifying geometric elements in the architecture around them. At the end of the tour children receive an origami crown as a reward.

3.2 Students aged 8 to 11 years: Analysis

Aim of the visit is teaching the visitors not only to observe and recognize some architectural shapes and their geometry, but also to analyze them.

Fig. 3. Observing visible bricks in the Portrait Gallery (photo by Centro Studi Venaria Reale)

Fig. 4. Body gaming and origami folding (photos by U. Comollo)

In addition to the previous activities, students will fold a complete origami 3D model of the Great Gallery, which includes the underlying Portrait Gallery, the overlying hall and the extrados, the "skin" of the building.

With regard to the architecture, the observation of a 3D model allows also to show what the eye does not necessarily capture, due to the inaccessibility of certain areas.

For example, during the visit, it happens to cross rooms which are vertically aligned, without noticing their actual relative positions; a 3D model enables people to read proportional relationships, to understand the role of floor slabs and to perceive the geometries of intrados and extrados.

In a specific case, in the Great Gallery, we could widen the offer, allowing the access to the attic; this would provide a virtual vertical section of the building, actually difficult to perceive by a not expert person, but immediately comprehensible through origami modeling.

Fig. 5. A complete origami 3D model (photo by F. Manino) of three different levels of Juvarra's Gallery (views by Centro Studi Venaria Reale)

The students of this group are also involved in a 2D treasure hunt, but in a richer and broader way than the younger children: they have a map to orient themselves in space, a list of figures to look for and they must be able to place the required figures on the map.

Children are surprised by the fact that, starting from a simple square sheet, it is possible to obtain various figures with simple folds. In the example we show how to get a rhombus and a parallelogram by the same square sheet.

3.3 Students aged 12 to 13 years: Abstraction

In addition to activities of the previous group (in particular: 2D treasure hunt and folding of a complete origami 3D model of the Great Gallery), these students are more involved in the study of surfaces from an architectural and mathematical point of view.

Fig. 6. 2D treasure hunt (photos by U. Comollo)

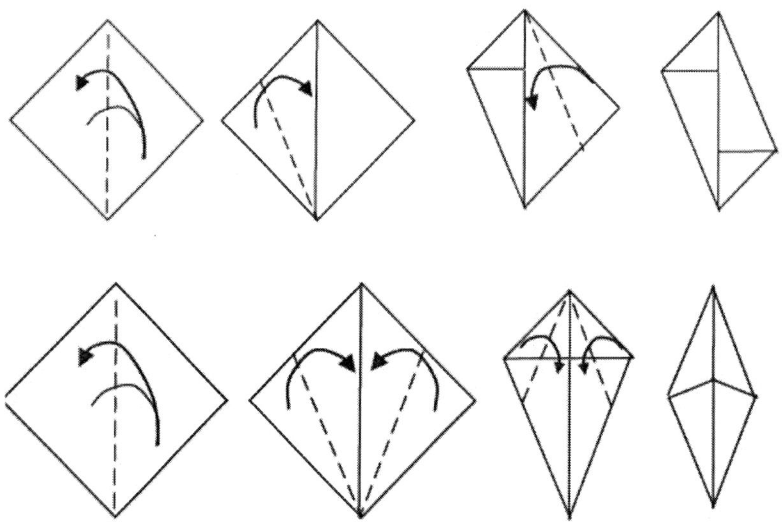

Fig. 7. How to fold a parallelogram and a rhombus

The definitions of translational cylinder and rotational surfaces are introduced, looking around during the visit. In a hall suitably chosen for its architectural configuration, a three-dimensional model will be shown to explain groin vaults and cloister vaults generated by the intersection of two cylinders.

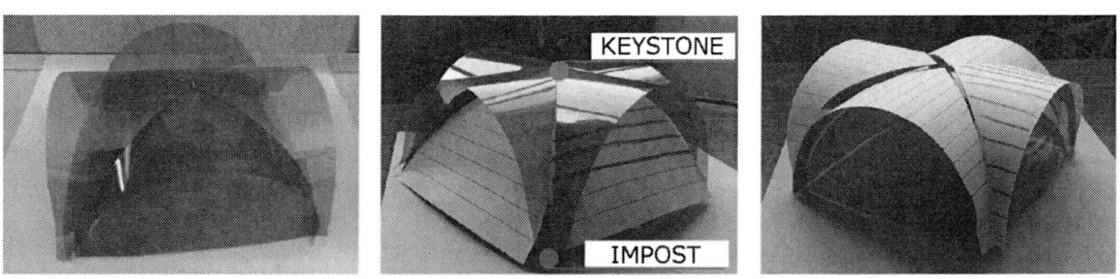

Fig. 8. Support structure for the visualization of groin and cloister vaults through the "dressing" of ruled surfaces (photos by the authors)

Then they are invited to the calculation of the aerolighting ratios and to play with the proportions of the main areas visited: a map with simplified shapes and approximated measures is distributed and students have to verify if the light area is at least 1/8 of the floor.

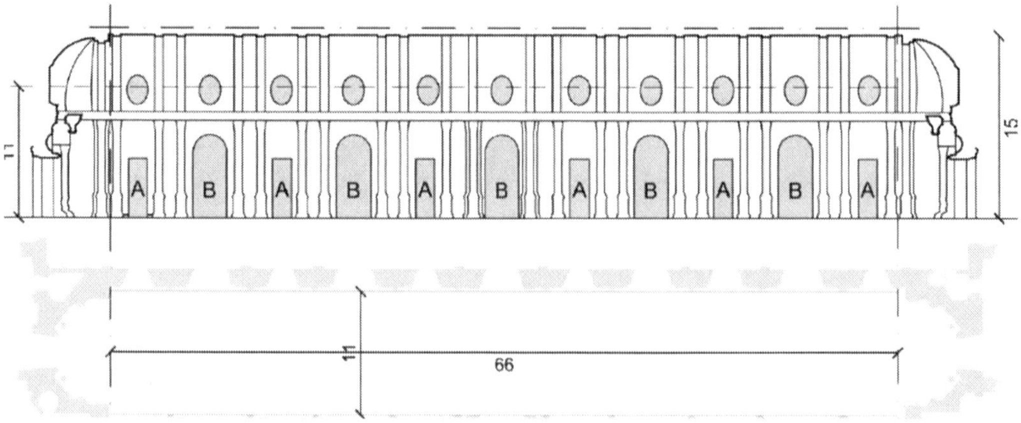

Fig. 9. Aerolighting ratios Sheet (graphic elaboration of a survey by Centro Studi Venaria Reale)

3.4 Students aged 14 to 16 years: Deduction

In agreement with van Hiele theory, students of this group are expected to look for relationships between properties of shapes and, in particular, they work on the concept of translational cylinder and, through direct experimentation with suitable educational materials, they study the intersection of a cylinder with a plane and its development, in order to understand the mathematical research that underlies an origami project; they become able to fold a 3D model of a barrel vault, and two cloister vaults (intersection of circular cylinders and prisms respectively).

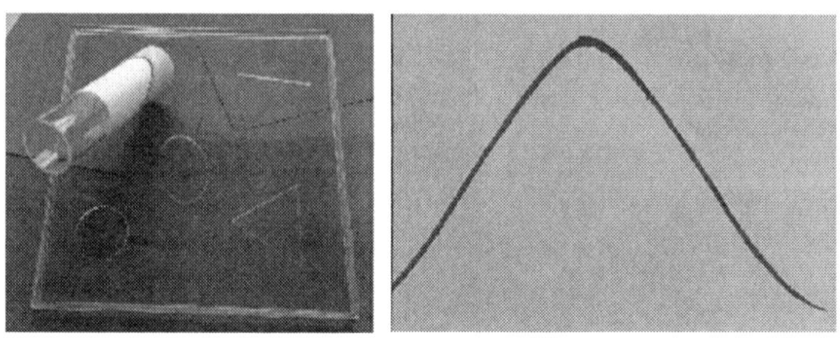

Fig. 10. Educational material to study the intersection of a cylinder with a plane (photos by the authors)

Instead of the 2D treasure hunt, a 3D treasure hunt is proposed, in which the tutor presents various types of vaults generated by cylinders intersections and the students have to mark on a map the vaults they recognize during the visit.

3.5 Students aged 17 to 19 years: Deduction and Rigor

Following van Hiele pattern, as in the previous cases, we can ask these students to deduce rigorously some mathematical fact about the architecture description.

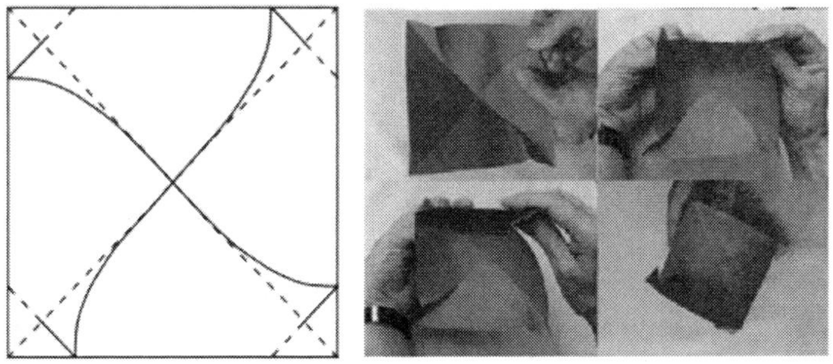

Fig. 11. How to fold an origami model of a cloister vault (model created thanks to collaboration of the company Bottega Fagnola, Torino [7], photos by the authors)

Fig. 12. Examples of origami models of vaults (see [3])

Suitable educational materials give them a way to experiment directly and design rigorously origami models of surfaces generated by intersection of cylinders, via the development of real surfaces.

These activities are supported by more sophisticated mathematical tools, such as topics in Differential Geometry: e.g. calculation of the length of a curve arc; beginning with a simple 3D model, the cylinders intersection curve is developed and in addition its equations are found, using a line integral (see for details [3]).

More detailed explanations from an architectural point of view are provided, for example the tutor remarks that the geometry of a vault becomes an element of recognition of an architect: every architect who worked at the Royal Residence has left his signature, choosing shapes, proportions, decorations that are his personal expression, but also expression of the age in which he lived; moreover from the mathematical point of view students are invited to look at sections of barrel vaults, recognizing the type of diretrix curves (ellipse, circle, polycentric curve) of a cylinder and they are also led to fold some conic as an envelope of tangents as

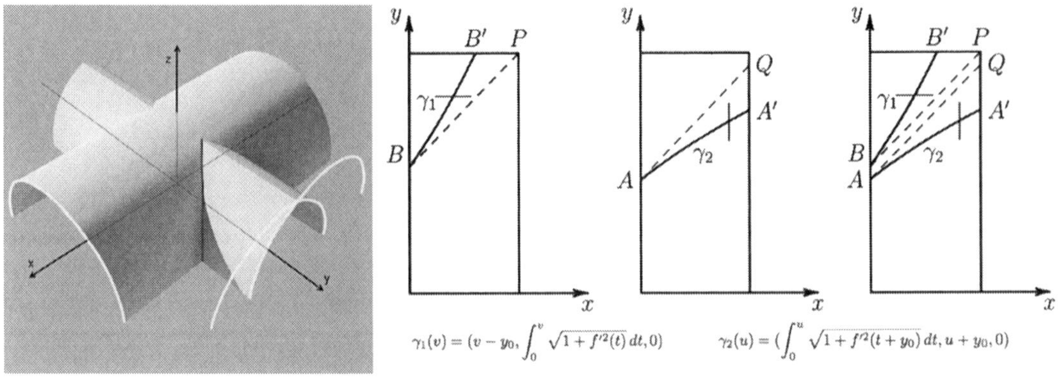

Fig. 13. Developments of the red curve in the first octant, as a curve on the first cylinder and as a curve on the second cylinder and their parametric representation

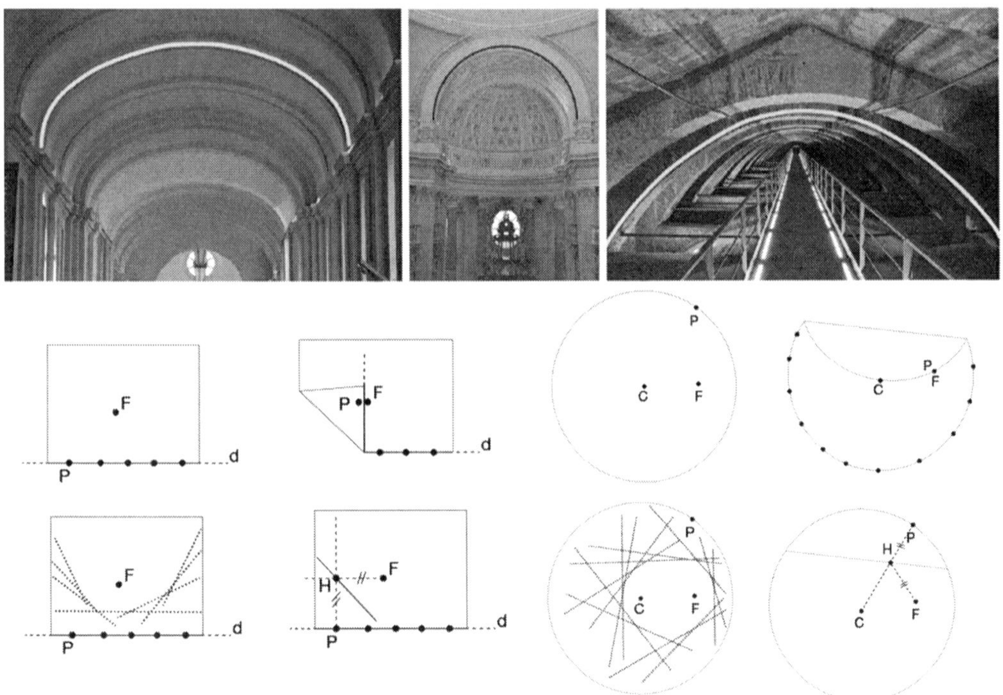

Fig. 14. Examples of barrel vault conic sections and how to fold some of them (photos by U. Comollo, drawings by V. Bosetto)

parabola and ellipse.

4 From experimentation to feasibility: conclusion and insights

The experiment done with schools has also led us to propose a tour adapted to families and adults, borrowing some materials from the project for schools. However planning these activities has allowed us to produce new and specific materials, which in turn has proved useful for schools.

For example, we produced a die-cut origami model cloister vault, in order that everyone can have the satisfaction of being able to fold it.

Fig. 15. Adult visitors during the tour, playing the treasure hunt (photo by U. Comollo)

The purposes of our project are not accomplished with the end of the visit.

In fact schools receive educational materials to continue and to deepen in their classroom the work begun at the Royal Residence, in the spirit of van Hiele model.

For example, a large white paper crease pattern of the Great Gallery with folding instructions, that kindergarten children may fold and color, folding instructions for a crown and a proportion game, crease pattern of various types of vaults.

But, most importantly, all participants begin to look with new eyes the reality that surrounds them and so they find out that everything around them is geometrically describable, although evidently with a certain degree of approximation due to the context.

References

[1] Crowley, M.L.: *The van Hiele model of the development of geometric thought*, In Learning and Teaching Geometry, K-12, 1987 Yearbook of the National Council of Teachers of Mathematics, Mary Montgomery Lindquist, Reston, Va.: National, 1987, pp. 1-16.

[2] Do Carmo, M.P.: *Differential geometry of curves and surfaces*, Pearson Education: Canada, 1976.

[3] Cumino, C., Frigerio, E. , Gallina, S., Spreafico, M. L., and Zich, U.: *Modeling Vaults in Origami: a Bridge between Mathematics and Architecture*, in Origami6: II. Technology, Art, Education, Proceeding of the sixth international meeeting on Origami Science, Mathematics and Education, Ed. AMS, 2015. AMS.

[4] Piaget, J.: *Avviamento al calcolo*, la Nuova Italia: Firenze, Italia, 1956.

[5] van Hiele, P.: *Structure and insight. A theory of Mathematics Education*, Academic Press: Orlando, USA, 1986.

[6] Volpiano, M., Zich, U.: *Scientific monitoring and documentation of the Venaria Reale restoration sites*, Proceedings of the XX International Symposium: International Cooperation to Save the World's Cultural Heritage, CIPA, 2005.

[7] URL: `http:www.bottegafagnola.it`

[8] URL: `https://flickrock.com/oschene`.

[9] URL: `http://www.lavenariareale.it`.

[10] URL: `http://www.lavenaria.it/web/it/educazione/scuole/progetti-insieme.html`.

Current address

Cumino Caterina, Associate Professor
Department of Mathematical Sciences, Politecnico di Torino
corso Duca degli Abruzzi 24, Torino, Italy
Tel. number: +39 011 090 7500, e-mail: caterina.cumino@polito.it

Spreafico Maria Luisa, Assistant Professor
Department of Mathematical Sciences, Politecnico di Torino
corso Duca degli Abruzzi 24, Torino, Italy
Tel. number: +39 011 090 7500, e-mail: maria.spreafico@polito.it

Zich Ursula, Assistant Professor
Department of Architecture and Design, Politecnico di Torino
viale Pier Andrea Mattioli 39, Torino, Italy
Tel. number: +39 011 090 4364, e-mail: ursula.zich@polito.it

Slovak University of Technology in Bratislava
Faculty of Mechanical Engineering
16th Conference on Applied Mathematics **APLIMAT 2017**
Proceedings

OPTIMALITY CONDITIONS FOR SCALAR L INEAR DELAYED DIFFERENTIAL EQUATIONS

DEMCHENKO Hanna (CZ), DIBLÍK Josef (CZ)

Abstract. In the contribution is considered a linear differential equation with a control function and a delay together with a problem on minimizing of an integral-type functional. The second Lyapunov's method is used to solve the problem.

Keywords: delayed differential equation, optimal control problem, Lyapunov-Krasovskii functional

Mathematics subject classification: Primary 34H15; Secondary 49J15

1 Introduction

The paper considers a problem of minimizing a functional

$$I = \min_u \int_{t_0}^{\infty} \left[\alpha x^2(t) + 2\beta x(t)x(t-\tau) + \gamma x^2(t-\tau) + \delta u^2(t)\right] \mathrm{d}t \qquad (1)$$

where $\alpha > 0$, $\alpha\gamma - \beta^2 > 0$, $\delta > 0$ for a linear differential equation with delay

$$x'(t) = ax(t) + bx(t-\tau) + cu(t), \ \ t \geq t_0 \qquad (2)$$

where $a, b, c \in \mathbb{R}$, $b \neq 0$, $\tau > 0$, $x\colon [t_0 - \tau, \infty) \to \mathbb{R}$ and $u\colon [t_0, \infty) \to \mathbb{R}$ is a control function. The existence of the improper integral in (1) is assumed, nevertheless, we note that their existence is guaranteed by the conditions of the below results.

To solve the problem, we are motivated by Malkin's approach, utilizing Lyapunov's second method applied to ordinary differential equations presented in [1]. We extend this approach to delayed differential equations in section 2 with the necessary modifications. Problem (2), (1) is considered in section 3. In spite of Malkin's approach making it possible to find optimal control functions for large classes of systems of ordinary linear differential equations and minimizing problems, the results that can be derived for the linear delayed differential equations considered are not so general and cover only narrow classes of problems.

2 Preliminaries

Consider a process, controlled by means of a function $u\colon [t_0,\infty) \to \mathbb{R}$, and assume that it can be represented by a differential equation of delayed type

$$x'(t) = f(t, x_t, u(t)), \quad t \geq t_0, \tag{3}$$

where $t_0 \in \mathbb{R}$, $f\colon D \to \mathbb{R}$,

$$D := \{(t, x_t, u) \in [t_0, \infty) \times C_\tau \times \mathbb{R},\ \|x\|_\tau < M,\ |u| < M\},$$

M is a positive constant, $C_\tau = C([-\tau, 0], \mathbb{R})$ is the space of continuous mappings from the interval $[-\tau, 0]$ into $\mathbb{R}$,

$$\|x(t)\|_\tau := \max_{\theta \in [-\tau, 0]} (|x(t+\theta)|), \quad t \geq t_0$$

and $x_t \in C_\tau$ is defined by $x_t(\theta) := x(t+\theta)$, $\theta \in [-\tau, 0]$.

We will assume that

1. $f(t, 0, 0) = 0$, $t \geq t_0$.
2. f is locally Lipschitzian in every bounded neighborhood of each point $(t^*, x, u) \in D$.

The problem under consideration is the following. We need to determine a function (control function) $u\colon [t_0, \infty) \to \mathbb{R}$ such that the zero solution $x(t) = 0$, $t \geq t_0$ of equation (3) is asymptotically stable and, for an arbitrary solution $x = x(t)$, $t \geqslant t_0 - \tau$ of equation (3) satisfying $|x| < M$, the integral

$$\int_{t_0}^{\infty} \omega\,(t, x_t, u(t))\,\mathrm{d}t \tag{4}$$

exists and attains a minimum value. Here, $\omega\colon D \to \mathbb{R}$ is a positive-definite functional.

Define an auxiliary functional $V\colon [t_0, \infty) \times C_\tau \to \mathbb{R}$. Below we assume that there exists the derivative $\mathrm{d}V(t, x_t)/\mathrm{d}t$ of functional $V(t, x_t)$ along the trajectories of system (3). Repeating the well-known definitions, we say that functional V is positive definite if there exists a continuous nondecreasing function w on $[0, \infty)$ which is zero at 0 and positive on $(0, \infty)$ such that

$$V(t, x_t) \geq w(\|x_t\|_\tau), \quad t \geq t_0$$

where x is assumed to be defined on $[t_0 - \tau, \infty)$. The functional V has an infinitesimal upper bound if there exists a continuous nondecreasing function W on $[0, \infty)$ which is zero at 0 and positive on $(0, \infty)$ such that

$$V(t, x_t) \leq W(\|x_t\|_\tau), \quad t \geq t_0.$$

Moreover, define an auxiliary functional $B\colon D_1 \to \mathbb{R}$,

$$D_1 := \{(V, t, x_t, u) \in \mathbb{R} \times [t_0, \infty) \times C_\tau \times \mathbb{R},\ \|x\|_\tau < M,\ |u| < M\}$$

by the formula

$$B\,(V, t, x_t, u) := \frac{\mathrm{d}V(t, x_t)}{\mathrm{d}t} + \omega(t, x_t, u(t)). \tag{5}$$

The following theorem was motivated by a similar theorem for non-delay systems [1, Theorem IV]. We omit its proof.

Theorem 1 *Assume that, for differential equation* (3)*, there exists a positive definite functional V having an infinitesimal upper bound and a vector-function $u_0\colon [t_0, \infty) \to \mathbb{R}$, $\|u_0(t)\| \le M$, $t \ge t_0$ such that*

i) Functional $\omega(t, x_t, u_0(t))$ is positive-definite for every $t \ge t_0$, $\|x_t\|_\tau < M$.

ii) Identity $B(V, t, x_t, u_0(t)) \equiv 0$ holds on $[t_0, \infty)$ for every solution $x\colon [t_0, \infty) \to \mathbb{R}$ of equation (3) *where $u = u_0$.*

iii) Inequality $B(V, t, x_t, u(t)) \ge 0$ holds on $[t_0, \infty)$ for every solution $x\colon [t_0, \infty) \to \mathbb{R}$ of equation (3) *and every function $u\colon [t_0, \infty) \to \mathbb{R}$ with $|u(t)| < M$, $t \in [t_0, \infty)$.*

Then, the function u_0 is a solution of the problem (3), (4) *and*

$$\int_{t_0}^{\infty} \omega(t, x_t, u_0(t))\mathrm{d}t = \min_u \left[\int_{t_0}^{\infty} \omega(t, x_t, u(t))\mathrm{d}t\right] = V(t_0, x_{t_0}).$$

3 Equations with single delay

Let us consider equation (2) together with a quality criterion (1). Let $t_0 = 0$, then

$$I = \min_u \int_0^{\infty} [\alpha x^2(t) + 2\beta x(t)x(t-\tau) + \gamma x^2(t-\tau) + \delta u^2(t)]dt. \tag{6}$$

Theorem 2 *If for optimal control problem* (2), (6) *a Lyapunov-Krasovskii functional is used in the form*

$$V(t, x_t) = hx^2(t) + d\int_{t-\tau}^{t} x^2(s)ds, \ h > 0, \ d > 0,$$

with $h = -\beta/b$ ($\beta b < 0$), $d = \gamma$ and

$$\delta(2ha + d + \alpha) - h^2c^2 = 0,$$

then an optimal stabilization control function $u_0(t)$ equals

$$u_0(t) = -\frac{hc}{\delta}x(t). \tag{7}$$

Equation (2) *with $u(t) = u_0(t)$, i.e.,*

$$x'(t) = ax(t) + bx(t-\tau) + cu_0(t)$$

is asymptotically stable and

$$V(t_0, x_{t_0}) = \int_{t_0}^{\infty} \left[\alpha x^2(t) + 2\beta x(t)x(t-\tau) + \gamma x^2(t-\tau) + \delta u_0^2(t)\right] \mathrm{d}t =$$

$$= \min_u \int_{t_0}^{\infty} \left[\alpha x^2(t) + 2\beta x(t)x(t-\tau) + \gamma x^2(t-\tau) + \delta u^2(t)\right] \mathrm{d}t.$$

Proof. We utilize Theorem 1. In accordance with condition $iii)$ of Theorem 1, we analyze the non-negativity of the expression B given by (5), i.e.,

$$B(V,t,x_t,u(t)) = 2hx(t)[ax(t)+bx(t-\tau)+cu(t)]+ \\ +d[x^2(t)-x^2(t-\tau)]+\alpha x^2(t)+2\beta x(t)x(t-\tau)+\gamma x^2(t-\tau)+\delta u^2(t) \geq 0.$$

Simplifying the last expression, we get

$$B(V,t,x_t,u(t)) = (2ha+d+\alpha)x^2(t)+(\gamma-d)x^2(t-\tau)+ \\ +(2hb+2\beta)x(t)x(t-\tau)+2hcx(t)u(t)+\delta u^2(t) \geq 0.$$

The conditions of non-negativity are the following:

$$2ha+d+\alpha > 0, \tag{8}$$

$$(2ha+d+\alpha)(\gamma-d)-(hb+\beta)^2 \geq 0, \tag{9}$$

$$\delta(2ha+d+\alpha)(\gamma-d)-h^2c^2(\gamma-d)^2-\delta(hb+\beta)^2 \geq 0. \tag{10}$$

For $u(t)=u_0(t)$, by $ii)$, we have

$$B(V,t,x_t,u_0(t)) = (2ha+d+\alpha)x^2(t)+(\gamma-d)x^2(t-\tau)+ \\ +(2hb+2\beta)x(t)x(t-\tau)+2hcx(t)u_0(t)+\delta u_0^2(t) = 0. \tag{11}$$

Looking for an extremum of the last expression, we get

$$B'_{u_0}(V,t,x_t,u_0(t)) = 2hcx(t)+2\delta u_0(t) = 0,$$

i.e.,

$$u_0(t) = -\frac{hc}{\delta}x(t),$$

which is the minimum of the function B because

$$B''_{u_0u_0}(V,t,x_t,u_0(t)) = 2\delta > 0.$$

For (11) to be valid, i.e.,

$$\left(2ha+d+\alpha-\frac{h^2c^2}{\delta}\right)x^2(t)+(\gamma-d)x^2(t-\tau)+(2hb+2\beta)x(t)x(t-\tau)=0$$

we obtain

$$\delta(2ha+d+\alpha)-h^2c^2=0, \\ d=\gamma, \\ h=-\frac{\beta}{b}.$$

So the left-hand side of (8) becomes

$$2ha+d+\alpha=\frac{h^2c^2}{\delta}>0,$$

and (9)-(10) are true because their left-hand sides are equal to zero.

We obtain an optimal control function

$$u_0(t) = -\frac{hc}{\delta}x(t).$$

□

Example 3 *Consider equation* (2) *with* $a = 1$, $b = -1$, $c = 3$, *i.e.,*

$$\dot{x}(t) = x(t) - x(t - \tau) + 3u(t)$$

with a quadratic quality criterion (6) *where* $\alpha = 2 > 0$, $\beta = 1$, $\gamma = 2$, $\delta = 1 > 0$ *(here* $\alpha\gamma - \beta^2 = 3 > 0$*), i.e.,*

$$I = \int_0^\infty [2x^2(t) + 2x(t)x(t-\tau) + 2x^2(t-\tau) + u^2(t)]\mathrm{d}t.$$

By formula (7), *the optimal stabilization control function*

$$u_0(t) = -\frac{hc}{\delta}x(t) = -3x(t)$$

exists and

$$\int_0^\infty [2x^2(t) + 2x(t)x(t-\tau) + 2x^2(t-\tau) + u_0^2(t)]\mathrm{d}t = x^2(t) + 2\int_{-\tau}^0 x^2(s)\mathrm{d}s.$$

Acknowledgement

The authors were supported by the Grant FEKT-S-14-2200 of Faculty of Electrical Engineering and Communication, Brno University of Technology and the first author was supported by the Visegrad/V4EaP Scholarship 51600029.

References

[1] Malkin, I.G.: *Theory of Stability of Motion, Second revised edition (Russian)*, Moscow, Nauka Publisher, 1966. 530 pp.

Current address

Demchenko Hanna, Mgr.
Faculty of Electrical Engineering and Communication, Brno University of Technology
Technicka 2848/8, Zabovresky, 61600, Brno, Ceska republika
Tel. number: +420 54114 6097, e-mail: xdemch02@stud.feec.vutbr.cz

Diblik Josef, prof. RNDr. DrSc.
Faculty of Electrical Engineering and Communication, Brno University of Technology
Technicka 2848/8, Zabovresky, 61600, Brno, Ceska republika
Tel. number: +420 54114 6064, e-mail: diblik@feec.vutbr.cz

Slovak University of Technology in Bratislava
Faculty of Mechanical Engineering
16th Conference on Applied Mathematics **APLIMAT 2017**
Proceedings

SOLVING A HIGHER-ORDER LINEAR DISCRETE EQUATION

DIBLÍK Josef (CZ), MENCÁKOVÁ Kristýna (CZ)

Abstract. In this paper we will find two linearly independent solutions of a linear discrete homogenous equation of the $(m + 2)$-nd order. These solutions are a special functions called delayed discrete cosine and delayed discrete sine. There is given a representation of an initial problem and a nonhomogenous linear equation.

Keywords: Delayed cosine, delayed sine, discrete equation

Mathematics subject classification: Primary 39A06; Secondary 39A05

1 Introduction

Below it is used following notation: for integers s, r, $s \leq r$ define $\mathbb{Z}_s^r := \{s, s+1, \dots r\}$. Similarly symbols $\mathbb{Z}_{-\infty}^r$, $\mathbb{Z}_s^\infty$ are defined.

In this paper we consider a linear discrete homogeneous equation of the order $(m + 2)$

$$\Delta^2 x(k) + b^2 x(k - m) = 0, \quad k \in \mathbb{Z}_0^\infty, \tag{1}$$

where $b \neq 0$ and $m \in \mathbb{N}$.

We will find two linearly independent solutions of (1) as a special functions called delayed discrete cosine and delayed discrete sine. A representation of a solution of Cauchy problem is given as well. Previously, a similar problem for discrete linear systems

$$\Delta x(k) = Bx(k - m) + f(k), \quad k \in \mathbb{Z}_0^\infty,$$

where m is a fixed integer, B is a constant $n \times n$ matrix, $f : \mathbb{Z}_0^\infty \to \mathbb{R}^n$, was considered in [1]. A fundamental matrix was constructed as a delayed discrete matrix exponential. Later, in the paper [2], there was consider a differential system of the second order with delay

$$\ddot{x}(t) + \Omega^2 x(t-\tau) = 0, \tag{2}$$

where $\tau > 0$ and Ω is a constant $n \times n$ matrix. A fundamental matrix for (2) was constructed as a special matrix functions called delayed matrix cosine and delayed matrix sine. Such special matrix functions served as a motivation for present investigation.

2 Delayed discrete cosine and sine

In this part there we define auxiliary discrete functions – delayed cosine, delayed sine and some of their properties are proved.

Definition 1. Delayed cosine is defined as:

$$\mathrm{Cos}_m bk := \begin{cases} 0 \quad \text{if} \quad k \in \mathbb{Z}_{-\infty}^{-m-1}, \\ 1 \quad \text{if} \quad k \in \mathbb{Z}_{-m}^{1}, \\ 1 - b^2 \cdot \binom{k}{2} \quad \text{if} \quad k \in \mathbb{Z}_2^{(m+2)+1}, \\ 1 - b^2 \cdot \binom{k}{2} + b^4 \cdot \binom{k-m}{4} \quad \text{if} \quad k \in \mathbb{Z}_{(m+2)+2}^{2(m+2)+1}, \\ \dots \\ 1 - b^2 \cdot \binom{k}{2} + b^4 \cdot \binom{k-m}{4} - b^6 \cdot \binom{k-2m}{6} + \dots + (-1)^l b^{2l} \cdot \binom{k-(l-1)m}{2l} \\ \text{if} \quad k \in \mathbb{Z}_{(l-1)(m+2)+2}^{l(m+2)+1}, \quad \ell = 0, 1, 2, \dots, \\ \dots \quad . \end{cases}$$

Definition 2. Delayed sine is defined as:

$$\mathrm{Sin}_m bk := \begin{cases} 0 \quad \text{if} \quad k \in \mathbb{Z}_{-\infty}^{-m-1}, \\ b \cdot \binom{k+m}{1} \quad \text{if} \quad k \in \mathbb{Z}_{-m}^{1}, \\ b \cdot \binom{k+m}{1} - b^3 \cdot \binom{k}{3} \quad \text{if} \quad k \in \mathbb{Z}_2^{(m+2)+1}, \\ b \cdot \binom{k+m}{1} - b^3 \cdot \binom{k}{3} + b^5 \cdot \binom{k-m}{5} \quad \text{if} \quad k \in \mathbb{Z}_{(m+2)+2}^{2(m+2)+1}, \\ \dots \\ b \cdot \binom{k+m}{1} - b^3 \cdot \binom{k}{3} + b^5 \cdot \binom{k-m}{5} + \dots + (-1)^l b^{2l+1} \cdot \binom{k-(l-1)m}{2l+1} \\ \text{if} \quad k \in \mathbb{Z}_{(l-1)(m+2)+2}^{l(m+2)+1}, \quad \ell = 0, 1, 2, \dots, \\ \dots \quad . \end{cases}$$

In the following theorem, there will be given basic properties of $\mathrm{Cos}_m bk$ and $\mathrm{Sin}_m bk$.

Theorem 1. *For* $\mathrm{Cos}_m bk$, $\mathrm{Sin}_m bk$ *and any* $k \in \mathbb{Z}$ *are hold:*

$$\Delta \mathrm{Cos}_m bk = -b \cdot \mathrm{Sin}_m b(k-m), \tag{3}$$

$$\Delta \mathrm{Sin}_m bk = b \cdot \mathrm{Cos}_m bk. \tag{4}$$

Proof. In the proof we will use well-known formula

$$\binom{p+1}{q} - \binom{p}{q} = \binom{p}{q-1}, \tag{5}$$

where p, q are whole numbers.

At the first we prove the formula (3). The proof is divided into three parts.
a) If $k \in \mathbb{Z}_{-\infty}^{-m-2}$:

$$\Delta \mathrm{Cos}_m bk = \mathrm{Cos}_m b(k+1) - \mathrm{Cos}_m bk = 0 - 0 = 0 = -b \cdot \mathrm{Sin}_m b(k-m).$$

For these k formula (3) obviously holds.

b) If $k \in \mathbb{Z}_{(l-1)(m+2)+2}^{l(m+2)}$ we get

$$\Delta \mathrm{Cos}_m bk = \mathrm{Cos}_m b(k+1) - \mathrm{Cos}_m bk =$$

$$= 1 - b^2 \cdot \binom{k+1}{2} + b^4 \cdot \binom{k+1-m}{4} - b^6 \cdot \binom{k+1-2m}{6} + \dots$$

$$+(-1)^l b^{2l} \cdot \binom{k+1-(l-1)m}{2l} - \left[1 - b^2 \cdot \binom{k}{2} + b^4 \cdot \binom{k-m}{4} - b^6 \cdot \binom{k-2m}{6} + \right.$$

$$\left. \dots + (-1)^l b^{2l} \cdot \binom{k-(l-1)m}{2l}\right] =$$

$$= 1 - 1 - b^2 \cdot \left[\binom{k+1}{2} - \binom{k}{2}\right] + b^4 \cdot \left[\binom{k-m+1}{4} - \binom{k-m}{4}\right] - b^6 \cdot \left[\binom{k-2m+1}{6} - \right.$$

$$\left. \binom{k-2m}{6}\right] + \dots + (-1)^l b^{2l} \cdot \left[\binom{k-(l-1)m+1}{2l} - \binom{k-(l-1)m}{2l}\right] =$$

[we use formula (5)]

$$= 0 - b^2 \cdot \binom{k}{1} + b^4 \cdot \binom{k-m}{3} - b^6 \cdot \binom{k-2m}{5} + \dots + (-1)^l b^{2l} \cdot \binom{k-(l-1)m}{2l-1} =$$

$$= -b \cdot \left[b \cdot \binom{k}{1} - b^3 \cdot \binom{k-m}{3} + b^5 \cdot \binom{k-2m}{5} + \dots \right.$$

$$\left. +(-1)^{l-1} b^{2l-1} \cdot \binom{k-(l-1)m}{2l-1}\right] =$$

$$= -b \cdot \left[b \cdot \binom{k-m+m}{1} - b^3 \cdot \binom{k-m}{3} + b^5 \cdot \binom{k-m-m}{5} + \dots \right.$$

$$+(-1)^{l-1}b^{2(l-1)+1}\cdot\binom{k-(l-2)m-m}{2(l-1)+1}\Bigg] =$$

$= -b\cdot \mathrm{Sin}_m b(k-m)$.
In this case formula (3) holds too.

c) Let $k = l(m+2)+1$. Then

$$\Delta \mathrm{Cos}_m bk = \mathrm{Cos}_m b(k+1) - \mathrm{Cos}_m bk =$$

$$= 1 - b^2\cdot\binom{k+1}{2} - b^4\cdot\binom{k+1-m}{4} + \dots$$

$$+(-1)^l b^{2l}\cdot\binom{k+1-(l-1)m}{2l} + (-1)^{l+1}b^{2(l+1)}\cdot\binom{k+1-lm}{2(l+1)}$$

$$-\left[1 - b^2\cdot\binom{k}{2} - b^4\cdot\binom{k-m}{4} + \dots + (-1)^l b^{2l}\cdot\binom{k-(l-1)m}{2l}\right] =$$

$$= 1 - 1 - b^2\cdot\left[\binom{k+1}{2} - \binom{k}{2}\right] + b^4\cdot\left[\binom{k-m+1}{4} - \binom{k-m}{4}\right] + \dots$$

$$+(-1)^l b^{2l}\cdot\left[\binom{k-(l-1)m+1}{2l} - \binom{k-(l-1)m}{2l}\right] + (-1)^{l+1}b^{2(l+1)}\cdot\binom{k-lm+1}{2(l+1)} =$$

[utilising formula (5)]

$$= -b^2\cdot\binom{k}{1} + b^4\cdot\binom{k-m}{3} + \dots$$

$$+(-1)^l b^{2l}\cdot\binom{k-(l-1)m}{2l-1} + (-1)^{2l+1}b^{2(l+1)}\cdot\binom{k-lm+1}{2(l+1)} =$$

$$= -b\cdot\left[b\cdot\binom{k}{1} - b^3\cdot\binom{k-m}{3} + \dots\right.$$

$$\left.+(-1)^{l-1}b^{2l-1}\cdot\binom{k-(l-1)m}{2l-1} + (-1)^l b^{2l+1}\cdot\binom{k-lm+1}{2l+2}\right] =$$

$$= -b\cdot\left[b\cdot\binom{k+m-m}{1} - b^3\cdot\binom{k-m}{3} + \dots\right.$$

$$\left.+(-1)^{l-1}b^{2l-1}\cdot\binom{k+m-lm}{2l-1} + (-1)^l b^{2l+1}\cdot\binom{k-lm+1}{2l+2}\right] = (*).$$

The last binomial coefficient can be decomposed by (5):

$$\binom{k-lm+1}{2l+2} = \binom{k-lm}{2l+1} + \binom{k-lm}{2l+2}$$
$$= \binom{k-m-lm+m}{2l+1} + \binom{l(m+2)+1-lm}{2l+2}$$

$$=\binom{k-m-(l-1)m}{2l+1}+\binom{lm+2l+1-lm}{2l+2}$$
$$=\binom{k-m-(l-1)m}{2l+1}+\binom{2l+1}{2l+2}=\binom{k-m-(l-1)m}{2l+1}+0$$
$$=\binom{k-m-(l-1)m}{2l+1}.$$

Then

$$(*)=-b\cdot\left[b\cdot\binom{k+m-m}{1}-b^3\cdot\binom{k-m}{3}+\dots\right.$$
$$\left.+(-1)^{l-1}b^{2l-1}\cdot\binom{k+m-lm}{2l-1}+(-1)^l b^{2l+1}\cdot\binom{k-m-(l-1)m}{2l+1}\right]=$$
$$=-b\cdot \mathrm{Sin}_m b(k-m).$$

So formula (3) holds in this case and we proved it for each $k\in\mathbb{Z}$.

Now we prove formula (4), again in three parts.
a) If $k\in\mathbb{Z}_{-\infty}^{-m-2}$

$$\Delta\mathrm{Sin}_m bk=\mathrm{Sin}_m b(k+1)-\mathrm{Sin}_m bk=0-0=0=b\cdot\mathrm{Cos}_m bk$$

and the formula obviously holds.

b) If $k\in\mathbb{Z}_{l(m+2)+2}^{(l+1)(m+2)}$ then

$$\Delta\mathrm{Sin}_m bk=\mathrm{Sin}_m b(k+1)-\mathrm{Sin}_m bk=$$
$$=b\cdot\binom{k+m+1}{1}-b^3\cdot\binom{k+1}{3}+b^5\cdot\binom{k+1-m}{5}+\dots$$
$$+(-1)^l b^{2l+1}\cdot\binom{k+1-(l-1)m}{2l+1}$$
$$-\left[b\cdot\binom{k+m}{1}-b^3\cdot\binom{k}{3}+b^5\cdot\binom{k-m}{5}+\dots+(-1)^l b^{2l+1}\cdot\binom{k-(l-1)m}{2l+1}\right]=$$

$$=b\cdot\left[\binom{k+m+1}{1}-\binom{k+m}{1}\right]-b^3\cdot\left[\binom{k+1}{3}-\binom{k}{3}\right]+b^5\cdot\left[\binom{k-m+1}{5}\right.$$
$$\left.-\binom{k-m}{5}\right]+\dots+(-1)^l b^{2l+1}\cdot\left[\binom{k-(l-1)m+1}{2l+1}-\binom{k-(l-1)m}{2l+1}\right]=$$

[we use formula (5)]

$$=b\cdot\binom{k+m}{0}-b^3\cdot\binom{k}{2}+b^5\cdot\binom{k-m}{4}+\dots+(-1)^l b^{2l+1}\cdot\binom{k-(l-1)m}{2l}=$$

$$= b \cdot \left[1 - b^2 \cdot \binom{k}{2} + b^4 \cdot \binom{k-m}{4} + \cdots + (-1)^l b^{2l} \cdot \binom{k-(l-1)m}{2l} \right] =$$

$= b \cdot \mathrm{Cos}_m bk.$
In this case formula (4) holds too.

c) Let $k = l(m+2)+1$ we get

$$\Delta \mathrm{Sin}_m bk = \mathrm{Sin}_m b(k+1) - \mathrm{Sin}_m bk =$$

$$= b \cdot \binom{k+m+1}{1} - b^3 \cdot \binom{k+1}{3} + b^5 \cdot \binom{k+1-m}{5} + \dots$$
$$+(-1)^l b^{2l+1} \cdot \binom{k+1-(l-1)m}{2l+1} + (-1)^{l+1} b^{2(l+1)+1} \cdot \binom{k+1-lm}{2(l+1)+1} -$$
$$- \left[b \cdot \binom{k+m}{1} - b^3 \cdot \binom{k}{3} + b^5 \cdot \binom{k-m}{5} + \cdots + (-1)^l b^{2l+1} \cdot \binom{k-(l-1)m}{2l+1} \right] =$$

$$= b \cdot \left[\binom{k+m+1}{1} - \binom{k+m}{1} \right] - b^3 \cdot \left[\binom{k+1}{3} - \binom{k}{3} \right] + b^5 \cdot \left[\binom{k-m+1}{5} \right.$$
$$\left. - \binom{k-m}{5} \right] + \cdots + (-1)^l b^{2l+1} \cdot \left[\binom{k-(l-1)m+1}{2l+1} - \binom{k-(l-1)m}{2l+1} \right]$$
$$+(-1)^{l+1} b^{2(l+1)+1} \cdot \binom{k-lm+1}{2(l+1)+1} =$$

[utilising formula (5)]

$$= b \cdot \binom{k+m}{0} - b^3 \cdot \binom{k}{2} + b^5 \cdot \binom{k-m}{4} + \dots$$
$$+(-1)^l b^{2l+1} \cdot \binom{k-(l-1)m}{2l} + (-1)^{l+1} b^{2l+3} \cdot \binom{k-lm+1}{2l+3} =$$

$$= b \cdot \left[1 - b^2 \cdot \binom{k}{2} + b^4 \cdot \binom{k-m}{4} + \dots \right.$$
$$\left. +(-1)^l b^{2l} \cdot \binom{k-(l-1)m}{2l} \right] + (-1)^{l+1} b^{2l+3} \cdot \binom{lm+2l+1-lm+1}{2l+3} =$$

$= b \cdot \mathrm{Cos}_m bk,$

because the last binomial coefficient

$$\binom{lm+2l+1-lm+1}{2l+3} = \binom{2l+2}{2l+3} = 0.$$

So formula (4) holds in the last case and for every $k \in \mathbb{Z}$ too.

□

Remark 1. From formulas (3), (4) follows that for any $k \in \mathbb{Z}$:

$$\Delta^2 \text{Cos}_m bk = -b^2 \cdot \text{Cos}_m b(k-m),$$

$$\Delta^2 \text{Sin}_m bk = -b^2 \cdot \text{Sin}_m b(k-m),$$

i.e. $\text{Cos}_m bk$ and $\text{Sin}_m bk$ are linearly independent solutions of equation (1).

Obviously, the expression

$$x(k) = C_1 \cdot \text{Cos}_m bk + C_2 \cdot \text{Sin}_m bk, \quad k \geq 2,$$

where C_1, C_2 are arbitrary constant, is a family of solutions of equation (1).

Theorem 2. *Solution of initial problem* (1), (6), *where*

$$x(k) = \varphi(k), k = -m, \ldots, 1, \tag{6}$$

is expressed by formula

$$x(k) = (\text{Cos}_m bk)\,\varphi(-m) + \\ +b^{-1}\left[(\text{Sin}_m bk)\,\Delta\varphi(-m) + \sum_{j=-m+1}^{0} \text{Sin}_m b(k-m-j)\cdot\Delta^2\varphi(j-1)\right], \tag{7}$$

where $k \in \mathbb{Z}_{-m}^{\infty}$.

Proof. According to Remark 1 it is easy to see that $x(k)$ from formula (7) is a solution of equation (1) for any $k \in \mathbb{Z}_2^{\infty}$.

Now we prove that formula (7) satisfy also conditions (6).

Let $k \in \mathbb{Z}_{-m}^{1}$. Then $\text{Cos}_m bk = 1$, $\text{Sin}_m bk = b \cdot \binom{k+m}{1}$ and

$$x(k) = \varphi(-m) + b^{-1}\cdot b\cdot\binom{k+m}{1}\cdot\Delta\varphi(-m) \\ +b^{-1}\cdot\sum_{j=-m+1}^{0} \text{Sin}_m b(k-m-j)\cdot\Delta^2\varphi(j-1).$$

Now we divide the proof into three cases.

a) For $k = -m$ we get

$$x(-m) = \varphi(-m) + \binom{0}{1}\Delta\varphi(-m) + b^{-1}\cdot\sum_{j=-m+1}^{0} \text{Sin}_m k(-2m-j)\Delta^2\varphi(j-1) \\ = \varphi(-m) + 0 + 0 = \varphi(-m).$$

b) If $k = -m+1$, we have

$$x(-m+1) = \varphi(-m) + \binom{1}{1}\Delta\varphi(-m) + b^{-1}\cdot\sum_{j=-m+1}^{0} \text{Sin}_m k(-2m+1-j)\Delta^2\varphi(j-1)$$

$$= \varphi(-m) + \varphi(-m+1) - \varphi(-m) + 0 = \varphi(-m+1).$$

c) Finaly let $k \in \mathbb{Z}^1_{-m+2}$. Then

$$x(k) = \varphi(-m) + \binom{k+m}{1} \cdot [\varphi(-m+1) - \varphi(-m)]$$

$$+b^{-1}\left[\sum_{j=-m+1}^{k-1} \mathrm{Sin}_m b(k-m-j)\Delta^2\varphi(j-1) + \sum_{j=k}^{0} \mathrm{Sin}_m b(k-m-j)\Delta^2\varphi(j-1)\right] = (*).$$

The second sum is composed of all null terms. So we have

$$(*) = \varphi(-m) - \binom{k+m}{1}\varphi(-m) + \binom{k+m}{1}\varphi(-m+1)$$

$$+b^{-1} \cdot \sum_{j=-m+1}^{k-1} \mathrm{Sin}_m b(k-m-j) \cdot \Delta^2\varphi(j-1) =$$

$$= \binom{1-k-m}{1}\varphi(-m) + \binom{k+m}{1}\varphi(-m+1) + b^{-1} \cdot \big[\mathrm{Sin}_m b(k-m+m-1)\Delta^2\varphi(-m)$$

$$+\mathrm{Sin}_m b(k-m+m-2)\Delta^2\varphi(-m+1) + \mathrm{Sin}_m b(k-m+m-3)\Delta^2\varphi(-m+2) + \ldots$$

$$+\mathrm{Sin}_m b(k-m-k+3)\Delta^2\varphi(k-4) + \mathrm{Sin}_m b(k-m-k+2)\Delta^2\varphi(k-3)$$

$$+\mathrm{Sin}_m b(k-m-k+1)\Delta^2\varphi(k-2)\big] =$$

$$= \binom{1-k-m}{1}\varphi(-m) + \binom{k+m}{1}\varphi(-m+1) + b^{-1} \cdot \big[\mathrm{Sin}_m b(k-1)\Delta^2\varphi(-m)$$

$$+\mathrm{Sin}_m b(k-2)\Delta^2\varphi(-m+1) + \mathrm{Sin}_m b(k-3)\Delta^2\varphi(-m+2) + \ldots$$

$$+\mathrm{Sin}_m b(-m+3)\Delta^2\varphi(k-4) + \mathrm{Sin}_m b(-m+2)\Delta^2\varphi(k-3)$$

$$+\mathrm{Sin}_m b(-m+1)\Delta^2\varphi(k-2)\big] =$$

$$= \binom{1-k-m}{1}\varphi(-m) + \binom{k+m}{1}\varphi(-m+1) + b^{-1} \cdot \left[b \cdot \binom{k-1+m}{1}\Delta^2\varphi(-m)\right.$$

$$+b \cdot \binom{k-2+m}{1}\Delta^2\varphi(-m+1) + b \cdot \binom{k-3+m}{1}\Delta^2\varphi(-m+2) + \ldots$$

$$\left.+b \cdot \binom{3}{1}\Delta^2\varphi(k-4) + b \cdot \binom{2}{1}\Delta^2\varphi(k-3) + b \cdot \binom{1}{1}\Delta^2\varphi(k-2)\right] =$$

$$= (1-k-m)\varphi(-m) + (k+m)\varphi(-m+1) + b^{-1} \cdot b \cdot \big[(k-1+m)\Delta^2\varphi(-m)$$

$$+(k-2+m)\Delta^2\varphi(-m+1) + (k-3+m)\Delta^2\varphi(-m+2) + \ldots$$

$$+3\Delta^2\varphi(k-4) + 2\Delta^2\varphi(k-3) + \Delta^2\varphi(k-2)\big] =$$

$$= (1-k-m)\varphi(-m) + (k+m)\varphi(-m+1)$$

$$
\begin{aligned}
&+(k-1+m)\left[\varphi(-m+2)-2\varphi(-m+1)+\varphi(-m)\right]\\
&+(k-2+m)\left[\varphi(-m+3)-2\varphi(-m+2)+\varphi(-m+1)\right]\\
&+(k-3+m)\left[\varphi(-m+4)-2\varphi(-m+3)+\varphi(-m+2)\right]+\dots\\
&+3\left[\varphi(k-2)-2\varphi(k-3)+\varphi(k-4)\right]+2\left[\varphi(k-1)-2\varphi(k-2)+\varphi(k-3)\right]\\
&+\left[\varphi(k)-2\varphi(k-1)+\varphi(k-2)\right]=\\
&=\varphi(-m)\left[1-k-m+k-1+m\right]+\varphi(-m+1)\left[k+m-2(k-1+m)+k-2+m\right]\\
&+\varphi(-m+2)\left[k-1+m-2(k-2+m)+k-3+m\right]\\
&+\varphi(-m+3)\left[k-2+m-2(k-3+m)+k-4+m\right]+\dots\\
&+\varphi(k-3)\left[4-2\cdot 3+2\right]+\varphi(k-2)\left[3-2\cdot 2+1\right]+\varphi(k-1)\left[2-2\right]+\varphi(k)=\\
&=\varphi(-m)\cdot 0+\varphi(-m+1)\cdot 0+\dots+\varphi(k-1)\cdot 0+\varphi(k)=\varphi(k).
\end{aligned}
$$

So we shown that $x(k)=\varphi(k)$ for any $k\in\mathbb{Z}_{-m}^{1}$.

□

Acknowledgement

The authors were supported by Grant FEKT-S-14-2200 of Faculty of Electrical Engineering and Communication, BUT.

References

[1] Diblík, J., Khusainov, D. Ya.: *Representation of solutions of linear discrete systems with constant coefficients and pure delay*, Advances in Difference Equations, Volume 2006, Article ID 80825, Pages 1–13.

[2] Khusainov, D. Ya., Diblík, J., Růžičková, M., Lukáčová, J.: *Representation of a solution of the Cauchy problem for an oscillating system with pure delay*, Nonlinear Oscillations, Volume 11, No. 2, 2008, Pages 261–270.

Current address

Diblík Josef, prof. RNDr. DrSc.
Department of Mathematics,
Faculty of Electrical Engineering and Communication, Brno University of Technology
Technická 2848/8, Žabovřesky, 61600, Brno, Česká republika
Tel. number: +420 54114 6064, e-mail: diblik@feec.vutbr.cz

Mencáková Kristýna, Mgr.
Department of Mathematics,
Faculty of Electrical Engineering and Communication, Brno University of Technology
Technická 2848/8, Žabovřesky, 61600, Brno, Česká republika
Tel. number: +420 54114 7608, e-mail: mencakova.k@fce.vutbr.cz

Slovak University of Technology in Bratislava
Faculty of Mechanical Engineering
16th Conference on Applied Mathematics **APLIMAT 2017**
Proceedings

SOLUTION OF WEAKLY DELAYED LINEAR DISCRETE SYSTEMS IN $\mathbb{R}^3$

DIBLÍK Josef (CZ), ŠAFAŘÍK Jan (CZ)

Abstract. In the paper we investigate weakly delayed linear discrete systems with constant delay

$$x(k+1) = Ax(k) + Bx(k-m), k = 0, 1, \dots, m,$$

in $\mathbb{R}^3$. Conditions for the system to be a weakly delayed system are given and the initial problem is explicitely solved in one of possible cases.

Keywords: Discrete system, weakly delayed system, linear system, initial problem

Mathematics subject classification: Primary 65D17; Secondary 53A04, 14Q05

1 Weakly delayed systems in $\mathbb{R}^3$.

In the paper, weakly delayed systems of linear discrete equations are investigated. For previous results we refer to [1]–[6].
Consider a linear system of difference equations

$$x(k+1) = Ax(k) + Bx(k-m), \quad k = 0, 1, \dots \tag{1}$$

where $x(k)$ is an unknown 3-dimensional vector-function, A and B are 3 by 3 constant matrices with elements a_{ij} and b_{ij}, $i, j = 1, 2, 3$ respectively, and $m \geq 1$ is a natural number.

We assume that (1) is a weakly delayed system in the sense of the following definition.

Definition 1. *System* (1) *is called weakly delayed if the characteristic equations for* (1) *and for the system without delay*

$$x(k+1) = Ax(k)$$

have identical roots, that is, if, for every $\lambda \in \mathbb{C} \setminus \{0\}$,

$$\det\left(A+\lambda^{-m}B-\lambda E\right)=\det\left(A-\lambda E\right),$$

where E is a 3 by 3 unit matrix.

In the paper we assume that the matrix A is given in the Jordan form and has the form

$$A\equiv\Lambda=\begin{pmatrix}\lambda_1 & 0 & 0\\ 0 & \lambda_2 & 0\\ 0 & 0 & \lambda_3\end{pmatrix} \tag{2}$$

and its eigenvalues λ_1, λ_2 and λ_3 are mutually different real numbers.

Applying Definition 1 to system (1) with $A=\Lambda$, we get conditions under which the system is weakly delayed.

Theorem 1. *System* (1) *is a weakly delayed system if and only if*

$$b_{11}=b_{22}=b_{33}=0, \tag{3}$$
$$b_{12}b_{23}b_{31}+b_{13}b_{21}b_{32}=0, \tag{4}$$
$$b_{12}b_{21}+b_{13}b_{31}+b_{23}b_{32}=0, \tag{5}$$
$$\lambda_3 b_{12}b_{21}+\lambda_2 b_{13}b_{31}+\lambda_1 b_{23}b_{32}=0. \tag{6}$$

For the proof, we refer to [6], Theorem 3.

The purpose of the paper is to give an explicit solution of the initial problem

$$x(0)=x_0=\begin{pmatrix}x_{0,1}\\ x_{0,2}\\ x_{0,3}\end{pmatrix},\ldots,x(m)=x_m=\begin{pmatrix}x_{m,1}\\ x_{m,2}\\ x_{m,3}\end{pmatrix} \tag{7}$$

to system (1) where $x_{i,j}$, $i=0,\ldots,m$, $j=1,2,3$, are real constants.

System (1) can be transformed into a system without delay. Below we describe this transformation. Taking into account the above properties of matrix B, we have

$$B=\begin{pmatrix}0 & b_{12} & b_{13}\\ b_{21} & 0 & b_{23}\\ b_{31} & b_{32} & 0\end{pmatrix} \tag{8}$$

and its elements satisfy (3)–(6). From these formulas, it obviously follows that B is a nilpotent matrix (all its eigenvalues are equal to zero). Below we assume that the geometrical multiplicity of the zero eigenvalue equals 1 or 2.

Below we investigate the system (1) in the form

$$x(k+1)=\Lambda x(k)+Bx(k-m)$$

with initial data (7). Define new dependent 3-dimensional vector functions $z^1, \ldots, z^m$ by formulas

$$\begin{array}{lcl} z^1(k) = x(k-1), & \Rightarrow & z^1(k+1) = x(k), \\ z^2(k) = x(k-2), & \Rightarrow & z^2(k+1) = x(k-1), \\ \vdots & & \\ z^m(k) = x(k-m), & \Rightarrow & z^m(k+1) = x(k-(m-1)) \end{array}$$

and, instead of (1), consider a new system without delay

$$\begin{array}{rcllllll} x(k+1) & = & \Lambda x(k) & & & & & +Bz^m(k), \\ z^1(k+1) & = & x(k), & & & & & \\ z^2(k+1) & = & & z^1(k), & & & & \\ z^3(k+1) & = & & & z^2(k), & & & \\ \vdots & & & & & \ddots & & \\ z^m(k+1) & = & & & & & z^{m-1}(k). & \end{array}$$

Unifying the notation of the dependent variables as

$$\begin{aligned} y_i(k) &= x_i(k), \ \ i = 1,2,3, \\ y_{j+3}(k) &= z_j^1(k), \ \ j = 1,2,3, \\ y_{j+6}(k) &= z_j^2(k), \ \ j = 1,2,3, \\ &\vdots \\ y_{j+3m}(k) &= z_j^m(k), \ \ j = 1,2,3, \end{aligned}$$

we have

$$y(k+1) = \mathcal{A}y(k), \ k \geq 0 \tag{9}$$

where $y(k) = (y_1(k), \ldots, y_{3m+3}(k))^T$ and

$$\mathcal{A} = \begin{pmatrix} \Lambda & \Theta & \ldots & \Theta & B \\ E & \Theta & \ldots & \Theta & \Theta \\ \Theta & E & \ldots & \Theta & \Theta \\ \vdots & \vdots & \ddots & \vdots & \vdots \\ \Theta & \Theta & \ldots & E & \Theta \end{pmatrix} \tag{10}$$

where Θ is a 3 by 3 zero matrix.

2 Solution of the system (9)

The initial data for the system (9), as seen from (7), are

$$y(0) = y_0 = (x_0, \ldots, x_m)^T.$$

We will transform system (9) using the transformation $y(k) = Sw(k)$ where S is a regular transient matrix and $w(k)$ is a new dependent $3(m+1)$-dimensional vector into a system with a matrix of the Jordan form. We get

$$Sw(k+1) = \mathcal{A}Sw(k)$$

or

$$w(k+1) = \gamma w(k) \tag{11}$$

where

$$\gamma = S^{-1} \mathcal{A} S$$

with the initial data for (11) being

$$w(0) = S^{-1} y(0).$$

Then, the solution of the system (11) is

$$w(k) = \gamma^k w(0), \quad k = 1, 2, 3, \ldots. \tag{12}$$

Below we will need the following auxiliary result (we omit its proof).

Theorem 2. *Let a matrix A be of the type* (2) *and let the entries of a matrix B satisfy* (3)–(6). *Then, the eigenvalues $\mu_i, i = 1, \ldots, 3m+3$ of the matrix $\mathcal{A}$ are $\mu_1 = \lambda_1$, $\mu_2 = \lambda_2$, $\mu_3 = \lambda_3$, $\mu_4 = \mu_5 = \cdots = \mu_{3m+3} = 0$.*

When using formula (12), it is necessary to consider the geometrical multiplicity of the zero eigenvalue of matrix B.

2.1 Case I - the geometrical multiplicity of B equals 1

Due to Theorem 2, we can assume that the transition matrix S is such that

$$\gamma = \gamma_1 = \begin{pmatrix} \Lambda_B^I & \Theta & \ldots & \Theta & \Theta \\ \Theta & E & \ldots & \Theta & \Theta \\ \vdots & \vdots & \ddots & \vdots & \vdots \\ \Theta & \Theta & \ldots & E & \Theta \\ \Theta & \Theta & \ldots & \Theta & \Lambda \end{pmatrix},$$

where

$$\Lambda_B^I = \begin{pmatrix} 0 & 1 & 0 \\ 0 & 0 & 1 \\ 0 & 0 & 0 \end{pmatrix}$$

is the Jordan form of the matrix B. Obviously, powers of γ are following:

$$\gamma_1^2 = \begin{pmatrix} (\Lambda_B^I)^2 & \Theta & \ldots & \Theta & \Theta \\ \Theta & E & \ldots & \Theta & \Theta \\ \vdots & \vdots & \ddots & \vdots & \vdots \\ \Theta & \Theta & \ldots & E & \Theta \\ \Theta & \Theta & \ldots & \Theta & \Lambda^2 \end{pmatrix},$$

where

$$(\Lambda_B^I)^2 = \begin{pmatrix} 0 & 0 & 1 \\ 0 & 0 & 0 \\ 0 & 0 & 0 \end{pmatrix}, \quad \Lambda^2 = \begin{pmatrix} \lambda_1^2 & 0 & 0 \\ 0 & \lambda_2^2 & 0 \\ 0 & 0 & \lambda_3^2 \end{pmatrix},$$

and for $k \geq 3$

$$\gamma_1^k = \begin{pmatrix} \Theta & \Theta & \dots & \Theta & \Theta \\ \Theta & E & \dots & \Theta & \Theta \\ \vdots & \vdots & \ddots & \vdots & \vdots \\ \Theta & \Theta & \dots & E & \Theta \\ \Theta & \Theta & \dots & \Theta & \Lambda^k \end{pmatrix}$$

where

$$\Lambda^k = \begin{pmatrix} \lambda_1^k & 0 & 0 \\ 0 & \lambda_2^k & 0 \\ 0 & 0 & \lambda_3^k \end{pmatrix}.$$

2.2 Case II - the geometrical multiplicity of B equals 2

Due to Theorem 2, we can assume that the transition matrix S is such that

$$\gamma = \gamma_2 = \begin{pmatrix} \Lambda_B^{II} & \Theta & \dots & \Theta & \Theta \\ \Theta & E & \dots & \Theta & \Theta \\ \vdots & \vdots & \ddots & \vdots & \vdots \\ \Theta & \Theta & \dots & E & \Theta \\ \Theta & \Theta & \dots & \Theta & \Lambda \end{pmatrix},$$

where

$$\Lambda_B^{II} = \begin{pmatrix} 0 & 0 & 0 \\ 0 & 0 & 1 \\ 0 & 0 & 0 \end{pmatrix}.$$

Since

$$(\Lambda_B^{II})^2 = \Theta,$$

the powers γ_2^k, for $k \geq 2$, are

$$\gamma_2^k = \begin{pmatrix} \Theta & \Theta & \dots & \Theta & \Theta \\ \Theta & E & \dots & \Theta & \Theta \\ \vdots & \vdots & \ddots & \vdots & \vdots \\ \Theta & \Theta & \dots & E & \Theta \\ \Theta & \Theta & \dots & \Theta & \Lambda^k \end{pmatrix}.$$

2.3 Solution of the problem (1), (7)

The solution of (9) is

$$y(k) = Sw(k) = S\gamma_i^k w(0), \quad k = 1, 2, 3, \dots$$

where $i = 1$ if the geometrical multiplicity of the zero eigenvalue of B equals 1 and $i = 2$ if the geometrical multiplicity of the zero eigenvalue of B equals 2. Using an auxiliary matrix

$$Q = (E, \underbrace{\Theta, \dots, \Theta}_{m}),$$

we can write the solution of the initial problem (1), (7) in the form

$$x(k) = QS\gamma_i^k w(0), \quad i = 1, 2, \quad k = 1, 2, 3, \ldots, \tag{13}$$

where

$$w(0) = S^{-1}x(0). \tag{14}$$

Therefore, the following theorem holds.

Theorem 3. *Let the matrix A have the form* (2) *where λ_1, λ_2, λ_3 are different real numbers, let matrix B have the form* (10)*, with its elements satisfying* (3)–(6)*. Then, the solution of the initial problem* (1)*,* (7) *is given by formula* (13) *where $i = 1$ if the geometrical multiplicity of the zero eigenvalue of B equals 1 and $i = 2$ if the geometrical multiplicity of the zero eigenvalue of B equals 2 and $w(0)$ given by* (14).

Remark 1. *The initial-value problem* (1)*,* (7) *is specified by $3(m+1)$ initial values* (7)*. In the case considered, the solution given by formula* (13) *depends, as it follows from* (12) *and formulas for γ_i^k, $i = 1, 2$, $k = 1, 2, 3, \ldots$, for $k \to \infty$, not on $3(m+1)$ parameters, but only on $3m$ parameters. It means that, among the initial $3(m+1)$ parameters (initial data), there are three unnecessary initial values.*

Acknowledgement

The authors were supported by the Grant FEKT-S-14-2200 of Faculty of Electrical Engineering and Communication, BUT.

References

[1] Diblík J., Halfarová H.: *Explicit general solution of planar linear discrete systems with constant coefficients and weak delays.* Adv. Difference Equ. 2013, Art. number: 50, doi:10.1186/1687-1847-2013-50, 1–29.

[2] Diblík J., Halfarová H.: *General explicit solution of planar weakly delayed linear discrete systems and pasting its solutions.* Abstr. Appl. Anal. 2014, doi:10.1155/2014/627295, 1–37.

[3] Diblík J., Khusainov D. Ya., Šmarda Z.: *Construction of the general solution of planar linear discrete systems with constant coefficients and weak delay.* Adv. Difference Equ. 2009, Art. ID 784935, 18 pp.

[4] Elaydi, S. N.: *An Introduction to Difference Equations*, Third Edition, Springer, 2005.

[5] Šafařík, J.: *Solution of a Weakly Delayed Difference System.* In Proceedings of the 22nd Conference STUDENT EEICT 2016. Brno: Vysoké učení technické v Brně, Fakulta elektrotechniky a komunikačních technologií, 2016. s. 763-767. ISBN: 978-80-214-5350- 0.

[6] Šafařík, J., Diblík, J., Halfarová, H.: *Weakly Delayed Systems of Linear Discrete Equations in $\mathbb{R}^3$.* In MITAV 2015 (Matematika, informační technologie a aplikované vědy), Post-conference proceedings of extended versions of selected papers. Brno: Univerzita obrany v Brně, 2015. s. 105-121. ISBN: 978-80-7231-436-2.

Current address

Diblík Josef, prof. RNDr. DrSc.
Faculty of Electrical Engineering and Communication,
Brno University of Technology in Brno
Technická 2848/8, 61600 Brno, Czech Republic
Tel. number: +420 54114 6064, e-mail: diblik@feec.vutbr.cz

Šafařík Jan, Mgr.
Faculty of Electrical Engineering and Communication,
Brno University of Technology in Brno
Technická 2848/8, 61600 Brno, Czech Republic
Tel. number: +420 54114 6097
E-mail: safarik.j@fce.vutbr.cz, xsafar19@stud.feec.vutbr.cz

Slovak University of Technology in Bratislava
Faculty of Mechanical Engineering
16th Conference on Applied Mathematics **APLIMAT 2017**
Proceedings

THE SOLUTIONS OF SECOND-ORDER NONHOMOGENUOUS LINEAR DIFFERENTIAL SYSTEMS WITH CONSTANT DELAYS

DIBLÍK Josef (CZ), SVOBODA Zdeněk (CZ)

Abstract. In the paper, representations of solutions to initial problems for n-dimensional nonhomogeneous second-order linear differential equations with delays

$$x''(t) - (A+B)x'(t-\tau) + BAx(t-2\tau) = f(t), \quad t \geq 0,$$

where A and B are $n \times n$ matrices, $f(t)$ is continuous n-dimensional vector-function and $\tau > 0$, are derived by means of special matrix delayed functions.

Keywords: representation of solutions, delay, matrix delayed functions

Mathematics subject classification: Primary 34K05; Secondary 34K065

1 Introduction

Recently a new formalization of the well-known method of steps in the theory of linear differential equations with constant coefficients and single delay was developed in [4, 5]. It uses so-called matrix function delayed matrix exponential, which is defined as a matrix polynomial on every step-interval. This formalization has been widely applied, e.g., in boundary-value problems, control and stability problems, modification to discrete equations was performed, as well as generalizations to the case of several delays, etc. (we refer, e.g., to [1, 2, 3]). The paper [6] deals with homogeneous second-order linear differential systems with constant delays. The results was derived by comparing solutions of the studied problem with solutions of the first-order linear system which comprise $2n$ equations. In the paper we study the initial problem for linear non-homogeneous system of second-order delayed differential equations

$$x''(t) - (A+B)x'(t-\tau) + BAx(t-2\tau) = f(t), \quad t \geq 0, \tag{1}$$

$$x^{(i)}(t) = \xi^{(i)}(t), \quad i = 0, 1, \quad t \in [-2\tau, 0] \tag{2}$$

where A, B are $n \times n$ matrices, $\tau > 0$, $f\colon [0,\infty) \to \mathbb{R}^n$ and the function $\xi\colon [-2\tau, 0] \to \mathbb{R}^n$ is twice continuously differentiable.

2 Delayed exponential of matrix

In this section we recall the basic notions and results about linear differential systems of the first order with constant coefficients and a constant delay. A delayed matrix exponential e_τ^{At}, $t \in \mathbb{R}$ is defined as

$$e_\tau^{At} = \sum_{s=0}^{\lfloor t/\tau \rfloor + 1} A^s \frac{(t-(s-1)\tau)^s}{s!} \tag{3}$$

where $t \in \mathbb{R}$ and $\lfloor \cdot \rfloor$ is the floor function. The delayed matrix exponential equals the unit matrix on $[-\tau, 0]$ and represents the fundamental matrix of a homogeneous linear system with a single delay

$$\dot{x}(t) = Ax(t-\tau).$$

In [4] is studied initial problem system

$$z'(t) = Az(t) + Bz(t-\tau) + G(t), \tag{4}$$

$$z(t) \equiv \varphi(t) \quad \text{for} \quad -\tau \leq t \leq 0 \tag{5}$$

with commuting matrices A, B and continuous vector function $G(t)$. Integral representation of solution $z = z(t)$ of initial problem (4), (5) in the form

$$z(t) = e^{A(t+\tau)} e_\tau^{B_1 t} \varphi(-\tau) + \int_{-\tau}^{0} e^{A(t-s)} e_\tau^{B_1(t-\tau-s)} [\varphi'(s) - A\varphi(s)] ds$$
$$+ \int_0^t e^{A(t-s)} e_\tau^{B_1(t-\tau-s)} G(s) ds \tag{6}$$

is given, where the matrix B_1 is defined as $B_1 = e^{-A\tau} B$. In the paper we utilize a special case of this result, specified by the condition $A = \Theta$ (the zero matrix). Then the integral representation (6) of solution $z = z(t)$ of initial problem (4), (5) reduces to

$$z(t) = e_\tau^{Bt} \varphi(-\tau) + \int_{-\tau}^{0} e_\tau^{B(t-\tau-s)} \varphi'(s) ds + \int_0^t e_\tau^{B(t-\tau-s)} G(s) ds. \tag{7}$$

3 Main result

We show that the equation (1) can be written as a system

$$x'(t) = Ax(t-\tau) + \quad y(t-\tau) + \int_0^t f(u)\, du\,, \tag{8}$$

$$y'(t) = By(t-\tau) + B \int_0^t f(u)\, du \tag{9}$$

where $t \geq 0$, or in the matrix form as

$$\begin{pmatrix} x'(t) \\ y'(t) \end{pmatrix} = C \begin{pmatrix} x(t) \\ y(t) \end{pmatrix} + F(t) \tag{10}$$

where

$$C = \begin{pmatrix} A & I \\ \Theta & B \end{pmatrix} \quad \text{and} \quad F(t) = \begin{pmatrix} \int\limits_0^t f(u)\,du \\ B\int\limits_0^t f(u)\,du \end{pmatrix}.$$

The first n components of the solution of the system given by (8), (9) is a solution of equation (1). From (8) it follows

$$y(t-\tau) = x'(t) - Ax(t-\tau) - \int\limits_0^t f(u)\,du \Rightarrow$$

$$\Rightarrow y(t-2\tau) = x'(t-\tau) - Ax(t-2\tau) - \int\limits_0^{t-\tau} f(u)\,du$$

and from (9) we have

$$y'(t-\tau) = By(t-2\tau) + B\int\limits_0^{t-\tau} f(u)\,du \tag{11}$$

Differentiating (8) and using (8)–(11), we derive

$$\begin{aligned} x''(t) =& Ax'(t-\tau) + y'(t-\tau) + f(t) \\ =& Ax'(t-\tau) + By(t-2\tau) + B\int\limits_0^{t-\tau} f(u)du + f(t) \\ =& Ax'(t-\tau) + B\left(x'(t-\tau) - Ax(t-2\tau) - \int\limits_0^{t-\tau} f(u)\,du \right) \\ &+ B\int\limits_0^{t-\tau} f(u)du + f(t) = (A+B)x'(t-\tau) - BAx(t-2\tau) + f(t). \end{aligned}$$

Now we define initial functions to the system (8), (9). Using (8) we compute (for $t = 0$)

$$y(-\tau) = x'(0) - Ax(-\tau)$$

and, for $-\tau \leq t \leq 0$, we derive

$$\begin{aligned} y(t) =& y(-\tau) + \int_{-\tau}^{t} y'(s)ds \\ =& y(-\tau) + \int_{-\tau}^{t} \left(By(s-\tau) + B\int_0^s f(u)\,du \right) ds \\ =& y(-\tau) + \int_{-\tau}^{t} \left(B\left(x'(s) - Ax'(s-\tau) - \int_0^s f(u)\,du \right) + B\int_0^s f(u)\,du \right) ds \\ =& x'(0) - Ax(-\tau) + \int_{-\tau}^{t} B(x'(s) - Ax'(s-\tau))ds. \end{aligned} \tag{12}$$

Using (2) and (12) we define:

$$\varphi(t) = \begin{pmatrix} x(t) \\ y(t) \end{pmatrix} = \begin{pmatrix} \xi(t) \\ \xi'(0) - A\xi(-\tau) + \int_{-\tau}^{t} B(\xi'(s) - A\xi(s-\tau))ds \end{pmatrix}$$

where $t \in [-\tau, 0]$.

Let us define an auxiliary matrix function

$$\mathrm{e}_\tau^{(A,B)t} = \sum_{s=0}^{\lfloor t/\tau \rfloor} \frac{(t-(s-1)\tau)^s}{s!} \sum_{i=0}^{s} A^{s-i} B^i, \quad t \in \mathbb{R}.$$

Below we need the powers of the matrix C. Relevant formula is

$$C^k = \begin{pmatrix} A^k & \sum_{i=0}^{k-1} A^{k-1-i} B^i \\ \Theta & B^k \end{pmatrix}, \quad k \geq 0.$$

Then from (3) we derive

$$\mathrm{e}_\tau^{Ct} = \begin{pmatrix} \mathrm{e}_\tau^{At} & \mathrm{e}_\tau^{(A,B)t} \\ \Theta & \mathrm{e}_\tau^{Bt} \end{pmatrix}.$$

Theorem 1 *The initial problem* (1), (2) *has the solution*

$$x(t) = \mathrm{e}_\tau^{At}\xi(-\tau) + e_\tau^{(A,B)t}(\xi'(0) - A\xi(-\tau)) + \int_\tau^0 \left(\mathrm{e}_\tau^{A(t-\tau-s)}\xi'(s) + \mathrm{e}_\tau^{(A,B)(t-\tau-s)} B\left(\xi''(s+\tau) - A\xi'(s-\tau)\right)\right) ds + \int_0^t \left(\mathrm{e}_\tau^{A(t-\tau-s)} + B\mathrm{e}_\tau^{(A,B)(t-\tau-s)}\right)\left(\int_0^s f(u)du\right) ds. \quad (13)$$

PROOF We apply formula (7) to the system (10). The solution is given as the sum of 3 summands. Now we compute each of summands separately: The first summand equals

$$e_\tau^{B_1 t}\varphi(-\tau) = \begin{pmatrix} \mathrm{e}_\tau^{At} & \mathrm{e}_\tau^{(A,B)t} \\ \Theta & \mathrm{e}_\tau^{Bt} \end{pmatrix} \begin{pmatrix} \xi(-\tau) \\ \xi'(0) - A\xi(-\tau) \end{pmatrix} = \begin{pmatrix} \mathrm{e}_\tau^{At}\xi(-\tau) + e_\tau^{(A,B)t}(\xi'(0) - A\xi(-\tau)) \\ e_\tau^{Bt}(\xi'(0) - A\xi(-\tau)) \end{pmatrix}.$$

The second summand takes the form:

$$\int_{-\tau}^0 \mathrm{e}_\tau^{C(t-\tau-s)}\varphi'(s)ds = \int_\tau^0 \begin{pmatrix} \mathrm{e}_\tau^{A(t-\tau-s)} & \mathrm{e}_\tau^{(A,B)(t-\tau-s)} \\ \Theta & \mathrm{e}_\tau^{B(t-\tau-s)} \end{pmatrix} \begin{pmatrix} \xi'(s) \\ B\left(\xi''(s+\tau) - A\xi'(s-\tau)\right) \end{pmatrix} ds = \begin{pmatrix} \int_\tau^0 \left(\mathrm{e}_\tau^{A(t-\tau-s)}\xi'(s) + \mathrm{e}_\tau^{(A,B)(t-\tau-s)} B\left(\xi''(s+\tau) - A\xi'(s-\tau)\right)\right) ds \\ \int_\tau^0 \mathrm{e}_\tau^{B(t-\tau-s)} B\left(\xi''(s+\tau) - A\xi'(s-\tau)\right) ds \end{pmatrix}.$$

Finally, the third summand is

$$\int_0^t e_\tau^{C(t-\tau-s)} F(s) ds =$$

$$\int_0^t \begin{pmatrix} e_\tau^{A(t-\tau-s)} & e_\tau^{(A,B)(t-\tau-s)} \\ \Theta & e_\tau^{B(t-\tau-s)} \end{pmatrix} \begin{pmatrix} \int_0^s f(u)\, du \\ B\int_0^s f(u)\, du \end{pmatrix} ds =$$

$$\begin{pmatrix} \int_0^t \left(e_\tau^{A(t-\tau-s)} + B e_\tau^{(A,B)(t-\tau-s)} \right) \left(\int_0^s f(u)\, du \right) ds \\ \int_0^t B e_\tau^{B(t-\tau-s)} \left(\int_0^s f(u)\, du \right) ds \end{pmatrix}.$$

The first n components of the vector

$$z(t) = \begin{pmatrix} x(t) \\ y(t) \end{pmatrix}$$

are components of the solution $x = x(t)$ of the equation (1), i.e.

$$\begin{aligned} x(t) = {} & e_\tau^{At}\xi(-\tau) + e_\tau^{(A,B)t}(\xi'(0) - A\xi(-\tau)) \\ & + \int_\tau^0 \left(e_\tau^{A(t-\tau-s)}\xi'(s) + e_\tau^{(A,B)(t-\tau-s)} B \left(\xi''(s+\tau) - A\xi'(s-\tau) \right) \right) ds \\ & + \int_0^t \left(e_\tau^{A(t-\tau-s)} + B e_\tau^{(A,B)(t-\tau-s)} \right) \left(\int_0^s f(u)\, du \right) ds \end{aligned}$$

and the formula (13) holds. □

Acknowledgment

The work was supported by the project FEKT-S-14-2200 of Brno University of Technology.

References

[1] A. Boichuk, J. Diblík, D. Khusainov, M. Růžičková, *Boundary value problems for delay differential systems,* Adv. Differ. Equ. (2010), Article ID 593834, 20 pages.

[2] M. Pospíšil, J. Diblík, M. Fečkan, *Observability of difference equations with a delay,* AIP Conference Proceedings, American Institute of Physics Inc., Volume 1558, 2013, 478–481, 11th International Conference of Numerical Analysis and Applied Mathematics 2013, ICNAAM 2013; Rhodes; Greece; 21 September 2013 through 27 September 2013,

[3] D. Khusainov, J. Diblík, M. Růžičková, Linejnie Dinamicheskie Sistemi s Posledejstvijem. Predstavlenije reshenij, Ustojchivost, Upravlenije, Stabilizacija (Linear Dynamical Systems with Aftereffect, Representation of Solutions, Control, Stabilization), *Informacionno-analiticheskoe agentstvo*, Kiev 2015. ISBN 978-617-571-114-9. (In Russian)

[4] D.Ya. Khusainov, G.V. Shuklin, *Linear autonomous time-delay system with permutation matrices solving,* Stud. Univ. Žilina, Math. Ser. **17** (2003) 101–108.

[5] D.Ya. Khusainov, G.V. Shuklin, *Relative controllability in systems with pure delay*. (English. Russian original) Int. Appl. Mech. **41**, No. 2, 210–221 (2005); translation from Prikl. Mekh., Kiev **41**, No. 2, 118–130 (2005).

[6] Z. Svoboda, *Representation of solutions of linear differential systems of second-order with constant delays*. Nonlinear Oscil. **19**, No 1, 129-141, 2016 ISSN 1562-3076.

Current address

Diblík Josef, prof. DrSc.
Department of Mathematics, Faculty of Electrical Engineering and Communication
Brno University of Technology
Technická 3058/10, 616 00, Brno, Czech Republic
Tel. number: +420 54114 7618, e-mail: diblik.j@fce.vutbr.cz

Svoboda Zdeněk, RNDr. CSc.
Department of Mathematics, Faculty of Electrical Engineering and Communication
Brno University of Technology
Technická 3058/10, 616 00, Brno, Czech Republic
Tel. number: +420 54114 6082, e-mail: svobodaz@feec.vutbr.cz

Slovak University of Technology in Bratislava
Faculty of Mechanical Engineering
16th Conference on Applied Mathematics **APLIMAT 2017**
Proceedings

ANALYTIC SOLUTION OF THE MULTI-RESONANT ELECTRICAL CIRCUIT UNDER IMPULSE SUPPLY

DOBRUCKÝ Branislav (SK), MARČOKOVÁ Mariana (SK), ŠTEFANEC Pavol (SK)

Abstract. The paper deals with analytical solution which make possible to estimate instantaneous state of dynamical system in any time instant of the multi-resonant converter filter after using Laplace-Carson transform to the input voltage. Results of theoretical analysis confirmed by numerical computer simulation are given in the paper.

Keywords: multi-resonant converter, inductor, capacitor, electrical circuit, dynamical system, Laplace-Carson transform, non-harmonic impulse supply, analytic method, numerical method

Mathematics Subject Classification: Primary 44A05; Secondary 65R10

1 Introduction

In this paper we deal with dynamical resonant circuit model. Multi-resonant converters are important part of the field of conversion and conditioning of electrical energy. One of them is multi-resonant converter known under acronym L1C1L2C2, [1]. The basic equivalent electrical scheme is depicted in Fig. 1without active electronic switches.

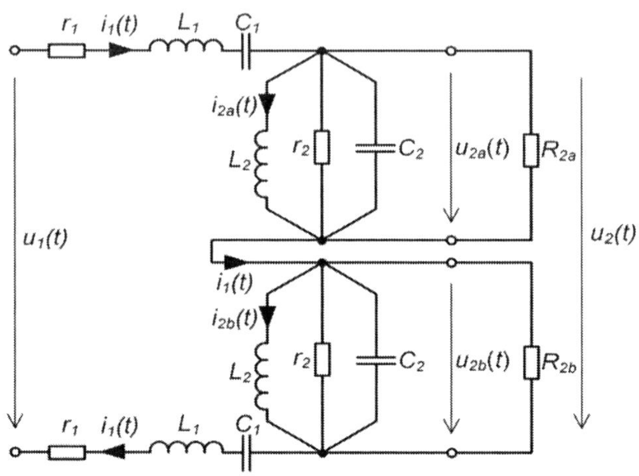

Fig. 1. Basic equivalent electrical scheme of L1C1L2C2 multi-resonant converter filter

The filtering part of the converter consists of 8 accumulative elements as inductors and capacitors ($2L_1 - 2L_2$, $2C_1 - 2C_2$). Its dynamical model can be presented by the system of 6 equations because of $2r_1$, $2L_1$ and $2C_1$ elements. Task of such a filter is to provide symmetrical harmonic output voltages $u_2(t)$ with neutral point with minimum total harmonic distortion. The input voltage $u_1(t)$ (as exciting function) is strongly non-harmonic one, actually rectangular with changing width of pulses.

2 Analytic method using Laplace-Carson transform

There are many mathematical methods which make possible to analyze electrical circuits as dynamical systems [2], [3]. Basically, solution of electrical circuit as dynamical system can be done in time or frequency domain. If focusing on time domain solution the possibilities are analysis or identification depending on availability of mathematical model and its parameters [7]. Generally, we can use analytic and (or) numerical solution for the analysing and investigating of dynamical system. In the next chapter we will deal with dynamical resonant circuit model.

In [5] and [6] investigation regarding to time domain after Laplace-Carson transform with parameter p has been used. After Laplace-Carson transform used to $u_1(t)$ we have input voltage in the form

$$U_1(p) = \frac{U_1^{T/2}(p)}{1 + \mathrm{e}^{-pT/2}}$$

where $U_1^{T/2}(p) = U(1 - \mathrm{e}^{-pT_s})$, T_s is width of input pulses and U is amplitude of input voltage.

Then, using Kirchhoff laws [5] one can write

$$U_1(p) = rI_1(p) + pLI_1(p) + \frac{1}{pC}I_1(p) + U_{2a}(p) + U_{2b}(p)$$

where

$$I_1(p) = \frac{1}{pL_2}U_{2a}(p) + pC_2U_{2a}(p) + \left(\frac{1}{r_2} + \frac{1}{R_{2a}}\right)U_{2a}(p)$$

and

$$I_1(p) = \frac{1}{pL_2}U_{2b}(p) + pC_2U_{2b}(p) + \left(\frac{1}{r_2} + \frac{1}{R_{2b}}\right)U_{2b}(p),$$

respectively, see Fig. 1.

By adapting

$$U_2(p) = U_{2a}(p) + U_{2b}(p) = \frac{2}{3}\frac{Y_{2b}(p)}{Y_{2a}(p)}\frac{1}{Z_1(p)}U_1(p)$$

where

$$Z_1(p) = r + pL + \frac{1}{pC}, \quad r = 2r_1,$$

$$Y_{2a}(p) = \frac{1}{R_a} + pC_2 + \frac{1}{pL_2}, \qquad \frac{1}{R_a} = \frac{1}{r_2} + \frac{1}{R_{2a}},$$

$$Y_{2b}(p) = \frac{1}{R_b} + pC_2 + \frac{1}{pL_2}, \qquad \frac{1}{R_b} = \frac{1}{r_2} + \frac{1}{R_{2b}}.$$

Taking into account equality of the parameters of both voltage loops $u_{2a}(t), u_{2b}(t)$ we get

$$U_{2a}(p) = U_{2b}(p) = \frac{U_2\,(p)}{2}.$$

Solution for different values of resistors ($R_a \neq R_b$) – rather complicated – is just shown as numerical one in Chapter 3.

Then, choosing $T_s = \frac{T}{2}$ for first half-period

$$U_{2a}(p) = U_{2b}(p) = \frac{U_1^{T/2}(p)}{2 + Z_1(p)Y_2(p)} = \frac{U\left(1 - \mathrm{e}^{-pT/2}\right)}{2 + Z_1(p)Y_2(p)} = U\frac{A(p)}{B(p)}$$

and using inverse transform [2-3], the $u_{2a}(t)$ and $u_{2b}(t)$ can be determined by

$$u_2^{0-T/2}(t) = U\sum_{k=1}^{N} \frac{A(p_k)}{p_k B'(p_k)} \mathrm{e}^{p_k t}$$

in the range $0 - \frac{T}{2}$ since $\frac{A(0)}{B(0)} = 0$ and

$$u_2^{0-T/2}(t) = 2u_{2a}^{0-T/2}(t) = 2u_{2b}^{0-T/2}(t).$$

By similar way transient component can be calculated

$$u_{trans}(t) = \sum_{k=1}^{N} U_1(p)\frac{A(p_k)}{p_k B'(p_k)} \mathrm{e}^{p_k t} = \sum_{k=1}^{N} U\frac{1 - \mathrm{e}^{-p_k T/2}}{1 + \mathrm{e}^{-p_k T/2}} \frac{A(p_k)}{p_k B'(p_k)} \mathrm{e}^{p_k t}$$

and steady-state component as

$$u_{steady}^{0-T/2}(t) = u_2^{0-T/2}(t) - u_{trans}^{0-T/2}(t).$$

Finally, for total voltage waveforms in entire time range

$$u_2(t) = u_{trans}(t) \pm \left[u_2^{0-T/2}(t) - u_{trans}^{0-T/2}(t)\right].$$

Results of simulation are shown in Fig. 2.

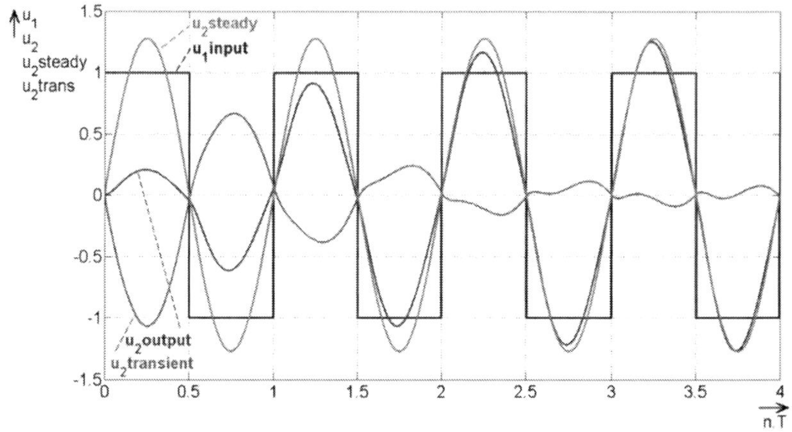

Fig. 2. Time waveforms of output voltage components

3 Verification by numerical computer simulation

For verification following dynamical model has been used (cf. Fig. 1)

$$\frac{di_1(t)}{dt} = -\frac{r_1}{L}i_1(t) - \frac{1}{L}u_C(t) - \frac{1}{L}u_{2a}(t) - \frac{1}{L}u_{2b}(t) + \frac{1}{L}u_1(t), \quad (1)$$

$$\frac{du_{2a}(t)}{dt} = \frac{1}{C_2}i_1(t) - \frac{1}{R_aC_2}u_{2a}(t) - \frac{1}{C_2}i_2(t), \quad (2)$$

$$\frac{du_{2b}(t)}{dt} = \frac{1}{C_2}i_1(t) - \frac{1}{R_bC_2}u_{2b}(t) - \frac{1}{C_2}i_2\,(t), \quad (3)$$

$$\frac{du_C(t)}{dt} = \frac{1}{C}i_1(t), \qquad \frac{di_{2a}(t)}{dt} = \frac{1}{L_2}u_{2a}(t), \qquad \frac{di_{2b}(t)}{dt} = \frac{1}{L_2}u_{2b}\,(t). \quad (4)-(6)$$

By time discretization (1) - (6) with step Δ one create discrete dynamical model [6]

$$\boldsymbol{x}_{k+1} = \boldsymbol{F}_\Delta \boldsymbol{x}_k + \boldsymbol{G}_\Delta \boldsymbol{u}_k,$$

where

$$\boldsymbol{x}_k = [i_1, u_{2a}, u_{2b}, u_C, i_{2a}, i_{2b}]^T \text{ and } \boldsymbol{u}_k = \left[\sqrt{2}U\sin\left[fix\left(4\frac{\Delta}{T}k\right)\frac{\pi}{2} + \frac{\pi}{4}\right], 0, 0, 0, 0, 0\right]^T$$

and $fix\left(4\frac{\Delta}{T}k\right)$ means integer part of $\left(4\frac{\Delta}{T}k\right)$ and $\boldsymbol{F}_\Delta$, $\boldsymbol{G}_\Delta$ are matrices of the system.

Determination of matrices $\boldsymbol{F}_\Delta$, $\boldsymbol{G}_\Delta$ is possible to obtain by using of:

- analytic method (suitable for low order system);
- numerical methods depending on their type, e.g. Euler direct explicit method and Euler indirect implicit method [8].

Results of verification worked-out by computer simulation are shown in Fig. 3a, Fig. 3b for different output values of R_a and R_b.

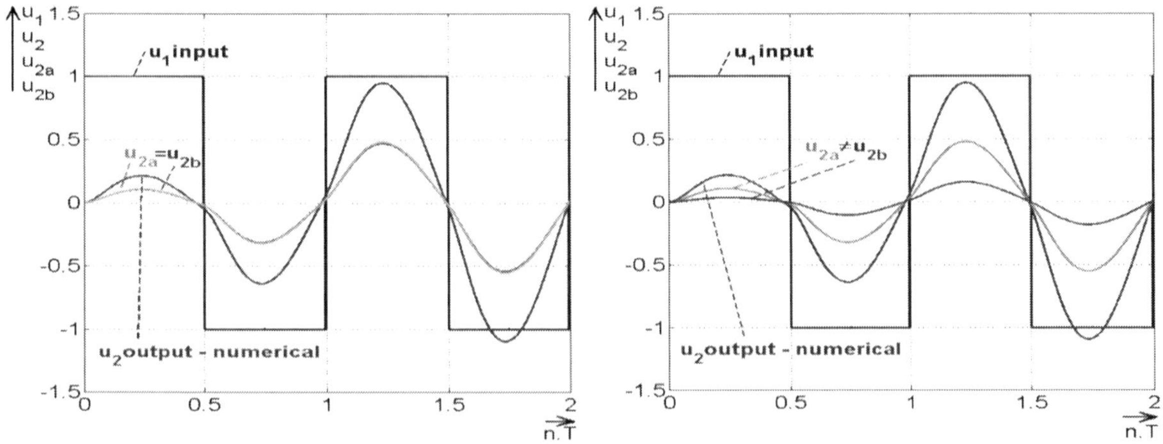

a) b)

Fig. 3. Time waveforms of output voltages u_2, u_{2a}, u_{2b}.
a) $R_a = R_b$, b) $R_a \neq R_b$ $(R_a = \frac{1}{3} R_b)$

Based on above equations and using new modifying method proposed in the work [10] it is possible to obtain the solution for any discrete time instant

$$\boldsymbol{x}_k = (\boldsymbol{F}_\Delta)^k \boldsymbol{x}_0 + \boldsymbol{G}_\Delta \sum_{l=k-1}^{0} (\boldsymbol{F}_\Delta)^l \{\boldsymbol{u}_{k-(l+1)}\} \qquad (7)$$

for $k \in < 0, \infty)$ where k is integer.

Verification results achieved by numerical computer simulation using the method (7) are not given in the paper.

4 Conclusion

Method introduced in the paper presents good mean for analytic solution of complex electrical circuits under periodic non-harmonic impulse supply. Worked-out results have confirmed that theoretical assumptions and verifying computer simulations are in very good agreement.

Acknowledgement

The first and the third authors acknowledge the Grant Agency for Research and Development (APVV) of Slovak Republic for supporting this work under the Grant No. 0314-12.

The first and the second authors were also supported by Russian Science Foundation through the Grant No. 14-49-00079.

References

[1] FRIVALDSKÝ, M., DOBRUCKÝ, B., PRAŽENICA, M., KOSCELNÍK, J. Multi-tank resonant topologies as key design factors for reliability improvement of power converter for power energy applications. *Electrical Engineering - Archiv für Elektrotechnik*, vol. 97, no. 4, 2015, 287-302.

[2] MANN, H. Using computers for electrotechnics designs (in Czech). SNTL, Prague, 1984.

[3] REKTORYS, K. et al., Overview of used mathematics, part 2, SNTL, Prague, 1988.

[4] MAYER, I. Methodological contribution of operator calculus application for transient solution under periodical non-harmonic supply (in Slovak). *Journal of Electrical Engineering*, vol. XXII, no. 2, 1971, 126-129.

[5] KNEPPO, Ľ. *Theoretical electrotechnics – Transient phenomena in electrical circuits, 2nd edition* (in Slovak). SVŠT Bratislava, 1973.

[6] UŠÁK, E. *Transient phenomena solution using integral transforms* (in Slovak). STU, Bratislava, 2011.

[7] DOBRUCKÝ, B., SOLÍK, I. Contribution to state analysis of power semiconductor systems. *Journal of Electrical Engineering*, vol. XXXXIII, no. 3, 1992, 65-71.

[8] STREJC, V. *State-space theory of linear discrete control* (in Czech). Academia, Prague, 1978, Chapter 7.

[9] MORAVČÍK, J. *Mathematical analysis 3* (in Slovak). Alfa, Bratislava, 1992.

[10] DOBRUCKÝ, B. Comparison of two methods for determination of instantaneous state of dynamical system with LCLC circuit. *International Journal of Circuits, Systems and Signal Processing*, vol. 10, 2016, ISSN 1998-4464, 384-389.

Current addresses

Dobrucký Branislav, prof. Ing. PhD.
University of Žilina
01026 Žilina, Slovak Republic
Tel. number: +421415131602, e-mail: branislav.dobrucky@fel.uniza.sk

Moscow Power Energetic Institute – National Research University
111 250 Moscow, Russia

Marčoková Mariana, doc. RNDr. CSc.
University of Žilina
01026 Žilina, Slovak Republic
Tel. number: +421415136205, e-mail: mariana. marcokova@fstav.uniza.sk

Moscow Power Energetic Institute – National Research University
111 250 Moscow, Russia

Štefanec Pavol, PhD. student
University of Žilina
01026 Žilina, Slovak Republic
Tel. number: +421415131606, e-mail: pavol.stefanec@fel.uniza.sk

Slovak University of Technology in Bratislava
Faculty of Mechanical Engineering
16th Conference on Applied Mathematics **APLIMAT 2017**
Proceedings

AN APPLICATION OF POLYNOMIAL FUNCTIONS OF DIFFERENT ORDER IN SMOOTHING OF MORTALITY CURVES: THE CASE STUDY OF BULGARIA AND CZECHIA FROM 1960 TO 2010

DOTLAČILOVÁ Petra (CZ), ŠIMPACH Ondřej (CZ)

Abstract. In this paper we use classical polynomial functions of 2nd, 3rd and 4th order for smoothing the mortality curves at advanced ages (60–110 years of life) in the case of Bulgaria and Czechia. Classical polynomial functions are good approach as a supplement to weighted moving averages and commonly used models for smoothing variance in the age-and-sex-specific death rates. Our results are compared with each other using adjusted index of determination and it will be shown that the quality of smoothing is at very good level in the case of Bulgaria and Czechia.

Keywords: polynomial function, mortality curve, age-specific death rate, Bulgaria, Czechia

Mathematics Subject Classification: Primary 90C30; Secondary 62H12

1 Introduction

Mortality at the advanced ages (60 years and above) becomes important topic for demographers and for analysts in the field of health and pension insurance and also for public policy planning [15]. Extensions of length of human life is influenced mainly by improvement in medicine and in healthcare. For our future it means, that the proportion of persons at higher ages will continue to increase (see e.g. papers by Fiala [7], [8] or Gavrilov, Gavrilova [9]). One cause of population ageing is decrease in mortality. But the empirical values of age-specific death rates show fluctuations, which are noticeable mainly at higher ages. That is the reason why is useful to smooth these values. At higher ages (60 years and above), where mortality has more natural character and variability in observations is higher, it is possible to use some of available functions (e.g. Gompertz-Makeham, Kanistö model, Coale-Kisker model, Thatcher model and others). On the other hand classical mathematical approaches as e.g. weighted moving averages or polynomial functions of different order could be used for smoothing of death rates as well.

The countries of Central and Eastern Europe is aging as it is in the case of developed Western European countries [13], [14]. At first, there was a significant improvement in a care of live born persons and infants, which caused the decrease of infant mortality and mortality rates during childhood. Later, mortality began to improve even at the advanced ages. Among reasons for this evolution could be included higher level of health care, standard of living, better living conditions or functional health and pension and social system. Another equally important reason could be more interest in a healthy lifestyle and also better environment in most countries (especially in industrial cities). Given to this evolution, it is important to have the best imagination about how long in average will live not only the youngest persons, but also the oldest ones.

In this paper we review the existing forms of levelling functions, which are used in mortality curves smoothing. After we describe the methodology of polynomial functions of 2nd, 3rd and 4th order and apply this approach on mortality curves of Bulgaria and Czechia for males and females separately, for the period from 1960 to 2010. Our results are compared with each other using adjusted index of determination and it is shown that the quality of smoothing is at very good level. Last sections of this paper discuss the results and concludes remarks with future challenges.

2 Models review

From the past analysis, performed e.g. by Boleslawski, Tabeau [1] or Burcin, Tesárková, Šídlo [3] it is evident that the level of mortality of younger persons is different in comparison with the oldest ones. Therefore, it is necessary to correct estimates of mortality at the highest ages. For this correction are used various types of models. Nowadays we can use several existing models which are used for smoothing and for estimating of unknown parameters. It is possible to obtain them by using the professional demographical software (e.g. DeRaS, see Burcin, Hulíková Tesárková, Kománek [2], Statistica, or IBM SPSS Statistics). Among the most famous are included Coale-Kisker model (see e.g. Coale, Kisker [4])

$$m_{x,t} = e^{ax^2+bx+c}, \tag{1}$$

where $m_{x,t}$ are age-and-sex specific death rates and x is age. Obtained model corresponds with an exponential quadratic function, where a, b and c are parameters of model. Next one is Thatcher model (see e.g. Thatcher, Kanistö, Vaupel [16])

$$\mu_{x,t} = \frac{z}{1+z} + \gamma, \tag{2}$$

where $z = \alpha e^{\beta x}$, α, β, γ are parameters of model, x is age (and μ_x is the intensity of mortality at exact age x)[1]. The other one is Kannistö model (Thatcher, Kanistö, Vaupel [16])

$$\mu_{x,t} = \frac{e^{[\theta_0+\theta_1(x-80)]}}{1+e^{[\theta_0+\theta_1(x-80)]}}, \tag{3}$$

where θ_0 and θ_1 are unknown parameters. Kanistö model is a special case of the logistic function, where logit transformation of death rates is expressed by linear function of age. The oldest model (but still very frequently used) is the Gompertz-Makeham function (Gompertz [10], Makeham [12])

[1] For the intensity of mortality is valid a simplistic equation $\mu(x+1/2) \doteq m_x$ for $x = 1,2,3,\ldots,59$ (see e.g. [7]).

$$\mu_{x,t} = a + bc^{x}, \tag{4}$$

where a, b and c are parameters. All these models are suitable for the elimination of fluctuations in age-specific death rates and their subsequent extrapolation (more e.g. in Ekonomov, Yarigin [6] or Dotlačilová, Šimpach, Langhamrová [5]). Another option is application of polynomial functions.

In this paper we use the polynomial functions of 2nd, 3rd and 4th order and estimation of parameters of regression functions will be based on the age range x = 60–85 years. We evaluate the significance of models by adjusted indices of determination $adj.R^2$ and this information will be one of the most important about the suitability of models. The other information about suitability can be comparison of smoothed and extrapolated values with theoretical expectations (according to Gompertz [10] theory about the law of mortality in a graphical form prepared by 3D Bicubic splines in *Statgraphics Centurion XVI* software).

3 Materials and Methods

For the purposes of mortality analysis we use demographic data from the "Human Mortality Database" (HMD [11]) for Bulgaria and Czechia, with annual frequency of observations. Let us denote the polynomial function of 2nd, 3rd and 4th order as

$$m_{x,t} = \begin{cases} \beta_0 + \beta_1 x_t + \beta_2 x_t^2 + \varepsilon_t \\ \beta_0 + \beta_1 x_t + \beta_2 x_t^2 + \beta_3 x_t^3 + \varepsilon_t \\ \beta_0 + \beta_1 x_t + \beta_2 x_t^2 + \beta_3 x_t^3 + \beta_4 x_t^4 + \varepsilon_t \end{cases}, \tag{5}$$

where $m_{x,t}$ are the age-specific mortality rates calculated by ratio

$$m_{x,t} = \frac{D_{x,t}}{E_{x,t}}, \tag{6}$$

where $D_{x,t}$ is the number of died x-years old persons in calendar year t, $E_{x,t}$ is the exposure to risk of x-years old persons in calendar year t, which is (commonly) estimated as a number of mid-year population x-years old in year t, β_0, β_1, β_2, β_3 and β_4 are parameters of polynomial functions and ε_t should be a residual element with characteristics of white noise process, where $E(\varepsilon_t) = 0$, $D(\varepsilon_t) = \sigma^2$, $cov(\varepsilon_t ; \varepsilon_t') = 0$ and $\varepsilon_t \approx$ N distribution, x is age and used range is $x \in < 0 ; 110 >$. (The estimation of parameters of polynomial functions will be based on the age range x = 60–85 years only), t is time where $t \in < 1960; 2010 >$ for the case of Bulgaria and Czechia. These polynomial functions we use for smoothing of age-and-sex specific death rates from 60 to 85 years. This interval is not obligatory, but there are some authors (e.g. Burcin, Tesárková, Šídlo [3]), who recommend it. For the advanced ages we perform the extrapolation and our recommendation for the end is $x = 110$.

4 Results

There are modelled the intensities of mortality using polynomial functions of the 2nd, 3rd and 4th order in the following part. We chose Bulgaria (as the population of Eastern Europe with higher levels of mortality) and Czechia (as the population of Central Europe with a lower mortality rates). The resulting outputs are fed sequentially presented using 3D Bicubic splines. In Appendix of this paper is presented the evaluation of obtained outputs using adjusted coefficients of determination.

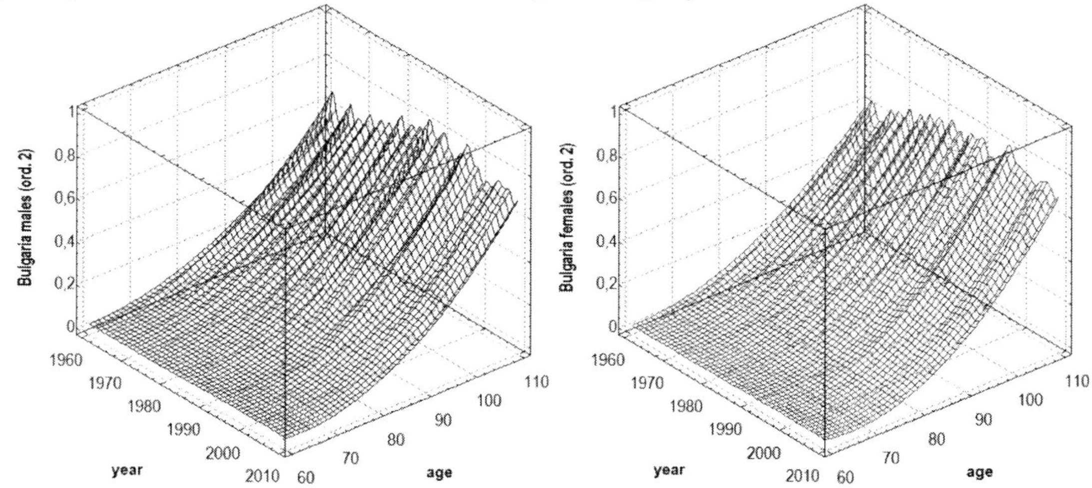

Fig. 1. Smoothing by polynomial function of the 2nd order for Bulgarian males (left) and females (right). Source: authors' calculations and construction

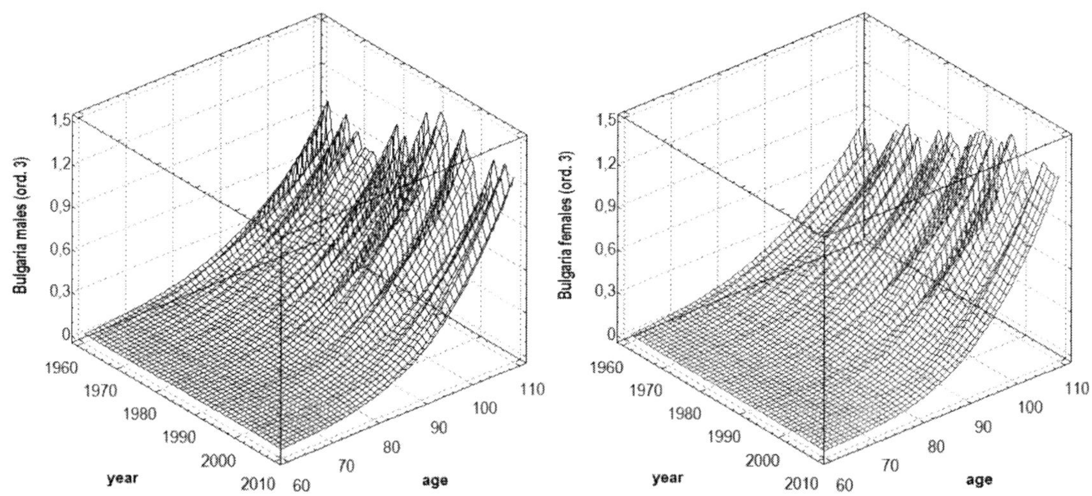

Fig. 2. Smoothing by polynomial function of the 3rd order for Bulgarian males (left) and females (right). Source: authors' calculations and construction

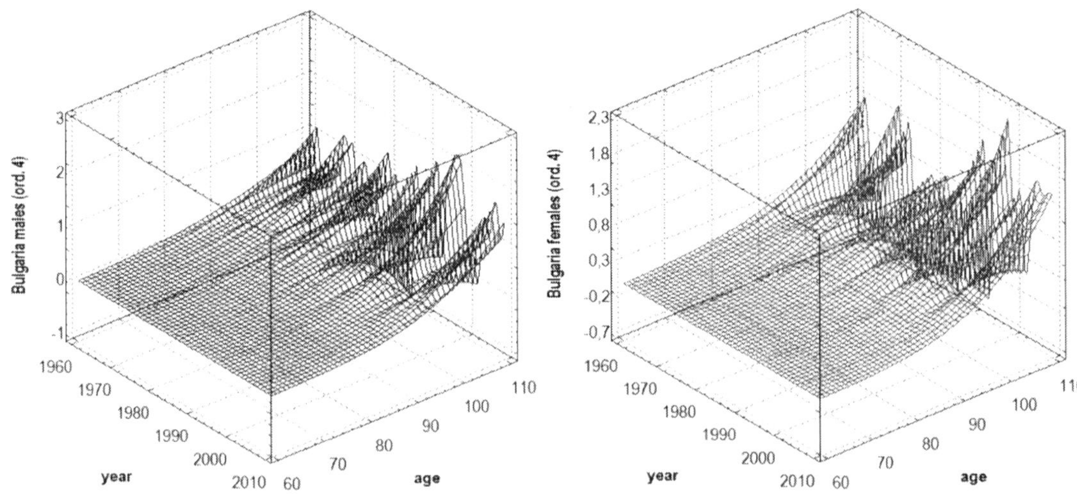

Fig. 3. Smoothing by polynomial function of the 4th order for Bulgarian males (left) and females (right). Source: authors' calculations and construction

4.1 Bulgarian population

Fig. 1 shows the modelled intensity of mortality for the Bulgarian population using a polynomial function of the 2nd order. From the results it is evident that at high ages (90+) we receive relatively smooth fitting. This leads to significant removal of random and systematic variations, which are present in empirical death rates, especially in ages 85+. Fig. 2 shows modelled death rates for the same population using a polynomial function of the 3rd order. From the obtained values we can conclude that the selected function provides a higher level of mortality (in comparison with 2nd order of polynomial function). Then, it is also obvious that the function of 3rd order does not provide fluent fitting (especially in the highest ages). Fig. 3 shows modelled intensity of mortality using a polynomial of 4th order. When applying this polynomial for Bulgarian population we receive the least fluent fitting. (This is especially true in the highest ages – 100+). Moreover, it also shows that the model is not applicable for ages 100+, because there is a decrease in the mortality law (Gompertz [10]). Comparison of models using the adjusted indices of determination is presented in Fig. 7, 8 and 9 in Appendix of this paper. The best results are generally in the case of function of the 4th order, but that cannot be used universally. The significance of the polynomial of the 3rd order is sufficient in most cases and we can claim that can be used universally.

4.2 Czech population

Following part is devoted to modelling of mortality of the Czech population. Fig. 4 shows modelled intensity of mortality obtained when applying a polynomial function of the 2nd order. Also here is evident that we receive relatively smooth fitting in the highest ages. Moreover, modelled intensity of mortality corresponds to the assumption that the mortality decreases from past to present. In Fig. 5 is graphically shown the intensity of mortality modelled by polynomial function of 3rd order. If we compare the results obtained by function of the 2nd and 3rd order, we find that the second one provides higher values of the intensity of mortality. Then, it is also possible to say that we receive for Czech males less continuous fitting in the highest ages (compared to the previous function). Last

Fig. 6 shows the modelling of intensity of mortality using a polynomial function of 4th order. It is obvious even in the case of the Czech population that this function fails in some years. There is a reverse decline of intensity of mortality (breaking down the extrapolated development to the opposite direction). Compared to the previous functions is thus applicable in selected years only. Comparison of models for the Czech population using the adjusted indices of determination is presented in Fig. 10, 11 and 12 in Appendix of this paper. Again in this case are the best results generally provided by function of the 4th order. Using this function is not possible in general. The significance of the polynomial of the 3rd order was sufficient in most cases during development of 1960–2010 and then we can claim that it can be universally used.

5 Discussion and Conclusion

The aim of this paper was to show different ways of smoothing mortality curve at ages 60+. As an analytical functions were chosen polynomial functions of the 2nd, 3rd and 4th order. By these functions there was modelled the intensity of mortality in Bulgarian and Czech population. Another goal was to assess the obtained fitting using by *adj.*R^2. Based on the results we can say that using a polynomial function of the 2nd and 3rd order brings relatively smooth fitting. A different situation arises when applying a function of the 4th order. This function is failing mainly in the case of Bulgarian population, and this misalignment is evident mainly in the highest ages (about 100 years).

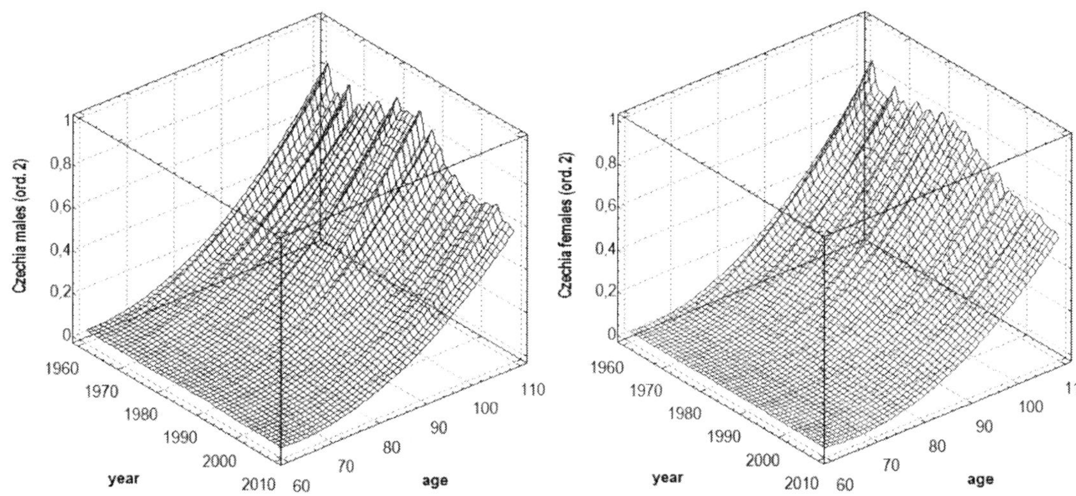

Fig. 4. Smoothing by polynomial function of the 2nd order for Czech males (left) and females (right).
Source: authors' calculations and construction

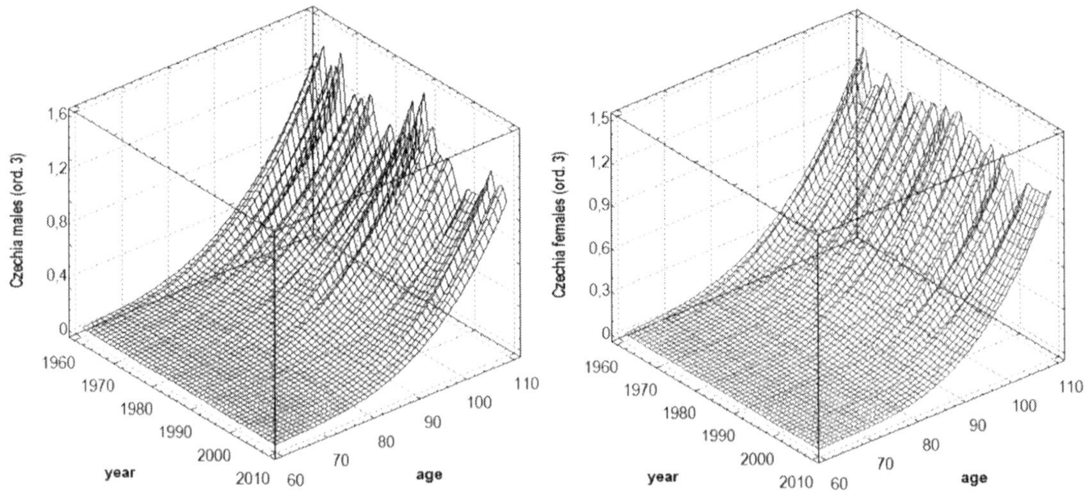

Fig. 5. Smoothing by polynomial function of the 3rd order for Czech males (left) and females (right).
Source: authors' calculations and construction

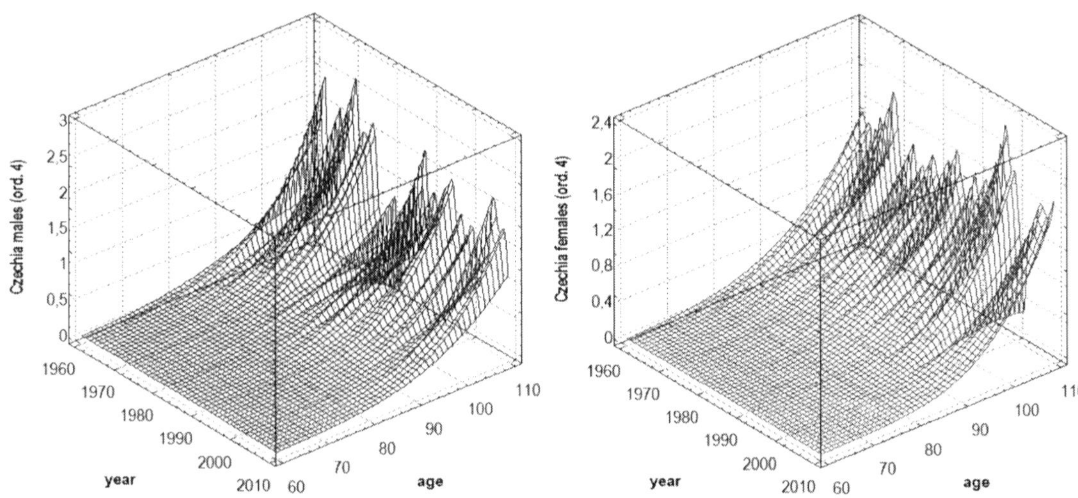

Fig. 6. Smoothing by polynomial function of the 4th order for Czech males (left) and females (right).
Source: authors' calculations and construction

A similar conclusion can be claimed in the case of the Czech population, where a polynomial function of 4th order works correctly in selected years only.

In the next part we dealt with the evaluation of selected models using the adjusted coefficient of determination. The evaluation was firstly carried out in Bulgarian population. Both in the case of males and females we can say, that for the modelling of intensity of mortality in the age range of 60–85 years is the most appropriate polynomial function of the 3rd or 4th order, but towards the present, as seems to be more appropriate polynomial function of the 3rd order. In the case of the Czech population we came to a similar conclusion. Here is the most appropriate model the polynomial function of the 3rd (or 4th, but not in general) order.

Acknowledgement

The authors gratefully acknowledge to the Czech Science Foundation project no. P402/12/G097 DYME – Dynamic Models in Economics for supporting this paper.

References

[1] BOLESLAWSKI, L., TABEAU, E. Comparing Theoretical Age Patterns of Mortality beyond the Age of 80. *In: Tabeau, E., Van Den Berg Jeths, A. and Heathcote, CH. (Eds.): Forecasting Mortality in Developed Countries: Insights from a Statistical, Demographic and Epidemiological Perspective*, 2001, pp. 127 – 155.

[2] BURCIN, B., HULÍKOVÁ TESÁRKOVÁ, K., KOMÁNEK, D. *DeRaS: software tool for modelling mortality intensities and life table construction*. Charles University in Prague, 2012, Prague. URL: http://deras.natur.cuni.cz

[3] BURCIN, B., TESÁRKOVÁ, K., ŠÍDLO, L. Nejpoužívanější metody vyrovnávání a extrapolace křivky úmrtnosti a jejich aplikace na českou populaci. *Demografie*, no. 52, vol. 2010, pp. 77 – 89.

[4] COALE, A., KISKER, E. Defects in Data on Old-age Mortality in the United States: New Procedures for Calculating Mortality Schedules and Life Tables at the Highest Ages. *Asian and Pacific Population Forum*, 1990, vol. 4, pp. 1 – 31.

[5] DOTLAČILOVÁ, P., ŠIMPACH, O., LANGHAMROVÁ, J. DeRaS versus MS Excel Solver in Levelling the Life Expectancy at Birth. *In: APLIMAT 2013*. Bratislava: Publishing House of STU, 2014, pp. 108 – 114.

[6] EKONOMOV, AL., YARIGIN, VN. The age dynamics of mortality and the Gompertz-Makeham law, *Zhurnal Obshchei Biologii*, 1989, vol. 50, no. 2, pp. 236 – 243.

[7] FIALA, T. Mortality in the Age from 20 to 70 Years in the Czech Republic in the Period 1950 to 1996. *Acta Oeconomica Pragensia*, 1999, vol. 7, no. 4, pp. 35 – 44.

[8] FIALA, T. Reducing the Consequences of Population Ageing for the Pension System by Raising the Retirement Age in the Czech Republic. *In: The 7th International Days of Statistics and Economics*. Slaný: Melandrium, 2013, pp. 398 – 407.

[9] GAVRILOV, L.A., GAVRILOVA, N.S. Mortality measurement at advanced ages: a study of social security administration death master file. *North American actuarial journal*, 2011, vol. 15, no. 3, pp. 432 – 447.

[10] GOMPERTZ, B. On the Nature of the Function Expressive of the Law of Human Mortality, and on a New Mode of Determining the Value of Life Contingencies. *Philosophical Transactions of the Royal Society of London*, vol. 1825, no. 115, pp. 513 – 585.

[11] HMD: Wilmoth, J.R., Shkolnikov, V, Barbieri, M. *The Human Mortality Database*, URL: http://www.mortality.org/

[12] MAKEHAM, W.M. On the Law of Mortality and the Construction of Annuity Tables. *The Assurance Magazine, and Journal of the Institute of Actuaries*, vol. 1860, no. 8, pp. 301 – 310.

[13] ŠIMPACH, O., LANGHAMROVÁ, J. Czech Household Computer Facilities as a Reliable Variable in a Life Expectancy Forecast Model up to the Year 2060. *In: IDIMT-2012 (ICT Support for Complex Systems)*. Linz: Trauner, 2012, pp. 143 – 152.

[14] ŠIMPACH, O., LANGHAMROVÁ, J. The Impact of ICT Growth on Households and Municipalities in the Czech Nuts-3 Regions: The Application of Cluster Analysis. *In: IDIMT-2014 (Networking Societies – Cooperation and Conflict)*. Linz: Trauner, 2014, pp. 63 – 70.

[15] ŠIMPACH, O., PECHROVÁ, M. Development of the Czech Farmers' Age Structure and the Consequences for Subsidy Policy. *Agris on-line Papers in Economics and Informatics*, 2015, vol. VII, no. 3, pp. 57 – 69.

[16] THATCHER, R.A., KANISTÖ, V., VAUPEL, J.W. *The Force of Mortality at Ages 80 to 120*. Odense University Press, 1998.

Appendix

Following part presents the results of adjusted determination indices for particular populations.

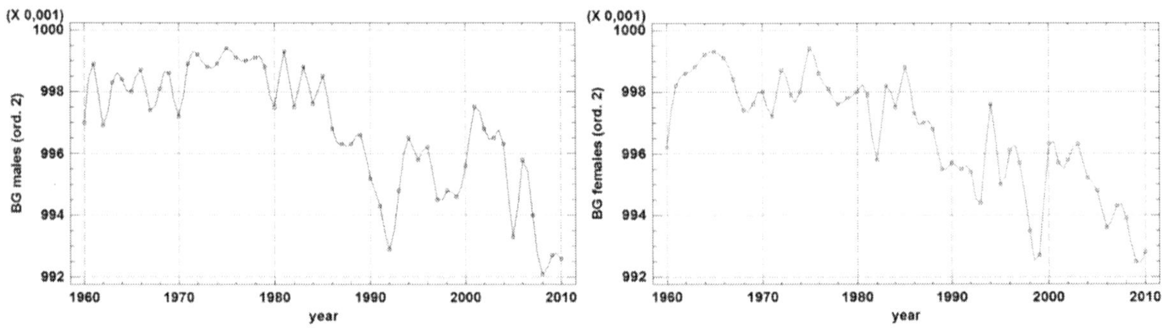

Fig. 7. Development of *adj.*R^2 for polynomial functions of the 2nd order for Bulgarian males (left) and females (right). Source: authors' calculations and construction

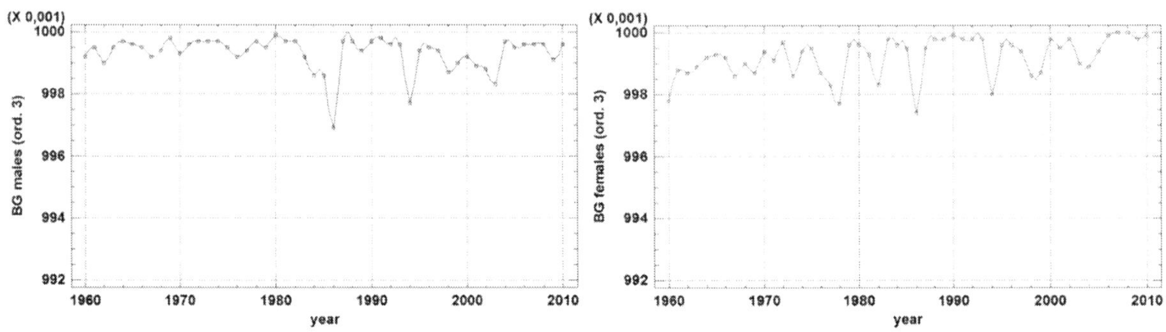

Fig. 8. Development of *adj.*R^2 for polynomial functions of the 3rd order for Bulgarian males (left) and females (right). Source: authors' calculations and construction

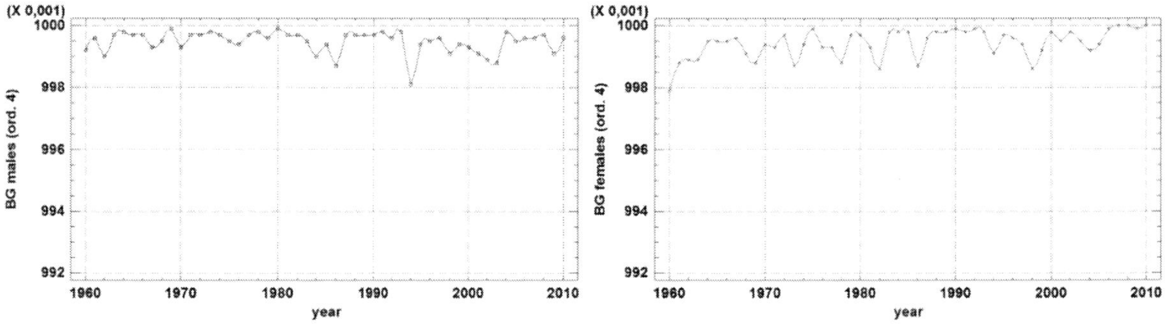

Fig. 9. Development of $adj.R^2$ for polynomial functions of the 4th order for Bulgarian males (left) and females (right). Source: authors' calculations and construction

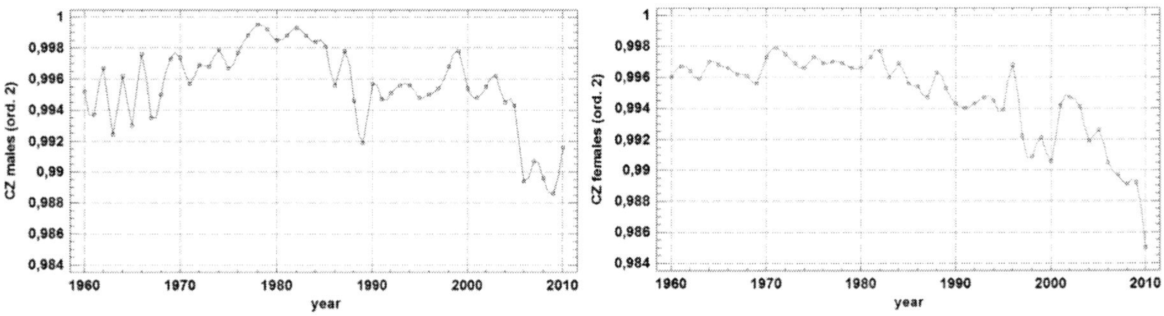

Fig. 10. Development of $adj.R^2$ for polynomial functions of the 2nd order for Czech males (left) and females (right). Source: authors' calculations and construction

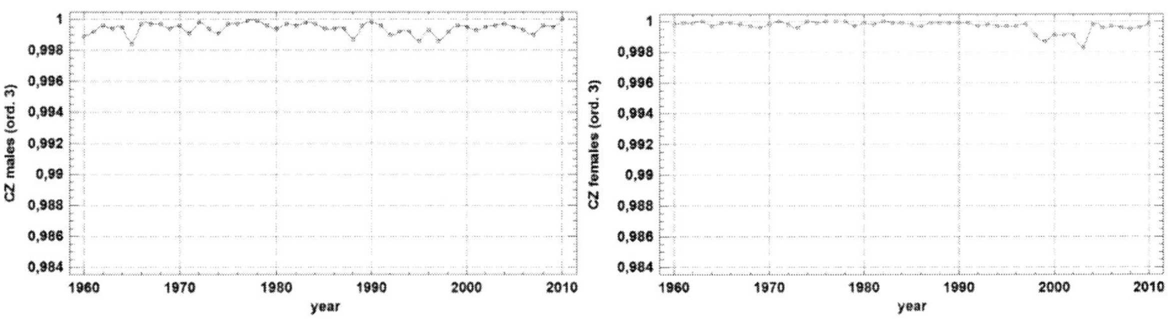

Fig. 11. Development of $adj.R^2$ for polynomial functions of the 3rd order for Czech males (left) and females (right). Source: authors' calculations and construction

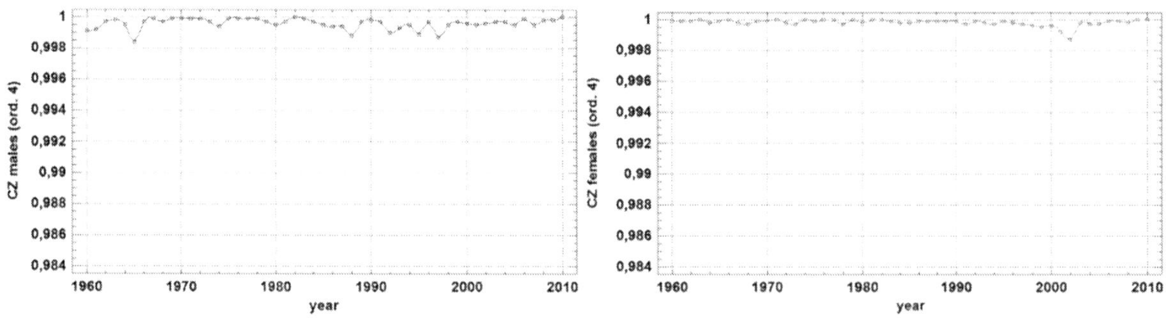

Fig. 12. Development of $adj.R^2$ for polynomial functions of the 4th order for Czech males (left) and females (right). Source: authors' calculations and construction

Current address

Dotlačilová Petra, Ing.
University of Economics in Prague, Faculty of Informatics and Statistics,
Department of Mathematics
W. Churchill sq. 4, 130 67 Prague 3, Czech Republic
Tel. number: +420 224 095 273, e-mail: petra.dotlacilova@vse.cz

Šimpach Ondřej, Ing. Ph.D.
University of Economics in Prague, Faculty of Informatics and Statistics,
Department of Statistics and Probability
W. Churchill sq. 4, 130 67 Prague 3, Czech Republic
Tel. number: +420 224 095 273, e-mail: ondrej.simpach@vse.cz

Slovak University of Technology in Bratislava
Faculty of Mechanical Engineering
16th Conference on Applied Mathematics APLIMAT 2017
Proceedings

ON THE CROSSING NUMBERS OF SEVERAL 8-VERTEX TREES WITH PATHS AND CYCLES

DRAŽENSKÁ Emília (SK)

Abstract. The crossing number of a simple graph G is the minimum number of edge crossings in any drawing of G in the plane. Garey and Johnson have proved that compute the crossing number for a given graph is a very difficult problem, it is NP–complete problem, in general. There are several classes of graphs for which crossing numbers have been studied. We extend these results by giving the exact values or upper bounds of crossing numbers of some graphs.

Keywords: graphs, drawings, crossing numbers

Mathematics subject classification: 05C10, 05C38

1 Introduction

Let G be a simple, undirected and connected graph with set of vertices V and set of edges E. A mapping that assigns a point in the plane for each vertex and for each edge a continuous curve between its two endpoints is called a *drawing* of the graph $G = (V, E)$. A *crossing* of two edges is the intersection of the interiors of the corresponding curves. The *crossing number*, $\mathrm{cr}(G)$, of a graph G is the minimum number of edge crossings in a drawing of G in the plane. The drawing with minimum number of crossings must be a *good drawing*, that means, each two edges have at most one point in common, which is either a commom end-vertex or a crossing and no three edges cross at the same point. Garey and Johnson proved that compute the crossing number for a given graph is NP-complete problem [6]. The problem of determining the crossing number of a given graph has been studied in graph theory and in computer science, VLSI-layout. Many researchers have focused on optimizing the VSLI circuit layout. Especially, the layout theory studied the efficiency of embedding the graphs in the plane according to VLSI rules. The main problem is minimizing the number of wire crossings in a circuit. C. Thompson introduced a graph-theoretic model for very large integration circuitry [20].

Cartesian products of two graphs are one of few graph classes for which the crossing numbers were studied. Cartesian product $G_1 \square G_2$ of graphs $G_1 = (V_1, E_1)$ and $G_2 = (V_2, E_2)$ is a graph with vertex set $V = V_1 \times V_2$ and two vertices (v_1, v_2) and (u_1, u_2) are adjacent in $G_1 \square G_2$ if and only if either $v_1 = u_1$ and v_2 is adjacent with u_2 in G_2 or $v_2 = u_2$ and v_1 is adjacent with u_1 in G_1.

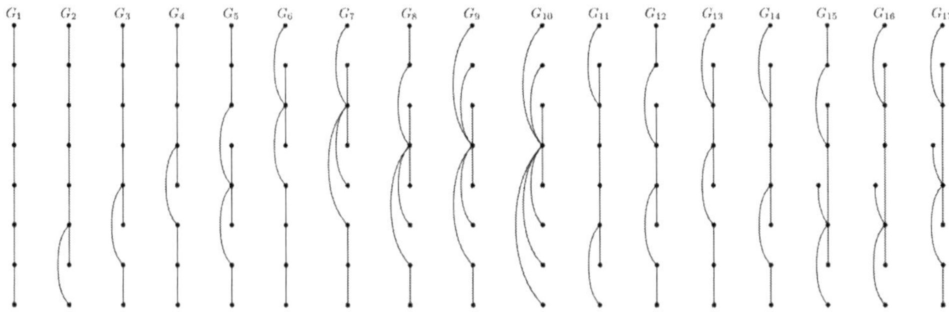

Fig. 1. Several eight vertex trees

Let C_n be the cycle with n edges, P_n be the paths with n edges and S_n be the star with n edges. A tree is a connected graph without cycle. The crossing numbers for the Cartesian products of some specific graphs G with cycles, paths and stars were determined. The crossing numbers of $G \square C_n$ were studied in [1], [3], [4], [7], [8], [10], [12], [11], and [15]. The crossing numbers of Cartesian products $G \square P_n$ are determined in [5], [9], [13], [14], [16], [19], and [21] . Besides of Cartesian product, there are join and strong products of two graphs for which exact values of crossing numbers were studied (see [17], [18]).

In this paper we extend these results by giving the exact values or upper bounds of the crossing numbers of Cartesian products for several trees on eigth vertices with paths P_n and with cycles C_n. There exist 23 connected graphs G with eight vertices and seven edges. We state the crossing numbers of $G \square P_n$ or $G \square C_n$ for 17 of them (see Fig. 1).

2 The crossing number of the Cartesian products with paths

Every graph $G_i \square P_n$ contains $n+1$ copies of the graph G_i. An uppper bound for the crossing number of the graph $G_i \square P_n$ is a number of crossings in a drawing of $G_i \square P_n$. On the other hand, if the graph $G_i \square P_n$ contains a subgraph for which the crossing number is known yet, we have a lower bound for the crossing number for the graph $G_i \square P_n$.

Theorem 2.1 *For* $n \geq 1$, $\mathrm{cr}(G_1 \square P_n) = 0$,

$$\mathrm{cr}(G_2 \square P_n) = \mathrm{cr}(G_3 \square P_n) = \mathrm{cr}(G_4 \square P_n) = n - 1,$$

$$\mathrm{cr}(G_5 \square P_n) = \mathrm{cr}(G_6 \square P_n) = 2(n - 1),$$

$$\mathrm{cr}(G_7 \square P_n) = \mathrm{cr}(G_8 \square P_n) = 4(n - 1),$$

$$\mathrm{cr}(G_9 \square P_n) = 6(n - 1),$$

$$\mathrm{cr}(G_{10} \square P_n) = 9(n - 1),$$

$$\mathrm{cr}(G_{11} \square P_n) = 2(n - 1).$$

Proof: It is easy to see, that the graph $P_m \square P_n$ is planar. As the graph G_1 is isomorphic to the path P_7, $\mathrm{cr}(G_1\square P_n) = 0$. The graph $G_{10}\square P_n$ is isomorphic to the graph $S_7\square P_n$. Bokal in [2] proved that $\mathrm{cr}(S_7\square P_n) = 9(n-1)$.

In Figs. 2(a), 2(b), and 2(c) one can find the drawings of the graphs $G_2\square P_n$, $G_3\square P_n$, and $G_4\square P_n$. The edges of $n-1$ copies of the graphs G_2, G_3, and G_4 are crossed once,

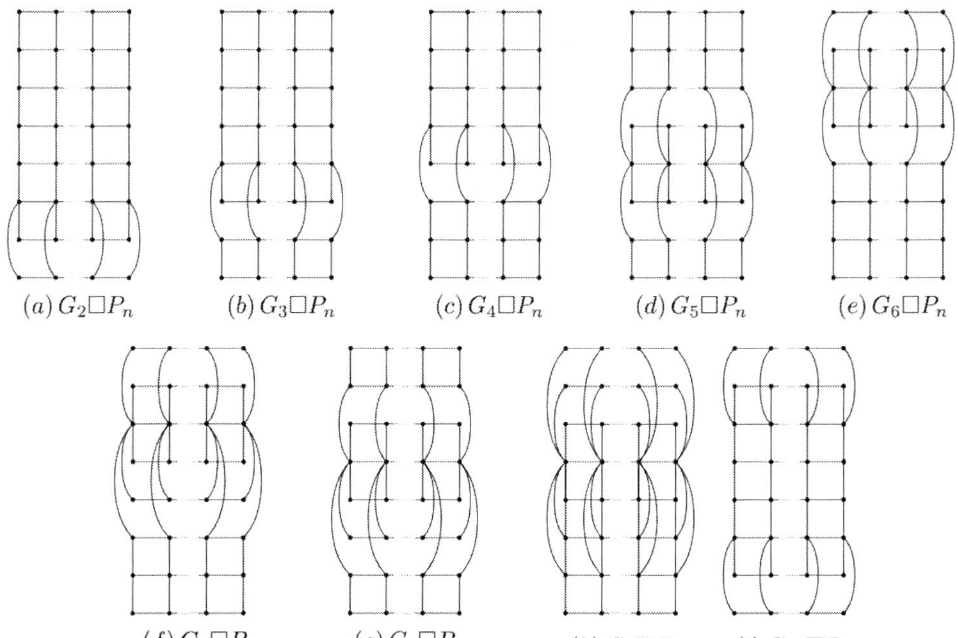

$(a)\ G_2\square P_n$ $(b)\ G_3\square P_n$ $(c)\ G_4\square P_n$ $(d)\ G_5\square P_n$ $(e)\ G_6\square P_n$

$(f)\ G_7\square P_n$ $(g)\ G_8\square P_n$ $(h)\ G_9\square P_n$ $(i)\ G_{11}\square P_n$

Fig. 2. Cartesian products $G_i\square P_n$ for $i = 2, \ldots 9, 11$

so the crossing number of $G_2\square P_n$, $G_3\square P_n$, and $G_4\square P_n$ is at most $n-1$. The drawings of the graphs $G_5\square P_n$, $G_6\square P_n$, and $G_{11}\square P_n$ in which the edges of $n-1$ copies of subgraphs isomorphic to G_5, G_6, and G_{11} are crossed twice are shown in Figs. 2(d), 2(e), and 2(i). So, $\mathrm{cr}(G_5\square P_n) \leq 2(n-1)$, $\mathrm{cr}(G_6\square P_n) \leq 2(n-1)$, and $\mathrm{cr}(G_{11}\square P_n) \leq 2(n-1)$. The drawings of the graphs $G_7\square P_n$ and $G_8\square P_n$ with $4(n-1)$ crossings are shown in Figs. 2(f) and 2(g). Thus, $\mathrm{cr}(G_7\square P_n) \leq 4(n-1)$ and $\mathrm{cr}(G_8\square P_n) \leq 4(n-1)$. There is a drawing of the graph $G_9\square P_n$ with $6(n-1)$ crossings (see Fig. 2(h)). So, $\mathrm{cr}(G_9\square P_n) \leq 6(n-1)$.

Now, we find the lower bounds of crossing numbers $\mathrm{cr}(G_i\square P_n)$. The graphs G_2, G_3, and G_4 contain the graph S_3 as a subgraph. So, the Cartesian products $G_2\square P_n$, $G_3\square P_n$, and $G_4\square P_n$ contain $S_3\square P_n$ as a subgraph. It was proved in [7] that $\mathrm{cr}(S_3\square P_n) = n-1$. It implies, that the crossing number of $G_2\square P_n$, $G_3\square P_n$, and $G_4\square P_n$ is at least $n-1$. The graphs $G_5\square P_n$, and $G_6\square P_n$ contain the graph $S_4\square P_n$ as a subgraph. As $\mathrm{cr}(S_4\square P_n) = 2(n-1)$ (see [8]), $\mathrm{cr}(G_5\square P_n) \geq 2(n-1)$, and also $\mathrm{cr}(G_6\square P_n) \geq 2(n-1)$. The graph $S_5\square P_n$ is a subgraph of the graph $G_7\square P_n$ and $G_8\square P_n$. The graph $G_9\square P_n$ contains the graph $S_6\square P_n$ as a subgraph. It was determined in [2] that $\mathrm{cr}(S_5\square P_n) = 4(n-1)$ and $\mathrm{cr}(S_6\square P_n) = 6(n-1)$. So, the lower bound $4(n-1)$ for the crossing numbers of $G_7\square P_n$ and $G_8\square P_n$ and $6(n-1)$ for $\mathrm{cr}(G_9\square P_n)$. As the graph G_{11} contains two vertex–disjoint subgraphs, both isomorphic to S_3, we obtain $\mathrm{cr}(G_{11}\square P_n) \geq \mathrm{cr}(S_3\square P_n) + \mathrm{cr}(S_3\square P_n) \geq 2(n-1)$.

The upper and lower bounds of crossing numbers for graphs $G_i\square P_n$ for $i = 2, \ldots, 9$ and 11 are the same. Thus, we get exact values of crossing numbers of corresponding graphs. □

Formatting Issue - Best Available Paper/Graphic

Theorem 2.2 *For* $n \geq 1$, $\mathrm{cr}(G_{12}\Box P_n) = \mathrm{cr}(G_{13}\Box P_n) = \mathrm{cr}(G_{14}\Box P_n) = 2(n-1)$.

Proof: The drawings of the graphs $G_{12}\Box P_n$, $G_{13}\Box P_n$, and $G_{14}\Box P_n$ with $2(n-1)$ crossings are shown in Figs. 3(a) – (c). Thus, $cr(G_i\Box P_n) \leq 2(n-1)$ for $i = 12, 13, 14$.

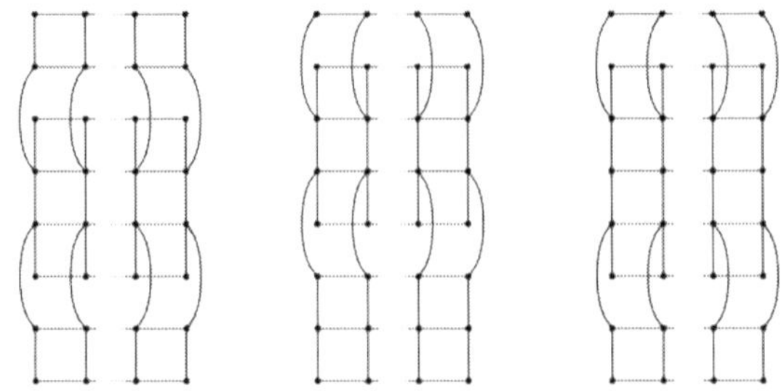

$(a)\, G_{12}\Box P_n$ $(b)\, G_{13}\Box CP_n$ $(c)\, G_{14}\Box P_n$

Fig. 3. Cartesian products $G_i\Box P_n$ for $i = 12, 13, 14$

Fig. 4. The graph T

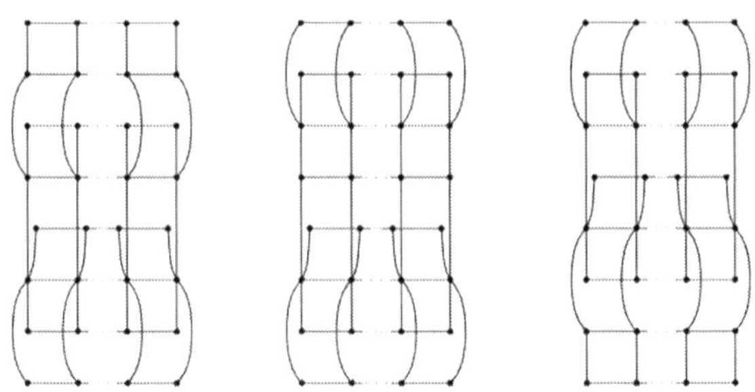

$(a)\, G_{15}\Box P_n$ $(b)\, G_{16}\Box P_n$ $(c)\, G_{17}\Box P_n$

Fig. 5. Cartesian products $G_i\Box P_n$ for $i = 15, 16, 17$

The graphs G_{12} and G_{13} contain the graph T (see Fig. 4) as a subgraph and the graph G_{14} contains the elementary subdivision of the graph T (subdivision of the edge which is incident to the vertices of degree three). So, the graph $T\Box P_n$ or its subdivision is a subgraph of the graphs $G_i\Box P_n$. It was published in [16], that $\mathrm{cr}(T\Box P_n) = 2(n-1)$. Hence, the crossing numbers of graphs $G_{12}\Box P_n$, $G_{13}\Box P_n$, and $G_{14}\Box P_n$ are at least $2(n-1)$. Thus, $\mathrm{cr}(G_{12}\Box P_n) = \mathrm{cr}(G_{13}\Box P_n) = \mathrm{cr}(G_{14}\Box P_n) = 2(n-1)$. □

Theorem 2.3 *For $n \geq 1$, $\mathrm{cr}(G_{15}\Box P_n) = \mathrm{cr}(G_{16}\Box P_n) = \mathrm{cr}(G_{17}\Box P_n) = 3(n-1)$.*

Proof: There are drawings of the graphs $G_{15}\Box P_n$, $G_{16}\Box P_n$, and $G_{17}\Box P_n$ (see Figs. 5 (a)–(c)) with $3(n-1)$ crossings. Thus, the crossing number for every graph $G_i\Box P_n$, $i = 15, 16, 17$ is at most $3(n-1)$. The graph S (see Fig. 6) is a subgraph of every graph G_i for $i = 15, 16$ and 17. So, the graph $S\Box P_n$ is a subgraph of the graphs $G_i\Box P_n$. As $\mathrm{cr}(S\Box P_n)$ = $3(n-1)$ (see [5]), $\mathrm{cr}(G_i\Box P_n) \geq 3(n-1)$. This completes our proof. □

Fig. 6. The graph S

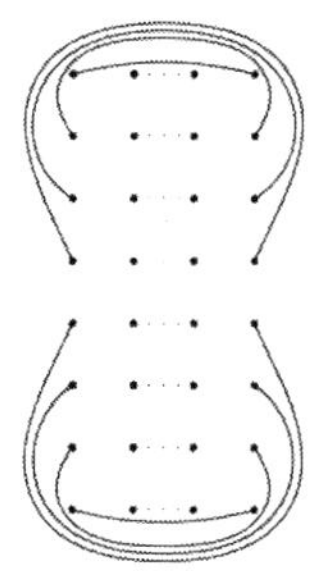

Fig. 7. Added edges

3 The crossing number of the Cartesian products with cycles

The graph $G_i\Box C_n$ contains n copies of the graph G_i. The graph $G_i\Box C_n$ is obtained by adding eight edges to the graph $G_i\Box P_{n-1}$, see Fig. 7.

The graph $G_1\Box C_n$ is planar. It was proved in [7] that $\mathrm{cr}(S_3\Box C_n) = n$ and in [8] $\mathrm{cr}(S_4\Box C_n) = n$ for $n \geq 6$. It was proved in the same papers that the crossing number of $S_3\Box C_n$ for $n = 3, 4, 5$ is 1, 2, 4, respectively and the crossing number of $S_4\Box C_n$ for $n = 3, 4, 5$ is 2, 4, 8, respectively. As the graphs G_2, G_3, and G_4 contain the star S_3 as a subgraph, the Cartesian product $S_3\Box C_n$ is a subgraph of graphs $G_2\Box C_n$, $G_3\Box C_n$, and $G_4\Box C_n$. Thus, $\mathrm{cr}(G_2\Box C_n) \geq n$, $\mathrm{cr}(G_3\Box C_n) \geq n$, and $\mathrm{cr}(G_4\Box C_n) \geq n$ for $n \geq 6$ and the lower bound for the crossing numbers of $G_2\Box C_n$, $G_3\Box C_n$, and $G_4\Box C_n$ for $n = 3, 4, 5$ is 1, 2, 4, respectively. Both graphs G_5 and G_6 contain the graph S_4 as a subgraph, thus $\mathrm{cr}(G_5\Box C_n) \geq 2n$ and $\mathrm{cr}(G_6\Box C_n) \geq 2n$ for $n \geq 6$ and 2, 4, 8 for $n = 3, 4, 5$, respectively. Thus, we receive lower bounds for the crossing numbers of the graphs $G_i\Box C_n$ $(i = 2, \ldots 6)$ for $n \geq 3$.

The graphs G_{11}, G_{12}, G_{13}, and G_{14} contain the graph T (see Fig. 4) or its subdivision as a subgraph. Thus, the Cartesian product $T\Box C_n$ or its subdivision is a subgraph of graphs $G_{11}\Box C_n$, $G_{12}\Box C_n$, $G_{13}\Box C_n$, and $G_{14}\Box C_n$. In [4], it was published that $\mathrm{cr}(T\Box C_3) \geq 2$, $\mathrm{cr}(T\Box C_4) \geq 4$, $\mathrm{cr}(T\Box C_5) \geq 8$, and $\mathrm{cr}(T\Box C_n) \geq 2n$ for $n \geq 6$. Thus, for $i = 11, 12, 13$, and 14, $\mathrm{cr}(G_i\Box C_3) \geq 2$, $\mathrm{cr}(G_i\Box C_4) \geq 4$, $\mathrm{cr}(G_i\Box C_5) \geq 8$, and $\mathrm{cr}(G_i\Box C_n) \geq 2n$ for $n \geq 6$ hold.

Let every copy of G_i in the drawing of $G_i \Box C_n$ has the same number of crossings, say k crossings. Then the graph $G_i \Box C_n$ has at most $k \cdot n$ crossings. So, we have the upper bound for the crossing numbers of every graph $G_i \Box C_n$ for $i = 2, \ldots 17$. We can reduce this upper bound for $\mathrm{cr}(G_i \Box C_n)$, $i = 2, \ldots 6, 11, \ldots 14$ and $n = 3, 4, 5$. Using drawing in Fig. 8 (a) we have $\mathrm{cr}(G_3 \Box C_5) \geq 4$. By deleting the edges of one copy of subgraphs isomorphic to G_3 which edges are crossed two times, we obtaind a drawing of subdivision of $G_3 \Box C_4$ with two crossings. And by removing the edges of one more copy of subgraph isomorphic to G_3 which edges are crossed once, we get a drawing of $G_3 \Box C_3$ with one crossing. So, $\mathrm{cr}(G_3 \Box C_4) \geq 2$ and $\mathrm{cr}(G_5 \Box C_5) \geq 1$. The drawing in Fig. 8 (b) shows that $\mathrm{cr}(G_5 \Box C_5) \geq 8$. By deleting the edges of one copy of subgraphs isomorphic to G_5 which edges are crossed four times, we obtaind a drawing of subdivision of $G_5 \Box C_4$ with four crossings. And by deleting the edges of one more copy of subgraph isomorphic to G_5 which edges are crossed twice, we get a drawing of $G_5 \Box C_3$ with two crossing. Thus, $\mathrm{cr}(G_3 \Box C_4) \geq 4$ and $\mathrm{cr}(G_5 \Box C_5) \geq 2$. We have $\mathrm{cr}(G_{14} \Box C_5) \geq 4$ (see Fig. 8 (c)). We delete the edges of one respektively we delete the edges of two copies of the subgraphs isomorphic to G_{14}. And we get subdivisions of $G_{14} \Box C_4$ respektively $G_{14} \Box C_4$ with four respektively two crossings. It is not difficult to find drawings of the graphs $G_i \Box C_5$, $G_i \Box C_4$, and $G_i \Box C_3$ for $i = 2, 4, 6, 11, 12, 13$. So,

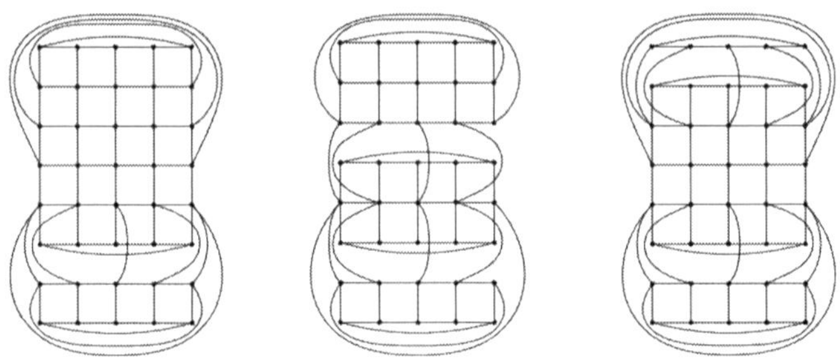

$(a)\ G_3 \Box C_5$ $(b)\ G_5 \Box C_5$ $(c)\ G_{14} \Box C_5$

Fig. 8. Cartesian products with C_5

we have the crossing number of the graphs $G_i \Box C_n$ for $n = 1, \ldots, 6, 11, 12, 13, 14$. For crossing number of graphs $G_i \Box C_n$ for $n = 7, 8, 9, 10, 15, 16, 17$ we have only upper bounds.

Theorem 3.1 *For* $n \geq 3$, $\mathrm{cr}(G_1 \Box C_n) = 0$.

For $n \geq 6$, $\mathrm{cr}(G_2 \Box C_n) = \mathrm{cr}(G_3 \Box C_n) = \mathrm{cr}(G_4 \Box C_n) = n$,

$$\mathrm{cr}(G_5 \Box C_n) = \mathrm{cr}(G_6 \Box C_n) = 2n,$$

$$\mathrm{cr}(G_{11} \Box C_n) = \mathrm{cr}(G_{12} \Box C_n) = \mathrm{cr}(G_{13} \Box C_n) = \mathrm{cr}(G_{14} \Box C_n) =$$

$$= 2n.$$

4 Acknowledgement

The paper was supported by the Slovak VEGA grant No. 1/0389/15.

Bibliography

[1] BEINEKE, L. W., RINGEISEN, R. D.: On the crossing numbers of products of cycles and graphs of order four, *J. Graph Theory* **4** (1980), 145–155.

[2] BOKAL, D.: On the crossing numbers of Cartesian products with paths, *J. Graph Theory B* **97** (2007), 381–384.

[3] DRAŽENSKÁ, E., KLEŠČ, M.: On the crossing numbers of $G\square C_n$ for graphs G on six vertices, *Discussiones Mathematicae Graph Theory*,**36** (2007), 109–119.

[4] DRAŽENSKÁ, E., KLEŠČ, M.: The crossing numbers of products of cycles with 6–vertex trees, *Tatra Mountains Math. Publ.*, **36** (2007), 191–197.

[5] DRAŽENSKÁ, E.: The crossing numbers of products of paths with 7-vertex trees, *Creative Mathematics and Informatics*, **23** (2014), 109–119.

[6] GAREY, M. R., JOHNSON, D. S.: Crossing number is NP-complete, *SIAM J. Algebraic and Discrete Methods* **4**, (1983) 312–316.

[7] JENDROL', S., ŠČERBOVÁ,M.: On the crossing numbers of $S_m \times P_n$ and $S_m \times C_n$, *Časopis pro pěstování matematiky* **107** (1982), 225–230.

[8] KLEŠČ, M.: On the crossing numbers of Cartesian products of stars and paths or cycles, *Mathematica Slovaca* **41** (1991), 113–120.

[9] KLEŠČ, M.: The crossing numbers of Cartesian products of paths with 5–vertex graphs, *Discrete Mathematics* **233** (2001), 353–359.

[10] KLEŠČ, M.: The crossing number of $K_{2,3} \times C_3$, *Discrete Mathematics* **251** (2002), 109–117.

[11] KLEŠČ, M.: Some crossing numbers of products of cycles, *Discussiones Mathematicae - Graph theory* **25** (2005), 197–210.

[12] KLEŠČ, M., KOCÚROVÁ, A.:The crossing number of products of 5-vertex graphs with cycles, *Discrete Mathematics* **307** (2007), 1395–1403.

[13] KLEŠČ, M., KRAVECOVÁ, D., PETRILLOVÁ, J.: On the crossing numbers of Cartesian products of paths with special graphs, *Carpathian Journal of Mathematics* **30** (2014), 317–325.

[14] KLEŠČ, M., PETRILLOVÁ, J.: On the optimal drawings of products of paths with graphs, *Acta Elektrotechnica et Informatica* **13** (2013), 56–61.

[15] KLEŠČ, M., PETRILLOVÁ, J.: The crossing number of $P_N^2\square C_4$, *Acta Electrotechnica et Informatica* **12** (2012), 42–46.

[16] KLEŠČ, M., PETRILLOVÁ, J.:The crossing numbers of products of paths with graphs of order six, *Discussiones Mathematicae - Graph theory* **33** (2013), 571–582.

[17] KLEŠČ, M., PETRILLOVÁ, J., Valo, M.: Minimal number of crossings in strong products of paths, *Carpathian Journal of Mathematics* **29** (2013), 27–32.

[18] KLEŠČ, M., Valo, M.: Minimum crossings in join of graphs with paths and cycles, *Acta Electrotechnica et Informatica* **12** (2012), 32–37.

[19] OUYANG, Z., WANG, J. and HUANG, Y.: The crossing number of the Cartesian product of paths with complete graphs, *Discrete Mathematics* **328** (2014), 71–78.

[20] THOMPSON, C. D.: "Area-time complexity for VLSI", *Proceedings of the eleventh annual ACM symposium on Theory of computing, ser. STOC 79*, (1979) 81–88.

[21] WANG, J., HUANG, Y.: The crossing number of $K_{2,4} \times P_n$, *Acta Math.Sci, Ser. A, Chin. Ed.* **28** (2008), 251–255.

Current address

Draženská Emília, RNDr. PhD.
Department of Mathematics and Theoretical Informatics,
Faculty of Electrical Engineering and Informatics, Technical University of Košice
Nemcovej 32, 042 00 Košice, Slovak Republic
Tel. number: +421556022445, e-mail: emilia.drazenska@tuke.sk

Slovak University of Technology in Bratislava
Faculty of Mechanical Engineering
16th Conference on Applied Mathematics **APLIMAT 2017**
Proceedings

MATHEMATICS TEACHING PROBLEMS AND ITS SOLUTION AT RIGA TECHNICAL UNIVERSITY

DZENITE Ilona (LV), VOLODKO Inta (LV)

Abstract. Mathematics is one of the basic subjects in all academic programs at Riga Technical University (RTU). At the same time it is one of the subjects which causes the most problems for many students. The main reasons are the students' insufficient knowledge of elementary mathematics, the students' disinterest and lack of motivation. In this article we present a description of a series of actions which are used in RTU for solving the above-mentioned problems.

Keywords: teaching of mathematics, elementary mathematics knowledge, motivation of students, e-learning

Mathematics Subject Classification: 97D60, 97D70, 97U50

1 Introduction

The Riga Technical University (RTU) is not only the oldest, but also the biggest Higher Education Institution in Latvia. There are more than 15 000 students studying in RTU this year. Its nine faculties have in total 137 academic programs of different levels.

Since all engineering sciences are based on mathematics, all RTU students study mathematics in the 1st year of their studies. However, higher mathematics studies are hampered by several factors such as: (i) The students' insufficient knowledge of elementary mathematics; (ii) The students' disinterest and lack of motivation. Let us consider each factor in detail.

2 The Students' Insufficient Knowledge of Elementary Mathematics

In order to check the knowledge level of elementary mathematics of newly accepted RTU students, every year, during the 1st tutorial on mathematics, the Professors of the Department of Engineering Mathematics give a small test for all the 1st year students. This test consists of 5 simple tasks involving: 1) operations with fractions, 2) calculation of a function value, 3) expression of a linear variable from a proportional equality, 4) basic properties of power functions, 5) basic properties of logarithmic functions. Each correctly solved task is awarded 2 points. The test evaluation is given a positive result if the student is awarded 4 or more points. The points distribution in 2016/2017 academic year is shown in Fig. 1.

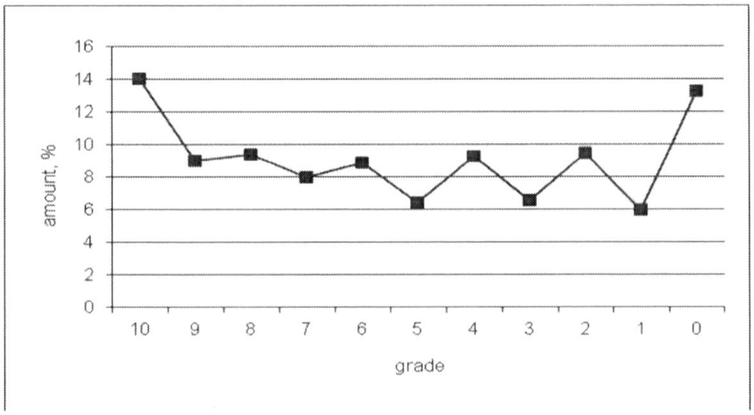

Fig. 1. The marks distribution for the elementary mathematics test in 2016/2017 a.y.

As can be seen from Fig. 1, only 14 % of students have solved all tasks in the test, but 35 % of students have not solved even 2 of 5 given tasks. This tendency remains the same from year to year, so that 35-55 % of students newly accepted into the university do not have the appropriate level of knowledge of elementary mathematics to study higher mathematics successfully. Fig. 2 shows the number of students who have failed the elementary mathematics test during the last 9 years.

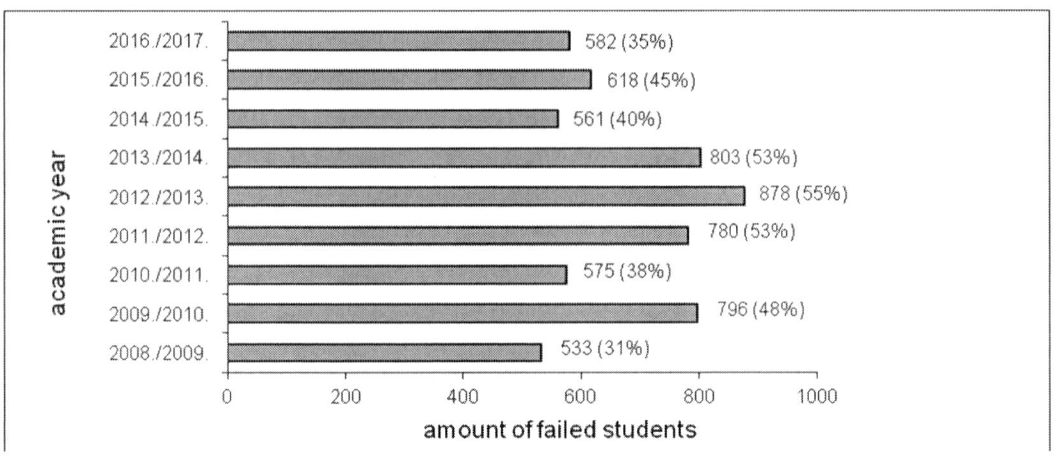

Fig. 2. The comparison of number of failed students during the last 9 years

From year to year the Professors of the Department of Engineering Mathematics have been searching for the solution to the problem: how to help weak students fill gaps in their knowledge of elementary mathematics so that they will be able to study the course of higher mathematics successfully. For this purpose:

1) Video lectures were created, freely available on the internet, in which the basics of elementary mathematics are explained in a simple and easily understandable form. The video lectures were made, processed and placed on the internet with technical support and assistance of colleagues from the RTU Information Technology Centre. There are 43 lectures available for students in

total. Each video lecture is short, lasting 5-18 minutes only, to ensure that students do not loose attention during the lecture. Thanks to these video lectures, students can repeat the elementary mathematics basics at the appropriate time, place and speed. These lectures are available not only for RTU students, but for everyone interested in the subject. The lectures can be used by schoolchildren both during their school studies and in preparation for university studies.

2) A free mathematical e-course based on MOOC (Massive Open Online Course) was created. The students of this course have access to lecture summaries, samples of task solutions with detailed explanations and self-assessment tests.

3) A new subject "Basic chapters of Elementary Mathematics" was created. At present this subject is taught only for part-time students, whilst for full-time students this subject is available as a subject of free choice. This subject lasts for 1 semester, 2 hours per week, and it helps students to learn elementary mathematics at a level which allows them to study higher mathematics successfully.

4) Starting from this academic year, for students who failed the elementary mathematics test, there are additional lectures on mathematics during the 1st semester.

3 The Students' Disinterest and Lack of Motivation

For many students mathematics seems not only difficult and complicated, but also redundant and unnecessary. The Professors of RTU Department of Engineering Mathematics try to solve this problem in the following two ways:

1) In order to make students interested in studying mathematics, Professors try to show them tasks that are connected with students' chosen specialization.

2) In order to motivate students to study during whole semester, the semester's final evaluation consists of not only the examination mark, but also of the work made during the semester, i.e. homework and classroom tests results. The semester's final mark is calculated by the formula 0,6*E+0,3*K+0,1*M, where E is the exam mark, K is the average mark of all classroom tests, M is the average mark of all homework, under the condition that the exam was passed successfully.

As the semester results show, this formula motivates those students who have a weak knowledge of mathematics, but who are diligent and work during the whole semester.

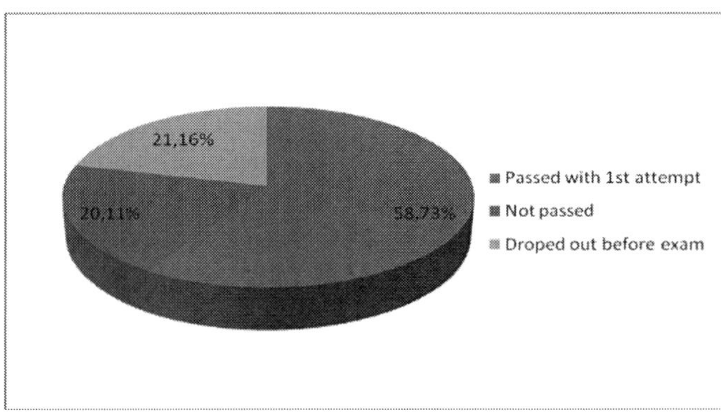

Fig. 3. The number of students that passed and failed in the 1st semester exam on mathematics

The following data was taken within only one RTU Faculty, the Faculty of Computer Science and Information Technology. In the faculty there are 443 students in their 1st year of studies. Summarizing the results of the 1st semester mathematics exam of 2015/2016 a.y., the following data was obtained (see Fig. 3). Slightly more than a half (58.73 %) of the 1st year students have passed the exam from the 1st attempt, 21.16 % of students dropped out before the 1st session. For the remaining 20.11 % of students who failed the exam, there is the possibility to retake and pass the exam before the next semester.

Fig. 4 shows the influence of the above mentioned formula on final results of students which passed the exam successfully. For most of them (73.87 %) the semester final mark is not changed, as those students showed the same knowledge as they did during the semester. For 16.22 % of students, the work during the semester helped them to increase the semester final mark by 1, while for the final 9.91% of students, who ignored the work during the semester, they have obtained a lower semester final mark than their exam mark.

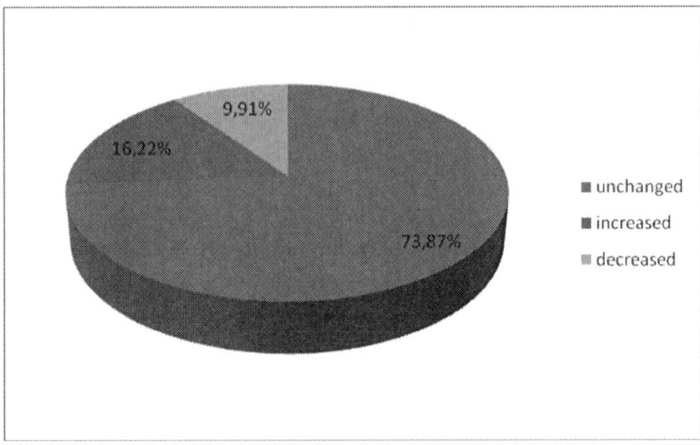

Fig. 4. The change of the semester final evaluation due to the application of the formula

4 E-learning Course – an Assistant for Students and Teachers

A good assistant for studies both for students and Professors is the portal ORTUS, created in 2007 by the staff of RTU Department of Information technology. Every year the portal is updated and improved. In ORTUS students can find all the necessary information they need for their studies. One part of ORTUS is dedicated to E-learning, which is based on the program MOODLE (Modular Object-Oriented Dynamic Learning Environment). In the E-learning section students have access to all subjects included in their chosen academic program.

In the E-environment of Mathematics subjects students can find the Professors' academic plans, timetable, the requirements for obtaining a successful assessment, summaries of lectures, samples of task solutions, homework tasks, tasks for self-study and self-assessment and aptitude tests. Here the Professors may inform students about upcoming tests, consultations and exams, as well as post the students' results after tests.

Students very widely use the above-mentioned portal. The following data is taken from the mathematics e-course of only one RTU Faculty, the Faculty of Computer Science and Information

Technology. Fig. 5 and 6 present the number of students who visited the E-course in the 1st and 2nd semesters, respectively.

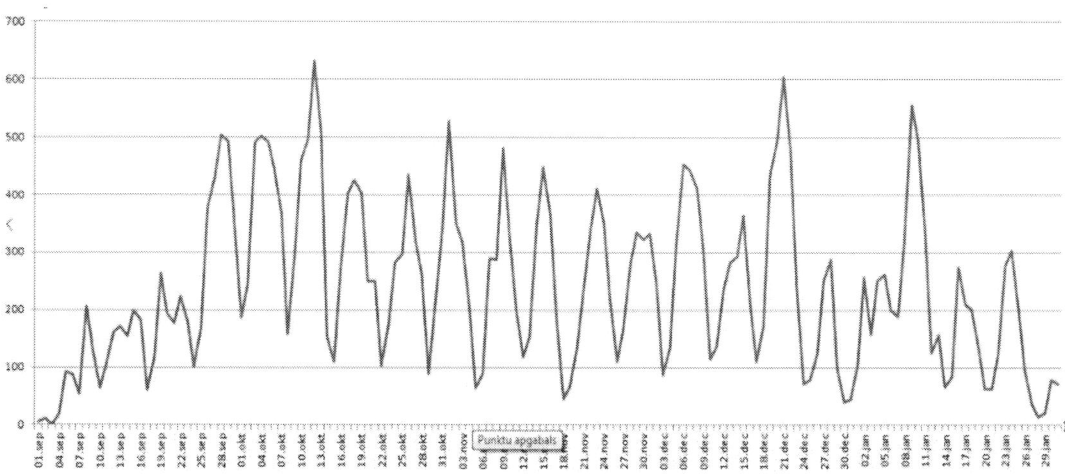

Fig. 5. The number of students' visits to E-course in the 1st semester

In the 1st semester the distribution of visits is quite smooth, except that there are less visits in the beginning and at the end of the semester. That can be explained very easily: in the beginning of the semester the 1st year students are not used to working with ORTUS materials, but by the end of semester most of students have already passed the exam. In the 2nd semester the situation is different: the ORTUS materials are used less during the semester, but the use of it increases during the examination period. The top use of ORTUS materials is on the day before the exam, and the second jump of it is on the day before the exam re-take.

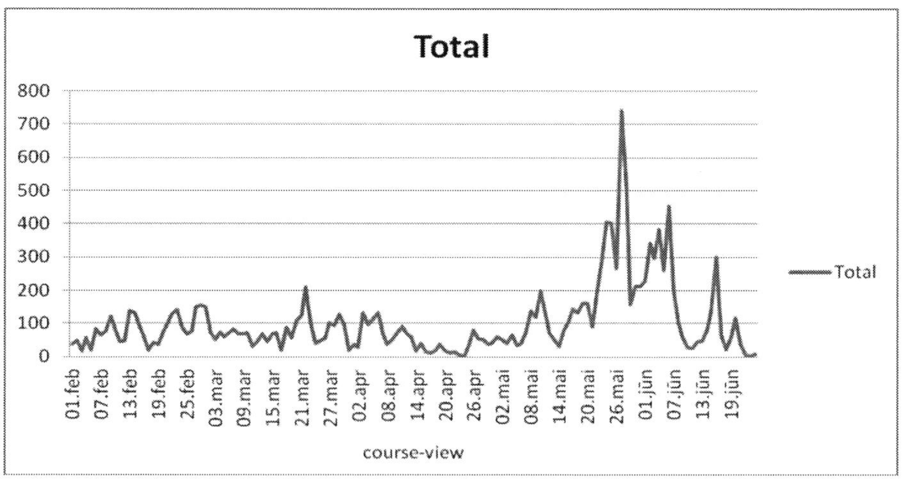

Fig. 6. The number of students' visits to E-course in the 2nd semester

The use of E-course facilitates studies and has the following benefits:

- Materials are available for students at any time and place they have internet access;
- Students may study at their own speed;

- Students have the opportunity to combine studies with work and more effectively manage their time;
- The use of technology stimulates studies and increases the students' interest;
- Professors have the opportunity to use information blocks of different types and sizes, not only texts, but also video lectures and interactive materials;
- Professors may reduce their workload, by substituting part of students' classroom and homework tests with online tests at the portal.

In our opinion, there is only one disadvantage: many students are not able study in the E-environment, they don't know how to work with work plans and course materials, and are not able to communicate with Professors.

The ORTUS portal has a feedback facility. Every semester through this portal, students take a survey, in which students express their thoughts about Professors and their teaching methods. Each Professor has access to the survey summary, which shows his/her assessment. This indicates any issues with the Professors' teaching, and gives them the opportunity to improve their working methods.

Modern youth are native users of computers and willingly use information technology. In order to make mathematics courses more attractive, the mathematical software MATHEMATICA and MATLAB were introduced. During the classes students learn how to use these packages to solve mathematical problems, similar to the tasks solved at tutorials, but with a higher degree of difficulty.

Despite all above mentioned actions, and taking into account that nowadays the amount of available information grows rapidly, university cannot give to students all necessary knowledge, skills and abilities that they will need in their later life. The student himself must be able to distinguish the important from the unimportant, and to learn how to optimize the relationship between knowledge and their own learning needs.

References

[1] ČERŅAJEVA, S., VOLODKO, I. Teaching Mathematics in Riga Technical University. *International Conference "Mathematical Competence Development in Higher Education Institutions"*, Jelgava, 2012, Latvia, p. 21.

[2] DZENITE, I., VOLODKO, I. The use of the package „Mathematica" in teaching of mathematics in Riga Technical University. In *Proceedings of the International Conference on Engineering and Mathematics*. Bilbao, Spain, 2009, June 17–19, p. 75–79.

[3] VOLODKO, I., ROŠČINA, I. Использование портала ORTUS для обучения математике в Рижском Техническом университете. *Некоторые актуальные проблемы современной математики и математического образования. Материалы научной конференции „Герценовские чтения - 2012"*, Санкт-Петербург, Библиотека Академии наук, 2012, April 16-21, p. 168–173, (Saint Petersburg, Russia, in Russian).

[4] VOLODKO, I., ROŠČINA, I. Использование пакета программ Mathematica9 при обучении математике в Рижском Техническом университете. *Труды международной*

научной конференции „Образование, наука и экономика в вузах и школах. Интеграция в международное образовательное пространство". Цахкадзор, Армения, 2014, March 24-29,p. 482–485, (Tsakhkadzor, Armenia, in Russian).

Current address

Dzenite Ilona, assistant professor
Department of Engineering Mathematics, Institute of Applied Mathematics,
Faculty of Computer Science and Information Technology, Riga Technical University
2 Daugavgrivas street, Riga, LV-1007, Latvia
Tel. number: (+371)67089527, e-mail: ilona.dzenite@rtu.lv

Volodko Inta, professor
Department of Engineering Mathematics, Institute of Applied Mathematics,
Faculty of Computer Science and Information Technology, Riga Technical University
2 Daugavgrivas street, Riga, LV-1007, Latvia
Tel. number: (+371)67089528, e-mail: inta.volodko@rtu.lv

Slovak University of Technology in Bratislava
Faculty of Mechanical Engineering
16th Conference on Applied Mathematics **APLIMAT 2017**
Proceedings

LINEAR TRIANGULATIONS OF POLYTOPES

FAILLA Gioia (IT)

Abstract. We study the linear triangulation of the initial simplicial complexes associated the second squarefree Veronese ring $K[V_2] \subset k[x_1, \ldots, x_n]$, for n=5.

Keywords: Computational Geometry, Algebraic Geometry, Simplicial Complexes, Gröbner bases

Mathematics Subject Classification: Primary 13P10; Secondary 13F55

Introduction

The most important term orders on the set of monomials of the polynomial ring S=K[x_1,..,x_n], K a field, are the lexicographic order and the reverse lexicographic order.

The linear order L is introduced in [11] (chap. 9) on the set $\underline{T}$ of r-indexed variables of the presentation K-algebra K[$\underline{T}$] of the rth-Veronese squarefree ring, the subring of S generated on K by all squarefree monomials in r variables $x_1, \ldots, x_n$. In the paper [1] the authors considered the universal Gröbner basis G for the toric deal I of K[$\underline{T}$], universal with respect to all linear orders on the set $\underline{T}$ of variables and the lexicographic order on the set of monomials of the polynomial ring K[$\underline{T}$] of the presentation. They proved that the degree of the binomials of the Gröbner basis wit respecth all linear orders of the variables $\underline{T}$ is 2 or 3.

In this paper we compute the initial complexes [10] in correspondence of the Gröbner basis G and we study the triangulations of the representing polytopes. Combinatoric properties of the triangulations are studied.

In particular, for n=5, the linear order yealds to initial simplicial complexes more complicated than for n=4, depending from the degree of the Gröbner basis of the toric ideal. Since such triangulations depend from the chosed term order, it is a fertile field to explore in the direction of varying the term order and to check properties whether the reverse lexicographic order is applied.

Starting from the founded polytopes in the paper, geometric realizations and new applications in different fields are obtained in [6], [7].

1 Notations and known results

Let K be a field and let R = $K[x_1, x_2, \ldots, x_n]$ be the polynomial ring in n variables $x_1, x_2, \ldots, x_n$. We denote the set of the power products of R by:

$$M^n = \{x_1^{a_1} \cdots x_n^{a_n} \mid a_i \in \mathbb{N}, i = 1, \ldots, n\}$$

where $x_1^{a_1} \cdots x_n^{a_n} = \mathbf{x}^{\mathbf{a}}$, $\boldsymbol{a} = (a_1, a_2, \ldots, a_n) \in \mathbb{N}^n$.

Definition 1.1. A term order on M^n is a total order < on M^n satisfying:
(1) $1 < \boldsymbol{x^b}$ for all $\boldsymbol{x^b} \in M^n$, $\boldsymbol{x^b} \neq 1$;
(2) If $\boldsymbol{x^a} < \boldsymbol{x^b}$, then $\boldsymbol{x^a x^c} < \boldsymbol{x^b x^c}$, for all $\boldsymbol{x^c} \in M^n$.

Definition 1.2. We define the lexicographical order on M^n, if $x_1 > x_2 > \ldots > x_n$ as follows:
for $\boldsymbol{a} = (a_1, a_2, \ldots, a_n)$, $\boldsymbol{b} = (b_1, b_2, \ldots, b_n)$
$\boldsymbol{x^a} < \boldsymbol{x^b} \Leftrightarrow \exists h \mid a_1 = b_1, \ldots, a_h = b_h, a_{h+1} < b_{h+1}\ h = 0, \ldots, n-1$

Definition 1.3. Let $G \subset$ R, $G = \{f_1, \ldots, f_s\}$ be a set of polynomials of R. G is an universal Gröbner basis for the ideal $(f_1, \ldots, f_s)$ if G is a Gröbner basis with respect to all term orders on the monomials of R.

Definition 1.4. Let $G \subset$ R, $G = \{f_1, \ldots, f_s\}$ be a set of polynomials. G is called a universal lexicographical Gröbner basis for $(f_1, \ldots, f_s)$ if G is a Gröbner basis with respect to the lexicographical order on the monomials of R and with respect to all orders of the variables $x_1, x_2, \ldots, x_n$.

Now we recall the definition of linear orders of variables introduced in [11].
Given two integers n and k such that $4 \leq 2k \leq n$, we consider the set of variables:

$$\mathbf{T} = \{T_{i_1 i_2 \ldots i_k} : 1 \leq i_1 < i_2 < \cdots < i_k \leq n\}.$$

Let A = (a_{ij}) be a non singular matrix of order k with entries in R

Definition 1.5. (Linear lexicographical order on $\boldsymbol{T}$)
Let $\boldsymbol{i} = (i_1, i_2, \ldots, i_k) \in \mathbb{Z}^k$ and $\boldsymbol{j} = (j_1, j_2, \ldots, j_k) \in \mathbb{Z}^k$. Then

$$\boldsymbol{i} <_{A(lex)} \boldsymbol{j} \text{ iff } \boldsymbol{i}A <_{lex} \boldsymbol{j}A$$

that is:

$$(i_1, i_2, \cdots, i_k)(a_{ij}) = (b_1, b_2, \cdots, b_k)$$

where $b_t = i_1 a_{1t} + i_2 a_{2t} + \cdots + i_k a_{kt}$, $\boldsymbol{i}A = (b_1, b_2, \cdots, b_k)$ and $\boldsymbol{J}A = (c_1, c_2, \cdots, c_k)$.
So $\boldsymbol{i}A >_{lex} \boldsymbol{j}A$ iff $(b_1, b_2, \cdots, b_k) >_{lex} (c_1, c_2, \cdots, c_k)$ that is $b_1 = c_1$, $b_2 = c_2, \ldots, b_l > c_l$,

for some $l = 1, \dots, k$.

Thus, we can define the linear lexicographical order of variables by:

$$T_{i_1 i_2 \dots i_k} >_{A(lex)} T_{j_1 j_2 \dots j_k} \text{ iff } (i_1, i_2, \dots, i_k) >_{A(lex)} (j_1, j_2, \dots, j_k).$$

Now we consider the set $\mathcal{A}_k$ of matrices of order k of the form $(\pm e_{i_j})$,$j = 1, \dots, k$, where the set $\{e_{i_1}, \dots, e_{i_k}\}$s a permutation of the set of canonical vectors$\{e_1, \dots, e_k\}$of $\mathbb{R}^k$, (See [11], chap.9).
Let R $=K[x_1, x_2, \dots, x_n]$ be the polynomial ring, K a field, with the standard graduation on the variables. Let V_k the r-th- Veronese square free ring of R, generated on K by all square free monomials of degree k of R. Let I be the toric ideal of $K[V_k]$, that is the kernel of the ring homomorphism: h: $K[T_1 \dots T_{\binom{n}{k}}] \mapsto K[V_k]$ such that $T_{i_1 \dots i_k} \mapsto x_{i1} \dots x_{ik}$. We recall the definition of linear order L on the variables $T_1 \dots T_{\binom{n}{k}}$.

Consider the presentation:

$$\varphi: K[\boldsymbol{T}] \longrightarrow K[V_k] \text{ induced by } T_{i_1 i_2 \dots i_k} \mapsto x_{i_1} \dots x_{i_k},$$

where $I = \ker(\varphi)$ is the toric ideal of $K[V_k]$. We call the monomial algebra $K[V_k]$ the squarefree *k*-Veronese subring of R. In [1], [11], for some values of k and *n*, $0 \leq 2k \leq n$, the authors give positive answer to the problem to obtain lexicographic Gröbner bases for linearly ordered variables.

Let be $K[V_2]$ be the squarefree second Veronese subring of the polynomial ring *or*, where $V_2$is the set of square-free monomials of or of degree 2.

Proposition 1.6. The short list by the matrices of $\mathcal{A}_2$ is the following

$$A_1 = \begin{pmatrix} 1 & 0 \\ 0 & 1 \end{pmatrix}, A_2 = \begin{pmatrix} -1 & 0 \\ 0 & 1 \end{pmatrix}, A_3 = \begin{pmatrix} 1 & 0 \\ 0 & -1 \end{pmatrix}, A_4 = \begin{pmatrix} -1 & 0 \\ 0 & -1 \end{pmatrix},$$

$$A_5 = \begin{pmatrix} 0 & 1 \\ 1 & 0 \end{pmatrix}, A_6 = \begin{pmatrix} 0 & 1 \\ -1 & 0 \end{pmatrix}, A_7 = \begin{pmatrix} 0 & -1 \\ 1 & 0 \end{pmatrix}, A_8 = \begin{pmatrix} 0 & -1 \\ -1 & 0 \end{pmatrix}$$

Proof. See ([1], prop 1.8).

Proposition 1.7. Let *i, j, k, l* be positive integers such that $1 \leq i < j < k < l \leq n$. Then the linear lexicographical orders on the variables of **T** induced by the matrices of $\mathcal{A}_2$ are given by the list:

$$T_{kl} > T_{jl} > T_{jk} > T_{il} > T_{ik} > T_{ij} \qquad \text{(by } A_1\text{)},$$

$$T_{il} > T_{ik} > T_{ij} > T_{jl} > T_{jk} > T_{kl} \qquad \text{(by } A_2\text{)},$$

$$T_{kl} > T_{jk} > T_{jl} > T_{ij} > T_{ik} > T_{il} \qquad \text{(by } A_3\text{)},$$

$$T_{ij} > T_{ik} > T_{il} > T_{jk} > T_{jl} > T_{kl} \quad \text{(by A}_4\text{)},$$

$$T_{kl} > T_{jl} > T_{il} > T_{jk} > T_{ik} > T_{ij} \quad \text{(by A}_5\text{)},$$

$$T_{kl} > T_{jk} > T_{jl} > T_{ij} > T_{ik} > T_{il} \quad \text{(by A}_6\text{)},$$

$$T_{il} > T_{jl} > T_{kl} > T_{ih} > T_{jk} > T_{ij} \quad \text{(by A}_7\text{)},$$

$$T_{ij} > T_{ik} > T_{jk} > T_{il} > T_{jl} > T_{kl} \quad \text{(by A}_8\text{)}.$$

After, we will extend systematically the above orders to all the variables $T_{ij}, 1 \leq i < j \leq n$.

Proof: See [1] (Prop 1.8).

Theorem 1.8. Let $K[V_2]$ be the squarefree secondVeronese subring of R = $K[x_1, x_2, \dots, x_n]$ and let I be the toric ideal of $K[V_2]$ = $K[T]$ /I. Then the following facts hold:

(1) The set $B_1 = \left\{\underline{T_{ij}T_{kl}} - T_{il}T_{jk}, \underline{T_{ik}T_{jl}} - T_{il}T_{jk}, 1 \leq i < j < k < l \leq n\right\}$ is a lexicographical reduced Gröbner basis for I with respect to the linear orders on the variables induced by the matrices A_1, A_4, A_5, A_8.

(2) The set $B_2 = \left\{T_{ij}T_{kl} - \underline{T_{il}T_{jk}}, T_{ij}T_{kl} - \underline{T_{ik}T_{jl}}, \underline{T_{ik}T_{jk}T_{lm}} - T_{ij}T_{km}T_{kl}, 1 \leq i < j < k < l < m \leq n\right\}$ is a reduced lexicographical Gröbner basis for I with respect to the linear orders on the variables induced by the matrices A_2, A_7.

(3) The set $B_3 = \left\{\underline{T_{ij}T_{kl}} - T_{il}T_{jk}, T_{ik}T_{jl} - \underline{T_{il}T_{jk}}, 1 \leq i < j < k < l \leq n\right\}$ is a lexicographical reduced Gröbner basis for I with respect to the linear orders on the variables induced by the matrices A_3 and A_6.

The underlined terms in each binomial are the initial monomials.

Proof: See [1], (Theorem 1.11).

Corollary 1.9. Let $K[V_2]$ = $K[T]/I$. The set $\{T_{ij}T_{kl} - T_{il}T_{jk}, T_{ik}T_{jl} - T_{il}T_{jk}, T_{ij}T_{kl} - T_{ik}T_{jl}, T_{ik}T_{jk}T_{lm} - T_{ij}T_{km}T_{kl}\ 1 \leq i < j < k < l < m \leq n\}$ of binomials of I is a Γ -universal lexicographic Gröbner basis of *I*.
Proof: See [1], (Corollary 1.12)

Example 1.10. If n=4, let T = $\{T_{12}, T_{13}, T_{14}, T_{23}, T_{24}, T_{34}\}$ be the set of variables for $K[V_2] \subseteq K[x_1, x_2, \dots, x_n]$.

The Γ -universal lexicographical square free quadratic Gröbner basis for I is of degree two and is the following

$$\{T_{24}T_{13} - T_{23}T_{14}, T_{34}T_{12} - T_{23}\ T_{14}, T_{12}T_{34} - T_{13}, T_{24}\}.$$

Example 1.11 If n=5, let $\underline{T} = \{T_{12}, T_{13}, T_{14}, T_{15}, T_{23}, T_{24}, T_{25}, T_{34}, T_{35}, T_{45}\}$ be the set of variables in the presentation of $K[V_2] \subset k[x_1, \dots, x_n]$ Then, the Γ-universal lexicographic squarefree Gröbner bases for *I* is of degree two and three and it is the following.

$$\{T_{12}T_{34} - T_{14}T_{23}, T_{13}T_{24} - T_{14}T_{23}, T_{12}T_{34} - T_{13}T_{24}, T_{13}T_{23}T_{45} - T_{12}T_{34}T_{35}\}.$$

2 Triangulation of initial complexes

We recall the definition of initial simplicial complex $in_{<}\Delta$ and triangulation of a set of lattice points.

Let I be any ideal of R= $K[x_1, \dots, x_n]$ and let < be a term order on the monomials of R.

Definition 2.1. The initial complex $\Delta_{<}(I)$ of I with respect to < is the simplicial complex on the vertex set $\{1,2, \dots, n\}$ whose Stanley- Reisner ideal is the radical of the monomial ideal $in_{<}(I)$.

Let $\mathcal{A} = \{\boldsymbol{a}_1, \dots, \boldsymbol{a}_i\}$, where $\underline{a}_i \in \mathbb{N}^n$, be a set $\mathcal{A}$ of lattice points. Assume that the set $\mathcal{A}$ spans an affine hyperplane. This means that the toric ideal $I_{\mathcal{A}}$ is a homogeneous ideal of the polynomial ring.

Definition 2.2. ([10], chapter 8) Let σ be a subset of lattice points, $\sigma \subseteq \mathcal{A}$ and *pos (σ)* be the polyhedral cone generated by σ, then one defines triangulation of $\mathcal{A}$ a collection Δ of subsets of $\mathcal{A}$, such that the set *{pos (σ), σ ∈ $\mathcal{A}$ }* is the set of cones of a simplicial fan whose support is the same as *pos ($\mathcal{A}$), supp {pos (σ), σ ∈Δ} = pos ($\mathcal{A}$),*

Let $K[V_k]$ be the kth−Veronese squarefree subring of R and let $\mathcal{A}$ be the set of lattice points of the semigroup under $K[V_k]$.
We will identify the given set of lattice points $\mathcal{A} = \{\boldsymbol{a}_1, a_2, \dots, \boldsymbol{a}_{\binom{n}{k}}\} \subset \mathbb{N}^n$ with the index set $\{1, \dots, t\}, t = \binom{n}{k}$.

Definition 2.3. Let < any linear order of the set of variables $T = \{T_{i_1 i_2 \dots i_k} : 1 \le i_1 < i_2 < \dots < i_k \le n\}$. Let $K[V_k] = K[T]/I_{\mathcal{A}}$. We call initial linear simplicial complex of I any initial simplicial complex with respect to the linear order < of the variables.

Definition 2.4. Let < any linear order of the set of variables $T = \{T_{i_1 i_2 \dots i_k} : 1 \le i_1 < i_2 < \dots < i_k \le n\}$. Let $K[V_k] = K[T]/I_{\mathcal{A}}$. We call initial linear lexicographical simplicial complex of $I_{\mathcal{A}}$ any initial linear simplicial complex of I with respect to the lexicographical order on the monomials of K[T].

For k = 2, we call the variables $\mathbf{T}_{i_1 i_2}$ of **T** with $\mathbf{T}_{ij}$, , $1 \le i < j \le n$.

Theorem 2.5.

Let $K[V_2]$ be the second squarefree Veronese ring. For n=5, the set $\mathcal{A}$ has the following linear lexicographic triangulations:

i) $\Delta^{(1)}$with maximal faces set given by
{{1,2,3,7,8,9},{6,7,8,9,10},{2,3,6,7,8,9}}.

ii) $\Delta^{(2)}$with maximal faces set given by
{{1,2,3,5,8,9}; {1,3,5,6,8,9}, {1,5,6,7,8,9}}.

iii) $\Delta^{(3)}$with maximal faces set given by
{{1,4,5,6,7,8,9,10},{1,2,3,4,5,8,9,10}}.

Proof: It is sufficient to consider only the linear orders associated to matrices A_1, A_2, A_3.
Let $\Delta^{(i)}$ the triangulation coming from the linear order on the variables and associated to the initial simplicial complex $\Delta_{<}^{(i)}$ $(I_{\mathcal{A}})$, i=1,2,3.

We have the following

1. Write the set $\mathcal{A}$ as an ordered set induced by the matrix A_1 on the variables

$$T_{kl} > T_{jl} > T_{il} > T_{jk} > T_{ik} > T_{ij}$$

Then the Gröbner basis of $I_{\mathcal{A}}$ is

$$\left\{\underline{T_{ij}T_{kl}} - T_{il}T_{jk}, \underline{T_{ik}T_{jl}} - T_{il}T_{jk}, 1 \le i < j < k < l \le 5\right\}$$

For

$$(i)\ \ i = 1, j = 2, k = 3, l = 4$$
$$T_{12} < T_{13} < T_{15} < T_{23} < T_{24} < T_{34}$$

with binomials

$$\underline{T_{12}T_{34}} - T_{14}T_{23}, \underline{T_{13}T_{24}} - T_{14}T_{23}$$

For

$$(ii)\ \ i = 1, j = 2, k = 3, l = 5$$
$$T_{12} < T_{13} < T_{15} < T_{23} < T_{25} < T_{35}$$

and binomials given by

$$T_{12}T_{35} - T_{15}T_{23}, T_{13}T_{25} - T_{15}T_{23}$$

For

$$(iii)\ \ i = 1, j = 2, k = 4, l = 5$$

$$T_{12} < T_{14} < T_{15} < T_{24} < T_{25} < T_{45}$$

and binomials

$$\underline{T_{12}T_{45}} - T_{15}T_{24}, \underline{T_{14}T_{25}} - T_{15}T_{24}$$

For

$$(iv) \quad i = 2, j = 3, k = 4, l = 5$$
$$T_{23} < T_{24} < T_{25} < T_{34} < T_{35} < T_{45}$$
$$\underline{T_{23}T_{45}} - T_{25}T_{34}, \underline{T_{24}T_{35}} - T_{25}T_{34}$$

Then we have the Gröbner basis

$$G_1 = \{\underline{T_{23}T_{45}} - T_{25}T_{34}, \underline{T_{24}T_{35}} - T_{25}T_{34}, \underline{T_{12}T_{45}} - T_{15}T_{24}, \underline{T_{14}T_{25}} - T_{15}T_{24}, \underline{T_{12}T_{34}} - T_{14}T_{23}, T_{13}T_{24} - T_{14}T_{23}, \underline{T_{12}T_{35}} - T_{15}T_{23}, \underline{T_{13}T_{25}} - T_{15}T_{23}\}$$

and $in(I_{\mathcal{A}}) = (T_{12}T_{34}, T_{13}T_{24}, T_{12}T_{35}, T_{13}T_{25}, T_{12}T_{45}, T_{14}T_{25}, T_{23}T_{45}, T_{24}T_{35})$.

Choose an order on the variables that is compatible with the previous orders, then we can order the set of lattice points as:

(1,1,0,0,0)< (1,0,1,0,0)< (1,0,0,1,0)< (1,0,0,0,1)< (0,1,1,0,0)< (0,1,0,1,0)< (0,1,0,0,1)<
< (0,0,1,1,0)< (0,0,1,0,1)< (0,0,0,1,1)

and

$$a_{10} < a_9 < a_8 < a_7 < a_6 < a_5 < a_4 < a_3 < a_2 < a_1$$

In the correspondence
$\underline{a}^i \longrightarrow \underline{x}^{\underline{a}_i} \longrightarrow z_i, \underline{x} = (x_1, \dots, x_n)$ $in(I_{\mathcal{A}})$ can be written as $in_<(I_{\mathcal{A}})$ $=(z_{10}z_3, z_9z_5, z_{10}z_2, z_9z_4, z_{10}z_1, z_8z_4, z_6z_1, z_5z_2)$ and the sets {3, 10}, {5, 9}, {2, 10}, {4, 9},{1, 10}, {4, 8},{1, 6}, {2, 5} do not belong to $\Delta_<^{(1)}(I_{\mathcal{A}})$.

It follows that the set of maximal faces of is $\Delta^{(1)}$ is:

{{1,2,3,7,8,9},{6,7,8,9,10},{2,3,6,7,8,9}}

2. Write the set $\mathcal{A}$ as an ordered set induced by the order A_3 on the variables

$$T_{kl} > T_{jl} > T_{jl} > T_{ij} > T_{ik} > T_{il}$$

Then the Gröbner basis of $I_{\mathcal{A}}$ is

$$\left\{\underline{T_{ij}T_{kl}} - T_{il}T_{jk}, T_{ik}T_{jl} - \underline{T_{il}T_{jk}}, 1 \leq i < j < k < l \leq 5\right\}$$

For

$$(i)\mathrm{i} = 1, j = 2, k = 3, l = 4$$
$$T_{14} < T_{13} < T_{12} < T_{24} < T_{23} < T_{34}$$

with binomials

$$\underline{T_{12}T_{34}} - T_{14}T_{23}, T_{13}T_{24} - \underline{T_{14}T_{23}}$$

For

$$(ii)\mathrm{i} = 1, j = 2, k = 3, l = 5$$
$$T_{15} < T_{13} < T_{12} < T_{25} < T_{23} < T_{35}$$

and binomials given by

$$\underline{T_{12}T_{35}} - T_{15}T_{23}, T_{13}T_{25} - \underline{T_{15}T_{23}}$$

For

$$(iii) \quad \mathrm{i} = 1, j = 2, k = 4, l = 5$$
$$T_{15} < T_{13} < T_{12} < T_{25} < T_{24} < T_{45}$$
$$\underline{T_{12}T_{45}} - T_{15}T_{24}, T_{14}T_{25} - \underline{T_{15}T_{24}}$$

For

$$(\mathrm{i}v) \quad \mathrm{i} = 2, j = 3, k = 4, l = 5$$
$$T_{15} < T_{14} < T_{23} < T_{35} < T_{34} < T_{45}$$
$$\underline{T_{23}T_{45}} - T_{25}T_{34}, T_{24}T_{35} - \underline{T_{25}T_{34}}$$

then we have the Gröbner basis

$$G_2 = \left\{\underline{T_{23}T_{45}} - T_{25}T_{34}, T_{24}T_{35} - \underline{T_{25}T_{34}}, T_{14}T_{25} - \underline{T_{15}T_{24}}, \underline{T_{12}T_{45}} - T_{15}T_{24}, T_{12}T_{34} - \underline{T_{14}T_{23}}, T_{13}T_{24} - T_{14}T_{23}, \underline{T_{12}T_{35}} - T_{15}T_{23}, T_{13}T_{25} - \underline{T_{15}T_{23}}\right\}$$

and $in_{<}(I_{\mathcal{A}}) = (T_{12}T_{34}, T_{14}T_{23}, T_{12}T_{35}, T_{15}T_{23}, T_{15}T_{24}, T_{12}T_{45}, T_{23}T_{45}, T_{25}T_{34})$.
Choose an order on the variables that is compatible with the previous orders:

$$T_{15} < T_{14} < T_{13} < T_{12} < T_{25} < T_{24} < T_{23} < T_{35} < T_{34} < T_{45}$$

Then we can order the set of lattice points as:

(1,0,0,0,1)< (1,0,0,1,0)< (1,0,1,0,0)< (1,1,0,0,0)< (0,1,0,0,1)< (0,1,0,1,0)< (0,1,1,0,0)<
< (0,0,1,0,1) <(0,0,1,1,0)< (0,0,0,1,1)

and

$$\boldsymbol{a}_{10} < \boldsymbol{a}_9 < \boldsymbol{a}_8 < \boldsymbol{a}_7 < \boldsymbol{a}_6 < \boldsymbol{a}_5 < \boldsymbol{a}_4 < \boldsymbol{a}_3 < \boldsymbol{a}_2 < \boldsymbol{a}_1$$

In the correspondence
$a_i \longrightarrow x^{a_i} \longrightarrow z_i, x = (x_1, \dots, x_n),\ in_<(I_{\mathcal{A}})$ can be written as:
$in_<(I_{\mathcal{A}})$=(z_7z_2 ,z_9z_4 ,z_7z_3 ,$z_{10}z_4$,$z_{10}z_5$,$z_{10}z_1$,z_4z_1 ,z_6z_2) and the sets {7,2}, {9,4},{7,3},{10,4}, {10,5},{10,1}, {4,1}, {6,2} dont belong to in $\Delta^{(2)}<(I_{\mathcal{A}})$

It follows that the set of maximal faces of $\Delta^{(2)}$ is:

{1,2,3,5,8,9},{1,3,5,6,8,9},{1,5,6,7,8,9}
The maximal faces do not have the same dimension.

3. Write the set $\mathcal{A}$ as an ordered set induced by the matrix A_2 on the variables

$$T_{il} > T_{ik} > T_{ij} > T_{jl} > T_{jk} > T_{kl}$$

Then the Gröbner basis of $I_{\mathcal{A}}$ is

$$\left\{T_{ij}T_{kl} - \underline{T_{il}T_{jk}}, T_{ij}T_{kl} - \underline{T_{ik}T_{jl}}\ , \underline{T_{ik}T_{jk}T_{em}} - T_{ij}T_{km}T_{ke}, 1 \leq i < j < k < l < m \leq 5\right\}$$

For

(i) i $= 1, j = 2, k = 3, l = 4$, m=5

$$T_{34} < T_{23} < T_{24} < T_{12} < T_{13} < T_{14}$$

with binomials

$$T_{12}T_{34} - \underline{T_{13}T_{24}}, T_{12}T_{34} - \underline{T_{13}T_{24}}, \underline{T_{13}T_{23}T_{45}} - T_{12}T_{35}T_{34},$$

Then we have the Gröbner basis

$$G_3 = \left\{T_{12}T_{34} - \underline{T_{14}T_{23}}, T_{12}T_{34} - \underline{T_{13}T_{24}}, \underline{T_{13}T_{23}T_{45}} - T_{12}T_{35}T_{34}\right\}.$$

Choose an order on the variables that is compatible with the previous order, then we can order the set of lattice points as:

(0,0,0,1,1)< (0,0,1,1,0)< (0,0,1,0,1)< (0,1,1,0,0) < (0,1,0,1, 0)< (0,1,0,0,1)< (1,1,0,0,0)<
< (1,0,1,0,0)< (1,0,0,1,0)< (1,0,0,0,1)

and $in_{<}(I_{\mathcal{A}}) = (T_{14}T_{23}, T_{13}T_{24}, T_{13}T_{23}T_{45})$

$$a_{10} < a_9 < a_8 < a_7 < a_6 < a_5 < a_4 < a_3 < a_2 < a_1$$

In the correspondence

$\underline{a_i} \longrightarrow \underline{x}^{\underline{a_i}} \longrightarrow z_i, \underline{x} = (x_1, \dots, x_n)$, $in_{<}(I_{\mathcal{A}})$ can be written as in
$in_{<}(I_{\mathcal{A}})$=(z_2z_7 ,z_3z_6 ,z_3z_7 z_{10}) and the sets {2,7}, {3,6}, {3, 7,10}, dont belong to in $\Delta_{<}^{(3)}(I_{\mathcal{A}})$

It follows that the set of maximal faces of $\Delta^{(3)}$ is:
{{1,4,5,6,7,8,9,10},{1,2,3,4,5,8,9,10}}

Remark 2.6. The kth-Veronese squarefree ring V_k is involved in several algebraic and geometric applications. In particular in the transportation problem [10] and in the higher transportation problem [3], [4], [5], [2] Segre products of two squarefree Veronese rings are considered, if we view in [3] and [10] the polynomial ring S =K[$x_1, \dots, x_n$] over the field K as the first squarefree Veronese ring of the same ring S. An open problem is to study the initial simplicial complexes of the Segre product of Veronese squarefree rings. Moreover, V_k is naturally linked to the class of mixed product ideals of the polynomial ring K[$x_1, \dots, x_n, y_1, \dots, y_n$] in two sets of variables introduced in [9], being K[V_k] the subalgebra of S generated on the field K by the minimal system of generators of I_k, the kth-squarefree squarefree ideal of K[$x_1, \dots, x_n$], which appears in all the mixed product ideals. For the ideals of this class, the simplicial complexes having such ideals as Stanley-Reisner ideals can be investigated ([8],[10]).

References

[1] G. D'Agui, G.Failla, Linear order of variables for complete graphs, Afrika mathematika, Série 3,Volume 20 (2008), 131-146.

[2] G. Failla, R. Utano, Connected graphs arising from products of Veronese varieties, Algebra Colloquium 23:2 (2016), 281-292.

[3] G. Failla, On the (1,1)-Segre model for business, Applied Mathematical Sciences, Vol.8 (2014), no.166, 8329-8336.

[4] G. Failla, Moves for Segre products of Veronese varieties, Rendiconti del Circolo Matematico di Palermo, Serie II, Suppl.80, 2008, 161-167.

[5] G. Failla, Projective toric varieties and associated fibers,Rendiconti del Circolo Matematico di Palermo, Serie II, Suppl.77, 2006, 267-280.
[6] V. Iorfida, G. Restuccia, Simplicial complexes and new applications, Applied Mathematical Sciences, Vol. 10, 2016, no. 34, 1653 – 1661.
[7] V. Iorfida, Simplicial complexes present in art and arts facilities. Aplimat 2015, 14th Conference on Applied Mathematics, Proceedings (2015), 84-94.
[8] M. Murgia, R. Utano, Algebraic invariants of ideals of mixed products, Bollettino di Matematica Pura e Applicata,Vol.VII, 2014, 43- 52.
[9] G. Restuccia, R. Villarreal, On the normality of monomial ideals of mixed products, Comm. Algebra 29 (8) (2001), 3571-3580.
[10] B. Sturmfels, Groebner bases and convex polytopes, Univ. Lectures Series, Amer. Math.Soc.,Vol.8 (1995).
[11] R. Villarreal, Monomial Algebras, Pure and Applied Mathematics, Marcel Dekker,Inc. New York, vol.238 (2001).

The reserch that led to the present paper was partially supported by the group GNSAGA of INDAM. The author wrote while being a member of the activity group AACC&AP-SIMAI.

Current address

Failla Gioia
DIIES, University of Reggio Calabria
via Graziella, salita Feo di Vito, Reggio Calabria, Italy
E-mail: gioia.failla@unirc.it

Slovak University of Technology in Bratislava
Faculty of Mechanical Engineering
16th Conference on Applied Mathematics **APLIMAT 2017**
Proceedings

PARAMETRIC PLANAR SECTIONS OF A POINT CLOUD WITH APPLICATIONS IN CULTURAL HERITAGE

FALCOLINI Corrado (IT)

Abstract. Given a discrete ("Point Cloud") representation of a surveyed object it is possible to analyse its three dimensional features with the help of several planar sections.
We survey a few applications of original algorithms to specific cases of interest in archaeology or architecture such as reconstruction of fragmented objects or symmetry detection.

Keywords: Point Clouds, Planar Sections, Symmetry Detection, Object Reconstruction

Mathematics Subject Classification: Primary 65K05, 53A04; Secondary 97N50

1 Introduction

Data sets coming from modern surveys (see [2], [4]) are basically made by a long list of coordinates of points detected on the surface of the object under study.

Each point can be associated to some local feature: color (photogrammetric survey), reflectance (an index of material reflection from laser scanner survey), orientation (average normal to the closest faces of the triangulated surface through the points) but point coordinates are often enough to evaluate its 3D geometrical model and feature.

The algorithms mentioned in this paper have as input the point cloud of a given fragment, or portion of a monument, and as output the best possible sectional curve on a given plane: with the help of several planar sections of an object it is then possible to analyse its three dimensional features.

Applications in Cultural Heritage presented in this paper run from the detection of an optimal fluting column model, in the virtual reconstruction of the Arch of Titus at the Circus Maximus in Rome (see [3], [8], [10]), to the contact probability of two possibly adjacent fragments, in the recomposition and restoration of the fragmented statue of S. Andrea at Stiffe, L'Aquila (see [1], [6], [9]) or the shape and tasselation of Borromini's San Carlino dome (see for example [5], [7], [12]).

2 Planar sections of a point cloud

Starting from the list of points over a column fragment, as in Fig.1, and for a given plane we test a segmentation procedure to approximate its curve of section. Starting as in Fig. 1a, we select all those points at a distance smaller than a certain threshold from the fixed plane and we project them (see Fig. 1c) on the plane. Shape thickness in Fig. 1d depends on plane orientation and on the fixed threshold.

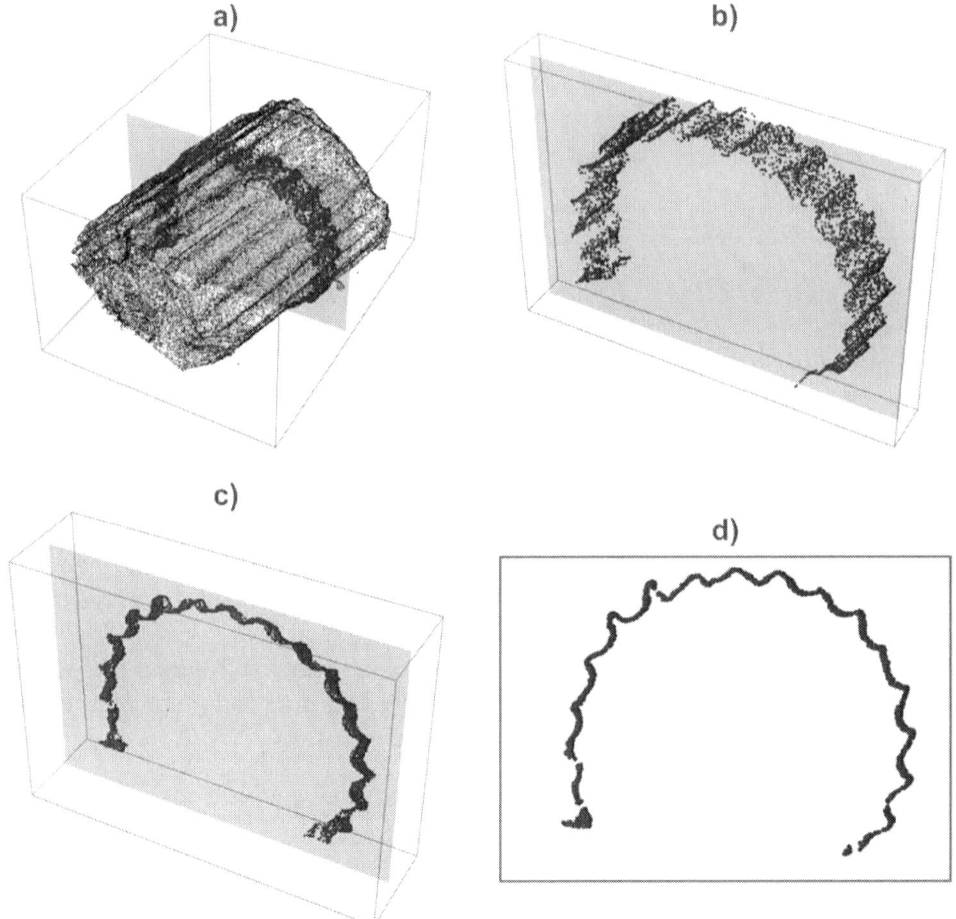

Fig. 1. Points on a given plane section of a surveyed column fragment: a) selected points at a distance from the given plane below a threshold; b) selected points in a box; c) projection of selected points on the given plane; d) planar representation of selected points: thickness depends on plane orientation and threshold

Decreasing the threshold we may look for a thinner curve, as regular as possible. Such a curve-like shape will be made of unordered points, difficult to connect but still useful for testing orientation and symmetry.

As a second step we evaluate an interpolating curve with the special feature of being parametric. Among several algorithms we choose to draw a polyline, ordering the selected points, or to apply a continuation method with an appropriate condition.

Another possible issue with a planar parametric curve as output (see [9]) is to find the border of a given set of points S on the plane: this could be found with a convex hull algorithm or with the algorithm shown in Fig. 2.

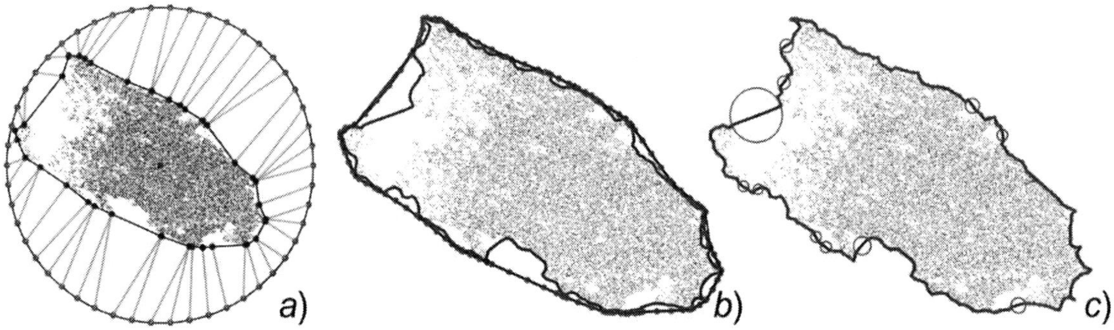

Fig. 2. a) Properly choosing the maximal circle of the circumscribed sphere connect the points of the set at minimal distance; b) take a uniform set of points on the interpolating parametric curve and iterates the procedure; c) corrections for straight portions of the limiting parametric curve

In order to detect a parametric border of S (small black points in Fig. 2) one can start with a uniform distribution of base points on a circle containing S, select for each base point the element of S at minimal distance (large blue points in Fig. 2a) and construct an interpolating parametric curve.

On this new curve take a uniform distribution of new base points and iterate the procedure: some corrections might be useful to prevent stright pieces of the limiting parametric curve which prevent to move closer to S. This algorithm works also in three dimensions.

3 Parametric curve of sections and inflection points

The first example refers to the application of an automatic procedure (see [7]) for the extraction of a piecewise regular parametric curve of section directly from the point cloud.

To detect a moulding profile of a surveyed fragment of the Arch of Titus at the Circus Maximus in Rome (see Fig. 3) we first have to select a relevant plane of section: if the point cloud is properly oriented we could choose a vertical plane which optimize points selection.

The correct orientation of the fragment, or that of the relevant plane of section, can be found analysing the normal components in the standard point cloud data: at any point of the cloud is associated an average normal to the closest faces of the corresponding triangulated surface. For instance it is possible to automatically select planar regions of the surface analyzing tstatistical properties of nearby normal directions and to check precise alignments using specific properties and symmetry.

We then look for a parametric curve fitting the selected points, like in Sec. 2, on the chosen plane (see Fig. 1).

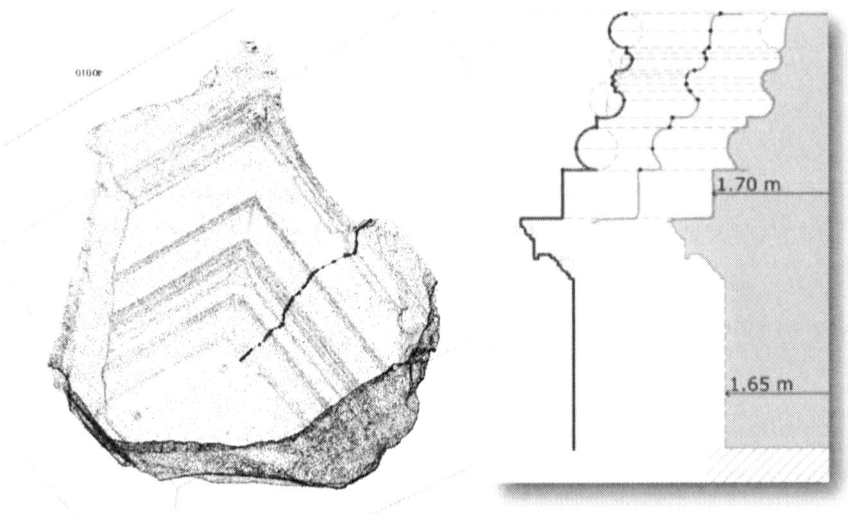

Fig. 3. Left: points on a molding section of a surveyed fragment. Right: molding interpolating curve of a column base section, inflection points detection and final geometrical reconstruction

Special points on the parametric curve, such as point of discontinuity in the direction of normal vectors and inflection points, allow an automatic geometrical moulding construction which can be translated in feature elements of the database and used in matching procedures.

As another example we show in Fig.4 a superposition of vertical parametric sections to check symmetries in coffers dimensions and placement in Borromini's San Carlino alle Quattro Fontane dome.

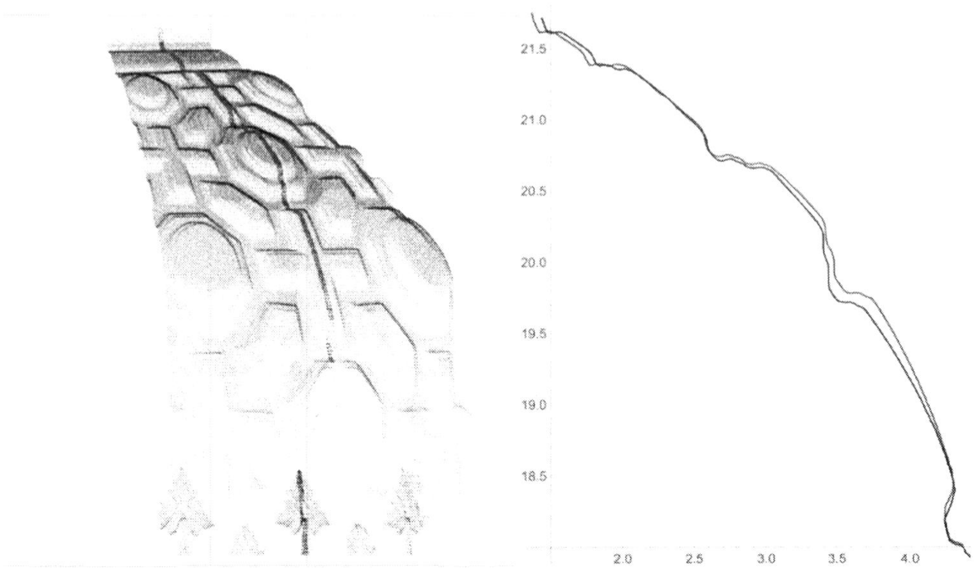

Fig. 4. Left: point cloud of a portion of S. Carlino's dome and selected points on a vertical section. Right: vertical sections comparison. The distance between two different curves can be measured using their parametric representation

Two different sectional curves represent the central molding profile of coffers placed in different positions on the dome. Curve profile gives informations on single coffers (octagons and crosses) size and its simmetric properties, but also on their mutual orientation and the geometrical shape of the dome intradox.

The distance between the two curves, and other geometrical features, can be measured using their parametric representation.

Comparison between the curves gives informations on simmetric properties of the dome and indirectly on its construction procedure.

4 Optimal fluted column model

From the survey of a ruined column drum (the same shown in Fig. 1), the second algorithm example shows how it is possible to find an optimal fluted column model (see [10]).

A simple n-fluted column section model can be easily constructed: take a circle C_0 of radius r_0 and a smaller circle C of radius r centered on a point of C_0 and call P one of the two intersection points of C with C_0 (see Fig. 5). The coordinates of the point P are then immediately computed as a solution of the system in Fig. 5. Also the angular distance θ between consecutive flutes is easy to compute and these formulas allow a parametric section model of n-fluted column of radius r_0 depending only on the size of r.

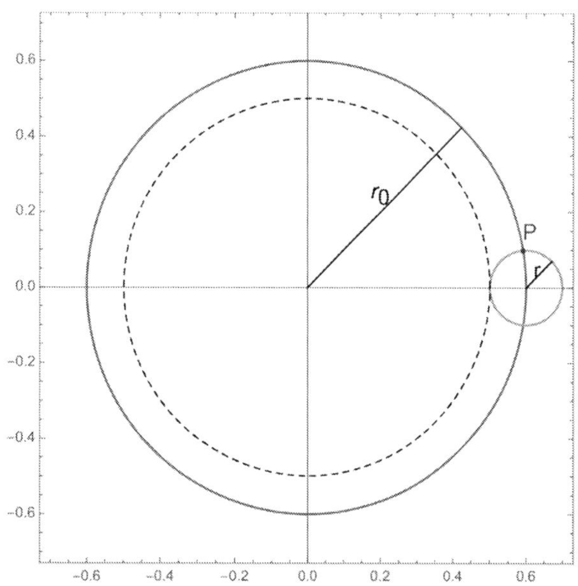

$$\begin{cases} x^2 + y^2 = r_0^2 \\ (x - r_0)^2 + y^2 = r^2 \end{cases}$$

$$x_P = r_0 - \frac{r^2}{2r_0}$$

$$\theta = \frac{2\pi}{n} - 2\arccos\frac{x_P}{r_0}$$

Fig. 5. n-fluted column section model: P coordinates and the angle θ between consecutive flutes are easily computed

The point cloud of the surveyed column fragment of Fig. 1 are shown in Fig. 6a projected on a vertictal plane. The projection is not completely orthogonal to the best cylinder containing the fragment thus the black border of the projection (see Fig. 6a) appears thicker than needed.

In Fig. 6b the same fragment is shown with a better orientation and cutting out the drum basis.

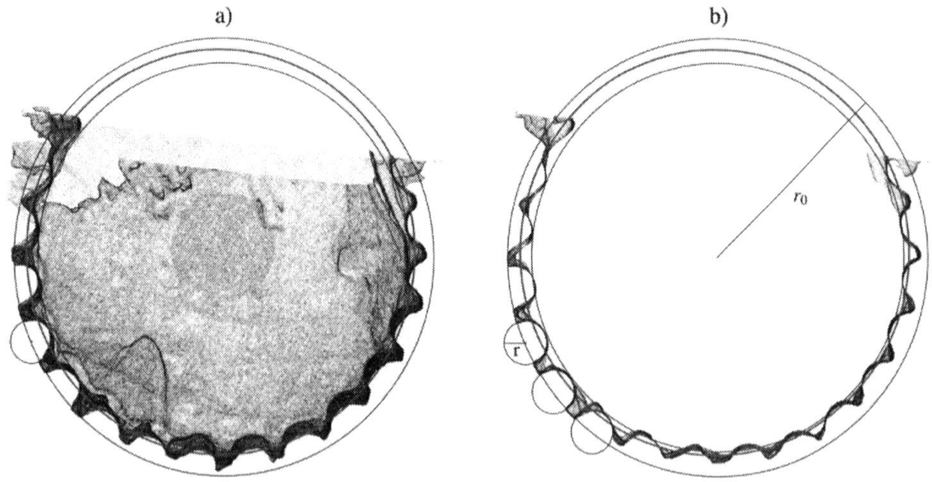

Fig. 6. a) Projected points of the drum on a plane: the projection is not completely orthogonal to the best cylinder containing the fragment; b) The same projection with a better orientation and cutting out drum basis. Radii r and r_0 can be estimated on a proper scale

Radii r and r_0 can then be estimated on a proper scale and the section model used to reproduce the original column, taking into account historical and architectural order information about the shape of a planar section containing column axis (entasis, tapering).

The n-fluted column section model of Fig. 5 agrees very well with the projected points of Fig. 6b whose border represents an optimal vertical section of a ruined fallen column fragment. Note that the size of the drum which has been projected is around 1 meter.

Comparing a cylindrical projection of the oriented point cloud with several section models, allows a very precise measure of column and flutings radii and then a possible drum reposition along the virtually reconstructed column.

5 Lines of fracture

The third application is for the broken statue reconstruction of S. Andrea at Stiffe, L'Aquila (see [9]) and shows the use of section curves to detect lines of fracture which could be tested for matching probability.

Points on a given planar section of a surveyed statue fragment (see Fig. 7 and Fig. 8) are obtained first as projection on the given plane (as in Sec. 2) and we look for a parametric curve which interpolates these points using an automatic procedure. Then we take uniformly spaced points on the parametric interpolating curve and use automatic inflection points detection to select special points depending on geometrical properties of the parametric representation (see Fig. 7).

These nodal points on many different parallel sections might line up to evidentiate a line of fracture (see Fig. 8) where the normals to the surface of the 3D-model of the surveyed fragment change suddenly direction: the problem of reconstructing a broken statue from its pieces can then be attacked by testing high matching probability checking corresponding curves and not only surfaces of contact.

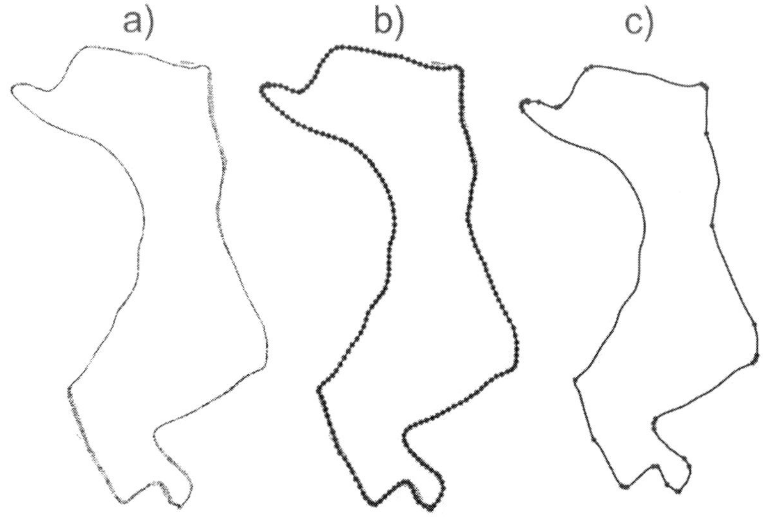

Fig. 7. Points on a given planar section of the surveyed statue fragment of Fig. 8: a) projection of selected points on the given plane; b) uniformly spaced points on an interpolating curve close to the selection; c) parametric curve and automatic inflection points detection

All measures have been optimized up to a certain error using the processed data and have been verified and improved after any new fragment addition or change in a suitable database.

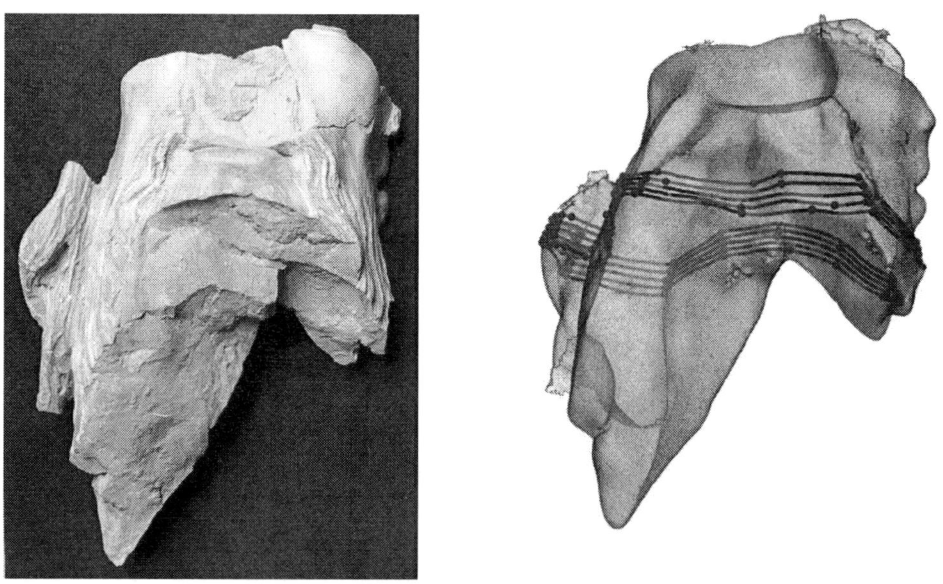

Fig. 8. Left: a fragment of the statue of S. Andrea at Stiffe, L'Aquila. Right: parallel planar section curves with some inflection points (the same as in Fig. 7c) lined up on lines of fracture.

Some of the original algorithms mentioned here have been also applied in different contexts.

6 Conclusions

Parametric planar sections of a point cloud are good starting point to analyze special features of a given surveyed architectural or archaeological object. Some interesting questions about the object at study can be partially answered using algorithms to construct the closest parametric curve to the points based mainly on point cloud coordinates.

References

[1] Huang Q.X. et al., Reassembling Fractured Objects by Geometric Matching, *ACM Trans. Graphics*, vol. 25, no. 3, 569-578, 2006.

[2] Dellepiane M. et al., Multiple Uses of 3D Scanning for the Valorization of an Artistic Site: The Case of Luni Terracottas, *Proc. Eurographics Italian Charter Conf.*, Eurographics, 2008.

[3] Thuswaldner B., Flory S., Kalasek R., Hofer M. Digital Anastylosis of the Octagon in Ephesos, *Journal on Computing and Cultural Heritage*, 2(1), 1–27, 2009.

[4] Scopigno R., Callieri M., Cignoni P., Corsini M., Dellepiane M., Ponchio F., Ranzuglia G, *3D models for cultural heritage: beyond plain visualization*. IEEE Computer, 44, 2011.

[5] Falcolini C., Vallicelli M. Modelling the vault of San Carlo alle Quattro Fontane, *Aplimat Journal of Applied Mathematics*, 4, 143-150, 2011.

[6] Palmas G., Pietroni N., Cignoni P., Scopigno R. A computer-assisted constraint-based system for assembling fragmented objects, In *Proc. of Digital Heritage* 2013 International Congress, vol. 1, IEEE, pp. 529–536 .

[7] Canciani M., Falcolini C., Saccone M., Spadafora G. From point cloud to Architectural models: algorithms for shape reconstruction. *ISPRS Archives,* XL-5/W1, 2013.

[8] Canciani, M., Falcolini, C., Buonfiglio, M., Pergola, S., Saccone, M., Mammı, B., Romito, G. Virtual Anastylosis of the Arch of Titus at Circus Maximus in Rome. International Journal of Heritage in the Digital Era, 3(2), 393–412, 2014.

[9] Canciani M., Capriotti G., D'Alessandro L., Falcolini C., Saccone M. The recomposition of fragmented objects: the case study of St. Andrea statue at Stiffe, L'Aquila, *Le vie dei Mercanti*, XIII Forum Internazionale di Studi, Capri, 2015.

[10] Falcolini C., Talamanca V. Modelli geometrici per nuvole di punti, In Atti del Convegno Mathematica Italia UGM, Napoli 28-29 maggio 2015, ISBN 978-88-96810-04-0.

[11] Falcolini C. Algorithms for Geometrical Models in Borromini's San Carlino alle Quattro Fontane. *Handbook of Research on Visual Computing and Emerging Geometrical Design Tools*, 26, 642–665, 2016

Current address

Falcolini Corrado, Associate Professor
Dipartimento di Architettura, Università Roma Tre
Via Madonna dei Monti, 40, 00184 Roma, Italy
Tel. number: +390657339621, e-mail: falco@mat.uniroma3.it

Slovak University of Technology in Bratislava
Faculty of Mechanical Engineering
16th Conference on Applied Mathematics **APLIMAT 2017**
Proceedings

UTILIZATION OF SOFTWARE MEANS AT SOLVING DIFFERENTIAL EQUATIONS IN TEACHING TECHNICAL SUBJECTS

FECHOVÁ Erika (SR)

Abstract. The present time brings a great surge in the use of information and communication means, the application of which can be found in almost all areas of our lives. They are important means of modernization in educational process. The paper deals with the possibility of using application software in educational process of natural science and technical subjects. Suitability of utilization of computer means is presented at solving the problem from electrical engineering – solving differential equations of second order and describing time dependencies.

Keywords: information and communication technologies, educational process, software means, MATLAB, Mathematica, MS Excel

Mathematics Subject Classification: 97M50

1 Introduction

The most distinctive feature of the present time is implementation of information and communication technologies into everyday lives of people. These changes influence not only private (spending free time, communication) and work spheres, but also educational process. One of the most important tasks in educational area is to work out such programs and methods, at which computers would become common work tools of a teacher and at the same time they would not eliminate development of creative thinking of a student. Nowadays the issue of using computers in educational process is very often discussed at various levels. As a result of those discussions, it can be said that computers form reliable and attractive environment for learning, provide positive feedback, help to create shapely correct text, respect individual requirements , pace, speed and skills, allow to return to the problem and start or finish work in various places, help students with specific disorders of learning and handicapped students to learn, make rich information sources available, comprehensibly present complex advancement of thoughts and relations by means of graphics, offer environment for development of students thinking [1], [2].

Information and communication technologies support non-traditional forms of education (e.g. e-learning) and can contribute to development of lifelong education that is inevitable for continuous renewing and gaining necessary knowledge and skills for life in digital world. Using information and communication technologies means possibility to improve learning and thinking in many ways [3], [4].

Information technologies present one of the factors, due to which mathematical education is continuously changed, transformed and modernized. Teaching natural science and technical subjects using mathematical apparatus at technical schools is not easy and the task of pedagogues is to make educational process interesting, extraordinary and attractive.

In the following part of the paper the possibility to make teaching the subject of electric engineering by means of computer support of MS Excel, Matlab and Mathematica programs more effective is presented by particular example of solving transient performance in RLC circuit.

2 Physical analysis of the problem

Problem assignment: Calculate and depict the course of voltage and current in the capacitor in RLC circuit with parameters $R = 5\,\Omega$, $L = 0{,}1\,\mathrm{H}$, $C = 100\,\mu\mathrm{F}$ connected to voltage $u = 10\,\mathrm{V}$, if voltage in the capacitor at the switch was $0\,\mathrm{V}$ and current flowing through the circuit at the moment of switching the circuit was zero [5].

Solution: It is a series electric circuit with *R*, *L*, *C* connected to harmonic voltage with initial conditions $i(t = 0\,\mathrm{s}) = 0$, $u_C(t = 0\,\mathrm{s}) = 0$ (Fig. 1).

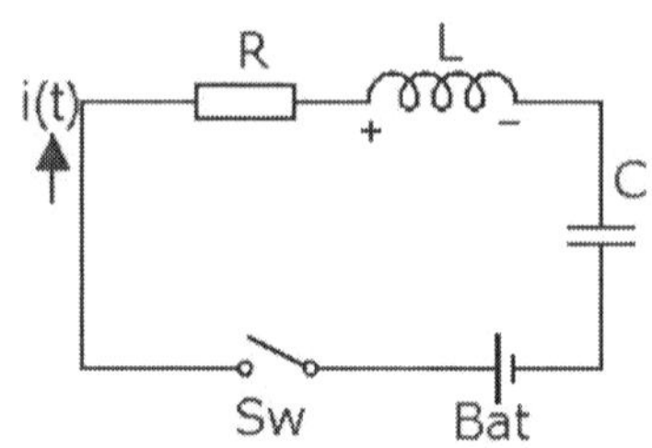

Fig. 1. RLC series circuit

Electric current, which flows through the circuit at given moment, can be determined by means of 2nd Kirchhoff's Law. For electric voltages in the circuit we get

$$u_L + u_C + u_R = u \tag{1}$$

where u_L, u_C, u_R present momentary values of voltage on the reel with *L* induction, capacitor with *C* capacity and resistance with *R* value.
For momentary values of voltage we have

$$u_L = L\frac{\mathrm{d}i}{\mathrm{d}t},\ u_C = \frac{1}{C}\int_0^t i\,\mathrm{d}t,\ u_R = Ri \tag{2}$$

and for the current flowing through capacitor we have

$$i = C\frac{\mathrm{d}u_c}{\mathrm{d}t} \tag{3}$$

If we substitute (2), (3) equations into (1) equation we get

$$\begin{aligned} & L\frac{\mathrm{d}i}{\mathrm{d}t} + \frac{1}{C}\int_0^t i\,\mathrm{d}t + Ri = u \\ & L\frac{\mathrm{d}}{\mathrm{d}t}\left(C\frac{\mathrm{d}u_C}{\mathrm{d}t}\right) + \frac{1}{C}\int_0^t C\frac{\mathrm{d}u_C}{\mathrm{d}t}\,dt + RC\frac{\mathrm{d}u_C}{\mathrm{d}t} = u \end{aligned} \tag{4}$$

We got linear integral-differential equation. For further solving it is suitable to transpose (4) equation into differential equation

$$LC\frac{\mathrm{d}^2u_C}{\mathrm{d}t^2} + RC\frac{\mathrm{d}u_C}{\mathrm{d}t} + u_C = u \tag{5}$$

It is a linear differential equation of second order with constant coefficients with right side. Such a differential equation can be solved either analytically or numerically. Analytic solution of this equation requires considerable mathematical knowledge and skills. On the other side numeric solution requires computer skills (or skills with application software).

General solution of the equation can be found as total of general solution of the appropriate linear differential equation with constant coefficients without right side and particular solution of linear differential equation with constant coefficients with right side [6].After substitution of numeric values into (5) equation we have differential equation

$$10^{-5}\frac{\mathrm{d}^2u_C}{\mathrm{d}t^2} + 5\cdot 10^{-4}\frac{\mathrm{d}u_C}{\mathrm{d}t} + u_C = 10$$

Analytic solution of the differential equation (expression of dependence of voltage on time) is as follows

$$u_c(t) = -10\cdot e^{-25t}\cos(25\sqrt{159}t) - \frac{10}{\sqrt{159}}e^{-25t}\sin(25\sqrt{159}t) + 10 \tag{6}$$

It is valid for dependence of current on time that

$$i = C\frac{\mathrm{d}u_C}{\mathrm{d}t} = 10^{-4}\frac{\mathrm{d}}{\mathrm{d}t}\left(-10\cdot e^{-25t}\cos(25\sqrt{159}t) - \frac{10}{\sqrt{159}}e^{-25t}\sin(25\sqrt{159}t) + 10\right)$$

After substitution we get

$$i = \frac{4}{\sqrt{159}}e^{-25t}\sin\left(25\sqrt{159}t\right) \tag{7}$$

For dependencies of voltage and current on time the graphs of dependencies given by (6) and (7) equations are described in MS Excel program (Fig. 2 and Fig. 3).

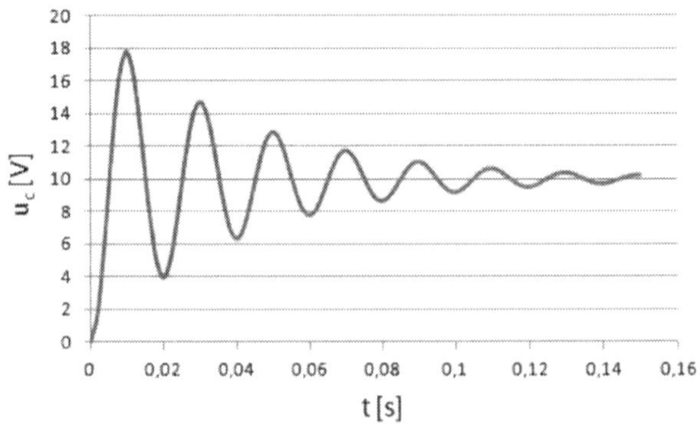

Fig. 2. Dependence of voltage on time

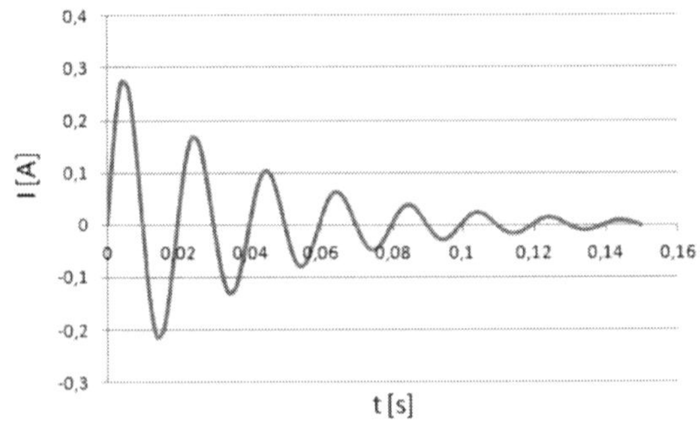

Fig. 3. Dependence of current on time

Analytic solution of the system of differential equations is difficult, it requires considerable mathematical knowledge and skills from the theory of solving differential equations and obviously does not lead to simple dependencies and results. Therefore the problem will be solved numerically by using MATLAB [7], [8].

3 Numerical solution in Matlab

Efficiency of educational process can be increased by application of some modern teaching methods. One of them is implementation of information and communication technologies into teaching such as utilization of means of MATLAB at solving differential equations.

MATLAB presents highly efficient language for technical calculations. It combines calculations, visualization and programming into simply usable environment. It is an interactive tool in which the basic data type is the field without necessity to declare its parameters. This property together with number of in-built functions enables relatively easy solution of many technical problems. In school environment MATLAB is a standard tool in teaching mathematics and other technical subjects, but it is also an efficient tool for research, development and data analysis [9], [10].

MATLAB is closer to the programming language compared to other similar products. From didactic point of view it is a suitable system, because it does not require complicated programming formulae and after relatively short time a beginner can manage to work in MATLAB. On the other side it presents a strong tool for experienced users. MATLAB does not have so many prearranged mathematical functions as for example MATHEMATICA. It does not have integrated properties such as MathCad. Support of symbolic calculations is not its standard part as it is at above mentioned products. However, it does not mean that MATLAB is depleted of these possibilities. It contains more than 500 simple or more complex mathematical functions implemented in the form of highly efficient and robust algorithms. From these functions it is possible to compose arbitrarily other functions. Sets of functions suitable for solution of a certain type of problems in MATLAB are called toolboxes. SIMULINK is an independent extension of MATLAB – solution of the system of nonlinear differential equations with a graphic entry of the system being solved. It enables graphically to observe dependencies of parameters at any connection point. It is used for simulation of dynamic behavior of the observed system. It is possible to use MATLAB in case of robust calculations, processing of extensive data files, work with large matrices and in cases when solution of the problem can be converted into vector and matrix operations. With regard to programming possibilities it is advantageous to use MATLAB also in case of branched or iterative algorithms of solution.

It is necessary to realize at the solution of differential equations of higher order in MATLAB that every differential equation of higher order can be transposed to the equivalent set of differential equations of first order with known initial conditions. At the problem solution it is suitable to transpose the differential equation of second order (5) to the set of differential equations of first order (8) as follows

$$\begin{aligned} \frac{\mathrm{d}u_C}{\mathrm{d}t} &= \frac{i}{C} \\ \frac{\mathrm{d}i}{\mathrm{d}t} &= \frac{u - u_C - R \cdot i}{L} \end{aligned} \tag{8}$$

Basic standard function for the solution of differential equations is *ode45* function, which syntax is: [*t*,*y*] = *ode*45 (*'name of the_function'*, *time_interval*, *initial_conditions*)

where *name of the_function* is reference to the function describing the set of differential equations, the parameter of *time_interval* is presented by the vector with two elements – initial time of solution t_0 and final time of solution t, the parameter of *initial_conditions* is presented by the vector of initial conditions y_0 from which we find $y(t_0) = y_0$. Two parameters are the output of the *ode45* function: t - the vector that contains instants of time, in which solution values are determined and y - the matrix containing its own solutions. To depict the current and voltage dependence on time, program writing in MATLAB is used, where the initial problem parameters, time and properties of depicted voltage and current dependences are given.

The following initial parameters are used at the problem solution: $R = 5\,\Omega$, $L = 10^{-1}$ H, $C = 10^{-4}$ F, $u = 10$ V.

```
function [] = RLC_obvod
U0 = 10; % V
R = 5; % ohm
C = 1e-4; % F
L = 1e-1; % H
Uc0 = 0; % V
I0 = 0; % A
t_konec = 1.5e-1;
% duc/dt = i/C
% di/dt = (U0 - uc - Ri)/L
function dxdt = dif_rce(t, x)
    dxdt = zeros(2,1);
    dxdt(1) = x(2)/C;
    dxdt(2) = (U0 - x(1) - R*x(2))/L;
end
[t, x] = ode45(@dif_rce, [0, t_konec], [Uc0, I0]);
uc = x(:,1);
i = x(:,2);
ur = R*i;
ul = U0 - uc - ur;
subplot(1, 2, 1);
plot(t, uc, 'r');
grid on;
title('The time dependence of voltage in the circuit ');
xlabel('t [s]'); ylabel('uc [V]');
subplot(1, 2, 2);
plot(t, i, 'b');
grid on;
title('The time dependence of current in the circuit ');
xlabel('t [s]'); ylabel('i [A]');
end
```

The result of the program initialization is depiction of time dependence of the voltage and current of the RLC series circuit (Fig. 4):

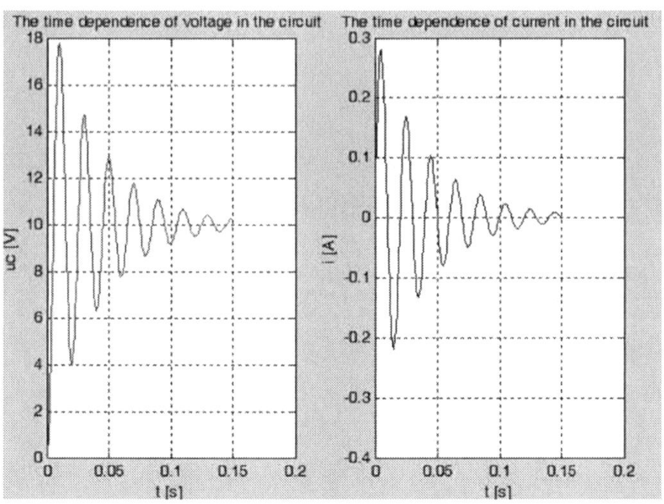

Fig. 4. Voltage and current flow in the RLC circuit

4 The problemsoluition in Mathematica Program

Mathematica is a computational software program used in scientific, engineering, and mathematical fields and other areas of technical computing, which we use to solve the problem. The parameters of the problem are substituted into the equation (4) and equation is written into the Mathematica program as $(10\text{^-}5)(\mathrm{d}(\mathrm{d}u/\mathrm{d}t)/\mathrm{d}t)+(5*10\text{^-}4)(\mathrm{d}u/\mathrm{d}t)+u=10$. After the program initialization we have the solution for voltage: $u_C(t) = C_1 e^{-25t}\cos\left(25\sqrt{159t}\right) + C_2 e^{-25t}\sin\left(25\sqrt{159t}\right) + 10$, where the solution of the equation is the sum of general solution of the appropriate equation without the right side u_c and optional particular solution u_P, t. j. $u_C(t) = u_c + u_P$. The C_1, C_2 constants are determined on the basis of the initial conditions. The result of the solution of our differential equation (6) after substitution of initial parameters in Mathematica program is as follows $u_C(t) = -10e^{-2500t}\cos\left(2500\sqrt{159t}\right) - \frac{10}{\sqrt{159}}e^{-2500t}\sin\left(2500\sqrt{159t}\right) + 10$.

The graph of voltage dependence can be found in Fig. 5.

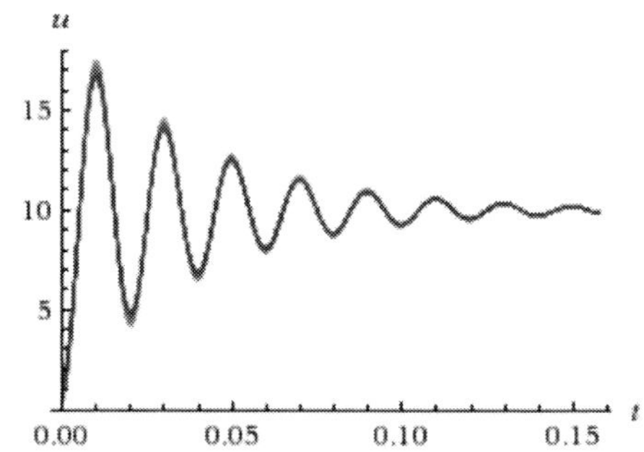

Fig. 5. Voltage characteristics of RLC circuit

For the current we have:

$$i(t) = C\frac{\mathrm{d}u_C}{\mathrm{d}t} = \frac{400e^{-2500t}\sin\left(2500\sqrt{159t}\right)}{\sqrt{159}} \tag{9}$$

The equation (9) is written into the Mathematica program as follows:

$$400*(e\text{^-}(2500*t))*\sin(2500*\mathrm{sqrt}(159)*t)/\mathrm{sqrt}(159) = i.$$

The graph of current dependence can be found in Fig. 6:

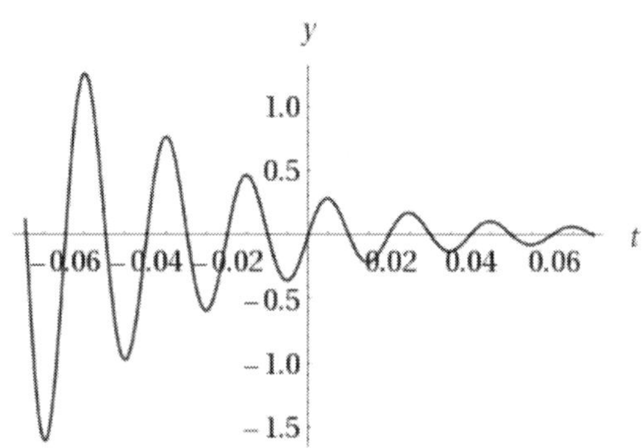

Fig. 6. Current dependence of RLC circuit

6 Conclusion

The presence of information and communication technologies in teaching has positive influence on efficiency of educational process and students accept them very positively. Based on the results it can be said that it is possible to facilitate and improve educational process by suitable combination of classic teaching methods and introduction of new elements using information and communication means. Gradual introduction and utilization of computer technique into teaching natural science and technical subjects at universities is one of the significant elements of modernizing technical and natural science education. The main goal of these modernizing tendencies is continual increase of educational level corresponding to present-day degree of knowledge in individual teaching disciplines with the respect to the age of a student and type of school studied. On the other side teaching these subjects with the support of computer technique cannot be understood in any case as a universal way how to solve the problems in educational system.

Acknowledgement

The paper was supported by the project KEGA 026TUKE-4/2016: „Implementation of modern information and communication technologies in education of natural science and technical subjects at technical faculties".

References

[1] KALAŠ, I. *Čo ponúkajú informačné a komunikačné technológie iným predmetom*. ŠPU Bratislava, 2001.

[2] TUREK, I. *Zvyšovanie efektívnosti vyučovania*. MC Bratislava, 1997.

[3] KIREŠ, M. Informačno-komunikačné technológie a potreba ďalšieho vzdelávania učiteľov. In: *Medzinárodná vedecko–odborná konferencia Fyzika v kontexte kultúry*. FHPV PU Prešov, 2001.

[4] BRESTENSKÁ, B. a kol. *Premena školy s využitím informačných a komunikačných technológií*. Elfa, s. r. o. Košice, 2010.

[5] HAJKO, V., DANIEL-SZABÓ J. *Základy fyziky*. Veda Bratislava, 1983.

[6] VAGASKÁ, A., MIŽÁKOVÁ, J. MATEMATIKA II. TU v Košiciach, Prešov, 2014. ISBN: 978-80-553-1670-3.

[7] HREHOVÁ, S. Approximations of the solution of the differential equation using the choosen application programs. In: *20. DIDMATTECH 2007*. Votobia, Olomouc, 2007: p. 323-326.

[8] VAGASKÁ, A. Computer support of applied mathematics in the engineering study programme. In: *6. konference o matematice a fyzice na vysokých školách technických*. Brno: Univerzita obrany, 2009: p. 309-316.

[9] KARBAN, P. Výpočty a simulace v programoch Matlab a Simulink. Computer Press Brno, a. s, 2007, 224 s. ISBN 80-2511-448-1.

[10] DUŠEK, F. MATLAB a SIMULINK, úvod do používaní. Univerzita Pardubice, 2002, ISBN 80-7194-273-1.

Current address

Fechová Erika, RNDr. PhD.
Technical University of Košice, Faculty of Manufacturing Technologies with a seat in Prešov, Department of Mathematics, Informatics and Cybernetics
Bayerova 1, 080 01 Prešov, Slovak Republic
Tel. number: 00421 55 602 6421, e-mail: erika.fechova@tuke.sk

Slovak University of Technology in Bratislava
Faculty of Mechanical Engineering
16th Conference on Applied Mathematics **APLIMAT 2017**
Proceedings

NEW TOOLS FOR MONGE PROJECTION IN GEOGEBRA

FERDIÁNOVÁ Věra (CZ), PORUBA Jakub (CZ)

Abstract. GeoGebra is a dynamic mathematics software for all levels of education that brings together geometry, algebra, spreadsheets, graphing, statistics and calculus in one easy-to-use package. Materials created in GeoGebra can be stored or shared on official website. Having a look at the official website of GeoGebra, we find out that not so much Tools have been created and shared yet. However, we can find some basic constructions there, but they often describe the same problemacy and are not included in GeoGebra software in general - e.g. construction of inscribed and circumscribed circles of a triangle. We decided to create a new tool that would help to improve teaching Monge projection.

Keywords: Monge projection, tools, GeoGebra

Mathematics subject classification: Primary 97G80, 51N05, 68N19; Secondary 00A99

1 Introduction and background of Monge projection

Monge projection is one of first projective methods, which students at secondary school and universities meet in subjects focused on constructive geometry or technical drawing. However, it can be said that Monge developed this method by accident. Thanks to his lower-class origin, when he enrolled at the military school in Meziéres, he was made to attend classes for poor students. These classes created various plaster models for classes of students of higher social status and during this process he found out this method that he named *geométrie descriptive*. This method simplified solving 3D space problems by constructions in plane – it was such a ground-breaking method that it was forbidden to publish it for 15 years – it became a military secret. Official teaching of this method was possible after the French Revolution, when Monge was called-up to teach at newly established École normale, which should educate 1500 young men as teachers as soon as possible (lots of teachers were executed during the French Revolution). In 1795, École normale is then the first school where the *geométrie descriptive* was brought to public. His knowledge was published in a book in

1798 in book named *Geométrie descriptive*. This book was rewritten couple of times and amended by notes of other authors. However, Monge there defines what the descriptive geometry is, sets up basic methods and principles of solving problems, deals with tangents and tangential planes, spheres and curved planes and their plane sections. In planned continuing, he wanted to deal with perspective, shades and statuary, but he never managed to finish this part thanks to his often participation on scientific voyages. Anyway, the principle of projection on projection planes was known even at Ancient times and in Middle Ages, it was used in perspective problems. Before Monge, Two orthogonal plane projections were used by Albrecht Dürer, who dealt with spatial curves, or Frézier, who used this planes in engineering and military cases. But Monge was the first one, who managed to organise the knowledge, simplify it and show, that basic geometric laws are enough for depicting this objects. [2] [6] [7] [8]

Nowadays, at the time of digital world, there is a need to improve teaching methods, because we can say that teaching Monge projection have not changed much from the time when Monge developed it (drawings are still common on sheets of paper). Several digital materials of constructions can be found od GeoGebra website, e.g. series of constructions by Jana Volná, Roman Hašek and others or a book by Tomáš Vysloužil that uses projection of 3D image and a plane. Thus we decided to create a new tool that would help to improve teaching Monge projection. As a software we chose GeoGebra.

2 GeoGebra

GeoGebra is a dynamic mathematics software for all levels of education that brings together geometry, algebra, spreadsheets, graphing, statistics and calculus in one easy-to-use package. It also is a rapidly expanding community of millions of users located in just about every country. GeoGebra has become the leading provider of dynamic mathematicsl software, supporting science, technology, engineering and mathematics (STEM) education and innovations in teaching and learning worldwide.[1] This software contains special areas for Geometry, Algebra and Spreadsheet, all of them are connected and fully dynamic. Interface of GeoGebra is user-friendly, you can use too many tools for construction. Main advantage of this software is in visualization of geometrical problems.

GeoGebra won these awards:

- Archimedes 2016: MNU Award in category Mathematics (Hamburg, Germany)
- Microsoft Partner of the Year Award 2015: Finalist, Public Sector: Education (Redmond, WA, USA)
- MERLOT Classics Award 2013: Multimedia Educational Resource for Learning and Online Teaching (Las Vegas, Nevada, USA)
- NTLC Award 2010: National Technology Leadership Award (Washington D.C., USA)
- Tech Award 2009: Laureat in the Education Category (San Jose, California, USA)
- BETT Award 2009: Finalist in London for British Educational Technology Award
- SourceForge.net Community Choice Awards 2008: Finalist, Best Project for Educators

- AECT Distinguished Development Award 2008: Association for Educational Communications and Technology (Orlando, USA)
- Learnie Award 2006: Austrian Educational Software Award (Vienna, Austria)
- eTwinning Award 2006: 1st prize for "Crop Circles Challenge" with GeoGebra (Linz, Austria)
- Trophées du Libre 2005: International Free Software Award, category Education (Soisson, France)
- Comenius 2004: German Educational Media Award (Berlin, Germany)
- Learnie Award 2005: Austrian Educational Software Award (Vienna, Austria)
- digita 2004: German Educational Software Award (Cologne, Germany)
- Learnie Award 2003: Austrian Educational Software Award (Vienna, Austria)
- EASA 2002: European Academic Software Award (Ronneby, Sweden) [1]

We can see, GeoGebra won too many awards in education sections and is constantly being improved.

3 Custom tools in GeoGebra

GeoGebra allows you to create your own construction tools based on an existing construction. Once created, your custom tool can be used both with a pointing device and as a command in the Input Bar. All tools are automatically saved in your GeoGebra file.[1] If user wants, he can export or import tools to .ggt files or share tools on GeoGebra website. Custom tools are easy to create, fully dynamic and function. One of disadvantages of Custom tools is immposibility of setting automatical labels of points, because GeoGebra has a function for automatical labels of points in default. It is a problem, if you need to use indexing of points - you must rewrite it manually. One of ideas for solving this problem is using a Script. It is possible to create new functions on one click, but it is not friendly for advanced users. Second option is using external software for renaming points, but once again, users need highest skill in programming.

3.1 How to create a new custom tool?

At first, we need completed constructions like a construction of square, triangle, circumscribe circle, etc. The best option for a tool is construction, which is simple, universal and often in use. We need to know what will be an input and output of the tool - we can see it in an example of square. Algorithm of creating a new tool on example of Square is also described in other part of this paper. Input for tool Square will be 2 points or vertices of one side of square and output will be the square. If we have a decision for basic task, we can start to create Custom Tool.

- In menu, open *Tools* and choose *Create new tool*

- The window *Create new tool* contains three bookmarks: *Output objects, Input objects and Name & Icons.*

- Specify the *Output Objects:* Click on the construction, e. g. a square, or select it from the drop down menu. Also specify edges of the square in *Output Objects* (1). Click to button *Next.*

- Specify the *Input Objects*: GeoGebra automatically specifies Input Objects, which usually are first objects in our construction. We can change default set by clicking on the points or select 2 points from the list in the drop down menu (1). If we have more objects, we can use icon *delete* on the right side of window with sign a cross. Click to button *Next.*

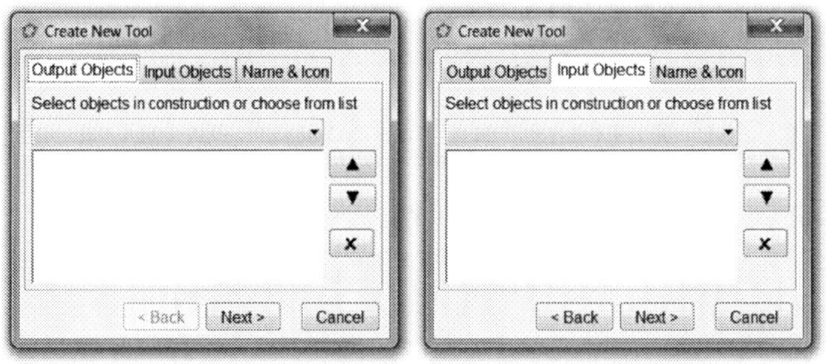

Fig. 1. Windows of Create new tool in GeoGebra

- Specify the *Name & Icon*: choose name of the tool and command name (2). Fill help tool, because function of this tool can be forgot or another user can use a tool, but it does not influence functioning of the tool. Check box is for visibility tool in toolbar. We can change picture of icon. Icon is a standard figure of file type PNG, BMP, JPG etc.

- Last step is click on button *Finish*

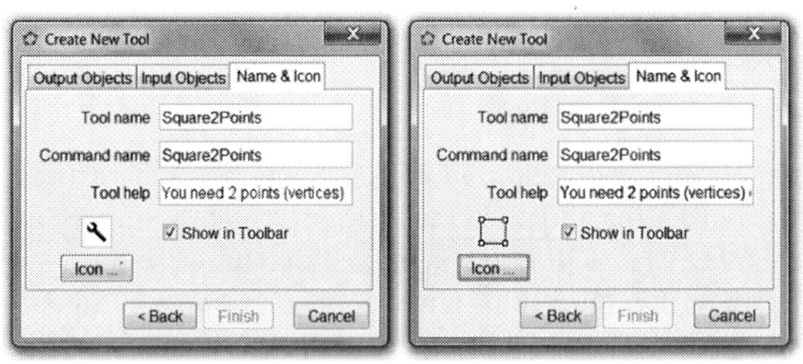

Fig. 2. Windows of Create new tool in GeoGebra

We created tool very simply and GeoGebra automatically saves the tool in our file. If we want to use the tool in another file we, can save it as GGT files through Manage tool.

4 Tools for Monge projection

We would like to present a few tools for Monge projection. We chose the basic constructions, which are often used in geometrical problems in Monge projection and help drawing the final construction.

When drawing drawings of basic constructions by default (a point, a straight line, a plane), we continue in following way:

- *a point:* If we want to make a projection of a point, we have to construct its first and second projections. It means that we project it perpendicularly to horizontal plane π and vertical plane ν, we call these projections related views of a point. But practically, if we are given point coordinates, we map the x coordinate on a folding line, plus y-coordinate and minus z-coordinate under the folding line on a line crossing the x coordinate of the point and parallel to the vertical axis , minus y-coordinate and plus z-coordinate by analogy above the folding line.

- *a straight line (given by points):* If the straight line is given by points, we continue in way of mapping points and connect correspoding first and second projections of points by line, which gives us first and second projection the straight line; if these projections cut the folding line, we map tracing points

- *a plane(given by a trace line):*Trace lines are given by intersection lines of this plane with horizontal and vertical plane. The intersection lines cut the folding line and the vertical axis. Thus, when the plain is given by a trace line, we are usually given the coordinates of these "cuts". Then the maping is simple - we map the x-coordinate on the folding line and y and z-coordinates on the vertical axis analogically to points. [3] [4] [5]

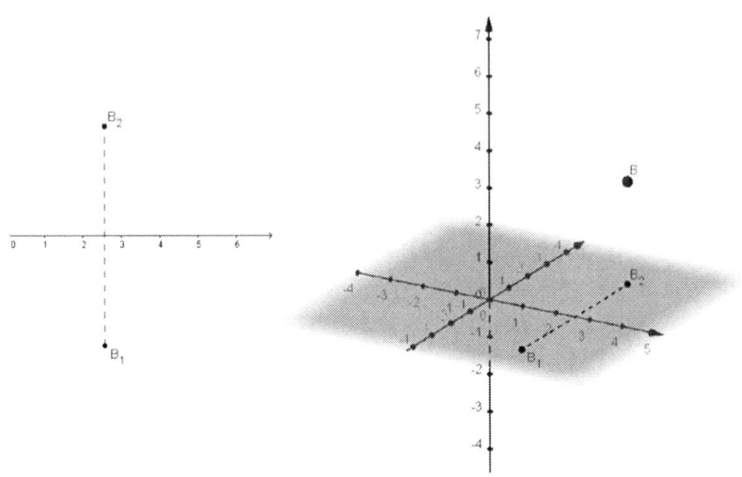

Fig. 3. Graphics and 3D Graphics windows in GeoGebra with visualization of a Point in Monge projection

We created three tools for these basic constructions. We combinated visualization in 3D space with 2D plane in GeoGebra. Description of "algorithm" contains input set objects (Input), steps of construction (Construction)and output set objects (Output).

Tool Point in Monge Projection (command PointMP[A]):
Input: New point $A = (x, y, z)$ or existing point in space
Construction:

```
A_1=(x(A),-y(A))

A_2=(x(A),z(A))

a=Segment[A_1,A_2]
```

Change property type line of segment and hide name of segment

Output: Segment a, points A_1, A_2
Description: Tool Point in Monge Projection (3) vizualizes point A from 3D Graphics window to Graphics window, it means point $A \in \mathbb{E}^3$ is shown in Monge projection. The first and second projection of point A lies on order line, which is represented by dashed segment without description.

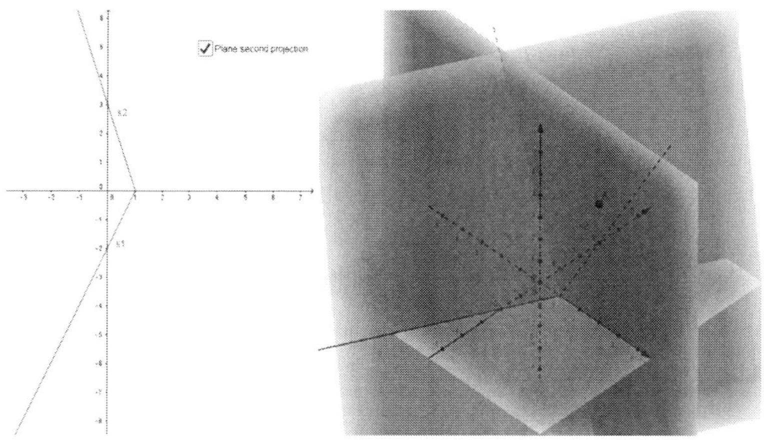

Fig. 4. Graphics and 3D Graphics windows in GeoGebra with visualization of a Plane in Monge projection

Tool Trace line of plane in Monge Projection (command TLPlaneMP[A]):
Input: New point $A = (x, y, z)$ or existing point in space, which coordinates represent coordinates of trace lines of plane α
Construction:

```
s_1=Ray[(x(A),0),(0,-y(A))]

s_2=Ray[(x(A),0),(0,z(A))]

α=Plane[(0,0,z(A)),s_1]

β=Plane[xAxis,zAxis]
```

A line p is intersection $\alpha,\ \beta$

Create checkbox for show of a plane β and a line p

Output: Graphic window trace lines of plane $s_1,\ s_2$, Graphic 3D window plane α
Description: Tool Trace line of plane in Monge projection (4) constructs trace lines of plane from coordinators of trace lines $\alpha \equiv (x, y, z)$,to simplyfy we use a point A, which represent steps of plane α. Tool creates trace lines of plane α in Graphics window and a real plane α in 3D Graphics window. Tool turns on the second projection plane β (standard description is ν, but a symbol is not in GeoGebra), its intersection with the plane α and a checkbox vizualization of the plane β and the line p.

Tool Line in Monge Projection (command LineMP[A,B]):
Input: New points $A = (x_A, y_A, z_A)$, $B = (x_B, y_B, z_B)$ or existing two points $A,\ B$ in space, which $A,\ B$ in line p
Construction:

```
PointMP[A]

PointMP[B]

p=Line[A_1,B_1]

q=Line[A_1,B_1]

N_1=Intersec[xAxis,p]

P_2=Intersec[xAxis,q]

p_1=Ray[N_1,A_1]

p_2=Ray[P_2,A_2]
```

Output: Rays $p_1,\ p_2$

Description: Tool Line in Monge projection constructs first and second projection of a line p, which points $A,\ B \in p$. Tool visualizes second projection of first projection trace point P and first projection of second projection trace point N.

These constructions are very simply and very often used in standard constructions in the Monge projection. Tools were uploaded on website GeoGebra and we will see reflection from users.

5 Conclusion

The Custom tool in teaching descriptive geometry seems to be useful. It can be used in all basic constructions we use as functions in algorithms. We have already prepared three basic tools for Monge projection, if it acquits well, the plan is to extend tools in way of searching horizontal and vertical lines of a plane etc. Created tools might be helpful for users, who are good at geometry but lose time by constructions of basic objects. We think, that one of main benefits is making the lesson faster, when students already acquire basic knowledge about constructions and need to solve more difficult tasks as plane sections of solids, etc.

6 Acknowledgement

The paper was supported by grant from SGS12/PřF/2016-2017: Geometric mechanics, optimization and number theory.

References

[1] URL: `http://www.geogebra.org`.

[2] PORUBA, J. *Geometrické útvary v Mongeově promítání* [online]. Ostrava, 2016 [cit. 2016-11-23]. Bakalářská práce. Ostravská univerzita, Přírodovědecká fakulta. Vedoucí práce Mgr. Věra Ferdiánová, Ph.D.

[3] Medek, V. a Píska, R. *Deskriptivní geometrie I.* 1. vyd. SNTL, 1966.

[4] Harant, M. a Lanta, O. *Deskriptivní geometrie: pro II. a III. ročník SVVŠ* 1. vyd. SPN, Praha, 1965.

[5] Chodorová, M. *Principy zobrazení.* [online]. Olomouc: KAG PřF UP v Olomouci. 3 s.

[6] Balvínová, M. *Život a dílo Gasparda Monge* [online]. Brno, 2009. Diplomová práce. Masarykova univerzita, Přírodovědecká fakulta. Vedoucí práce Josef Janyška.

[7] Carlbom, I. and Paciorek, J. Planar Geometric Projections and Viewing Transformations In. *ACM Computing Surveys* [online]. 1978, Vol. 10, No. 4, pg. 465-502.

[8] The Editors of Salem P. *Great Mathematicians* [e-book]. Ipswich, Mass: Salem Press; 2013.

Current address

Ferdiánová Věra, Mgr. Ph.D.
Department of Mathematics, Faculty of Science, University of Ostrava
30. dubna 22, 703 01 Ostrava, Czech Republic
E-mail: vera.ferdianova@osu.cz

Poruba Jakub, Bc.
Department of Mathematics, Faculty of Science, University of Ostrava
30. dubna 22, 703 01 Ostrava, Czech Republic

Slovak University of Technology in Bratislava
Faculty of Mechanical Engineering
16th Conference on Applied Mathematics **APLIMAT 2017**
Proceedings

FIRST ORDER DIFFERENTIAL EQUATIONS INDUCED BY THE INFINITE SERVERS QUEUE WITH POISSON ARRIVALS TRANSIENT BEHAVIOR PROBABILITY DISTRIBUTION PARAMETERS STUDY AS TIME FUNCTIONS

FERREIRA Manuel Alberto M. (PT)

Abstract. The M|G|∞ queue system state transient probabilities, considering the time origin at the beginning of a busy period, mean and variance monotony as time functions is studied. These studies, for which results it is determinant the hazard rate function service time length, induce the consideration of two differential equations, one related with the mean monotony study and another with the variance monotony study, which solutions lead to some particular service time distributions, for which those parameters present specific behaviors as time functions.

Keywords: M|G|∞, queue, transient probabilities, service time length, hazard rate function, first order differential equations, exact differential equations

Mathematics Subject Classification: 60K35

1 Introduction

The M|G|∞ queue customers arrive at this queue system according to a Poisson process at rate λ and receive a service which time length is a positive random variable with distribution function $G(.)$ and mean α. Upon their arrival, they find immediately an available server. Each customer service time length is independent from the other customers' services time length and from the arrivals process. The traffic intensity is a no dimensional parameter given by $\rho = \lambda\alpha$.

Call $N(t)$ the number of occupied servers (or the number of customers being served) at instant t, in a M|G|∞ queue system.

Be $p_{1'n} = P[N(t) = n | N(0) = 1']$, $n = 0,1,2,\ldots$, meaning $N(0) = 1'$ that the time origin is an instant at which a customer arrives at the system, jumping the number of customers from 0 to 1. That is: an instant at which a busy period begins. Then, see (Ferreira and Andrade, 2009),

$$p_{1'0}(t) = p_{00}(t)G(t)$$
$$p_{1'n}(t) = p_{0n}(t)G(t) + p_{0n-1}(t)(1 - G(t)),\ n = 1,2,... \tag{1.1}$$

where

$$p_{0n}(t) = \frac{\left(\lambda \int_0^t [1 - G(v)]dv\right)^n}{n!} e^{-\lambda \int_0^t [1-G(v)]dv}, \quad n = 0,1,2,..., \tag{1.2}$$

being here the time origin an instant at which the queue system is customers empty, see again (Ferreira and Andrade, 2009).

It is easy to check that

$$\lim_{t \to \infty} p_{1'n}(t) = \frac{\rho^n}{n!} e^{-\rho}, \quad n = 0,1,2,... \tag{1.3}$$

such as $\lim_{t \to \infty} p_{0n}(t) = \frac{\rho^n}{n!} e^{-\rho}, \quad n = 0,1,2,...$

Denoting $\mu(1',t)$ the distributions given by (1.1) mean value:

$$\mu(1',t) = 1 - G(t) + \lambda \int_0^t [1 - G(v)]dv \tag{1.4}$$

And denoting $V(1',t)$ the distribution defined by (1.1) variance, it is obtained

$$V(1',t) = \lambda \int_0^t [1 - G(v)]dv + G(t) - G^2(t). \tag{1.5}$$

The main target is to study $\mu(1',t)$ and $V(1',t)$ monotony as time functions. Then two first order differential equations, induced by this study, are considered allowing very interesting conclusions. In this work are continued and deepened the works of Ferreira (2005 and 2014).

2 Study of $\mu(1',t)$ as time function

Call *h(t)* the service time length hazard rate function. It is given by, see Ross (1983),

$$h(t) = \frac{g(t)}{1-G(t)} \tag{2.1}$$

where $g(t) = \frac{dG(t)}{dt}$. It is interpreted as the rate at which the services end.

Proposition 2.1

If $G(t) < 1, t > 0$ continuous and differentiable and

$$h(t) \le \lambda,\ t>0 \tag{2.2}$$

$\mu(1',t)$ is non- decreasing.

Dem.:

It is enough to note, according to (1.4), that $\frac{d}{dt}\mu(1',t)=(1-G(t))(\lambda-h(t))$. ■

Obs.:

-The expression (2.2) meaning is that if the rate at which the services end is lesser or equal than the customers arrivals rate, $\mu(1',t)$ is non- decreasing.

-For the M|M|∞ system, that is M|G|∞ queue with exponentially distributed service times length, (2.2) is equivalent to

$$\rho \ge 1 \tag{2.3}$$

- $\lim_{t\to\infty} \mu(1',t)=\rho$, as expected. ■

Defining $\beta(.)$ as

$$\beta(t) = h(t) - \lambda \tag{2.4}$$

The first order differential equation (2.4) solution is the following collection of service time distributions:

$$G(t)=1-(1-G(0))e^{-\lambda t-\int_0^t \beta(u)du}, t\ge 0, \frac{\int_0^t \beta(u)du}{t} \ge -\lambda \tag{2.5}$$

So, evidently,

Proposition 2.2

If $\beta(t) = 0$

$$G(t)=1-(1-G(0))e^{-\lambda t},\quad t\ge 0 \tag{2.6}$$

and $\mu(1',t)=1-G(0)=\rho,\quad t\ge 0.$ ■

Proposition 2.3

If $\beta(t) = -\lambda$

$$G(t) = 1\,, t \ge 0 \tag{2.7}$$

and $\mu(1',t)=0, \quad t \geq 0.$ ■

For some particular service time length distributions:

-Deterministic with value α (**M|D|∞** system)

$$\mu(1',t)=\begin{cases}1+\lambda t, t<\alpha \\ \rho, t\geq \alpha\end{cases} \tag{2.8}$$

-Exponential (**M|M|∞** system)

$$\mu(1',t)=\rho+(1-\rho)e^{-\frac{t}{\alpha}} \tag{2.9}$$

$$-\ G(t)=1-\frac{\left(1-e^{-\rho}\right)(\lambda+\beta)}{\lambda e^{-\rho}\left(e^{(\lambda+\beta)t}-1\right)+\lambda}, t\geq 0, -\lambda\leq\beta\leq\frac{\lambda}{e^{\rho}-1}$$

$$\mu(1't)=\frac{\left(1-e^{-\rho}\right)(\lambda+\beta)}{\lambda e^{-\rho}\left(e^{(\lambda+\beta)t}-1\right)+\lambda}+\rho-\log\left(1+\left(e^{\rho}-1\right)e^{-(\lambda+\beta)t}\right) \tag{2.10}$$

For this collection of service time distributions the busy period time length is exponentially distributed with an atom at the origin, see (Ferreira and Andrade, 2009) and Ferreira (2016 and 2016a):

$$B^{\beta}(t)=1-\frac{\lambda+\beta}{\lambda}\left(1-e^{-\rho}\right)e^{-e^{-\rho}(\lambda+\beta)t}, t\geq 0, -\lambda\leq\beta\leq\frac{\lambda}{e^{\rho}-1} \tag{2.11}$$

Note:

-For a M/G/∞ queue, if the service time length distribution function belongs to the collection, see again (Ferreira and Andrade, 2009) and Ferreira (2016 and 2016a),

$$G(t)=1-\frac{1}{\lambda}\frac{\left(1-e^{-\rho}\right)e^{-\lambda t-\int_0^t\beta(u)du}}{\int_0^{\infty}e^{-\lambda w-\int_0^w\beta(u)du}dw-\left(1-e^{-\rho}\right)\int_0^t e^{-\lambda w-\int_0^w\beta(u)du}dw},$$

$$t\geq 0, -\lambda\leq\frac{\int_0^t\beta(u)du}{t}\leq\frac{\lambda}{e^{\rho}-1},$$

which is solution for the Riccati equation

$$\frac{dG(t)}{dt} = -\lambda G^2(t) - (\beta(t) - \lambda)G(t) + \beta(t),$$

the busy period time length distribution function is

$$B(t) = \left(1 - (1 - G(0))\left(e^{-\lambda t - \int_0^t \beta(u)du} + \lambda \int_0^t e^{-\lambda w - \int_0^w \beta(u)du} dw\right)\right) *$$

$$* \sum_{n=0}^{\infty} \lambda^n (1 - G(0))^n \left(e^{-\lambda t - \int_0^t \beta(u)du}\right)^{*n}, -\lambda \leq \frac{\int_0^t \beta(u)du}{t} \leq \frac{\lambda}{e^{\rho} - 1}.$$

If $\beta(t) = \beta$ (constant) it results (2.11).■

3 Study of $V(1',t)$ as time function

Proposition 3.1

If $\frac{1}{2} < G(t) < 1, t > 0$ continuous and differentiable and

$$h(t) \leq \frac{\lambda}{2G(t)-1} \qquad (3.1)$$

$V(1',t)$ is non-decreasing.

Dem.:
According to (1.5), $\frac{d}{dt} V(1',t) = \lambda(1 - G(t)) + g(t) - 2G(t)g(t) = \lambda(1 - G(t)) + g(t)(1 - 2G(t)) = (1 - G(t))\,(h(t)(1 - 2G(t)) + \lambda)$. So that $\frac{d}{dt} V(1',t) \geq 0$ it is enough to have $(h(t)(1 - 2G(t)) + \lambda) \geq 0 \Leftrightarrow h(t)(2G(t) - 1) - \lambda \leq 0 \Leftrightarrow h(t) \leq \frac{\lambda}{2G(t)-1}$ since $G(t) > \frac{1}{2}$.■

Obs.:

- $\lim_{t \to \infty} V(1',t) = \rho$, as expected.■

Define $\beta(.)$ as $\beta(t) = h(t) - \frac{\lambda}{2G(t)-1}$. It results the following differential equation in G(.):

$$\frac{dG(t)}{dt} = \left(\beta(t) + \frac{\lambda}{2G(t)-1}\right)(1 - G(t)), t \geq 0 \qquad (3.2)$$

Give (3.2) the following form

$$\left(\lambda+\beta(t)(2G(t)-1)\right)dt+\left(2-\frac{1}{1-G(t)}\right)dG=0 \quad (3.3)$$

that can be turned in

$$\left(\lambda+\beta(2G(t)-1)\right)dt+\left(2-\frac{1}{1-G(t)}\right)dG=0 \quad (3.4)$$

for $\beta(t)=\beta\ (const.)$.So $\frac{\partial}{\partial G}\left(\lambda+\beta(2G-1)\right)=2\beta$ and $\frac{\partial}{\partial t}\left(2-\frac{1}{1-G}\right)=0$.Then

it is possible to consider the following situations:

A) $\boldsymbol{\beta=0}$

Equation (3.2) is an exact differential equation which solution is

$$\frac{1-G(t)}{1-G(0)}e^{2(G(t)-G(0))}=e^{-\lambda t}, t\geq 0 \quad (3.5)$$

Proposition 3.2

If $G(.)$ is implicitly defined as

$$\frac{1-G(t)}{1-G(0)}e^{2(G(t)-G(0))}=e^{-\lambda t}, t\geq 0 \quad (3.6)$$

$V(1',t)=\rho, t\geq 0.$ ■

Obs.:

-The density associated to (3.6) is

$$g(t)=-\frac{\lambda e^{-\lambda t}(1-G(0))}{(1-2G(t))e^{2(G(t)-G(0))}}. \quad (3.7)$$

-After (3.7), denoting S the associated random variable, it is easy to see that, with $G(0)>\frac{1}{2}$,

$$\frac{(1-G(0))n!e^{-2(1-G(0))}}{\lambda^n}\leq E\left[S^n\right]\leq\frac{(1-G(0))n!}{(2G(0)-1)\lambda^n}, n=1,2,\ldots\ . \ ■ \quad (3.8)$$

B) $\boldsymbol{\beta\neq 0}$

Now equation (3.2) is not an exact differential equation but, using the integrating factor

$$\mu = \frac{1}{\beta(2G(t)-1)+\lambda} \tag{3.9}$$

that is a solution for the differential equation

$$\frac{d\ln\mu}{dG} = \frac{\frac{\partial}{\partial t}\left(2-\frac{1}{1-G}\right)-\frac{\partial}{\partial G}\left(\lambda+\beta(2G-1)\right)}{\lambda+\beta(2G-1)} = \frac{0-2\beta}{\lambda+\beta(2G-1)} = -\frac{2\beta}{\beta(2G-1)+\lambda} \tag{3.10}$$

it becomes an exact differential equation which solution is

$$\left(\frac{1-G(t)}{1-G(0)}\right)^{\frac{1}{\beta+\lambda}} \left|\frac{2\lambda G(t)-\beta+\lambda}{2\lambda G(0)-\beta+\lambda}\right|^{\frac{\lambda}{\beta(\beta+\lambda)}} = e^{-t}\ . \tag{3.11}$$

It is interesting to make $\beta = \lambda$ in (3.11). In fact it becomes

$$\left(\frac{1-G(t)}{1-G(0)}\right)\left(\frac{G(t)}{G(0)}\right) = e^{-2\lambda t}$$

that lead to

$$G(t) = \frac{1+\sqrt{1-4(1-G(0))G(0)e^{-2\lambda t}}}{2}\ . \tag{3.12}$$

It is easy to check that in (3.12) $G(\infty) = 1$ and $G(0) = \frac{1}{2}$.

Note:

- Cases A) and B) may be resolved using the separation variables technic, maybe easier. The proceeding used comes from an attempt to get solutions for the general case with $\beta(t)$ non constant. But, in this situation, the second member of (3.10) does not depend only on G what makes this proceeding to determine the integrating factor unviable.

- $G(t) = 1, t \geq 0$ is a trivial solution of equation (3.2). And, for this case, it is easy to check after (1.5) that $V(1',t) = 0, t \geq 0$. ■

For some particular service time length distributions:

-Deterministic with value α (**M|D|∞** system)

$$V(1',t) = \begin{cases} \lambda t, t < \alpha \\ \rho, t \geq \alpha \end{cases}\ . \tag{3.13}$$

-Exponential (**M|M|∞** system)

$$V(1',t)=\rho\left(1-e^{-1/\alpha}\right)+e^{-\frac{t}{\lambda}}+e^{-\frac{2t}{\alpha}}, \quad (3.14)$$

$$-G(t)=1-\frac{\left(1-e^{-\rho}\right)(\lambda+\beta)}{\lambda e^{-\rho}\left(e^{(\lambda+\beta)t}-1\right)+\lambda}, t\geq 0, -\lambda\leq\beta\leq\frac{\lambda}{e^{\rho}-1}$$

$$V(1',t)=\rho-\log\left(1+\left(e^{\rho}-1\right)e^{-(\lambda+\beta)t}\right)+\frac{\left(1-e^{-\rho}\right)(\lambda+\beta)}{\lambda e^{-\rho}\left(e^{(\lambda+\beta)t}-1\right)+\lambda}+$$

$$\left(\frac{\left(1-e^{-\rho}\right)(\lambda+\beta)}{\lambda e^{-\rho}\left(e^{(\lambda+\beta)t}-1\right)+\lambda}\right)^2. \quad (3.15)$$

4 Conclusions

It was possible to study $\mu(1',t)$ and $V(1',t)$, as time functions, playing there an important role the service time length hazard rate function. This circumstance leads to differential equations (2.4) and (3.2). The first was completely solved and the second only in some particular cases. As consequence, some service time length distributions, for which those parameters have particular behaviors, as time functions, were determined.

References

[1] CARRILLO, M. J.(1991), " Extensions of Palm's Theorem: A Review ". *Management Science*, 37 (6), 739-744.

[2] FERREIRA, M. A. M (1988), *Redes de filas de espera*. Master dissertation presented at IST, UL, Lisboa.

[3] FERREIRA , M. A. M. (1998), " Aplicação da Equação de Riccati ao Estudo do Período de Ocupação do Sistema M|G|∞". *Revista de Estatística*, 1 (1º Quadrimestre) , 23-28.

[4] FERREIRA, M. A. M. (2005), " Differential Equations Important in the M|G|∞ Queue System Transient Behaviour and Busy Period". *APLIMAT 2005- 4th International Conference on Applied Mathematics 2005, Proceedings 2005,* Bratislava, 119-132.

[5] FERREIRA, M. A. M. (2007), "M|G|∞ System Parameteres for a Particular Collection of Service Time Distributions". *6h tInternational Conference APLIMAT 2007,* Bratislava; Slovakia; 6 February 2007 through 9 February 2007; 131-137.

[6] FERREIRA, M. A. M. (2013), "Modelling and Differential Costs Evaluation of a Two Echelons Repair System through Infinite Servers Nodes Queuing Networks". *Applied Mathematical Sciences,* 7(109-112), 5567-5576.

[7] FERREIRA, M. A. M. (2014), "Differential Equations Arising in the M|G|∞ Queue Transient Behavior Study". *International Journal of Academic Research*, 6 (1), 5-10.

[8] FERREIRA, M. A. M. (2016), " Results and Applications in Statistical Queuing Theory". *APLIMAT 2016-15th Conference on Applied Mathematics 2016, Proceedings 2016*, Bratislava, 362-375.

[9] FERREIRA, M. A. M. (2016a), " Some Achievements from a Statistical Queuing Theory Research". *Acta Scientiae et Intellectus,* 2(1),70-83.

[10] FERREIRA, M. A. M., ANDRADE, M. (2009), "The Ties Between the M|G|∞ Queue System Transient Behavior and the Busy Period". *International Journal of Academic Research*, 1(1), 84-92.

[11] FERREIRA, M. A. M., ANDRADE, M. (2009a), " M|G|∞ Queue System Parameters for a Particular Collection of Service Time Distributions". *AJMCSR-African Journal of Mathematics and Computer Science Research*, 2(7), 138-141.

[12] FERREIRA, M. A. M., ANDRADE, M. (2010), "M|G|∞ System Transient Behavior with Time Origin at the Beginning of a Busy Period Mean and Variance". *9th International Conference APLIMAT 2010, Proceedings*, 883-890.

[13] FERREIRA, M. A. M., ANDRADE, M. (2012), " Busy Period and Busy Cycle Distributions and Parameters for a Particular M|G|∞ Queue System". *American Journal of Mathematics and Statistics,* 2(2), 10-15. DOI: 10.5923/j.ajms.20120202.03.

[14] FERREIRA, M. A. M., ANDRADE, M., FILIPE, J. A. (2008), " The Riccati Equation in the M|G|∞ System Busy Cycle Study". *Journal of Mathematics, Statistics and Allied Fields,* 2(1).

[15] FERREIRA, M. A. M., ANDRADE, M., FILIPE, J. A. (2009), " Networks of Queues with Infinite Servers in Each Node Applied to the Management of a Two Echelons Repair System". *China-USA Business Review*, 8 (8), 39-45.

[16] FERREIRA, M. A. M., FILIPE, J. A.(2015),"Occupation Study of a Surgery Service through a Queues Model. *Applied Mathematical Sciences,* 9(93-96), 4791-4798.

[17] FERREIRA, M. A. M., FILIPE, J. A., Coelho, M. (2015)," A Queue Model for Idle Motorcars Management Considering an Energy Sources'New Scenario".In *Filipe, J. A., Nagopoulos, N., Rontos, K., Vavouras, I., "Towards a Society with more Fair Economy or an Economy with a more Social Face: The Contribution of Scientific Social Knowledge to the Alternative Models of Socioeconomic Development"*, 143-152.

[18] HERSHEY, J. C., WEISS, E. N., MORRIS, A. C. (1981), " A Stochastic Service Network Model with Application to Hospital Facilities". *Operations Research* 29, 1-22.

[19] KELLY, F. P.(1979) . *Reversibility and Stochastic Networks*. New York, John Wiley & Sons.

[20] ROSS, S. (1983). *Stochastic Processes.* New York, Wiley.

[21] TAKÁCS, L. (1962). An Introduction to Queueing Theory. Oxford University Press.

Current address

Ferreira Manuel Alberto M., Professor Catedrático
INSTITUTO UNIVERSITÁRIO DE LISBOA (ISCTE-IUL)
BRU – IUL, ISTAR-IUL
Av. das Forças Armadas, 1649-026 Lisboa, Portugal
Tel. number: + 351 21 790 37 03. FAX: + 351 21 790 39 41,
E-mail: manuel.ferreira@iscte.pt

Slovak University of Technology in Bratislava
Faculty of Mechanical Engineering
16th Conference on Applied Mathematics **APLIMAT 2017**
Proceedings

PREFERENCES RELATIONS AND CONSUMER THEORY

FERREIRA Manuel Alberto M. (PT), FILIPE José António (PT)

Abstract. In this paper an interesting application of mathematics in economics is presented: the formulation of the theory of consumer basic problem, grounded on the concept of preferences relation and operationalized with optimization tools.

Keywords: Preferences relation, optimization, consumer theory, demand

Mathematics Subject Classification: MSC 2010: 49K20

1 Introduction

This work target is to present a mathematical formulation of the consumer theory as rigorous as possible. This is done using mainly the concept of preferences relation and then optimization mathematical tools.

So, in the next section, the preferences relation concept and the respective axioms are presented and the model to be used is outlined.

In Section 3 is established the consumer theory and some important consequent results are shown. Through Section 4 it is made a function $u(\cdot)$, which plays a key role in this work, short study.
In the end there is a short section of conclusions and bibliography on these subjects.

2 Preferences Relations

In order to use mathematical tools to deal with the concept of consumer basket, consider a non-empty convex closed subset X in $\mathbb{R}_+{}^n$ where is defined a preferences relation, symbolized $\succcurlyeq$, as follows:

Definition 2.1
For any x, y and z belonging to X,
i) $x \succcurlyeq y \vee y \succcurlyeq x$ (Completeness),
ii) $x \succcurlyeq y \wedge y \succcurlyeq z \Rightarrow x \succcurlyeq z$ (Transitivity)

iii) For any $y \in X$, $\{x \in X: x \succcurlyeq y\}$ and $\{x \in X: y \succcurlyeq x\}$ are closed subsets (Continuity),
iv) $x \geq y$ (that is: $x_i \geq y_i, i = 1,2,..n$) and $x \neq y \Longrightarrow x \succ y$ (Strong Monotonicity).■

Note:

- $x \succ y$ if $x \succcurlyeq y$ and $y \succcurlyeq x$ is false,
- In economics language *X* is the set of a consumer possible baskets and $\succcurlyeq$ represents that consumer preferences lengthwise this set of baskets.

Definition 2.2

$x, y \in X$ are indifferent, and it is denoted $x \sim y$, when $x \succcurlyeq y$ and $y \succcurlyeq x$.■

Definition 2.3[1]

A basket $x \in X$ is said to be redundant when there is $y \in X$ such that $x \geq y, y \neq x,$ and $x \sim y$. ■

To present some results, to the Definition 1.1 must be added the following two axioms:

v) $x, y \in X, x \neq y$ and x, y not redundant $x \sim y \Longrightarrow (1-t)x + ty \succ x, \forall t \in [0,1]$.■

Note:

-That is: the convex linear combination of two different, not redundant and indifferent baskets is preferred to one of them isolated.

vi) $\forall x \in X, \forall \varepsilon > 0, \exists y \in X$ such that $\|x - y\| < \varepsilon$ and $y \succ x$.■

Note:

-That is: in any neighborhood of a basket there is at least one other basket preferable to it.

Theorem 2.1

Supposing that $\succcurlyeq$ fulfills the axioms *i), ii), iii)* and *iv)* in Definition 2.1,if $X = \mathbb{R}_+{}^n$ there is a continuous function $u: X \to \mathbb{R}$ such that $u(x) \geq u(y)$ if and only if $x \succcurlyeq y, \forall x, y \in X$.

Dem:

Call **1** the vector with the whole *n* coordinates equal to 1. For $x \in X$ define $A = \{t \in \mathbb{R}_+ : t\mathbf{1} \succcurlyeq x\}$ and $B = \{t \in \mathbb{R}_+ : x \succcurlyeq t\mathbf{1}\}$. According to *iv) A* and *B* are nonempty and by *iii)* are both closed since the function $\varphi: \begin{matrix} \mathbb{R}_+ \to \mathbb{R}_+{}^n \\ t \to t\mathbf{1} \end{matrix}$ is continuous and, here, $A = \varphi^{-1}(\{y \in X: y \succcurlyeq x\})$ and $B = \varphi^{-1}(\{y \in X: x \succcurlyeq y\})$. After *i)* it results $A \cup B = \mathbb{R}_+$and by the $\mathbb{R}_+$ connexity[2] there is $t \in \mathbb{R}$ fulfilling $t\mathbf{1} \sim x$. Again by *iv)* this *t* is unique. So define $u: X \to \mathbb{R}$ as $u(x) = t$. Besides $u^{-1}([t_0, \infty]) = \{x \in X: u(x) \geq t_0\} = \{x \in X: x \succcurlyeq u^{-1}(t_0)\}$ and $u^{-1}([0, t_0]) = \{x \in X: u^{-1}(t_0) \succcurlyeq x\}$ are closed and then it results that *u* is continuous[3].■

Note:

-More generally it is also true that "If $X \subset \mathbb{R}^n$ is a connex set and $\succcurlyeq$ is a preferences relation in *X* fulfilling *i), ii)* and *iii)*, there is a continuous function $u: X \to \mathbb{R}$ such that $u(x) \geq u(y) \Longleftrightarrow x \succcurlyeq y, \forall x, y \in X$", see Debreu (1959),

---

[1]See Simonsen (1989).

[2] A set is connex if it cannot be represented as the union of two separable sets both non-empty.

[3] Because in this case the whole closed sets in $[0, \infty)$ will have inverse image closed. For details see (Cysne and Moreira, 1997), page 95.

-This theorem shows that under axioms *i), ii), iii)* and *iv)* it is possible to determine a numerical scale for the consumer preferences what is extremely important in Decision Theory.

3 Consumer Theory

Begin by a formulation, among many equivalent, of the theory of consumer basic problem:

Problem 3.1
Call r the consumer income and $p = (p_1, p_2, \dots, p_n)$ the n goods vector prices. If $u: X \to \mathbb{R}$ represents the consumer preferences, the preferences maximization problem can be mathematically formulated as

$$\begin{matrix} \max\ u(x) \\ s.to\ \langle p, x\rangle \leq r \end{matrix} \quad (3.1),$$

being supposed that $A = \{x \in X: \langle p, x\rangle \leq r\}$, the opportunities set, is non empty.■

Note:

-If u continuous and $p \in \mathbb{R}_{++}{}^n$, so being A compact, the Weierstrass theorem guaranties that (3.1) is viable. ■

Definition 3.1
The indirect utility function $v(p, r)$ is defined through the problem:

$$\begin{matrix} v(p, r) = \max\ u(x) \\ s.to\ \langle p, x\rangle \leq r, x \in X \end{matrix} \quad (3.2).\ ■$$

Note:

-It is easy to check that if x is a solution for (3.2) $\langle p, x\rangle = r$ since either *vi)* or *iv)* are satisfied for $X = \mathbb{R}_+{}^n$,

- x is not redundant,
- If, in addition, *v)* is satisfied x is unique.

Definition 3.2
The vector only solution x for the problem (3.2), given p and r: $x(p, r)$ is the Marshallian demand.■
Note:

-$x_i(p, r)$ is the Marshallian demand for the good $i, i = 1,2, \dots, n$.

Theorem 3.1
If a consumer preferences satisfy *i), ii), iii)* and *vi)*, the Marshallian demand $x_i: \mathbb{R}^n_{++} \times \mathbb{R}_+ \to \mathbb{R}$ is continuous $\forall i = 1,2, \dots, n$.

Dem:
See Simonsen (1989).■

4 The Function u(·)

Consider the Problem 3.1 following version:

$$\begin{array}{c} \max\ u(x) \\ s.to\ \langle p, x\rangle = r \end{array} \quad (4.1).$$

It is a constrained optimization problem. The Lagrange's multipliers method may be applied in its solution, see for instance (Ferreira and Amaral, 2002). The Lagrangean function is

$$\mathcal{L}(x_1, x_2, \dots, x_n, \lambda) = u(x_1, x_2, \dots, x_n) + \lambda(r - x_1p_1 - x_2p_2 - \cdots - x_np_n) \qquad (4.2).$$

The first order conditions are

$$\begin{gathered} \frac{\partial u}{\partial x_1} - \lambda p_1 = 0 \\ \frac{\partial u}{\partial x_2} - \lambda p_2 = 0 \\ \vdots \\ \frac{\partial u}{\partial x_n} - \lambda p_n = 0 \\ x_1p_1 + x_2p_2 + \cdots + x_np_n = r. \end{gathered}$$

Using the first and the last conditions it is obtained, $x_1 \neq 0$,

$$\frac{\partial u}{\partial x_1} - \lambda\left(\frac{Y}{x_1} - \frac{x_2}{x_1}p_2 - \cdots - \frac{x_n}{x_1}p_n\right) = 0$$

or

$$x_1\frac{\partial u}{\partial x_1} + \lambda x_2p_2 + \cdots + \lambda x_np_n = \lambda r.$$

As $\lambda p_i = \frac{\partial u}{\partial x_i}, i = 1, 2, \dots, n$, the former equation becomes

$$x_1\frac{\partial u}{\partial x_1} + x_2\frac{\partial u}{\partial x_2} + \cdots + x_n\frac{\partial u}{\partial x_n} = \lambda r$$

and noting that

$$\lambda = \frac{1}{p_1}\frac{\partial u}{\partial x_1}$$

it assumes the form

$$\left(x_1 - \frac{Y}{p_1}\right)\frac{\partial u}{\partial x_1} + x_2\frac{\partial u}{\partial x_2} + \cdots + x_n\frac{\partial u}{\partial x_n} = 0 \qquad (4.3).$$

It is a first order homogeneous partial derivatives equation, (Ferreira and Amaral, 2005). Solving it:

$$u(x_1, x_2, \ldots, x_n) = F\left(\frac{r}{p_1 + p_2 + \cdots + p_n}\right) \qquad (4.4).$$

Note:

-If u_1 and u_2 solve (4.3), $\alpha_1 u_1 + \alpha_2 u_2, \alpha_1, \alpha_2 \in \mathbb{R}$ also solve it:

$$\left(x_1 - \frac{r}{p_1}\right)\left(\alpha_1\frac{\partial u_1}{\partial x_1} + \alpha_2\frac{\partial u_2}{\partial x_1}\right) + x_2\left(\alpha_1\frac{\partial u_1}{\partial x_2} + \alpha_2\frac{\partial u_2}{\partial x_2}\right) + \cdots + x_n\left(\alpha_1\frac{\partial u_1}{\partial x_n} + \alpha_2\frac{\partial u_2}{\partial x_n}\right) = \alpha_1\left(\left(x_1 - \frac{r}{P_1}\right)\frac{\partial u_1}{\partial x_1} + x_2\frac{\partial u_1}{\partial x_2} + \cdots + x_n\frac{\partial u_1}{\partial x_n}\right) + \alpha_2\left(\left(x_1 - \frac{r}{P_{x_1}}\right)\frac{\partial u_2}{\partial x_1} + x_2\frac{\partial u_2}{\partial x_2} + \cdots + x_n\frac{\partial u_2}{\partial x_n}\right) = \alpha_1 0 + \alpha_2 0 = 0 \quad .$$

So a linear combination of solutions is also a solution,

-$F(.)$ is any differentiable function,

-$r = x_1 P_{x_1} + x_2 P_{x_2} + \cdots + x_n P_{x_n}$,

-It is easy to check that (4.4) is a solution for (4.3) substituting directly,

-The expression (4.4) evidences the utility functional dependency from the whole goods and the income.

Some Examples:

One particular concretization of (4.4) may be

$$u(x_1, x_2, \ldots, x_n) = \alpha^{\frac{r}{p_1 + p_2 + \cdots + p_n}} \quad (4.5)$$

for which $U(1, 1, \ldots, 1) = \alpha$. That is α is a standard utility: the value of the utility when unitary quantities of every good are used.■

Defining $z_i = \alpha^{x_i}, i = 1,2, \ldots, n$, (4.5) becomes

$$u(x_1, x_2, \dots, x_n) = z_1{}^{\alpha_1} z_2{}^{\alpha_2} \dots z_n{}^{\alpha_n} \quad (4.6)$$

where $\alpha_i = \frac{p_i}{p}, i = 1,2, \dots, n$ and $p = \sum_{i=1}^{n} p_i$ are the standardized prices of each good. Formally, (4.6) is a Cobb-Douglas function. That is: in terms of the standard utility base exponential of each quantity, (4.5) assumes the Cobb-Douglas utility form.■

Another example is

$$u(x_1, x_2, \dots, x_n) = \beta \frac{r}{p_1 + p_2 + \dots + p_n} \qquad (4.7)$$

being now β the standard utility. ■

The expression (4.7) may be written as

$$u(x_1, x_2, \dots, x_n) = \beta \sum_{i=1}^{n} \alpha_i x_i \quad (4.8)$$

with α_i defined above. So, in this case, the utility is given by a linear combination of the quantities, which weights are the standardized prices, multiplied by the standard utility.■

And, finally, as a last example

$$u(x_1, x_2, \dots, x_n) = \gamma \, ln \frac{r}{p_1 + p_2 + \dots + p_n} \qquad (4.9)$$

for which $U(1, 1, \dots, 1) = 0$ and $U(e, e, \dots, e) = \gamma$. Evidently

$$u(x_1, x_2, \dots, x_n) = ln(\sum_{i=1}^{n} \alpha_i x_i)^{\gamma} \quad (4.10).$$■

5 Concluding Remarks

In this paper, economic concepts were presented through a rigorous formulation, allowing that some economic results were demonstrated in the mathematical sense. Plays here a fundamental role the concept of preferences relation that is axiomatically formulated.

References

[1] BORCH, K. (1974), *The Mathematical Theory of Insurance*. Lexington, Mass.: Lexington Books, D.C. Heath and Co..

[2] BORCH; K. (1990), *Economics of Insurance*. Edited by Aase K. K. and Sandmo A., Amsterdam, North-Holland.

[3] CHAVAGLIA, J., FILIPE, J. A., FERREIRA, M. A. M. (2016), "Neuroeconomics and reinforcement learning an exploratory analysis", *APLIMAT 2016-15th Conference on Applied Mathematics 2016, Proceedings 2016*, Bratislava, 527-534.

[4] CYSNE, R. P., MOREIRA, H. A. (1997), *Curso de Matemática para Economistas*. São Paulo, Editora Atlas, S. A..

[5] DEBREU,G. (1959), *Theory of value and axiomatic analysis of economic equilibrium*. Monograph, 17, Cowles Foundation.

[6] FERREIRA, M. A. M. (1991), "Equações com derivadas parciais e funções de utilidade", *Revista de Gestão*, XI-XII, 63.

[7] FERREIRA, M. A. M. (2015), "The Hahn-Banach theorem in convex programming", *APLIMAT 2015-14th Conference on Applied Mathematics 2015, Proceedings 2015*, Bratislava, 283-291.

[8] FERREIRA, M. A. M. (2015a), "The minimax theorem as Hahn-Banach theorem consequence", *Acta Scientiae et Intellectus*, 1 (1), 58-65.

[9] FERREIRA, M. A. M., AMARAL, I. (2002), *Cálculo Diferencial em* $\mathbb{R}^n$. Edições Sílabo-5ª Edição, Lisboa.

[10] FERREIRA, M. A. M., AMARAL, I. (2005), *Integrais Múltiplos e Equações Diferenciais*. Edições Sílabo-5ª Edição, Lisboa.

[11] FERREIRA, M. A. M., ANDRADE, M. (2011), "A note on partial derivatives equations and utility functions", *Journal of Economics and Engineering*, 2(1), 23-24.

[12] FERREIRA, M. A. M., FILIPE, J. A. (2014), "Convex sets strict separation in Hilbert spaces", *Applied Mathematical Sciences*, (61-64), 3155-3160. **DOI**: 10.12988/ams.2014.44257

[13] FERREIRA, M. A. M., FILIPE, J. A. (2015), "A ratio to evaluate harvest procedures management in an economic system where resources dynamics is ruled by an ornsteein-uhlenbeck process", *Applied Mathematical Sciences*, 9(25-28), 1213-1219. **DOI**: 10.12988/ams.2015.410871

[14] FERREIRA, M. A. M., MATOS, M. C. (2014), "Convex sets strict separation in the minimax theorem", *Applied Mathematical Sciences*, 8(33-36), 1781-1787. **DOI**: 10.12988/ams.2014.4271

[15] FERREIRA, M. A. M., FILIPE, J. A., ANDRADE, M. (2013), "A note on partial derivatives equations and utility functions (Revisited)", *Journal of Economics and Engineering*, 2(1), 11-13.

[16] FERREIRA, M. A. M., ANDRADE, M., MATOS, M. C., FILIPE, J. A., COELHO, M. (2012), "Kuhn-Tucker's Theorem-the Fundamental Result in Convex Programming Applied to Finance and Economic Sciences", *International Journal of Latest Trends in Finance & Economic Sciences*, 2 (2), 111-116.

[17] FIGUEIRA, J., FERREIRA, M. A. M. (2006), "Resources dynamics from the point of view of the regenerated Ornstein-Uhlenbeck model", *5th International Conference APLIMAT 2006, Proceedings*, 433-436.

[18] FILIPE, J. A., FERREIRA, M. A. M. (2013), "Chaos in humanities and social sciences: An approach", *APLIMAT 2013-12th Conference on Applied Mathematics 2013, Proceedings 2013*, Bratislava.

[19] NEUMÄRKER, B. (2007), "Neuroeconomics and the Economic Logic of Behavior", *Analyse & Kritik*, 29, 60-85.

[20] ROBERTS, B., SCHULZE, D. L. (1973), *Modern Mathematics and Economic Analysis*. New York: W. W. Norton & Company, Inc..

Current address

Ferreira Manuel Alberto M., Professor Catedrático
INSTITUTO UNIVERSITÁRIO DE LISBOA (ISCTE-IUL)
BRU – IUL, ISTAR-IUL
Av. das Forças Armadas, 1649-026 Lisboa, Portugal
Tel. number: + 351 21 790 37 03. Fax: + 351 21 790 39 41,
E-mail: manuel.ferreira@iscte.pt

Filipe José António, Professor Auxiliar
INSTITUTO UNIVERSITÁRIO DE LISBOA (ISCTE-IUL)
BRU – IUL, ISTAR-IUL
Av. das Forças Armadas, 1649-026 Lisboa, Portugal
Tel. number: + 351 21 790 34 08, Fax: + 351 21 790 39 41,
E-mail: jose.filipe@iscte.pt

Slovak University of Technology in Bratislava
Faculty of Mechanical Engineering
16th Conference on Applied Mathematics **APLIMAT 2017**
Proceedings

APPROACH OF THE DYNAMICAL SYSTEMS USING THE TRANSFER FUNCTION

FLOREA Olivia (RO)

Abstract. In this paper it is studied the behavior of a dynamical system using the transfer function. This study is used in the theory of automatic regulation. The behavior of some dynamical systems is analyzed using Simulink environment from Matlab software.

Keywords: transfer function, Laplace transform, dynamical system, Simulink

Mathematics subject classification: Primary 44A10; Secondary 93A30

1 Introduction

We consider two functions $x(t)$ and $y(t)$ that belong to a vectorial space. If the function $x(t)$ is an input function, and the function $y(t)$ is an output function, than the ratio between the image of the output function through the Laplace transform and the image of the input function through the Laplace transform is the transfer function:

$$W(s) = \frac{Y(s)}{X(s)} = \frac{\mathcal{L}[y(t)](s)}{\mathcal{L}[x(t)](s)} \tag{1}$$

Next we will consider a movement equation having the form:

$$a\ddot{q} + b\dot{q} + cq = Q(t) \tag{2}$$

where $Q(t)$ is the disturbing force. Using the non-dimensionalization this equation can be written in the following form:

$$T_2^2\ddot{y} + T_1\dot{y} + y = kx \tag{3}$$

Applying the Laplace transform on the equation (3) we have:

$$T_2^2\left(s^2Y(s) - sy(0) - \dot{y}(0)\right) + T_1\left(sY(s) - y(0)\right) + Y(s) = kX(s) \tag{4}$$

From this final relation it can be deduced the functional equation:

$$Y(s) = \frac{kX(s) + T_1 y(0) + T_2^2\left(sy(0) + \dot{y}(0)\right)}{s^2T_2^2 + sT_1 + 1} \tag{5}$$

The dynamic transfer function can be written under the ratio from (1) in the case when the initial conditions are null:

$$y(0) = \dot{y}(0) = 0$$

Thus, from (5) we have:

$$W(s) = \frac{Y(s)}{X(s)} = \frac{k}{s^2T_2^2 + sT_1 + 1} \tag{6}$$

The ratio (6) does not depends on the rule through the non-dimensionalization was realized. To obtain the original output function $y(t)$ we will use the inverse transformation. If the initial conditions are different from zero then the changing unknown function can be realized:

$$z = y - \dot{y}_0 t - y_0, \text{ where } y_0 = y(0), \dot{y}_0 = \dot{y}(0) \tag{7}$$

and thus it can be determined $Y(s)$ function by the initial conditions:

$$Y(s) = W(s)X(s) + \frac{T_2^2(sy_0 + \dot{y}_0)}{s^2T_2^2 + sT_1 + 1} \tag{8}$$

With the help of function from (8) it can be analyzed the behavior of dynamical systems. In the theory of automatic regulation the transfer function is often met. In the case of a dynamical system having the following form:

$$\ddot{y} + 2\gamma\dot{y} + \lambda^2 y = kx \tag{9}$$

where the initial conditions are null, it can be obtained:

$$s^2Y(s) + 2\gamma sY(s) + \lambda^2 Y(s) = kX(s)$$

that can lead to the expression of the transfer function:

$$W(s) = \frac{k}{s^2 + 2\gamma s + \lambda^2} \tag{10}$$

2 Frequency characteristics

If the disturbing force is harmonic, then it can be represented as follows:

$$Q(i\omega) = A_1 e^{i(\omega_1 t + \phi_1)} = A_1(i\omega)e^{i\omega_1 t} \tag{11}$$

where $A_1 e^{i\phi_1} = A_1(i\omega)$ and A_1 is the amplitude of the harmonic function, ω_1 is the frequency and ϕ_1 is the phase. The output frequency function can be written in the following form:

$$q(i\omega) = A_2 e^{i(\omega_2 t + \phi_2)} = A_2(i\omega)e^{i\omega_2 t} \tag{12}$$

where $A_2 e^{i\phi_2} = A_2(i\omega)$. The frequency transfer function will be the ratio between the output and input frequency amplitudes, respectively:

$$W(i\omega) = \frac{A_2(i\omega)}{A_1(i\omega)} \tag{13}$$

The oscillator equation (9) where t is replacing by $i\omega$ will be:

$$\ddot{y}(i\omega) + 2\gamma\dot{y}(i\omega) + \lambda^2 y(i\omega) = kx(i\omega) \tag{14}$$

In the example considered above y is the displacement function and x is the disturbing force. The expression of y function by the complex amplitude of the harmonic oscillations is:

$$y(i\omega) = A e^{i(\omega t+\phi)} = A(i\omega)e^{i\omega t}; \quad x(i\omega) = e^{i\omega t} \tag{15}$$

Using the derivatives with respect to time, we have:

$$\dot{y}(i\omega) = i\omega A(i\omega)e^{i\omega t}; \ddot{y}(i\omega) = -\omega^2 A(i\omega)e^{i\omega t}$$

Replacing in (15) we obtain the expression of the amplitude:

$$A(i\omega) = \frac{k}{-\omega^2 + 2\gamma i\omega + \lambda^2} \tag{16}$$

But, $A(i\omega) = U(\omega) + iV(\omega)$. Finding of the two components of the complex amplitude can be realized through the amplification of the ratio with the denominator conjugate, that will lead to:

$$U(\omega) = k\frac{-\omega^2 + \lambda^2}{(-\omega^2 + \lambda^2)^2 + 4\gamma^2\omega^2}; V(\omega) = \frac{-2\gamma k\omega}{(-\omega^2 + \lambda^2)^2 + 4\gamma^2\omega^2}$$

The intensity of the complex amplitude of harmonic oscillator is:

$$A(\omega) = \sqrt{U(\omega)^2 + V(\omega)^2} = \frac{k}{\sqrt{(-\omega^2 + \lambda^2)^2 + 4\gamma^2\omega^2}}$$

The initial phase of the harmonic oscillator is:

$$\phi = \text{arctg}\frac{V}{U} = \text{arctg}\frac{-2\gamma\omega}{\lambda^2 - \omega^2}$$

therefore the solution of equation (15) is:

$$y(i\omega) = \frac{k}{\sqrt{(\lambda^2 - \omega^2)^2 + 4\gamma^2\omega^2}} e^{i(\omega t - \phi)}$$

or, using the Euler's relation: $e^{i\theta} = \cos\theta + i\sin\theta$ we have:

$$y(i\omega) = \frac{k\cos(\omega t - \phi)}{\sqrt{(\lambda^2 - \omega^2) + 4\gamma^2\omega^2}} + i\frac{k\sin(\omega t - \phi)}{\sqrt{(\lambda^2 - \omega^2) + 4\gamma^2\omega^2}} \tag{17}$$

For $x = \cos\omega t$ we have $y = \frac{k\cos(\omega t - \phi)}{\sqrt{(\lambda^2 - \omega^2) + 4\gamma^2\omega^2}}$ and for $x = \sin\omega t$ we have $y = \frac{k\sin(\omega t - \phi)}{\sqrt{(\lambda^2 - \omega^2) + 4\gamma^2\omega^2}}$.

Constructing the characteristics $A(\omega)$ and $\phi(\omega)$ we are interested in an area where the frequencies are big. The characteristic graphic of the amplitude A of the disturbing oscillations show that the amplitude decreases when the frequencies of harmonic perturbations ω tend to λ. In the case when $\omega = \lambda$ it is obtained the resonance frequency.

Another example is considered in the dynamical oscillator:

$$\ddot{y} + \lambda^2 y = k \sin \lambda t \tag{18}$$

Applying the Laplace transform we have:

$$Y(s) = \frac{sy_0 + \dot{y}_0}{s^2 + \lambda^2} + k\frac{\lambda}{(\lambda^2 + s^2)^2} \tag{19}$$

with the solution:

$$y(t) = y_0 \cos \lambda t + frac\dot{y}_0 \lambda \sin(\lambda t) + k\frac{\lambda t \cos \lambda t - \sin \lambda t}{2\lambda^2} \tag{20}$$

3 Multiphysics simulations using Simulink

Every system is characterized by a functional dependence between the variation in time of the output $y(t)$ and the variation in time of the input $x(t)$. This connection of dynamic regime can be expressed through a differential equation obtained based on the physical laws that characterize the functionality of some elements of the system. For a linear system input/output, the differential equation has generally the form:

$$a_n y^{(n)}(t) + a_{n-1}y^{(n-1)} + \ldots + a_0 y(t) = b_m x^{(m)}(t) + b_{m-1}x^{(m-1)} + \ldots + b_0 x(t) \tag{21}$$

where the coefficients a_i, b_j with $i = \overline{1,n}, j = \overline{1,m}$ have a physical meaning, and the condition that the system can be physically realized is $n \geq m$. Using the Laplace transform to the above equation with null initial conditions, the transfer function of the system is:

$$W(s) = \frac{Y(s)}{X(s)} = \frac{b_m s^m + b_{m-1}s^{n-1} + \ldots + b_1 s + b_0}{a_n s^n + a_{n-1}s^{n-1} + \ldots + a_1 s + a_0}$$

For a single-output system, the input and output of the block are scalar time-domain signals. To model this system in Simulink we have to enter a vector for the numerator coefficients of the transfer function in the Numerator coefficients field and a vector for the denominator coefficients of the transfer function in the Denominator coefficients field.

If we consider the following differential equation:

$$y'' + 3y' + 2y = 2x' - 3x$$

where the input signal is considered a step function, a sine wave function and a ramp function, respectively. The transfer function corresponding to the above differential equation is:

$$W(s) = \frac{2s - 3}{s^2 + 3s + 2}$$

From the category continuous we will insert the block Transfer Fcn. Into the numerator we will insert the vector $[2 - 3]$ and into the text box from denominator we will insert the vector $[132]$. The Simulink scheme is:

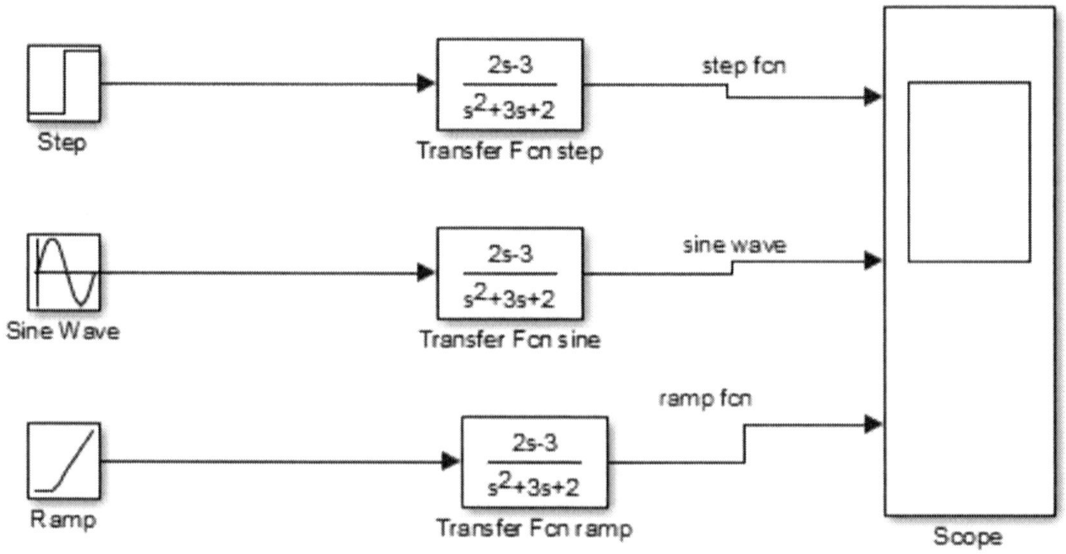

The graphical representation of all three cases is:

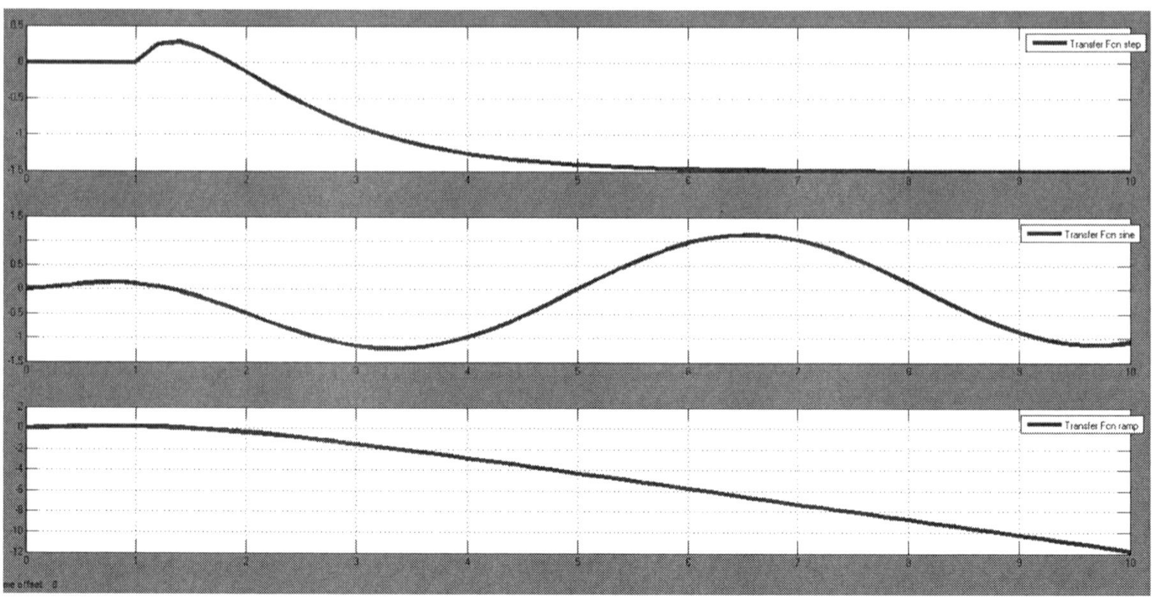

References

[1] Learning MATLAB, Matlab Simulink Student version, The MathWorks, Inc, 2004.

[2] Learning Simulink, Matlab Simulink Student version, The MathWorks, Inc, 2004.

[3] Jesper Bank, A Harmonic-Oscillator Design Method- ology Based on Describing Functions, Chalmers Reproservice Goteborg, Sweden, 2006

Current address

Florea Olivia, Associate Professor
Faculty of Mathematics and Computer Science,
Transilvania University of Braşov, Romania
E-mail: olivia.florea@unitbv.ro

Slovak University of Technology in Bratislava
Faculty of Mechanical Engineering
16th Conference on Applied Mathematics **APLIMAT 2017**
Proceedings

CONTINUITY OF A FUNCTION OF SEVERAL VARIABLES AND A GENERALIZATION OF ONE COUNTEREXAMPLE

FULIER Jozef (SK)

Abstract. The submitted paper discusses the issue of teaching university students about the limits and continuity of a function of several variables. Our attempt was to find a cohesive view on one group of interesting counterexamples related to the function which was studied in detail by the well-known Italian mathematician G. Peano in 1884. We provide a proof of a simple proposition which is further broadly applicable in proposing analogous counterexamples with other functions. Also, we generalize the range of validity of a noteworthy counterexample.

Keywords: education in mathematics, a function of several variables, limits, continuity, counterexamples

Mathematics Subject Classification: Primary 97B50, 97B40; Secondary 97A30

1 Introduction

In order to develop certain mathematical culture and build up professional insight it is very important to include also elements of multivariable calculus into the primary and secondary school mathematics teacher education. The concept of a function of several variables can be qualitatively understood from simple examples in everyday life. The temperature in a room may vary from point to point. A point in space can be defined by an ordered triple of numbers which are coordinates of the point in certain coordinate system, e.g. (x, y, z). Measurements of the temperature at every point from a set D in space assign a real number T (the temperature) to every point from set D. The dependence of T on coordinates of the point is indicated by writing $T = T(x, y, z)$. It is worth noting that functions of several variables have their origin in geometry (e.g. curves depending on parameters; *Leibniz,* 1694) and in physics. A famous problem throughout the 18th century was the calculation of the movement of a vibrating string (*d'Alembert,* 1748). The position of a string $u(x, t)$ is actually a function of variable x, the space coordinate, and of variable t, the time. According to *Hairer* and *Wanner* (2008) an important breakthrough for the systematic study of several variables, which occurred around the middle of the 19th century, was the idea of denoting pairs (then n −tuples)$(x_1, x_2) =: \boldsymbol{x}$, $(x_1, x_2, \dots, x_n) =: \boldsymbol{x}$ by a single letter and of considering them as new mathematical objects. They were called "extensive Grősse" by *Grassmann* (1844, 1862), "complexes" by *Peano* (1888), and "vectors" by *Hamilton* (1853), (*Hairer-Wanner*, 2007).

2 Limits, continuity of a function of several variables, and the teaching of these concepts

In mathematics teacher education at the Constantine the Philosopher University in Nitra students are expected to master various topics of multivariable calculus, such as topology of $n-$ dimensional space, limits and continuity, partial derivatives and differentiability, maxima and minima of functions of several variables $f: D \longrightarrow \mathbb{R}, D \subset \mathbb{R}^n$, with emphasis on $n = 2, n = 3$.
In the following we deal only with limits and continuity of a function of several variables. For the sake of completeness we give their definitions and basic theorems.

Definition 1. (*Limit of a Function of Several Variables*). Let f be a function of several variables whose domain is a set D in the Euclidean space E_n. Let $\boldsymbol{x_0}$ be a limit point of D. Then the limit of $f(\boldsymbol{x})$ as $\boldsymbol{x} \to \boldsymbol{x_0}$ is said to be a number L if for every number $\varepsilon > 0$ there exists a corresponding number $\delta > 0$ such that if $\boldsymbol{x} \in D$ and $0 < \|\boldsymbol{x} - \boldsymbol{x_0}\| < \delta$, then $|f(\boldsymbol{x}) - b| < \varepsilon$ (for $\boldsymbol{a} = (a_1, a_2, \dots, a_n)$ is $\|a\| = \sqrt{a_1^2 + a_2^2 + \cdots + a_n^2}$), which is denoted by $\lim_{\boldsymbol{x} \to \boldsymbol{x_0}} f(\boldsymbol{x}) = L$.

Definition 2. (*Continuity*). Function f of several variables with domain D is said to be continuous at point $\boldsymbol{x_0} \in D$ if $\lim_{\boldsymbol{x} \to \boldsymbol{x_0}} f(\boldsymbol{x}) = f(\boldsymbol{x_0})$. Function f is said to be continuous on D if it is continuous at every point from D.

A huge advantage is that these definitions are formally almost identical to definitions of limits and continuity of functions of one variable (for $n = 1$). Formal differences consist only in the new symbols which are present in two inclusions $D \subset \mathbb{R}^n, \boldsymbol{x_0} \in \mathbb{R}^n$ and in distance $\|\boldsymbol{x} - \boldsymbol{x_0}\|$. The lecturer generally emphasizes to students that they do not need to "re-learn" new definitions, but recall the "old" definitions and "update" them in their memory with new symbols. This, however, does not help us to avoid fundamental problems with limits and continuity of functions of several variables, but at least it eliminates partially the fear of the unknown, especially among low-achieving students. Naturally, the lecturer must bear in mind that the practical determination of limits and continuity of functions of several variables is not as easy as it might seem at the first glance. We can also use a new way of teaching calculus to prospective mathematics teachers. This "upgrade" can be based on employment of modern technology (software *Mathematica, Maple, WinPlot*) into the educational process instead of only chalk-and-blackboard method with a very strong "definition-theorem-proof" model of lessons.

The definition of the limit gives only the criterion for whether number L is the limit of $f(\boldsymbol{x})$ as $\boldsymbol{x} \to \boldsymbol{x_0}$. In practice, however, a possible value of the limit is typically unknown. Some examinations are needed to make a "reasonable" guess of a possible value of the limit.
Here a method to investigate limits is outlined that might be helpful. In what follows, the limit point is often set to the origin $\boldsymbol{x_0} = (0, 0, \dots, 0)$. This is not a limitation because one can always move the origin of the coordinate system to any particular point by shifting the values of the argument, for example $\lim_{(x,y) \to (a,b)} f(x, y) = \lim_{(x,y) \to (0,0)} f(x + a, y + b)$.

A general strategy to investigate limits of functions of several variables involves four main steps (Shabanov, 2012):

Step 1: *Continuity argument.*
The simplest scenario in examining the limit happens when the function f in question is continuous at the limit point $\lim_{\boldsymbol{x} \to \boldsymbol{x_0}} f(\boldsymbol{x}) = f(\boldsymbol{x_0})$.

Step 2: Composition rule.
Theorem A. *(Composition rule for limits).* Let $g(t)$ be a function of one variable and let h be a continuous function of several variables such that $h(\boldsymbol{x}) \to t_0 = h(\boldsymbol{x_0})$ *as* $\boldsymbol{x} \to \boldsymbol{x_0}$. Suppose that the function f is the composition $f(\boldsymbol{x}) = g(h(\boldsymbol{x}))$. *Then* $lim_{\boldsymbol{x} \to \boldsymbol{x_0}} f(\boldsymbol{x}) = lim_{t \to t_0} g(t)$.

Step 3: Limits along curves.
For functions of several variables, there are infinitely many paths along which the limit point can be approached. They include straight lines and paths of any other shape, in contrast to the one-variable case. Nevertheless, a similar result holds for multivariable limits, that is, *if the limit exists, then it should not depend on the curve which the limit point may be approached along*. If there are two curves along which the limits do not coincide, then the multivariable limit does not exist.
A curve in the Euclidean space is a set of points $\boldsymbol{x}(t) = (x_1(t), x_2(t), \ldots, x_n(t))$, where $x_k(t)$, $k = 1, 2, \ldots, n$, are continuous functions of a variable $t \in [a, b]$.

Definition. (*Limit along a curve*). Let $\boldsymbol{x_0}$ be a limit point of domain D of function f. Let $\boldsymbol{x}(t) = (x_1(t), x_2(t), \ldots, x_n(t)$ be a curve C in D such that $\boldsymbol{x}(t) \to \boldsymbol{x_0}$ as $t \to t_0$ (*or* $t \to t_0^+$). The function $F(t) = f(\boldsymbol{x}(t))$ defines the values of f on the curve C. The limit $lim_{t \to t_0} F(t) = lim_{t \to t_0} f(x_1(t), x_2(t), \ldots, x_n(t))$ ($\lim_{t \to t_0^+} F(t)$) is called the limit of f along the curve C if it exists.

Theorem B. (*Independence of the limit on a curve through the limit point*). If the limit of $f(\boldsymbol{x})$ exists *as* $\boldsymbol{x} \to \boldsymbol{x_0}$, then the limit of f along any curve leading to $\boldsymbol{x_0}$ from within the domain of f exists and does not depend on the curve.

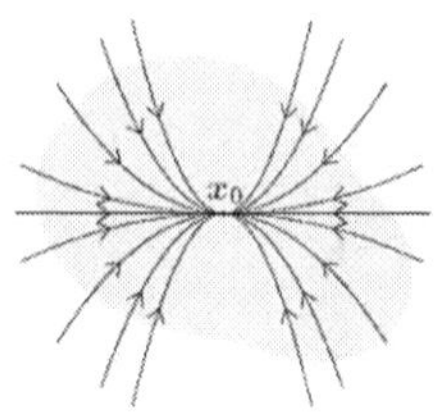

Fig. 1. Different ways of approaching the point $\boldsymbol{x_0}$

A direct consequence of this theorem is the following result, which is very useful in many practical applications.

Corollary B. (*Criterion for nonexistence of the limit*). Let f be a function of several variables on D. If there is a curve $\boldsymbol{x}(t)$ in D such that $\boldsymbol{x}(t) \to \boldsymbol{x_0}$ as $t \to t_0^+$ and the limit $lim_{t \to t_0^+} f(\boldsymbol{x}(t))$ does not exist, then the multivariable limit $\lim_{\boldsymbol{x} \to \boldsymbol{x_0}} f(\boldsymbol{x})$ does not exist either. If there are two curves in D leading to $\boldsymbol{x_0}$ such that the limits of f along them exist but do not coincide, then the multivariable limit $\lim_{\boldsymbol{x} \to \boldsymbol{x_0}} f(\boldsymbol{x})$ does not exist.

Step 4: *Applying the squeeze principle:* If Steps 1 and 2 do not apply to the multivariable limit in question, then a "reasonable" guess of a possible value of the limit is helpful. This is the outcome of Step 3. If limits along a family of curves (e.g. straight lines) happen to be the same number L, then this number is the sought-after "reasonable" guess. The definition of the multivariable limit or *the squeeze principle* can be used in order to prove or disprove that L is the multivariable limit.

Theorem C. *(The squeeze theorem).* If $\lim_{x\to x_0} g(x) = \lim_{x\to x_0} h(x) = L$ and the inequality $g(x) \le f(x) \le h(x)$ holds for all x in punctured neighbourhood of x_0, then $\lim_{x\to x_0} f(x) = L$.

Example 1. Find the limits or prove that $lim_{(x,y)\to(0,0)} \wp(x,y)$ does not exist. By $\wp(x,y)$ we denote

$$\wp(x,y) = \frac{2xy}{x^2+y^2}, \quad or \quad \wp(x,y) = \begin{cases} \frac{2xy}{x^2+y^2} & \text{if } (x,y) \neq (0,0) \\ 0 \quad \text{if} & (x,y) = (0,0). \end{cases}$$

Solution.
Step 1. The continuity argument does not apply.
Step 2. No substitution exists to transform the two-variable limit to a one-variable limit.
Step 3. The function $\wp(x,y) = \frac{2xy}{x^2+y^2}$ has both coordinate limits equal to zero at the origin: $lim_{x\to 0} \wp(x,0) = \lim_{x\to 0} \frac{0}{x^2} = 0 = lim_{x\to 0} \wp(0,y) = \lim_{y\to 0} \frac{0}{y^2}$. However, it does not have the general limit, since the partial limit along straight lines $y = kx, k \in \mathbb{R}$, depends on the chosen line: $lim_{x\to 0} \wp(x,kx) = \lim_{x\to 0} \frac{2kx^2}{x^2+k^2x^2} = \frac{2k}{1+k^2}$, for example it holds: $lim_{x\to 0} \wp(x,x) = 1 \neq 0$. Therefore, if a function approaches zero, as a point approaches the origin along any of the coordinate axes, then its limits may not be equal to zero. In addition, if we put $\wp(0,0) = 0$, then this statement can be rephrased as follows: If a function is continuous with respect to each of variables at some point, then it may not be continuous at this point.

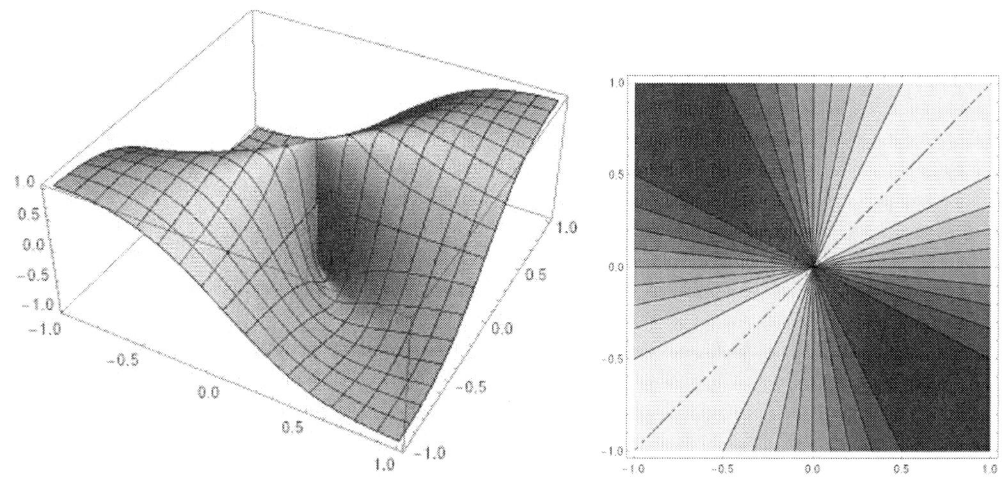

Fig. 2. Computer-generated (Mathematica) graph and level curves of Peano function $z = \wp(x,y)$

***Remark* 1.** This remarkable function $\wp(x,y)$ was studied by the great Italian mathematician *Giuseppe Peano* (1858 - 1932) in chapter *"Annotazioni" al trattato di calcolo del* (see Genocchi & *Peano*, 1884). *Peano* is perhaps most widely known as a pioneer of symbolic logic and a promoter of the axiomatic method, but he considered his work in analysis most important. As early as in the "*Calculus*" of 1884 Peano gave many counterexamples to commonly accepted notions in mathematics, but his most famous example was the space-filling curve published in 1890. This curve is given by continuous parametric functions and goes through every point in a square as the parameter ranges over some interval. He also constructed an example of a function whose partial derivatives do not commute $(f(x,y) = \frac{xy(x^2-y^2)}{x^2+y^2}$, for $(x,y) \neq (0,0)$, $f(0,0) = 0)$ and the important (perhaps less known) is also an example of the function $\wp$. Note that G. Peano in his *Annotazioni* introduced an example of this function as a counterexample to the defective reasoning proposed by L. A. Cauchy (1789 - 1857) in his famous work *Cours d'analyze* (1821). In that work Cauchy identified the continuity of function $f : z = f(x,y)$ with the so-called *separate continuity* with respect to both variables (by the continuity of the two functions of one variable: $g_y(x) = f(x,y), h_x(y) = f(x,y)$). (See also *Maestro di controesempi*).

Example 2. Find the limits or prove that $lim_{(x,y)\to(0,0)} h(x,y)$ does not exist. By $h(x,y)$ we denote

$$h(x,y) = \begin{cases} \dfrac{2x^2y}{x^4+y^2}, & \text{if } (x,y) \neq (0,0), \\ 0, & \text{if } (x,y) = (0,0). \end{cases}$$

Solution. For any straight line $y = Ax,\ A \in \mathbb{R},\ A \neq 0$ passing through the origin, we have $lim_{x\to 0} h(x,Ax) = \lim\limits_{x\to 0} \frac{2Ax^3}{x^4+A^2x^2} = \lim\limits_{x\to 0} \frac{2Ax}{x^2+A^2} = 0 = h(0,0)$, and for the straight line $x = 0$ the limit is also zero $lim_{y\to 0} h(0,y) = \lim\limits_{y\to 0} \frac{0}{y^2} = 0$. However, along any parabola $y = Kx^2, K \in \mathbb{R}$, the result is different

$$lim_{x\to 0} h(x,Kx^2) = \lim_{x\to 0} \frac{2Kx^4}{x^4+K^2x^4+} = \frac{2K}{1+K^2}.$$

For example for $K = 1$ we have $lim_{y\to 0} h(x,Kx^2) = 1 \neq 0$. Therefore, if a function is continuous along any straight line passing through a chosen point, then it may not be continuous in this point.

It can be easily verified that $h(x,y) = \wp(x^2,y)$ and also that an analogous result can be obtained for function $h^*(x,y) = \wp(x,y^2)$.
It must be acknowledged that nowadays Peano function $\wp(x,y)$ and also the function $h(x,y)$, $(h^*(x,y))$ can be found in any textbook of a multivariable calculus, and we believe that every prospective mathematics teacher should be familiar with them.

In the following we show how the function

$$\wp_0(x,y) = \begin{cases} \frac{2xy}{x^2+y^2}, & \text{if } (x,y) \neq (0,0) \\ 0, & \text{if } (x,y) = (0,0) \end{cases}, \tag{1}$$

can be applied in the construction of new counterexamples or in order to design functions with interesting properties, which can enrich the existing collection books of advanced problems in mathematical analysis.

3 Main results

Theorem 1. Let g: $y = g(x)$ be any continuous non-constant function in some neighbourhood $U(0)$ of 0 with property: $\lim\limits_{x\to 0} g(x) = 0$ and let

$$F: F(x,y) := \wp_0(g(x),y) = \frac{g(x)y}{g^2(x)+y^2}, x \in U(0).$$

Then the limit

$$lim_{(x,y)\to(0,0)} F(x,y) = \lim_{(x,y)\to(0,0)} \frac{g(x)y}{g^2(x)+y^2}$$

does not exist. This means that function F is neither continuous, nor differentiable at (0,0), although it has interesting properties:

1. It is continuous on $U(0) \times (-\infty, \infty)$ with respect to variable x and with respect to variable y.

2. For any continuous function $h = h(x)$ on $U(0)$ with property:

   a) $\lim_{x\to 0} h(x) = 0$,

   b) $\lim_{x\to 0} \frac{g(x)}{h(x)} = 0$, *or* $\lim_{x\to 0} \frac{h(x)}{g(x)} = 0$,

   any partial function $F(x, Ah(x)), A \in \mathbb{R}$ of F is a continuous function on whole $U(0)$ and $\lim_{x\to 0} F(x, Ah(x)) = 0$,

3. Any of the partial functions $F(x, Kg(x)), K \in \mathbb{R}, x \in U(0), x \neq 0$ of F is the constant function $y = \frac{2K}{K^2+1}$, which implies that $lim_{x\to 0} F(x, g(x))$ exists, but it is dependent on *K*.

4. $\frac{\partial F(0,0)}{\partial x} = \frac{\partial F(0,0)}{\partial y} = 0.$

Proof. Let functions satisfy assumptions of the theorem. Proposition 1 is trivial.
2. Then $\lim_{x\to 0} g(x) = \lim_{x\to 0} h(x) = 0$ and for example $\lim_{x\to 0} \frac{g(x)}{h(x)} = 0$. For any curve $y = Ah(x),\ A \in \mathbb{R}$, $A \neq 0$ passing through the origin, we have

$$lim_{x\to 0} F(x, Ah(x)) = \lim_{x\to 0} \frac{2Ag(x)h(x)}{g^2(x)+A^2h^2(x)} = \lim_{x\to 0} \frac{2A\left(\frac{g(x)}{h(x)}\right)}{\left(\frac{g(x)}{h(x)}\right)^2 + A^2} = \lim_{x\to 0} \frac{2A\left(\frac{h(x)}{g(x)}\right)}{1+A^2\left(\frac{h(x)}{g(x)}\right)^2} = 0 = F(0,0).$$ For

$A = 0$ the proof is trivial. This means that any partial function H is continuous on the whole $U(0)$.
3. However, along any curve $y = Kg(x), K \in \mathbb{R}, x \in U(0)$ we have for $x \neq 0$
$F(x, Kg(x)) = \frac{2Kg^2(x)}{g^2(x)+K^2g^2(x)} = \frac{2K}{K^2+1} = const.$ Then the limit $lim_{x\to 0} F(x, Kg(x)) = \frac{2K}{K^2+1}$ depends on *K*. This means that $\lim_{(x,y)\to(0,0)} F(x,y)$ does not exist.
4. Finally, we have $\frac{\partial F(0,0)}{\partial x} = \lim_{x\to 0} \frac{F(x,0)-F(0,0)}{x} = \lim_{x\to 0} \frac{0}{x} = 0 = \frac{\partial F(0,0)}{\partial y}$.

Remark 2. It can be easily verified that in the formulation of Theorem 1 neighbourhood $U(0)$ of point 0 may be replaced by the right neighbourhood $U(0+)$, respectively by the left neighbourhood $U(0-)$ of this point, or even by $U(a)$ of arbitrary point $a \in \mathbb{R}$.

By the above theorem it is not demanding to solve the following problem.

Example 3. Let $\alpha \in \mathbb{R},\ \alpha > 0$ and let $F: F(x,y) = \wp_0(|x|^\alpha, y)$. Then the limit

$$\lim_{(x,y)\to(0,0)} F(x,y) = \lim_{(x,y)\to(0,0)} \frac{2|x|^\alpha y}{|x|^{2\alpha} + y^2}$$

does not exist. This means that function F is neither continuous, nor differentiable at $(0,0)$, although it has interesting properties:

1. It is continuous in $\mathbb{R} \times \mathbb{R}$ with respect to variable x and with respect to variable y.
2. Any of its partial functions $F\left(x, A|x|^\beta\right)$, $A \in \mathbb{R}$ is a continuous function for $\beta > 0, \beta \neq \alpha$.
3. Any of its partial functions $F(x, K|x|^\alpha), K \in \mathbb{R}, x \neq 0$ is the constant function $y = \frac{2K}{K^2+1}$, which implies that $lim_{x\to 0} F(x, K|x|^\alpha)$ exists, but it is dependent on K.
4. $\frac{\partial F(0,0)}{\partial x} = \frac{\partial F(0,0)}{\partial y} = 0.$

Example 4. Let $\alpha \in \mathbb{R},\ \alpha > 0$ and $g(x) = |x|^\alpha \sin\frac{1}{x}, g(0) = 0$. Let $F: F(x,y) = \wp_0(g(x), y)$.

Then the limit $\quad lim_{(x,y)\to(0,0)} F(x,y) = \lim_{(x,y)\to(0,0)} \frac{2y|x|^\alpha \sin\frac{1}{x}}{|x|^{2\alpha}\sin^2\frac{1}{x} + y^2}$

does not exist. This means that function F is neither continuous, nor differentiable at $(0,0)$, although it has interesting properties:

1. It is continuous in $\mathbb{R} \times \mathbb{R}$ with respect to variable x and with respect to variable y.
2. Any of its partial functions $F\left(x, A|x|^\beta\right)$, $A \in \mathbb{R}$ is a continuous function for $0 < \beta < \alpha$.
3. Any of its partial functions $F\left(x, K|x|^\alpha \sin\frac{1}{x}\right), K \in \mathbb{R}, x \neq 0$ is the constant function $y = \frac{2K}{K^2+1}$, which implies that $lim_{x\to 0} F\left(x, K|x|^\alpha \sin\frac{1}{x}\right)$ exists, but it is dependent on K.
4. $\frac{\partial F(0,0)}{\partial x} = \frac{\partial F(0,0)}{\partial y} = 0.$

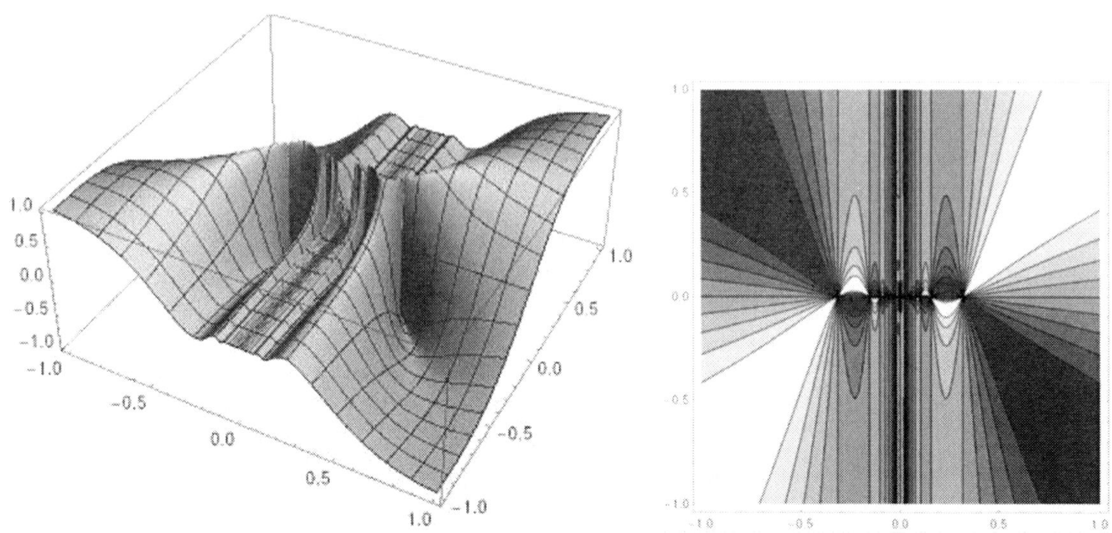

Fig. 3. Computer-generated (Mathematica) graph and level curves of function $F: F(x,y) = \wp_0(g(x), y)\, for\ g(x) = |x|^2\, sin\frac{1}{x}, g(0) = 0$

Remark 3. It is worth noting that in the creation of more demanding problems of this type we may "complicate" the definition of the function $F: F(x,y) := \wp(g(x),y) = \frac{2g(x)y}{g^2(x)+y^2}$ by multiplying it by a "harmless" factor, e.g. by function $f: z = f(x,y)$ with property $\lim_{[x,y]\to[0,0]} f(x,y) = b \in \mathbb{R}$, $b \neq 0$.

Finally, we formulate an assertion related to one interesting function which appears in many counterexamples of mathematical analysis. It is a truly remarkable function, and it is defined as follows:

$$g: g(x) = \begin{cases} e^{-1/x^2}, & \text{if } x \neq 0, \\ 0, & \text{if } x = 0. \end{cases} \tag{2}$$

This function is infinitely differentiable at $x = 0$, and it has all derivatives equal to zero at that point. Consequently, the Taylor series of g about $x = 0$ is also zero. However, g is not a zero function ($g(0) = 0, g(x) > 0$ for $x \neq 0$), therefore it is not equal to its Taylor series around the origin.

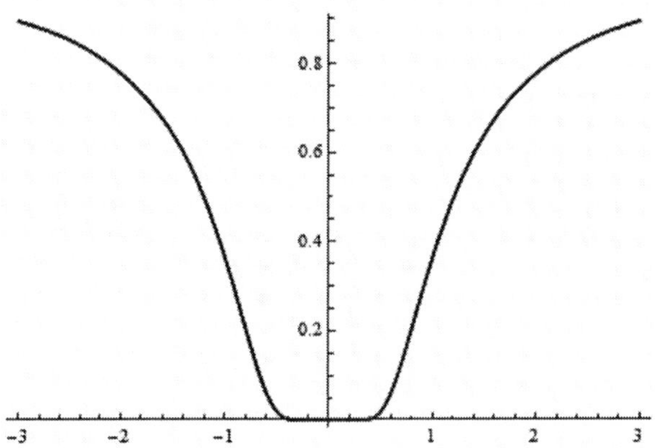

Fig. 4. The graph of function $y = e^{-1/x^2}$

Remark 4. It is worth noting that in the monograph (GELBAUM - OLMSTED, 2003, pp. 116-117) it is stated:

$$f(x,y) = \begin{cases} \dfrac{e^{-\frac{1}{x^2}} y}{e^{-\frac{2}{x^2}} + y^2} & \text{if } x \neq 0, \\ 0 & \text{if } x = 0. \end{cases},$$

and let C be an arbitrary curve through the origin and of the form $x^m = (y/c)^n$ or $y = cx^{\frac{m}{n}}$ where m and n are relatively prime positive integers and c is a nonzero constant ($x \geq 0$ in case n is even). Then if the point (x,y) is permitted to approach $(0,0)$ along C, we have

$\lim_{x\to 0} f(x, cx^{m/n}) = \lim_{x\to 0} \frac{ce^{-\frac{1}{x^2}} x^{-m/n}}{e^{-\frac{2}{x^2}} x^{-m/n} + c^2} = 0$. (cf. Example 10, Chapter 3). In spite of the fact that the limit of $f(x,y)$ as (x,y) approaches the origin along an arbitrary algebraic curve of type

$y = cx^{m/n}$ is zero, the function $f(x, y)$ is discontinuous at (0,0) since there are points of the form $\left(a, a^{-\frac{1}{a^2}}\right)$ arbitrarily near (0,0) at which f has the value $\frac{1}{2}$."

We generalize this result in the following theorem.

Theorem 2. Let $\mathcal{F}: \mathcal{F}(x, y) = \wp_0(g(x), y)$. Then the limit

$$\lim_{(x,y)\to(0,0)} \mathcal{F}(x, y) = \lim_{(x,y)\to(0,0)} \frac{2e^{-1/x^2} y}{e^{-2/x^2} + y^2}$$

does not exist. This means that function $\mathcal{F}$ is neither continuous, nor differentiable at (0,0), although it has interesting properties:

1. It is continuous in $\mathbb{R} \times \mathbb{R}$ with respect to variable x and with respect to variable y.
2. Any of its partial functions $\mathcal{F}(x, Ax^\alpha), x \geq 0, A \in \mathbb{R}, \alpha \in \mathbb{R}, \alpha > 0$, is a continuous function (from the right) since

$$\lim_{x\to 0^+} \mathcal{F}(x, Ax^\alpha) = \lim_{x\to 0^+} \frac{2e^{-1/x^2} Ax^\alpha}{e^{-2/x^2} + (Ax^\alpha)^2} = 0 = \mathcal{F}(0,0).$$

3. Any of its partial functions $\mathcal{F}(x, Kg(x)), K \in \mathbb{R}, \ x \in \mathbb{R}, x \neq 0 \ of \mathcal{F}$ is the constant function $y = \frac{2K}{K^2+1}$, which implies that $lim_{x\to 0} \mathcal{F}(x, g(x))$ exists, but it is dependent on K.
4. $\frac{\partial F(0,0)}{\partial x} = \frac{\partial F(0,0)}{\partial y} = 0.$

Proof. The propositions 1 and 4 are trivial.

2. Just consider that the limit $\lim_{x\to 0^+} \frac{g(x)}{x^\alpha} = \lim_{x\to 0^+} \frac{e^{-1/x^2}}{x^\alpha}$ for all $\alpha \in \mathbb{R}, \ \alpha > 0$.

In (KLUVÁNEK – MIŠÍK – ŠVEC, p.112) it is proved that $\lim_{x\to 0^+} \frac{e^{-1/x^2}}{x^k} = \lim_{x\to 0^+} \frac{1}{x^k e^{1/x^2}} = 0$ for every $k = 0,1,2,\ldots$ Let $\alpha \in \mathbb{R}, \ \alpha > 0$ be given; choose $k \in \mathbb{N}, k > \alpha$, then $\lim_{x\to 0^+} \frac{g(x)}{x^\alpha} = \lim_{x\to 0^+} \frac{e^{-1/x^2}}{x^\alpha} = \lim_{x\to 0^+} \frac{e^{-1/x^2}}{x^k} x^{k-\alpha} == \lim_{x\to 0^+} \frac{e^{-1/x^2}}{x^k} . \ \lim_{x\to 0^+} x^{k-\alpha} = 0$ for all $\alpha \in \mathbb{R}, \ \alpha > 0$.

Let C be an arbitrary curve through the origin and of the form $\mathrm{y} = \mathrm{A}x^\alpha$ where $\mathrm{A} \in \mathbb{R}, A \neq 0 \ \alpha \in \mathbb{R}$, $\alpha > 0$. Then if the point (x, y) is permitted to approach $(0, 0)$ along, we have

$$\lim_{x\to 0^+} \mathcal{F}(x, Ax^\alpha) = \lim_{x\to 0^+} \frac{2e^{-1/x^2} Ax^\alpha}{e^{-2/x^2} + (Ax^\alpha)^2} = \lim_{x\to 0^+} \frac{2A \frac{e^{-1/x^2}}{x^\alpha}}{\left(\frac{e^{-1/x^2}}{x^\alpha}\right)^2 + A^2} = 0 = \mathcal{F}(0,0).$$

It is also apparent that in case $\alpha \in \mathbb{N}; \ \ \alpha = \frac{m}{n}, \ \ m, n \in \mathbb{N}, (m, n) = 1$, while m, n are odd, then it is possible to replace the right-hand limit $x \to 0^+$ for the limit $x \to 0$.

3. For $x \neq 0, K \in \mathbb{R}$ we have

$$\mathcal{F}(x, Kg(x)) = \mathcal{F}\left(x, Ke^{-1/x^2}\right) = \frac{2e^{-1/x^2}Ke^{-1/x^2}}{e^{-2/x^2}+\left(Ke^{-1/x^2}\right)^2} = \frac{2K}{K^2+1}, \text{ therefore for } K \neq 0$$

$$\lim_{x\to 0^+} \mathcal{F}(x, Kg(x)) = \mathcal{F}\left(x, Ke^{-1/x^2}\right) = \frac{2K}{K^2+1} \neq 0,$$

which implies that $\lim_{(x,y)\to(0,0)} \mathcal{F}(x,y) = \lim_{(x,y)\to(0,0)} \frac{2e^{-1/x^2}y}{e^{-2/x^2}+y^2}$ does not exists.

4 Conclusion

Searching for and proposing counterexamples belong to features of good mathematics teaching at secondary schools and, in particular, at universities, as it positively develops not only the students' critical thinking, but it also contributes to the development of their creativity. In this paper we are focused on the issue of teaching university students about limits and continuity of functions of several variables. Although teaching about this topic is not the most challenging part of mathematical analysis education at universities, in the initial stages of this education students often formulate incorrect propositions. The source of their incorrect formulations, usually associated with various students' misconceptions of functions of several variables, is that students uncritically shift their conceptions of functions of one variable on multivariable functions. For such cases it is useful for a university lecturer to have appropriate counterexamples at hand. Most of such counterexamples can be found in quality textbooks and specialized monographs, such as *Counterexamples in Analysis* (GELBAUM - OLMSTED, 2003). Naturally, lecturers can propose new counterexamples by themselves, or modify the counterexamples already proposed by someone else. In this paper we aimed to find a cohesive view on one group of counterexamples related to the function which was thoroughly studied in 1884 by the famous Italian mathematician G. Peano, in form

$$\wp_0(x,y) = \begin{cases} \frac{2xy}{x^2+y^2} & \text{if} \quad (x,y) \neq (0,0), \\ 0 & \text{if} \quad (x,y) = (0,0). \end{cases}$$

We have provided a proof of a simple proposition (Theorem 1) by means of which analogous counterexamples can be proposed. These functions with their properties can enrich current collection of tasks and problems in books of mathematical analysis. Also, by a tiny step we have generalized one already known counterexample.

Acknowledgement

The author gratefully acknowledges Mgr. Zuzana Naštická for technical proof-reading.

References

[1] CAUCHY, A.L. *Cours d'Analyse algébrique*. Oeuvres series 2, vol. III, 1821

[2] BOURCHTEIN, A. - BOURCHTEIN, L. *Counterexamples: From Calculus to the Beginnings of Analysis*. CRC Press Taylor & Francis Group LLC, 2015.

[3] FULIER, J. O jednej zaujímavej triede funkcií viac premenných II. In: *Acta Mathematica 16*, Nitra, UKF Nitra 2013, s. 68-75.

[4] FULIER, J. –VRÁBEL, P. *Diferenciálny počet*. Nitra, UKF Nitra 1997, 2007, 304 s.

[5] GELBAUM, B.R. – OLMSTED, J.M.N. *Counterexamples in Analysis*. Dover Publications, Inc. Mineola, New York, 2003.

[6] GENOCCHI, A. – PEANO, G. *Calcolo differeziale e principia di calcolo integrale "pubblicato con aggiunte dal Dr. Giuseppe Peano"*.Torino 1884.

[7] HAIRER, E. - WANNER, G. *Analysis by Its History*. Springer 2008. ISBN 978-0-387-77031-4, 377 p.

[8] KENNEDY, H.C. *Selected works of Giuseppe Peano*. London, George Allen & UMWIN LTD, University Toronto Press 1973.

[9] KLUVÁNEK, I. - MIŠÍK, L. - ŠVEC, M. *Matematika II*. SVTL Bratislava, 1961, 855 s.

[10] PEANO, G. *"Annotazioni" al trattato di calcolo del*. Opere scelte I, 1884, s. 47-73.

[11] SHABANOV S. *Concepts in Calculus III, Multivariable Calculus*. Univ. Press of Florida, 2012.

[12] STACHO, M. – BRANICKÁ. *Nové trendy vo vyučovaní matematiky na fakulte PEDAS Žilinskej Univerzity v Žiline*. In: 6. konference o matematice a fyzice na vysokých školách technických: s mezinárodní účastí : sborník příspěvků : Brno, 10. září 2009. Část 2. - Brno: Univerzita obrany, 2009. - ISBN 978-80-7231-667-0. - S. 259-263

[13] VARGA, M. - NAŠTICKÁ, Z. *Multiple Proofs*. In Balko, Ľ., Szarková, D., Richtáriková, D. (Eds.) Proceedings of the 15th Conference on Applied Mathematics APLIMAT 2016, 2016, pp. 1070-1077. Bratislava, Slovakia: Slovak University of Technology, Faculty of Mechanical Engineering. ISBN 978-80-227-4531-4

[14] URL: *http://matematica.unibocconi.it/sites/default/files/GP_2.pdf* (*Maestro di controesempi*)

Current address

Fulier Jozef, prof. RNDr. CSc.
Department of Mathematics, Faculty of Natural Sciences,
Constantine the Philosopher University in Nitra
Tr. A. Hlinku 1, SK-949 74 Nitra, Slovak Republic
Tel. number: +421 37 6408 707, e-mail: jfulier@ukf.sk

Slovak University of Technology in Bratislava
Faculty of Mechanical Engineering
16th Conference on Applied Mathematics **APLIMAT 2017**
Proceedings

CONVENTIONALIZED DISTORTIONS IN UPPER-PALAEOLITHIC CAVE ART: CALCULATIONS OF THE KEYSTONE EFFECT

GATTON Matt (US), ACKERMAN Michael (US), ACKERMAN Jennifer (US) BROCK Walter (US), BROCK Dylan (US)

Abstract. Upper-Palaeolithic cave artworks sometime display conventionalized distortions: sliver-thin heads and enormous protruding bellies. Prehistorians have grappled with the reasons for these distortions, given that no animal in the archeological record was so disfigured, and have produced a number of explanations: pregnancy, bloating and death. Countervailing arguments — How, exactly, can a male horse be pregnant? — largely dismantled these explanations decades ago, but no viable explanation was offered as a replacement, so the question was largely abandoned. During the course of recent experiments with reconstructions of simple, animal-hide, Palaeolithic tents, a potential explanation arose. When light passed through small holes in the animal hides and fell upon non-parallel surfaces within the darkened tents it caused projected images to distort (the keystone effect) in a manner similar to the conventionalized distortions seen in the cave artworks. The optical distortions of a horse documented inside a camera obscura are presented side-by-side with Paleolithic artwork. This study measures the keystone distortion evident in the famous 'Second Chinese Horse' in the Axial Gallery of la Grotte de Lascaux (Dordogne, France).

Keywords: Protruding bellies, small heads, habitation sites, animal-hide tents, keystone effect, camera obscura, image projection, Lascaux, and zoomorph

Mathematics Subject Classification: Mathematics and Art—00A66

1 Introduction

Palaeolithic cave art includes figures that adhere to conventionalized distortions such as extremely small heads and enormous, often bowed, bodies. A prime example is the 'Second Chinese Horse' in the Axial Gallery, Lascaux (Dordogne, France) (Fig. 1). In this work we compare the distorted depiction to the actual morphology of the ancestral horse, the Przewalski's horse, which once ranged across Eurasia (fig. 2). The Przewalski's horse is more compact and zebra-like than the lean saddlebreds to which we are accustomed today.

Fig. 1. Second Chinese Horse, Axial Gallery, Lascaux (Dordogne, France) (redrawn by author)

Fig. 2. Przewalski's horse (drawn by author)

The Lascaux Horse is outlined in black, has a black mane, and its flank is comprised of a yellowish 'M'-shape on top and a white underbelly and legs. Its large body is out of proportion to its small, thin head. The base of the neck is as thick as the body, but tapers off to accommodate the head.

There are fifty-eight figurative representations in the Axial Gallery, of which nineteen, all horses, display the conventionalized distortion. It is not just the paintings in the Axial Gallery that display the distortions, the Nave is dominated by the 'Great Black Cow', which is surrounded by and overlaps a herd of horses, all of which follow the distortion convention. The Passageway holds an engraving of a horse in the center of the composition that follows the same conventions, as does another engraving known as the 'Bearded Horse' at the nexus of the Passageway, Nave, and Apse. Similarly, the engraved Horse with Claviform Signs in the Apse and another horse on the entablature of the entry to the Passageway are additional examples. While most of the renderings in Lascaux do not follow the distortion convention, a significant minority does, constituting a running theme.

The shape of the horses in Lascaux follows a template regardless of the scale of the depiction — big, medium, or small — they are all of one design. First they were outlined with a drawn or engraved line, or both, then filled in with color with a spitting technique, where pigments were sprayed onto the surface. A mask, sometimes called a mark or stencil, kept the pigment in the intended area. The visual effect is quite similar to airbrush. Details were added with brushes and pads. In terms of the distortions, the important fact is the shape of the animal existed in the artist's head before any pigment was applied. In some artmaking processes, production is an act of discovery itself as the artist makes it up as s/he goes along. This was not the case in Lascaux. The shape came first.

The distortion convention extends beyond Lascaux, even if it is most prevalent there. Most of the figures at the cave of Altamira (Santander, Spain) are proportional, but there is an outline of a horse, part of the black sequence in the inner gallery, that follows the distortion convention. To name but a few, other distorted images include: a black outline of a megaloceros in the Megaloceros Gallery of Chauvet (Ardèche, France); two very large, dotted, overlapping horses in the Big Chamber, three superimposed 'antelopes', and two horses in the panel of the lioness, in the Le Combel Gallery of the cave of Pech Merle (Lot, France); an aurochs in the central panel at Grotte Chabot (Gard, France); an engraved horse and ibexes from Pair-non-Pair (Gironde, France); a pair of ibex from Cougnac (Lot, France); horse figures from Le Portel (France); and the black outline of a horse from Grotte de Mayenne-Sciences (Pays de la Loire, France). The distortions are not limited to wall art, as a small bone found at the cave of Mas d'Azil (Ariège, France) contains an engraving of a horse that follows the conventions, as do several engravings of animals on small flat stones from the cave of Parpalló (València, Spain).[1]

2 Explanations

These characteristic distortions have long puzzled researchers, because no animal in the fossil record is known to have been so disfigured, so different explanations were produced: pregnancy, bloating, and death. An early, and obvious, explanation for the enlarged bellies was that the animals were pregnant. This notion was dismissed, however, because many of the animals were male, with penile shafts clearly represented (Leroi-Gourhan 1966, 39; Leroi-Gourhan 1967, 310; Bandi 1968, 16; Baffier 1984, 148, 149; Bahn and Vertut 1997, 185). Also, the animals were depicted with different seasonal coats, and herd animals adhere to annual breeding and calving cycles that enhance the survivability of the young. Offspring are born in the late spring and summer when food is plentiful. Even if the distorted animals had been all female, late-term pregnancy was not a year round occurrence.

Hans-Georg Bandi saw the protruding bellies as evidence that the animals were merely well-fed (1968, 16). Certain plant materials can cause digestive bloating, so perhaps the impossibly ballooned stomachs in the artworks were an overemphasis of the effects of the animal's diet (Bahn and Vertut 1997, 185). The weakness of this explanation was the drooping stomachs appear on animal depictions year round, as evidenced by their seasonal coats. Animals do indeed pack on weight in the fall when food is plentiful, but shed weight over the course of the winter when food is scarce. The herd thins in the winter, collectively and individually, as starvation looms. The idea of year round over-full bellies contravenes the natural rhythm of plant and animal biology.

P. A. Leason investigated the odd characteristics of certain Palaeolithic cave artworks — dangling lower extremities, non-weight bearing limbs, drooping tongues, and limp postures — and concluded the depicted animals were dead (1939). Animal carcasses splayed out on the ground. Drawn before butchered, the art of the hunter. In Leason's view, the emphasis in some figures on the underside of the belly was result of a viewing angle of an artist standing over a dead beast (1939, 59). W.H. Riddell countered many of Leason's arguments, but passed over the viewpoint argument (1940). A viewpoint above a dead animal causes the belly to balloon upward, not downward.

The larger point is neither pregnancy, nor bloating, nor death causes an animal's skull to compress to microcephalic proportion. With no plausible explanation for the conventionalized distortions, the matter was written off as a matter of style, which is a fairly unapproachable construct, because of its inherent subjectivity. The schemes of broad stylistic criteria applied to Palaeolithic art have been, according to Romain Pigeaud, exercises that have led nowhere (2007, 410). But styles, in the sense of a set of communicative characteristics, do not just come out of thin air. They are evidence of the visual language of the time and place, which springs from the life experiences of the artists, whose concerns, understandings, and mental constructs are embedded in and translated through the mode of production, the materials, and techniques, which further influence the look of the art. Art making processes follow a logic — decision-making incumbent in the task — even though the sources and steps taken are not always immediately evident.

3 Experiments

Palaeolithic people lived as small bands of hunter-gatherers. Their habitation sites usually consisted of a number of tents, fire pits, and rubbish pits. The tents came in several forms, including yurt-like, teepee-like, and A-frame. The structural framework of the tents was made either with wooden posts and branches in less harsh climates or mammoth bones and tusks on the frozen steppe where trees did not grow. The framework was covered with animal hides. Anchor stones held down the animal-hide covering and kept the posts firmly in place.

Spaces enclosed by opaque materials like animal hides, tend to be dimmer than the sunlit outside world. The contrast between interior and exterior illumination levels is the basis of the camera obscura principle, where light rays pass from a bright area through a small hole into a dim area projecting an inverted image (Gatton 2009, 146) (Fig. 3). The image appears like an upside-down movie. The effect seems like magic, but is actually a matter of optical physics.

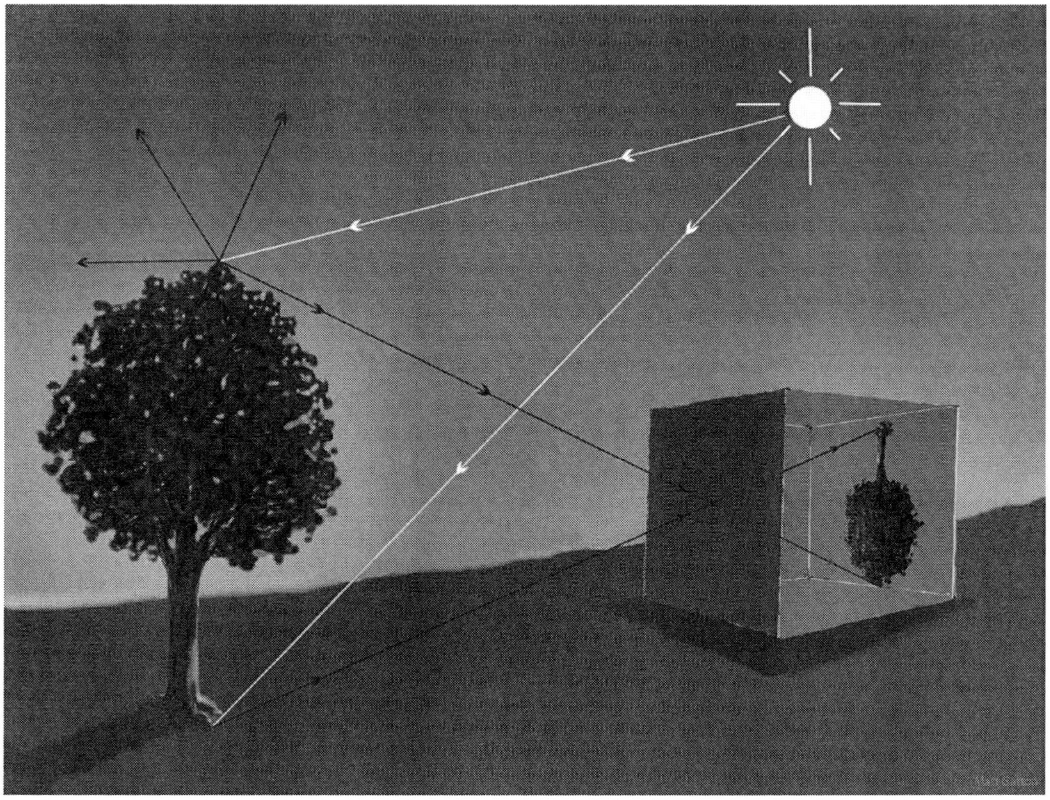

Fig. 3. Camera obscura diagram. The image is inverted because light rays travel in straight lines (paleo-camera.com)

The first author explored the suitability of Palaeolithic tents/shelters for coincidental camera obscura formation during field research in 2005 and 2006. The first set of experiments took place on the bison ranch in the USA. A small, rudimentary bison hide teepee was erected next to bison grazing pasture. A hole in the hides allowed light to enter and project images of bison milling about the field into the tent. The second set of experiments occurred at a rock overhang in the USA. A tree branch framework was leaned against the overhang, enclosing a room-size space, and covered with bison hides. A horse was stationed outside and its image was projected through a rough hole in the hides and into the interior space. The third set of experiments took place at the Musée du Malgré-Tout in Belgium, where archaeologists Pierre Cattelain and Claire Bellier had reconstructed a variety of Palaeolithic structures. A smattering of small holes was present in the hide coverings of each tent. Images were projected inside the Gönnersdorf tent reconstruction onto a rabbit hide.

During the course of the tent reconstruction experiments, the distortion of images on tilted surfaces was observed, so another set of experiments was devised to explore the possibility further. A camera obscura was set up on a farm in the USA. (At this point, it had already been established that Palaeolithic tents could act as random camera obscuras, so in the interest of efficiency and transportability, a tent was not reconstructed, and the back of a box truck was used as the enclosed 'tent' space. It does not matter whether the light passes through a rough hole in wood, sheet metal, or leather. The effect is the same.) A hole was drilled in the side of the box truck, forming a makeshift camera obscura. A horse was stationed outside, to the 'hole' side of the truck. Inside the camera obscura, a hand-sized flat rock was held near the hole, so as to 'catch' the projected image of the horse. (The floors of Palaeolithic tents were covered with such flat 'paver' stones, which were sometimes picked up and used as an engraving surface.) The stone was held waist high, and to keep it still, secured against the body. The bottom of the stone was tilted away from the body, angling the face of the stone slightly upward for easier viewing. From this downward looking perspective, the upside-down image appeared right side up. The more the stone was tilted, the more the image of the horse distorted, causing the animal's head to reduce in size and the stomach to protrude downward. The stone was pivoted up and down, and the horse morphed back and forth between a regular horse and a 'Lascaux' horse (Figs. 4-5).

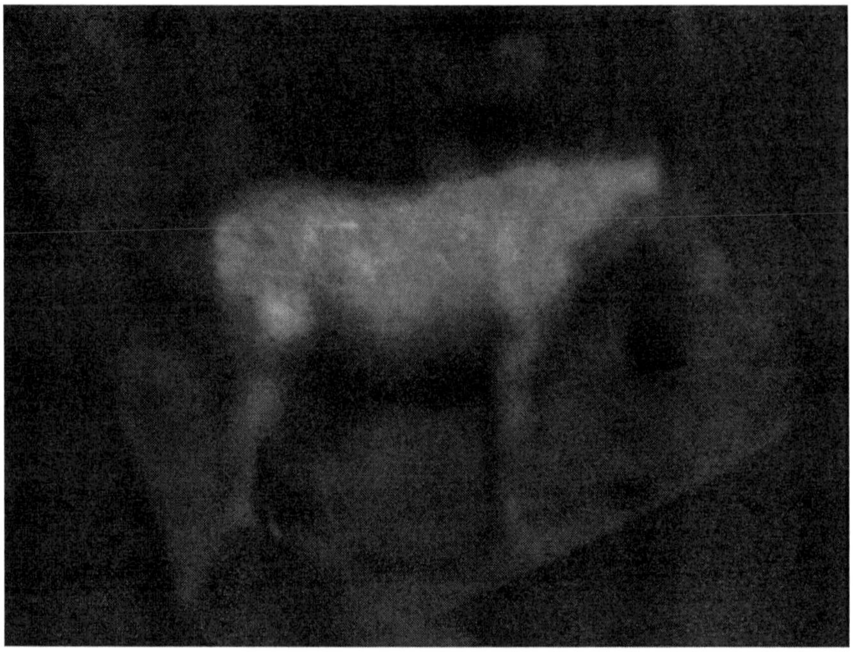

Fig. 4. Image of horse projected on parallel stone. No distortion (image rotated 180°)

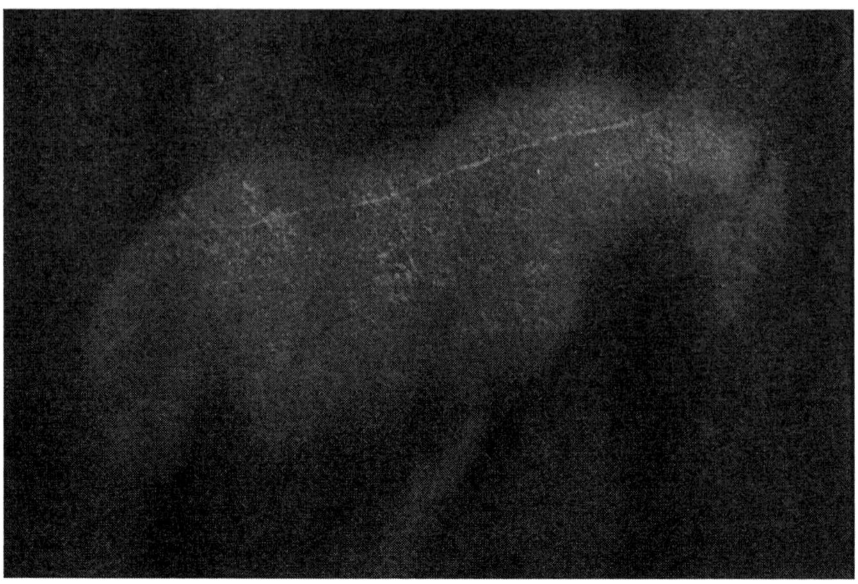

Fig. 5. Image of horse projected on tilted stone. Increased distortion (image rotated 180°)

Optically the distortion is known as the keystone effect, where the image plane is tilted out-of-parallel with the aperture plane. Increases in the tilt of the image plane cause increases in distortion. The keystone effect is generally considered from a rectilinear framework, where a rectangle is altered, either vertically or horizontally, into a symmetric trapezoid, because of the manner in which modern projectors cast framed images onto flat walls (Fig. 6). In a room size camera obscura there is no rectangular frame around the projected image; the image fills the interior space, covering nearly 180°, and the stones can be tilted, not just in vertical or horizontal planes, but in any angular plane. The rectangular understanding of the keystone effect, while valid, is not evident in the context of simple camera obscura projections on irregularly shaped stones.

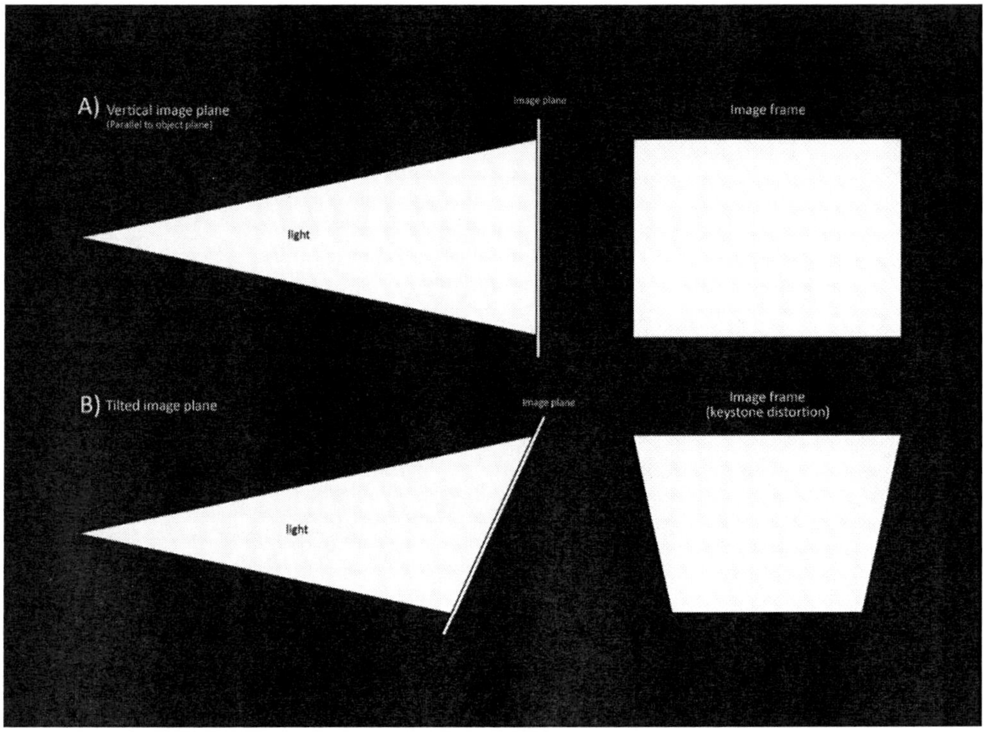

Fig. 6. Keystone distortion diagram

Compare the image of the horse projected on the tilted stone to the ‘Second Chinese Horse’ in the Axial Gallery, Lascaux (Fig. 7). The peculiar head-to-torso ratio in the cave artwork is evident in the projected image. This is not to say there were projected images in the deep recess of the cave. As it is completely dark with no sunlight in the cave, the cave artworks visually replicated an experience that happened in the tents.

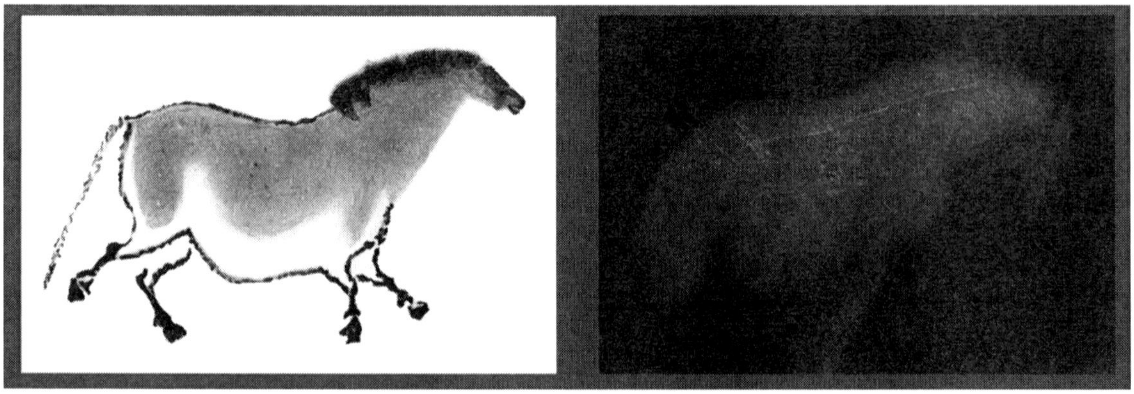

Fig. 7. Comparison.
On left: 'Second Chinese Horse' in the Axial Gallery, Lascaux (Dordogne, France)
On right: Image of horse projected on tilted stone (image rotated 180°) (paleo-camera.com)

4 Calculations

The purpose of this section is to provide mathematical evidence of the keystone effect in the Lascaux horse and the projected image on the tilted stone, relative to the baseline Przewalski's horse. For each image, we measured the torso thickness from the low point on the back to the low point on the stomach, the head length from behind the ear to the tip of the muzzle, and the maximum head thickness measured from the brow to the jowl (Fig. 8).

(A)

(B)

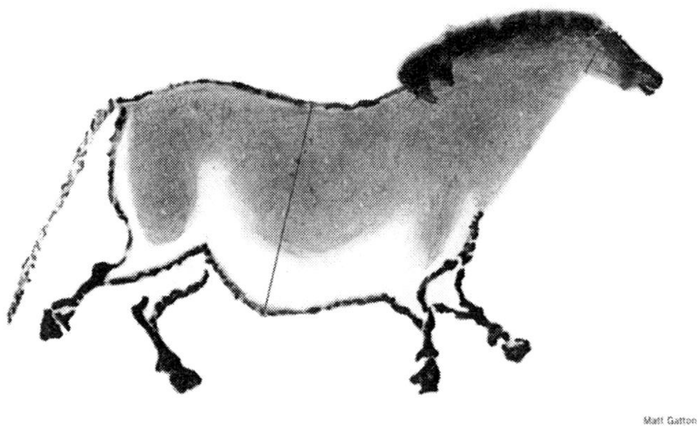

(C)

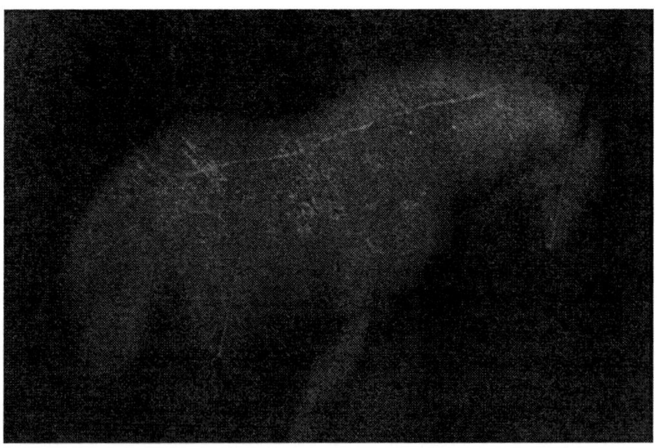

Fig. 8. Measurements of torso thickness and head thickness and head length for (A) the Przewalski's horse, (B) the Lascaux Horse, and (C) the distorted projected image (image rotated 180°)

The measurements, taken in millimeters, do not matter per se as it is the ratios of the measurements that reflect the distortion. We calculated the ratio of the Lascaux horse and projected image lengths relative to the corresponding Przewalski's horse lengths. The ratios appear in Table 1.

	Lascaux horse	Projected Image
Ratio of torso thickness to Przewalski's horse	1.4730	1.7703
Ratio of head length to Przewalski's horse	0.7077	0.9077
Ratio of max head thickness to Przewalski's horse	0.6622	0.7297

Tab. 1. Comparison of anatomical ratios to Przewalski's horse

We would expect the ratios for the torso thickness to both be significantly greater than 1.0, and indeed this is the case. Thus, the ratio reflects the elongated stomachs from the keystone effect. Both the head length and thickness ratios reflect the shrinking of the head to a significantly smaller size. These ratios quantify the visual distortion created by the keystone effect in these images, whether created from the artist's recollection or projected onto the tilted stone.

Furthermore, for each image, calculating the ratio of the head length to the torso thickness and the ratio of the maximum head thickness to the torso thickness allows us to compare the degree of distortion between the Lascaux horse and the projected image. The calculations appear in the table Table 2.

	Przewalski's horse	Lascaux horse	Projected Image
Ratio of head length to torso thickness	0.8784	0.4220	0.4504
Ratio of max head thickness to torso thickness	0.5000	0.2248	0.2061

Tab. 2. Comparison of ratios of head measurements to torso thicknesses for each horse

Interestingly, the ratios for the Lascaux horse and the projected image are fairly close (Fig. 9). Thus, the projected image was a good replication of the distortion observed in the Lascaux horse, which supports the premise that although the Lascaux horse was drawn from memory, it still reflected the keystone effect from the artist's original observations. In all four ratios for the Lascaux horse and projected image, the torso thickness has a greater effect on the value of the ratio when compared to the corresponding ratio for Przewalski's horse. This comparison is another indication of the keystone effect on these two images. If the keystone effect were not in place, then those ratios would more closely match the ratios for Przewalski's horse and would not be such a close match to each other.

Fig. 9. Lascaux horse and tilted-projected image overlapped

5 Discussion

Prehistorian Abbé André Glory drew upon the Siberian ethnological work of Chimkiévitch, Schrenk, Khangalov, Anokhin and Potanin, and proposed that some Palaeolithic animal artworks may not be copies of real animals per se, as much as renderings of spirits that took the form of animals: zoomorphs (1964, 346). The difficulty is determining when a figure is a real animal, alive or dead, or a zoomorph. If images of animals projected inside the tents were interpreted as spirits, then the distortion convention might provide some indication whether a depicted animal is real or a spirit.

Palaeolithic people spent the overwhelming majority of their lives out-in-the-world and only occasionally made art-making forays into the deep cave (Guthrie 2005, 36; based on Sieveking 1979, 102). The visual syntax of the culture was developed within the context of campsite living. Unfortunately, any paintings made in the habitation sites were erased by the elements, but fortunately, the deep cave artworks were preserved, which provide an impression of campsite life. The subject matter of cave artwork is life outside, the pursuit of large animals in the fields. The manner of depicting the animals reflects the way Palaeolithic people experienced, encountered, and interacted with the beasts. The distortion convention was woven into the artistic fabric of the culture, bearing witness to a source outside in the day-to-day living areas, the tents.

References

[1] BANDI, H.G. Art quaternaire et zoologie. In *Simposio de Arte Rupestre, Barcelona 1966*. 1968. 13-19.

[2] BAFFIER, D. Les caractères sexuels secondaires des mammifères dans l'art pariétal paléolithique Franco-Cantabrique. In Bandi etal, *La Contribution de la Zoologie et de l'ethologie à l'interprétation de l'art des peuples chasseurs préhistoriques, 3e colloque de la Society Suisse des sciences humaines, Sigriswil 1979*. Fribourg: Editions Universitaires, 1984. 143-154.

[3] BAHN, P. *Pyrenean Prehistory: A Palaeoeconomic Survey of the French Sites*. Warminster: Aris & Phillips, 1985.

[4] BAHN, P. and VERTUT, J. *Journey through the Ice Age*. Berkeley: University of California Press, 1997.

[5] BONILLA, V.V. *Arte Paleolítico de la Cova del Parpalló: Estudio de la collección de plaquetas y cantos grabados y pintados, Vol. I and Vol. II*. Servei d'investigació Prehistòrica: Diputació de València, 1994.

[6] GATTON, M.; CARREON, L.; CAWEIN, M.; BROCK, W.; and SCOTT, V. The Camera Obscura and the Origin of Art: The case for image projection in the Paleolithic. In the *Official Proceedings of the XV World Congress of the Union Internationale des Sciences Préhistoriques et Protohistoriques (UISPP) 35, Lisbon 2006. Global State of the Art--SO7 (Giriraj Kumar and Robert Bednarik, eds.)* BAR S2108. Oxford: Archaeopress, 2010.

[7] GATTON, M. First Light: Inside the Palaeolithic camera obscura. In *Acts of Seeing: Artists, Scientists and the History of the Visual -- a volume dedicated to Martin Kemp (Assimina Kaniari and Marina Wallace, eds.)*. London: Zidane, 2009.

[8] GATTON, M. Excerpts from — The Camera Obscura and the Origin of Art: The case for image projection in the Palaeolithic, first presented at the University of Louisville. In <*Paleo-camera.com*> October 6, 2005.

[9] GERNSHEIM, H. and GERNSHEIM, A. *The History of Photography: From the Camera Obscura to the Beginning of the Modern Era*. New York: McGraw-Hill, 1969.

[10] GLORY, A. L'énigme de l'art quaternaire peut-elle être résolue par la théorie du culte des Ongones? In *Revue des Sciences Religieuses*, 38, 4, 1964. 337-388.

[11] GUTHRIE, R. D. *The Nature of Paleolithic Art*. Chicago: University of Chicago Press, 2005.

[12] LEASON, P.A. A New View of the Western European Group of Quaternary Cave Art. In *Proceedings of the Prehistoric Society* 5, 1939.

[13] LEROI-GOURHAN, A. *Treasures of Prehistoric Art* [Norbert Guterman, translator]. New York: Harry N. Abrams, 1967.

[14] LEROI-GOURHAN, A. Réflexions de méthode sur l'art paléolithique. In *Bulletin de la Société Préhistorique Française, 63, 1966*. 35-49.

[15] LOMMEL, A. *The World of Early Hunters: Medicine-men, Shamans, and Artists*. London: Evelyn, Adams, and Mackay, 1967.

[16] PERRY, D. *Life above the Jungle Floor*. New York: Simon and Schuster, 1988.

[17] PIGEAUD, R. Determining style in Palaeolithic cave art: a new method derived from horse images. In *Antiquity*, 81, 312, June 2007. 409-422.

[18] RENNER, E. *Pinhole Photography: Rediscovering a Historic Technique (3rd ed)*. Oxford: Focal Press, 2004.

[19] RIDDELL, W.H. Dead or Alive? In *Antiquity* 14, 1940. 154-162.

[20] SIEVEKING, A. *The Cave Artists*. London: Thames and Hudson, 1979.

Current address

Gatton Matt, MFA
Paleo-camera
1900 Gladstone Avenue, Louisville KY 40205, USA
1 (502) 974-2079
E-mail: Gatton@paleo-camera.com

Ackerman Michael, PhD
Department of Mathematics, Bellarmine University
Louisville KY 40205, USA
1 (502) 272-8125
E-mail: mackerman@bellarmine.edu

Ackerman Jenifer, MA
Jefferson Community and Technical College
109 E. Broadway, Louisville KY 40202, USA
1 (502) 213-5145
E-mail: Jackerman0006@kctcs.edu

Brock Walter, MFA
Independent Filmmaker
Louisville KY, USA

Brock Dylan

Notes

[1] See stones: 16004, 16033, 16036 A, 16079 A, 16093, 16116 A, 16116 B, 16229 A, 16330 A, 16330 B, 16341, 16343, 16358 B, 16361, 16341, 17365 A, 18889, 19010, 19016 A, 19150 B, 19347 A, 19349 B, 19357 A, 19360 A, 19362 A, 19659, 19782, 20216, 20127, 20236, 20369, 20393, 20565, 20664 A, 20672, 21084, 21126 B, 21473, 21587 B, and 21597 (in Bonilla 1994).

Slovak University of Technology in Bratislava
Faculty of Mechanical Engineering
16th Conference on Applied Mathematics **APLIMAT 2017**
Proceedings

SHEPHERDSON'S THEOREMS FOR FRAGMENTS OF OPEN INDUCTION

GLIVICKÁ Jana (CZ), GLIVICKÝ Petr (CZ)

Abstract. By a well-known result of Shepherdson, models of the theory IOpen (a first order arithmetic containing the scheme of induction for all quantifier free formulas) are exactly all the discretely ordered semirings that are integer parts of their real closures. In this paper we prove several analogous results that provide algebraic equivalents to various fragments of IOpen.

Keywords: weak arithmetics, open induction, linear induction, real closed fields

Mathematics subject classification: Primary 03C62, 03F30; Secondary 06F25

1 Introduction

Peano arithmetic (PA) is a first-order theory which attempts to axiomatize the structure $\underline{\mathbb{N}} = \langle \mathbb{N}, 0, 1, +, \cdot, \leq \rangle$ of natural numbers (called the standard model of PA). Its axiomatics consists of elementary properties of $0, 1, +, \cdot, \leq$ in $\mathbb{N}$ and of the scheme of infinitely many axioms of induction:

$$(\varphi(0, \overline{y}) \mathbin{\&} (\forall x)(\varphi(x, \overline{y}) \rightarrow \varphi(x + 1, \overline{y})) \rightarrow (\forall x)\varphi(x, \overline{y})), \tag{1}$$

one for every formula $\varphi(x, \overline{y})$ of the language $L = \langle 0, 1, +, \cdot, \leq \rangle$ with a distinguished variable x (see section 2.3 for details and references).

Models of PA other than the standard model $\underline{\mathbb{N}}$ (called nonstandard models) are very complicated structures – one of the explications of this fact being the classical theorem of Tennenbaum [6] stating that in any (countable) nonstandard model of PA neither $+$ nor $\cdot$ are recursive ("algorithmically describable") functions.

That is one of the reasons why theories weaker than PA (also called fragments of PA) are studied. There are two main ways how to weaken PA. The first one is to restrict the language

This paper was processed with contribution of long term institutional support of research activities by Faculty of Informatics and Statistics, University of Economics, Prague. The study was supported by the Charles University, project GA UK No. 270815. The work was supported by the grant SVV-2016-260336.

by leaving out either $+$ (together with $\leq$) or $\cdot$. In the former case the resulting theory is called Skolem arithmetic (SA), in the latter one we obtain Presburger arithmetic (Pr). The second possibility how to weaken PA is to restrict the amount of induction present by including the axioms of induction (1) only for formulas φ from certain prescribed set. We will be particularly interested in the theory IOpen which contains axioms (1) just for quantifier free (also called open) formulas φ (see the Preliminaries section for details).

In [5] Shepherdson proved that there is a recursive nonstandard model of IOpen (in contrast to PA, where no such model exists by the Tennenbaum's theorem). Shepherdson's model is constructed as an integer part of certain real closed field (see the Preliminaries section for the precise definitions). Implicitly, even more is proved: a discretely ordered semiring $\mathcal{M}$ is a model of IOpen if and only if $\mathcal{M}$ is an integer part of the real closure $\mathrm{R}(\mathcal{M})$ of $\mathcal{M}$.

In this paper we prove several analogous results – versions of the Shepherdson's theorem – for other fragments of PA in place of IOpen that correspond to various algebraic properties of (extensions of) their models.

2 Preliminaries

2.1 Discretely ordered rings and their extensions

A discretely ordered ring is a structure $\mathcal{R} = \langle R, 0, 1, +, -, \cdot, \leq \rangle$ such that $\langle R, 0, 1, +, -, \cdot \rangle$ is a commutative ring, $\leq$ is a linear ordering on R such that 1 is the least positive element, and $\leq$ respects $+$ and $\cdot$ in the following way:

$$a \leq b \rightarrow a + c \leq b + c, \qquad 0 \leq a, b \rightarrow 0 \leq a \cdot b,$$

for all $a, b, c \in R$.

A nonnegative part of a discretely ordered ring in the language without $-$ is called a discretely ordered semiring. We denote the semiring corresponding to the ring $\mathcal{M}$ by $\mathcal{M}^+$.

Further on, $\mathcal{M}$ always denotes a discretely ordered ring. (Such an $\mathcal{M}$ necessarily contains negative elements and therefore, strictly speaking, can not be a model of arithmetical theory T. If we say that $\mathcal{M}$ is a model of T, which we denote by $\mathcal{M} \models T$, we mean that $\mathcal{M}^+$ is.)

We define $\mathrm{F}(\mathcal{M})$ as the fraction field of $\mathcal{M}$ and $\mathrm{R}(\mathcal{M})$ as the unique, up to isomorphism (by the Artin-Schreier theorem [4, Theorem B.14]), ordered real closure of $\mathrm{F}(\mathcal{M})$ that preserves the ordering of $\mathrm{F}(\mathcal{M})$. By $\frac{\mathcal{M}}{\mathbb{N}}$ we denote the ordered ring of all formal fractions of the form m/n, where $m \in M$ and $0 \neq n \in \mathbb{N}$.

Let $\mathcal{R}$ be a ring. A discretely ordered subring I of $\mathcal{R}$ is called an integer part of $\mathcal{R}$ (denoted by $I \subseteq^{IP} \mathcal{R}$) if for every $r \in R$, there is $i \in I$ such that $r - 1 < i \leq r$.

We call $m \in M$ the integer part of $r \in \mathrm{R}(\mathcal{M})$ if $r - 1 < m \leq r$. (Note that every $r \in \mathrm{R}(\mathcal{M})$ has an integer part $m \in M$ if and only if $\mathcal{M} \subseteq^{IP} \mathrm{R}(\mathcal{M})$.)

We say that an ordered ring $\mathcal{R}'$ is a dense subring of an ordered ring $\mathcal{R}$, and denote it by $\mathcal{R}' \subseteq^{d} \mathcal{R}$, if $\mathcal{R}'$ is a subring of $\mathcal{R}$ and for every $q < r$ from R there is $r' \in R'$ such that $q < r' < r$.

2.2 Integer-parts-of-roots property

Let $f(x)$ be a definable unary function on $\mathcal{M}$. By $\mathrm{IPR}(f)$ (integer-parts-of-roots) we denote the following formula:

$$(a < b \,\&\, f(a) \leq y < f(b)) \rightarrow (\exists x)(a \leq x < b \,\&\, f(x) \leq y < f(x+1)).$$

The intended meaning of $\mathcal{M} \models \mathrm{IPR}(f)$ can be expressed in vague terms as "existence of integer parts for all f-roots of values $y \in M$" or, in other words, "existence of integer parts of all values $f^{-1}(y)$, where f^{-1} is an inverse function of f, i.e. a function such that $f(f^{-1}(y)) = y$, for all $y \in M$".

If $\mathcal{F}$ is a set of definable unary functions on $\mathcal{M}$, we write $\mathrm{IPR}(\mathcal{F})$ for the scheme $\{\mathrm{IPR}(f); f \in \mathcal{F}\}$.

2.3 Arithmetical theories

Now we define several arithmetical theories that we use in this paper. Robinson arithmetic (Q) is a basic theory of arithmetic in the language $L = \langle 0, 1, +, \cdot, \leq \rangle$. It's axioms are just elementary properties of the symbols from the language. For our purposes the precise axiomatics is not important. We refer the reader to [1, page 28, Definition 1.1].

Peano arithmetic is the extension of Q by the scheme (1) of induction for all L-formulas $\varphi(x, \overline{y})$ with a distinguished variable x (see e.g. [2] for a detailed list of axioms and basic properties of PA).

When we extend Robinson arithmetic just by the scheme (1) for all quantifier free (also called open) formulas, we get the arithmetic of open induction (IOpen). Its models are exactly all the discretely ordered semirings that satisfy the induction axioms (1) for all quantifier free formulas φ.

The theory $\mathrm{IOpenLin}$ (open linear induction) is the extension of Robinson arithmetic by the induction scheme (1) for all linear formulas $\varphi(x, \overline{y})$ with a distinguished variable x. Here, we say that $\varphi(x, \overline{y})$ is linear if in every occurrence of $\cdot$ in φ at least one of the two factors is y_i, for some i.

Presburger arithmetic (Pr) is the theory of the structure $\langle \mathbb{N}, 0, 1, +, \leq \rangle$. It can be explicitly axiomatized as the theory in the language $L^+ = \langle 0, 1, +, \leq \rangle$ containing the following axioms:

$$\begin{array}{llll} \text{(A1)} & 0 \neq z+1, & \text{(A2)} & x+1 = y+1 \rightarrow x = y, \\ \text{(A3)} & x+0 = x, & \text{(A4)} & x+(y+1) = (x+y)+1, \end{array}$$
$$(\mathrm{D}_{\leq})\; x \leq y \leftrightarrow (\exists z)(x + z = y),$$

and the scheme of induction (1) for all formulas of the language L^+.

3 Results

We describe how the relations between $\mathcal{M}$, $\mathrm{F}(\mathcal{M})$ and $\mathrm{R}(\mathcal{M})$ translate to certain forms of induction in $\mathcal{M}$. The Shepherdson's result on the relation between IOpen and integer parts of real closures [5] can be reformulated in the following way:

Theorem 1 *The following are equivalent for any discretely ordered ring $\mathcal{M}$:*

1. $\mathcal{M} \subseteq^{IP} R(\mathcal{M})$,
2. *all roots $r \in R(\mathcal{M})$ of polynomials $p \in M[x]$ have integer parts in M,*
3. $\mathcal{M} \models \mathrm{IPR}(M[x])$,
4. $\mathcal{M}^+ \models \mathrm{IOpen}$.

For Presburger arithmetic a similar theorem easily follows from well known properties of Pr:

Theorem 2 *Let $\mathcal{M}$ be a discretely ordered ring. Then the following are equivalent:*

1. $\mathcal{M} \subseteq^{IP} \frac{M}{\mathbb{N}}$,
2. *all fractions m/n with $m \in M$ and $0 \neq n \in \mathbb{N}$ have integer parts in M,*
3. $\mathcal{M} \models \mathrm{IPR}(\{n(x); n \in \mathbb{N}\})$, *where $n(x) = x + \ldots + x$ with n summands,*
4. $\mathcal{M}^+ \models \mathrm{Pr}$.

Theorems 3 and 5 state analogous results for other arithmetical theories, providing equivalents to the following algebraic properties: $\mathcal{M} \subseteq^{IP} \mathrm{F}(\mathcal{M})$ and $\mathrm{F}(\mathcal{M}) \subseteq^{d} \mathrm{R}(\mathcal{M})$.

Theorem 3 *Let $\mathcal{M}$ be a discretely ordered ring. Then the following are equivalent:*

1. $\mathcal{M} \subseteq^{IP} F(\mathcal{M})$,
2. *all fractions m'/m with $m', m \in M$ and $0 < m$ have integer parts in M,*
3. $\mathcal{M} \models \mathrm{IPR}(\{m(x); m \in M\})$, *where $m(x) = m \cdot x$,*

*3. $\mathcal{M} \models (\forall n, k \neq 0)(\exists n')(\exists 0 \leq l < k)(n = n' \cdot k + l)$,

4. $\mathcal{M}^+ \models \mathrm{IOpenLin}$.

Proof. Clearly, $1 \Leftrightarrow 2$ and $3 \Leftrightarrow$ *3. Further:

- $1 \Rightarrow 4$: Let $\varphi(x, \bar{y})$ be a quantifier free linear formula, $\bar{m} \in M$. We prove that the induction axiom for $\varphi(x, \bar{m})$ holds in $\mathcal{M}$. $\varphi(x, \bar{m})$ is equivalent to a boolean combination of formulas of the form $x \cdot u \geq v$, for some $u, v \in M$. Suppose that $\mathcal{M}^+ \models \varphi(0, \bar{m}) \wedge \neg(\forall x)\varphi(x, \bar{m})$. The set $\{x \in M^+; \mathcal{M}^+ \models \varphi(x, \bar{m})\}$ is a finite union of intervals with endpoints of the form $\frac{v}{u}$, for some $u, v \in M$, or $\pm\infty$. By condition 1, for any $u, v \in M$, there exists $w \in M$ such that $w = \lfloor \frac{v}{u} \rfloor$. This implies that $\mathcal{M}^+ \models (\exists w)(\varphi(w, \bar{m}) \wedge \neg\varphi(w + 1, \bar{m}))$.
- $4 \Rightarrow$ *3: Fix $n, k \neq 0 \in M^+$. $\mathcal{M} \models 0 \cdot k \leq n \wedge \neg(\forall n')(n' \cdot k \leq n)$, however $n' \cdot k \leq n$ is an open linear formula. Condition 4 implies $\mathcal{M}^+ \models (\exists n')(n' \cdot k \leq n < (n' + 1) \cdot k)$; such an n' and $l := n - n' \cdot k$ satisfy the condition *3.

- *3 ⇒ 1: Let $\frac{n}{k} \in$ F($\mathcal{M}$) and $n', l \in M$ satisfying the condition *3 for n, k. Then $\frac{n}{k} = \frac{n' \cdot k + l}{k} = n' + \frac{l}{k}$ and $\left|n' - \frac{n}{k}\right| = \frac{l}{k} < 1$.

□

Lemma 4 *Let $r \in R(\mathcal{M})$ satisfy $0 < r < 1$. Then there is some $m \in M^+$ such that $0 < \frac{1}{m} < r$.*

Proof. Suppose for contradiction that for every $m \in M^+$ it is $0 < r < \frac{1}{m}$. Let $f(x) \in$ F($\mathcal{M}$)$[x]$ be of the least degree such that r is the root of $f(x)$. Then the absolute coeficient of $f(x)$ is 0 (if not then it can be expressed as a linear combination of powers of r which is easily a contradiction) and $f(x) = g(x) \cdot x$ for some $g(x) \in$ F($\mathcal{M}$)$[x]$. This contradicts the minimality of degree of $f(x)$. □

Notation:

$$I_x^l \varphi(x, \bar{y}) = I_x^l \varphi := ((\forall u < l)\varphi(u, \bar{y}) \wedge (\forall v)(\varphi(v, \bar{y}) \rightarrow \varphi(v + l, \bar{y}))) \rightarrow (\forall x)\varphi(x, \bar{y})$$

eIOpen is an axiom schema $(\exists l > 0)(I_x^l \psi(\frac{x}{l}, \bar{y}))$, where $\psi(x, \bar{y})$ is an open formula.

If we rewrite $I_x^l \psi(\frac{x}{l}, \bar{y})$ we get the following formula

$$((\forall u < l)\psi(\frac{u}{l}, \bar{y}) \wedge (\forall v)(\psi(\frac{v}{l}, \bar{y}) \rightarrow \psi(\frac{v}{l} + 1, \bar{y}))) \rightarrow (\forall x)\psi(\frac{x}{l}, \bar{y}).$$

Theorem 5 *Let $\mathcal{M}$ be a discretely ordered ring. Then the following are equivalent:*

1. *F($\mathcal{M}$) $\subseteq^d$ R($\mathcal{M}$),*
2. *$(\forall r \in R(\mathcal{M}))(\exists f \in F(\mathcal{M}))(r < f < r + 1)$,*
3. *$(\forall r \in R(\mathcal{M}))(\exists l, m \in M)(m \leq lr < m + l)$ [i.e. $\frac{m}{l} \leq r < \frac{m}{l} + 1$],*
4. *$\mathcal{M}^+ \models$ eIOpen.*

Proof. 1 ⇒ 4: Let $\psi(x, \bar{y})$ be an open formula, $\bar{a} \in M^+$; in the next, $\psi(x)$ means $\psi(x, \bar{a})$. $\psi(x)$ can be written as a boolean combination of fomulas of the form $p(x) \geq 0$, where $p(x) \in$ F($\mathcal{M}$)$[x]$. $\psi(x)$ can change its truth value only at roots of these polynomials.

Suppose there is some $f \in$ F($\mathcal{M}$) such that $0 \leq f < 1$ and $\neg\psi(f)$. As $f = \frac{u}{l}$ for some $u < l \in M^+$, we have $\neg\psi(\frac{u}{l})$ and $I_x^l \psi(x)$ holds.

In the next, we suppose $\psi(x)$ holds on $[0, 1) \cap$ F($\mathcal{M}$). Let $r \in$ R($\mathcal{M}$) be the largest root of some polynomial from $\psi(x)$ such that $\psi(x)$ holds on $[0, r) \cap$ F($\mathcal{M}$) (if there is not such an r, $\psi(x)$ holds on all $f \in$ F($\mathcal{M}$); in particular, $I_x^l \psi(x)$ holds for any $l > 0$). Let $r' \in$ R($\mathcal{M}$) be the smallest root of some polynomial from $\psi(x)$ greater then r (or $r' := +\infty$, if such a root does not exist). By 1, there exists some $f \in$ F($\mathcal{M}$) with $1 \leq r < f < r'$, $f < r + 1$. Let us take $m, l \in M^+$ such that $f - 1 = \frac{m}{l}$. Then $0 < \frac{m}{l} < r < \frac{m}{l} + 1 < r'$. Then, by the choice of r, r', it holds $\psi(\frac{m}{l})$ and $\neg\psi(\frac{m}{l} + 1)$. That implies $I_x^l \psi(\frac{x}{l})$.

4 ⇒ 3: $r \approx r'$ denotes that there is no $f \in$ F($\mathcal{M}$) between r and r'. Let us fix $r \geq 1$, the case for $0 \leq r < 1$ is trivial, $r < 0$ symmetric. Note that it suffices to find some $m, l \in M$

such that 3 holds for m, l and any $r' \approx r$. Let $p(x) \in \mathrm{F}(\mathcal{M})[x]$ be such that $p(r) = 0$. By (repeateadly) differentiating $p(x)$, we arrive at some polynomial $p'(x)$ such that it has only one, simple root $r' \approx r$. We may suppose that $p'(x)$ is increasing at r'. Choose $f, f' \in \mathrm{F}(\mathcal{M})$ so that r' is the only root of $p'(x)$ between them.

Consider $\psi(x) := (x \leq f') \wedge (x \leq f \vee p'(x) \leq 0)$. Obviously, $\psi(x)$ holds on $[0, r']$ and does not hold on $(r', +\infty)$. By 4, there exists $l \in M$, $l > 0$ such that $I^l_x \psi(x)$. We chose $r \geq 1$, thus also $r' \geq 1$. Therefore $\psi(\frac{u}{l})$ holds for every $u < l$. On the other hand, $\neg(\forall x)\psi(\frac{u}{l})$. From $I^l_x \psi(x)$, we get $m \in M$ such that $\psi(\frac{m}{l})$ and $\neg\psi(\frac{m}{l} + 1)$. This implies $\frac{m}{l} \leq r' < \frac{m}{l} + 1$ and we are done.

$3 \Rightarrow 2$: Easy.

$2 \Rightarrow 1$: Let $r < r' \in \mathrm{R}(\mathcal{M})$. By Lemma 4, we may fix $0 < k \in M$ such that $\frac{1}{k} < r' - r$. Then $kr < kr + 1 < kr'$. By 2, there exists $f \in \mathrm{F}(\mathcal{M})$ for which $kr < f < kr + 1 < kr'$ holds. This implies $r < \frac{f}{k} < r'$. Since $\frac{f}{k} \in \mathrm{F}(\mathcal{M})$ we are done. □

As clearly $(\mathcal{M} \subseteq^{IP} \mathrm{F}(\mathcal{M}) \,\&\, \mathrm{F}(\mathcal{M}) \subseteq^{d} \mathrm{R}(\mathcal{M})) \rightarrow \mathcal{M} \subseteq^{IP} \mathrm{R}(\mathcal{M})$, we get the following:

Corollary 6 $\mathrm{IOpen} = \mathrm{IOpenLin} + \mathrm{eIOpen}$.

Proof. The inclusion $\subseteq$ follows from the above implication. The other inclusion is trivial as all instances of the induction scheme in IOpenLin and eIOpen are just special cases of open induction. □

Similarily from the inclusion $\mathrm{IOpen} \supseteq \mathrm{IOpenLin} + \mathrm{eIOpen}$, we get:

Corollary 7 $(\mathcal{M} \subseteq^{IP} F(\mathcal{M}) \,\&\, F(\mathcal{M}) \subseteq^{d} R(\mathcal{M})) \leftrightarrow \mathcal{M} \subseteq^{IP} R(\mathcal{M})$.

4 Questions

The obvious similarity of Theorems 1, 2, 3 (and 5) suggests that there may be a common generalization. We state this as a rather vague

Open question 8 *Is there a truly general Shepherdson's theorem, i.e. a theorem such that Theorems 1, 2, 3 (and 5) are its special cases?*

The theory IOpen is the strongest fragment of PA that we have a Shepherdson's type theorem for. Recently, however, Shepherdson's original ideas were generalized by Kołodziejczyk [3] and used to construct interesting models of the theory T^0_2 (an arithmetic with sharply bounded induction – see [3, Section 2] for the precise definition) that is stronger than IOpen. Therefore it seems interesting to ask:

Open question 9 *Are there variants of the Shepherdson's theorem for theories stronger than* IOpen*? In particular is there such a variant for the theory T^0_2?*

References

[1] Hájek, P., Pudlák, P.: *Metamathematics of first order arithmetic*, Perspectives in Mathematical Logic, Springer-Verlag, Berlin, 1993.

[2] Kaye, R.: *Models of Peano arithmetic*, Oxford University Press, Oxford, 1991.

[3] Kołodziejczyk, L. A.: *Independence results for variants of sharply bounded induction*, Ann. Pure Appl. Logic, vol. 162, (2011), pp. 981–990.

[4] Marker, D.: *Model theory: An introduction*, Graduate Texts in Mathematics, Springer-Verlag, New York, 2002.

[5] Shepherdson, J. C.: *A non-standard model for a free variable fragment of number theory*, Bull. Acad. Polon. Sci. Sér. Sci. Math. Astronom. Phys., vol. 12, (1964), pp. 79–86.

[6] Tennenbaum, S.: *Non-archimedean models for arithmetic*, Not. Amer. Math. Soc., vol. 6, (1959), p. 270.

Current addresses

Glivická Jana, Mgr.
Department of Theoretical Computer Science and Mathematical Logic,
Faculty of Mathematics and Physics, Charles University
Malostranské náměstí 25, 118 00 Praha 1, Czech Republic
Tel. number: +420 951 554 245, e-mail: jana.glivicka@gmail.com

Glivický Petr, RNDr. PhD.
Department of Mathematics, Faculty of Informatics and Statistics,
University of Economics, Prague
Ekonomická 957, 148 00 Praha 4, Czech Republic
Tel. number: +420 224 094 246, e-mail: petrglivicky@gmail.com

Slovak University of Technology in Bratislava
Faculty of Mechanical Engineering
16th Conference on Applied Mathematics **APLIMAT 2017**
Proceedings

ENHANCEMENT EFFECTS OF A LEARNING MANAGEMENT SYSTEM ON STUDENT LEARNING

GRIRA Sofiane (AE)

Abstract. This paper describes an investigation of the enhancement effects of using MyMathLab (MML) in math courses for engineering students.In this paper we compared the student's results with and without using MyMathlab and how MyMathLab's features improved student's performance in math courses. In addition, the benefits and difficulties of using MyMathLab were discussed. Finally, we examined students responses to the uses of MyMathLab.

Keywords: MyMathLab, online assessment,learning management system

Mathematics Subject Classification: Primary 97U50; Secondary 97U70

1 Introduction

For the last thirty years there has been considerable debate over teaching methods and approaches that might stimulate student-centered learning in undergraduate math courses. Many universities are still using no technology to deliver math courses. It may be due to a broad range of factors related to faculty preparation and training, and the impact of technology in student learning and performance. One of the major challenge while teaching engineering students math courses was how can you make them motivated? Usually students do not feel comfortable asking questions in class, do not come often to office hours and go rarely to the learning support center to seek help especially the weak students. After years of teaching math It was obvious to me that teaching math like the way we were taught when we were students will not help us to achieve our goals in this oriented technology life. Students use technology and digital tools in every part of their lives and very few for learning.They have never really been pushed or asked to use that same technology inside the classroom. As technology inextricably penetrates into our daily lives, it entails different opportunities and challenges for teaching and learning to meet students' needs. In a traditional delivery classroom, an instructor gives lecture, uses blackboard and chalk and students listen and take notes. he monitors the activities of each student, identify their problems and gets solution to it. As for exams, it has always been conducted with the use of paper and pencil. With the evolution of technology where everyone has technology devices as computers, tablets or smartphone, integrating technology was understandable and expectable. In nowadays, The use of technology is in every aspect of our life, we all depend on technology and we use various technologies to accomplish specific tasks in our lives.

Unfortunately, instructional technologies have not been utilized to their full potential. The aim of this paper is to study the effect of introducing MyMathLab in math courses over student's performance. It investigates the effectiveness of computer-assisted learning. A comparison between student's performance in a paper and pencil exams with an online exams using MyMathLab will be given.

2 Design of the study

After teaching for five years math courses without use of any technology except the use of the projector and slides and where the homework and exams were paper based, we implemented MyMathLab, a management learning system for the majority of the math courses (calculus 1,2 and 3, linear algebra, and statistics) taken by engineering students. Over the period of six years students were using MyMathLab for studying and doing their homework. All the proctored exams were administrated online using MyMathLab. MyMathLab was used in conjunction with the conventional lecture method of instruction.

3 Use of MyMathLab in Classroom

MyMathLab was used as a mean of communicating with students and administering and grading exams. E-books were used to teach instead of using slides. Students using their laptops or smartphone can access the study plan in MyMathLab which consists of sets of problems and solve them. Students will get an immediate feedback and know if their answer was correct or not which will help the instructor to identify the students who are struggling.

4 Use of MyMathLab Outside Classroom

Students using MyMathLab could view streamlined video lectures connected with any topic in the book and also do sample homework problems and sample tests on any topic with immediate feedback on their work to help them learn. Also from time to time and especially before exams an online office hours is scheduled using liveclassrom chat where students and instructor use a real-time interactive whiteboard to interact. Moreover, a multimedia library is available which lets you access the multimedia resources related to your course. By linking to all multimedia from one place, the multimedia library makes it easy to identify and locate the resources needed by students.

5 Methodology

The participants for this research were university Engineering students who enrolled in Calculus and linear algebra courses at Abu Dhabi University.1148 students were selected. Half of them did use MyMathLab in their study as well as to do their homework. All exams taken by these students were online proctored exams using MyMathLab. The other half of students did not use MyMathLab and all their homework and exams were paper-based. All the courses had the same syllabus and were taught by different instructors. The grades used in the study are the final grades obtained by students in a math course.

6 Results and Discussion

One of the most important features of MML is the study plan.Study Plan is designed for students to improve their skills by practicing problems and getting help to achieve the course objectives. After completing each assigned question, Study Plan will generate questions that focus on each learning objective that the student have struggled with. Without the aid of technology, it is difficult, if not impossible, for faculty to ascertain how much time a student is actually spending on task. MML provides detailed tracking features, thereby enabling faculty to reliably and automatically monitor studentsâ€™ time and intervene at the first sign of trouble.

To study the effect of the study plan on student's performance we checked the correlation between the objectives mastered by a student and his grade on a calculus course. Fig. 1 shows that the correlation is strong (r = 0.91).

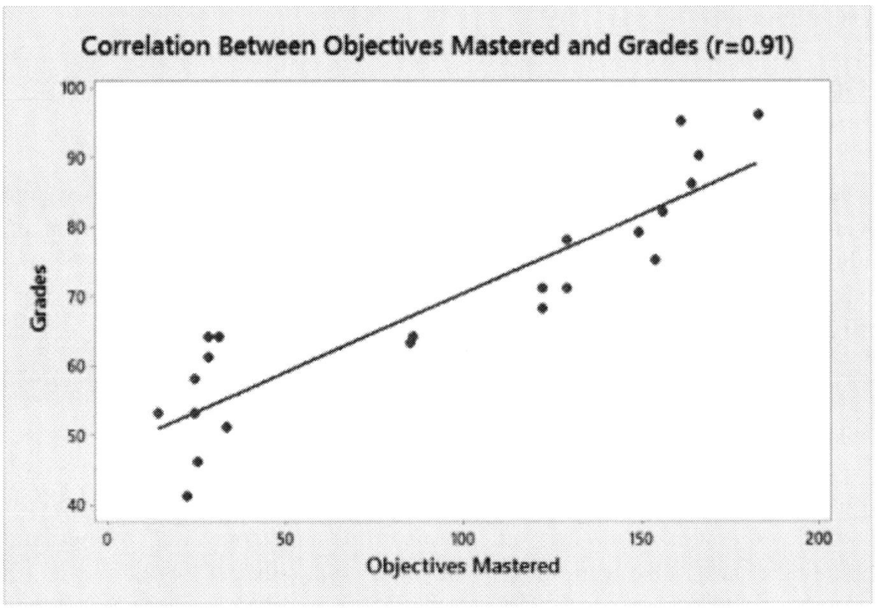

Fig. 1

It is obvious that working on the study plan and mastering the objectives had a positive impact on student's grades. Study plan helps students monitor their own progress, letting them see at a glance exactly which topics they need to practice. Students are motivated by the guided solutions, by the ability to immediately see what they did wrong. Students are taking responsibility for and ownership of their learning. Fig. 2 compares the grades of students who used MML versus those who did not.

Two sample T hypothesis test:

Difference	Sample Diff.	Std. Err.	DF	T-Stat	P-value
$\mu_1 - \mu_2$	0.18	0.073000644	1133.5096	2.4657317	0.0069

Sample Statistics:

Sample	n	Mean	Std. Dev.
Population 1	574	2.23	1.3
Population 2	574	2.05	1.17

Statistical Results (StatCrunch)

We noticed that the number of students getting A has almost doubled. Statistical results shows the result of a hypothesis testing comparing the average grades of students with and without using MML. We remarked that P-value (0.0069) was small which implied that the average grades of students using MML is greater than the average grades of students who did not use MML. We can say that MML had helped many students to excel and become excellent students.

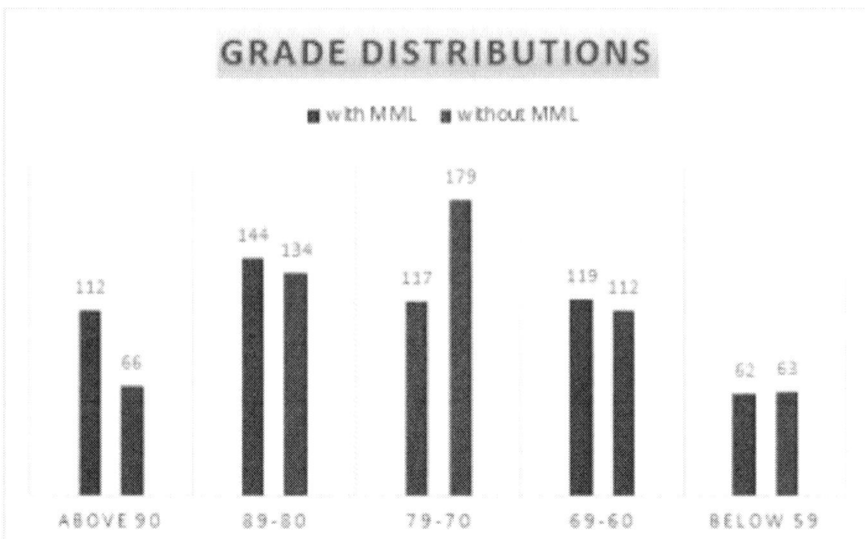

Fig. 2

On the other hand Fig. 2 shows that the number of students who failed the course has almost remain the same. This can be due to many factors. One of them is the initial transition from a traditional lecture-based format to the student-centered Web format was a challenge for some students. Also some students are not familiar with the use of technology in their courses which requires some technology skills. Moreover, it is only in math courses where students were introduced to MML and were given online exams. This new way of learning and assessment was challenging for weak and slower learning students.

7 Student's feedback

A survey was conducted for students who used MML.

Questions	Strongly agree	Agree	Neither agree nor disagree	Disagree	Strongly disagree
MML provided adequate practice to understand the concepts	58%	38%	4%	0%	0%
MML adequately prepared me for exams	60%	28%	10%	0%	2%
MML allowed me to get a better grade in the course	53%	23%	13%	10%	1%
My understanding of the course increased as a result of using MML	50%	35%	13%	2%	0%
I would recommend MML to another student taking this course	58%	30%	12%	0%	0%
I preferred online exam using MML over paper and pencil exams	68%	15%	5%	8%	4%

The survey showed that 85% strongly agree/agree that MML increased the understanding of the course. 76% strongly agree/agree that MML help them to get better grade.Also 83% of students preferred online exam using MML over paper and pencil exams. We can conclude from the above table that students agreed that MML helped them to learn and understand math. Regarding grades 76% agreed that MML allowed them to have better grades . Moreover 88% will recommend this online system to other students too.

8 Limitation

This study was solely based on a survey where all the respondents were from one university. Thus, the results cannot be generalized to students in other universities. An additional weakness of this study is that is focused solely on students studying non-remedial mathematics courses and did not examine the impact of MML on remedial courses.

9 Conclusion

The study has shown that there is a strong positive correlation between the objectives mastered from the study plan in MML with the final grade. Consequently, Study Plan helped students to improve their mathematics skills as well as their grades for the majority of students. Unlimited practice offered by online assessments motivated students to do more mathematics exercises, thereby increasing not just students' success but mastery. The study also has shown that the passing rate did not increase much by using MML. More study should be done focusing on weak and slower learning students. It should also include the effects of MML on students taking remedial courses. In overall, students were satisfied with the use of MML in their studies.

References

[1] M. D. Speckler. *Making the Grade, Version 4: How to Improve Your MyMathLab Implementation through Personalization + Best Practices.* Boston, USA: Pearson, pp. 2-66, 2010.

[2] HONG, K.S. , LAI K. W., HOLTON D. *Students' satisfaction and perceived learning with a web-based course.* Educational Technology & Society. 6 (1): 116-124, 2003.

[3] RAINES, J. M., CLARK, L. M. *A brief overview on using technology to engage students in mathematics.* Current Issues in Education, 14(2) , 2011.

[4] TRILLO R., ILLARRI S., LOPEZ J. R., BRISABOA N. R. *Development of an online assessment system to track the performance of students.* Proc. 3rd International Conference on Web Information System and Technologies, Spain, pp. 452-457 , 2007.

[5] BORMAN, D., SLEIGH, A. *An evaluation of the use of interactive approaches and integrated on-line resources* . Teaching Mathematics Applications, 30, 166-177 , 2011.

[6] BASHAM, J., SMELTZER, A., PIANFETTI, E. *An integrated framework used to increase preservice teacher NETS-T ability.* Journal of Technology and Teacher Education, 257-276 , 2013.

[7] WESTOVER, J. H., WESTOVER, J. P *Teaching hybrid courses across disciplines: Effectively combining traditional learning and e-learning pedagogy.* International Journal of Information and Education Technologies, 4(1), 93-96 , 2014.

Current address

Grira Sofiane, PhD
College of Arts and Sciences, Al Ain Campus, 1790, Abu Dhabi University, United Arab Emirates
Tel. number: +97137090719, e-mail: grira.sofiane@adu.ac.ae

Slovak University of Technology in Bratislava
Faculty of Mechanical Engineering
16th Conference on Applied Mathematics **APLIMAT 2017**
Proceedings

STATISTICAL ANALYSIS OF THE LEVEL OF MATHEMATICS AT SLOVAK LOWER SECONDARY SCHOOLS

GUNČAGA Ján (SK), JUREČKOVÁ Mária (SK)

Abstract. The aim of this article is to analyse and compare the achieved level of mathematical education at Slovak lower secondary schools in recent years and to point to the tendencies in the development of the level of teaching mathematics. The results of the analysis will help identify the quality of the mathematical competencies of pupils and the causes of the lack of interest of pupils in secondary schools in technology and natural sciences.

Keywords: statistical analysis, teaching mathematics, evaluation

Mathematics Subject Classification: 97B10, 97C70, 97D60

1 Introduction

In educational practice there are different ways of measuring pupils' knowledge [1]. Despite the fact that there are different opinions of the forms of examining and marking students, it is indisputable that the verification of the achieved knowledge is an important component of the educational process [2]. It not only provides important information on the level of achieved abilities to a student and his/her parents, but it is also a form of control for a teacher who can, on the basis of them, evaluate the efficiency of the teaching methods used. Regularly carried out in Slovakia by the National Institute for Certified Educational Measurements in Bratislava (NICEM), centrally implemented testing is irreplaceable in this process. The aim of the nation-wide testing is to objectively compare the performances of pupils within the tested subjects and, based on the results, to give schools feedback about their level as compared to other schools in Slovakia.

In our paper we would like to focus on the comparison of the results of nation-wide T9 tests in mathematics used in the years 2011-2016. The tests in mathematics measured the pupils' skills of applying basic mathematical principles and procedures in both the mathematical context and real life. Our goal is a statistical analysis of the results of standardized tests from the period of last six years to identify the direction of the level of mathematical education of pupils at lower secondary schools. Gathering a sufficient quantity of data to identify existing shortcomings in the mathematical capabilities of our pupils can help identify the causes of the lack of interest in technical and natural scientific study programs.

2 Sources and methodology of research

The source of the data is an anonymized database provided by the NICEM. It contains information on the results of the tests of 251,757 pupils of the ninth grade of lower secondary schools across Slovakia. In the years 2011-2016 there were 45,381, 43,485, 41,774, 42,007, 40,880 and 38,230 pupils respectively. For our analysis, we used a point-based evaluation of all tasks in the tests in the selected period as well as the information on the sex of pupils and the founders of the schools that they attended (state, private, church). Besides the overall evaluation, we also paid attention to the results achieved in the four categories to which the problems were divided. The division reflects the standard of competences that pupils are supposed to attain in individual sectors [11]. The thematic units were not the same in all the years of testing. In 2013, however, they were unified to the following categories:

T1 Numbers, variables and arithmetic operations with numbers,
T2 Relations, functions, tables, charts,
T3 Geometry and measurement,
T4 Combinatorics, probability, statistics, logic, reasoning, evidence.

In each of the selected years a test contained 20 tasks to verify the ability to use mathematical thinking for solving various problems of everyday life. There was the same number of items with the context of real life and mathematical context. However, the number of tasks in the above mentioned categories was not within individual tests the same and we also had to take into account a different number of tasks in individual categories within the selected years [11], [12], [13], [14], [15]. The statistical analysis was carried out in the statistical program IBM SPSS.

3 The results of the analysis

In the introduction we will pay attention to the overall results of the tests and their distribution in individual selected years. Initial information can be obtained from the chart (Fig. 1) that indicates how the mean of gained points in T9 tests developed in individual years.

From the chart (Fig. 1) it is evident that after the gradual increase of the selected parameter in 2011-2013, a rapid decrease in the success rate of pupils came in 2014 and in 2016 children even failed to achieve the level of 2011. Based on comparing the medians it can be also stated that the year 2013 was most successful in T9 testing in mathematics. In 2013 half of the tested pupils obtained at least 13 points (of the total number of 20 points) and 25% of pupils had at least 17 points. The distribution of the total number of the obtained points in individual years is also interesting (Fig. 2).

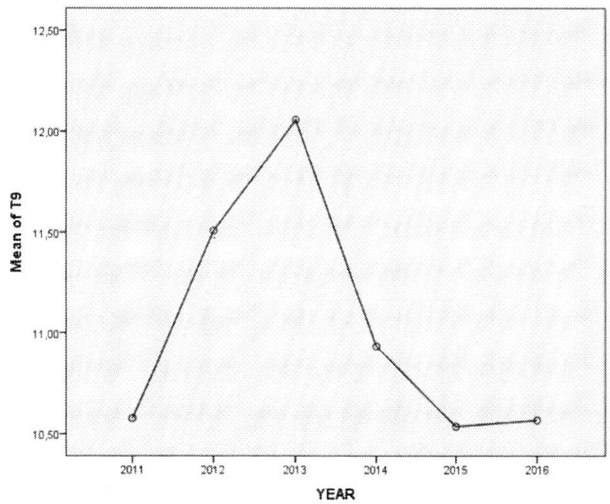

Fig. 1. Mean of the results of T9 testing

In 2013 there was a significant skewness (-0,311) in the distribution of the selected feature towards higher values. It was a great progress as compared to 2011 (skewness = 0,358) when the lower point levels were more frequent. In 2016 the frequency of obtained points shows the highest symmetry with the median of 11 points and with the average number of points (10.6 points). Based on the above mentioned charts there is a natural indication that the results of T9 tests vary in individual years of the selected period. In regard to the results of the test of normality (Kolmogorov-Smirnov test), that in all above mentioned years refused a normal distribution of the number of obtained points, we used Kruskal-Wallis test [3], [4]. Output of the test in SPSS (Tab. 1) confirmed our hypothesis.

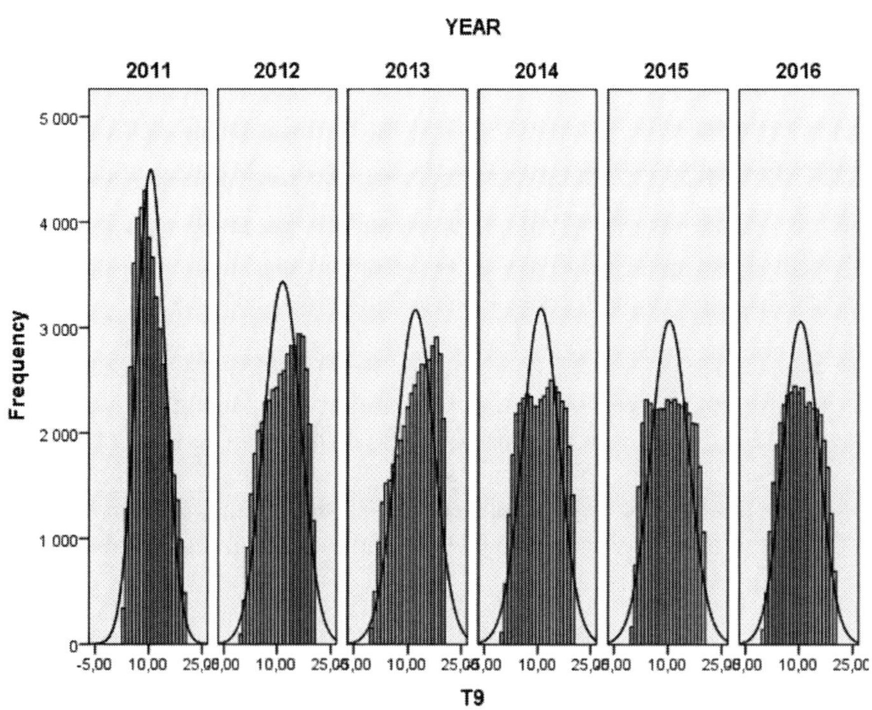

Fig. 2. Distribution of results in T9 testing

Hypothesis Test Summary

	Null Hypothesis	Test	Sig.	Decision
1	The distribution of T9 is the same across categories of YEAR.	Independent-Samples Kruskal-Wallis Test	,000	Reject the null hypothesis.

Asymptotic significances are displayed. The significance level is ,05.

Tab. 1. Kruskal-Wallis test for the distribution of T9 across the years 2011-2016

The above mentioned results reflect the consequences of the introduction of the new State Educational Program in 2011 as well as the changes in the number of lessons in the subject of mathematics.

According to 1997 curriculum, in the subject of mathematics the number of lessons in the 5th, 6th and 7th grades of primary schools was 5 lessons per week, 165 lessons per year. In the 8th grade there were 4 lessons per week, 132 lessons per year. In the 9th grade there were again 5 lessons per week and 165 lessons per year. After the introduction of the State Educational Program ISCED 2, the number of lessons in individual subjects is determined by the framework curriculum. Within the meaning of the school law, the number of lessons is defined only for the whole stage of education and not according to grades, as it was in the old curriculum. The allotment of the number of lessons to individual grades falls into a school's authority. The same applies to the allotment of curriculum to individual grades. In this case, however, the curriculum must be age-appropriate. Comparing the numbers of mandatory mathematics lessons, according to 1997 curriculum there was a total number of 24 lessons in the 5th to 9th grades. According to [17] it is 19 lessons, which is a decrease of more than 20 percent. If we compare the mathematics curricula of 1997 and 2010, in terms of the content the curriculum was reduced almost in all thematic units and there were shifts from lower to higher grades as well. Even if 31 optional lessons fall within a school's authority, not always 5 lessons are devoted to mathematics, which would compensate the above mentioned decrease.

In the further analysis we focused on the successfulness in T9 tests depending on sex. The chart (Fig. 3), which depicts the histogram of points in 2011 to 2016 in the case of boys (0) and girls (1), indicates that there are not significant differences in the distribution.

The differences will show when we compare the successfulness of boys and girls in individual categories T1-T4. Based on the output of the Mann-Whitney test (Tab.2) it can be stated that the results are comparable only in the case of a T9 test as a whole and in the category of geometry. In other categories the differences are statistically significant. Compared to boys, girls achieve worse results only in the category T4 (Combinatorics, probability, statistics, logic, reasoning, evidence).

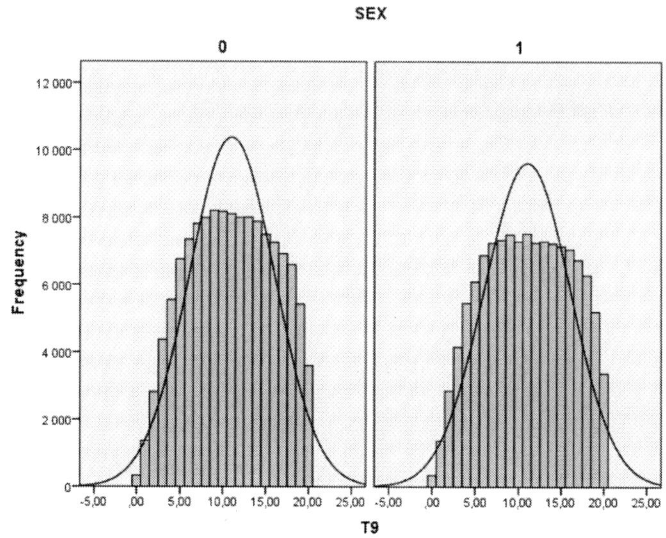

Fig. 3. Distribution of T9 across categories sex

Hypothesis Test Summary

	Null Hypothesis	Test	Sig.	Decision
1	The distribution of T9 is the same across categories of SEX.	Independent-Samples Mann-Whitney U Test	,426	Retain the null hypothesis.
2	The distribution of PT3 is the same across categories of SEX.	Independent-Samples Mann-Whitney U Test	,492	Retain the null hypothesis.
3	The distribution of PT1 is the same across categories of SEX.	Independent-Samples Mann-Whitney U Test	,000	Reject the null hypothesis.
4	The distribution of PT2 is the same across categories of SEX.	Independent-Samples Mann-Whitney U Test	,000	Reject the null hypothesis.
5	The distribution of PT4 is the same across categories of SEX.	Independent-Samples Mann-Whitney U Test	,000	Reject the null hypothesis.

Asymptotic significances are displayed. The significance level is ,05.

Tab. 2. Mann-Whitney U test for the distribution of T9 and T1-T4 across categories sex

In the selected period 94.24% of the tested pupils attended state schools (1), 1.46% private schools (2) and 4.3% church schools (3). Based on the results of the six-year T9 testing in mathematics we made box plots (Fig. 3) that indicate certain differences in the successfulness in the case of state (1), private (2) and church schools (3). The Kruskal-Wallis test confirmed that these differences were statistically significant. The long-term evaluation showed that the pupils of church schools achieved better results in T9 testing than the children from state and private schools whose results are comparable.

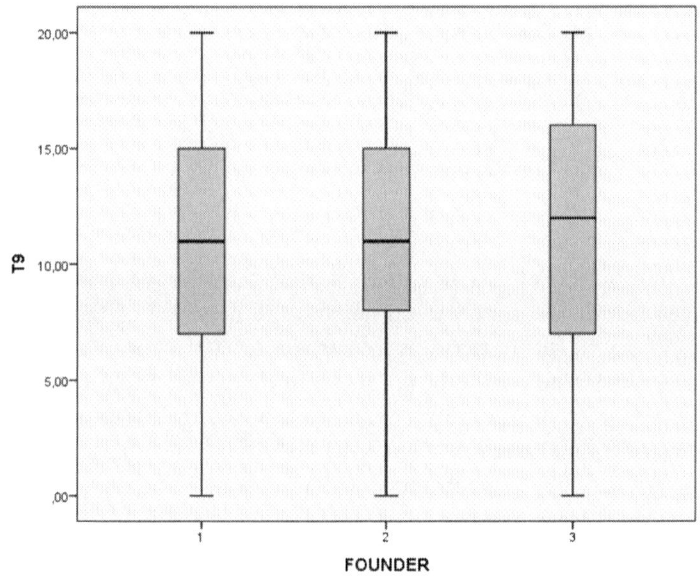

Fig. 4. Boxplots of T9 across categories founders

In all the years, Geometry and measurement (T3) problems prevailed in individual mathematics tests. In the final report published by NICEM in 2012 it was stated that the pupils of the 9th grade of lower secondary school had underdeveloped spatial imagination, they could not decompose composite geometric solid figures into simpler ones and they lacked the capability of planning a strategy of solving problems [13]. Similarly, Ruthven in [5] also pays special attention to the research of the geometric drawing and graphic representation of solids. In regard to the fact that geometry problems were clearly identifiable in all selected years, it was possible to monitor the level of lower secondary school geometry in a longer term perspective. The number of geometry problems was not the same in each year. Therefore, we analysed the percentage of success in this area (PT3). The following charts (Fig. 5, Fig. 6) bring the development of average successfulness (Fig. 5) and boxplots in individual years (Fig. 6).

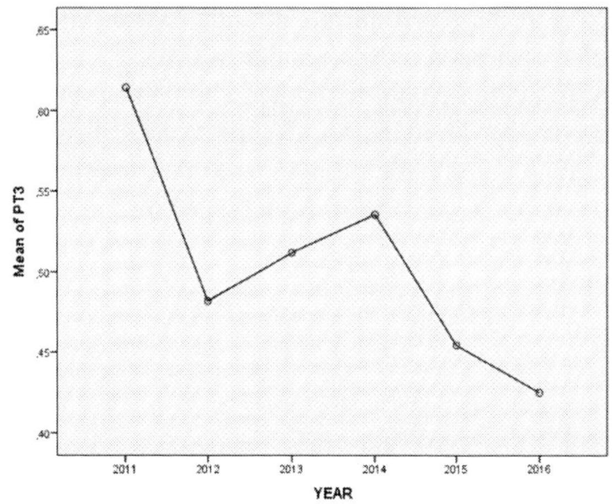

Fig. 5. Mean of mathematics results in geometry

Based on the above mentioned charts it can be stated that the year 2012 really brought a rapid decrease in the success rate in geometry even despite the increase of the total successfulness in T9 tests. There has been a very negative trend in this area after 2014. In 2016 the level of 2012 was not even reached. It is also illustrated by the boxplots from the above mentioned years. In 2013, 25% of ninth graders had success rate in geometry of at least 83%. In 2016 such successfulness was not even reached by 10% of ninth graders. In 2014 to 2016 one quarter of ninth graders did not overcome the success rate of 20% in geometry.

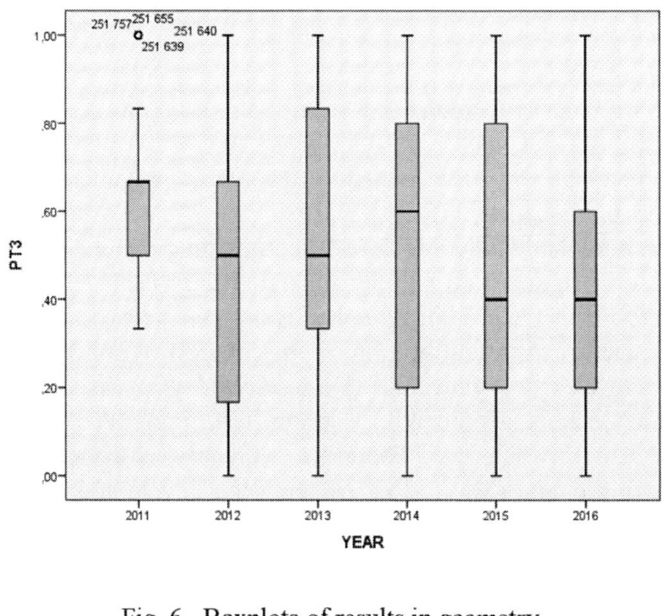

Fig. 6. Boxplots of results in geometry

The impact of the reduction of curriculum is thus evident in the field of geometry as well. In 1997 curriculum approximately 200 lessons were allotted to geometry in the 5th to 9th grades. In the new curriculum (2010) there are about 170 geometry lessons in these grades, which is about a 15 percent decrease as compared to the previous number.

4 Conclusion

The above mentioned results that constitute a part of the depth analysis of T9 tests in 2011-2016 show a very negative trend in terms of the quality of the mathematical education at lower secondary schools. We conclude it not only from the overall test results but, surprisingly enough, also from a partial area in which we supposed an opposite trend.

Recently, high-quality freely available geometric software for teaching mathematics providing an interactive environment for pupils has been developed at schools. For instance, Cabri Geometry is purely didactic software distributed to all schools involved in the project Infovek. These programs allow to study geometric relations in a dynamic way and to experiment with forms and objects, and thus help develop imagination. It turns out, however, that the use of this kind of software is either insufficient or inefficient.

The decrease of the number of mathematics lessons at the second stage of primary education, the need for rectification of the problematic mathematics teaching at the first stage and many times an insufficient qualification of the teachers of mathematics at lower secondary schools may be a reason for this situation.

A pupil creates his/her relationship to mathematics as early as at primary school. Therefore, it is very important exactly at this stage of education to work intensively on its proper formation. After all, not only in Slovakia but also in the neighbouring countries, there is a lack of interest in the study of those study programs that are based on the knowledge of mathematics. It is, therefore, necessary to increase the number of mathematics lessons within the new State educational program ISCED 2 as well as the Framework curriculum for primary schools. This would create a space for a greater use of modern educational tools that can make a subject more attractive and teaching it more efficient (see also [17], [18]). Undoubtedly, however, it is also necessary to pay attention to the preparation of teachers for the use of information technologies in educational process [8]. In addition, the results of T9 statistical surveys should be also seen in a broader international context as they provide an opportunity for comparison with analogous surveys in other countries [7].

Acknowledgement

The authors would like to thank the National Institute for Certified Educational Measurements in Bratislava (NICEM) for providing the database of the results of T9 tests. The paper was written with the support of the grant VEGA1/0440/15 "*Geometric Conceptions and Misconceptions of Preschool and School Age Children*".

References

[1] GUTIERREZ, A., JAIME, A. *On the assessment of the van Hiele levels of reasoning*. In: Focus on Learning Problems in Mathematics, Vol. 20, Nr.2/3, pp. 27-46.

[2] BRINDLEY, S. et all. *Master Class in Mathematics Education International Perspectives on Teaching*. London, New York: Bloomsbury Academic, 2014. ISBN 978-1-4411-7975-3.

[3] ŘEZÁNKOVÁ, H. *Analýza dat z dotazníkových šetření*, Profesional publishing Praha, 2010, ISBN 978-80-7431-019-5.

[4] AGRESTI, A. *Categorical Data Analysis*, Second Edition, John Wiley&Sons, Hoboken 2002

[5] RUTHVEN, K. *Mathematical technologies as a vehicle for intuition and experiment: a foundational theme of the ICMI, and a continuing preoccupation*. In: Interantional Journal for the History of Mathematics Education, Vol. 3, Nr. 2, pp. 91-102.

[6] ŽILKOVÁ, K. *Teória a prax geometrických manipulácií v primárnom vzdelávaní.* Praha: Powerprint, 2013, pp 115. ISBN 978-80-87415-84-9.

[7] MULLIS, I., MARTIN, M., GONZALEZ, E., & CHROSTOWSKI, S. (2004). *TIMMS 2003 International Mathematics Report*. Boston : TIMSS & PIRLS International Study Center, Boston College.

[8] SIMONKA, Z., MOJŽIŠOVÁ, E. *Derive, Mathematica, Cabri Geometry a MS Excel v školstve.* In: Informačné a komunikačné technológie v matematickom vzdelávaní na EU v Bratislave: 22.-24.6.2009, Bratislava: Katedra matematiky Fakulty hospodárskej informatiky Ekonomickej univerzity v Bratislave, 2009, pp. 23-27.

[9] GREBEŇ, R., MOLNÁR, J.: *Geometric terminology: How do teachers see it?.* In: *Mathematica V, Scientific Issues.* Ružomberok, 2015, pp. 21-28. VERBUM, ISBN 978-80-561-0296-1

[10] *Štátny vzdelávací program Matematika (Vzdelávacia oblasť: Matematika a práca s informáciami) Príloha ISCED 2.* Bratislava: Štátny pedagogický ústav, 2010, In: http://www.statpedu.sk/sites/default/files/dokumenty/statny-vzdelavaci-program/matematika_isced2.pdf

[11] [T11] *Testovanie 9-2011 Priebeh, výsledky analýzy.* In: http://www.nucem.sk/documents//26/testovanie_9_2011/Sprava_T9_2011_v2.3.pdf

[12] [T12] *Testovanie 9-2012 Priebeh, výsledky analýzy.* In: http://www.nucem.sk/documents//26/testovanie_9_2012/Sprava_T9_2012_v2.3.pdf

[13] [T13] *Testovanie 9-2013 Priebeh, výsledky analýzy.* In: http://www.nucem.sk/documents//26/testovanie_9_2013/Sprava_T9_2013_v2.3.pdf

[14] [T14] *Testovanie T9-2014 – priebeh, výsledky a analýzy.* In: http://www.nucem.sk/documents//26/testovanie_9_2014/Sprava_T9_FINAL_2.pdf

[15] [T15] *Testovanie T9-2015 – priebeh, výsledky a analýzy.* In: http://www.nucem.sk/documents//26/testovanie_9_2015/Sprava-T-9-FINAL_2015.pdf

[16] *Rámcový učebný plán pre základné školy s vyučovacím jazykom slovenským* In: *http://www.statpedu.sk/sites/default/files/dokumenty/statny-vzdelavaci-program/rup2.pdf*

[17] REPOVSKÝ, M., KUBIŠ, T. et all. *Zbierka uvoľnených úloh z testovania matematickej a čitateľskej gramotnosti pre 2. stupeň základných škôl a 1. -4. ročník osemročných gymnázií.* Bratislava: NÚCEM, 2012, ISBN 978-80-970261-8-9.

[18] PICHANIČOVÁ, I., POLGÁRYOVÁ E. et all. *Zbierka úloh pre vzdelávací stupeň ISCED 2: Matematika, Slovenský jazyk a literatúra, Slovenský jazyk a slovenská literatúra, Maďarský jazyk a literatúra.* Bratislava: NÚCEM, 2013, ISBN 978-80-89638-04-8.

Current address

Gunčaga Ján, Doc. PaedDr. PhD.
Catholic University in Ružomberok
Hrabovská 1, 034 01 Ružomberok, Slovak Republic
Tel. number: 0908220253, e-mail: jan.gunčaga@ku.sk

Jurečková Mária, Doc. RNDr. CSc.
Catholic University in Ružomberok
Hrabovská 1, 034 01 Ružomberok, Slovak Republic
Tel. number: 0911543924, e-mail: maria.jureckova@ku.sk

Slovak University of Technology in Bratislava
Faculty of Mechanical Engineering
16th Conference on Applied Mathematics **APLIMAT 2017**
Proceedings

INSIGHT INTO THE ART OF SOLVING A PRACTICAL GEOMETRIC TASK

HAŠKOVÁ Simona (CZ)

Abstract. The article presents and compares three approaches to the problem solving of a practical geometric task: the standard mathematical approach, the experimental solution and the mathematical "*art*" solution; the last two stated illustrate the unconventional and inventive approach to the solution of the problem. The step-by-step solutions are presented in the form of a case study and serve as an inspiration to the teachers of a linear algebra or geometric class.

Keywords: geometric task, algebraic solution, graphic and experimental solution, the "art" of problem solving, economic solution

Mathematics Subject Classification: Primary 14G99; Secondary 14P99

1 Introduction

The goal of mathematics education is more than teaching students to imitate standard problem-solving techniques. Its role should lie in encouraging students to think independently and creatively, so that they are able to approach problems in multiple ways. The creative ability is closely linked with the inventive ability in the sense of "*art*" to find a solution from a different perspective on the problems. As in other fields the word "*art*" in mathematical solutions has a similar meaning; a mere knowledge of technique is not enough for either the creation of an original work of art or for the genuine solution of a mathematical problem. The „*art*" of solving problems rests in the ability to find the points of view, which *simplify* and *clarify* the solution along with the utilization of the specifics of given problems (see [1], [3]).

Cognitive mathematical "*art*" abilities can be purposefully developed by training just as artistic abilities can be cultivated [2]. And this is a challenge for teachers! Here we describe a challenging, engaging, and realistic carpet-runner problem that may be solved in multiple ways in the form of a case study. Within it we show the procedure for a standard mathematical solution, an experimental and graphic solution and a mathematical "*art*" solution arising from reframing the problem – in other words, arising from taking a different perspective on the problem.
Even though the performed solutions require creative thinking together with sophisticated mathematical tools, this problem may be shared with students at almost any level, for

example anywhere in the calculus sequence, in a linear algebra or geometry class and, with appropriate methodological support, possibly even in a precalculus class (see [12], [8]).

A general algorithm of step-by-step guidance for solving these kinds of tasks is provided by G. Polya's classic work How to Solve It [9], and we follow its precepts. We concretize his general proposals in the solution of a geometric task consisting of the calculation of the right length of a runner placed across a rectangular room in a specified manner (diagonally from one corner to the opposite corner). This task cannot be solved by geometric construction. Therefore, it is not easy to obtain algebraic relations for the standard analytical (numerical) solution. As we shall see, experimentation is also not an effective method here.

The point in the mathematical "*art*" solution lies in the fact that everything that can be geometrically constructed can also be calculated. Thus, the given task (to find a suitable runner for the room) is replaced with a dual task (to find a suitable room for a given runner), which is solvable by a simple geometric construction and this geometric construction can be easily analytically described. From an analytical description of the solution to a dual task we then derive the solution to the original problem.

The presented solutions to the carpet-runner problem should attract attention to the phenomena of flexibility in thinking (also in this topic in [14]). Flexible thinking stands behind the creation of "*inventions*" in various social and technical fields and mathematics education can significantly contribute to it as shown in [11] and [13].

2 Entering the Problem

Imagine a rectangular room with a width **a** and a length **b**, for which we want to buy a carpet – a runner width **c**, which is also a rectangle (see Fig. 1).

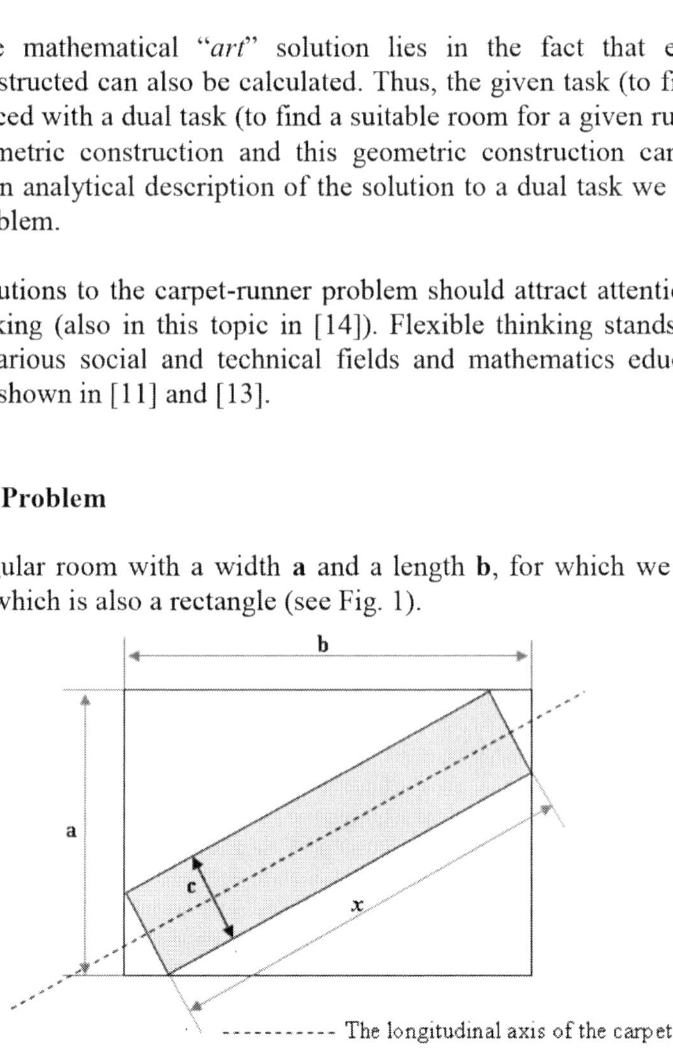

Fig. 1. Location of the carpet in the room

We want to place the carpet diagonally in the room so that each of its corners touches one of four walls. First, we have to determine what length x it should have. We can determine the required length of the runner by **calculation**, by reasoning **graphically** or by **experimental** investigation. Let us start with a standard calculation. After we get the proper length of the runner, we will deal with the practical problem of laying it in the room.

2.1 The Standard Solution

In a classical way to the solution we proceed from the known dimensions of a room for which we look for a carpet of an appropriate length. The Fig. 2 shows the result of the correctly situated carpet in the rectangular room. We see two pairs of similar right-angled triangles in it. One pair has a constant hypotenuse **c**, and the second pair has a variable hypotenuse x dependent on the dimensions of the room. For both pairs the Pythagorean Theorem is valid:

$$z^2 + v^2 = c^2, \tag{1}$$

$$(b - z)^2 + (a - v)^2 = x^2. \tag{2}$$

From the similarity of the triangles mentioned it follows (see Fig. 2) that

$$z / c = (a - v) / x, \tag{3}$$

$$v / c = (b - z) / x, \tag{4}$$

$$z / v = (a - v) / (b - z). \tag{5}$$

The last relation (5) is the ratio of the previous two and does not bring anything new. Therefore, it is discarded from our further considerations, which leaves us with the four conditions for the three variables z, v and x. Any triad of these conditions enable us to identify the correct length x of the carpet. To find the simplest and most economical solution we choose the trinity of the simplest formulas, the triad (1), (3) and (4), which is also referred to in Fig. 2. The last two terms (3) and (4) can be (after multiplying by the product $x \cdot c$ and after simplification) converted to a system of two linear equations in the variables v and z

$$c \cdot v + x \cdot z = a \cdot c, \tag{6}$$

$$x \cdot v + c \cdot z = b \cdot c. \tag{7}$$

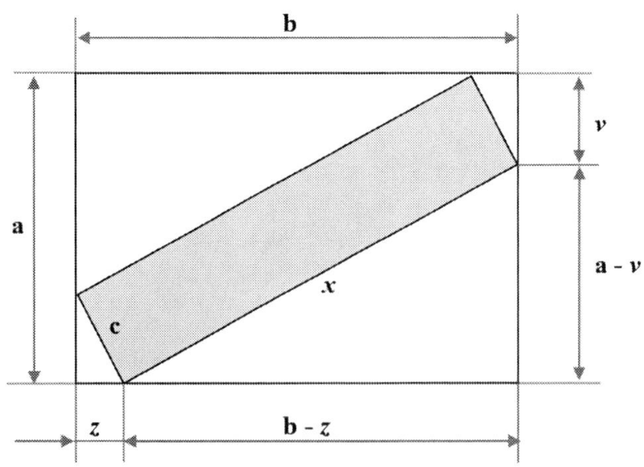

$$z^2 + v^2 = c^2$$

$$z / c = (a - v) / x$$

$$v / c = (b - z) / x$$

Fig. 2. The outline and the conditions of the standard solution

After applying, for example, the elimination method we obtain

$$y = (b \cdot c \cdot x - a \cdot c^2) / (x^2 - c^2) = c \cdot (b \cdot x - a \cdot c) / (x^2 - c^2), \tag{8}$$

$$z = (a \cdot c \cdot x - b \cdot c^2) / (x^2 - c^2) = c \cdot (a \cdot x - b \cdot c) / (x^2 - c^2). \tag{9}$$

Substituting the results (8) and (9) in the condition (1) (the Pythagorean Theorem) and after dividing by $\mathbf{c^2}$ we obtain

$$(a \cdot x - b \cdot c)^2 / (x^2 - c^2)^2 + (b \cdot x - a \cdot c)^2 / (x^2 - c^2)^2 = 1. \tag{10}$$

The condition (10) is entirely for the variable $\boldsymbol{x}$ (all other parameters are constants). Multiplying both sides of this equation by the expression $\mathbf{(x^2 - c^2)^2}$ we obtain

$$(a \cdot x - b \cdot c)^2 + (b \cdot x - a \cdot c)^2 = (x^2 - c^2)^2, \tag{11}$$

which after multiplying and after further simplification gives

$$x^4 + (-b^2 - a^2 - 2c^2) \cdot x^2 + 4 \cdot a \cdot b \cdot c \cdot x - c^2 \cdot (a^2 + b^2) + c^4 = 0. \tag{12}$$

Thus, the sought correct length of our carpet is determined by the solution of this polynomial equation lying in the interval $\mathbf{0 < x < \sqrt{a^2 + b^2}}$. The solution can be calculated by any iteration method. To solve this problem we can also use the effective tools that mathematics offers. Although the result of the solution will be the same, these tools, however, do not often help us to fully understand the nature of the problem.

For instance, linear algebra uses Cramer´s Rule operating with the determinants of various matrices assembled over a system of equations to solve the system of linear equations. A variety of mathematical computer programs can also help us with the adjustment of the relations to find their zero points as shown e.g. in ([4] and ([5]). For example, Fig. 3 shows the behaviour of the value of our polynomial on the left side of the equation (12) depending on the variable x for $\mathbf{a = 3}$, $\mathbf{b = 9}$, $\mathbf{c = 1}$, which leads to $\mathbf{P(x) = x^4 - 92 \cdot x^2 + 108 \cdot x - 89}$ enumerated and drawn by the mathematical program Maple. We seek roots $\boldsymbol{x > 0}$ of $\mathbf{P(x) = 0}$, which occurs at the point (9, 0). Thus, in a room with a length of 9 m and a width of 3 m, our 1 m wide runner is 9 m long.

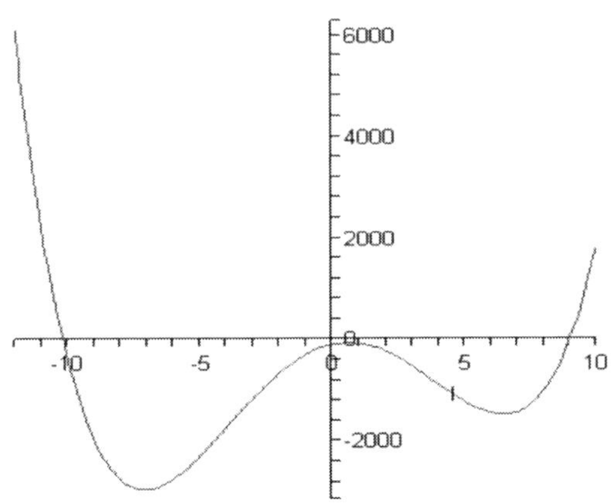

Fig. 3. Behaviour of function $\mathbf{P(x)}$

We now have the correct length of the runner and we should place it in the room. However, this is not a trivial matter. Even if there are two people to place it on the floor, one on each side of the runner and they cooperate to rotate and shift it to the required position, it will not be an easy task, as the 9 m long and 1 m wide runner will bend and twist when being handled. For one person this strategy of "trial and error" appears to be impossible.

Yet, one person can easily handle this problem if he pre-calculates the final value of the variable z or v (see Figure 2). For instance in case of **a** = **3**, **b** = **9**, **c** = **1** and x = **9** it applies z = **0.225** (the value is obtained after substitution **a**, **b**, **c** and x to the expression (9)).

If the person starts to lay the rolled runner, one of the accessible corners of the runner must touch the longer wall of the room at a distance of 22.5 cm from its corner. Then he rotates the rolled runner so that its second accessible corner touches the side wall and unrolls it. If he has counted, measured, and manipulated the runner correctly, the opposite side of it touches the opposite walls at the right points.

2.2 The Experimental and Graphic Solution to the Task

Let us try to find the right solution to this task through other means than calculation. If we do not want to calculate the length of the carpet (either because we simply do not want to or because we do not know how) and if money is no object, we can try to find it experimentally.

It is obvious that the runner cannot be longer than the diagonal of the room. Then we could buy a runner of the diagonal length with the intention of cutting off the excess length after placing it in the room. Such a procedure would not be very wise for two reasons: firstly, the wider the runner the more the correct length will vary from the diagonal; secondly, due to the problem of situating it, which is mentioned above. Only by chance we might be able to "hit" the correct angle of the longitudinal axis with a diagonal of the room so that the runner side, formed by cutting (with its ends touching the walls), would also be perpendicular to the longitudinal axis of the runner (the runner should be a rectangle), see Fig. 1.

The more rational approach to solving the problem of the runner length is provided by a graphical model of the situation. We draw the layout of the room to an appropriate scale and lay the strip width of the runner to the same scale diagonally on it. The points where its edges intersect the room walls are the imaginary touch points of the runner with the walls. By suitable shifting and turning of the strip we make sure that both connecting lines of the touch points are perpendicular to the longitudinal axis of the strip (see Fig. 1). Once this has been accomplished, we draw them on the strip. The mutual distance of the points, which is parallel to the longitudinal axis of the runner on the given scale, shows the sought length of the runner.

Why have we played with the strip of paper in the above case and not tried to draw a rectangular runner directly in a rectangular room? This is because to hit the right length and position of the inner rectangle by direct plotting is much more difficult. And this is the heart of the problem! One thesis of graphic solution to geometric problems says that *what can be easily and accurately drawn, can also be easily calculated* ([6], [7]). Here we demonstrate a case of a graphic solution that is not easy to draw accurately.

2.3 A Solution to the Same Task From a Different Perspective

R. L. Ackoff [1] would therefore advise us to try to look at it from a different point of view, from which the right solution can be drawn accurately and easily, and to utilize this specificity in the solution procedure. B. Siskind in his strategy 5 says [10]: "*Change the causal relation, clarify the desired result, and formulate what you want to achieve.*"

To change the causal relation means not to seek a runner for the room, but a room for the given runner. And this is an easily solvable graphic task. It suffices to lead the base line (that is not parallel to any side of the runner) through the arbitrary top of the runner; then from the neighbouring tops to run perpendiculars to it, and through the opposite top to run a parallel with the base line. The intersections of four lines indicate the position of the corners and sides of the room depending on angle α (see Fig. 4). It is advantageous to express angle α by means of the value of some of its trigonometric function, e.g. $\sin \alpha = y$. From the Pythagorean Theorem for the length of catheti of the right–angled triangles we get the relations attributed to the arrows of the outline in Fig. 4.

Now we know the constants **c** and x. The task is to calculate from these constants and the variable y the width **a** and the length **b** of the room; and this is a simple task. The results are given on the right side of the outline in Figure 4. However, from these results we also deduce what we originally wanted: From any equation stated next to the outline in Fig. 4 we make x independent and by substituting this expression in the other equation we get an equation only for y.

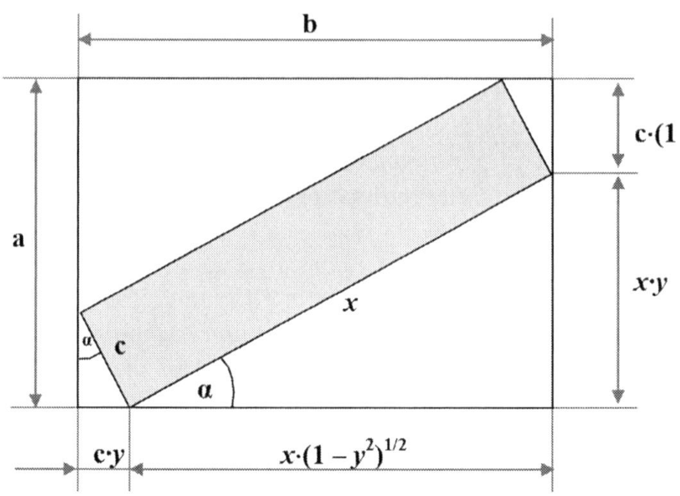

$$c\cdot(1-y^2)^{1/2} + x\cdot y = a$$
$$c\cdot y + x\cdot(1-y^2)^{1/2} = b$$
$$0 \le y < 1$$

Fig. 4. The solution to the problem of finding the right length of the carpet from a different point of view

Here we can see a complete prescription for calculating the length x of the runner from the known values **a**, **b** and **c**:

$$x = (b - c\cdot y) / (1 - y^2)^{1/2}, \tag{13}$$

$$(c\cdot(1 - 2\cdot y^2) + b\cdot y) / (1 - y^2)^{1/2} = a, \quad 0 \le y \le (1/2)^{1/2}. \tag{14}$$

The relation for x (13) is an elementary modification of the equation $\mathbf{c} \cdot y + x \cdot (\mathbf{1} - y^2)^{1/2} = \mathbf{b}$; the condition for y was formed by substituting the right side of (13) for x to the equation $\mathbf{c} \cdot (\mathbf{1} - y^2)^{1/2} + x \cdot y = \mathbf{a}$ and by converting its left side to a common denominator, i.e. (14). The value y can then be obtained by an iteration method.

For $\mathbf{a} = \mathbf{3}$, $\mathbf{b} = \mathbf{9}$ and $\mathbf{c} = \mathbf{1}$ a simple bisection iteration stabilizes after just few steps at the value $y = \mathbf{0.225}$. Substituting this value in the equation (13) we get $x = \mathbf{9}$. As $\mathbf{c} = \mathbf{1}$ then $\mathbf{c} \cdot y = y = \mathbf{0.225}$, the location of the runner in the room is the same as in the standard solution.

Regarding the problem of the correct placement of the runner, it is worth mentioning that while the data necessary for it does not cost anything in this case, in the standard approach we have to find it out additionally from a rather complicated relation. As we can see, the calculation procedure in this latter case is much easier.

3 Final Discussion and Evaluation

Let's ask how the solution procedures of the problem differ from one another? A standard method of calculation directly formulates and calculates what we want to know. It is based on three conditions for the three variables, one major variable x (independent) and two auxiliary variables v and z, mutually bound by the Pythagorean Theorem. From two of these conditions it separates by elimination both auxiliary variables and by substituting them in the binding relation it forms a relation for just the main variable x, which is solved by the iterative method.

We do not consider as wise the direct experimental solution of the problem by cutting off the excess length of the runner. A better solution is based on the use of fundamental analogy, in which the source analogue of a solved situation is a model consisting of the outline of the room and a strip of paper representing the runner. The benefit of this model is the realization that it is possible to move towards the desired outcome by a method of "*trial and error*" and that it can, although with difficulty, be drawn accurately. From there came the motivation to change our perspective on the problem in order to make this difficulty disappear.

We succeeded by reversing the perspective by 180 degrees (using Strategy 5 as described in [10] that inter alia says: "*Look for the unusual approaches. An instinctive and at the first glance obvious approach may not be the best one*"), which resulted in a different point of view on the problem. Within this approach we formulated equations for something other than what we initially wanted. However, from these relations it was easy to deduce it. The resulting effect was a simpler, clearer, and more economical solution to the original problem. Unlike the standard approach, in the new approach the number of variables and conditions dropped from three to two. The technical complexity of derivation was significantly reduced and the whole process became more transparent and rationalized. We can conclude that the portion of "*art*" resulting, in this case, from the reversed perspective on the problem, paid off.

References

[1] ACKOFF, R., L., VERGARA, E. Creativity in problem solving and planning: A review. *European Journal of Operational Research*, 7(1), 1981: 1-13, ISSN 03772217.

[2] BURTON, L. The teaching of mathematics to young children using a problem solving approach. *Educational studies in mathematics*, *11*(1), 1980: s. 43-58.

[3] ERNEST, P. The problem-solving approach to mathematics teaching. *Teaching mathematics and its applications*, *7*(2), 1988: s. 82-92.

[4] HAŠEK, R. Systems of Computer Algebra and Dynamic Geometry as Tools of Mathematical Investigation. *International Journal for Technology in Mathematics Education*, *20*(3), s. 2013: 103-108.

[5] HAŠEK, R., ZAHRADNÍK, J. Study of Historical Geometric Problems by Means of CAS and DGS. *International Journal for Technology in Mathematics Education*, *22*(2), 2015: s. 53-58.

[6] JIANG, Y., TIAN, F., WANG, H., ZHANG, X., WANG, X., DAI, G. Intelligent understanding of handwritten geometry theorem proving. *Proceedings of the 15th International Conference on Intelligent User Interfaces - IUI '10*, 2010: s. 199-128, doi:10.1145/1719970.1719988.

[7] LOURIDAS, S., E., RASSIAS, M. T. *Problem-Solving and Selected Topics in Euclidean Geometry.* The Spirit of the Mathematical Olympiads, 2013: s. 1-235, DOI: 10.1007/978-1-4614-7273-5.

[8] MONAGHAN, J., POOL, P., ROPER, T., THRELFALL, J. Open-start mathematics problems: an approach to assessing problem solving. *Teaching Mathematics and its Applications*, *28*(1), 2009: s. 21-31.

[9] POLYA, G. *How to solve it.* Princeston University Press, 2nd ed., 1957, ISBN 10:0691023565.

[10] SISKIND, B. *Bumblebees Can't Fly: A Practical Guide to Making Everything Work.* Stoddard, Toronto, 2001, ISBN 0773733000.

[11] STEHEL, V., & VOCHOZKA, M. The Analysis of the Economical Value Added in Transport. *NAŠE MORE, Znanstveno-stručni časopis za more i pomorstvo*, *63*(3 Special Issue), 2016: s. 185-188.

[12] TURNER, E. E., DRAKE, C., MCDUFFIE, A. R., AGUIRRE, J., BARTELL, T. G., FOOTE, M. Q. Promoting equity in mathematics teacher preparation: A framework for advancing teacher learning of children's multiple mathematics knowledge bases. *Journal of Mathematics Teacher Education*, *15*(1), 2012: s. 67-82.

[13] VEIDE, Z., STROZHEVA, V. The Visualizations Methods of Geometrical Forms in teaching of Civil Engineering Students. *Environment. Technology. Resources. Proceedings of the International Scientific and Practical Conference*, Vol. 2, 2015: s. 312-316).

[14] ZEIDMANE, A., DUKA, V. Students' Perceptions on Practical Problem Solving in Mathematics in E-environment. *International Journal of Engineering Pedagogy*, *4*(2), 2014: s. 82-87.

Current address

Hašková Simona, Ing. Ph.D.
School of Expertness and Valuation, Institute of Technology and Business in České Budějovice
Okružní 517/10, 37001 České Budějovice, Czech Republic
Tel. number: +420 380 070 220; e-mail: haskovas@post.cz

Slovak University of Technology in Bratislava
Faculty of Mechanical Engineering
16th Conference on Applied Mathematics **APLIMAT 2017**
Proceedings

PLUG-IN PROBLEM IN DISCRETE GNSS-PIM ALGORITHMS

HECKENBERGEROVA Jana (CZ)

Abstract. The aim of the contributed paper is to find a solution of plug-in problem in Discrete GNSS-PIM algorithms. Global Navigation Satellite Systems - Position Integrity Monitoring(GNSS-PIM) algorithms determine train position and verify its safeness. Plug-in problem has to be solved, when covariance matrix of GNSS positions $\mathbf{\Sigma}$ is unknown and we would like to use its estimate instead. Confidence area for variance components σ_1^2, σ_2^2 and insensitivity region for the test in a given alternative are derivated first as a necessary part of the plug-in problem solution. If the confidence area is a subset of insensitivity region, then the estimate of covariance matrix can be considered as precise enough to be use in the Discrete GNSS-PIM algorithm instead of a real covariance matrix.

Keywords: Discrete GNSS-PIM algorithm; estimate of covariance matrix components; confidence area; insensitivity region

Mathematics subject classification: Primary 62F03; Secondary 62F10, 62F25, 62J10

1 Notation

R^k	k-dimensional linear space
$\|\cdot\|$	euclidian norm
$A_{m,n}$	matrix A with $m \times n$ dimensions
A^+	Moore-Penrose pseudoinverse of matrix A
$\mathcal{M}(A)$	column space generated by matrix A
$\mathcal{O}(A)$	orthogonal complement of space $\mathcal{M}(A)$
P_A	projection operator on $\mathcal{M}(A)$
M_A	projection operator on $\mathcal{O}(A)$
$\mathbf{Y}$	observation vector
$E_{\beta,\vartheta}(\mathbf{Y}),\ E(\mathbf{Y}\|\beta,\vartheta)$	mean value of observation vector $\mathbf{Y}$ with parameters β and ϑ
$\Sigma(\vartheta) = \mathrm{var}_{\vartheta}(\mathbf{Y})$	covariance matrix with parameter ϑ
$\mathcal{C}(y)$	confidence area given by realization yof observation vector $\mathbf{Y}$
$\mathbf{Y} \sim N_n(\boldsymbol{\mu}, \boldsymbol{\Sigma})$	n-dimensional observation vector $\mathbf{Y}$ has normal probability distribution with mean value $\boldsymbol{\mu}$ and covariance matrix $\boldsymbol{\Sigma}$
$\chi^2_r(0)$	random variable with central χ^2 distribution with r degrees of freedom
$\chi^2_r(\delta; 1-\alpha)$	$(1-\alpha)$-quantile of random variable with noncentral χ^2 distribution with r degrees of freedom and non-centrality parameter δ
H_0	null hypothesis
H_a	alternative hypothesis
α	significance level
κ_α	power of the test in a given alternative H_a
$\mathcal{T}_{\kappa_\alpha}$	threshold domain for given power of the test κ_α
$\mathcal{N}_{H_a}$	insensitivity region for test in a given alternative

2 Introduction

Modern satellite systems (GPS-USA, GLONASS-Russia, GALILEO-EU) are successfully used in car GPS navigations, air transportation, geodesy, agriculture. It seems advantageous to use satellite navigation (Global Navigation Satellite Systems-GNSS) for railway transportation (eg. view of driven trace of the train, determinate current train position, verify safeness of train position). Reasons of this application are evident - to increase safety, capacity and effectiveness and to reduce running costs of train operation.

The GNSS Position Integrity Monitoring (GNSS-PIM) algorithms monitor the position integrity of the train GNSS receiver [5]. The aim of these algorithms is to verify that GNSS position of the train corresponds to the given train track and to find out that obtained GNSS position complies with safety requirements - Safety Integrity Level (SIL). The GNSS-PIM algorithms could be divided into analytical [4], parametrical

[7], [8] and discrete [9] according to the train track definition. Another division could be made by the track character (linear, arch or cubic). Algorithms development begins from straight line tracks; arch and cubic algorithms are explored afterwards using linearization of gained nonlinear statistical models. If supposed track is discretely defined by polygon of points, then GNSS-PIM algorithms is called discrete. The greatest advantage compared to others GNSS-PIM algorithms is that knowledge of the track character is not demanded.

3 Problem Formulation

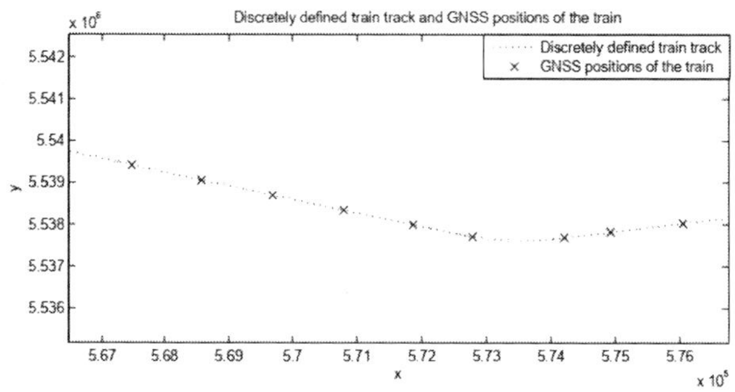

Fig. 1. Discretely defined train track and GNSS positions of the train

Let us suppose that the train track is defined by polygon of discrete, timely equidistant, (x,y)-coordinate points of track axis. Position of train locomotive is monitored with GNSS and it is given also by (x,y)-coordinates as shown in Fig. 1. The aim of GNSS-PIM algorithms is to verify that GNSS train position corresponds to the given train track, and to find out that obtained GNSS position complies with safety requirements - Safety Integrity Level (SIL) - for given probability of missed detection (P_{MD}).

GNSS position integrity monitoring is provided by comparing of obtained Protection Level (PL) with Horizontal Alert Limit (HAL). In statistical point of view, the PL is radius of the threshold domain, which is limited by probability of the second kind error, and HAL is for given SIL already defined by user. If PL value is same or lower than HAL, then GNSS position has enough integrity and it could be regarded as safe. If PL is greater than HAL, the GNSS position does not satisfy safety requirements and must not be used for following safety related applications.

Let $\mathbf{Y}$ be matrix of discretely defined track, and α, $P_{\mathrm{M}D}$, κ_α, HAL be known constants. Let Z_j be the GNSS position of train reference point and let us assume that

$$Z_j \sim N_2(\boldsymbol{\mu}_j, \Sigma),$$

where $\boldsymbol{\mu}_j$ is an unknown first order parameter and Σ is covariance matrix.

Discrete GNSS-PIM algorithms solve these two situations:

- **GNSS-PIM algorithms with known covariance matrix**
  Let be supposed that covariance matrix Σ is known. The goal of GNSS-PIM is to verify safeness of GNSS position Z_j. If null hypotheses $H_0 : E(Z_j) \in \mathbf{Y}$ is refused, than GNSS position is not with given probability on the train track. In this case red light is shining on signal semaphore. When H_0 can not be refused, Protection Level is evaluated and moreover if P$L <$ HAL, than GNSS position is safety and the green light occurs on semaphore. Otherwise when P$L \geq$ HAL, algorithm is not able to verify safeness of the GNSS train position and emergency sign (orange) is shown on the signal lights. For more detailed description see [9, 10].

- **GNSS-PIM algorithms with an unknown covariance matrix**
  Let be supposed that covariance matrix Σ is unknown. The aim of algorithm is evaluation of unbiased covariance matrix estimate $\widehat{\Sigma}$. Afterwards it is necessary to find out, whether $\widehat{\Sigma}$ could be used instead of the real covariance matrix, therefore so called Plug-in Problem has to be solved. When this statement is confirmed GNSS-PIM algorithms with known covariance matrix is utilized to verify GNSS position safeness.

4 Discrete GNSS-PIM algorithms with unknown covariance matrix

Let us assume that covariance matrix Σ is diagonal

$$\Sigma = \begin{pmatrix} \sigma_1^2, & 0 \\ 0, & \sigma_2^2 \end{pmatrix} = \sigma_1^2 \begin{pmatrix} 1, & 0 \\ 0, & 0 \end{pmatrix} + \sigma_2^2 \begin{pmatrix} 0, & 0 \\ 0, & 1 \end{pmatrix} = \sigma_1^2 \mathbf{V}_1 + \sigma_2^2 \mathbf{V}_2,$$

where σ_1^2, σ_2^2 are unknown variance components and $\mathbf{V}_1 = \begin{pmatrix} 1, & 0 \\ 0, & 0 \end{pmatrix}$,

$\mathbf{V}_2 = \begin{pmatrix} 0, & 0 \\ 0, & 1 \end{pmatrix}$ are known positive semi-definite matrices.

This structure of covariance matrix Σ follows physical characteristics of GNSS measurements, where position error is concentrated on North-South and West-East directions (axes directions) [3], [11].

4.1 Estimate of covariance matrix with unknown variance components

Derivation of variance component estimates $\widehat{\sigma}_1^2$, $\widehat{\sigma}_2^2$ is based on a comparison of projection residual vectors that are given by movement on two straight train tracks with different deflections from the x axis. It is necessary to use unitary direction of train tracks, which can be easily described with track deflection α. Geometrical representation of all used terms is shown in Fig. 2.

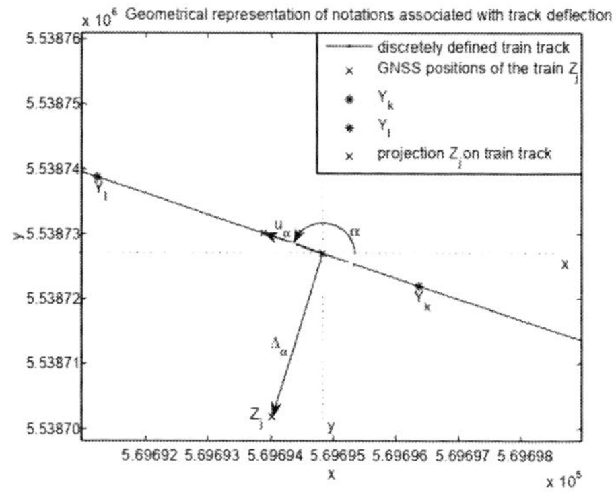

Fig. 2. Geometrical representation of unitary track direction u_α, track deflection α and projection residual vector $\mathbf{\Delta}_\alpha$

Theorem 4.1
Let $\mathbf{Y}$ be matrix of discretely defined train track, let Z_j be GNSS position of reference point and let k, l be indexes of two nearest points of discretely defined track $\mathbf{Y}$.
Let be supposed that

$$Z_j \sim N_2(\boldsymbol{\mu}_j, \Sigma),$$

where parameters $\boldsymbol{\mu}_j$ and Σ are unknown. Let

$$\vec{u} = \frac{\mathbf{Y}_l - \mathbf{Y}_k}{\sqrt{(\mathbf{Y}_l - \mathbf{Y}_k)'(\mathbf{Y}_l - \mathbf{Y}_k)}}.$$

be unitary direction of track axis,
then

$$\widetilde{Z}_j = P_u Z_j + M_u \mathbf{Y}_k \sim N_2(P_u \boldsymbol{\mu}_j + M_u \mathbf{Y}_k, P_u \Sigma P_u)$$

and

$$\mathbf{\Delta}_j = M_u(Z_j - \mathbf{Y}_k) \sim N_2(M_u(\boldsymbol{\mu}_j - \mathbf{Y}_k), M_u \Sigma M_u).$$

Let covariance matrix be diagonal $\Sigma = \begin{pmatrix} \sigma_1^2, & 0 \\ 0, & \sigma_2^2 \end{pmatrix}$, *where variance components* σ_1^2, σ_2^2 *are unknown and let GNSS position* Z_j *corresponds to train track* $\mathbf{Y}$, *so* $E(Z_j) \in \mathbf{Y}$,
then

$$\boldsymbol{\Delta}_j \sim N_2(0, \sigma_1^2 \mathbf{V}_{u1} + \sigma_2^2 \mathbf{V}_{u2}),$$

where $\mathbf{V}_{u1}, \mathbf{V}_{u2}$ *are known positive semi-definite matrices*

$$\begin{aligned} \mathbf{V}_{u1} &= \{M_u\}_{\cdot 1}\{M_u\}_{1\cdot} \\ \mathbf{V}_{u2} &= \{M_u\}_{\cdot 2}\{M_u\}_{2\cdot} \end{aligned}$$

and $\{M_u\}_{i\cdot}$ *is i-th row (resp.* $\{M_u\}_{\cdot j}$ *is j-th columb) of projection matrix* M_u.

Let us suppose that track direction u_α is given by deflection α from x axis

$$u_\alpha = \begin{pmatrix} \cos\alpha \\ \sin\alpha \end{pmatrix},$$

then projection of GNSS train position on the train track is defined by

$$\widetilde{Z}_\alpha = P_{u_\alpha} Z_j + M_{u_\alpha} \mathbf{Y}_k,$$

projection residual vector

$$\boldsymbol{\Delta}_\alpha = M_{u_\alpha}(Z_j - \mathbf{Y}_k)$$

and

$$\Delta_\alpha = \|\boldsymbol{\Delta}_\alpha\|$$

is magnitude of projection residual vector $\boldsymbol{\Delta}_\alpha$. Supposing that presumptions of previous theorem are valid, magnitude Δ_α of projection residual vector $\boldsymbol{\Delta}_\alpha$ has normal distribution

$$\Delta_\alpha \sim N(0, \sigma_1^2 \sin^2\alpha + \sigma_2^2 \cos^2\alpha).$$

Theorem 4.2 ***Unbiased estimator of covariance matrix*** Σ
Let $\mathbf{Y}_1, \mathbf{Y}_2$ *be discretely defined straight train tracks with deflections* α_1, α_2 *and let* $Z_{\alpha_1,1}, \ldots, Z_{\alpha_1,m_1}, Z_{\alpha_2,1}, \ldots, Z_{\alpha_2,m_2}$ *be measured GNSS train positions on these tracks, that*

$$\begin{aligned} &\forall i = 1, \ldots, m_1 \quad Z_{\alpha_1,i} \sim N_2\left(\boldsymbol{\mu}_{\alpha_1,i}, \Sigma\right) \quad \textit{and} \quad E(Z_{\alpha_1,i}) \in \mathbf{Y}_1 \\ &\forall j = 1, \ldots, m_2 \quad Z_{\alpha_2,j} \sim N_2\left(\boldsymbol{\mu}_{\alpha_2,j}, \Sigma\right) \quad \textit{and} \quad E(Z_{\alpha_2,j}) \in \mathbf{Y}_2, \end{aligned}$$

where means are unknown and covariance matrix $\Sigma = \begin{pmatrix} \sigma_1^2, & 0 \\ 0, & \sigma_2^2 \end{pmatrix}$ *is diagonal with unknown variance components* σ_1^2, σ_2^2.
Let $\sin(\alpha_1 + \alpha_2) \cdot \sin(\alpha_1 - \alpha_2) \neq 0$, *then unbiased point estimators of variance components* $\widehat{\sigma}_1^2, \widehat{\sigma}_2^2$ *are in form*

$$\widehat{\sigma}_1^2 = \frac{\frac{1}{m_1}\sum_{j=1}^{m_1}\Delta_{\alpha_1,j}^2\cos^2\alpha_2 - \frac{1}{m_2}\sum_{j=1}^{m_2}\Delta_{\alpha_2,j}^2\cos^2\alpha_1}{\sin(\alpha_1+\alpha_2)\sin(\alpha_1-\alpha_2)},$$

$$\widehat{\sigma}_2^2 = \frac{\frac{1}{m_2}\sum_{j=1}^{m_2}\Delta_{\alpha_2,j}^2\sin^2\alpha_1 - \frac{1}{m_1}\sum_{j=1}^{m_1}\Delta_{\alpha_1,j}^2\sin^2\alpha_2}{\sin(\alpha_1+\alpha_2)\sin(\alpha_1-\alpha_2)}.$$

Proof: Unbiased estimators of variance components are solution of equation system

$$S_{m_1}^2 = \frac{1}{m_1}\sum_{j=1}^{m_1}\Delta_{\alpha_1,j}^2 = \widehat{\sigma}_1^2\sin^2\alpha_1 + \widehat{\sigma}_2^2\cos^2\alpha_1$$

$$S_{m_2}^2 = \frac{1}{m_2}\sum_{j=1}^{m_2}\Delta_{\alpha_2,j}^2 = \widehat{\sigma}_1^2\sin^2\alpha_2 + \widehat{\sigma}_2^2\cos^2\alpha_2.$$

□

4.2 Plug–in problem

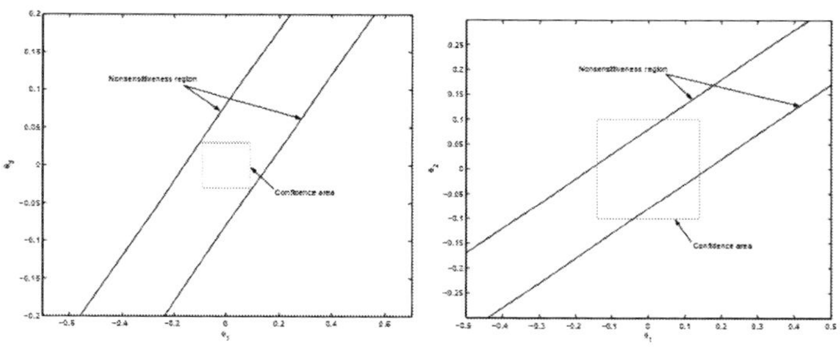

Fig. 3. Mutual positions of confidence area and insensitivity region

Plug-in problem has to be solved, when real covariance matrix of GNSS position Σ is unknown and we would like to use its estimate $\widehat{\Sigma}$ instead. If confidence area of variance components σ_1^2, σ_2^2 is a subset of insensitivity region for test of GNSS position integrity in a given alternative as shown on Fig. 3 (left), then estimation of covariance matrix $\widehat{\Sigma}$ can be considered precise enough to be used in Discrete GNSS-PIM algorithm instead of an unknown real covariance matrix Σ. Otherwise as in Fig. 3 (right) replacement of covariance matrix Σ can caused unexpected errors in results of GNSS-PIM algorithm.

Insensitivity region for variance components is area, where components has minimal impact on given feature. Deeper description of insensitivity region in general linear models can be found in [1, 2, 6].

Theorem 4.3 ***Confidence area for variance components $\boldsymbol{\sigma}_1^2, \boldsymbol{\sigma}_2^2$***
Let all presumptions of theorem 4.2 be satisfied.
$(1-2\alpha)$*-confidence area for variance components $\boldsymbol{\sigma}_1^2, \boldsymbol{\sigma}_2^2$ is given by*

$$\mathcal{C}_{\sigma_1^2,\sigma_2^2} = \left\{ u = \begin{pmatrix} u_1 \\ u_2 \end{pmatrix} \in R^2 : d_{\sigma_1^2} \le u_1 \le h_{\sigma_1^2}, d_{\sigma_2^2} \le u_2 \le h_{\sigma_2^2} \right\}$$

where α is significance level,

$$d_{\sigma_1^2} = \frac{a_{\alpha_1}\cos^2\alpha_2 - b_{\alpha_2}\cos^2\alpha_1}{\sin(\alpha_1+\alpha_2)\sin(\alpha_1-\alpha_2)}, \quad h_{\sigma_1^2} = \frac{b_{\alpha_1}\cos^2\alpha_2 - a_{\alpha_2}\cos^2\alpha_1}{\sin(\alpha_1+\alpha_2)\sin(\alpha_1-\alpha_2)},$$
$$d_{\sigma_2^2} = \frac{a_{\alpha_2}\sin^2\alpha_1 - b_{\alpha_1}\sin^2\alpha_2}{\sin(\alpha_1+\alpha_2)\sin(\alpha_1-\alpha_2)}, \quad h_{\sigma_2^2} = \frac{b_{\alpha_2}\sin^2\alpha_1 - a_{\alpha_1}\sin^2\alpha_2}{\sin(\alpha_1+\alpha_2)\sin(\alpha_1-\alpha_2)}$$

and

$$a_{\alpha_1} = \frac{\sum_{j=1}^{m_1}\Delta_{\alpha_1,j}^2}{\chi_{m_1}^2(0,1-\frac{\alpha}{2})}, \qquad b_{\alpha_1} = \frac{\sum_{j=1}^{m_1}\Delta_{\alpha_1,j}^2}{\chi_{m_1}^2(0,\frac{\alpha}{2})},$$
$$a_{\alpha_2} = \frac{\sum_{j=1}^{m_2}\Delta_{\alpha_2,j}^2}{\chi_{m_2}^2(0,1-\frac{\alpha}{2})}, \qquad b_{\alpha_2} = \frac{\sum_{j=1}^{m_2}\Delta_{\alpha_2,j}^2}{\chi_{m_2}^2(0,\frac{\alpha}{2})}.$$

Theorem 4.4 ***Insensitivity region for the test in given alternative***
Let $\mathbf{Y}$ be matrix of discretely defined straight train track and let GNSS position of train $Z_j \sim N_2(\boldsymbol{\mu}_j, \boldsymbol{\Sigma})$, where mean value $\boldsymbol{\mu}_j$ is unknown and $\boldsymbol{\Sigma}$ is diagonal covariance matrix with unknown variance components $\boldsymbol{\sigma}_1^2, \boldsymbol{\sigma}_2^2$.
Let k, l be indexes of two nearest track points and

$$\boldsymbol{\Delta}_j \sim N_2(\boldsymbol{\nu}, M_{\mathbf{Y}_l-\mathbf{Y}_k}\boldsymbol{\Sigma}M_{\mathbf{Y}_l-\mathbf{Y}_k})$$

be projection residual vector belonging to position Z_j.
Let be supposed that

$$\boldsymbol{\vartheta}_0 = \begin{pmatrix} \vartheta_1 \\ \vartheta_2 \end{pmatrix} = \begin{pmatrix} \widehat{\boldsymbol{\sigma}}_1^2 \\ \widehat{\boldsymbol{\sigma}}_2^2 \end{pmatrix}$$

is unbiased estimator of variance components and

$$\boldsymbol{\Sigma}(\boldsymbol{\vartheta}_0) = \begin{pmatrix} \widehat{\boldsymbol{\sigma}}_1^2, & 0 \\ 0, & \widehat{\boldsymbol{\sigma}}_2^2 \end{pmatrix} = \widehat{\boldsymbol{\sigma}}_1^2\mathbf{V}_1 + \widehat{\boldsymbol{\sigma}}_2^2\mathbf{V}_2$$

is unbiased estimator of variance matrix $\boldsymbol{\Sigma}$.

Let the null hypothesis

$$H_0 : \; E(Z_j) \in \mathbf{Y}$$

be tested against alternative

$$H_a : \; E(Z_j) \notin \mathbf{Y},$$

then insensitivity region of variance components for test in given alternative $\mathcal{N}_{H_a}$ is in form

$$\begin{aligned}\mathcal{N}_{H_a} &= \Big\{\boldsymbol{\vartheta}_0 + \boldsymbol{\delta\vartheta} \in R^2 : \\ &(\boldsymbol{\delta\vartheta} + \delta_{\min}(t^2A - aa')^+a)'\,(t^2A - aa')\,(\boldsymbol{\delta\vartheta} + \delta_{\min}(t^2A - aa')^+a) \leq \\ &\leq \delta_{\min}\,(1 + a'(t^2A - aa')^+a)\Big\},\end{aligned}$$

where

$$\{a\}_i = Tr[\mathbf{V}_i U^+] + \boldsymbol{\nu}'(M_{\mathbf{Y}_l-\mathbf{Y}_k}\boldsymbol{\Sigma}(\boldsymbol{\vartheta}_0)M_{\mathbf{Y}_l-\mathbf{Y}_k})^+\mathbf{V}_i U^+\boldsymbol{\nu},$$

$$\{A\}_{i,j} = 2Tr[\mathbf{V}_i(M_{\mathbf{Y}_l-\mathbf{Y}_k}\boldsymbol{\Sigma}(\boldsymbol{\vartheta}_0)M_{\mathbf{Y}_l-\mathbf{Y}_k})^+\mathbf{V}_j U^+] + 4\boldsymbol{\nu}'U^+\mathbf{V}_i U^+\mathbf{V}_j U^+\boldsymbol{\nu},$$

and $U = (M_{\mathbf{Y}_l-\mathbf{Y}_k}\boldsymbol{\Sigma}(\boldsymbol{\vartheta}_0)M_{\mathbf{Y}_l-\mathbf{Y}_k})$,

$t \in \langle 3; 5\rangle$ *is selected parameter,*
$\delta_{\min}$ *is solution to equation*

$$P\{\chi_1^2(\delta_{\max}) - \delta_{\min} \geq \chi_1^2(0, 1-\alpha)\} = 1 - P_{MD} - \epsilon = \kappa_\alpha - \epsilon,$$

where α *is chosen significance level,*
P_{MD} *is given probability of undetectable error,*
$\kappa_\alpha = 1 - P_{MD}$ *is corresponding power of the test,*
ϵ *is given acceptable fall of* κ_α *and*
$\delta_{\max}$ *is parameter of noncentrality satisfying*

$$P\{\chi_1^2(\delta_{\max}) \geq \chi_1^2(0, 1-\alpha)\} = 1 - P_{MD} = \kappa_\alpha.$$

Proof: Insensitivity region for variance components for the test in given alternative $\mathcal{N}_{H_a}$ is defined by

$$\mathcal{N}_{H_a} = \left\{\boldsymbol{\vartheta}_0 + \boldsymbol{\delta\vartheta} \in R^2 : P\left(T(\boldsymbol{\vartheta}_0 + \boldsymbol{\delta\vartheta}) \geq \chi_1^2(0, 1-\alpha)\right) \geq 1 - P_{MD} - \epsilon\right\}.$$

Test statistics $T(\boldsymbol{\vartheta}_0 + \boldsymbol{\delta\vartheta})$ is linearized by the help of Taylor expansion. Then approximation with mean value and standard deviation is applied and after few statistical operations insensitivity region $\mathcal{N}_{H_a}$ is in form

$$\mathcal{N}_{H_a} = \left\{\boldsymbol{\vartheta}_0 + \boldsymbol{\delta\vartheta} \in R^2; \boldsymbol{\delta\vartheta}'\left(t^2A - aa'\right)\boldsymbol{\delta\vartheta} + 2\delta_{\min}a'\boldsymbol{\delta\vartheta} \leq \delta_{\min}^2\right\}.$$

Last step of this proof is to rewrite $\mathcal{N}_{H_a}$ as quadratics form. □

5 Conclusion

Global Navigation Satellite Systems have a lot of advantages and their usability in railway transportation seems to be reasonable. It is also necessary to develop algorithms for train position determination and safeness verification. This is the aim of GNSS Position Integrity Monitoring (GNSS-PIM) algorithms. Unfortunately variances of the actual errors in determination of train position are usually unknown and must be estimated. This fact influences heavily the procedure of confidence domain determination and power functions of statistical tests. Plug-in problem has to be solved, when covariance matrix of GNSS positions is unknown and we would like to use its estimation instead. Unbiased covariance matrix estimation, confidence area and insensitivity region are derived in this paper as most important parts of Plug-in problem solution.

Utilization of other inertial sensors and measure techniques (e.g. odometers, gyroscopes and Doppler radar), make the set of results heterogeneous. To solve this development of other special numerical algorithms will be needed. Only satisfactory solution of mentioned problems gives chances for utilization of GNSS-PIM algorithms in the practice. Afterwards they are able to increase effectiveness, safety and capacity of train operation.

6 Acknowledgement

This research has been supported by the institutional funds of the University of Pardubice, Czech Republic.

References

[1] BOHÁČOVÁ, H. Insensitivity region for variance components in general linear model. *Acta Universitas Palackianae, Fac. rer. nat., Mathematica* (47) 2008, pg. 7–22, ISBN: 978-80-244-1819-3.

[2] BOHÁČOVÁ, H., HECKENBERGEROVÁ, J. Properties of the Insensitivity Region for a Linear Function of the Fixed Effects Parameters in the Mixed Linear Model. *Forum Statisticum Slovacum* 7/2008, pg. 7-12, Slovenská štatistická a demografická spoločnosť, Bratislava 2008, ISSN: 1336-7420.

[3] BOORE, K., AKOS, D.M., BERTELSEN, N., RINDER, P., JENSEN, S.H. A Software-Defined GPS and GALILEO Receiver, A Single-Frequency Approach. *Applied and Numerical Harmonic Analysis* (Series Editor John J. Benedetto), Birkhäuser Boston-Basel-Berlin 2007, ISBN: 978-0-8176-4390-4.

[4] DVOŘAKOVÁ, J. *Statistické metody pro identifikaci polohy vlaku.* KMAaAM Faculty of Science, UP Olomouc, 2004. (master thesis in Czech)

[5] DVOŘAKOVÁ, J., MOCEK, H., MAIXNER, V. Statistical Approach to the Train position Integrity Monitoring. *2nd International Conference Proceedings "Reliability, safety and diagnostics of transport structures and means 2005"*, pg. 79 - 86, Univerzita Pardubice, Pardubice 2005, ISBN 80-7194-769-5.

[6] FIŠEROVÁ, E., KUBÁČEK, L., KUNDEROVÁ, P. *Linear Statistical Models: Regularity and Singularities*. Academia, Praha 2007, ISBN 978-80-200-1562-4.

[7] HECKENBERGEROVÁ, J. Parametrické algoritmy pro ověření GNSS polohy vlaku. *Infotrans 2007 – conference proceeding*, pg. 103-108, Univerzita Pardubice, Pardubice 2007, ISBN: 978-80-7194-989-3.(in Czech)

[8] HECKENBERGEROVÁ, J. GNSS train position verification by the help of parametrical PIM algorithms. *Journal of Applied Mathematics*, pg. 397-408, Slovak Univerzity of Technology in Bratislava, Bratislava 2008, ISSN: 1337-6365.

[9] HECKENBERGEROVÁ, J., BOHÁČOVÁ, H. Introduction to the Discrete GNSS Position Integrity Monitoring Algorithms. *Forum Statisticum Slovacum* 7/2008, pg. 19-24, Slovenská štatistická a demografická spoločnosť, Bratislava 2008, ISSN: 1336-7420.

[10] HECKENBERGEROVÁ, J. *Statistické problémy metod pro určovaní polohy - spolehlivost a úroveň integrity*. KMAaAM Faculty of Science, Palacký University in Olomouc 2009.(doctoral thesis in Czech)

[11] MERVART, L. *Základy GPS*. Vydavatelství ČVUT, Praha 1993.(in Czech)

Current address

Heckenbergerova Jana, Mgr. Ph.D.
University of Pardubice
Studentska 95, 532 10 Pardubice, Czech Republic
E-mail: Jana.Heckenbergerova@upce.cz

Slovak University of Technology in Bratislava
Faculty of Mechanical Engineering
16th Conference on Applied Mathematics **APLIMAT 2017**
Proceedings

ON 4-PLANAR MAPPINGS ONTO ALMOST QUATERNIONIC HERMITIAN SPACES

HINTERLEITNER Irena (CZ)

Abstract. In this paper we study 4-planar mapping of almost quaternionic spaces. This mapping generalize geodesic, quasigeodesic and holomorphically projective mappings of Riemannian and Kählerian spaces. We obtained the new fundamental equations of 4-planar mappings onto almost quaternionic Hermitian spaces.

Keywords: 4-planar mapping, almost quaternionic Hermitian space, almost quaternionic structure

Mathematics subject classification: Primary 53B35; Secondary 53B05, 53B30, 53B35

1 Introduction

4-planar and 4-quasiplanar mappings of almost quaternionic spaces have been studied in [13], [21] and [19]. These mappings generalize geodesic, quasigeodesic and holomorphically projective mappings of Riemannian and Kählerian spaces, see [2], [3], [5], [6], [7], [12], [15], [16], [17], [18], [22], [23], [24], [25], [26], [27], [28], [32], [33].

Almost quaternionic structures were studied by many authors for example [1], [11], [20]. Generalizations of the above introduced mappings were studied in [8], [9], [10], [29], [31], [30].

4-planar mappings of almost quaternionic manifolds were studied by Kurbatova, see [13, 14]. Related problems were studied in the paper [4]. In the above papers the conditions for "preservation" of covariant derivatives of quaternionic structures were supposed.

We obtained the new fundamental equations of 4-planar mappings onto almost quaternionic Hermitian spaces.

2 Almost quaternionic and quaternionic Hermitian and Kähler manifolds

Under an *almost quaternionic* space we understand a differentiable manifold M_n with the almost complex structures $\overset{1}{F}$ and $\overset{2}{F}$ defined on it, satisfying

$$\overset{1}{F} \cdot \overset{1}{F} = -I; \quad \overset{2}{F} \cdot \overset{2}{F} = -I; \quad \overset{1}{F} \cdot \overset{2}{F} + \overset{2}{F} \cdot \overset{1}{F} = 0. \tag{1}$$

In the local coordinate system x the conditions (1) have the following form

$$\overset{1}{F}{}^h_\alpha \overset{1}{F}{}^\alpha_i = -\delta^h_i; \quad \overset{2}{F}{}^h_\alpha \overset{2}{F}{}^\alpha_i = -\delta^h_i; \quad \overset{1}{F}{}^h_\alpha \overset{2}{F}{}^\alpha_i + \overset{2}{F}{}^h_\alpha \overset{1}{F}{}^\alpha_i = 0,$$

where δ^h_i is the Kronecker symbol, see e.g. [1], [2].

The tensor $\overset{3}{F}{}^h_i \equiv \overset{1}{F}{}^\alpha_i \overset{2}{F}{}^h_\alpha$ is a further almost complex structure. The relations among the tensors $\overset{1}{F}$, $\overset{2}{F}$ and $\overset{3}{F}$ are the following

$$\overset{1}{F}{}^h_i = \overset{2}{F}{}^\alpha_i \overset{3}{F}{}^h_\alpha = -\overset{3}{F}{}^\alpha_i \overset{2}{F}{}^h_\alpha; \quad \overset{2}{F}{}^h_i = \overset{3}{F}{}^\alpha_i \overset{1}{F}{}^h_\alpha = -\overset{1}{F}{}^\alpha_i \overset{3}{F}{}^h_\alpha; \quad \overset{3}{F}{}^h_i = \overset{1}{F}{}^\alpha_i \overset{2}{F}{}^h_\alpha = -\overset{2}{F}{}^\alpha_i \overset{1}{F}{}^h_\alpha. \tag{2}$$

Each pair chosen from the three structures $\overset{1}{F}$, $\overset{2}{F}$ and $\overset{3}{F}$ determines an *almost quaternionic structure*. The tensors ${}^*\overset{1}{F}, {}^*\overset{2}{F}, {}^*\overset{3}{F}$ and $\overset{1}{F}, \overset{2}{F}, \overset{3}{F}$ define the same almost quaternionic structure if ${}^*\overset{\sigma}{F} = \sum_{\rho=1}^{3} \alpha_\rho \overset{\rho}{F}$ where α_ρ are some functions.

An almost quaternionic manifold A_n is called a *quaternionic Hermitian manifold*, if there exists a metric g such that $(g, \overset{s}{F})$, $s = 1, 2, 3$, are Hermitian spaces, so that

$$g(X, \overset{s}{F}X) = 0$$

for any $X \in TA_n$. In the following ∇ is an affine connection on A_n.

Moreover, if $\nabla \overset{s}{F} = 0$, then A_n is called a *quaternionic Kähler manifold.*

Let $A_n(\nabla, \overset{1}{F}, \overset{2}{F}, \overset{3}{F})$ be an almost quaternionic Hermitian space with affine connection ∇ without torsion and with almost quaternionic structures $(\overset{1}{F}, \overset{2}{F}, \overset{3}{F})$.

Definition 1 *A curve ℓ in A_n which is given by the equation $\ell = \ell(t)$, $\lambda = d\ell/dt$ $(\neq 0)$, $t \in I$, where t is a parameter is called* 4-planar, *if under the parallel translation along the curve, the tangent vector λ belongs to the four-dimensional distribution*

$$D = Span\{\lambda, \overset{1}{F}\lambda, \overset{2}{F}\lambda, \overset{3}{F}\lambda\},$$

that is, it satisfies

$$\nabla_t \lambda = a(t)\lambda + b(t)\overset{1}{F}\lambda + c(t)\overset{2}{F}\lambda + d(t)\overset{3}{F}\lambda,$$

where $a(t)$, $b(t)$, $c(t)$ and $d(t)$ are some functions of the parameter t.

Particularly, in the case $b(t) = c(t) = d(t) = 0$, a 4-planar curve is a geodesic. Evidently, a 4-planar curve with respect to the structure $(\overset{1}{F}, \overset{2}{F}, \overset{3}{F})$ is 4-planar with respect to the structure $({}^*\overset{1}{F}, {}^*\overset{2}{F}, {}^*\overset{3}{F})$, too.

3 4-planar mappings

Consider two almost quaternionic manifolds with affine connections without torsion A_n and $\bar{A}_n$ with connections ∇ and $\bar{\nabla}$, respectively. Let an almost quaternionic structure $(\overset{1}{F}, \overset{2}{F}, \overset{3}{F})$ be defined on A_n.

Definition 2 *A diffeomorphism f: $A_n \to \bar{A}_n$ is called a* 4-planar mapping, *if it maps any 4-planar curve in A_n onto a 4-planar curve in $\bar{A}_n$.*

Assume a 4-planar mapping f: $A_n \to \bar{A}_n$. Since f is a diffeomorphism, we can suppose local coordinate charts on M or $\bar{M}$, respectively, such that locally f: $A_n \to \bar{A}_n$ maps points onto points with the same coordinates, and $\bar{M} = M$.

A manifold A_n admits a 4-planar mapping onto $\bar{A}_n$ if and only if the following equations [19]:

$$\bar{\nabla}_X Y = \nabla_X Y + \sum_{s=0}^{3} \left\{ \underset{s}{\psi}(X)\overset{s}{F}Y + \underset{s}{\psi}(Y)\overset{s}{F}X \right\} \tag{3}$$

hold for any tangent fields X, Y and where $\underset{s}{\psi}$ are differential forms; $\overset{0}{F} = \mathrm{Id}$.

If $\underset{s}{\psi} \equiv 0$ $(s = 0, 1, 2, 3)$ then f is *affine*.

Beside these facts it was proved [19] that the quaternionic structure of A_n and $\bar{A}_n$ is preserved; for this reason we can suppose that $\overset{s}{\bar{F}} = \overset{s}{F}$. This was a priori assumed in the definition and results by Kurbatova [13, 14].

Equation (3) in the common coordinate system x with respect to the mapping, has the following form

$$\bar{\Gamma}^h_{ij}(x) = \Gamma^h_{ij}(x) + \sum_{s=0}^{3} \underset{s}{\psi}_{(i}\, \overset{s}{F}{}^h_{j)} \tag{4}$$

where Γ^h_{ij} and $\bar{\Gamma}^h_{ij}$ are components of ∇ and $\bar{\nabla}$, $\underset{s}{\psi}_i(x)$ are components of $\underset{s}{\psi}$, $(i\,j)$ denotes a symmetrization without division by 2.

Finally we will suppose that the space $A_n(\nabla, \overset{1}{F}, \overset{2}{F}, \overset{3}{F})$ is mapped onto the (pseudo-) Riemannian space $\bar{V}_n(\bar{g})$. A mapping f: $A_n \to \bar{V}_n$ is 4-planar if and only if the metric tensor $\bar{g}_{ij}(x)$ satisfies the following equations:

$$\bar{g}_{ij,k} = \sum_{s=0}^{3} \left(\underset{s}{\psi}_k\, \bar{g}_{\alpha(i}\, \overset{s}{F}{}^{\alpha}_{j)} + \underset{s}{\psi}_{(i}\, \bar{g}_{j)\alpha}\, \overset{s}{F}{}^{\alpha}_{k} \right) \tag{5}$$

where the comma is the covariant derivative in A_n (see [19]).

4 4-planar mapping theory for $A_n \to \bar{H}_n$

Assume the quaternionic manifold $A_n = (M, \nabla, F)$ with affine connection ∇ and structures $F = (\overset{1}{F}, \overset{2}{F}, \overset{3}{F})$; and the quaternionic Hermitian manifold $\bar{H}_n = (\bar{M}, \bar{g}, \bar{F})$ with metrics $\bar{g}$ and structures $\bar{F} = (\overset{1}{\bar{F}}, \overset{2}{\bar{F}}, \overset{3}{\bar{F}})$, and the Levi-Civita connection $\bar{\nabla}$.

We further assume that A_n admits a 4-planar mapping onto $\bar{H}_n$. Then we can consider $\bar{M} = M$ and $\overset{s}{\bar{F}} = \overset{s}{F}$ for $s = 1, 2, 3$. These assumptions are correlated with results in [21]. Because $\bar{H}_n$ is an almost quaternionic Hermitian space, we have

$$\bar{g}(X, \overset{s}{F}X) = 0, \quad s = 1, 2, 3. \tag{6}$$

From that we can simplify formula (5) as follows:

$$\bar{g}_{ij,k} = 2\psi_k\, \bar{g}_{ij} + \sum_{s=0}^{3} \left(\underset{s}{\psi}_i\, \bar{g}_{j\alpha}\, \overset{s}{F}{}^{\alpha}_{k} + \underset{s}{\psi}_j\, \bar{g}_{i\alpha}\, \overset{s}{F}{}^{\alpha}_{k} \right). \tag{7}$$

Here and in the following $\psi_k \equiv \underset{0}{\psi}_k$.

Formula (7) is a system of partial equations in covariant derivatives with unknown functions $\bar{g}_{ij}(x)$ and $\underset{s}{\psi}_j(x)$, $s = 0, 1, 2, 3$. This means that the number of unknown functions in the system (7) is

$$\frac{1}{2}\, n \cdot (n+1) + 4\, n.$$

Now we prove that the number of unknown functions will be less then $3\, n$.

We proved the following theorem

Theorem 1 *Let a quaternionic space A_n admit a 4-planar mapping onto an almost quaternionic Hermitian space $\bar{H}_n$. Then in the fundamental equations* (7) *the functions $\underset{s}{\psi}_i(x)$, $s = 1, 2, 3$, are expressed from the unknown functions $\bar{g}_{ij}(x)$ and $\psi_i(x)$ $(\equiv \underset{0}{\psi}_i(x))$.*

Proof. Let $A_n(\nabla, \overset{1}{F}, \overset{2}{F}, \overset{3}{F})$ admit a 4-planar mapping onto $\bar{H}_n(\bar{g}, \overset{1}{F}, \overset{2}{F}, \overset{3}{F})$, therefore equations (7) holds.

Formulas (6) in local coordinates have the following form

$$\bar{g}_{i\alpha} \overset{s}{F}{}^{\alpha}_{k} + \bar{g}_{k\alpha} \overset{s}{F}{}^{\alpha}_{i} = 0, \quad s = 1, 2, 3. \tag{8}$$

After contracting with F^k_j we get

$$\bar{g}_{\alpha\beta} \overset{s}{F}{}^{\alpha}_{i} \overset{s}{F}{}^{\beta}_{j} = \bar{g}_{ij}, \quad s = 1, 2, 3. \tag{9}$$

By differentiating formulas (9) along x^k we obtain

$$\bar{g}_{\alpha\beta,k} \overset{s}{F}{}^{\alpha}_{i} \overset{s}{F}{}^{\beta}_{j} = \bar{g}_{ij,k} - \bar{g}_{\alpha\beta} (\overset{s}{F}{}^{\alpha}_{i} \overset{s}{F}{}^{\beta}_{j})_{,k}, \quad s = 1, 2, 3.$$

Now we exclude $\bar{g}_{ij,k}$ from the last formula with help of equations (7)

$$\begin{aligned} &\sum_{s=0}^{3} \left(\underset{s}{\psi}_i\, \bar{g}_{j\alpha}\, \overset{s}{F}{}^{\alpha}_{k} + \underset{s}{\psi}_j\, \bar{g}_{i\alpha}\, \overset{s}{F}{}^{\alpha}_{k} \right) = \bar{g}_{\alpha\beta} (\overset{s}{F}{}^{\alpha}_{i} \overset{s}{F}{}^{\beta}_{j})_{,k} + \\ &\sum_{s=0}^{3} \left(\underset{s}{\psi}_\alpha\, \overset{s}{F}{}^{\alpha}_{i} \cdot \overset{s}{F}{}^{\beta}_{j}\, \bar{g}_{j\alpha}\, \overset{s}{F}{}^{\alpha}_{k} + \underset{s}{\psi}_\alpha\, \overset{s}{F}{}^{\alpha}_{j} \cdot \overset{s}{F}{}^{\beta}_{j}\, \bar{g}_{i\alpha}\, \overset{s}{F}{}^{\alpha}_{k} \right). \end{aligned} \tag{10}$$

Taking account of expressions (2), (8) and (9) for $s=1$ we obtain

$$\begin{aligned}&(\psi_i - \underset{1}{\psi}_i)\cdot \bar g_{jk} + (\psi_j - \underset{1}{\psi}_j)\cdot \bar g_{ik} + \bar g_{\alpha\beta}(\overset{1}{F}{}^{\alpha}_{i}\overset{1}{F}{}^{\beta}_{j})_{,k} + \\ &\sum_{s=1}^{3}\left(\overset{s}{\theta}_i \cdot \bar g_{j\alpha}\overset{s}{F}{}^{\alpha}_{k} + \overset{s}{\theta}_j \cdot \bar g_{i\alpha}\overset{s}{F}{}^{\alpha}_{k}\right) = 0\end{aligned} \tag{11}$$

where $\overset{s}{\theta}_i$, $s=1,2,3$, are certain covectors.

We symmetrized (11) with respect to the indices i,j,k. Because $\bar g_{j\alpha}\overset{s}{F}{}^{\alpha}_{k}$ are skew-symmetric, we obtain

$$(\psi_i - \underset{1}{\psi}_i)\cdot \bar g_{jk} + (\psi_j - \underset{1}{\psi}_j)\cdot \bar g_{ki} + (\psi_k - \underset{1}{\psi}_k)\cdot \bar g_{ij} + \bar g_{\alpha\beta}(\overset{1}{F}{}^{\alpha}_{(i}\overset{1}{F}{}^{\beta}_{j})_{,k)} = 0$$

where round brackets denote symmetrization without division.

After contraction with the inverse metric tensor $\bar g^{jk}$ we have

$$\underset{1}{\psi}_i = \psi_i + \frac{1}{3\,(n-2)}\,\bar g_{\alpha\beta}(\overset{1}{F}{}^{\alpha}_{(i}\overset{1}{F}{}^{\beta}_{j})_{,k)}\,\bar g^{jk}$$

and analogically follows

$$\underset{s}{\psi}_i = \psi_i + \frac{1}{3\,(n-2)}\,\bar g_{\alpha\beta}(\overset{s}{F}{}^{\alpha}_{(i}\overset{s}{F}{}^{\beta}_{j})_{,k)}\,\bar g^{jk}$$

for $s=2,3$.

This proves the theorem.

Acknowledgment

The paper was supported by the project No. LO1408 "AdMaS UP - Advanced Materials, Structures and Technologies", supported by the Ministry of Education, Youth and Sports under the "National Sustainability Programme I" of the Brno University of Technology.

References

[1] Alekseevsky, D. V., Marchiafava, S.: *Transformations of a quaternionic Kähler manifold*, C. R. Acad. Sci. Paris Sér. I Math., vol. 320, no. 6, (1995), pp. 703–708.

[2] Beklemishev, D. V.: *Differential geometry of spaces with almost complex structure*, in *Geometry 1963*, VINITI, Moscow, 1965, pp. 165–212.

[3] Domashev, V. V., Mikeš, J.: *Theory of holomorphically projective mappings of Kählerian spaces*, Mat. Notes, vol. 23, (1978), pp. 160–163.

[4] Hinterleitner, I.: *4-planar mappings of quaternionic Kähler manifolds*, in *Geometric methods in physics*, Trends Math., Birkhäuser/Springer, Basel, 2013, pp. 187–193.

[5] Hinterleitner, I., Mikeš, J.: *On F-planar mappings of spaces with affine connections*, Note Mat., vol. 27, no. 1, (2007), pp. 111–118.

[6] Hinterleitner, I., Mikeš, J.: *Geodesic mappings and Einstein spaces*, in *Geometric methods in physics*, Trends Math., Birkhäuser/Springer, Basel, 2013, pp. 331–335.

[7] Hinterleitner, I., Mikeš, J., Stranská, J.: *Infinitesimal F-planar transformations*, Russ. Math., (2008), pp. 13–18.

[8] Hrdina, J.: *Almost complex projective structures and their morphisms*, Arch. Math. (Brno), vol. 45, no. 4, (2009), pp. 255–264.

[9] Hrdina, J.: *Remarks on F-planar curves and their generalizations*, Banach Center Publ., vol. 93, (2011), pp. 241–249.

[10] Hrdina, J., Slovák, J.: *Morphisms of almost product projective geometries*, in *Differential geometry and its applications*, World Sci. Publ., 2008, pp. 253–261.

[11] Jukl, M., Juklová, L., Mikeš, J.: *Some results on traceless decomposition of tensors*, J. Math. Sci., vol. 174, (2011), pp. 627–640.

[12] Kurbatova, I. N.: *HP-mappings of H-spaces*, Ukrain. Geom. Sb., vol. 27, (1984), pp. 75–82.

[13] Kurbatova, I. N.: *4-quasi-planar mappings of almost quaternion manifolds*, Sov. Math., vol. 30, (1986), pp. 100–104.

[14] Kurbatova, I. N.: *On diffeomorphisms of almost quaternion manifolds*, Mat. Stud., vol. 40, no. 1, (2013), pp. 96–103.

[15] Mikeš, J.: *Holomorphically projective mappings of Kähler spaces*, Ukrain. Geom. Sb., vol. 23, (1980), pp. 90–98.

[16] Mikeš, J.: *Special F-planar mappings of affinely connected spaces onto Riemannian spaces*, Mosc. Univ. Math. Bull., vol. 49, (1994), pp. 18–24, 97.

[17] Mikeš, J.: *Geodesic mappings of affine-connected and Riemannian spaces*, J. Math. Sci. (New York), vol. 78, no. 3, (1996), pp. 311–333.

[18] Mikeš, J.: *Holomorphically projective mappings and their generalizations*, J. Math. Sci. (New York), vol. 89, no. 3, (1998), pp. 1334–1353.

[19] Mikeš, J., Bělohlávková, J., Pokorná, O.: *On special 4-planar mappings of almost Hermitian quaternionic spaces*, in *Quaternionic structures in mathematics and physics*, Univ. Studi Roma "La Sapienza", Rome, 1999, pp. 265–271.

[20] Mikeš, J., Jukl, M., Juklová, L.: *The decomposition of tensor spaces with quaternionic structure*, in *APLIMAT 2007, 6th Int. Conf., I*, 2007, pp. 217–222.

[21] Mikeš, J., Němčíková, J., Pokorná, O.: *On the theory of the 4-quasiplanar mappings of almost quaternionic spaces*, Rend. Circ. Mat. Palermo (2) Suppl., vol. 54, (1998), pp. 75–81.

[22] Mikeš, J., Shiha, M., Vanžurová, A.: *Invariant objects by holomorphically projective mappings of Kähler spaces*, in *8th Int. Conf. Proc. APLIMAT (Bratislava)*, 2009, pp. 439–444.

[23] Mikeš, J., Sinyukov, N. S.: *Quasiplanar mappings of spaces with affine connection*, Sov. Math., vol. 27, (1983), pp. 63–70.

[24] Mikeš, J., Vanžurová, A., Hinterleitner, I.: *Geodesic mappings and some generalizations*, Palacky Univ. Olomouc, 2009.

[25] Petrov, A.: *Simulation of physical fields*, Grav. and Theor. of Relativity, Kazan State Univ. Press, Kazan, vol. 4–5, (1968), pp. 7–21.

[26] Prvanović, M.: *Holomorphically projective transformations in a locally product space*, Math. Balkanica, vol. 1, (1971), pp. 195–213.

[27] Sinyukov, N. S.: *Geodesic mappings of Riemannian spaces*, Nauka, Moscow, 1979.

[28] Sinyukov, N. S., Kurbatova, I. N., Mikeš, J.: *Geodesic mappings of Riemannian spaces*, Odessk. Univ., Odessa, 1985.

[29] Stanković, M., Zlatanović, M., Velimirović, L.: *Equitorsion holomorphically projective mappings of generalized Kählerian space of the second kind*, Int. Electron. J. Geom., vol. 3, (2010), pp. 635–653.

[30] Stanković, M. S., Minčić, S. M., Velimirović, L. S.: *On equitorsion holomorphically projective mappings of generalized Kählerian spaces*, Czechoslovak Math. J., vol. 54(129), no. 3, (2004), pp. 701–715.

[31] Stanković, M. S., Zlatanović, M. L., Velimirović, L. S.: *Equitorsion holomorphically projective mappings of generalized Kählerian space of the first kind*, Czechoslovak Math. J., vol. 60(135), no. 3, (2010), pp. 635–653.

[32] Tachibana, S., Ishihara, S.: *On infinitesimal holomorphically projective transformations in Kählerian manifolds*, Tôhoku Math. J. (2), vol. 12, (1960), pp. 77–101.

[33] Yano, K.: *Differential geometry on complex and almost complex spaces*, New York, 1965.

Current address

Hinterleitner Irena, Mgr. PhD.
Institute of Mathematics and Descriptive Geometry,
Faculty of Civil Engineering, Brno University of Technology
Žižkova 17, 602 00 Brno, Czech republic
Tel. number: +420 541 147 612, e-mail: hinterleitner.i@fce.vutbr.cz

Slovak University of Technology in Bratislava
Faculty of Mechanical Engineering
16th Conference on Applied Mathematics **APLIMAT 2017**
Proceedings

DERIVATIVES IN PHYSICS IN NOVEL APPEARANCE – – SEVERAL APPLICATIONS FOR MATH LECTURES

HLAVÁČOVÁ M. Irena (CZ)

Abstract. The paper proposes a collection of various interesting and important applications of derivatives in physics. Apart from the most common applications in mechanics, namely instantaneous velocity and acceleration, less known extreme search is introduced not only in mechanics but also in optics and quantum mechanics. Some not so common views on electrodynamics problems are enclosed in the survey. Examples from mechanics are presented in detail in order to emphasize novelty in their solution. Other examples are given in a brief form – either only the most important step is presented (application of the Faraday's law) or they are only named to remind their key ideas.

Keywords: application of derivatives, physical problems, time, space, min-max problems

Mathematics Subject Classification: Primary 70-01, 78-01; Secondary 70B05, 70B15.

1 The role of derivatives in technical disciplines namely in physics

Contemporary research in educational practice states that considered in terms of an individual learner, the essence of a constructivist view of conceptual change is, that it is the learner who must recognize his or her conceptions, evaluate these conceptions, decide whether to reconstruct the conceptions, and, if they decide to reconstruct, whether to review and restructure other relevant aspects of their understanding in ways that lead to consistency [1]. Unfortunately, nowadays there are too many scholars attending various types of institutes, colleges as well as universities in Czech Republic, who have never even thought about consistency in their knowledge. No matter how lamentable this situation is, it is impossible for their teachers to change it, as they are not able to change their minds. Nevertheless, there exists another important concept in education process – metacognition. It refers to the knowledge, awareness, and control of one's own learning. Metacognitive knowledge refers to knowledge of the nature and processes of learning, personal learning characteristics, effective learning strategies and where to use them. Metacognitive awareness includes perceptions of the purpose of the current activity and personal progress through the activity. This paper aims to help math teachers to strengthen their impact on metacognitive knowledge of their students as far as calculus, or more exactly derivatives, is concerned.

According to recent research mathematical understanding of students can be improved to a great extent if proper representations are used to manifest mathematical conception [2]. Some authors already pointed out that usually there is lack of good examples from everyday life used in teaching calculus [3], [4]. This situation may have risen from the fact that mathematicians often concentrate on abstractions and they should not be so proficient in specialized fields.

There are lots of problems in human life in which mathematical description of the situation or the process requires calculus and namely derivatives. Some of them are closely connected with mathematic, e.g. geometry or statistics, others are more independent. In differential geometry the curvature and torsion of the curve are calculated from the formulas using the first and the second derivatives with respect to curves parameter [5]. On the other hand, in various applications of statistics [6] the first and the second derivatives are applied for identification of extremes most often. We could certainly find derivations in other fields of human scientific activity – in biology, pharmacy and chemistry. Physics stand aside among these, its interconnection and liaison with mathematic lead to much more diverse application of calculus.

It is generally considered that satisfactory training in physics at technical colleges requires implementation of derivatives and integrals. This is really true in many cases, but sometimes the students do not understand this requirement. They are forced to solve relatively simple problems in rather complicated manner, without having understood the reason or having perceived any benefit. Having passed their math exam they suspect their lecturers both in mathematics and physics of having just ganged up to make their higher education more complicated and almost impassable. I believe that this pessimistic conclusion may be proved to be false if the choice of examples of use is made more convincing. This paper aims to introduce some frequent or important applications of derivatives in physics and to highlight the most interesting of them.

Let us start with something we should call "physical meaning" of derivative. The definition of derivative of the quantity b according to a is given by the following formula [7]

$$\frac{\mathrm{d}b}{\mathrm{d}a} = \lim_{\Delta a \to 0} \frac{\Delta b}{\Delta a} \tag{1}$$

It expresses how much the quantity b changes with infinitesimally small change of the quantity a. Three basic different physical meanings can be found depending on the choice of the independent quantity a. We shall discuss all of them and explain their importance on practical examples and solved problems.

2 Time Derivative in Mechanics

In case that quantity a is time, we usually say that derivative expresses instantaneous rate of change of the quantity b. This meaning is the most common one and unfortunately in minds of most college graduates the only one. So let us start with this type of application. In mechanics there are two generally known applications of the rate of change of something: First of all the rate of change of the position vector represents *instantaneous velocity* and secondly the rate of change of instantaneous velocity is called *acceleration*. Both these quantities can be defined and used either in three eventually two dimensions as vectors on in one dimension problems.

Let us start our trip around the physical world of derivatives with three different examples of exploitation of the energy conservation law to determine the equation of motion. The mathematic formula of mechanical energy conservation of a rigid body or a particle is commonly known in the following form

$$E_p + E_k = konst. \tag{2}$$

Even the most ignorant students should be acquainted with bit more exact expression valid in the homogenous gravity field with constant gravity acceleration g in minus y direction

$$mgh + \frac{1}{2}mv^2 = konst. \tag{3}$$

Supposing the mass remains constant enables us to divide the whole equation by m. Derivation of this modified equation with respect to time leads to

$$g\frac{dh}{dt} + v\frac{dv}{dt} = gv_y(t) + va = 0 \tag{4}$$

in which $v_y(t)$ stands for vertical component of instantaneous velocity.

This equation should be used for any rigid body moving in the homogenous gravity field. We shall use it in the following three practical examples presenting "description of the problem" followed by its solution and finally by evaluation of its suitability for "domestication" of derivatives in students' minds.

2.1 Projectile motion

Example:
A gun located on a cliff 49 m high fires a shell with a muzzle velocity of 490 m·s^{-1} at an angle of elevation of $\theta = 37°$. Find the kinematic description of motion of the shell: the time of flight, the horizontal distance the shell will travel, and the velocity with which the shell will strike the ground.

$g = 9{,}81$ m·s^{-2}
$h_0 = 49$ m
$u = 490$ m·s^{-1}
$\theta = 37°$

$t = ?$
$l = ?$
$v = ?$

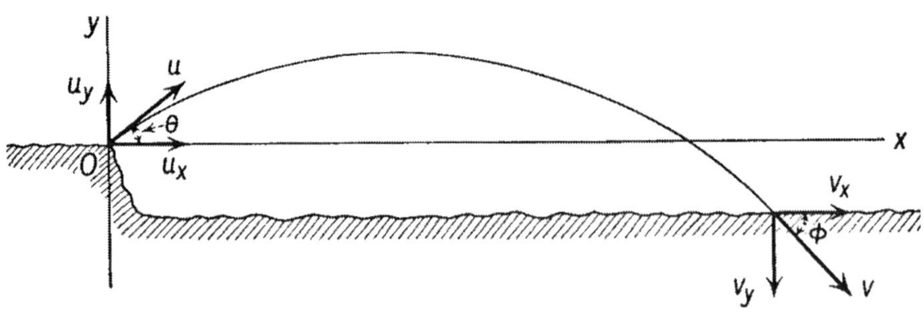

Fig. 1. Projectile motion [6]

Solution:

The first example can be classified as a plane motion, it should be solved in x and y directions independently. In vertical direction, we should specify the equation (4) for this problem in the form

$$g \cdot \frac{\mathrm{d}y}{\mathrm{d}t} + v_y(t)\frac{\mathrm{d}v_y}{\mathrm{d}t} = g \cdot \frac{\mathrm{d}y}{\mathrm{d}t} + v_y(t)(-g) = 0 \tag{5}$$

Consulting the Fig. 1, as far as the initial value of vertical velocity is concerned, leads to

$$y = h_0 + u_y t - \frac{1}{2}gt^2 \Rightarrow \frac{\mathrm{d}y}{\mathrm{d}t} = u_y - gt \Rightarrow v_y = u_y - gt \tag{6}$$

The time of flight is given by the condition that y coordinate is equal to zero.

$$0 = h_0 + u_y t - \frac{1}{2}gt^2 \Rightarrow t_{1,2} = \frac{-u_y \pm \sqrt{u_y^2 + 2gh_0}}{-g} \Rightarrow t_2 = \frac{u_y}{g} + \sqrt{\left(\frac{u_y}{g}\right)^2 + 2\frac{h_0}{g}} \tag{7}$$

In horizontal direction, taking into account Fig. 1 as far as the initial value of horizontal velocity is concerned we should specify the equation (4) for this problem in the form

$$\frac{\mathrm{d}v_x}{\mathrm{d}t} = 0 \Rightarrow v_x(t) = u_x \tag{8}$$

The horizontal distance the shell will travel is then given by product of horizontal velocity and time calculated in equation (7)

$$l = u_x \cdot \left[\frac{u_y}{g} + \sqrt{\left(\frac{u_y}{g}\right)^2 + 2\frac{h_0}{g}}\right] \tag{9}$$

The velocity with which the shell will strike the ground is calculated using equation (3) both at the beginning of motion when the y coordinate of the shell is h_0 and in time t_2 when it is equal to zero

$$mgh_0 + \frac{1}{2}u^2 = \frac{1}{2}v^2 \Rightarrow v = \sqrt{u^2 + 2gh_0} \tag{10}$$

Finally we should determine the "elevation" angle knowing that its tangent is the ratio of vertical and horizontal components of final velocity

$$\tan\Phi = -\frac{u_y^2 + 2gh_0}{u_x^2} \Rightarrow \Phi = \arctan\left(-\frac{u_y^2 + 2gh_0}{u_x^2}\right) \tag{10}$$

Substituting given numerical values leads to the following results: the time of flight is 60 s, the shell will travel 23592 m and it hits the ground with the velocity 491 $\mathrm{m \cdot s^{-1}}$ falling with the angle -37.2°.

Evaluation:

In my opinion this problem represents a wrong choice of representation of derivative usefulness. The solution is very complicated and it could be achieved much more easily with simple knowledge of basic linear acceleration equations. The application of derivative seems to be purely theoretical abstraction and it should attract only few students. Although it may be used to demonstrate applicability of derivatives in calculus training, general application of this way of solution would just confirm the conviction of most students that physics is incomprehensible science and that derivatives and calculus as a whole are just one way how to make description of common and simple processes more difficult.

2.2 A box sliding down the inclined plane

Example:

A box is sliding down the inclined plane with the slope of $\theta = 37°$. Neglecting friction find the acceleration of its motion.

$g = 9{,}81\ \mathrm{m{\cdot}s^{-2}}$
$\theta = 37°$
$a = ?$

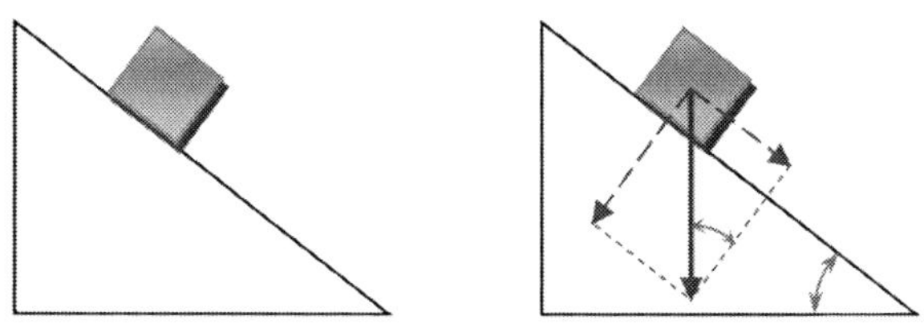

Fig. 2. Sliding down the inclined plane

Solution:

We can express the height h of the slope in terms of its length s and the angle of inclination θ and then rewrite the equation (4) for this problem in the form

$$g \cdot \frac{\mathrm{d}s \cdot \sin\theta}{\mathrm{d}t} + va = g\sin\theta \cdot \frac{\mathrm{d}s}{\mathrm{d}t} + va = 0 \tag{11}$$

We can substitute the first derivative of displacement s by velocity v and having divided both sides of the equation by velocity leads immediately to the final result

$$g\sin\theta \cdot v + va = 0 \Rightarrow a = -g\sin\theta \tag{12}$$

This result is in perfect agreement with the result coming from dynamics analysis of the problem based on decomposition of gravity force (see Fig. 2 b). Substituting given numerical values leads to the conclusion that acceleration of the box is – 5.90 $\mathrm{m{\cdot}s^{-2}}$.

Evaluation:

This problem can be easily solved also by decomposition of gravity force into tangential and normal components. Nevertheless, the students are rarely acquainted with inclined plane and they often interchange the force components. Presented way of calculation may be clearer for them.

2.3 A heavy hollow sphere rolling down the hill

Example:
A hollow sphere with radius R is rolling down the inclined plane with the slope of $\theta = 37°$ without any slipping. Neglecting friction find the acceleration of its motion. The moment of inertia of the hollow sphere is 2/3 mR^2 [3].

$g = 9{,}81\ \mathrm{m{\cdot}s^{-2}}$
$\theta = 37°$
$J = \frac{2}{3}mR^2$
$a = ?$

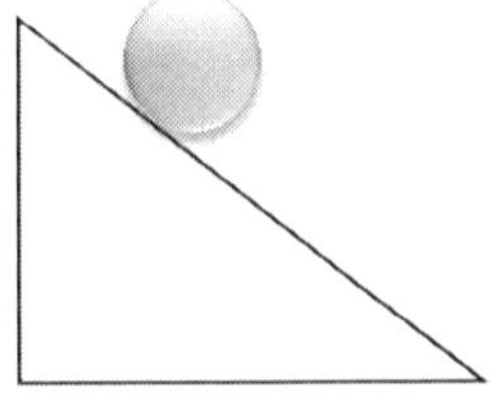

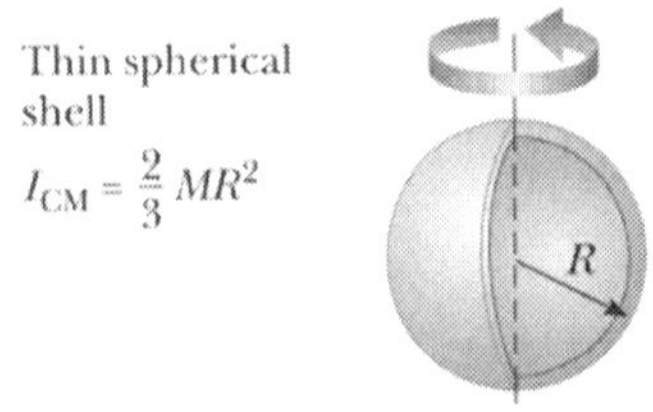

Fig. 3. Rolling down the inclined plane; moment of inertia of a thin spherical shell [3]

Solution:
This time it is necessary to return to the equation (3) and modify the expression for kinetic energy. The kinetic energy of a rolling body consists of two parts – kinetic energy of the translational motion and kinetic energy of the rotational motion.

$$mgh + \frac{1}{2}mv^2 + \frac{1}{2}J\omega^2 = konst. \tag{13}$$

Substituting the moment of inertia by the proper formula and ω by v/R leads to the simplified expression

$$mgh + \frac{1}{2}mv^2 + \frac{1}{2}\left(\frac{2}{3}mR^2\right)\left(\frac{v}{R}\right)^2 = mgh + \frac{5}{6}mv^2 = 0\,, \tag{14}$$

Expressing the height h of the slope in terms of its length s and the angle of inclination θ, dividing the equation by m and derivation with respect to time leads to

$$g\sin\theta\frac{ds}{dt} + \frac{5}{3}v\frac{dv}{dt} = g\sin\theta\cdot v + \frac{5}{3}v\cdot a = 0 \tag{15}$$

We can again substitute the first derivative of displacement s by velocity v. Having divided both sides of the equation by velocity we come to the final result

$$g\sin\theta + \frac{5}{3}a = 0 \Rightarrow a = -\frac{3}{5}g\sin\theta \tag{16}$$

which means that the acceleration of the hollow sphere rolling downhill is $-3{,}42\ \mathrm{ms^{-2}}$.

Evaluation:
Rolling is much more difficult to understand than sliding, therefore there is no other simple way to solve the problem.

2.4 Tangential and centripetal acceleration as time derivative in polar coordinates

Cartesian coordinates of a moving particle should be expressed in polar coordinates in the form

$$x = r\cos\varphi; y = r\sin\varphi \Rightarrow \vec{r} = r\cos\varphi\vec{i} + r\sin\varphi\vec{j} \tag{17}$$

Calculation of the first derivative with application of the chain rule gives us instantaneous velocity

$$\vec{v} = \frac{d\vec{r}}{dt} = \left(\frac{dr}{dt}\cos\varphi - r\frac{d\varphi}{dt}\sin\varphi\right)\vec{i} + \left(\frac{dr}{dt}\sin\varphi + r\frac{d\varphi}{dt}\cos\varphi\right)\vec{j} = v_x\vec{i} + v_y\vec{j} \tag{18}$$

The rate of change of the angle φ is called *angular velocity* ω, we can rewrite the instantaneous velocity in the form

$$\vec{v} = \frac{d\vec{r}}{dt} = \left(\frac{dr}{dt}\cos\varphi - r\omega\sin\varphi\right)\vec{i} + \left(\frac{dr}{dt}\sin\varphi + r\omega\cos\varphi\right)\vec{j} \tag{19}$$

Let us think about the special type of movement when the distance of a particle from the origin remains constant, i.e. circular motion of a particle. In such case instantaneous velocity simplifies

$$\vec{v} = \frac{d\vec{r}}{dt} = (-r\omega\sin\varphi)\vec{i} + r\omega\cos\varphi\vec{j} = r\omega(-\sin\varphi, \cos\varphi) \tag{20}$$

Another differentiating by time brings

$$\vec{a} = \frac{d\vec{v}}{dt} = \left(-r\frac{d\omega}{dt}\sin\varphi - r\omega^2\cos\varphi\right)\vec{i} + \left(r\frac{d\omega}{dt}\omega\cos\varphi - r\omega^2\sin\varphi\right)\vec{j} \tag{21}$$

The rate of change of the angular velocity ω is called *angular acceleration* α. The equation will be rewritten in vector components to be better arranged. In this arrangement we can see at once, that the direction of the first part of the acceleration is the same as direction of velocity (Fig.4), while the direction of the second part of the acceleration is opposite to the direction of position vector

$$\vec{a} = r\alpha(-\sin\varphi, \cos\varphi) + (-r\omega^2)(\cos\varphi, \sin\varphi) = \vec{a}_t + \vec{a}_c \tag{22}$$

Evaluation:

Using polar coordinates we succeeded to introduce four kinematic quantities: angular velocity ω, angular acceleration α, tangential acceleration a_t and centripetal acceleration a_c. Although there exist at least two other ways to derive tangential acceleration, namely differentiation of the product of magnitude and direction, or differentiation of the cross product of angular velocity and position vector, I have not met any other way how to introduce simply and clearly the formula for calculation of centripetal acceleration.

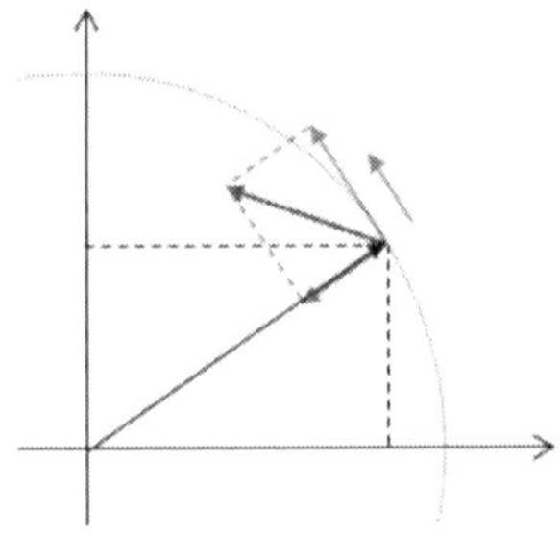

Fig. 4. Tangential and centripetal acceleration

2.5 Force impulse and displacement effect of force

The next example of uncommon use of derivatives in physics will be the last one chosen from mechanics. Application of differentials on practical problem in mechanics of continuum will help us to present another point of view on work and energy. Dealing with energy at school students of all grades are uniformly taught that there is no work done when the body is moving horizontally. At the same time Newton's first law of dynamics states: a body moving at a constant velocity in an inertial frame of reference is not acted upon by a nonzero force, otherwise its velocity certainly changes. Both these statements seem to be in contradiction with our everyday life, and therefore, they represent the cornerstones of a general conception of many people, that physics is a mysterious science, based on special laws, which have to be learned and exactly used when passing the exam but which in fact have nothing in common with the real world.

Example:
Coal is falling from the hopper onto a horizontal conveyor belt, which transports it to control and further processing. Determine the power of the conveyor engine where the mass flow rate of coal falling from the hopper is 375 kg·s^{-1}, the belt material moves at speed of 3 m·s^{-1}.

$Q_m = 375$ kg·s^{-1}
$u = 3$ m·s^{-1}
$P_{total} = ?$

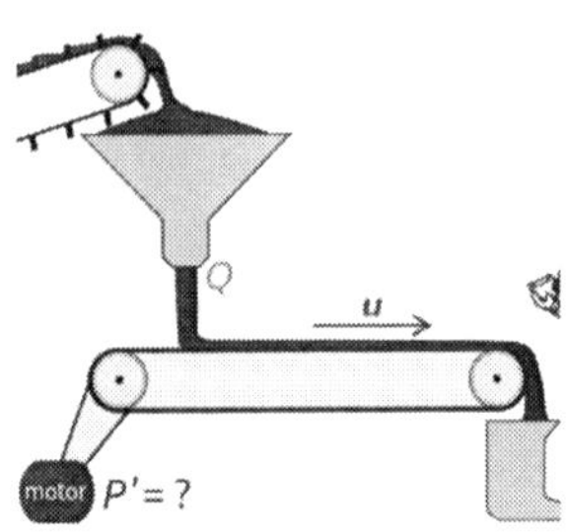

Fig. 5. Practical example of force-displacement and force-time calculus [9]

Solution:
The coal falls on conveyor belt downwards, therefore it has no horizontal component of velocity. There is mass of coal dm which falls down on the belt each time interval dt, it gains momentum dmu thanks to conveyor motion. This instantaneous change of momentum is equal to the action of

the force F that drives the conveyor. The change of coal momentum is equal to the impulse of the force F.

$$F\mathrm{d}t = \mathrm{d}mu \Rightarrow F = \frac{\mathrm{d}m}{\mathrm{d}t}u = Q_m u \tag{23}$$

The work which must be done when accelerating the coal from zero to velocity u is given by dot product of the force and displacement. Power which is necessary to sustain the constant movement of the conveyor belt is defined as time derivative of the work done.

$$\mathrm{d}W = \vec{F}\mathrm{d}\vec{s} = \vec{F}\vec{u}\mathrm{d}t \Rightarrow P = \frac{\mathrm{d}W}{\mathrm{d}t} = Fu = Q_m u^2 \tag{24}$$

It should be mentioned that the contradiction to basic physical laws cited above is only superficial. The coal must be first accelerated, then it goes on moving with constant velocity and there is no more force necessary.
The power of the conveyor engine is 3375 W.

Evaluation:
This practical logistic and mining problem which can be found in [9] in the Czech version of the collection of Solved Problems in Physics developed by Department of Physics Education, Faculty of Mathematics and Physics, Charles University in Prague may help to explain and understand the difference as well as connection between physical theory and real action. It represents less usual application of derivative – it was used primarily for the process description. The students should perceive that a problem may occur, in which the Newton‘s law has to be written in differential form, i.e. the rate change of momentum is equal to the force impulse.

3 Time Derivative in Electrodynamics - Electromagnetic induction

The Faraday's law states and production of electric energy certifies that voltage is induced on a conductive loop when changing magnetic flux through the loop. The induced voltage is proportional to the flux temporal change, in fact, it is the first time derivative of the magnetic flux

$$U = -\frac{\mathrm{d}\Phi}{\mathrm{d}t} = -\frac{\mathrm{d}}{\mathrm{d}t}\int \vec{B}\cdot\mathrm{d}\vec{S} \tag{25}$$

The minus sign indicates the orientation of the induced voltage. The law itself can be derived from the Lorentz force acting on a charged particle which is moving with non-zero velocity component perpendicular to magnetic induction $\vec{B}$. Although this derivation is wonderful it is not short and clear enough to be welcome by students, they usually prefer casual attitude with change of area and without calculus at all. There are, however, two very simply applications of Faraday law which should demonstrate quickly and easily the behavior of a coil and capacitor in the alternating circuit.

3.1 A coil carrying an alternating current

Magnetic flux is directly proportional to the current, which is passing through the coil: $\Phi = Li$, the constant of proportionality is the inductance L. If we let harmonic current $i = I\sin\omega t$ to pass through the coil, the induced voltage will be

$$u_i = -\frac{d\Phi}{dt} = -\ \omega I\cos\omega t \tag{26}$$

The function $-\cos\omega t$ is the sine function shifted in time by $-\frac{\pi}{2}$, i.e. $-\cos\omega t = \sin\left(\omega t - \frac{\pi}{2}\right)$

By differentiating we discovered, that voltage on the coil anticipates the current by half π, therefore in time relations voltage on the coil anticipates the current by a quarter of a period.

3.2 A capacitor carrying an alternating current

If we join capacitor to the source of harmonic voltage, the change of the amount of charge on capacitor plates will be harmonic as well: $q = Q\sin\omega t$. The current which is bringing the charge to the capacitor is expressing the rate of charge delivery to the plates of the capacitor, i.e. the rate of charge change in time

$$i = \frac{dq}{dt} = \omega Q\cos\omega t \tag{26}$$

The function $\cos\omega t$ is the sine function shifted in time by $+\frac{\pi}{2}$, tj. $\cos\omega t = \sin\left(\omega t + \frac{\pi}{2}\right)$.

By differentiating we discovered, that voltage in the capacitor delays after the current by half π, therefore in time relations voltage in the capacitor delays after the current by a quarter of a period.

4 Space Derivatives

In case that quantity a is a space coordinate, it expresses local rate of change of the quantity b along the increasing coordinate a. When used in three independent space coordinates a_1, a_2, a_3 (using partial derivatives in such case) it enables to find the vector of maximum growth of the quantity b in the point where the derivatives are calculated, well known as operator gradient.

On the contrary if the coordinates are not treated individually, they are not separated, representing special direction in space, but used (multiplied) together they form either infinitesimal area dS or infinitesimal volume dV leading to definition of local area density or space density of the quantity b. These are, however, necessary in integral calculus, which is not discussed in this paper.

If the force acting on an object is a function of position only, it is said to be a conservative force, and it can be represented by a potential energy function which for a one-dimensional case satisfies the derivative condition

$$F = -\frac{dV}{dx} \qquad (25)$$

Potential energy is energy which results from position or configuration. An object may have the capacity for doing work as a result of its position in a gravitational field (gravitational potential energy), an electric field (electric potential energy), or a magnetic field (magnetic potential energy).

It may also have elastic potential energy as a result of a stretched spring or other elastic deformation. Its differentiating with respect to x enables to find the source of oscillation motion – the elastic force. Nevertheless I do not recommend anybody to persuade students to find elastic force in free oscillations by differentiating elastic potential energy. It is much more likely that they know or even understand the elastic force from their personal experience, while they have definitely no idea, how to express elastic potential energy.

A magnetic dipole moment in a magnetic field will possess potential energy which depends upon its orientation with respect to the magnetic field. Its differentiating with respect to angular displacement θ enables to find the source of its rotation – the magnetic torque.

5 Derivatives in Min-Max Problems

One of the great powers of calculus is in the determination of the maximum or minimum value of a function. Take $b(a)$ to be a function of a. Then the value of a for which the derivative of $b(a)$ with respect to a is equal to zero corresponds to a maximum, a minimum or an inflexion point of the function $b(a)$. This knowledge is very often used in physics and so I will close my survey by citation of four examples which are using it:

Fermats principle (Light follows the path of least time) applied on the light beam reflected from a planar interface leads to the *Law of Reflection*, the same law applied on a light beam passing through a planar interface between two materials with different refractive index leads to the *Law of Refraction*. It should be emphasized that the principle is dealing with the least time therefore the derivative applied in the refraction case is derivative of time with respect to x.

From the classical mechanics the *maximum height* calculation in *projectile motion* should be mentioned as an example of min-max problem.

Quantum mechanics, although usually not taught in common undergraduate courses, will close our exhibition of physical applications of derivatives. *Wien's displacement law* states that the spectral radiance of black body radiation per unit wavelength, peaks at the wavelength λ_{max} given by: $\lambda_{max} = b/T$, where T is the absolute temperature in kelvin and b is a constant of proportionality called Wien's displacement constant, equal to 2.8977729×10^{-3} m K. It should be found, if we differentiate Plank's law, explaining the spectral-energy distribution of radiation emitted by a blackbody.

6 Conclusions

There are a lot of problems in physics, which should be solved with the involvement of derivatives. Not all of them are suitable to attract students to try to use them. The first example presented in this survey was on purpose chosen to demonstrate the main important fact: proper example should be clear and simple enough so that any student will understand it and, moreover, it must offer either faster or easier way or, if possible, the only way, how to solve the problem.

Acknowledgement

The paper was not supported by any grant. The author gratefully acknowledges her colleagues teaching mathematics at The Institute of Technology and Business for supporting her work.

References

[1] GUNSTONE, R.F., MITCHELL, I.J. Metacognition and Conceptual Change. In *Teaching Science for Understanding*, 2005, ISBN: 978-012498360-1, pp. 133-163.

[2] PAPE, S.J., TCHOSHANOV, M.A. The Role of Representation(s) in Developing Mathematical Understanding. *Theory Into Practice*, Vol. 40 , Iss. 2, 2001: pp. 82-83, ISSN: 0040-5841.

[3] KLEPANCOVÁ M., VARGA, M. Limity okolo nás. In *didZA 7: Nové trendy vo vyučovaní matematiky a informatiky na základných a stredných a vysokých školách: zborník príspevkov zo 7. žilinskej didaktickej konferencie s medzinárodnou účasťou*, 2010, ISBN 978-80-554-0216-1, s. 1-4.

[4] VARGA, M. Derivácia okolo nás. In *Učme aplikovať matematiku: zborník z vedeckého seminára Učme aplikovať matematiku organizovaného Katedrou matematiky dňa 17. októbra 2007* / Ondrej Šedivý, Dušan Vallo. - Nitra: UKF, 2008. s. 93-97, ISBN 978-80-8094-290-8.

[5] MIKEŠ, J., STEPANOVA, E. et al. *Differential Geometry of Special Mappings*. 1. vyd. Olomouc: Palacky University, Olomouc, 2015. Monografie. ISBN 978-80-244-4671-4, 566 s.

[6] ČEJKA, J., BARTUŠKOVÁ, P., BARTUŠKA, L. The Application of Mathematical Methods in Transport and Logistic Area. In *APLIMAT 2016, 15th Conference on Applied Mathematics*, Proceedings, 2016, ISBN 978-80-227-4140-8, s. 225-235.

[7] KNICHAL,V., BAŠTA, A., PIŠL, M.,REKTORYS, K. *Matematika I*. SNTL, 1965, 544 s.

[8] URL: http://slideplayer.com/slide/4063669/

[9] URL: http://reseneulohy.cz/cs/fyzika

Current address

Hlaváčová Irena M., doc. Ing. Ph.D.
The Institute of Technology and Bussiness,
Department of Informatics and Natural Sciences
Okružní 517/10, 370 01 České Budějovice, Czech Republic
Tel. number: + 420 387 842 185, e-mail: 19512@mail.vstecb.cz

Slovak University of Technology in Bratislava
Faculty of Mechanical Engineering
16th Conference on Applied Mathematics **APLIMAT 2017**
Proceedings

MINIMAL GENERATING SETS FOR ASSOCIATED IDEALS OF MONOMIAL CURVES

HOLEŠOVÁ Michaela (SK)

Abstract. We deal with associated prime ideals of monomial curves in afinne space. We search conditions for concrete descriptions of minimal generating sets for associated prime ideals of monomial curves. We show some results also about associated prime ideals as set-theoretic complete intersections.

Keywords: Minimal generating set, monomial curve, associated prime ideal

Mathematics subject classification: Primary 14H45, 14M10; Secondary 14A05

1 Introduction

Let $n_1, n_2, \ldots, n_h, h \geq 3$ be positive integers, $(n_1, n_2, \ldots, n_h) = 1$ and it is a minimal set of generators for the numerical semigroup $H = \langle n_1, n_2, \ldots, n_h \rangle$. Let K be an arbitrary field, $R = K[x_1, x_2, \ldots, x_h]$ the polynomial ring in h variables over K. Let $C = C(n_1, n_2, \ldots, n_h)$ be a monomial curve in affine space A^h over K having parameterization $x_i = t^{n_i}$, $i = 1, 2, \ldots, h$. The ideal P of all polynomials $f(x_1, x_2, \ldots, x_h) \in R$ such that $f(t^{n_1}, t^{n_2}, \ldots, t^{n_h}) = 0$, for transcendental t over K, is the associated prime ideal of local ring $R_{(x_1, x_2, \ldots, x_h)}$ of the monomial curve C. The height of a prime ideal P, ht(P), is the supremum of the lenghts r of all prime ideal chains $p_0 \subseteq p_1 \subseteq \ldots \subseteq p_r = P$. Let P be the corresponding ideal in R with its height ht(P)=$h - 1$.
The ideal I is called a set-theoretic complete intersection (s.t.c.i., for short), if there are s=ht(I) elements $g_1, g_2, g_3, ... g_s$, such that rad(I)=rad($g_1, g_2, ..., g_s$). For example the associated prime ideal P of a monomial curve C in A^4 is a s.t.c.i., if P=rad(g_1, g_2, g_3) and also a monomial curve C is a s.t.c.i. (more information can be found in [10]).

The general problem of whether all associated ideals of monomial curves (or monomial curves) are s.t.c.i. is still open. In connection with this problem the description of the minimal generating set for associated prime ideal of a monomial curve is examined. There are nevertheless some partial results in this direction.
E. Kunz [10] showed that every monomial curve in 3-dimensional affine space is a s.t.c.i and minimal generating sets of these ideals are

(1)$M = \{F_1 = x_1^{r_1} - x_3^{r_3}, F_2 = x_2^{r_2} - x_1^{\alpha_{21}}x_3^{\alpha_{23}}\}, 0 < \alpha_{21} < r_1$, or
(2)$M = \{F_1 = x_1^{r_1} - x_2^{\alpha_{12}}x_3^{\alpha_{13}}, F_2 = x_2^{r_2} - x_1^{\alpha_{21}}x_3^{\alpha_{23}}, F_3 = x_3^{r_3} - x_1^{\alpha_{31}}x_2^{\alpha_{32}}\}, 0 < \alpha_{ij} < r_j, \{i, j\} \subset \{1, 2, 3\}$.

Monomial curves are called monomial Gorenstein curves when the numerical semigroup H is symmetric. $H = \langle n_1, n_2, n_3, n_4 \rangle$ is symmetric, that is if $c-1 \notin H$, but $c-1+j \in H, j \geq 1$, then for all $s \in H$, $(c-1) - s \notin H$.
In 4-dimensional affine space A^4, H. Bresinsky proved (see [2]), that if numerical semigroup H is symmetric, then the monomial Gorenstein curve $C(n_1, n_2, n_3, n_4)$ and its the associated prime ideal is s.t.c.i. with a minimal generating set

(1)$M = \{F_1 = x_1^{r_1} - x_2^{r_2}, F_2 = x_3^{r_3} - x_1^{\alpha_{31}}x_2^{\alpha_{32}}, F_3 = x_4^{r_4} - x_1^{\alpha_{41}}x_2^{\alpha_{42}}x_3^{\alpha_{43}}\}, r_j \geq 2$ $(j = 1, \ldots, 4), \alpha_{3i} \geq 0, \alpha_{4j} \geq 0$ $(i = 1, 2, j = 1, 2, 3)$, or
(2)$M = \{F_1 = x_1^{r_1} - x_2^{r_2}, F_2 = x_3^{r_3} - x_4^{r_4}, F_3 = x_1^{\alpha_1}x_2^{\alpha_2} - x_3^{\beta_3}x_4^{\beta_4}\}, r_j \geq 2$ $(j = 1, \ldots, 4)$, $\alpha_1 + \alpha_2 \geq 2, \beta_3 + \beta_4 \geq 2, \alpha_i \geq 0$ $(i = 1, 2), \beta_j \geq 0$ $(j = 3, 4)$, or
(3)$M = \{F_1 = x_1^{r_1} - x_3^{\alpha_{13}}x_4^{\alpha_{14}}, F_2 = x_2^{r_2} - x_1^{\alpha_{21}}x_4^{\alpha_{24}}, F_3 = x_3^{r_3} - x_1^{\alpha_{31}}x_2^{\alpha_{32}}, F_4 = x_4^{r_4} - x_2^{\alpha_{42}}x_3^{\alpha_{43}}\}, r_1 = \alpha_{21} + \alpha_{31}, r_2 = \alpha_{32} + \alpha_{42}, r_3 = \alpha_{13} + \alpha_{43}, r_4 = \alpha_{14} + \alpha_{24}, 0 < \alpha_{ij} < r_j$, $1 \leq i \leq 4$, $1 \leq j \leq 4$.

W. Gastinger in [6] proved, that associated prime ideals of monomial curves in A^4 are s.t.c.i., if minimal generating sets of these ideals have four generators. The set M can be

(1)$M = \{F_1 = x_1^{r_1} - x_2^{r_2}, F_2 = x_1^{\gamma_1}x_2^{\gamma_2} - x_3^{\gamma_3}x_4^{\gamma_4}, F_3 = x_3^{r_3} - x_1^{\alpha_1}x_2^{\alpha_2}x_4^{\alpha_4}, F_4 = x_4^{r_4} - x_1^{\beta_1}x_2^{\beta_2}x_3^{\beta_3}\}, \gamma_i = \alpha_i + \beta_i$ $(i = 1, 2)$, $r_3 = \gamma_3 + \beta_3, r_4 = \alpha_4 + \gamma_4$, a $0 < \alpha_i < r_i$ $(i = 2, 4)$, $0 < \beta_i < r_i$ $(i = 2, 3)$, $\alpha_1 > 0, \beta_1 > 1, (\alpha_1, \beta_1) \not\subseteq r_1 Z$;
(2) $M = \{F_1 = x_1^{r_1} - x_2^{r_2}, F_2 = x_1^{\gamma_1}x_2^{\gamma_2} - x_3^{\gamma_3}x_4^{\gamma_4}, F_3 = x_3^{r_3} - x_1^{\alpha_1}x_4^{\alpha_4}, F_4 = x_4^{r_4} - x_1^{\beta_1}x_2^{\beta_2}x_3^{\beta_3}\}$, $\gamma_1 = \alpha_1 + \beta_1, \gamma_2 = \beta_2, r_3 = \gamma_3 + \beta_3, r_4 = \alpha_4 + \gamma_4, 0 < \alpha_4 < r_4, 0 < \beta_i < r_i$ $(i = 2, 3)$, $\alpha_1 > 0, \beta_1 > 1, (\alpha_1, \beta_1) \not\subseteq r_1 Z$;
(3) $M = \{F_1 = x_1^{r_1} - x_2^{r_2}, F_2 = x_1^{\gamma_1}x_2^{\gamma_2} - x_3^{\gamma_3}x_4^{\gamma_4}, F_3 = x_3^{r_3} - x_1^{\alpha_1}x_4^{\alpha_4}, F_4 = x_4^{r_4} - x_2^{\beta_2}x_3^{\beta_3}\}$, $\gamma_1 = \alpha_1, \gamma_2 = \beta_2, r_3 = \gamma_3 + \beta_3, r_4 = \alpha_4 + \gamma_4, 0 < \alpha_i < r_i$ $(i = 1, 4), 0 < \beta_i < r_i$ $(i = 2, 3)$, $\alpha_1 > 0, \beta_1 > 1, (\alpha_1, \beta_1) \not\subseteq r_1 Z$;
(4)$M = \{F_1 = x_1^{r_1} - x_2^{\alpha_{12}}x_3^{\alpha_{13}}x_4^{\alpha_{14}}, F_2 = x_2^{r_2} - x_1^{\alpha_{21}}x_3^{\alpha_{23}}, F_3 = x_3^{r_3} - x_1^{\alpha_{31}}x_4^{\alpha_{34}}, F_4 = x_4^{r_4} - x_1^{\alpha_{41}}x_2^{\alpha_{42}}\}, 0 < \alpha_{ij} < r_j, 1 \leq i \leq 4, 1 \leq j \leq 4$ $r_1 = \alpha_{21} + \alpha_{31} + \alpha_{41}, r_2 = \alpha_{12} + \alpha_{42}, r_3 = \alpha_{13} + \alpha_{23}, r_4 = \alpha_{14} + \alpha_{34}$.

2 Minimal generating set in 4-dimensional polynomial ring R

Let $M \subseteq P$ be a set of binomials with one coefficient 1, the other -1 and two terms relatively prime to each other.

When a binomial term $\prod_{i=1}^4 x_i^{\gamma_i} - \prod_{i=1}^4 x_i^{\vartheta_i} \in P$, where $\gamma_i\vartheta_i = 0$, $i = 1, 2, 3, 4$, it is clear that $\sum_{i=1}^4 \gamma_i n_i = \sum_{i=1}^4 \vartheta_i n_i$.

We have basically two types of binomial terms of the associated prime ideal P of the monomial curve $C = C(n_1, n_2, n_3, n_4)$

i) $x_i^{\gamma_i}x_j^{\gamma_j} - x_k^{\gamma_k}x_l^{\gamma_l}, \{i, j, k, l\} = \{1, 2, 3, 4\}, \gamma_i\gamma_j\gamma_k\gamma_l \neq 0$ or
ii) $x_i^{r_i} - x_j^{\alpha_{ij}}x_k^{\alpha_{ik}}x_l^{\alpha_{il}}, \{i, j, k, l\} = \{1, 2, 3, 4\}, r_i \neq 0$.

We denote the binomial term $x_i^{r_i} - x_j^{\alpha_{ij}}x_k^{\alpha_{ik}}x_l^{\alpha_{il}} \in P$ either by $(x_i^{r_i})$ if r_i is minimal or by $(x_i^{r_i}, x_j^{r_j})$ if $x_j^{r_j} - x_i^{\alpha_{ji}}x_k^{\alpha_{jk}}x_l^{\alpha_{jl}} \in P$ with r_j minimal and $\alpha_{ji} = r_i, \alpha_{jk} = \alpha_{jl} = 0$.

Every generating set of P contains for each i $(i = 1, 2, 3, 4)$ at least one polynomial $x_i^{r_i} - x_j^{\alpha_{ij}} x_k^{\alpha_{ik}} x_l^{\alpha_{il}}$ with r_i minimal. We also denote polynomial $x_i^{r_i'} - x_j^{\alpha_{ij}} x_k^{\alpha_{ik}} x_l^{\alpha_{il}}$ by $(x_i^{r_i'}(k,l))$ if r_i' is minimal with respect to the condition either $\alpha_{ik} \neq 0$ or $\alpha_{il} \neq 0$.

We define as H. Bresinsky a set B in three mutually exclusive cases

- For binomials $(x_s^{r_s}), s = i, j, k, l, \{i, j, k, l\} = \{1, 2, 3, 4\}$ be with at least two non-zero exponents $\alpha_{sh}, h \in \{i, j, k, l\} - \{s\}$, let $B = \{(x_i^{r_i}), (x_j^{r_j}), (x_k^{r_k}), (x_l^{r_l})\}$.

- Let $(x_i^{r_i}, x_j^{r_j}) \in P$, but $(x_k^{r_k}, x_l^{r_l}) \notin P$. Then either $B = \{(x_i^{r_i}, x_j^{r_j}), (x_k^{r_k}), (x_l^{r_l})\}$ or $B = \{(x_i^{r_i}, x_j^{r_j}), (x_k^{r_k}), (x_l^{r_l}), (x_j^{r_j'}(k,l))\}$,

- $B = \{(x_i^{r_i}, x_j^{r_j}), (x_k^{r_k}, x_l^{r_l})\} \cup C,\ C \subseteq \{(x_j^{r_j'}(k,l)), (x_l^{r_l'}(i,j))\}$.

We write $x_i^{\gamma_{i1}} x_j^{\gamma_{j1}} \nless x_i^{\gamma_{i2}} x_j^{\gamma_{j2}}$ if either $\gamma_{i1} > \gamma_{i2}$ and $\gamma_{j1} < \gamma_{j2}$ or both the inequalities are reversed. For binomials we write $x_i^{\gamma_{i1}} x_j^{\gamma_{j1}} - x_k^{\gamma_{k1}} x_l^{\gamma_{l1}} \nless x_i^{\gamma_{i2}} x_j^{\gamma_{j2}} - x_k^{\gamma_{k2}} x_l^{\gamma_{l2}}$ if $x_i^{\gamma_{i1}} x_j^{\gamma_{j1}} \nless x_i^{\gamma_{i2}} x_j^{\gamma_{j2}}$ and $x_k^{\gamma_{k1}} x_l^{\gamma_{l1}} \nless x_k^{\gamma_{k2}} x_l^{\gamma_{l2}}$. We next define a set $D_{ij}, i \neq j, \{i, j\} \subset \{1, 2, 3, 4\}$ as $D_{ij} = \{f = x_i^{\gamma_i} x_j^{\gamma_j} - x_k^{\gamma_k} x_l^{\gamma_l}, \{k, l\} \subset \{1, 2, 3, 4\} - \{i, j\}, \gamma_h < r_h$ for the polynomials $(x_h^{r_h'}(k,l))$ if $h \in \{i, j\}$; for the polynomials $(x_h^{r_h'}(i,j))$ if $h \in \{k, l\}$ and for each binomial term $f' = x_i^{\gamma_{i1}} x_j^{\gamma_{j1}} - x_k^{\gamma_{k1}} x_l^{\gamma_{l1}} \in P, f' \neq f$ is $f' \nless f\}$.

The definition of sets B, D_{ij} gives simply

Corollary 2.1.
i) If $B = \{(x_1^{r_1}), (x_2^{r_2}), (x_3^{r_3}), (x_4^{r_4})\}$ and $x_i^{\gamma_i} x_j^{\gamma_j} - x_k^{\gamma_k} x_l^{\gamma_l} \in D_{ij}$, $\{i, j, k, l\} = \{1, 2, 3, 4\}$, then $\gamma_h < r_h,\ h = 1, 2, 3, 4$.
ii) Let $B = \{(x_i^{r_i}, x_j^{r_j}), (x_k^{r_k}), (x_l^{r_l})\}$. If $x_i^{\gamma_i} x_k^{\gamma_k} - x_j^{\gamma_j} x_l^{\gamma_l} \in D_{ik}$, or $x_i^{\gamma_i} x_l^{\gamma_l} - x_j^{\gamma_j} x_k^{\gamma_k} \in D_{il}$, $\{i, j, k, l\} = \{1, 2, 3, 4\}$, then $\gamma_h < r_h,\ h = i, j$. If $x_i^{\gamma_i} x_j^{\gamma_j} - x_k^{\gamma_k} x_l^{\gamma_l} \in D_{ij}$, $\{i, j, k, l\} = \{1, 2, 3, 4\}$, then $\gamma_h < r_h,\ h = k, l$.
iii) Let $B = \{(x_i^{r_i}, x_j^{r_j}), (x_k^{r_k}), (x_l^{r_l}), (x_j^{r_j'}(k,l))\}$. If $x_i^{\gamma_i} x_k^{\gamma_k} - x_j^{\gamma_j} x_l^{\gamma_l} \in D_{ik}$, or $x_i^{\gamma_i} x_l^{\gamma_l} - x_j^{\gamma_j} x_k^{\gamma_k} \in D_{il}$, $\{i, j, k, l\} = \{1, 2, 3, 4\}$, then $\gamma_h < r_h,\ h = i, j$. If $x_i^{\gamma_i} x_j^{\gamma_j} - x_k^{\gamma_k} x_l^{\gamma_l} \in D_{ij}$, $\{i, j, k, l\} = \{1, 2, 3, 4\}$, then $\gamma_j < r_j', \gamma_h < r_h,\ h = k, l$.
iv) Let $B = \{(x_i^{r_i}, x_j^{r_j}), (x_k^{r_k}, x_l^{r_l})\} \cup C,\ C \subseteq \{(x_j^{r_j'}(k,l)), (x_l^{r_l'}(i,j))\}$. If $x_i^{\gamma_i} x_k^{\gamma_k} - x_j^{\gamma_j} x_l^{\gamma_l} \in D_{ik}$, or $x_i^{\gamma_i} x_l^{\gamma_l} - x_j^{\gamma_j} x_k^{\gamma_k} \in D_{il}$, $\{i, j, k, l\} = \{1, 2, 3, 4\}$, then $\gamma_h < r_h,\ h = 1, 2, 3, 4$.

If $x_i^{\gamma_i} x_j^{\gamma_j} - x_k^{\gamma_k} x_l^{\gamma_l} \in D_{ij}$, $\{i, j, k, l\} = \{1, 2, 3, 4\}$, then $\gamma_h < r_h',\ h = j, l$.

In [1], H. Bresinsky gives the following theorems.
Theorem 2.1. $M = B \cup D_{ij} \cup D_{ik} \cup D_{il}, \{i, j, k, l\} = \{1, 2, 3, 4\}$ is a minimal generating set for the associated prime ideal $P = P(n_1, n_2, n_3, n_4)$ of the monomial curve in A^4.

Theorem 2.2. Let $M = B \cup D_{ij} \cup D_{ik} \cup D_{il}, \{i, j, k, l\} = \{1, 2, 3, 4\}$ a minimal generating set for the associated prime ideal P. If $x_i^{\gamma_{i1}} x_j^{\gamma_{j1}} - x_k^{\gamma_{k1}} x_l^{\gamma_{l1}}, x_i^{\gamma_{i2}} x_j^{\gamma_{j2}} - x_k^{\gamma_{k2}} x_l^{\gamma_{l2}}$ are two different elements from a set D_{ij}, then there exists a uniquely determined element f in D_{ik} or D_{il}

such that

$$f = x_i^{\gamma_{i2}-\gamma_{i1}} x_k^{\gamma_{k1}-\gamma_{k2}} - x_j^{\gamma_{j1}-\gamma_{j2}} x_l^{\gamma_{l2}-\gamma_{l1}} \text{ for } \gamma_{i1} < \gamma_{i2},\ \gamma_{j1} > \gamma_{j2},\ \gamma_{k1} > \gamma_{k2},\ \gamma_{l1} < \gamma_{l2},$$
$$f = x_i^{\gamma_{i2}-\gamma_{i1}} x_l^{\gamma_{l1}-\gamma_{l2}} - x_j^{\gamma_{j1}-\gamma_{j2}} x_k^{\gamma_{k2}-\gamma_{k1}} \text{ for } \gamma_{i1} < \gamma_{i2},\ \gamma_{j1} > \gamma_{j2},\ \gamma_{k1} < \gamma_{k2},\ \gamma_{l1} > \gamma_{l2},$$
$$f = x_i^{\gamma_{i1}-\gamma_{i2}} x_l^{\gamma_{l2}-\gamma_{l1}} - x_j^{\gamma_{j2}-\gamma_{j1}} x_k^{\gamma_{k1}-\gamma_{k2}} \text{ for } \gamma_{i1} > \gamma_{i2},\ \gamma_{j1} < \gamma_{j2},\ \gamma_{k1} > \gamma_{k2},\ \gamma_{l1} < \gamma_{l2},$$
$$f = x_i^{\gamma_{i1}-\gamma_{i2}} x_k^{\gamma_{k2}-\gamma_{k1}} - x_j^{\gamma_{j2}-\gamma_{j1}} x_l^{\gamma_{l1}-\gamma_{l2}} \text{ for } \gamma_{i1} > \gamma_{i2},\ \gamma_{j1} < \gamma_{j2},\ \gamma_{k1} < \gamma_{k2},\ \gamma_{l1} > \gamma_{l2}.$$

We can find in [6], Lemma 7.1, the next property of minimal generating set for a prime ideal P.

Lemma 2.1. Let $g_j = \prod_{i=1}^{4} x_i{}^{\gamma_i}, j = 1, \ldots, h$ be a monomial term in R. Let $M = \{x_1^{r_1} - g_1, x_2^{r_2} - g_2, g_3 - g_4, \ldots, g_{h-1} - g_h\}$ be a minimal generating set for the associated prime ideal P of the monomial curve in A^4. If $x_2 \mid g_1$ and $x_1 \mid g_2$, then there is an integer k, $3 \leq k \leq h$ with $g_k = x_1^{\delta_1} x_2^{\delta_2}$.

In [9] we proved additional conditions of minimal generating set for a prime ideal P.

Theorem 2.3. Let $B = \{(x_1^{r_1}), (x_2^{r_2}), (x_3^{r_3}), (x_4^{r_4})\} \subseteq M$, where M is a minimal generating set for the associated prime ideal P of a monomial curve in A^4. If $x_i^{r_i} - x_j^{\alpha_{ij}} x_k^{\alpha_{ik}} x_l^{\alpha_{il}} \in B$ and $\alpha_{ij} \neq 0$, $\{i, j, k, l\} = \{1, 2, 3, 4\}$, then $\alpha_{ji} < r_i$.

Theorem 2.4. Let $B = \{(x_i^{r_i}, x_j^{r_j}), (x_k^{r_k}), (x_l^{r_l}), (x_j^{r_j'}(k, l))\}$ and $B \subseteq M$, $\{i, j, k, l\} = \{1, 2, 3, 4\}$, where M is a minimal generating set for the associated prime ideal P of a monomial curve in A^4.
If $\alpha_{kl} \neq 0$, then $\alpha_{lk} < r_k$. If $\alpha_{ju} \neq 0$, then $\alpha_{uj} < r_j'$, $u = k, l$.
If $\alpha_{lk} \neq 0$, then $\alpha_{kl} < r_l$. If $\alpha_{uj} \neq 0$, then $\alpha_{ju} < r_u$, $u = k, l$.

Theorem 2.5. If $B = \{(x_i^{r_i}, x_j^{r_j}), (x_k^{r_k}, x_l^{r_l})\} \cup C \subseteq M$,

$$C \subseteq \{(x_j^{r_j'}(k, l)) = x_j^{r_j'} - x_i^{\alpha_{ji}} x_k^{\alpha_{jk}} x_l^{\alpha_{jl}}, (x_l^{r_l'}(i, j)) = x_l^{r_l'} - x_i^{\alpha_{li}} x_j^{\alpha_{lj}} x_k^{\alpha_{lk}}\},$$

$\{i, j, k, l\} = \{1, 2, 3, 4\}$ and M is a minimal generating set for the associated prime ideal P of a monomial curve in A^4. Then $\alpha_{ji} < r_i$ and $\alpha_{lk} < r_k$.

Theorem 2.6. Let P be the associated prime ideal of a monomial curve in A^4. Let $B = \{(x_1^{r_1}), (x_2^{r_2}), (x_3^{r_3}), (x_4^{r_4})\}$ and $M = B \cup D_{12} \cup D_{13} \cup D_{14}$ be a minimal generating set for the ideal P, $x_i^{\gamma_i} x_j^{\gamma_j} - x_k^{\gamma_k} x_l^{\gamma_l} \in D_{ij}$, $\{i, j, k, l\} = \{1, 2, 3, 4\}$. If $(x_s^{r_s}) \in B$, $s \in \{k, l\}$, $(x_s^{r_s}) = x_s^{r_s} - N_s$ and $N_s = x_i^{\alpha_{si}} x_j^{\alpha_{sj}}$, then either $\alpha_{si} > \gamma_i \wedge \alpha_{sj} < \gamma_j$ or $\alpha_{si} < \gamma_i \wedge \alpha_{sj} > \gamma_j$.

When we use permutation of index $(i, j, k, l) \to (k, l, i, j)$ in Theorem 2.6 we receive following

Corollary 2.2. Let P be the associated prime ideal of a monomial curve in A^4. Let $B = \{(x_1^{r_1}), (x_2^{r_2}), (x_3^{r_3}), (x_4^{r_4})\}$ and $M = B \cup D_{12} \cup D_{13} \cup D_{14}$ be a minimal generating set for the ideal P, $x_i^{\gamma_i} x_j^{\gamma_j} - x_k^{\gamma_k} x_l^{\gamma_l} \in D_{ij}$, $\{i, j, k, l\} = \{1, 2, 3, 4\}$.

If $(x_s^{r_s}) \in B$, $s \in \{i, j\}$, $(x_s^{r_s}) = x_s^{r_s} - N_s$ and $N_s = x_k^{\alpha_{sk}} x_l^{\alpha_{sl}}$, then either $\alpha_{sk} > \gamma_k \wedge \alpha_{sl} < \gamma_l$ or $\alpha_{sk} < \gamma_k \wedge \alpha_{sl} > \gamma_l$.

Using these results, we can find other conditions for the set $M = B \cup D_{12} \cup D_{13} \cup D_{14}$. We denote the number of elements of the set D_{ij} as $|D_{ij}|$, $\{i, j\} \subset \{1, 2, 3, 4\}$ and the number of elements of the set M as $\mu(P)$.

Theorem 2.7. If $M = \{(x_i^{r_i}, x_j^{r_j}), (x_k^{r_k}, x_l^{r_l})\} \cup D_{ij} \cup D_{ik} \cup D_{il}, \{i,j,k,l\} = \{1,2,3,4\}$ is a minimal generating set for the associated prime ideal P of the monomial curve in A^4 and $\mu(P) = 3$, then $D_{ij} \cup D_{ik} \cup D_{il} = \{x_i^{\gamma_i} x_j^{\gamma_j} - x_k^{\gamma_k} x_l^{\gamma_l}\}$.

Proof: Assume that $D_{ij} = \emptyset$ and $D_{ik} = \{x_i^{\omega_i} x_k^{\omega_k} - x_j^{\omega_j} x_l^{\omega_l}\}$. By Corollary 2.1 iv) is $\omega_h < r_h, h = i,j,k,l$ holds. Hence

$$x_i^{r_i-\omega_i} x_l^{\omega_l} - x_j^{r_j-\omega_j} x_k^{\omega_k} \in P \subseteq P + (x_j, x_k) \subseteq (x_i^{r_i}, x_j, x_k, x_l^{r_l})$$

$x_i^{r_i-\omega_i} x_l^{\omega_l} \in (x_i^{r_i}, x_j, x_k, x_l^{r_l})$ and $\omega_l < r_l, r_i - \omega_i < r_i$, it is impossible.
When $D_{ij} = \emptyset$ and $D_{il} = \{x_i^{\omega_i} x_l^{\omega_l} - x_j^{\omega_j} x_k^{\omega_k}\}$, we have

$$x_i^{r_i-\omega_i} x_k^{\omega_k} - x_j^{r_j-\omega_j} x_l^{\omega_l} \in P \subseteq P + (x_j, x_l) \subseteq (x_i^{r_i}, x_j, x_k^{r_k}, x_l), \quad \text{where}$$

$\omega_h < r_h, h = i,j,k,l$. Such a binomial cannot be in the prime ideal P due to $D_{ij} \neq \emptyset$.

Remark 2.1. H. Bresinsky in [3] dealt with ideal complete intersections and showed that monomial curves in A^4 are ideal complete intersections if and only if its associated prime ideals have minimal generating sets $M = \{(x_i^{r_i}, x_j^{r_j}), (x_k^{r_k}, x_l^{r_l}), x_i^{\gamma_i} x_j^{\gamma_j} - x_k^{\gamma_k} x_l^{\gamma_l}\}$ or $M = \{(x_i^{r_i}, x_j^{r_j}), (x_k^{r_k}), (x_l^{r_l})\}$.

Theorem 2.8. Let P be the associated prime ideal of a monomial curve in A^4, $\mu(P) > 3$. If $B = \{(x_i^{r_i}, x_j^{r_j}), (x_k^{r_k}, x_l^{r_l})\}$, $\{i,j,k,l\} = \{1,2,3,4\}$, $D_{ik} = \emptyset$ or $D_{il} = \emptyset$, then $M = B \cup D_{ij} \cup D_{ik} \cup D_{il}, \{i,j,k,l\} = \{1,2,3,4\}$ is not a minimal generating set of the ideal P.

Proof: If $M = B \cup D_{ij} \cup D_{ik} \cup D_{il}$, $B = \{(x_i^{r_i}, x_j^{r_j}), (x_k^{r_k}, x_l^{r_l})\}$, $\{i,j,k,l\} = \{1,2,3,4\}$ is the minimal generating set of the ideal P and $D_{il} = \emptyset$, then is $D_{ij} \neq \emptyset$ and $D_{ik} \neq \emptyset$.
In case that $D_{ik} = \emptyset$ is $D_{ij} \neq \emptyset$ and $D_{il} \neq \emptyset$. When $x_i^{\omega_i} x_k^{\omega_k} - x_j^{\omega_j} x_l^{\omega_l} \in D_{ik}$, using Corollary 2.1 $\omega_h < r_h, h = i,j,k,l$ hold and

$$x_i^{r_i-\omega_i} x_l^{\omega_l} - x_j^{r_j-\omega_j} x_k^{\omega_k} \in P \subseteq P + (x_j, x_k) \subseteq (x_i^{r_i}, x_j, x_k, x_l^{r_l}),$$

thus $x_i^{r_i-\omega_i} x_l^{\omega_l} \in (x_i^{r_i}, x_j, x_k, x_l^{r_l})$ and $\omega_l < r_l, r_i - \omega_i < r_i$, which is a contradiction. When $x_i^{\omega_i} x_l^{\omega_l} - x_j^{\omega_j} x_k^{\omega_k} \in D_{il}$, we have again $\omega_h < r_h, h = i,j,k,l$, $x_i^{r_i-\omega_i} x_k^{\omega_k} - x_j^{r_j-\omega_j} x_l^{\omega_l} \in P \subseteq P + (x_j, x_l) \subseteq (x_i^{r_i}, x_j, x_k^{r_k}, x_l)$, which is a contradiction and the set M cannot be minimal generating set for the associated prime ideal of a monomial curve in A^4.

Theorem 2.9. Let P be the associated prime ideal of a monomial curve in A^4, $\mu(P) > 3$. If $B = \{(x_i^{r_i}, x_j^{r_j}), (x_k^{r_k}, x_l^{r_l})\}$, $\{i,j,k,l\} = \{1,2,3,4\}$, $|D_{ik}| = |D_{il}| = 1$, then $M = B \cup D_{ij} \cup D_{ik} \cup D_{il}, \{i,j,k,l\} = \{1,2,3,4\}$ is not a minimal generating set of the ideal P.

Proof: If $M = B \cup D_{ij} \cup D_{ik} \cup D_{il}$, $B = \{(x_i^{r_i}, x_j^{r_j}), (x_k^{r_k}, x_l^{r_l})\}$, $\{i,j,k,l\} = \{1,2,3,4\}$ is the minimal generating set of the ideal P and $|D_{ik}| = |D_{il}| = 1$, for $\mu(P) \geq 7$ is $|D_{ij}| \geq 3$. From Theorem 2.2 is $|D_{ik}| > 1$ or $|D_{il}| > 1$, it is contradiction. Next we suppose that $7 > \mu(P) > 3$.
Let $x_i^{\omega_i} x_k^{\omega_k} - x_j^{\omega_j} x_l^{\omega_l} \in D_{ik}$ and $x_i^{\delta_i} x_l^{\delta_l} - x_j^{\delta_j} x_k^{\delta_k} \in D_{il}$. Using Corollary 2.1 $\omega_h < r_h, \delta_h < r_h$, $h = i,j,k,l$ hold. We have

$$1) \quad x_i^{r_i-\omega_i} x_l^{\omega_l} - x_j^{r_j-\omega_j} x_k^{\omega_k} \in P,$$
$$2) \quad x_l^{r_l-\omega_l} x_i^{\omega_i} - x_k^{r_k-\omega_k} x_j^{\omega_j} \in P,$$
$$3) \quad x_i^{r_i-\delta_i} x_k^{\delta_k} - x_j^{r_j-\delta_j} x_l^{\delta_l} \in P,$$
$$4) \quad x_k^{r_k-\delta_k} x_i^{\delta_i} - x_l^{r_l-\delta_l} x_j^{\delta_j} \in P.$$

The condition 1) gives
$x_i^{r_i-\omega_i}x_l^{\omega_l} - x_j^{r_j-\omega_j}x_k^{\omega_k} \in P + (x_j, x_k) \subseteq (x_i^{r_i}, x_j, x_k, x_l^{r_l}, x_i^{\delta_i}x_l^{\delta_l})$,
$x_i^{r_i-\omega_i}x_l^{\omega_l} \in (x_i^{r_i}, x_j, x_k, x_l^{r_l}, x_i^{\delta_i}x_l^{\delta_l})$ and

$$x_i^{r_i-\omega_i}x_l^{\omega_l} - x_j^{r_j-\omega_j}x_k^{\omega_k} \in P + (x_i, x_l) \subseteq (x_i, x_j^{r_j}, x_k^{r_k}, x_l, x_j^{\delta_j}x_k^{\delta_k}),$$

thus $x_j^{r_j-\omega_j}x_k^{\omega_k} \in (x_i, x_j^{r_j}, x_k^{r_k}, x_l, x_j^{\delta_j}x_k^{\delta_k})$. Clearly,

$$r_i - \omega_i \geq \delta_i \wedge \omega_l \geq \delta_l,\ r_j - \omega_j \geq \delta_j \wedge \omega_k \geq \delta_k.$$

By analogy, we get the following conditions

From 2)	$r_l - \omega_l \geq \delta_l \wedge \omega_i \geq \delta_i,\ r_k - \omega_k \geq \delta_k \wedge \omega_j \geq \delta_j,$
From 3)	$r_i - \delta_i \geq \omega_i \wedge \delta_k \geq \omega_k,\ r_j - \delta_j \geq \omega_j \wedge \delta_l \geq \omega_l,$
From 4)	$r_k - \delta_k \geq \omega_k \wedge \delta_i \geq \omega_i,\ r_l - \delta_l \geq \omega_l \wedge \delta_j \geq \omega_j.$

Hence is $\omega_h = \delta_h,\ r_h \geq 2\omega_h,\ h = i,j,k,l,\ x_i^{\omega_i}x_k^{\omega_k} - x_j^{\omega_j}x_l^{\omega_l},\ x_i^{\omega_i}x_l^{\omega_l} - x_j^{\omega_j}x_k^{\omega_k} \in P$. Combination of these binomials gives $x_i^{\omega_i}(x_k^{2\omega_k} - x_l^{2\omega_l}) \in P$, $x_l^{\omega_l}(x_i^{2\omega_i} - x_j^{2\omega_j}) \in P$.
We know that P is a prime ideal and therefore is $x_k^{2\omega_k} - x_l^{2\omega_l} \in P$, $x_i^{2\omega_i} - x_j^{2\omega_j} \in P$. Moreover, is $x_k^{\omega_k} - x_l^{\omega_l} \in P$, $x_i^{\omega_i} - x_j^{\omega_j} \in P$, but r_h, $h = i,j,k,l$ must be minimal, a contradiction. Thus the set M cannot be a minimal generating set for the associated prime ideal of a monomial curve in A^4.

Theorem 2.10. Let P be the associated prime ideal of a monomial curve in A^4, $\mu(P) > 3$. If $B = \{(x_i^{r_i}, x_j^{r_j}), (x_k^{r_k}, x_l^{r_l})\}$, $\{i,j,k,l\} = \{1,2,3,4\}$, $|D_{ik}| = 2, |D_{il}| = 1$, then $M = B \cup D_{ij} \cup D_{ik} \cup D_{il}$, $\{i,j,k,l\} = \{1,2,3,4\}$ is not a minimal generating set of the ideal P.

Proof: If $M = B \cup D_{ij} \cup D_{ik} \cup D_{il}$, $B = \{(x_i^{r_i}, x_j^{r_j}), (x_k^{r_k}, x_l^{r_l})\}$, $\{i,j,k,l\} = \{1,2,3,4\}$ is the minimal generating set of the ideal P, $\mu(P) \geq 9$, then $|D_{ij}| \geq 4$, but from Theorem 2.2 $|D_{ik}| > 2$ or $|D_{il}| > 1$, a contradiction. Next, we can assume that $9 > \mu(P) > 3$.
Denote $f_1 = x_i^{\omega_i}x_k^{\omega_k} - x_j^{\omega_j}x_l^{\omega_l} \in D_{ik}$, $f_2 = x_i^{\gamma_i}x_k^{\gamma_k} - x_j^{\gamma_j}x_l^{\gamma_l} \in D_{ik}$ and
$g = x_i^{\delta_i}x_l^{\delta_l} - x_j^{\delta_j}x_k^{\delta_k} \in D_{il}$. Using Corollary 2.1 is $\omega_h < r_h, \gamma_h < r_h, \delta_h < r_h$, $h = i,j,k,l$. This gives

a)	$x_i^{r_i-\omega_i}x_l^{\omega_l} - x_j^{r_j-\omega_j}x_k^{\omega_k} \in P,$
b)	$x_l^{r_l-\omega_l}x_i^{\omega_i} - x_k^{r_k-\omega_k}x_j^{\omega_j} \in P,$
c)	$x_i^{r_i-\delta_i}x_k^{\delta_k} - x_j^{r_j-\delta_j}x_l^{\delta_l} \in P,$
d)	$x_k^{r_k-\delta_k}x_i^{\delta_i} - x_l^{r_l-\delta_l}x_j^{\delta_j} \in P,$
e)	$x_i^{r_i-\gamma_i}x_l^{\gamma_l} - x_j^{r_j-\gamma_j}x_k^{\gamma_k} \in P,$
f)	$x_l^{r_l-\gamma_l}x_i^{\gamma_i} - x_k^{r_k-\gamma_k}x_j^{\gamma_j} \in P.$

The same reasons as in the proof of Theorem 2.9, these conditions implies for the exponents of binomials in the set M that

$$\begin{array}{ll} \text{a)} & r_i - \omega_i \geq \delta_i \wedge \omega_l \geq \delta_l,\ r_j - \omega_j \geq \delta_j \wedge \omega_k \geq \delta_k, & (1) \\ \text{b)} & r_l - \omega_l \geq \delta_l \wedge \omega_i \geq \delta_i,\ r_k - \omega_k \geq \delta_k \wedge \omega_j \geq \delta_j, & (2) \\ \text{e)} & r_i - \gamma_i \geq \delta_i \wedge \gamma_l \geq \delta_l,\ r_j - \gamma_j \geq \delta_j \wedge \gamma_k \geq \delta_k, & (3) \\ \text{f)} & r_l - \gamma_l \geq \delta_l \wedge \gamma_i \geq \delta_i,\ r_k - \gamma_k \geq \delta_k \wedge \gamma_j \geq \delta_j, & (4) \\ \text{c)} & r_i - \delta_i \geq \omega_i \wedge \delta_k \geq \omega_k,\ r_j - \delta_j \geq \omega_j \wedge \delta_l \geq \omega_l \text{ or} & (5) \\ & r_i - \delta_i \geq \gamma_i \wedge \delta_k \geq \gamma_k,\ r_j - \delta_j \geq \gamma_j \wedge \delta_l \geq \gamma_l, & (6) \\ \text{d)} & r_k - \delta_k \geq \omega_k \wedge \delta_i \geq \omega_i,\ r_l - \delta_l \geq \omega_l \wedge \delta_j \geq \omega_j \text{ or} & (7) \\ & r_k - \delta_k \geq \gamma_k \wedge \delta_i \geq \gamma_i,\ r_l - \delta_l \geq \gamma_l \wedge \delta_j \geq \gamma_j. & (8) \end{array}$$

First assume inequalities (5) and (1), we obtain $\omega_u = \delta_u,\ u = k, l$ and from (3) is $\gamma_u \geq \omega_u,\ u = k, l$. By the definition of the set D_{ik}, the following inequalities

$$(\omega_i > \gamma_i \wedge \omega_k < \gamma_k) \wedge (\omega_j > \gamma_j \wedge \omega_l < \gamma_l) \tag{9}$$

have to be satisfied. The inequalities (2), (4), (7) gives $\omega_s = \delta_s,\ s = i, j$ and $\gamma_s \geq \omega_s,\ s = i, j$, which is contrary to the inequalities (9), thus (5) and (7) cannot be both true. The inequalities (8), (2), (4) gives that $\gamma_s = \delta_s,\ \omega_s \geq \gamma_s,\ s = i, j$. From (1) and (3), (2) and (4), we have that $r_h \geq \gamma_h + \omega_h,\ h = i, j, k, l$.

If $|D_{ik}| = 2$, in accordance with Theorem 2.2 the binomial

$$x_i^{\omega_i - \gamma_i} x_l^{\gamma_l - \omega_l} - x_j^{\omega_j - \gamma_j} x_k^{\gamma_k - \omega_k} \in P \tag{10}$$

belong into to the set D_{il}, hence $\omega_s = 2\gamma_s,\ s = i, j,\ 2\omega_u = \gamma_u,\ u = k, l$. With respect to all conditions for exponents of binomials in the set M

$$x_l^{\gamma_l} f_2 - x_k^{\gamma_k} x_l^{\gamma_l - \omega_l} g = x_j^{\gamma_j} x_l^{\gamma_l - \omega_l} (x_k^{\gamma_k + \omega_k} - x_l^{\gamma_l + \omega_l}) \in P.$$

Since P is a prime ideal, $x_k^{\gamma_k + \omega_k} - x_l^{\gamma_l + \omega_l} \in P$ and $r_u = \gamma_u + \omega_u = 3\omega_u,\ u = k, l$. The generator $(x_k^{r_k}, x_l^{r_l})$ can also be described by

$$(x_k^{\omega_k} - x_l^{\omega_l})(x_k^{2\omega_k} + x_k^{\omega_k} x_l^{\omega_l} + x_l^{2\omega_l}),$$

such that $x_k^{\omega_k} - x_l^{\omega_l} \in P$, but the exponent r_l is minimal, a contradiction. Thus (5) and (8) cannot be both true.

Next, we assume the inequalities (6), (1), (3), we get $\gamma_u = \delta_u,\ \omega_u \geq \gamma_u,\ u = k, l$ and by the definition of the set D_{ik}, the following inequalities have to be satisfied

$$(\omega_k > \gamma_k \wedge \omega_i < \gamma_i) \wedge (\omega_l > \gamma_l \wedge \omega_j < \gamma_j). \tag{11}$$

Hence, the inequalities (6) and (8) can not be met, because from (2), (4), (8) is $\gamma_s = \delta_s,\ s = i, j\ \omega_s \geq \gamma_s,\ s = i, j$ and it is contradiction with inequalities (11). Next we suppose that the inequalities (5), (8) are false, now $\omega_s = \delta_s,\ s = i, j$ and with inequalities (10) is $\omega_u = 0,\ u = k, l$, it is contradiction. Thus, the set M cannot be a minimal generating set for the associated prime ideal P.

Theorem 2.11. Let P be the associated prime ideal of a monomial curve in A^4, $\mu(P) > 3$. If $B = \{(x_i^{r_i}, x_j^{r_j}), (x_k^{r_k}, x_l^{r_l})\}$, $\{i,j,k,l\} = \{1,2,3,4\}$, $|D_{ik}| = 1, |D_{il}| = 2$, then $M = B \cup D_{ij} \cup D_{ik} \cup D_{il}, \{i,j,k,l\} = \{1,2,3,4\}$ is not a minimal generating set of the ideal P.

Proof: Analogously to the proof of Theorem 2.10.

Corollary 2.3. Let P be the associated prime ideal of a monomial curve in A^4, $\mu(P) > 3$. If $B = \{(x_i^{r_i}, x_j^{r_j}), (x_k^{r_k}, x_l^{r_l})\}$, $\{i,j,k,l\} = \{1,2,3,4\}$, $D_{il} = \emptyset$ or $D_{ik} = \emptyset$ or $|D_{ik}| = |D_{il}| = 1$ or $|D_{ik}| = 2 \wedge |D_{il}| = 1$ or $|D_{ik}| = 1 \wedge |D_{il}| = 2$, then $M = B \cup D_{ij} \cup D_{ik} \cup D_{il}, \{i,j,k,l\} = \{1,2,3,4\}$ is not a minimal generating set of the ideal P.

Remark 2.2. Previous results give for example that if $\mu(P) = 5$ then the set $M = \{(x_i^{r_i}, x_j^{r_j}), (x_k^{r_k}, x_l^{r_l})\} \cup D_{ij} \cup D_{ik} \cup D_{il}, \{i,j,k,l\} = \{1,2,3,4\}$ cannot be a minimal generating set of the ideal P.

When we use ideas and above properties, we can prove the following theorem about minimal generating sets for associated prime ideals of monomial curves, which excludes some descriptions of the prime ideal P.

Theorem 2.12. Let P be the associated prime ideal of a monomial curve in A^4, $\mu(P) \geq 5$. If

1) $B = \{(x_i^{r_i}, x_j^{r_j}), (x_k^{r_k}, x_l^{r_l})\} \cup C$, where $\emptyset \neq C \subset \{(x_j^{r'_j}(k,l)), (x_l^{r'_l}(i,j))\}$, $D_{il} = \emptyset$ or $D_{ik} = \emptyset$ or $|D_{ik}| = |D_{il}| = 1$,

2) $B = \{(x_i^{r_i}, x_j^{r_j}), (x_k^{r_k}, x_l^{r_l}), (x_j^{r'_j}(k,l)), (x_l^{r'_l}(i,j))\}$, $D_{ik} = \emptyset$ or $D_{il} = \emptyset$,

3) $B = \{(x_i^{r_i}, x_j^{r_j}), (x_k^{r_k}), (x_l^{r_l})\}$, $D_{ij} = \emptyset$ or $D_{ik} = \emptyset$ alebo $D_{il} = \emptyset$,

4) $B = \{(x_i^{r_i}, x_j^{r_j}), (x_k^{r_k}), (x_l^{r_l}), (x_j^{r'_j}(k,l))\}$, $D_{ik} = \emptyset$ or $D_{il} = \emptyset$,

then $M = B \cup D_{ij} \cup D_{ik} \cup D_{il}, \{i,j,k,l\} = \{1,2,3,4\}$ is not a minimal generating set of the ideal P.

Remark 2.3. The inducted propositions give necessary conditions on a minimal set of generators for an associated prime ideal P of a monomial curve in A^4. We suppose in [9] that this minimal generating set has five generators. We presented in [9] that if P is the associated ideal of a monomial curve in A^4 and $\mu(P) = 5$, then the minimal generating set M is

$$\begin{aligned} M = \quad & \{x_i^{r_i} - x_j^{\alpha_{ij}} x_k^{\alpha_{ik}} x_l^{\alpha_{il}}, x_j^{r_j} - x_i^{\alpha_{ji}} x_k^{\alpha_{jk}} x_l^{\alpha_{jl}}, x_k^{r_k} - x_i^{\alpha_{ki}} x_l^{\alpha_{kl}}, \\ & x_l^{r_l} - x_i^{\alpha_{li}} x_j^{\alpha_{lj}}, x_i^{\gamma_i} x_k^{\gamma_k} - x_j^{\gamma_j} x_l^{\gamma_l}\}, \end{aligned}$$

where $\{i,j,k,l\} = \{1,2,3,4\}$ and the only one of the following cases occurs:

I) only α_{il} or α_{ji} can be zero and all exponents must satisfy next equations:

$$\begin{aligned} r_i &= \alpha_{ji} + \alpha_{ki} + \alpha_{li}, r_j = \alpha_{ij} + \alpha_{lj}, r_k = \alpha_{ik} + \alpha_{jk}, r_l = \alpha_{il} + \alpha_{jl} + \alpha_{kl}, \\ \omega_i &= \alpha_{ji} + \alpha_{li}, \omega_j = \alpha_{ij}, \omega_k = \alpha_{jk}, \omega_l = \alpha_{il} + \alpha_{kl}, \end{aligned} \tag{12}$$

II) $\alpha_{jl} = 0$, only α_{il} can be zero and it is $u \in N$, that all exponents must satisfy next equations:

$$\begin{aligned} & r_i = \alpha_{li} + \alpha_{ji} + (u+1)\alpha_{ki}, r_j = \alpha_{lj} + \alpha_{ij}, r_k = \alpha_{ik} + \omega_k, \\ & r_l = \alpha_{il} + (u+1)\alpha_{kl}, \alpha_{jk} + \alpha_{ik} = (u+1)r_k, \\ & \omega_i = \alpha_{li} + \alpha_{ji} + u\alpha_{ki}, \ \omega_j = \alpha_{ij}, \\ & \omega_k = r_k - \alpha_{ik} = \alpha_{jk} - ur_k, \ \omega_l = \alpha_{il} + \alpha_{kl}. \end{aligned} \tag{13}$$

We proved in [9] that if P is the associated prime ideal of a monomial curve C in A^4 and its a minimal generating set is the set M having five generators and noted above, then this prime ideal P (monomial curve C) is a set-theoretic complete intersection.

Example 2.1. a) We know that minimal generating sets for the associated prime ideals P of monomial Gorenstein curves (non ideal-theoretic intersection) in A^4 belongs to case I) and $\alpha_{il} = 0 \wedge \alpha_{ji} = 0$ (see [2],[3]).

b) In [8] we can find two groups of monomial curves in A^4

$$C(4s-1, 4s, 4s+1, 6s) \qquad \text{and}$$
$$C(4s+1, 4s+2, 4s+3, 6s+3), \ s \in N, \ s \geq 2.$$

Minimal generating sets of their associated prime ideals are

$$\{x_i^{s+1} - x_j x_k {x_l}^{s-1}, x_j^2 - x_i x_k x_l, x_k^2 - x_i x_l, x_l^{s+l} - x_i^{s-1} x_j, x_i^s x_k - x_j {x_l}^s\}$$

and

$$\{x_i^{s+2} - x_j x_k {x_l}^{s-1}, x_j^2 - x_i x_k x_l, x_k^2 - x_i x_l, x_l^{s+l} - x_i^s x_j, x_i^{s+1} x_k - x_j {x_l}^s\}$$

for $(i, j, k, l) = (3, 4, 2, 1)$ and $(i, j, k, l) = (1, 4, 2, 3)$. So we can say that these ideals are from the case I) and $\alpha_{il} \neq 0 \wedge \alpha_{ji} \neq 0$. Thus these curves and their associated prime ideals are s.t.c.i.

References

[1] BRESINSKY, H. *Binomial generating sets for monomial curves, with applications in A^4*. Rend.Sem.Mat., Univers. Politecn. Torino, 46, 1988, pp. 353-370.

[2] BRESINSKY, H. *Monomial Gorenstein curves in A^4 as set-theoretic complete intersections.* Manuscripta Math, 27, 1979, pp. 353-358.

[3] BRESINSKY, H. *Symmetric semigroup of integers generated by 4 elements.* Manuscripta Math, 17, 1975, pp. 205-219.

[4] COX, D., LITTLE, J., O'SHEA, D. *Using Algebraic Geometry.* Graduate Texts in Mathematics, Volume 185, Springer - Verlag, 1998.

[5] EISENBUD, D., EVANS,E.G., Jr. *Every algebraic set in n-space is the intersection of n hypersurfaces.* Inventiones. math., 19: 380-383, 1973.

[6] GASTINGER, W. *Über die Verschwindungsideale monomialer Kurven.* Dissertation zur Erlangung des Doktorgrades der Naturwissenschaften, Landshut, 1989.

[7] HOLEŠOVÁ, M. *Associated ideals of monomial curves in A^4*, PhD. thesis [slovak], FMFI UK Bratislava, 2005.

[8] HOLEŠOVÁ, M. *About monomial curves $C(2k-1, 2k, 2k+1, 3k)$*.[slovak] Acta Facultatis Paedagogicae Universitatis Tyrnaviensis, Trnavská univerzita, 2004, pp. 5-12.

[9] HOLEŠOVÁ, M. *Asociated prime ideals of monomial curves with five generators as set-theoretic complete intersections.* Studies of the University of Žilina, Mathematical Series Vol. 25/2011, pp. 1-8.

[10] KUNZ, E. *Einführung in die kommutative Algebra und algebraische Geometrie.* Vieweg, Braunschweig, 1980.

Current address

Holešová Michaela, RNDr. PhD.
Department of Structural Mechanics and Applied Mathematics,
Faculty of Civil Engineering, University of Žilina
Univerzitná 8215/1, 010 26 Žilina, Slovak republic
Tel. number: +42141 513 6275, e-mail: michaela.holesova@fstav.uniza.sk

Slovak University of Technology in Bratislava
Faculty of Mechanical Engineering
16th Conference on Applied Mathematics **APLIMAT 2017**
Proceedings

ILL-POSED PROBLEMS. APPLICATION TO THE TIME SERIES ANALYSIS

HORSKÝ Richard (CZ)

Abstract. Many problems in mathematics and applied science take the form of an operator equation. The operator is usually a matrix or a difference, differential or an integral operator. It turns out that a great number of such problems are ill-posed which means that the solution either does not exist or it is not unique or it is unstable. In time series analysis the ill-posedness is demonstrated by the non-stationarity of a stochastic process. The typical example of such a process is the random walk.

Keywords: ill-posed problem, regularization, symmetrization, random walk

Mathematics Subject Classification: Primary 39A50, 65J20; Secondary 47B39

1 Ill-posed problems

The definition of the well-posed problem was introduced by J. Hadamard in 1902 when he studied what types of boundary conditions should be used for various types of differential equations. Hadamard's concept reflected the idea that any mathematical model of physical phenomena should have all the following properties:

A. There exists a solution to the problem (existence).
B. There is only one solution to the problem (uniqueness).
C. The solution depends continuously on the data (stability).

A problem which does not satisfy at least one of these conditions was called *ill-posed* (improperly posed) problem. Hadamard also gave the example of ill-posed problem, which is today taken as classical one: the Cauchy problem for the Laplace equation. It became clear later that many branches of natural sciences in general involve ill-posed problem, e.g. computer tomography (see [5], p.183) and others. The great development in research of methods for solving ill-posed problems began in 1950'. Here must be mentioned the names of outstanding Russian mathematicians: A. N. Tikhonov, M. M. Lavrentiev, V. K. Ivanov and also their disciples, e.g. V. A. Morozov. They contributed much to the development of the theory and techniques which became a new, remarkable and important area in numerical analysis (e.g. [7]). The foundations of approximate methods for solving ill-posed problems were laid by A. N. Tikhonov. He generalized the classical Hadamard's concept of well-posedness of the problem.

It is usual that the problems of mathematical physics and analysis are formulated as the problem of solving the equation

$$Ax = b\,, \tag{1}$$

where $A: V \to W$ is an operator defined on a normed space V with values in a normed space W, b is a given element in W. Tikhonov was the first to invoke a topological theorem providing sufficient conditions for the problem (1) to be well-posed in his new sense: *Let A be a continuous operator on a non-empty compact set M in V, and suppose A is injective. Then its inverse A^{-1} considered on $A(M)=N$ is continuous.* In Tikhonov's viewpoint a certain compact set plays a basic role. Generalizations of this theorem to metric and topological spaces were obtained by Ivanov. As to the ill-posed problems Tikhonov suggested a new method which he called *regularization*. This method is based on the idea that the minimum deviation of the values Ax from the specified right hand side b is stabilized by means of some non-negative convex functional (see below).

1.1 Some examples of ill-posed problems

1) If the spaces V and W in the definition of the operator A in (1) are both equal to the euclidean n-dimensional space then the operator A is represented by a square matrix $\mathbf{A}$. It is well-known that in this case A is always bounded, i.e. continuous (w.r.t. any norm in V and W resp.). The well-posedness of (1) is violated if $\mathbf{A}$ is singular. Then A is neither injective nor surjective. However, even if $\mathbf{A}$ is nonsingular the system (1) can be problematic in the numeric viewpoint. In this context the notion of the condition number of the matrix is introduced: $\mathrm{cond}\mathbf{A} = \|\mathbf{A}\|\|\mathbf{A}^{-1}\|$. Let b_δ be a perturbation of b in (1), $Ax^*=b$, $Ax_\delta= b_\delta$. Then $x^*-x_\delta=A^{-1}(b- b_\delta)$. For the relative error of the solution it holds $\frac{\|x^* - x_\delta\|}{\|x^*\|} \le \mathrm{cond}\mathbf{A}\frac{\|b - b_\delta\|}{\|b\|}$. One can see from this estimate that a large condition number can result in a considerable change in the solution even if the relative change in the right hand side is small.

2) Suppose the normed spaces V and W are Banach spaces and A is a linear, bounded and injective operator with a closed range $R(A)$. Then its inverse A^{-1} is also bounded (see [9], p.176). Thus the equation (1) is well-posed. The typical example of this situation is the Fredholm integral equation of the second kind (see [8], p.88). The equation (1) may be ill-posed. It happens if either the operator A is not bounded below (see [6], p.73) or the right hand side of the equation (1) is not in the range $R(A)$. The important case of such an operator is a compact operator in the infinite dimensional space. Let us recall that A is the *compact operator* if it transforms any bounded set in V onto a relatively compact set (its closure is compact) in W. The compact operator is a natural generalization of the operator in finite dimensional spaces. If V or W are spaces of finite dimension then any linear operator $A: V \to W$ is compact. If $V = W$ then the compact operators form a closed ideal $\mathcal{L}_C(V)$ in the space of all bounded linear operators $\mathcal{L}(V)$ the domain of which is the Banach space V and their range is a subspace of V. If the Banach space V is of infinite dimension then

a compact operator A may not have bounded inverse operator: if A^{-1} were bounded then the identity $I = A^{-1}A$ should be a compact operator since $\mathcal{L}_C(V)$ is an ideal. However the identity I on V is compact if and only if the space V is of a finite dimension. The typical example of a compact operator is the Fredholm integral operator of the first kind, for instance in the form:

$$y(t) = Ax(t) = \int_0^1 k(t,s)x(s)ds. \tag{2}$$

In the formula (2) $k(t,s)$ is the given function lying in the space $C([0,1]\times[0,1])$ (kernel of the integral operator (2)), $x, y \in V = W = C[0,1]$, where the space $C(K)$ of all continuous functions on the compact set K equipped with a supremum norm is the Banach space (e.g. see [6], p.30).

3) Differentiation and integration are two problems which are inverse to each other. Even if the symbolic differentiation is much simpler than symbolic integration it may be surprising the differentiation is ill-posed in the settings considered below. The differentiation turns out to be the more delicate problem from the numerical point of view. We define the operator

$$y(t) = Ax(t) = \int_0^t x(s)ds, \tag{3}$$

where $t \in [0,1]$, $x \in C[0,1]$. The spaces $V = C[0,1]$, $W = C[0,1] \cap C^1(0,1)$ are regarded as normed linear spaces, both with the supremum norm $\|\cdot\|_\infty$, W is a subspace of V. The inverse operator A^{-1} to the operator (3) is defined on the space $R(A)$ of those functions in W satisfying the condition $y(0) = 0$ and it is the derivative. Let us assume that we are given certain function $b \in R(A)$ and noisy data b_δ for which $\|b - b_\delta\|_\infty \le \delta$. If we have a sequence of functions $b_{\delta,n}(t) = b(t) + \delta \sin \frac{nt}{\delta}$, $\delta \in (0,1)$, $n = 1,2,3,\ldots$ then it is obviously $\|b - b_{\delta,n}\|_\infty \le \delta$. On the other hand for the solutions of (1) with operator (3) we obtain $\|A^{-1}b - A^{-1}b_{\delta,n}\|_\infty = n$ since $A^{-1}b - A^{-1}b_{\delta,n} = \frac{db}{dt} - \frac{db_{\delta,n}}{dt} = n\cos\frac{nt}{\delta}$. Hence, the error in the solution increase without bound as $n \to \infty$ although the error in the data is bounded by δ. This shows that the equation (1) with the operator (3) is ill-posed.

1.2 General concept of regularization

Suppose again the general problem (1), where $A: V \to W$, V and W are normed linear spaces, and A is a bounded linear and injective operator with inverse which is not bounded, i.e. (1) is ill-posed problem. This is a situation which is usual in applications and we met it in the examples with the operators (2) and (3) respectively. The problem (1) is instable and the idea is to approximate this problem by a well posed one. The procedure for this replacement is called *regularization method*.

First we deal with approximations of the inverse operator A^{-1} (which is not bounded). A *regularization family for A* is a collection of bounded linear operators $R_\alpha : W \to V$, $\alpha > 0$, for which $\lim_{\alpha \to 0+} \|R_\alpha Ax - x\| = 0$ for all $x \in V$. A regularization family may not be uniformly bounded. In fact, there is a sequence of positive numbers $\alpha_n \to 0$ such that $\|R_{\alpha_n}\| \to \infty$. If there were a constant c such that for all $\alpha > 0$, we would have $\|R_\alpha\| \le c$, then for any $y \in R(A)$ the following inequalities would hold:

$$\|A^{-1}y\| \le \|A^{-1}y - R_\alpha y\| + \|R_\alpha y\| \le \|x - R_\alpha Ax\| + \|R_\alpha\|\|y\| \le \|x - R_\alpha x\| + c\|y\| \quad (4)$$

The first term on the right hand side in (4) tends to zero as $\alpha \to 0$. However, it means that $\|A^{-1}y\| \le c\|y\|$ which is a contradiction with A^{-1} is not bounded. In other words, the product $R_\alpha A$ converges to identity I pointwise, not uniformly.

Second the right hand side of the equation (1) contains a noise. We could watch the effect of this noise in the third example in the previous section. Suppose that $\boldsymbol{b_\delta}$ is a perturbation of b in (1), $\|b - b_\delta\|_\infty \le \delta$. We then define

$$x_{\alpha,\delta} = R_\alpha b_\delta. \quad (5)$$

The vector (5) is called the *regularized solution to (1)*.

Let $x^* = A^{-1}b$ is the exact solution to the problem (1). Now we derive the fundamental estimate for the error of the regularized solution

$$\|x^* - x_{\alpha,\delta}\| \le \|x^* - R_\alpha b\| + \|R_\alpha b - R_\alpha b_\delta\| \le \|x^* - R_\alpha Ax^*\| + \|R_\alpha\|\delta. \quad (6)$$

As α tends to zero the first term in the right hand side in (6) converges to zero (*regularization effect*) while the second term grows to infinity (*ill-posedness effect*). Thus two competing effects enter (6). These effects forces us to make a trade off between accuracy and stability. The basic question is how to choose the value of the parameter α which is dependent on the given $\delta > 0$ (and also on $b_\delta \in W$).

Let us return to the third example in the previous section to illustrate this. It is usual to approximate the derivative by the *central difference quotients*

$$R_h b(t) = \frac{b(t+h) - b(t-h)}{2h}, \quad h > 0.$$

We can notice that the role of the parameter $\alpha > 0$ plays here the difference step $h > 0$. A Taylor expansion of the function b yields $\|A^{-1}b - R_h b\|_\infty \le \frac{h}{2}\|A^{-2}b\|_\infty$ suppose that b is continuously differentiable of the second order (the operator of the second derivative is here denoted as A^{-2}). For noisy data the total error can be estimated as follows:

$$\|A^{-1}b - R_h b_\delta\|_\infty \le \|A^{-1}b - R_h b\|_\infty + \|R_h b - R_h b_\delta\|_\infty \le \frac{h}{2}\|A^{-2}b\|_\infty + \frac{\delta}{h} \quad (7)$$

In this estimate we have split the total error into an *approximation error*, which tends to 0 as $h \to 0$ and a *data noise error*, which explodes to infinity as $h \to 0$. If we want to obtain a good approximation it is necessary to balance these two error terms by a suitable choice of the discretization parameter h. The minimum of the right hand side in (7) is attained at $h=\left(2\|A^{-2}b\|_{\infty}^{-1}\right)^{1/2}\delta^{1/2}$. With this step h the total error is of the order $\|A^{-1}b-R_h b_\delta\|_{\infty}=O\left(\delta^{1/2}\right)$. It can be shown that for higher order difference schemes the convergence rate is always smaller than $O(\delta)$. It is well-known that the operator of derivative A^{-1} is closed but not bounded. The convergence rate $O\left(\delta^{1/2}\right)$ reflects the fact that stability is restored due to a-priori information

$$\|A^{-2}b\|_{\infty}\leq C \tag{8}$$

for some constant C. The inequality (8) guarantees the continuity of the operator of derivative A^{-1} on the set of functions b satisfying (8). It follows from

$\|A^{-1}b\|_{\infty}\leq\|A^{-1}b-R_h b\|_{\infty}+\|R_h b\|_{\infty}\leq\frac{h}{2}\|A^{-2}b\|_{\infty}+\frac{\|b\|_{\infty}}{h}\leq 2\sqrt{C\|b\|_{\infty}}$ if we set $h=\sqrt{\frac{\|b\|_{\infty}}{C}}$.

This last example exposes the typical properties of ill-posed problems:

- Amplification of high frequency errors.
- Restoration of stability by a-priori information
- Trade-off between accuracy and stability in the choice of the discretization parameter

1.3 Least square solution as the generalized solution

The exact solution to the operator equation is usually called the classical solution. However it need not exist. If $b\notin R(A)$ one can try to look for a solution in a generalized sense. It is natural to require that such a solution minimizes the residual norm $\|Ax-b\|$. The following notions and considerations are performed when both spaces V and W are Hilbert spaces: The vector $x_0\in V$ (if it exists) for which

$$\|Ax_0-b\|=\inf_{x\in V}\|Ax-b\| \tag{8}$$

is called the *least square solution* or the *generalized solution to the equation (1) in the sense of the least squares* (GSLS). However it need not exist. The sufficient condition for the existence of the GSLS is that the range *R(A)* is a closed subspace of W. It will be the case if for example the space *R(A)* is of a finite dimension. On the other hand if *R(A)* is a dense subspace of W different from W there is always an element $b\in R(A)$ such that the equation (1) has no GSLS.

It is the known fact that $x_0\in V$ is a GSLS of (1) if and only if it is the solution to the *normal form of the equation (1)*

$$A^*Ax=A^*b, \tag{9}$$

where A^* is the adjoint operator to the operator A. The optimalization problem (8) is thus equivalently reformulated as the linear equation (9). If GSLS exists it need not be unique. All the

solution of (9) form a linear set $L_b = \{x \in V : A^* Ax = A^* b\} = x_0 + N(A)$, where $N(A)$ is the kernel of the operator A. To assure the uniqueness of the solution to (9) we impose some additional condition. It is usually the requirment to the least norm of an element within the set L_b: if $L_b \neq \emptyset$ there exists a unique element $\tilde{x} \in L_b$ for which $\|\tilde{x}\| = \inf_{x \in L_b} \|x\|$. We can notice that the new problem (9) is an operator equation with a self-adjoint positive operator. That is why we talk about the symmetrization of the original problem (1).
This acces to the problem (1) is at the bottom of the Tikhonov regularization method for solving the ill-posed problems. We used it in application in the last section.

2 The spaces of stochastic sequences and non-stationary processes

The non-stationary time series are generated by non-stationary stochastic processes. These processes can be understood as a solutions of operator equations which are classified as ill-posed problems. We will introduce an application of the Tikhonov regularization method to the essential problem the solution of which is well-known random walk. The other non-stationary models in Box Jenkins methodology involve this case. In time series analysis the classical access to stabilization is the differencing of the time series. It can be also taken as a certain type of regularization.
The stochastic process is defined as a mapping $\mathbf{X} : T \to L_2$, where the set T (time domain) is here set of all integers and $L_2 = L_2(\Omega, \pi)$ is the space of all functions (random variables) defined almost everywhere on the measurable space Ω, that are square integrable over the space Ω with respect to a probability measure π. It means that this mapping is in fact a (both sided) sequence. However we adopt slightly different modification of this current definition which consists in the assumption that the sequence is actually one sided of the type

$$\mathbf{X} = (X_t, X_{t-1}, \ldots) = (X_{t-k}), k \in \mathbf{N}_0, t \in T. \tag{10}$$

If we extend the stochastic sequences (10) by zeros to time points $t+1$, $t+2$,... we get the both sided stochastic sequence from the original definition given above.
Basic characteristics of the stochastic sequences are as follows:

- the function of means $\mu_t = EX_t$;
- the function of variances $\sigma_t^2 = DX_t = EX_t^2 - \mu_t^2$;
- the autocovariance function $C(t,s) = cov(X_t, X_s) = E(X_t - \mu_t)(X_s - \mu_s)$.

The definitions of the basic notions as the *stationary process,* the *white noise process* and the *general linear process,* are given for instance in [1] or in [4]. Let us recall that the stationary process is such a process for which the functions of means and variances are constant and the values of its autocovariance function depend only on the time lag of its terms. The sequences (10) are elements of certain spaces. Instead of the Banach space of all bounded stochastic sequences $\ell_\infty(L_2)$ (see [4]) we will consider now the space $\ell_2(L_2)$, which is the space of all square summable stochastic sequences. The framework of the Hilbert space $\ell_2(L_2)$ will be more suitable for our purposes.

2.1 Stochastic sequence as an element of Hilbert coordinate space

All the terms of the stochastic sequence (10) are elements of the Hilbert space $L_2 = L_2(\Omega, \pi)$. Their norms in L_2 are $\|X_{t-k}\|_2^2 = EX_{t-k}^2 = \sigma_{t-k}^2 + \mu_{t-k}^2$. If (10) is the centered (with zero mean) and stationary sequence then its terms lie on the sphere with radius σ^2 in L_2. The space of all bounded stochastic sequences $\ell_\infty(L_2)$ with the norm $\|\mathbf{X}\|_\infty = \sup_{k\geq 0}\sqrt{EX_{t-k}^2}$ was used in [4]. It was natural since this space contains all stationary sequences and it is the Banach space. However in what follows we will employ another space. The space $\ell_2(L_2)$ of all stochastic sequences (10) which satisfy $\sum_{k=0}^{\infty} EX_{t-k}^2 < \infty$ is the Hilbert space with respect to the norm derived from the inner product

$$\langle \mathbf{X}, \mathbf{Y}\rangle = \sum_{k=0}^{\infty} E(X_{t-k}Y_{t-k}), \text{ i.e. the norm } \|\mathbf{X}\| = \left(\sum_{k=0}^{\infty} EX_{t-k}^2\right)^{1/2}.$$

Unfortunately the space $\ell_2(L_2)$ contains no stationary stochastic sequences as well as the constant sequences (except the zero sequence). In spite of this apparent drawback this space has more comfortable structure for our intentions than the $\ell_\infty(L_2)$. We may overcome this deficiency if we slightly modify the concepts of stationarity of the process and white noise respectively. As to the mean we will consider only centered sequences. We allow the sequence to be stationary as far as to the past as one wishes then the norm (variance) of its terms has to fall to zero (otherwise the series of variances may not be convergent).

2.2 Lag operator and its spectral properties

The lag operator is defined for example in [4]. We will take this operator in the Hilbert space $\ell_2(L_2)$ and thus

$$B: \ell_2(L_2) \to \ell_2(L_2),\ B(X_{t-k}) = (X_{t-1-k}) \tag{11}$$

It is a bounded operator with the norm equal to one, i.e. the same value when this operator is defined on the space $\ell_\infty(L_2)$ (see [4]).
The spectral properties of the lag operator in the space $\ell_\infty(L_2)$ are summarized in [4], the section 2.2, Proposition 1. We can recall that the spectrum of the lag operator is the closed unit circle in the complex plane and all elements of this circle are the eigenvalues of this operator. In the case of (11) it is almost the same thing. There is only one difference: if λ is a complex unit, i.e. $|\lambda| = 1$ then the operator $(\lambda I - B)^{-1}$ exists but it is not bounded. It follows that the operator $(\lambda I - B)^{-1}$ may not be defined on the whole Hilbert space $\ell_2(L_2)$ but only on some proper subspace of this space. In fact the space $R(\lambda I - B)$ is a proper subspace of $\ell_2(L_2)$ which is dense in $\ell_2(L_2)$. If we take for example $\lambda = 1$ the corresponded eigenvector in $\ell_\infty(L_2)$ is $\mathbf{E} = (1,1,1,\ldots)$ but obviously $\mathbf{E} \notin \ell_2(L_2)$.

There is an important class of stochastic sequences which are the solutions to the stochastic difference equation of the type

$$\Phi(B)\mathbf{X} = \Theta(B)\mathbf{W}, \tag{12}$$

where $\Phi(B)$ and $\Theta(B)$ are polynomials in B of the degree p and q respectively, and $\mathbf{W}$ is a white noise. As was shown in [4] the operator equation (12) is the well-posed problem if and only if all the roots of the polynomial $\Phi(z)$ are outside the closed unit circle in the complex plane. In this case the solution to (12) is well-known ARMA(p,q) process. On the other hand if $\Phi(1) = 0$ then $\Phi(B)$ is either not invertible or it has inverse which is not bounded. In any case (12) is then an ill-posed problem. In terminology of Box Jenkins methodology the solution to (12), if it exists, is of the class ARIMA(p,d,q), where d is the multiplicity of the unit as a root of the $\Phi(z)$. Such a process is the *non-stationary stochastic process.*

3 The random walk as the solution to an ill-posed problem

If we set in (12) $\Phi(B) := I - B$, i.e. $\Phi(B)$ is the difference operator, and $\Theta(B) := I$ respectively, then we get the essential form of the ill-posed equation (12), i.e.

$$(I - B)\mathbf{X} = \mathbf{W}. \tag{13}$$

The equation (13) is the stochastic difference equation of the first order. We will deal with the problem of the regularization of its solution in the Hilbert space $\ell_2(L_2)$ by the Tikhonov regularization method. The solution to the operator equation (13), if it exists, is called the *random walk process.*

3.1 The least square solution to the equation $(I - B)\mathbf{X} = \mathbf{W}$

The adjoint operator to the difference operator $(I - B)$ is obviously $I - B^*$ and the normal form (9) of the equation (13) is

$$(I - B - B^* - B^*B)\mathbf{X} = (I - B^*)\mathbf{W}. \tag{14}$$

For simple writings let us set $A = (I - B)$. It follows from the relations among the kernels and the ranges respectively of mutually inverse operators (see [9], p. 238) that we have $N(A^*A) = N(A) = \{o\}, \overline{R(A^*A)} = \overline{R(A^*)} = \ell_2(L_2), R(A^*A) \subseteq R(A^*) \subset \ell_2(L_2)$. If we add the spectral properties of the adjoint operator B^* we see that the equation (14) is an ill-posed problem as well as (13). Thus the GSLS to (13) need not exist for some right side (from the space $\ell_2(L_2)$).

3.2 Tikhonov regularization on the equation $(I - B)\mathbf{X} = \mathbf{W}$

The Tikhonov regularization consists in the regularization of the normal form of the given problem. We keep the denotation from the previous section, i.e. $A = I - B$. The basic idea of the Tikhonov regularization is to add some positive multiple of the identity αI to the operator A^*A. This way we get a bounded linear bijection $A^*A+\alpha I$ on the space $\ell_2(L_2)$. The regularized form of the equation (14) is

$$\left(I-B-B^*-B^*B+\alpha I\right)\mathbf{X}=\left(I-B^*\right)\mathbf{W}. \tag{15}$$

The equation (15) is well-posed. The collection of operators $R_\alpha=\left(A^*A+\alpha I\right)^{-1}A^*$ is a regularization family for the operator A^*A. As in the section 1.2 we replace in (15) the white noise $\mathbf{W}$ by its perturbation $\mathbf{W}_\delta$, $\|\mathbf{W}-\mathbf{W}_\delta\|\le\delta,\ \delta>0$. Then we get the regularized solution $\mathbf{X}_{\alpha,\delta}=R_\alpha\mathbf{W}_\delta$.

Let us describe the fundamental estimate for the error of the Tikhonov method. First we estimate the norm $\|R_\alpha\|$. For this we use the polar decomposition of a bounded linear operator (see [6], p.234) and the Riesz functional calculus (see [6], p.87). By polar decomposition $A=U|A|$, where U is an isometry, in our case even unitary operator, and $|A|$ is the square root of the operator $T:=A^*A$ which is self-adjoint. Hence we have $R_\alpha=(T+\alpha I)^{-1}|A|U$ and if we denote $f(T)=(T+\alpha I)^{-1}|A|$ we obtain $\|R_\alpha\|\le\|f(T)\|$. To compute the norm of the self-adjoint operator $f(T)$we employ the Riesz functional calculus

$$\|f(T)\|=\sup_{\lambda\in[0,4]}|f(\lambda)|=\sup_{\lambda\in[0,4]}\frac{\sqrt{\lambda}}{\lambda+\alpha}=\frac{1}{2\sqrt{\alpha}}. \tag{16}$$

The upper bound 4 in the interval in (16) is obtained from the fact that $\|A\|=\|I-B\|=2$ (the spectrum of the difference operator is only translated spectrum of the lag operator, which follows from the spectral mapping theorem, and $\|T\|=\|A\|^2$ (see [9], p.239)).

Second we estimate the regularization effect $\|\tilde{\mathbf{X}}-R_\alpha A\tilde{\mathbf{X}}\|$. Here we will suppose that $\tilde{\mathbf{X}}=A^*\mathbf{Z}$ for some $\mathbf{Z}\in\ell_2(L_2)$, i.e. the GSLS (the exact solution to (14)) belongs to the range of the adjoint operator A^*. This can be interpreted as a certain requirment on smoothness of the GSLS. For comfortable writing set $G_\alpha:=(T+\alpha I)^{-1}$. We derive

$$\begin{aligned}&\|R_\alpha A\tilde{\mathbf{X}}-\tilde{\mathbf{X}}\|=\|G_\alpha T\tilde{\mathbf{X}}-\tilde{\mathbf{X}}\|=\|G_\alpha(T\tilde{\mathbf{X}}-\alpha\tilde{\mathbf{X}}-T\tilde{\mathbf{X}})\|=\alpha\|G_\alpha\tilde{\mathbf{X}}\|=\alpha\|G_\alpha A^*\mathbf{Z}\|\le\\&\le\alpha\|R_\alpha\|\|\mathbf{Z}\|\le\frac{\alpha}{2\sqrt{\alpha}}\|\mathbf{Z}\|=\frac{\sqrt{\alpha}}{2}\|\mathbf{Z}\|.\end{aligned} \tag{17}$$

Finally we join the two estimations (16) and (17) and thus get the fundamental estimation for the error of the Tikhonov method:

$$\|\tilde{\mathbf{X}}-\mathbf{X}_{\alpha,\delta}\|\le\left\|\left(A^*\right)^{-1}\tilde{\mathbf{X}}\right\|\frac{\sqrt{\alpha}}{2}+\frac{\delta}{2\sqrt{\alpha}}. \tag{18}$$

The first term on the right hand side in (18) reflects the regularization effect while the second one the ill-posedness effect. We have to balance the parameters α and δ in such a manner so that $\frac{\delta}{\sqrt{\alpha}} \to 0$ if $\alpha \to 0_+$ and, at the same time, $\delta \to 0_+$. The rule for the choice of the value of the parameter α may be so called Morozov discrepancy principle. It consists in taking $\alpha = \alpha(\delta)$ such that $\left\| A\mathbf{X}_{\alpha,\delta} - b_\delta \right\| = \delta$. This choice guarantees the balanced approximation and data noise errors. The numerical computation of α can be carried out by Newton´s method.

The biggest drawback of Tikhonov regularization is the repeated matrix manipulation, computation of the inverses $G_\alpha := (T + \alpha I)^{-1}$.

References

[1] ARLT, J. *Moderní metody modelování ekonomických časových řad.* Grada Publishing, 1999, ISBN 80-7169-539-4, 307 s.

[2] DHRYMES, J.P. *Distributed Lags. Problems of Estimation and Formulation,* North Holland, Amsterdam, 1985.

[3] HORSKÝ, R. The lag operator and its spectral properties. *Mathematical Methods in Economics 2013* Proceedings Part I (2013), 285–290.

[4] HORSKÝ, R. Algebra of linear filters in time series analysis. *15th Conference on Applied Mathematics* , Aplimat 2016, Bratislava

[5] KLEBANOV, L., KOZUBOWSKI, T.J., RACHEV, S.T.: Ill-Posed Problems in Probability and Stability of Random Sums, Nova Science Publishers, Inc., New York, 2006.

[6] LUKEŠ, J. *Zápisky z funkcionální analýzy.* Univerzita Karlova v Praze. Nakladatelství Karolinum. 2012.

[7] MOROZOV, V.A.: Methods for Solving Incorrectly Posed Problems, Springer Verlag, New York, 1984.

[8] NAIR, M.T. (2009): Linear Operator Equations, Approximation and Regularization. World Scientific, ISBN: 978-981-283-565-9.

[9] TAYLOR, A.E. *Úvod do funkcionální analýzy.* Academia Praha, 1973.

Current address

Horský Richard, RNDr. Ph.D.
University of Economics, Department of mathematics
Ekonomická 957, 148 00 Praha 4, Czech Rupblic
E-mail: rhorsky@vse.cz

Slovak University of Technology in Bratislava
Faculty of Mechanical Engineering
16th Conference on Applied Mathematics **APLIMAT 2017**
Proceedings

CREATION OF COMPUTER MODEL VIA TOOLS OF MS EXCEL

HREHOVA Stella (SK), VAGASKA Alena (SK)

Abstract. Creation of various types of models in education has an irreplaceable position. Currently we can use various computer applications to create the desired model. The simulation models developed by software resources are mainly used to determine the expected behaviour of the modelled reality. The article describes the Excel tools that were used to create the model which describes the static characteristics of the artificial muscle.

Keywords: pneumatic artificial muscle, dependence, MS Excel, VBA

Mathematics Subject Classification: 97M50

1 Brief description of artificial muscles

McKibben muscle is an actuator which converts pneumatic (or hydraulic) energy into mechanical form by transferring the pressure applied on the inner surface of its bladder into the shortening tension. To find the relationship of the tension, length, and pressure a theoretical approach and several experiments will be analyzed with simplified modelling [1]. Pneumatic artificial muscles (PAM) allow us to design and construct the so-called low-cost actuators with very good power / weight ratio. When we compare industrial artificial muscle with other actuators, artificial muscle has several advantages:

- At first it is its low weight converted to tensile force. This ensures its design, which is mainly formed from lightweight membrane. It can be extremely strong, despite of its weight and flexibility, and it is able to build up the force of several thousand [N].
- PAM needs to develop at the same force less power than the pneumatic cylinder and the initial force during contraction can be up to several times greater.
- There is easy replacement and low maintenance along with its ability to use this technology in dusty and explosive environments.

McKibben artificial muscle is the most common type of artificial muscles which is produced. From the structural point it is a very simple device which consists from the inner layer, outer layer and terminals [2].

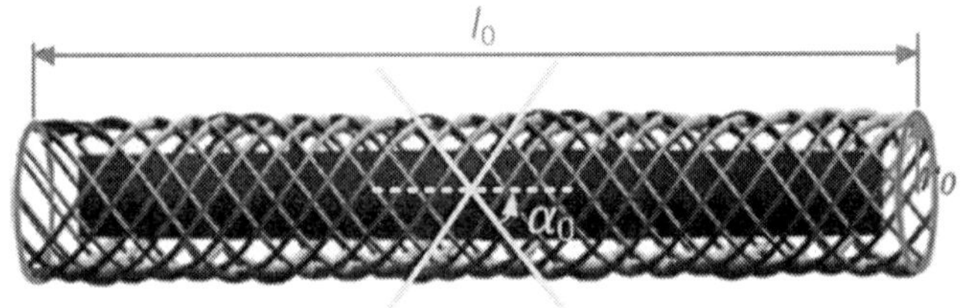

Fig 1. Scheme of McKibben artificial muscle

The pressurization of pneumatic artificial muscle means shortening of its length from the initial length l_0 to the length l and it is increasing of its radius from r_0 to r (diameter from d_0 to d). It also changes the angle between the fibers′ braid from α_0 to α. All these changes have resulted in a force F, which is transmitted by other important components of the pneumatic artificial muscles [3].

1.1 Description of chosen characteristics

The format of the model was simplified and linearized from theoretical results, and the parameters were optimized with experimental data.

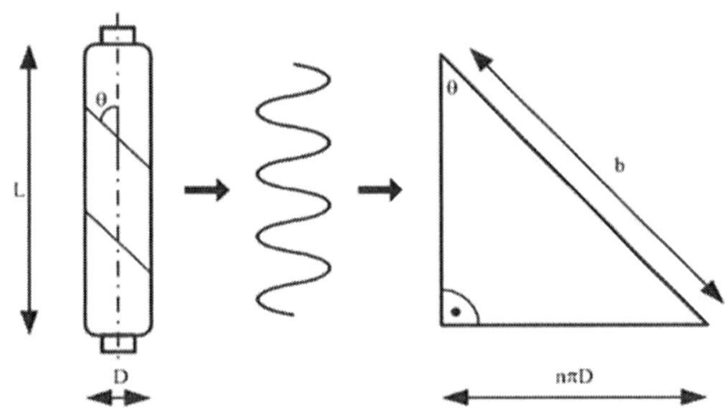

Fig. 2. The geometry of the actuator

Geometric constants of the system are b fibre length and n number of a fibre orbs provided inextensibility of fibres of braiding.
To determine the values of individual static characteristics the following mathematical formulas [4] will be used:

1. Tension force

$$F = p\frac{\left(3L^2 - b^2\right)}{4\pi n^2} \tag{1}$$

where $L = b\cos\alpha$

It results from the relation that the tension force is directly proportional to the air pressure inside the muscle and it is the function of the angle of fibres.

2. Volume

$$V = \frac{\left(b^2 L - L^3\right)}{4\pi n^2} \tag{2}$$

3. Contraction size of pneumatic muscle depends on the air pressure in the muscle and muscle load.

$$\kappa = 1 - \sqrt{\frac{F\,4\pi n^2}{p3L_0^2} + \frac{b^2}{3L_0^2}} \tag{3}$$

2 Used tools of MS Excel

Visual Basic for Application (VBA) is an integration of the Microsoft's event-driven programming language Visual Basic with Microsoft Office application. By running Visual Basic we can build customized solutions and programs.It allows the creation of user-defined functions and the automation of specific computer processes and calculations. The following picture represents the basic window of VBA.

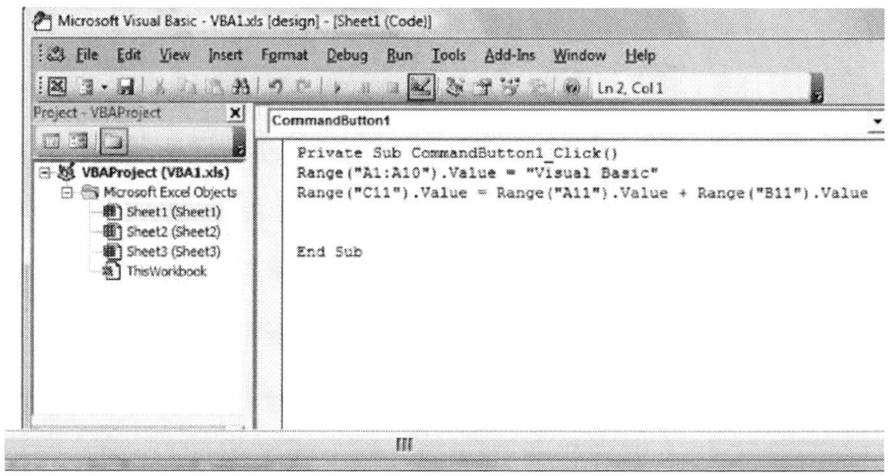

Fig. 3. The basic window of VBA

In this environment the code is created for the command buttons. Within each code there will be calculated the values needed to create the required charts. The user selects the task by either display "Dependence contraction on pressure", or display "Dependence force on contraction". After pressing the appropriate button there will be generated a new graph for the specified values. Load force is changing between 100 and 500 N and pressure is changing between 100 and 500 kPa. After change of the input data, a new graph will be showed in this same presentation.

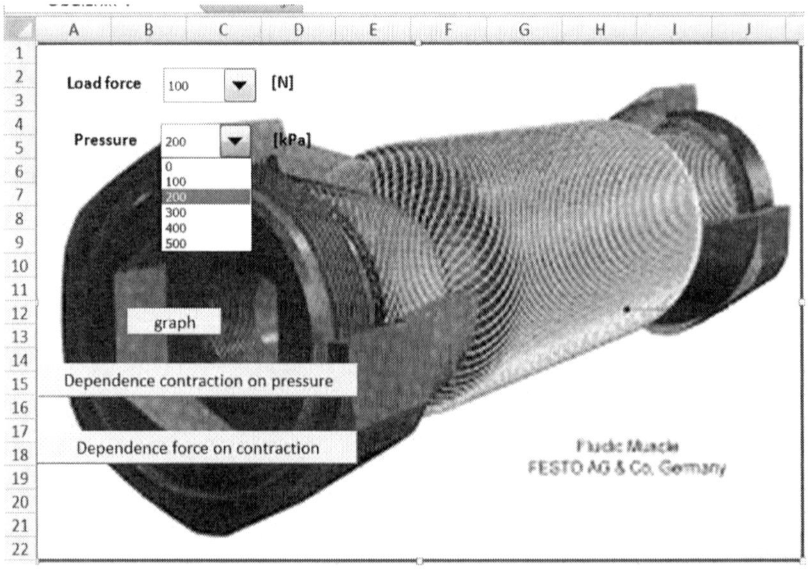

Fig. 4. User settings to show a graph

We used the opportunity to create a macro for rendering graphs. Via macros, each step is gradually entered directly in VBA. It is possible to directly call that macro by its name. We can also edit the code, respectively supplement it by other commands. The next figure represents the final graph of dependence contraction on pressure.

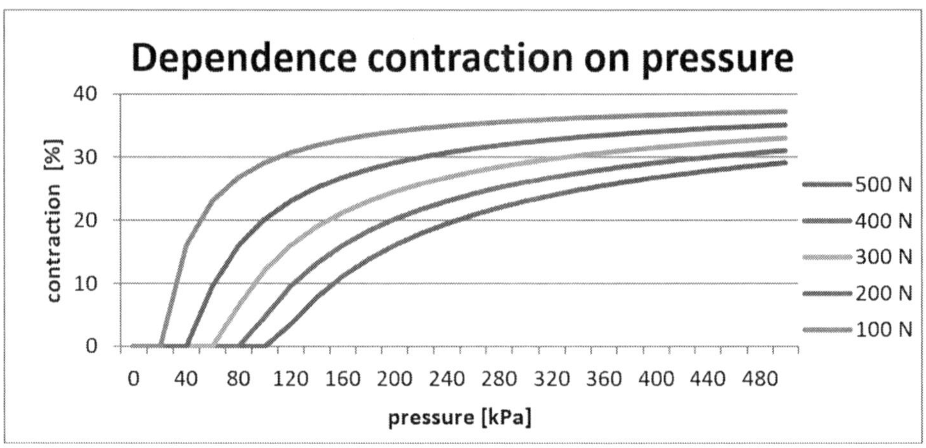

Fig. 5. Graph dependence contraction on pressure

If we change the input data the data forming the graph will be automatically changed. For this reason it is necessary to press F9 for the current data series in the chart.

=SERIES(;Hárok1!$H$2:$H$18;Hárok1!$I$2:$I$18;1)

This ensures that the change of the input data does not automatically change the actual graph.

To show the dependence force on contraction we can change the value of pressure.

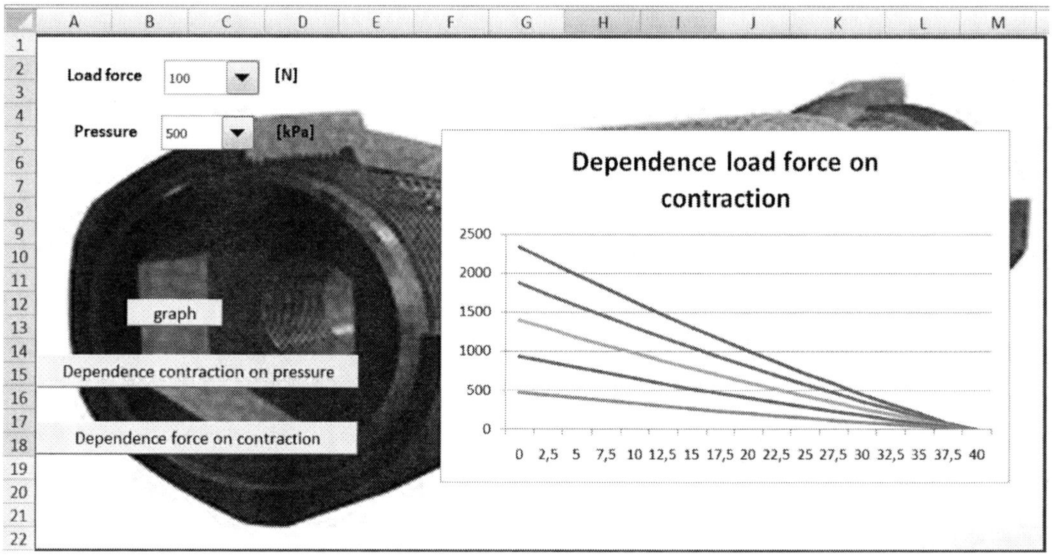

Fig. 6. Graph dependence contraction on pressure

2.1 User form

We create a simple user interface and we insert the required values there. After pressing the "Solve" button results are displayed for the input parameters.

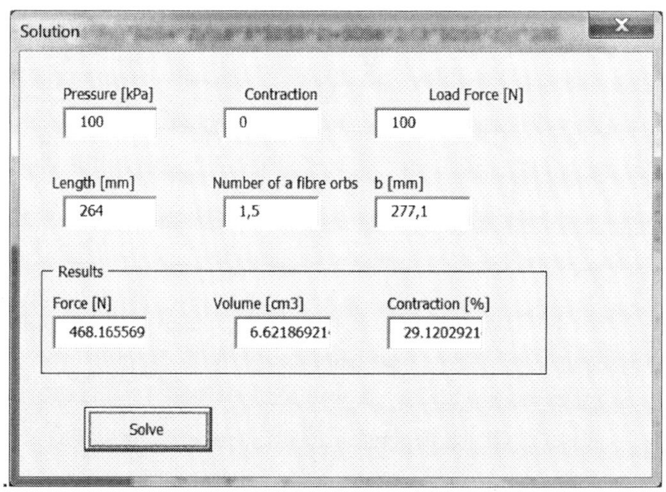

Fig. 7. User interface

Conclusion

In this contribution was used the tools of MS Excel such as an appropriate way to analyze some static characteristics of artificial muscle. There were used the base programming command of VBA and there was described a manner of how to create a simple user interface. We can analyze the static characteristics of artificial muscle according to their mathematical formulas. It is an useful information if we want to use an artificial muscle in dynamic system.

Acknowledgement

The research work is supported by the Project KEGA 026TUKE-4/2016. „Implementation of Modern Information and Communication Technologies in Education of Natural Science and Technical Subjects at Technical Faculties."

References

[1] HANNAFORD, B., CH.P.CHOU. Measurement and Modeling of McKibben Pneumatic Artificial Muscles. Online *www.dtic.mil/cgi-bin/GetTRDoc?AD=ADA299458*

[2] VAGASKÁ, A. Mathematical description and static characteristics of the spring actuator with pneumatic artificial muscle. *Applied Mechanics and Materials* Vol. 460, 2014, pp. 65-72, ISSN 1662-7482

[3] KOPEČNÝ, L. ŠOLC, F. McKibben's pneumatic muscle in robotics. AT&P Journal - Robotics, 2/2003, pp. 62-64. URL: http://www.atpjournal.sk/buxus/docs/atp-2003-2-62.pdf

[4] PITEĽ, J. et al: Pneumatic artificial muscles: modeling, simulation, control. Technical university of Kosice, 2015, ISBN 978-80-553-2164-6

Current address

Hrehova Stella, Ing. PhD.
Technical University of Kosice, Faculty of Manufacturing Technologies with a seat in Presov,
Department of Mathematics, Informatics and Cybernetics
Bayerova 1, 080 01 Prešov, Slovak Republic
Tel. number: +421 55 6026426, e-mail: stella.hrehova@tuke.sk

Vagaska Alena, PaedDr. PhD.
Technical University of Kosice, Faculty of Manufacturing Technologies with a seat in Presov,
Department of Mathematics, Informatics and Cybernetics
Bayerova 1, 080 01 Prešov, Slovak Republic
Tel. number: +421 55 6026430, e-mail: alena.vagaska@tuke.sk

Slovak University of Technology in Bratislava
Faculty of Mechanical Engineering
16th Conference on Applied Mathematics **APLIMAT 2017**
Proceedings

STATISTICAL EVALUATION OF THE IMPACT OF ECOLOGICAL FACTORS ON THE ACTIVE AND EXCHANGE SOIL REACTION

HUDEC Michal (SK), PORUBCOVÁ Lýdia (SK), NOGA Henryk (PL), FESZTEROVÁ Melánia (SK)

Abstract. The work is focused on the monitoring of changes the active and the exchange soil reaction. These characteristics were observed on grass and forest ecosystem in Kremnické and Štiavnické vrchy Mts., in the identified soil types in 2011 - 2014 years. The obtained results were evaluated and graphically processed using the statistical software. Analysis of the variance of pH values without and with interactions from the ecological factors was realized through two - factor analysis (ANOVA), with using Tukey´s HSD - test.

Keywords: soil reaction (active and change), Kremnické and Štiavnické mountains, parametric test

Mathematics Subject Classification: Primary 62F, Secondary 92E

1 Introduction

Soil is a product of nature and is one of the most valuable human assets. Soil as a part of the geographical environment maintains the life of plants, animals and humans. It is a crucial natural resource and the part of the economic (production base) and eco-social potential (social support) of each country. Land management can not completely avoid the impact on the soil and the environment (Hanes, 1995). Negative effects of industrial technologies on soils are most frequently demonstrated in the damage of physical condition of the soil, degraded soil chemistry, disturbance of the soil condition (Zrubec, 1990). Soil quality integrates biological, chemical and physical components and processes of land connecting it with the environment in the country (Bedrna 2002a). The quality of soil depends on the part of natural soil composition (natural soil quality), which is a function of soil-forming factors (e. g. parent rock), but also on changes due to land use by humans (dynamic soil quality) (Vilček, Hronec, Bedrna, 2005). Ecological factors are natural phenomena; it is the relationship of the living to live or living to the inanimate. These factors can be positive but also have negative effects on the environment (Bedrna 2002a).

In connection with the context of soil it is important not only to monitor changes in land management and vegetation, but also changes of the chemical properties of soil (chemical composition of the mineral constituent of soil, the organic layer of soil, the sorption capacity of soil and the character of sorption complex, soil reaction and buffering capacity soils, the carbonate content in the soil, the oxidation - reduction potential of soil, the chemical composition of soil solution) (Šlepetienė et al., 2010; Hudec, Hreško, 2012). The land use in practice can have a permanent effect on the distribution of nutrients in the soil and can substantially affect the natural ecosystem recovery after land abandonment (Lal, 2000; Fraterrigo et al., 2003; Aleksandrowicz - Trzcińska, 2005). The extent to which a country can be seamlessly utilized depends on the physical, chemical and biological soil quality (Ryan et al., 2008).

The soil reaction significantly affects the properties of soils and it is one of the important indicators of soil fertility (Zaujec et al., 2009). It directly affects the solubility of many substances, the nutrients accessibility, adsorption and desorption of ions, the biochemical reaction, and the quality of the humus content, soil structure, and other properties (Hanes, 1995). The reaction of soil solution is a basic agrochemical parameter (Kováčik, 2013). It is involved in many soil-forming processes and affects the vital activity of soil organisms (Hanes, 1995). Soil reaction is the result of the capacity of soil release into the soil solution, and it also binds hydrogen and hydroxide ions in the soil and forms a solution of various ratios of the amounts (Bielek, 2014). The reaction of the soil solution of different soils varies over a wide range of pH = 3.5 to 9 exceptionally more (Kováčik, 2013). Natural (background) pH soils in Slovakia are not detectable in the present time, because of the applied liming acid soils for decades (environmental factor) practically on most of their areas (Bielek, 2014). Moreover, the acidity of our soils is relatively strongly influenced by acid precipitation, acid fertilizers and other mostly anthropic influences (Némethová, Dubcová, Kramáreková, 2014). Largely on the value of soil reaction geology has influence that affects it greatly.

Soil reaction depending on the factors that determine it, are divided as the current (pH_{H2O}) and potential. Within potential exchange (pH_{KCl}) and hydrolytic (H) acidity are distinguished (Bielek, 2014). Current which is also called as active acidity is determined by the concentration of free hydrogen ions in an aqueous extract (Bielek, 2014), or it is determined by the amount of acids and the bases of organic and inorganic origin located in the soil solution, and their degree of dissociation of CO_2 in the soil solution (Kováčik, 2013).

The paper highlights the importance of monitoring exchange values and active soil reaction as one of the main indicators of soil fertility. It is approaching the possibility of using parametric tests in the evaluation of selected characteristics of the soil. The practical part is to employ methods of statistical evaluation of ecological factors at pH values in soil samples. In terms of ecological factors that affect the values of pH in soil samples, we observed: elevation, slope and shape of the relief, season (autumn, spring), climatic conditions (rainfall), biota at selected locations. The obtained results were evaluated and graphically processed using the statistical software STATISTICA at significance level $\alpha = 0.05$. The variability of the active and exchange soil reaction between habitats was assessed using F - test (ANOVA) at a significance level of $\alpha = 0.05$. Analysis of the variance of pH values without and with interactions from the land use factor (type of habitat), soil type, location and season was realized through two - factor analysis (ANOVA), with using Tukey´s HSD test for the detection of contrasts values at the significance level $\alpha = 0.05$.

2 Material and Methods

The study was located in Kremnické and Štiavnické mountains. (Fig. 1) There selected area was 5 x 5 m on each location and it soil sampling from 27 different sampling sites was conducted over the years between 2011 - 2014. Kremnické and Štiavnické vrchy Mountains are volcanic mountains that are located in central Slovakia, part of the Western Carpathians province, sub - province Inner Western Carpathians and the Slovak Central Mountains. The localization of monitoring places of the soil samples according to Mazút (1980) were incorporated into three height degrees of altitude (foothill, mountain, higher mountain). The soil sampling places (S1 - S27) were assigned to the localities with the nearest meteorological station. We got an overview of climatic conditions for sampling places through the Slovak Hydrometeorological Institute (SHMI) from the following meteorological stations: Kremnické Bane (sample numbers S1 - S3), Žiar nad Hronom (sample numbers S4 - S19) and Banská Štiavnica (sample numbers S20 - S27).

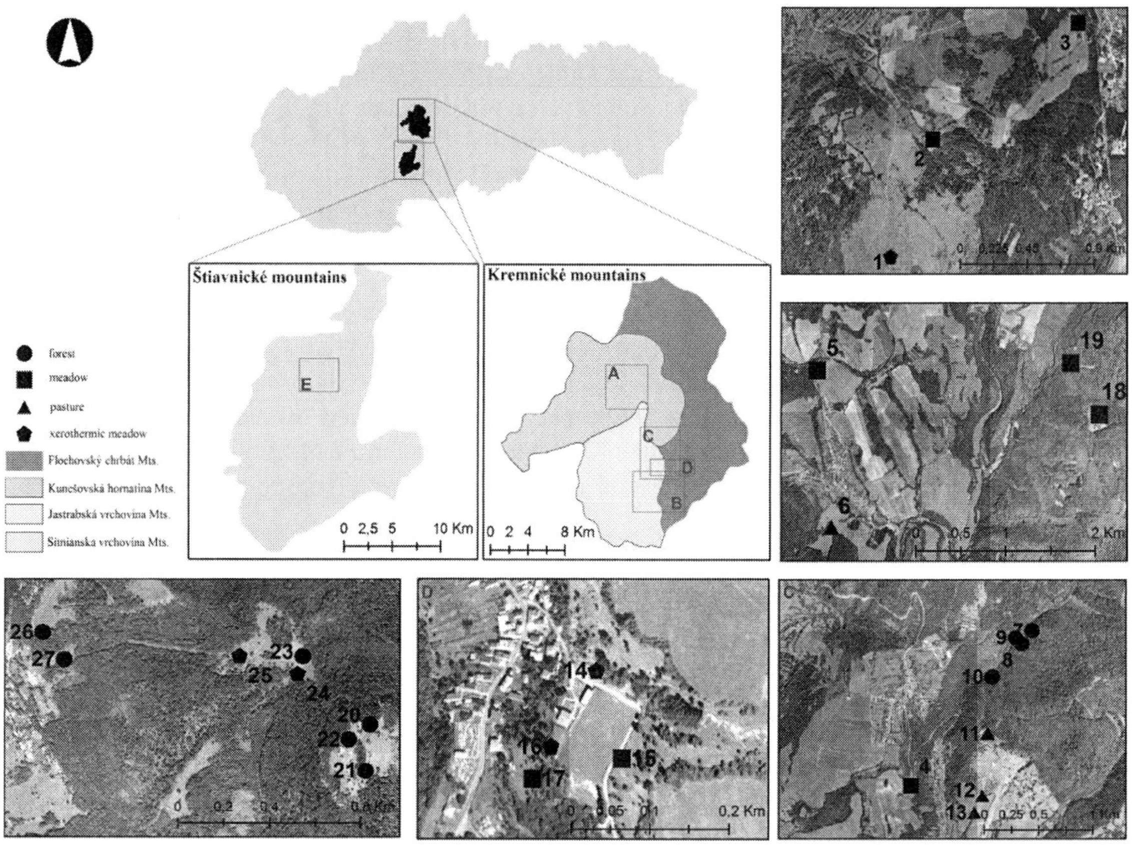

Fig. 1. The soil sampling places according to the land use

The analyzed soil samples were collected by a probe (open borrow pits) from a depth of about 1 m in the autumn of 2011, 2012, 2013 and the spring of 2012, 2013 and 2014 (Hudec, Feszterová, 2015). The opened probe is used for soil for determination of soil characteristics; such as soil profile description, its depth, the stratigraphy (the arrangement of genetic horizons) and morphological characteristics. The location of soil borrow pits, which collected soil samples were situated so that represented different altitudes, the impact of macro- and microrelief and land use (mown meadow, pasture, forest). Soil samples were collected from 5 main soil types (Andosol, Cambisol, Fluvisol, Leptosol, Distric Planosol) of the selected horizons.

Soil was air-dried at room temperature and sieved (< 2 mm), and the sample was taken using standard procedures. Visible plant remnants were removed manually. Soil reaction was determined in distilled water as active soil reaction (pH_{H2O}) and KCl as an exchange of soil reaction (pH_{KCl}) in a solution of 1 $mol.L^{-1}$. The ratio of soil and solution was 1 : 2.5 (VAN REEUWIJK, 2002).

3 Results and Discussion

Soil reaction has between individual habitats variability values. According to Islam and Weil (2000), the soil pH values, of natural forest, meadows and cultivated soil are extremely different. The variability of the active soil reaction (pH_{H2O}) and exchange soil reaction (pH_{KCl}) were assessed using the F - test (ANOVA) on the significant level of $\alpha = 0.05$ ($F = 5.993$, $P = 0.001$ and $P = 7.643$, $p = 0.000$). Gündoğan et al. (2011) found out statistically significant differences in values of soil reaction (pH_{H2O}), while grassland has higher pH values than the forest ($p < 0.05$). Authors' findings Gundoğan et al. (2011) have not been confirmed for the exchange of soil reaction for the greatest variability and the maximum value is already in the soil of the forest.

Soil reaction is affected by rainfall, especially its amount and chemical composition (ecological factors). The amount of rainfall that reaches the soil is affected by a range of other slope reliefs. When assessing the statistical variability of soil reaction values, considering the slope relief, we found a significant difference between the values of the active soil reaction ($F = 7.260$, $p = 0.000$) at the significance level $\alpha = 0.05$. Hudec and Feszterová (2013) have not been observed in their work affect of the slope relief on soil reaction values. The shape of relief had a statistically significant effect only on the sorption properties of soil. The slope relief, as one of the main indicators impacts the mostly on landscape. It is the main factor of erosion conditions affecting the movement direction of runoff water and material down the slope. This factor is very important to assess whether the material at the site is dispersed or concentrated (Koleda et al., 2012).

Development of active and exchange soil reaction values for different categories of erosion material in those seasons soil sampling is illustrated in Tab. 1 a Tab. 2. Testing contrasts of pH soil values which interact with erosion material, and a soil cover were not found in a statistically significant interaction between the soil pH values ($F = 2.185$, $p = 0.141$).

Soil erosion	2011	2012		2013		2014
	Autumn	Spring	Autumn	Spring	Autumn	Spring
Weak	5.99±1.10	5.92±0.54	6.01±0.30	5.93±0.51	5.08±1.53	5.59±0.85
Moderate weak	5.73±0.31	5.67±0.44	6.30±0.38	5.52±0.23	5.29±0.36	5.16±0.08
Severe	5.82±0.46	5.97±0.28	6.14±0.07	5.93±0.35	5.35±0.40	5.50±0.27
Moderate severe	5.77±0.49	5.41±0.10	5.75±0.29	5.66±0.22	5.42±0.28	5.67±0.22

Tab. 1. The development of active soil reaction values ± standard deviation (SD) for different categories erosion material in those seasons soil sampling

Soil erosion	2011	2012		2013		2014
	Autumn	Spring	Autumn	Spring	Autumn	Spring
Weak	4.75±1.01	4.78±0.51	4.62±0.39	4.71±0.56	4.06±1.28	4.40±0.76
Moderate weak	5.10±0.57	4.69±0.37	4.58±0.18	4.47±0.14	4.14±0.33	4.15±0.25
Severe	4.73±0.31	5.06±0.40	4.68±0.09	4.84±0.47	4.38±0.48	4.63±0.29
Moderate severe	4.46±0.22	4.48±0.10	4.49±0.02	4.53±0.21	4.35±0.12	4.43±0.10

Tab. 2. The development of exchange soil reaction values ± standard deviation (SD) for different categories outflow material in these seasons soil sampling

The development the active soil reaction values in different season habitats studied is shown on Fig. 2. This development is decreasing for all types of habitats. The average values of active soil reaction in seasons are described in Tab. 3.

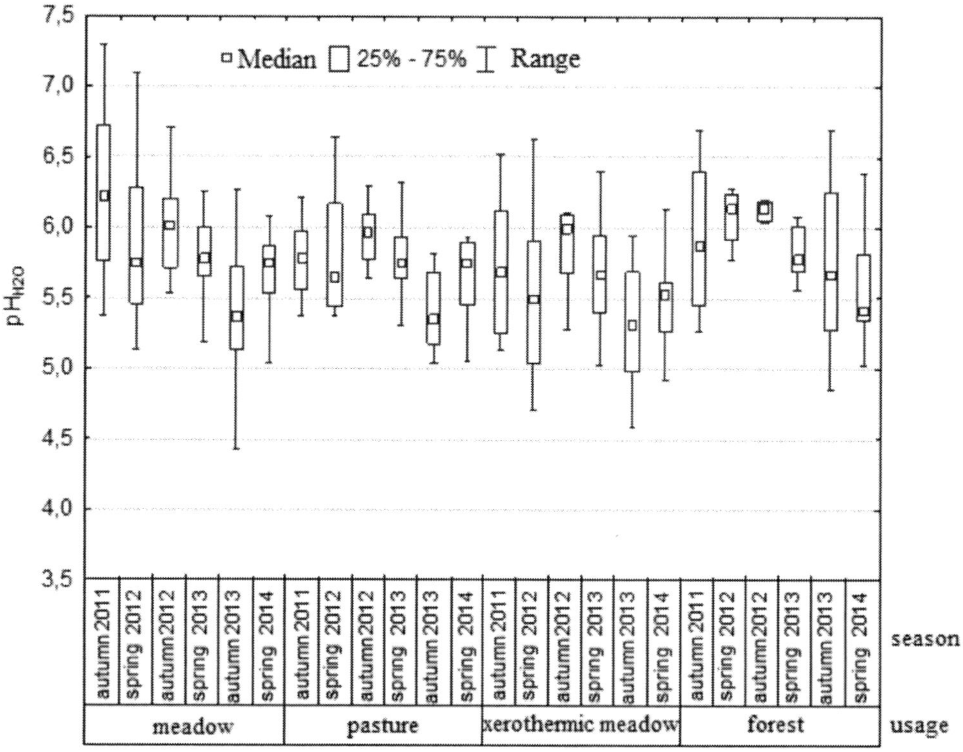

Fig. 2. The difference of pH_{H2O} values depending on the habitat use and season

Year	2011	2012		2013		2014
Biotope	Autumn	Spring	Autumn	Spring	Autumn	Spring
Meadow	6.06±1.21	5.90±0.52	6.00±0.32	5.96±0.55	5.47±0.54	5.71±0.38
Pasture	5.80±0.27	5.84±0.40	5.99±0.25	5.79±0.26	5.39±0.24	5.63±0.27
Xerothermic meadow	5.57±0.77	5.53±0.60	5.97±0.47	5.66±0.43	5.29±0.45	5.49±0.29
Forest	5.93±0.49	6.04±0.27	6.12±0.07	5.89±0.31	5.74±0.59	5.57±0.40

Tab. 3. The average values ± SD of active soil reaction on the monitored habitats

The monitored soil samples, according to the values of active soil reaction, (Čurlík, Šurina, 1998) were ranked among from the weakly acidic to strongly acidic (mown meadow), from moderately acidic to strongly acidic (pasture, xerothermic meadow) and from weakly acidic to moderately acidic (forest). A significant difference in values between the individual seasons is in the autumn of 2012, when the analysis values significantly increased in comparison with the spring of 2012. This could also be affected among other things by climatic conditions (ecological factor). The year 2012 was relatively warm and dry based on the data from the Slovak Hydrometeorological Institute (a total of 734.1 mm of rain fell at three locations as to meteorological station from the beginning of April 2012 to the end of September 2012). On the contrary, in the autumn of 2013 there was a decline in the value of active soil reaction compared to the level in the spring of 2013. The reason for increasing the quantity of precipitation fell on all sites (total 1162.4 mm rainfall at 3 locations according to meteorological station from the beginning of April 2013 to the end of September 2013). Impact of rainfall and soil-forming processes on pH values confirms its work even Bedrna (2002b). According to Fatubarina and Olojugba (2014), the distribution of soil pH falls into the depths of the soil profile during the dry season due to low precipitation. There is no movement of cations from the top layer, which are bound to the soil organic matter. Fatubarin and Olojugba (2014) argue that soil pH value have tendency to increase depending on the depth of the soil sample collection, due to the vertical movement or relocation dissolved cations and because of low organic matter content.

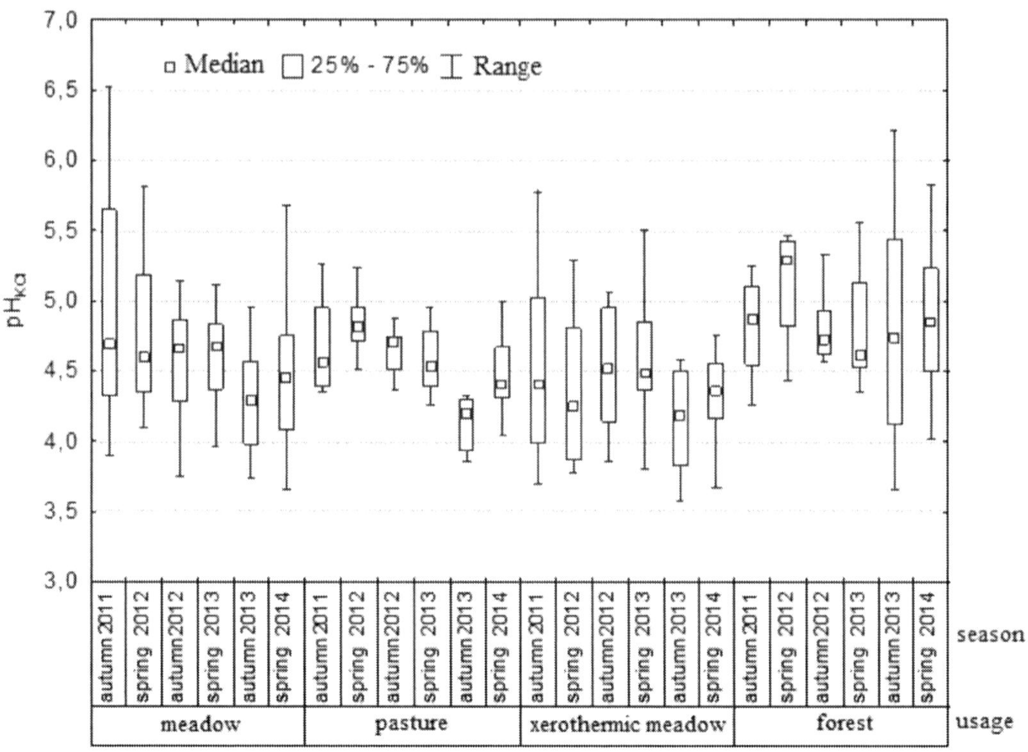

Fig. 3. Differentiation of pH_{KCl} depending on habitat use and season

In case of exchange soil reaction, the differentiation in the individual seasons is processed in Fig. 3 and the average values are shown in Tab. 4.

Year	2011	2012		2013		2014
Biotope	**Autumn**	**Spring**	**Autumn**	**Spring**	**Autumn**	**Spring**
Meadow	6.06±1.21	5.90±0.52	6.00±0.32	5.96±0.55	5.47±0.54	5.71±0.38
Pasture	5.80±0.27	5.84±0.40	5.99±0.25	5.79±0.26	5.39±0.24	5.63±0.27
Xerothermic meadow	5.57±0.77	5.53±0.60	5.97±0.47	5.66±0.43	5.29±0.45	5.49±0.29
Forest	5.93±0.49	6.04±0.27	6.12±0.07	5.89±0.31	5.74±0.59	5.57±0.40

Tab. 4. Average values ± SD exchange soil reaction on the monitored habitats

The values of soil reaction (active and exchange) were statistically evaluated by the analysis of variance without interaction depends on the soil type (ecological factor) (Tab. 5a - 5b).

	SS	Degrees of Freedom	MS	F	p
Soil type	13.837	4	3.459	15.100	0.000

$p < 0.05$, SS = Sum of Square Errors, MS = Mean Equared Errors, F = F - value, p = p – value

Tab. 5a. The variance analysis of pH_{H2O} values without interactions depending on the soil type factor

	SS	Degrees of Freedom	MS	F	p
Soil type	15.262	4	3.815	16.260	0.000

$p < 0.05$, SS = Sum of Square Errors, MS = Mean Equared Errors, F = F - value, p = p – value

Tab. 5b. The variance analysis of pH_{H2O} values without interactions depending on the soil type factor

Regarding the analysis of variance, we found out that the values of pH are actively affected by the factor, which is assessed on the significant level $\alpha = 0.05$. Diversity of the obtained results may be caused not only by variables like altitude, climate conditions, type of vegetation, but especially by different types of soil. Analysis of variance values of soil reaction with interaction, depending on factor of location (Žiar nad Hronom, Kremnické Bane, Banská Štiavnica) into which individual locations of soil sampling were assigned and factor of season soil sampling are processed in tables 6a - 6b.

	SS	Degrees of Freedom	MS	F	p
Locality	10.679	2	5.339	26.110	0.000
Season	12.763	5	2.553	12.480	0.000
Locality*Seaso	3.766	10	0.377	1.840	0.052

$p < 0.05$, SS = Sum of Square Errors, MS = Mean Equared Errors, F = F - value, p = p – value

Tab. 6a. The variance analysis of pH_{H2O} values with interactions depending on the factor of locality and season

	SS	Degrees of Freedom	MS	F	p
Locality	30.390	2	15.195	81.240	0.000
Season	3.954	5	0.791	4.230	0.001
Locality*Seaso	2.497	10	0.250	1.340	0.210

$p < 0.05$, SS = Sum of Square Errors, MS = Mean Equared Errors, F = F - value, p = p – value

Tab. 6b. The variance analysis of pH_{KCl} values with interactions depending on the factor of locality and season

Bossuyt and Hermy (2001) found out that the pH of soil is influenced by geographic region ($p < 0.05$). It is confirmed by our statistical results by using analysis of variance values pH_{H2O} with interactions in Tab. 6a ($F = 26.110$, $p = 0.000$) and values pH_{KCl} with interactions in Tab. 6b ($F = 81.240$, $p = 0.000$). Concerning the analysis of variance with interactions, there was not found out statistically significant interaction between location and season for active and exchange soil reaction. We found out statistically significant differences between Žiar nad Hronom and locations of Kremnické Bane and Banská Štiavnica by testing the impact of location on active soil reaction with Tukey´s HSD - test. The probable reason is that the places of soil sampling of location Kremnické Bane and Banská Štiavnica are mostly at higher altitudes. These places of soil sampling are, on the basis of vertical division, classified in the montane and oreal stage, while the soil sampling place in Žiar nad Hronom to the submontane and montane stage.

Locality	Žiar nad Hronom	Kremnické Bane	Banská Štiavnica
Žiar nad Hronom		0.000	0.000
Kremnické Bane			0.220
Banská Štiavnica			

$p < 0.05$

Tab. 7. The testing of contrasts values pH_{H2O} on the level of meteorological station location factors (Tukey´s HSD test)

For the completing the analysis of variance values of soil reaction (active and exchange) by testing levels of contrasts of factor season, we used Tukey's method of multiple comparing averages, which we implemented in the program STATISTICA. Having tested contrasts of values active soil reaction from a factor of season, we found out statistically significant differences between the other seasons and season of autumn 2013 for $p < 0.05$ (Tab. 8), when were recorded the lowest values of soil reaction $pH_{H2O} = 5.46 \pm 0.51$. (Tab. 9)

Year		**2011**	**2012**		**2013**		**2014**
Season		**Autumn**	**Spring**	**Autumn**	**Spring**	**Autumn**	**Spring**
Autumn	**2011**		0.362	1.000	0.405	0.000	0.000
Spring	**2012**			0.486	1.000	0.000	0.131
Autumn	**2012**				0.539	0.000	0.000
Spring	**2013**					0.000	0.061
Autumn	**2013**						0.209
Spring	**2014**						

$p < 0.05$

Tab. 8. The testing of contrasts values pH_{H2O} on the season factor level (Tukey´s HSD test)

Season	2011	2012		2013		2014
	Autumn	Spring	Autumn	Spring	Autumn	Spring
pH_{H2O}	5,93±0,98	5,86±0,51	6,01±0,32	5,88±0,48	5,46±0,51	5,64±0,36
pH_{KCl}	4,74±0,90	4,77±0,49	4,61±0,35	4,69±0,51	4,36±0,55	4,47±0,43

Tab. 9. Average values ± SD pH_{H2O} and pH_{KCl} in the seasons of soil sampling

The reason was that in 2013 rainfall fell more than in 2012, which can affect the acidity of the soil by its chemical composition. Fatubarin and Olojugba (2014) found out statistically significant differences between the soil pH in each month by testing the influence of rainfall season on soil pH during the 2 years.

Testing levels of contrast factor of season and locality were also carried out for values exchange soil reaction (pH_{KCl}), which are processed in Tab. 10 and in Tab. 11. In comparison with active soil reaction there were recorded minor differences in testing contrast values pH_{KCl}. Differences were found out between the season of the autumn 2012 and the autumn 2013. (Tab. 10) In the autumn of 2013 there were found out significantly lower values in comparison with the autumn of 2012. (Tab. 9) Difference of pH_{KCl} values was found out between the seasons of the spring 2012 and the spring 2014 (Tab. 10).

Year		2011	2012	2012	2013	2013	2014
Season		Autumn	Spring	Autumn	Spring	Autumn	Spring
Autumn	2011		0.993	0.226	0.682	0.000	0.003
Spring	2012			0.559	0.952	0.000	0.021
Autumn	2012				0.955	0.099	0.722
Spring	2013					0.004	0.159
Autumn	2013						0.808
Spring	2014						

$p < 0.05$

Tab. 10. The testing of contrasts values pH_{KCl} on the level of season factor (Tukey´s HSD test)

Locality	Žiar nad Hronom	Kremnické Bane	Banská Štiavnica
Žiar nad Hronom		0.000	0.000
Kremnické Bane			0.220
Banská Štiavnica			

$p < 0.05$

Tab. 11. The testing of contrasts values pH_{KCl} on the level of meteorological stations locality season factor (Tukey´s HSD test)

4 Conclusion

Functioning of meadow and forest ecosystems involves important processes influencing nutrient cycling, water cycle as well as the degradation of toxic substances. Regarding the analysis of variance, we found out that the pH values of the active soil reaction are influenced by factors (use of land and type of soil) on the significant level α = 0.05. We found out significant differences in the values of the active soil reaction, whereby the grassland has higher pH than forest ($p < 0.05$). The biggest variability and maximum value of active soil reaction was found out in the soil of the forest. Soil reaction was influenced among other things by slope of relief, season and locality. Soil of meadow and forest ecosystems is a future in mitigation of global climate change and therefore we must protect these natural resources.

Acknowledgement

The paper was supported by grant UGA entitled "*The content of labile nitrogen in soil samples of selected area*".

References

[1] ALEKSANDROWICZ – TRZCIŃSKA, M. 2005. Stan mikoryz sosny zwyczajnej w uprawie założonej na gruncie porolnym. In *Sylwan,* vol. 149, no. 2, 200: pp. 42-49.

[2] BEDRNA, Z. *Environmentálne pôdoznalectvo*. Bratislava : Slovenská akadémia vied v Bratislave, 2002a, ISBN 80-224-0660-0, 352 s.

[3] BEDRNA, Z. Acidifikácia pôd Belianskych Tatier. In *Oecologia Montana, International Journal of Mountain Ecology*, vol. 11, no. 1-2, 2002b: pp. 13-15. ISSN 1210-3209.

[4] BIELEK, P. *Kompendium praktického pôdoznalectva*. Nitra : SPU, 2014. ISBN 978-80-552-1155-8, 244 s.

[5] BOSSUYT, B., HERMY, M. Influence of land use history on seed banks in European temperate forest ecosystems: a review. In *Ecography.*, vol. 24, no. 2. 2001: pp. 225-238, ISSN 0906-7590.

[6] ČURLÍK, J., ŠURINA, B. *Príručka terénneho prieskumu a mapovania pôd*. Bratislava : Výskumný ústav pôdnej úrodnosti, 1998, ISBN 80-85361-37-X, 134 s.

[7] FATUBARIN, A., OLOJUGBA, M. R. Effect of rainfall season on the chemical properties of the soil of a Southern Guinea Savanna ecosystem in Nigeria. In *Journal of Ecology and the Natural Environmen,* vol. 6, no. 4, 2014: pp. 182-189, ISSN 2006-9847.

[8] FRATERRIGO, J. M., TURNER, M. G., PEARSON, S. M. The effect of prior land use on the spatial distribution of soil chemical properties in southern Appalachian forests. In *18th Annual general meeting*, Alberta, Canada, 2003.

[9] GÜNDOĞAN, R., ÖZYURT, H., AKAY, A. E. The effects of land use on properties of soils developed over ophiolites in Turkey. In *Int. J. Forest, Soil and Erosion,* vol. 1, no. 1, 2011: pp. 36-42, ISSN 2251-6387,

[10] HANES, J. *Antropogénne vplyvy na vlastnosti poľnohospodárskych pôd*. Nitra : Vysoká škola poľnohospodárska v Nitre, 1995 ISBN 80-7137-238-2, 89 s.

[11] HUDEC, M., HREŠKO, J. Obsah humusu na pôdach kamenných miest Kremnických vrchov. In *Ekologické štúdie,* roč. 3, č. 2, 2012: pp. 20-27.

[12] HUDEC, M., FESZTEROVÁ, M. Spatial variability in chemical properties of eutric cambisol, In Acta fytotechnica et Zootechnica, roč. 16, č. 2, 2013: pp. 11-20, ISSN 1336-9245.

[13] ,HUDEC, M., FESZTEROVÁ, M. *Humifikácia na neovulkanitoch Kremnických a Štiavnických vrchov*. Nitra : UKF, 2015, ISBN 978-80-558-0866-6, 143 s.

[14] ISLAM, K. R., WEIL, R. R. Land use effects on soil quality in a tropical forest ecosystem of Bangladesh. In *Agriculture, Ecosystems and Environment*, 79, 2000: pp. 9-16.

[15] KOLEDA, P., HUDEC, M., FESZTEROVÁ, M. GIS a jeho využitie na polohovú charakteristiku pri sledovaní procesu humifikácie v pôde Kremnických vrchov. In Zborník príspevkov, Nováková, M., Sviček, M. (Eds.): *Environmentálne indexy a indikátory ako nástroje analýzy a hodnotenia stavov a procesov v krajine*, 2012, ISBN 978-80-89128-97-6, pp. 105-113.

[16] KOVÁČIK, P. 2013. *Agrochémia a výživa rastlín*. Nitra : SPU v Nitre, 2013, ISBN 978-80-552-1012-4, 179 s.

[17] LAL, R. Soil management in developing countries. In *Soil Sci*, 165, 2000: pp. 57-72.

[18] MAZÚR, E. Hypsografické stupne 1 : 500 000. In MAZÚR E. et al. (Eds.) *Atlas SSR,* 1980, s. 38-39.

[19] NÉMETHOVÁ, J., DUBCOVÁ, A., KRAMÁREKOVÁ, H. The impacts of the European Union's common agricultural policy on agriculture in Slovakia. In *MORAVIAN GEOGRAPHICAL REPORTS*, vol. 22, no. 4, 2014: pp. 51-64, ISSN 1210-8812,

[20] RYAN, J., MASRI, S., IBRIKÇI, H., SINGH, M., PALA, M., HARRIS, H. C. Implications of Cereal-Based Crop Rotations, Nitrogen Fertilization, and Stubble Grazing on Soil Organic Matter in a Mediterranean-Type Environment. In *Turk J Agric For*, 32, 2008: pp. 289-297.

[21] ŠLEPETIENĖ, A., ŠLEPETYS, J., LIAUDANSKIENĖ, I. Chemical composition of differently used Terric Histosols. In *Zemdirbystė – Agriculture*, vol. 97, 2010: pp. 25-32..

[22] VAN REEUWIJK, L. P. *Producer for Soil Analysis*. International Soil reference and Information Centre, Wageningen, 2002.

[23] VILČEK, J., HRONEC, O., BEDRNA, Z. *Environmetána pedológia*. Nitra, Bratislava: SPU, VÚPOP, 2005, ISBN 80-8069-501-6. 299 s.

[24] ZAUJEC, A., CHLPÍK, J., NÁDAŠSKÝ, J., POLÁKOVÁ, N., TOBIÁŠOVÁ, E. *Pedológia a základy geológie*. Nitra: SPU, 2009, ISBN 978-80-552-0207-5, 399 s.

[25] ZRUBEC, F. Všestranná starostlivosť o úrodnosť pôdy alebo „zúrodňovanie„ pôdy? In *Pôda najcennejší zdroj,* Bratislava: VÚPÚ, 1990, pp. 39-45.

Current address

Hudec Michal, PaedDr. PhD.
Piaristické gymnázium sv. Jozefa Kalazanského
Piaristická 6, 949 01 Nitra, Slovak Republic
Tel. number: +421 907615135, e-mail: michal.hudec81@gmail.comI

Porubcová Lýdia, Ing.
Faculty of Natural Sciences, Constantine the Philosopher University
Tr. A. Hlinku 1, 949 74 Nitra, Slovak Republic
Tel. number: +421 6408 653, e-mail: lydia.porubcova@ukf.sk

Noga Henryk, prof. hab. Dr. PhD.
Faculty of Mathematics, Physics and Technical Science, The Pedagogical University of Cracow
Podchorążych 2, 31-015 Cracow, Poland
Tel. number: +48 604895789, e-mail: senoga@cyf-kr.edu.pl

Feszterová Melánia, doc. Ing. PhD.
Faculty of Natural Sciences, Constantine the Philosopher University
Tr. A. Hlinku 1, 949 74 Nitra, Slovak Republic
Tel. number: +421 903456414, e-mail: mfeszterova@ukf.sk

Slovak University of Technology in Bratislava
Faculty of Mechanical Engineering
16th Conference on Applied Mathematics **APLIMAT 2017**
Proceedings

COMPOSITIONS OF POLYTOPES TRIANGULATED BY LINEAR ORDERS

IORFIDA Vincenzo (IT)

Abstract. We consider convex polytopes arising from Veronese squarefree rings and we study their triangulations realized starting by initial complexes linked to different term orders. We utilize their geometric realization to obtain new models in different fields.

Keywords: Computational Geometry, Algebraic Geometry, Simplicial Complexes

Mathematics Subject Classification: AMS_01A99, 13P10, 13755

Introduction

Consider term orders on the set of monomials of the polynomial ring S=K[$x_1, x_2, \dots, x_n$], K a field. The linear order L is introduced in [9] (chap.9) on the set $\underline{T}$ of r indexed variables appearing in the rth-Veronese squarefree ring, the subring of S generated on K by all squarefree monomials of S of degree r in n variables, useful in many subjects of combinatorics and convex geometry.

In the papers [1], the authors considered the universal Gröbner basis G for the toric ideal I of K[$\underline{T}$], the algebra associated to the second Veronese squarefree ring, with respect to all linear orders on the set $\underline{T}$ of variables and the lexicographic order on the set of monomials of $K[\underline{T}]$. They proved that the degree is 2 for n=4 and 3 for n>4 and they compute the initial complexes with respect to those term orders and the triangulations of the polytopes arising from the initial complexes.

In this paper, we construct the polytopes that are the geometric realizations of the initial complexes. By using their triangulations, one can obtain new interesting configurations, composing the triangulated polytopes, for applications in the field of engineering, or business and architecture (see [4], [5])

More precisaly, for n=4, we utilize octahedrons for obtaining new models that offer multidisciplinary approaches in the teaching of scientific disciplines.

N. 1 Notations, basic concepts and known results.

Here and throughout this paper, $\mathbb{R}$ denotes the field of real numbers and $\mathbb{R}^+$denotes the non–negative real numbers.

Definition 1.1. A polyhedron is a finite intersection of closed half-spaces in $\mathbb{R}^n$. Thus a polyhedron P can be written as $P = \{\boldsymbol{x} \in \mathbb{R}^n : A \leq \boldsymbol{x} \leq \boldsymbol{b}\}$, (1.1)

where A is a matrix with *n* columns. If **b**=0 then there exist vectors $\mathbf{u}_1, ..., \mathbf{u}_m \in \mathbb{R}^n$ such that
$P = pos\ (\{\mathbf{u}_1, \dots, \mathbf{u}_m\}) := \{\boldsymbol{\lambda}_1\mathbf{u}_1 + \dots + \boldsymbol{\lambda}_m\mathbf{u}_m : \boldsymbol{\lambda}_1, \dots, \boldsymbol{\lambda}_m \in \mathbb{R}^+\}$ (1.2)

Definition 1.2. A polyhedron of the form (1.2) is called a polyhedral cone.

Definition 1.3. A polyhedron Q which is bounded is called a polytope and it is the convex hull of a finite set of points

$$Q = conv\ (\{\mathbf{v}_1, \dots, \mathbf{v}_m\}) := \left\{\sum_{i=1}^{m} \lambda_i \mathbf{v}_i : \lambda_1, \dots, \lambda_m \in \mathbb{R}^+, \sum_{i=1}^{m} \lambda_i = 1\right\}$$

Example 1.4. (*the cube*)

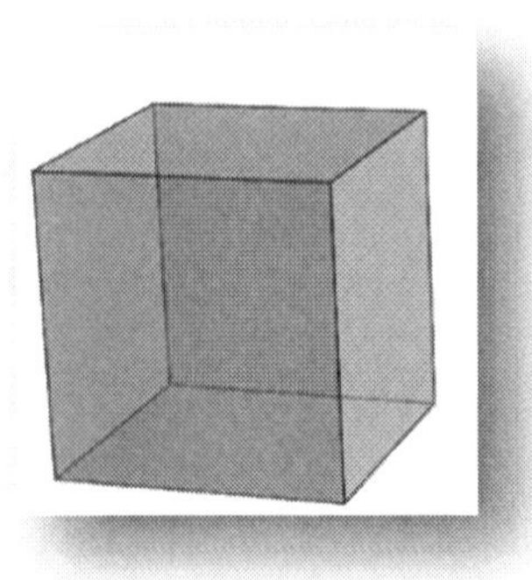

Example 1.5. (*the 1- tetrahedron*)

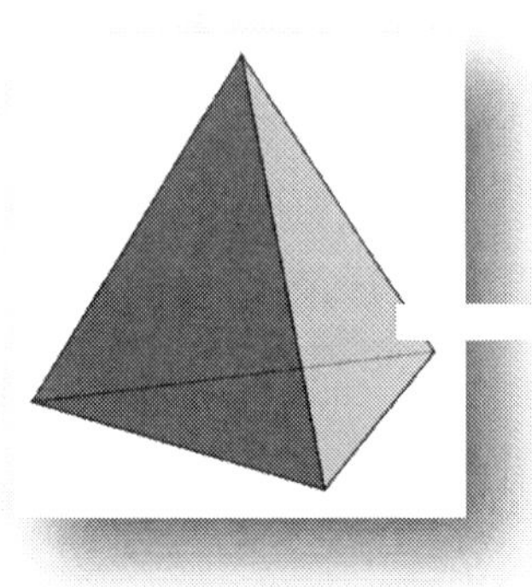

Example 1.6. (*the 1- octahedron*)

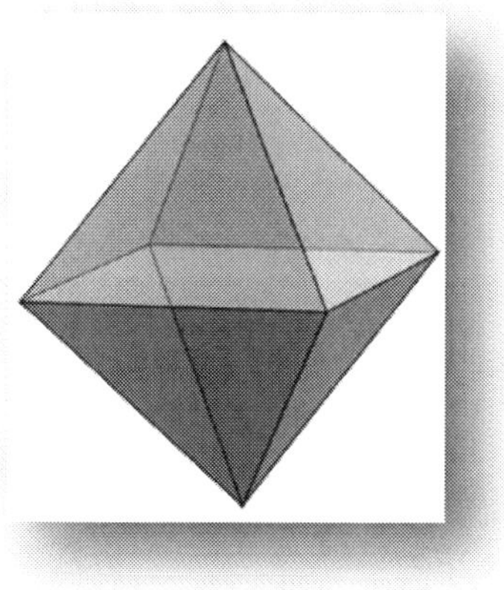

Remark 1.7. For n=3, the points (1,1,0), (1,0,1), (0,1,1) are affinely independent and we obtain as convex hull the full plane $\mathbb{P}^2$.

Let K be a field (that will be $\mathbb{Q}$ or $\mathbb{R}$) and let R = $K\,[x_1, x_2, \dots, x_n]$ be the polynomial ring in n variables $x_1, x_2, \dots, x_n$. The set of the power products of R is denoted by:

$$M^n = \{x_1^{a_1} \cdots x_n^{a_n} \,|a_i \in \mathbb{N}, i = 1, \dots, n\}$$

where $x_1^{a_1} \cdots x_n^{a_n} = \mathbf{x}^{\mathbf{a}},\ \boldsymbol{a} = (a_1, a_2, \dots, a_n) \in \mathbb{N}^n$.

Definition 1.8. We define a term order on M^nas a total order < on M^n satisfying:
(1) 1 <$\boldsymbol{x}^{\boldsymbol{b}}$ for all $\boldsymbol{x}^{\boldsymbol{b}} \in M^n$, $\boldsymbol{x}^{\boldsymbol{b}} \neq 1$;
(2) If xa $\boldsymbol{x}^{\boldsymbol{a}}$<$\boldsymbol{x}^{\boldsymbol{b}}$, then $\boldsymbol{x}^{\boldsymbol{a}}\boldsymbol{x}^{\boldsymbol{c}}$< $\boldsymbol{x}^{\boldsymbol{b}}\boldsymbol{x}^{\boldsymbol{c}}$, for all $\boldsymbol{x}^{\boldsymbol{c}} \in M^n$

Definition 1.9. We define the lexicographical order on $M^n$with order of variables x_1>x_2> . . . >x_n as follows:
for $\boldsymbol{a} = (a_1, a_2, \dots, a_n)$, $\boldsymbol{b} = (b_1, b_2, \dots, b_n)$
$\boldsymbol{x}^{\boldsymbol{a}} < \boldsymbol{x}^{\boldsymbol{b}} \Leftrightarrow \exists h \mid a_1 = b_1, \dots, a_h = b_h, a_{h+1} < b_{h+1} h = 0, \dots, n-1$

Definition 1.10. Let $G \subset$ R, $G = \{f_1, \dots, f_s\}$ a set of polynomials. G is an universal Gröbner basis for the ideal $(f_1, \dots, f_s)$ if G is a Gröbner basis with respect to all term orders on the monomials of R.

Definition 1.11. Let $G \subset$ R, $G = \{f_1, \dots, f_s\}$ a set of polynomials. G is said a universal lexicographical Gröbner basis for $(f_1, \dots, f_s)$ if G is a Gröbner basis with respect to the lexicographical order on the monomials of R and with respect to all orders of the variables $x_1, x_2, \dots, x_n$.

Let Γ = $\{<_i\}$ a set of orders of the variables $x_1, x_2, \dots, x_n$ with |Γ| < n!.

Definition 1.12. Let $G \subset R$, $G = \{f_1, \dots, f_s\}$ be a set of polynomials. We say that G is a Γ−universal lexicographical Gröbner basis for $(f_1, \dots, f_s)$if G is a Gröbner basis with respect to the lexicographical order on the monomials of $\mathcal{R}$ and with respect to all orders of Γ on the variables $x_1, x_2, \dots, x_n$.

Now we recall the definition of linear orders of variables introduced in [9].
Given two integers n and k such that $4 \leq 2k \leq n$ we consider the set of variables:

$$\mathbf{T} = \{T_{i_1 i_2,\dots,i_k}: 1 \leq i_1 < i_2 < \cdots < i_k \leq n\}.$$

Let $A = (a_{ij})$ be a non singular matrix of order k with entries in R.

Definition 1.13. (Linear lexicographical order in $\boldsymbol{T}$)
Let $\boldsymbol{i} = (i_1, i_2, \dots, i_k) \in \mathbb{Z}^k$ and $\boldsymbol{j} = (j_1, j_2, \dots, j_k) \in \mathbb{Z}^k$. Then

$$\boldsymbol{i} <_{A(lex)} \boldsymbol{j} \text{ iff } \boldsymbol{i}A <_{lex} \boldsymbol{j}A$$

that is:

$$(i_1, i_2, \cdots, i_k)(a_{ij}) = (b_1, b_2, \cdots, b_k)$$

where$b_t = i_1 a_{1t} + i_2 a_{2t} + \cdots + i_k a_{kt}$, $\boldsymbol{i}A = (b_1, b_2, \cdots, b_k)$ and $\boldsymbol{J}A = (c_1, c_2, \cdots, c_k)$.
So $\boldsymbol{i}A >_{lex} \boldsymbol{j}A$ iff $(b_1, b_2, \cdots, b_k) >_{lex} (c_1, c_2, \cdots, c_k)$ that is $b_1 = c_1, b_2 = c_2, \dots, b_l > c_l$
For some $l = 1, \dots, k$.

Thus, we can define the linear lexicographical order of variables by:

$T_{i_1 i_2,\dots,i_k} >_{A(lex)} T_{j_1 j_2,\dots,j_k}$ iff $(i_1, i_2, \dots, i_k) >_{A(lex)} (j_1, j_2, \dots, j_k)$.

Now, consider the set $\mathcal{A}_k$of matrices of order k of the form $(\pm e_{i_j})$, $j = 1, \dots, k$, where the set $\{e_{i_1}, \dots, e_{i_k}\}$ is a permutation of the set of canonical vectors $\{e_1, \dots, e_k\}$ of $\mathbb{R}^k$. See [9].

Now consider the presentation:
$\varphi: K[\boldsymbol{T}] \longrightarrow K[V_k]$ induced by $T_{i_1 i_2 \dots i_k} \mapsto x_{i_1}, \dots, x_{i_k}$,
where $I = \ker(\varphi)$ is the toric ideal of $K[V_k]$. We call the monomial algebra $K[V_k]$ the squarefree kth-Veronese subring of R. In [1], [9], for some values of k and n, $0 \leq 2k \leq n$, the authors give positive answer to the problem to obtain lexicographic Gröbner bases for linearly ordered variables.

Let be $K[V_2]$ be the squarefree second-Veronese subring of the polynomial ring or, where V_2 is the set of square-free monomials of or of degree 2.

Proposition 1.14. The short list by the matrices of $\mathcal{A}_2$is the following

$$A_1 = \begin{pmatrix} 1 & 0 \\ 0 & 1 \end{pmatrix}, A_2 = \begin{pmatrix} -1 & 0 \\ 0 & 1 \end{pmatrix}, A_3 = \begin{pmatrix} 1 & 0 \\ 0 & -1 \end{pmatrix}, A_4 = \begin{pmatrix} -1 & 0 \\ 0 & -1 \end{pmatrix},$$
$$A_5 = \begin{pmatrix} 0 & 1 \\ 1 & 0 \end{pmatrix}, A_6 = \begin{pmatrix} 0 & 1 \\ -1 & 0 \end{pmatrix}, A_7 = \begin{pmatrix} 0 & -1 \\ 1 & 0 \end{pmatrix}, A_8 = \begin{pmatrix} 0 & -1 \\ -1 & 0 \end{pmatrix}$$

Proof: See [1](Prop 1.8.)

Proposition 1.15. Let *i, j, k, l* be positive integers such that $1 \leq i < j < k < l \leq n$. Then the linear lexicographical orders on the variables of **T** induced by the matrices of $\mathcal{A}_2$ are given by the list:

$$T_{kl} > T_{jl} > T_{jk} > T_{il} > T_{ik} > T_{ij} \qquad \text{(by } A_1\text{)},$$

$$T_{il} > T_{ik} > T_{ij} > T_{jl} > T_{jk} > T_{kl} \qquad \text{(by } A_2\text{)},$$

$$T_{kl} > T_{jk} > T_{jl} > T_{ij} > T_{ik} > T_{il} \qquad \text{(by } A_3\text{)},$$

$$T_{ij} > T_{ik} > T_{il} > T_{jk} > T_{jl} > T_{kl} \qquad \text{(by } A_4\text{)},$$

$$T_{kl} > T_{jl} > T_{il} > T_{jk} > T_{ik} > T_{ij} \qquad \text{(by } A_5\text{)},$$

$$T_{kl} > T_{jk} > T_{jl} > T_{ij} > T_{ik} > T_{il} \qquad \text{(by } A_6\text{)},$$

$$T_{il} > T_{jl} > T_{kl} > T_{ih} > T_{jk} > T_{ij} \qquad \text{(by } A_7\text{)},$$

$$T_{ij} > T_{ik} > T_{jk} > T_{il} > T_{jl} > T_{kl} \qquad \text{(by } A_8\text{)}.$$

After, we will extend systematically the above orders to all the variables $T_{ij}, 1 \leq i < j \leq n$.

Proof: See [1](Prop 1.8.)

Theorem 1.16. Let $K[V_2]$ be the square free second-Veronese subring of R = $K[x_1, x_2, \ldots ,x_n]$ and let I be the toric ideal of $K[V_2]$= $K[T]$ /I. Then the following facts hold:

(1) The set $B_1 = \{\underline{T_{ij}T_{kl}} - T_{il}T_{jk}, \underline{T_{ik}T_{jl}} - T_{il}T_{jk}, 1 \leq i < j < k < l \leq n\}$ is a lexicographical reduced Gröbner basis for I with respect to the linear orders on the variables induced by the matrices A_1, A_4, A_5, A_8.

(2) The set $B_2 = \{T_{ij}T_{kl} - \underline{T_{il}T_{jk}}, T_{ij}T_{kl} - \underline{T_{ik}T_{jl}}, \underline{T_{ik}T_{jk}T_{lm}} - T_{ij}T_{km}T_{kl}, 1 \leq i < j < k < l < m \leq n\}$; is a reduced lexicographical Gröbner basis for I with respect to the linear orders on the variables induced by the matrices A_2, A_7.

(3) The set $B_3 = \{\underline{T_{ij}T_{kl}} - T_{il}T_{jk}, T_{ik}T_{jl} - \underline{T_{il}T_{jk}}, 1 \leq i < j < k < l \leq n\}$ is a lexicographical reduced Gröbner basis for I with respect to the linear

orders on the variables induced by the matrices A_3 and A_6.
The underlined terms in each binomial are the initial monomials.

Proof: See [1](Theorem 1.11.)

Corollary 1.17. Let $K[V_2]= K[T]/I$. The set $\{T_{ij}T_{kl} - T_{il}T_{jk}, T_{ik}T_{jl} - T_{il}T_{jk}, T_{ij}T_{kl} - T_{ik}T_{jl}, T_{ik}T_{jk}T_{lm} - T_{ij}T_{km}T_{kl}\ 1 \leq i < j < k < l < m \leq n\}$ ofbinomials of I, is a Γ-universal lexicographic Gröbner basis of *I*.

Proof: See [1](Corollary 1.12.)

Example 1.18. If n=4, let T $= \{T_{12}, T_{13}, T_{14}, T_{23}, T_{24}, T_{34}\}$ be the set of variables for $K[V_2] \subseteq K[x_1, x_2, ... ,x_n]$.
The Γ-universal lexicographical square free quadratic Gröbner basis for I is of degree twe and is the following

$$\{T_{24}T_{13} - T_{23}T_{14}, T_{34}T_{12} - T_{23}\ T_{14}, T_{12}T_{34} - T_{13}, T_{24}\}.$$

N.2 Triangulation and compositions.

We recall the definition of initial simplicial complex $in_{<}\Delta$ and triangulation of a set of lattice points.
Let I be any ideal of R= $K[x_1, ..., x_n]$ and let < be a term order on the monomials of R.

Definition 2.1. The initial complex $\Delta_{<}(I)$ of I with respect to < is the simplicial complex on the vertex set $\{1,2, ..., n\}$ whose the Stanley- Reisner ideal is the radical of $in_{<}(I)$.

Let A = $\{\boldsymbol{a}_1, ..., \boldsymbol{a}_i\}$, where $\underline{a}_i \in \mathbb{N}^n$

Definition 2.2. ([9], chap.8).
Let σ be a subset of lattice points, $\sigma \subseteq \mathcal{A}$ and *pos (σ)* be the polyhedral cone generated by σ, then one defines triangulation of $\mathcal{A}$ a collection Δ of subsets of $\mathcal{A}$,such that the set *{pos (σ), σ ∈ $\mathcal{A}$ }* is the set of cones of a simplicial fan whose support is the same as *pos ($\mathcal{A}$)*,
supp {pos (σ), σ ∈Δ} = pos ($\mathcal{A}$).

Let $K[V_k]$ be the kth–Veronese square free subring of R and let $\mathcal{A}$ be the set of lattice points of the semigroup under $K[V_k]$.
We will identify the given set of lattice points $\mathcal{A} = \{\boldsymbol{a}_1, \boldsymbol{a}_2, ... ,\boldsymbol{a}_{\binom{n}{k}}\} \subset \mathbb{N}^n$ with the index set$\{1, ..., t\}, t = \binom{n}{k}$.

Definition 2.3. Let < be any linear order of the set of variables $T = \{T_{i_1 i_2 \dots i_k} : 1 \leq i_1 < i_2 < \dots < i_k \leq n\}$. Let $K[V_k] = K[T]/I_{\mathcal{A}}$. We call initial linear simplicial complex of I any initial simplicial complex with respect to the linear order < of the variables.

Definition 2.4. Let < be any linear order of the set of variables $T = \{T_{i_1 i_2 \dots i_k} : 1 \leq i_1 < i_2 < \dots < i_k \leq n\}$. Let $K[V_k] = K[T]/I_{\mathcal{A}}$. We call initial linear lexicographical simplicial complex of $I_{\mathcal{A}}$ any initial linear simplicial complex of I with respect to the lexicographical order on the monomials of K[T].
For k = 2, we call the variables $T_{i_1 i_2}$ of I with T_{ij}, $1 \leq i < j \leq n$.

Theorem 2.5. Let $K[V_2] = K[T]/I_{\mathcal{A}}$. Then we have:
i): The ideal $I_{\mathcal{A}}$ has three linear lexicographical initial complexes $\Delta_<^{(1)}(I_{\mathcal{A}})$, $\Delta_<^{(2)}(I_{\mathcal{A}})$ and $\Delta_<^{(3)}(I_{\mathcal{A}})$ corresponding respectively to B_1, B_2, B_3.
ii): The linear lexicographical initial complexes $\Delta_<^{(1)}(I_{\mathcal{A}})$ and $\Delta_<^{(2)}(I_{\mathcal{A}})$ have a Stanley-Reisner ideal of degree two.
iii): The linear lexicographical initial complex $\Delta_<^{(3)}(I_{\mathcal{A}})$, has a Stanley-Reisner ideal of degree two for n = 4 and of degree three for n > 4.

Proof: See [1](Corollary 2.5)

Theorem 2.6. (Geometric realization)

Let $K[V_2]$ be the second Veronese free ring of $K[x_1, \dots, x_n]$. For n=4 we have the following facts:
(1) $\Delta_<^{(1)}(I_{\mathcal{A}})$, is an octahedron with set of vertices T= {(0,0,1,1) >(0,1,0,1)> (0,1,1,0)> (1,0,0,1) (1,0,1,0)>(1,1,0,0) }, with the following labeling of the vertices

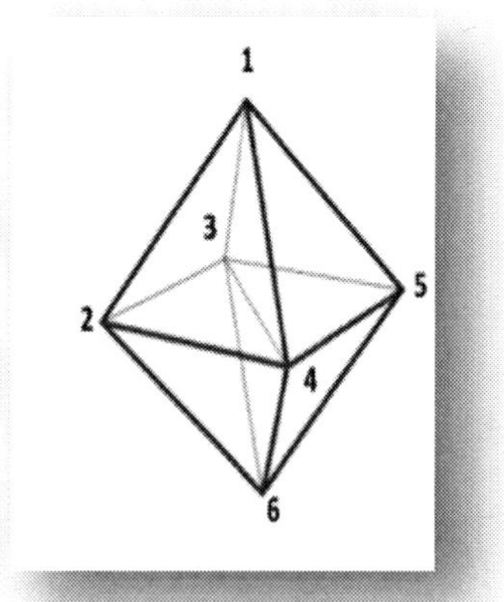

and triangulation $\Delta^{(1)}$with a maximal faces set {{1, 2, 3, 4}, {3, 4, 5, 6}, {1, 5, 3, 4}, {5, 6, 3, 4}}.

(2) $\Delta_<^{(2)}(I_{\mathcal{A}})$, is an octahedron with set of vertices T= {(0, 0, 1, 1)>(0, 1, 1, 0)> (0, 1, 0, 1)> (1, 1, 0, 0)> (1, 0, 1, 0)> (1, 0, 0, 1)}, with the following labeling of the vertices

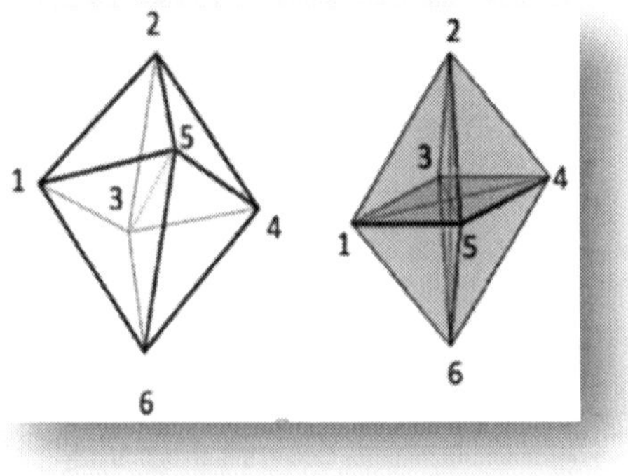

and triangulation $\Delta^{(2)}$ with a maximal faces set {{1, 2, 3, 5}, {3, 4, 5, 6}, {2, 3, 4, 5}, {1, 3, 5, 6}}

(3) $\Delta_{<}^{(3)}$ ($I_{\mathcal{A}}$), is an octahedron with set of vertices T= {(1, 0, 0, 1)> (1, 0, 1, 0)>(1, 1, 0, 0)> (0, 1, 0, 1)> (0, 1, 1, 0)> (0, 0, 1, 1} with the following labeling of the vertices

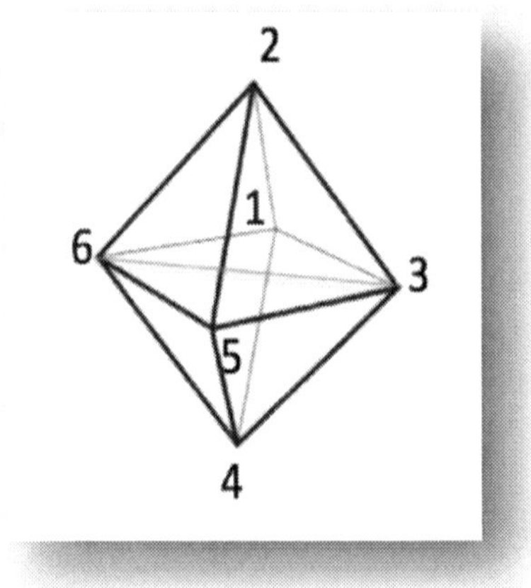

and triangulation $\Delta^{(3)}$ with a maximal faces set {{1, 3, 4, 6}, {3, 4, 5, 6}, {2, 3, 5, 6}, {1, 2, 3, 6}}

Proof: Let $\Delta^{(i)}$ be the regular triangulation coming from the linear order on the variables and associated to the simplicial complex $\Delta_{<}^{(i)}(I_{\mathcal{A}})$, i= 1,2,3.

(a) Consider the linear initial simplicial complex$\Delta_{<}^{(1)}(I_{\mathcal{A}})$. Write the set $\mathcal{A}$ as an ordered set

$\mathcal{A}$ = {(0, 0, 1, 1), (0, 1, 0, 1), (0, 1, 1, 0), (1, 0, 0, 1), (1, 0, 1, 0), (1, 1, 0, 0)},

with ($\boldsymbol{a}_1 > \boldsymbol{a}_2 > ... > \boldsymbol{a}_6$). In the correspondence

$\boldsymbol{a}_i \longrightarrow \boldsymbol{x}^{a_i} \longrightarrow z_i, \quad \boldsymbol{x} = (x_1, \dots, x_n)$

$G_1 = \left\{\underline{z_1 z_6} - z_3 z_4 \text{ , } \underline{z_2 z_5} - z_3 z_4 \right\}$and the sets {1, 6}, {2, 5} $\notin \Delta_{<}^{(1)}(I_{\mathcal{A}})$

Since {1, 6}, $\notin \Delta_{<}^{(1)}(I_{\mathcal{A}})$ the square {1, 3, 6, 4} is not a face of the initial complex $\Delta_{<}^{(i)}(I_{\mathcal{A}})$ an since {2, 5} $\notin \Delta_{<}^{(1)}(I_{\mathcal{A}})$ the square {2, 3, 5, 4} is not a face of $\Delta_{<}^{(1)}(I_{\mathcal{A}})$. It follows that we have 4 maximal faces of $\Delta^{(1)}(I_{\mathcal{A}})$: {2, 3, 5, 4}, {1, 5, 3, 4}, {2, 6, 3, 4}, {5, 6, 3, 4} that are 4 tetrahedrons.

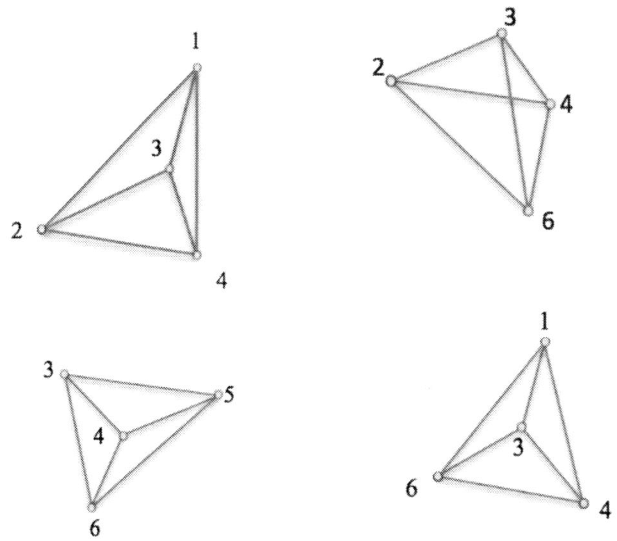

It follows that the set of maximal faces of $\Delta(1)$ is given by:
$\{\{1, 2, 3, 4\}, \{1, 5, 3, 4\}, \{2, 6, 3, 4\}, \{5, 6, 3, 4\}\}$.

(b) Consider the linear initial simplicial complex $\Delta_<(I_{\mathcal{A}})$. Write the set $\mathcal{A}$ as an ordered set

$\mathcal{A} = \{(0, 0, 1, 1), (0, 1, 1, 0), (0, 1, 0, 1), (1, 1, 0, 0), (1, 0, 1, 0), (1, 0, 0, 1)\}$,

with $(\underline{a}_1 > \underline{a}_2 >, \dots, > \underline{a}_6)$. In the correspondence

$\mathbf{a}_i \longrightarrow \mathbf{x}^{a_i} \longrightarrow z_i, \quad \mathbf{x} = (x_1, \dots, x_n)$

$G_2 = \left\{ \underline{z_1 z_4} - z_2 z_6 \;,\; \underline{z_3 z_5} - z_2 z_6 \right\}$ and the sets $\{1, 4\}, \{2, 6\} \notin \Delta_<^{(2)}(I_{\mathcal{A}})$

Since $\{1, 4\}, \notin \Delta_<^{(2)}(I_{\mathcal{A}})$, the square $\{1, 3, 4, 5\}$ is not a face of the initial complex $\Delta_<^{(2)}(I_{\mathcal{A}})$ an since $\{2, 6\} \notin \Delta_<^{(2)}(I_{\mathcal{A}})$ the square $\{2, 3, 5, 6\}$ is not a face of $\Delta_<^{(2)}(I_{\mathcal{A}})$. It follows that the set of maximal faces of $\Delta^{(2)}$ is given by:

$\{\{1, 2, 3, 5\}, \{3, 4, 5, 6\}, \{2, 3, 4, 5\}, \{1,3,5, 6, \}\}$.

We have 4 maximal faces of $\Delta^{(2)}(I_{\mathcal{A}})$ that are 4 tetrahedrons.

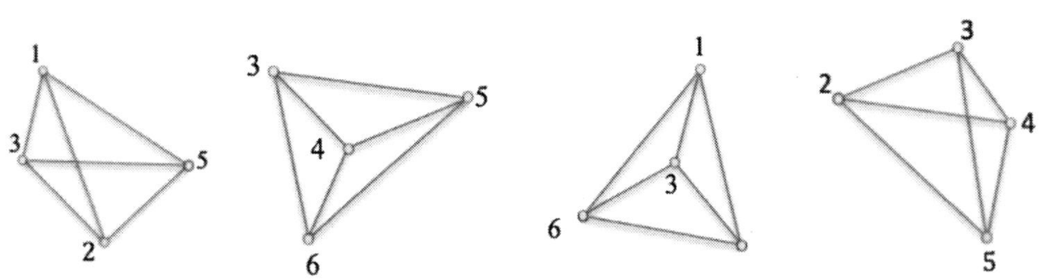

(c) Consider the linear initial simplicial complex $\Delta_<^{(3)}(I_{\mathcal{A}})$. Write the set $\mathcal{A}$ as an ordered set

$\mathcal{A} = \{(1,0,0,1), (1,0,1,0), (1,1,0,0), (0,1,0,1), (0,1,1,0), (0,0,1,1)\}$,
with $(\boldsymbol{a}_1 > \boldsymbol{a}_2 > \ldots > \boldsymbol{a}_6)$.

In the correspondence $\mathbf{a_i} \longrightarrow \mathbf{x}^{\mathbf{a_i}} \longrightarrow \mathrm{z_i}$, $\quad \mathbf{x} = (x_1, \ldots, x_n)$
$\mathrm{G}_3 = \left\{\underline{\mathrm{z_3z_6}} - \mathrm{z_1z_5}\ ,\ \underline{\mathrm{z_2z_4}} - \mathrm{z_3z_6}\right\}$ and the sets $\{1, 6\}, \{2, 5\} \notin \Delta_{<}^{(1)}(\mathrm{I}_{\mathcal{A}})$
Since $\{1, 5\}, \notin \Delta_{<}^{(1)}(\mathrm{I}_{\mathcal{A}})$ the square $\{1, 3, 5, 6\}$ is not a face of the initial complex $\Delta_{<}^{(3)}(\mathrm{I}_{\mathcal{A}})$ an since $\{2, 4\} \notin \Delta_{<}^{(3)}(\mathrm{I}_{\mathcal{A}})$ the square $\{1, 2, 5, 4\}$ is not a face of $\Delta_{<}^{(3)}(\mathrm{I}_{\mathcal{A}})$. It follows that we have 4 maximal faces of $\Delta^{(3)}(I_{\mathcal{A}})$ are $\{1, 3, 4, 6\}, \{3, 4,5, 6\}, \{2, 3, 5, 6\}, \{1, 2, 3, 6\}$, that is 4 tetrahedrons.

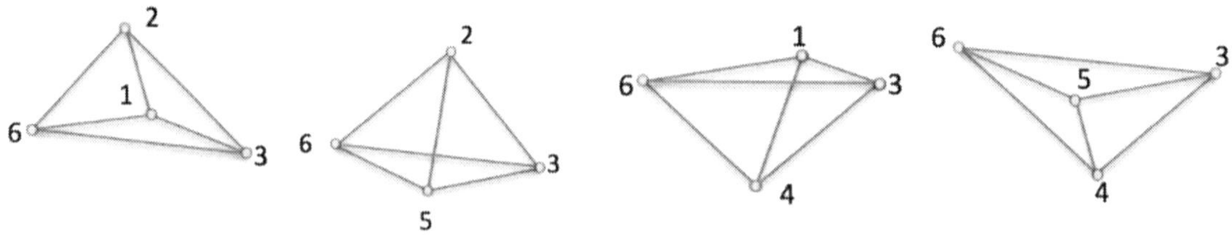

Remark 2.7. The kth-Veronese squarefree ring V_k has a very important role in many problems of algebraic and geometric combinatorics. In particular in some transportation problems considered in [7], [2], [3] the Segre product of two squarefree Veronese rings is considered. Moreover V_k is linked, naturally to the class of mixed product ideals in two sets of variables introduced in [6]. In the simplicial complexes having as Stanley- Reisner ideals such ideals can by studied (see [8] and [9]). It is an open problem to study the geometric realization of the simplicial complexes associated to this class.

Now we consider some compositions of tetrahedrons:

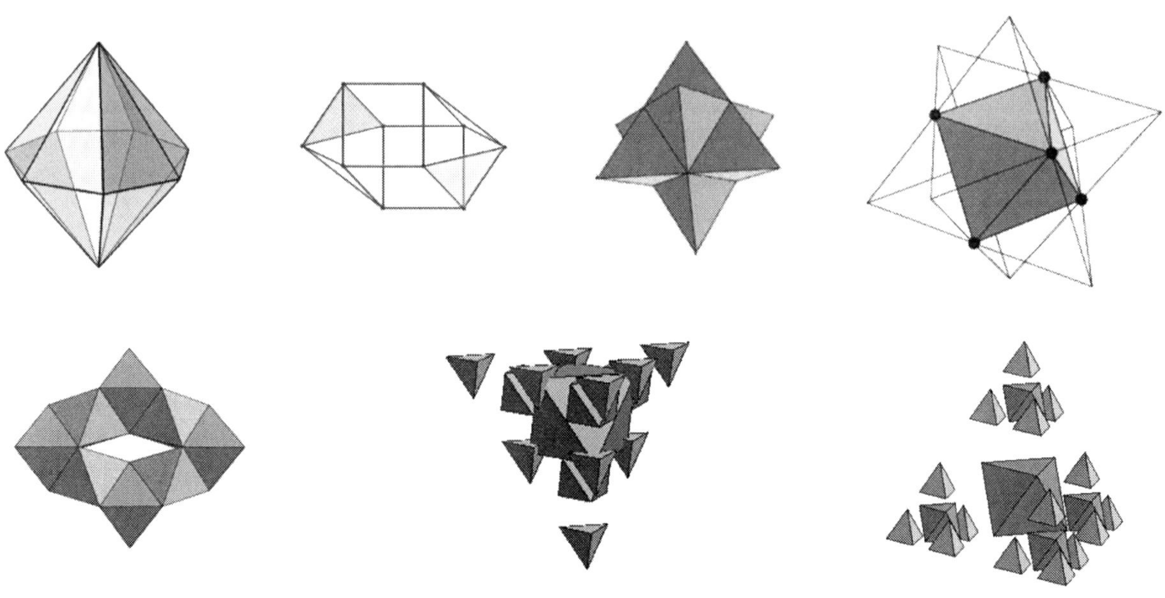

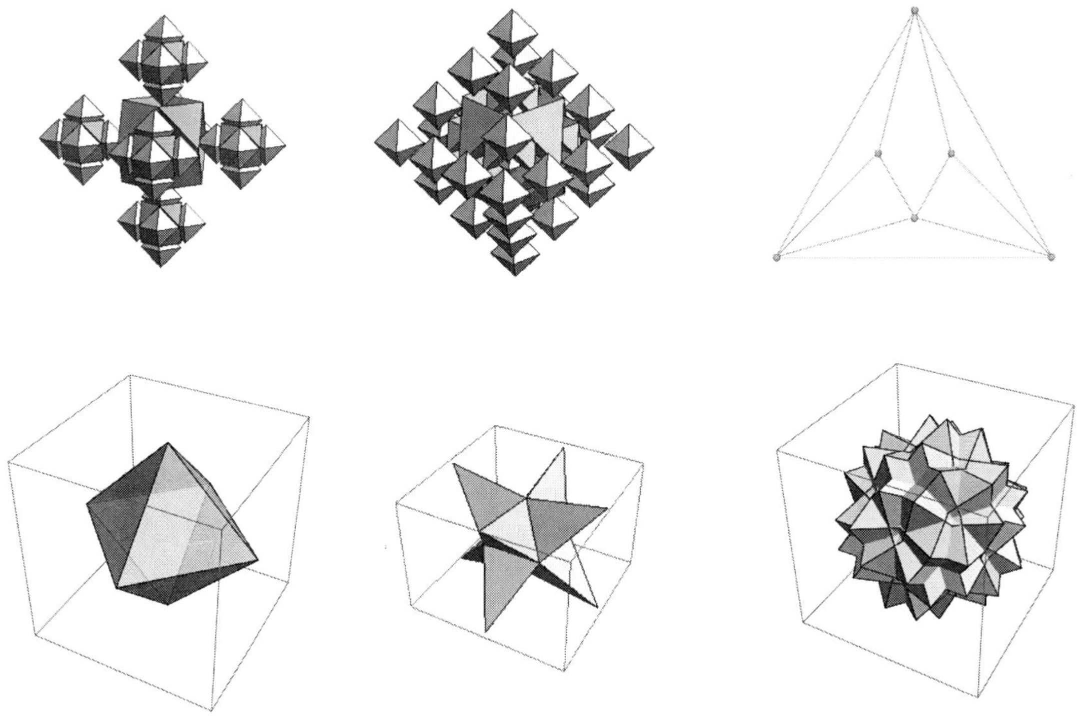

The Geometric figures are software created with Mathematica and Wolfram 3d-XplorMath-J.

References

[1] G.D'Agui-G.Failla Afrika mathematika,Série 3,Volume 20, 2008, 131 -146.

[2] G. Failla,On the (1,1) –Segre model for business, Applied Mathematical SciencesVol.8, (2014), no.166, pp. 8329-8336.

[3] G. Failla-R. Utano, Connected graphs arising from products of Veronese varieties,Algebra Colloquium 23 (2016), no.2, 281-292.

[4] V.Iorfida, Simplicial complexes present in art and arts facilities. Aplimat 2015, 14th Conference on Applied Mathematics, Proceedings (2015), pp.84-94.

[5] V.Iorfida-G. Restuccia, Simplicial complexes and new applications, Applied Mathematical Sciences, Vol. 10, 2016, no. 34, 1653 – 1661.

[6] G. Restuccia, R. Villarreal- On the normality of monomial ideals of mixed products, Comm. Algebra 29 (8) (2001) pp. 3751-3580.

[7] B. Sturmfels, Gröbner bases and convex polytopes, Univ.Lectures Series,Amer. Math.Soc.,Vvol.8,(1995).

[8] R. Utano, M. Murgia -Algebraic invariants of ideals of mixed products, Bollettino di Matematica Pura e Applicata,Vol.VII,2014, pp. 43- 52

[9] R. Villarreal, Monomial Algebras, Pure and Applied Mathematics,Marcel Dekker, Inc. New York,vol.238(2001).

The reserch that led to the present paper was partially supported by the group GNSAGA of INDAM The author wrote while being a member of the activity group AACC&AP-SIMAI.

Current address

Iorfida Vincenzo, Ph.D.
School "Archimedes University" of Calabria, Campus of Arcavacata
Via P. Bucci, 87036 Rende CS, Italy

Associazione Nazionale Mathesis, Sezione di Serra San Bruno, Italy
and Laboratorio per la Comunicazione Scientifica, Universita della Calabria
Ponte Bucci, Cubo 30b, 87036 Arcavacata di Rende CS, Italy
E-mail: unilmat@gmail.it

Slovak University of Technology in Bratislava
Faculty of Mechanical Engineering
16th Conference on Applied Mathematics APLIMAT 2017
Proceedings

COMPUTER ASSISTED SEARCH FOR GRAPHS WITH PRESCRIBED DEGREES AND CYCLE STRUCTURE

JAJCAYOVÁ B. Tatiana (SK), MASÁR Martin (SK)

Abstract. The excess of a k-regular graph G of girth g is defined to be the difference between G's order and the well-known Moore bound, and k-regular graphs of girth g and minimal excess are called (k, g)-cages.
In this paper we introduce theoretical results, as well as computer experiments and algorithms to produce and analyze data of cycle spectra of Moore graphs and graphs close to Moore graphs, i.e. graphs with small excess. The results of these experiments where used to improve lower bounds for (k, g)-graphs, and to study a structure of subgraphs induced by excess vertices.

Keywords: graph, cage, Moore bound, excess, lower bound, computer assisted search

Mathematics subject classification: Primary 05C30, 05C35, 05C38; Secondary 05C85

1 Introduction

We use the term (k, g)-*graph* to denote a finite, simple k-regular graph of girth g, where the girth g of a graph is the length of its shortest cycle. Examples of k-regular graphs include n-cycle, C_n, with $k = 2$, $g = n$, complete graph, K_n, with $k = n - 1$, $g = 3$, complete bipartite graph, $K_{n,n}$, with $k = n$, $g = 4$, Petersen graph with $k = 3$, $g = 5$, Heawood graph with $k = 3$, $g = 6$, Tutte graph with $k = 3$, $g = 8$.

The existence of (k, g)-graphs for any pair of parameters (k, g), k-degree, g-girth, with $k \geq 2$ and $g \geq 3$ was shown by Erdös and Sachs in 1963 [4, 14].

A (k, g)-*cage* is a smallest k-regular graph of girth g. We will denoted its order by $n(k, g)$. The orders $n(k, g)$ have been determined only for very limited sets of parameters (k, g) [6].

The so-called *Moore bound* is a natural lower bound on the order $n(k, g)$ of k-regular cages of girth g. This lower bound is obtained from an iterative construction of a graph without

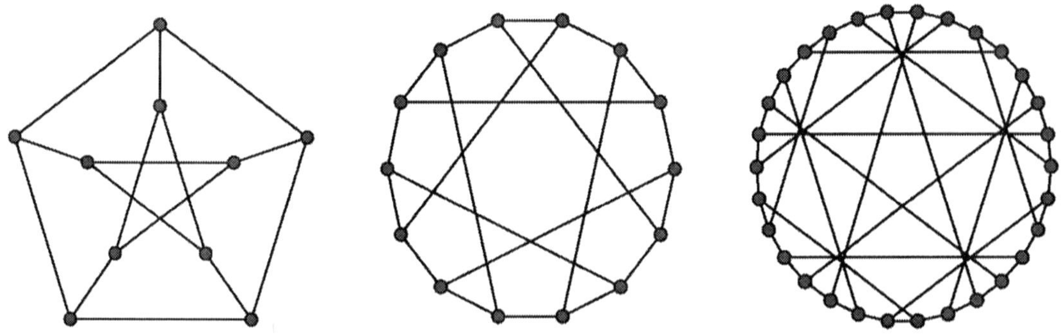

Fig. 1. Petersen graph, Heawood graph, Tutte graph

introducing cycles of length shorter than g, that is, by building a "base" Moore tree up to certain hight, and only later adding "horizontal" edges to close cycles, so that all cycles are of length at least the girth.

Here is the example of buiding $(3,5)$-cage with *odd* girth 5 and $(3,6)$-cage with *even* girth 6:

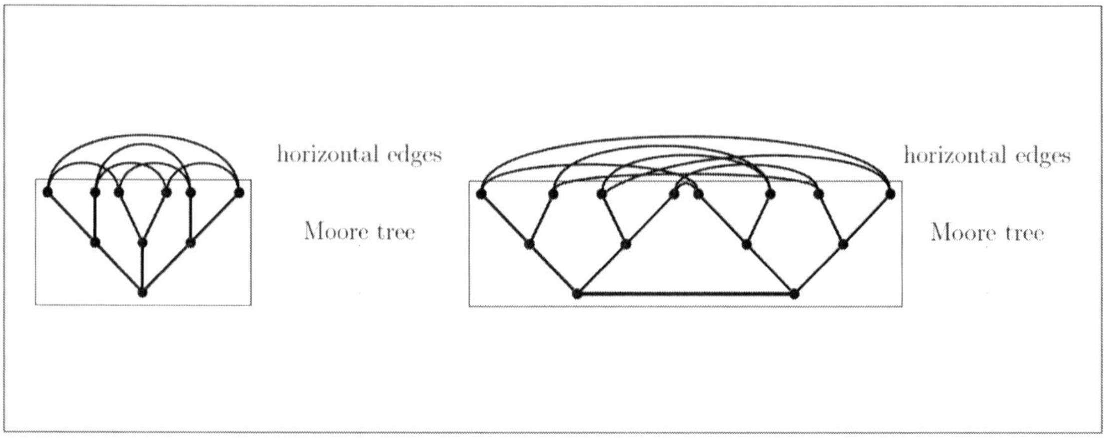

Fig. 2. The Moore tree and horizontal edges in the $(3,5)$-cage and $(3,6)$-cage.

The precise form of the Moore bound depends on the parity of g:

$$n(k,g) \geq M(k,g) = \begin{cases} 1+k+k(k-1)+...+k(k-1)^{(g-3)/2}, & g \text{ odd} \\ 2(1+(k-1)+...+(k-1)^{(g-2)/2}), & g \text{ even} \end{cases} \quad (1)$$

Graphs whose orders equal the Moore bound are called *Moore graphs* and are very rare. They are known to exist if and only if

$k = 2$ and $g \geq 3$: cycles;

$g = 3$ and $k \geq 2$: complete graphs;

$g = 4$ and $k \geq 2$: complete bipartite graphs;

$g = 5$ and $k = 2, 3, 7$: the 5-cycle, the Petersen graph, the Hoffman-Singleton graph;

$g = 6, 8$, or 12: symmetric generalized n-gons of order $k - 1$;

with the existence of the $(57, 5)$-graph of order matching the Moore bound still unresolved [1, 3, 6].

2 Counting cycles close to girth

The above clasification has one (bad) practical consequence: there is no hope for completing the construction from just building the Moore tree and adding horizontal edges for any other cases than those listed. To obtain a cage, we will need to introduce extra vertices. The difference between the order n of a (k, g)-graph G and the Moore bound $M(k, g)$ is called the *excess* e of the graph G, $e = n - M(k, g)$. Determining the excess of (k, g)-cages is equivalent to determining $n(k, g)$. As mentioned above, the exact values $n(k, g)$ are not known for the majority of parameter pairs (k, g).

Outside the cases for which the classification of the parameter pairs of the Moore graphs asserts the existence of a graph whose order matches the Moore bound, the obvious lower bound for the order of a (k, g)-cage is the value of the Moore bound plus one, $M(k, g) + 1$, when k is even, and the value of the Moore bound plus two, $M(k, g) + 2$, when k is odd (with the second statement following from the fact that odd degree regular graphs must be of even order, and the fact that both Moore bounds are even for odd k). It is almost universally believed that the excess of the majority of cages is significantly bigger than 2. The following table lists some results from the dynamic survey [6] maintained by G. Exoo and R.Jajcay, and illustrates that in reality there are huge gaps between known lower and upper bounds for known (k, g)-graphs, with the excess quickly becoming a multiple of the Moore bound. Looking at the table one is almost tempted to state that for each excess $e > 0$, there exist parameters (k, g) with the property that the excess of any (k, g)-graph G exceeds e. The existence of such parameter pairs for arbitrarily large e has already been established for the (much more restricted) family of vertex-transitive (k, g)-graphs [2, 7], but has only been conjectured for the case of general cages.

Girth g	Lower Bound	Upper Bound	# of Cages	Due to
5	10	10	1	Petersen
6	14	14	1	Heawood
7	24	24	1	McGee
8	30	30	1	Tutte
9	58	58	18	Brinkmann-McKay-Saager
10	70	70	3	O'Keefe-Wong
11	112	112	1	McKay-Myrvold; Balaban
12	126	126	1	Benson
13	202	272		McKay-Myrvold; Hoare
14	258	384		McKay; Exoo
15	384	620		Biggs
16	512	960		Exoo
17	768	2176		Exoo
18	1024	2560		Exoo
19	1536	4324		Hoare, H(47)
20	2048	5376		Exoo
21	3072	16028		Exoo
22	4096	16206		Biggs-Hoare, S(73)
23	6144	49326		Exoo
24	8192	49608		Bray-Parker-Rowley
25	12288	108906		Exoo
26	16384	109200		Bray-Parker-Rowley
27	24576	285852		Bray-Parker-Rowley
28	32768	415104		Bray-Parker-Rowley
29	49152	1143408		Bray-Parker-Rowley
30	65536	1227666		Biggs-Hoare, S(313)
31	98304	3649794		Bray-Parker-Rowley
32	131072	3650304		Bray-Parker-Rowley

To construct new cages requires two hard steps: a really nice construction, and a proof of non-existence of better (smaller) graphs.

R.Jajcay, S.Filipovski and the first author studied graphs of odd order and excess $1, 2$ and 3 [9], and graphs of even girth and excess 4 in [8]. There they analyzed structure of graphs of even girth and excess 4:

Lemma 2.1 ([8]) *Let $k \geq 6$, $g = 2m \geq 6$. Let G be a (k, g)-graph of excess 4, u, v be two adjacent vertices in G, and $X_f = \{w_1, w_2, w_3, w_4\}$ be the four excess vertices with respect to the edge $f = \{u, v\}$. The induced subgraph $G[w_1, w_2, w_3, w_4]$ is isomorphic to $2K_2$ (two disjoint copies of K_2) or $\mathcal{P}_3$ (a path of length 3).*

So we knew, how the excess vertices of the graph with, say, $k = 3$ and $g = 6$ would be arranged, but we were not sure such graph existed. And if, how many non-isomorphic $(3, 6)$-

graphs with excess 4 were there. Here came useful computer search and experiments done by the second author, that not only shown existence of such graph, but found all non-isomorphic graphs with prescribed properties.

3 Computer Assisted Search

In this section we will describe algorithms and their implementations in experiments designed to test graphs for required structure, and to create new graphs with prescribed properties.

3.1 Graph representation

We first have to create a representation of a graph in the program. We have chosen Java, and thus we have to split the graph into separate objects.

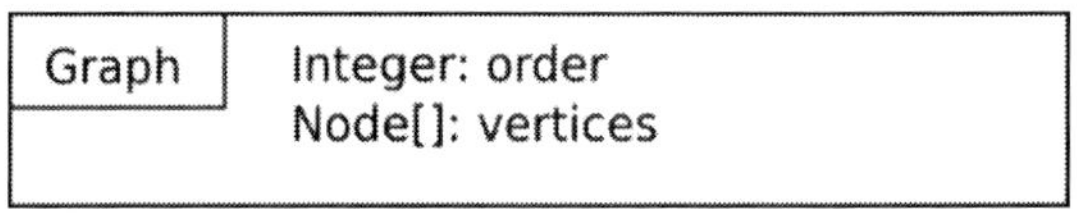

Fig. 3. Data structure of a graph

The graph itself is represented as an array of objects, and an integer representing the order of each node.

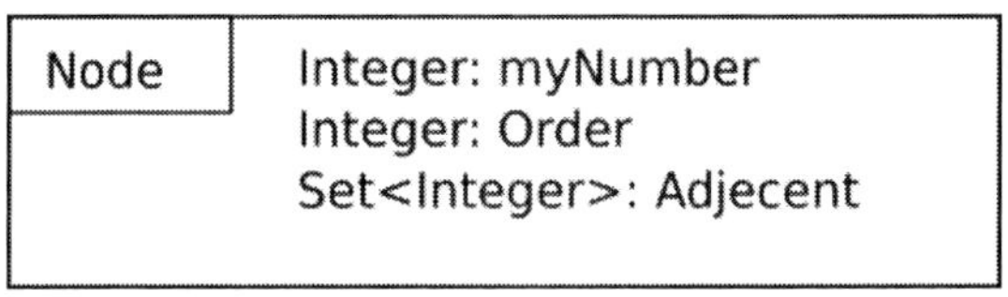

Fig. 4. Data structure of a vertex.

A vertex of a graph is an object containing an id of the vertex, which is the same as the position of the vertex in the array of vertices, an integer representing the order of the vertex, and a set of ids of adjacent vertices.

Finally, we have a GraphFactory class that creates the graphs using static functions. It contains functions to build a single graph, all graphs and the Moore tree for a given number of vertices, girth and order of the graph.

To simplify operations, we have an edge object containing two integers corresponding to two adjacent vertices ids. It is used to compare two edges, to store used edges in a cycle finding algorithms and to aid in debugging.

3.2 Algorithms

In this section, we will describe two basic algorithms: Graph construction and Identifying cycles. We also discuss their possible future improvements.

3.2.1 Graph construction

To construct a graph, we started by a naive algorithm mimicking a human approach to a construction of a Moore graph. Later, we added several computer-specific optimizations.

Fig. 5. Creating a Moore tree for a $(3, 5)$-cage

First, we build a Moore tree. For graph with excess $e \geq 1$, there will be several (excess) vertices outside the tree. Due to the linear complexity of this operation, it is desirable to speed up the complex algorithm.

Next, we have to connect unsaturated vertices - to add horizontal edges. The complexity of this step is $O(2^n)$. To add an horizontal edge, we do a depth-first algorithm connecting a vertex with a lower id to a vertex with a higher id. This is possible, since we work with undirected graphs. It also saves about half of the computation time.

While we are adding new edges connecting unsaturated vertices, we need to check for cycles. We do this by using several layers of checks.

0.) Reserve a place in computer memory for a graph and initialize a data structure for a given number of vertices. Also create an array for used edges.

1.) While a graph contains an unsaturated vertex, do:

  Select an unsaturated vertex u.

  For every other unsaturated vertex $v \neq u$ check if the edge u, v is already in the graph. If at least one of the vertices u or v is isolated, go to 4.)

2.) If edge (u, v) is new, do:

3.) For length up to the girth of the graph starting from the new edge, see if one of these is true in order:

  3.1.) Is the length is less than 3?

  3.2.) Is the length less than the girth and does not create a cycle?

  3.3.) Is the length more than the girth?

4.) Accept the edge as correct

  In other cases, we create a cycle that is shorter than the given girth, and therefore unacceptable.

We also need to deal with isomorphism of graphs. The graph isomorphism problem is one of problems in computational complexity theory belonging to NP, but not known to belong to either P nor NP-complete subclasses. There are several algorithms that could be used for our purposes. We have chosen the VF2 algorithm [15], because its memory requirements are smaller than of other graph isomorphism algorithms, and therefore can work with big graphs. It is as follows:
If we want to know if two graphs G and G' are isomorphic.

1.) If all vertices are bijectively mapped and all edges are mapped to edges and all non-edges are mapped to non-edges, return a true result. Otherwise proceed.

2.) Try to map an unused node from the first graph to an unused node in the second graph.

3.) Check if the mapped vertices match up to the vertices in the second graph. If all of the mapped ones match, go to 1.). If the result is true, return true, otherwise proceed.

4.) Repeat steps 2 and 3 for all available vertices. If all possible mapping have been used and not one of them was correct, return false.

Once we have done these steps, we obtain all non-isomorphic graphs for the given number of vertices, girth and degree. If we only want a single true result, we can skip checking for isomorphism. Note, that there are more optimal approaches, such as the one created in [11], but our algorithm served for our purposes well.

3.2.2 Identifying cycles

Compared to graph construction, finding cycles is both simpler and faster. We can choose suboptimal methods while suffering negligible performance losses in the program.

We started with a simple depth-first search. The optimization we used initially is once again based on human naive approach. We used a recursive implementation that is based on a starting edge, vertex or going through the entire graph. As we work on small graphs, this algorithm is fast enough for now. For larger graphs this will need to be addressed differently.

For a more efficient approach in case of large graphs with thousands of vertices, we need to use Paton's algorithm [13]. The algorithm is an improvement on the base cycles method. The basis of this approach is building a spanning tree on the graph. Once we have done so, we start adding other edges for the graph to the spanning tree, therefore creating cycles. Here we give a brief overview of the later method:

For every connected component of the graph:

1.) For every vertex of the component selected in the previous step

2.) If the stack is not empty, take out the last vertex in the stack

   2.1.) If the vertex was already covered in the spanning tree, try another one, otherwise proceed.

   2.2.) If it was not covered, check all edges that were not previously used. If they contain a new vertex, add it to the spanning tree and the stack. If it contains the same vertex, then add a self loop to the set of the found cycles. If the edge connects to an already found vertex, we have found a cycle.

3.) If the stack is not empty, go back to 2.) Otherwise choose another connected component.

4.) Return the set of all cycles.

3.3 Experimental results for $(3,6)$-graph with excess 4.

Here, we provide the visualization of a result of graph construction for a graph with order 3, girth 6 and excess $e = 4$, i.e. on 18 vertices. This was used to experimentally check the results of a recent paper on Moore graphs with an excess of 4 [8]. As said above, a general structure of such graphs could be deduced from theoretical results in [8], but we were not sure if such graphs existed.

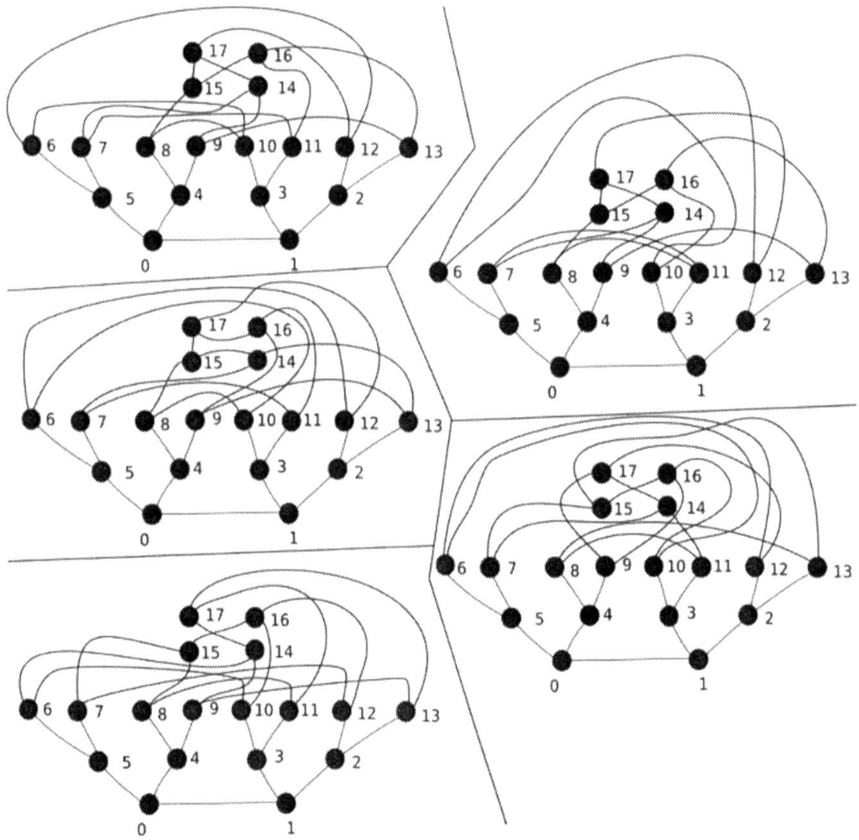

Fig. 6. (3,6) + 4 result

4 Concluding notes

As we said earlier, to construct new cages requires to come up with some smart construction, and a proof of non-existence of a smaller graph. Brute force approaches are doomed to be unsuccessful, because number of graphs to be searched grows exponentially. Just how bad is it is illustrated by the next table where we list number of vertices vs. number of graphs

4	5	6	7	8	9	10	11
11	34	156	1044	12346	274668	12005168	1018997864

So it is clear that computer experiments and computer searches must be combined with clever constructions and employment of deep theoretical results. Therefore it make sense to try to improve lower bounds for (k,g)-graphs (as in [1],[5],[8],[9]), or observe some properties of cages, that could be helpful. In fact, all the known Moore graphs are highly symmetric - translated into language of graphs, they are all vertex transitive. So studies of symmetries may help. On the other hand, $(57, 5)$-graph, the only graph for which the existence of the Moore graph is still unresolved, is known to be *non*-vertex transitive (if it exists) [10]. Note that algorithms and experiments introduced in the previous section, together with some additional onces from [12], can be used to test graphs for several given properties, as well as for building graphs with prescribed properties and cycle structure.

References

[1] E. Bannai and T. Ito, On finite Moore graphs, J. Fac. Sci. Tokyo, Sect. 1A, 20 (1973) 191-208.

[2] N. L. Biggs. Excess in vertex-transitive graphs. Bull. London Math. Society 14 (1982) 52-54.

[3] R. M. Damerell, On Moore graphs, Proc. Cambridge Phil. Soc. 74 (1973) 227-236.

[4] P. Erdős and H. Sachs, Reguläre Graphen gegebener Taillenweite mit minimaler Knotenzahl, Wiss. Z. Uni. Halle (Math. Nat.) 12 (1963) 251-257.

[5] L. Eroh and A. Schwenk, Cages of girth 5 and 7, Congr. Numer. 138 (1999) 157-173.

[6] G. Exoo and R. Jajcay, Dynamic cage survey, Electron. J. Combin., Dynamic Survey 16, September 2008.

[7] S. Filipovski and R. Jajcay. On the excess of vertex-transitive graphs of given degree and girth. Submitted for publication.

[8] T. B. Jajcayová, S. Filipovski and R. Jajcay. Improved lower bounds for the orders of even girth cages The Electronic Journal of Combinatorics 23 (2016).

[9] T. B. Jajcayová, S. Filipovski and R. Jajcay. Counting cycles in graphs with small excess. Accepted for publication in Lecture Notes of Seminario Interdisciplinare di Matematica.

[10] M. Mačaj and J. Širáň, Search for properties of the missing Moore graph, Linear Algebra and its Applications 432 (2010) 2381-2398.

[11] M. Meringer, Fast Generation of Regular Graphs and Construction of Cages. Journal of Graph Theory 30, 137-146, 1999.

[12] M.Masár, Counting cycles in Moore graphs, Bachelor thesis Comenius University, (2016).

[13] K. Paton, An algorithm for finding a fundamental set of cycles for an undirected linear graph, Comm. ACM 12 (1969), pp. 514-518.

[14] H. Sachs, Regular graphs with given girth and restricted circuits, J. London Math. Soc. 38 (1963) 423-429.

[15] L. P. Cordella, P. Foggia, C. Sansone, and M. Vento. A (sub)graph isomorphism algorithm for matching large graphs. IEEE PAMI, 26(10):1367–1372, 2004

Current address

Jajcayová B. Tatiana, RNDr. PhD.
Comenius University, Bratislava, Slovakia
Supported by VEGA 1/0577/14 and VEGA 1/0474/15
E-mail: tatiana.jajcayova@fmph.uniba.sk

Masár Martin
Comenius University, Bratislava, Slovakia
E-mail: martin.masar@uniba.sk

Slovak University of Technology in Bratislava
Faculty of Mechanical Engineering
16th Conference on Applied Mathematics **APLIMAT 2017**
Proceedings

ASYMPTOTIC NORMALITY OF PARAMETER ESTIMATES FOR STOCHASTIC DIFFERENTIAL EQUATION OF SECOND ORDER

JANÁK Josef (CZ)

Abstract. We introduce (scalar) stochastic differential equation of second order with two unknown parameters and its solution, which is Ornstein–Uhlenbeck process. Based on ergodicity, we derive the strong consistent estimators of these parameters. Then we study the asymptotic normality of these estimators.

Keywords: Ornstein–Uhlenbeck process, parameter estimation, asymptotic normality

Mathematics subject classification: Primary 62M05; Secondary 60G35, 93E10

1 Introduction

In this article, we are improving and extending some previous results on the parameter estimation for stochastic differential equation of second order, which were previously published in [1]. Namely, we are focusing on asymptotic normality of estimators, which is the main result of this paper.

The following section introduces the stochastic differential equation for the harmonic oscillation, its rewriting and its solution, that is the Ornstein–Uhlenbeck process. Using the observation of the trajectory $\{X_t^{x_0}, 0 \le t \le T\}$ and ergodicity of the Ornstein–Uhlenbeck process, strong constistent estimators of unknown parameters are proposed. The proof of Theorem 1 may be found in [2]. The proof of Lemma 1 is only a matter of computation.

In third section, the asymptotic normality of these estimators is derived. We are using Itô's formula to obtain different form of quantity $Y_T = \int_0^T \|X_t\|_{\mathbb{R}^2}^2 \, dt$ (on which our estimators are based on) and then central limit theorem for the stochastic integral (which may be found in [3]). The results are summarized in Theorem 2.

2 Parameter estimation and strong consistency

Consider the following (scalar) stochastic differential equation

$$\ddot{x} + 2a\dot{x} + bx = \sigma\dot{\beta}_t, \tag{1}$$

with initial values $X_0 = x_1$ and $\dot{X}_0 = x_2$. Let $a > 0$, $b > 0$ are unknown real parameters, $\sigma > 0$ is known and $\dot{\beta}_t$ is the formal time derivative of the standard Brownian motion.

We may rewrite this equation in the form

$$dX_t = AX_t\,dt + \Phi dB_t, \quad X_0 = x_0, \tag{2}$$

if we set

$$X_t = \begin{pmatrix} X_t^1 \\ X_t^2 \end{pmatrix} \in \mathbb{R}^2,\ A = \begin{pmatrix} 0 & 1 \\ -b & -2a \end{pmatrix} \in \mathbb{M}_{2\times 2},\ \Phi = \begin{pmatrix} 0 & 0 \\ 0 & \sigma \end{pmatrix} \in \mathbb{M}_{2\times 2},$$

$$B_t = \begin{pmatrix} B_t^1 \\ B_t^2 \end{pmatrix} \in \mathbb{R}^2,\ x_0 = \begin{pmatrix} x_1 \\ x_2 \end{pmatrix} \in \mathbb{R}^2,$$

where $B_t^2 = \beta_t$, B_t^1 is an independent Brownian motion and $\mathbb{M}_{2\times 2}$ stands for the space of all 2×2 matrices with real entries, equipped with the operator norm $\|\cdot\|_{\mathcal{L}(\mathbb{R}^2)}$.

If we denote $S(t) := e^{At}$ a strongly continuous semigroup on $\mathbb{R}^2$, then the solution $(X_t^{x_0}, t \geq 0)$ to (2) is defined by the mild form

$$X_t^{x_0} = S(t)x_0 + Z_t, \quad t \geq 0, \tag{3}$$

where $(Z_t, t \geq 0)$ is the convolution integral

$$Z_t = \int_0^t S(t-u)\Phi\,dB_u. \tag{4}$$

The solution $(X_t^{x_0}, t \geq 0)$ to the equation (2) is called the Ornstein–Uhlenbeck process.

Theorem 1. *If the semigroup $(S(t), t \geq 0)$ is exponentially stable, i.e. there exist constants $M > 0$ and $\rho > 0$ such that for all $t \geq 0$*

$$\|S(t)\|_{\mathcal{L}(\mathbb{R}^2)} \leq Me^{-\rho t}, \tag{A1}$$

then there is a Gaussian centered limiting measure $\mu_\infty^{(a,b)} = N(0, Q_\infty^{(a,b)})$ for $(X_t^{x_0}, t \geq 0)$, such that

$$w^* - \lim_{t\to\infty} \mu_t^{x_0} = \mu_\infty^{(a,b)}$$

for each initial condition $x_0 \in \mathbb{R}^2$, where $\mu_t^{x_0} =$ Law $(X_t^{x_0})$ and Law $(\cdot)$ denotes the probability distribution. The covariance matrix $Q_\infty^{(a,b)}$ has the following form

$$Q_\infty^{(a,b)} = \int_0^\infty S(t)\Phi\Phi^\top S^\top(t)\,dt. \tag{5}$$

The eigenvalues of matrix A are $\lambda_1 = -a + \sqrt{a^2 - b}$, $\lambda_2 = -a - \sqrt{a^2 - b}$. If $a^2 < b$, then the real parts of eigenvalues of matrix A are negative and condition (A1) holds true. If we denote $\alpha = -a$, $\beta = \sqrt{b - a^2}$, then we may write $\lambda_1 = \alpha + i\beta$, $\lambda_2 = \alpha - i\beta$.

Lemma 1. The fundamental system $(S(t), t \geq 0)$ for the equation (1) has the following form

$$S(t) = \begin{pmatrix} e^{\alpha t}\left(\cos(\beta t) - \frac{\alpha}{\beta}\sin(\beta t)\right) & \frac{1}{\beta}e^{\alpha t}\sin(\beta t) \\ e^{\alpha t}\left(-\beta - \frac{\alpha^2}{\beta}\right)\sin(\beta t) & \frac{1}{\beta}e^{\alpha t}(\alpha\sin(\beta t) + \beta\cos(\beta t)) \end{pmatrix}.$$

The covariance matrix $Q_\infty^{(a,b)}$ of the limiting measure $\mu_\infty^{(a,b)}$ equals to

$$\begin{aligned} Q_\infty^{(a,b)} &= \int_0^\infty S(t)\Phi\Phi^\top S^\top(t)\,dt \\ &= \int_0^\infty \frac{\sigma^2}{\beta^2}\begin{pmatrix} q_{11}(t) & q_{12}(t) \\ q_{21}(t) & q_{22}(t) \end{pmatrix} dt \\ &= \frac{\sigma^2}{\beta^2}\begin{pmatrix} \frac{\beta^2}{4ab} & 0 \\ 0 & \frac{\beta^2}{4a} \end{pmatrix} = \sigma^2 \begin{pmatrix} \frac{1}{4ab} & 0 \\ 0 & \frac{1}{4a} \end{pmatrix}, \end{aligned} \tag{6}$$

where

$$\begin{aligned} q_{11}(t) &= e^{2\alpha t}\sin^2(\beta t), \\ q_{12}(t) &= q_{21}(t) = e^{2\alpha t}\sin(\beta t)(\alpha\sin(\beta t) + \beta\cos(\beta t)), \\ q_{22}(t) &= e^{2\alpha t}(\alpha\sin(\beta t) + \beta\cos(\beta t))^2. \end{aligned}$$

For the parameter estimation, we are interested in the trace of the covariance matrix $Q_\infty^{(a,b)}$. From the expression (6), it follows, that

$$\text{Tr}\, Q_\infty^{(a,b)} = \frac{b+1}{4ab}\sigma^2. \tag{7}$$

Since the Ornstein–Uhlenbeck process X_t is ergodic in $\mathbb{R}^2$ (see Example 2.1. in [5]), we may use Birkhoff's theorem. Namely

$$\lim_{T\to\infty} \frac{1}{T}\int_0^T \|X_t^{x_0}\|_{\mathbb{R}^2}^2\,dt = \int_{\mathbb{R}^2} \|y\|_{\mathbb{R}^2}^2\,d\mu_\infty^{(a,b)}(y) = \text{Tr}\, Q_\infty^{(a,b)}, \tag{8}$$

for any initial contidion $x_0 \in \mathbb{R}^2$ (see Theorem 4.9. in [4]).

If we denote $Y_T := \frac{1}{T}\int_0^T \|X_t^{x_0}\|_{\mathbb{R}^2}^2\,dt$, then (based on (7)) some strongly constistent estimates of parameters a and b may be proposed. If we know the true value of the parameter b, then the strongly consistent estimator of the parameter a is

$$\hat{a}_T = \frac{b+1}{4bY_T}\sigma^2. \tag{9}$$

Similarly, if we know the true value of the parameter a, then

$$\hat{b}_T = \frac{\sigma^2}{4aY_T - \sigma^2} \tag{10}$$

is the strong consistent estimator of the parameter b.

3 Asymptotic normality of estimators

In this section, we are interested in asymptotic normality of estimators (9) and (10), i. e. we will study, if $\sqrt{T}\,(\hat{a}_T - a)$ or $\sqrt{T}\left(\hat{b}_T - b\right)$ converge to some normal distribution. To this aim, define a linear operator $P : \mathbb{R}^2 \to \mathbb{R}^2$ by

$$Px = P\begin{pmatrix} x_1 \\ x_2 \end{pmatrix} = \begin{pmatrix} b + \frac{4a^2}{b+1} & \frac{2a}{b+1} \\ \frac{2a}{b+1} & 1 \end{pmatrix}\begin{pmatrix} x_1 \\ x_2 \end{pmatrix}, \quad \forall x = \begin{pmatrix} x_1 \\ x_2 \end{pmatrix} \in \mathbb{R}^2. \tag{11}$$

Let us denote $\langle \cdot, \cdot \rangle$ and $\|\cdot\|$, the Euclidean scalar product and the norm in $\mathbb{R}^2$. The properties of matrix P are summarized in the following Lemma 2.

Lemma 2. The matrix P is symmetric and

$$\langle Px, Ax \rangle = -\frac{2ab}{b+1}\|x\|^2, \quad \forall x = \begin{pmatrix} x_1 \\ x_2 \end{pmatrix} \in \mathbb{R}^2. \tag{12}$$

Proof. It is evident, that $P = P^\top$, so P is symmetric. Moreover, for every $x = \begin{pmatrix} x_1 \\ x_2 \end{pmatrix} \in \mathbb{R}^2$, we have

$$\begin{aligned}
\langle Px, Ax \rangle &= \left\langle \begin{pmatrix} b + \frac{4a^2}{b+1} & \frac{2a}{b+1} \\ \frac{2a}{b+1} & 1 \end{pmatrix}\begin{pmatrix} x_1 \\ x_2 \end{pmatrix}, \begin{pmatrix} 0 & 1 \\ -b & -2a \end{pmatrix}\begin{pmatrix} x_1 \\ x_2 \end{pmatrix} \right\rangle \\
&= \left\langle \begin{pmatrix} bx_1 + \frac{4a^2}{b+1}x_1 + \frac{2a}{b+1}x_2 \\ \frac{2a}{b+1}x_1 + x_2 \end{pmatrix}, \begin{pmatrix} x_2 \\ -bx_1 - 2ax_2 \end{pmatrix} \right\rangle \\
&= bx_1x_2 + \frac{4a^2}{b+1}x_1x_2 + \frac{2a}{b+1}x_2^2 - \frac{2ab}{b+1}x_1^2 - \frac{4a^2}{b+1}x_1x_2 - bx_1x_2 - 2ax_2^2 \\
&= -\frac{2ab}{b+1}x_1^2 + \left(\frac{2a}{b+1} - 2a\right)x_2^2 = -\frac{2ab}{b+1}\|x\|^2.
\end{aligned}$$

□

Let us define the function $g : \mathbb{R}^2 \to \mathbb{R}$ by

$$g(x) = \langle Px, x \rangle, \quad \forall x \in \mathbb{R}^2. \tag{13}$$

The application of Itô's formula to the function $g(X_t)$ yields

$$dg(X_t) = 2\langle PX_t, dX_t \rangle + \frac{1}{2}\operatorname{Tr}\left(2P\Phi\Phi^\top\right) dt. \tag{14}$$

The second term may be simplified via following calculation

$$\frac{1}{2}\operatorname{Tr}\left(2P\Phi\Phi^\top\right) = \operatorname{Tr}\begin{pmatrix} 0 & \frac{2a}{b+1}\sigma^2 \\ 0 & \sigma^2 \end{pmatrix} = \sigma^2.$$

The expression (14) may be now written in the following way

$$\begin{aligned}
dg(X_t) &= 2\langle PX_t, AX_t \rangle\, dt + 2\langle PX_t, \Phi dB_t \rangle + \sigma^2\, dt \\
&= -\frac{4ab}{b+1}\|X_t\|^2\, dt + 2\langle PX_t, \Phi dB_t \rangle + \sigma^2\, dt.
\end{aligned}$$

After integrating previous formula over the interval $(0, T)$ and after some algebraic operations, we will arrive to

$$Y_T = \frac{1}{T}\int_0^T \|X_t\|^2\, dt = -\frac{b+1}{4abT}\left(\langle PX_T, X_T\rangle - \langle PX_0, X_0\rangle\right) + \frac{b+1}{2abT}\int_0^T \langle PX_t, \Phi dB_t\rangle + \frac{b+1}{4ab}\sigma^2. \tag{15}$$

This is an alternative formula for the quantity Y_T, which was introduced in Section 2. Also note, that we are simplifying notation here by writing X_t instead of $X_t^{x_0}$.

3.1 Asymptotic normality of the estimator $\hat{a}_T$

Let us recall, that the estimator of the parameter a has the form (9), that is

$$\hat{a}_T = \frac{b+1}{4bY_T}\sigma^2.$$

Using this formula and formula (15) for Y_T, we are able to compute the following

$$\begin{aligned}
\sqrt{T}\left(\hat{a}_T - a\right) &= \sqrt{T}\left(\frac{b+1}{4bY_T}\sigma^2 - a\right) = \sqrt{T}\frac{(b+1)\sigma^2 - 4abY_T}{4bY_T} \\
&= \frac{\sqrt{T}}{4bY_T}\left(\frac{b+1}{T}\left(\langle PX_T, X_T\rangle - \langle PX_0, X_0\rangle\right) - \frac{2(b+1)}{T}\int_0^T \langle PX_t, \Phi dB_t\rangle\right) \\
&= \frac{b+1}{4bY_T}\frac{1}{\sqrt{T}}\left(\langle PX_T, X_T\rangle - \langle PX_0, X_0\rangle\right) - \frac{b+1}{2bY_T}\frac{1}{\sqrt{T}}\int_0^T \langle PX_t, \Phi dB_t\rangle .
\end{aligned} \tag{16}$$

Using Chebyshev's inequality, it is easy to show, that the first term $\frac{1}{\sqrt{T}}\left(\langle PX_T, X_T\rangle - \langle PX_0, X_0\rangle\right) \to 0$ in probability as $T \to \infty$. We may also write

$$\begin{aligned}
Z_T &:= \frac{1}{\sqrt{T}}\int_0^T \langle PX_t, \Phi dB_t\rangle = \frac{1}{\sqrt{T}}\int_0^T \sum_{n=1}^{2} \langle PX_t, e_n\rangle\, d\langle e_n, \Phi B_t\rangle \\
&= \frac{\sigma}{\sqrt{T}}\int_0^T \langle PX_t, e_2\rangle\, d\beta_t,
\end{aligned} \tag{17}$$

where we have used, that $\Phi B_t = \begin{pmatrix} 0 \\ \sigma\beta_t \end{pmatrix}$.

Define the operator $\tilde{P} : \mathbb{R}^2 \to \mathbb{R}$ by

$$\tilde{P}x = \begin{pmatrix} \frac{2a}{b+1} & 1 \end{pmatrix}\begin{pmatrix} x_1 \\ x_2 \end{pmatrix} = \frac{2a}{b+1}x_1 + x_2, \quad \forall x = \begin{pmatrix} x_1 \\ x_2 \end{pmatrix} \in \mathbb{R}^2. \tag{18}$$

By the central limit theorem for the stochastic integral (see Proposition 1.22. in [3]), Z_T converges weakly to a Gaussian distribution with a zero mean and variance given by

$$\begin{aligned}\lim_{T\to\infty}\frac{\sigma^2}{T}\int_0^T \langle PX_t, e_2\rangle^2\,dt &= \lim_{T\to\infty}\frac{\sigma^2}{T}\int_0^T \left(\tilde{P}X_t\right)^2 dt = \sigma^2\,\mathbb{E}\left(\tilde{P}X_\infty\right)^2 \\ &= \sigma^2\,\mathrm{Var}\left(\tilde{P}X_\infty\right) = \sigma^2\,\mathrm{Var}\left(\frac{2a}{b+1}X_\infty^1 + X_\infty^2\right) \\ &= \frac{4a^2}{(b+1)^2}\frac{\sigma^4}{4ab} + \frac{\sigma^4}{4a} = \frac{a\sigma^4}{b(b+1)^2} + \frac{\sigma^4}{4a},\end{aligned} \tag{19}$$

where $X_\infty = \begin{pmatrix} X_\infty^1 \\ X_\infty^2 \end{pmatrix}$ is an $\mathbb{R}^2$–valued Gaussian random variable with zero mean and covariance matrix $Q_\infty^{(a,b)}$. (It has the invariant distribution.) We have also used the fact, that random variables X_∞^1 and X_∞^2 are independent.

Since the multiplicative factor $-\frac{b+1}{2bY_T}$ of Z_T in (16) converges to $-\frac{2a}{\sigma^2}$ as $T \to \infty$, we have the following results

$$Z_T \overset{\text{as.}}{\sim} N\left(0, \frac{a\sigma^4}{b(b+1)^2} + \frac{\sigma^4}{4a}\right),\ T\to\infty, \tag{20}$$

$$\sqrt{T}\left(\hat{a}_T - a\right) \overset{\text{as.}}{\sim} N\left(0, \frac{4a^3}{b(b+1)^2} + a\right),\ T\to\infty. \tag{21}$$

3.2 Asymptotic normality of the estimator $\hat{b}_T$

Let us recall, that the estimator of the parameter b has the form (10), that is

$$\hat{b}_T = \frac{\sigma^2}{4aY_T - \sigma^2}.$$

Using this formula and formula (15) for Y_T, we are able to compute the following

$$\begin{aligned}\sqrt{T}\left(\hat{b}_T - b\right) &= \sqrt{T}\left(\frac{\sigma^2}{4aY_T - \sigma^2} - b\right) = \frac{\sqrt{T}}{4aY_T - \sigma^2}\left(\sigma^2 - 4abY_T + b\sigma^2\right) \\ &= \frac{\sqrt{T}}{4aY_T - \sigma^2}\left(\frac{b+1}{T}\left(\langle PX_T, X_T\rangle - \langle PX_0, X_0\rangle\right)\right. \\ &\qquad \left. - \frac{2(b+1)}{T}\int_0^T \langle PX_t, \Phi dB_t\rangle\right) \\ &= \frac{b+1}{4aY_T - \sigma^2}\frac{1}{\sqrt{T}}\left(\langle PX_T, X_T\rangle - \langle PX_0, X_0\rangle\right) \\ &\qquad - \frac{2(b+1)}{4aY_T - \sigma^2}\frac{1}{\sqrt{T}}\int_0^T \langle PX_t, \Phi dB_t\rangle.\end{aligned} \tag{22}$$

As before, the term $\frac{1}{\sqrt{T}}\left(\langle PX_T, X_T\rangle - \langle PX_0, X_0\rangle\right) \to 0$ in probability as $T \to \infty$ and

$$Z_T \overset{\text{as.}}{\sim} N\left(0, \frac{a\sigma^4}{b(b+1)^2} + \frac{\sigma^4}{4a}\right),\ T\to\infty.$$

The multiplicative factor $-\frac{2(b+1)}{4aY_T-\sigma^2}$ of Z_T in (22) converges to $-\frac{2b(b+1)}{\sigma^2}$ as $T \to \infty$, which brings us to the following result

$$\sqrt{T}\left(\hat{b}_T - b\right) \overset{as.}{\approx} N\left(0,\ 4ab + \frac{b^2(b+1)^2}{a}\right),\ T \to \infty. \tag{23}$$

We may summarize the results in the following theorem.

Theorem 2. *The estimators $\hat{a}_T$ and $\hat{b}_T$ are asymptotically normal, i. e.*

$$\sqrt{T}\left(\hat{a}_T - a\right) \overset{as.}{\approx} N\left(0,\ \frac{4a^3}{b(b+1)^2} + a\right),\ T \to \infty,$$
$$\sqrt{T}\left(\hat{b}_T - b\right) \overset{as.}{\approx} N\left(0,\ 4ab + \frac{b^2(b+1)^2}{a}\right),\ T \to \infty.$$

4 Conclusion

We have introduced the stochastic differential equation of second order (which is used for modeling of stochastic oscillation) and its solution, which is Ornstein–Uhlenbeck process. Based on ergodicity, the strong consistent estimators of unknown parameters have been derived.

In the second part of the paper, we have shown, that these estimators are asymptotically normally distributed with their means and variances stated in Theorem 2.

5 Acknowledgement

The paper was supported by the GAČR Grant no. 15–08819S entitled "Stochastic Processes in Infinite Dimensional Spaces".

The paper was supported by KMAT VŠE.

References

[1] JANÁK, J.: Fractional Brownian Motion, *Proceedings of the 15th Conference on Applied Mathematics Aplimat 2016*, Slovak University of Technology, Bratislava, 557–567, 2016.

[2] JANÁK, J.: *Geometrical Fractional Brownian Motion*, Bachelor Thesis, Faculty of Mathematics and Physics, Charles University in Prague, 2007.

[3] KUTOYANTS, Y. A.: *Statistical Inference for Ergodic Diffusion Processes*, Springer London, 2004.

[4] MASLOWSKI, B., POSPÍŠIL, J.: Ergodicity and parameter estimates for infinite–dimensional fractional Ornstein–Uhlenbeck process, *Applied Mathematics and Optimization*, Vol. 57, Issue 3, 401–429, 2007.

[5] MASLOWSKI, B., SEIDLER, J.: Probabilistic approach to the strong Feller property, *Probability Theory and Related Fields*, Vol. 118, Issue 2, 187–210, 2000.

Current address

Janák Josef, Mgr.
Department of Mathematics, University of Economics in Prague
Ekonomická 957, 148 00 Prague 4, Czech Republic
E-mail: janj04@vse.cz

Slovak University of Technology in Bratislava
Faculty of Mechanical Engineering
16th Conference on Applied Mathematics **APLIMAT 2017**
Proceedings

FRACTALS, ROBOTS AND MATHEMATICS EDUCATION

JANČAŘÍK Antonín (CZ), ČADEK Jan (CZ)

Abstract. The aim of this paper is to present an interdisciplinary project that was developed within the frame of a diploma thesis [1]. The first author of this paper is the supervisor of the thesis, the second author is the author of the thesis. The goal of the developed interdisciplinary project was to verify whether the topic of fractal geometry can be used in teaching mathematics at upper secondary schools. For these ends robotics kit LEGO Mindstorms NXT was used.

Keywords: fractal, robot, LEGO, mathematics education, LOGO, microwords

Mathematics Subject Classification: Primary 97B10, 97U50; Secondary 28A80

1 Introduction

There are a number of topics and areas that are common for mathematics and information technologies. In the following text we present the area of fractal geometry. This area, although belonging to the field of mathematics, is of strongly interdisciplinary nature. It is related to geography, art and philosophy ([2]). The paper shows that the topic can also be used in lessons of information technology on upper secondary school level. The project described in the paper was created by an undergraduate student – an in-service teacher. The project allowed him to connect two disciplines he studies at the Faculty of Education. And the connection is made in an area that he personally finds very attractive.

2 Fractals

Fractals, or more precisely fractal geometry, is one of the areas of modern twentieth century mathematics that are still being studied and developed. The central concept of the theory – fractal – is not rigorously defined. Different authors approach the concept differently. For example one of the founders of fractal geometry, Benoît Mandelbrot, uses a relatively broad definition of the concept: “Fractal is a set or geometrical figures whose Hausdorff-Besicovitch dimension is (sharply) greater than its topological dimension” [3, p. 8] (comp. [4]). This definition is slightly problematic as it does not consider those objects which are perceived as fractals despite the fact that their fractal dimension in various parts differs or cannot be calculated.

Fig. 1. Fourth iteration of Koch's snowflake, Sierpiński triangle, Mandelbrot's set

The basic concept when introducing pupils to fractals is the concept of self-similarity – i.e. invariance with respect to rescaling [3, p. 87]. This is a unique property when one part of an object looks similar as the whole object when rescaled (different magnification). Mandelbrot introduces the concept of self-similarity as internal homothety of the studied object [5, p. 28]. It can be divided into three types [3, p. 88]:

Exact self-similarity – an exactly self-similar set is created by iteration of the same motive by rescaling, rotation or translation. Its parts always look the same whatever the scale is, the objects display regularity. Examples are Sierpiński triangle, Koch's snowflake etc.

Statistic self-similarity – reflects the fact that natural fractals (coastlines, leaves, snowflakes etc.) are never utterly regular, i.e. they are not exactly self-similar. Randomness comes in play in their generation.

Quazi self-similarity – is the property of objects that are generated by deterministic algorithms with no randomness but that are not exactly self-similar. Examples of these are Julia sets of Mandelbrot set.

2.1 Lindenmayer's systems

Although there are many different fractal objects, we can distinguish between several types of fractals that have the same basic properties. This results in the systematic nature of fractal geometry. Each type of fractals is specific by being suitable for solution of some set of problems. The types differ in the way they are generated.

The most important for the here presented work are the so called L-systems (Lindenmayer's systems) named after the Hungarian biologist Aristid Lindenmayer (1925 – 1989). Sometimes they are referred to as Logo-like systems (with reference to turtle geometry). These systems were chosen for the project because the goal of the work was to allow pupils to draw fractals with the help of a robot. And L-systems are growing more and more popular in computer graphics.

The deterministic context-free L-system is an ordered tuple formed by a finite alphabet of symbols (the geometrical meaning of each of them is stated in advance), by a finite set of transcription rules (which defines conversion of a symbol to a word) and by axiom (determines the initial state of the system). Typical examples of fractals generated by L-systems are Koch's snowflake, Peano and Hilbert curves or Pythagoras tree.

Apart from deterministic L-systems there are also stochastic (random) L-systems that in generating an object use a specific element of randomness (e.g. randomisation of length of a step or of the angle of rotation), so the created fractal is always somewhat different. This allows us to create naturally looking tree vegetation etc.

Fig. 2. Grass modelled using stochastic L-systems in 3D

2.2 Fractals in curricular documents

In the Czech Republic, fractals are included neither in Framework Education Programme for Elementary Education (RVP ZV) [6] nor in Framework Education Programme for Secondary General Education (RVP G) [7]. However, this does not imply topics from fractal geometry could not be included in lessons. The objectives defined for different educational areas, or more precisely the educational content in these areas allow introduction of fractals in several subjects, namely mathematics, physics, informatics but also art or geography. Fractal geometry can become an important element in building cross-curricular links.

When looking for resources abroad, we found a rich source of inspiration in Turkey, where fractals are integral part of the common educational content. Turkish educational programme for lower secondary schools uses fractals in geometry in the educational area Patterns and Tessellation [8]. It is also included in the curriculum for upper secondary schools where fractals are not only used in geometry (pupils should be able to create fractal objects, explain their construction, determine their length or area at a specific level of iteration [9] – one can come across e.g. Sierpiński triangle, Koch and dragon curves, Pythagoras tree) but also in algebra, namely in the educational area Complex numbers [8] where Mandelbrot and Julia sets are used to illustrate multiplication and division of complex numbers and in the educational area Arithmetic progressions and geometric series [8] used e.g. for determination of the area of a Sierpiński triangle.

Another source used in the project were materials available from a website of the American non-profit organization Fractal Foundation. Its founder, Jonathan Wolfe, was the first to include fractals successfully in lessons on an American elementary school Alameda Elementary School (Portland, state Oregon) in 2001.

3 Research methodology

For implementation of the teaching experiment, the author decided to use the robotic kit LEGO Mindstorms Education NXT for work with fractals. The experiment was planned into 9 lessons within which pupils were introduced to fractals and subsequently tested on the topic. Pupils' results on the test were compared to performance of Turkish pupils on a similar test. Both qualitative and quantitative methods were used for evaluation of the experiment.

3.1 Stages of the experiment

The whole project was conducted in three stages which are further divided into smaller blocks (see Tab. 1.).

Stage	Activity	Number of lessons
Preparatory	Algorithmic game Lighbot	2
	The environments L-systems NXT Generator	1
Main	Search for rules generating fractals	2
	Lecture and discussion	1
	Creative production of fractals	1
Concluding	Test focusing on theory and algorithmic problems	1
	Recognition of fractals, project evaluation	1

Tab. 1. Stages of the experiment

The goal of the preparatory stage of the experiment was to unify individual pupils' algorithmic thinking to the same basic level and to introduce them to the environment L-systems NXT Generator which was programmed for the need of the project.

The first two lessons of the main stage of the experiment were devoted to the worksheet Generating L-systems. The worksheet consists of four problems that were designed to guide the pupils to find experimentally the rules for drawing basic mathematical fractals: Koch curve, C curve, Cantor set and Sierpiński triangle formed by one polygonal chain.

The third lesson of the main stage was conducted as an explanatory lecture followed by a discussion. The lesson was opened by a presentation that lead pupils to a discussion whose goal was to make them realize the limits of Euclidean geometry when describing natural objects. The pupils came to an agreement that it was necessary to look for a more powerful tool that would enable them to define the complex structures that surround them in a simpler way.

In the last lesson of the main stage the pupils were again working on a worksheet. The goal of the activities was to summarize previous activities into one integral whole. This was planned to allow the pupils apply the so far gained experience with fractals in creation of new objects.

The concluding stage of the project consisted of a test and evaluation of the teaching experiment with fractal geometry from the pupils' perspective. A thirty minute test combined opened questions from the above mentioned Turkish research with three algorithmic problems. The algorithmic problems were thematically related to the project and were taken from the contest Bobřík informatiky (Beaver of Informatics), category Kadet. In the competition these particular problems had been classified as problems of the highest difficulty. In the last lesson the pupils filled in a knowledge quiz and a questionnaire in which they evaluated the project, use of technology in lessons etc.

3.2 Theoretical background for teaching

As L-system and turtle geometry were used in the teaching experiment. Thus it was natural to come out of two theoretical backgrounds: constructionism and the environment of microworlds.

3.2.1 Constructionism

Constructionism shares its fundamental idea with constructivism. This is the idea that learning occurs when constructing knowledge structures. Above that, constructionists are convinced that this happens most conveniently in those environments in which the learner is actively involved in creation of the surrounding world, albeit building a sand castle on a beach or creating the theory of the universe ([10]). This idea is underlain by the conviction that the most efficient learning comes out of human activity, i.e. we learn best by doing. In contrast to Piaget's constructivism, constructionism puts emphasis on pupils' individual preferences and individual learning styles. Constructionists believe that if construction of one's own knowledge is connected to some creative activity whose outcome is a product, it is more efficient because the learner is creating the world immediately around them. The final product is not as important as the process of its production. Learners often learn without actually being aware of it ([11]). That is why the test included problems of algorithmic nature, despite the fact that the project itself had not focused on them.

3.2.2 Microworlds

Microworld is the world created by a computer. Its generation is guided by specific, rigorously defined rules. Examples of microcosms used in the Czech Republic are turtle geometry in the programme Logo ([12]) or Robot Karel ([13]). In both of these microworlds pupils draw picture using a programmed path. Although pupils may perceive such work in the microworlds as a game, it is a very efficient way of learning ([14]). It has been proved ([15]) that work in microworlds supports development both of geometric and algorithmic thinking.

The environment of microworlds is typical for teaching informatics and programming. However, it can also be come across in teaching mathematics. One of the interesting applications of microworlds in mathematics is the use of small programmable robots ([16]). And this is what we did in our experiment.

3.3 Preparatory technical stage

The author of this text kept construction different versions of robots until he managed to assemble a model that could move precisely enough and that would really enable pupils to draw fractals. Simultaneously the environment L-systems NXT Generator was programmed in Imagine Logo for the needs of the project. In its essence it is a very simple applet that gives pupils some freedom in creation of fractal figures but at the same time does not distract them by providing too many possibilities provided by Imagine Logo as such.

Fig. 3. Final version of the model of robotic turtle

Using this applet pupils could not only experiment with fractal systems on the monitor but they could directly control the robot who plotted everything on a sheet of paper, following the assigned parameters. The plan of the model as well as the environment L-systems NXT Generator currently available at the author of this text.

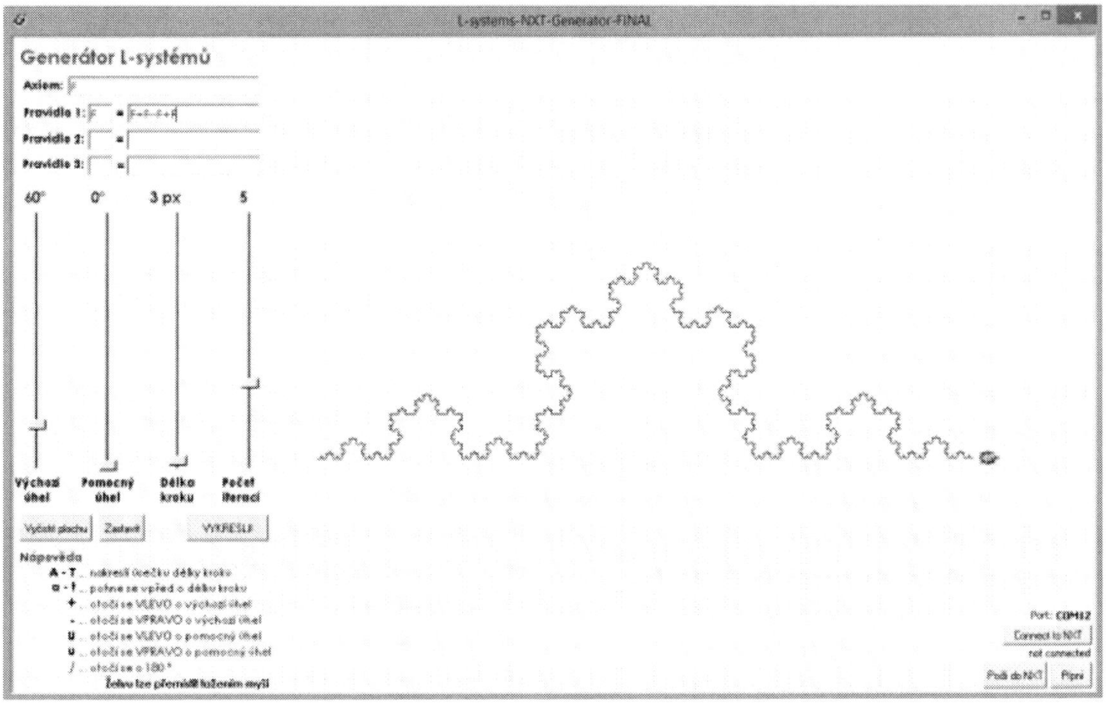

Fig. 4. Graphical user interface of the applet L-systems NXT Generator

4 Results of the experiment

The final test was sat by 18 pupils. The test took 30 minutes. The pupils were working on it on a computer. The test contained seven questions. The first five questions were open questions, the last two were closed multiple choice items. The last three items were taken from the contest Beaver of Informatics. The results of the test are presented in the following table.

	Correct answer	Wrong answer	Incomplete/no answer
Definition of a fractal	89%	11%	0%
Examples of fractals in nature	78%	0%	22%
Sketch of a fractal	34%	33%	28%
Creating an L-system according to the assignment	50%	22%	28%
Plotting robot (Beaver of Informatics)	100%	0%	0%
Robotic bee (Beaver of Informatics)	94%	6%	0%
Plotting robots (Beaver of Informatics)	94%	6%	0%

Tab. 2. Results of the test

A part of the follow-up quiz was naming concrete fractals and deciding whether a particular object is or is not a fractal. The results of naming objects are presented in the following table. The results in telling whether an object is a fractal or not is presented in the next section where they are compared with results of study conducted in Turkish schools.

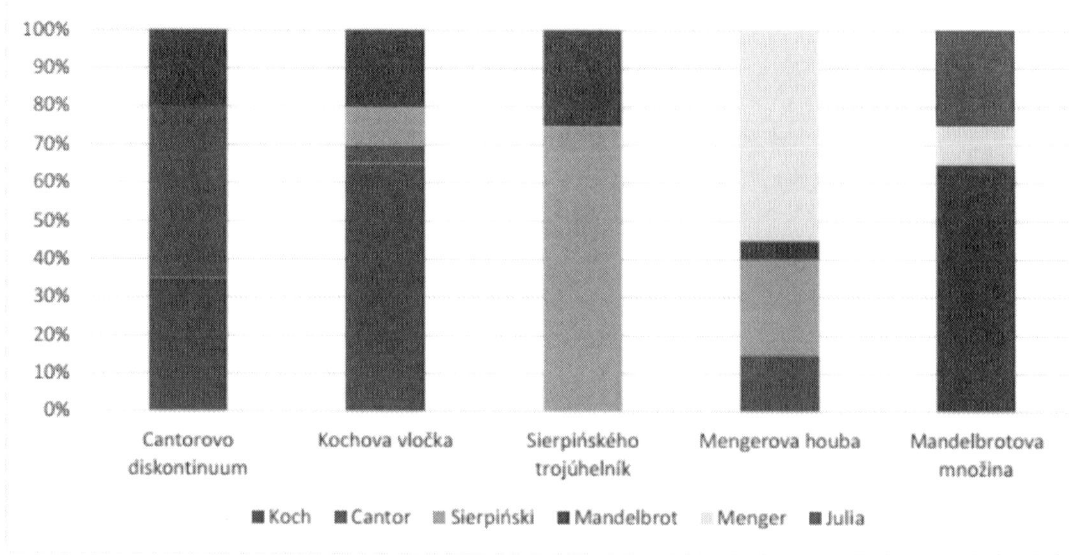

Fig. 5. Results of the quiz

4.1 Comparison of results of Czech and Turkish pupils

Comparison of results of Czech and Turkish pupils [17] shows that Czech pupils achieve much better results when replicating formal knowledge of the definition of a fractal (see Tab. 4). However, they find it difficult to construct an object that would be a fractal following the definition (see Tab. 5).

The following table shows the comparison of Czech and Turkish pupils' recognition of specific figures in the following test.

Picture	Class	Is fractal	Is not fractal	I don't know
	8 - TUR **9 - CZE** 10 - TUR	70,6 % **70 %** 96,6 %	29,4 % **25 %** 3,4 %	0 % **5 %** 0 %
	8 - TUR **9 - CZE** 10 - TUR	29,4 % **0 %** 3,4 %	70,6 % **90 %** 96,6 %	0 % **10 %** 0 %
	8 - TUR **9 - CZE** 10 - TUR	76,5 % **55 %** 58,6 %	23,5 % **40 %** 41,4 %	0 % **5 %** 0 %
	8 - TUR **9 - CZE** 10 - TUR	26,5 % **20 %** 0 %	73,5 % **80 %** 100 %	0 % **0 %** 0 %
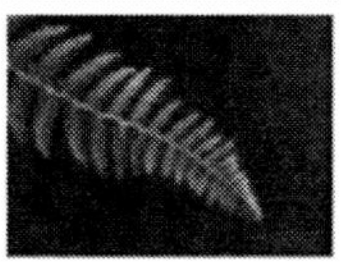	8 - TUR **9 - CZE** 10 - TUR	100 % **95 %** 100 %	0 % **0 %** 0 %	0 % **5 %** 0 %
	8 - TUR **9 - CZE** 10 - TUR	29,4 % **5 %** 3,4 %	67,6 % **85 %** 96,6 %	3 % **10 %** 0 %
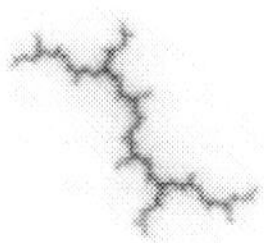	8 - TUR **9 - CZE** 10 - TUR	73,5 % **90 %** 72,5 %	26,5 % **10 %** 24,1 %	0 % **0 %** 3,4 %
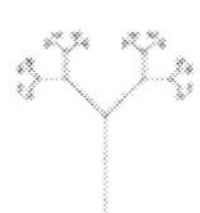	8 - TUR **9 - CZE** 10 - TUR	100 % **85 %** 89,7 %	0 % **10 %** 10,3 %	0 % **0 %** 0 %
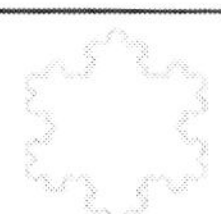	8 - TUR **9 - CZE** 10 - TUR	55,9 % **95 %** 79,3 %	44,1 % **0 %** 20,7 %	0 % **5 %** 0 %

Tab. 3. Results of the quiz

Class	Correct answer	Incorect answer	Incomplete response
8-TUR	64,70%	14,70%	20,60%
9-CZE	88,89%	11,11%	0,00%
10-TUR	44,80%	31,00%	24,20%

Tab. 4. Definition of a fractal

Class	Correct answer	Incorect answer	Incomplete response
8-TUR	70,60%	29,40%	0,00%
9-CZE	38,59%	33,33%	27,78%
10-TUR	51,70%	44,80%	3,50%

Tab. 5. Application of definition

5 Conclusion

The results of the experiment, especially the results in solving more difficult algorithmic tasks from the Beaver of Informatics contest (Czech localization of the Bebras competition) show that introduction of fractal geometry with the use of the Logo language could support development of algorithmic and mathematical thinking of pupils.

Observations carried out during individual activities and feedback from pupils indicate that fractals are attractive, interesting and motivating enough for pupils. Of course this does not imply that fractal geometry should immediately enter Framework education programmes and become compulsory. However, this experiment gives an insight into how Czech pupils are able to grasp this area.

Acknowledgement

The paper was supported by grant from Charles University in Prague entitled "P15 - Schools and the Teaching Profession in the Context of Increasing Demands on Education".

References

[1] ČADEK, J. *Didaktické využití stavebnice LEGO Mindstorms ve výuce matematiky se zaměřením na fraktály* (Diploma thesis). PedF UK v Praze, 2016.

[2] WINITZKY DE SPINADEL, V.M. Mathematics and artistic design. *In APLIMAT 2016 - 15th Conference on Applied Mathematics*, Bratislava, 2016, 1139-1146.

[3] ZELINKA, I. *Fraktální geometrie: Principy a aplikace*. Praha: BEN, 2006. ISBN: 80-7300-193-4.

[4] KOŤÁTKOVÁ STRÁNSKÁ, P., MAREK, J. Clustering of paintings by using of the fractal dimension, . *In APLIMAT 2015 - 14th Conference on Applied Mathematics*, Bratislava, 2015, 500-506.

[5] MANDELBROT, B. *Fraktály: Tvar, náhoda a dimenze*. Praha: Mladá fronta, 2003. ISBN: 80-204-1009-0.

[6] *Rámcový vzdělávací program pro základní vzdělávání*. Praha: MŠMT, 2016 [22/11/2016]. URL: http://www.nuv.cz/uploads/RVP_ZV_2016.pdf.

[7] *Rámcový vzdělávací program pro gymnázia*. Praha: VÚP v Praze, 2007 [22/11/2016]. ISBN: 978-80-87000-11-3. URL: http://www.nuv.cz/file/159_1_1/

[8] *Ortaöğretim Matematik (9, 10, 11 Ve 12. Sınıflar) Dersi Öğretim Programi*. Ankara: MEB, 2011 [22/11/2016]. URL: http://ttkb.meb.gov.tr/program2.aspx.

[9] *Ortaöğretim Geometri Dersi 9–10. Sınıflar: Öğretim Programi*. Ankara: MEB, 2010 [03/07/2016]. URL: http://tinyurl.com/zq555f9

[10] PAPERT, S., HAREL, I. Situating constructionism. *Constructionism*, 1(11), 1991.

[11] JANČAŘÍK, A.. *Vybrané teorie učeni a jejich projekce do využívání ICT ve výuce matematiky*. Praha: UK v Praze, Pedagogická fakulta, 2013. 188 . ISBN 978-80-7290-766-3.

[12] ŠTĚPÁNEK, P. Co je LOGO? *Pokroky matematiky, fyziky a astronomie*, 26(2), 1981, 98-105.

[13] BERGIN, J., ROBERTS, J., PATTIS, R., STEHLIK, M. Karel++: A Gentle Introduction to the Art of Object-Oriented Programming. New York: John Wiley & Sons, Inc., 1996.

[14] RIEBER, L. P. Seriously considering play: Designing interactive learning environments based on the blending of microworlds, simulations, and games. Educational Technology Research and Development, 44(2), 1996, 43-58.

[15] LABORDE, C., KYNIGOS, C., HOLLEBRANDS, K., STRASSER, R. Teaching and learning geometry with technology. In A. Gutierrez & P. Boero (Eds.), Handbook of research on the psychology of mathematics education: Past, present and future, Rotterdam, The Netherlands: Sense Publishers, 2006, 275-304.

[16] HIGHFIELD, K. Playing with simple robotics: a tool for mathematics learning in the early years. In Semt 13, Proceedings: Tasks and Tools in Elementary Mathematics, Prague, 2013.

[17] KARAKUS, F. a I. KARATAS. Secondary School Students' Misconceptions about Fractals. In: *Journal of Education and Human Development*. Madison: American Research Institute for Policy Development, 2014, s. 241-50 [03/07/2016]. ISSN: 2334-296X. URL: http://jehdnet.com/journals/jehd/ Vol_3_No_3_September_2014/19.pdf

Current address

Jančařík Antonín, Doc. RNDr. Ph.D.
Charles University, Faculty of Education
M. Rettigové 4, 116 39 Prague 1, Czech Republic
Tel. number: +420 221 900 251, e-mail: antonin.jancarik@pedf.cuni.cz

Čadek Jan, Mgr.
Charles University, Faculty of Education
M. Rettigové 4, 116 39 Prague 1, Czech Republic
E-mail: honza.cadek@seznam.cz

Slovak University of Technology in Bratislava
Faculty of Mechanical Engineering
16th Conference on Applied Mathematics **APLIMAT 2017**
Proceedings

TOLERANCE CHARTS IN STATISTICAL PROCESS CONTROL

JANIGA Ivan (SK), GABKOVA Jana (SK), RICHTARIKOVA Daniela (SK)

Abstract. The article analyzes the process of tightening screw fasteners. Tolerance diagrams are used in a process control, because in this case, they are more appropriate than the control charts which are used by the firm.

Keywords: tolerance chart, control chart

Mathematics Subject Classification: 62P30

1 Introduction

The article analyzes the process of tightening screws for fastening the two sides of the boot lid when it is being installed in the car. A well-known automotive company which we will call ABC Ltd. since it does not wish to be named, provided real data on the regulation of the production process.

ABC Ltd. makes a sample of size $k = 6$ observations every other day for two months. The total number of samples is 30. From the sample data it constructs an average control chart ($\bar{X}$ −diagram) and range control charts (R −diagram) with unknown basic parameters. ABC Ltd. recalculate the upper control limit (UCL) and lower control limit (LCL) every two months. Thus, the company always sets the process of tightening the screw connections after two months for regulation but they never actually regulate. We consider this a "luxury" but the company can afford it only because the specification limits are very wide, i.e. from 14 Nm to 34 Nm.
This article will first present a more precise analysis of each regulation diagram of two month samples for one year. Then the data is applied to tolerance charts, which are also statistical tools of process control. The results of the two analyses are compared.

2 Analysis of control charts

Data from the control of the production process, provided by ABC Ltd., are once again analyzed, in this case from calculations in STATGRAPHICS Centurion software. The calculated parameters in the control charts are identical to those provided by ABC Ltd. The results are shown in Tab. 1.

Average control chart				
Months	Left side		Right side	
	UCL – LCL	**CL**	**UCL – LCL**	**CL**
Jan-Feb	5,59	20,57	4,06	23,24
Mar-Apr	5,18	20,69	4,98	22,96
May-June	7,96	23,52	5,36	23,90
July-Aug	10,16	24,92	3,78	25,39
Sep-Oct	5,61	23,33	4,81	24,51
Nov-Dec	4,65	22,57	4,44	24,94

Tab. 1. Range of control limits every two months during one year

In the table it can be seen that the range between the upper and lower control limits in average control charts change every two months. On the left side of the boot lid ranges vary from 4,65 to 10,16 Nm but on the right side only from 3,78 Nm to 5,36 Nm. The center lines on the left side of the boot lid vary from 20,57 to 24,92 Nm and on the right side from 23,24 to 25,39 Nm. The values of individual ranges on the left side fluctuate more than on the right side. This is better shown in Fig. 1 and Fig. 2.

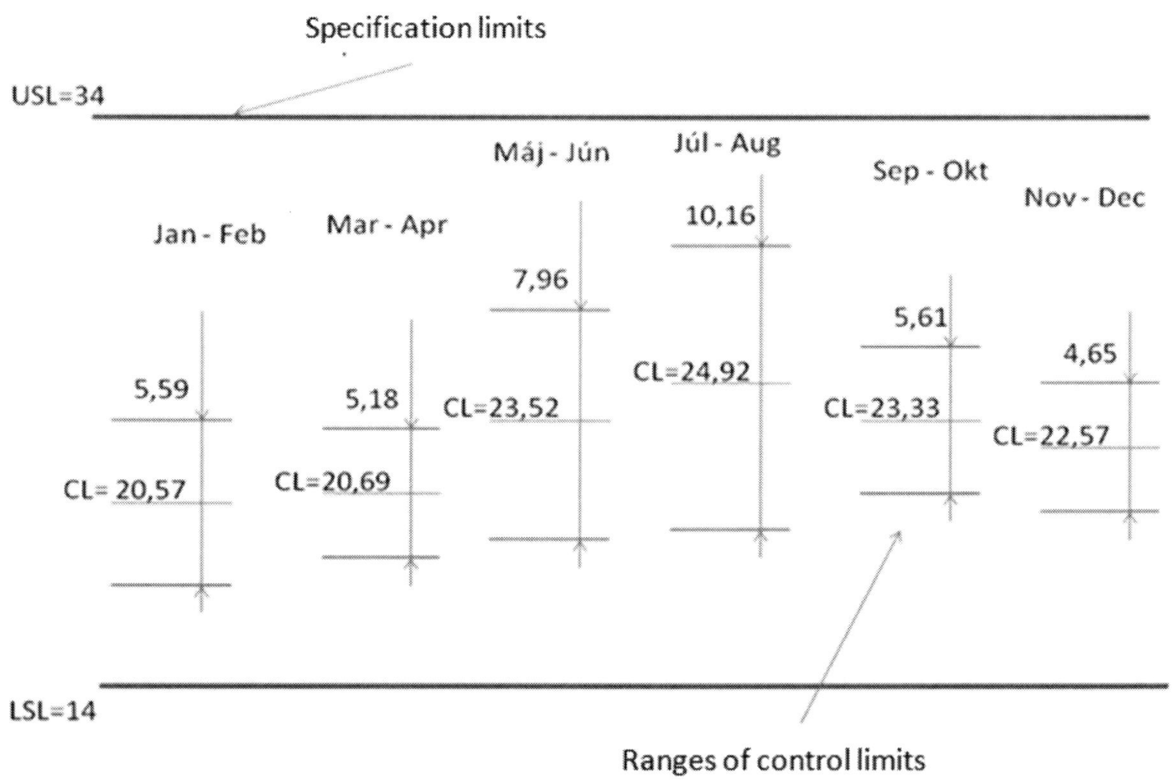

Fig. 1. Comparison of control limits with the specification limits – the left side of the boot lid

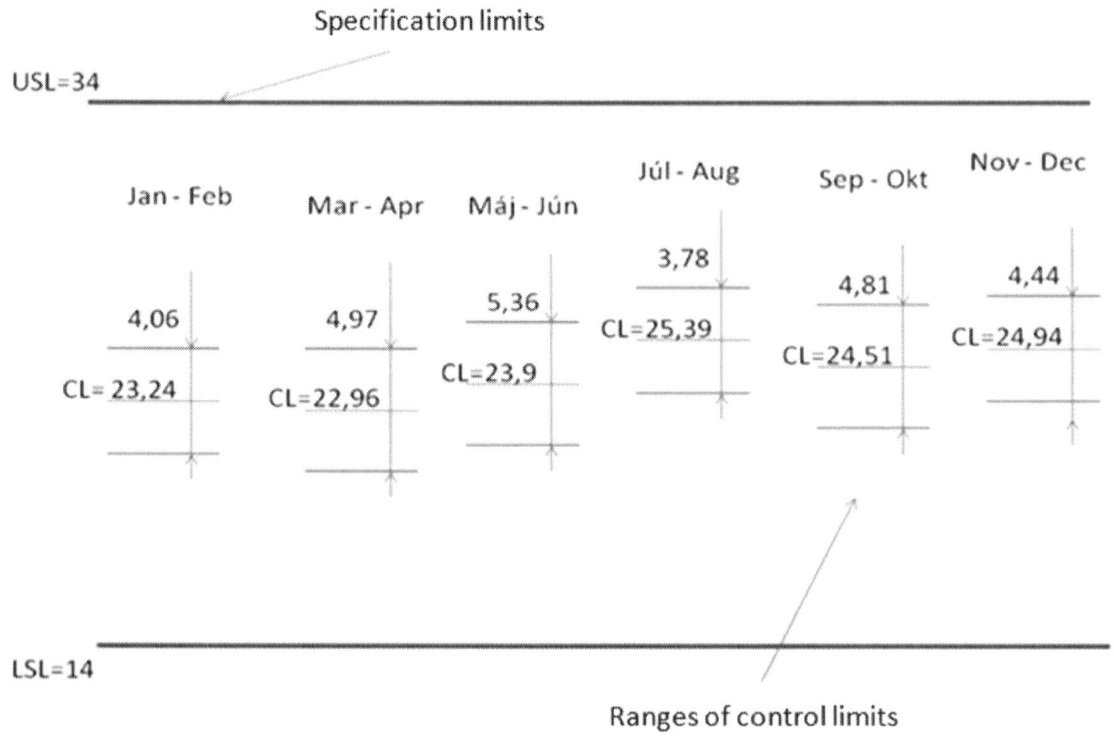

Fig. 2. Comparison of control limits with the specification limits – the right side of the boot lid

3 Tolerance diagram with data from ABC Ltd.

Tolerance diagram shows the six measured values in each of 30 samples in the range of specification limits from 14 Nm to 34 Nm, as can be seen in Fig. 3.

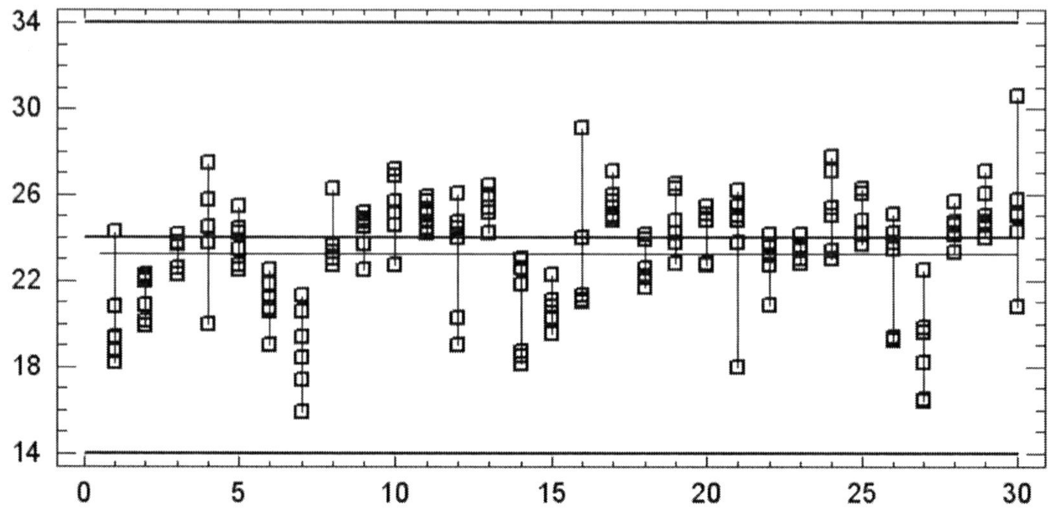

Fig. 3. Tolerance diagram per Jan–Feb – the right side of the boot lid

The figure shows that all measured values are within the specifications with a sufficient margin.

Tolerance diagrams were applied to the data from the control of the production process and found that all meet the specifications in addition to data from the regulation in July-Aug - left side of the boot lid. This problem and other problems will be discussed in the third part of the article.

4 Tolerance diagrams versus control charts

In this section, we discuss two issues. The first concerns the regulation of the right side of the boot lid in the months of January-February.

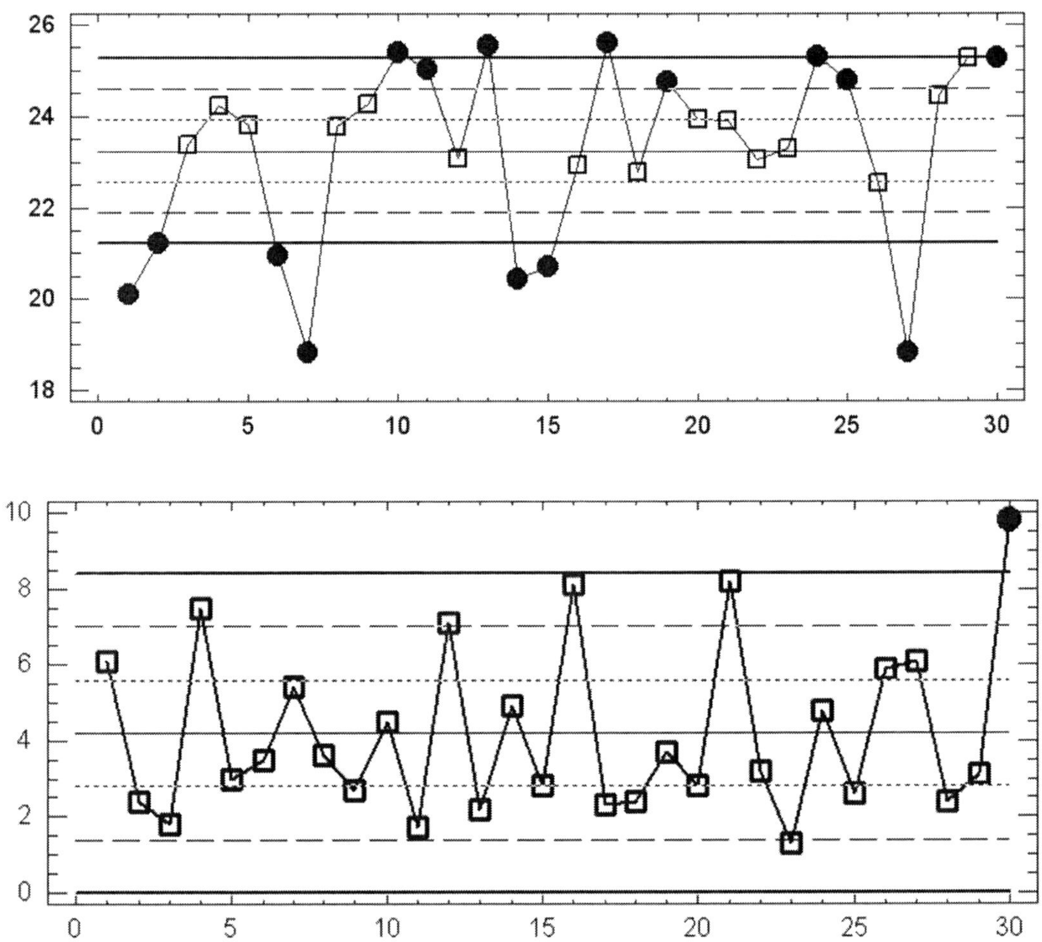

Fig. 4. Average control charts and range control charts per Jan–Feb – the right side of the boot lid

The tolerance diagram (Fig. 3), indicating individual measurements, shows the stability of observations within the specification limits.The average control chart, on the other hand, is strongly unstable within the control limits.

The second problem concerns the regulation of the left side of the boot lid in the months from July to August.

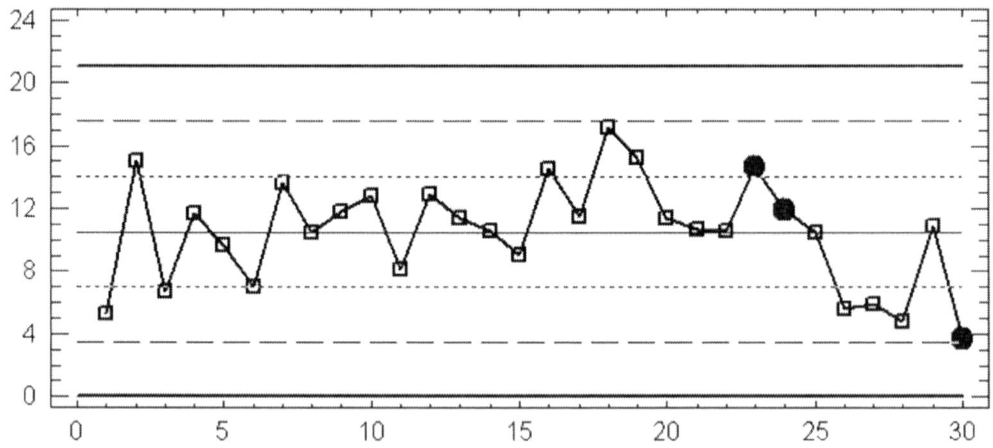

Fig. 5. Range control charts per Jul–Aug – the left side of the boot lid

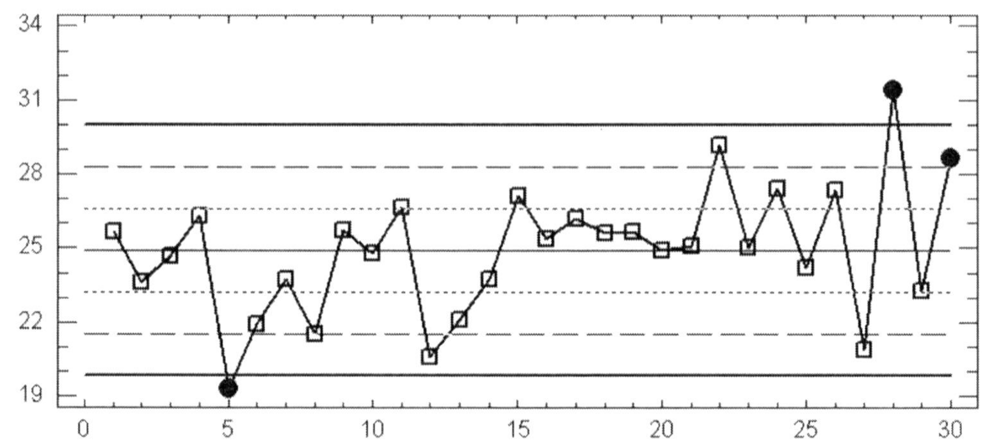

Fig. 6. Average control chart per Jul–Aug – the left side of the boot lid

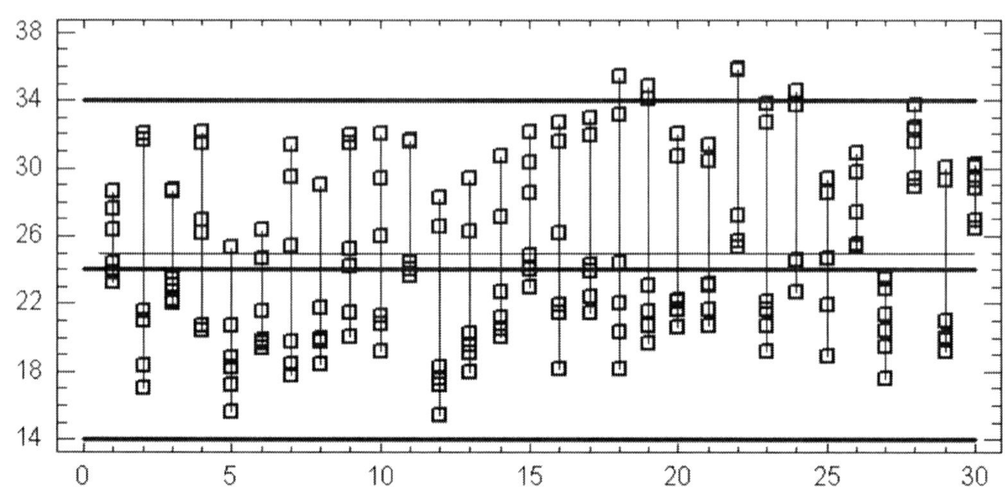

Fig. 7. Tolerance chart per Jul–Aug – the left side of the boot lid

In the range chart (Fig. 5) three samples occurred - numbers 23, 24 and 30- which point at instability despite the average values being within the specification limits.

The average control chart (Fig. 6) contains two outlying averages: one below the lower control limit - LCL (sample number 5)- and the other above the upper control limit - UCL (sample number 28), which point at the instability of the process. Sample number 30 indicates an assignable cause.

All three values (5, 28 and 30) show that the process of tightening the screw fastening is out of control. Also in this diagram, the value of all averages are inside the specification limits, i.e. from 14 Nm to 34 Nm.

The tolerance diagram (Fig. 7) presents five measurements, samples 18, 19, 22 and 24, beyond the upper specification limit of 34 Nm. This indicates that higher torque values than the prescribed specifications have been used in tightening the screws. The control charts in Fig. 5 and Fig. 6 the facts do not signal this reality.

The charts for the averages and ranges use average values in subgroups, while the tolerance charts use individual values. Based on the tolerance chart, we found that in the period analyzed products had failed to conform to specifications. The torque values for individual measurements exceeded the specifications limits in four samples, with serial numbers 18, 19, 22 and 24. Only chart range indicated instability in sample number 24. This fact was due to a distorted view of samples 18, 19, 22 and 24 in the control charts, working with average values only. The averages measured in the above samples are influenced by the "great weight" of individual values under the center line or just above the center line. Therefore, the average values of samples 18, 19, 22 and 24 are evaluated as satisfactory, although some values of these samples are located above the upper specification limit.

5 Conclusion

The paper presents findings revealed on the basis of the tolerance chart for the left side of the boot lid in July and August. In these months five values were found which fell outside the specification limits, although these values were not found in the analysis of control charts for average and range. The reason why these values were not found in control charts for averages and ranges is that they are working with average values in the samples. Because most of the values in these samples were located either below nominal value or slightly above nominal value, their weight prevented average values in the averages and ranges control charts from deviating above the upper specification limit. The fact that the 5 values in samples 18, 19, 22, and 24 can be detected only by the tolerance chart is that the tolerance chart uses individual values. It follows that when a specification range is wide, the Shewhart control chart is appropriate to supplement the tolerance diagrams. Thus, it is possible to prevent situations which produce nonconforming products even though the control charts signal neither values that are beyond control limits nor instability.

References

[1] GRANGER, C.W. – NEWBOLD, P. 1974. Spurious Regression in Econometrics. In: Journal of Econometrics, č. 2, 1974, s. 111 – 120.

[2] GREEN, W. H. 1997. Econometric Analyses. London: Prentice – Hall, 1997. 1076 s. ISBN 0-13-7246659-5.

[3] KANDEROVÁ, M. 2004. Metódy prognózovania sezónnosti v ekonomických časových radoch. In: Acta Facultatis Aerari PubliciI. 2004, č. 1. Banská Bystrica : Fakulta financií UMB, 2004, s. 53 – 60.

Current address

Janiga Ivan, doc. RNDr., PhD.
Institute of Mathematics and Physics
Faculty of Mechanical Engineering, Slovak University of Technology in Bratislava
Nám. slobody 17, 812 31 Bratislava, Slovak Republic
E-mail: ivan.janiga@stuba.sk

Gabková Jana, RNDr., PhD.
Institute of Mathematics and Physics
Faculty of Mechanical Engineering, Slovak University of Technology in Bratislava
Nám. slobody 17, 812 31 Bratislava, Slovak Republic
E-mail: jana.gabkova@stuba.sk

Richtáriková Daniela, RNDr., PhD.
Institute of Mathematics and Physics
Faculty of Mechanical Engineering, Slovak University of Technology in Bratislava
Nám. slobody 17, 812 31 Bratislava, Slovak Republic
E-mail: daniela.richtarikova@stuba.sk

Slovak University of Technology in Bratislava
Faculty of Mechanical Engineering
16th Conference on Applied Mathematics **APLIMAT 2017**
Proceedings

POLYNOMIALS WITH THE SCHUR PROPERTY AND THEIR CHARACTERIZATIONS

Jánský Jiří (CZ)

Abstract. The paper discusses a distribution of the zeros of the polynomial

$$p(\lambda) = \lambda^k - \alpha\lambda^{k-m} - \beta$$

with respect to the unit circle. This problem is of a theoretical as well as practical importance which motivated many authors to formulate necessary and sufficient conditions guaranteeing the location of all the zeros of $p(\lambda)$ inside the unit circle. Using various approaches, different forms of conditions ensuring this property have been obtained. We give here an overview and comparisons of these conditions, and show equivalence between some of them.

Keywords: polynomial, zero location, difference equation, stability

Mathematics subject classification: Primary 26C10; Secondary 39A30, 39A06

1 Introduction

We deal with the problem of characterization of the Schur property of the polynomial

$$p(\lambda) = \lambda^k - \alpha\lambda^{k-m} - \beta \quad (1)$$

with real coefficients α, β and integer $k > m > 0$. We say that $p(\lambda)$ has the Schur property if all its zeros are located inside the unit circle in the complex plane. This problem has a long history and its investigations are connected with many papers [4, 7, 8, 9, 12].

The common property of conditions derived in these papers is that they are not explicit. These conditions either require solving of some auxiliary nonlinear equations ([4, 7, 8]), or they formulate appropriate regions (where (1) has the Schur property) in the (α, β)-plane via their boundaries formed by lines and certain parametric curves ([8, 12]). Recently, another type of conditions guaranteeing the Schur property for (1) has appeared in [1, 2, 3]. These conditions are of a different type and their equivalence with the results stated in [4, 8] is not trivial.

The paper is focused on comparisons of various necessary and sufficient conditions stated in [4, 8, 1], which assure that polynomial (1) has the Schur property. Besides these comparisons, we make a proof of equivalence of appropriate conditions stated in [8, 1].

2 Some known results and their comparisons

We recall here the relevant assertions where integers k and m are regarded as coprimes. We emphasize that this assumption can be imposed without loss of generality.

At first we present criterion stated in [4]. It provides a direct generalization of the famous result [7].

Theorem 2.1 *Let α, β be real numbers such that $\alpha \neq 0$ and let k odd and m even be positive coprime integers such that $k > m$. The polynomial (1) has the Schur property if and only if either*

$$0 < \alpha < 1, \quad |\beta| < 1 - \alpha\,,$$

or

$$-1 < \alpha < 0, \quad |\beta| < \min_{\phi \in S}(\alpha^2 - 2\alpha\cos(m\phi) + 1)^{1/2}\,,$$

where S is the solution set of

$$\frac{\sin((k-m)x)}{\sin(kx)} = \frac{1}{\alpha}, \qquad x \in (0, \pi)\,.$$

Note that the paper [4] involves also formulations where (1) with k, m odd and k even, m odd has the Schur property. Since these criteria are formally very close to the assertion of Theorem 2.1, we do not state them explicitly and refer to [4].

This criterion can be utilized especially when entries k, m and at least one of coefficients α, β are fixed. In the remaining cases, its usage seems to be problematic. The following example illustrates application of Theorem 2.1 to (1) with given α, k, and m.

Example 2.2 *Let $k = 5$, $m = 2$ and $\alpha = 0.8$. We analyze the Schur property for the polynomial*

$$q_5(\lambda) = \lambda^5 + 0.8\lambda^3 - \beta \tag{2}$$

with respect to β.

Theorem 2.1 requires to solve the nonlinear equation

$$\frac{\sin(3x)}{\sin(5x)} = \frac{1}{-0.8}, \qquad x \in (0, \pi)\,.$$

We have to use appropriate numerical method to obtain the solution set

$$S = \{0.7591, 1.4161, 1.7255, 2.3824\}.$$

Then, Theorem 2.1 yields the condition

$$|\beta| < \min(0.3406, 1.3130, 0.3406, 1.3130)\,,$$

i.e. (2) has the Schur property if and only if $\beta \in (-0.3406, 0.3406)$. This interval is depicted in the Fig. 1.

The next lemma from the elementary number theory will allow us to present alternative conditions.

Lemma 2.3 *Let $k > m$ be coprime positive integers. Then there exists a couple of positive integers (j, s) such that*

$$|mj - ks| = 1, \quad j < k, \quad s \text{ odd}. \tag{3}$$

If m is odd, then this couple is unique; if m is even, then there are precisely two such couples (in one of which j is even and in the other odd).

Now we can formulate the following conditions derived in [8].

Theorem 2.4 *Let α, β be real numbers, $k > m$ be coprime positive integers and let (j, s) be a couple of positive integers satisfying (3). The polynomial (1) has the Schur property if and only if the couple (α, β) is the internal point of the finite domain bounded by the following lines and curves:*

(i) $\alpha + \beta = 1$;

(ii) $\alpha = \frac{\sin(k\omega)}{\sin((k-m)\omega)}, \quad \beta = -\frac{\sin(m\omega)}{\sin((k-m)\omega)}$;

(iii) $(-1)^m\alpha + (-1)^k\beta = 1$;

(iv) $\alpha = (-1)^m \frac{\sin(k\omega)}{\sin((k-m)\omega)}, \quad \beta = -(-1)^k \frac{\sin(m\omega)}{\sin((k-m)\omega)}$

where ω varies between $j\pi/k$ and $s\pi/m$. If m is even, then any of two couples (j, s) satisfying (3) can be taken.

Theorem 2.4 describes the boundary of domain in the (α, β)-plane, where (1) has the Schur property. Hence, this formulation enables an easy plotting of this domain, but it is not optimal from the analytical viewpoint.

Example 2.5 *We put $k = 5$, $m = 2$ and analyze the Schur property for the polynomial*

$$q_5(\lambda) = \lambda^5 - \alpha\lambda^3 - \beta$$

with respect to α and β. Using Lemma 2.3, we can set $j = 2$ and $s = 1$. Then Theorem 2.4 yields the domain for coefficients (α, β) where $q_5(\lambda)$ has the Schur property. This domain is depicted in the Fig. 1.

There is an issue with the use of Theorem 2.4 consisting in its geometrical stylization. For example, let

$$q_5(\lambda) = \lambda^5 + 0.8\lambda^3 - 0.25.$$

Then it is not a quite straightforward matter to decide whether the point $(-0.8, 0.25)$ lies inside the domain depicted in the Fig. 1. Therefore, we present the assertion stated in [1], which yields explicit conditions on α, β, k, m ensuring the Schur property.

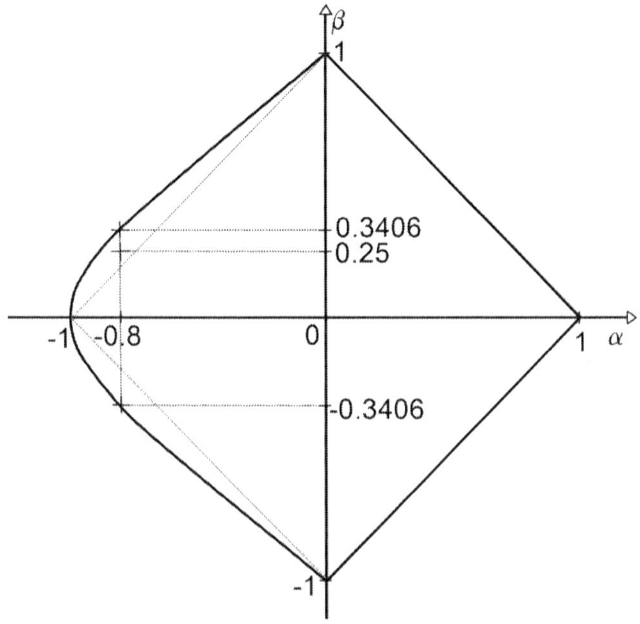

Fig. 1. The domain in the (α, β)-plane, where $q_5(\lambda)$ has the Schur property

Theorem 2.6 *Let α, β be real numbers and $k > m$ be coprime positive integers. The polynomial (1) has the Schur property if and only if any one of the following conditions holds:*

(C1) $|\alpha| + |\beta| < 1$;

(C2) $|\alpha| + |\beta| \geq 1, \quad |\alpha| - 1 < |\beta| < 1, \quad \alpha^k \beta^m < 0$ *and*

$$k \arccos \frac{1 + \alpha^2 - \beta^2}{2|\alpha|} + m \arccos \frac{1 - \alpha^2 + \beta^2}{2|\beta|} < \pi \tag{4}$$

The system of conditions presented in Theorem 2.6 is explicit and does not require any additional operations when discussing the Schur property of a given polynomial (1). In particular, using this criterion it is very easy to decide whether (1) with all fixed entries has the Schur property or not. Beside this direct test for (1), it can provide effective conditions on k, m. Notice also that (C1) is independent of m, k, while (C2) depends on these parameters. In the case $|\alpha| + |\beta| = 1$, (C2) is reduced to $\alpha^k \beta^m < 0$.

Example 2.7 *We consider (1) with $\alpha = -0.8$, $\beta = 0.25$, $m = 2$ and investigate the Schur property of*

$$q_k(\lambda) = \lambda^k + 0.8\lambda^{k-2} - 0.25$$

with respect to integer $k > 2$. A direct application of Theorem 2.6 yields that $q_k(\lambda)$ has the Schur property if and only if

$$k < \frac{\pi - m \arccos \frac{1-\alpha^2+\beta^2}{2|\beta|}}{\arccos \frac{1+\alpha^2-\beta^2}{2|\alpha|}} = 11.98 \quad \textit{and} \quad k \textit{ is odd},$$

i.e. when $k \in \{3,5,7,9,11\}$.

Note that deduction of this conclusion from conditions of Theorems 2.1 and 2.4 is difficult.

The boundary of the domain (4) in the (α,β)-plane is given implicitly by

$$k \arccos \frac{1+\alpha^2-\beta^2}{2|\alpha|} + m \arccos \frac{1-\alpha^2+\beta^2}{2|\beta|} = \pi \, .$$

Our next aim is to show that this curve is equivalent to that stated in Theorem 2.4. More precisely, we are going to prove equivalence of Theorems 2.4 and 2.6.

3 Equivalence of Theorems 2.4 and 2.6

Let m, k be such that $mj - ks = 1$. Then $\omega \in (s\pi/m, j\pi/k)$, hence

$$k\omega \in (ks\pi/m, j\pi), \quad m\omega \in (s\pi, mj\pi/k), \quad (k-m)\omega \in ((j-s)\pi-\pi/m, (j-s)\pi-\pi/k) \, .$$

This implies $\sin m\omega < 0$ (s is always odd) and

$$\begin{aligned} &\sin(k\omega) > 0, \quad \sin((k-m)\omega) < 0 \qquad (\text{if } j \text{ is odd}), \\ &\sin(k\omega) < 0, \quad \sin((k-m)\omega) > 0 \qquad (\text{if } j \text{ is even}). \end{aligned}$$

Analogously, if $mj - ks = -1$ then $\omega \in (j\pi/k, s\pi/m)$ and

$$k\omega \in (j\pi, ks\pi/m), \quad m\omega \in (mj\pi/k, s\pi), \quad (k-m)\omega \in ((j-s)\pi+\pi/k, (j-s)\pi+\pi/m) \, .$$

Consequently, $\sin m\omega > 0$ and

$$\begin{aligned} &\sin(k\omega) < 0, \quad \sin((k-m)\omega) > 0 \qquad (\text{if } j \text{ is odd}), \\ &\sin(k\omega) > 0, \quad \sin((k-m)\omega) < 0 \qquad (\text{if } j \text{ is even}). \end{aligned}$$

Further, we analyze the form of parametric curves (ii) and (iv). In both the cases it holds

$$\begin{aligned} \frac{1+\alpha^2-\beta^2}{2|\alpha|} &= \frac{\sin^2((k-m)\omega) + \sin^2(k\omega) - \sin^2(m\omega)}{2|\sin(k\omega)\sin((k-m)\omega)|} \\ &= \frac{(\sin(k\omega)\cos(m\omega) - \cos k\omega \sin(m\omega))^2 + \sin^2(k\omega) - \sin^2(m\omega)}{2|\sin(k\omega)\sin((k-m)\omega)|} \\ &= \frac{\sin^2(k\omega)(1+\cos^2(m\omega)) - \sin^2(m\omega)(1-\cos^2(k\omega))}{2|\sin(k\omega)\sin((k-m)\omega)|} \\ &\quad - \frac{2\sin(k\omega)\cos(m\omega)\sin(m\omega)\cos(k\omega)}{2|\sin(k\omega)\sin((k-m)\omega)|} \\ &= \frac{2\sin^2(k\omega)\cos^2(m\omega) - 2\sin(k\omega)\cos(m\omega)\sin(m\omega)\cos(k\omega)}{2|\sin(k\omega)\sin((k-m)\omega)|} \\ &= \frac{\sin(k\omega)\cos(m\omega)(\sin(k\omega)\cos(m\omega) - \sin(m\omega)\cos(k\omega))}{|\sin(k\omega)\sin((k-m)\omega)|} \\ &= \cos(m\omega)\ \operatorname{sgn}\left[\sin(k\omega)\sin((k-m)\omega)\right] \end{aligned}$$

and

$$\begin{aligned}
\frac{1-\alpha^2+\beta^2}{2|\beta|} &= \frac{\sin^2((k-m)\omega)-\sin^2(k\omega)+\sin^2(m\omega)}{2|\sin(m\omega)\sin((k-m)\omega)|} \\
&= \frac{(\sin(k\omega)\cos(m\omega)-\cos(k\omega)\sin(m\omega))^2-\sin^2(k\omega)+\sin^2(m\omega)}{2|\sin(m\omega)\sin((k-m)\omega)|} \\
&= \frac{-\sin^2(k\omega)(1-\cos^2(m\omega))+\sin^2(m\omega)(1+\cos^2(k\omega))}{2|\sin(m\omega)\sin((k-m)\omega)|} \\
&\quad -\frac{2\sin(k\omega)\cos(m\omega)\sin(m\omega)\cos(k\omega)}{2|\sin(m\omega)\sin((k-m)\omega)|} \\
&= \frac{2\sin^2(m\omega)\cos^2(k\omega)-2\sin(k\omega)\cos(m\omega)\sin(m\omega)\cos(k\omega)}{2|\sin(m\omega)\sin((k-m)\omega)|} \\
&= \frac{\sin(m\omega)\cos(k\omega)(\sin(m\omega)\cos(k\omega)-\sin(k\omega)\cos(m\omega)}{|\sin(m\omega)\sin((k-m)\omega)|} \\
&= -\cos(k\omega)\ \operatorname{sgn}\left[\sin(m\omega)\sin((k-m)\omega)\right].
\end{aligned}$$

Taking into account the previous sign discussions, we can observe that if j is odd then

$$\cos(m\omega)=\frac{-1-\alpha^2+\beta^2}{2|\alpha|}, \qquad \cos(k\omega)=\frac{-1+\alpha^2-\beta^2}{2|\beta|}, \tag{5}$$

while if j is even then

$$\cos(m\omega)=\frac{-1-\alpha^2+\beta^2}{2|\alpha|}, \qquad \cos(k\omega)=\frac{1-\alpha^2+\beta^2}{2|\beta|}. \tag{6}$$

Now we again distinguish two sign cases of expression $mj-ks$. First let $mj-ks=1$. Then

$$\cos(m\omega)=\frac{-1-\alpha^2+\beta^2}{2|\alpha|}, \quad m\omega\in(s\pi, mj\pi/k)$$

if and only if

$$m\omega=s\pi+\arccos\frac{1+\alpha^2-\beta^2}{2|\alpha|}. \tag{7}$$

Similarly,

$$\cos(k\omega)=(-1)^j\frac{1-\alpha^2+\beta^2}{2|\beta|}, \quad k\omega\in(ks\pi/m, j\pi)$$

if and only if

$$k\omega=j\pi-\arccos\frac{1-\alpha^2+\beta^2}{2|\beta|} \tag{8}$$

(we emphasize that the previous equivalency holds for both cases j odd and j even). Comparing (7), (8) and eliminating parameter ω we get

$$k\left(s\pi+\arccos\frac{1+\alpha^2-\beta^2}{2|\alpha|}\right)=mj\pi-m\arccos\frac{1-\alpha^2+\beta^2}{2|\beta|},$$

i.e.

$$k\arccos\frac{1+\alpha^2-\beta^2}{2|\alpha|}+m\arccos\frac{1-\alpha^2+\beta^2}{2|\beta|}=\pi \tag{9}$$

by use of $mj - ks = 1$.

Analogously we can discuss the case $mj - ks = -1$. Then

$$\cos(m\omega) = \frac{-1-\alpha^2+\beta^2}{2|\alpha|}, \qquad m\omega \in (mj\pi/k, s\pi)$$

if and only if

$$m\omega = s\pi - \arccos\frac{1+\alpha^2-\beta^2}{2|\alpha|}. \tag{10}$$

Further,

$$\cos(k\omega) = (-1)^j\frac{1-\alpha^2+\beta^2}{2|\beta|}, \qquad k\omega \in (j\pi, ks\pi/m)$$

if and only if

$$k\omega = j\pi + \arccos\frac{1-\alpha^2+\beta^2}{2|\beta|}. \tag{11}$$

Similarly as in the previous part, the relation (9) follows from (10), (11) with respect to $mj - ks = -1$.

Note, that (5), (6) implies

$$-1 \leq \frac{-1-\alpha^2+\beta^2}{2|\alpha|} \leq 1, \qquad -1 \leq \frac{1-\alpha^2+\beta^2}{2|\beta|} \leq 1,$$

i.e.

$$|\alpha| + |\beta| \geq 1, \qquad |\alpha| - 1 \leq |\beta| \leq 1 + |\alpha|. \tag{12}$$

Thus the cyclometric functions in (9) are well defined if and only if (12) holds.

Considering the above sign analysis of $\sin(m\omega)$, $\sin(k\omega)$ and $\sin((k-m)\omega)$ we can conclude that the domain where (1) has the Schur property is bounded by:

(a) the straight lines $\alpha + \beta = 1, \quad (-1)^m\alpha + (-1)^k\beta = 1$;

(b) the curve (9) under $\alpha^k\beta^m < 0, \quad |\alpha| + |\beta| \geq 1, \quad |\alpha| - 1 \leq |\beta| \leq 1 + |\alpha|$.

First, let $|\alpha| + |\beta| < 1$. Then we immediately get (C1) (see Fig. 2).

Assuming $|\alpha| + |\beta| \geq 1$, we complete the verification of validity of the part (C2). First notice that the curve (9) is extending the domain $|\alpha| + |\beta| < 1$. Indeed, if $|\alpha| + |\beta| = 1$ then

$$\frac{1+\alpha^2-\beta^2}{2|\alpha|} = \frac{1-\alpha^2+\beta^2}{2|\beta|} = 1$$

and the lefthand-side of (9) is zero for arbitrary k and m. In other words, points of the straight lines $|\alpha| + |\beta| = 1$, $\alpha^k\beta^m < 0$ are internal points of domain (where (1) has the Schur property) for any k and m. Thus the polynomial (1) has the Schur property if nad only if

$$(4), \quad \alpha^k\beta^m < 0, \quad |\alpha| + |\beta| \geq 1, \quad |\alpha| - 1 \leq |\beta| \leq 1 + |\beta|.$$

To conclude the proof of (C2) we note that inequalities $|\alpha| - 1 < |\beta| < 1$ follows from (4).

Thus we can completely describe the problem of finding domain where (1) has the Schur property in terms of conditions presented in parts (C1) and (C2).

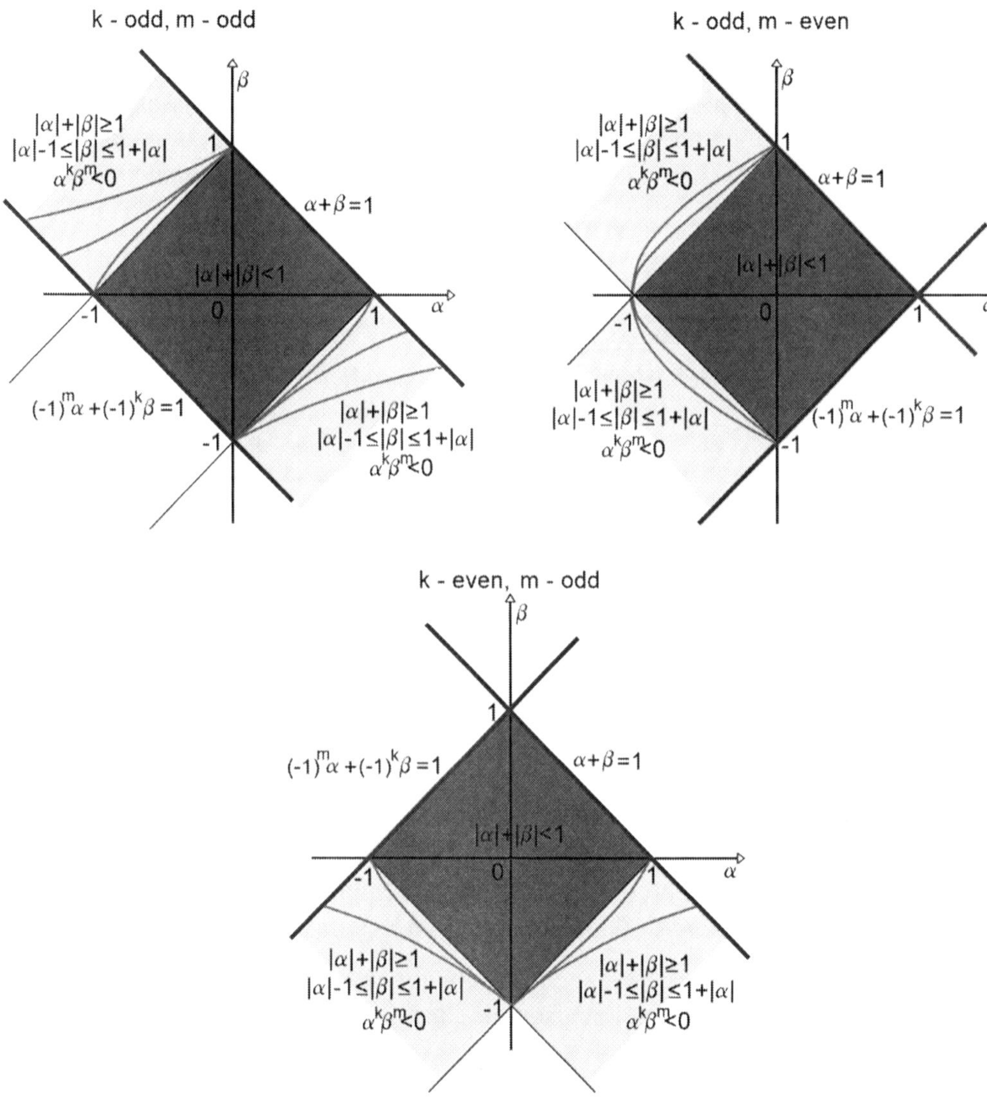

Fig. 2. Boundaries of domains in the (α, β)-plane, where (1) has the Schur property. The curves (9) (i.e. the curves (ii), (iv)) are plotted in green colour for various k, m

4 Final remarks

We have surveyed and compared conditions ensuring that (1) has the Schur property. Moreover, we have shown equivalence between two types of conditions. The presented conditions have applications to stability theory of linear difference equations. In particular, (1) is the characteristic polynomial of the linear trinomial difference equation

$$x(n) = \alpha x(n-m) + \beta x(n-k) \tag{13}$$

which serves as a model of some problems from population biology and as a discretization of some delay differential equations. Using some elementary properties of theory of linear time-invariant difference equations we get that conditions of Theorems 2.1, 2.4 and 2.6 are necessary and sufficient conditions for asymptotic stability of the zero solution of (13).

As the last remark we note that the study of polynomials and locations of their zeros is connected with many other mathematical problems in various branches of mathematics see [5, 6, 10, 11, 13]. Generalization and application of theorems presented in this paper will be a subject matter of the future research.

Acknowledgements

The author was supported within the project "VÝZKUMFVT (DZRO K-217)" supported by the Ministry of Defence, Czech Republic

References

[1] J. Čermák, J. Jánský, *Explicit stability conditions for a linear trinomial delay difference equation*, Appl. Math. Lett. 43 (2015), 56–60.

[2] J. Čermák, P. Tomášek, *On delay-dependent stability conditions for a three-term linear difference equation*, Funkcial. Ekvac. 57 (2014), 91–106.

[3] J. Čermák, *Stabiliy conditions for delay difference equations: a survey and perspectives*, Tatra mountains, 63 (2015)

[4] F. Dannan, *The asymptotic stability of* $x(n+k)+ax(n)+bx(n-\ell)=0$, J. Difference Equ. Appl. 10 (2004), 589–599.

[5] Hasilová, K. *Iterative Method for Bandwidth Selection in Kernel Discriminant Analysis*, In: 32nd International Conference Mathematical Methods in Economics MME2014. Olomouc: Palacký University, Olomouc, 2014, 263-268.

[6] I. Horová, J. Koláček, K. Hasilová, *Full bandwidth matrix selectors for gradient kernel density estimate*. Computational Statistics & Data Analysis, 2013, vol. 57, no. 1, p. 364-376.

[7] S.A. Kuruklis, *The asymptotic stability of* $x_{n+1}-ax_n+bx_{n-k}=0$, J. Math. Anal. Appl. 188 (1994) 719–731.

[8] M.M. Kipnis, R.M. Nigmatullin, *Stability of the trinomial linear difference equations with two delays*, Autom. Remote Control 65 (11) (2004) 1710–1723.

[9] M.M. Kipnis, R.M. Nigmatullin, *Stability of some difference equations with two delays*, Autom. Remote Control 64 (5) (2003) 782–790.

[10] V. Talhofer, A. Hofmann, S. Hoskova-Mayerova, *Application of Fuzzy Membership Function in Mathematical Models for Estimation of Vehicle Trafficability in Terrain*, APLIMAT 2015 - 14th Conference on Applied Mathematics, Proceedings, Slovak University of Technology in Bratislava, (2015), 711–719.

[11] R. Ameri, S. Hoskova-Mayerova, *Fuzzy Continuous Polygroups*, APLIMAT 2016 - 15th Conference on Applied Mathematics, Proceedings, Slovak University of Technology in Bratislava, (2016), 13–19.

[12] V.G. Papanicolaou, *On the asymptotic stability of a class of linear difference equations*, Math. Mag. 69 (1) (1996) 34-43.

[13] Potůček, R. *The sum of the series of reciprocals of the quadratic polynomials with double non-positive integer root*, APLIMAT 2016 - 15th Conference on Applied Mathematics, Proceedings, Slovak University of Technology in Bratislava, (2016), 919-925.

Current address

Jánský Jiří, Ing. RNDr. PhD.
Katedra matematiky a fyziky,
Fakulta vojenských technologií, Univerzita obrany
Kounicova 65, 662 10 Brno, Czech republic
Tel. number: +420 736 215 145, e-mail: jiri.jansky@unob.cz

Slovak University of Technology in Bratislava
Faculty of Mechanical Engineering
16th Conference on Applied Mathematics **APLIMAT 2017**
Proceedings

CONTROL CHARTS FOR PROCESSES WITH FLUCTUATING MEAN

JAROŠOVÁ Eva (CZ)

Abstract. Control charts for processes in which the variation of sample averages arise due to other causes than the inherent within-sample variation are examined. The control chart with extended limits and modified control chart are applied on a real industrial process and the operating characteristic and average run length are calculated. The analysis is based on the random-effect model.

Keywords: control chart with extended limits, modified control chart, random-effect model, operating characteristic, average run length

Mathematics Subject Classification: 62P30, 62-07

1 Introduction

Some modern industrial processes show behaviour that does not correspond to the classical assumptions on which the conventional Shewhart chart is based. One of them is the assumption of constant process mean. When samples are taken from a process, the observed within-sample variation is small in comparison with the distance of specification limits, however, the variation of sample averages is substantially larger than it would correspond to the random differences between items within samples. If all observations lie well within given specification limits, some fluctuation of the process mean is allowed unless non-conforming units are produced.

In ISO 21747 several so called time-dependent process models are distinguished and processes with the non-constant mean are classified as type C processes. In a more general meaning, a process may be considered in control if the process mean fluctuates between determined limits. To distinguish special causes from the allowed mean fluctuation, adjusted control limits that are wider than the classical three-sigma limits are constructed.

Two approaches can be used to adjust the X-bar control limits. The modified control chart and acceptance control chart, see e.g. [3], [6], [7], [9] are derived from specification limits, the control chart with extended limits in [1] or [2] is based only on process observations. The first two approaches contradict the aim of continuous process improvement by reducing its variation, however, it can be argued that once a high level of process capability ($C_{pk} > 2$, which is the aim of the Six Sigma methodology), has been achieved, the control limits can be relaxed. The third approach gives a scope to some improvement; major shifts from the reference value are signalled regardless how far from the specification limits they might be.

Information regarding to the properties of the charts with relaxed limits, such as the operating characteristic or average run length, is not available. Especially the properties of the extended control chart should be known when a process is to be improved.

The aim of the paper is to examine the performance of control charts with relaxed limits that have been designed for a real industrial process. Representatives of both approaches were chosen, namely the control chart with extended limits and the modified control chart.

2 The random-effect model

If samples are taken from a process the mean of which is not constant and changes randomly, observation x_{ij} in sample j can be represented by the model

$$x_{ij} = \mu_0 + a_j + \varepsilon_{ij} \tag{1}$$

where μ_0 is the grand process mean, $a_j \sim N(0, \sigma_A^2)$ is the random effect of sample j, and $\varepsilon_{ij} \sim N(0, \sigma^2)$ is the random error effect. The instantaneous distribution is normal with the mean $E(x_{ij}) = \mu_0 + a_j$ and variance σ^2, the outcoming distribution is also normal with the mean μ_0 and the variance given by $\sigma_x^2 = \sigma_A^2 + \sigma^2$.

The null hypothesis $H_0: \sigma_A^2 = 0$ corresponds to the assumption of constant mean μ_0. When k samples of size n are taken, test statistic $F = MSA/MSE$ has the F distribution with $k - 1$ and $k(n - 1)$ degrees of freedom. Sums of squares

$$SSA = n\sum_{j=1}^{k}(\bar{x}_j - \bar{\bar{x}})^2 \qquad SSE = \sum_{j=1}^{k}\sum_{i=1}^{n}(x_{ij} - \bar{x}_j)^2 \tag{2}$$

measure the between-sample and within-sample variation, respectively, and $SST = SSA + SSE$ is the total sum of squares (Tab. 1).

Source of variation	Sum of Squares	Df	Mean Square	F-ratio
Samples	SSA	$k - 1$	$MSA = SSA/(k - 1)$	MSA/MSE
Residual	SSE	$k(n - 1)$	$MSE = SSE/(k(n - 1))$	
Total	SST	$kn - 1$		

Tab. 1. ANOVA table

Rejection of H_0 implies that the variation of sample averages is significant and consequently, the classical control chart is unusable in this case. The variance components can be estimated by equating *MSA* and *MSE* to their expected values

$$E(MSA) = n\sigma_A^2 + \sigma^2 \qquad E(MSE) = \sigma^2 \tag{3}$$

and this yields

$$\hat{\sigma}_A^2 = \frac{MSA - MSE}{n} \qquad \hat{\sigma}^2 = MSE \tag{4}$$

The variance of the outcoming distribution is estimated by means of

$$\hat{\sigma}_x^2 = \hat{\sigma}_A^2 + \hat{\sigma}^2 \tag{5}$$

By taking square roots of the variances in (4) and (5) slightly biased estimates arise, nevertheless, they are used in further analysis.

3 Capability and performance indices

The performance of industrial processes is evaluated by indices which express the ability to produce within desirable specifications. The most common indices C_p and C_{pk} are given by the formulas [5]

$$C_p = \frac{USL - LSL}{6\sigma} \qquad C_{pk} = \min(C_{pU}, C_{pL}) \tag{6}$$

where

$$C_{pU} = \frac{USL - \mu}{3\sigma} \qquad C_{pL} = \frac{\mu - LSL}{3\sigma} \tag{7}$$

Their construction assumes the normal distribution of a quality characteristic with constant mean μ and standard deviation σ. If the process mean fluctuation is considered inherent to the process, the use of performance indices P_p and P_{pk} is recommended by the Automotive Industry Action Group. Standard deviation σ, which measures the within-sample variation, is replaced by the overall standard deviation σ_x. Both the capability and performance indices are implemented in statistical software products such as Minitab or Statgraphics. Several estimators of σ and σ_x can be chosen; the estimates based on the random-effect model are used in this study.

4 The control chart with extended limits

Limits for the true process mean are located at distance Δ from the centre line and the control limits are constructed at the three-sigma distance outwards from the limits for the process mean. They are

$$LCL = \bar{\bar{x}} - \frac{3\hat{\sigma}}{\sqrt{n}} - \Delta \qquad UCL = \bar{\bar{x}} + \frac{3\hat{\sigma}}{\sqrt{n}} + \Delta \tag{8}$$

where $\bar{\bar{x}}$ is the total average, σ denotes the within-sample standard deviation and n is the sample size. Based on the random-effect model the choice $\Delta = 1.5\,\hat{\sigma}_A$ is recommended in [2].

The ability of the chart to detect a shift in the mean is described by its operating-characteristic function and average run length (*ARL*). The operating characteristic curve describes the relationship between the probability of not detecting a shift from μ_0 to $\mu = \mu_0 + m\sigma_A$ on the first subsequent sample

$$\beta = P(LCL \le \bar{x} \le UCL \mid \mu = \mu_0 + m\sigma_A) \tag{9}$$

It seems reasonable for shifts in the grand mean to be expressed in multiples of σ_A instead of σ in the extended control chart. To calculate β, the distribution function of normal distribution $N(\mu, \sigma_A^2)$ is considered.

If the averages plotted on the chart can be considered independent, then the number of samples taken until the first point exceeds a control limit is a random variable with the geometric distribution G(1–β). The average run length is the expected value of this variable

$$ARL = \frac{1}{1-\beta} \tag{10}$$

5 The modified control chart

The modified control chart is suitable for so called six-sigma processes, i.e. processes with the fraction nonconforming lower than 3.4 ppm. Differently from the previous chart, where the extension is based on the observed mean's variation, the control limits of the modified chart are derived from the allowable process variation given by specification limits *USL* and *LSL*, and from the acceptable limit of a fraction non-conforming p_A (see e.g. [6], [7], [8] or [9]).

For the process fraction non-conforming to be less than p_A, the true process mean must be in the interval (μ_L; μ_U), where

$$\mu_L = LSL + u_{1-p_A}\sigma \qquad \mu_U = USL - u_{1-p_A}\sigma \tag{11}$$

The normal within-sample distribution is assumed but no assumptions are needed for sample mean distribution. Then the upper and lower control limits are located $\pm 3\sigma/\sqrt{n}$ from limits (11)

$$LCL = LSL + u_{1-p_A}\hat{\sigma} - \frac{3}{\sqrt{n}}\hat{\sigma} \qquad UCL = USL - u_{1-p_A}\hat{\sigma} + \frac{3}{\sqrt{n}}\hat{\sigma} \tag{12}$$

Values $u_{1-p_A} = 3$ or $u_{1-p_A} = 4.5$ are used in the references mentioned above. Since it may be cumbersome to choose a specific value p_A of the order less than 10^{-6}, the formula for the capability index C_{pk} can be used. Considering the region above the centre line and the process mean in its utmost value μ_U; the probability of exceeding USL is

$$P(X > USL) = 1 - \Phi\left(\frac{USL - \mu_U}{\sigma}\right) = 1 - \Phi(3C_{pU}) = p_A \tag{13}$$

Consequently $u_{1-p_A} = 3C_{pU}$. Probability $P(X < LSL)$ is approximately zero for $\mu = \mu_U$. Similarly, for the region below the centre line we get $u_{1-p_A} = 3C_{pL}$. The choice of $C_{pk} = 2$ should meet the aim of the Six Sigma approach, however, another preferable value of C_{pk} may be considered.

6 Application on a moulding process

In one of the manufacturing processes in Laird company frames from tin-coated cold-rolled steel are moulded. Subgroups of 5 frames are taken from the process and the frame length and other dimensions are measured optically by Micro-Vu Excel 4520 with the scale resolution of 0.5 μm. The process mean fluctuates over time, however, individual observations lie within the specified tolerance limits, which are USL= 35.1 mm and LSL = 34.9 mm. The conventional three-sigma control limits for sample averages are unusable, see Fig. 1, where the Shewhart chart based on 28 samples of size 5 is displayed. While the R chart indicates that the within-sample variation is in control, most of the points in the X-bar chart lie outside the control limits and obviously, all of them cannot be considered the signals of special causes.

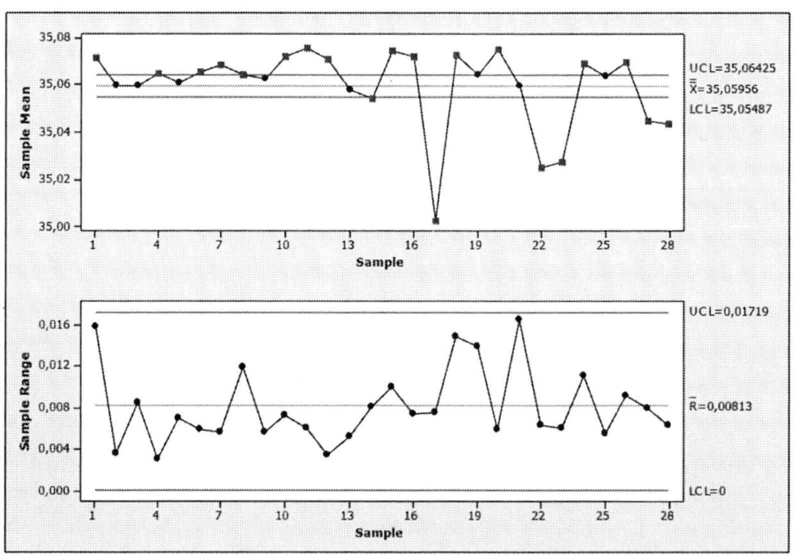

Fig. 1. Shewhart X-bar and R charts

Two approaches with different aims are applied:

1. The focus is on the statistical control of the process, which means reduction of variation and continuous process improvement. The extended control chart is constructed and its properties are examined.
2. As the process capability is sufficiently high, the role of a control chart is the process monitoring and the modified control chart is used. Unless the expected fraction nonconforming exceeds a specified level, no signals should occur.

6.1 Control chart with extended limits

Since the conventional X-bar chart cannot be used to identify outlying averages, outliers in the data were identified by the Grubbs test with the significance level of 0.05. As a result, samples number 17, 22, and 23 containing outliers were excluded from the further analysis.

The differences between the process levels at different times are significant at the 0.001 level, as follows from the ANOVA table (Tab 2)

Source of variation	Sum of Squares	Df	Mean Square	*F*-ratio	P-value
Subgroups	0.008785	24	$3.66\cdot10^{-4}$	28.46	0.000
Residual	0.001286	100	$1.29\cdot10^{-5}$		
Total	0.010071	124			

Tab. 2. ANOVA table after 3 samples were excluded

The probability plot of residuals from ANOVA in Fig. 2 and the p-value of 0.229 of the Anderson-Darling test do not contradict to the assumption of the normal instantaneous (within-sample) distribution of the process.

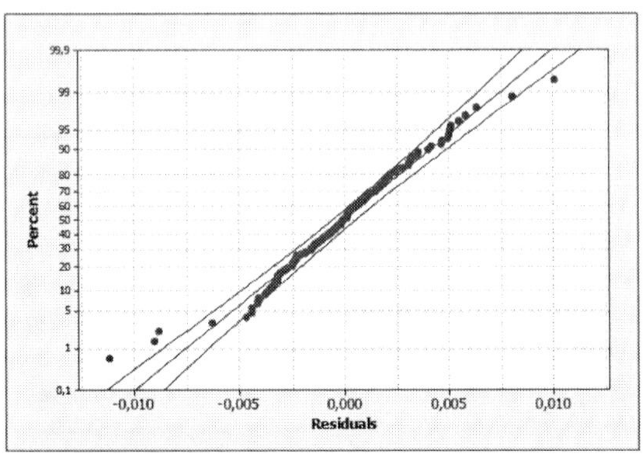

Fig. 2. Normal probability plot for residuals

The total average is $\bar{\bar{x}} = 35.0645$, the estimated variance components obtained through ANOVA are $\hat{\sigma}_A^2 = 7.06\cdot10^{-5}$ and $\hat{\sigma}^2 = 1.29\cdot10^{-5}$. The extended control limits according to Eq. (3) are UCL = 35.082, LCL = 35.047 (Fig. 3).

The trial control limits obtained from the gathered data help to decide whether the process was in control over the time period of collecting the data. Since two last points below the lower limit indicate that some special cause might exist, further two samples were dropped to determine process characteristics under relatively stable conditions.

The new total average based on 115 observations is $\bar{\bar{x}} = 35.066$, estimated variance components are $\hat{\sigma}_A^2 = 3.54\cdot10^{-5}$ and $\hat{\sigma}^2 = 1.33\cdot10^{-5}$, and the extended control limits according to Eq. (3) are UCL = 35.080, LCL = 35.052. The resulting X-bar chart with original 28 sample averages is displayed in Fig. 4.

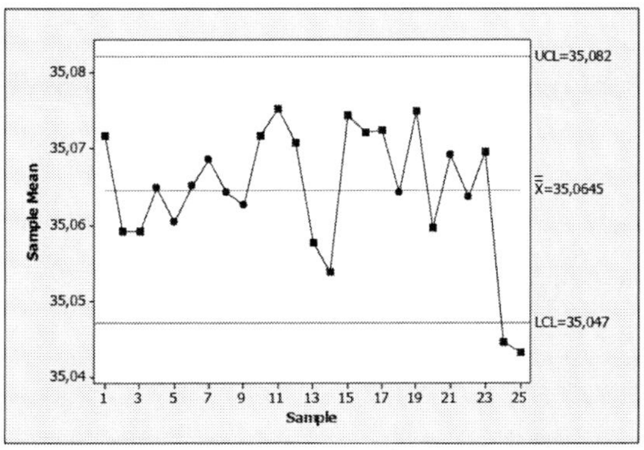

Fig. 3. Control chart with trial extended limits

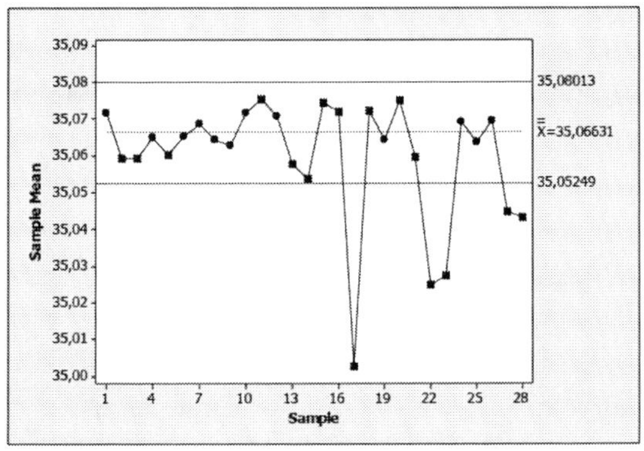

Fig. 4. Control chart with revised extended limits

The assumption of a normal instantaneous distribution was verified above (Fig. 2), normal probability plots for sample averages and for the outcoming distribution (both without 5 excluded samples) displayed in Fig. 5 and Fig. 6 indicate that both the sample averages and individual observations follow approximately a normal distribution (p-values of the Anderson-Darling test are 0.303 and 0.116).

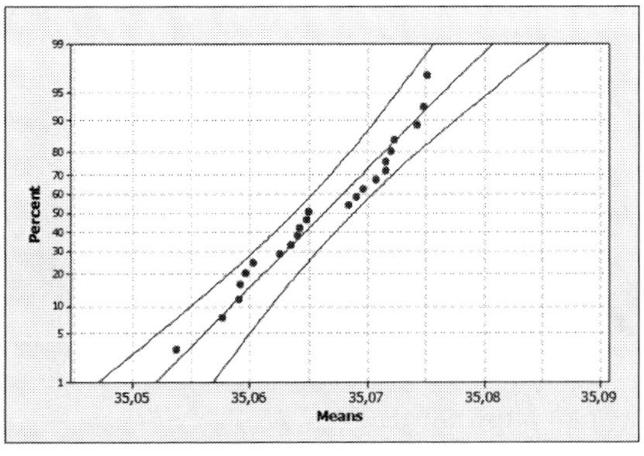

Fig. 5. Normal probability plot for sample averages

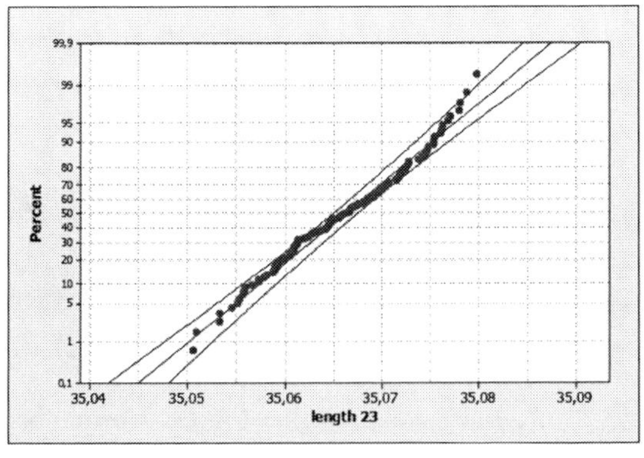

Fig. 6. Normal probability plot for observations

For different shifts $\delta = m\sigma_A$ off the centre line values of the operating characteristic β and ARL were calculated. To assess the effect on the process ability to meet specification limits, indices C_{pU} and P_{pU} as well as the expected fraction nonconforming were estimated (Tab. 3). The following parameter estimates were used

$$\bar{\bar{x}} = 35.066 \qquad \hat{\sigma}_A = 5.95 \cdot 10^{-3} \qquad \hat{\sigma} = 3.65 \cdot 10^{-3} \qquad \hat{\sigma}_x = 6.98 \cdot 10^{-3}$$

As can be seen from Fig. 7, indices C_{pU}, which correspond to the red curve, do not reflect the real ability to yield output within the specification limits.

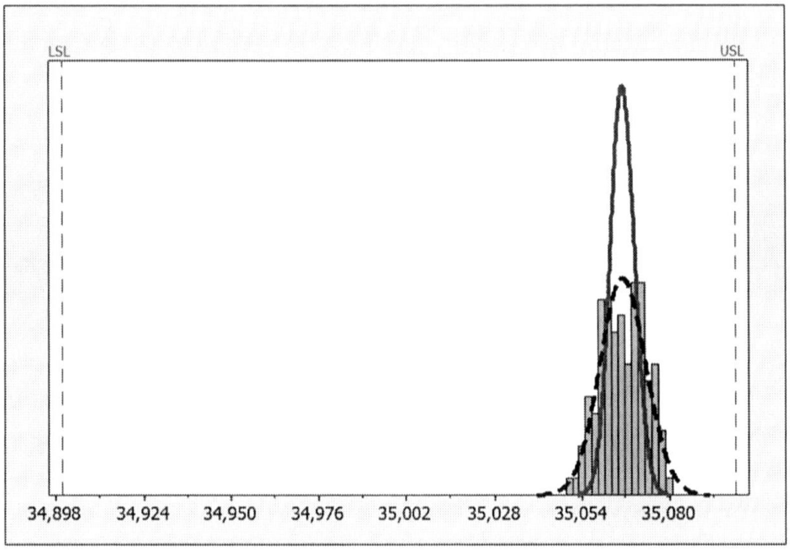

Fig. 7. Capability plot

To compare the relative quickness of detection a shift with the conventional Shewhart chart, the utmost right column was included. In this case the shifts of $m\sigma$ are considered.

m	β	ARL	$\hat{C}_{pU}$	$\hat{P}_{pU}$	$\hat{p}_{out,\sigma_x}$	ARL_{Sh}
0	0.97986	49.7	3.08	1.61	$6.92 \cdot 10^{-7}$	370.4
0.5	0.96354	27.4	2.80	1.47	$5.38 \cdot 10^{-6}$	33.4
1	0.90677	10.7	2.53	1.33	$3.52 \cdot 10^{-5}$	4.5
1.5	0.79491	4.9	2.26	1.18	$1.93 \cdot 10^{-4}$	1.6
2	0.62695	2.7	1.99	1.04	$8.95 \cdot 10^{-4}$	1.1
2.5	0.43008	1.8	1.72	0.90	$3.50 \cdot 10^{-3}$	1.0
3	0.24946	1.3	1.45	0.76	$1.16 \cdot 10^{-2}$	1.0

Tab. 3. Properties of the extended control charts

As $ARL(0)$ is more than seven times shorter than that of the conventional Shewhart chart, the risk of false signal is quite high, i.e. $ARL(0) = 1 - \beta(0) = 0.02$. The shifts which can be detected reasonably quickly with the extended chart are of $2.5\sigma_A$ or more, while the conventional chart is able to detect shifts of 1.5 σ or more. Since the geometric distribution is skewed, it may take a long time than the fall of $\hat{P}_{pU}$ from 1.61 to 1.18 corresponding to the upward shift of $1.5\sigma_A$ is detected. To assure a sufficient quality level, the process should be adjusted towards the centre of the specification limits.

6.2 Modified control chart

Since only within-sample variation is examined, none of the samples was excluded.

Source of variation	Sum of Squares	Df	Mean Square	*F*-ratio	P-value
Subgroups	0.039375	27	$1.46 \cdot 10^{-3}$	119.95	0.000
Residual	0.001362	112	$1.22 \cdot 10^{-5}$		
Total	0.040737	139			

Tab. 4. ANOVA table for all 28 samples

The estimated pooled standard deviation is $\hat{\sigma} = 0.0035$. For chosen $C_{pk} = 2$ and $u_{1-p_A} = 6$ the upper and lower control limits become $UCL = 35.084$ and $LCL = 34.916$ (Fig. 8)

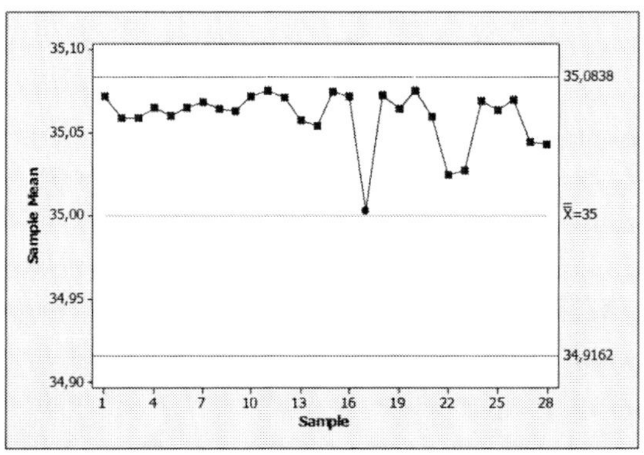

Fig. 8. Modified control chart

The modified control limits do not depend on the actual process location but *ARL* is affected by the grand mean. The distance between the upper control limit and the total average is larger than it was at the extended control chart, which means that values of *ARL* will be higher. Although it is a welcome change in $ARL(0)$ since it increased to 681.3, the detection of shifts is slower than before. After the shift of $1.5\sigma_A$ the fraction nonconforming will correspond to 1.18, which is the value of P_{pU}, instead of the chosen value of 2 before the process is adjusted. It follows that the choice of $u_{1-p_A} = 6$ is not sufficient for the given process mean.

7 Conclusion

It should be noted that other ways of calculating the extended control limits are possible; in [1] the observed sample average variation is considered. However, the method based on the ANOVA model is preferable since it makes use of a larger amount of information.

As it was mentioned earlier, each of the charts described in the paper serves to different aims. Considering the data from the moulding process with C_{pk} = 3.08 and P_{pk} = 1.61 it is pointless to endeavour to improve the process and therefore the modified control chart seems to be more suitable in this case. The extended control chart was shown as an illustration of steps that are needed to be performed in the case of less capable processes.

It can be concluded that in in this study the properties of the extended control chart are much worse than the properties of the conventional Shewhart chart if the shifts expressed in the same multiples of σ_A for the former chart and σ for the latter chart are considered. The low value of $ARL(0)$ is mainly due to the choice $\Delta = 1.5\,\sigma_A$. However, the decrease of ARL with the increasing shift in the mean is relatively slow; the shift of $1.5\sigma_A$ will be detected after 4.7 samples on average in comparison with 1.6 samples in the conventional chart if samples of size 5 are considered. This study confirmed that some scepticism about the use of these charts is justified as some assignable causes may not be detected or are detected too late (see e.g. [10]). Nevertheless, control charts with better properties have not been published so far.

References

[1] BISSELL, D. *Statistical Methods for SPC and TQM.* Chapman & Hall, 1994, ISBN 0-412-39440-5, 373 p.

[2] DIETRICH, E., SCHULZE, A. *Statistical Procedures for Machine and Process Qualification.* Hanser, 2010, ISBN 978-3-446-42249-0, 396 p.

[3] HILL, D.: Modified Control Limits. *Applied Statistics*, 1956, vol. 5, no. 1, p. 12–19.

[4] ISO 21747: Process performance and capability statistics for measured quality characteristics. ISO 2006

[5] KOTZ, S., JOHNSON, N. L.: Process capability indices: A Review, 1992-2000. *Journal of Quality Technology*, 2002, vol. 34, no. 1, p. 2–19

[6] MITRA, A.: *Fundamentals of Quality Control and Improvement.* John Wiley & Sons, 2008, ISBN 978-0-470-22653-7, 700 p.

[7] MONTGOMERY, D. C. *Introduction to Statistical Quality Control.* J. Wiley & Sons, 2013, ISBN 978-80-247-5355-3, 734 p.

[8] OTT, E. R.: *Process Quality Control. Troubleshooting and Interpretation of Data.* ASQ Quality Press, 2005, ISBN 978-0873896559, 628 p.

[9] WADSWORTH, H. M., STEPHENS, K.S., GODFREY, A.B. *Modern Methods for Quality Control and Improvement.* John Wiley & Sons, 2002, ISBN: 978-0-471-29973-8, 704 p.

[10] WHEELER, D. J.: *Advanced Topics in Statistical Process Control.* 2nd ed. SPC Press, 2004, ISBN 978-0-945320-63-0, 470 p.

Current address

Jarošová Eva, doc. Ing. CSc.
ŠKODA AUTO University
Na Karmeli 1457, 293 01 Mladá Boleslav, Czech Republic
Tel. number:+420 732 469 892, e-mail: jarosova@is.savs.cz

Slovak University of Technology in Bratislava
Faculty of Mechanical Engineering
16th Conference on Applied Mathematics **APLIMAT 2017**
Proceedings

FAST ALGORITHM FOR FINDING CLIQUE NUMBER OF SCALE-FREE NETWORKS AND SPARSE GRAPHS

JURSA Andrej (SK)

Abstract. In the graph theory the maximum clique problem is NP hard. There are many algorithms attempting to solve it. In this paper we present a new algorithm specially targeted for solving maximum clique problem in scale-free networks and sparse graphs. We use the general algorithm developed by Östergård in 2002 as a base for our new algorithm. In our algorithm the search process has been improved by an effective initial heuristic. First, preliminary clique size is computed and then the graph is reduced by the k-core decomposition, where k is the size of the preliminary clique. Subsequently, greedy algorithm for coloring is used upon induced subgraph of selected vertex. Our new algorithm solves maximum clique problem for an arbitrary graph, but together with the above mentioned and some other, less significant pruning techniques, the algorithm performs exceptionally well on scale-free networks and sparse graphs. Its functionality has been tested on many real graphs as well as randomly generated graphs.

Keywords: clique number, scale-free networks, sparse graphs, k-core decomposition, graph coloring

Mathematics subject classification: Primary 05C85; Secondary 05C69, 05C82

1 Introduction

This paper has been inspired by our studies of functional brain networks [8]. In these studies, our aim is to find an appropriate set of statistical measures able to distinguish among three sets of functional brain networks of young healthy participants, elderly healthy participants and elderly participants with Alzheimer disease. One of these measures is a maximum clique.

We realize that using a general algorithm for this task may lead to very long computation times. But due to the scale-free structure of functional brain networks, we can implement several pruning techniques to make the search process significantly faster. There exist randomized or heuristic algorithms like [9] or [11] which can give solution quickly, but for our study we need an algorithm which can give an exact solution in the shortest possible time. For solving problems of maximum clique in graphs there are many other studies like [6, 3, 7]. Our new algorithm is based on the exact algorithm created by Östergård [10] and it is improved version of our previous algorithm [5]. Our algorithm is usable on all types of simple undirected graphs but it works best on scale-free networks and other types of sparse graphs.

This paper is organized as follows. In the next section we will describe some preliminaries and definitions. In section three we will show basic idea of our algorithm. The fourth section is devoted to the original Östergårds maximum clique algorithm. The fifth section describes the details of our new algorithm. The sixth section summarizes our results and achievements on various types of graphs. Last section describes the conclusion of our work.

2 Preliminaries

Our functional brain networks are undirected simple graphs, therefore we limit ourselves to this type of graphs. Undirected graph $G = (V, E)$ consists of V set of vertices and E set of edges. We say that two vertices are adjacent, if there is an edge between them. The degree k of the vertex v is a number of edges connecting v to its adjacent neighbors. For the vertex v we define two sets: an open set of neighbors $N(v) = \{u | u \in V \land (u, v) \in E\}$ and a close set of neighbors $N[v] = N(v) \cup \{v\}$.

Simple unweighted graph does not have loops and multiple edges. The density of the such simple unweighted and undirected graph is defined as

$$D = \frac{2\,|E|}{|V|\,(|V| - 1)} \tag{1}$$

A clique in a simple undirected graph is a subset of vertices $V' \subseteq V$ such that

$$\forall v \in V', \forall u \in V' : u \neq v \implies (v, u) \in E \tag{2}$$

The maximum clique is the largest clique in the graph. The size of the maximum clique is a clique number $\omega(G)$ of the graph G.

Scale-free networks in our case are simple undirected graphs having power law degree distribution, which means that a fraction $P(k)$ of vertices having the degree k scales as

$$P(k) \propto C.k^{-\gamma} \tag{3}$$

Here γ is a scaling exponent typically in the range $2 < \gamma < 3$ and C is a constant. For the given graph G the degree distribution is easily measured and visualized in a log-log plot (Fig 1 (a)).

The k-core decomposition of the graph G forms an induced subgraph H, in which vertices have the degree at least k. The k-core decomposition of the sparse graph removes large amount of vertices with the degree lower than k. In Fig. 1 (b) the red area represents all

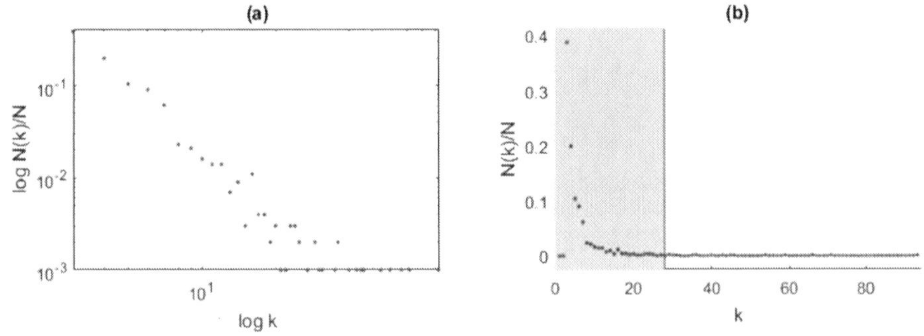

Fig. 1. Scale-free network degree distribution in the log-log plot (a) and vertices potentially removed by k-core decomposition, where $k = 0.3 \cdot k_{max}(G)$. Here k_{max} is a maximum degree in the network (b)

vertices removed from the scale-free network due to the k-core decomposition. Our algorithm is using k-core decomposition for reducing set of vertices composed from neighbors of starting node. This is one of improvements which leads to shorter processing times.

We also using pruning technique of vertex coloring by greedy color assignment in descending order of the vertex degrees. This coloring is used on induced subgraph of starting vertex and its neighbors upon k-core decomposition. By comparing the chromatic number against best clique number known during computation this allows the algorithm to decide if it is reasonable to branch over selected subset of vertices or not.

3 Main idea of algorithm

The main very basic skeleton of our algorithm is following:

1. Run heuristic $\rightarrow$ preliminary clique size (max).
2. Iterate n times from higher to lower indexes:
   a) Select vertex v_i and extract induced subgraph consisting of $N[v_i]$ vertices $\rightarrow G_{v_i}$.
   b) Decompose G_{v_i} using max-core decomposition $\rightarrow$ decomposed graph $G_{v_i}^{max}$.
   c) If $\left|V(G_{v_i}^{max})\right| > max \wedge \chi(G_{v_i}^{max}) > max$ then branch over $G_{v_i}^{max}$ and possibly find new max.

As a result verified by many tests we have created algorithm which performs very effective on most graphs but it performs best on scale-free networks or other sparse graphs.

4 Revision of Östergårds algorithm

The original algorithm of Östergård [10] processes an indexed graph in witch the vertices are arbitrary indexed from 1 to N. Here N denotes number of nodes. The algorithm starts with vertex having highest index and proceeds from the highest to the lowest one. During this process an array called $c[i]$ is maintained, in which i is the index of already processed vertex. Each item of array c stores maximum clique size in which vertex v_i participates. Currently highest clique size is stored in the global variable *max*.

The values of array c changes as follows:

- $c[i] = c[i+1]$ - if the processing of vertex i does not change the value of *max*
- $c[i] = c[i+1] + 1$ - otherwise

For each v_i the algorithm uses the set $S_i = \{v_j | i \leq j \leq |V|\}$. This set contains all nodes with the indices $\geq i$. S is important in order to pick the neighbors of vertex v_i with the higher index than i.

The algorithm consists of two procedures. The first one is MAIN in which the algorithm is iterating over all vertices in descending order of their indices. The vertex selected in each iteration becomes a root vertex v_i and its neighborhood is selected for branching. The neighborhood of root vertex is called U and is determined as $S_i \cap N(v_i)$ so it takes only neighbors with higher indices. When U is known the algorithm runs the second procedure called CLIQUE.

The second procedure CLIQUE obtains two parameters. The first one is U, the set of neighbors of the root vertex. The second parameter is called $size$ which contains the size of a clique formed over root vertex and its neighborhood. First call of CLIQUE procedure starts with $size = 1$ because the root vertex is the first vertex of the clique. The CLIQUE procedure is branching over U and searching for $size > max$.

In the **Algorithm 1** the pseudocode of the Östergårds algorithm is presented. The algorithm is starting in procedure MAIN (line 24) where the global variable max is initialized to zero and then the iteration over vertices starts. One can see that each time the procedure CLIQUE is finished, the max value is stored in $c[i]$. There is also a global variable $found$ which is set to $false$ in each iteration.

Procedure CLIQUE is searching for clique size by branching over U. When U is empty (line 2) and also $size > max$ (line 3) the variable max is changed and the variable $found$ is set to $true$. This causes return of the control into MAIN procedure.

Branching over U is done in a loop. In each iteration the new vertex v_i is selected while i is the minimum index in U. There are two pruning tests:

1. if $size + |U| \leq max$ the algorithm knows that there is not enough vertices in U to construct clique with $size > max$
2. if $size + c[i] \leq max$ then by choosing vertex v_i as next one for branching does not help to find a clique with $size > max$

Algorithm 1 Östergårds original algorithm.

Input: Simple undirected graph G.
Output: Clique number $\omega(G)$ in global variable *max*.

1: **procedure** CLIQUE(U, *size*)
2:     **if** $|U| = 0$ **then**
3:         **if** *size* $>$ *max* **then**
4:             *max* $\leftarrow$ *size*
5:             *found* $\leftarrow$ **true**
6:         **end if**
7:         **return**
8:     **end if**
9:     **while** $U \neq \emptyset$ **do**
10:         **if** *size* $+ |U| \leq$ *max* **then**
11:             **return**   ▷ Nothing better can be found.
12:         **end if**
13:         $i \leftarrow \min\{j | v_j \in U\}$
14:         **if** *size* $+ c[i] \leq$ *max* **then**
15:             **return**   ▷ Nothing better can be found.
16:         **end if**
17:         $U \leftarrow U \setminus \{v_i\}$
18:         CLIQUE($U \cap N(v_i)$, *size* $+ 1$)
19:         **if** *found* $=$ **true then**
20:             **return**   ▷ Higher clique size was found, return to main loop.
21:         **end if**
22:     **end while**
23: **end procedure**
24: **procedure** MAIN
25:     *max* $\leftarrow 0$
26:     **for** $i \leftarrow |V|$ **downto** 1 **do**
27:         *found* $\leftarrow$ **false**
28:         CLIQUE($S_i \cap N(v_i)$, 1)
29:         $c[i] \leftarrow$ *max*   ▷ Store clique size for next iteration, this is used for pruning.
30:     **end for**
31: **end procedure**

In the process of branching on the same level, U is reduced in size in each iteration (line 17). When one of the above mentioned tests passes for some $v_i \in U$ then it will pass also for all $v_j \in U$ where $j > i$.

Branching itself is done by calculating new neighbor set as intersection $U \cap N(v_i)$ where v_i is the vertex chosen during branching on current level. The CLIQUE procedure is called recursively with new neighbor set and with value of the parameter *size* incremented by one.

This algorithm can be used for arbitrary simple undirected graph and it finds an exact maximum clique size.

5 New algorithm

Our new algorithm is using the original Östergårds CLIQUE subroutine (from **Algorithm 1**). It is separated into two phases. The first phase is using greedy function [4, 11] to find quickly some preliminary clique number of the input graph. We call this heuristic phase. The second phase is responsible for finding the exact clique number of the input graph, we refer to this one as the search phase.

The search phase is using preliminary solution found in the heuristic phase to improve it or to approve it, in case the heuristic phase have found the real clique number. To achieve this, the search phase incorporates the greedy coloring algorithm with k-core decomposition, both performed on the neighbor set of selected root vertex. The same principles as in the original Östergårds algorithm are present, the graph is processed from the the highest index to the lowest one and the neighbor set is selected just for the vertices with higher indices that the selected root vertex has.

5.1 The heuristic phase

The first heuristic phase is mainly using the greedy vertex selection to iterate over the vertices of the input graph. The vertices wich have degree high enough are selected as root vertices. The algorithm searches in the neighborhood by greedy selection (procedure CLIQUEHEURISTICSTARTINVERTEX from **Algorithm 2**).

To determine the root vertices, the algorithm uses degree threshold which is set to 95% of the maximum degree in the graph. This value was found by experimenting with multiple settings and it is sufficient to select a relatively small subset of vertices for scale-free networks, for which this algorighm is optimized.

In the pseudocode of **Algorithm 2** the first function which is called is CLIQUEHEURISTIC. This function takes the graph G and determines a lower limit for the degree $k_{threshold}$ (line 15). This threshold is used to select that vertices, which have degree higher or equal to $k_{threshold}$. For each such root vertex it calls CLIQUEHEURISTICSTARTINVERTEX function, which builds the clique by greedy selection.

In the CLIQUEHEURISTICSTARTINVERTEX function the first vertex is put into the $Clique$ set and its neighbors are put into the Nb set. In the process the algorithm picks a vertex v from Nb set which has highest degree. This vertex is then put into $Clique$ set and Nb is reduced to $Nb \cap N(v)$. This process is repeated until Nb is empty. There is one pruning incorporated. The heuristic keeps the highest size of the found clique, so it is able to stop early if the size of the $Clique$ and of the Nb is lower or equal the known size.

The result of the heuristic search is stored to the global variable max. In some cases this can be the real clique number of the input graph. But in typical case the result is arbitrary close to the real clique number. There is no approximation factor for this algorithm, that's why we call it heuristic.

Having the preliminary solution it is wise to decompose graph and reduce sizes of its vertex and edge sets. To do that the simple k-core decomposition is used, where the value of k is equal to the value of max. For scale-free networks or other sparse graphs this can greatly reduce the search space for the next phase.

The k-core decomposition algorithm is presented in **Algorithm 3**. This function uses queue

Algorithm 2 Preliminary clique size heuristic function.

Input: Simple undirected graph G.

Output: Preliminary clique size in graph G.

```
function CLIQUEHEURISTICSTARTINVERTEX(G, i, lastMax)
    Clique ← {v_i}, Nb ← N(v_i)                 ▷ Initialize clique and neighbor sets.
    while true do
        if |Clique| + |Nb| ≤ lastMax then       ▷ Higher clique size can not be found.
            return lastMax
        end if
        if |Nb| = 0 then                        ▷ All possible vertices are processed.
            return |Clique|
        end if
        j ← index of vertex w_j ∈ Nb which have maximum degree in Nb
        Clique ← Clique ∪ {v_j}, Nb ← Nb ∩ N(v_j) ▷ Update clique and neighbor sets.
    end while
end function
function CLIQUEHEURISTIC(G)
    k_threshold ← ⌊0.95 · k_max⌋
    heurMax ← 0
    for i ← 1 to |V| do
        if GETDEGREE(G, v_i) ≥ k_threshold then
            max ← CLIQUEHEURISTICSTARTINVERTEX(G, i, heurMax)
            heurMax ← max {max, heurMax}
        end if
    end for
    return heurMax
end function
```

of vertices, which does not contain duplicates. First, it iterates over all vertices in graph and add each vertex with degree lower than k into the end of this queue. Then it polls vertices from the start of this queue and tests if they have degree lower than k. If this test passes for the current vertex it adds each neighbor of that vertex to the end of the queue (without duplicates) and then deletes the current vertex. As a result it returns a reduced graph, where each vertex have degree at least k or this returned graph can be empty.

The k-core decomposition is not the last step in this phase. Just before the algorithm steps into the search phase it re-indexes the graph nodes according to degree in the decomposed remnant of the input graph. This guarantees that vertex with lower index have the lower degree as well. It is useful because in the search phase the algorithm iterates over vertex indices from the higher to the lower one. Only neighbors of the root vertices with higher index as the root vertex has are selected. As a result of such procedure, the vertices with the higher indices have only small neighbors sets to be searched.

This algorithm returns the clique number of input graph. If one wants to know the vertex indices of the nodes in maximum clique of the original graph, one can store them in the process and use mapping table from re-indexed values to the initial ones.

The final step of the heuristic phase is to compute degeneracy of the reduced graph. It is also

Algorithm 3 k-core decomposition of graph G.

Input: Simple undirected graph G, desired value k.
Output: k-core decomposition of graph G.

function COMPUTEKCORE(G, k)
    $queue \leftarrow$ empty vertex queue   ▷ This queue can't contain duplicates.
    for all $v \in V(G)$ **do**
        if GETDEGREE(G, v) $< k$ **then**
            ADDLAST($queue, v$)
        end if
    end for
    while $queue$ is not empty **do**
        $u \leftarrow$ POLLFIRST($queue$)
        if GETDEGREE(G, u) $< k$ **then**
            ADDLASTALL($queue$, NOTINQUEUE($queue, N(u)$))   ▷ By deletion of u from G degree of each neighbor of u will be lowered by 1, so test it again.
        end if
        DELETEVERTEX(G, u)
    end while
    return G
end function

the degeneracy of the input graph as well. This is done by computation of higher and higher k-core decomposition where the degeneracy of the graph d is the last d-core decomposition. Following $(d + 1)$-core is empty graph. Having the value of d we know there is no greater clique than $d + 1$ clique.

5.2 The search phase

The search phase is the main phase of our algorithm. Thanks to the heuristic phase the preliminary max is already known. Also the reduced graph is available for processing with re-indexed vertices. The search phase iterates over the vertex set of the reduced graph. The iterations run from the higher index of the root vertex to the lower one. For each selected root vertex with index i, and its neighborhood composed just from the neighboring vertices with higher indices than i, the algorithm computes an induced subgraph and then performs several tasks on it. The induced subgraph is called the root subgraph.

First operation on the root subgraph is another k-core decomposition due to the knowledge of current max ($k = max$). This can yield to a reduced root subgraph which can have number of vertices lower or equal to max. If this case occurs it is clear that algorithm can't find a new clique of size $max + 1$ here, so it just set the $c[i] = max$ and continues to the next root vertex. Otherwise it performs the next operation.

Algorithm 4 Greedy algorithm used to find minimum colors needed to color vertex i and it's neighborhood.

Input: Simple undirected graph G, index of vertex i.
Output: Suboptimal number of minimum colors needed to color v_i and it's neighborhood.

function GETMINCOLORS(G, i)
    $colors \leftarrow 1, colorMap[v_i] \leftarrow 1$
    $Neighbors \leftarrow N(v_i), AllNodes \leftarrow Neighbors \cup \{v_i\}$
    for all unprocessed $u \in Neighbors$ having highest degree **do** ▷ Iterate over neighbors u of vertex v_i.
        $N_u \leftarrow N(u) \cap AllNodes$ ▷ Assemble set of neighbors of u within neighbors of v_i.
        $minColor \leftarrow 1, UsedColors \leftarrow \{\}$
        for all $w \in N_u$ **do** ▷ Iterate over all neighbors w of vertex u ...
            if $colorMap[w]$ is already set **then** ▷ ... and collect all colors already used.
                $UsedColors \leftarrow UsedColors \cup \{colorMap[w]\}$
            end if
        end for
        for $color \leftarrow 1$ **to** $colors + 1$ **do**
            if $color \notin UsedColors$ **then** ▷ New unused minimum color is found.
                $minColor \leftarrow color$
                break
            end if
        end for
        $colorMap[u] \leftarrow minColor$
        $colors \leftarrow \max(colors, minColor)$ ▷ Remember the highest color number used.
    end for
    return $colors$
end function

If the reduced root subgraph is sufficiently large then the greedy graph coloring is performed (**Algorithm 4**). Clearly the selected index i is the one, which have higher or equal degree as the rest of the vertices. So the vertex v_i gets the color 1, and the rest of the vertices from the reduced root subgraph will take colors with higher numbers. First all these remaining vertices are put into sorted list by their degree so that the first vertex have the highest degree. Note that the degrees are computed from the reduced root subgraph not from the input graph, so they are smaller and accurate. The greedy coloring iterates over this sorted list picking vertices one by one and attaching lowest possible color numbers with constraint that the two vertices connected by an edge can't have the same color. This process does not find the optimal coloring, but the minimum number of colors needed is not higher than number returned by this procedure.

It is known that the chromatic number $\chi(G) \geq \omega(G)$ [13]. According to this fact, if the greedy coloring function returns number of colors lower or equal to max, then branching over the reduced root subgraph is not necessary. We know that the exact minimum number of colors is lower or equal to this approximate result. If approximate result is lower or equal to the max, then algorithm sets again the $c[i] = max$ and continues with the next root vertex.

If the reduced root subgraph passes the coloring test then the algorithm determines set of neighbors of root vertex. It is equal to the vertex set of reduced root subgraph without vertex v_i. Then it calls Östergårds CLIQUE subroutine to branch over this set. As a result the process of branching may find $max(t+1) = max(t)+1$ where t represents time, which will be also new maximum clique size for the input graph found in the process of the search phase. When CLIQUE ends then $c[i] = max$ is set.

When all vertices in reduced graph are processed by the search phase, the clique number is in the global max variable. The pseudocode for our algorithm can be seen on **Algorithm 5**.

Algorithm 5 New fast algorithm for finding clique number of simple undirected graphs.

Input: Simple undirected graph G.
Output: Clique number $\omega(G)$ in variable max.

procedure MAIN
  $max \leftarrow$ CLIQUEHEURISTIC(G) ▷ From **Algorithm 2.**
  $G_{reduced} \leftarrow$ COMPUTEKCORE(G, max) ▷ From **Algorithm 3.**
  $d \leftarrow$ GETDEGENERACY($G_{reduced}$)
  $G_{sorted} \leftarrow$ SORTVERTICES($G_{reduced}$) ▷ Will sort graph vertices according to degree.
  for $i \leftarrow \max \arg V(G_{sorted})$ **downto** $\min \arg V(G_{sorted})$ **do**
    if $max = d+1$ **then**
      break ▷ Can't find clique with higher size than degeneracy plus one.
    end if
    $G_{root} \leftarrow$ GETINDUCED(G_{sorted}, $N[v_i] \cap S_i$) ▷ Where v_i is root vertex.
    $G_{reducedRoot} \leftarrow$ COMPUTEKCORE(G_{root}, max)
    if $|V(G_{reducedRoot})| > max$ **then**
      if GETMINCOLORS($G_{reducedRoot}$, i) $> max$ **then** ▷ From **Algorithm 4.**
        CLIQUE($V(G_{reducedRoot}) \setminus \{v_i\}$, 1) ▷ From **Algorithm 1.**
      end if
    end if
    $c[i] \leftarrow max$
  end for
end procedure

---

6 Tests and results

In this section we will discuss about tests of our new algorithm and compare them to original Östergårds algorithm. Sometimes we will refer to Östergårds algorithm as the old algorithm. At the end of this section we will discuss about possible future improvements to our new algorithm.

6.1 Tests of our algorithm

Our algorithm have been tested on several different datasets:

- Real data:
  - FMRi data of functional brain networks.

  - Simple undirected graphs from the database on https://snap.stanford.edu/data/.

- Artificial data:
  - Some graphs of DIMACS dataset.
  - Barabási-Albert networks [1].
  - Watts–Strogatz networks [12].
  - Bernoulli graphs [2].

We have implemented our algorithm in Java 1.8 programming language and all tests were performed on computer with Intel core i7 4790K@4.0GHz processor with 32GB of DDR3 RAM. The implementations of our algorithm as well as the Östergårds algorithm are single-threaded.

Table 1 contains the results for tests of our algorithm while table 2 contains comparative results for our new algorithm and original Östergårs algorithm as well as number of calls to CLIQUE subroutine. There are several graphs denoted by name in column G. First graph „C125.9" is from DIMACS dataset. It is the smallest graph from this dataset and it is not representing scale-free network or sparse graph. Having just 125 vertices with 6963 edges it have density around 0.898, our algorithm needed 0.006 seconds in heuristic phase and 1513 second in search phase to determine clique number of 34. This value of clique number is also known from DIMACS dataset website so both results matches. The real computation time is 4 times better than the original Östergårds algorithm needed, which was 6309 seconds. From table 2 we can also see that CLIQUE subroutine was called just for 60 root vertices thanks to heuristic phase and prunings in search phase. Whole virtual search tree has over one billion nodes, which is caused by high density of graph.

Graphs denoted as „BG_n_q" represents random graphs generated by Wolfram's Mathematica Bernoulli graph distribution method, where n is number of vertices and $p = \frac{q}{100}$ is probability of edge between any two vertices. The density D is proportional to p here. As it can be seen the algorithm is able to solve clique number for these graphs but the computational time is rising rapidly with higher density. These times needed for our algorithm are still lesser compared with Östergårds algorithm. Graph „BG_500_40" needed 25 seconds to finish and graph „BG_500_50" needed 404 seconds to finish by old algorithm. As we can see the number of root CLIQUE calls for root vertices are less than 500 vertices in both graphs, which means that some root vertices were skipped from branching. The k-core decomposition in the heuristic phase returns always the same graph, because each vertex has degree higher than $\omega(G)$.

The graphs called „facebook", „as-skitter", „Email-Enron", „CA-AstroPh", „CA-CondMat", „CA-HepPh" and „CA-HepTh" represents real scale-free networks from Stanford snap datasets. One may see that the „facebook" network have highest density than other graphs from this dataset and the time needed to compute the maximum clique is aproximatelly 268 seconds for just 4039 vertices and 88234 edges. On the other hand „as-skitter" having 1696415 vertices and 11095298 edges and very low density is processed in only 42 seconds. Time of the initialization phases for this graph is 18.4 seconds due to the large amount of vertices. We may say that number of vertices and the edges density is the key factor of the algorithm speed.

G	$\lvert V\rvert$	$\lvert E\rvert$	D	d	Preliminary $\omega_p(G)$	Heuristic phase (s)	Final $\omega(G)$	$\omega_p(G)/\omega(G)$ ratio	Search phase (s)
C125.9	125	6963	0.8984516129	102	31	0.006	34	0.91176	1513.053
BG_500_40	500	49820	0.3993587174	175	9	0.032	11	0.81818	8.924
BG_500_50	500	62441	0.5005290581	223	11	0.046	13	0.84615	239.734
BG_1000_40	1000	199909	0.4002182182	361	10	0.251	12	0.83333	910.971
BG_1000_50	1000	250302	0.5011051051	461	12	0.257	15	0.80000	40141.835
facebook	4039	88234	0.0108199635	115	7	0.059	69	0.10145	268.018
as-skitter	1696415	11095298	0.0000077109	111	37	18.437	67	0.55224	41.716
Email-Enron	36692	183831	0.0002730976	43	17	0.208	20	0.85000	0.818
CA-AstroPh	18771	198050	0.0011242248	56	23	0.227	57	0.40351	0.509
CA-CondMat	23133	93439	0.0003492312	25	4	0.103	26	0.15385	0.200
CA-HepPh	12006	118489	0.0016441731	238	239	0.116	239	1.00000	0.078
CA-HepTh	9875	25973	0.0005327481	31	5	0.032	32	0.15625	0.050
BA_300_20	300	5790	0.1290969900	20	20	0.003	21	0.95238	0.006
BA_1000_20	1000	19790	0.0396196196	20	17	0.015	21	0.80952	0.030
BA_10000_20	10000	199790	0.0039961996	20	20	0.281	21	0.95238	0.372
BA_100000_20	100000	1999790	0.0003999620	20	20	4.055	21	0.95238	6.740
BA_250000_50	250000	12498725	0.0003999608	50	51	34.514	51	1.00000	25.357
awith_09	7558	102091	0.0035748773	50	23	0.114	29	0.79310	0.712
awith_37	6218	73880	0.0038223046	49	33	0.070	36	0.91667	0.336
awith_02	7529	66494	0.0023463649	51	28	0.088	41	0.68293	0.346
awith_40	5677	17200	0.0010675720	12	11	0.019	12	0.91667	0.014
awout_13	7416	139482	0.0050730283	118	62	0.215	68	0.91176	16.531
awout_14	6849	172969	0.0073757698	137	80	0.252	94	0.85106	6.626
awout_04	8000	183940	0.0057488436	91	45	0.203	55	0.81818	2.871
awout_15	5439	20276	0.0013710523	40	36	0.042	38	0.94737	0.050
young_26	8183	211232	0.0063098303	166	103	0.195	124	0.83065	25.972
young_17	7553	134565	0.0047182467	132	84	0.110	94	0.89362	19.550
young_16	7032	131887	0.0053350197	147	88	0.128	108	0.81481	3.897
young_34	7131	93643	0.0036835396	106	66	0.070	89	0.74157	0.693
WS_500_25_50	500	25000	0.2004008016	88	13	0.020	16	0.81250	0.560
WS_500_40_50	500	25000	0.2004008016	88	8	0.016	11	0.72727	0.333
WS_5000_25_50	5000	250000	0.0200040008	87	13	0.205	16	0.81250	4.815
WS_5000_40_50	5000	250000	0.0200040008	86	8	0.244	11	0.72727	2.483
WS_10000_25_50	10000	500000	0.0100010001	87	14	0.424	16	0.87500	10.320
WS_10000_40_50	10000	500000	0.0100010001	86	9	0.478	12	0.75000	4.907

Tab. 1. Results of test runs of our algorithm for different datasets. Times are in seconds. G is name of graph, $|V|$ is number of vertices, $|E|$ is number of edges, D is density of edges, d is degeneracy of graph, $\omega_p(G)$ is preliminary clique number known from heuristic phase, $\omega(G)$ is final clique number of graph. Heuristic phase time denotes time needed to traverse through graph to find preliminary clique number. Search phase time denotes time needed to branch over all remaining vertices

G	Our algorithm	Östergårs algorithm	$\|V\|$	Root CLIQUE calls	Total CLIQUE calls
C125.9	1513.059s	6309.648s	125	60	1034571801
BG_500_40	8.956s	24.761s	500	424	3945976
BG_500_50	239.780s	403.817s	500	434	70312920
BA_300_20	0.009s	0.016s	300	1	21
BA_1000_20	0.035s	0.035s	1000	5	101
BA_10000_20	0.653s	0.410s	10000	1	23
BA_100000_20	10.795s	6.235s	100000	1	21
BA_250000_50	59.871s	89.586s	250000	0	0
facebook	268.077s	> 30.000m	4039	81	103038963
as-skitter	60.143s	> 30.000m	1696415	32	1445753
Email-Enron	1.026s	8.203s	36692	13	5521
CA-AstroPh	0.736	0.453s	18771	34	1377
CA-CondMat	0.303	0.041s	23133	22	354
CA-HepPh	0.194	0.360s	12006	0	0
CA-HepTh	0.082	0.010s	9875	27	513
awith_09	0.826s	0.768s	7558	24	12646
awith_37	0.406s	0.615s	6218	3	381
awith_02	0.434s	1.857s	7529	14	1932
awith_40	0.033s	0.005s	5677	1	12
awout_13	16.746s	> 30.000m	7416	42	2723352
awout_14	6.878s	> 30.000m	6849	39	528298
awout_04	3.074s	40.102s	8000	30	142715
awout_15	0.092s	0.147s	5439	2	76
young_26	26.167s	> 30.000m	8183	29	5013089
young_17	19.660s	> 30.000m	7553	19	5388977
young_16	4.025s	> 30.000m	7032	26	367461
young_34	0.763s	> 30.000m	7131	24	2803
WS_500_25_50	0.580s	2.272s	500	233	247814
WS_500_40_50	0.349s	0.606s	500	219	97173
WS_5000_25_50	5.020s	9.924s	5000	1878	1808610
WS_5000_40_50	2.727s	3.095s	5000	1615	402399
WS_10000_25_50	10.744s	21.482s	10000	3571	3384796
WS_10000_40_50	5.385s	4.971s	10000	2940	595681

Tab. 2. Comparison of real run times for our new algorithm and original Östergårs algorithm. Computations where times are > than some value were stopped prematurely. Table also shows number of CLIQUE calls for each test instance by our algorithm

„CA-HepPh" is an example of a graph, where heuristic function finds maximum clique size and the algorithm ends immediately in search phase because the clique size found is higher than degeneracy of the graph. The comparison between new and old algorithm shows that on the graphs prefixed with „CA-" the computation times are very similar, in much cases the old algorithm is perform better. The „facebook" graph have highest density of these data, for which our algorithm takes 268 seconds to finish computation (4.47 minutes), while old algorithm needs more than 30 minutes until we have stopped the computation. On the contrary „as-skitter" graph have smallest density from all tested graphs but having a large number of vertices and edges, our algorithm still finish it in one minute but the computation with old algorithm needs more than half hour too until we have stopped the computation. We can say that in general case our new algorithm performs better. Also there is no exact boundary of number of vertices, edges and the density to say when it is better to run new or old algorithm. We can also see that for each of this test instance the number of root calls to CLIQUE subroutine for each root vertex is very small according to number of vertices in these graphs. The after-heuristic k-core decomposition and prunings in search phase help to reduce number branching, but still in some cases the size of virtual search trees are very high – over 100 millions for „facebook" graph – which is relative to the density of reduced graph after heuristic phase.

Graphs denoted as „BA_n_m" represents artificial scale-free networks generated from Barabási-Albert model, where n is number of vertices in network and m is number of edges connected to network with each new vertex. The total computational times are rising with number of vertices and edges. We can see that for each of tested graph $m = d$ and the $\omega(G) = m+1 = d+1$. This is just coincidence and we can't expect based of this observation that this holds for all instances of generated networks by this model. The times needed to process these data by both algorithms are very similar, for „BA_100000_20" the Östergårs algorithm performs better while for „BA_250000_50" our algorithm ends in shorter time. We can see that our algorithm is able to find clique size in the heuristic phase very close to the real clique number and in the search phase the number of root CLIQUE calls is always very close to zero. By the observation of these instances we know that k-core decomposition is not reducing the input graph at the end of the heuristic phase. Based on this fact the number of skipped branchings over root vertices are possible thanks to the prunings in the search phase. If we observe the number of vertices, edges or densities of each instance, we can't say where it is better to use our new algorithm and where to use the old one.

Next graphs prefixed with „awith_", „awout_" and „young_" represents some of our real data for scale-free functional brain networks for old participants with Alzheimer disease, old participants without Alzheimer disease and young participants without Alzheimer disease. Graphs for group with disease have low densities, the computational times are similar for both algorithms but in average they are shorter for our new algorithm. Group of graphs for old participants without disease shows that while they have similar size of vertices sets as „awith_" graphs, their density is higher and computation time needed for old algorithm is substantially higher. The last group of data is the young health participants group, which have computation time needed by our new algorithm in several seconds while old algorithm needed more than half of hour for each graph until we have terminated the computation. We can see that for real complex networks our algorithm performs very good against old one. Similar to the Barabási-Albert random graphs these instances needed very small number of root CLIQUE calls. On the contrary the k-core decomposition on tdhe end of the heuristic phase leads to reduced graph, this helped to reduce number of root calls as well as prunings in the search phase.

Last graphs are named „WS_n_q_k" representing networks generated by Watts-Strogatz model of small world networks. Parameter n represents number of vertices, $p = \frac{q}{100}$ is rewiring probability and k represents starting regularity of graph, which are $2k$-regular on start. For examples with 500 vertices the graphs have very high density 0.2, each of this graph is processed faster by our new algorithm than by the old algorithm. With higher number of vertices the density of graphs is decreasing and the computational times are increasing for both, our algorithm and the old one. Just in one tested case the Östergårds algorithm performs better, in average our new algorithm is faster. These test instances shows that the ratio between the number of root CLIQUE calls and the number of vertices is related to the rewiring probability, where it is decreasing with increase of this probability.

6.2 Results of tests

Our new algorithm was tested on wide range of graph instances. Half of the graphs presented in section 6.1 are artificial generated graphs, the rest of graphs are real data of FMRi functional brain networks or data from Stanford dataset. The performance of our algorithm was compared to performance of the Östergårds algorithm for each test instance. Both algorithms are implemented in Java programming language so the times needed to solve each instance can be compared.

In several cases the times needed to solve instance of test graph was similar for both algorithms and almost always these times was under one second. We think that this is caused by specific combination of properties of these graphs. It seems that the graph structure with the size of graph or with the density of edges in the graph as well as the ordering of vertices during processing by Östergårds algorithm can sometimes leads to faster computation time than our new algorithm is able to achieve. In these cases the heuristic searching of clique size, renumbering of graph, k-core decompositions and colorings of vertices can slow down the performance of our new algorithm. This was observed almost always when computational times were under one second for both algorithms. On the contrary, when the times needed to solve test instance by our algorithm was greater than one second only in two cases the Östergårds algorithm achieved times better than our one.

In most cases for computational times greater than one second we can see that new algorithm performs better. Thanks to new pruning techniques incorporated into it, for test cases of sparse graphs we have been able to achieve times of several seconds by our algorithm while the old one needs substantially more time. The performance of our algorithm have proved to be better for dense graphs as well.

For our algorithm we have also count the number of CLIQUE calls from root vertices. This number was always under the number of vertices in graphs. We have not count this for Östergårds algorithm because it is not doing any pruning before first CLIQUE call for each root vertex so it always branch over all root vertices in the graph. From the table 2 we can see that for the graphs that represents scale-free networks this number is very close to zero and substantially smaller than number of vertices in that graphs. This means that our pruning techniques is doing great part in decision for each root vertex to skip branching over it neighborhood, leading to short computational times. These techniques were also successful for small-world networks test cases. For dense graphs the number of root CLIQUE is closer to number of vertices, meaning that our pruning techniques helps to skip branching just for smaller set of root vertices, but still are able to help made the algorithm faster.

As the result from tests we can say that for most general cases the new algorithm will solve clique number of arbitrary simple undirected graph faster than Östergårds algorithm.

6.3 Future improvements

Our pruning techniques highly depends on speed of decomposition and coloring as well as on the approximation factor of coloring. By improving these techniques the algorithm can perform better. Having solution of coloring closer to optimum can disallow branching over vertices of more root subgraphs and save a lot of time. The heuristic phase also need some improvements. From some results there is clear that sometimes it takes several seconds to finish which can be improved as well.

For now the algorithm is working in single computation thread. Because of the use of $c[i]$ array during computation this needs from the algorithm to finish branching over root subgraphs belonging to root vertices with higher i parameter to be able to branch over root subgraph of root vertex j where $\forall i : j > i \wedge i$ finished $\rightarrow j$ can finish branching. Doing this in multiple computation threads means that just the thread processing root subgraph for root vertex with lowest index can finish without possible waiting for another threads. If we find the solution on how to overcome this limitation, then the algorithm can perform significantly better on computers with multiple logical CPUs.

7 Conclusion

Our new algorithm is as well as Östergårds able to find exact clique number of any kind of simple undirected graph. It performs best on the scale-free networks and other sparse graphs. The improvements was achieved by incorporating several improvements. The greedy search for maximum clique lower bound helps the algorithm to start the branching process over reduced graph and with some preliminary knowledge. It uses k-core decomposition widely over the search phase as well as at the end of the heuristic phase.

For each of searched root vertex the algorithm extracts induced subgraph reduced by k-core decomposition where $k = max$. This is the first important pruning technique. Second important pruning technique is greedy computation of chromatic number of this reduced subgraph, which is then compared with known max. These two prunings allows the algorithm to branch just over the vertices which have the right property to possibly be a part of the maximum clique. Branching itself is done by original Östergårds CLIQUE subroutine while the process is following the original principle to iterate from highest vertex index to lowest one. To improve this process the input graph is reordered according to degree at the end of heuristic phase.

The algorithm is also using one early stopping criterion. During the search phase if the max clique number meets the input graph degeneracy $d + 1$ it is clear that there can't be found higher clique size and the algorithm terminates. For some input graphs the heuristic phase can find exact clique number and the search phase is just verifying if there is no better result. In this case if this clique number is also equal to $d + 1$ the algorithm will terminate right before first searched root subgraph.

By the tests that we carried out we have show than in most cases our new algorithm performs better than original Östergårds algorithm. In some cases it is slower due to additional work done against the old algorithm. Pruning techniques like computation of preliminary clique size, extraction of induced subgraph, k-core decomposition and greedy coloring takes some extra time, wich may for some inputs slow down the performance. In most cases these pruning techniques will save time by disallowing to branch over root subgraphs without potential to find better solution. In most general cases of simple undirected graphs our algorithm can find solution substantially faster than Östergårds algorithm.

References

[1] Albert, R., Barabási, A.-L.: *Statistical mechanics of complex networks*, Reviews of Modern Physics, vol. 74, (2002), pp. 47–97, doi:10.1103/RevModPhys.74.47.

[2] Butts, C. T.: *Bernoulli graph bounds for general random graphs*, Sociological

Methodology, vol. 41, (2011), pp. 299–345, ISSN 00811750, 14679531, URL: http://www.jstor.org/stable/41336925.

[3] Carmo, R., Züge, A.: *Branch and bound algorithms for the maximum clique problem under a unified framework*, Journal of the Brazilian Computer Society, vol. 18, no. 2, (2012), pp. 137–151, ISSN 1678-4804, doi:10.1007/s13173-011-0050-6, URL: http://dx.doi.org/10.1007/s13173-011-0050-6.

[4] Eblen, J. D., Phillips, C. A., Rogers, G. L., Langston, M. A.: *The maximum clique enumeration problem: Algorithms, applications and implementations*, in *Proceedings of the 7th International Conference on Bioinformatics Research and Applications*, ISBRA'11, Springer-Verlag, Berlin, Heidelberg, 2011, ISBN 978-3-642-21259-8, pp. 306–319, URL: http://dl.acm.org/citation.cfm?id=2009164.2009194.

[5] Jursa, A.: *Fast algorithm for finding maximum clique in scale-free networks*, in *ITAT 2016 : Information Technologies - Applications and Theory*, CreateSpace Independent Publishing Platform, September 2016, ISBN 978-1-5370-1674-0, pp. 212–217.

[6] Konc, J., Janezic, D.: *An improved branch and bound algorithm for the maximum clique problem*, MATCH Communications in Mathematical and in Computer Chemistry, URL: http://www.sicmm.org/~konc/%C4%8CLANKI/MATCH58(3)569-590.pdf.

[7] Li, C.-M., Quan, Z.: *An efficient branch-and-bound algorithm based on maxsat for the maximum clique problem*, in *Proceedings of the Twenty-Fourth AAAI Conference on Artificial Intelligence*, AAAI'10, AAAI Press, 2010, pp. 128–133, URL: http://dl.acm.org/citation.cfm?id=2898607.2898628.

[8] Markošová, M.: *Dynamic model of functional brain networks*, in *ITAT 2016 : Information Technologies - Applications and Theory*, CreateSpace Independent Publishing Platform, September 2016, ISBN 978-1-5370-1674-0, pp. 226–231.

[9] Nehéz, M.: *Analysis of the randomized algorithm for clique problems*, in *Proceedings of the 15th Conference on Applied Mathematics APLIMAT 2016*, SUT Publishing, Bratislava, Slovak Republic, Feb. 2016, pp. 867–875.

[10] Östergård, P. R. J.: *A fast algorithm for the maximum clique problem*, Discrete Appl. Math., vol. 120, no. 1-3, (2002), pp. 197–207, ISSN 0166-218X, doi:10.1016/S0166-218X(01)00290-6, URL: http://dx.doi.org/10.1016/S0166-218X(01)00290-6.

[11] Pattabiraman, B., Patwary, M. M. A., Gebremedhin, A. H., Liao, W., Choudhary, A. N.: *Fast algorithms for the maximum clique problem on massive sparse graphs*, CoRR, vol. abs/1209.5818, URL: http://arxiv.org/abs/1209.5818.

[12] Watts, D. J., Strogatz, S. H.: *Collective dynamics of'small-world'networks.*, Nature, vol. 393, no. 6684, (1998), pp. 409–10.

[13] Weisstein, E.: *Clique Number*, http://mathworld.wolfram.com/CliqueNumber.html, from MathWorld–A Wolfram Web Resource.

Current address

Jursa Andrej, Mgr.
Department of Applied Informatics,
Faculty of Mathematics, Physics and Informatics,
Comenius University in Bratislava
Mlynská dolina, 824 48 Bratislava, Slovak Republic
E-mail: andrej.jursa@fmph.uniba.sk

Slovak University of Technology in Bratislava
Faculty of Mechanical Engineering
16th Conference on Applied Mathematics **APLIMAT 2017**
Proceedings

NUMBER OF THEORETICAL PLATES IN A DISTILLATION COLUMN

KABÁT Juraj (SK), GUŽELA Štefan (SK), PECIAR Peter (SK)

Abstract. The number of theoretical plates (NTP) is the basic process parameter of every plate column. All the methods that deal with NTP are based on the idea that every plate represents one theoretical equilibrium stage. This paper is focused on the Lewis-Matheson method. The calculation of NTP through the Lewis-Matheson method is demonstrated by the solving of the multi-component mixture separation problem.

Keywords: distillation, rectification, theoretical plate, plate column, Lewis-Matheson

Mathematics Subject Classification: Primary 97M50; Secondary 65Y99

1 Introduction

Many authors have been dealing with the NTP of distillation column calculations. In general, these calculations can be divided into two main groups: design and rating methods [2]. Design methods are focused on the calculation of the NTP needed to obtain the necessary purity and composition of a distillate and a bottom product. Rating methods are used to determine the composition of the material streams of the existing column.

From another point of view [2, 6] the methods can be divided into: quick methods based on empirical functions (Gilliland correlation, Smith-Brinkley method), graphical methods (Hengstebeck method), and rigorous methods (Lewis-Matheson method, Thiele-Gheddes method).

The complexity of a calculation can be estimated by the number of processed mixture components. The NTP of a binary mixture problem can be calculated precisely by the analytical or graphical methods, for example a graphical McCabe-Thiele or Fenske method [4, 7].

While solving a multicomponent mixture problem the analytical calculation cannot be used anymore. The methods for solving these problems are usually based on numerical and iterative calculations, as will be demonstrated in this paper.

2 Design parameters of the simulated problem

The design parameters and specifications of the problem solved in this article can be seen in Tab. 1. These parameters are necessary for NTP calculation through the Lewis-Matheson method.

Composition of feed and distillate			Process and operation parameters	
Compound	Feed mass fraction $\bar{x}_{Fi}$	Distillate flow rate [kg/s]		
Methane C_1	0.095	0.417	overall feed flow rate	4.39 kg/s
Ethane C_2	0.0616	0.27	operational pressure	2MPa
Propane C_3	0.2511	1.085	operational reflux ratio	1,5
Butane C_4	0.2251	0.017	feed temperature	78°C
Pentane C_5	0.1808	0		
Hexane C_6	0.1864	0		
$\sum$	1	1.789		

Tab. 1. Design and process specifications of the simulated problem

The distillate flow rates of the compounds C_5 and C_6 shown in Tab. 1 are equal to zero. These given values are related to a light key compound (LKC) and a heavy key compound (HKC). According to the definition [2], the LKC and HKC in this case are represented by propane and butane respectively. The physical abstraction of the LKC and the HKC significantly simplifies the calculation [2].

The equilibrium state data of the mixture compounds are given by the vapour-liquid (V-L) distribution ratio K_i. These ratios can be generally given in the form of a graph, table or equation. The vapour-liquid distribution ratios used in this paper are calculated through a 3rd degree polynomial equation as follows:

$$K_i = \frac{y_i}{x_i} = a + b.T + c.T^2 + d.T^3. \qquad (1)$$

Eq. 1 represents an approximation of the experimental data published in [8]. The temperature *t* is substituted in °C. The values of the coefficients *a, b, c* and *d* used in Eq. 1 can be seen in Tab. 2. A comparison of the approximated equations and the experimental data is shown in Fig. 1.

Compound	a	b	c	d
C_1	7.180	$2.912 .10^{-2}$	$7.959 .10^{-4}$	$-2.153 .10^{-6}$
C_2	1.405	$4.013 .10^{-3}$	$2.979 .10^{-4}$	$-6.727 .10^{-7}$
C_3	$4.272 .10^{-1}$	$4.582 .10^{-3}$	$9.174 .10^{-5}$	$-1.373 .10^{-7}$
C_4	$1.299 .10^{-1}$	$8.786 .10^{-4}$	$8.811 .10^{-5}$	$-2.361 .10^{-7}$
C_5	$4.790 .10^{-2}$	$3.240 .10^{-4}$	$4.727 .10^{-5}$	$-6.897 .10^{-8}$
C_6	$1.812 .10^{-2}$	$5.830 .10^{-4}$	$2.678 .10^{-5}$	$-1.721 .10^{-8}$

Tab. 2. Coefficients for Eq. 1 which can be used in the following range:
temperature 10÷150°C, pressure 2MPa

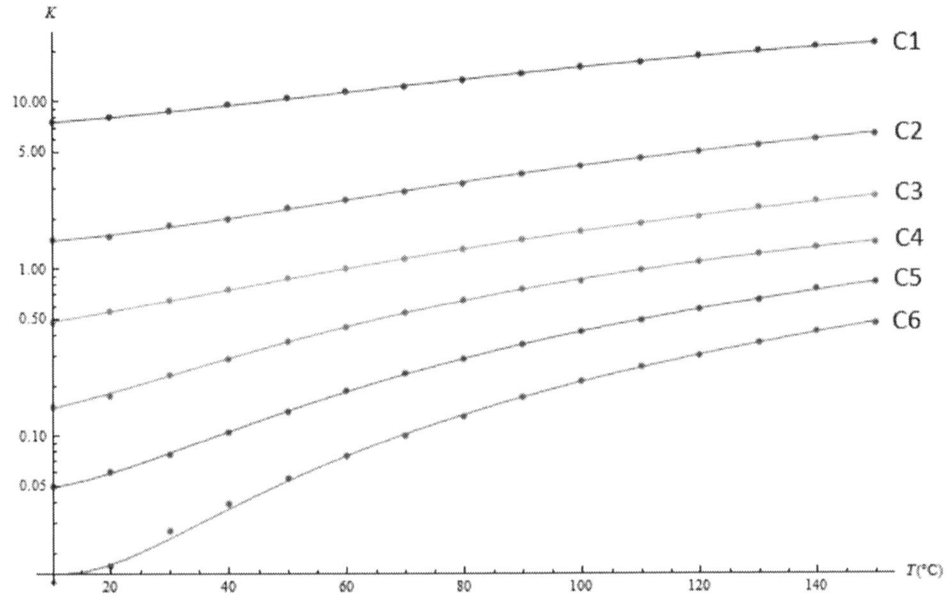

Fig. 1. Dependency of vapour-liquid distribution ratios on temperature for 2MPa pressure. The lines represent Eq. 1 and the points show the experimental data [8]

3 Calculation of the NTP by the Lewis-Matheson method

The method is based on an iterative calculation of the vapour and liquid composition in each stage by estimating the bubble and dew point temperatures. The calculation is divided into two parts: the determination of the NTP in the enriching (rectifying) and stripping section of a column. The plates are numbered according to Fig. 2. This calculation considers a reboiler and a partial condenser as two theoretical stages, therefore the NTP has to be decreased by 2.

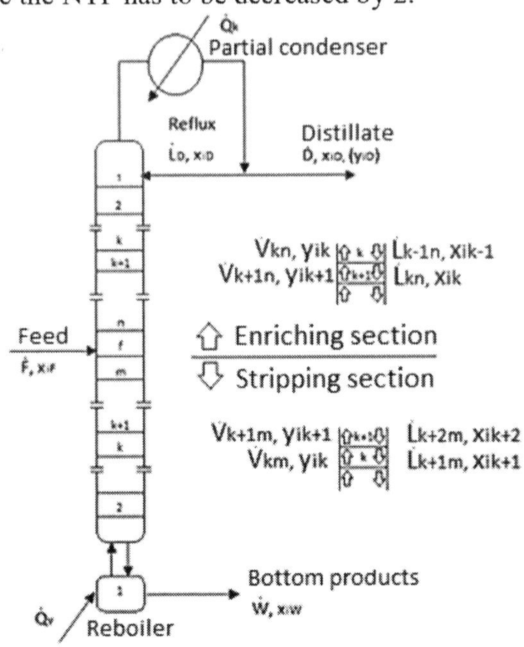

Fig. 2. Scheme of a plate column

Algorithm of the Lewis-Matheson method:

1. The calculation of an enriching section starts with an estimation of the dew point vapour temperature leaving a partial condenser . According to the estimated temperature, the V-L distribution ratios of every compound are determined through Eq. 1.

2. Based on the V-L distribution ratios, the liquid composition is calculated by the following equation:

$$x_i = \frac{y_i}{K_i} \tag{2}$$

3. If the estimated dew point temperature was correct, the liquid composition calculated in the second point has to meet the condition given by the equation as follows:

$$\sum_i x_i = \sum_i \frac{y_i}{K_i} = 1. \tag{3}$$

4. After the condition given by Eq. 3 is accomplished, the calculation continues with a determination of the composition of the vapour on the following plate according to Eq. 4 [2]:

$$y_{i,k+1} = \frac{\dot{L}_{kn}}{\dot{V}_{k+1n}} x_{ik} + \frac{\dot{D}}{\dot{V}_{k+1n}} x_{iD} \tag{4}$$

   The values $y_{i,k+1}$ given by Eq. 4 are necessary for estimating the temperature and the composition of the liquid stream leaving the next stage (stage number 1).

5. The temperature in stage number one is estimated and specified using the steps 1-3. Next, Eq. 4 is used for calculation of the vapour stream composition.

6. Steps 1-5 are repeated until the LKC/HKC ratio in the vapour phase exceeds the LKC/HKC ratio in the vapour part of the feed (the value used in this paper is 1.85). This step concludes the calculation of the enriching section.

7. Point 7 starts the stripping section calculation. The bubble point temperature of a bottom product is estimated according to a known liquid composition. Next, the V-L distribution ratios K_i and the vapour phase composition $y_i$of the components are calculated by using Eq. 1.

8. Estimated bubble point temperature is checked by the following condition:

$$\sum_i y_i = \sum_i K_i x_i = 1. \tag{5}$$

9. If the condition given by Eq. 5 is correct, the liquid phase composition on the next stage is calculated by Eq. 6.

$$x_{i,k+1} = \left(y_{ik} + \frac{\dot{W}}{\dot{V}_{km}} x_{iw} \right) \frac{\dot{V}_{km}}{\dot{L}_{k+1m}} \tag{6}$$

10. Steps 7-9 are repeated until the LKC/HKC ratio in the liquid phase exceeds the LKC/HKC in the liquid part of the feed (the value used in this paper is 0.90). Step 10 concludes the stripping section calculation.

3.1 Material balance of the column

The material flow rates represent further data needed for the Lewis-Matheson method calculation. The material streams leaving the column are calculated by the following equations:

$$\dot{F} = \dot{W} + \dot{D}, \tag{7}$$

$$\dot{F}.z_{Fi} = \dot{W}.x_{Wi} + \dot{D}.x_{Di}\,. \tag{8}$$

The material balance of the column is shown in Tab. 3.

Compound	Feed		Distillate		Bottom product	
	z_{Fi}	$\dot{f}_i$ *[mol/s]*	x_{Di}	$\dot{d}_i$*[mol/s]*	x_{Wi}	$\dot{w}_i$*[mol/s]*
C_1	0,2666	25,9941	0,4341	25,9941	0	0
C_2	0,0922	8,99315	0,15	8,99315	0	0
C_3	0,2564	24,9978	0,411	24,6048	0,0105	0,393
C_4	0,1744	17,0061	0,0049	0,293	0,4445	16,7131
C_5	0,1128	11,0009	0	0	0,2925	11,0009
C_6	0,0974	9,4963	0	0	0,2525	9,4963
$\sum$	1	97,4883	1	59,8675	1	37,6208

Tab. 3. The material balance of the column

4 Computer programming of the calculation

The algorithm described in the chapter 3 was programmed in Wolfram Mathematica software. The input values needed for a successful calculation are show in Fig. 3. These values are: the equations of the V-L distribution ratios (KC1 [t_]...KC6 [t_]), the composition of the distillate (xWC1...xWC6), the flow rates of the vapour and liquid streams in both sections of the column (Vn, Ln, Vm, Lm) and the LKC and HKC molar flow rates in the vapour and liquid part of the feed (vFC3, vFC4, lFC3, lFC4).

```
KC1[t_] = 7.180659340659343` + 0.029121866695395918` t + 0.0007958818142641693` t^2 - 2.1524635495223842`*^-6 t^3;
KC2[t_] = 1.4049743589743575` + 0.004013012042423835` t + 0.00029787132801838644` t^2 - 6.726579520697155`*^-7 t^3;
KC3[t_] = 0.42716483516483544` + 0.004582264957264944` t + 0.00009174342814048717` t^2 - 1.373177476118663`*^-7 t^3;
KC4[t_] = 0.12994871794871796` + 0.000878646544087719` t + 0.00008811364289305471` t^2 - 2.360901625607514 8`*^-7 t^3;
KC5[t_] = 0.04790109890109886 5` - 0.0003240119765855102` t + 0.00004726621417797 8966` t^2 - 6.897310206133782`*^-8 t^3;
KC6[t_] = 0.01811501831501818` - 0.0005829950142450129` t + 0.00002677565898 1541336` t^2 - 1.7209234120998955`*^-8 t^3;
xDC1 = 0.43414261629312734`; xDC2 =0.14998207353647824`; xDC3 = 0.4109885389110591`; xDC4 = 0.004886771259335255`;
xDC5 = 0; xDC6 = 0;
xWC1 = 0; xWC2 = 0; xWC3 = 0.010450437613097583`; xWC4 = 0.4444652857888889`; xWC5 = 0.2925476614367136`;
xWC6 = 0.2525366151612999`;
Vn = 149.66870163568186`; Ln = 89.80122098140912`;
Vm = 80.8419296118737`; Lm = 118.46279460038872`;
De = 59.86748065427275`; W = 37.603648126372455`;
vFC3 = 18.852741545696002`; vFC4 = 10.189788140177706`;
lFC3 = 6.145081435459179`; lFC4 = 6.816286754845352`;
```

Fig. 3. The input values for the programmed algorithm in Wolfram Mathematica interface

The program continues calculating the NTP of the enriching section as shown in Fig. 4. The initial data are represented by the distillate compounds mole fractions, which are equal to the mole fraction of the vapour phase leaving the partial condenser. These data are read by the "For" cycle which continues to the "Do" cycle. The "Do" cycle repeatedly estimates the temperature in the predefined range and interval length and calculates the V-L distribution ratios. If the condition given by Eq. 3 is fulfilled, the vapour and liquid compositions in the stage are calculated. Then the initial conditions of the "For" cycle are redefined according to the previous iteration. The calculation continues to the next stage until the termination condition given by the LKC/HKC ratio in the vapour phase is exceeded. This concludes the enriching section calculation.

The form of the stripping section calculation is analogous to the enriching one. The only differences are the equations calculating the stream composition and the termination condition, which is given by the LKC/HKC ratio in the liquid phase.

```
For[k = 1; yC1 = xDC1; yC2 = xDC2; yC3 = xDC3; yC4 = xDC4; yC5 = xDC5; yC6 = xDC6; ratio = 2, ratio > vFC3/vFC4,
 k++; ratio = ratiov; yC1 = yC1v; yC2 = yC2v; yC3 = yC3v; yC4 = yC4v; yC5 = yC5v; yC6 = yC6v, Print["NUMBER OF THEORETICAL STAGE: ", k];
 Do[{sum = yC1/KC1[i] + yC2/KC2[i] + yC3/KC3[i] + yC4/KC4[i] + yC5/KC5[i] + yC6/KC6[i]; If[Abs[sum - 1] < 0.001,
   Print["Σxi=", yC1/KC1[i] + yC2/KC2[i] + yC3/KC3[i] + yC4/KC4[i] + yC5/KC5[i] + yC6/KC6[i], "  T=", i, "°C"];
   Print["VAPOUR MOLE FRACTION: ", " yC1=", yC1, " yC2=", yC2, " yC3=", yC3, " yC4=", yC4, " yC5=", yC5, " yC6=", yC6, " ...Σyi=",
    yC1 + yC2 + yC3 + yC4 + yC5 + yC6];
   Print["VAPOUR MOLAR FLOW RATE [mol/s]: ", " vC1=", Vn*yC1, " vC2=", Vn*yC2, " vC3=", Vn*yC3, " vC4=", Vn*yC4, " vC5=", Vn*yC5, " vC6=",
    Vn*yC6, " ...Σvi=", Vn*(yC1 + yC2 + yC3 + yC4 + yC5 + yC6)];
   Print["LIQUID MOLE FRACTION: ", " xC1=", yC1/KC1[i], " xC2=", yC2/KC2[i], " xC3=", yC3/KC3[i], " xC4=", yC4/KC4[i], " xC5=", yC5/KC5[i], " xC6=", yC6/KC6[i],
    " ...Σxi=", yC1/KC1[i] + yC2/KC2[i] + yC3/KC3[i] + yC4/KC4[i] + yC5/KC5[i] + yC6/KC6[i]];
   Print["LIQUID MOLAR FLOW RATE [mol/s]: ", " lC1=", Ln*yC1/KC1[i], " lC2=", Ln*yC2/KC2[i], " lC3=", Ln*yC3/KC3[i], " lC4=", Ln*yC4/KC4[i], " lC5=",
    Ln*yC5/KC5[i], " lC6=", Ln*yC6/KC6[i], " ...Σli=", Ln*(yC1/KC1[i] + yC2/KC2[i] + yC3/KC3[i] + yC4/KC4[i] + yC5/KC5[i] + yC6/KC6[i])];
   Print["LKC/HKC RATIO=", (Vn*yC3)/(Vn*yC4)]; yC1v = Ln/Vn*yC1/KC1[i] + De/Vn*xDC1; yC2v = Ln/Vn*yC2/KC2[i] + De/Vn*xDC2; yC3v = Ln/Vn*yC3/KC3[i] + De/Vn*xDC3;
   yC4v = Ln/Vn*yC4/KC4[i] + De/Vn*xDC4; yC5v = Ln/Vn*yC5/KC5[i] + De/Vn*xDC5; yC6v = Ln/Vn*yC6/KC6[i] + De/Vn*xDC6;
   Print["NEXT STAGE VAPOR MOLE FRACTION: ", " yC1=", yC1v, " yC2=", yC2v, " yC3=", yC3v, " yC4=", yC4v, " yC5=", yC5v, " yC6=", yC6v];
   ratiov = (Vn*yC3)/(Vn*yC4); Print[""]]}, {i, 0, 100, 0.1}]]
```

Fig. 4. Enriching part calculation in Wolfram Mathematica software interface

The program output data can be seen in Fig. 5. These are the composition and flow rates of the vapour and liquid stream in each stage, as well as the stage temperature.

```
NUMBER OF THEORETICAL STAGE: 1
Σxi=0.999743  T=13.5°C
VAPOUR MOLE FRACTION:  yC1=0.434143 yC2=0.149982 yC3=0.410989 yC4=0.00488677 yC5=0 yC6=0 ...Σyi=1.
VAPOUR MOLAR FLOW RATE [mol/s]:  vC1=64.9776 vC2=22.4476 vC3=61.5121 vC4=0.731397 vC5=0. vC6=0. ...Σvi=149.669
LIQUID MOLE FRACTION:  xC1=0.0562831 xC2=0.0992088 xC3=0.813182 xC4=0.0310689 xC5=0. xC6=0. ...Σxi=0.999743
LIQUID MOLAR FLOW RATE [mol/s]:  lC1=5.05429 lC2=8.90907 lC3=73.0247 lC4=2.79002 lC5=0. lC6=0. ...Σli=89.7781
LKC/HCK RATIO=84.1023
NEXT STAGE VAPOR MOLE FRACTION:  vC1=0.207427 vC2=0.119518 vC3=0.652305 vC4=0.020596 vC5=0. vC6=0.
```

Fig. 5. Program output data for the first stage

5 Calculation results

The program calculated NTP as follows: 8 theoretical plates in enriching part and 11 theoretical plates in stripping part of the column. The temperature and concentration profiles are shown in Fig. 6 and 7 respectively. These are not connected in the feed stage, a phenomenon caused by ignoring the influence of C_5 and C_6 compounds in the enriching section and C_1 and C_2 compounds in the stripping section of the column. While these zero concentrations are in accord with the LKC and HKC concept, this represents only a physical abstraction. All the compounds entering the column in a feed are distributed to both sections. But some compounds effects on the equilibrium state of a mixture can be ignored because they achieve very low concentration values.

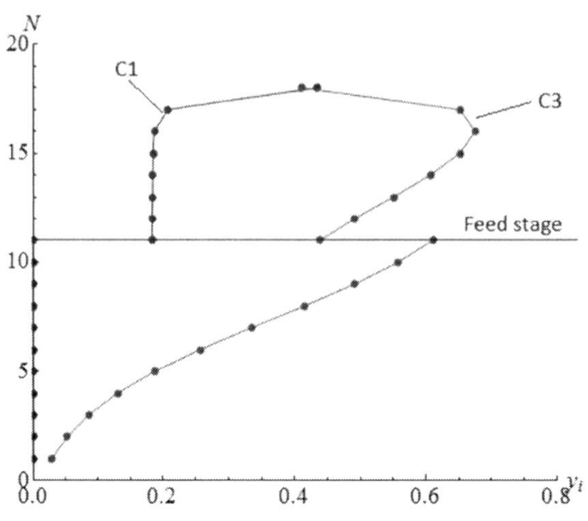

Fig. 6. Concentration profile of C_1 and C_3 in the vapour phase

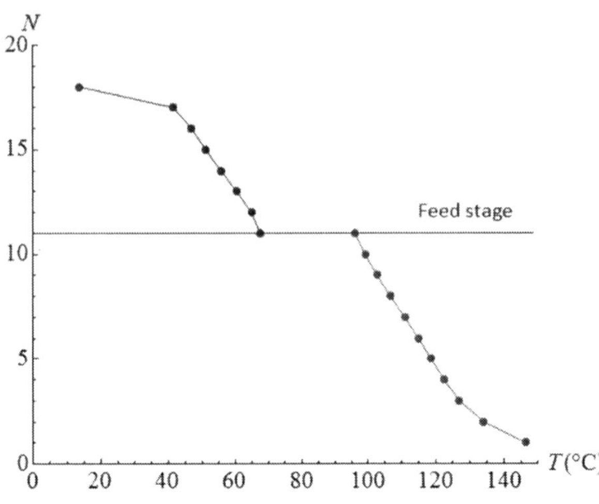

Fig. 7. Temperature profile of the column

The discrepancy of the concentration and temperature profiles shown in Fig. 6 and 7 respectively is unacceptable. The profiles have to be connected and continuous, and so the first calculation needs to be corrected. The correction of the first iteration can be made using the new estimation of the distillate composition, as follows [2]:

$$\Delta \dot{d}_i = -\frac{x_{iFD} - x_{iFW}}{\frac{x_{iFD}}{\dot{d}_{ip}} + \frac{x_{iFW}}{\dot{w}_{ip}}} \tag{9}$$

The values x_{iFD} and x_{iFW} represent a molar fraction on a feed stage calculated for an enriching and stripping section respectively. According to the change of distillate flow rate $\Delta \dot{d}_i$ the material balance is recalculated through Eq. 10 and 11 and the whole algorithm described in the chapter 3 is repeated.

$$\dot{d}_i = \dot{d}_{ip} + \Delta \dot{d}_i \tag{10}$$

$$\dot{w}_i = \dot{w}_{ip} - \Delta \dot{d}_i \tag{11}$$

We have to realize that after the first iteration the values x_{iFD}(for C_5 and C_6) and x_{iFW}(for C_1 and C_2) are equal to zero. Therefore in Fig. 9 the zero value would be divided by zero. So before the correction through Eq. 9 can be used, one iteration where the values x_{iFD} and x_{iFW} of the zero compounds are only estimated has to be done. It is very difficult to suggest an estimated value that will generally lead to the correct results. But the Lewis-Matheson method belongs among design methods, so NTP should not change after the iteration calculated by the estimated values. The new correction of the distillate composition in this paper was estimated as follows:$\Delta \dot{d}_{C1} = -10^{-13} mol/s$, $\Delta \dot{d}_{C2} = -10^{-13} mol/s$, $\Delta \dot{d}_{C5} = 10^{-13} mol/s$ a $\Delta \dot{d}_{C6} = 10^{-13} mol/s$.

The values of the LKC and HKC were not corrected because the non-key compounds of the mixture (C_1, C_2, C_5 and C_6) influence the equilibrium state the most [8]. The change of key component values is recommended when equilibrium temperature discrepancy is only a few °C.

The composition discrepancy in the feed stage ($\Delta x_{Fi} = x_{iFD} - x_{iFW}$) after every iteration and the temperatures in the feed stage calculated for both sections of the column are shown in Fig. 8 and Fig. 9 respectively.

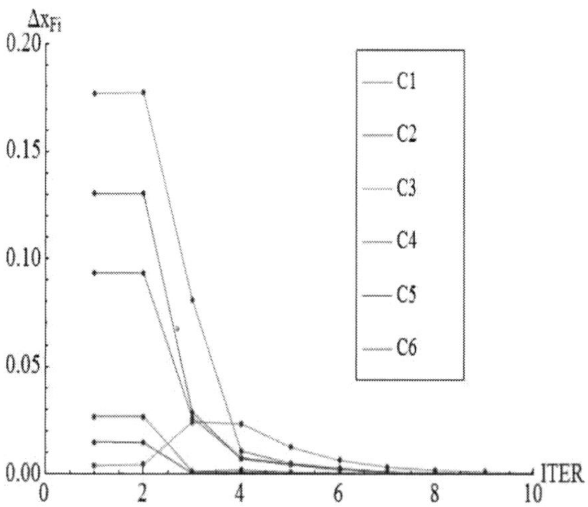

Fig. 8. Composition discrepancy in the feed stage after every iteration

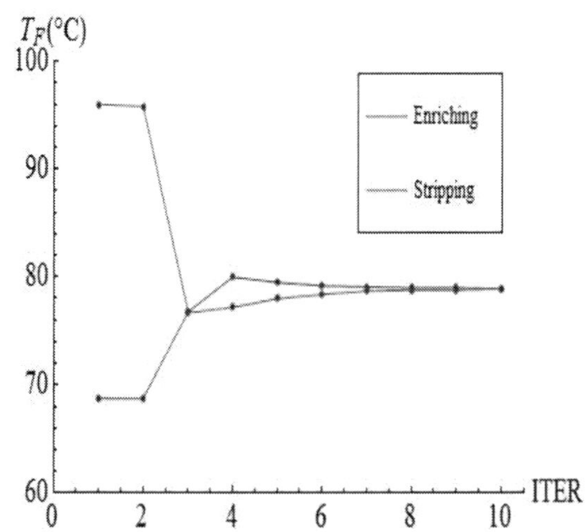

Fig. 9. Temperature in the feed stage calculated for both sections of the column

6 Conclusion

In the paper, NTP estimation by the Lewis-Matheson method was described. The algorithm started with the necessary input data and the first iteration results. These were followed by correction of the concentration and temperature profiles. The whole algorithm was programmed in Wolfram Mathematica software. The program work was demonstrated by solving a multi-component mixture problem. The correction led the calculation to more accurate and continuous concentration and temperature profiles. The simulation could describe a real process even more accurately if the operation conditions in the column were considered as non-isobaric. If the real conditions are taken into consideration, every plate represents a pressure drop which depends highly on the chosen type of a plate (bubble-cap, sieve, valve…). If the plate type was known and the equations describing the V-L distribution ratios were dependent on pressure as well, the simulation results would be closer to a real operation conditions.

Nomenclature

C_1	methane	x_i	liquid phase mole fraction
C_2	ethane	$\bar{x}_i$	mass fraction
C_3	propane	y_i	vapour phase mole fraction
C_4	butane	z_i	feed mole fraction
C_5	pentane	**Subscripts**	
C_6	hexane	D	distillate
$\dot{D}$	distillate molar flow rate (mol.s^{-1})	f	feed stage
$\dot{d}_i$	distillate *i*-compound molar flow rate (mol.s^{-1})	F	feed
$\dot{F}$	feed molar flow rate (mol.s^{-1})	FD	calculation of enriching part
$\dot{f}_i$	feed *i*-compound molar flow rate (mol.s^{-1})	FW	calculation of stripping part
HKC	heavy key compound	i	general mixture compound
K_i	vapour liquid distribution ratio	k	general stage number
$\dot{L}_i$	liquid phase molar flow rate (mol.s^{-1})	m	stripping part index
LKC	light key compound	n	enriching part index
N	number of theoretical plate	p	previous iteration
T, t	temperature (°C)	W	bottom products
$\dot{V}_i$	vapour phase molar flow rate (mol.s^{-1})		
$\dot{W}$	bottom product molar flow rate (mol.s^{-1})		
$\dot{w}_i$	bottom prod. *i*-compound molar flow rate (mol.s^{-1})		

Acknowledgement

This article was created with the support of the Ministry of Education, Science, Research and Sport of the Slovak Republic within the Research and Development Operational Programme for the project "University Science Park of STU Bratislava", ITMS 26240220084, co-funded by the European Regional Development Fund.

References

[1] COUPER, J. R., PENNY, W. R., FAIR, J. R., WALAS, S. M. *Chemical process equipment.* Elsevier, 2005, ISBN 0-7506-7510-1

[2] DOJČANSKÝ, J. *Separačné procesy I: Výpočtové metódy v difúznych procesoch.* Vydavateľstvo STU, 1995, ISBN 80-227-0807-0

[3] DOJČANSKÝ, J., LONGAUER, J. *Chemické inžinierstvo II.* Vydavateľstvo MALÉ CENTRUM, 2000, ISBN 80-967064-8-9

[4] CHRIAŠTEĽ, L., GUŽELA, Š. *Difúzne pochody I.* Nakladateľstvo STU v Bratislave, 2011, ISBN 987-80-227-3439-4

[5] KABÁT, J. *Etážová rektifikačná kolóna.* Ústav procesného inžinierstva, SjF STU BA, 2016, Diplomová práca

[6] KISTER, H. Z. *Distillation design.* 1992, ISBN 0-07-034909-6

[7] PECIAR, P., FEKETE, R., PECIAR, M. *Procesné strojníctvo II.* Vydavateľstvo STU, 2016, ISBN 978-80-227-4540-6

[8] VERMOUZEK, Č. *Difuzní pochody chemického a potravinářského průmyslu.* SNTL- Státní nakladatelství technické literatury, Praha 1983

Current address

Kabát Juraj, Ing.
Institute of Process Engineering, Faculty of Mechanical Engineering,
Slovak University of Technology in Bratislava
Nám. slobody 17, 812 31 Bratislava 1, Slovak Republic
Tel. number: 02 57296 526, e-mail: juraj.kabat@stuba.sk

Guželа Štefan, doc. Ing. PhD.
Institute of Process Engineering, Faculty of Mechanical Engineering,
Slovak University of Technology in Bratislava
Nám. slobody 17, 812 31 Bratislava 1, Slovak Republic
Tel. number: 02 57296 535, e-mail: stefan.guzela@stuba.sk

Peciar Peter, Ing. PhD.
Institute of Process Engineering, Faculty of Mechanical Engineering,
Slovak University of Technology in Bratislava
Nám. slobody 17, 812 31 Bratislava 1, Slovak Republic
Tel. number: 02 57296 588, e-mail: peter.peciar@stuba.sk

Slovak University of Technology in Bratislava
Faculty of Mechanical Engineering
16th Conference on Applied Mathematics **APLIMAT 2017**
Proceedings

SEGMENTED CONSTRUCTION OF PLANAR B-SPLINE CURVES

KAŇKA Miloš (CZ)

Abstract. This work deals with the construction of planar B-spline curves that are constructed piecewise, i.e., segment-wise, in comparison to their final (complete) graph. The task for segmented construction of planar B-spline curves was assigned by Swedish company MIR in Enskede. During the cutting process with laser or water beam a situation sometimes occours in which a course of the whole curve (cutting track) is not desired. When a course of only a part of the curve is preferred.

Keywords: B-spline base functions, Cox-de Boor recursive formula, B-spline curves clamped, open, closed.

Mathematics subject classification: Primary 65D07; Secondary 93A30

1 Introduction

Let $p \geq 1$ be an integer. For $n \geq p$ let us consider $n+1$ "control" points in $\mathbb{R}^2$

$$P(0),\ P(1),\ \dots,\ P(n), \tag{1}$$

which are, except for some exceptions, distinct. For $m = n + p + 1$, let us further consider the $(m+1)$-dimensional "nodal" vector

$$U = (U(0), U(1), \dots, U(m)), \tag{2}$$

the components, so-called nodals, which are real numbers, form a non-decreasing sequence (with the exception of a stationary sequence). For example, for

$$p = 2, \quad n = 9 \tag{3}$$

we might speak about the following control points

$$\begin{array}{llll} P(0) = (1,5), & P(1) = (3,8), & P(2) = (10,6), & P(3) = (12,3), \\ P(4) = (7,1), & P(5) = (6,3), & P(6) = (3,2), & P(7) = (5,5), \\ P(8) = (3,5), & P(9) = (0,2), & & \end{array} \tag{4}$$

and, for $m = n + p + 1 = 9 + 2 + 1 = 12$, the $m + 1 = 12 + 1 = 13$-dimensional nodal vector

$$U = (0, 0, 0, 5, 10, 10, 10, 10, 15, 20, 25, 25, 25). \quad (5)$$

In this case, the nodal $0 = U(0) = U(1) = U(2)$ has multiplicity $k(0) = 3$, the nodal $5 = U(3)$ has multiplicity $k(5) = 1$, the nodal $10 = U(4) = U(5) = U(6) = U(7)$ has multiplicity $k(10) = 4$, and so on, at last, the nodal $25 = U(10) = U(11) = U(12)$ has multiplicity $k(25) = 3$.

With the help of the integer $sk \geq 1$, we express the total number $sk + 1$ of distinct nodals (components) of the vector (2). For the case of (3)–(5) there is $sk = 5$.

We assign to every $i = 0, 1, \ldots, sk$ exactly one nodal denoted as $V(i)$ in such a way that $V(0), V(1), \ldots, V(sk)$ forms an increasing sequence. Hence, for the case of (3)–(5),

$$0 = V(0) < 5 = V(1) < 10 = V(2) < 15 = V(3) < 20 = V(4) < 25 = V(5).$$

For $X = U(p)$, let us denote by the symbol i_0 the number from the sequence $i = 0, 1, \ldots, sk$ such that $X = V(i_0)$. Similarly, for $Y = U(m - p)$ we denote by i_1 such number from the sequence $i = 0, 1, \ldots, sk$ that $Y = V(i_1)$. For the case of (3)–(5) it holds for $X = U(p) = U(2) = 0$ that $X = 0 = V(0)$, hence $i_0 = 0$, for $Y = U(m - p) = U(10) = 25$ we have $Y = 25 = V(5)$, hence $i_1 = 5$.

For $j = 0, 1, \ldots, n$, let us denote $N_{p,j} = N_{p,j}(u)$ the B-spline base functions of degree p, where $u \in \langle U(0), U(m))$, and for r, s integers such that $0 \leq r \leq s \leq sk$ we use the more accurate notation $N_p(j; V(r), V(s))(u)$, if $N_{p,j}(u)$ is non-zero in the interval $\langle V(r), V(s))$, and zero everywhere else. Specifically, for the case of (3)–(5), we get with the help of the Cox - de Boor recursive formula:

$$N_{0,j} = \begin{cases} 1 & \text{for } U(j) \leq u < U(j+1), \\ 0 & \text{otherwise }, \end{cases} \quad (6)$$

$$N_{p,j} = \frac{u - U(j)}{U(j+p) - U(j)} N_{p-1,j} + \frac{U(j+p+1) - u}{U(j+p+1) - U(j+1)} N_{p-1,j+1};$$

during the calculations, the meaningless expressions are equal to zero. Then we get the following results:

$$
\begin{aligned}
N_{2,0} &= N_2(0;\langle 0,5)) = 0.04u^2 - 0.4u + 1 \quad \text{for } u \in \langle 0,5),\\
N_{2,1} &= N_2(1;\langle 0,10)) = \begin{cases} -0.06u^2 + 0.4u & \text{for } u \in \langle 0,5),\\ 0.02u^2 - 0.4u + 2 & \text{for } u \in \langle 5,10),\end{cases}\\
N_{2,2} &= N_2(2;\langle 0,10)) = \begin{cases} 0.02u^2 & \text{for } u \in \langle 0,5),\\ -0.06u^2 + 0.8u - 2 & \text{for } u \in \langle 5,10),\end{cases}\\
N_{2,3} &= N_2(3;\langle 5,10)) = 0.04u^2 - 0.4u + 1 \quad \text{for } u \in \langle 5,10),\\
N_{2,4} &= N_2(4;\langle 10,10)) = 0 \quad \text{for every } u,\\
N_{2,5} &= N_2(5;\langle 10,15)) = 0.04u^2 - 1.2u + 9 \quad \text{for } u \in \langle 10,15),\\
N_{2,6} &= N_2(6;\langle 10,20)) = \begin{cases} -0.06u^2 + 1.6u - 10 & \text{for } u \in \langle 10,15),\\ 0.02u^2 - 0.8u + 8 & \text{for } u \in \langle 15,20),\end{cases}\\
N_{2,7} &= N_2(7;\langle 10,25)) = \begin{cases} 0.02u^2 - 0.4u + 2 & \text{for } u \in \langle 10,15),\\ -0.04u^2 + 1.4u - 11.5 & \text{for } u \in \langle 15,20),\\ 0.02u^2 - u + 12.5 & \text{for } u \in \langle 20,25),\end{cases}\\
N_{2,8} &= N_2(8;\langle 15,25)) = \begin{cases} 0.02u^2 - 0.6u + 4.5 & \text{for } u \in \langle 15,20),\\ -0.06u^2 + 2.6u - 27.5 & \text{for } u \in \langle 20,25),\end{cases}\\
N_{2,9} &= N_2(9;\langle 20,25)) = 0.04u^2 - 1.6u + 16 \quad \text{for } u \in \langle 20,25).
\end{aligned} \tag{7}
$$

The number of base functions $N_{2,j}$ is given by the number $n + 1 = 9 + 1 = 10$, and we say that they belong to the nodals vector (5).

2 Parametrized equation of a B-spline of order p

The parametrized equation of a B-spline of order p, that is given by the control points (1), nodal vector (2), and its corresponding base B-splines of order p, is defined for $u \in \langle U(0), U(m)) \neq \emptyset$ by

$$
B(u) = \sum_{j=0}^{n} N_{p,j}(u) \cdot P(j). \tag{8}
$$

The curve consists of sk segments. For example, for the case of (3)–(5), for $u \in \langle U(9), U(10)) = \langle 20, 25) \subset \langle U(0), U(12))$, only $N_{2,7}$, $N_{2,8}$, and $N_{2,9}$ are non-zero (see (7)), while according to (8) the coordinates $x(u)$, $y(u)$ of $B(u)$ are for $u \in \langle 20, 25)$:

$$
\begin{aligned}
x(u) &= 5 \cdot (0.02u^2 - u + 12.5) + 3 \cdot (-0.06u^2 + 2.6u - 27.5)\\
&\quad + 0 \cdot (0.04u^2 - 1.6u + 16) = -0.08u^2 + 2.8u - 20,\\
y(u) &= 5 \cdot (0.02u^2 - u + 12.5) + 5 \cdot (-0.06u^2 + 2.6u - 27.5)\\
&\quad + 2 \cdot (0.04u^2 - 1.6u + 16) = -0.12u^2 + 4.8u - 43.
\end{aligned} \tag{9}
$$

For $u \in \langle 20, 25\rangle$, the equations (9) are obviously the parametrized equations of the last fifth segment of the constructed B-spline of order 2 which is closed for $u \in \langle 20, 25\rangle$.

3 The most used B-spline curves

In application at MIR company, the so-called "clamped", or "open", or "closed" B-spline curves of order p, with non-empty domain interval $\langle U(p), U(m-p)\rangle$ are used the most frequently. From a clamped curve we demand that the initial and end point of the vector (2) have multiplicity $p+1$; from an open curve we demand the same multiplicity not to be greater than p and at last, from a closed curve we demand that $P(i) = P(n-p+i+1)$, $i = 0, 1, \ldots, p-1$, for the control points (1) and that the coordinates of the vector (2) form an arithmetic sequence. Hence, the case (3)–(5) corresponds with the construction of a clamped B-spline curve of order 2.

In all the three aforementioned cases the B-spline curve "copies" the "control polygonal trail" with vertices (1). It is characteristic for a clamped curve that at $P(0)$ (where it has its beginning) and $P(n)$ (where it ends) it touches the segments $P(0)P(1)$ and $P(n-1)P(n)$, respectively, of the polygonal trail. For a closed curve its initial and end points coincide.

For a clamped open, or closed B-spline curve of order p, defined on the interval $\langle U(p), U(m-p)\rangle$, or on $\langle U(p), U(m-p)\rangle$ (because it makes sense only to consider the case when $\langle U(p), U(m-p)\rangle$ is a non-empty set), the following holds: if it holds, for q integer from $1, 2, \ldots, i_1 - i_0 - 1$ (where $i_1 - i_0 \geq 2$, see Subsection 1), that the multiplicity $k \geq 1$ of the nodal $V(i_0+q)$ is greater or equal than $p+1$, then the (closed) q-th and $(q+1)$-th segments of the curve do not meet; if $k < p+1$, then they do meet. For example, for the case of (3)–(5), where $i_0 = 0$ and $i_1 = 5$, it holds for $q = 4$ from the sequence $1, 2, \ldots, i_1 - i_0 - 1 = 4$ that the multiplicity of the nodal $V(i_0+q) = V(4) = 20$ is $k = 1 < p+1 = 2+1 = 3$, therefore the fourth (closed) segment, for which $u \in \langle V(3), V(4)\rangle = \langle 15, 20\rangle$, is connected to the fifth (closed) segment, where $u \in \langle V(4), V(5)\rangle = \langle 20, 25\rangle$. For $q = 2$, the multiplicity of the nodal $V(2) = 10$ is $k = 4 \geq p+1 = 2+1 = 3$, hence the second and third segments of the curve are not connected. The total number of (closed) segments of the constructed B-spline curve of order 2 is equal to $i_1 - i_0 = 5$.

Literature:

a) Spline-functions and spline-curves, theory: [1], [3], [6], [8], [10], [11]

b) Spline-functions for statistical purposes: [4], [7]

c) Elements of difference calculus: [9]

4 Segmented construction of planar B-spline curve

Let us consider a clamped, or open, or closed B-spline of order $p \geq 1$ defined on $\langle U(p), U(m-p)\rangle$. We describe the so-called "segmented" construction of these curves.

Let Q be an integer, let $1 \leq Q \leq i_1 - i_0$ and $1 \leq g = \lfloor \frac{i_1 - i_0}{Q} \rfloor$ be integers such that $i_1 - i_0 = g \cdot Q + h$, for some integer $0 \leq h < Q$. For the case of (3)–(5), where $i_1 - i_0 = 5$, it follows that for $Q = 2$ there is $g = \lfloor \frac{5}{2} \rfloor = 2$, $h = 5 - 2 \cdot 2 = 1$. Or, for $Q = 2$ we get $g = \lfloor \frac{5}{5} \rfloor = 1$ and $h = 5 - 1 \cdot 5 = 0$.

1. In the first step of the mentioned segmented construction of the B-spline curve we choose from the set of all base functions $N_p(j; V(r), V(s))$, for $j = 0, 1, \ldots, n$, those ones for which the intersection $\langle V(r), V(s)\rangle \cap \langle V(i_0), V(i_0 + Q)\rangle$ is non-empty, $r < s$, or, for $r = s$, the nodal $V(r) = V(s)$ lies in the closure of the interval $\langle V(i_0), V(i_0 + Q)\rangle$:

$$N_{p,j_1},\ N_{p,j_1+1},\ \ldots,\ N_{p,j_1+w_1}. \tag{10}$$

We assign accordingly the control points to the sequence (10)

$$P(j_1),\ P(j_1 + 1),\ \ldots,\ P(j_1 + w_1), \tag{11}$$

$n_1 + 1 = w_1 + 1$ in number, hence $n_1 = w_1$, then, for $m_1 = n_1 + p + 1$, we choose an $(m_1 + 1)$-dimensional nodal vector

$$U_1 = (U(j_1),\ U(j_1 + 1),\ \ldots,\ U(j_1 + w_1)) \tag{12}$$

from the vector (2). We will denote by B_1 the B-spline of order p that was created with the help of the control points (11), the nodal vector (12), and the corresponding B-splines (10) and call it *the first segment* of the constructed B-spline of order p.

For the case of (3)–(5) and chosen $Q = 2$ the base functions $N_2(j; V(r), V(s))$, $r < s$, for which the intersection $\langle V(r), V(s)\rangle \cap \langle V(i_0), V(i_0 + Q)\rangle$ is non-empty, or, for $r = s$, the nodal $V(r) = V(s)$ lies in the closure of the interval $\langle 0, 10\rangle$, are the following:

$$N_{2,0},\ N_{2,1},\ N_{2,2},\ N_{2,3},\ N_{2,4}. \tag{13}$$

In a corresponding way we assign to this sequence the sequence of control points.

$$P(0),\ P(1),\ P(2),\ P(3),\ P(4), \tag{14}$$

altogether $n_1 + 1 = w_1 + 1 = 5$ in number, hence $n_1 = w_1 = 4$, and for $m_1 = n_1 + p + 1 = 4 + 2 + 1 = 7$ we choose from the nodal vector (5) the $7 + 1 = 8$-dimensional nodal vector

$$U_1 = (U(0), U(1), \ldots, U(7)) = (0, 0, 0, 5, 10, 10, 10, 10). \tag{15}$$

We denote by B_1 the B-spline curve of order 2 constructed with the help of the control points (14), the nodal vector (15), and with the corresponding B-spline functions (13), the domain of which is $\langle 0, 10\rangle$ and belongs to a further not specified category (with respect to the aforementioned clamped, open, or closed B-splines) and call it the first segment of the constructed B-spline curve of order 2.

2. In the second step (if it is realizable, i.e., $2Q \leq i_1 - i_0$) we proceed analogously as in the first step, that is, from the sequence of base functions $N_p(j; V(r), V(s))$ we subsequently pick all the base functions for which, if $r < s$, the intersection $\langle V(r), V(s)\rangle \cap \langle V(i_0 + Q), V(i_0 + 2Q)\rangle$ is non-empty or the nodal $V(r) = V(s)$, for $r = s$, lies in the closure of the interval $\langle V(i_0 + Q), V(i_0 + 2Q)\rangle$:

$$N_{p,j_2},\ N_{p,j_2+1},\ \ldots,\ N_{p,j_2+w_2}. \tag{16}$$

In a corresponding way, we assign, to the sequence (16), the sequence of control points.

$$P(j_2),\ P(j_2 + 1),\ \ldots, P(j_2 + w_2), \tag{17}$$

altogether $n_2 + 1 = w_2 + 1$ in number, hence $w_2 = n_2$, and for $m_2 = n_2 + p + 1$ we choose from (2) the $(m_2 + 1)$-dimensional nodal vector

$$U_2 = (U(j_2), U(j_2 + 1), \ldots, U(j_2 + w_2)). \quad (18)$$

We denote by B_2 the B-spline of order p that was formed with the help of the control points (17), the nodal vector (18), and the corresponding B-splines (16) and call it *the second segment* of the constructed B-spline of order 2.

For the case of (3)–(5) and the chosen $Q = 2$ we have the following base functions $N_2(j; V(r), V(s))$, for which the intersection $\langle V(r), V(s)\rangle \cap \langle V(i_0+Q), V(i_0+2Q)\rangle = \langle V(2), V(4)\rangle = \langle 10, 20\rangle$ is non-empty, if $r < s$, or the nodal $V(r) = V(s)$, for $r = s$, lies in the closure of the interval $\langle 10, 20\rangle$:

$$N_{2,4},\ N_{2,5},\ N_{2,6},\ N_{2,7},\ N_{2,8}. \quad (19)$$

In a corresponding way, we assign, to this sequence, the sequence of control points

$$P(4),\ P(5),\ P(6),\ P(7),\ P(8), \quad (20)$$

altogether $n_2 + 1 = w_2 + 1 = 5$ in number, hence $n_2 = w_2 = 4$, and for $m_2 = n_2 + p + 1 = 4 + 2 + 1 = 7$ we choose from (5) the $7 + 1 = 8$-dimensional nodal vector

$$U_2 = (U(4), U(5), \ldots, U(11)) = (10, 10, 10, 10, 15, 20, 25, 25). \quad (21)$$

We denote by B_2 the B-spline of order 2 that was formed with the help of the control points (20), the nodal vector (21), and the corresponding B-splines (19) and call it the second segment of the constructed B-spline of order 2.

It follows from the equation $i_1 - i_0 = g \cdot Q + h$ that there exist $g + \mathrm{sgn}(h)$ segments B_1, B_2, ..., B_g, $B_{g+\mathrm{sgn}(h)}$ of the constructed B-spline curve of order p. Above in Step 1 and 2 we described the procedure how to obtain B_1 and B_2, the other segments up to the g-th are obtained analogously. If $h > 0$, then it remains to consturct the last $(g + 1)$-th segment, when from the sequence of base functions $N_p(j; V(r), V(s))$ we subsequently choose those for which the intersection $\langle V(r), V(s)\rangle \cap \langle V(i_0 + gQ), V(i_1)\rangle$ is non-empty, if $r < s$, or the nodal $V(r) = V(s)$, for $r = s$, lies in the closure of the interval $\langle V(i_0 + gQ), V(i_1)\rangle$:

$$N_{p,j_{g+1}},\ N_{p,j_{g+1}+1},\ \ldots,\ N_{p,j_{g+1}+w_{g+1}}. \quad (22)$$

In a corresponding way, we assign, to the sequence (22) the sequence of control points

$$P(j_{g+1}),\ P(j_{g+1} + 1),\ \ldots, P(j_{g+1} + w_{g+1}), \quad (23)$$

altogether $n_{g+1} + 1 = w_{g+1} + 1$ in number, hence $n_{g+1} = w_{g+1}$, and for $m_{g+1} = n_{g+1} + p + 1$ we choose from (2) the $(m_{g+1} + 1)$-dimensional nodal vector

$$U_{g+1} = (U(j_{g+1}), U(j_{g+1} + 1), \ldots, U(j_{g+1} + m_{g+1})). \quad (24)$$

The B-spline of order p that was formed with the help of the control points (23), the nodal vector (24), and the corresponding B-splines (22) is the last $(g + 1)$-th segment B_{g+1} of the constructed B-spline of order p.

For the case of (3)–(5) and the chosen $Q = 2$, it follows from $i_1 - i_0 = gQ + h$, i.e. $5 = 2 \cdot 2 + 1$, that there exist $g + \mathrm{sgn}(h) = 2 + 1 = 3$ segments of the constructed B-spline of order 2: B_1, B_2, B_3. Since B_1 and B_2 have already been contructed above, it is necessary to construct only the segment B_3. The base functions $N_2(j; V(r), V(s))$, for which the intersection $\langle V(r), V(s)\rangle \cap \langle V(i_0 + gQ), V(i_1)) = \langle V(0 + 2 \cdot 2), V(5)) = \langle V(4), V(5)) = \langle 20, 25)$ is non-empty, if $r < s$, or the nodal $V(r) = V(s)$, for $r = s$, lies in the closure of the interval $\langle 20, 25)$:

$$N_{2,7},\ N_{2,8},\ N_{2,9}. \tag{25}$$

In a corresponding way, we assign, to this sequence the sequence of control points

$$P(7),\ P(8),\ P(9), \tag{26}$$

altogether $n_3 + 1 = w_3 + 1 = 3$ in number, hence $n_3 = w_3 = 2$, and for $m_3 = n_3 + p + 1 = 2 + 2 + 1 = 5$ we choose from (5) the $5 + 1 = 6$-dimensional nodal vector

$$U_3 = (U(7), U(8), \ldots, U(12)) = (10, 15, 20, 25, 25, 25). \tag{27}$$

The B-spline of order 2 with domain $\langle 20, 25)$ that was formed with the help of the control points (26), the nodal vector (27), and the corresponding B-splines (25), and belongs to a further not specified category, is the last segment B_3 of the constructed B-spline of order 2.

Let us note that for the segments B_i, $i = 1, 2, \ldots, g$, the variable of the corresponding parametric equations lies in the interval $\langle V(i_0 + (i-1)Q), V(i_0 + iQ))$, while for B_{g+1} (if it exists) in the interval $\langle V(i_0 + gQ), V(i_1))$. The particular segments of the constructed B-spline curve can be considered closed.

Example 1

For $n = 9$ let us consider in $\mathbb{R}^2$ the following $n + 1 = 9 + 1 = 10$ control points

$$P(0),\ P(1), \ldots,\ P(9). \tag{28}$$

For $p = 3$, there will be $m = n + p + 1 = 9 + 3 + 1 = 13$ and let there be the $m + 1 = 13 + 1 = 14$-dimensional nodal vector

$$U = (U(0), U(1), \ldots, U(13)) = (1, 2, 2, 2, 4, 6, 6, 8, 8, 8, 10, 12, 12, 12). \tag{29}$$

We will proceed as in Section 1. The number $sk \geq 1$ that gives the total number $sk + 1$ of mutually different nodals of (29) is equal to $sk = 6$. For $i = 0, 1, \ldots, sk = 6$, the nodals $V(i)$ are the following:

$$1 = V(0) < 2 = V(1) < 4 = V(2) < 6 = V(3) < 8 = V(4) < 10 = V(5) < 12 = V(6).$$

For $X = U(p) = U(3) = 2$, there is $X = 2 = V(1)$, i.e. $i_0 = 1$, further, for $Y = U(m - p) = U(10) = 10$ there is $Y = 10 = V(5)$, i.e. $i_1 = 5$.

The B-splines $N_{3,j}$ of order 3, $j = 0, 1, \ldots, n = 9$ corresponding to the nodal vector (29) are the following (we use the notation introduced above, however, to save space we do not give their particular analytic expressions):

$$\begin{array}{llll} N_3(0; \langle 2, 4)), & N_3(1; \langle 2, 6)), & N_3(2; \langle 2, 6)), & N_3(3; \langle 2, 8)), \\ N_3(4; \langle 4, 8)), & N_3(5; \langle 6, 8)), & N_3(6; \langle 6, 10)), & N_3(7; \langle 8, 12)), \\ N_3(8; \langle 8, 12)), & N_3(9; \langle 8, 12)). & & \end{array} \tag{30}$$

In the spirit of Section 3 the constructed B-spline of order 3 will be open, as the multiplicity of the initial and end nodals of (29) is not greater than $p = 3$; its domain will be the interval $\langle U(p), U(m-p)) = \langle U(3), U(10)) = \langle 2, 10)$, eventually its closure $\langle 2, 10\rangle$. The curve will consist of $i_1 - i_0 = 5 - 1 = 4$ (closed) segments. Since, for example, for $q = 1$, chosen from the sequence $1, 2, 3 = i_1 - i_0 - 1$, the multiplicity of the nodal $V(i_0 + q) = V(1+1) = V(2) = 4$ is equal to $k = 1 < p + 1 = 4$ the first (closed) segment with domain $\langle V(i_0 + q - 1), V(i_0 + q)\rangle = \langle V(1), V(2)\rangle = \langle 2, 4\rangle$ is connected onto the second segment with domain $\langle V(2), V(3)\rangle = \langle 4, 6\rangle$. Or for $q = 2$, chosen from the sequence $1, 2, 3 = i_1 - i_0 - 1$, the multiplicity of the nodal $V(i_0 + q) = V(1+2) = V(3) = 6$ is equal to $k = 2 < p + 1 = 4$, therefore the second (closed) segment of the curve is connected onto the third (closed) segment. We easily verify that (for $q = 3$) the third (closed) segment is connected onto the fourth (closed) segment. The curve is continuous.

We proceed further as in Section 4. Let, for example, be $Q = 3$. We then have

$$g = \lfloor (i_1 - i_0)/Q \rfloor = \lfloor 4/3 \rfloor = 1 \tag{31}$$

and it holds that $i_1 - i_0 = 4 = g \cdot Q + h = 1 \cdot 3 + h = 3 + h$, hence $h = 1$. In the first step of the construction of the corresponding B-spline of order 3, which consists of $g + \text{sgn}(h) = 1 + \text{sgn}(1) = 1 + 1 = 2$ segments B_1, B_2, we choose from the sequence of base functions $N_3(j; < V(r), V(s)))$, where $j = 0, 1, \ldots, 9$ (see (30)), subsequently all the base functions for which the intersection $\langle V(r), V(s)\rangle \cap \langle V(i_0), V(i_0 + Q)) = \langle V(1), V(4)) = \langle 2, 8)$ is non-empty, if $r < s$, or the nodal $V(r) = V(s)$, for $r = s$, lies in the closure of $\langle V(1), V(4)) = \langle 2, 8)$:

$$N_{3,0},\ N_{3,1},\ N_{3,2},\ N_{3,3},\ N_{3,4},\ N_{3,5},,\ N_{3,6}. \tag{32}$$

In a corresponding way, we assign, to the sequence (32), the sequence of control points

$$P(0),\ P(1),\ P(2),\ P(3),\ P(4),\ P(5),\ P(6), \tag{33}$$

altogether $n_1 + 1 = w_1 + 1 = 7$ in number, hence $n_1 = w_1 = 6$, and for $m_1 = n_1 + p + 1 = 6 + 3 + 1 = 10$ we choose from (29) the $m_1 + 1 = 10 + 1 = 11$-dimensional nodal vector

$$U_1 = (U(0), U(1), \ldots, U(10)) = (1, 2, 2, 2, 4, 6, 6, 8, 8, 8, 10). \tag{34}$$

The B-spline of order 3 with domain $\langle 2, 8)$ that was formed with the help of the control points (33), the nodal vector (34), and the corresponding B-splines (32), and belongs to the category of open B-spline curves, is the first segment B_1 of the segment-wise construction of the given B-spline of order 3.

As $h = 1 > 0$, it is necessary to construct the second (and last) segment $B_2 = B_{g+\text{sgn}(h)} = B_{1+1}$. We choose from $N_3(j; < V(r), V(s)), j = 0, 1, \ldots, 9$, the sequence of B-spline base functions (see (30)), subsequently all the base functions for which the intersection $\langle V(r), V(s)\rangle \cap \langle V(i_0 + gQ), V(i_1)) = \langle V(1 + 1 \cdot 3), V(5)) = \langle V(4), V(5)) = \langle 8, 10)$ is non-empty, if $r < s$, or the nodal $V(r) = V(s)$, for $r = s$, lies in the closure of $\langle 8, 10)$:

$$N_{3,6},\ N_{3,7},\ N_{3,8},\ N_{3,9}. \tag{35}$$

In a corresponding way, we assign, to the sequence (35), the sequence of control points

$$P(6),\ P(7),\ P(8),\ P(9), \tag{36}$$

altogether $n_2 + 1 = w_2 + 1 = 4$ in number, hence $n_2 = w_2 = 3$, and for $m_2 = n_2 + p + 1 = 3 + 3 + 1 = 7$ we choose from (29) the $m_2 + 1 = 7 + 1 = 8$-dimensional nodal vector

$$U_2 = (U(6), U(7), \ldots, U(13)) = (6, 8, 8, 8, 10, 12, 12, 12). \tag{37}$$

The B-spline of order 3 with domain $\langle V(i_0+gQ), V(i_1)) = \langle V(4), V(5)) = \langle 8, 10)$ that was created with the help of the control points (36), the nodal vector (37), and the corresponding B-splines (35), and belongs to the category of open B-spline curves, is the second (and last) segment B_2 of the segment-wise construction of the given B-spline of order 3.

Example 2

Let $p = 2$. For $n = 9$, let us consider in $\mathbb{R}^2$ the following $n + 1 = 9 + 1 = 10$ control points:

$$\begin{array}{llll} P(0) = (1, 11), & P(1) = (12, 11), & P(2) = (12, 2), & \\ P(3) = (8, 2), & P(4) = (8, 8), & P(5) = (4, 8), & \\ P(6) = (4, 3), & P(7) = (1, 3), & P(8) = (1, 11), & P(9) = (12, 11). \end{array} \tag{38}$$

For $i = 0, 1, \ldots, p-1 = 2-1 = 1$, it holds that $P(i) = P(n-p+i+1)$: $P(0) = P(9-2+0+1) = P(8)$, $P(1) = P(9-2+1+1) = P(9)$. Further, for $m = n+p+1 = 9+2+1 = 12$ let us consider the $m + 1 = 12 + 1 = 13$-dimensional nodal vector

$$U = (U(0), U(1), \ldots, U(12)) = (0, 1, 2, 3, 4, 5, 6, 7, 8, 9, 10, 11, 12), \tag{39}$$

the components of which form an arithmetic sequence. Obviously, it will be a construction of a closed B-spline of order 2 with domain $\langle U(p), U(m-p)) = \langle U(2), U(10)) = \langle 2, 10)$.

We proceed as in Section 1. The number $sk \geq 1$, which gives the total number $sk + 1$ of mutually different nodals (components) of the vector (39), is equal to $sk = 12$. For $i = 0, 1, \ldots, sk = 12$, there is $V(i) = i$.

For $X = U(p) = U(2) = 2$, there is $X = 2 = V(2)$, i.e., $i_0 = 2$, for $Y = U(m-p) = U(10) = 10$ there is $Y = 10 = V(10)$, i.e., $i_1 = 10$. The constructed B-spline of order 2 will consist of $i_1 - i_0 = 10 - 2 = 8$ (closed) segments. For a chosen number q from the sequence $1, 2, \ldots, i_1 - i_0 - 1 = 8 - 1 = 7$, the nodal $V(i_0 + q) = V(2 + q) = 2 + q$ has multiplicity $k = 1 < p+1 = 3$, what means that the q-th (closed) segment is connected onto the $(q + 1)$-th (closed) segment. The trail of the curve is continuous.

Considering the facts from the general theory of B-spline curves, we can see that the following holds for the B-splines $N_{2,j}$, $j = 0, 1, \ldots, n = 9$, corresponding to the nodal vector (39):

$$\begin{aligned} N_{2,j} &= N_2(j; \langle j, j+3)) = \\ &= \begin{cases} \frac{1}{2}(u-j)^2 & \text{for } j \leq u < j+1, \\ \frac{1}{2}(-2u^2 + 2u(2j+3) - (2j^2 + 6j + 3)) & \text{for } j+1 \leq u < j+2, \\ \frac{1}{2}(u - (j+3))^2 & \text{for } j+2 \leq u < j+3. \end{cases} \end{aligned} \tag{40}$$

For example,

$$N_{2,7} = N_2(7; \langle 7, 10)) = \begin{cases} \frac{1}{2}(u^2 - 14u + 49) & \text{for } 7 \leq u < 8, \\ \frac{1}{2}(-2u^2 + 34u - 143) & \text{for } 8 \leq u < 9, \\ \frac{1}{2}(u^2 - 20u + 100) & \text{for } 9 \leq u < 10. \end{cases}$$

We proceed in accordance with Section 4. Let, for example, $Q = 4$. There is then $g = \lfloor (i_1 - i_0)/Q \rfloor = \lfloor 8/4 \rfloor = \lfloor 2 \rfloor = 2$ and it holds that $i_1 - i_0 = 8 = g \cdot Q + h = 2 \cdot 4 + h = 8 + h$, hence $h = 0$. In the first step of the construction of the considered B-spline of order 2, which consists of $g + \operatorname{sgn}(h) = 2 + \operatorname{sgn}(0) = 2 + 0 = 2$ segments B_1, B_2, we choose from the sequence of base functions $N_2(j; < V(r), V(s)), j = 0, 1, \ldots, n = 9$ (see (40)) subsequently those for which the intersection $\langle V(r), V(s)) = \langle r, s) \cap \langle V(i_0), V(i_0 + Q)) = \langle i_0, i_0 + Q) = \langle 2, 2 + 4) = \langle 2, 6)$ is non-empty, if $r < s$, or the nodal $V(r) = r = s = V(s)$, for $r = s$ lies in the closure of the interval $\langle 2, 6)$:

$$N_{2,0},\ N_{2,1},\ N_{2,2},\ N_{2,3},\ N_{2,4},\ N_{2,5}. \tag{41}$$

In a corresponding way, we assign, to the sequence (41), the sequence of control points

$$P(0),\ P(1),\ P(2),\ P(3),\ P(4),\ P(5), \tag{42}$$

altogether $n_1 + 1 = w_1 + 1 = 6$ in number, hence $n_1 = w_1 = 5$, and for $m_1 = n_1 + p + 1 = 5 + 2 + 1 = 8$ we choose from (39) the $m_1 + 1 = 8 + 1 = 9$-dimensional nodal vector

$$U_1 = (U(0), U(1), \ldots, U(8)) = (0, 1, 2, 3, 4, 5, 6, 7, 8). \tag{43}$$

The B-spline of order 2 with domain $\langle V(i_0), V(i_0 + Q)) = \langle 2, 6)$ that was created with the help of the control points (42), the nodal vector (43), and the corresponding B-splines (41) is the first segment B_1 of the given B-spline of order 2 constructed segment-wise.

In the second step which is feasible as $2Q = 2 \cdot 4 = 8 \leq i_1 - i_0 = 8$ we choose from the sequence of base functions (40) subsequently those for which the intersection $\langle r, s) \cap$ $\cap \langle V(i_0 + Q), V(i_0 + 2Q)) = \langle V(2 + 4), V(2 + 2 \cdot 4)) = \langle V(6), V(10)) = \langle 6, 10)$ is non-empty, if $r < s$, or the nodal $V(r) = r = s = V(s)$, for $r = s$ lies in the closure of the interval $\langle 6, 10)$:

$$N_{2,4},\ N_{2,5},\ N_{2,6},\ N_{2,7},\ N_{2,8},\ N_{2,9}. \tag{44}$$

In a corresponding way, we assign, to the sequence (44), the sequence of control points

$$P(4),\ P(5),\ P(6),\ P(7),\ P(8),\ P(9), \tag{45}$$

altogether $n_2 + 1 = w_2 + 1 = 6$ in number, hence $n_2 = w_2 = 5$, and for $m_2 = n_2 + p + 1 = 5 + 2 + 1 = 8$ we choose from (39) the $m_2 + 1 = 8 + 1 = 9$-dimensional nodal vector

$$U_2 = (U(4), U(5), \ldots, U(12)) = (4, 5, 6, 7, 8, 9, 10, 11, 12). \tag{46}$$

The B-spline of order 2 with domain $\langle V(i_0 + Q), V(i_0 + 2Q)) = \langle 6, 10)$ that was created with the help of the control points (45), the nodal vector (46), and the corresponding B-splines (44) is the second and last segment B_2 of the given B-spline of order 2 constructed segment-wise.

Lastly, we attach the parametric equations of B_1 and B_2, computed according to Section 2.

B_1 :

$$x(u) = \begin{cases} \frac{1}{2}(-11u^2 + 66u - 75) & \text{for } 2 \leq u < 3, \\ -2u^2 + 12u - 6 & \text{for } 3 \leq u < 4, \\ 2u^2 - 20u + 58 & \text{for } 4 \leq u < 5, \\ -2u^2 + 20u - 42 & \text{for } 5 \leq u < 6, \end{cases}$$

$$y(u) = \begin{cases} \frac{1}{2}(-9u^2 + 36u - 14) & \text{for } 2 \le u < 3, \\ \frac{1}{2}(9u^2 - 72u + 148) & \text{for } 3 \le u < 4, \\ 3u^2 - 24u + 50 & \text{for } 4 \le u < 5, \\ -3u^2 + 36u - 100 & \text{for } 5 \le u < 6, \end{cases}$$

B_2 :

$$x(u) = \begin{cases} 2u^2 - 28u + 102 & \text{for } 6 \le u < 7, \\ \frac{1}{2}(-3u^2 + 42u - 139) & \text{for } 7 \le u < 8, \\ \frac{1}{2}(3u^2 - 54u + 245) & \text{for } 8 \le u < 9, \\ \frac{1}{2}(11u^2 - 198u + 893) & \text{for } 9 \le u < 10, \end{cases}$$

$$y(u) = \begin{cases} \frac{1}{2}(-5u^2 + 60u - 164) & \text{for } 6 \le u < 7, \\ \frac{1}{2}(5u^2 - 80u + 326) & \text{for } 7 \le u < 8, \\ 4u^2 - 64u + 259 & \text{for } 8 \le u < 9, \\ -4u^2 + 80u - 389 & \text{for } 9 \le u < 10. \end{cases}$$

5 Conclusion

Author of this paper made a computer program which can solve mathematical and constructive part of segmented construction of planar B-spline curves of 1st, 2nd and 3rd order. With help of this program, the specific tasks can be easily solved.

References

[1] Bézier, P. (1972). *Numerical Control: Mathematics and Applications*. John Wiley & Sons Ltd, New York.

[2] De Boor, C. (1972). On calculating with B-splines. In: *Journal of Approximation Theory*, 6, 50–62.

[3] Bőhmer, K. (1974). *Spline-Funktionen*. Teubner, Stuttgart.

[4] Kaňka, M. (2015). Segmented Regression Based on B-splines with Solved Examples. *Statistics and Economy Journal* 95 (4), 47–66.

[5] Kaňka, M. (2016). Segmented Regression Based on Cut-off Polynomials. *Statistics and Economy Journal* 96 (2), 60–72.

[6] Makarov, V. L., and Chlobystov, V. V. (1983). *Splajn-approximacija funkcij*. Vysšaja škola, Moscow.

[7] Meloun, M., and Militký, J. (1994). *Statistické zpracování experimentálních dat*. PLUS, s.r.o., Prague.

[8] Spät, H. (1966). *Spline Algorithmen zur Konstruktion glatter Kurven und Flächen*, 4th Ed. Oldenbourg, München-Wien.

[9] Schrutka, L. (1945). *Leitfaden der Interpolation*. Springer-Verlag, Wien.

[10] Vasilenko, V. A. (1983). *Splajn-funkcii: tĕorija, algoritmy, programmy*. Nauka, Novosibirsk.

[11] Wang, X., Wu, Z., Shen, J., Zhang, T., Mou, X., Zhou, M. (2016). *Repairing the cerebral vascular through blending Ball B-spline curves with G(2) continuity*. Neurocomputing, Amsterdam, 768-777.

Current address

Kaňka Miloš, Doc. RNDr. CSc.
Department of Mathematics, University of Economics
Ekonomická 957, 14800 Prague, Czech Republic
E-mail: kanka@vse.cz

Slovak University of Technology in Bratislava
Faculty of Mechanical Engineering
16th Conference on Applied Mathematics **APLIMAT 2017**
Proceedings

EWMA BASED LTPD VARIABLES SAMPLING PLANS AND COST MODELS

KASPŘÍKOVÁ Nikola (CZ)

Abstract. The calculation of the recently proposed unknown standard deviation case LTPD acceptance sampling plans minimizing the mean inspection cost per lot of the process average quality, when the design of the variables sampling plans is based on the exponentially weighted moving average statistic is recalled. Then the economic efficiency of the plans is discussed using cost model considering difference in variable and fixed cost.

Keywords: Sampling plans, LTPD, EWMA, cost optimization, inspection by variables

Mathematics subject classification: 62D99

1 Introduction

Sampling inspection is a basic tool of industry statistics. The acceptance sampling plans are designed and used to improve the economic efficiency of business process. There exist many types of such acceptance sampling plans. New rectifying lot tolerance percent defective (LTPD) plans for sampling inspection by variables based on the exponentially weighted moving average statistic have been recently introduced. [6] addresses the case of known standard deviation and the plans for the unknown standard deviation case are introduced in [5]. This paper recalls the design of these plans and focuses on new approach to the comparative economic efficiency evaluation of the plans. The R software extension package [8] which implements the algorithms for the calculation of the plans and which has been published on the Comprehensive R Archive Network can be used for the calculation of the plans.

The LTPD sampling plans minimizing the mean inspection cost per lot of process average quality when the remainder of rejected lots is inspected were originally designed by Dodge and Romig for the inspection by attributes - see [2]. Plans for the inspection by variables and for the inspection by variables and attributes (all items from the sample are inspected by variables, the remainder of rejected lots is inspected by attributes) were then proposed and it was shown that these plans are in many situations more economical than the corresponding

Dodge-Romig attribute sampling plans. The LTPD plans for the inspection by variables and attributes have been introduced in [10], using an approximate calculation of the plans. Exact plans, using non-central t distribution in the calculation of the operating characteristic, have been reported in [11] and implemented in the LTPDvar R package [7]. The operating characteristics used for these plans are discussed in [3] and [4]. It has been shown that these plans are in many situations superior to the original attribute sampling plans and similar results have been obtained for the AOQL plans – the analysis is provided in [8] and in [9]. An economic model considering difference in variable cost is used in [8] and [9].

The plans proposed in [5] reflect the recent development in the field of the design of the acceptance sampling plans and makes use of the exponentially weighted moving average (EWMA) statistic for the inspection by variables. The EWMA statistic has often been used in the process monitoring applications, nevertheless its application in the design of the acceptance sampling plans is rather new (see [13], and [1]).

The recently proposed LTPD plans for the inspection by variables and attributes which are designed to use the EWMA statistics in the decision procedure are addressed in the present paper and the discussion of the economic performance of such plans is provided. The case of the unknown standard deviation is considered in this paper. The structure of the paper is as follows: the LTPD plans for the inspection by attributes minimizing the mean inspection cost per lot of the process average quality when the remainder of the rejected lots is inspected are recalled first, then the LTPD plans for the inspection by variables and attributes using EWMA statistic are calculated using the LTPDvar package and the evaluation of the comparative economic efficiency is shown, using a simple economic model which considers differences in both variable and fixed costs.

1.1 LTPD sampling plans for the inspection by attributes

First we recall the attribute sampling plans. Dodge and Romig in [2] consider the case, where each inspected item is to be classified as either good or non-conforming (acceptance sampling by attributes) and they search for the sampling plans which minimize the mean number of the items inspected per lot of the process average quality

$$I_s = N - (N - n) \cdot L(\bar{p}; n; c) \tag{1}$$

under the condition

$$L(p_t; n; c) \leq \beta, \tag{2}$$

where $L(p, n, c)$ is the operating characteristic (the probability of accepting a submitted lot with the proportion defective p when using the plan (n, c) for acceptance sampling),
N is the number of the items in the lot (an input parameter),
$\bar{p}$ is the process average proportion defective (an input parameter),
p_t is the lot tolerance proportion defective (an input parameter, $P_t = 100p_t$ is the lot tolerance per cent defective, denoted LTPD),
n is the number of items in the sample $(n < N)$,
c is the acceptance number (the lot is rejected when the number of defective items in the sample is greater than c).

Condition (2) provides a guarantee for the consumer that the lots of the unsatisfactory quality level, with the proportion defective p_t are going to be accepted only with the specified probability β at most (the so-called consumer's risk). The standard value 0.1 is used for the consumer's risk in [2].

1.2 Sampling plans for the inspection by variables

We recall the recently introduced LTPD plans for the inspection by variables and attributes - see [5]. The plans have been designed in [5] under the following assumptions: The measurements of a single quality characteristic X are independent and identically distributed normal random variables with parameters μ and σ^2. We consider the unknown σ case. For the quality characteristic X, either an upper specification limit U (the item is defective if its measurement exceeds U), or a lower specification limit L (the item is defective if its measurement is smaller than L), is given.

For the design of the rectifying LTPD plans minimizing the mean inspection cost per lot of the process average quality we shall use a procedure based on the EWMA statistic. The procedure is as follows: draw a random sample of n items from the lot and compute the sample mean $\bar{x}$, sample standard deviation s and the statistic Z at time t as

$$Z_t = \lambda\bar{x} + (1-\lambda)Z_{t-1}, \tag{3}$$

where λ is a smoothing constant (between 0 and 1). Accept the lot if

$$\frac{U - Z_t}{\sigma} \geq k \quad \text{or} \quad \frac{Z_t - L}{\sigma} \geq k. \tag{4}$$

The operating characteristic is (see e.g. [13] or [1])

$$L(p) = \Phi(u_{1-p}c_4 - k)\sqrt{\frac{1}{\frac{\lambda}{n(2-\lambda)} + k^2(1-c_4{}^2)}}, \tag{5}$$

where

$$c_4 = \sqrt{(2/(n-1))}\frac{\Gamma(n/2)}{\Gamma((n-1)/2)}, \tag{6}$$

The plan parameters (n, k) are to be determined so that the plan has optimal economic characteristics and satisfies the requirement (2), when (5) is used as the operating characteristic.

Regarding the economic optimality, Klufa in [10] uses the economic model, described in more detail in [8] and searches for the acceptance plan (n, k), minimizing the mean inspection cost per lot of the process average quality C_{ms} under the condition (2). We are going to use this model for the design of the plans too. The inspection cost per lot, assuming that the remainder of the rejected lots is inspected by attributes (the inspection by variables and attributes), is $n\,c_m^*$, with the probability $L(p; n, k)$, and $[n\,c_m^* + (N-n)\,c_s^*]$ with the probability $[1-L(p; n, k)]$, where c_s^* is the cost of the inspection of one item by attributes, and c_m^*

is the cost of the inspection of one item by variables. The mean inspection cost per lot of the process average quality is then

$$C_{ms} = n \cdot c_m^* + (N - n) \cdot c_s^* \cdot [1 - L(\bar{p}; n, k)]. \tag{7}$$

Let us denote

$$c_m = \frac{c_m^*}{c_s^*}. \tag{8}$$

Instead of C_{ms} we shall look for the acceptance plan (n, k) minimizing

$$I_{ms} = n \cdot c_m + (N - n) \cdot [1 - L(\bar{p}; n, k)] \tag{9}$$

(both functions C_{ms} and I_{ms} have a minimum for the same acceptance plan $(C_{ms} = I_{ms} \cdot c_s^*)$) under the condition (2).

For the discussion of the c_m parameter, see [8]. The value of this parameter must be estimated based on the real cost calculation in practice. Usually it is $c_m > 1$.

2 Calculation and evaluation of the plans

The LTPD acceptance sampling plan based on the EWMA model for sampling inspection by variables when the remainder of rejected lots is inspected by attributes, as implemented in [7] will be calculated in the example below. The solution for the unknown σ case will be searched for, using the operating characteristic given by (5). The resulting sampling plan will then be evaluated with regard to the economic characteristics and compared with the corresponding LTPD attributes sampling plan as discussed in [2].

Example. A lot with $N = 1000$ items is considered in the acceptance procedure. The lot tolerance proportion defective is given to be $p_t = 0.01$, and the process average quality is $\bar{p} = 0.001$. It is known that the cost of the inspection of one item by variables is 2 times higher than the cost of the inspection of one item by attributes, so parameter c_m equals 2. Find the LTPD acceptance sampling plan for sampling inspection by variables when remainder of rejected lots is inspected by attributes with the EWMA statistic with smoothing constant $\lambda = 0.7$.

The plan can be calculated using the functions available in the LTPDvar package [7] for the R software [12], see the documentation of the package for a more detailed description. The implementation is based on the following principles: the k corresponding to particular sample size n is obtained using the condition (2) and the cost-optimizing n is searched for using numerical methods, where the approximate solution, which is calculated first, is used for setting the bounds of the intervals to be searched in.

The resulting plan may be obtained using the following call in the LTPDvar package:

```
planLTPD(N=1000, pbar=0.001, pt=0.01,cm=2,method="ewma2",

lam=0.7)
```

and the output of the calculation is as follows.

```
An object of class "ACSPlan"
Slot "n":
[1] 66

Slot "k":
[1] 2.656216
```

So the resulting solution is $n = 66$, $k = 2.656216$.

The corresponding Dodge-Romig attributes sampling plan is $n = 250$, $c = 0$.

For the comparison of these two plans - the EWMA plan for inspection by variables and attributes ($n = 66$, $k = 2.656216$) and Dodge-Romig plan ($n = 250$, $c = 0$) from an economic point of view we use a simple ecnomic model of cost difference (measured in the c_s^* units) in which we consider the difference of the mean inspection costs per lot of the process average quality of the particular plans and the effect of the difference in fixed costs:

$$c = I_{ms}(66, 2.656216) - I_s(250, 0) + f, \tag{10}$$

where f is the increase in fixed cost (measured in c_s^*) in case that we opt for the plan for the inspection by variables and attributes. The decision to implement the plan for the inspection by variables and attributes may bring higher fixed cost in some cases in practice in comparison with performing just the attribute inspection due to presumably more difficult preparation of the inspection procedure.

The I_{ms} function is implemented in the LTPDvar package too and (for plan $n = 66$, $k = 2.656216$) may be called as shown below.

```
Ims(n= 66, k=2.656216, N= 1000, pbar=0.001, cm=2,
type="ewma2", lam=0.7)
```

In this case we obtain $I_{ms} = 166.9973$. The cost measure c then represents the cost difference (measured in the c_s^* units) in the mean inspection cost per lot of the process average quality in case that the EWMA based plan for the inspection by variables and attributes is used in place of the corresponding attribute sampling plan. The guidelines for the application and interpretation of the measure are as follows: in case that the resulting c value is negative, then one should preferably choose the plan for the inspection by variables and attributes.

Let's suppose that in our example situation $f = 50c_s^*$, then we get $c = -151$, which shows that there can be expected savings in the mean inspection cost per lot of the process average quality if the EWMA plan is used in place of the corresponding attribute inspection plan.

3 Conclusion

The EWMA-based unknown standard deviation case LTPD plans for the sampling inspection by variables when the remainder of the rejected lots is inspected by attributes minimizing the mean inspection cost per lot of the process average quality in many cases bring significant economic advantages over the corresponding attribute inspection plans. The improvement in economic efficiency can be assessed using simple model considering difference in variable and fixed cost and the LTPDvar package for the R computing environment can be conveniently used for the calculation of the optimal acceptance sampling plan.

4 Acknowledgment

This paper has been produced with contribution of long term institutional support of research activities by Faculty of Informatics and Statistics, University of Economics, Prague.

References

[1] ASLAM, M., AZAM, M., JUN, C. A new lot inspection procedure based on exponentially weighted moving average, *International Journal of Systems Science* **46**, 2015, 1392–1400.

[2] DODGE, H. F., ROMIG, H. G. *Sampling Inspection Tables: Single and Double Sampling*. John Wiley, New York, 1998.

[3] JENNETT, W. J., WELCH, B. L. The Control of Proportion Defective as Judged by a Single Quality Characteristic Varying on a Continuous Scale, *Supplement to the Journal of the Royal Statistical Society* **6**, 1939, 80–88.

[4] JOHNSON, N. L., WELCH, B. L. Applications of the Non-central t distribution. *Biometrika* 31: 1940, 362 – 389.

[5] KASPRIKOVA, N.. Rectifying LTPD Plans Minimizing Mean Inspection Cost Per Lot of Process Average Quality Using EWMA Statistics. *Journal of Testing and Evaluation* **4**. 45:2017, DOI: 10.1520/JTE20150259.

[6] KASPRIKOVA, N. Calculation of the LTPD Variables Sampling Plans in R Computing Environment. *15th Conference on Applied Mathematics (Aplimat 2016)*, Bratislava, Slovenská technická univerzita, 2016.

[7] KASPRIKOVA, N. *LTPDvar: LTPD and AOQL plans for acceptance sampling inspection by variables*. R package, 2015, URL http://CRAN.R-project.org/package=LTPDvar.

[8] KASPRIKOVA, N., KLUFA, J. AOQL sampling plans for inspection by variables and attributes versus the plans for inspection by attributes, *Quality technology and quantitative management* **6**, 2015, 133–142.

[9] KASPRIKOVA, N., KLUFA, J. Calculation of LTPD single sampling plans for inspection by variables and its software implementation. In: International Days of Statistics and Economics, Prague, 2011.

[10] KLUFA, J. Acceptance sampling by variables when the remainder of rejected lots is inspected. *Statistical Papers*, 35, 1994, 337 – 349.

[11] KLUFA, J. Exact calculation of the Dodge-Romig LTPD single sampling plans for inspection by variables. *Statistical Papers*, 51(2), 2010, 297–305.

[12] R CORE TEAM. *R: A language and environment for statistical computing*. R Foundation for Statistical Computing, Vienna, Austria, 2015. URL http://www.R-project.org.

[13] SANKLE, R., SINGH, J. R. Single sampling plans for variables indexed by aql and aoql under EWMA model. *Journal of Reliability and Statistical Studies*, Vol. 5, Issue 2, 2012, 11–20.

Current address

Kaspříková Nikola, Ph.D.
University of Economics in Prague, Department of Mathematics
Nám. W. Churchilla 4, 130 67 Praha, Czech Republic
Tel. number: +420 224 09 4247, e-mail: nb33@tulipany.cz

Slovak University of Technology in Bratislava
Faculty of Mechanical Engineering
16th Conference on Applied Mathematics **APLIMAT 2017**
Proceedings

IDENTIFICATION OF SEA TURTLES USING IMAGE PROCESSING AND NEURAL ARBITRATION

KHASHMAN Adnan (CY), CARSTEA G. Claudia (RO), OYEDOTUN Oyebade K. (CY)

Abstract. This paper presents a novel neural network application in a rarely explored area related to marine life and the tracking of the mysterious sea turtles. Preservation of marine life is an important integral part of realizing a self sustaining ecosystem, especially the endangered species. Governmental and non-governmental organizations hope that with the availability of relevant information, planning and management, environmental policies could be structured to align with emerging challenges attributed to the ecosystem. In this work, we propose the identification of the eight species of sea turtles using image data obtained from a public database and neural network. Image processing techniques and pattern averaging are applied to images, on which feedforward neural networks are trained. Furthermore, we investigate particle swarm optimization of the designed feedforward neural network, and present comparative analysis based on training time, identification rate, and run time.

Keywords: intelligent identification, image processing, neural network, particle swarm optimization, sea turtles

Mathematics Subject Classification: Primary 97R40, 82C32; Secondary 62M45

1 Introduction

Sadly, we humans have been engaging various ecosystem dwindling activities; with diverse and often uncontrolled subsequences on the environment, thus, many animal species have recently been officially declared as endangered [1]. The preservation of our ecosystem relies on our ability to monitor and control these endangering activities as well as the life species which are under threat of extinction. Although, many governmental and non-governmental agencies are rising to the challenge by creating massive awareness campaigns, strict environmental rules, anti-poaching laws etc., the situation is still unsettled and the damage to the ecosystem continues [2].

In order to recover and realize a stable ecosystem, it is vital that animal species can be quickly and correctly identified, especially the endangered species. One example of these species is the mysterious sea turtles (Chelonioidea turtles) which are an important part of marine life; contributing immensely to coral life and aquatic ecosystem. Generally, sea turtles are of eight species, namely, '*Caretta caretta, Chelonia agassizii, Chelonia mydas, Dermochelys coriacea, Eretmochelys imbricate, Lepidochelys kempii, Lepidochelys olivacea and Natator depressus*' [3]. The identification of sea turtle species is complex and difficult due to the fact that all sea turtles have more or less the same overall morphology (physical appearance); and hence, finer details are required for the identification of the different species. Some of the features that can be used for the identification of sea turtles include shell or back patterns, shell or back textures, beak shapes, etc. With the identification of sea turtle species, continuing to remain largely elusive, even to a human expert; we attempt to contribute to the global efforts for preserving such gradually disappearing species by developing an effective and intelligent visual identification system to identify the eight turtle species. Such a system will have to suffice on such a hard identification task, which is can be even more complex when sea turtles are conceived in ocean deeps harboring various marine coral lives. i.e. environmental clutters make identification much more complex.

Artificial neural networks have been successfully utilized in various difficult identification or recognition tasks over a wide range of application fields [4]-[12]. Therefore, we propose in this paper the use of a neural network model for the task of identifying the species of the sea turtles. Random images of the eight species of the sea turtles are obtained by searching online free sources and to form our experiments database. These images undergo basic image processing and pattern averaging processes and the result features for each image are used in the neural network arbitration. The neural network we choose is based on the popular back propagation learning algorithm (BPNN) due to its simplicity and efficiency in learning. The weights optimization of the designed BPNN is achieved using particle swarm optimization (PSO). The completed intelligent system will also be scale-invariant as well as rotational-invariant. This is due to the turtle images being randomly selected.

The paper is organized as follows: in section 2 we describe the database analysis and preparation for neural processing. In section 3 we present the BPNN identification model and describe its design and topology. In section 4 we discuss the experimental results and finally, in section 5 we conclude this work.

2 Database Description and Analysis

In this section, we describe the eight species of sea turtles with image examples[3][13][14]. We also apply image processing and pattern averaging techniques in order to minimize the amount of data used in processing. The extracted feature vectors from each image will be used for training and testing the proposed neural network model. The eight species of Chelonioidea sea turtles are shown in Fig. 1.

The color images of sea turtles undergo several basic image processing operations including conversion from RGB color to grayscale images using Equation 1. Fig. 2 shows the grayscale images of the turtles shown in Fig. 1.

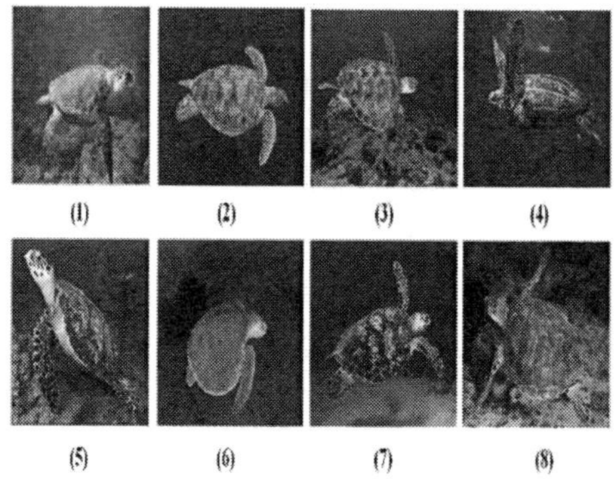

Fig. 1. Color images of the eight sea turtle species: (1) Caretta caretta (2) Chelonia agassizii (3) Chelonia mydas (4) Dermochelys coriacea (5) Eretmochelys imbricate (6) Lepidochelys kempii, (7) Lepidochelys olivacea (8) Natator depressus

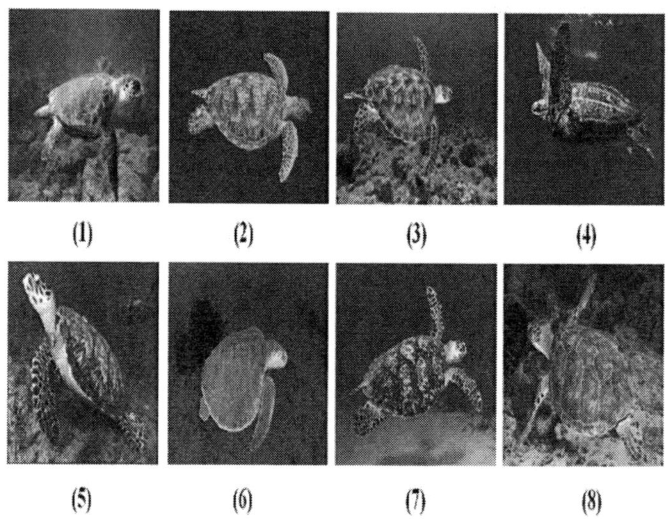

Fig. 2. Grayscale images of the 8 sea turtles species that are shown in Fig. 1

$$f'(x,y) = \frac{(R+G+B)}{3}, \qquad (1)$$

where R, G, B are the red, green and blue color channels intensities, respectively; f'(x,y) is the transformed grayscale intensity value [15].

The grayscale images are then resized to 16×16 pixels images using pattern averaging method as defined in Equations 2 & 3.

$$Seg_i = \frac{\sum p[x,y]}{D}, \qquad (2)$$

where Seg is the segment index, p is the pixel value and D is the total number of pixels per segment. D itself can be computed using Equation 3.

$$D = \frac{TP_x.TP_y}{S}, \qquad (3)$$

where TP is the x and y dimension of the image and S is the total number of segments [7].

In order to expand and enlarge the database; which is necessary in this particular application due to the general lack of turtle images, the original images are reprocessing in two ways:

1- by adding Gaussian noise with zero mean and standard deviation in the range 1% to 5%.

2- By adding salt & pepper noise of density in the range 1% to 5%.

This allows the identification systems to learn and approximate noisy images of the sea turtles, such that at test time, systems can be observed to show robustness to noise. Fig. 3 shows the Gaussian noise-corrupted images of sea turtles with 0 mean and 2% standard deviation. Whereas, Fig. 4 shows the salt & pepper noise-corrupted images of the sea turtles using 2% density.

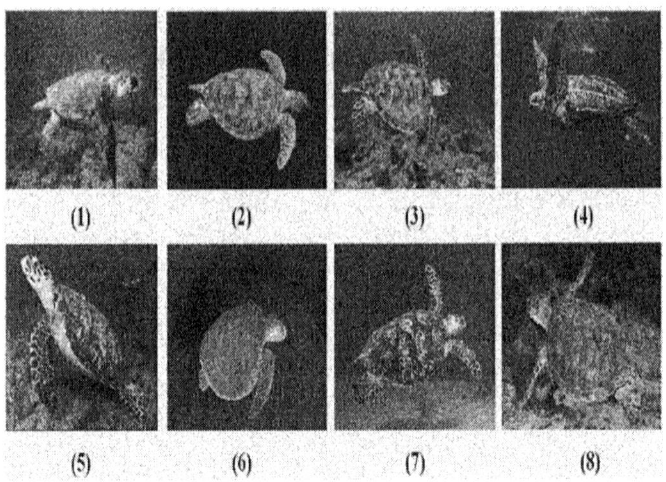

Fig. 3. Images of sea turtles species with added Gaussian noise of 0 mean and 2% standard deviation

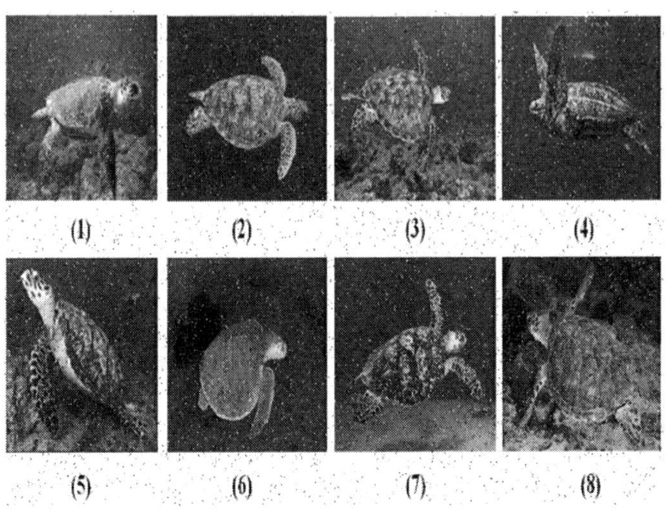

Fig. 4. Images of sea turtles with added salt & pepper noise of density 2%

As a result of reprocessing the database image by corrupting them with two types of noise, we have a total of 400 images available for training and testing our neural network model BPNN. The distribution of available data is 50 images per class. i.e. 50 images for 8 sea turtle classes which results in a total of 400 images. The images pixel values are normalized values in the range 0 to 1, which is more appropriate for BPNN input layer. The normalization of pixel values is achieved using Equation 4, where *N.P* is the value of normalized pixel.

$$N.P = \frac{Pixel\ value}{Range\ of\ pixel\ values} \tag{4}$$

3 Neural Network Arbitration

A back propagation neural network (BPNN) was designed for the identification of the considered sea turtles. The network input layer has 256 input neurons that receive the 16×16 pixels feature vector of each image. The suitable number of hidden neurons was determined through experimental trails. The number of output neurons is 8. i.e. the number of output classes for sea turtles. Fig. 5 shows the designed back propagation neural network.

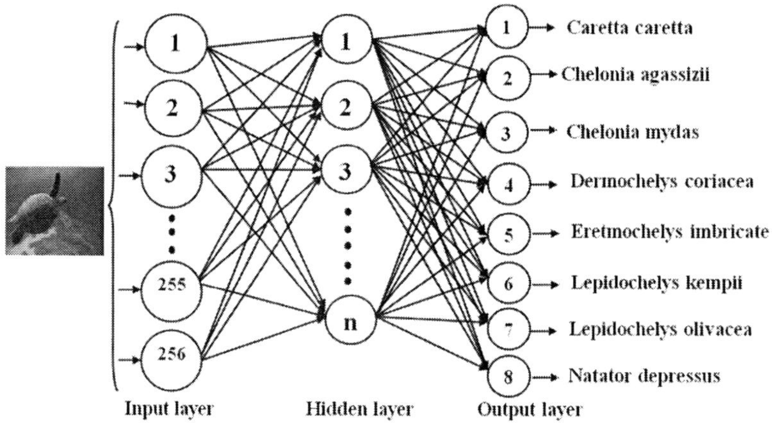

Fig. 5. Neural network topology

3.1 Training of neural networks

A. Back propagation learning algorithm

For training the designed neural network, 30 images (of 256 pixels each) per class are used; since, there are 8 classes, a total of 240 images (60% of available data) are used for training the network. Training parameters of the back propagation network for different number of hidden neurons are shown in Tab. 1.

Network parameter	BPNN1	BPNN2	BPNN3
Training samples	240	240	240
Hidden neurons	25	30	35
Learning rate (η)	0.008	0.008	0.008
Momentum rate (α)	0.9	0.9	0.9
Iterations	5788	4500	3313
Training time (s)	126	132	146
Obtained Error	0.249	0.248	0.246

Tab. 1. Training parameters for BPNN models

B. Particle swarm optimization (PSO) algorithm

The particle swarm optimization (PSO) algorithm is used to learn the optimal weights for the designed feedforward neural network. PSO relies on the swarm intelligence approach to obtaining weights which optimizes the network for the described task. Particles (potential solutions) fly around stochastically in solution space, in attempt to find better solutions iteratively [16] [17]. Fitness values are used to rank solutions achieved by each of the particles. Also, particles are capable of influencing neighboring particles towards the optimal solution. Positions and velocities of particles are updated after every iteration; particles flock towards other particles with better or higher fitness values after each iteration [18]. Furthermore, all particles keep track of global and local fitness value histories during heuristic search of optimal weights. In PSO, the particles' positions represent the solutions to the domain specific problem [19]; hence the achievable positions of particles are constrained or clamped to the range {-1, 1}, which is the desirable range for designed feedforward neural network weights. Also, PSO less often get trapped in local optima, since, PSO does not rely on the local gradients errors as obtains in the back propagation algorithm.

The same training data (240 samples) used in the back propagation optimized neural network have been used in the particle swarm optimized neural network. The idea is to observe any significant improvement in performance of the feedforward network as a factor of the approach used for weights optimization. Also, the number of neurons in the hidden layer is varied as done with the back propagation; hence, performances of networks can be observed as a function of the number of hidden neurons. Tab. 2 shows the training parameters for the PSO approach. It will be seen from Tab. 2 that PSO1 achieved the lowest error value at 0.247 in 91.1 seconds.

Network parameter	PSO1	PSO2	PSO3
Training samples	240	240	240
Hidden neurons	25	30	35
Number of particles	5	5	5
Iterations	299	299	299
Training time (s)	91.1	107.5	118.2
Obtained Error	0.247	0.248	0.248

Tab. 2. PSO training parameters

3.2 Testing of neural networks

The trained networks are tested on reserved data that are not part of the training data. 20 images per class are used for testing the trained networks; hence, a total of 160 images or samples as test data, since there are 8 classes of sea turtles. The networks obtained using back propagation algorithm and PSO, respectively. Equation 5 can be used to evaluate identification rate, IR.

$$IR = \frac{\text{Number } of\ correctly\ classified\ samples}{Total\ number\ of\ test\ samples} \tag{5}$$

A. Back propagation learning algorithm

The three models of BPNN are tested with the above described test data. The performance of the networks is based on achieved identification rates. Tab. 3 shows the identification rates achieved on the training and test data. It will be seen from Tab. 3 that the range of overall identification rate for training data is roughly 60%-64%. Also, it will be seen from Tab. 3 that BPNN3 achieved the best identification rates on training and testing data.

Network parameter	BPNN1	BPPN2	BPNN3
Training samples	240	240	240
IR on training data	65.83%	66.67%	68.33%
Testing samples	160	160	160
IR on testing data	53.13%	55.00%	56.88%
Overall IR	**60.75%**	**62.00%**	**63.75%**
Runtime (s)	0.24	0.27	0.31

Tab. 3. Identification rates (IR) for BPNNs

B. Particle swarm optimization learning algorithm

The particle swarm optimized (PSO) neural networks were also implemented with the training and test data. The identification rates for the PSO networks are shown in Tab. 4.

Network parameter	PSO1	PSO2	PSO3
Training samples	240	240	240
IR on training data	67.50%	67.08%	67.08%
Testing samples	160	160	160
IR on test data	56.52%	53.75%	55.00%
Overall IR	**63.00%**	**61.75%**	**62.25%**
Runtime (s)	0.12	0.15	0.17

Tab. 4. Identification rates (IR) for PSO networks

From Tab. 4, it will be seen that the PSO networks have an overall identification rate range of 61%-63%. Also, it can be seen that PSO1 achieved the highest identification rates of both the training and testing data.

4 Results and Discussion

Feedforward neural networks with one hidden layer have been designed in this work. The neural networks have been trained using both the back propagation learning algorithm and particle swarm optimization algorithm. The designed networks are of different number of hidden neurons; constructive and heuristic approach has been used to determine the suitable number of hidden neurons for optimum performance or optimization.

The back propagation optimized network with best performance, BPNN3, required a training time of 146secs (3313 iterations) to converge to an error value of 0.246. The simulation of BPNN3, shows that an overall identification rate of 63.75% is achieved in a runtime of 0.31secs.

Furthermore, the particle swarm optimized networks are trained on the same data used in the BPNN models. The PSO network with best performance, PSO1, required a training time of 91.1secs (299 iterations) to converge to a MSE value of 0.247. The simulation of PSO1 shows that an overall identification rate of 63.00% is achieved in a runtime of 0.12secs.

The faster training time observed in the PSO networks as compared to the BPNN models can be associated with the fact that weights optimization of PSO networks do not rely on the back propagation of error gradients; metaheuristic and stochastic search of solution space is used in obtaining optimal weights for the PSO networks. In contrast, BPNNs learn weights through back propagation of local gradients, hence more likely to get stuck in local optima. However, it can be seen from Tab. 3 & 4 that the best BPNN network, BPNN3, slightly outperforms the best PSO network, PSO1, on overall identification rate.

Also, it can be seen that the PSO networks have lesser runtimes for simulation as compared to the BPNNs.

The task described within this work, identification of sea turtles, is not a trivially achievable task; considering the database that have been used in training the networks. The identification of sea turtles using trivially processed images makes the task even more challenging for the neural networks. Significantly impacting factors that affect the learning of sea turtles identification include highly varying backgrounds, lighting conditions, different captured views or parts of sea turtles.

5 Conclusions

In this paper we presented a novel intelligent system that is based on image processing and neural network arbitration with two algorithms: the back propagation learning algorithm (BPNN) and the particle swarm optimization algorithm (PSO). The developed intelligent system is implemented in a rare field of application that is of utmost important to the environment and the preservation of ecosystems.

Our application is specifically to identify the often hard to recognize sea turtles. There are eight different types of these shy marine life creatures and differentiating them by their physical appearance is a challenging task. That is because of their similar morphological shapes. However, such a complex task is ideal for the non-linear neural networks and hence we developed the identification system.

The obtained results are considered to be successful as detailed in the previous section. This is more valid when considering the image database that we used in our experiments; these were random and comprise different scales and rotations of the turtles. Therefore, our scale-invariant rotational-invariant neural model is considered to be efficient. Future, work will focus on further enlarging the database in order to improve the identification rates.

References

[1] WILCOVE D. S., Endangered *species management: the US experience, Conservation Biology for All,* pp. 220-235. 2010.

[2] MESSER K., *The poacher's dilemma: the economics of poaching and enforcement, Endangered Species* UPDATE, 17(3), 2000. pp. 50-56.

[3] *Marine Turtles*, Web Accessed Monday, September 28, 2015. http://www.starfish.ch/reef/marine-turtles.html

[4] KHASHMAN A., *Application of an emotional neural network to facial recognition. Neural Computing and Applications,* 2009; 18(4): 309-320.

[5] KHASHMAN A, DIMILILER K., *Comparison criteria for optimum image compression,* Proceedings of the IEEE International Conference on 'Computer as a Tool' (EUROCON 2005). 2005; Serbia & Montenegro.

[6] KHASHMAN A, *IBCIS: Intelligent blood cell identification system. Progress in Natural Science,* 2008; 18(10): 1309-1314.

[7] KHASHMAN A., *Blood Cell Identification using Emotional Neural Networks*, J. Inf. Sci. Eng. 2009; 25(6): 1737-1751.

[8] KHASHMAN A., *Face Recognition Using Neural Networks and Pattern Averaging*, International Symposium on Neural Networks, China (2006) 98-103.

[9] KHASHMAN A. & AL-ZGOUL E., *Image Segmentation of Blood Cells in Leukemia Patients,* 4th WSEAS International Conference on Computer Engineering and Applications (CEA'10), The Harvard Inn, Cambridge, MA, USA, 27-29 January 2010.

[10] KHASHMAN A, DIMILILER K. *Neural networks arbitration for optimum DCT image compression*, Proceedings of the IEEE International Conference on 'Computer as a Tool' (EUROCON 2007). 2007; Warsaw, Poland.

[11] KHASHMAN A, NWULU NI., *Support vector machines versus back propagation algorithm for oil price prediction,* Advances in Neural Networks. 2011: 530-538.

[12] OYEDOTUN O.K., EBENEZER O.O., HELWAN A., KHASHMAN A., *Hybrid Auto Encoder Network for Iris Nevus Diagnosis Considering Potential Malignancy*, Proceedings of the 3rd IEEE International Conference on Advances in Biomedical Engineering (ICABME'2015), Beirut, Lebanon, Sep. 16-18, 2015.

[13] *MarineBio, Hawksbill Sea Turtles, Eretmochelys imbricate, MarineBio Conservation Society.* Web. Accessed Monday, August, 2016. http://marinebio.org/species.asp?id=164

[14] *Sea World, Parks and Entertainment, Scientific Classification*, Web Accessed Monday, August 28, 2016. http://seaworld.org/animal-info/animal-infobooks/sea-turtles/scientific-classification/

[15] KUMAR T., & VERMA, K., *A Theory Based on Conversion of RGB image to Gray image,* International Journal of Computer Applications, 7(2), 2010. pp. 7-10

[16] ABDULLAH A.G, SURANEGARA G.M, HAKIM D.L., *Hybrid PSO-ANN Application for Improved Accuracy of Short Term Load Forecasting*, WSEAS Transactions On Power Systems, vol.9, 2014. pp. 446-451

[17] RINI D.P, SHAMSUDDIN S.M, YUHANIZ S.S., *Particle Swarm Optimization: Technique, System and Challenges*, International Journal of Computer Applications, vol.14 (1), 2011. pp. 19-27

[18] BAI Q., *Analysis of particle swarm optimization algorithm,* Computer and information science, 3(1), 2010. p180.

[19] KENNEDY J., EBERHART R.C., *Particle swarm optimization*, in Proc. IEEE Int. Conference on Neural Networks, Piscataway, NJ, 1995. pp. 1942–1948

Current address

Khashman Adnan, Prof.
European Centre for Research and Academic Afairs (ECRAA),
Lefkosa, North Cyprus
Engineering Faculty, Final International University
Girne, Mersin 10, Turkey.
E-mail: adnan.khashman@ecraa.com

Carstea Claudia G., Assist.Prof.
Transilvania University of Brasov,Department of Mathematics and Informatics
Iuliu Maniu nr.50, Brasov, Romania
E-mail: carstea.claudia@unitbv.ro

Oyedotun Oyebade K., Assist.Prof.
European Centre for Research and Academic A_airs,
Lefkosa, Mersin-10, North Cyprus, Turkey
E-mail: oyebade.oyedotun@ecraa.com

Slovak University of Technology in Bratislava
Faculty of Mechanical Engineering
16th Conference on Applied Mathematics **APLIMAT 2017**
Proceedings

MULTIPLE CHOICE QUESTION TESTS IN MATHEMATICS FOR ENTRANCE EXAMS

KLŮFA Jindřich (CZ)

Abstract. The causes of differences of average number of points in the test in mathematics between test variants in entrance examinations are studied in present paper. The analysed data are the results of 4614 students in mathematics in 2013 at University of Economics in Prague. Different statistical methods (test of hypothesis, ANOVA, correlation, regression) were used for this analyse. Calculations were performed using MS Excel.

Keywords: Entrance examinations, test variants, mathematics, statistical methods

Mathematics Subject Classification: Primary 97A99; Secondary 62F03

1 Introduction

For entrance examinations at university are frequently used the multiple choice question tests. Advantages of these tests: the multiple choice question tests are objective (there is clearly no impact of any subjective factor in evaluation) and results can be evaluated quite easily for large number of students. On the other hand, a student can obtain certain number of points in the test purely by guessing the right answers. This problem was analysed in Zhao (2006), Premadasa (1993), Klůfa and Kaspříková (2012). In Klůfa (2012) it was shown that risk of success of students with lower performance levels at University of Economics in Prague is negligible.
The difficulty of test variants for students is poorly measured. If we use the multiple choice question tests, the difficulty of test variants depends also on variants of answers from which the correct answer is chosen, e.g. the example

$$\text{“combinatorial number } \binom{7}{3} \text{ is} \quad \text{(a) 21,} \quad \text{(b) 35,} \quad \text{(c) 49”}$$

is more difficult than the example

$$\text{“combinatorial number } \binom{7}{3} \text{ is} \quad \text{(a) negative number,} \quad \text{(b) zero,} \quad \text{(c) positive number”.}$$

This problem shall be analysed for the test variants in mathematics which are used for the entrance examinations at University of Economics in Prague.

Test variants in mathematics are generated from a database created by the Department of Mathematics. The tests in mathematics at University of Economics have 10 questions for 5 points and 5 questions for 10 points (100 points total) – see Klůfa and Eliášová (2013). Questions are independent. Each question has 5 answers (one answer is correct), wrong answer is not penalized. The number of points in the test in mathematics can be

$$0, 5, 10, 15, 20, 25, 30, 35, 40, 45, \ldots, 90, 95, 100.$$

These tests are used to the three faculties of the Prague University of Economics (Faculty of Informatics and Statistics, Faculty of Finance and Accounting, Faculty of Business Administration). Analysis of the entrance examinations in mathematics in 2013 is provided in this paper (similar problems are solved in Brožová and Rydval (2013), Klůfa (2013), Kohanová (2012), Hrubý (2013), Kaspříková (2012), Mošna (2013), Klůfa (2015), Kučera, Svatošová and Pelikán (2015)). These results will be used to further improve of the preparation of test variants in the coming years.

2 Entrance exams at the Prague University of Economics

The analysed data are the results of 4614 students in mathematics in 2013. Seventeen test variants were used for the entrance examinations in mathematics in 2013 (these test variants we can find in Klůfa and Eliášová (2013)). Part of one of them is on Fig. 1.

matematika – A7

Instrukce k testu:

Z uvedených odpovědí je právě jedna správná.
Příklady č.1 až 10 jsou za 5 bodů.
Příklady č.11 až 15 jsou za 10 bodů.

1. Číslo $\log_{\sqrt{7}}\left(|1-2\sqrt{7}|-|\sqrt{7}-1|\right)$ je rovno číslu:
   a) -1 b) 0 c) $\sqrt{7}$ d) $\frac{1}{2}$
   e) jiná odpověď
2. Na sázence je 25 čísel. Kolika způsoby lze označit tři z nich.
   a) 13 800 b) 750 c) 2 300 d) 625
   e) jiná odpověď
3. Číslo $\log_{\frac{1}{8}}\left(\frac{1}{2}\right)$ je rovno číslu:
   a) -3 b) $-\frac{1}{3}$ c) 3 d) $\frac{1}{3}$
   e) jiná odpověď
4. Kvadratická rovnice $x^2+px+q=0$ má jeden kořen $x_1 = 2+\sqrt{2}\,i$. Součet $p+q$ je
   a) 5 b) 4 c) 6 d) 2
   e) jiná odpověď
5. Množina všech reálných čísel, pro která platí $\log_{\frac{1}{2}} x > 2$ je rovna množině:
   a) $(\frac{1}{2},+\infty)$ b) $(0,\frac{1}{2})$ c) $(\frac{1}{4},+\infty)$ d) $(0,\frac{1}{4})$
   e) jiná odpověď
6. Množina všech reálných čísel, pro která platí $\left(\frac{5}{6}\right)^x < \frac{6}{5}$, je rovna množině:
   a) $(-\infty,-1)$ b) $(0,+\infty)$ c) $(0,1)$ d) $(-1,+\infty)$
   e) jiná odpověď
7. Množina všech reálných čísel, pro která platí $x^2-5x>0$, je rovna množině:
   a) $(-5,0)$ b) $(0,5)$ c) $(-\infty,0)\cup(5,+\infty)$ d) $(-\infty,-5)\cup(0,+\infty)$
   e) jiná odpověď

Fig. 1. Part of the variant A7 in mathematics in 2013

The test variants have 15 questions, each question has 5 answers – see Fig. 1. Between these answers is the answer "e) another answer". If e) is correct answer, to mark the letter e may be for the students more complicated than to mark the letters a, b, c, d. For example solution of the question 1 in the test variant A7 (see Fig. 1) is

$$\log_{\sqrt{7}}((-1+2\sqrt{7})-(\sqrt{7}-1))=\log_{\sqrt{7}}(\sqrt{7})=1,$$

i.e. the correct answer is e). The question 1 in the test variant A7 with variants of answers e.g.

a) 1 b) 0 c) $\sqrt{7}$ d) $\frac{1}{2}$ e) another answer

may be easier for students. Therefore we shall study the dependence of the number of points in the test in mathematics (denoted Y) on the number of answers e) in tests variants (denoted X).

3 Results

The number of answers e) in tests variants which were used for the entrance examinations at University of Economics in Prague in 2013 is 1, 2, 3, 4 and 5. Now we shall compare distributions of number of points in the test in mathematics in test variants with 1, 2, 3, 4 and 5 answers e) - see Fig. 2 and Tab. 1.

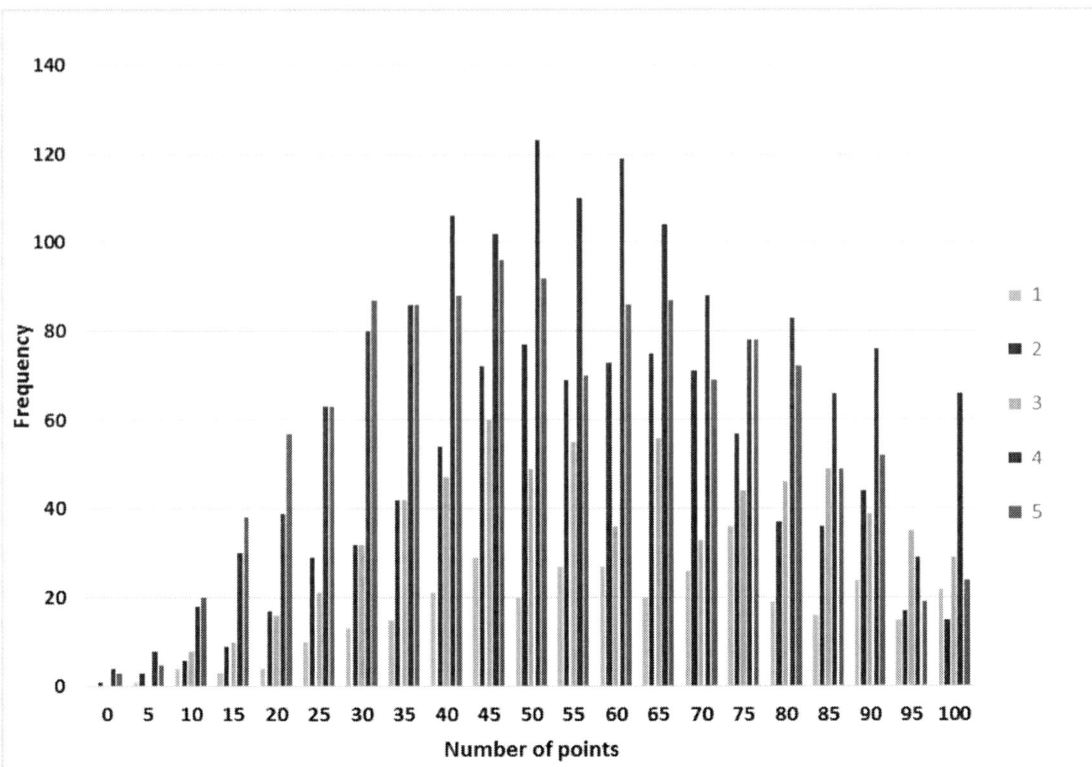

Fig. 2. Distribution of number of points in test in mathematics in 2013 in test variants with 1, 2, 3, 4 and 5 answers e) (histogram)

Number of answers e)	Frequency n_i	Average number of points	Variance
1	352	62.699	506.655
2	836	57.829	419.742
3	707	60.184	528.578
4	1478	56.712	532.850
5	1241	52.961	527.494

Tab. 1. Distribution of number of points in test in mathematics in 2013 in test variants with 1, 2, 3, 4 and 5 answers e) (source: own calculation)

Now we shall test null hypothesis

$$H_0: \mu_1 = \mu_2 = \mu_3 = \mu_4 = \mu_5, \quad (1)$$

where $\mu_1, \mu_2, \mu_3, \mu_4, \mu_5$ is mean number of points in mathematics in test variants with 1, 2, 3, 4 and 5 answers e), i.e. mean number of points in test variants with 1, 2, 3, 4 and 5 answers e) is the same.

To verify the validity of the hypothesis we use ANOVA. In the first step we verify assumption of this method by Bartlett's test (see e.g. Anděl (1978)). The hypothesis "variance of number of points in mathematics in test variants with 1, 2, 3, 4 and 5 answers e) is the same" is not rejected at 1% significance level, assumption of ANOVA can be considered to have been met.

Results of ANOVA we got with MS Excel (Marek, 2013) – see Tab. 2. Since

$$F = 19.422 > 2.374,$$

the null hypothesis (1) is rejected at 5% significance level. There are some differences between the test variants with 1, 2, 3, 4 and 5 answers e), the differences between average number of points in test variants with 1, 2, 3, 4 and 5 answers e) in Tab. 1 are statistically significant.

Source of variability	Sum of Squares	Degrees of freedom	Fraction	F	p value	F crit
Test variants	39486	4	9871.58	19.422	7.31E-16	2.374
Rezidual	2342608	4609	508.27			
Sum	2382094	4613				

Tab. 2. Results of ANOVA (source: own calculation)

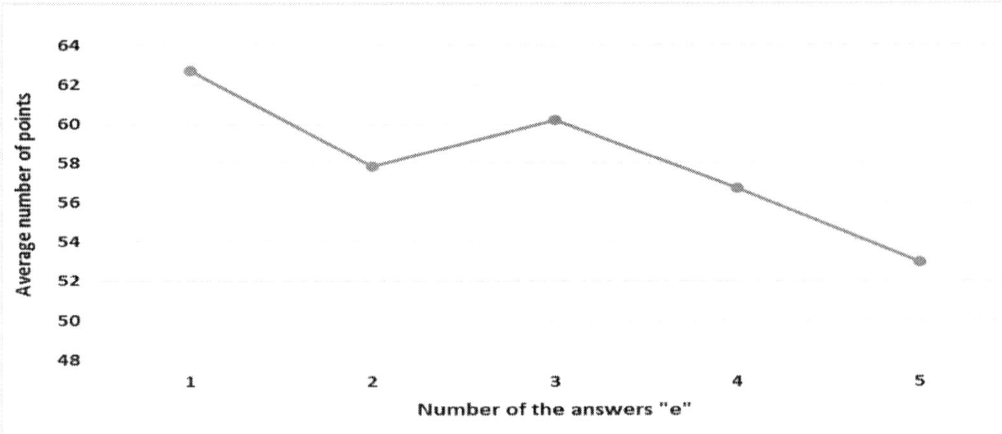

Fig. 3. Dependence of the average number of points in the test in mathematics on the number of answers e) in tests variants

From results of ANOVA it seems that Y=the number of points in the test in mathematics depends on X=the number of answers e) in tests variants. From Fig. 3 it seems that dependence is linear, therefore we shall compute the correlation coefficient r and regression line $y = a + bx$. Using MS Excel (Marek, 2013) we have

$$r = -0.112$$

From this result follows that the dependence of the number of points in the test in mathematics on the number of answers e) in tests variants is weak or the dependence does not exist. Therefore we shall test null hypothesis

$$H_o: \rho=0, \tag{2}$$

i.e. the number of points in the test in mathematics does not depend on the number of answers e) in tests variants. We shall compute - see e.g. Rao (1973)

$$t = \frac{r}{\sqrt{1-r^2}}\sqrt{n-2} = -7.66$$

Critical value of t distribution for $n-2 = 4612$ degrees of freedom and significance level $\alpha = 0.01$ is $t_{0.01}(4612) = 2.58$. Since

$$|t| > 2.58,$$

the null hypothesis (2) is rejected at 1% significance level, i.e. the dependence of the number of points in the test in mathematics on the number of answers e) in tests variants exists. Using MS Excel we obtain furthermore regression line

$$y = 63.974 - 2.009x$$

From this regression line follows that expected the number of points in mathematics in the test variant with 0 answers e) is 63.974 and if the number of answers e) in tests variants increases by 1, then the expected the number of points in mathematics decreases by 2.009.

4 Conclusion

The differences between number of points in mathematics in test variants, which were used for the entrance examinations in mathematics at University of Economics in Prague in 2013, were analysed in present paper. The differences may arise due to the varying difficulty of the test variants, but also because of the different level of knowledge of students who write these variants. From Klůfa (2016) it follows that differences between results of tests variants may arise due to different level of knowledge of the students who wrote these variants. On the other hand, the difficulty of test variants for students is poorly measured. This problem was solved in the present paper. From results of the paper it follows that the number of points in the test in mathematics depends on the number of answers "e) another answer" in tests variants, but the dependence is very weak.

Acknowledgement

This paper was processed with contribution of long term institutional support of research activities by Faculty of Informatics and Statistics, University of Economics, Prague.

References

[1] ANDĚL, J. *Matematická statistika.* SNTL/ALFA, Prague, 1978.

[2] BROŽOVÁ, H., RYDVAL, J. Analysis of the exam test quality. In *Efficiency and Responsibility in Education, Proceedings of the 10th International Conference*, Prague, 2013, pp. 47–54.

[3] HRUBÝ, M. Extended elements of the question objects. In *Efficiency and Responsibility in Education, Proceedings of the 10th International Conference*, Prague, 2013, pp. 204–209.

[4] KASPŘÍKOVÁ, N. Data analysis of students' performance. In *Efficiency and Responsibility in Education, Proceedings of the 9th International Conference*, Prague, 2012, pp. 213–218.

[5] KLŮFA, J. Tests from probability point of view. In *Efficiency and Responsibility in Education, Proceedings of the 9th International Conference*, Prague, 2012, pp. 229–233.

[6] KLŮFA, J., KASPŘÍKOVÁ, N. Multiple Choice Question Tests for Entrance Examinations – A Probabilistic Approach, *Journal on Efficiency and Responsibility in Education and Science*, vol. 5, no. 4, 2012: pp. 195–202.

[7] KLŮFA, J., ELIÁŠOVÁ, L. *Matematika – příprava k přijímacím zkouškám na VŠE.* Ekopress, Prague, 2013.

[8] KLŮFA, J. Comparison of entrance examinations in mathematics. In *Efficiency and Responsibility in Education, Proceedings of the 10th International Conference*, Prague, 2013, pp. 270–275.

[9] KLŮFA, J. Dependence of the Results of Entrance Examinations on Test Variants. *Procedia - Social and Behavioral Sciences*, vol. 174, 2015: pp. 3565–3571.

[10] KLŮFA, J. Analysis of the differences between results of test variants. In *Efficiency and Responsibility in Education, Proceedings of the 13th International Conference,* Prague, 2016, pp. 279–285

[11] KOHANOVÁ, I. Analysis of university entrance test from mathematics. *Acta Didactica Universitatis Comenianae Mathematics,* vol. 12, 2012: pp.31-46.

[12] KUČERA, P., SVATOŠOVÁ, L., PELIKÁN, M. University study results as related to the admission exam results. In *Efficiency and Responsibility in Education, Proceedings of the 12th International Conference*, Prague, 2015, pp. 318–324.

[13] MAREK, L. *Statistika v příkladech.* Professional Publishing, Prague, 2013.

[14] MOŠNA, F. E-learning and interactive dynamical figures for teaching of mathematics. In *Efficiency and Responsibility in Education, Proceedings of the 10th International Conference*, Prague, 2013, pp. 449–454.

[15] PREMADASA, I. A reappraisal of the use of multiple-choice questions. *Medical Teacher,* vol. 15, no. 2-3, 1993: pp. 237-242.

[16] RAO, C.R. *Linear Statistical Inference and Its Applications.* John Wiley, New York, 1973.

[17] ZHAO, Y. How to design and interpret a multiple-choice-question test: A probabilistic approach. *International Journal of Engineering Education*, vol. 22, no. 6, 2006: pp. 1281-1286.

Current address

Klůfa Jindřich, Prof. RNDr. CSc.
University of Economics
W. Churchill Sq. 4, 13067 Prague 3, Czech Republic
Tel. number: +420224094244, e-mail: klufa@vse.cz

Slovak University of Technology in Bratislava
Faculty of Mechanical Engineering
16th Conference on Applied Mathematics **APLIMAT 2017**
Proceedings

GEOMETRY EDUCATION IN HISTORICAL CONTEXT

KMEŤOVÁ Mária (SK)

Abstract. The paper is aimed to analyse our way of geometry teaching in a historical context. We deal with questions: How was the geometry curriculum influenced by new developments in science, mathematics and geometry itself? How can we partially compensate for the newly reduced geometry teaching hours and content of curriculum by using appropriate computer programs for creative thinking and investigation in geometry? Finally, we would like to emphasise the importance of geometry education for its impact on natural scientific and philosophical view of the world.

Keywords: Geometry education, geometry in sciences, mathematics curriculum, dynamic geometry software

Mathematics Subject Classification: 97G10, 97D20, 97B50

1 Introduction

Geometry has a long history intimately connected with the development of mathematics. Euclid's Elements, considered first axiomatic construction of geometry, captured ancient mathematics usually formulated in the language of geometry. The geometry is therefore an integral part of our culture, being a vital component of numerous aspects of life and science from astronomy to architecture, from biology to physics. What is more, geometry appeals to our visual, aesthetic and intuitive senses [1].

Euclidean geometry formed the fundamentals of geometry studies and geometry teaching for centuries. The scientific developments in geometry implied some changes in teaching of the subject. This effect was not the same in different countries and different school systems. Some of the countries were very modern and applied the new approaches immediately (for example a strong influence of Bourbaki school in France), some of them remained on the historically proven way of teaching for longer time. Now we can compare some results – for example by PISA tests. However, this comparison is not unambiguous because of many other effects, they can show the differences in deeper understanding of geometric relationships and capability of application by students of the subject in everyday situations [2]. The results show, that many students are not able to interpret basic geometric relations correctly in practical life problems [2, 3].

2 Geometry in Sciences

In this section we would like to pick up some milestones of development of geometry as a science, which could influence the geometry teaching to a great extent.

2.1 Big Leaps in History of Geometry

The beginnings of geometry trace back to Mesopotamian, Babylonian and ancient Egypt history. They were focused on the practical calculation of specific tasks, the constellation of celestial bodies, volumes and surfaces required in preparing various tools, buildings and other needs for human activities. The first mathematical theory was created according to the many specific tasks of the same type by generalization. The ancient Greeks were the first to introduce an abstract logical method to systemise previously known facts in mathematics. Their aim was to create a logically consistent foundation of mathematics. At that time several Elements of Mathematics were created, the first of them was written by Hippocrates, the other fell into oblivion under the shade of the Euclid's Elements [4]. Euclid's Elements gained widespread recognition as such a systematisation of mathematical knowledge which in terms of logical rigor remained unsurpassable for nearly 2,000 years. "The extraordinary brilliance with which the Greeks succeeded in erecting geometry as on exact science puts the geometrical achievements of other ancient civilizations in the shadow, and it has inevitably biased our historical judgements in favour of the Greek produce."[5]

New axiomatic systems fulfilling the growing demand for mathematical strictness were created at the end of the 19th century (Pasch, Peano). Euclid's Elements remained the only original source of school education geometry for a long time. Hilbert's axiomatic construction of geometry was first published only in 1899.

Even though the first hints of coordinates emerged in ancient Greek mathematics, the real analytical geometry supported by algebra and coordinate system arose only in the 17th century. Until that period the arithmetic and geometry evolved separately. At that time works of Pierre Fermat (1601-1665) and René Descartes (1596-1650) arose, which introduced analytical method into geometry almost simultaneously. Primacy belongs to Descartes, because his work was published 42 years earlier than the Fermat's one. Descartes was as a philosopher so enthusiastic of his new method that expressed a vision that in the future every problem of mechanics, physics, natural science, but also of human life could be interpreted and solved as an arithmetic problem in the language of equations [6].

The Pythagoreans were the first to understand the difference between the real performing and the thought experiment in mathematics. Eudoxus' method of exhaustion allowed getting arbitrarily close to the point in which we know that it can be never achieved. This geometric idea led to the discovery of irrational numbers in mathematics and also to the method that creates the basis of Dedekind's theory and the infinitesimal calculus. The idea of a never achievable point is similar to the idea of a point at infinity.

While the remaining part of mathematics struggled with definition of (actual or potential) infinity, the geometry captured it by mapping of parallel lines into a plane. First it was shown in several woodcuts of Albrecht Dürer (1471-1528) where he shows a method for creating a perspective 2-dimensional map of the 3-dimensional object. The most cited of them is the woodcut "*A man drawing a lute*" from 1525. Central projection illustrated on Dürer's pictures forms the fundamental idea of projective geometry. In central projection there is a correspondence between points of the object and points of the image, which is established by associating each point of the object to the point of intersection of the image plane with the line containing the object point and the eye. For example, an artist adds a vanishing point to the picture to a pair of railroad tracks that disappear off into the distance. A vanishing point, in general, is that point in a picture at which two parallel lines in the scene appear to meet [7].

The important concept of a point at infinity occurred independently to the German astronomer Johann Kepler (1571-1630) and the French architect Girard Desargues (1591-1661). Kepler (in his *Paralipomena in Vitellionem*, 1604) declared that a parabola has two foci, one of which is infinitely distant in both of two opposite directions, and that any point on the curve is joined to this "blind focus" by a line parallel to the axis. Desargues declared (in 1639) that parallel lines have a "common end" at an infinite distance [8]. Then the groundwork was laid to derive projective space from ordinary space by adding a common point at infinity for all lines parallel to each other and adding a common line at infinity for all planes parallel to a given plane.

Jean Victor Poncelet (1788-1867) fought in Napoleon's Russian campaign (1812) until the Russians took him prisoner. As a prisoner at Saratoff on the Volga (1812-1814) he still had the vigour of spirit to implement a great work, he decided to reconstruct the whole science of geometry. The result was his epoch-making "*Traité des propriétés projectives des figures*", published eight years later, in 1822 [9]. In this work the power of central projection was first made prominent. His leading idea was the study of projective properties, and as a foundation principle he introduced the ratio today known as cross-ratio [10]. The discovery of the principle of duality was also claimed by Poncelet. This principle of geometric reciprocation has been greatly elaborated and has found its way into modern algebra and elementary geometry [10].

The concept of infinity was not such a big problem in geometry as in other parts of mathematics. J.C.F. Gauss (1777-1855) wrote about infinity: "I protest against the use of an infinite quantity as something completed, which is never permissible in mathematics." [11]. A half century later a mathematician, theoretical physicist and philosopher H. Poincaré (1854-1912) stated: "Actual infinity does not exist." (At that time geometry had an actual infinity in the form of the "meeting point" of parallel lines.)

J. F. Cantor (1845-1918) was the most significant mathematician who defended actual infinities. He introduced set theory opposed to the general attitude coining the theory of actual infinity. Cantor's discoveries about infinite sets were just part of a deeper philosophical revolution that affected all branches of mathematics, not just the set theory. D. Hilbert (1862-1943) said on the set theory: "Nobody shall ever expel us from one of the most vigorous and fruitful branches of mathematics, a paradise created by Cantor." [12].

Geometry was changed and evolved, specialisations arose, which were responsive to the diverse theoretical and practical challenges. Many branches we did not mention, however they create remarkable part of modern geometry (non-Euclidean geometry, differential geometry, algebraic geometry, etc.) Today classification distinguishes more than 50 geometric disciplines [13]. Such development was not an end in itself, geometry provides the basis for the natural sciences.

2.2 Geometry and Physics

The great advantage of Descartes's analytic geometry was that it translated geometrical problems into algebra thus opening up the solution of complicated geometrical problems with analytic manipulations. But the goal of mathematicians was not only to obtain new results with the help of formal operations. Sometimes a new approach of a problem can be of greater importance than new formal results. This happened with the discoveries of David Hilbert (1862 -1943). He put geometry in the service of analysis by a remarkable isomorphism between the properties of linear differential and integral operators and the geometrical theory of quadratic surfaces. Than J. von Neumann laid mathematically unshakable foundations of quantum mechanics using operators of linear Hilbert space [5].

One day C.F. Gauss wrote down in his scrapbook "Large geometric discovery". A sentence, which he called "Theorema egregium" (excellent theorem) enabled a new view of space in which we live. Gauss figured out that the main curvature (which is thereafter referred to as Gaussian curvature) is an invariant intrinsic property of each area. This discovery (published in 1828) is the germ of the Riemannian geometry and Einstein's general relativity at the same time [5].

In 1917 A. Einstein published his theory of gravity, which provides a convincing argument that our physical space shall be closed in the sense that when a line is continuously lengthened then finally it gets back to the starting point. Maybe it is contrary to our experience, but our analytical capabilities (thanks to Gauss) allow detailed study of such space.

H. Minkowski in 1907 showed that Einstein's two postulates from 1905, which are demolishing Newton's concept of absolute time and absolute space, have rather geometric than physical characteristics [4]. The unification of space and time into a single four-dimensional (Minkowski) geometry has a large contribution to the successful completion of Einstein's ideas.

Today's physics still struggles to understand the dimensionality (macroscopic and microscopic dimensions) of a space where the physical actions happen. The fact that our space has three macroscopic dimensions was adopted at the time of Ptolemy. Modern science still lacks a deeper theory substantiated evidence of the origin of this number [14]. Moreover, string theory requires at least ten dimensions [15].

Time is even much more mysterious then space. There is a sharp contrast between our perception of time and what the modern physical theories show us [14]. All of these "mysteries" of modern science are closely related to the geometry and its deep understanding.

3 Geometry Teaching and the Curricula

At the time of the philosopher Plato, geometry was considered to be the basis of scientific research. Plato did not allow anyone to enter his Academy, who did not have at least basic knowledge of geometry. Euclid's Elements were not only the foundations of mathematics but nearly 2,000 years also the guidebook for teaching of geometry. Some countries have kept this tradition for a surprisingly long time, even at a time when many other countries modernised the geometry teaching. Not every modernisation, however, can be viewed as successful.

Current analytic geometry naturally includes also the concept of vector. Historically the concept of vector in relation to points was introduced by German geometer Möbius in 1828 as the subtraction of any two points. The key properties of this subtraction in affine space are the so-called Weyl´s axioms (after the German mathematician, theoretical physicist and philosopher Hermann Klaus Hugo Weyl (1885 –1955)), which create the base of an axiomatic system of analytic Euclidean geometry. This new approach allowed us to express all geometry in the language of linear algebra. In the early 20th century, it might seem that synthetic geometry is outdated, in some curricula for mathematics teachers also the analytic geometry was replaced by linear algebra with a short summary of geometric interpretation of the results. The secondary school mathematics curriculum prefers that part of mathematics, which has a practical advantage of easier testing of knowledge on tasks with numerical results. A result of a non-trivial construction or other interesting geometric tasks when the solution contains an unexpected idea, can only be appreciated by a well-prepared teacher.

Since we wanted to include newer and newer things progressively into mathematics curricula, geometry was pushed into the background. Starting from the end of the 20th century more and more statistics replaced other parts, mainly geometry in mathematics curricula, because statistics was considered to be a basis for numeracy and its application in practise.

Since geometry itself specialised in such a scale, that it is hard to follow by school curricula, now it is required from each specific sector to prepare their specialists at universities on the part of geometry that they need. Under pressure because of decreasing teaching hours of mathematics, different parts of the geometry program are being moved gradually into higher and higher grades, and therefore - logically – the element at the end drops off. An example of this situation is the study of symmetries in plane, which could serve as a basis for motivational games for 10-11year olds, but this part has already moved into the final year geometry program in primary school (pushing out trigonometry). The result of this development is that descriptive geometry gradually disappeared not only from curricula at the secondary schools but also at the most university study programmes at technical universities, where it was once regarded to be the queen of all technical disciplines (history of this undesirable development see in [23]). And finally, recently, the missing analytic geometry in the secondary school studies has also disappeared. If students will need them in their professional training, they must study them in the first year of their university studies as a part of special preparation. But on the other hand, again there is a strong pressure of specialised vocational subjects, so there is not enough time to study general geometry. This is valid also to the technical colleges as well as to the teacher training in mathematics.

Not all math teachers today have heard of differential geometry and the large geometric discovery of Gauss, which we have mentioned previously. They also do not have to meet during their study with projective geometry, with the idea of the extended Euclidean space, and do not have to use his abstract imagination for detecting the relations among subspaces of the multidimensional space. They are thus deprived of that knowledge, important not only for future teachers, but above all, for the future generations of students as well. Teacher without this (and other) knowledge conveys a completely different view of the world of geometry, as if they did have this knowledge.

Similar concerns are not entirely new. The next quote from [16] according to [1] is quite topical and maybe characterises the last 20-30 years of geometry teaching in the UK also: "It is a shame that higher institutions that prepare secondary school teachers to send them back to the high schools knowing no more about the subject [geometry] they are to teach than when they left high school. In fact, they know less, for they have had several years to forget what they once knew". In that context the question arises: What does the teacher of mathematics need to know from geometry to transfer the subject to a higher level? There are two types of knowledge domains: subject matter knowledge (which includes key facts, concepts, principles, and explanatory frameworks of a discipline), and pedagogical content knowledge (which consists of an understanding of how to present specific topics in ways appropriate to the students being taught) [1]. A range of studies has analysed teachers' mathematical knowledge, but only a very few of them have looked at teachers' knowledge of geometry. From the research literature it follows that many beginning teachers of mathematics and, possibly, a range of experienced teachers, do not possess the sorts of repertoires of subject-matter knowledge and pedagogical content knowledge of geometry which would enable them to teach geometrical topics as well as they might. This indicates that attention could be usefully paid both to the initial education of teachers of mathematics and to their continuing professional development [1].

4 Geometry Education of Mathematics Teachers

The research mentioned suggests how important good quality mathematics teacher education is. It can influence a generation of students.

The compulsory geometry program for math teachers at our university now contains subjects which follow basic parts of the secondary school geometry program completed with a structural view of each part, and also other important parts of traditional Euclidean geometry which are not included in primary and secondary school geometry programs right now. The optional subjects are special courses on projective geometry, differential geometry, non-Euclidean geometry and so on, which have the aim to complete a view of the whole science of geometry as mentioned before.

However it seems to be nearly ideal, the praxis shows different results. Again, there is not enough time to acquire the required high level of knowledge for each student. But there is a possibility to compensate for the missing teaching hours by using dynamic geometry software. That way we can achieve not only better level of subject matter knowledge but equally, better level of pedagogical content knowledge – here it means how to present and solve specific problems by dynamic geometry support on computer [17, 18].

Utilization of dynamic geometry software is ideal for solving locus problems in plane geometry. For example there is a possibility to develop and understand conic sections as loci, without any algebraic symbolism, using geometric constructions.

Today computer software packages also allow 3D visualisation and manipulation with objects in virtual 3D space. Besides the most popular Cabri and GeoGebra software there are Euler3D, SketchUp, Blender, and finally Leonardo3D, which is different from the previous ones by the possibility of creating a spatial object directly in space in front of the computer screen. (Refer to the picture from [19].)

Fig. 1. Working with Leonardo3D

5 Current Problems of Geometry Teaching

5.1 What to Teach

There are quite different views on what is and what is not to be trained in geometry generally (now we omit the special needs for technical studies). The design of the geometry curriculum remains a critical question.

According to [13] the aims of teaching geometry can be briefly summarised as follows:

- To develop spatial awareness, geometrical intuition and the ability to visualise;
- To provide a breadth of geometrical experiences in 2 and 3 dimensions;
- To develop knowledge and understanding of and the ability to use geometrical properties and theorems;
- To encourage the development and use of conjecture, deductive reasoning and proof;
- To develop skills of applying geometry through modelling and problem solving in real world contexts.

Reaching these goals can be achieved in different ways and usually the curriculum is quite mixed according to special geometry disciplines (plane geometry, constructions, spatial shapes, area and volume measurements, transformations etc.). It is very important to apply the different disciplines for students at the right time, when they have more interest in, and are more receptive to it. For example, it is a well-known phenomenon that students are not equally receptive to spatial relationships at every age [6].

5.2 How to Teach Geometry

Probably the most prevalent way to teach geometry is to inform students of the properties associated with plane or solid shapes, requiring them to learn the properties and then to complete exercises which show that they have learned the facts. Such an approach can mean that little attempt is made to encourage students to make logical connections and explain their reasoning. To have a good knowledge of geometrical facts, students should develop their spatial thinking and geometrical intuition by a variety of approaches. For example, some facts can be introduced informally, others developed deductively or found through exploration. It is very important to give opportunity to students to investigate and develop some relationships.

Dynamic geometry software GeoGebra has proved itself to be an excellent tool for experimentation and investigation in geometry [20]. It is very intuitive to use, skills are easy to acquire, and last but not least, it is freely available. Work with GeoGebra in geometry education allows us direct manipulation with basic objects as points, lines and circles. They provide a model of Euclidean geometry which offers feedback through „dragging" the objects and seeing whether the constructions or hypotheses are correct. Students are able to generate ample empirical evidence for geometric theorems in ways which would have been difficult if not impossible before [21] (working on paper).

To teach geometry effectively to students of any age or ability, it is important to ensure that students understand the concepts they are learning and the steps that are involved in particular processes. The concept of genetic development can show the best order of possible investigation area. More influential authors of mathematical didactic literature acquired this concept. Now we accept that the relation between ontogenesis and phylogenesis is universally recognised to be much more complex than was originally believed, i.e. that the development of mathematical knowledge of one person repeats, to some extent, the development of mathematics in the history of science. The specificity of this pedagogical use of history is that it interweaves our knowledge of past conceptual developments with the design of classroom activities, the goal of which is to enhance the students' development of mathematical thinking [22].

Of course, historical development is not always straightforward, sometimes closely related discoveries were born with a surprisingly large time gap. For example Pappus theorem coming from the 4th century is a special case of Pascal's theorem from the 17th century, and its dual version (Brianchon theorem) arose only in the 19th century. In teaching we do not separate them. In such cases we use rather thematic than genetic (or historical) parallel. The genetic approach therefore should not be regarded strictly as historic, but more as a natural, practical and logical path to knowledge.

The genetic approach gives us the answer to the question, how we can explain the meaning and importance of a particular mathematical discovery. In teaching, we should take into account

a) Prior knowledge, experience and the level of thinking of students;

b) Partial problems that students can solve independently, without teacher´s support;

c) An abstract consideration should be preceded by intuitive or heuristic considerations, supported by a sufficient amount of particular models [6].

If a student acquires his geometric knowledge through his own activity, the quality of his knowledge will be deeper, longer-lasting and better quality.

5.3 Why to Teach Geometry

Given the above consideration of the aims of teaching geometry, it is possible to say why it should be included in the school mathematics curriculum in sufficient extent.

The study of geometry contributes to helping students to develop the skills of visualisation, critical thinking, intuition, perspective, problem-solving, conjecturing, deductive reasoning, logical argument and proof. Geometric representations can be used to help students make sense of other areas of mathematics: fractions and multiplication in arithmetic, the relationships between the graphs of functions and graphical representations of data in statistics. Spatial reasoning is important in other curriculum areas. Geometry is a rich source of opportunities for developing notions of proof. Visual images, particularly those, which can be manipulated on the computer screen, invite students to observe and conjecture generalisations. Proving conjectures requires students to understand how the observed images are related to one another and are linked to fundamental building blocks [13].

The author of article [24] is convinced by his experience in teaching geometry to mathematics and engineering students that geometry ultimately helps to save time in education and problem solving processes. "Besides these more or less practical effects, geometry helps to build a (natural scientific and philosophical) view of the world and allows us to get deeper insight into a lot of works of fine art. So geometry is a general educational value."

6 Conclusion

Geometry is often seen only as a part of mathematics helping to solve technical problems. This view neglects a strong influence of geometry on sciences, philosophy and art throughout history. Reducing the number of teaching hours in school geometry curriculum in recent decades seems to have negative effects not only on technical training but also the training of mathematics teachers themselves. New computer tools bring new opportunities and new trends into teaching and learning geometry – investigation, experimentation, manipulation with virtual objects, but also geometric visualisation of non-geometric relationships. This way geometry can become an inspiratory and creative basis for mathematics, sciences and philosophy.

Acknowledgement

The paper was supported by grant from Slovak Grant Agency KEGA no. 016UKF-4/2016.

References

[1] JONES, K. Teacher Knowledge and Professional Development in Geometry. In: *Rowland, T. (Ed.) Proceedings of the British Society for Research into Learning Mathematics* 20(3), November 2000, pp. 109-114.

[2] URL: http://www.oecd.org/pisa/.

[3] VIDERMANOVÁ, K., VALLO, D. Practical Geometry Tasks as a Method for Teaching Active Learning in Geometry. In: *Procedia - Social and Behavioral Sciences: WCES 2014, 6th World Conference on Educational Sciences,* Vol. 191 (2015), ISSN 1877-0428, pp. 1796-1800.

[4] RIBNYIKOV, K. A. *A matematika története.* Tankönyvkiadó, Budapest, 1974.

[5] LÁNCZOS, K. Space through the Ages – Hungarian translation: A geometirai térfogalom fejlődése, Gondolat, Budapest, 1976, ISBN 963 280 206 3.

[6] HEJNÝ, M. a kolektív. *Teória vyučovania matematiky 2.* SPN Bratislava 1990. ISBN 80-08-01344-3.

[7] PENNA, M.A., PATTERSON, R.R. *Projective Geometry and its Applications to Computer Graphics*, Prentice-Hall, 1986, ISBN 0-13-730649-0.

[8] COXETER, H.S.M. Projective geometry, Second edition, University of Toronto Press, 1974, Hungarian translation, Gondolat, Budapest, 1986, ISBN 963-281-678-1.

[9] COXETER, H.S.M. *The real projective plane*, Third Edition, 1993, Springer-Verlag New York, ISBN 0-387-97889-5.

[10] LAKSA, A. 2013. *An Insight into Geometric Modelling and Curves and Surfaces,* 2013, available on http://fagweb.hin.no/ala/geomod/ERBSbok_main.pdf.

[11] URL: http://math.stackexchange.com/questions/141130/what-did-gauss-think-about-infinity.

[12] URL: http://www-history.mcs.st-andrews.ac.uk/Quotations/Hilbert.html.

[13] JONES, K. Issues in the Teaching and Learning of Geometry. In *Linda Haggarty (Ed.), Aspects of Teaching Secondary Mathematics: perspective and practice,* RoutledgeFalmer, London, 2002, ISBN 0-415-26641-6, pp. 121 -139.

[14] SANIGA, M. Geometry of Time and Dimensionality of Space. In: *The Nature of Time: Geometry, Physics and Perception.* NATO Science Series II. Vol. 95, Kluwer Academic Publishers, 2003.

[15] POLCHINSKI, J. *String theory.* Cambridge University Press, 2005, ISBN 13-978-0-521-63303-1.

[16] O'TOOLE, A. L. *College geometry for secondary school teachers.* American Mathematical Monthly, 1941, 48(5), pp. 319-326.

[17] RUMANOVÁ, L., VALLO, D., ĎURIŠ, V. *Didactical Phenomena of Unusual Geometry Tasks in Teaching of Stereometry.* In: Procedia - Social and Behavioral Sciences : WCLTA 2014 - 5th World Conference on Learning, Teaching and Educational Leadership, Prague, October 2014, vol. 186, ISSN 1877-0428, pp. 354-358.

[18] KMEŤOVÁ, M. *Kružnicová inverzia a Steinerov reťazec.* Proceedings of Symposium on Computer Geometry SCG 2010, Kočovce, Vol. 19, s. 57-62, ISBN 978-80-227-3364-9.

[19] URL: http://geek-mag.com/posts/240404/ Interview with the creator of a 3D mouse Leonardo.

[20] URL: http://www.geogebra.org/.

[21] HOYLES, C., JONES, K. Proof in Dynamic Geometry Contexts. In: *C. Mammana and V. Villani (Eds), Perspectives on the Teaching of Geometry for the 21st Century*. Dordrecht: Kluwer, 1998, pp. 121-128. ISBN: 0-792-34990-3.

[22] FURINGHETTI, F., RADFORD, L. Historical Conceptual Developments and the Teaching of Mathematics: from Phylogenesis and Ontogenesis Theory to Classroom Practice, in: *Handbook of International Research in Mathematics Education,* LEA Publishers, London, 2002, ISBN 0-8058-3371-4, pp. 631-654.

[23] VELICHOVÁ, D., *Deskriptívna geometria na Strojníckej fakulte STU v Bratislave - Quo Vadis?*, Deskriptívna geometria na slovenských univerzitách včera a dnes, Žilina 2014, SR, ISBN 978-80-554-0968-9, pp. 86-96.

[24] WEIS, G. *Is geometry a cultural or a technical need?* 24. Conference on Geometry and Computer Graphics 2004, Proceedings, pp 1-13. ISBN 80-248-0581-2.

Current address

Kmeťová Mária, doc. RNDr. PhD.
Department of Mathematics, Faculty of Natural Sciences,
Constantine the Philosopher University in Nitra
Tr. A. Hlinku 1, 949 74 Nitra, Slovak Republic
E-mail: mkmetova@ukf.sk

Slovak University of Technology in Bratislava
Faculty of Mechanical Engineering
16th Conference on Applied Mathematics **APLIMAT 2017**
Proceedings

ANALYSIS OF EFFICIENCY OF ENTRANCE EXAMINATION BASED ON PERMEATION IN THE FIRST YEAR OF STUDY

KONEČNÁ Petra (CZ), BUJOK Petr (CZ)

Abstract. Article extends partial outcomes received in year 2016, in which the author tries to answer the question, based on the monitoring of the admission procedure strategy, interest in studies and permeation in the first year of study in selected undergraduate Bachelor's degree programmes, whether the entrance examination is able to guarantee reduction in academic failure rates and increasing number of successful graduates. In this extending article authors process needed data of all Bachelor's degree programmes and put them through more-detailed analysis. Based on results of analysis, they also try to classify study programmes according to permeation as well as other common features.

Keywords: entrance examination, tertiary education, academic failure rates, statistics

Mathematics Subject Classification: Primary 62P25, 97A10; Secondary 97B10

1 Introduction

Entrance examination is considered to be efficient, whether its process and form ensure reduction in academic failure rates, which leads to higher permeation during the years of study, better study results, successful graduates in standard length of study and increase of ratio of graduates of all enrolled students in total. It is influenced by lots of parameters, not only by process and form of entrance examination – e.g. difficulty of chosen study, acceptance in preferred study next year, attractive job offer, social conditions, preference of starting a family life. The entrance examination itself is significantly influenced by success in first year of study. Thus we focused on permeation in first year of study – we studied the ratio of students meeting all conditions for enrolling at the second year of study on the total number of enrolled students at first year of study.

In pilot research in year 2016, we focused on trends in permeation in first year of study of Bachelor's degree programmes in years 2006 – 2013, with special aim at mathematical programmes and comparing them with all Bachelor's degree programmes in total as well as with selected groups of natural science's programmes. Even though we worked with partial results based on simple comparing of data of selected programmes, findings are following:

- The fact whether an applicant took or did not take the entrance examination has a clear impact on increased failure rate particularly in the first year of study considering the overall evaluation of all study programmes. This, however, cannot be applied to individual programmes.
- Entrance examination is perceived as a tool for verifying the applicants' prerequisites for university study. However, choice of the a posteriori approach within the admission procedure strategy does not actually mean that students who do not pass the entrance examination do not have the ability to succeed in the first year of study.
- High failure rate is typical for combined and distance forms of studies regardless of the fact, if the entrance examination was exempted or not
- Resolution of problem of efficiency of entrance examination as a follow-up to increasing a rate of successful graduates from university studies thus will not be easy as it might be seen. Fast and simplified solutions might be counterproductive especially for programmes with low number of applicants and graduates, including mathematical programmes. To find an answer to the question, where the entrance examination could be applied as an appropriate tool to reduce the academic failure rate and where not, we would have to work with data from all study programmes and use it for a thorough statistical analysis. [2]

Thus we decided to monitor trends in permeation in first year of study in particular Bachelor's degree programmes in years 2006 – 2014[1], to compare them with average permeation of all Bachelor's degree programmes in particular years, as well as to find links with the form of entrance examination and divide the programmes into groups with similar characteristics.

2 Admission procedure strategy in years 2006 - 2014

Regulating the number of first-year students, thus controlling the number of accepted applicants, can be managed in two basic ways [4].

1. A priori way, which means that the acceptance or rejection is decided before entering the university; that decision is usually made on the basis of the entrance examination results or other criteria evaluation.

2 A posteriori way, which means that at the beginning of their studies all applicants meeting legal requirements are admitted and those who fail to meet the study requirements during their university studies are made to leave.

There were changes in admission procedure strategy at Faculty of Science of University of Ostrava during the years 2006 – 2014, which were based on previously mentioned ways of controlling the number of accepted applicants. An overview of years, in which a priori and a posteriori way was chosen, is shown in Tab. 1.

[1] [1] Chosen period is maximal possible, because all data for the year 2015 will be available on 31st December 2016

	Academic Year								
Programme code	2006 2007	2007 2008	2008 2009	2009 2010	2010 2011	2011 2012	2012 2013	2013 2014	2014 2015
AM	+	+	+	+	-	-	-	-	-
AME	+	+	+	+	-	-	-	-	-
IP	+	+	+	+	-	-	+	+	+
CH	+	+	+	+	-	+	+	+	+
IN	+	+	+	+	-	-	-	-	+
AI	+	+	+	+	-	-	+	+	+
SBE	+	+	+	+	+	+	+	+	+
PKG	+	+	+	+	+	+	+	+	+
BF	+	+	+	+	-	-	-	-	-
OKB	+	+	+	+	+	+	+	+	+
AEB	+	+	+	+	+	+	+	+	+
EXB	+	+	+	+	+	+	+	+	+
GRB	+	+	+	+	+	+	+	+	+
FGB	+	+	+	+	+	+	-	-	-
KGB	+	+	+	+	-	-	-	-	-
2OB	+	+	+	+	+	+	+	+	+

Tab. 1. Overview of taking (+) and not taking (-) the entrance examination

There were several reasons – a lower interest of applicants in some study programmes in relation to its difficulty and lower attractiveness, similar strategy in admission procedure in other universities, an attempt to make study in selected programmes more attractive in way that an applicant can focus on secondary school's leaving exam and do not have to take other tests.

3 Permeation in first year of study in years 2006 - 2014

By the permeation in first year of study we mean the ratio of students meeting all conditions for enrolling at the second year of study on the total number of enrolled students at first year of study. There are two basic reasons (thus creating two groups) why students do not enrol at the second year of study. First group is made by students whose study was terminated in first year of their study on their own request. The second group is made by students whose study was terminated due to not meeting the study requirements – in general [2]: receiving minimal number of credits in first and second semester, value of weighted grade point average in the first year of study may not exceed the amount of 3.4. a student also cannot have twice registered and unsuccessfully passed mandatory subject. If students do not meet all mentioned requirements, he/she is sent a resolution of termination of his/her study. However, due to statute of limitation, is the termination of study at the second group processed during the winter semester. Thus, to make the research relevant, the data of first and the second group must be collected up to different limit dates.
Permeation in first year of study is given by following formula:

$$S = \frac{N - F_1 - F_2}{N}, \tag{1}$$

where N is number of enrolled student in first year of study, F_1 is number of students of first group (up to 31st August next year) and F_2 is number of students of second group (up to 31st December next year).

With regard to the fact, which was proved in previous research – that there is unequivocally low permeation in combined and distance forms of study regardless of way of entrance examination, we decided to analyse only data of Bachelor's degree programmes in full-time form of study.

3.1. Descriptive statistics – results

In first phase, we evaluated data in the same way as in pilot research. Total permeation in Bachelor's study programmes is continually decreasing during the years, which was also confirmed by academic year 2014/2015.

In most cases, permeation is below the average at programmes without taking an entrance exam. There are, however, programmes, in which entrance exam was taken, but it did not influence the success in first year of study enough.

According to values in Tab. 2 we can identify programmes, in which taking entrance exam has probably positive effect on permeation – those are Applied Informatics (AI), Social geography programmes (GRB, PKG) and two-major undergraduate study programmes. (2OB BC). In last five years, this trend can be seen even in Chemistry (CH). We can also identify programmes whose permeation is low regardless of the fact, whether the entrance exam is taken or not – those programmes are mathematical and physical ones (AM, AME, BF), physical geography programmes (FGB, KGB) and Informatics (IN). Surprising is, that there is a huge number of study programmes, where the permeation is low even though the entrance exam is taken (SBE, OKB, AEB, EXB).

	Academic Year								
Code of programme	2006/2007	2007/2008	2008/2009	2009/2010	2010/2011	2011/2012	2012/2013	2013/2014	2014/2015
AM	66,67%	33,33%	50,00%	100,00%	50,00%	77,78%	38,46%	41,18%	30,00%
AME	40,00%	88,89%		80,00%	66,67%	70,00%	50,00%	30,00%	44,44%
IP		100,00%	58,33%	57,89%	54,55%	68,42%	40,00%		
CH	72,22%	46,67%	55,56%	72,22%	41,67%	74,19%	54,17%	75,00%	63,64%
IN	60,38%	71,43%	59,09%	75,00%	56,00%	55,17%	53,33%	46,30%	28,57%
AI			78,95%	70,59%	54,76%	60,71%	72,41%	85,37%	68,09%
SBE	77,27%	57,58%	76,47%	63,16%	63,33%	61,76%	59,26%	72,00%	50,00%
PKG	88,24%	75,00%	61,82%	87,23%	72,09%	71,15%	63,16%	73,17%	61,76%
BF	50,00%	100,00%	0,00%		30,77%	0,00%	36,84%	34,62%	43,48%
OKB	72,73%	61,54%	34,21%	52,27%	63,33%	65,52%	56,76%	47,27%	51,52%
AEB	79,31%	71,43%	61,11%	62,50%	58,82%	66,67%	70,59%	50,00%	60,00%
EXB	75,00%	46,67%	45,00%	58,97%	66,67%	69,70%	56,00%	58,70%	60,61%
GRB	61,76%	90,63%	83,33%	74,29%	82,22%	78,57%	69,57%	71,43%	68,18%
FGB	75,00%	64,29%	62,50%	60,00%	62,86%	65,71%	66,67%	37,50%	52,78%
KGB	78,26%	63,64%	45,45%	64,29%	37,50%	72,73%	38,89%	63,64%	51,85%
2OB	68,48%	53,27%	73,17%	74,07%	66,67%	67,48%	66,04%	65,29%	59,79%
All bachelor programmes	71,25%	64,49%	63,36%	68,59%	60,35%	66,45%	60,04%	59,78%	56,05%

Tab. 2. Percentage of students who successfully completed the first year of study, out of all enrolled students. Red values are below the average value of total permeation in Bachelor's programmes. Dark boxes are cases, in which entrance examination was not taken

Based on permeation in first year of study, we tried to classify study programmes and thus create groups with similar permeation in study. We were also interested, if the grouped programmes also have other common characteristics.

For the first comparison, we used parameters for arithmetic mean, median and trimmed mean, which we received by deleting the highest and the lowest value of the monitored period.

	mean	trimmed mean	median	min	max
AM	0,54157	0,51059	0,5	0,3	1
AME	0,5875	0,58519	0,58333	0,3	0,88889
IP	0,63199	0,59799	0,58114	0,4	1
CH	0,61703	0,62666	0,63636	0,41667	0,75
IN	0,56141	0,57386	0,56	0,28571	0,75
AI	0,70125	0,7015	0,70588	0,54762	0,85366
SBE	0,64537	0,64795	0,63158	0,5	0,77273
PKG	0,72625	0,71947	0,72093	0,61765	0,88235
BF	0,36963	0,32617	0,35729	0	1
OKB	0,56127	0,56887	0,56757	0,34211	0,72727
AEB	0,64492	0,64445	0,625	0,5	0,7931
EXB	0,59701	0,59615	0,58974	0,45	0,75
GRB	0,75553	0,7537	0,74286	0,61765	0,90625
FGB	0,60811	0,62115	0,62857	0,375	0,75
KGB	0,5736	0,57212	0,63636	0,375	0,78261
2OB	0,66029	0,66702	0,66667	0,53271	0,74074

Tab. 3. Values of Parameters for particular study programmes

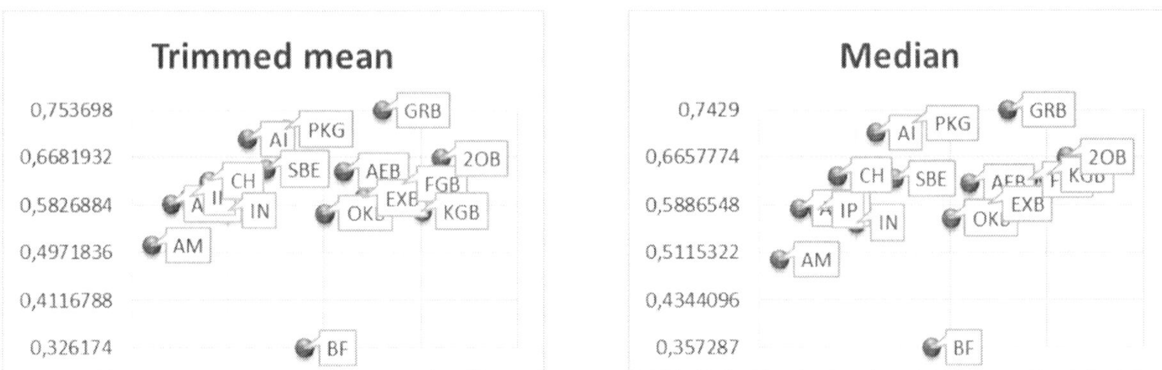

Fig. 1

In both classifications, we can see that the lowest permeation is an extreme between other programmes and relates only to study programme Biophysics (BF). Values of permeation are on level of combined and distance study programmes, which show high failure rate. Low permeation can be also observed in other programmes, most of them are mathematical ones (AM, AME) or those with more mathematical subjects (IN, KGB). All those programmes are included in a posteriori way of entrance examination in long term period. Surprising is, that strongly different programme also belongs to category 2-3 - Landscape protection and formation (OKB), where entrance exam is always taken. Category 5 is the same according to both classifications and includes social geography programmes (PKG, GRB) which stand on border between natural and humanistic sciences, but this group also includes Applied Informatics. (AI). All those programs took entrance examinations, with exception of two years at (AI), which unequivocally showed decrease of permeation, but later, when entrance examination shown again, the permeation rapidly raised again. Thus there is visible a strong influence of a priori way. The group 4 is in both cases created by experimental natural science's programmes. Those balance on average level between all Bachelor's degree programmes, often even below the average, even though these programmes, with a few exceptions, undergo entrance examination. But its influence is not visible.

Trimmed mean	category	median
BF	1	BF
	2	AM
AM, IN, OKB, KGB	3	IN, OKB, AME, IP
AME, IP, CH, SBE, AEB, EXB, FGB, 2OB	4	CH, SBE, AEB, EXB, FGB, KGB, 2OB
AI, PKG, GRB	5	AI, PKG, GRB

Tab. 4. Classification of Bachelor's degree programmes into groups according to trimmed mean and median

However, methods and parameters of descriptive statistics cannot be sufficiently exact, thus the data were analysed by another method.

3.2 Cluster analysis - results

Cluster analysis belongs between methods of multidimensional data, whose aim is to group a set of objects in such a way that objects in the same group are similar to each other than to those in different groups - clusters. Clusters are created in a way not to overlap mutually but each element must be put into any of the cluster. There are two basic approaches – hierarchical and non-hierarchical. In hierarchical approach, each element creates a cluster at first, and gradually those elements cluster from small ones to one only. On the contrary, non-hierarchical approach tries to cluster elements into given number of clusters. To find out similarity between study programmes, we used an algorithm k-means [1]. The aim of k-means is to arrange elements into clusters in way that variance of elements into clusters is the least. Finding minimal variances in all clusters is a difficult task that cannot be solved in polynomial time. However, for most tasks the k-means can find acceptable solutions. More details about this method can be found in guide of method in programme NCSS, in which this analysis was made [3].

Input of the algorithm *k*-means is, except of elements data matrix, the given number of clusters – *k*. It is evident that for a small number of cluster the variance of elements is larger than for a larger number. Common output of k-means method is coefficient of ratio of variance of elements to total variance of elements. It is evident, that this coefficient decreases with rising number of cluster and is recommended to choose k, for which the transition of ratio of variances is distinctive in comparison to *k*-1.

There was suggested a simple procedure, how to estimate a suitable number of cluster k that the number of clusters was not high and the ratio of variance inside clusters to variance without clusters was the smallest. By transformation of both criteria to the scale 0-100 was, thanks to Pythagoras' theorem, calculated a simple criterion, which should be as smallest as possible in ideal conditions (low number of cluster and simultaneously small variance in clusters). In criterion, var_k symbolises variance of elements put into k clusters, var is a total variance of elements without clusters and k_{max} symbolises maximal number of clusters for given data. The best value of the criterion was for $k = 5$.

$$krit = \sqrt{(var_k/var \cdot 100)^2 + (k/k_{max} \cdot 100)^2} \quad (2)$$

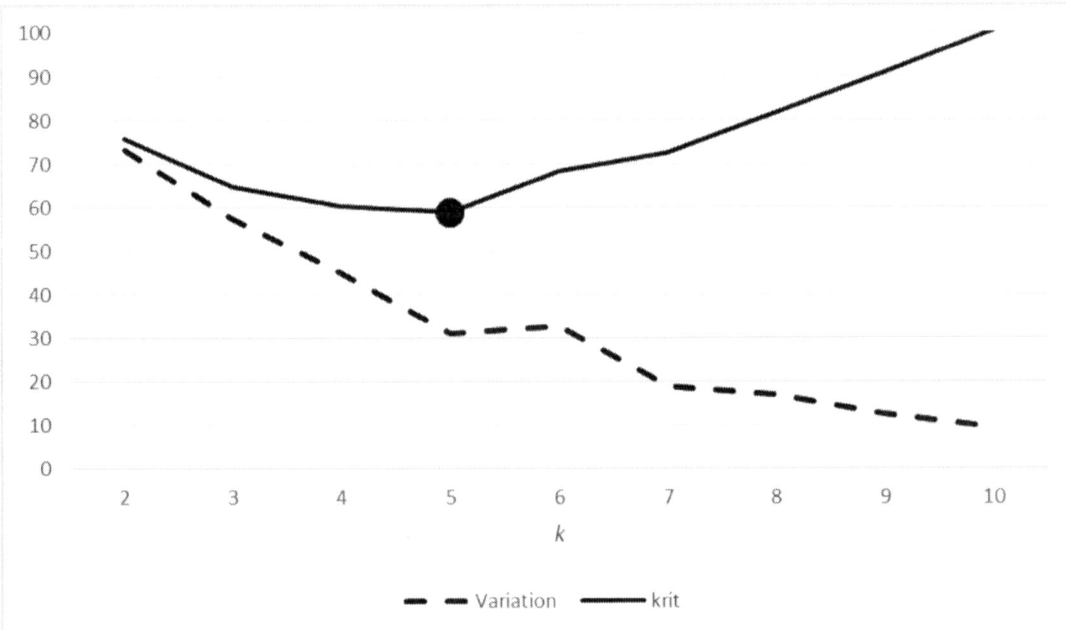

Fig. 2

By application of algorithm *k*-means on permeation of Bachelor's degree programmes we got following clusters:

1. BFB
2. AMB
3. AMEB, INB, IPB
4. AEB, FGB, EXB, CHB, KGB, OKB
5. PGB, GRB, SBB, AIB, two-major undergraduate study programmes

4 Conclusion

With the use of descriptive statistics and cluster analysis, we can divide Bachelor's degree programmes into four groups, which have not only similar permeation, but other characteristics as well.

The first group is created by the study programme Biophysics, which has extremely low permeation between Bachelor's degree programmes, it is on level of combined and distance forms. There is a posteriori way chosen in case of entrance examination. It is a programme with a low number of students.

Second group is created by programmes, which, according to previously described classifications, represent second and third category with a low permeation. Those are all mathematical programmes and Informatics, which includes a huge number of difficult mathematical subjects. There is also a posteriori way in case of entrance examination, as well as the number of students is low.

These two groups show a low permeation. However, strict implementation of a priori way, obligatory entrance examination, seem not to have expected effect. In years, in which entrance exam was taken, the permeation did not rise much. Moreover, these programmes have a low number of applicants and students, which means that there is a low effect on the total faculty parameter of permeation.

Third group is created by chemistry, physical geography, biological and ecological programmes[2], let's say experimental ones. Those programmes show rather average, in some years even below the average, permeation, even though most of these programmes have in most years a priori way in case of entrance examination. It might be considered, if current form of entrance examination should be adjusted to be one of efficient tools for verification of presumptions of university studies' applicants. Those programmes belong to ones that have the highest number of applicants and students, thus the low permeation strongly influences the total value of permeation in Bachelor's degree programmes of the faculty, which should be one of monitored parameters for evaluation the quality of university.

Fourth group is created by social geography programmes and Applied Informatics. Those programmes show high permeation, which is also caused by a priori way in case of entrance examination, which is an efficient tool here. Those programmes have a high number of students, thus there is no need to modify the a priori way.

References

[1] HARETIGAN, J. A., WONG M. A. Algorithm AS 136: A K-Means Clustering Algorithm. *Journal of the Royal Statistical Society*. Series C (Applied Statistics), vol. 28, no. 1, 1979, pp. 100–108.

[2] KONEČNÁ, P. Entrance Examination as a Guarantee to Reduce Academic Failure Rates and Increase Number of Successful Graduates – Yes or No? *Proceedings of 15th Conference of Applied Mathematics,* Bratislava, 2016, 978-80-227-4531-4, s. 680 – 689.

[3] NCSS 11 Statistical Software (2016). NCSS, LLC. Kaysville, Utah, USA, ncss.com/software/ncss.

[4] POTŮČEK, M. *Sociální politika*. Praha, Sociologické nakladatelství, 1995, ISBN 80-85850-01-X, s. 142.

[2] In case of Cluster analysis, one of biological programmes (SBE) was put into category with the highest permeation. However, based on results of descriptive statistics, considering the type of programme, we decided to put it into the third category.

Current address

Konečná Petra, Ph.D.
Department of Mathematics, Faculty of Science, University of Ostrava
30. dubna 22, 70103 Ostrava, Czech Republic
Tel. number: +420597092103, e-mail: petra.konecna@osu.cz

Bujok Petr, Ph.D.
Department of Informatics and Computers, Faculty of Science, University of Ostrava
30. dubna 22, 70103 Ostrava, Czech Republic
Tel. number: +420597092176, e-mail: petr.bujok@osu.cz

Slovak University of Technology in Bratislava
Faculty of Mechanical Engineering
16th Conference on Applied Mathematics **APLIMAT 2017**
Proceedings

SYMMETRY IN ELEMENTARY EDUCATION WITH THE USE OF DIGITAL TECHNOLOGIES AND MANIPULATIONS

KOREŇOVÁ Lilla (SK)

Abstract. Education of geometry in elementary education has gone over a lot of changes in the past decades in Slovakia, reducing the contents and changing some concepts. Building geometric concepts in elementary education is a very important process. To avoid formalism in education it is a good thing to increase not just the motivation of students, but more importantly to substitute transmissive educational process with a constructivist approach. Digital technologies certainly aid to deeper understanding. In this paper we present a few proposals of methodics for the topic Axial Symmetry using digital technologies, manipulative activities and controlled exploration. We also present the results of the research of this method.

Keywords: Education of geometry, axial symmetry, GeoGebra, elementary education, manipulative techniques, digital technologies, constructivism

Mathematics Subject Classification: Primary 97C30, 97D40; Secondary 97U50, 97U60

1 Introduction

The education of school mathematics has changed significantly in the past decades. Not just in the Europe, but all over the world. Places that did not change the methods, forms and also the contents of education of school mathematics have louder students and also parents against mathematics as a taught subject.

In the 70s education of mathematics was influenced with 'New Math' reform that can be characterized with a structurally-formalistic direction, in the 80s didactic theories emphasizing visualisation and geometry emerged to the front in teaching mathematics. Another outstanding direction is orienting school mathematics to its application in real-life situations (so-called Real-life Math). The 90s are well known for the rise of computers and intelligent calculators into the educational process. (Ambrus, 2002)

In the past three decades various psychological approaches have been incorporated into the education of mathematics, such as:

Deductive (behavioural) approach is characterized in a direct transfer (transmission) of knowledge from the teacher to the student.

Inductive (constructivist) approach in teaching is very different from deductive approach in many aspects. Cognitive evolution and process of education is defined as (Kostrub, 2008):

- Always start with the already achieved degree of progress of the student,
- Make sure of constructing meaningful teaching,
- Make possible the performing of the students' own meaningful educational process,
- Influencing the students in such a way, that they can themselves modify the schemes of knowledge,
- Creating and maintaining rich relations between newly gained knowledge and pre-existing schemes of knowledge.

Currently we live in the era, where we indispensably need a computer, the internet, mobile and other means of ICT (Information and Communications Technologies), therefore we are a part of the digital world. This generation of students was already born into this world and it is natural to them. Students of primary education are "digital aboriginals" as defined by Marc Prensky. Thankfully this fact is being leveraged in many schools to increase the quality of the educational process.

The theoretical framework for integrating technologies into the education process was developed by Mishra, P. and Koehler, M. J., and published as Technological Pedagogical Content Knowledge, known abbreviated also as TPACK (Mishra-Koehler, 2006).

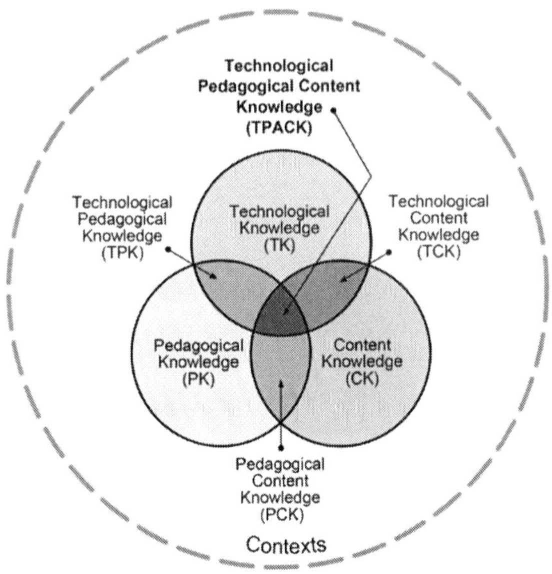

Fig. 1. TPACK

The TPACK model expresses the degree of teachers understanding digital technologies and the degree of their effective use in achieving educational goals (Fig. 1). The authors of TPACK are claiming, that it exceeds expert knowledge, pedagogy, technologies and is different in expertise, that the teacher gains by studying specific disciplines. According to Mishra and Koehler, teachers must understand how technologies, pedagogy and content work together and create a form of knowledge that exceeds above all of the three components combined. Teaching with technologies requires a flexible framework that explains how can new and fast changing technologies be used and integrated into new pedagogic approaches and areas of educational content (Mishra - Koehler, 2008).

Tablets became very popular in the past years, thanks to mobile operators and coupled with mobile internet are very attractive to students. There are many projects abroad and in Slovakia that focus on early adoption of m-learning into schools.

GeoGebra
We will focus on GeoGebra as it is an Open-source software, freely accessible to teachers and students, it dynamically grows and is adequate for PC, notebooks and tablets.
GeoGebra can be categorized into dynamic geometry systems, but the will of the developers community is to connect also arithmetic and algebraic inputs with visualisation. Therefore, it is appropriate to consider categorizing it into CAS (Computer Algebra Systems). The very name GeoGebra came into existence as an aggregation of the words GEOmetry and alGEBRA.

State education program (SEP) and geometry
In Slovakia, according to the latest SEP standards, axial symmetry is introduced from the first year of elementary school on the level of propaedeutics. The student should be able to complete the drawing of an axially symmetric picture. In secondary education, axial and central symmetry are introduced only in the 9th grade. Many teachers, however, start teaching it from the 5th grade, where students start to intuitively imagine symmetry.

2 Digital resources for teaching axial symmetry in primary education

In the next part, we are going to focus on ideas of methodics for teaching axial symmetry in the first grade of elementary schools.

According to the SEP this topic should be approached from a propaedeutic level. For improving the motivation of students, the teacher can use digital materials appropriate for interactive whiteboards. Websites and applets can be used as m-learning, therefore can be used on tablets or smartphones (individually or in pairs). We can find such a portal in Slovakia called Planéta vedomostí (Planet of knowledge). Teaching section called Zobrazenia (Symmetry) contains several nice activities about axial symmetry. The topic is introduced intuitively using a game with mirrors (Fig. 2).

In the next part, we provide some links to web pages that are suitable as didactic games, where mostly the task of the student is to fill in the second part of the image, so that the image will be symmetrical, or the student has to place an axis in a manner that it would become the symmetry axis. The meaning of the word axial symmetry is introduced as a "mirror image" (Fig. 4 to 8).

Fig. 2

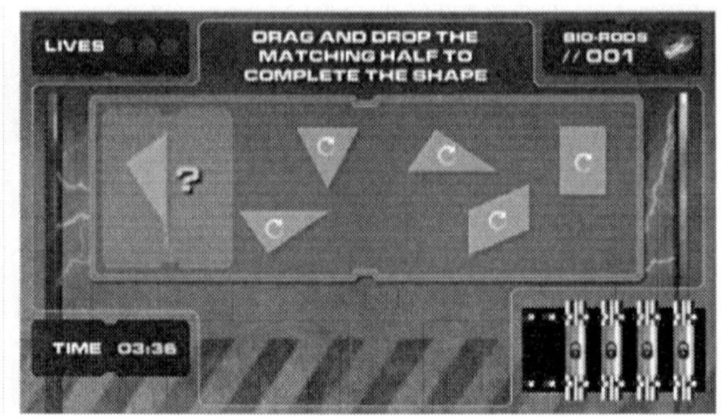

Fig. 3
http://www.topmarks.co.uk/Flash.aspx?a=activity19

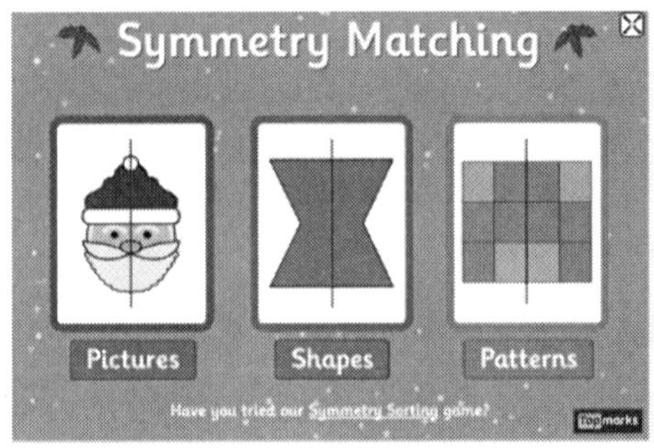

Fig. 4
http://www.topmarks.co.uk/symmetry/symmetry-matching

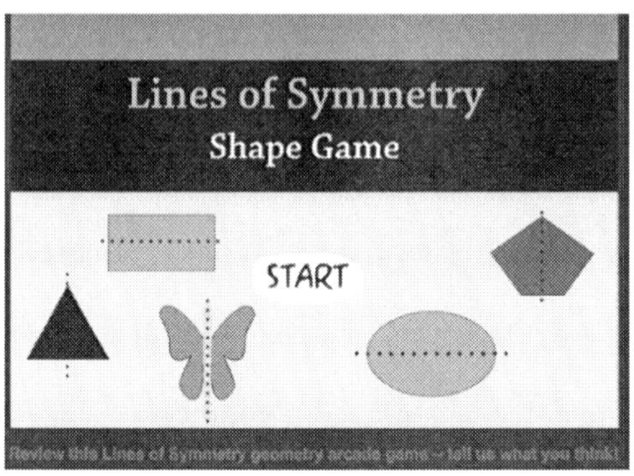

Fig. 5
http://www.sheppardsoftware.com/mathgames/geometry/shapeshoot/SymmetryLinesShapesShoot.htm

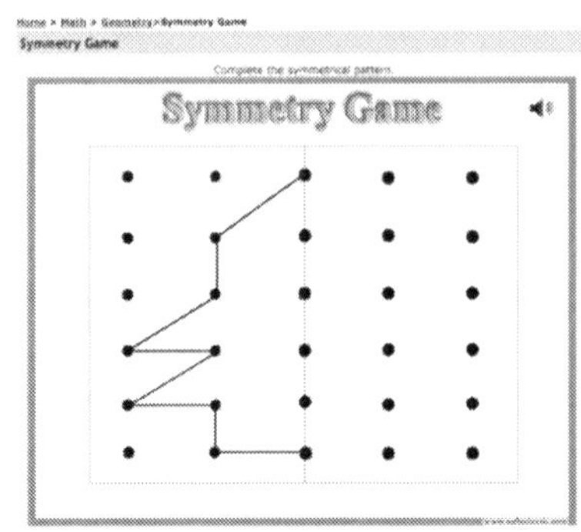

Fig. 6
http://www.softschools.com/math/geometry/symmetry_game/

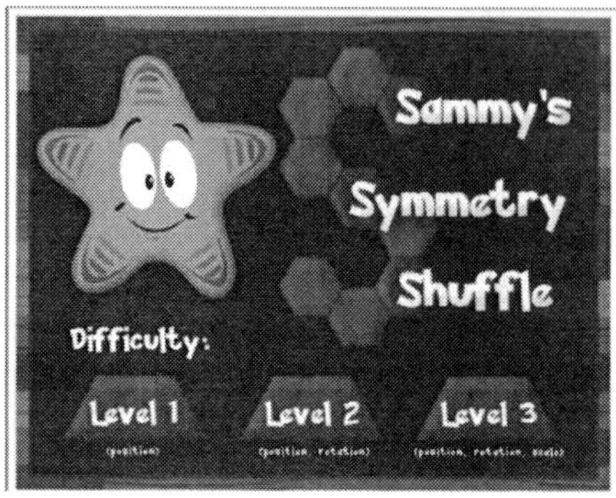

Fig. 7
http://www.fuelthebrain.com/games/sammys-symmetry-shuffle/

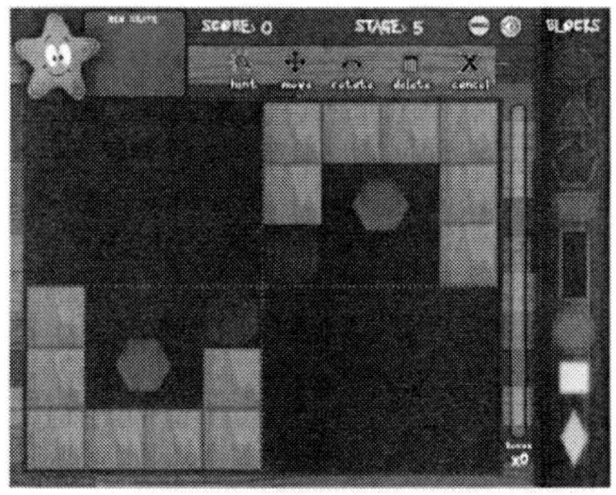

Fig. 8
http://www.fuelthebrain.com/games/sammys-symmetry-shuffle/

3 Survey

The ideas for teaching axial symmetry were tested in four classrooms at the Elementary school Hlboká cesta in Bratislava, between October and December in 2016. Students of the 4th and the 5th grade were involved (the 5th grade is considered as a transitional grade between primary and secondary education). The activities took place during several classes and were focused mainly to manipulative techniques with digital technologies. The students used paper, scissors, strings, cubes, drawing aids, tablets or smartphones and interactive whiteboards. Education was focused on constructivist principles. We were using controlled exploration. The tasks of activities were achieved in the dimension of cognitive process, understanding, application, analysis and evaluation. During manipulative activities we used Bruner's theory, meaning students went from inactive representation, through iconic to symbolic. During research we used the methodology of qualitative research.

In the next chapter we present the specific activities with description of the research.

Activity 1: Mirror theater

This is a group activity. Aids: mirror, pairs of identical geometric shapes, blocks, squared paper with a drawn line approximately in the middle of the sheet
The activity consists of these sub-activities:

- A pair of students played a game, where one of the students placed a geometric shape on the squared sheet of paper, then the other one placed the second shaped in such a way, that it would be the first object's mirror image. After that they verified their correctness using a mirror.
- One of the students constructed a building using blocks and the other one constructed the buildings mirrored version. Again, the results were verified using a mirror.
- A group of students created an imaginary line in the classroom (two students spanned a string). Some of the students created a "live image" on one side and the other group tried to create the same "live image" in such a way that it would be a mirrored image. The third group evaluated if it is the correct image.

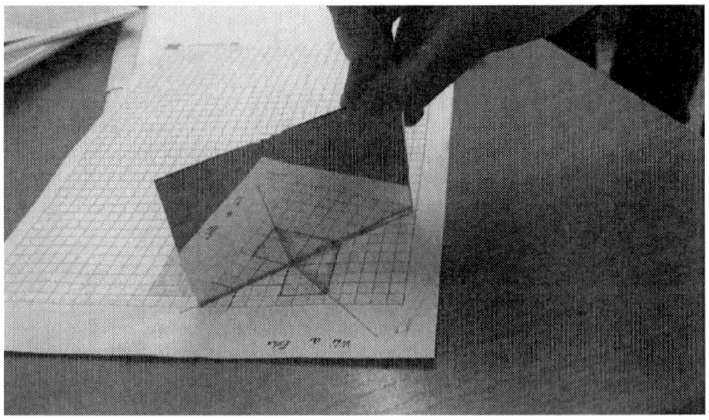

Fig. 9

Activity 2: Drawing with transilumination and scission

Aids: paper, drawing aids, window.

Students were drawing on paper. First an arbitrary line and then different geometric shapes. After that they folded the paper alongside the drawn line. They placed the sheet of paper on a window and re-drew the shapes on the other side of the paper. After they completed the drawing they verified with a mirror if it is the mirrored image. Next, they had to connect the vertices of the original shape with the vertices of the mirrored shape and were supposed to find out, how it would be possible to create the image without folding the sheet of paper.

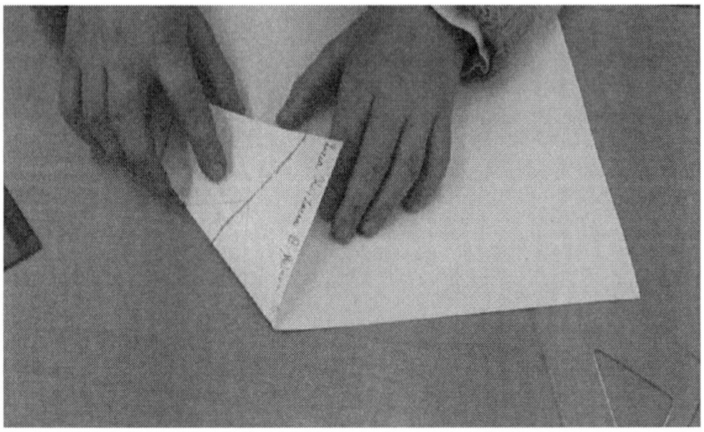

Fig. 10

In this activity, we observed a significant difference between classes, especially in pace and skill. None of the pairs of students were able to determine how to construct the mirrored image without folding the paper in half and re-drawing the picture on a window.

Fig. 11

Activity 3: Drawing symmetric shapes on a mobile in GeoGebra

Aids: tablet or smartphone, interactive whiteboard, GeoGebrabook available at https://www.geogebra.org/m/sxhCscUZ#material/k7ePsWjM

During this activity, students were supposed to draw in shapes in GeoGebra on a squared sheet that would be axially symmetrical. Some students worked on an interactive whiteboard. This GeoGebrabook has 11 applets that have increasing difficulty.

This activity significantly motivated the students as they had immediate feedback. Interestingly, some girls placed a mirror on their smartphone to help themselves to draw the image.

Fig. 12

Fig. 13

Activity 4: Create an educational video for your classmates

In this activity, pairs of students were supposed to write a scenario about axial symmetry and they shoot a video. The aim of the video was to explain to their classmates what axial symmetry is, what are the characteristics of axially symmetric shapes and how do we draw them.

This was assigned as homework for weekend. On Monday, the students presented their videos. Some pairs just read out definitions from school books or Wikipedia. Most of the students described axial symmetry as a mirrored image and to construct such an image they would use a mirror or squared paper or a window. Only 20% of students described the construction using perpendicular lines and equal distances from the axis. Despite that, after several classes of exercises they completed the test with 87% success rate.

4 Conclusion

Education of Axial Symmetry with the proposed method, using a constructivist approach and manipulative techniques and digital technologies lead to greater motivation of the students for the given topic, as well as to activating higher cognitive functions. The disadvantage is material and time demand (sometimes it is hard to acquire tablets for the entire class). For future, we would like to continue with the given research and create an electronic methodic material for teaching the given topic.

Acknowledgement

The paper was supported by grant from Grant APVV -15-0378 Optimalizácia výučbových materiálov z matematiky na základe analýzy súčasných potrieb a schopností žiakov mladšieho školského veku.

References

[1] AMBRUS, A. *Nemzetközi tendenciák a matematika oktatásban*. Budapest, 2002. https://dl.dropboxusercontent.com/u/100162898/ambrus/tendenc.pdf

[2] ČAVOJSKÝ, I. *Uplatnenie informačných a komunikačných technológií vo výučbe*. Zborník MMK 2012, roč. 3 [elektronický zdroj] Hradec Králové, Magnanimitas, 2012, s. 3276-3280 [CD-ROM]

[3] GUNČAGA, J. *GeoGebra in Mathematical Educational Motivation*. Annals. Computer Science Series, 2011. http://anale-informatica.tibiscus.ro/download/lucrari/9-1-07-Guncaga.pdf

[4] HAŠEK, R. *Několik příkladů užití programu GeoGebra ve výuce*. Sborník příspěvků 4. konference Užití počítačů ve výuce matematiky, 5. – 7. 11. 2009, České Budějovice, 2009, s. 65 – 74.

[5] HEJNÝ, M., NOVOTNÁ, J. a STEHLÍKOVÁ, N. *Dvacet pět kapitol z didaktiky matematiky*. Praha, Univerzita Karlova v Praze – Pedagogická fakulta, 2004.

[6] HERENDINÉ-KÓNYA, E., *A matematika tanítása alsó tagozaton*. Nemzedékek Tudása Tankönyvkiadó, Budapest, 2013.

[7] KALAŠ, I. a MORAVČÍK, M. *Digitálne technológie v materskej škole 1*. Bratislava, Metodicko—pedagogické centrum, 2012.

[8] KOSTRUB, D. *Dieťa/žiak/študent – učivo – učiteľ, didaktický alebo bermudský trojuholník?* Prešov, Rokus, 2008.

[9] KOSTRUB, D. *Základy kvalitatívnej metodológie*. UK Bratislava, 2016

[10] LORENZ, M. *Kde nechala škola díru: m-learning aneb Vzdělávání pro záškoláky*. ProInflow 2010, Praha, 2010. vyd.: 53 - 57. http://pro.inflow.cz/kde-nechala-skola-diru-m-learning-aneb-vzdelani-pro-zaskolaky

[11] MISHRA, P., KOEHLER, M. J. *Introducing Technological Pedagogical Content Knowledge*. Washington, DC, AACTE Committee on Innovation and Technology, 2008.

[12] MISHRA, P., KOEHLER, M. J. *Technological Pedagogical Content Knowledge: A new framework for teacher knowledge*. 2006, Teachers College Record

[13] PARTOVÁ, E. *Metódy a formy využívania pedagogického softvéru vo vyučovaní*. Olomouc, Univerzita Palackého, 2004.

[14] PARTOVÁ, E. *Vyučovanie matematiky pomocou moderných technológií*. Bratislava, Univerzita Komenského v Bratislave, 2011.

[15] PRENSKY, M. „Shaping Tech for the Classroom." 2005. edutopia. http://www.edutopia.org/ adopt-and-adapt-shaping-tech-for-classroom

[16] PŠENÁKOVÁ, I., HORVÁTH, R., SZABÓ T. "Interaktívna tabuľa vo vzdelávaní budúcich pedagógov." Edukacja-Technika-Informatyka 7.1 ,2001, 224-229.

[17] STOFFOVÁ, V., et al. Didaktika informatiky v príprave budúcich učiteľov informatiky a informačnej výchovy. Edukacja-Technika-Informatyka, 2016, 7.1: 230-242.

[18] STOFFOVÁ, V., ŠTRBO, M.: *Educational technologies to support language teaching* Proceedings of XXIX. DidMatTech 2016, Eötvös Loránd Tudományegyetem, Budapest, 2016, http://didmattech.inf.elte.hu/proceedings/

[19] SIMONKA, Zs. „Kreslenie grafov elementárnych funkcií a ich transformácií s využitím IKT (projekty a štúdie vo svete)“. Zborník APLIMAT 2008 - 7th International Conference on Applied Mathematics. STU. Bratislava, 2008.

[20] VELIKOVA, E. „Geo-Gebra: *Technology can advance creative mathematical activities in the classroom.* 2011. http://www.igmcg.org/files/newsletter_1.pdf

[21] VÁSÁRHELYI, É. *A számítógép a matematika oktatásában*. Budapest, 2015. http://dl.dropbox.com/u/100162898/vasar/szgep.pdf

[22] WINTSCHE, G. *Egy kísérlet nem kísérlet. A játéktól a kutatásig*. Budapest, Varga Tamás Tanítványainak Emlékalapítványa, 2014, s. 90-93.

[23] ŽILKOVÁ, K. *Školská matematika v prostredí IKT*. Bratislava, Univerzita Komenského Bratislava, 2009.

Current addres

Koreňová Lilla, doc. PaedDr. PhD.
Comenius University in Bratislava, Faculty of Education
Račianska 59, 813 34 Bratislava, Slovak Republic
Tel. number: 02/50 222 151, e-mail: korenova@fedu.uniba.sk

Slovak University of Technology in Bratislava
Faculty of Mechanical Engineering
16th Conference on Applied Mathematics **APLIMAT 2017**
Proceedings

DYNAMIC MODEL OF TAX ASSIGNMENT FOR YIELD TAX ON PERSONAL INCOME FROM DEPENDENT ACTIVITIES

KOŤÁTKOVÁ STRÁNSKÁ Pavla (CZ)

Abstract. The fundamental changes that accompany the allocation of taxes were due to the changing political situation or a reflection of the political cycle and a non-system approach in the creation of modifications. The aim of the paper is to create a dynamic model. The assembled dynamic model will be a tool for specifying the expected proceeds from the selected tax revenues (e.g. yield tax on personal income from dependent activities). Assembled model will be simulated on a change to the selected criteria.

Keywords: dynamic model, tax assignment, yield tax on personal income from dependent activities

Mathematics Subject Classification: Primary 60A05; Secondary 005C62

1 System dynamics and dynamic model

System science is a field of science dealing with systems. System science can be broken down in system theory (e.g. the issue of Cybernetics), and systemic application (e.g. the issue of system analysis, operational analysis, and others).

A system can be defined as a set of interacting elements, which together form a single unit ensuring the overall purpose. (FORRESTER, 1989)

One branch of system science is called "system dynamics", which belongs to the modern system approaches. System dynamics is a type of "systems thinking" and a methodology for solving real systemic problems. Systemic thinking is a modern approach with a direct link to system dynamics. Thanks to its existence we are able to model the real world. (MILDEOVÁ, 2008)
Visualization in system dynamics is used to express relationships and links between elements using graphical diagrams.

1.1 Mathematical basis of system dynamics of accumulation and flow-diagram

Interdependence of the individual is through the use of a system of differential equations. In the system dynamics are applied the so-called systems of nonlinear differential equations that are used to create and solution models.

Let us have an m-dimensional vector X fixed variables m_k–dimensional vector constant k, then m-dimensional system of differential equations is in the form (1)

$$\mathrm{x}' = f(t, \mathrm{x}, \mathrm{k}) \ , \tag{1}$$

where $f = (f_1, \ldots, f_m)'$ are functions defined on set $G \subseteq \mathrm{R}^{m+m_k+1}$.

Let there be i-the and j-the fixed variable, then their equations are of the form (2)

$$\begin{aligned} x_i' &= f_i(t, x_1, \ldots, x_i, x_j, \ldots, x_m, k_1, \ldots, k_{m_k}) \\ x_j' &= f_j(t, x_1, \ldots, x_i, x_j, \ldots, x_m, k_1, \ldots, k_{m_k}) \end{aligned} \ , \tag{2}$$

where dependence on the individual variables need not necessarily be non-constant. Function f_i where dependence on the individual variables need not necessarily be non-constant x_i and is called as the "flow". In the model that contains the elemental there is $x' = f(x)$ (see Fig. 1). (BURIANOVÁ, 2007)

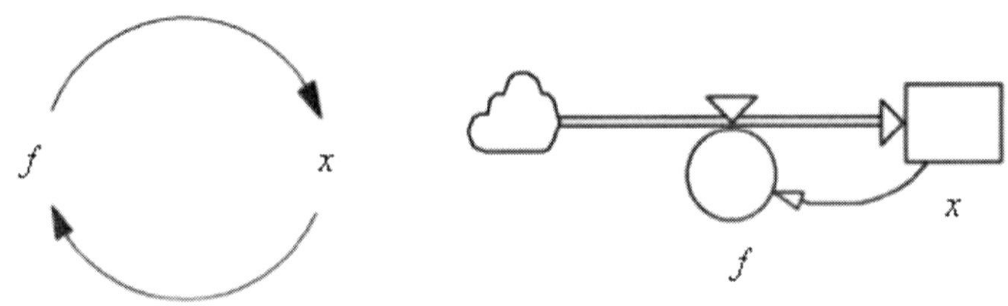

Fig. 1. Basic loop

If $\frac{\partial f}{\partial x} > 0$, then occurs to the elemental loop with positive polarity (see Fig. 2a), i.e., that the loop is reinforcing loop in the model. Otherwise, talks about the loop with a negative polarity, loop buffer (see Fig. 2b).

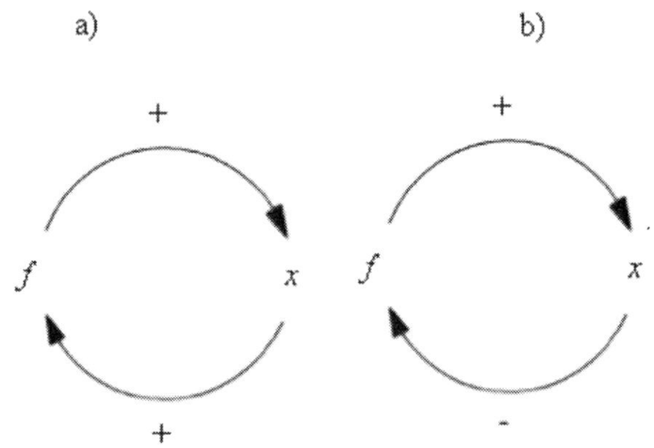

Fig. 2. Reinforcing loop and buffer loop

The reinforcing loop

It is assumed loop with positive polarity, ie. reinforcing loop with one fixed variable $x(t)$ with single flow $x' = f(x)$ a inequalities $\frac{\partial f}{\partial x} > 0$.

If

$$x' = f(x)$$
$$x'' = f'(x) \cdot x' = f(x) \cdot f'(x) = f(x) \cdot \frac{\partial f}{\partial x} \quad (5)$$

Let us have $x' > 0$, respectively $f > 0$, then $x(t)$ is increasing. From the term $\frac{\partial f}{\partial x} > 0$ implies, that $x'' = f(x) \cdot \frac{\partial f}{\partial x} > 0$. Then $x(t)$ for $x' > 0$ is an increasing convex function.

Conversely, if $x' < 0$, respectively $f < 0$, then $x(t)$ is degreacing. From the term $\frac{\partial f}{\partial x} > 0$ implies, that $x'' = f(x) \cdot \frac{\partial f}{\partial x} < 0$. Then $x(t)$ pro $x' < 0$ is degrecing concave function.

Buffer loop

It is believed the same case, a fixed variable $x(t)$ with single flow $x' = f(x)$ a inequalities $\frac{\partial f}{\partial x} < 0$.

If

$$x' = f(x)$$
$$x'' = f'(x) \cdot x' = f(x) \cdot f'(x) = f(x) \cdot \frac{\partial f}{\partial x} \quad (6)$$

Let us have $x' > 0$, respectively $f > 0$, then $x(t)$ is increasing. From the term $\frac{\partial f}{\partial x} < 0$ implies, that $x'' = f(x) \cdot \frac{\partial f}{\partial x} < 0$. Than for $x' > 0$ is increasing concave function. And conversely.

1.2 Dynamic model

The term "model" in system science is considered as a representative or real object. A dynamic model is a model that includes time factors and time effects.

The dynamic system can be understood as a set of interrelated elements whose values evolve over time and influence each other. It is a system where all variables are in time t, which is an element of a given set of time moments denoted by T.

Models in system dynamics are divided into physical and abstract models. Under the physical model it is possible to imagine a material, subject or realization model. Abstract models are formal or ideological and are further subdivided into mathematical and simulation models. Abstract mathematical models are based on mathematical relationships and equations. Abstract simulation models are expressed by means of programming.

A dynamic model is a useful tool to understand the relationships and links between elements in the real world as it is possible to simulate the functioning of the system, interactively affect the behavior of the model, and suggest improvements to the model. (BURIANOVÁ, 2007)

Definition of the simulation (KINDLER, 1980): "*The simulation is a research technique which essentially substitutes the investigated dynamical system with a simulator which is experimented on to obtain information about the originally analyzed dynamic system.*" (KŘIVÝ, 2001)

2 Dynamic model of tax assignement

Dynamic model consists of four dynamic sub-models were compiled using the simulation program Vensim PLE, which are interconnected through causal relationships and characterize the proposed model of tax assignment (next as TA).

These dynamic models include: model of the population in the municipality (focusing on the observation of changes in population due to the natural increase of population in the municipality); model of population in the Czech Republic (focused on observation of changes in population due to the natural increase of the population in the Czech Republic); model depicting the methodology for calculating tax revenues from shared taxes, which is redistributed to municipalities from the national gross yield of 23,58%; model depicting the methodology for calculating taxable income from taxes on personal income from dependent activities (next as PIDA).

The simulation examined the behavior of the model with a fixed time horizon of one year. The simulation of the developed model was applied to the selected municipality, and also performed in the simulation program Vensim PLE, which yielded two outputs:

- predicted behavior of the model from a situation of constant baseline parameters;

- examining model behavior due to changes in baseline parameters.

The default values of input parameters of the selected municipalities were data derived from national and regional statistics for the years 2014 and 2015. The results of the simulations found a slightly increasing trend for tax from personal income, which is supported by the increasing number of pupils and students.

2.1 Graphical visualization of dynamic model TA for yield PIDA

The amount of the tax revenue accruing to the specific municipality is affected by three criteria, where the most important criterion is the number of inhabitants.
The following figure (see Fig. 3) the calculation methodology captures tax revenue PIDA in graphical form as one of shared taxes.

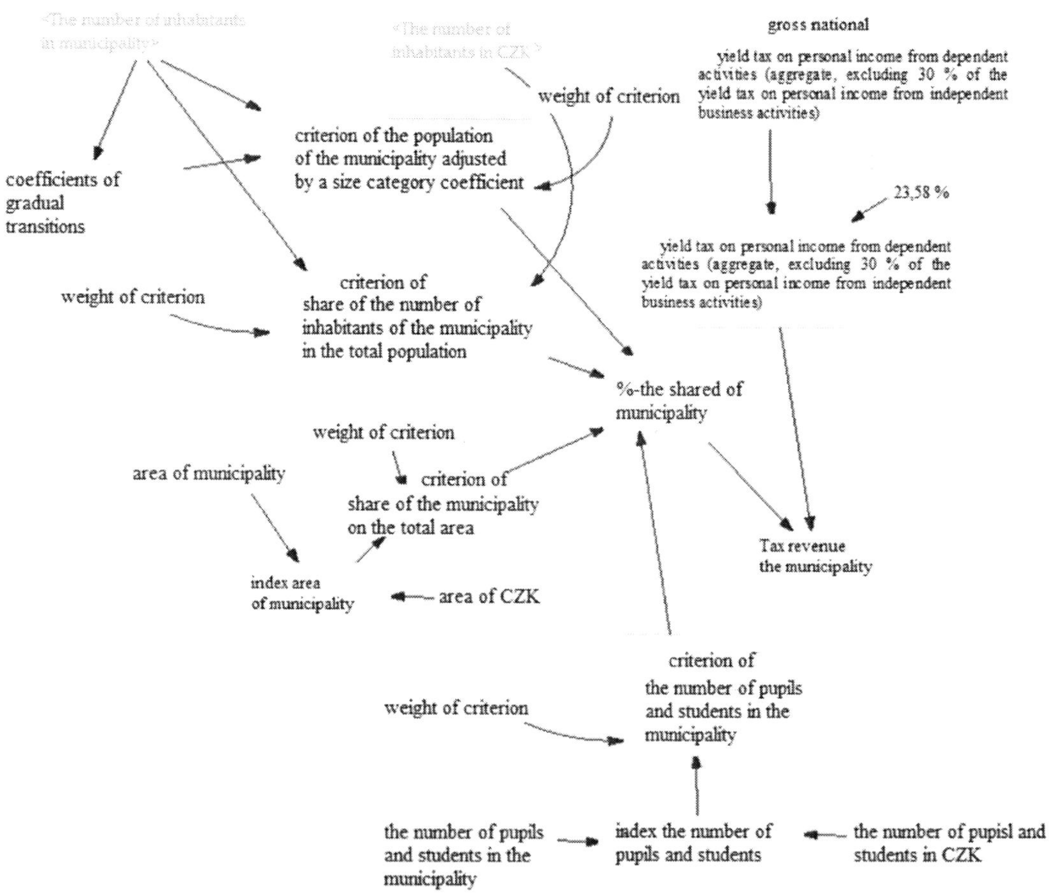

Fig. 3. Model methodology for calculating shared tax revenue from yield tax on the personal income from dependent activities

The issue of the revenue from this tax is not only from the point of view of the budget of the municipality, but also for the state budget. Tax revenue from tax on the yield rate personal income from dependent activities particularly affects the tax rate, then the possibility of tax deductions and tax rebates.

2.2 Simulation - effect of changes in te values of the weights of pupils and students on TA

The simulation shown in the figure below, has demonstrated a slight increase in the value of the monitored parameters. The increasing trend has remained essentially unchanged.

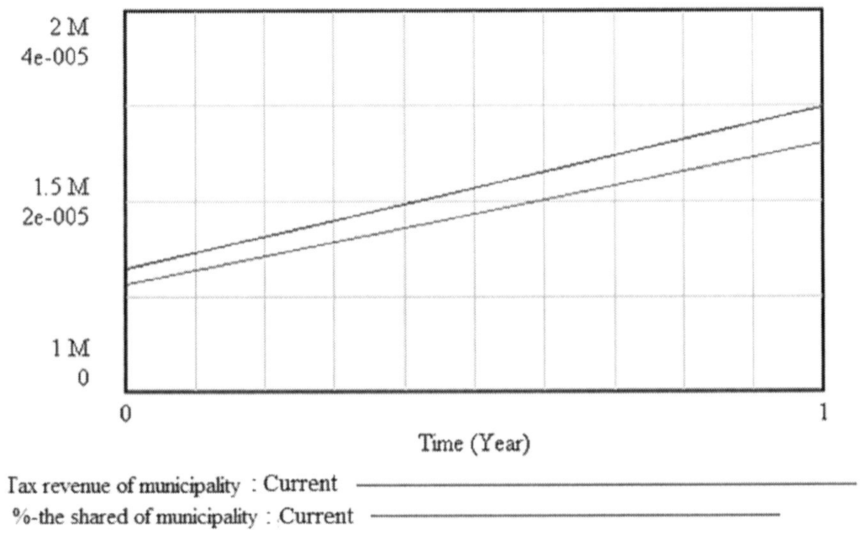

Fig. 4. Effect of changing the values of the weights of the criteria on the overall shared tax revenue of the municipality

From the above figure (see Fig. 4), it is evident that the municipality draws positive effects from the existence of the school facility.

3 Conclusion

Revenues from taxes, which are redistributed to municipalities from the state budget, make up the largest share of budget revenues, and should respect their individual needs and preferences. Due to the low fiscal autonomy of municipalities, revenues from taxes represent an unstable income and municipalities often do not know how much money they receive from taxes. They can only surmise and base their planning on previous years and in particular the year preceding. Municipality expenditure has largely a current expenditure character and in contrast to income the volume of expenditure is known, especially if it has a mandatory character. Municipalities must have mandatory expenditure under the law.

The key legislation that governs the allocation of resources within the system of public budgets is Act No. 243/2000 Coll. on the budgetary allocation of taxes (called also Tax assignment). Budgetary allocation of taxes in the Czech Republic has developed since 1993.

The result of the simulation is made to the fact that the criterion of the number of pupils and students to help the municipality to get to the budget funds from the proceeds of the taxes due to the increase in the value of the percentage.

Assembled model is a good tool for future amendment or editing for the TA. This model can take advantage of the municipality in budget planning for next year. Simulation will be done a particular image on the future developments in tax revenues.

References

[1] MILDEOVÁ, S., VOJTKO, V. Systémová dynamika. 2. přepracované vydání. Praha: Oeconomica, 2008. ISBN 978-80-245-1448-2.

[2] BURIANOVÁ, E. Simulace dynamických modelů s využitím metod systémové dynamiky. In Zborník príspevkov z vedeckého seminára s medzinárodnou účasťou. Nitra: Fakulta prídrodných vied UKF – KI, 2007, s. 19-29. ISBN978-80-8094-167-3.

[3] FORRESTER, J. W. The Beginning of Systém Dynamics. Studgart: Banquet Talk at the international meeting of the Systém Dynamics Society. 1989.

[4] JÍLEK, M. Fiskální decentralizace, teorie a empirie. 1. vyd. Praha: ASPI – Wolters Kluwer, 2008, s. 428. ISBN 978-80-7357-355-3

[5] KINDLER, E. Simulační programovací jazyky. Praha: SNTL, 1980.

[6] KOŤÁTKOVÁ STRÁNSKÁ, P. Analysis of financing the Municipalities in the Czech Republic. In International journal of mathematical models and methods in applied sciences, 2012, vol. 6, no. 8, pp. 917-925. ISSN 1998-0140.

[7] KŘIVÝ, I., KINDLER, E. Simulace a modelování. Ostrava: Ostravská univerzita – Přírodovědecká fakulta, 2001, s. 146.

[8] The Act no. 243/2000 Sb., o rozpočtovém určení výnosů některých daní územním samosprávným celkům a některým státním fondům (zákon o rozpočtovém určení daní)

[9] The Act no. 128/2000 Sb., o obcích

Current address

Koťátková Stránská Pavla, Ing. Ph.D.
Masaryk Institute of Advanced Studies, MIAS School of Business Institution address including Department of Economic studies, Czech Technical University in Prague
Kolejní 2637/2A, 160 00 Praha 6, Czech Republic
Tel. +420 602 611 505, e-mail: pavla.kotatkova.stranska@cvut.cz

Slovak University of Technology in Bratislava
Faculty of Mechanical Engineering
16th Conference on Applied Mathematics APLIMAT 2017
Proceedings

APPLYING DISCRETE MATHEMATICS IN SITUATIONAL INFORMATICS TASKS - IT STUDENTS' PERSPECTIVE

KOTLASOVÁ Michala (CZ), VANÍČEK Jiří (CZ)

Abstract. The paper reports on findings of a research among IT students focused on the impact of passing an undergraduate course of discrete mathematics on students' ability to recognize concepts, principles and methods used in discrete mathematics in situational informatics tasks. The performance in solving such problems, ability to find discrete mathematics in these tasks and also the use of knowledge gained in the course of discrete mathematics were studied and are discussed in the article.

Keywords: discrete mathematics, graph theory, situational informatics tasks, Beaver of informatics

Mathematics Subject Classification: Primary 97K20, 97K30, Secondary 97Q30

1 Discrete mathematics and its teaching

The effort to improve mathematics education has recently lead to introduction of new teaching methods and approaches such as inquiry based education, augmented reality, problem based education, building schemes [1]. Another innovative approach is supplementing the contents of mathematics by application tasks, whose solution makes the learner link theoretical and abstract environments of the learned content to the real world. This may result in an increase in motivation but also provides a natural environment in which the learner finds space to develop their competence to mathematize real situation.

The course of discrete mathematics in the training of future bachelors in informatics (both future teachers and informatics experts) encompasses a relatively wide spectrum of mathematics topics from combinatorics and combinatory methods through finitary relations to graph theory. Especially two areas, basic combinatorics and graph operations, are essential for any further studies in the field of information technology but also in practical application of informatics knowledge.

Theories in the areas of discrete mathematics have attracted attention of a number of prominent mathematicians. For example hypergraphs, namely Ramsey theory, have been explored by Rödl [2], mapping of Algorithmic Graph Theory for example by Golumbic [3]; these authors primarily focus on verification of theories. Another research area of great potential is how knowledge from discrete mathematics can be applied in various scientific disciplines, not just mathematics.

Although the previous paragraph may seem to suggest that in teaching we are able to find applications of discrete mathematics only on theoretical level, the contrary is true. Teaching informatics subjects nowadays asks for knowledge of discrete mathematics as the cornerstone in solving of informatics tasks. This can be illustrated by teaching computer networks, namely routing. Here we use Dijkstra's algorithm to look for the best way to optimal computer network throughput. As Deo [4] and other publications focusing on practical applications of discrete mathematics in computer science show, knowledge in this area is essential for deeper understanding of principles of algorithmization, contact and electrical networking etc. Foulds [5] shows in his publication that we often come across the use of graph theory. Interpreting a problem by a graph is not only used in algorithmization but also in networking, especially in the area of computer networks in linking nodes using the least amount of connections, maintaining network functionality while removing some nods, in looking for the shortest paths in networks, circuit layouts, Hamiltonian circuits.

Already this short list of publications signals the importance of knowledge of discrete mathematics for informatics as a scientific discipline. Less attention has so far been paid by the research community to teaching of informatics at universities. In consequence to developments in the used teaching methods as well as to changes in the target group of education, we must ask whether teaching of discrete mathematics with its stable methods and procedures does not need to undergo some change. Research conducted by Haviger [6], focusing on analysis of teaching of graph theory and combinatory algorithms and its subsequent optimization among other found that a task's success rate increases if a graph in the assignment is visualized in contrast to the situation when it is assigned in the form of a matrix. The success rate is provably influenced by solvers' motivation and interest in the task. The effort to innovate teaching methods in order to increase success rate in solving model problems using discrete mathematics is the starting point of the here presented research.

2 Beaver of Informatics contest

The Bebras project [www.bebras.org] can undoubtedly be regarded as one of the most important milestones in the area of reintroduction of informatics content into schools across Europe over the past few years [7]. In contrast to "official" ways such as definition of national curricula, taxonomy of learning objectives, outcomes and contents, the form of an informatics contest designated for average eager pupils interested in technology approaches the issue from the other pole: it looks for suitable topics and problems, suitable situations, didactical transformations of the original informatics topic.). Recently the contest runs in more than 40 countries with more than 1 million contestants of age from 9 to 19 yearly so it is one of biggest world contributors of introducing of computer science content into schools.

This cluster of national contests introduces the fundamental concepts of computer science to all students, beginning at the elementary school level. These contests react to the need to develop computational thinking; this concept was defined by J.Wing [8] and is very closely connected to other universal fundamental concepts that go beyond contemporary technology. The shift from the concept of digital literacy to computational thinking can be perceived not only as broadening of the educational area but also as a parallel of transformation of transmisive teaching methods to approaches developing pupils' critical thinking and ability to solve problems.

Beaver of informatics uses so-called situational informatics tasks. These tasks that are not based neither on a practical activity, nor on project work and creativity. This kind of task is not "develop a program, create an algorithm, propose a solution to the problem". In a situational task solvers emerge into a described situation in which they must grasp, get to understand the used concepts and terms, find an informatics principle the task is based on, solve the problem using cognitive and thinking skills and select the right answer from the offered choices [9]. The contest tasks always have informatics background but are set in situations and environments from the contestants' lives. This allows the contestants to imagine and visualize the situation and to use their experience. Part of these tasks are from topic included in discrete mathematics.

As an example, we present here two typical situational contest tasks for high school students connected with discrete mathematics:

Exemplary task Building a chip *(author Valentina Dagienė)*
A small chip is composed of a grid of contacts (marked as dots in Fig. 1). Some are already connected (marked as line segments). Connectors are always only between adjacent contacts, horizontally or vertically. We want to connect S and R with a continuous sequence of connectors, which do not touch any already connected contacts.

Question: How many different ways are there to connect S and R with the least possible number of connectors?
Set of answers: A) 5 B) 13 C) 15 D) 16

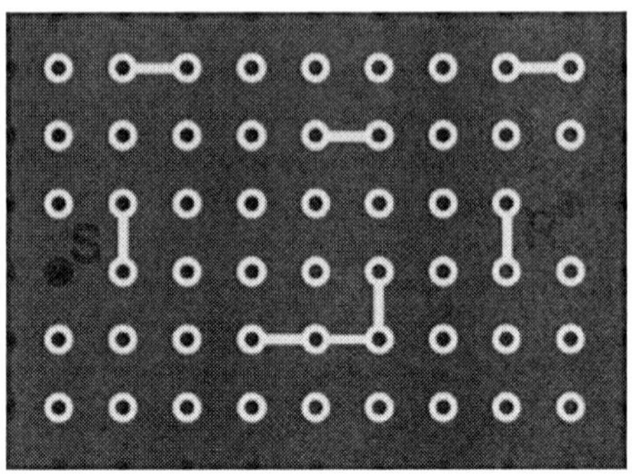

Fig. 1. Illustrative picture to the exemplary task *Building a chip*

Exemplary task Graph of a map *(author Michael Königsmayr)*
Maps can be easily pictured as graphs. In such a graph every node is a country and the lines between the nodes mean that they border each other. The picture shows a graph of a map with seven countries (Fig. 2).

Question: Can you find a map that matches the graph?

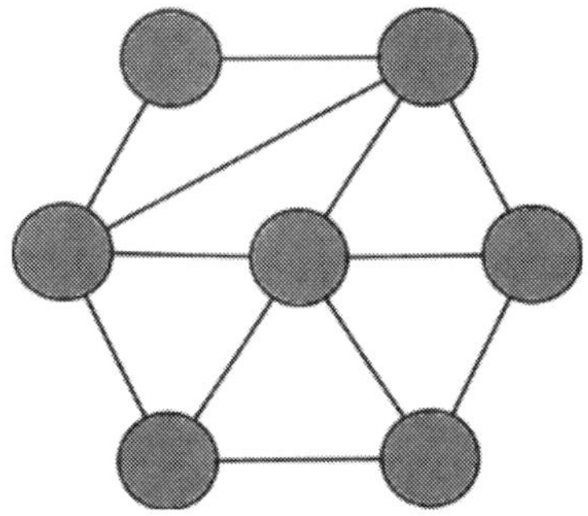

Fig. 2. Graph of a map with seven countries

Set of answers

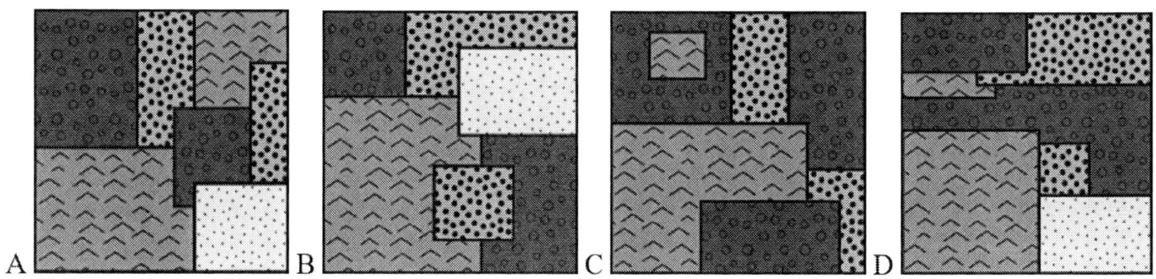

Fig. 3. Set of answers to the exemplary task *Graph of a map*

For comparison, discrete mathematics in an undergraduate university course uses abstract tasks. The learner may fail to see their connection to reality or potential application in informatics. A typical task looks like this [10]:

1. We are selecting three out of five marbles of different colours (red, blue, yellow, green, white). How many possibilities do we have if we definitely want to select white or green marble?

2. Let A = {1,2,3,4} and R = {(1,1), (3,3), (4,4), (2, 3), (3, 2), (1,2), (2,1)} be a relation on a set A. Find the properties of the assigned relation R (reflexivity, symmetry, antisymmetry, transitivity)

Our department guarantees and organizes the contest Beaver of informatics. Thus it was very convenient to try use the contest tasks in teaching of discrete mathematics (both to increase motivation and to set the solved problems into meaningful situations, allowing mathematization of a real problem).

3 Method and design

The question we wanted to inquire into was whether students who had finished the course of discrete mathematics were good enough in the area to recognize discrete mathematics tasks when disguised in the attire of a situation and whether they then could solve such tasks using methods acquired in the course of discrete mathematics.

We constructed a test from a selection of tasks from past years of the Beaver of informatics (Czech localization of the Bebras contest, www.ibobr.cz). The topics of the tasks corresponded to the contents of the course of discrete mathematics. The tasks were from the areas of propositional calculus, combinatory models and principles (combinations, logical trees), graphs (looking for the shortest path), trees (binary trees) and graph skeletons (looking for minimal graph skeletons). When selecting the tasks we were guided by applications of different topics in other areas of informatics. Apart from application of graphs and looking for the shortest path, tasks dealing with other topics were selected. For example:

- *Rodinný strom (Family tree, author Zoltán Molnár)* representing creation of a logical tree. It is based on a logical programming language. Systems using logical programming allow the user to simplify decision making and search for data according to given criteria. Logical programming is much needed in the field of artificial intelligence.

- *Jak se dostat k číslu 100 (Descend the Tree, author Troy Vasiga)* concerns an alternative way of coding numbers using two values, i.e. binary expression of natural numbers. Binary representation of data is the cornerstone of computer work.

- *Trezor (Electronic Lock, author Volodymyr Ksiondzyk)* is a task focusing on the use of elements running on the principle of logical operators. These are called logical gates in informatics.

The constructed test was an online test. The testees went through 10 items and selected from a choice of 4 answers (text or pictorial), recorded their answer to a text box or manipulated with objects in an applet, e.g. dragged them to the right places, marked them etc. While sitting the test the testees could go back to previous tasks, change their answers, but also could skip a question and not answer it at all.

The test with tasks related to discrete mathematics was assigned to 19 first year undergraduate students of Information Technology and e-Learning and students of Informatics for future teachers who had not enrolled in the course of discrete mathematics yet (referred to as no DM group in the following text) and to 19 students who had finished the course of discrete mathematics (after DM group). Their results on the test were compared. The results were also compared to the results of upper secondary school students in the previous years of the Beaver of Informatics.

Having completed the test, the students were asked to fill in a questionnaire asking about the relation of the test items to the course of discrete mathematics. The questionnaire had three sections of questions. The first section was about the testee's opinion on the test as a whole. The answers were given on a five-point scale and the items were asking about the testee's satisfaction with results of the test and their ability to relate the test thematically to a specific informatics discipline (programming, operation systems, principles of computers etc.). We were primarily interested if the respondents would relate the tasks to discrete mathematics. In the second section we asked explicitly whether the assignments in the test reminded them of tasks taught in the course of discrete mathematics and whether procedures known from the course helped them solve the tasks. The third section focused on different procedures in individual tasks used or noticed by the respondents.

The questionnaire was filled in by all the tested students. The findings from this survey are an important source of information about how test items were solved by the students. These findings were compared to partial results from the test.

Having completed the questionnaire, the students were informed about their own results. This was followed by a joint discussion on this type of tasks. The findings from this discussion were included into an analysis of the sources of wrong solutions in some test items.

4 Outcomes of the questionnaire survey

The individual test items involved discrete mathematics both in problems from real life and in problems from informatics environment. All students were asked in the questionnaire whether they would relate the assigned tasks to some informatics topic (programming, algorithmization, computer networks ...), or to discrete mathematics. Their opinions differed depending on the year of their studies, or more precisely on whether they had already finished the course of discrete mathematics. In case of no DM group, knowledge of different informatics disciplines was not expected. However, the students did show some knowledge in the area.

The question "Into which subjects would you place this type of task?" was a multiple choice item. The Graph 1 shows the frequency of selection of the different answers.

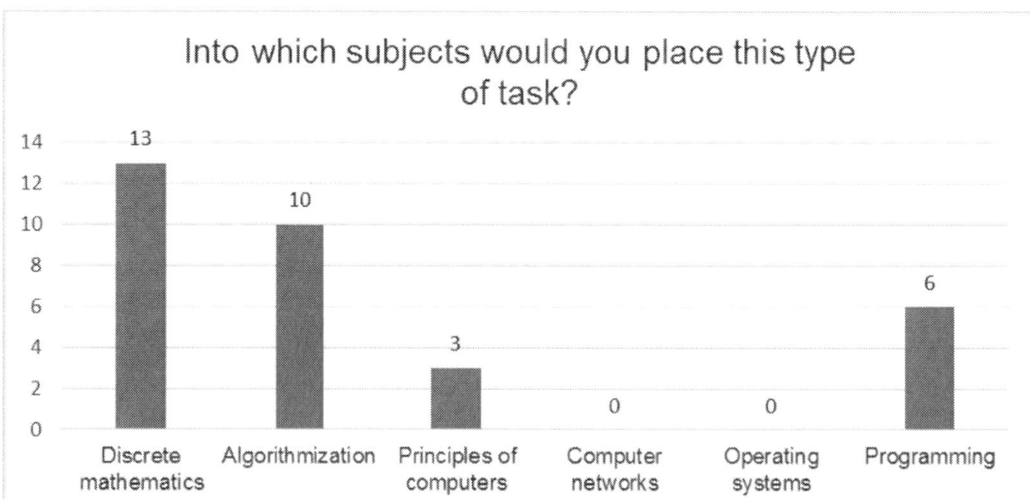

Graph 1. Frequency of answers to question "Into which subjects would you place this type of task?" no DM group

13 students from no DM group placed the tasks into discrete mathematics without actually knowing the content of the discipline. It seems the students chose the answer because of the word mathematics, not because of the adjective discrete. We presume the students observed some mathematical principles in the problems although they were not able to name them. Other students related the problems to the area of algorithmization or programming, none of them to the technical area.

Let us now look at the same question as answered by after DM group. Again, predominant was relating the tasks to discrete mathematics.

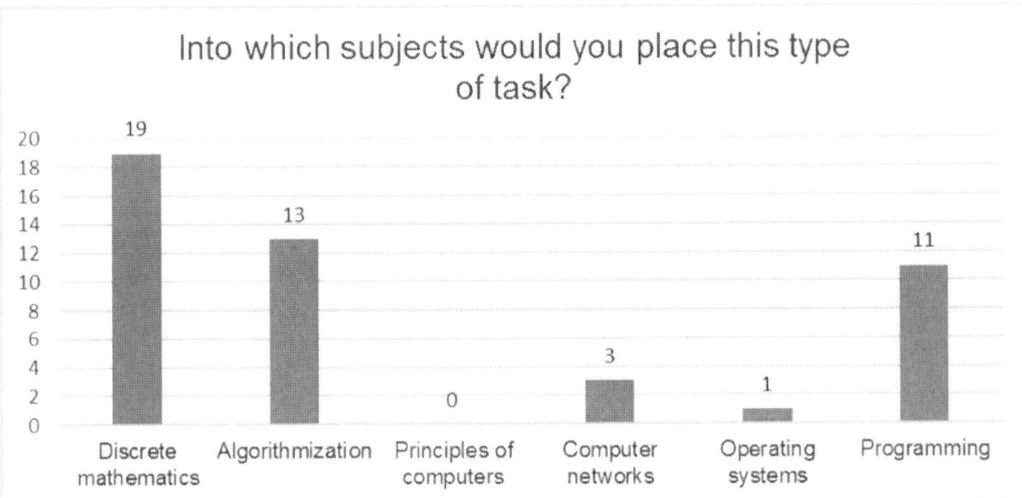

Graph 2. Frequency of answers to question "Into which subjects would you place this type of task?" after DM group

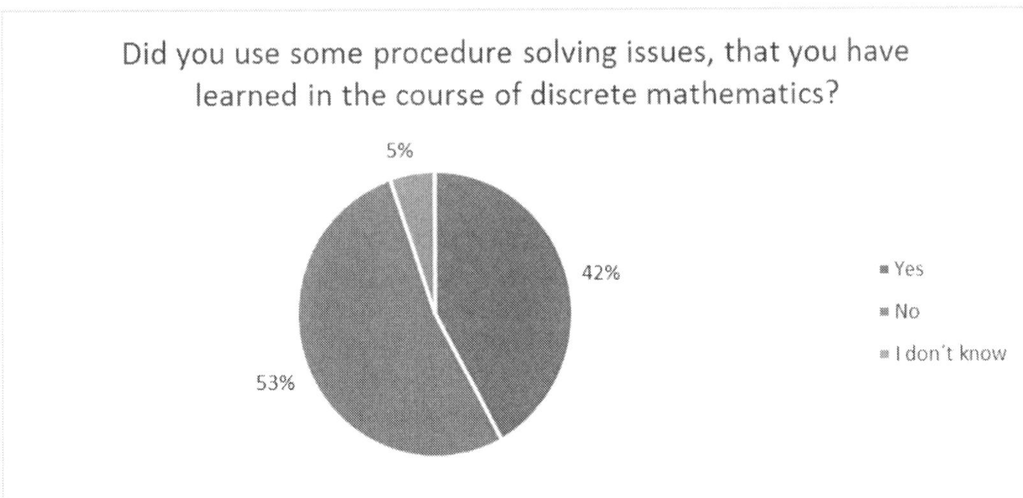

Graph 3. Frequency of answers to question "Did you use some procedure solving issues, that you have learned in the course of discrete mathematics?" after DM group

There is another question that in case of after DM group shows whether the students were able to recognize procedures or structures introduced in discrete mathematics. It is logical not to present results of no DM group here as they had not participated in the course. The specific topics mentioned were the following: evaluation of graphs, looking for optimal path, skeletons of graphs, descending a tree. Some students stated that they tried to apply the selected operation but either were not successful (the result was not correct) or they did not manage to finish the solution (due to lack of knowledge)

In the survey the students were asked in which tasks they recognized the given area and possibly used it for its solution. Within the frame of the research we compared students' answers to test results, see the chapter Analysis of test results and questionnaire outcomes.

5 Results of testing based on tasks from an informatics contest

In our research we asked whether students after having finished the course of discrete mathematics would become more successful in solving the tasks or not. Within the scope of the whole group we could not observe improved performance.

Name of task	*Link to the original task in Czech language*	*no DM group*	*after DM group*	*Beaver contest*
Sada karet	https://goo.gl/Ws6Wfp	5 %	16 %	
Mosty přes ostrovy	https://goo.gl/c5FTk3	10 %	11 %	4 %
Cesty na hřiště	https://goo.gl/aOzbh6	25 %	32 %	18 %
Jak se dostat k číslu 100	https://goo.gl/1QQpUW	30 %	26 %	28 %
Nejlepší překlad	https://goo.gl/aeaDK6	45 %	32 %	31 %
Špióni	https://goo.gl/nXut18	50 %	32 %	30 %
Výroba čipu	https://goo.gl/XDP4Ms	50 %	47 %	45 %
Trezor	https://goo.gl/g1vyH4	60 %	79 %	54 %
Rodinný strom	https://goo.gl/UMf0iS	70 %	58 %	50 %
Průzkum řeky	https://goo.gl/1zgXEs	85 %	68 %	77 %

Tab. 1. Overview of success rate in test items in the studied groups

An overview of tasks included in the test and success rate of all three groups is presented in Tab 1. In each item we provide a link to its full assignment in original as they were presented to students (the original assignments were in English and translated to Czech which could have slightly changed the meaning). In the chapter presenting analysis of test results the assignments are in English. The tasks are presented in the table in the order from the lowest success rate in no DM group. For comparison we also present results in the national round of the Beaver of Informatics where the tasks had been solved by upper secondary students (the column Beaver contest).

The data in the table do not imply that after DM group would be more successful in more tasks than no DM group. The collected data could lead to the conclusion that having finished the course of discrete mathematics has no impact on success on the test. However, these are only indicative results as there might be other factors at play. For example, experience from previous contests (i.e.

familiarity with the process and form of testing), former education in informatics at upper secondary school, the level of logical thinking and other.

6 Analysis of test results and questionnaire outcomes

Having assessed the success rate in the test we took a closer look at individuals' performance in different tasks. A student's performance was compared to their answers in the questionnaire. Let us now present some findings. We studied in more details those tasks in whose solution students were rather unsuccessful. Using answers in the questionnaires we tried to reconstruct the reasons for this low success rate. We also present comparison with tasks focusing on the same area or with tasks that students placed into this area.

The test item with lowest success rate was the problem *Sada karet (Card match, author Troy Vasiga)*. It was labelled as a difficult task also in the competition. With respect to discrete mathematics it represents application of a combinatory principle – logical tree.

Task Sada karet

Each of the nine cards below has three attributes:

- the shape of the objects (triangle, square or circle)
- the number of objects (1, 2 or 3 objects)
- the fill style of the objects (empty, striped, solid)

A group is a collection of three cards such that the three cards satisfy all of the following three conditions:

- the objects on the three cards either all have the same shape, or all have a different shape
- all three cards have the same number of objects, or there is different number of objects on each card
- the objects on the three cards have the same fill style, or the objects on the cards all have a different fill style.

Consider the following nine cards (numbered from 1 to 9):

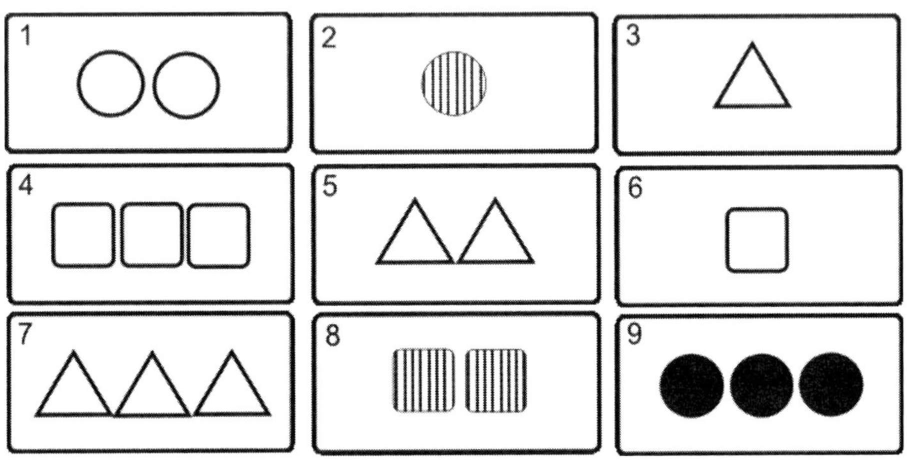

Fig. 4. Illustrative picture to the exemplary task *Sada karet*

Cards 1, 2, 9 form a group because:

- the objects have the same shape (circle)
- there is a different number of objects on each card
- the fill style of the objects on each card is different

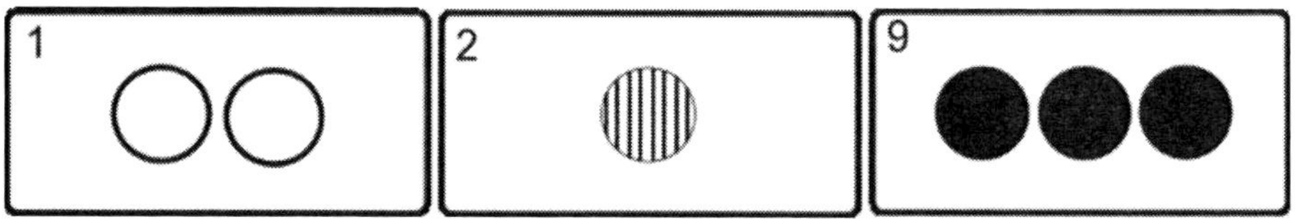

Fig. 5. Illustrative picture to the exemplary task *Sada karet*

Each group is described by list of increasing numbers of cards, it is impossible describe the group 1, 2, 9 as 2, 9, 1.

Question: Allowing a card to appear in multiple groups, how many groups can be formed with the cards above (including the group 1, 2, 9)?
Set of answers: A) 3 B) 4 C) 5 D) 6

Let us discuss the correct solution. All the conditions about card properties must be fulfilled at the same time. Let us start from the first, arbitrarily selected card. In each subsequent step we make a decision according to the conditions about which card we can use. At the same time the order of cards does not matter. As soon as a card is used and placed into the solution, we do not make a solution with reverse order of the same series of cards. The solution can be illustrated by a graph, namely a logical tree, see Fig. 6.

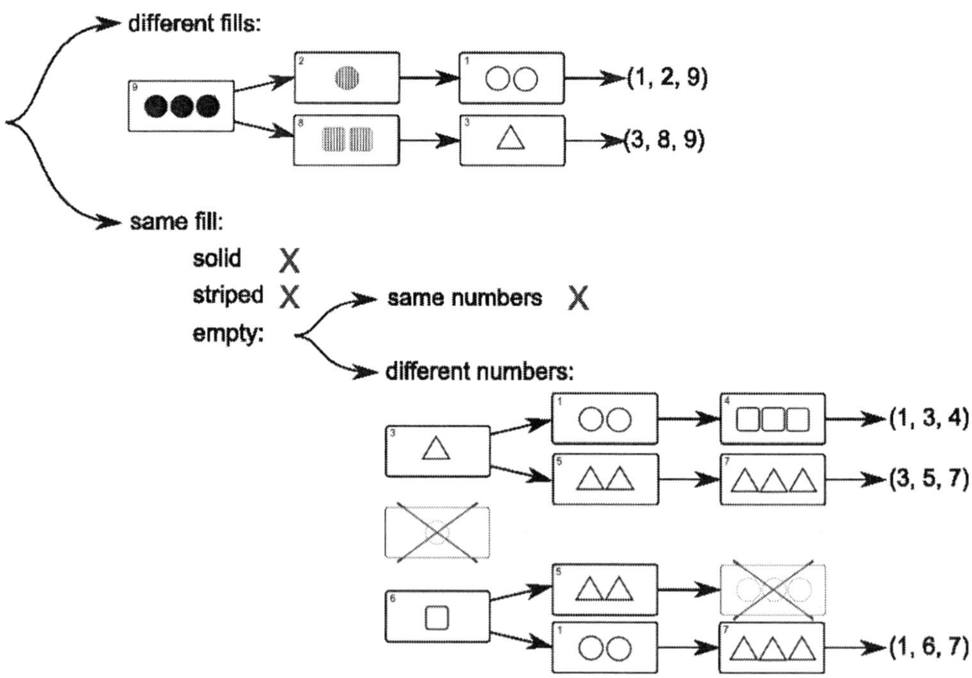

Fig. 6. Logical tree in a solution of the task *Sada karet*

This is a clearly visible example of a logical tree. Based on previous steps we determine the following possible solutions. I.e. we start from an initial decision (the root of the tree) and continue selecting until the last card in the set (the leaf). The number of leaves represents the number of possible solutions, the number of sets that can be created. In its essence a logical tree is an example of combinatory principles.

This is one of the few problems that caused less trouble to after DM group. The questionnaire answers imply that students from this group mostly did not use any area or topic from discrete mathematics and did not recognize it in the task. A minority of respondents mentioned the topic of combinatorics or combination. We can infer that most students used logical reasoning or chose the right answer at random.

If we consider the set of offered topics in the relevant item in the questionnaire we can presume that a possible cause of poor performance in this task was mistaking the task for its numerical variant. We presume the students came out of typical problems using combinations, which is characteristic for problem solving in discrete mathematics. What they did was selecting three cards from a collection of all available cards while the order of selection did not matter. However, combinations do not bring the desired result. Combination does not fulfil the given conditions for a set of cards. It seems the students replaced the question from the assignment by their own question that they expected to be asked in this type of a task.

The test included one more task from the area of logical trees with a different assignment, *Rodinný strom (Family tree, author Zoltán Molnár)*. The goal of this problem is to construct a family tree from the given relationships. Each relationship is described by the command `parent(A,B)`, which means that A is the parent of B. From the following relationships the contestant has to find all grandparents of Zoltan among beavers:

```
parent(Thomas, Peter).
parent(Ann, Eve).
parent(Reka, George).
parent(Margaret, Ann).
parent(Ann, Zoltan).
parent(Peter, George).
parent(Charles, Zoltan).
parent(Margaret, Peter).
parent(Thomas, Ann).
parent(Charles, Eve).
```

The students selected the correct answer from 4 possible combinations of names. The solution can again be illustrated by a logical tree as shown in Fig. 7.

As far as the number of correct answers is concerned, this was one of the most successful test items. It is another task in which having finished the course of discrete mathematics did not improve performance, since no DM group was more successful. Although the success rate in the task was marked in both groups, a logical tree was not mentioned in the subsequent questionnaire by any of the students. The answers suggest the students solved the problem by logical reasoning. The task seems to have been regarded as too simple to ask for construction of a logical tree. Thus we can infer that the use of models from discrete mathematics is likely to be motivated by the level of difficulty of the model situation.

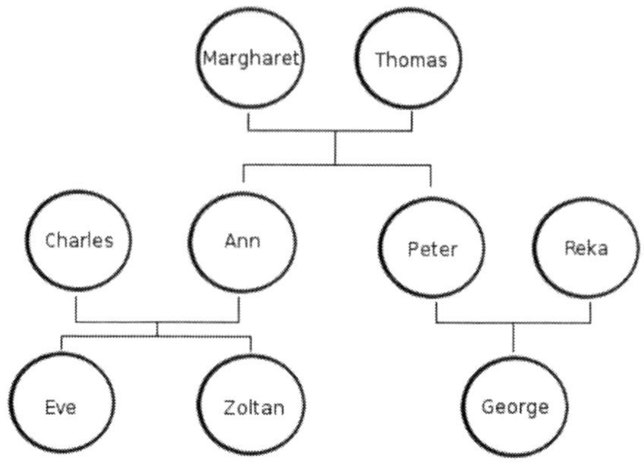

Fig. 7. Logical tree in the solution of the task *Rodinný strom*

The use of two thematically similar tasks was deliberate. The aim was to explore whether students were able to recognize the topic in two different situations. The difference in success rates in both tasks and the difference in reactions in the questionnaire suggests this area is hard to be recognized by students in problems of application nature. Use of similar problems and tasks in the course of discreet mathematics could prove to be helpful for their better insight.

Although students failed to see logical trees in the previous two tasks, the questionnaire analysis shows that they were quite often able to find a logical tree in the task *Jak se dostat k číslu 100 (Descend the Tree, author Troy Vasiga)*. However, in case of this problem the sought principle was not a logical tree but a binary tree.

Task Jak se dostat k číslu 100

Your friend writes down all of the integers starting from 0 in the following way:

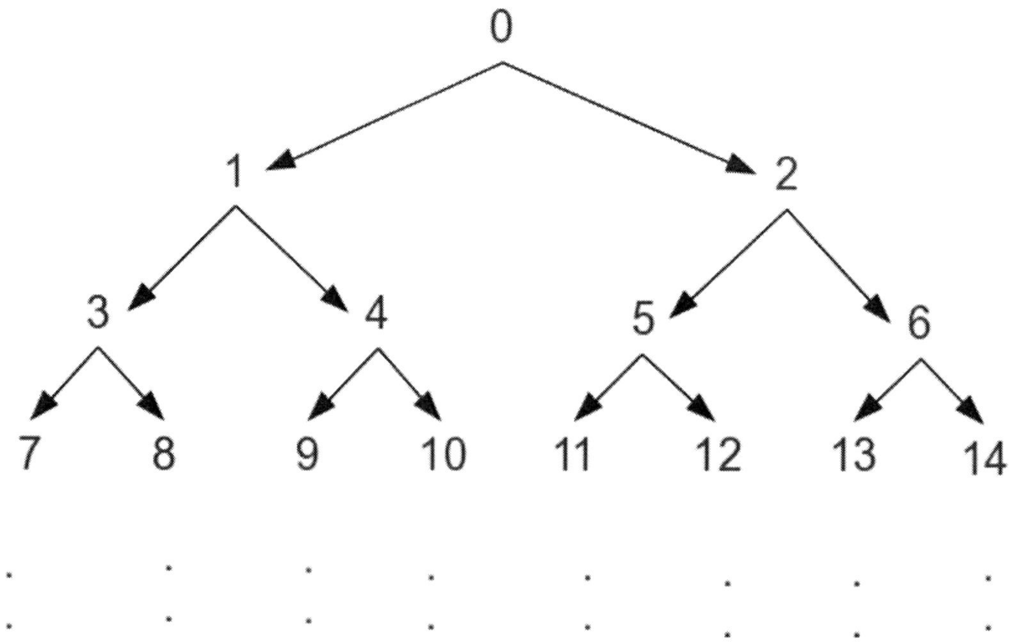

Fig. 8. Illustrative picture to the exemplary task *Jak se dostat k číslu 100*

Specifically, below every number there are two numbers: one on the left and one on the right. For example, below 3, the number 7 is on the left, and the number 8 is on the right. The numbers can be read in increasing order from top row to bottom row and from left-to-right within a row. Notice that we can get from 0 to 11 by going right (R), left (L) then left (L).

Question: Starting at 0, what sequence of left (L) and right (R) moves will end up at 100?
Set of answers: (A) LRRLRR (B) RLLRLR (C) RRLLRL (D) LLRRLL

The goal of the task is to determine how we will go through a tree to get to number 100 if all vertices are numbered progressively from the root to leaves in levels from left to right. This represents a typical example of a binary tree. The problem tests understanding of regularity of such a tree and ability to determine the position of a number. Students were selecting from four possible paths through the tree.

Correct answer is (B). Explanation: Notice that all even numbers are arrived at by a (R) move. Therefore, 100 must end with an (R). The parent of 100 is floor(100-1)/2 = 49. Using a similar argument, 49 (which is odd) must be arrived by a (L) move, and the parent of 49 is 24. Similarly, 11, 5, 2, 0 is the parent chain, which causes moves to be RLLRLR. No other path can lead to the number 100. In fact you only need to find the last two directions, and then check the others to be sure.

Binary tree as relevant for this task was mentioned only by one student in the questionnaire. The rest of the answers were referring to a logical tree. This shows that students do not make a sufficient distinction between a binary and a logical tree, in other words they do not understand the structures well enough or do not have the terminology anchored. This topic is very hard to be recognized by students in practical problems, which can result in having troubles when solving problems that require higher degree of expertise. This again makes us conclude that it might be useful to show the mentioned structures in different model situations.

The assignments of the above presented tasks do not represent the typical situations that would be discussed in the course of discrete mathematics. One of the typical problems whose alternatives were worked on in the course of discrete mathematics is the problem *Mosty přes ostrovy (Build the bridges!, author Ivo Blöchliger)*. This problem is an application of looking for the minimal skeleton of a graph with values.

Task Mosty přes ostrovy
Mr. Beaver must build bridges connecting all islands and both river banks directly or indirectly (see Fig. 9). No bridge is built yet. Next to each possible bridge is the number of work days it takes to build it. You can click on any bridge to build it or remove it.

Question: How many work days are needed at least? Write integer number.

In this task, we are to solve so-called minimum spanning tree problem and look for the skeleton of a graph that does not include any circular connection of edges and at the same time includes all vertices of the graph. When we construct the skeleton we achieve connection of all the demanded islands and at the same time discover that there is only one path to each of the islands. However, finding the skeleton does not necessarily fulfil the demand on the shortest possible time for construction. Thus we must look for a minimal skeleton of the graph. When solving the problem, we can use one of greedy searches, for example Kruskal's algorithm. Its principle is construction of a non-decreasing sequence of masses of the various edges. Then edges with the lowest mass are gradually selected and included in the skeleton. If the edge forms a circle with the previous ones, it has to be removed from the selection. The process is over when we select n – 1 edges of the skeleton. The students were asked to record the solution of this task as a numerical value of the sum of the shortest time to build the bridges.

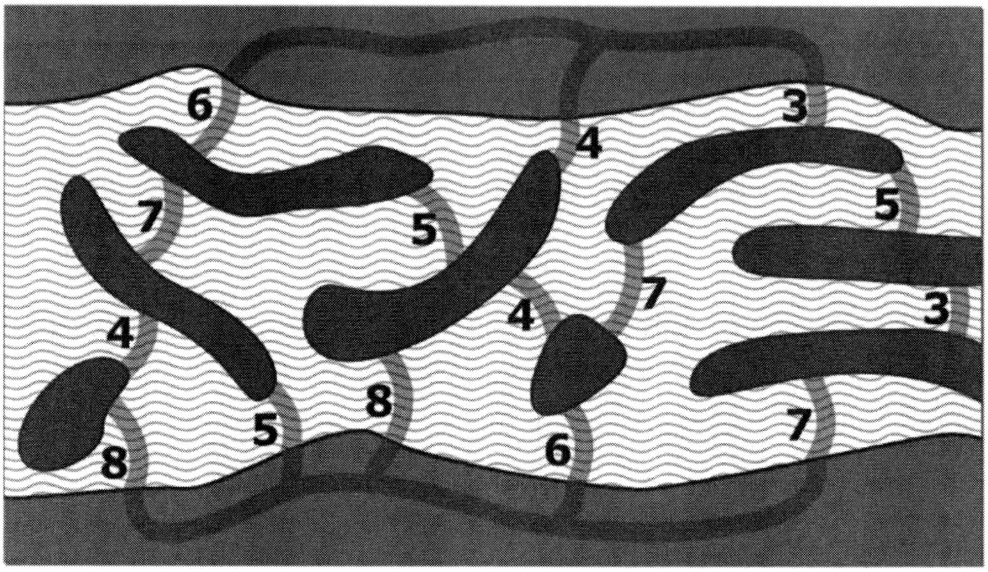

Fig. 9. Illustrative picture to the exemplary task *Mosty přes ostrovy*

This task was one of the least successful tasks on the test. Only one respondent mentioned in the questionnaire the use of a skeleton of a graph. And this respondent did not solve the task correctly. The other answers in the questionnaire indicate that students mistook this topic for the topic of looking for the shortest path. However, the latter is not applicable for the solution of this task. When looking for the shortest path we look for the shortest possible connection between two graph nods (islands here) but when constructing it we do not have to go through all the nods, which is required in this task.

For comparison we present an assignment of analogical task used in the course of DM.

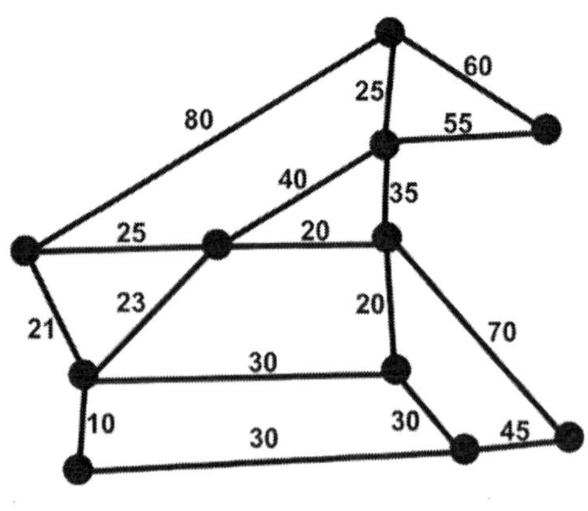

Fig. 10. Assignment of a graph with values for making the minimal skeleton

As it is visible in Fig. 10, this is an analogical case to looking for the cheapest connection of nods in a computer network. Although the assignment of this task is markedly similar to tasks that had been solved in the course of discrete mathematics, those area seems to be ungraspable for students in practical tasks.

7 Conclusion

The conducted survey shows that the impact of the undergraduate course of discrete mathematics on recognition of principles and methods of this discipline in situational tasks is negligible. The differences in performance and recognition of a task as a discrete mathematics task could be observed between after DM group and no DM group only in some tasks. At the same time, we cannot determine what aspect or criterion would allow us to find a type of task in which students who had finished the course of discrete mathematics would outperform other students. The outcomes from the questionnaire survey do not imply that students who had completed the course of discrete mathematics would use the learned procedures for solving the problems in any greater degree. On the contrary, in majority of cases they either did not recognize them or used them incorrectly.

One of the reasons may be the deliberate difference between situational Bebras contest tasks and school model tasks. The respondents failed to see the relation and then their knowledge of DM did not help them solve the problems. While the contestants in Beaver of Informatics are expected to use heuristic approaches, intelligence and logic and they are not expected to know the methods specific for the discipline, after DM students are expected to look for discrete mathematics in application tasks and to use the learnt methods for their solution. It seems after DM students failed to grasp the concepts and methods of DM within the course of discrete mathematics on sufficiently high level of Bloom's taxonomy.

It would be interesting to find out whether use of similar situational tasks within the course of discrete mathematics would result in better performance of students who had finished the course when solving situational and application tasks and also in their insight into whether a task is related to discrete mathematics, its concepts, principles and solving procedures. This would be very useful, taking into account that bachelor studies in IT should have practical focus and the theoretical nature of the subject may not be in line with this focus. Moreover, theoretical nature of the subject may also not suit students, their learning styles, thinking and motivation. The following stage of this research could focus on definition and search for situational tasks of application nature that would make discrete mathematics more accessible to students.

Acknowledgement

The paper was supported by grant GAJU No. 121/2016/S entitled "Problem solving as one of key competencies in mathematics and computer science education".

References

[1] HEJNÝ, M. *Exploring the Cognitive Dimension of Teaching Mathematics through Scheme-oriented Approach to Education.* Orbis Scholae 6(12), 41-55, 2012.

[2] DELLAMONICA, D, S. LA FLEUR and V. RÖDL. On induced Ramsey numbers for k - uniform hypergraphs. *Random Structures & Algorithms* [online]. 2016, **48**(1), 5-20 [cit. 2016-11-22]. DOI: 10.1002/rsa.20580. ISSN 10429832. Available from: http://doi.wiley.com/10.1002/rsa.20580

[3] GOLUMBIC, M. Ch. *Algorithmic graph theory and perfect graphs.* New York: Academic Press, 1980. ISBN 0122892607.

[4] DEO, N. *Graph theory with applications to engineering & computer science.* Dover edition. Mineola, New York: Dover Publications, Inc., 2016. ISBN 0486807932.

[5] FOULDS, L. R. *Graph theory applications.* Springer Science & Business Media, 2012.

[6] HAVIGER, J. *Analýza a návrh optimalizace výuky teorie grafů a kombinatorických algoritmů (Analysis and design of optimalization of graph theory and combinatoric algorithms education)* [online]. Hradec Králové, 2010 [cit. 2016-11-13]. Doctoral theses. University of Hradec Králové, UHK, Faculty of sciences. Thesis supervisor Eva Milková. Available from: <http://theses.cz/id/2e6p6g/>.

[7] DAGIENĖ, V. and S. SENTANCE. *It's Computational Thinking! Bebras Tasks in the Curriculum* [online]. s. 28 [cit. 2016-11-22]. DOI: 10.1007/978-3-319-46747-4_3. Available from: http://link.springer.com/10.1007/978-3-319-46747-4_3

[8] WING, J. M. *Computational thinking.* Communications of the ACM, 49 (3), 33-35, 2006.

[9] VANÍČEK, J. *What Makes Situational Informatics Tasks Difficult?* [online]. s. 90 [cit. 2016-11-22]. DOI: 10.1007/978-3-319-46747-4_8. Dostupné z: http://link.springer.com/10.1007/978-3-319-46747-4_8

[10] NÝDL, V. *Diskrétní matematika v příkladech (Discrete mathematics in examples).* České Budějovice: Jihočeská univerzita v Českých Budějovicích, Pedagogická fakulta, 2006. ISBN 80-7040-889-8.

Current address

Kotlasová Michala, Mgr.
Jihočeská univerzita v Českých Budějovicích, Pedagogická fakulta
Jeronýmova 10, 371 15 České Budějovice, Czech Republic
Tel. number: +420 387 773 078, e-mail: kotlasovam@pf.jcu.cz

Vaníček Jiří, doc. PaedDr. Ph.D.
Jihočeská univerzita v Českých Budějovicích, Pedagogická fakulta
Jeronýmova 10, 371 15 České Budějovice, Czech Republic
Tel. number: +420 387 773 074, e-mail: vanicek@pf.jcu.cz

Slovak University of Technology in Bratislava
Faculty of Mechanical Engineering
15th Conference on Applied Mathematics **APLIMAT 2016**
Proceedings

COMPARING THE EFFECTS OF CONSTANTS FOR GALLIUM NITRIDE BASED ON THE CHARGE OF A THIN MEMBRANE

KOVÁČ Tomáš (SK), HORVÁT František (SK), ČEKAN Michal (SK), HUČKO Branislav (SK), DZUBA Jaroslav (SK), VANKO Gabriel (SK)

Abstract. The contribution focuses on the comparison of constants in the compliance matrix of a mechanical model based on magnitude of the resulting charge when a thin piezoelectric membrane of circular cross section made from aluminum gallium nitride is loaded. The article compares 15 sets of varying mechanical constants used in numerical analysis when determining the magnitude of the charge. The constants were obtained from different sources in world databases. The numerical results of each set were compared with experimental results and conclusions were drawn.

Keywords: piezoelectric, gallium nitride, constitutive equations, median, variance, deviation

Mathematics Subject Classification: Primary 74D10; Secondary 62J10; Secondary 60F10

1 INTRODUCTION

The most common configuration of thin Al-GaN membranes is a circular section intended for the measurement of pressure (sensor). In order to subject pressure to the membrane, it must be fixed around its circumference. In PECIAR et. al. [4] the methods for calibrating a membrane intended to measure pressure are discussed. A spike introduces a load over a given area on the membrane and a reference value is obtained for the membrane [3]. The numerical computation used in the design and verification of such a sensor requires the consideration of variations in the mechanical properties of Al-GaN and the resulting charge that they produce. Therefore this article first deals with gathering the material constants from numerous publications worldwide, implementing them into the numerical analysis, and then finally comparing results with experimental measurements.

2 INVESTIGATED MODEL

The numerical model is based on a thin membrane used in experimental measurements. The configuration and geometrical properties of the membrane can be seen in Fig. 1. It is made of a piezoelectric AlGaN layer over GaN. The ratio of each corresponding layer is 99.9524% GaN overlaid with 0.0476 % AlGaN. The membrane is circular with a diameter of 750 μm. The membrane is placed over a ceramic substrate measuring 350 x 350 μm. the thickness of each layer is 4.22 μm and 0.02 μm for GaN and AlGaN respectively. Once a charge is created a thin 2D layer of electron gas is produced between the layers and acts as a transfer for current across the membrane. This layer has an effect on the magnitude of the piezoelectric charge which occurs on the surface of the AlGaN layer.

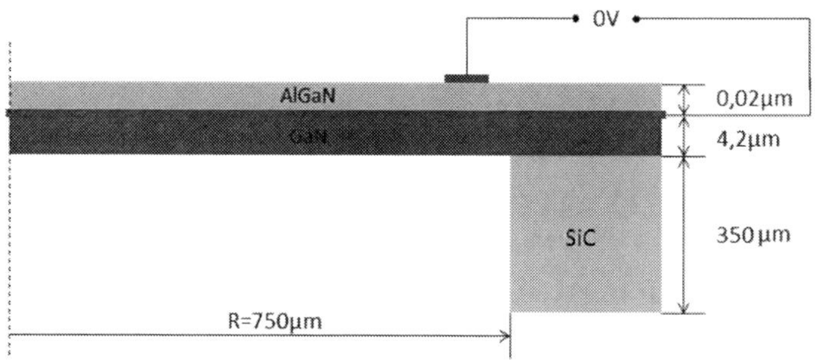

Fig. .1. Dimensions of the equivalent model of the membrane

3 GENRAL DESCRIPTION OF PIEZOELECTRIC MATERIALS

The solution of piezoelectric problems requires a physical and electrical description of the material. These can be divided into two general equations. Mechanical properties are given by equation (1) while the electric properties are given with respect to dielectrics in equation (2).

$$S = \frac{\sigma}{E_x} \quad [\frac{\mathrm{m}}{\mathrm{m}}; \frac{\mathrm{N}}{\mathrm{m}^2}; \frac{\mathrm{N}}{\mathrm{m}^2}] \tag{1}$$

$$D = \varepsilon . E \quad [\frac{\mathrm{C}}{\mathrm{m}^2}; \frac{\mathrm{F}}{\mathrm{m}}; \frac{\mathrm{V}}{\mathrm{m}}] \tag{2}$$

where: S is the deformation, σ is stress, E_x is Young's modulus, D is the electric displacement, ε is the permittivity of the dielectric and E is the electric field. Because the piezoelectric effect deals with both, it is necessary to combine the electric and mechanical properties into an electro mechanical description given by (3).

$$S = \frac{\sigma}{E_z} + d.E \quad [\frac{\mathrm{m}}{\mathrm{m}}; \frac{\mathrm{N}}{\mathrm{m}^2}; \frac{\mathrm{N}}{\mathrm{m}^2}; \frac{\mathrm{m}}{\mathrm{V}}; \frac{\mathrm{N}}{\mathrm{m}^2}] \tag{3}$$

$$D = d.\sigma + \varepsilon^{\sigma}.E \quad [\frac{\mathrm{C}}{\mathrm{m}^2}; \frac{\mathrm{C}}{\mathrm{m}^2}; \frac{\mathrm{N}}{\mathrm{m}^2}; -; \frac{\mathrm{N}}{\mathrm{m}^2}] \tag{4}$$

where d – piezoelectric deformation constant with respect to electric field E. In the absence of mechanical stress, d in equation (4) represents the electric charge per unit stress over area D provided that the electric field is zero giving the permittivity ε^{σ} of the material subject to constant stress. Equation (3) describes the inverse piezoelectric effect. The membrane is subject to an external electric field which results in the total strain S. Equation (4) describes the direct piezoelectric effect. Mechanical stress applied to the body in combination with the piezoelectric constant and presence of electric field, the result of which creates a charge and it's corresponding displacement D.

AlGaN generally is described as an anisotropic piezoelectric material. In describing the material constants, it is necessary to use tensor formulation based on the conventions in Fig. 2

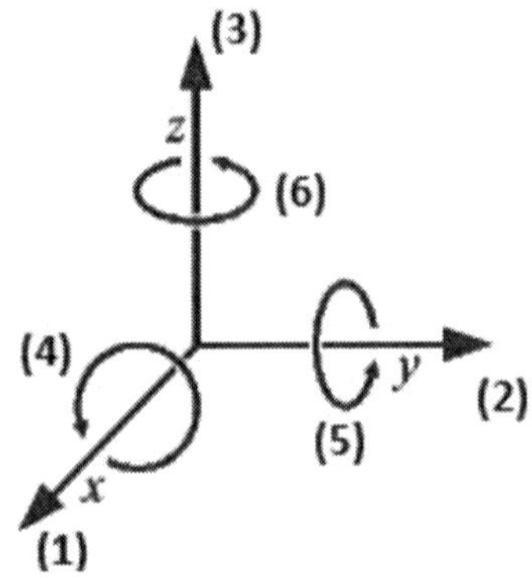

Fig. 2. Cartesian coordinate conventions utilized in the tensor description

where: 1 - load in the *x* axis, 2 – load in *y* axis, 3 – load in *z* axis, 4 - shear in *yz*, 5 - shear in *xz* and 6 – shear in *xy*

to coincide with the direction of polarization (3) and (4) can be rewritten as:

$$S_i = s_{ij}.\sigma_j + d_{mi}.E_m \tag{5}$$

$$D_m = d_{mj}.\sigma_j + \varepsilon_{mk}.E_k \tag{6}$$

where: indexes i, j = 1,2,3,...,6 and m, k = 1,2,3 and represent the individual orientations in the coordinate system.

In cases where the material is polarized in the *z* direction, the material is understood as transversely isotropic. A material polarized in such a way simplifies equations (5) and (6). Further description requires the application into matrix form which a finite element software such as ANSYS can easily formulate.

4 FINITE ELEMENTS AND MESHING

In solving for the piezoelectric study, it was necessary to define each element to calculate the mechanical and piezoelectric properties. Mechanical properties were defined by 8 node, PLANE 183 elements. The element is suitable for axisymmetric problems. The AlGaN layer was represented by PLANE 223 elements. This is the electrically active layer. It consists of the same 2D 8 node element which defines 4 degrees of freedom at each node. The properties allow for solutions with large deformations caused by piezoelectricity. 2D elements PLANE183 and PLANE223 are represented in Fig. 3.

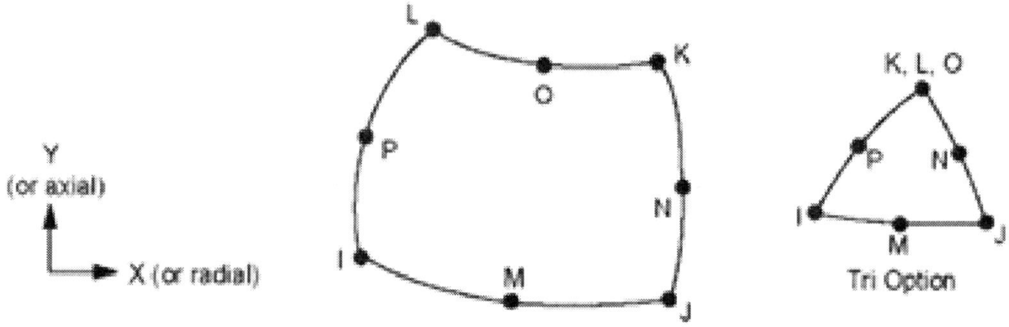

Fig. 3. Geometry of the 2-D elements PLANE183 and PLANE223

The finite element model was constructed as an axisymmetric problem with mapped meshing. To accommodate midplane effects a minimum of three elements over any given thickness was used. The meshed model can be observed in Fig. 4.

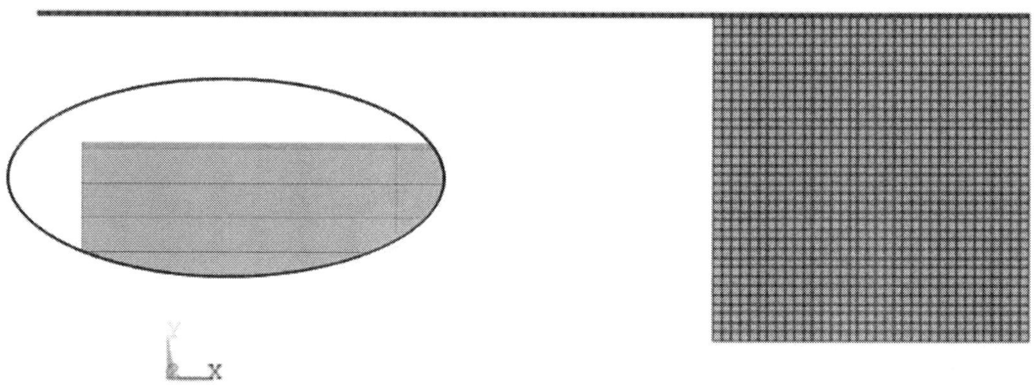

Fig. 4. Finite element mesh of the membrane model with edge detail (left bubble)

5 COMPARING THE NUMERICAL CALCULATIONS

Investigating the effects of changing mechanical constants of GaN and observing its behavior was based on the results of numerical evaluation. Primarily, the evaluation concentrated on the resulting change in charge calculated through the 2DEG layer. In the numerical analysis, input constant $C_{11} - C_{44}$ of the compliance matrix where changed while constant C_{66} was calculated according to relation (7).

$$C_{66} = \frac{C_{11} - C_{12}}{2} \tag{7}$$

The constants are based on findings in 15 different publications which have formulated these constants based on experimental evaluation. Their results can be observed in Tab. 1.

p. č.	C11	C12	C13	C22	C33	C44	C66	Ref.
1	367	135	103	367	405	95	116	[8]
2	390	145	106	390	398	105	122.5	[9]
3	374	106	70	374	379	101	134	[10]
4	365	135	114	365	381	109	115	[11]
5	373	141	80.4	373	387	93.6	116	[12]
6	396	144	100	396	392	91	126	[13]
7	369	94.2	66.7	369	397	118	137.4	[14]
8	377	160	114	377	209	81.4	108.5	[15]
9	370	145	110	370	390	90	112.5	[16]
10	351	142	106	351	397	93	104.5	[17]
11	334	132	99	334	372	86	101	[18]
12	386.3	159.8	141.3	386.3	391.2	115	113.25	[19]
13	396	144	64	396	476	91	126	[20]
14	373	141	80	373	387	94	116	[21]
15	374.2	141.4	98.1	374.2	388.6	98.3	116.4	[22]

Tab. 1. Input constants based on findings from various authors

The axisymmetric finite element model was created in ANSYS Multiphysucs. The loading and boundary conditions can be observed in Fig. 5. The outer surface was fixed while axisymmetry was defined at the center of the model. Using the coupling function, electric charge was possible to calculate. A pre-stress is applied in the first load step based on displacement to simulate the clamping of the membrane and results in an in-plane stress of 40 MPa. The second load step applies pressure perpendicularly to the membrane itself at magnitudes of 1, 5, 10 and 20 kPa.

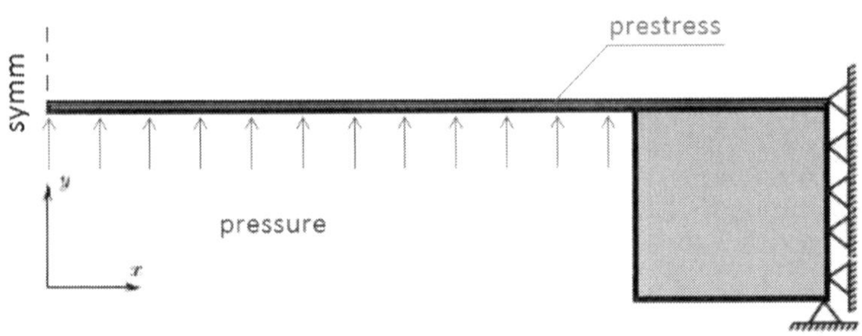

Fig. 5. Boundary and loading conditions in the model

6 COMPARISON OF RESULTS

Sixty numerical simulations were performed utilizing the varying material constants of gallium nitride and loading conditions discussed in parts 4 and 5 of this article. Each set of material constants was numerically calculated for four loading conditions. The goal of each simulation was to determine the magnitude of the electrical charge. The results of all fifteen sets of material constants and their corresponding charge are displayed in Tab. 2. Experimental data was also provided and can be seen at the end of Tab. 2. Graphical representation of these results can be seen in Fig. 6.

Simulation data	p [kPa]	1	5	10	20
1	charge Q [pC]	1,7918	8,2695	14,78	24,599
2		1,6905	7,8456	14,108	23,62
3		1,7312	8,019	14,392	24,075
4		1,8538	8,5295	15,197	25,226
5		1,6994	7,8803	14,157	23,665
6		1,6526	7,6842	13,847	23,23
7		1,7634	8,1566	14,618	24,432
8		1,9398	8,8802	15,742	26
9		1,7906	8,2631	14,767	24,564
10		1,8676	8,5816	15,267	25,279
11		1,9604	8,9625	15,862	26,142
12		1,8192	8,3861	14,969	24,885
13		1,5665	7,316	13,248	22,33
14		1,6985	7,8765	14,151	23,656
15		1,7409	8,0563	14,441	24,095
Experimental data		1,0388	5,6272	13,7016	25,2063

Tab. 2. Values obtained from experimental measurements and simulation results

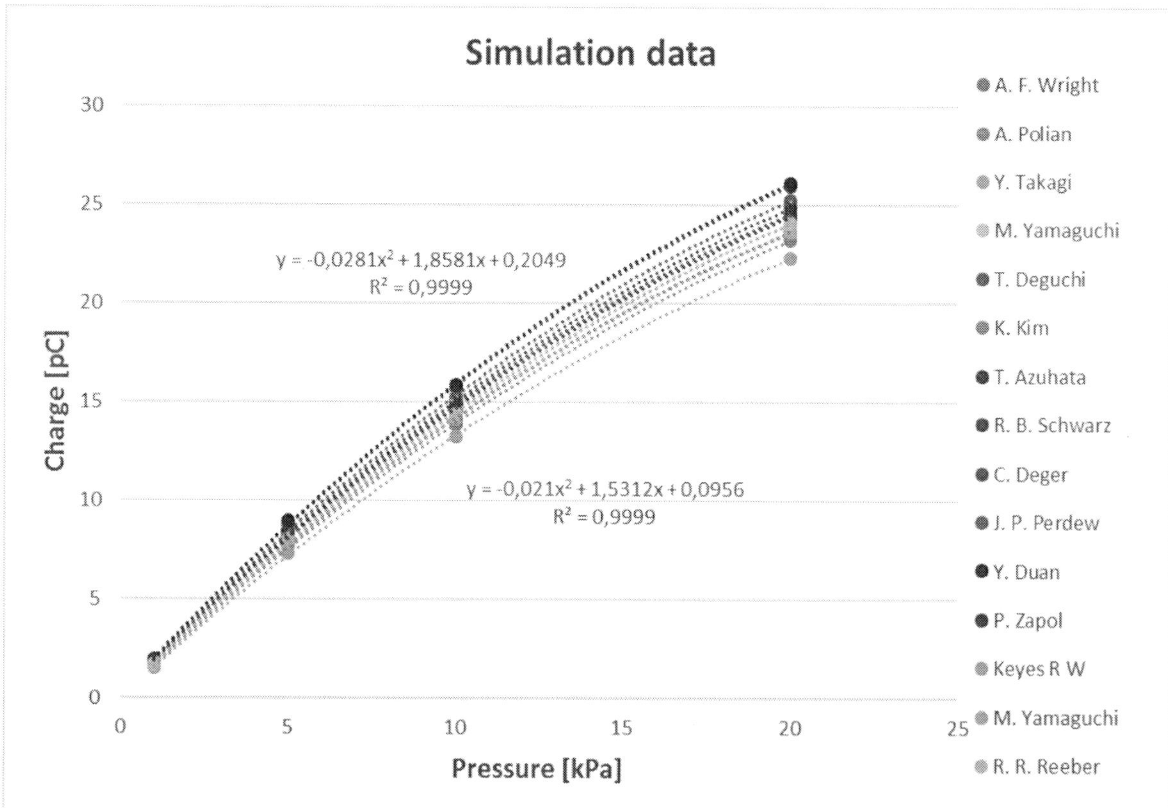

Fig. 6. Graphical representation of the simulation results

Fig. 6 above shows the magnitude of charge which was numerically calculated by ANSYS Multiphysics for four different load cases. Between each point a 2nd degree polynomial regression curve was overlaid onto the data. To maintain clarity in the graph, only the edge curve equations are shown. From these results, it was shown that the maximum charge occurring at a load of 1, 5, 10, 20 kPa were 1.9604, 8.9625, 15.862, and 26.142 pC respectively. The minimum charge was 1.5665, 7.316, 13.248, 22.33 [pC]. The median value for each load case was 1.77108, 8.180467, 14.6364 and 24.38653 [pC] respectively. The coefficients of variation for each load case were 1 [kPa] – 0.59829 [%], 5 [kPa] – 2.258427 [%], 10 [kPa] – 3.178549 [%] and 20 [kPa] – 4.074412 [%]. These values are more clearly displayed in Tab. 3 below.

pressure [kPa]	relative maxima of the charge value [pC]	relative minima of the charge value [pC]	mean [pC]	coefficients of variation [%]
1	1.9604	1.5665	1.77108	0.59829
5	8.9625	7.316	8.180467	2.258427
10	15.862	13.248	14.6364	3.178549
20	26.142	22.33	24.38653	4.074412

Tab. 3. statistical values of the results

It is worth mentioning that as the load increases, so does the deviation of the charge value. Based on the number of simulations and statistically calculated results, each set of material constants has a relatively small effect on the results. The median values in Tab. 3 where compared with experimental results which can be seen in Fig. 7.

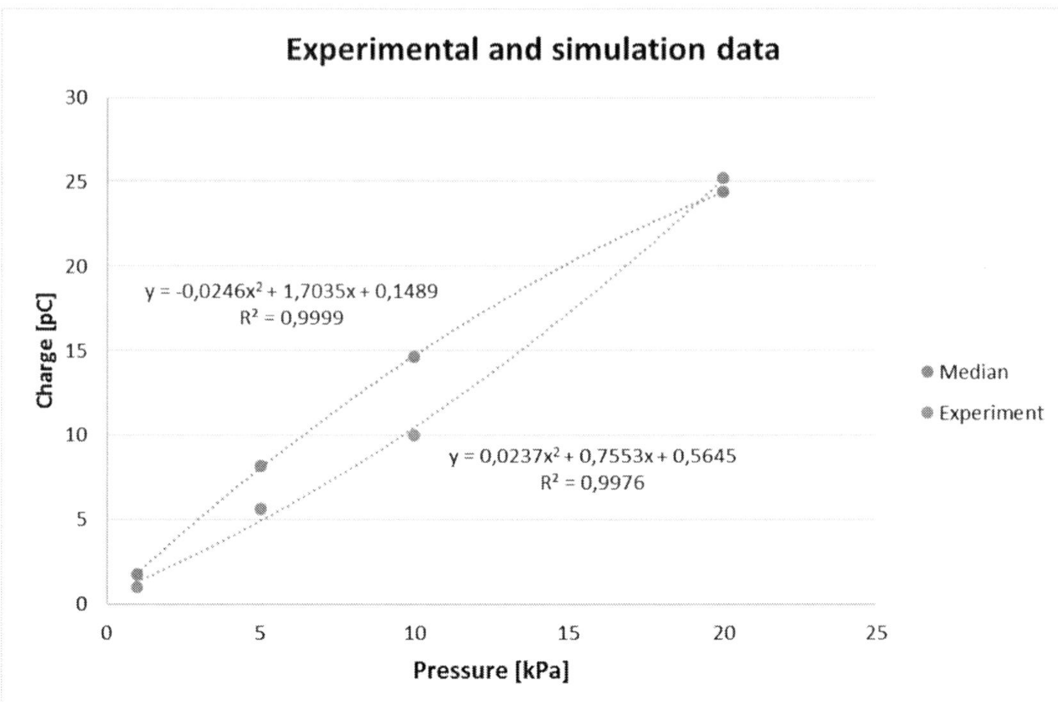

Fig. 7. Graphical representation of the median values (simulated) and experimental results

In Fig. 7 above it can be observed that the charge at load 5 and 10 kPa is quite different while at 1 and 20 kPa the results correspond quite well. This difference may be caused by simplifications in the model, mechanical and electric properties and/or even imperfections in the experimental test bench due to the very small dimensions. Future publications will focus on obtaining more data from experiments and perfecting the experimental process followed by further investigation of the material model and further investigation into the finite element model.

Acknowledgement

This work was supported by the Scientific Grant Agency of the Ministry of Education, Science, Research and Sport of the Slovak Republic and Slovak Academy of Sciences under the contract No. 1/0712/14 Micro-electromechanical (MEMS) energy harvesting system for applications in medicine.

References

[1] KERMANI M., MOALLEM M., PATEL R. *Applied Vibration Suppresion Using Piezoelectric Materials*, New York: Nova Scienece Publishers, Inc., 2008, ISBN-13: 978-1-60021-896-5, 176 p.

[2] URL: http://bme240.eng.uci.edu/students/06s/yuhsianh/Piezoelctric%20effect.htm

[3] PECIAR, P., PECIAR, M., FEKETE, R. *Mobilný univerzálny merací systém tlakov v partikulárnych materiáloch a spôsob kalibrácie tlakových snímačov : úžitkový vzor č. 7283.* Banská Bystrica Úrad priemyselného vlastníctva SR, 2015. 12 s.

[4] PECIAR, P., FEKETE, R. Calibration of the pressure sensor with a ceramic membrane. In *Annals of Faculty of Engineering Hunedoara*. 2014, Tome 12, Fasc. 2, s. 193-196. ISSN 1584-2665.

[5] ERHART, J. *Základy piezoelektřiny pro aplikace*, Brno: Ústav automatizace a měřici techniky VUT v Brne, 2011. URL: http://www.crr.vutbr.cz/system/files/brozura_06_1112.pdf

[6] URL: http://www.ansys.com/Products/Simulation+Technology/Fluid+Dynamics/Specialized+Products/ANSYS+Polyflow/Features/Online+Help+&+Documentation

[7] ERTURK, A., INMAN, D. J. *Piezoelectric energy harvesting.* First edition. Wiley Publication, 2011, ISBN: 978-0-470-68254-8, 402 p.

[8] A. F. Wright, J. *Appl. Phys.* 82, 2833 (1997).

[9] A. Polian, M. Grimsdich, and I. Grzegory, *J. Appl. Phys.* 79, 3343 (1996).

[10] Y. Takagi, M. Ahart, T. Azuhata, T. Sota, K. Suzuki, and S. Nakamura, *Physica B* 219/20, 547 (1996).

[11] M. Yamaguchi, T. Yagi, T. Azuhata, T. Sota, K. Suzuki, S. ChiChibu, and S. Nakamura, *J. Phys. C* 9, 241 (1997).

[12] T. Deguchi, D. Ichiryu, K. Toshikawa, K. Sekiguchi, T. Sota, R. Matsuo, T. Azuhata, T. Yagi, S. Chichibu, and S. Nakamura, *J. Appl. Phys.* 86, 1860 (1999).

[13] K. Kim, W. R. L. Lambrecht, and B. Segall, *Phys. R*ev. B 53, 16310 (1996); *Phys. Rev.* B 56, 7018 (1997).

[14] T. Azuhata, T. Sota, and K. Suzuki, *J. Phys. Condens. Matter* 8, 3111 (1996).

[15] R. B. Schwarz, K. Khachaturyan, and E. R. Weber, *Appl. Phys. Lett.* 70, 1122 (1997).

[16] C. Deger, E. Born, H. Angerer, O. Ambacher, M. Stutzmann, J. Hornsteiner, E. Riha, and G. Fischerauer, *Appl. Phys. Lett.* 72, 2400 (1998).

[17] J. P. Perdew, A. Ruzsinszky, G. I. Csonka, O. A. Vydrov, G. E. Scuseria, L. A. Constantin, X. Zhou, and K. Burke, *Phys. Rev.* Lett. 100, 136406 (2008)

[18] Y. Duana, G. Tangb, L. Qin, and L. Shi, *Phys. J.* B 66, 211–215 (2008)

[19] Peter Zapol, Ravindra Pandey and Julian D Gale, *J. Phys.: Condens. Matter* 9 (1997) 9517–9525.

[20] Keyes R W 1962 *J. Appl. Phys.* 33 3371

[21] M. Yamaguchi, T. Yagi, T. Sota, T. Deguchi, K.Shimada, S. Nakamura, *J. Appl. Phys.* 85,8502 (1999).

[22] R. R. Reeber and K. Wang, *MRS Internet J. Nitride Semicond. Res.* 6, 3 (2001).

Current address

Kováč Tomáš, Ing.
Slovak University of Technology
Nám. slobody 17, 812 31 Bratislava, Slovak Republic
Tel. number: +421904090168, e-mail: tomas.kovac@stuba.sk

Horvát František, Ing.
Slovak University of Technology
Nám. slobody 17, 812 31 Bratislava, Slovak Republic
Tel. number: +421949317313, e-mail: frantisek.horvat@stuba.sk

Čekan Michal, Ing. PhD.
Slovak University of Technology
Nám. slobody 17, 812 31 Bratislava, Slovak Republic
Tel. number: +421902781182, e-mail: michal.cekan@stuba.sk

Hučko Branislav, doc. Ing., PhD.
Slovak University of Technology
Nám. slobody 17, 812 31 Bratislava, Slovak Republic
Tel. number: +421907893278, e-mail: branislav.hucko@stuba.sk

Dzuba Jaroslav, Ing. PhD.
Institute of Electrical Engineering SAS
Dúbravská cesta 9, 841 04 Bratislava, Slovak Republic
Tel. number: +421-2-59222739, e-mail: jaroslav.dzuba@savba.sk

Vanko Gabriel, Ing.
Institute of Electrical Engineering SAS
Dúbravská cesta 9, 841 04 Bratislava, Slovak Republic
Tel. number: +421-2-5922 2739, e-mail: gabriel.vanko@savba.sk

Slovak University of Technology in Bratislava
Faculty of Mechanical Engineering
16th Conference on Applied Mathematics **APLIMAT 2017**
Proceedings

MATHEMATICAL MODEL OF SERUM PSA – URINARY SARCOSINE DEPENDENCY AND ITS IMPACT ON PROSTATE TUMOR GEOMETRY

KOVÁR Martin (CZ), HEGER Zbyněk (CZ), CERNEI Natalia (CZ), PŘÍBORSKÝ Jan (CZ)

Abstract. Coming out from the well-known Brunn-Minkowski Theorem and its immediate consequence, the isoperimetric inequality, we attempt to reveal the geometrical information regarding the prostate tumor, hiding behind the dependency between the two biomarkers – serum prostate specific antigen (PSA) and urinary sarcosine.

Keywords: Sarcosine, PSA, prostate cancer, tumor geometry, isoperimetric inequality, linear regression, Pearson correlation coefficient

Mathematics subject classification: Primary 92C50, 92C40, 92C99; Secondary 52A40, 26D15, 62J05, 62J10

1 Introduction

The study of relationships between various biomarkers, in particular between prostate specific antigen (PSA) and urinary sarcosine levels, potentially may improve the chance of early diagnostics of prostate cancer and the precision of the patient's diagnosis during the watchful observation period. However, no significant (linear) correlation between urinary sarcosine levels and serum PSA has been discovered so far. In this paper we will attempt to give an explanation of this fact. We will demonstrate that there is a strong, but nonlinear relationship between the biomarkers mentioned above. Moreover, coming out from the well-known Brunn-Minkowski Theorem and its immediate consequence, the isoperimetric inequality [5], we claim that the relationship carries some additional, geometrical or morphological information about the prostate tumor, which, as we hope, may help in getting more accurate information about the actual disease development.

Because of complexity of the general cancerous process it is very difficult to determine precisely the physical and geometrical parameters of a tumor. However, for simplicity, we consider that still there exist some effective values of these quantities. Our conjecture, relying on certain physical and biochemical reasons, expects the urinary sarcosine level be proportional to the tumor effective volume, while the PSA level depends proportionally on its effective surface.

These phenomena are connected with the differences in biomarkers nature. Increased secretion of PSA is due to a disorganization of prostate cells within the surface regions of tumor. In contrast, sarcosine is an oncometabolic product, resulting from unknown alterations in sarcosine-metabolic pathway. These occur during malignant transformation and are transmitted through clonogenic growth to all cells within a tumor mass. It worth noting that despite PSA, which is large secreted glycoprotein, sarcosine is small imino acid, which can be simply released from cells and can pass through the extracellular matrix to accumulate in urine of subjects [3].

Recall that the classical isoperimetric inequality states that a sphere has the smallest surface area per given volume. Taking a set $M \subseteq \mathbb{R}^n$ with surface area $\mathrm{surf}(M)$ and volume $\mathrm{vol}(M)$, it follows

$$\mathrm{surf}(M) \geq n \cdot \mathrm{vol}(M)^{\frac{n-1}{n}} \, \mathrm{vol}(B_n(1))^{\frac{1}{n}}, \tag{1}$$

where $B_n(r) \subseteq \mathbb{R}^n$ is the ball of the radius $r > 0$ in $\mathbb{R}^n$. The equality holds if $M = B_n(r)$. We take $n = 3$ for the 3-dimensional Euclidean space and apply the inequality (1) to the effective surface and effective volume of the prostate tumor. The most general conditions under which the isoperimetric inequality holds are still a subject of research (see, for instance [1], [6], [7] or [8] for some recent results and generalizations), however the existence of a finite Lebesgue measure of the closure $\mathrm{cl}\, M$ of M with respect to the Euclidean topology, ensuring validity of (1) by [5], we consider being sufficient for our purpose. Then

$$\mathrm{surf}(M) \geq 3 \cdot \mathrm{vol}(M)^{\frac{2}{3}} \, \mathrm{vol}(B_3(1))^{\frac{1}{3}},$$

and since $\mathrm{vol}(B_3(1)) = \frac{4}{3}\pi$, it subsequently follows

$$\mathrm{surf}(M) \geq 3 \cdot \mathrm{vol}(M)^{\frac{2}{3}} \left(\frac{4}{3}\pi\right)^{\frac{1}{3}},$$

$$\mathrm{surf}(M)^{\frac{1}{2}} \geq \sqrt{3} \cdot \mathrm{vol}(M)^{\frac{1}{3}} \left(\frac{4}{3}\pi\right)^{\frac{1}{6}},$$

$$\frac{\mathrm{surf}(M)^{\frac{1}{2}}}{\mathrm{vol}(M)^{\frac{1}{3}}} \geq \sqrt[3]{6\sqrt{\pi}}. \tag{2}$$

Denoting by p the PSA serum concentration level and by s the urinary sarcosine concentration level, the quantity $k = \frac{\sqrt{p}}{\sqrt[3]{s}} = p^{\frac{1}{2}} \cdot s^{-\frac{1}{3}}$ should be proportional to the expression $\mathrm{surf}(M)^{\frac{1}{2}} \cdot \mathrm{vol}(M)^{-\frac{1}{3}}$ from the left side of (2) and constant if the prostate tumor of a selected patient does not change its shape (although, we admit that it may change its size). On the other hand, changes of k should point to certain, usually more dramatic development of the disease – for instance a metastatic process, usually increasing the overall effective surface of a tumor with respect to its effective volume. Due to the isoperimetric inequality (1) and its consequence (2), we suggest to call k as the *coefficient of (a prostate tumor) roundness*.

For the first sight, the verification of the described dependency between p and s by a measurement seems to be a trivial task, but the reality is unfortunately very different. The stable period of the disease for a single patient is usually not long enough, so only a few measurements (say, approximately, 3 up to 6) of PSA and sarcosine concentration levels could be taken. That gives no sufficient data for any serious statistical analysis. In addition, the proportional coefficients between p and the effective surface or between s and the effective volume of a prostate tumor may depend on prostate cancer type (originating from acinar, ductal, transitional, squamous or small cells), differentiation or other metabolic parameters and so they may vary from one patient to another [4]. Consequently, the data of different patients are not directly comparable, so some additional workaround is necessary. We will deal with this problem in one of the next sections. For the used statistical notions, the reader is referred to [9].

2 Organization of measurements

2.1 Chemicals

Sarcosine standard and other chemicals were purchased from Sigma-Aldrich (St. Louis, MO, USA) in ACS purity, unless noted otherwise.

2.2 Volunteers and sample collection

For purpose of this study, urine and whole blood specimens from patients suffering from prostate cancer (n = 13, before any oncological treatment applied), were used. Average age of patients was 58.40 years. In all cases subjects were diagnosed with acinar adenocarcinoma. The specimens were collected from participants with signed informed consent. The blood (venipuncture from antecubital vein) and urine were sampled to adequate sterile tubes and immediately processed for analyses.

2.3 Preparation of urinary and blood specimens for sarcosine and PSA analyses

For sarcosine analysis, 500 µl of urine was pipetted into the 96-well evaporation plate Deepwell plate 96 (Eppendorf AG, Hamburg, Germany) and evaporated by nitrogen blow-down evaporator Ultravap 96 with spiral needles (Porvair Sciences limited, Leatherhead, United Kingdom). After this procedure sample was diluted with 500 µl of dilution buffer and subsequently used for analysis by ion-exchange chromatography (IEC). In case of tPSA and fPSA, whole blood was collected into anticoagulant-(ethylenediaminetetraacetic acid)-treated tubes and centrifuged for 10 min at 2000 g at 4 °C. The resulting plasma was immediately analyzed using the immunoenzymometric assay.

2.4 Estimation of sarcosine amount in urinary specimens

Sarcosine in urine was determined by using IEC with Vis detection after post-column derivatization with ninhydrin (AAA-400, Ingos, Prague, Czech Republic).

2.5 Determination of tPSA (immunoenzymometric assay)

The levels of tPSA were analyzed using an automated analyzer AIA 600 II (Tosoh Bioscience, Tokyo, Japan) utilizing the ST-AIA-PACK PSAII testing cups. The analyses were performed according to the manufacturer's instructions.

3 Data and their analysis

The measured patient data are seen from the following table. For them we use the standard matrix notation, that is, p_{ij} is the PSA value and s_{ij} is the urinary sarcosine value of the i-th patient from the j-th measurement. The number of all measurements for the i-th patient we denote by n_i.

	tPSA [μg/l]						Sarcosine [μmol/mmol crea]					
Patient	1	2	3	4	5	6	1	2	3	4	5	6
1	5.24	6.12	6.25	7.51			3.25	10.80	6.89	11.20		
2	3.24	3.57	4.21	4.99	5.16		0.98	1.65	2.41	2.51	1.98	
3	8.66	10.11	12.56				6.88	7.10	11.90			
4	3.64	4.16	5.55	6.10			3.51	3.01	6.50	8.10		
5	5.97	6.13	6.55	6.99	7.25		11.80	18.10	21.00	17.10	25.60	
6	7.59	7.79	7.88	8.01	8.16		7.81	8.98	15.10	20.10	17.60	
7	3.31	4.15	4.21	4.25			0.10	0.10	0.35	0.15		
8	2.38	2.42	3.84	5.11	5.21		3.64	1.05	2.64	1.99	3.51	
9	5.31	5.66	5.91				6.10	8.12	25.13			
10	1.95	2.16	3.55	3.88			0.10	0.39	0.66	2.50		
11	3.54	3.69	3.87	4.51	4.79		3.40	1.50	2.20	5.10	3.50	
12	6.11	6.39	6.01	6.48	6.22	7.01	6.03	8.70	4.80	7.80	9.90	11.50
13	8.25	6.53	7.61				9.90	5.30	9.00			

Table 1.

If the prostate tumor of the i-th patient is stable and develops only slowly during the observation period in which the measurements are taken, it keeps approximately the same shape and so the ratio

$$\frac{\sqrt{p_{ij}}}{\sqrt[3]{s_{ij}}}$$

should be approximately constant with respect to $j \in \{1, 2, \ldots, n_i\}$. Of course, in the real case the ratio is not constant due to many influences, including the measurement error, random changes of the patient's metabolism and other. However, taking a large number n_i of measurements, the random deviations will approximately cancel each other, so the mean value

$$k_i = \frac{1}{n_i} \sum_{j=1}^{n_i} \frac{\sqrt{p_{ij}}}{\sqrt[3]{s_{ij}}} \tag{3}$$

should be close to the actual value of the coefficient k of tumor roundness of the i-th patient. Unfortunately, as we explained in the introductory section, in reality n_i cannot be very large (in our data set it varies from 3 to 6), so the relative error of the mean value could be relatively significant. That may be really observed from the next table, which may be considered as an illustration of this phenomenon.

Patient no.:	1	2	3	4	5	6	7	8	9	10	11	12	13
Mark of graph point	○	●	□	■	△	▲	▽	▼	★	▷	◁	◆	∞
Roundness coefficient k	1.3	1.68	1.58	1.29	0.98	1.2	3.78	1.45	1.09	2.16	1.42	1.28	1.38
Relative error of k	0.14	0.08	0.04	0.06	0.07	0.13	0.16	0.2	0.21	0.3	0.13	0.09	0.06

Table 2.

The problem of small number of measurements could be got over by a certain kind of transformation, normalizing and linearizing the data of individual patients, allowing joining them into a one large single file. Consider the transformation given by the following relationships:

$$x_{ij} = k_i \cdot \sqrt[3]{s_{ij}}, \qquad y_{ij} = \sqrt{p_{ij}} \tag{4}$$

Denoting $z_{ij} = \frac{y_{ij}}{x_{ij}}$, it holds

$$\sum_{j=1}^{n_i} z_{ij} = \sum_{j=1}^{n_i} \frac{y_{ij}}{x_{ij}} = \sum_{j=1}^{n_i} \frac{\sqrt{p_{ij}}}{k_i \cdot \sqrt[3]{s_{ij}}} = \frac{1}{k_i} \cdot \sum_{j=1}^{n_i} \frac{\sqrt{p_{ij}}}{\sqrt[3]{s_{ij}}} = n_i \tag{5}$$

by (3), so

$$\frac{1}{n_i} \sum_{j=1}^{n_i} z_{ij} = 1 \tag{6}$$

Similarly, if we denote by $m = \sum_{i=1}^{n} n_i$ the number of all measured pairs of PSA and urinary sarcosine for all patients, we get

$$\frac{1}{m} \sum_{i=1}^{n} \sum_{j=1}^{n_i} z_{ij} = \frac{1}{m} \sum_{i=1}^{n} n_i = 1 \tag{7}$$

This shows that the transformation (4) converts our problem to the verification that the random quantities

$$y = \sqrt{p}, \qquad x = k \cdot \sqrt[3]{s} \tag{8}$$

are approximately proportional to each other with the coefficient 1, independently on the individually chosen patient. We may do it in a standard way, for instance, graphically, with help of the linear regression and the method of the least squares, or by calculating the Pearson correlation coefficient between both transformed random quantities. Note that in our case, $m = 56$.

Recall that the Pearson correlation coefficient ρ of the random quantities x, y is defined as

$$\rho(x, y) = \frac{\operatorname{cov}(x, y)}{\sigma(x)\sigma(y)}, \tag{9}$$

where cov is the covariance and σ is the standard deviation. In our notation, one may easily prove that (9) has a form

$$\rho(x,y) = \frac{m\left(\sum_{i=1}^{n}\sum_{j=1}^{n_i} x_{ij}y_{ij}\right) - \left(\sum_{i=1}^{n}\sum_{j=1}^{n_i} x_{ij}\right)\cdot\left(\sum_{i=1}^{n}\sum_{j=1}^{n_i} y_{ij}\right)}{\sqrt{m\left(\sum_{i=1}^{n}\sum_{j=1}^{n_i} x_{ij}^2\right) - \left(\sum_{i=1}^{n}\sum_{j=1}^{n_i} x_{ij}\right)^2}\cdot\sqrt{m\left(\sum_{i=1}^{n}\sum_{j=1}^{n_i} y_{ij}^2\right) - \left(\sum_{i=1}^{n}\sum_{j=1}^{n_i} y_{ij}\right)^2}}.$$

After a short calculation (implemented in Wolfram Mathematica and webMathematica software) we get

$$\rho(x,y) \approx 0.832 \tag{10}$$

Note that in medical and biological sciences, a value of ρ above 0.8 is usually interpreted as a confirmation of very high, strong dependence between two random quantities, close to the linear dependence.

We also may interpose a best-fitting straight line through the points (x_{ij}, y_{ij}) corresponding to our measurements. The straight line should respect the theoretical data of a healthy subject. The median value P_0 of the PSA level of a healthy men population slightly depends on the age and it is approximately 0.7 between 40 and 50 years, 0.9 between 50 and 60 years, and it is about 1.4 for the age above 60. On the other hand, the concentration of urinary sarcosine of men without the prostate cancer should be very small, usually unmeasurably close to 0. In this case, the coefficient k of prostate tumor roundness is undefined. However, using a limit transition we may calculate the (x_0, y_0)-point corresponding to that situation and the transformation given by (4). While it is clear that $y_0 = \sqrt{P_0}$, for the x-coordinate we successively get

$$x_0 = \lim_{s\to 0} k\cdot\sqrt[3]{s} = \lim_{s\to 0}\frac{\sqrt{p}}{\sqrt[3]{s}}\sqrt[3]{s} = \lim_{s\to 0}\sqrt{p} = \sqrt{P_0}, \tag{11}$$

where we suppose k and p depending continuously on s if $s \in (0,\infty)$. Thus the desired regression line goes through the point $(\sqrt{P_0}, \sqrt{P_0})$ and its slope a we may determine by the method of the least squares.

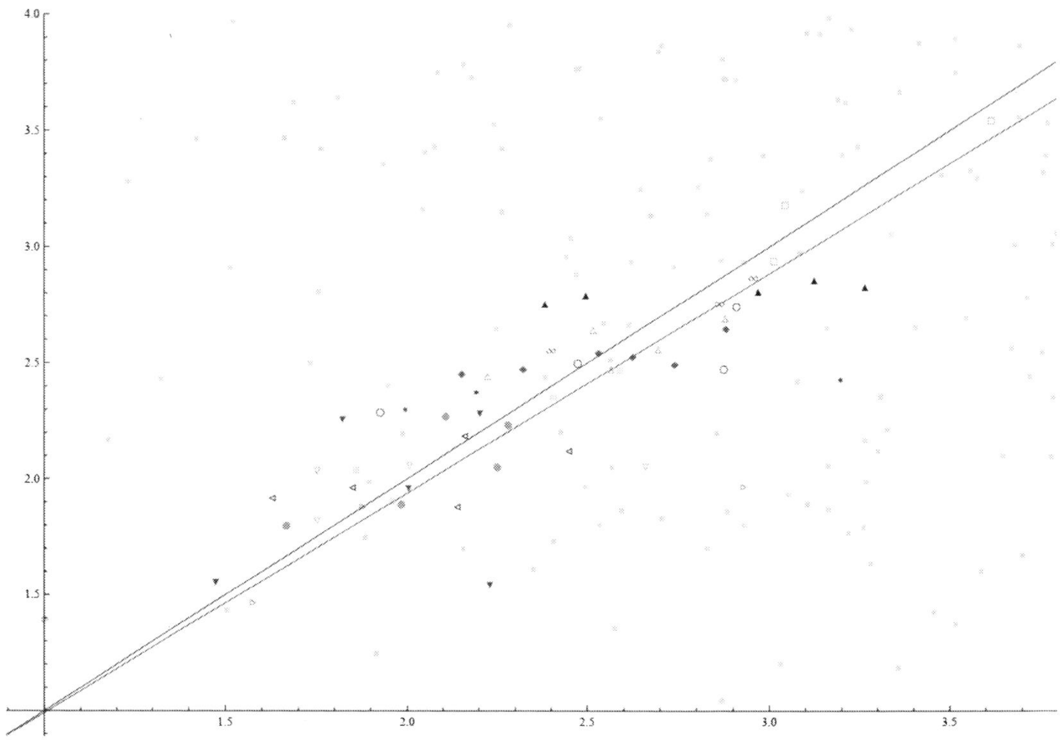

Fig. 1.

The regression line has the equation

$$y = \sqrt{P_0} + a(x - \sqrt{P_0}) \tag{12}$$

and we want to minimize the expression

$$S = \sum_{i=1}^{n}\sum_{j=1}^{n_i}(y(x_{ij}) - y_{ij})^2 = \sum_{i=1}^{n}\sum_{j=1}^{n_i}(\sqrt{P_0} + a(x_{ij} - \sqrt{P_0}) - y_{ij})^2 \tag{13}$$

with respect to a. Then, putting

$$\frac{\partial S}{\partial a} = 0,$$

we subsequently get

$$a = \frac{\sum_{i=1}^{n}\sum_{j=1}^{n_i}(y_{ij} - \sqrt{P_0})(x_{ij} - \sqrt{P_0})}{\sum_{i=1}^{n}\sum_{j=1}^{n_i}(x_{ij} - \sqrt{P_0})^2}. \tag{14}$$

Applying the software Wolfram Mathematica at our data and (14), after a short calculation one may realize that

$$a \approx 0.95, \tag{15}$$

which is not very far from the theoretical value 1, following from (8). On Fig. 1, the regression line is red, while the theoretical dependency, represented by the axis of the first and the third quadrants and having the slope equal to 1, is blue.

Finally, we may calculate the standard deviation $\sigma(z)$ of the mean value of the ratio $z = \frac{y}{x}$. Note that the mean value of the random quantity z is always 1, as it follows from (7), so $\sigma(z)$ is also an estimation of the relative error of the mean value. It follows that

$$\sigma(z) = \sqrt{\frac{1}{m}\sum_{i=1}^{n}\sum_{j=1}^{n_i}(z_{ij}-1)^2} \approx 0.127. \tag{16}$$

This corresponds with the precision of our measurements of PSA and urinary sarcosine well, being standardly of order of several percent. The figure also contains pseudo-randomly generated data with $m = 200$, which are marked by small grey squares. The pseudo-randomly generated data are processed by the same algorithm as the measured data and their Pearson correlation coefficient is approximately 0.002.

A webMathemtica applet, realizing the complete calculation and allowing to switch data of individual selected patients on and off, is available from the following URL:
`http://matika.umat.feec.vutbr.cz/software/webmath/Prostate.jsp`

4 Closing remarks

Due to differences in metabolism of individual patients, the difference in their cancer type, differentiation of cancer cells and overall development of the disease, the values of k are not directly comparable between patients, nor any concrete conclusion regrading the tumor shape and morphology is possible yet. For that purpose, some further, much larger study, which would include a simultaneous observation of many biological and medicine parameters, including MRI or PET/CT tests, would be necessary for the calibration of k. No matter how such research would be interesting and useful, it could be relatively complicated, long-termed and perhaps also expensive.

On the other hand, there is a number of natural conjectures concerning the utility of k, which can be clinically tested very easily and under low costs. The observation of k of a single patient may give a valuable additional information, which is not available if only one of the biomarkers, PSA or urinary sarcosine, is observed. For instance, the simultaneous growth of PSA, urinary sarcosine and k is probably a much worse sign of the disease development than the case, in which both biomarkers are slightly growing, but k remains approximately constant. Recall that growth of k probably means that the surface of the whole tumor body prevails over its volume, which could mean that the cancerous process is spreading. On the contrary, decreasing k probably means that tumor is getting more "compact", round, with a smaller surface, and consequently its spreading capabilities are getting less potent.

There is also an interesting question whether the urinary sarcosine as a potential prostate tumor marker has a "grey zone" for low values, similarly as PSA. Since

$$\frac{\partial k}{\partial s} = -\frac{1}{3} \cdot p^{\frac{1}{2}} \cdot s^{-\frac{4}{3}}$$

and so

$$\lim_{s \to 0+} \frac{\partial k}{\partial s} = -\infty,$$

for urinary sarcosine concentration s close to 0 the error of determination of k as a coefficient of proportionality between $\sqrt{p}$ and $\sqrt[3]{s}$ is very high, and so both biomarkers become practically independent for a healthy men population. This is in a contrast with respect their strong dependency for the pathological case. This fact means, besides other things, that the "grey zone" of PSA does not automatically mean that urinary sarcosine as a prostate tumor marker behaves in the same way, although both biomarkers become closely related if the prostate cancer is already present.

The work is supported by the project FEKT-S-14-2200 of Brno University of Technology.

Bibliography

[1] Boroczky, K. J., Lutwak, E., Yang, D., Zhang, G., *The log-Brunn–Minkowski inequality*, Advances in Mathematics 231 (3-4) (2012), 1974–1997, ISSN: 0001-8708.

[2] Catalona, W. J., *Dr. Catalona's Recommended Guidelines for PSA Screening*, URL: `http://drcatalona.com/quest/quest_fall06_1.asp`.

[3] Cernei, N., Heger, Z., Gumulec, J., Zitka, O., Masarik, M., Babula, P., Eckschlager, T., Stiborova, M., Kizek, R., Adam, V., *Sarcosine as a Potential Prostate Cancer Biomarker-A Review*, International Journal of Molecular Sciences 14 (7) (2013), 13893-13908, ISSN: 1422-0067

[4] Chang, A., J., Autio, K., A., Roach, M., Scher, H., I., *High-risk prostate-cancer-classification and therapy*. Nature Reviews Clinical Oncology, 11 (6) (2014), 308-323, ISSN: 1759-4774.

[5] Feder, H., *Geometric Measure Theory*, Springer-Verlag, Berlin, 1969, pp. 676, ISBN: 978-3-5406-0656-7.

[6] Liu, J., *Generalizations of the Brunn-Minkowski inequality*, Linear Algebra and Its Applications 508 (2016), 206-213, ISSN: 0024-3795.

[7] Lutwak, E., Yang, D., Zhang, G., *The Brunn-Minkowski-Firey inequality for non-convex sets*, Advances in Applied Mathematics 48 (2012), 407–413, ISSN: 0196-885.

[8] Milman, E., Rotem , L., *Complemented Brunn-Minkowski inequalities and isoperimetry for homogeneous and non-homegenous measures*, Advances in Mathematics 262 (2014), 867-908, ISSN: 0001-8708.

[9] Shay, G., *Introduction to Probability with Statistical Applications*, Birkhauser, Boston, 2007, pp. 313, ISBN: 978-0-8176-4497-0.

Current address

Kovár Martin, doc. RNDr. Ph.D.
Department of Mathematics, Faculty of Electrical Engineering and Communication
Brno University of Technology
Technická 8, 616 00 Brno, Czech Republic
E-mail: kovar@feec.vutbr.cz

Heger Zbyněk, Mgr. Ph.D.
Department of Chemistry and Biochemistry, Mendel University in Brno
Zemědělská 1, 613 00 Brno, Czech Republic
E-mail: zbynek.heger@mendelu.cz

Central European Institute of Technology, Brno University of Technology
Technická 10, 616 00 Brno, Czech Republic
E-mail: zbynek.heger@ceitec.vutbr.cz

Cernei Natalia, Mgr. Ph.D.
Department of Chemistry and Biochemistry, Mendel University in Brno
Zemědělská 1, 613 00 Brno, Czech Republic
E-mail: qqcernei@node.mendelu.cz

Central European Institute of Technology, Brno University of Technology
Technická 10, 616 00 Brno, Czech Republic
E-mail: natalia.cernei@ceitec.vutbr.cz

Příborský Jan, Ing.
Department of Mathematics, Faculty of Electrical Engineering and Communication
Brno University of Technology
Technická 8, 616 00 Brno, Czech Republic
E-mail: priborsky.jan@seznam.cz

Slovak University of Technology in Bratislava
Faculty of Mechanical Engineering
16th Conference on Applied Mathematics **APLIMAT 2017**
Proceedings

SOME PROPERTIES OF CAUSAL STRUCTURES

Martin Kovár (CZ), Jan Příborský (CZ)

Abstract. We introduce and study a new notion, *causal venue*, which is a slight modification of causal site, defined by J. D. Christensen and L. Crane in 2005. We omit the assumption of anti-reflexivity of the causal relation and the necessity of the presence of the least element in the spatial relation. As we show and discuss, our modification of the studied algebraic causal structure potentially solves some limitations of its original version.

Keywords: Causal site, causal venue, partial order and its modifications, binary relations and their properties

Mathematics subject classification: Primary 06A06, 08A02. Secondary 83C99, 83C45

1 Preliminaries and Introduction

In this paper we slightly generalize some results of M. Kovár and A. Chernikava stated in Section 3 of [2] regarding causal sites. We introduce a new notion, a *causal venue*. Some properties of causal sites will now follow as natural consequences from the properties of the causal venues. Note that a causal site is a general algebraic structure capturing the point-set behavior of the light cones in general relativity, but its limitation is that it does not cover the case of Lorentzian manifolds which are not globally hyperbolic. The anti-reflexivity of the causal relation $\prec$ puts a very strong limitation on the construction of a causal site, as it is shown in [2]. In this paper we attempt to prove similar results even if the condition of anti-reflexivity is omitted.

Throughout this paper by a relation we always mean a binary relation. We say that a relation $\prec$ on a set X is *linear* if for every $a, b \in X$, $a \neq b$, it holds $a \prec b$ or $b \prec a$. Note that this notion slightly differs from the widely used notion of a *total relation*, in whose definition the assumption of $a \neq b$ is omitted.

Recall that a *causal site* $(\mathcal{S}, \sqsubseteq, \prec)$ defined by J. D. Christensen and L. Crane in [1] is a set $\mathcal{S}$ of *regions* equipped with two binary relations $\sqsubseteq$, $\prec$, where $(\mathcal{S}, \sqsubseteq)$ is a partial order having the binary suprema $\sqcup$ and the least element $\bot \in \mathcal{S}$, and $(\mathcal{S} \smallsetminus \{\bot\}, \prec)$ is a strict partial order (i.e., anti-reflexive and transitive), linked together by the following axioms, which are satisfied for all regions $a, b, c \in \mathcal{S}$:

(i) $b \sqsubseteq a$ and $a \prec c$ implies $b \prec c$,

(ii) $b \sqsubseteq a$ and $c \prec a$ implies $c \prec b$,

(iii) $a \prec c$ and $b \prec c$ implies $a \sqcup b \prec c$.

(iv) There exits $b_a \in \mathcal{S}$, called *cutting of a by b*, such that

(1) $b_a \prec a$ and $b_a \sqsubseteq b$;

(2) if $c \in \mathcal{S}$, $c \prec a$ and $c \sqsubseteq b$ then $c \sqsubseteq b_a$.

Omitting the condition of anti-reflexivity of the relation $\prec$ and the existence of the least element in the partially ordered set $(\mathcal{S}, \sqsubseteq)$, we introduce a new notion, a *causal venue*. We say that $c \in \mathcal{S}$ is a *causa-sui element*, if $c \prec c$. Further, $L \subseteq \mathcal{S}$ is called *a causal anti-chain*, if there are no $a, b \in L$, $a \neq b$, such that $a \prec b$ or $b \prec a$.

2 Main results

As we already mentioned in the introductory section, now, besides other things, we reformulate and generalize some results concerning causal sites, stated in [2].

Lemma 2.1. *Let $(\mathcal{S}, \sqsubseteq, \prec)$ be a causal venue with $L \subseteq \mathcal{S}$ linearly ordered by the relation $\sqsubseteq$. Then L is a causal anti-chain or it contains a causa-sui element.*

Proof. Suppose that L is not a causal anti-chain. Then there exist $a, b \in L$, $a \neq b$, such that $a \prec b$ or $b \prec a$. Without loss of generality, we may choose the denotation of the elements a, b such that $b \sqsubseteq a$. If $a \prec b$, Then, by the axiom (i), it should be $b \prec b$. Similarly, if $b \prec a$, by the axiom (ii) we get again that $b \prec b$. Hence, $b \in L$ is a causa-sui element. □

Although the axioms of a causal site fit at the most usual models of space-time (as it is shown in [2]) well, the following example illustrates that even very simple causal relationships cannot be realized as a causal site but they might be realized as a causal venue.

Example 2.1. Consider a system $\mathcal{S} = \{a, b\}$ (perhaps, a very simple abstract "universe") alternately changing its state from a collapsed state a to an expanded state b and vice-versa. It is natural to put $a \sqsubseteq b$.

Taking a source of inspiration from the philosophy of *Tao* [3], one may say that the state a already contains a "germ" of b and conversely, b already has a "germ" of a in itself. This may be expressed by the causal relation $\prec$ as $a \prec b$ and $b \prec a$. Finally, from transitivity of the causal relation $\prec$ we get also $a \prec a$, $b \prec b$. These relationships are also seen from Figure 1, which is a combination of the Hasse diagram of $(\mathcal{S}, \sqsubseteq)$ and the one-set arrow diagram of $(\mathcal{S}, \prec)$.

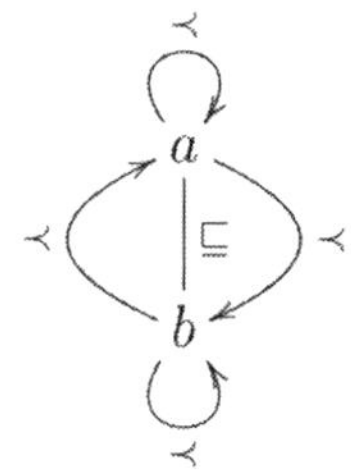

Figure 1.

More precisely, in terms of binary relations, we have $\sqsubseteq = \{(a,a),(a,b),(b,b)\}$ and $\prec = \{(a,a),(a,b),(b,b),(b,a)\} = \mathcal{S} \times \mathcal{S}$. Obviously, $(\mathcal{S}, \sqsubseteq, \prec)$ is a causal venue with $\bot = b$, $a_b = a$, and $b_a = b$, although it is not a causal site, since it holds $a \prec a$.

Corollary 2.1. *Let $(\mathcal{S}, \sqsubseteq, \prec)$ be a causal site with $L \subseteq \mathcal{S} \setminus \{\bot\}$ linearly ordered by the relation $\sqsubseteq$, $|L| \geq 2$. Then L is a causal anti-chain with respect to $\prec$.*

Proof. Since $L \subseteq \mathcal{S} \setminus \{\bot\}$, $\bot \notin L$, so, by the definition of a causal site, L does not contain a causa-sui element. Then L is a causal anti-chain by Lemma 2.1. □

Lemma 2.2. *Let $(\mathcal{S}, \sqsubseteq, \prec)$ be a causal venue with the least element $\bot$ with respect to $\sqsubseteq$. Then $\bot$ is a causa-sui element of $\mathcal{S}$ with respect to $\prec$.*

Proof. Let $a \in \mathcal{S}$ be an arbitrary element. By the axiom (iv), there exists $a_a \in \mathcal{S}$ such that $a_a \prec a$ and $a_a \sqsubseteq a$. Then, by the axiom (ii), it follows $a_a \prec a_a$. Since $\bot \sqsubseteq a_a$ by the assumption, by the axiom (i) it holds $\bot \prec a_a$. Considering $\bot \sqsubseteq a_a$ again, by axiom (ii) we finally get $\bot \prec \bot$. □

Corollary 2.2. *Let $(\mathcal{S}, \sqsubseteq, \prec)$ be a causal site. Then $\bot$ is the least and causa-sui element of $\mathcal{S}$ with respect to $\prec$.*

Proof. Following the argumentation the proof of Lemma 2.2 again, $a_a \prec a_a$ implies $a_a = \bot$, since the relation $\prec$ is anti-reflexive on the set $\mathcal{S} \setminus \{\bot\}$. □

Corollary 2.3. *Let $(\mathcal{S}, \sqsubseteq)$ be a linearly ordered set with the least element $\bot$. Then there exists a unique binary relation $\prec$ on $\mathcal{S}$, such that $(\mathcal{S}, \sqsubseteq, \prec)$ is a causal site. In this relation, $\mathcal{S} \setminus \{\bot\}$ is a causal anti-chain and $\bot$ is the least element of $(\mathcal{S}, \prec)$.*

Proof. It is not difficult to check the axioms of the causal site constructed as suggested in the corollary statement.

Suppose conversely, that $(\mathcal{S}, \sqsubseteq, \prec)$ is a causal site, in which there are $a, b \in \mathcal{S} \setminus \{\bot\}$, $a \neq b$, $a \prec b$. Then $L = \{a, b\} \subseteq \mathcal{S} \setminus \{\bot\}$ is not anti-chain, $|L| \geq 2$, which contradicts to Corollary 2.1. □

Corollary 2.4. *Suppose that $\prec$ is a linear transitive relation on $\mathcal{S}$, $|\mathcal{S}| \geq 3$, having the only one element $\perp$ satisfying $\perp \prec \perp$. Then there is no binary relation $\sqsubseteq$ such that $(\mathcal{S}, \sqsubseteq, \prec)$ is a causal venue.*

Proof. Suppose conversely, that $(\mathcal{S}, \sqsubseteq, \prec)$ is a causal venue with $\prec$ being a strictly linear order on $\mathcal{S} \setminus \{\perp\}$, $|\mathcal{S}| \geq 3$. There exist $b, c \in \mathcal{S} \setminus \{\perp\}$, $b \neq c$. Since $(\mathcal{S}, \sqsubseteq)$ has binary suprema by the definition of a causal venue, there exist $a = b \sqcup c \in \mathcal{S} \setminus \{\perp\}$. Clearly, $b, c \sqsubseteq a$. Since $b \neq c$, at least one of the elements b, c differs from a. Let L be the two-element set containing a and that element from $\{b, c\}$, different from a. Then L is a chain with respect to $\sqsubseteq$ as well as for $\prec$, so by Lemma 2.1 L should contain a causa-sui element. But this is not possible since $\perp$ is the only element with $\perp \prec \perp$. □

Theorem 2.1. *Let $(\mathcal{S}, \sqsubseteq, \prec)$ be a causal venue and let $C \subseteq \mathcal{S}$ be the set of all its causa-sui elements. Denote $N(x) = \{y | \, y \in \mathcal{S}, y \not\prec x \not\prec y\}$. Then for every distinct $x_1, x_2, \ldots x_k \in \mathcal{S} \setminus C$, where $k \geq 2$,*

$$\bigcap_{i=1}^{k} N(x_i) \neq \varnothing.$$

Proof. Let $x_1, x_2, \ldots x_k \in \mathcal{S} \setminus C$ and denote $s = x_1 \sqcup x_2 \sqcup \cdots \sqcup x_k$. We will show that

$$s \in \bigcap_{i=1}^{k} N(x_i) \neq \varnothing.$$

It is clear that $x_i \sqsubseteq s$ for every $i = 1, 2, \ldots, k$. Suppose, for some $i = 1, 2, \ldots, k$, that it also holds $x_i \prec s$ or $s \prec x_i$. Because of anti-reflexivity of the relation $\prec$ on $\mathcal{S} \setminus C$, the equality $x_i = s$ is possible only for $s \in C$. But this would contradict to the choice of $x_1, x_2, \ldots x_k$. Therefore $s \in C$ is not possible, and so $x_i \neq s$. Then $L = \{x_i, s\} \subseteq \mathcal{S} \setminus C$, $|L| = 2$ is linearly ordered by $\sqsubseteq$ and it does not contain any causa-sui element (because all such elements are in C). Now, by Lemma 2.1, L is an anti-chain with respect to $\prec$. But this is a contradiction to our assumption that $x_i \prec s$ or $s \prec x_i$. Hence, $s \in \bigcap_{i=1}^{k} N(x_i) \neq \varnothing$. □

Note that the previous theorem essentially establishes that the collection $\{N(x) | x \in \mathcal{S} \setminus C\}$ is centered (or, in other terms, has the *finite intersection property*, [2]) if $\mathcal{S} \setminus C$ contains at least two distinct elements. It is easy to see that this condition need not be necessarily satisfied – the causal venue in Example 2.1 has $\mathcal{S} \setminus C = \varnothing$. However, such causal venues has a very simple structure.

Taking $C = \{\perp\}$ in a causal site, we directly obtain a similar result presented in [2] as Theorem 3.1, now following as a corollary.

Let $(\mathcal{S}, \sqsubseteq, \prec)$ be a causal venue. Two elements $a, b \in \mathcal{S}$ are said *causally connected* if $a \prec b$. Then we also say that a *causally precedes* b.

We will close the paper with a thought, in addition having a rather philosophical background. Although the existence of the least and causa-sui element of a general causal venue is not ensured, which is in a contrast with Corollary 2.2, we still may prove its relaxed form.

Proposition 2.1. *Let $(\mathcal{S}, \sqsubseteq, \prec)$ be a causal venue. If there is an element $s \in \mathcal{S}$, which causally precedes every non-causa-sui element, then s is a causa-sui element.*

Proof. Let $(\mathcal{S}, \sqsubseteq, \prec)$ be a causal venue and let N be its set of all non-causa-sui elements. Suppose that there is some $s \in \mathcal{S}$, such that $s \prec x$ for every $x \in N$. However, it is not possible $s \in N$ since it would lead to $s \prec s$, contradicting with $s \in N$. Then $s \notin N$, which implies $s \prec s$. □

The previous proposition points out some kind of exceptionality of the element s, causally preceding the non-causa-sui subjects, and its very simple proof incorporates some ideas of the well-known Russell's set-theoretical antinomy. The result does not depend on the relation $\sqsubseteq$, and, of course, it also perfectly holds in causal sites, where, however, the idea of exceptionality is hidden by the causal site definition. In our opinion, any good algebraic model of causality should be universal and it should not presume or prefer any concrete philosophical conception of Universe. On the other hand, Proposition 2.1 seems to gently indicate that the definition of a causal venue (or a causal site, of course) admits of also the creationist conceptions of reality.

The work is supported by the project FEKT-S-14-2200 of Brno University of Technology.

References

[1] Christensen J. D., Crane L., *Causal Sites as Quantum Geometry*, J. Math. Phys. 46 (2005), 122502-122523.

[2] Kovár M., Chernikava A., *The framework, Causal and Co-compact Structure of Space-time*, arXiv:1311.2986 [math-ph], 1-34.

[3] Lao-Tzu, *Tao Te Ching: An Illustrated Journey*, Frances Lincoln (2009), pp. 96, ISBN: 978-0711229648.

Current address

Kovár Martin, doc. RNDr. PhD.
Department of Mathematics, Faculty of Electrical Engineering and Communication,
Brno University of Technology
Technická 8, 616 00 Brno, Czech Republic
E-mail: kovar@feec.vutbr.cz

Příborský Jan, Ing.
Department of Mathematics, Faculty of Electrical Engineering and Communication,
Brno University of Technology
Technická 8, 616 00 Brno, Czech Republic
E-mail: priborsky.jan@seznam.cz

Slovak University of Technology in Bratislava
Faculty of Mechanical Engineering
16th Conference on Applied Mathematics **APLIMAT 2017**
Proceedings

INTEGRATING THE HISTORY OF ALGEBRA IN THE MATHEMATICS CLASS

KRČMÁROVÁ Monika (SK), FULIER Jozef (SK)

Abstract. The paper deals with the use of history of algebra in mathematics classroom. We describe our experience with the integrating of the history of mathematics. We used the single and double false position method in the historical problems to solve linear equations in the high school. We present also some examples of solving quadratic equations by completing the square from historical point of view. They could be used in the classroom too.

Keywords: algebra, history, teaching, equations, linear, quadratic, integrating history

Mathematics Subject Classification: 97D40, 97H30

1 Introduction

Using[1] history of mathematics in the teaching of mathematics has a long tradition. There are several studies about this topic last years, working groups on history of mathematics and teaching, projects focused on using history in mathematic class and many activities from this field. The Institutes, working groups or societies produce reprints of original resources, materials for the mathematical classrooms, they group teachers and researches, or they organize seminars.

For example in France it works the IREM (*Instituts de Recherche sur l'Enseignement des Mathématiques*). In Canada it is the CSHPM/SCHPM (*The Canadian Society for History and Philosophy of Mathematics /La Société Canadienne d'Histoire et de Philosophie des Mathématiques)*. In Great Britain it works The British BSHM (*British Society for the History of Mathematics*). It is interested not only in the research of the history of mathematics but also in the using of history in mathematics education. The BSHM arranged several „HIMED" (History of Mathematics in Education) meetings (they bring together mathematicians, historians and teachers; Wilson-Flood, 2013). The HPM (*The International Study Group on the Relations between the History and Pedagogy of Mathematics)* is the international group affiliated to the *International*

[1] Siu and Tzanakis (Siu–Tzanakis, 2004) think that instead of "using" history of mathematics in teaching would be better to use the world "integrating", or *"permeating"*.

Commission on Mathematical Instruction (ICMI). Their goal is to combine the history of mathematics with the teaching and learning of mathematics. HPM group has many activities. Organizing conferences that are satellite meetings to the *International Congress on Mathematical Education* (ICME), organized by ICMI every four years (for example in 2016: ICME-13 in Hamburg, Germany, satellite meeting in Montpellier (France)). Another activity of the HPM Group is the organization of ESU- *The European Summer University on the History and Epistemology in Mathematics Education* (the last, seventh ESU was held in July, 2014, in Copenhagen, Denmark) [18]. Also in the Czech republic there is a long tradition of (today called) International Conference of History of Mathematics (the next conference in 2018 will be the 38.annual).

There are several activities and materials focused on the using history of algebra in teaching mathematics. The MAA´s free online journal *Convergence (Where Mathematics, History, and Teaching Interact* (founding editors: Victor J. Katz and Frank Swetz, [21]) contains resources for using history of mathematics in the teaching (articles about history - also about history of algebra), historical problems, historical images, translation of original sources, classroom activities, projects, modules, quotations, collections of portraits and snapshots of mathematicians that we may use in the classroom).

We could mention also some studies, books and collection of materials for mathematics class (also the school algebra): *Using History to Teach Mathematics: An International Perspective* (Katz, 2000); The ICMI Study book on "*History in Mathematics Education*", edited by J. Fauvel and J. van Maanen (2000), *Vita Mathematica: Historical Research and Integration with Teaching* (Calinger, 1996), *Learn from the Masters* (edited by Swetz, Fauvel, Bekken, Johannsson, Katz, 1995). The MAA´s *Historical Modules for the Teaching and Learning of Mathematics* (editors V.Katz and K. D. Michalowicz) contains 11 modules including for example the Linear Equation Module. The course „*Study the Masters: Using Primary Historical Sources in Mathematics Teaching*" (organised by Joint Mathematics Meetings; Boston, 2012, led by D. Otero and D. Pengelley).

There are many questions associated with the idea of integrating the history in the teaching of mathematics. The question *Why* should be history of mathematics integrated in mathematics education is enough discussed in many papers and works. Puig and Rojano (Puig-Rojano, 2004) even think that there is no need to further discuss the necessity or usefulness of studying the history of mathematics for mathematics education. We think that the difficult question of problem is how *exactly* to use history of math in mathematic class. Siu (Siu, 2000) thinks that there are as many different ways to integrate history of mathematics into classroom teaching as there are teachers (depending on teacher style, preferences about historical topics).

Special part of this problem and the main key to the successful and active use of history in the teaching of mathematics is the use of history in the teacher training. Many authors argue (Kimličková, 2012; Winicki, 2000; Šedivý-Fulier, 2004) that the requirement for the success in the using history in the mathematic classroom is the adequate preparation the future math teachers in this area. Therefore, it is necessary not only to prepare ready-use materials for in-service teachers but also to focus on the future teachers and their education in history of mathematics and possibilities how and why to work with history in the classroom.

2 Historical Topics

2.1 Solving linear equations using the single (double) method of false position

The first word problems that lead to linear equations appeared in the Egyptian Rhind and Moscow papyrus (1650 B.C., 1850 B.C.), in the Kahun papyrus, old Babylonian clay tablets and the Chinese Nine Chapters on the Mathematical Art. They were solved by false position method or „direct division" with no symbolic or any form of equation that we know today.

The single false position method (the method of surplus and deficiency, the simple false position method, the method of false position, regula falsi). The main principle of the method (to solve the equation $ax = b$ or $ax + by = c$) can be simplified in three words „guess-check-adjust" (Katz-Michalowitz, 2004): we have to guess a solution of the equation, check if it is right and adjust it in the case it isn't (we have this method more described in (Krčmárová, 2013). In the classroom we can solve for example the historical problems R24-R29 from the Rhind mathematical papyrus using the method of false position (applied once or twice (see the problem 29). This kind of problems are called the „*aha*" problems (*aha* as the unknown quantity).

Problem R27. *A quantity and its 1/5 added together become 21. What is the quantity?*

Solution: the problem leads to the equation $x + \frac{1}{5}x = 21$.

1. Guess: for example 5;
2. Check the guess: $5+\frac{1}{5}.5 = 6 \neq 21$;
3. Adjust the solution: $\frac{21}{6} = 3,5$. 21 is 3,5-times bigger then 6, so we have to multiply our guess 3,5-times, so the right answer is $5.3,5 = 17,5$.

Problem R28. *A quantity and its 2/3 are added together and from the sum 1/3 of the sum is subtracted, and 10 remains. What is the quantity?*

Problem R29. *A quantity and its 2/3 are added together and 1/3 of the sum is added; then 1/3 of this sum is taken and the result is 10. What is the quantity?*

(Problem 29 leads to the equation $x + \frac{2}{3}x - \frac{1}{3}\left(x + \frac{2}{3}x\right) = 10$. First we have to substitute $x + \frac{2}{3}x = y$ a then to use the method two times; the solution $x = 9$). This problem is appropriate for higher grades (it depends; we solved it with the students aged 17 years (they knew the method of substitution from other mathematics classes, but it was still difficult for them to understand).

The double false position method (the second false position method, *regula duorum falsorum (positionum), hisab al-Chata'ayn*) was first described in Nine Chapters on the Mathematical Art (Book 7: *Jiuzhang Suanshu, Ying Buzu* method: *ying* = „excess", too much and *buzu* = „deficit", not enough).

The first eight problems deal directly with excess and deficit, in the other problems the excess nor deficit are not mentioned in text. The method was used when the single method didn't work (equations of type $ax+b=0$ or a system of linear equations).
We have to solve the equation:

$$10x-6=0$$

Solution:

1. Guess two solutions, for example $g_1 = 1$, $g_2 = 2$.
2. Compute the „errors" (e_1, e_2) : $g_1 = 10.1-6=4$, $g_2 = 10.2-6=14$.
3. Complete the solution using the formula:

$$x = \frac{g_1e_2 - g_2e_1}{e_2 - e_1}, \qquad (1)$$

so $$x = \frac{1.14-2.4}{14-4} = \frac{6}{10} = \underline{\underline{\frac{3}{5}}} = x.$$

We solved with the students some equations of type $ax+b=0$. Then we solved some problems from Nine Chapters (Problems 1-8 from the 7th book). All these problems deal with purchasing some good (gold, sheep, dog or pig) and lead to system of two equations. We show the solution on the next problem from Nine Chapters.

Problem YB1

A number of people are going to purchase goods. If everyone pays 8, there is an excess of 3; if everyone pays 7, there is a deficit of 4. How many people are involved and how much do the goods cost?

We can arrange the numbers in the grid like the Chinese did:

$$\begin{matrix} 8 & 3 \\ 7 & 4 \end{matrix} \qquad (2)$$

The Chinese had more rules to calculate the result (according to type of a problem: if it is an „excess-excess problem", „excess-deficit problem" and so on (Juškevič, 1978)). We write the „deficit of 4" as a negative number -4.

If we use the formula (1) (or see the Fig. 1): $\frac{8.(-4)-7.3}{-4-3} = \frac{-53}{-7} = \underline{\underline{\frac{53}{7}}}$. That is not a solution we are searching. It is the intercept for the left column in the Fig. 1:

$$\begin{matrix} \lceil 8 & & & 3 \\ 7\ \square & & \square & -4\ \uparrow \end{matrix}$$

Fig. 1

And similarly, according to Fig. 2 the result $\frac{3.7-8.(-4)}{7-8}=\frac{53}{-1}=-53$ is the intercept for the right column in the next picture (Fig. 2):

8 3

↑7 □ □ –4

Fig. 2

If we rewrite the results in the grid, we have the (Tab. 1).

So if no one pays anything, it will be the deficit of 53, so the goods count 53. If everyone pays $\frac{35}{7}$, there is enough money, so there are $53 \div \frac{53}{7} = 7$ people involved.

paid pro person	excess/deficit
8	3
7	-4
$\frac{53}{7}$	0
0	-53

Tab. 1

Problem YB5. *A number of people are going to purchase gold. If everyone pays 400, there is an excess of 3400; if everyone pays 300, there is an excess of 100. How many people are involved and how much does the gold cost?* [Answer: 33; 9800]

Problem YB7. *A number of people are going to purchase a dog. If everyone pays 5, there is a deficit of 90; if everyone pays 50, there is just enough. How many people are involved and how much does the dog cost?* [Answer: 2; 100]

2.2 Solving the quadratic equations

2.2.1 Geometric algebra

The development of *geometric algebra* is often attributed to classical Greek mathematics (The Book II of Euclid's *Elements* contains 14 problems from geometric algebra). The ideas of geometric algebra have appeared in Egypt, Mesopotamia and Asia several centuries before.

The main principle is using geometrical concepts and proofs as the underpinnings of algebraic techniques (Allaire-Bradley, 2001). Numbers are presented like lines of segments of a given lengths, quadratic quantities like the areas of squares and cubic quantities like volumes.

We focused on the use of geometric algebra to solve quadratic equations (because of our future using in high schools). The importance of using geometric algebra can be understood from a historical and current point of view:

a) the historical significance: the symbolical algebra was not developed, nor the general idea of equation

b) the current significance: suitable use of the geometric algebra in mathematics education (for its geometric clearness)

The use of geometric algebra in mathematics may have several advantages. Jahnke thinks (Jahnke, 1997) that the current school algebra is closely linked with arithmetic (it is reinforced by the impact of technology and computers) and some algebra formulas serve only as a prescription, only as a help in a calculation. Therefore, it may occur the loss of more "structural" information contained in the formulas – so here could help the clearness of geometry.

But the purpose of geometric algebra is not to replace the symbolic approach but to offer alternative perspectives (Allaire-Bradley, 2001). They also write that their classroom experience suggests that a historical approach always makes the subject more appealing to the typical student.

We think it may be useful at school at first to offer the students the methods of solving the quadratic equations in the past (before the using the formula based on discriminant), for example thegeomteric algebra methods, discuss about the advantages or disadvantages the method (clearness, only positive roots, loss of some solutions etc).

2.2.2 Al-Khwarizmi's Solution of Quadratic Equations

In The *Compendious Book on Calculation by Completion and Balancing* (*Al-kitāb al-mukhtaṣar fī ḥisāb al-ğabr wa'l-muqābala*) Al-Khwarizmi defined "squares" (x^2, from Arabic māl), "roots" (bx, džízr) and "numbers" (regular constants, numbers, *dirham* (it also meant a type of coin), and he identified six types of quadratic equations:

1.) Squares equal roots ($ax^2 = bx$)

2.) Squares equal number ($ax^2 = c$),

3.) Roots equal number ($bx = c$),

4.) Squares and roots equal number ($ax^2 + bx = c$),

5.) Squares and number equal roots ($ax^2 + c = bx$),

6.) Roots and number equal squares ($bx + c = ax^2$).

Other types of quadratic equations (he considered only positive coefficients *a, b, c*) he adjusted to the one of 6 types. He used the operations: *al-jabr*, *al-muqabala*, *al-radd* and *al-ikmāl.* The operation Al-jabr, „restoration", eliminates the negative terms in the equation by adding the same positive quantity to both sides of an equation (restoring of deficiency). Al-radd (reduction) vs.al-ikmāl (completion) means dividing the equation by the positive number bigger (vs. smaller) then 1 to get the equation with the coefficient $a = 1$. Al-muquabala (the opposition or the compensation) means the removal of identical quantities from both sides. At first he studied the equation with the coefficient $a = 1$.

For the last three types of equations he gave the geometric proof. He founded just the positive solutions. This part can be used in high school classes, especially the fourth type of equation (because it is not difficult).

Type 4.) Squares and roots equal number ($ax^2 + bx = c$),
Problem 1. *What must be the square which, when increased by ten of its own roots, amounts to 39 dirhem?*

Solution:
This problem and its solution is the best known problem from Al-Khwarizmi´s types of quadratic equations and for its simplicity it is optimal for using it in the mathematic classroom.
We have to solve the equation: $x^2 + 10x = 39$
Al-Khwarizmi writes the solution:

„*You halve the number of roots, which in the present instance yields five. This you multiply by itself; the product is 25. Add this to 39; the sum is 64. Now take the root of this which is eight, and subtract from it half the number of the roots, which is five; the remainder is three. This is the root of the square which you sought for.*" (Katz, 1998, p.245)

The geometric way to find the positive solution is described on the Fig. 3. We begin with a square of area x^2 and the rectangle with the sides x and 10. Then we halve the rectangle in two rectangles, each having one side of 5, and the other side x and replace one of them. The total shaded area is $x^2 + 10x$, which is the expression on the left side of the equation. The total area is therefore 39. We get the diagram called *gnomon*[2]. We have to add the small square with the area of 25 to it (we complete the square) to get the whole square with the sides $(5+x)$. It has the area $39+25=64$, so the one side of a square is $8=5+x$, so $x = 3$, which is one solution of the equation.

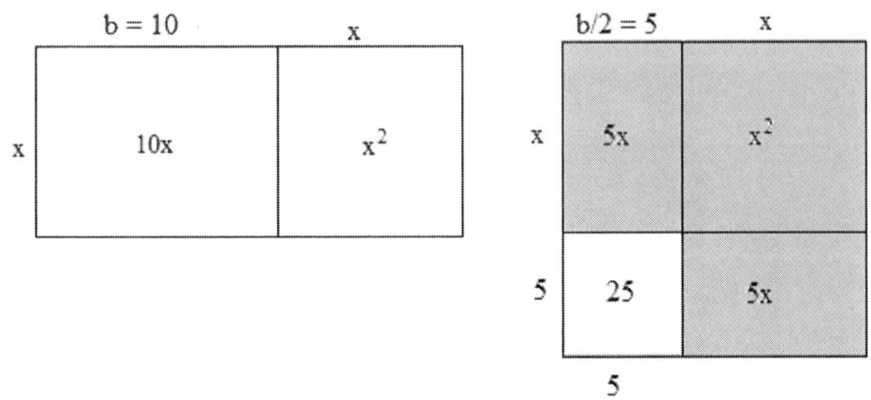

Fig. 3

[2] A shape which, when added to a figure, yields another figure similar to the original [17]. Gnomon can be a square with a little corner missing.

Problem 2. *Two squares and ten roots added together amounts 48 dirhem.*
(Positive solution is 3. At first we have to use the al-radd operation.)

Problem 3. *Half of the square and five roots added together amouns to 28 dirhem.*
(Positive solution is 4. At first we have to use the al-ikmāl operation.)

Type 5.) Squares and number equal roots ($ax^2 + c = bx$)

Problem 4. „A square and 21 dirhem together amounts ten roots.
Solution:
We solve the equation $x^2 + 21 = 10x$.

Al-Khwarizmi writes the solution:

„Halve the number of the roots. It is 5. Multiply this by itself and the product is 25. Subtract from this the 21 added to the square term and the remainder is 4. Extract its square root, 2, and subtract this from half the number of roots, 5. There remains 3. This is the root you wanted, whose square is 9. Alternatively, you may add the square root to half the number of roots and the sum is 7. This is then the root you wanted and the square is 49". [19]
So in this case Al-Khwarizmi knew that there are two solutions.
To get the two positive solutions we have to draw two pictures (see Fig. 4, Fig. 5).

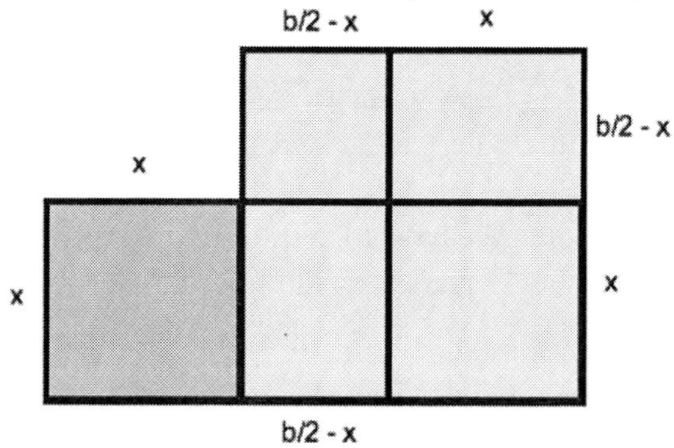

Fig. 4. Solving the equation $ax^2 + c = bx$ $(\frac{b}{2} > x)$

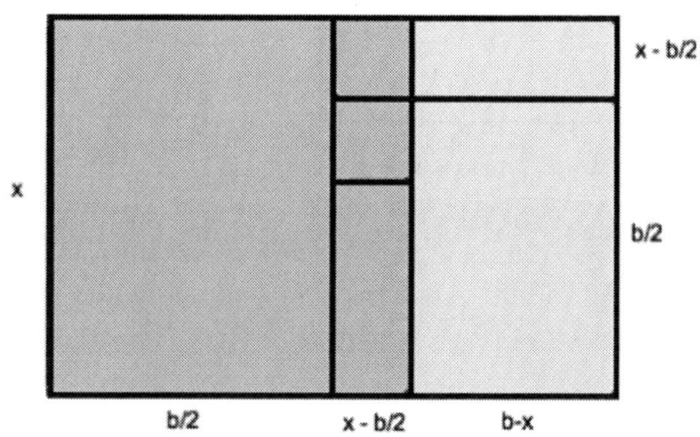

Fig. 5. Solving the equation $ax^2 + c = bx$ $(\frac{b}{2} < x)$

Al-Khwarizmi knew that only the type 5.) has two (positive) solutions.

In the case that $D<0$ (where $D = \left(\frac{b}{2}\right)^2 - c$) Al-Khwarizmi states that the task is impossible.

3 Methods

In 2014 and 2016 we carried out the small experiments with the prepared historical materials with the students of high school. We focused on the part Linear equations (the method of false position-single and double and some historical problems). At the beginning of the experiment we made a questionnaire survey (with 73 students of all grades of economic secondary school (business college) in spring 2014 and 8 students of fourth grade of grammar school in spring 2016. It contained both closed and open questions. It focused on the relationship the students to mathematics and to the history of mathematics. We wanted to find out what relationship have the students to mathematics and to its history in general, and what is their experience with integrating the history in the classroom last years. If they would like to learn something about history of mathematics and what they understand by history of mathematics. We also wanted to know if they think the history of mathematics in the class could make the math more interesting for them. We tried to find out what are their ideas about solving equations in the past, when did people solve the linear equations. After this questionnaires we made some historical materials and tried them in the math (spring 2014, business college, second grade, 17 students, solving historical problems and the method of false and double false position). We wanted to gain experience how to work with history and with the students (we just observed the student and did some recommendations for our next work).

In spring 2016 we did some similar historical activities with the students from the grammar school (fourth grade = the last). The scheme of the lesson looked as follows:

1.) Reading material – Pictures, about the Rhind Mathematical Papyrus, about The Nine Chapters on the Mathematical Art

2.) The principle of the method, one problem solved by teacher

3.) Problems to solve in students working groups (3-4 student in the group)

4.) Discussion – about results, methods

We tried to discuss for what kind of equations doesn´t the method work, why the method works – it was in the case of Rule of double false position difficult (we showed only the single false position method[3]).

At the end of the lessons with the historical activities the students wrote the questionnaires - what they liked about the lessons, their perceptions. It included closed questions with the rating scale 1-5 and also opened questions about the lessons.

[3] The picture based on simple proportions.

4 Results

One question included in the first questionnaire was about the history of solving equations: When people have already solved the linear equations (linear problems)? The most student´s answers were about 12.-19.century (almost the half of the student didn´t even write anything), a very few of all students thought that it was earlier (about 12 student wrote it was perhaps before Christ). They were surprised when they realized how long ago it really was.

Some detailed results of questionnaires (from the business college, for first and fouth grade) we discussed in (Krčmárová, 2013).

In all cases (of grades and school) the results of the questionnaires were very similar. 78% of students think that they heard about history of mathematics in mathematics class just a little. Almost 55% of students declared the mostly negative relationship to mathematics at school (relationship to the subject History was rated a bit better). They don´t like mathematics for example because of the many formulas and procedures they have to know, but they don´t see their purpose in the real life. But the most of students (especially after the historical activities in class) think that their relationship to mathematics would be better if the teacher would integrate the elements of the history of mathematics to the class and the real-life problems.

In spite of some negative comments about their relationship to mathematics, the students were curious about the upcoming historical activities. They asked when it will be and what will be the subject of historical activities, they didn´t know what to expect from it. They asked if anything from these history will be on the exam (in the first questionnaire there was a question if they would like to have anything from the history on the exam: 81% of them answered that they don´t expect better results of the exam with history in it; so they´d rather have standard exams as usual).

We had two lessons with the historical activities, on the different days, in the second semester of the school year (but - based on our previous experiences - not at the end of semester/year. Because at the end of the schoolyear the students are no longer motivated enough. They had almost no interest in teaching (first grade, business college).

The students kept attention during the lesson and they were more interested then normally. Most of them appreciated that they work in groups or they are doing something different than usual. What they didn´t like was the part when the teacher explained the formulas (like in the typical math class). Some of them were glad that they can also discuss something about the solution, but on the other side it was the most problematic part of the lesson, because for some students it was difficult (or they are not accustomed to discuss), so some student were not very interested (but it was almost the end of the lesson). As a teacher I found it interesting to discuss with the students about the methods: why we don´t use them today, what were the problems with using them and so on. In the second questionnaire all of the student (except one) changed a bit their opinion about using history in the mathematic class and the effect to their relationships to mathematics. On the same question like in the first questionnaire they asked more positive about the chance of enhancing their relation to math through history. We think that most of the students we were working with, didn´t have any or had very little experience with the using the history of mathematics in the class. Perhaps that´s why they mostly first answered negatively. Every teacher has also to consider (according to type of students, school, etc.) how long historical activities he would like to integrate to the mathematic class and how often. We thing that shorter activities but more frequently are better for motivation.

5 Discussion

We showed some opportunities of using some historical methods and problems in the mathematical class and our experience with the historical activities with students at the grammar school. Of course we would like to continue and try the historical activities focused on the quadratic and cubic equations with the secondary school students and especially with the university students (future teachers).

References

[1] ALLAIRE, P., R., BRADLEY, R., E. *Geometric Approaches to Quadratic Equations from Other Times and Places,* In: Mathematics Teacher, v94, n4, p.308-13,319, April 2001, ISSN-0025-5769.

[2] HUDEČEK, Jiří. *Matematika v devíti kapitolách.* Praha: Matfyzpress, 2008, ISBN 978-80-7378-046-3.

[3] JAHNKE, H., N. *Zur geometrischen Deutung der quadratischen Gleichung* [online], [cit.2016-12-3.]. https://www.uni-due.de/imperia/md/content/didmath/ag_jahnke/wt97_1.pdf.

[4] JUŠKEVIČ, A. P. *Dějiny matematiky ve středověku*, Praha, Academia, 1978.

[5] KATZ, V., J. *A History of Mathematics: An Introduction,* Addison-Wesley, 1998, ISBN 0321016181.

[6] KATZ, V., MICHALOWITZ, K. D. (eds). *Historical Modules for the Teaching and Learning of Mathematics (Module 7: Linear Equations*). 2004, Electronic ISBN: 9780883857410.

[7] KIMLIČKOVÁ, Z. *História ako motivačný nástroj pre podporu rozvoja matematického myslenia v sekundárnom vzdelávaní* [dizertačná práca], Bratislava, 2012.

[8] KRČMÁROVÁ, M. *História na hodine matematiky na strednej škole.* In: 8.didaktická konferencia: zborník príspevkov z 8.didaktickej konferencie s medzinárodnou účasťou, Dubnica nad Váhom : DTI, 2014. - ISBN 978-80-89732-01-2.

[9] KRČMÁROVÁ, M. *Využitie niektorých historických metód riešenia lineárnych rovníc vo vyučovaní matematiky*, In: Acta Mathematica 16: zborník príspevkov z 11. nitrianskej matematickej konferencie, 1. vyd. - Nitra: UKF, 2013. - ISBN 978-80-558-0365-4, S. 76-81.

[10] PUIG, L., ROJANO, T. *The history of algebra in mathematics education.* In: K. Stacey, H. Chick, and M. Kendal (Eds.), The teaching and learning of algebra: The 12th ICMI study (pp. 189-224). Boston / Dordrecht / New York / London: Kluwer Academic Publishers, 2004.

[11] SIU, M–K. *Historical support for particular subjects.* In: J. Fauvel & J. van Maanen (Eds.), History in mathematics education: An ICMI book, (pp. 241 – 290). Dordrecht, The Netherlands: Kluwer Academic Publishers, 2000.

[12] SIU, M.-K, TZANAKIS, C. *History of Mathematics in Classroom Teaching - Appetizer? Main Course? Or Dessert?* [online]. In Mediterranean Journal for Research in Mathematics Education, Volume 3, No. 1-2, pp. 5-10., 2004 [cit. 2012-7-30]. http://hkumath.hku.hk/~mks/J-MedEdexcerpt.pdf.

[13] ŠEDIVÝ, O. - FULIER, J. *Úlohy a humanizácia vyučovania matematiky*, Univerzita Konštantína Filozofa v Nitre, Nitra 2004, 245 strán, ISBN 80-8050-700-7.

[14] TALSMA, G. *A Little History of Some Familiar Mathematics* [online], Mathematics and Statistics Department, Calvin College, 2011, [cit. 2016-12-1] http://www.calvin.edu/~tals/images.380/history.1st.yr.sem.113011.pdf.

[15] WILSON, R.-FLOOD, R. *The British Society for the History of Mathematics*, In: Newsletter Of The European Mathematical Society, March 2013, Issue 87, ISSN 1027-488X.

[16] WINICKI, G. *The Analysis of Regula Falsi as an Instance for Proffesional Development of Elementary School Teachers*. In: Using History to Teach Mathematics. An International Perspective. MAA, ISBN: 0-88385-163-6, USA, 2000.

[17] URL: http://mathworld.wolfram.com/Gnomon.html.

[18] URL: http://www.clab.edc.uoc.gr/HPM/about%20HPM.htm.

[19] URL: http://www-history.mcs.st-and.ac.uk/Extras/Al-Khwarizmi_quadratics.html.

[20] URL: http://www.raikas.net/alkh.html.

[21] URL: http://www.maa.org/press/periodicals/convergence.

Current address

Fulier Jozef, prof. RNDr. CSc.
Department of Mathematics, Faculty of Natural Sciences,
Constantine the Philosopher University in Nitra
94974 Nitra, Slovak Republic
Tel. number: 037 6408 707, e-mail: jfulier@ukf.sk

Krčmárová Monika, Mgr.
Department of Mathematics, Faculty of Natural Sciences,
Constantine the Philosopher University in Nitra
94974 Nitra, Slovak republic
Tel. number: 037 6408 709, e-mail: monika.krcmarova@ukf.sk

AUTHOR INDEX

AUTHOR INDEX

AUTHOR INDEX

Institute of Mathematics and Physics, Faculty of Mechanical Engineering
Slovak University of Technology in Bratislava
Namestie Slobody 17
812 31 Bratislava 1
Slovak Republic

ISBN 978-1-5108-3698-3

CURRAN ASSOCIATES INC.
proceedings.com

16th Conference on Applied Mathematics (APLIMAT 2017)

Bratislava, Slovak Republic
31 January – 2 February 2017

Volume 2 of 2

STU
SjF

16th Conference on Applied Mathematics (APLIMAT 2017)

Bratislava, Slovak Republic
31 January – 2 February 2017

Volume 2 of 2

ISBN: 978-1-5108-3698-3

Printed from e-media with permission by:

Curran Associates, Inc.
57 Morehouse Lane
Red Hook, NY 12571

Some format issues inherent in the e-media version may also appear in this print version.

Printed by Curran Associates, Inc. (2017)

Institute of Mathematics and Physics, Faculty of Mechanical Engineering
Slovak University of Technology in Bratislava
Namestie Slobody 17
812 31 Bratislava 1
Slovak Republic

Phone: +421 2 572 96 111
Fax: +421 2 529 25 749

www.stuba.sk

Additional copies of this publication are available from:

Curran Associates, Inc.
57 Morehouse Lane
Red Hook, NY 12571 USA
Phone: 845-758-0400
Fax: 845-758-2633
Email: curran@proceedings.com
Web: www.proceedings.com

STU
SjF
SLOVAK UNIVERSITY OF
TECHNOLOGY IN BRATISLAVA
FACULTY OF MECHANICAL
ENGINEERING

Aplimat

16th CONFERENCE ON APPLIED MATHEMATICS

APLIMAT 2017 PROCEEDINGS

January 31 - February 2, 2017

Bratislava, Slovak Republic

SLOVAK UNIVERSITY OF TECHNOLOGY IN BRATISLAVA

Slovenská asociácia strojných inžinierov SASI

ZVÄZ SLOVENSKÝCH VEDECKOTECHNICKÝCH SPOLOČNOSTÍ

Vydavateľstvo Spektrum STU
Bratislava

Proceedings, 16th Conference on Applied Mathematics Aplimat 2017
First edition

Editors: Dagmar Szarková, Peter Letavaj, Daniela Richtáriková, Monika Prašílová

ISBN 978-80-227-4650-2

Slovak University of Technology in Bratislava
Faculty of Mechanical Engineering
16th Conference on Applied Mathematics **APLIMAT 2017**
Proceedings

The purpose of Aplimat Conference is to promote discussions and interactions between researchers and practitioners focused on disciplinary, interdisciplinary and transdisciplinary issues, ideas, concepts, theories, methodologies and applications. We are particularly interested in fostering the exchange of concepts, research ideas and other results which could contribute to the academic area and also benefit business and the industrial community.

Conference main topics

- Algebra and geometry and their applications
- Differential equations, dynamic systems and their applications
- Mathematics and art, section dedicated to the memory of Mauro Francaviglia
- Modeling and simulation in engineering and scientific computations
- New trends in mathematical education
- Statistical methods in technical and economic sciences and in practise
- Mathematical Aspects of Network Science and Applications
- Mathematics and Physics in Contemporary Science and Technique

List of reviewers

BAŠTINEC Jaromír, Czech Republic
BERÁNEK Jaroslav, Czech Republic
BOHÁČ Zdeněk, CZech republic
BUIKIS Māris, Latvia
CAPANNA Alessandra, Italy
CARKOVS Jevgenijs, Latvia
CARRANTI Andrea, Italy
CHALMOVIANSKÝ Pavel, Slovakia
CHVALINA Jan, Czech Republic
COCCHIARELLA Luigi, Italy
ČERMÁK Jan, Czech Republic
DOLEŽALOVÁ Jarmila, Czech Republic
FALCOLINI Corrado, Italy
FERREIRA Manuel A. M., Portugal
FILPA Andrea, Italy
FRONC Marek, Slovakia
Ferdinand HAVELKA, Slovakia
HOŠKOVÁ-MAYEROVÁ Šárka, Czech Republic
JANIGA Ivan, Slovakia
JUKL Marek, Czech Republic
KOLÁŘOVÁ Edita, Czech Republic
KOLYSHKIN Andrej, Latvia
KREML Pavel, Czech republic
LETAVAJ Peter, Slovakia
LINNERTOVÁ Dagmar, Czech Republic
LORENZI Marcella G., Italy
MAGDOLEN Ľuboš, Slovakia
MAGRONE Paola, Italy
MARIN Marin, Romaia
MIKEŠ Josef, Czech Republic
MOUČKA Jiří, Czech Republic
NEHÉZ Martin, Slovakia
NUCCI Lucia, Italy
PĂUN Marius, Romania
RICHTÁRIKOVÁ Daniela, Slovakia
RŮŽIČKOVÁ Miroslava, Slovakia
SIVÝ Július, Slovakia
SPADAFORA Giovanna, Italy
SVOBODA Zdeněk, Czech Republic
ŠMARDA Zdeněk, Czech Republic
TEDESCHINI-LALLI Laura, Italy
ÚRADNÍČEK Juraj, Slovakia
VALA Jiří, Czech Republic
VELICHOVÁ Daniela, Slovakia
VOLNÁ Jana, Czech Republic
VOLNÝ Petr, Czech Republic

CONTENTS

First Edition

Bratislava 2017

Slovak University of Technology in Bratislava
Faculty of Mechanical Engineering
16th Conference on Applied Mathematics **APLIMAT 2017**
Proceedings

STATISTIC OF QUASI-PERIODIC SIGNAL WITH RANDOM PERIOD - FIRST APPLICATION ON VOCAL CORDS OSCILLATION

KROUPA Lukáš (CZ), VÁVRA František (CZ), NOVÝ Pavel (CZ)

Abstract. This paper will introduce problem of statistics of quasi-periodic signal in relation to detection of the vocal cords pathology from audio recording. Distribution function of period lengths and its relation to distribution function of immediate frequencies is defined and application on vocal cord diagnostic by classification of periods and frequencies to common (normal) and anomalous is devised.

Keywords: vocal cords, quasi-periodic signal, early diagnosis, random period, statistics, probability model

Mathematics subject classification: Primary 62P10; Secondary 62F10, 62F25

1 Introduction

Quasi-periodic signal is assumed to be continuous bounded signal which repeatedly crosses given level (e.g. zero level) in direction from below to above this level. It's apparent this definition is loose and heuristic, but obvious when applied. See figure 1.

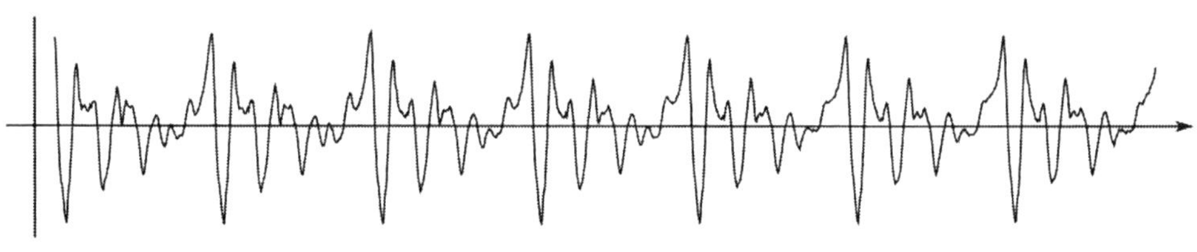

Fig. 1. Example of quasi-periodic signal

The problem of random periods is mainly studied in field of signal processing (for example see [2]). For our problem of vocal cords diagnostics from audio recording[1] we can define

[1]For problem definition and details see [1]

three sub-problems. First, description and classification of length of individual periods and representation in frequency domain. Second, obtaining of amplitude (shape) parameters of individual oscillations. Third problem is "measurement" of deviation of real oscillation from it's estimated ideal shape. The first sub-problem is subject of this paper. We will assume, we can measure period of oscillation[2]. We have random selection of observed oscillation lengths (periods) with given scope, that is the distance between consecutive transition from below to above the zero level. We will focus on signals, where one "period" is composed of sum of components[3], therefore the assumption of asymptotic normal probability distribution is reasonable (see [3], page 376).

2 Distribution function of frequency and it's relation to distribution function of period lengths

Distribution function and density function of period length $T > 0[sec]$ will be denoted as $G_t(x)$, $g_t(x)$ and assume the values are known. Than distribution function of frequency $f = \frac{1}{T} > 0[1/sec]$ will be

$$G_f(x) = P(f \le x) = P(\frac{1}{T} \le x) = P(T \ge \frac{1}{x}) = 1 - P(T \le \frac{1}{x}) = 1 - G_t(\frac{1}{x})$$

, in summary $G_f(x) = 1 - G_t(\frac{1}{x})$ from which the density is

$$g_f(x) = \frac{1}{x^2} g_T(\frac{1}{X}) \tag{1}$$

For the first two moments of frequency f probability description, following would be valid if these moments exist[4]:

$$E(f) = \int_0^{+\infty} \frac{x}{x^2} g_T(\frac{1}{x}) dx = \int_0^{+\infty} \frac{1}{x} g_T(\frac{1}{x}) dx = \int_0^{+\infty} \frac{1}{y} g_T(y) dy = E_T(\frac{1}{T})$$

summarized as $E(f) = E(\frac{1}{T})$,

$$E(f^2) = \int_0^{+\infty} \frac{x^2}{x^2} g_T(\frac{1}{x}) dx = \int_0^{+\infty} g_T(\frac{1}{x}) dx = \int_0^{+\infty} \frac{1}{y^2} g_T(y) dy = E_T(\frac{1}{T^2})$$

and accordingly $\sigma^2(f) = \sigma_T^2(\frac{1}{T})$. To avoid problems caused by existence or non-existence of moments, quantile replacement of these moments is used.

For this reason and because of the normality we will assume continuous distribution function $f(x)$ for $x > 0$. We can infer that $G_T(x)$ for $x > 0$ is continuous as well. Onward we will denote $z_p \stackrel{def}{=} G_z^{-1}(p)$ for $0 < p < 1$ and z_p is p quantile of random variable z, i.e. $z_p = inf\{x \in R : p \le G_z(x)\}$.

[2]For description of signal pre-processing and period length detection see [1], page 23

[3]That is the case for acoustic signal produced by vocal cords.

[4]Transformation of period distribution to frequency distribution might lead to distribution without existence of some of the moments. Even for uniform distribution of periods on interval $(0, T_{max})$ or often used exponential distribution of period duration. Non-existence of moments might not be just academical issue. Fundamentals of this problems are in relation $f = \frac{1}{T}$

3 Quantiles and relation of $G_T(T_p) = p$ and $G_f(f_p) = p$

Let $0 < p < 1$ and $G_T(T_p) = p, G_f(f_p) = p$ from which

$$p = 1 - G_T(\frac{1}{f_p}) \Rightarrow \frac{1}{f_p} = G_T^{-1}(1-p) = T_{1-p}$$

, summarized as

$$f_p = \frac{1}{T_{p-1}} \tag{2}$$

By that we have obtained relation between quantiles of period length and immediate frequency. It might be difficult to get (estimate) one of both f_p, T_{1-p} quatiles in some cases (statistical, estimative). Hence this simple inference, which connects quatiles f_p and T_{1-p}. If following applies

$$G_T(T_p) = G_f(f_p) = 1 - G_T(\frac{1}{f_p} \Rightarrow G_T(\frac{1}{f_p}) = 1 - G_t(T_p) \Rightarrow \frac{1}{f_p = G_T^{-1}(1 - G_T(T_p))}$$

, again summarized:

$$f_p = \frac{1}{G_T^{-1}(1 - G_T(T_p))} \tag{3}$$

then medians $T_{0,5}$, $f_{0,5}$ can be used as replacement values. To describe variability the quartile ranges $T_{0,75} - T_{0,25}$, $f_{0,75} - f_{0,25}$ can be used, or alternatively the width of zone equivalent to σ zone above the median for normal distribution (for technical comparisons), i.e. $T_{0,84134} - T_{0,5}$ and $f_{0,84134} - f_{0,5}$. Obviously this form of variability comparison might be problematic for unsymmetrical distributions. Another alternative might be width of $\pm 3\sigma$ zone in symmetrical position of normal distribution $\frac{T_{0,99865} - T_{0,00135}}{6}$, $\frac{f_{0,99865} - f_{0,00135}}{6}$. These values are equal to value of σ in case of normal distribution. Consequently statistical inference is now possible.

4 Application on vocal cords oscillation

Fundamental prerequisite for vocal cord diagnostics based on audio signal is classification of oscillations (in this case length of their periods and immediate frequencies) to common (standard) and anomalous. We assume phonation recording of vowel "a" ("ááá..."). This process can be divided in several steps:

1. Using statistical methods create probability description of period length.
   a) Elimination of identical values by addition of noise. (We are working with model of continuous random variables.)
   b) Normality test of period lengths set. See appendix 7 for one of possible methods.
   c) Exclusion of observations from this set, which do not conform to this assumption.
   d) Creation of point or eventually range parameter estimations from this reduced set.
2. Create probability model of frequency from model of period lengths.
3. Based on given level of significance $0 < \alpha < 1$ form zones in which common values occur.
4. Further diagnostic comparisons and decisions.

5 Example of process

Observed values:

Inx	Minimum	Average	Sampled median	Maximum	StD. sampled	Skewness - Excel	Kurtosis - Excel	Count	Observed period lengths [sample count]																			
2	164	170	170	180	4,4	0,70	0,11	20	180	173	178	171	174	173	174	170	171	169	170	168	169	165	167	168	165	165	164	166

Tab. 1. Observed period lengths

Values after randomization and reduction by normality tests.

Inx	Minimum	Average	Sampled median	Maximum	StD. sampled	Skewness - Excel	Kurtosis - Excel	Count	Observed period lengths [sample count]																			
2	164	169	169	178	3,7	0.45	-0,23	19		173,13	177,84	170,55	174,08	172,64	174,15	169,92	170,71	168,56	170,48	167,71	169,42	164,91	167,23	168,39	164,95	165,14	163,97	166,31

Tab. 2. Values after randomization and reduction

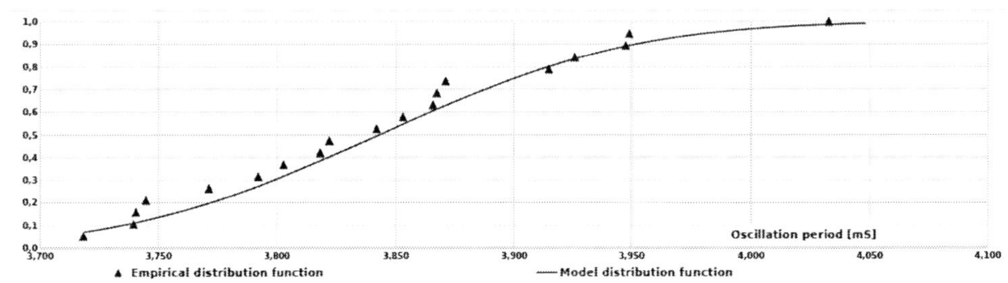

Fig. 2. Comparison of empirical data and model distribution function

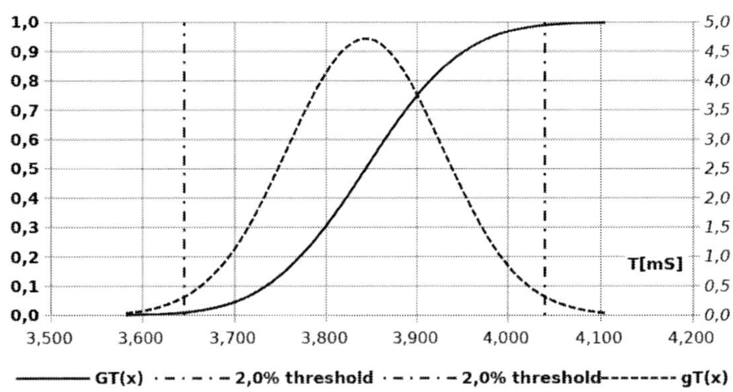

Fig. 3. Modelled distribution functions and density functions for period lengths

Bounds for common or anomaly test and quantile parameters.

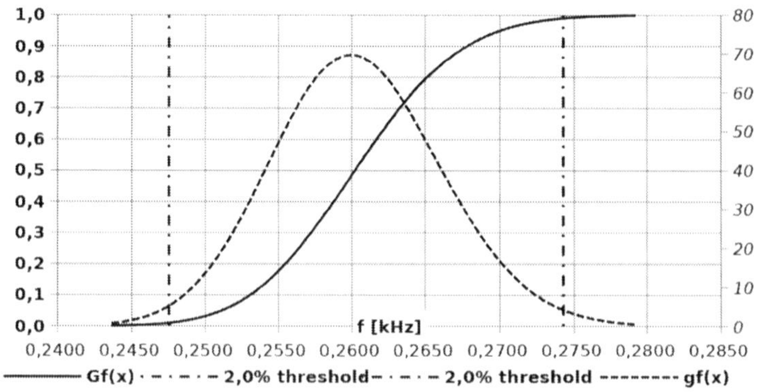

Fig. 4. Modelled distribution functions and density functions for immediate frequencies

6 Conclusion, discussion of results and further work

We assume probability model of oscillation for partial-diagnostic from standardized phonation. Research in this field implies probability model of immediate frequency resemble normal probability distribution, even though it's analytical expression is significantly different. For example density:

$$g_f = \frac{1}{x^2} g_T\left(\frac{1}{x}\right) = \frac{1}{x^2} \frac{1}{\sqrt{2\pi}\sigma} e^{-\frac{(\frac{1}{x}-\mu)^2}{2\sigma^2}} = \frac{1}{x^2} \frac{1}{\sqrt{2\pi}\sigma} e^{\frac{(1-\mu x)^2}{2(\sigma x)^2}}$$

This holds not only for presented data but for all data available. This leads to following questions:

- Is this the case only for our data, or is it general fact?
- Does it confirm intuitive approaches, when normal distribution is used for both random variables?
- If this is general fact, define scope and how this could be utilized?
- If this is general fact, what is the physiological, physical and acoustical basis?

It is definite description and probability restriction (Bonferroni inequality) for our modified algorithm of normality tests (for example related to Jarque-Bera test) with respect to identification of outlier observations. Parametric and non-parametric descriptions of ideal oscillation shape (in given domain that would be mainly more or less randomly distorted harmonic signals). Degree of "deviation" from estimated ideal shape, a measure with diagnostic value. Joint distribution of triplets period (frequency), shape parameters and degree of deviation.

References

[1] Kroupa, L.: *Systém vyhodnocení parametrů jednoho kmitu hlasivek*, Diploma thesis, University of West Bohemia, Pilsen 2015

[2] URL: `http://web.stanford.edu/group/cioffi/doc/book/`, `http://web.stanford.edu/group/cioffi/`.

[3] Rényi, A.: *Teorie pravděpodobnosti*, Book, ACADEMIA, Prague 1972

[4] Hátle, J., Likeš, J.,: *Základy počtu pravděpodobnosti a matematické statistiky.*, Book, SNTL, Prague 1974

[5] Likeš, J., Machek, J.,: *Matematická statistika*, Book, SNTL, Prague 1983

[6] Jílek, M.: *Statistické toleranční meze*, Book, SNTL, Prague 1988

7 Appendix: Normality test

Based on skewness and kurtosis[5]

Sampled skewness:

$$A_3 = \frac{\sqrt{n}\sum_{i=1}^{n}(x_i - \overline{x})^3}{[\sum_{i=1}^{n}(x_i - \overline{x})^2]^{\frac{3}{2}}} - 3$$

with assumption of sample selection from normal distribution is:

$$E(A^3) = 0$$

and

$$\sigma^2(A_3) = \frac{6(n-2)}{(n+1)(n+3)}$$

Sampled kurtosis:

$$A_4 = \frac{n\sum_{i=1}^{n}(x_i - \overline{x})^4}{[\sum_{i=1}^{n}(x_i - \overline{x})^2]^2} - 3$$

with assumption of sample selection from normal distribution is:

$$E(A_4) = -\frac{6}{n-1}$$

and

$$\sigma^2(A_4) = \frac{24n(n-2)(n-3)}{(n+1)^2(n+3)(n+5)}$$

. Asymptotically has the A_3 distribution[6] $N = (0, \frac{6(n-2)}{(n+1)(n+3)})$ and A_4 distribution $N = (-\frac{6}{n+1}, \frac{24n(n-2)(n-3)}{(n+1)^2(n+3)(n+5)})$. Therefore the hypothesis of normality on level of confidence α is rejected, if: $\frac{|A_3|}{\sqrt{\sigma^2(A_3)}} > u_{1-\frac{\alpha}{2}}$ or $\frac{|A_4+\frac{6}{n+1}|}{\sqrt{\sigma^2(A_4)}} > u_{1-\frac{\alpha}{2}}$ If the hypothesis is rejected by at least one of stated tests, observation by which removal will lead to improvement of affected test has to be found, i.e. the test will draw closer to critical value or eventually surpass this value. This

[5]See for example [5], page 138.

[6]Convention $N = (\mu, \sigma^2)$

observation is removed from further statistical processing. This process is repeated until the normality hypothesis is accepted by both tests or until the cardinality of the set decreases under acceptable threshold. This is "brute force" algorithm which repeatedly iterates over whole set. Detailed description of individual steps and decisions exceeds scope of this paper. We expect it would be published including study of premises needed for effective function.

Current address

Kroupa Lukáš, Ing.
Department of Computer Science and Engineering,
Faculty of Applied Sciences, University of West Bohemia
Univerzitní 8, 306 14 Pilsen, Czech Republic
E-mail: kroupal@kiv.zcu.cz

Vávra František, doc. Ing. CSc.
Department of Mathematics,
Faculty of Applied Sciences, University of West Bohemia
Univerzitní 8, 306 14 Pilsen, Czech Republic
Tel. number: +420 377 63 2615, e-mail: vavra@kma.zcu.cz

Nový Pavel, Ing. PhD.
Department of Computer Science and Engineering,
Faculty of Applied Sciences, University of West Bohemia
Univerzitní 8, 306 14 Pilsen, Czech Republic
Tel. number: +420 377 63 2411, e-mail: novyp@kiv.zcu.cz

Slovak University of Technology in Bratislava
Faculty of Mechanical Engineering
16th Conference on Applied Mathematics **APLIMAT 2017**
Proceedings

THE SMART GREENHOUSE: PHOTONIC CRYSTALS TO CONTROL INDOOR TEMPERATURES

LALATTA COSTERBOSA Cecilia (IT), ROCCETTI CAMPAGNOLI Beatrice (IT), GIUSTO Paolo (IT), COMORETTO Davide (IT)

Abstract. Smart Greenhouse is the prototype of a greenhouse, built in compliance with parameters of the *Parametric Hybrid Wall* (PHW) method (Aplimat 2016 *The smart greenhouse*). The aim of the project is to achieve the optimization of the greenhouse indoor condition by means of a semi-automated control. A device able to independently manage the micro temperatures exploiting the combination of a biodegradable elastic polymer integrated to a Photonic Crystals film.

Keywords: Light, Photonic Crystals, Refraction Index, Responsive Surfaces, Parametric Hybrid Walls (PHW), Greenhouses

Mathematics Subject Classification: Primary 74M05

1 PHW Method, Molecola and The Smart Greenhouse background

The Smart Greenohuse belongs to the Molecola Project.

Molecola was developed following the publication of a Master Theses project called *Parametric Hybrid Wall* (PHW), presented for the first time at Aplimat in 2014. PHW is a methodological process that lead to a prototype of interactive and dynamic structures for Architecture. The PHW method correlates Architecture and IT (Information Technology), with the goal of achieving easily assembled and manageable structural units, equipped with artificial intelligence, in order to respond to different surrounding factors. The prerogative of the PHW is the production of systems by rapid prototyping tools such as 3D printing, CNC, and LaserCut in an Open Source environment and Open Hardware (therefore easily accessible), ready for a low-cost market.

1.1 About the Smart Greenhouse

Molecola explores various areas of development, always implying the use of PHW method. Smart Greenhouse is the final product, obtained declining Molecola to the agricultural sector, with particular reference to the greenhouses and the Precision Farming.

Smart Greenhouse is therefore the prototype of a greenhouse, built according to the parameters of the PHW method. Its objective is the management and the optimization of the internal conditions of a greenhouse by means of a semi-automated control.

The expected results of this research are mainly related to energy saving purposes and the use of renewable sources, in order to sensitize farmers on the issue of climate change, inducted by the use of polluting energy sources and the waste of water

Moreover, this project focuses on innovative greenhouses able to monitor and optimize crops temperatures.
The project is still under development. Assumptions made on the functionality of the mechanical structure are based on studies made with the first five PHW prototypes developed between November 2013 and October 2016. As for the behaviour of Photonic Crystals (PC) and the polymeric materials which integrates PC, and their reaction on the mechanical structure, the project is still based on theoretical suppositions.

2 The Smart Greenhouse Configuration

The Smart Greenhouse is made of modular elements that can be assembled. These modules are in turn composed by two layers. The outermost (A), called *skin* is a functional material showing: transparency, reflectivity, elasticity, and in general responsivity to sun-light stimuli. The underlying (B) is instead a structural skeleton holding, once assembled, the whole structure as well as the skin.

2.1 The *Skin*

In this phase, the project focuses on (A). and in particular on the implementation of all-polymer planar photonic crystal structures into the skin. Photonic crystals are nano-composite media where materials having different refractive index are assembled into an highly ordered and periodical structure. The typical periodicity of few hundredths of nanometers. The own nature of photonic crystals allows the material to reflect selected light wavelengths. In the simplest case, the reflectance spectral range can be easily tuned and engineered by changing material composition (i.e. the refractive index) and layers thicknesses. Therefore, by tuning the system into the near infrared spectral region, photons inducing heating can be prevented to overcome the skin thus reducing inner temperature. Moreover, it is possible to create reflection in the visible spectral range thus providing bright structure color as often observed in nature, both in the biological (i.e. Morpho Rethenor Butterflyies) and mineral world (opal gemstones) [see for instance Organic and Hybrid Photonic Crystals, edited by D. Comoretto, Springer 2015]. This opportunity, obtained through the photonic crystal film, can result in better control of internal climatic conditions of the greenhouse without the need of artificial climate control systems.

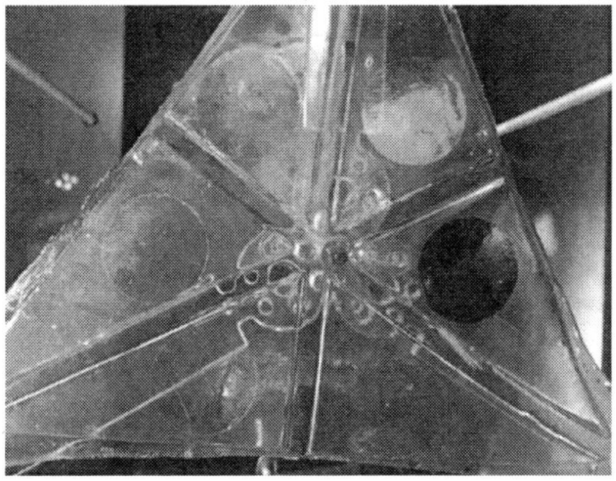

Fig. 1. Prototype of a module of the Skin, with PC integrated

3 Photonic Crystals

Photonic crystals (PC) are materials whose dielectric function (or refractive index) is periodic in one, two or three (1D, 2D or 3D) dimensions in the space. When light travel through them, it is coherently scattered at each interface. When a constructive interference among scattered light beams occur, light does propagate thorough the structure. However, when destructive interference occurs, propagation through the PC is forbidden and light is backward diffracted giving rise to a strong reflection signal. We call the spectral range of reflected light photonic band gap (PBG), in analogy to the electronic structure of crystalline solids. Since the interference conditions depends on the light incidence/observation angle, strong iridescence is associated with the photonic band gap thus providing unusual chromatic effects to photonic crystal structures. Nowadays, PC are widely used for different technological applications as sensing, photovoltaics, and lasing.

3.1 Photonic Crystals Photonic Band Gap

In this work we would like to introduce the 1D PC, well known as distributed Bragg reflectors (DBR) or dielectric mirrors (Fig. 2). In these composite systems the dielectric function is periodic

$$\varepsilon(x,y,z)=\varepsilon(x,y,z+iD) \quad (1)$$

where i is an integer and D is the period, i.e., the length of the unit cell. The periodic modulation of the refractive index ($n=\sqrt{\varepsilon}$) allows the generation of bands (i.e. an ensemble of photon energies) allowed/forbidden to propagation of photons (Fig. 2). These latter are called PBG. If the PBG energy is equal to the energy of visible spectrum (1.65-3.1 eV correspondent to ~400-750 nm), the result is a "structural color". In this case the color comes from backward diffraction phenomena driven by the electronic structure of the material as usually observed in dyes or pigments. Notice that fully transparent materials when properly assembled into a PC structure give rise to bright color.

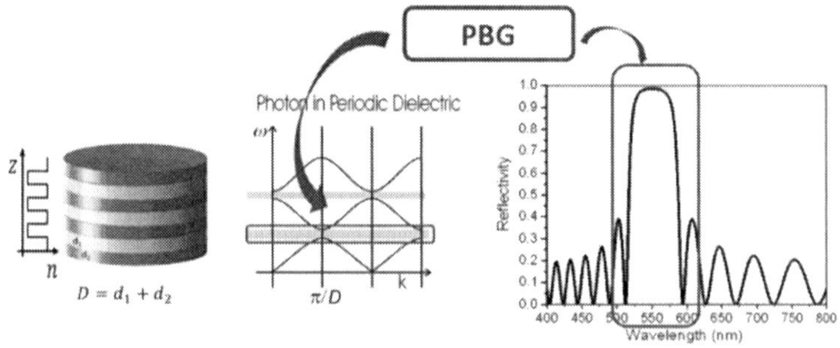

Fig. 2. Disposition of the two dieletric materials

3.1.1 Controlling PC's Colours

As a DBR is made of planar stacks of two dielectric materials, the position of the PBGs center can be easily calculated from the well-known Bragg-Snell's law (neglecting the absorbance of materials):

$$m\lambda = 2D\sqrt{n_{eff}^2 - sin^2\,\theta} \tag{2}$$

where m is the diffraction order, λ the diffracted wavelength, D the period, θ the angle of incidence of the light and n_{eff} the effective refractive index of the PC defined as the average refractive index of composing materials weighted over their volume fraction:

$$n_{eff}^2 = n_1^2\left(\frac{d_1}{d_1 + d_2}\right) + n_2^2\left(\frac{d_2}{d_1 + d_2}\right) \tag{3}$$

Where the number 1 and 2 are referred to low and high refractive index materials, respectively. By modifying the thickness and/or the refractive index of the two materials it is possible to obtain all colors in the visible spectrum.

According to Fig. 4, the color of the system, i.e. the center of the PBG λ_{PBG} depends on the observation angle. This provide to structural color two properties never observed in standard dyes or pigments:

1) The color depends on the observation angle and becomes more blueish for high detection angles or by tilting the sample.
2) For large area PC, different points of the sample are focused by our eye at different angles thus providing a strong iridescence to the sample.

An additional opportunity in order to control the color of the system is to tune the photonic band gap spectral width (W). For instance, in Fig. 2, the PBG spans form about 520 to 590 nm, i.e. form the green to the orange. The width can be tuned by the refractive index of composing materials according to

$$W = \frac{hc}{\lambda_{PBG}} \frac{n_1 - n_2}{n_1 + n_2} \quad (4)$$

When the dielectric contrast (i.e. the refractive index difference, n_1-n_2) is reduced the spectral purity is high, while when large additive mixture of spectrally adjacent color becomes easier. We notice that all these structures obtained by polymer materials can be peeled off from the substrate where are grown thus giving rise to free-standing structures. These PCs, being made by commodity polymers, have also all mechanical properties typical of such materials. Then they can be bent, rolled up or stretched by simply selecting the suitable polymer and can be easily embedded into the skin.

A final comment on the tunability and responsivity of all-polymer PCs. According to the formula so far reported, the spectral position of the PBG can be tuned in suitable spectral ranges. For instance, with thick layers (hundredths of nanometers), the PBG can be tuned into the near infrared. At such wavelengths, sun irradiation gives rise to heating. The use of PCs suitably tuned can be used as a screen to prevent un-wanted building heating. Moreover, polymer properties (refractive index or thickness) can be modified by exposure to different chemical environments and/or to pollutants. Swelling induced by analyte adsorption might induce a change in color of the structure thus opening the way for chromatic sensors. Then, color is not only an aesthetic property but could become an active feature of the structure integrated into buildings.

4 Innovation

Design is taking advantages from Chemistry. This link leads to a new perspective in the world of materials: for the first time Photonic Crystals are not seen only as colourful films to be used for aesthetic purposes in the field of design, but they become a turning point in the efficiency of the materials involved in the greenhouse's structure.

5 Advantages

The combination of PC with elastomeric polymers, produces notable advantages when integrated into building structures: (i) ability to adapt to climate changes, (ii) energy efficiency, (iii) environmental sustainability, (iv) innovative techniques related to natural resources. From Design and Technologies side: (v) Dynamic Shapes concerning natural and memory form materials can be envisaged and realized.

6 Final considerations

The considerations we just made, do open a new scenario on energy saving, around the subject of air conditioning systems. The expected results of this research are in fact essentially related to energy saving and the use of renewable sources with the intention to sensitize farmers on the issue of climate change linked to the abuse of polluting energy sources and waste of water

Acknowledgements

The authors gratefully acknowledge Professor Alessandra Capanna from La Sapienza University of Rome and Marcella Giulia Lorenzi, Aplimat Maths and Art section Chair Person. Work in Genoa has been supported by Italian Ministry of University, Research and Instruction thorough the "Progetti di Ricerca di Rilevante Interesse Nazionale 2010-2011" Program (2010XLLNM3) and by University of Genoa.

References

[1] YUHUI, L., *Advanced Functional Materials* Volume 23, Issue 6, Feb. 11, 2013, pp 660–672

[2] MABEY, R., *Fencing Paradise: Exploring the Gardens of Eden*, Eden Project Books, London 2005, ISBN 1-903919- 31-2

[3] PEARMANN, H., WHALLEY, A. *The Architecture of Eden*. With a foreword by Sir Nicholas Grimshaw.: Eden Project Books, London 2003, ISBN 1-903919-15-0

[4] BALDWIN, J. *The Pillow Dome*, The Buckminster Fuller Institute 1985. Retrieved 9 February 2014

[5] COMORETTO, D. *Organic and Hybrid Photonic Crystals.* Springer (New York), 2016 ISBN 978-3-319-16579-0

[6] CECCHINI, C. *Plastiche: I Materiali del possibile.* ALINEA, Firenze 2004, ISBN 88-8125-874

Current Address

Lalatta Costerbosa Cecilia, PhD candidate
Dep. PDTA (Planning, Technologies for Architecture and Design)
La Sapienza University of Rome
Via Flaminia 72, 00196, Roma, Italy
Tel. number (+39) 3409172448, e-mail: c.lalatta.costerbosa@gmail.com

Roccetti Campagnoli Beatrice, Graduated in Architecture
Dep.t DiAP La Sapienza University of Rome, Faculty of Architecture
Via Flaminia 359, 00196 Roma, Italy
e-mail: beatriceroccetti@gmail.com

Giusto Paolo, Master student
Dep. Chemistry and Industrial Chemistry University of Genova
Via Dodecaneso 31, 16146 Genova, Italy
e-mail: paologiusto89@gmail.com

Comoretto Davide, Associate Professor
Dep. Chemistry and Industrial Chemistry University of Genoa
Via Dodecaneso 31, 16146 Genova, Italy
Tel. number (+39) 010 3538736, e-mail: davide.comoretto@unige.it

Slovak University of Technology in Bratislava
Faculty of Mechanical Engineering
16th Conference on Applied Mathematics **APLIMAT 2017**
Proceedings

THE "GEOMETRY OF MUSIC" VISUAL ART OF ANDY GILMORE

LORENZI Marcella Giulia (IT)

Abstract. Starting from the parallel developments of Mathematics and Art during the centuries, and the possibilities offered nowadays by computer graphics to create geometric visuals, in this paper we will deal with the works of a contemporary artist, Andy Gilmore, specifically focusing on his technique of incorporating mathematics and nature into the structure of his designs, while being inspired by music. Finally, we will examine some his rendering of new geometric models of reality.

Keywords: mathematics and art, geometrical inspiration, music and geometry, complexity, physics and representation of models

Mathematics Subject Classification: Primary 00A67

"Mighty is geometry; joined with art, irresistible"
Euripides

1 Introduction

Going back in time, looking at primordial forms of Art – those that belong to prehistoric paintings in the caves and to proto-artistic drawings in Pleistocene and subsequent pottery – we understand that the concept of "geometric form" slowly arouse from the observation of the forms already existing in Nature and hidden in the structures through which our mind tends to perceive the "Order of Nature".

Most of primordial forms of Art do in fact contain those "primordial forms" that were later encoded in Euclidean Geometry: in two dimensions straight lines, triangles and other more complicated "polygons" (that account for the structure nowadays called the "linear and affine structure of Euclidean Space") as well as circles (that account for the structure that we call the "metric structure of Euclidean Space"), together with their three-dimensional extensions known as polyhedrons and spheres.[21]

In other papers we dealt with the general subject of "form persistence" [12] and we addressed explicitly how the simple forms of Geometry – and more generally, Mathematics - have again become in the XX Century a fresh source of inspiration for Art, in a kind of "backward travel in time" towards a renewed aesthetics of "simple forms": circles, straight lines, triangles *("senza i quali mezi è impossibile a intenderne umanamente parola"* as Galileo said; , as well as squares) and other geometric constructions [21].

As is well known, the turn of the XX Century (more specifically, the years between the last two decades of XIX Century and the first two decades of the new Century) saw great revolutions in thought, related to new understandings of the physical world, of Technology, of the Psychology of vision and perception, along with the artistic sensibility that refers to these new issues.

A new age was thence born: the age of Motion, Light, Dynamism, Perception and Visualization, as well as of the new ways of putting them in relation with the way in which brain reacts to external stimuli. [...] *"Thanks to the informatics innovation we are rapidly approaching an <<always on>> society, in which most inhabitants of rich countries will be continuously connected with the network [...] the <<globalvillage>> theorized at the beginning of the sixties by Marshall McLuhan is becoming a reality. [...] In the meantime visualization has assumed an increasingly important role: the language of images has become the true international jargon. Techniques as Photography, Cinema, digital and computer-aided reproductions [....] telescopes as Hubble and electronic microscopes, as well as (obviously) Television and Internet, have invaded scientific research and our daily life, drastically changing our ways of perceiving. [...] Frontline artists have begun to use mainly and at the same time several techniques, neither one representing the only way of expression: from time to time the same artist can pass from the production of video to pictures, sculptures, performances".* [21]

Without understanding their mathematical and technological roots it is not possible to focus appropriately the new forms of Art that developed in the Century that just passed by and opened a new age of artistic and scientific sensibility...[21]

Among the large number of artists operating within this framework, in the present work we will describe the work of a very up-to-edge, appreciated and prestigious artist in the field of Digital Visual art, Andy Gilmore.

2 The "geometry of music" visual art of Andy Gilmore

"A master of color and geometric composition, Andy Gilmore's work is often characterized as kaleidoscopic and hypnotic, though it could just as well be described as visually acoustic, his often complex arrangements referencing the scales and melodies in music." [11]

Gilmore's journey toward this masterful style followed a long and winding road. He characterizes his pursuit of education as a tangled web, noting the bookstores, bakeries, and theaters where he worked were of more importance than the schools he attended in pursuit of his BFA, which he received from SUNY Empire State College. After finishing school, Gilmore relocated to San Clemente, California, where he took up work for FourStar Distribution, resizing print ads for C1RCA Footwear and Forum Snowboards. It would serve as his introduction to the computer and lay the foundation for a career in digital art and design.

Andy Gilmore has an impressive reputation, his extensive list of clients includes Wired Magazine, The New York Times, Fast Company, Wallpaper*, Nike, Warp Records, New Scientist, Warp

Records, The New York Times, Cool Hunter, Ghostly International and McCann Erickson. He currently lives and works in Rochester, New York. [11]

His digital work can be identified through the synchronization of form and color, producing mesmerizing kaleidoscopic and geometric artworks, images of bright swirling colors, intricately interwoven lines that morphing into sleek harmonius forms.

Kaleidoscope is word coming from the greek words greek "kalos" (beautiful) and "eidos" (image). Andy states his love for music as a huge influence in his work, particularly in regards to the physics of sound and acoustic [23]: his work is an exploration of the physical properties of sound and light put to form, he experiments with the proportions, perception and properties of sound and light.[8] So his designs transcend straight-forward aesthetic beauty, their forms inhabiting a realm of mathematical complexity, containing a certain geometric musicality.

In his pieces, Gilmore explores the physical properties of tone – the resonance of a single string creating a range of vibrations and sound waves, both audible and inaudible – and mirrors that relationship visually, producing complex images with overtones swimming through undertones, melodies dancing with accompaniment.[3]

Fig. 1. Andy Gilmore. All the images released under Creative commons licence, via spacecollective.org or pinterest.com. All the artworks can be purchased at www.agilmore.com

The "musicality" of his works is more than a coincidence – Andy Gilmore is also a musician. Deeply interested in the physics of sound and acoustics, his work reflects just that.

In His Own Words: "Music is at the center of my artistic life. As a musician I have always been fascinated by the harmony and the physics of sound—in a sense defining music as waveforms, waveforms whose properties and proportions define our scales, from which we write our melodies, in which we weave our emotions and memories into songs." [23] Playing guitar, he started to see that just as in geometry you have a certain grid, or unit of measurement that defines a form, an instrument has a system as well that has its base in proportions and its pitch relates to colour and frequency. At the same time, geometry made sense to hime because it was a way of controlling proportions. In many ways he looked at it as the equivalent of playing scales on an instrument, it was essentially a way of refining his intuitive proportional aesthetic relationships.

Fig. 2. Andy Gilmore, Creative commons

In addition to music as an experience he has been very interested in the theories, methods and language of music—serialism, minimalism, spectralism, indeterminacy, improvisation, tonality/atonality, silence/noise, raga and rhythm—these have always informed the language that he applied to design and ultimately shaped his work. [22]

Indeed he affirms that his interests musically were rooted in 20th Century American experimental music and harmonic theories. There were a number of composers studying these harmonic systems and they used these grids to determine the harmonic ratios. These grids and were exactly what he was working on geometrically, and realised that if he assigned values to each individual piece of this geometric form, then he could apply the same harmonic principles. [8]

"From that system, it becomes something else. There's nothing arbitrary about the way a plant grows or the way seeds are distributed in an apple. They make numbers much more real to me. The world is rooted in numbers and there are a lot of recurring forms.[...] All of us started off as a single cell that became two cells and then four and eight and it keeps branching out. This process is as much rooted in my body and my mind as it is in all of us. There is this creative principle at the root of everything. In our material world you can lose sight of the fact that we are part of process that's been going on for billions of years. [8]

Fig. 3. Andy Gilmore, Creative commons

If for Bach music was the sound of maths, for the artist Andy Gilmore, graphics are the geometry of music. Chromatic vortices capable of hypnotising the eye and bewitching it with deep digital notes.

The constant creative oscillation between graphic design, illustration and music has harmonised to produce complex digital graphics, that are ultramodern yet at the same time retro, a mix of genres that are synthesised in an original abstract vision. [4]

The process to create his artworks is rather simple, with the inspiration of patterns and numbers, bringing with awareness those relationships and the numbers that exist in nature into his work. [11] Working many hours a day in a dimly lit studio, "Generally, I build a form/pattern in Illustrator line by line, piece by piece—values are then assigned across the form to create the illusion of depth and movement. I essentially then use Photoshop to develop the image as if it were film. This all can take days or hours [...] Overall, it's a very Gestalt approach to image making—in that I just allow the suggestion of form to guide me along."

Fig. 4. Andy Gilmore, Creative commons

As stated in a previous paper in collaboration with E. Conversano, M. Francaviglia, and L. Tedeschini Lalli, "Our mind tends to recognize shapes and forms in the world: perception is transformed into proportions and geometrical shapes, that artists and architects have continuously used to produce their artworks, from cavern paintings to modern Art. Forms have been recognized in Nature, used because of their aesthetic value or functionality. They have been first understood at the emotional level and later elaborated more consciously. Some of them, more easily reproducible and recognizable, have pervaded the visual symbolic communication and have been linked to Semantics". [12]

Fig. 5. Andy Gilmore, Creative commons

Composed primarily of bold and bright colors on black backgrounds, his geometric compositions are a perfect synchronisation of form and colour. [22]

Playing with levels, blend modes and masks, he often exploits also moiré patterns.

"In mathematics, physics, and art, a moiré pattern (/mwɑːrˈeɪ/; French: [mwaˈʁe]) or moiré fringes are large scale interference patterns that can be produced when an opaque ruled pattern with transparent gaps is overlaid on another similar pattern. For the moiré interference pattern to appear, the two patterns are not completely identical, they must be displaced, rotated, etc. or have different, but similar pitch. Moiré patterns appear in many different situations. In printing, the printed pattern of dots can negatively interfere with the image. In television and digital photography, a pattern on an object being photographed can interfere with the shape of the light sensors to generate unwanted artifacts". [33]

A video looking into the artist's creative process was released by the design house / online store Ghostly International, with which Gilmore has collaborated creating album covers for artists releasing music through the brand's record label.

"The scope of Gilmore's compositions is very broad, ranging from soothing, almost-kaleidoscopic geometric patterns to bold, fractal-inspired loops. Although often inspired by patterns both in nature or urban environments, there's a certain rhythm prevalent throughout his work; Ghostly International fittingly describes it as "visually acoustic." Whatever you make of his creations, watching Gilmore work while listening to his rationale is a rare treat." [28]

Fig. 6 a.-6 b. Andy Gilmore, Creative commons

3 Art for visualization of complexity and physics

In order to understand any complex system, it becomes vital to understand its most basic components. Theoretically, anything can be broken down into a set of discrete smaller parts, whose predictable behavior will then help inform us as to the nature of the whole. This idea is the foundation of Western thought and most, if not all, progress since the ancient Greeks. In the physical sciences, these parts are particles and forces. In the social sciences they are ideas and people. And lastly, and perhaps most strangely, in math and art they are the point and line. These two fields have only recently begun a beautiful courtship empowered by technology. Both rely on collections of points forming lines and shapes, numbers and functions being interchangeably writing or represented as geometry. Likewise in both, the simple components are recombined and elaborated upon to create new works. The simplest component to have shape is the polygon, making them the fundamental building brick of all art (pointillism aside) but also truly inspiring.

"The thing that humans most hate about complexity is that it's infinite: it can never be fully understood. And because of that, there will always be limits — serious limits — to the models of reality that scientists and philosophers put forward. And similar limits on our ability to control or even intervene in any reliable, useful way in these systems." [24]

The newest model of reality is one that comes closer to previous models to embracing complexity and unknowingness instead of fighting to simplify and fully understand it. It's a geometric model, and it's seriously shaking up the scientific establishment, because it does away with the need for the concepts of space and time entirely, seeing them as human mental constructs that are not essential to the nature of reality at all.

Fig. 7. Andy Gilmore, Creative commons

Andy Gilmore was recently engaged to use his art to represent models of the universe.
For example, he created visuals for the New Scientist and their 'Entangled Universe' feature: "Entangled universe: Could wormholes hold the cosmos together? Weird connections through space-time might make reality real, giving us a promising new route to a theory of everything." [10].

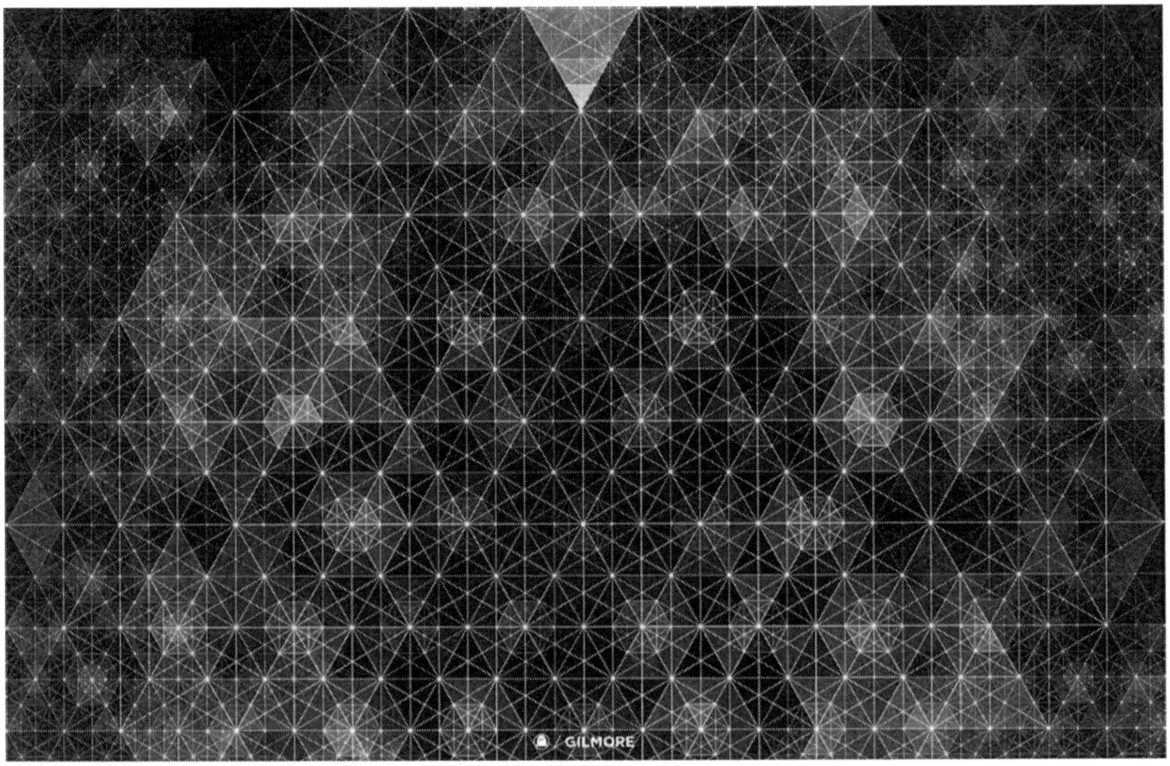

Fig. 8. Andy Gilmore, Creative commons

In addition, he made a rendering of an *amplituhedron*, a jewel-shape new geometric model of reality. It's a model that reflects the complexity of the universe, appreciates it, rather than trying to explain and understand it. "Indeed, the problematic issues of black holes and the big bang and the impossibility of testing the newest theories of subatomic particles are rendered by this model, not solved by it. They are seen for what they are — errors in our simplistic human modelling of reality that are inherent in the very scientific language and tools we use to create them."

Here's a quote from Nima Arkani-Hamed, a professor at the Institute for Advanced Study, one of the new model's developers, talking about how "change" is represented in this new model:

"In a sense, we would see that change arises from the structure of the object," he said. "But it's not from the object changing. The object is basically timeless." Change is emergent, in other words, and emergence is the closest our language can come to expressing the concept of complexity. [1]

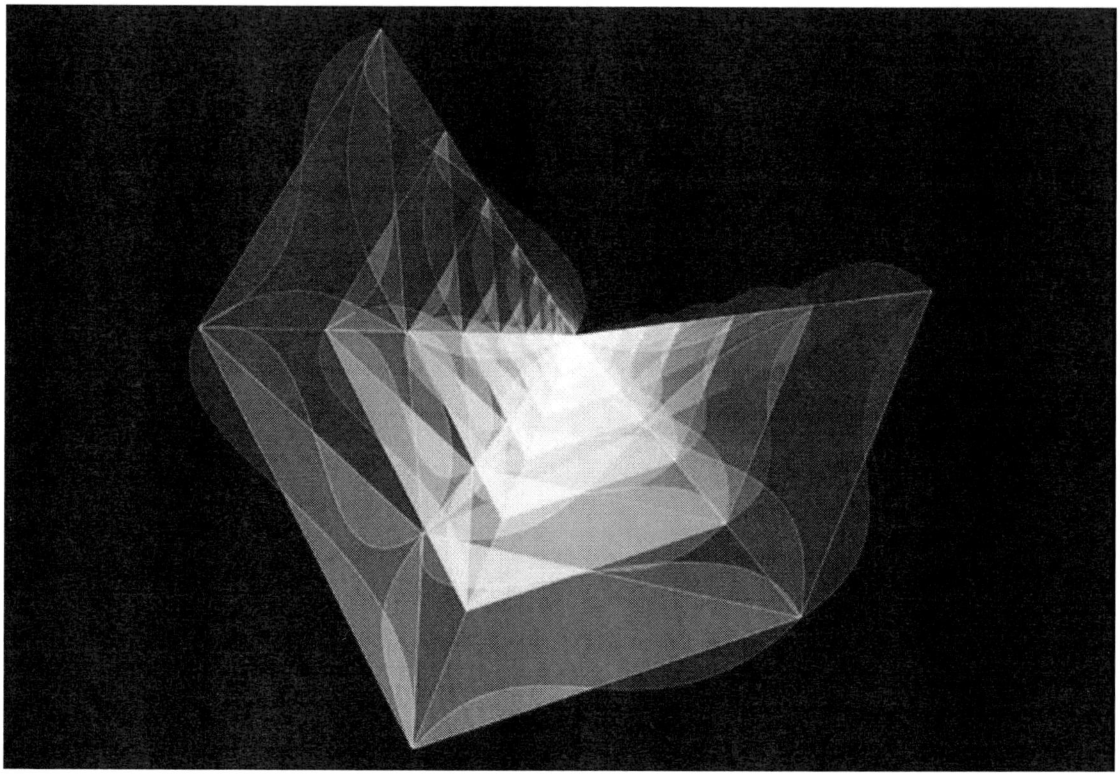

Fig. 9. Andy Gilmore, *amplituhedron*

4 Conclusions

More than a participant in the rising trend of digital art, Andy Gilmore is an innovator, dreaming up works that, without computers, could not exist. "The beauty of such simple shapes is endless. Their abstract form allows for infinite possibilities; complexity or simplicity. Andy Gilmore leverages this unique ability in his kaleidoscopic works. It's easy to simply appreciate them as wonderful pieces in and of themselves but the more I stare the more I begin to see shapes appear, like cloud watching. His sketches are even more fascinating as they reveal a pointillist nature to his method and a very precise control of symmetry".[6]

It may not look simple, but the colors had to be carefully placed in the image. A great artists makes their medium look easy. It doesn't matter if they use paint, paper, cement, clay, marble, or keyboards and mouses. It's the art of technology. This is perfect for this day and age. Art should also depict an era. However these images are created, they are beautiful. They force one to ask oneself what is art, what is decoration, and what is the difference?

The artist either has a deep understanding of color, or has a mathematical algorithm that captures the essence of color. Artists have ever sought that deep understanding by any means available to them. [2] Finally, his ability to create artworks served to render also new models of reality.

As Albert Einstein stated in Principles of research: "The Artist, but also the Scientist, substitute the empirical world with a self-created one, aiming to transcendence".

Acknowledgements

This paper was written under the auspices of the INDAM – GNASAGA research group.

This paper is dedicated to the memory of Prof. Mauro Francaviglia, co-organizer of the Math&Art section from the beginning. I wish to thank Daniela Velichova and Monica Kovachova and all the staff of Aplimat and all the participants to this session.

I'm very grateful to Alessandra Capanna for her precious help for the entire conference section process and her heartily support.

Last but not least, I thank from the bottom of my heart Kristof Fenyvesi, and his collaborators, for his wonderful Bridges Math Art and Experience Workshop exhibition in memoriam of Mauro Francaviglia.

References

[1] ANANTHASWAMY A., Entangled universe: Could wormholes hold the cosmos together? https://www.newscientist.com/article/mg22830460-400-entangled-universe-could-wormholes-hold-the-cosmos-together/ 4 November 2015 © Copyright Reed Business Information Ltd.

[2] Andy Gilmore, Http://Artistaday.com, http://www.agilmore.com.

[3] Andy Gilmore "Black Math" a futuristic, almost alien feel of graphic works, http://www.shearyadi.com/myworld/andy-gilmore-black-math-a-futuristic-almost-alien-feel-of-graphic-works/.

[4] Andy Gilmore's musical geometries _ Lancia Trendvisions, https://s3-eu-west-1.amazonaws.com/adv-vault/upload/lancia/musica-per-i-vostri-occhi/LTVs-ANDY-GILMORE-02.jpg.

[5] ASHTON, D., Contrasts Of Form: Geometric Abstract Art, 1910-1980. Arts Magazine 60.(1985): 18-20. Art Full Text (H.W. Wilson). Web. 2 Sept. 2013.

[6] BENCHIRLIN, The Maya Method, in The Sketching Mechanism http://www.themechanism.com/voice/tag/andy-gilmore/.

[7] BRAY C., Andy Gilmore, http://www.peopleofprint.com/solo-artist/andy-gilmore/ April 22, 2014.

[8] BREED, A conversation between Rob Alderson and Andy Gilmore about his eclectic influences, from music to nature. http://www.breedlondon.com/rob-alderson-andy-gilmore/ © 2015 - 2016 Breed London.

[9] BREED, Andy Gilmore, http://www.breedlondon.com/artists/andy-gilmore/ © 2015 - 2016 Breed London.

[10] BREED, New Scientist, Andy Gilmore creates visuals for the New Scientist and their 'Entangled Universe' feature. http://www.breedlondon.com/new-scientist-andygilmore/© 2015 - 2016 Breed London.

[11] CARIDAD P., Andy Gilmore, A Visual Adventure In Geometry and Light http://www.theghostlystore.com/collections/andy-gilmore https://www.visualnews.com/2011/08/12/a-visual-adventure-in-geometry-and-light/ August 12, 2011.

[12] CONVERSANO, E., FRANCAVIGLIA, M., LORENZI, M. G., TEDESCHINI LALLI L., (2012) Geometric Forms that Persist in Art and Architecture in Bridges 2011 pp 463–466.
[13] CROW D., Visible Signs (Second Edition): An Introduction to Semiotics in the Visual Arts.
[14] DI SANZO, A., Ghostly International Presents: Andy Gilmore. Opere psichedeliche per una vita immersa nell'arte. http://www.goodshortfilms.it/en/time/0-5/ghostly-international-presents-andy-gilmore.
[15] DOBLE R., LORENZI M. G., FRANCAVIGLIA M., Experimenting & Beauty. Scientific Method & Insight in Mathematics and Photography, Bridges math art conference, 2012.
[16] DYLAN D., Sacred Geometry – The Art of Andy Glimore, http://www.mindcontrolindustries.com/geometric-wizardry-the-art-of-andy-glimore/ November 3, 2013.
[17] FRANCO, A. M. , Geometric Abstraction, American Art 26.2 (2012): 34-41. Art Full Text (H.W. Wilson). Web. 2 Sept. 2013.
[18] FULLEYLOVE R., Andy Gilmore: Activities of the Inanimate, http://www.itsnicethat.com/articles/andy-gilmore-2, 09 May 2012.
[19] GHOSTLY INTERNATIONAL, Andy Gilmore, http://ghostly.com/artists.
[19] GRANDE J. K., Art Nature Dialogues: Interviews with Environmental Artists. Albany: State University of New York Press, 2004.
[20] HATCHELL J., What is a 'standard candle'?
[21] LORENZI M. G., FRANCAVIGLIA M., The role of mathematics in contemporary art at the turn of the millennium. Aplimat 2011, Journal of applied mathematics, Vol. IV, N. IV, 2011.
[22] MELT POINCARÉ, Andy Gilmore, http://visualmelt.com/Andy-Gilmore.
[23] ONARTO, Artist Spotlight: Andy Gilmore, https://onarto.com/andy-gilmore-geometric-kaleidoscopic-artwork/.
[24] POLLARD D., Embracing Complexity: Seeking Appreciation Instead of 'Understanding' http://howtosavetheworld.ca/2013/10/08/embracing-complexity-seeking-appreciation-instead-of-understanding/ October 8, 2013.
[25] PONZELLI, M., Le caleidoscopiche geometrie di Andy Gilmore, http://www.organiconcrete.com/2013/01/18/le-caleidoscopiche-geometrie-di-andy-gilmore/ 18 gennaio 2013.
[26] POTTAGE J., Geometrical Investigations: Illustrating the Art of Discovery in the Mathematical Field. Reading, Mass: Addison-Wesley, 1983.
[27] SALA, N., CAPPELLATO, G., (2004), Architetture della complessità. La geometria frattale tra arte, architettura e territorio, Franco Angeli, Milano.
[28] SOUPPOURIS A., Seeing with geometry: inside the mind of Andy Gilmore, The Verge http://www.theverge.com/2013/8/13/4617076/andy-gilmore-ghostly-international-video August 13, 2013.
[29] SPACECOLLECTIVE, Andy Gilmore's project, http://spacecollective.org © 2006-2016, American Scenes Inc.
[30] THE GHOSTLY STORE BLOG, In Process with Andy Gilmore, http://blog.ghostly.com/post/20957306212/in-process-with-andy-gilmore
[31] The Work of Andy Gilmore http://thetripatorium.com/pictures/detail/the_work_of_andy_gilmore

[32] VOGT P., Andy Gilmore, http://www.witness-this.com/art/andy-gilmore January 21, 2015

[33] Wikimedia, Moiré

Current addresses

Lorenzi Marcella Giulia, Dr. PhD
University of Calabria - CEL
Via Savinio, Ed. Polifunzionale, 87036 Arcavacata di Rende (CS), Italy
Tel. +39 (0)984-49.3440, e-mail: marcella.lorenzi@unical.it

Slovak University of Technology in Bratislava
Faculty of Mechanical Engineering
16th Conference on Applied Mathematics **APLIMAT 2017**
Proceedings

COMPARISON OF DIFFERENT WAYS HOW TO DETERMINE THE OPTIMAL NUMBER OF CLUSTERS IN CLUSTER ANALYSIS

LÖSTER Tomáš (CZ)

Abstract. Cluster analysis is a multivariate statistical method which main aim is to classify objects into groups called clusters. Classification of objects into groups is a natural requirement of various scientific disciplines. There are a lot of methods which can be used to classification of objects into clusters, in the current scientific literature. The main aim of this article is to compare selected ways, which can be used to determine the optimal number of clusters. We used generated files for evaluation of coefficients. These files were generated in different conditions. It can be said, that CHF coefficient is one of the best coefficient for finding the optimal number of clusters. Euclidean distance measure is more useful in different conditions in comparison with Mahalanobis distance measure in clustering process.

Keywords: clustering, evaluating of clustering, methods, coefficients for determining the optimal number of clusters, distance measures

Mathematics Subject Classification: Primary 62H30; Secondary 91C20

1 Introduction

The basic issue of many scientific disciplines is the modernly so called taxonomy: to organize objects in groups. They may be represented by customers, patients, cars, documents, etc. To create groups there could be used different statistical methods and procedures. If the object (observation) is included into the existing group, there is used the discrimination analysis, if the object is classified into classes that may not be known in advance, there is used the cluster analysis. This is a multivariate statistical method that aim is to create groups of objects, which are called clusters. The objective of cluster analysis is to find objects within one cluster that are as similar as possible and objects from two different clusters should be as least similar as possible. Cluster analysis is very often used statistical method, see e.g. [3], [4], [8], [11], [13]. Very often it is used to regions classification. Authors very often used wages to describe regions. The problem of wages and poverty is described e.g. in [1], [2], [16]. Other demographic variables, which are very often considered in cluster analysis, are described in [9], [10], [14], etc. In the current scientific literature, there are many cluster analysis methods that are used to classify objects. These methods and procedures can be categorized according to various criteria see e.g. [3], [12].They can be divided into hierarchical and non-hierarchical. They can also be divided into traditional or new approaches.

Among best known methods we can include, for example, the nearest neighbor method, furthest neighbor method, average method, centroid method or Ward's method. Detailed description of these methods can be found, for example in [3]. Methods differ, apart from other facts, in the time of formation and also in the way of clustering. Some of them minimize the distance between objects, which means the distance among the closest and furthermost objects. Ward's method solves the principle of clustering based on minimizing the intra-group variability, see [3]. Clustering methods may be combined with various measures of distances. The most famous include Euclidean distance measure or Mahalanobis distance measure, see for example [3]. The result of these combinations is a large number of options that can be used to classify objects.

Part of the cluster analysis is very often the number of clusters determination in which the objects may be distributed. To determine the number of clusters (groups of objects), there is a number of criteria and procedures. As noted above, various techniques for clustering may involve a different distribution of objects into clusters. The issues of determining the number of clusters, in the case of quantitative variables, are the subject of many scientific papers, including for example [4], [5], [6], [7], etc. However, in the current literature there is no clear rule to determine the use of concrete coefficients in different conditions. Various authors create and modify their coefficients and sometimes the coefficients are compared with the existing ones. Among the selected criteria for determining the number of clusters can be included Davies-Bouldin's (DB) index, Dunn's index, RMSSTD index, CHF index and PTS index. These coefficients are already applied e.g. in SYSTAT software and they are commonly used for determining the number of clusters.

For clustering can be used a lot of software products. Among the best-known and most frequently used can be included e.g. IBM SPSS, SAS, STATISTICA, S-PLUS, SYSTAT, STATGRAPHICS, etc. In these are implemented especially the traditional clustering methods, including possible methods of decomposition. Some of them, such as the SAS system do not allow to user to select the appropriate combination of clustering methods and distance measures. The system offers a combination like that. When determining the number of clusters can be used only selected software products. But only the selected coefficients are there implemented. Another widespread software products, such as STATISTICA or STATGRAPHICS does not contain coefficient for determining the number of clusters. Tab. 1 summarizes the evaluation and the location of selected coefficients listed above for determining the number of clusters.

Coefficient	Founded extreme	Software
CHF index	maximum	SAS system LE, SYSTAT
PTS index	maximum	SAS system LE, SYSTAT
RMSSTD	minimum	SYSTAT
Davies-Bouldin (DB)	minimum	SYSTAT
Dunn's index	maximum	SYSTAT

Tab. 1. Overview of selected coefficients in software products

2 Evaluation of coefficients for determining the number of clusters

For the evaluation of a selected coefficient was necessary to generate a sufficient amount of data files that satisfy the same conditions, in order to the objective evaluation of the ability of the coefficient for the determination of optimal number of clusters. The input parameters of the generator are: the number of objects in each of the clusters (equal in all clusters), the number of generated clusters, the number of variables which characterize the objects, variability within the cluster and separation of clusters. The generator principle is as follows: firstly, in random *m*-dimensional space (according to number of selected variables) place the centroids of *k* clusters. Next, on the basis of the separation of clusters (the distance between the centroids of clusters) and on the basis of variability within the clusters, place randomly the selected numbers of objects around the centroids of clusters. For the purpose of examining selected coefficients there were created three groups of files.

Group 1

The clusters in this group of files are well separated. Within this group of files were generated 2,3,4,5,10,20 clusters. For every number of clusters were genereted 10 files, i.e. the total number in this group is 60 data files. Tab. 2 shows the number of correctly set clusters (in %) at each of the methods using Eclidean distance measure. It is evident that the succes rate (in %) of RMSSTD, CHF, D-B an Dunn coefficient is 100 %.

Method/coefficient	RMSSTD	CHF	PTS	D-B	Dunn
Nearest neighbour	100,00	100,00	60,00	100,00	100,00
Farthest neighbour	100,00	100,00	48,33	100,00	100,00
Centroid method	100,00	100,00	60,00	100,00	100,00
Average distance	100,00	100,00	53,33	100,00	100,00
Ward's method	100,00	100,00	48,33	100,00	100,00

Tab. 2. Number of correctly set clusters (in %) – Euclidean distance measure

Table 3 shows the number of correctly set clusters (in %) at each of the method using Mahalanobis distance measure. It is evident that succes rate of all coefficients is lower then 100 %.

Method/coefficient	RMSSTD	CHF	PTS	D-B	Dunn
Nearest neighbour	83,33	98,33	60,00	65,00	96,67
Farthest neighbour	75,00	65,00	18,33	65,00	63,33
Centroid method	73,33	73,33	48,33	75,00	85,00
Average distance	75,00	83,33	63,33	65,00	95,00
Ward's method	68,33	75,00	48,33	65,00	95,00

Tab. 3. Number of correctly set clusters (in %) – Mahalanobis distance measure

Table 4 shows comparisons of success rate (in %) of Euclidean and Mahalanobis distance measure in clustering. All values were calculated as the difference between success rate of Euclidean and Mahalanobis distance measure. It is evident that success rate is higher for almost all coefficients when using Euclidean distance measure.

Method/coefficient	RMSSTD	CHF	PTS	D-B	Dunn
Nearest neighbour	16,67	1,67	0,00	35,00	3,33
Farthest neighbour	25,00	35,00	30,00	35,00	36,67
Centroid method	26,67	26,67	11,67	25,00	15,00
Average distance	25,00	16,67	-10,00	35,00	5,00
Ward's method	31,67	25,00	0,00	35,00	5,00

Tab. 4. Difference in the success rate (in %) of Euclidean and Mahalanobis distances

Group 2

For this group of files is characteristic, that clusters are touched but not significantly overlapped. Within this group of files were generated three clusters. Total number is 20 data files in this group. Table 5 shows the number of correctly set clusters (in %) at each of the methods using Euclidean distance measure.

Method/coefficient	RMSSTD	CHF	PTS	D-B	Dunn
Nearest neighbour	70,00	70,00	65,00	70,00	55,00
Farthest neighbour	75,00	85,00	60,00	60,00	55,00
Centroid method	75,00	75,00	55,00	50,00	60,00
Average distance	75,00	85,00	55,00	60,00	70,00
Ward's method	75,00	80,00	50,00	60,00	50,00

Tab. 5. Number of correctly set clusters (in %) – Euclidean distance measure

Table 6 shows the number of correctly set clusters (in %) at each of the methods using Mahalanobis distance measure.

Method/coefficient	RMSSTD	CHF	PTS	D-B	Dunn
Nearest neighbour	40,00	55,00	55,00	5,00	20,00
Farthest neighbour	10,00	20,00	20,00	0,00	5,00
Centroid method	10,00	15,00	15,00	15,00	20,00
Average distance	55,00	80,00	65,00	30,00	40,00
Ward's method	50,00	70,00	80,00	50,00	35,00

Tab. 6. Number of correctly set clusters (in %) – Mahalanobis distance measure

Table 7 shows comparisons of success rate (in %) of Euclidean and Mahalanobis distance measure in clustering. All values were calculated as the difference between success rate of Euclidean and Mahalanobis distance measure. It is evident that success rate is higher for almost of coefficients in using Euclidean distance measure.

Method/coefficient	RMSSTD	CHF	PTS	D-B	Dunn
Nearest neighbour	30,00	15,00	10,00	65,00	35,00
Farthest neighbour	65,00	65,00	40,00	60,00	50,00
Centroid method	65,00	60,00	40,00	35,00	40,00
Average distance	20,00	5,00	-10,00	30,00	30,00
Ward's method	25,00	10,00	-30,00	10,00	15,00

Tab. 7. Difference in the success rate (in %) of Euclidean and Mahalanobis distances

Group 3

Within this group of files were generated 20 data files. There were generated three clusters in each data files. The clusters in this group of files may be partially or significantly overlapped. Tab. 8 shows the number of correctly set clusters (in %) at each of the methods using Euclidean distance measure. It is evident that success rate (in %) is very low.

Method/coefficient	RMSSTD	CHF	PTS	D-B	Dunn
Nearest neighbour	30,00	10,00	0,00	10,00	0,00
Farthest neighbour	50,00	0,00	0,00	0,00	0,00
Centroid method	10,00	0,00	0,00	20,00	20,00
Average distance	10,00	10,00	0,00	0,00	10,00
Ward's method	10,00	20,00	0,00	10,00	10,00

Tab. 8. Number of correctly set clusters (in %) – Euclidean distance measure

Table 9 shows the number of correctly set clusters (in %) at each of the methods using Mahalanobis distance measure.

Method/coefficient	RMSSTD	CHF	PTS	D-B	Dunn
Nearest neighbour	20,00	10,00	20,00	0,00	0,00
Farthest neighbour	40,00	40,00	40,00	30,00	0,00
Centroid method	30,00	0,00	0,00	0,00	0,00
Average distance	0,00	50,00	10,00	20,00	0,00
Ward's method	0,00	50,00	40,00	60,00	20,00

Tab. 9. Number of correctly set clusters (in %) – Mahalanobis distance measure

Table 10 shows comparisons of Euclidean and Mahalanobis distance measure in clustering. All values were calculated as difference between success rate of Euclidean and Mahalanobis distance measure.

Method/coefficient	RMSSTD	CHF	PTS	D-B	Dunn
Nearest neighbour	10,00	0,00	-20,00	10,00	0,00
Farthest neighbour	10,00	-40,00	-40,00	-30,00	0,00
Centroid method	-20,00	0,00	0,00	20,00	20,00
Average distance	10,00	-40,00	-10,00	-20,00	10,00
Ward's method	10,00	-30,00	-40,00	-50,00	-10,00

Tab. 10. Difference in the success rate (in %) of Euclidean and Mahalanobis distances

3 Conclusion

Cluster analysis is a multivariate statistical method which aim is to classify objects into groups called clusters. The objective of cluster analysis is to find objects within one cluster that are as similar as possible and objects from two different clusters should be as least similar as possible. The output of cluster analysis very often contains the number of clusters into which the objects are classified. This information is not known in advance in most real cases. To determine the optimal number of clusters there are many ways (coefficients) that again can be combined with different methods and different measures of distances.

Based on the results of three groups of files were compiled conclusins to Tab. 11, which contains average succes rate (in %) for Euclidean distance.

Method/coefficient	RMSSTD	CHF	PTS	D-B	Dunn
Nearest neighbour	66,67	60,00	41,67	60,00	51,67
Farthest neighbour	75,00	61,67	36,11	53,33	51,67
Centroid method	61,67	58,33	38,33	56,67	60,00
Average distance	61,67	65,00	36,11	53,33	60,00
Ward's method	61,67	66,67	32,78	56,67	53,33

Tab. 11. Average success rate (in %) – Euclidean distance measure

The average succes rate (in %) for Mahalanobis distance measure is in Tab. 12.

Method/coefficient	RMSSTD	CHF	PTS	D-B	Dunn
Nearest neighbour	47,78	54,44	45,00	23,33	38,89
Farthest neighbour	41,67	41,67	26,11	31,67	22,78
Centroid method	37,78	29,44	21,11	30,00	35,00
Average distance	43,33	71,11	46,11	38,33	45,00
Ward's method	39,44	65,00	56,11	58,33	50,00

Tab. 12. Average success rate (in %) – Mahalanobis distance measure

In Tab. 13 are displayed differences between average succes rate (in %) for both distances. It is evident that success rate is higher for almost of coefficients in using of Euclidean distance measure.

Method/coefficient	RMSSTD	CHF	PTS	D-B	Dunn
Nearest neighbour	18,89	5,56	-3,33	36,67	12,78
Farthest neighbour	33,33	20,00	10,00	21,67	28,89
Centroid method	23,89	28,89	17,22	26,67	25,00
Average distance	18,33	-6,11	-10,00	15,00	15,00
Ward's method	22,22	1,67	-23,33	-1,67	3,33

Tab. 13. Difference Average success rate (in %) of Euclidean and Mahalanobis distances

Graphical comparison of differences between success rate of Euclidean and Mahalanobis distance measure is evident from the Fig. 1.

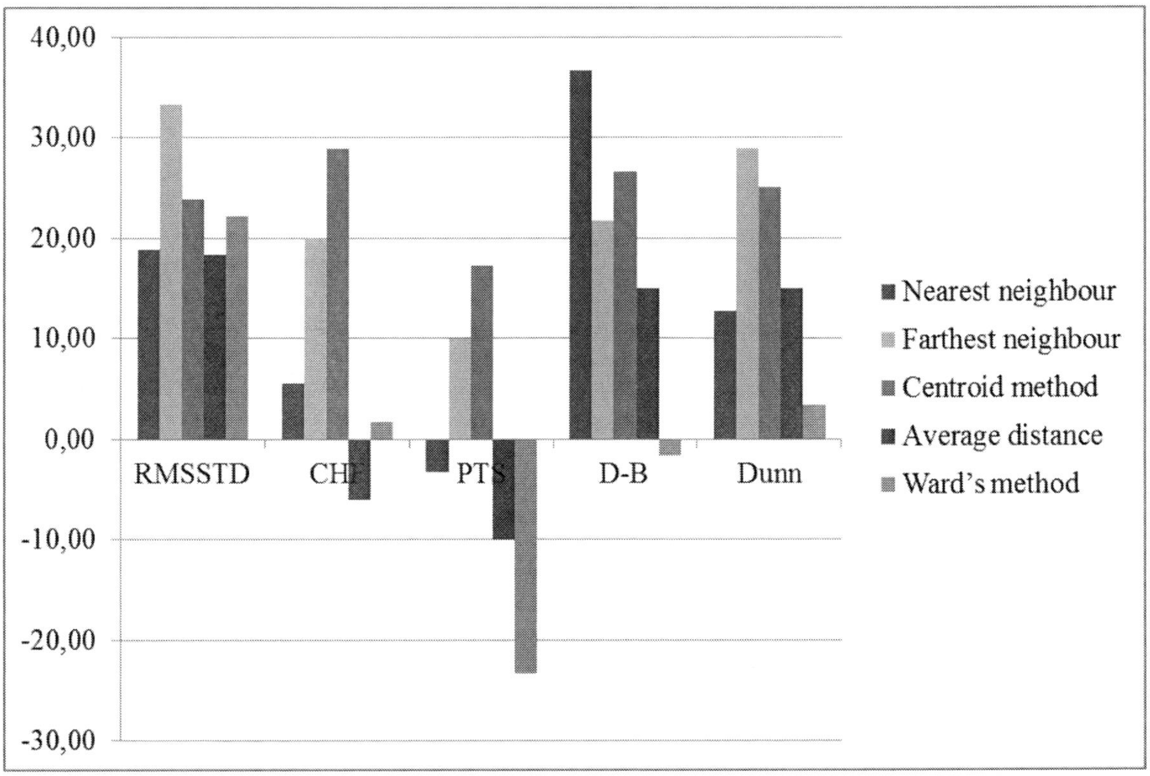

Fig. 1. Comparison of differences between success rate of Euclidean and Mahalanobis distance (Eucledian – Mahalanobis)

In conclusion we can say, that the CHF coefficient is more successful in determining the number of clusters compared with the other coefficients, which are noted e.g. in [6]. The more are the individual clusters overlapped, (i.e. that the separation rate decreases), the less its success rate. Best results are achieved in connection with the average distance method, and Ward's method. The success rate in use of other methods is low. The lowest success of CHF coefficient was achieved in connection with centroid method. When we compare the CHF coefficient in connection with the squares of the Euclidean distance measures at the same group of files, it is obvious that the use of square Euclidean distances extent always leads to a higher success of this coefficient, rather than using Mahalanobis distance measure.

Acknowledgement

This paper was processed with contribution of long term institutional support of research activities number IP400040 by Faculty of Informatics and Statistics, University of Economics, Prague, Czech Republic.

References

[1] BÍLKOVA, D. *Modelling of income and wage distribution using the method of l-moments of parameter estimation* . In Loster Tomas, Pavelka Tomas (Eds.), *International Days of Statistics and Economics* (pp. 40-50), 2011, ISBN 978-80-86175-77-5.

[2] BÍLKOVA, D. *Development of wage distribution of the czech republic in recent years by highest education attainment and forecasts for 2011 and 2012*. In Loster Tomas, Pavelka Tomas (Eds.), *6th International Days of Statistics and Economics* (pp. 162-182), 2012, ISBN 978-80-86175-86-7.

[3] GAN, G., MA CH., WU J.: *Data Clustering Theory, Algorithms, and Applications*, ASA, 2007, Philadelphia.

[4] HALKIDI, M., VAZIRGIANNIS, M.: *Clustering validity assessment: Finding the optimal partitioning of a data set*, Proceedings of the IEEE international conference on data mining, 2001, (pp. 187-194).

[5] LÖSTER, T. *The Evaluation of CHF coefficient in determining the number of clusters using Euclidean distance measure*. In: The 8th International Days of Statistics and Economics., 2014, (pp. 858–869), ISBN 978-80-87990-02-5.

[6] LÖSTER, T. *The Evaluation of CHF coefficient in determining the number of clusters using Mahalanobis distance measure*. In: *14th Conference on Applied Mathematics – Aplimat 2015 [CD]*. Bratislava, 03.02.2015 – 05.02.2015. Bratislava : Slovak University of Technology, 2015, s. 546–554. ISBN 978-80-227-4314-3.

[7] LÖSTER, Tomáš. Evaluation of Coefficients for Determining the Optimal Number of Clusters in Cluster Analysis on Real Data Sets. In: The 9th International Days of Statistics and Economics (MSED 2015) [online]. Praha, 10.09.2015 – 12.09.2015. Slaný : Melandrium, 2015, s. 1014–1023. ISBN 978-80-87990-06-3.

[8] LÖSTER, Tomáš. Evaluation of Coefficients for Determining the Optimal Number of Clusters in Cluster Analysis on Real Data Sets. In: *The 9th International Days of Statistics and Economics (MSED 2015)*. [online] Praha, 10.09.2015 – 12.09.2015. Slaný : Melandrium, 2015, s. 1014–1023. ISBN 978-80-87990-06-3.

[9] MAKHALOVÁ, E. 2015. The Fuzzy Clustering Problems and Possible Solutions *The 9th International Days of Statistics and Economics*. Praha, 2015. s. 1052–1061. ISBN 978-80-87990-06-3.

[10] MEGYESIOVA, S., & LIESKOVSKA, V. *Recent population change in europe*. In Loster Tomas, Pavelka Tomas (Eds.), *International Days of Statistics and Economics*, (pp. 381-389), 2011, ISBN 978-80-86175-77-5.

[11] MELOUN, M., MILITKÝ, J., HILL, M.: Počítačová analýza vícerozměrných dat v příkladech, Academia, 2005, Praha.

[12] ŘEZANKOVA, H., HÚSEK, D., SNÁŠEL, V.: *Shluková analýza dat,* 2. vydání, Professional Publishing, 2009, Praha.
[13] ŘEZANKOVA, H., & LOSTER, T. Shlukova analyza domacnosti charakterizovanych kategorialnimi ukazateli. *E+M. Ekonomie a Management, 16*(3), (pp 139-147), 2013, ISSN: 1212-3609.
[14] STANKOVIČOVÁ, I., VOJTKOVÁ, M.: Viacrozmerné štatistické metódy s aplikáciami, Ekonómia, 2007, Bratislava.

Current address

Löster Tomáš, Ing. Ph.D.
University of Economics, Prague, Dept. of Statistics and Probability
W. Churchill sq. 4, 130 67 Prague 3, Czech Republic
E-mail: tomas.loster@vse.cz

Slovak University of Technology in Bratislava
Faculty of Mechanical Engineering
16th Conference on Applied Mathematics **APLIMAT 2017**
Proceedings

MODELING OF TURBULENT WIND FLOW AROUND HIGH-RISE BUILDING VERIFIED BY RESULTS FROM TEST IN WIND TUNNEL LABORATORY

MACAK Marek (SK), HUBOVA Olga (SK), KONECNA Lenka (SK)

Abstract. The main purpose of this paper is determination of the accuracy of CFD simulation of turbulent wind flow around high-rise building. The external wind pressures coefficients obtained by calculation model were compared with the results from tests carried out in SvF STU Wind Tunnel Laboratory. Short description of the wind tunnel laboratory, experimental measurements, and used measuring devices are also mentioned. However, the largest space of this paper is devoted to created 3D calculation model and used method for its solution.

Keywords: CFD simulation, turbulent wind flow, experimental measurements, BLWT, high-rise building, external wind pressures

Mathematics Subject Classification: Primary 76-05, 76-04; Secondary 35Q35

1 Introduction

Computer simulation is one from the three possibilities how to determine the influence of wind and its effects on investigated objects. Especially, this method is suitable for the objects with atypical shapes, where required information about the wind effects could not be found in valid standards [1]. This method is faster and cheaper than experimental measurements made in wind tunnel laboratory or in-situ measurements. However, it is very important to use right input parameters describing the wind flow (laminar or turbulent) around the object. Therefore, it can be said that this method gives "approximate results". In the case of atypical structures or arrangement of several structures, the best solution is using the combination of both methods, i.e. CFD simulation and the tests in wind tunnel, [2].

As it was mentioned above, experimental measurements are technically and financially demanding [3]. Therefore, the main purpose of this paper was to create 3D model of atypical structure with sufficient accuracy, which could be used for next steps in our research. In this case, the high-rise building with triangular cross-section was chosen. The photos of similar buildings designed by architects all over the world are depicted in Fig. 1.

Fig. 1. High-rise buildings with triangular cross-sections – a) high-rise building in Chicago [4], b) Flatrion building in New York [5], c) Lincoln Plaza in Dallas [6], d) Panorama City in Bratislava [7]

2 Description of the Investigated High-Rise Building

The cross-section of the analysed high-rise building had the shape of equilateral triangle with the length of the edge 52.2 m and total height equalled to 114.3 m (34-storey building with considered structural height 3.2 m for the typical floor and 4.5 m for the ground floor). The ground plan of the structure was the same in all levels. At the top of the structure, the flat roof was designed. In this case, the building was located in large city, such as Bratislava. This type of city can be classified by the terrain category between III and IV (closely to IV).

3 Wind Tunnel Laboratory, Created 3D Model of Structure, Used Measuring Devices

Experimental measurements were done in SvF STU Wind Tunnel Laboratory in Bratislava. It is classified as Boundary Layer Wind Tunnel with opened circuit scheme. There are two operating spaces used for the investigation of the uniform wind flow (with the fluctuation less than 5 %) and turbulent wind flow. It is universal wind tunnel, which could be used due to its parameters (its dimensions, parameters of the wind flow) in various branches, i.e. civil and mechanical engineering, chemical industry, ecology, etc. It was confirmed by set of the tests, that this laboratory provides the results with the same accuracy as other well-known laboratories. More information about this wind tunnel laboratory can be found in [8-10].

3D model of structure in scale 1 : 300 was created from plexiglass. The length of edge was 17.3 cm and total height of the model was 38.1 cm. Pressure taps were located in three levels – in the height of 36.8 cm (A), 20 cm (B) and 3 cm (C), see Fig. 2a. There were16 pressure taps in each measured level. They were located only on two edges (nos. 1-7 and nos. 8-16) because of the symmetry of the model. The model was placed on the rotating table which was used for the simulating of various wind directions (from 0° to 120°, with the angle of rotation 15°). All measurements were repeated for 4 various wind velocities, see Table 1. Reference velocities were measured in total height of the model (without the model) by ALMEMO probe. Average temperature during the tests was 24.1°C, barometric pressure was 99 800 Pa and air density was 1.16457 kg/m^3.

16-channel pressure scanner SCANIVALVE DSA 3217/16 was used for the measurements of the external pressures. Prandtl probe, located in front of rear operating space, was used as a reference probe. The terrain category between III and IV was created by using plastic film with the height of elements equalled to 20 mm and rectangular barrier with the height of 150 mm.

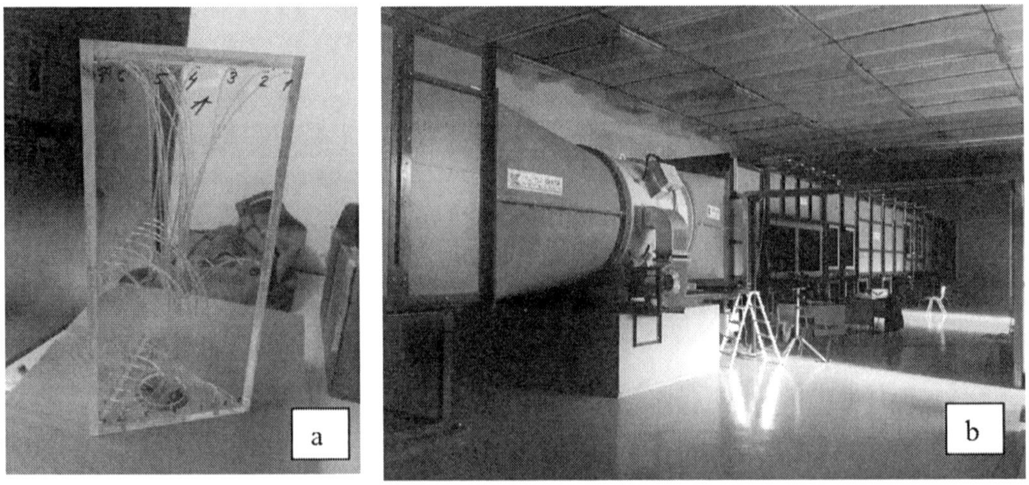

Fig. 2. a) 3D model of the structure used during the tests, b) wind tunnel SvF STU in Bratislava

Reference wind velocity in the top of the model (m/s)	6.975	9.440	12.505	15.130
Reynolds number (-)	8.04e3	1.09e4	1.44e4	1.74e4

Tab. 1. Reference velocities and Reynolds numbers

$$\mathrm{Re} = dv/\nu\,. \tag{1}$$

Reynolds number was calculated by using Eq. 1, where d is characteristic dimension of the structure [m], v is reference wind velocity [m/s] and ν is the value of kinematic viscosity [m^2/s] depending on the air temperature and barometric pressure. From the Table 1 it is evident that calculated Reynolds number for the velocity v = 6.975 m/s is less than the value recommended for this type of structure by valid standard [1] - Re = 1e4. This velocity should not be used for the calculation of external wind pressure coefficients needed for the design of structure.

4 Numerical Model

4.1 Wind Flow Modeling

Wind speed in bottom part of the atmosphere should be represented by logarithmic wind profile Eq. 2 or power low wind profile Eq. 3, [11].

$$v(z) = \frac{v^*}{\kappa}\ln\frac{z+z_0}{z_0}\,, \tag{2}$$

$$\frac{v(z)}{v_{ref}} = \left(\frac{z}{z_{ref}}\right)^{\alpha}, \tag{3}$$

where $v(z)$ is mean wind velocity at height z, v^* is shear velocity calculated by Eq. 4, z_0 is aerodynamic roughness height, κ is von Karman constant ($\kappa = 0.42$), v_{ref} is the known wind speed at reference height $z_{ref,}$, α is coefficient derived from the stability of athmosphere.

$$v^* = \frac{v_{ref}\kappa}{\ln\frac{z_{ref}+z_0}{z_0}}. \tag{4}$$

4.2 Turbulence Modeling

Turbulence models were well described by several authors in detail in the cited references and therefore, only their brief description is presented here.

Standard k-ε model is widely used despite its known limitations. It is a two equation model. It means that it includes two extra transport equations to represent the turbulent properties of the flow. The first transported variable is turbulent kinetic energy k with equation

$$\frac{\partial(\rho k)}{\partial t} + \frac{\partial(\rho k u_i)}{\partial x_i} = \frac{\partial}{\partial x_i}\left[\frac{\mu_t}{\sigma_k}\frac{\partial k}{\partial x_j}\right] + P_k + P_b + \rho\varepsilon - Y_M + S_k. \tag{5}$$

The second transported variable in this case is the turbulent dissipation ε with equation

$$\frac{\partial(\rho k)}{\partial t} + \frac{\partial(\rho k u_i)}{\partial x_i} = \frac{\partial}{\partial x_i}\left[\frac{\mu_t}{\sigma_k}\frac{\partial k}{\partial x_j}\right] + C_{1\varepsilon}\frac{\varepsilon}{k}(P_k + C_{3\varepsilon}P_b) - C_{2\varepsilon}\rho\frac{\varepsilon^2}{k} + S_\varepsilon. \tag{6}$$

Modifications have been introduced to improve its performance: the RNG (Re-Normalisation Group) model and the Realizable model [12].

Standard k–ω model incorporates modifications for low-Reynolds number effects, compressibility, and shear flow spreading. The sensitivity of the solutions to values for k and ω outside the shear layer were improved.

The first transported variable is turbulent kinetic energy k with equation

$$\frac{\partial k}{\partial t} + \frac{\partial k u_i}{\partial x_i} = \tau_{ij}\frac{\partial u_i}{\partial x_j} - \beta^* k\omega + \frac{\partial}{\partial x_j}\left[\upsilon + \sigma\upsilon_T\frac{\partial k}{\partial x_j}\right]. \tag{7}$$

The second transported variable is the specific dissipation ω with equation

$$\frac{\partial\omega}{\partial t}+\frac{\partial\omega u_i}{\partial x_i}=\alpha\frac{\varepsilon}{k}\tau_{ij}\frac{\partial u_i}{\partial x_j}-\beta\omega^2+\frac{\partial}{\partial x_j}\left[\upsilon+\sigma\upsilon_T\frac{\partial k}{\partial x_j}\right]. \tag{8}$$

Modifications have been also introduced in the BSL (Baseline) model and the SST (Shear Stress Transport) model [13, 14].

4.3 Numerical Method

For the solution, finite volume method implemented under program ANSYS Fluent [15, 16], was chosen. Fluent offers three variants of this turbulence model: Standard, RNG and Realizable; and four types of near-wall modeling: standard wall functions, scalable wall functions, non-equilibrium wall functions, enhanced wall treatment and user defined wall functions. In second case, Fluent offers two variants of turbulence model: Standard and SST.

4.3.1 Computation Domain

Size of the computational domain for coarse grid was 7×4.5×2 m^3 (l×w×h) according to recommended maximum value of block ratio 3% [17, 18]. The distance from the triangle object to the inlet, sides and top of the domain was at least five times the height of the building, and the distance from the outlet was at least eleven times the height (according to the guidelines given by [17, 18]).

4.3.2 Boundary Conditions

The inlet boundary conditions of the domain are defined by the vertical profiles (Eqs. 2 and 3 in Section 4.1). For the simulation of the wind flow developed in SvF STU Wind Tunnel Laboratory in Bratislava, it was recommended to use logarithmic wind profile calculated by Eq. 2 with aerodynamic roughness height in tunel scale z_0 equalled to 0.0023. Aditional inputs for *k*-ε model were: the equation for turbulent kinetic energy *k* and the equation for turbulence dissipation rate ε - Eqs. 9 and 10.

$$k=\frac{u^{*2}}{\sqrt{C_\mu}}, \tag{9}$$

$$\varepsilon(z)=\frac{u^{*3}}{\kappa(z+z_0)}, \tag{10}$$

where $C_\mu = 0.09$ is a model constant. The *k*-ω model inputs are based on turbulent kinetic energy *k* (Eq. 9) and specific turbulence dissipation rate ω defined by Eq. 11.

$$\omega(z)=\frac{\varepsilon(z)}{k}. \tag{11}$$

The outlet boundary was defined as pressure outflow and the side and upper boundary as zero gradient or symmetry. The bottom boundary was simulated as a rough wall with adapted standard wall functions [17]. For rough wall, the method proposed in [19] was used.

$$k_S = \frac{E \cdot z_0}{c_s}, \tag{12}$$

where k_s is the roughness height of wall function, c_s is the roughness constant of wall function and $E = 9.79$ is the log law of the wall constant. In our case, c_s was set to one and k_s was defined in computed form by Eq. 12.

4.3.3 Computational Grid

Mesh was created by the size function set on adaptive. The element size on surface of the object was 0.01 m with soft behaviour and inflation for these surfaces was applied with 5 layers with growth rate of 10%. All other settings were left on default. In the case of turbulent boundary layer, dimensionless quantity y^+ defined by Eq. 13 has to be considered.

$$y^+ = \frac{v^* \Delta y}{\upsilon}, \tag{13}$$

where υ is the local kinematic viscosity of the fluid and Δy is the distance of first cell from the nearest wall. For standard wall functions with log-law layer it is recommended to keep $30 < y^+ < 300$, [20].

4.4 Result Transformation

External wind pressures were measured for different wind directions and for various wind speeds. For better presentation of the results, these measured values were recalculated to the values of external wind pressure coefficients using Eq. 14. The pressure coefficient is defined as a dimensionless number which describes the relative pressure and its values are independent from wind speed if the following condition is satisfied: Reynolds number *Re* has to be greater than 1e4.

$$c_{pe}(x) = \frac{p(x) - p_{ref}}{0.5 \rho v_{ref}^2}, \tag{14}$$

where C_{pe} is external wind pressure coefficient [-], $p_{(x)}$ is the value of wind pressure in measuring point located on the investigated model [Pa], p_{ref} is static pressure of undisturbed flow [Pa], ρ is air density [kg/m^3] and v_{ref} is reference wind speed [m/s].

5 Comparison of the Results

For the determination of the accuracy of 3D model solved by CFD simulation, only two angles of rotation were chosen - 0° and 90°. External wind pressure coefficients obtained from tests were compared with the values calculated by CFD simulation - see Figs. 3-5 (0°) and Figs. 6-8 (90°). As it was mentioned above (see Tab. 1), only Reynolds number calculated by the value of wind velocity equalled to 6.975 m/s, did not meet the recommended value 1e4. Other calculated Reynolds numbers were greater than recommended value. Therefore, in the presented graphs, only results from CFD simulation solved for one chosen wind velocity (v = 12.505 m/s) are shown.

In Tables 2 and 3, the percentage average errors calculated as a difference between the solution by calculation model (v = 12.505 m/s) and experimental measurements (v = 12.505 m/s) are mentioned. From the comparison of the results, good coincidence was obtained. Hence, we can say that created 3D model and setting of the CFD simulation are good. In some pressure taps, the difference in values was very small (e.g. see Figs. 3, 4, 8). However; we can see in Figs. 5, 6, 7 that there are some points where greater differences between the values from tests and CFD simulation are evident. Better setting of input parameters into the CFD simulation (the characteristics of wind flow) and using smaller size of elements in created 3D model (this is a problem of the capacity of used computer), can help to improve the results.

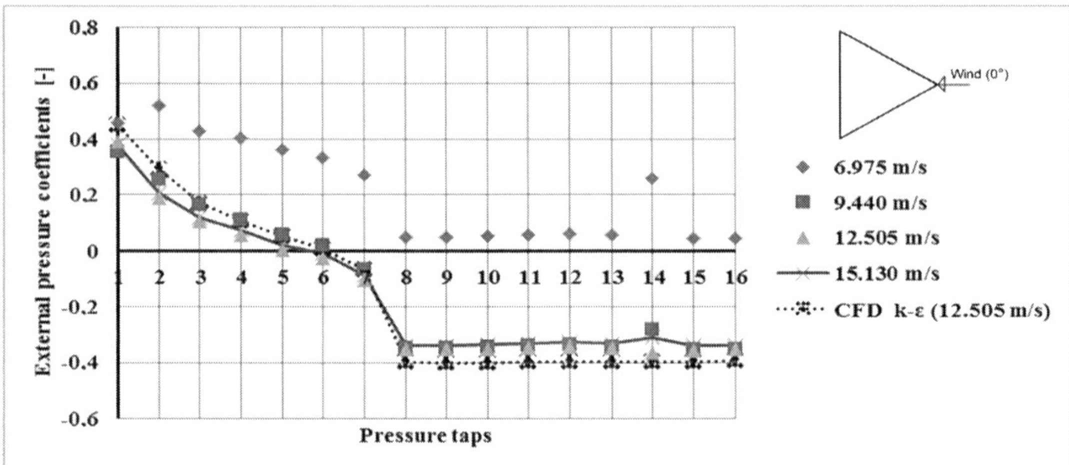

Fig. 3. Comparison of external wind pressure coefficients - wind direction 0°, level A

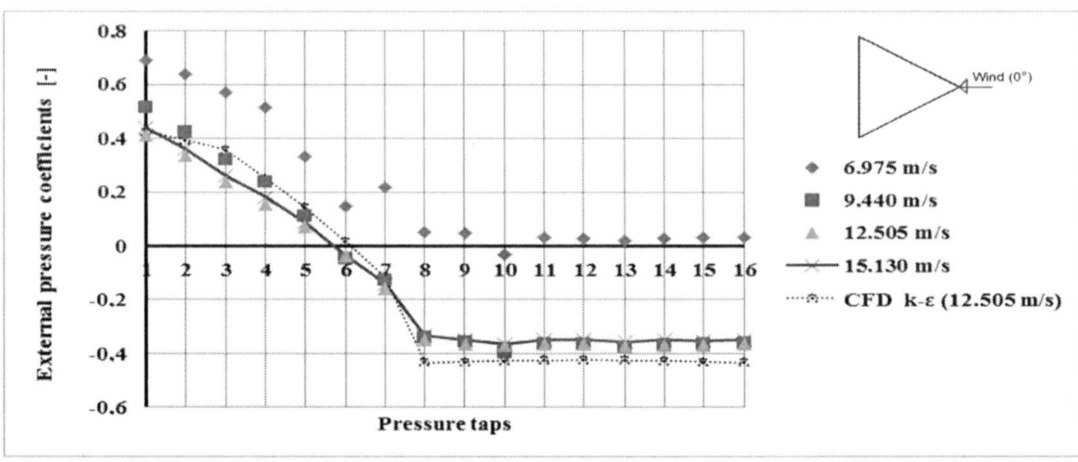

Fig. 4. Comparison of external wind pressure coefficients - wind direction 0°, level B

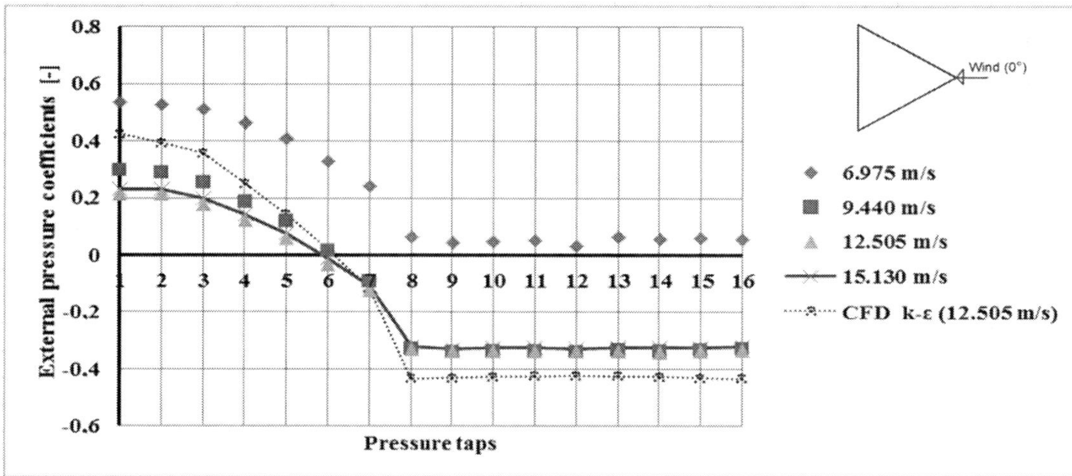

Fig. 5. Comparison of external wind pressure coefficients - wind direction 0°, level C

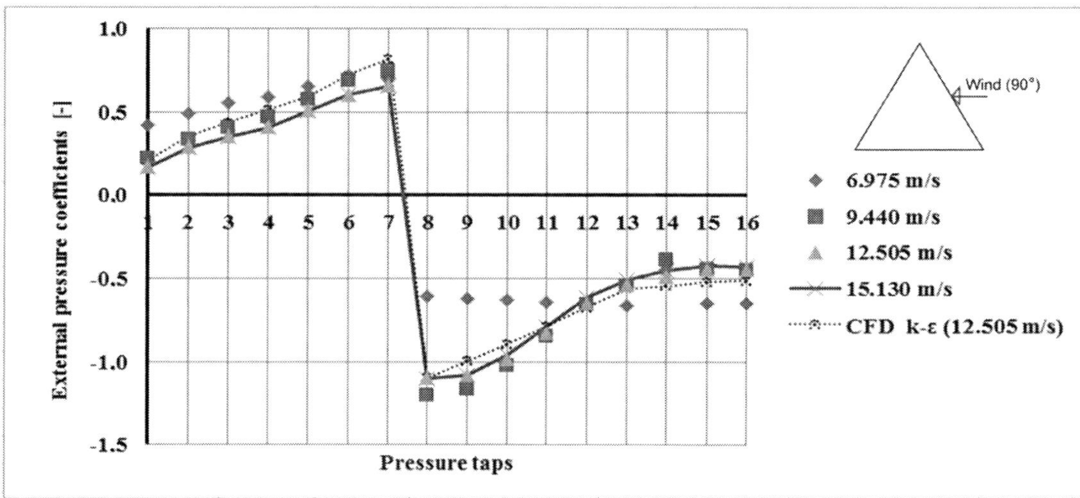

Fig. 6. Comparison of external wind pressure coefficients - wind direction 90°, level A

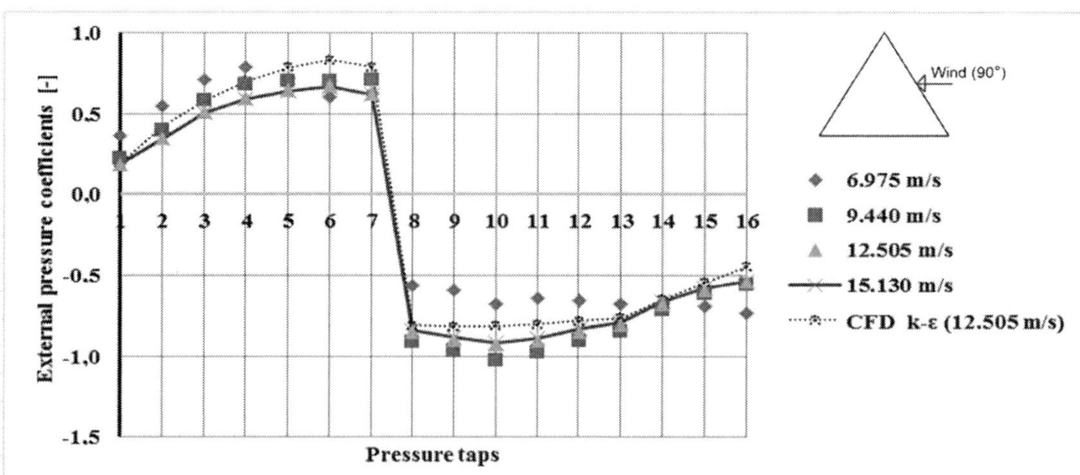

Fig. 7. Comparison of external wind pressure coefficients - wind direction 90°, level B

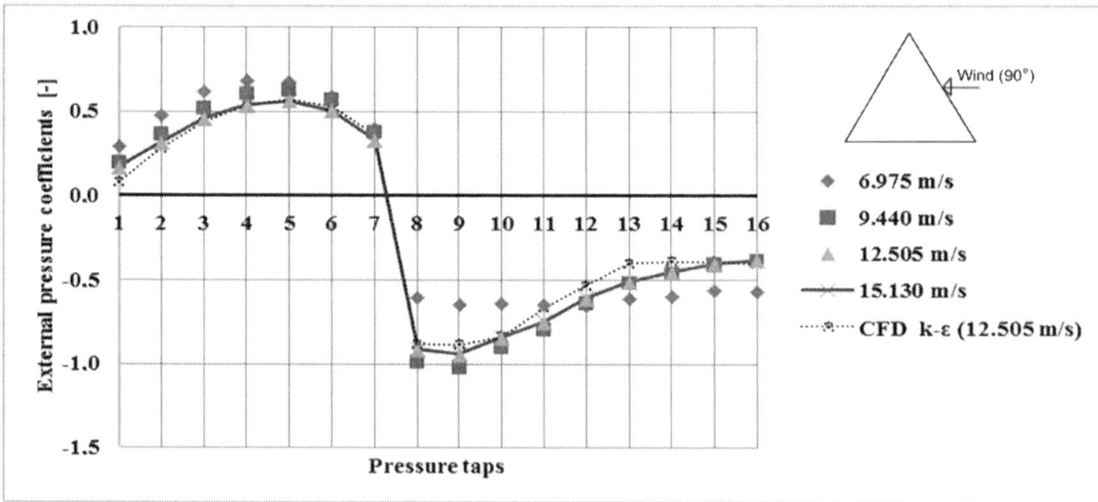

Fig. 8. Comparison of external wind pressure coefficients - wind direction 90°, level C

Model	Level A	Level B	Level C	Average error
Standard k-ε model	2.53%	21.69%	12.05%	**12.09%**
Standard k-ω model	14.76%	27.25%	7.90%	16.63%

Tab. 2. Percentage average errors for different turbulence models - wind direction 0°

Model	Level A	Level B	Level C	Average error
Standard k-ε model	19.98%	11.60%	9.06%	**13.55%**
Standard k-ω model	16.58%	13.90%	13.90%	14.79%

Tab. 3. Percentage average errors for different turbulence models - wind direction 90°

5 Conclusion

The results from experimental measurements carried out in wind tunnel were used for the verification of the CFD simulation. They helped to set created 3D model and some other important input parameters needed for the calculation. From the comparison of both analysed models it is evident that k-ε model provides better results than k-ω model. Therefore this model will be used for future investigation – the influence of rounding of corners on the values of external wind pressure coefficients.

Acknowledgement

This paper was created with the support of the Ministry of Education, Science, Research and Sport of the Slovak Republic within the Research and Development Operational Programme for the project "University Science Park of STU Bratislava", ITMS 26240220084 and the Scientific Grant Agency - VEGA grants No. 1/0544/15, 1/0265/16. Authors thank to the Slovak University of

Technology for the support - financial sources No. 1619 and No. 1651. This paper was created with the support of the TU1304 COST action "WINERCOST".

References

[1] STN EN 1991-1-4 Eurocode 1, Actions on structures. Part 1-4: General actions, wind actions, SUTN, 2007. (in Slovak)

[2] H. HEMIDA, et al., Experimental investigation of wind flow above the roof of a high-rise building, in: Trends and challenges for wind energy harvesting, March 30–31, 2015, Coimbra, Portugal, pp. 25-34.

[3] CERMAK,J.E. et al., Wind tunnel studies of buildings and structures, ASCE manuals and reports on engineering practice, No. 67, American Society of Civil Engineers, 2009. ISBN 978-0-7844-0319-8.

[4] URL:https://blog.buzzmath.com/2013/10/29/did-chicago-steal-buzzcitys-mathematicalgenius/.

[5] URL:https://www.google.sk/search?q=trojuholn%C3%ADkov%C3%A9+budovy&biw=739&bih=633&source=lnms&tbm=isch&sa=X&ved=0ahUKEwio6-b46bnQAhWIDMAKHZVSBz4Q_AUIBigB#imgrc=v4guFgte8BZr1M%3A.

[6] URL: http://www.rickrubens.com/DallasFtWorth.htm.

[7] URL: https://dennikn.sk/162225/mali-to-byt-najvyssie-budovy-ceska-a-slovenska-dopadlo-to-takto-fotogaleria/.

[8] HUBOVA, O., KONECNA., L., Experimental determination of wind flowing around a building configuration. In: World Multidisciplinary Civil Engineering-Architecture-Urban Planning Symposium , June 12-16, 2017, Prague, Czech Republic.

[9] HUBOVA, O., KONECNA, L., Comparison of experimental determination of wind pressure distribution in steady and turbulent wind flow. In: *Proceedings of 15-th International Scientific Conference VSU'2015*, Sofia, Bulgaria, Vol. I, pp. 90-95. ISSN 1314-071X.

[10] SKRUCANY, T., et al., Influence of aerodynamic trailer devices on drag reduction measured in a wind tunnel. *Maintenance and Reliability*, 18(1), 2016, pp. 151-154.

[11] LAUNDER, B.E. AND SPALDING, D.B., The numerical computation of turbulent flows, Comput. Methods Appl. Mech. Eng., 3 (1974) 269-289.

[12] WILCOX, D.C., Multiscale model for turbulent flows. In: AIAA 24th Aerospace Sciences Meeting. American Institute of Aeronautics and Astronautics, 1986. ISBN: 978-1-84821-001-1.

[13] WILCOX, D.C., Formulation of the k-omega Turbulence Model Revisited. In: AIAA Journal, Vol. 46, No. 11, pp. 2823-2838, 2008. DOI: 10.2514/1.36541.

[14] WILCOX, D.C., Turbulence Modelling for CFD. 3rd edition. La Canada CA: DCW Industries, Inc. 2006. DOI: 10.1017/S0022112095211388.

[15] ANSYS Inc., ANSYS Fluent Theory Guide, Release 17.2 (2016).

[16] ANSYS Inc., ANSYS Fluent User's Guide, Release 17.2 (2016).

[17] FRANKE J. et al., Best practice guideline for the CFD simulation of flows in the urban environment, COST Office Brussels (2007).

[18] TOMINAGA Y., AIJ guidelines for practical applications of CFD to pedestrian wind environment around buildings, J Wind Eng Ind Aerod 2008. 96 (10-11) 1749-1761.

[19] BLOCKEN, B., STATHOPOULOS, T., CARMELIET, J., CFD simulation of the atmospheric boundary layer–wall function problems. Atmospheric Environment 41(2):238-252. Elsevier (2007)

[20] URL: http://www.cfd-online.com/Wiki/Dimensionless_wall_distance_(y_plus)

Current address

Macak Marek, Researcher
Slovak University of Technology in Bratislava, Faculty of Civil Engineering,
Department of Mathematics and Descriptive Geometry
Radlinskeho 11, SK-810 05 Bratislava, Slovak Republic
E-mail: marek.macak@stuba.sk

Hubova Olga, Associate professor
Slovak University of Technology in Bratislava, Faculty of Civil Engineering,
Department of Structural Mechanics
Radlinskeho 11, SK-810 05 Bratislava, Slovak Republic
E-mail: olga.hubova@stuba.sk

Konecna Lenka, Researcher
Slovak University of Technology in Bratislava, Faculty of Civil Engineering,
Department of Structural Mechanics
Radlinskeho 11, SK-810 05 Bratislava, Slovak Republic
E-mail: lenka.konecna@stuba.sk

Slovak University of Technology in Bratislava
Faculty of Mechanical Engineering
16th Conference on Applied Mathematics **APLIMAT 2017**
Proceedings

MATEMARTIAMO: MATHS & ART COMPETITION

MAGNAGHI-DELFINO Paola (IT), NORANDO Tullia (IT)

Abstract. The **FDS** Laboratory (**F**ormation, **D**idactics and Experimental Teaching, **S**cience Communication) promotes large-scale initiatives, as awards, games, television programs and conferences, for the dissemination and the "demystification" of the difficulties of mathematics.
In 2015-2016 FDS proposed to the students of the last year of the Italian high schools and the freshmen of Politecnico di Milano to test their artistic skills (Visual Arts, Literature and Photography) in the contest MatemArtiAmo **(Maths&Art)**. All artworks were published on the Facebook FDS page and public voted them. The works that had better interpreted the theme of the relationship between math and arts were shown in an exhibition at the Politecnico di Milano.

Keywords: Art, Mathematics, Contest

Mathematics Subject Classification: AMS_97M80

Are there rocking chairs in Heaven
where little babies go?
to Davide

Introduction

The authors are engaged in the interdisciplinary research initiatives of the **FDS** (**F**ormation, **D**idactics, **and Science** Communication) Laboratory of the Mathematics Department of the Politecnico of Milano. The most important aim of the FDS laboratory has always been to contribute to the revaluation of the role which mathematics naturally play in the liberal arts.

Historically mathematics always constituted an integral part of the liberal arts, in all meanings, which this term has had through the centuries. Nowadays in the Italian secondary schools, the students still believe that there are two different learning fields which have alternative interests and goals: the scientific and the humanistic studies and careers.

The connection between Mathematics and Figurative Arts is an argument that only recently had been investigated by Italian mathematical researchers and popularisers as Bruno D'Amore, Michele Emmer, Mauro Francaviglia, Piergiorgio Odifreddi, among the others

However, high school's teachers, who generally do not exhibit any bias towards the other teaching arguments, develop each of the cultural trails in a separate way, nor they feel the lack of connections.

For many years, as FDS active members, we did many interdisciplinary projects with high school students of different orientations and we are more and more convinced that these experiences contributes significantly to increase of the wealth of skills of the students. A chance to experience their creativity by producing an artwork in which they express mathematical laws is the first step towards the enjoyment of mathematics that we want students to feel.

Among all the projects we did, the authors realised many educational projects oriented to connect learning experiences of Maths to other educational experiences in the Humanities and in the Art world [12]. The aim of these interdisciplinary projects is mostly to reach students that cannot otherwise be reached and we try to transform their opinions or prejudices about the role of Maths and Arts in the real world. Some of the projects we did provide new challenges for those students already considered successful.

The idea to hold a contest on the theme of Art and Mathematics was born in one of our FDS meetings and strengthened through the exchange of experiences had in Tortona during the event of award ceremony of the Competition "Adotta Scienza e Arte nella tua classe". [8]
In 2015-2016, FDS Laboratory published the competition MatemArtiAmo (Math&Arts) for pupils of secondary schools and freshmen of Politecnico di Milano in order to let them experience mathematical concepts by way of artworks creation. The contest provided six sections: Painting, Graphics, Photography, Literature, Sculpture and Video. The rules are simple:

- Individual students or groups of up to three students can participate to the contest. Each student may submit only one work to a single section.
- All submitted artworks were posted on the FDS Facebook page and judged by the public by means of "I *like*" preferences applied on its original photo. The authors of the artworks highly appreciated by the FDS and the popular juries were exhibited in the show "MatemArtiAmo: mathematics becomes art".

We received one hundred or so artworks, especially in the section of Painting and Graphics.
Most of the artworks' authors went to secondary schools of almost all Italian regions and, in our opinion, some of their artworks are very interesting.

The artworks were exhibited at the Politecnico di Milano and many of them were awarded during the ceremony that concluded the exhibition.

In this paper, we describe production and operating methods of the contest and we describe the content and the meaning of some prize winning artworks.

1 Mathematics and Figurative Arts

Galileo said that nature is an open book if you only know the language in which it is written, and this language is math. Plato in the *Philebus* attributes the absolute beauty to geometric figures (curves and straight lines constructed by ruler, square and compasses) because the beauty of the natural ones is relative. Doubts about the figurative art, which in the Bible is theological, in Plato are because nature is considered the projection of the real world (*Ideas world*) and so the figurative art is the copy of the copy of the ideas. The constructive restriction is philosophical: the tools allow building curves and straight lines always equal to themselves. Galileo used the same topic of Plato but applied to figurative arts. Many scholars wrote treatises about the artistic theory: we recall the Renaissance painters, sculptors and architects as Piero Della Francesca, Leonardo da Vinci and Leon Battista Alberti and mathematicians as Luca Pacioli. In modern times, Vasilij Kandinskij (*Concerning the spiritual in art, point and line to plane* [11]) goes back to the same tools of Galileo: straight lines and curves can be the alphabet of the artworks. In all figurative arts the same ideas were expressed in the manifesto titled "*The Basis of Concrete Art*" (1930) signed by Theo van Doesburg, or in the artistic movement *The Style*, also known as *Neoplasticism*, whose principal member was Piet Mondrian and the architect Gerrit Rietveld.

In the paintings, mathematics can be into the artworks in two ways: or in the figures or in the structure. In many well-known authors, the mathematics can be easily found. Some examples are the *Pointillism* (Gino Severini *The black cat*, George Seurat *La Parade de Cirque*), *Cubism* (Pablo Picasso *Three Musicians*, *Woman with flower*), Kazimir Malevic *Peasant Woman with Buckets and Child*, Vincent Van Gogh *The starry night*, Salvador Dali *Galatea of the Spheres*, Josef Albers *Homage to the square*.

The other aspect concerns mathematics as structure of the work; clearly, we can consider all the painters, which use the rules of the perspective. The principal mathematical topics are the theory of proportions, the similarity and, for this reason, the golden ratio. As regards Symmetry, in particular, Leonardo da Vinci studied the central symmetry and Brunelleschi, Alberti and Piero Della Francesca studied the three ways of the axonometric theory. Examples of studies of perspective is the *Leda Atomica*. In this painting, Dalì left the drawing with a frame: circle with pentagon and star inscribed.

In the sculpture, the artist Max Bill further promoted the ideas associated with *Concrete Art*; we remember in particular his artworks inspired by the Möbius ribbon.

Umberto Boccioni, one of the most prominent artist s among the Italian Futurists, not only developed the movement's theories but also produced some significant *Futuristic Sculptures* as the well-known *Unique Forms of Continuity in Space*. Boccioni borrowed from the geometric forms to evoke crashing and sounds that accompany the depicted movement.

In the architecture, the Bauhaus, the Rationalism and the works and treatises of Le Corbusier highlight the strictly connection between the beauty canon and the mathematical geometry and proportions. In Italy, latest works of Santiago Calatrava, Massimiliano Fuksas and Renzo Piano show an undeniable inspiration to many different geometric shapes, that technology now help to use.

In the artworks presented to the contests, we can see personal developments of the ideas we have mentioned here.

1.1 Point and Line to Plane

	*Essence* by Martina Pellizzoni Author's Comment According to the Buddhist tradition, a Mandala represents, with its geometric shape, the process by which the cosmos is formed from its centre. In Buddhism, the Universe and inner self coincide, so the symbolism of the Mandala recounts a trip, which refers to an inner growth. The painting consists of several colours, in each of which an important meaning is hidden. Technique: acrylic on wood.

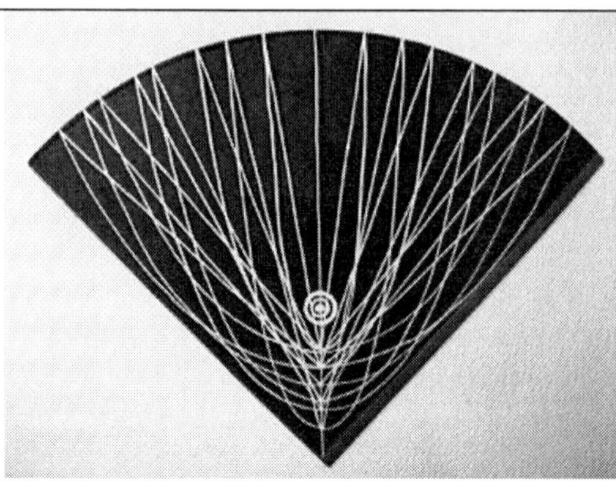	*The function is the aim* by Ada Rosito Author's Comment My work is an abstract composition, consisting of catenaries overlapping quadratic functions. In the middle, there is a round shape, which intersects the vertical axis, consisting of concentric circles: it is the goal, the dream that a person never stops thinking of, the focal point of his life. In particular, I chose to surround this round shape by catenaries because they possess a mysterious elegance. Material: white glossy lacquer on tempered glass, with a background of black construction paper.

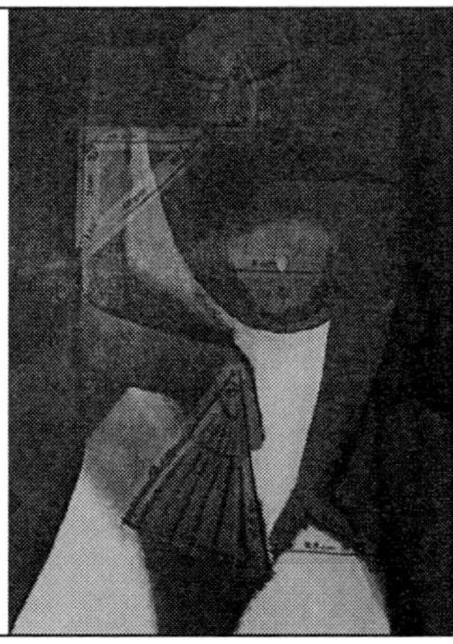	*Woman with triangle* by Laura Stella Author's Comment My work repurposes a painting by Picasso, *Woman with a fan*, tempera on canvas, accompanied by the measures of its fundamental geometric forms.

1.2 Golden Section and Fibonacci series

	*Phi Mask* by Chiara Brambilla Author's Comment Dr Stephen Marquardt has theorized that there are proportions of the human face that correspond to visual beauty's ideal and that this human archetype can be mapped mathematically using the Golden Ratio, thus defining it the "Phi Mask". "I'm fascinated by the interplay between the flatness of the surface and the plasticity of the face," said Chuck Close, from whom I was inspired. In my artwork, the Golden Section is suggested by the proportion of the canvas and by the golden background.
	*My spiral mind* by Elisa Moretti Author's Comment Italian mathematician Leonardo of Pisa, known as Fibonacci, invented the famous series that we know as Fibonacci sequence. Fibonacci sequences appear in biological settings such as branching in trees, arrangement of leaves on a stem, etc. Approximate logarithmic spirals can occur in nature; golden spirals are one special case of these logarithmic spirals. The Fibonacci spiral approximates the golden spiral created by drawing circular arcs connecting the opposite corners of squares in the Fibonacci tiling. The Fibonacci spirals inspired my painting.

1.3 Spirals

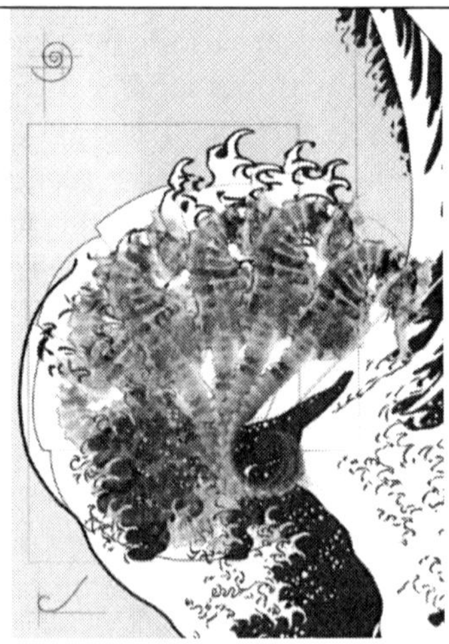

Kanagawa spiral by Laura Foppiani
Author's Comment
The work depicts the construction of a golden spiral by means natural elements. Starting from the origin of the spiral, I positioned seahorses, whose tail follows the shape of the spiral. I decided to insert in the background the *Kanagawa great wave*, retaining the original light color in background.

Reality spiral by Anna Nutarelli
Author's Comment
The logarithmic spiral is the structure of some shells, is the form in which we see a spiral stairway from below and the shape of some galaxies. In my artwork, the design of the spiral turns in these familiar figures. The work represents the requirement of the human being to take part in the world around him: they describe the world with mathematical rules, which become basic to satisfy and to broaden his knowledge.

1.4 Mathematics as Structure

One letter can make the difference? by Martina Praticò
Author's Comment
Symmetric and asymmetric parts in the work were compared on the same surface. In this print were represented the symmetry and asymmetry as if they embodied two opposite emotions. The first is very calm, regular, abstract and austere and the second irregular, shaken and outside the box. The difference is enhanced by the colour of the background: the two figures appear as one the negative of the other. In the upper right, you can read the letter "a" that justifies the title of the artwork: I thought that this single letter, when it precedes the word "symmetry", completely change the meaning of an object or a situation.

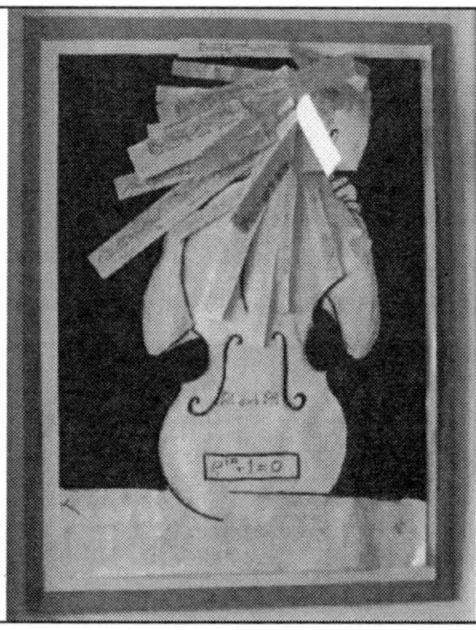

Untitled by Bijay Polotti - Francesco Romagnoli
Author's Comment
The human figure is represented as a tool that exudes art. Cutting in the violin sounding board remember the integral's symbol. I represented a combination between music, which is also expressed through rhythm and quantity, and mathematics. The blackboard in the background evokes the mathematics teaching and learning. The man, in prominence compared to the black background, symbolizes a mediator and an instrument of connection between art and mathematics.

Binary System by Martina Merlo
Author's Comment
My artwork is explicitly inspired by *Not to be Reproduced* (*La reproduction interdite*) painted by René Magritte in 1937 that is currently owned by the Museum Boijmans Van Beuningen in Rotterdam. The work depicts a man standing in front of a mirror, but whereas the book on the mantelpiece is reflected correctly, the man can see only the back of his head.
In my painting, I include two units, an order and the next higher order. I have an individual, his double and a surreal deception...
How to get back to the departure? 0 or 1: true or false?

Man's Harmony by Francesca Colturani
Author's Comment
The human being is the centre of the Universe: the wonderful expression of a mysterious harmony. In my painting, the man is both the protagonist and the prisoner of the order around him. The miracle of the man is the discovery that he is the perfect outcome of the same harmony that is in the world.
I created my artwork in the physics laboratory of my school. A black, square plate was made vibrating through a sinusoidal wave format a given, gradually increasing frequency. A green powder, sprinkled onto the plate, will come to rest only at the nodes of the predominant mode of vibration of the plate, which renders the nodes visible as green lines. As the frequency increased, it excited modes of vibration of ever-higher order, making intricate patterns consisting of increasingly larger numbers of lines.
The normal modes of oscillation of the metallic plate reworked and overlapping each other and so generated the world and the muscles of the man into the space.

1.5 Möbius ribbon

	∞^{∞} by Chiara Magni Author's Comment The structure consists of four faces of a cube, covered by mirrors. Inside I suspended a Möbius strip in turn reflective. It does not represent the infinite, but the infinite path, so how do you create an infinite journey from reflection between the mirrors. Infinity is a characteristic feature of mathematics, abstract essence; but infinity (or almost infinity) is also the journey of discoveries that science can perform, also thanks to mathematics.

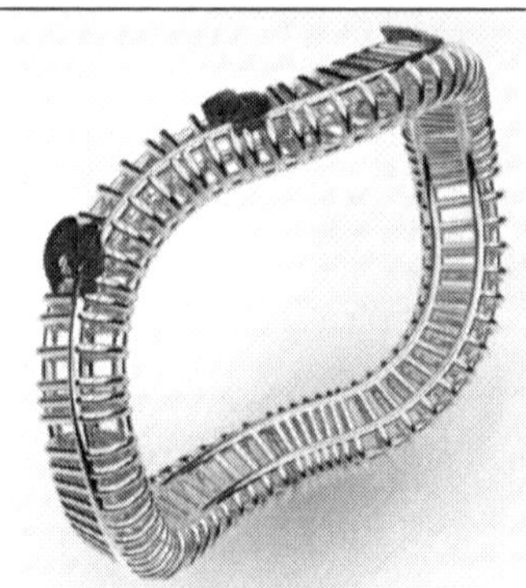	*Cage of numbers* by Luca Porco Author's Comment A Faraday cage is a mesh of electrically conducting material, which blocks electromagnetic waves, and electrostatic fields by letting the free charges move to counter the external fields. I imagined that a cage retains the prime numbers as a magnetic cage. Therefore, I realized a bracelet with wire and coloured hard stone facings.

1.6 New Style in Sculpture

	*Conic sections* by Arianna Parmeggiani Author's Comment The sections of a cone crossed by inclined planes are special mathematical curves of four kinds: circle, ellipse, parabola and hyperbola. I imagined breaking the ring's shape with inclined planes and created a jewel's model composed of five sectors, each one in a different material.

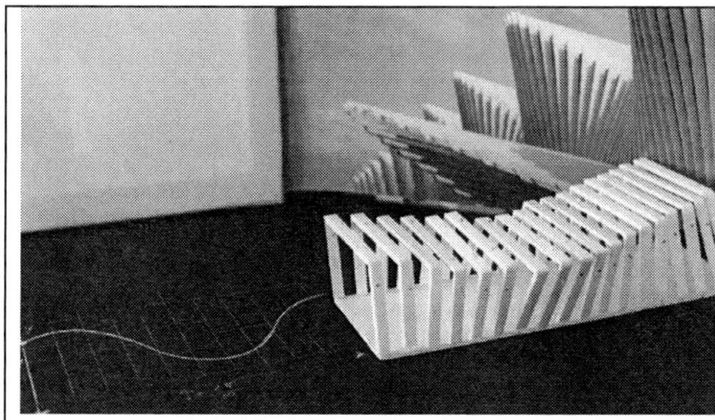	*Sine Bridge* by Daniele Mantovani Author's Comment Model of a bridge inspired by the concept of sine wave, which became part of the urban context. I designed the structure to avoid sudden changes in brightness and besides it is windbreak. The various forms of coverage have inclination and different length due the choice of changing gradually the cosine of angle while maintaining the height of the structure. Materials: foamcore, pins.

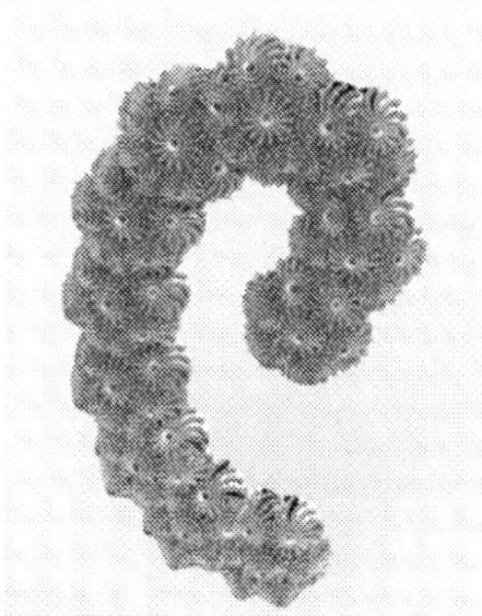	*Network of relationships* by Vanessa Bersano Author's Comment I created my artwork using a 3D modelling software. The shape of the 14 Karat yellow gold earring is composed of elements characterized by a fractal geometry that are repeated to form a rounded and lengthened body.

2 Mathematics and Literature

Many writers have enthusiasm for mathematics just as many mathematicians love the world of literature, despite the wall that seems to stand between the two disciplines. At a glance, mathematics and literature have no reciprocity as subjects and no one to date has been successful in both fields. We consider an exceptional occurrence that Bertrand Russell, one of the greatest philosophers and mathematicians of the 20th century, received the Nobel Prize for literature in 1950.

If we exclude the obvious influence of science on the minor literature genres of scientific popularisation and sci-fi, the relation between mathematics and literary works seem to be secondary and exceptional, even though on various occasions it has been the case that a literary work has contained interesting observations pertaining to mathematics.

Love stories are dynamic processes in which the feelings change over time, starting from a state of indifference. For this reason, they can be explain with the formal structure of the theory of dynamical systems. Sergio Rinaldi, professor of Politecnico di Milano, worked on several famous loving affairs, as for example, the love story between Scarlett and Rhett in *Gone with the wind*, the cyclical story between Petrarch and Laura De Sade described in the *Canzoniere*. Rinaldi also examined love stories that tend towards schemes that are more complex, like that is described in *Jules and Jim*, a tale that became famous in 1962 after the movie of François Truffaut.

The few writers, who have ventured headlong into high-level mathematics, as Lewis Carroll, Thomas Pynchon, David Foster Wallace, have been most inventive in both the sentences they constructed and the stories they created.

Italo Calvino, which grew up in a family of Italian scientists, wrote that *the scientific and poetic attitude coincide: both are attitudes towards research and design, discovery and invention*; starting in the late 1960s he changed direction of his novels using combinatorics and metanarratives in novels as *The castle of crossed destinies* [5], *Invisible cities* [4] and *Palomar* [6].

Ernest Hemingway wrote to Maxwell Perkins, in 1945 that *the laws of prose writing are as immutable as those of flight, of mathematics and of physics.*

The very best passages of Hemingway have the mathematical complexity of a fractal: a seemingly simple formula that, in its recurrence, causes slight but crucial changes over time, as for example the famous retreat from Caporetto in *A Farewell to Arms* [9].

At the contest, we received few literary works, some of them are short proses, and others are poems. *Theophany* by Emanuele Lo Presti [A1] is a talk by numbers to humanity, too haughty to recognize that their achievements are due to the sciences. In response to the blind arrogance of the human race, the numbers they throw the man, by means a monologue characterized sometimes by strong sarcasm, his ephemeral condition and his helplessness in front of Nature, dominated by scientific and mathematical laws.

Poets have been more conversant with mathematics, probably because they have to pay attention to the numerical qualities of words when working in meter, forced to consider the form and even physical shape of what they write, not just its meaning. Wordsworth praised *poetry and geometric truth* for *their high privilege of lasting life.*

The French poet Paul Valéry once expressed his great admiration for mathematics, saying *I worship this most beautiful subject of all and I do not care that my love remains unrequited.*

A poem is a text: the author wants to send a message to his readers. In order to do this, the writer use a code with the linguistic signs by that code. A linguistic sign is composed of two inseparable elements: a signifier and a signified. In the poetic text, these two aspects of a linguistic sign usually overlap.

However, precisely because poetry uses linguistic marks, you can analyse separately aspects of meaning and significance and with this, disassemble the poetic text. A poem dismounted is a text that you can interpret.

In the poem *Love code* [A2], Elisa D'Elia starts from the need to create a parallelism between the arithmetic rules and the love's reasons. She uses two different linguistic signs and, in addition, she plays with the readers, taking advantage of the versatility of numeric symbols.

Recently also comics had begun to deal with math. As with traditional media, even in comic strips mathematics can appear through the symbolic approach or the structural approach.

Marco Abate, mathematician at Pisa University, examined [14] as an example of the first case the comic story *Three times Zero*, an episode of the comic series *Dylan Dog*[1]. Mathematics enters the story in two ways. The first is a quote by the famous "butterfly effect" of chaotic dynamical systems, the real interest in mathematics is however elsewhere. One of the key characters of the story is a mathematician, obviously eccentric, that has discovered that *three times zero equals three*. The use in the history of this sentence is purely symbolic, evocative, and metaphoric; it is not important to its real meaning, but only the effect, which results in the reader's mind, impressions, images and sensations that surround the environment.

Mario Abate has projected and wrote the text to the comic strip *The Lemma of Levemberg* [2], having in mind two main objectives: to use realistic mathematics as characters and to tell the story using a structural approach, i.e. by inserting real mathematical concepts inside the story.

The story is based on the theory of chaotic dynamical systems. To explain what a chaotic dynamical system is in understandable terms to a non-specialist reader, Abate used analogies with phenomena of everyday life, in particular of love life.

David Osenda, the author of the comic story *Last lecture at Göttingen* [13], is one of the most startling revelations of the 2008 edition of the *Festival della Matematica* in Rome. A story set in Germany during the Nazi dictatorship, where an old Jewish professor and a young man talking about Cantor's theory of infinity. David Osenda succeeds admirably with its beautiful tables to transpose in comics a mathematical theorem.

Luana Caselli, researcher in Neuroscience, Luca Iaboli, emergency physician, Grace Lancon, Marco Madoglio, writer and designer, set up a good teamwork and created the graphic novel *Doctor G* [7]. Through scientific investigation of crimes, Doctor G, professor of statistics at University of Berlin and police consultant, analyses cases and famous judicial stories, trying to document and explain the events through the method of scientific research. Each survey is taken as a pretext to examine a theme with health implications and the possible methodological approaches and techniques to try to quantify the extent of the damage and the probability of occurrence.

In the 2015-16 edition of the contest, unexpectedly, three students of the Politecnico Design School sent to the contest a comic strip. We received the following attempt

[1] Dylan Dog (text by Tiziano Sclavi, design by Bruno Brindisi) became in a short time one of the most sold Italian comics. Dylan Dog is a private investigator *sui generis*, specializing in investigations with horrific slopes, so he is known as the "Nightmare Investigator".

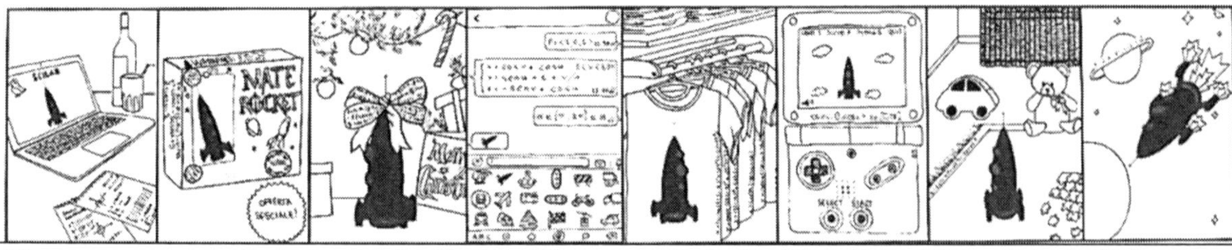

Give space to your imagination by Beatrice Bazzan – Federica Borrello – Gianna Maria Strianese
Authors' Comment
We wanted to demonstrate with our graphic work that, starting from any object or better from an idea, our imagination can wander to the point of being able to produce a multiplicity of situations that do not necessarily imply direct access to the object itself. We used the program Scilab that allows you to create images with 3D effect by using mathematical formulas. We created a rocket and we entered it into a sequence of eight figures that form the comic strip. When you think of a rocket, you immediately think that it is destined to travel to space. However, why do you not make it a toy rocket or a print on a t-shirt for teenagers or a funny emoticon for the modern media? It is all about to give space to your imagination.

In the second edition of the contest, we decided to enhance this form of artistic expression, introducing a new section: Meme.

3 Mathematics and Graphic Arts

The term Graphic Art usually refers to the arts that rely more on line or tone than on colour, especially drawing and the various forms of engraving. Graphic art further includes photography and computer graphics. It also encompasses drawn plans and layouts for interior and architectural designs. The next major change in graphic arts came when the personal computer was invented in the Twentieth century. Powerful computer software enables artists to manipulate images in a much faster and simpler way than the skills of board artists prior to the 1990s. With quick calculations, computers easily recolor, scale, rotate, and rearrange images.
The students presented at the contest artworks that they created in different ways.

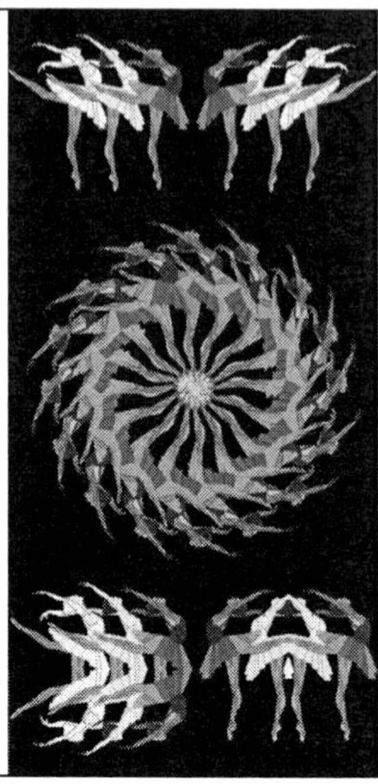

Geometric Dance by Vanessa Fisicaro
Author's Comment
Something magical happens when you dance, everything lines up, the world is in perfect sync and that is when you feel fully alive. This is a phrase, taken from the movie *Step up 3*, which expresses well the work I wanted to achieve.
Analysing the concept of symmetry associated with the theme, Art and Mathematics, I wanted to enhance its presence in dance, considered nowadays the art of movement.
My artwork consists of three panels in which the main subject is a dancer that I repeated several times using different kinds of symmetry: horizontal, vertical and central. The geometric shapes and colours of the dancers create a strict structure.

Blow by Giorgia Tripoli
Author's Comment
This work represents the morphology of the face. I drew the face of a girl with mathematical geometric proportions. According to recent studies, all the faces, regardless of race, gender and age, are consistent with the Golden Ratio. So I breached down the face using this criterion.

What is the number of Marilyn?
by Giulio Cigno-Membo
Author's Comment
The artwork consists of graphic elaboration of the photograph of Marilyn Monroe. I obtained the image from the decomposition of the original one by means of a program in C++. I prepared numerical chromatic points and placed two squares in bright colours, pleasantly in proportion to the rest of the picture.

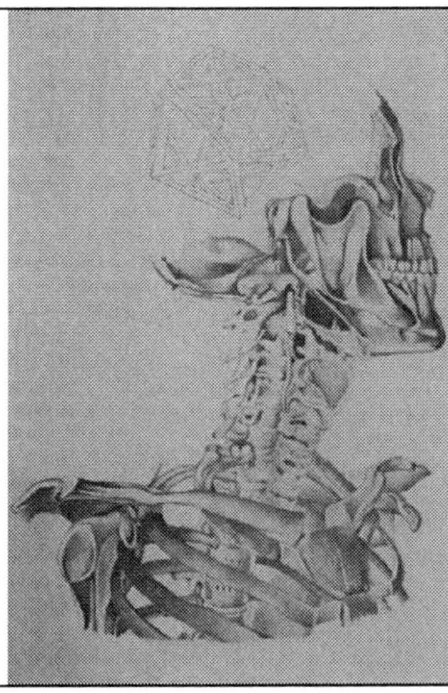

Abstraction by Mirko Cassago
Author's Comment
A Man, only because of the concrete contact with reality, is able to abstract what surrounds him and understand thoroughly the deepest meaning of materiality. Only by creating models, inspired by the practical experience, the human being is able to grasp the true essence of nature.

GuerMaturità by Michele Caruso - Dalila de Dominicis - Raffaele Pierro
Authors' Comment
The research of the relationship between art and mathematics led us to the study of the Golden Section. Our work is inspired by the painting *Guernica*. We focused on the study of this painting by Pablo Picasso for its powerful message and we decided to take a photograph based on it but making also a liberal use of irony. Via Photoshop, we enriched our work by various effects and graphic elements. Our artwork presents our difficult reality of students, which is enclosed in the Golden Ratio, although it may seem disastrous.

4 Mathematics and others Visual Media

The biographical approach is the simplest, and the most widely used in movies just like in literature. Many movies focuses on the person of the mathematician as such rather than the mathematical aspect of his work. In most cases, the artists narrate the life of an existed mathematician (for example in *A Beautiful Mind* [M6] or *Death of a Neapolitan Mathematician* [M2]); other times they focus on a fictional mathematician mainly on his psychological aspects (for example in *Good Will Hunting* or *Proof* [M4]).

Mathematical terminology is used more for suggestions that offers that for its scientific content, as in *21 Black Jack* [M7] or *Cube* [M3]. Another example is the Argentine film *Möbius* [M5], in which topological terminology is used to symbolize complexity beyond comprehension of the human mind.

Less usual the brave Disney's choice to introduce mathematical concepts in cartoons with amazing results, as in *Donald in Mathmagic Land* [M1], directed by Hamilton Luske.
The structural approach is more ambitious: it consists in inserting mathematical aspects in the structure of the work as such and not only for their symbolic value, as in [M8]. This can take place both within the plot, as more or less crucial elements for the development of the story, even in the formal composition of the work. A good example of the latter case is the drama work *Infinities* by Luca Ronconi and J.D. Barrow and *Numbers* (stylized NUMB3RS), the American crime drama television series created by Nicolas Falacci and Cheryl Heuton, that ran on CBS from January 23, 2005, to March 12, 2010.

Among the video presented to the contest, we found examples of both approaches.
In *The Banach-Tarski Paradox*[2], the author, Giovanni Affatato, presents the theorem in ironic way and after he uses doubling the ball as pretext to represent the dual essence of mathematics. In the

[2] The **Banach–Tarski paradox** is a theorem in set-theoretic geometry, which states the following: given a solid ball in a 3-dimensional space, there exists a decomposition of the ball into a finite number of disjoint subsets, which can be put back together in a different way to yield two identical copies of the original ball. The reason the Banach–Tarski theorem is called a paradox is that it contradicts basic geometric intuition.

first part of the video, he identifies the clarity and the transparency that are the basis of all logical-mathematical processes. In the second part, he highlights the extraordinary ability of the mathematics to amaze us even when we explore artistic languages, in which each element, like symmetry and fractal iterations, find new definition.

In *Math&Magic*, Laura Baldo represents some famous mathematical rules by means of magic tricks and illusions. A simple mathematical formula indeed may surprise if applied to particular objects with the right gestures and timings. The illusion games in the video are due to the courtesy of the magician Flavio Cyrano Negro.

Valeria Vergani in *Analytic Cubism* presents her geometric analysis of *Fabrica in Horta de Ebro*, painted by Pablo Picasso in 1909, during the period of the Cubism. Using GeoGebra, she reproduces the lines of the painting by means of the connection between points that are the intersections of right lines in the coordinate plane.

The quoted videos are published online in the following website:
www.fds.mate.polimi.it/matemartiamo

Conclusions

In the real life, students have gained experience of transformations and mathematical symmetries, mathematics are all around them and are waiting for being identified. Nevertheless, students in general do not love mathematics, perhaps due to the way the school shows it.

In our opinion, MatemArtiAmo gives the students an opportunity to learn techniques, use tools, and prove theorems while they build their artistic work with their hands and with their minds. The art making experience achieves value of cultural experience; by means of it, the students enhance both scientific and humanistic knowledge, using their personal skills.

Due the success of the first edition, FDS decided to launch a second edition of the contest MatemArtiAmo. We increased the sections and now we have Painting, Graphics, Meme, Photography, Literature, Video, Sculpture, Fashion and Jewellery Design. We extended the competition to students of the last three years of Secondary School and to all the students in the Politecnico di Milano.

Appendix A1

TEOFANIA (*Theophany*) by Emanuele Lo Presti

"Abbiamo vinto noi. Non moriremo, non ce ne andremo.

Nessuna lapide vedrà mai il nostro nome, il nostro cielo non conoscerà tramonto.

Siamo stati noi a portarvi fin qui, noi a condurvi per mano.

Senza di noi avreste brancolato nel buio: sareste potuti soccombere nel momento stesso in cui avete mandato giù il primo respiro.

Chi di voi ha mostrato gli astri agli Assiri, il giro dei pianeti agli antichi? Chi illuminò Galileo, o con quale inchiostro Copernico scrisse le sue poesie? Chi ha svelato per voi le leggi del firmamento, chi vi ha dato le chiavi con cui vi vantate di poter aprire ogni porta?

Siete forse perfetti come noi siamo, o potete allungare la vostra destra verso l'infinito? Potete toccarlo? Quanto veloce sa correre verso l'orizzonte il più rapido tra i cavalli dei vostri armenti? Sarà come la più lenta delle testuggini dell'Oceano di fronte a noi.

Hai forse tu parlato al crepuscolo, e gli hai detto quando abbracciare le selve? Dov'eri quando abbiamo scritto le leggi del mondo, quando abbiamo parlato alle stagioni mostrando loro il loro corso? Hai suggerito tu alle rondini il loro rifugio, indicato al leone la sua preda? Oppure hai guardato nel volto della tigre, e non ne hai temuto la forza? Hai tu soffiato i monsoni sulle Indie? Non ricordiamo di averti visto sul Parnaso. Ti ha forse Apollo donato la sua Arpa? Non cantavi con Orfeo nei boschi, dov'eri quando la bella Euridice gli fu tolta?

Rispondici, se puoi. Rispondici, e non ti interrogheremo ancora, non crederemo più che tu non sia grande come noi. Alla nostra voce rispondono i mari, per noi corre la luce.

Non saranno più copiose di noi le gocce di acqua del Giordano, nessun monte potrà mai superarci. Non ci conterai tu, né tuo figlio né tuo nipote, fino alla decima generazione, perché nemmeno le mura di Ilio potranno contenerci. Noi siamo la lingua attraverso la quale ha voce questo libro: niente esiste all'infuori di noi.

Noi siamo i Numeri."

Appendix A2

Love code (*Amore in codice*) by Elisa D'Elia

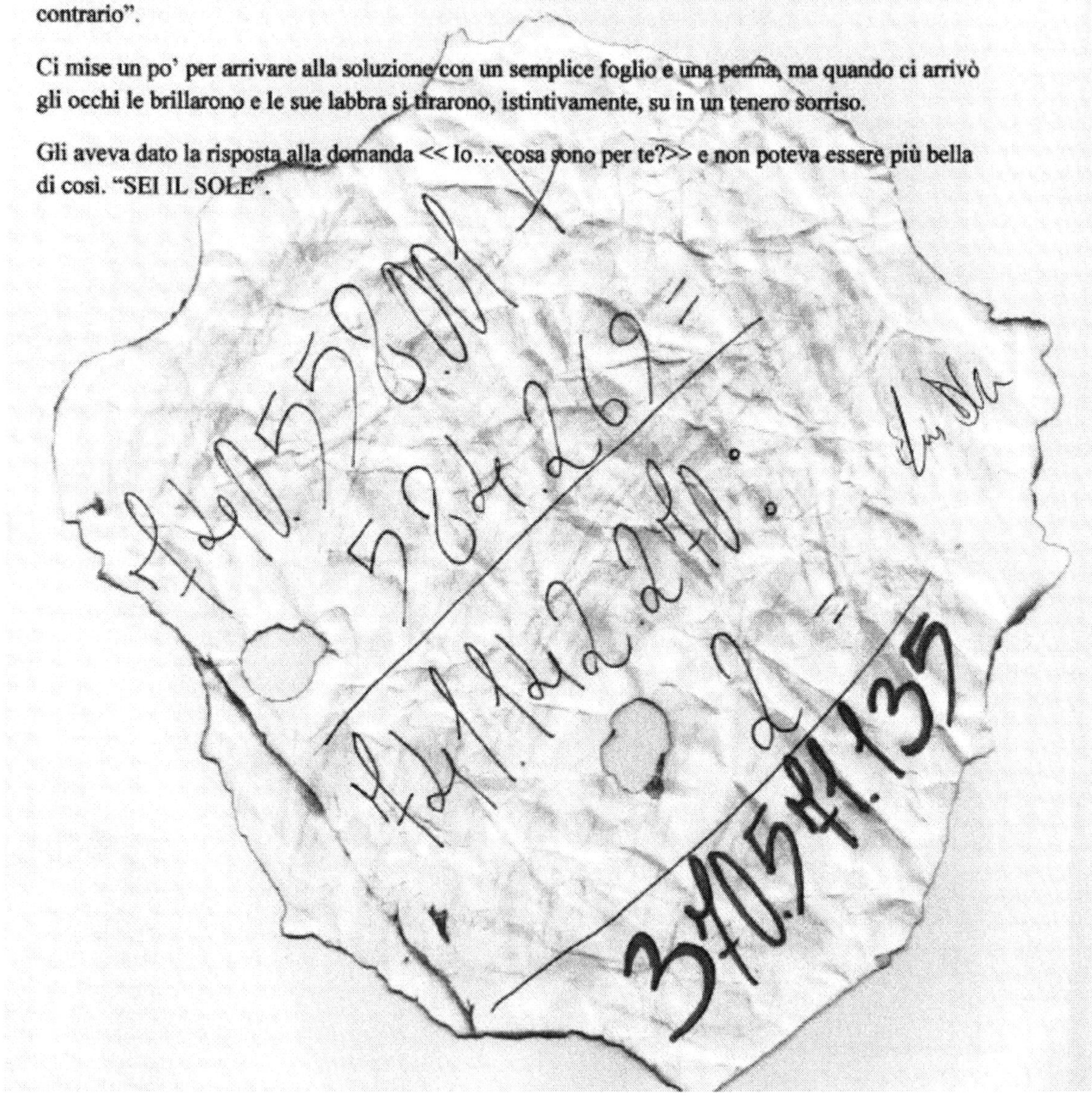

Rimase lì, con quel foglietto sgualcito in mano.

Pensava a lui e a quel giorno in cui glielo diede,

scritto con il loro solito codice, che utilizzavano sempre, ma ora più complesso.

Per arrivare ai numeri che gli interessavano doveva trovarsi la soluzione.

"Settecentoquarantamilionicinquecentocinquantottomilauno più cinquecentoottantaquattromiladuecentosessantanove il risultato diviso due e leggere la soluzione al contrario".

Ci mise un po' per arrivare alla soluzione con un semplice foglio e una penna, ma quando ci arrivò gli occhi le brillarono e le sue labbra si tirarono, istintivamente, su in un tenero sorriso.

Gli aveva dato la risposta alla domanda << Io… cosa sono per te?>> e non poteva essere più bella di così. "SEI IL SOLE".

Acknowledgements

The authors want to thank

- Anna Rho and Laura Guarino, EventiMate of Politecnico di Milano.Without their help, this adventure would have been impossible.

- Domenico Brunetto, doctoral candidate at Politecnico di Milano, for his will to try innovative ways and the precious help in technological difficulties.

- Marcella Giulia Lorenzi, for her friendly encouragement.

- Giulio Magli, full professor at Politecnico di Milano, for his convinced support.

References

[1] ABATE, M. Scrivere Matematica nel fumetto. In *Matematica e Cultura* 2003, Ed. M. Emmer, Springer Italia, Milano, 2004, pp. 19-29, ISBN 978-88-470-0210-4.

[2] ABATE, M., NATALI, S. Il Lemma di Levemberg. Lazarus Ledd Extra 3, Star Comics, Perugia, 1969.

[3] AUBURN, D. Proof, Faber and Faber, New York, 2001.

[4] CALVINO, I. Le città invisibili.(Invisibles Cities) Einaudi, Torino, 1972.

[5] CALVINO, I. Il castello dei destini incrociati (The Castle of Crossed Destines) Einaudi, Torino, 1973.

[6] CALVINO, I. Palomar.(Mr.Palomar) Einaudi, Torino, 1983.

[7] CASELLI, L., IABOLI, L., LOBACCARO, G., MADOGLIO, M. Doctor G – A Graphic Novel. 2016, URL: www.l-inkproject.com/doctor-g/

[8] FABBRI, F., PAROLINI, G., SARTORI, R., BOCCARDI, B., GIURGOLA, G. Adotta arte e scienza nella tua classe. A project of mathematics and physics popularisation in the italian middle and high schools, Art and Mathematics International Conference Aplimat, Bratislava, 2012

[9] HEMINGWAY, E, M. A Farewell to Arms, Scribnur, 1929.

[10] KANDINSKIJ, W. The Art of Spiritual Harmony, Sadler T.H. translation, 1946 URL: https://archive.org/details/onspiritualinart00kand

[11] KANDINSKIJ, W. Point and Line to Plane, Dearstyne H. and Rebay H. translation, 1947 URL: https://archive.org/details/pointlinetoplane00kand

[12] MAGNAGHI-DELFINO, P., NORANDO, T., Alessandro Mazzucotelli: an Artistic and Educational Project, Aplimat Journal of Applied Mathematics, Bratislava, 2016 , 771-791

[13] OSENDA, D. Last Lecture in Göttingen. 001 Edizioni, 2009

URL: http://www.001edizioni.com/

[14] SCLAVI, T.; BRINDISI, B. Three times Zero. Dylan Dog 125, Sergio Bonelli Editore, Milano, 1997

Movies

[M1] Donald in Mathmagic Land. Directed by LUSKE, H., USA, 1959 URL: https://www.youtube.com/watch?v=U_ZHsk0-eF0

[M2] Morte di un matematico napoletano. Directed by MARTONE, M., Italia, 1992

[M3] Cube. Directed by NATALI, V., Canada, 1997

[M4] Good Will Hunting- Genio ribelle. Directed by VAN SANT, G., USA, 1998

[M5] Moebius. Directed by MOSQUERA, G., Argentina, 1998

[M6] A Beautiful Mind. Directed by HOWARD, R., USA, 2001

[M7] 21 Black Jack. Directed by LUKETIC, R., USA, 2008

[M8] Le ruban de Moebius. Directed by BLONDEAU, E., France, 2011
URL: https://www.youtube.com/watch?v=wfkxEsKrOyo

Current address

Magnaghi-Delfino Paola, Researcher
Dipartimento di Matematica, Politecnico di Milano
piazza Leonardo da Vinci 32, 20133 Milano (MI), Italy
Tel. number: +390223994515, e-mail: paola.magnaghi@polimi.it

Norando Tullia, Assistant Professor
Dipartimento di Matematica, Politecnico di Milano
piazza Leonardo da Vinci 32, 20133 Milano (MI), Italy
Tel. number: +390223994533 e-mail: tullia.norando@polimi.it

MATEMARTIAMO
Concorso d'arte & matematica

Slovak University of Technology in Bratislava
Faculty of Mechanical Engineering
16th Conference on Applied Mathematics **APLIMAT 2017**
Proceedings

FOLDING CUBIC ROOTS: MARGHERITA PIAZZOLLA BELOCH'S CONTRIBUTION TO ELEMENTARY GEOMETRIC CONSTRUCTIONS

MAGRONE Paola (IT), TALAMANCA Valerio (IT)

Abstract. In this paper we describe the contributions made by Margherita Piazzolla Beloch to the field of elementary geometric construction. In particular we examine her solution of the classical Greek problem of doubling the cube via paper folding. The solution was achieved by creating a new fold which in turn enabled Piazzolla Beloch to construct, also by paper folding, the root of any given cubic polynomial.

Keywords: Paper folding, projective geometry, roots of cubic polynomial, elementary geometric constructions

Mathematics subject classification: Primary 51M15; Secondary 51N15

1 Introduction

Paper folding, or origami, has been known for centuries as a fine art form, a way to transform a piece of paper into a stunning three dimensional object. In the last 30-40 years it has been understood that paper folding can also be an important scientific and technological tool. One of the pioneers of the application of paper folding to geometrical problems is Margherita Piazzolla Beloch. In her book *Lezioni di matematica complementare, la matematica elementare vista dall'alto* Piazzolla Beloch devotes the last two chapters to elementary solutions of classical geometric problems. It is probable that her deep interest in elementary mathematics, a common thread among Italian algebraic geometers at the beginning of the '900 (cf. [7]), is one of the main reason in Piazzolla Beloch's appreciation and usage of paper folding. Mainly because paper folding is a powerful but simple tool to perform geometric constructions. Following Piazzolla Beloch, see [21, p.353], we call a geometric problem a *problem of third degree* if its resolution requires finding a root of a cubic polynomial irreducible over the rational numbers. Gauss was the first one to state, even though without proof (see [6]), that angle trisection and cube doubling are not solvable by ruler and compass. The first published proof is due to M.L. Wantzel (cf. [30]) who actually proved that every problem of the third degree cannot be solved by ruler and compass. In her book, Piazzolla Beloch after describing and investigating classical solutions via mechanical instruments and higher degree curves turns her attention to a new method: paper folding.

Margherita Piazzolla Beloch's source of information about paper folding was the book written by Sundara Row [23] and an article of Rupp [26]. She published her reflections about paper folding in a series of articles ([18], [19], and [20]), which appeared several years before her lecture notes [21]. Her fundamental contribution was the discovery of what is now often called[1] Beloch's fold: *Given two point P_1 and P_2 and two lines r_1 and r_2 then, when it exists, we can fold the line reflecting P_1 onto r_1 and P_2 onto r_2.* (Fig. 1)

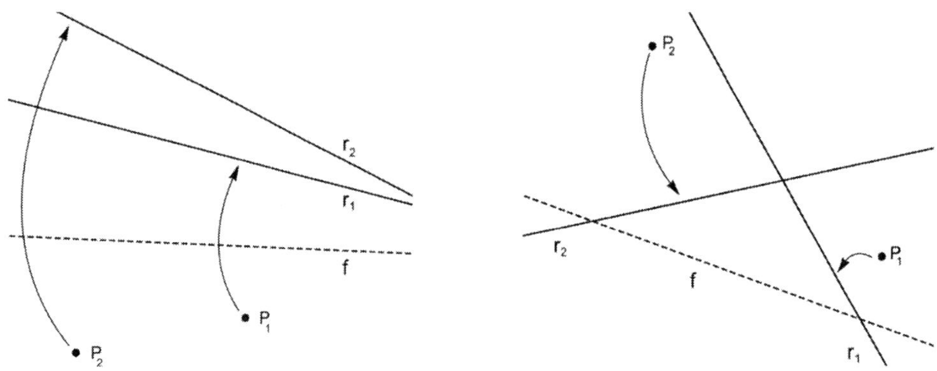

Fig. 1. Two instances of Piazzolla Beloch's fold

It was this fold that enabled Piazzolla Beloch to show that by using paper folding it is possible to double the cube and, with the aid of Lill's graphical method, to solve all cubic equations. As we shall see in section 7, in all the applications that Piazzolla Beloch considered it was evident that one could perform the above fold. Nonetheless it is interesting to establish a sufficient condition for when it is possible to execute Piazzolla Beloch's fold. We provide such a sufficient condition by means of (synthetic) projective geometry, see also [8] where similar conditions are established by means of analytic geometry.

This paper is organized as follows: section 2 contains a short biography of Margherita Piazzolla Beloch. In section 3 we examine her writings around solutions via mechanical instruments and higher degree curves of classical geometric problems. Section 4 encompasses a brief history of the application of paper folding to geometric problems. In section 5 we introduce Piazzolla Beloch's fold as well as the other six basic folds of paper folding; section 6 contains our deduction of the sufficient condition for executing Piazzolla Beloch's fold, based on synthetic projective geometry and Bezout's theorem. In the last section we exhibit three different approaches (graphical, geometrical and algebraic) to show that Piazzolla Beloch's fold can be used to construct a solution to any given cubic equations. The graphical approach, via Lill's method, was the one used originally by Piazzolla Beloch, and has been revisited several times since then. The other two, although quite straightforward, have not received the same kind of attention.

[1]Most authors who are native English speakers mistakenly think that Piazzolla is her middle name.

2 A few biographical notes about Margherita Piazzolla Beloch

The original source of information about Margherita Piazzolla Beloch life is the introduction of the guide [5] describing the "Collection Montesano", a private collection of offprints owned by Domenico Montesano which was acquired by Piazzolla Beloch on behalf of the library of the Mathematics department of Ferrara University.

Margherita Piazzolla Beloch was born in Frascati, near Rome, in 1879. She graduated in mathematics from Sapienza Università di Roma, with "dignità di stampa"[2], in 1908, under the supervision of Guido Castelnuovo. Her thesis, entitled "On birational transformations of space", was indeed published in the Annali di Matematica Pura ed Applicata, [17], one of the oldest math journal in Italy. Her first official academical appointment was as an assistant to the chair of "Descriptive Geometry" at the University of Pavia, and then in Palermo. In 1927 she became full professor of geometry at the University of Ferrara. In Ferrara she gave courses on many topics, such as descriptive geometry, higher geometry, complementary mathematics, superior mathematics.

In 1955 she retired but continued to be very active publishing widely for several years. In the mid '60 she personally prepared for publication a selection of her works [22], which was finally published in 1967. It consists of over 50 research papers which were divided in the following three topics: algebraic geometry, projective topology, and photogrammetry. The latter field was particularly dear to her. As stated in the preface of [22] it was the Italian Society of Photogrammetry and Topography that in 1961 proposed the publication of her selected works, to show their gratitude and affection to the scientist. As she herself acknowledges this was her "preferred field of study".

3 Classical problems: geometric construction and mechanical devices

Of the many geometrical problems studied by the ancient Greek only three gained enduring fame: doubling a cube, trisecting an angle, and squaring the circle. They owe their celebrity chiefly to the fact that they withstood every attempt to be solved by ruler and compass for centuries, until, in the nineteenth century, it was proven that was impossible to solve them with only ruler and compass, see [11] for a detailed account.

Piazzolla Beloch's interest in these matters was more on the positive side so to say: she was interested on which mechanical devices one needed to solve a particular problem or alternatively which curve of higher degree was needed to obtain the desired construction. We now briefly review the content, relevant to our investigation, of the two chapters of [21] dedicated to geometric construction and mechanical devices. After giving a detailed construction of the cissoid of Diocles, the conchoid of Nicomedes, and the Quadratrix of Hippias, she used them to solve the above mentioned classical problems. Specifically: doubling a cube can be solved by using either the cissoid or the conchoid, trisecting an angle is solved by the conchoid, and squaring the circle require the use of the Quadratrix.

She also remarks that any third degree problem can be reduced, by the use of ruler and compass, to either doubling a cube or trisecting an angle. Since both these problems are solvable by the conchoid, she concludes that every third degree problem is solvable by the combined use of ruler, compass and the conchoid.

[2] worthy of publication

Having settled these matters she takes a slightly different point of view. Suppose we fix the set of tools we are allowed to use (where tools mean both mechanical devices and curves). We then call the *frame of problems* of the given set of tools all the geometric problems that set solves. So for example the frame of problems of ruler, compass and conchoid of Nicomedes contains all the problem of degree less or equal to three. If two sets of tools have the same frame of problems they are called *equivalent*. For example ruler and compass is equivalent to the ruler and a fixed circle completely drawn. She then determines the frame of problems for several mechanical (hinged) devices, with special regard to those constructed by Alfred B. Kempe in [10].

The next set of tools she examines is paper folding. Firstly she remarks that several geometric constructions are more readily done with paper folding than ruler and compass e.g. drawing a line through a point perpendicular to a given line takes only one fold while one has to use the compass three time and the ruler once. She then continues showing how to bisect an angle, find the intersection of a given circle and a given line and lastly find the intersection of two given circles by paper folding. As a consequence paper folding can solve all the problems that ruler and compass can. She then shows that paper folding can actually solve all third degree problems. We examine this latter topic in greater details in sections 5, 6, and 7.

4 A few remarks on the history of geometric constructions via paper folding

As remarked by Piazzolla Beloch in [20], the first author to draw attention to the use of paper folding in the solution of classical geometric problems was Felix Klein in [11]

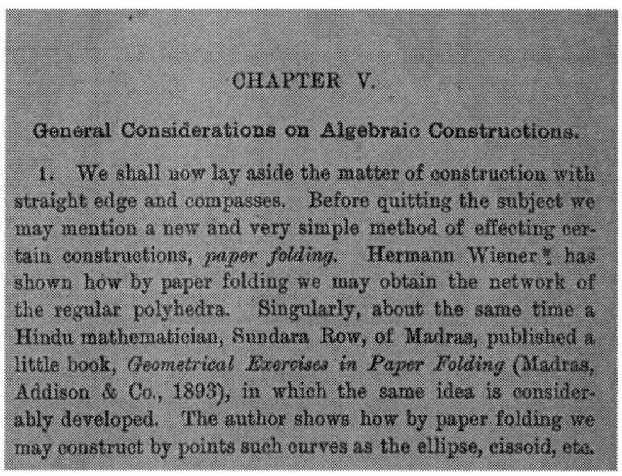

CHAPTER V.

General Considerations on Algebraic Constructions.

1. We shall now lay aside the matter of construction with straight edge and compasses. Before quitting the subject we may mention a new and very simple method of effecting certain constructions, *paper folding*. Hermann Wiener* has shown how by paper folding we may obtain the network of the regular polyhedra. Singularly, about the same time a Hindu mathematician, Sundara Row, of Madras, published a little book, *Geometrical Exercises in Paper Folding* (Madras, Addison & Co., 1893), in which the same idea is considerably developed. The author shows how by paper folding we may construct by points such curves as the ellipse, cissoid, etc.

Fig. 2. Klein's reference to Wiener's work and Sundara Row's book

In his brief remark Klein mentions the two forefathers of the application of paper folding to geometric problems: Herman Wiener and Sundara Row. While Wiener's contribution to paper folding has been largely forgotten, Sundara Row's book has become a classic and has enjoyed a flurry of re-printings (Fig.3)

Fig. 3. A selection of covers of reprints of Sundara Row's book

The version of *Geometric exercises in paper folding* by Sundara Row that can be found nowadays is the one edited by W.W. Beman and D.E. Smith. They also became interested in the work of Sundara Row thanks to Klein. In fact the first sentence of the preface of their edition of the book reads "*Our attention was first attracted to Sundara Row's book by a reference in Klein's Vorträge über ausgewählte Fragen der Elementargeometrie*", a text which they translated from German into English. Beman and Smith in their edition of Sundara Row's book added several pictures of actual folded paper, see Fig. 4 as an example , in order to replace some of the illustrations present in the original version, which were mere line drawings.

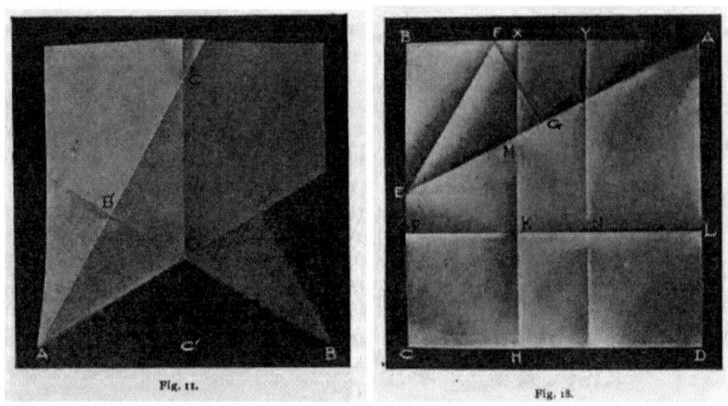

Fig. 4. Two pictures taken from the book by Sundara Row

It is only in the recent article by Michael Friedman [3] that Herman Wiener work on paper folding is examined. As both the book of Sundara Row and the article by Wiener on paper folding appeared in 1893, Friedman asserts that 1893 may be thought of as the first year that saw a modern mathematical treatment of the problem of determining what kind of object can be produced by folding paper. Wiener exhibited some paper objects during an exhibition on mathematical tools and models in 1892 in Munich and wrote a short note [31] for the catalogue of the exhibition, which is admittedly hard to follow being without the illustrations which were present at the exhibition. On the other hand, as Friedman points out, there are some interesting observation in the paper but a general lack of mathematical rigor: for example he gives the instructions to construct a regular pentagon by folding and knotting a strip of paper (Fig. 5), but does not supply a proof of the regularity (a proof can be found in [15]).

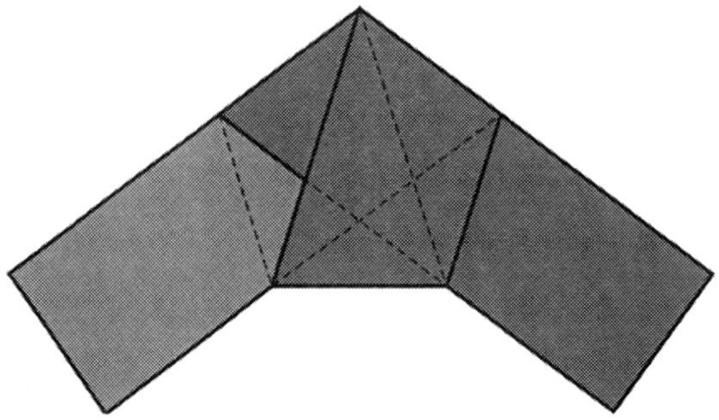

Fig. 5. The construction of a regular pentagon by Wiener

It must be noted that also Piazzolla Beloch contributions went unnoticed for several year as mathematicians generally lost interest in the subject. It was only in the late '80 that, thanks to the efforts of Huzita Humizaki with the aid of Benedetto Scimemi, there was a revival of interest in paper folding. The first International meeting on Origami Science and Technology, organized by Humizaki, was held in Ferrara in December 1988 [1]. It is rather peculiar that even though the first pages of the proceeding of this meeting host a reproduction of Piazzolla Beloch's article on paper folding many of the subsequent authors were not aware of her contributions. The Origami meeting have now become a regular event, the last of which was held at Tokyo university in 2014 and its proceedings are published by the American Mathematical Society [16].

5 Margherita Piazzolla Beloch innovative folding

Margherita Piazzolla Beloch seems to have been the first one to discover that problems of the third degree can be solved by means of paper folding. At that time the applications of paper folding to geometry were still in their infancy and a list of basic folds was not available. We now have such a complete list of basic foldings (see 6), sometimes improperly called axioms. It has to be noted that some foldings were discovered several times over, see the introduction of [9] and [13], for more details.

There is still one basic fold: Piazzolla Beloch's fold. This fold appeared first in print in the short note [18], an extract from the lecture notes for her course *Matematiche complementari* containing the new result about paper folding. Let us recall its definition: *Given two point P_1 and P_2 and two lines r_1 and r_2 then, when it exists, we can fold the line reflecting P_1 onto r_1 and P_2 onto r_2.* (cf. Fig. 1). To see that, when possible, one can actually perform this fold we quote B. Scimemi (see [28]) "taking advantage of translucency, first superimpose P_1 on r_1, then let P_1 run along r_1 and meanwhile pay attention to the motion of the point P_2, waiting until it comes to lie on the line r_2".

The folds O1-O5 were already extensively used by Sundara Row in [23]. Fold O7 became

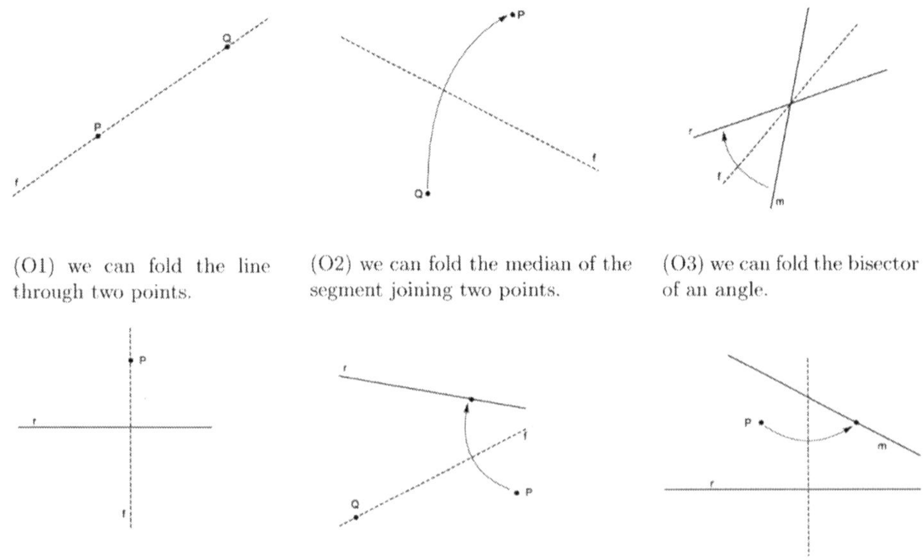

(O1) we can fold the line through two points.

(O2) we can fold the median of the segment joining two points.

(O3) we can fold the bisector of an angle.

(O4) we can fold the line perpendicular to a given line passing through a given point.

(O5) Given two point P and Q and a line r, we can fold the line passing through Q and reflecting P onto r.

(O7) Given a point P and two lines r and m we can fold the line perpendicular to r reflecting P ont m

Fig. 6. Six of the seven basic paper foldings

widely known only after Koshiro Hatori announced its discovery in 2002, but Justin (see [12]) already listed it as one of his seven basic fold, 13 years earlier.

Piazzolla Beloch introduced O6 to solve the following problem:*Given two points P_1 and P_2 and two lines r_1 and r_2 construct a square with vertices X, Y, W, and Z so that X and Y lie on r_1 and r_2 and P_1 (respectively P_2) lies on the line joining X and Z (respectively Y and W)* (Fig. 7):

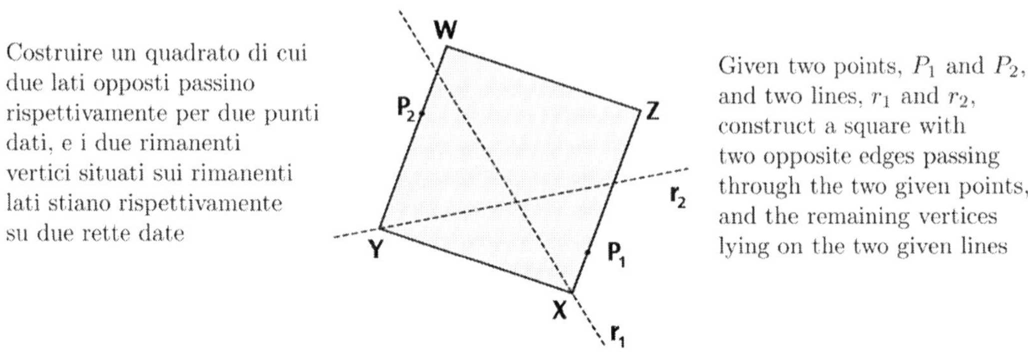

Fig. 7. Piazzolla Beloch's square

This square construction is the key to construct solutions of third degree problems by paper folding. To show why fold O6 enables us to construct the square in Fig. 7, we have to give a geometrical interpretation of fold O5 and O6. As was already known to Sundara Row the folding line in O5 is the tangent line, in Q, to the parabola, $\mathcal{P}_P^r$, having r has directrix and

P has focus. In fact as Piazzolla Beloch remarks in [18] one can reconstruct a parabola by folding its tangents in the following way [3]: use the bottom edge of a sheet of paper as the directrix of the parabola, and mark the focus at a given distance from it. Then fold the paper on itself, without moving the focus making sure that the bottom edge of the sheet passes through the focus.

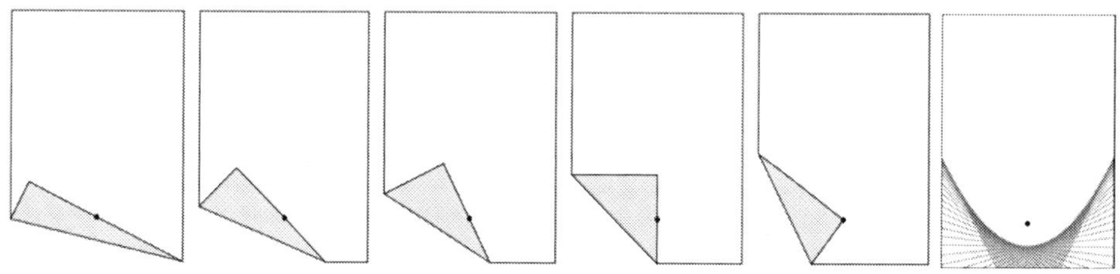

Fig. 8. Folding the tangents of a parabola

The folding line produced by fold O6 is then a common tangent to $\mathcal{P}_{P_1}^{r_1}$ and $\mathcal{P}_{P_2}^{r_2}$. To show way fold O6 enable us to construct the square of Fig. 7, we argue along the lines of [18]: consider the parabola $\mathcal{P}_1$ (respectively $\mathcal{P}_2$) having focus in P_1 (resp. P_2) and whose tangent line in the vertex is r_1 (resp. r_2). Then we perform fold O6 and the folding line, call it f, is tangent to both $\mathcal{P}_1$ and $\mathcal{P}_1$. Let X (respectively Y) be the intersection of f and r_1 (resp. r_2), see Fig. 9

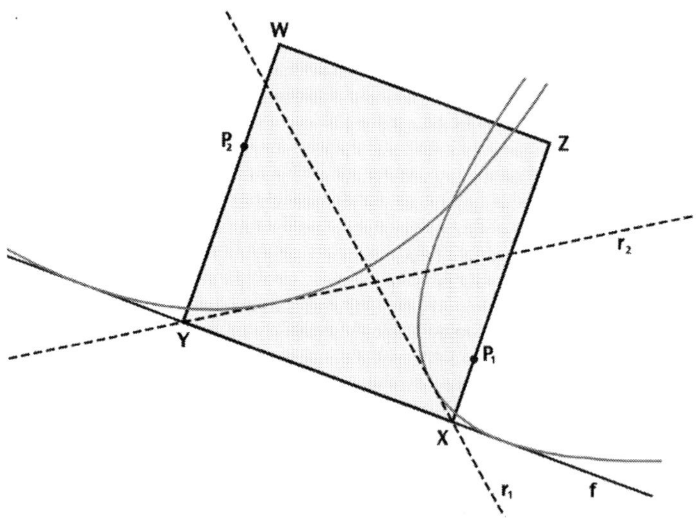

Fig. 9. The underlying parabolae in Piazzolla Beloch's square

The line passing through X (resp. Y) and perpendicular to f passes through the focus of $\mathcal{P}_1$ (resp. $\mathcal{P}_2$) which is P_1, (resp. P_2). Therefore the square having the segment $\overline{XY}$ as an edge is the desired square.

[3]"*Prendendo cioè l'orlo (rettilineo) di un foglio di carta come direttrice della parabola, e segnando il fuoco alla data distanza da questa, basta tener fermo il fuoco e ripiegare la carta su se stessa in modo che l'orlo ripiegato venga a passare per il fuoco.*" [18, p.187]

6 Piazzolla Beloch's fold and projective duality

As we already remarked is not always possible to perform fold O6 and most probably Piazzolla Beloch was well aware of this. Moreover, as we shall see in the last section, all the instance of fold O6 that she needed to perform in her investigation were clearly possible. Even though several authors seem unaware that fold O6 can not be performed in all cases, already in [8] there is a discussion, in terms of analytic geometry of when it is possible, see also [2] for an approach closer to the one presented here. We would like now to give a simple criterion for when two parabolae do have a common tangent in terms of projective geometry and projective duality in the plane. Projective duality was well known to Margherita Piazzolla Beloch, who extensively worked on projective algebraic curves. We briefly recall the basic facts about projective duality, referring the interested reader to [24].

So let $\mathbb{A}^2(\mathbb{R})$ denote the (affine, real) plane and $\mathbb{P}_2(\mathbb{R}) = \mathbb{A}^2(\mathbb{R}) \cup \ell^\infty$ the projective real plane obtained from $\mathbb{A}^2(\mathbb{R})$ adjoining ℓ^∞ the "line at infinity" whose points represent the directions of lines in $\mathbb{A}^2(\mathbb{R})$. Formally points in $\mathbb{P}_2(\mathbb{R})$ are represented by three homogeneous coordinates $P[X : Y : Z]$ and $\mathbb{A}^2(\mathbb{R})$ correspond to the open subset given by $Z \neq 0$. It is straightforward to check that all parabolae in the affine plane are actually tangent to the line at infinity, the tangency point being the direction of the symmetry axis of the parabola.

Now we come to projective duality in the plane. The dual projective plane $\mathbb{P}_2^*(\mathbb{R})$ is the set of all lines in $\mathbb{P}_2(\mathbb{R})$, and points in $\mathbb{P}_2(\mathbb{R})$ correspond to lines in $\mathbb{P}_2^*(\mathbb{R})$. Given a conic $\mathcal{C} \in \mathbb{P}_2(\mathbb{R})$ its dual conic $\mathcal{C}^*$ is the set of all tangent lines to $\mathcal{C}$. It is readily seen that $\mathcal{C}^*$ is a conic as well. Therefore the common tangents to two parabolae $\mathcal{C}_1$ and $\mathcal{C}_2$ are the point of intersections of the dual conics $\mathcal{C}_1^*$ and $\mathcal{C}_2^*$.

By Bézout's theorem [4, section 5.3], there are exactly 4 points of intersection if counted with the appropriate multiplicity. In this simple case the multiplicity is one unless $\mathcal{C}_1^*$ and $\mathcal{C}_2^*$ have the same tangent line at the point of intersection, in which case the multiplicity is 2. This means that there is a line which not only is tangent to both $\mathcal{C}_1$ and $\mathcal{C}_2$, but also that the point of tangency is the same.

Now let us go back to the problem of finding the common tangent to two given parabolae $\mathcal{C}_1$ and $\mathcal{C}_2$. As we already remarked the line at infinity is tangent to both $\mathcal{C}_1$ and $\mathcal{C}_2$, thus $\mathcal{C}_1^*$ and $\mathcal{C}_2^*$ always intersect in the point Q_0 of $\mathbb{P}_2^*(\mathbb{R})$ corresponding to the line at infinity of $\mathbb{P}_2(\mathbb{R})$. It follows that the remaining points of intersection can be found by solving the cubic equation obtained by eliminating the appropriate variable, unless the multiplicity of Q_0 is two, which happens if and only if $\mathcal{C}_1$ and $\mathcal{C}_2$ have parallel symmetry axes. If $\mathcal{C}_1$ and $\mathcal{C}_2$ have parallel symmetry axes we are left with a quadratic equation which may or may not have real solutions, and both cases do happen. If $\mathcal{C}_1$ and $\mathcal{C}_2$ do not have parallel symmetry axis then the common tangents in $\mathbb{A}^2(\mathbb{R})$, different from the line at infinity, are either 3 or 1 and both cases do happen (10).

7 Folding solutions of cubic equations

In this section we will show in three different ways, graphic, geometric, and algebraic, how Margherita Piazzolla Beloch's folding enables us to construct solution of cubic equations by paper folding. As is clear from her writings she was aware of all of them but only wrote down the details for the graphical one. The graphical method is based on Lill's method (see [14], [11]) to find real roots of polynomials (with real coefficients) of every degree.

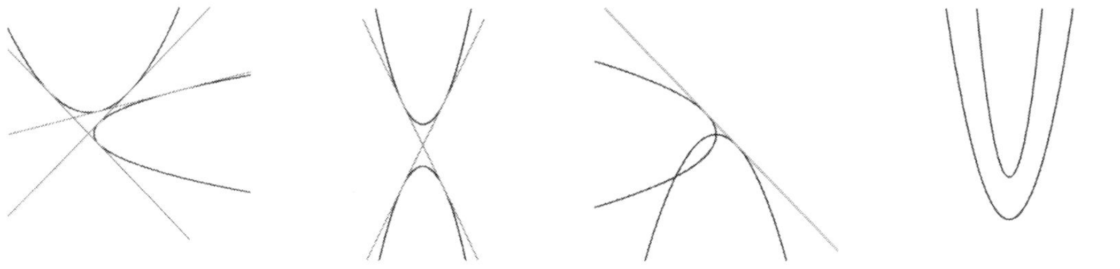

(a) Three common tangents (b) Two common tangents (c) One common tangent (d) No common tangent

Fig. 10. Common tangents to two parabolae

Lill's method was used by Margherita Piazzolla Beloch in her original solution of cubic equations by paper foldings in [20]. Since then, a few authors dwell on the subject, notably Benedetto Scimemi [27] and Thomas C. Hull [9]. Therefore we will be rather brief on Lill's method and concentrate on the application of Piazzolla Beloch's fold. Being a graphical method we will relay heavily on illustrations.

Lill's method, which we describe here only for cubic polynomials, goes as follows (we use the convention of [27]): given a cubic polynomial $P(X) = a_3x^3 + a_2x^2 + a_1x + a_0$ we construct a 4 edge polygonal chain $OA_3A_2A_1A_0$, with only right angles, where the starting point is the origin O, the length of the segment ending in A_i is $|a_i|$ and in A_i the turn is made clockwise if $a_ia_{i-1} \geq 0$ and counterclockwise if $a_ia_{i-1} \leq 0$. As usual a picture is worth a thousand words so we refer the reader to Fig. 11. To find a root Lill proceeds as follows: draw a line from the origin, forming an angle of θ radiants with the x-axis and whenever it hits a edge of the Lill's polygon, or its continuation, it bounces off on the line orthogonal to r, choosing the direction that ensures that it will hit the next edge of the Lill's polygon (or its continuation). Since we are in degree 3, the Lill's polygon has at most 4 edges and we have to perform only two reflections. If after the second reflection we end up with a line which passes through the point P_0 then $\tan(\theta)$ is a root of the cubic polynomial associated to the polygon. Also in this case it is much more illuminating to look at a picture so we refer the reader to Fig. 11.

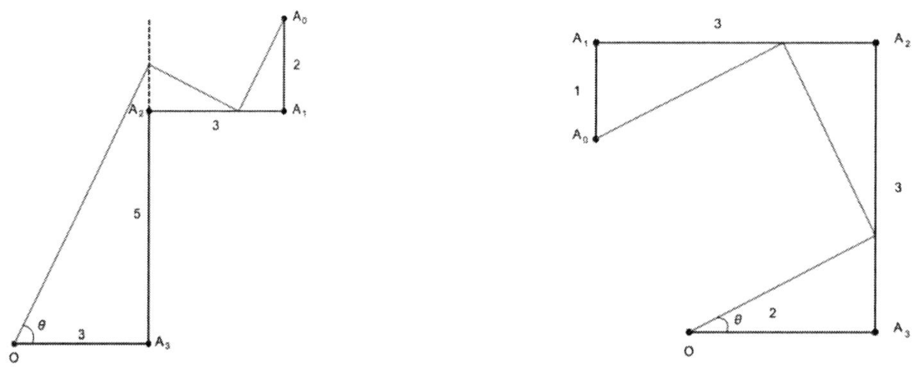

(a) $\tan(\theta) = 2$ is a root of $3x^3 - 5x^2 - 3x + 2$ (b) $\tan(\theta) = \frac{1}{2}$ is a root of $2x^3 - 3x^2 + 3x - 1$

Fig. 11. Finding roots of cubic polynomials via Lill's polygons

Piazzola Beloch's square construction shows that we can always find, by paper folding, such an angle θ. To see this suppose first that all the coefficients of $P(x)$ are not zero. Then we can apply Piazzola Beloch square construction to the points O and A_0 and the lines r_1 joining A_3 and A_2, and r_2 joining A_2 and A_1. Since r_1 and r_2 are perpendicular, hence not parallel, we are sure that we can perform Piazzolla Beloch's fold and construct a square with O on one edge or on its continuation, A_0 on the parallel edge and two vertices on r_1 and r_2. Therefore the line passing trough the origin on which the edge of the square lies is the line required to find the root of the cubic polynomial by Lill's method.

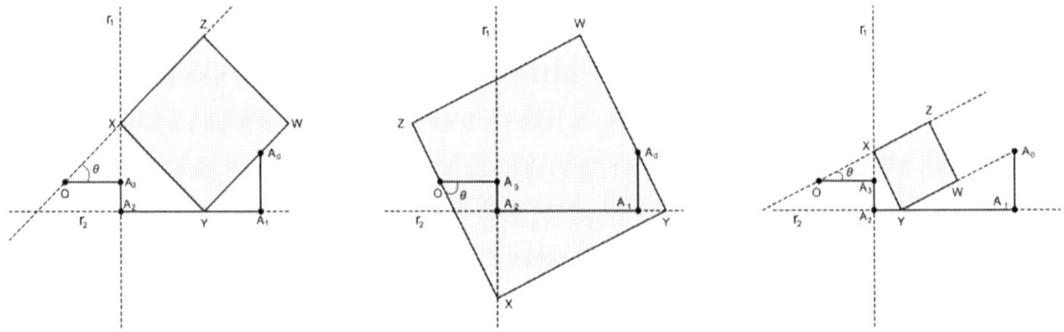

(a) $\alpha = 1$ (b) $\alpha = -2$ (c) $\alpha = \frac{1}{2}$

Fig. 12. Piazzolla Beoch' s squares correspondig to the three real roots of $x^3 + \frac{x^2}{2} - \frac{5}{2}x + 1$

It is possible to apply Piazzolla Beloch's construction also if some of the coefficients vanishes, one has just to be careful in the choice of the lines r_1 and r_2. Namely one line will be the one connecting the two vertices of the polygon and the other one has to be choose perpendicular to the previous one passing to the vertex corresponding to the coefficient preceding the missing one. One argues similarly if there are two missing coefficients. Note that we only need to deal with polynomials of the following form $x^3 + a_2x^2 + a_0$, $x^3 + a_1x + a_0$ and $x^3 + a_0$. As before, everything is clearer if one looks at a picture and so we refer the reader to Fig. 13.

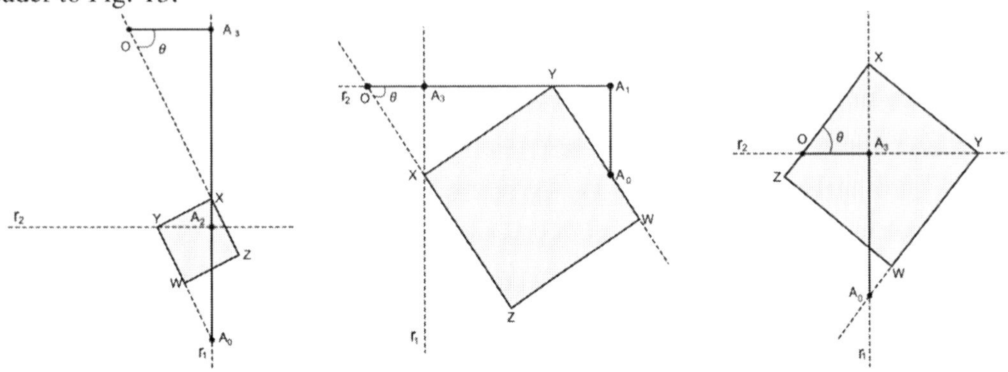

(a) $P(x) = x^3 + \frac{7}{3}x^2 - \frac{4}{3}$ (b) $P(x) = x^3 - \frac{13}{4}x - \frac{3}{2}$ (c) $P(x) = x^3 - 2$

Fig. 13. Piazzolla Beloch's square construction for polynomials with missing terms

It has to be noticed that in order to construct the cubic root of a given number, one does not need Lill's method but it is enough to use Piazzolla Beloch's square construction directly as shown in [18].

Next we turn to geometry. Let $P(x) = a_0 + a_1x + a_2x^2 + x^3$ be a generic monic cubic polynomial. It suffices to show that we can find its real roots via paper folding. To do so we have to exhibit two parabolae whose common tangents have slopes that are precisely the real roots of $P(x)$. Piazzolla Beloch's square construction suggests that should be enough to search among parabolae that have symmetry axis either horizontal or vertical. So we consider the following two families of parabolae

$$\mathcal{P}_c : y = x^2 + c \qquad \mathcal{P}_{a,b} : x = ay^2 + b.$$

A straightforward computation shows that the slopes of the lines that are common tangents to $\mathcal{P}_c$ and $\mathcal{P}_{a,b}$ are the roots of $t^3 - 4bt^2 - 4c + \frac{1}{a}$. The self-evident choice of a, b, c shows that we can recover the polynomial $P(x)$.

We come at last to the algebraic approach. Our construction is based on Cardano's formula for solving the cubic equations. First of all a general cubic polynomial $P(x) = a_0 + a_1x + a_2x^2 + x^3$ can be reduced, by a linear change of variable, to a polynomial of the form $x^3 + px + q$. Next Cardano's formula gives us a real root for this type of polynomials. Namely:

$$x = \sqrt[3]{-\frac{q}{2} + \sqrt{\frac{q^2}{4} + \frac{p^3}{27}}} + \sqrt[3]{-\frac{q}{2} - \sqrt{\frac{q^2}{4} + \frac{p^3}{27}}}.$$

Since it is possible by paper folding to construct, without the aid of Lill's polygon, both the square root and the cubic root of a given quantity, it follows, again, that by using paper folding we can construct the root of cubic polynomial.

Acknowledgments We would like to thank Benedetto Scimemi for sharing his views and memories around paper foldings, Corrado Falcolini and Laura Tedeschini Lalli for several discussion on the subject of paper folding, Marcello Liberato who brought to our attention the book of Sundara Row, and Ciro Ciliberto for lending us his copy of the selected works of Margherita Piazzolla Beloch. We are grateful to Laura Tedeschini Lalli for passing on to us the idea of Mario Fiorentini to inquire more into the figure of his former colleague Margherita Piazzolla Beloch.

References

[1] *Proceedings of the first international meeting on Origami Science and Technology*, Ferrara, Italy, 6-7 December 1989, H. Huzita editor. (Printed by the Physics Department "Galileo Galilei" of the Università di Padova).

[2] Alperin, R. C.: *A mathematical theory of origami constructions and numbers*, New York J. Math., Vol 6 (2000).

[3] Friedman, M.: *Two beginnings of geometry and folding: Hermann Wiener and Sundara Row*, Bulletin of the British Society for the History of Mathematics, 2015 published on-line at `http://dx.doi.org/10.1080/17498430.2015.1045700`.

[4] Fulton, W.: *Algebraic curves*, Benjamin Press, New York, also available on line at `http://www.math.lsa.umich.edu/ wfulton/CurveBook.pdf`

[5] Gambini G., Pepe L.: *La raccolta Montesano di opuscoli nella biblioteca dell'istiuto matematico dell'università di Ferrara*, available online at `http://dm.unife.it/comunicare-la-matematica/filemat/pdf/rcmont.pdf`.

[6] Gauss, C. F.: *Disquisitiones Arithmeticae, english translation*, English Edition, Springer Verlag New York Berlin Heidelberg Tokyo 1986. Latin original version available at http://edoc.hu-berlin.de/ebind/hdok2/h284_gauss_1801/pdf/h284_gauss_1801.pdf

[7] Giacardi, L.: *The Italian School of Algebraic Geometry and Mathematics Teaching in Secondary Schools. Methodological Approaches, Institutional and Publishing Initiatives*, International Journal for the History of Mathematics Education 5, 1, (2010), 1-19.

[8] Huzita, H., Scimemi, B.: *Algebra of Paper Folding (Origami)* in [1]

[9] Hull, T. C.: *Solving cubics with creases: the work of Beloch and Lill*, Amer. Math. Monthly, Vol. 118, 2011.

[10] Kempe, A. B.: *How to draw a straight line; a lecture on linkages*, MacMillan and Co., London, 1877.

[11] Klein, F. *Elementarmathematik vom höheren Standtpunkt aus*, second edition, Teubner, Leipzig, 1911; Jbuch 42, 109

[12] Justine, H.: *Resolution par le pliage de l'equatione de troisieme degre et applicationes geometriques*, in [1]

[13] Lang, R.: *Origami and Geometric Constructions*, available on line at: http://www.langorigami.com/files/articles/origami_constructions.pdf

[14] Lill, E. M.: *Résolution graphique des équations numériques de tous degrés à une seule inconnue, et description d'un instrument inventé dans ce but.* Nouv. Ann. de Math. (2), Vol 6 (1867) pp. 359-362

[15] Maekawa, J.: *Introduction to the study of tape knots* in Origami 5: Fifth International Meeting of Origami Science, Mathematics, and Education, Boca Raton P. Wang-Iverson, R. J. Lang, and M. Yim (eds), A K Peters/CRC Press, 2011, 395-403.

[16] *Origami*^{6} K. Miura, T. Kawasaki, T. Tachi, R. Uehara, R. J. Lang, P. Wang-Inverson (eds), AMS, Providence R.I., 2015

[17] Piazzolla Beloch, M.: *Sulle Trasformazioni birazionali dello spazio*, Annali di Matematica pura ed applicata, Serie III, t. XVI, pp. 27-68.

[18] Piazzolla Beloch, M.: *Alcune applicazioni del metodo del ripiegamento della carta di Sundara Row*, Atti Acc. Scienze Mediche e Naturali di Ferrara, 1934

[19] Piazzolla Beloch, M.: *Sul metodo del piegamento della carta per la risoluzione dei problemi geometrici*, Periodico di Matematiche, Serie IV, 41, p. 104.

[20] Piazzolla Beloch, M.: *Sulla risoluzione dei problemi di 3° e 4° grado col metodo del ripiegamento della carta*, in Scritti matematici offerti a Luigi Berzolari, 1936.

[21] Piazzolla Beloch, M.: *Lezioni di Matematica Complementare*, redatte dal Prof. Egidio Orzalesi, Pubblicazioni dell'Istituto di Geometria dell'Università di Ferrara, 1953.

[22] Piazzolla Beloch, M.: *Opere Scelte*, CEDAM, Padova 1967.

[23] Row, S.: *Geometric Exercises in Paper Folding*, third edition (revised and edited by Wooster W. Beman and David E. Smith), The open court publishing company, Chicago 1917.

[24] Richert-Gebert, J.:*Perspectives on Projective Geometry*, Springer-Verlag, 2011.

[25] Runge, C.: Graphical methods. No. 4. Columbia University Press, 1912.

[26] Rupp, M. C. *On a trasformation by paper foldings*, Amer. Math. monthly, vol XXXI (1924) p.432.

[27] Scimemi, B.: *Un metodo grafico per la ricerca degli zeri dei polinomi*, Archimede 39 (1987), no. 4, 163-178.

[28] Scimemi, B.: *Algebra e geometria piegando la carta* in Matematica: gioco ed apprendimento, (ed. B. d'Amore), Apeiron, Bologna,1990.

[29] Strickland, E.: Scienziate d'Italia, Diciannove vite per la ricerca, Roma, Donzelli Editore, 2011, pp. 33-35.

[30] Wantzel, M. L.: *Recherches sur les moyens de reconnaître si un Problème de Géométrie peut se résoudre avec la règle et le compas*, Journal de Mathématiques Pures et Appliquees, 1837, 366-372.

[31] Wiener, H.: *Herstellung der platonischen koörper ans papierstreifen*, in Walther von Dyck (ed), *Katalog mathematischer und mathematisch-physikalischer modelle, apparate und instrumente*, München, 1893, 52-54.

Current address

Magrone Paola, Dr. researcher
Dipartimento di Architettura
Via Madonna dei Monti 40 I-00146 Rome, Italy
E-mail: magrone@mat.uniroma3.it

Talamanca Valerio, Dr.
Dipartimento di Matematica e Fisica Università Roma Tre
L.go San Leonardo Murialdo 1, 00146 Roma
Tel. number: +39 0657338247, e-mail: valerio@mat.uniroma3.it

Slovak University of Technology in Bratislava
Faculty of Mechanical Engineering
16th Conference on Applied Mathematics **APLIMAT 2017**
Proceedings

MATHEMATICAL LITERACY SUPPORTED BY COMPUTER TOOLS

MAJOVSKA Renata (CZ)

Abstract. Modern technologies affect all areas of human life but mathematical education resists their influence. In this paper we present some experiences with using mathematical programs, online calculators and interactive tools at the Faculty of Economics VŠB-TU in Ostrava. The special attention is paid to the knowledge engine Wolfram|Alpha. We carried out the research to assess the appropriateness of the use of electronic products in teaching. The results of this research are presented.

Keywords: Mathematics, ICT, online calculator, Wolfram|Alpha

Mathematics Subject Classification: Primary 97U70; Secondary 97U50

1 Introduction

New knowledge, information and technologies change our habits, professional and everyday life. There is a great shortage of technically educated experts while graduates of some humanitarian branches, as well as some graduates of faculties of economics have begun to have problems at the labor market in the Czech Republic. Mathematical education at secondary and university levels often relies on routine memorization without understanding and we can even hear that mathematics is an unimportant and unnecessarily difficult subject. Fortunately, students′ opinions have been changing in recent years. Most of them consider mathematics as an important subject though still very demanding, and they are afraid of mathematics. Many students at faculties of economics are convinced that mathematical competence and literacy open the door to productive futures. Problems with mathematics and the poor level of mathematical knowledge of students coming from secondary schools have been discussed for several years now (PISA, 2012). Many teachers and experts hold discussions about the content and methods which teachers should use. Scientists and teachers think over how to make mathematics more interesting, more objective and more accessible. They believe that mathematics can be taught in such a way that it forces students to think, solve problems around us and think about relation between mathematics with other subjects. On the other hand, many teachers believe that quality teaching means to give students a great amount of knowledge at the highest professional level. Then encyclopedic teaching predominates, teachers examine facts not skills. However, the results are paradoxically the very opposite.

Many students come to the university with poor knowledge, they often do not know the basic concepts such as the function, or they know the definitions but they are not able to imagine what the definitions mean and are used for. They know mechanical procedures for solving some standardized tasks, but they do not know what they have really solved and why (Stavy, Tirosh, 2000). We can also see a significant lack in basic arithmetic and geometric skills, the students cannot count either with low or high numbers, with decimal numbers, fractions or with percentages. They cannot draw a simple sketch or graph. They do not know modern and applied topics such as matrices, probability and statistics, mathematical logic, etc. Students are not able to use the calculator, they have not seen any online calculator or mathematical program at the Internet.

Classical mathematical education uses static educational methods, which do not enable deeper view of the dynamics and philosophy of solving problems. Lessons miss the elements of constructivism, critical thinking, problem solving and interactivity (Piaget, 2002). Students absorb subject matters passively without a real understanding and therefore they are not able to apply their knowledge in other subjects. Only students who understand the fundamental principles of mathematical concepts very well can apply mathematics effectively. Cognitive is the synonym of the terms as knowing, thought or reasonable. The accentuation of cognitive methods and the infiltration of information and communication technologies into mathematical education cause principal changes which bring education closer to students and fight against so-called "math phobia". The cognitive methods in mathematical education should be supported with the tools which increase their clearness and comprehension.

The implementation of modern information and communication technology into the education brings many advantages. New learning methods help students to overcome their difficulties with mathematics and present mathematics as the powerful tool for problem solving in the real word. The modern lecture supported with new elements increases the imagination and creativity of students and simplifies the mathematical content. ICT can be the connecting element between the mathematical theory and its applicability to practice. It enables students to approach education more actively. Students can work in a heuristic cognitive way. The utilization of modern technology in lectures represents a new approach to education. In spite of these advantages computers cannot replace the person of the teacher in the educational process.

2 Teacher and Cognitive Mathematical Education

In order cognitive technologies can play the positive role in mathematical education we have to find not only the methods of their including in the temporary mathematical curriculum but firstly we should create a new curriculum suitable for present requirements that would prepare teachers for applying those methods. Teachers gradually lose the role of the source of information, they rather become helpmates and guides. Many teachers are not able to adjust to this new role. Some university mathematicians have tended to attitude that at any level students have to learn a set of basic skills and concepts without their utilization or application, and unfortunately without understanding. Thus students pass the courses of 'pure' mathematics without understanding why and what they learn.

The input of cognitive technologies has not met with positive public acceptance generally. We can find many opponents, oldish as well as young ones, of modern trends in education mainly among mathematical teachers. Mathematicians are afraid of immoderate application of technology in education. They believe that students will concentrate on routine control of computer programs without thinking about topical mathematical problem. To help teachers to modify their teaching methods they should be equipped with a library of portable, interactive, computer and web based support packages suitable for the new interactive e-learning classroom environment. These support packages should help teachers to understand how, when and where to use technology tools and resources in their classrooms.

3 New Mathematical Courses

At the Faculty of Economics VŠB-Technical University in Ostrava we have tried to help students to overcome their mathematical troubles for several years. Firstly, in the year 2003, we started with new clearer study materials without animation and interactive computer support but also without precise difficult definitions, see Figure 1.

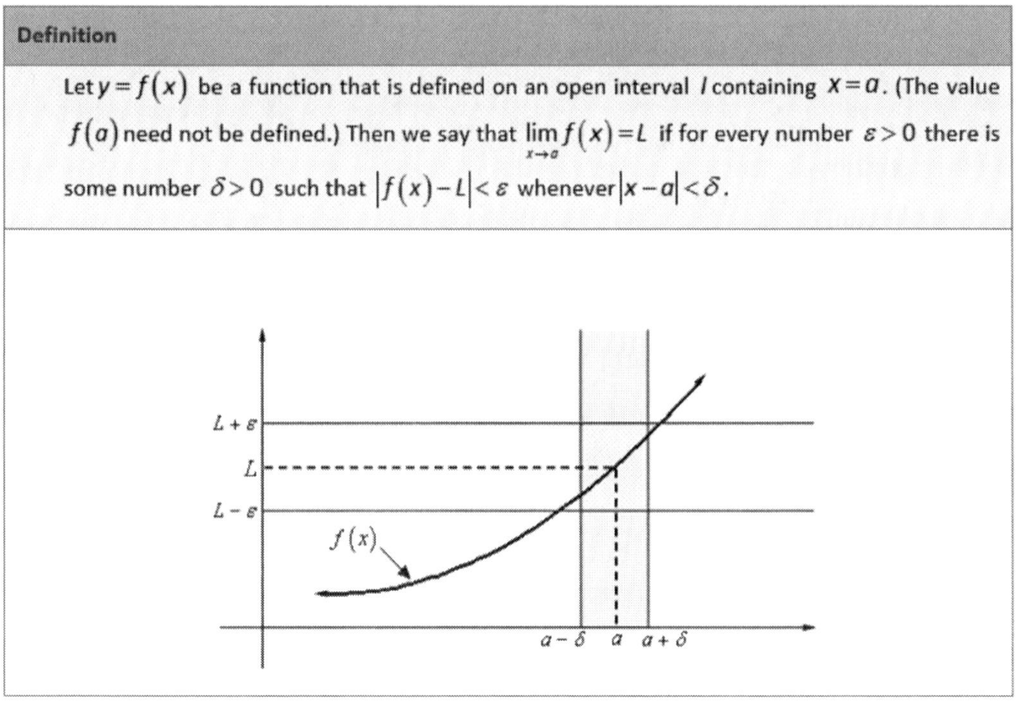

Fig. 1. A part of study material

In the year 2007 LMS Moodle was installed at our university and new Computer Algebras Systems (CAS), online calculators and other interactive tools started to occur at the Internet. A small group of teachers has begun to create interactive mathematical courses which fundamentally change the features of classic lectures. To improve the students´ knowledge, to strengthen their ability to apply mathematical methods to their economic study and specially to help to overcome their fear of mathematics we implemented study materials with interactive applets, access to suitable online calculators and open source mathematical programs, see Figure 2.

Exercise

Given the matrices $A=\begin{pmatrix}1&0&0\\1&1&0\\1&1&1\end{pmatrix}$, $B=\begin{pmatrix}0&1&1\\1&0&0\\0&0&1\end{pmatrix}$ and $C=\begin{pmatrix}1&0&0\\0&1&0\\1&0&1\end{pmatrix}$ solve the equation $X\cdot A+2B=3C$ for unknown matrix X . Use online calculator http://matrixcalc.org/en/.

For more information and examples see http://www.vitutor.com/alg/matrix/matrices_index.html.

Fig. 2. A part of new study material with access to online calculator

All study materials, applets and mathematical computer programs are available in LMS Moodle not only for students but for other teachers, see Figure 3.

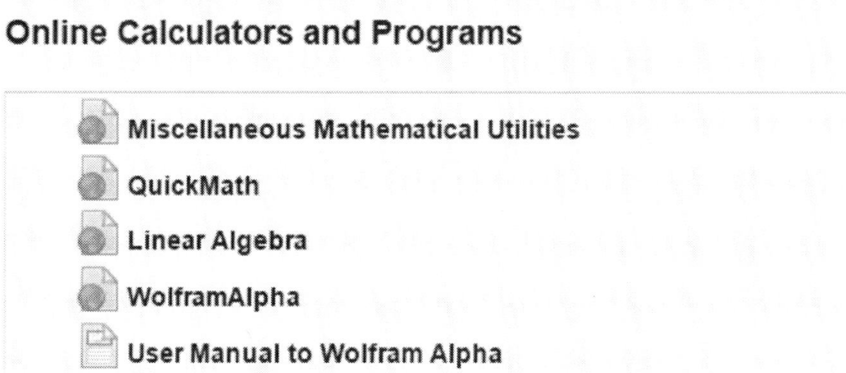

Fig. 3. Access to online calculators in LMS Moodle

4 Wolfram|Alpha in Mathematical Education

In the last two years students are recommended to use Wolfram|Alpha. It is a part of the sophisticated program called Mathematica and it is characterized as the knowledge and intelligence engine. The access to broad computational resources of Wolfram|Alpha is free. The free Wolfram|Alpha service is only available for ad hoc, personal, non-commercial use. Wolfram|Alpha LLC is continually working to improve and expand this product, creating new algorithms for computation, linguistic analysis, data comprehension and computational aesthetics to provide smart and attractive results. It introduces a fundamentally new way to acquire knowledge. This tool answers immediately by doing dynamic computations based on an enormous collection of built-in data, algorithms and methods. The Wolfram| Alpha search box accepts natural language input in keyword, phrase or sentence formats, as well as mathematical equations. The results are computed and presented dynamically with maximum clarity.

We observed that only a small number of students uses Wolfram|Alpha or other online calculators for problem solving and doing exercises. We found out that they are not able to write down the instructions into the search box properly in spite of the fact that notating any mathematical formulas is very easy and intuitive. Even if students make a mistake in the command the program can cope with it. For example, a student can write "derivat" instead of "derivative" and the software gives the correct result with the advice how to enter the notation correctly next time, see Figure 4.

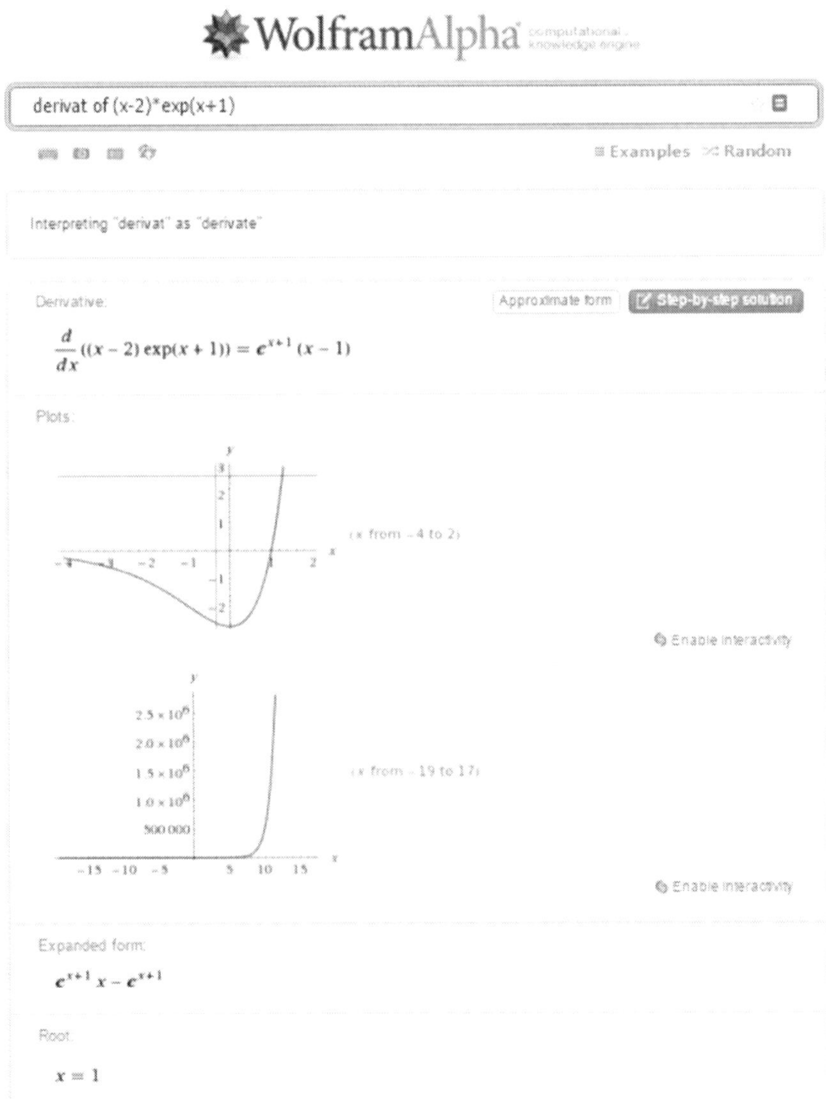

Fig. 4. Wolfram|Alpha environment

Students could start with Wolfram|Alpha by watching Stephen Wolfram's introduction screencast available from https://www.wolfram.com/broadcast/ or browsing the web pages *Examples by Topic*, see Figure 5. But it did not help very much.

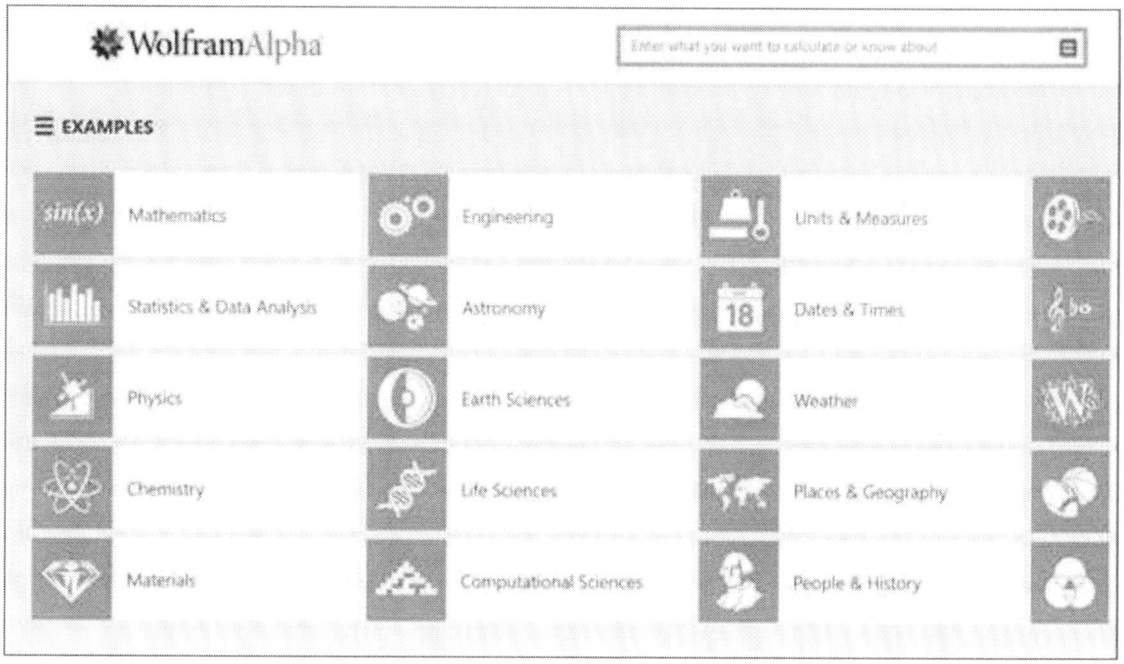

Fig. 5. Examples by Topic Wolfram|Alpha environment

To help students to overcome the initial reluctance to use Wolfram|Alpha we prepared a brief guide comprising instructions how to use it. Students can find many examples with precise notations of formulas, functions, commands etc. in this guide. Every example includes the command and the link to Wolfram|Alpha environment, where the solution is immediately displayed, see Figure 6.

Find the local extremes of the function $y=x\cdot e^{\frac{-x^2}{2}}$ (note: the function values in extremes are calculated, the graph of function is drawn):	
local extrema of x*exp(-x^2/2)	*online evaluation*

Fig. 6. A part of Wolfram|Alpha guide

As we can see in the Figure 4, Wolfram|Alpha gives more than the result of our demand. We wanted to find the first derivative and the program draws the graph of the first derivative, expands it and calculates the roots.

5 Effect of Modern Technologies on Mathematical Education

In the academic year 2014/2015 we realized a research. The aim of our research was to explore whether using open source mathematical programs, applets and online calculators in mathematical education could change students' attitude towards mathematics and improve their mathematical knowledge. We chose the sample among students of the first grade of the bachelor study program. The students were divided into two groups. The first one was the control group consisting of 76 students and the second one was the experimental group comprised of 72 students. The experimental group consisted of volunteers who attended special extra lessons supported by

Wolfram|Alpha, open source computer programs, applets and online calculators. Lectures and seminars were conducted by the same teacher in both groups. We used a questionnaire to explore changes in students' attitudes towards mathematics and the process of learning. The questionnaire has two parts; the first one in the beginning of experiment and the second one in the end, and a didactic test with six tasks to examine the level of the students′ knowledge. Students from both groups solved six tasks in which they should find the domain of the function, draw the graph of the function of one real variable, find the limit of the function, examine intervals of the monotonicity of the function, intervals of concavity and convexity, and solve one matrix equation. The students had the same conditions, i.e. they had the access to the Internet and were informed about the possibility to use any online calculators, computer programs and applets during writing the test in both groups. More details about this experiment can be found in Majovska (2015). Our results respond to the results of larger researches (Koedinger et al., 2010), (Luu, Freeman, 2011), (De Witte et al., 2014).

6 Conclusion

In this paper we wanted to point out that a clearer and simpler mathematical theory and implementation of appropriate modern information and communication technologies not only improves students' results in mathematics but these tools can change students′ attitude towards mathematics. We verified that students began to use online calculators and mathematical programs more frequently if suitable electronic tools were prepared for them and their teachers helped them to overcome initial troubles with using new technologies. We were very surprised by our students′ positive changes in opinions on usefulness of mathematics for their economic studies. The students who used ICT in special seminars mostly evaluated mathematics as rather useful for economic studies. Therefore, it should be the stimulus to teachers to pay more attention to these opinions and take greater responsibility for mathematical teaching.

References

[1] DE WITTE, K., HAELERMANS, C., ROGGE, N. The effectiveness of a computer-assisted math learning program. *Journal of Computer Assisted Learning*, vol. 31, No. 4, 2015, pp. 314-329, ISSN 0266-4909, doi: 10.1111/jcal.12090.

[2] KOEDINGER, K. R., MCLAUGHLIN, E. A., HEFFERNAN, N. T. A quasi-experimental evaluation of an on-line formative assessment and tutoring system. *Journal of Educational Computing Research*, vol. 43, No. 4, 2010, pp. 489-510, ISSN: 0735-6331.

[3] LUU, K., FREEMAN, J. G. An analysis of the relationship between information and communication technology (ICT) and scientific literacy in Canada and Austria. *Computers & Education*, vol. 56, No. 4, 2011, pp. 1072-1082, ISSN: 0360-1315.

[4] MAJOVSKA, R. Open source programmes as the support in mathematical education. In *Proceedings of the 12th Conference ERIE 2015*, Prague, 2015, ISBN 978-80-213-2560-9, ISSN 2336-744X, pp. 341–347.

[5] PIAGET, J. *The Construction of Reality in the Child*. Routledge. London, 2002, ISBN 978-0465014071, 386 p.

[6] PISA. *PISA 2012 Assessment and Analytical Framework: Mathematics, Reading, Science, Problem Solving and Financial Literacy*. Paris OECD. Retrieved from http://www.oecd.org/pisa/pisaproducts/PISA%202012%20framework%20e-book_final.pdf.

[7] STAVY, R., TIROSH, D. *How Students (Mis-)Understand Science and Mathematics*. New York: Teachers College Press, 2000, ISBN 978-0807739587, 127 p.

Current address

Majovská Renata, PaedDr. PhD.
Faculty of Economics VŠB -Technical University
Sokolská třída 33, 702 00 Ostrava, Czech Republic
Tel. number: +420 597 322 506, e-mail: renata.majovska@vsb.cz

MATHEMATICAL LINGUISTICS AND GREEK TRANSLATION OF THE "RAVEN" BY E.A.POE

MAREK Jaroslav (CZ), GKOSDI Maria (GR), DELIGIANNAKI Katherine (GR)

Abstract. The poem Raven of E. A. Poe is not only exciting story about meeting an ill man and a raven, about meeting a live with a disappointment, but a great occasion for comparing a texts. Translations of Raven exists in every language across whole world probably. In this article values of linguistics characteristics will be calculated. Our attention has been given to looking for similarities and differences among four Greek translation of Raven and of original text.

Keywords: linguistic characteristics, frequencies of letters, aggregation, alliteration, E. A. Poe Raven, translations of poem Raven

Mathematics Subject Classification: Primary 91F20, 03B65; Secondary 68T50

1 Introduction

Text analysis includes a large number of phenomena, that can be observed on different text units. The frequency of letters, length of words, length of sentences can be tracked. For poetry phonic phenomena (measuring the Euphoniums) are under investigation. The euphony is given by arrangement and subsequent repetition of sounds (especially vowels) in verse. In this paper, we will measure values of aggregation and alliteration. Linguistics are often interesting about Raven, because many writers translated E. A. Poe text to every language probably. Therefore calculations will be demonstrated on one original and four translation into the Greek language by Kostas Ouranis (T1) , G.B. Ioannidis (T2), Kaisaras Emannouil (T3), Ilias Polyxronakis (T4).

1.1 Philosophy of Raven

"The Philosophy of Composition" is an 1846 essay written by American writer Edgar Allan Poe that elucidates a theory about how good writers write when they write well. He concludes that length, "unity of effect" and a logical method are important considerations for good writing. He also makes the assertion that "the death... of a beautiful woman" is "unquestionably the most poetical topic in the world". Poe uses the composition of his own poem "The Raven" as an example. The three central elements of Poe's philosophy of composition are: length, method, unity of effect,

In the essay, Poe traces the logical progression of his creation of "The Raven" as an attempt to compose "a poem that should suit at once the popular and the critical taste." He claims that he considered every aspect of the poem. For example, he purposely set the poem on a tempestuous evening, causing the raven to seek shelter. He purposefully chose a pallid bust to contrast with the dark plume of the bird. The bust was of Pallas in order to evoke the notion of scholar, to match with the presumed student narrator poring over his "volume[s] of forgotten lore." No aspect of the poem was an accident, he claims, but is based on control by the author.

Even the term "Nevermore," he says, is based on logic following the "unity of effect." The sounds in the vowels in particular, he writes, have more meaning than the definition of the word itself. He had previously used words like "Lenore" for the same effect.

The raven itself, Poe says, is meant to symbolize Mournful and Never-ending Remembrance. This may imply an autobiographical significance to the poem, alluding to the many people in Poe's life who had died.

1.2 Greek translation

In this article five text of Poe's Raven, one original and four translation into the Greek language by Kostas Ouranis (1890-1953), Kaisaras Emannouil (1902-1970), G.B. Ioannidis (1919), Ilias Polyxronakis, has been analyzed.

Then, methought, the air grew denser, perfumed from an unseen censer
Swung by seraphim whose foot-falls tinkled on the tufted floor.
"Wretch," I cried, "thy God hath lent thee — by these angels he hath sent thee
Respite — respite and nepenthe, from thy memories of Lenore;
Quaff, oh quaff this kind nepenthe and forget this lost Lenore!"
Quoth the Raven "Nevermore."

Tab. 1. Part No 14 from poem "The Raven" by E. A. Poe

Τότε ο αγέρας φάνηκε σα να 'ταν μυρωμένος
από 'να θυμιατήριο αόρατο που αγγέλοι
και Σεραφείμ το κούναγαν και τ' αλαφρά τους πόδια
ακούγονταν στο μαλακό χαλί της κάμαράς μου.
«Ναυαγισμένε» φώναξα, «αναβολή σου στέλνει
με τους αγγέλους, ο Θεός και μαύρη λησμοσύνη
για τη χαμένη αγάπη σου την όμορφη Λεονόρα.
Πιες απ' το μαύρο το πιοτό της Λήθης και λησμόνα
εκείνην όπου χάθηκε «. Και το Κοράκι είπε:
«Ποτέ από δω και πια!».

Tab. 2. Part No 14 from translation of poem "The Raven" by Kostas Ouranis

Ξάφνου ως νά νοιωσα μου εφάνη γύρω μου άκρατο λιβάνι
Και πλημμύρα να μου φτάνει σύννεφο η θεία του καπνιά .
- 'Αθλιε , φώναξα , στοχάσου , που ο Θεός στέλνει κοντά σου .
Αγγέλους να σου σταλάσουν νηπενθές για λησμονιά ,
Πιέστο , ω , πιέστο , τη Λεωνόρα να ξεχάσεις μ' απονιά ,

Και μου λέει : – Ποτέ πια !

Tab. 3. Part No 14 from translation of poem "The Raven" by G. B. Ioannidis

Τότε, σαν πιο βαρύς, μου εφάνη ο αγέρας
από' να θυμιατήρι αθέατο ευώδα
στα χέρια Σεραφείμ, που, όπως γλιστρούσαν,
τα πόδια τους κροτούσαν στο ταπέτο.
"Δύστυχε, ο Θεός σου - ανάκραξα - σου στέλνει,
μ' αυτά τα Σεραφείμ του σου προσφέρει
ο γλυκό νηπενθές και να μπορέσει
να λησμονήσεις την Ελεωνόρα!
Ω, πιέ το , και λησμόμησε γαι πάντα
τη νεκρη σου, χαμένη Ελεωνόρα!"
Π ο τ έ π ι ά το κορακι μού αποκρίθει.

Tab. 4. Part No 14 from translation of poem "The Raven" by Kaisaras Emannouil

Κατόπιν μου φάνηκε να πυκνώνει ο αέρας,
αρωματιζόμενος από κάποιο αόρατο θυμιατήρι
που το έσειε ένα Σεραφείμ και τα βήματά του
κουδούνιζαν στο πυκνό δάπεδο.
«Φουκαρά» είπα με φωνή μεγάλη,
«ο Θεός έχει προσφέρει σε εσένα —
δια μέσω αυτών των αγγέλων, έχει προσφέρει σε εσένα
Ανακούφιση — ανακούφιση και νηπενθές από τις αναμνήσεις της Λενόρ!
Πίνε, ω, πίνε με γουλιές μεγάλες αυτό το ευγενικό το νηπενθές
και ξέχασε αυτήν την απολεσθείσα Λενόρ».
«Ποτέ πια», είπε το Κοράκι.

Tab. 5. Part No 14 from translation of poem "The Raven" by Ilias Polixronakis

In the following text linguistics characteristics will be discussed with the aim to compare texts. Firstly, the reader will be familiar with the aggregation and alliteration.

2 Aggregation and Alliteration

2.1 Aggregation

Aggregation attempts to measure, whether sentences (or in this manuscript verse) lying closer to each other, are more phonic similar than those whose distance is greater. Usually estimates of unknown parameters a and b of aproximation function $y = a \cdot x^{-b}$ are determined. This function is solution of differential equation

$$\frac{dy}{y} = -b\frac{dx}{x}, \tag{1}$$

where dx/x is the relative change in distance, dy/y is relative change of similarity, b is coefficient of proportionality. It is valid that $b > 0$.

The similarity of verses is calculated so that for a pair of verses two pairs of sets are created: sets A_i, A_j and B_i, B_j, where i is number of first verse and j is index of second verse. The set A_i is formed by arranged phoneme from first verse, in the set A_j are phonemes from next verse. Pairs of consecutive phonemes are saved in sets B_i and B_j.

If we want to explore the aggregation, for example of the first two verses of G. B. Ioannidis translation.

Δώδεκα έδειχνεν η ώρα, μεσονύχτι, όπως και τώρα
Κι ήμουν βυθισμένος ώρα σε βιβλία αλλοτινά,

Once upon a midnight dreary, while I pondered, weak and weary,
Over many a quaint and curious volume of forgotten lore

First we assemble set $A_1 = \{\alpha,\delta,\varepsilon,\eta,\iota,\kappa,\mu,\nu,o,\pi,\rho,\sigma,\tau,\upsilon,\chi,\omega\}$ and set $A_2 = \{\alpha,\beta,\varepsilon,\eta,\theta,\iota,\kappa,\lambda,\mu,\nu,o,\rho,\sigma,\tau,\upsilon,\omega\}$. Both sets have number of elements equal to 16.
Set B_1 contains following pairs of phonemes from first verse:

$B_1 = \{\alpha\varepsilon,\alpha\iota,\alpha\mu,\delta\varepsilon,\delta\omega,\varepsilon\delta,\varepsilon\iota,\varepsilon\kappa,\varepsilon\nu,\varepsilon\sigma,\eta\omega,\iota o,\iota\tau,\iota\chi,\kappa\alpha,\mu\varepsilon,\nu\varepsilon,\nu\eta,\nu\upsilon,o\pi,\pi\omega,\rho\alpha,\sigma\omega,\tau\iota,\tau\omega,\chi\nu,\chi\tau,\omega\delta,\omega\nu,\omega\rho,\omega\varsigma\}$
and set
$B_2 = \{\alpha\alpha,\alpha\lambda,\alpha\varsigma,\beta\iota,\beta\lambda,\beta\upsilon,\varepsilon\beta,\varepsilon\nu,\eta\mu,\theta\iota,\iota\alpha,\iota\beta,\iota\eta,\iota\nu,\iota\sigma,\kappa\iota,\lambda\iota,\lambda\lambda,\lambda o,\mu\varepsilon,\mu o,\nu\alpha,\nu\beta,\nu o,o\varsigma,o\tau,o\upsilon,\rho\alpha,\sigma\varepsilon,\sigma\mu,\tau\iota,\upsilon\theta,\upsilon\nu,\omega\}$.
First set contains 31 elements and second set contain 34 pairs of phoneme.

Formula for calculating of verses similarities number i and j is

$$S_{i,j} = 100 \cdot \left(\frac{|A_i \cap A_j|}{|A_i| \cdot |A_j|} + \frac{|B_i \cap B_j|}{|B_i| \cdot |B_j|} \right). \tag{2}$$

where $X_i \cap X_j$ – intersection of X_i and X_j, $|X_i|$ is number of elements of set X_i.
After substitution of calculated values into formula

$$S_{1,2} = 100 \cdot \left(\frac{13^2}{16 \cdot 16} + \frac{5^2}{31 \cdot 34} \right) \cong 68 .$$

For the estimation of aggregation more pairs of verses should be calculated for different distances $k = 1, 2, \ldots, n$ and each shift can be characterized by averages of measured values $\overline{S}_i$. The values obtained by previously described procedure are often approximated by function $y = a \cdot x^{-b}$.

2.2 Alliteration

Alliteration is a literary technique, where the same letter repeats at the beginning of words in verse. We have made the analysis of the poem translation "Raven" by Kostas Ouranis. Following sentence can be an illustration:

*δεν είσαι **κάνας άνανδρος**, αρχαϊκό **κοράκι**,*
*που **κατοικείς** στις **πένθιμες ακρογιαλιές** της **Νύχτας**;*
*Στ' όνομα της **Πλουτωνικής** της **Νύχτας**, τ' όνομά σου!»*
***Και** το **κοράκι απάντησε**: «**Ποτέ** από 'δω **και πια**».*

From the perspective of text analysis is not conclusive, if words beginning with the same letter lies in the verse immediately behind, and among them there are one or more words starting with a different letter. Thanks to this alliteration occurs in poetry quite often. It is interesting to examine, whether it is a random event or a characteristic of poem or author. The basis for calculating of alliteration are probabilities of occurrence of individual letters p_i ($i = 1,2,\dots,k$), where k is number of letters in the text). Estimates of these probabilities can be determined from a larger database of texts, optionally directly from the examined text. The probability that x-times multiple alliteration appears in verse, where x words begins with letter i and all other words begin with other letters, is given by formula

$$P(X_i = x) = \binom{n}{x} \cdot p_i^x \cdot q_i^{n-x} \tag{3}$$

where n – is total number of words in the analysis verse, $\binom{n}{x}$ – combinational number, that is calculated by $\frac{n!}{n! \cdot (n-x)!}$, p_i – probability of occurring of letter i, $q_i = 1 - p_i$.

For determination of alliteration it is not important only calculation of multiple x-times alliteration in the analyses verse, but also probabilities of occurring more extreme event $X_i \geq x$ given by

$$P(X_i \geq x) = \sum_{r=x}^{n} \binom{n}{r} \cdot p_i^r \cdot q_i^{n-r}, \tag{4}$$

Coefficient KA for characterization of alliteration is defined by

$$KA = \begin{cases} 100 \cdot (\alpha - P(X \geq x)), & \alpha > P(X \geq x), \\ 0, & \alpha \leq P(X \geq x). \end{cases} \tag{5}$$

Situation would be different in the case of multiple alliteration, e.g. if two or more alliteration exists in verse, for instance in the first verse of original Raven: „*And my soul **from** out that shadow that lies **floating** on the **floor**"*, where total number of words is $n = 13$, two of which start with letter „s", two words start with „o" and three words with „f" (the remaining six words start with any other letters). Probability of triple alliteration is given by multinomial distribution

$$P(X_1 = x_1, X_2 = x_2, X_3 = x_3, X_4 = v) = \frac{n!}{x_1!x_2!x_3!v!} \cdot p_1^{x_1} \cdot p_2^{x_2} \cdot p_3^{x_3} \cdot (1 - p_1 - p_2 - p_3)^v, \qquad (6)$$

where $v = n - x_1 - x_2 - x_3$.
As in the case of simple alliteration calculation of $P(X_1 = y_1, X_2 = y_2, X_3 = w)$, it is necessary for calculation of coefficient *KA*, where $w = n - y_1 - y_2$, $x_1 \le y_1 \le n$, $x_2 \le y_2 \le n$ a $y_1 + y_2 \le n$. For comparison alliteration among different poems (for example, a sample of poems by one author) alliteration can be determined as mean of coefficients KA calculated for n verses in the poem.

3 Results

Measured basic characteristics of original text and of four Greek translation of Raven are presented at Tab. 1, Tab. 2, Tab. 3, Tab. 4, and Tab. 5.

aggregation	T1	T2	T3	T4	Original
all stanzas	34,519	38,455	34,375	44,316	

Tab. 1. Aggregation

alliteration	T1	T2	T3	T4	Original
1	2,436	1,5	2,377	4,115	3,948
2	2,174	3,181	1,957	4,684	3,273
3	2,649	2,009	1,268	4,998	1,693
4	1,989	2,919	1,768	3,914	2,495
5	3,117	3,214	2,138	4,369	4,927
6	2,941	2,86	1,059	4,202	4,118
7	2,684	3,689	2,197	4,994	4,13
8	2,887	4,257	3,256	4,452	3,227
9	3,018	3,421	2,888	4,163	2,995
10	3,152	4,832	2,451	4,963	4,286
11	3,894	3,27	2,771	4,971	4,684
12	3,269	3,27	2,543	4,79	3,824
13	2,487	3,137	1,754	4,696	4,48
14	3,201	4,115	2,446	4,799	3,211
15	2,15	4,654	2,459	4,792	3,7
16	3,778	4,577	2,343	4,798	3,994
17	3,131	4,838	2,676	4,82	4,04
18	3,468	4,444	3,269	4,7	4,161
mean	2,913	3,566	2,312	4,623	3,733

Tab. 2. Alliteration

frequencies	T1	T2	T3	T4	original	
A	10,72%	10,46%	10,52%	9,98%	E	12,77%
O	8,84%	7,51%	8,54%	8,87%	T	9,14%
T	7,65%	6,95%	7,61%	7,71%	N	7,70%
Σ	6,48%	6,20%	6,43%	6,72%	O	7,64%
I	6,11%	8,59%	6,37%	6,26%	A	7,02%
N	5,35%	6,51%	5,85%	6,14%	R	7,00%
E	5,28%	5,82%	5,50%	6,72%	H	5,97%
Π	4,78%	3,79%	4,51%	5,04%	S	5,71%
K	4,44%	3,61%	3,61%	3,89%	L	4,68%
P	3,88%	4,33%	4,33%	3,57%	M	3,34%
Sum	63,53%	63,77%	63,27%	64,90%		70,97%

Tab. 3. Measured basic characteristics

typical words	T1	T2	T3	T4	original
number of words/different words	1004/478	891/454	1254/574	1364/614	1073/484
το (the)	3,5%	1,7%	2,3%	4,0%	5,2%
και (and)	3,0%	2,1%	2,0%	3,6%	2,8%
να (of)	2,5%	2,1%	1,6%	1,8%	1,9%
μου (my)	2,3%	3,8%	2,5%	2,2%	2,2%
Που (where)	2,1%	1,4%	1,7%	2,3%	1,4%
Από (from)	1,9%	0,5%	0,5%	1,2%	1,9%
Λενορ (Lenor)	0,4%	0,5%	0,4%	0,6%	0,7%
Κοράκι (raven)	1,0%	0,5%	0,8%	0,7%	0,9%
Πουλί (bird)	0,7%	0,1%	0,4%	0,9%	0,8%
ποτε-πια (nevermore)	1,0%	1,2%	0,9%	0,8%	0,7%

Tab. 4. Number of words, typical words.

length of words	T1	T2	T3	T4	Original
1	3,98%	4,71%	10,37%	1,91%	4,38%
2	16,63%	16,27%	13,00%	12,61%	14,07%
3	25,70%	24,02%	22,17%	26,39%	15,94%
4	11,55%	13,92%	12,76%	14,66%	24,23%
5	12,65%	12,68%	11,08%	9,53%	11,56%
6	10,36%	11,11%	9,81%	9,75%	9,32%
7	7,67%	6,51%	6,78%	7,99%	9,97%
8	6,27%	4,83%	6,62%	6,60%	4,66%
9	1,79%	3,48%	3,59%	4,40%	3,63%
10	1,99%	1,35%	1,91%	2,49%	1,30%
Mean	4,259	4,240	4,200	4,444	4,438

Tab. 5. Length of words

4 Conluding remarks

Measured values have prepared a possibility for comparing of analysed texts. Characteristics of analysed poems are very close. However, achieving and exceeding of Poe's alliteration in the translation by Ilias Polyxronakis is very surprising, because Poe was trying to reach large values. Translation by G. B. Ioannidis has a very low frequency of words raven, and bird. The average word length is comparable in all translations.

Acknowledgement

The paper was supported by the internal support of University of Pardubice.

References

[1] WIMMER, G., ALTMANN, G., HŘEBÍČEK, L., ONDREJOVIČ, S., WIMMEROVÁ, S., Úvod do analýzy textov. VEDA, Bratislava, 2003, 345 pages.

[2] ANDRES, J., BENESOVA, M. Fractal analysis of Poe's Raven, Glottometrics, no. 21, 2011: pp. 73-98.

[3] ANDRES, J., BENESOVA, M. Fractal analysis of Poe's Raven II, Journal of Quantitative Linguistics, no. 19, vol. 4, 2012: pp. 301-324.

Current address

Marek Jaroslav, Mgr. Ph.D.
University of Pardubice, Faculty of Electrical Engineering and Informatics,
Department of Mathematics and Physics
Studentská 95, 532 10 Pardubice, Czech Republic
Tel. number: +420 466 037 219, e-mail: jaroslav.marek@upce.cz

Deligiannaki Katherine, Bc.
University of Pardubice, Faculty of Electrical Engineering and Informatics,
Department of Mathematics and Physics
Studentská 95, 532 10 Pardubice, Czech Republic
Tel. number: +420 466 037 219, e-mail: katerinad92@hotmail.gr

Gkosdi Maria, Bc.
University of Pardubice, Faculty of Electrical Engineering and Informatics,
Department of Mathematics and Physics
Studentská 95, 532 10 Pardubice, Czech Republic
Tel. number: +420 466 037 219, e-mail: mariag92@hotmail.gr

Slovak University of Technology in Bratislava
Faculty of Mechanical Engineering
16th Conference on Applied Mathematics **APLIMAT 2017**
Proceedings

EFFECTS OF RULE CHANGES AND THREE-POINT SYSTEM IN NHL

MAREK Patrice (CZ)

Abstract. There are two main reasons for changing rules in the ice hockey. The first reason is a safety of players and spectators and the second reason is an attractiveness of matches. This paper studies effects of rule changes that were made because of the second reason, e.g., allowing a two-line pass, narrowing a neutral zone, overtime in the case of a tie match. Rule changes are analyzed from two perspectives – the first one is a distribution of goals scored in a match and the second one is a relative number of ties after a 60-minute regulation time. All seasons since the last big expansion of the NHL in 1979 are used in this analysis. The second part of this paper is dedicated to study of the three-point system that is often named as a cure for a high number of ties in the NHL. This system was earlier introduced in the world's most important ice hockey leagues, i.e., in the Czech Republic, Finland, Germany, Russia, Switzerland and Sweden and its effect in these leagues is analyzed.

Keywords: Rules, three-point system, distribution of goals, relative number of ties, ice hockey

Mathematics subject classification: Primary 62-07; Secondary 62P99

1 Introduction

Ice hockey is a dynamic sport that has evolved during time. One reason is an advance in technology, e.g., composite sticks, goaltender masks (introduced in the National Hockey League (NHL) by Jacques Plante during the 1959–60 season), or the availability of information about every single player. A change caused by an advance in technology is usually slow and with no clear starting point (in 1970s, there still were some goaltenders without a mask).

On the contrary rule changes take effect rapidly and it is easy to identify their starting point as the rules are changed only between seasons. There are two main reasons to change rules:

- The first reason is a safety of players and spectators.

- The second reason attractiveness of matches, e.g., how many goals are scored.

This paper analyses effects of rule changes made because of the second reason, e.g., allowing a two-line pass, narrowing a neutral zone, overtime in a case of a tie.

Although data from the first NHL season (1917–1918) can be obtained it is necessary to maintain some consistency of data and therefore, only results since the 1979–1980 season are used. The main reason to start with 1979–1980 season is the league expansion from 17 to 21 teams in 1979. It was the last big expansion of the NHL and since then maximum of two teams joined the NHL in a single season. Since the 2000–2001 season NHL consists of 30 teams.

Full list of main rule changes in the NHL since the 1979–1980 season are given in Section 2 and effects of these rule changes on the distribution of goals is analyzed in Section 3.1 and their effect on the relative number of ties is analyzed in Section 3.2. In the case of ties the same approach as in [13] is used. However a different method is used in the case of the distribution of goals. This method does not require an assumption about the Poisson distribution of goals.

Section 4 studies effects of the three-point system that was earlier introduced in the Czech Republic, Finland, Germany, Russia, Switzerland and Sweden and it is often named as a cure to high amount of ties.

2 List of Rule Changes

Following list contains major rule changes in the NHL since the 1979–1980 season. Mentioned seasons are those when a new rule took effect and therefore, some change in the distribution of goals, relative number of ties or both can be expected in the given season.

(A) 1983–1984 season: Five-minute overtime period was introduced if the score was tied after the end of a regulation time. If either team scored during the overtime period, the game ended and the scoring team received two points and the losing team received no point. If no goal was scored during the overtime, each team received one point [6, pp. 178].

(B) 1999–2000 season: Number of skaters in the overtime was reduced from five to four, plus a goalkeeper, and both teams automatically received one point for games that went into the overtime, i.e., if either team scored during the overtime period, the game ended and the scoring team received two points and the losing team received one point. If no goal was scored then each team received one point [6, pp. 178-9].

(C) 2003–2004 season: Maximum length of goaltenders' pads was reduced to 38 inches [15, pp. 11].

(D) 2005–2006 season: If no goal was scored during the overtime period then a shootout was used to decide the game's winner, i.e., the team that receives the extra point [6, pp. 179-180]. The NHL salary cap was introduced [6, pp. 124]. The neutral zone was reduced from 54 feet to 50 feet, center red line was eliminated for two-line passes, size of goaltenders' equipment was reduced by approximately 11 percent and some more minor changes to rules were made [15, pp. 11].

(E) 2010–2011 season: Goaltenders' pads shall not exceed 11 inches. Size of pads is set to be anatomically proportional [14, pp. 18].

(F) 2013–2014 season: Pads size was reduced due to a change in the calculation of an anatomically proportional size [16, pp. 19].

All analyzed seasons are later depicted in Fig. 1 and Fig. 2; vertical lines in these figures indicate rule changes from the previous list.

3 Analysis of Rule Changes

Historical match results from the NHL can be obtained from numerous online sources. Data for this analysis were obtained from a website [5] that offers all results since the 1942–1943 season. As explained before, only results from the 1979–1980 season to 2014–2015 season are used. The season 2005–2006 is used in the analysis as a subsequent season for the 2003–2004 season due to the cancellation of the 2004–2005 season. Next, only regular season matches are used in the analysis since playoff matches are played with different rules for an overtime.

3.1 Distribution of Goals

The first part of the analysis concerns influence of rule changes to the distribution of goals. Important motivation for rule changes is to influence a mean value of goals; however, using the approach described bellow it is possible to detect more changes, e.g., increase of variance. Distribution of goals is usually considered to be Poisson or negative binomial, i.e., with the variance similar to the mean value (see [12]). Therefore, in the most of cases, a significant change of the distribution results also in a significant change of the mean value. This is the reason why an average number of scored goals per a game in each season is used for a brief comparison, see Fig. 1.

First, a contingency table can be used to display an observed number of matches with given total goals scored in a match in two consecutive seasons. An example for the 2003–2004 and 2005–2006 seasons is shown in Tab. 1. As explained before, the 2004–2005 season was canceled and therefore, these two seasons are considered as consecutive.

Total number of goals in a match	0	1	2	3	4	5	6	7
2003–2004	10	33	118	155	235	192	185	132
2005–2006	4	16	62	89	182	195	210	161

Total number of goals in a match	8	9	10	11	12	13	14	15
2003–2004	87	43	25	11	2	1	1	0
2005–2006	141	83	54	17	5	7	3	1

Tab. 1. Observed number of matches with given total number of goals in a match for seasons 2003–2004 and 2005–2006

Following hypotheses are formulated and the chi-square test of homogeneity (for details see [17, pp. 602-7]) is used to test whether there is a significant change in a distribution between two consecutive seasons[1]:

- H_0 : Distributions of total number of goals scored in the regulation time in both seasons are identical.
- H_1 : Distribution of total number of goals scored in the regulation time in one season differs from distribution of total number of goals scored in the regulation time in the second season.

P-value for the chi-square test of homogeneity for data presented in Tab. 1 is less than 0.001 and therefore, H_0 can be rejected and H_1 can be accepted on a traditional significance level 0.05. The positive effect (i.e., an increased mean value of goals scored in the regulation time) can be clearly seen in Tab. 1 and it can be also demonstrated by sample statistics of both seasons shown in Tab. 2.

	Average	Sample variance
2003–2004	5.019	4.998
2005–2006	5.941	5.543

Tab. 2. Sample statistics obtained for seasons 2003–2004 and 2005–2006

Six major rule changes were presented in Section 2. Rule changes (A) and (B) contain no reason for a change in the distribution of goals. On the other hand rule changes (C)–(F) contain a reduction of goaltenders' equipment and some other changes that can result in a change of the distribution of goals so that a mean value of goals rises. Fig. 1 shows average numbers of total goals scored per a match in each season with highlighted rule changes.

Tab. 3 contains p-values for the chi-square test of homogeneity for each two consecutive seasons, e.g., first p-value 0.863 is valid for the test between seasons 2014–2015 and 2013–2014. As expected, for the first rule change (A) it is not possible to reject hypothesis about identical distribution in both seasons. The second expectation, that rule change (B) will have no effect, is not confirmed as the null hypothesis is rejected and alternative hypothesis that distributions in both seasons differs from each other is accepted.

Expectation for rule changes (C)–(F) was that they can result in a change of the distribution because all of them contain reduction of goaltenders' equipment. Null hypothesis can be rejected only for the rule change (D). For this rule change the alternative hypothesis can be accepted, i.e., that distributions in seasons 2003–2004 and 2005–2006 differs from each other. The 2005–2006 season is the last season where the null hypothesis can be rejected and therefore, no significant change between distributions can be seen since then.

Results contain many rejections of the null hypothesis for seasons with no significant change in rules. This can be seen mainly in 1990s when average number of goals dropped by two

[1]We emphasize that we do not test any compound hypothesis but we want to test each change individually. Compound hypothesis can be used for seasons with no significant change but we rather choose to present results for each pair of season. The Bonferroni correction can be used in test of compound hypothesis about change of distribution in years with no significant rule change with result that null hypothesis can be rejected and alternative hypothesis that distributions differs can be accepted.

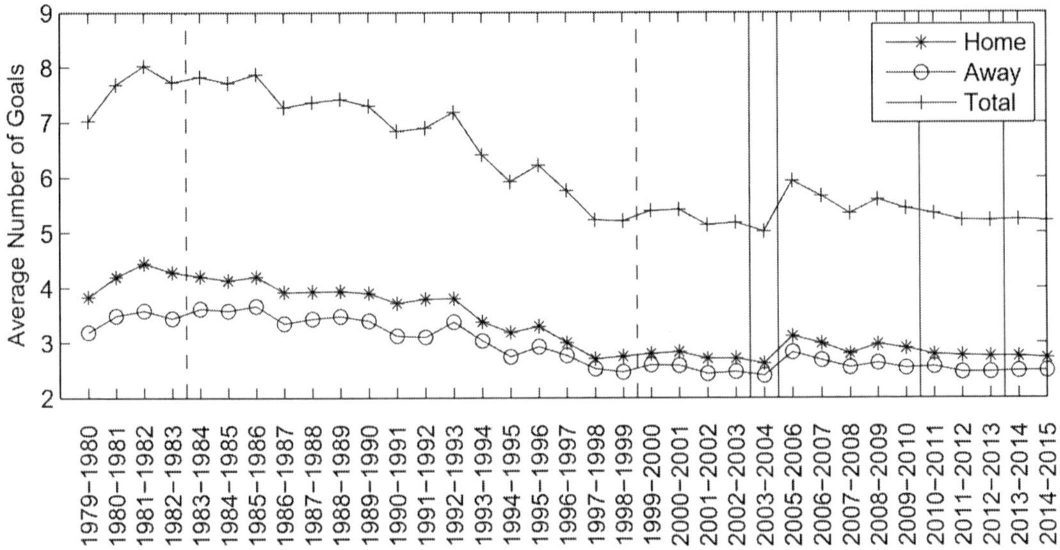

Fig. 1. Average number of total goals scored per a match in seasons (the dashed vertical line is used for rule changes where no effect is expected and the full vertical line is used for rule changes where increase is expected)

goals per game. From this point of view, 1990s can be seen as a transformation of ice hockey in the NHL.

3.2 Relative Number of Ties

Second part of the analysis studies an influence of rule changes to the relative number of ties in the regulation time of a match. This part is slightly connected with the previous one because it can be expected that the fewer goals in a match is scored the more tied matches are recorded.

Season	P-value	Avg	s^2	Season	P-value	Avg	s^2
2014–2015	0.863	5.214	4.892	1995–1996	0.094	6.226	6.697
2013–2014$_{(F)}$	0.274	5.239	4.651	1994–1995$^\bullet$	0.001	5.931	6.045
2012–2013	0.632	5.218	5.130	1993–1994$^\bullet$	<0.001	6.417	6.504
2011–2012	0.855	5.224	5.033	1992–1993$^\bullet$	0.012	7.188	7.454
2010–2011$_{(E)}$	0.135	5.344	5.145	1991–1992	0.346	6.899	6.760
2009–2010	0.056	5.436	5.071	1990–1991$^\bullet$	0.003	6.846	6.354
2008–2009	0.175	5.596	5.803	1989–1990	0.816	7.302	8.242
2007–2008	0.172	5.346	5.143	1988–1989	0.978	7.421	7.798
2006–2007	0.100	5.663	5.428	1987–1988	0.919	7.367	7.515
2005–2006$^\bullet_{(D)}$	<0.001	5.941	5.543	1986–1987$^\bullet$	0.010	7.274	7.155
2003–2004$_{(C)}$	0.394	5.019	4.998	1985–1986	0.368	7.870	8.080
2002–2003	0.975	5.182	5.046	1984–1985$^\bullet$	0.026	7.717	7.600
2001–2002	0.158	5.139	4.912	1983–1984$_{(A)}$	0.115	7.825	8.814
2000–2001	0.512	5.415	5.519	1982–1983	0.412	7.730	8.186
1999–2000$^\bullet_{(B)}$	0.047	5.394	4.774	1981–1982	0.257	8.025	8.027
1998–1999	0.059	5.213	5.354	1980–1981$^\bullet$	<0.001	7.687	7.514
1997–1998$^\bullet$	<0.001	5.225	4.646	1979–1980	0.538	7.026	6.536
1996–1997$^\bullet$	0.003	5.765	5.738				

Tab. 3. Table contains p-values for the chi-square test of a homogeneity and average (Avg) and sample variance (s^2) of goals per a game in the given season. Symbol $\bullet$ indicates rejection of a null hypothesis with a level of significance 0.05 and (A)–(F) indicates rule changes described in Section 2.

The test for a difference in two population proportions (as defined in [17, pp. 486-491]) is used to compare relative number of ties in each two consecutive seasons. Given the number of matches in the first season n_1 and in the second season n_2 and the number observed ties in the first season t_1 and in the second season t_2 it is easy to obtain the maximum likelihood estimate of success probability p (i.e., a probability of a tie) of Bernoulli distribution in a given season as $\hat{p}_i = \frac{t_i}{n_i}, i = 1, 2$.

The null hypothesis is that there is no difference between parameters p_1 and p_2 and one-, or two-sided alternative hypothesis can be formulated. In this case, following hypotheses are tested[2].

- H_0 : Relative number of ties in the regulation time in both seasons are identical, i.e., $p_1 = p_2$.
- H_1 : Relative number of ties in the regulation time in one season differs from the relative number of ties in the regulation time in the second season, i.e., $p_1 \neq p_2$.

[2]We emphasize that we do not test any compound hypothesis but we want to test each change individually. Compound hypothesis can be used for seasons with no significant change but we rather choose to present results for each pair of season. The Bonferroni correction can be used in test of compound hypothesis about change of relative number of ties in years with no significant rule change with result that null hypothesis cannot be rejected.

In some seasons, appropriate one-sided alternative hypothesis can be used. These situations will be described later.

First, Equation 1 is used to estimate a common population proportion p under the assumption that null hypothesis is true.

$$\hat{p} = \frac{n_1\hat{p}_1 + n_2\hat{p}_2}{n_1 + n_2} = \frac{t_1 + t_2}{n_1 + n_2} \tag{1}$$

Next, a z-statistic is computed as shown in Equation 2. If the null hypothesis is true then z has a standard normal distribution.

$$z = \frac{\hat{p}_1 - \hat{p}_2}{\sqrt{\hat{p}(1-\hat{p})(\frac{1}{n_1} + \frac{1}{n_2})}} \tag{2}$$

Finally, z is compared with an appropriate quantile and a final decision about hypotheses is made.

One-sided alternative hypothesis is used for the rule change (B) where teams can be motivated to maintain a tie to obtain one secured point for a game that continues to the overtime, i.e., alternative hypothesis that relative number of ties in the regulation time increased in the second season is used. On the contrary, rule changes (C)–(F) can result in lowering amount of ties (because of their motivation to increase the number of scored goals) and therefore, one-sided alternative hypothesis that relative number of ties in the regulation time decreased in the second season is used. Fig. 2 shows relative number of ties in each season with highlighted rule changes.

Tab. 4 contains p-values of tests for each two consecutive seasons. There is only one pair of seasons, 2001–2002 and 2002–2003, where the null hypothesis can be rejected and the alternative hypothesis that a relative number of ties in the regulation time in one season differs from a relative number of ties in the regulation time in the second season can be accepted. In a case that 0.1 significance level is used then the null hypothesis can be rejected for two more rule changes – the rule change (B), a motivation to continue to the overtime that increased relative number of ties and the rule change (D), a significant pack of rule changes that lowered the relative number of ties. Next two rejections of the null hypothesis can be seen in 1990s.

Given these results, it can be seen that efforts to affect a relative number of ties are more problematic than efforts to affect a distribution of goals. This part also confirms two main results from Section 3.1, i.e., the rule change (D) can be considered as significant and 1990s can be considered as some transformation of ice hockey in the NHL.

4 Three-Point System

Three points for a win in the regulation time is often named as a cure to a high amount of ties. Many articles in newspapers claim that this change cured a soccer and therefore, it can cure the ice hockey. LeBrun [7] and Bondy [1] can be named as examples of articles

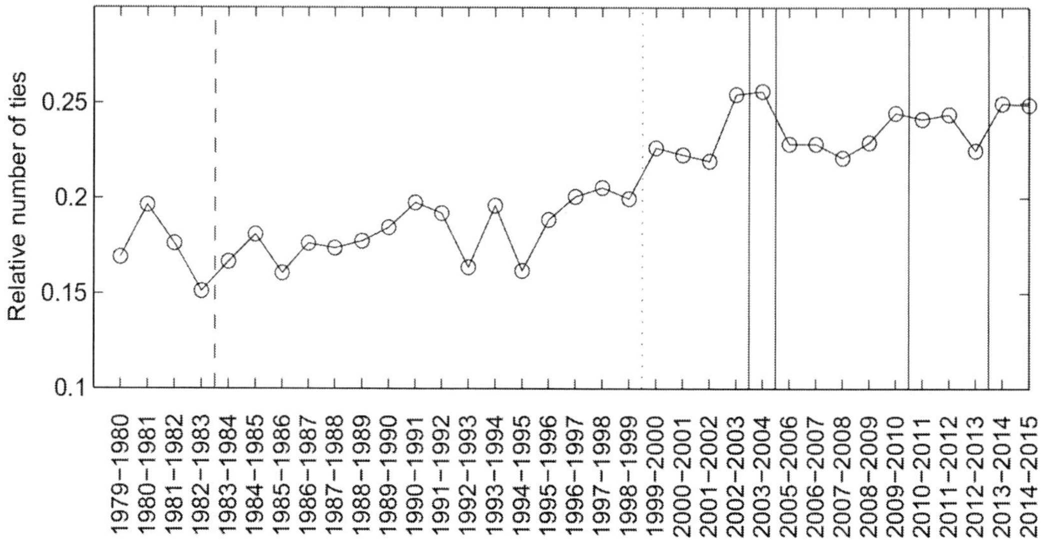

Fig. 2. Relative number of ties in seasons (a dashed vertical line is used for a rule change where no effect is expected, a dotted vertical line is used for a rule change where an increase is expected and a full line is used for rule changes where a decrease is expected)

where an effect on the number of ties and fairness is discussed. The rest of this paper will answer only the first question, i.e., whether the three-point system can lower the number of ties and the fairness of the point-systems is not examined. The analysis is not focused on a study of the effect of the three-point system in the soccer (as named sources suggest) but its effect to the world's most important ice hockey leagues is analyzed – in the Czech Republic, Finland, Germany, Russia, Switzerland and Sweden. In each league the first season that uses the three-point system is found and the same hypotheses as in Section 3.1 and Section 3.2 are tested.

4.1 The Czech Republic

The 2000–2001 season is the first season when the three-point system was used in the Czech Republic. Summary of data obtained from [2] is in Tab. 5.

Same approach like in Section 3.1 is used to test whether this change had some effect on the distribution of goals. P-value in this case is 0.163 and therefore, the null hypothesis that distributions of total number of goals scored in the regulation time are identical in both seasons cannot be rejected.

The effect of the three-point system should be a reduction of amount of ties and therefore, the same approach like in Section 3.2 is used with a one-sided alternative hypothesis, i.e., that a relative number of ties in the regulation time is lower in the second season. P-value of this test is 0.608 and therefore, the null hypothesis that a relative numbers of ties in the regulation time in both seasons are identical cannot be rejected.

Season	P-value	Ties	Season	P-value	Ties
2014–2015	0.963	0.249	1995–1996	0.167	0.189
$2013–2014_{(F)}$ (x,–)	0.890	0.250	1994–1995•	0.079	0.162
2012–2013	0.343	0.225	1993–1994•	0.055	0.196
2011–2012	0.888	0.244	1992–1993	0.107	0.164
$2010–2011_{(E)}$ (x,–)	0.425	0.241	1991–1992	0.770	0.192
2009–2010	0.368	0.245	1990–1991	0.495	0.198
2008–2009	0.629	0.229	1989–1990	0.704	0.185
2007–2008	0.664	0.221	1988–1989	0.847	0.177
2006–2007	1.000	0.228	1987–1988	0.898	0.174
$2005–2006^{\bullet}_{(D)}$ (x,–)	0.055	0.228	1986–1987	0.397	0.176
$2003–2004_{(C)}$ (x,–)	0.537	0.256	1985–1986	0.270	0.161
2002–2003••	0.041	0.254	1984–1985	0.440	0.181
2001–2002	0.846	0.220	$1983–1984_{(A)}$	0.386	0.167
2000–2001	0.828	0.223	1982–1983	0.166	0.151
$1999–2000^{\bullet}_{(B)}$ (–,x)	0.060	0.226	1981–1982	0.287	0.176
1998–1999	0.737	0.200	1980–1981	0.146	0.196
1997–1998	0.788	0.205	1979–1980	—	0.169
1996–1997	0.477	0.201			

Tab. 4. Table contains p-values for a test of a hypothesis that a relative number of ties in the regulation time in both seasons are identical along with an observed relative number of ties in the given season. (–,x) or (x,–) indicates that a one-sided alternative hypothesis was used, i.e., a relative number of ties is higher or lower. A symbol •• indicates rejection of the null hypothesis with a level of significance 0.05 and symbol • indicates rejection of the null hypothesis with a level of significance 0.10. Appropriate rule changes described in Section 2 are indicated by (A)–(F).

	1999–2000	2000–2001
Number of matches	364	364
Number of ties	74	77
Relative number of ties	0.203	0.212
Average number of goals	5.742	5.448
Sample variance	6.104	5.460

Tab. 5. Summary for seasons in the Czech Republic

4.2 Finland

The 2004–2005 season is the first season when the three-point system was used in Finland. Summary of data (with exactly same number of ties and goals scored in both seasons) obtained from [10] is in Tab. 6.

Tests described in Section 4.1 were used with the identical findings, i.e., the null hypothesis that distributions of total number of goals scored in the regulation time are identical in both

	2003–2004	2004–2005
Number of matches	364	364
Number of ties	82	82
Relative number of ties	0.225	0.225
Average number of goals	5.255	5.255
Sample variance	5.171	5.447

Tab. 6. Summary for seasons in Finland

seasons cannot be rejected (p-value is 0.516) and the second null hypothesis that relative numbers of ties in the regulation time in both season are identical cannot be rejected (p-value is 0.500).

4.3 Germany

The 1998–1999 season is the first season when the three-point system was used in Germany. Summary of data obtained from [9] is in Tab. 7.

	1997–1998	1998–1999
Number of matches	342	364
Number of ties	63	72
Relative number of ties	0.184	0.198
Average number of goals	6.313	6.657
Sample variance	6.133	6.226

Tab. 7. Summary for seasons in Germany

Tests described in Section 4.1 were used with the identical findings, i.e., the null hypothesis that distributions of total number of goals scored in the regulation time are identical in both seasons cannot be rejected (p-value is 0.654) and the second null hypothesis that relative numbers of ties in the regulation time in both season are identical cannot be rejected (p-value is 0.677).

4.4 Russia

The 1999–2000 season is the first season when the three-point system was used in Russia. Summary of data obtained from [8] is in Tab. 8.

	1998–1999	1999–2000
Number of matches	462	380
Number of ties	84	70
Relative number of ties	0.182	0.184
Average number of goals	4.894	4.939
Sample variance	5.258	5.856

Tab. 8. Summary for seasons in Russia

Tests described in Section 4.1 were used with the identical findings, i.e., the null hypothesis that distributions of total number of goals scored in the regulation time are identical in both seasons cannot be rejected (p-value is 0.923) and the second null hypothesis that relative numbers of ties in the regulation time in both season are identical cannot be rejected (p-value is 0.536).

4.5 Switzerland

The 2006–2007 season is the first season when the three-point system was used in Switzerland. Summary of data obtained from [11] is in Tab. 9.

	2005–2006	2006–2007
Number of matches	264	264
Number of ties	47	47
Relative number of ties	0.178	0.178
Average number of goals	5.920	6.159
Sample variance	5.579	4.956

Tab. 9. Summary for seasons in Switzerland

Tests described in Section 4.1 were used with the identical findings, i.e., the null hypothesis that distributions of total number of goals scored in the regulation time are identical in both seasons cannot be rejected (p-value is 0.651) and the second null hypothesis that relative numbers of ties in the regulation time in both season are identical cannot be rejected (p-value is 0.500).

4.6 Sweden

The 1998–1999 season is the first season when the three-point system was used in Sweden. Summary of data obtained from [4] and [3] is in Tab. 10.

	1997–1998	1998–1999
Number of matches	276	300
Number of ties	54	61
Relative number of ties	0.196	0.203
Average number of goals	5.609	5.760
Sample variance	5.926	6.611

Tab. 10. Summary for seasons in Sweden

Tests described in Section 4.1 were used with the identical findings, i.e., the null hypothesis that distributions of total number of goals scored in the regulation time are identical in both seasons cannot be rejected (p-value is 0.978) and the second null hypothesis that relative numbers of ties in the regulation time in both season are identical cannot be rejected (p-value is 0.591).

5 Conclusion

Major rule changes in the NHL since 1979 were presented and their effect was tested from two perspectives – their effect on the distribution of goals and the relative number of ties.

The rule change (A) – five-minute overtime for a game that ends by a tie in the regulation time – had no effect from both perspectives, as expected. For the rule change (B) – one point guaranteed in the overtime – the expectation was that it will have no effect on distribution; however the p-value of the test is 0.047 and therefore, it is possible to reject the null hypothesis that distribution is identical before and after the change. Sample statistics indicates that the main change was in the variance of goals – the average number of goals per game increased from 5.231 to 5.394 and the sample variance decreased from 5.354 to 4.774. Expected change can be seen for the relative number of ties where the p-value of the test that relative number of ties in both seasons is identical against the alternative hypothesis that the relative number of ties increased is 0.06, i.e., the null hypothesis was rejected on the level of significance 0.1 (the observed relative number of ties increased from 0.200 to 0.226).

The expectation for rule changes (C)–(F), usually some equipment reduction, was that distribution of goals will be changed and relative number of ties will be decreased. The null hypothesis that distribution is identical before and after the change was rejected only for the biggest rule change (D) that included a two-line pass legalization and a reduction of a neutral zone. The p-value was less than 0.001 and the change can be illustrated by a significant increase in an average number of goals scored per one game – from 5.019 to 5.941. The p-value of the test that relative number of ties in both seasons is identical against the alternative hypothesis that a relative number of ties declined is 0.055, i.e., the null hypothesis was rejected on the level of significance 0.1 (observed relative number of ties decreased from 0.256 to 0.228).

From the results of tests in other seasons we can formulate a hypothesis that the NHL undergone a transformation in 1990s – the average number of goals decreased and the relative number of ties increased.

The last part of this paper studied the three-point system that is usually named as a cure for high amount of ties. Effects of the three-point system were analyzed for the world's most important leagues – in the Czech Republic, Finland, Germany, Russia, Switzerland and Sweden. Null hypotheses were not rejected and therefore, it is not possible to prove a significant effect of the three-point system on the distribution of goals and the relative number of ties. Thus, another changes should be done to increase the number of scored goals and to decrease the relative number of ties. Good inspiration can be found in the rule change (D) where it is possible to show its significant effect. More radical changes can be inspired by rugby union bonus points system, i.e., teams in the ice hockey can be rewarded by bonus points for scoring some amount of goals and for losing by no more than a specified margin. This can maintain a motivation to score even when the result of a game looks clear.

References

[1] Bondy, F.: *The NHL's current point system just doesn't add up*, NY Daily News, `http://www.nydailynews.com/sports/hockey/nhl-current-point-system-doesn-add-article-1.139711`.

[2] eSports s.r.o.: *Zápasy: Tipsport extraliga*, 2015, `http://www.hokej.cz/tipsport-extraliga/zapasy`.

[3] FoG Data ab: *Online-reports 1997-98*, 2015, `http://historical.stats.swehockey.se/history/Games_Hist_Total.asp?qsGameTitle=Elitserien&qsSeasonId=1&qsSerieId=90001&qsShowTableLink=1&qsSerieNamn=Elitserien&qsLayout=1004`.

[4] FoG Data ab: *Online-reports 1998-99*, 2015, `http://historical.stats.swehockey.se/history/Games_Hist_Total.asp?qsGameTitle=Elitserien&qsSeasonId=30&qsSerieId=90001&qsShowTableLink=1&qsSerieNamn=Elitserien&qsLayout=1004`.

[5] Hockey History: *Hockey History - Birthplace, Ancestors, NHL*, 2015, `http://hockey-history.com/nhl/historical-results/`.

[6] Kahane, L. H., Shmanske, S.: *The Oxford Handbook of Sports Economics, Volume 1: The Economics of Sports*, Oxford University Press, New York, 2012.

[7] LeBrun, P.: *Time to revisit three-point system*, ESPN: The Worldwide Leader in Sports, `http://sports.espn.go.com/nhl/columns/story?columnist=lebrun_pierre&id=4706309`.

[8] LiveSport s.r.o.: *Hockey russia - superliga results*, 2015, `http://www.sportstats.com/hockey/russia/superliga/results/`.

[9] LiveSport.cz: *Del - archiv výsledků, tabulky*, 2015, `http://www.livesport.cz/hokej/nemecko/del/archiv/`.

[10] LiveSport.cz: *Liiga - archiv výsledků, tabulky*, 2015, `http://www.livesport.cz/hokej/finsko/liiga/archiv/`.

[11] LiveSport.cz: *Nla - archiv výsledků, tabulky*, 2015, `http://www.livesport.cz/hokej/svycarsko/nla/archiv/`.

[12] Maher, M. J.: *Modelling association football scores*, Statistica Neerlandica, vol. 36, no. 3, (1982), pp. 109–118.

[13] Marek, P.: *Effects of rule changes in nhl*, in *Proceedings of the 5th International Conference on Mathematics in Sport*, 2015, pp. 115–120.

[14] The National Hockey League: *Official Rules 2010–2011*, Dan Diamond and Associates, Inc., Canada, 2010.

[15] The National Hockey League: *Official Guide & Record Book 2013*, Dan Diamond and Associates, Inc., Canada, 2012.

[16] The National Hockey League: *Official Rules 2013–2014*, Dan Diamond and Associates, Inc., Canada, 2013.

[17] Utts, J. M., Heckard, R. F.: *Mind on Statistics*, Cengage Learning, Stamford, 2015.

Current address

Marek Patrice, Ing. PhD.
European Centre of Excellence NTIS – New Technologies for Information Society,
Faculty of Applied Sciences, University of West Bohemia
Technická 8, 306 14 Plzeň, Czech Republic
Tel. number: +420 377 632 686, e-mail: patrke@kma.zcu.cz

Slovak University of Technology in Bratislava
Faculty of Mechanical Engineering
16th Conference on Applied Mathematics **APLIMAT 2017**
Proceedings

ANALYTICALLY SOLVABLE MODELS OF THE SCALE FREE GROWING NETWORK WITH HIERARCHY

MARKOŠOVÁ Mária (SK), BORIS Rudolf (SK), ČAJÁGI Martin (SK), NÁTHER Peter (SK)

Abstract. It is well known, that the preferential node attachment leads to the scale free network structure. Less is, however, known about the local processes leading to the hierarchical node organization. Our previous studies have shown, that adding new edges in a certain pattern creates complex scale free network with hierarchy. These studies were preferably numerical. In this paper we look for a minimal analytically solvable model of growing scale free hierarchical network. We define a minimal number of local processes leading to the hierarchical scale free structure and derive dynamical equations by which the evolution of network is driven.

Keywords: scale free network, hierarchy

Mathematics Subject Classification: Primary 60K20; Secondary 90B12

1 Introduction

Theoretical, as well as numerical results on growing complex networks show, that their global structure is strongly influenced by the local processes of their growth. Pioneering work on this subject has been done by Barabási and Albert [1]. These authors have shown, both numerically and analytically, that the growth of a network by the preferential node attachment leads to the scale free structure of the network.

Numerically, ***preferential attachment*** means this:

1. The simulation starts from an arbitrary small network.
2. Each time unit one node is injected and labelled by the time of its addition s.
3. In general, a new node brings m new links. These links are added to the older nodes and the probability of this addition is proportional to the older node degree (preferential linking).

These processes are mathematically tractable and for great t lead to the easily solvable integro – differential equation

$$\frac{\partial k(s,t)}{\partial t} = m\frac{k(s,t)}{\int\limits_{s=1}^{t} k(s,t)ds} \tag{1}$$

Here $k(s,t)$ denotes an average degree of the node s at time t. In (1) $\int\limits_{s=1}^{t} k(s,t)ds = 2mt$ because each edge adds to the degree of the two nodes and $k(s,t)/2mt$ is the probability of the one edge linking . The solution of (1) is $k(s,t) \propto \left(\frac{t}{s}\right)^{\beta}$ with $\beta = \frac{1}{2}$. From (1) one gets the degree distribution

$$P(k) \propto k^{-\gamma}\ ,\gamma = 3.0\,, \tag{2}$$

where

$$\gamma = 1 + \frac{1}{\beta} \tag{3}$$

is the scaling exponent [2] . This power law distribution is a typical sign of the complex ***scale free*** networks [2, 3].

Variations of the local processes described above lead to the different final network structure [2, 3, 4, 5]. One can build a model such as (1), which shows in detail, how the local dynamics modifies global network properties. On the other hand, having a real network and measuring its properties, such as the degree distribution (2) for example, one can, with a help of the suitable model, guess the network history and its local dynamics [2, 3, 4, 5]. This way the structure of internet [5, 6], www network [5, 6], collaboration network [6] etc. has been explained and various other properties were analysed. Ravász and Barabási have recently noticed that the scale free structure in real networks is often accompanied by the hierarchical node organization [7]. This cannot be completely explained by the preferential node linking. They therefore suggested a model, in which the node addition was, somewhat artificially, replaced by the addition of certain five node pattern. Numerical simulations of the Ravász - Barabási model have revealed, that such process leads to the hierarchical scale free structure with the power law degree distribution (2), and, moreover, with the similar distribution of an average clustering coefficients $C(k)$ of nodes with the degree k [7]

$$C(k) \propto k^{-\delta}\,. \tag{4}$$

In (4) δ is another scaling exponent. Clustering coefficient of certain node is given as a ratio of the existing edges among the k node neighbours to the number of edges given if the k neighbours create a complete graph [7] (7). As has been shown by Barabási and Ravász, distributions (2, 4) are typical for the scale free network with the node hierarchy.

Our own studies have shown [8], that the Ravász – Brabási process of network growth is not the only one creating distributions (2, 4) and that the pattern addition can be understood in a more general and a more natural way. Namely:

1. The growth of the network starts from an arbitrary small graph.

2. Each time unit one node is injected and labelled by the time of its addition s.
3. In general, a new node brings m new links. These links are added to the older nodes by a specific way:
   a) The first of the m edges links to the randomly chosen node.
   b) m-1 edges link to the neighbours of the selected node.

The irregular "pattern" is thus given by the local configuration of a newly added edges and one node on the top.

Numerical simulations of the described processes [8] have found, that the resulting network has power law distributions (2) and (4) suggesting thus a scale free hierarchical structure.

This network is also analytically tractable, but only for $m=2$ [8]. In this case, our network can be transformed to the analytically solvable Vásquez model of surfers revealing and modifying the unknown network by jumping to the randomly chosen node and subsequent walking along the edges [9]. Vásquez model simulates the movement of surfers through the www network, in which a surfer can discover certain www page either by searching engine (jumping) or following hyperlinks on already discovered page (walking). The surfer simultaneously modifies the www net by adding a hyperlink to his own page from time to time [9]. We have found more complete analytical solution of the Vásquez model [10]. We have shown how the model behaves in two different regions of the parameter space. For certain parameter values we related the Vásquez model to the other known models described detaily in [7].

In this paper our motivation is to find a minimal analytically solvable model having distributions (2) and (4). By another words, we are looking for the simple growth processes leading to the scale free hierarchical network structure. We do not require realistic counterpart of our model(s) at the moment. Such course is not meaningless from the theoretical point of view. To give another example - in reality there are not too many networks growing by the pure preferential linking (as states Barabási and Albert [1]), but understanding, that the scale free structure is a result of the preferential linking is very important for analysing the real networks structure.

In this paper we suggest two simple minimal models A and B, which have a scale free hierarchical structure. We present numerical simulations of these models. We show that the models are analytically tractable. We found the distribution $P(k)$, $C(k)$ for the model A analytically and compare them to the measured ones. The analysis of the model B is still in progress. We calculated the distributions $P(k)$, $C(k)$ for the model B as well, but they somewhat differ from the measured ones. This signalizes, that there is something missing in our analysis. Measurements have been done on numerically simulated networks.

2 The model A

The model utilizes the knowledge, that the scale free structure is a result of the preferential linking and the power law degree distribution of $C(k)$ (4) is a result of local edge linking. The dynamics of the absolutely minimal model A is as follows:

1. The growth of the network starts from an arbitrary small graph.

2. Each time unit one node is injected and labelled by the time of its addition s.
3. A new node brings $m=2$ new links. Then an older edge is chosen randomly and the two links are added to the nodes at both ends of the selected edge.

Edge selection ensures, that two of the new links are added to the neighbour nodes. It also ensures a preferential linking, because, if certain node has a great degree, it is more probable, that its edge will be selected.

The model has been studied both numerically and analytically, as described in the next subsections.

2.1 Analytical calculations

The analytical calculation of the degree distribution is easy. Above mentioned processes are captured by the equation

$$\frac{\partial k(s,t)}{\partial t} = \frac{k(s,t)}{2t}, \tag{5}$$

which is the equation (1) with $m = 2$ and thus

$$k(s,t) \propto \left(\frac{t}{s}\right)^{\frac{1}{2}}, \tag{6}$$

and the degree distribution is given by (2). On the right hand side of (5) is the probability of choosing one of the edges of the node s from 2t network edges (we omit the initial small net).

Calculation of the $C(k)$ distribution is also rather straightforward. As the new node comes, there are two possibilities of its attachment:

a) A chosen edge is not one of the edges attached to the node s. In this case the clustering coefficient of the node s is not changed $C(s,t+1) = C(s)$, where

$$C(s) = \frac{2E(s,t)}{k(s,t)[k(s,t)-1]}. \tag{7}$$

In (7) $E(s,t)$ is the number of edges among the neighbours of the node s.

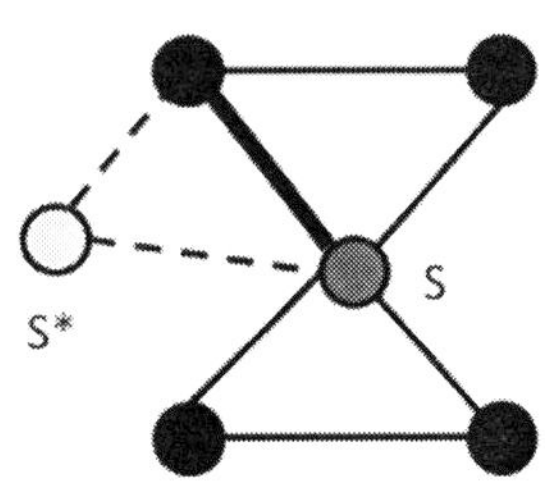

Fig. 1. Model A: A new node s^* links itself to the network changing the clustering coefficient of the node s by changing its degree and number of edges among its neighbours. The selected edge is the thick one

b) A chosen edge is one of the edges attached to the node s. Then the situation is depicted on Fig. 1. The probability of this event is

$$p = \frac{k(s,t)}{2t}, \tag{8}$$

because 2t is the number of edges in the network at time t and

$$\begin{aligned} C(s,t+1) &= \frac{2(E(s,t)+1)}{[k(s,t)+1]\,k(s,t)} = \\ &= C(s,t)\frac{k(s,t)-1}{k(s,t)+1} + \frac{2}{k(s,t)[k(s,t)+1]} \end{aligned} \tag{9}$$

Therefore the following equation holds

$$\begin{aligned} C(s,t+1) &= (1-p)C(s,t) + \\ &+ p\left\{C(s,t)\frac{k(s,t)-1}{k(s,t)+1} + \frac{2}{k(s,t)[k(s,t)+1]}\right\}. \end{aligned} \tag{10}$$

From the equation (10) one gets (for great t) a differential equation

$$\frac{\partial C(s,t)}{\partial t} = \frac{1}{t}\frac{1}{k(s,t)+1}[1-k(s,t)C(s,t)] \tag{11}$$

Solving (11) using (6) one gets

$$C(k) \propto \frac{ln(k)}{k}. \tag{12}$$

2.2 Numerical simulations

The model has been numerically simulated for N=50000 nodes. Both distributions P(k), C(k) are depicted on Fig. 2. Numerical results are in a good accordance with the analytically calculated distributions, having exponents $\gamma = 3.0$ and $\delta = 0.98$ which is slightly below 1 as expected from (12).

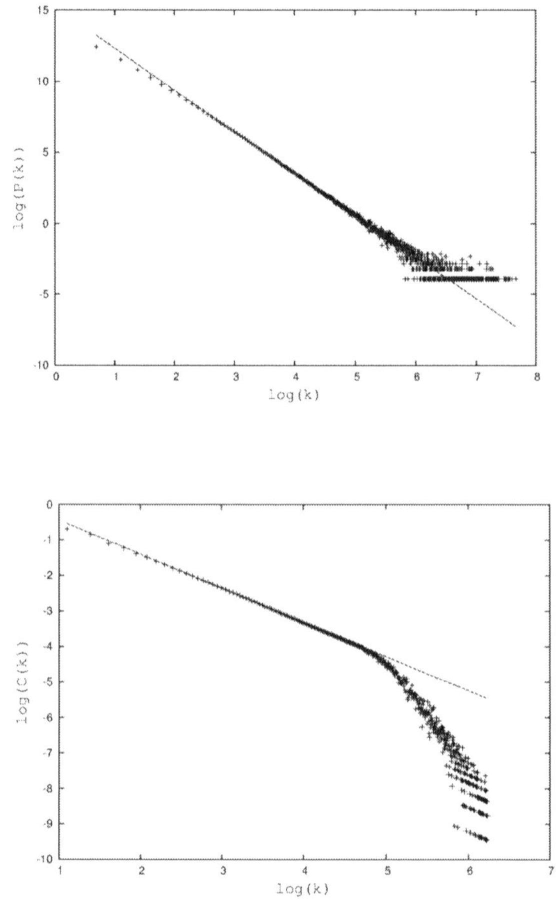

Fig. 2. Degree ($P(k)$, a) and clustering coefficient ($C(k)$, b) distributions for the model A constructed from the simulated network. Slopes of the solid lines define scaling exponents $\gamma = 3.0$ and $\delta = 0.98$

3 The model B

The model B is based on these dynamical processes:

1. The growth of the network starts from an arbitrary small graph.
2. Each time unit one node is injected and labelled by the time of its addition s.
3. A new node brings three new links. These links are added to the older nodes by a specific way:
   a) The first one to a randomly chosen node .
   b) Two of them to the end nodes of a randomly chosen edge.

3.1 Analytical calculations

First, let us calculate the degree distribution. The new node brings three new links which are added to the older ones as mentioned above.

The network dynamics is described by the equation

$$\frac{\partial k(s,t)}{\partial t}=\frac{1}{t}+\frac{k(s,t)}{3t}, \tag{13}$$

where the first term on the right hand side is the probability of a node selection and the second one the probability of an edge selection. It is because there are t nodes and 3t edges in the system at time t (we omit the initial small network for great t). The solution of (13) is

$$k(s,t)\propto\left(\frac{t}{s}\right)^{\beta}, \tag{14}$$

where

$$\beta=\frac{1}{3}. \tag{15}$$

Using the formula (3) we get the scaling exponent of the degree distribution (2)

$$\gamma=4.0. \tag{16}$$

To calculate analytically the distribution is more complicated for the model B, because there are several possibilities how to change the clustering coefficient of certain node s as the network grows (Fig. 3). Each of five depicted situations, which we take into account occur with different probabilities. The six - th situation having the probability $p_6=1-\sum_{i=1}^{5}p_i$ does not change the clustering coefficient of the node s, because the chosen edge is not related to the node s. Thus

$$C(s,t+1)=\sum_{i=1}^{6}p_iC_i(s,t+1). \tag{17}$$

Here

$$p_1=\frac{1}{t-2}\frac{3t-E(s,t)-a(s,t)-k(s,t)}{3t},\quad C_1(s,t+1)=C(s,t)\frac{k(s,t)-1}{k(s,t)+1},$$

$$p_2=\frac{1}{t-2}\frac{a(s,t)}{3t},\quad C_2(s,t+1)=C(s,t)\frac{k(s,t)-1}{k(s,t)+1}+\frac{2}{k(s,t)(k(s,t)+1)},$$

$$p_3=\frac{1}{t-2}\frac{E(s,t)}{3t},\quad C_3(s,t+1)=C(s,t)\frac{k(s,t)-1}{k(s,t)+1}+\frac{4}{k(s,t)(k(s,t)+1)},$$

$$p_4=\frac{k(s,t)}{3t}\frac{k(s,t)-1}{t-2},\quad C_4(s,t+1)=C(s,t)\frac{k(s,t)-1}{k(s,t)+1}+\frac{4}{k(s,t)(k(s,t)+1)},$$

$$p_5 = \frac{k(s,t)}{3t}\frac{t-1-k(s,t)}{t-2}, \quad C_5(s,t+1) = C(s,t)\frac{k(s,t)-1}{k(s,t)+1} + \frac{2}{k(s,t)(k(s,t)+1)},$$

where $a(s,t) = \sum_{i\in\{nn\}} k(i,t) - E(s,t) - k(s,t)$ and the summation includes all nearest neighbours $\{nn\}$ of the node s.

With a help of the equation (17) we get for great t

$$\frac{\partial C(s,t)}{\partial t} = \frac{(-2tk(s,t)+8k(s,t)-6t-2)}{3t(t-2)(k(s,t)+1)}C(s,t) + \frac{2\bar{a}(s,t)+2t+2k(s,t)-6}{3t(t-2)(k(s,t)+1)} \tag{18}$$

Giving $a(s,t) = k(s,t)\bar{a}(s,t)$, and using the exact solution of (13) which is

$$k(s,t) = 6\left(\frac{t}{s}\right)^{\frac{1}{3}} - 3 ,$$

together with the equation (7), the solution of the equation (18) is approximated by

$$C(k) \propto k^{-1}. \tag{19}$$

3.2 Numerical simulations

The model B has been simulated numerically for the network with 800000 nodes. We use far greater number of nodes for the model B because of the slow numerical convergence of the exponents γ and δ with the network size to the analytically calculated values. We have measured scaling exponents γ and δ for the simulated network and found, that even for the great network they does not match exactly the calculated ones. Measured exponent $\gamma = 3.95$ is quite in accordance with the calculated (16), but $\delta \cong 0.83$ is worse approximation of the calculated scaling exponent $\delta = 1.0$ (19).

4 Summary

We have studied the two models of the growing network with the local dynamical rules. For both models we made numerical simulations, as well as analytical calculations of the two measurable distributions P(k) (2), C(k) (4). The model A is completely analytically solvable, and the numerical results converge well and are in a good accordance with the analytical ones. This cannot be told about the numerical results of the model B. Both scaling exponents converge slowly to the analytically calculated value even for the great network.

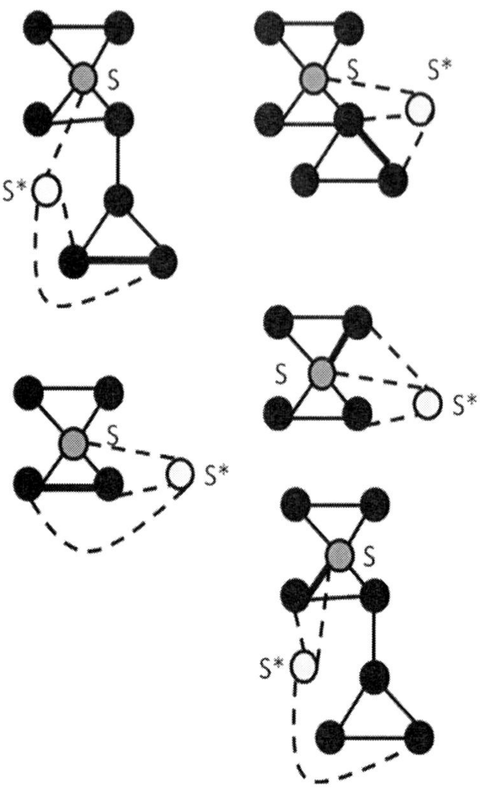

Fig. 3. Model B. A new node s* links itself by three new edges to the network. All of these five linking possibilities change the clustering coefficient of the node s. The selected edge is the thick one

To find the distribution (4) for the model B analytically is more difficult, because there are several local processes changing the clustering coefficient (7) of the defined node *s*. They also occurred with different probabilities. Moreover, probabilities of these processes include the variable $a(s,t)$ which we do not know exactly. We only estimate that it does not grow more quickly than t and support our estimate by the numerical simulations. If our estimate is not correct, this can partly explain the discrepancy between analytically calculated and measured distribution (4). We suppose that the reason of the slow convergence of the equation (18) may lie in the second term of its solution which is $\approx \frac{ln(k(s,t))}{k^2(s,t)}$.

Nevertheless, our simple models are analytically tractable. Numerical simulations of the model A have shown that our analytical method is promising. For both models A and B also another parameters, suggested recently, can be studied [11]. The analysis of the model B and improvement of the match of the analytical and numerical results is still in progress.

References

[1] BARABÁSI, A.L., ALBERT, R, Emergence of scaling in random networks, *Science* vol. 286, 1999, pp. 3616.

[2] DOROGOVTSEV, S.N., MENDES, J.F.F., Evolution of networks, *Advanced Phys.* vol. 51, 2002, pp. 1079.

[3] ALMEES, E., BARABÁSI, A.L. Barabási, Power laws, scale free networks and genome biology, *Springer*, US 2006
[4] MARKOŠOVÁ, M., "Komplexné siete", in *Umelá inteligencia a kognitívna veda II*, STU Bratislva, 2010, pp. 321.
[5] ALBERT, R., BARABÁSI, A. L., Statistical mechanics of complex networks, *Rec. Modern. Phys* vol. 74, 2002, pp. 47.
[6] COSTA, L. and others, Analysing and modelling real world phenomena with complex networks, *Advances in Physics,* vol. 60, 2011, pp. 329.
[7] RAVÁSZ, E., BARABÁSI, A. L., Hierarchical organization in complex networks, *Phys. Rev.* E, vol. 67, 2003. pp. 026112.
[8] NÁTHER, P., MARKOŠOVÁ, M., RUDOLF, B.,Hierarchi in the growing scale free network with local rules, *Physica A*, vol. 388, 2009, pp. 5053.
[9] VÁSQUEZ, A., Growing networks with local rules, *Phys. Rev.*E, vol. 67, 2003, pp. 056104.
[10] RUDOLF, B., MARKOŠOVÁ, M., ČAJÁGI, M., TIŇO, P., Distributions and scaling in the connecting nearest neighbours models *Phys. Rev.* E, vol. 85, 2012, pp 026114.
[11] MONES, E., VICSEK, L., VICSEK, T., Hierarchy measure for complex networks, cond.mat/1202.0191, 2012

Acknowledgement

The paper was supported by Tatrbanka research grant (2016) .

Current address

Markošová Mária, Doc. PhD.
Department of Applied Informatics, Faculty of Mathematics, Physics and Informatics UK
Mlynská dolina, 842 48 Bratislava 4, Slovak Republic
E-mail: markosova@ii.fmph.uniba.sk

Boris Rudolf, Doc. PhD.
Department of Mathematics, Faculty of Electrical Engineering and Information Technology STU
Ilkovičova 3, 812 19 Bratislava 4, Slovak Republic
E-mail: boris.rudolf@stuba.sk

Čajági Martin
Department of Applied Informatics, Faculty of Mathematics, Physics and Informatics UK
Mlynská dolina, 842 48 Bratislava 4, Slovak Republic
E-mail: martin.cajagi@gmail.com

Náther Peter, PhD.
Department of Applied Informatics, Faculty of Mathematics, Physics and Informatics UK
Mlynská dolina, 842 48 Bratislava 4, Slovak Republic
E-mail: nather@ii.fmph.uniba.sk

Slovak University of Technology in Bratislava
Faculty of Mechanical Engineering
16th Conference on Applied Mathematics **APLIMAT 2017**
Proceedings

REMARKS ON THE LOCAL INTERSECTION MULTIPLICITY OF PLANE CURVES

MAŤAŠOVSKÝ Alexander (SK)

Abstract. The two dimensional version of Bézout's theorem states the global number of common points of two projective algebraic plane curves. This result says nothing about the connection between the multiplicity of a point C in $U \cap V$ and the multiplicities of a point C on U and of a point C on V. Denote r resp. s the multiplicity of common point C on U resp. V. Then, there is the following inequality for multiplicities $i(U, V; C) \geq r \cdot s$. The aim of the paper is a characterization of equality. In particular, we give an equation for improving the inequality to equality.

Keywords: local intersection multiplicity, Bézout's theorem

Mathematics subject classification: Primary 14C17; Secondary 14B05, 13D40

1 Introduction

The basic roots for the creation of the many roads of the intersection theory are represented by classical Bézout's theorem. The original theorem of Bézout discussing the number of solutions of n polynomials in n variables was published in 1764. Its two dimensional version states that two plane algebraic curves of degrees m and n intersect in at most $m \cdot n$ different points, unless they have infinitely may points in common.

It took more than three centuries of hard work to master the Bézout's theorem and to attach an integer to each intersection point measuring the "multiplicity" of the intersection, such that the sum of multiplicities is $m \cdot n$. One way to define such a multiplicity is a resultant of two polynomials. It is also possible to define the above multiplicity by theory of infinitely near singularities (A. Weil). In a view of our comment below, we accent that the starting point for the local algebra definition of above multiplicity has been the observation, that the intersection multiplicity at the origin $O = (0, 0)$ of two affine algebraic curves $f(x, y) = 0$, $g(x, y) = 0$ may be defined to be the degree of the field extension $k((x, y))/k((f, g))$, where $k((x, y))$ is the field of quotients of the ring of power series in x and y with coefficients in the base field k and where $k((f, g))$ is the field of quotients of the ring which can be expressed as power series in f and g. From there C. Chevalley was able to define the multiplicity of a local ring with respect to a system of parameters and then the general notion of intersection multiplicity (P. Samuel).

2 Local intersection multiplicity and Bézout's theorem

Let X, Y be algebraic varieties of the projective space $\mathbb{P}_k^n$ over an algebraic closed field k. If C is an irreducible component of $X \cap Y$ with $\dim(C) = \dim(X \cap Y)$, we denote by $i(X, Y; C)$ the local intersection multiplicity of X and Y along C defined by A. Weil (see [9]). Let us assume that one variety, say Y, is a complete intersection. Then we know that the multiplicity $i(X, Y; C)$ is given by the Samuel multiplicity of the ideal $I(Y)$ in the local ring $R = (k[x_0, x_1, \ldots, x_n]/I(X))_{I(C)}$, so

$$i(X, Y; C) = e_0(I(Y) \cdot R; R)$$

see [9, Chap. II., §7b]. The Samuel multiplicity e_0 is defined as follows.

Let $(R, \mathfrak{m})$ be a Noetherian local ring with $\dim(R) = d$. For any $\mathfrak{m}$-primary ideal $\mathfrak{q}$ in R the R-module $R/\mathfrak{q}^n$ is of finite length for all $n \in \mathbb{N}$. For large n, this length function becomes a polynomial (Hilbert-Samuel polynomial) which can be written as

$$\ell_R(R/\mathfrak{q}^n) = e_0(\mathfrak{q}; R)\frac{n^d}{d!} + \text{terms of lower degree.}$$

The coefficient $e_0(\mathfrak{q}; R)$ is called the Samuel multiplicity of $\mathfrak{q}$ in R. Recall that the length $\ell_R(R/\mathfrak{q})$ of the R-module $R/\mathfrak{q}$ is the number s of ideals in a maximal chain of the $\mathfrak{m}$-primary ideals in R of the form

$$\mathfrak{q} = \mathfrak{q}_0 \subsetneq \mathfrak{q}_1 \subsetneq \cdots \subsetneq \mathfrak{q}_s = \mathfrak{m}.$$

The following theorem says how to calculate the Samuel multiplicity e_0.

Theorem 2.1. *Let* $(R, \mathfrak{m})$ *be a Noetherian local ring with* $\dim(R) = d$ *then*

a) if $\mathfrak{q}_1$, $\mathfrak{q}_2$ *are* $\mathfrak{m}$*-primary ideals in* $(R, \mathfrak{m})$ *and* $\mathfrak{q}_1 \subseteq \mathfrak{q}_2$ *then* $e_0(\mathfrak{q}_1; R) \geq e_0(\mathfrak{q}_2; R)$,

b) if $\mathfrak{q}$ *is an* $\mathfrak{m}$*-primary ideal in* $(R, \mathfrak{m})$ *then for any* $s \in \mathbb{N}$ *the multiplicity* $e_0(\mathfrak{q}^s; R) = s^d e_0(\mathfrak{q}; R)$,

c) if $\mathfrak{q}_1 = (a_1, a_2, \ldots, a_d) \cdot R$, $\mathfrak{q}_2 = (b_1, a_2, \ldots, a_d) \cdot R$ *are parameter ideals in* $(R, \mathfrak{m})$ *then the ideal* $\mathfrak{q} = (a_1 b_1, a_2, \ldots, a_d) \cdot R$ *is also a parameter ideal in* $(R, \mathfrak{m})$ *and the Samuel multiplicity* $e_0(\mathfrak{q}; R) = e_0(\mathfrak{q}_1; R) + e_0(\mathfrak{q}_2; R)$,

d) for any parameter ideal $(a_1, a_2, \ldots, a_d) \cdot R$ *in* $(R, \mathfrak{m})$ *and for any* $m \in \mathbb{N}$, *the multiplicity* $e_0\big((a_1^m, a_2, \ldots, a_d) \cdot R; R\big) = m e_0\big((a_1, a_2, \ldots, a_d) \cdot R; R\big)$,

e) for any parameter ideal $(a_1, \ldots, a_d) \cdot R$ *in* $(R, \mathfrak{m})$ *and for any* $s \in \mathbb{N}$, *the multiplicity* $e_0\big((a_1^s, \ldots, a_d^s) \cdot R; R\big) = e_0\big((a_1, \ldots, a_d)^s \cdot R; R\big) = s^d e_0\big((a_1, \ldots, a_d) \cdot R; R\big)$.

Proof. [5, Ch. 7.]. □

Example 1. Let $A = k[x, y]$ be a polynomial ring over an algebraic closed field k, $R = k[x, y]_{(x,y)\cdot k[x,y]}$ be the localization of A with respect to maximal ideal $(x, y) \cdot A$. Suppose that $\mathfrak{q} = (a, b) \cdot R$ denotes a parameter ideal in R. Then, there are among others the following computations for multiplicities

$$e_0((a^m - b^n, a^k - b^l) \cdot R; R) = e_0(\mathfrak{q}; R) \cdot \min\{ml, nk\}, \text{ provided } ml \neq nk,$$
$$e_0((a^m, b^n, a^k b^l) \cdot R; R) = e_0(\mathfrak{q}; R) \cdot \min\{mn, ml + nk\}.$$

When $A = k[x, y, z]$, $R = k[x, y, z]_{(x,y,z)\cdot k[x,y,z]}$ and $\mathfrak{q} = (a, b, c) \cdot R$ is a parameter ideal in R, we have

$$e_0((a^m - b^n, a^k - c^l, b^r - c^s) \cdot R; R) = e_0(\mathfrak{q}; R) \cdot \min\{mlr, nks\},$$
$$e_0((a^m, b^n, c^l, a^k b^r c^s) \cdot R; R) = e_0(\mathfrak{q}; R) \cdot \min\{mnl, mns + mlr + nlk\}.$$

The statements of the previous example follow from [2].

2.1 The two dimensional version of Bézout's theorem

Let U defined by the equation $F(x_0, x_1, x_2) = 0$ and V defined by the equation $G(x_0, x_1, x_2) = 0$ be two algebraic projective plane curves, where F, G are homogeneous polynomials of degrees m and n respectively in polynomial ring $k[x_0, x_1, x_2]$ over an algebraic closed field k. Suppose that the curves U and V have no common components. The classical Bézout's theorem says the following.

Theorem 2.2. *There is a positive integer $i(U, V; C)$ for all common points C of U and V such that*

$$m \cdot n = \sum_{C \in U \cap V} i(U, V; C),$$

where the sum is over all common points C of U and V.

Proof. We present here a short and very simple proof with the use of the notion of degree of homogeneous ideal (for the properties of deg() see [7]). We know that $\deg(F) \cdot R = m$, $deg(G) \cdot R = n$ and

$$\deg(F, G) \cdot R = m \cdot n$$

since the ideal $(F, G) \cdot R$ is homogeneous of dimension zero. Let

$$(F, G) \cdot R = \mathfrak{q}_1 \cap \mathfrak{q}_2 \cap \cdots \cap \mathfrak{q}_s$$

be a primary decomposition of the ideal $(F, G) \cdot R$, where $\mathfrak{q}_i$ is $\mathfrak{p}_i$-primary ideal defining a point C_i so

$$U \cap V = \{C_1, C_2, \ldots, C_s\}.$$

By the same argument, the ideal $(F, G) \cdot R$ is zero-dimensional, we have

$$\deg(F, G) \cdot R = \sum_{i=1}^{s} \ell_R(\mathfrak{q}_i),$$

where $\ell_R(\mathfrak{q}_i)$ is the length of the factor module $k[x_0, x_1, x_2]/\mathfrak{q}_i$. In order to finish the proof, we define $i(U, V; C_i) \overset{\text{def}}{=} \ell_R(\mathfrak{q}_i)$. □

The Bézout's theorem states, that two algebraic projective curves U and V of degrees m and n (without common components) intersect exactly in $m \cdot n$ points when each common point C is counting as many times as its multiplicity $i(U, V; C)$. What is the connection between this multiplicity and the Samuel intersection multiplicity?

As the local ring $k[x_0, x_1, x_2]_{\mathfrak{p}_i}$ is a Cohen-Macaulay ring (for details see [10, App. 4.]), we have

$$\ell_R(\mathfrak{q}_i) = e_0((F, G) \cdot k[x_0, x_1, x_2]_{\mathfrak{p}_i}; k[x_0, x_1, x_2]_{\mathfrak{p}_i}).$$

Corollary 2.3. *By the notion and hypothesis listed in front of the theorem above, we have*

$$\deg(F) \cdot \deg(G) = \sum_{C \in U \cap V} e_0((F, G) \cdot k[x_0, x_1, x_2]_{I(C)}; k[x_0, x_1, x_2]_{I(C)}),$$

where the sum is over all common points C of U and V, and $I(C)$ denotes the associated prime ideal of the point C.

Therefore the classical Theorem of Bézout states the global number of common points of two projective algebraic plane curves. This result says nothing about the connection between the multiplicity of C on $U \cap V$ and the multiplicities of C on U and C on V. In the following we investigate the mentioned connection. We are motivated by the classical result of B. Bydžovský related to this problem. In his work [3, Chap. 11. 134. f.], B. Bydžovský published the following result on the multiplicity of the common point of intersection of two curves.

Theorem 2.4. *The point which is on the curve U with the multiplicity r, on the curve V with the multiplicity s and in which the curves have no common tangent lines, is exactly $r \cdot s$-multiple point on the intersection. If moreover these curves have t common tangent lines in this point, the intersection point in $U \cap V$ has the multiplicity at least $r \cdot s + t$, or strictly speaking the multiplicity*

$$i(U, V; C) = r \cdot s + t + \varrho,$$

where ϱ is a non-negative integer depending on U, V and C.

Let us give one motivated example related to this number. Since the multiplicity $i(U, V; C)$ is a local property, we can study this problem in an affine plane over a field k. Let $f(x, y)$ and $g(x, y)$ are polynomials in the polynomial ring $k[x, y]$ and U, V are affine algebraic curves in affine plane defining by the equations $f(x, y) = 0$, $g(x, y) = 0$ respectively. Let us assume that the point C is the origin $O = (0, 0)$ (by a suitable change of the coordinate system). We denote by $R = k[x, y]_{(x,y) \cdot k[x,y]}$ the local polynomial ring. Then we have

$$i(U, V; C) = e_0((f, g) \cdot R; R).$$

Let $f^\star(x, y)$ denote the homogeneous part of $f(x, y)$ of lowest degree (similar $g^\star(x, y)$) and $h(x, y)$ be the greatest common divisor of $f^\star(x, y)$ and $g^\star(x, y)$. It is known, that the equations

$f^\star(x, y) = 0$ define the pencil of tangent lines to U at C,

$g^\star(x, y) = 0$ define the pencil of tangent lines to V at C,

$h(x, y) = 0$ define the pencil of common tangent lines to U and V at C.

With the notion above, we have

$$\deg(f^{\star}(x,y)) = r, \quad \deg(g^{\star}(x,y)) = s, \quad \deg(h(x,y)) = t.$$

Example 2. Let $U = V(f) \subseteq \mathbb{A}_k^2$ and $V = V(g) \subseteq \mathbb{A}_k^2$ the two plane curves defined by

$$f(x,y) = x^3 + y^3 - 3xy, \qquad g(x,y) = x^2 + y^2 - 2ry,$$

where $f(x,y)$, $g(x,y) \in k[x,y]$ and $r \in k$ is a constant (k is an algebraically closed field). In this example set $k = \mathbb{C}$. The curve U is the "folium cartesium", while the curve V is the circle with center $S = [0,r]$ and radius r (see Fig. 1).

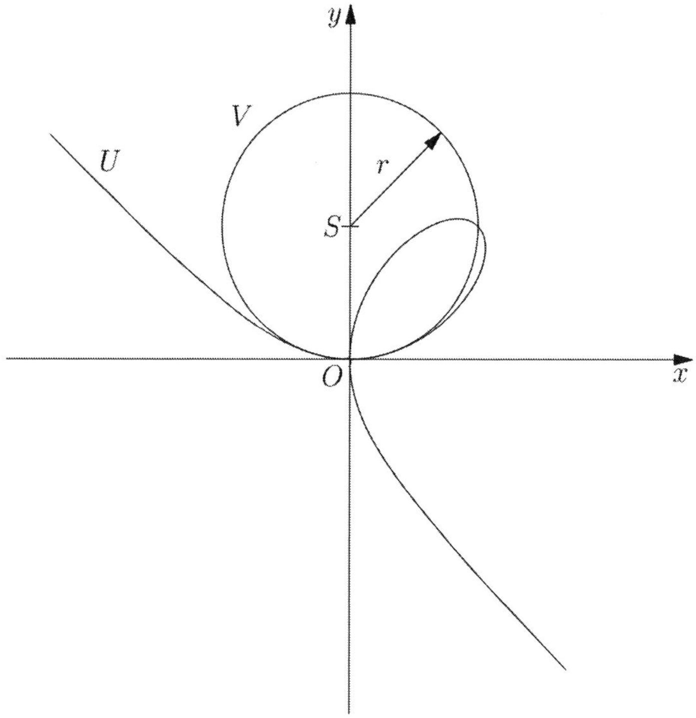

Fig. 1. The intersection of curves U and V at O.

We have that $O = (0,0) \in U \cap V$. We consider the local intersection multiplicity at the origin. To this end, put $R = k[x,y]_{(x,y)\cdot k[x,y]}$. Because U and V do not have a component in common, the polynomials f, g form a system of parameters of R. Then by using the Theorem 2.1 the local intersection multiplicity $i(U,V;O) = e_0((f,g)\cdot R; R)$ is given by

$$e_0((f,g)\cdot R; R) = \begin{cases} 3 & \text{if } r \neq 0 \text{ and } r \neq \frac{3}{2}, \\ 4 & \text{if } r = 0, \\ 5 & \text{if } r = \frac{3}{2}. \end{cases}$$

Note that for $r = 0$ the curve V consists of one (real) point, but it is the union of two (conjugate) complex lines intersecting in the origin.

With the notions above for the numbers of the intersection of the "folium cartesium" with the circle we have

e_0	r	s	t	ϱ
3	2	1	1	0
4	2	2	0	0
5	2	1	1	2

Here e_0 denotes the local intersection multiplicity and from Theorem 2.4 it is equal to $r \cdot s + t + \varrho$. Let us remark, that if the local intersection multiplicity $e_0((f,g)\cdot R; R) = 5$, then the curves U and V have at the point O the same curvature which is equal to $\frac{2}{3}$.

For $r = \frac{3}{2}$, there exists one more common point with coordinates $\left[\frac{3}{2}, \frac{3}{2}\right]$ and with multiplicity 1. If $r = 0$, then there are two more intersection points with multiplicity 1. Their coordinates are $\left[-\frac{3}{2} - \frac{3}{2}i, -\frac{3}{2} + \frac{3}{2}i\right]$, $\left[-\frac{3}{2} + \frac{3}{2}i, -\frac{3}{2} - \frac{3}{2}i\right]$. In other cases there are three more intersection points with multiplicity 1, one point has real coordinates and the other two have complex coordinates, which we have omitted.

3 Computation of correction term

Denote the sum $t + \varrho$ from Theorem 2.4 as $\operatorname{corr}(U, V; C)$, which is also a non-negative integer depending on curves U, V and on intersecting point C. The non-negative integer $\operatorname{corr}(U, V; C)$ is called the *correction term* which makes equality between the local intersection multiplicity and product of multiplicities of intersecting point on each curves U, V. In this section we compute this correction term using homological algebra methods.

Definition 3.1. Let $M_\bullet$ denote a bounded complex of R-modules

$$M_\bullet \colon 0 \longrightarrow M_n \xrightarrow{\varphi_n} M_{n-1} \xrightarrow{\varphi_{n-1}} \cdots \xrightarrow{\varphi_2} M_1 \xrightarrow{\varphi_1} M_0 \longrightarrow 0.$$

Since $\operatorname{Im}(\varphi_{k+1}) \subseteq \operatorname{Ker}(\varphi_k)$ there exists a factor module $\operatorname{Ker}(\varphi_k)/\operatorname{Im}(\varphi_{k+1})$, which is called k^{th} *homology module* denoted by $H_k(M_\bullet)$ for all $k = 0, \ldots, n$.

Suppose that for all k the homology module $H_k(M_\bullet)$ is of finite length. Then define *Euler-Poincaré characteristic* of $M_\bullet$

$$\chi(M_\bullet; R) = \sum_{k=0}^{n} (-1)^k \ell_R\big(H_k(M_\bullet)\big).$$

Theorem 3.1. *Let $A = k[x, y]$ be the polynomial ring over algebraic closed field k with $\dim(A) = 2$ and $R = k[x, y]_{(x,y)\cdot k[x,y]}$ be the localization of A with maximal ideal $\mathfrak{m} = (x, y) \cdot k[x, y]$. Let f, g denote a system of parameters of R. Choose c, $d \in \mathbb{N}$ such that $f \in \mathfrak{m}^c \setminus \mathfrak{m}^{c+1}$ and $g \in \mathfrak{m}^d \setminus \mathfrak{m}^{d+1}$. For all $n \gg 0$ we have:*

a) $\ell_R\big(\mathfrak{m}^n/(f\mathfrak{m}^{n-c}, g\mathfrak{m}^{n-d}) \cdot R\big)$ is a constant,

b) $e_0\big((f, g) \cdot R; R\big) = c \cdot d + \ell_R\big(\mathfrak{m}^n/(f\mathfrak{m}^{n-c}, g\mathfrak{m}^{n-d}) \cdot R\big)$.

Proof. Define the following sequence of complexes:

$$0 \longrightarrow C_\bullet \xrightarrow{\varphi_\bullet} K_\bullet(f, g; R) \xrightarrow{\psi_\bullet} L_\bullet(f, g; R) \longrightarrow 0, \qquad (1)$$

i.e.

$$\begin{array}{ccccccccc}
 & & 0 & & 0 & & 0 & & \\
 & & \downarrow & & \downarrow & & \downarrow & & \\
0 & \longrightarrow & \mathfrak{m}^{n-c-d} & \xrightarrow{\varphi_3} & A & \xrightarrow{\psi_3} & A/\mathfrak{m}^{n-c-d} & \longrightarrow & 0 \\
 & & \downarrow{\scriptstyle \alpha_2} & & \downarrow{\scriptstyle \beta_2} & & \downarrow{\scriptstyle \gamma_2} & & \\
0 & \longrightarrow & \mathfrak{m}^{n-c} \oplus \mathfrak{m}^{n-d} & \xrightarrow{\varphi_2} & A \oplus A & \xrightarrow{\psi_2} & A/\mathfrak{m}^{n-c} \oplus A/\mathfrak{m}^{n-d} & \longrightarrow & 0 \\
 & & \downarrow{\scriptstyle \alpha_1} & & \downarrow{\scriptstyle \beta_1} & & \downarrow{\scriptstyle \gamma_1} & & \\
0 & \longrightarrow & \mathfrak{m}^{n} & \xrightarrow{\varphi_1} & A & \xrightarrow{\psi_1} & A/\mathfrak{m}^{n} & \longrightarrow & 0 \\
 & & \downarrow & & \downarrow & & \downarrow & & \\
 & & 0 & & 0 & & 0 & &
\end{array}$$

Complex $K_\bullet(f,g;R)\colon 0 \longrightarrow A \xrightarrow{\beta_2} A \oplus A \xrightarrow{\beta_1} A \longrightarrow 0$ is a Koszul complex with maps

$$\beta_2\colon r \mapsto (rg, rf),$$
$$\beta_1\colon (r_1, r_2) \mapsto r_1 f - r_2 g.$$

Complex $C_\bullet$ is a subcomplex of Koszul complex and $L_\bullet(f,g;R)$ is a factorcomplex. The details of Koszul's complexes see in [4]. The maps φ_i are natural injections and ψ_i are canonical homomorphisms for $i = 1, 2$. The maps α and γ are defined as follows:

$$\alpha_2\colon u \mapsto (ug, uf),$$
$$\alpha_1\colon (v_1, v_2) \mapsto v_1 f - v_2 g,$$
$$\gamma_2\colon u + \mathfrak{m}^{n-c-d} \mapsto (ug + \mathfrak{m}^{n-c}, uf + \mathfrak{m}^{n-d}),$$
$$\gamma_1\colon (v_1 + \mathfrak{m}^{n-c}, v_2 + \mathfrak{m}^{n-d}) \mapsto v_1 f - v_2 g + \mathfrak{m}^{n}.$$

It is clear that the rows of diagram above make exact sequences, then (1) is a short exact sequence of complexes. There exists a long exact sequence (see [6, Theorem 3.3])

$$0 \longrightarrow H_2(C_\bullet) \longrightarrow H_2\big(K_\bullet(f,g;R)\big) \longrightarrow H_2\big(L_\bullet(f,g;R)\big) \longrightarrow$$
$$\longrightarrow H_1(C_\bullet) \longrightarrow H_1\big(K_\bullet(f,g;R)\big) \longrightarrow H_1\big(L_\bullet(f,g;R)\big) \longrightarrow$$
$$\longrightarrow H_0(C_\bullet) \longrightarrow H_0\big(K_\bullet(f,g;R)\big) \longrightarrow H_0\big(L_\bullet(f,g;R)\big) \longrightarrow 0.$$

From the properties of Koszul's complex and from the fact that $f^\star$ is a $k[x,y]$-regular element follows

$$H_1\big(K_\bullet(f,g;R)\big) = H_2\big(K_\bullet(f,g;R)\big) = 0$$

(see e.g. [8, Ch. IV.]). Recall that the $f^\star$ denotes the non-zero homogeneous part of polynomial f of lowest degree. Then $H_2(C_\bullet) = 0$. For the second homology module $H_2\big(L_\bullet(f,g;R)\big) = \mathrm{Ker}(\gamma_2)$. An element $u + \mathfrak{m}^{n-c-d} \in \mathrm{Ker}(\gamma_2)$ if and only if $uf \in \mathfrak{m}^{n-d}$ and $ug \in \mathfrak{m}^{n-c}$. Therefore

$$\mathrm{Ker}(\gamma_2) = \big((\mathfrak{m}^{n-d} :_R f) \cap (\mathfrak{m}^{n-c} :_R g)\big)/\mathfrak{m}^{n-c-d}.$$

Because $f^\star$ is a $k[x,y]$-regular element $\mathfrak{m}^{n-d} :_R f = \mathfrak{m}^{n-c-d}$ and then

$$H_2\big(L_\bullet(f,g;R)\big) = 0.$$

Therefore $H_1(C_\bullet) = 0$. The homology module $H_0(C_\bullet) = \mathfrak{m}^n/(f\mathfrak{m}^{n-c}, g\mathfrak{m}^{n-d}) \cdot R$ and

$$H_0\big(K_\bullet(f,g;R)\big) = H_0\big(L_\bullet(f,g;R)\big) \simeq R/(f,g) \cdot R.$$

The length of these homology modules are finite and we have a long exact sequence of homology modules then the following equality

$$\begin{aligned} 0 &= \ell_R\big(R/(f,g) \cdot R\big) - \ell_R\big(R/(f,g) \cdot R\big) \\ &\quad + \ell_R\big(\mathfrak{m}^n/(f\mathfrak{m}^{n-c}, g\mathfrak{m}^{n-d}) \cdot R\big) - \ell_R\Big(H_1\big(L_\bullet(f,g;R)\big)\Big) \end{aligned}$$

holds. Therefore

$$\ell_R\Big(H_1\big(L_\bullet(f,g;R)\big)\Big) = \ell_R\big(\mathfrak{m}^n/(f\mathfrak{m}^{n-c}, g\mathfrak{m}^{n-d}) \cdot R\big).$$

It is known by Serre and Auslander Buchsbaum (see [1] resp. [8]) that the Euler-Poincaré characteristic of the Koszul complex coincides with the multiplicity e_0. It means for $n \gg 0$

$$\begin{aligned} \chi\big(L_\bullet(f,g;R)\big) &= \sum_{i=0}^{2} (-1)^i \ell_R\Big(H_i\big(L_\bullet(f,g;R)\big)\Big) \\ &= \ell_R\big(R/(f,g) \cdot R\big) - \ell_R\big(\mathfrak{m}^n/(f\mathfrak{m}^{n-c}, g\mathfrak{m}^{n-d}) \cdot R\big). \end{aligned}$$

Moreover, $\chi\big(L_\bullet(f,g;R)\big) = c \cdot d \cdot e_0(\mathfrak{m};R)$ then

$$\ell_R\big(R/(f,g) \cdot R\big) = c \cdot d \cdot e_0(\mathfrak{m};R) + \ell_R\big(\mathfrak{m}^n/(f\mathfrak{m}^{n-c}, g\mathfrak{m}^{n-d}) \cdot R\big).$$

Since the local ring R is regular, $\ell_R\big(R/(f,g) \cdot R\big) = e_0\big((f,g) \cdot R;R\big)$ and $e_0(\mathfrak{m};R) = 1$. Finally,

$$e_0\big((f,g) \cdot R;R\big) = c \cdot d + \ell_R\big(\mathfrak{m}^n/(f\mathfrak{m}^{n-c}, g\mathfrak{m}^{n-d}) \cdot R\big).$$

This completes the proof. □

Let us recall the fact when the elements $f^\star$, $g^\star$ make a system of parameters in $k[x,y]$. Then $\mathfrak{m}^n = f\mathfrak{m}^{n-c} + g\mathfrak{m}^{n-d}$ and from this $\ell_R\big(\mathfrak{m}^n/(f\mathfrak{m}^{n-c}, g\mathfrak{m}^{n-d}) \cdot R\big) = 0$. Therefore the multiplicity $e_0\big((f,g) \cdot R;R\big) = c \cdot d$.

Acknowledgement

This paper was supported by Cultural and Educational Grant Agency of the Slovak Republic under the KEGA grant No. 047STU-4/2016.

References

[1] AUSLANDER, M. – BUCHSBAUM, D. A. Codimension and multiplicity. *Ann. of Math.* (2) 68, 1958, pp. 625–657.

[2] BOĎA, E. On the Formulas for Calculation of Multiplicity. In *Zborník vedeckých prác z medzinárodnej vedeckej konferencie "Matematika a jej aplikácie v inžinierskom vzdelávaní"*, Nitra, 2006, ISBN 80-8069-708-6, pp. 20–25.

[3] BYDŽOVSKÝ, B. *Úvod do algebraické geometrie.* Praha : JČMF, 1948.

[4] EISENBUD, D. *Commutative Algebra with a View Toward Algebraic Geometry.* New York : Springer, 1995. ISBN 978-0-387-94269-8.

[5] NORTHCOTT, D. G. *Lessons on Rings Modules and Multiplicities.* New York : Cambridge University Press, 1968. ISBN 978-0-521-09807-6.

[6] OSBORNE, M. S. *Basic Homological Algebra.* New York : Springer, 2000. ISBN 0-387-98934-X.

[7] SAMUEL, P. M'ethodes d'algèbre abstraite en géométrie algébrique. In *Ergebnisse der Mathematik und ihrer Grenzgebiete*, 1955.

[8] SERRE, J. P. *Local Algebra.* Berlin : Springer, 2000. ISBN 978-3-540-66641-7.

[9] WEIL, A. *Foundations of algebraic geometry.* Providence : American Mathematical Society, 1946, ISBN 0-8218-1029-4.

[10] ZARISKI, O. – SAMUEL, P. *Commutative Algebra. Vol. II.* Princeton : Van Nostrand, 1960. ISBN 0-387-90171-X.

Current address

Maťašovský Alexander, RNDr. PhD.
Institute of Information Engineering, Automation and Mathematics,
Faculty of Chemical and Food Technology,
Slovak University of Technology in Bratislava
Radlinského 9, 812 37 Bratislava
Tel. number: +421 259 325 342, e-mail: alexander.matasovsky@stuba.sk

Slovak University of Technology in Bratislava
Faculty of Mechanical Engineering
16th Conference on Applied Mathematics **APLIMAT 2017**
Proceedings

PERSPECTIVE AND PROPORTIONS IN THE MONTEFELTRO ALTARPIECE OF PIERO DELLA FRANCESCA

MELE Giampiero (IT)

Abstract. The aim of this research is to highlight the architect's creative skills while designing an artefact that refers to the divine model. Perspective representation is the means by which everything is seen in proportional relationships, producing an artificial image that filters reality through another system of metric relations.
After a brief investigation of the historical context of the Holy Conversation, the paper focuses on the study of proportion and perspective.
Piero's painting is a rectangle composed of 6x10 square modules, whose sides are equal to 8/10 of the Urbino foot (38.3541 cm). The horizon line is placed at a height of 4 modules, and the square 6x6 acts as the basic unit for a Latin cross church plan whose nave's area is equal to three 6x6 squares, and whose transept is composed of one 6x6 square and two 4x6 rectangles. An additional 4x6 rectangle produces a space containing the apse with lacunars. The viewpoint of the perspective is 15 modules away from the picture plane.
The study of the geometrical construction and its relationship with the measure of the architecture and its perspective representation, allows us to formulate hypotheses and discuss some doubtful conclusions previously set by historians.
The research is a part of the National Research project PRIN *"Architectural Perspectives: digital preservation, dissemination and study"*[1].

Keywords: geometry, perspective, proportions, measure, Piero della Francesca, Montefeltro Altarpiece

Mathematics Subject Classification: Primary 51N99, 51N05; Secondary 68U01

1 The Montefeltro Altarpiece. A brief history of research on the theme of perspective.

The Montefeltro Altarpiece, also called Pala Montefeltro or Brera Madonna (1472-1474), housed in the Pinacoteca di Brera in Milan, is one of the masterpieces of Piero della Francesca whose perspective construction has already been studied in depth, trying to bridge the gap derived from the scarcity of direct sources, and with the aim to unveil the origins and show the consequences of this model in Renaissance painting.

The importance of the role of Piero della Francesca in the development of perspective in painting is well known. Indeed, the transition from empirics and the medieval *"perspectiva communis or naturalis"* to the rigorous Renaissance *"perspectiva artificialis or pingendi"*, is one of the basis for the development of Italian Renaissance painting, and Piero della Francesca, together with Filippo Brunelleschi, and Leon Battista Alberti, is one of the main pioneers of this revolution.

Since the Montefeltro Altarpiece is one of the most meaningful early examples of perspective in painting, many scholars made comparisons with the first models of perspective, the concrete buildings, and Piero's treatise *De Prospectiva Pingendi*. In this field of research, we can identify three main lines of investigations: first the search for beautiful proportions in classical architecture represented by light and colour[2], second the connections between the painting and Piero della Francesca' theoretical study of perspective[3], and, last, the typology of the represented architecture (the majority of scholars agreeing that the painted apse is part of a centrally planned temple).

Given the extent of studies about the perspective and the architecture in the Montefeltro Altarpiece, it is necessary to draw an update review of what has been written and proposed on this topic in recent years, so to contextualize the present research in the history of the studies on Piero della Francesca's work.

The scientific debate started in the Sixties of the XX Century, when Meiss and Jones, followed by Shearman[4], stated that the space where the Holy Conversation takes place, should be considered to be a cross plan, whose distance from the crossing is directly proportional to the depth of the choir.

Welliver[5], instead, asserts that the interior corresponds to a three niches structure, the middle one being larger than the side ones. The perspective framework determines the proportion, in which the basic measure would be equal to the height of the viewer's eye, who is looking towards Jesus. [Welliver 1973:19-20]

This interpretation is rejected by Davis, Snyder and Maltese[6] who believe that the group is placed in front of the crossing. The represented perspective – considered to be incorrect by Davis and Snyder - derives from a model used by Piero, and reproduced in two of the most meaningful perspective projects by Donato Bramante: the Prevedari engraving, and the false choir of Santa Maria presso San Satiro in Milan.

Another interesting comparison between the Brera Altarpiece and other masterpieces, was made by Battisti[7] who calculated the number of panels depicted in the niche and compared them with the Pantheon, indicating that the design has been reused in the Prevedari engraving: *"...Bramante corrected the architectural solution of the Urbino Altarpiece in two ways: he put at the center of the arc, not an empty space of panels, but the solid moulded band, starting the arc, resting on two-sided pillars, with a half-coffer instead of a whole one.* " [Battisti 1986:230]

Borsi[8], comparing the painting with the work of Filarete, considers the architectural plan to be a variant of the cross in the square, and its design would be similar to Masaccio's Trinity in Santa Maria Novella. This interpretation is followed by Bruschi, who states: *"The space represented consists of 4 rectangular arms covered by barrel vaults, forming a cross around a central space square, the apse is a square with sides in ratio of 1:1.04719"*. [Bruschi 1992:291]

Agreeing with Davis and Snyder, Bruschi too compares the architecture of the painting with the ones by Donato Bramante's in *the Prevedari* engraving, in the fake choir of Santa Maria presso San Satiro and, subsequently, redesigned for the project of the new St. Peter in Rome. [Bruschi 1996:219-221]

Further inquiries on the architecture of the Brera Altarpiece are conducted by Bertelli[9] who notices a discrepancy between the floor and the span of the architecture, and between the real distance between the viewer's viewpoint and the painted scene, due perhaps to the absence of a plan at the base of the painting.

The research on the architecture of the Brera Altarpiece continues with Di Teodoro[10] who assumes that the painted space is *"a cruciform space with a coffered barrel-vaulted presbytery of a semi-elliptical apse - whose two arms are suggested by the arches orthogonal to the picture plane - and a square cross vault "*. [Di Teodoro 1996:29]

A particular focus in the analysis of the light in the painting is present in the essays by Bruschi, Battisti, Kemp[11] who agree in pointing out that the lighting and perspective model developed by Piero for his Altarpiece, influenced Bramante's *Prevedari engraving*, and fake choir of *Santa Maria presso San Satiro*.

Further considerations on the architecture of the Brera Altarpiece focus on the apse. In particular Ceriana[12] considers the apse to be derived from Brunelleschi and Alberti models, similarly to those carried out by Francesco di Giorgio Martini and Donato Bramante in Santa Maria presso San Satiro. The architecture is the result of a combination of regular solids, which leads naturally to think of a flat roof, as it would be in the Prevedari engraving, and in a drawing derived from it, in the Kupferstichkabinett collection in Berlin, but also in the School of Athens (1509-1511) by Raphael in the Vatican. [Ceriana 1997:118-120,154 n. 46] In fact evidences of quotas, incisions of adapted cartons at different points, as well as changes during the assembly phase have been found in the painting. According to the author, the orthogonal lines were first outlined, then the diagonal lines, while the barrel vault would be designed with nails, horse hair, wooden rods and paper strips to mark the measurements (width and height). [Ceriana 1997: 123-126]

Camerota[13] focuses on the relation between the observer and the painting and considers the distance of the observer's eye to be equal to the height of an equilateral triangle, whose side would be equal to the side of a square, producing a 6:7 proportion. The same would happen in the choir of San Satiro, between the point of view of an observer and the height of the square. In San Satiro - whose purpose is to create an optical angle of 60° *"... the line from the floor instead of going up towards the vanishing point remains horizontal"*. [Camerota 2001:27]

Finally, it is important to underline that the studies on the painting by Piero della Francesca carefully inquired the connections between the art work and the theory explained in his treatise, De Prospectiva Pingendi (1472-80 ca.). The impact of Piero's treatise on his fellows is pointed out by Luca Pacioli (1445c -1517) who praised Piero, in the *Summa de Arithmetica*[14] and, later, in *De Divina Proportione*[15]. Vasari too in his vast historical survey of artists insisted on the double talent of Piero della Francesca who was both a mathematician and artist.

The studies on the relations between theory and art in Piero's work are many and well documented. Particularly remarkable are the studies by Nicco Fasola, Venturi, and Battisti[16]; followed by the researches by Dalai Emiliani [1984:7-16], Field[17], Kemp [1990: 41-43], Chastel[18], Sorci[19], Mussini, and Grasselli[20].

Among different findings in this field of research, it is relevant to notice that the central perspective does not assume that the represented objects correspond to the real size, as resulting from the drawing, which is reproduced in scale. In fact, several times in the treatise, Piero showed how to apply it. The point of view could also vary, according to what the painter portrayed; it depended on the size of the angle, which in turn, depended on the distance from where it would have been observed.

Indeed, the retinal perception does not match to the perspective view; the latter is limited and finite in comparison with reality. Later in the Renaissance, Leonardo da Vinci was the one who noticed this characteristic of vision and developed a natural alternative, using light and shadows to correct the distortion[21]. Conceivably Piero represented what was being observed and, consequently, from his drawn interpretation, we expect to perceive approximately the original location, nowadays shifted by the distortion of the interior design of the Gallery.

2 The Perspective measure "de corpi chubi"

Unveiling the geometrical construction in relationship with the measure of the architecture and perspective, allows us to formulate hypotheses and oppose results to some doubtful conclusions previously set by historians. The theory of Renaissance perspective technique has already stated considerations on the design method of space representation[22]. However, the analytical study on other subjects and the use of digital tools allows us to highlight a set of assumption around the measurement, the squared grids and the cubic space, a way by which the Renaissance treatises themselves allude as an empirical way[23]. For this reason, the research conducted on the Montefeltro Altarpiece, and its comparison with some exercises listed in Piero's treatise, unveils an invisible knowledge, while understanding and demonstrating the method used by the artist.

The question that arises is: given a cube of known dimensions divided into sub cubes, the representation framework corresponds with one of its faces, the view point **A** is located at a certain distance and height, which is proportional to its edge. What is the approach in the projected surfaces on the framework with the grid? This problem is shown in (Fig. 1).

It shows a cube whose faces are divided into 36 (6 x 6) smaller squares. The cube is composed by 36 x 6= 216 cubic units. If we place the view point **A** at a distance from the cube[24], equal to the length of the edge (6 units), and 2 units high from the bottom line, we can see that the projection on the framework of the edge opposite to the ground floor- of the upper face of the cube, starting from the top, coincides with the second horizontal grid line of the picture. Furthermore, the projection of the edge, opposite to the picture of the cube's face coinciding with the ground floor, corresponds to the first horizontal line of this grid. If we divide in two equal parts the distance of the view point (Fig. 1, top left), it emerges the coincidence of the opposite edges belonging to the orthogonal square between the perspective and the grid. The view point can vary according to the distance from the image plane, so that it can redouble, or else it is the height from the ground floor, that can vary, according to the identified modular grid.

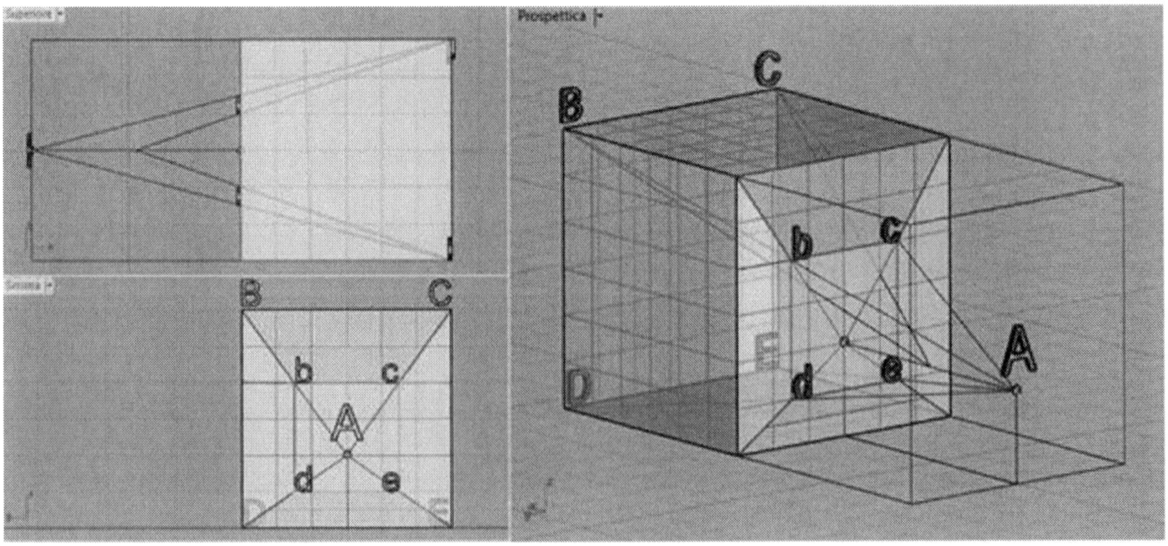

Fig. 1. Scheme of the cube's "accidental" perspective with edge equal to 6 units

In this exercise we have chosen a basic 6x6 square matrix. The view point is placed at a distance to the image plane, equal to the cube edge and respect to the ground plane the height is equal to half edge. In so doing, we recall – in a way – the same exercise of the quadrangles in Leon Battista Alberti's *De Pictura*[25]. This particular position of the viewpoint (3 braccia fiorentine) is a sort of paradigm for many scholars' analysis of the Renaissance perspective, but the Altare Montefeltro is a particular case in which the representation of the cube in perspective leads to the representation of a room , in which the picture plan and his elevation are contained in the same pattern of a square matrix.

The grid on the image plane is at the same time that of the plan and of the elevation of the cubic room. The idea of a space based on a cubical module and its bi-dimensional drawing, are under the same modular scheme: the square grid. Icnografphia and Ortographia can be traced on the same basic module, which allows us to link the number and measure in perspective.

The architectural drawing from square modules can easily be delineated in perspective with simple geometric patterns, easily expressible through numbers associated to a mesh.

Moreover, If the position of viewpoint is well studied, the coincidence between the perspective projection and the square mesh facilitates the construction of the perspective drawing,

The same premises are the basis of the studies of the Holy Conversation by Piero della Francesca and of Santa Maria at San Satiro.

The purpose is to identify the basic grid of the perspective drawing, and in so doing deducing the real measure of the architectural space portrayed in the picture. The digital image of the Altarpiece[26], imported to the CAD and associated to the measures of the painting, allows us to trace the basic squared module (measures of the rectangle in the perspective output: 172 cm bottom, 173 cm upper and 251 cm lateral)[27]. The module is the same both in the architectural context and in the geometrical construction of the perspective. The identification of the main point A made possible to place the horizon line. This line is the basis of the square in which is inscribed the arch of the architectural context behind the group of figures, which corresponds to the plane of the transition between the crossing span of the transept-nave and the apse (Fig. 2). Starting from the division into 4 equal parts of this square's side, we got the basic module, which is equal to 30,7995 cm.

The unit of measure in use in Urbino during the Renaissance was the "Piede agrimensorio" (agrimensorio food), equal to 38.3541 cm, that could be divided into 10 Once (Ounce)[28]. We can explain the measure identified in the Montefeltro Altarpiece only if we consider the value of the basic module equal to 8/10 of the Urbino "Piede Agrimensorio" (8 x 38.3541/10 = 30,680 cm) (Fig. 3).

It is interesting to note that if we divide the painting surface[29] into the historical square unit[30]; the result nearly corresponds to 30 square feet[31]: this cannot be a coincidence.

We can assume that Piero della Francesca agreed with the Duke of Urbino a payment quantified per square foot painted and the evidence of the data-outcomes validates the conjecture that the intention was just to create a painting of 30 square feet[32]. In fact, if we express the framework' side dimensions of the wooden painting-support in tenths of the Urbino "Piede agrimensorio", (45 and 66 plus 2/3 tenths of feet) the area is exactly 30 square feet. Moreover, the measurement of the width of the painting obtained through the survey corresponds to the theoretical one described above, conveyed in centimeters, whereas the height is different by about 4,7 cm.

Fig. 2. Recognize the main point of the quadrangular A let us to discover the basic module

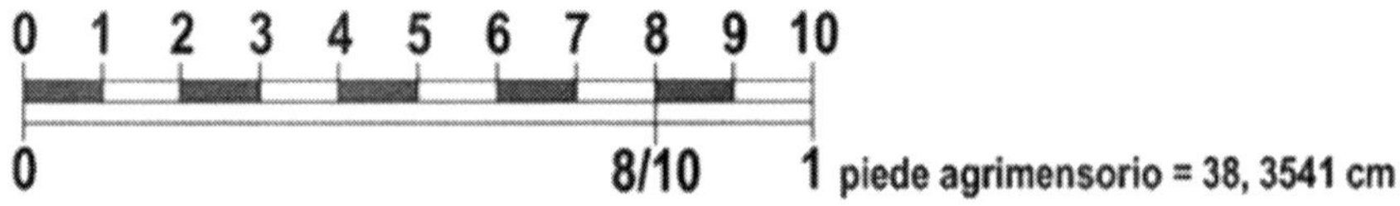

Fig. 3. Agrimensorio Urbino Foot

Some scholars state about a possible decrease of the wooden panel because of a sort of a posteriori decision, but the measurements here taken into consideration do not confirm the hypothesis. If the panel had trimmed – assumption supported by the presence of traces of pigment on the borders – it would be probably when Piero handed his masterpiece to the Duke with the purpose to put the painting on another bigger frame, forming the surface for the paid dimension of the painting. Clarified the question about the historical metrics units[33], let's analyze the structure of the perspective of the architecture depicted as a background; the aim is to understand the construction of the drawing and to find out the dimensions of the architectural space represented in it. In order to reveal the unknown drawing of the altarpiece it is necessary to resort to the exercises included in De prospectiva pingendi with particular reference to figures XII, XIII, XVa, XVI, XVII, XXIII and XLIII published in the Nicco Fasola ediction.

The first six exercises in the first book concern *"de puncti de linee et superficie piane"*, and in the last one, the XLIII, in the second book, we read *"diremo de corpi chubi, de pilastri quadri, de colonne tonde et de più facce"*. In the exercise XII, we read: *"Da l'occhio dato nel termine posto il piano asignato degradare"*, that means: given the view point and the framework, let's identify the perspective on the assigned plane. The figure, in profile, shows all the terms of the projection but omits the distances reported in the text, that are: the finite size of 20 arms of the plane, the distance of 10 arms of the view point from the framework, the height of 3 arms of the view point from ground floor, and the decrease of 2 arms of the height of the plane in the perspective, according to the data parameters (2/3 AD) (Fig.4 A). In this figure, it is important to note the ratio between the measure of the plane *"da digradare"* (that is "to decrease") with that of the distance of the view point from the framework (2:1), and the height of the plane in perspective decreased in function of a hypothetical grid of 1 x 1 arms.

This method is the same used in the subsequent exercise *"il piano degradato in quadro riducere"*, in which is described how to put into perspective the quadrangular plane of 20 arms. If the square in the figure represents both the framework and the plane to put into perspective, one can find that - even if it is not explicitly said - if the ratio between the distances of the view point from the framework and the side of the square are 10:20, the height of the plane projected on the framework is equal to 2 arms. (Fig.4 B) The third exercise, "la superficie quadrata diminuita, in più parti equali devisa, quelle divisioni in quadrati producere" expresses the existence of a square grid in which is divided an overturned plane, below the ground line, and it is not perceivable the coincidence among the framework, the plane, the reduced plane in perspective and the grid (Fig. 4 C).

The next two figures (XVI, XVII) consider the octagon inscribed in the square; the latter is rotated as to obtain a hexadecagon[34]. The aim of the exercise is to approximate the circumference into a polygon inscribed in a square, whose vertices are dependent on its related coordinates, in order to make clear that in perspective the circumferences must be reduced to sequence of points. The last exercise of the first book exemplifies a rectangle obtained by 5 squares drawn in perspective with

the method previously exposed. The purpose is probably that to show how the square matrix planes can be reduced in proportion to the framework (Fig. 5).
In relation to the exercise of the second book, we read: *"sopra del piano degradato una volta in crociera sopra a muraglia quadrata ponere"* (Fig. 6). The author describes in minute details how to construct a perspective; even if he does not explain the fundamentals of the construction, he explains - in the handbook style - step by step, the procedure that we should follow to obtain the perspective of a square bay on quadrangular pillars, roofed by a cross vault.

Two are the reasons why this exercise is very important: the first is about the identification of a square in plan and its double in elevation. In this case, the latter is not placed on the framework but on a plane parallel to it, as in the Montefeltro Altarpiece. The second is about the method – with particular reference to the method of the 'hexadecagon' described before - to draw arches in perspective, two of which are situated on orthogonal planes respect to the framework, and the other two on planes which are parallel to the framework. This method to set the perspective drawing is analogous to that proceeded for the background architecture shown in the Montefeltro Altarpiece. Piero's obsessions for square plans and *corpi chubi* solids, could be explained with the idea of describing a mathematical space that recognize in the square the generator. The architectural space so conceived for the background of the Holy Conversation, was subsequently depicted in perspective. To construct the perspective drawing without points situated outside the framework, it is essential to know the relationships between the cubes that compose it and the picture that is obtained on the framework according to the grid.

Thanks to the studies on the grid method which Piero della Francesca exposed in the exercises quoted above, we can resume the questions about the Holy Conversation, in order to identify the distance of the view point from the framework, therefore. it is possible to "reconstruct" the architecture shown in it the painting.

A careful observation of the structure of the painting helped to identify, the main point A and the horizon line, in addition to the square previously described.

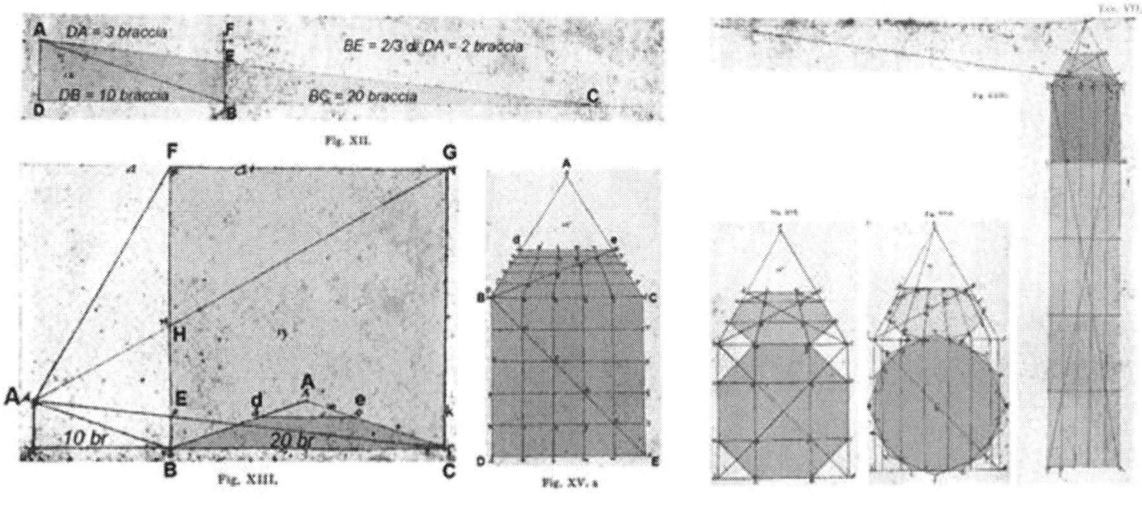

Fig. 4, 5. figure XII, XIII, XV A, XVI, XVII, XXIII from Niccola Fasola edition of the De prospectiva pingendi by Piero della Francesca, integrated with color uniform and carachters able to a better understanding of the author exercises

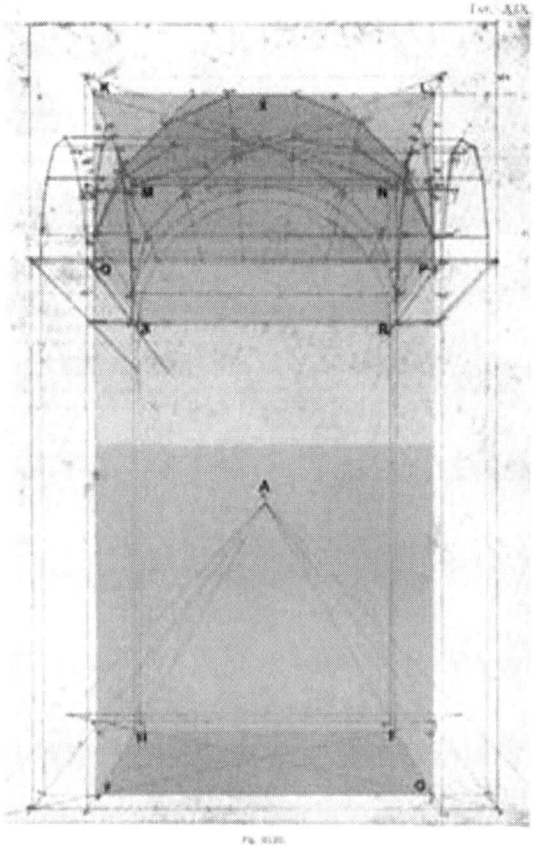

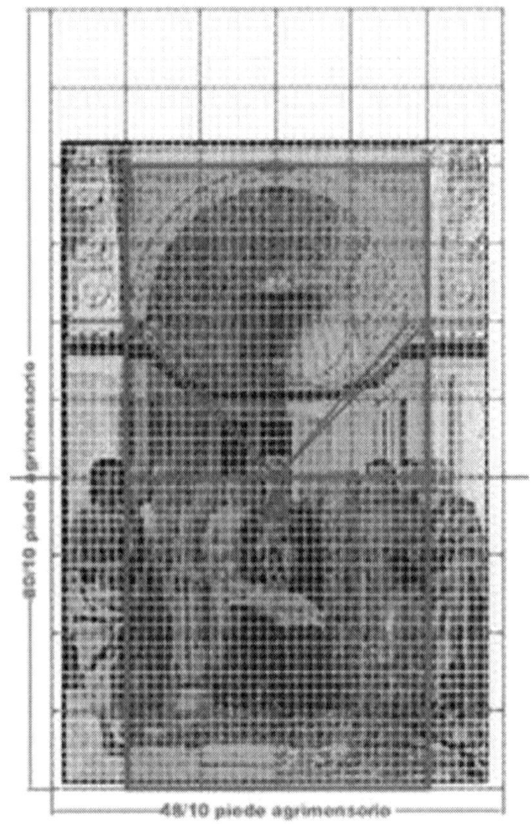

Fig. 6, 7. Figure XLIII from the treatise De perspectiva pingendi.
The disclosed grid superimposed on the painting

If another square of the same size (4x4 modules)[35], as the previous one is placed below the horizon line, we identified the ground line. (Fig. 7)

At that point, the square plan of the bay situated in the foreground has been determined, directly in perspective, at the height of the impost block of the arches of the tax-aisle transept crossing bay. This procedure has disclosed the dimension of the side of the square, of 6 x 6 modules which, in turn, allowed to find the rules of that perspective. The 1,2,3,4 square, divided in half at points 5 and 6[36] (Fig. 8 Scheme A). Taking into account the position of points 7 and 8, points 5 and 6, joined with the point A, make us identify the real dimension of the architecture and of the grid. The distance between point 7 and 8 is the side of a new square, equal to 6 modules, that is the real size of the plan of the bay.

It is also noteworthy that the intersection of the two isosceles triangles identified by the diagonals of the two halves of this 6x6 square, correspond to the reduced height of the identified plane at the impost of the arches of the bay (Fig. 8 Scheme B). Moreover, two of these four diagonals are also the two edge sides of the square located at the height of arches. The lines joining the points 2,3,9,10 outline precisely this square plane (Fig. 8 Scheme C).

Now we have a complete vision both of the grid and of the general problem. What we need is to identify the distance of the view point from the framework, and to find it, we should just follow the instructions provided by the author in the exercise of Figure XIII of the De Prospectiva Pingendi.
In fact, if we join the point 9 with points 11, 7 and 12, and if these two segments are extended to the horizon we can get the distance of the point A to the framework (Fig. 9), equal to (48 x2)/10 Urbino Agrimensorio foot. Consequently, the distance of the view point from the framework is equal 2 times the side of the square of the plant of the bay.

At this point, we can find out all the measures of the architecture that were not modulated on the real scale, but in a suitable scale to the perspective drawing.

It becomes simple as well to find out the depth of the transept and to understand that the grid located on the framework is also the grid of the architectural space both in plan and elevation.

To find out the depth of the apse it is enough to draw the straight line that starts from point 5 to point A. This line intersects the major side of the rectangle of the grid in point 13. This 48/10 foot height, is the depth of the apse impost plane. The width of the apse, in plan, is minor and to evaluate its real measure it is enough to extend to the framework the straight line which goes from the main point to impost point at the intrados of the arch. This value is 3/10 of foot, therefore, the width is (48 – 6)/10 = 42/10 feet. (Fig. 10 scheme 1).

As we can see from the grid, the height at the impost is (7 x 8) / 10 = 56/10 feet and the total height of the impost of the "tambour" is 80 /10 feet. We could go into detail of the description of the geometrical constructions of the painting, but we refer to a more extensive dissertation on the Alterpiece. However, we should write down some consideration on the identified scheme. As we can see from the images, the perspective drawing, conceived by Piero della Francesca is placed in a larger rectangle than the painted support. This does not mean that painting borders were cut, but, as we have previously speculated, the reason is in the economic agreements with the Duke of Urbino. Piero knew that by choosing the basic unit of tenth foot he could reach that given surface, if the board lengths was such that their product gives that particular result (30 square feet)

One must not forget the expertise of the painter in "numeracy"; it is not by chance that he wrote his first treatise, "De abacus", on the commercial calculation.

The resulting measures of the present study allow us not only to construct a plan and elevation scheme of the architecture, which Piero della Francesca conceived for the setting of the Holy Conversation, but also to gain a more critical consciousness of the typology of the church. If we think at the choir and transept drawn in the Altarpiece, with the measures that now we know, it could even be easy to complete the composition of the whole church with the choir so designed by Piero. By using a sequence of squares of 48/10 x 48/10, it is enough to add a square to the right, and another one to the left of the first, in order to complete the transept, and to add other three squares below the transept to complete the aisle. The diagram that comes out is very close to what Leon Battista Alberti designed in Mantova, for the church of Sant'Andrea (Fig. 10), although there are different details.

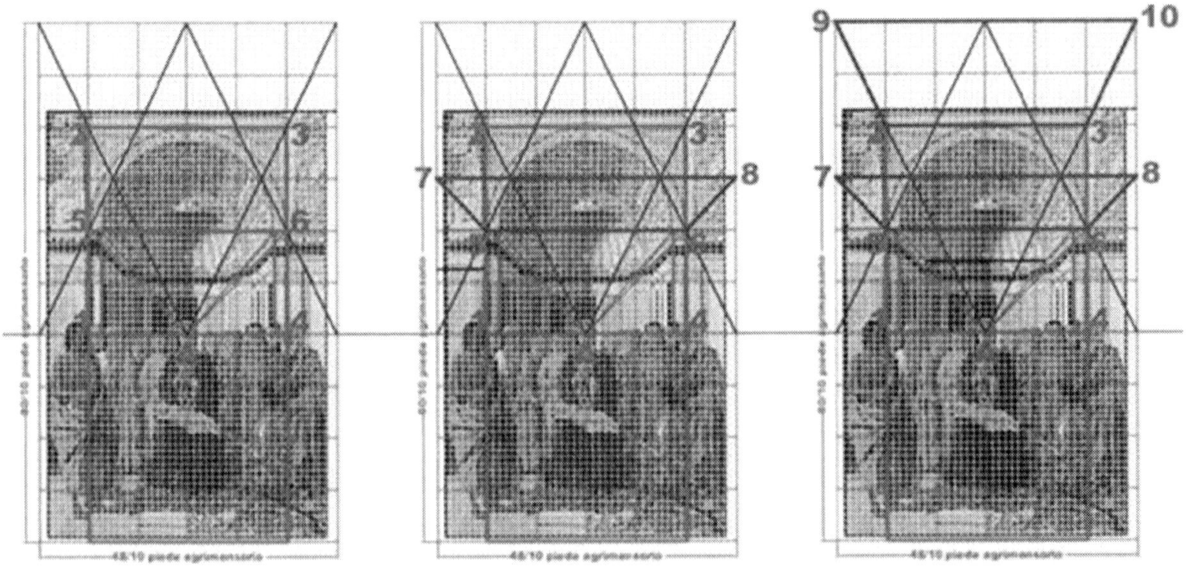

Fig. 8. Geometric patterns proportional to the grid, which identify the geometric elements of perspective

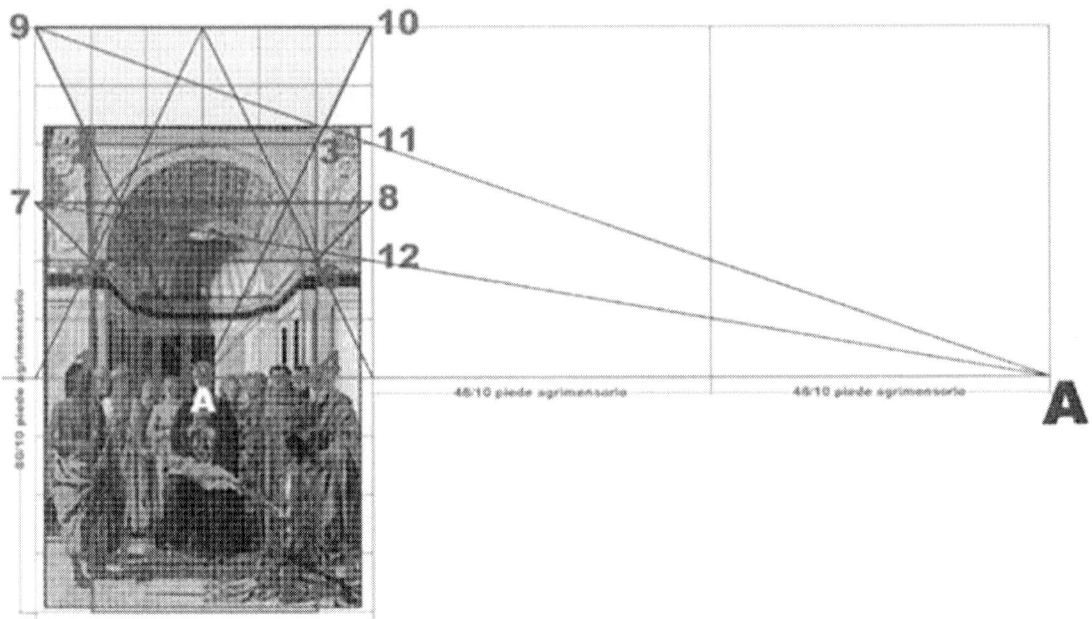

Fig. 9. Geometric patterns proportional to the grid, which identify the geometric elements of perspective

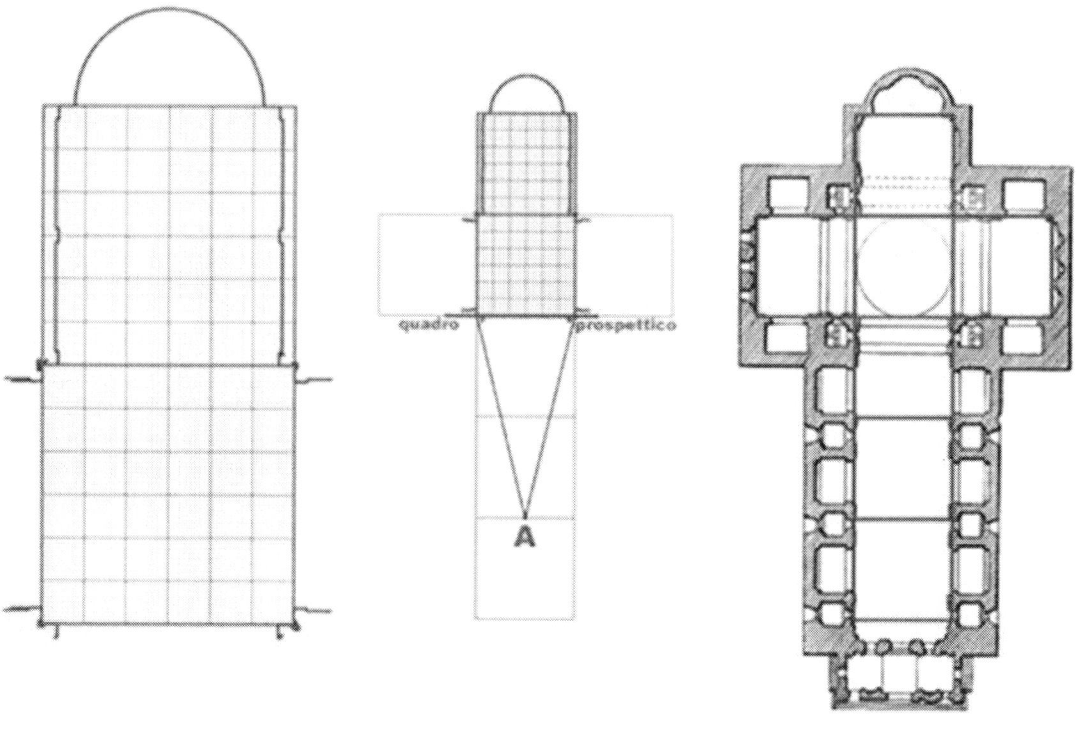

Fig. 10. 1) Scheme of the plant of the choir proceeds from the perspective of Brera Altarpiece. 2) Hypothesis of the church plan represented in perspective with the location of the perspective square and the viewer's point. 3) Plan of Sant'Andrea in Mantua

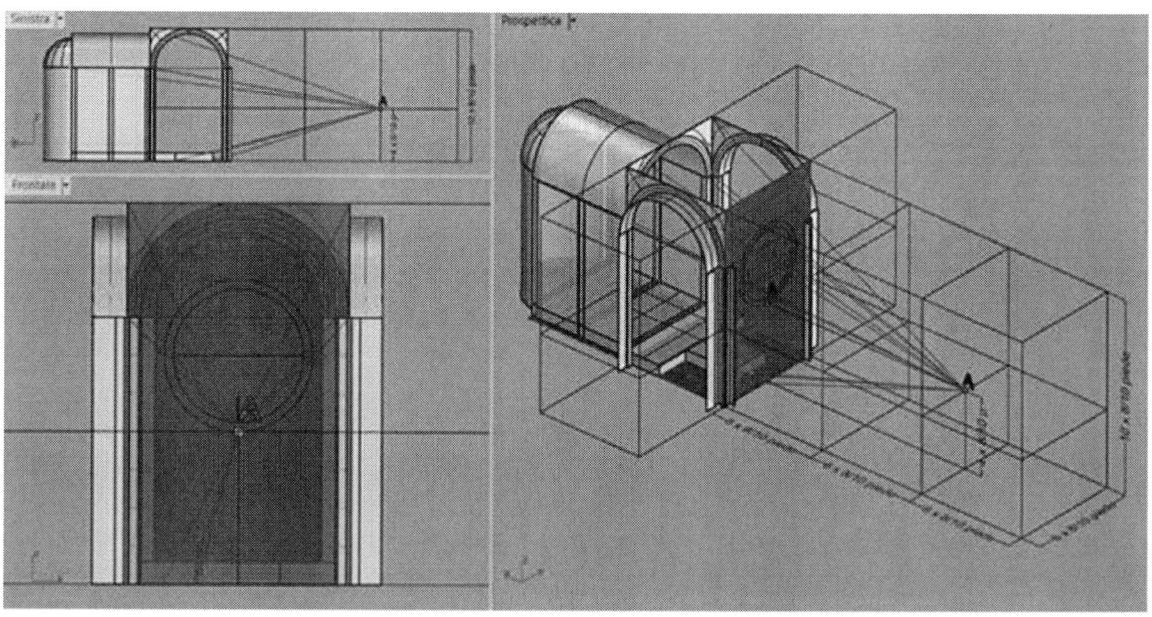

Fig. 11. 3D Model of architecture represented in the Brera Altarpiece. It is shown the viewer's point and the framework, with the assumptions volumetric of the church

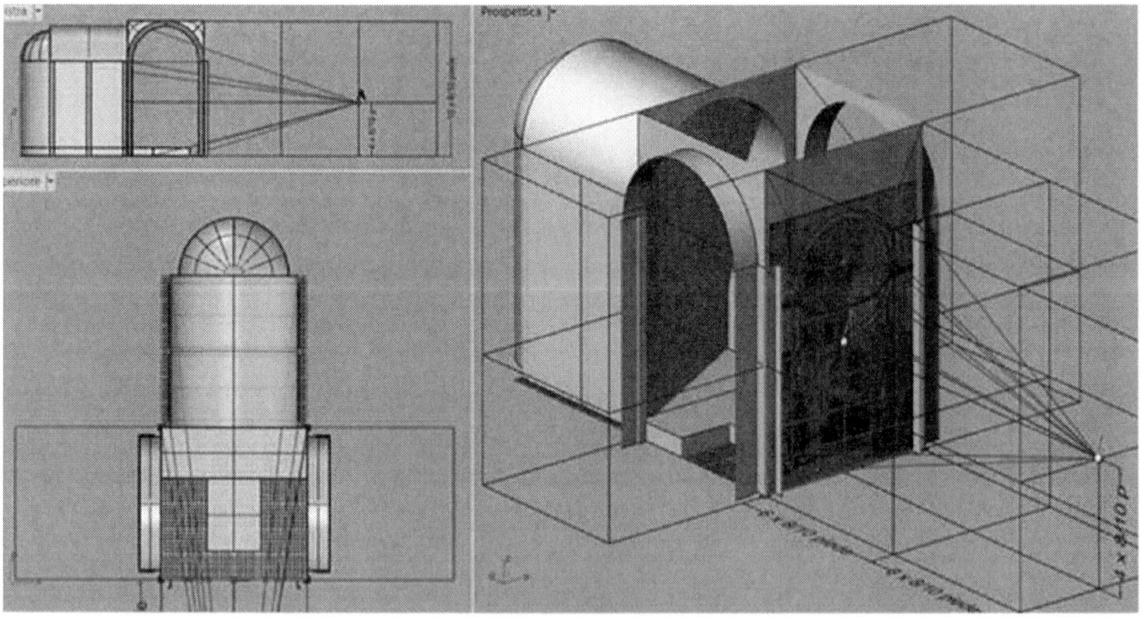

Fig. 12. 3D Model of architecture represented in the Brera Altarpiece. The viewer's point and framework with the metric image applied to the representation framework.
At the top, the basement on which stands the Virgin and all the figures

However what is useful for the purposes of our demonstration is that some Renaissance church, square matrix based, are particularly suitable to be performed in perspective. The debate about the position of the personages inside the architectural space has been solved by a virtual three-dimensional model, which count all measures and the right position of the view point. (Fig. 11)

It was possible to apply the calculated image[37] of the painting on the representation framework and verify both the correspondence between the painted image and the three-dimensional model. Furthermore, it was possible to find the right position of the personages, thanks to the basement, located under the feet of the Virgin.

The dimension of the basement, in fact, have been derived from the perspective, combining the projecting rays and the image points with the center of projection and extending these lines until the intersection the horizontal plane directly in the virtual space in 3D (Fig.12). From the calculations, the position of the Virgin is exactly in the center of that bay, which intersects the transept and the nave. Under the basement are three figures, St. John the Baptist, St. Jerome and behind them, St. Bernardine from Siena; on the left, St. John the Evangelist, St. Francis and St. Peter Martyr. Behind the Virgin, on the basement are four angels, and in the foreground of the basement a knelling figure: the Duke Federico from Montefeltro.

3 Conclusions

In conclusion, what is notable in the Sacred Conversation is that the idea of space, generated through the number and measurements, is functional to the representation in perspective, and the square grid is a useful tool for its construction. The geometry and arithmetic, in their historical meaning, play an important role both in the definition of space both in its transformation to obtain its image. The square grid achieved by the module explains the essence of the project in the Renaissance age, as well as its representation, both when it is used for figurative-pictorial purposes, both when needed to surprise with spectacular scenographic effects.

Notes

[1] National Research Program 2010-2011, National Coordinator is Prof. Riccardo Migliari and Milan Research Unit coordinator is Prof. Michela Rossi.

[2] WITTKOWER, Rudolf 1953. Brunelleschi and "Proportion" in perspective, 'Journal of the Courtauld and Warburg Institutes', 275-291; MARZOLI, Carla 1951. Studi sulle proporzioni in Nona Triennale di Milano, cat. Exh.Triennale Milano, 114-134; BRUSCHI, Arnaldo 1969. Bramante Architetto, Bari 71-73; MEISS, Millard 1971. La Sacra Conversazione di Piero della Francesca, 'Quaderni di Brera I', Firenze; BATTISTI, Eugenio 1974. Bramante, Piero e Pacioli ad Urbino, 'Studi Bramanteschi', 267-283; MALTESE, Corrado 1974. Architettura Ficta 1472 circa, in Studi Bramanteschi, atti del congresso internazionale Milano-Urbino-Roma, 1970:283-292; FOCILLON, Henry 1992. Piero della Francesca, intr. Emiliani A., Parma; HALL, Peter 1992. Colour and Meaning. Practice and Theory in Renaissance Painting, Cambridge, 79-84; KEMP, Martin 1995. In the Light of Dante. Meditations on natural and divine light in Piero della Francesca, Raphael and Michelangelo, 'Ars naturam adiuvans' Festschrift für Matthias Winner, 160-177; KEMP, Martin 1996. New Light on old theories: Piero' studies of the transmission of light, in Piero della Francesca tra arte e scienza, 333-345.

[3] VAGNETTI, Luigi 1975. Riflessioni sul De prospective pingendi, in Commentari, XXVI, 19-22, 32-35; Daly Davis M. 1977. Piero della Francesca's mathematical treatises: the 'Trattato d'abaco' and 'Libellus de quinque corporibus regularibus', Longo Editore, Ravenna; DALAI EMILIANI, Marisa 1980. La prospettiva rinascimentale: codificazioni e trasgressioni, Atti del Convegno internazionale, Milano, Castello Sforzesco 11- 15 ottobre 1977, Firenze: Centro Di; DALAI EMILIANI, Marisa 1984. Figure rinascimentali dei poliedri platonici. Qualche problema di storia e autografia, in MARANI, Pietro C. (c.by), Fra Rinascimento, Manierismo e Realtà. Scritti in memoria di Anna Maria Brizio, Firenze 7-16; DALAI EMILIANI Marisa 1996. Piero della Francesca, Milano; BESOMI, Ottavio & SORCI, Alessandra 2004. Edizione nazionale degli scritti di Piero della Francesca: criteri di studio per il Trattato dell'Abaco e il De Prospectiva pingendi, Atti del Seminario, Stiftung Bibliothek Werner Oechslin Einsiedln, 21-24 sett. 2003, 'Scholion', 3, 150-159.

[4] MEISS Milard, JONES Theodore G. 1966. Once again Piero della Francesca's Montefeltro Altarpiece, 'The Art Bulletin', XLVIII, 2, 203-207; SHEARMAN John 1968.The Logic and Realism of Piero della Francesca, in 'Festschrift Ulrich Middeldorf, herausgegeben von A. Kosegarten und P. Tigler', Berlin 180-186.

[5] WELLIVER Warman.1973. The Simbolic Architecture of Domenico Veneziano and Piero della Francesca, in 'The Art Quarterly, XXXVI, 1-30.

[6] DAVIS E.M., SNYDER Thimoty D.1974. Piero della Francesca's Madonna of Urbino: a Further Examination, 'Gazette des Beaux-Arts', LXXV, 193-212; MALTESE Corrado1974. Architectura "ficta" 1472 circa, "Studi Bramanteschi", Roma 283-292.

[7] BATTISTI Eugenio 1986. E' possibile identificare in Piero della Francesca uno stile di corte?, in Federico da Montefeltro. Le arti..., Roma, 223-232; Battisti E. 1992. Piero della Francesca (1971), DALAI EMILIANI Marisa c.by, Milano 2 voll., I, 263-268; II, 504-513; 590-591.

[8] BORSI Franco1989. Bramante, Electa:10, 49, 51-54.

[9] BERTELLI, Carlo 1991. Piero della Francesca, La forza divina della pittura, Milano 131-150; 212-213. BERTELLI, Carlo 2012. Piero un pittore per due amici, Skirà, Milano

[10] DI TEODORO Francesco P. 1996. La Sacra Conversazione di Piero della Francesca, Tea Spa, Milano

[11] BATTISTI, Eugenio 1986. E' possibile identificare in Piero della Francesca uno stile di corte?, in Federico da Montefeltro. Le arti, Roma :230; Bruschi 1990:15 n.2; KEMP, Martin 1990. The science of Art. Optical Themes in Western Art from Brunelleschi to Seurat, New Haven, London.

[12] CERIANA, Matteo 1997, Sull'architettura dipinta della pala, in 'Nuovi Studi oltre il Restauro', TREVISANI, Filippo, DAFFRA, Manuela (c. by) Milano, 115-149.

[13] CAMEROTA Filippo 2001. Donato Bramante: ricerche, proposte, rilettura, a.c. DI TEODORO Francesco P. Accademia Raffaello, Urbino.

[14] PACIOLI, Luca 1523. Summa de Arithmetica geometria Proportioni et proportionalità, Casa Editrice Le Lettere, Tuscolano, Paganino Paganini, 1523, Distinctio Sexta, Tractatus Primus, Articulus Secundus, c. 68v.

[15] PACIOLI, Luca 1509. De divina proportione (Brixiae, Paganius Paganinus, 1509), rist. anast. Maslianico (Como). Dominioni 1967, Pars Prima, c. 33r.
[16] NICCO FASOLA, Giusta 1985. (c.by) Piero della Francesca, De Perspectiva pingendi, Casa Editrice Le Lettere, Firenze; Venturi L. 1954. Piero della Francesca, Genève 13-14; Battisti E. 1986 cit.
[17] According to Field without metrics data, it is impossible to reproduce the parts that are proportional. FIELD Judith V. 2005. Piero della Francesca. A Mathematician's Art, New Haven, London.
[18] CLERGEAU, Marie F. 1992. De prospectiva pingendi à la peinture de Piero: quel lien? In Piero della Francesca tra arte e scienza, Atti del Convegno Arezzo 8-10 ottobre 1992, DALAI EMILIANI, Marisa & CURZI Valter (c.by), Venezia: 65.
[19] SORCI, Alessandra 2001. La forza delle linee. Prospettiva e stereometria in Piero della Francesca, Firenze.
[20] MUSSINI Massimo, GRASSELLI Luigi 2008. Piero della Francesca. De Perspectiva Pingendi, Aboca Museum, Sansepolcro.
[21] MARANI Pietro C. 2010. 9 febbraio 1498. Il «Cenacolo» svelato, in I giorni di Milano, Laterza, Roma-Bari: 69-95; MARANI Pietro C. 2010. Leonardiana. Studi e saggi su Leonardo da Vinci, Skira, 207-222.
[22] SPAGNESI Gianfranco, PADS Mario, MANDELLI Emma 1984. Raffaello: l'architettura "picta," percezione e realtà, exhib. cat. Rome , Palazzo Venezia , January 14-February 4, Multigrafica Venezia; Bartoli Maria Teresa 1984. La Scuola d'Atene e il metodo prospettico di Raffaello, 'Bollettino degli ingegneri della Toscana' 7 /8, Florence.
[23] See for instance: GRYSON Cecil 1980. (c.by) Alberti Leon Battista, De pictura, Laterza Bari.
[24] The cube face is at the same time the representation framework.
[25] The problem starts from a square belonging to a geometral plane of 6x6 'Florentine arms' - and put the framework in equivalence of one square's side and the observer's point of view, at a distance from the panel equal to the square's side (6 arms) and a height from geometral equal to half of the side (3 arms). ""dove io debbo dipingere scrivo uno quadrangolo di retti angoli grande quanto io voglio.....e quivi determino quanto mi piaccino nella mia pittura uomini grandi; e divido la lunghezza di questo uomo in tre parti, quali a me ciascuna sia proporzionale a quella misura si chiama braccio, però che commisurando uno comune uomo si vede essere quasi tre braccia; e con queste braccia segno la linea di sotto la qual giace nel quadrangolo in tante parti quante ne riceva; ed émmi questa linea medesima proporzionale a quella ultima quantità quale prima mi traversò innanzi. Poi dentro questo quadrangulo in un punto, dove a me paia, fermo uno punto il quale occupi quello luogo dove il razzo centrico ferisce, e per questo chiamo punto centrico.... Segno dritte linee da esso a ciascuna divisione posta nella linea del quadrangolo che giace, quali segnate linee a me dimostrino in che modo, quasi persino in infinito, ciascuna traversa quantità segua alterandosi" ALBERTI Leon Battista, De pictura, Book I chapter 19.
[26] As downloaded from the Brera Gallery digital archive's website. Even if it was not " high resolution," it was enough to identify the related rules in the perspective view.
[27] The measurements were taken from the publication by TREVISANI, DAFFRA 1997:269. Here in addition to the general measures of the painting are presented in a systematic way the measurements of the horizontal axes that built the painting support.
[28] MARTINI Angelo 1883. Manuale di metrologia, ossia misure, pesi e monete in uso attualmente e anticamente presso tutti i popoli, Loescher, Torino.
[29] The surface of the painting is 4.3423 m = 1.73 mx 2.51 m.
[30] One square foot is equal to 0.147104 m.
[31] 30 square feet are equal to 4.41312 mq the measurements produce a surface slightly lower respect to the theoretical surface, which is equal to 4.3423 square meters.
[32] 45 cm x 3.83541 = 172.59 cm it is the theoretical measure corresponding with the measurement taken during the restoration of the painting. It is 172 at the bottom and 173 cm at the top; (66 +2/3) x 3.83541 cm = 255.69 cm. The theoretical measure differs about 4 cm from the real measure (the measurement in height was detected by TREVISANI, DAFRA 2006:269 and equal to 251 cm).
[33] to examine in deep different metric units, that has been taken into consideration by previous papers and books, please see the essay written by TREVISANI, DAFFRA 1997: 51.
[34] The hexadecagon is obtained by reversing the top center of the figure of the octagon on the diagonal of the square.
[35] The double square reminds the exercise in the De Prospective Pingendi in Fig. XLIII.
[36] The two points determine the segment impost-arc transition between the nave and transept apse.
[37] The image with the overlaid grid in the CAD environment was printed virtually and imported into the three-dimensional model to obtain the 3D grid vector, drawn on the image. In this way, we found the perfect correspondence between image and metric 3D model.

References

[1] WITTKOWER, Rudolf 1953. Brunelleschi and "Proportion" in perspective, 'Journal of the Courtauld and Warburg Institutes'

[2] MARZOLI, Carla 1951. Studi sulle proporzioni in Nona Triennale di Milano, cat. Exh.Triennale Milano

[3] BRUSCHI, Arnaldo 1969. Bramante Architetto, Bari

[4] MEISS, Millard 1971. La Sacra Conversazione di Piero della Francesca, 'Quaderni di Brera I', Firenze

[5] BATTISTI, Eugenio 1974. Bramante, Piero e Pacioli ad Urbino, 'Studi Bramanteschi'

[6] MALTESE, Corrado 1974. Archittetura Ficta 1472 circa, in Studi Bramanteschi, atti del congresso internazionale Milano-Urbino-Roma, 1970

[7] FOCILLON, Henry 1992. Piero della Francesca, intr. Emiliani A., Parma

[8] HALL, Peter 1992. Colour and Meaning. Practice and Theory in Renaissance Painting, Cambridge

[9] KEMP, Martin 1995. In the Light of Dante. Meditations on natural and divine light in Piero della Francesca, Raphael and Michelangelo, 'Ars naturam adiuvans' Festschrift für Matthias Winner

[10] KEMP, Martin 1996. New Light on old theories: Piero' studies of the transmission of light, in Piero della Francesca tra arte e scienza

[11] VAGNETTI, Luigi 1975. Riflessioni sul De prospective pingendi, in Commentari, XXVI

[12] DALY DAVIS M. 1977. Piero della Francesca's mathematical treatises: the 'Trattato d'abaco' and 'Libellus de quinque corporibus regularibus', Longo Editore, Ravenna

[13] DALAI EMILIANI, Marisa 1980. La prospettiva rinascimentale: codificazioni e trasgressioni, Atti del Convegno internazionale, Milano, Castello Sforzesco 11- 15 ottobre 1977, Firenze: Centro Di

[14] DALAI EMILIANI, Marisa 1984. Figure rinascimentali dei poliedri platonici. Qualche problema di storia e autografia, in MARANI, Pietro C. (c.by), Fra Rinascimento, Manierismo e Realtà. Scritti in memoria di Anna Maria Brizio, Firenze

[15] DALAI EMILIANI Marisa 1996. Piero della Francesca, Milano

[16] BESOMI, Ottavio & SORCI, Alessandra 2004. Edizione nazionale degli scritti di Piero della Francesca: criteri di studio per il Trattato dell'Abaco e il De Prospectiva pingendi, Atti del Seminario, Stiftung Bibliothek Werner Oechslin Einsiedln, 21-24 sett. 2003, 'Scholion', 3

[17] MEISS Milard, JONES Theodore G. 1966. Once again Piero della Francesca's Montefeltro Altarpiece, 'The Art Bulletin', XLVIII, 2,

[18] SHEARMAN John 1968.The Logic and Realism of Piero della Francesca, in 'Festschrift Ulrich Middeldorf, herausgegeben von A. Kosegarten und P. Tigler', Berlin

[19] WELLIVER Warman.1973. The Simbolic Architecture of Domenico Veneziano and Piero della Francesca, in 'The Art Quarterly, XXXVI

[20] DAVIS E.M., SNYDER Thimoty D.1974. Piero della Francesca's Madonna of Urbino: a Further Examination, 'Gazette des Beaux-Arts', LXXV

[21] MALTESE Corrado1974. Architectura "ficta" 1472 circa, "Studi Bramanteschi", Roma

[22] BATTISTI Eugenio 1986. E' possibile identificare in Piero della Francesca uno stile di corte?, in Federico da Montefeltro. Le arti..., Roma

[23] BORSI Franco1989. Bramante, Electa
[24] BERTELLI, Carlo 1991. Piero della Francesca, La forza divina della pittura, Milano
[25] BERTELLI, Carlo 2012. Piero un pittore per due amici, Skirà, Milano
[26] DI TEODORO Francesco P. 1996. La Sacra Conversazione di Piero della Francesca, Tea Spa, Milano
[27] BATTISTI, Eugenio 1986. E' possibile identificare in Piero della Francesca uno stile di corte?, in Federico da Montefeltro. Le arti, Roma
[28] KEMP, Martin 1990. The science of Art. Optical Themes in Western Art from Brunelleschi to Seurat, New Haven, London.
[29] CAMEROTA Filippo 2001. Donato Bramante: ricerche, proposte, rilettura, a.c. Di Teodoro Francesco P. Accademia Raffaello, Urbino
[30] PACIOLI, Luca 1523. Summa de Arithmetica geometria Proportioni et proportionalità, Casa Editrice Le Lettere, Tuscolano, Paganino Paganini, 1523
[31] PACIOLI, Luca 1509. De divina proportione (Brixiae, Paganius Paganinus, 1509), rist. anast. Maslianico (Como). Dominioni 1967, Pars Prima
[32] NICCO FASOLA, Giusta 1985. (c.by) Piero della Francesca, De Perspectiva pingendi, Casa Editrice Le Lettere, Firenze
[33] FIELD Judith V. 2005. Piero della Francesca. A Mathematician's Art, New Haven, London.
[34] CLERGEAU, Marie F. 1992. De prospectiva pingendi à la peinture de Piero: quel lien? In Piero della Francesca tra arte e scienza, Atti del Convegno Arezzo 8-10 ottobre 1992, DALAI EMILIANI, Marisa & CURZI Valter (c.by), Venezia
[35] SORCI, Alessandra 2001. La forza delle linee. Prospettiva e stereometria in Piero della Francesca, Firenze.
[36] MUSSINI Massimo, GRASSELLI Luigi 2008. Piero della Francesca. De Perspectiva Pingendi, Aboca Museum, Sansepolcro
[37] MARANI Pietro C. 2010. 9 febbraio 1498. Il «Cenacolo» svelato, in I giorni di Milano, Laterza, Roma-Bari
[38] MARANI Pietro C. 2010. Leonardiana. Studi e saggi su Leonardo da Vinci, Skira
[39] SPAGNESI Gianfranco, PADS Mario, MANDELLI Emma 1984. Raffaello: l'architettura "picta," percezione e realtà, exhib. cat. Rome , Palazzo Venezia , January 14-February 4, Multigrafica Venezia; Bartoli Maria Teresa 1984. La Scuola d'Atene e il metodo prospettico di Raffaello, 'Bollettino degli ingegneri della Toscana' 7 /8, Florence.
[40] GRYSON Cecil 1980. (c.by) Alberti Leon Battista, De pictura, Laterza Bari.
[41] MARTINI Angelo 1883. Manuale di metrologia, ossia misure, pesi e monete in uso attualmente e anticamente presso tutti i popoli, Loescher, Torino

Current address

Mele Giampiero, Associate Professor
Università degli studi eCampus
Via degli Isimbardi, 10, 22060 Novedrate, Italy
Tel. number:+393895903667, e-mail: giampiero.mele@uniecampus.it

Slovak University of Technology in Bratislava
Faculty of Mechanical Engineering
16th Conference on Applied Mathematics **APLIMAT 2017**
Proceedings

ON CONFORMAL MAPPINGS ONTO COMPACT EINSTEIN SPACES

MIKEŠ Josef (CZ), GUSEVA Nadezhda (RU), HINTERLEITNER Irena (CZ)

Abstract. This paper is devoted to conformal mappings of special (pseudo-) Riemannian spaces onto Einstein spaces. On the basis of studying the fundamental linear equations which were obtained by J. Mikeš, M.L. Gavril'chenko and E.I. Gladyscheva we found new results of conformal mappings of compactess n-dimensional pseudo-Riemannian manifolds onto Einstein manifolds.

Keywords: Conformal mapping, Einstein space, compact space, Ricci form

Mathematics subject classification: Primary 53C25; Secondary 53C24

Introduction

This paper is devoted to conformal mappings of special (pseudo-) Riemannian spaces onto Einstein spaces.

On the basis of studying the fundamental linear equations which were obtained by J. Mikeš, M.L. Gavril'chenko and E.I. Gladyscheva [14] we found new results of conformal mappings of compactess n-dimensional pseudo-Riemannian manifolds onto Einstein manifolds.

Conformal mappings of n-dimensional Riemannian spaces V_n were studied in many papers, see for example [2, 3, 6, 5, 10, 11, 12, 16, 17]. We assume that the metrics g of V_n under study are of arbitrary signature, i.e., V_n is either a proper Riemannian space or a pseudo-Riemannian space. Conformal mappings have applications in the general theory of relativity (see, e.g., [1, 4, 5, 16]).

In 1920 H. Brinkmann [1] started with research of conformal mapping of (pseudo-) Riemannian manifolds V_n onto Einstein spaces $\bar{V}_n$. He obtained the fundamental system of differential equations in covariant derivatives of Cauchy type with respect to $(n + 1)$ unknown functions. This problem is stated in detail in A. Z. Petrov's monograph [16].

As we said above, in [13, 14] J. Mikeš, M.L. Gavril'chenko and E.I. Gladyscheva found the mentioned fundamental system of differential equations in linear form, see [9], [15](pp. 112–116), [12] (pp. 242–246). This system was precisely studied in paper [7].

The number p ($\leq n+2$) of substantial parameters on which depends the general solutions is called a *degree of mobility of a Riemannian space with respect to conformal mappings onto Einstein spaces*.

In [7, 8, 9, 14] were found lacunas in the distribution of mobility degrees of Riemannian spaces with respect to conformal mappings onto Einstein spaces.

In the above mentioned papers, it is assumed that the geometric objects in question are of rather high differentiability class. Our paper is devoted to the study of minimal conditions on the differentiability of the geometric objects under conformal mappings of V_n onto Einstein spaces. As it is known, by means of the choice of a coordinate system one can decrease the differentiability class to the "minimal level". What is more, the smoothness of the function defining a conformal mapping influences immediately and in an "inpenetrable" way the smoothness of the metric of a space conformal to an Einstein space. In the paper we obtain the basic system of the problem in question in closed linear form under minimal requirements on the differentiability of the metrics of spaces that are in conformal correspondence. All considerations are local. We assume that the dimensions of the spaces under consideration is greater than two.

1 Basic facts on conformal mappings

Consider a Riemannian space V_n with metric g of arbitrary signature. If in terms of a coordinate chart (U, x) the components $g_{ij}(x)$ of the metric belong to the class C^r, we will write $V_n \in C^r$. We assume that $r = 2, 3, \ldots, \infty, \omega$, where C^∞ is the class of functions possessing continuous partial derivatives of any order, C^ω is the class of real analytic functions.

In $V_n \in C^1$ there are defined the Christoffel symbols of the first and the second types

$$\Gamma_{ijk} = 1/2\,(\partial_i g_{jk} + \partial_j g_{ik} - \partial_k g_{ij}) \quad \text{and} \quad \Gamma^h_{ij} = g^{h\alpha}\Gamma_{ij\alpha},$$

where g^{ij} are the components of the inverse of the matrix $\|g_{ij}\|$ and $\partial_i = \partial/\partial x^i$.

In $V_n \in C^2$, there are defined the Riemann and the Ricci tensors and the scalar curvature

$$R_{hijk} = g_{h\alpha}R^\alpha_{ijk},\ \ R^h_{ijk} = \partial_j\Gamma^h_{ik} - \partial_k\Gamma^h_{ij} + \Gamma^\alpha_{ik}\Gamma^h_{\alpha j} - \Gamma^\alpha_{ij}\Gamma^h_{\alpha k},\ \ R_{ij} = R^\alpha_{i\alpha j},\ \ R = R_{\alpha\beta}g^{\alpha\beta}.$$

Note that in many works (e.g., [17, 18]) the Ricci tensor is defined with the opposite sign. Further, in V_n , there are defined the Weyl tensor of conformal curvature C

$$C^h_{ijk} = g^{h\alpha}C_{\alpha ijk},\ \ C_{hijk} = R_{hijk} - g_{hj}L_{ik} - g_{ik}L_{hj} + g_{hk}L_{ij} + g_{ij}L_{hk},$$

and the tensors

$$L_{ij} = \frac{1}{n-2}\,(R_{ij} - \frac{R}{2(n-1)}\,g_{ij}),\ \ L^h_i = g^{h\alpha}L_{\alpha i} \quad \text{and} \quad L_{ijk} = L_{ij,k} - L_{ik,j}.$$

Here and in what follows the comma stands for the covariant derivative with respect to the Levi-Civita connection of V_n. If $V_n \in C^r$, then

$$\Gamma^h_{ij}, \Gamma_{ijk} \in C^{r-1};\ \ R^h_{ijk}, R_{hijk}, R_{ij}, R, C^h_{ijk}, C_{hijk}, L_{ij}, L^h_i \in C^{r-2} \quad \text{and} \quad L_{ijk} \in C^{r-3}.$$

Two Riemannian spaces V_n and $\bar{V}_n$ are in a *conformal correspondence* if, in a coordinate system (U, x) which is common with respect to a mapping between V_n and $\bar{V}_n$, their metric tensors g_{ij} and $\bar{g}_{ij}$ are related as follows [6, 5, 15, 12, 16, 17]:

$$\bar{g}_{ij}(x) = e^{2\sigma(x)}g_{ij}(x), \tag{1}$$

where σ is an invariant.

As it is known, for spaces V_n and $\bar{V}_n$ which are in a conformal correspondence, the following relations hold:

$$\begin{aligned}
\bar{\Gamma}^h_{ij} &= \Gamma^h_{ij} + \delta^h_i \sigma_j + \delta^h_j \sigma_i - \sigma^h g_{ij};\\
\bar{R}^h_{ijk} &= R^h_{ijk} + \delta^h_k \sigma_{ij} - \delta^h_j \sigma_{ik} + \sigma^h_k g_{ij} - \sigma^h_j g_{ik} + \Delta_1 \sigma \, (\delta^h_k g_{ij} - \delta^h_j g_{ik});\\
\bar{R}_{ij} &= R_{ij} - (n-2)\sigma_{ij} - (\Delta_2 \sigma + (n-2)\Delta_1 \sigma) g_{ij};\\
\bar{C}^h_{ijk} &= C^h_{ijk}.
\end{aligned}$$

Here Γ^h_{ij} $(\bar{\Gamma}^h_{ij})$ are the Christoffel symbols, R^h_{ijk} $(\bar{R}^h_{ijk})$ the Riemann tensors, R_{ij} $(\bar{R}_{ij})$ the Ricci tensors, and C^h_{ijk} $(\bar{C}^h_{ijk})$ the conformal curvature tensors of V_n and $\bar{V}_n$, respectively. In addition, δ^h_i is the Kronecker's delta, $\Delta_1 \sigma = g^{\alpha\beta}\sigma_\alpha \sigma_\beta$ and $\Delta_2 \sigma = g^{\alpha\beta}\sigma_{,\alpha\beta}$ are the first and the second Beltrami operators, respectively,

$$\sigma_i = \sigma_{,i}, \; \sigma_{ij} = \sigma_{i,j} - \sigma_i \sigma_j, \; \sigma^h = \sigma_\alpha g^{\alpha h}, \; \sigma^h_i = \sigma_{\alpha i} g^{\alpha h}.$$

The above formulas hold when V_n and $\bar{V}_n \in C^2$ are considered in a common coordinate system with respect to the mapping between V_n and $\bar{V}_n$.

2 Conformal mappings onto Einstein spaces

In [14] it is proved that a Riemannian space V_n admits a conformal mapping onto an Einstein space $\bar{V}_n$ if and only if in V_n there exists a solution of the following system of linear homogeneous differential equations in covariant derivatives of Cauchy type with respect to the functions $u(x)$ and $s(x)$ (> 0):

$$s_{,ij} = u\, g_{ij} - s\, L_{ij}. \tag{2}$$

In this case, $s = e^{-\sigma}$, and (1) takes the form

$$\bar{g}_{ij} = s^{-2} g_{ij}.$$

These conditions are fulfilled for minimal requirements on the smoothness class of the functions under consideration, i.e., when $s \in C^2$ and u is a continuous function. Then obviously V_n and $\bar{V}_n \in C^2$.

In [14], as a matter of fact, under the condition that V_n and $\bar{V}_n \in C^3$, it is proved that a Riemannian space V_n admits a conformal mapping onto an Einstein space $\bar{V}_n$ if and only if in V_n there exists a solution to the following closed system of linear homogeneous differential equations in covariant derivatives of Cauchy type with respect to the functions $u(x)$, $s(x)$ (> 0) and a vector $s_i(x)$:

$$\text{(a) } s_{,i} = s_i; \quad \text{(b) } s_{i,j} = u\, g_{ij} - s\, L_{ij}; \quad \text{(c) } u_{,i} = -s_\alpha L^\alpha_i. \tag{3}$$

In [7] we have obtained the following theorem.

Theorem 1. *A Riemannian space $V_n \in C^r$, $r > 2$, admits a conformal mapping onto an Einstein space $\bar{V}_n \in C^2$ if and only if in V_n there exists a solution to closed system of linear homogeneous differential equations in covariant derivatives of Cauchy type* (3) *with respect to the functions $u(x)$, $s(x)$ (> 0) and a vector $s_i(x)$. In this case, $\bar{V}_n \in C^r$.*

3 Modification of a certain Theorem by Hopf

After modification of a well known Hopf Theorem about the existence of the solutions of differential equations in partial derivative on manifolds [18] (pp. 26) we prove the new results about conformal mappings onto Einstein spaces.

In advance we modify the well known Hopf Theorem.

Theorem 2. *Let $\varphi \in C^2$ be a function on a connected compact manifold M_n without boundary. Then if for every point $P_0 \in M_n$ there exists the coordinate neighbourhood $U(x^1, x^2, \dots, x^n) \in M_n$ and in this neighbourhood there exists the continuous functions $A^{ij}(x)$ and $B^i(x)$ of $P(x) \in U$ such, that on the whole U the following inequality holds*

$$A^{ab}(x)\,\frac{\partial^2\varphi}{\partial x^a \partial x^b} + B^a(x)\,\frac{\partial\varphi}{\partial x^a} \geq 0 \quad (resp.\ \ \leq 0) \tag{4}$$

where $A^{ab}(x)z_a z_b$ is a positive form then φ is constant on M_n, and "$\geq$" or "$\leq$" in inequalities (4) *for all coordinates neighbourhood are the same.*

Here and later we suppose that the studied manifolds are connected without boundary.

Obviously, in (pseudo-) Riemannian manifolds and manifolds with affine connection the partial derivative in formula (4) may be replaced by the covariant derivative.

We point out that in Theorem 2 we do not suppose that the functions $A^{ij}(x)$ and $B^i(x)$ define in all coordinate neighborhoods, geometric objects which are determined "in the whole" of M_n, in such a way as required, as for example in a theorem by K. Yano and S. Bochner [18] (p. 26).

Proof. For a compact manifold M_n we can choose a finite set of coordinate neighbourhoods U, such that the unification of these coordinate neighbourhoods covered M_n and in each of them the condition (4) holds (for definition take with sign "$\geq$").

Because φ is smooth and M_n is compact, the function φ on M_n reaches its maximum at the point $P_0 \in U_0$, where U_0 is one of the neighbourhoods chosen earlier. Then $\varphi(P) \leq \varphi(P_0)$ for each points $P \in U_0$ and on U_0 all the conditions of Hopf's Theorem hold, and we have $\varphi(P) = \varphi(P_0)$ for any point $P \in U_0$.

Further consider a coordinate neighbourhood U_1 which has an intersection with U_0. Evidently, $\varphi(P) \leq \varphi(P_0)$ for any point $P \in U_1$. From this $\varphi(P) = \varphi(P_0)$ for any point $P \in U_1$. Analogically we exhausted any coordinate neighbourhood U. Because the number of them is finite and M_n is connected, we verify that $\varphi(P) = \varphi(P_0)$ for any point $P \in M_n$. The theorem is proved. □

4 Conformal mappings of pseudo-Riemannian manifolds onto Einstein spaces

We prove the following theorem.

Theorem 3. *Let V_n be a compact pseudo-Riemannian manifold. If the Ricci tensor constitutes a positive (or negative) form, then a conformal mapping V_n onto Einstein spaces is only homothetic.*

Proof. Let V_n be a compact pseudo-Riemannian manifold which admits a conformal mapping onto Einstein space $\bar{V}_n$.

We note that we can rewrite equation (2) in the form

$$s_{,ij} = \alpha\, g_{ij} - \frac{s}{n-2}\, R_{ij} \tag{5}$$

where α is a function on V_n and R_{ij} are components of the Ricci tensor. We suppose that the Ricci form Ric is positive (or negative).

For any point P_0 of V_n there exists a coordinate neighbourhood $U^*(x)$ such that

$$R_{ij}(P_0) = \pm diag(1, 1, \ldots, 1) \quad \text{and} \quad g_{ij}(P_0) = \pm diag(g_1, g_2, \ldots, g_n)$$

where $g_1 > 1,\ g_2 < 0,\ g_i \neq 0$.

We put

$$A(x) \stackrel{\text{def}}{=} -\frac{1}{g_1}\,(B(x) \cdot g_2 + g_3 + \cdots + g_n).$$

We choose the smooth positive function $B(x)$ suche that $A(x) > 0$.

Therefore

$$A^{ij}(x) \stackrel{\text{def}}{=} diag(A(x), B(x), 1, 1, \ldots, 1)$$

gives rise to a positive form $A^{ij}(x) z_\alpha z_\beta$ on the neighbourhood U^*.

In the point P_0 holds

$$g_{ij}(P_0) \cdot A^{ij}(P_0) = 0 \quad \text{and} \quad R_{ij}(P_0) \cdot A^{ij}(P_0) > 0 \quad (\text{resp.} \ < 0).$$

From (5) it follows that

$$s_{,ij}(P_0) \cdot A^{ij}(P_0) < 0 \quad (\text{resp.} \ > 0).$$

It implies the existence of the coordinate neighbourhood $U(x)$ in which for any point $P \in U$

$$s_{,ij}(P) \cdot A^{ij}(P) < 0 \quad (\text{resp.} \ > 0)$$

holds.

Finally, because $s_{,ij} = \partial_{ij} s - \partial_\alpha s\, \Gamma^\alpha_{ij}$ we have that from Theorem 2 follows that the function s is constant, i.e. the conformal mapping is homothetic. □

The last Theorem implies

Theorem 4. *A compact non-Einstein pseudo-Riemannian manifold with the Ricci tensor constituting a positive (or negative) form does not admit conformal mapping onto Einstein spaces.*

ACKNOWLEDGMENTS. The paper was supported by the grant IGA Faculty of Science 2014016 Mathematical Structures of the Palacky University and the project No. LO1408, AdMas UP—Advanced Materials, Structures and Technologies (supported by the Ministry of Education, Youth and Sports under the National Sustainability Programme I), Brno University of Technology.

References

[1] Brinkmann, H. W.: *Einstein spaces which are mapped conformally on each other*, Math. Ann., vol. 94, no. 1, (1925), pp. 119–145.

[2] Chudá, H., Mikeš, J.: *Conformally geodesic mappings satisfying a certain initial condition*, Arch. Math. (Brno), vol. 47, no. 5, (2011), pp. 389–394.

[3] Chudá, H., Shiha, M.: *Conformal holomorphically projective mappings satisfying a certain initial condition*, Miskolc Math. Notes, vol. 14, no. 2, (2013), pp. 569–574.

[4] Denisov, V. I.: *Special conformal mappings in general relativity*, Ukrain. Geom. Sb., vol. 28, (1985), pp. 43–50.

[5] Eisenhart, L. P.: *Non-Riemannian geometry*, AMS, Providence, 1990.

[6] Eisenhart, L. P.: *Riemannian geometry*, Princeton University Press, 1997.

[7] Evtushik, L. E., Hinterleitner, I., Guseva, N. I., Mikeš, J.: *Conformal mappings onto Einstein spaces*, Russian Math., vol. 60, no. 10, (2016), pp. 5–9.

[8] Evtushik, L. E., Kiosak, V. A., Mikeš, J.: *The mobility of Riemannian spaces with respect to conformal mappings onto Einstein spaces*, Russ. Math., vol. 24, no. 8, (2010), pp. 29–33.

[9] Mikeš, J.: *Holomorphically projective mappings and their generalizations*, J. Math. Sci. (New York), vol. 89, no. 3, (1998), pp. 1334–1353.

[10] Mikeš, J., Berezovski, V. E., Stepanova, E., Chudá, H.: *Geodesic mappings and their generalizations*, J. Math. Sci. (N.Y.), vol. 217, no. 5, (2016), pp. 607–623.

[11] Mikeš, J., Chudá, H., Hinterleitner, I.: *Conformal holomorphically projective mappings of almost Hermitian manifolds with a certain initial condition*, Int. J. Geom. Methods Mod. Phys., vol. 11, no. 5, (2014), pp. 1450 044, 8.

[12] Mikeš, J., et al.: *Differential geometry of special mappings*, Palacky Univ. Press, Olomouc, 2015.

[13] Mikeš, J., Gavril'chenko, M. L., Gladysheva, E. I.: *Conformal mappings onto einstein spaces*, in *Proc. Kazan Univ.*, 1992, p. 64.

[14] Mikeš, J., Gavril'chenko, M. L., Gladysheva, E. I.: *Conformal mappings onto Einstein spaces*, Mosc. Univ. Math. Bull., vol. 49, no. 3, (1994), pp. 10–14.

[15] Mikeš, J., Vanžurová, A., Hinterleitner, I.: *Geodesic mappings and some generalizations*, Palacky Univ. Press, Olomouc, 2009.

[16] Petrov, A. Z.: *Einstein spaces*, Pergamon Press, New York, 1969.

[17] Sinyukov, N. S.: *Geodesic mappings of Riemannian spaces*, Nauka, Moscow, 1979.

[18] Yano, K., Bochner, S.: *Curvature and Betti numbers*, Princeton Univ. Press, 1953.

Current address

Mikeš Josef, prof. RNDr. DrSc.
Department Algebra and Geometry, Faculty of Natural Science, Palacky University
17. listopadu 12, 77146 Olomouc, Czech republic
Tel. number: +420 731 579 395, e-mail: josef.mikes@upol.cz

Guseva Nadezda Ivanovna, doc. CSc.
Department of Geometry, Faculty of Mathematics,
Moscow State Pedagogical University
Malaya Pirogovskaya 1, 119882 Moscow, Russia
Tel. number: +996 773 829 671, e-mail: ngus12@mail.ru

Hinterleitner Irena, Mgr. PhD.
Institute of Mathematics and Descriptive Geometry,
Faculty of Civil Engineering, Brno University of Technology
Žižkova 17, 602 00 Brno, Czech republic
Tel. number: +420 774 239 488, e-mail: hinterleitner.i@fce.vutbr.cz

Slovak University of Technology in Bratislava
Faculty of Mechanical Engineering
16th Conference on Applied Mathematics **APLIMAT 2017**
Proceedings

ARE MATHEMATICS A SOURCE OF ARTISTIC CREATION? INCURSION INTO MAPS PROPERTIES

MIRA Christian (FR)

Abstract. The article object is only to show how certain models of dynamic systems may generate strange shapes, some reminding possible known forms, or communicating an "artistic" feeling. In this context the properties of such models are essential for understanding the mathematical sense of generated pictures. This gives the opportunity of an incursion into the maps properties, through their phase plane singularities.

Keywords: applied mathematics, dynamical systems, maps, chaotic dynamics, mathematics and art relation

Mathematics Subject Classification: 37C05, 37C70, 37D45, 37E30, 34C15, 34D45

1 Introduction

A clear answer to the question raised by the article title will not be given in this text. Indeed the evaluation of the artistic quality of an image depends on the observer, on his culture. For the pictures illustrating this text, the judgment on their possible aesthetic value is left to the reader. The object is only to show how certain equations of dynamic systems can generate strange shapes, some reminding possible known forms, or communicating an "artistic" feeling. In this context the properties of such models are essential for understanding the mathematical sense of generated pictures. This gives the opportunity of an incursion into the maps properties, through their phase plane singularities.

As far as I am concerned, at the beginning the revelation of such shapes was an exhibition of "*stochastic*" (i.e. now *chaotic*, at that time this adjective did not exist) "attractive" images, created by some map equations. This exhibition took place in the framework of the international colloquium "*Points Mappings and Applications*" (French title "*Transformations Ponctuelles et leurs Applications*", Sept. 10-14, 1973), essentially devoted to discrete models (maps) of nonlinear dynamical systems, which took place in the "*Laboratoire d'Automatique et d'Analyse des Systèmes*" (now called "*Laboratoire d'Analyse et d'Architecture des Systèmes*", Centre National de la Recherche Scientifique). Having been in charge with the colloquium presentation [1], at its start I announced the exhibition by quoting the Birkhoff's papers which deal with the laws of aesthetic [2]-[6], and I read a Poincaré's text extracted from a "*Notice sur Halphen*" (Journal de l'Ecole

Polytechnique 60, 137-161, 1890, also cf. [7]). In this text Poincaré deals with aesthetic feelings communicated to mathematicians through some of their results:

> *"Le savant digne de ce nom, le géomètre surtout, éprouve en face de son œuvre la même impression que l'artiste; sa jouissance est aussi grande et de même nature. Si je n'écrivais pour un public amoureux de la Science, je n'oserais pas m'exprimer ainsi; je redouterais l'incrédulité des profanes. Mais ici, je puis dire toute ma pensée. Si nous travaillons, c'est moins pour obtenir ces résultats positifs auxquels le vulgaire nous croit uniquement attachés, que pour ressentir cette émotion esthétique et la communiquer à ceux qui sont capables de l'éprouver".*

This extract speaks about the artistic emotion felt by a mathematician when a line of his researches succeeds, sensation out-of-reach for a non specialist. I took the liberty of saying that now, in a different form, the exhibited images may reveal such feelings to non mathematicians (cf. [1], page 27), this due to the new possibilities offered by numerical simulations. In particular, I risked this statement on the basis of the five, and three wings *"mythic bird"* pictures (Figures. 1-3, and 4 on the left). These pictures are generated by the map equation written below the Fig. 1 image and that of the left cover of Fig. 4. They are due to a hidden fractal (at that time, this word did not exist) substratum (FS), made up of infinitely many sets of unstable periodic points with period tending toward infinity, whose limit sets are organized in classes (or types) of order 1,2,...,∞, in the Pulkin'sense ([8] and pp. 99-102 of [36]), i.e. the structure of the unstable sets is now called *fractal*. When the unstable cycles are saddle points, their unstable manifolds also belong to this "hidden" fractal organization. The "aesthetic" quality of Fig. 1 inspired the French artist Jean Dieuzaide who created the Fig. 2 tapestry. With two other equivalent images, this picture was chosen for illustrating the cover of four books (Figs. 3 & 4), and that of one journal (Fig. 5).

Oiseau mythique

$\alpha = 0.009 \qquad \mu = \cos(4.\pi/5)+0.008$

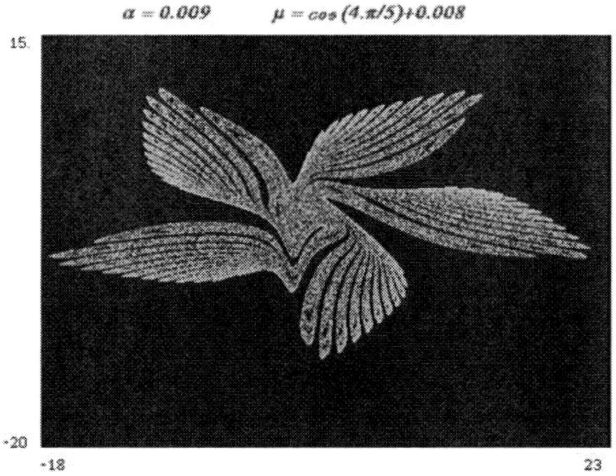

$$\begin{cases} x_{n+1} = y_n + \alpha.(1-0.05y_n^2).y_n + \mu.x_n + \dfrac{2.(1-\mu).x_n^2}{1+x_n^2} \\ y_{n+1} = -x_n + \mu.x_{n+1} + \dfrac{2.(1-\mu).x_{n+1}^2}{1+x_{n+1}^2} \end{cases}$$

Fig. 1. Left: *"Stochastic"* (now *"chaotic"*, at the colloquium time this word did not exist) attractor created by the above map

Fig. 2. Right: tapestry carried out by the French artist Jean Dieuzaide, which decorates the access hall to the manager office of the *Laboratoire d'Analyse et d'Architecture des Systèmes* (CNRS Toulouse)

Fig. 3. The five wings "Mythic Bird", on the left, as cover of the Cépadues book published in 1980, on the right as cover of the World Scientific book (2000) whose picture illustrates the chapter 8 of the book [9]

Fig. 4. On the left the three wings "*Mythic Bird*", in the cover of the World Scientific Ed. published in 1987. On the right "*Strange owls*" cover of this 1996 World Scientific book. The picture of a basin is created from a family of cubic noninvertible maps of $Z_1 < Z_3 >$ type (in this book pp. 437-444)

Fig. 5. In the journal *Prikladnaja Nelineinaja Dinamika* (Nonlinear Applied Dynamics, Journal of the Saratov University), the Fig. 4 "*Strange owls*" have different humorous interpretations. On the left, as a "*walrus*" in the journal cover. On the right, the journal back cover sees a "*walrus*" bifurcation as giving rise to a couple of *gooses* (associated with the upper yellow image of the Fig. 4 right cover). These pictures are related to the paper "*O basins porojdaemuik dvumernuik neobratimuimi otobrajenjami*" (About the basins generated by two-dimensional noninvertible maps), T. 4, N° 2, 1996, pp. 40-71)

It appears that an originality of the 1973 Toulouse colloquium was this exhibition of chaotic images, generated by nonlinear maps. Their mathematical sense has been given in the colloquium proceedings, from the article [10] devoted to maps which can generate different types of pictures, in particular "aesthetic" ones. Two years after, the same exhibition, entitled "Morphogénèse et Mathématiques", was organized by Marcel Barthes, Director of the "Alliance Française Rio de Janeiro-Centre" in the "Centre Culturel de la Maison de France" (May 8 to 30, 1975). Some years later, the book [11] was published and illustrated by twelve pictures of the colloquium exhibition. Pages 309-318 of this book show the *homoclinic*, and *heteroclinic* structures at the origin of the above hidden fractal structure (FS). They also give the mathematical sense of such pictures. The same year the book [12], in a more complete form, also deals with the mathematical sense of such structures (pp. 104-233). Page 186 of this book reproduces the Fig. 1, and another "aesthetic" ones exhibited in the colloquium framework (pp. 156-157). In fact a large variety of other maps families can generate interesting pictures such those of the right part of Fig. 4.

It is important to underline that the pictures exhibited in the colloquium framework were not due to a research of equations generating aesthetical effects. Indeed, at that time I worked in a laboratory of engineering researches (LAAS Toulouse), and Igor Gumowski occupied a position of Senior Physicist in the "*European Organization for Nuclear Research*" in Geneva. In this framework he had particularly to study the problem of "*stochastic*" (i.e. now "*chaotic*") instability in accelerators and storage rings. Such instabilities appear to increase with the amount of non-linearity. Physically this amount increases with the self-fields, and the latter increase with beam intensity. As a first step the purpose was to understand fundamental properties of invertible maps (conservative, or not), and noninvertible maps (not very studied at that time) as discrete dynamical systems models. A second step was to obtain rough qualitative models of "real" discrete systems from experimental observation of their behaviors. It was the case of the Fig. 1 map with a bounded nonlinearity (caption below Fig.1: $x^2/(1+x^2)$, $\alpha > 0$ being a damping term, if $\alpha = 0$ the map is conservative), the aim being to approach the qualitative behaviors of storage rings and accelerators (dynamics of the longitudinal motion of particles in an accelerator) [13]-[15], also pp. 148-177 of the book [12]. The third step (toward a "quantitative" model) was an approach via a more elaborated model (cf. pp. 326-328 of the above book [11].

In 1986 the book "*The Beauty of Fractals*" by H.O. Peitgen & P.H. Richter (Springer Verlag) was published. Being unaware of the 1973 exhibition and of the two above books (1980), this book is essentially devoted to Julia-Fatou sets and Mandelbrot sets, i.e. generated by Dim2 maps defined by two functions satisfying the Cauchy-Riemann conditions (thus issued from a Dim1 map of a complex variable). This book was a best seller which contributed to the popularization of mathematics as tool of artistic creation, which was not the case of the two above books, and the proceedings of the 1973 conference. The preface "*Frontiers of Chaos*" of the book "*The Beauty of Fractals*" begins with a section entitled "*Science and art*".

The maps family, defined below by the caption below the Fig. 1 picture, has inspired several authors who have found a subject of articles with either purpose of "aesthetical" creation, or use of its mathematical properties for applications (cf. some of them [16]-[22]). They have presented this maps family under the title "*Gumowski- Mira map*". A web search based on operative words "Gumowski Mira map" shows other examples.

The maps family also was object of a Sony sound recording with heading "*Gumowski Mira map*", played by members of the NHK Symphony Orchestra, as part of the CD "*Chaos zone / Kohsei Morimoto*" (1990) [23]. I don't know if the Birkoff's aesthetic theory applied to music (cf. [5] Part

III The diatonic chords, Harmony, & Melody) would be interesting for measuring the artistic quality of this record.

The first part of the article is devoted to an attempt of classifying two-dimensional maps with a potential "artistic vocation", either as discrete models, or resulting from a Poincaré' section of an ordinary differential equation (ODE). The classification is made via five criteria: (a) Nature of the picture as an aesthetic object (attractor, basin, parameter plane structure, and its bifurcation sets), (b) Picture properties (fractal or non-fractal image, connected, or multiply connected, or non-connected image, ...), (c) Nature of the two-dimensional dynamical system which generates the picture (ODE, conservative, quasi-conservative map, invertible and non-invertible maps, ...), (d) Properties of the functions defining the dynamical system (analyticity, smoothness, continuity, ...)., (e) Colored picture (with or without shade variation, ...).

The second part is devoted to the Birkhoff's mathematical theory of aesthetics, and the limitations of its formula of aesthetic measure for most of the pictures of this article.

The third part deals with conservative and almost conservative maps.

The fourth section concerns several types of two-dimensional invertible maps.

It is followed by a fifth part related to various forms of two-dimensional noninvertible maps

Independently of the topic defined by the title, this text may give a general information on two-dimensional maps properties.

2 An attempt of classifying maps with a potential "artistic vocation"

First we note that the solutions of ordinary differential equations (ODE) are not defined by the classical transcendental functions of Mathematical Analysis. These solutions are generally obtained via the Poincaré method of surface section, which leads to a map T constructed by numerical integration, a fixed point being a "*main*" periodic solution of period τ, a period k cycle being a $1/k$ *subharmonic* of the "main" periodic solution [24], [25]. If the ODE is smooth, the map T is invertible. Nevertheless, the surface section method does not reduce completely the problem to a map one. Indeed the solution behavior inside a time interval τ play an important role, because generally it gives rise to well defined regions of a parameter space associated with rank m *higher harmonics resonances*, or *synchronizations*, m/k *fractional harmonics resonances*, or *synchronizations*, and also in the parameter space to *isoordinal* (same period k) sequences of regions of period k stable cycles, which differ by m, $m = 1,2,3,...$ [24].

Now consider a picture generated by a two-dimensional map T (the hypothesis of this paper). Without knowledge of its origin, as a first step, with some experience of the matter, a simple observation of the image frequently may permit to guess its nature: either attractor, or conservative situation, or basin with, or without, attractor, or image of a parameter plane representing regions where T has the same qualitative behavior. From this observation, a second step may discern some picture properties such as: fractal type, connected, non-connected, multiply connected image, focal points. Sometimes, it is possible to go further. The following classification of aesthetic objects is proposed:

- A. Nature of the picture as an aesthetic object
    - (a) Attractors (phase plane),
    - (b) Non attracting stationary states generated by a conservative map,
    - (c) Basins,
    - (d) Basins with attractor,
    - (e) Parameter plane.

- B. Picture properties (maps)
    - (a) Non fractal,
    - (b) Visible fractal,
        - -1. The boundary is fractal, but not the global image
        - -2. The global image is fractal.
    - (c) Hidden fractal (substratum made up of infinitely many sets of unstable periodic points with period tending toward infinity, different classes (in the Pulkin's sense ([8] and pp. 99-102 of [36]) of limit sets and related unstable manifold of saddles points in the attractor case)
    - (d) Visible focal points for an attractor (the inverse map T^{-1} has a cancelling denominator)
    - (e) Visible focal points for a basin (the map T has a cancelling denominator),
    - (f) Connected image,
    - (g) Multiply connected image,
    - (h) Non-connected image.
    - (i) Mixed : multiply connected and non-connected.

- C. Nature of the two-dimensional dynamical system which generates the picture :
    - (a) Ordinary Differential Equation (map obtained via a Poincaré section),
    - (b) Conservative map,
    - (c) Quasi-conservative map (imbedding of a conservative map into a dissipative one, with an imbedding parameter α such that $|\alpha|$ is small).
    - (d) Invertible map,
    - (e) Noninvertible map with a critical curve and its associated class.
    - (f) Noninvertible map with a critical point,

- D. Properties of the functions defining the dynamical system
    - (a) Analytic,
    - (b) Smooth
    - (c) Piecewise smooth

- (d) Continuous
- (e) Piecewise continuous
- (f) Satisfy the Cauchy-Riemann conditions (giving rise to Julia-Fatou sets and Mandelbrot sets)
- (g) Close to the case (f)

- E. Colored picture (different algorithms of coloring with different aesthetical results)
  - (a) Gradual variation of a same color shade according an algorithm (Ljapunov's exponents, distance from the boundary of a basin, or an attractor)
  - (b) Without shade variation
  - (c) With three-dimensional rendering.
  - (d) Without three-dimensional rendering.

- F. Non colored picture (or only two colors)
  - (a) For an attractor: shade variation due to a density variation of the iterated points
  - (b) Without shade variation

In the case of a noninvertible map (Ce) its class is defined by a phase space divided into open regions Z_i, each point of a Z_i having i distinct inverses. Class examples: maps class generating two regions Z_0–Z_2, or class associated with three regions Z_1–Z_3–Z_1, or with three regions Z_0–Z_2–Z_4. So the one-dimensional map $x' = x^2 - \lambda$ belongs to the Z_0–Z_2 class. The map $x' = \lambda x - x^3$, $\lambda > 0$, belongs to the Z_1–Z_3–Z_1 class, if $\lambda \leq 0$ this map is invertible. The map $x' = x^4 - \lambda x^2 + \mu$, $\lambda > 0$, belongs to the Z_0–Z_2–Z_4 class. More complex classes of two dimensional noninvertible maps are defined pp. 113-183 of the book [26], with consideration of a "foliated" phase plane.

Each picture can be identified via a sequence of symbols whose number depends on the image. So the Fig. 1 picture is that of an attractor, generated by a Dim2 invertible map, identified via the sequence of symbols AaBcCdDaEb (*F* is absent because it concerns non colored picture). The "*strange owls*" on the right part of Fig 4 are images of a multiply connected basins associated with the sequence AcBb2CeDaFb (the symbol E is absent because the image is non colored). This identification of maps with an "artistic vocation" might concern mathematicians interested by the relation between art and mathematics (?). In this text, it will not be used below.

3 Birkhoff's mathematical theory of aesthetics

3.1. Elements of the theory

Birkhoff was intellectually the greatest Poincaré's disciple, taking over the technics and problems of Poincaré and carrying on. So the results about fixed points, or periodic orbits, of saddle (hyperbolic) type, or center (elliptic) type, *homoclinic* and *heteroclinic* structures, unstable centers, *instability rings*, the concepts of *signature*, for the two-dimensional surface transformations, have given some of the greatest advances in this field. The Birkhoff's qualitative theory reaches its highest point with the long paper, written in French language, "*Nouvelles recherches sur les systèmes dynamiques*"

(Memoriae Pont. Acad. Sci. Novi Lincei, 1935, 53, vol. 1, pp. 85-216). As well as such researches, Birkhoff was interested by the notion of *aesthetic*. In a paper presented at the International Mathematical Congress at Bologna in 1928 [3] he proposed a mathematical theory of aesthetics in the following terms:

> "The typical aesthetic experience may be regarded as compounded of three successive phases: (1) a preliminary effort, necessary of attention, which is necessary for the act of perception, and which increases in proportion to what we shall call the complexity (C); (2) the feeling of value or aesthetic measure (M) which rewards this effort; and finally (3) a realization that the object is characterized by a certain harmony, symmetry, or order (O), more or less concealed, which seems necessary to the aesthetic effect."

Thus there arises almost at once the question of determining in a given case to what extent this aesthetic measure is simply the effect of the density of the relations of order, compared with the complexity. And so it seems natural to propose such a formula as $M = O/C$.

In this paper Birkhoff gives preliminary indications of the application of this formula to polygonal forms, tiles and vases. He remarks that the formula suggests how it is possible to find new forms which possesses a high aesthetic measure M, and gives the example of the Fig. 6 vase for which $M = 1$.

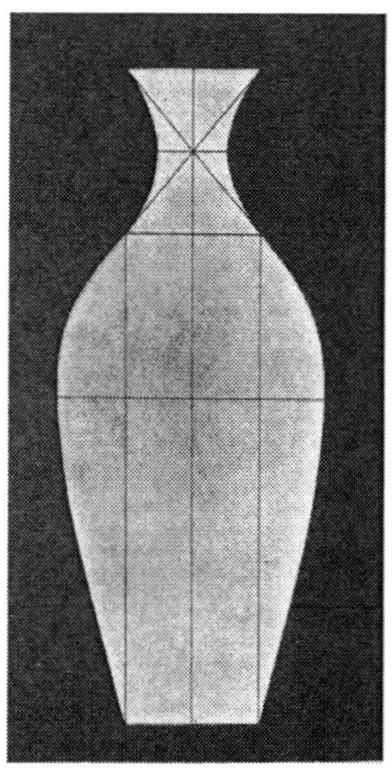

Fig. 6. Vase with $M = 1$, the highest possible value (according to Fig. 4 of the paper [3])

In paper [4] Birkhoff specifies :

> "This analysis of the aesthetic experience suggests that the aesthetic feelings arise primarily because of an unusual degree of harmonious interrelationship within the object. More definitely, if we regard, M, O, and C, as measurable variables, we are led to write $M = O/C$, and thus to embody in a formula the conjecture that the aesthetic measure may be regarded as determined by the density of order relations in the aesthetic object. The well known aesthetic demand for « unity in variety » is evidently closely connected with this formula. The definition of the beautiful as that which exhibits the greatest number of ideas in the shortest time, given by the Dutchman Hemsterhuis in the eighteenth century, is also of an analogous nature. If we admit the validity of such a formula, the following mathematical formulation of

the fundamental aesthetic problem may be made: within each class of aesthetic objects, to define the order O and the complexity C so that their ratio $M = O/C$ yields the aesthetic measure of the different objects of the class."

A psychological justification of such a formula may be given. The act of perception of an aesthetic object involves certain automatic motor adjustments, and the feeling of effort or tension which always accompanies perception appears as a summational effect due to the tensions which accompany the various automatic adjustments. Thus if A, B, C, ... represent these adjustments, each with its index of tension a, b, c, ..., and if these respective adjustments are made r, s, t, ... times, we may consider the sum $C = ra + sb + tc + \ldots$ to represent the complexity. On the other hand, the order O corresponds to certain associations called into play by the act of perception. An instance of such an association would be that of symmetry. More precisely, if associations of various types L, M, N, ... take place, each with its index of tone of feeling, l, m, n, ... and if these occur u, v, w, ... times, then we may regard the total tone of feeling (positive or negative) to be $O = ul + vm + wn + \ldots$ representing the order O of the object.

According to our theory it is the intuitive estimate of the amount of order, O, inherent in the aesthetic object, as compared with its complexity, C, from which arises the feeling of the aesthetic measure, M, of the object considered; the fundamental formula embodies this conjecture.

An instructive analogy is the following. Among business enterprises those are regarded as most successful in which the annual profit p is largest in comparison to the investment i; thus the ratio p/i is regarded as the economic measure of success. Now in the aesthetic experience, the effort of attention measured by C corresponds to the investment, and the order O corresponds to the profit. By analogy it is the ratio O/C which represents the aesthetic measure M.

It is obvious that any such mathematical theory can only represent the facts for certain simple classes of aesthetic objects in which the aesthetic factors involved are essentially mathematical or formal. These formal factors comprehend symmetry, balance, repetition, sequence, etc. On the other hand there are connotative factors which are obviously beyond the reach of any such theory. An instance would be the associations involved in the meaning of a beautiful poem.

Thus the quantitative application of the mathematical theory will be limited to that part of the aesthetic effect which is due to the formal elements of order of simple type. However, the formula can be held to be qualitatively applicable to the complete aesthetic experience. [4]

Remark. Birkhoff does not take into account the eventual coloring of an aesthetical object, also related to a certain harmony, symmetry, balance, repetition, more or less concealed for the aesthetic experience, which takes part in the amount of order O. Nevertheless, in many cases colored pictures (for example those of basins related to several attractors) are essential to clarify the mathematical sense of regions with different qualitative properties, secondarily giving rise to interesting aesthetic effects. He was interested by the problem of map (in a geometrical sense) coloring (Collected Mathematical Papers, Vol. 3, five papers collected under the name "*The four color problem*", 1-144), but not directly related to his aesthetic theory.

3.2 Application to pictures generated by a dynamical system

In the above text, Birkhoff mentions limitations of his theory, saying: "*this theory can only represent the facts for certain simple classes of aesthetic objects in which the aesthetic factors involved are essentially mathematical or formal. These formal factors comprehend symmetry, balance, repetition, sequence, etc.*". It is the case of Figs.1-4, and another fractal pictures, for which his formula cannot be directly applied. Nevertheless, considering the Birkhoff studies of *polygonal forms* related to his formula [27], perhaps it would be possible to try the formula use via such a form circumscribed to a "fractal" picture created by a map. In this case, a degree of liberty exists in the choice of the length ratio of one side of the rectangle frame of the image over its width, which can permit to optimize the M value. It remains the question of the aesthetic feeling due the complex structure inside the domain limited by the polygonal form. For pictures created from piecewise linear continuous, or discontinuous map the formula could be directly applied.

A first example is given by the parameter plane with non fractal connected pictures (AeBcfCeDeEb) of Fig. 7, generated by the family of piecewise continuous piecewise linear maps family with the parameter vector $(a,b,\lambda_1,\lambda_2)$:

$$x' = \begin{cases} T_1 x = a + \lambda_1 x & (x<0) \\ T_2 x = b + \lambda_2 x & (x>0) \end{cases} + y \qquad (3.1)$$
$$y' = \beta x$$

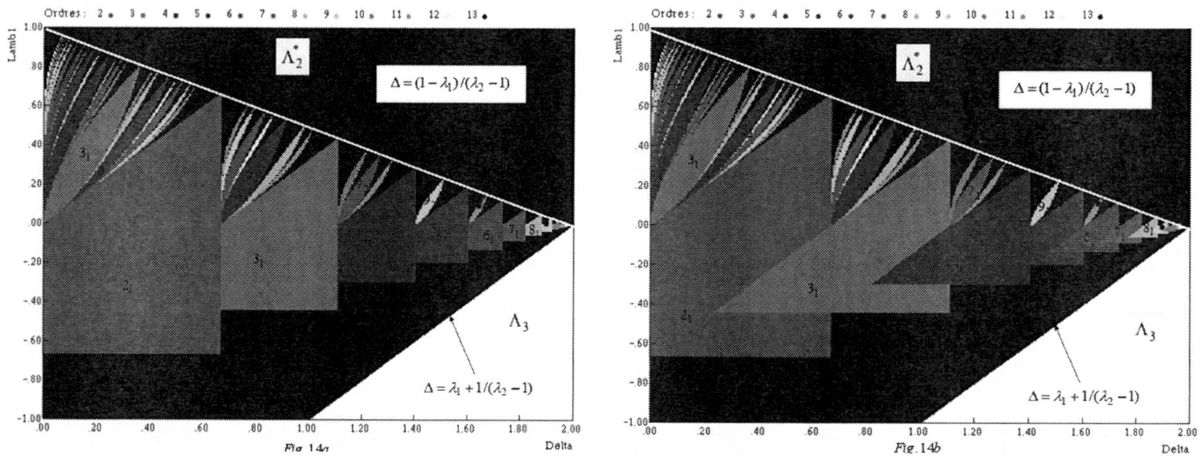

Fig. 7. Map (3.1), $\Delta = |a/b|$, $b = -2$, $\lambda_2 = 1.5$, $\beta = 0$. Global parameter plane (Δ,λ_1), $-1 < \lambda_1 < 1$, obtained from two different numerical scannings. The comparison of Figs. (left) and (right) permits to see that the sheets of the plane overlap only for $-1 < \lambda_1 < 0$. For $0 \le \lambda_1 < 1$ one has the Leonov's "*boxes in files*" structure, with only one sheet. The domains of stable cycles are denoted by r_i, $i = 1,2$, r being the cycle period. The white region corresponds to diverging orbits. The black one corresponds to a stable behavior in the Poisson sense

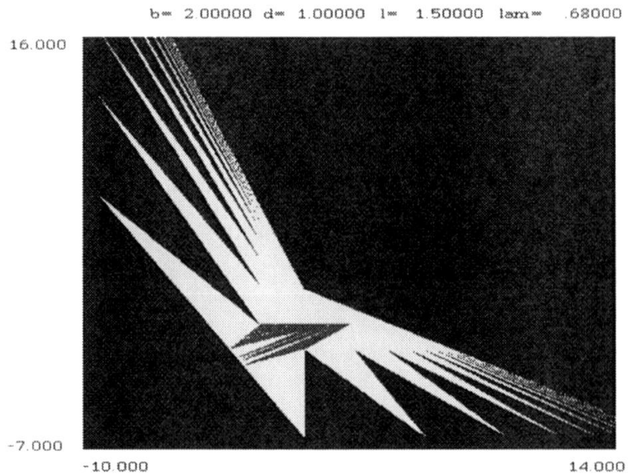

Fig. 8. Map (3.1), $a = 2$, $b = -2$, $\lambda_1 = 1$, $\lambda_2 = 1.5$, $\beta = 0.68$, phase plane (x,y) representation. The attractor and its basin are respectively red and yellow colored. The blue region is that of diverging orbits

Fig. 8 gives the situation in the phase plane (x,y) with the pair attractor and its basin (AdBcfDeEb)

Fig. 9 (similar to a "blew up symmetric quartz") is a second example. It represents a connected attractor with hidden fractal (AaBcf), generated by the family of continuous, piecewise linear noninvertible maps T of (CeDe) Z_1–Z_3–Z_1 class, with a shade variation due to a continuous change of the iterated points density (Fa):

$$\left.\begin{aligned} x' &= (1-\lambda)x + y \\ y' &= \begin{cases} y - \lambda x/5, & -1/2 \le x \le 1/2, \\ y - 2\lambda x + 0.9, & x \ge 1/2, \\ y - 2\lambda x - 0.9, & x \le 1/2, \end{cases} \end{aligned}\right\} \tag{3.2}$$

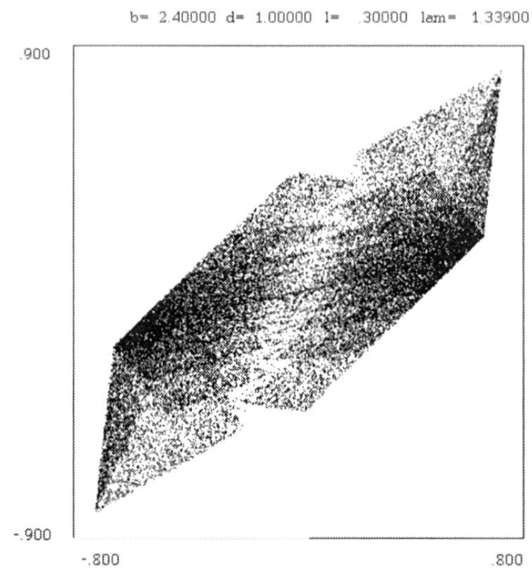

Fig. 9. Map (3.2), $\lambda = 2.4$. "Blew up symmetric quartz", the shade variation due to a continuous change of the iterated points density

The Z_1–Z_3–Z_1 noninvertible map is defined from the three regions bounded by the *critical curve* made up of the two lines $L \cup L'$, $y = x \pm (1/2 - 0.4\lambda)$ first rank images of the lines $L_{-1} \cup L'_{-1}$, $x = \pm 1/2$. The external attractor boundary is made up of two segments of $L \cup L'$ and their images belonging to L_{-1} $L_n \cup L'_n$ = $T^n(L \cup L')$, $n = 1, 2, 3$, (more details are given in sec. 6, and chap. 4 of the book [26], and below sec. 6.1), the neighborhood of these segments are regions with higher density of iterated points. The internal segments with neighborhood of upper density of iterated points belongs to segments $L_r \cup L'_r$, $r = 3, 4, 5$.

4 Mathematical factors at the pictures origin. Conservative maps and almost conservative maps

4.1 Conservative maps

A conservative map is invertible, has analytic data, its Jacobian determinant J is such that $|J| = 1$. According to Birkhoff [28], the phase plane picture, around a reference fixed point O of type center (i.e. its multipliers, or eigen-values, are $S_{1,2} = \exp(\pm j\varphi)$, $j^2 = -1$, $\varphi \neq 2\pi p/q$, p, q = integers), is given by Fig. 10. The center O is first surrounded by a family of regular invariant closed curves, then by one or more island structures, defined by the stable and unstable manifolds of a period k saddle cycle, limiting a domain of closed invariant curves related to a period k center. This situation is followed by one or more *instability rings* defined by closed invariant curves related to a period k center, and intersecting stable and unstable manifolds of a period k saddle S_i (here $i = 1,2,3$), giving rise to *homoclinic* points. Each *instability ring* is bounded by at least two smooth closed invariant curves. More details are given in chapter 3 of the book [11] and chapter 5 of the book [11].

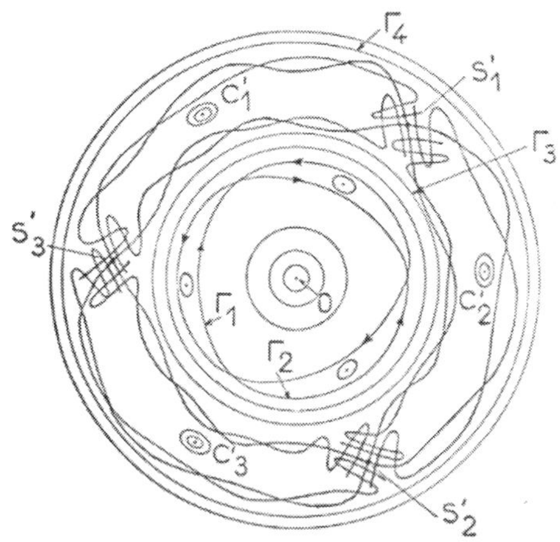

Fig. 10. Phase plane according to Birkoff

In the framework of the 1973 Toulouse colloquium, conservative pictures (Figs. 11-14) were exhibited from various initial conditions associated with the bounded quadratic map T:

$$x' = y + F(x)$$
$$y' = -x + F(x')\; F(x) = \mu x + 2(1-\mu)\, x^2/(1+x^2), \quad -1 < \mu < 1 \tag{4.1}$$

Here $\delta = 1 - \mu$ is a sort measure of the strength of the nonlinearity. The fixed points of (4.1) are: $O(0;0)$ with multipliers $S_{1,2} = \exp(\pm j\varphi) = \mu + j\sqrt{1-\mu^2}$, $j^2 = -1$, $\mu = \cos\varphi$, $0 < \varphi < \pi$, and the point ($x = 1$; $y = 0$), singular point of type "cusp" with multipliers $S_1 = S_2 = +1$. The point O is a center, except possibly $\phi \neq 2\pi\, p\,/\,q$, p, q = integers.

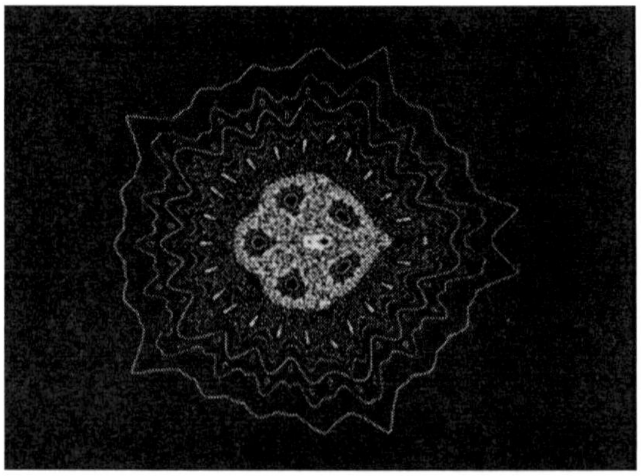

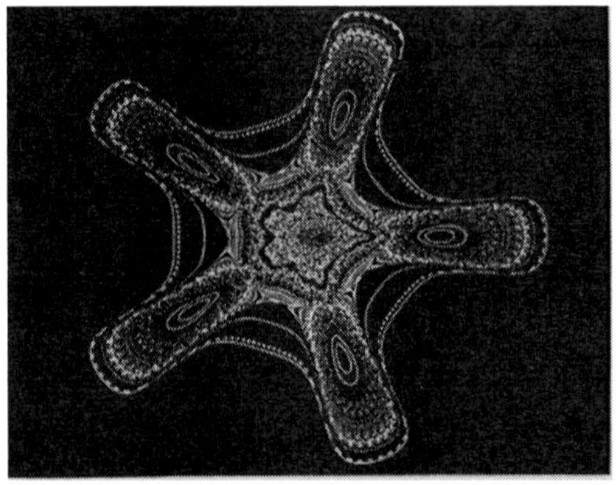

The saddle cycles, with positive multipliers (eigenvalues), having the same period k, and the same rotation number, are such that these multipliers increase simultaneously with δ. Such a period k cycle, saddle, or center, with rotation number r, is designated by saddle (or center) k/r. It is related to the Poincaré's rotation number N by the relation $N=r/k$. Let s_n be the distance between the fixed point (0;0) and the image, or preimage, of an initial point (x_0, y_0) after n iterations. For non bounded quadratic nonlinearities ($F(x)=\mu\, x + (1 - \mu)\, x^2$) saddles k/r satisfy asymptotic relations given in pp. 112-116 of [12], and pp. 276-281, 307-309 of [11], with a new property with respect to Fig. 10 (pp. 121-122 of [12], pp. 273 of [11]). This additional feature consist in regions of intersecting *instability rings* (bounded chaotic region due to *homoclinic* and *heteroclinic* points), *region of diffusion* (s_n grows in the mean, but less than an exponential rate), and an unbouded *region of stochastic instability* (s_n grows exponentially). When $\delta=1-\mu$ increases, the regions with *chaotic* (*stochastic* at the colloquium time) behavior increase.

A bounded nonlinearity adds another new feature: the distinction between *internal*, *external*, and *mixed* cycles with respect to the intersecting stable and unstable manifolds of the cusp point (1;0) (cf. pp. 149-159 of [12], and pp. 311-318 of [11]), and the presence of bounded regions of local *stochastic instability* which separate two *diffusion domains*, giving rise to an oscillation between two long *chaotic* evolutions communicating via a fast *chaotic transient*. These new features associated with the asymptotic relations satisfied by the saddle cycles, play a fundamental role in the Figs. 11-14 structure, and their eventual "aesthetic" aspect".

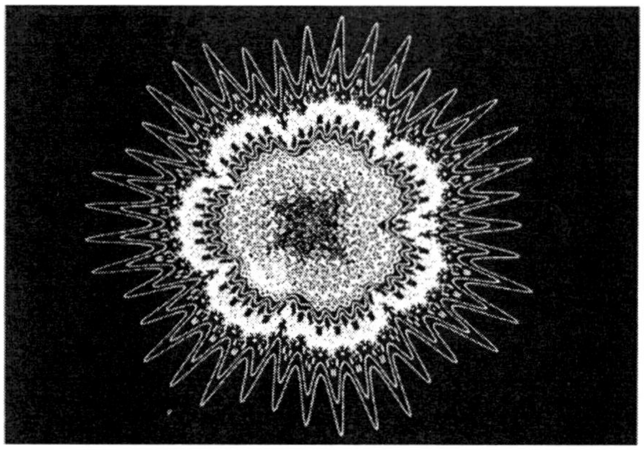

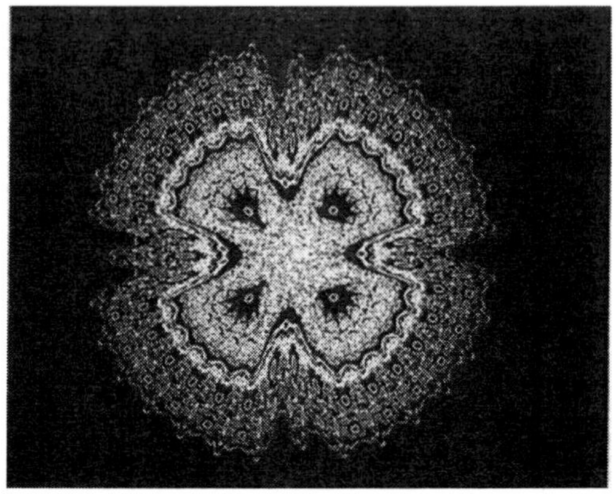

Fig. 11 - 12. Map (4.1), $\mu = -0.2$. Map (4.1), μ=0.05, near $\cos(2\pi/4) = 0$

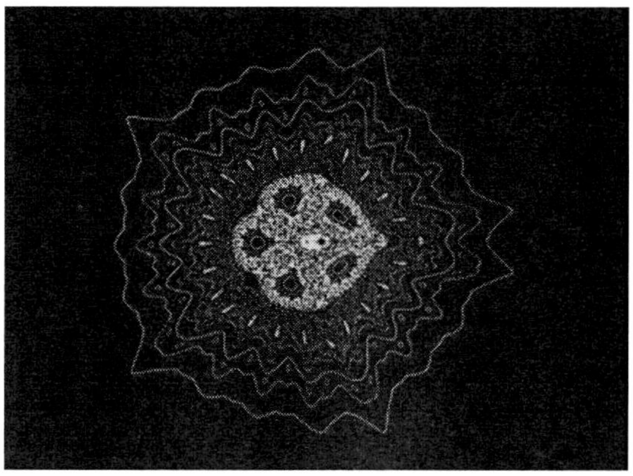

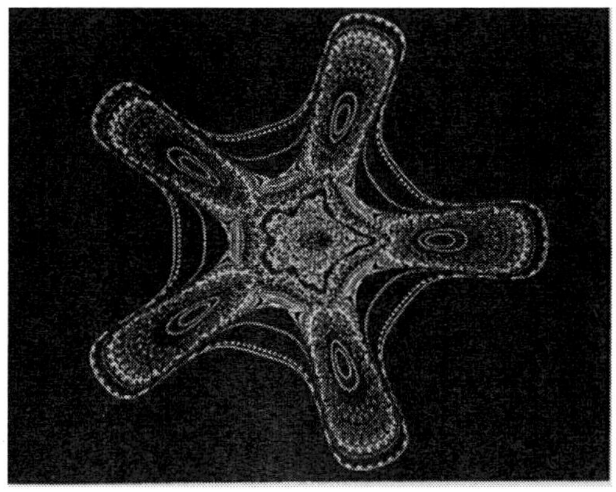

Fig. 13 - 14. Map (4.1), $\mu = 0.25$. Map (4.1), $\mu = 0.3$, near $\cos(2\pi/5) \simeq 0.3090169944$

Another family of bounded map T, giving rise to interesting pictures, is of exponential type:

$$\begin{aligned} x' &= y + F(x) \\ y' &= -x + F(x'), \quad F(x) = \mu x + (1-\mu)\, x^2 \exp[-(x^2-1)/4], \quad -1 < \mu < 1 \end{aligned} \tag{4.2}$$

A particular feature of this map is the existence of a third fixed point $(x_1;0)$ independently of the μ value, $x_1 \approx 1.8742631665$ being the root of $x^2\exp[-(x^2-1)/4] = 1$. This point is also a center. The (4.2) cycles are organized into four distinct families: *internal* ones surrounding the main center (0;0), *internal* ones surrounding the secondary center $(x_1;0)$, *external* ones surrounding both (0;0) and $(x_1;0)$, *mixed* ones. Otherwise the qualitative properties of (4.2) are similar to those of (4.1).

4.2 Almost conservative maps

The imbedding of the maps (4.1) and (4.2) into a maps family by introduction of a small non linear damping term (the Jacobian determinant J having a constant sign) gives examples for obtaining maps with an "aesthetic" vocation.

A first example related to (4.1) is:

$$
\begin{aligned}
x' &= y+\alpha y(1-\sigma y^2)+F(x) \\
y' &= -x+F(x'), \qquad F(x) = \mu x+2(1-\mu)\, x^2/(1+x^2), \qquad -1<\mu < 1
\end{aligned}
\tag{4.3}
$$

The damping term is $\alpha y(1-\sigma y^2)$. For $\alpha \neq 0$, the mathematical properties of the pictures generated by (4.3) are profoundly modified with respect to the case $\alpha = 0$ (cf. chapter 4 of [12], and pp. 375-377 of [11]). As for the "aesthetical" aspect Figs. 1 and 15 give examples of such pictures.

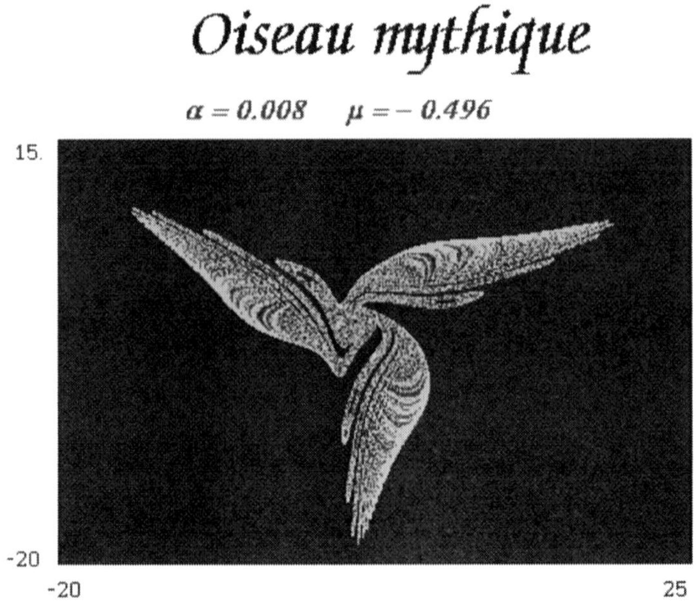

Fig. 15. Map (4.3), $\alpha = 0.008$, $\sigma = 0.05$, $\mu = -0.496 \simeq \cos(2\pi/3)$. Three wings mythic bird

Fig. 16. Map (4.3), $\alpha = 0.005$, $\sigma = 1.0$, $\mu = 0.125$. Strange sea mollusk (exhibited for the 1973 colloquium)

Other types of embedding, using various forms of nonlinear damping,

$$x' = y + \alpha y[1-\sigma(y^2+3x^2)] + F(x)$$
$$y' = -x + F(x'), \qquad F(x) = \mu x + 2(1-\mu)x^2/(1+x^4)), \qquad -1 < \mu < 1 \tag{4.4}$$

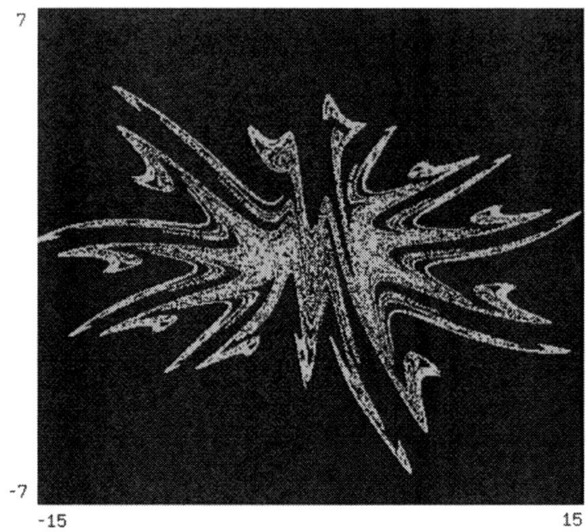

Fig. 17. Map (4.4), $\alpha = 0.000999$, $\sigma = 0.49859$, $\mu = -0.95$. Two headed mythic bird

Another example, obtained by embedding the "exponential map" (4.2) into a maps family by introduction of a small non linear damping term, is given by:

$$x' = y + \alpha y(1-\sigma y^2) + F(x)$$
$$y' = -x + F(x'), \qquad F(x) = \mu x + (1-\mu)x^2\exp[-(x^2-1)/4], \qquad -1 < \mu < 1 \tag{4.5}$$

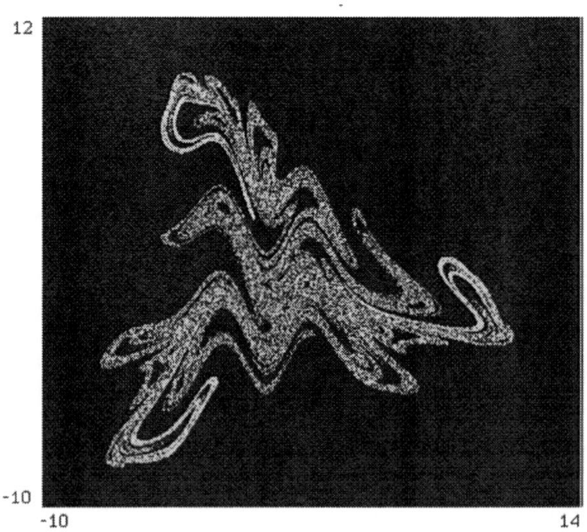

Fig. 18. Map (4.5), $\alpha = 0.005$, $\sigma = 0.45$, $\mu = -0.42$. "Exponential" mythic bird 1

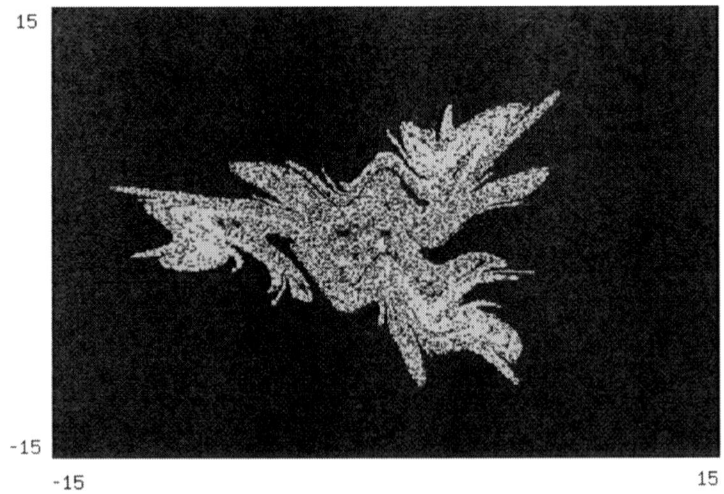

Fig. 19. Map (4.4), $\alpha = 0.0083$, $\sigma = 0.1$, $\mu = -0.38$. "Exponential" mythic bird 2

5 Two-dimensional invertible maps

5.1 Invertible maps T without cancelling denominator, and inverse T^{-1} without cancelling denominator

An attractor A, numerically generated by a two-dimensional invertible map, is essentially "moulded" by the unstable manifold of saddle fixed points, or saddle period k cycles. The simplest case showing this property is that of the Henon's map, not specially known for generating aesthetic pictures

$$x' = 1 - ax^2 + y$$
$$y' = bx \tag{5.1}$$

So, Fig. 20a represents the "oscillating" unstable manifold of the saddle fixed point q_2

$$(x = [-e + \sqrt{e^2 + 4a}\,/(2a), \quad y = bx, e = 1 - b),$$

support of the classical Henon' strange attractor A, Fig. 20b the boundary of its basin, and Fig. 21 (numerical scanning of the (x, y) phase plane) the Hénon' strange attractor A (without evident aesthetic effect).

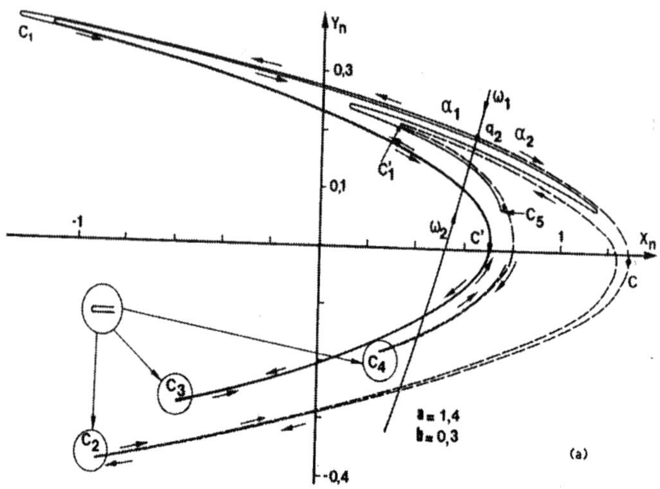

(a)

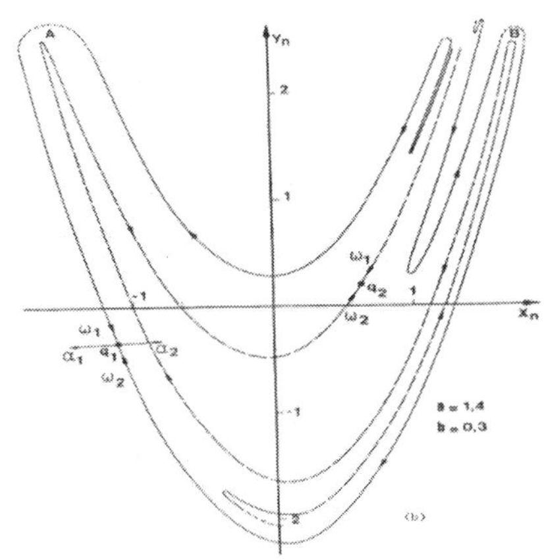

(b)

Fig. 20. Map (4.1) (according to pp. 292-293 of [23]), phase plane (x,y).
(a) Unstable manifold $W^u(q_2) = \alpha_1(q_2) \cup \alpha_2(q_2)$ of the saddle point fixed q_2, basis of the map attractor. The arc $\omega_1 \cup \omega_2$ belongs to the stable manifold of q_2.
(b) Stable manifold $W^s(q_1) = \omega_1(q_1) \cup \omega_2(q_1)$ of the saddle fixed point q_1, boundary of the attractor basin, and stable manifold $W^s(q_2) = \omega_1(q_2) \cup \omega_2(q_2)$ of the saddle fixed point q_2.

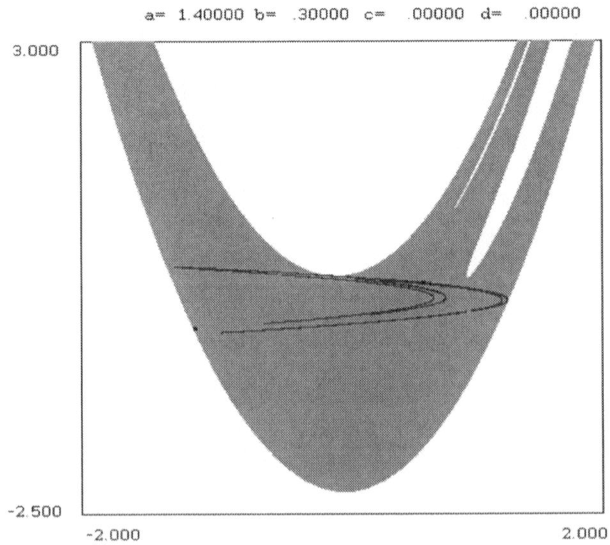

Fig. 21. Map (4.1). Attractor (red colored set) and its basin (brown colored area).
The saddle fixed points q_1 and q_2 are blue colored.
The two green points are those of an unstable period 2 cycle.
The white area is that of diverging orbits

Fig. 20 shows the existence of infinitely many *homoclinic* points $[\alpha_1(q_2) \cup \alpha_2(q_2) \cap [\omega_1(q_2) \cup \omega_2(q_2)]$, limit points of unstable period k cycles, $k \to \infty$. Indeed, a Birkhoff theorem states that, in the general analytic case, an arbitrarily small neighborhood of a *homoclinic* point contains infinitely many unstable cycles, and an other *homoclinic* point (p. 349 of [28]). This means that the *A* properties correspond to a hidden fractal, defined by the sequence AaBcCdDaEb.

An example of a more aesthetic picture, "moulded" by the unstable manifold of saddle fixed points, or saddle period k cycles, is given by the non autonomous two-dimensional non-autonomous differential equation (ODE):

$$dx/dt = E \sin \omega t - \alpha f(x) - y$$
$$dy/dt = \beta f(x), \quad f(x) = ax + bx^3 \tag{5.2}$$

From the map T constructed by numerical integration, this ODE generates the Fig. 22 attractor (AaBcCaDaFb). For more mathematical details see sec. 3.4.2 of [25])

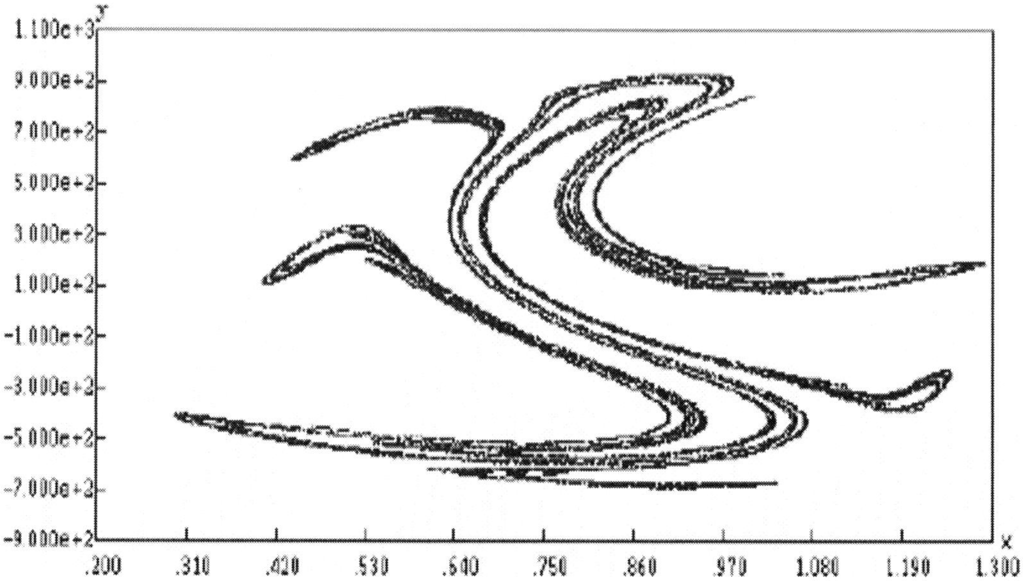

Fig. 22. ODE (5.2), according to Fig. 27 of [25]. Chaotic attractor obtained for $\alpha = 18, \beta = 10^6, a = 0.01, b = 0.375, \omega = 236.12, E = 2315.08, x_0 = 0.82633, y_0 = 729.802$

5.2 Invertible maps with vanishing denominator.

In order to characterize the qualitative behaviors generated by a two-dimensional map T in presence of a vanishing denominator in its equation, it is necessary to introduce the concepts of *focal point* and *prefocal curve* [29]-[33]. Roughly speaking a *prefocal curve* δ_Q is a set of points such that the inverse map T^{-1} converts δ_Q into a single point Q, called a *focal point*, $T^{-1}(\delta_Q) = Q$. Such points may belong to the basin boundary of an attractor.

Let T the map is given by:

$$x' = y$$
$$y' = y(1 - ay) - bx + cx/(y + \sqrt{d}\,)^2 \tag{5.3}$$

T is invertible if $bc < 0$. Figure 23 represents the basin of the stable period 2 cycle made up of the point ($x\sim1.89$, $y\sim9.53$) and ($x\sim9.5$, $y\sim1.95$). The focal point is $Q(0;-\sqrt{3})$, i.e. the point for which the numerator $N(x,y)\equiv cx$, and the denominator $D(x,y)\equiv(y + \sqrt{d}\,)^2$ are equal to zero. The prefocal curve is $x = -\sqrt{d}$.

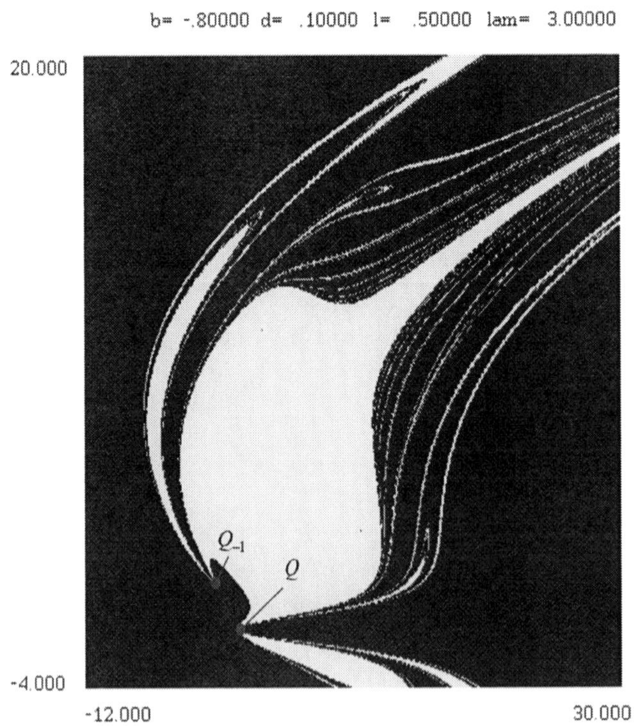

Fig. 23. Map (5.3). Basin (yellow colored) of the stable period 2 cycle, $a = 0.1$, $b = -0.8$, $c = 0.5$, $d = 3$.
The focal point is $Q(0;-\sqrt{d}\,)$. Its rank one preimage is the point Q_{-1},
and the rank n preimages Q_{-n} belong to the basin boundary.
The blue area corresponds to diverging orbits

5.3 Invertible maps the inverse of which has a vanishing denominator

Let T be an invertible map. If the inverse T^{-1} has a vanishing denominator, a focal point of T^{-1} are called *knot*.An example is given by the inversible maps family T:

$$
\begin{aligned}
x' &= ax[1 + y + f(x)] - b \\
y' &= c[1 + y + f(x)]
\end{aligned}
\tag{5.4}
$$

Its inverse inverse T^{-1} is:

$$
\begin{aligned}
x &= c(x' + b)/(ay') \\
y &= y'/c - f[c(x' + b)/(ay')]
\end{aligned}
$$

Consider the Jacobian determinant J of T, $y = -f(x)$ is the curve (γ) for which $J = 0$. A *knot* point is defined by two properties. (a) The image $T(\gamma)$ is the "basic" *knot* point $K(-b;0)$. (b) The "basic" knot is the focal point K ($x' = -b; y' = 0$) of the inverse map T^{-1}, i.e. numerator and denominator of T^{-1} cancel. The attractor knots are K and its rank n images K_{-n}, $n = 1,2,3,....$ Figure 24 represents the attractor generated by T for $f(x) = -x(x - dx^2)$, $a = 1.795$, $b = 0.6$, $c = -0.559$, $d = -0.2$.

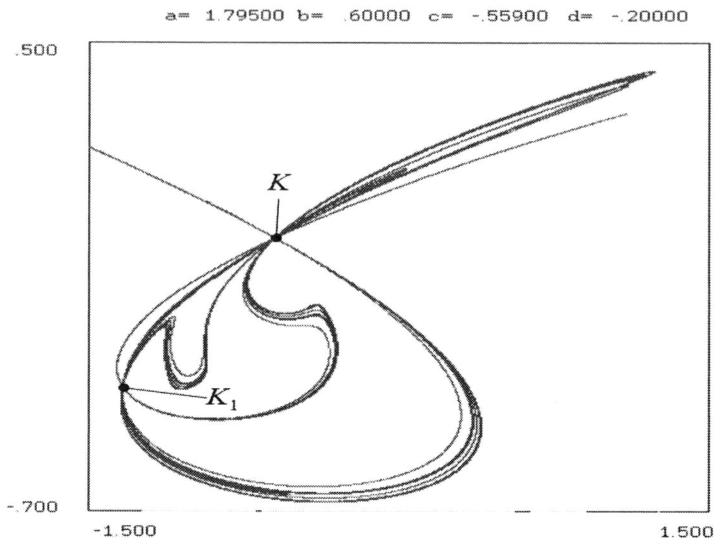

Fig. 24. Map (5.4), $f(x) = -x(x-dx^2)$, $a = 1.795$, $b = 0.6$, $c = -0.559$, $d = -0.2$. K is the "basic" *knot*, K_1 its first rank image

Remark. From the point of view of an "aesthetic" attractor A research, the most interesting images are generally obtained when the picture complexity increases. The experience shows that this occurs when the map parameters are closed to the bifurcation which destroys the attractor. This bifurcation takes place when the A "external" boundary (an unstable manifold of a saddle) has a contact with its basin boundary (a stable manifold of a saddle). It is the case of Figs. 25 and 26.

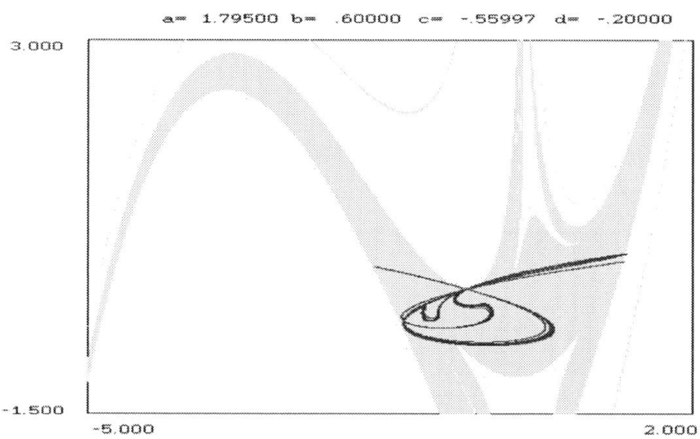

Fig. 25. Map (5.4), $a = 1.795$, $b = 0.6$, $c = -0.55983$, $d = -0.2$. Attractor and its basin (yellow colored), when the parameters are closed to the bifurcation value $c = c_b = -0.55983...$, which destroys the attractor ($c < c_b$) giving rise to a *strange repeller* [34]. The white area is that of diverging orbits

A second example with another form of attractor is:

$$
\begin{aligned}
x' &= \frac{1-a}{a}x(b-x) + \frac{y(x-b)}{a} + c \\
y' &= x,
\end{aligned}
\tag{5.5}
$$

the inverse T^{-1} of which being:

$$x = y'$$

$$y = \frac{a(x'-c)-(1-a)y'(b-y')}{y'-b}$$

The "basic" *knot* is defined as for (5.4): focal point $K(x = c,\ y = b)$ of the inverse map T^{-1}, and $T(J = 0)$ is the point K. The attractor knots are K and its rank n images K_{-n}, $n = 1,2,3,\ldots$.

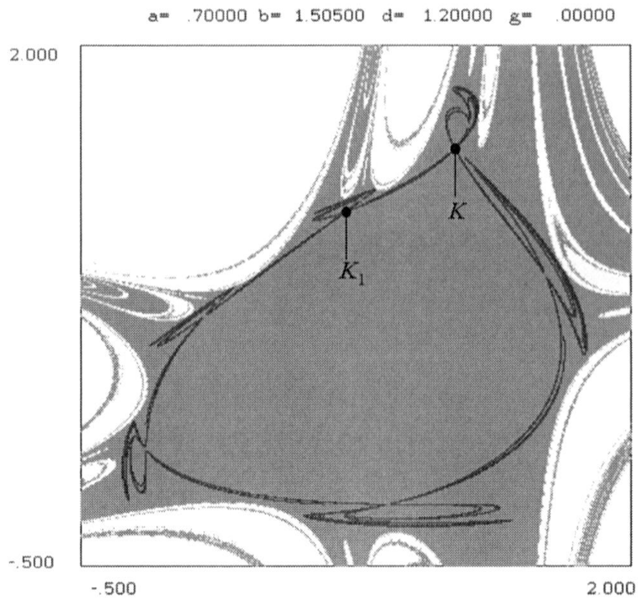

Fig. 26. Map (5.5), $a = 0.7$, $b = 1.505$, $c = 1.2$. Attractor with *knots*, and its basin (brown colored). K is the "basic" *knot*, K_1 its first rank image. The rank n knots $n = 1,2,\ldots,8$ images of K are the green colored points

5.4 Examples of basins generated by an ordinary differential equation

The non autonomous two-dimensional non-autonomous differential equation is the above Duffing type (5.2) with a nonlinear damping [24] and [25].

An eventual artistic feeling may result of the association of differently colored basins of attractors, with a spiraling form these basins, the contrast and the luminosity of the image.

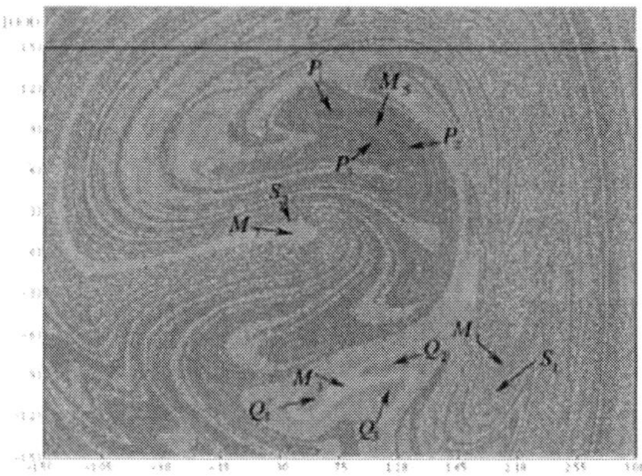

Fig. 27. ODE (5.2). Basins of six attractors obtained for the point (ω = 489,495 and E = 1176.561): four stable fixed points M_i, i = 1, 3, 5, 7, and two stable period three cycles P_j and Q_j, j = 1, 2, 3. S_1 and S_7 are saddles associated with M_1 and M_7

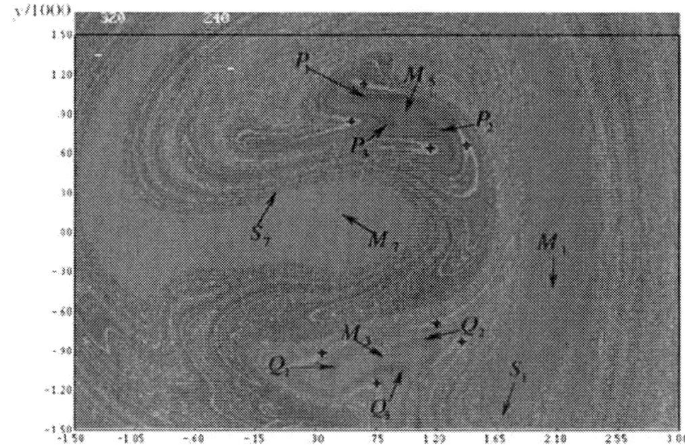

Fig. 28. ODE (5.6). Another choice of parameter point (ω,E) with ω = 480 and E = 1176.561. Two new basins appear related to two stable period 4 cycles, associated with the six basins of Fig. 27. The points of each stable period four cycles are denoted by four small pointed stars

6 Two-dimensional noninvertible maps

Two-dimensional noninvertible maps $X' = TX$ have been quickly presented above at the end of sec. 2. They are the object of the book [35], a very short survey being given in [34]. These two last references provide the details of a too long topic to be developed in this article. Such maps offer a large range of models with a potential "artistic vocation" (some of the figures of [35]): attractors, diversity of basins, parameter plane situations.

6.1 Map singularities and bifurcations specific to noninvertible maps

With respect to invertible maps, noninvertible maps T introduce a singularity of a different nature: the *critical set*. The rank-one *critical set CM* is the geometrical locus of points X having at least two coincident rank-one preimages. Such preimages are located on a set CM_{-1}, the set of merging (or coincident) of rank-one preimages. The set CM satisfies the relations $T^{-1}(CM) \supseteq CM_{-1}$ and $T(CM_{-1}) = CM$. A rank–q critical set CM_{q-1} is given by the rank–q image $CM_{q-1} = T^{q}(CM)$, $CM_0 \equiv CM$. If dim$X = p = 1$, CM is a rank-one *critical point* C. If dim$X = p = 2$, CM is a rank-one critical curve LC.

Such new singularities play a fundamental role in the attractors and basins structure, and their bifurcations. It is the case of "*contact bifurcations*", resulting from the meeting of two singularities of different nature: an invariant manifold (or set) by T or T^{-1} with a *critical set*. This situation generally gives rise to global bifurcations, which may be related to *homoclinic* and *heteroclinic* bifurcations. Most of results obtained till now concern the general class of maps T of the line (Dim$X = 1$) and of the plane (Dim$X = 2$). For plane maps (the topic of this paper) the critical set CM generally becomes a critical curve LC (for *Ligne Critique* in French). In exceptional cases the critical set may include isolated points. This may occur when the inverse map T^{-1} has a vanishing denominator (cf. sec. 5.2 and 5.3). When the two functions defining T satisfy the Cauchy-Riemann conditions (case of one-dimensional maps of the complex variable, related to a *Julia set*), the set LC in the plane is generally made up of isolated points.

In general the critical curve LC is made up of several branches separating the plane into open regions, whose points have different numbers of rank-one preimages. So the plane R^2 can be subdivided into open regions Z_i ($R^2 = \bigcup_i \overline{Z}_i$, $\overline{Z}_i$ being the closure of Z_i), each point of Z_i having i distinct rank-one preimages, $i = 0,1,...N$. Generally the boundaries of the regions Z_i are branches of the rank-one critical curve LC, locus of points such that at least two determinations of T^{-1} are merging. The locus of these "*coincident first rank preimages*" is a curve LC_{-1}, called rank-one curve of merging preimages. If T is smooth LC_{-1} is a set of points for which the Jacobian determinant of T vanishes. If the map is nonsmooth LC_{-1} belongs to the set for which the noninvertible map T is not smooth. The curve LC satisfies the relations $T^{-1}(LC) \supseteq LC_{-1}$ and $T(LC_{-1}) = LC$. Exceptionally non critical lines in the above sense can separate regions with distinct rank-one preimages (cf. [34]).

The simplest case is that of maps in which LC (made up of only one branch) separates the plane into two open regions Z_0 and Z_2. The corresponding maps are said of (Z_0–Z_2) type (or *class*). In more complex cases a classification of noninvertible maps from the structure of the set of Z_i regions can be made (cf. chapter 3 of [35]).

The *stable set* $W^S(p^*)$ of a saddle p^* is backward invariant $T^{-1}[W^S(p^*)] = W^S(p^*)$. It is mapped into itself by T, $T[W^S(p^*)] \subseteq W^S(p^*)$. It is invariant if T is invertible, while for a noninvertible map it may be strictly mapped into itself. When T is noninvertible with $p = 2$, $W^S(p^*)$ may be non-connected and made up of infinitely many closed curves, passing through the increasing rank preimages of p^*. An equivalent property holds for higher dimensions $p > 2$.

When the map T is continuous and noninvertible, the invariant unstable set $W^u(p^*)$ of a saddle point p^* is connected, but self intersections may occur (so that it may not be a manifold), which cannot happen for invertible maps. These self intersections are different from those due to the existence of *knots*. The role of critical sets CM_q and sets CM_{-1} of merging preimages is again essential in the understanding of the formation of self intersections of the unstable set of a saddle fixed point, and properties of invariant closed curves.

It is worth noting that the bifurcations organization, for example in a parameter plane, may be very different from that given by invertible maps. Indeed with respect to the invertible case, the noninvertiblity also adds new bifurcation structures in a parameter plane due to the presence of new co–dimension 2 points related to cycles with real multipliers $S_1 = +1$ and $S_2 = -1$.

6.2 Absorbing areas. Chaotic areas. Bifurcations

Critical curves permit to define the essential notions of *absorbing area*, and *chaotic area*. Roughly speaking an absorbing area (d') is a region bounded by arcs of critical curves of finite, or infinite, rank LC_{-n}, $n = 0,1,2,...r$, $LC_0 \equiv LC$, such that the successive images of all points of a neighborhood $U(d')$, from a finite number of iterations, enter into (d') and cannot get away after entering. Except for some bifurcation cases, a chaotic area (d) is an *invariant* absorbing area, containing a fractal set made up of infinitely many repelling cycles with increasing period, and their limit sets (in the Pulkin'sense, cf. pp. 99-102 of [36]), giving rise to orbits with the property of sensitivity to initial conditions. Its boundary ∂d is made up of LC_n arcs. A chaotic area (d) is not always a strange attractor in a mathematical sense (cf. the sec. "Important remark" of [34]). A chaotic area may be periodic of period k, i.e. a k-cyclic chaotic area is constituted by k nonconnected chaotic areas invariant by T^k . An extended notion of absorbing area and chaotic area, that of *mixed absorbing area*, *mixed chaotic area* (sec. 4.9 of [35]. For non invertible maps, as for invertible ones, when the inverse map has a vanishing denominator, a chaotic attractor may present "points of focalization", called *knots*.

The role of critical curves is fundamental in the definition of bifurcations leading to the destruction, or to a sudden and qualitative modification of absorbing areas and chaotic areas. In particular such modifications concern: transitions "*simply connected chaotic area* → *doubly connected chaotic area*" (or "*annular area*"), "*nonconnected chaotic area* → *doubly connected chaotic area*". A chaotic area (d) is destroyed via a bifurcation resulting from the contact of its boundary ∂d with the boundary $\partial D(d)$ of the basin $D(d)$ of (d). Then, as soon as it is destroyed, (d) turns into a *strange repeller*, fractal set made up of repelling cycles with increasing period, and their limit sets. It is contained into a region of *chaotic transient* before a regular convergence toward another attractor.

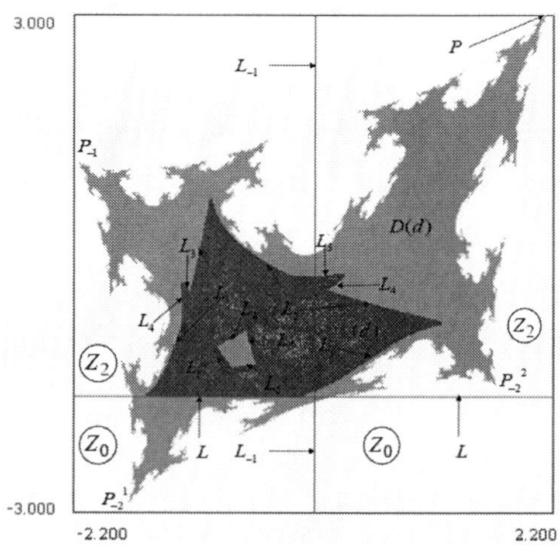

Fig. 29. Map (6.1). The boundary of the chaotic area (d) is made up of critical arcs L_n, $n = 1,2,...,6$. The basin boundary $\partial D(d)$ of the chaotic area (d) is fractal, but the basin itself is nonfractal

An example is that of the Kawakami's map

$$x' = ax + y, \qquad y' = x^2 + b \tag{6.1}$$

This map gives many interesting situations (type of attractors, basins, bifurcations) shortly described in [34], and widely in chapter 5 of [35]. Fig. 29 shows one of these situations.

Another example (chosen for the 1973 Toulouse colloquium) is given by Fig. 30 (map equation below the image). The lines with higher density of iterated points are critical arcs.

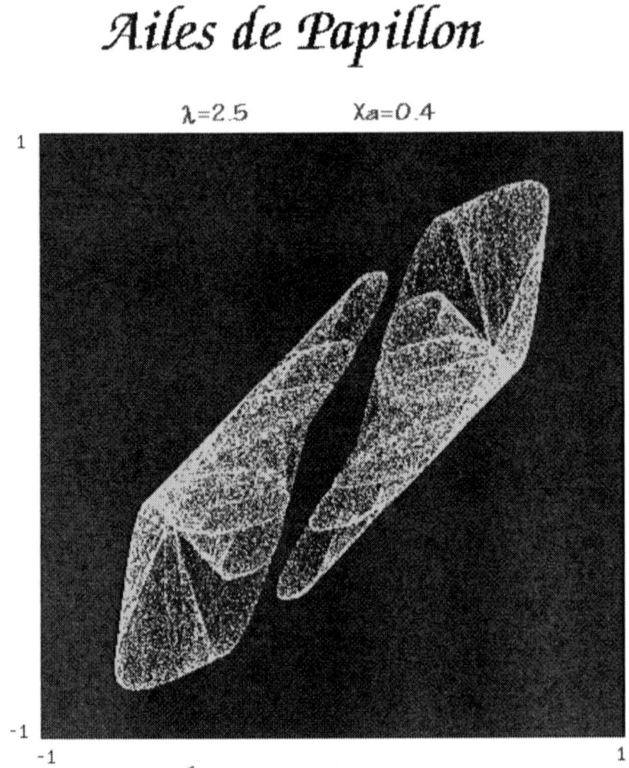

$$\begin{cases} x_{n+1} = (1-\lambda).x_n + y_n \\ y_{n+1} = y_n - f(x_n) \end{cases}$$

$$\text{avec:} \begin{cases} f(x) = 0.2\,\lambda x & \text{pour } 0 < x < x_A \\ f(x) = 0.45\,\lambda \dfrac{(x^2 - x_A^2 - x + x_A)}{(0.5 - x_A)} + 1.1\,\lambda(x - x_A) + y_A & \text{pour } x_A < x < 1 - x_A \\ f(x) = 2.\lambda(x + x_A - 1) + y_B & \text{pour } x > 1 - x_A \\ f(-x) = -f(x) & \end{cases}$$

$$y_A = 0.2\,\lambda x_A \qquad y_B = 1.1\,\lambda(1 - 2x_A) + y_A$$

Fig. 30

6.3 Basins and their bifurcations

Colored basins generated by noninvertible maps offer a large variety of "aesthetic" pictures.

Let D be the basin of an attracting set A, generated by a noninvertible map T. D is invariant under backward iteration T^{-1} of T, but not necessarily invariant by T. The basin D and its boundary ∂D satisfy the relations:

$$T^{-1}(D) = D, \quad T(D) \subseteq D, \quad T^{-1}(\partial D) = \partial D, \quad T(\partial D) \subseteq \partial D$$

Here, the strict inclusion holds iff D contains points of a Z_0 region (i.e. with no real preimage). As said above D may simply connected as in the invertible case, but also non connected, and multiply connected. Its boundary ∂D may contain repulsive sets related to the presence of *strange repellers SR*. Such a fractal unstable set SR is made up of infinitely many unstable cycles with increasing period, their limit sets of increasing class in the Pulkin' sense, the preimages of increasing rank of all these points. This situation gives rise to *fractal* basin boundaries (or fuzzy boundaries) separating the domain of influence of different attractors, and chaotic transients toward a defined attractor.

Critical curves play a basic role in the bifurcations of type "*simply connected basin* ↔ *non–connected basin*". It is the same for bifurcations of type "*simply connected basin* ↔ *multiply connected basin*", and for the destruction of the basin of a chaotic area. These bifurcations always result from the contact of a basin boundary with an arc of critical curve. They are generated by the same mechanism, but in many different ways. One of this way is related to the destruction of the basin of chaotic area (d) after contact of its boundary ∂d with the boundary $\partial D(d)$ of the basin $D(d)$.

The following cubic map gives interesting images of multiply connected and non connected basins (Figs. 31 & 32):

$$\begin{aligned} x' &= y \\ y' &= x(a + by + cx^2) + dy \end{aligned} \qquad (6.2)$$

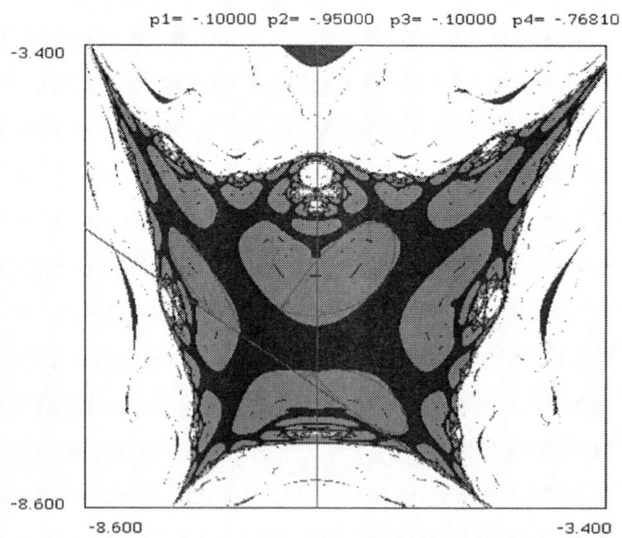

Fig. 31. Map (6.2). $a = -0.1$, $b = 0.95$, $c = -0.1$, $d = -0.7681$. The brown area is the fractal nonconnected basin of a period 2 self-intersecting attracting closed curve.
The blue fractal multiply connected region is the basin of a stable fixed point.
The red domain is the fractal nonconnected domain of another attractor.
The white part is the region of diverging orbits

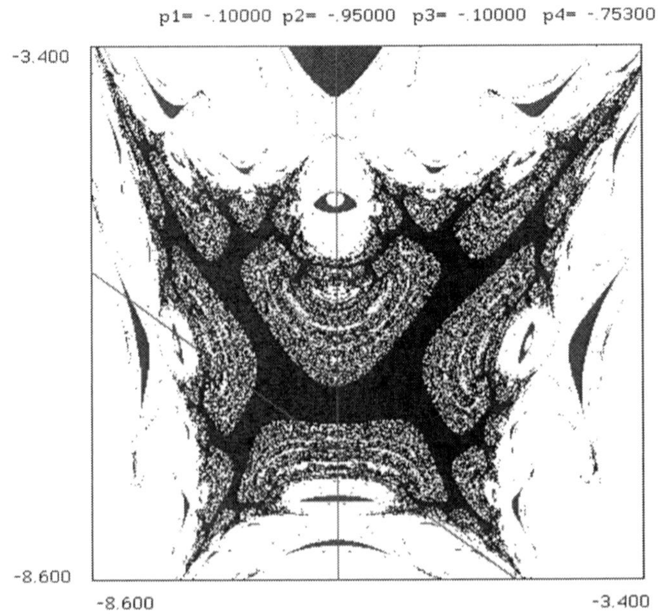

Fig. 32. Map (6.2). $a = -0.1$, $b = 0.95$, $c = -0.1$, $d = -0.753$.
The blue fractal multiply connected region is the basin of a stable fixed point.
The red domain is the fractal nonconnected domain of another attractor, it has penetrated white holes inside a hole of the blue part. The fractal white part is the region of diverging orbits

The Kawakami's map (6.1) provides another type of fractal basins

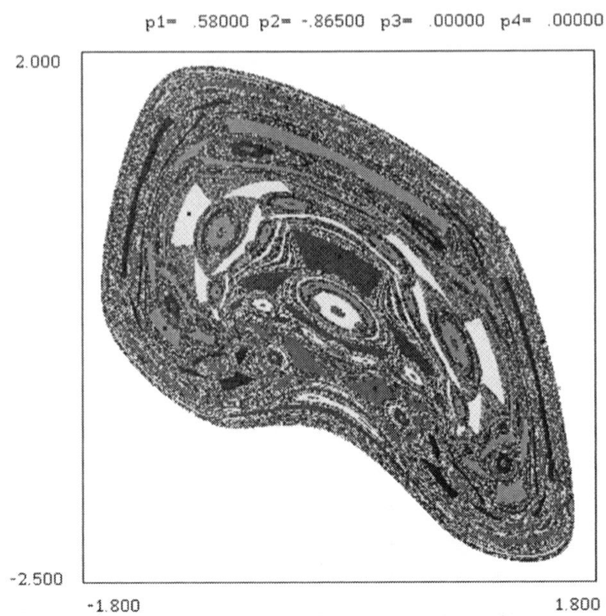

Fig. 33. Map (6.1). $a = 0.58$, $b = -0.865$. Basins (differently colored) of 6 attractors.
These basins are of different types: multiply connected, nonconnected, mixed (multiply-nonconnected)

7 Parameter plane generated by invertible and noninvertible maps

From an "artistic" point of view, this section does not differentiate them. From a mathematical point of view the noninvertibility adds new features (cf. sec. 6).

7.1 Two types of programs used for a parameter plane scan

Pictures of parameter plane can be obtained via two programs of scanning with different aesthetic effects, and different mathematical information.

The first program is related to a scanning based on the detection of stable cycles, each period being associated with a well defined color. The aesthetic effect is essentially to the picture geometry and the colors association, but without any relief effect due to a continuous change of shade (second program), in particular for regions of chaos. It is used for the Henon's map giving rise to Figs. 34-36.

The second program is a scanning based on the largest nonzero Lyapunov's exponent μ of each plane point. Negative exponents $\mu < 0$ (stable cycles for the associated map) are grey shading colored, the color white indicates the most negative values (two branches of a topographic saddle cf. p. 353 of [36]), the color black is related to $\mu = 0$. Positive exponents, $\mu > 0$ (chaotic behavior), are related to a continuous shade change in the yellow-red scale, from lighter yellow shades in the neighborhood of the crossroad area boundary to somber red denoting a stronger chaos (more positive values of μ). The aesthetic effect is essentially due to the picture geometry and a relief effect due to a continuous change of shade, in particular for regions of chaos. From a mathematical point of view, the important disadvantage is the absence of identification of the period of stable cycles, but from an aesthetical point of view the effect is more interesting. It is used for the ordinary differential equation studied in [39] and [40] (obtained via a Poincaré section), cf. Figs. 37-39.

7.2 Hénon's map: crossroad areas and spring areas

For a better understanding of the bifurcation mechanisms, the parameter plane must be considered as made up of sheets, each sheet being associated with a well defined cycle $(k; j)$, where k is the cycle period, and j an integer characterizing the permutation way of the cycle points via k successive iterations. When k increases the number of possible permutation ways increases exponentially. This parameter plane is called *foliated*, and can be associated with a qualitative "artificial" three-dimensional representation of the sheets. Covering areas of several sheets, associated with stable cycles (or more complex attractors), gives rise the *multistability phenomenon* for a given point of the parameter plane.

For the Henon's map, a study of the bifurcation curves organization in the parameter plane (a,b), and of the three-dimensional foliated representation of some of its parts, is given pp. 313-395, of the book [36]. An overview presents the topic in [37]. A complete knowledge of the bifurcations organization implies the identification of the sheets "geometry" in this three-dimensional auxiliary qualitative space. This identification amounts to seeing how to pass continuously from a sheet to another one following a continuous path of the parameter plane, i.e. to knowing the possible communications between sheets. Three basic communication structures are: the *crossroad area*, the *saddle area* and the *spring area* (cf. [36], [37], and their references from the years 1982).

The notion, and the structure, of *crossroad area* (1982, in French *zone échangeur*) was rediscovered in 1993 under the now popular name *shrimp* [40]. As for the *spring area* (1982, in French *zone source*) seems having been discovered in 2007 (Fig. 4b of [40]), but without any appellation. From the years 1982 it has appeared that a *crossroad area* is a very widespread structure met in parameter plane patterns of dynamical systems, and the *spring area* structure at a lesser extent. These *crossroad areas* and *spring areas* are organized in a fractal way according the *box-within-a-box* fractal structure of the Dim1 Myrberg quadratic map T: $y' = x^2 - \lambda$ (cf. a survey in [37]), and more completely in chapters 4 and 6 of [36].

More particularly sections 3.2.1 and 3.2.2 provide the details about the fractal *box-within-a-box* organization of the *crossroad areas* of Fig. 35, and the same organization for the Fig. 36 *spring area*.

The cycle period is $K=5k$

So red is $K=5$

blue is $K=10$

yellow is $K=20$

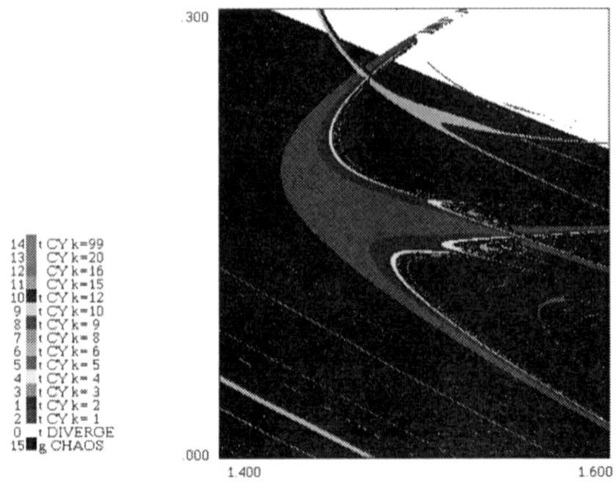

Fig. 34. Henon's map (5.1). Component of the fractal organization of the parameter plane (a,b), with view of a crossroad area, $1.4 \leq a \leq 1.6$, $0.0 \leq b \leq 0.3$. With respect to the colors scale ($k = 1,2,...$) on the left, the cycle period of numerous coexisting crossroad areas is $K = 5k$. So $k = 1$(red) corresponds to a stable cycle of period $k = 5$, $k = 2$(blue) to a stable cycle of period $k = 10$, resulting from a period doubling of the period 5 cycle

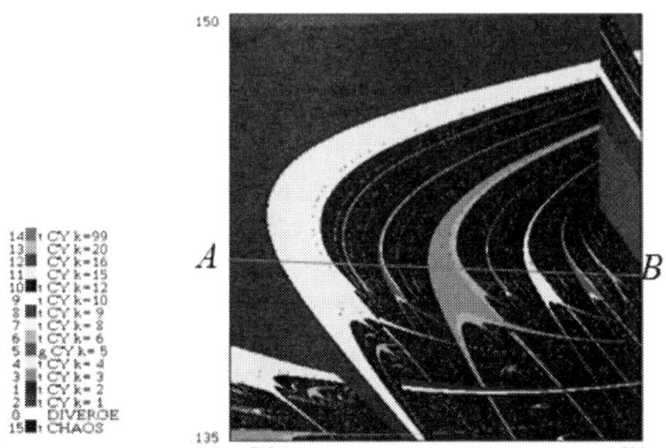

Fig. 35. Henon's map (5.1). Enlargement of a part of Fig. 34, with many coexisting *crossroad areas* ($K = 6k$) organized with a box-within-a-box structure. So $k = 1$(red) corresponds to a stable cycle of period $k = 6$, $k = 2$ (blue) to a stable cycle of period $k = 12$, ..., with view of the fractal *box-within-a-box* structure of the *crossroad areas*, $k = 6$ (light blue)$\rightarrow$ $k = 30$, etc.

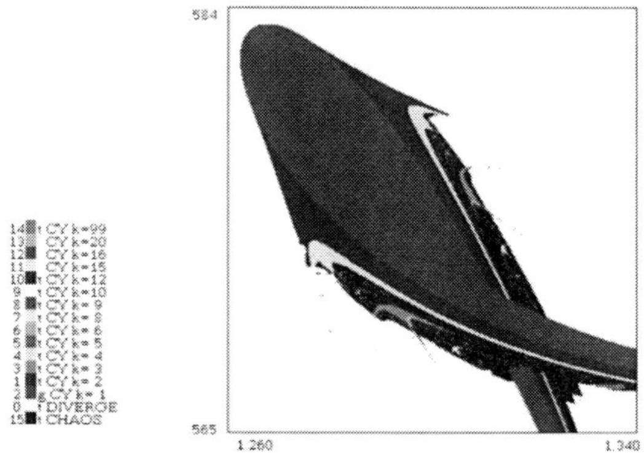

Fig. 36. Henon's map (5.1). Element of the parameter plane (a,b) with view of a *spring area* $1.26 \leq a \leq 1.340$, $0.565 \leq b \leq 0.584$. With respect to the colors scale ($k = 1,2,...$) on the left, the cycle period is $K = 6k$. So $k = 1$(red) corresponds to a stable cycle of period $k = 6$, $k = 2$ (blue) to a stable cycle of period $k = 12$, ...

7.3 Ordinary differential equations: crossroad areas, spring areas, and isoordinal cascades

The publication [40] deals with the ODE model of optically injected semiconductor lasers. Figure 37 represents a *crossroad area* obtained via a parameter plane scanning based on the second program of sec. 7.1. 32.

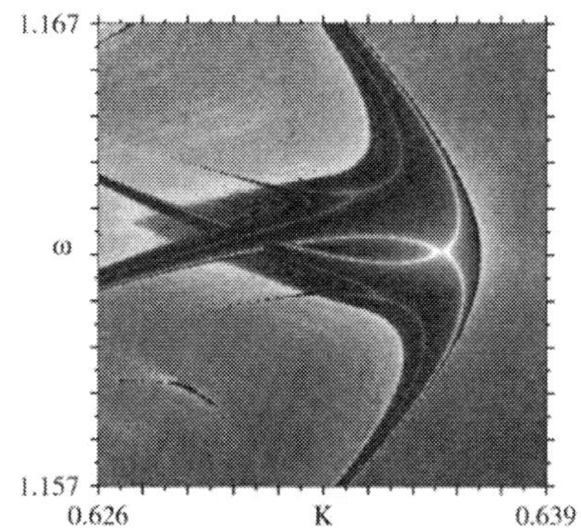

Fig. 37. Crossroad area (According to Fig. 1c of [40]), compared with Fig. 36. Parameter plane scanning based on the largest nonzero Lyapunov's exponent μ of each plane point for the ordinary differential equation studied in [40] (obtained via a Poincaré section). Negative exponents $\mu < 0$ (stable cycles for the associated map) are grey shading colored, the color white indicates the most negative values (called *topographic saddle* p. 353 of [36]), the color black is related to $\mu = 0$. Positive exponents, $\mu > 0$ (chaotic behavior), are related to a continuous shade change in the yellow-red scale, from lighter yellow shades in the neighborhood of the crossroad area boundary to somber red denoting a stronger chaos (more positive values of μ)

The solutions generated by an ordinary differential equation are generally obtained via the Poincaré method of surface section, which leads to a map T constructed by numerical integration. For such a map a fixed point is a "*main*" periodic solution of period τ, a period k cycle is a $1/k$ *subharmonic* of the "main" periodic solution [24], [25]. If the ODE is smooth, the map T is invertible.

Nevertheless, the surface section method does not reduce completely the problem to a map one. Indeed the solution behavior inside a time interval τ play an important role, because generally it gives rise to well defined regions of a parameter space associated with rank m *higher harmonics resonances*, or *synchronizations*, *m/k fractional harmonics resonances*, or *synchronizations* [24], [25]. The article [38] provides a survey about this topic.

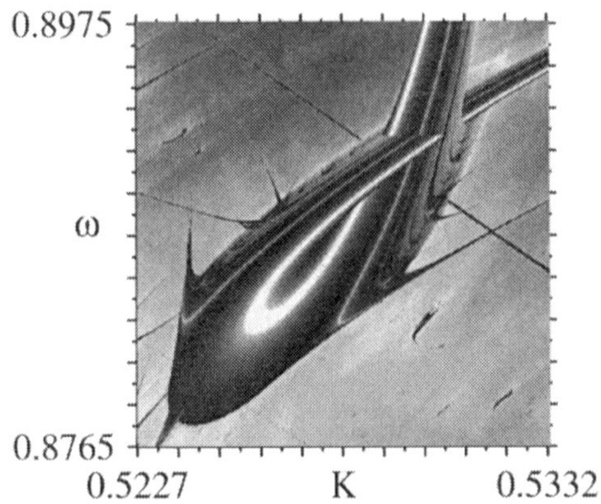

Fig. 38. Spring area (According to Fig. 4b of [40]), compared with Fig. 36. Parameter plane scanning based on the largest nonzero Lyapunov's exponent μ of each plane point for the ordinary differential equation studied in [40] (obtained via a Poincaré section)

With respect to maps as models of discrete systems, in a *foliated* parameter space, periodic solutions of ODEs generally give rise to a sequence of *communication structures* (or *cells*) ${}^{m}S_{j}^{k}$, $m = 1, 2, 3, \ldots$, forming an *isoordinal cascade*, i.e. its elements have the same basic period k, the same index j, but they differ in the integer m (cf. [24] and the survey [38], and their references). The paper [24] provides a three-dimensional qualitative view of such an *isoordinal cascade*, and also Fig.1 of [38] for the ODE (5.2) (cf. also the article references). Figure 39 represents such a cascade obtained via a parameter plane scanning based on the largest nonzero Lyapunov's exponent μ of each plane point for the ordinary differential equation studied in [40].

8 Conclusion

As said in Sec. 1, the judgment on the possible aesthetic value of the pictures illustrating this text is left to the reader. The object is also an incursion into the maps properties which generate such images, for understanding the mathematical sense of generated images. Except for images containing polygonal forms (sec. 3), it appears that most of the pictures of this text cannot be evaluated by the Birkhoff's mathematical theory of aesthetics. Perhaps, some mathematicians motivated by the relations between mathematics and art will find an interest in perfecting this theory? In particular about the colored pictures, the Birkoff's aesthetic theory applied to music (cf. [5], Part III The diatonic chords, Harmony, & Melody) might be associated with wave lengths of rainbow colors (visible spectrum)?

References

[1] MIRA, C. "*Exposé d'Introduction*". Colloque International du CNRS n° 229 Transformations Ponctuelles et Applications (Toulouse Sept. 1973), p. 19-27. Proceedings: Editions du CNRS, Paris, 1976.

[2] BIRKHOFF, G.D. *Science and spiritual perspective*. Century Mag., June 1929, vol. 118, 156-165; or Collected Mathematical Papers, Dover Publications, 1968, Inc., New York, Vol. III, 278-287.

[3] BIRKHOFF, G.D. *Quelques éléments mathématiques de l'art*. Atti Congr. Intern. di Math., Bologna, vol. 1, 1929, 315-333.

[4] BIRKHOFF, G.D. *A mathematical theory of aesthetics*. Scientia, September 1931, 133-146.

[5] BIRKHOFF, G.D. *A mathematical theory of aesthetics and its applications to poetry and music* - A course of Lectures: I. Relation of the theory to earlier aesthetic theories, II. The musical quality in poetry, III. The diatonic chords, IV. Harmony, V. Melody. Rice Institute Pamphlet, vol.19, 1932, 189-342.

[6] BIRKHOFF, G.D. *On drawings composed of uniform straight lines*. Jn. D. Math. (Liouville), S. 9, vol. 19, 1940, 221-2386.

[7] POINCARÉ, H. *Œuvres*. Tome II. Gauthier-Villars, Paris, 1916.

[8] PULKIN, C.P. *Oscillating iterated sequences* (in Russian). Dokl. Akad. Nauk SSSR, vol. 76, n. 6, 1950, 1129-1132.

[9] MIRA, C. *I. Gumowski and a Toulouse research group in the "prehistoric" times of chaotic dynamics*. In *The Chaos Avant-Garde. Memories of the Early Days of Chaos Theory*. Editors Ralph Abraham & Yoshisuke Ueda. World Scientific Series on Nonlinear Science (Editor Leon O. Chua). Series A, vol. 39, 2000.

[10] BERNUSSOU, J., LIU HSU, MIRA, C. *Quelques exemples de solutions stochastiques bornées dans les récurrences autonomes du deuxième ordre*. Colloque International du CNRS n. 229 Transformations Ponctuelles et Applications (Toulouse Sept. 1973), pp. 195-226.

[11] GUMOWSKI, I., MIRA, C. *Dynamique Chaotique. Transformations ponctuelles. Transition Ordre - Désordre*. Cépadues Editions, Toulouse, 1980.

[12] GUMOWSKI, I., MIRA, C. *Recurrences and discrete Dynamic systems*. Lecture Notes in Mathematics n. 809, Springer-Verlag Berlin, 1980.

[13] GUMOWSKI, I., MIRA, C. *Boundaries of stochasticity domains in Hamiltonian systems*. Proc. 8th Conf. on High-Energy Accelerators. CERN Genève, 1971, 374-376.

[14] GUMOWSKI, I. *Some properties of large amplitude solutions of conservative dynamic systems*. Part 1: "Quadratic and cubic non-linearities". Part 2 "Bounded nonlinearities". Rapport CERN/SI/Int. BR/72-1, Genève, 1972.

[15] GUMOWSKI, I. *Stochastic effects in longitudinal phase space*. Paper contributed to the 9th International Conference on High Energy Particle Accelerators. Standford, 2-7 May 1974.

[16] SAHA, L.M., DAS, M.K., MRIDULA BUDHRAJA, M. *Characterization of Attractors in Gumowski-Mira Map Using Fast Lyapunov Indicator*. J. Forma, Letter, 21, 2006, 151-158.

[17] AMBIKA, K., AMBIKA G. *Synchronisation schemes for Gumowski-Mira maps*. URL: http://www.ncnsd.org/proceedings/proceeding06/paper/122.pdf.

[18] *The strange beauty of the Gumowski-Mira Attractor*. URL: http://petervandernoord.nl/blog/?p=289

[19] BEN MAALLEM, H., RICHARD, P., FERRIER, J.-L., LABIB, A. *Using Gumowski-Mira Maps for Artistic Creation*. Proceedings 2th Generative Art Conference GA2009, pp. 308-315.

[20] KULENOVIĆ, M.R.S., NURKANOVIĆ, Z., PILAV, E. *Birkhoff Normal Forms and KAM Theory for Gumowski-Mira Equation*. Hindawi, Scientific World Journal, vol. 2014 (2014).

[21] HANCHINAMANI, G., KULKARNI, L. *An Efficient Image Encryption Scheme Based on Quintuple Encryption Using Gumowski-Mira and Tent Maps...* International Journal of Contents, vol. 11, n. 4, 2015, 56-69.

[22] KOTARSKI, W., GDAWIEC, K., LISOWSKA, A. *On Gumowski-Mira Aesthetic Superfractal Forms*. URL: http://kgdawiec.bplaced.net/badania/pdf/cacs_2010.pdf

[23] URL: https://searchworks.**stanford.edu/view**/3301410

[24] KHAMMARI, H., MIRA, C., CARCASSÈS, J.-P. *Behavior of harmonics generated by a Duffing type equation with a nonlinear damping. Part I.* Int. J. Bifurcation Chaos, vol. 15, n. 10, 2005, 3181-3221.

[25] KHAMMARI, H., MIRA, C. *Behavior of harmonics generated by a Duffing type equation with a nonlinear damping. Part II.* Int. J. Bifurcation Chaos, vol. 19, n. 4, 2009, 1227-1254.

[26] MIRA, C., GARDINI, L., BARUGOLA, A., CATHALA, J.C. *Chaotic Dynamics in two-dimensional noninvertible maps*. World Scientific, Series A on Nonlinear Sciences, 20, 1996.

[27] BIRKHOFF, G.D. *Polygonal forms*. Sixth Yearbook of Nat. Council of Teachers of Math., 1931, 175-195.

[28] BIRKHOFF, G.D., SMITH, P.A. *Structure analysis of surface transformations*. Journal de Math. (Liouville), S. 9, vol. 7, 1926, 345-379.

[29] MIRA, C. *Singularités non classiques d'une récurrence et d'une équation différentielle*. Compte Rendus Acad. Sci. Paris, Série I, vol. 292, 1981, 146-151.

[30] BISCHI, G.I., GARDINI, L., MIRA, C. *Plane maps with denominator. Part I. Some generic properties*. Intern. J. of Bif. and Chaos vol. 9, n. 1, 1999, 119-153.

[31] BISCHI, G.I., GARDINI, L., MIRA, C. *Plane maps with denominator. Part II. Noninvertible maps with simple focal points*. Intern. J. of Bif. and Chaos vol. 13, n. 8, 2003, 2253-2277.

[32] BISCHI, G.I., GARDINI, L., MIRA, C. *Plane maps with denominator. Part III. Noninvertible maps with simple focal points* .Intern. J. of Bif. and Chaos vol. 15, n. 2, 2005, 451-496.

[33] BISCHI, G.I., GARDINI, L., MIRA, C. *Maps with vanishing denominator*. Scholarpedia article, 2 (9):3277 (2007). URL: http://www.scholarpedia.org/article/Maps_with_vanishing_denominators

[34] MIRA, C. *Noninvertible Maps*. Scholarpedia article, 2 (9): 2328 (2007). URL: http://www.scholarpedia.org/article/Noninvertible_maps

[35] MIRA, C., GARDINI, L., BARUGOLA, A., CATHALA, J.C. *Chaotic Dynamics in two-dimensional noninvertible maps*. World Scientific, Series A on Nonlinear Sciences, 20, 1996.

[36] MIRA, C. *Chaotic dynamics. From the one-dimensional endomorphism to the two-dimensional diffeomorphism*. World Scientific, Singapore, 1987.

[37] MIRA, C. *"Shrimp" fishing, or searching for foliation singularities in the parameter plane? Part I: basic elements of the parameter plane*. Research Gate (2016). URL: https://www.researchgate.net/publication/302960218_Shrimp_fishing_or_searching_for_foliation_singularities_of_the_parameter_plane_Part_I_basic_elements_of_the_parameter_plane_foliation

[38] MIRA, C. *"Shrimp" fishing, or searching for foliation singularities in the parameter plane? Part II: Non autonomous ordinary differential equations. Harmonic, higher harmonic, subharmonic, and fractional harmonic nonlinear resonances.* Research Gate (2016). URL: https://www.researchgate.net/publication/309430434_Shrimp_fishing_or_searching_for_foliation_singularities_in_the_parameter_plane_Part_IIa_nonautonomous_ordinary_differential_equations_Harmonic_higher_harmonic_sub_harmonic_and_fractional_harmonic_nonl

[39] GALLAS, J.A.C. *Structure of Parameter Space of the Hénon map.* Physical Review Letters, vol. 70, n. 18, 1993, 2714-2721.

[40] BONATO, C., GALLAS, J.A.C. *Accumulation horizons and period adding in optically injected semiconductor lasers.* Phys. Rev. E, vol. 75, 2007, Rapid communications, 055204(R)

Current address

Mira Christian, Phd
Due to retirement no affiliation to an institution
E-mail: christian.mira@sfr.fr

Slovak University of Technology in Bratislava
Faculty of Mechanical Engineering
16th Conference on Applied Mathematics **APLIMAT 2017**
Proceedings

DIFFERENCES IN CLASSROOM PRACTICES IN ORDINARY AND A CLIL MATHEMATICS LESSON

MORAOVÁ Hana (CZ), NOVOTNÁ Jarmila (CZ)

Abstract. The goal of the paper is to show the specifics of communication and practices in a CLIL mathematics lesson in contrast to a mathematics lesson conducted in a mother tongue. The reported research is conducted within the frame of the international Lexicon Project. A CLIL lesson must have dual focus (mathematical and language objectives) and thus we must expect some significant differences in classroom practices and communication patterns. For these reasons, the authors of the paper use Czech Lexicon to describe several CLIL lessons and look for what specific aspects of a CLIL lesson do not have a corresponding term in the Lexicon. Lexicon is a book of reference, a lexicon that provides the needed terminology as well as examples of what is meant by different terms used for the description of classroom practices in a mathematics lesson. The aim of the study is to point to the fact that teachers who face the situation of teaching mathematics through a foreign language should be familiar with practices that will be different from a traditional mathematics lesson and thus it will need more of their and their pupils' attention. Moreover, pupils must be allowed extra time to get used to these new practices and patterns to develop a new classroom culture – the culture of a CLIL mathematics lesson. The results of the presented analysis illuminate the specifics of teaching mathematics in CLIL, and thus they are useful both for CLIL teachers and teacher trainers.

Keywords: CLIL, mathematics, classroom practices, classroom management, communication, stages of a lesson

Mathematics Subject Classification: Primary 97D40, 97C50; Secondary 97D80

1 Introduction

In 1966, the Council of Europe adopted a basic programme document on education entitled "Teaching and learning. Towards the learning society". One of the declared primary objectives of education is mastery of three languages of European Union. The list of methods for achievement of this objective includes teaching of a selected subject through a foreign language – CLIL (Content and Language Integrated Learning).

CLIL is a broad term that refers to teaching of a non-language subject in a foreign language (e.g. mathematics in English). Knowledge and mastery of the contents of the selected subject is developed through a foreign language and a foreign language is practiced and developed thanks to the content of the particular subject. The term CLIL can be used already when at least 25 % of the lesson is conducted in the foreign language. There are two objectives in each lesson – one objective is connected to the topic and the subject, the other to the foreign language. [7], [5]

The CLIL methodology is learner-centred. Pupils get actively involved in the lessons, the teacher uses activating methods. Pupils are encouraged to communicate and collaborate. The teacher uses a variety of classroom management techniques. A lot of non-verbal means of communication are used in the classroom. Emphasis is put on selection of interesting, practical and engaging topics.

Hofmannová and Novotná [3] monitor the teacher's role when teaching mathematics through English to Czech pupils and study the extent of the teacher's verbal and non-verbal support in their pupils' cognitive development. When a subject is taught through a foreign language, a much greater degree of responsiveness and more adaptability than in regular classrooms. Sometimes teachers in regular classrooms find it much harder to respond to their pupil's different learning styles [4]. In a foreign language lesson the pupil must try much harder to understand, both when the subject matter is presented and when grasping problem assignments etc. In a CLIL classroom, the teacher's language production, based on their general competence in the foreign language as well as knowledge of terminology specific for the subject, is likely to be adapted to the pupils' level. [7]

The paper focuses on the specifics of CLIL lessons when compared to lessons conducted in the teacher's and most pupils' mother tongue. We come out of the Czech Lexicon (see 2.2) created for teaching mathematics in the mother tongue. We show what areas are specific and unique for situations when mathematics is taught through a foreign language. The text is limited to teaching mathematics through English. However, we are convinced that the presented results can be useful also for other subjects and also for other foreign languages.

2 Lexicon project

2.1 Lexicon in general

In the year 2014, David Clarke created an international community from nine countries (Australia, Chile, China, Czech Republic, Finland, France, Germany, Japan, USA) to solve the project named *The Lexicon Project: Analysing pedagogical naming systems from different cultures to reconceptualise classroom practice and advance educational theory* [1]. The main goal of the project was to create a database of terms for classroom research on international level. Cultural specificity of classrooms poses challenges for international comparative research. Educational theories, research and descriptions of practice are currently framed in English, which names some aspects of the classroom but ignores key aspects named in other languages. It limits the capacity to access, connect and adapt the wisdom of other cultures. [2] The Lexicon Project initiates cross-cultural dialogue to identify pedagogical terms from selected Asian and European educational communities and use them as analytical tools to categorise, interrogate and enrich our classroom practice, classroom research, and educational theory.

2.2 Czech Lexicon

In [9], the process of creating the Czech Lexicon and its final form, as it was subject to national evaluation, are presented. The Czech Lexicon contains about 100 lexical items. It includes general didactical terms more than "mathematical didactical vocabulary". The structure of the Czech Lexicon is based on the following large categories: Classroom Management, Introductory Communication, Explanation of New Topic, Revision of Previously Taught Topic, Solving of a Problem, Checking Individual Work, Institutionalisation, Summary, Non-mathematical Social Interaction, Assessment, Concluding the Lesson, Individual Consultation with a Pupil.

Czech Lexicon is highly stratified. For each term, there are several not necessarily disjoint sub-terms. It has the form of a table of eight columns. The content of these columns is illustrated in Table 1, which is a part of Czech lexicon – section Explanation of a new topic. The term is structured as shown in the third column of Table 1:

In Local Language (where applicable)		In English		User	Video Examples		
Pedagogical Term	Description	Pedagogical Term	Description		Lesson	Time Stamp IN	Time Stamp OUT
Výklad nového učiva	U vykládá nové učivo	Explanation of new topic	T explains a new topic				
• linking	U upozorňuje na souvislosti s dříve naučeným	linking	T draws attention to links with previously acquired knowledge	R	Australia	00:27:25	00:27:32
• za použití pomůcek		using aids					
○ s oporou o kreslený obrázek	U vykládá a současně kreslí na tabuli	using a drawing	U explains and at the same time makes illustrative drawing on the board	T, R	China Finland	00:23:56 00:11:12	00:36:18 00:14:30
○ s použitím projektoru, interaktivní tabule	U vykládá a prezentuje na interaktivní tabuli, resp. na projektoru	using a projector, a smart board	T explains and presents on a smart board or a projector	T, R	China	00:03:55	00:05:00

Tab. 1. Sample of Czech Lexicon

2.3 Methodology of creation of Czech Lexicon

The process of creation of Czech Lexicon is described in detail in [8]. As no similar study could be found, qualitative research design was used. The materials used for creation of Czech Lexicon were mathematics classroom video recordings of one Czech mathematics lesson and eight video recordings of lessons from other countries (of other project teams). The main methodological approach used was open and axial coding of these episodes.

At first, open coding was used [9]. The Czech team members named the events they were observing. They divided the lesson into shorter episodes they found distinct. They defined the beginning and the end of each episode. The episodes were labelled by a code. As expected, there were huge differences in approaches to description of the phenomena. Having finished individual work, all team members were shown the coding of other team members. The team met and different codes were grouped into larger codes using axial coding. The result of this process was a list of codes – the first draft of the Czech Lexicon. This was followed by verification of the Czech Lexicon. Verification had the form of coding foreign lessons – two members of the team were always coding two lessons. The results of their coding were compared and the list of codes extended and restructured.

3 Our research: CLIL – What terms are not in Czech Lexicon

Unlike a usual mathematics lesson a CLIL lesson has a dual focus – learning the subject matter and acquiring new language, vocabulary and language structures. The question we ask here is how well a CLIL lesson with its specific communication and practices can be described using the terminology from Czech Lexicon and what terms are missing in the Lexicon. While the Lexicon should provide all terminology needed for description of any mathematics lesson, it may lack terminology needed for description of how language goals are achieved in the course of the lesson. The Lexicon does not describe activities typical for teaching English as a foreign language and activities typical for a situation when learners are not very proficient and confident in the language of instruction and the teacher is not a native speaker.

We believe that if we manage to find what terms are missing in the Czech Lexicon with respect to the progress of a CLIL lesson, we will be able to name the areas that are specific for a CLIL lesson and that are likely to need to be paid more attention by the teacher as they challenge the mathematics teachers' ad their pupils' beliefs about what it means to be conducting and following a mathematics lesson. The missing areas are exactly those that will have to be addressed by the teacher if the pupils are to feel confident and safe in a mathematics CLIL lesson.

3.1 Methodology of supplementing Czech Lexicon for teaching mathematics in CLIL

The paper answers the following research question: What are the terms suitable for description of a lesson of mathematics conducted in CLIL that are not included in Czech Lexicon for mathematics education?

The methodology for supplementing the Czech Lexicon for mathematics CLIL lessons is similar to the methodology of creating described in 2.3. The authors of the paper analysed, coded and described CLIL lessons. They were members of the team that created the original Czech Lexicon.

The analysed material consisted of four video recordings of mathematics CLIL lessons. The difference between creation of the original Lexicon and this survey is that the coding was conducted only by the two authors of this paper – a specialist in didactics of mathematics and a specialist in didactics of a foreign language (cooperation of maths and language educators proved to be very valuable already in precious research.). Due to the size of the team the discussion on how to sequence the video recordings and describe the different activities using fitting terms was done only by two people, which made it much easier to come to an agreement on a term and discussions were much briefer.

3.2 Language activities

A CLIL lesson unlike a regular mathematics lesson conducted in mother tongue has many stages that are adapted from an EFL lesson. The observed lessons all started with a lead-in language activity – the teacher was telling a story from ancient China, the teacher activated the needed vocabulary by a game, the pupils were asked to think about superstitions and fairy tales, a short video in English was shown to the pupils. These are language focused ice-breakers that are used to make the pupils involved in the lesson and make the atmosphere in the classroom relaxed and open. They also serve for activation of the needed language structures.

Another type of activity that could be observed in the CLIL lessons was the use of pair work communication activities. The activities have two goals – practicing mathematics but also communicating in English. These pair work activities are often rooted in the real world and mathematics is of "secondary" importance. They are not a separate item in the Czech Lexicon. The goal of pair work in an ordinary mathematics lesson is not to make pupils speak.

More attention is paid to activities in which the teacher checks understanding not only of mathematical concepts, but also of the language of assignments. Pupils are asked to reformulate, to state in their own words, to show, to draw or even to translate (which is not used in regular mathematics lessons in Czech).

Typical for a CLIL lesson is a section in which the teacher verifies knowledge of the needed vocabulary – not necessarily by a test but by other types of activities (matching words and definitions, gap-filling, creating definitions, guessing the word from its definition, games).

3.3 Code switching

In linguistics, code-switching occurs when a speaker alternates between two or more languages in the context of a single conversation. For code switching intermediate level of competence is needed. It is an interim stage of production where the speaker switches from their mother tongue to the foreign language as it corresponds to the ′inner speech′. Presence of these phenomena can be observed e.g. when in a less teacher controlled pair or group work while solving a problem the pupils switch to their mother tongue and switch back to English for collective feedback. [7]

The area of code-switching is not included in the Czech Lexicon as it never occurs in a regular mathematics lesson. The Lexicon includes terms in the area of switching into symbolical language, use of pictures, illustrations and other visual aids including electronic devices. Otherwise it is silent in this area and CLIL teachers will have to think of how to managing these situations in a lesson.

3.4 Scaffolding

To put it briefly, scaffolding is anything the teacher (and not only the teacher but also the author of a learning material, a parent or a pupil) does to make learning easier – both as far as language and the subject itself are concerned. "Scaffolding" is more important in CLIL than in regular lessons [12]. In CLIL the teacher must build double scaffolding: one for the language, the other for the subject itself. [11]

The following are strategies used as scaffolding in CLIL. Italics is used to show those activities that are already included in the Czech Lexicon. The examples clearly point at types of scaffolding activities that are not typical for a mathematics lesson conducted in Czech.

a) on verbal level: use of

- synonyms,
- antonyms,
- hyponyms and hypernyms,
- categories – "use of semantic clusters"
- *examples and non-examples in context,*
- *questions,*
- associations,
- mnemonic aids,
- *reformulation,*
- change in pace of speaking,
- translation

etc.;

b) on non-verbal level: use of

- *graphic representations and symbols,*
- *visualization of concepts,*
- *visual aids and models,*
- mimics and gestures (to help pupils understand unknown words),
- elements of dramatization and motion (to help pupils understand unknown words),
- *use of IT*

etc.

[6], [11]

Other methods and strategies are for example:

- motivation through appropriate context, interesting activities, short texts (developing reading comprehension skills),
- *dividing? the task into subtasks to facilitate solution,*
- *activating prior knowledge,*
- *creating mind maps,*
- *brainstorming,*
- choice of language means (including work with a dictionary, vocabulary lists, explanatory phrases),
- adapting a text – shortening, structuring, personifying, emphasising important words, summarizing, work with key words,
- *questions the teacher asks to guide and help,*
- *growing aware of the goals,*

- *search for logical links, for connections to real world, cross-curricular approaches*,
- *positive teacher's feedback*,
- peer teaching (pupils help each other and learn from each other),
- use of media and internet (e.g. short YouTube videos in English, CLIL websites, websites with mathematics materials in English).

[11], [12]

4 Concluding remarks

Work on Czech Lexicon showed that Czech terminology in the area of mathematics education seems to be unsettled. The need to cultivate subject didactic discourse is seen as one of the prerequisites to emancipation of subject didactics [10: 449]. This only highlights the importance of creation of Czech Lexicon and the need of its circulation.

The Lexicon project continues with new activities. National groups validate national lexicons within the national professional community. All teams under the leadership of the Australian team compare national lexicons. An international lexicon of terms for the activities of the mathematics classroom, together with an illustrative library of video material is created.

The here presented survey shows that Czech Lexicon cannot be used exhaustively for a mathematics CLIL lesson due to its different patterns of communication and classroom practices. There are whole areas of doing things in a CLIL lesson that do not occur in an ordinary mathematics lesson and thus they are excluded from the Lexicon. The point of this research is not to conclude that the Czech Lexicon must be adapted, the point is to show which areas of a CLIL lesson need more attention both of the teacher and the pupils as they are not in line with general beliefs about what it means to do mathematics at school.

The goal of the paper is to trigger discussion in pedagogical community on Czech terminology used in didactics not only of mathematics but also of the CLIL community. Teachers who face the situation of teaching mathematics through a foreign language must be well aware of those practices that are different and require more attention. Also pupils must be expected to need extra time to get used to these new practices and patterns to develop a new classroom culture – the culture of a CLIL mathematics lesson. The results of the analysis do not only illuminate the specifics of teaching mathematics in CLIL, but are useful both for CLIL teachers and teacher trainers.

Acknowledgement

The paper was partly supported by a project DP140101361 *The Lexicon Project: Analysing pedagogical naming systems from different cultures to reconceptualise classroom practice and advance educational theory* supported by the Australian Research Council.

References

[1] CLARKE, D. Comparative Research in Mathematics Education: Boundary Crossing and Boundary Creation. In *PME 39 Proceedings, vol. 2,* Hobart, 2015, s. 169-176.

[2] CLARKE, D., MESITI, C., O'KEEFE, C. et al. Addressing the Challenge of Legitimate International Comparisons of Classroom Practice. *International Journal of Educational Research*, ročník 46, č. 5, 2007: s. 280-293, ISSN 0883-0355.

[3] HOFMANNOVÁ, M., NOVOTNÁ, J. Implementation of Content and Language Integrated Learning (CLIL) in Mathematics Classroom. In *PME 26 Proceedings, vol. 1,* Norwich, 2002, s. 346.

[4] MARSH, D., LANGÉ, G. (Eds.). *Using Languages to Learn and Learning to Use Languages.* University of Jyväskylä, 2000. Czech text: Část 6. HOFMANNOVÁ, M., NOVOTNÁ, J. Jak využít jazyky k učení a jak se naučit používat jazyky.

[5] MORAOVÁ, H.. Výuka matematiky metodou CLIL – jak sladit jazykové a matematické cíle výuky. In *Dva dny s didaktikou matematiky,* Praha, 2014, ISBN 978-80-7290-801-1, s. 103–106.

[6] NOVOTNÁ, Jarmila. *CLIL – Neverbální prostředky komunikace a různé formy reprezentace.* [online]. 2010 [cit. 20.8.2016]. <http://clanky.rvp.cz/clanek/o/z/9749/CLIL---NEVERBALNIPROSTREDKY-KOMUNIKACE-A-RUZNE-ORMYREPREZENTACE.html/>..

[7] NOVOTNÁ, J., HOFMANNOVÁ, M. Cizí jazyk jako nástroj při výuce matematiky. In *8. setkání učitelů matematiky všech typů a stupňů škol,* Praha, 2002, s. 225-230.

[8] NOVOTNÁ, J., MORAOVÁ, H., HOŠPESOVÁ, A., ŽLÁBKOVÁ, I., BUREŠ, J. How do we understand each other when we describe classroom activities – Lexicon. In *ERIE 2016 Proceedings,* Praha, 2016, s. 440-447.

[9] STRAUSS, A., CORBIN, J. *Basics of qualitative research: techniques and procedures for developing grounded theory.* SAGE Publications, 1998.

[10] STUCHLÍKOVÁ, I., JANÍK, T. et al. *Oborové didaktiky, vývoj – stav – perspektivy.* Brno, 2015, ISBN 978-80-210-7769-0.

[11] ŠTEFLÍČKOVÁ, A. *Diagnostika obtíží žáků při výuce CLIL.* Praha, 2012.

[12] TEJKALOVÁ, . *Postavte žákům lešení aneb jak na scaffolding v hodinách CLIL a nejen tam* [online]. 2010 [cit. 20.8.2016]. <http://clanky.rvp.cz/clanek/o/z/9797/POSTAVTE-ZAKUM-LESENI-ANEBJAK-NA-SCAFFOLDING-V-HODINACH-CLIL-A-NEJEN-TAM.html/>.

Current address

Moraová Hana, PhDr.
Univerzita Karlova, Pedagogická fakulta
M. Rettigové 4, 116 39 Praha 1, Czech Republic
E-mail: hana.moraova@pedf.cuni.cz

Novotná Jarmila, Prof. RNDr. CSc.
Univerzita Karlova, Pedagogická fakulta
M. Rettigové 4, 116 39 Praha 1, Czech Republic
E-mail: jarmila.novotna@pedf.cuni.cz

Slovak University of Technology in Bratislava
Faculty of Mechanical Engineering
16th Conference on Applied Mathematics **APLIMAT 2017**
Proceedings

STATISTICAL TOOLS OF EXCEL

MOŠNA František (CZ)

Abstract. Use of computer technologies has become an integral part of statistics education at all levels of schools. This paper shows how the widespread program MS Excel can be used for teaching/learning of statistics. Statistical calculations in Excel are enabled through following three main options – statistical functions, add-in Analysis ToolPak and graphical tools. In one example relating to correlation it is shown that Excel can also be used for calculating of multiple and partial correlation and regression. Excel remains a good tool not only for scientific work but also for teaching statistics in schools and universities. It can substitute statistical programs even in relatively advanced topics.

Keywords: statistics teaching, Excel, correlation, regression

Mathematics Subject Classification: Primary 97K70, Secondary 97U70

1 Introduction

e-Learning can be understood in the narrow sense of the word as a use of internet technologies for gaining knowledge [3] or in the broader sense as educational process using information and communication technologies working with data in electronic form.

The use of computer technologies has become an integral part of statistics education at all levels of schools [4]. While the calculators are mainly used at primary schools, students of secondary school are often working with widespread program MS Excel. The use of this program continues at universities or colleges and in any research or other scholarly activities. Besides Excel, there is a plenty of other useful special statistical programs like

- SPSS – the most widespread program of the company IBM, its advantage is easy handling and clarity of outputs,
- STATISTICA – competitive program from Statsoft, providing plenty of handy and useful tools with spreadsheet and graphical outputs [5],
- R – the most widely used freeware program for professionals, it is necessary to study manual and know something for using it [2],
- the package of statistical tools in MATHEMATICA – from Wolfram,

- other useful programs like Statgraphic, SAS, NCSS etc.

However, Excel has a privileged position among them because of its broad and general expansion.

2 Access to statistical tools in Excel

Roughly speaking, in Excel there are following three types of tools for statistical calculations and presentations.

a) **Statistical functions:** Excel contains a lot of statistical functions that can be used in formulas or tables. Statistical functions are being developed. For instance, since version 2010 Excel has implemented so called functions "with dots" – such as T.DIST or T.INV, giving the values of the distribution function of Student's distribution and quantiles of Student's distribution, which is actually the inverse function of the distribution function. Finally, these functions correspond to their abbreviations. Moreover, the old functions "without dots" such as TDIST or TINV are preserved, too. The function TINV and similar ones give somewhat different values, so called "critical values". Their origin is historical and their definitions are a little confusing. They are fortunately replaced by quantiles corresponding to T.INV etc. in new textbooks. We have e.g. T.INV(p;n)=TINV(2*(1-p);n).

b) **Graphical tools:** Tools associated with graphs are very comfortable, natural and effective, especially for descriptive statistics or regression models.

c) **Add-in programs Analysis ToolPak:** In Excel we can take advantage of add-in programs for statistical calculations. They are part of basic installation but for their use it is necessary to carry out their activation. Analysis ToolPak is very useful for the calculations. On the other hand, some errors or problems are registered. Translations into national languages are on very low level. For instance, in procedure *Descriptive statistics* (in Czech version since 2007) some error message occurs on ninth row instead of „Range", which describes the difference between maximum and minimum value (in version 2003 it was correct). Another question, it is not entirely clear why Fisher's two-sample F-test does not have two-tailed version (contrast to the Student's t-test).

As we noted, Excel can be used at the secondary school level mainly for working out of characteristics, charting, and working with models concerning descriptive statistics. Offers of Excel are more fully used at university level or in research [6]. Let us look at one example of the use of advanced techniques in Excel. We will deal with multiple and partial correlation and multiple linear regression.

3 Correlation and regression

Correlation ranges from -1 to 1, and it represents a certain kind of characteristic of dependence between random variables. When this rate is close to 1, it means that the variables are directly proportional. Conversely, when the correlation approaches -1, the dependence between these variables is indirectly proportional. Correlation around 0 indicates independence of variables. We use correlation coefficient for independence testing. Excel enables to perform the classical methods using so called Pearson correlation coefficient denoted by r. But Excel does not include so called non-parametric methods, e.g. Spearman coefficient r_S or Kendal coefficient τ. The properties of

Excel tables can be used for independence tests of categorical variables in contingency tables (cross tabs), too.

Classical Pearson correlation coefficient is accessible

- through the functions CORREL or
- in Analysis ToolPak parts *Correlation* and *t-test: Paired Two Sample of Mean.*

Correlation coefficient is closely related to linear regression $y=a+bx$. We are usually interested in estimates for the regression coefficients a, b and also so called coefficient of determination R^2. It expresses the ratio (in percents) which is explained by the investigated model and the rest remains to randomness. It holds $R^2=r^2$.

Estimates of regression coefficients and coefficient of determination are accessible

- through the functions INTERCEPT for a, SLOPE for b etc.
- in Analysis ToolPak part *Regression* or
- via graphs.

But if we need to study, for instance, multiple or partial correlation, there are no suitable special statistical functions, graphical tools or add-in procedures in Excel. This can be considered as a notable disadvantage of this program compared with other ones.

Nevertheless, Excel can be used for the calculation of multiple correlation and multiple linear regression, too. Only some formulas are necessary [1] and a bit of invention.

For simplicity, we will introduce one example of the multiple correlation and estimates for regression. It expresses the dependence of the random variable Z on the variables X and Y and linear regression $z=a+b_1x+b_2y$.

4 Example – multiple and partial correlation

We have average monthly rainfall or precipitation (in mm per month), mould infestation (in percent) and hectare potato yields (in tonnes per hectar) at 10 experimental fields.

Precipitation	36	40	30	25	41	23	35	34	44	42
mould infestation	40	35	35	25	45	20	30	35	45	40
Yields	160	200	140	130	170	150	150	200	230	195

a) We are interested in how yields depend on the average monthly rainfall and relative mould infestation.
b) Furthermore, we are interested in how yields depend on mould infestation without affecting precipitation.

Assume that the data are realization of random variables $X_1, \ldots, X_{10}$, $Y_1, \ldots, Y_{10}$ and $Z_1, \ldots, Z_{10}$ with normal distribution. Data are written in table (Fig. 1) together with their squares, products and sums in the last row.

	A	B	C	D	E	F	G	H
1	X	Y	Z	Xsquare	Ysquare	XY	XZ	YZ
2	36	40	160	1296	1600	1440	5760	6400
3	40	35	200	1600	1225	1400	8000	7000
4	30	35	140	900	1225	1050	4200	4900
5	25	25	130	625	625	625	3250	3250
6	41	45	170	1681	2025	1845	6970	7650
7	23	20	150	529	400	460	3450	3000
8	35	30	150	1225	900	1050	5250	4500
9	34	35	200	1156	1225	1190	6800	7000
10	44	45	230	1936	2025	1980	10120	10350
11	42	42	195	1764	1764	1764	8190	8190
12	350	352	1725	12712	13014	12804	61990	62240

Fig. 1. Data

As we know, the dependency measure is the correlation coefficient. At first we calculate Pearson correlation coefficients r_{XY}, r_{XZ} and r_{YZ} using Analysis ToolPak *Correlation* (see Fig. 2).

	X	Y	Z		
X	1			**rZ,(X,Y) =**	**0,79377**
Y	0,90172	1			
Z	0,77653	0,62906	1	**rY,Z.X =**	**-0,26117**

Fig. 2. Correlation coefficients

Multiple correlation coefficient can be calculated by formula

$$r^2_{Z,(X,Y)} = \frac{r^2_{XZ} + r^2_{YZ} - 2r_{XY}r_{XZ}r_{YZ}}{1 - r^2_{XY}} , \tag{1}$$

so $r^2_{Z,(X,Y)} = 0.630072$ and $r_{Z,(X,Y)} = 0.793771$, (see Fig. 2).

Also partial correlation is a quite interesting concept. It expresses the dependence of variable Z on Y without effect of X and we can calculate it by formula

$$r_{Y,Z.X} = \frac{r_{YZ} - r_{XY}r_{XZ}}{\sqrt{1 - r^2_{XY}}\sqrt{1 - r^2_{XZ}}} . \tag{2}$$

In our example, the ordinary correlation coefficient $r_{YZ} = 0.629064$ between Y and Z is positive which indicates the positive influence. It is rather peculiar, because mould infestation can hardly improve the yield of potatoes. This dependency is only apparent and is a consequence of the fact that both the yield and mould infestation are positively influenced by falls. The correct information about this relationship is given just by partial correlation coefficient $r_{Y,Z.X} = -0.26117$, (see Fig. 2). It shows us that the influence of mould infestation for yields is negative in real, which is logical.

To obtain an estimate of the regression coefficients a, b_1 and b_2 we use system of linear equations corresponding to the matrix $(A|b)$

$$\begin{pmatrix} n & \Sigma x_k & \Sigma y_k \\ \Sigma x_k & \Sigma x_k^2 & \Sigma x_k y_k \\ \Sigma y_k & \Sigma x_k y_k & \Sigma y_k^2 \end{pmatrix} \begin{pmatrix} a \\ b_1 \\ b_2 \end{pmatrix} = \begin{pmatrix} \Sigma z_k \\ \Sigma x_k z_k \\ \Sigma y_k z_k \end{pmatrix}, \qquad (3)$$

where necessary elements are calculated in table on Fig. 1.

Because Excel enables calculations with matrices, we can obtain inverse matrix A^{-1} (by function INVERSE) and product $A^{-1}\ b$ (by function MATRIXPRODUCT) which gives the required regression coefficients.

	10	350	352	1725		z = 47,988 + 5,041 x - 1,475 y	
(A|b) =	350	12712	12804	61990			
	352	12804	13014	62240		R2 =	0,63007
	2,76997	-0,08894	0,01258		a		47,9883
A-1 =	-0,08894	0,01158	-0,00899		b1	A-1 . B =	5,04082
	0,01258	-0,00899	0,00858		b2		-1,47491

Fig. 3. Multilinear regression

We have $a = 47.988$, $b_1 = 5.041$, $b_2 = -1.475$ and the coefficient of determination R^2=0.63007 (see Fig. 3).

5 Conclusion

Excel is a very effective tool for descriptive statistics as well as for hypothesis testing which provides the basis for important decisions. As we can see from the above example, Excel can help us even in quite difficult situations. Excel is not primarily statistical program. There exists a lot of mistakes, problem etc. Nevertheless Excel remains a good tool for scientific work but also for teaching statistics in schools and universities.

References

[1] BAILEY, N.T.J. *Statistical Methods in Biology*, Cambridge: University Press, 1995.

[2] DALGAARD, P. *Introductory Statistics with R*, Springer, 2008.

[3] HORTON, W., HORTON, K. *e-Learning Tools and Technologies*, Wiley, 2003.

[4] JANČAŘÍK, A., NOVOTNÁ, J. From a textbook to an e-learning course (e-Learning or e-book?). In: *Proceedings of 10th International conference "Models in Developing Mathematics Education"*. Dresden, 2009, p. 272-286.

[5] LEWICKI, P., HILL, T. *Statistics: Methods and Application*, Statsoft, 2007.

[6] REMENYI, D., ONOFREY, G., ENGLISH, J. *An introduction to statistics using Microsoft Excel*, Academic Publishing, 2009.

Current address

Mošna František, RNDr. Ph.D.
Czech University of Life Sciences
Kamýcká 129, 16521 Praha 6, Czech Republic
Tel. number: +0420 224 383 276, e-mail: mosna@tf.czu.cz

Slovak University of Technology in Bratislava
Faculty of Mechanical Engineering
16th Conference on Applied Mathematics APLIMAT 2017
Proceedings

SEISMIC RESPONSE OF ELEVATED LIQUID STORAGE TANKS

MUSIL Miloš (SK), SIVÝ Martin (SK), CHLEBO Ondrej (SK), PROKOP Ondrej (SK)

Abstract. The paper deals with the seismic analysis of elevated liquid storage tanks with the aim to compute the response of the structure to earthquake loading. In the seismic analysis, the unfavourable mode of oscillation is computed by the analytical and numerical approach. The computation of the response to the seismic event is performed by spectrum analysis with El Centro earthquake response spectrum plot and results are compared by time-history analysis using El Centro accelerogram. All analyses are carried out by FE method in software ANSYS.

Keywords: elevated storage tank, seismic excitation, spring-mass model, Eurocode 8

Mathematics Subject Classification: Primary 74H45, 70J10

1 Introduction

Large capacity tanks and vessels are mostly used as a storage for liquids (non-toxic, non-flammable liquids to highly flammable, toxic chemicals with explosive nature) or bulk materials. Tanks can be found almost in each industry sector, e.g., nuclear, chemical, food, etc. In addition to storage function, vessels are essential components in transmission and processing systems (e.g., heat exchangers, evaporators, drying devices, cooling system, etc.) [1].

In today's global industry development and improving the efficiency of production, emphasis must be more and more laid to increasing the safety of facilities using new advanced technical resources that will ensure effective prevention of the large industrial accidents. In addition to standard requirements for devices design, there are some qualitative methods of hazard identification (HAZOP, DOW index) to prevent negative consequences in ordinary operation [2].

One of the most installed types of the tank of various sizes and shapes is elevated liquid storage tank. It is often used in industrial facilities, nuclear power plants and military bases. This type of tanks consists of a supporting structure and a vessel. The supporting structure can be assumed as a braced frame, a multi-column assembly or an axisymmetric pedestal and the vessel can be of various shapes (e.g., cylindrical, conical, spherical, etc.).

During operation tanks can be subjected to different static and dynamic loading. One of the most critical external dynamic events affecting safety and operation efficiency of the tank-liquid system is the seismic excitation. The vulnerability of the structures to ground motions has been demonstrated by almost every major past and recent earthquake (e.g. El Centro, Fukushima). In regions prone to strong ground shaking, negative consequences may lead to damage, ruptures of storage tanks or failures of tank accessories. If the industrial facilities store flammable or toxic liquids, disastrous effects, such as uncontrolled fire, explosion or toxic dispersion can arise.

Elevated tanks are vulnerable to dynamic loading due to the relatively small resistance of the supporting structure. The most specific destructions of this type of tank-liquid system are e.g., disruptions of slender brace due to their inability to resist against the compression, disruptions of the turnbuckles used in the braces, disruptions of the vertical brittle brace, a plastic joint formation in one of the beams at the highest level of the structure and the disruption in the welded joint of piping due to the lack of free movement for the largely displacing tank [3].

Due to negative effects which can be developed in storage tanks, it was an effort to describe tank response during a seismic event to prevent pollution, radiation or loss of human lives. Based on analytical results of G. W. Housner [4] and A. S. Veletsos [5] and from experimental research, different provisions for storage tanks have been developed and some of them were adopted in standards dedicated to seismic resistance (e.g., API, Eurocode 8, etc.) [6].

2 Simplified procedure

When an elevated storage tank, which is not completely filled with a liquid, is subjected to a horizontal earthquake excitation, staging, tank walls and liquid are subjected to a horizontal acceleration. In the model, the liquid may be substituted by two masses – impulsive and convective. The impulsive part represents effects of the lower portion of the liquid which moves in unison with the structure. It is represented by an impulsive mass m_i rigidly connected to the tank walls, located at a height h_i or h_i' above the tank bottom. The upper liquid region with the free surface correspondents to the convective part which causes the sloshing effects. This portion is described by a mass m_c, connected to the walls through a spring of stiffness k_c, located at the height h_c or h_c'.

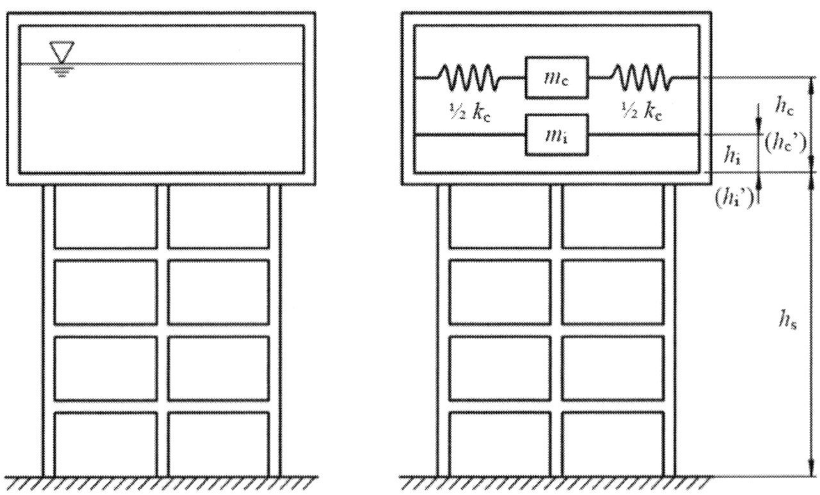

Fig. 1. Elevated storage tank and simplified model

Parameters introduced in Fig. 1 are used in calculations of basic seismic characteristics, such as base shear, overturning moment and hydrodynamic pressure [7, 8]. Procedure for parameters calculation of cylindrical storage tanks can be found in [9]. The dependency of these parameters to slenderness ratio γ is shown in Fig. 2.

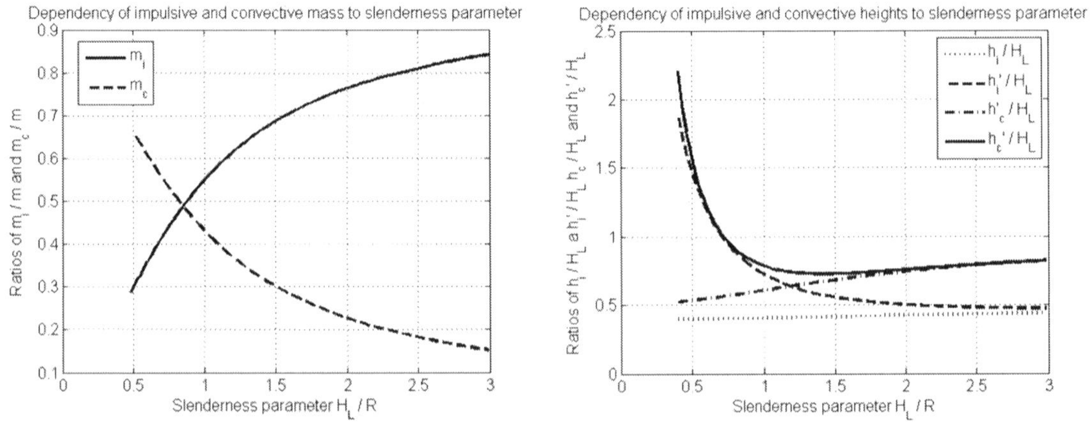

Fig. 2. Dependency of equivalent model's parameters to slenderness parameter

If a closed elevated tank is empty or full of liquid, it is essential to assume this structure as a one-mass structure. However, elevated tanks are almost never filled with the liquid but have a free surface; hence, they can be idealized by a two-mass model (Fig. 3a).

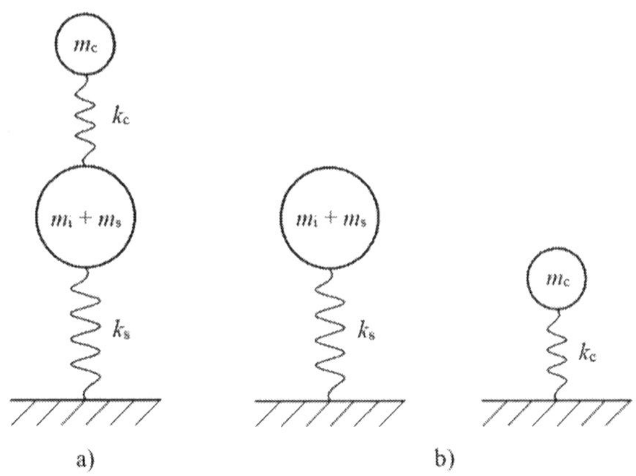

Fig. 3. Two-mass idealization of elevated liquid storage tank

Two-mass idealization was proposed by G. W. Housner [4] and is used in the most of the international codes [9, 10]. The model has two degrees of freedom corresponding to masses m_i and m_c (Fig. 3a). Structural mass m_s should be added to impulsive mass. It includes the mass of the tank and the portion of the mass of staging. The sum of impulsive and structural mass should be connected to the ground by a spring representing the staging stiffness k_s (it acts like a lateral spring).

The response of this simplified model can be obtained by elementary structural dynamics. Due to large differences between natural frequencies, the system can be treated as two uncoupled SDOF systems – one representing the impulsive and structural mass behaving as an inverted pendulum with lateral stiffness k_s and the other representing the convective mass with a spring of stiffness k_c (Fig. 3b).

3 Seismic response of elevated liquid storage tanks

For seismic analysis of structures and equipment following methods are acceptable [11]:

- Response spectrum method (spectrum analysis)
- Method of direct time integration of the motion equations system (time-history analysis)
- Equivalent static method (static analysis)

Following analyses for determination of seismic response are carried out on the model of the flexible (steel) elevated storage tank with geometric parameters shown in Fig. 4a. The tank contains liquid (water) with the height of free surface H_L from floor slab. Tank base is fixed to the rigid foundation.

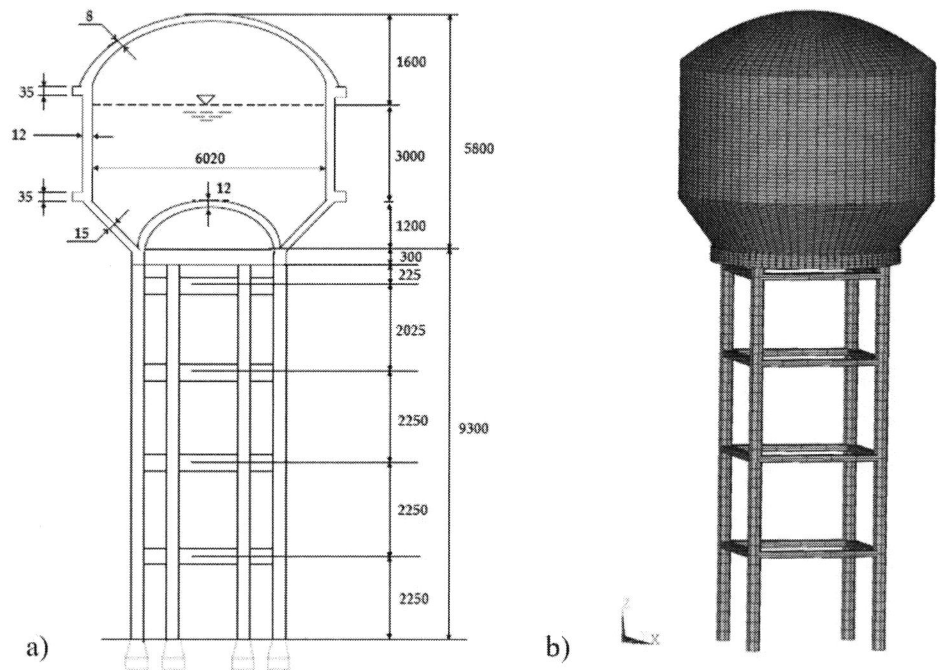

Fig. 4. Drawing and FE model of elevated liquid storage tank

The seismic analysis of elevated storage tank containing liquid is a problem involving fluid-structure interaction. This interaction can be investigated by using different (classical and modern) approaches, such as mentioned the added mass concept, Lagrangian or Eulerian methods, etc. These analyses can be performed by the analytical methods or by finite element method. Software like ANSYS enables to model the liquid using elements based on Lagrangian or Eulerian formulation. Due to longer computational time in time-history analysis, the liquid in FE model (Fig. 4b) is modeled assuming the added mass concept.

The aim of the paper is to determine the seismic response to earthquake event in the form of displacements of the structure with liquid. Therefore, the liquid does not have to consist of impulsive and convective mass (sloshing is not investigated) and can be modeled only as one mass with the total mass equal to the liquid mass in the model. It is not reasonable to attach the mass to a single point on the shell wall since this practice will result to misrepresentation of the mode of oscillation and response as well. A simple solution is to add the liquid mass uniformly to the shell wall to the height of the liquid free surface. Then the investigated FE model of elevated tank consists of beam elements (four column staging), shell elements (vessel) and mass elements (liquid).

For elevated tanks, staging components should be designed for the critical direction of the seismic excitation. It follows different components of staging may have different critical directions [10]. Fig. 5 represents the critical direction of the seismic force for the typical four column frame type, in which a) correspondents to the critical direction for shear force in a brace and b) represents the critical direction for axial force in a column.

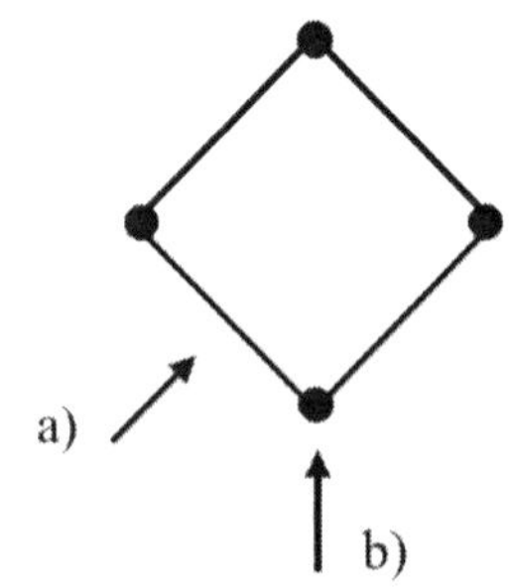

Fig. 5. Critical direction of the seismic excitation

As the seismic excitation, acceleration time-history of El Centro earthquake (California, 1940) with PGA 3.417 m/sec^2 at 2.14 sec and respective response spectrum are applied (Fig. 6).

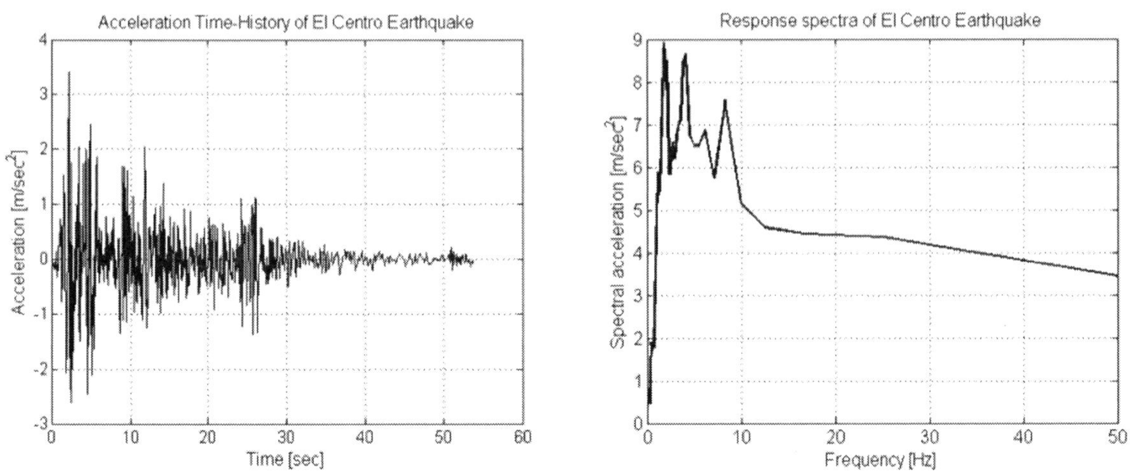

Fig. 6. Acceleration time-history and response spectrum of El Centro earthquake

3.1 Impulsive mode of oscillation

Modal analysis is the first analysis which should be carried out with the aim to determine dynamic properties of investigated tank-liquid system. Standards concerning with seismic resistance usually contain the procedure for analytical calculation of the impulsive natural frequency f_i when the unfavourable response of the liquid storage tank occurs. This response is described by liquid oscillation in unison with the oscillation of the tank. Using simplified procedure [10] this frequency can be calculated as follows

$$f_{\mathrm{i}} = \frac{1}{2\pi}\sqrt{\frac{k_{\mathrm{S}}}{m_{\mathrm{i}}+m_{\mathrm{S}}}} = \frac{1}{2\pi}\sqrt{\frac{g}{\Delta}} \tag{1}$$

where Δ is the deflection of the center of gravity of the tank when a lateral force of magnitude $(m_{\mathrm{i}} + m_{\mathrm{s}})g$ is applied at the center of gravity of tank.

Applying (1), impulsive natural frequency is 1.25 Hz. The first impulsive mode of oscillation obtained from modal analysis using ANSYS is presented in Fig. 7. Respective impulsive response occurs at 1.18 Hz.

Fig. 7. Impulsive mode of oscillation

3.2 Spectrum analysis

Response spectrum method (spectrum analysis) is the significant method for seismic analysis of the structures and equipment. It is mainly used instead of time-history analysis to compute the response of structures. Spectrum analysis can be employed if the seismically loaded structure has the character of linear, proportionally damped system, the seismic excitation does not correlate in each direction and the first resonant frequency is lower than 35 Hz.

The structure is usually modeled as MDOF system on the base of finite element method [12]. The equation of motion for this structure under earthquake ground motion $\ddot{u}_{\mathrm{g}}(t)$ can be written as

$$\mathbf{M}\ddot{\mathbf{u}}(t) + \mathbf{C}\dot{\mathbf{u}}(t) + \mathbf{K}\mathbf{u}(t) = -\mathbf{M}\mathbf{d}\ddot{u}_{\mathrm{g}}(t) \tag{2}$$

where $\mathbf{d}$ is a vector of excitation direction (direction cosines).

For computation of the response, results from modal analysis must be known which are obtained by solving following eigenvalue problem

$$\begin{aligned}(\mathbf{K} - \boldsymbol{\omega}^2\mathbf{M})\boldsymbol{\phi} &= 0 \\ \det(\mathbf{K} - \boldsymbol{\omega}_i^2\mathbf{M}) &= 0\end{aligned} \tag{3}$$

The solution of (3) gives n frequencies and respective modal vectors $\boldsymbol{\phi}_i$, where n is the number of DOFs of the investigated structure. The modal vectors have the following orthogonal properties

$$\boldsymbol{\phi}_i^{\mathrm{T}}\mathbf{M}\,\boldsymbol{\phi}_j = 0 \qquad \boldsymbol{\phi}_i^{\mathrm{T}}\,\mathbf{K}\,\boldsymbol{\phi}_j = 0 \qquad i \neq j \tag{4}$$

The modal vectors are often normalized such that

$$\boldsymbol{\phi}_i^{\mathrm{T}}\,\mathbf{M}\,\boldsymbol{\phi}_i = 1 \qquad \boldsymbol{\phi}_i^{\mathrm{T}}\,\mathbf{K}\,\boldsymbol{\phi}_i = \omega_i^2 \tag{5}$$

It is often assumed that the damping matrix $\mathbf{C}$ has orthogonality properties similar to those of $\mathbf{M}$ and $\mathbf{K}$ and that the damping ratio can be defined for each mode

$$\boldsymbol{\phi}_i^{\mathrm{T}}\,\mathbf{C}\,\boldsymbol{\phi}_j = 0 \;\text{ for }\; i \neq j; \quad \boldsymbol{\phi}_i^{\mathrm{T}}\,\mathbf{C}\,\boldsymbol{\phi}_i = 2\omega_i\xi_i \tag{6}$$

The response of the structure is represented in terms of a linear superposition of mode shapes

$$\mathbf{u}(t) = \textstyle\sum_{i=1}^{n}\boldsymbol{\phi}_i \mathrm{w}_i\,(t) \tag{7}$$

where $\mathrm{w}_i(t)$ are called normal coordinates and are functions of the time variable t.
Substitution of (7) in (2) and pre-multiplication by $\boldsymbol{\phi}^{\mathrm{T}}$

$$\boldsymbol{\phi}^{\mathrm{T}}\mathbf{M}\boldsymbol{\phi}\ddot{\mathbf{w}}(t) + \boldsymbol{\phi}^{\mathrm{T}}\mathbf{C}\boldsymbol{\phi}\dot{\mathbf{w}}(t) + \boldsymbol{\phi}^{\mathrm{T}}\mathbf{K}\boldsymbol{\phi}\mathbf{w}(t) = -\boldsymbol{\phi}^{\mathrm{T}}\mathbf{M}\mathbf{d}\ddot{u}_{\mathrm{g}}(t) \tag{8}$$

Application of the properties (4) to (6) gives

$$\ddot{\mathrm{w}}_i(t) + 2\omega_i\xi_i\dot{\mathrm{w}}_i(t) + \omega_i^2\mathrm{w}_i(t) = -\gamma_i\ddot{u}_{\mathrm{g}}(t) \tag{9}$$

in which γ_i is called the participation factor for mode i and is given by

$$\gamma_i = \boldsymbol{\phi}_i^{\mathrm{T}}\mathbf{M}\mathbf{d} \tag{10}$$

Maximum displacement vector in the ith mode can be written

$$\mathbf{u}_{i\,\max} = \frac{\gamma_i\, S_{\mathrm{A}i}}{\omega_i^2}\boldsymbol{\phi}_i \tag{11}$$

where $S_{\mathrm{A}i}$ represents spectral acceleration for ith frequency.

Given displacement vector (11) represents the maximum value of any response of interest. The overall response of the system is calculated by combining the maximum modal responses specified by one of the mode combination methods (SRSS, CQC, GRP, etc.). General rule for modal responses combination can be defined as

$$R = \pm\sqrt{\varepsilon_{i,j} R_i R_j} \tag{12}$$

where R_i, R_j are the ith and jth modal response respectively and $\varepsilon_{i,j}$ is a combination matrix whose shape and values are based on the chosen modal response combination.

For computation of seismic response of investigated elevated storage tank, single-point response spectrum analysis is performed. The tank-liquid system is excited with respective response spectrum in the x-direction (the critical direction for axial force in a column). The seismic acceleration response spectrum of the El Centro acceleration time-history with 5 % damping ratio (Fig. 6) is applied.

Based on results from modal analysis, GRP method is used for the combination of maximum modal responses. This combination rule is used due to closely spaced modes of oscillation. Figure 8 presents the overall response of the tank-liquid system to the El Centro response spectrum. The maximum response of the liquid-structure system is 7.6e-2 m.

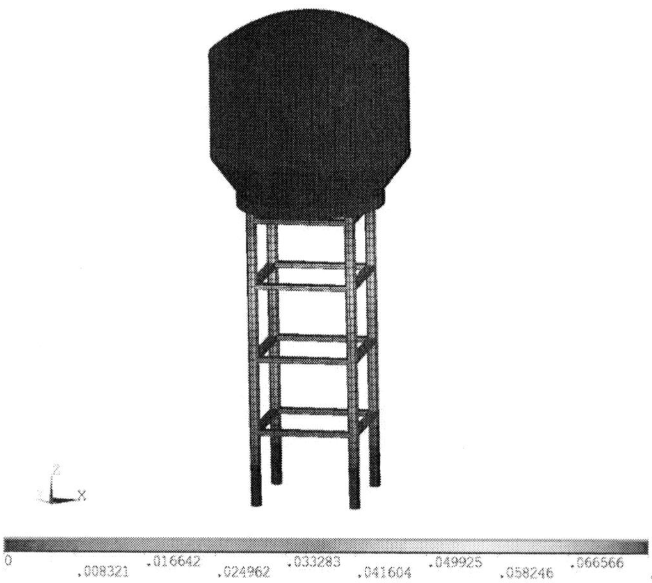

Fig. 8. Maximum seismic response of liquid storage tank to the correspondent response spectra

3.3 Time-History analysis

This analysis uses the method of integrating the equations of motion. Time-History analysis is a technique used to determine the dynamic response of a structure under the action of any time-dependent loads (solution in time domain). This method is universal and can be used for various structure (e.g. rigid body, linear proportionally damped systems). If the structure shows a significant nonlinear behaviour, it is non-proportionally damped, or excitation from different directions shows a significant correlation, it is necessary to use this method. For computation of time-dependent responses, various numerical integration methods are used. One of the methods, that is mostly used and is available in ANSYS, is Newmark's implicit method [13].

$$\mathbf{M\ddot{u}}(t) + \mathbf{C\dot{u}}(t) + \mathbf{Ku}(t) = \mathbf{F}(t) \tag{13}$$

The equation of motion (13) can be rewritten as

$$\mathbf{M\ddot{u}}_{t+\Delta t} + \mathbf{C\dot{u}}_{t+\Delta t} + \mathbf{Ku}_{t+\Delta t} = \mathbf{F}_{t+\Delta t} \tag{14}$$

where $\mathbf{\ddot{u}}_{t+\Delta t}$ is the nodal acceleration vector, $\mathbf{\dot{u}}_{t+\Delta t}$ the nodal velocity vector, $\mathbf{u}_{t+\Delta t}$ the nodal displacement vector and $\mathbf{F}_{t+\Delta t}$ the applied load. All vectors are assumed at time $t + \Delta t$.

In the equation of motion (14), for convenience, only force $\mathbf{F}(t)$ on the right side is considered. In the case of the seismic excitation, the force $\mathbf{F}(t)$ is replaced by $-\mathbf{M}\ddot{u}^{g}_{t+\Delta t}$.

Displacement and velocity are obtained as follows

$$\mathbf{u}_{t+\Delta t} = \mathbf{u}_t + \mathbf{\dot{u}}_t\Delta t + [(1/2 - \alpha)\mathbf{\ddot{u}}_t + \alpha\mathbf{\ddot{u}}_{t+\Delta t}]\Delta t^2 \tag{15}$$

$$\mathbf{\dot{u}}_{t+\Delta t} = \mathbf{\dot{u}}_t + [(1 - \delta)\mathbf{\ddot{u}}_t + \delta\mathbf{\ddot{u}}_{t+\Delta t}]\Delta t \tag{16}$$

where $\mathbf{\ddot{u}}_t$ represents the nodal acceleration vector, $\mathbf{\dot{u}}_t$ the nodal velocity vector, $\mathbf{u}_t$ the nodal displacement vector. All vectors are assumed at time t. α, δ represent Newmark's integration parameters which influence accuracy and stability of algorithm (average acceleration method, linear acceleration method). Introducing (15) and (16) to (14) the following equation is obtained

$$\begin{aligned}\left(\frac{1}{\alpha\Delta t^2}\mathbf{M} + \frac{\delta}{\alpha\Delta t}\mathbf{C} + \mathbf{K}\right)\mathbf{u}_{t+\Delta t} = \mathbf{F}_{t+\Delta t} + \mathbf{M}\left(\frac{1}{\alpha\Delta t^2}\mathbf{u}_t + \frac{1}{\alpha\Delta t}\mathbf{\dot{u}}_t + \left(\frac{1}{2\alpha} - 1\right)\mathbf{\ddot{u}}_t\right) + \\ + \mathbf{C}\left(\frac{\delta}{\alpha\Delta t}\mathbf{u}_t + \left(\frac{\delta}{\alpha} - 1\right)\mathbf{\dot{u}}_t + \frac{\Delta t}{2}\left(\frac{\delta}{\alpha} - 2\right)\mathbf{\ddot{u}}_t\right)\end{aligned} \tag{17}$$

First, the unknown $\mathbf{u}_{t+\Delta t}$ is computed using (17), then two unknowns are obtained using (18)

$$\begin{aligned}\mathbf{\dot{u}}_{t+\Delta t} &= \frac{\delta}{\alpha\Delta t}(\mathbf{u}_{t+\Delta t} - \mathbf{u}_t) - \left(\frac{\delta}{\alpha} - 1\right)\mathbf{\dot{u}}_t - \frac{\Delta t}{2}\left(\frac{\delta}{\alpha} - 2\right)\mathbf{\ddot{u}}_t \\ \mathbf{\ddot{u}}_{t+\Delta t} &= \frac{1}{\alpha\Delta t^2}(\mathbf{u}_{t+\Delta t} - \mathbf{u}_t) - \frac{1}{\alpha\Delta t}\mathbf{\dot{u}}_t - \left(\frac{1}{2\alpha} - 1\right)\mathbf{\ddot{u}}_t\end{aligned} \tag{18}$$

To achieve correct results, the choice of integration step is important and depends on the integration method. The general requirement for the integration step is to be at maximum 1/20 of the modal period which is significant for the response of a given component if the Newmark's method is used. If earthquake ground motion is applied, the integration step should not be shorter than the interval of the input acceleration data which is usually 0.01 or 0.005 sec [14].

Acceleration time-history of the seismic excitation (Fig. 6) is applied to the elevated tank-liquid system in the *x*-direction which is the critical direction for axial force in a column. Seismic responses are computed at every time step.

Figure 9 shows only the maximum response of investigated model to the seismic excitation and presents the time-varying displacements of the system in a node in which maximum response occurs. Maximum response with the peak value of 7.9e-2 m occurs at 6.28 sec.

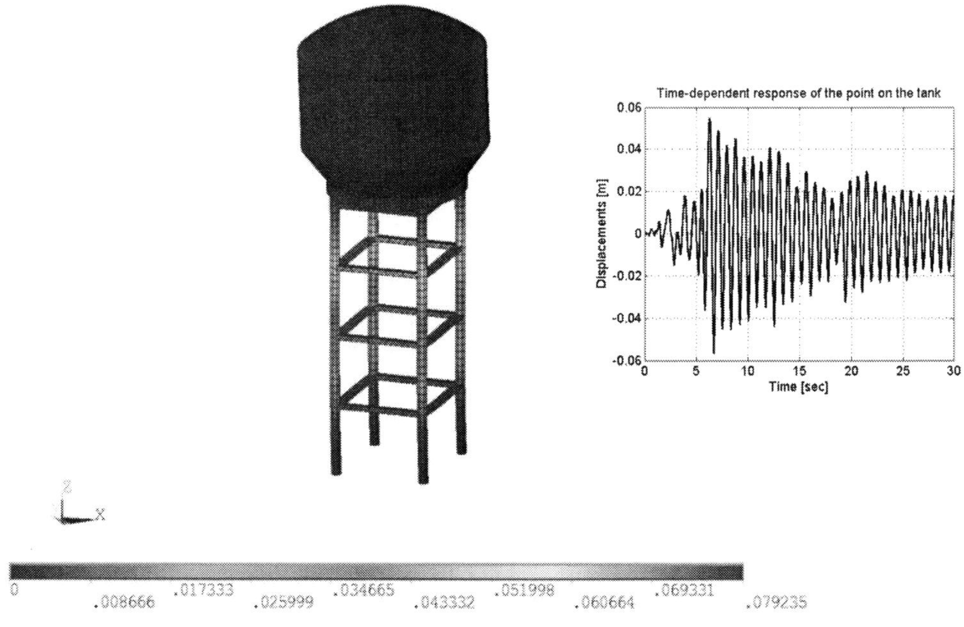

Fig. 9. Maximum seismic response of liquid storage tank to earthquake excitation El Centro

4 Conclusion

The paper was dedicated to performing a seismic analysis of a model of the elevated liquid storage tank with the aim to compute its response to given seismic excitation (El Centro). Seismic analysis can be assumed as one of the analyses which should be carried out to ensure satisfactory performance of tanks, especially in earthquake prone regions.

The basic concept for evaluating the equivalent simplified model of elevated tanks used in international standards was introduced. This procedure is valid for circular or rectangular vessels. For other tank shapes, e.g. intze, truncated conical shape, slenderness parameter can correspondent to that of the equivalent circular tank of the same volume and radius equal to the radius at top liquid level; mass, height and stiffness parameters of the equivalent circular tank shall be used.

The situation when liquid oscillates in unison with the tank represents an unfavorable response to the system. International standards provide some provisions for calculation of the respective impulsive frequency of that response. It can be noted, for shaft type staging, in addition to the flexural deformation, shear deformation may be included in the calculation of lateral stiffness of staging.

Seismic response to earthquake loading was computed by FE software ANSYS using spectrum and time-dependent analysis. When considering only overall response for linear behaved structure, method of response spectrum is advantageous due to a less computational effort. The benefit of time-history analysis is in its generality. The disadvantage is longer computational time compared to other methods and problems with the convergence of the solution. The results between each solution of dynamic analyses represented good conformity.

Acknowledgement

The paper was supported by the STU Grant scheme for Support of Young Researchers entitled "Design and Optimization of the Measuring Equipment for Vibration Tests."

References

[1] ÚRADNÍČEK, J., HORNIAKOVÁ, A., MUSIL, M., KENÍŽ, J. An Applied Research of Seismic and Vibration Resistance of the Equipment with Assessment of Seismic and Vibration Effects. In *EuroNoise 2015: Proceedings of the 10th European Congress and Exposition on Noise Control Engineering*, Maastricht (Netherlands), 2015.

[2] PECIAR, P., FEKETE, R., PECIAR, M. *Process Engineering II*. Bratislava: Vydavateľstvo STU, 2016, ISBN 978-80-227-4540-6.

[3] ADIBI, M., ESHGHI, S. Seismic Performance of Eight Elevated Water Tanks During Silakhor, Iran Earthquake of 31 March 2006. In *The 14th World Conference on Earthquake Engineering*, Beijing (China), 2008.

[4] HOUSNER, G. W. The Dynamic Behavior of Water Tanks. *Bulletin of the Seismological Society of America*, Vol. 53, No. 2, 1963, pp. 381-387.

[5] VELETSOS, A. S. *Seismic Response and Design of Liquid Storage Tanks. Guidelines for the Seismic Design of Oil and Gas Pipeline Systems*. Technical Council on Lifeline Earthquake Engineering (ASCE), 1984.

[6] SIVÝ, M., MUSIL, M. Procedure for Seismic Analysis of Liquid Storage Tanks using FEM Approach and Analytical Models. In *Advances in Mechanism Design II: Proceedings of the 12th International conference on the Theory of Machines and Mechanisms*, Liberec (Czech Republic), 2017, pp. 213-219, ISBN 978-3-319-44086-6.

[7] MALHOTRA, P. K., WENK, T., WIELAND, M. Simple Procedure for Seismic Analysis of Liquid Storage Tanks. *Structural Engineering*, Vol. 10, No. 3, 2010, ISSN 1016-8664.

[8] SIVÝ, M. Seismic Resistance of Storage Tanks Containing Liquid in accordance with Principles of Eurocode 8 Standard. *Journal of Mechanical Engineering - Strojnícky časopis*, Vol. 66. No. 2, 2016, ISSN 0039-2472.

[9] EUROCODE 8 – *Design of Structures for Earthquakes Resistance: Part 4 - Silos, tanks and pipelines*. European Committee for Standardization, Brussels, 1998.

[10] IITK-GSDMA. *Guidelines for Seismic Design of Liquid Storage Tanks. Provisions with Commentary and Explanatory Examples*. Indian Institute of Technology, Kanpur, 2007.

[11] PNM34080183 – *Requirements on Evaluation of Seismic Resistance of Structures, Systems and Components of Mochovce NPP*, 2009.

[12] GUPTA, A. K. *Response Spectrum Method in Seismic Analysis and Design of Structures*. Boston: Blackwell Scientific Publications, 1990, ISBN 08-654-2115-3.

[13] ANSYS, Inc. *ANSYS Mechanical APDL Theory Reference*. Canonsburg (USA), 2016.

[14] FARDIS, M. N., CARVALHO, E. C., FAJFAR, P., PECKER, A. *Seismic Design of Concrete Buildings to Eurocode 8*. Boca Raton, FL: CRC Press, Taylor & Francis Group, 2015, ISBN 978-1-466-55974-5.

[15] RICHTÁRIKOVÁ, D. Enhancing Ability to Identify and Use Mathematical Concepts. In *Proceedings of the 18th SEFI Mathematics Working Group Seminar*, European Society for Engineering Education, Brussels (Belgium), 2016, pp. 208-213, ISBN 978-2-87352-013-7.

Current address

Musil Miloš, prof. Ing. CSc.
Institute of Applied Mechanics and Mechatronics, Faculty of Mechanical Engineering, Slovak University of Technology
Nám. slobody 17, 812 31 Bratislava, Slovak Republic
Tel. number: +421 (0)2 57296389, e-mail: milos.musil@stuba.sk

Sivý Martin, Ing.
Institute of Applied Mechanics and Mechatronics, Faculty of Mechanical Engineering, Slovak University of Technology
Nám. slobody 17, 812 31 Bratislava, Slovak Republic
Tel. number: +421 (0)2 57296479, e-mail: martin.sivy@stuba.sk

Chlebo Ondrej, Ing. PhD.
Institute of Applied Mechanics and Mechatronics, Faculty of Mechanical Engineering, Slovak University of Technology
Nám. slobody 17, 812 31 Bratislava, Slovak Republic
Tel. number: +421 (0)2 57296376, e-mail: ondrej.chlebo@stuba.sk

Prokop Ondrej, Ing.
Institute of Applied Mechanics and Mechatronics, Faculty of Mechanical Engineering, Slovak University of Technology
Nám. slobody 17, 812 31 Bratislava, Slovak Republic
Tel. number: +421 (0)2 57296479, e-mail: prokop.ondrej6@gmail.com

Slovak University of Technology in Bratislava
Faculty of Mechanical Engineering
16th Conference on Applied Mathematics **APLIMAT 2017**
Proceedings

ANALYSIS OF VIDEO FILES USING INFORMATION ENTROPY

NÁHLÍK Tomáš (CZ)

Abstract. The real-time experiments, especially biological experiments, provides us huge amount of data. They can be of different type; some of them are images, or video-files. Problem with biological experiment is that it cannot be repeated under exactly the same condition. Once the experiment is done we need to mine all possible data. But it can happen that the quality of data is poor or is changed during the experiments. Then it is necessary to enhance or improve obtained data.

Keywords: Image Processing, Video Processing, Data Mining, Information Entropy

Mathematics Subject Classification: Primary 94A17, 94A08; Secondary 68U10

1 Introduction

Due the amount of data provided by biological experiments the storing of them is difficult. Usually it is the compromise of needed data and the capacity of the storage medium. Sometime the users are forced to use some type of compression. In case of image the files are converts to one video file. Or instead of taken images the video camera is used. The video file is for manual (human) analysis better than the series of images, because it can be viewed in a fraction of time. But for some automatic analysis it is problematic. The video files have usually lower resolution and use some type of compression.

It can happen that during the experiment the optical condition are changed. The intensity of light can change due day-night cycle, or the sample can move little bit and will be defocused. In this case the techniques of image processing should be used.

One of these techniques is based on measuring of the information content of the image. This method is called information entropy calculation and it is based on Shannon formula.

$$H = -\sum_i p_i \log_2(p_i) \tag{1.1}$$

This Shannon's entropy is the limit case of Rényi entropy.

$$H_\alpha(X) = \frac{1}{1-\alpha}\log(\sum_{i-1}^{n} p_i^\alpha) \quad (1.2)$$

2 Materials and Methods

Formulas for entropy calculations must be changed to support the images (histograms of images) as an input. Instead of calculation of probability of some event we need to calculate the information content of the image.

Principle of calculation of information carried by the image point (PIG – Point Information Gain) was described in a few articles [1, 2, 3]. It can be calculated as simple difference of the information calculated from the whole image and from the image without that particular pixel.

$$PIG_{\alpha,x,y} = \frac{1}{1-\alpha}\ln\left(\sum_{i-1}^{n} p_{i,x,y}^\alpha\right) - \frac{1}{1-\alpha}\ln(\sum_{i-1}^{n} p_i^\alpha) \quad 2.1)$$

Where α is Rényi parameter, $p_{i,x,y}$ and p_i are probabilities of occurrence of given intensity in the image without and with the examined point.

It is also possible to calculate the information content or value of the whole image (PIE – Point Information gain Entropy).

$$PIE_\alpha = \sum_{j=0}^{m}\sum_{i=0}^{n} PIG_{\alpha,x,y} \quad 2.2)$$

Where m is number of points in x direction and n is number of points in y direction and $PIG_{\alpha,x,y}$ is calculated according the equation mentioned above.

There is also option to calculate the changes of information between two subsequent images in the series. This is called PDG – Point Divergence Gain.

$$PDG_{\alpha,x,y} = \frac{1}{1-\alpha}\log_2\frac{\sum_{i=1}^{n} p_i^\alpha}{\sum_{i=1}^{n} p_{i,x(l+1),y(l+1)}^\alpha} \quad 2.3)$$

where l is index of the current image in the series. So this means the PDG value is calculated such that you calculate information of the image and then the pixels from this image are one by one change for the pixels of the next image and the information content is recalculated.

All these values can be calculated in two different ways – as semantic or as syntactic information. In case of semantic information we take the histogram of the whole image, in the case of syntactic information we take the histogram just from the neighborhood (for example all pixels from the same line and same column where the examined point is – we called it cross calculation). The cross methods also helps to eliminate effect such as uneven light.

There is software called Entropy Calculator which can provide calculation of information content of the images based on different settings. It is designed for calculation point information gain (PIG) and derived quantities [3] of single image or series of images. On Fig.1 it is shown that it has many different settings. It can works with different types of images – colored (RGB) images, greyscale or black and white images or RAW images. Images can also have different bit depth – 1 bit (Black-

White images), 8 bits, 12 bits or 16 bits. When raw images are used proper Bayer mask of the camera can easily selected. It is possible also to crop the images or to omit intensities with exceptionally low occupancy (designed for removing of defect camera pixels). In the settings of Entropy calculation there are also different parameters to set up. PDG or PIG can be selected. For calculation of syntactic entropy different neighborhood can be selected – cross, rectangle, elliptic.

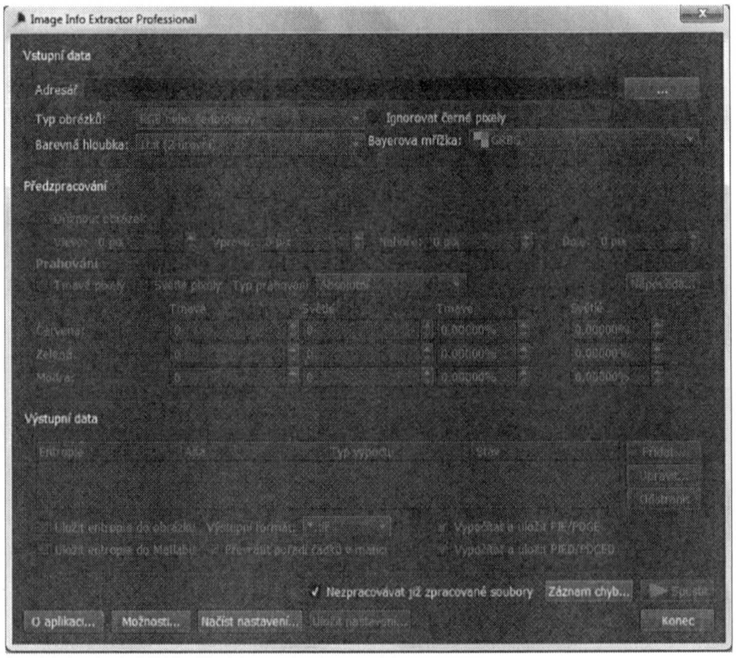

Fig. 1. GUI of Entropy Calculator

Video which was used for testing was video of *Paratrypanosoma confusum*. In this video sequence was necessary to analyze the development and movement of the flagellum. But unfortunately during the video sequence the flagellum oscillates in the *Z* axis. Standard methods for image improvements do not help because of the optical properties of flagellum.

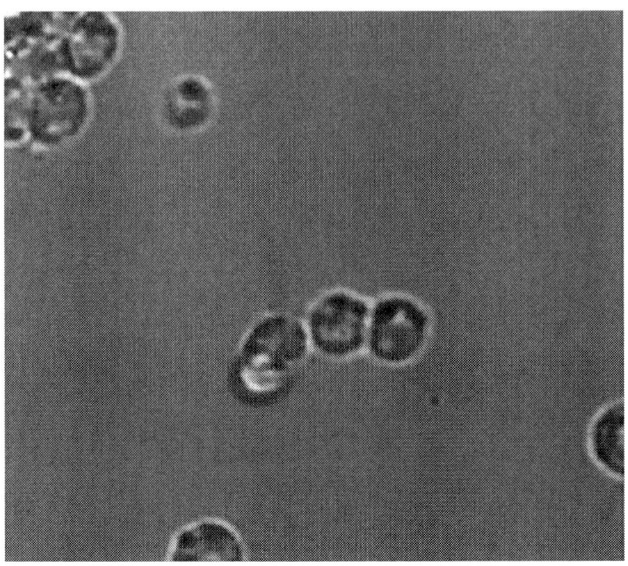

Fig. 2. Original image of *Paratrypanosoma confusum*

3 Results and Discussion

Images in computers are stored in form of matrixes full of numbers. And video files are only sequences of images. It is possible to use different mathematical tools for analysis, enhancement and processing of the images and video files. The analysis of video files is almost the same as the analysis of separated images. The only thing is that in video files some type of compression is usually used. This means that analysis of video files needs some more processing steps.

On the Fig. 3 there are examples of different processing of the original image (Fig.2) using information entropy calculation. Images A and C are visualization of syntactic information with different value of parameter α (1.5 resp. 0.9999) and B and D show semantic information. α parameter is 1.5 resp. 0.9999.

You can see that different method of calculation enhance different part of the image (Fig. 3). Selection of proper method depends on the expecting results. It can be shown that parameter α higher than 1 enhance more probable values (in our case the background) and values lower than 1 works with more rare point (intensities – in our case cells). If the Rényi parameter is 1 or almost one it handles all pixels (intensity values) equally.

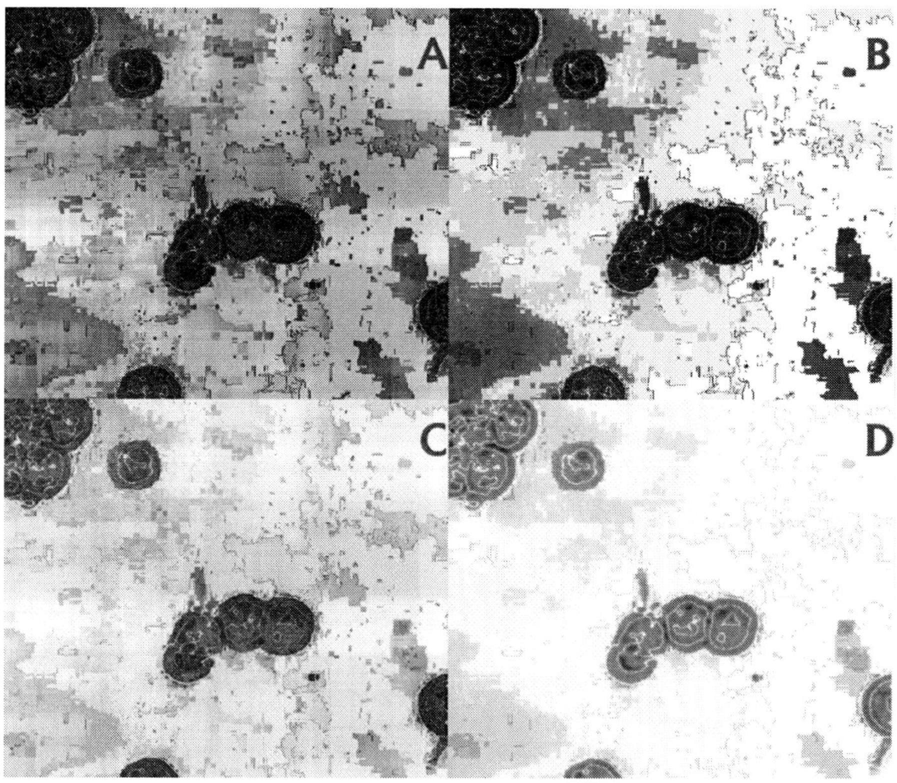

Fig. 3. Entropy images of *Paratrypanosoma confusum*
A – Cross method, α=1.5
B – Whole image method, α=1.5
C - Cross method, α=0.9999
D – Whole image method, α=0.9999

Our analysis of the video of the *P. confusum* shows that for this experiment is better to use α close to 1 because the optical properties of the flagellum are somewhere between the properties of background and the properties of the cells.

Also this analysis enhanced the cell wall and even the inner parts of the cells. The mathematical operations help biologists to understand and analyze their experiments. This analysis can be done automatically and faster than manual analysis of images one by one.

References

[1] ŠTYS, D., URBAN, J., VANĚK, J., CÍSAŘ, P. *Analysis of biological time-lapse microscopic experiment from the point of view of the information theory.* Micron, 42(4):360-365, 2011.

[2] DALIBOR, Š., VANĚK, J., NÁHLÍK, T., URBAN, J., CÍSAŘ, P. *The cell monolayer trajectory from the system state point of view.* Molecular biosystems, 7(10):2824-2833, 2011.

[3] RYCHTARIKOVA, R., NAHLIK, T., SMAHA, R., URBAN, J., STYS, D.,Jr, CISAR, P., STYS, D. *Multifractality in imaging: Application of information entropy for observation of inner dynamics inside of an unlabeled living cell in bright-field microscopy.* In ISCS 2014: Interdisciplinary Symposium on Complex Systems, pages 261-267. Springer, 2015.

Current address

Náhlík Tomáš, Mgr. Ph.D.
The Institute of Technology and Business in České Budějovice
Okružní 517/10, 370 01 České Budějovice, Czech Republic
Tel. number: +420 737 931 802, e-mail: nahlik@mail.vstecb.cz

Slovak University of Technology in Bratislava
Faculty of Mechanical Engineering
16th Conference on Applied Mathematics **APLIMAT 2017**
Proceedings

A CLASS OF HYPERLATTICES INDUCED BY QUASI-ORDERED SEMIGROUPS

NOVÁK Michal (CZ), CRISTEA Irina (SLO), KŘEHLÍK Štěpán (CZ)

Abstract. In this paper we study hyperstructure generalizations of lattices. We give conditions for EL–semihypergroups to be H_v–semilattices, hypersemilattices or hyperlattices. We also study some of their special elements.

Keywords: Join hyperlattice, hyperlattice, hypersemilattice, H_v-semilattice, quasi-ordered semigroup

Mathematics subject classification: Primary 06B99; Secondary 06F05

1 Introduction

The concept of a *lattice* is one of the important concepts of abstract algebra with many significant applications in logic, topology, IT and other algebra-related areas. It can be approached from two directions: either as an order-theoretic concept or as an algebraic structure with two binary operations. However, in many applications where lattices are used, *multi-valued aspects* play an important role. In these cases, results of the *hyperstructure theory*, based on the idea that a result of an operation, a *hyperoperation*, is a subset of the carrier set (instead of its element), are used.

In this paper we study hyperstructure generalizations of lattices. The notion of *hyperlattice* is not new as it was introduced in 1977 by Konstantinidou and Mittas [12]. The notion of *ordering* is also a topic well established in the hyperstructure theory. Originally, whenever ordering was involved, hyperstructures constructed from ordered sets were studied or special hyperoperations making use of ordering were constructed. Among other concepts this includes the study of quasi-order hypergroups [2, 3, 5] or EL–hyperstructures (see [18, 19] which give a partial theoretic background for results obtained at random since late 1990s) or the use of Varlet's [23] interval lattice hyperoperation. Lattices were often studied as tools for constructing hyperstructures, both binary and n–ary; see e.g. [14, 15]. In 2009, Heidari and Davvaz [8] made an important move in the area of ordering in hyperstructure theory by introducing the concept of *ordered hyperstructures*, in which ordering is defined on the hyperstructure itself. This, combined with numerous results obtained by Konstantinidou and Seramifidis (for references see the list in [6]; most papers written in French), enhanced the study of hyperstructure generalizations of lattice-like structures such as *join / meet hyperlattices*, *hypersemilattices* or H_v*–semilattices* [16, 21, 22] and enabled to view these hyperstructures from the two perspectives common in lattices.

In our paper we link the study of one class of hyperstructures, EL*–hyperstructures* [18, 19, 20], to concepts studied by Dehghan Nezhad, Rasouli, Soltani Lashkenari and Davvaz [16, 21, 22] and Xiao and Zhao [25]. For the time being we study lattice-like hyperstructures from the perspective of sets endowed with one or two hyperoperations.

2 A construction of hyperstructures from quasi-ordered semigroups

We are going to use the following natural construction first presented in a detailed discussion in [4] (though it can be traced back to [3]), applied in a number of papers including [9] and studied from the theoretical point of view in [18, 19]. Notice that in the following text we work with *principal ends*, i.e. for an arbitrary $a \in (S, \leq)$ we set $[a)_\leq = \{x \in S; a \leq x\}$. Associativity of the hyperoperation "$*$" defined on S means that $a * (b * c) = (a * b) * c$ for all $a, b, c \in S$; an associative hypergroupoid is called a *semihypergroup*. The following lemma and its add-ons regarding hypergroups and join spaces are often referred to as the *Ends lemma*.

Lemma 1. *([4], Theorem 1.3, p. 146) Let $(S, \cdot, \leq)$ be a partially ordered semigroup. The binary hyperoperation $* : S \times S \to \mathcal{P}^*(S)$, where $\mathcal{P}^*(S)$ is the system of all nonempty subsets of S, defined by*

$$a * b = [a \cdot b)_\leq \tag{1}$$

is associative. The semi-hypergroup $(S, *)$ is commutative if and only if the semigroup $(S, \cdot)$ is commutative.

Hyperstructures $(S, *)$, in this case semihypergroups, constructed by means of (1) are called EL*–hyperstructures*. As mentioned in Section 4, the concept can be dualized. It can also be adjusted to suit the study of sets $(S, \leq)$ which are lattices or modified in order to decrease the number of assumptions and simultaneously increase applicability of the concept. For details see [13, 17].

Remark 1. *Since the antisymmetry of relation $\leq$ is needed only to prove that commutativity of the hyperoperation implies commutativity of the single-valued operation, Lemma 1 is often quoted assuming quasi–ordered semigroups.*

3 H_v–semilattices and hypersemilattices

Konstantinidou and Mittas in [12] introduced the concept of a *hyperlattice* as a generalization of the concept of a lattice. Later, Xiao and Zhao in [25] introduced and studied the concept of a *hypersemilattice* as a multivalued analogy of a semilattice. Dehghan Nezhad and Davvaz in [16] presented numerous results on hypersemilattices and especially H_v*–semilattices*, which are the natural weakening of the strong hypersemilattice concept and are members of the important family of H_v*–structures*, the study of which was initiated by Vougiouklis (for a basic overview and applications see [24]). In this section we relate EL–hyperstructures to hypersemilattices and H_v–semilattices.

Definition 1. *([16], Definition 3.1) Let L be a nonempty set with a binary hyperoperation $*$ on L such that, for all $a, b, c \in L$, the following conditions hold:*

1. $a \in a * a$ *(idempotency)*
2. $a * b = b * a$ *(commutativity)*
3. $(a * b) * c \cap a * (b * c) \neq \emptyset$ *(weak associativity)*

Then $(L, *)$ is called an H_v–semilattice. When in the condition 3 we have equality, then $(L, *)$ is called a hypersemilattice.

First of all we show the connection between EL–(semi)hypergroups and hypersemilattices / H_v–semilattices.

Theorem 1. *Let $(L, *)$ be a hyperstructure constructed from a quasi-ordered semigroup $(L, \cdot, \leq)$ by defining*

$$a * b = [a \cdot b)_{\leq} = \{x \in L; a \cdot b \leq x\}. \qquad (2)$$

1. *If "$\cdot$" is commutative and $(L, \cdot)$ is not a group, then the condition that for all $a \in L$ there holds $a \cdot a \leq a$ is equivalent to the fact that $(L, *)$ is a hypersemilattice.*
2. *If "$\cdot$" is not commutative and "$\leq$" is antisymmetric, then $(L, *)$ is neither a hypersemilattice nor an H_v–semilattice.*
3. *If $(L, \cdot)$ is a non-trivial group, then $(L, *)$ is neither a hypersemilattice nor an H_v–semilattice.*

Proof. Given the construction presented in Section 2, we have that (2) constructs a semihypergroup, i.e. validity of condition 3 of Definition 1 (in its strong associative version) is secured by default. Therefore, the question of whether (2) gives rise to hypersemilattices, is for *commutative* "$*$" equivalent to the question of validity of condition in statement 1. Moreover, for hyperoperation (2), the condition "$a \in a * a$ for all $a \in L$" rewrites as "$a \cdot a \leq a$ for all $a \in L$".

If $(L, \cdot)$ is a semigroup which is not a group, this has no special implications and we obtain statement 1.

However, if $(L, \cdot)$ is a group, then this is equivalent to $a \leq u$ for all $a \in L$, where u is the neutral element of $(L, \cdot)$. This means that u is the greatest element of $(L, \leq)$. Yet $a \leq u$ is in a quasi-ordered group equivalent to $u \leq a^{-1}$ for all $a \in L$, which is possible only if $u = a^{-1}$. Yet since this should hold for all $a \in L$, there is $L = \{u\}$ and we obtain statement 3.

Finally, if "$\leq$" is antisymmetric, then $(L, \cdot, \leq)$ is a partially ordered semigroup and commutativity of $(L, *)$ is equivalent to commutativity of $(L, \cdot)$ and we obtain statement 2. □

Remark 2. *Further on, in Corollary 2 and in Corollary 3 we assume that $(L, *)$ is a semihypergroup constructed from a commutative semigroup $(L, \cdot)$. Notice that this means that $(L, *)$ is commutative. However, commutativity of $(L, *)$ might be secured also for non-commutative carrier semigroups $(L, \cdot)$ – in case $\leq$ is a quasi-ordering which is not partial* *ordering. Of course, only such semigroups $(L, \cdot)$ which are not groups are relevant in this respect.*

Corollary 1. *In Theorem 1, an idempotent quasi-ordered semigroup $(L, \cdot, \leq)$ always creates a hypersemilattice.*

Proof. Obvious since $\leq$ is reflexive. □

Example 1. *The set union and set intersection are idempotent operations. Therefore, if we regard $(\mathcal{P}^*(S), \cap)$ or $(\mathcal{P}^*(S), \cup)$, then condition 1 of Definition 1 holds in both cases. Moreover, both $(\mathcal{P}^*(S), \cap, \subseteq)$ and $(\mathcal{P}^*(S), \cup, \subseteq)$ are quasi-ordered semigroups which are not groups. Therefore, if we set $A * B = \{X \subseteq \mathcal{P}^*(S); A \cap B \subseteq X\}$ or $A * B = \{X \subseteq \mathcal{P}^*(S); A \cup B \subseteq X\}$, then $(\mathcal{P}^*(S), *)$ is in both cases a hypersemilattice. The same is true if we consider $\mathcal{P}(S) = \mathcal{P}^*(S) \cup \{\emptyset\}$ in either of the cases.*

Further on, we will call the hypersemilattice constructed from a quasi-ordered semigroup using (2) an *EL–hypersemilattice.*

The results of Theorem 1 and Corollary 1 may be worded also in terms of *idempotent elements* and *idempotent sets.*

Definition 2. *([1], Definitions 2.13 and 2.16) An element a of a semihypergroup $(S, *)$ is called an idempotent if $a \in a * a$. The semihypergroup $(S, *)$ is called an idempotent semihypergroup if all its elements are idempotent. A nonempty subset A of a semihypergroup $(S, *)$ is called an idempotent subset if $A \subseteq A * A$.*

Corollary 2. *Let $(L, *)$ be a semihypergroup constructed from a quasi-ordered commutative semigroup $(L, \cdot, \leq)$ by (2). Then $(L, *)$ is a hypersemilattice if and only if it is an idempotent semihypergroup.*

Proof. Obvious, since $(L, *)$ is both commutative and associative thanks to Lemma 1 and the condition $a \in a * a$ for all $a \in L$, equivalent to $a \cdot a \leq a$ is present in definitions of both concepts. □

Corollary 3. *Let $(S, *)$ be a commutative semihypergroup constructed from a quasi-ordered semigroup $(S, \cdot, \leq)$ by (2). Then if for every $s \in S$ there holds $s \cdot s \leq s$, then every subset A of $(S, *)$ is idempotent.*

Proof. Obvious thanks to Corollary 2, also Corollary to [16], Proposition 3.1. □

Remark 3. *Notice that as far as the case of regular EL–semihypergroups $(S, *)$ is concerned (where* regular *means that $(S, *)$ has an identity and every element of S has an inverse, see [1]), [18], Corollary 2, links idempotent subsemihypergroups of $(S, *)$ to hyperideals of $(S, *)$.*

Example 2. *Denote the closed interval of real numbers $\langle 0, 1\rangle$ by L. Obviously, $(L, \cdot, \leq)$, where "$\cdot$" is the usual multiplication and "$\leq$" the usual ordering of real numbers, is a quasi-ordered semigroup, which is not a group. Also obviously, the condition $a \cdot a \leq a$ for all $a \in L$ holds in L. Therefore, $(L, *)$, where "$*$" is defined by (2), is a hypersemilattice. It is also an example of a hypersemilattice constructed from a non-idempotent quasi-ordered semigroup.*

Remark 4. *In [16], Proposition 3.3, the authors take the aditive partially ordered group of all integers* $(\mathbb{Z},+,\leq)$*, define the hyperoperation in a way similar to* (2) *by*

$$k * l = \{u \in \mathbb{Z}; k + l \leq 2u\}$$

for all $k, l \in \mathbb{Z}$*, and show that* $(\mathbb{Z}, *)$ *is an* H_v*-semilattice which is not a hypersemilattice. Notice that the choice of* double sum *instead of* sum *in the definition of the hyperoperation is crucial as it secures idempotency of* $*$ *(even though at a price of breaking its strong associativity). On the other hand, choosing* sum *instead of* double sum*, i.e. converting the hyperoperation to* (2) *(thus regarding an* EL*–hyperstructure), would secure strong associativity of* $*$ *yet prevent* $(\mathbb{Z}, *)$ *from being an* H_v*–semilattice or a hypersemilattice as* $(\mathbb{Z},+,\leq)$ *is a partially ordered group. In general, number domains with operations of sum or product of numbers are problematic as a basis for construction of* EL*–hypersemilattices as they themselves or their important subsets are mostly groups. For an application of* $(\mathbb{Z}, *)$ *defined as an* EL*–hypergroup see [9], Example 1.*

Also notice, that introducing the concept of double-sum *in [16] (or any other multiple of the sum) is possible only because* $(\mathbb{Z},+,\cdot)$ *is a ring as* $k + l \leq 2u$ *is a definition that relies on two single-valued operations, i.e. cannot be generalized so easily as definition* (2).

In [17] modified EL–hyperstructures are studied. Motivated by simplification of assumptions and increasing applicability of the concept, the hyperoperation (2) is – among others – modified to include elements a, b, i.e. we have $a *_m b = \{a,b\} \cup [a \cdot b)_{\leq}$ (thus $(L, *)$ becomes an *extensive* hyperstructure). In this case there is $a \in a *_m a$ for all $a \in L$ by default, i.e. axiom 1 of Definition 1 always holds. Also, again by default, axiom 3 of the definition is secured for this hyperoperation. Moreover, it can be easily proved that hyperoperation (2) of Theorem 2 is associative if and only if hyperoperation (3) is associative (the rather straightforward proof is included in [17]). As a result we have the following theorem and corollary.

Theorem 2. *Let* $(L, *_m)$ *be a hyperstructure constructed from a quasi-ordered semigroup* $(L,\cdot,\leq)$*, where* $\leq$ *is not antisymmetric, by defining*

$$a *_m b = \{a, b\} \cup [a \cdot b)_{\leq} = \{x \in L; a \cdot b \leq x\}. \tag{3}$$

1. *If "*$\cdot$*" is commutative and* $(L,\cdot)$ *is not a group, then* $(L, *_m)$ *is a hypersemilattice.*
2. *If* $(L,\cdot)$ *is a non-trivial group, then* $(L, *_m)$ *is neither a hypersemilattice nor an* H_v*–semilattice.*

Proof. Obvious. □

Corollary 4. *Let* $(L, *_m)$ *be a hyperstructure constructed from a partially-ordered semigroup* $(L,\cdot,\leq)$ *which is not a group, where "*$*_m$*" is defined using* (3)*. Then* $(L, *_m)$ *is a hypersemilattice if and only if "*$\cdot$*" is commutative.*

Proof. Idempotency and weak associativity are obvious. Since $(L,\cdot,\leq)$ is a partially ordered semigroup, validity of the remaining axiom of Definition 1, commutativity of the hyperoperation "$*_m$", is equivalent to the commutativity of the single-valued operation. □

Let us call the hypersemilattice constructed from a quasi-ordered semigroup using (3) an *mEL–hypersemilattice.*

The list of important elements of H_v–semilattices and hypersemilattices includes *absorbent* and *fixed* elements.

Definition 3. *([16], Definition 3.4) Let $(L, *)$ be an H_v–semilattice. An element $a \in L$ is called an absorbent element of L if it satisfies $c \in a * c$ for all $c \in L$. An element $b \in L$ is called a fixed element of L if it satisfies $b * c = \{b\}$ for all $c \in L$.*

Remark 5. *Notice that, for commutative semihypergroups, the definition of* fixed *elements is equivalent to the definition of* zero scalar elements, *which are used in classification of* simple *semihypergroups. For details cf. [18] or [10, 11].*

The following theorem lets us identify the absorbent and fixed elements easily. Notice that due to Theorem 1 the study of groups $(L, \cdot)$ is irrelevant, i.e. the cancellation law cannot be used in Theorem 3, statement 1.

Theorem 3. *Let $(L, *)$ be a hypersemilattice constructed from a quasi-ordered semigroup $(L, \cdot, \leq)$ using hyperoperation $*$ defined by* (2). *Then:*

1. *$a \in L$ is an absorbent element of $(L, *)$ if and only if $a \cdot c \leq c$ holds for all $c \in L$.*
2. *If $a \in L$ is an absorbent element of $(L, *)$, then all elements $b \in L$ such that $b \leq a$ are absorbent.*
3. *$b \in L$ is a fixed element of $(L, *)$ if and only if b is a maximal element of $(L, \leq)$ such that $b \cdot c = c \cdot b = b$ holds for all $c \in L$.*
4. *If $(L, \cdot, \leq)$ is non-trivial, then no element of L can be simultaneously absorbent and fixed element of $(L, *)$.*

Proof. Statement 1 is obvious.

In statement 2 suppose an arbitrary element $b \in L$ such that $b \leq a$. In a quasi-ordered semigroup $(L, \cdot, \leq)$ this means that, for all $c \in L$, there is $b \cdot c \leq a \cdot c$. Yet $a \in L$ is an absorbent element of $(L, *)$. Thus $a \cdot c \leq c$ for all $c \in L$. From transitivity we get $b \cdot c \leq c$ for all $c \in L$, i.e. b is also absorbent.

As regards statement 3: according to the definition, $b \in L$ is a fixed element of $(L, *)$ if for all $c \in L$ there holds $b * c = \{b\}$. The hyperoperation $*$ is defined using (2), i.e. $\{b\} = b*c = \{x \in L; b\cdot c \leq x\}$ for all $c \in L$. Moreover, in hypersemilattices, commutativity is assumed, i.e. in our case there must be $b * c = c * b$ for all $c \in L$. However, since $\leq$ is reflexive, two simultaneous conclusions are equivalent with this: $b \cdot c = c \cdot b = b$ must hold for all $c \in L$ and on top of that, b must be a maximal element of $(L, \leq)$.

In statement 4, if a is absorbent, then $a \cdot c \leq c$ for all $c \in L$. If a is fixed, then $a \cdot c = a$ for all $c \in L$. Thus $a \leq c$, for all $c \in L$, i.e. a is the smallest element of $(L, \leq)$. Yet as follows from statement 3, fixed elements are also maximal ones. Thus we get a contradiction or $(L, \cdot, \leq)$ is trivial. ☐

Theorem 4. *Let $(L, *)$ be a hypersemilattice constructed from a quasi-ordered semigroup $(L, \cdot, \leq)$ using hyperoperation $*$ defined by (2). Denote $AE(L)$ the set of all absorbent elements of $(L, *)$. Then:*

1. *$(AE(L), \cdot)$ is a subsemigroup of $(L, \cdot)$.*
2. *If $(L, \cdot)$ is a monoid, then $AE(L)$ is always non-empty.*
3. *$(AE(L), *)$, where $*$ is defined in the same way as in $(L, *)$ yet restricted on elements of $AE(L)$, is a hypersemilattice.*

Proof. Suppose that $a, b \in L$ are absorbent elements of $(L, *)$. Then $b \cdot c \leq c$ for all $c \in L$. In a quasi-ordered semigroup this implies $a \cdot b \cdot c \leq a \cdot c$ for all $c \in L$. However, since also $a \in L$ is absorbent, we get from transitivity that $a \cdot b \cdot c \leq c$ for all $c \in L$, which means that also $a \cdot b$ is absorbent.

Further, denote u the neutral element of $(L, \cdot)$, if it exists. Obviously, since $\leq$ is reflexive, $u \cdot c = c \leq c$ for all $c \in L$, which means that u is absorbent.

Finally, the proof of statement 3 is obvious. □

Remark 6. *Notice that in [16], p. 384, plain text, an H_v–subsemilattice of $(L, *)$ is defined as an arbitrary non-empty subset $M \subseteq L$ such that $a * b \in \mathcal{P}^*(M)$, i.e. such that $M * M = M$. However, in our case of $(AE(L), *)$ we have a different situation as the equality $AE(L) * AE(L) = AE(L)$ is all but self-evident. $(AE(L), *)$ can be – but equally well need not be – a subsemihypergroup of $(L, *)$. For details cf. [20], where a detailed discussion of subhyperstructures of EL–hyperstructures is included, or Example 3. Notice that this has also important influence on the study of hyperideals of the hypersemilattices in question (or hyperstructures based on* (2) *in general).*

Example 3. *If we continue with Example 2, where we considered $L = \langle 0, 1 \rangle$ with the usual multiplication and ordering of real numbers, then we see that all elements of L are absorbent and no element of L is fixed.*

If we denote by $L(Q)$ the set of all rational numbers from $L = \langle 0, 1 \rangle$, we get an example of what has been discussed in Remark 6: $(L(Q), \cdot)$ is a subsemigroup of $(L, \cdot)$, thus we may use $(L(Q), \cdot, \leq)$ to define the hyperoperation using (2)*. However, e.g. $\frac{1}{4} * \frac{1}{5} = \{x; \frac{1}{20} \leq x\}$ yet we have to decide whether we take $x \in L$ or $x \in L(Q)$ as both of these choices are equally justifiable and meaningful. Only based on answer to this question can we consider cases such as fractions of π and thus decide in what relation $(L(Q), *)$ to $(L, *)$ is.*

Example 4. *If we continue with Example 1, where we considered the power set of an arbitrary set S, and examine the set intersection, then we see that $A \cap C \subseteq C$ for all $C \in \mathcal{P}(S)$ holds for all $A \in \mathcal{P}(S)$ while condition $B \cap C = B$ for all $C \in \mathcal{P}(S)$ holds for $B = \emptyset$ only.*

On the other hand, if we examine the set union, we see that $A \cup C \subseteq C$ for all $C \in \mathcal{P}(S)$ holds for $A = \emptyset$ only while $B \cup C = B$ for all $C \in \mathcal{P}(S)$ holds for $B = S$ only.

Thus in $(\mathcal{P}(S), *)$, constructed from $(\mathcal{P}(S), \cap, \subseteq)$, all elements are absorbent and none is fixed – because $\emptyset$ is not a maximal element of $(\mathcal{P}(S), \subseteq)$. However, if we construct $(\mathcal{P}(S), *)$ from $(\mathcal{P}(S), \cup, \subseteq)$, then only $\emptyset$ is absorbent and only S is fixed.

4 Duality

It can be easily proved that if in Lemma 1 we use hyperoperation $*_d : S \times S \to \mathcal{P}(S)$ defined by

$$a *_d b = (a \cdot b]_{\geq} = \{x \in S; x \leq a \cdot b\} \tag{4}$$

instead of (1), the lemma remains valid. This means that the results of Section 3 are valid also for the hyperoperation $*_d$ and its modification $a *_{dm} b = \{a, b\} \cup (a \cdot b]_{\geq} = \{a, b\} \cup \{x \in S; x \leq a \cdot b\}$. For more details concerning duality and modifications of the hyperoperation in the context of matrices, see [13], in which the connection to Varlet's [23] interval hyperoperation $a \diamond b = \{x \in L \mid a \wedge b \leq x \leq a \vee b\}$, for all $a, b \in L$, where L is a lattice, is also provided.

5 Hyperlattices

Hyperlattices of Konstantinidou and Mittas [12] are sets H with one hyperoperation "$\bigvee$" and one single-valued operation "$\wedge$" under the following Definition 4. In this way they are also defined in [6].

Definition 4. *([6], p. 151) Let H be a set, "$\bigvee$" a hyperoperation on H and "$\wedge$" an operation. We say that $(H, \bigvee, \wedge)$ is a hyperlattice if the following conditions are satisfied, for all $a, b, c \in H$:*

1. $a \in a \bigvee a$ *and* $a \wedge a = a$
2. $a \bigvee b = b \bigvee a$ *and* $a \wedge b = b \wedge a$
3. $(a \bigvee b) \bigvee c = a \bigvee (b \bigvee c)$ *and* $(a \wedge b) \wedge c = a \wedge (b \wedge c)$
4. $a \in [a \bigvee (a \wedge b)] \cap [a \wedge (a \bigvee b)]$
5. $a \in a \bigvee b \Rightarrow b = a \wedge b$

In the course of time other approaches have emerged and further concepts have been defined. The most prominent of these include *total hyperlattices*, also known as *superlattices*, which are structures with two hyperoperations. Also, some authors such as Soltani Lashkenari, Rasouli and Davvaz [21, 22] make distinction between *join* and *meet hyperlattices*, which they consider to be hyperstructures with one hyperoperation and one single-valued operation in the sense of Definition 4 yet such that only axioms 1–4 hold. (In their terminology the above definition defines the *join hyperlattice*.) If moreover axiom 5 of Definition 4 holds, they call the join hyperlattice a *strong join hyperlattice*. Naturally, join and meet hyperlattices are mutual duals, i.e. dually, a *strong meet hyperlattice* may be defined. In this section we will adopt this terminology and study strong join hyperlattices (defined by Definition 4) and join hyperlattices (defined by axioms 1–4 of Definition 4).

Suppose now that the hyperoperation "$\bigvee$" is the hyperoperation constructed on a quasi-ordered semigroup $(L, \cdot, \leq)$ using formulas (2). In other words, let us study structures $(L, *, \wedge)$.

First of all, we explain the meaning of Definition 4 in this context of ours. Since axioms 1–3 (with respect to "$\bigvee$", i.e. with respect to "$*$") are defining axioms of hypersemilattices, we can use results of Section 3. In axiom 4, every element $a \in L$ must be, for all $a, b \in L$, in the intersection of $a \bigvee(a \wedge b)$ and $a \wedge (a \bigvee b)$. Yet since "$\bigvee$", i.e. "$*$" in our case, is defined as

$$a \bigvee b = a * b = [a \cdot b)_{\leq} = \{x \in L \mid a \cdot b \leq x\}, \tag{5}$$

we have that $a \in a \bigvee(a \wedge b)$ is equivalent to $a \in [a \cdot (a \wedge b))_{\leq}$ which is equivalent to $a \cdot (a \wedge b) \leq a$, for all $a, b \in L$. On the other hand, the fact that $a \in a \wedge (a \bigvee b)$ is equivalent to $a \in a \wedge [a \cdot b)_{\leq}$ which is equivalent to $a \in \{a \wedge x \mid x \in [a \cdot b)_{\leq}\}$, i.e. $a \in \{a \wedge x \mid a \cdot b \leq x\}$. This means that, for all $a, b \in L$, there exists an element $x \in L$ such that $a = a \wedge x$, where $a \cdot b \leq x$. Axiom 5 is in our context equivalent to the fact that $a \cdot b \leq a \Rightarrow b = a \wedge b$, for all $a, b \in L$.

As can be seen in the following examples, hyperoperation (5) can be used to construct both join hyperlattices and strong join hyperlattices.

Example 5. *Let $L = \mathbb{N}$, operation "$\cdot$" be defined as minimum of natural numbers and operation "$\wedge$" be defined as maximum of natural numbers. Finally, $\leq$ is the usual ordering of natural numbers by size. In this case, for all $a, b \in \mathbb{N}$, $a \bigvee b = a * b = \{x \in \mathbb{N} \mid \min\{a, b\} \leq x\}$ and $a \wedge b = \max\{a, b\}$. Obviously, $(\mathbb{N}, *)$ is a hypersemilattice and $(\mathbb{N}, *, \wedge)$ is a join hyperlattice (for all $a \in \mathbb{N}$ the desired $x \in \mathbb{N}$ equals a) which is not strong.*

Example 6. *Suppose that $L = \mathcal{P}^*(S)$, where $\mathcal{P}^*(S)$ is the system of all non-empty subsets of an arbitrary set S. For an arbitrary subsets $A, B \subseteq \mathcal{P}^*(S)$ define that*

$$A \bigvee B = A * B = \{X \in \mathcal{P}^*(S) \mid A \cup B \subseteq X\},$$

i.e. the operation "$\cdot$" is defined as union of subsets of S and ordering is defined as inclusion, and $A \wedge B = A \cap B$ for all $A, B \subseteq \mathcal{P}^*(S)$. Then obviously, $A \cup (A \cap B) \subseteq A$ for all $A \subseteq \mathcal{P}^*(S)$ and for $X = S$ we have that $A = A \cap X$, where $A \cup B \subseteq X$ for all $A, B \subseteq \mathcal{P}^*(S)$. Thus axiom 4 of Definition 4 holds and $(\mathcal{P}^*(S), *, \wedge)$ is a join hyperlattice. Moreover, $A \cup B \subseteq A$ is, for all $A, B \subseteq \mathcal{P}^*(S)$, equivalent to the fact that $B \subseteq A$ and in this case $B = A \cap B$. Therefore, the join hyperlattice $(\mathcal{P}^*(S), *, \wedge)$ is strong.

Therefore, we can ask: *If we want $(L, *, \wedge)$ to be a join hyperlattice, how shall the single-valued operations "$\cdot$" and "$\wedge$" be linked?*

Theorem 5. *Let $(L, *)$ be an EL–hypersemilattice, i.e. a hyperstructure constructed from a single-valued quasi- / partially ordered semigroup $(L, \cdot, \leq)$ using theorems of Section 3 and hyperoperation (5), and let "$\wedge$" be an idempotent, commutative and associative operation on L.*

1. *If, for all $a, b \in L$, there is $a \cdot (a \wedge b) \leq a$ and $a \cdot b \leq a$, then $(L, *, \wedge)$ is a join hyperlattice.*
2. *If, for all $a \in L$, there is $a \cdot (a \wedge b) \leq a$ and $a = a \wedge 1$, where 1 is the greatest element of $(L, \leq)$, then $(L, *, \wedge)$ is a join hyperlattice.*

Proof. Since $(L, *)$ is a hypersemilattice and we assume that "$\wedge$" is an idempotent, commutative and associative operation on L, we verify axiom 4 of Definition 4 only.

1. As has been mentioned above, the left-hand side of the axiom is equivalent to the condition $a \cdot (a \wedge b) \leq a$. In the right-hand-side it is enough, for all $a \in L$, to set $x = a$ and we obtain that, for all $a, b \in L$, there must be $a = a \wedge a$ (which is true because "$\wedge$" is idempotent), where $a \cdot b \leq a$ (which is what we suppose).

2. The left-hand side of the axiom is equivalent to the condition $a \cdot (a \wedge b) \leq a$. In the right-hand side we assume that $a = a \wedge 1$ and since 1 is the greatest element of $(L, \leq)$, the fact that $a \cdot b \leq 1$ holds, for all $a, b \in L$, trivially.

□

Remark 7. *Example 5 is an example of a join hyperlattice constructed using Theorem 5, statement 1. However, as Example 6 suggests, the converse of this condition is not true as $A \cup B \subseteq A$ does not hold for all $A, B \in \mathcal{P}^*(S)$. Notice that the condition of statement 2 holds in $(\mathcal{P}^*(S), *, \wedge)$ from Example 6.*

Example 7. *Suppose the partially ordered semigroup $(\mathbb{N}, \gcd, |)$, where* gcd *stands for the greatest common divisor of natural numbers and "$|$" is the usual relation of divisibility. By Corollary 1, $(N, *)$, where*

$$a * b = \{x \in \mathbb{N} \mid \gcd\{a, b\}|x\},$$

for all $a, b \in \mathbb{N}$, is a semihyperlattice. Set "$\bigvee$" = "$*$" and $a \wedge b = \min\{a, b\}$. Since there is $\gcd\{a, b\}|a$ and $\gcd\{a, \min\{a, b\}\}|a$, for all $a, b \in \mathbb{N}$, we have that $(\mathbb{N}, *, \min)$ is a join hyperlattice. Since e.g. $\gcd\{16, 24\} = 16 \not\Rightarrow 24 = \min\{16, 24\}$, the join hyperlattice is not strong.

If we use the modification of the EL–hyperoperation (3), i.e. assume that for all $a, b \in L$ there is

$$a \bigvee b = a *_m b = \{a, b\} \cup [a \cdot b)_{\leq} = \{a, b\} \cup \{x \in L; a \cdot b \leq x\}, \tag{6}$$

we have $a \in a \bigvee (a \wedge b)$ by default. As a result, no special link between operations "$\cdot$" and "$\wedge$" is needed under conditions of Theorem 5.

Corollary 5. *Let $(L, *_m)$ be an EL–hypersemilattice, i.e. a hyperstructure constructed from a single-valued quasi- / partially ordered semigroup $(L, \cdot, \leq)$ using theorems of Section 3 and hyperoperation (6), and let "$\wedge$" be an idempotent, commutative and associative operation on L. If, for all $a, b \in L$, there is $a \cdot b \leq a$, then $(L, *_m, \wedge)$ is a join hyperlattice.*

Proof. Obvious because in Theorem 5, statement 1, condition $a \cdot (a \wedge b) \leq a$ of the left-hand side of axiom 4 becomes reduntant and on the right-hand side of axiom 4 it is enough to set $x = a$ for all $a \in L$. □

Acknowledgement

The work was supported by the project FEKT-S-14-2200 of Brno University of Technology.

References

[1] S. Chattopadhyay, A semihypergroup associated with a Γ–semigroup, Analele Ştiinţ. Univ. Al. I. Cuza Iaşi – Ser. Noua – Mat., LVI(1) (2010), 209–225.

[2] I. Chajda, Š. Hošková, A factorization of quasiorder hypergroups, Comment. Math. Univ. Carolin. 45(4) (2004), 573–581.

[3] J. Chvalina, Commutative hypergroups in the sense of Marty and ordered sets, Gen. Alg. and Ordered Sets, Proc. Inter. Conf. Olomouc, (1994), 19–30.

[4] J. Chvalina, Functional Graphs, Quasi-ordered Sets and Commutative Hypergroups, Masaryk University, Brno, 1995 (in Czech).

[5] J. Chvalina, L. Chvalinová, State hypergroups of automata, Acta Math. et Inform. Univ. Ostraviensis 4(1) (1996), 105–120.

[6] P. Corsini, V. Leoreanu, Applications of Hyperstructure Theory, Kluwer Academic Publishers, Dodrecht – Boston – London, 2003.

[7] P. Corsini, Prolegomena of Hypergroup Theory, Aviani Editore, 1993.

[8] D. Heidari, B. Davvaz, On ordered hyperstructures, U.P.B. Sci. Bull. Series A, 73(2) (2011), 85–96.

[9] Š. Hošková, J. Chvalina, Discrete transformation hypergroups and transformation hypergroups with phase tolerance space, Discret. Math. 308(18) (2008), 4133–4143.

[10] H. M. Jafarabadi, N. H. Sarmin, M. R. Molaei, Completely simple and regular semi hypergroups, Bull. Malays. Math. Sci. Soc., 35(2) (2012), 335–343.

[11] H. M. Jafarabadi, N. H. Sarmin, M. R. Molaei, Simple semi hypergroups, Australian Journal of Basic and Applied Sciences, 5(4), (2011), 51–55.

[12] M. Konstantinidou, J. Mittas, An introduction to the theory of hyperlattices, Math. Balcanica 7 (1977), 187–193.

[13] Š. Křehlík, M. Novák, From lattices to H_v–matrices. Accepted in An. Şt. Univ. Ovidius Constanta, XXIV(3) (2016).

[14] V. Leoreanu-Fotea, I. G. Rosenberg, Hypergroupoids determined by lattices, European J. Combin. 31 (2010), 925–931.

[15] V. Leoreanu-Fotea, I. G. Rosenberg, Join spaces determined by lattices, J. Mult.-Valued Log. Soft Comput., 16(1-2) (2009), 421–432.

[16] A. Dehghan Nezhad, B. Davvaz, An Introduction to the Theory of H_v–Semilattices, Bull. Malays. Math. Sci. Soc. (2)32(3) (2009), 375–390.

[17] M. Novák, Š. Křehlík, EL–hyperstructures revisited (submitted)

[18] M. Novák, On EL–semihypergroups, European J. Combin., 44 (2015), 274–286.

[19] M. Novák, Some basic properties of EL–hyperstructures, European J. Combin., 34 (2013), 446–459.

[20] M. Novák, The notion of subhyperstructure of "Ends lemma"–based hyperstructures, Aplimat – J. of Applied Mathematics, 3(II) (2010), 237–247.

[21] S. Rasouli, B. Davvaz, Lattices derived from hyperlattices, Commun. Algebr., 38 (2010), 2720–2737.

[22] A. Soltani Lashkenari, B. Davvaz, Complete join hyperlattices, Indian J. Pure Appl. Math, 46(5) (2015), 633-645.

[23] J. V. Varlet, Remarks on distributive lattices, Bull. de l'Acad. Polonnaise des Sciences, Serie des Sciences Math., Astr. et Phys., 23 (1975), 1143-1147.

[24] T. Vougiouklis, Hyperstructures and their Representations, Monographs in Mathematics, Hadronic Press, 1994.

[25] Y. Xiao, B. Zhao, Hypersemilattices and their ideals, J. Shaanxi Normal Univ. Nat. Sci. Ed. 33(1) (2005), 7–10.

Current address

Novák Michal, RNDr. Ph.D.
Department of Mathematics, Faculty of Electrical Engineering and Communication,
Brno University of Technology
Technická 10, 616 00 Brno, Czech Republic
Tel. number: +420541146080, e-mail: novakm@feec.vutbr.cz

Cristea Irina, Assist. Prof. Ph.D.
Center for Systems and Information Technologies, University of Nova Gorica
Vipavska cesta 13, 5000 Nova Gorica, Slovenia
Tel. number: +38653315232, e-mail: irinacri@yahoo.co.uk, irina.cristea@ung.si

Křehlík Štěpán, Mgr. Ph.D.
Department of Applied Mathematics and Computer Science,
Faculty of Economics and Administration, Masaryk University
Lipová 41a, 602 00 Brno – Pisárky, Czech Republic
Tel. number: +420549495894, e-mail: Stepan.Krehlik@econ.muni.cz

Slovak University of Technology in Bratislava
Faculty of Mechanical Engineering
16th Conference on Applied Mathematics **APLIMAT 2017**
Proceedings

PARAMETERS INFLUENCING WORD PROBLEMS DIFFICULTY

NOVOTNÁ Jarmila (CZ), VONDROVÁ Naďa (CZ)

Abstract. The paper presents the methodology and preliminary results of a 3-year project GAČR 16-06134S *Context problems as a key to the application and understanding of mathematical concepts*. The project aims to identify linguistic, psychological and mathematical parameters of context problems influencing their difficulty, describe those influences on primary school pupils' success and the relationships between levels of linguistic and mathematical competence at different ages. Via linguistic, psychological and mathematical analyses of word problems in textbooks, international mathematical surveys and existing research results, about 71 linguistic psychological and structural parameters influencing word problems difficulty were identified. Seven parameters have been selected for the first stage of research: Pa. Formulation of numerical information in words or in numerals, Pb. (Mis)leading triads of numbers, Pc. Functional relationships, Pd. Order of information, Pe. Presence of false implication, Pf. Length of the text, Pg. Several levels of communication. Pairs of word problems differing in only one of the parameters have been created. They will be used with pupils of Grade 3 through to 9 from four Prague primary schools to find out how exactly the parameters influence the success rate of pupils and their solving strategies and mistakes. After presenting a theoretical framework consisting of word problems and related research, the paper concentrates on the methodology of research and presents illustrations of pairs of word problems differing in one parameter. For each parameter, we hypothesise how it will influence the solution and success rates. Finally, possible uses and implications of future research results are given.

Keywords: Word problems, linguistic parameters, mathematical parameters, psychological parameters, context

Mathematics Subject Classification: Primary 97D40, 97C50; Secondary 97D80.

1 Introduction

In our previous work, we asked Czech teachers what areas of school mathematics are critical for their pupils. One of the most frequent ones were word problems [18]. As the main causes, they mentioned pupils' insufficient reading skills, tendency to give up on common sense when solving word problems, inability to imagine the situation, inability to mathematise the situation, select relevant information from the text, etc. Our interviews with pupils confirmed what the teachers said.

We decided to look into the problem more deeply within project GAČR *Context problems as a key to the application and understanding of mathematical concepts*. Thus, its main aim is to identify causes of lasting problems of Czech pupils with the solution of word problems and investigate possibilities of their overcoming. In particular, to identify those linguistic and structural parameters of the word problems that influence their difficulty (depending on age), determine how the influence of these parameters changes with the age of pupils, beginning with Grade 3 and ending with Grade 9 of the primary school, and find out whether there is the correlation between the levels of linguistic literacy in the Czech language and of mathematical literacy for pupils of different age groups.

In order to investigate the influence of various parameters on pupils' success in solving word problems, the parameters to be dealt with had to be identified. This paper presents the process of selecting linguistic, mathematical and psychological parameters that aspire to have this impact.

2 Theoretical background

Word problems are defined in different ways, there is no agreed-upon meaning. For example, Resnick and Ford [19] see word problems as one important part of context-embedded problems: "Context-embedded problems, exemplified by word problems, draw upon the processes of language understanding and strategy seeking as well as upon basic computational skills." (p. 95) Odvárko et al. [15] understand word problems in school mathematics as such problems in which objects, phenomena and situations (with their diverse properties and relationships) from various non-mathematical domains occur. For Kuřina [10], word problems are problems where usually a certain real situation is described and the task of the solver is to determine answers to asked questions. A more detailed definition is given by Verschaffel, Greer and De Corte [23]. They characterize word problems as verbal descriptions of problem situations wherein one or more questions are raised the answer to which can be obtained by the application of mathematical operations to numerical data available in the problem statement.

In this paper, in accordance with the project GAČR, we characterise word problems as problems which include some context (real, real-like or imaginary) within which some numerical data are given and a question (questions) is posed for pupils to solve using their mathematical knowledge and out-of-school experience.

Throughout the whole history of mathematics, mathematical problems serve to carry information, to practise mathematics, to diagnose the level of acquiring mathematical skills, etc. Word problems hold a prominent place within them mainly because most life situations are described in words. Word problems constitute one of a few school mathematics domains which require mathematization of situations described in words and the transformation of a mathematical solution back to the context of the problem [14]. There is a general agreement on the importance of word problems. For example, Blum and Niss [2] mention, among others, that they develop the ability to activate mathematics to extra-mathematical situations and lead to a rich and comprehensive picture of mathematics in all its facets, as a science, as a field of activity in society and culture. Verschaffel, Greer and De Corte [23] stress the application and motivation functions of word problems on the one hand and the thought-provoking and concept-formation functions on the other.

Given the importance of word problems, it is alarming that the area of word problems has been identified as an area of concern in many countries because of the pupils' difficulties in solving them. Teachers report that solving word problems belong among those in which pupils have the poorest performance [7], [18]. Studies with pupils confirm this assertion (e.g., [22], [14]). Word problems are often feared by pupils and contribute to their negative attitude to mathematics [18]. There may be various causes of word problems difficulty and researchers from different fields have been studying them. We will mention some examples only.

The formulation of word problem assignment is fundamental [5]. In fact, as Richard [20, p. 228] underlines, "it is completely unjustifiable to consider that the formulation is secondary, that there is a structure and content in a problem [...]. The formulation is a part of the problem as essential as the expressed relationships, as forth as it has a determining role in the construction of the representation of the problem." Thus reading skills are of capital importance. For example, Korhonen, Linnanmäki and Aunio [9] established a close link between reading and mathematics performance and showed that no such link is between spelling and mathematics performance. The link between linguistic and mathematics competence was suggested by Pape [17] for Grade 6 and 7 pupils. Vilenius-Tuohimaa, Aunola and Nurmi [25] showed that performance on word problems of pupils from Grade 4 was strongly related to performance in reading comprehension. However, even after controlling for the level of technical reading involved, performance in word problems was still related to reading comprehension, suggesting that both of these skills require overall reasoning abilities. For Grade 3 pupils, Grimm's study [6] showed that early reading comprehension is associated with changes in the applications and conceptual understanding of mathematics as opposed to performing mathematical operations.

There are many ways to mistranslate word problems (e.g., [3], [21]). MacGregor and Stacey [11] showed that it is not the syntax of the relational wording only which leads to error. Errors occur even for problems in which the statement could be directly translated into an equation. Other studies determined that relational information used in word problems can sometimes facilitate problem solving if this relational information is consistent with pupils' out of-classroom experience and if this experience can be used to create non-algebraic solution strategies (e.g., [8], [12]).

It is taken as self-evident that contexts play an important role in pupils' understanding and success in word problems, however, the evidence of their influence is rather anecdotal and speculative [1]. For example, Palm [16] found an impact of authenticity on both the presence of 'real life' considerations in the solution process and on the proportion of pupils' written solutions that were really affected by these considerations.

The GAČR project will focus on the developmental aspect. Studies investigating the development in the solving procedures of pupils of different ages (through the primary school) and in their text comprehension are sparse. For example, Morales, Shute and Pellegrino [13] showed that unlike 5th and 6th graders, 3rd graders were more strongly influenced by the surface structure of word problems than by any underlying conceptual similarities. Davis-Dorsey, Ross and Morrison [4] looked into the impact of both personalizing mathematical word problems and rewording them for explicitness: the 2nd graders benefited only when personalization and rewording were used in combination. The 5th graders benefited consistently from personalization but realized no advantages from rewording. A comprehensive survey of studies investigating the influence of reformulations of word problems on pupils' success in solving them is provided in [24]. We did not find any relevant studies dealing with the influence of particular grammatical phenomena as understood by linguistics on the understanding of word problems texting.

As far as the authors know, in the Czech Republic there are no research studies on the influence of linguistic knowledge and skills on the solution of mathematical word problems. The ones which are most closely linked to this topic are those which are based on overall results of international comparative studies such as PISA, TIMSS and PIRLS and do not deal with mathematical and language phenomena on the level of concrete word problems.

3 Our research – methodology

Our project is unique at least in the Czech context that the research is organised in the cooperation of researchers in linguistic, mathematical education and psychology. This enables us to investigate the problem from different standpoints.

3.1 Identification of parameters

In the first stage, we identified word problems in which Czech pupils had bad results in international surveys TIMSS and PISA and also in the Czech testing organised by Kalibro. These problems were analysed from the point of their linguistic, mathematical and psychological features. Gradually, a list of parameters hypothesised to influence the difficulty of word problems originated. It has been used for the analysis of other word problems from the surveys above and mathematical textbooks as well.

During the whole team meetings, the researchers from the above fields negotiated which parameters, according to their expertise, influence the difficulty of word problems. Some of them were accepted as "obvious". For example, the team agreed that the familiarity of context or the presence of units of measures influence the difficulty of the problem. For others, the situation was not so clear and it was decided that these parameters are worth looking into (see section 4). It was not an easy task, as the final list included 71 parameters, out of which 44 were of linguistic nature and 14 were mathematical. Finally, we decided to test these parameters in the first main test: Pa. Formulation of numerical information in words or in numerals, Pb. (Mis)leading triads of numbers, Pc. Functional relationships, Pd. Order of information, Pe. Presence of false implication, Pf. Length of the text, Pg. Several levels of communication (see section 4 for detail). For each parameter, pairs of word problems were made which only differ in the parameter in question.

3.2 Participants

An opportunity sample of pupils has been made. The total population of Grade 3 through to 9 from four primary schools takes part in our research in order to get a sufficient number of pupils. As we had to eliminate the influence of intervening variables other than those which we follow we needed schools of a similar type (homogeneous environment). We used reports by the Czech School Inspection and website of schools to choose primary schools whose characteristics are: no specialisation, attended by children from their immediate surroundings, with a varied socio-economic background, with the percentage of children from abroad not exceeding the average for the whole Czech Republic (which is according to PISA 2012 2,8 % and according to the Czech Statistical Office 2 %), in the area of outer Prague. Altogether, there are 1 700 pupils, the total in each grade ranges from 178 pupils in Grade 9, to 320 pupils in Grade 4.

All the pupils underwent initial testing in mathematics and Czech language via written tests in October 2016 to balance their mathematical and linguistic competences. Based on their results in the tests, each class was divided into two halves of the same ability. (The process of this division is beyond the scope of the paper.)

In the next stage in February 2017 (after piloting the first main test), participating pupils will be given a test comprising 4 to 6 word problems (depending on their age). Pairs of word problems differing in one parameter will be distributed to two versions of the test (see below) given to the alternate halves of the class. This will enable us to see in what way the investigated parameter influences pupils' solving strategy and success rate. Some parameters will be investigated in one or two grades only, while others can be used in tests for all involved grades.

In the analysis of data (pupils' written solutions), we will use a mixture of qualitative and quantitative methods. Prior to the statistical analyses of data, a qualitative analysis will be made in order to identify relevant phenomena to code for in pupils' solutions. At least some word problems used in the main testing stage will be rather complex, the pupils' solving strategies will be varied and pupils can fail in different stages of their solutions. Thus a detailed analysis of at least some pupils' solutions must be done prior to enumerating the presence or not of some phenomena.

We plan to follow each class for two years. The pupils will write four main tests within the two years which will enable us to investigate the influence of more parameters and their combination on the difficulty of word problems and also to consider pupils' development in solving word problems.

As stated above, we are now at the stage of finalising the first main test. It will focus on 7 parameters which will be illustrated in the next section.

4 Results and their discussion

In this section, we present pairs of word problems for 7 parameters which we plan to test in the first main test. For each pair, differences in pupils' success rate expected by us are given and reasons are suggested. The problems serve as illustrations only. In the test, some parameters will be tested across grades to see whether they are age dependent. Thus, more pairs of word problems differing in the same parameter but of various difficulty are being formulated for different grades.

Pa. Formulation of numerical information in words or in numerals.

Version A: In a bike race, Lucy will cover twenty kilometres in one hour. Kate will cover four kilometres less during the same time. How many kilometres will Kate cover in two hours?

Version B: In a bike race, Lucy will cover 20 kilometres in one hour. Kate will cover 4 kilometres less during the same time. How many kilometres will Kate cover in two hours?

Note: In version B, the number of kilometres is written in numerals rather than words while the values associated with time are left in words. It was much discussed whether they should be changed into numerals, too, but then we decided not to do it for the sake of authenticity. Pupils are familiar with such records from everyday world, too. Values given in numerals attract attention of pupils more than values in words that is why we expect higher success rate for version B. The question is whether it will be true for older pupils, too.

Pb. (Mis)leading triads of numbers.

Version A: A scale of a map says that 1 centimetre on the map represents 4 km in reality. The distance between two towns on the map is 8 centimetres. How many kilometres are between the two towns?

Version B: A scale of a map says that 1 centimetre on the map represents 4 km in reality. The distance between two towns on the map is 9 centimetres. How many kilometres are between the two towns?

Note: Version A was used in TIMSS for grade 4 pupils and their success rate was 67.2 %. Considering the fact that many pupils chose incorrect results of 2 and 16, we hypothesise that this was caused by the presence of numbers which lead to what we call a "good" triad of numbers (8-4-2). That is why we use number 9 instead of 8 in version B and want to see how it will affect the solution.

Pc. Functional relationships (inspired by [12]).

Version A: On each pallet, there are 270 books for children and 45 books for adults. How many books for adults are altogether on the pallets which contain 4 860 books for children?

Version B: On each pallet, there are 270 books packed into 45 boxes. How many boxes are on the pallets which contain altogether 4 860 books?

Note: In version B, there is a clear functional relationship between books and boxes – books are put into boxes, thus there are more books than boxes. There is no such relationship in version A; there is no apparent reason why there should be a link between the number of the two types of books. We hypothesise that the functional relationship might help pupils when solving the problem.

Pd. Order of information.

Version A: The Machs made a three-day bike trip. The first day, they covered 32 km, the second one by 8 km more than the first day, the third day by 11 km less than the second day. How many kilometres did they cover the third day?

Version B: The Machs made a three-day bike trip. The third day they covered by 11 km less than the second day. The first day they covered 32 km, the second day by 8 km more than the first day. How many kilometres did they cover the third day?

Note: In version A, the calculation can be made in the order in which the values are presented. Each intermediate result is the input for the next step. The last intermediate result reached after using all the information is the final result of the problem. In version B, there is a change in the order of information and the solver must analyse the problem to find the first value which he/she should start with. That is why we expect a higher success rate for version A.

Pe. Presence of false implication.

Version A: Veronika and Petra gathered white dead nettles. After drying them, they got 1.5 kg of flower. In the collecting centre, they got 750 CZK for it. How much did Petra get and how much did Veronika get, if Petra gathered by 25 % of flower more than Veronika?

Version B: Veronika and Petra gathered white dead nettles. After drying them, they got 1.5 kg of flower. In the collecting centre, they got 750 CZK for it. Petra gathered by 25 % of flower more than Veronika. How much did Petra get and how much did Veronika get?

Note: This is a linguistic parameter. It is hypothesized that pupils might be misled by the false implication in version A. From the linguistic point of view, it is an imprecise expression (quite common in word problems, though) because this false implication is not of "if A, then B" kind. We have not succeeded in getting rid of the second parameter which is changed in version B (order of information). Due to the nature of the Czech language, if we wanted to use the information "if Petra gathered by 25 % of flower more than Veronika" before the question (as is the case in version B), it would sound strange. If we changed the order of these two pieces of information in version B, the question would not be at the end of the problem as is the case in version A.

Pf. Length of the text.

Version A: Each participant of a ski trip is to pay 2 100 CZK. The teacher, who was to collect the total of 42 000 CZK from the pupils, has only 25 200 CZK so far. How many pupils have not brought the money yet?

Version A: A class goes on a ski trip. Each participant of the trip is to pay 2 100 CZK. Money for the trip are collected by the teacher from the pupils. He was to collect the total of 42 000 CZK from them. Some pupils have not brought the money yet, the teacher has only 25 200 CZK so far. How many pupils have not brought the money yet?

Note: The order of information is the same in both versions. However, version B includes "non-mathematical" text which is not necessary for the solution. It can be expected that for some pupils, mainly the ones with reading problems, it will be difficult to find values relevant for the solution of the problem in the text. On the other hand, the additional text provides more detail about the situation and thus, it may help some pupils to imagine the situation.

Pg. Several levels of communication

Version A: A pen costs 1 zed more than a pencil. It is possible to buy 2 pens and 3 pencils for 17 zeds. How many zeds are needed to buy 1 pen and 2 pencils?

Version B: Joe knows that a pen costs 1 zed more than a pencil. His friend bought 2 pens and 3 pencils for 17 zeds. How many zeds will Joe need to buy 1 pen and 2 pencils?

Version C: A pen costs 1 CZK more than a pencil. It is possible to buy 2 pens and 3 pencils for 17 CZK. How many crowns are needed to buy 1 pen and 2 pencils?

Note: Version B was used in TIMSS 2007 test for Grade 8 and the Czech pupils only reached 25 % success rate. By using versions A and B in our research, we plan to see first, whether the second level of communication represented by "Joe knows that" adds difficulty to the problem and second, what mistakes pupils made which caused this low success rate. Version C looks into possible problems pupils might have with an imaginary currency rather than the currency they are familiar with.

5 Concluding remarks

As stated above, the research outlined in the paper is unique in that it connects mathematical, linguistic and psychological views of word problems. It transpires that it is not possible to strictly distinguish the three views, some parameters sit in their intersections. Moreover, when creating pairs of problems, we realised that it was not easy to make two problems which only differ in one parameter. Often more parameters change (e.g., length of the text, order of information) which is caused by linguistic features of the Czech language – for example, one of the variant for the parameter reads strange if we do not change other features of the sentence. Thus a very fine grained analysis is necessary in order to get a clear distinction between the versions of the word problem. The cooperation of researchers from the three fields is really necessary to achieve this goal as they separately tend to neglect features of word problems which are outside the field of their expertise.

The goal of the presented research is not to simplify word problems for pupils. The goal is to find out which parameters influence the difficulty of their solution. It is important information for textbook authors, authors of mathematical tests and teachers, too. Word problems are used for testing pupils' mathematical competence in both national and international studies and thus such information is needed. Moreover, if the influence of some linguistic parameters is confirmed, results will be valuable for the teaching non only of mathematics but also of the Czech language. Recommendations will be given to teachers of the Czech languages what phenomena are important for pupils in order to understand not only a general text but a text of word problems, too. Even though the literature on word problems is abundant, there is a shortage of systematic studies which investigate the whole age range of primary pupils and look into the link between mathematical and linguistic competences. We hope to fill this gap.

Acknowledgement

The authors gratefully acknowledge the Scientific Grant Agency of Czech Republic for financially supporting this work under the Grant No GAČR 16-06134S.

References

[1] BESWICK, K. Putting context in context: An examination of the evidence for the benefits of 'Contextualised' Tasks. *International Journal of Science and Mathematics Education*, ročník 9, č. 2, 2011: 367–390.

[2] BLUM, W., NISS, M. Applied Mathematical Problem Solving, Modelling, Applications, and Links to other Subjects – State, Trends and Issues in Mathematics Instruction. *Educational Studies in Mathematics*, ročník 22, 1991.

[3] CAPRARO, M. M., JOFFRION, H. Algebraic equations: Can middle-school students meaningfully translate from words into mathematical symbols? *Reading Psychology*, ročník 27, 2006: 147–164.

[4] DAVIS-DORSEY, J., ROSS, S.M., MORRISON, G.R. The role of rewording and context personalization in the solving of mathematical word problems. *Journal of Educational Psychology*, ročník 83, č. 1, 1991: 61–68.

[5] DE CORTE, E., VERSCHAFFEL, L., DE WIN, L. Influence of rewording verbal problems on children's problem representations and solutions. *Journal of Educational Psychology*, ročník 77, č. 4, 1985: 460–470.

[6] GRIMM, K.J. Longitudinal associations between reading and mathematics achievement. *Developmental Neuropsychology*, ročník 33, č. 3, 2008: 410–426.

[7] HOFFER, T.B., et al. *Final Report on the National Survey of Algebra Teachers for the National Math Panel*. Chicago, (2007.

[8] KOEDINGER, K.R., NATHAN, M.J. The real story behind story problems: Effects of representations on quantitative reasoning. *Journal of the Learning Sciences*, ročník 13, 2004: 129–164.

[9] KORHONEN, J., LINNANMÄKI, K., AUNIO, P. Language and mathematical performance: A comparison of lower secondary school students with different level of mathematical skills. *Scandinavian Journal of Educational Research*, ročník 56, č. 3, 2012: 333–344.

[10] KUŘINA, F. *Umění vidět v matematice*. Praha, 1989.

[11] MACGREGOR, M., STACEY, K. Cognitive models underlying students' formulation of simple linear equations. *Journal for Research in Mathematics Education*, ročník 24, 1993: 217–232.

[12] MARTIN, S.A., BASSOK, M. Effects of semantic cues on mathematical modeling: Evidence from word-problem solving and equation construction tasks. *Memory & Cognition*, ročník 33, 2005: 471–478.

[13] MORALES, R.V., SHUTE, V.J., PELLEGRINO, J.W. Developmental differences in understanding and solving simple mathematics word problems. *Cognition and Instruction*, ročník 2, 1985: 41–57.

[14] NOVOTNÁ, J. *Study of solving word problems in teaching of mathematics. From atomic analysis to the analysis of situations*. Saarbrücken, 2010. ISBN 978-3-8433-8631-9.

[15] ODVÁRKO, O. et al. *Metody řešení matematických úloh*. Praha, 1990.

[16] PALM, T. "Impact of authenticity on sense making in word problem solving." *Educational Studies in Mathematics,* ročník 67, č. (1), 2008: 37-58.

[17] PAPE, S.J. Middle school children's problem-solving behavior: A cognitive analysis from a reading comprehension perspectives. *Journal for Research in Mathem. Education*, ročník 35, č. 3, 2004: 187–219.

[18] RENDL, M., VONDROVÁ, N. et al. *Kritická místa matematiky na základní škole očima učitelů*. Praha, 2013. ISBN 978-80-7290-723-6

[19] RESNICK, L.B., FORD, W.W. *The Psychology of Mathematics for Instruction*. Lawrence Erlbaum Assoc., 1981.

[20] RICHARD, J.-F. La construction de la représentation du problème. *Psychologie Française*, ročník 29, č. 21, 1984: 226–230.

[21] SEBRECHTS, M.M., et al. Using algebra word problems to assess quantitative ability: Attributes, strategies, and errors. *Cognition and Instruction*, ročník 14, 1996: 285–343.

[22] VAN DOOREN, W., et al. Just answering ... or thinking? Contrasting pupils' solutions and classifications of missing-value word problems. *Mathematical Thinking & Learning*, ročník 12, č. (1), 2010: 20–35.

[23] VERSCHAFFEL, L., GREER, B., DE CORTE, E. *Making Sense of Word Problems*. Sweets & Zeitlinger Publ., 2000.

[24] VICENTE, S., ORRANTIA, J., VERSCHAFFEL, L. Influence of situational and mathematical information on situationally difficult word problems. *Studia psychologica*, řočník 50, č. 4, 2008: 337–356.

[25] VILENIUS-TUOHIMAA, P.M., AUNOLA, K., NURMI, J-E. The association between mathematical word problems and reading comprehension. *Educational Psychology*, ročník 28, č. 4, 2008: 409–426.

Current address

Novotná Jarmila, Prof. RNDr. CSc.
Univerzita Karlova, Pedagogická fakulta
M. Rettigové 4, 116 39 Praha 1, Czech Republic
E-mail: jarmila.novotna@pedf.cuni.cz

Vondrová Naďa, Doc. RNDr. Ph.D.
Univerzita Karlova, Pedagogická fakulta
M. Rettigové 4, 116 39 Praha 1, Czech Republic
E-mail: nada.vondrova@pedf.cuni.cz

Palm, T. (2008). "Impact of authenticity on sense making in word problem solving." Educational Studies in Mathematics 67(1): 37-58.

Slovak University of Technology in Bratislava
Faculty of Mechanical Engineering
16th Conference on Applied Mathematics **APLIMAT 2017**
Proceedings

MATHEMATICAL TASKS WITH APPLICATIONS AND THEIR PRESENTATION VIA INFORMATION TECHNOLOGY

ORSZÁGHOVÁ Dana (SK), HORNYÁK GREGÁŇOVÁ Radomíra (SK), TÓTHOVÁ Darina (SK)

Abstract. In this paper we focused on the interconnection of three areas: how to link together knowledge in mathematics, economics and computer science? To use mathematics, economics and informatics in the one complex task - this is the difficult problem for students who enter the university in the first year. Students have to solve tasks that are formulated in the study subject mathematics and some problems appear in the topic of the function with two variables. We also present the results of the questionnaire survey about the skill level of university teachers to use IT tools.

Keywords: Mathematics, applied problems, economics, information technology

Mathematics Subject Classification: 97D40, 46N10, 68U01

1 Introduction

The main aim of the study of mathematics at the economic and technical faculties is to provide students with the adequate mathematical apparatus and theoretical methods. At the Faculty of Economics and Management (Slovak University of Agriculture in Nitra) in the winter term the content of mathematics course is: elementary functions with basic features, domains, graphs, inverse functions; differential calculus of the function with one real variable, limits and asymptotes, derivatives of functions, investigation of the features of functions; differential calculus of functions of two real variables, local and constrained extremes of functions of two variables and economic applications. The decisions in planning, production and sale are conditioned by information and methods needed for solving of economic problem. Mathematics and its methods represent proper mean for solving these tasks and for coming to the correct answer. The mathematical knowledge and mathematical methods are used in the other specialized subjects in applied form. Mathematics is very useful means for solving applied tasks of economics study subjects and for solving tasks in the professional life after finishing the university study. Solving of applied tasks is the important way to show students the practical application of mathematical methods and also argued that without knowledge of the general mathematical theory cannot be solved applied tasks. In this frames we will discuss these topics: samples of applied mathematical-economic problems, graphical visualization of used math functions and results of conducted survey about usage of information technology in economical courses.

Education in the information society is based on access to knowledge and information, which are then processed by a person or institution. The use of electronic documents and developing courses were associated with many issues that are the subject solutions. The Portal of Universities was created to draft of the association EUNIS Slovakia and supported by the Ministry of Education, Science, Research and Sport SR in the central development projects. The portal is accessible from URL: https://www.portalvs.sk as a gateway to information for prospective students, university students and the general public. The Portal currently includes several modules. The module "E-Learning" is designed especially for university teachers to gain an overview which courses are developed at Slovak universities and to what extent. University teachers can establish cooperation with the authors of these courses, in order to avoid unnecessary work –making duplicate courses.

2 Material and methods

While working on this paper, we used basic qualitative and quantitative scientific methods, e.g. analysis, synthesis, deduction, comparison, descriptive statistic and mathematics calculation methods.

Graphical approach is suitable for demonstration of various mathematical and economic functions, characteristics and mutual relations. Graphic interpretation of the function represents its properties exactly. The most frequently used production function is Cobb-Douglas production function

$$f : z = A \cdot x^{\alpha} \cdot y^{1-\alpha},\ 0 < \alpha < 1, \tag{1}$$

where A, α are constants, z is a total volume of a production, x, y are inputs (production factors), capital x and work y.

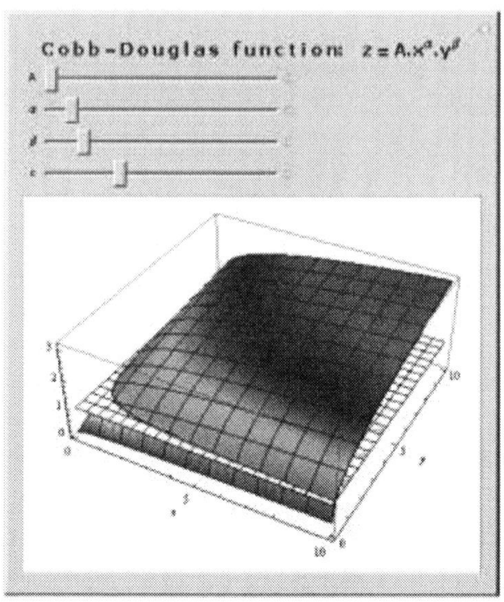

Fig. 1. Graph of the Cobb-Douglas production function. Source: Fajkus, 2014

The Cobb-Douglas function at the Fig. 1 (drawn in Mathematica software) shows a panel which appears when the third module is started. A graph of a Cobb-Douglas function is already displayed. Moving the appropriate sliders the user may change all of the three constants A, α and $\beta = 1 - \alpha$. To display different isoquants, the user has to change the value of another constant c. From the mathematical point of view, the equation $z = c$ in 3D graphically represents a plane. The intersection of this plane with the graph of the Cobb-Douglas function is in economy called an isoquant and in the output window is displayed in a red colour. Using the mouse the whole picture can be rotated, so both the function and the isoquant can be viewed from different angles. This program was written in the Mathematica software (Fajkus, 2014).

There are a lot of economic branches where we can apply mathematical methods. The creation of mathematics models for the prediction of future growth or decrease allow, for example, to evaluate the development of the debt service of the municipalities by special parameters (Papcunová and Gecíková, 2011). Via the Pearson correlation coefficient it was confirmed the dependence between total incomes and capital expenditures of municipalities (Hornyák Gregáňová, Országhová and Papcunová, 2015). Next area for applying of mathematics and statistics is corporate social responsibility (CSR). We can find also a multidimensional approach and techniques for CSR assessment and evaluation of sustainability reporting practices (Habek, 2014).

Factual aspects of the topics taught within the specialized courses in the field of accounting allows students of economical faculties orient themselves in the issue of depreciation of fixed assets and its effect on earnings entrepreneurs (Ferenczi Vaňová and Krajčírová and Váryová and Košovská, 2015). As Semrádová et al. (2010) stated 'in the preparation of courses and learning materials in a virtual environment it is necessary to consider "why" and "what instruments" to use. An advantage of virtual environment is the possibility of flexible options, instant updates, which are set in the didactic process and readily available for pupils and students'.

The process of a complex integration of IT to subjects requires the vast demands on teachers because it requests:

a) to utilize new technologies adequately and productively for acquiring of subject's educational aims;
b) to take care of adequate development of information literacy of teachers' students, to teach them to utilize IT solely and effectively (Bošák, Kubliha and Labaš, 2010).

3 Results

3.1 Functions with two variables in applied problems

In this part we present task about optimization – conditioned extremes of a function of two variables. Except the main function there is given another function - condition, which is important for the calculation of conditioned extremes. The Lagrange method of calculation of conditioned extremes uses auxiliary function, so called Lagrange function

$$F(x,y) = f(x,y) + \lambda \cdot g(x,y) \tag{2}$$

The calculations are based on partial derivatives. The calculation itself is easy to understand; more difficult is to imagine the graph of the given function and the constraint function. Graphs of functions of an optimization problem are shown in Fig. 2 which was created in the Mathematica software. This mathematical problem has applications in economics. For example: consumers make decisions on what to buy and the constraint is their income.

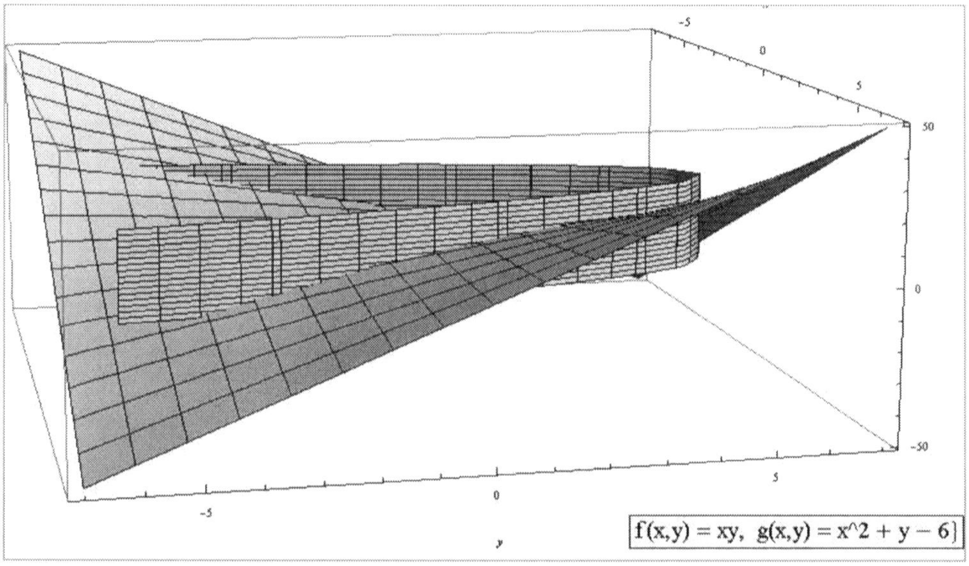

Fig. 2. Graph of the function $f(x, y) = xy$ and constraint function $g(x, y) = x^2 + y - 6$. Source: authors

We can formulate example for optimization problem in applied form:
The fishing company sells two different kinds of fish: Rainbow Trout and Brook Trout. Based on the cost analysis it was identified the function of total cost $TC: f(x, y) = 6x^2 + 10y^2 - xy + 30$, where x is the number of Rainbow Trout and y is the number of Brook Trout. Find the minimum cost of the company if they want to sell 30,000 pieces of fish. It implies that the constraint is $g(x, y): x + y = 30{,}000$.

3.2 Questionnaire survey

In connection with the implementation of information technology into university education new requirements are given on the professional erudition of university teachers from the informatics branch (Országhová, 2010).

In this part we present the results of realized questionnaire survey which was conducted at the Faculty of Economics and Managements (FEM), Slovak University of Agriculture in Nitra within the years 2012 - 2015. Respondents of this survey (in each academic year) were students of the 1st study year in bachelor study programs at the Faculty of Economics and Managements. We were interested in the assessment of information competence of teachers from the students' point of view. The survey was focused on teachers of this faculty – how they apply IT in their taught topics (all taught subjects, not only mathematics). At the end of summer semester students have experience with their teachers from lectures and seminars. They were asked to answer the question: if they consider their teachers as capable in the area of information technologies and their utilization in the education. The questionnaire was anonymous; it was distributed and collected on seminars from mathematics by the teacher from the Department of Mathematics at FEM.

The numbers of participated respondents/students were: 50 respondents in the year 2012; 60 respondents in the year 2013, 64 respondents in the year 2014 and 68 respondents in year 2015.

Year	Very capable	Averagely capable	Less capable	Frequency
2012	32	62	6	50
2013	42	53	5	60
2014	52	42	6	64
2015	49	46	5	68

Tab. 1. Results of the survey about the capability level of teachers. Source: authors

Total results of the questionnaire survey related to the teachers' capability in the information competence within the year 2012 – 2015 are summed up in the Tab. 2. Graphical presentation of total results is on the Fig. 3. From the graph follows that changes has occurred between levels "very capable" and "averagely capable". We can assume that teachers know and control the tools of IT in professional courses. In mathematics graphical interpretation of problems via IT tools is a suitable support for understanding of some topics by students.

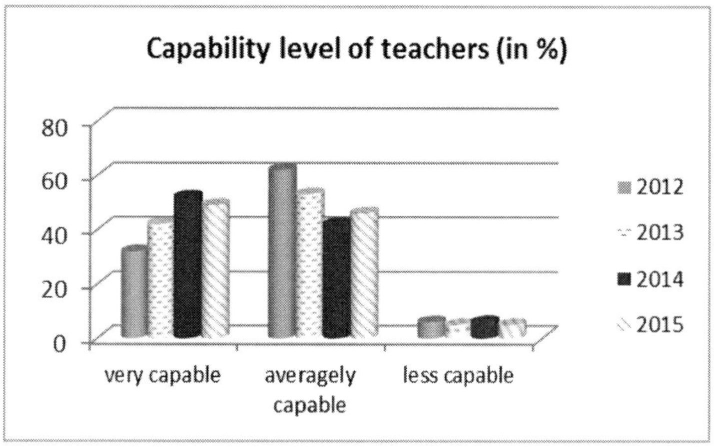

Fig. 3. Graphical interpretation of the survey results. Source: authors

Based on the results of obtained students' answers from the questionnaire survey it arose that year after year the majority of students consider their teachers as very capable in the area of IT. It might be stated that information competence is the part of knowledge of the increasing number of university teachers and its utilization in the education is still spreading whether in the form of lecture presentation or electronic educational materials used for individual and additional study (see also Hornyák Gregáňová, 2015).

4 Conclusion

Mathematical apparatus of the functions with one or more real variables are proper means for economic analysis, formulation of relations in business market and production sphere by quantitative methods. In this contribution we analyzed relations within four basic factors: mathematics, economics, informatics and mutual application. Plotting of graphs of functions belongs to tasks for which it is suitable to use graphical interpretation by means of IT. In economic applications the functions represent economic functions such as costs, profits, consumptions or else.

New kinds of smart and intelligent technology change the process of knowledge presentation and acquiring new skills. Present system of education gives a new requirement for students and teachers – the communication and computer literacy. We can notice these features of the IT application in mathematical education:

- Some kinds of mathematical tasks cause to students difficulty; therefore it is useful and effective to use link between mathematics, economics and computer science. This is the important reason to improve interdisciplinary relations and to show students the comprehensibility and applicability of mathematics.
- Changes in the technical support of education, creation and usage of electronic educational courses. The use of digital technology is most beneficial in the motivational teaching phase.
- Application of graphical tools in mathematics improves the imagination of students, develops their creativity and increases their motivation to study mathematics and to acquire new knowledge.
- In the present system of education a university lecturer cannot educate students without the knowledge of information technology tools, which are affecting the educational process in various ways. We have found that majority of teachers use information technology tools. Students can identify teachers' level of abilities and evaluate their different levels.

Acknowledgement:

This contribution was created with the financial support of EUNIS - Slovakia, Association for University Information Systems (http://www.eunis.sk).

References

[1] BOŠÁK, O., KUBLIHA, M. and LABAŠ, V. Support conception of new competences for IT application in material research and education (Koncepcia podpory nových kompetencií pre uplatňovanie IT v materiálovom výskume a vzdelávaní). In *Trendy ve vzdělávání 2010* (conference proceedings). [CD-ROM]. Olomouc: Votobia, 2010, s. 314-317. ISBN 978-80-87244-09-0.

[2] FAJKUS, M. IT-supported teaching of Microeconomics. In Proceedings of IX. international conference on applied business research (ICABR 2014). Talca, Chile, October 6-10, 2014. P.174-180. [Online]. Retrieved August 3, 2016 from http://www.icabr.com/fullpapers/icabr2014.pdf.

[3] FERENCZI VAŇOVÁ, A., KRAJČÍROVÁ, R., VÁRYOVÁ, I. and KOŠOVSKÁ, I. Depreciation influence of fixed assets on accounting result and corporate income tax base pursuant accounting and tax legislation in the Slovak Republic. In *Finance and Performance of Firms in Science, Education and Practice:* proceedings of the 7th International Scientific Conference. Zlín: Tomas Bata University in Zlín, Faculty of Management and Economics, 287-296 [Online]. ISBN 978-80-7454-482-8. Retrieved August 10, 2016 from http://ufu.utb.cz/konference/sbornik2015.pdf.

[4] HABEK, P. Evaluation of sustainability reporting practices in Poland. *Quality and Quantity,* Volume 48, Issue 3, 2014, Pages 1739-1752.

[5] HORNYÁK GREGÁŇOVÁ, R. Information competence – one of the key competences of mathematics teachers. In *Conference proceedings of eLearning Conference and Competition 2015*. Faculty of Informatics and Management, University of Hradec Králové: Gaudeamus, 2015, pp. 82-84.

[6] HORNYÁK GREGÁŇOVÁ, R., ORSZÁGHOVÁ, D. and PAPCUNOVÁ, V. The debt service of the municipalities and its impact on the financial management of the municipalities in Slovakia. In *Conference proceedings of Hradec Economic Days 2015*. Hradec Králové, 2015. p. 216-222. ISBN 978-80-7435-549-3. Retrieved August 3, 2015 from http://fim.uhk.cz/hed/images/KOMPLET_1_AG.pdf.

[7] ORSZÁGHOVÁ, D. The use of information technology in mathematics education. *Aplimat – Journal of Applied Mathematics. STU, Bratislava*, *3*(3), 147-154.

[8] PAPCUNOVA, V. and GECÍKOVÁ, I. Eeconomic crisis from the perspective of local self-government. In *Conference proceedings: 14th international colloquium on regional sciences.* Boretice, Czech Republic, June 22-24, 2011, Pages: 175-180.

[9] PortalVS.sk (2006 - 2015). Retrieved October 19, 2016 from https://www.portalvs.sk.

[10] SEMRÁDOVÁ, I. et al. *Reflections on the Exploitation of a Virtual Study Environment.* University of Hradec Králové, 1st ed., 121 p., 2010, printed by Miloš VOGNAR Publishing House-M & V, Hradec Králové. ISBN 978-80-86771-43-4.

Current address

Országhová Dana, Doc. RNDr. CSc.
Department of Mathematics, Faculty of Economics and Management,
Slovak University of Agriculture in Nitra
949 76 Nitra, Slovak Republic
Tel. number: +421 37 6414181, e-mail: dana.orszaghova@uniag.sk

Hornyák Gregáňová Radomíra, Mgr. PhD.
Department of Mathematics, Faculty of Economics and Management,
Slovak University of Agriculture in Nitra
949 76 Nitra, Slovak Republic
Tel. number: +421 37 6414180, e-mail: radomira.greganova@uniag.sk

Tóthová Darina, RNDr. PhD.
Centre of Information Technologies, Faculty of Economics and Management,
Slovak University of Agriculture in Nitra
949 76 Nitra, Slovak Republic
Tel. number: +421 37 6414129, e-mail:. darina.tothova@uniag.sk

Slovak University of Technology in Bratislava
Faculty of Mechanical Engineering
16th Conference on Applied Mathematics **APLIMAT 2017**
Proceedings

EXAMPLES OF GOOD PRACTICES OF EXPERIENTIAL LEARNING AND STUDENTS' OPINIONS ON EXPERIENTIAL LEARNING

PÁLENÍKOVÁ Kitti (SK), NAŠTICKÁ Zuzana (SK), VITÉZOVÁ Zuzana (SK), VALLO Dušan (SK), SMIEŠKOVÁ Edita (SK), RUMANOVÁ Lucia (SK), ZÁHORSKÁ Júlia (SK), KLIMENTOVÁ Lucia (SK)

Abstract. The present research is focused on constructivist approach in mathematics instruction referred to as experiential learning. The research presents findings of a questionnaire survey concerning with students' attitudes towards particular workshops they took part in, and popularity of science subjects. The workshops were designed following the experiential learning theory, with emphasis on learners' active work. The workshops' topics were chosen with intention to show students where geometry learnt at school can be applied, and to make them experiment and observe actively in order to find patterns for solutions of linear Diophantine equations.

Keywords: experiential learning theory; manipulatives; students' opinions; popularity of science subjects

Mathematics Subject Classification: 97D40

1 Introduction

The structure of every school lesson depends on various factors, such as the number of students in class, their level of knowledge, teacher's enthusiasm and attitudes to new methods and strategies in education etc. (Vidermanová & Klepancová, 2012). A typical school mathematics lesson usually starts with the teacher's lecture in which s/he explains the new subject matter or an algorithm. Then, students normally solve a few math tasks and exercises with the teacher, and after that they are assigned homework or sit a test. Such kind of lessons can be assessed by students as "out of date" lessons. Goldin (1990) remarks that it takes less time to state a well-established method than it does to guide students to its discovery and the stated method appears to take care of the set of problems at hand. Then, naturally, the prevailing emphasis on skills tests influences many teachers toward the short-term goal of teaching rote procedures. Goldin (1990) indicates the following possible reasons: (i) teachers with the least mathematical preparation (not all) see mathematics only as such a set of rules and procedures; some are insecure with their own mathematical ability, (ii) college professors of unusually high mathematical ability reason extraordinarily rapidly and tacitly; thus, they may themselves be unaware of the complexity of their pattern-recognition, visualization, and problem-solving heuristics, and may describe their reasoning in terms of its overt product an efficient and effective procedure; (iii) many teachers, at all levels, are over-concerned with students' efficiency in arriving at solutions to problems.

On the other hand, it turns out that current students have different behaviour in comparison to students in the past. Students are willing to learn only things which they consider to be necessary in life. However, sometimes students do not know what job they would like to do as adults. It is very difficult to attract students' attention and catch their interest in topics of an unpopular school subject. Therefore, nowadays teachers need to look for and implement new teaching approaches and methods. There are many "new" educational theories on how to improve students' attitudes to mathematics, motivate them, and make mathematics instruction more attractive. One of the approaches teachers can choose from is an approach focused on active work of students. For example using ICT during instruction can help students visualize the given problem and, thus, improve the quality of geometry teaching (Vallo & Rumanová, 2013). Results of Jenisová et al. (2013) indicate that incorporating new experiments based on inquiry approach into the teaching process can meet with a positive response and can develop scientific as well as mathematical competence. From the many constructivist approaches we decided for experiential learning theory. In the following sections we discuss experiential learning and workshops designed in context of this approach.

2 Experiential Learning Theory

Experiential learning is considered to be an appropriate approach that can help learners understand mathematics and remember the achieved knowledge for a longer period. Experiential mathematics education is a constructivist educational approach. The main characteristics of constructivist teaching and learning are active learning and exploration (Kuřina, 2002). According to Silberman (2007) in an experiential lesson learners are involved in concrete activities which provide the learners with an opportunity to experience learning content. Activities should be based on experiences from real world. During such instruction various teaching methods and techniques can be applied, such as games, competitions, excursions, role-play, problem solving or cooperative learning and the like.

Kolb (1984) defined the experiential learning theory as "the process whereby knowledge is created through the transformation of experience. Knowledge results from the combination of grasping and transforming experience" (Kolb & Kolb, 2009, p. 298). The experiential learning theory model (Fig. 1) illustrates two dialectically related modes of *grasping* experience: Concrete Experience (CE) and Abstract Conceptualization (AC), and two dialectically related modes of *transforming* experience: Reflective Observation (RO) and Active Experimentation (AE). Experiential learning is a process in which knowledge is constructed through the so-called creative tension among the four learning modes. In an idealized form, such learning is a process running in a cycle or a spiral where the learner "touches all the bases — experiencing, reflecting, thinking, and acting", recursively and as a response to the learning situation and content. The basis for observations and reflections are immediate experiences. Reflections are adapted and purified into abstract concepts which can initialize new implications for action. These implications can be further actively tested and thus serve as guides in creating new experiences (Kolb & Kolb, 2009).

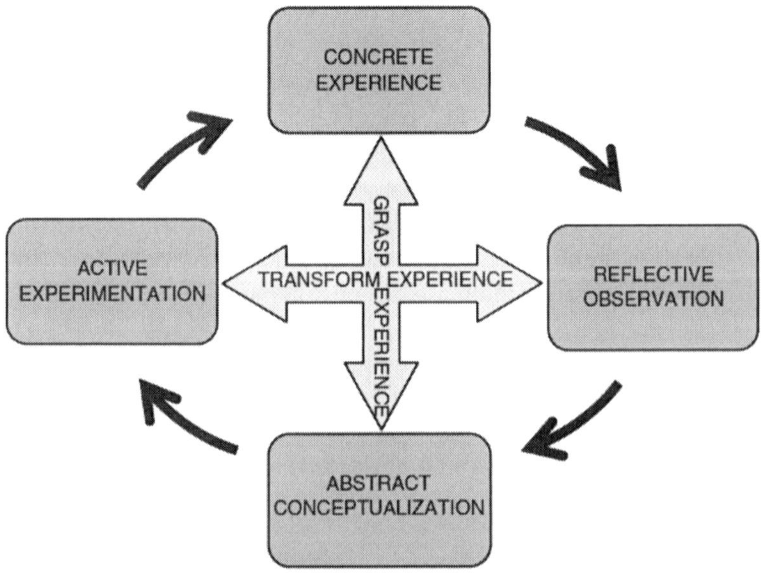

Fig. 1. The Experiential Learning Cycle (Kolb & Kolb, 2009, p. 299)

Naturally, not all learners are able to explore and discover regularities, patterns and rules by themselves. For efficient use of this approach it is recommended for teachers to prepare supporting questions for students. During students experimentation teachers' question posing can be useful, especially if used when the students really need it rather than for faster completion of the problem solving activity. Kirchner et al. (2006) in their analysis of constructional approaches during instruction notice that minimal guidance does not work.

3 Organized workshops following the Experiential Learning Theory

For experiential learning it is recommended to provide learners with opportunities to learn outside the classroom (Qualters, 2010). This can be done in various ways, such as visiting museums, expositions, amusement parks, or organizing educational workshops. In May 2016 three mathematical workshops were organized in the atrium of the Constantine the Philosopher University in Nitra, Slovakia, in order to show the participants various applications of mathematics in practice. The workshops were attended by 70 students at the age of 15-16 from a grammar school in Nitra. The students worked in small groups of 4-5. Each group took part in all the three workshops. During the workshops students faced unconventional tasks which were designed to develop creativity, logical thinking and spatial imagination. In accordance with the recommendations, the attending students also appreciated that even mathematical problems can be solved outside, literally *in the fresh air*. As explained below, in the workshops several manipulatives (objects which can be touched and moved) were applied, which is also recommended for experiential learning (Hartshorn & Boren, 1999).

Workshop *Constructions in use*

According to Vidermanová & Vallo (2015) students often encounter problems such as measuring or marking out an area. These are problems associated with surveying topographical areas, which are based on real needs and require measuring distances, marking out areas in the field. Forty years ago the Slovak mathematics curriculum included many surveying tasks: marking out line segments and shapes, measuring out line segments, distances and angles and specific problems related to measuring distances through an obstacle etc. These topics were reduced over time, and, eventually, they completely disappeared from the mathematics curriculum.

In workshop *Constructions in use* students were asked to measure the base of a pool having the following shapes: (i) square (side length 2 m), (ii) rectangle (side lengths 2 m x 3.5 m; Fig. 2a) and (iii) regular hexagon (side length of 2 m, Fig. 2b). The second task was to measure distance between two points (marked by the teacher) through an obstacle. The third task was to measure the height of a building (picked by the teacher).

Fig. 2a

Fig. 2b

Fig. 2. Students' work and outcome from *Constructions in use* workshop

Tasks of this type provide students with opportunities to recognize the practical use of symmetry (reflection, rotation). Their other benefits are stated in Vidermanová & Vallo (2015).

Workshop *Diophantus*

The inspiration for this workshop comes from the Bruce Willis's and Samuel S. Jackson's attempt to defuse a bomb in *Die Hard with a Vengeance* (1995). Though, such an introduction seems to be interesting for boys only; therefore, in order to make the theme gender-free we introduced the task as follows: *Sometimes it happens that you do not have a cup or a vessel with the specific volume you need. For example, one effervescent tablet with vitamin C is recommended to be dissolved in 4 dcl of water. However, you only have a 3 dcl cup and a 5 dcl cup. How can you measure 4 dcl using only these two cups?* (see Fig. 3a) This problem is known worldwide as *the water puzzle*. This task mathematically leads to solving a linear Diophantine equation.

The introductory task was followed by three analogous tasks: (1) *You have to measure exactly 2 dcl of water, but you only have a 3 dcl cup and a 4 dcl cup. How can you do it?* (2) *You have to measure exactly 3 dcl of water, but you only have a 4 dcl cup and a 3 dcl cup. How can you do it?* (3) *You have to measure exactly 2 dcl of water, but you only have a 5 dcl cup and a 4 dcl cup. How can you do it?*

We asked students to describe their solutions by words and to write their solutions using mathematical notation (Fig. 3b). The students' solutions will be subjected to detailed analysis in further research.

Fig. 3a

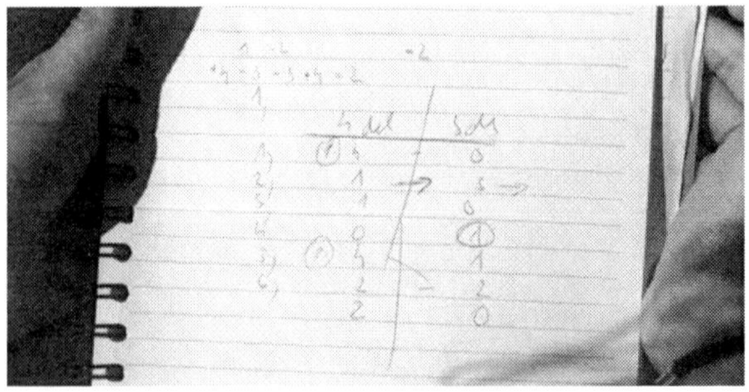

Fig. 3b

Fig. 3. Students' work during *Diophantus* workshop

Workshop *Ornaments*

The introduction of the workshop was focused on providing students with information about creation of gardens and flower beds. After the theoretical presentation, including the presentation of mostly used patterns, the students were asked to propose their own geometrical design of a garden or a flower bed. They seemed to be intrigued by this form of *Land Art*. Then, students were asked to use the presented patterns in certain ratio to draw their own designs. After that they continued outside on a grass area where they marked out their own designs using plastic cutlery and twine (Fig. 2a, 2b).

Fig. 4a

Fig. 4b

Fig. 4. Students' outcomes from *Ornaments* workshop

4 Students' opinions on the workshops

After the workshops all the seventy students participated in an attitude questionnaire survey.
The questionnaire consisted of 5 items focused on the following areas:

- marks achieved by students in mathematics classes;
- students' favourite science subjects;
- students' opinions on the experiential learning – about understanding and retaining knowledge from workshops; their interest in participating in further subject-related activities.

Results of the questionnaire survey

The first item of the questionnaire was focused on the marks which the students achieved in mathematics. The grading system usually used at Slovak schools contains five marks: 1 = excellent; 2 = laudable; 3 = good; 4 = pass; 5 = fail (Kiselova, 2012). The students' average mark achieved in mathematics is 2.07. Frequencies of particular marks are shown in Tab. 1 (marks 4 and 5 were not stated by students).

Mark in Maths	Mark 1	Mark 2	Mark 3
Frequency	11	43	16
Percent	15.71%	61.43%	22.86%

Tab. 1. Students' marks in Maths

The second item asked students to arrange (in prepared rows, see Fig. 5) mathematics and the following subjects: physics, chemistry, biology, geography, from the most favourite to the least favourite.

the most favourite ☺				the least favourite ☹

Fig. 5. Rows for students' answer

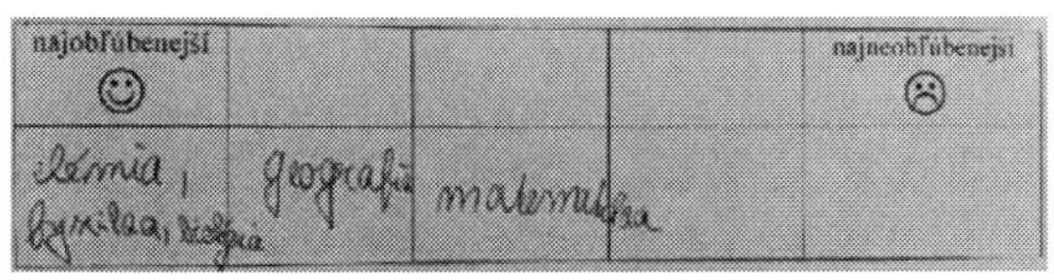

Fig. 6. Example of a student's arrangement

In order to evaluate the popularity of the subjects we proceeded as follows: the first written subject was assigned 5 points; the second subject was assigned 4 points; etc., finally, the least favourite subject was assigned 1 point. Based on the total sum for all subject the following popularity chart was obtained: the most favourite subject is biology, after that geography, chemistry, mathematics and the last is physics. In case students wrote more than one subject in a row, the same point value was assigned to all of the written subjects (Fig. 6; in that case biology, chemistry and physics got 5 points; geography got 4 points and mathematics got 3 points).

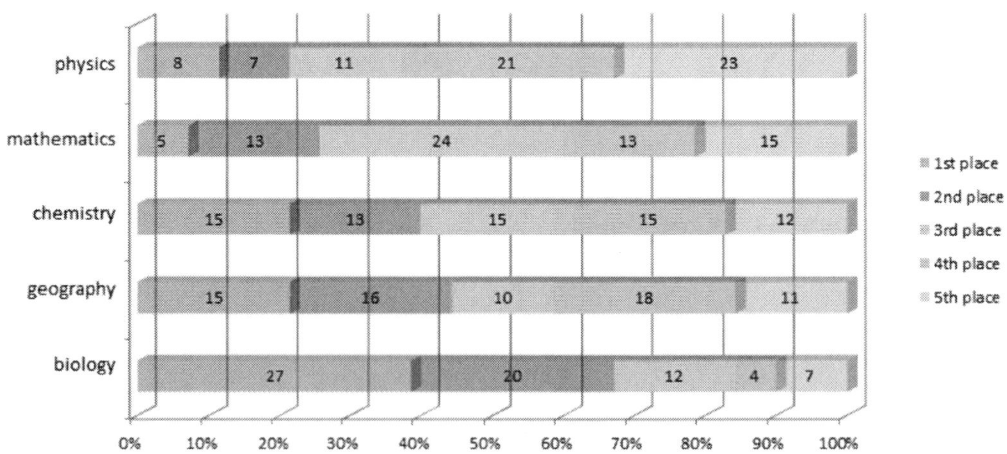

Fig. 7. Popularity of subjects

In Fig. 7 the frequencies of students' arrangements of each place for all subjects are listed. Mathematics is reported as the most favourite subject by as few as 5 students.

We believe, based on these results, that students could get motivated to solve mathematical tasks and problems through interdisciplinary relations with favourite subjects (e.g. interdisciplinary projects). The popularity of all school subjects at different kinds of schools and levels of education will be the subject of further research.

The third item comprised two statements focused on the students' need to *experience* problem-solving. Students were asked to mark how they feel about the following statements:

(1) *I can better understand a solution of a mathematical problem if I experience it.*
strongly agree – agree – neutral – disagree – strongly disagree

(2) *I can better memorize a solution of a mathematical problem if I experience it.*
strongly agree – agree – neutral – disagree – strongly disagree

Statement (1)	strongly agree	agree	neutral	disagree	strongly disagree
Frequency	26	26	13	2	3
Percent	37.14%	37.14%	18.57%	2.86%	4.29%

Tab. 2. Students' agreement with the first statement

Statement (2)	strongly agree	agree	neutral	disagree	strongly disagree
Frequency	33	25	7	2	3
Percent	47.14%	35.71%	10.00%	2.86%	4.29%

Tab. 3. Students' agreement with the second statement

We dare say it seems that students' enthusiasm influenced their instant (dis)agreement with the statements. One student declared that he could not answer because he had never took part in such a kind of an activity before (he marked neutral).

The fourth questionnaire item was the following question: *Which of the realised workshops do you like most and why*? Apart from four students who liked all workshops equally, each student marked one of the workshops. Altogether 54.29% of the students favoured workshop *Diophantus* the most.

Workshop	Constructions in use	Diophantus	Ornaments	All workshops
Frequency	10	34	22	4
Percent	14.29%	48.57%	31.43%	5.71%

Tab. 4. Popularity of the workshops

Only a few students stated a reason for their choice of the workshop they favoured the most. We want to point out that 9 out of 11 students with mark 1 in mathematics tagged *Diophantus* as the most favoured workshop (Tab. 5). The reason that students stated for liking *Diophantus* the most was the need to use their logical thinking. In case of *Ornaments*, students appreciated the need to use their creativity. Those students who tagged *Constructions in use* stated that these problems are in a great way related to the real world.

		Workshop		
		Constructions in use	Diophantus	Ornaments
Mark in Maths	Mark 1	0	9	2
	Mark 2	7	22	18
	Mark 3	7	7	6

Tab. 5. Popularity of the workshops in relation to students' marks in mathematics

Finally, we asked students about their interest to participate in similar activities in future. Not surprisingly, 90 % of students expressed their will to participate in further activities following experiential learning theory.

5 Conclusion

Constructivist approaches to students' learning and teaching can be effective provided that the focus of students' work is on their active involvement in problem solving. Among the benefits of these approaches we can list that they can be organized in or outside the classroom, various teaching aids can be applied (usual as well as unusual), and collaboration with peers is valuable. Experiential learning processes are likely to teach students to work systematically, communicate effectively in order to be able to draw the best possible conclusions, and to develop their reasoning skills.

On the other hand, the experiential learning approach is, naturally, time-consuming. Teachers have to prepare problems very carefully. The selected problems should be solvable in groups and in reasonable time period. Too difficult problems can demotivate students. In addition, the teacher needs to observe students' progress in problem-solving and help them by asking relevant questions. If it is possible, we recommend co-teaching of at least two teachers who guide the students' work. When assessment of students' work is necessary, additional methods are required, such as video or audio recordings of group work.

The questionnaire survey findings suggest that the workshops following the experiential learning theory were perceived as engaging by students. The teachers reported no problems with students' behaviour. We plan to process further mathematical and interdisciplinary topics by the principles of various constructivist approaches.

Acknowledgement

The organized workshops for students and the paper was supported by national project KEGA 016UKF-4/2016 *The Implementation of Constructivist Approaches to Mathematics Teaching with a Focus on Active Acquisition of Knowledge by Pupils in the Context of Bilingual Education* supported by Ministry of Education, Science, Research and Sport of the Slovak Republic, an institutional project UGA VI/16/2014 *Rozvíjanie špecifického matematického myslenia a schopností riešiť matematické problémy prostredníctvom objavného vyučovania a adekvátna príprava učiteľov* and by grant from the Slovak Research and Development Agency under the contract No. APVV-15-0368 entitled *Practice in the centre of the subject field didactics, subject field didactics in the centre of preparation for practice.*

References

[1] GOLDIN, G.: *Epistemology, constructivism, and discovery learning in mathematics*. In: Journal for Research in Mathematics Education. Monograph Constructivist Views on the Teaching and Learning of Mathematics, Vol. 4 (Chap. 3), 1990. p. 31-47. ISBN0873533003.

[2] HARTSHORN, R.; BOREN, S. Experiential Learning of Mathematics: Using Manipulatives. ERIC ED 321967. 1990. p. 1-7. Retrieved from: http://www.eric.ed.gov/PDFS/ED321967.pdf

[3] JENISOVÁ, Z.; BRANIŠA, J.; MELUŠOVÁ, J. *Implementation of Inquiry-based Learning Supported by Digital Technologies in Courses of Professional Development for Chemistry Teachers*. In: 2013 International Conference on Advanced ICT and Education (ICAICTE-13). Atlantis Press, 2013. p. 237 – 241. ISBN 978-90786-77-79-6

[4] KIRSCHNER, P. A.; SWELLER, J.; CLARK, R. E. Why minimal guidance during instruction does not work: An analysis of the failure of constructivist, discovery, problem-based, experiential, and inquiry-based teaching. In: *Educational psychologist*, 2006, Vol. 41. No. 2: p. 75-86. ISSN: 1532-6985.

[5] KISELOVA, D. *System of education in Slovakia*. Comenius University, 2012. Retrieved from https://www.jfmed.uniba.sk/fileadmin/jlf/Pracoviska/ustav-cudzich-jazykov/EDUCATIONAL_SYSTEM_IN_SLOVAKIA.pdf

[6] KOLB, A. Y.; KOLB, D. A. The learning way meta-cognitive aspects of experiential learning. In: *Simulation & Gaming*, 2009, Vol. 40, No. 3: p. 297-327.

[7] KUŘINA, F. O matematice a jejím vyučování. In: *Obzory matematiky, fyziky a informatiky*. 2002. Roč. 31, č. 1, s. 1-8. ISSN 1335-4981.

[8] QUALTERS, D. M. (ed.) *Experiential Education: Making the Most of Learning Outside the Classroom: New Directions for Teaching and Learning*, 2010. No. 124. 128 p. ISBN: 978-0-470-9

[9] SILBERMAN, M. *The handbook of experiential learning*. San Fransisco: John Wiley & Sons, 2007. 400 p. ISBN 978-0-7879-8258-4.

[10] VALLO, D.; RUMANOVÁ, L. Skladanie osových súmerností vo vyučovaní elementárnej geometrie. In: *Matematika v primárnej škole. Rôzne cesty, rovnaké ciele*. Prešov: Prešovská univerzita, 2013. s. 261-265. ISBN 978-80-555-0765-1.

[11] VIDERMANOVÁ, K.; KLEPANCOVÁ, M. Students' Investigation of Platonic Solids - they are Really only Five. In: Billich M. (ed.) *Mathematica IV*, Ružomberok: Catholic University in Ružomberok. p. 191 – 198. ISBN 978-80-8084-954-2.

[12] VIDERMANOVA, K.; VALLO, D. Practical Geometry Tasks as a Method for Teaching Active Learning in Geometry. In: *Procedia-Social and Behavioral Sciences*, 2015, Vol. 191. p. 1796-1800. ISSN 1877-0428.

Current address

Páleníková Kitti, RNDr. PhD.
Department of Mathematics, Faculty of Natural Sciences,
Constantine the Philosopher University in Nitra
Tr. A. Hlinku 1, SK-949 74 Nitra, Slovak Republic
Tel. number: +421 37 6408 691, e-mail: kpalenikova@ukf.sk

Naštická Zuzana, Mgr.
Department of Mathematics, Faculty of Natural Sciences,
Constantine the Philosopher University in Nitra
Tr. A. Hlinku 1, SK-949 74 Nitra, Slovak Republic
Tel. number: +421 37 6408 696, e-mail: zuzana.nasticka@gmail.com

Vitézová Zuzana, Mgr. PhD.
Základná škola s materskou školou Čeľadice 87
Čeľadice 87, SK-951 03 Čeľadice, Slovak Republic
E-mail: zuzana.vitezova@gmail.com

Vallo Dušan, RNDr. PhD.
Department of Mathematics, Faculty of Natural Sciences,
Constantine the Philosopher University in Nitra
Tr. A. Hlinku 1, SK-949 74 Nitra, Slovak Republic
Tel. number: +421 37 6408 690, e-mail: dvallo@ukf.sk

Smiešková Edita, PaedDr.
Department of Mathematics, Faculty of Natural Sciences,
Constantine the Philosopher University in Nitra
Tr. A. Hlinku 1, SK-949 74 Nitra, Slovak Republic
Tel. number: +421 37 6408 703, e-mail: edita.smieskova@ukf.sk

Rumanová Lucia, PaedDr. PhD.
Department of Mathematics, Faculty of Natural Sciences,
Constantine the Philosopher University in Nitra
Tr. A. Hlinku 1, SK-949 74 Nitra, Slovak Republic
Tel. number: +421 37 6408 693, e-mail: lrumanova@ukf.sk

Záhorská Júlia, PaedDr. PhD.
Department of Mathematics, Faculty of Natural Sciences,
Constantine the Philosopher University in Nitra
Tr. A. Hlinku 1, SK-949 74 Nitra, Slovak Republic
Tel. number: +421 37 6408 693, e-mail: jzahorska@ukf.sk

Klimentová Lucia, Mgr.
Department of Mathematics, Faculty of Natural Sciences,
Constantine the Philosopher University in Nitra
Tr. A. Hlinku 1, SK-949 74 Nitra, Slovak Republic
Tel. number: +421 37 6408 703, e-mail: lucia.klimentova@ukf.sk

Slovak University of Technology in Bratislava
Faculty of Mechanical Engineering
16th Conference on Applied Mathematics **APLIMAT 2017**
Proceedings

TROPICAL GEOMETRY

PĂUN Marius (RO)

Abstract. Starting from a simple problem of flow we construct the context in which the Tropical Geometry is working and we present some applications of this algebraic geometry. Low level examples constitute a very good model of how difficult and modern mathematics can be used in teaching.

Keywords: Semifields, Tropical geometry

Mathematics subject classification: Primary 14T97

1 Introduction

Teaching higher algebra in school is always a challenge. If for groups and fields the reasons they are studied for are quite easy to explain, the more sophisticated structures as free semigroups, semirings, moduloids or algebroids are difficult to introduce and more difficult to understand. But a simple application can make things clearer. Let us see how a simple bidirectional flow problem can be the starting point of the construction of an algebraic structure. Consider a bidirectional flow motion of some kind of information and a set of filters. (Fig.1.) One can combine the action of the set of filters only in two ways: sequential or parallel. (Fig.2.)
By direct observation one can see that the sequential composition of filters is commutative, associative and admits a "neutral" element the unlimited filter (or we can call it the zero filter).

The parallel composition of filters is also associative, has a "neutral" element the blocking filter, is commutative, and for any filter "a" the filter with the same dimension but with opposite flow is the inverse of the filter "a".
Can those two combinations of filters be correlated? A simple observation leads us to the conclusion that the parallel combination is distributive with respect to the serial one and the serial one is not distributive with respect to the parallel combination.

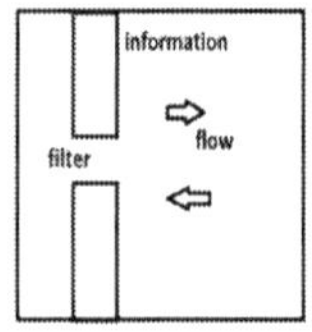

Fig. 1. Field of information with bidirectional flow

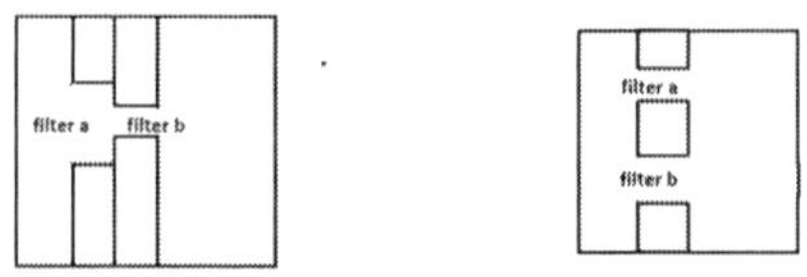

(a) sequential (b) parallel

Fig. 2. Combinations of filters

2 Semirings, semifields, moduloids, algebroids

Definition 1 *(**Semirings**) A nonempty set K is called a semiring if it is endowed with two internal operations $\oplus$ and $\otimes$ with the properties:*

1. $\oplus$ is associative, admits neutral elemnt denoted ε and is commutative.

2. $\otimes$ is associatve, admits neutral element e and it is distributive with respect to $\oplus$.

3. $\varepsilon \otimes e = e \otimes \varepsilon = \varepsilon$

Definition 2 *(**Semifields**) The semiring K is a semifield if $K^* = K \setminus \{\varepsilon\}$ is a group with respect to $\otimes$.*
The semifield is:
- commutative if $\otimes$ is commutative.
- idempotent if $a \oplus a = a \ \forall a \in K$.

Definition 3 *(**Idempotent moduloids**) An unempty set $\mathcal{M}$ is called a moduloid over an idempotent semiring $K(\oplus, \otimes, \varepsilon, e)$ if it is endowed with an internal operation denoted also $\oplus$, associative, commutative, with neutral element ε, and with an external operation $K \times \mathcal{M} \longrightarrow \mathcal{M}$ indicated by simple juxtaposition, with the properties:*

1. $\alpha(x \oplus y) = \alpha x \oplus \alpha y \quad \forall\, x,\ y \in \mathcal{M}$ *and* $\alpha \in K$
2. $(\alpha \oplus \beta)\, x = \alpha x \oplus \alpha y \quad \forall \alpha,\ \beta \in K$ *and* $x \in \mathcal{M}$
3. $\alpha(\beta\, x) = (\alpha \otimes \beta)\, x \quad \forall \alpha,\ \beta \in K$ *and* $x \in \mathcal{M}$
4. $e\, x = x \quad \forall\, x \in \mathcal{M}$
5. $\varepsilon\, x = \varepsilon \quad \forall\, x \in \mathcal{M}$

If K is a semifield then $\mathcal{M}$ is a kind of vectorial space.

The simplest example of such a kind of moduloid is the K^n with the addition defined componentwise, $\varepsilon = (\varepsilon, ..., \varepsilon)$ and the external operation defined also componentwise.
On $\mathcal{M} \times \mathcal{M} \longrightarrow K$ we can define an usual dotproduct. If $x = (x_1, ..x_n)$ *and* $y = (y_1, ..., y_n)$ then

$$x \odot y = \oplus_{i=1}^{n}(x_i \otimes y_i)$$

Definition 4 ***(Idempotent algebroids)*** *we say that the idempotent moduloid $\mathcal{M}$ is an idempotent algebroid if we have defined on it an internal product, denoted also by $\otimes$, associative, with neutral element denoted also e and distributive with respect to $\oplus$.*

The most usual idempotent algebroid we shall work with is the algebroid $K^{n\times n}$ of the square matrices with elements in K. The sum is defined componentwise and the internal product is defined as follows: If $A = (a_{ij}),\ B = (b_{ij}),\ C = (c_{ij}), i,\ j = 1, .., n$ then

$$C = A \otimes B \Leftrightarrow c_{ij} = \oplus_{k=1}^{n}(a_{ik} \otimes b_{kj})$$

The zero matrix is $\varepsilon = \begin{pmatrix} \varepsilon & \dots & \varepsilon \\ . & \dots & . \\ \varepsilon & \dots & \varepsilon \end{pmatrix}$ and the unit matrix is $e = \begin{pmatrix} e & \dots & \varepsilon \\ . & \dots & . \\ \varepsilon & \dots & e \end{pmatrix}$

3 Tropical geometry

3.1 The semifield $\mathcal{R}_{min}$

Let us consider $\mathcal{R}_{min} = R \cup \{\infty\}$. On it we define two internal operations $\oplus$ and $\otimes$ as follows:

$$x \oplus y = min(x, y) \quad and \quad x \otimes y = x + y$$

It is easy to see that $\mathcal{R}_{min}(\oplus, \otimes)$ is an idempotent commutative semifield with the "zero" element $\varepsilon = \infty$ and the "unity" element $e = 0$. It is the "algebraic model" of the bidirectional flow with filters introduced at the beginning.

Observation 1 *1. There is no symmetry for the addition on $\mathcal{R}_{min}(\oplus, \otimes, \varepsilon, e)$. For any $a \neq \varepsilon \in \mathcal{R}_{min}$ there is no $b \in \mathcal{R}_{min}$ susch as $a \oplus b = \varepsilon$ and $a \oplus b = c \oplus b$ does not necessary imply that $a = c$.*

2. *The zero element is concellative with respect to $\otimes$ i.e. $a \otimes \varepsilon = \varepsilon \otimes a = \varepsilon \quad \forall a \in \mathcal{R}_{min}$*

3. *There are no zero divisors i.e. $a \otimes b = \varepsilon \Rightarrow \ a = \varepsilon \ or \ b = \varepsilon$*

4. $a \oplus e = \begin{cases} 0 & for\ x \geq 0 \\ a & for\ x < 0 \end{cases}$

5. *$(x \oplus y)^2 = x^2 \oplus y^2$ holds true because $2min(x, y) = min(2x, x+y, 2y) = min(2x, 2y)$ and by a simple induction we obtain that $(x \oplus y)^n = x^n \oplus y^n$.*

Observation 2 *$\mathcal{R}_{min}$ is an ordered semifield if we take $P = \mathcal{R}_{min}{}_{+}^{*}$. The natural order relation is defined as $x \leq y \Leftrightarrow x \oplus y = x$.*

As for the field R, we can define on $\mathcal{R}_{min}$ an "inverse" of the multiplicative operation, the "division", as follows $x \oslash y = x - y$.
We can introduce also the "max" operation on $\mathcal{R}_{min}$ as follows:
$max : \mathcal{R}_{min} \times \mathcal{R}_{min} \longrightarrow \mathcal{R}_{min}, \quad max(x, y) = (x \otimes y) \oslash (x \oplus y)$.

Observation 3 *Using the properties of the $\oplus$, $\otimes$ and $\oslash$ we can prove that the next relations hold true:*

- $-max(a, b) = -a \oplus -b$
- $-(a \oplus b) = max(-a, -b)$
- $max(a, max(b, c)) = max(max(a, b), c)$
- $max(a, b \oplus c) = max(a, b) \oplus max(a, c)$
- $a \oplus max(b, c) = max(a \oplus b, a \oplus c)$.

The idempotent moduloid and the idempotent algebroid we are working with are ${\mathcal{R}_{min}}^n$ and ${\mathcal{R}_{min}}^{n \times n}$.
Here are some examples of computations in the structures we mentioned before.
$5 \oplus 3 = 3; \quad -2 \oplus -6 = -6; \;\; 2 \otimes 5 = 7; \quad -3 \otimes -1; 1 \otimes 2 \oplus 5 = 3 \oplus 5 = 3.$
$\varepsilon \oplus 1 = 1; \;\; \varepsilon \oplus e = e; \;\; \varepsilon \otimes 5 = \varepsilon; \;\; \varepsilon \otimes e = \varepsilon; \;\; e \oplus 2 = e; \;\; e \oplus -3 = -3.$
$5^2 = 5 \otimes 5 = 2^5 = 10 = 1^10; \;\; e^3 = 3^0 = 0;$
$3 \oslash 2 = 1; \;\; (3 \otimes 5) \oslash (3 \oplus 5) = \; max(3, 5) = 5; \;\; (3 \oplus 6)^2 = 3^2 \oplus 6^2 = 6;$
$e \oslash 3 = -3; \;\; 4 \oslash e = 4; \;\; \varepsilon \oslash a$ has no sense because the operation $\oslash$ is defined on ${\mathcal{R}_{min}}^*$
$\sqrt{4} = 2; \;\; \sqrt{8} = 4; \;\; \sqrt[3]{12} = 4.$
$(2, 1, 3) \oplus (1, 4, 2) = (1, 1, 2); \;\; 3(1, 2, 3) = (3 \otimes 1, 3 \otimes 2, 3 \otimes 3) = (4, 5, 6);$
$(1, 2, 3) \odot (2, 4, 1) = 1 \otimes 2 \oplus 2 \otimes 4 \oplus 3 \otimes 1 = 3 \oplus 6 \oplus 4 = 3.$
$\begin{pmatrix} 1 & 2 \\ 3 & 4 \end{pmatrix} \oplus \begin{pmatrix} 4 & 3 \\ 1 & -1 \end{pmatrix} = \begin{pmatrix} 1 & 2 \\ 1 & -1 \end{pmatrix}$
$\begin{pmatrix} 1 & 2 \\ 3 & 4 \end{pmatrix} \otimes \begin{pmatrix} 4 & 3 \\ 1 & -1 \end{pmatrix} = \begin{pmatrix} 3 & 3 \\ 5 & 3 \end{pmatrix}$
$2 \begin{pmatrix} 1 & 2 \\ 3 & 4 \end{pmatrix} = \begin{pmatrix} 2 \otimes 1 & 2 \otimes 2 \\ 2 \otimes 3 & 2 \otimes 4 \end{pmatrix} = \begin{pmatrix} 3 & 4 \\ 5 & 6 \end{pmatrix}.$

The algebraic structure $\mathcal{R}_{min}$ is called "tropical semiring". The name "tropical" was given by the french mathematicians in the honor of the brasilian mathematician Imre Simon who first used $\mathcal{R}_{min}$ algebra in optimization theory.
One have to define over $\mathcal{R}_{min}$ linear, affine, rational, momomial and polynomial functions. The zero sets of systems of multivariate polinomials are algebraic varieties in the algebraic tropical geometry. The functions we considered are piecewise-linear so the tropical geometry is a piecewise-linear algebraic geometry.

3.2 Linear and affine functions and equations

Definition 5 ***(Linear function)*** *A function $f : \mathcal{R}_{min} \longrightarrow \mathcal{R}_{min}$ is called a linear function if there exists a constant $a \in \mathcal{R}_{min}$ such that $f(x) = a \otimes x \quad \forall\, x \in \mathcal{R}_{min}$.*

Observation 4

- *The graphic of a linear function is a straight line parallel to the first bisectrix. The intersection to the vertical axis is the point* (e, a).
- *We have to emphasize that* $f(x) = x$ *is the linear function* $f(x) = 0 \otimes x$ *and not the function* $f(x) = 1 \otimes x$.

Definition 6 ***(Affine functions)*** *The function* $f : \mathcal{R}_{min} \longrightarrow \mathcal{R}_{min}$ *is called an affine function if there exist* $a,\ b \in \mathcal{R}_{min}$ *such that* $f(x) = a \otimes x \oplus b, \quad \forall x \in \mathcal{R}_{min}$.

The graphics of linear and affine functions are presented in Fig.3. (a) and (b). Since there is

(a) Linear (b) Affine

Fig. 3. Linear and affine functions

no symmetry with respect to $\oplus$, the general form of the affine equation is

$$a \otimes x \oplus b = a' \otimes x \oplus b'.$$

1. If $(a - a')(b - b') < 0$ then we have an unique solution $x_0 = (b \oplus b') \oslash (a \oplus a')$.
2. If $(a - a')(b - b') = 0$ then the solution is not unique. $x_0 > b \oslash (a \oplus a')$ or $x_0 < (b \oplus b') \oslash a$.
3. In rest the equation has no solutions.

3.3 Systems of linear equations

The general form of a linear system is:

$$A \otimes x \oplus b = C \otimes x \oplus d.$$

where $A,\ C \in {\mathcal{R}_{min}}^{n \times n};\ \ b,\ d \in {\mathcal{R}_{min}}^{n}$.

Observation 5 *The system can be simplified to a canonical performing this replacements:*

- $a_{ij} = \varepsilon$ *if* $a_{ij} > c_{ij}$.
- $c_{ij} = \varepsilon$ *if* $c_{ij} > a_{ij}$.
- $b_i = \varepsilon$ *if* $b_i > d_i$.
- $d_i = \varepsilon$ *if* $d_i > b_i$.

There are only two types of systems of linear equations for which the theory is well developed: $A \otimes x = b$ and $x = A \otimes x \oplus b$. For the first type, there is no special method of solving the system or to write down directly the solution. We can obtain first a weak solution. If this solution verifies the system then this is the unique solution. If it does not verify the system then this weak solution is minimal.

Theorem 1 *Let us consider $\overline{\mathcal{R}_{min}} = \mathcal{R}_{min} \cup \{-\infty\}$. If the $n \times n$ matrix A and the n vector b have components in $\overline{\mathcal{R}_{min}}$ then the minimal weak solution of the system $A \otimes x = b$ is given componentwise by:*

$$e \oslash x_j = \oplus_i((e \oslash b_i) \otimes a_{ij})$$

The proof is the same as in the case of "max-plus Algebra" and can be find in [1].
Examples:
If we consedier the system:

$$\begin{pmatrix} 2 & 3 \\ 4 & 5 \end{pmatrix} \otimes \begin{pmatrix} x \\ y \end{pmatrix} = \begin{pmatrix} 6 \\ 7 \end{pmatrix}$$

then we compute
$(-6,\ -7) \otimes \begin{pmatrix} 2 & 3 \\ 4 & 5 \end{pmatrix} = (-4,\ -3)$ and the weak solution is $\begin{pmatrix} 4 \\ 3 \end{pmatrix}$. Cecking directly in the system we observe that:

$$\begin{pmatrix} 2 & 3 \\ 4 & 5 \end{pmatrix} \otimes \begin{pmatrix} 4 \\ 3 \end{pmatrix} = \begin{pmatrix} 6 \\ 8 \end{pmatrix} \geq \begin{pmatrix} 6 \\ 7 \end{pmatrix}$$

meaning that $\begin{pmatrix} 4 \\ 3 \end{pmatrix}$ is a minimal weak solution.
If we take the system:

$$\begin{pmatrix} 1 & 3 \\ 2 & 1 \end{pmatrix} \otimes \begin{pmatrix} x \\ y \end{pmatrix} = \begin{pmatrix} 5 \\ 4 \end{pmatrix}$$

and perform similar calculations we obtain:
$(-5,\ -4) \otimes \begin{pmatrix} 1 & 3 \\ 2 & 1 \end{pmatrix} = (-4,\ -3)$ the weak solution is $\begin{pmatrix} 4 \\ 3 \end{pmatrix}$ and this is the solution of the system.

$$\begin{pmatrix} 1 & 3 \\ 2 & 1 \end{pmatrix} \otimes \begin{pmatrix} 4 \\ 3 \end{pmatrix} = \begin{pmatrix} 5 \\ 4 \end{pmatrix}$$

We can associate to every directed graph with n nodes, labeled from 1 to n, a matrix D_G with positive elements d_{ij} representing the length of the directed edge (i, j). If (i, j) is not an edge then $d_{ij} = \infty$. D_G is not necessary a symmetric matrix.
In $\mathcal{R}_{min}$ algebra holds true:

Theorem 2 *In this conditions, the (n-1)-th power of the matrix D_G has in the entrance d^*_{ij} the length of the shortest path from node i to node j.*

Using just the definition of the matrix multiplication in ${\mathcal{R}_{min}}^{n\times n}$ the proof of the theorem is immediate.
Now, considering $A \in {\mathcal{R}_{min}}^{n\times n}$ one can define the matrix

$$A^* = e \oplus A \oplus A^2 \oplus .. \oplus A^n \oplus ..$$

This is the matrix of minimum weigth of all path of any length from node j to node i.

Observation 6 *If* $a_{ij} > 0$ $\forall i, j$ *then* $A^* = e \oplus A \oplus .. \oplus A^{n-1}$.

Theorem 3 *If the entries of the matrix A satisfy* $a_{ij} \geq 0$ *then the system* $x = A \otimes x \oplus b$ *has a solution* $x = A^* \otimes b$.
If $a_{ij} > 0$ *then the solution is unique.*

Proof If $A^* \otimes b$ exists then it is a solution of the system. Indeed:

$$A^* \otimes b = (e \oplus A \otimes A^*) \otimes b = A \otimes (A^* \otimes b) \oplus b.$$

The proof of the existence and uniqueness of the solution is the same as the proof of Theorem 3.17 in [1].
Example: Let us consider the system

$$\begin{pmatrix} x \\ y \end{pmatrix} = \begin{pmatrix} 1 & 2 \\ 3 & 4 \end{pmatrix} \otimes \begin{pmatrix} x \\ y \end{pmatrix} \oplus \begin{pmatrix} 3 \\ 7 \end{pmatrix}$$

Then $A^2 = \begin{pmatrix} 2 & 3 \\ 4 & 5 \end{pmatrix}$ and $A^* = \begin{pmatrix} 0 & \infty \\ \infty & 0 \end{pmatrix} \oplus \begin{pmatrix} 1 & 2 \\ 3 & 4 \end{pmatrix} \oplus \begin{pmatrix} 2 & 3 \\ 4 & 5 \end{pmatrix} = \begin{pmatrix} 0 & 2 \\ 3 & 0 \end{pmatrix}$.
The solution of the system is:

$$\begin{pmatrix} x \\ y \end{pmatrix} = \begin{pmatrix} 0 & 2 \\ 3 & 0 \end{pmatrix} \otimes \begin{pmatrix} 3 \\ 7 \end{pmatrix} = \begin{pmatrix} 3 \\ 6 \end{pmatrix}$$

3.4 Polynomials in one variable

As usual, we shall define a tropical polynomial as a finite linear combination of tropical monomials. A tropical monomial of degree $k \geq 0$ is a product $a \otimes X^k$. This monomial has associate the tropical function $p(x) = a \otimes x^k$ and the graph of this monomial in the (x, y) plane is a straight line with the slope k and the intersection to the vertical axis equal to a. Thus, to any tropical polynomial we can associate a tropical polynomial function which is continous, piecewise-linear and concave. The correspondence between polynomials and their associate polynomial function is not biunivoque, for example $X^2 \oplus 3 \otimes X \oplus 2$ has the same associate polynomial function as $X^2 \oplus 2$.
Let us consider the general quadratic tropical polynoimial $p(X) = a \otimes X^2 \oplus b \otimes X \oplus c$. Every monomial of $p(X)$ has a linear graph; $a \otimes X^2$ is represented by the line $y = 2x + a$, $b \otimes X$ has $y = x + b$ and c has $y = c$ as representations. In order that all the three lines contribute to the graph of $p(x)$ is necessary that $b - a \leq c - b$. (Figure 4.) The points $x_1 = b - a$ and $x_2 = c - b$ are called the tropical roots of tropical polynomial function $p(x)$. Then, the tropical polynomial function $p(x)$ admits a linear factorization

$$p(x) = a \otimes [x \oplus (b - a)] \otimes [x \oplus (c - b)].$$

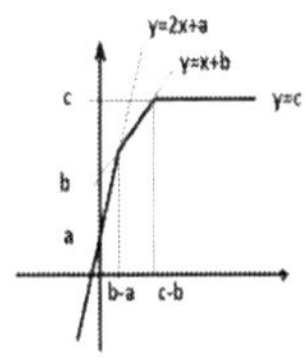

Fig. 4. Field of information with bidirectional flow

It is easy to see that if $b - a \geq \frac{c-a}{2}$ then the decomposition is $p(x) = a \otimes [x \oplus (\frac{c-a}{2})]^2$.

For tropical polynomial functions the Fundamental Theorem of Algebra holds true and $\mathcal{R}_{min}[X]$ is tropicaly algebraic closed. It means that to every polynomial $p(X) \in rmi[X]$ there exists an equivalent tropical polynomial having the same tropical polynomial function and this function can be linear factored.

4 Conclusions

Tropical Geometry is a relative new field of mathematics with a spectacular evolution. The applications in solving problems in discrete events are many and efficient. We can mention here some domain where this theory can be applied:Dynamic Programming, Parallel Computation, Finite Capacity Queing Systems, Traffic Problems, Timed Event Graphs and so on. From the didactical point of view, in terms of teaching, the introductory part of this theory can be a good exercice of improvement of the capacity of abstract thinking for the young students and a very good example of how modern mathematics is really usefull.

References

[1] Baccelli F., Cohen G., Olsder G. j., Quadrat J-P., *Syncronization and Linearity. An Algebra for Discrete Event Systems*, Wiley, 1992, (ISBN 0 471 93609 X).

[2] Maclagan d., Sturmfels B.,*Introduction to tropical geometry*, AMS Graduate studies in Mathematics, vol 161, 2009

Current address

Păun Marius, Conf. univ. dr.
Dept. of Mathematics and Computer Sciences,
Faculty of Mathematics and Computer Sciences,
Transilvania University of Brasov
Iuliu Maniu str. 50, 500090 Brasov, Romania
Tel. +40744631409, e-mail: m.paun@unitbv.ro

Slovak University of Technology in Bratislava
Faculty of Mechanical Engineering
16th Conference on Applied Mathematics **APLIMAT 2017**
Proceedings

INVERTIBILITY OF GRAPHS WITH A UNIQUE PERFECT MATCHING

PAVLÍKOVÁ Soňa (SK), ŠEVČOVIČ Daniel (SK)

Abstract. In this paper we investigate invertibility of graphs with a unique perfect matching, i.e. graphs having a unique 1-factor. We recall the new notion of the so-called negatively invertible graphs investigated by the authors in the recent paper. It is an extension of the classical definition of an inverse graph due to Godsil. We characterize all graphs with a unique perfect matching on $m \leq 6$ vertices with respect to their positive and negative invertibility. We show that negatively invertible graphs exhibit properties like selfinvertibility which cannot be observed within the class of positively invertible non-bipartite graphs with a unique perfect matching.

Keywords: Integrally invertible graph, positively an negatively invertible graphs, 1-factor, unique perfect matching

Mathematics subject classification: Primary 05C50; Secondary 15A09

1 Introduction and preliminaries

The main purpose of this paper is to investigate invertibility of graphs with a unique perfect matching (graphs having a unique 1-factor) (c.f. Pavlíková *et al.* [10, 9]). Following our recent paper [12], we recall the new notion of the so-called negatively invertible graphs. It is an extension of the classical definition of an inverse graph due to Godsil [4]. Our goal is to characterize all graphs with a unique perfect matching on $m \leq 6$ vertices with respect to their positive and negative invertibility. To our best knowledge, the concept of negative invertibility of graphs is a novel approach in the theory of graph inverses when compared to the classical concept of (positive) invertibility of graphs due to Godsil.

According to the result due to Sachs (c.f. [2, p. 32]), it is known that, for a bipartite graph G having a unique perfect matching, the adjacency matrix A is integrally invertible. The aim of this paper is to investigate a larger class of graphs which are not necessarily bipartite but still have a unique 1-factor and an integrally invertible adjacency matrix.

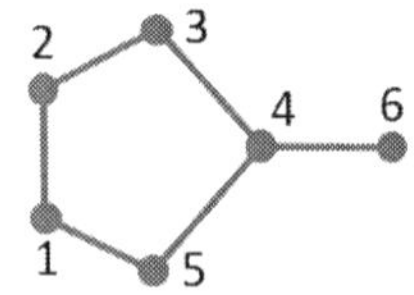

Fig. 1. A graph representing the structural graph of a chemical molecule of fulvene

Let $G = (V, E)$ be an undirected graph with the set of vertices V and the set of edges E. By A_G we denote its adjacency matrix. Conversely, if A is a $\{0,1\}$-symmetric matrix then G_A denotes the graph with the adjacency matrix A. The spectrum $\sigma(G)$ of an undirected graph G consists of eigenvalues of its adjacency symmetric matrix A_G, i.e. $\sigma(G) = \sigma(A_G) = \{\lambda, \lambda$ is an eigenvalue of $A_G\}$ (c.f. [3, 2]). If the spectrum does not contain zero then the adjacency matrix A is invertible.

Clearly, a graph G_A has the integral inverse matrix A^{-1} if and only if the determinant $\det(A) = \pm 1$ (c.f. Kirkland and Akbari [6]). The inverse matrix A^{-1} however may contain positive as well as negative entries. In fact, it is well known that the inverse matrix of the adjacency matrix of a graph is nonnegative if and only if the graph is a union of isolated edges (c.f. Harary [5]). Nevertheless, the concept of an inverse graph is indeed based on the inverse of its adjacency matrix. In [4] Godsil defined a graph to be invertible if the inverse of its (non-singular) adjacency matrix is diagonally similar (c.f. [13]) to a non-negative matrix.

Definition 1 (Godsil [4]) *An undirected vertex-labeled graph G_A is called positively invertible (or invertible in Godsil's sense) if inverse of its adjacency matrix is signable to a nonnegative integer matrix, i.e. there exists a diagonal matrix D^A containing ± 1 elements and such that all elements of the matrix $D^A A^{-1} D^A$ are nonnegative integers only. The inverse graph $H = G_A^{-1}$ is defined by the adjacency matrix $A_H = D^A A^{-1} D^A$.*

Unfortunately, the class of positively invertible graphs does not contain many important examples of graphs having integral inverse matrix. Among them there is a graph representing the chemical organic molecule of fulvene (see Fig. 1). The inverse of its adjacency matrix is integral but it is not signable to a nonnegative matrix.

The structural graph of fulvene however plays an important role in computational chemistry as it represents a chemical molecule. For instance, spectral properties of inverse graphs are important in order to compute the so-called binding energies of molecular orbitals.

Motivated by this shortcoming of positive invertibility of a graph, in the recent paper [12], Pavlíková and Ševčovič introduced the new concept of the so-called negative invertibility of a graph with integral inverse of its adjacency matrix.

Definition 2 (Pavlíková and Ševčovič [12]) *An undirected vertex-labeled graph G_A is called negative invertible if the inverse matrix of its adjacency matrix is signable to a nonpositive integer matrix, i.e. there exists a diagonal matrix D^A containing ± 1 elements and such that all elements of the matrix $D^A A^{-1} D^A$ are nonpositive integers only. The inverse graph $H = G_A^{-1}$ is defined by the adjacency matrix $A_H = -D^A A^{-1} D^A$.*

The adjacency matrix A of the fulvene graph and its inverse matrix A^{-1} are as follows:

$$A = \begin{pmatrix} 0 & 1 & 0 & 0 & 1 & 0 \\ 1 & 0 & 1 & 0 & 0 & 0 \\ 0 & 1 & 0 & 1 & 0 & 0 \\ 0 & 0 & 1 & 0 & 1 & 1 \\ 1 & 0 & 0 & 1 & 0 & 0 \\ 0 & 0 & 0 & 1 & 0 & 0 \end{pmatrix}, \qquad A^{-1} = \begin{pmatrix} 0 & 0 & 0 & 0 & 1 & -1 \\ 0 & 0 & 1 & 0 & 0 & -1 \\ 0 & 1 & 0 & 0 & -1 & 1 \\ 0 & 0 & 0 & 0 & 0 & 1 \\ 1 & 0 & -1 & 0 & 0 & 1 \\ -1 & -1 & 1 & 1 & 1 & -2 \end{pmatrix}.$$

The matrix A^{-1} is signable to a nonpositive matrix by the diagonal matrix $D^A = diag(1, 1, -1, -1, -1, 1)$.

Both positive as well as negatively invertible graphs have the following useful property. If $G = G_A$ is positively invertible then $A_H = D^A A^{-1} D^A$ is the adjacency matrix of the inverse graph $H = G^{-1}$. As $A = D^A A_H^{-1} D^A$ we have the identity $G = (G^{-1})^{-1}$. This property also remains true in the case of negatively invertible graphs. Indeed, if $G = G_A$ is negatively invertible then $A_H = -D^A A^{-1} D^A$ is the adjacency matrix of the inverse graph $H = G^{-1}$. As $A = -D^A A_H^{-1} D^A$ we again obtain $G = (G^{-1})^{-1}$.

In the next section we recall the complete list of graphs with a unique perfect matching on $m \leq 6$ vertices discovered by Pavlíková and Ševčovič [12]. In addition, we present their inverse graphs and maximal subgraphs with a unique perfect matching. Interestingly enough, there are no positively invertible non-bipartite graphs with a unique perfect matching that are selfinvertible. On the other hand, there exist two negatively invertible graphs which are selfinvertible.

2 Connected graphs with a unique 1-factor

In this section we present the complete list of invertible graphs on $m \leq 6$ vertices with a unique 1-factor. Recall that a graph G has a unique 1-factor if there exists a unique subgraph spanning G and such that each vertex has the degree 1. Notice that any graph having a 1-factor should have even number of vertices.

For $m = 2$ the graph K_2 is the unique connected graph and it has a unique 1-factor. It is a positively and negatively invertible bipartite graph.

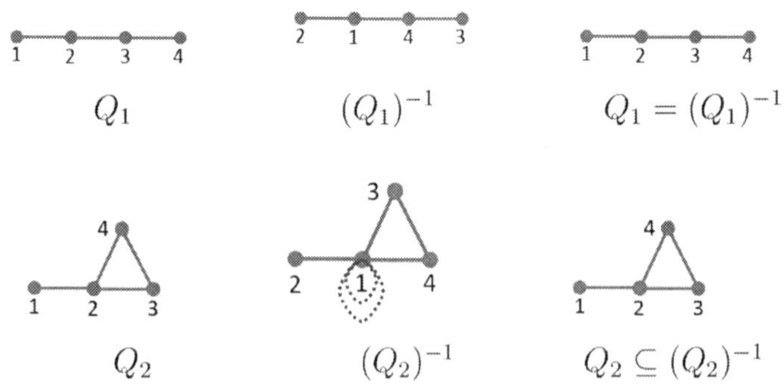

Fig. 2. Invertible graphs on four vertices with a unique perfect matching, their inverse graphs and maximal subgraphs with a unique perfect matching

For $m = 4$ there are two connected graphs Q_1, Q_2 with a unique 1-factor shown in Fig. 2. The graph Q_1 is bipartite and selfinvertible. The graph Q_2 is positively invertible but it is not selfinvertible.

Next we analyse the connected graphs on $m = 6$ vertices with a unique 1-factor. We have the following theorem.

Theorem 1 *[12] There exist 20 undirected connected graphs on $m = 6$ vertices with a unique 1-factor shown in Figures 3,4,5,6,7. All of them have invertible adjacency matrices. Except of the graph H_{19} they are integrally invertible.*

There are three bipartite graphs H_1, H_2, and H_6 which are simultaneously positively and negatively invertible. There are twelve graphs

$$H_3, H_4, H_7, H_8, H_9, H_{13}, H_{14}, H_{15}, H_{16}, H_{17}, H_{18}, H_{20},$$

which are positively invertible. The three graphs H_5, H_{10}, and H_{12} are negatively invertible. The integrally invertible graph H_{11} is neither positively nor negatively invertible. The graphs H_8, and H_{18} are iso-spectral but not isomorphic.

Recall that the proof of this result is based on the well-known Kotzig's theorem. It states that a graph with a unique 1-factor should contain a bridge belonging to the 1-factor sub-graph. Using this characterization one can construct all 20 connected graphs on 6 vertices having a unique 1-factor. We refer the reader to [12] for details.

Tab. 1. The family of graphs on 6 vertices with a unique 1-factor and their signability. Source: [12]

Graph	invertibility
H_1	positively and negatively invertible (bipartite)
H_2	positively and negatively invertible (bipartite)
H_3	positively invertible
H_4	positively invertible
H_5	negatively invertible
H_6	positively and negatively invertible (bipartite)
H_7	positively invertible
H_8	positively invertible
H_9	positively invertible
H_{10}	negatively invertible
H_{11}	noninvertible with integral inverse of the adjacency matrix
H_{12}	negatively invertible
H_{13}	positively invertible
H_{14}	positively invertible
H_{15}	positively invertible
H_{16}	positively invertible
H_{17}	positively invertible
H_{18}	positively invertible
H_{19}	noninvertible with nonintegral inverse of the adjacency matrix
H_{20}	positively invertible

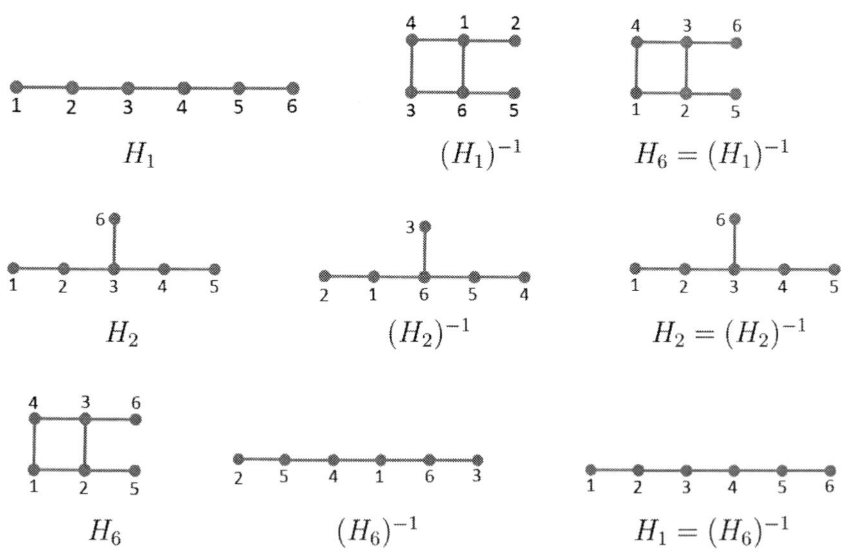

Fig. 3. Positively and negatively invertible (bipartite) graphs with a unique perfect matching, their inverses and maximal subgraphs with a unique perfect matching

2.1 Positively and negatively invertible (bipartite) graphs with a unique perfect matching on $m = 6$ vertices

Recall that a graph G_B is called bipartite if the set of vertices can be divided into two disjoint subsets such that every edge connects a vertex from the first subset into the one from the second subset.

Theorem 2 (Pavlíková and Ševčovič [12]) *Let G be an integrally invertible graph. Then G is a bipartite graph if and only if G is simultaneously positively and negatively invertible.*

In [8] McLeman and McNicholas considered graphs which can be obtained from a given connected graph by attaching pendant edges to each of vertices. They proved that such graphs are selfinvertible, i.e. $H^{-1} = H$. An example illustrating their result is the graph H_2 which is obtained from a path with three vertices by adding pendant edges to each of them.

In Fig. 3 we show all three bipartite graphs on $m = 6$ vertices with a unique perfect matching. According to the previous theorem they are simultaneously positive as well as negative invertible.

2.2 Positively invertible graphs with a unique perfect matching on $m = 6$ vertices

In this section we present the complete list of twelve non-bipartite graphs with a unique perfect matching on $m = 6$ vertices which are positively invariant (see Fig. 4,5). Interestingly, none of them is selfinvertible and their inverse graphs contain vertices with loops with even multiplicity. The inverse graphs $(H_4)^{-1}, (H_8)^{-1}, (H_{13})^{-1}, (H_{15})^{-1}$ contain multiple edges. There are four graphs denoted by H_9, H_{13}, H_{14}, and H_{18} such that they are maximal subgraphs of their inverse graphs, i.e. $H_i \subseteq (H_i)^{-1}, i = 9, 13, 14, 18$. There are two pairs of graphs (H_i, H_j) such that $H_i \subseteq (H_j)^{-1}$ and $H_j \subseteq (H_i)^{-1}$ for pairs $(i, j) \in \{(3, 7), (15, 17)\}$.

2.3 Negatively invertible graphs with a unique perfect matching on $m = 6$ vertices

In this section we present the complete list of all three non-bipartite graphs with a unique perfect matching on $m = 6$ vertices which are negatively invariant. The graphs H_5, and H_{12} are selfinvertible. Recall that according to the results due to McLeman and McNicholas [8] a graph which is obtained from a given connected graph by attaching pendant edges to each of vertices is selfinvertible. Among negatively invertible graphs with unique perfect matching on $m = 6$ vertices as an example illustrating this property one can present the graph H_5. In contrast to positively invertible bipartite graphs there exists the graph H_{12} which is selfinvertible but it cannot be obtained by means of the construction due to McLeman and McNicholas.

2.4 Noninvertible graphs with a unique perfect matching on $m = 6$ vertices

In this section we present two remaining graphs on $m = 6$ vertices which are not invertible. The first graph is denoted by H_{11} and it has the adjacency matrix which is integrally invertible. Nevertheless, A^{-1} is neither positively nor negatively signable, so H_{11} is not invertible. The graph H_{19} has the adjacency matrix with $\det(A) = 3$ and, as a consequence, A^{-1} is not integer matrix. Hence H_{19} is noninvertible as well.

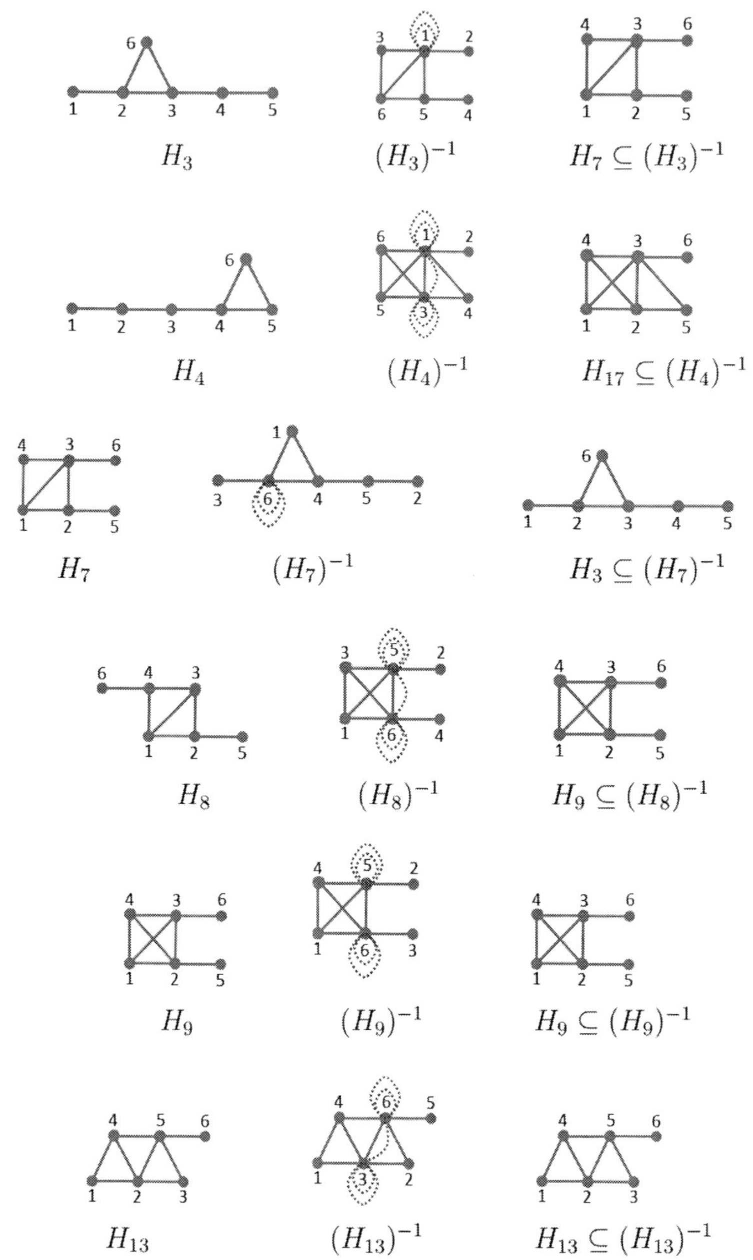

Fig. 4. Positively invertible graphs with a unique perfect matching, their inverses and maximal subgraphs with a unique perfect matching (1. part)

3 Conclusions

In this note we investigated invertible graphs with a unique perfect matching on $m \leq 6$ vertices. We recalled classical concept of graph inversion due to Godsil which we referred to as positive invertibility of a graph. We also recalled the new concept of negative invertibility proposed by the authors in [12]. By inspecting properties of the inverse graphs we showed that negatively invertible graphs exhibits properties like selfinvertibility which cannot be observed within the class of positively invertible non-bipartite graphs with a unique perfect matching on $m \leq 6$ vertices.

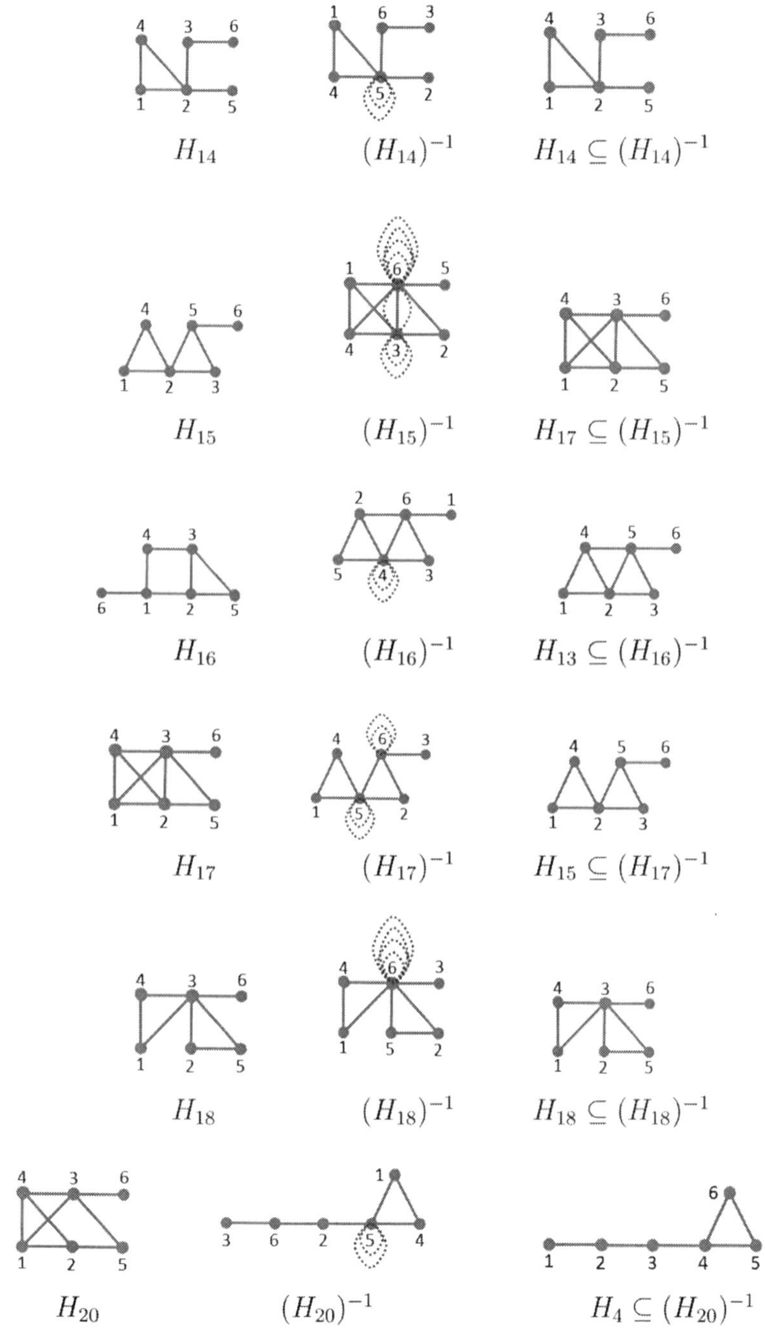

Fig. 5. Positively invertible graphs with a unique perfect matching, their inverses and maximal subgraphs with a unique perfect matching (2. part)

Acknowledgements

The research was supported by the APVV Research Grant 0136-12 (SP) and VEGA grant 1/0780/15 (DŠ).

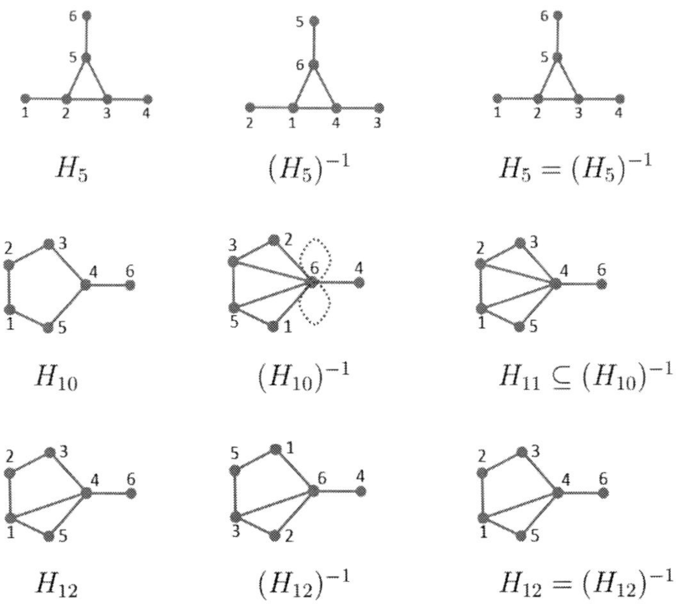

Fig. 6. Negatively invertible graphs with a unique perfect matching, their inverses and maximal subgraphs with a unique perfect matching

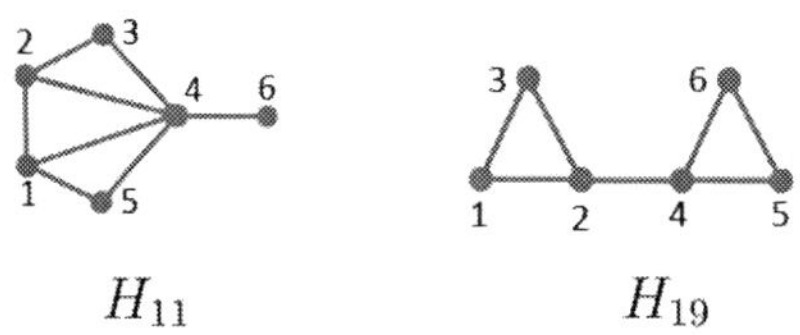

Fig. 7. Noninvertible graphs with a unique perfect matching: H_{11} has integral inverse of its adjacency matrix but it is neither positively nor negatively invertible, H_{19} is not integrally invertible

References

[1] R. B. Bapat, and E. Ghorbani, Inverses of triangular matrices and bipartite graphs, Linear Algebra and its Applications 447 (2014), 68-73.

[2] D. Cvetković, M. Doob, and H. Sachs, *Spectra of graphs - Theory and application*, Deutscher Verlag der Wissenchaften, Berlin, 1980; Academic Press, New York, 1980.

[3] D. Cvetković, I. Gutman, and S. Simić, On selfpseudo-inverse graphs, Univ. Beograd Publ. Elektrotehn. Fak. Ser. Mat. fiz., No. 602 – 633, (1978), 111–117.

[4] C. D. Godsil, Inverses of Trees, Combinatorica 5 (1985) 1, 33–39.

[5] F. Harary, and H. Minc, Which nonnegative matrices are self-inverse? Math. Mag. (Math. Assoc. of America) 49 (1976) 2, 91–92.

[6] S. J. Kirkland, and S. Akbari, On unimodular graphs, Linear Algebra and its Applications 421 (2007), 3–15.

[7] S. J. Kirkland, and R. M. Tifenbach, Directed intervals and the dual of a graph, Linear Algebra and its Applications 431 (2009), 792–807.

[8] C. McLeman, and E. McNicholas, Graph Invertibility, Graphs and Combinatorics (2014) 30, 977-1002

[9] S. Pavlíková, *Graphs with a unique* 1*-factor and matrices*, PhD Thesis (in Slovak), Comenius University, Bratislava, 1994, www.iam.fmph.uniba.sk /studium/phd/pavlikova

[10] S. Pavlíková, and J. Krč-Jediný, On the inverse and dual index of a tree, Linear and Multilinear Algebra 28 (1990), 93–109.

[11] S. Pavlíková, A note on inverses of labeled graphs. to appear: Australasian Journal on Combinatorics 2016.

[12] S. Pavlíková, and D.Ševčovič, On a Construction of Integrally Invertible Graphs and their Spectral Properties, submitted, 2016.

[13] T. Zaslavsky, Signed graphs, Discrete Applied Math. 4 (1982), 47–74.

Current address

Pavlíková Soňa, RNDr. CSc.
Institute of Information Engineering, Automation, and Mathematics,
FCFT, Slovak Technical University
812 37 Bratislava, Slovak Republic
E-mail: sona.pavlikova@stuba.sk

Ševčovič Daniel, Prof. RNDr. CSc.
Institute of Mathematics and Physics, Faculty of Mechanical Engineering,
Slovak University of Technology in Bratislava
Nám. slobody 17, 812 31 Bratislava, Slovak Republic
E-mail: sevcovic@fmph.uniba.sk

Slovak University of Technology in Bratislava
Faculty of Mechanical Engineering
16th Conference on Applied Mathematics **APLIMAT 2017**
Proceedings

ON SEMISYMMETRIC PROJECTIVE EUCLIDEAN SPACES

PEŠKA Patrik (CZ), MIKEŠ Josef (CZ), SABYKANOV Almazbek Asanovich (KG)

Abstract. In this paper we study n-dimensional semisymmetric projective Euclidean spaces. These spaces are generalization of symmetric spaces for which the covariant derivative of curvature tensor is vanishing. In our paper, we find new properties of semisymmetric projective Euclidean spaces.

Keywords: Semisymmetric spaces, projective Euclidean spaces

Mathematics subject classification: Primary 53B05; Secondary 53B99

1 Introduction

This paper is devoted to n-dimensional semisymmetric projective Euclidean spaces.

Semisymmetric spaces are spaces with affine connection ∇ in which the curvature tensor R satisfies the following condition

$$R \circ R = 0. \tag{1}$$

These spaces are generalization of *symmetric spaces* for which the covariant derivative of the curvature tensor is vanishing

$$\nabla R = 0. \tag{2}$$

P.A. Shirokov begin to study symmetric spaces [19, 20, 21] and later É. Cartan [2]. They implicitly have started to study the conditions (2) as integrability conditions of (1). The notion semisymmetric was explicitly introduced in a paper by Sinyukov [22]. In his paper he studied geodesic mappings of symmetric and semisymmetric spaces, and later in this field continues Mikeš who obtained new results, see [5, 7, 10, 16, 11, 12, 13, 14, 15, 17, 23, 24].

The geometry of symmetric and semisymmetric spaces plays an important role in Riemannian manifolds and their generalizations, see [1, 4, 8, 9, 18, 25, 26]. A great concern in semisymmetric spaces had Nomizu hypothesis [18] which was disproved later [27].

Projective Euclidean spaces were investigated in many different ways. These spaces are geodesically equivalent to Euclidean spaces. Components of affine connection of symmetric projective Euclidean spaces were obtained by P.A. Shirokov [20], see [19, 21].

In our paper, we find new properties of semisymmetric projective Euclidean spaces.

2 Special spaces with affine connection

Let $A_n = (M, \nabla)$ be an n-dimensional manifold M with affine connection ∇. In local coordinate system x with respect to a chart (U, φ), $\varphi = (x^1, \dots, x^n)$, the components $\Gamma^h_{ij}(x)$ characterize the connection ∇.

The torsion of a connection ∇ on M is a tensor field S of type $(1, 2)$ and given by

$$S(X, Y) = \nabla_X Y + \nabla_Y X - [X, Y],$$

where $[,]$ denotes the Lie bracket: $[X, Y]f = X(Yf) - Y(Xf)$, and X, Y are tangent vectors.

In a local coordinates, we obtain the following formula for the components of S

$$S^h_{ij} = \Gamma^h_{ij} - \Gamma^h_{ji}.$$

A connection ∇ is called *torsion-free* if $S \equiv 0$ and i.e. $\Gamma^h_{ij} = \Gamma^h_{ji}$.

In A_n there are defined the *curvature* and *Ricci tensors*:

$$R(X, Y)Z = \nabla_X(\nabla_Y Z) - \nabla_Y(\nabla_X Z) + \nabla_{[X,Y]} Z$$

and

$$\mathrm{Ric}(Z, Y) = \mathrm{trace}\{X \mapsto R(X, Y)Z\}.$$

In local coordinate system, we have

$$R^h_{ijk} = \partial_j \Gamma^h_{jk} - \partial_k \Gamma^h_{ij} + \Gamma^\alpha_{ik}\Gamma^h_{\alpha j} - \Gamma^\alpha_{ij}\Gamma^h_{\alpha k} \quad \text{and} \quad \mathrm{Ric}_{ij} = R_{ij} = R^\alpha_{i\alpha j}.$$

Further, the *Weyl tensor of projective curvature* is defined as

$$\begin{aligned} W^h_{ijk} \;=\; & R^h_{ijk} - \tfrac{1}{n-1}\,(\delta^h_k R_{ij} - \delta^h_j R_{ik}) + \\ & \tfrac{1}{n+1}\,(\delta^h_i R_{[jk]} + \tfrac{1}{n+1}\,(\delta^h_k R_{[ij]} - \delta^h_j R_{[ik]})), \end{aligned}$$

where δ^h_i is the Kronecker symbol and $[jk]$ denotes antisymmetrization.

A space A_n with symmetric Ricci tensor (i.e. $R_{ij} = R_{ji}$) is called e*quiaffine* and in this case the projective Weyl tensor is simplified to

$$W^h_{ijk} = R^h_{ijk} - \frac{1}{n-1}\,(\delta^h_k R_{ij} - \delta^h_j R_{ik}).$$

A space A_n is called *flat* (or *affine*), if there exists a coordinate system x for which $\Gamma^h_{ij}(x) = 0$. It is known, that the tensor criterion for these spaces is that the curvature and torsion tensors are vanishing, i.e. $S = 0$ and $R = 0$. In a natural way, we can implement the Euclidean metric, thus we call them *Euclidean* E_n.

3 Projective Euclidean spaces

In this part, we remind known facts about geodesic mappings and projective Euclidean spaces, see [3, 6, 12, 23].

A diffeomorphism f between spaces A_n and $\bar{A}_n$ with affine connection ∇ and $\bar{\nabla}$ is called *geodesic mapping* if any geodesic in A_n is mapped onto a geodesic in $\bar{A}_n$.

In a common coordinate system x with respect to f, the necessary and sufficient condition of geodesic mapping f: $A_n \to \bar{A}_n$ is the Levi-Civita equation:

$$\bar{\Gamma}^h_{ij} = \Gamma^h_{ij} + \delta^h_i \psi_j + \delta^h_j \psi_i \,, \tag{3}$$

where Γ^h_{ij} and $\bar{\Gamma}^h_{ij}$ are components of the affine connections on A and $\bar{A}_n$, ψ_i are components of a linear form and δ^h_i is the Kronecker symbol.

The Curvature, the Ricci and the Weyl projective tensors of A_n and $\bar{A}_n$ are connected by the following formulas:

$$\bar{R}^h_{ijk} = R^h_{ijk} + \delta^h_i \psi_{[kj]} + \delta^h_k \psi_{ij} - \delta^h_j \psi_{ik}, \tag{4}$$

$$\bar{R}_{ij} = R_{ij} + (n-1)\, \psi_{ij} + \psi_{[ij]},$$

$$\bar{W}^h_{ijk} = W^h_{ijk},$$

where

$$\psi_{ij} = \psi_{i,j} - \psi_i\, \psi_j, \tag{5}$$

" , " denotes the covariant derivative with respect to ∇ and the square bracket denotes alternation of indices.

The space A_n is *projective Euclidean* if A_n admits a geodesic mapping onto a Euclidean space. It is known [3, 12, 15, 23, 21], a space A_n $(n > 2)$ is projective Euclidean if and only if $W^h_{ijk} = 0$.

Evidently from (4) follows that A_n is projective Euclidean if and only if the curvature tensor has the following form

$$R^h_{ijk} = \delta^h_i \psi_{[jk]} + \delta^k_j \psi_{ik} - \delta^h_k \psi_{ij}, \tag{6}$$

where ψ_{ij} is a tensor.

The tensor ψ_{ij} necessarily satisfies the conditions

$$\psi_{ij,k} = \psi_{ik,j}.$$

The Ricci tensor of projective a Euclidean space has form

$$R_{ij} = -(n-1)\, \psi_{ij} + \psi_{[ij]}. \tag{7}$$

4 Semisymmetric projective Euclidean spaces

As we mentioned above, semisymmetric spaces are defined by the condition (1). Formula (1) can be written in the following form [25, 26, 27]:

$$R^h_{\alpha jk}R^\alpha_{ilm} + R^h_{i\alpha k}R^\alpha_{jlm} + R^h_{ij\alpha}R^\alpha_{klm} - R^\alpha_{ijk}R^h_{\alpha lm} = 0.$$

These conditions are algebraic polynomials of second order in the components of the curvature tensor R. On the base of the Ricci identities, they can be simplified into the following

$$R^h_{ijk,[lm]} = 0. \tag{8}$$

In the next lemmas, we are going to formulate when a projective Euclidean space is semisymmetric.

Lemma 1. *A projective Euclidean space A_n is semisymmetric if and only if the tensor ψ_{ij} satisfies the following condition*

$$\psi_{ij,[lm]} = 0. \tag{9}$$

Proof. If (9) is valid in a projective Euclidean space A_n, then from formula (6) follows condition (8), i.e. A_n is semisymmetric.

Conversely, let a projective Euclidean space A_n be semisymmetric, i.e. the curvature tensor satisfies (8), then from formula (6) we obtain

$$\delta^h_i\psi_{[jk],[lm]} + \delta^h_j\psi_{ik,[lm]} - \delta^h_k\psi_{ij,[lm]} = 0. \tag{10}$$

After contracting (10) with respect to the indices h and i, we have $\psi_{[jk],[lm]} = 0$, and after contracting the same formula with respect to the indices h and k, we obtain (9). □

Now, the following lemma holds.

Lemma 2. *A projective Euclidean space A_n is semisymmetric if and only if the tensor ψ_{ij} is symmetric.*

Proof. Let A_n be a projective Euclidean space.

The Ricci identity for the tensor ψ_{ij} can be written in the following form

$$\psi_{ij,[lm]} = \psi_{\alpha j}R^h_{ilm} + \psi_{i\alpha}R^\alpha_{jlm}. \tag{11}$$

Substituting (6) to (11), we obtain

$$\psi_{ij,[lm]} = 2\psi_{ij}\psi_{[lm]} + \psi_{im}\psi_{[lj]} - \psi_{il}\psi_{[mj]}. \tag{12}$$

Firstly, we suppose that tensor ψ_{ij} is symmetric. Then from (12) we have $\psi_{ij,[lm]} = 0$ and from Lemma 1 follows that A_n is semisymmetric.

On the other hand, if A_n is semisymmetric then from Lemma 1 follows $\psi_{ij,[lm]} = 0$, and from formula (12) we obtain the following equation

$$2\psi_{ij}\psi_{[lm]} + \psi_{im}\psi_{[lj]} - \psi_{il}\psi_{[mj]} = 0. \tag{13}$$

Alternating (13) with respect to the indices i and l we have

$$\psi_{ij}\psi_{[lm]} + \psi_{lj}\psi_{[im]} = 0.$$

Thus from the last formula, we obtain

$$\psi_{ij} \cdot \psi_{[lm]} = 0.$$

It evidently follows that $\psi_{[lm]} = 0$, i.e. the tensor ψ_{ij} is symmetric.

Finally, lemma 2 is proved. □

On the base of Lemma 2 we get the following theorems.

Theorem 1. *A projective Euclidean space A_n is semisymmetric if and only if A_n is equiaffine.*

Proof. From (7) follows that the tensor ψ_{ij} is symmetric if and only if the Ricci tensor is symmetric. This is the criterion of an equiaffine connection ∇, see [12] (p. 85). □

Theorem 2. *Components of a semisymmetric projective Euclidean space A_n have in a projective coordinate system x the following form:*

$$\Gamma^h_{ij} = \delta^h_i \psi_j + \delta^h_j \psi_i, \tag{14}$$

where $\psi_i = \partial_i \psi$ and ψ is a function.

Proof. From Lemma 2 follows that the tensor ψ_{ij} is symmetric and from (5) follows that the vector field ψ_i is locally gradient, i.e. there exists a function ψ that $\psi_i = \partial_i \psi$. If A_n is Euclidean with affine coordinates then formula (14) follows from formula (3). □

Remark. Spaces of constant curvature are special (pseudo-)Riemannian manifolds for which the Riemannian tensor has the following form

$$R^h_{ijk} = K \cdot (\delta^h_j g_{ik} - \delta^h_k g_{ij}), \tag{15}$$

where K is a constant and g_{ij} is a metric tensor.

It is known, how E. Beltrami proved that these spaces are projective Euclidean.

From formula (15) elementary follows that $\nabla R = 0$, and also $R \circ R = 0$, i.e. spaces of constant curvature are special symmetric and semisymmetric spaces.

Every two dimensional (pseudo-) Riemannian space V_2 has a Riemannian tensor in form (15) with Gaussian curvature K (in general it is not constant). It is clear to see from formula (15) that in V_2 it follows $R \circ R = 0$, i.e. every V_2 is semisymmetric.

References

[1] Boeckx, E., Kowalski, O., Vanhecke, L.: *Riemannian manifolds of conullity two*, World Sci. Publ., NJ, 1996.

[2] Cartan, É.: *Les espaces riemanniens symétriques*, Verhandlungen Kongress Zürich, vol. 1, no. 3, (1932), pp. 152–161.

[3] Eisenhart, L. P.: *Fields of parallel vectors in a Riemannian geometry*, Trans. Amer. Math. Soc., vol. 27, no. 4, (1925), pp. 563–573.

[4] Evtušik, L., Lumiste, Ü., Ostianu, N., Širokov, A.: *Differential-geometric structures on manifolds*, in *Probl. in geometry, Vol. 9*, VINITI, Moscow, 1979, p. 248.

[5] Hinterleitner, I., Mikeš, J.: *Geodesic mappings onto Weyl manifolds*, in *APLIMAT 2009, Proc. Conf.*, Bratislava, 2009, pp. 423–430, URL: `WOS:000272790800046`.

[6] Hinterleitner, I., Mikeš, J.: *Geodesic mappings and differentiability of metrics, affine and projective connections*, Filomat, vol. 29, no. 6, (2015), pp. 1245–1249.

[7] Hinterleitner, I., Mikeš, J., Peška, P.: *On F_2^ε-planar mappings of (pseudo-) Riemannian manifolds*, Arch. Math. (Brno), vol. 50, no. 5, (2014), pp. 287–295.

[8] Kaigorodov, V. R.: *Riemannian spaces* $^*K_n^s$, Trudy Geometr. Sem. VINITI, vol. 5, (1974), pp. 359–373.

[9] Lumiste, Ü.: *Semisymmetric submanifolds*, J. Math. Sci., vol. 7, no. 2, (1994), pp. 1609–1623.

[10] Mikeš, J.: *Geodesic mappings of special Riemannian spaces*, in *Topics in differential geometry, Vol. I, II (Debrecen, 1984)*, vol. 46, North-Holland, Amsterdam, 1988, pp. 793–813.

[11] Mikeš, J.: *Geodesic mappings of affine-connected and Riemannian spaces*, J. Math. Sci., New York, vol. 78, no. 3, (1996), pp. 311–333.

[12] Mikeš, J., et al.: *Differential geometry of special mappings*, Palacky Univ. Press, Olomouc, 2015.

[13] Mikeš, J., Radulović, Ž.: *On geodesic and holomorphically projective mappings of generalized-recurrent spaces*, Publ. Inst. Math. (Beograd) (N.S.), vol. 59(73), (1996), pp. 153–160.

[14] Mikeš, J., Radulović, Ž., Haddad, M.: *Geodesic and holomorphically projective mappings of m-pseudo- and m-quasisymmetric Riemannian spaces*, Izv. Vyssh. Uchebn. Zaved. Mat., , no. 10, (1996), pp. 30–35.

[15] Mikeš, J., Vanžurová, A., Hinterleitner, I.: *Geodesic mappings and some generalizations*, Palacky Univ. Press, Olomouc, 2009.

[16] Mikeš, J.: *On geodesic and holomorphic-projective mappings of generalized m-recurrent Riemannian spaces*, Sib. Mat. Zh., vol. 33, no. 5, (1992), p. 215.

[17] Mikeš, J., Berezovski, V., Stepanova, E., Chudá, H.: *Geodesic mappings and their generalizations*, J. Math. Sci., New York, vol. 217, no. 5, (2016), pp. 607–623.

[18] Nomizu, K.: *On hypersurfaces satisfying a certain condition on the curvature tensor*, Tôhoku Math. J. (2), vol. 20, (1968), pp. 46–59.

[19] Shirokov, A. P.: *P. A. Shirokov's work on the geometry of symmetric spaces*, J. Math. Sci., New York, vol. 89, no. 3, (1998), pp. 1253–1260.

[20] Shirokov, P. A.: *Projectively Euclidean symmetric spaces*, Trudy Sem. Vektor. Tenzor. Analizu, vol. 8, (1950), pp. 73–81.

[21] Shirokov, P. A.: *Selected investigations on geometry*, Kazan Univ. Press, 1966.

[22] Sinyukov, N. S.: *On geodesic mappings of Riemannian spaces onto symmetric Riemannian spaces*, Dokl. Akad. Nauk SSSR (N.S.), vol. 98, (1954), pp. 21–23.

[23] Sinyukov, N. S.: *Geodesic mappings of Riemannian spaces*, Nauka, Moscow, 1979.

[24] Sobchuk, V. S., Mikeš, J., Pokorná, O.: *On almost geodesic mappings π_2 between semisymmetric Riemannian spaces*, Novi Sad J. Math., vol. 29, no. 3, (1999), pp. 309–312.

[25] Szabó, Z. I.: *Structure theorems on Riemannian spaces satisfying $R(X, Y) \cdot R = 0$. I. The local version*, J. Diff. Geom., vol. 17, no. 4, (1982), pp. 531–582.

[26] Szabó, Z. I.: *Structure theorems on Riemannian spaces satisfying $R(X,Y) \cdot R = 0$. II. Global versions*, Geom. Dedicata, vol. 19, no. 1, (1985), pp. 65–108.

[27] Takagi, H.: *An example of Riemannian manifolds satisfying $R(X, Y) \cdot R = 0$ but not $\nabla R = 0$*, Tôhoku Math. J. (2), vol. 24, (1972), pp. 105–108.

Current address

Peška Patrik, Mgr.
Department of Algebra and Geometry, Faculty of Natural Science, Palacky University
17. listopadu 12, 77146 Olomouc, Czech republic
Tel. number: +420 739 142 782, e-mail: patrik_peska@seznam.cz

Mikeš Josef, prof. RNDr. DrSc.
Department of Algebra and Geometry, Faculty of Natural Science, Palacky University
17. listopadu 12, 77146 Olomouc, Czech republic
Tel. number: +420 731 579 395, e-mail: josef.mikes@upol.cz

Sabykanov Almazbek Asanovich, PhD.
Department of Algebra, Geometry, Topology, and the Teaching of Higher Mathematics,
Faculty of Mathematics, Informatics and Cybernetics, Kyrgyz National University
Abdymomunova 238, 720033 Bishkek, Kyrgyzstan
Tel. number: +996 773 829 671, e-mail: almazbek.asanovich@mail.ru

Slovak University of Technology in Bratislava
Faculty of Mechanical Engineering
16th Conference on Applied Mathematics **APLIMAT 2017**
Proceedings

REACTION TIME OF SIMPLE CATEGORIZATION OF SUBCLASSES OF FUNCTIONS

PILOUS Derek (CZ), JANDA David (CZ)

Abstract. In the last decades, there is a tendency to use general psychological methods of research in mathematics education. In our contribution, we present mostly quantitative results of our first study focused on measuring reaction time and accuracy of respondents in a specially developed task – the categorization of mathematical objects, namely one subclass of class of functions. Particular hypotheses and interpretations of the process of decision in this task, based on the statistical analysis, are proposed.

Keywords: reaction time, categorization, functions

Mathematics Subject Classification: Primary 97C30; Secondary 97I20

1 Introduction

In mathematics education, we can observe and describe behaviour of students and teachers and their mutual interactions within mathematics or rather, we can focus on the description of cognitive processes during learning and teaching. In the latter area of research, it is necessary to develop new methods of observing these cognitive processes as objectively as possible because it is impossible (yet) to observe or describe them directly. The methods used for the observation of mathematical concept development (representation, etc.) in the last decades include, among others, observations through measuring neural activity, eye-tracking or measuring reaction (response) time of some operations. In our research, we will use the third one, with a goal to describe several aspects which are connected to the concept of function, its development and its usage in a very simple cognitive task. It deals with a decision making process whether something is or is not an instance of a category created as one (specially developed) subset of set of functions, made according to the mathematical definition.

Our experiment was conducted with 22 respondents from an average high school and in this contribution, the method and selected results will be presented together with the potential and limitations of the method. Finally, some implications for mathematics education will be given in the form of some hypotheses even though the number of respondents was rather small.

2 Theoretical framework

2.1 Theories of concept representation and development

There are several influential concepts describing the quality of cognitive structure and mental processes of students in mathematics education. For example, Duffin and Simpson [2] distinguish between *natural* and *alien learning experience* in accordance to how an individual responds to them. Similarly, Skemp [7] describes *relational* and *instrumental understanding* to capture the quality of processing concept by an individual or Tall and Vinner introduce *concept image* and *concept definition* [9].

The nature of the above theories and notions is such that they qualitatively differentiate between mental processes of individuals – they create categories of these processes. In other words, in the whole spectrum of an individual's mental activities, we could find a process which could be categorized either in the first or the second category. This also shows the way in which these categories (and adequate notions) can be developed – researchers observe phenomena which, in a special context, are so characteristic or representative that there is a reason to name and describe them. But on the other hand, we cannot say that these notions describe disjoint categories and it is very rarely possible to categorize a process as a clear instance of only the first or the second category.

Intentionally, we did not use any concrete notions in previous paragraphs. Each theory and its notions are formulated in a different mathematical, psychological and educational context and that is the reason why it can enrich our description of reality in different ways even if their meanings overlap. Meanwhile, we have to be aware that we are talking about a very complex system – human brain – thus all these notions are only artificially created constructs, which should help us to better describe how our cognitive system works in a specific context. We mention this because a general quality of mental processes, not the concrete one, is behind our ideas.

These theoretical concepts (as concept image or concept definition) are useful because of their wide range. Our long-term goal is to find the way to say something new about them and through that, also something new about learning and the use of mathematical concepts by students.

Our idea of accomplishing it lies in the method of measuring reaction time of respondents. But for this purpose, it was necessary to find a task which is very simple (to learn, realize, repeat, etc.) and universal (it will be possible to use it generally, for different concepts – functions, numbers, geometrical figures, etc.). Thus we chose the concept of categorization – a decision whether something is or is not an instance of a (precisely) defined category.

Nevertheless, before it is possible to provide some statements about the connection of our proposed method with some theoretical concepts, it is necessary to describe what actually happens if an individual is doing our test. This is the main goal of this contribution.

2.2 Reaction time in mathematics education research

It is possible to find several applications of measuring reaction time in mathematics education. For example, Babai et al. [1] used the measuring of reaction time to distinguish intuitive and non-intuitive rules (in framework of Theory of intuitive rules [8]) – a similar kind of duality we described above. The task was to compare the areas and perimeters of two rectangles. The authors track reaction time with connection to intuitive rule "if A is bigger, then B is bigger". They showed that the reaction time of respondents was really faster when comparing those figures, which were in accordance to this rule.

Oberstainer et al. [4] used reaction time in their study on natural number bias with connection to fractions. They found out that even experienced mathematicians show tendency to bias their judgments by relations of natural numbers involved and explain it by different types of processes which are conducted for different fractions.

In our study, the measuring of RT was used in a task of simple categorization of mathematical objects, in which students are asked to decide if the presented object is an instance of the given category (created for this purpose) or not. It is clear that in this context, we need methodology very different from the studies above. Moreover, the usage of reaction times as a research tool in mathematics education is still new and its properties and methods of analyzing data far from settled. For these reasons, we consider methodological results of our study to be at least as important as conclusions for the studied phenomenon itself.

3 Methodology

The sample of the study is a group of twenty two students, 16 and 17 years old, from one class of an average high school in Prague. The study was realized in June 2016, in the same year when functions were taught in mathematics lessons. All the data gathered were used in the analysis.

The reaction time experiment consisted of several parts. The first one was instructive. The concept of tall function was presented to students through its definition ("Every function whose maximum is greater than the absolute value of its minimum is called *tall function.*") and some examples and non-examples[1] (see Fig. 2), and we also pointed to some inherent critical properties such as the existence of minimum and maximum (without which the function cannot fulfil the definition). We further revised the rules of graph interpretation, such as the meaning of full dot (a point included in the graph of function) and empty dot (a point excluded from the graph of function) or the presumed course of graphs as they continue beyond drawing. All demonstrated examples and non-examples differed from those used in the categorization task itself.

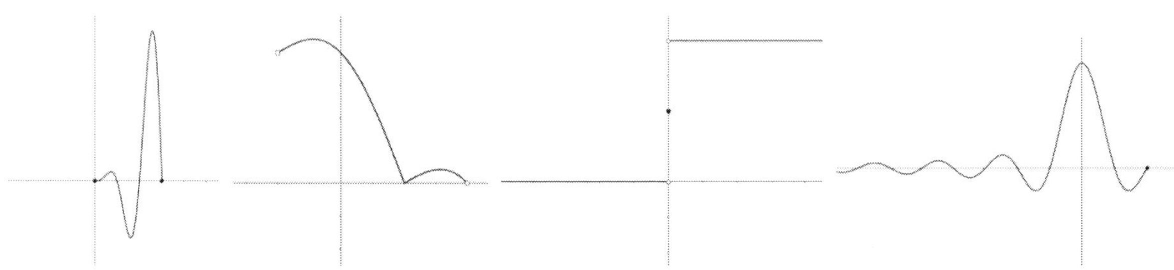

Fig. 1. Figures used as examples of tall function

[1] We use the word non-example for a particular object that is not an instance of the category in question.

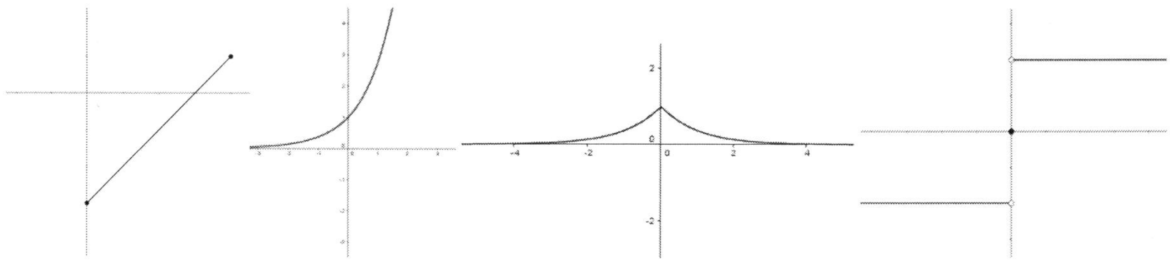

Fig. 2. Figures used as non-examples of tall function

The second part consisted of a control test, whose aim was to measure general reaction time and to set the respondents' focus on task. It was done by five letters "A"[2] or "N" displayed in a random order on the screen; respondents were asked to press the appropriate key.

The third part included the first categorization test. It was separated from the previous part by 10 seconds lasting screen where instructions were repeated. Next, twenty figures, ten examples and ten non-examples of tall function, were presented in a random order. Each figure was displayed until the respondent reacted. After that, there was one second delay before the next figure was displayed. Once the participant reacted, there was no possibility to reverse his or her decision. During this part, the participants were unaware that their reaction time was recorded.

After the first round of the categorization test, the participants were informed that not only accuracy but also speed matters. The whole test was repeated in exactly the same way as in the second and third parts of the experiment. The same set of twenty figures was used but their order was randomized again.

3.1 Pre-observations

While developing the final form of the test, we made several pre-tests and gathered information from a small group of participants including ourselves. First, almost all the participants considered the awareness of the fact that reaction time is recorded very stressful. This is the reason why we included two test trials in the main experiment – without and with the information that the reaction time is measured.

Second, many respondents mentioned, and we ourselves observed, a phenomenon which we call *blackout* – it is the situation when the respondent suddenly, often after a string of rapid and (at least subjectively) accurate responses, without a clear cause or warning, loses focus and needs to "restart". This happens regardless of the respondent's mathematical competency or any other parameter we have been able to observe.

4 Results

As described in methodology, our collected data consist of reaction times and accuracy (in the form of numbers 0/1 for wrong/right categorizations) for 20 examples and non-examples (hereinafter

[2] These keys are used because the word *ano* means *yes* and *ne* means *no* in Czech and are also conveniently placed on the keyboard.

only „examples") from 22 participants in two trials. A common problem for most tests and comparisons we wanted to perform on reaction times (like finding a relationship between reaction time and accuracy or determining a statistical distribution of reaction times) was an uneven difficulty of tasks. As we expected, some figures were more difficult to categorize than others and it took more time on average for respondents to decide about them. To control for the task's difficulty, we needed to relate reaction times to the difficulty of respective tasks. However, it cannot be determined a priori and thus, we approximated it by mean of reaction times for a given example and normalized reaction times for that example by dividing them by that mean – for reasons described below, we used geometric mean. Such normalized data will be called example-normalized reaction times here.

Let us start with the most important finding in terms of methodology which determined all further analysis. From the nature of reaction times themselves – (theoretically) unbounded positive values influenced by many factors – we expected that they might be distributed at least approximately log-normally. This assumption was confirmed beyond doubt. We used the Shapiro-Wilk test [6], considered the best test of normality of distribution as it has the best power for a given significance [5]. This test is known for its sensitivity to sample size [3], refuting normality of most large samples, even when they pass other normality tests; for that reason, the Shapiro-Wilk test is not recommended for large samples. However, the test of moderately large sample of 440 logarithmized reaction times from the first trial (in which the respondents were unaware of reaction time measuring) showed p-value of 0.319, considerably higher than chosen alpha level of 0.05. The test of logarithmized example-normalized reaction times showed (as expected, since we have not a controlled set of examples for the normality of distribution of difficulty in any way) even higher p-value of 0.399. Thus, we can conclude that example-normalized reaction times and even reaction times themselves of respondents unaware of reaction time measuring in our case are log-normally distributed and consider it safe to assume the same for any reaction time data of unaware respondents (to the extend of normality of tasks' difficulties).

However, the awareness of time measuring changes the situation quite a lot. The same tests on (logarithmized) data from the second trial show a dramatic difference: p-value for reaction times is abysmal $1.12 \cdot 10^{-5}$ and although it dramatically rises for example-normalized reaction times to 0.033, it still does not exceed the chosen alpha level, let alone come anywhere near values for the first trial. Such a low p-value can be a result of oversensitivity of the Shapiro-Wilk test, as the same data pass Kolmogorov-Smirnov test (p-value 0.196) and also visual control by Q-Q plot suggests that distribution of at least logarithmized example-normalized reaction times from the second trial is very close to normal. Still, the histogram shows leaning to left which is confirmed by z-value of skewness 2.09, and a huge difference from otherwise comparable data from the first trial stands out. Thus, we conclude that time stress – even largely self-imposed, as we did not instruct the respondents to prefer speed before accuracy nor set fixed time for completing the test – influences categorization process in a way which significantly changes the distribution of reaction times (rather than just shrink them proportionally).

The conclusion about the approximate log-normality especially of example-normalized reaction times led us to use methods connected with normal distribution not on the original data but on their logarithms. That is also why we have chosen geometric mean as central tendency of original reaction times as it corresponds to the arithmetic mean of their logarithms.

Besides the distribution of the whole datasets, we also examined the distribution of logarithmized times for individual respondents and examples. These small samples mostly passed the Shapiro–Wilk test; those which did not were considered good candidates for in-depth qualitative research. Because their frequency is higher (approximately twice) than the probability of false refuting normality by the test, we hope that at least some of them carry real and extractable information about categorization process. An interesting fact,which seemingly contradicts findings about the log-normality of trials, is that the average logarithmized reaction times of individual respondents are normally distributed according to the Shapiro-Wilk test only in the second trial and not in the first one (p-value 0.882 vs. 0.046).

Let us proceed to basic quantitative characteristics of the data. The geometrical mean of all reaction times was 6 891 ms in the first trial and 1 829 ms in the second, decreasing by ratio more than 3.75. Interesting and probably very important is a significant difference between the average time for examples and non-examples in both trials, latter being faster: 8 478 vs. 5 602 ms in the first trial and 2 141 vs. 1 562 ms in the second one. Our explanation is as follows: to confirm that a given function fulfils the definition of tall function, several conditions must be met concurrently, but to refute it only one unfulfilled condition suffices; hence checking conditions ends sooner on average in the latter case. In case of means, we would get the same results by calculating the arithmetic mean of logarithmized values and exponentializing the result. This does not hold for standard deviation: in the first trial, it is as high as 5 831 ms when calculated directly, while calculating the standard deviaton of logarithmized values and exponentializing the result yields much lower value of 1 921 ms. For the second trial, the situation is reversed, respective numbers being 1 332 and 1 752 ms. The correlation between the average reaction times for individual examples in the first and second trial is 0.765, the same correlation for logarithmized times is 0.783. These values are higher than correlations for either individual respondents or whole datasets, with the exception of one respondent in the original, non-logarithmized times.

For methodological reasons, especially assuming the learning period in the beginning of the test, the progression of average reaction time in the course of the test was of high interest to us. We found that in both trials, the first task was on average the slowest (in both non-normalized and normalized times) and after that the reaction times for the following several figures were significantly decreasing (four in the first trial, one in the second one). We believe that this fact can be caused by a combination of two effects – a specific time-consuming „activation" of respondent for work and big amount of learning from the first examples. Interestingly enough, the decrease was followed by a similarly sharp increase in the next two figures, ending in the second highest average time (for both trials). Then there was a decrease again followed by steady minor fluctuations around a more or less constant value for the rest of the test (again in both trials and both non-normalized and normalized times). These findings are less statistically significant than the ones above. We need to confirm them on a larger sample of respondents and in case of their real significance, interpret them by qualitative research.

Let us turn our attention to accuracy. The average accuracy was 76.82 % in the first trial and 74.55 % in the second. As we can see, reaction times dropped under time stress considerably, while accuracy remained almost the same. It raises a question whether respondents could memorize examples and react in the second trial not based on decision process but on simple association. While we cannot completely rule out such influence, we consider its large role unlikely for several reasons. First, it is quite improbable that respondents could memorize such complex data as graphs of functions in such a short time. Second, the interviews with respondents ruled out the possibility of active memory record, as they were mostly not able to recall even immediately the preceding graph, and our own experience while taking the test shows that even we, who are much more familiar with graphs of functions than the respondents, were able to recognize passively at most one or two most „impressive" graphs. Third and most important, the casewise change of accuracy between trials (the ratio of examples categorized once correctly and once incorrectly by the same respondent) was quite high – 10.5 %. Given the assumed stability of decision process (in the sense that it reaches the same results with the same data), such a ratio is more consistent with new decisions on most items than with remembering them from before.

The average accuracy for individual respondents ranges from 0.25 to 1 in both trials. The correlation between them is 0.928. Two respondents in the first trial and three in the second one achieved accuracy smaller (but only one in each case significantly) than 0.5, which corresponds to random answers. The average accuracy in individual examples ranged from 0.455 to 1 in the first trial and from 0.409 to 0.955 in the second one, the correlation between them being 0.877.

We found a moderately strong negative correlation between the average reaction times and the average accuracy for individual examples, –0.67 for the original data and –0.64 for the logarithmized ones in the first trial and –0.43 for the original data and –0.45 for the logarithmized ones in the second trial. These result show that the difficulty of examples manifests itself in both reaction time and accuracy.

5 Discussion

Generally, there is evidence that decision making processes in cases in which respondents are/are not aware of reaction time measuring differ, even when their accuracy is almost the same. Self-imposed time stress significantly influences the distribution (not just the overall average) of reaction times; in its absence, distribution is log-normal with high fidelity. Distributions for individual respondents and tasks are also mostly log-normal; the exception to the rule can be a good criterion for selecting cases for qualitative research. Moreover, decision making (at least in this particular case) is not dominantly guided by time spending nor is error tolerance; both criteria are used comparably regardless of time stress. As for the methodology of reaction time measuring, using the same medium sized set of data as complex as graphs of functions two times in the row with a minimal pause does not seem to affect results; the average accuracy does not increase and the casewise change of decision is quite high.

Acknowledgement

The authors gratefully acknowledge project PRVOUK P15 *The school and teaching profession in the context of growing educational demands* for financially supporting this work.

References

[1] BABAI, R., et al. Intuitive rules in science and mathematics: A reaction time study. *International Journal of Mathematical Education in Science and Technology*, 2006, 37.8: 913-924.

[2] DUFFIN, Janet M.; SIMPSON, Adrian P. A search for understanding. *The Journal of Mathematical Behavior*, 2000, 18.4: 415-427.

[3] FIELD, Andy. *Discovering statistics using SPSS (3rd ed.).* Los Angeles [i.e. Thousand Oaks, Calif.]: SAGE Publications, 2009. p. 143.

[4] OBERSTEINER, Andreas, et al. The natural number bias and magnitude representation in fraction comparison by expert mathematicians. *Learning and Instruction*, 2013, 28: 64-72.

[5] RAZALI, N. M.; WAH, Y. B. Power comparisons of Shapiro-Wilk, Kolmogorov-Smirnov, Lilliefors and Anderson-Darling tests. *Journal of Statistical Modeling and Analytics*, 2011, 2(1): 21-33.

[6] SHAPIRO, S. S.; WILK, M. B. An Analysis of Variance Test for Normality (Complete Samples). *Biometrika,* 1965, 52(3/4): 591-611.

[7] SKEMP, R. Instrumental understanding and relational understanding. *Mathematics Teaching*, 1976, 77: 20-26.

[8] STAVY, Ruth; TIROSH, Dina. *How students (mis-) understand science and mathematics: Intuitive rules.* Teachers College Press, 2000.

[9] TALL, David; VINNER, Shlomo. Concept image and concept definition in mathematics with particular reference to limits and continuity. *Educational studies in mathematics*, 1981, 12.2: 151-169.

Current address

Pilous Derek, Mgr. PhD.
Department of Mathematics and Mathematical Education, Faculty of Education, Charles University
Magdalény Rettigové 4, 116 39 Praha 1, Czech Republic
E-mail: Derek.Pilous@seznam.cz

Janda David, Mgr.
Secondary School for the Administration of the EU
Lipí 1911, 193 00 Praha 9 - Horní Počernice, Czech Republic
and
Department of Mathematics and Mathematical Education, Faculty of Education, Charles University
Magdalény Rettigové 4, 116 39 Praha 1, Czech Republic
E-mail: Djanda@seznam.cz

Slovak University of Technology in Bratislava
Faculty of Mechanical Engineering
16th Conference on Applied Mathematics **APLIMAT 2017**
Proceedings

AN ANALYSIS OF ERRORS IN ALGEBRAIZATION OF A MATH WORD PROBLEM AND THE ROSNICK-CLEMENT PHENOMENON

PLOTHOVÁ Lucia (SK), NAŠTICKÁ Zuzana (SK)

Abstract. The purpose of this research was to examine the basic algebraic skills of students, and to identify errors they make when solving a simple math word problem. The research was conducted among 103 first graders of a secondary comprehensive school aged 15-16 and 135 first-grade university students. The research instrument was a short three-item test. Despite proper acquisition of the algorithm for solving linear equations, in algebraization of the math word problem altogether 35.93% of the secondary school students and as much as 72.59% of the university students were unsuccessful. The reversal error occurred in 27.18% of the incorrect secondary school students' solutions and in 35% of the incorrect university students' solutions of the word problem. The students' low success rates could be caused by their miscomprehension of the texting of the word problem, which is associated with their reading comprehension skills. The reversal error occurred due to mechanical procedures employing no deeper thinking. Many of the errors can be remedied by teaching students to verify and interpret their results back into the non-mathematical language.

Keywords: algebraic thinking, math word problem, linear equation, reversal error

Mathematics Subject Classification: 97H20, 97H30

1 Introduction

The main objective of mathematics education at primary and secondary schools is to make students learn how to employ mathematics and mathematical thinking in their future daily lives, i.e to mathematize real-life situations in order to achieve an objective solution. The manner and the content of education have to correspond with this objective (ŠPÚ, 2015c). One of the crucial points in this effort is to systematically develop students' algebraic thinking.

According to Knuth (2005) algebraic reasoning depends on an understanding of a number of key ideas, of which equivalence and variable are, arguably, two of the most fundamental. The ubiquitous presence of the equal sign symbol in mathematics at all levels highlights its important role in mathematics (Kieran, 1992). Several research studies indicate that many students do not perceive the equal sign as a symbol expressing equivalence (i.e. a symbol which denotes the relation between two variables), but rather as a way to express the result of arithmetic operations (Knuth, 2005).

In this paper we are concerned with the analysis of errors made by students in solutions of a math word problem whose mathematical model is a simple linear equation. We primarily focus on the identification of Rosnick-Clement phenomenon, i.e. the reversal error, in the incorrect students' solutions of the math word problem. Our effort is to explain possible causes of the errors made by students in their solutions and to propose a remedial strategy.

2 Algebraic thinking

Algebraic thinking is one of the essential components of mathematical thinking. It can be viewed in two manners. Algebraic thinking can be considered as a transition from arithmetic to algebra in which its characteristic properties are generality and abstraction. The second view on algebraic thinking is considering it to be an indispensable requirement for the investigation of mathematical functions.

Experts whose research is focused on algebraic thinking define it by characterizing its various aspects. According to Kaput (2008) there are two main aspects of algebraic thinking: (i) making generalizations and expressing those generalizations in increasingly, conventional symbol systems, and (ii) reasoning with symbolic forms, including the syntactically guided manipulations of those symbolic forms. In terms of the first aspect, generalizations are produced, justified and expressed in various ways. The second aspect refers to the association of meanings to symbols and to the treatment of symbols independently of their meaning (Chimoni, 2015). Kaput (2008) suggests that these two aspects of algebraic thinking denote reasoning processes which are considered to flow through varying degrees throughout three strands of algebraic activity: (i) generalized arithmetic, (ii) functional thinking, and (iii) the application of modelling languages for describing generalizations.

While thinking algebraically to solve problems, various thinking habits come into play, such as doing-undoing, building rules to represent functions, and abstracting from computation (Cai, 2005). Algebraic thinking comprises the construction and representation of patterns and regularities, deliberate generalization, and active exploration and conjecture (NCTM, 1993). By these assumptions, algebraic thinking cannot be developed passively; quite the contrary, in order to develop and improve algebraic thinking, learners have to be actively involved in mathematical activities. Kieran (2004) suggests that for proper development of learners' algebraic thinking it is important to focus (i) on relations rather than on mere calculation of a numerical answer, (ii) on both operations and their inverses, and on the mutually related idea of doing and undoing, (iii) on representing as well as solving a problem, not only on merely solving it, (iv) on numbers as well as letters, not just on numbers alone, and (v) to focus anew on the meaning of the equal sign. Rephrasing (iv), focusing on numbers as well as letters includes (a) working with letters that can be either unknowns, variables, or parameters, (b) accepting unclosed literal expressions as responses, and (c) comparing expressions for equivalence based on properties rather than on numerical evaluation. It is, thus, not the mere use of mathematical notation which constitutes the algebraic thinking of learners, but the manner in and extent to which they are able to manipulate the general (Radford, 2008).

2.1 Algebraic thinking in primary and secondary education

Algebraic thinking of learners ought to be developed gradually, starting as early as in pre-primary education. It is an important prerequisite of successful mastering of all content areas related to algebra. Also, it is crucial for learners to acquire such knowledge which they are likely to need in their further education and future daily life (ŠPU, 2015c).

In primary education algebraic thinking is mostly based on long-term propaedeutics of concepts such as variable, equality and inequality, in form of various mathematical tasks and exercises (ŠPU, 2015a). At this stage algebraic thinking is developed when learners are asked to fill in a number in an equality where the missing number is represented by an object (e.g. a blot, a piece of fruit, a flower, a planar figure), and by solving the so-called *sum pyramid puzzles* or other mathematical puzzles such as *"I'm thinking of a number. If I add 20 and take away 10, I get a number which is by 5 greater than 45. What number am I thinking of?"*

In lower secondary education the significant proportion of propaedeutics of equalities, inequalities and variables persists in the development of algebraic thinking. Algebraic thinking is further developed through curricular content about ratio, direct and inverse proportion, algebraic expressions, and the graphical representation of linear dependences. By the end of the 8th grade students should be able to compose a simple algebraic expression with a variable based on a word description of a certain situation, express a particular unknown from a simple formula, and solve simple math word problems leading to a linear equation without particular formalization. By the end of lower secondary education students should be able to solve linear equations with a multiple occurrence of one variable, and solve context-based tasks by means of linear equations (ŠPÚ, 2015b). The result of mathematics education should be learners who are able to use mathematical notation properly in order to represent relations, and who understand letters in mathematical notations as numbers, and who can do useful rearrangements of simple algebraic expressions.

As students proceed from lower to upper secondary educational level, the have to master the use of mathematical notation in a much broader extent. At secondary school students should be able to use a wider range of mathematical tools and achieve a higher level of formalization of mathematical knowledge, including use of symbols and ability to abstract and generalize. Their algebraic thinking is developed mainly through curricular content about equations, functions, sequences, and analytical geometry. Inter alia, students should be able to express simple verbally described relationships by means of variables, constants, equalities and inequalities, provide relevant reasoning of solution strategies to various types of equations, and in simple cases choose an appropriate representation of given relations between two quantities (ŠPÚ, 2015c).

As shown above, students start encountering and experiencing algebraization as early as they start attending primary school, and their algebraic thinking is further developed until the very end of their secondary education.

2.2 The reversal error

In 1980 U.S. mathematics education researchers Peter Rosnick and John Clement conducted a research focused on the "translation" of a verbal text into the language of formulas and backwards. The research sample consisted of students from various age categories and from a variety of study programs. Besides other things, they were asked to solve the following math word problem:

Write an equation using the variables S and P to represent the following statement: "There are six times as many students as professors at this university" Use S for the number of students and P for the number of professors (Rosnick, 1980).

Fig. 1. The texting of the math word problem by Rosnick & Clement

Based on the analysis of students' solutions, Rosnick and Clement found that around 37 % of the students were unsuccessful in solving the task. As much as two thirds of the students made the so-called reversal error (i.e. as a results they stated $6S = P$). Such an error is usually referred to as the Rosnick-Clement phenomenon. Rosnick and Clement learned, however, that such an error cannot be remedied by a short instruction. The researchers informally interviewed some of the students who made the reversal error, the interviews lasting from 45 to 60 minutes each. The interviewer applied one of the following instructional strategies in each of the interviews:

1. Simply telling the students that the reversal is incorrect.
2. Telling the student that the variable should be thought of as "number of students," not "students."
3. Pointing out (with pictures) that since "students" is a bigger group than "professors", one must multiply the professors by six to create an equality.
4. Asking the students to test the equations by "plugging in" numbers.
5. Asking the students to draw graphs and/or tables.
6. Specifically showing the students how to set up a proportion to solve the problem.
7. Demonstrating a correct solution to the students, using an analogous problem.

After the interviews, students sat another test with an analogous task. Although during the interviews it seemed that students could be able to solve such a task, it turned out that most of the students made the reversal error again (Rosnick, 1980).

The follow up research conducted by J. Lochhead brought no better results. It confirmed the occurrence of the reversal error in most of the respondents whose solution to the given task was incorrect (Lochhead, 1980). Further research in this issue was carried out by the Mathematical Institute of the University in Klagenfurt in Austria. They found that the reversal error can be observed in all age groups above 13 years (Fischer, 1992).

3 Research sample and methods

The research was conducted at a town comprehensive secondary school and at a university in Slovakia. The research sample consisted of 103 first graders of a secondary comprehensive school aged 15-16 and 135 first-grade university students of four bachelor study programs (Pre-school and Elementary Education, Marketing Communication and Advertising, Applied Informatics, Mathematical and Informatical Methods in Economy).

The purpose of the research was to examine the basic algebraic skills of the respondents, i.e. their ability to translate a word problem into an algebraic equation, and to identify errors they make when solving a simple math word problem. The research instrument was a short test consisting of three tasks (see Fig. 2).

***Task* 1** Solve the equation $5x = 7$.
***Task* 2** Solve the equation $4x = 2x + 2$.
***Task* 3** Let M be the number of men and Z the number of women present in university/school canteen. There are three times as many women as men in the canteen. Express the dependency using variables M and Z.

Fig. 2. The texting of the test

Tasks 1 and 2, which require trivial solutions of linear equations based on lower secondary school curricular content, were included in the test in order to show that the respondents had had acquired the algorithm for solving simple linear equations. Task 3 served for identification of errors made by respondents, in particular the reversal error. The respondents completed the test in around 8 minutes.

3 Results

In this section we present results obtained by means of the content analysis of students' solutions of the test. In particular, we focus on identification of the reversal error in students' solutions, but also on identification of other errors made by students in their solutions of Task 3.

	Secondary school students		University students	
	incorrect (%)	reversal error (%)	incorrect (%)	reversal error (%)
Task 1	2.91	-	3.70	-
Task 2	3.88	-	2.22	-
Task 3	35.93	**27.18**	72.59	**35.00**

Tab. 1. The un-success rates of the respondents

Neither the secondary comprehensive school students, nor the university students had serious difficulties when solving Task 1 and 2 (see Tab. 1). Those few students who were unsuccessful in solving the two tasks made errors related to incorrect performance of arithmetic operations. This indicates that students had had acquired the algorithm for solving simple linear equations. Despite proper acquisition of the algorithm, many of the students had difficulties with algebraization of the math word problem, requiring a linear equation analogous to the two equations in Task 1 and 2, which is suggested by high un-success rates of students in Task 3. As shown in Tab. 1, there is a significant difference in success rates of the secondary school students and the university students in Task 3. In addition, all secondary school students at least attempted to solve Task 3, whereas as much as 17% of university students did not even try (most of them added a brief comment, saying "I don't know"). In solving the math word problem in Task 3 altogether 35.93% of the secondary school students were unsuccessful, and the reversal error (i.e. 3Z=M) occurred in 27.18% of the incorrect secondary school students' solutions. As much as 72.59% of the university students made an error in algebraization in their solutions of Task 3, and the reversal error occurred in 35% of the incorrect university students' solutions of the task.

Secondly, we were interested in the un-success rates of university students with respect to the type of secondary school which they had graduated at.

	University students who graduated at comprehensive secondary schools		University students who graduated at secondary vocational schools	
	incorrect (%)	reversal error (%)	incorrect (%)	reversal error (%)
Task 1	0.00	-	4.17	-
Task 2	0.00	-	2.09	-
Task 3	57.78	26.67	77.08	16.67

Tab. 2. The un-success rates of the university students who graduated at secondary comprehensive or vocational schools

	University students who graduated at secondary business schools		University students who graduated at secondary pedagogical schools	
	incorrect (%)	reversal error (%)	incorrect (%)	reversal error (%)
Task 1	3.85	-	6.25	-
Task 2	7.09	-	0.00	-
Task 3	**81.82**	**38.46**	66.67	18.75

Tab. 3. The un-success rates of the university students who graduated at secondary business or pedagogical schools

In accordance with what was hypothesized, university students who had graduated at comprehensive secondary schools were the most successful in solving Task 3 in comparison to students who had graduated at other types of secondary school. Surprisingly, there was as much as 57.78% of comprehensive secondary school graduates whose solution of Task 3 was wrong. Secondary business school graduates were the least successful university students in algebraization of the math word problem, as only 18.18% of them managed to solve Task 3. Also, in this group of university students the reversal error occurred the most frequently. In case of secondary vocational school graduates the occurrence of the reversal error was quite low in comparison with the occurrence of other types of errors. The results of the analysis also show that the occurrence of the reversal error in secondary comprehensive school students and secondary comprehensive school graduates is almost the same (27.18% and 26.67% respectively), though the occurrence of correct algebraization of the math word problem is by almost 22% lower in the secondary comprehensive school graduates.

Next, the analysis is focused on other types of errors which occurred in the respondents' solutions of Task 3. The second most frequent error was solution in form of an inequality (henceforth the *inequality error*, see Fig. 3). This type of error was made by 17% of university students. None of the secondary school students made the inequality error.

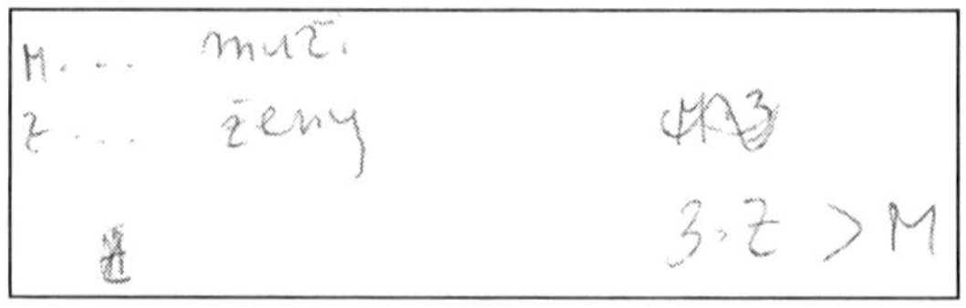

Fig. 3. A sample of a student solution of Task 3 containing the reversal error and the inequality error.

One of the possible causes of the presence of the inequality error in students' solutions can be the misinterpretation of the input information. The texting of Task 3 asks solvers to express the dependency between the number of women and the number of men by means of an equation. The inequality notation represents the algebraization (i.e. mathematization) of another situation, namely "triple the number of women is greater than the number of men". Another cause of the inequality error can be the misconception of "equation". Students might have understood it as a general mathematical notation which serves for expression of a dependence between variables.

Some students (7%) mathematized Task 3 by means of an expression which represents the total number of people in the canteen (henceforth the *sum error*, see Fig. 4). Although they correctly expressed the number of women and men, they failed in understanding the task (to express the dependence).

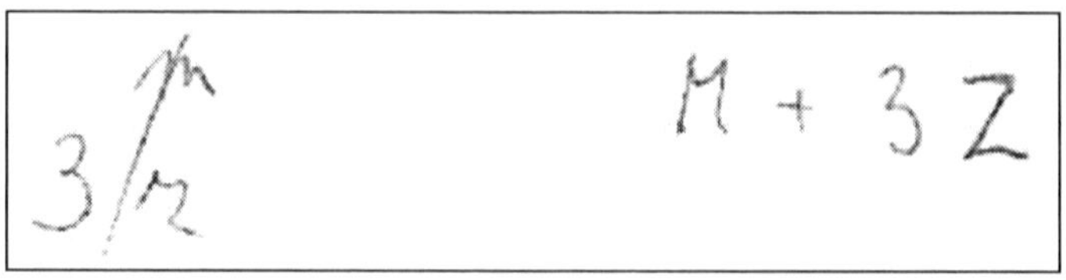

Fig. 4. A sample of a student solution of Task 3 containing the reversal error and the sum error.

The most serious error which occurred in the solutions seems to be the students' conclusion that the sum of the women and men in the canteen should be equal to zero (henceforth the *zero error*, see Fig. 5). The zero error was made by 8% of all students. They did not realize that if such an equation were to hold, either the number of women or the number of men in the canteen would be negative.

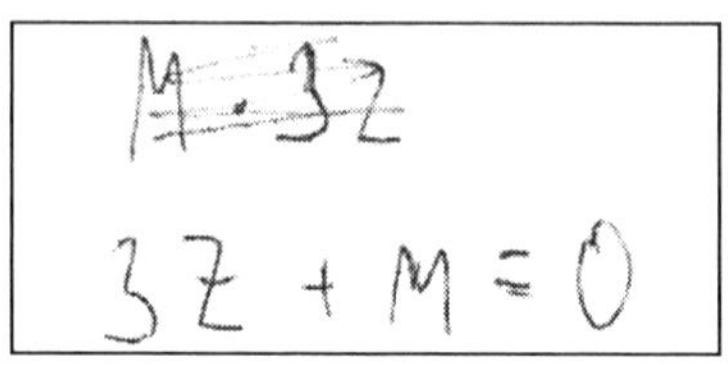

Fig. 5. A sample of a student solution of Task 3 containing the reversal error and the zero error.

An interesting fact is that in several solutions variables x and y were used instead of variables Z and M which were, actually, required by the texting of Task 3 (Figs. 6 and 7). This can be caused by the generally prevailing use of symbols x, y and z for variables or unknowns in mathematics lessons, i.e. the students had probably encountered only this notation for variables and unknowns when composing or solving equations.

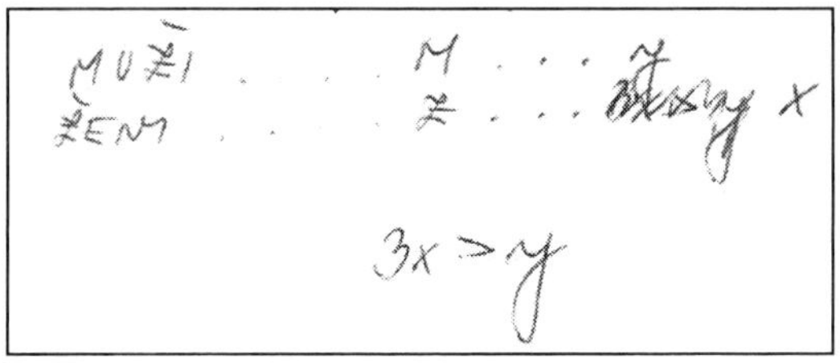

Fig. 6. A sample of a student solution of Task 3 containing the *x* and *y* notation, the reversal error and the inequality error.

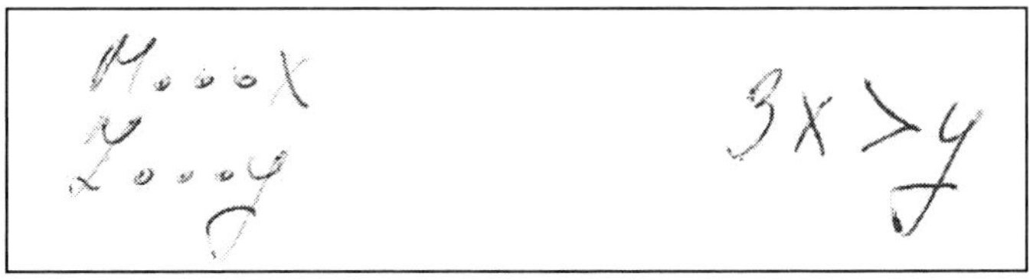

Fig. 7. A sample of a student solution of Task 3 containing the *x* and *y* notation and the inequality error.

4 Discussion

The analysis of students' solutions has shown that students had no serious difficulties with solving given linear equations. In other words, they had had acquired the algorithm for solving linear solutions. If any mistakes were made in solutions of the given linear equations, the mistakes were related to arithmetic operations performed improperly. As far as the mathematization of Task 3 is concerned, the respondents achieved unsatisfying results. The mathematical model of the situation described in Task 3 was properly developed by 64.07% of secondary school students and as few as 27.41% of university students. One of the possible reasons of low success rates of students in solving Task 3 can be miscomprehension of the texting of the word problem. The issue of (mis)comprehension of the word problem is connected with the level of one's reading comprehension skills, whose development is often undervalued and neglected.

Another cause of the high proportion of incorrect solutions of Task 3 are errors which were made in the mathematization process. The most frequent error in solutions of Task 3 was the so-called reversal error. The reversal error occurs when the solver literally translates the texting of a math word problem into mathematical notation, without realizing the important nuances. We believe that such a way of solving math word problems is mechanical and employs no deeper thinking.

The secondary comprehensive school students, whose algebraic thinking is expected to be developed less in comparison to secondary comprehensive school graduates (see subsection 2.1), achieved higher success rates in solving Task 3. This can indicate that the manner in which algebraic thinking is developed at school does not lead to long-term acquisition of quality algebraic thinking. However, the high success rates of secondary school students in comparison with university students who had graduated at the same type of secondary school can be simply caused by the fact that around the time when the research was carried out the students were just learning curricular content about linear equations. The frequency of the occurrence of the reversal error in secondary comprehensive school students and in secondary comprehensive school graduates differed just subtly (27.18% and 26.67% respectively). Surprisingly, as much as 81.82% of secondary business school graduates did not manage to solve Task 3, although their algebraic thinking is supposed to be developed both in mathematics lessons and in the economic subject lessons.

The second most frequent error made by students was proposing the mathematical model of given situation in form of an inequality. Most likely the reason was that students literally transcribed the given situation by means of mathematical notation, without deeper thinking. Also, it turns out that the concept of equation (several students thought of the sum of women and men equal to zero) and the meaning of the equal sign are often misunderstood (the equal sign is often understood only as a sign after which the final result comes).

In general, one type of notation for variables in algebraization of math word problems prevails. This was shown in those students' solutions of Task 3 where letters x and y were used instead of M and Z which was originally required by the texting of the task. That is a consequence of students having encountered mostly x and y as symbols for variables and unknowns.

We believe that many of the errors made by students in their attempts to mathematize situation given in the math word problem can be remedied by teaching students to verify and interpret their results back into the non-mathematical language. Verification and interpretation of mathematical result expressed by means of mathematical notation can at least suggest the solver that something is wrong, if the interpretation e.g. includes a negative number of people in a room. We believe verification and interpretation of mathematical result should be paid more attention in school instruction in general.

Acknowledgement

The presented research was supported by institutional projects UGA VII/14/2016 *Improving key competences of mathematical modelling in secondary mathematics education,* supported by Constantine the Philosopher University in Nitra. The authors gratefully acknowledge Gymnázium Andreja Vrábla in Levice for continual cooperation.

References

[1] CAI, J., et al. The development of students' algebraic thinking in earlier grades. *ZDM Mathematics Education*, vol. 37, no. 1, 2005: p. 5-15, ISSN 1863-9704.

[2] FISCHER, R., MALLE, G. *Človek a matematika*. SPN, 1992, ISBN 80-08-01309-5, 336 p.

[3] CHIMONI, M., PITTA-PANTAZI, D. Connections between algebraic thinking and reasoning processes. In *Proceedings of the 9th Congress of the European Society for Research in Mathematics Education*, Prague, 2015, p. 398-404.

[4] KAPUT, James J. What is algebra? What is algebraic reasoning. In *Algebra in the early grades*, New York: Routledge, 2008, p. 5-17, ISBN 978-0-8058-5473-2.

[5] KIERAN, C. The learning and teaching of algebra. In *Handbook of research on mathematics teaching and learning*, New York: Macmillan, 1992, p. 390-419, ISBN 978-1-59311-598-2.

[6] KIERAN, C. Algebraic thinking in the early grades: What is it? *The Mathematics Educator*, vol. 8, no.1, 2004: p.139-151, ISSN 0218-9100.

[7] KNUTH, E., et al. Middle school students' understanding of core algebraic concepts: Equivalence & Variable1. *ZDM Mathematics Education*, vol. 37, no. 1, 2005: p. 68-76, ISSN 1863-9704.

[8] LOCHHEAD, J. Faculty interpretations of simple algebraic statements: The professor's side of the equation. *The Journal of Mathematical Behavior*, vol. 3, no. 1, 1980: p. 29-37, ISSN 0732-3123.

[9] NCTM. Algebra for the Twenty-First Century: Proceedings of the August 1992 Conference. Reston: NCTM, 1992.

[10] RADFORD, L. Iconicity and contraction: A semiotic investigation of forms of algebraic generalizations of patterns in different contexts. *ZDM Mathematics Education*, vol. 40, no. 1, 2008: p. 83-96, ISSN 1863-9704.

[11] ROSNICK, P., CLEMENT, J. Learning without understanding: The effect of tutoring strategies on algebra misconceptions. *The Journal of Mathematical Behavior*, vol. 3, no.1, 1980: p. 3-27, ISSN 0732-3123.

[12] ŠPÚ. Štátny vzdelávací program pre 1. stupeň základnej školy; príloha *Matematika*. 2015a. Retrieved from http://www.statpedu.sk/sites/default/files/dokumenty/inovovany-statny-vzdelavaci-program/matematika_pv_2014.pdf.

[13] ŠPÚ. Štátny vzdelávací program pre 2. stupeň základnej školy; príloha *Matematika*. 2015b. Retrieved from http://www.statpedu.sk/sites/default/files/dokumenty/inovovany-statny-vzdelavaci-program/matematika_nsv_2014.pdf.

[14] ŠPÚ. Štátny vzdelávací program pre gymnáziá so štvorročným a päťročným vzdelávacím programom; príloha *Matematika*. 2015c. Retrieved from http://www.statpedu.sk/sites/default/files/dokumenty/inovovany-statny-vzdelavaci-program/matematika_g_4_5_r.pdf.

Current address

Plothová Lucia, Mgr.
Department of Mathematics, Faculty of Natural Sciences,
Constantine the Philosopher University in Nitra
Tr. A. Hlinku 1, SK-949 74 Nitra, Slovak Republic
Tel. number: +421 37 6408 722, e-mail: lucia.plothova@ukf.sk

Naštická Zuzana, Mgr.
Department of Mathematics, Faculty of Natural Sciences,
Constantine the Philosopher University in Nitra
Tr. A. Hlinku 1, SK-949 74 Nitra, Slovac Republic
Tel. number: +421 37 6408 696, e-mail: zuzana.nasticka@gmail.com

AN ANALYSIS OF STUDENTS' USE OF MATHEMATICAL MODELS IN SOLVING TASKS WITH REAL-LIFE CONTEXT

PLOTHOVÁ Lucia (SK), PÁLENÍKOVÁ Kitti (SK), NAŠTICKÁ Zuzana (SK), FLORKOVÁ Michaela (SK), ŠUBOVÁ Lenka (SK)

Abstract. Despite the general tendency to emphasize the importance of mathematics in solving daily-life problems, students' difficulties with such tasks still widely persist. Mathematics education should be focused on making students learn how to mathematize real-life situations by means of mathematical models. Also, it seems wise to incorporate team work in mathematics education, for while solving problems in groups students have to communicate and provide proper arguments. In this research study we present results of a content analysis conducted on students' solutions of two mathematical tasks related to gas consumption in households and gas distribution costs. The analysis was focused on how 204 students, aged 16-18, working in 51 groups of four, mathematically modelled the given real-life situations. Their solutions' were analysed in terms of four main stages of the modelling cycle by Kaiser (real situation, real world model, mathematical model, and mathematical results). The results of the analysis indicate that students experienced serious difficulties at all four stages of the modelling cycle. Most students failed at the very first stage of the cycle, i.e. they misunderstood the real-life situation, its conditions and consequences. We assume this is closely associated with the reading comprehension skills of the students. Also, the interpretation of the obtained mathematical results was absent in students' solutions, which suggests that students neglected proper verification of their mathematical model. Although we expected students to go through the stages of the modelling cycle sequentially, their solutions demonstrated that students often skipped some of the stages.

Keywords: mathematical modelling, real-life problems, mathematical literacy, problem solving

Mathematics Subject Classification: 97M10; 97M30

1 Introduction

One of the most important goals of mathematics education in school environment is to make students learn how to use correct mathematical language, i.e. terms and symbols, and, especially, to mathematize real-life situations by means of mathematical models. The teaching of mathematics should develop students' functional, cognitive and metacognitive abilities. At the end of the secondary level of education students should (i) be able to reflect on the process of their own learning and thinking when acquiring and processing new knowledge and information, and to apply various learning strategies; (ii) be able to critically evaluate information and information sources, and to creatively process and make practical use of information; (iii) apply mathematical thinking in order to solve practical problems in daily situations; (iv) use mathematical models of logical and spatial thinking, and use various mathematical forms of representation, such as formulas, models, statistics, diagrams, graphs, tables (ŠPÚ, 2015).

A suitable choice of organizational forms and methods of instruction facilitates development of many other competences useful in life, such as ability to collaborate and communicate effectively. It is also necessary to encourage these competences in solving problems, e. g. by incorporating team work task and projects in education, since while solving problems in groups students have to communicate, collaborate and provide proper arguments to support their reasoning.

Zakaria and Iksan (2007) emphasize the need for changes in science and mathematics teaching. Teachers should give less emphasis on students' acquisition of information, presenting scientific and mathematical knowledge through lecture, asking for recitation of acquired knowledge and working alone. More emphasis should be given on students understanding of a particular concept, guiding students in active learning, providing opportunities for discussion and elaboration and encouraging them to work with peers and teachers.

In relation to these requirements on educational objectives, we further focus on the concepts of collaborative learning, mathematical literacy, problem-solving skills, and mathematical modelling of mathematical tasks with real-life context.

1.1 Collaborative mathematics learning

The positive influences of collaborative work of students on their learning outcomes are supported by several researches. According to Zakaria's (2003) research about cooperative learning affects on students' achievement and problem solving skills, the cooperative group instruction showed significantly better results in mathematics achievement and problem solving skills. The effect size was moderate and therefore practically meaningful. He also found that students in the cooperative learning group had a favourable response towards group work. Zakaria concluded that the utilization of cooperative learning methods is a preferable alternative to traditional instructional method. Other research published by Springer et al (1999) suggest that small-group learning is effective in undergraduate science, mathematics, engineering and technology (SMET) courses and programs, and that widespread implementation of small-group learning in undergraduate SMET education should be given more support. Students who learn in small groups generally demonstrate greater academic achievement, express more favourable attitudes toward learning, and persist through SMET courses or programs to a greater extent than their more traditionally taught counterparts. The reported effects are relatively large in research on educational innovation and have a great deal of practical significance. Yackel at al (1991) described their research focused on the construction of classroom norms for cooperation and on the learning opportunities that arise in the course of interactions when children work in small groups to complete mathematical activities.

1.2 Mathematical literacy, problem-solving, and math tasks with real-life context

For the sake of reaching learning goals in mathematics education it is advised to teachers to adapt not only their teaching approach and methods, but also the teaching content. In mathematics education it seems to be essential to spend reasonable time and attention to application of students' acquired knowledge in solving mathematical tasks related to real-life contexts. We agree with Marja van den Heuvel-Panhuizen (2003) who emphasizes that students should learn mathematics by developing and applying mathematical concepts and tools in daily-life problem situations that make sense to them. Despite the general tendency to emphasize the importance of mathematics in solving daily-life problems, students' difficulties with such tasks still widely persist. This is also demonstrated by the Programme for International Student Assessment (PISA) international tests which focus on real-life problems that are in accordance with the definition of mathematical literacy.

The OECD Programme for International Student Assessment (PISA), founded in 1997, represents a commitment by the governments of OECD countries to monitor the outcomes of education systems, in terms of students' achievement, within a common, internationally agreed framework. They evaluate education systems worldwide by testing the skills and knowledge of 15-year-old students. Since 2000, PISA has been testing students worldwide in the key subjects: reading, mathematics and science.

OECD define mathematical literacy as „an individual's capacity to formulate, employ and interpret mathematics in a variety of contexts" which "includes reasoning mathematically and using mathematical concepts, procedures, facts and tools to describe, explain and predict phenomena" (PISA 2015 Mathematics Framework, p. 65). According to this definition mathematical literacy "assists individuals to recognize the role that mathematics plays in the world and to make the well-founded judgements and decisions needed by constructive, engaged and reflective citizens" (ibid).
PISA also regularly introduces new tests to assess students' skills in other areas relevant to modern life, such as creative problem solving and financial literacy (tested for the first time in 2012) and collaborative problem solving (testing will begin in 2015). As far as problem-solving is concerned, according to OECD mathematical literacy "frequently requires devising strategies for solving problems mathematically. This involves a set of critical control processes that guide an individual to effectively recognize, formulate and solve problems. This skill is characterized as selecting or devising a plan or strategy to use mathematics to solve problems arising from a task or context, as well as guiding its implementation. This mathematical capability can be demanded at any of the stages of the problem-solving process." (PISA 2015 Mathematics Framework, p. 69)

In creative problem solving Slovak students achieved average result of 483 points. Their performance is significantly lower than the average students' results in OECD countries, which is 500 points.

In Fig 1. a sample task from PISA pilot tests is shown. The task is focused on students' ability to work with mathematical models. The mathematical model to the real-life situation is already included in the texting of the task, and students are expected to understand and work with the given mathematical model, i.e. to solve the related tasks. Mathematics teachers, however, should teach students not only to work with given models, but also to propose mathematical models by themselves. Following these requirements, we have designed two mutually related tasks about gas distribution in households (see section 2, Fig. 3).

POWER OF THE WIND

Zedtown is considering building some wind power stations to produce electricity.

The Zedtown Council gathered information about the following model.

Model:	E-82
Height of tower:	138 metres
Number of rotor blades:	3
Length of one rotor blade:	40 metres
Maximum speed of rotation:	20 rotations per minute
Price for construction:	3 200 000 zeds
Turnover:	0.10 zeds per kWh generated
Maintenance cost:	0.01 zeds per kWh generated
Efficiency:	Operational 97% of the year

Note: kilowatt hours (kWh) is a measure of electrical energy.

Fig. 1a. Texting of task from PISA pilot test 2012 (OECD, 2013)

Question 2: POWER OF THE WIND PM922Q02

Zedtown wants to estimate the costs and the profit that would be created by constructing this wind power station.

Zedtown's mayor proposes the following formula for estimating the financial gain, F zeds, over a number of years, y, if they build the E-82 model.

$$F = \underbrace{400\,000\,y}_{\text{Profit from the yearly production of electricity}} - \underbrace{3\,200\,000}_{\text{Costs of building the wind power station}}$$

Based on the mayor's formula, what is the minimum number of years of operation required to cover the cost of construction of the wind power station?

A 6 years
B 8 years
C 10 years
D 12 years

Fig. 1b. Question to task from PISA pilot test 2012 (OECD, 2013)

1.3 Mathematical models and modelling in mathematics education

Mathematics enables us to describe phenomena occurring in nature and the surrounding world by means of mathematical models. The models allow simplifying real situations. Mathematical models often describe very intricate situations just fragmentarily and with certain inaccuracy, which sometimes causes that rigorously obtained results and values considerably differ from reality. This, however, cannot be taken as a serious disadvantage, because it is the simplifying effect of mathematical models which, in fact, makes them useful. The usefulness of a mathematical model consists in highlighting certain conditions of the situation in question (Fisher, 1992).

According to Lesh et al (2002) models are conceptual systems (consisting of elements, relations, operations and rules governing interactions) that are expressed using external notation systems and that are used to construct, describe, or explain the behaviours of other systems. A mathematical modelling focuses on structural characteristics of the relevant systems. These models reside inside the minds of the students and are embodied in the equations, diagrams, computer programs, or other representational media. Their huge advantage is that for a concrete situation several models can be created (like in the gas consumption related task designed for the purposes of this research, see section 2). In addition, one model can be applied in various situations. For instance, a linear equation can serve for modelling the total weight of a vehicle which is dependent on the number of people in the vehicle, and also for modelling the total costs for electric power as a dependence on the power consumption.

The process of designing a model itself, i.e. leading from a real situation to a mathematical model, or the whole problem-solving process, or again any manner of connecting the real world with mathematics, is referred to as mathematical modelling (Blum 1993). The particular stages of mathematical modelling are illustrated by the following diagram (Fig. 2) representing the modelling cycle.

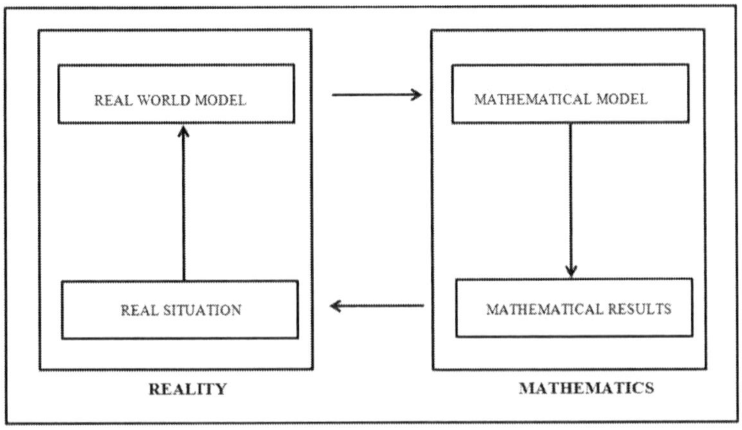

Fig. 2. Kaiser's modelling cycle (Ferri, 2006)

The starting point of the modelling cycle is a *real-life situation* which represents certain problem to be solved. Understanding the situation is the very first prerequisite for creating a *real world model.* At this stage an important role is played by idealization and simplification of the problem. When reified, the real model is most commonly shown in the form of a simple drawing (diagram) in which the most important information, conditions and relations are included. The next step is to mathematize the real situation, which serves for translating between the real world and mathematics, i.e. the pieces of information provided by the real model are transformed to the language of mathematics. Thus, the *mathematical model* of the real situation is formed. This stage closes the transition from reality to mathematics. Applying appropriate mathematical methods and techniques mathematical results are obtained which, consequently, have to be translated back to the language of common life. An essential part of this process is the verification and interpretation of mathematical result with respect to the real-life situation.

Mathematical models and modelling are distinctively represented in technical and natural sciences. Mathematical modelling plays an irreplaceable role in the teaching and learning of mathematics. It is said to help students better understand the world. Support mathematics learning (motivation, concept formation, comprehension, retaining). Mathematical models contribute to development of various mathematical competencies and appropriate attitudes, and to formation of an adequate picture of mathematics (Blum, Ferri 2009). The main objective of the teaching of mathematical modelling is to provide students with opportunities to experience and acquire skills which are necessary for problem-solving, because they will continue to encounter them in their everyday lives (Burkhart, 1989). Students need to be shown that mathematics is not just a heap of formulas and algorithms; quite the contrary, mathematics is present everywhere around us and is broadly applicable. Mathematical modelling is a source of relevant arguments to questions often asked by students, such as *'Where can I apply this in real life?'* and *'What do I need this for?'*

2 Research sample, methods and materials

The research was conducted at a town comprehensive school in Nitra, Slovakia. The research sample consisted of 204 students aged 16-18 who worked in 51 groups of four. The co-operating school also provides bilingual instruction, therefore, the sample partially consisted of bilingually educated students. More specifically, students worked in 27 Slovak speaking groups and 24 English speaking groups. Students were provided worksheets in the main language of their instruction. Yet, the possible differences between bilingually and monolingually educated students are not within scope of this study.

We were interested in *how* the students modelled the real-life situation (content) stated in the worksheet, and *what models* they used in solutions of tasks. These tasks were a part of a larger worksheet oriented on financial literacy of students. The first part of the worksheet was focused on the possibilities of housing and mortgage loans for young people; however, it is not subjected to the analysis presented in this research study. We analyse two tasks related to gas consumption in households and gas distribution costs (see Fig. 3). For solutions of the two tasks the students needed approximately 20 minutes.

Many house owners use gas in their households (for heating, cooking, water heating etc.). A real estate owner can choose from several gas consumption rates. Company PPS offers four rates. The price for gas distribution is calculated by the fixed monthly rate and the consumed gas rate (depending on the amount of the consumed gas).

Price list of gas distribution for households:

Gas rate label	Fixed monthly rate (€/month)	Consumed gas rate (€/kWh)
D1	2.11	0.0577
D2	4.98	0.0413
D3	7.75	0.0394
D4	36.43	0.0547

**The prices include VAT.*

TASK 1: Determine the annual gas consumption amount at which D3 rate would be more favourable for the Smiths than D2 rate.

TASK 2: Suggest the payment method for gas consumption for one year which would be the most advantageous for the Smiths. Use an algebraic expression.

Fig. 3. Texting of the gas related tasks

The tasks used in this research can be mathematically modelled in several ways. Below we provide an overview of expected solutions for subsequent references in section *Results*.

TASK 1

A: Solution applying an inequality

In the following solution we demonstrate the particular stages of the modelling cycle. The starting point is the description of the *real situation* in the task, including a table. Based on detailed analysis of the provided information we move to formation of the real model. The description of the situation itself indicates that the total annual costs for gas consumption consists of the fixed monthly rate (henceforth FMR), and the consumed gas rate (CGR) which depends on the actual amount of consumed gas. It is assumed that at certain amount of consumed gas the rate (or tariff) D3 should be more favourable than D2 tariff. In other words, the total annual costs for gas consumption should be lower for tariff D3 than for D2. After recognition of the dependences, the *real model* is developed which allows the mathematization of the real situation. Let variable x denote the amount of gas consumed per one year; then the wished *mathematical model* is

$$12.4,98+0,0413x>12.7,75+0,0394x, \quad (1)$$

where the expression on the left represents the total costs for annual gas consumption for tariff D2, and the expression on the right represent the total costs for annual gas consumption for tariff D3. The real situation can be also modelled step by step, i.e. first, the mathematical models for the total costs for annual gas consumption for particular tariffs need to be developed. Then, the final model, properly respecting the total costs comparison between tariffs D2 and D3, needs to be proposed. The fact that D3 is required to be more favourable than D2 is expressed by the inequality sign denoting that the total costs for tariff D3 are wished to be lower than the total costs for tariff D2. After that, the *mathematical result* is obtained by means of particular rearrangements of the inequality (1).

$$0,0019x>33,24, \quad (2)$$

which implies

$$x>17\,494,74. \quad (3)$$

The final stage of the modelling cycle is the *interpretation* of the mathematical result. The inequality (1) holds if variable x assumes values greater than 17 494,74. In terms of the real situation it means that tariff D3 is more favourable (than D2) for the Smith family if the total amount of consumed gas per one year in their household is greater than 17 494,74 kWh.

In the following methods of mathematical modelling of the situation only brief solutions are shown.

B: Solution applying an equality supported by further reasoning

In this case, an equality is applied in order to determine the amount of consumed gas for which the total costs are the same for both tariffs D2 and D3. So,

$$12.4,98+0,0413x=12.7,75+0,0394x, \quad (4)$$

Hence,

$$x=17\,494,74. \quad (5)$$

If the total amount of yearly consumed gas is lower than 17 494,74 kWh, then tariff D2 is more favourable. If the total amount of yearly consumed gas is greater than 17 494,74 kWh, then tariff D3 is more favourable.

C: Solution applying the graphical method

The linear functions which are plotted in Fig. 4 represent the total costs for annual gas consumption for tariff D2, $f(x)=12.4,98+0,0413x$, and for tariff D3 $g(x)=12.7,75+0,0394x$.

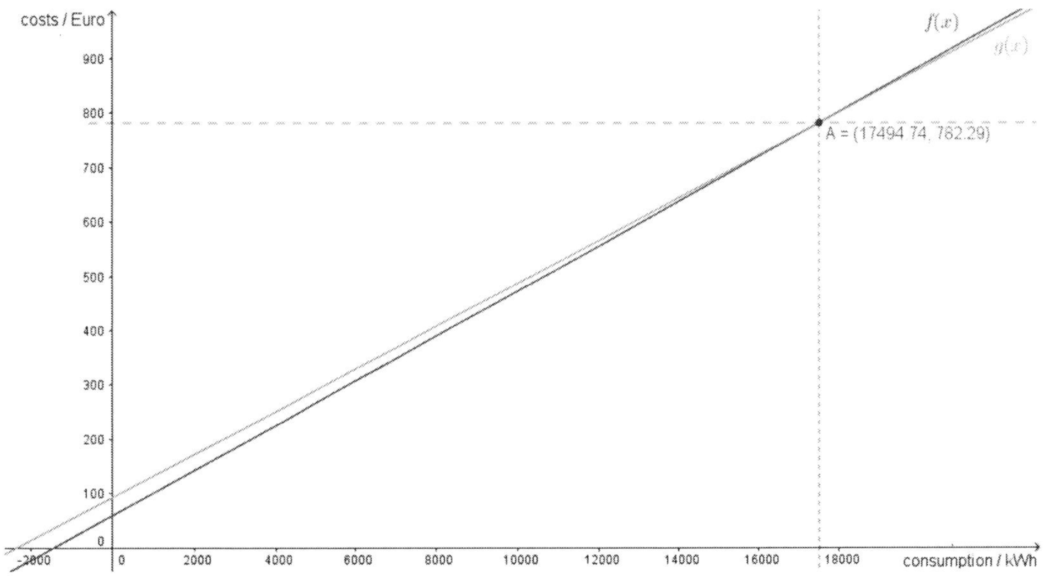

Fig. 4. Graphical solution of gas related Task 1

A disadvantage of the graphical method is that without any proper software it is not possible to determine the exact value from the graph. Actually, the result is in form of a non-terminating and non-recurring decimal number, so still the result has to be rounded to some extent. As for the graphical method, the result is represented as the intersection point of the graphs of the two linear functions, where the x-coordinate represents the annual gas consumption amount and the y-coordinate represents the corresponding costs. The coordinates of the intersection point can be calculated as in the solution applying the equality (see equations (4) and (5)).

TASK 2

When solving Task 2 it is necessary to consider the gas consumption in relation to specific household. Naturally, during warmer months in a year far less gas is consumed than in colder months, when besides cooking and water heating extra gas is used for heating the whole real estate. Based on detailed analysis of the information provided in the table, tariff D4 can be considered as absolutely irrelevant for common households, in comparison to the other three tariffs. Therefore, the most advantageous way of combining the three tariffs during a year can be expressed by means of the following mathematical model:

$$(2{,}11i+0{,}0577x)+(4{,}98j+0{,}0413y)+(7{,}75k+0{,}0394z), \tag{5}$$

where variables i, j, k denote the number of months for particular tariff (assuming that during one month only one tariff can be applied), and x, y, z denote the amount of gas consumed in the corresponding months.

Note: Although it is possible to develop a theoretical model of situation when the three tariffs are combined in order to minimize the total costs, it is, unfortunately, not possible in real life, since the companies distributing gas to households allow their customers to change the tariff only once in twelve months.

Secondly, we applied an additional research method, one-item questionnaire survey, in order to determine the level of students‘ awareness of household costs. Students were asked to circle one option from *deep – average – limited*, in order to complete the following statement according their own opinion:

I think that my awareness of our household costs is...

3 Results

In this section we present the results obtained by means of content analysis of the students' solutions. The analysis was focused primarily on how students modelled the given situations.

Task 1

Altogether 67% of the student groups tried to solve Task 1. Then, 58 % of those groups who at least tried to solve Task 1 did not understand the real situation, and, consequently, managed to develop neither the real model, nor the mathematical model of the situation. Their misconception of the problem was mostly displayed in the formation of the mathematical model. For instance, some students calculated the gas consumption as the difference between the FMR and the CGR for both tariffs, some students concluded that D3 tariff is (always) more favourable based on comparison of the CGR for both tariffs, and there were also students who divided the difference between the annual fixed rates by the factor of CGR for tariff D3. A special group consists of students whose mathematical model, besides annual fixed rate, included also the CGR multiplied by the number of hours in one year. This was probably caused by the students' confusion due to the measurement units of gas consumption, kilowatt-hours (Fig. 5). This also indicates that students lack relevant knowledge from Physics.

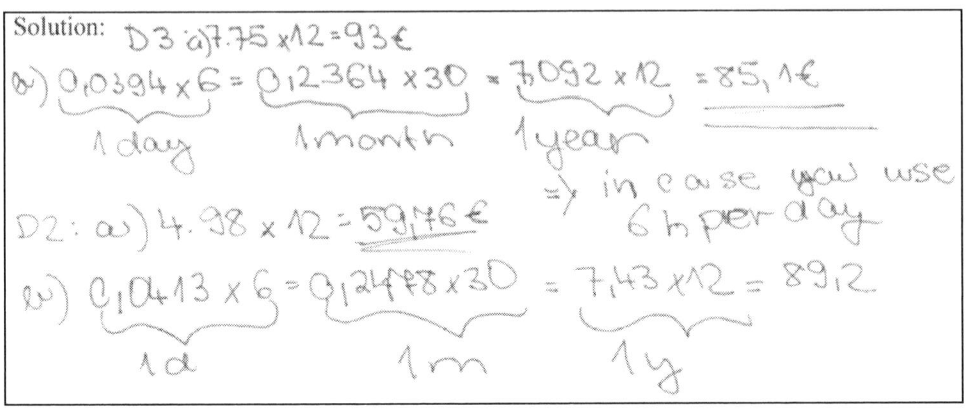

Fig. 5. A sample of students‘ solution. Students obviously lacked relevant Physics knowledge, and were confused by the measurement unit of gas consumption

Only 26 % of the student groups who attempted to solve Task 1, modelled the situation using an inequality. In two cases students modelled the situation by means of an equality. No student group applied the graphical method. The rest of the groups used other approaches to the modelling of the situation. In total, only four student groups managed to develop a correct mathematical model of the real situation, three of them obtained the correct mathematical result, and only one group correctly interpreted the obtained mathematical result in terms of the real-life situation.

Next, we analyze the *types of models* which students decided to apply in their solutions to Task 1.

Mathematical model in form of inequality

One of the incorrect considerations when developing the mathematical model in form of an inequality was the assumption that the total costs of gas consumption for tariff D3 should be greater than for tariff D2, which contradicted the requirement of D3 being the more favourable tariff.

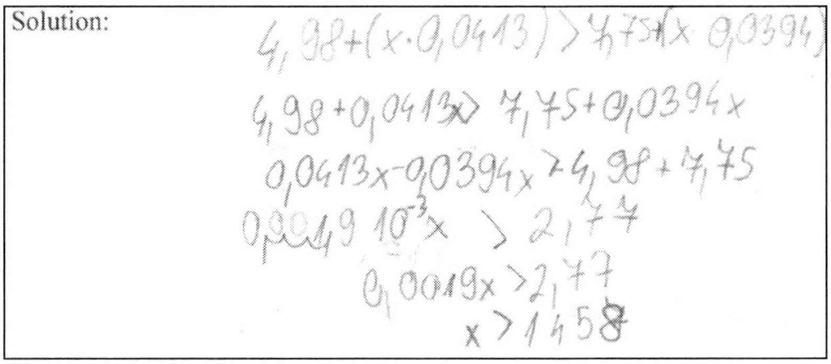

Fig. 6. A sample of students' solution. Instead of the annual gas consumption students considered the monthly gas consumption

As much as 44% of the student groups who used the inequality model in their solutions actually modelled the monthly gas consumption instead of the annual one (Fig. 6). It can be concluded that these students also understood the real situation and managed to develop the mathematical model for the more favourable offer. Yet, they underestimated the importance of verification and interpretation of the mathematical result. Considering the monthly gas consumption instead of the annual one can be referred to as a lapse or a mistake, not an error.

Mathematical model in form of equality

When developing the model in form of an equality students did not verified the obtained mathematical results. In most cases the mathematical model described the monthly costs of gas consumption instead of the annual one, similarly to students' solutions in form of an inequality. One of these student groups modelled the situation correctly. In their interpretation of the mathematical result they took into consideration the requirement on D3 to be more favourable in relation to the annual gas consumption (Fig. 7). Their whole solution is, however, not captured in Fig. 7, since these students accidentally wrote their notes on some other piece of paper, not the worksheet.

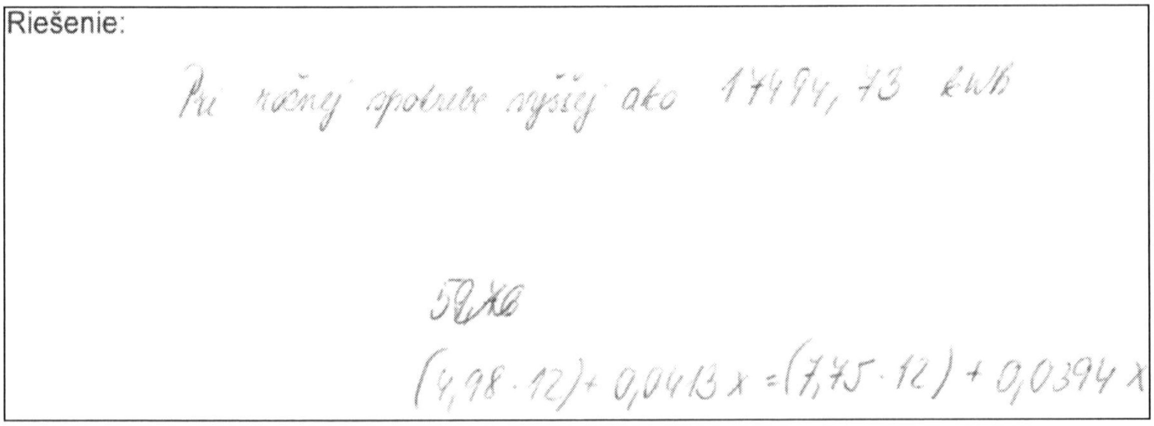

Fig. 7. A sample of correct students' solution applying an equality

Other approaches to the modelling of the real situation

A few student groups modelled the problem for one particular value of the annual gas consumption (Fig.8). These students failed at the third stage of the modelling cycle. The understood the real situation and managed to develop the real model. They also correctly concluded that the mathematical model should include the value of the fixed rate for twelve months and the CGR. However, they digressed from the task when they calculated the total costs for a particular value of the consumed gas, and, actually, modelled a situation for a different question, namely '*Which of the tariffs D2 and D3 is more favourable for the Smiths, if their annual gas consumption is 1 000 000 kWh?*'

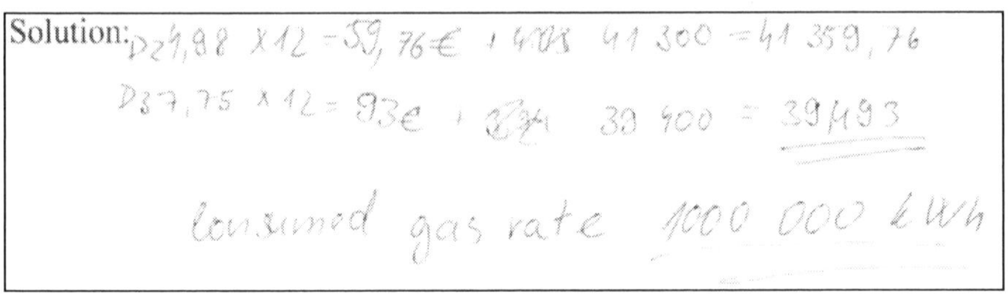

Fig. 8. A sample of students' solution. Although students fully understood the real situation and developed a correct real model, they digressed from the task

A few student groups tried to solve the task by trial-and-error method. As captured in Fig. 9, students developed a mathematical model for monthly costs (not annual) for tariffs D2 and D3. They gradually calculated the monthly costs for specific values of variable x which represented the amount of the consumed gas, and obtained the total monthly costs for both tariffs. When they found a value of x for which the monthly costs for tariff D3 were lower than for tariff D2, they finished the procedure. From this they concluded that tariff D3 is more favourable at the amount of consumed gas greater than 1500 kWh. In terms of modelling the total annual costs, the students developed an incorrect model. They did not verify the obtained mathematical results. They found only one of many possible values for which the (monthly) costs are lower for tariff D3 than D2. Moreover, by trial-and-error method they could just hardly obtain an accurate result.

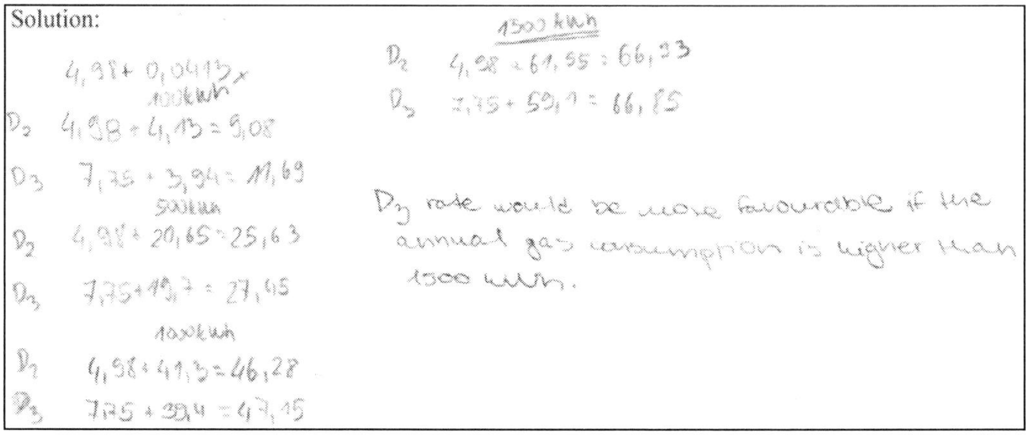

Fig. 9. A sample of students' solution. Students digressed from the task and considered monthly costs. In addition, they applied trial-and-error method, which did not help them obtain the wished result.

Task 2

Modelling the situation in Task 2 turned out to be demanding for students, which is also indicated by the fact that only 27 % of the groups attempted to solve the task. None of the groups managed to develop a correct mathematical model of the situation. However, their attempts to develop a model include also a few correct ideas. As captured in Fig. 10 and 11, students took into account that the amount of the gas people generally consume in their households varies during a year. That is, the students understood the real situation. Developing the real model was dependent on how students modelled the situation in the previous task. The errors which first occurred in their solution to Task 1, re-appeared in solution to Task 2. When creating the mathematical model, all student groups picked specific number of months in which the particular tariffs were applied. Also, they correctly concluded that tariff D4 is in general unfavourable for a common household, and in their mathematical models they considered only tariffs D1, D2 and D3. Students' failure at developing the mathematical models is most likely caused by the fact that students did not realize that the value of the amount of consumed gas is different for each of the three tariffs, which should be expressed by three different variables.

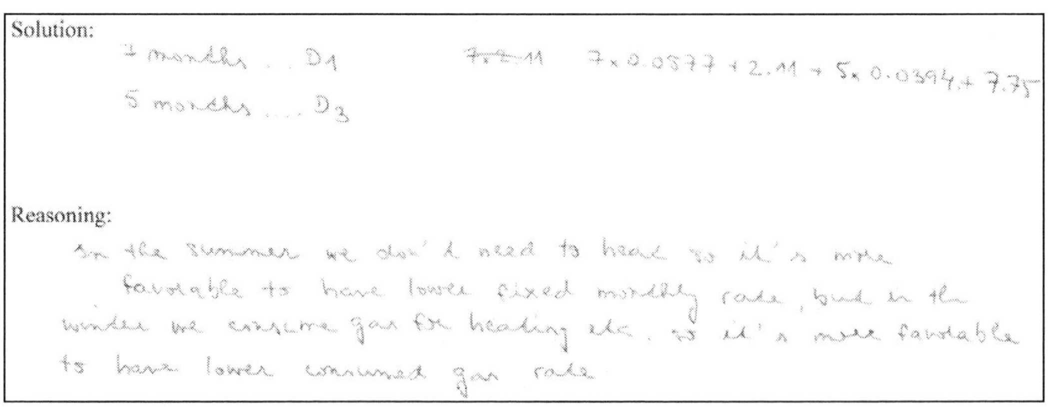

Fig. 10. A sample of students' solution. Although students' model is over-simplified, it shows that they understood the conditions of the real situation

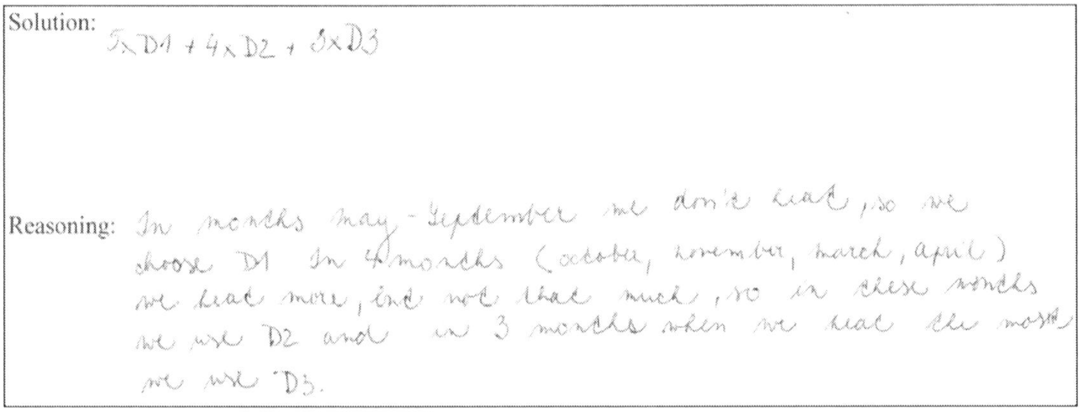
Solution: 5×D1 + 4×D2 + 3×D3

Reasoning: In months may - September we don't heat, so we choose D1. In 4 months (october, november, march, april) we heat more, but not that much, so in these months we use D2 and in 3 months when we heat the most we use D3.

Fig. 11. A sample of students' solution. The errors which first occurred in their solution to Task 1, re-appeared in solution to Task 2.

4 Discussion

The low success rate of students' solutions is in most cases caused by failures in the first stage of the modelling cycle, i.e. misconception of the real situation. The misconception can be a consequence of the students' low reading comprehension skill, or their inattention which goes hand in hand with their want to solve the tasks as quickly as possible. Reading comprehension skill is essential in most (if not all) school subjects, and is especially crucial in natural sciences, as it is the sciences where context-based problems often occur which require attentive reading.

In connection with the first stage of the modelling cycle and the reading comprehension, several students' mistakes arise from their difficulties when looking for relevant information in tables and charts and its interpretation. For instance, some students believed that the total annual costs of gas consumption at certain tariff (Task 1) could be calculated as the difference of fixed monthly rates, or the difference (in some case the product) of a fixed monthly rate and its corresponding consumed gas rate. This implies that students misunderstood the real situation. Tasks and problems that require work with information provided in form of a table or chart also belong to an area of problems which are challenging for Slovak students, as shown in the results of PISA measurements (NUCEM, 2012).

Incidentally, although the comparison of bilingually and monolingually educated students' performance is not covered in the scope of this research study, we dare to remark that bilingual student groups showed more modelling and reasoning skills. This might also be associated with the reading comprehension skill and the so called *selective attention*, i.e. an enhanced ability to selectively attend to information and inhibit misleading cues (Bialystok, 2001; Moschkovich, 2005), which are said to be better developed in bilinguals than monolinguals.

Some students showed correct understanding of the real situation in their solutions; yet, they considered the *monthly* costs instead of the *annual* costs (see Fig. 6, though, those can also be just cases of students' *lapses*, not necessarily *error* in comprehension), or they disregarded the requirement on D3 tariff to be *more favourable* (some students did not realize that if D3 is to be *more favourable*, then the total costs for D3 rate should be *smaller than* the total costs for D2 rate, and used improper inequality sign).

Applying the trial-and-error method, students could obtain the correct solution to Task 1 by randomly picked value relatively close to the wished result, or after time-demanding calculations, which would probably discourage them from trying enough values to get a satisfactorily accurate result. The trial-and-error method would have been effective enough if students had applied suitable software, which would have made their calculation faster. Still, this way they would have only been able to get certain interval including the more accurate value.

Another influential flaw in the students' solutions was the absence of an answer, i.e. interpretation of the mathematical result back into the context of the real situation. It is generally known that in school lessons teachers together with students usually interpret the mathematical results orally, yet, students rarely write down the interpretations. In work life, however, it is often required to record, explain and report solution strategies and the obtained results and interpretations in written form.

The high number of student groups who did not even try to solve the two tasks can be a consequence of the students' lack of interest in the context of the tasks. We believe that although the tasks are closely related to the needs of common people in real life, to teenaged students the issue does not seem to be relevant and interesting enough. The questionnaire survey, which was conducted as an additional research method, indicates that 30% of the students think that their awareness of their household costs is limited; as much as 52% of the students thinks their awareness of the households costs is average; and only 18% considers their awareness of the households costs to be deep.

To sum up, while solving the tasks students experienced serious difficulties at all four stages of the modelling cycle. The misconception of the real situation led to inaccurately developed model, or an inaccurately developed model led to incorrect results. We expected that students would go through the stages of the modelling cycle sequentially. Surprisingly, their solutions demonstrate skipping some stages. In further research we suggest monitoring the students' modelling routes in terms of the Blum's (2005) 7-staged cycle, which could enable us to analyze in more detail how students think while modelling real situations.

5 Conclusion

As already mentioned, in international assessments of mathematical literacy Slovak students achieve results below the average of OECD countries, and their achievement has long been decreasing. In order to improve the current state it is necessary to identify its causes. In conclusion, we dare to name some of the causes.

Firstly, it is necessary to take into account that students' interest in mathematics generally decreases, despite global efforts of teachers and educational policy makers to incorporate more tasks and problems with real-life context into the educational process. However, it seems natural that teenaged students are not very interested in problems related to household and the planning of future financial investments, although the knowledge which they might acquire through such tasks is of high importance for their future lives. We assume that in order to catch students' interest in similar types of problems it would be useful to include such topics in context-based mathematical tasks which are close to the students' spare time activities, such as social networks, cinematography, popular music, computer games etc.

Another influential factor related to the incorporation of context-based task into the teaching of mathematics is the insufficient number of mathematics lessons at many primary and secondary schools. Unfortunately, teachers usually prefer to teach mathematics with focus on the development of students' algorithmic skills, mainly due to preparation for various national tests and university entrance exams.

Basically, mathematics teachers can be, figuratively speaking, divided into two main groups. The first group includes teachers who, based on their professional and life experience, are aware of the importance of solving context-based tasks with students; for the above stated reasons, however, they decide not to. The second group represents the teachers who have never encountered context-based mathematical tasks within their own education, and due to various reasons and despite their life experience they do not arrive at the conclusion that they should prepare students for certain problem situations. In relation to this, a question arises if designing of such tasks (i. e. original, non-routine and new to the students; Doorman et al, 2007) should not be included in university teacher training programs which educate future mathematics teachers. Wubbels et al (1997) warn that preparation of teachers for education using real-life problems is no simple task. It requires the search for techniques and strategies in teacher education which induce important changes in the student teachers' conceptions of mathematics and mathematics teaching, conceptions which are deeply engrained.

Acknowledgement

The research study was supported by national project KEGA 016UKF-4/2016 *The Implementation of Constructivist Approaches to Mathematics Teaching with a Focus on Active Acquisition of Knowledge by Pupils in the Context of Bilingual Education* supported by Ministry of Education, Science, Research and Sport of the Slovak Republic, and institutional projects UGA VII/14/2016 *Improving key competences of mathematical modelling in secondary mathematics education,* and UGA VII/11/2016 *Improving the quality of bilingual Slovak and English mathematics instruction in lower secondary education with emphasis on the development of mathematical competences* supported by Constantine the Philosopher University in Nitra. The authors gratefully acknowledge Gymnázium Golianova 68 in Nitra for continual cooperation.

References

[1] BIALYSTOK, E. *Bilingualism in Development: Language, Literacy and Cognition.* Cambridge: Cambridge University Press, 2001, ISBN 0-521-63231-5, 288 p.

[2] BINNS, B., BURKHARDT, H., GILLESPIE, J., SWAN, M. B.: Mathematical modelling in the School Classroom: Developing Effective Support for Representative Teachers. In: *Applications and Modelling in Learning and Teaching mathematics*. Chichester: Ellis Horwood Limited, 1989, ISBN 0-7458-0355-5, p. 136 – 143.

[3] BLUM, W. Mathematical modelling in mathematics education and instruction. In: *Teaching and learning mathematics in context,* 1993, p. 3 – 14. Retrieved from http://citeseerx.ist.psu.edu/viewdoc/download?doi=10.1.1.580.7314&rep=rep1&type=pdf

[4] BLUM, W. Mathematical modelling in mathematics education and instruction. In: *Teaching and learning mathematics in context*, 1993, p. 3-14.

[5] BLUM, W., LEISS, D. „Filling Up "-the problem of independence-preserving teacher interventions in lessons with demanding modelling tasks. In: *CERME 4–Proceedings of the Fourth Congress of the European Society for Research in Mathematics Education*, Sant Feliu de Guíxols, 2005, ISBN 84-611-3282-3, p. 1623-1633.

[6] BLUM, W., FERRI, R. B. Mathematical modelling: Can It Be Taught And Learnt? *Journal of Mathematical Modelling and Application*, vol. 1, no. 1, 2009: p. 45 – 58, ISSN 2178-2423.

[7] DOORMAN, M. et al. Problem solving as a challenge for mathematics education in The Netherlands. *ZDM Mathematics Education*, vol. 39, no. 5, 2007: p. 405 – 418, ISSN 1863-9704.

[8] FISCHER, R., MALLE, G. *Človek a matematika*. SPN, 1992, ISBN 80-08-01309-5, 336 p.

[9] LESH, R., CARMONA, G., POST, T. Models and modelling. In: *Proceedings of the Annual Meeting [of the] North American Chapter of the International Group for the Psychology of Mathematics Education*, Georgia, 2002, ISBN 0-88119-002-0, p. 89-98.

[10] FERRI, R. B. Theoretical and empirical differentiations of phase in the modelling process. *ZDM International Journal on Mathematics Education*, vol. 38, no. 2, 2006: p. 86 – 95, ISSN 1863-9690.

[11] MOSCHKOVICH, J. N. Using two languages when learning mathematics. *Educational Studies in Mathematics*, vol. 64, 2005: p. 121-144, ISSN 1573-0816.

[12] NUCEM. *PISA 2012 Národná správa Slovensko*. 2015. Retrieved from http://www.nucem.sk/documents//27/medzinarodne_merania/pisa/publikacie_a_diseminacia/1_narodne_spravy/N%C3%A1rodn%C3%A1_spr%C3%A1va_PISA_2012.pdf

[13] NUCEM. *Zbierka uvoľnených úloh z testovania matematickej a čitateľskej gramotnosti pre 2. Stupeň základných škôl a 1. – 4. Ročník osemročných gymnázií*. 2012: Retrieved from http://www.nucem.sk/documents/26/testovanie_9_2013/Zbierka_uloh_2012_v58_fin.pdf

[14] OECD. *PISA 2012 Released Mathematical Items*. May 2013. Retrieved from http://www.oecd.org/pisa/pisaproducts/pisa2012-2006-rel-items-maths-ENG.pdf

[15] OECD. *PISA 2015 Assessment and Analytical Framework: Science, Reading, Mathematic and Financial Literacy*. 2016. DOI: http://dx.doi.org/10.1787/9789264255425-en

[16] SPRINGER, L., STANNE, M. E., DONOVAN, S. S. Effects of small-group learning on undergraduates in science, mathematics, engineering, and technology: A meta-analysis. *Review of educational research*, vol. 69, no. 1, 1999: p. 21 – 51, ISSN 1935-1046.

[17] ŠPÚ. Štátny vzdelávací program pre gymnáziá; príloha *Matematika*. 2015. Retrieved from http://www.statpedu.sk/sites/default/files/dokumenty/inovovany-statny-vzdelavaci-program/matematika_g_4_5_r.pdf

[18] VAN DEN HEUVEL-PANHUIZEN, Marja. The didactical use of models in realistic mathematics education: An example from a longitudinal trajectory on percentage. *Educational Studies in Mathematics*, vol. 54, no. 1, 2003: p. 9 – 35, ISSN 1573-0816.

[19] WUBBELS, T., KORTHAGEN, F., BROEKMAN, H. Preparing teachers for realistic mathematics education. *Educational Studies in Mathematics*, vol. 32, no. 1, 1997: p 1-28. ISSN 1573-0816.

[20] YACKEL, E., COBB, P., WOOD, T. Small-group interactions as a source of learning opportunities in second-grade mathematics. *Journal for research in mathematics education*, vol. 22, no. 5, 1991: p. 390 - 408. ISSN 1945-2306.

[21] ZAKARIA, E. *Kesan Pembelajaran Koperatif Ke Atas Pelajar Matrikulasi*. Tesis Doktor Falsafah. Fakulti Pendidikan, Universiti Kebangsaan Malaysia, 2003.

[22] ZAKARIA, E., IKSAN, Z Promoting Cooperative Learning in Science and Mathematics Education: A Malaysian Perspective. *Eurasia Journal of Mathematics, Science & Technology Education*, vol. 3, no. 1, 2007: p. 35-39. ISSN: 1305-8223.

Current address

Plothová Lucia, Mgr.
Department of Mathematics, Faculty of Natural Sciences,
Constantine the Philosopher University in Nitra
Tr. A. Hlinku 1, SK-949 74 Nitra, Slovak Republic
Tel. number: +421 37 6408 722, e-mail: lucia.plothova@ukf.sk

Páleníková Kitti, RNDr. PhD.
Department of Mathematics, Faculty of Natural Sciences,
Constantine the Philosopher University in Nitra
Tr. A. Hlinku 1, SK-949 74 Nitra, Slovak Republic
Tel. number: +421 37 6408 691, e-mail: kpalenikova@ukf.sk

Naštická Zuzana, Mgr.
Department of Mathematics, Faculty of Natural Sciences,
Constantine the Philosopher University in Nitra
Tr. A. Hlinku 1, SK-949 74 Nitra, Slovak Republic
Tel. number: +421 37 6408 696, e-mail: zuzana.nasticka@gmail.com

Florková Michaela, Mgr.
Department of Mathematics, Faculty of Natural Sciences,
Constantine the Philosopher University in Nitra
Tr. A. Hlinku 1, SK - 949 74 Nitra, Slovak Republic
Tel. number: +421 37 6408 709, e-mail: michaela.florkova@ukf.sk

Šubová Lenka
Department of Mathematics, Faculty of Natural Sciences,
Constantine the Philosopher University in Nitra
Tr. A. Hlinku 1, SK - 949 74 Nitra, Slovak Republic
Tel. number: +421 37 6408 691, e-mail: lenka.subova2@student.ukf.sk

Slovak University of Technology in Bratislava
Faculty of Mechanical Engineering
16th Conference on Applied Mathematics **APLIMAT 2017**
Proceedings

METHOD TO PROGRAMMATICALLY GENERATE A LARGE NUMBER OF ASSIGNMENTS FOR STUDENTS AND TO ASSESS THEIR RESULTS IN MATLAB

POLCEROVÁ Marie (CZ), POLCER Jaroslav (CZ), POLCER Pavel (CZ), POLCEROVÁ Lenka (CZ)

Abstract. This paper presents a method of generating large number of different individual assignments for students, their fully automated electronic assessment and its role and possibility in computer-aided teaching of mathematics at technical higher education institutions. This paper describes methodology and the specific realization, including programs written in MATLAB and examples of inputs and outputs. Also this article summarizes current experience with this method and discusses evaluations in mathematics.

Keywords: e-assessment, mathematics, MATLAB, interactive computer programs

Mathematics Subject Classification: 97Q70

1 Introduction

Introduction of computer technology into teaching of mathematics at technical universities has a global character [17], [18], [22], [16], [33], [36], [37], [38], [39]. Using Computer algebra systems (CAS) in computer-aided teaching of mathematics has its pros and cons and brings new problems [43], [49], [15], [4], [13], [24], [20], [23], [59]. Evaluation of students that is not only objective, but also quick, cheap and without a huge workload for teachers [27], [29], [45], [30], [32], [58], [51], [52] is imperative for universities with high numbers of students [10], [9].

One of the main goals of computer aided teaching of mathematics at technical higher education institutions is to ensure that the future engineers are able to present the results of their work at the appropriate graphical level (according to International standard [28]) without mathematical errors [47], [30]. The objective of the subject mathematics is to make students learn and to apply the acquired knowledge and mathematical skills in technical subjects [21], [4]. Students can use any Computer Algebra System (CAS) to solve and check their tasks [7], [8], [19], [20], [50].

Our faculty selected program MATLAB, because it is applicable not only in mathematics (Symbolic Math Toolbox), but also in physics (Simulink) and in chemistry (MATLAB, Comsol). This decision was made in 2000 and our faculty (Faculty of Chemistry) received funding in 2001 to purchase this program. Program MATLAB was selected by faculty management with regard to needs of both general subjects (such as mathematics and physics) and specialized subjects (such as physical chemistry and chemical engineering). Teachers of mathematics first considered programs Maple and Mathematica, because they wanted to improve content of exercises from mathematics by adding analytical (symbolic), numerical and graphical solutions. Teachers of physics requested MATLAB instead, because they needed to approximate experimental data and MATLAB has functions (polyfit, interp1, spline, interp1q, etc.) exactly for this purpose. MATLAB was also needed for doctoral students to simulate physical phenomena (using Simulink). Teachers of physical chemistry and chemistry engineering clearly preferred MATLAB, because it included pre-made packages with ode solvers (ode23, ode45, etc.) for solving differential equations. They also need Simulink and Comsol (which is based on MATLAB) to simulate chemical phenomena. Teachers of mathematics were also satisfied with choice of MATLAB because since version 5.3 it contains the same symbolic core (Symbolic Math Toolbox) as Maple V Release 5 [41].

As a result students work with unified software interface in all their subjects [48], [9], [21], [51]. Program MATLAB allows students to combine symbolic, numerical and graphical solutions, which reduces the probability that their calculations are incorrect [41], [50].

2 Methodology

The computer-aided teaching of mathematics is done in a specialized classroom with 24 computers with adequate hardware and software. During the semester students solve individually assigned practical verbal tasks involving numerical and graphical solutions using program MATLAB [57]. Text in the first task DU1 is the same for all students except for numerical values, which are different for each student. Tasks DU2, DU5 and DU6 have dozens of variants of text and each text variant has further different numerical values. In case of DU3 and DU4 every student has a completely different task [49]. We attach great importance to the first task because students first encounter the new software interface. Given this fact, the task should not be too difficult for them, it should rather be motivational [3], [6], [12], [40], but it should also reflect necessity for students to learn the basics.

First it is necessary that students learn the way elementary functions are implemented in MATLAB and how to use them correctly [4]. Then they have to learn how to save the calculated results as variables and to display them with the required number of decimal places [3].

In the first lessons of mathematics students learn not only new elementary functions such as antitrigonometric functions, but at the same time, they also repeat other functions. We consider this introduction to elementary functions to be a basic knowledge and the main purpose of these first lessons is to make every student learn these basic skills. We created task that contains a total of 12 simple exercises requiring use of these basic skills. If we gave the same task to all students, only one or few students would solve it and others would just copy the results which is not intended. Therefore, we used MATLAB as a programming language and we created a program (script **`gen_du1.m`**), that generates an individual assignment for each student and which is available on webpage ROBOT (link http://www.fch.vutbr.cz/~polcerova/eng/robot.html). In this case each of these tasks contains the same text, but different values, generated using a random number generator. An example of a generated assignment is in appendix 1.

After generating assignments we post an Excel table at e-learning portal of our faculty (system Moodle) which, among other things, contains a list of all students and number of their assignment for task DU1 [58]. At the same time we give all students access to directory **`ASSIGNMENT`** which contains the generated tasks. During lessons, after interpretation and demonstration of subject matter, every student starts to solve his/her own task.

Because it is not possible for one teacher to immediately check every task of every student and because each student wants immediate feedback [9], [44], [12] we created an e-assessment program ROBOT, which checks and evaluates all submitted tasks and saves the results of assessment in a table every 30 seconds. This allows teacher to focus attention on those students who have trouble solving the task.

As mentioned before, we consider this introduction to elementary functions to be a basic knowledge, so each student has to solve this task within a week. We even kept program ROBOT running after the lessons so students can repeatedly check their results. Some students took this task as a game and kept trying until they reached maximum number of points.

After the deadline for submission the teacher checks and evaluates all the tasks using the program ROBOT and only processes and publishes results on the e-learning portal Students immediately receive their grading.

3 Assessment

Task DU1 (Appendix 1) is only one of six tasks (DU1-DU6) that each student has to submit during the semester. Task DU1 consists of 12 exercises which can be divided in four categories. Exercises no. 1, 10 and 12 represent the first category which is focused on the calculation of the functional value of a polynomial and calculation of algebraic expression with powers, roots and absolute value. Exercises no. 2, 3, 4 and 5 represent the second category which is focused on the calculation of the functional value of trigonometric functions of a specified angle in degrees, minutes and seconds. Exercises no. 6, 7, 8 and 9 represent the third category which is focused on the calculation of angle within a specified quadrant (in degrees, minutes and seconds) from given values of trigonometric functions. Exercise no. 11 represents the last, fourth, category which is focused on the calculation of the logarithms with different bases.

3.1 Verification and evaluation (e-assessment) by program ROBOT

Program ROBOT consists of two scripts (**`assess_du1.m`**, **`robot.m`**) and one user function (**`assess_du1_help.m`**). To start the program ROBOT further requires one text file (**`tables_for_asses.txt`**) which was generated by the script **`gen_du1.m`** and one **`.csv`** file (**`names.csv`**), which was created by the teacher (surname, name and login of all students). Program ROBOT (after running script **`robot.m`**) first identifies student's submited script by the greatest match with correct results and checks student's name using the **`.csv`** file (**`names.csv`**). After that program checks all 60 variables (Tab. 1) in every task using previously correctly calculated values of these variables (**`tables_for_assess.txt`**) and also evaluates the accuracy of the calculation based on the actual content found.

If the variable is numeric (integer or real), the program checks if it is the same as the correctly calculated value of the variable. If they are the same (full match) then evaluation is 1, otherwise it is 0. If the variable is text, the program again checks for a full match. If there is a full match with the correct text variable then the evaluation is 1. If there is not a match then the program converts the text variable to a numeric variable. If this numerical value and correct value differ only in the last required decimal place then the evaluation is 0.5. If there is greater difference, the evaluation is 0.

When the program checks the entire task (all 60 variables) it moves the file to directory **ASSESSED**, appends evaluation in the table **table.txt** (feedback for students) and to the **.csv** file **table_for_excel.csv** (for evaluation by the teacher). The script **assess_du1.m** was originally written in language which uses a comma instead of a dot as decimal separator. Because we do not know future users, we left commands for replacing dots by commas (at last lines 567 and 569) although they are currently "irrationally" replacing dots by dots.

Number of exercise	Variable INTEGER	Variable REAL	Variable STRING	Quantity
1		`a`	`as`	2
2	`s1, m1, v1`	`b, c`	`bs, cs`	7
3	`s2, m2, v2`	`d, e`	`ds, bs`	7
4	`s3, m3, v3`	`f, g`	`fs, gs`	7
5	`s4 ,m4, v4`	`h, m`	`hs, ms`	7
6	`degrees1, minutes1, seconds1`	`radians1`		4
7	`degrees2, minutes2, seconds2`	`radians2`		4
8	`degrees3, minutes3, seconds3`	`radians3`		4
9	`degrees4, minut4, seconds4`	`radians4`		4
10	`n`		`ns`	2
11		`p1, p2, p3, p4`	`p1s, p2s, p3s, p4s`	8
12		`p5, p6`	`p5s, p6s`	4
Total	25	19	16	60

Tab. 1. Variables

3.2 Our evaluation criteria

Students can obtain a maximum of 3 points from the task DU1 (it is only one task from six). At least one exercise from each of the four categories (mentioned above) has to be correct otherwise student will not get any points.

If you write down all the variables (Tab. 1), that student has to attach values to, you can see that there is a total of 60 variables. If the student meets the requirement that at least one exercise from each of the four categories is correct then it is sufficient to divide number of correct answers by 20 to get students assessment (60/20 = 3 points for classification). Because every teacher asked for slightly different criteria or different number of points, the output of the e-assessment program ROBOT includes spreadsheet in **.csv** format which contains assessment for each individual variable. For every correctly attached value student gets 1 point, for partially correct value (usually the wrong number of decimal places for text variables) student receives 0.5 points and for wrong value 0 points are given. Each teacher can than copy data from this spreadsheet to assessment they are used to and add their own criteria or do other modifications (Tab. 2).

1	Number assignment	Group	11th exercise								12th exercise				Sum	Con. 1	Con. 2	Con. 3	Con. 4	Total assessment	Date
2			p1	p1s	p2	p2s	p3	p3s	p4	p4s	p5	p5s	p6	p6s							
22	20	PO	0	0	0	0	0	0	0	0	0	0	0	0	34	1	1	1	0	0	16/02/2015
23	21	CT7	1	1	1	1	1	1	1	1	1	1	1	1	60	1	1	1	1	3	18/02/2015
24	22	SO	1	1	1	1	1	1	1	1	1	1	1	1	60	1	1	1	1	3	18/02/2015
25	23	SO	1	1	1	1	1	1	1	1	1	1	1	1	24	1	0	0	1	0	19/02/2015
26	24	CT16	1	1	1	1	1	1	1	1	1	1	1	1	60	1	1	1	1	3	19/02/2015
27	25	CT16	1	1	1	1	1	1	1	1	1	1	1	1	57.5	1	1	1	1	2.875	16/02/2015
28	26	PO	1	1	1	1	1	1	1	1	1	1	1	1	60	1	1	1	1	3	19/02/2015
29	27	CT7	1	1	1	1	1	1	1	1	1	1	1	1	59	1	1	1	1	2.95	16/02/2015
30	28	PO	0	0	0	0	0	0	0	0	0	0	0	0	0	0	0	0	0	0	16/02/2015
31	29	CT7	1	1	1	1	1	1	1	1	1	1	1	1	50	1	1	1	1	2.5	16/02/2015
32	30	CT7	1	1	1	1	1	1	1	1	1	1	1	1	60	1	1	1	1	3	12/02/2015
33	31	CT7	1	1	1	1	1	1	1	1	1	1	1	1	60	1	1	1	1	3	13/02/2015
34	32	PO	1	1	1	1	1	1	1	1	1	1	1	1	60	1	1	1	1	3	11/02/2015
35	33	CT7	1	1	1	1	1	1	1	1	1	1	1	1	60	1	1	1	1	3	17/02/2015
36	34	CT7	1	1	1	1	1	1	1	1	1	1	0	0	56	1	1	1	1	2.8	18/02/2015
37	35	CT16	1	1	1	1	1	1	1	1	1	1	1	1	60	1	1	1	1	3	18/02/2015
38	36	PO	1	1	1	1	1	1	1	1	1	1	1	1	60	1	1	1	1	3	13/02/2015
39	37	CT7	0	0	0	0	0	0	0	0	0	0	0	0	10	0	0	1	0	0	16/02/2015
40	38	CT16	1	1	1	1	1	1	1	1	1	1	1	1	60	1	1	1	1	3	17/02/2015
41	39	PO	1	1	1	1	1	1	1	1	1	1	1	1	59	1	1	1	1	2.95	17/02/2015
42	40	SO	1	1	1	1	1	1	1	1	1	1	1	1	57	1	1	1	1	2.85	16/02/2015
43	41	PO	1	1	1	1	1	1	1	1	1	1	1	1	60	1	1	1	1	3	16/02/2015
44	42	CT16	1	1	1	1	1	1	1	1	1	1	1	1	60	1	1	1	1	3	18/02/2015
45	43	SO	1	1	1	1	1	1	1	1	1	1	1	1	52.5	1	0	1	1	0	16/02/2015
46	44	PO	1	1	1	1	1	1	1	1	1	1	1	1	60	1	1	1	1	3	16/02/2015
47	45	PO	1	1	1	1	1	1	1	1	1	1	1	1	60	1	1	1	1	3	16/02/2015
48	46	CT7	1	1	1	1	1	1	1	1	1	1	1	1	60	1	1	1	1	3	19/02/2015
49	47	PO	1	1	1	0.5	1	1	1	1	0	0	1	1	41	1	0	1	1	0	11/02/2015
50	48	PO	1	1	1	1	1	1	1	1	1	0.5	1	1	56	1	1	1	1	2.8	16/02/2015
51	49	PO	1	1	1	1	1	1	1	1	1	1	1	1	56	1	1	1	1	2.8	11/02/2015
52	50	CT16	1	1	1	1	1	1	1	1	1	1	1	1	60	1	1	1	1	3	19/02/2015

Tab. 2. Results

4 Realization

4.1 Student's part

Students log into the computers in the specialized classroom using their usernames and passwords which were given to them when they were admitted to our faculty. They open the Excel table with assignment which contains a list of all students and number of their assignment for DU1. Then they open their own assignment and run MATLAB 6p1 Release12.1. After that students solve first exercise according to the teacher's instructions and save a script they created as **username.m** (based on surname). If they want to check their results by program ROBOT they copy the script to directory **C/PCM/submission/DU1** and wait until it is checked. Students find out that task was checked when it disappears form the directory. Then they change directory to **C/PCM** and open a file with updated assessment (**table.txt**). They immediately see (at the end of the file) which of the variables are correct, partially correct, incorrect or missing (Tab. 3). They correct errors in the script and run the checking procedure again until they are satisfied with the result or until the deadline for submission as recommended by Nicol and Macfarlane-Dick [44].

```
Assessment run 09-Jun-2016  08:06:57
Assessment surname1.m version 08-Jun-2015 14:31:36
surname1   assignment 00000007  points 58.0
Points: 1-30 1 1 1 1 1 1 1 1 1 1 1 1 1 1 1 1 1 1 1 1 1 1 1 1 1 1 1 1 1 1
Points:31-60 1 1 1 1 1 1 1 1 1 1 1 1 1 0 0 1 1 1 1 1 1 1 1 1 1 1 1 1 1 1
Excercise  1.: a    1, as    1
Excercise  2.: s1    1, m1    1, v1    1, b    1, bs    1, c    1, cs    1
Excercise  3.: s2    1, m2    1, v2    1, d    1, ds    1, e    1, es    1
Excercise  4.: s3    1, m3    1, v3    1, f    1, fs    1, g    1, gs    1
Excercise  5.: s4    1, m4    1, v4    1, h    1, hs    1, m    1, ms    1
Excercise  6.: degrees1   1, minutes1   1, seconds1   1, radians1   1
Excercise  7.: degrees2   1, minutes2   1, seconds2   1, radians2   1
Excercise  8.: degrees3   1, minutes3   1, seconds3   1, radians3   1
Excercise  9.: degrees4   1, minutes4  ##, seconds4   0, radians4   1
Excercise 10.: n    1, ns    1
Excercise 11.: p1    1, p1s    1, p2    1, p2s    1, p3    1, p3s    1, p4    1, p4s    1
Excercise 12.: p5    1, p5s    1, p6    1, p6s    1

Assessment surname2.m version 09-Jun-2016 07:06:47
surname2   assignment 00000014  points 60.0
Points: 1-30 1 1 1 1 1 1 1 1 1 1 1 1 1 1 1 1 1 1 1 1 1 1 1 1 1 1 1 1 1 1
Points:31-60 1 1 1 1 1 1 1 1 1 1 1 1 1 1 1 1 1 1 1 1 1 1 1 1 1 1 1 1 1 1
Excercise  1.: a    1, as    1
Excercise  2.: s1    1, m1    1, v1    1, b    1, bs    1, c    1, cs    1
Excercise  3.: s2    1, m2    1, v2    1, d    1, ds    1, e    1, es    1
Excercise  4.: s3    1, m3    1, v3    1, f    1, fs    1, g    1, gs    1
Excercise  5.: s4    1, m4    1, v4    1, h    1, hs    1, m    1, ms    1
Excercise  6.: degrees1   1, minutes1   1, seconds1   1, radians1   1
Excercise  7.: degrees2   1, minutes2   1, seconds2   1, radians2   1
Excercise  8.: degrees3   1, minutes3   1, seconds3   1, radians3   1
Excercise  9.: degrees4   1, minutes4   1, seconds4   1, radians4   1
Excercise 10.: n    1, ns    1
Excercise 11.: p1    1, p1s    1, p2    1, p2s    1, p3    1, p3s    1, p4    1, p4s    1
Excercise 12.: p5    1, p5s    1, p6    1, p6s    1

Assessment surname3.m version 09-Jun-2016 07:24:45
surname3   assignment 00000036  points 53.5
Points: 1-30 1 1 1 1 1 1 1 1 1 1 1 1 1 1 1 1 1 1 1 1 1 1 1 1 1 1 1 1 1 1
Points:31-60 1 1 0 1 1 1 0 1 1 1 1 1 0 0 1 1 1 1 1 1 1 1 1 1 1 1 1 0.5 0 0
Excercise  1.: a    1, as    1
Excercise  2.: s1    1, m1    1, v1    1, b    1, bs    1, c    1, cs    1
Excercise  3.: s2    1, m2    1, v2    1, d    1, ds    1, e    1, es    1
Excercise  4.: s3    1, m3    1, v3    1, f    1, fs    1, g    1, gs    1
Excercise  5.: s4    1, m4    1, v4    1, h    1, hs    1, m    1, ms    1
Excercise  6.: degrees1   1, minutes1   1, seconds1   0, radians1   1
Excercise  7.: degrees2   1, minutes2   1, seconds2   0, radians2   1
Excercise  8.: degrees3   1, minutes3   1, seconds3   1, radians3   1
Excercise  9.: degrees4  ##, minutes4   0, seconds4   1, radians4   1
Excercise 10.: n    1, ns    1
Excercise 11.: p1    1, p1s    1, p2    1, p2s    1, p3    1, p3s    1, p4    1, p4s    1
Excercise 12.: p5    1, p5s 0.5, p6    0, p6s    0
```

```
Assessment surname4.m version 09-Jun-2016 07:25:10
surname4    assignment 00000050  points 32.0
Points: 1-30 1 1 0 0 0 1 1 1 1 0 0 0 1 1 1 1 0 0 0 1 1 1 1 0 0 0 1 1 1 1
Points:31-60 1 1 1 1 1 1 1 1 1 1 1 1 0 0 0 0 1 1 0 0 0 0 0 0 0 0 0 0 0 0
Excercise  1.: a     1, as    1
Excercise  2.: s1    0, m1    0, v1    0, b    1, bs    1, c    1, cs    1
Excercise  3.: s2    0, m2    0, v2    0, d    1, ds    1, e    1, es    1
Excercise  4.: s3    0, m3    0, v3    0, f    1, fs    1, g    1, gs    1
Excercise  5.: s4    0, m4    0, v4    0, h    1, hs    1, m    1, ms    1
Excercise  6.: degrees1    1, minutes1    1, seconds1    1, radians1    1
Excercise  7.: degrees2    1, minutes2    1, seconds2    1, radians2    1
Excercise  8.: degrees3    1, minutes3    1, seconds3    1, radians3    1
Excercise  9.: degrees4   ##, minutes4   ##, seconds4   ##, radians4   ##
Excercise 10.: n     1, ns    1
Excercise 11.: p1   ##, p1s   ##, p2   ##, p2s   ##, p3   ##, p3s   ##, p4   ##, p4s   ##
Excercise 12.: p5   ##, p5s   ##, p6   ##, p6s   ##
```

Tab. 3. Assessment

4.2 Teacher's part

The teacher first copies m-file **gen_du1.m** into a MATLAB work directory (**WORK**) and runs it. This creates a directory **AssignmentDU1** which contains 600 different assignments for students. If the teacher wants to generate assignments for a different number of students or change (update) the year, he/she opens **gen_du1.m** and modifies academic year (at line 1) and number of assignments (at line 6). Simultaneously the m-file creates three other text files. Text file **assigDU1.txt** contains all generated tasks which allows for archiving or printing of the assignments and giving them to students on paper. Text file **resultDU1.txt** contains the results of all the generated tasks which, if printed, allows teacher to check the results without using a computer. Third text file **tables_for_assess.txt** contains a table with results of all the generated assignments and it is used to check the results via assessment program ROBOT.

The teacher has to download full name and login of all students from faculty information system for which the assignments will be generated. The teacher creates a file **name.csv** (semicolon separated) which contains three columns: surname, name and login. This file is needed to run the program ROBOT.

The teacher creates a directory **PCM** on a network drive **C** in such a way that students can only read files in this directory. Then the teacher creates a subdirectory **SUBMISSIONS** where students can only copy their scripts and see their presence and size (they can not save, delete, rewrite or read files here). Because students submit not only the tasks DU1-DU6 but also the semester work etc. to this directory there are created corresponding subdirectories. The subdirectory for this partial task is named **DU1**.

Before a lesson the teacher copies these five files into MATLAB work directory (**WORK**): **assess_du1.m**, **assess_du1_help.m**, **names.csv**, **tables_for_assess.txt**, **robot.m**. Students submit their work to the directory **C/PCM/SUBMISSIONS/DU1**. As mentioned above, students can copy their script here and see its presence and size, but they are not able to save files, rename, delete, open or modify them. This prevents one student to damage someone else's file. In the directory **P/PCM** students can only read files, which means, that they can view **table.txt**, but they can not modify it, delete it, etc.

After that the teacher just runs the program ROBOT (script **robot.m** on webpage ROBOT is set to run until 10:18 pm). The program checks the scripts that students continuously submit to the directory **C/PCM/SUBMISSIONS/DU1** and the results are continuously updated in the text file **table.txt** (at the end of the file). Students find their name with the assessment and see which variable is incorrect (0), which is correct (1), which is only partially correct (0.5) and which is missing (##) after opening the table. Once ROBOT checks the script it moves it to the directory **ASSESSED** which is created in the MATLAB work directory (**WORK**). More recent attempt of the same student overwrites the previous attempt of this student so after lesson there is only the latest and presumably best attempt of each student. The teacher does not have to copy all 60 items into a spreadsheet because file **table_for_excel.csv** is created. It is sufficient as assessment for some teachers. It contains number of assignment, actual name of the script, name it was supposed to have, date and time of submission, total number of points gained and 60 columns with points for variables.

We copy this data into an Excel file **assess.xls** where are formulas representing above mentioned assessment criteria (Tab. 2). If the teacher wants to stop the program earlier than set he/she has to press Ctrl + C in the MATLAB command line. If the teacher wants to run the script again he/she writes "robot" in the command line again. This is useful when different teachers or groups of students are using the same classroom etc. The teacher stops the program ROBOT, moves the directory **ASSESSED** and files **table_for_excel.csv** and **table.txt** to his/her drive and the e-assessment is done. The program ROBOT repeats the assessment using file **assess_du1.m** every 30 seconds. Because there are 24 students in the classroom, program usually waits until there is any file in the directory **C/PCM/SUBMISSIONS/DU1**. Even so, some students are able to get their script checked more than 30 times during a lesson. If the teacher has different number of students, he/she can set a different checking period.

5 Experience

Task DU1 has met with a very positive response by both students and teachers at our faculty. Students greatly appreciated the fact that they have almost immediate feedback which multiple authors such as Blyth and Labovic [9] or Sangwin [51] consider very important. Some students even abused this immediate feedback and instead of trying the script on their local computer, they immediately copied it in the evaluation directory. We had to modify the program for this reason so in case of a corrupt file, the program ROBOT is not terminated earlier than set. Currently, if students do not use command "clear all" the program works very reliably.

In order to gain feedback from students we asked them to complete a questionnaire. In this questionnaire students reported that they enjoyed this work, that they learned a lot, that they could finish the work at home when the program ROBOT was kept running and that they find it as a truly objective assessment etc. Teachers positively evaluated mainly huge reduction of workload and the opportunity to focus on those students who encountered difficulties and they can quickly archive the results of the evaluation and work of students. Overall, in the survey, both sides reported that not only the immediate feedback is beneficial but also that they do not need any papers. They also reported that it motivates some students and that the assessment is independent on the human factor etc.

At the same time teachers and students stated that this task is uncompromising in terms of compliance with MATLAB syntax [58], [51]. If the students submitted the script incorrectly or submitted a corrupt script, they got 0 points even when they previously had 58 items correctly.We kept intermediate assessment for this reason and if the student accidentally submitted wrong or corrupt script for final evaluation we checked whether a better assessment was previously achieved. We consider the file containing list of names (**`names.csv`**) necessary because we always find students who are trying to cheat instead of doing proper work. If someone would like to cheat by solving someone else's assignment, he/she would not succeed because assignment is tied to a particular student and assessment program recognizes this and tells the teacher which student does assignment truly belong to. We also consider setting rights in individual directories very important because no student could open, rename, move, copy, or delete the script no matter who created it (as mentioned before). Task DU1 met with a very positive response and when some students saw (after deadline for submission) that even their classmates (less skilled in mathematics) were able to gain full points, they wanted to improve their assessment. This option was provided to them, but the overall rating was then calculated as arithmetic mean of their previous and new results and they could not achieve full points.

The disadvantage of this assessment is that students do not know why particular variable is wrong and what mistake they made. However this task is so simple and divided into such small steps that feedback through symbols (1, 0, 0.5, #) was sufficient. Corbalan et al. [12] mentioned that: "(...) *for students of a technical domain, feedback provided on all the problem-solving steps is more effective and it is clearly preferred by learners and more motivating than feedback provided on the final solution step only.*" In this study they suggest that: "(...) *electronic environments should incorporate step-wise guidance which support students understand the correctness or incorrectness of their solutions.*"

6 Discussion

The programs used to generate assignments and check the results are simple, so everyone can manage to use them. This is why all used (and mentioned) programs are free for use on the webpage ROBOT. Example DU1 (used in this paper) generates different tasks by changing values in assignment but not text of the task. How about other tasks?

There are tasks (on our faculty DU2, DU4, DU6) which have only dozens of different variants differing in both values and text. Here we encounter the problem, how to check verbal responses of students and their graphs without forcing them to write results to a prescribed template [58], [22], [45]. We think that students in mathematics should be able to properly express themselves using proper terminology and symbols. This is the reason why we require composition of students work to be at an appropriate level. As mentioned in introduction our faculty selected MATLAB as CAS system and that is why we only generate assignments for students and results for teachers using program MATLAB. The evaluation itself is done by the teacher who can assess the accuracy of the results, their justification, graphic outputs and overall formal quality of the document [47], [48].

Our use of MATLAB is similar to the way Blyth and Labovic [9] of University of Melbourne use Maple. They consider as a unique feature of their work that their students use only Maple and external CAA (Computer Assisted Assessment) package is not required. Blyth and Labovic [9] have very good experiences with this approach to assessing students. "*They [students] would like more of these automatically marked assignments. They appear to be particularly engaged with these automatically (and semi-automatically) marked assignments and behave in a manner akin to computer game players intent on completing (or 'clocking') the game.*" They [9] also noted that creating these Maple files is time consuming but benefit of reduction of assessment time outweighs this effort.
Heck et al. [24] of University of Amsterdam and Utrecht University use Mathematica empowered applets for learning and practising algebra and calculus at secondary school level. They discuss that: „*Mathematica can be used as back engine to evaluate students' answers and provide appropriate feedback. These applets provide scoring facilities, are fully integrated in a virtual learning environment in which answers are stored, and can be edited by teachers to create their own exercises and tests.*"

Neither Blyth and Labovic [9] or Heck et al. [24] use CAS to generate tasks or assignments for students. They only use these systems to check the results. While we found attempts to use Maple and Mathematica for the purpose of electronic assessment we did not find any such attempts made in MATLAB. We then consider our use of MATLAB for e-assessment as useful contribution because MATLAB is used for teaching of numerical methods of solving of differential equations and other engineering applications on many universities worldwide, for example United Arab Emirates University [1], North Dakota State University [2], University of Minho [11], University of Connectcut [14], Ben-Gurion University of the Negev [14], [53], Technical University Budapest [35], DeVry University [42], University of Pittsburgh at Johnstown [46], Tel-Aviv University [53] and Francis Marion University [56].

Pacheco-Venegas et al. [45] assert that "*the simplest form of Computer Aided Assessment (CAA) systems are those in which students have to choose the correct answer from a number of options (multiple choice questions)*" and that "*the feedback provided only refers to the final solution and disregards the steps followed in the process.*" We use this form of CAA in mathematics only for training tests. Students have these training tests with multiple choice questions on e-learning (system MOODLE). If they select wrong answer there will be a comment on possible mistakes and why it may be wrong after assessment. In exam test we do not require only numerical accuracy, which is very well controllable using e-assessment, but also correct procedure, which may be different for each student. For tasks (on our faculty DU3 and DU4) where task for every student is completely different, we use database of entries and results, which each teacher generates on his/her own. For the reasons mentioned above we do not use e-assessment for these tasks.

Pacheco-Venegas et al. [45] assert that "*recent research has focused on the study of systems that can provide a more thorough assessment*" and that "*however, most CAA systems do not provide convenient user interfaces necessary to input the required mathematical symbols.*" Many other authors are trying to minimize disadvantages of e-assessment mentioned above. For example Schvinning et al. [54] "*consider the domain of mathematics and introduce a general framework that allows localising mistakes precisely and offers many options in exercise and feedback design*". Shirai and Fukui [55] "*present a web-based drill system named DIGITAL-WORK which assists learners in mastering some basic formulae using a new interactive math input method*". Jones et al. [31] "*report a study that tested an alternative approach to assessment, called comparative judgement, which may represent a superior method for assessing open-ended questions that encourage a range of unpredictable responses*". Holmes [26] "*suggests that carefully designed assessments can be used to increase student engagement and learning, and, as a result, contribute to improving the quality of the overall student experience*". Lan et al. [34] "*study the problem of automatically grading the kinds of open response mathematical questions that figure prominently in STEM (science, technology, engineering, and mathematics) courses*". Araújo et al. [5] "*present two algorithms for exercise generation admitting smart assessment*". Hettiarachchi et al. [25] "*proposes an e-assessment system that supports a formative assessment model that goes beyond simple types of questions and introduces an interactive dynamic environment for both skill and knowledge assessment in an online education*". However we believe it is not yet possible to use fully automated e-assessment for seminar work, bachelor thesis etc. on our faculty.

7 Conclusion

We agree with Pacheco-Venegas et al. [45] that "*formative assessment allows the students to contextualize and regulate their academic work. In practice, the teacher's heavy workload, or the large number of students attending class, makes it difficult to provide appropriate and timely assessment and feedback.*" For these and other reasons (lack of specialists and specialized classrooms, scarcity of economic means etc.) we produced above mentioned method to programmatically generate a large number of individual assignments of task DU1 for students and to provide fully automatic check and evaluation of their results. This e-assesment provides to students sufficient feedback immediatelly which motivates students to achieve better results. For our best knowledge this is the only attempt to use MATLAB in way of generating assignments and their results, and in way of assessment of student's work which is done in MATLAB as well.

Appendix 1

```
Assignment no. 3
Create m-file (script) which:
 1. Calculates the value of a polynomial p = -4x^9+7x^6-6x^5-1x^4-4x^3+3x^2-8 for x = 0.64. The
    script should also save this result as variable a, but display it on the monitor using text
    variable as showing eight decimal places (rounded up).
 2. Finds base angle beta (angle lies in the first quadrant) to given angle alpha and saves degrees
    as variable s1, minutes as variable m1, seconds as variable v1 and in radians as variable b, but
    displays it on the monitor using text variable bs showing six decimal places (rounded up). The
    script should also calculate using angle beta cotangent of angle alpha. It should also save this
    value as variable c, but display it on the monitor using text variable cs showing six decimal
    places (rounded up). Angle alpha = 1084 degrees 14 minutes 24 seconds.
 3. Finds base angle beta (angle lies in the first quadrant) to given angle alpha and saves degrees
    as variable s2, minutes as variable m2, seconds as variable v2 and in radians as variable d, but
    displays it on the monitor using text variable ds showing six decimal places (rounded up). The
    script should also calculate using angle beta tangent of angle alpha. It should also save this
    value as variable e, but display it on the monitor using text variable es showing six decimal
    places (rounded up). Angle alpha = 2983 degrees 7 minutes 17 seconds.
 4. Finds base angle beta (angle lies in the first quadrant) to given angle alpha and saves degrees
    as variable s3, minutes as variable m3, seconds as variable v3 and in radians as variable f, but
    displays it on the monitor using text variable fs showing six decimal places (rounded up). The
    script should also calculate using angle beta cosine of angle alpha. It should also save this
    value as variable g, but display it on the monitor using text variable gs showing six decimal
    places (rounded up). Angle alpha = 4545 degrees 56 minutes 32 seconds.
 5. Finds base angle beta (angle lies in the first quadrant) to given angle alpha and saves degrees
    as variable s4, minutes as variable m4, seconds as variable v4 and in radians as variable, but
    displays it on the monitor using text variable hs showing six decimal places (rounded up). The
    script should also calculate using angle beta sine of angle alpha. It should also save this
    value as variable m, but display it on the monitor using text variable ms showing six decimal
    places (rounded up). Angle alpha = 5345 degrees 59 minutes 41 seconds.
```

6. Finds angle alpha in degrees, minutes and seconds to given value x1=cot(alpha) if you know, that angle alpha lies in the first quadrant. The script should also save degrees as variable degrees1, minutes as variable minutes1 and seconds as variable seconds1 (round seconds to integers). It should also save angle alpha in radians as variable radians1 and display everything on the monitor. Value x1 = 5.8644.
7. Finds angle alpha in degrees, minutes and seconds to given value x2=tan(alpha) if you know, that angle alpha lies in the second quadrant. The script should also save degrees as variable degrees2, minutes as variable minutes2 and seconds as variable seconds2 (round seconds to integers). It should also save angle alpha in radians as variable radians2 and display everything on the monitor. Value x2 = -6.0087.
8. Finds angle alpha in degrees, minutes and seconds to given value x3=cos(alpha) if you know, that angle alpha lies in the third quadrant. The script should also save degrees as variable degrees3, minutes as variable minutes3 and seconds as variable seconds3 (round seconds to integers). It should also save angle alpha in radians as variable radians3 and display everything on the monitor. Value x3 = -0.5386.
9. Finds angle alpha in degrees, minutes and seconds to given value x4=sin(alpha) if you know, that angle alpha lies in the third quadrant. The script should also save degrees as variable degrees4, minutes as variable minutes4 and seconds as variable seconds4 (round seconds to integers). It should also save angle alpha in radians as variable radians4 and display everything on the monitor. Value x4 = -0.6707.
10. Finds the cube roots of the sum of given values a, b. The script should also save this result as variable n, but display it on the monitor using text variable ns showing nine decimal places (rounded up). Value a = -0.9663, value b = -2.8330.
11. Finds the natural logarithm of the number a, saves it as variable p1, but displays it on the monitor using text variable p1s showing five decimal places (rounded up). The script should also find the common logarithm of the number a, save it as variable p2, but display it on the monitor using text variable p2s showing seven decimal places (rounded up). It should also find the binary logarithm of the number a, save it as variable p3, but display it on the monitor using text variable p3s showing ten decimal places (rounded up). The script should also find the logarithm of number a to base z, save it as variable p4, but display it on the monitor using text variable p4s showing six decimal places (rounded up). Value a = 4.2859, value of base z = 8.
12. Finds the square root of the sum multiplied by k of exp(a) and the absolute value of b. The script should also save this result as variable p5, but display it on the monitor using text variable p5s showing seven decimal places (rounded up). It should also save as variable p6 the k-th power of the sum of exp(b) and the absolute value of a. The script should also display this result on the monitor using text variable p6s showing five decimal places. Value a = 1.7266, value b = -6.2173, multiple k = 11.

Name this m-file your login (for example novakj.m) and copy to disk C/PCM to a directory SUBMISSION to a subdirectory DU1. Use only MATLAB commands explained in class or in a textbook. After running the script, all variables must contain the required values. Do not use command clear all.

Acknowledgement

The authors would like to express their gratitude to the Faculty of Chemistry of Brno University of Technology in Brno for the provided facilities and opportunity to implement this project.

References

[1] AHMED, W. K. Advantages and Disadvantages of Using MATLAB/ode45 for Solving Differential Equations in Engineering Applications. *International Journal of Engineering*, 7(1), 2013, s. 25–31.

[2] ALIMI, M. H., MAHINFALAH, M., NAZARI, G. H. MATLAB in engineering education. In *Proceedings of the 2001 ASME Design Engineering Technical Conference and Computers and Information in Engineering Conference*, 9–12 September 2001, Pittsburgh, PA; United States, Volume 1, Code 62084, s. 461–465.

[3] ANDRADE-ARÉCHIGA, M., LÓPEZ, G., LÓPEZ-MORTEO, G. Assessing effectiveness of learning units under the teaching unit model in an undergraduate mathematics course. *Computers & Education,* 59(2), 2012, s. 594-606.

[4] ANDRADE-ARÉCHIGA, M., LÓPEZ, G., PULIDO, J. Mathematics education and information technologies in emerging economies. *Creative Education*, 4(10), 2013, s. 40-47.

[5] ARAÚJO, I., BRITO, I., MACHADO, G. J., PEREIRA, R. M. S., ALMEIDA, J. J., SMIRNOV, G. New algorithms for smart assessment of math exercises. *Advances in Intelligent Systems and Computing*, 353, 2015, s. 1221-1230.

[6] ARDITO, C., COSTABILE, M. F., MARSICO, M. D., LANZILOTTI, R., LEVIALDI, S., ROSELLI, T. et al. An approach to usability evaluation of e-learning applications. *Universal Access in the Information Society*, 4(3), 2005, s. 270-283.

[7] ARTIGUE, M. Learning Mathematics in a CAS Environment: The Genesis of a Reflection about Instrumentation and the Dialectics between Technical and Conceptual work. *International Journal of Computers for Mathematical Learning*, 7, 2002, s. 245-274.

[8] BEESON, M., NICAUD, J. F. MathXpert: un logiciel pour aider les élèves á apprendre les mathématiques par l'action. *Sciences et techniques éducatives*, 9 (1-2), 2002, s. 37-62.

[9] BLYTH, B., LABOVIC, A. Using Maple to implement eLearning integrated with computer aided assessment. *International Journal of Mathematical Education in Science and Technology*, 40(7), 2009, s. 975-988.

[10] BUBENÍK, F. Příspěvek k testování znalostí v matematických předmětech. In *6th Conference on Mathematics and Physics at Technical Universities*, Brno, Czech republic, (1), 2009, ISBN 978-80-7231-667-0, s. 65-70.

[11] CARNEIRO, F., LEÃO, C. P., TEIXEIRA, S. F. C. F. Teaching Differential Equations in Different Environments: A First Approach. *Computer Applications in Engineering Education*, 18, 2010, s. 555-562.

[12] CORBALAN, G., PAAS, F., CUYPERS, H. Computer-based feedback in linear algebra: effects on transfer performance and motivation. *Computers & Education*, 55(2), 2010, s. 692-703.

[13] COSTABILE, M., DE MARSICO, M., LANZILOTTI, R., PLANTAMURA, V., ROSELLI, T. On the usability evaluation of e-learning applications. In *Proceedings of the 38th Annual Hawaii International Conference on System Sciences*, 6b-6b, 2005.

[14] CUTLIP, M. B. SHACHAM, M. Integration of Numerical Problem Solving into the Chemical Engineering Curriculum. In *Proceedings of the 2003 American Society for Engineering Education Annual Conference & Exposition, Copyright American Society for Engineering Education*, 8.754.1–8.754.5, 2003.

[15] DE LA VILLA, A., LOIS, A., MILEVICICH, L., DEL REY, A. M., RODRIGUEZ, G. New trends in computer math teaching. *Proceedings of the Frontiers in Mathematics and Science Education Research Conference*, 1-3 May 2014, Famagusta, North Cyprus.

[16] DRIJVERS, P., TROUCHE, L. Handheld technology for mathematics education: flashback into the future. *ZDM Mathematics Education*, 42, 2010, s. 667-681.

[17] European Commission/EACEA/Eurydice. *The European Higher Education Area in 2015: Bologna Process Implementation Report.* Luxembourg: Publications Office of the European Union, 2015, ISBN 978-92-9201-847-4.

[18] *European Conference Modern Mathematical Methods in Engineering*. The Department of Mathematics and Descriptive Geometry, VŠB – Technical University of Ostrava, Czech republic, 2015, annual.

[19] EVANS, M., NORTON, P., LEIGH-LACANCASTER, D. *Mathematical methods computer algebra system (CAS) pilot study examinations 2004*, 2005, s. 329-336.

[20] FLYNN, P. CAS in Australia: a brief Perspective. In *The Fifth CAME Symposium.* Pécs, Hungary, 2007.

[21] GARCÍA, A., GARCÍA, F., RODRÍGUEZ, G., DE LA VILLA, A. Small projects: Learning and Assessing Competencies: New challenges for Mathematics in Engineering Degrees in Spain. *Proceedings of 16th SEFI MWG Seminar*. Salamanca, Spain, 2012.

[22] GARCÍA, A., GARCÍA, F., DEL REY, A. M., RODRÍGUEZ, G., DE LA VILLA, A. Changing assessment methods: new rules, new roles. *Journal of Symbolic Computation*, 2014, s. 61-62, s. 70-84.

[23] GRANIĆ, A. Experience with usability evaluation of e-learning systems. *Universal Access in the Information Society*, 7(4), 2008, s. 209-221.

[24] HECK, A., BOON, P., VAN VELTHOVE, W. Mathematica empowered applets for learning school algebra and calculus. In *Proceedings of the 9th International Mathematica Symposium*, 2008.

[25] HETTIARACHCHI, E., HUERTAS, M. A., MOR, E. E-assessment system for skill and knowledge assessment in computer engineering education. *International Journal of Engineering Education*, 31(2), 2015, s. 529-540.

[26] HOLMES, N. Student perceptions of their learning and engagement in response to the use of a continuous e-assessment in an undergraduate module. *Assessment and Evaluation in Higher Education*, 40(1), 2015, s. 1-14.

[27] *International conference on e-Assessment and computer supported assessment and qualifications, the e-Assessment Question Conference and Exhibition*. The focus for the conference, London, Great Britain, 2015, annual, http://www.e-assess.co.uk.

[28] International standard. ISO 80000-2. *Quantities and units – Part 2: Mathematical signs and symbols to be used in the natural science and technology, first edition*, Published in Switzerland, 2009, ISO 80000-2:2009(E).

[29] JISCinfoNet. *Effective Use of VLEs: e–Assessment*, JISCinfoNet applied infoKit, 2006, http://tools.jiscinfonet.ac.uk/downloads/vle/eassessment-printable.pdf.

[30] JONES, I. Computer-aided assessment questions in engineering mathematics using MapleTA. *International Journal of Mathematical Education in Science and Technology*, 39(3), 2008, s. 341-356.

[31] JONES, I., MATTHEW, I. The problem of assessing problem solving: can comparative judgement help? *Educational Studies in Mathematics*, 89(3), 2015, s. 337-355.

[32] KAJETANOWICZ, P., WIERZEJEWSKI, J. Application of computer algebra systems in automatic assessment of math skills. Teaching, *Mathematics and Computer Science*, 6(2), 2008, s. 395-408.

[33] LAGRANGE, J. B., ARTIGUE, M., LABORDE, C., TROUCHE, L. Technology and mathematics education: multidimensional overview of recent research and innovation. In: F. K. S. Leung (Ed.), *Second international handbook of mathematics education*, 1, Dordrecht: Kluwer Academic Publisher, 2003, s. 237-270.

[34] LAN, A. S., VATS, D., WATERS, A. E., BARANIUK, R. G. Mathematical language processing: Automatic grading and feedback for open response mathematical questions. *L@S 2015 - 2nd ACM Conference on Learning at Scale*, 2015, s. 167-176.

[35] LÁNG-LÁZI, M., DIÓSPATONYI, I., PETZ, D., VICZIÁN, ZS., Fetter, GY. Computer and multimedia in chemical engineering education. *Computer and Chemical Engineering Supplement*, 1999, s. 637–640.

[36] LIMNIOU, M., SMITH, M. The role of feedback in e-assessments for engineering education. *Education and Information Technologies*, 17(4), 2012.

[37] LOIS, A., MILEVICICH, L., RODRÍGUEZ, G., DE LA VILLA, A. Perspectiva de las TIC'S en la educación superior en Iberoamérica. P. Lestón (Ed), *Acta Latinoamericana de Matemática Educativa*, 24, México: Comité Latinoamericano de Matemática Educativa, 2011, s. 1170-1178.

[38] LOIS, A., MILEVICICH, L., RODRÍGUEZ, G., DE LA VILLA, A. (2013). Enseñar Matemática: un reto en el nuevo paradigma tecnológico. R. Flores (Ed), *Acta Latinoamericana de Matemática Educativa*, 26, México: Comité Latinoamericano de Matemática Educativa , 2013, s. 1859-1866.

[39] LOIS, A., MILEVICICH, L., RODRÍGUEZ, G., DE LA VILLA, A. La revolución tecnológica en la enseñanza de las matemáticas: el nuevo paradigma ¿es una oportunidad de cambio o un simple engaño? R. Flores (Ed), *Acta Latinoamericana de Matemática Educativa*, 26, México: Comité Latinoamericano de Matemática Educativa, 2013, s. 1867-1876.

[40] LOPEZ-MORTEO, G., LÓPEZ, G. Computer support for learning mathematics: a learning environment based on recreational learning objects. *Computers & Education*, 48(4), 2007, s. 618-641.

[41] MATLAB creative team. *MATLAB The Language of Technical Computing*. USA: Natick, The MathWorks, Inc., 2000.

[42] MUQRI, M. R., MUQRI, H., EK CHNG, S. Leveraging the power of java and Matlab to solve ODEs. In *Proceedings of the 121st ASEE Annual Conference and Exposition*, 15–18 June 2014, Indianapolis, United States.

[43] *National Conference Using of computers in education of mathematics which is accredited by the Ministry of Education of the Czech Republic as part of teacher training*. Department of Mathematics. Faculty of Education. University of South Bohemia in České Budějovice, 2015, biennial.

[44] NICOL, D. J., MACFARLANE-DICK, D. Formative assessment and self-regulated learning: a model and seven principles of good feedback practice. *Studies in Higher Education*, 31(2), 2006, 199-218.

[45] PACHECO-VENEGAS, N. D., LÓPEZ, G., ANDRADE-ARÉCHIGA, M. Conceptualization, development and implementation of a web-based system for automatic evaluation of mathematical expressions. *Computer @ Education*, 88, 2015, s. 15-28.

[46] PIETRYGA, F. W. Solving Differential Equations Using MATLAB/Simulink. In *Proceedings of the 2005 American Society for Engineering Education Annual Conference & Exposition,* Copyright American Society for Engineering Education, 2005.

[47] POLCEROVÁ, M. Computer technique seminar in the bachelor's program at the Faculty of Chemistry of Brno, University of Technology. In *Aplimat 2005 - 4th International Conference*, Bratislava, Slovakia, 2005, ISBN 80-969264-3-8, s. 357-363.

[48] POLCEROVÁ, M. Computer Exercises in Teaching Mathematics, In *13th SEFI European Seminar on Mathematics in Engineering Education*, MWG Mathematics Working Group, Proceeding Buskerud University College, Kongsberg, Norway, 2006, ISSN 1476-1378, s. 38-41.

[49] POLCEROVÁ, M. *MATLAB Počítačová cvičení z matematiky*. Faculty of Chemistry, Brno University of Technology. Czech republic, 2010, ISBN 978-80-214-4236-8.

[50] POLCEROVÁ, M. Use of MATLAB in teaching, In *MITAV 2014 Mathematics Information Technology and applied sciences*, University of defence, Brno, Czech republic, 2014, ISBN 978-80-7231-961-9, s. 211-216.

[51] SANGWIN, C. J. Assessing elementary algebra with STACK. *International Journal of Mathematical Education in Science and Technology*, 38(8), 2007, s. 987-1002.

[52] SANGWIN, C. J. *Computer aided assessment of mathematics*. Oxford University Press, 2013.

[53] SHACHAM, M., BRAUNER, N., POZIN, M. Application of feedback control principles for solving differential-algebraic systems of equation in process control education. *Computers & Chemical Engineering*, 20, 1996, s. 1329–1334. S0098-1354(96)00228-1.

[54] SCHWINNING, N., SCHYPULA, M., STRIEWE, M., GOEDICKE, M. Concepts and realisations of flexible exercise design and feedback generation in an e-assessment system for mathematics. *CEUR Workshop Proceedings*, 1186, 2014.

[55] SHIRAI, S., FUKUI, T. Development and evaluation of a web-based drill system to master basic math formulae using a new interactive math input method. *Lecture Notes in Computer Science (including subseries Lecture Notes in Artificial Intelligence and Lecture Notes in Bioinformatics)*, 8592 LNCS, 2014, s. 621-628.

[56] SZURLEY, D. Projects with applications of differential equations and MATLAB. In *Electronic Proceedings of the Twenty-third Annual International Conference on Technology in Collegiate Mathematics*, 17–20 March 2011, Denver, Colorado, 2011, Paper S022, s. 234–243.

[57] The MathWorks, Inc. *MATLAB version 6.1.0.450 Release 12.1* [software]. License Brno University of Technology: PC Windows NT4 and higher, 128 MB RAM; 250 MB disc; resolution 800×600, 256 colours, 2001.

[58] TOWNSLEY, L. Testing across precalculus sections using WebAssign. In *Electronic Proceedings of the Twenty-first Annual International Conference on Technology in Collegiate Mathematics*, 2009.

[59] WICK, D. Free and open-source software applications for mathematics and education. In *Electronic Proceedings of the Twenty-first Annual International Conference on Technology in Collegiate Mathematics*. Louisiana, New Orleans, 2009, s. 300-304.

Current address

Polcerová Marie, RNDr. Ph.D.
Institute of Physical and Applied Chemistry, Faculty of Chemistry,
Brno University of Technology
Purkyňova 118, 612 00 Brno, Czech Republic
Tel. number: +420 541 149 487, e-mail: polcerova@fch.vutbr.cz

Polcer Jaroslav, Ing.
ROmiLL, spol. s r.o.
Tkalcovská 799/14,602 00 Brno, Czech Republic
Tel. number: +420 518 714 404, e-mail: polcer@romill.cz

Polcer Pavel, Bc
Faculty of Mechanical Engineering,
Brno University of Technology
Technická 2896/2, 616 69 Brno, Czech Republic,
Tel. number: +420 728 419 672, e-mail: 153046@vutbr.cz

Polcerová Lenka, Mgr.
Department of Anthropology, Faculty of Science,
Masaryk University
Kotlářská 267/2, 602 00 Brno, Czech Republic
Tel. number: +420 549 494 966, e-mail: polcerova@mail.muni.cz

Slovak University of Technology in Bratislava
Faculty of Mechanical Engineering
16th Conference on Applied Mathematics **APLIMAT 2017**
Proceedings

THE USAGE OF STATISTICAL METHODS FOR EVALUATION OF NITRATE NITROGEN CONTENT

PORUBCOVÁ Lýdia (SK), TIRPÁKOVÁ Anna (SK), FESZTEROVÁ Melánia (SK)

Abstract. The article shows the impacts of agro - technical practices on changes of nitrate nitrogen content. From the tilth practices this effect described two methods of cultivation: medium and shallow depth tilth. In cultivation method we used three kinds of fertilization (without fertilizer – O, balanced fertilization with NPK fertilizers – PH, balanced fertilization with NPK fertilizers after harvest residues incorporating – PZ). The statistical analysis of the obtained data was performed by means of Wilcoxon two - sample test.

Keywords: Wilcoxon two - sample test, nitrate nitrogen, soil tillage system, depth ploughing

Mathematics Subject Classification: Primary 62G08; *JEL classis:* C40

1 Introduction

Agriculture is one of the most extensive human activities in nature, which can have positive as well as negative impact on the quality of the natural environment. For improving the efficiency of agricultural activity it is very important to determine fertilizers doses with respect to soil analysis findings. Maintaining adequate application of mineral and organic fertilizers, agricultural managers and farmers seek to ensure sufficient nutrient content in soil and subsequent nutrient usability by crops.

One of the essential biogenic elements is nitrogen (Javoreková et al., 2008), which in comparison with other elements has the most intensive effects on crop life processes. Its importance follows from the requirement of nitrogen in organic matter formation as well as in human nutrition (Torma, 2005). Nitrogen present in soil undergoes various biological transformations (Mikkelsen, 2012). Transformations of nitrogen in soil (Javoreková et al., 2008) are a crucial factor affecting crop yield. Therefore, it is inevitable to understand these processes and to influence them in favour of man and the environment. The changes of nitrogen forms and content in soil depend not only on the soil properties, which condition microbial processes of mineralization and its immobilization in soil, but also on soil fertilization. This also is one of the reasons why it is necessary to adjust the fertilization to soil order and type. It is, however, essential to realize that high fertilization intensity may cause nitrogen abundance. This fact is related to the risk of nitrite and nitrate accumulation in soil, which affects nitrogen accumulation in crops. Since excessive nitrogen content in crops meant for consumption can influence consumers in a negative manner, in present paper we are focused on changes of nitrate nitrogen content in soil.

1.1 The Changes of Nitrogen Content in Soil

According to Bujnovský (2002), the representation of the main nitrogen resources in its overall entry to soil is as follows: industrial fertilizers 38.7 % (up to 53.2 %), animal production fertilizers 29.2 % (ranging from 21.6 % to 37.3 %), atmospheric deposition 17.6 % (ranging from 11.2 % to 32.3 %), and biological nitrogen fixation 12.4 % (from 5.3 % to 29.8 %). One of the most extensive nitrogen resources is the atmosphere, containing approximately 78 % molecular nitrogen (Brown, 2008). Based on previous ionization and formation of NH_4^+ and NO_3^- around 10 – 40 kg.ha^{-1} of nitrogen enters soil annually (Richter, Hlušek, 1994). Post-harvest residues are decomposed in soil and thus contribute to formation of organic soil matter, which contains 5 % nitrogen. From the immobilized nitrogen in organic matter around 1 – 3 % is released annually, and it thus becomes receivable by plants (Barbarick, 2006).

Nitrogen is also present in the Earth's crust as a component of organic matter and humus (Killpack, Buchholz, 1993). Soil tillage practices have a major impact on soil carbon storage and cropping sustainability, due to their effect on soil aggregation, organic residue decomposition rate, oxygen-carbon dynamics, microbial abundance and diversity, nitrogen mineralization and nutrient availability (Valboa et al., 2015).

Papini et al. (2007) report that nitrogen content and dynamics in soil is affected by various tillage systems. In reduced tillage higher nitrogen content was found than in shallow and medium depth ploughing, and differences were also found in the lower tissue of samples (Papini et al., 2007). Soil tillage also has an effect on interaction relationships with climatic conditions (Demo, 1990). Processes of mineralization and nitrification can be also affected by means of agro-technologies in form of soil aeration. According to Demo (1990) agro - technologies influence these processes more significantly in case of heavier soils than lighter ones.

From the ecological, productivity – related and economical perspective, it is very important to possess knowledge related to factors such as tillage, fertilization, soil - ecological and climatic conditions. Their variability in soil changes during the vegetative period. These factors influence the dynamics of nitrate nitrogen content in soil and soil processes.

The aim of the present experiment was to determine the impacts of agro-technical practices on the changes of nitrate nitrogen content in soil. Various statistical methods were used to assess the impact of mutual relationships between agro-technical practices and fertilization with industrial fertilizers in combination with incorporation of post-harvest residues in order to make the use of such fertilizers as efficient as possible. The aim of this article is to describe and the use of non-parametric test in evaluation of nitrate nitrogen content. The statistical analysis of the obtained data was performed by means of the Wilcoxon two – sample test.

2 Material and Methods

The study was located on research - experimental base of Slovak University of Agriculture in Nitra (SUA) – Dolná Malanta. (Fig. 1) Dolná Malanta is 45 km away from the SUA campus, and its altitude is 175 – 180 m above sea level. The site of the research - experimental base has a flat-land character with a slight southern inclination. Geographically, the site is situated in the western part of Žitavská pahorkatina whose characteristic triangular shape is bounded by the mountain range of Tribeč and rivers Nitra and Žitava.

Fig. 1. Dolná Malanta – the research-experimental base of Slovak University of Agriculture in Nitra (Slovak Republic)

Soils of the researched site evolved from originally proluvial solifluction sediments. The research-experimental base is situated at the boundary between the loess sediments of Žitavská pahorkatina and the hillside sediments of Tribeč mountain range. The site of the research-experimental base of SUA in Nitra, Dolná Malanta – is a very warm agro-climatic area, where the average temperature during the vegetative period under consideration is 16.3°C, and a very dry agro-climatic subarea with average annual precipitation being 561 mm (Špánik, Repa, Šiška, 2002).

As mentioned above, dynamics and values of $N\text{-}NO_3^-$ content in soil profile were monitored in two consecutive vegetative periods (2009/2010, 2010/2011) of winter wheat (*Triticum aestivum*, variety Verita). As a preceding crop the red clover (*Trifolium pratense*) was used. In the experiment two manners of basic soil tillage were applied: medium depth ploughing (0.20 – 0.25 m) – henceforth B1, and shallow ploughing (0.15 – 0.20 m) – henceforth B2. For each of the two tillage manners three fertilization variants were applied: without fertilizer (henceforth 0), industrial fertilizer containing nitrogen-phosphorus-potassium-based (NPK) fertilizers (henceforth PH), and industrial fertilizer with NPK fertilizers after incorporating post-harvest residues of the preceding crop (henceforth PZ). Both PH and PZ variant were balanced fertilization variants. Fertilizing winter wheat (variety Verita) with nitrogen follows from the demands and biological specifics of the variety, the soil fertility rate, the applied sowing practice, and soil conditions.

In variants PH and PZ ammonium nitrate with dolomite (LAD) was used for fertilization. Phosphorus was added in form of triple superphosphate (2009) and single phosphate (2010). In 2009 potassium was not used for fertilization, and in 2010 potassium salt was applied for fertilization (60 %).

The analyzed soil samples were taken from two depths, 0.00 – 0.3 m and 0.30 – 0.6 m. Soil samples were taken in fortnight intervals during the whole vegetative period. Variant repetition was quadruple. In both manners of basic tillage the experiment was based on the method of perpendicularly divided blocks. Fertilization variants were arbitrarily distributed within the particular tillage manners, so that the influence of soil heterogeneity was eliminated. Soil samples for $N\text{-}NO_3^-$ analysis were taken by means of standard procedures. Visible plant remnants were removed by hand. Nitrate nitrogen ($N\text{-}NO_3^-$) content was determined by the colour method (Radov et al., 1985). The obtained data were statistically processed with the use of STATISTICA software.

3 Results and Discussion

We analysed the values of nitrate nitrogen ($N\text{-}NO_3^-$) content in soil samples. The soil samples were collected during vegetation periods 2009/2010 and 2010/2011 in two soil tillage manners (B1, B2) and three fertilization variants (0, PH, PZ) in two depths (0.00 – 0.3 m; 0.30 – 0.6 m) (Porubcová, 2013). During the two vegetation periods the lowest recorded value of $N\text{-}NO_3^-$ content was 0.67 $mg.kg^{-1}$. The soil samples were collected on April 20, 2010; in B2 tillage 0 fertilization variant in depth 0.30 – 0.6 m. The highest concentration 21.76 $mg.kg^{-1}$ was recorded in B1 tillage. The soil samples were collected on October 19, 2009; 0 fertilization variant in depth 0.00 – 0.30 m. The average $N\text{-}NO_3^-$ content in vegetation periods under consideration was 4.9439 $mg.kg^{-1}$.

3.1 Nitrate Nitrogen Content at Medium Depth Ploughing

During the vegetation periods under consideration (2009/2010 and 2010/2011) at B1 tillage the highest recorded value of $N\text{-}NO_3^-$ content was 21.76 $mg.kg^{-1}$ in 0 fertilization variant in depth 0.0-0.3 m (sampling date: October 19, 2009). The lowest $N\text{-}NO_3^-$ content 0.82 $mg.kg^{-1}$ was recorded in 0 fertilization variant in depth 0.0 – 0.3 m (sampling date: April 20, 2010). The average value of $N\text{-}NO_3^-$ content during the two vegetation periods was 4.5714 $mg.kg^{-1}$.

In 0 fertilization variant the lowest $N\text{-}NO_3^-$ content was 0.82 $mg.kg^{-1}$ in depth 0.0 – 0.3 m (April 20, 2010). The highest value of $N\text{-}NO_3^-$ content was 21.76 $mg.kg^{-1}$ and was recorded in depth 0.0 – 0.3 m (October 19, 2009). The average value of $N\text{-}NO_3^-$ content was 4.54 $mg.kg^{-1}$.

In PH fertilization variant the lowest value of $N\text{-}NO_3^-$ content 0.85 $mg.kg^{-1}$ was recorded in depth 0.0 – 0.3 m (April 20, 2010). The highest $N\text{-}NO_3^-$ content was 16.48 $mg.kg^{-1}$ and was recorded in depth 0.0 – 0.3 m (October 5, 2009). The average value of $N\text{-}NO_3^-$ content was 4.52 $mg.kg^{-1}$.
In PZ fertilization variant the lowest $N\text{-}NO_3^-$ content 0.88 $mg.kg^{-1}$ was recorded in depth 0.3 – 0.6 m (April 20, 2010). The highest $N\text{-}NO_3^-$ content 14.22 $mg.kg^{-1}$ was recorded in depth 0.0 – 0.3 m (October 19, 2009). The average value of $N\text{-}NO_3^-$ content was 4.65 $mg.kg^{-1}$.

3.2 Nitrate Nitrogen Content at Shallow Depth Ploughing

During the vegetation periods under consideration (2009/2010 and 2010/2011) at B2 tillage the highest recorded $N\text{-}NO_3^-$ content was 15.12 $mg.kg^{-1}$ at PZ fertilization in depth 0.0 – 0.3 m (sampling date: November 2, 2009). The lowest $N\text{-}NO_3^-$ content 0.67 $mg.kg^{-1}$ was recorded in 0 fertilization variant in depth 0.3 – 0.6 m (sampling date: April 20, 2010). The average N-NO_3^- content during the two consecutive vegetation periods was 5.3166 mg.kg^{-1}.

In 0 fertilization variant the lowest value of N-NO_3^- content was 0.67 mg.kg^{-1} and was recorded in depth 0.3 – 0.6 m (April 20, 2010). The highest N-NO_3^- content was 14.87 mg.kg^{-1} and was recorded in depth 0.0 – 0.3 m (November 2, 2009). The average value of N-NO_3^- content was 4.3933 mg.kg^{-1}.

In PH fertilization variant the lowest value of N-NO_3^- content 1.44 mg.kg^{-1} was recorded in depth 0.00 – 0.3 m (June 17, 2011). The highest N-NO_3^- content 13.89 mg.kg^{-1} was recorded in depth 0.0 – 0.3 m (October 19, 2009). The average value of N-NO_3^- content was 5.4337 mg.kg^{-1}.
In PZ fertilization variant the lowest value of N-NO_3^- content was 2.06 mg.kg^{-1} and was recorded in depth 0.30 – 0.6 m (April 11, 2011). The highest N-NO_3^- content 15.12 mg.kg^{-1} was recorded in depth 0.0 – 0.3 m (November 2, 2009). The average value of N-NO_3- content was 6.1225 mg.kg^{-1}.

3.3 Analysis of results

The first issue we addressed by the experiment was the question if there are any significant differences in N-NO_3^- content in soil samples at various depths of ploughing B1 (N-NO_3^-) and B2 (N-NO_3^-). Since parametric test use prerequisites were not met, in order to verify statistical significance of the differences we applied a nonparametric statistical method, namely Wilcoxon two - sample test. Nonparametric Wilcoxon two - sample test allows testing hypothesis about equivalence of two distributions. Assume that $(X_1, X_2, \dots, X_m)$ and $(Y_1, Y_2, \dots, Y_n)$ are two mutually independent random samples from continuous distributions. Assume that $(X_1, X_2, \dots, X_m)$ is a random sample from a population with continuous distribution function F, and $(Y_1, Y_2, \dots, Y_n)$ is a random sample independent on the latter one from a population with continuous distribution function G. We test null hypothesis $H_0 : F = G$ against alternative hypothesis $H_1 : F \neq G$.

Here, the observed variables are X, Y, where X is the N-NO_3^- content in soil samples at B1 ploughing and Y is the N-NO_3^- content in soil samples at B2 ploughing. The first sample set consists of soil samples at B1 ploughing and the second sample set comprises soil samples at B2 ploughing. The recorded values of N-NO_3^- content in soil samples under consideration are the realizations of the two mutually independent random samples from continuous distributions. The tested hypothesis is null hypothesis H_0: *the distributions of the random samples are equivalent*. Against the null hypothesis we pose alternative hypothesis H_1: *the distributions of the random samples are different*. In order to test if hypothesis H_0 holds we applied Wilcoxon two-sample test with the use of software STATISTICA.

Providing the software with input data, for Wilcoxon two-sample test we obtain the following output: the value of test criterion Z and the value of probability p. The test can be interpreted by p value, which is the probability of error which is made in case the null hypothesis is rejected. If the value of probability p is low enough ($p < 0.05$ or $p < 0.01$), the tested hypothesis is rejected (at significance level 0.05 or 0.01). Otherwise, the tested hypothesis cannot be rejected.

Fertilization variant 0 - without fertilizer, control variant

Firstly, we tested the statistical significance of differences between ploughing manners B1 and B2 in N-NO_3^- content in soil samples taken from various depths. The analyzed soil samples were taken from two depths, 0.00 – 0.3 m and 0.30 – 0.6 m. The soil samples were in fertilization variant 0. Having tested the statistical significance of differences between ploughing manners B1 and B2 in N-NO_3^- content in soil samples in fertilization variant 0, considering samples taken from depth 0.0-0.3 m, the obtained value of test criterion was $Z = 0.670139$ and probability p value $p = 0.502770$. Since the obtained p value is greater than 0.05, the null hypothesis cannot be rejected, i.e. the difference between ploughing manners B1 and B2 with respect to the observed variable – N-NO_3^- content in soil samples in fertilization variant 0, considering soil samples taken from depth 0.00 – 0.3 m – is not statistically significant. The recorded differences between ploughing manners B1 and B2 in N-NO_3^- content in soil samples without fertilization taken from depth 0.00 – 0.3 m are not statistically significant. This is illustrated in Fig. 2

Using software STATISTICA, for soil samples taken from depth 0.3 – 0.6 m we obtained the following results: the value of test criterion was $Z = 0.564227$, and probability p value was $p = 0.572600$. Since the obtained p value is greater than 0.05, the null hypothesis cannot be rejected, i.e. the difference between ploughing manners B1 and B2 with respect to the observed variable – N-NO_3^- content in soil samples without fertilization, considering soil samples taken from depth 0.3 – 0.6 m – is not statistically significant. This is illustrated in Fig. 3.

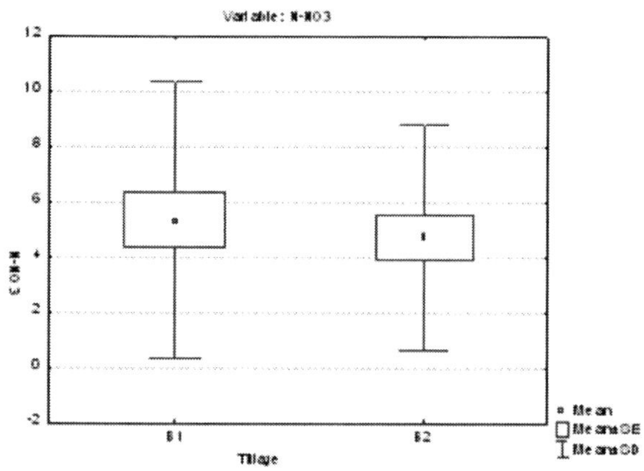

Fig. 2. The average value of N-NO_3^- content in soil samples between ploughing manners B1 and B2 in fertilization variant 0 (depth 0.00 – 0.3 m)

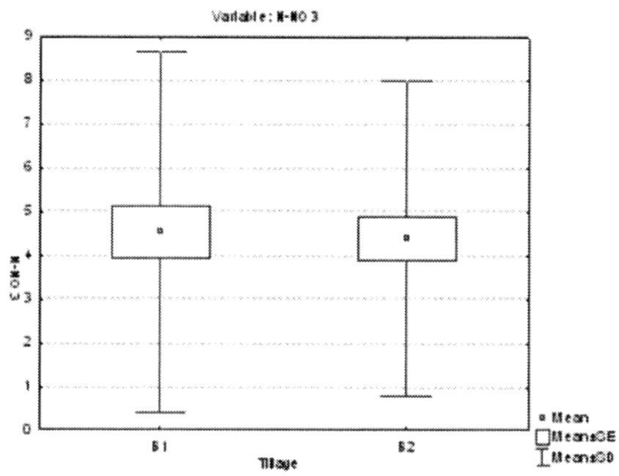

Fig. 3. The average value of N-NO_3^- content in soil samples between ploughing manners B1 and B2 in fertilization variant 0 (depth 0.30 – 0.6 m)

Fertilization variant with NPK fertilizers (PH)

Analogously, we tested the statistical significance of differences between ploughing manners B1 and B2 in N-NO_3^- content in soil samples with balanced fertilization (PH variant). The analyzed soil samples were taken from two depths (0.00 – 0.3 m, 0.30 – 0.6 m).

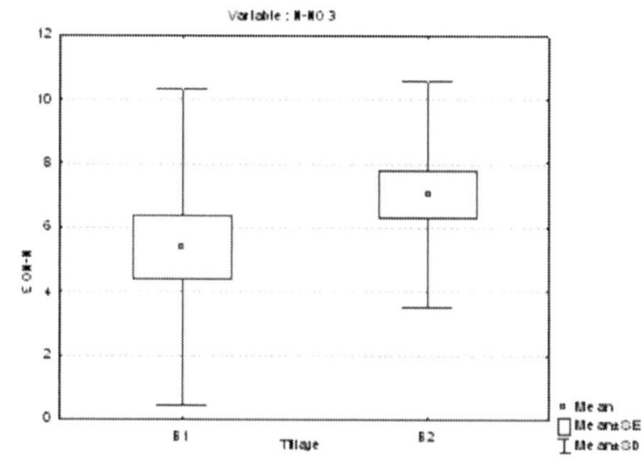

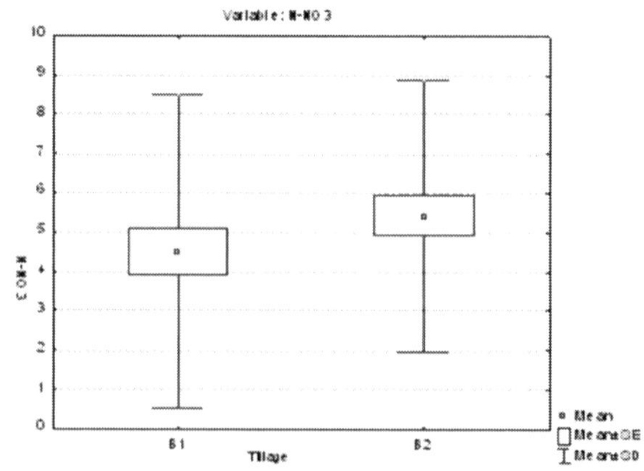

Fig. 4. The average value of $N\text{-}NO_3^-$ content in soil samples between ploughing manners B1 and B2 in fertilization variant PH (depth 0.0 – 0.3 m)

Fig. 5. The average value of $N\text{-}NO_3^-$ content in soil samples between ploughing manners B1 and B2 in fertilization variant PH (depth 0.3 – 0.6 m)

Having tested the statistical significance of differences between ploughing manners B1 and B2 in $N\text{-}NO_3^-$ content in soil samples taken from depth 0.00 – 0.3 m, the obtained value of test criterion was $Z = -2.28878$ and probability p value $p = 0.022093$ and from depth 0.30 – 0.6 m, the obtained value of test criterion was $Z = -2.02609$ and probability p value $p = 0.042756$.
Both the obtained p value is smaller than 0.05, the null hypothesis is rejected (at significance level = 0.05), i.e. the difference between ploughing manners B1 and B2 with respect to the observed variable – $N\text{-}NO_3^-$ content in soil samples in PH fertilization variant, considering soil samples taken from depth 0.00 – 0.3 m and from 0.30 – 0.6 m – is statistically significant. This is illustrated in the following figures (Fig. 4 – 5).

Fertilization variant with NPK fertilizers after incorporating post-harvest residues (PZ)

Thirdly, we tested the statistical significance of differences between ploughing manners B1 and B2 in $N\text{-}NO_3^-$ content in soil samples with balanced fertilization after incorporating post-harvest residues of the preceding crop (PZ variant).The analyzed soil samples were taken from two depths (0.0 – 0.3 m, 0.3 – 0.6 m).

For test the statistical significance of differences between ploughing manners B1 and B2 we applied Wilcoxon two-sample test with the use of software STATISTICA. The obtained value of test criterion from depth 0.0 – 0.3 m, was $Z = -2.03104$ and probability p value $p = 0.042252$ and from depth 0.3 – 0.6 m, the obtained value of test criterion was $Z = -2.28988$ and probability p value $p = 0.022029$.

Both the obtained p value is smaller than 0.05. The null hypothesis is rejected (at significance level = 0.05), i.e. the difference between ploughing manners B1 and B2 with respect to the observed variable – $N\text{-}NO_3^-$ content in soil samples with balanced fertilization after incorporating post-harvest residues of the preceding crop (PZ variant). Considering soil samples taken from depth 0.0 – 0.3 m and from 0.3 – 0.6 m – is statistically significant. This is illustrated in the following figures (Fig. 6 – 7).

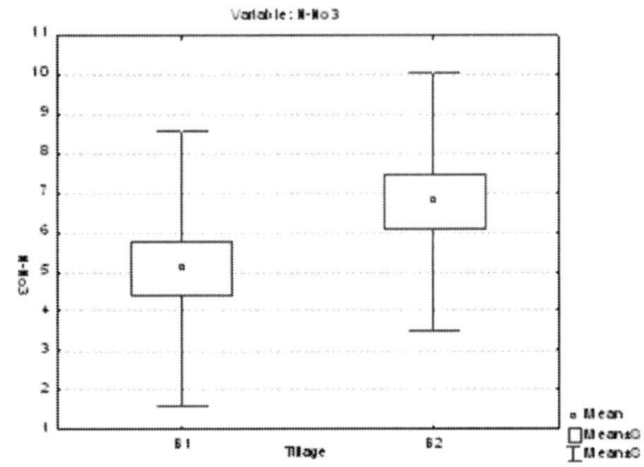

Fig. 6. The average value of $N\text{-}NO_3^-$ content in soil samples between ploughing manners B1 and B2 in fertilization variant PZ (depth 0.0 – 0.3 m)

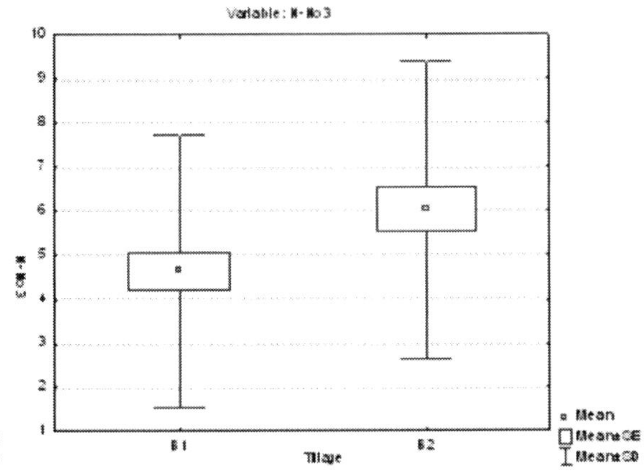

Fig. 7. The average value of $N\text{-}NO_3^-$ content in soil samples between ploughing manners B1 and B2 in fertilization variant PZ (depth 0.3 – 0.6 m)

4 Conclusion

If nitrogen supplies in soil get exhausted too early in the vegetation period, it can lead to decrease in quality and quantity of the crop yield. In case of nitrogen shortage plants "broach their reserves" which results in protein decomposition. As a consequence, plants get older faster, in some cultivated crops their vegetation period gets shorter, their seeds and fruits are not matured enough. On the other hand, nitrogen overabundance causes that plants grow luxuriantly with many deep green leaves. This may result in over-thickening of the plant cover. In over-thickened cover plants get weaker and more disease-prone. Their root system gets exhausted, its growth falls behind the growth of the aboveground parts, it does not branch out, nor does it grow in depth, which makes the plant lose one of its possibilities to resist drought.

In $N\text{-}NO_3^-$ dynamics in several combinations of the observed factors (tillage system, fertilization variant, the depth from which soils samples were taken) certain interdependence was identified between the autumn and the spring maximum in soil. If a considerable autumn maximum had been reached, then the following spring maximum was less noticeable. This relation was identified in the first observed depth from which the soil samples were taken (0 – 0.3 m). Contrariwise, in the second observed depth (0.3 – 0.6 m) the spring maximum was more dominant. Partially, this was caused by $N\text{-}NO_3^-$ seeping into deeper soil levels which resulted in the transport of $N\text{-}NO_3^-$ during winter months from the first observed depth, where it supported the autumn maximum, to the second observed depth, supporting the spring maximum. The statistical evaluation of $N\text{-}NO_3^-$ content indicates statistically very significant influence (at significance level = 0.05) of three observed factors: the depth from which the soil samples were taken, the sampling date, and the vegetation period.

Acknowledgement

The paper was supported by grant UGA entitled "The content of labile nitrogen in soil samples of selected area".

References

[1] BARBARICK, K. A. Nitrogen Sources and Transformations. In Colorado State University Cooperative Extension 2/96 [online], 2006 [cit. 2013-03-04]. Available on internet: <http://www.ext.colostate.edu/ pubs/crops/00550.html>.

[2] BIELEK, P. *Dusík v poľnohospodárskych pôdach Slovenska.* Autoreferát doktorandskej práce. Nitra: Slovenská poľnohospodárska univerzita, 1997, 50 p.

[3] BROWN, B. A. *Burley Tobacco Information.* In N.C. Cooperative Extension Publication. Situation and Outlook. 2008, 376 p.

[4] BUJNOVSKÝ, R. *Efektívne hnojenie pôdy a poľných plodín.* Nitra: ÚVTIP – NOI, 2002, ISBN 80-89088-05-8, 42 p.

[5] DEMO, M. et al. *Obrábanie pôdy.* Nitra: VŠP, 1995, ISBN 80-7137-255-2, 315 p.

[6] DOLAN, M. S., CLAPP, C. E., ALLMARAS, R. R., BAKER, J. M., MOLINA, J. A. E. Soil organic carbon and nitrogen in a Minnesota soil as related to tillage, residue and nitrogen management. In *Soil and Tillage Research*, vol. 89, 2006: pp. 221-231.

[7] GALUŠČÁKOVÁ, Ľ. *Vplyv pestovateľských zásahov na zmeny obsahu anorganického dusíka v pôde.* Dizertačná práca. Nitra: SPU, 2007, 197 p.

[8] IVANIČ, J., HAVELKA, B., KNOPP, K. *Výživa rastlín a hnojenie.* Bratislava: Príroda, 1979, 361 p.

[9] JAVOREKOVÁ, S. et al. *Biológia pôdy v agroekosystémoch.* 1. vyd. Nitra: SPU, 2008, ISBN 978-80-552-0007-1, 349 p.

[10] KILLPACK, S. C., BUCHHOL, Z. D. Nitrogen in the Environment: What is Nitrogen? [online] University of Missouri: 1993 [cit. 2013-02-08]. Available on internet: <http://extension.missouri.edu/p/WQ251>.

[11] MALHI, S. S., LEMKE, R., WANG, Z. H., CHHABRA, B. S. Tillage, nitrogen and crop residue effects on crop yield, nutrient uptake, soil quality and greenhouse gas emissions. In *Soil Till Res.*, vol. 90, no. 1-2, 2006, ISSN 0167-1987.

[12] MIKKELSEN, R. Where does nitrogen fertilizer come from?. In Plant Nutrition TODAY [online], Fall 2012, no. 3 [cit. 2013-02-27]. Available on internet: <http://www.ipni.net/ ipniweb/pnt.nsf/5a4b8be72a35cd46852568d9001a18da/fe44a1a01b8e627606257a69007205c e!OpenDocument>.

[13] ONDRIŠÍK, P., ČERNÝ, I. Changes in the content of inorganic nitrogen in the soil during the growing season of winter wheat in dependence on fertilisation. In *Acta fytotechnica et zootechnica,* roč. 5, č. 3, s. 64-67.

[14] ONDRIŠÍK, P., SOVIŠOVÁ, M., URMINSKÁ, J. The effect of different soil tillage and fertilization methods on the Dynamics of inorganic nitrogen in soil. In *Acta fytotechnica et zootechnica,* vol. 2, 2012: pp. 47-51.

[15] ONDRIŠÍK, P., URMINSKÁ, J., KANTOR, M. The dynamic of inorganic nitrogen in the soil during the growing season of winter wheat in dependence on agrotechnical measures. In *Acta fytotechnica et zootechnica,* vol. 7, no. 1, 2004: pp. 14-19. ISSN 1335-8X.

[16] PAPINI, R., VALBOA, G., PIOVANELLI, C., BRANDI, G. Nitrogen and phosphorus in a loam soil of central Italy as affected by 6 years of different tillage systems. In *Soil and Tillage Research,* 92, 2007: pp. 175-180.

[17] PORUBCOVÁ, L. *Dynamika anorganického dusíka v pôde v závislosti od agrotechnických opatrení.* Diplomová práca. Nitra: SPU, 2013. 78 p.

[18] RADOV, A. C., PUSTOV, I. V., KOROĽKOV, A. V. *Praktikum po agrochimiji.* Moskva: Agropromizdat, 1985.

[19] REDDY, K. R., DELAUNE, R. D. *Biogeochemistry of Wetlands: Science and Applications.* CRC Press, Boca Raton, FL. 2008.

[20] RICHTER, R., HLUŠEK, J. *Výživa a hnojení rostlin* (I. obecná část). Brno: Vysoká škola zemědělská, 1994, ISBN 80-7157-138-5, 177 p.

[21] SMATANA, J., CIGĽAR, J., POSPIŠIL, R., TÝR, Š. Vplyv obrábania pôdy na dynamiku anorganického dusíka v integrovanej sústave hospodárenia. *In 2. Majstrovstvá Slovenskej republiky v orbe Maňa,* Nitra : SPU, 1997, ISBN 80-7137-394-X, pp. 53-56.

[22] SPAGNI, A., MARSILI-LIBELLI, S. Nitrogen removal via nitrite in a sequencing batch reactor treating sanitary land fill leachate. In *Bioresource Technology*, 100, 2009: pp. 609-614.

[23] ŠOLTÝSOVÁ, B. 1998. Sezónne zmeny pôdnych živín v závislosti na diferencovanom spôsobe spracovania pôdy. In *Trvalo udržateľný rozvoj poľnohosp. výroby na reg. úrovni.* 2. diel. Michalovce, 12.-13. 11. 1998. Michalovce : OVÚA, pp. 315-320.

[24] ŠPÁNIK, R., REPA, Š., ŠIŠKA, B. *Agroklimatické a fenologické pomery Nitra (1991 - 2000).* Nitra: SPU, 2002, ISBN 80-7137-987-5, 39 p.

[25] TORMA, S. Dusík – nenahraditelný prvek v půdě a rostlině. In *Agro,* roč. 10, č. 1, 2005: pp. 27 – 29.

[26] VALBOA, G., LAGOMARSINO, A., BRANDI, G., AGNELLI, A. E., SIMONCINI, S., PAPINI, R., VIGNOZZI, N., PELLEGRINI, S. Long-term variations in soil organic matter under different tillage intensities. In *Soil and Tillage Research,* vol. 154, 2015: pp. 126–135. doi:10.1016/j.still.2015.06.017.

Current address

Porubcová Lýdia, Ing.
Constantine the Philosopher University
Tr. A. Hlinku 1, 949 74 Nitra, Slovak Republic
Tel. number: 00 421 6408 653, e-mail: lydia.porubcova@ukf.sk

Tirpáková Anna, prof. RNDr. CSc.
Constantine the Philosopher University
Tr. A. Hlinku 1, 949 74 Nitra, Slovak Republic
Tel. number: 00 421 6408 699, e-mail: atirpakova@ukf.sk

Feszterová Melánia, doc. Ing. PhD.
Constantine the Philosopher University
Tr. A. Hlinku 1, 949 74 Nitra, Slovak Republic
Tel. number: 00 421 903456414, e-mail: mfeszterova@ukf.sk

Slovak University of Technology in Bratislava
Faculty of Mechanical Engineering
16th Conference on Applied Mathematics **APLIMAT 2017**
Proceedings

THE SUM OF THE SERIES OF RECIPROCALS OF THE QUADRATIC POLYNOMIALS WITH ONE NEGATIVE AND ONE POSITIVE INTEGER ROOT

POTŮČEK Radovan (CZ)

Abstract. This contribution deals with the series of reciprocals of the quadratic polynomials with one negative and one positive integer root. We derive the formula for the sum of this series and verify it using the computer algebra system Maple 16. This contribution can be an inspiration for teachers of mathematics whose are teaching the topic Infinite series.

Keywords: Sequence of partial sums, telescoping series, harmonic number, CAS Maple

Mathematics subject classification: Primary 40A05; Secondary 65B10.

1 Introduction and basic notions

This contribution, which is a follow-up to author's papers [1], [2], and [3] dealing with the sums of the series of reciprocals of quadratic polynomials with different positive integer roots, double non-positive integer root, and different negative integer roots. We derive the formula for the sum of this series and verify it by some examples evaluated using the basic programming language of the computer algebra system Maple 16. One of interesting papers concerning verification tests of mathematical model is the paper [4]. A co-author of this article is as well a co-author of another interesting paper [5].

Reciprocals of some numbers or functions belong to interesting part of number theory and mathematical analysis. It is well-known that, for example, the nth harmonic number, is the sum of the reciprocals of the first n positive integers, the sum of the reciprocals of all the non-zero triangular numbers is 2, the sum of the reciprocals of the square numbers (the Basel problem) is $\pi^2/6$, the sum of the reciprocals of the cubes of positive integers is called Apéry's constant, and equals approximately 1.2021. Very interesting is the Kempner series, which is the sum of the reciprocals of all positive integers not containing the digit "9" in base 10. This series converges, specifically to approximately 22.9207. This contribution, which deals with reciprocals of the quadratic polynomials with one negative and one positive integer root, can be an inspiration for teachers of mathematics whose are teaching the topic Infinite series or as a subject matter for work with talented students.

Now, let us recall the basic terms. For any sequence $\{a_k\}$ of numbers the associated *series* is defined as the sum

$$\sum_{k=1}^{\infty} a_k = a_1 + a_2 + a_3 + \cdots .$$

The *sequence of partial sums* $\{s_n\}$ associated to a series $\sum_{k=1}^{\infty} a_k$ is defined for each n as the sum

$$s_n = \sum_{k=1}^{n} a_k = a_1 + a_2 + \cdots + a_n .$$

The series $\sum_{k=1}^{\infty} a_k$ *converges* to a limit s if and only if the sequence of partial sums $\{s_n\}$ converges to s, i.e. $\lim_{n\to\infty} s_n = s$. We say that the series $\sum_{k=1}^{\infty} a_k$ has a *sum* s and write $\sum_{k=1}^{\infty} a_k = s$.

The *telescoping series* is any series where nearly every term cancels with a preceding or following term, so its partial sums eventually only have a fixed number of terms after cancellation. The method of changing series whose terms are rational functions into telescoping series is that of transforming the rational functions by the method of partial fractions.

For example, the series $\sum_{\substack{k=1\\k\neq 2}}^{\infty} \frac{1}{(k+1)(k-2)}$ has the general nth term

$$a_n = \frac{1}{(n+1)(n-2)} = \frac{C}{n+1} + \frac{D}{n-2}.$$

After removing the fractions we get the equation $1 = C(n-2) + D(n+1)$. For $n = -1$ we get $C = -1/3$ and for $n = 2$ we obtain $D = 1/3$, so we have

$$a_n = \frac{-1}{3(n+1)} + \frac{1}{3(n-2)} = \frac{1}{3}\left(\frac{1}{n-2} - \frac{1}{n+1}\right).$$

After that we arrange the terms of the nth partial sum $s_n = a_1 + a_3 + a_4 + \cdots + a_n$ in a form where can be seen what is cancelling. Then we find the limit of the sequence of the partial sums s_n in order to find the sum s of the infinite telescoping series as $s = \lim_{n\to\infty} s_n$. In our case we get

$$s_n = \frac{1}{3}\left[\left(\frac{1}{-1} - \frac{1}{2}\right) + \left(\frac{1}{1} - \frac{1}{4}\right) + \left(\frac{1}{2} - \frac{1}{5}\right) + \left(\frac{1}{3} - \frac{1}{6}\right) + \left(\frac{1}{4} - \frac{1}{7}\right) + \cdots\right.$$
$$\cdots + \left(\frac{1}{n-6} - \frac{1}{n-3}\right) + \left(\frac{1}{n-5} - \frac{1}{n-2}\right) + \left(\frac{1}{n-4} - \frac{1}{n-1}\right) +$$
$$\left. + \left(\frac{1}{n-3} - \frac{1}{n}\right) + \left(\frac{1}{n-2} - \frac{1}{n+1}\right)\right] = \frac{1}{3}\left(\frac{1}{3} - \frac{1}{n-1} - \frac{1}{n} - \frac{1}{n+1}\right).$$

Hence $s = \frac{1}{3}\lim_{n\to\infty}\left(\frac{1}{3} - \frac{1}{n-1} - \frac{1}{n} - \frac{1}{n+1}\right) = \frac{1}{9}$, as $\lim_{n\to\infty}\frac{1}{n+c} = 0$ for any integer c.

The *nth harmonic number* is the sum of the reciprocals of the first n natural numbers:

$$H_n = 1 + \frac{1}{2} + \frac{1}{3} + \cdots + \frac{1}{n} = \sum_{k=1}^{n}\frac{1}{k},$$

H_0 being defined as 0. First ten values of the harmonic numbers are stated in the following table:

n	1	2	3	4	5	6	7	8	9	10
H_n	1	$\frac{3}{2}$	$\frac{11}{6}$	$\frac{25}{12}$	$\frac{137}{60}$	$\frac{49}{20}$	$\frac{363}{140}$	$\frac{761}{280}$	$\frac{7\,129}{2\,520}$	$\frac{7\,381}{2\,520}$

Tab. 1. First ten values of the harmonic numbers

Basic information about harmonic numbers can be found e.g. in the web-sites [6] or in [7], interesting information are included e.g. in the paper [8].

2 The sum of the series of reciprocals of the quadratic polynomial with one negative and one positive integer root

Let us consider the series of reciprocals of the normalized quadratic polynomials of the form $k^2 - (a+b)k + ab = (k-a)(k-b)$ with two integer roots $a < 0$, $b > 0$, i.e. the series

$$\sum_{\substack{k=1\\k\neq b}}^{\infty}\frac{1}{(k-a)(k-b)} = \sum_{\substack{k=1\\k\neq b}}^{\infty}\frac{1}{(k-(-A))(k-b)} = \sum_{\substack{k=1\\k\neq b}}^{\infty}\frac{1}{(k+A)(k-b)}, \tag{1}$$

where $A = -a$, and let us determine its sum $s(a,b)$.

We differentiate three following cases:

1. Let us assume that $a = -b$ and $b > 1$, i.e. $A = b > 1$. Then the series (1) has the form

$$s(-b,b) = \sum_{\substack{k=1\\k\neq b}}^{\infty}\frac{1}{(k+b)(k-b)} = \sum_{k=1}^{b-1}\frac{1}{(k+b)(k-b)} + \sum_{k=b+1}^{\infty}\frac{1}{(k+b)(k-b)}. \tag{2}$$

Because the nth term $(n\neq b)$ of the series (2) is

$$a_n = \frac{1}{(n+b)(n-b)} = \frac{1}{2b}\left(\frac{1}{n-b} - \frac{1}{n+b}\right) = \frac{-1}{2b}\left(\frac{1}{n+b} - \frac{1}{n-b}\right) = \frac{-1}{2b}\left(\frac{1}{b-n} + \frac{1}{b+n}\right),$$

then the nth partial sum of the series (2), $n > b$, is

$$s_n(-b,b) = \frac{-1}{2b}\left[\sum_{k=1}^{b-1}\left(\frac{1}{b-k}+\frac{1}{b+k}\right)+\sum_{k=b+1}^{n}\left(\frac{1}{k+b}-\frac{1}{k-b}\right)\right] =$$

$$= \frac{-1}{2b}\left(\sum_{k=1}^{b-1}\frac{1}{b-k}+\sum_{k=1}^{b-1}\frac{1}{b+k}+\sum_{k=b+1}^{n}\frac{1}{k+b}-\sum_{k=b+1}^{n}\frac{1}{k-b}\right) =$$

$$= \frac{-1}{2b}\left[\left(\frac{1}{b-1}+\frac{1}{b-2}+\cdots+\frac{1}{2}+\frac{1}{1}\right)+\left(\frac{1}{b+1}+\frac{1}{b+2}+\cdots+\frac{1}{2b-2}+\frac{1}{2b-1}\right)+\right.$$

$$\left.+\left(\frac{1}{2b+1}+\frac{1}{2b+2}+\cdots+\frac{1}{n+b-1}+\frac{1}{n+b}\right)-\left(\frac{1}{1}+\frac{1}{2}+\cdots+\frac{1}{n-b-1}+\frac{1}{n-b}\right)\right] =$$

$$= \frac{-1}{2b}\left[H_{2b-1}-\frac{1}{b}+\left(\frac{1}{n-b+1}+\frac{1}{n-b+2}+\cdots+\frac{1}{n+b-1}+\frac{1}{n+b}\right)-H_{2b}\right] =$$

$$= \frac{-1}{2b}\left[-\frac{1}{b}-\frac{1}{2b}+\left(\frac{1}{n-b+1}+\frac{1}{n-b+2}+\cdots+\frac{1}{n+b-1}+\frac{1}{n+b}\right)\right].$$

Because for arbitrary c integer is

$$\lim_{n\to\infty}\frac{1}{n+c} = 0, \tag{3}$$

we have

$$s(-b,b) = \lim_{n\to\infty} s_n(-b,b) = \frac{-1}{2b}\left(-\frac{1}{b}-\frac{1}{2b}\right) = \frac{1}{2b}\left(\frac{1}{b}+\frac{1}{2b}\right) = \frac{1}{2b}\cdot\frac{2+1}{2b} = \frac{3}{4b^2}.$$

In case $a = -b = -1$, i.e. $A = b = 1$, the series (1) has the form

$$s(-1,1) = \sum_{k=2}^{\infty}\frac{1}{(k+1)(k-1)}. \tag{4}$$

Because the nth term ($n \geq 2$) of the series (4) is

$$a_n = \frac{1}{(n+1)(n-1)} = \frac{1}{2}\left(\frac{1}{n-1}-\frac{1}{n+1}\right) = \frac{-1}{2}\left(\frac{1}{n+1}-\frac{1}{n-1}\right),$$

then the nth partial sum of the series (4), $n > 1$, is

$$s_n(-1,1) = \frac{-1}{2}\sum_{k=2}^{n}\left(\frac{1}{k+1}-\frac{1}{k-1}\right) = \frac{-1}{2}\left(\sum_{k=2}^{n}\frac{1}{k+1}-\sum_{k=2}^{n}\frac{1}{k-1}\right) =$$

$$= \frac{-1}{2}\left[\left(\frac{1}{3}+\frac{1}{4}+\cdots+\frac{1}{n}+\frac{1}{n+1}\right)-\left(\frac{1}{1}+\frac{1}{2}+\cdots+\frac{1}{n-2}+\frac{1}{n-1}\right)\right] =$$

$$= \frac{-1}{2}\left(\frac{1}{n}+\frac{1}{n+1}-\frac{1}{1}-\frac{1}{2}\right) = \frac{-1}{2}\cdot\frac{2(n+1)+2n-2n(n+1)-n(n+1)}{2n(n+1)} =$$

$$= \frac{-1}{2}\cdot\frac{2n+2+2n-2n^2-2n-n^2-n}{2n(n+1)} = \frac{-1}{2}\cdot\frac{-3n^2+n+2}{2n(n+1)} = \frac{3n^2-n-2}{4n^2+4n}.$$

Since for arbitrary $c \neq 0, d, e, f \neq 0, g, h$ real is $\lim\limits_{n\to\infty} \dfrac{cn^2+dn+e}{fn^2+gn+h} = \dfrac{c}{f}$, we have

$$s(-1,1) = \lim_{n\to\infty} s_n(-1,1) = \lim_{n\to\infty} \frac{3n^2-n-2}{4n^2+4n} = \frac{3}{4},$$

so the formula $s(-1,1) = \dfrac{3}{4}$ is a special case of the formula $s(-b,b) = \dfrac{3}{4b^2}$ for $b = 1$.

2. Let us assume that $|a| = -a = A > b$, i.e. $A = b + (A-b) = b + \ell$, where integer $\ell > 0$, hence $a = -b - \ell$. Then the series (1) has the form

$$s(-b-\ell,b) = \sum_{k=1}^{b-1} \frac{1}{(k+b+\ell)(k-b)} + \sum_{k=b+1}^{\infty} \frac{1}{(k+b+\ell)(k-b)}. \tag{5}$$

Because the nth term $a_n = \dfrac{1}{(n+b+\ell)(n-b)}$, $n \neq b$, of the series (5) is

$$\frac{1}{2b+\ell}\left(\frac{1}{n-b} - \frac{1}{n+b+\ell}\right) = \frac{-1}{2b+\ell}\left(\frac{1}{n+b+\ell} - \frac{1}{n-b}\right) = \frac{-1}{2b+\ell}\left(\frac{1}{b-n} + \frac{1}{b+n+\ell}\right),$$

then the nth partial sum of the series (5), $n > b$, is

$$\begin{aligned}
s_n(-b-\ell,b) &= \frac{-1}{2b+\ell}\left[\sum_{k=1}^{b-1}\left(\frac{1}{b-k} + \frac{1}{b+k+\ell}\right) + \sum_{k=b+1}^{n}\left(\frac{1}{k+b+\ell} - \frac{1}{k-b}\right)\right] = \\
&= \frac{-1}{2b+\ell}\left(\sum_{k=1}^{b-1}\frac{1}{b-k} + \sum_{k=1}^{b-1}\frac{1}{b+k+\ell} + \sum_{k=b+1}^{n}\frac{1}{k+b+\ell} - \sum_{k=b+1}^{n}\frac{1}{k-b}\right) = \\
&= \frac{-1}{2b+\ell}\left[\left(\frac{1}{b-1} + \frac{1}{b-2} + \cdots + \frac{1}{1}\right) + \left(\frac{1}{b+1+\ell} + \frac{1}{b+2+\ell} + \cdots + \frac{1}{2b-1+\ell}\right) + \right. \\
&\quad \left. + \left(\frac{1}{2b+1+\ell} + \frac{1}{2b+2+\ell} + \cdots + \frac{1}{n+b+\ell}\right) - \left(\frac{1}{1} + \frac{1}{2} + \cdots + \frac{1}{n-b}\right)\right] = \\
&= \frac{-1}{2b+\ell}\left[H_{2b-1+\ell} - H_{b+\ell} + H_{b-1} + \left(\frac{1}{n-b+1} + \frac{1}{n-b+2} + \cdots + \frac{1}{n+b+\ell}\right) - H_{2b+\ell}\right] = \\
&= \frac{-1}{2b+\ell}\left[-\frac{1}{2b+\ell} - H_{b+\ell} + H_{b-1} + \left(\frac{1}{n-b+1} + \frac{1}{n-b+2} + \cdots + \frac{1}{n+b+\ell}\right)\right].
\end{aligned}$$

For the sum

$$s(-b-\ell,b) = \lim_{n\to\infty} s_n(-b-\ell,b),$$

by (3), we have,

$$s(-b-\ell,b) = \frac{1}{2b+\ell}\left(\frac{1}{2b+\ell} + H_{b+\ell} - H_{b-1}\right) = \frac{(2b+\ell)\big(H_{b+\ell} - H_{b-1}\big) + 1}{(2b+\ell)^2},$$

or, for $-b-\ell = a$, whence $2b+\ell = b + (b+\ell) = b - a$, we have

$$s(a,b) = \frac{(b-a)\big(H_{-a} - H_{b-1}\big) + 1}{(b-a)^2}.$$

3. Let us assume that $|a| = -a = A < b$, i.e. $A = b - (b - A) = b - \ell$, where integer $\ell > 0$, hence $a = -b + \ell$. Then the series (1) has the form

$$s(-b+\ell, b) = \sum_{k=1}^{b-1} \frac{1}{(k+b-\ell)(k-b)} + \sum_{k=b+1}^{\infty} \frac{1}{(k+b-\ell)(k-b)}. \tag{6}$$

Because the nth term $a_n = \dfrac{1}{(n+b-\ell)(n-b)}$, $n \neq b$, of the series (6) is

$$\frac{1}{2b-\ell}\left(\frac{1}{n-b} - \frac{1}{n+b-\ell}\right) = \frac{1}{\ell-2b}\left(\frac{1}{n+b-\ell} - \frac{1}{n-b}\right) = \frac{1}{\ell-2b}\left(\frac{1}{b-n} + \frac{1}{b+n-\ell}\right),$$

then the nth partial sum of the series (6), $n > b$, is

$$\begin{aligned}
s_n(-b+\ell, b) &= \frac{1}{\ell-2b}\left[\sum_{k=1}^{b-1}\left(\frac{1}{b-k} + \frac{1}{b+k-\ell}\right) + \sum_{k=b+1}^{n}\left(\frac{1}{k+b-\ell} - \frac{1}{k-b}\right)\right] = \\
&= \frac{1}{\ell-2b}\left(\sum_{k=1}^{b-1}\frac{1}{b-k} + \sum_{k=1}^{b-1}\frac{1}{b+k-\ell} + \sum_{k=b+1}^{n}\frac{1}{k+b-\ell} - \sum_{k=b+1}^{n}\frac{1}{k-b}\right) = \\
&= \frac{1}{\ell-2b}\left[\left(\frac{1}{b-1} + \frac{1}{b-2} + \cdots + \frac{1}{1}\right) + \left(\frac{1}{b+1-\ell} + \frac{1}{b+2-\ell} + \cdots + \frac{1}{2b-1-\ell}\right) + \right. \\
&\quad \left. + \left(\frac{1}{2b+1-\ell} + \frac{1}{2b+2-\ell} + \cdots + \frac{1}{n+b-\ell}\right) - \left(\frac{1}{1} + \frac{1}{2} + \cdots + \frac{1}{n-b}\right)\right] = \\
&= \frac{1}{\ell-2b}\left[H_{2b-1-\ell} - H_{b-\ell} + H_{b-1} + \left(\frac{1}{n-b+1} + \frac{1}{n-b+2} + \cdots + \frac{1}{n+b-\ell}\right) - H_{2b-\ell}\right] = \\
&= \frac{1}{\ell-2b}\left[-\frac{1}{2b-\ell} - H_{b-\ell} + H_{b-1} + \left(\frac{1}{n-b+1} + \frac{1}{n-b+2} + \cdots + \frac{1}{n+b-\ell}\right)\right].
\end{aligned}$$

For the sum $s(-b+\ell, b) = \lim_{n\to\infty} s_n(-b+\ell, b)$, by (3), we have

$$s(-b+\ell, b) = \frac{1}{2b-\ell}\left(\frac{1}{2b-\ell} + H_{b-\ell} - H_{b-1}\right) = \frac{(2b-\ell)\big(H_{b-\ell} - H_{b-1}\big) + 1}{(2b-\ell)^2},$$

or, for $-b+\ell = a$, whence $2b - \ell = b + (b-\ell) = b - a$, we have

$$s(a, b) = \frac{(b-a)\big(H_{-a} - H_{b-1}\big) + 1}{(b-a)^2}.$$

Altogether, we derived this statement:

Theorem 1 The series $\displaystyle\sum_{\substack{k=1 \\ k\neq b}}^{\infty} \frac{1}{(k-a)(k-b)}$, where $a < 0$ and $b > 0$ are integers, has the sum $s(a, b)$ in the form

$$s(a, b) = \frac{(b-a)\big(H_{-a} - H_{b-1}\big) + 1}{(b-a)^2}, \tag{7}$$

where H_n is the nth harmonic number.

Corollary 1 From Theorem 1 it evidently follows:

1. $s(-b+1,b) = \dfrac{(2b-1)\big(H_{b-1}-H_{b-1}\big)+1}{(2b-1)^2} = \dfrac{1}{(2b-1)^2}.$

2. $s(-b,b) = \dfrac{2b\big(H_b-H_{b-1}\big)+1}{(2b)^2} = \dfrac{2b\cdot\frac{1}{b}+1}{4b^2} = \dfrac{3}{4b^2}.$

Example 1 Using i) nth partial sum, ii) formula 2. from Corollary 1 calculate the sum $s(-5,5)$ of the series

$$\sum_{\substack{k=1\\k\neq 5}}^{\infty} \frac{1}{(k+5)(k-5)}. \tag{8}$$

i) The series (8) has the sum

$$s(-5,5) = \sum_{\substack{k=1\\k\neq 5}}^{\infty} \frac{1}{(k+5)(k-5)} = \sum_{k=1}^{4} \frac{1}{(k+5)(k-5)} + \sum_{k=6}^{\infty} \frac{1}{(k+5)(k-5)}.$$

Because the nth term $(n\neq 5)$ of the series (8) is

$$a_n = \frac{1}{(n+5)(n-5)} = \frac{1}{2\cdot 5}\left(\frac{1}{n-5}-\frac{1}{n+5}\right) = \frac{-1}{10}\left(\frac{1}{n+5}-\frac{1}{n-5}\right) = \frac{-1}{10}\left(\frac{1}{5-n}+\frac{1}{5+n}\right),$$

then the nth partial sum $s_n(-5,5)$, where $n>5$, of the series (8) is

$$\begin{aligned}
s_n(-5,5) &= \frac{-1}{10}\left[\sum_{k=1}^{4}\left(\frac{1}{5-k}+\frac{1}{5+k}\right)+\sum_{k=6}^{n}\left(\frac{1}{k+5}-\frac{1}{k-5}\right)\right] = \\
&= \frac{-1}{10}\left(\sum_{k=1}^{4}\frac{1}{5-k}+\sum_{k=1}^{4}\frac{1}{5+k}+\sum_{k=6}^{n}\frac{1}{k+5}-\sum_{k=6}^{n}\frac{1}{k-5}\right) = \\
&= \frac{-1}{10}\left[\left(\frac{1}{4}+\frac{1}{3}+\frac{1}{2}+\frac{1}{1}\right)+\left(\frac{1}{6}+\frac{1}{7}+\frac{1}{8}+\frac{1}{9}\right)+\right. \\
&\left.+\left(\frac{1}{11}+\frac{1}{12}+\cdots+\frac{1}{n+4}+\frac{1}{n+5}\right)-\left(\frac{1}{1}+\frac{1}{2}+\cdots+\frac{1}{n-6}+\frac{1}{n-5}\right)\right] = \\
&= \frac{-1}{10}\left[H_9-\frac{1}{5}+\left(\frac{1}{n-4}+\frac{1}{n-3}+\cdots+\frac{1}{n+4}+\frac{1}{n+5}\right)-H_{10}\right] = \\
&= \frac{-1}{10}\left[-\frac{1}{5}-\frac{1}{10}+\left(\frac{1}{n-4}+\frac{1}{n-3}+\cdots+\frac{1}{n+4}+\frac{1}{n+5}\right)\right].
\end{aligned}$$

By (3) we have

$$s(-5,5) = \lim_{n\to\infty} s_n(-5,5) = \frac{-1}{10}\left(-\frac{1}{5}-\frac{1}{10}\right) = \frac{1}{10}\cdot\frac{3}{10} = \frac{3}{100}.$$

ii) By the formula formula 2. from Corollary 1, i.e.

$$s(-b,b) = \frac{3}{4b^2},$$

we get the sum $s(-5,5)$ much more easily:

$$s(-5,5)=\frac{3}{4\cdot 5^2}=\frac{3}{100}.$$

Example 2 Using i) nth partial sum, ii) formula (7) calculate the sum $s(-8,5)$ of the series

$$\sum_{\substack{k=1\\k\neq 5}}^{\infty}\frac{1}{(k+8)(k-5)}. \tag{9}$$

i) The series (9) has the sum $s(-8,5)$ in the form

$$s(-5-3,5)=\sum_{\substack{k=1\\k\neq 5}}^{\infty}\frac{1}{(k+5+3)(k-5)}=\sum_{k=1}^{4}\frac{1}{(k+5+3)(k-5)}+\sum_{k=6}^{\infty}\frac{1}{(k+5+3)(k-5)}.$$

Because the nth term $a_n=\dfrac{1}{(n+8)(n-5)}$, $n\neq 5$, of the series (9) is

$$\frac{1}{8-(-5)}\left(\frac{1}{n-5}-\frac{1}{n+8}\right)=\frac{-1}{13}\left(\frac{1}{n+8}-\frac{1}{n-5}\right)=\frac{-1}{13}\left(\frac{1}{5-n}+\frac{1}{8+n}\right),$$

then the nth partial sum $s_n(-8,5)$, where $n>5$, of the series (9) is

$$\begin{aligned}
s_n(-8,5)&=\frac{-1}{13}\left[\sum_{k=1}^{4}\left(\frac{1}{5-k}+\frac{1}{8+k}\right)+\sum_{k=6}^{n}\left(\frac{1}{k+8}-\frac{1}{k-5}\right)\right]=\\
&=\frac{-1}{13}\left(\sum_{k=1}^{4}\frac{1}{5-k}+\sum_{k=1}^{4}\frac{1}{8+k}+\sum_{k=6}^{n}\frac{1}{k+8}-\sum_{k=6}^{n}\frac{1}{k-5}\right)=\\
&=\frac{-1}{13}\left[\left(\frac{1}{4}+\frac{1}{3}+\frac{1}{2}+\frac{1}{1}\right)+\left(\frac{1}{9}+\frac{1}{10}+\frac{1}{11}+\frac{1}{12}\right)+\right.\\
&\left.+\left(\frac{1}{14}+\frac{1}{15}+\cdots+\frac{1}{n+7}+\frac{1}{n+8}\right)-\left(\frac{1}{1}+\frac{1}{2}+\cdots+\frac{1}{n-6}+\frac{1}{n-5}\right)\right]=\\
&=\frac{-1}{13}\left[H_{12}-H_8+H_4+\left(\frac{1}{n-4}+\frac{1}{n-3}+\cdots+\frac{1}{n+7}+\frac{1}{n+8}\right)-H_{13}\right]=\\
&=\frac{-1}{13}\left[-\frac{1}{13}-H_8+H_4+\left(\frac{1}{n-4}+\frac{1}{n-3}+\cdots+\frac{1}{n+7}+\frac{1}{n+8}\right)\right].
\end{aligned}$$

From (3) we have

$$\begin{aligned}
s(-8,5)&=\lim_{n\to\infty}s_n(-8,5)=\frac{-1}{13}\left(-\frac{1}{13}-H_8+H_4\right)=\frac{-1}{13}\left(-\frac{1}{13}-\frac{1}{8}-\frac{1}{7}-\frac{1}{6}-\frac{1}{5}\right)=\\
&=\frac{1}{13}\left(\frac{1}{13}+\frac{1}{8}+\frac{1}{7}+\frac{1}{6}+\frac{1}{5}\right)=\frac{7769}{141960}\doteq 0.054727.
\end{aligned}$$

ii) By the formula (7) from Theorem 1 we get the sum $s(-8,5)$ much more easily:

$$s(-8,5)=\frac{[5-(-8)](H_{-(-8)}-H_{5-1})+1}{[5-(-8)]^2}=\frac{13\cdot\dfrac{533}{840}+1}{169}=\frac{7769}{840\cdot 169}=\frac{7769}{141960}.$$

Example 3 Using i) nth partial sum, ii) formula (7) calculate the sum $s(-2,5)$ of the series

$$\sum_{\substack{k=1\\k\neq 5}}^{\infty} \frac{1}{(k+2)(k-5)}. \tag{10}$$

i) The series (10) has the sum $s(-2,5)$ in the form

$$s(-5+3,5) = \sum_{\substack{k=1\\k\neq 5}}^{\infty} \frac{1}{(k+5-3)(k-5)} = \sum_{k=1}^{4} \frac{1}{(k+5-3)(k-5)} + \sum_{k=6}^{\infty} \frac{1}{(k+5-3)(k-5)}.$$

Because the nth term $a_n = \dfrac{1}{(n+2)(n-5)}$, $n \neq 5$, of the series (9) is

$$\frac{1}{2-(-5)}\left(\frac{1}{n-5} - \frac{1}{n+2}\right) = \frac{-1}{7}\left(\frac{1}{n+2} - \frac{1}{n-5}\right) = \frac{-1}{7}\left(\frac{1}{5-n} + \frac{1}{2+n}\right),$$

then the nth partial sum $s_n(-2,5)$, where $n > 5$, of the series (10) is

$$\begin{aligned}
s_n(-2,5) &= \frac{-1}{7}\left[\sum_{k=1}^{4}\left(\frac{1}{5-k} + \frac{1}{2+k}\right) + \sum_{k=6}^{n}\left(\frac{1}{k+2} - \frac{1}{k-5}\right)\right] = \\
&= \frac{-1}{7}\left(\sum_{k=1}^{4}\frac{1}{5-k} + \sum_{k=1}^{4}\frac{1}{2+k} + \sum_{k=6}^{n}\frac{1}{k+2} - \sum_{k=6}^{n}\frac{1}{k-5}\right) = \\
&= \frac{-1}{7}\left[\left(\frac{1}{4}+\frac{1}{3}+\frac{1}{2}+\frac{1}{1}\right) + \left(\frac{1}{3}+\frac{1}{4}+\frac{1}{5}+\frac{1}{6}\right) + \right.\\
&\quad \left. + \left(\frac{1}{8}+\frac{1}{9}+\cdots+\frac{1}{n+1}+\frac{1}{n+2}\right) - \left(\frac{1}{1}+\frac{1}{2}+\cdots+\frac{1}{n-6}+\frac{1}{n-5}\right)\right] = \\
&= \frac{-1}{7}\left[H_6 - H_2 + H_4 + \left(\frac{1}{n-4}+\frac{1}{n-3}+\cdots+\frac{1}{n+1}+\frac{1}{n+2}\right) - H_7\right] = \\
&= \frac{-1}{7}\left[-\frac{1}{7} - H_2 + H_4 + \left(\frac{1}{n-4}+\frac{1}{n-3}+\cdots+\frac{1}{n+1}+\frac{1}{n+2}\right)\right].
\end{aligned}$$

According (3) we have

$$\begin{aligned}
s(-2,5) &= \lim_{n\to\infty} s_n(-2,5) = \frac{-1}{7}\left(-\frac{1}{7} - H_2 + H_4\right) = \frac{-1}{7}\left(-\frac{1}{7}+\frac{1}{3}+\frac{1}{4}\right) = \\
&= \frac{1}{7}\left(\frac{1}{7}-\frac{1}{3}-\frac{1}{4}\right) = -\frac{37}{588} \doteq -0.062925.
\end{aligned}$$

ii) By the formula (7) from Theorem 1 we get the sum $s(-2,5)$ immediately:

$$s(-2,5) = \frac{[5-(-2)](H_{-(-2)} - H_{5-1}) + 1}{[5-(-2)]^2} = \frac{7\cdot\left(-\frac{7}{12}\right)+1}{49} = \frac{-37}{12\cdot 49} = \frac{-37}{588}.$$

3 Numerical verification

We solve the problem to determine the values of the sum

$$s(a,b)=\sum_{\substack{k=1\\k\neq b}}^{\infty}\frac{1}{(k-a)(k-b)}$$

for $a=-10,-9,\ldots,-1$ and $b=1,2,\ldots,10$. We use on the one hand an approximative direct evaluation of the sum

$$s(a,b,t)=\sum_{\substack{k=1\\k\neq b}}^{t}\frac{1}{(k-a)(k-b)},$$

where $t=10^8$, using the basic programming language of the computer algebra system Maple 16, and on the other hand the formula (7) for evaluation the sum $s(a,b)$.

We compare 10^2 pairs of these ways obtained sums $s(a,b,10^8)$ and $s(a,b)$ to verify the formula (7). We use following simple procedures `hnum`, `rp2anegbpos` and two **for** statements:

```
hnum:=proc(h)
    local i,s; s:=0;
    if h=0 then s:=0 else
    for i from 1 to h do
        s:=s+1/i;
    end do;
    end if;
end proc:

rp2anegbpos:=proc(a,b,n)
    local k,sab,sumab; sumab:=0;
    sab:=((b-a)*(hnum(-a)-hnum(b-1))+1)/((b-a)*(b-a));
    print("n=",n,"s(",a,b,")=",evalf[20](sab));
    for k from 1 to n do
        if k<>b then
        sumab:=sumab+1/((k-a)*(k-b))
        else sumab:=sumab+0;
        end if;
    end do;
    print("sum(",a,b,")=",evalf[20](sumab),"diff=",evalf[20](abs(sumab-sab)));
end proc:

for i from -10 to -1 do
    for j from 1 to 10 do
        rp2anegbpos(i,j,100000000);
    end do;
```

```
end do;
```

The approximative values of the sums $s(a,b)$ for $a = -1, -2, \ldots, -10$, $b = 1, 2, \ldots, 10$, rounded to 3 decimals and obtained by these procedures are written into the following table 2:

$a\backslash b$	1	2	3	4	5	6	7	8	9	10
−1	0.750	0.111	−0.063	−0.127	−0.153	−0.163	−0.166	−0.165	−0.162	−0.158
−2	0.611	0.188	0.040	−0.028	−0.063	−0.082	−0.093	−0.099	−0.102	−0.104
−3	0.521	0.207	0.083	0.020	−0.016	−0.038	−0.052	−0.061	−0.067	−0.071
−4	0.457	0.208	0.104	0.047	0.012	−0.010	−0.025	−0.036	−0.043	−0.048
−5	0.408	0.204	0.114	0.062	0.030	0.008	−0.007	−0.018	−0.026	−0.032
−6	0.370	0.197	0.118	0.072	0.042	0.021	0.006	−0.005	−0.013	−0.020
−7	0.340	0.189	0.119	0.077	0.049	0.030	0.015	0.004	−0.004	−0.010
−8	0.314	0.182	0.119	0.081	0.055	0.036	0.022	0.012	0.004	−0.003
−9	0.293	0.175	0.118	0.083	0.058	0.041	0.028	0.017	0.009	0.003
−10	0.275	0.168	0.116	0.083	0.061	0.044	0.032	0.022	0.014	0.008

Tab. 2. The sums $s(a,b)$ for $a = -1, -2, \ldots, -10$, $b = 1, 2, \ldots, 10$

Computation of 100 couples of the sums $s(a,b,10^8)$ and $s(a,b)$ took about 8 hour and 35 minutes. The absolute errors, i.e. differences $|s(a,b) - s(a,b,10^8)|$, are all almost exactly equal to $1 \cdot 10^{-8}$.

4 Conclusion

We dealt with the sum of the series of reciprocals of the quadratic polynomials with two integer roots $a < 0$, $b > 0$, i.e. with the series

$$\sum_{\substack{k=1\\k\neq b}}^{\infty} \frac{1}{(k-a)(k-b)}.$$

We derived that the sum $s(a,b)$ of this series is given by the formula

$$s(a,b) = \frac{(b-a)\big(H_{-a} - H_{b-1}\big) + 1}{(b-a)^2},$$

where H_n is the nth harmonic number. We verified this result by computing 100 various sums by using the CAS Maple 16.

We also stated two simple cases of the sum $s(a,b)$:

$$s(-b+1,b) = \frac{1}{(2b-1)^2}, \qquad s(-b,b) = \frac{3}{4b^2}.$$

The series of reciprocals of the quadratic polynomials with one negative and one positive integer root so belong to special types of infinite series, such as geometric and telescoping series, which sums are given analytically by means of a simple formula.

References

[1] Potůček, R.: *The sum of the series of reciprocals of the quadratic polynomial with different positive integer roots.* In: Matematika, informační technologie a aplikované vědy (MITAV 2016). Univerzita obrany, Brno, 2016, p. 32-43. ISBN 978-80-7231-464-5.

[2] Potůček, R.: *The sum of the series of reciprocals of the quadratic polynomials with double non-positive integer root.* In: Aplimat — 15th Conference on Applied Mathematics 2016, Proceedings. Vydavateľstvo STU, Bratislava, Slovak Republic, 2016, p. 919-925. ISBN 978-80-227-4531-4.

[3] Potůček, R.: *The sum of the series of reciprocals of the quadratic polynomials with different negative integer roots.* Ratio Mathematica – Journal of Foundations and Applications of Mathematics, 2016, vol. 30, no. 1/2016, p. 59-66. ISSN 1592-7415.

[4] Talhofer, V., Hošková-Mayerová, Š.: *Verification tests of mathematical model of passage through terrain.* In: Aplimat — 15th Conference on Applied Mathematics 2016, Proceedings. Vydavateľstvo STU, Bratislava, Slovak Republic, 2016, p. 1010-1025. ISBN 978-80-227-4531-4.

[5] Hošková-Mayerová, Š., Hofmann, A.: *Development of Applications in the Analysis of the Natural Environment.* In: Aplimat — 15th Conference on Applied Mathematics 2016, Proceedings. Vydavateľstvo STU, Bratislava, Slovak Republic, 2016, p. 467-481. ISBN 978-80-227-4531-4.

[6] URL: `https://en.wikipedia.org/wiki/Harmonic_number`

[7] URL: `http://mathworld.wolfram.com/HarmonicNumber.html`

[8] Benjamin, A. T., Preston, G. O., Quinn, J. J.: *A Stirling Encounter with Harmonic Numbers.* Mathematics Magazine 75 (2), 2002, p. 95–103, [online], URL: `https://www.math.hmc.edu/~benjamin/papers/harmonic.pdf`

Current address

Potůček Radovan, RNDr. Ph.D.
Department of Mathematics and Physics, Faculty of Military Technology,
University of Defence
Kounicova 65, 662 10 Brno, Czech Republic
Tel. number: + 420 973 443 056, e-mail: Radovan.Potucek@unob.cz

Slovak University of Technology in Bratislava
Faculty of Mechanical Engineering
16th Conference on Applied Mathematics **APLIMAT 2017**
Proceedings

USAGE OF ARTIFICIAL INTELLIGENCE AND SPECTRAL ANALYSIS FOR PREDICTING THE BEHAVIOR OF STOCK PRICES

POZDÍLKOVÁ Alena (CZ), MAREK Jaroslav (CZ), TODT Jaroslav (CZ)

Abstract. In this paper methods of artificial intelligence and spectral analysis to build an algorithm for predicting the behavior of stock prices are applied. Spectral decomposition of a time series was calculated using known methods based on Fourier transformation. The results obtained from periodogram analysis simply provide information about periodicities. Significance analysis was not performed and we worked with four frequencies. This spectral information is then used in clustering of data. Comparison of behavior of price oscillation in clusters was carried out. The presented contribution aims to describe a new algorithm for predicting the behavior of stock prices. The clustering algorithm is based on spectral analysis and SOM. The whole procedure is tested on selected time sections of Dow Jones Industrial Averages, where the algorithm is performed. Results of analysis and final discussion, presented in the Case Study, show that the new method successfully signalizes the trend of stock market prices.

Keywords: Stock market price, SOM, spectral analysis, artificial intelligence

Mathematics Subject Classification: Primary 62J02, 62J05; Secondary 62B15

1 Introduction

Time series analysis can help us for prediction of future prices of shares. Prediction of fund prices is more difficult, because we have to consider construction of initial and final fund prices. Frequently predicting the future value of securities, especially stock prices, is a constantly developing area. In this area, neural networks and unsupervised systems can be used. One particularly interesting class of an unsupervised system is based on competitive learning, in which the output neurons compete amongst themselves to be activated, with the result that only one is activated at any one time. This activated neuron is called a winning neuron. Such competition can be induced/implemented by having lateral inhibition connections (negative feedback paths) between the neurons. The result is that the neurons are forced to organize themselves. For obvious reasons, such a network is called a Self Organizing Map (SOM).

2 Periodogram

Given a real-valued time series $X_1, X_2, \dots X_n$ of length n, consider the function

$$I(\lambda) = \frac{1}{2\pi n}\left|\sum_{i=1}^{n} X_t e^{-it\lambda}\right|^2, -\pi \le \lambda \le \pi \tag{1}$$

Function $I(\lambda)$ is called periodogram of sequence $X_1, X_2, \dots X_n$.

The periodogram $I(\lambda)$ is a key tool in harmonic analysis. If the data contains strong periodic components, these will cause peaks in the periodogram at the frequencies of oscillation. Thus, it is tempting to attribute peaks in the periodogram to systematic periodic components in the underlying mechanism which generated the data.

For every fix n function $I(\lambda)$ is random variable. Therefore $I(\lambda), -\pi \le \lambda \le \pi$, forming random process. Each implementation of the process is a continuous function. If the sequence length n is small compared with the length of the actual period

$$T_j = \frac{2\pi}{\lambda_j}, \tag{2}$$

where $\lambda_1, \dots, \lambda_p$ are mutually different numbers from interval $(-\pi, \pi)$, this period will appear more like trend. In contrast, very short period also will not be to recognize them correctly. It is clear that the shortest periodicity detectable it has a length T = 2. According to (2.2) it corresponds to the frequency $\lambda = \pi$, which is called Nyquist frequency.

The calculation can be easily performed with the formula

$$I(\lambda) = \frac{1}{4\pi} A^2(\lambda) + \frac{1}{4\pi} B^2(\lambda), \tag{3}$$

where

$$A(\lambda) = \sqrt{2/N} \sum_{t=1}^{N} X_t \cos x\lambda, \quad B(\lambda) = \sqrt{2/N} \sum_{t=1}^{N} X_t \sin x\lambda. \tag{4}$$

Fisher's test allows us to assess the statistical significance of an observed periodogram peak. Fisher's g-statistic is the ratio of the maximum periodogram value to the sum of all periodogram values.

3 Self-Organizing Feature Maps (SOM)

The self-organized feature map is a multidimensional scaling method to project m-dimensional space to a discrete k-dimensional output space. The map (the output space) is usually a two dimensional grid.

Let consider m-dimensional input vector $z_i, i = 1, \dots, n$.

The winning neuron is found by computing the Euclidean distance from each codebook vector to the input vector, and selecting the neuron closest to the input vector.

$$\|w_{mn} - x_p\| = \min_{\forall k,j}\left\{\|w_{kj} - x_p\|^2\right\}. \tag{5}$$

The stages of the SOM algorithm can be summarized as follows:

1. Initialization — Choose at random values for the initial weight vectors w_j .
2. Sampling — Draw a sample training input vector x from the input space.
3. Matching — Find the winning neuron $I(x)$ with weight vector closest to input vector.
4. Updating— Apply the weight update equation $\Delta w_{ji} = \eta(t) T_{j,I(x)}(t)(x_i - w_{ji})$. We consider epoch t dependent learning rate $\eta(t) = \eta_0 e^{-t/\tau_\eta}$
5. Continuation — keep returning to step 2 until the feature map stops changing.

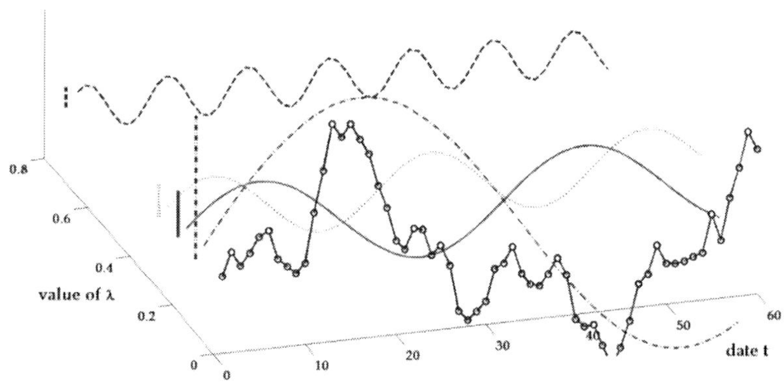

Fig. 1. Spectral analysis

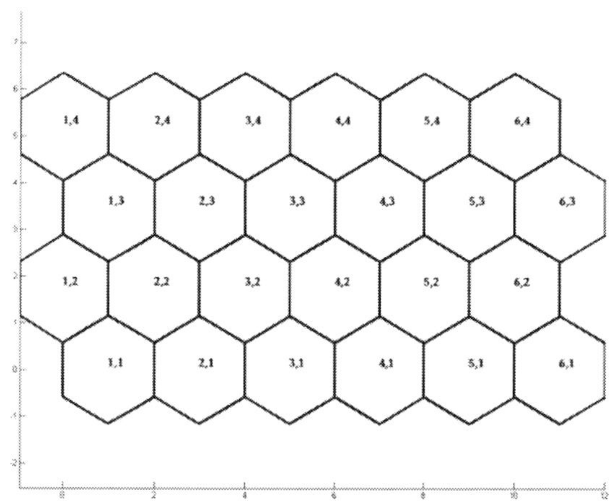

Fig. 2. Neurons

The Fig. 2 shows the frequencies of neurons in SOM algorithm. Neuron with higher frequencies cover the most part of data, but we will be interested in nodes with higher mean value of growth rate.

4 Algorithm

We will aply periodogram and SOM based strategy on test data.

Every window can be explained by the first largest, second largest, third largest and fourth largest periodogram coordinates. So $x_i = (\lambda_1^i, \lambda_2^i, \lambda_3^i, \lambda_4^i)$. A concrete example is used to illustrate the spectral decomposition. Dashed lines in Fig. 1 show the location of the hidden frequencies λ^i.

For all windows we can find four largest values of λ, so we will get matrix X of size $N \times 4$. We learn map in SOM algorithm. We will find a relative growth (mean return value) for all initialized neurons. Mean return value will be $\frac{x_{n+1}}{x_n}$.
We store the mean return value for all neurons to use as historical parameters for a future forecasting.

We find neuron, on which appears a new analyzed window. The present window can be classified as follows:

Overbought: This scan reveals stocks with a relative growth (mean return value) above upper bound, once this criterion is met, a bearish signal is triggered.

Oversold: This scan reveals stocks with a relative growth below lower bound, once this criteria is met, a bullish signal is triggered.

Uninteresting: Neuron was characterized by unattractive relative growth between lower bound and upper bound.

Unknown: Neuron was not initialized yet.

Only the first class we will use as a buying signal.

5 Experiment and results

The markets data that we will use for testing are daily data from NASDAQ and NYSE. We will use 12 years historical market daily data 15 selected companies from the Dow Jones Industrial Average: Apple, IBM, Microsoft, VISA, Wal/Mart, Pfizer Inc., Procter & Gamble, General Electric, JPMorgan Chase, Verizon communications, Goldman Sachs BDC, Merck & Co, Nike Inc., Intel and Walt Disney('AAPL','IBM','MSFT','V','WMT','PFE','PG','GE','JPM','VZ','GS','MRK','NKE', 'INTC','DIS'). Firstly we optimized neural network on the whole historical data set. Clustering was carried out for the parameters: the size of a neural network 24, learning rate 0.01, radius learning 5, the length of period 0.01, and lower bound of signal -0.005, upper bound of signal 0.005.

Eventually we used algorithm for prediction.
Specific formulas for calculation of characteristics will be provided in following paragraph. Input parameters will be denoted as follows: n is number of trades, n_{loss} is number of loosing trades, n_{win} is number of winning trades.

$$n = n_{loss} + n_{win} \tag{6}$$

The win ratio is ratio of number of winning trades to number of all trades.

$$W_R = \frac{n_{win}}{n} \tag{7}$$

The profit loss ratio is ratio of mean value of positive price changes to mean value of negative price changes.

$$PL_R = \frac{profit}{n_{win}} / \frac{loss}{n_{loss}} \tag{8}$$

The profit factor is equal to the ratio of the sum of positive price changes to sum of negative price changes.

$$P_F = \frac{profit}{loss} \tag{9}$$

The profitability of a trading system is defined as the ratio of winning trades to total trades and ranges between 0 and 1 or, as a percent value, between 0 and 100.

$$P_{rule} = \frac{P_F}{PL_R} \tag{10}$$

In Tab. 1. several win lose characteristics for better idea of the values of monitored parameters will been presented.

Market Title	Trades	Win Ratio	P/L Ratio	Profit Factor	Percent Profit	Duration
GOOGL	1141	0.8238	7.4285	34.7405	12.59%	22
MSFT	743	0.7160	2.2188	5.5943	3.17%	20
KO	509	0.7799	9.4592	33.5295	17.85%	35

Tab. 1. Win lose statistics calculated in the period 2010–2015, source: own

Date	Price	Operation	Percent Profit	Duration	Type
29.12.2000	56,4	purchase	-0.069148936	5	reduce losses
8.2.2001	57,6	purchase	0.034722222	2	time frame
11.4.2001	52,3	purchase	0,056405354	13	limitation of profit
3.7.2001	55,75	purchase	-0,052017937	6	sale
27.7.2001	53	purchase	-0,03	20	time frame
5.9.2001	50,6	purchase	-0,160079051	6	sale
2.10.2001	37,35	purchase	0,097456493	2	limitation of profit
3.1.2002	51,9	purchase	-0,010597303	7	sale
25.1.2002	51,91	purchase	0,069158158	6	limitation of profit
27.2.2002	54,05	purchase	0,092506938	5	limitation of profit
10.5.2002	56	purchase	0,059821429	13	limitation of profit
19.7.2002	43,7	purchase	-0,083524027	3	sale
29.12.2000	56,4	purchase	-0.069148936	5	reduce losses
8.2.2001	57,6	purchase	0.034722222	2	time frame
11.4.2001	52,3	purchase	0,056405354	13	limitation of profit

Tab. 2. Example of signals: Volkswagen, source: own

6 Comparison of technical analysis and neural networks

In this chapter neural networks and technical analysis will be compared. Indicator Rate of change will be used. Rate of change (ROC) is simple technical analysis indicator showing the difference between today's closing price and the close N days ago. The number of trading days was set to 12 to minimize the Day of the Week Effect. The Day of the Week Effect reflects the fact that stock prices during the week move in a particular pattern. On Monday stock quotes regularly falling, the highest positive changes occur Wednesday and Friday. On Tuesday, the rate is nearly unchangeable and Thursday slightly positive.

These methods were applied to data from 15 companies of Dow Jones industrial average. Analysis is done for previous 12 years. Example of computed results is in the following Table 3.

Compared stocks are of the following companies: Apple, IBM, Microsoft, VISA, Wal/Mart, Pfizer Inc., Procter & Gamble, General Electric, JPMorgan Chase, Verizon communications, Goldman Sachs BDC, Merck & Co, Nike Inc., Intel and Walt Disney.

APPLE	Win Ratio	P/L Ratio	Number of trades	Win Ratio	P/L Ratio	Number of trades
2005	0,69	1,98	90	0,46	1,61	167
2006	0,85	2,25	88	0,5	0,9	163
2007	0,84	1,67	82	0,33	1,01	156
2008	0,85	1,31	155	0,44	0,76	160
2009	0,86	1,37	71	0,43	1,18	141
2010	0,85	0,99	107	0,4	1,05	124
2011	0,8	3,16	49	0,67	1,27	106
2012	0,76	1,84	25	0,32	1,23	147
2013	0,81	2,81	58	0,59	1,03	133
2014	0,45	6,86	123	0,57	15,23	97
2015	0,5	13,0	2	0,51	0,97	88
2016	0,72	3,4	36	0,49	0,74	100

Tab. 3. Comparison: Apple, source: own

Following Table is comparison of average values of Win Ratio and P/L Ratio for both used methods – clustering of significant periods and rate of change. Comparison is made for all 15 companies.

Company	Average Win Ratio (clustering of significant periods)	Average Win Ratio (rate of change)	Average P/L Ratio (clustering of significant periods)	Average P/L Ratio (rate of change)
Apple	0,75	0,48	3,39	2,25
IBM	0,84	0,51	3,75	1,01
Microsoft	0,81	0,49	5,90	1,06
VISA	0,75	0,55	4,45	1,08
Wal-Mart	0,84	0,58	5,38	1,97
Pfizer Inc.	0,78	0,56	2,15	1,36
Procter & Gamble	0,8	0,45	1,82	1,52
General Electric	0,81	0,56	2,02	5,09
JPMorgan Chase	0,83	0,55	1,76	1,21
Verizon com.	0,81	0,54	2,78	1,07
Goldman Sachs	0,84	0,47	2,43	1,08
Merck & Co.	0,83	0,53	1,98	1,04
Nike Inc.	0,75	0,49	2,74	1,24
Intel	0,80	0,48	7,42	1,12
Walt Disney	0,81	0,5	3,35	1,17

Tab. 4. Comparison of Average Win Ratio and Average P/L Ratio, source: own

All results are illustrated by following Graphs. Fig. 3 shows Win Ratio for all companies using clustering of significant periods, Fig. 4 shows Win Ratio using rate of change, Fig. 5 shows P/L

Ratio for all companies using clustering of significant periods, Fig. 4 shows P/L Ratio using rate of change. Fig. 7, Fig. 8, Fig. 9 and Fig. 10 show boxplot analysis for all previous situations.

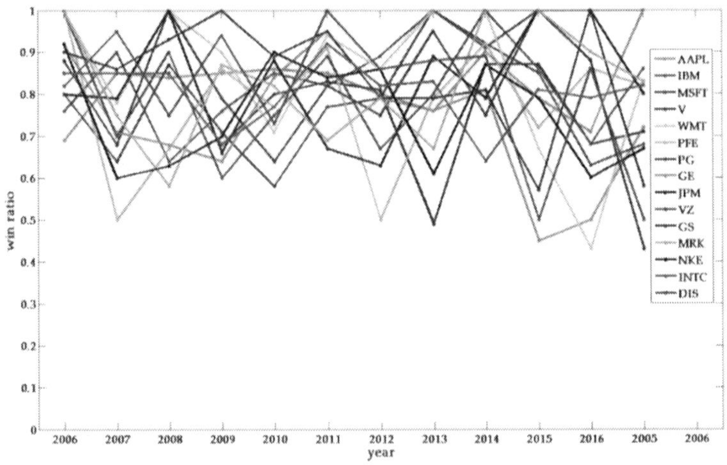

Fig. 3. Win Ratio using clustering of significant periods, source: own

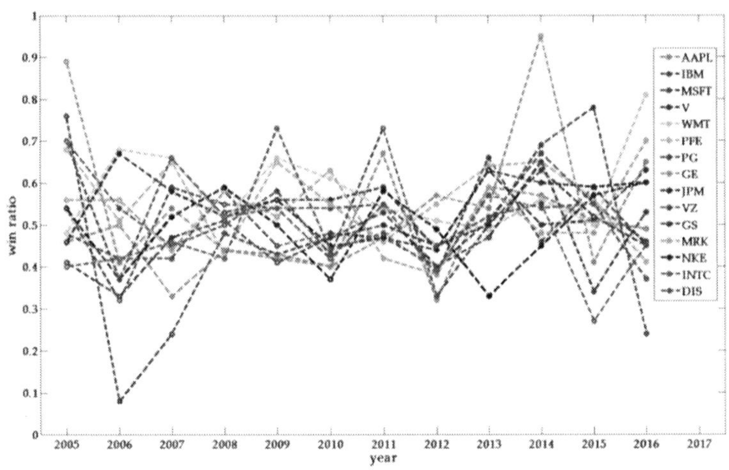

Fig. 4. Win Ratio using rate of change, source: own

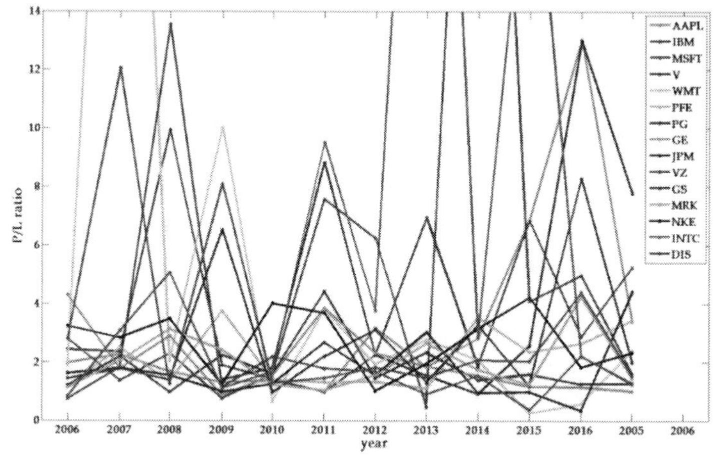

Fig. 5. P/L Ratio using clustering of significant periods, source: own

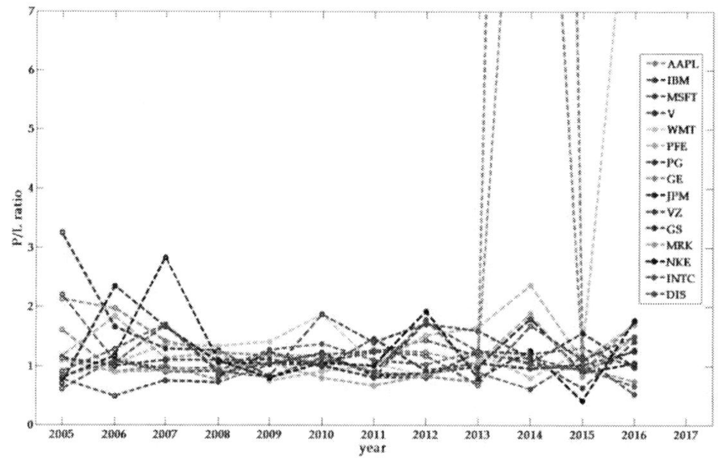

Fig. 6. P/L Ratio using rate of change, source: own

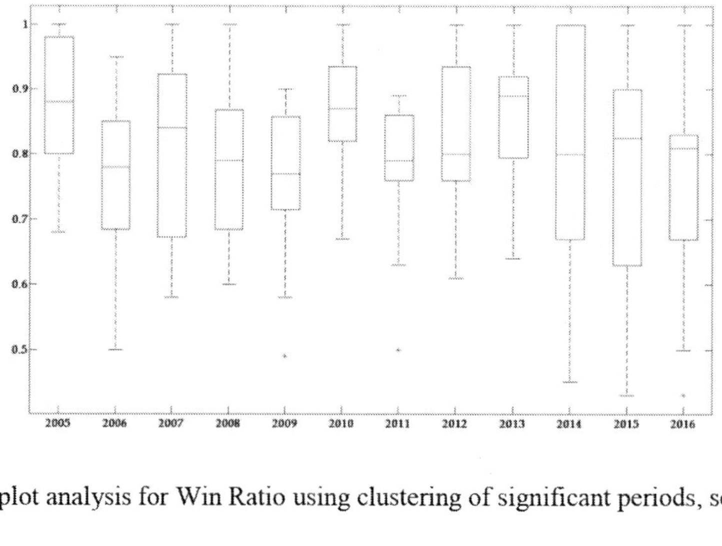

Fig. 7. Boxplot analysis for Win Ratio using clustering of significant periods, source: own

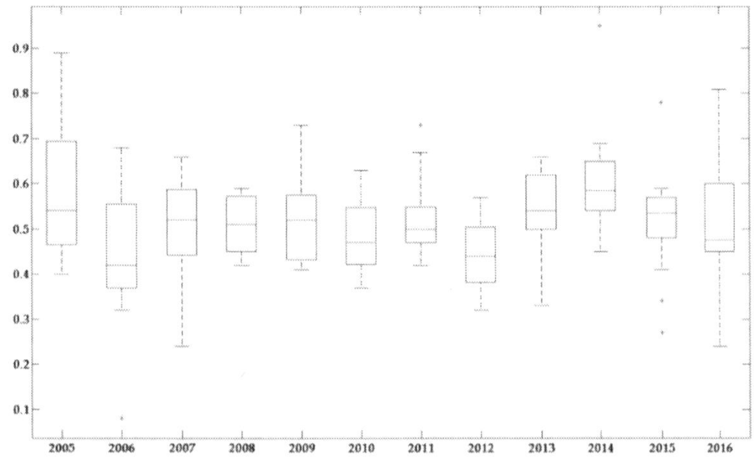

Fig. 8. Boxplot analysis for Win Ratio using rate of change, source: own

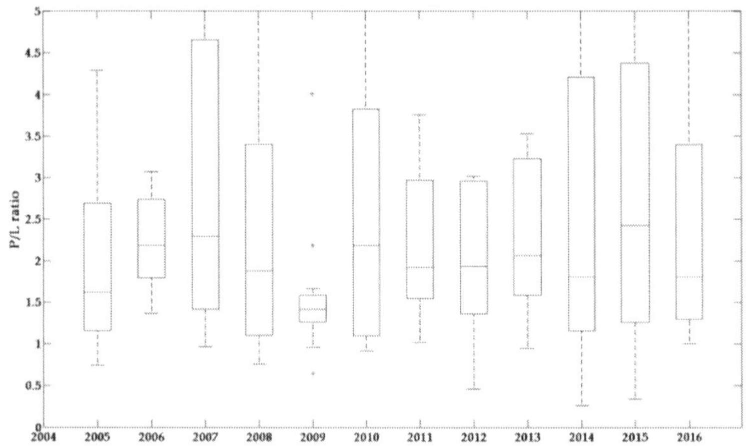

Fig. 9. Boxplot analysis for P/L Ratio using clustering of significant periods, source: own

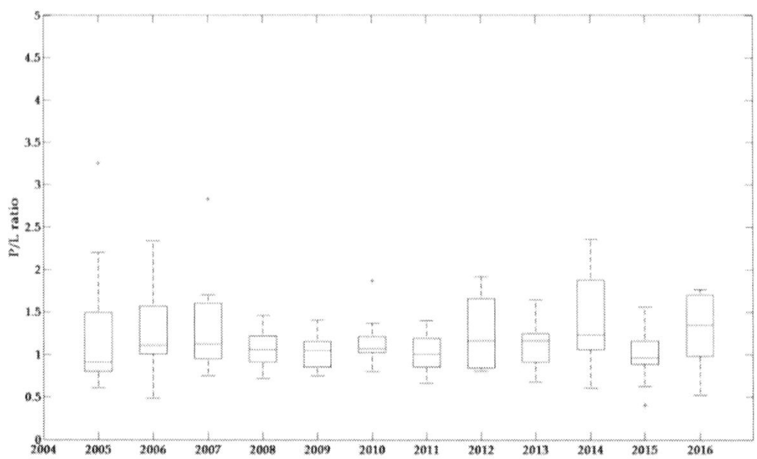

Fig. 10. Boxplot analysis for P/L Ratio using rate of change, source: own

From Table 4 can be seen, then all computed values of average win ratio and average p/l ratio are bigger wih using clustering of significant periods.

7 Conclusion

Knowledge of the expected annual interest rate of share and risks for financial investing on a stock exchange is rarely at our disposal. The main goal was to use periodogram and SOM for prediction of the stock market. Analysis of win loss statistics show that the described method can be used for prediction of future stock market prices and for successful financial speculations.

It was demonstrated that the percentage of profitable trades were very interesting. This is consistent with the expected power and utility of hidden periods and neural networks.

From comparison of technical analysis and neural networks, it can be seen that all computed values of average win ratio and average p/l ratio are bigger when using clustering of significant periods. Usage of neural networks gives better results than technical analysis.

Acknowledgements

This contribution has been supported by the Ministry of Education, Youth and Sports of the Czech Republic project No. CZ.1.07/ 2.3.00/30.0058, and partly supported by SGS project and institutional support of the University of Pardubice, Czech Republic.

References

[1] SCHUSTER, A. On the investigation of hidden periodicities with application to a supposed 26 day period of meteorological phenomena. Terrestrial Magnetism, 3, 1898, pp. 13-41.

[2] NEELY, Ch., WELLER, P. and DITTMAR, R. Is Technical Analysis in the Foreign Exchange Market Profitable?: A Genetic Programming Approach [online]. Journal of Financial and Quantitative Analysis, 1997, http://research.stlouisfed.org/wp/1996/96-006.pdf

[3] YAO, J., TAN, C. and POH, L. H. Neural Networks for Technical Analysis: A Study on KLCI. International Journal of Theoretical and Applied Finance, vol. 2, No. 2, 1999, pp. 221-241.

[4] TAYLOR, D. J. Machine Learning, Neural and Statistical Classification. Ellis Horwood, 1994, ISBN 0-13-106360-X.

[5] BISHOP, C. M. Neural Networks for Pattern Recognition. Oxford University Press, 1995, ISBN 0-19-853864-2.

[6] RIPLEY, B. D. Pattern Recognition and Neural Networks. Cambridge University Press, 1996, ISBN 0-521-46086-7

[7] The Profitability Rule. URL: http://www.priceactionlab.com/Literature/profitability.pdf, online, [cit. 20. 11. 2015].

[8] FISHER, R. A. Tests of significance in harmonic analysis. Proc. Roy. Statist. Soc., Ser. A, 125 (1929), pp. 54–59.

Current address

Pozdílková Alena, Mgr. PhD.
Department of Mathematics and Physics, University of Pardubice,
Faculty of Electrotechnics and Informatics
Studentská 85, 530 02 Pardubice, Czech Republic
E-mail: alena.pozdilkova@upce.cz

Marek Jaroslav, Mgr. PhD.
Department of Mathematics and Physics, University of Pardubice,
Faculty of Electrotechnics and Informatics
Studentská 85, 530 02 Pardubice, Czech Republik
E-mail: jaroslav.marek@upce.cz

Todt Jaroslav, Ing.
Department of Mathematics and Physics, University of Pardubice,
Faculty of Electrotechnics andInformatics
Studentská 85, 530 02 Pardubice, Czech Republic
E-mail: st44009@student.upce.cz

Slovak University of Technology in Bratislava
Faculty of Mechanical Engineering
16th Conference on Applied Mathematics **APLIMAT 2017**
Proceedings

FREE FORM DEFORMATION METHODS – THE THEORY AND PRACTICE

PROCHAZKOVA Jana (CZ)

Abstract. Solid modeling is one of the essential techniques used in common 3D studios (Maya, Rhinoceros, Blender). The editing or modification of the solids is often done using free-form deformation (FFD). In our article, we present two different techniques of the FFD. Firstly, we outline Sederberg's scheme based on Bernsteins polynomials that is the headstone of the method. Secondly, we describe the method NURBS FFD that is more complex and is independent on the shape of the object hull.

Keywords: Free form deformations, spline, NURBS, Lagrange function

Mathematics subject classification: Primary 68U05; Secondary 65D17

1 Introduction

The editing of geometric objects leads to a novel development in this field. One of the key methods is the FFD technique. The basic idea is to enclose an object into a hull. The object is then transformed via the hull deformation. The mathematical background encompases e.g. Bernsteins polynomials [5] or B-spline functions [8, 9].

Solid modeling itself covers many representation schemes. The commercial geometric modelers usually use boundary representation (B-rep) because of its flexibility and geometric complexity. Data structure forms combinatorial maps: vertices, edges, faces as parts of manifold faces or Non-Uniform B-Spline surfaces (NURBS).

Surface (polygonal) mesh models are similar to the boundary representation; nevertheless, they only use simple surface mesh that consist of vertices and edges (e.g. trinagular mesh, quadrilaterals, convex polygon meshes).

We have to mention a constructive solid geometry (CSG) as an important scheme of the solid representation. This method creates a complex surface or an object using Boolean operators to combine simpler objects, e.g sphere, cone or cube, see [1].

The demand of the technical practice to have powerful and easy to manipulate tool for this purpose has started the development of free-form modeling (FFD) tools. The demands on the solid manipulation are independent on the scheme of solid representation: (1) to select a point on the solid surface; (2) to place the point to a target position to form required shape; (3) to work with these points effortlessly.

The development of the deformation tools lead to improvements in many other areas. The FFD can be used in the methods of the medical image registration (detection process of breast cancer, heart diseases), gesture recognition and microscopy [2], [3].

1.1 Historical perspective of FFD

One of the first deformation methods is Barr technique [4] that is based on the hierarchical transformations. It covers basic operations as bending, twisting, stretching and taper operators. The method recomputes the normal vectors of transformed surfaces using special transformation matrices.

The term Free-Form Deformation (FFD) was firstly published by Sederberg and Perry [5] in the year 1986. The main idea is the insertion of the solid representation to the regular parallelpiped; subsequent transformation recomputes the position of the solid points with respect to the modified points of the lattice. The brief overview of method is in section 2.1.

The extension of Sederberg's technique was presented in 1990 by Coquillart [6]. It is based on usage of irregular lattice that extends the applicability of the algorithm. Subsequently, the FFD was modified e.g. by de Casteljau algorithm [7], general spline [8], NURBS [9].

In our article, we present two different techniques of the FFD: Sederberg's scheme (sec. 2.1) based on Bernsteins polynomials that is the foundation of the FFD techniques and the method NURBS FFD (sec. 2.2) that is more complex and is independent on the shape of the object hull. We also provide algorithms of both methods.

2 Free form deformation

2.1 Sederberg's technique

FFD technique presented in [5] is based on the tensor product trivariate Bernestein polynomial. Every object point $\mathbf{X}$ has (s, t, u) coordinates in the coordinate system on a parallelpiped region:

$$\mathbf{X} = \mathbf{X}_0 + s\mathbf{S} + t\mathbf{T} + u\mathbf{U} \tag{1}$$

Let P_{ijk}, $i = 0, \ldots, l$, $j = 0, \ldots, m$, $k = 0, \ldots, n$ are control points on a lattice. The deformation is represented by a movement of the control points P_{ijk} from their latticial positions. New position of point $\mathbf{X}_{def}$ is computed by:

$$\mathbf{X}_{def} = \sum_{i=0}^{l} \binom{l}{i} (1-s)^{l-i} s^i \left[\sum_{j=0}^{m} \binom{m}{j} (1-t)^{m-j} t^j \left(\sum_{k=0}^{n} \binom{n}{k} (1-u)^{n-k} u^k \mathbf{P}_{ijk} \right) \right] \tag{2}$$

where (s, t, u) are parameters in Eq.(1).

The algorithm is easy to program and the psedocode can be written as:

Algorithm psedocode

Input: parallelpiped points, parameters (s, t, u)
Output: new coordinates of the point $\boldsymbol{X}$ for parameters (s, t, u)

for $i = 0, \ldots, l$ **do**
  for $j = 0, \ldots, m$ **do**
    for $k = 0, \ldots, n$ **do**
      $X = X + \text{binom}(l, i) * \text{pow}(1 - s, l - i) * \text{pow}(s, i) *$
      $\text{binom}(m - 1, j) * \text{pow}(1 - t, m - j) * \text{pow}(t, j) *$
      $\text{binom}(n, k) * \text{pow}(1 - u, n - k) * \text{pow}(u, k) * \boldsymbol{P}_{ijk};$
    end for
  end for
end for

The result of one latticial position change is visible in Fig. 1. Note that function `binom` is the computation of binomial number and function `pow`(a, b) is power a^b.

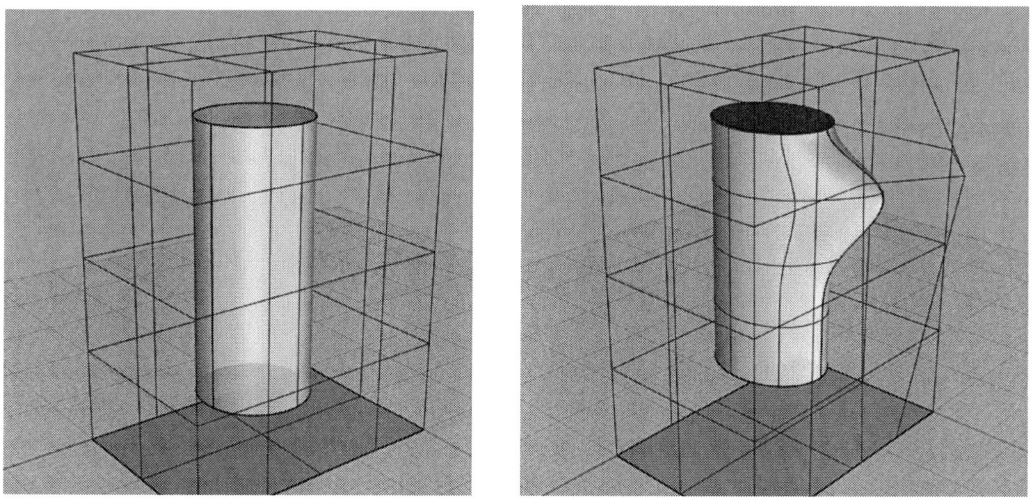

(a) (b)

Fig. 1. Sederbergs technique FFD: a) before and b) after transformation

2.2 NURBS FFD

Generalization of Sederberg's first FFD is method based on NURBS. NURBS volume consists of the regular net of control points $\mathbf{P}_{ijk}$ that are connected to knot vectors and basis B-spline functions. The mathematical expression can be written as:

$$Q(u, v, w) = \sum_{i,j,k=0}^{l,m,n} \mathbf{P}_{ijk} R_{ijk}(u, v, w), \quad 0 \leq u, v, w \leq 1 \tag{3}$$

where

$$R_{ijk}(u, v, w) = \frac{w_{ijk} B_i^p(u) B_j^q(v) B_k^r(w)}{\sum_{i,j,k=0}^{l,m,n} w_{ijk} B_i^p(u) B_j^q(v) B_k^r(w)} \tag{4}$$

Let w_{ijk} be the weights of corresponding control points P_{ijk} and $B_i^p(u)$, $B_j^q(v)$, $B_k^r(w)$ are basis functions of degree p, q, r defined on knot vectors $\mathbf{u} = (u_0, u_1, \ldots, u_{l+p+1})$, $\mathbf{v} = (v_0, v_1, \ldots, v_{m+q+1})$ and $\mathbf{w} = (w_0, w_1, \ldots, w_{n+r+1})$ (in Eq. (4)).

Main idea of this algorithm is replace the parallelpiped region in Sederbergs scheme by general NURBS volume. The arbitrary object $\mathcal{O}$ is covered by NURBS volume $Q(u, v, w)$. Nevertheless, the NURBS FFD do not move the points of the lattice like the Sederberg's scheme but we work with movement of the object points.

2.2.1 Computation for one point movement

Let $\mathbf{S} = \mathcal{O}(u_s, v_s, w_s)$ be a source point of the object $\mathcal{O}$ and $\mathbf{T}$ be a target point. The values (u_s, v_s, w_s) are the parameters of the point in the NURBS control net. Our task is the computation of a new NURBS control net $\mathbf{P}_{ijk} + \boldsymbol{\delta}_{ijk}$ that satisfies:

$$\mathbf{T} = \sum_{i,j,k=0}^{l,m,n} (\mathbf{P}_{ijk} + \boldsymbol{\delta}_{ijk}) R_{ijk}(u_s, v_s, w_s) = \mathbf{S} + \sum_{i,j,k=0}^{l,m,n} \boldsymbol{\delta}_{ijk} R_{ijk}(u_s, v_s, w_s). \tag{5}$$

The variables $\boldsymbol{\delta}_{ijk}$ express the shift of the NURBS control points $\mathbf{P}_{ijk}$. This demand leads us to optimization problem (e.g. solved in [9]) that can be solved by the method of Lagrange multipliers. We search the minimum:

$$\min \sum_{i,j,k=0}^{l,m,n} \|\boldsymbol{\delta}_{ijk}^2\| \tag{6}$$

Lagrange function is:

$$\mathbf{L} = \sum_{i,j,k=0}^{l,m,n} \|\boldsymbol{\delta}_{ijk}^2\| + \lambda \left(\mathbf{T} - \mathbf{S} - \sum_{i,j,k=0}^{l,m,n} \boldsymbol{\delta}_{ijk} R_{ijk}(u_s, v_s, w_s) \right), \tag{7}$$

where $\boldsymbol{\lambda} = (\lambda_1, \lambda_2, \lambda_3$ is Lagrange multipliers and operation $\|.\|$ is Euclidean norm.

We set:

$$\frac{\partial \mathbf{L}}{\partial \lambda_1} = \frac{\partial \mathbf{L}}{\partial \lambda_2} = \frac{\partial \mathbf{L}}{\partial \lambda_3} = 0 \quad \text{and} \quad \frac{\partial \mathbf{L}}{\partial \delta_{ijk}^x} = \frac{\partial \mathbf{L}}{\partial \delta_{ijk}^y} = \frac{\partial \mathbf{L}}{\partial \delta_{ijk}^z} = 0 \tag{8}$$

Consequently, we get the equations (in vector form)

$$\mathbf{T} = \mathbf{S} + \sum_{i,j,k=0}^{l,m,n} \boldsymbol{\delta}_{ijk} R_{ijk}(u_s, v_s, w_s) \tag{9}$$

$$\boldsymbol{\delta}_{ijk} = \frac{\lambda}{2} R_{ijk}(u_s, v_s, w_s), \quad \text{for} \quad 0 \le i \le l, \quad 0 \le j \le m, \quad 0 \le k \le n. \tag{10}$$

We substitute $\boldsymbol{\delta}_{ijk}$ in Eq. (9) by the term in Eq. (10) and we express $\frac{\lambda}{2}$ as:

$$\frac{\lambda}{2} = \frac{\mathbf{T} - \mathbf{S}}{\sum_{i,j,k=0}^{l,m,n} R_{ijk}^2(u_s, v_s, w_s)} = \frac{\boldsymbol{\delta}_{ijk}}{R_{ijk}(u_s, v_s, w_s)} \tag{11}$$

The expression $\boldsymbol{\delta}_{ijk}$ gives the explicit solution of the point movements:

$$\boldsymbol{\delta}_{ijk} = \frac{R_{ijk}(u_s, v_s, w_s)}{\sum_{i,j,k=0}^{l,m,n} R_{ijk}^2(u_s, v_s, w_s)}(\mathbf{T} - \mathbf{S}) \tag{12}$$

New positions of the NURBS control points $\mathbf{P}_{ijk} + \boldsymbol{\delta}_{ijk}$ generates new shape $\mathcal{O}'$ of the object $\mathcal{O}$. This deformation ensures that point $\mathbf{T}$ is on the object $\mathcal{O}'$ in desired position .

Algorithm psedocode

Input: NURBS volume, source point S, target point T
Output: array $\boldsymbol{\delta}$

1 Compute (u_s, v_s, w_s) for source point S (in [11])
2 Compute basis polynomials $\mathbf{R}(u_s, v_s, w_s)$ with Eq.(4)
3 Compute result array $\boldsymbol{\delta}$ with Eq.(12)
4 Recompute NURBS volume $\mathbf{P}_{new} = \mathbf{P} + \boldsymbol{\delta}$

2.2.2 The composition of transformations

In this part, we introduce direct computation method for the FFD in case we need a transformation of a point set. There are three possible approaches: sequential (one transformation followed by another one), composite computation [9] or multilevel approach[8]. In our case, we choose the composite computation because of the simplicity of the algorithm and lower time-consuming.

Let $\mathbf{S}_f$, $f = 1, \ldots, h$ are origin points on an object $\mathcal{O}$ with parameters (u_f, v_f, w_f) and $\mathbf{T}_f$ are the target points. The Eq. 13 holds:

$$\mathbf{T}_f = \sum_{i,j,k=0}^{l,m,n} (\mathbf{P}_{ijk} + \boldsymbol{\delta}_{ijk}) R_{ijk}(u_f, v_f, w_f) = \mathbf{S}_f + \sum_{i,j,k=0}^{l,m,n} \boldsymbol{\delta}_{ijk} R_{ijk}(u_f, v_f, w_f) \tag{13}$$

where $f = 1, \ldots, h$.

Lagrange function is (similarly to the one point FFD):

$$\mathbf{L} = \sum_{i,j,k=0}^{l,m,n} \|\boldsymbol{\delta}_{ijk}\|^2 + \sum_{f=1}^{h} \boldsymbol{\lambda_f} \left(\mathbf{T}_f - \mathbf{S}_f - \sum_{i,j,k=0}^{l,m,n} \boldsymbol{\delta}_{ijk} R_{ijk}(u_f, v_f, w_f) \right) \tag{14}$$

where $\boldsymbol{\lambda_f} = (\lambda_{f1}, \lambda_{f2}, \lambda_{f3})$, $f = 1, \ldots, h$ are Lagrange multipliers. We get:

$$\mathbf{T}_f = \mathbf{S}_f - \sum_{i,j,k=0}^{l,m,n} \boldsymbol{\delta}_{ijk} R_{ijk}(u_f, v_f, w_f) \tag{15}$$

and

$$\boldsymbol{\delta}_{ijk} = \sum_{f=1}^{h} \frac{\boldsymbol{\lambda_f}}{2} R_{ijk}(u_f, v_f, w_f) \tag{16}$$

$0 \leq i \leq l, 0 \leq j \leq m, 0 \leq k \leq n.$

The solution of the system of linear equations determines all the shift vectors $\boldsymbol{\delta}_{ijk}$.

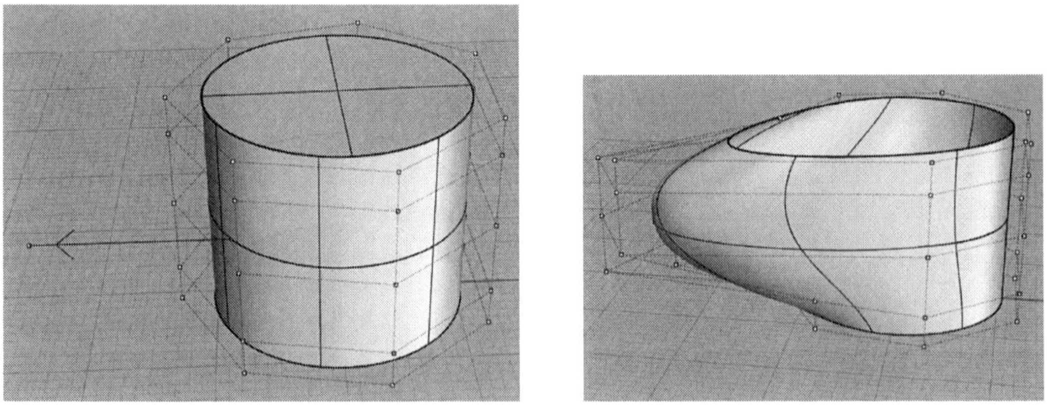

(a) (b)

Fig. 2. NURBS FFD: a) before and b) after transformation

3 Conclusion

The FFD methods are common tool for in the technical practice. Common commercial applications contain the FFD technique to modify the models. Nonetheless, the the FFD techniques are widely even in other areas: medical research, general image registration, gesture recognition, microscopy etc. We described original approach that was proposed by Sederberg and in Sec. 2.2 the modern NURBS FFD technique is presented. We also provided the algorithms pseudocode and influence of the transformations on the basic shapes.

Acknowledgment

This work was supported by Project LO1202 by financial means from the Ministry of Education, Youth and Sports under the National Sustainability Programme I.

References

[1] Foley, J.D.: *Computer Graphics: Principles and Practice* (1996)

[2] Mattes, D., Haynor, D.R., Vesselle, H., Lewellen, T.K., Eubank W.: PET-CT Image Registration in the Chest Using Free-form Deformations IEEE Transactions on Medical Imaging, 22(1), 2003.

[3] Rueckert, D., Sonoda L.I., Hayes C., Hill D.L.G., Leach M.O., Hawkes D.J.: Non-rigid registration using free-form deformations: application to breast MR images. IEEE Transactions on Medical Imaging, 18(7) (1999) 12–21.

[4] Barr, A. H.: Global and Local Deformations of Solid Primitives, Proceedings of SIGGRAPH '84, Computer Graphics 18, 3 (1984), 21-30.

[5] Sederberg, T. W., Parry, S. R.: Free-Form Deformation of Solid Geometric Models, Proceedings of SIGGRAPH '86, Computer Graphics 20, 4 (1986), 151-159.

[6] Coquillart, S.: Extended Free-Form Deformation: A sculpturing Tool for 3D Geometric Modeling, Proceedings of SIGGRAPH '90, Computer Graphics 24, 4 (1990), 187-196.

[7] Chang, Y.and Rockwood, A. P., A Generalized de Casteljau Approach to 3D Free-form Deformation, Proceedings of SIGGRAPH '94, Computer Graphics (July 1994), 257-260.

[8] Lee, S., Wolberg, G., Chwa, K.-Y., Shin, S.Y.: Image Metamorphosis with Scattered Feature Constraints, IEEE Transactions on Visualization and Computer Graphics 2, 4 (1996), 337-354.

[9] Hu, S.M., Tai, Ch.L.: Direct manipulation of FFD: Efficient explicit solutions and decomposible multiple point constraints, Visual Computer 17 (2001), 370–379.

[10] Hsu, W.M., Hughes, J.F., Kaufman, H.: Direct Manipulation of Free-Form Deformations, Computer Graphics 26, 2 (1992), 177-184.

[11] Piegl, L., Tiller, W.: The NURBS Book. Springer-Verlag New York, 1997.

Current address

Procházková Jana, Mgr. Ph.D.
Institute of Mathematics, Faculty of Mechanical Engineering,
Brno University of Technology
Technická 2, 616 69 Brno, Czech republic
E-mail: prochazkova.j@fme.vutbr.cz

Slovak University of Technology in Bratislava
Faculty of Mechanical Engineering
16th Conference on Applied Mathematics **APLIMAT 2017**
Proceedings

POLYNOMIAL CALIBRATION WITH USE OF LINEARIZED MODEL WIHT ERRORS IN VARIABLES, AND KENWARD ROGER TYPE OF APPROXIMATION

RÁBOŇOVÁ Petra (CZ)

Abstract. In the contribution we focuse on polynomial calibration. We estimate parameters of transformation function and obtained confidence region for transformation curve.

Keywords: Polynomial calibration, transformation curve, transformation function, MINQUE method, replicated model

Mathematics subject classification: Primary 62F25, 62F10; Secondary 62P30

Calibration is a set of tasks which gives the relationship between a reference device and a calibrated device (if some special conditions are fullfilled). This relationship is described by the transformation function and represented by the transformation curve. In the contribution we assume that the transformation curve is a polynomial of degree k. In the contribution we assume that the transformation curve is a polynomial of degree k. Therefore, we focus on the polynomial calibration. We use a method based on linearized errors-in-variables model and Kenward-Roger type approximation for obtaining the transformation curve. Calibration process can be divided into two parts: 1) creation of the calibration model (calibration of the device), 2) application (use) of the calibration model (measuring by calibrated device). In the contribution we estimate parameters of the transformation function and their confidence region by the mentioned methods and we also show the use of the transformation curve in the measuring process.

1 Creation of calibration model

We assume that we have m different objects. Each of these objects we measure by two different measuring devices (device A and device B), and we repeat the measurement n times. We assume that device A is less precise than device B. It is assumed that measured values on both devices are realizations of random variables with normal distribution, and for each of m objects, values measured errorlessly by device A are $\mu = (\mu_1, \ldots, \mu_m)$, and by device B are $\nu_i = a_0 + a_1\mu_i \cdots + a_k\mu_i^k$, where $a_0, a_1, \cdots, a_k \in \mathbb{R}$, $i = 1, 2, \ldots, m$. The function $\nu_i = a_0 + a_1\mu_i \cdots + a_k\mu_i^k$ is called transformation function. Next assumption is that measurements realised by device A and device B are independent and $X_{ij} \sim N(\mu_i, \sigma_x^2)$ (X_{ij} is normally distributed with mean μ_i and dispersion σ_x^2), $Y_{ij} \sim N\left(a_0 + a_1\mu_i + \cdots a_k\mu_i^k, \sigma_y^2\right)$, $i = 1, \cdots, m, j = 1, \cdots, n$. The aim is to estimate the parameters $\left(a_0, a_1 \cdots, a_k, \sigma_x^2, \sigma_y^2,\right)'$.

1.1 Replicated model

In the paper, a replicated model will be used for estimating of transformation curve. At firts we suppose that by the devices A and B we realize only one measurement on each object. In this case we have a vector of random variables-measurements by device A as $\mathbf{X} = (X_1, \cdots, X_m)$'. For each $i = 1, \cdots, m$ is $X_i \sim N(\mu_i, \sigma_x^2)$. $\mathbf{Y} = (Y_1, \cdots, Y_m)'$ measurements are obtained by the device B, where $Y_i \sim N\left(\nu_i, \sigma_y^2\right)$. We denote $\mu = (\mu_1, \cdots \mu_m)$, $\nu = (\nu_1, \cdots \nu_m)$ as in the introduction and suppose that measurements are not correlated. So we can write the measurement model as:

$$\begin{pmatrix} \mathbf{X} \\ \mathbf{Y} \end{pmatrix} \sim N\left[\begin{pmatrix} \mu \\ \nu \end{pmatrix}, \begin{pmatrix} \sigma_x^2\mathbf{I_m} & \mathbf{0}_m \\ \mathbf{0}_m & \sigma_y^2\mathbf{I_m} \end{pmatrix}\right].$$

If we denote ν and μ^b as $\nu = a_0\mathbf{1_m} + a_1\mu \cdots + a_k\mu^k$, $\mu^b = \left(\mu_1^b, \ldots, \mu_m^b\right)'$, $b = 1, 2, \cdots, k$, $\mathbf{1_m} = (1, \ldots, 1)' \in \mathbb{R}^m$, we obtain a linear regression model with nonlinear constraints on parameters.

$$\nu = \mathbf{1_m}a_0 + \mu a_1 + \cdots + \mu^k a_k$$

Using Taylor expansion in values $a_{10}, \cdots, a_{k0}$, $\mu_0 = (\mu_{10}, \cdots, \mu_{m0})$, $\mu_0^b = \left(\mu_{10}^b, \cdots, \mu_{m0}^b\right)$, $\delta\mu_i = \mu_i - \mu_{i0}$ for $i = 1, \ldots, m$, $\delta\mu = (\delta\mu_1, \ldots, \delta\mu_m)'$ and neglecting the terms of the second and higher order we obtain the linear regression model with linear constraints

$$\begin{pmatrix} \mathbf{X} - \mu_0 \\ \mathbf{Y} \end{pmatrix} \sim N\left[\begin{pmatrix} \delta\mu \\ \nu \end{pmatrix}, \begin{pmatrix} \sigma_x^2\mathbf{I_m} & \mathbf{0}_m \\ \mathbf{0}_m & \sigma_y^2\mathbf{I_m} \end{pmatrix}\right]$$

$$\left(diag\left(a_{10}\mathbf{1_m} + \cdots + ka_{k0}\mu_0^{k-1}\right), -\mathbf{I_m}\right)\begin{pmatrix} \delta\mu \\ \nu \end{pmatrix} + \left(\mathbf{1_m}, \mu_0, \cdots, \mu_0^k\right)\begin{pmatrix} a_0 \\ a_1 \\ \vdots \\ a_{k0} \end{pmatrix} = \mathbf{0},$$

($diag$ is the diagonal matrix with elements of vector $\left(a_{10}\mathbf{1_m} + \cdots + ka_{k0}\mu_0^{k-1}\right)$ on the diagonal). If we repeat measurement by devices A and B n times, we obtain the replicated model:

$$\begin{pmatrix} \mathbf{X^1} - \mu_0 \\ \mathbf{Y^1} \\ \vdots \\ \mathbf{X^n} - \mu_0 \\ \mathbf{Y^n} \end{pmatrix} \sim N \left[\mathbf{1_n} \otimes \begin{pmatrix} \delta\mu \\ \nu \end{pmatrix}, \ \mathbf{I_n} \otimes \begin{pmatrix} \sigma_x^2 \mathbf{I_m} & \mathbf{0}_m \\ \mathbf{0}_m & \sigma_y^2 \mathbf{I_m} \end{pmatrix} \right] \tag{1}$$

with linear constraints on parameters

$$\left(diag\left(a_{10}\mathbf{1_m} + \cdots + k a_{k0}\mu_0^{k-1}\right), -\mathbf{I_m}\right) \begin{pmatrix} \delta\mu \\ \nu \end{pmatrix} + \left(\mathbf{1_m}, \mu_0, \cdots, \mu_0^k\right) \begin{pmatrix} a_0 \\ \vdots \\ a_k \end{pmatrix} = \mathbf{0},$$

where $\otimes$ is the Kroneker's product of matrices.
This model is a special case of the incomplete indirect measurement model with constraints of the II. type to parameters of the 1. order.

1.2 Estimate of the 1. order parameters

In this section we focuse on the estimate of transformation function paramaters $\begin{pmatrix} a_0 \\ \vdots \\ a_k \end{pmatrix}$ and estimate of vector $\begin{pmatrix} \delta\mu \\ \nu \end{pmatrix}$. For this purpose we use the procedure describe in [4].
According to [4, page 129], let $Y \sim (\mathbf{X}\beta_1, \Sigma)$ be a model with contstrains $\mathbf{b} + \mathbf{B}_1\beta_1 + \mathbf{B}_2\beta_2 = \mathbf{0}$, where $r(\mathbf{X}_{n,k_1}) = k_1 < n$, $r(\mathbf{B}_{1,(q,k_1)}, \mathbf{B}_{2,q,k_2}) = q < k_1 + k_2$, $r(\mathbf{B}_2) = k_2 < q$ and Σ is a positive definite matrix, then BLUE (best linear unbiased estimation) of vector $\begin{pmatrix} \beta_1 \\ \beta_2 \end{pmatrix}$ is:

$$\begin{pmatrix} \widehat{\widehat{\beta}}_1 \\ \widehat{\widehat{\beta}}_2 \end{pmatrix} = - \begin{pmatrix} \mathbf{C}^{-1}\mathbf{B}_1'\mathbf{Q}_{11} \\ \mathbf{Q}_{21} \end{pmatrix} \mathbf{b} + \begin{pmatrix} \mathbf{I} - \mathbf{C}^{-1}\mathbf{B}_1'\mathbf{Q}_{11}\mathbf{B}_1 \\ -\mathbf{Q}_{21}\mathbf{B}_1 \end{pmatrix} \widehat{\beta}_1,$$

where $\mathbf{C} = \mathbf{X}'\Sigma^{-1}\mathbf{X}$ and $\hat{\beta}_1 = \mathbf{C}^{-1}\mathbf{X}'\Sigma^{-1}\mathbf{X}$.

$$var(\widehat{\widehat{\beta}}_2) = -\mathbf{Q}_{22},$$

$$\begin{pmatrix} \mathbf{B}_1\mathbf{C}^{-1}\mathbf{B}_1' & \mathbf{B}_2 \\ \mathbf{B}_2' & \mathbf{0} \end{pmatrix}^{-1} = \begin{pmatrix} \mathbf{Q}_{11} & \mathbf{Q}_{12} \\ \mathbf{Q}_{21} & \mathbf{Q}_{22} \end{pmatrix}.$$

Now we apply the procedure on the model (1). We denote $\mathbf{B}_2 = \left(\mathbf{1}_m; \mu_0; \cdots; \mu_0^k\right)$, $\mathbf{B}_1 = (\mathbf{S}; -\mathbf{I}_m)$, $\mathbf{S} = diag\left(a_{10}\mathbf{1_m} + \cdots + ka_{k0}\mu_0^{k-1}\right)$, $\mathbf{A}_1 = \mathbf{B}_1\mathbf{C}^{-1}\mathbf{B}_1' = \frac{1}{n}(\sigma_x^2\mathbf{SS} + \sigma_y^2\mathbf{I})$, $\overline{\mathbf{X}} = \frac{1}{n}\left(\sum_{i=1}^n X_{1i}, \ldots, \sum_{i=1}^n X_{mi}\right)'$, $\overline{\mathbf{Y}} = \frac{1}{n}\left(\sum_{i=1}^n Y_{1i}, \ldots, \sum_{i=1}^n Y_{mi}\right)'$.
Firstly, we compute matrices $\mathbf{Q}_{11}, \mathbf{Q}_{21}, \mathbf{Q}_{22}$ for our model with use of [5, page 65] and $\hat{\beta}_1$.

$$\mathbf{C}^{-1} = \left[(\mathbf{1}_n \otimes \mathbf{I}_{2m})' \left[\mathbf{I}_n \otimes \begin{pmatrix} \sigma_x^2\mathbf{I}_m & \mathbf{0}_m \\ \mathbf{0}_m & \sigma_y^2\mathbf{I}_m \end{pmatrix}^{-1}\right] (\mathbf{1}_n \otimes \mathbf{I}_{2m})\right]^{-1} =$$
$$= \left[(\mathbf{1}_n' \cdot \mathbf{1}_n) \otimes \left(\begin{pmatrix} \sigma_x^2\mathbf{I}_m & \mathbf{0}_m \\ \mathbf{0}_m & \sigma_y^2\mathbf{I}_m \end{pmatrix}^{-1}\right)\right]^{-1} = \frac{1}{n}\begin{pmatrix} \sigma_x^2\mathbf{I}_m & \mathbf{0}_m \\ \mathbf{0}_m & \sigma_y^2\mathbf{I}_m \end{pmatrix}$$

$$\mathbf{Q}_{11} = \left(\mathbf{B}_1\mathbf{C}^{-1}\mathbf{B}_1'\right)^{-1} - \left(\mathbf{B}_1\mathbf{C}^{-1}\mathbf{B}_1'\right)^{-1}\mathbf{B}_2\left(\mathbf{B}_2'\left(\mathbf{B}_1\mathbf{C}^{-1}\mathbf{B}_1'\right)^{-1}\mathbf{B}_2\right)^{-1}\mathbf{B}_2'\left(\mathbf{B}_1\mathbf{C}^{-1}\mathbf{B}_1'\right)^{-1}$$

$$\mathbf{Q}_{12} = \left(\mathbf{B}_1\mathbf{C}^{-1}\mathbf{B}_1'\right)^{-1}\mathbf{B}_2\left(\mathbf{B}_2'\left(\mathbf{B}_1\mathbf{C}^{-1}\mathbf{B}_1'\right)^{-1}\mathbf{B}_2\right)^{-1},$$

$$\mathbf{Q}_{12} = \mathbf{Q}_{21}',$$

$$\mathbf{Q}_{22} = -\left(\mathbf{B}_2'\left(\mathbf{B}_1\mathbf{C}^{-1}\mathbf{B}_1'\right)^{-1}\mathbf{B}_2\right)^{-1},$$

$$\widehat{\beta}_1 = \begin{pmatrix} \frac{\sigma_x^2}{n}\mathbf{I}_m & \mathbf{0}_m \\ \mathbf{0}_m & \frac{\sigma_y^2}{n}\mathbf{I}_m \end{pmatrix} (\mathbf{1}_n \otimes \mathbf{I}_{2m})' \left(\mathbf{I}_n \otimes \begin{pmatrix} \sigma_x^2\mathbf{I}_m & \mathbf{0}_m \\ \mathbf{0}_m & \sigma_y^2\mathbf{I}_m \end{pmatrix}^{-1}\right) \begin{pmatrix} \mathbf{X}^1 - \mu_0 \\ \mathbf{Y}^1 \\ \vdots \\ \mathbf{X}^n - \mu_0 \\ \mathbf{Y}^n \end{pmatrix} =$$
$$= \begin{pmatrix} \frac{\sigma_x^2}{n}\mathbf{I}_m & \mathbf{0}_m \\ \mathbf{0}_m & \frac{\sigma_y^2}{n}\mathbf{I}_m \end{pmatrix} \left(\mathbf{1}_n' \otimes \begin{pmatrix} \sigma_x^2\mathbf{I}_m & \mathbf{0}_m \\ \mathbf{0}_m & \sigma_y^2\mathbf{I}_m \end{pmatrix}^{-1}\right) \begin{pmatrix} \mathbf{X}^1 - \mu_0 \\ \mathbf{Y}^1 \\ \vdots \\ \mathbf{X}^n - \mu_0 \\ \mathbf{Y}^n \end{pmatrix} =$$
$$= \frac{1}{n}(\mathbf{1}_n' \otimes \mathbf{I}_{2m}) \begin{pmatrix} \mathbf{X}^1 - \mu_0 \\ \mathbf{Y}^1 \\ \vdots \\ \mathbf{X}^n - \mu_0 \\ \mathbf{Y}^n \end{pmatrix} = \begin{pmatrix} \bar{\mathbf{X}} - \mu_0 \\ \bar{\mathbf{Y}} \end{pmatrix}$$

Therefore:

$$\begin{pmatrix} \widehat{\delta\mu} \\ \widehat{\nu} \end{pmatrix} = \left(\mathbf{I}-\mathbf{C}^{-1}\mathbf{B}_1'\mathbf{Q}_{11}\mathbf{B}_1\right)\widehat{\beta}_1 = \begin{pmatrix} \bar{\mathbf{X}}-\mu_0 \\ \bar{\mathbf{Y}} \end{pmatrix} - \frac{1}{n}\begin{pmatrix} \sigma_x^2\mathbf{I}_m & \mathbf{0}_m \\ \mathbf{0}_m & \sigma_y^2\mathbf{I}_m \end{pmatrix} \times$$

$$\times \begin{pmatrix} \mathbf{S} \\ -\mathbf{I}_m \end{pmatrix} \mathbf{Q}_{11}\left(\mathbf{S};-\mathbf{I}_m\right)\begin{pmatrix} \bar{\mathbf{X}}-\mu_0 \\ \bar{\mathbf{Y}} \end{pmatrix} =$$

$$= \begin{pmatrix} \bar{\mathbf{X}}-\mu_0 \\ \bar{\mathbf{Y}} \end{pmatrix} - \begin{pmatrix} \frac{\sigma_x^2}{n}\mathbf{S}\mathbf{Q}_{11}\left(\mathbf{S}\left(\bar{\mathbf{X}}-\mu_0\right)-\bar{\mathbf{Y}}\right) \\ -\frac{\sigma_y^2}{n}\mathbf{Q}_{11}\left(\mathbf{S}\left(\bar{\mathbf{X}}-\mu_0\right)-\bar{\mathbf{Y}}\right) \end{pmatrix}$$

$$\widehat{\mu}-\mu_0 = \bar{\mathbf{X}}-\mu_0-\tfrac{\sigma_x^2}{n}\mathbf{S}\mathbf{Q}_{11}\left(\mathbf{S}\left(\bar{\mathbf{X}}-\mu_0\right)-\bar{\mathbf{Y}}\right)$$

$$\widehat{\mu} = \bar{\mathbf{X}}-\tfrac{\sigma_x^2}{n}\mathbf{S}\mathbf{Q}_{11}\left(\mathbf{S}\left(\bar{\mathbf{X}}-\mu_0\right)-\bar{\mathbf{Y}}\right)$$

$$\widehat{\nu} = \bar{\mathbf{Y}}+\tfrac{\sigma_y^2}{n}\mathbf{Q}_{11}\left(\mathbf{S}\left(\bar{\mathbf{X}}-\mu_0\right)-\bar{\mathbf{Y}}\right)$$

$$\begin{pmatrix} \widehat{a}_0 \\ \vdots \\ \widehat{a}_k \end{pmatrix} = -\mathbf{Q}_{21}\mathbf{B}_1\widehat{\beta}_1$$

$$var\begin{pmatrix} \widehat{a}_0 \\ \vdots \\ \widehat{a}_k \end{pmatrix} = var(\widehat{\widehat{\beta}_2}) = -\mathbf{Q}_{22} = \left(\mathbf{B}_2'\mathbf{A}_1^{-1}\mathbf{B}_2\right)^{-1}.$$

1.3 Estimate of variance matrix

To estimate the parameters of the 1. order, firstly we need to estimate the covariance matrix. For estimating $\left(\sigma_x^2,\sigma_y^2\right)'$ MINQUE procedure is proposed (see [4, page 99]). Before use of MINQUE procedure, it is necessary to convert the model with constrains to a model without constrains (see [4, page 130]).
Firstly we convert the model $Y \sim (\mathbf{X}\beta_1, \Sigma)$ with contstrains $\mathbf{b}+\mathbf{B}_1\beta_1+\mathbf{B}_2\beta_2=\mathbf{0}$ to the model $\mathbf{Y}-\mathbf{X}\beta_{1,0} \sim (\mathbf{X}\mathbf{K}_1\lambda, \Sigma)$ without contstrains where $\lambda \in \mathbb{R}^{k_1+k_2-q}$, $\beta_0 = \begin{pmatrix} \beta_{1,0} \\ \beta_{2,0} \end{pmatrix}$ is arbitrary vector for which $\mathbf{b}+\mathbf{B}_1\beta_{1,0}+\mathbf{B}_2\beta_{2,0}=\mathbf{0}$, $\begin{pmatrix} \mathbf{K}_1 \\ \mathbf{K}_2 \end{pmatrix}$ is $(k_1+k_2)\times(k_1+k_2-q)$ a matrix with property $\mathcal{M}\left[\begin{pmatrix} \mathbf{K}_1 \\ \mathbf{K}_2 \end{pmatrix}\right] = Ker(\mathbf{B}_1,\mathbf{B}_2)$ and $\begin{pmatrix} \beta_1 \\ \beta_2 \end{pmatrix} = \begin{pmatrix} \beta_{1,0} \\ \beta_{2,0} \end{pmatrix} + \begin{pmatrix} \mathbf{K}_1 \\ \mathbf{K}_2 \end{pmatrix}\lambda$.

Then we use the MINQUE procedure (see [5,p.100]) to estimate a vector of variance components ϑ in the model wihtout constrains.

$$\hat{\vartheta} = \left[\mathbf{S}_{(M_L\Sigma_0 M_L)+}\right]^{-1}\begin{pmatrix} F_1 \\ \vdots \\ F_p \end{pmatrix},$$

$$\mathbf{var}\left(\hat{\vartheta}|\vartheta_0\right) = 2\left[\mathbf{S}_{(M_L\Sigma_0 M_L)+}\right]^{-1},$$

where

$$F_i = \mathbf{Z}'\left(\mathbf{M_L}\mathbf{\Sigma}_0\mathbf{M_L}\right)^+\mathbf{V}_i\left(\mathbf{M_L}\mathbf{\Sigma}_0\mathbf{M_L}\right)^+\mathbf{Z} \quad \text{for} \quad i = 1, \cdots, p \text{ and } j = 1, \cdots, p.$$

$\left\{S_{(\mathbf{M_L}\mathbf{\Sigma}_0\mathbf{M_L})^+}\right\}_{ij} = Tr\left(\left(\mathbf{M_L}\mathbf{\Sigma}_0\mathbf{M_L}\right)^+\mathbf{V}_i\left(\mathbf{M_L}\mathbf{\Sigma}_0\mathbf{M_L}\right)^+\mathbf{V}_j\right)$,
$\mathbf{L} = \mathbf{X}\mathbf{K}_1$, $\mathbf{Z} = \mathbf{Y} - \mathbf{X}\beta_{1,0}$, $\mathbf{V}_i$ are known positive semidefinite matrices such that $\mathbf{\Sigma}(\vartheta) = \sum_{i=1}^{p}\vartheta_i\mathbf{V}_i$ and $\vartheta_i > 0$ are variance components $i = 1, \cdots, p$ and $\mathbf{\Sigma}_0$ is a variance matrix where unknown variance components are replaced by initial values $\vartheta_0, \mathbf{M_L} = \mathbf{I} - \mathbf{L}(\mathbf{L}'\mathbf{L})^{-1}\mathbf{L}'$.

Now we apply the procedure on model (1). As initial values $\left(\sigma_{x0}^2, \sigma_{y0}^2\right)'$ are proposed $\sigma_{x0}^2 = \frac{1}{nm}\sum_{i=1}^{m}\sum_{j=1}^{n}\left(X_{ij} - \overline{X}_i\right)^2$, $\sigma_{y0}^2 = \frac{1}{nm}\sum_{i=1}^{m}\sum_{j=1}^{n}\left(Y_{ij} - \overline{Y}_i\right)^2$.

$$\mathbf{Z} = \begin{pmatrix} \mathbf{X}^1 - \mu_0 \\ \mathbf{Y}^1 \\ \vdots \\ \mathbf{X}^n - \mu_0 \\ \mathbf{Y}^n \end{pmatrix}, \qquad \mathbf{L} = (\mathbf{1}_n \otimes \mathbf{I}_{2m})\mathbf{K}_1 = \mathbf{1}_n \otimes \mathbf{K}_1,$$

$$\mathbf{V}_1 = \begin{pmatrix} \mathbf{I}_m & \mathbf{0}_m \\ \mathbf{0}_m & \mathbf{0}_m \end{pmatrix}, \mathbf{V}_2 = \begin{pmatrix} \mathbf{0}_m & \mathbf{0}_m \\ \mathbf{0}_m & \mathbf{I}_m \end{pmatrix}$$

$$\begin{pmatrix} \widehat{\sigma_x^2} \\ \widehat{\sigma_y^2} \end{pmatrix} = \left[\mathbf{S}_{(M_L\Sigma_0 M_L)+}\right]^{-1}\begin{pmatrix} F_1 \\ F_2 \end{pmatrix}$$

$$F_1 = \sigma_{x0}^{-4}\left(\sum_{k=1}^{n}\left(\mathbf{X}^\mathbf{k} - \overline{\mathbf{X}}\right)'\left(\mathbf{X}^\mathbf{k} - \overline{\mathbf{X}}\right) + n\left(\overline{\mathbf{X}} - \widehat{\mu}\right)'\left(\overline{\mathbf{X}} - \widehat{\mu}\right)\right),$$

$$F_2 = \sigma_{y0}^{-4}\left(\sum_{k=1}^{n}\left(\mathbf{Y}^\mathbf{k} - \overline{\mathbf{Y}}\right)'\left(\mathbf{Y}^\mathbf{k} - \overline{\mathbf{Y}}\right) + n\left(\overline{\mathbf{Y}} - \widehat{\nu}\right)'\left(\overline{\mathbf{Y}} - \widehat{\nu}\right)\right),$$

$$S_{(\mathbf{M_L}\mathbf{\Sigma}_0\mathbf{M_L})^+} = (n-1)\,\mathbf{S}_{\Sigma_{00}^{-1}} + S_{(\mathbf{M}_{\mathbf{K}_1}\mathbf{\Sigma}_{00}\mathbf{M}_{\mathbf{K}_1})^+}$$

$$\mathbf{S}_{\Sigma_{00}^{-1}} = m\begin{pmatrix} \sigma_{x0}^{-4} & 0 \\ 0 & \sigma_{y0}^{-4} \end{pmatrix},$$

$$\mathbf{S}_{(\mathbf{M}_{\mathbf{K}_1}\mathbf{\Sigma}_{00}\mathbf{M}_{\mathbf{K}_1})^+} = \frac{1}{n^2}\begin{pmatrix} Tr\{\mathbf{S}\mathbf{Q}_{11}\mathbf{S}\mathbf{S}\mathbf{Q}_{11}\mathbf{S}\} & Tr\{\mathbf{Q}_{11}\mathbf{S}\mathbf{S}\mathbf{Q}_{11}\} \\ Tr\{\mathbf{Q}_{11}\mathbf{S}\mathbf{S}\mathbf{Q}_{11}\} & Tr\{\mathbf{Q}_{11}\mathbf{Q}_{11}\} \end{pmatrix}.$$

The covariance matrix of $(\sigma_x^2, \sigma_y^2)'$ is

$$\mathbf{var}\left(\begin{pmatrix} \widehat{\sigma_x^2} \\ \widehat{\sigma_y^2} \end{pmatrix}|\sigma_{x0}^2, \sigma_{y0}^2\right) = 2\left[(n-1)\,\mathbf{S}_{\Sigma_{00}^{-1}} + S_{(\mathbf{M}_{\mathbf{K}_1}\mathbf{\Sigma}_{00}\mathbf{M}_{\mathbf{K}_1})^+}\right]^{-1}.$$

For calculation the confidence region for the vector $(a_0, a_1 \cdots, a_k)'$ the Kenward and Roger method can be used (see [2, page 983-188]).
By [4]

$$\begin{pmatrix} \widehat{a_0} \\ \vdots \\ \widehat{a_k} \end{pmatrix} \approx N \left[\begin{pmatrix} a_0 \\ \vdots \\ a_k \end{pmatrix}, \left(\mathbf{B'}_2\mathbf{A}_1^{-1}\mathbf{B}_2\right)^{-1} \right].$$

2 Aplication of calibration model

In the previous section are estimated parameters of the transformation function. In this section, we derive the confidence region for the transformation function and descrite the range of values, which can be obtained on device B with predetermined probability $1 - \alpha$, if we measure the concrete value on device A.

2.1 Confidence region for transformation curve

Using the Kenward and Roger approximation (see[3, page 985]), we obtain an approximate covariance matrix $\widehat{\Sigma_\mathbf{A}}$.

$$\widehat{\Sigma_A} = \widehat{\Sigma} + 2\widehat{\Sigma}\{\sum_{x,y}\sum_{x,y} V_{ij}\left(\mathbf{W_{ij}} - \mathbf{P_i}\widehat{\Sigma}\mathbf{P_j}\right)\}\widehat{\Sigma},$$

where

$$\widehat{\Sigma} = \left(\mathbf{B'}_2\mathbf{A}_1^{-1}\mathbf{B}_2\right)^{-1}$$

$$\mathbf{V} = \begin{pmatrix} V_{xx} & V_{xy} \\ V_{yx} & V_{yy} \end{pmatrix} = 2\left[(n-1)\,\mathbf{S}_{\Sigma_{00}^{-1}} + S_{(\mathbf{M_{K_1}}\Sigma_{00}\mathbf{M_{K_1}})^+}\right]^{-1},$$

$$\mathbf{P}_x = -\frac{1}{n}\mathbf{B'}_2\mathbf{A}_1^{-1}\mathbf{S}\mathbf{S}\mathbf{A}_1^{-1}\mathbf{B}_2,$$

$$\mathbf{P}_y = -\frac{1}{n}\mathbf{B'}_2\mathbf{A}_1^{-1}\mathbf{A}_1^{-1}\mathbf{B}_2,$$

$$\mathbf{W}_{xx} = \frac{1}{n^2}\mathbf{B'}_2\mathbf{A}_1^{-1}\mathbf{S}\mathbf{S}\mathbf{A}_1^{-1}\mathbf{B}_2(\mathbf{B'}_2\mathbf{A}_1^{-1}\mathbf{B}_2)^{-1}\mathbf{B'}_2\mathbf{A}_1^{-1}\mathbf{S}\mathbf{S}\mathbf{A}_1^{-1}\mathbf{B}_2,$$

$$\mathbf{W}_{xy} = \frac{1}{n^2}\mathbf{B'}_2\mathbf{A}_1^{-1}\mathbf{S}\mathbf{S}\mathbf{A}_1^{-1}\mathbf{B}_2(\mathbf{B'}_2\mathbf{A}_1^{-1}\mathbf{B}_2)^{-1}\mathbf{B'}_2\mathbf{A}_1^{-1}\mathbf{A}_1^{-1}\mathbf{B}_2,$$

$$\mathbf{W}_{yx} = \frac{1}{n^2}\mathbf{B'}_2\mathbf{A}_1^{-1}\mathbf{A}_1^{-1}\mathbf{B}_2(\mathbf{B'}_2\mathbf{A}_1^{-1}\mathbf{B}_2)^{-1}\mathbf{B'}_2\mathbf{A}_1^{-1}\mathbf{S}\mathbf{S}\mathbf{A}_1^{-1}\mathbf{B}_2,$$

$$\mathbf{W}_{yy} = \frac{1}{n^2}\mathbf{B'}_2\mathbf{A}_1^{-1}\mathbf{A}_1^{-1}\mathbf{B}_2(\mathbf{B'}_2\mathbf{A}_1^{-1}\mathbf{B}_2)^{-1}\mathbf{B'}_2\mathbf{A}_1^{-1}\mathbf{A}_1^{-1}\mathbf{B}_2,$$

According to [2, page 986] the approximative $(1-\alpha)$-confidence region for the vector $(a_0, \cdots, a_k)'$ is:

$$C^2_{(1-\alpha)} = \left\{ \begin{pmatrix} a_0 \\ \vdots \\ a_k \end{pmatrix} : \begin{pmatrix} \widehat{a_0} - a_0 \\ \vdots \\ \widehat{a_k} - a_k \end{pmatrix}' \widehat{\Sigma_A}^{-1} \begin{pmatrix} \widehat{a_0} - a_0 \\ \vdots \\ \widehat{a_k} - a_k \end{pmatrix} \leq \frac{(k+1) \cdot F_{k+1,v}(1-\alpha)}{\lambda} \right\},$$

where $F_{k+1,v}$ is the $(1-\alpha)$ quantile of Fisher-Schnedecor distribution with $k+1$ and v degrees of freedom (we obtain v and λ by the use of the Kenward and Roger method, see [1]). If we suppose, that the errorless measurement by the device A is x, we would like obtain the $(1-\alpha)$-confidence region for $\nu = a_0 + a_1 x + \cdots + a_k x^k$. We denote $\mathbf{L}_x = \left(1, x, \cdots, x^k\right)'$. According to [2] we obtain:

$$A_1 = \sum_{x,y}\sum_{x,y} V_{ij} Tr\{\mathbf{\Gamma}\widehat{\Sigma}\mathbf{P}_i\widehat{\Sigma}\} Tr\{\mathbf{\Gamma}\widehat{\Sigma}\mathbf{P}_j\widehat{\Sigma}\},$$

$$A_2 = \sum_{x,y}\sum_{x,y} V_{ij} Tr\{\mathbf{\Gamma}\widehat{\Sigma}\mathbf{P}_i\widehat{\Sigma}\mathbf{\Gamma}\widehat{\Sigma}\mathbf{P}_j\widehat{\Sigma}\},$$

$$\mathbf{\Gamma} = \mathbf{L}_x(\mathbf{L}_x'\hat{\Sigma}\mathbf{L}_x)^{-1}\mathbf{L}_x',$$

$$g = \frac{2A_1 - 5A_2}{3A_2},$$

$$B = \frac{1}{2}(A_1 + 6A_2),$$

$$c_1 = \frac{2A_1 - 5A_2}{25A_2 - 4A_1},$$

$$c_2 = \frac{-2A_1 + 8A_2}{25A_2 - 4A_1},$$

$$c_3 = \frac{-2A_1 + 14A_2}{25A_2 - 4A_1},$$

$$E^* = (1 - A_2)^{-1},$$

$$V^* = 2\left(\frac{1 + c_1 B}{(1 - c_2 B)^2 (1 - c_3 B)}\right),$$

$$\rho = \frac{V^*}{2E^{*2}},$$

$$v = 4 + \frac{3}{\rho - 1},$$

$$\lambda = \frac{v}{E^*(v-2)}.$$

$$P\left\{\widehat{a_0} + \widehat{a_1}x + \cdots + \widehat{a_k}x^k - \sqrt{\frac{F_{1,v}(1-\alpha)\cdot\left(\mathbf{L}'_x\widehat{\boldsymbol{\Sigma}_A}\mathbf{L}_x\right)}{\lambda}} \le a_0 + a_1x + \cdots + a_kx^k \le \right. \quad (2)$$

$$\left. \le \widehat{a_0} + \widehat{a_1}x + \cdots + \widehat{a_k}x^k + \sqrt{\frac{F_{1,v}(1-\alpha)\cdot\left(\mathbf{L}'_x\widehat{\boldsymbol{\Sigma}_A}\mathbf{L}_x\right)}{\lambda}}\right\} = 1 - \alpha.$$

If we compute this confidence interval for all x in a given range we get a confidence region around the calibration curve $\widehat{a_0} + \widehat{a_1}x + \cdots + \widehat{a_k}x^k$.

In case that the true value measured using device A is μ_x, we calculate the confidence region for $\nu_x = a_0 + a_1x + \cdots + a_kx^k$. Let x be the measured value i.e. the realization of random variable $X \sim N(\mu_x, \sigma_x^2)$.

According to [3, p.72-73]

$$\frac{X - \mu_x}{s} \sim t_u$$

where t_u is a student distribution with u degrees of freedom, s^2 is an estimator of σ_x^2 independent of X and $\frac{u \cdot s^2}{\sigma_x^2} \sim \chi_u^2$. It means $D(s^2) = \frac{2\sigma_x^4}{u}$. If we replace s^2 by estimator $\widehat{\sigma_x^2}$, then $u = \frac{2\widehat{\sigma^4}}{D(\widehat{\sigma_x^2})} = \frac{2\widehat{\sigma_x^4}}{\{var\left(\begin{pmatrix}\widehat{\sigma_x^2}\\ \widehat{\sigma_y^2}\end{pmatrix} | \sigma_{x0}^2, \sigma_{y0}^2\right)\}_{1,1}}$.

The $(1-\gamma)$ confidence interval for μ_x is

$$P\left\{X - \widehat{\sigma_x}t_u\left(1 - \frac{\gamma}{2}\right) \le \mu_x \le X + \widehat{\sigma_x}t_u\left(1 - \frac{\gamma}{2}\right)\right\} = 1 - \gamma.$$

Where $t_u\left(1 - \frac{\gamma}{2}\right)$ is $\left(1 - \frac{\gamma}{2}\right)$ quantile of the student t distribution with u degrees of freedom. Returning to measuring by device B, we calculate the confidence interval for $\nu_x = a_0 + a_1\mu_x + \cdots + a_k\mu^k$ for the measured x. We denote

$$\mathbf{x_1} = \begin{pmatrix} 1 \\ x - \widehat{\sigma_x} t_u \left(1 - \frac{\gamma}{2}\right) \\ \vdots \\ \left(x - \widehat{\sigma_x} t_u \left(1 - \frac{\gamma}{2}\right)\right)^k \end{pmatrix},$$

$$\mathbf{x_2} = \begin{pmatrix} 1 \\ x + \widehat{\sigma_x} t_u \left(1 - \frac{\gamma}{2}\right) \\ \vdots \\ \left(x + \widehat{\sigma_x} t_u \left(1 - \frac{\gamma}{2}\right)\right)^k \end{pmatrix}.$$

Utilizing the Bonferroni's inequality (see [6, page 24]), supposing the function $\widehat{a_0} + \widehat{a_1}\mu_x + \cdots + \widehat{a_k}\mu_x^k$ is increasing, we obtain at least $(1 - \alpha - \gamma)$ confidence interval for $\widehat{a_0} + \widehat{a_1}\mu_x + \cdots + \widehat{a_k}\mu_x^k$:

$$P\left\{\mathbf{x_1}'\widehat{\mathbf{\Theta}} - \sqrt{\frac{\mathbf{x_1}'\widehat{\mathbf{\Sigma}_A}\mathbf{x_1} \cdot F_{1,u}(1-\alpha)}{\lambda}} < a_0 + a_1\mu_x + \cdots + a_k{\mu_x}^k < \right. \tag{3}$$
$$\left. < \mathbf{x_2}'\widehat{\mathbf{\Theta}} + \sqrt{\frac{\mathbf{x_2}'\widehat{\mathbf{\Sigma}_A}\mathbf{x_2} \cdot F_{1,u}(1-\alpha)}{\lambda}}\right\} = 1 - \alpha - \gamma,$$

where $\widehat{\mathbf{\Theta}} = (\hat{a}, \hat{a}_1, \cdots, \hat{a}_k)'$. If the function is decreasing, then

$$P\left\{\mathbf{x_2}'\widehat{\mathbf{\Theta}} - \sqrt{\frac{\mathbf{x_2}'\widehat{\mathbf{\Sigma}_A}\mathbf{x_2} \cdot F_{1,u}(1-\alpha)}{\lambda}} < a_0 + a_1\mu_x + \cdots + a_k{\mu_x}^k < \right. \tag{4}$$
$$\left. < \mathbf{x_1}'\widehat{\mathbf{\Theta}} + \sqrt{\frac{\mathbf{x_1}'\widehat{\mathbf{\Sigma}_A}\mathbf{x_1} \cdot F_{1,u}(1-\alpha)}{\lambda}}\right\} = 1 - \alpha - \gamma.$$

References

[1] WIMMER, Gejza, Rudolf PALENČÁR, Viktor WITKOVSKÝ a Stanislav ĎURIŠ. *Vyhodnotenie kalibrácie meradiel, Štatistické metódy pre analýzu neistôt v metrológii.* Slovenská technická univerzita. Bratislava, 2015. ISBN 978-80-227-4374-7.

[2] KENWARD, Michael, G., ROGER, James H. *Small Sample Inference for Fixed Effects from Restricted Maximum Likelihood*, Biometric, Volume 53, Issue 3 (Sep.,1997), 983-997.

[3] ANDĚL, Jiři. *Základy matematické statistiky.* 1. vyd. Praha: Matfyzpress, 2005, 358 s. ISBN 8086732401.

[4] KUBÁČEK, Lubomír a Ludmila KUBÁČKOVÁ. *Statistika a metrologie.* 1. vyd. Olomouc: Univerzita Palackého, 2000, 307 s. ISBN 8024400936.

[5] ANDĚL, Jiři. *Matematická statistika.* 2. vyd. Praha: SNTL - Alfa, 1985.

Current address

Ráboňová Petra, Mgr.
Department of Mathematics and Statistics -Faculty of Science, Masaryk University
Kotlářská 2, 611 37 Brno
Department of Applied Mathematics and Computer Science,
Faculty of Economics and Administration, Masaryk University
Lipová 41a, 602 00 Brno
Tel. number: +420 549 49 6073, e-mail: 324037@mail.muni.cz

Slovak University of Technology in Bratislava
Faculty of Mechanical Engineering
16th Conference on Applied Mathematics **APLIMAT 2017**
Proceedings

NUMERICAL COMPUTING OF SINGULAR INTEGRALS IN INTEGRAL EUQATIONS

RAK Josef (CZ)

Abstract. This paper shows three possible methods to numerically compute a special singular integral in an integral equation of the second kind. The methods are isolation of the singularity, singularity subtraction and substitution. All methods are compared with their error behavior and numerical examples. The main aim is to show, that the subtraction method, which is usable in multidimensional problems, has good error behavior.

Keywords: numerical integration, singular integral, integral equation

Mathematics subject classification: Primary 65R20, 45B05; Secondary 65Z05

1 Introduction

Many phenomena in physics and engineering are described by integral equations with a singular kernel function. There are many methods for computing singular integrals. Three of them are described in this paper. The first is isolation of the singularity, where the singular integrand is changed by a finite function. The next way is singularity subtraction. It is used when the primitive function of the singular part of the integrand can be computed analytically. Then the singular function is weakened in the neighborhood of the singular point. In several cases the coordinate transformation can be used. After transformation we obtain a non-singular integrand which can be easily computed by common numerical integration rules. This method is called substitution.

This paper compares the speed convergence of the methods mentioned above. Let us have an integral equation of the second kind. It has following form

$$y(x) - \lambda \int_a^b x(t)k(x,t)dt = g(x), \tag{1}$$

where $\lambda \neq 0$ and the kernel function $k(x,t)$ is singular if $x = t$. Let the kernel function $k(x,t)$ have a singular factor $|x-t|^{-\alpha}$, where $\gamma \in (0,1)$. It means that the kernel function is of the form

$$k(x,t) = \frac{f(t)}{|x-t|^\alpha}. \tag{2}$$

For future use, let us sign the singular factor as

$$s(x,t) = \frac{1}{|x-t|^\alpha}. \tag{3}$$

Without the loss of generality we can assume that $a = 0$ and $b = 1$. Otherwise the coordinate transformation can be made. So the singular integral to be computed is of the form

$$I(x) = \int_0^1 \frac{f(t)}{|x-t|^\alpha} dt. \tag{4}$$

Without the loss of generality we can also assume that $x \in [0,1]$. If not the integral is not singular and can be easily calculated.

2 Isolation of the singularity

Let us have a numerical integration rule of the form:

$$\int_a^1 v(x)dx \approx \sum_{j=1}^{n} \omega_j v(x_j) \tag{5}$$

where $x_j \in D$ are called the node points and ω_j weights. Since the kernel function is singular when $x = t$ the integration rule cannot be applied directly.

Let us approximate the singular factor of the kernel function $s(x,t)$ by a finite function $s_\varepsilon(x,t)$, which coincide with $s(x,t)$ outside the certain neighborhood of $x = t$

$$s(x,t) \approx s_\varepsilon(x,t) = \begin{cases} \frac{1}{|x-t|^\alpha}, |x-t| \geq \varepsilon \\ \frac{1}{\varepsilon^\alpha}, |x-t| < \varepsilon. \end{cases} \tag{6}$$

Then the common numerical integration rule (5) can be used and integral (4) can be approximated by

$$I(x) \approx I_1(x) = \sum_{j=1}^{n} \omega_j f(x_j) s_\varepsilon(x, x_j). \tag{7}$$

3 Singularity subtraction

Another way is to weaken the singularity by following steps. First let us rewrite (2) into the form as in [1].

$$k(x,t) = \frac{f(t)}{|x-t|^\alpha} = \frac{f(t)-f(x)}{|x-t|^\alpha} + \frac{f(x)}{|x-t|^\alpha} = s(x,t)[f(t)-f(x)] + \frac{f(x)}{|x-t|^\alpha} \tag{8}$$

By changing $s_\varepsilon(x,t)$ and $s(x,t)$ in the first element of (8) we get an approximation of (4):

$$I(x) \approx I_2(x) = \sum_{j=1}^{n} \omega_j [f(x_j) - f(x)] s_\varepsilon(x, x_j) + f(x) \int_0^1 \frac{1}{|x-t|^\alpha} dt. \tag{9}$$

Also here the common numerical integration rule can be used. The integral in (9) is computed analytically or by special integration rule.

4 Substitution

The final method described in this paper is substitution. It is based on the transformation of the coordinates (see [2]). We assumed that $x > 0$. So we can split the integral $I(x)$ into a sum of two integrals

$$I(x) = \int_0^1 \frac{f(t)}{|x-t|^\alpha} dt = \int_0^x \frac{f(t)}{(x-t)^\alpha} + \int_x^1 \frac{f(t)}{(t-x)^\alpha} \tag{10}$$

Using substitution $u = x - t$ the left integral in (10) can be written as

$$\int_0^x \frac{f(t)}{(x-t)^\alpha} = \int_0^x \frac{f(x-u)}{u^\alpha}. \tag{11}$$

Using substitution $u = t - x$ the right integral in (10) can be written as

$$\int_0^x \frac{f(t)}{(x-t)^\alpha} = \int_0^{1-x} \frac{f(x+u)}{u^\alpha}. \tag{12}$$

Finally using substitution $\tau = u^{1-\alpha}$ the (11) can be written as a non-singular integral

$$\int_0^x \frac{f(x+u)}{u^\alpha} = \frac{1}{1-\alpha} \int_0^{x^{1-\alpha}} f\left(x - \tau^{\frac{1}{1-\alpha}}\right) d\tau \tag{13}$$

and the (12) can be written as a non-singular integral

$$\int_0^{1-x} \frac{f(x-u)}{u^\alpha} = \frac{1}{1-\alpha} \int_0^{(1-x)^{1-\alpha}} f\left(x + \tau^{\frac{1}{1-\alpha}}\right) d\tau. \tag{14}$$

From (11), (12), (13) and (14) the integral $I(x)$ have the non-singular form

$$I(x) = \frac{1}{1-\alpha} \left[\int_0^{x^{1-\alpha}} f\left(x - \tau^{\frac{1}{1-\alpha}}\right) d\tau + \int_0^{(1-x)^{1-\alpha}} f\left(x + \tau^{\frac{1}{1-\alpha}}\right) d\tau. \right] \tag{15}$$

and common numerical integration rule can be used.

5 Numerical tests

In this section let us test all methods described in previous chapters. The results were calculated by Maple [3] software.

Let us choose $\alpha = \frac{1}{2}$. It is the most common singular factor because it is the Euclidean distance. We will compare methods from previous chapters for functions $f(x) = e^x$ and $f(x) = \sqrt{x}$ at points $x = 0.1$ and $x = 1$.

Let us start with the first method - isolation of the singularity. Since the integrand is only a continuous function we can not expect a good error behavior and it is not necessary to use too exact numerical integration rule. Here a compound mid-point rule which has an error $\mathcal{O}(n^{-2})$ was used. Since the kernel function is Hölder continuous with constant $\frac{1}{2}$, the expected error is $\mathcal{O}(\sqrt{n}^{-1})$. The columns labeled "ratio" in following tables give the ratio of successive errors. In this case the value in the column ratio is expected to be $\sqrt{2} \approx 1.41$.

The following tables shows results of $I_1(x)$ (isolation of the singularity) at points $x = 0.1$ and $x = 1$ when $f(x) = e^x$ and $f(x) = \sqrt{x}$.

n	$I(0.1) = 3.616031448$			$I(1) = 4.060156938$		
	$I_1(0.1)$	error	ratio	$I_1(1)$	error	ratio
10	3.200041237	0.415990211	-	3.535650417	0.524506521	-
20	3.318869286	0.297162162	1.399876109	3.690801962	0.369354976	1.420060795
40	3.405087646	0.210943802	1.408726681	3.799559688	0.260597250	1.417340267
80	3.466661994	0.149369454	1.412228514	3.876099135	0.184057803	1.415844619
160	3.510358575	0.105672873	1.413508025	3.930085526	0.130071412	1.415051933
320	3.541296260	0.074735188	1.413964102	3.968210323	0.091946615	1.414640572
640	3.563182398	0.052849050	1.414125476	3.995150799	0.065006139	1.414429720
1280	3.578660684	0.037370764	1.414181685	4.014194202	0.045962736	1.414322659

Tab. 1. Isolation of the singularity, $f(x) = e^x$

n	$I(0.1) = 1.287607577$			$I(1) = 1.570796327$		
	$I_1(0.1)$	error	ratio	$I_1(1)$	error	ratio
10	1.172875273	0.114732304	-	1.380492123	0.190304204	-
20	1.204268197	0.083339380	1.376687756	1.435880646	0.134915681	1.410541774
40	1.227891017	0.059716560	1.395582398	1.475274400	0.095521927	1.412405353
80	1.245106342	0.042501235	1.405054700	1.503209257	0.067587070	1.413316586
160	1.257458265	0.030149312	1.409691704	1.522989950	0.047806377	1.413766829
320	1.266254910	0.021352667	1.411969381	1.536986787	0.033809540	1.413990755
640	1.272497020	0.015110557	1.413095957	1.546887491	0.023908836	1.414102301
1280	1.276918589	0.010688988	1.413656466	1.553889562	0,016906765	1.414157942

Tab. 2. Isolation of the singularity, $f(x) = \sqrt{x}$

We can see that the error was, as it was expected. However the convergence is really slow, so this method is not usable. To get accuracy 0.01 more than 1000 node points are needed.

Now let us test the second method. If the function $f(x)$ is Lipschitz continuous, the function $[f(t) - f(x)]|x - t|^{-\alpha}$ is a Hölder continuous function with constant $\frac{3}{2}$. So the expected error for $f(x) = e^x$ is $\mathcal{O}(\sqrt{n^{-3}})$ and the value in the column ratio is expected to be $\sqrt{8} \approx 2.83$. If the function is only continuous the expected error is $\mathcal{O}(\sqrt{n-1})$. So for function $f(x) = \sqrt{x}$ the value in the column ratio is expected to be $\sqrt{2} \approx 1.42$. Also here the compound midpoint integration rule was used. The improper integral was (9) computed analytically by Maple [3].

The following tables show results of $I_2(x)$ (singularity subtraction) at points $x = 0.1$ and $x = 1$ when $f(x) = e^x$ and $f(x) = \sqrt{x}$.

n	$I(0.1) = 3.616031448$ $I_2(0.1)$	error	ratio	$I(1) = 4.060156938$ $I_2(1)$	error	ratio
10	3.615899379	0.000132069	-	4.055053294	0.005103644	-
20	3.615996754	0.000034694	3.806681271	4.058333725	0.001823213	2.799258233
40	3.616022212	0.000009236	3.756388047	4.059508583	0.000648355	2.812059751
80	3.616029020	0.000002428	3.803953871	4.059926936	0.000230002	2.818910270
160	3.616030819	0.000000629	3.860095389	4.060075457	0.000081481	2.822768498
320	3.616031287	0.000000161	3.906832299	4.060128093	0.000028845	2.824787658
640	3.616031408	0.000000040	4.024999980	4.060146733	0.000010205	2.826555610
1280	3.616031438	0.000000010	3.999999867	4.060153324	0.000003614	2.823741007

Tab. 3. Singularity subtraction, $f(x) = e^x$

n	$I(0.1) = 1.287607577$ $I_2(0.1)$	error	ratio	$I(1) = 1.570796327$ $I_2(1)$	error	ratio
10	1.291866724	0.004259147	-	1.571569763	0.000773436	-
20	1.289286680	0.001679103	2.536560890	1.571088026	0.000291699	2.651486635
40	1.288246720	0.000639143	2.627116310	1.570904253	0.000107926	2.702768564
80	1.287845433	0.000237856	2.687100599	1.570835726	0.000039399	2.739308104
160	1.287694760	0.000087183	2.728238303	1.570810574	0.000014247	2.765424300
320	1.287639197	0.000031620	2.757210626	1.570801445	0.000005118	2.783704572
640	1.287618962	0.000011385	2.777338603	1.570798157	0.000001830	2.796721312
1280	1.287611654	0.000004077	2,792494481	1.570796979	0.000000652	2.806748466

Tab. 4. Singularity subtraction, $f(x) = \sqrt{x}$

We can see that the method has good results that correspond to the theory. In the case of the function $f(x) = \sqrt{x}$ the estimate was too pessimistic. The reason is simple. Numerical integration rules applied to a function which is only continuous, have the modulus of the continuity in the error bound.

The last method was substitution. It transformed a singular integral into non-singular one. Also here, a compound mid-point rule was used. The reason for using such rule was to compare this method with second method - the singularity subtraction. Following tables show results of $I_3(x)$ (substitution) at points $x = 0.1$ and $x = 1$ when $f(x) = e^x$ and $f(x) = \sqrt{x}$. When the function $f(x)$ is $\mathcal{C}^2[0,1]$ the expected error is $\mathcal{O}(n^{-2})$, so the value in the column ratio for function $f(x) = e^x$ should be 4. If the function is only continuous the expected error is $\mathcal{O}(\sqrt{n^{-1}})$. So for the function $f(x) = \sqrt{x}$ the value in the column ratio is expected to be $\sqrt{2} \approx 1.42$.

n	$I(0.1) = 3.616031448$ $I_3(0.1)$	error	ratio	$I(1) = 4.060156938$ $I_3(1)$	error	ratio
10	3.612225691	0.003805757	-	4.061824576	0.001667638	-
20	3.615078189	0.000953259	3.992364089	4.060573664	0.000416726	4.001761349
40	3.615793022	0.000238426	3.998133592	4.060261110	0.000104172	4.000364781
80	3.615971835	0.000059613	3.999563853	4.060182980	0.000026042	4.000153598
160	3.616016537	0.000014911	3.997920997	4.060163442	0.000006504	4.003997540
320	3.616027730	0.000003718	4.010489510	4.060158572	0.000001634	3.980416158
640	3.616030515	0.000000933	3.984994639	4.060157354	0.000000416	3.927884615
1280	3.616031217	0.000000231	4.038961040	4.060157066	0.000000128	3.250000000

Tab. 5. Substitution, $f(x) = e^x$

n	$I(0.1) = 1.287607577$ $I_3(0.1)$	error	ratio	$I(1) = 1.570796327$ $I_3(1)$	error	ratio
10	1.287436948	0.000170629	-	1.576205716	0.005409389	-
20	1.287621592	0.000014015	12.17474134	1.572715294	0.001918967	2.818906735
40	1.287631067	0.000023490	0.596636867	1.571475931	0.000679604	2.823654658
80	1.287620509	0.000012932	1.816424374	1.571036806	0.000240479	2.826043022
160	1.287613302	0.000005725	2.258864629	1.570881385	0.000085058	2.827235533
320	1.287609890	0.000002313	2.475140510	1.570826405	0.000030078	2.827914090
640	1.287608466	0.000000889	2.601799775	1.570806963	0.000010636	2.827942836
1280	1.287607911	0,000000334	2.661676646	1.570800089	0.000003762	2.827219564

Tab. 6. Substitution, $f(x) = \sqrt{x}$

Also here, the theory correspond with the results. The estimate for function $f(x) = \sqrt{x}$ was also too pessimistic. The reason is the same as in case of the previous method. Note that, when putting $f(x) = 1$ this method can be also used for computing the improper integral in (9).

6 Conclusion

We can see that the first method - isolation of the singularity has a very slow convergence and is not usable.

So let us compare only the results of second and the third method in following tables for the same functions $f(x)$. They are showed in tables 7 and 8. We can not say which of the methods is better. But both are usable. If we compare the time needed to compute value in case of 10^5 node points on Intel(R) Core(TM) i3-3220@3.30Ghz computer we need 5.2 seconds for the subtraction method and 5.7 second for the substitution method. So the subtraction method is a little bit faster. Note that some time - circa 0.2 for computing a singular integral analytically was spend by Maple.

The main aim of this paper was comparison of subtraction and substitution method. The subtraction method has a big advantage. It is usable in multidimensional problems because there is no changing of coordinates needed. In the case of the substitution method, it will be a big problem to get analytical formula equivalent to (15). As we saw in one dimensional case the subtraction method has similar error behavior as the substitution method.

n	$I(0.1) = 3.616031448$ $I_2(0.1)$	$I_3(0.1)$	error	ratio	$I(1) = 1.570796327$ $I_2(1)$	$I_3(1)$	error	ratio
10	0.000132069	0.003805757	-	-	0.005103644	0.001667638	-	-
20	0.000034694	0.000953259	3.81	3.99	0.001823213	0.000416726	2.80	4.00
40	0.000009236	0.000238426	3.76	4.00	0.000648355	0.000104172	2.81	4.00
80	0.000002428	0.000059613	3.80	4.00	0.000230002	0.000026042	2.82	4.00
160	0.000000629	0.000014911	3.86	4.00	0.000081481	0.000006504	2.82	4.00
320	0.000000161	0.000003718	3.91	4.01	0.000028845	0.000001634	2.82	3.98
640	0.000000040	0.000000933	4.02	3.98	0.000010205	0.000000416	2.83	3.93
1280	0.000000010	0.000000231	4.00	4.04	0.000003614	0.000000128	2.82	3.25

Tab. 7. Method comparison, $f(x) = e^x$

n	$I(0.1) = 1.287607577$				$I(1) = 4.060156938$			
	$I_2(0.1)$	$I_3(0.1)$	error	ratio	$I_2(1)$	$I_3(1)$	error	ratio
10	0.004259147	0.000170629	-	-	0.000773436	0.005409389	-	-
20	0.001679103	0.000014015	2.54	12.17	0.000291699	0.001918967	2.65	2.82
40	0.000639143	0.000023490	2.63	0.60	0.000107926	0.000679604	2.70	2.82
80	0.000237856	0.000012932	2.69	1.82	0.000039399	0.000240479	2.74	2.83
160	0.000087183	0.000005725	2.73	2.26	0.000014247	0.000085058	2.77	2.83
320	0.000031620	0.000002313	2.76	2.48	0.000005118	0.000030078	2.78	2.83
640	0.000011385	0.000000889	2.78	2.60	0.000001830	0.000010636	2.80	2.83
1280	0.000004077	0.000000334	2.79	2.66	0.000000652	0.000003762	2.81	2.83

Tab. 8. Method comparison, $f(x) = \sqrt{x}$

References

[1] KANTOROVICH L. V. , KRYLOV V. I. Approximate Methods of Higher Analysis. Interscience, New York, 1958,

[2] AMARI S., BORNEMANN J. Efficient Numerical Computation of Singular Integrals with Applications to Electromagnetics, 1958, IEE Trans. on Atennas and Propag, vol 43, no 11, pp-1342-1348, 1995

[3] Maple. http://www.maplesoft.com.

Current Address

Rak Josef, RNDr. Ph.D.
Univerzita Pardubice, Fakulta elektrotechniky a informatiky
Studentská 95, 532 10 Pardubice, Czech Republic
Tel. number: +420 466 036 726, e-mail: josef.rak@upce.cz

Slovak University of Technology in Bratislava
Faculty of Mechanical Engineering
16th Conference on Applied Mathematics **APLIMAT 2017**
Proceedings

APPLICATION OF FUZZY CLUSTERING ON DIRECT MARKETING CASE

RIHOVA Elena (CZ), RIHA David (CZ)

Abstract. There are many marketing campaigns made by banks, nowadays. However, every of those campaigns has just one and only goal. This goal is a success, i.e. the client subscribes the deposit. In this paper we are trying to determinate the success of the campaign with fuzzy clustering help (if the campaign was sucessfull or not). Real world data were collected from a Portugauese direct bank marketing campaign in 2012.

Keywords: fuzzy clustering, indices, marketing

Mathematics Subject Classification: 62H30, 62A86

1 Introduction

Cluster analysis is a multivariate statistical classification method, implying different methods and procedures. Clustering methods can be divided into hard and fuzzy; the latter one provides a more precise picture of the information by clustering objects than hard clustering. But *in practice,* the optimal number of clusters is not known *a priori*, and therefore it is necessary to determine the optimal number of clusters. To solve this problem, the validity indices help us. However, there are many different validity indices to choose from.

Each of these indices has its strengths and weaknesses. In this research, several coefficients for estimating the optimal number of clusters (for fuzzy clustering techniques) will be examined. Those coefficients are: Dunn's coefficient (PC), modified Dunn's coefficient (PC_{mod}), and the Xie-Beni (XB) index. All those indices will be applied on direct marketing case.

2 Fuzzy Sets and Fuzzy *K*-means clustering

The hard clustering methods restrict that each object of the data set belongs to exactly one cluster, which does not happen in real life. Fuzzy set theory proposed by L. Zadeh in 1965 gave an idea of uncertainty of belonging, which was described by a membership function. The concept of fuzzy sets expands to the classical concept the classical concept of a set. One of the fundamental assuming that membership function can take any value in the range [0, 1], and not only the values 0 or 1, like in the concept of classical sets.

The classical set theory is built on the fundamental concept of "set" of which an individual is either a member or not a member. In other words, when one asks the question, is this element a member of that set. The answer is either "yes" or "no." For instance, in probability and statistics, one may ask us a question like, what is the probability of this element being a member of that set. In this case an answer could be like "The probability for this element to be a member of this set is 70%." But the final outcome is still either "it is" or "it is not" a member of this set. It means, that the chance for one to make a correct prediction "it is a member of the set" is 70%, which does not mean that element has 70% membership in the set and in the meantime it possesses 30% non-membership.

That is to say, in the classical set theory, it is not allowed that an element is in a set and not in the set at the same time. There are many real - world application cases cannot be described and handled by the classical set theory. On the contrary, fuzzy set theory accepts partial memberships, and, therefore, in a sense generalizes the classical set theory to some extent.

Fuzzy set theory was proposed by Lofti A. Zadeh in his 1965 paper entitled none other than: *Fuzzy Sets.* According Zadeh, the notion of a fuzzy set provides a convenient object of departure for the construction of a conceptual frame-work which parallels in many respects the framework used in the case of ordinary sets, but is more general than the latter and, potentially, may prove to have a much wider scope of applicability, particularly in the fields of object classification and information processing. Essentially, such a framework provides a natural way of dealing with problems in which the source of imprecision is the absence of sharply defined criteria of class membership rather than the presence of random variables (Zadeh, 1965). To sum up the Zadeh's words we can say, that the sense of vagueness rather than the lack of knowledge about the value of a parameter. Let's begin the discussion about fuzzy sets with several basic definitions.

Let X be a space of objects with generic objects, denoted by x, thus $X = \{x\}$. Zadeh in his work in 1965s. writes, „a fuzzy set U in X characterized by a membership function $f_U(x)$ which associates with each object in X a real number in the interval $\langle 0, 1\rangle$, with the value of $f_U(x)$at x representing „the grade of membership" of x in U. Thus, the nearer the value of $f_U(x)$to unity, the higher the grade of membership of x in U. When U is a set in the ordinary sense of the term, its membership function can take on only two values 0 and 1, with $f_U(x) = 1$or 0 according x does or does not belong to U". Roughly, with the help of fuzzy sets theory we can use in machine learning such definition like „very low", „low", „medium", „high" and „very high", not like in traditional binary machine language, where we have only use „low" and „high". The significance of fuzzy variables is that they facilitate gradual transitions between states and, consequently, possess a natural capability to express and deal with observation and measurement uncertainties. Traditional variables, which we may refer to as crisp variables, do not have this capability.

For instance, in Fig.1.1 there is the The probability, that a client will subscribe a term deposit within a range $\langle T_1, T_2\rangle$, which is characterized as a fuzzy variable and it is contrasted in Fig.1.2 with comparable no fuzzy variable.

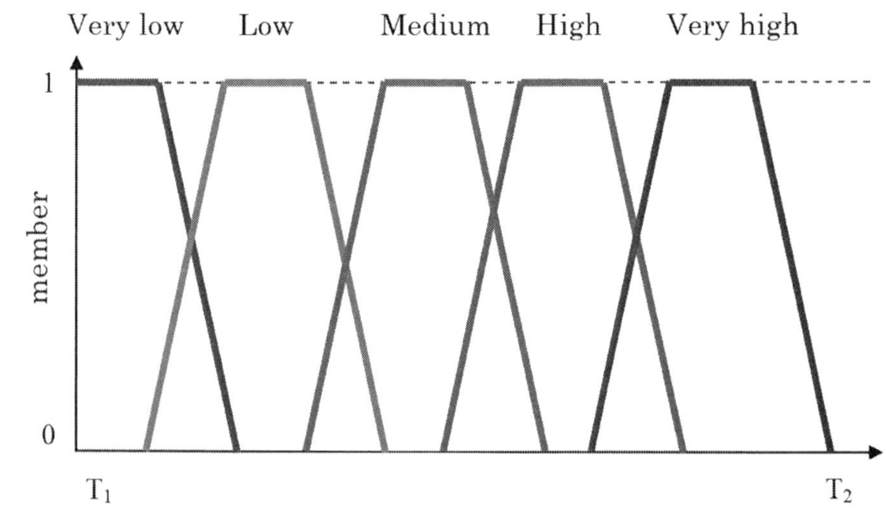

Fig. 2.1. Client will subscribe a term deposit as fuzzy variable

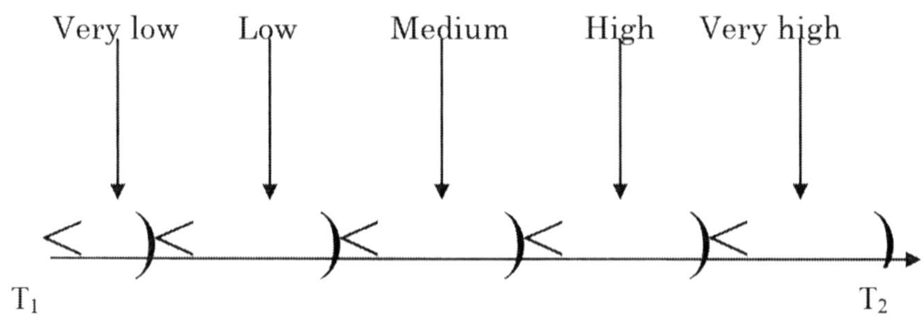

Fig. 2.2. Client will subscribe a term deposit as crisp variable

Several definitions involving fuzzy sets are obvious extensions of the corresponding definitions of ordinary, no fuzzy sets, below are introduced those definitions for fuzzy sets:

1. **Emptiness.** A fuzzy set U is empty if and only if its membership function $f_U(x)$ is identically zero on X, i.e., $f_U(x) = 0$.

2. **Equal.** Two fuzzy sets U_1 and $U_2$are equal, written as $U_1 = U_2$, if and only if their membership functions are identical, i.e., $f_{U1}(x) = f_{U2}(x),\ \forall x \in X$.

3. **Complementation.** The complement of a fuzzy set U, denoted by $\overline{U}$, is defined by:

$$f_{U'}(x) = 1 - f_U(x), \forall x \in X. \tag{2.1}$$

4. **Containment.** A fuzzy set U_1 is said to be contained in a fuzzy set U_2 (or, equivalently, U_1 is a subset of U_2, if and only if $f[U_1(x)] \leq f[U_2(x)]\ \forall\ x \in X$. In symbols:

$$U_1 \subset U_2 \Leftrightarrow f_{U_1} \leq f_{U_2}. \quad (2.2)$$

5. **Union.** The union of two fuzzy sets U_1 and U_2 is again a fuzzy set U_3, written as $U_3 = U_1 \cup U_2$, whose membership function is defined by

$$f_{U_3}(x) = \max\{f_{U_1}(x), f_{U_2}(x)\}, x \in X, \quad (2.3)$$

where $f_{U1}(x)$ and $f_{U2}(x)$, are membership functions of U_1 and U_2, respectively. Let V stand for maximum. Then the union can be represented as:

$$f_{U_3}(x) = f_{U_1}(x) \vee f_{U_2}(x). \quad (2.4)$$

Like the union of ordinary sets, the union $\cup$of fuzzy sets has the associative optionality, i.e., $U_1 \cup (U_2 \cup U_3) = (U_1 \cup U_2) \cup U_3$.It can be shown that the union of fuzzy set U_1 and $U_2$is the smallest fuzzy set containing both $U_1$and U_2.

6. **Intersection.** The intersection of two fuzzy sets $U_1$and U_2 is again a fuzzy set U_4, written as $U_4 = U_1 \cap U_2$, whose membership function is defined as:

$$f_{U_4}(x) = \min[f_{U_1}(x), f_{U_2}(x)], \forall x \in X, \quad (2.5)$$

where $f_{U1}(x)$ and $f_{U2}(x)$, are membership functions of U_1 and U_2, respectively.

„Based on the operations of union, intersection, and complementation, it is easy to extend many of the basic identities that hold for ordinary sets to fuzzy sets, such us the De Morgan laws:

$$(U_1 \cup U_2)' = U_1' \cap U_2', \quad (2.6)$$

$$(U_1 \cap U_2)' = U_1' \cup U_2', \quad (2.7)$$

and the distributive laws:

$$U_3 \cap (U_1 \cup U_2) = (U_3 \cap U_1) \cup (U_3 \cap U_2), \quad (2.8)$$

$$U_3 \cup (U_1 \cap U_2) = (U_3 \cup U_1) \cap (U_3 \cup U_2), \quad (2.9)$$

where U_1, U_2 and U_3 are fuzzy sets.
Let's list in brief the advantages of fuzzy sets compared to others:

1. The ability to operate with fuzzy input data and with the values, which cannot be found exactly,
2. the possibility of formalizing the fuzzy evaluation criteria ("very low", "low", "medium", "high" and "very high"),
3. the possibility the qualitative assessments for input data as well as for output results,

4. the possibility to operate not only the level of data values, but the level of data confidence.

In this regard, has become one of the most advanced techniques, which are

widely used in all the world, beginning with computer science and statistics and ending management and medicine. The emergence and the blossoming of fuzzy set theory takes place in the era of information technology, mainly in fuzzy data mining and fuzzy clustering.

Let's start by considering, what is it fuzzy k-means clustering. We can see some differences in comparison with k-means clustering (hard clustering). Moreover, one should note here: k-means clustering algorithm is a starting object for the fuzzy extensions (Löster, 2012).

In the probabilistic fuzzy k-means algorithm the centers of every cluster are driven apart, it means includes a part of the object's membership values, hence thus leaves less that may attract other cluster centers.

These algorithms are based on objective functions J, which are mathematical criteria that quantify the goodness of cluster models that comprise prototypes and data partition. Objective functions serve as cost functions that have to be minimized to obtain optimal cluster solutions (Řezanková, 2010). Thus, for each of the following cluster models the respective objective function expresses desired optional ties of what should be regarded as "best" results of the cluster algorithm.

This algorithm is based on objective functions J_h, which is a mathematical criteria that quantify the goodness of cluster models that comprise prototypes and data partition. A large family of fuzzy clustering algorithms is based on minimization of the fuzzy k-means functional formulated as (Dunn, 1974; Bezdek, 1981):

$$J = \sum_{i=1}^{n} \sum_{j=1}^{k} u_{ij}^{q} \left\| x_i - k_j \right\|^2 , \tag{2.10}$$

where n is the total number of objects in a given data set and k is the number of clusters, $X = \{x_1, x_2, \ldots, x_n\} \subset R^n$ are the feature data and cluster centers. $U = [u_{ij}]_{kxn}$ is a fuzzy partition matrix composed of the membership grade of object x_i to each cluster h. $|| x_i - C_h||$ is the Euclidean norm between x_i and $C_{h,}$ the weighting exponent q is called fuzzifier which can have influence on the clustering performance of *FKM*. The fuzzifier $q \in (1, +\infty)$, the greater q, the fuzzier the clustering is (Oliveira et al., 2007).

The main advantage of fuzzy k-means clustering (*FKM*) is, it allows gradual memberships of objects to clusters measured as degrees in [0, 1]. This gives the flexibility to express that objects can belong to more than one cluster.

Fuzzy partitioning is carried out through an iterative optimization of the objective function shown above, with the update of membership u_{ij} and the cluster centers k_h (larger membership values indicate higher confidence in the assignment of the object to the cluster). The membership degrees have to be chosen according the following update formula that is independent of the chosen distance measure (Bezdek, 1981; Pedrycz, 2005):

$$u_{ij} = \frac{1}{\sum_{h=1}^{k} \left(\frac{\|x_i - C_h\|}{\|x_i - C_k\|} \right)^{\frac{2}{q-1}}} = \frac{D_{ij}^{-\frac{2}{q-1}}}{\sum_{h=1}^{k} d_{kj}^{-\frac{2}{q-1}}}. \quad (2.11)$$

The above equation clearly shows the relative character of the probabilistic membership degree: it depends not only on the distance of the object x_i to cluster j, but also on the distance between this object and other clusters (Oliveira, 2007).

This iteration will stop when $max_{ij}\ \{|u_{ij}^{(k+1)} - u_{ij}^{(k)}|\} < \varepsilon$, where ε is a termination criterion between 0 and 1, whereas k are the iteration steps.

As Bezdek (Bezdek, 1981) pointed out: in the case of the basic k-means model the cluster center vectors serve as prototypes, while an inner product norm induced metric is applied as distance measure; the derivations of J with the centers:

$$c_h = \frac{\sum_{i=1}^{n} u_{ij}^q x_i}{\sum_{i=1}^{n} u_{ij}^q}. \quad (2.12)$$

Eqs. (3.2) and (3.3) constitute an iterative optimization procedure. The goal is to iteratively improve a sequence of sets of fuzzy clusters until no further improvement in J is possible.

The *FKM* algorithm is executed in the following steps:

***Step* 1**: Given a pre-selected number of cluster k, a chosen value of q, initialize memberships u_{ij} of x_i belonging to cluster h such that

$$\sum_{j=1}^{k} u_{ij} = 1,\ \forall i \in \{1,...,n\}, \quad (2.13)$$

$$\sum_{i=1}^{n} u_{ij} > 0,\ \forall j \in \{1,...,k\}. \quad (2.14)$$

The condition Eq. (2.5) avoids the trivial solution of minimization problem, i.e. $u_{ij} = 0,\ \forall i, j$ (Oliveira, 2007). As Oliveira (Oliveira, 2007) pointed out: the norma-lization constraint Eq.(2.4) leads to a distribution of the weight of each object over the different clusters.

Step 2: Calculate the fuzzy cluster centroid c_h for $h = 1, 2.... , k$ using Eq. (2.3).

Step 3: Employ Eq. (2.2) to update the fuzzy membership u_{ij}.

Step 4: If the improvement in J is less than a certain threshold ε, then halt; otherwise go to step 2.

FKM attempts to find the most characteristic object in each cluster, which can be considered as the "centroid" of the cluster and, then, the grade of membership for each object in the clusters. Such aim is achieved by minimizing the objective function. (Wanga, Zhang, 2007)

Compared with the hard *k*-means clustering it is quite intensive to its initialization and it is not likely to get stuck in an undesired local minimum of its objective function in practice. The membership degrees for a given datum formally resemble the probabilities of its being a member of the corresponding cluster (Oliveira, 2007). Fig. 2.3 shows a probabilistic fuzzy classification of a two-dimensional symmetric data-set with two clusters.

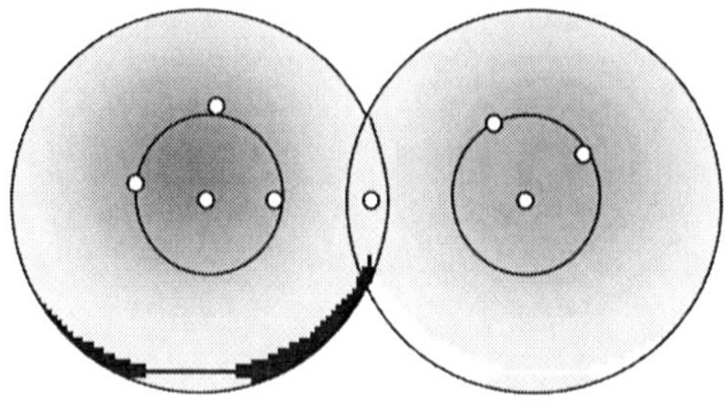

Fig. 2.3. Probabilistic fuzzy classification of a two-dimensional symmetric data-set with two clusters

The blue scale indicates the strength of belonging objects x to clusters k: where the darker shading indicates a high degree of membership for object close to the cluster center c and light shading in otherwise. The centers of clusters can have a degree equal to 1, while objects close to boundaries have the membership degrees smaller than 1.

Objects on class boundaries can be classified as undetermined, because they are equidistant from the both cluster centers.

3 Some cluster validity indices applied on Bank Marketing case

There are many cluster validity indexes that have been proposed in the literature for evaluating fuzzy clustering techniques. All those indices can be divided into three groups:

1) validity indices involving only the membership values,

2) validity indices involving the membership values and the data set,

3) other approaches.

In current research will be considered indices of the first and second groups.

Dunn's index or Partition coefficient (*PC*)

Bezdek attempted to define a performance measure based on minimizing the overall content of pair wise fuzzy intersection in *U*, the partition matrix. He proposed cluster validity index for fuzzy clustering: partition coefficient (*PC*). The index was defined as

$$PC = \frac{1}{n}\sum_{j=1}^{k}\sum_{i=1}^{n} u_{ij}^2 . \qquad (3.1)$$

The PC index indicates the average relative amount of membership sharing done between pairs of fuzzy subsets in U, by combining into a single number, the average contents of pairs of fuzzy algebraic products. The index values range in $\langle 1/k,\ 1\rangle$, where k is the number of clusters. In general, we find an optimal cluster number k^* by solving $\max_{2\le k\le n-1}$ to produce the best clustering performance for the data set X.

Modified Dunn's index (PC_{mod})

The next validity index was proposed by Dave (Dave, 1992) as a modification of the previous one:

$$PC_{\mathrm{mod}}(k) = 1 - \frac{k}{k-1}(1 - PC(k)) . \qquad (3.2)$$

This index can take values $\langle 0,1\rangle$, where k^* is the optimal number of clusters. This cluster number k^* is defined by solving of:

$$\max_{2\le k\le n-1} PC_{\mathrm{mod}}(k) . \qquad (3.3)$$

When the variability in clusters is small, this modified Dunn's coefficient PC_{mod} usually determined the number of clusters correctly (Řezanková, Húsek, 2012).When the cluster variability is greater, the normalized Dunn's coefficient usually achieved its highest value for the highest possible number of clusters. (Řezanková, Húsek, 2012)

Xie and Beni (*XB*) validity index

Those index was proposed in 1991 by Xie and Beni (Xie, Beni, 1991) with $q = 2$ and modified by Pal and Bezdek (Pal, Bezdek, 1995) was defined as:

$$XB = \frac{J_m(u,v)/n}{Sep(v)} = \frac{\sum_{i=1}^{k}\sum_{j=1}^{n} u_{ij}^2 \left\| x_j - x_i \right\|^2}{n.\min_{i,j}\left\| v_i - v_j \right\|^2} \qquad (3.4)$$

Like we can observe, this index includes two components: compactness in the numerator and separation, which is represented in denominator. Small value of compactness is evidence of a good partition, and high value of separation is evidence of a good partition. The optimal number of clusters k^* can be find by solving $\min_{2\le k\le n-1} XB$ to produce the best clustering performance for the data set X.

Unfortunately, this index has tendency to monotonically decrease with increasing number of clusters.

The case study

The data is related with direct marketing campaigns (phone calls) of a Portuguese banking institution. The classification goal is to predict if the client will subscribe a term deposit. Here are two groups of clients: which are subscribe a term deposit or do not. This data set contains 4119 examples with 7 numerical variables. The attribute information is following:

1. The age of the client;
2. Balance: average yearly balance, in euros;
3. Last contact day of the month;
4. Last contact duration;
5. Number of contacts performed during this campaign and for this client;
6. Number of days that passed by after the client was last contacted from a previous campaign;
7. Number of contacts performed before this campaign and for this client.

To find validity indices, an extensive comparison of some of the abovementioned indices was conducted with the probabilistic fuzzy *k*-means algorithm on an artificial number and a number of well-known data sets. In all experiments the distance function used was Euclidian distance. Those indices were calculated: Dunn's index, modified Dunn's index, the Xie-Beni index.

The name of the coefficient / The number of clusters	PC	PC_{mod}	XB
2	**0.914568**	**0.908977**	**0.897774**
3	0.897447	0.887474	0.887423
4	0.887456	0.860101	0.871211
5	0.871236	0.854931	0.866674
6	0.651278	0.671236	0.856554
7	0.613213	0.600092	0.831214

Tab. 1. The dataset Bank Marketing (source: author's calculation)

All those indices work correct, and estimate the optimal number of clusters right (the optimal number of clusters 2). We can assume, that in the case with small number of clusters all indices work correct, such assume was made in the research and publicized in dissertation thesis Evaluation of Fuzzy Clustering Results (Rihova, 2016). High variability does not affect the results, just only in case the small number of clusters. And finally, the number of variables does not affect the results (Bezdek, 1995).

4 Conclusion

The validation of clustering structures is the most difficult and frustrating part of cluster analysis. That's why the issue of the definition of the indices, which would be good for data with large variability and a big number of clusters, is still under discussion.

We plan to study those indices in data sets with various characteristics to explain their behavior.

References

[1] Bezdek, J.C. (1981). *Pattern Recognition With Fuzzy Objective Function Algorithms.* New York: Plenium Press.

[2] Dave, R.N., Bhaswan, K. (1992) Adaptive fuzzy *c*-shells clustering and detection of ellipses. *IEEE Transactions of Neural Networks,* vol.3, no.5, p: 643–662.

[3] Dunn, J.C. (1974) A fuzzy relative of the ISODATA process and its use in detecting compact, well-separated clusters, Journal of Cybernetics, vol.3, no.3, p.: 32–57.

[4] Löster, T. (2012). Kritéria pro hodnocení výsledků shlukování se známým zařazením do skupin založená na konfuzní matici, *Forum Statisticum Slovacum*, sv. 8, No. 7, 85—89 p.

[5] Oliveira, J.W. (2007) *Advances in Fuzzy Clustering and its Applications*. London: , John Wiley & Sons, Ltd.

[6] Pal N. R., Bezdek J. C. (1995). On cluster validity for fuzzy C-means model, *IEEE Trans. Fuzzy Systems* 3 (3), 370–379.

[7] Pedrycz, W. (2005). *Knowelarge – Based Clustering: From Data to Information Granules.* New Yoork: J.Wiley&Son Inc. Holboken.

[8] Řezanková H., Húsek D. (2012). Fuzzy Clustering: determining the number of clusters, *Computational Aspects of Social Networks* (CASoN).

[9] Řezanková H., Húsek. D., LÖSTER, R.(2010): Clustering with Mixed TypeVariables and Determination of Cluster Numbers, In: *CNAM and INRIA*, Paris, p.: 1525-1532.

[10] Říhová, E. (2016) Evaluating of fuzzy clustering results

[11] Wang W., Zhang Y. (2007). On fuzzy cluster validity indices, *Fuzzy Sets and Systems, No.* 158.

[12] Xie X. L., Beni G. (1991). A validity measure for fuzzy clustering, *IEEE Trans. Pattern Anal. Mach. Intell.* 13, 841–847.

[13] Zadeh L. A. (1965). Fuzzy Sets, *Information and Control*, No.

The internet resource: the database of the real datasets

http://archive.ics.uci.edu/ml/datasets.html?format=&task=cla&att=num&area=&numAtt=&numIns =&type=&sort=nameUp&view=table –

Current address

Říhová Elena, Mgr. Ph.D.
University College of Business in Prague
Spálená 76/14, 110 00 Prague 1, Czech Republic
E-mail: rihova.elena@gmail.com

Říha David, Ing. MBA, Ph.D.
University of Economics, Prague
W. Churchill Sq. 4, 130 67 Prague 3, Czech Republic
E-mail: david.riha@vse.cz

Slovak University of Technology in Bratislava
Faculty of Mechanical Engineering
16th Conference on Applied Mathematics **APLIMAT 2017**
Proceedings

DESCRIPTION OF MEMBRANE ELEMENTS FOR PIEZOELECTRIC MATERIALS

RÓKOVÁ Simona (SK), HORVÁT František (SK), ČEKAN Michal (SK), HUČKO Branislav (SK), KOVÁČ Tomáš (SK)

Abstract. Most FEA software does not offer elements which incorporate the properties of thin piezoelectric membranes within their libraries. The need to implement piezoelectric materials into mechanical designs has become of increasing importance. This contribution formulates a thin membrane element based on shape function and material properties of AlGaN and PZT materials and compares them to generic Shell 181 finite elements. The element formulation shows promising results for the use of more suitable elements for the description of thin membranes in finite element analysis.

Keywords: FEA, shell elements, piezoelectric material, material properties

Mathematics Subject Classification: Primary 74S05; Secondary 74G06

1 Introduction

There is currently an increasing interest in the use of piezoelectric materials within advanced structures in an effort to transform them into "smart" adaptive structures. Piezoelectric materials can be used as sensors that can measure physical quantities such as strain, or actuators which respond to an applied voltage to cause deformation or damping [1][6][7]. Real objects such as storage tanks, roofing, and aircraft fuselages often can be represented by shells. However in very thin structures (membranes) such as those often found in biomechanics, this may not be the best formulation. This article deals with the formulation of a dedicated membrane element description in terms of the finite element method that incorporates the properties of some common piezoelectric materials such as lead zirconate titanate (PZT) and aluminum gallium nitride AlN and gallium nitride GaN.

2 Characteristics of piezoelectric materials

Piezoelectricity is the property of some materials to develop an electric charge on their surface when mechanical stress is applied to them. The direction of positive polarization is usually chosen to be identical with the *Z*-axis of a rectangular system of crystallographic axes *X*, *Y*, *Z*. The directions of *X*, *Y*, *Z* are represented by *1*, *2*, *3* respectively and the shear about the axes by *4*, *5* and *6*, respectively, see fig. 1 [2].

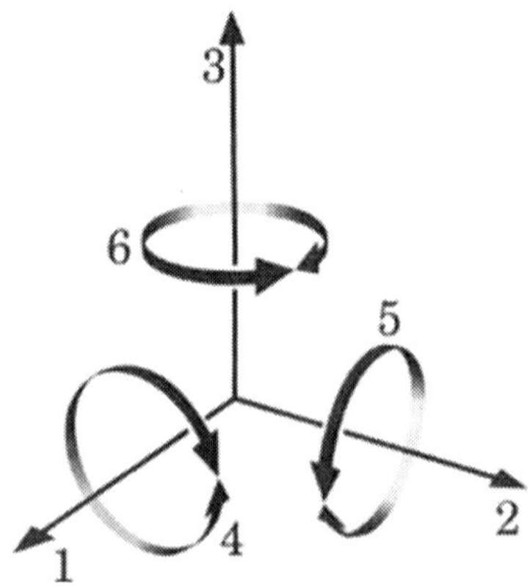

Fig. 1. Axes and directions of deformations

In general, poled piezoelectric materials are transversely isotropic.[1] The plane of isotropy is defined as the 12-plane (or the *xy*-plane). The piezoelectric material therefore exhibits symmetry about the 3-axis (the *z*-axis), which is the poling axis of the material. The field variables are the stress components (T), strain components (S), electric field components (E), and the electric displacement components (D^e). The standard form of the piezoelectric constitutive equations can be given in four different forms by taking either two of the four field variables as the independent variables.. [2]

$$S_{ij} = s^E_{ijkl}T_{kl} + d_{kij}E_k \tag{2.1}$$

$$D^e_i = d_{ikl}T_{kl} + \varepsilon^T_{ik}E_k \tag{2.2}$$

(2.1) and (2.2) are the preferred form of the piezoelectric constitutive equations, where the superscripts E and T denote that the respective constants are evaluated with constant electric field and stress, respectively. Utilizing the conventions set in fig. 1, expressions (2.1) and (2.2) can be expanded into matrix form which are not shown for the purpose of brevity.

3 Formulation of the finite element equations

The following form of the shape function is considered:

$$\begin{aligned} N_1 &= \frac{1}{4}(1-\xi)(1-\eta) \\ N_2 &= \frac{1}{4}(1+\xi)(1-\eta) \\ N_3 &= \frac{1}{4}(1+\xi)(1+\eta) \\ N_4 &= \frac{1}{4}(1-\xi)(1+\eta) \end{aligned} \tag{3.1}$$

[1]Special classes of orthotropic materials are those that have the same properties in one plane (e.g. the *x-y* plane) and different properties in the direction normal to this plane (the *z*-axis). Such materials are called transverse isotropic, and they are described by 5 independent elastic constants, instead of 9 for fully orthotropic. [20]

Shape functions are not direct functions of x and y in the global coordinates of the system. The coordinates ξ and η refer to the natural coordinates of the element. To be able to do strain and stress calculations, the derivatives with respect to the global coordinates are necessary [4].

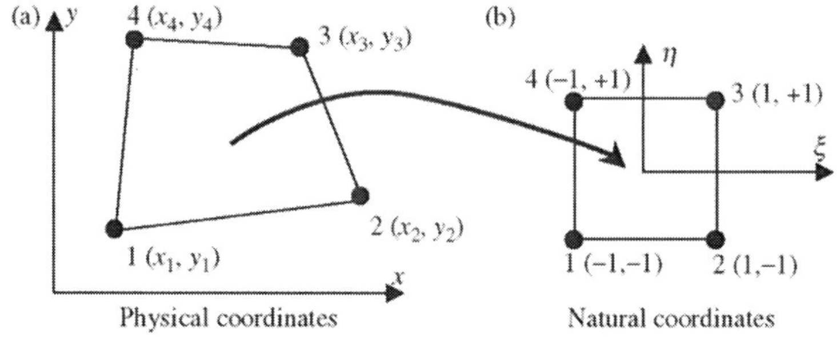

Fig. 2. Q4 element in global and natural coordinates [4]

To obtain the relationship between Cartesian and natural coordinates, the Jacobian operator is used.

$$\begin{bmatrix} dx \\ dy \end{bmatrix} = \begin{bmatrix} \frac{\partial x}{\partial \xi} & \frac{\partial x}{\partial \eta} \\ \frac{\partial y}{\partial \xi} & \frac{\partial y}{\partial \eta} \end{bmatrix} \begin{bmatrix} d\xi \\ d\eta \end{bmatrix} = J^T \begin{bmatrix} d\xi \\ d\eta \end{bmatrix}$$

$$\begin{bmatrix} d\xi \\ d\eta \end{bmatrix} = \begin{bmatrix} \frac{\partial \xi}{\partial x} & \frac{\partial \xi}{\partial y} \\ \frac{\partial \eta}{\partial x} & \frac{\partial \eta}{\partial y} \end{bmatrix} \begin{bmatrix} dx \\ dy \end{bmatrix} = J^{-T} \begin{bmatrix} dx \\ dy \end{bmatrix} \tag{3.2}$$

To compute the entries of the J at any quadrilateral location, geometry is considered:

$$x = \sum_{i=1}^{n} x_i N_i^e$$

$$y = \sum_{i=1}^{n} y_i N_i^e \tag{3.3}$$

where n is the number of nodes and e represents the element. After differentiation with respect to the quadrilateral coordinates:

$$\frac{\partial x}{\partial \xi} = \sum_{i=1}^{n} x_i \frac{\partial N_i^e}{\partial \xi} \tag{3.4}$$

The same procedure as in Equation (3.4) is made for partial derivatives of x with respect to η and partial derivatives of y with respect to ξ and η. In matrix for, the Jacobian matrix is written:

$$J = \begin{bmatrix} \frac{\partial N_1^e}{\partial \xi} & \cdots & \frac{\partial N_n^e}{\partial \xi} \\ \frac{\partial N_1^e}{\partial \eta} & \cdots & \frac{\partial N_n^e}{\partial \eta} \end{bmatrix} \begin{bmatrix} x_1 & y_1 \\ \cdots & \cdots \\ x_n & y_n \end{bmatrix} \tag{3.5}$$

Partials of the shape functions with respect to the global coordinates are present in the strain-displacement matrix B that appears in the computation of the element stiffness matrix. The calculation of those partials is done by computing J [4].

To solve Equation (3.5), the following must first be solved:

$$\begin{bmatrix} \frac{\partial N_i^e}{\partial x} \\ \frac{\partial N_i^e}{\partial y} \end{bmatrix} = \begin{bmatrix} \frac{\partial \xi}{\partial x} & \frac{\partial \eta}{\partial x} \\ \frac{\partial \xi}{\partial y} & \frac{\partial \eta}{\partial y} \end{bmatrix} \begin{bmatrix} \frac{\partial N_i^e}{\partial \xi} \\ \frac{\partial N_i^e}{\partial \eta} \end{bmatrix} \tag{3.6}$$

the Jacobian relates the natural and global coordinates. In heavily distorted elements the Jacobian is inverse, indicating self-intersecting geometry and should not be allowed [4]. After this calculation, the stiffness matrix can be obtained by:

$$K = \int B^T C B dV \tag{3.7}$$

Because the strain-displacement matrix B depends on the element natural coordinates, the element volume is given by:

$$dV = h \; detJ \; d\xi \; d\eta \tag{3.8}$$

Where $detJ$ is the determinant of the Jacobian matrix and h is the thickness of the shell.
It is not possible in many cases to calculate the integral; therefore the numerical computation by Gauss quadrature takes place [4].

Let F be substitute into the right side of Equation (3.9). Then:

$$\begin{aligned} F &= B^T C B \; detJ \\ K &= h \int F d\xi \; d\eta \\ K &= \sum_{i,j,k} F_{i,j} \alpha_{i,j} \end{aligned} \tag{3.9}$$

Where $F_{i,j}$ is a matrix dependent on the natural points. Integration points and integration weights depend on the type of integration the user performs [4]. In the quadrilateral 4-noded element a *2 x 2* numerical integration is chosen.

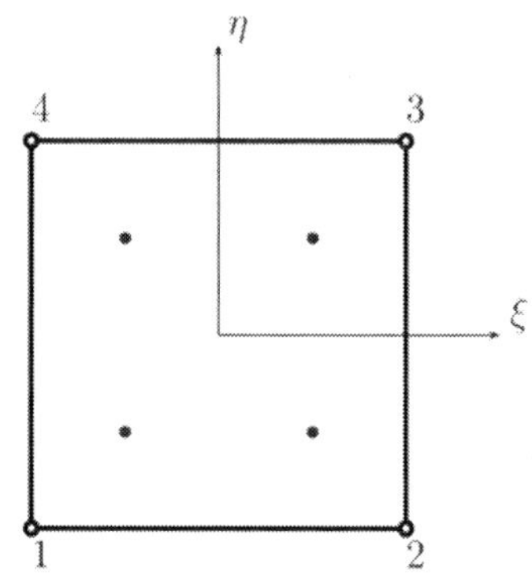

Fig. 1. Two Gauss point integration (ξ, η = +- 1/sqrt(3)) [4]

The Gauss quadrature in two dimensions replaces the integration by a summation over the Gauss points:

$$\iint_{-1}^{1} F(\xi,\eta)\, d\xi\, d\eta = \sum_{i=1}^{p}\sum_{j=1}^{q} w_i w_j F(\xi_i,\eta_j) \tag{3.10}$$

where p, q are the number of integrating points in the direction of the natural coordinates and w_i, w_j are the corresponding weights. Using the first four orders of Gauss quadrature we can proceed to calculate the stiffness.

$$\iint_{-1}^{1} F(\xi,\eta)\, d\xi\, d\eta = \sum_{i=1}^{p}\sum_{j=1}^{q} w_i w_j F(\xi_i,\eta_j) \tag{3.11}$$

$$K_e = h\sum_{i=1}^{2}\sum_{j=1}^{2} B^T CB\, detJ\; w_i w_j \tag{3.12}$$

4 Numerical analysis and results

There are two analyses performed within this article. The first calculates the deflection of the shell/membrane element created analytically within Matlab using the priciples discussed above. The second is an analysis of the shell element available within the Ansys element library.

A *5 x 5 mm* square membrane with a thickness of *0,25 mm* was modeled in both Matlab and Ansys meshed with *10* elements along the *x* and *y* directions. This discretized shell is loaded in the *–z* direction with a pressure of *10 Pa*. The material, piezo and permittivity properties of the element are defined for AlGaN and PZT-5H.

The Ansys analysis utilized the general SHELL181 element suitable for analyzing thin to moderately thick shell structures as recommended by its documentation. SHELL181 is a four node element with 6 DOF at each node, translations in the x, y, and z directions, and rotations about the x, y, and z axes [5]. Formulation is based off of the logarithmic strain and true stress measures. Element kinematics allow for finite membrane strains (stretching). However, the curvature changes within an increment are assumed to be small [5].

Calculated deflections of the AlGaN and PZT-5H membranes from Matlab and Ansys are shown in Fig. 4, Fig. 5, Fig. 6 and Fig. 7, respectively. The right side of the figures displays the Matlab results, the scale of deflection are shown in meters.

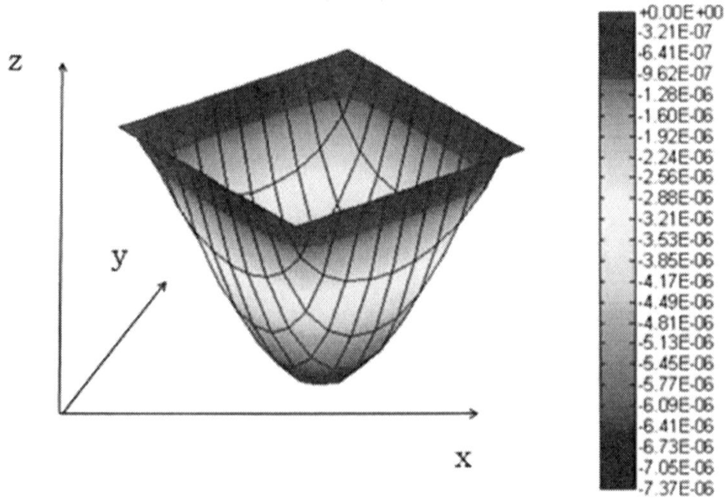

Fig. 4. Deflection of the AlGaN membrane (Matlab)

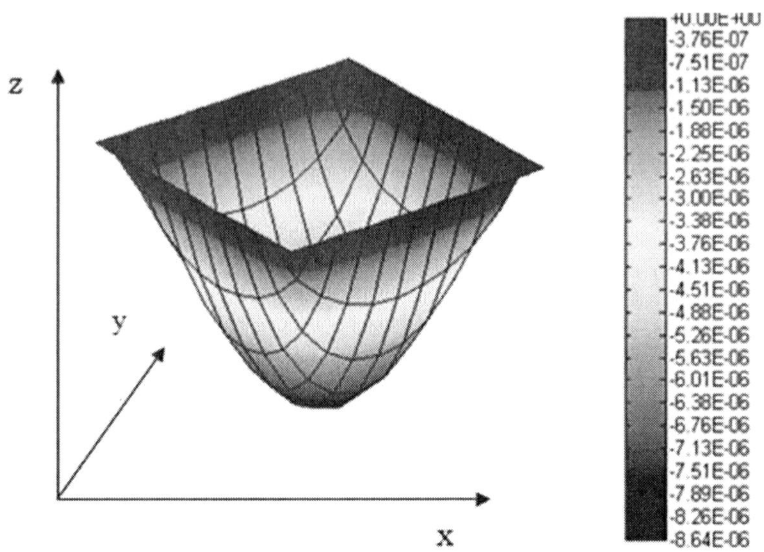

Fig. 5. Deflection of the PZT-5H membrane (Matlab)

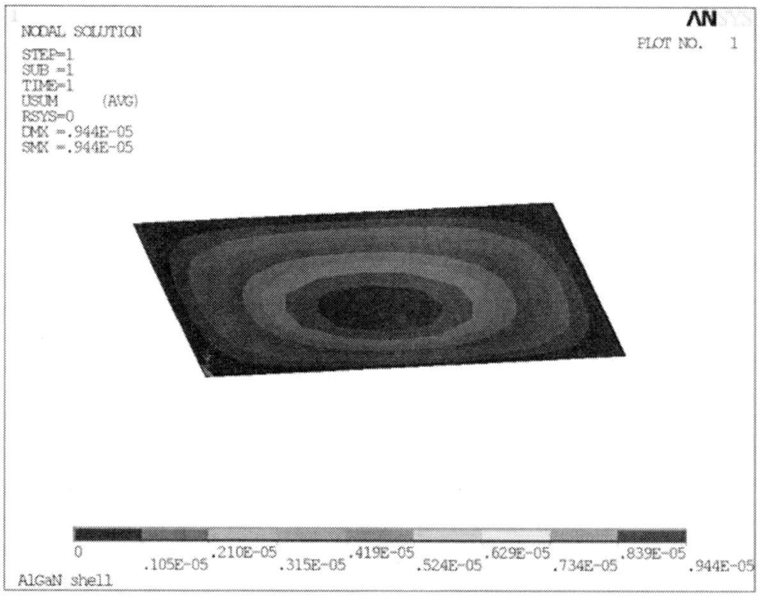

Fig. 6. Deflection of the AlGaN membrane (Ansys)

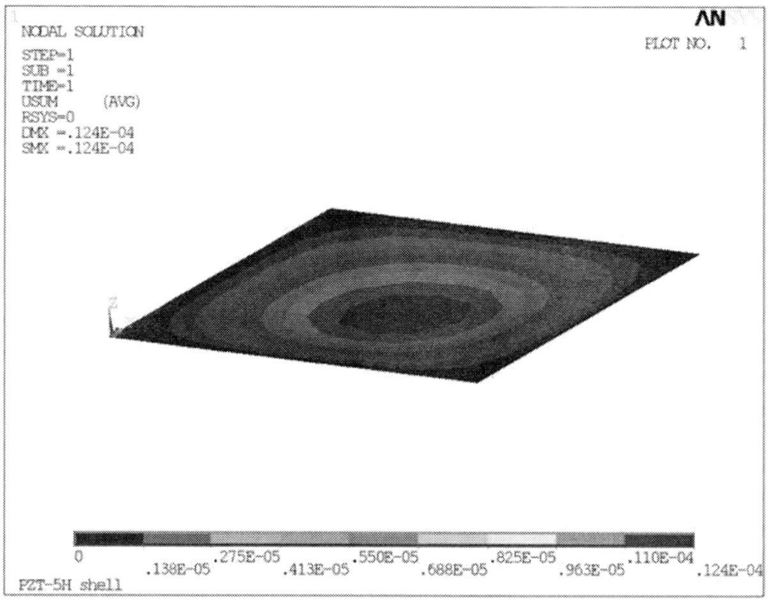

Fig. 7. Deflection of PZT-5H membrane (Ansys)

5 Conclusions and discussion

At the first stage of this article, the shell element was formulated using the usual method of defining shape functions, substituting into the constitutive equations, and thus obtaining the element matrices. Comparison of the analytical solution with the solution from FEM software was carried out afterwards.

Tab. 1 represents Element Types (ET) and Materials (MAT) with their corresponding results. Aluminum Gallium Nitride shows greater deflections in comparison with the stiffer PZT material in all cases of the used elements. The "black box" phenomenon of the Ansys software doesn't provide the exact equations used to solve the problem. Therefore it can only be assumed that the Shell181 and analytical membrane have different result because of their differing mathematical formulation of the element chosen. Another reason of the different behavior may be that in the Matlab code, only the membrane is taken into account which means that only in-plane stiffness's were present. In shell elements, not only are the in-plane effects considered, but the plate stiffness is also considered. The question now remains which formulation is more suitable for piezoelectric membranes. To prove this, experiments are being prepared in cooperation with the Slovak academy of science on PZT and AlGaN wafers. Future works will focus on comparing the experimental results with the formulation within this paper.

MAT/ET	*MATLAB max. deflection [m]*	*ANSYS max. deflection [m]*
	Analytical Membrane	**SHELL181**
AlGaN	7.37E-06	9.44E-06
PZT-5H	8.64E-06	12.4E-06

Tab. 1. Comparison of the analysis results

Acknowledgement

This work was supported by the Scientific Grant Agency of the Ministry of Education, Science, Research and Sport of the Slovak Republic and Slovak Academy of Sciences under the contract No. 1/0712/14 Micro-electromechanical (MEMS) energy harvesting system for applications in medicine "

References

[1] VINSON, J.R, *Plate and Panel Structures of Isotropic, Composite and Piezoelectric Materials, Including Sandwich Construction.* First Edition. s.l. : Springer Netherlands, 2005. p. 424. ISBN 978-1-4020-3111-3.

[2] ERTURK, A., INMAN, J.D. *Piezoelectric Energy Harvesting.* s.l. : Wiley, 2011. ISBN: 978-0-470-68254-8.

[3] RAO, S. *The Finite Element Method in Engineering.* s.l. : Elsevier Science & Technology Books , 2004. ISBN: 0750678283.

[4] FERREIRA, A.J.M. *MATLAB Codes for Finite Element Analysis. Solid Structures.* Porto : Springer, 2009. ISBN 978-1-4020-9200-8.

[5] ANSYS 14.0 Documentation.

[6] RADOŠ, R., Et Al., *Metodika merania parametrov elektrického vozidla.* In Zborník príspevkov 42. medzinárodnej konferencie katedier dopravných, manipulačných, stavebných

a poľnohospodárskych strojov : Kočovce, 6.-7. 9. 2016. 1. vyd. Bratislava : Vydavateľstvo STU, 2016, S. 163-168. ISBN 978-80-227-4584-0.

[7] FIĽA, M., Et Al., *Support of the State in Implementation of Innovations in Enterprise Sector in the Slovak Republic.* In: Journal on Law, Economy and Management. -- ISSN 2048-4186. -- Vol. 5/2015, no. 1, s. 13-18.

Current address

Róková Simona, Ing.
Faculty of Mechanical Engineering,
Slovak University of Technology in Bratislava
Námestie slobody 17, 812 31 Bratislava, Slovak Republic
e-mail: simonarokova@gmail.com

Horvat František, Ing.
Faculty of Mechanical Engineering,
Slovak University of Technology in Bratislava
Námestie slobody 17, 812 31 Bratislava, Slovak Republic
E-mail: frantisek.horvat@stuba.sk

Čekan Michal, Ing. PhD.
Faculty of Mechanical Engineering,
Slovak University of Technology in Bratislava
Námestie slobody 17, 812 31 Bratislava, Slovak Republic
E-mail: michal.cekan@stuba.sk

Hučko Branislav, doc. Ing. PhD.
Faculty of Mechanical Engineering,
Slovak University of Technology in Bratislava
Námestie slobody 17, 812 31 Bratislava, Slovak Republic
E-mail: simonarokova@gmail.com

Kováč Tomáš, Ing.
Faculty of Mechanical Engineering,
Slovak University of Technology in Bratislava
Námestie slobody 17, 812 31 Bratislava, Slovak Republic
E-mail: branislav.hucko@stuba.sk

Slovak University of Technology in Bratislava
Faculty of Mechanical Engineering
16th Conference on Applied Mathematics **APLIMAT 2017**
Proceedings

ASSESSING A TEACHER'S COMPETENCE FOR IMPLEMENTATION OF INQUIRY BASED MATHEMATICS EDUCATION FROM A DISCUSSION OF OPEN GEOMETRICAL SITUATIONS

ROUBÍČEK Filip (CZ)

Abstract. This study is focused on the possibility to assess and develop primary teachers' abilities to implement some principles of inquiry based mathematics education into their teaching practice. The requirements on a primary teacher and their competence are assessed through a joint discussion of open geometrical situations presented through Concept Cartoons. Teachers' written comments and notes and verbal statements expressed in the discussion of four geometrical problems were analysed and presented.

Keywords: inquiry based mathematics education, primary school teacher, classroom mathematical discussion, Concept Cartoon, geometrical problems

Mathematics Subject Classification: Primary 97C70; Secondary 97G40

1 Introduction

Contemporary approaches to mathematical education put emphasis on pupils' (or students') active role and on the use of methods that support better understanding of mathematical concepts, develop the ability to apply knowledge and also boost pupils' (or students') interest in mathematics. One of these approaches is inquiry based education, which is primarily used in sciences but also in mathematics (cf. [1], [7]). The aim of inquiry based education is to allow pupils (or students) to use procedures known from scientific inquiry in their learning ([4]).

Prerequisite to pupils' inquiry is creation of a learning environment that a) allows autonomous and cooperative learning, b) allows the use of methods such as modelling, experimenting and analysing, c) gives sufficient space for mathematical discussion, asking questions, formulation of conjectures, justification and sharing of solving procedures and ideas (cf. [5]). If implementation of inquiry based education is to be successful, the teacher must have solid knowledge of the mathematical content but he/she also must be able to encourage pupils, to support them and guide them through the solving process, be flexible in reacting correctly to their proposals, ideas and findings. As the main obstacle in implementation of inquiry based mathematics education it is often seen the fact that the teachers have no hands-on experience with this approach from the perspective of pupils ([3]).

The goal of the here reported study is to get insight into how primary teachers are able to implement some principles of inquiry based education when solving open geometrical problems. The core of the study is a qualitatively designed empirical survey. The data were collected from analysis of teachers' statements on posed geometrical situations presented by Concept Cartoons (cf. [8]) and their reflections of pupils' responses (from the viewpoint of mathematical content, relations, obstacles in understanding) during a joint discussion.

2 Concept Cartoon as a stimulus to mathematical discussion

The presented requirements on teachers and their competence can be assessed through a mathematical discussion triggered by simulated didactical situations. Concept Cartoons seem to be a convenient tool for this. They were developed as a tool for learning and teaching science ([6]) and lately designed for mathematics ([2]). It is an educational tool based on the choice of a right statement and its justification. Concept Cartoon (see Fig. 1) is a picture showing a group of children in a bubble-dialogue who present alternative viewpoints on their pictured mathematical situation.

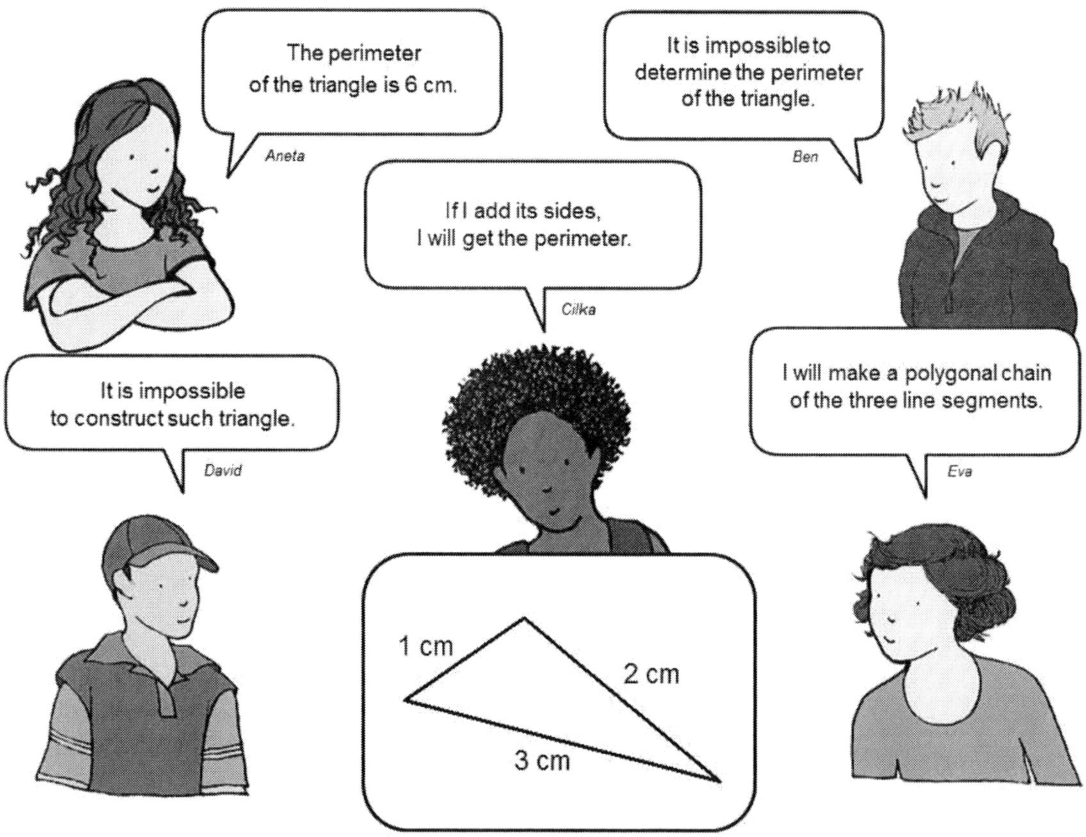

Fig. 1. The first concept cartoon – the perimeter of a triangle

The situation described in Concept Cartoon is intentionally conceived as open in the sense of open-approach: their starting or goal situation is not exactly given, they have multiple levels of grasping ([9]). Although the solved problem is only suggested by a picture and a few pupils' statements, it is possible to determine what the children were commenting on. However, it is impossible to infer what they reacted upon. The pupils' statements allow several interpretations and make teachers ask questions. Thus they are didactically interesting stimuli for a joint discussion. They also give the opportunity to reflect on pupils' ideas and considerations and on the possible ways to correction of the uncovered misconceptions and misunderstandings.

A group of 15 primary teachers were given four geometrical didactical situations in the form of a Concept Cartoon. They were asked to do the following:

Comment how differently children spoke about the given situation.

How would you react to their comments?

Which of them are wrong and why?

2.1 Description of Concept Cartoons with geometrical content

When creating the Concept Cartoons, authentic children's statements were used. Designed Concept Cartoons contain five different statements which are marked by names of children (A – Aneta, B – Ben, C – Cilka, D – David, E – Eva). These statements were recorded when solving similar geometrical problems. Many of the statements in the speech bubbles seem to be vague – the statements are inaccurate, incomplete, equivocal – they require clarification or additional information. The situations were all connected to selected areas of school geometry. They involved knowledge from plane geometry, solid geometry, knowledge of calculations and constructions.

2.1.1 The first Concept Cartoon – the perimeter of a triangle (CC1)

The first Concept Cartoon (Fig. 1) concerned the concept of perimeter of a geometrical figure, its determination and the question of existence of a triangle of given lengths of sides (triangle inequality).

2.1.2 The second Concept Cartoon – the area of a square (CC2)

The second Concept Cartoon (Fig. 2) was about determination of the area of a square using a square gird. It included several procedures for determination of the area, or for calculating the area of a polygon in a square grid:

a) calculating the area using the length of its sides,

b) assembling the polygon from unit squares,

c) extending the polygon with another figure to make it a polygon of a known area,

d) splitting the polygon into several polygons of a known area.

The square was depicted in a rotated position. Thus it was expected to evoke a rhombus, which is a relatively common pupils' misconception.

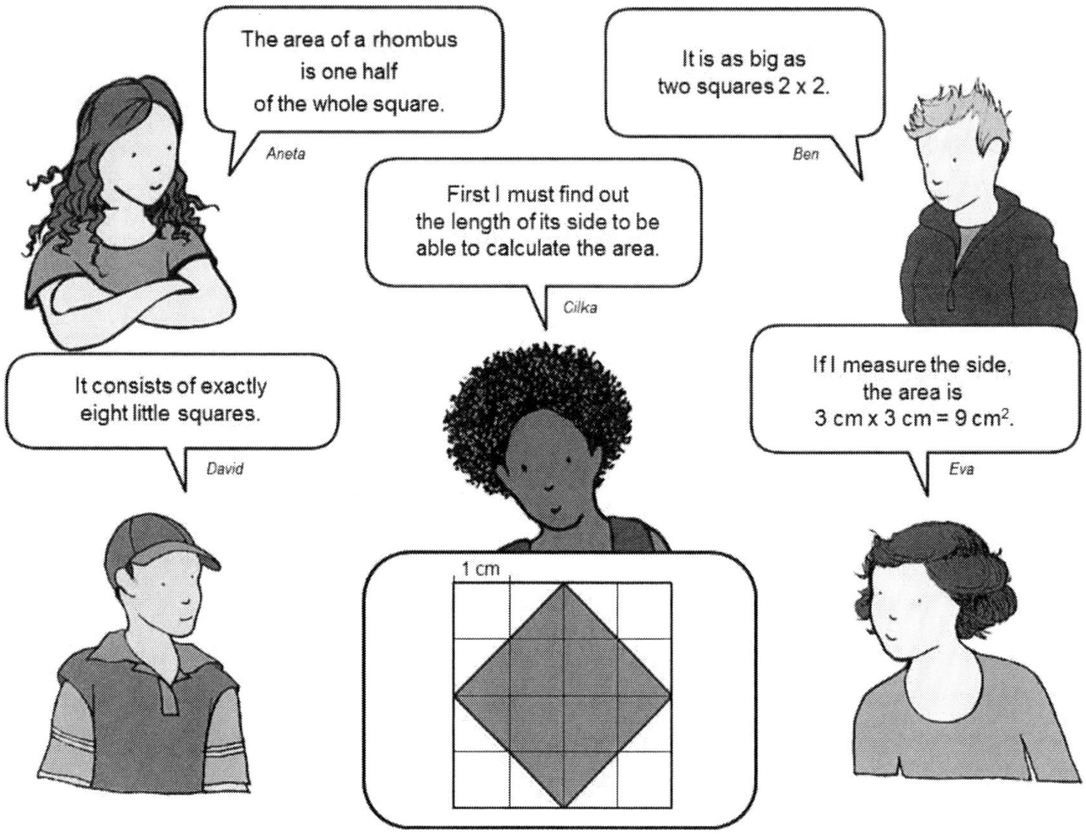

Fig. 2. The second Concept Cartoon – the area of a square

2.1.3 The third Concept Cartoon – orthogonal projection of solids (CC3)

The third Concept Cartoon (Fig. 3) focused on recognition of solids (prism, pyramid, cone) from their orthogonal projection (front view) and the use of correct terminology when describing solids and their properties. The aim was to determine which solids have a triangular projection.

2.1.4 The fourth Concept Cartoon – the volume of a box (CC4)

The fourth Concept Cartoon (Fig. 4) focused on introduction of volume of a prism (box). It was based on filling in the space with unit cubes. It looked for relationships between dimensions of a prism and its volume. The task was to compare the size of two prisms of given dimensions.

Fig. 3. The third Concept Cartoon – orthogonal projection of solids

Fig. 4. The fourth Concept Cartoon – the volume of a box

2.2 Mathematical discussion

The different Concept Cartoons were discussed from three different points of view:

a) from the point of view of their geometrical content, or of the problem and its solution,
b) from the point of view of the teacher's reaction to the different statements,
c) from the point of view of re-education of the discovered misconceptions.

This means that in each Concept Cartoon the teachers first described the geometrical problem that they reconstructed from the picture and the pupils' statements and assessed correctness of the different statements. Then they reflected on how they would react to the different pupils' statements, what questions they would ask or what explanations they would provide. Finally they proposed how the discovered misconceptions could be eliminated and what problems they would offer to their pupils to guide them to discovery of their mistakes or to help them understand the contents better.

When assessing correctness of the statements in the speech bubbles, it turned out that some teachers judge the statements out of the context of the solved task or out of the context of other pupils' statements. Thus two groups of assignments were provided in case of some statements. Some teachers perceived the statements as correct, others as incorrect. It sometimes turned out that even the teacher was not really sure about the problem or that they perceive the concept in a narrower sense.

2.2.1 Geometrical content of Concept Cartoons

In case of the first Concept Cartoon some teachers thought the geometrical problem was about the universal procedure for determination of perimeter of a triangle. Others thought it was about existence of the triangle of the given lengths (i.e. verification of triangle inequality). Not all the teachers realized that both of these perspectives have to be combined. It also turned out that they understood the concept of perimeter of a triangle as the length of its boundary (i.e. a number), not as the boundary itself (i.e. a polygonal line made up of three line segments).

The geometrical situation described in the second Concept Cartoon is familiar to the teachers. They solve similar problems with their pupils very often. Thus their idea of its solution was very pronounced. The teachers were able to see the different methods of determining the area of the square in a square grid. However, sometimes they were not able to grasp the idea that made the pupils answer in the way they did.

When analysing the geometrical situation described in the third Concept Cartoon, not all the teachers were familiar with the concept of orthogonal projection. They were convinced that a triangle cannot be an orthogonal projection of a cone as the base of the cone cannot be a straight line. This misconception affected their assessment of the individual statements. They also discussed the issue of the position of a solid, or the position of its base.

In case of the fourth Concept Cartoon teachers discussed what is actually meant by the size of a box, whether it is only related to volume or also to its surface. The context implies that in case of filling in a box with cubes, the point is to determine the volume. However, also other questions can be considered: *How large wrapping paper do you need to wrap the boxes? How long piece of string is needed to tie up the boxes?* In these cases the teachers compared the areas of the faces or the sum of lengths of edges.

2.2.2 Teacher's reaction to the different statements

If this was a discussion in their lesson, the teachers would react to these statements either by drawing attention to the mistake and explanation, or by inquiring and asking *How do you mean this? Why do you think so?* Some teachers used the sequence of questions:

CC1-A *How did you come to that? (I added the lengths of the sides.) And are these really the sides of a triangle? What condition do sides of a triangle have to meet?*

CC1-B *It is possible to determine the perimeter of every triangle. Why do you think it is not possible here?*

CC1-C *Yes, this holds for every triangle. But is this a triangle? (Yes, there is a triangle in the picture.) And have you heard anything about triangle inequality?*

In other words some of the teachers would tell the children the correct solution, others would only hints at how to proceed. Some reactions show that the teacher tends to correct pupils' mistakes rather than to guide them to better understanding. One of the teachers immediately points out a mistake or asks such questions that do not let the pupil rethink and correct their reasoning. Another teacher asks for verification or justification and using a model or counterexample guides the pupil to discovery that their original statement is not true.

CC1-A *The calculation is correct. Now try to construct the triangle.*

CC1-C *Yes, and why? Change one of the items to make it possible to construct it.*

CC4-A *It will be so in this case. Which of the boxes has a larger base? Will this always hold? What else does the size of a box depend on? Does it depend on its height? Let us look for cases when it holds and when it does not.*

CC4-B *Let us find out whether this is true. How do we determine the size of a box? How many cubes with the side of 10 cm can we put in the box? Let us model it with cubes.*

CC4-C *Your statement seems to be true. Do you think the size of a box can also be influenced by its third dimension?*

CC4-D *Great. Show us how you proceeded. Show us how you reached 80 cubes.*

CC-4E *Let us check the size of the boxes by modelling them from cubes.*

Although transmissive or instructive approaches were more common among the teachers than constructivist approaches, many of them tried to grasp the pupils' thinking, get to understand their statement and discover the source of their misunderstanding.

2.2.3 Re-education of the discovered misconceptions

The discovered misconceptions served as the basis for the teachers' discussion on what problems they would present to their pupils to help those pupils who do not understand yet to understand and those who have grasped the concept to deepen and develop their knowledge.

For clarification of the concepts of triangle inequality and perimeter of a triangle they proposed to use the following problem:

> *Find out how many different triangles you can make from line segments of lengths 1 cm, 2 cm, 3 cm, 4 cm, 5 cm.* Model the triangles (for example using wooden straws), consider either scalene triangles (i.e. each line segment is only used once) or also isosceles and equilateral triangles (i.e. line segments can be used more times). *Determine their perimeters, the smallest and the greatest possible perimeter.*

As far as the environment of the area of a polygon was concerned, they would pose the following task:

> *Find out how many squares (i.e. squares of different areas) can be created in a limited square grid (e.g. on a geoboard). Solve the same problem for a rectangle and a right triangle.*

To clarify knowledge of solids, their properties and projections, the teachers suggested to show models of different solids, to create models of solids from construction kits and other materials. An open problem could be formulated as follows:

> *Which solid can/cannot be seen as a) a square, b) a circle?*

They also proposed to use models (for example a set of cubes) in the case of introduction to a prism. The teachers would for example pose the following problem:

> *Propose the dimensions of two boxes whose size is equal but dimensions are different.*

2.3 Comments on the discussion

If the teachers want to guide pupils to individual discovery, they should never transmit ready-made knowledge or procedures. On the contrary, they should encourage their pupils to experiment, to propose procedures and verify them, to formulate conclusions and justify them. This can be achieved if the teacher asks more or less guiding questions, comes with aimed challenges, asks for justification etc. Another good stimulus, not only for the pupil but also their classmates, might be summarization of what the pupil has said and hinting at how he/she should proceed in his/her considerations. If the teacher only assesses correctness of the statement without giving any guidance, the pupil may not be sufficiently activated.

Examples of various reactions to children's statements:

CC1-A *You would be right if the figure was a triangle.*

CC1-A *At the first sight the perimeter of the triangle seems to be 6 cm. It seems you have added the sides from the illustration. However, have you tried to construct the triangle? Can such triangle exist?*

CC2-A *I can't see a rhombus, could you help me?*

CC2-A *This is an interesting result. Can you explain how have you reached it? How many* cm^2 *would it be?*

CC2-C *Can we determine the area in a different way in this case?*

CC2-E *Is your measuring accurate? Check your measurements. Can you determine the area of the figure without measuring?*

CC2-E *No, the side measures a little less than 3 cm. This mistake will occur if you measure inaccurately. When we determine the area using a square grid, we will avoid this mistake.*

CC3-A *Yes, it could be a pyramid. But Ben is also right.*

CC3-A *Are you sure it must be a pyramid?*

CC3-E *Look at the solids and say which of them have one triangular face.*

CC4-A *Do you mean these specific boxes or does it generally hold that you can put more cubes into a box with a larger base? What would it be if the box on the left was 10 cm lower?*

CC4-B *Does the size depend only on the height?*

CC4-C *Do you think two dimensions are enough to compare sizes of boxes?*

CC4-C *Are you sure you can tell this from your calculation? If I were you I would recalculate it.*

CC4-E *What are you calculating? The length of three sides or the volume of the box?*

The teacher's reaction to a pupil's statement may not always be appropriate – the teacher responds out of context, disregards what the pupil says, uses the pupil's statement imprudently (without asking for clarification or verification), asks misleading questions to make the pupil follow their own procedure. This might be caused either by their different understanding of the statement than was meant by the pupil or by their rejection of the pupil's alternative procedure or unusual idea. Sometimes it happens that a teacher regards a correct result as a wrong one because they have not come across a similar solution before or because the solution does not correspond to the teacher's idea of how it should be. A teacher's inappropriate reaction may cause confusion on the pupil's side or his/her refusal to communicate with the teacher.

The reaction to statement the CC2-D *Yes. What is the correct unit of area? Specify your answer. What is the area of 1 square?* shows that the teacher wants the pupil to express the area in cm^2. The teacher does not regard giving area in the number of square units as a good enough answer.

For example the reaction to statement CC4-C *And also because* $5 \times 4 \times 4 > 6 \times 4 \times 3$. indicates the teacher's failure to understand the pupil. Maybe the teacher does not consider the pupil's omission of one dimension (which is the same for both boxes in this case) as possible with respect to the general formula for calculation of volume of a prism (block) $V = abc$. Also the reaction to statement CC4-E *No, because* $V = abc$ *and* $S = 2(ab + bc + ac)$. is not good. A pupil who reasons like this cannot be expected to be able to make use of this teacher's explanation to discover the mistake in his/her own reasoning.

It may also happen that a teacher makes a mistake because his/her conception or ideas are incomplete or wrong. This may be the case in the following reaction to statement CC3-C *Look at a cone and a pyramid. Do they look the same if you view them from the front? ... We see that a cone is more round. It would have a triangular shape only in case it were its cross-section.*

In pupils' geometrical discourse we often come across inaccurate use of geometrical terminology, confusion of some words (edge – side, side – face) and use of words from everyday language for description of geometrical situations and objects (a spiky solid). While we tolerate this inaccuracy in case of pupils, the teacher must use their language accurately. The teacher's reaction to statement CC4-E *How do we calculate the area of a solid?* is worse than merely misleading.

In a spontaneous discussion the teacher should primarily follow the idea the pupil is trying to communicate and then comment on inaccurate formulations and terminological mistakes. It is important that the pupils be warned their statement can be understood differently and they should get the chance to clarify what they mean. The teacher's reaction *Is the word "spiky" a geometrical term?* fails to respond to the essence of the idea communicated in CC3-C and may cause the pupil not to be willing to communicate any further.

3 Conclusion

Mathematical discussion on Concept Cartoons was based on written notes in advance prepared by the teachers. They were given the set of four Concept Cartoons two weeks before the discussion. Thus they had the chance to get familiar with the geometrical situations and to think their ideas over. What they presented in the discussion was not a spontaneous reaction that would be typical for a classroom situation. Although the discussion was simulated and planned, it showed the problems and misconceptions that may occur in the teachers' lessons and thus showed their competence to react to pupils' proposals appropriately. It enabled us to observe how they assess correctness of pupils' ideas, how they ask questions, how they diagnose and possibly re-educate pupils' misconceptions. The discussion allowed some of the teachers to clarify the relations between some concepts and to realize that the described situation could be regarded from different points of view. They also had the chance to reflect on their teaching style when confronted with opinions of other teachers.

The analysis of individual teachers' reactions (both written comments and notes and verbal statements expressed in the discussion) allowed us to infer indirectly whether the teacher applies the principles of inquiry based education in their lessons (whether they are used to them) or whether they are ready to start incorporating the principles into their lessons. The ability to guide pupils to independent discovery is affected not only by the teacher's skill to implement the principles of inquiry based education in their lessons but also by their beliefs. The use of Concept Cartoons with open geometrical situations is one of the ways to triggering a fruitful mathematical discussion but also of finding out whether a teacher has the skills and abilities needed for successful implementation of inquiries into their teaching.

Acknowledgement

The paper was supported by grant from Czech Science Foundation (project No. 14-01417S) and by RVO 67985840.

References

[1] ARTIGUE, M. et al. *Learning through inquiry. The Fibonacci Project Resources*, 2011, http://www.fibonacci-project.eu.

[2] DABELL, J. KEOGH, B., NAYLOR, S. *Concept Cartoons in Mathematics Education*. Sandbach: Millgate House Education, 2008.

[3] DORIER, J.-L., GARCÍA, F.J. Challenges and opportunities for the implementation of inquiry based learning in day-to-day teaching. *ZDM Mathematics Education*, 45, 2013, pp. 837-849.

[4] DORIER, J.-L., MAASS, K. Inquiry-based mathematics education. In S. Lerman (Ed.), *Encyclopedia of mathematics Education*. Dordrecht: Springer, 2014, pp. 300-304.

[5] HOŠPESOVÁ, A., SAMKOVÁ, L., TICHÁ, M., ROUBÍČEK, F. How Primary School Teachers Posed Problems for Inquiry Based Mathematics Education. In Novotná, Moraová, *International Symposium Elementary Maths Teaching SEMT '15. Proceedings*. Prague: Charles University in Prague, Faculty of Education, 2015, s. 156-165.

[6] KEOGH, B., NAYLOR, S. Learning in science: another way in. *Primary Science Review*, 26, 1993, pp. 22-23.

[7] SAMKOVÁ, L., HOŠPESOVÁ, A., ROUBÍČEK, F., TICHÁ, M. Badatelsky orientované vyučování matematice. (In Czech: Inquiry based mathematics teaching). *Scientia in educatione (sciED)*, 2015, 6 (1), pp. 91-122.

[8] SAMKOVÁ, L., HOŠPESOVÁ, A. Using Concept Cartoons to investigate future teachers' knowledge. In Krainer, Vondrová, *Proceedings of the Ninth Congress of the European Society for Research in Mathematics Education*. Praha: Charles University in Prague, Faculty of Education and ERME, 2015, s. 3241-3247.

[9] SAMKOVÁ, L., TICHÁ, M. Developing open approach to mathematics in future primary school teachers. *ERIE*, 2016, pp. 494-501.

Current address

Roubíček Filip, PhDr. Ph.D.
Institute of Mathematics CAS
Žitná 25, 115 67 Praha 1, Czech Republic
Tel. number: +420 222 090 726, e-mail: roubicek@math.cas.cz

Slovak University of Technology in Bratislava
Faculty of Mechanical Engineering
16th Conference on Applied Mathematics APLIMAT 2017
Proceedings

HOMOGENIZATION ON DOMAINS WITH HOLES

ROZEHNALOVÁ Petra (CZ), FRANCŮ Jan (CZ)

Abstract. Homogenization is a mathematical technique which can be use for mathematical modeling of mechanical behavior of materials with fine structure (composite material, finely perforated material etc.). In this paper we describe homogenization based on unfolding operator, we briefly state properties of such operator and present an example of its usage.

Keywords: Homogenization, perforated domain, torsion, two-scale convergence

Mathematics subject classification: 41A60, 35J15

1 Introduction

Theory of homogenization was developed to describe behaviour of composite materials. In the broad sense by homogenization we mean replacing heterogeneous material by an equivalent homogeneous one. Mathematical approach consists in considering a sequence of problems with a material with more and more refined structure. This approach was first introduced by J.B. Keller (1973) and developed by I. Babuška (1975). More about homogenization can be found in the monograph [1] or in the textbook [3].

Other problems for which a similar approach can be used are boundary valued problems defined on a perforated domain. Let Ω be a domain in $\mathbb{R}^n$ and let it be perforated by holes. This perforated domain will be denoted by Ω_ε^*. We shall study Laplace or more generally elliptic equation on the Ω_ε^* with different types of boundary conditions. The solution of such a problem let be denoted by u^ε. Let us consider a sequence of the perforated domains (and associated differential problems) in which for $\varepsilon \searrow 0$ the number of holes is increasing and their size is decreasing.

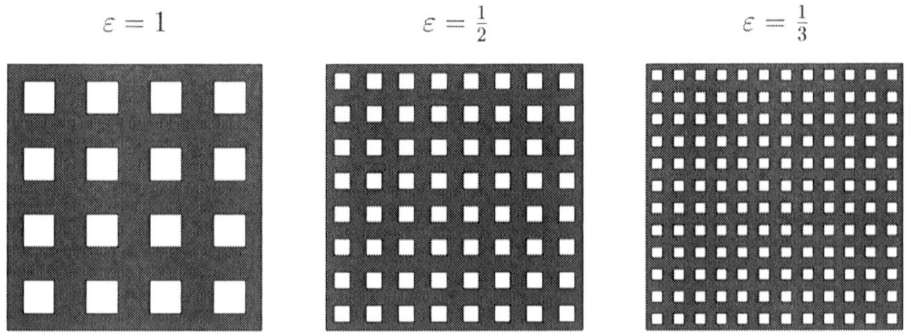

Fig. 1. Sequence of Ω_ε^*

The principal questions are: How does the sequence of the solutions u^ε behaves when the parameter ε tends to zero? Does the limit homogenized solution exist? If yes, how can it be characterized? Does it depends on the manner of behaviour of holes?

Our approach how to solve this problem is to use an *unfolding operator for periodically perforated domains* $\mathcal{T}_\varepsilon^*$ to transform functions u_ε, resp. ∇u_ε defined on Ω_ε^* to the functions $\mathcal{T}_\varepsilon^*(u_\varepsilon)$, resp. $\mathcal{T}_\varepsilon^*(\nabla u_\varepsilon)$ defined on the fixed domain.

In the following we will introduce definition, basic properties of unfolding operator and conditions under which $\{\mathcal{T}_\varepsilon^*(u_\varepsilon)\}$, resp. $\{\mathcal{T}_\varepsilon^*(\nabla u_\varepsilon)\}$ converge weakly or strongly. Finally we will show application of this theory to partial differential problem describing torsion of the rod (for derivation of the model see [4] or [5]).

2 Unfolding operator in perforated domains

Domain with holes Let us begin with defining a domain with holes. Let Ω be a bounded domain (connected open set) in $\mathbb{R}^N$ with Lipschitz boundary. We consider a descending sequence $E = \{\varepsilon_k\}_{k=0}^{\infty}$ of positive numbers, such that $\varepsilon_k \searrow 0$ as $k \to \infty$ (we call it a scale). For each ε, let $\{T_\varepsilon^j\}_{j=1}^{n(\varepsilon)}$ be a system of disjoint bounded domains in $\mathbb{R}^N$ with Lipschitz boundary representing the "holes".

Part of Ω occupied by material is denoted by Ω_ε^* and it is defined as Ω without holes T_ε^j, i.e. $\Omega_\varepsilon^* = \Omega \setminus T_\varepsilon$, where $T_\varepsilon = \bigcup_{j=1}^{n(\varepsilon)} \overline{T_\varepsilon^j}$. We assume that Ω_ε^* is a multiply connected set. Furthermore we denote by Ω_ε^i, $i = 1, \dots, m(\varepsilon)$ the "interior holes", they are such sets $\overline{T_\varepsilon^j}$ that are completely inside Ω_ε^* and do not intersect the boundary $\partial\Omega$, i.e. sets $\overline{T_\varepsilon^j} \subset \Omega$. Their union is denoted by $\Omega_\varepsilon^{\text{int}}$ and $\Omega_\varepsilon^{\text{int}} = \bigcup_{i=1}^{m(\varepsilon)} \Omega_\varepsilon^i$. Let the holes which are intersecting boundary be denoted by $\Omega_\varepsilon^{\text{ext}}$, i.e. $\Omega_\varepsilon^{\text{ext}} = \left(T_\varepsilon \setminus \Omega_\varepsilon^{\text{int}}\right) \cap \Omega$, and $\partial_{\text{ext}}\Omega_\varepsilon^*$ represent the exterior boundary of Ω_ε^* and be defined as $\partial_{\text{ext}}\Omega_\varepsilon^* = \partial\Omega_\varepsilon^* \setminus \partial\Omega_\varepsilon^{\text{int}}$.

Periodically perforated domain Till now the holes T_ε are distributed in very general manner. In sequel we define a periodical distribution of the holes.

Let us define the *reference cell* Y in $\mathbb{R}^N$ by setting $Y = \langle 0, l_1) \times \dots \times \langle 0, l_N)$, where $l_1, \dots, l_N$ are given positive numbers. Let $T \subset Y$ be open bounded set with smooth boundary. This set represents a hole in Y. The part of the reference cell Y occupied by a material is denoted by Y^* and $Y^* = Y \setminus \overline{T}$.

Then the set T_ε is defined as a union of translates and scaled images of T, so

$$T_\varepsilon = \bigcup_{k\in\mathcal{K}} (\varepsilon (T + k)), \quad \text{where} \quad \mathcal{K} = \left\{k \in \mathbb{R}^N \mid k = \xi \cdot (l_1, l_2, \dots, l_N), \xi \in \mathbb{Z}^N\right\}$$

Definition 1 (Unfolding operator for perforated domains)**.** The operator $\mathcal{T}_\varepsilon^*$ maps a function $u : \Omega_\varepsilon^* \to \mathbb{R}$ to a function $\mathcal{T}_\varepsilon^*(u) : \Omega \times Y \to \mathbb{R}$, and is defined as follows:

$$\mathcal{T}_\varepsilon^*(u)(x,y) = \begin{cases} u\left(\varepsilon\left[\frac{x}{\varepsilon}\right]_Y + \varepsilon y\right) & \text{for } (x,y) \in \widehat{\Omega}_\varepsilon \times Y^*, \\ u(x) & \text{for } (x,y) \in \Lambda_\varepsilon^* \times Y, \\ 0 & \text{otherwise.} \end{cases} \tag{1}$$

The following basic properties of the unfolding operator for perforated domains can be proven (proofs of all theorems in this paper can be found in [6]). Proper definition of used functional spaces, as $L^p(\Omega)$, $W^{1,p}(\Omega)$, $L^2(\Omega, H^1_{\text{per}}(Y))$, can be found in e.g. [3].

Theorem 2. *Let $\mathcal{T}_\varepsilon^*$ be the unfolding operator for perforated domains. Then for each $\varepsilon > 0$ we have:*

(i) The unfolding $\mathcal{T}_\varepsilon^*$ is multiplicative, i.e. for the functions $u, v : \Omega_\varepsilon^* \to \mathbb{R}$ holds $\mathcal{T}_\varepsilon^*(uv) = \mathcal{T}_\varepsilon^*(u)\mathcal{T}_\varepsilon^*(v)$.

(ii) The operator $\mathcal{T}_\varepsilon^*$ is linear, i.e. for $\alpha, \beta \in \mathbb{R}$ and $u, v : \Omega_\varepsilon^* \to \mathbb{R}$ there is $\mathcal{T}_\varepsilon^*(\alpha u + \beta v) = \alpha\mathcal{T}_\varepsilon^*(u) + \beta\mathcal{T}_\varepsilon^*(v)$.

(iii) The operator $\mathcal{T}_\varepsilon^*$ conserves the integral, i.e. for $u \in L^1(\Omega_\varepsilon^*)$ holds $\iint_{\Omega\times Y} \mathcal{T}_\varepsilon^*(u)(x,y)\,\mathrm{d}x\,\mathrm{d}y = |Y| \int_{\Omega_\varepsilon^*} u(x)\,\mathrm{d}x$.

(iv) The unfolding operator $\mathcal{T}_\varepsilon^* : L^p(\Omega_\varepsilon^*) \to L^p(\Omega \times Y)$, where $p \in \langle 1, \infty)$, conserves the norm in the sense that for every $u \in L^p(\Omega_\varepsilon^*)$ holds $\|\mathcal{T}_\varepsilon^*(u)\|_{L^p(\Omega\times Y)} = |Y|^{\frac{1}{p}} \|u\|_{L^p(\Omega_\varepsilon^*)}$.

Thus $\mathcal{T}_\varepsilon^*$ is bounded and for it's norm holds $\|\mathcal{T}_\varepsilon^*\|_{\mathcal{L}(L^p(\Omega_\varepsilon^*), L^p(\Omega\times Y))} = |Y|^{\frac{1}{p}}$.

(v) The unfolding $\mathcal{T}_\varepsilon^*$ is continuous operator from $L^p(\Omega_\varepsilon^*)$ to $L^p(\Omega \times Y)$, where $p \in \langle 1, \infty)$.

The unfolding operator can be used to define so called two-scale convergence.

Definition 3 (Two-scale convergence for perforated domains)**.** Let $\mathcal{T}_\varepsilon^*$ by the unfolding operator for perforated domains, $E = \{\varepsilon\}$ be a scale, $\{u_\varepsilon\}$ be a sequence in $L^p(\Omega_\varepsilon^*)$ and $u_0 \in L^p(\Omega \times Y)$, $p \in \langle 1, \infty)$. A sequence $\{u_\varepsilon\}$ is said to *strongly (resp. weakly) two-scale converge* to u_0 in $L^p(\Omega)$ with respect to the scale E if the sequence $\{\mathcal{T}_\varepsilon^*(u_\varepsilon)\}$ converges to u_0 strongly (resp. weakly) in $L^p(\Omega \times Y)$.

Let us list some properties of two-scale convergence.

Theorem 4. *Let $\{u_\varepsilon\}$ be a sequence in $L^p(\Omega_\varepsilon^*)$ and $u_0 \in L^p(\Omega \times Y)$, $p \in \langle 1, \infty)$.*

(i) Any constant sequence $\{u\} \in L^p(\Omega)$ *strongly two-scale converges to itself, i.e.* $\mathcal{T}_\varepsilon^*(u) \to u_0$ *strongly in* $L^p(\Omega \times Y)$*, where*

$$u_0(x,y) = \begin{cases} u(x) & [x,y] \in \Omega \times Y^*, \\ 0 & \textit{otherwise.} \end{cases}$$

(ii) Any sequence $\{u_\varepsilon\} \in L^p(\Omega_\varepsilon^*)$ *two-scale converging (strongly or weakly) in* $L^p(\Omega)$ *is bounded in* $L^p(\Omega_\varepsilon^*)$.

(iii) If the two-scale limit u_0 *exist, then is unique as an element of* L^p*-spaces.*

(iv) If $\{u_\varepsilon\}$ *strongly two-scale converges to* u_0 *in* $L^p(\Omega)$*, then it weakly two-scale converges to the same limit.*

(v) For $p \in (1,\infty)$*. If* $\{u_\varepsilon\}$ *weakly two-scale converges to* u_0 *in* $L^p(\Omega)$*, i.e.* $\mathcal{T}_\varepsilon^*(u_\varepsilon) \rightharpoonup u_0$ *weakly in* $L^p(\Omega \times Y)$*, Then* $\widetilde{u}_\varepsilon \rightharpoonup u^*$ *weakly in* $L^p(\Omega)$*, where* $u^*(x) = \frac{1}{|Y|}\int_{Y^*} u_0(x,y)\,\mathrm{d}y$.

Unfolding operator $\mathcal{T}_\varepsilon^*$ and gradients Consider a function $u \in W^{1,p}(\Omega)$. Then it is straightforward that

$$\mathcal{T}_\varepsilon^*(\nabla u) = \begin{cases} \frac{1}{\varepsilon}\nabla_y \mathcal{T}_\varepsilon^*(u) & \text{on } \widehat{\Omega}_\varepsilon \times Y^*, \\ \nabla u = \nabla \mathcal{T}_\varepsilon^*(u) & \text{on } \Lambda_\varepsilon^* \times Y, \\ 0 & \text{otherwise.} \end{cases} \tag{2}$$

Now we will state the main result about convergence of the unfolded sequence of gradients $\{\mathcal{T}_\varepsilon^*(\nabla u_\varepsilon)\}$.

Theorem 5. *Let a sequence* $\{u_\varepsilon\}$ *be bounded in* $W^{1,p}(\Omega_\varepsilon^*)$*, for* $p \in (1,\infty)$*. i.e.* $\|u_\varepsilon\|_{W^{1,p}(\Omega_\varepsilon^*)} \leq C$*. Then, there exist a subsequence (still denoted* $\{u_\varepsilon\}$*) and functions* $u_0 \in W^{1,p}(\Omega)$ *and* $u_0^* \in L^p(\Omega; W^{1,p}_{\mathrm{per}}(Y))$ *such that*

(i) $\mathcal{T}_\varepsilon^*(u_\varepsilon) \rightharpoonup u_0$ *weakly in* $L^p(\Omega; W^{1,p}(Y))$*,*

(ii) $\mathcal{T}_\varepsilon^*(\nabla u_\varepsilon) \rightharpoonup \nabla u_0 + \nabla_y u_0^*$ *weakly in* $L^p(\Omega \times Y)$*, i.e.* ∇u_ε *converges two-scale weakly in* $L^p(\Omega)$.

Moreover, $\mathcal{M}_Y^*(u_0^*) = 0$*, where* $\mathcal{M}_Y^*(u_0)(x) = \frac{1}{|Y^*|}\int_{Y^*} u_0(x,y)\,\mathrm{d}y$.

3 Application

The following boundary problem was derived in [5] and describes torsion of prismatic bar.

$$\begin{cases} -\nabla \cdot (A^\varepsilon \nabla u_\varepsilon) = f & \text{in } \Omega_\varepsilon^*, \\ u_\varepsilon = 0 & \text{on } \partial_{\mathrm{ext}}\Omega_\varepsilon^*, \\ u_\varepsilon = \text{const.} & \text{on } \partial\Omega_\varepsilon^i; \quad i = 1,\dots,m(\varepsilon), \\ \displaystyle\int_{\partial\Omega_\varepsilon^i} A^\varepsilon(x)\frac{\partial u_\varepsilon}{\partial n}(x)\,\mathrm{d}x = \int_{\Omega_\varepsilon^i} f(x)\,\mathrm{d}x & i = 1,\dots,m(\varepsilon), \end{cases} \tag{3}$$

where $f \in L^2(\Omega)$, $A^\varepsilon(x)$ is a matrix function such that the problem (3) has a unique weak solution (for proper definition see for example [2]), vector n is outward-pointing unite normal (i.e. on inner boundary, n is directed towards the interior of the holes), $m(\varepsilon)$ denotes number of interior holes.

Let us introduce the vector space

$$\mathcal{S}_\varepsilon(\Omega) = \left\{ v \in H_0^1(\Omega), \text{ s.t. } v = 0 \text{ in } \overline{\Omega_\varepsilon^{\text{ext}}},\ v = \text{const. in } \overline{\Omega_\varepsilon^i},\ i = 1, \ldots, m(\varepsilon) \right\}$$

with the norm $\|v\|_{\mathcal{S}_\varepsilon(\Omega)} = \|\nabla v\|_{[L^2(\Omega_\varepsilon^*)]^2}$.

The weak formulation of the problem (3) is: Find $u_\varepsilon \in \mathcal{S}_\varepsilon(\Omega)$ such that

$$\int_{\Omega_\varepsilon^*} A^\varepsilon(x)\,\nabla u_\varepsilon(x) \cdot \nabla v(x)\,\mathrm{d}x = \int_\Omega f(x)\,v(x)\,\mathrm{d}x, \qquad \forall v \in \mathcal{S}_\varepsilon(\Omega). \tag{4}$$

The following theorem states homogenization result for the problem (4).

Theorem 6. *Let u_ε be the solution of the Problem (4). Assume that $\mathcal{T}_\varepsilon(A^\varepsilon) \to A$ a. e. in $\Omega \times Y^*$ for some matrix $A = A(x, y)$. Then, there exist $u_0 \in H_0^1(\Omega)$ and $u_0^* \in L^2(\Omega, H^1_{\text{per}}(Y^*))$ such that*

$$\begin{aligned} &\|u_\varepsilon - u_0\|_{L^2(\Omega_\varepsilon^*)} \to 0, && \\ &\mathcal{T}_\varepsilon(u_\varepsilon) \rightharpoonup u_0 && \text{weakly in } L^2(\Omega, H^1(Y^*)), \\ &\mathcal{T}_\varepsilon(\nabla u_\varepsilon) \rightharpoonup \nabla u_0 + \nabla_y u_0^* && \text{weakly in } L^2(\Omega \times Y^*), \end{aligned} \tag{5}$$

and the pair (u_0, u_0^*) is the unique solution of the problem: Find $u_0 \in H_0^1(\Omega)$ and $u_0^* \in L^2(\Omega, H^1_{\text{per}}(Y^*))$ such that

$$\frac{1}{|Y|} \iint\limits_{\Omega \times Y^*} A(x,y)\,[\nabla u_0(x) + \nabla_y u_0^*(x,y)] \cdot [\nabla \Psi(x) + \nabla_y \Phi(x,y)]\,\mathrm{d}x\,\mathrm{d}y =$$

$$= \int_\Omega f(x)\,\Psi(x)\,\mathrm{d}x, \quad \forall \Psi \in H_0^1(\Omega),\ \forall \Phi \in L^2(\Omega, H^1_{\text{per}}(Y^*)). \tag{6}$$

4 Numerical examples

Let $x = (x_1, x_2) \in \Omega$ and $y = (y_1, y_2) \in Y$, where Ω is a simple domain in $\mathbb{R}^2$ and $Y = \langle 0, l_1) \times \langle 0, l_2)$, l_1, l_2 are real positive numbers. Vector function y^c has the form $y^c = (\,y_1^c,\ y_2^c\,)$. Furthermore, let us suppose that A is a function only in variable y, i.e. $A(x, y) = A(y)$.

We would like to solve the problem, derived in the Theorem 6:

$$\begin{cases} \text{Find } u_0 \in H_0^1(\Omega) \text{ and } u_0^* \in L^2(\Omega, H^1_{\text{per}}(Y)) \text{ such that} \\ \dfrac{1}{|Y|} \displaystyle\iint\limits_{\Omega \times Y^*} A(y)\,[\nabla u_0(x) + \nabla_y u_0^*(x,y)] \cdot [\nabla \Psi(x) + \nabla_y \Phi(x,y)]\,\mathrm{d}x\,\mathrm{d}y = \int\limits_\Omega f(x)\,\Psi(x)\,\mathrm{d}x, \\ \forall \Psi \in H_0^1(\Omega),\ \forall \Phi \in L^2(\Omega, H^1_{\text{per}}(Y)),\ \text{s. t. } \Phi + y^c \cdot \nabla \Psi \text{ is constant in } y \text{ on } \Omega \times T, \\ \mathcal{M}_Y(u_0^*) = 0, \\ u_0^* = -y^c \cdot \nabla u_0 \ \text{ on } \Omega \times T. \end{cases} \tag{7}$$

We will look for u_0, u_0^* in two steps. At first, we will compute auxiliary functions denoted $\hat{\chi}_1, \hat{\chi}_2$ and subsequently, using them, we will find homogenized solutions u_0, u_0^*.

Let us choose $\Psi(x) \equiv 0$ as a test function in (7). We suggest function u_0^* in the form

$$u_0^*(x,y) = -\hat{\chi}_1(y)\,\frac{\partial u_0}{\partial x_1}(x) - \hat{\chi}_2(y)\,\frac{\partial u_0}{\partial x_2}(x).$$

Then, (7) takes the form

$$\iint\limits_{\Omega\times Y^*} A\left[\frac{\partial u_0}{\partial x_1}\frac{\partial \Phi}{\partial y_1} + \frac{\partial u_0}{\partial x_2}\frac{\partial \Phi}{\partial y_2}\right] \mathrm{d}x\,\mathrm{d}y =$$

$$= \iint\limits_{\Omega\times Y^*} A\left[\frac{\partial \hat{\chi}_1}{\partial y_1}\frac{\partial u_0}{\partial x_1}\frac{\partial \Phi}{\partial y_1} + \frac{\partial \hat{\chi}_2}{\partial y_1}\frac{\partial u_0}{\partial x_2}\frac{\partial \Phi}{\partial y_1} + \frac{\partial \hat{\chi}_1}{\partial y_2}\frac{\partial u_0}{\partial x_1}\frac{\partial \Phi}{\partial y_2} + \frac{\partial \hat{\chi}_2}{\partial y_2}\frac{\partial u_0}{\partial x_2}\frac{\partial \Phi}{\partial y_2}\right] \mathrm{d}x\,\mathrm{d}y.$$

From this we see that the problem (7) is fulfilled when the auxiliary function $\hat{\chi}_i$, $i = 1, 2$, satisfies

$$\int\limits_{Y^*} A\,\frac{\partial \Phi}{\partial y_i}\,\mathrm{d}y = \int\limits_{Y^*} A\,\nabla\hat{\chi}_i \cdot \nabla\Phi\,\mathrm{d}y, \quad \forall\,\Phi \in L^2(\Omega, H^1_{\mathrm{per}}(Y)), \text{ s. t. } \Phi \text{ is constant in } y \text{ on } T.$$

Rewriting this, we derive the following problem

$$\left\{\begin{array}{l} \text{Find } \hat{\chi}_i \in H^1_{\mathrm{per}}(Y) \text{ such that} \\ \displaystyle\int\limits_{Y^*} A\,\nabla\left(\hat{\chi}_i - y_i\right) \cdot \nabla\Phi\,\mathrm{d}y = 0, \quad \forall\,\Phi \in L^2(\Omega, H^1_{\mathrm{per}}(Y)), \text{ s. t. } \Phi \text{ is constant in } y \text{ on } T, \\ \mathcal{M}_Y(\hat{\chi}_i) = 0, \\ \hat{\chi}_i = -y_i^c \text{ on } T. \end{array}\right. \tag{8}$$

Now, let us choose as a test function in (7) a function

$$\Phi(x,y) = -\varphi_1(y)\,y^c \cdot \Psi(x),$$

where $\Psi \in \mathcal{D}(\Omega)$ and φ is Y-periodic function which $\varphi_1|_Y \in \mathcal{D}(Y)$, $\varphi_1 \equiv 1$ on T.

Then, (7) takes the form

$$\frac{1}{|Y|}\iint\limits_{\Omega\times Y^*} A\left[\nabla u_0 + \nabla_y\left(-\hat{\chi}_1\frac{\partial u_0}{\partial x_1} - \hat{\chi}_2\frac{\partial u_0}{\partial x_2}\right)\right]\cdot\left[\nabla\Psi - \nabla_y\left(\varphi_1\,y^c\cdot\Psi\right)\right]\mathrm{d}x\,\mathrm{d}y = \int\limits_{\Omega} f\,\Psi\,\mathrm{d}x.$$

Simple computations yield the problem

$$\left\{\begin{array}{l} \text{Find } u_0 \in H^1_0(\Omega) \text{ such that} \\ \displaystyle\frac{|Y^*|}{|Y|}\int\limits_{\Omega} \mathcal{A}\,\nabla u_0 \cdot \nabla\Psi\,\mathrm{d}x\,\mathrm{d}y = \int\limits_{\Omega} f\,\Psi\,\mathrm{d}x, \quad \forall\Psi \in H^1_0(\Omega). \end{array}\right. \tag{9}$$

Where matrix $\mathcal{A}$ is given by $\mathcal{A} = (a_{ij})_{i,j=1,2}$

$$\begin{aligned} a_{11} &= \int_{Y^*} A\left[\left(1-\frac{\partial\hat\chi_1}{\partial y_1}\right)\left(1-\frac{\partial(y_1^c\varphi_1)}{\partial y_1}\right)+\frac{\partial\hat\chi_1}{\partial y_2}\frac{\partial(y_1^c\varphi_1)}{\partial y_2}\right]dy,\\ a_{12} &= -\int_{Y^*} A\left[\left(1-\frac{\partial\hat\chi_1}{\partial y_1}\right)\frac{\partial(y_2^c\varphi_1)}{\partial y_1}+\frac{\partial\hat\chi_1}{\partial y_2}\left(1-\frac{\partial(y_2^c\varphi_1)}{\partial y_2}\right)\right]dy,\\ a_{21} &= -\int_{Y^*} A\left[\left(1-\frac{\partial\hat\chi_2}{\partial y_2}\right)\frac{\partial(y_1^c\varphi_1)}{\partial y_2}+\frac{\partial\hat\chi_2}{\partial y_1}\left(1-\frac{\partial(y_1^c\varphi_1)}{\partial y_1}\right)\right]dy,\\ a_{22} &= \int_{Y^*} A\left[\left(1-\frac{\partial\hat\chi_2}{\partial y_2}\right)\left(1-\frac{\partial(y_2^c\varphi_1)}{\partial y_2}\right)+\frac{\partial\hat\chi_2}{\partial y_1}\frac{\partial(y_2^c\varphi_1)}{\partial y_1}\right]dy. \end{aligned}$$

In the sequel, we present numerical results of the homogenization of torsion problem. We assume $\Omega = (0,1)\times(0,1)$, reference cell $Y = \langle 0,1)\times\langle 0,1)$, reference hole $T = \left(\frac{1}{4},\frac{3}{4}\right)\times\left(\frac{1}{4},\frac{3}{4}\right)$. Torsion problem is obtained for $A(y) = 1$, $f(x) = -2$.

In the first step, by solving problem (8) we get two auxiliary functions $\hat\chi_1$ (on Figure 2) and $\hat\chi_2$. In the second step the problem (9) is solved to obtain the homogenized solution. A comparison of functions u_ε and homogenized solution u_0 is on Figure 4. Graph of function $u_{1/4}$ is on Figure 3. The numerical results are obtained by finite element method implemented in MATLAB.

Periodic boundary conditions Boundary conditions prescribed in the formulation (8) can be replaced, in our case, by Dirichlet and homogenous Neumann condition. Let us remind that in our model problem the reference cell $Y = \langle 0,l_1)\times\langle 0,l_2)$ and reference hole $T = \left(\frac{l_1-a}{2},\frac{l_1+a}{2}\right)\times\left(\frac{l_2-b}{2},\frac{l_2+b}{2}\right)$, where $0 < a < l_1$ and $0 < b < l_2$. So perforated reference cell Y^* is symmetric with respect to axes $y_2 = \frac{l_2}{2}$ and $y_1 = \frac{l_1}{2}$.

If a function $\hat\chi_1(y_1,y_2)$ is a solution of the problem (8), then also the function $-\hat\chi_1(l_1 - y_1, y_2)$ is a solution of the same problem. Indeed, if a function $w = w(y_1,y_2)$ belongs to $L^2(\Omega, H^1_{\text{per}}(Y))$ then also the function $-w(l_1-y_1,y_2)$ belongs to the same space. Let us choose it in (8) as a test function Φ. If a function $\hat\chi_1$ solves the problem (8), then we can derive:

$$\int_0^{l_2}\int_0^{l_1}\nabla\left(\hat\chi_1(y_1,y_2)-y_1\right)\cdot\nabla\left[-w(l_1-y_1,y_2)\right]\,1_{Y^*}\,\mathrm{d}y_1\,\mathrm{d}y_2 \left|\begin{array}{c} l_1-y_1=t\\ -\mathrm{d}y_1=\mathrm{d}t\\ y_1=0\Rightarrow t=l_1\\ y_1=l_1\Rightarrow t=0\end{array}\right| =$$

$$\begin{aligned} &= \int_0^{l_2}\int_{l_1}^{0}\nabla\left(\hat\chi_1(l_1-t,y_2)-(l_1-t)\right)\cdot\nabla w(t,y_2)\,1_{Y^*}\,\mathrm{d}t\,\mathrm{d}y_2 =\\ &= -\int_0^{l_2}\int_0^{l_1}\nabla\left(\hat\chi_1(l_1-t,y_2)+t\right)\cdot\nabla w(t,y_2)\,\mathrm{d}t\,1_{Y^*}\,\mathrm{d}y_2 =\\ &= \int_0^{l_2}\int_0^{l_1}\nabla\left(-\hat\chi_1(l_1-t,y_2)-t\right)\cdot\nabla w(t,y_2)\,\mathrm{d}t\,1_{Y^*}\,\mathrm{d}y_2 = 0. \end{aligned}$$

Thus, the function $-\hat{\chi}_1(l_1 - y_1, y_2)$ also solves the problem (8). From the uniqueness of the solution of the problem (8) we get $\hat{\chi}_1(y_1, y_2) = -\hat{\chi}_1(l_1 - y_1, y_2)$, which means that $\hat{\chi}_1$ "odd in variable y_1 with respect to point $y_1 = \frac{l1}{2}$". Finally, since this symmetry holds and the function $\hat{\chi}_1 \in H^1_{\text{per}}(Y)$ we can prescribe to the boundary $y_1 = 0$ and $y_1 = l_1$ Dirichlet conditions.

Similar reasoning leads to Neumann conditions on the boundary $y_2 = 0$ and $y_2 = l_2$. Indeed, let us choose in (8) as a test function $\Phi = w(y_1, l_2 - y_2)$, which belongs to $L^2(\Omega, H^1_{\text{per}}(Y))$. If a function $\hat{\chi}_1$ solves the problem (8), then we can derive:

$$\int_0^{l_1}\int_0^{l_2} \nabla\left(\hat{\chi}_1(y_1, y_2) - y_1\right)\cdot\nabla w(y_1, l_2 - y_2)\, 1_{Y^*}\, \mathrm{d}y_2\, \mathrm{d}y_1 \left|\begin{array}{l} l_2 - y_2 = t \\ -\,\mathrm{d}y_2 = \mathrm{d}t \\ y_2 = 0 \Rightarrow t = l_2 \\ y_2 = l_2 \Rightarrow t = 0 \end{array}\right| =$$

$$= -\int_0^{l_1}\int_{l_2}^{0} \nabla\left(\hat{\chi}_1(y_1, l_2 - t) - y_1\right)\cdot\nabla w(y_1, t)\, 1_{Y^*}\, \mathrm{d}t\, \mathrm{d}y_2 =$$

$$= \int_0^{l_1}\int_0^{l_2} \nabla\left(\hat{\chi}_1(y_1, l_2 - t) - y_1\right)\cdot\nabla w(y_1, t)\, 1_{Y^*}\, \mathrm{d}t\, \mathrm{d}y_2$$

Thus, the function $\hat{\chi}_1(y_1, l_2 - y_2)$ also solves the problem (8). Since the solution of (8) is unique it means that $\hat{\chi}_1(y_1, y_2) = \hat{\chi}_1(y_1, l_2 - y_2)$, i.e. the function $\hat{\chi}_1$ is "even in variable y_2 with respect to axis $y_2 = \frac{l2}{2}$". Finally, because this symmetry holds and the function $\hat{\chi}_1 \in H^1_{\text{per}}(Y)$ we can prescribe to the boundary $y_2 = 0$ and $y_2 = l_2$ homogenous Neumann conditions.

So, the boundary conditions for function $\hat{\chi}$ are: $\hat{\chi}_1(y_1, 0) = c, \hat{\chi}_1(y_1, l_2) = c, \frac{\partial\hat{\chi}_1}{\partial y_2}(0, y_2) = 0, \frac{\partial\hat{\chi}_1}{\partial y_2}(l_2, y_2) = 0$, where constant c is determined from the condition $\mathcal{M}_Y(\hat{\chi}_i) = 0$. For the function $\hat{\chi}_2$ the reasoning is analogical.

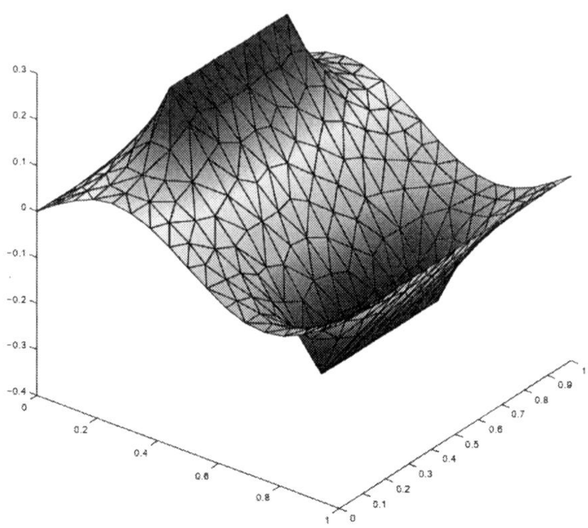

Fig. 2. Auxiliary function $\hat{\chi}_1$

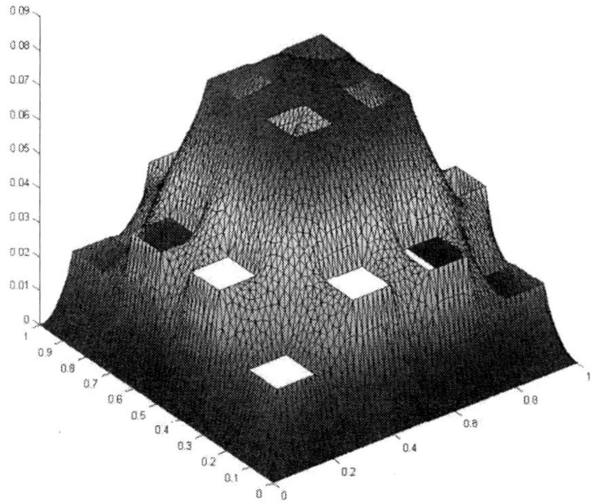

Fig. 3. Graph of function $u_{1/4}$

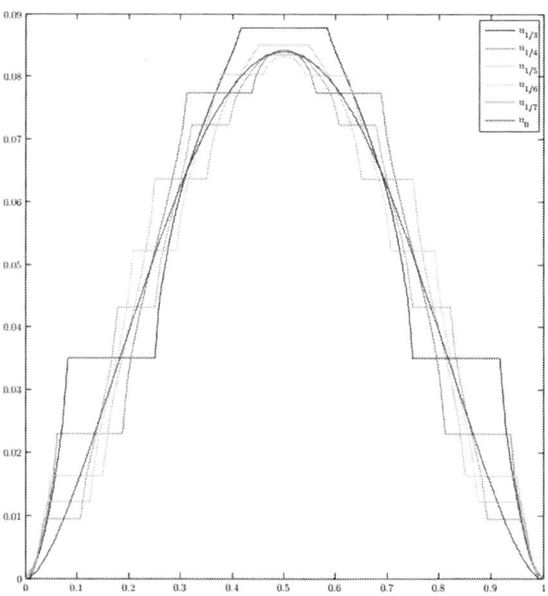

Fig. 4. Diagonal cuts of functions u_ε, for $\varepsilon = {}^1/_3, {}^1/_4, {}^1/_5, {}^1/_6, {}^1/_7$, and of homogenized solution u_0

Acknowledgements

This research has been supported by the project of specific university research of Brno University of Technology, No. FAST-S-16-3385, in cooperation with the project LO1408 AdMaS UP (Advanced Materials, Structures and Technologies, Ministry of Education, Youth and Sports of the Czech Republic, National Sustainability Programme I).

References

[1] Bensoussan, A., Lions, J.-L., Papanicolaou, G.: *Asymptotic Analysis for Periodic Structures*, Studies in Mathematics and Its Applications, North-Holland Publishing Company, 1978, ISBN 9780444851727.

[2] Cioránescu, D., Damlamian, A., Griso, G.: *The periodic unfolding method in homogenization*, SIAM Journal on Mathematical Analysis, vol. 40, no. 4, (2008), pp. 1585–1620, ISSN 0036-1410.

[3] Cioránescu, D., Donato, P.: *An Introduction to Homogenization*, Oxford lecture series in mathematics and its applications, Oxford University Press, 1999, ISBN 9780198565543.

[4] Franců, J., Nováčková, P., Janíček, P.: *Torsion of a non-circular bar*, Engineering Mechanics, vol. 19, no. 1, (2012), pp. 45–60, ISSN 1805-4633.

[5] Franců, J., Rozehnalová, P.: *Torsion of a bar with holes*, Engineering Mechanics, vol. 22, no. 1, , pp. 3–24, ISSN 1805-4633.

[6] Rozehnalová, P.: *Homogenization in Perforated Domains*, Ph.D. thesis, Brno University of Technology, 2016.

Current address

Rozehnalová Petra, Ing. PhD.
Institute of Mathematics and Descriptive Geometry, Faculty of Civil Engineering,
Brno University of Technology
Veveří 95, 602 00 Brno, Czech Republic
E-mail : rozehnalova.p@fce.vutbr.cz

Franců Jan, prof. RNDr. CSc.
Institute of Mathematics, Faculty of Mechanical Engineering,
Brno University of Technology
Technická 2, 616 69 Brno, Czech Republic
E-mail: francu@fme.vutbr.cz

Slovak University of Technology in Bratislava
Faculty of Mechanical Engineering
16th Conference on Applied Mathematics APLIMAT 2017
Proceedings

ON GLOBAL GEODESIC MAPPINGS OF QUADRICS OF REVOLUTION

RÝPAROVÁ Lenka (CZ), MIKEŠ Josef (CZ)

Abstract. In this paper we study quadrics of revolution. Quadrics of revolution (except circular cylinder) admit nontrivial geodesic mappings and deformations in which they remain rotational. Surfaces (except circular cylinder and sphere) obtained in this geodesic deformation are no longer quadrics.

Keywords: geodesic mapping, geodesic deformation, quadric of revolution, cylinder, cone, hyperboloid, paraboloid

Mathematics subject classification: Primary 53B20; Secondary 53A05, 53C22

1 Indroduction

In 1865, E. Beltrami begin to study geodesic mappings of spaces of constant curvature [1]. Geodesic mappings of surfaces in Euclidean space $\mathcal{E}_3$ were studied by U. Dini [2]. The general properties of the theory of geodesic mappings of Riemannian spaces have been studied by Levi–Civita [5] and others, see [10, 11, 12, 14]. Many results of general theory of geodesic mappings and deformations can be found in [10, 11, 12, 13, 14].

Geodesic deformations of surfaces of revolution were studied in [9]. A globally geodesic deformation of a sphere was constructed in [8]. In the paper [3] global geodesic mappings and transformations of an ellipsoid were constructed. It was proved that many Riemannian spaces do not have trivial geodesic mappings [6, 10, 11, 12, 13, 14]. For these global mappings the condition of compactness is an especially strong one [4, 7, 10, 15].

In this paper we present geodesic deformations of quadrics of revolution. We study geodesic mappings that deform quadrics to surfaces of revolution, which are no longer quadrics.

2 Surfaces of revolution

A diffeomorphism between surfaces $\mathcal{S}_2$ and $\bar{\mathcal{S}}_2$ is called a *geodesic mapping* if any geodesic on $\mathcal{S}_2$ maps onto a geodesic on $\bar{\mathcal{S}}_2$.

A *geodesic deformation* is a deformation of $\mathcal{S}_2$ which preserves geodesics.

Let $\mathcal{S}_2$ be a surface of revolution in Euclidean space $\mathcal{E}_3$ which is in Cartesian coordinates (x, y, z) given by the equations

$$x = r(w) \cos t, \quad y = r(w) \sin t, \quad z = z(w),$$

where $w \in \langle w_1, w_2 \rangle,\ t \in \langle 0, 2\pi)$. The parameters w_1 and w_2 can reach values $\pm\infty$.

The metric of the surface $\mathcal{S}_2$ has the form

$$\mathrm{d}s^2 = a(w)\,\mathrm{d}w^2 + b(w)\,\mathrm{d}t^2, \tag{1}$$

where $a(w)$ and $b(w)$ are non zero functions and where

$$a(w) = r'^2(w) + z'^2(w) \quad \text{and} \quad b(w) = r^2(w).$$

Let us suppose that the surface $\mathcal{S}_2$ maps geodesically onto a surface of revolution $\bar{\mathcal{S}}_2$, such that parallels and meridians on $\mathcal{S}_2$ map onto parallels and meridians on $\bar{\mathcal{S}}_2$. From this follows that the equations of the surface $\bar{\mathcal{S}}_2$ have the following form

$$x = \bar{r}(w) \cos t, \quad y = \bar{r}(w) \sin t, \quad z = \bar{z}(w),$$

where w is the same parameter as on the surface $\mathcal{S}_2$ and $t \in \langle 0, 2\pi)$.

The metric of the surface $\bar{\mathcal{S}}_2$ has the following form

$$\mathrm{d}\bar{s}^2 = \left(\bar{r}'^2(w) + \bar{z}'^2(w)\right)\mathrm{d}w^2 + \bar{r}^2(w)\,\mathrm{d}t^2.$$

It was proved by Mikeš [9], see also [10] (p. 302), that the surface $\mathcal{S}_2$ with the metric (1) maps geodesically onto the surface $\bar{\mathcal{S}}_2$ with the metric

$$\mathrm{d}\bar{s}^2 = \frac{p\,a(w)}{(1 + qb(w))^2}\,\mathrm{d}w^2 + \frac{p\,b(w)}{1 + qb(w)}\,\mathrm{d}t^2,$$

where p and q are real parameters, t and w are coordinates.

The surface $\bar{\mathcal{S}}_2$ can be obtained from the surface $\mathcal{S}_2$ using these transformations:

$$\bar{r}(w) = \frac{r(w)}{\sqrt{1 + qr^2(w)}}, \tag{2a}$$

$$\bar{z}(w) = \int_{w_0}^{w} \sqrt{\frac{z'^2 + qr^2\left(r'^2 + z'^2\right)}{(1 + qr^2)^3}}\,\mathrm{d}\tau, \tag{2b}$$

where q is a parameter. The coordinate w is the same as on the surface $\mathcal{S}_2$. The above mentioned functions $\bar{r}$ and $\bar{z}$ have to fulfill the condition of smoothness at the poles $w = w_1$ and $w = w_2$, where $\bar{r} = 0$, namely $\dfrac{\mathrm{d}\bar{r}}{\mathrm{d}w} = \pm 1$ and $\dfrac{\mathrm{d}\bar{z}}{\mathrm{d}w} = 0$.

The mapping is a nontrivial geodesic mapping if $q \neq 0$ and $b \not\equiv 0$, and the transformation is nontrivial if $b \not\equiv 0$.

We note that a projective deformation of an n-dimensional sphere was constructed in [8] and geodesic deformations of an ellipsoid of revolution were studied in [3].

Therefore, in the following part of the paper, we study the above mentioned transformation on surfaces of revolution: circular cone, circular cylinder, one-sheet hyperboloid of revolution, two-sheet hyperboloid of revolution and paraboloid of revolution. We restrict our study to $q \geq 0$.

3 Geodesic mapping of a circular cylinder

Let $\mathcal{S}_2$ be a circular cylinder given by the parametric equations

$$x = k\cos t, \quad y = k\sin t, \quad z = w,$$

where k (> 0) is its radius. The metric of the surface $\mathcal{S}_2$ has the form $\mathrm{d}s^2 = \mathrm{d}w^2 + k^2\,\mathrm{d}t^2$.

Using (2), the surface $\bar{\mathcal{S}}_2$ is characterized by the following equations

$$\bar{r}(w) = \frac{k}{\sqrt{1+qk^2}} \quad \text{and} \quad \bar{z}(w) = \int_{w_1}^{w} \sqrt{\frac{1+qk^2}{(1+qk^2)^3}}\,\mathrm{d}\tilde{w} = \frac{1}{1+qk^2}(w - w_1).$$

Evidently, $\bar{r}(w)$ is a constant, therefore surface $\bar{\mathcal{S}}_2$ is also a circular cylinder. Moreover, from above follows, that the geodesic transformation is trivial.

4 Geodesic mapping of a circular cone

Let $\mathcal{S}_2$ be a circular cone given by the parametric equations

$$x = w\cos t, \quad y = w\sin t, \quad z = w,$$

Now let $w > 0$, thus we are considering the top part of the cone. For the value $w = 0$ we get the wertex of the cone which is a singular point.

The equations of a generalized circular cone can be written in the following form

$$x = aw\cos t, \quad y = aw\sin t, \quad z = cw,$$

where $a,\, c \in \mathbb{R}$. We choose $a = c = 1$. Any other circular cone can be obtained from the considered one by an affine transformation of $\mathcal{E}_3$.

The metric of the circular cone $\mathcal{S}_2$ has the form $\mathrm{d}s^2 = 2\,\mathrm{d}w^2 + w^2\,\mathrm{d}t^2$.

Using the transformation equations (2) we get

$$\bar{r}(w) = \frac{w}{\sqrt{1+qw^2}} \quad \text{and} \quad \bar{z}(w) = \int_{0}^{w} \sqrt{\frac{1+2q\tilde{w}^2}{(1+q\tilde{w}^2)^3}}\,\mathrm{d}\tilde{w}. \qquad (3)$$

We can choose the lower bound of the integral as 0. For $q = 0$ is $\bar{\mathcal{S}}_2 \equiv \mathcal{S}_2$. If $q > 0$ then $\bar{r}(w) > 0$.

The metric of the surface $\bar{\mathcal{S}}_2$ has the following form

$$\mathrm{d}\bar{s}^2 = \frac{2}{(1+qw^2)^2}\,\mathrm{d}w^2 + \frac{w^2}{1+qw^2}\,\mathrm{d}t^2.$$

We consider the following limits of equations (3) for $w \to \infty$

$$\bar{r}_{\max} = \lim_{w\to\infty} \bar{r}(w) = \lim_{w\to\infty} \frac{w}{\sqrt{1+qw^2}} = \frac{1}{\sqrt{q}}$$

and

$$\bar{z}_{\text{max}} = \lim_{w\to\infty} \bar{z}(w) = \lim_{w\to\infty} \int_0^w \sqrt{\frac{1+2q\tilde{w}^2}{(1+q\tilde{w}^2)^3}}\, \mathrm{d}\tilde{w} \le \lim_{w\to\infty} \int_0^w \frac{\sqrt{2}}{1+q\tilde{w}^2}\, \mathrm{d}\tilde{w} =$$

$$= \sqrt{\frac{2}{q}} \lim_{w\to\infty} \arctan\left(\sqrt{q}\, w\right) = \frac{\pi}{\sqrt{2q}}\,.$$

The transformation of the surface $\mathcal{S}_2$ onto $\bar{\mathcal{S}}_2$ is not trivial. The circular conical surface deforms onto the surface, which is no longer a quadric. For this mapping all geodesics (except "parallels") map onto incomplete geodesics on the surface $\bar{\mathcal{S}}_2$.

Let $\bar{\mathcal{S}}_2$ be an image of the cone $\mathcal{S}_2$ in non trivial geodesic deformation.

Proposition 1 *A circular conical surface admits a nontrivial global geodesic transformation in which it remains rotational. The surface $\bar{\mathcal{S}}_2$ is no longer a quadric.*

5 Geodesic mapping of an one-sheet hyperboloid of revolution

Let $\mathcal{S}_2$ be a one-sheet hyperboloid of revolution given by the parametric equations

$$x = \cosh w\, \cos t, \quad y = \cosh w\, \sin t, \quad z = \sinh w,$$

The metric of the surface $\mathcal{S}_2$ has the form

$$\mathrm{d}s^2 = (\sinh^2 w + \cosh^2 w)\, \mathrm{d}w^2 + \cosh^2 w\; \mathrm{d}t^2.$$

Using transformations (2) we get

$$\bar{r}(w) = \frac{\cosh w}{\sqrt{1+q\cosh^2 w}} \tag{4}$$

and

$$\bar{z}(w) = \int_0^w \frac{\cosh\tilde{w}}{1+q\cosh^2\tilde{w}} \cdot \sqrt{\frac{1+q\,(2\cosh^2\tilde{w}-1)}{1+q\cosh^2\tilde{w}}}\, \mathrm{d}\tilde{w}. \tag{5}$$

Notice that the functions $\bar{r}(w)$ and $\bar{z}(w)$ are monotonically increasing for $w \ge 0$ and $q > 0$. Firstly, we consider the limit of (4) for $w \to \infty$

$$\bar{r}_{\text{max}} = \lim_{w\to\infty} \frac{\cosh w}{\sqrt{1+q\cosh^2 w}} = \frac{1}{\sqrt{q}}.$$

Instead of an explicit solution of the integral (5) let us consider

$$\bar{z}_{\text{max}} = \lim_{w\to\infty} \int_0^w \frac{\cosh\tilde{w}}{1+q\cosh^2\tilde{w}} \cdot \sqrt{\frac{1+q\,(2\cosh^2\tilde{w}-1)}{1+q\cosh^2\tilde{w}}}\, \mathrm{d}\tilde{w}.$$

The limit exists and we can estimate its value in this manner

$$\bar{z}_{\text{max}} \le \lim_{w\to\infty} \int_0^w \frac{\cosh\tilde{w}}{1+q\cosh^2\tilde{w}} \sqrt{2}\, \mathrm{d}\tilde{w} \le \frac{1}{q} \lim_{w\to\infty} \int_0^w \frac{\cosh\tilde{w}}{\cosh^2\tilde{w}}\, \mathrm{d}\tilde{w} =$$

$$= \frac{1}{q} \lim_{w\to\infty} \int_0^w \frac{1}{\cosh\tilde{w}}\, \mathrm{d}\tilde{w} = \frac{\pi}{2q}.$$

From this follows that

$$\bar{z}_{\max} \leq \frac{\pi}{2q} \quad \text{and similarly} \quad \bar{z}_{\min} \geq -\frac{\pi}{2q}.$$

The geodesics on the one-sheet hyperboloid of revolution are complete. Evidently, the geodesics on the surface $\bar{S}_2$ are incomplete (except "parallels").

Let $\bar{S}_2$ be an image of the one-sheet hyperboloid S_2 in non trivial geodesic deformation.

Proposition 2 *A circular hyperboloid of one sheet admits a nontrivial geodesic transformation in which it remains rotational. The surface $\bar{S}_2$ is no longer a quadric.*

6 Geodesic mapping of a two-sheet hyperboloid of revolution

Let S_2 be a two-sheet hyperboloid of revolution given by the equations

$$x = \sinh w \cos t, \quad y = \sinh w \sin t, \quad z = \cosh w.$$

Note that the above mentioned equations describe only the top part of the hyperboloid. The metric of the hyperboloid has the following form

$$\mathrm{d}s^2 = (\cosh^2 w + \sinh^2 w)\, \mathrm{d}w^2 + \sinh^2 w \ \mathrm{d}t^2.$$

Using formulas (2), we obtain the transformation equations for the hyperboloid

$$\bar{r}(w) = \frac{\sinh w}{\sqrt{1 + q \sinh^2 w}}, \tag{6}$$

$$\bar{z}(w) = \int_{w_0}^{w} \frac{\sinh \tilde{w}}{1 + q \sinh^2 \tilde{w}} \cdot \sqrt{\frac{1 + q\,(2 \sinh^2 \tilde{w} + 1)}{1 + q \sinh^2 \tilde{w}}}\, \mathrm{d}\tilde{w}. \tag{7}$$

For the pole $w_1 = 0$ where $\bar{r}(w_1) = 0$, the smoothness conditions are fulfilled, therefore $\frac{\mathrm{d}\bar{r}}{\mathrm{d}w}(w_1) = 1$ and $\frac{\mathrm{d}\bar{z}}{\mathrm{d}w}(w_1) = 0$.

For $q = 0$ formulas (6) and (7), when $w_0 = 0$, have the solution $\bar{r} = \sinh w$, $\bar{z} = \cosh w$. For $q \approx 0$ these solutions are $\bar{r} \approx \sinh w$ and $\bar{z} \approx \cosh w$, which can be verified by integration of Taylor series.

Similarly as on the one-sheet hyperboloid of revolution, we study limits of (6) and (7) for $w \to \infty$

$$\bar{r}_{\max} = \lim_{w \to \infty} \frac{\sinh w}{\sqrt{1 + q \sinh^2 w}} = \frac{1}{\sqrt{q}}$$

and

$$\bar{z}_{\max} = \lim_{w \to \infty} \int_{1}^{w} \frac{\sinh \tilde{w}}{1 + q \sinh^2 \tilde{w}} \cdot \sqrt{\frac{1 + q\,(2 \sinh^2 \tilde{w} + 1)}{1 + q \sinh^2 \tilde{w}}}\, \mathrm{d}\tilde{w} \leq$$

$$\leq \lim_{w \to \infty} \int_{1}^{w} \frac{\sqrt{2} \sinh \tilde{w}}{1 + q \sinh^2 \tilde{w}}\, \mathrm{d}\tilde{w} \leq \lim_{w \to \infty} \int_{1}^{w} \frac{1}{q \sinh \tilde{w}}\, \mathrm{d}\tilde{w} =$$

$$= -\frac{1}{q} \ln \left(\tanh \left(\frac{1}{2} \right) \right) \approx \frac{1}{q} \cdot 0,77194.$$

Evidently, from the calculated limits, the two-sheet hyperboloid of revolution deforms to a surface of revolution which is bounded in space (as in example above). Note, that all geodesics on $\bar{S}_2$ (except "parallels") are incomplete.

Let $\bar{S}_2$ be an image of the hyperboloid S_2 in non trivial geodesic deformation.

Proposition 3 *A circular hyperboloid of two sheets admits a nontrivial geodesic mapping in which it remains rotational. The surface $\bar{S}_2$ is no longer a quadric.*

7 Geodesic mapping of a paraboloid of revolution

Let S_2 be a paraboloid of revolution given by the equations

$$x = w\cos t, \quad y = w\sin t, \quad z = w^2,$$

The metric of the paraboloid has the form $\mathrm{d}s^2 = (1+4w^2)\,\mathrm{d}w^2 + w^2\,\mathrm{d}t^2$.

The transformation equations for this paraboloid are

$$\bar{r}(w) = \frac{w}{\sqrt{1+qw^2}} \quad \text{and} \quad \bar{z}(w) = \int_0^w \frac{\tilde{w}}{1+q\tilde{w}^2}\cdot\sqrt{\frac{4+q(1+4\tilde{w}^2)}{1+q\tilde{w}^2}}\,\mathrm{d}\tilde{w}. \tag{8}$$

We verify if the smoothness conditions are satisfied for the pole $w = w_1$ where $\bar{r} = 0$. Evidently, for the pole $w_1 = 0$ applies $\frac{\mathrm{d}\bar{z}}{\mathrm{d}w}(w_1) = 0$ and further $\frac{\mathrm{d}\bar{r}}{\mathrm{d}w}(w_1) = 1$, thus both conditions are fulfilled.

As for the previous surfaces we find limits of (8) for $w \to \infty$

$$\bar{r}_{\max} = \lim_{w\to\infty} \frac{w}{\sqrt{1+qw^2}} = \frac{1}{\sqrt{q}},$$

$$\bar{z}_{\max} = \lim_{w\to\infty} \int_0^w \frac{\tilde{w}}{1+q\tilde{w}^2}\cdot\sqrt{\frac{4+q\,(1+4\tilde{w}^2)}{1+q\tilde{w}^2}}\,\mathrm{d}\tilde{w} \leq \lim_{w\to\infty} \frac{2}{2q}\ln|1+qw^2| = +\infty.$$

From the calculated limits and transformation equations (8) it follows that the circular paraboloid S_2 maps non trivially geodesically onto the surface $\bar{S}_2$ which is no longer a quadric.

Proposition 4 *A circular paraboloid admits a nontrivial geodesic deformation in which it remains rotational, but the obtained surface is no longer a quadric.*

Finally, the results can be summarized in form of a theorem.

Theorem 1 *Quadrics of revolution (except a circular cylinder) admit nontrivial geodesic mappings and deformations, in which they remain rotational. Surfaces (except circular cylinder and sphere) obtained in these geodesic deformations are no longer quadrics.*

References

[1] Beltrami, E.: *Teoria fondamente degli spayii di curvatura constante*, Annali di Mat. (2), vol. 2, (1868), pp. 232–255.

[2] Dini, U.: *Sopra un problema che si presenta nella teoria generale delle rappresentazione geografiche di una superficie su di un'altra*, Annali di Mat. (2), vol. 3, (1869), pp. 269–294.

[3] Hinterleitner, I.: *On global geodesic mappings of ellipsoids*, AIP. Conf. Proc., vol. 1460, (2012), pp. 180–184.

[4] Hinterleitner, I.: *Geodesic mappings on compact Riemannian manifolds with conditions on sectional curvature*, Publ. Inst. Math., vol. 94(108), (2013), pp. 125–130.

[5] Levi-Civita, T.: *Sulle trasformazioni dello equazioni dinamiche.*, Annali di Mat. (2), vol. 24, (1896), pp. 255–300.

[6] Mikeš, J.: *Geodesic mappings of special Riemannian spaces*, Colloq. Math. Soc. János Bolyai, vol. 46, (1988), pp. 793–813.

[7] Mikeš, J.: *On concircular global vector in compact riemannian spaces*, Dep. in Ukr. NIINTI, vol. 615, (1988), pp. 1–19.

[8] Mikeš, J.: *On the existence of n-dimensional compact Riemannian spaces that admit nontrivial projective transformations "in the large"*, Dokl. Akad. Nauk SSSR, vol. 305, no. 3, (1989), pp. 534–536.

[9] Mikeš, J.: *On existence of nontrivial global geodesic mappings of n-dimensional compact surfaces of revolution*, in *Differential geometry and its applications (Brno, 1989)*, World Sci. Publ., Teaneck, NJ, 1990, pp. 129–137.

[10] Mikeš, J., et al.: *Differential geometry of special mappings*, Palacky Univ. Press, Olomouc, 2015.

[11] Mikeš, J., Vanžurová, A., Hinterleitner, I.: *Geodesic mappings and some generalizations*, Palacky Univ. Press, Olomouc, 2009.

[12] Petrov, A. Z.: *New methods in the general relativity theory*, Nauka, Moscow, 1966.

[13] Radulovich, Z., Mikeš, J., Gavrilchenko, M. L.: *Geodesic mappings and deformations of Riemannian spaces*, CID, Podgorica; Odessa Univ. Publ., Odessa, 1997.

[14] Sinyukov, N. S.: *Geodesic mappings of Riemannian spaces*, Nauka, Moscow, 1979.

[15] Vrănceanu, G.: *Global properties of Riemannian spaces with constant affine connection*, Stud. si Cerc. Mat. Acad. RPR, vol. 14, (1963), pp. 7–22.

Current address

Rýparová Lenka, Mgr.
Department of Algebra and Geometry, Faculty of Science, Palacký University
17. listopadu 12, 771 46 Olomouc, Czech republic
Tel. number: +420 774 131 516, e-mail: lenka.ryparova01@upol.cz

Mikeš Josef, prof. RNDr. DrSc.
Department of Algebra and Geometry, Faculty of Science, Palacký University
17. listopadu 12, 771 46 Olomouc, Czech republic
Tel. number: +420 731 579 395, e-mail: josef.mikes@upol.cz

Slovak University of Technology in Bratislava
Faculty of Mechanical Engineering
16th Conference on Applied Mathematics **APLIMAT 2017**
Proceedings

HIGH SCHOOL OF BUSINESS PUPILS′ BACKGROUND AND MATHEMATICAL COMPETENCES

ŠABÍKOVÁ Henrieta (SK)

Abstract. Pupils′ background and their mathematical competences could be considered as basic factors that influence their high school study success. Based on previous teaching experience, characteristics of pupils′ background include origin (town or village), gender and number of pupils in their last class at the elementary school. Pupils′ mathematical competences could be characterized by their results of the entry exam in mathematics and by their math grade point average at the elementary school. The structure of pupils′ background and their mathematical competences will be analysed at the general and bilingual section of high school of business. This analysis should be initial resource of the experiment focused to bilingual English-Slovak mathematics education at high school of business and expected pupils results in mathematics.

Keywords: High school of business. Experiment. Bilingual and general mathematics education. Pupils′ background. Mathematical competences

Mathematics Subject Classification: Primary 97B10; Secondary 97C60

1 Intoduction

Pupils′ background and their mathematical competences could be considered as basic factors that influence their high school study success. Based on previous teaching experience, characteristics of pupils′ background include origin (town or village), gender and number of pupils in their last class at the elementary school. Pupils′ mathematical competences could be characterized by their results of the entry exam in mathematics and by their math grade point average at the elementary school. The structure of pupils′ background, their mathematical competences based on analysis at the general and bilingual section of high school of business will be compared in the contribution.

2 Pupils′ background

Based on previous teaching experience, pupils′ background include gender because of different problem solution ways, origin because of different educational system, number of pupils in their last class at the elementary school because of different learning possibilities.

2.1 Gender – Boys or girls

The first graphs illustrate proportion of boys and girls at the bilingual and at the general section of high school of business. It can be seen aproximentally equivalent distribution . The boys are less than quarter of all students, the girls are in the majority.

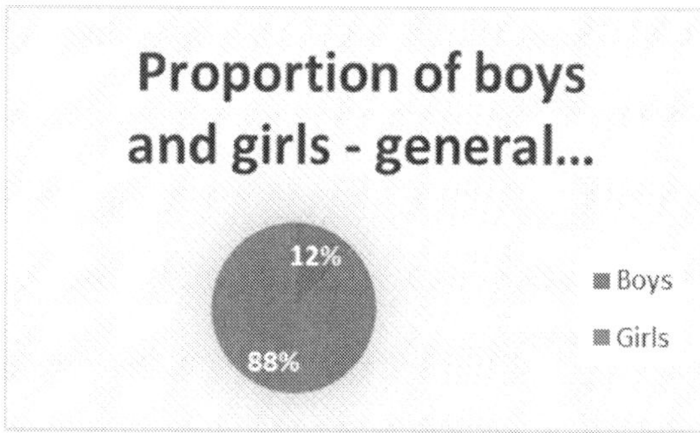

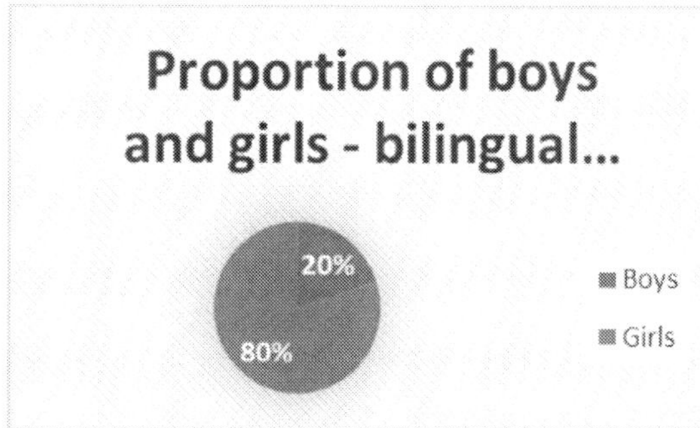

2.2 Origin – town or village

The second characteristic is pupils´ origin at the high school of business. By origin it is mean wheather the pupils are from town or village elementary school. The 60 % of bilingual and 65% of general education pupils came from village elementary school. Again we can talk about conformity in this charakteristic.

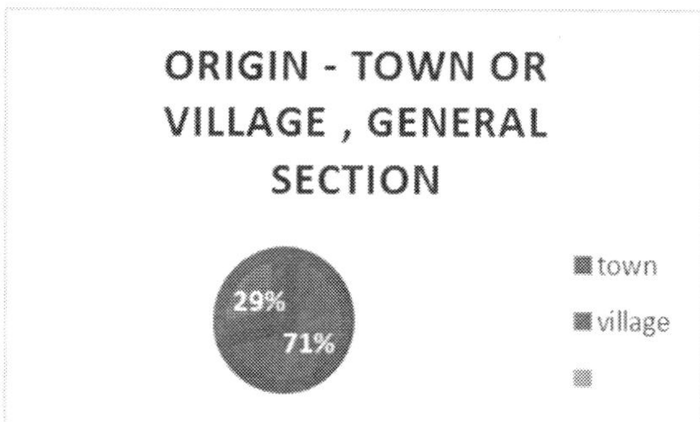

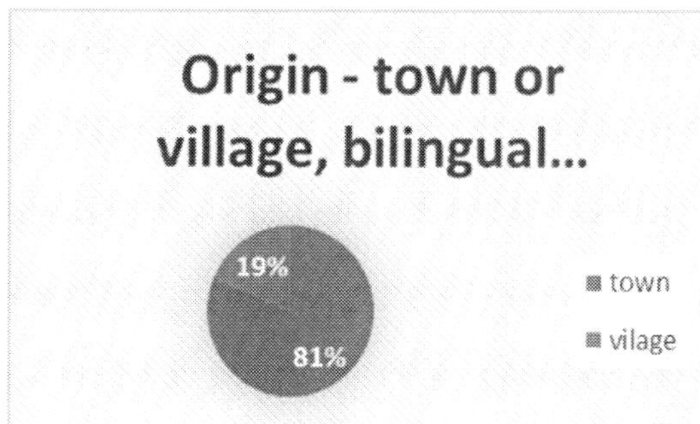

2.3 Number of pupils in the last class at the elementary school

The third question of analysis is: How many pupils were in the last class at the elementary school? The pupils were put into four groups. Bilingual pupils were average in the 23 members class and general education pupils were average in the 18 members class in the last year at the elementary school. This characteristic indicate andvantage of general education pupils at the high school of business.

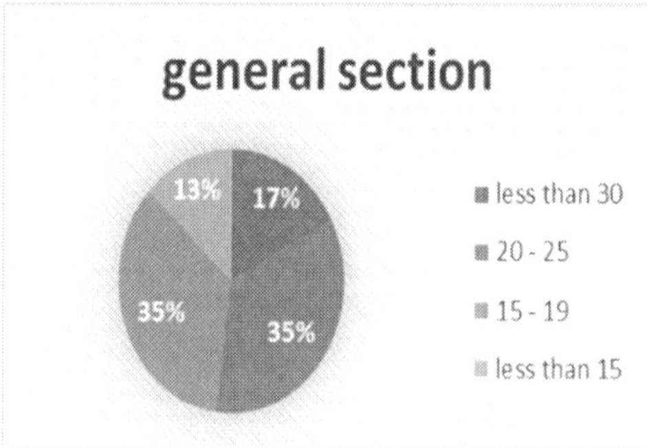

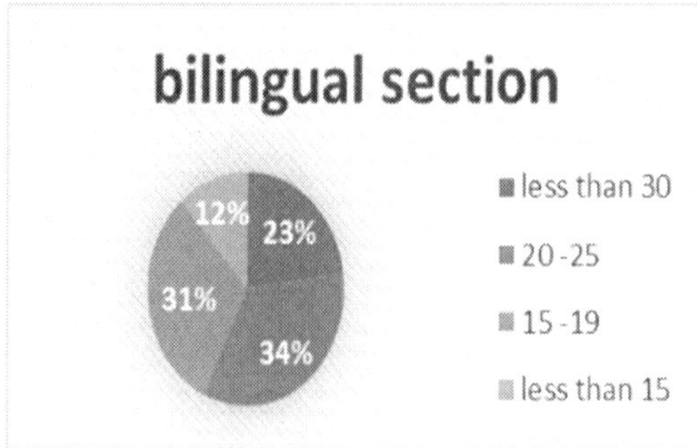

3 Mathematical competences

Pupils´competences could be characterized by their math grade average at the elementary school and by the results of the entry exam in mathematics.

3.1 Pupils´math grade point average at the elementary school

The majority of pupils achieved grade point average 2. Exactly bilingual education pupils obtained 2,30 and the general education pupils obtained 2,49. Seeing that difference is minimum, mathematical competences can be expected at the same level.

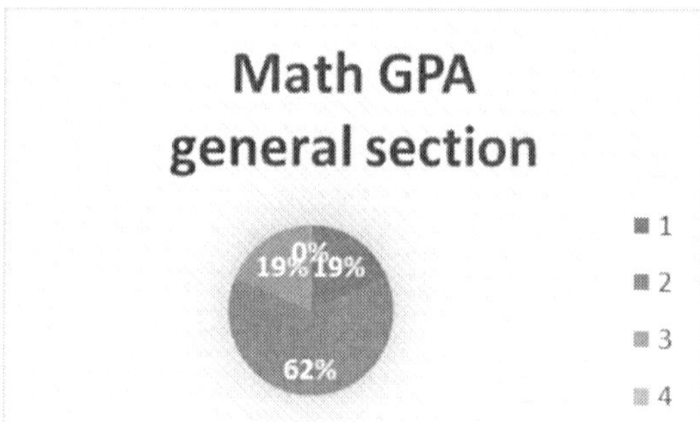

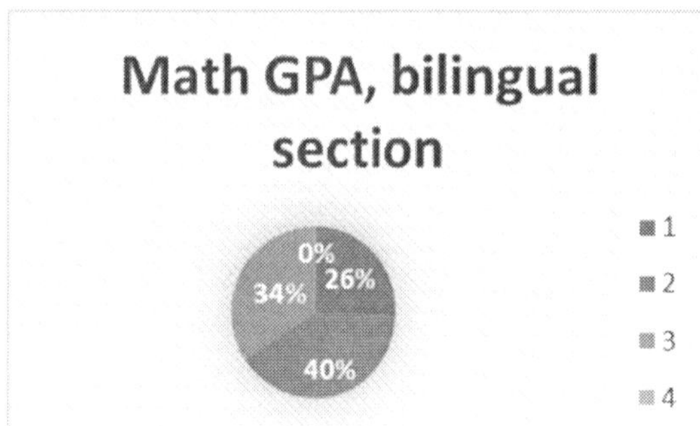

3.2 The results of the entry exam at the high school of business

Pupils mathematical competences could be also characterized by the results of the entry exam in mathematics. Pupils were classified from one to five (one as the best). The majority of bilingual and also general education students obtained grade 2. The results of bilingual and general section are very similar, 72,14% for bilingual education pupils and 68,59% for general education pupils.

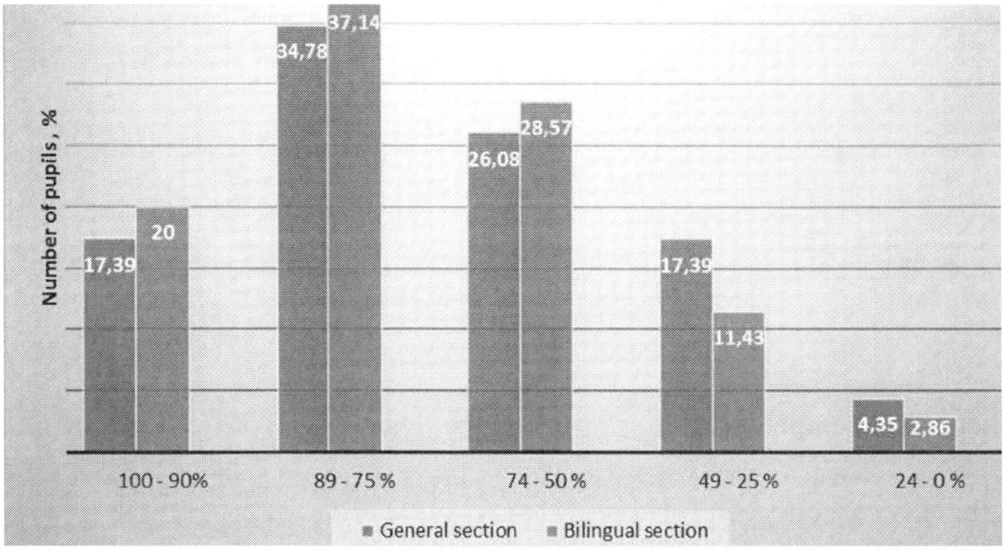

4 Conclusion

The profile of our high school´s pupil could be created based on these characteristics Average pupil in the bilingual (general) class is the girl, who came from village elementary school, in the last year she was in the class with less than 25(20) pupils, her math grade average was 2,30 (2,49) and her result of the entry exam is 72,14% (68,59).

This analysis should be initial resource of the experiment focused to bilingual English - Slovak mathematics education at high school of business and expected pupils´ results in mathematics.

Acknowledgement

The article and its presentation were supported by the project "Improving Quality of Higher Education Based on Development of Multilateral Institutional Cooperation"; EEA grants; project number EHP-11/SK06-IV-01-008/2014.

References

[1] BALL, P. What is CLIL [online] [cit. 2012-1-3]. One stop English, 2000 - 2012. Dostupný z WWW: <http://www.onestopenglish.com/support/methodology/teaching approaches/what-is-clil/156604.article>.

[2] BIALYSTOK, E. (2001). *Bilingualism in development: Language, literacy, and cognition.* Cambridge: Cambridge University Press.

[3] BONESRONNIG, H. (2003): *Class Size Effects on Student Achievement in Norway: Patterns and Explanations.* Southern Economic Journal, 69(4), 952-965.

[4] CLARKSON Philip, GALBRAITH P.: (1992). *Bilingualism and Mathematical Learning: Another Perspective.* In: *Journal for Research in Mathematics Education,* 23(1), 34-44.

Current address

Šabíková Henrieta, PaedDr.
Constantine the Philosopher University in Nitra, Slovak Republic
Tr. A. Hlinku 1, 979 74 Nitra, Slovak Republic
E-mail: henrieta.sabik@gmail.com

Slovak University of Technology in Bratislava
Faculty of Mechanical Engineering
16th Conference on Applied Mathematics **APLIMAT 2017**
Proceedings

OBSERVING HOW FUTURE PRIMARY SCHOOL TEACHERS REASON AND GENERALIZE: THE CASE OF NUMBER TRIANGLES AND CONCEPT CARTOONS

SAMKOVÁ Libuše (CZ), TICHÁ Marie (CZ)

Abstract. The contribution focuses on the possibility to use an educational tool called Concept Cartoons as a diagnostic instrument in problem solving and problem posing activities of future primary school teachers. The aim of the presented study is to observe which aspects of future primary school teachers' knowledge related to reasoning and generalization could be investigated through Concept Cartoons that are based on a substantial learning environment called "Number triangles".

Keywords: future primary school teachers, reasoning, generalization, Concept Cartoons, substantial learning environment, number triangles

Mathematics Subject Classification: Primary 97E50; Secondary 97F99, 97B50

"In general, mathematics can be described as the search for structures and patterns that bring order and simplicity to our universe." [6: p. 1]

1 Introduction

In our contribution at last year's APLIMAT conference [22] we focused on views of proof of future primary school teachers, and introduced Concept Cartoons as a form of classroom intervention aiming to promote in future primary school teachers the awareness of the need of proper argumentation, and lead them towards deductive argumentation. Now we shall present Concept Cartoons in a slightly different role, as a diagnostic tool in problem solving and problem posing activities of future primary school teachers. We shall again focus on aspects related to proving, namely on reasoning and argumentation that appear as an integral part of abstraction and generalization practices.

The art of abstraction and generalization is one of the most important for formation and development of mathematics [14], its cultivation is often emphasized among requirements on mathematics education. The starting point at pre-school and primary school levels is often grounded in informal looking for patterns and coherence. Such approach, just more formal, then constitutes mathematics at higher levels, giving it its power by bringing clarity to relations between objects and processes from diverse fields of mathematics or other sciences [6].

For many years we have been exploring the potential of discovering patterns and coherence in building the concept of arithmetic of natural numbers at primary school level, especially in the development of functional approach to problem solving [10, 11]. We understand such activities as crucial in mathematics education.

The aim of our study is to observe which aspects of future primary school teachers' knowledge related to reasoning and generalization could be investigated through Concept Cartoons during problem solving and problem posing. The role of the study is to get insight into the problematics, so that we choose qualitative design.

2 Background

2.1 Generalization, reasoning and patterns

Polya [20] characterizes generalization as "passing from the consideration of a given set of objects to that of a larger set, containing the given one" [20: p. 12]. To get a more precise characterization, we shall follow the construction presented and used by Ciosek [2]. In her study she combined two theoretical models: Dörfler's theory of generalization [4], and Krygovska's types of generalizations of theorems [13]. Dörfler describes generalization as a theoretical model of processes, with an initial stage called constructive abstraction. On the other hand, Krygovska's typology of generalization of statements consists of four types of generalization: through induction, through generalizing the reasoning, through unifying specific cases, and through perceiving recurrence. In our study we shall focus on the process of constructive abstraction, and on generalization through generalizing the reasoning. The constructive abstraction can be characterized as follows:

- The starting point is an **action** or system of actions (material, imagined or symbolic). Elements of this action are certain objects (material or ideal).

- Course of this action direct one's attention to some relation between the elements of the actions. This relation proves to be steady when the actions are repeated. They are called **invariant** (or schema) of the action.

- Stating invariants need a symbolic description; one has to introduce symbols for the elements of the actions, and then describe invariants stated by means of these symbols. This stating of invariants and their symbolic description have the character of a process of abstraction. It is constructive abstraction because what is abstracted is constituted by the action. [2: pp. 39-40]

The symbols used for the elements of the actions may be of various natures: verbal, iconic, geometric, or algebraic [4].

The generalization through generalizing the reasoning can be characterized this way:

> One notices that the reasoning carried out in a single case will remain correct in a different setting or minor modifications will be needed only to get a more general result. This often happens as the result of "**variation of constants**" or spontaneously resulting in the analysis of the proof. [2: p. 40]

The generalization process is closely related to patterns. In mathematics, a pattern may be described as "any predictable regularity, usually involving numerical, spatial or logical relationships" [17: p. 34]. The art of recognizing patterns plays an important role in the process of variation of constants; it helps to identify steady relations (invariants) of actions. Experiences in discovering and generalizing patterns help to develop analogical reasoning [5], to deepen conceptual understanding [19], etc.

Drawing on our experiences, we see pattern generalization as a four-step process: the solver

(i) conducts various inquiries with the observed elements, in order to look for commonalities;

(ii) finds a commonality;

(iii) explains the commonality;

(iv) draws general conclusions on the basis of the explanation.

We see the process as not linear, at any stage difficulties might appear that would force the solver to return to one of the previous stages. The solver may also return to the first stage in order to look for another generalization to compare. Similar models are presented e.g. by Rivera and Becker [21].

As valuable in our study we consider the project *mathe 2000* – project aimed to adopt an understanding of mathematics as the science of dynamic patterns [27], patterns that are not ready-made nor static, patterns that could be "explored, continued, re-shaped, and invented locally by the learners themselves" [27: p. 2]. Such a model allows including the method of releasing parameters into the generalization process.

2.2 Patterns in the primary school classroom

Children meet patterns in their everyday life from early years, even very simple experiments confirm that kindergarten children are able to discover and make use of them. Such patterns have usually the form of sequences of several real objects or pictures, and children are either asked to continue the sequence, or to fill in a missing component. The first typical patterns are geometrical; they use shapes, colours, or positions. These patterns are either repeating or growing (see Fig. 1), so that the solution procedure is based on identifying the repeating or the growing part. Although trivial and intuitive, such geometrical discoveries boost children's ability to generalize, their ability to perceive, observe, and experiment.

Some of the geometrical patterns can be also treated as arithmetical, when the relation between the geometrical figure and numeracy is taken into account. For instance, patterns from Fig. 1b and 1c can be considered as geometrical (when focusing on changes in shapes and their relations) as well as arithmetical (when focusing on changes in amounts of cubes and their relations). After meeting with numbers and operations on numbers, also patterns that do not use geometry in the assignment may be considered, e.g. patterns given by sequences 1, 2, 2, 3, 3, 3, 4, 4, 4, 4 or 3, 6, 9, 12, 15, 18, 21 or 1, 2, 3, 5, 8, 13, 21.

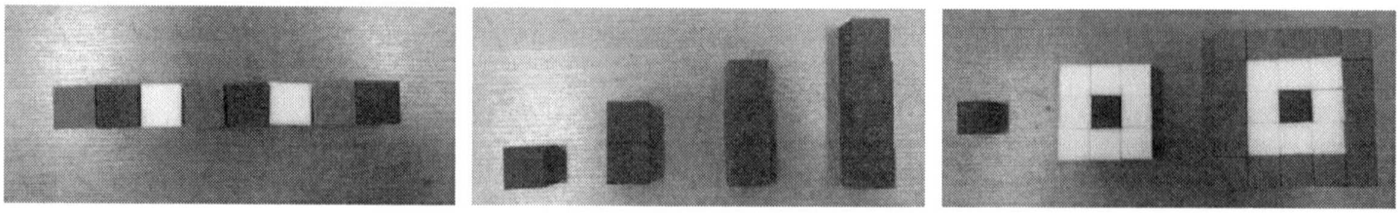

Fig. 1. Geometrical patterns: (a) repeating, (b) growing, (c) growing

We believe that mathematics education should be based on such natural abilities of children, and that development of these abilities could naturally lead the children to rudiments of argumentation, i.e. to asking questions, reasoning about phenomena and relations, formulating conjectures and conclusions, raising objections, generalizing.

The question is how educate teachers in order to provide these experiences. To prepare future teachers for implementing such activities in their teaching, we have to train them not only to apply the activities in the classroom, but also to be fully aware of how the activities could contribute to the pupils' development.

For this purpose we look for learning environments that could offer both pupils and future teachers the space for activities that would stimulate and develop their ability to perceive patterns in various phenomena, leading them this way towards the development of ability to draw general conclusions. We have drawn inspiration from the work of researchers involved in projects *mathe 2000* [27] and *NaDiMa* [12], and turn our attention to substantial learning environments.

According to Wittmann [26], substantial learning environment is "a teaching/learning unit with the following properties:

(1) It represents central objectives, contents and principles of teaching mathematics at a certain level.

(2) It is related to significant mathematical contents, processes and procedures *beyond* this level, and is a rich source of mathematical activities.

(3) It is flexible and can be adapted to the special conditions of a classroom.

(4) It integrates mathematical, psychological and pedagogical aspects of teaching mathematics, and so it forms a rich field for empirical research." [26: pp. 365-6]

Substantial learning environments employed in our teacher training are e.g. an arithmetical environment "Who hits 50?" [8], or geometrical environments "Way", "Room" and "Mosaic" [25]. In this particular study we shall use an arithmetical environment called "Number triangles".

2.3 Number triangles

Substantial learning environment called "Number triangles" or "Arithmogons" is based on work of McIntosh and Quadling [15], its detailed elaboration can be found e.g. in [12].

A number triangle consists of a triangle with three interior and three exterior fields containing natural numbers, where each exterior field holds the sum of the numbers that lay in its two adjacent interior fields; numbers in interior fields may also be replaced by counters (Fig. 2).

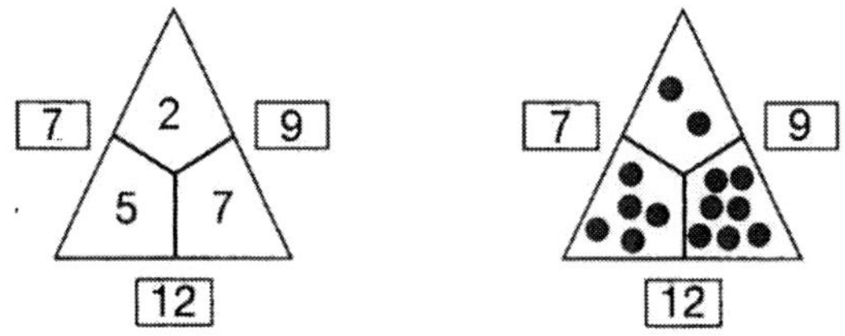

Fig. 2. Number triangles: (a) with numbers, (b) with counters [26]

The tasks are assigned with three numbers missing, the difficulty of the task depends on where lay the missing numbers. The basic version of the task is assigned with all the exterior numbers missing, i.e. the task deserves only three addition operations to solve. Other versions miss a combination of interior and exterior numbers, such tasks require both addition and subtraction operations to solve. The most difficult version is assigned with all interior numbers missing, in this case the solution cannot be obtained easily just by direct calculations. It may be found by trial-and-error method, by systematic variation of numbers, by using algebraic terms, or by using pattern oriented reasoning. General conclusions in this environment may refer to descriptions of solution procedures to particular versions of the task, to relations between the sum of interior numbers and the sum of exterior numbers, to the role that parity of particular numbers plays in the diagram, to the conditions of existence of a solution to particular versions of the task, etc. For detailed overview see e.g. [12: pp. 11-14]. We may also create tasks based on sequences of several number triangles with a pattern in the assignment, and ask the solver to find the pattern and use it in order to continue the sequence (Fig. 3).

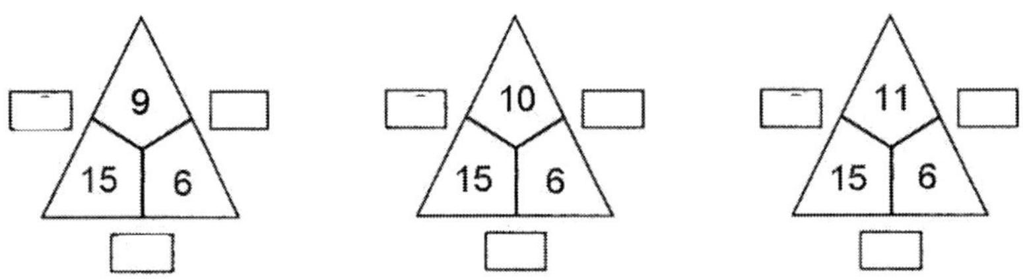

Fig. 3. Sequence of number triangles with a pattern

As follows from the previous paragraph, the environment of number triangles allows us to pose problems of different difficulty (Fig. 4). From the point of view of the generalization activities we consider valuable that number triangles enable the solvers to change and release various parameters of the task.

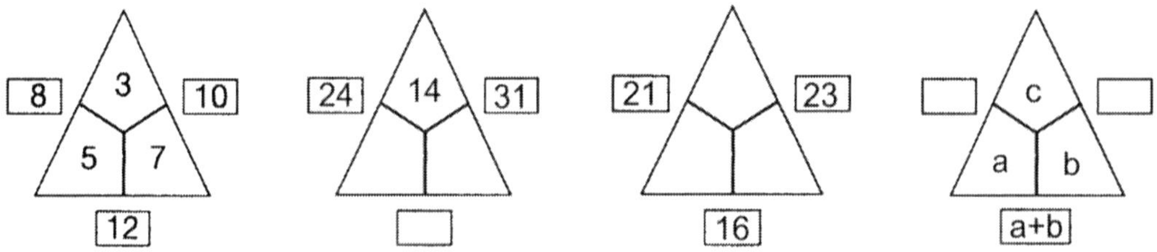

Fig. 4. Tasks of different difficulty

Among the number triangles that are often presented as samples (e.g. in [12]), we choose one that is very special in its composition: it is a triangle with 3, 5 and 7 as the interior numbers, and thus with 8, 10 and 12 as the exterior numbers. Such composition is specific from two reasons:

- first, the interior numbers are all odd, and the exterior numbers are all even;
- second, the interior numbers are consecutive odd numbers in one direction (counter clockwise), and the exterior numbers are consecutive even numbers in the other direction (clockwise).

This arrangement opens a wide space for the search for patterns. It offers easily seen but singular patterns, as well as patterns that require deeper consideration. Among the easily seen patterns we include e.g. the pattern "all the interior numbers are odd, exterior even". Such a conclusion cannot be generalized to other cases, the pattern is singular. Among the patterns that require deeper consideration of the context of the task, of other possible arrangements of numbers in the triangle, we include e.g. the pattern "the sum of exterior numbers is always even, because it equals twice the sum of the interior numbers". Such a conclusion is general. The opportunity of having both easily seen and not-easily seen patterns together in the task allows us to distinguish solvers that tend to apply only surface reasoning without looking for further relations.

2.4 Concept Cartoons

An educational tool called Concept Cartoons was developed by Keogh and Naylor [9] as a way to improve science education at primary school level, lately they have expanded also to other school subjects including mathematics [3]. Each Concept Cartoon is a picture showing a situation well-known to children from school or everyday reality, and a group of children in a bubble-dialog. Individual children in the picture present alternative viewpoints on the pictured situation, regardless of their correctness; the correctness may also be unclear or conditional. For a sample of a Concept Cartoon related to the "Number triangles" environment see Fig. 5.

When working with Concept Cartoons, pupils or students have to choose all children that are right, and justify their choice in classroom discussion. This is the way how Concept Cartoons can support teaching and learning: by generating discussion, provoking cognitive conflict, stimulating investigation and argumentation, promoting learners' involvement and motivation [18]. We used Concept Cartoons this way in the previous contribution [22], as an instrument that promotes in future primary school teachers the awareness of the need of proper argumentation, and leads them towards deductive argumentation.

Fig. 5. Concept Cartoon; template of children and empty bubbles taken from [3], number triangle and texts added

Now we present Concept Cartoons in a new role, as a diagnostic instrument in future primary school teachers' education, both in problem solving and problem posing related to substantial learning environments.

Recently we have already used Concept Cartoons as a diagnostic instrument to detect changes in future primary school teachers' open approach to mathematics, both during problem solving and problem posing [24].

3 Our Short Study

As an instrument in our study we apply Concept Cartoons related to substantial learning environments, namely to "Number triangles".

The aim of the study is to observe which aspects of future teachers' knowledge related to reasoning and generalization could be investigated through this instrument during problem solving and problem posing.

Our study is a pilot part of an ongoing qualitative research, and its role is to get insight into the problematics.

The participants of the study were future primary school teachers, students of the third or fourth year of five-year master degree program at the Faculty of Education, who attended the optional university course on didactics of mathematics. During the course, among others, the students get acquainted with Concept Cartoons as motivational and educational tool in mathematics education, with various tasks of application character (e.g. with unequal partition word problems, [23]), and with several substantial learning environments (e.g. "Who hits 50?").

Data collection took place twice during the course:

At first, we aimed to observe reasoning related to problem solving and to judging someone else's opinions. We assigned the participants a worksheet with a bubble-dialog task on "Number triangles". The worksheet consisted of two diverse parts: the first part presented an example of a number triangle with all numbers filled in, and asks some substantial questions about the triangle – the questions in the form "What if ...?" (Fig. 6a); the second part presented several alternatives about the triangle to judge (to decide whether the alternatives are right, and justify the decision; Fig. 6b). We asked the participants to respond to the worksheet in written form.

At second, we aimed to observe reasoning related to problem posing. We asked future teachers to choose a substantial learning environment, pose a task in this environment, pose their own Concept Cartoon related to the task, and elaborate the work as a written homework.

Fig. 6. Worksheet assigned to the participants, translated from Czech: (a) the first part, (b) the second part

Collected data were processed qualitatively, using open coding [16].

4 Results – problem solving

When observing future teachers' answers related to problem solving and judging, we revealed five diverse categories of responses which can be denoted as "awareness of the need of argumentation", "type of presented argumentation", "releasing parameters", "terminological correctness", and "mathematical content".

4.1 Awareness of the need of argumentation

These responses are related to the awareness of the need of proper argumentation, i.e., whether the respondent wants to offer arguments in his/her responses, whether he/she feels the inner need for argumentation (cf. [28]). Some of the respondents displayed lacks in this area, they presented only yes/no answers without any attempt to verify, or unjustified assumptions, e.g.

> Bubble: The sum of the exterior numbers equals the sum of the interior numbers.
> Response A: No.
> Response B: I think that the sum of the exterior numbers does not equal the sum of all the interior numbers.

4.2 Type of presented argumentation

This category area comprises of responses related to the type of argumentation used by the respondent. Our classification follows up the proof schemes framework established by Harel and Sowder [7].

Some of the students used empirical arguments, inductive proof schemes based on several cases, e.g.

> Bubble: The sum of the three exterior numbers may be even, but also odd.
> Response: 8 + 12 + 10 = 30 (even), 8 + 11 + 11 = 30 (even), 13 + 11 + 8 = 21 (even). Only even numbers appear to me!

Some of the respondents made attempts of deductive proof schemes, but did not succeeded. They usually displayed weaknesses in anticipation of results on objects used in their reasoning. These weaknesses mainly appeared as not noticing that a change in one interior number causes changes in two exterior numbers, and that a change in one exterior number causes changes in one interior and one exterior number, e.g.

> Bubble: The sum of the three exterior numbers may be even, but also odd.
> Response: Yes, if the numbers inside the triangle are diverse – not only even, not only odd, e.g. 3, 6, 7 or 8, 6, 3.

On the other hand, some of the participants anticipated results of operations on objects properly, e.g.

Bubble:	What happens if I replace 10 by 11?
Response:	Then I have to increase either number 3 or number 7 by 1. But I must not forget to increase also the result in the rectangle that is related to this number.

We also found students who correctly anticipated results and offered general deductive arguments:

Bubble:	If I put odd numbers to all exterior fields, then ... the numbers in interior fields will be also odd.
Response:	If I put an odd number, and split it to two numbers, then one of the numbers must be even and one odd. So, the bubble is not true.

4.3 Releasing parameters

One of the respondents successfully released the number domain in order to find a case that would solve the problem in the bubble:

Bubble:	So – to get all exterior numbers odd, I have to put inside...
Response:	... not integers (see Fig. 7)

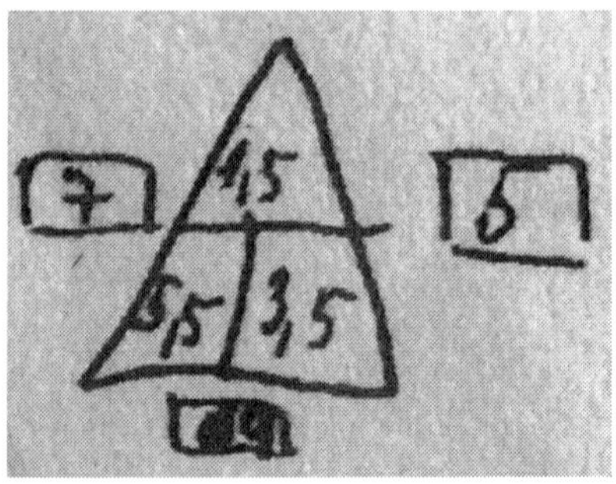

Fig. 7. Triangle with "not integers", excerpt from the worksheet

We welcome such reasoning because it enables to proceed with the generalization process to further stages, from constructive abstraction to extensional generalization [4].

On the contrary, we also found responses with reasoning too tied to cases given in the assignment, displays of non-dynamic thinking:

Bubble:	What happens if I replace 10 by 11?
Response A:	The rule about sum of adjacent inner numbers will not hold any longer.
Response B:	Then the result will not come out – that 2 adjacent numbers inside the triangle equal the number in the square beside them. I think we cannot do it.

4.4 Terminological correctness

Too often we revealed in the data terminological weaknesses such as awkward, inaccurate, unclear, ambiguous formulations or statements of rather narrative character. For instance they used *distance* instead of *difference*, and *perimeter* instead of *sum of exterior numbers*:

Bubble: What happen if I replace 3 by 4?
Response: The distance of the numbers in the triangle changes: 5-4 by 1, 7-4 by 3.

Bubble: So, to get all exterior numbers odd... I have to put inside...
Response: ... there must be at least one even number inside, to get the perimeter odd.

Some of the respondents used in their responses the term *result* without any further identification of the particular operation to which the result belongs; there was a response where *result* referred to the sum of two lower interior numbers, and also a response where *result* referred to the rule which establishes the triangle:

Bubble: What happen if I replace 3 by 4?
Response: If I replace 3 by 4, then in the lower part of the pyramid I will get numbers 6 and 4. The result will be 10 (thus the even number).

Bubble: What happen if I replace 10 by 11?
Response: Then the result will not come out – that 2 adjacent numbers inside the triangle equal the number in the square beside them.

4.5 Mathematical content

Another category of responses relates to mathematical content. We revealed here three different content areas that appeared significantly in data: logic (quantifiers and conjunctions), divisibility of numbers, and arithmetical operations. Displays of weaknesses in mathematical content were often combined from more content areas together, e.g. logic together with divisibility:

Response A: 30 is divisible by 2 as well as by 3, so that it might be even as well as odd.

Response B: If at least one of the interior numbers is even, then outside the triangle I will have one odd number, and the sum of exterior numbers will be odd.

5 Results – problem posing

The respondents posed their Concept Cartoons in various environments, one of them as the environment choose number triangles (Fig. 8). We shall focus on her Concept Cartoon closely.

We can see that there are many analogies with the Concept Cartoon that we assigned to her previously (that in Fig. 6). First, the Concept Cartoon is based on a number triangle that has all numbers filled, all interior numbers are odd, and all exterior numbers are even, i.e. the same attributes of the triangle as previously. Second, some of the texts in bubbles are modification of the previous texts; the "What-if-not" strategy of problem posing [1] was applied:

- in the bottom left bubble, the quantifier was changed, the verb *might be* was replaced by *is always*;

- in the top center bubble, the parity was changed, both the adjectives *odd* were replaced by *even*;

- in the top left bubble, both the quantifier and the parity were changed, to get an alternative that is correct.

The top right bubble is quite new, it presents an innovative view on the situation that was not discussed previously. The view is not correct, but is really plausible, and allows to distinguish solvers that tend to apply only surface reasoning.

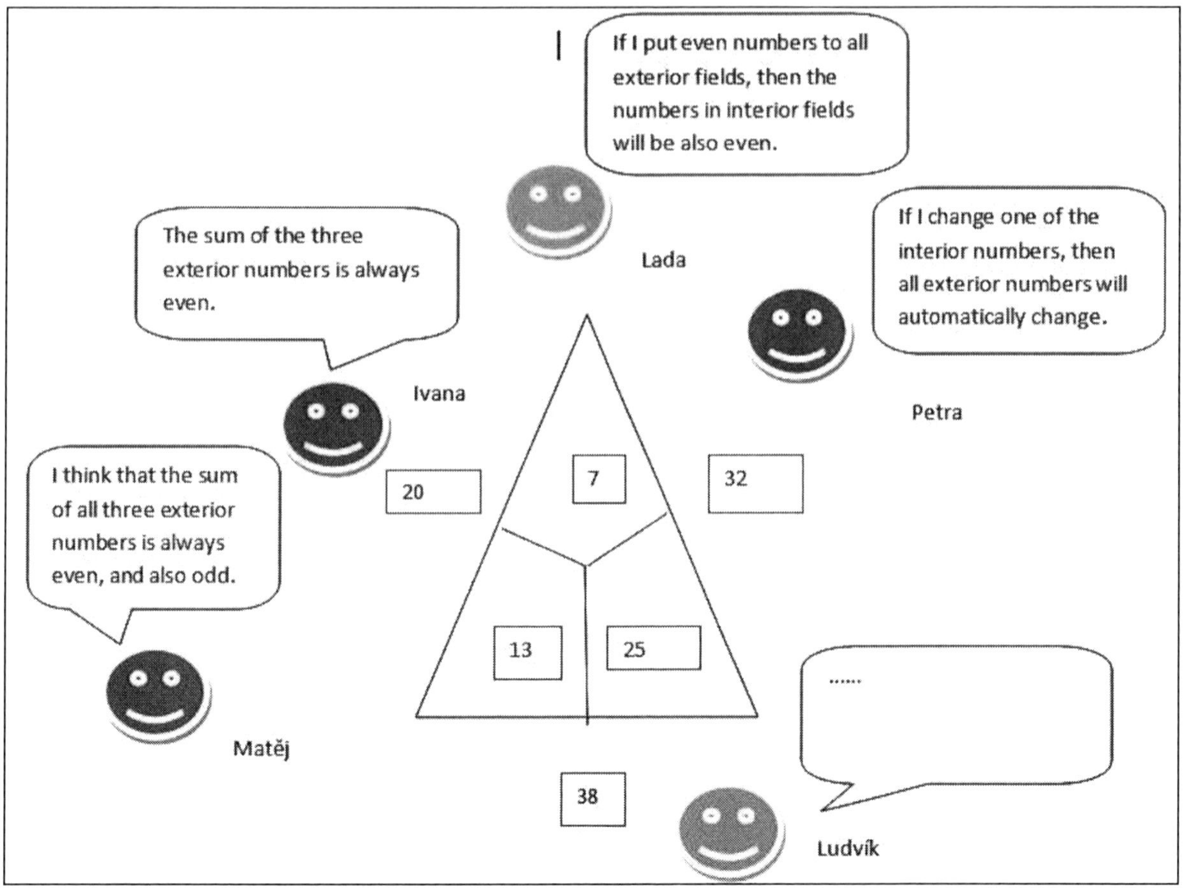

Fig. 8. Concept Cartoon posed by the respondent, translated from Czech

The Concept Cartoon was accompanied by a descriptive text, where the author presented correct solutions – new versions of her bubble texts that were modified as to be correct. Here we revealed her weakness in mathematical content when she stated that "The triangle always contains odd numbers as well as even numbers." Too tied to the context of the triangles that have odd numbers inside and even numbers outside, she was not able to realize that there might be triangles with all numbers even. All other modifications to correct versions were done properly.

6 Conclusion

In our study we showed Concept Cartoons in an innovative way, as a diagnostic tool when working with substantial learning environments. We used them in problem solving as well as in problem posing, in both cases Concept Cartoons proved their value. Once again (e.g. as in [23]) we confirmed our belief that while working with Concept Cartoons, the students (future primary school teachers) started to realize how important is to get in touch with various methods of argumentations.

We also presented how Concept Cartoons may help to observe how future primary school teachers reason and generalize, namely during the constructive abstraction stage and during generalization through generalizing the reasoning stage of the problem solving process. Concept Cartoons allow us to observe closely also the way how future primary school teachers pose problems, in that case from two complementary perspectives: on one hand we may observe how they pose sets of problems on the same topic – in fact, each Concept Cartoon is a combination of diverse tasks, where each of the bubbles represents one task, and all of the tasks are related to one common theme. At the same time we may see Concept Cartoon as a whole, as one complex task, and observe how future primary school teachers are able to handle the complexity. We see Concept Cartoons as really beneficial, and we shall continue in using them in future primary school teachers' education.

Acknowledgement

The paper was supported by grant from Czech Science Foundation no. 14-01417S entitled "Enhancing mathematics content knowledge of future primary teachers via inquiry based education", and by RVO 67985840.

References

[1] BROWN, S., WALTER, M. *The art of problem posing*. Philadelphia, PA: Franklin Institute Press, 1983.

[2] CIOSEK, M. Generalization in the process of defining a concept and exploring it by students. In B. Maj-Tatsis, K. Tatsis (Eds.), *Generalization in mathematics at all educational levels*. Rzeszów: Wydawnictwo Uniwersytetu Rzeszowskiego, 2012: pp. 38-56.

[3] DABELL, J., KEOGH, B., NAYLOR, S. *Concept Cartoons in Mathematics Education (CD-ROM)*. Sandbach: Millgate House Education, 2008.

[4] DŐRFLER, W. Forms and means of generalization in mathematics. In A. Bishop (Ed.), *Mathematical knowledge: its growth through teaching*. Mahwah, NJ: Erlbaum, 1991: pp. 63-85.

[5] ENGLISH, L. D. Promoting the development of young children's mathematical and analogical reasoning. In L. D. English (Ed.), *Mathematical and analogical reasoning of young learners*. Mahwah, NJ: Lawrence Erlbaum, 2004.

[6] GRIFFITHS, P. A. Mathematics at the turn of the millennium. *The American Mathematical Monthly*, vol. 107, no. 1, 2000: pp. 1-14.

[7] HAREL, G., SOWDER, L. Toward a comprehensive perspectives on the learning and teaching of proof. In F. Lester (Ed.), *Second handbook of research on mathematics teaching and learning*. Charlotte: NCTM, 2007: pp. 805-842.

[8] HOŠPESOVÁ, A., TICHÁ, M. One possible way of training teachers for inquiry based education. In B. Ubuz, C. Haser, M. A. Mariotti (Eds.), *Proceedings of the Eighth Congress of European Research in Mathematics Education (CERME 8)*. Antalya: Middle East Technical University, 2013: pp. 3105-3114.

[9] KEOGH, B., NAYLOR, S. Learning in science: another way in. *Primary Science Review*, vol. 26, 1993: pp. 22-23.

[10] KOMAN, M., TICHÁ, M. K rozvíjení funkčního přístupu k řešení úloh na základní škole. *Matematika a fyzika ve škole*, vol. 20, 1990: pp. 374-381.

[11] KOMAN, M., TICHÁ, M. *Basic research in didactics of mathematics in Czechoslovakia – Comments to the development of the functional and algorithmic approaches to problem solving*. Praha: MÚ ČSAV, preprint, 1988.

[12] KRAUTHAUSEN, G., SCHERER, P. (EDS.) *Ideas for Natural Differentiation in Primary Mathematics Classrooms, Vol. 1: Arithmetical Environment*. Rzeszów: Wydawnictwo Uniwersytetu Rzeszowskiego, 2010.

[13] KRYGOWSKA, A. Z. *Zarys dydaktyki matematyki, część 3*. Warszawa: WSiP, 1979.

[14] KUŘINA, F., PŮLPÁN, Z. *Podivuhodný svět elementární matematiky*. Praha: Academia, 2006.

[15] MCINTOSH, A., QUADLING, D. Arithmogons. *Mathematics Teaching*, vol. 70, 1975: pp. 18-23.

[16] MILES, M. B., HUBERMAN, A. M. *Qualitative data analysis. A sourcebook of new methods*. London: SAGE, 1994.

[17] MULLIGAN, J., MITCHELMORE, M. Awareness of pattern and structure in early mathematical development. *Mathematics Education Research Journal*, vol. 21, no. 2, 2009: pp. 33-49.

[18] NAYLOR, S., KEOGH, B. Concept Cartoons: What have we learnt? *Journal of Turkish Science Education*, vol. 10, no. 1, 2013: pp. 3-11.

[19] PITTA-PANTAZI, D., GREY, E., CHRISTOU, C. Elementary school students' mental representations of fractions. In M. J. Høines, A. Fuglestad (Eds.), *Proceedings of the 28th annual conference of the International Group for the Psychology of Mathematics Education (Vol. 4)*. Bergen: Bergen University College, 2004: pp. 41-48.

[20] POLYA, G. Induction and analogy in mathematics. Volume I of mathematics and plausible reasoning. New Jersey: Princeton University Press, 1954.

[21] RIVERA, F. D., BECKER, J. Algebraic reasoning through patterns. *Mathematics Teaching in the Middle School*, vol. 15, no. 4, 2009: pp. 212-221.

[22] SAMKOVÁ, L., TICHÁ, M. Developing views of proof of future primary school teachers. In Ľ. Balko, D. Szarková, D. Richtáriková (Eds.) *Proceedings, 15th Conference on Applied Mathematics Aplimat 2016*. Bratislava: STU, 2016, pp. 987-998.

[23] SAMKOVÁ, L., TICHÁ, M. Investigating future primary teachers' grasping of situations related to unequal partition word problems. In C. Sabena, B. Di Paola (Eds.), *Quaderni di Ricerca in Didattica (Mathematics), n. 25, Suppemento n. 2. Proceedings CIEAEM 67, Teaching and learning mathematics: resources and obstacles.* Palermo: G.R.I.M., 2015: pp. 295-303.

[24] SAMKOVÁ, L., TICHÁ, M. On the way to develop open approach to mathematics in future primary school teachers. *Journal on Efficiency and Responsibility in Education and Science*, vol. 9, no. 2, 2016: pp. 37-44.

[25] TICHÁ, M., HOŠPESOVÁ, A. Teacher competences prerequisite to natural differentiation. In M. Pytlak, T. Rowland, E. Swoboda (Eds.), *CERME 7 Proceedings of the Seventh Congress of the European Society for Research in Mathematics Education.* Rzeszow: University of Rzeszow, 2011: pp. 2888-2897.

[26] WITTMANN, E. CH. Mathematics education as a "design science". *Educational Studies in Mathematics*, vol. 29, no. 4, 1995: pp. 355-374.

[27] WITTMANN, E. CH. *Mathematics as the science of patterns – a guideline for developing mathematics education from early childhood to adulthood.* Plenary lecture presented at the international colloquium "Mathematical learning from early childhood to adulthood", Mons, Belgium, July 2005.

[28] ZASLAVSKY, O., NICKERSON, S. D., STYLIANIDES, A. J., KIDRON, I., WINICKI-LANDMAN, G. The need for proof and proving: Mathematical and pedagogical perspectives. In G. Hanna, M. de Villiers (Eds.), *Proof and proving in mathematics education, The 19th ICMI Study.* Dordrecht: Springer, 2012: pp. 215-230.

Current address

Samková Libuše, RNDr. Ph.D.
Katedra matematiky, Pedagogická fakulta Jihočeské univerzity
Jeronýmova 10, 371 15 České Budějovice, Czech Republic
Tel. number: +420 387 773 091, e-mail: lsamkova@pf.jcu.cz

Tichá Marie, Mgr. CSc.
Kabinet pro didaktiku matematiky, Matematický ústav AV ČR
Žitná 25, 115 67 Praha 1, Czech Republic
Tel. number: +420 222 090 726, e-mail: ticha@math.cas.cz

Slovak University of Technology in Bratislava
Faculty of Mechanical Engineering
16th Conference on Applied Mathematics **APLIMAT 2017**
Proceedings

MATHEMATICAL EDUCATION WITH WORKSHEETS

SCHREIBEROVÁ Petra (CZ), VOLNÝ Petr (CZ)

Abstract. We present our practical experiences with using of worksheets during education of Mathematics at VŠB - Technical University of Ostrava. We prepared worksheets for the students. These worksheets cover all main mathematical courses in the bachelor study program. We asked the students to answer a few questions concerning the worksheets. The feedback from students is presented in the paper and we also discuss our practical experiences with the worksheets.

Keywords: Mathematics, education, worksheets, statistical analysis

Mathematics subject classification: Primary 97A99; Secondary 97D40, 97K80

1 Introduction

In general, the study of science disciplines (especially mathematics) is treated as something very difficult and not easily understandable. It is obvious that there is a lack of interest and a low motivation for pupils and students to study mathematics, both in the primary as well as secondary and tertiary education. Their basic knowledge of secondary-school mathematics is very limited when they start a study at a university and therefore we need to focus on the repetition of basic mathematical principles during the first years of the education.

Currently, pupils and students are familiar to work with computers, tablets and smartphones which in the most cases replace the classic book. We try to attract students in learning mathematics and we try to adjust the exercises and lectures with the use of modern technologies. We want to respond to development in a field of information and communication technologies and we want to use it to increase the knowledge, motivation and activity in exercises and homework.

2 Worksheets

At our department the worksheets have been created in the last years to support the education of mathematics in the first two academic years. Worksheets concern Mathematics I (or

Bachelor Mathematics I), the content of which is the introduction to linear algebra, differential calculus and analytic geometry in 3-dimensional space, and Mathematics II (or Bachelor Mathematics II), where we focus on the integral calculus of functions of one variable, differential calculus of functions depending on two and more real variables and ordinary differential equations. Recently, the worksheets for Mathematics III (Statistics) have been created.

The worksheets consist of three different blocks. The first block is devoted to the theory and we use these worksheet throughout the lectures. A part of an every single sheet is a blank space where a student can make his own notes.

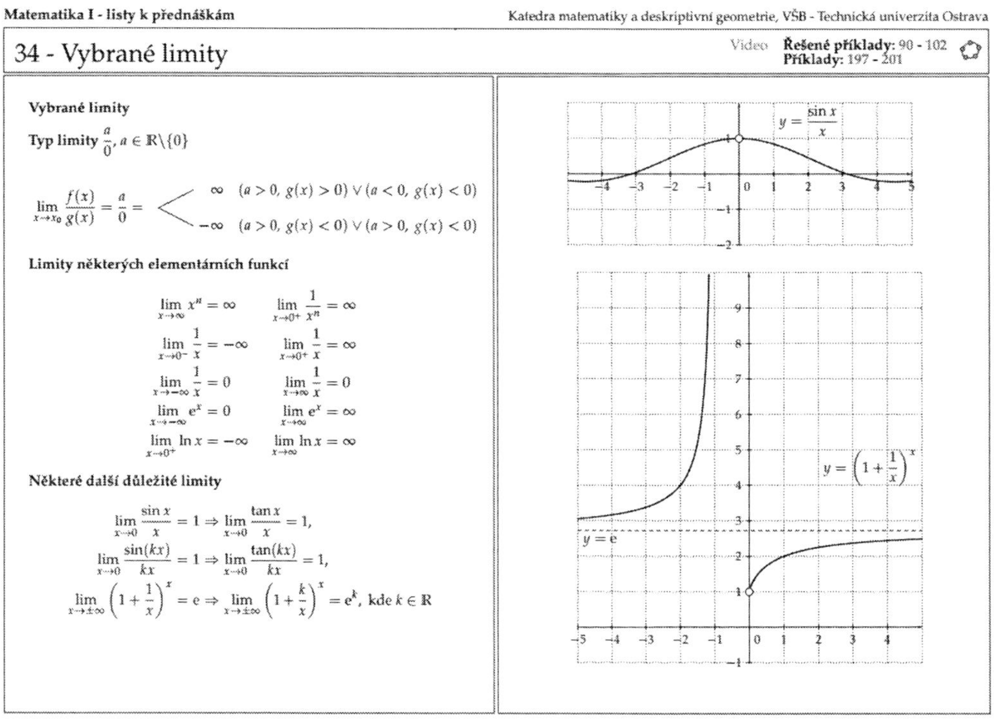

Matematika I - listy k přednáškám — Katedra matematiky a deskriptivní geometrie, VŠB - Technická univerzita Ostrava

34 - Vybrané limity — Video — Řešené příklady: 90 - 102 — Příklady: 197 - 201

Vybrané limity

Typ limity $\frac{a}{0}$, $a \in \mathbb{R}\setminus\{0\}$

$$\lim_{x\to x_0}\frac{f(x)}{g(x)} = \frac{a}{0} = \begin{cases} \infty & (a>0,\ g(x)>0) \vee (a<0,\ g(x)<0) \\ -\infty & (a>0,\ g(x)<0) \vee (a>0,\ g(x)<0) \end{cases}$$

Limity některých elementárních funkcí

$$\lim_{x\to\infty} x^n = \infty \qquad \lim_{x\to 0^+}\frac{1}{x^n} = \infty$$
$$\lim_{x\to 0^-}\frac{1}{x} = -\infty \qquad \lim_{x\to 0^+}\frac{1}{x} = \infty$$
$$\lim_{x\to -\infty}\frac{1}{x} = 0 \qquad \lim_{x\to \infty}\frac{1}{x} = 0$$
$$\lim_{x\to -\infty} e^x = 0 \qquad \lim_{x\to \infty} e^x = \infty$$
$$\lim_{x\to 0^+} \ln x = -\infty \qquad \lim_{x\to \infty} \ln x = \infty$$

Některé další důležité limity

$$\lim_{x\to 0}\frac{\sin x}{x} = 1 \Rightarrow \lim_{x\to 0}\frac{\tan x}{x} = 1,$$
$$\lim_{x\to 0}\frac{\sin(kx)}{kx} = 1 \Rightarrow \lim_{x\to 0}\frac{\tan(kx)}{kx} = 1,$$
$$\lim_{x\to\pm\infty}\left(1+\frac{1}{x}\right)^x = e \Rightarrow \lim_{x\to\pm\infty}\left(1+\frac{k}{x}\right)^x = e^k,\ \text{kde } k \in \mathbb{R}$$

The second block contains worksheet with solved examples. These examples provide a support for student homework. The third part of the worksheets is used for the practical classes and contains unsolved examples and problems. A typical worksheet consist of a task, a collection of hints and a blank space for solutions. The list of hints enables and helps a student to solve the given problem.

These three blocks are connected together by the hyperlinks throughout the text. The worksheets also contain direct links which lead students to our web pages where they can find the commented videos concerning problems in mathematics and the interactive tools supporting the education. The videos were created with help of the interactive board and the tools are of the form of GeoGebra applets.

GeoGebra is one of the best free software for the interactive education of mathematics. This year, the GeoGebra institute of Ostrava was established at our university. One of its main purposes is to cover all the activities concerning GeoGebra and its usage during the education. The institute also provides GeoGebra courses for teachers of mathematics and descriptive geometry.

On the next two figures we present worksheets with solved and unsolved examples.

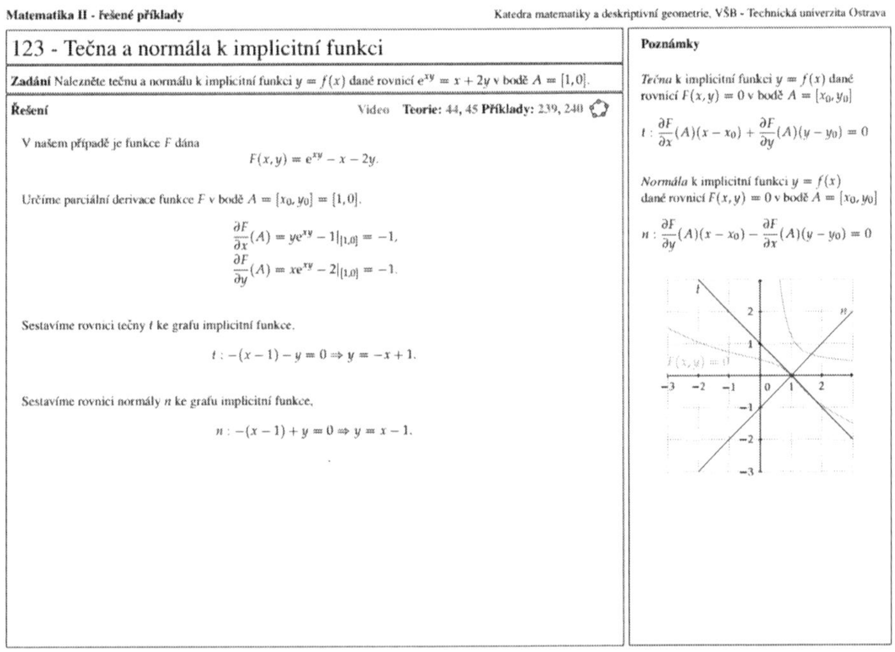

Matematika II - řešené příklady

Katedra matematiky a deskriptivní geometrie, VŠB - Technická univerzita Ostrava

123 - Tečna a normála k implicitní funkci

Zadání Nalezněte tečnu a normálu k implicitní funkci $y = f(x)$ dané rovnicí $e^{xy} = x + 2y$ v bodě $A = [1, 0]$.

Řešení Video **Teorie:** 44, 45 **Příklady:** 239, 240

V našem případě je funkce F dána

$$F(x, y) = e^{xy} - x - 2y.$$

Určíme parciální derivace funkce F v bodě $A = [x_0, y_0] = [1, 0]$.

$$\frac{\partial F}{\partial x}(A) = ye^{xy} - 1|_{[1,0]} = -1,$$
$$\frac{\partial F}{\partial y}(A) = xe^{xy} - 2|_{[1,0]} = -1.$$

Sestavíme rovnici tečny t ke grafu implicitní funkce.

$$t : -(x - 1) - y = 0 \Rightarrow y = -x + 1.$$

Sestavíme rovnici normály n ke grafu implicitní funkce,

$$n : -(x - 1) + y = 0 \Rightarrow y = x - 1.$$

Poznámky

Tečna k implicitní funkci $y = f(x)$ dané rovnicí $F(x, y) = 0$ v bodě $A = [x_0, y_0]$

$$t : \frac{\partial F}{\partial x}(A)(x - x_0) + \frac{\partial F}{\partial y}(A)(y - y_0) = 0$$

Normála k implicitní funkci $y = f(x)$ dané rovnicí $F(x, y) = 0$ v bodě $A = [x_0, y_0]$

$$n : \frac{\partial F}{\partial y}(A)(x - x_0) - \frac{\partial F}{\partial x}(A)(y - y_0) = 0$$

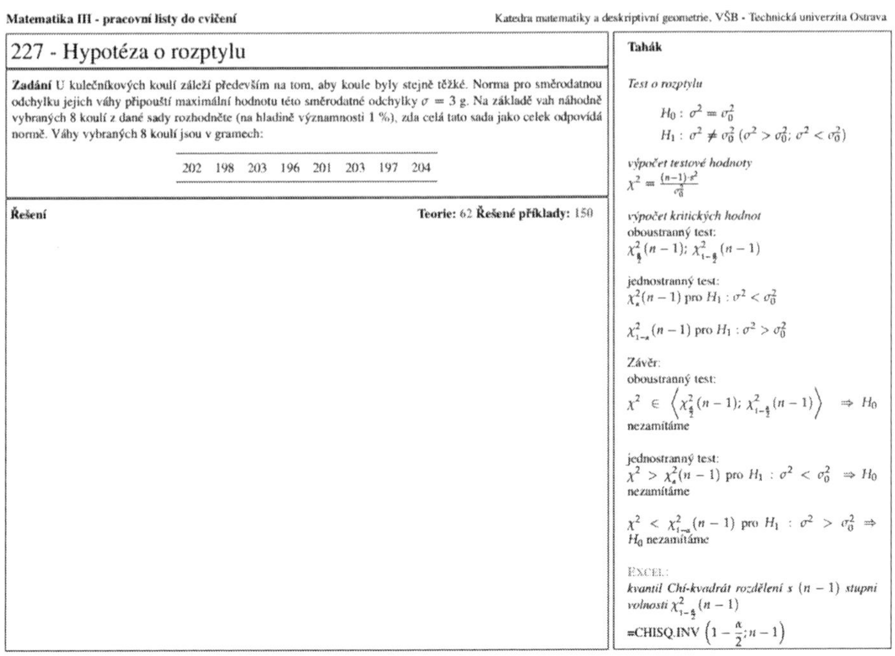

Matematika III - pracovní listy do cvičení

Katedra matematiky a deskriptivní geometrie, VŠB - Technická univerzita Ostrava

227 - Hypotéza o rozptylu

Zadání U kulečníkových koulí záleží především na tom, aby koule byly stejně těžké. Norma pro směrodatnou odchylku jejich váhy připouští maximální hodnotu této směrodatné odchylky $\sigma = 3$ g. Na základě vah náhodně vybraných 8 koulí z dané sady rozhodněte (na hladině významnosti 1 %), zda celá tato sada jako celek odpovídá normě. Váhy vybraných 8 koulí jsou v gramech:

202	198	203	196	201	203	197	204

Řešení **Teorie:** 62 **Řešené příklady:** 150

Tahák

Test o rozptylu

$$H_0 : \sigma^2 = \sigma_0^2$$
$$H_1 : \sigma^2 \neq \sigma_0^2 \; (\sigma^2 > \sigma_0^2; \sigma^2 < \sigma_0^2)$$

výpočet testové hodnoty

$\chi^2 = \frac{(n-1) \cdot s^2}{\sigma_0^2}$

výpočet kritických hodnot

oboustranný test:

$\chi^2_{\frac{\alpha}{2}}(n-1); \chi^2_{1-\frac{\alpha}{2}}(n-1)$

jednostranný test:

$\chi^2_{\alpha}(n-1)$ pro $H_1 : \sigma^2 < \sigma_0^2$

$\chi^2_{1-\alpha}(n-1)$ pro $H_1 : \sigma^2 > \sigma_0^2$

Závěr:

oboustranný test:

$\chi^2 \in \left\langle \chi^2_{\frac{\alpha}{2}}(n-1); \chi^2_{1-\frac{\alpha}{2}}(n-1) \right\rangle \Rightarrow H_0$ nezamítáme

jednostranný test:

$\chi^2 > \chi^2_{\alpha}(n-1)$ pro $H_1 : \sigma^2 < \sigma_0^2 \Rightarrow H_0$ nezamítáme

$\chi^2 < \chi^2_{1-\alpha}(n-1)$ pro $H_1 : \sigma^2 > \sigma_0^2 \Rightarrow H_0$ nezamítáme

EXCEL:

kvantil Chí-kvadrát rozdělení s $(n-1)$ stupni volnosti $\chi^2_{1-\frac{\alpha}{2}}(n-1)$

=CHISQ.INV $\left(1 - \frac{\alpha}{2}; n - 1\right)$

The figure below shows the example of the GeoGebra tool for finding local extrema of a function.

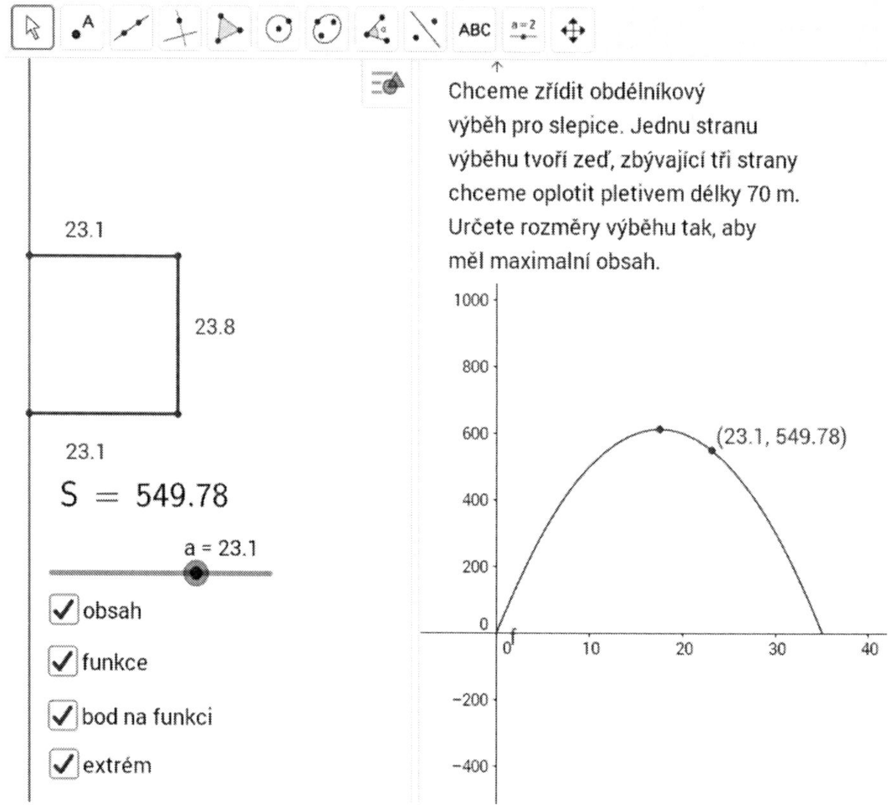

3 Practical experiences

In the years 2014 and 2015 (winter semester), we used worksheets during the lectures and tutorials of Mathematics I held at the Faculty of Civil Engineering, VŠB-Technical University of Ostrava. In the years 2013 and 2015 (summer semester), we used worksheets during lectures and tutorials of Mathematics II held at the Faculty of Civil Engineering and the Faculty of Mechanical Engineering.

At the end of the course, we let students to fill out two questionnaires because we would like to have a feedback from students which would allow us to improve our worksheets and the work with them. One of them was about unsolved problems used in tutorials and the second one contained questions about a theoretical part (the theory with solved examples) used during lectures. The both questionnaires are presented below - the first one contains questions about worksheets used during lectures and the second one relates of questions for using worksheets during tutorials.

The questionnaires were evaluated separately for Mathematics I and Mathematics II.

Mathematics, worksheets for lectures

Evaluation: 1-definitely yes, 5-definitely not

Faculty: Gender: M/F

		1 2 3 4 5
1.	Did you use worksheets during a lecture?	□ □ □ □ □
2.	Did you use worksheets during homeworks?	□ □ □ □ □
3.	Were worksheets useful for your studies?	□ □ □ □ □
4.	Was the number of examples sufficient?	□ □ □ □ □
5.	Was the pace of learning suitable for you?	□ □ □ □ □
6.	Were worksheets well-arranged?	□ □ □ □ □
7.	Did worksheets help you to gain better understanding in mathematics?	□ □ □ □ □
8.	Would you recommend worksheets to other students?	□ □ □ □ □
9.	Your commentary to worksheets	

Mathematics, worksheets for tutorials

Evaluation: 1-definitely yes, 5-definitely not

Faculty: Gender: M/F

		1 2 3 4 5
1.	Did worksheets help you to gain better understanding in mathematics?	□ □ □ □ □
2.	Was the style of worksheets acceptable?	□ □ □ □ □
3.	Was the number of examples sufficient?	□ □ □ □ □
4.	Was the pace of learning suitable for you?	□ □ □ □ □
5.	Did you consider the difficulty of the examples in worksheets sufficient?	□ □ □ □ □
6.	Were worksheets helpful for you?	□ □ □ □ □
7.	Would you recommend worksheets to other students?	□ □ □ □ □
8.	Were the hints useful for you?	□ □ □ □ □
9.	Your commentary to worksheets	

3.1 Results for Mathematics I

The number of students who filled out questionnaires was 192. The average rating of worksheets was 1.77, the most frequent response was 1 (49 %).

The part of worksheets for tutorials of Mathematics I was reviewed by 78 students, half of them were women. 56 % of responses were 1, 32 % at 2 and only 1 % of the evaluation was 5. The average rating was 1.59 with the standard deviation 0.79. Most questions were evaluated better than the average, i.e. 1.

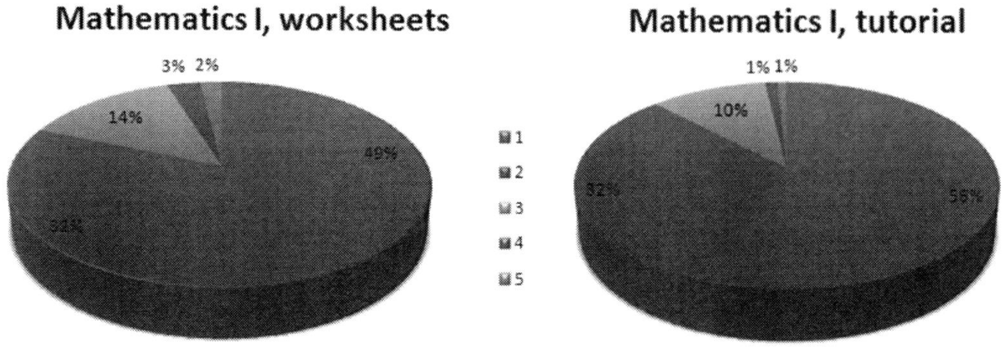

The best average score was at the question No. 1 (in the graph below it is No. 8 - Benefits of worksheets), 52 students answered 1 and 18 from 78 students marked 2. The majority of the students considers worksheets as very beneficial. The largest variance in responses was related to a difficulty of the unsolved problems and pace of learning.

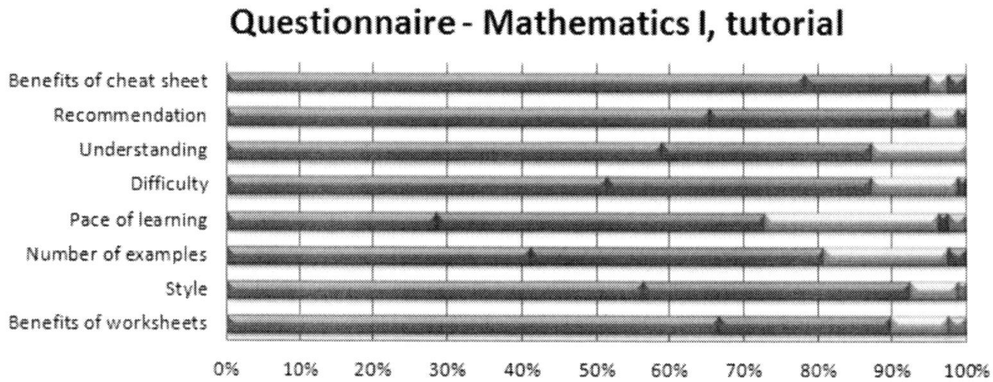

	Benefits of worksheets	Style	Number of examples	Pace of learning	Difficulty	Understanding	Recommendation	Benefits of cheat sheet
1	52	44	32	22	40	46	51	61
2	18	28	31	34	28	22	23	13
3	6	5	13	18	9	10	3	2
4	0	0	2	1	1	0	1	2
5	2	1	0	2	0	0	0	0

The statistical sample for the evaluation of worksheets used in lectures included answers from 114 students. Half of them were women (same as in the previous case). The evaluation of worksheets for the theory was slightly worse (45 % - 1, 32 % -2, 7 % - 4 or 5). The total average was 1.88 with a standard deviation 1.

The worst rating was on the question 4 (number of examples) where the most frequent response (45 cases) was 3 (neither satisfied nor dissatisfied). The question about benefits of

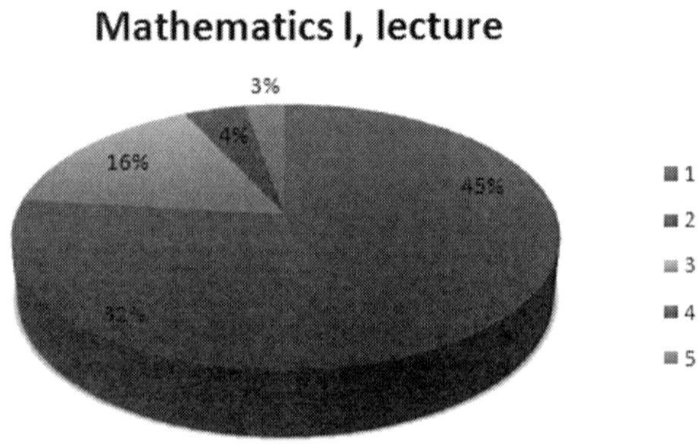

worksheets for the study had the best evaluation where 100 students said that worksheets for the lecture are very useful. The most varied answers were for questions about the number of solved examples. There prevailed the answers 2 and 3.

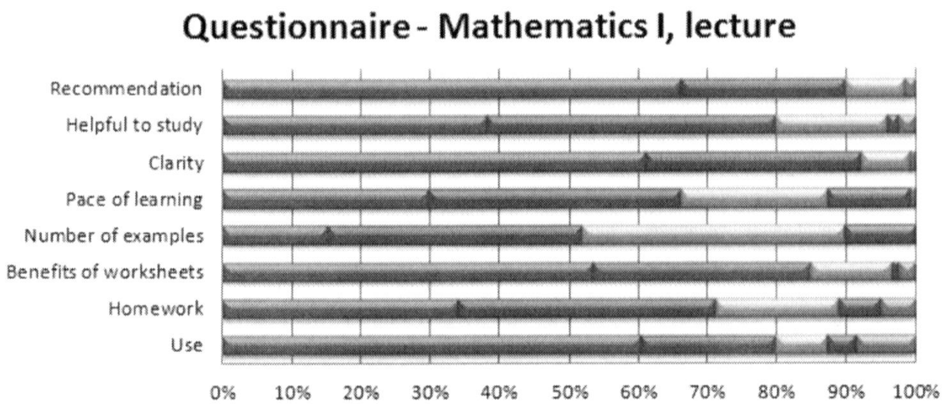

	Use	Homework	Benefits of worksheets	Number of examples	Pace of learning	Clarity	Helpful to study	Recommendation
1	71	40	63	18	35	70	45	78
2	23	44	37	43	43	36	49	28
3	9	21	14	45	25	8	19	10
4	5	7	1	12	14	0	2	0
5	10	6	3	0	1	1	3	2

3.2 Results for Mathematics II

Worksheets for Mathematics II were rated by 287 students (years 2013 and 2015). The average rating was 1.94 and the most questions were answered by 1 (41 %, answers 4 or 5 occurred in 8 % only).

An evaluation of worksheets for exercises were done by 186 students (65 % women) with the average rating of 1.68 and the standard deviation of 0.84 (means the answer 1 or 2 were most common). 51 % (762) answers was 1 and only 1 % of answers was 5.

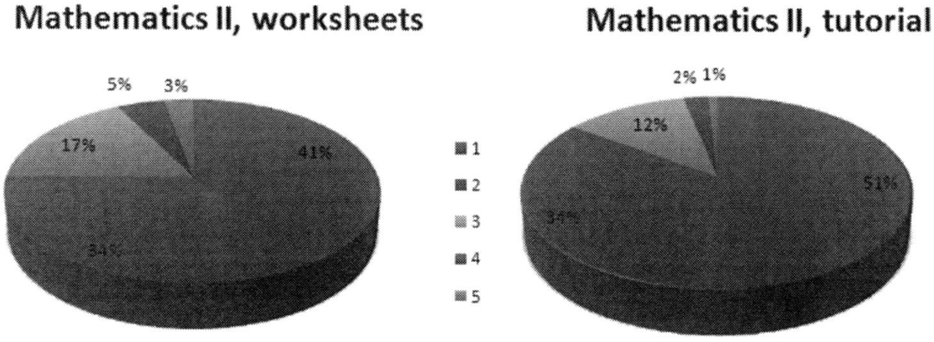

A formal style and benefits of cheat sheets were rated as the most valuable. Like in Mathematics I, students see the biggest problem in a pace of learning (15 students were dissatisfied with the pace of learning).

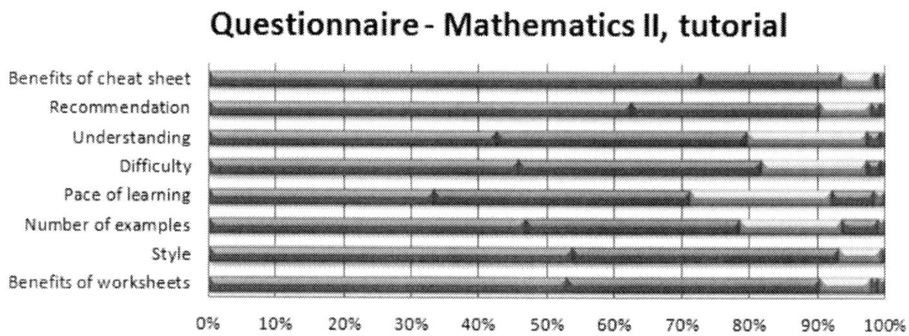

	Benefits of worksheets	Style	Number of examples	Pace of learning	Difficulty	Understanding	Recommendation	Benefits of cheat sheet
1	98	100	87	62	85	79	116	135
2	69	73	59	70	67	69	52	39
3	14	12	28	39	29	33	14	9
4	2	0	10	12	4	4	3	1
5	2	1	2	3	1	1	1	2

Worksheets for the lectures in Mathematics II were rated by 101 students (70 % women). The most common answers were in between 1 and 3 (89 % of total answers). The average rating was 2.12 and the standard deviation was 1.11.

The formal setup of the theoretical part was rated as the best one (78 % of students marked answer 1 or 2), valuable for study (only 7 % believes that the worksheets bring nothing valuable) and over 98 % of students recommends the worksheets for their colleagues. The answers 4 or 5 were mostly used for rating of worksheet usage during lectures (19 % of students did not use the worksheets as well). The biggest variance in the answers could be observed for a question about number of solved examples (one third of students answered that the number of examples is sometimes suitable, sometimes not).

If we compare worksheets for tutorials and lectures in Mathematics I and Mathematics II for

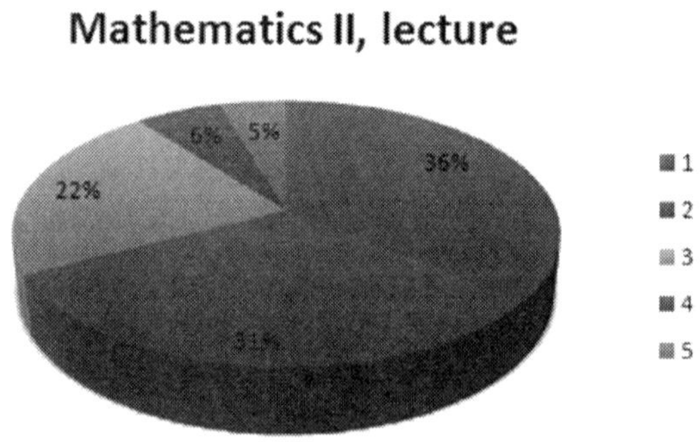

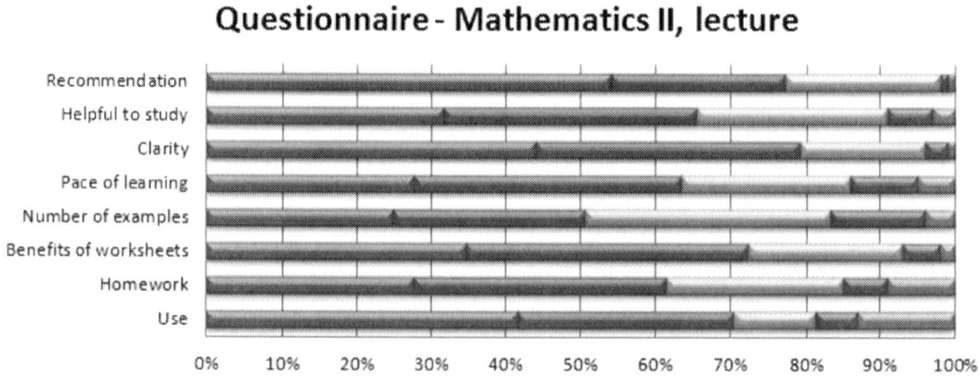

	Use	Homework	Benefits of worksheets	Number of examples	Pace of learning	Clarity	Helpful to study	Recommendation
1	42	28	35	25	28	44	32	54
2	29	34	38	26	36	35	34	23
3	11	24	21	33	23	17	26	21
4	6	6	5	13	9	3	6	1
5	13	9	2	4	5	1	3	1

the period of 2013-2015, we can argue at the 0.05 level of significance (p-value $= 1.2 \cdot 10^{-10}$; p-value $= 1.41 \cdot 10^{-22}$) that students evaluated benefit and use of worksheets on the exercises better than on the lectures. This conclusion is not a surprise for us. We can see the direct consequence with the form of an education through the lectures and the practical classes. The attitude of students is more less passive on the lectures, while in the tutorials the students are much more active and they work more independently.

4 Conclusion

We obtained the positive feedback from students. With worksheets and their use in teaching students were very satisfied, which was reflected by higher score achieved by them. For the interest, we present some interesting student comments:

- Mathematics I - Tutorial

  the need of more blank space in worksheets; more detailed hints; great materials; more difficult examples; less difficult examples.

- Mathematics I - Lecture

more solved examples; well arranged; clear; using all the time; helpful; lack of difficult examples; timesaving; I did not use them and now I feel that it was a mistake; great idea; more blank space.

- Mathematics II - Tutorial

  super; I did not have to write everything; convenient; more examples; useful; helpful for the study.

- Mathematics II - Lecture

  more applications; simplification of the work; more simple definitions; more complicated definitions; brief and accurate; add crosswords; more examples; I remember more; used only at home; perfect for the learning; well understood due to worksheets; improve education; good tool.

There is missing one comment which is the most interesting and the most frequent.

There is a lack of results in the worksheets for tutorials.

Our group discussed this issue and we have decided to not provide the students with results of unsolved examples. We would appreciate the opinion from the readers.

Is necessary to have unsolved examples complemented by the results?

From the point of view of a teacher, one can state that the education complemented by the worksheets is more effective and provides students standardized notes.

We are also going to prepare our worksheets in English version. There is a lack of such materials in the correspondence to the education of the foreign students.

Acknowledgments

Both of the authors appreciate the support of their department.

References

[1] DLOUHÁ, D., HAMŘÍKOVÁ, R., MORÁVKOVÁ, Z., TUŽILOVÁ, M.: *Matematika I: Pracovní listy*, VŠB - TU Ostrava, 2014, ISBN 978-80-248-3323-1.

[2] DUBOVSKÝ, V., JAROŠOVÁ, M., KRČEK, J., KRČKOVÁ, J., SCHREIBEROVÁ, P., VOLNÝ, P.: *Matematika III: Pracovní listy*, VŠB - TU Ostrava, 2016, ISBN 978-80-248-3875-5.

[3] *GeoGebra* [online]. 2013 [cit. 2013-12-09]. `http://www.geogebra.org/cms/cs/`

[4] GERGELITSOVÁ, Šárka. *Počítač ve výuce nejen geometrie: průvodce Geogebrou.* 1. vyd. Praha: Generation Europe, 2011, 247 s. ISBN 978-809-0497-436.

[5] HEBÁK, Petr a Jiří HUSTOPECKÝ. *Průvodce moderními statistickými metodami.* SNTL Praha, 1990, 293 s.,ISBN 80-03-00534-5.

[6] MORÁVKOVÁ, Z., PALÁČEK, R., SCHREIBEROVÁ, P., VOLNÝ, P.: *Matematika II: Pracovní listy*, VŠB - TU Ostrava, 2014, ISBN 978-80-248-3324-8.

[7] PAVELKA, Lubomír a Jarmila DOLEŽALOVÁ. *Pravděpodobnost a statistika.* Ostrava: VŠB - Technická univerzita, 2003, ISBN 80-7078-976-X.

[8] REKTORYS, Karel. *Přehled užité matematiky I.* 7. vyd. Praha: Prometheus, 2009, xxxii, 720 s. ISBN 978-80-7196-180-21.

[9] *Studijní opory* [online]. [cit. 2013-12-09]. `http://www.studopory.vsb.cz/`

Current address

Schreiberová Petra, Ing. PhD.
Department of Mathematics and Descriptive Geometry, VŠB-TU Ostrava
17. listopadu 15, 708 33 Ostrava-Poruba, Czech Republic
Tel. number: +420 597 324 152, e-mail: petra.schreiberova@vsb.cz

Volný Petr, RNDr. PhD.
Department of Mathematics and Descriptive Geometry, VŠB-TU Ostrava
17. listopadu 15, 708 33 Ostrava-Poruba, Czech Republic
Tel. number: +420 597 324 152, e-mail: petr.volny@vsb.cz

Slovak University of Technology in Bratislava
Faculty of Mechanical Engineering
16th Conference on Applied Mathematics **APLIMAT 2017**
Proceedings

THE PRINCIPLES OF RANDOM MATRIX THEORY AND THEIR APPLICATION TO THE PORTFOLIO CHOICE

ŠEDIVÁ Blanka (CZ), ŤOUPAL Tomáš (CZ)

Abstract. In this paper, we use random matrix theory to analyse eigenvalues and see if there is a presence of pertinent information by using Marčenko–Pastur distribution. Thus, we analyse cross correlations between price fluctuations of different stocks using methods of random matrix theory. Moreover, we try to clean correlation matrix from noisy elements to see if the gap between predicted risk and realized risk would be reduced. The cleaned correlation matrix was used in portfolio optimization problem.

Keywords: Random matrix theory, portfolio analysis, correlation matrix

Mathematics subject classification: Primary 60B20; Secondary 91G10, 90C31

1 Introduction

There are many articles about optimal portfolio in the science of mathematics. Investment decisions are governed by weighing risk vs. reward, that is possible loss against expected return. Markowitz' classical portfolio theory [5] assumes that the underlying stochastic process is multivariate normal with known returns and covariances. In practice, these parameters have to be determined from observations on the market. Since the number of observations is necessarily limited, empirically determined parameters will always contain a certain measurement error. Here, we analyse cross correlations between stocks by applying concepts and methods of random matrix theory, developed in the context of complex quantum systems where the precise nature of the interactions between subunits are not known.

2 Data acquisition and data processing

For our analysis we used the time series of daily closing prices of stocks available mainly on *NASDAQ*, which is the biggest stock market in the USA with over 3900 assets from about 39 countries. The financial instruments that can be traded here besides stocks are also options and futures. As a preferred server of datasets was chosen server *yahoo.finance.com*. We extract from this database time series of prices (the time series of prices have been adjusted for stock splits and dividends) of the 100 largest stocks by market capitalization on the starting date January 3, 2000.

3 Theoretical background

Suppose we have a set of N quantities (in our situation the quantities are stock returns of financial assets)and we denote the random vector $\mathbf{A} = (A_1, A_2, \ldots, A_N)$ of random returns on the individual assets. The distribution of vector $\mathbf{A}$ is characterized by vector of expected value with elements $\mathbb{E}A_i = r_i$ and by covariance matrix $\mathbf{V}$ which i, j^{th} element is the covariance between the A_i^{th} and the A_j^{th} random variables. The elements on diagonal σ_i^2 of matrix $\mathbf{V}$ represent variances of asset i. The correspondent correlation matrix is denoted by $\mathbf{C}$.

3.1 Markowitz portfolio theory

The Markowitz's theory of optimal portfolio is focused on the problem to find optimal weight of each assets such that overall portfolio provides the best return for a fixed level of risk, or conversely the smallest risk for a given overall return [5]. More precisely, the average return R_p of a portfolio P of N assets is defined as $R_p = \sum_{i=1}^{N} w_i r_i$ where w_i is the amount of capital invested in the assets i and $\mathbb{E}A_i = r_i$ are expected returns of the individual assets. Similarly the risk of a portfolio P can by associated with the total variance $\sigma_P^2 = \sum_{i,j=1}^{N} w_i V_{ij} w_j$ or in alternative form $\sigma_P^2 = \sum_{i,j=1}^{N} w_i \sigma_i C_{ij} \sigma_j w_j$ where σ_i^2 is the variance of asset i and $\mathbf{C}$ is the correlation matrix. The optimal portfolio which minimizes σ_P^2 for a given value of R_p can be easily found introducing a Lagrange multiplier and leads to a linear problem where the matrix $\mathbf{C}$ has to be invertible [3],[2].

The resulting from a Markowitz optimization scheme, which gives the portfolio with the minimum risk for a given return $R_P = \sum w_i r_i$

$$w_i \sigma_i = R_p \frac{\sum_j C_{ij}^{-1} r_j / \sigma_j}{\sum_{i,j} r_i / \sigma_i C_{ij}^{-1} r_j / \sigma_j} \tag{1}$$

By redefining w_i as $w_i \sigma_i$ the σ_i is absorbed in r_i and w_i and the equations (3) can be write in matrix notation

$$w_C = R_p \frac{C^{-1} r}{r^T C^{-1} r} \tag{2}$$

and the corresponding risk of the portfolio over the period using this construction is

$$\sigma_P^2 = \frac{R_P^2}{r^T C^{-1} r} \tag{3}$$

3.2 Empirical correlation

At the practical application the theoretical expectation values $\mathbb{E}A_i$ as theoretical covariation matrix $\mathbf{C}$ are unknown and it is necessary to estimate its from the realization of the of the $i-$th quantity $(i = 1, \ldots, N)$ at time t where $(t = 1, \ldots, T)$. We denote r_i^t the realization of f the $i-$th quantity at time t and we will assumed in the following to be demeaned and standardized. The normalized $T \times N$ matrix of returns will be denoted as $\mathbf{X} : X_{ti} = r_i^t/\sqrt{T}$. The simplest way to characterize the correlations between these quantities is to compute the Pearson estimator of the correlation matrix.

$$E_{ij} = \frac{1}{T}\sum_t r_i^t r_j^t, \tag{4}$$

or in matrix form $\mathbf{E} = \mathbf{X}^T\mathbf{X}$. where $\mathbf{E}$ will denote the empirical correlation matrix (i.e. on a given realization), that one must carefully distinguish from the "true" correlation matrix $\mathbf{C}$ of the underlying statistical model. One of the main point of the Marčenko-Pastur result is to characterize the difference between $\mathbf{E}$ and $\mathbf{C}$. The resulting empirical covariances $\mathbf{E}$ will fluctuate from sample to sample.

From equation (2) is obvious that stability of Markowitz model strongly depends on input data and estimator $\mathbf{E}$ of theoretical correlation matrix $\mathbf{C}$ which are used for asset mean return estimations and the dominant role for stability is given by quality of estimation of the covariance matrix.

3.3 Random matrix theory

For a set of N different assets, the correlation matrix contains $N(N-1)/2$ entries, which must be determined from N time series of length T. If T is not very large compared to N, we can expect that the determination of covariances is noisy and therefore that the empirical correlation matrix is to a large extent random. Because a covariance matrix is positive semidefinite, that the structure of it can by describe by real eigenvalues and corresponding eigenvectors. Eigenvalues of the covariance matrix that are small (or even zero) correspond to portfolios of stocks that have non-zero returns, but extremely low or vanishing risk; such portfolios are invariably related to estimation errors resulting from insufficient data. One of the approaches used to eliminate the problem of small eigenvalues in the estimated covariance matrix is the so-called random matrix technique. Random matrix theory (RMT), first developed by authors such as Dyson [4] and Mehta [9] for physical application, but there are also many results of interest in a financial context [7], [1], [10].

The spectral properties of $\mathbf{C}$ may be compared to those of random correlation matrix. As described by [7], [10] and others, if $\mathbf{R}$ is any matrix defined by $\mathbf{R} = (1/T)\,\mathbf{X}^T\mathbf{X}$, where $\mathbf{X}$ is an $T \times N$ matrix which elements are i.i.d random variables with mean zero and fixed variance σ^2, than in the limit $T, N \to \infty$ keeping ratio $Q = T/N \geq 1$ constant, the density of eigenvalues of $\mathbf{R}$ is given by

$$P(\lambda) = \frac{Q}{2\pi\sigma^2}\frac{\sqrt{(\lambda_{max} - \lambda)(\lambda - \lambda_{min})}}{\lambda}, \quad \lambda_{min} \leq \lambda \leq \lambda_{max}, \tag{5}$$

where the maximum and minimum eigenvalues are given by

$$\lambda_{max/min} = \sigma^2\left(1 \pm \sqrt{\frac{1}{Q}}\right)^2. \tag{6}$$

The distribution $P(\lambda)$ are known as the Marčenko-Pastur density [8] and the theoretical maximum and minimum eigenvalues determined the bounds for random matrix. If the eigenvalues of matrix are beyond these bounds, it is said that they deviate from random.

3.4 Cleaned correlation matrix

If we apply this theoretical background of RMT to the correlation matrix we can separate the noise and non-noise parts of $\mathbf{E}$. We cleaned the matrix by following procedure:

(1) to construct the empirical correlation matrix as (4),

(2) separate the noisy eigenvalues from non-noisy eigenvalues as (5),

(3) to keep the non-noisy eigenvalues the same and to take average of the noisy eigenvalues,

(4) to replace each eigenvalue associated with the noisy part by average of the eigenvalues,

(5) to reconstruct correlation matrix. The simple repair mechanism, based on the spectral decomposition of the correlation matrix, is described for example in [6].

4 Simulation studies

For a simulation study we used the geometric Brownian motion stock price model for the simulation of the time series by the following relation:

$$dS_t = F(t, S_t)dt + G(t, S_t)dW, \tag{7}$$

where $F(t, S_t)$ is the drift part of the stock price and the second term $G(t, S_t)$ is the shock part of stock price.

The number of generated quantities was $N = 15$ and the length of time series was $T = 60$. The simulated time series correspondent with stock prices and are correlated with random matrix $\mathbf{C}$.

Figure 1 present one simulated trajectories and the correlation matrix **C** as a heat matrix visualisation.

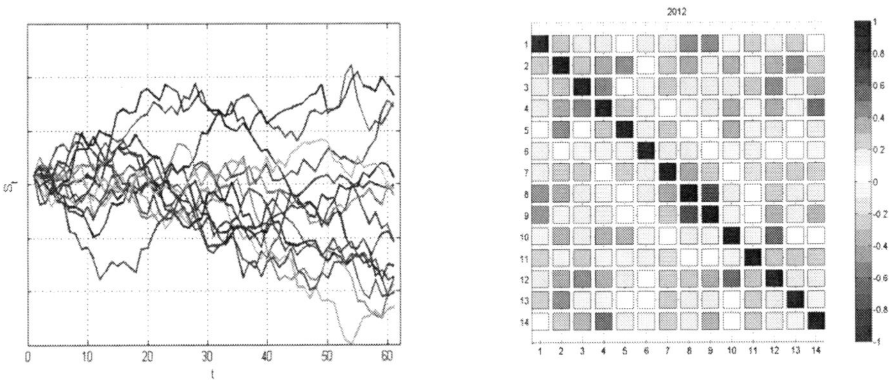

Fig. 1. The simulated trajectories and the heat matrix map of random correlation matrix

Figure 2 shows the theoretical probability density function of eigenvalues of matrix **C** for several value of parameters and the empirical histogram of eigenvalues for simulated datasets.

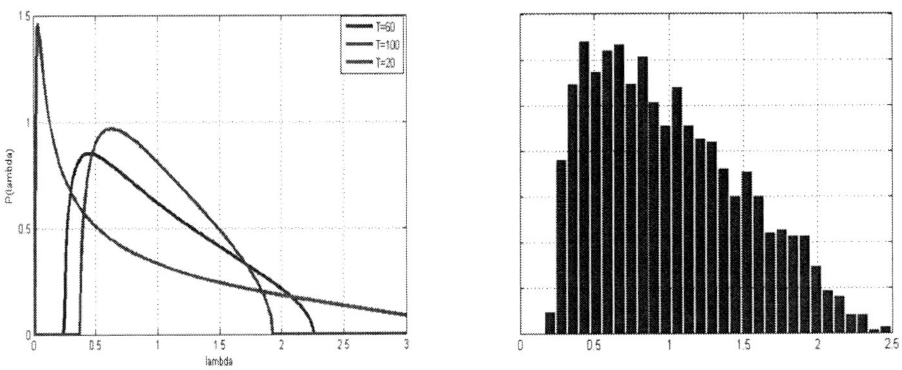

Fig. 2. Theoretical pdf of Marčenko-Pastur distribution of eigenvalues

5 Empirical results

For better understanding of the data we applied the method described in subsection 3.4. Firstly we constructed the empirical measured correlation matrix **E** by using several different numbers of observations (T) and we analysed the distribution of the eigenvalues. We compared the empirical distribution of the eigenvalues of the correlation matrix with the theoretical prediction given by (5) based on assumption that the correlation matrix is purely random. The results are summarized in Table 1.

From these results we can see that the important information about asset mutual connections is carried by 3 to 7% of eigenvalues of the correlation matrix. By increasing the number of observations on which is based the correlation matrix estimation, slightly increases the

# observation	% of $\lambda < \lambda_{min}$	% of $\lambda_{min} < \lambda < \lambda_{max}$	% of $\lambda_{max} < \lambda$
15	7%	91%	3%
30	20%	75%	5%
50	33%	60%	7%
100	54%	38%	8%
150	74%	20%	7%
200	77%	19%	8%

Tab. 1. Comparing eigenvalues of empirical correlation matrix and Marčenko-Pastur bounders

number of non-random correlations, but also increases the instability of the correlation matrix since its eigenvalues are very small. As an optimal number of observations in this case seems to be number between 50 and 100.

Figure 3 shows the results of our experiments on the data with 50 observations used for estimation of the correlation matrix. There is histogram of eigenvalues of the empirical correlation matrix and for comparison we plotted the density of (5) for $Q = 3.8462$ and $\sigma = 0.660$. A better fit can be obtained with a smaller value of $\sigma = 0.462$ (dashed blue line).

This result is in accordance with similar articles, for example [7] or [10] and shows problems related to correct market risk estimation.

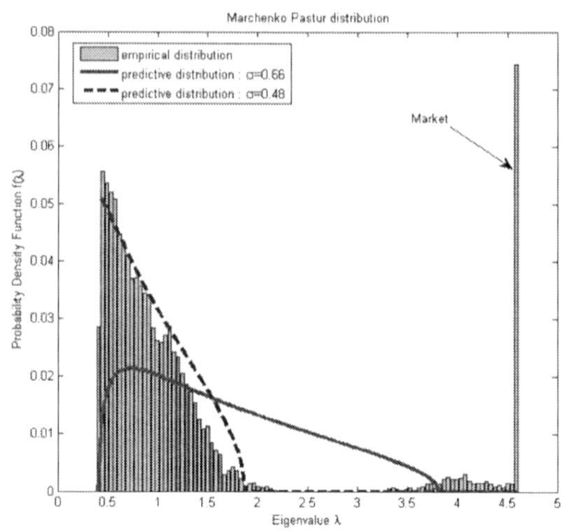

Fig. 3. Empirical and prediction distribution of eigenvalues

6 Conclusion

The goal of this article was to develop an advance approach for stability treatment of portfolio optimization. As we expected the traditional Markowitz model is very sensitive on input data and using the weighting we obtained different results. To conclude, we could say that RMT allows us to analyse in detail the correlation structure of a portfolio of equities. In this effect, Marčenko–Pastur distribution presented the theoretical interval of RMT predictions to observe which eigenvalues are deviating by plotting their empirical distribution. These deviating eigenvalues might contain important information about market and they represents about 3 to 7% of the studied eigenvalues of NASDAQ. This result is in accordance with similar articles, for example [7] or [10] and shows problems related to correct market risk estimation.

By observing the largest eigenvector components distribution, we see that there is a sharp asymmetry in the left, which means that the market reacts more to bad events than good events. Portfolio managers should consider this when they construct their portfolios. In addition, the cleaning procedure of correlation matrix reduced slightly the gap between predicted and realized risks. This procedure could be helpful for practitioners by reducing their errors of predictions.

Acknowledgements

This article was supported by the European Regional Development Fund (ERDF), project "NTIS - New Technologies for the Information Society", European Centre of Excellence, CZ.1.05/1.1.00/02.0090.

References

[1] Daly, J., Crane, M., & Ruskin, H. J. (2008). *Random matrix theory filters in portfolio optimisation: a stability and risk assessment.* Physica A: Statistical Mechanics and its Applications, 387(16), 4248-4260.

[2] Danielsson, J. (2011). *Financial risk forecasting: the theory and practice of forecasting market risk with implementation in R and Matlab* (Vol. 588). John Wiley & Sons.

[3] Dupačová, J.: Markowitzuv model optimální volby portfolia - předpoklady, data, alternativy *(n.d.). Retrieved April 25, 2016* URL: `http://msekce.karlin.mff.cuni.cz/ dupacova/downloads/Markowitz.pdf`

[4] Dyson, Freeman J. (1970).*Correlations between eigenvalues of a random matrix.* Comm. Math. Phys. 19, no. 3, 235–250.

[5] Elton, E. J., Gruber, M. J., Brown, S. J., & Goetzmann, W. N. (2009). *Modern portfolio theory and investment analysis.* John Wiley & Sons.

[6] Gilli, M., Maringer, D., & Schumann, E. (2011). *Numerical methods and optimization in finance.* Academic Press.

[7] Laloux, L., Cizeau, P., Potters, M., & Bouchaud, J. P. (2000). *Random matrix theory and financial correlations.* International Journal of Theoretical and Applied Finance, 3(03), 391-397.

[8] Marčenko, V. A., & Pastur, L. A. (1967). *Distribution of eigenvalues for some sets of random matrices.* Mathematics of the USSR-Sbornik, 1(4), 457.

[9] Mehta, M. L. (2004). *Random matrices* (Vol. 142). Academic Press.

[10] Potters, M., Bouchaud, J. P., & Laloux, L. (2005). *Financial applications of random matrix theory: Old laces and new pieces.* arXiv preprint physics/0507111.

Current address

Šedivá Blanka, RNDr. PhD.
Department of Mathematics, Faculty of Applied Sciences, University of West Bohemia
Univerzitni 8, 306 14 Pilsen, Czech Republic
Tel. number: +420 377 63 2618, e-mail: sediva@kma.zcu.cz

Ťoupal Tomáš, Ing. PhD.
Department of Mathematics, Faculty of Applied Sciences, University of West Bohemia
Univerzitni 8, 306 14 Pilsen, Czech Republic
Tel. number: +420 377 63 2674, e-mail: ttoupal@kma.zcu.cz

Slovak University of Technology in Bratislava
Faculty of Mechanical Engineering
16th Conference on Applied Mathematics **APLIMAT 2017**
Proceedings

THE FIBONACCI NUMBERS FOR THE MOLECULAR GRAPHS OF TWO TYPES OF BENT HEXAGONAL CHAINS

SEIBERT Jaroslav (CZ), ZAHRÁDKA Jaromír (CZ)

Abstract. The Fibonacci number of an undirected graph $G = (V, E)$ is given by the number of subsets U of V such that no two vertices in U are adjacent. This number is one of the most popular topological index in chemistry, which is called as the Merrifield-Simmons index there. Hexagonal chains are the graph representations of an important subclass of benzenoid molecules. In this contribution we follow our previous results on the Fibonacci number of the linear hexagonal chains. We obtain exact formulas for the Fibonacci numbers of two types of bent hexagonal chains.

Key words: Fibonacci number, simple graph, decomposition theorem, difference equation, hexagonal chain, linear chain, bent chain

Mathematics Subject Classification: Primary 12H10, 39A10; Secondary 11B39

1 Introduction

In this contribution, we consider undirected simple graphs (without loops of multiple edges). A subset S of vertex set of a graph G is called independent if none two vertices of S are adjacent in G. Other graph terminology and notation is taken from the book [2].

The Fibonacci numbers F_n are defined by the second order recurrence $F_{n+2} = F_{n+1} + F_n$, with $F_0 = 0, F_1 = 1$; the Lucas numbers L_n satisfy the same recurrence but with the initial terms $L_0 = 2, L_1 = 1$. These numbers appear – often with unexpected connections – in various areas of mathematics. In particular, we highlight the following connection to discrete mathematics: the total number of subsets of linearly ordered set $\{1, 2, \ldots, n\}$ such that no two elements are adjacent is equal to the Fibonacci number F_{n+2}. Based on this fact, Prodinger and Tichy introduced in [7] the Fibonacci number of a graph:

Definition 1. Let $G = (V, E)$ be a simple connected graph. The Fibonacci number $f(G)$ of G is defined as the number of all independent subsets of V.

Note that this graph invariant is also known as the Merrifield-Simmons index, and is one of important topological indices in chemical graph theory because of its correlation with boiling points (see [5]).

We presented in [8] a computer program for calculating of the Fibonacci number of a given connected graph with small number of vertices, also calculating the Fibonacci numbers of particular graphs being the chains of cycles of length four in which the neighboring circuits have a common edge or vertex. Other authors found the Fibonacci numbers of another classes of graphs. Yeh [11] computed algorithmically the Fibonacci numbers of lattice product graphs. Alameddine [1] found upper and lower bounds for the Fibonacci numbers of maximal outerplanar graphs on n vertices. Gutman [3] reported some results on extremal hexagonal chains.

The aim of this paper is to derive the exact formulae for the Fibonacci number of graphs two specific families of hexagonal systems. Recall that a hexagonal system is a connected plane graph without cut-vertices in which all inner faces are hexagons (that is, 6-vertex cycles), such that two hexagons are either disjoint or have exactly one common edge, and none three hexagons share a common edge. Two hexagons of a system may have either two common vertices (if they are adjacent) or none (if they are not adjacent). A hexagonal chain with n hexagons, $n \geq 2$, possesses two terminal hexagons and $n-2$ hexagons that have two neighbors. A hexagon adjacent to exactly two other hexagons possesses two vertices of degree 2. If these two vertices are adjacent, then the hexagon is angularly annelated, if these two vertices are not adjacent, then it is linearly annelated. The hexagonal systems we consider contain one or two angularly annelated hexagons, with lengths of linear subchains being the same; the derivations of their Fibonacci numbers is contained in Section 3, while Section 2 contains useful general and particular results used in these derivations.

2 Preliminary results

In this section, we review the general results for calculating the Fibonacci numbers of graphs and recall our results on the Fibonacci numbers of linear hexagonal chains from [9]. The basic results were derived by Prodinger and Tichy in [7]; another results of these authors can be found in [5].

Theorem 1. If G_1, G_2 are disjoint graphs, then $f(G_1 \cup G_2) = f(G_1) f(G_2)$.

Theorem 2.

(a) $f(P_n) = F_{n+2}$, where P_n is a path with n vertices,

(b) $f(C_n) = L_n$, where C_n is a cycle with n vertices.

Further, we will use the following decomposition theorems (see [6]) :

Theorem 3. Let G be a graph with at least two vertices and v be its arbitrary vertex. Then

$$f(G) = f(G-v) + f(G-N[v]),$$

where $G-v$ is the subgraph of G obtained by deletion of the vertex v and $G-N[v]$ is the subgraph of G obtained by deletion of the vertex v and all the vertices adjacent to v.

Theorem 4. Let vertices u,v be adjacent in a graph G. Then

(a) $f(G)=f(G-\{u,v\})+f(G-N[u])+f(G-N[v])$, where $G-\{u,v\}$ is the subgraph of G obtained by deletion of the vertices u and v of G,

(b) $f(G)=f(G-uv)-f(G-(N[u]\cup N[v]))$, where $G-uv$ is the subgraph of G obtained by deletion of the edge uv of G and $G-(N[u]\cup N[v])$ is the subgraph of G obtained by deletion of the vertices u ,v and all the vertices adjacent to them.

Consider now the graphs S_n, ${}^1S_n$, ${}^2S_n$ and ${}^3S_n$ of Fig. 1 (the members of $\{S_n\}$ are also known as acenes).

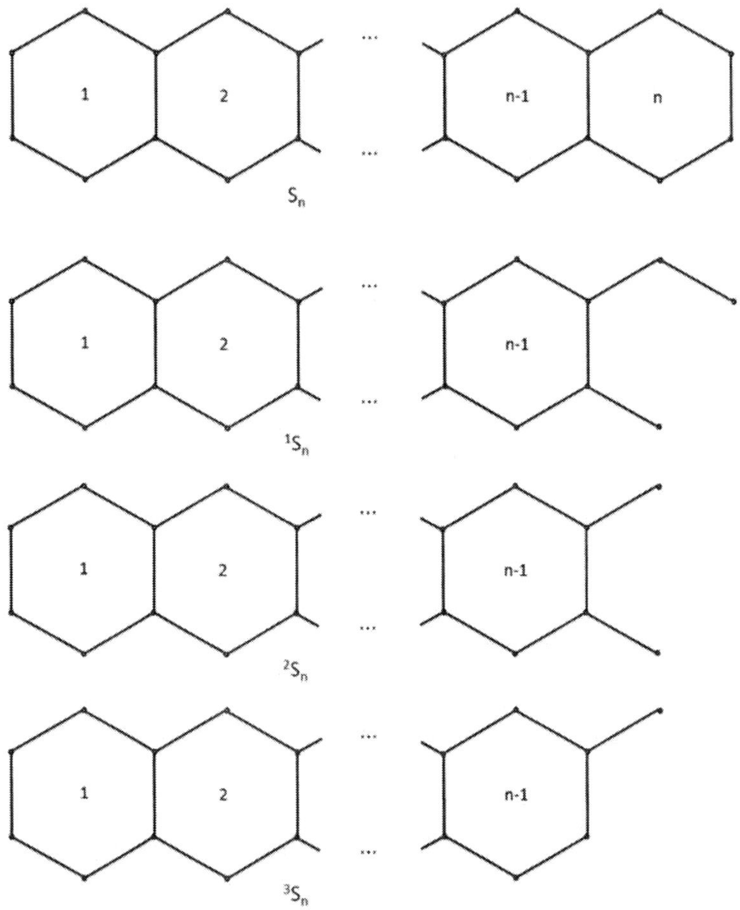

Fig. 1

Let us denote their Fibonacci numbers $f(S_n)=a_n$, $f({}^1S_n)=b_n$, $f({}^2S_n)=c_n$, $f({}^3S_n)=d_n$ for abbreviation. We derived relations (1)-(4) using Theorem 3.

Lemma 1. The terms of sequences $\{a_n\}$, $\{b_n\}$, $\{c_n\}$, $\{d_n\}$ satisfy the following system of difference equations

$$a_n = b_n + d_n \tag{1}$$

$$b_n = c_n + d_n \tag{2}$$

$$c_n = d_n + 2c_{n-1} + d_{n-1} \quad (3)$$

$$d_n = a_{n-1} + c_{n-1} + d_{n-1}. \quad (4)$$

Theorem 5. The formula $f({}^3S_n) = d_n = \frac{1}{\gamma - \delta}\left[(31-5\delta)\gamma^{n-1} - (31-5\gamma)\delta^{n-1}\right]$ holds for any positive integer n, where $\gamma = \frac{7+\sqrt{33}}{2}$, $\delta = \frac{7-\sqrt{33}}{2}$.

Proof. The system of identities of Lemma 1 can be transformed to the homogeneous linear difference equation of the second order with constant coefficients $d_{n+1} - 7d_n + 4d_{n-1} = 0$ and the initial conditions $d_1 = f(P_3) = 5, d_2 = f(C_6) + f(P_5) = 31$. The solution of this difference equation gives

$$d_n = \frac{31-5\delta}{\gamma(\gamma-\delta)}\gamma^n + \frac{31-5\gamma}{\delta(\gamma-\delta)}\delta^n$$

which finishes the proof.

Corollary. The following expressions for the Fibonacci numbers of the given graphs hold

$$f(S_n) = a_n = \frac{1}{2(\gamma-\delta)}\left[(31-5\delta)(1+\gamma)\gamma^{n-1} - (31-5\gamma)(1+\delta)\delta^{n-1}\right],$$

$$f({}^1S_n) = b_n = \frac{1}{2(\gamma-\delta)}\left[(31-5\delta)(-1+\gamma)\gamma^{n-1} - (31-5\gamma)(-1+\delta)\delta^{n-1}\right],$$

$$f({}^2S_n) = c_n = \frac{1}{2(\gamma-\delta)}\left[(31-5\delta)(-3+\gamma)\gamma^{n-1} - (31-5\gamma)(-3+\delta)\delta^{n-1}\right].$$

Proof. It is easy to obtain the relation $a_n = \frac{1}{2}(d_{n+1} + d_n)$ from identities (1)-(4). The remaining results we get from identities $b_n = a_n - d_n$ and $c_n = a_n - 2d_n$.

Remark. The above mentioned formulas for a_n, b_n, c_n can be rewritten to the form

$$a_n = \frac{1}{\gamma-\delta}\left[(18\gamma-12)\gamma^{n-1} - (18\delta-12)\delta^{n-1}\right],$$

$$b_n = \frac{1}{\gamma-\delta}\left[(13\gamma-8)\gamma^{n-1} - (13\delta-8)\delta^{n-1}\right],$$

$$c_n = \frac{1}{\gamma-\delta}\left[(8\gamma-4)\gamma^{n-1} - (8\delta-4)\delta^{n-1}\right]$$

using the obvious equalities $\gamma + \delta = 7$ and $\gamma\delta = 4$.

We have calculated the Fibonacci numbers of the graphs $f(S_n) = a_n$, $f({}^1S_n) = b_n$, $f({}^2S_n) = c_n$, $f({}^3S_n) = d_n$ for integers $1 \le n \le 10$ by using the system MATLAB. The obtained values are arranged in Table 1.

n	a_n	b_n	c_n	d_n
1	18	13	8	5
2	114	83	52	31
3	726	529	332	197
4	4 626	3 371	2 116	1 255
5	29 478	21 481	13 484	7 997
6	187 842	136 883	85 924	50 959
7	1 196 982	872 257	547 532	324 725
8	7 627 506	5 558 267	3 489 028	2 069 239
9	48 604 614	35 418 841	22 233 068	13 185 773
10	309 722 274	225 698 819	141 675 364	84 023 455

Tab. 1

3 Two types of molecular graphs containing the linear chains of hexagons

First, we will deal with the hexagonal chain $S_{n,n}$ which is consisting of two linear chains of the same length $n \geq 2$. This chain has just two angularly annelated hexagons. It is possible to use the results from the previous section.

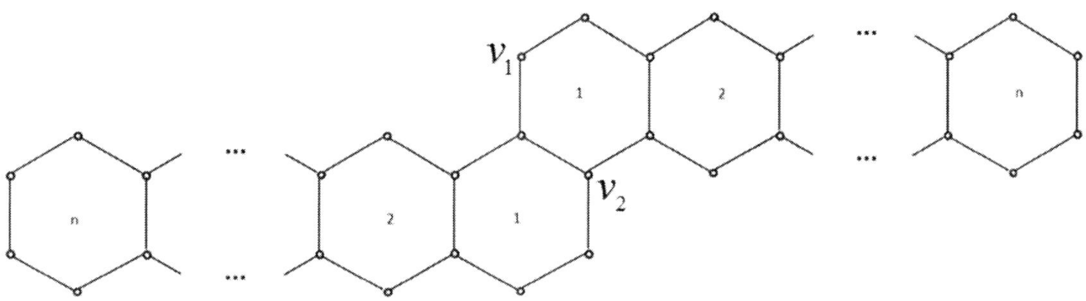

$S_{n,n}$

Fig. 2

Let us denote $f(S_{n,n}) = A_n$ for simplification and we prove the following Lemma.

Lemma 2. The terms of sequence $\{A_n\}$ satisfy the following relations

$$A_n = a_n(2c_{n-1} + d_{n-1}) + 2b_n d_{n-1} + d_n^2 + a_{n-1}^2, \qquad (5)$$
$$A_n = b_n d_n + a_{n-1}^2 + d_n(a_{n-1} + 2b_{n-1} + c_{n-1}), \qquad (6)$$

where $n \geq 2$.

Proof: These relations can be derived by repeatedly using Theorem 3. The vertex v_1 is chosen as the initial vertex for derivation of relation (5) and the vertex v_2 for derivation of relation (6). The simple (but lengthy) details of the proof are left to readers.

Theorem 6. The Fibonacci number of the graph $S_{n,n}$ can be written in the form

$$A_n = \frac{1}{(\gamma-\delta)^2}\left[(25827\gamma-16212)\gamma^{2n-4}+(25827\delta-16212)\delta^{2n-4}+432\cdot 4^{n-2}\right]$$

for any positive integer n.

Proof. We will use relation (5). Then for any positive integer $n \geq 2$ the following holds

$$\begin{aligned}
A_n &= a_n(2c_{n-1}+d_{n-1})+2b_n d_{n-1}+d_n^2+a_{n-1}^2 = \\
&= \frac{1}{4(\gamma-\delta)^2}\Big[\big((31-5\delta)(\gamma+\gamma^2)\gamma^{n-2}-(31-5\gamma)(\delta+\delta^2)\delta^{n-2}\big)\big(2(31-5\delta)(-3+\gamma)\gamma^{n-2}- \\
&\quad -2(31-5\gamma)(-3+\delta)\delta^{n-2}+2(31-5\delta)\gamma^{n-2}-2(31-5\gamma)\delta^{n-2}\big) + \\
&\quad +\big(2(31-5\delta)(-\gamma+\gamma^2)\gamma^{n-2}-2(31-5\gamma)(-\delta+\delta^2)\delta^{n-2}\big)\big(2(31-5\delta)\gamma^{n-2}-2(31-5\gamma)\delta^{n-2}\big)+ \\
&\quad +\big(2(31-5\delta)\gamma\gamma^{n-2}-2(31-5\gamma)\delta\delta^{n-2}\big)^2+\big((31-5\delta)(1+\gamma)\gamma^{n-2}-(31-5\gamma)(1+\delta)\delta^{n-2}\big)^2\Big] = \\
&= \frac{1}{4(\gamma-\delta)^2}\Big[(31-5\delta)^2(2\gamma^3+7\gamma^2-6\gamma+1)\gamma^{2(n-2)}+(31-5\gamma)^2(2\delta^3+7\delta^2-6\delta+1)\delta^{2(n-2)}- \\
&\quad -(31-5\gamma)(31-5\delta)(2\gamma^2\delta+14\gamma\delta+2\gamma\delta^2-6\gamma-6\delta+2)\gamma^{n-2}\delta^{n-2}\Big].
\end{aligned}$$

Using the obvious identities $\gamma^2-7\gamma+4=0$, $\delta^2-7\delta+4=0$, $\gamma\delta=4$, $\gamma+\delta=7$ we have

$$\begin{aligned}
A_n &= \frac{1}{4(\gamma-\delta)^2}\Big[(861-135\delta)(133\gamma-83)\gamma^{2(n-2)}+(861-135\gamma)(133\delta-83)\delta^{2(n-2)}- \\
&\quad -(48\gamma+48\delta+1392)\gamma^{n-2}\delta^{n-2}\Big] = \\
&= \frac{1}{4(\gamma-\delta)^2}\Big[(114513\gamma+11205\delta-143283)\gamma^{2n-4}+(114513\delta+11205\gamma-143283)\delta^{2n-4}+1728\cdot 4^{n-2}\Big]= \\
&= \frac{1}{4(\gamma-\delta)^2}\Big[(103308\gamma+11205(\gamma+\delta)-143283)\gamma^{2n-4}+(103308\delta+11205(\gamma+\delta)-143283)\delta^{2n-4}+ \\
&\quad +1728\cdot 4^{n-2}\Big]=\frac{1}{4(\gamma-\delta)^2}\Big[(103308\gamma-64848)\gamma^{2n-4}+(103308\delta-64848)\delta^{2n-4}+1728\cdot 4^{n-2}\Big]
\end{aligned}$$

and the proof is over for any positive integer $n \geq 2$.

Moreover, we can show that the formula is valid even for $n=1$. Thus

$$\begin{aligned}
A_1 &= \frac{1}{(\gamma-\delta)^2}\left[(25827\gamma-16212)\gamma^{-2}+(25827\delta-16212)\delta^{-2}+432\cdot 4^{-1}\right] = \\
&= \frac{1}{(\gamma-\delta)^2}\left[(25827\gamma-16212)\frac{\delta^2}{16}+(25827\delta-16212)\frac{\gamma^2}{16}+108\right] =
\end{aligned}$$

$$=\frac{1}{33}\left[\frac{180789}{8}\gamma\delta-\frac{25827}{4}(\gamma+\delta)-\frac{28371}{4}(\gamma+\delta)+8214\right]=114=a_2=f(S_2),$$

as $S_{1,1}$ is the linear chain of two hexagons.

We have also calculated the Fibonacci numbers of the graphs $S_{n,n}$ for the positive integers $n \leq 10$ by using of the system MATLAB.

n	A_n
1	114
2	4 509
3	182 613
4	7 413 261
5	301 014 981
6	12 222 974 397
7	496 325 599 989
8	20 153 781 566 829
9	818 363 832 870 693
10	33 230 456 635 551 261

Tab. 2

For now, we will deal with the sequence of the graphs $\overline{S}_{n,n}$ (see Fig. 3). Let us denote $f(\overline{S}_{n,n})=B_n$ for simplification and we prove the following Lemma.

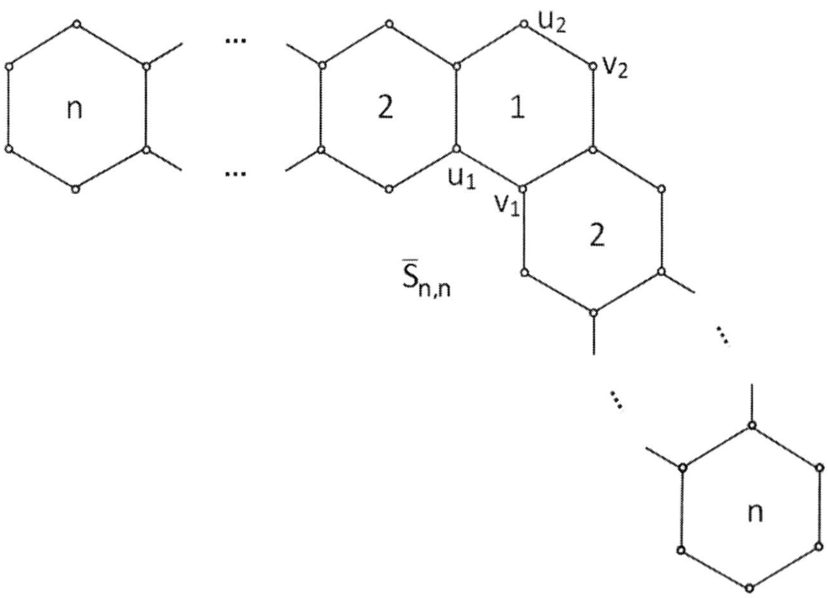

Fig. 3

Lemma 3. The terms of sequence $\{B_n\}$ satisfy for $n \geq 2$ the following relation

$$B_n = a_{n-1}^2 + 2a_{n-1}b_{n-1} - 3d_{n-1}^2. \quad (7)$$

Proof: This relation can be derived by using of the both statements of Theorem 4. The edge u_1v_1 (see Fig. 3) is chosen as the initial edge for the using of statement (b) and further the vertices u_2, v_2 for statement (a) used on the graph $\overline{S}_{n,n}$ - u_1v_1. The simple details of the proof are again left to readers.

Theorem 7. The Fibonacci number of the graph $\overline{S}_{n,n}$ can be written in the form

$$B_n = \frac{1}{(\gamma-\delta)^2}\left[(4107\gamma - 2580)\gamma^{2n-4} + (4107\delta - 2580)\delta^{2n-4} + 72\cdot 4^{n-2}\right]$$

for any positive integer n.

Proof. We will use relation (7). Then for any positive integer $n \geq 2$

$$B_n = a_{n-1}^2 + 2a_{n-1}b_{n-1} - 3d_{n-1}^2 =$$

$$= \frac{1}{(\gamma-\delta)^2}\left\{\left[(18\gamma-12)\gamma^{n-2} - (18\delta-12)\delta^{n-2}\right]^2 + \right.$$

$$+ 2\left[(18\gamma-12)\gamma^{n-2} - (18\delta-12)\delta^{n-2}\right]\left[(13\gamma-8)\gamma^{n-2} - (13\delta-8)\delta^{n-2}\right] -$$

$$\left. - 3\left[(31-5\delta)\gamma^{n-2} - (31-5\gamma)\delta^{n-2}\right]^2\right\} =$$

$$= \frac{1}{(\gamma-\delta)^2}\left[\left(792\gamma^2 - 75\delta^2 - 1032\gamma + 930\delta - 2547\right)\gamma^{2n-4} + \right.$$

$$\left. + \left(792\delta^2 - 75\gamma^2 - 1032\delta + 930\gamma - 2547\right)\delta^{2n-4} + (-1434\gamma\delta + 102\gamma + 102\delta + 5094)4^{n-2}\right].$$

By using of the relations $\gamma^2 = 7\gamma - 4,\ \delta^2 = 7\delta - 4,\ \gamma\delta = 4$ we obtain

$$B_n = \frac{1}{(\gamma-\delta)^2}\left[(4512\gamma + 405\delta - 5415)\gamma^{2n-4} + (4512\delta + 405\gamma - 5415)\delta^{2n-4} + (102\gamma + 102\delta - 642)4^{n-2}\right]$$

and

$$B_n = \frac{1}{(\gamma-\delta)^2}\left[(4107\gamma + 405(\gamma+\delta) - 5415)\gamma^{2n-4} + (4107\delta + 405(\gamma+\delta) - 5415)\delta^{2n-4} + \right.$$

$$\left. + (102(\gamma+\delta) - 642)4^{n-2}\right],$$

which gives the desired formula for $n \geq 2$ as $\gamma + \delta = 7$.

But it is easy to show that this expression of B_n also holds for $n = 1$. Then

$$B_1 = \frac{1}{(\gamma-\delta)^2}\left[(4107\gamma-2580)\gamma^{-2} + (4107\delta-2580)\delta^{-2} + 72\cdot 4^{-1}\right] =$$

$$= \frac{1}{(\gamma-\delta)^2}\left[(4107\gamma-2580)\frac{\delta^2}{16} + (4107\delta-2580)\frac{\gamma^2}{16} + 18\right] =$$

$$= \frac{1}{33}\left[\frac{28749}{8}\gamma\delta - \frac{4107}{4}(\gamma+\delta) - \frac{4515}{4}(\gamma+\delta) + 1308\right] = 18 = a_1 = f(S_1),$$

as $\bar{S}_{1,1}$ is the circuit with six vertices.

We have again calculated the Fibonacci numbers of the graphs $\bar{S}_{n,n}$ for positive integers $n \le 10$ by using of the system MATLAB.

n	B_n
1	18
2	717
3	29 037
4	1 178 757
5	47 863 293
6	1 943 530 293
7	78 918 910 893
8	3 204 578 788 197
9	130 125 027 446 877
10	5 283 852 863 531 157

Tab. 3

4 Concluding remarks

Many topological indices of graphs are closely connected with selected physico-chemical characteristics of the respective compounds. One of the most famous and widely studied topological indices is the Fibonacci number of graph, also known as Merrifield-Simmons index. This index is correlated with the boiling points of hydrocarbons; therefore, the knowledge of the Fibonacci number of a molecular graph is interesting for chemists. The classes of molecular graphs which are investigated in this contribution include some common hydrocarbons, as naphthalene ($S_{1,1}$), chrysene ($S_{2,2}$), phenanthrene ($\bar{S}_{2,2}$), pentaphene ($\bar{S}_{3,3}$) and so on. The methods of calculation of the Fibonacci numbers of graphs are possible to use for other classes of the molecular graphs. The main problem lies in a rather complicated solution of system of difference equations, to which the decomposition theorems lead.

Acknowledgment. The authors gratefully acknowledge the anonymous referee for his useful remarks and comments.

References

[1] ALAMEDDINE, A. F. *Bounds on the Fibonacci Number of a Maximal Outerplanar Graph.* Fibonacci Quarterly, **36** (1998), No. 2, 206-210.

[2] DIESTEL, R. *Graph Theory*, Springer, 2006.

[3] GUTMAN, I. *Extremal hexagonal chains.* Journal of Mathematical Chemistry, **12** (1993), 197–210.

[4] KIRSCHENHOFER, P., PRODINGER, H., TICHY, R. F. *Fibonacci numbers of graphs II.* Fibonacci Quarterly, **21** (1983), No. 2, 16-21.

[5] MERRIFIELD, R. E., SIMMONS, H. E. *Topological Methods in Chemistry*, Wiley, New York, 1989.

[6] PEDERSEN, A. S., VESTERGAARD, P. D. *The number of independent sets in unicyclic graphs.* Discrete Applied Mathematics, **152** (2005), 246-256.

[7] PRODINGER, H., TICHY, R. F. *Fibonacci numbers of graphs.* Fibonacci Quarterly, **20** (1982), No. 1, 16-21.

[8] SEIBERT, J., TROJOVSKY, P. *Recurrences and generating functions for Fibonacci numbers of graphs with Mathematica.* Proceedings of the Conference Informatika a Algoritmy, Presov 2002, 36-41.

[9] SEIBERT, J., ZAHRADKA, J. *Fibonacci numbers of graphs corresponding to a type of hexagonal chains.* Proceedings of 13th International Conference APLIMAT 2014. Slovak University of Technology, Bratislava, 356-363.

[10] VAGNER, S., GUTMAN, I. *Maxima and minima of the Hosoya index and the Merrifield-Simmons index. A survey of results and techniques.* Acta Appl. Math., **112** (2010), 323-346.

[11] YEH, L. *Fibonacci numbers of product graphs.* The Journal of Combinatorial Mathematics and Combinatorial Computing, **32**(2000), 223-229.

Current address

Seibert Jaroslav, Doc. RNDr. CSc.
Institute of Mathematics and Quantitative Methods,
Faculty of Economics and Administration, University of Pardubice
532 10 Pardubice, Czech Republic
Tel. number: +420 466 036 019, e-mail: jaroslav.seibert@upce.cz

Zahrádka Jaromír, RNDr. PhD.
Institute of Mathematics and Quantitative Methods,
Faculty of Economics and Administration, University of Pardubice
532 10 Pardubice, Czech Republic
Tel. number: +420 466 036 047, e-mail: jaromir.zahradka@upce.cz

Slovak University of Technology in Bratislava
Faculty of Mechanical Engineering
16th Conference on Applied Mathematics **APLIMAT 2017**
Proceedings

USING SEMANTIC DIFFERENTIAL FOR MEASURING CHANGES IN UNDERSTANDING SELECTED CONCEPTS BY RESPONDENTS

ŠIMANDL Václav (CZ), ŠERÝ Michal (CZ), DOBIÁŠ Václav (CZ)

Abstract. The article focuses on the analysis of semantic differential data based on the assessment of the global similarity of keywords. This method finds out the mutual distances of keywords in the semantic space. The calculation of this distance, creation of corresponding matrix of distances and application of cluster analysis for graphical representation of the keywords similarity in the form of dendrograms are described in the article. Using the comparison of the pretest and posttest dendrograms, changes in perceiving of selected keywords by respondents were identified.

Keywords: Semantic differential, multidimensional data, cluster analysis, dendrogram, attitudes, keyword, connotative meaning

Mathematics Subject Classification: 62P15, 62P25

1 Introduction

An inseparable part of the education is the assessment of the applied teaching methods and procedures. As it is necessary to select the optimum teaching method depending on the topic of lesson, it is also necessary to consider the suitable method of assessment of this lesson. While some methods of assessment of lesson results deal primarily with measuring changes in the pupil's knowledge (e.g. didactic tests), others concern rather the skills (e.g. intelligence tests) and other methods of assessment deal with attitudes or keyword understanding (e.g. semantic differential or Q-methodology) [1].

This article deals with the semantic differential method used for investigation of attitudes and keyword understanding. This method – although from the viewpoint of data collection it works primarily with the meaning of words – is based on the mathematical foundation and uses the method of statistic processing of multidimensional data. In the following text, the semantic differential method will be explained from the viewpoint of application of specific statistical methods in the scope of pedagogical research.

1.1 Pedagogical and psychological basis of semantic differential

In the educational process, the pupils and students acquire a number of knowledge and ideas about the things and phenomena creating the base of pupil's keyword understanding. Průcha, Walterová and Mareš define the keywords as generalized ideas of things and phenomena expressed by one or more terms of the natural or formal language [2]. Acquiring the knowledge of the abstract keywords is understood by renowned psychologists (Piaget, Vygotsky) as development of thinking and language. As constructivism accentuates the role of preconcept (spontaneous concept) in the cognitive process, the teaching performed in constructivist way corresponds to this. Jodelet characterizes preconcept as "reference system", in the scope of which the transformation, integration and acquisition of new or different information or representation takes place [3], [4]. The keywords have a denotative meaning for a certain individual (real object, class of objects: e.g.: *grandmother = old person*) and a connotative meaning (individual, emotive e.g.: *grandmother = granny Jane*) [5].

Attitudes express the relationship of an individual to a given object. Průcha, Walterová and Mareš define the attitudes as evaluating relationships taken by an individual towards the surrounding world and acquired during spontaneous learning in the family and in other social environment [2]. One of the targets of the pedagogical process must be necessarily influencing the attitudes of pupils or students in a certain direction. More methods may be used for investigating the attitudes, as e.g. Bogardus social distance scale or Q-methodology. Unlike other methods, semantic differential does not measure directly the attitudes, but it focuses on investigating the connotative meaning of keywords.

Thanks to this, the semantic differential may investigate also respondent's attitudes concerning certain keywords. For this reason, semantic differential may be considered a suitable research tool in the areas of teaching focusing on the change of attitudes or behaviour, as e.g. ecological education, computer security lessons or constructivist teaching. As for non-pedagogical sciences, semantic differential has been used for a long time, frequently and with success in marketing [6], [7], [8].

1.2 Semantic differential data analysis methods

The semantic differential (SD) is designed to measure individual psychological meanings of certain keywords used by particular people by placing these keywords in a so-called semantic space created by means of bipolar adjective pairs (e.g. necessary-useless) and the respective value scales [1]. Semantic space may be used also for measuring changes in understanding the meaning of these keywords [9]. Using this method, it is also possible to work with groups of respondents. It can be found out whether a certain group of respondents perceives the selected keywords in a homogenous way or if certain groups differ from one another in this aspect [10].

The method of semantic differential brings a quite extensive collection of data. From the mathematical viewpoint these data may be processed using methods of multidimensional statistical analysis. The data analysis can be carried out on three basic levels [11]:

- Answers analysis on the level of individual adjectives,
- Answers analysis on the level of individual dimensions of affective meaning,

- Answers analysis based on assessment of global similarity.

The answers analysis on the level of individual adjectives is implemented using the methods of descriptive statistics. Averaging of the resulting point scale for a selected group of respondents and the frequency of answers are used. In case of investigating the differences between respondent groups, the differences between averages may be tested (e.g. t-test).

The answers analysis on the level of individual dimensions can be carried out. It is based on the aggregation of data of individual adjectives scales into the low-dimensional space (usually 3 or 2 dimensions) e.g. by VARIMAX method [1]. The factor analysis may be carried out using some statistic programs, e.g. the program R [12].

The following text treats the answers analysis based on the assessment of global similarity. This analysis is carried out using the visualization of distance (associative) matrix. Euclidean distance metrics is usually applied to the creation of the distance matrix. By applying other metrics, another aspect in perceiving keywords may be accentuated during the analysis [13] (see below).

Applying semantic differential will be demonstrated on the example of our research [14]. In this, the changes in understanding selected e-safety keywords by university students were measured after lessons and compared with the state before lessons (see [14]). In the scope of this research, two groups of respondents were created. While the students included into the experimental group took part in the lessons, the students of control group did not take part in them. Two measurements of understanding the selected keywords were carried out for both groups. The first measurement (pretest) was implemented before lessons, the second measurement (posttest) was implemented after them.

2 Making up SD questionnaire

SD questionnaires were applied for measuring individual meanings of keywords. The questionnaires were made up of twelve seven-point rating scales with outer points representing pairs of adjectives of opposite meaning (bipolar adjectives) [1]. Respondents made judgements about their perception of fifteen keywords and recorded them on the scales by choosing a particular point on these scales [1]. Using this, an individual meaning was allocated to each keyword, perceived by the assessing subject [10].

During the choice of assessed keywords, it is necessary to focus on keywords concerning the researched issue [15]. The following keywords were chosen in our case: *Privacy*, *Email*, *Backup*, *Password*, *Ulož.to* (the Czech equivalent of RapidShare.com; often used for the illegal distribution of works protected by copyright law), *Facebook*, *Virus* and *Loss*. To be able to understand the connotative meaning of the above mentioned keywords in a wider context, the list was extended by the keywords exceeding the research issue and characterizing, in a certain way, the respondent himself [13]. These keywords concern the respondent, his view on the socially recognized values and education. Their choice was inspired by Pöschl's research [16] and the following words were selected as keywords: *Me*, *Fear*, *Knowledge*, *Life*, *Work*, *Money* and *Teacher*.
During the choice of bipolar adjectives it was proceeded from the character of assessed keywords [1]. As initial base, the original list of fifty assessment scales of C. Osgood was used [1] and adjectives from the Šerý's research were added to it [13]. Such pairs of adjectives were selected,

enabling to achieve the equal representation in dominant factors: Assessment, Potency and Activity. The selected pairs of adjectives are: good – bad; useful – useless; essential – superfluous; easy – complicated; heavy – light; problem – problem-free; thick – thin; strong – weak; dynamic – static; hot – cold; fast – slow; active – passive.

3 SD questionnaires data analysis

Pre-processing and analysis of data acquired by SD questionnaire survey were implemented applying software tools developed by us [13].

3.1 The part of data analysis carried out using automated techniques

Data analysis was based on the method of global evaluation of keyword similarity, which allows us to judge the extent to which keywords can be considered semantically similar or different, as claimed by Ferjenčík [11]. The Euclidean distance was used to calculate this so-called D-statistic [15]. The distance D_{ij} between keywords i and j was calculated according to this equation

$$D_{ij} = \sqrt{\sum_{k=1}^{12} \left(x_i(k) - x_j(k)\right)^2} \tag{1}$$

where $x_i(k)$ is the value of the kth scale (i.e. the chosen value for the kth pair of adjectives) for keyword i (see [15]).

Other possibilities for the calculation of D-statistics are represented by the use of Manhattan or Chebyshev metrics [13]; a different aspect of perceiving keywords may be accentuated by them. The influence of different metrics application may be illustrated on the comparison of Euclidean and Manhattan metrics. If we are interested in extreme differences in answers scales, it is more suitable to use the Manhattan metrics than Euclidean metrics. A model example may be adduced (see Fig. 1). Let us have two keywords, P_1 and P_2 and two seven-point assessment scales S_1 a S_2.

A

	P_1	P_2
S_1	1	7
S_2	7	1

B

	P_1	P_2
S_1	1	7
S_2	4	4

Fig. 1. Model situation

Status A represents the situation with the maximum distance of P_1 and P_2 keywords in any metrics. Let us designate it as 100 %. Status B represents the situation in which one keyword is perceived maximally differently and the other one identically. In this case, Euclidean metrics has the value 0.707 of maximum distance while Manhattan has only the value 0.5 of maximum distance.

The smaller the distance D_{ij}, the closer the given keywords i and j are in meaning and vice versa ([15]). The calculated values of distance D were then recorded in a symmetrical D-matrix [1]. To calculate the global assessment of keywords by group of respondents, the appropriate D-matrix was calculated for each respondent separately and then a final D-matrix with average distances was created.

In our opinion, the analysis of global similarity of keywords may be considered a certain form of cluster analysis. The keywords assessed in SD may be considered objects determined for clustering and the distances in D-matrix may be considered the characteristic based on which the objects are clustered. During the analysis, the agglomerative clustering method was used [17]. Each object at the beginning of this algorithm is regarded as a cluster. In each following step the two clusters nearest each other are joined into a single one. This process of gradual clustering gradually groups all objects into one big cluster [17]. During clustering, the method of group average was used for the calculation of the distance between clusters. This process of gradual clustering is expressed graphically using dendrograms. In the Fig. 2, the posttest dendrogram of experimental group is shown for illustration.

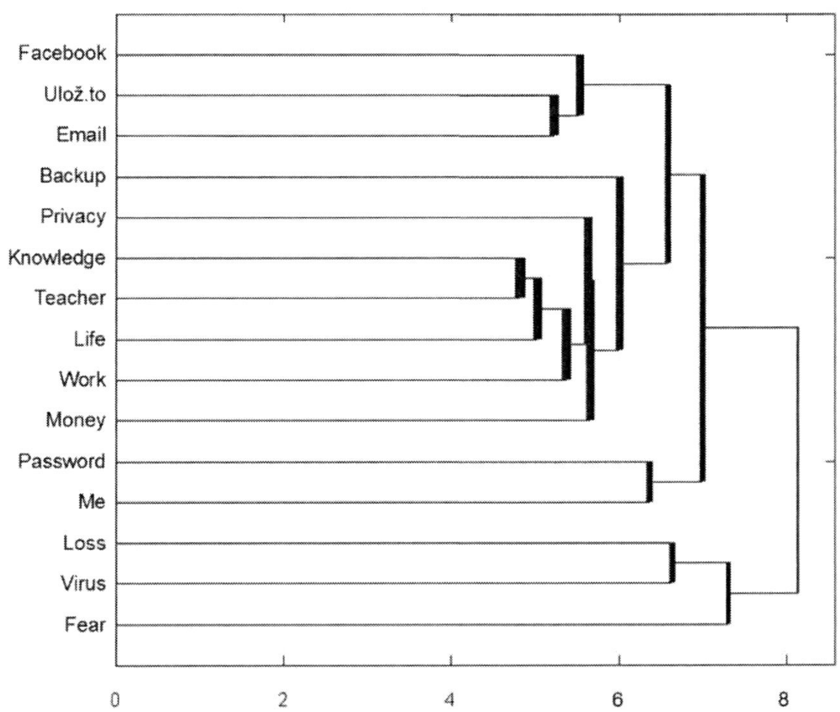

Fig. 2. Posttest dendrogram of experimental group

By creating dendrograms, the automated part of data analysis carried out using software tools was finished and the next analysis was based on the individual assessment of acquired graphical outputs.

3.2 Analysis of outputs

The analysis of graphical outputs was based on the analysis of occurred keywords clusters. We are interested in the position of keywords from the area of technical e-safety; their mutual position as well as position towards other keywords is important. The sensitivity of this method is influenced naturally by the size of sample (number of respondents). On the basis of the global design of our experiment, the analysis up to the level of ten clusters was carried out. The split into the selected number of clusters is done by the determination of limit distance in the analysed dendrogram. In Fig. 3, the imaginary cut (vertical line in this figure) cuts six branches of the dendrogram, creating six separate clusters of keywords.

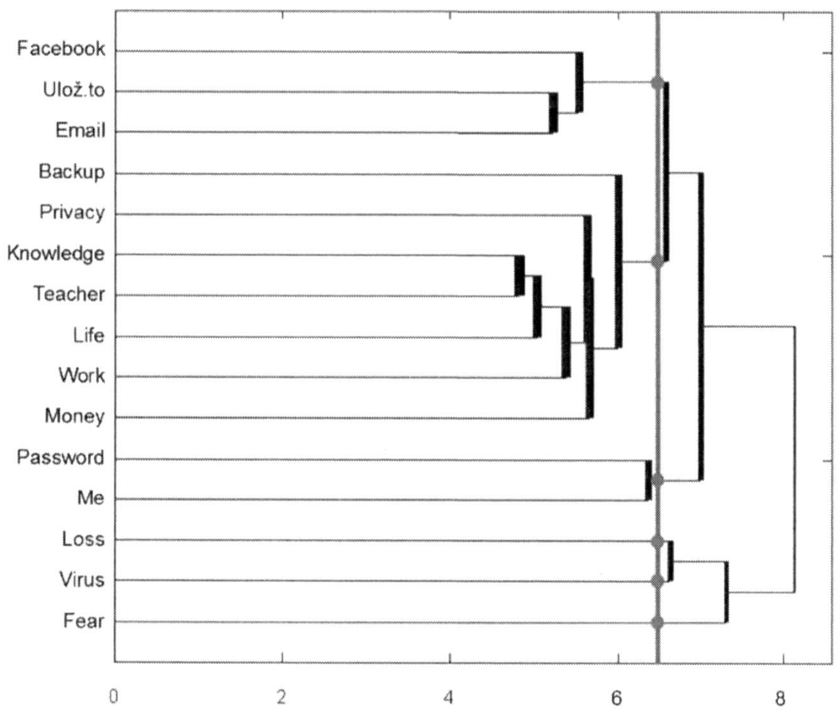

Fig. 3. The example of investigation with what other keywords the individual keywords create cluster in the dendrogram, if the dendrogram is split into six clusters. For example: *Password* keyword clusters with the keyword *Me*; *Facebook* keyword clusters with keywords *Ulož.to* and *Email*; *Virus* keyword represents an independent cluster

To enable us to determine changes in perception of keyword, the pretest dendrogram of a certain group of respondents was compared with the posttest dendrogram of this group, observing any possible inter-cluster movement of a particular keyword. The following situations can be considered inter-cluster movement of keywords:

- Certain keywords in the pretest dendrogram are in the same cluster, given a certain number of clusters, but in the posttest dendrogram they are in various clusters, given the same number of clusters (in the scheme in Fig. 4, this movement is expressed in the direction From one another).

- Certain keywords in the pretest dendrogram are in various clusters, given a certain number of clusters, but in the posttest dendrogram they are in one cluster, given the same number of clusters (in the scheme in Fig. 4, this movement is expressed in the direction To one another). The example of this movement is shown on a particular pair of dendrograms in Fig. 5.

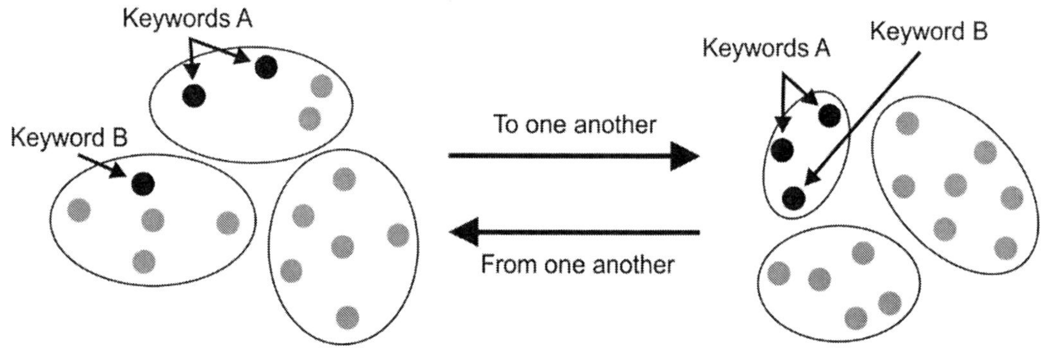

Fig. 4. Schematic example of inter-cluster movement of keywords, given three clusters. While in the diagram at the left side *Keywords A* and *B* create two various clusters, in the diagram at the right side they create one cluster

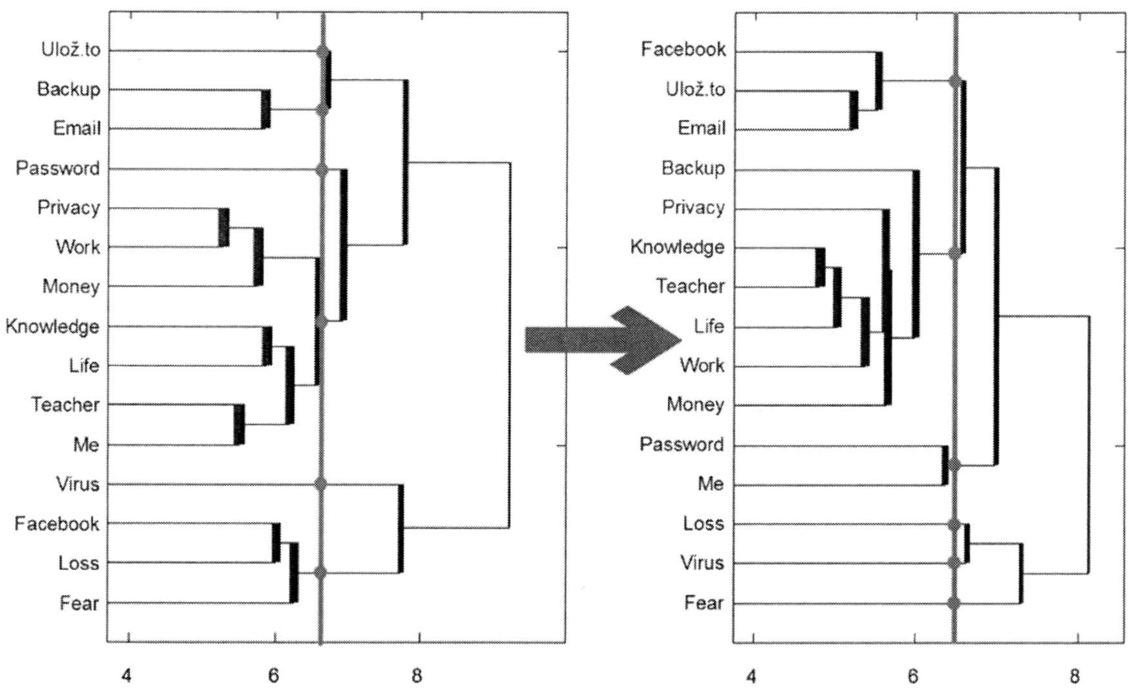

Fig. 5. Inter-cluster movement of keywords *Password* and *Me*, given six clusters. While these keywords are in various clusters in the pretest dendrogram (on the left), they are in one cluster in the posttest dendrogram (on the right)

To distinguish the cases in which the inter-cluster movement of keywords was caused not by lessons but by other factors (especially maturing of respondents), the list of inter-cluster movements created for the experimental group of respondents was compared with the inter-cluster movements in the control group. It was investigated if the same inter-cluster movement took place also in the control group. If a movement occurred in the control group as well, the given movement was excluded from the list of movements in the experimental group.

The identification of inter-cluster keyword movements caused probably not due to the effect of teaching but of abnormalities in the experimental group was subsequently carried out. The inter-cluster movements with a different initial state in the experimental group than in the control group were searched for. These were movements where, given a certain number of clusters in the pretest dendrogram of the experimental group, the investigated keywords were in one cluster, but in the pretest dendrogram of the control group, these keywords were in various clusters given the same number of clusters. Analogically these were the movements where, given a certain number of clusters in the pretest dendrogram of the experimental group, the investigated keywords were in various clusters, whereas in the pretest dendrogram of the control group, these keywords were in one cluster given the same number of clusters. These movements were removed from the list.

To find the inter-cluster keyword movements caused probably not due to the effect of lessons but of abnormalities in the experimental group, one more procedure was applied. In this case, the inter-cluster movements with the same resulting state in the experimental group as in the control group were searched for. These were the movements where, given a certain number of clusters, the investigated keywords were in various clusters in the posttest dendrogram not only in the experimental group but also in the control group. Analogically these were movements where, given a certain number of clusters, the investigated keywords were in one cluster in the posttest dendrogram not only in the experimental group but also in the control group. These movements were removed from the list.

Using this procedure such a state was achieved that the resulting list of inter-cluster movements of keywords contained only such movements where the initial deployment of investigated keywords among clusters was the same as in the control group and the resulting deployment of investigated keywords among clusters was different than in the control group. The list of such inter-cluster movements of keywords is stated in Tab. 1, where in the lines of the table:

- The direction To one another means that in the pretest dendrogram, the sets *Keywords A* and *Keywords B* created two various clusters while in the posttest dendrogram they created one cluster[1].

- Direction From one another means that in the pretest dendrogram the sets *Keywords A* and *Keywords B* created one cluster while in the posttest dendrogram they created two various clusters.

[1] This comparison was done given the disintegration of the keyword set to a certain number of clusters which was the same for the pretest dendrogram as well as posttest dendrogram. This number of clusters is stated in the column Number of clusters.

Keywords A	Keywords B	Direction	Number of clusters
Passwords	*Me*	To one another	5
Password, Me	*Privacy, Work, Life, Money, Knowledge, Teacher*	From one another	4
Virus	*Loss*	To one another	4
Facebook	*Ulož.to, E-mail*	To one another	10

Tab. 1. Inter-cluster movements of keywords for the selected type of teaching

4 Interpretation of acquired data

In the following text, the interpretation of data resulting from the SD questionnaires analysis using dendrograms is illustrated. As an example, it is possible to use our research [14] in which the changes in understanding selected e-safety keywords by university students after lessons in comparison with the state before lessons were measured. The data from the Tab. 1 will serve as a source of data for interpretation. The results are interpreted from the viewpoint how the selected type of lessons influenced the perception of selected e-safety keywords by the respondents. As movements showing fulfilment of teaching goals, only certain inter-cluster movements of keywords are designated. They are such inter-cluster movements which give the hope that the changes of attitudes included in them will result in the required change of attitudes or behaviour of respondents in the target area (in our case e-safety).

The movement of the keyword *Password* closer to the keyword *Me* can be explained in such a way that the respondents started to perceive these keywords as more similar to each other. It leads us to believe that respondents have become more aware that passwords are the key to their person and data. Choosing an appropriate password and keeping it secure are vital to staying safe when using ICT. This may be closely connected with lessons in which respondents participated. The mentioned change in perceiving the keywords *Password* and *Me* is moreover supported by distancing of these two keywords from the cluster of keywords *Privacy, Work, Life, Money, Knowledge* and *Teacher*. This change is perceived by us as positive. It gives hope that lessons have led to respondents' greater awareness of the fact that their password is an element that keeps their person and data secure (and that they should adopt a more consistent approach to choice and use of password than prior to lessons).

The keyword *Virus* moving closer to the keyword *Loss* can be explained in such a way that the respondents started to perceive these keywords as mutually more similar. It leads us to believe that respondents became more aware that a virus brings the risk of a certain loss. This may be connected with the lessons in which the respondents participated. This change is perceived by us as positive. It gives hope that lessons have led to respondents' greater awareness of the fact that a virus is a negative element resulting in a certain loss (and that they need to take antivirus protection more seriously than prior to lessons).

The movement of the keyword *Facebook* closer to the keywords *Ulož.to* and *Email* can be explained in such a way that the respondents started to perceive these concepts as more similar. It leads us to believe that respondents became more aware of what various services had in common – the possibility for users to share information and data. Nevertheless, as the lessons were not primarily focused on differentiating between various types of cloud services, we do not consider the established change in perceiving the concerned keywords for achieving the target of the used type of lessons. Therefore this change is perceived by us as neutral.

5 Conclusion

Semantic differential is a method designed for measuring individual, connotative meaning of keywords perceived by respondents. Although this method is used quite often, we did not succeed to find, in the scope of our research, the sources dealing with the process of data analysis consisting in the evaluation of the global similarity of keywords. For this reason, we described the possible use of this method for investigating changes in understanding selected keywords by respondents. Specifically it was the analysis and the subsequent data interpretation in the scope of semantic differential.

As the method of data analysis described by us uses the mathematic basis, the individual steps of data analysis are unambiguous. Various researchers should acquire the same results, when applying the same methodology and the same metrics. However, as follows from the previous chapter, interpretation of changes established by the comparison of dendrograms may be subjective. The reason is that from a certain point of view, this interpretation of data represents the qualitative research. A certain weakness of the described method may be seen in the subjective data interpretation. This weakness, however, may be minimized by the techniques determined for increasing the reliability and trustworthiness of qualitative research as is e.g. the audit of colleagues.

References

[1] CHRÁSKA, M. *Metody pedagogického výzkumu: Základy kvantitativního výzkumu*. Praha: Grada, 2007, ISBN 978-80-247-1369-4, 265 p.

[2] PRŮCHA, J., WALTEROVÁ, E., MAREŠ, J. *Pedagogický slovník.* 7., aktualiz. a rozš. vyd. Praha: Portál, 2013, ISBN 978-80-262-0403-9.

[3] JODELET, D. Représentation Sociale: phénomènes, concept et théorie. In *Psychologie Sociale.* 1984: p. 357–378.

[4] HEJNÝ, M., NOVOTNÁ, J., VONDROVÁ, N. (eds.). *Dvacet pět kapitol z didaktiky matematiky*. Praha: Univerzita Karlova v Praze, Pedagogická fakulta, 2004, ISBN 80-7290-189-3.

[5] MILL, J. S. System of Logic. In *The Collected Works of John Stuart Mill*, ed. J. M. Robson. Toronto: University of Toronto Press, volume VII, VIII, 1974.

[6] SHARPE, L. K., ANDERSON, W. T. Concept-Scale Interaction in the Semantic Differential. *Journal of Marketing Research*. Volume 9, issue 4, 1972: p. 432–434.

[7] STRUNGĂ, A. C. Osgood's semantic differential: a review of the Romanian social sciences literature. *Social Sciences and Education Research Review*. Volume 2, issue 2, 2014: p. 22–28, ISSN 2392-9683.

[8] KOZÁK, V., FUSEK, P. Vnímání značky pivovaru XYZ. *Kvasný průmysl*. Volume 51, issue 5, 2005: p. 166–168.

[9] CHRÁSKA, M. Změny v sémantickém prostoru studentů pedagogické fakulty. *Pedagogika* [online]. Volume 45, issue 1, 1995: p. 71–76 [cit. 2016-02-01]. Available from: http://pages.pedf.cuni.cz/pedagogika/?attachment_id=3118&edmc=3118

[10] PELIKÁN, J. *Základy empirického výzkumu pedagogických jevů*. Praha: Karolinum, 2004, ISBN 80-7184-569-8, 270 p.

[11] FERJENČÍK, J. *Úvod do metodologie psychologického výzkumu: Jak zkoumat lidskou duši*. Praha: Portál, 2000, ISBN 80-7178-367-6, 255 p.

[12] URL: http://www.r-project.org

[13] ŠERÝ, M. *Použití sémantického diferenciálu při hodnocení výuky na ZŠ* [online]. České Budějovice, 2013 [cit. 2016-02-01]. Doctoral dissertation. Jihočeská univerzita v Českých Budějovicích, Pedagogická fakulta, Katedra informatiky. Available from: http://theses.cz/id/m7rifg/Disertacni_prace_2013_v_28.pdf

[14] ŠIMANDL, V. *Kompetence učitelů ICT v oblasti technické e* bezpečnosti [online]. České Budějovice, 2016 [cit. 2016-10-01]. Doctoral dissertation. Jihočeská univerzita v Českých Budějovicích, Pedagogická fakulta, Katedra informatiky. Available from: http://theses.cz/id/1ndn00/simandl-dp9b5.pdf

[15] KERLINGER, F. N. *Základy výzkumu chování: Pedagogický a psychologický výzkum*. Praha: Academia, 1972, 708 p.

[16] PÖSCHL, R. *Vnímání významu matematiky a fyziky středoškolskými studenty* [online]. Praha, 2005 [cit. 2016-02-01]. Master's thesis. Univerzita Karlova v Praze, Matematicko-fyzikální fakulta, Katedra didaktiky fyziky. Available from: https://kdf.mff.cuni.cz/vyzkum/materialy/vnimani_vyznamu_M_a_F.pdf

[17] MELOUN, M., MILITKÝ, J. *Kompendium statistického zpracování dat: Metody a řešené úlohy*. Praha: Academia, 2006, ISBN 80-200-1396-2, 982 p.

Current address

Šimandl Václav, Mgr. Ph.D.
Faculty of Education, University of South Bohemia in České Budějovice
Jeronýmova 10, 371 15 České Budějovice, Czech Republic
Tel. number: +420 387 773 078, e-mail: simandl@pf.jcu.cz

Šerý Michal, Ing. Ph.D.
Faculty of Education, University of South Bohemia in České Budějovice
Jeronýmova 10, 371 15 České Budějovice, Czech Republic
Tel. number: +420 387 773 103, e-mail: kyklop@pf.jcu.cz

Dobiáš Václav, Mgr.
The Institute of Technology and Business in České Budějovice
Okružní 517/10, 370 01 České Budějovice, Czech Republic
Tel. number: +420 387 842 186, e-mail: dobias@mail.vstecb.cz

Slovak University of Technology in Bratislava
Faculty of Mechanical Engineering
16th Conference on Applied Mathematics **APLIMAT 2017**
Proceedings

DETERMINISTIC AND STOCHASTIC EXTRAPOLATION OF DEATH RATES IN THE CZECH REPUBLIC WITH AN IMPACT ON PROBABILITY OF DYING AND TABULAR NUMBER OF DEATHS

ŠIMPACH Ondřej (CZ), LANGHAMROVÁ Jitka (CZ)

Abstract. The aim of this paper is to compare the results of deterministic (linear regression) and stochastic (autoregressive integrated moving averages) extrapolations of logarithms of age-and-sex-specific death rates of the Czech population with impact on key characteristics of mortality tables – probability of dying and tabular number of deaths. The result is that the optimized stochastic ARIMA models provide more realistic results about the future development of the analysed characteristics and they are much less biased. Deterministic models have problems in applications on variable data, which can be expected especially in the highest ages.

Keywords: deterministic models, stochastic models, mortality, death rates, projection

Mathematics Subject Classification: Primary 90C30; Secondary 62H12

1 Introduction

Information about the future development of the age-and-sex specific structure of the population is very important for the central and regional administration. Prospective trend estimates of the number and structure of the population have their demographic importance, especially shows the prospective population consequences of the current population development. In practice, the population projections are either based on common mathematical extrapolation methods or methods of demographic population projections (Booth et al. [5]). The individual results depend on the procedure for calculating of demographic projection. Population structure and all derived characteristics are strongly determined by applied scenario and also by the model, which was used for extrapolation the future age-and-sex-specific demographic characteristics (Erbas et al. [11]). Future development for individual demographic processes can have an optimistic or pessimistic nature, resulting to the population projection in optimistic or pessimistic variant.

The aim of this paper is to focus on deterministic (which was used in the past and still uses) and stochastic (alternative) extrapolating of death rates, and compare the results between these approaches. The database is available for the Czech Republic at the age of 0–100+ completed years of life and the period from 1920 to 2012. We pay the attention to the differences that provides deterministic and stochastic model and compare the impacts of their applications.

2 Materials and Methods

For the description of mortality development is most often used an indicator known as life expectancy, (as a key output of the mortality tables), but for analytical purposes may be used other characteristics of mortality, such as e.g. age-and-sex specific death rates, probability of dying and the tabular number of deaths. Nowadays, the development of mortality for the oldest persons (and thus the development of life expectancy and other characteristics resulting from mortality tables) enters to the forefront of research interest (see e.g. paper by Bell [3]). The extension of human life is still in progress. It is important to note that for the examining of mortality at the highest ages it is difficult to use an empirical data, because of significant variability. It is caused by the different development of mortality trend in the highest age groups compared to younger ones. Probability of dying and tabular number of deaths is obtained as an output indicator from mortality tables. The calculation is carried out in several steps. First, we calculate the age-and-sex specific death rates as

$$m_{x,t} = \frac{M_{x,t}}{\bar{S}_{x,t}}, \tag{1}$$

where $M_{x,t}$ is the number of deaths at the exact age x and $\bar{S}_{x,t}$ is the mid-year number of living. Between the specific mortality rate and the intensity of mortality (which is considered in similar studies and analysis, see e.g. paper by Bogue, Anderton, Arriaga [4], Fiala [12] or Dotlačilová, Šimpach [9], [10]) is valid the followed relationship

$$m_{x,t} \approx \mu\left(x + \frac{1}{2}\right). \tag{2}$$

Next, we will calculate the probability of death for 0-year-old person as

$$q_{0,t} = \frac{M_{0,t}}{\alpha N_t^v + (1-\alpha) N_{t-1}^v} \quad \text{for } x = 0, \tag{3}$$

where $M_{0,t}$ is the number of deaths at the age 0, α is the proportion of lower elementary file of died persons (approximately 0.85) and N_t^v, respectively N_{t-1}^v is the number of live births in year t, respectively in year t-1 (Keyfitz [14]). The probability of dying

$$q_{x,t} = 1 - p_{x,t}, \tag{4}$$

is valid for $x > 0$. The calculation of the probability of surviving is given by formula

$$p_{0,t} = 1 - q_{0,t} \quad \text{for } x = 0, \tag{5}$$

and by formula

$$p_{x,t} = e^{-m_{x,t}} \text{ for } x > 0. \tag{6}$$

Next part of the calculation relates to tabular (i.e. imaginary) population. First, we select the initial number of live births in tabular population: $l_{0,t}$ = 100 000. Based on knowledge of the probability of surviving, we are able to calculate the number of survivors in the further exact ages by

$$l_{x+1,t} = l_{x,t} p_{x,t}, \tag{7}$$

where $l_{x,t}$ is the number of survivors at the exact age x from the default file of 100 000 live births of tabular population. The number of deaths of tabular population is given by

$$d_{x,t} = l_{x,t} q_{x,t}. \tag{8}$$

In the following steps it possible to calculate the number of lived years ($L_{x,t}$), the number of remaining years of life ($T_{x,t}$) and life expectancy ($e_{x,t}$), but because these characteristics are not used in the paper, they will not be explained here.

Extrapolation approaches applied on the mortality development, which are based on a deterministic principles are still used in the conditions of the Czech Republic. We can mention e.g. study by Fiala [12] or Koschin et al. [15]. Especially the study by Fiala [12] provides information that $\ln(m_{x,t})$ is for each age x approximately linear and can be modelled by linear or polynomial function (according to significance of parameters). According to this evidence we can conclude that if we find the appropriate constant (intercept) (β_0) and slope (β_1) of the regression line, we can easily perform linear extrapolation to the future as

$$\ln(m_{x,t}) = \beta_0 + \beta_1 t + \varepsilon_t, \tag{9}$$

where t is the calendar year and predict the future development of mortality rates. This approach is simple, however, because the database of the Czech Republic is quite variable, and the nature of individual time series of mortality is volatile, residuals of these regressions can be autocorrelated and the slope of the regression line (especially in the lowest and highest age groups, where is present the highest variability in the data) may not be optimal for achieving the desired results.

Another approach which was used e.g. by Arltová et al. [2], Šimpach [19], [20] or Šimpach, Pechrová [21] is based on Box and Jenkins [6] methodology, where the linear regression model (9) is replaced by autoregressive integrated moving averages model ARIMA, which can be written as

$$\ln m_{x,t} = \beta_0 + \sum_{i=1}^{p} \phi_i \ln m_{x,t-i} + \sum_{j=1}^{q} \theta_j \varepsilon_{t-j}, \tag{10}$$

where the autoregressive part is

$$\ln m_{x,t} = \beta_0 + \sum_{i=1}^{p} \phi_i \ln m_{x,t-i} + \varepsilon_t. \tag{11}$$

This is the model of order p AR(p), where ϕ_i (i = 1, ..., p) are the parameters, β_0 represents the constant, and ε_t is a pure white noise error term, where $E(\varepsilon_t) = 0$, $D(\varepsilon_t) = \sigma^2$, $cov(\varepsilon_t\ ;\ \varepsilon_t') = 0$ and $\varepsilon_t \approx$ N (normal) distribution (Mélard, Pasteels [16] or Ord, Lowe [17]. Moving average model of order q is formulated as

$$\ln m_{x,t} = \varepsilon_1 + \sum_{j=1}^{q} \theta_j \varepsilon_{t-j}\ , \tag{12}$$

where θ_j (j = 1, ..., q) are the parameters and ε_t is a white noise.

Followed by the above mentioned procedure it is obvious, that in the first step we estimate 101 linear regression models for males (age 0–100+ years of life) and 101 models for females with use of database of logarithms of age-specific death rates (see also Stauffer [18]). Then, these models are used for extrapolation up to the year 2100, (in the same way as it performs e.g. CZSO [8]). In the second step there are calculated new 2 × 101 models using an automatized process for time series modelling in Statgraphics Centurion XVI. ARIMA models with optimally selected parameters (Akaike [1]) are used to extrapolate the logarithms of age-specific death rates, also up to the year 2100. In the third step, there are the empirical and predicted death rates recalculated to probabilities of dying and table number of deaths using an algorithm of mortality tables.

3 Results

Estimated intercepts (β_0) and slopes (β_1) of the regression lines for logarithms of age-specific death rates of male and female population in the Czech Republic are presented in Tab. 3 in Appendix. Extrapolation is connected to empirical data, and the results are presented in Fig. 1 (left) for males and in Fig. 2 (left) for females. Optimized forms (see also [16] or [17]) of ARIMA models for these logarithms are shown in Tab. 1 below (for males) and in Tab. 2 below (for females).

Age	Model	Age	Model	Age	Model	Age	Model	Age	Model	Age	Model	Age	Model	Age	Model
0	(0,2,1) c	13	(0,1,1) c	26	(0,1,1) c	39	(0,1,1) c	52	(0,1,1) c	65	(0,1,1) c	78	(0,1,1) c	91	(0,1,2) c
1	(0,1,1) c	14	(0,1,1) c	27	(0,1,1) c	40	(2,2,1)	53	(0,1,1) c	66	(0,1,1) c	79	(1,1,1) c	92	(2,1,0) c
2	(2,2,1)	15	(0,1,1) c	28	(0,1,1) c	41	(0,1,1) c	54	(0,1,1) c	67	(0,1,1) c	80	(0,1,1) c	93	(0,1,1) c
3	(0,1,1) c	16	(0,1,1) c	29	(0,1,1) c	42	(0,1,1) c	55	(0,1,1) c	68	(0,1,1) c	81	(0,1,1) c	94	(0,1,2) c
4	(0,1,1) c	17	(0,1,1) c	30	(0,1,1) c	43	(0,1,1) c	56	(2,1,0) c	69	(0,1,1) c	82	(0,1,1) c	95	(1,1,1) c
5	(0,1,1) c	18	(0,1,1) c	31	(2,2,1)	44	(0,1,1) c	57	(0,1,1) c	70	(0,1,1) c	83	(1,1,1) c	96	(0,1,2) c
6	(0,1,1) c	19	(0,1,1) c	32	(0,1,1) c	45	(0,1,1) c	58	(0,1,1) c	71	(0,1,1) c	84	(0,1,1) c	97	(0,1,1) c
7	(0,1,1) c	20	(0,1,2) c	33	(0,1,1) c	46	(0,1,1) c	59	(0,1,1) c	72	(0,1,1) c	85	(0,1,1) c	98	(0,1,1) c
8	(0,1,1) c	21	(0,1,1) c	34	(0,1,1) c	47	(2,1,0) c	60	(0,1,1) c	73	(1,1,0) c	86	(0,1,1) c	99	(0,1,2) c
9	(0,1,1) c	22	(0,1,1) c	35	(0,1,1) c	48	(1,1,0) c	61	(0,1,1) c	74	(1,1,0) c	87	(0,1,1) c	100+	(0,1,1) c
10	(0,1,1) c	23	(0,1,1) c	36	(0,1,1) c	49	(0,1,1) c	62	(0,1,1) c	75	(2,1,0) c	88	(0,1,1) c		
11	(0,1,1) c	24	(0,1,1) c	37	(0,1,1) c	50	(2,1,0) c	63	(0,1,1) c	76	(0,1,1) c	89	(0,1,2) c		
12	(0,1,1) c	25	(2,1,0) c	38	(0,1,1) c	51	(1,1,0) c	64	(0,1,1) c	77	(0,1,1) c	90	(2,1,0) c		

Tab. 1. ARIMA (p, d, q) models with or without constant for logarithms of male's age-specific death rates in range of 0–100+ years in the Czech Republic. Source: authors' calculations and construction.

Age	Model	Age	Model	Age	Model	Age	Model	Age	Model	Age	Model	Age	Model	Age	Model
0	(0,2,1) c	13	(0,1,1) c	26	(0,1,1) c	39	(0,1,1) c	52	(0,1,1) c	65	(1,1,0) c	78	(2,1,2) c	91	(1,1,1) c
1	(0,1,1) c	14	(1,1,1) c	27	(0,1,1) c	40	(2,1,0) c	53	(0,1,1) c	66	(1,1,0) c	79	(0,1,1) c	92	(2,1,0) c
2	(1,1,1) c	15	(0,1,1) c	28	(0,1,1) c	41	(1,1,0) c	54	(0,1,1) c	67	(0,1,1) c	80	(0,1,1) c	93	(0,1,1) c
3	(2,1,0) c	16	(2,1,0) c	29	(0,1,1) c	42	(0,1,1) c	55	(0,1,1) c	68	(0,1,1) c	81	(0,1,1) c	94	(2,1,1) c
4	(2,1,0) c	17	(0,1,1) c	30	(0,1,1) c	43	(0,1,1) c	56	(0,1,1) c	69	(0,1,1) c	82	(0,1,1) c	95	(1,0,0) c
5	(0,1,1) c	18	(0,1,2) c	31	(0,1,1) c	44	(0,1,1) c	57	(0,1,1) c	70	(0,1,1) c	83	(0,1,1) c	96	(1,0,0) c
6	(1,1,1) c	19	(0,1,1) c	32	(0,1,1) c	45	(0,1,1) c	58	(2,1,0) c	71	(1,1,0) c	84	(0,1,1) c	97	(0,0,2) c
7	(0,1,1) c	20	(0,1,1) c	33	(0,1,1) c	46	(0,1,1) c	59	(1,1,1) c	72	(2,1,0) c	85	(0,1,1) c	98	(0,0,2) c
8	(2,1,0) c	21	(0,1,1) c	34	(0,1,1) c	47	(0,1,1) c	60	(0,1,1) c	73	(1,1,0) c	86	(0,1,1) c	99	(1,1,0) c
9	(2,1,0) c	22	(2,1,0) c	35	(2,1,0) c	48	(0,1,1) c	61	(0,1,2) c	74	(2,1,0) c	87	(0,1,1) c	100	(0,1,1) c
10	(0,1,1) c	23	(0,1,1) c	36	(0,1,1) c	49	(1,1,1) c	62	(1,1,0) c	75	(2,1,0) c	88	(0,1,1) c		
11	(0,1,1) c	24	(1,1,0) c	37	(0,1,1) c	50	(0,1,1) c	63	(1,1,0) c	76	(0,1,1) c	89	(1,1,1) c		
12	(1,1,1) c	25	(0,1,1) c	38	(0,1,1) c	51	(0,1,1) c	64	(0,1,1) c	77	(0,1,1) c	90	(2,1,0) c		

Tab. 2. ARIMA (p, d, q) models with or without constant for logarithms of female's age-specific death rates in range of 0–100+ years in the Czech Republic. Source: authors' calculations and construction.

Forecasts of death rates by ARIMA models are connected to empirical data, and the results are presented in Fig. 1 (right) for males and in Fig. 2 (right) for females. It is clear that the lower values of logarithms of these rates provide ARIMA models (in comparison with deterministic approach of linear regression). According to diagnostic tests (Jarque, Bera [13]) and comparison with other results (see e.g. Šimpach [20]) and theoretical expectations, these parameters are not deviated and probably provide more optimistic results.

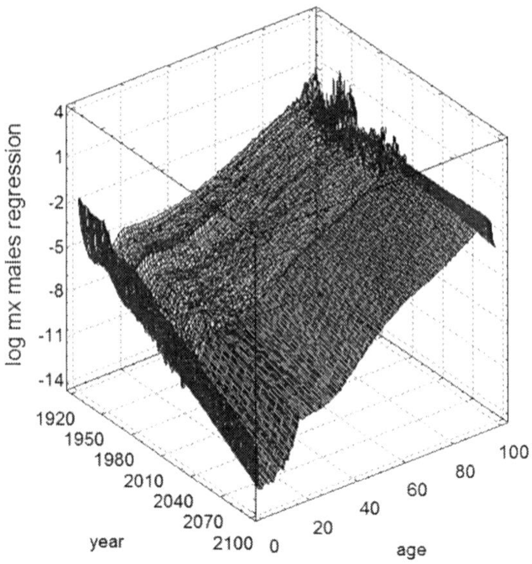

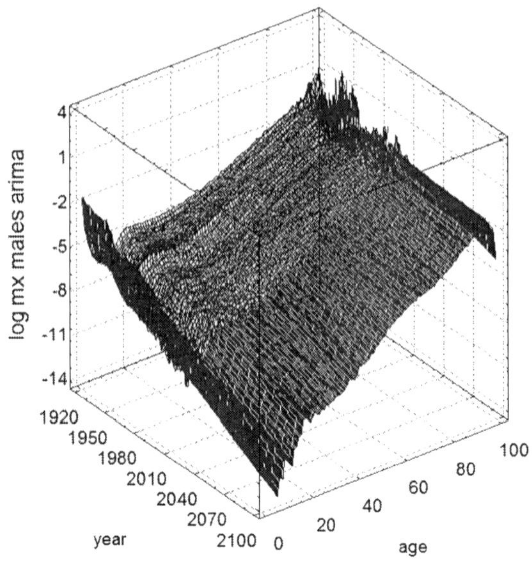

Fig. 1. Empirical values of age-specific death rates in logarithms of Czech males from 1920 to 2012 with attached extrapolated values up to the year 2100 according to individual linear regression models (left) and ARIMA models (right). Source: authors' calculations and construction

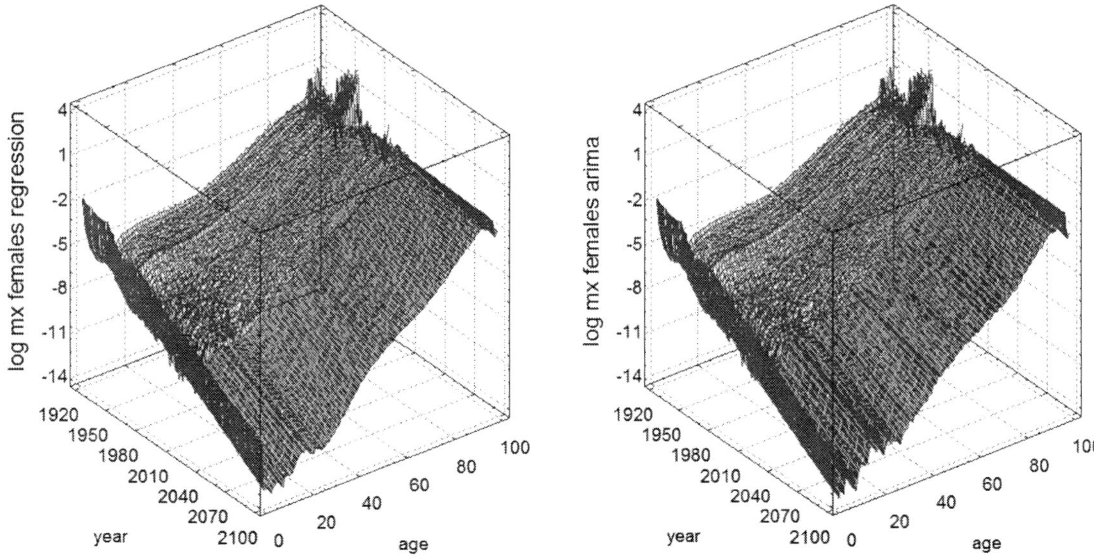

Fig. 2. Empirical values of age-specific death rates in logarithms of Czech females from 1920 to 2012 with attached extrapolated values up to the year 2100 according to individual linear regression models (left) and ARIMA models (right). Source: authors' calculations and construction

Larger differences between the results of linear regression and stochastic ARIMA models can be seen after recalculation of age-specific death rates to the probability of dying (Fig. 3 and 4). Due to the extreme values in empirical data of logarithms of death rates there is a bias in the highest ages and unreasonable escalation of probability of dying in the age of 90+. ARIMA models were not biased and provide more realistic results (see mainly the differences in highest age groups). In the case of linear models and population of males is the situation worse – the estimates are more biased.

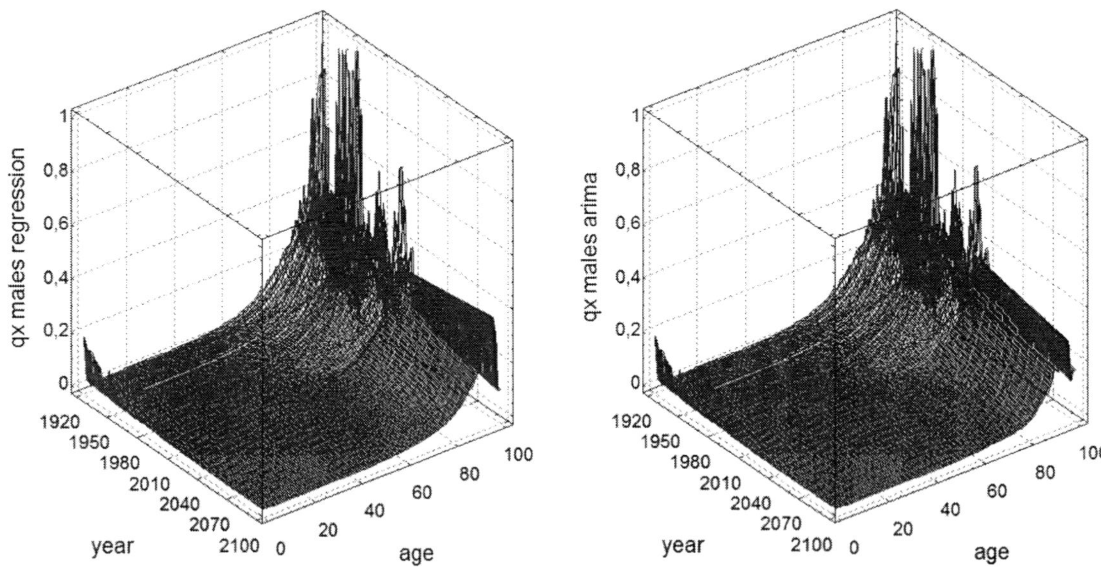

Fig. 3. Empirical values of age-specific probabilities of dying of Czech males from 1920 to 2012 with attached extrapolated values up to the year 2100 according to linear regression models (left) and ARIMA models (right) applied on log. of age-specific death rates. Source: authors' calculations and construction

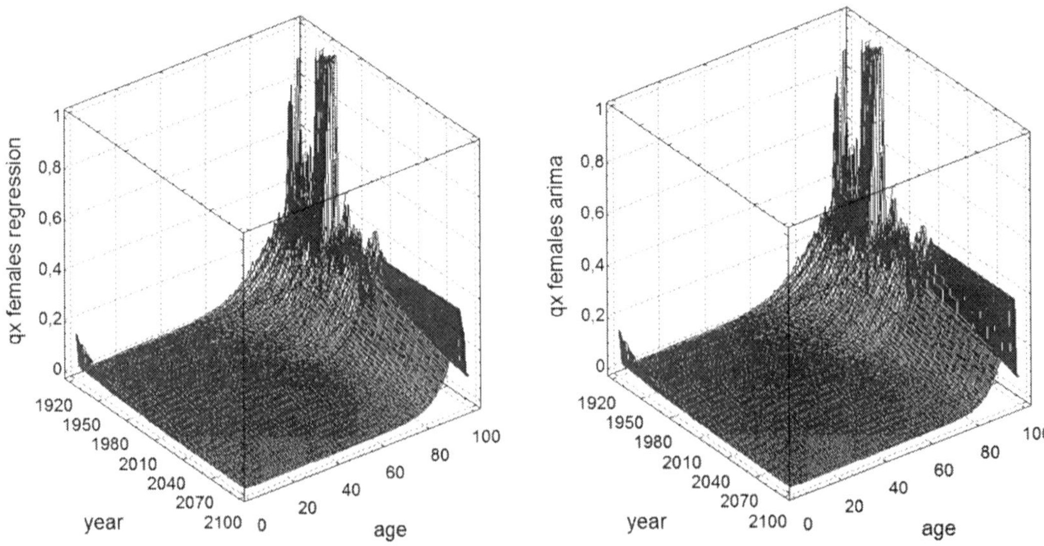

Fig. 4. Empirical values of age-specific probabilities of dying of Czech females from 1920 to 2012 with attached extrapolated values up to the year 2100 according to linear regression models (left) and ARIMA models (right) applied on log. of age-specific death rates. Source: authors' calculations and construction

In the last part of analysis we recalculated and extrapolated the empirical data of death rates to the table number of deaths on the basis of mortality tables algorithm. Not only that stochastic models predict moving of the mode of the deceased persons to the higher ages, but also sharper distribution of these numbers – i.e. that the concentration around the mode is denser. Much more important is this conclusion in the case of female population. While the tabular number of deaths in the case of the male population starts significantly growing from 62 years, for female population starts this increasing from 78 years. In all cases, the predicted values are continuously connected to the empirical data.

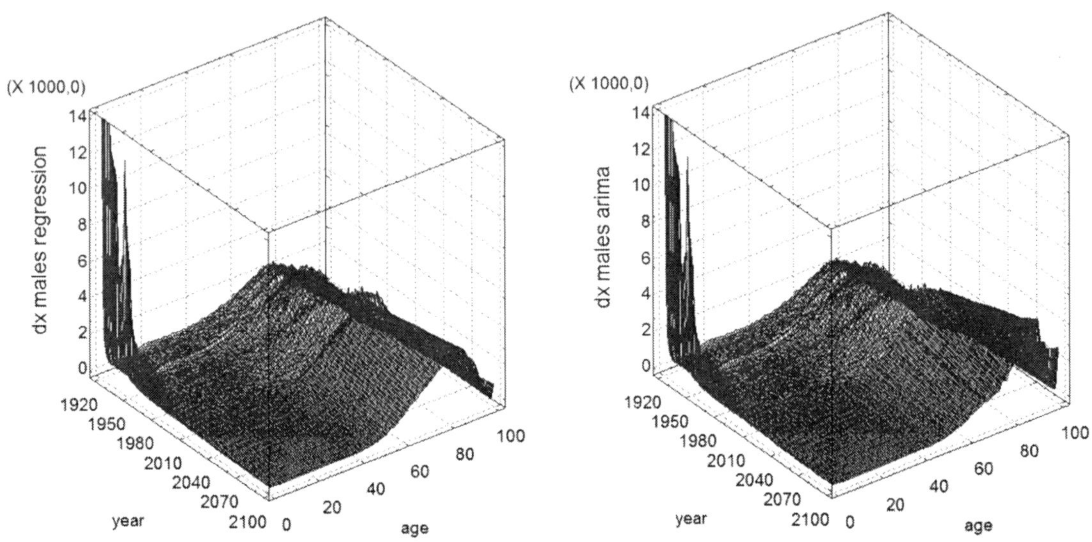

Fig. 5. Empirical values of age-specific tabular number of deaths of Czech males from 1920 to 2012 with attached extrapolated values up to the year 2100 according to linear regression models (left) and ARIMA models (right) applied on log. of age-specific death rates. Source: authors' calculations and construction

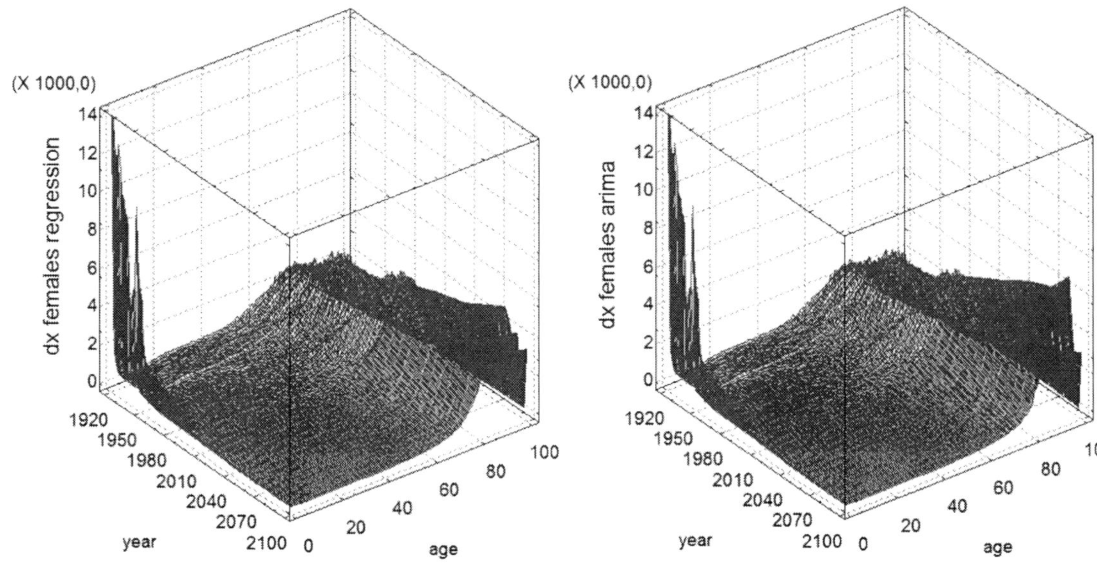

Fig. 6. Empirical values of age-specific tabular number of deaths of Czech females from 1920 to 2012 with attached extrapolated values up to the year 2100 according to linear regression models (left) and ARIMA models (right) applied on log. of age-specific death rates. Source: authors' calculations and construction

4 Discussion and Conclusion

Demographic projections of possible future population development are the basic information that is used to provide key information about the potential development of mortality (or fertility, migration and other demographic processes). Each projection is based on assumptions, which to a certain extent may or may not occur. The aim of this paper was to compare the results of deterministic extrapolation of logarithms of age-specific death rates of the Czech population with the results of extrapolation using stochastic ARIMA models of Box-Jenkins [6] methodology. We used data from the CZSO [7] database in the age range of 0–100+ completed years of life for each sex and because we used two types of methodological approach, there was designed $2 \times 2 \times 101 = 404$ individual models. On the basis of these models we made an extrapolation of logarithms of age-specific death rates up to the year 2100. Using by mortality tables algorithm the death rates were recalculated to the probability of dying and table numbers of deaths.

Results showed that extreme values, which contain analysed time series of logarithms of death rates, (mainly in the case of lowest and highest age groups), caused deflection of the estimated parameters of linear regression models, and thus the deflection of extrapolation of these rates to the future. Results are situated in the range in which it should not be placed. Stochastic models are much more robust for these extreme values. The extrapolation is not deflected, and the results are more relevant. Therefore it is obvious that the individual time series of logarithms of age-specific death rates contain a stochastic trend, which is explainable using by ARIMA models with optimized structure of parameters.

From the forecasted results of probabilities of dying and table number of deaths it is clear, that the linear regression models and stochastic ARIMA models are usable for age range approximately of

2–60 years. For ages 0, 1, and 61+ can be used stochastic ARIMA models only, because these models are robust in the case of strong variability and extreme values.

Acknowledgement

The authors gratefully acknowledge to the Czech Science Foundation project no. 15-13283S "Projection of the Czech Republic Population According to Educational Level and Marital status" for supporting this paper.

References

[1] AKAIKE, H. Information theory and an extension of the maximum likelihood principle. *In: Petrov, B. N., Csáki, F.: 2nd International Symposium on Information Theory*, Tsahkadsor, Armenia, USSR, September 2-8, 1971, Budapest: Akadémiai Kiadó, pp. 267 – 281.

[2] ARLTOVÁ, M., LANGHAMROVÁ, JI., LANGHAMROVÁ, JA. Development of life expectancy in the Czech Republic in years 1920-2010 with an outlook to 2050. *Prague Economic Papers*, 2013, vol. 22, no. 1, pp. 125 – 143.

[3] BELL, W. R. Comparing and assessing time series methods for forecasting age-specific fertility and mortality rates. *Journal of Official Statistics*, 1997, vol. 13, no. 3, pp. 279 – 303.

[4] BOGUE, D. J., ANDERTON, D. L., ARRIAGA E. E. Population models, Projections and Estimates. *Readings in Population Research Methodology*, 1993, vol. 5, United Nations Social Development Center, Chicago.

[5] BOOTH, H., TICKLE, L., SMITH, L. Evaluation of the variants of the Lee-Carter method of forecasting mortality: a multi-country comparison. *New Zealand Population Review*, 2005, 31(1), pp. 13 – 34.

[6] BOX, G. E. P., JENKINS, G. *Time series analysis: Forecasting and control.* San Francisco, Holden-Day, 1970, 537 p.

[7] CZSO. Obyvatelstvo podle Sčítání lidu, domů a bytů, Česká republika a kraje, *[DVD] Sčítání lidu, domů a bytů 2011*, Praha: Český statistický úřad, 2012.

[8] CZSO. *Projekce obyvatelstva České republiky do roku 2100, [on-line]*. Available at: http://www.czso.cz/csu/2013edicniplan.nsf/publ/4020-13-n_2013, [cited: 10.11.2015].

[9] DOTLAČILOVÁ, P., ŠIMPACH, O. Polynomial functions and smoothing of mortality rates: The Czech Republic and Slovakia during their independent development after their separation. *In: Aplimat 2015: 14th Conference on Applied Mathematics*. Bratislava: Slovak University of Technology, 2015, pp. 221 – 231.

[10] DOTLAČILOVÁ, P., ŠIMPACH, O. Smoothing of mortality curve: The usage of moving averages and Gomperzt-Makeham function in the case of the Czech Republic. *In: Aplimat 2016: 15th Conference on Applied Mathematics*. Bratislava: Slovak University of Technology, 2016, pp. 262 – 273.

[11] Erbas, B., ULLAH, S., HYNDMAN, R. J, SCOLLO, M., ABRAMSON, M. Forecasts of COPD mortality in Australia: 2006-2025. *BMC Medical Research Methodology*, vol. 2012, pp. 12 – 17.

[12] FIALA, T. Dva přístupy modelování vývoje úmrtnosti v populační projekci a jejich aplikace na populaci ČR. *Forum Statisticum Slovacum*, 2006, no. 4, pp. 44 – 55.

[13] JARQUE, C. M., BERA, A. K. Efficient tests for normality, homoscedasticity and serial independence of regression residuals. *Economics Letters*, 1980, vol. 6, no. 3, pp. 255 – 259.

[14] KEYFITZ, N. Experiments in the projection of mortality. *Canadian Studies in Population*, 1991, vol. 18, no. 2, pp. 1 – 17.

[15] KOSCHIN, F. ET AL. *Prognóza lidského kapitálu obyvatelstva České republiky do roku 2050*. Praha: Oeconomica. 2007, 105 p.

[16] MÉLARD, G., PASTEELS, J. M. Automatic ARIMA Modeling Including Intervention, Using Time Series Expert Software. *International Journal of Forecasting*, 2000, vol. 16, pp. 497 – 508.

[17] ORD, K., LOWE, S. Automatic Forecasting. *The American Statistician*, 1996, vol. 50, no. 1, pp. 88 – 94.

[18] STAUFFER, D. Simple tools for forecasts of population ageing in developed countries based on extrapolations of human mortality, fertility and migration. *Experimental Gerontology*, 2002, vol. 37, no. 8-9, pp. 1131 – 1136.

[19] ŠIMPACH, O. Fertility of Czech Females Could Be Lower than Expected: Trends in Future Development of Age-Specific Fertility Rates up to the Year 2050. *Statistika*, 2015, vol. 95, no. 1, pp. 19 – 35.

[20] ŠIMPACH, O. *Statistické metody v demografickém prognózování – Statistical methods in demographic forecasting [disertační práce – PhD thesis]*. Prague: University of Economics Prague, 2016, 176 p.

[21] ŠIMPACH, O., PECHROVÁ, M. Development of the Czech Farmers' Age Structure and the Consequences for Subsidy Policy. *Agris on-line Papers in Economics and Informatics*, 2015, vol. VII, no. 3, pp. 57 – 69.

Appendix

Tab. 3 shows the estimates of intercepts and slopes of 2 × 101 individual linear regression models.

MALES						FEMALES					
Age	**Intercept**	**Slope**	**Age**	**Intercept**	**Slope**	**Age**	**Intercept**	**Slope**	**Age**	**Intercept**	**Slope**
0	88,5109	-0,0468	51	9,6868	-0,0073	0	89,4360	-0,0474	51	24,9175	-0,0153
1	93,1204	-0,0505	52	9,0396	-0,0069	1	95,6204	-0,0518	52	24,6025	-0,0151
2	83,7540	-0,0460	53	8,5326	-0,0066	2	86,0081	-0,0473	53	23,5246	-0,0145
3	77,0660	-0,0427	54	7,0759	-0,0058	3	83,1380	-0,0459	54	22,8473	-0,0141
4	72,7697	-0,0406	55	6,9301	-0,0057	4	83,7996	-0,0464	55	22,2736	-0,0138
5	73,7969	-0,0412	56	6,7031	-0,0055	5	78,5892	-0,0438	56	22,1037	-0,0137
6	69,9993	-0,0393	57	6,0191	-0,0051	6	73,8460	-0,0414	57	21,1151	-0,0131
7	71,9836	-0,0404	58	5,6661	-0,0049	7	70,4510	-0,0397	58	21,1456	-0,0131
8	63,8499	-0,0363	59	5,5056	-0,0048	8	72,4563	-0,0408	59	20,5336	-0,0127
9	59,8880	-0,0343	60	5,4730	-0,0047	9	74,7851	-0,0421	60	21,2484	-0,0130
10	55,9073	-0,0323	61	4,6096	-0,0042	10	66,0924	-0,0376	61	20,7178	-0,0127
11	59,6766	-0,0342	62	4,4959	-0,0041	11	64,8665	-0,0370	62	21,6291	-0,0131
12	52,2472	-0,0304	63	4,4290	-0,0040	12	66,2089	-0,0377	63	21,0899	-0,0128
13	53,2289	-0,0309	64	4,7992	-0,0042	13	65,2777	-0,0372	64	21,4866	-0,0130
14	48,1625	-0,0282	65	5,0595	-0,0043	14	62,9035	-0,0359	65	20,9945	-0,0127
15	47,7204	-0,0279	66	5,0000	-0,0042	15	58,3992	-0,0335	66	21,0429	-0,0126
16	43,2882	-0,0255	67	5,0075	-0,0042	16	58,9243	-0,0337	67	21,1877	-0,0126
17	40,0813	-0,0238	68	5,2287	-0,0042	17	57,1798	-0,0328	68	20,9037	-0,0124
18	33,7400	-0,0204	69	5,3565	-0,0042	18	60,7097	-0,0345	69	20,7745	-0,0123
19	37,1909	-0,0222	70	5,9055	-0,0045	19	58,7712	-0,0335	70	20,9959	-0,0124
20	36,9525	-0,0220	71	5,7020	-0,0043	20	61,6050	-0,0349	71	20,8525	-0,0123
21	36,7412	-0,0219	72	6,4518	-0,0047	21	65,4689	-0,0369	72	20,7576	-0,0121
22	36,6589	-0,0218	73	6,8503	-0,0048	22	66,3813	-0,0374	73	20,2750	-0,0118
23	35,9527	-0,0215	74	6,9074	-0,0048	23	63,9931	-0,0361	74	19,8773	-0,0116
24	35,3041	-0,0211	75	6,8421	-0,0047	24	64,5962	-0,0364	75	18,9277	-0,0110
25	36,0789	-0,0216	76	6,9110	-0,0047	25	65,5761	-0,0369	76	18,5266	-0,0108
26	34,4657	-0,0207	77	7,1497	-0,0048	26	66,9680	-0,0376	77	18,3157	-0,0106
27	34,2254	-0,0206	78	7,1250	-0,0047	27	66,2216	-0,0372	78	16,9967	-0,0099
28	34,1289	-0,0206	79	7,3828	-0,0048	28	63,2096	-0,0357	79	16,8866	-0,0098
29	33,0804	-0,0200	80	7,3361	-0,0048	29	63,2161	-0,0357	80	16,0546	-0,0093
30	33,0810	-0,0200	81	7,2687	-0,0047	30	60,4873	-0,0342	81	14,8764	-0,0087
31	30,9401	-0,0189	82	6,8742	-0,0044	31	58,8892	-0,0334	82	14,2864	-0,0083
32	30,7952	-0,0188	83	7,4319	-0,0047	32	55,8598	-0,0318	83	13,6961	-0,0079
33	29,9794	-0,0183	84	7,3120	-0,0046	33	54,7079	-0,0312	84	13,0070	-0,0075
34	29,9375	-0,0183	85	6,4634	-0,0041	34	51,4006	-0,0295	85	12,4514	-0,0072
35	29,4114	-0,0180	86	6,1661	-0,0039	35	49,8194	-0,0287	86	11,9430	-0,0069
36	26,7796	-0,0166	87	5,5515	-0,0035	36	48,0991	-0,0277	87	10,2922	-0,0060
37	25,5494	-0,0160	88	5,9894	-0,0037	37	46,0115	-0,0266	88	9,5985	-0,0056
38	23,3905	-0,0148	89	5,1864	-0,0033	38	44,0519	-0,0256	89	8,8090	-0,0052
39	22,7481	-0,0145	90	4,2055	-0,0028	39	41,6228	-0,0243	90	7,2629	-0,0044
40	22,3299	-0,0142	91	2,9028	-0,0021	40	39,8129	-0,0234	91	5,2910	-0,0033
41	20,8742	-0,0134	92	1,8045	-0,0015	41	37,1529	-0,0220	92	5,0704	-0,0032
42	19,6673	-0,0128	93	1,1032	-0,0011	42	34,3563	-0,0205	93	4,5636	-0,0029
43	17,8824	-0,0118	94	1,7199	-0,0014	43	32,6876	-0,0196	94	5,4427	-0,0033
44	17,0943	-0,0114	95	3,8211	-0,0024	44	31,1862	-0,0188	95	11,8602	-0,0064
45	16,6479	-0,0111	96	1,5766	-0,0013	45	30,9918	-0,0187	96	-2,0406	0,0006
46	14,1438	-0,0098	97	-4,9009	0,0020	46	28,3308	-0,0173	97	1,7633	-0,0013
47	13,2588	-0,0093	98	-1,6089	0,0003	47	27,3924	-0,0168	98	6,1229	-0,0035
48	12,5806	-0,0089	99	15,4148	-0,0084	48	26,6438	-0,0163	99	20,6265	-0,0109
49	11,0328	-0,0080	100	19,2148	-0,0105	49	26,3721	-0,0162	100	9,0942	-0,0053
50	10,6577	-0,0078				50	25,7097	-0,0158			

Tab. 3. Estimated parameters (intercepts and slopes) of linear regression models for logarithms of males' and females' age-specific death rates in range of 0–100+ years in the Czech Republic.
Source: authors' calculations and construction.

Current address

Šimpach Ondřej, Ing. Ph.D.
University of Economics in Prague, Faculty of Informatics and Statistics,
Department of Statistics and Probability
W. Churchill sq. 4, 130 67 Prague 3, Czech Republic
Tel. number: +420 224 095 273, e-mail: ondrej.simpach@vse.cz

Langhamrová Jitka, doc. Ing. CSc.
University of Economics in Prague, Faculty of Informatics and Statistics,
Department of Demography
W. Churchill sq. 4, 130 67 Prague 3, Czech Republic
Tel. number: +420 +420 224 09 4315, e-mail: jitka.langhamrova@vse.cz

Slovak University of Technology in Bratislava
Faculty of Mechanical Engineering
16th Conference on Applied Mathematics **APLIMAT 2017**
Proceedings

CRYSTAL, MOLECULAR AND ELECTRONIC PROPERTIES OF 3-ACETYL-1-(3,5-DIMETHYLPHENYL)THIOUREA

SIVÝ Július (SK), VRÁBEL Viktor (SK), KRÁLIK Marián (SK), GOWDA B. Thimme (IN)

Abstract. We reported here the structures of new thiourea-derivatives, namely 3-acetyl-1-(3,5-dimethylphenyl)thiourea, $C_{11}H_{14}N_2OS$, which crystallizes as racemate in the space group P21/c. The C== S and C==O bonds are anti to each other and the N—H bonds are also anti to each other. The dihedral angle between the phenyl ring and the side chain (N1, C7, N2, C8, O1, S1) is 3.09 (1)°. The bond lengths of the carbonyl group C8=O1 is 1.224 (2) Å somewhat longer than typical carbonyl bonds. This may be due to the fact that atom O1 participates in hydrogen bonds. The crystal structure is slightly stabilized by the intra- and intermolecular N—H···O···H—N hydrogen bonds as H-atoms donor, link the molecules into infinite zigzag chains along the *c* axis. Calculations of the NBO analyses (DFT/B3LYP/6-311G, single point geometry) were performed.

Keywords: CCDC 1501555, crystal structure, DFT/B3LYP/6-311G, *R* factor, single crystal, T = 298 K, thiourea, X-ray diffraction, *wR* factor

Mathematics Subject Classification: 74E15; 37N20

1 Comments

Thiourea and its derivatives are known to exhibit a wide variety of biological activities. As part of studies on the substituent effects on the structures and other aspects of *N*-(aryl)-amides (Gowda *et al.*, 2001; *N*-(aryl)-methanesulfonamides (Gowda *et al.*, 2007) and *N*-chloroarylsulfonamides (Gowda & Ramachandra, 1989). Based on these facts and in continutation of our interest in developing simple and efficient outs for the synthesis of novel thiourea derivatives, we report here the crystal and molecular structure of the title compound (Fig.1).

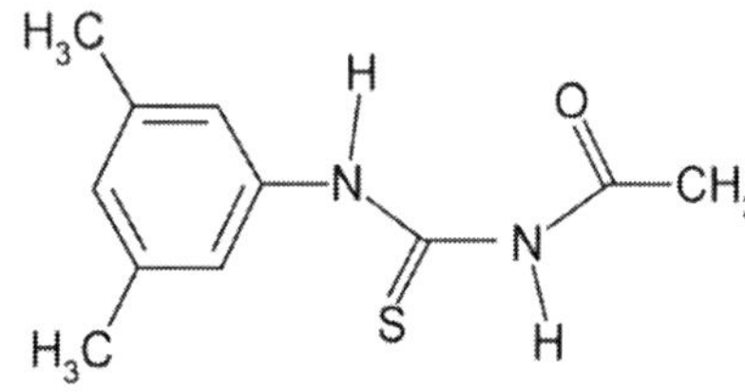

Fig. 1. Scheme of 3-Acetyl-1-(3,5-dimethylphenyl)thiourea

2 Experimental

3-Acetyl-1-(3,5-dimethylphenyl)thiourea (**I**) was synthesized by adding a solution of acetyl chloride (0.10 mol) in acetone (30 ml) dropwise to a suspension of ammonium thiocyanate (0.10 mol) in acetone (30 ml). The reaction mixture was refluxed for 30 min. After cooling to room temperature, a solution of 3,5-dimethylaniline (0.10 mol) in acetone (10 ml) was added and refluxed for 3 h. The reaction mixture was poured into acidified cold water. The precipitated title compound (**I**) was recrystallized to constant melting point from acetonitrile. The purity of the compound was checked and characterized by its infrared spectrum. Prism like colourless single crystals used in X-ray diffraction studies were grown in acetonitrile solution by slow evaporation of the solvent at room temperature. 3-Acetyl-1-(2-nitrophenyl)thiourea was synthesized by adding a solution of acetyl chloride (0.10 mol) in acetone (30 ml) dropwise to a suspension of ammonium thiocyanate (0.10 mol) in acetone (30 ml). The reaction mixture was refluxed for 30 min. After cooling to room temperature, a solution of 4-methylaniline (0.10 mol) in acetone (10 ml) was added and refluxed for 3 h. The reaction mixture was poured into acidified cold water. The precipitated title compound was recrystallized to constant melting point from acetonitrile. Prism like yellow single crystals used in X-ray diffraction studies were grown in acetonitrile solution by slow evaporation of the solvent at room temperature.

2.1 Refinement

Hydrogen atoms were placed in calculated positions with C—H distances of 0.93 Å (C-aromatic), 0.96 Å (*C*-methyl) and constrained to ride on their parent atoms. The amide H atom was visible in a difference map and refined with the N—H distance restrained to 0.860 (2) Å. The U_{iso}(H) values were set at $1.2U_{eq}$ (C-aromatic) or $1.5U_{eq}$ (*C*-methyl). Friedel pairs were merged. Refinement on F^2, least-squares matrix: full, $R[F^2 > 2\sigma(F^2)] = 0.049$, $wR(F^2) = 0.151$, $S = 1.00$, 2197 reflections, 139 parameters, 0 restraint, primary atom site location: structure-invariant direct methods, secondary atom site location: difference Fourier map, Hydrogen site location: inferred from neighbouring sites, H atoms treated by a mixture of independent and constrained refinement, $\Delta\rho_{max} = 0.34$ eÅ^{-3}, $\Delta\rho_{min} = -0.29$ eÅ^{-3}, Monoclinic, P2$_1$/c, $a = 4.8380(3)$, $b = 28.327(2)$, $c = 8.9622(6)$ (Å), $\beta = 102.095(5)$ (°).

2.1.1 Computing details

Data collection: *CrysAlis CCD* (Oxford Diffraction, 2009); cell refinement: *CrysAlis CCD*; data reduction: *CrysAlis RED* (Oxford Diffraction, 2009); program(s) used to solve structure: *SHELXS*

(Sheldrick, 2015), *Sir2014* (Burla *et al.*, 2014); program(s) used to refine structure: *SHELXL* (Sheldrick, 2015); molecular graphics: *DIAMOND* (Brandenburg, 1999); software used to prepare material for publication: *SHELXL* (Sheldrick, 2015), ShelXle (Hübschle, 2011), *PLATON* (Spek, 2009), *OLEX2* (Dolomanov *et al.*, 2009), *Gaussian 09* (Frisch *et al.*, 2013).

3 Results

We present here the crystal, molecular and electronic structure of one novel thiourea derivative, 3-acetyl-1-(3,5-dimethylphenyl)thiourea (**I**). Compound (**I**) crystallizes in monoclinic system, centrosymmetric space group $P2_1/c$. The dihedral angle between the phenyl ring and the side chain (N1, C7, N2, C8, O1, S1) is 3.093 (12)°. This is confirmed by the size of the torsion angle with respect to the phenyl ring C2—C1—N1—C7 of -177.26 (2)°. Crystallographic data for the structure reported in this paper is available from the Cambridge Crystallographic Data Centre, CCDC 1501555. The adjacent N—H bond is *syn* to the *ortho*-methyl group, compared to the *anti* conformation observed with respect to the *ortho*- and *meta*-methyl groups in the benzene ring. Furthermore, the conformations of the C=S and the C=O bonds are also *anti* to each other. The hydrogen atom of the NH attached to the phenyl ring and the amide oxygen are involved in intramolecular hydrogen bonding. The crystal structure is stabilized by the intramolecular N1—H101···O1 and two intermolecular N2—H102···O1^{i}, O1···(H102—N2)^{ii} hydrogen interactions as H-atoms donor, link the molecules into infinite C(4) (Bernstein *et al.,* 1995) zigzag chains along the *c* axis (Figs. 2., 3. and Tab. 4.). The bond lengths of the carbonyl group C8=O1 is 1.224 (2) Å somewhat longer than typical carbonyl bonds. This may be due to the fact that atom O1 participates in intra- and intermolecular hydrogen bonds. Atom N1 (N2) is sp^2-hybridized, as evidenced by the sum of the valence angles around it (360.0°). These data are consistent with conjugation of the lone-pair electrons on N1 (N2) with the adjacent carbonyl, similar to what is observed for amides.

The electronic structure of the molecule of the title compound was calculated by the *Ab Initio* method (Frisch, *et al.*, 2004), DFT/B3LYP/6-311G basis set was involved into calculation. Calculation of the electronic structure of a compound provides several indices which characterize the distribution of electron density in the molecule and the multiplicity of atomic bonds. The net charges give a picture of the distribution of electron density in the molecule and the values of the Wiberg bond indices enable one to estimate the multiplicity of individual atomic bonds. The net charge distribution in the molecule indicates that the large positive charges are localized at the atoms C7 (0.417) and C8 (0.815), whereas the negative net charges are located on the oxide and nitrogen atoms O1 (–0.706), N1 (–0.577) and N2 (–0.649) . This charge distribution and the spatial arrangement (geometry) of the molecule govern its biological activity and are important for the overall stabilization of the crystal structure. It follows from the Wiberg index value for the bond N1—C7 (I_W = 1.247) is not a pure single bond, but indicate the character of partial single or conjugated bonds. That π-electrons are delocalized in the region of the O1, C8, N2 atoms. The other bonds of the molecule have the character of single bonds (Tab. 4.). The results of these calculations are in good agreement with the experimental values of the bond lengths found by the X-ray structure analyses. It was found by evaluating experimental interactions of the structure (**I**) in the solid phase in this work that the intra- and intermolecular interactions are important in the parts of the molecule (**I**), (Tab. 1., Tab. 2.).

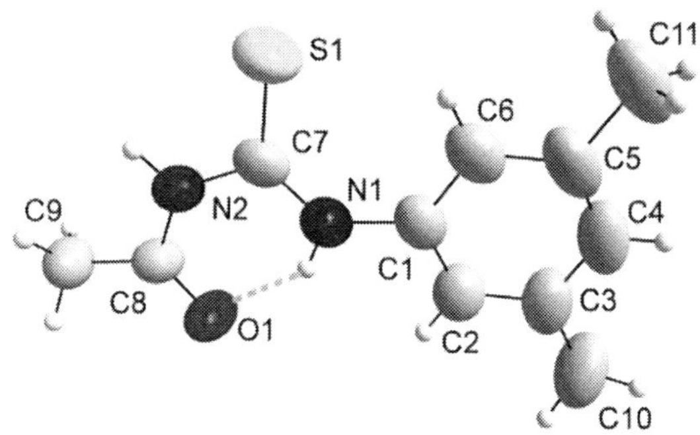

Fig. 2. Molecular structure of the title compound (**I**) with the atomic numbering scheme. Displacement ellipsoids are drawn at the 50 % probability level. H atoms are represented as small spheres of arbitrary radii

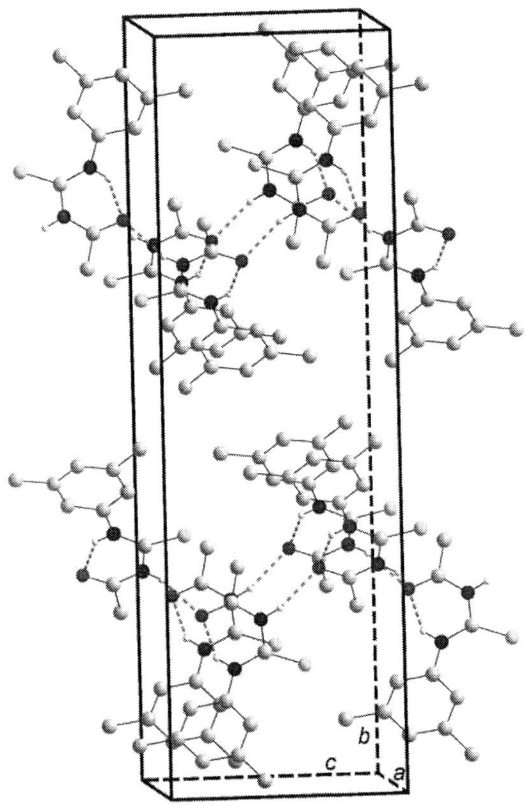

Fig. 3. Molecular packing view of the title compound (**I**). Molecular links along *c*-axis are generated by N—H · · · O hydrogen bonds which are shown by dashed lines. The rest of H atoms have been omitted for clarity

D—H···*A*	D—H	H···*A*	*D*···*A*	*D*—H···*A*
C6—H6···S1	0.93	2.58	3.245(3)	129
N1—H101···O1	0.86	1.90	2.642(2)	143
C9—H9*C*···O1[i]	0.96	2.70	3.256(3)	117
N2—H102···O1[i]	0.86	2.07	2.896(2)	161

Symmetry code: (i) +x, 1/2-y, -1/2+z; (ii)+x, 1/2-y, 1/2+z.

Tab. 1. Hydrogen-bond geometry (Å, °) for (**I**)

Atoms	Charge, *q*	Bond	I_w	Atoms	Charge, *q*
C1	0.230	N1-C1	0.984	N2	-0.649
C2	-0.247	N1-C7	1.247	H102	0.371
C3	0.081	C7=S1	1.508	C8	0.815
C4	-0.242	C7-N2	1.036	O1	-0.706
C5	0.080	C8-N2	1.095	S1	-0.306
C6	-0.238	C8=O1	1.644	C9	-0.561
N1	-0.577			C10	-0.500
H101	0.388			C11	-0.495
C7	0.417				

Tab. 2. Selected values of the net charges at individual atoms and Wiberg bonding indices I_w

Acknowledgement

This work was supported by the Grant Agency of the Slovak Republic (Grant No. 1/0371/16). This contribution is also the result of the project: Research Center for Industrial Synthesis of Drugs, ITMS 26240220061, supported by the Research & Development Operational Programme funded by the ERDF. JS, VV and MK thank the Structural Funds, Interreg IIIA, for financial support in purchasing the diffractometer. BTG thanks the University Grants Commission, Government of India, New Delhi, for a special grant under the UGC–BSR one-time grant to faculty.

References

[1] BRANDENBURG, K. 1999, Diamond. Version 2.1c. Crystal Impact GbR Bonn, Germany.

[2] BURLA, M. C., CALIANDRO, R., CARROZZINI, B., CASCARANO, G. L., GIACOVAZZO, C., MALLAMO, M., MAZZONE, A., POLIDORI, G. 2016, Sir2014, ver. 14.10. In preparation.

[3] DOLOMANOV, O. V., BOURHIS, L. J., GILDEA, R. J., HOWARD, J. A. K., PUSCHMANN, H. 2009, OLEX2. In *Journal of Applied Crystallography* **42**, 339-341.

[4] FRISCH, M. J. *et al.* 2013, Gaussian09. Gaussian. INC., Wallingford CT.

[5] GOWDA, B. T., RAMACHANDRA, P. 1989, In *J. Chem. Soc. Perkin Trans.* **2**, p. 1067–1071.
[6] GOWDA, B. T., PAULUS, H., FUESS, H. 2001, In *Z. Naturforsch. Teil A*, **56**, 386–394.
[7] GOWDA, B. T., FORO, S., FUESS, H. 2007, In *Acta Cryst.* E**63**, o2597.
[8] HÜBSCHLE, Ch. B., SHELDRICK, G. M., DITTRICH, B. 2011, In *J. App. Cryst.*, **44**, 1281–1284.
[9] SHELDRICK, G. M. 2015, In *Acta Cryst. C***71**, 3-8.
[10] SPEK, A. L. 2009, In *Acta Cryst.* D**65**, 148–155.

Current address

Sivý Július, doc. RNDr. PhD.
Institute of Mathematics and Physics,
Faculty of Mechanical Engineering, Slovak University of Technology
Nám. slobody 17, SK-812 31 Bratislava, Slovak Republic
E-mail: julius.sivy@stuba.sk

Vrábel Viktor, doc. Ing. PhD.
Institute of Analytical Chemistry,
Faculty of Chemical and Food Technology, Slovak University of Technology
Radlinského 9, SK-812 37 Bratislava, Slovak Republic
E-mail: viktor.vrabel@stuba.sk

Králik Marián, doc. Ing. PhD.
Institute of Manufacturing Systems, Environmental Technology and Quality Management,
Faculty of Mechanical Engineering, Slovak University of Technology
Nám. slobody 17, SK-812 31 Bratislava, Slovak Republic
E-mail: marian.kralik@stuba.sk

Gowda B. Thimme, Prof. PhD.
Department of Chemistry, Mangalore University
Mangalagangotri-574 199 IND, Mangalore, India
E-mail: gowdabt@yahoo.com

Slovak University of Technology in Bratislava
Faculty of Mechanical Engineering
16th Conference on Applied Mathematics **APLIMAT 2017**
Proceedings

COMPARISON OF ROBUST ESTIMATES

SLÁDEK Václav (CZ)

Abstract. The aim of this paper is to analyze properties of robust estimates which estimate the mean value and variability of a distribution and to compare properties of these estimates. This will provide background for selection which robust estimate is better estimate of an unknown parameter in specific situation. Properties of estimates are estimated by robust bootstrap. Considered robust estimates have relatively similar properties. If we dispose of a large sample, our selection can be based only on the mean value of estimate.

Keywords: robust estimates, properties of estimates, robust bootstrap, simulations

Mathematics Subject Classification: Primary 62G35; Secondary 62F40

1 Introduction

Estimation of an unknown reality based on limited information is one of the main aims of Statistics. An estimate should be also accurate estimate of an unknown reality. For example, we estimate an unknown parameter of distribution of random variable or dependence between two or more variables etc. Estimate is based on values of a random sample. This random sample should be representative sample of reality. This requirement of representativeness is fulfilled in most cases but sometimes the sample includes one or more observations which are different in values than others. These observations are called outliers. They are the reason why classic estimates, like sample mean, underestimate or overestimate the theoretical value of an unknown parameter θ. Robust estimates try to solve this problem. Examples of robust estimates can be sample median, winsored or trimmed mean. The question is: Which of these robust estimates choose? Decision is made on the basis of estimates properties. We want to have an estimate which is unbiased, consistent and efficient. Sometimes properties could be derived analytical, sometimes must be derived by simulations. Monte Carlo simulation is very useful method for estimation the properties. The main requirement of this method is knowledge of distribution of random variable. If the distribution is unknown, we use bootstrap method.

2 Robust estimates

2.1 Robustness and robust estimates

We know several definitions of robustness. Robustness can be defined as insensitivity to deviations from set of assumptions. Robust estimates keep their good properties even in case of deviations from assumptions, classic estimates do not keep them.

We know several characteristics which measure robustness. The first is influence function. It expresses the impact of the new observations to estimate. Good estimate is estimate whose value does not change by adding a new observation. The second most know characteristic is breakdown point. It is based on replacing several observations in a random sample by observations which are adverse, for example infinite. We are looking for count of replaced observations which leads to infinite value of estimate.

Robust estimate are estimates which are insensitive to deviations from set of assumptions. We know three basic categories of them: M-estimates, L-estimates and R-estimates. M-estimates are estimates which minimalize function which can be written as

$$\sum_{i=1}^{n} \rho(x_i, \theta) \rightarrow min, \tag{2.1}$$

where θ is unknown parameter and $\rho(\cdot)$ is suitable chosen function. L-estimates are based on values of ordered random sample, so called ordered statistics. L-estimates are linear combination of ordered statistics. They are very useful for description of distribution of random variable. Jurečková in [3] writes that general form of L-estimates can be written as

$$T_n = \sum_{i=1}^{n} c_{ni} h(X_{n:i}) + \sum_{j=1}^{k} a_j h^* \left(X_{n:[np_j]+1}\right), \tag{2.2}$$

where T_n is an estimate of an unknown parameter θ, c_{ni} and a_j are the coefficients are h and h^* the functions. R-estimates are relatively similar to L-estimates. They are based on ordered random sample too, but they use the rank of observations instead the value of observations. The rank of observation can be settled by formula

$$R_i = \sum_{j=1}^{n} I[X_j \leq X_i], i = 1, \dots, n, \tag{2.3}$$

where R_i is rank of i-th observation in sample. You can find more information in [2] and [4].

2.2 Robust measures of central tendency

Measure of central tendency of variable characterize level of distribution. The classic example of these measures is the mean value. Simply said, these measures inform us about position of data in defined space. Most of robust measures are based on quantiles, statistics which divide the vector of values into two parts in a set ratio.

The well-known robust measure is sample median $\tilde{x}_{0,5}$. Its value split an ordered random sample into two halves. It is often used to estimate the mean value of distribution or in description statistics. The sample median is consistent but biased estimate.

Trimean is used to estimate the mean value of distribution like sample median. But it is not often used. Insensitivity to outliers is its advantage. It is relatively more efficient estimate then sample median. It can be computed as weighted arithmetical mean of quartiles

$$trimean = 0{,}25\tilde{x}_{0,25} + 0{,}5\tilde{x}_{0,5} + 0{,}25\tilde{x}_{0,75}. \tag{2.4}$$

Winsorized mean is a characteristic which is based on an ordered sample. Extreme values (from both sides) of this sample are replaced by nearest "right" value. It is need to set k, a ratio of extreme values in sample, $0 < k < 0{,}5$. Then we set k-th and $(1-k)$-th quantile (y_k, y_{1-k}). Values lower than y_k are replaced by y_k and values greater than y_{1-k} are replaced by y_{1-k}. Formula of winsorized mean is following

$$\bar{x}_{w,k} = \frac{(nk+1)y_k + \sum_{i=nk+2}^{n-nk-1} y_i + (nk+1)y_{1-k}}{n}. \tag{2.5}$$

Next robust measure of central tendency is trimmed mean. Its construction is similar to winsorized mean. We do not replace the extreme values, we trim them. Formula is following

$$\bar{x}_{t,k} = \frac{1}{n-2nk} \sum_{i=nk+1}^{n-nk} y_i. \tag{2.6}$$

Sen weighted mean is the last robust measure which we mention. It is modification of trimmed mean. It is trimmed mean where each value has own weight which depends on the number of trimmed values, the rank of observation in sample, the ratio of trimmed observations and size of sample

$$\bar{x}_{nk} = \binom{n}{2nk+1}^{-1} \sum_{i=nk+1}^{n-nk} \binom{i}{nk}\binom{n-i+1}{nk} y_{i+1}. \tag{2.7}$$

2.3 Robust measures of variability

Measures of variability describe volatility of random variable. How its values are near or distant its mean value. Classic measures of variability are usually based on the mean value of variable, but it is unknown and it is often estimated by sample mean. But sample mean is sensitive to outliers. Robust measures are usually based on robust estimate of the mean value (sample median etc.).

The first robust measure is the mean absolute deviation from sample mean. Its computation is based on absolute distance between values in sample from sample mean. Its formula can be written as

$$s_a = \frac{1}{n}\sum_{i=1}^{n}|x_i - \bar{x}|. \tag{2.8}$$

Next measure is the mean absolute deviation from sample median. Its formula is different from previous thanks replacing sample mean by sample median

$$s_m = \frac{1}{n}\sum_{i=1}^{n}|x_i - \tilde{x}_{0,5}|. \tag{2.9}$$

Last measure which we mention is median absolute deviation. Its formula can be written as

$$MAD = medián|x_i - \tilde{x}_{0,5}|. \tag{2.10}$$

3 Robust bootstrap

Bootstrap is usually used for numeric estimating of properties of estimate. This method is based on regenerating random samples, so-called bootstrap samples $\boldsymbol{x}^* = (x_1^*, \dots, x_n^*)$. These bootstrap samples are generated from an origin sample which is from unknown distribution. The bootstrap sample has length n and empirical distribution function $\hat{F}(x)$ where each observation has probability of choosing equal $1/n$. Bootstrap sample are made by sampling with replacement from origin sample, so in $\boldsymbol{x}^*$ can be only values which are in origin sample. Some of these observations will appear in $\boldsymbol{x}^*$ only once, some of them more than once and some of them do not appear at all. We compute estimate of unknown parameter θ, which is denoted as $\hat{\theta}^*$, from each bootstrap sample. If we have enough $\hat{\theta}^*$ we can estimate distribution of estimate which is called bootstrap distribution of estimate.

Bootstrap has the same problem as classic estimates. It is sensitive to outliers in sample. Bootstrap samples are generated from origin sample with replacing, so the number of outliers in them could be greater than in origin sample. It can cause convergence some $\hat{\theta}^*$ to infinity. These bootstrap replications have negative impact on bootstrap distribution of estimate. This is the reason for using robust bootstrap.

There are a few robust bootstrap methods. All of them are focused on values adjustment in origin sample. Singh´s winsorized bootstrap is the most well-known method. Winsorized bootstrap is based on winsorizing the extreme values from origin sample. Assessment of extreme values is really important. Let us denote the ratio of extreme values in sample z. If we overestimate z we will lose a lot of information. But if we underestimate z then bootstrap distribution will not be accurate. Singh made a recommendation on z in [7]. We will use this method for estimating properties of estimates.

The second well-known method was published by Amado and Pires. It is based on weighted values in origin sample. Weights are assigned using robust standardized influence function. These weights are probability of choosing of each observation.

3.1 Comparison of classic and robust bootstrap

In this section we compare bootstrap distributions obtained by winsorized and classic bootstrap. Then we compare their results with the theoretical distribution of estimate. Theoretical distribution is approximated by Monte Carlo simulation. Let us consider a random sample x which is generated from normal distribution with parameters μ and σ^2. We will estimate the mean value using robust estimates. Let us consider sample with 40 and 80 observations and normal distribution $N(5,1)$, ratio of extreme value 0.1 and z has value 0.1. Extreme values in sample will be generated from the distribution $N(15,2)$. We will construct 2000 Monte Carlo simulations and 2000 bootstrap replications B. Considered robust estimates will be $\tilde{x}_{0,5}$ and $\bar{x}_{w,0.1}$.

Tab. 1 contains the mean value of sample median and winsorized mean. The theoretical mean values of estimates are in column *MC*, the classic bootstrap mean values are in column *CB* and the winsorized bootstrap mean values are in column *WB*.

n		*MC*	*CB*	*WB*
40	$\tilde{x}_{0,5}$	5.135	5.052	5.052
	$\bar{x}_{w,0.1}$	5.322	5.719	5.302
80	$\tilde{x}_{0,5}$	5.143	5.266	5.266
	$\bar{x}_{w,0.1}$	5.345	5.832	5.474

Tab. 1. Comparison of estimated mean values of robust estimates and their theoretical values

Estimated mean values of sample median are equal. If we estimate its distribution by winsorized bootstrap, we get the same distribution when we use classic bootstrap. The reason is probably maximal breakdown point of median. The second reason is that we do not winsorized values in origin sample which have impact on value of median. Fig. 1 shows difference between classic bootstrap and winsorized bootstrap distribution of $\bar{x}_{w,0,1}$.

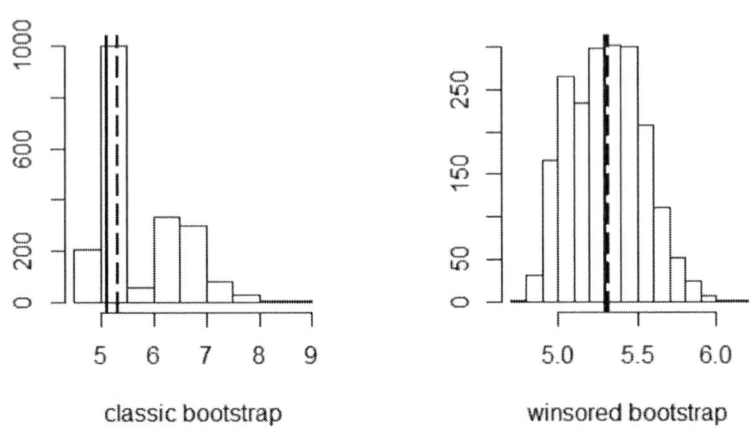

Fig. 1. Estimated distribution of $\bar{x}_{w,0,1}$ (classic and winsorized bootstrap)

4 Comparison of properties of estimates

In this section we will compare robust estimates of the mean value and the standard deviation of considered distributions. Also we will compare them with sample mean $\bar{x}$ and sample standard deviation s. Let us consider two scenarios:

1. Random variable with normal distribution $N(7,4)$. Outliers will be generated from distributions $N(0,1)$ and $N(20,4)$. The number of outliers from both distributions are identical.

2. Random variable with Poisson distribution $P(8)$. Outliers will be generated only from distribution $P(1)$.

For all scenarios we consider random samples with 20, 50, 100 and 500 observations. Ratio of outliers is 10 % and 20 %. We will compare these properties of estimates:

- the mean value,
- the standard deviation,
- the mean squared error.

4.1 The mean value of estimates

The mean value of estimate is one of the most important property of estimate. Tab. 2 contains results for normal distribution. The theoretical mean value of distribution is 7 and standard deviation is 2. It is obvious that the mean value of robust estimates converges to the theoretical value with increasing n. Difference between 10 % and 20 % of outliers is minimal. All robust estimates underestimate the theoretical value of distribution in sample with 20 observations. In section 2.3 we mentioned that sample median is biased estimate, we can see that its mean value is lower that theoretical mean value. Sample mean is biased for all n. None robust estimates (except $\tilde{x}_{0.5}$) systematically overestimate or underestimate the theoretical mean value. Fig. 2 shows case of 20 % outliers.

	$k = 0.1$				$k = 0.2$			
	$n = 20$	$n = 50$	$n = 100$	$n = 500$	$n = 20$	$n = 50$	$n = 100$	$n = 500$
$\bar{x}$	6.767	7.759	7.376	7.328	7.519	7.797	7.598	7.623
$\tilde{x}_{0,5}$	5.930	6.786	6.775	6.940	6.242	6.718	6.748	6.901
trimean	6.059	7.122	6.985	7.017	6.376	7.046	6.951	7.022
$\bar{x}_{w,k}$	6.502	7.351	7.069	7.043	6.717	7.246	7.114	7.054
$\bar{x}_{t,k}$	6.360	7.333	7.040	7.043	6.404	7.174	6.979	7.026
$\bar{x}_{nk}$	6.261	7.254	6.878	6.979	6.393	7.063	6.828	6.927
s	3.986	4.288	4.046	3.911	5.648	5.454	5.160	5.191
s_a	2.397	2.892	2.527	2.499	3.622	3.737	3.416	3.436
s_m	2.279	2.795	2.450	2.475	3.288	3.627	3.291	3.361
MAD	1.605	2.874	2.103	2.279	1.793	3.224	2.613	2.594

Tab. 2. The mean value of estimate, normal distribution, $k = 0.1$ and $k = 0.2$

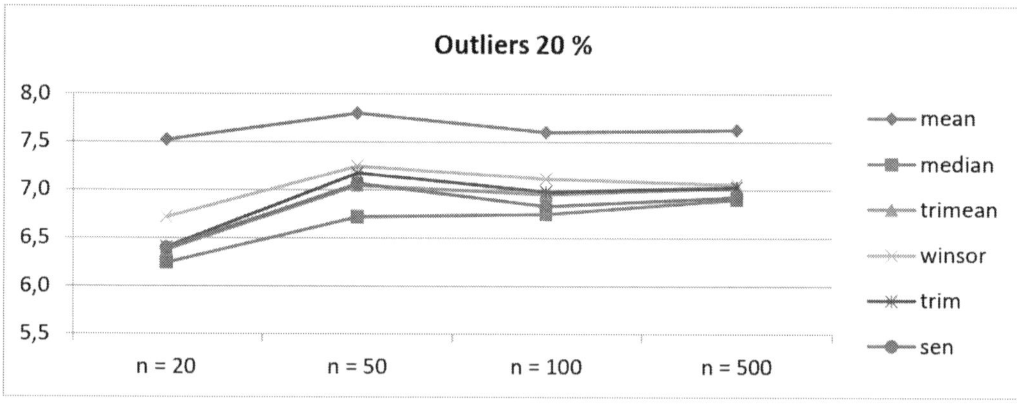

Fig. 2. The mean value of estimates of the mean value, normal distribution, $k = 0.2$

The mean value of estimates of standard deviations is different. Everyone considered estimate overestimates the theoretical value and their convergence is slower than convergence of estimates of mean value. We get accurate estimates for $k = 0.1$ then for $k = 0.2$. The most accurate estimate of variability is MAD. Fig. 3 shows situation of estimates of theoretical standard deviation of distribution for $k = 0.1$.

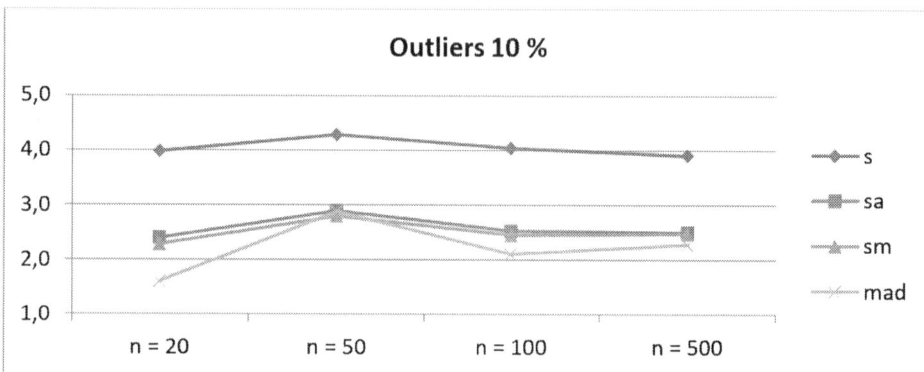

Fig. 3. The mean value of estimates of the standard deviation, normal distribution, $k = 0.1$

We estimate only one unknown parameter in the second scenario with Poisson distribution. This parameter λ is the mean value and also the variance of distribution. So every considered robust estimates is estimate of λ or estimate of $\sqrt{\lambda}$. Its true value in this scenario is equal 8. It is difficult to find systematic behaviour of estimates. Maybe it is caused by discrete distribution. The possible values of estimates are limited.

None estimates of the mean value fast converges to the theoretical value. It is probably caused by non-symmetrical layout of outliers (extreme values are only on one side) and considered robust estimates need symmetrical layout of outliers. The robust estimates of standard deviation give accurate estimates than estimates of the mean value. But only for small k and for small n (20 or 50 observations). These robust estimates do not converge to the theoretical value of λ with increasing n.

	$k = 0.1$				$k = 0.2$			
	$n = 20$	$n = 50$	$n = 100$	$n = 500$	$n = 20$	$n = 50$	$n = 100$	$n = 500$
$\bar{x}$	7.150	7.220	7.580	7.266	6.350	6.300	6.790	6.566
$\tilde{x}_{0,5}$	7.000	7.000	7.500	7.000	6.000	6.500	7.000	7.000
trimean	7.250	7.250	7.750	7.250	6.313	6.500	7.000	6.750
$\bar{x}_{w,k}$	7.420	7.360	7.660	7.282	6.640	6.220	6.960	6.636
$\bar{x}_{t,k}$	7.313	7.350	7.725	7.353	6.667	6.500	7.000	6.727
$\bar{x}_{nk}$	7.200	7.264	7.610	7.331	6.514	6.552	6.942	6.854
s	3.200	3.412	3.739	3.463	3.573	3.803	4.120	3.878
s_a	2.365	2.638	3.020	2.751	2.685	2.980	3.367	3.173
s_m	2.350	2.620	3.020	2.738	2.650	2.980	3.350	3.138
MAD	2.224	2.965	3.707	2.965	2.965	3.707	4.448	4.448

Tab. 3. The mean value of estimate, Poisson distribution, $k = 0.1$ and $k = 0.2$

4.2 Standard deviations of estimates

The second important property of estimate is its variability, which is measured by standard deviation. The convergence of standard deviation to zero with increasing n was proved in many papers, books etc. We will consider on speed of convergence for each estimate and we will compare them between themselves.

	$k = 0.1$				$k = 0.2$			
	$n = 20$	$n = 50$	$n = 100$	$n = 500$	$n = 20$	$n = 50$	$n = 100$	$n = 500$
$\bar{x}$	0.398	0.420	0.242	0.115	0.523	0.410	0.273	0.142
$\tilde{x}_{0,5}$	0.432	0.563	0.208	0.125	0.569	0.634	0.242	0.143
trimean	0.488	0.408	0.214	0.107	0.623	0.476	0.255	0.121
$\bar{x}_{w,k}$	0.455	0.449	0.279	0.123	0.663	0.460	0.306	0.147
$\bar{x}_{t,k}$	0.468	0.395	0.249	0.110	0.616	0.474	0.289	0.127
$\bar{x}_{nk}$	0.450	0.427	0.244	0.115	0.589	0.529	0.272	0.129
s	0.259	0.373	0.162	0.087	0.335	0.223	0.166	0.095
s_a	0.283	0.292	0.154	0.076	0.377	0.226	0.172	0.092
s_m	0.268	0.283	0.153	0.076	0.353	0.228	0.171	0.092
MAD	0.550	0.482	0.310	0.119	0.753	0.597	0.395	0.172

Tab. 4. Standard deviation of estimates, normal distribution, $k = 0.1$ and $k = 0.2$

Tab. 4 contains values of standard deviations of considered estimates. It is obvious that for same sample size the value of variability is more or less equal for each robust estimates of the mean value. If we doubled numbers of outliers, the variability increases, but not too much. The highest increase is in situation of 20 observations. Standard deviation of $\bar{x}$ is smallest for all estimates of the mean value in situation of 20 observations.

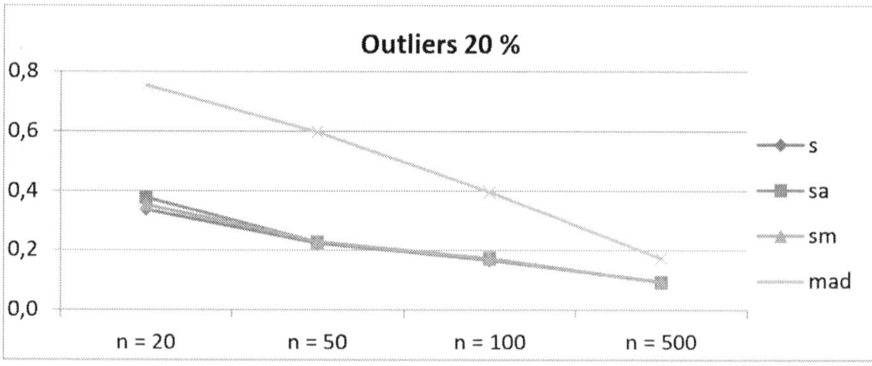

Fig. 4. Standard deviations of estimates of variability, normal distribution, $k = 0.2$

Variability of estimates of standard deviations gives more interesting results than the mean value. Variability of s, s_a and s_m are similar but variability of MAD is much greater even for $n = 500$. Absolute gap between MAD and others estimates decreases with increasing n, but relative gap stays on more or less similar level. From previous subsection we know that MAD has the mean value nearest to estimated parameter but it has the highest variability.

Tab. 5 contains results for scenario with Poisson distribution. It seems that variability of estimates of the mean value is slightly higher than first scenario. Variability decreases and converges to zero with increasing n. The convergence is relatively fast. The lowest variability has classic estimate of the mean value, sample mean. Robust estimate with lowest variability for $k = 0.1$ is trimmed mean and for $k = 0.2$ is sample median.

Variability of estimates of standard deviation is very similar to the previous scenario. But only for estimates s, s_a a s_m. Variability of MAD is much higher and its convergence to zero is distinctly.

	$k = 0.1$				$k = 0.2$			
	$n = 20$	$n = 50$	$n = 100$	$n = 500$	$n = 20$	$n = 50$	$n = 100$	$n = 500$
$\overline{x}$	0.489	0.456	0.346	0.148	0.690	0.473	0.374	0.154
$\tilde{x}_{0,5}$	0.832	0.603	0.545	0.350	0.810	0.573	0.543	0.167
trimean	0.669	0.554	0.451	0.246	0.860	0.678	0.607	0.231
$\overline{x}_{w,k}$	0.540	0.543	0.411	0.188	0.936	0.722	0.591	0.229
$\overline{x}_{t,k}$	0.575	0.515	0.398	0.165	0.832	0.597	0.516	0.210
$\overline{x}_{nk}$	0.587	0.477	0.417	0.170	0.787	0.513	0.449	0.180
s	0.254	0.288	0.198	0.085	0.423	0.253	0.177	0.073
s_a	0.267	0.278	0.198	0.084	0.433	0.288	0.206	0.087
s_m	0.277	0.280	0.200	0.086	0.428	0.284	0.205	0.089
MAD	0.715	0.750	0.714	0.259	1.187	0.952	0.947	0.587

Tab. 5. Standard deviation of estimates, Poisson distribution, $k = 0.1$ and $k = 0.2$

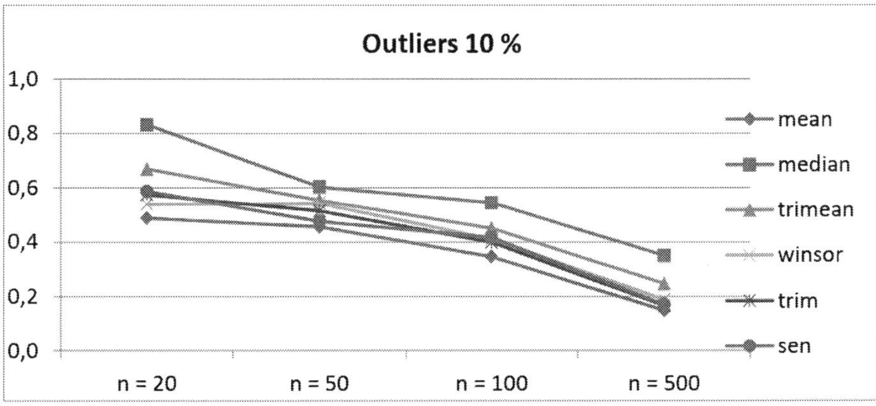

Fig. 5. Standard deviations of estimates of the mean value, Poisson distribution, $k = 0.1$

4.3 Mean squared error of estimates

The last property of estimates, which we analyse, is mean squared error, MSE. It is defined as sum of square bias and variability of estimate. Minimal MSE is often criteria which is required. It can be assumed that the value of MSE will converge to zero with increasing n and will increase with increasing k from previous results.

Tab. 6 contains results of analysis for scenario with normal distribution. Robust estimates of the mean value have similar value of MSE, absolute differences are small. The highest MSE has sample mean for both k. But it is caused by its bias not its variability. The "best" robust estimates are Sen weighted mean and trimmed mean.

MSE for standard deviation has different values, they are much higher. The highest MSE has classic estimate s. Robust estimates have multiple smaller MSE than s. The smallest value has MAD. Even if we have the largest sample size, its value is considerably smaller than the value of other robust estimates.

	$k = 0.1$				$k = 0.2$			
	$n = 20$	$n = 50$	$n = 100$	$n = 500$	$n = 20$	$n = 50$	$n = 100$	$n = 500$
$\bar{x}$	0.231	0.223	0.155	0.113	0.695	0.541	0.368	0.420
$\tilde{x}_{0,5}$	0.203	0.352	0.043	0.016	0.325	0.451	0.060	0.021
trimean	0.259	0.176	0.046	0.012	0.401	0.234	0.066	0.015
$\bar{x}_{w,k}$	0.208	0.202	0.078	0.015	0.442	0.212	0.094	0.022
$\bar{x}_{t,k}$	0.219	0.156	0.062	0.012	0.386	0.224	0.083	0.016
$\bar{x}_{nk}$	0.204	0.182	0.060	0.013	0.353	0.281	0.074	0.017
s	4.829	1.732	2.809	1.696	11.383	6.474	5.882	4.196
s_a	1.010	0.351	0.463	6.487	3.310	1.756	1.555	1.021
s_m	0.904	0.313	0.393	0.227	2.514	1.549	1.311	0.879
MAD	0.305	0.250	0.096	0.014	0.588	0.356	0.162	0.031

Tab. 6. Mean squared error of estiamtes, normal distribution, $k = 0.1$ and $k = 0.2$

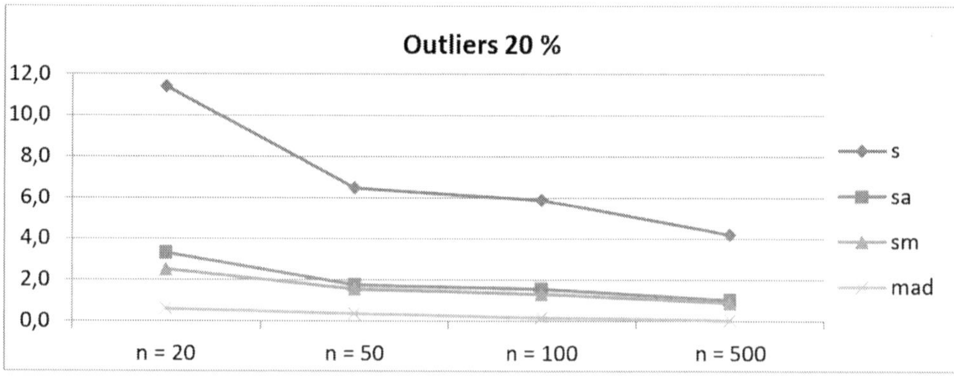

Fig. 6. *MSE* of estimates of standard deviation, normal distribution, $k = 0.2$

Tab. 7 contains results for second scenario. The highest *MSE* for all n and k in previous scenario has sample mean. This does not fit for second scenario with Poisson distribution. Sample mean has the smallest value for all n and k. *MAD* is not estimate with smallest *MSE* for this situation. Even classic estimate s has small value for majority of cases n and k. The "best" estimates are s_a and s_m.

	$k = 0.1$				$k = 0.2$			
	$n = 20$	$n = 50$	$n = 100$	$n = 500$	$n = 20$	$n = 50$	$n = 100$	$n = 500$
$\bar{x}$	0.313	0.209	0.120	0.022	0.500	0.229	0.141	0.035
$\tilde{x}_{0,5}$	0.694	0.383	0.297	0.145	0.927	0.331	0.306	0.029
trimean	0.450	0.310	0.204	0.062	0.762	0.472	0.374	0.053
$\bar{x}_{w,k}$	0.292	0.305	0.173	0.035	0.942	0.522	0.369	0.060
$\bar{x}_{t,k}$	0.330	0.268	0.160	0.027	0.721	0.359	0.269	0.044
$\bar{x}_{nk}$	0.350	0.227	0.174	0.029	0.620	0.263	0.202	0.032
s	1.236	0.132	0.095	0.047	0.510	0.300	0.168	0.208
s_a	0.421	0.093	0.060	0.018	0.288	0.169	0.097	0.081
s_m	0.488	0.096	0.066	0.018	0.307	0.191	0.097	0.084
MAD	0.512	0.576	0.512	0.070	1.410	1.017	0.898	0.436

Tab. 7. Mean squared error of estimates, Poisson distribution, $k = 0.1$ and $k = 0.2$

5 Conclusions

If we focus only on the mean values of estimates, we can generally say, that considered classic and robust estimates are not good estimates in considered situations. Only *MAD* gives us accurate results than others estimates. Robust estimates of the mean value of normal distribution give accurate results. Their values are accurate than values of sample mean. Robust estimates are inaccurate if we estimate unknown parameter of Poisson distribution. It is probably caused by non-symmetrical layout of outliers in random sample (extreme values are only on one side) and considered robust estimates need symmetrical layout of outliers. Robust estimates of variability overestimate the theoretical value of λ, robust estimates of the mean value underestimate it.

Variability of sample mean is in general smaller than variability of robust estimates. It also fits for estimates of standard deviation. Variability of all considered estimates converge to zero with increasing n. Values of se are similar from specific size of random sample. This does not fit for sample median and MAD in situation of Poisson distribution. The most variable measure of variability is MAD and the most variable measure of central tendency is sample median for majority considered n and k.

Value of MSE converges to zero with increasing n for all considered estimates. Generally, we can say that all robust estimates have smaller MSE than sample mean and sample standard deviation. High value of s is caused by its bias. The greatest impact on MSE has bias of estimates than their variance. Robust estimates have very similar value for all n and k. Sample mean has lowest MSE only for Poisson distribution.

Reference

[1] AMADO C., BIANCO A., BOENTE G., PIRES A. Robust Bootstrap: An Alternative To Bootstrapping Robust Estimators. *REVSTAT – Statistical Journal*, ročník 12, č. 2, 2014, ISSN 1645-6726.

[2] BERAN R.: Robust Location Estimates. *The Annals of Statistics*, ročník 5, č. 3, 1977, ISSN 0090-5364

[3] EFRON B., TIBSHIRANI R. An Introduction to the Bootstrap. Florida: CRC Press LLC, 1998. ISBN 0-412-04231-2

[4] JUREČKOVÁ J.: Robustní statistické metody. Praha: Nakladatelství Karolinum, 2001. ISBN 80-246-0259-8

[5] MALÁ I.: Vybrané kapitoly z teorie pravděpodobnosti a matematické statistiky. Praha: Oeconomica, 2011. ISBN 978-80-245-1829-9

[6] SHAO J.: Bootstrap estimation of the asymptotic variances of statistical functionals. Annals of the Institute of Statistical Mathematics, ročník 42, č. 4, 1990, ISSN 0020-3157

[7] SINGH K. Breakdown Theory for Bootstrap Quantiles. *The Annals of Statistics*, ročník 26, č. 5, 1998, ISSN 0090-5364

Current address

Sládek Václav, Ing.
Nám. Winstona Churchilla 4, 130 67 Praha 3, Czech Republic
Tel. number: +420 604 464 895, e-mail: xslav20@vse.cz

Slovak University of Technology in Bratislava
Faculty of Mechanical Engineering
16th Conference on Applied Mathematics **APLIMAT 2017**
Proceedings

OPTIMISATION OF VEHICLE ENERGY CONSUMPTION BY ALGORITHMIC GRAPH THEORY

ŠLAUKA Michal (SK), LIEDL Tomáš (SK), MILESICH Tomáš (SK)

Abstract. This publication describes the methodology of the energy consumption optimisation. Optimisation uses algorithm graph theory and Bellman-Ford method for search the shortest path. The shortest path means the lower consumption energy by a vehicle during transportation from point A to point B. As input to the algorithm is real measured data from the vehicle in the circuit defined by WTC cycle conditions.

Keywords: energy consumption, algorithm graph theory, the shortest path problem, optimisation, transportation, electric vehicle

Mathematics Subject Classification: Primary 00A69; Secondary 97K30

1 Introduction

The biggest negative aspect of the electric vehicles is their range. In comparison with commonly used conventional vehicles, it causes a decrease in their attractiveness to the extent of overshadowing all their benefits. The range of the electric vehicle directly depends on the consumption of electricity. In order to function, all electronic accessories as contained in the vehicle use the identical energy source as the powertrain. Air conditioning and heating unit take away in average 4 to 6kW of the battery. As input to the algorithm is real measured data from the vehicle in the circuit defined by WTC cycle conditions. This circuit contains the city, out of city and mountain round. The vehicle for optimisation is full electric vehicle VW E-Up!.

Data for this application is received from vehicle's communication network CAN. The main tracked data from a vehicle was actual speed, throttle and brake pedal position, actual torque and powertrain energy consumption. Tracked was also GPS position and road elevation. The measurement was repeated with four drivers every pass circuit one time. Optimisation uses algorithm graph theory and Bellman-Ford method for search the shortest path. The shortest path means the lower consumption energy by a vehicle during transportation from point A to point B. In this case is a circuit. Graph assembled using the principle of two conditions. Every minimum two drivers must be in the same place (means GPS position) and must have the same speed. If these two conditions are fulfilled there is the node for graph and in the directed edge is driver's energy consumption. The aim of the created algorithm is finding optimal speed profile for selected circuit with regard to minimising energy consumption.

2 Creation of the optimisation algorithm of the graph theory

Optimal power consumption required for the completion of the determined route can be acquired by means of optimisation algorithm of the graph theory, finding the best possible route pass. By splitting the route into certain sections and further searching for the best route pass of each one of them based on the consumption of a certain driver, the optimal route pass can be measured, however, the criteria have to be followed. The first criterion is the actual splitting of the route into the sections. Seeing that any of the drivers did not ultimately travel the identical distance, the route cannot be divided depending on that distance driven.

The result of the different pass of the turn´s radius and avoiding the route obstacles is the variation in the distance travelled. The shortest distance travelled by the driver B was precisely 86,657 km, then, in succession, driver A 86,795 km, the driver D 86,822 km and ultimately driver C 86,865 km. These determined variations would cause the inaccuracies in further searching for the optimal pass of the distance determined, because in the real environment each driver would be present on a different route, considering their identical distance travelled. With respect to the vehicle position in the coordinate system x, y, it is, therefore, necessary to divide the measured route into the sections. In the Fig. 1 the route splitting into 86 sections in the coordinate system x, y is displayed.

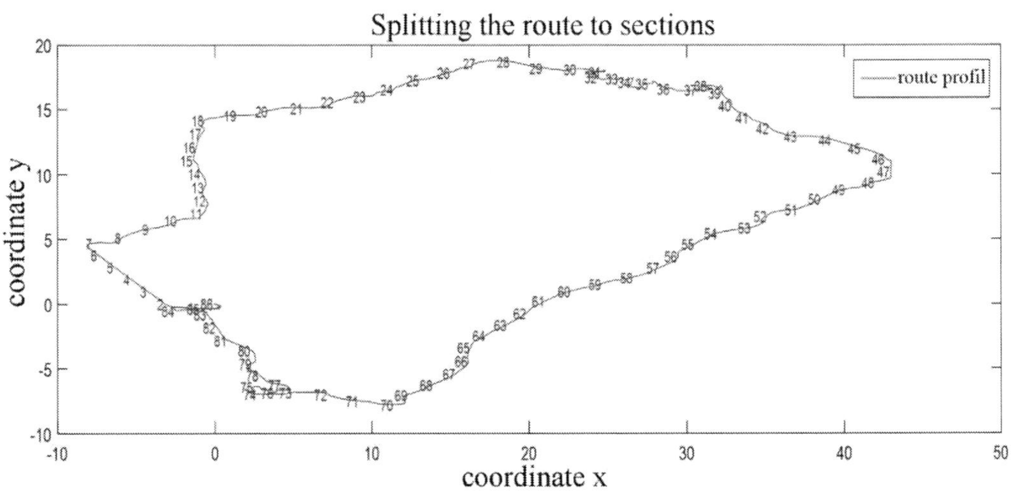

Fig. 1. Splitting the route to sections

The second criterion is a creating the continuity of the speed profile of the drivers´ pass between the respective sections, i.e. to find the curve speed intersections of the individual drivers in the given section. In the graph, the curve speed intersection presents a node. The number of graph nodes depends on the combination of creating possible pairs of the drivers´ curve speed intersections. For the pass of four drivers is the number of possible combinations without being repeated:

$$\binom{n}{k} = \frac{\prod_{i=n-k+1}^{n} i}{k} = \frac{n!}{k!(n-k)!} = \frac{4!}{2!(4-2)!} = 6 \tag{1}$$

The drivers´ combinations could also be expressed by the matrix of combinations:

$$\begin{matrix} A \\ B \\ C \\ D \end{matrix} \begin{bmatrix} 0 & 0 & 0 & 0 \\ 1 & 0 & 0 & 0 \\ 2 & 3 & 0 & 0 \\ 4 & 5 & 6 & 0 \end{bmatrix} \tag{2}$$

The matrix of combination is represented by the layers of the combinations assembled in the next scheme:

Layer							Intersection
1	ABCD						(6)
2	ABC	BCD					(3)
3	AB CD	AC BD	AD BC				(2)
4	AB	AC	AD	BC	BD	CD	(1)

The graph nodes are to be found in the layers, wherein the fourth layer, all combinations are present, i.e. from each driver each combination in every layer. Several connections, or more precisely paths, are created between the existing nodes, with a minimum of two for two drivers, while the paths present a size of the energy consumption.

3 Representation of the incidence and adjacency matrices of directed graph

It's represented by incidence matrix M = (m_{ij}) of type n x n where $n = |V|$is a number of the digraph's edge and its members are defined by the formula:

$$m_{ij} = \begin{cases} c\big((i,j)\big) & if\ (i,j) \in H \\ \infty & else \end{cases} \tag{3}$$

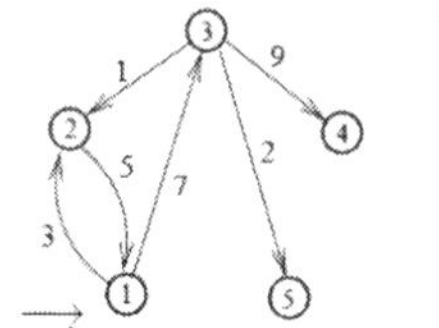

$\overrightarrow{G'_2} = (V_2, H_2, c_2)$

	1	2	3	4	5
1	-	3	7	-	-
2	5	-	-	-	-
3	-	1	-	9	2
4	-	-	-	-	-
5	-	-	-	-	-

Fig. 2. Adjacency matrix of digraph $\overrightarrow{G'_2}$

The adjacency matrix of a digraph is non-symmetric by main diagonal of the matrix. Representation by adjacency matrix or incidence matrix doesn't save a digital memory of the computing device. It needs to fill n^2 memory spaces and it's not addicted to a number of the edges. This method is suitable for algorithms which need to search existence of one and another one edge often. It's suitable for store digraph or subdigraph. It's suitable for few edge graph and for algorithms based on symmetric nodes search. [2]

4 Creating the digraph from measured data

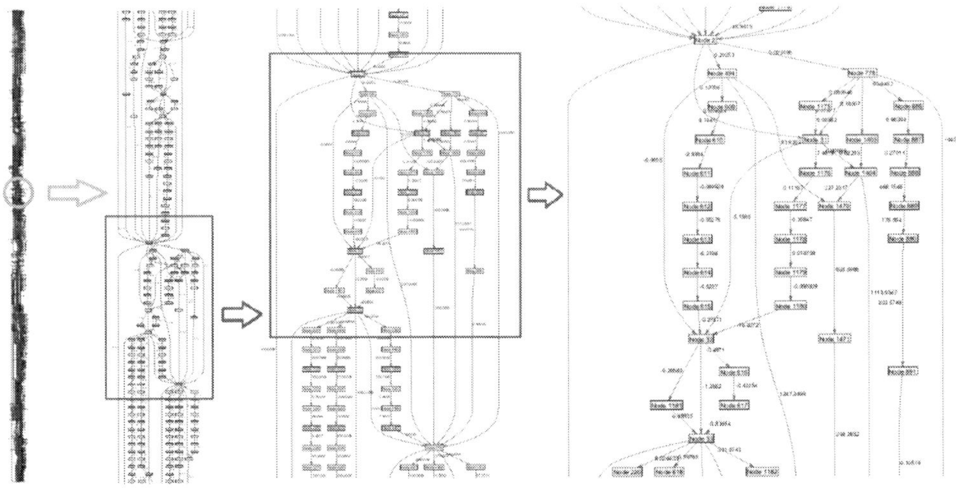

Fig. 3. The graph 1 visualisation in the diagram form for the consumption

The algorithm is going to be created, which keeps only routes with the smallest value of the consumption and assigns to them appertaining drivers. The algorithm result is the graph 1 for the consumption and the graph 2 for the drivers. These graphs are not the elements of graphic visualisation the values of the graph 1 and 2, but they are graphs from the theory of graphs, which includes the positions of particular notes and their paths.

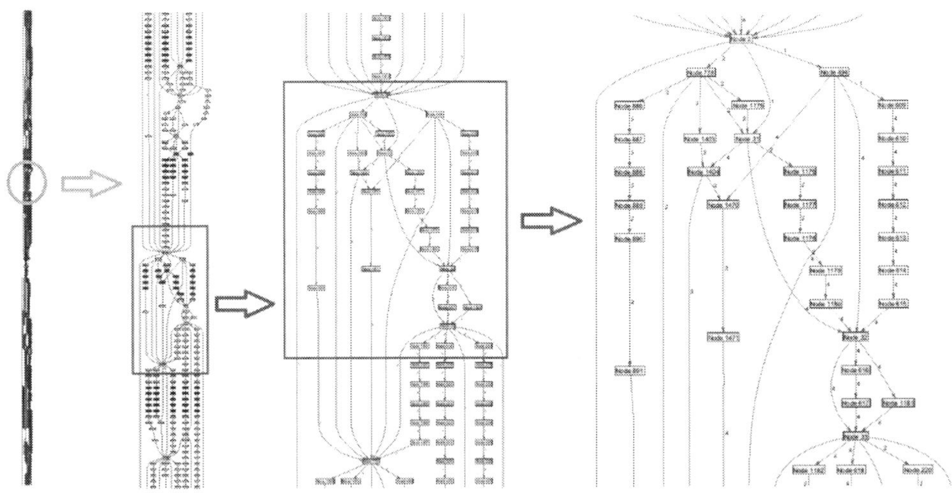

Fig. 4. The graph 2 visualisation in the diagram form for the drivers

The graph 1 visualisation in the diagram 1 form for the consumption is displayed in Fig. 3. According to a big quantity of the nodes and paths, the diagram is blind. Also, it is difficult to compute while drawing and viewing. It has a function only for the optical visualisation or a control of the right function of created algorithm. The graph 2 visualisation in the diagram 2 form for drivers is displayed in Fig. 4. It is identical to diagram 1, but the same notes are located on another place. The connection of particular paths is remaining. Instead of values, the consumption has these paths displayed descriptions of the single drivers.

5 The shortest path

May $\mu(u,v)$ is (u - v)-path in weighted graph $G = (V,H,c)$. Section distance $\mu(u,v)$ or also the length of the section is summary of its edges weights. Every edge weight is added as much as it appears in the section. Mark distance of the section $\mu(u,v)$ is $d(\mu(u,v))$. When all definiton as distance oriented move and path is the special case of the section. Simplified distance $d(\mu(u,v))$ section $\mu(u,v)$ is summary of edges weights, e. i.

$$d(\mu(u,v)) = \sum_{h\in\mu(u,v)} c(h). \tag{4}$$

The shortest ($u - v$)-path in weighted digraph $\vec{G} = (V,H,c)$ is oriented ($u - v$)-path $\mu(u,v)$ with minimum weight. Searching the shortest path in digraph is frequently used application with a practical sense. Often searching the shortest path is often part of other algorithms. Algorithms for searching the shortest (u-i)-path for fixed u assigned two markers for nodes. Upper estimated distance for the best (u-i)-path found is represented by sign $t(i)$ and $x(i)$ is its last node (Fig. 5).

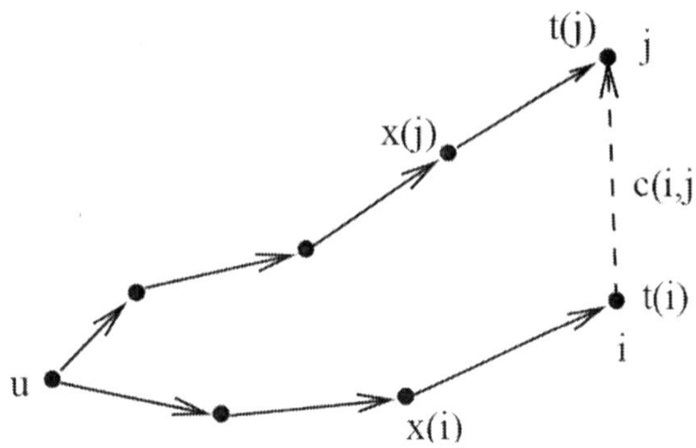

Fig. 5. Description of the algorithm function for searching the shortest path

During computation algorithm get (u-i)-path $\mu(u,v)$ with at best distance $t(i)$ with last node $x(i)$ and also the (u-j)-path $\mu(u,j)$ with at best distance $t(j)$ with last node $x(j)$. If (i,j) is oriented edge , i.e. $(i,j) \in H$ then

$$t(j) > t(i) + c(i,j) \tag{5}$$

Path $\mu(u,j)$ extended by edge (i,j) to node j with weight at best $t(i) + c(i,j)$ had less weight as estimate $t(j)$ present the shortest ($u - j$)-path. When node i is just the last node improved ($u - j$)-path then $t(j) := t(i) + c(i,j)$, $x(j) := i$. This path improving method is part of every algorithm for searching shortest path [1].

Application of the predefined algorithm which is contained in the user interface of the Matlab software. By applicating of the shortest path problem between two nodes of the digraph it obtains the path of the lowest energy consumption. For the correct shortest path search in the digraph is recommended to use a numerical method called Bellman-Ford. The figure of the shortest path digraph is described in Fig. 6 [3].

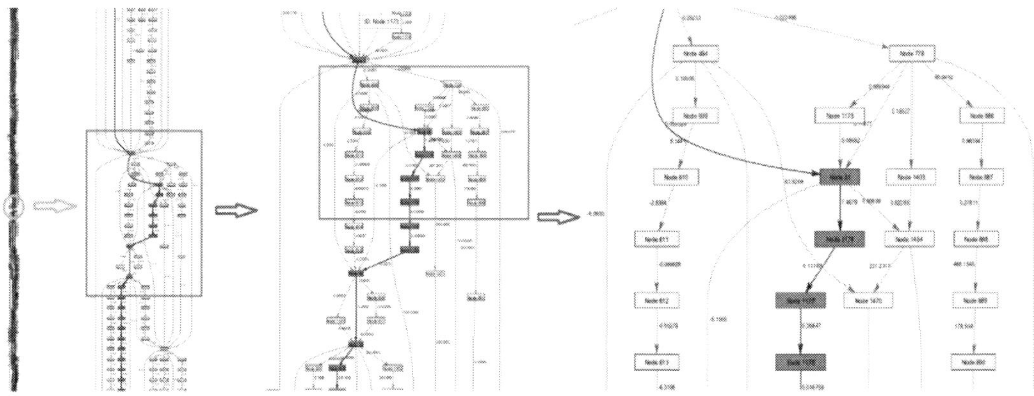

Fig. 6. The graph 1 visualisation in the diagram form displays the shortest path (red)

The last step to achieve the optimal pass of the measured route or more precisely optimisation of energy consumption is creating an algorithm to identifying paths selected by the method of shortest path for the speed profiles and each driver consumptions in selected section. Each driver's curves of the speed, energy consumption and torque for selected section are displayed in Fig. 7. Also, these curves are displayed for the optimal driver. In the displayed section 31 to 32km of measured route driver Opt. is copying driver's C speed profile. After approx. 200 meters driver Opt. starts copying driver B. The result of the driver Opt. is a curve of the energy consumption with smallest final value for selected section. [4]

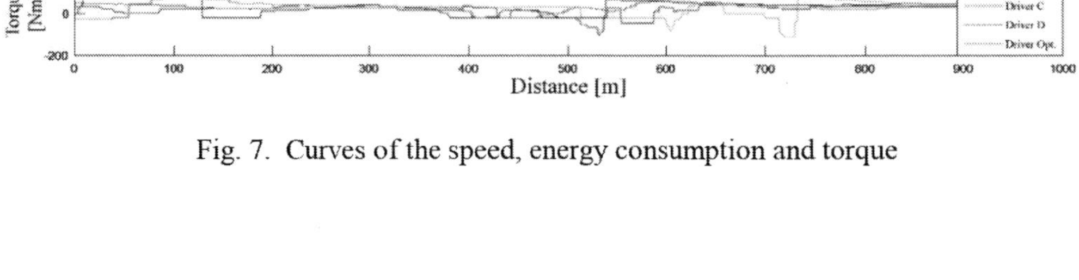

Fig. 7. Curves of the speed, energy consumption and torque

6 Conclusion

Final optimal curves of the speed and energy consumption for the whole measured route is displayed in Fig. 8. With help of the optimisation algorithm of the theory of graphs with using the method of the shortest path, the average energy consumption reached 9.65 kWh/100km. By comparison with the lowest value of average energy consumption measured before optimisation improvement was 3.5%, or more precisely 5 km.

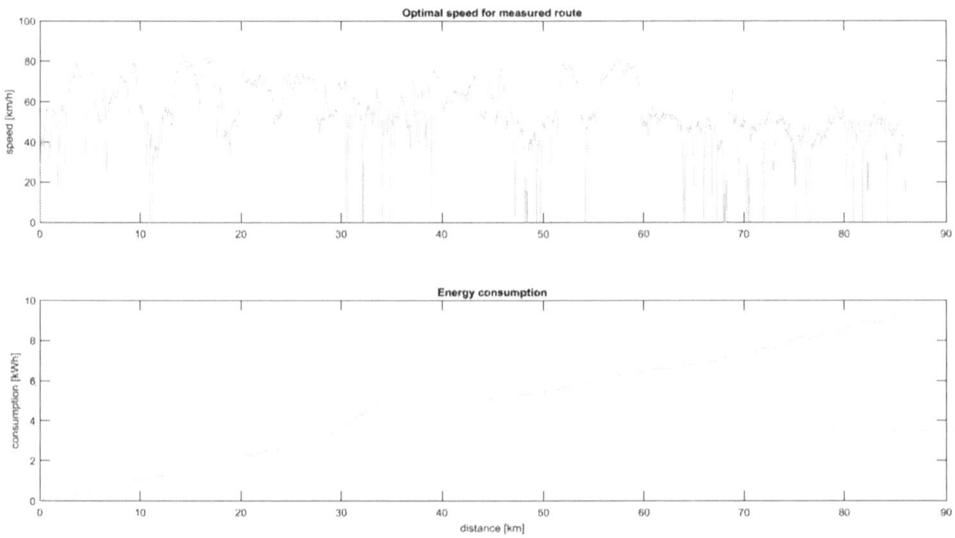

Fig. 8. Curves of the speed, energy consumption and torque

The vehicle completed the route in the most economic mode with deactivated air conditioning, heating, ventilation and all assistant systems of the vehicle. The special tires with a lower rolling-resistance inflated to a maximal allowable pressure were used. Each driver completed the measured route with compliance of the specified criteria, but also with the target reach the lowest value of the average energy consumption. The correctness of the optimisation and ability of the vehicle reaching specified energy consumption will be validated with help of simulation model and also with an experiment in real traffic. This method can bring the improvement up to 7% energy save with the best-experienced driver with combination more crossing in the same circuit.

Acknowledgement

This paper was created with support of VEGA agency within VEGA 1/044/15 project: Výskum možností uplatnenia metód generatívneho konštruovania pri vývoji modulov mobilných pracovných strojov and with support of APVV agency within APVV-15-0524 project: Výskum platformy modulov vybranej skupiny mobilných pracovných strojov, ich optimalizácia metódami generatívneho konštruovania.

References

[1] NOCEDAL, J.; WRIGHT, S. J. Numerical Optimalization. Springer. 651 s., 1999, ISBN 0-387-98793-2

[2] PALÚCH, S. Algoritmická teória grafov. Žilina: Žilinská univerzita v EDIS vydavateľstve ŽU. 2008. 272 s.

[3] MOŤOVSKÝ, M.; GULAN, L.; MAGDOLEN, Ľ. Optimalizácia mechanických vlastností mobilného pracovného stroja. In Zborník príspevkov 42. medzinárodnej konferencie katedier dopravných, manipulačných, stavebných a poľnohospodárskych strojov : Kočovce, 6.-7. 9. 2016. 1. vyd. Bratislava : Vydavateľstvo STU, 2016, s. 157-162. ISBN 978-80-227-4584-0.

[4] IZRAEL, G.; BUKOVECZKY, J., GULAN, L. Utilisation of modeling, stress analysis, kinematics optimisation, and hypothetical estimation of lifetime in the design process of mobile working machines. In Scientific Proceedings Faculty of Mechanical Engineering STU Bratislava : Vol. 19/2011. s.23-28. ISSN 1338-1954.

Current address

Šlauka Michal, Ing.
Slovak University of Technology in Bratislava, Faculty of Mechanical Engineering,
Institute of Transport Technology and Engineering Design
Nám. Slobody 17, 812 31 Bratislava, Slovak Republic
Tel. number: 02/57296429, e-mail: michal.slauka@stuba.sk

Liedl Tomáš, Ing.
Slovak University of Technology in Bratislava, Faculty of Mechanical Engineering,
Institute of Transport Technology and Engineering Design
Nám. Slobody 17, 812 31 Bratislava, Slovak Republic
Tel. number: 02/57296400, e-mail: tomas.liedl@stuba.sk

Milesich Tomáš, Ing. PhD.
Slovak University of Technology in Bratislava, Faculty of Mechanical Engineering,
Institute of Transport Technology and Engineering Design
Nám. Slobody 17, 812 31 Bratislava, Slovak Republic
Tel. number: 02/57296458, e-mail: tomas.milesich@stuba.sk

Slovak University of Technology in Bratislava
Faculty of Mechanical Engineering
16th Conference on Applied Mathematics **APLIMAT 2017**
Proceedings

SEMI-ANALYTICAL APPROACH TO SOLVING PARTIAL DIFFERENTIAL EQUATIONS

ŠMARDA Zdeněk (CZ)

Abstract. In the paper, an analytic solution of partial differential equations is deduced with the help of the differential transform method (DTM). General formalae of $(n+1)$ - dimensional DTM are given. To illustrate the capability and efficiency of the method two examples are solved. The method can be easily applied to many initial and boundary value problems and is capable of reducing the size of computational work.

Keywords: differential transformation method, partial differential equations, initial value problems

Mathematics subject classification: Primary 35A35; Secondary 35A02, 65M15

1 Introduction

It is known that partial differential equations and systems of partial differential equations have many applications in engineering, fluid mechanics, biology, hydrodynamics and physics (for example plasma physics, solid state physics, fluid mechanics), such as Korteweg-de Wries (KdV) equation, mKdV equation, RLW equation, Sine-Gordon equation, Boussinesq equation, Burgers equation, etc. Finding accurate and efficient methods for solving partial differential equations has long been an active research undertaking. Analytical methods commonly used to solve partial differential equations are very restricted and numerical techniques involving discretization of the variables on the other hand gives rise to rounding off errors. In recent years many researchers have applied the semi-analytical methods like the homotopy perturbation method (HPM) [1], [4], the variational iteration method (VIM) [3], [8], the Adomian decomposition method (ADM) [2], [7], the differential transformation method (DTM) [5] to solving initial and boundary value problems for partial differential equations.

In the paper certain initial value problems for partial differential equations are solved using DTM. General formulae of $(n+1)$ - dimensional DTM are given as well. Efficiency and reliability of the presented method is discussed in comparison with HPM and VIM.

2 Differential transformation method

The basic idea of DTM was introduced by Zhou [9] who applied this method to solve linear and nonlinear initial value problems in electric circuit analysis. DTM is a technique that uses Taylor series for the solution of differential equations in the form of a polynomial. It is different from the high-order Taylor series method, which requires symbolic computation of the necessary derivatives and integrals of the data functions. DTM leads to a system of recurrent algebraic equations for obtaining an analytic series solutions of functional equations and overcomes the difficulties of traditional methods.

Suppose that the function $u(\mathbf{x}, t)$, $\mathbf{x} = (x_1, x_2, \dots, x_n)$ is analytic at the point $(\bar{\mathbf{x}}, \bar{t})$, $\bar{\mathbf{x}} = (\bar{x_1}, \bar{x_2}, \dots, \bar{x_n})$. Then $u(\mathbf{x}, t)$ can be represented by the Taylor series

$$u(\mathbf{x}, t) = \sum_{k_1=0}^{\infty} \cdots \sum_{k_n=0}^{\infty} \sum_{h=0}^{\infty} \frac{1}{k_1! \dots k_n! h!} \left[\frac{\partial^{k_1+\dots k_n+h} u(\mathbf{x}, t)}{\partial x_1^{k_1} \dots \partial x_n^{k_n} \partial t^h} \right] \left(\prod_{i=1}^{n} (x_i - \bar{x}_i)^{k_i} \right) (t-\bar{t})^h. \quad (1)$$

The differential transform $U(\mathbf{k}, h)$ at the point $(\bar{\mathbf{x}}, \bar{t})$ is defined as follows:

$$U(\mathbf{k}, h) = \frac{1}{k_1! \dots k_n! h!} \left[\frac{\partial^{k_1+\dots k_n+h} u(\mathbf{x}, t)}{\partial x_1^{k_1} \dots \partial x_n^{k_n} \partial t^h} \right].$$

The differential inverse tranform of $U(\mathbf{k}, h)$ is defined by $u(\mathbf{x}, t)$ in the form

$$u(\mathbf{x}, t) = \sum_{k_1=0}^{\infty} \cdots \sum_{k_n=0}^{\infty} \sum_{h=0}^{\infty} U(\mathbf{k}, h) \left(\prod_{i=1}^{n} (x_i - \bar{x}_i)^{k_i} \right) (t - \bar{t})^h.$$

>From the definition we can easily prove the following formulae:

Lemma 1 *If*

$$u(x_1, \dots, x_n, t) = \frac{\partial g(x_1, \dots, x_n, t)}{\partial x_i}, 1 \le i \le n$$

then

$$U(k_1, \dots, k_n, h) = (k_i + 1) G(k_1, \dots, k_i + 1, \dots, k_n, h).$$

>From here it is easy to deduce a formula for the partial derivative with respect to the variable t.

Lemma 2 *If*

$$u(x_1, \dots, x_n, t) = \frac{\partial^{r+s} g(x_1, \dots, x_n, t)}{\partial x_i^r \partial x_j^s}, \ 1 \le i \ne j \le n,$$

then

$$U(k_1, \dots, k_n, h) = \frac{(k_i + r)!}{(k_i)!} \frac{(k_j + s)!}{(k_j)!} G(k_1, \dots, k_i + r, \dots, k_j + s, \dots, k_n, h).$$

Lemma 3 *If*

$$u(x_1, \dots, x_n, t) = x_1^{h_1} x_2^{h_2}, \dots x_n^{h_n} t^m,$$

then

$$U(k_1, \dots, k_n, h) = \delta(k_1 - h_1)\delta(k_2 - h_2) \dots \delta(k_n - h_n)\delta(h - m),$$

where $\delta(.)$ is the Kronecker function.

In applications we usually use two or three-dimensional DTM. To illustrate efficiency of the presented method we will apply DTM to initial value problems for partial differential equations of two variables. Though general formulae of DTM for functions of $(n+1)$-variables seem to be complicated, obtained recurrent algebraic equations are easily solvable using software Maple or Matlab.

Lemma 4 *If*

$$u(x,t) = f_1(x,t) f_2(x,t),$$

then

$$U(k,h) = \sum_{r=0}^{k} \sum_{s=0}^{h} F_1(r, h-s) F_2(k-r, s).$$

Lemma 5 *If*

$$u(x,t) = \frac{\partial f_1(x,t)}{\partial x} \frac{\partial f_2(x,t)}{\partial t},$$

then

$$U(k,h) = \sum_{r=0}^{k} \sum_{s=0}^{h} (r+1)(k-r+1) F_1(k-r, h-s) F_2(k-r+1, s).$$

3 Application

Example 1 Consider the Klein-Gordon type partial differential equation

$$\frac{\partial^2 u}{\partial t^2} = \frac{\partial^2 u}{\partial x^2} - u \tag{2}$$

subject to the initial conditions

$$u(x,0) = \mathrm{e}^{-x} + x, \quad \frac{\partial u}{\partial t}(x,0) = 0. \tag{3}$$

Applying DTM to equation (2) we get

$$(h+1)h+2)U(k,h+2) = (k+1)(k+2)U(k+2,h) - U(k,h). \tag{4}$$

>From initial conditions we obtain

$$U(k,0) = \begin{cases} 0, & k=1 \\ \frac{(-1)^k}{k!}, & k = 0,2,3,\ldots \end{cases} \qquad U(k,1) = 0, \; k = 0,1,2,\ldots$$

Solving recurrent equations (4) we have

$$k = 0, h = 0 : 2U(0,2) = 2U(2,0) - U(0,0) \Rightarrow U(0,2) = 0.$$
$$k = 1, h = 0 : 2U(1,2) = 6U(3,0) - U(1,0) \Rightarrow U(1,2) = -\frac{1}{2!}.$$
$$k = 0, h = 1 : 6U(0,3) = 2U(2,1) - U(0,1) \Rightarrow U(0,3) = 0.$$
$$k = 1, h = 1 : 6U(1,3) = 6U(3,1) - U(1,1) \Rightarrow U(1,3) = 0.$$
$$k = 2, h = 0 : 2U(2,2) = 12U(4,0) - U(2,0) \Rightarrow U(2,2) = 0.$$
$$k = 0, h = 2 : 12U(0,4) = 2U(2,2) - U(0,2) \Rightarrow U(0,4) = 0.$$
$$k = 3, h = 0 : 2U(3,2) = 20U(5,0) - U(3,0) \Rightarrow U(3,2) = 0.$$
$$k = 1, h = 2 : 12U(1,4) = 6U(3,2) - U(1,2) \Rightarrow U(1,4) = \frac{1}{4!}.$$
$$\vdots$$

Generally we get

$$U(1,h) = \begin{cases} 0, & h = 1,3,5,\dots \\ \frac{(-1)^h}{h!}, & h = 2,4,6,\dots \end{cases} \quad U(k,h) = 0, k = 0,2,3,4,\dots,\ h = 0,2,3,4,\dots$$

The solution of the given initial value problem has the form

$$u(x,t) = \sum_{k=0}^{\infty}\sum_{h=0}^{\infty} U(k,h)x^k t^h = \left(1 + \frac{1}{2!}x^2 - \frac{1}{3!}x^3 + \dots \frac{(-1)^k}{k!}x^k + \dots\right)$$
$$+ x\left(-\frac{1}{2!}t^2 + \frac{1}{4!}t^4 + \frac{(-1)^h}{2h)!}t^{2h} + \dots\right)$$
$$= \left(1 - x + \frac{1}{2!}x^2 - \frac{1}{3!}x^3 + \dots \frac{(-1)^k}{k!}x^k + \dots\right)$$
$$+ x\left(1 - \frac{1}{2!}t^2 + \frac{1}{4!}t^4 + \frac{(-1)^h}{2h)!}t^{2h} + \dots\right) = \mathrm{e}^{-x} + x\cos t.$$

The same initial value problem was solved by Jafari and Aminataei [4] and Mishra and Nagar [6] using the homotopy perturbation method and He-Laplace method respestively. But computational processes in both papers require complicated symbolic calculations of iterative approximations and a suitable selection of a initial guess.

Example 2 Consider the nonlinear nonhomogeneous partial differential equations

$$\frac{\partial^2 u}{\partial t^2} = \frac{\partial^2 u}{\partial x^2} - u^2 + x^2 t^2 \tag{5}$$

subject to the initial conditions

$$u(x,0) = 0, \ \ \frac{\partial u}{\partial t}(x,0) = x. \tag{6}$$

Taking the differential transform of (5), leads to

$$(h+1)h+2)U(k,h+2) = (k+1)(k+2)U(k+2,h)$$
$$- \sum_{r=0}^{k}\sum_{s=0}^{h} U(r,h-s)U(k-r,s) + \delta(k-2)\delta(h-2). \tag{7}$$

>From initial conditions (6) we have

$$U(k,0)=0,\ k=0,1,2,\ldots,\quad U(k,1)=\begin{cases}1, & k=1,\\ 0, & k=0,2,3,4,\ldots\end{cases}$$

Then from recurrent system (7) it follows

$$\begin{aligned}
&k=0,h=0: 2U(0,2)=2U(2,0)-U^2(0,0)\Rightarrow U(0,2)=0.\\
&k=1,h=0: 2U(1,2)=6U(3,0)-2U(1,0)U(0,0)\Rightarrow U(1,2)=0.\\
&k=0,h=1: 6U(0,3)=2U(2,1)-2U(0,1)U(0,0)\Rightarrow U(0,3)=0.\\
&k=1,h=1: 6U(1,3)=6U(3,1)-U(0,1)U(1,0)-2U(0,0)U(1,1)\\
&\qquad\qquad -U(1,0)U(1,1)\Rightarrow U(1,3)=0.\\
&k=2,h=0: 2U(2,2)=12U(4,0)-2U(0,0)U(2,0)-U^2(1,0)\Rightarrow U(2,2)=0.\\
&k=0,h=2: 12U(0,4)=2U(2,2)-2U(0,2)U(0,0)-U^2(0,1)\Rightarrow U(0,4)=0.\\
&\qquad\qquad\vdots
\end{aligned}$$

Generally we obtain

$$U(k,h)=\begin{cases}1, & k=1,h=1,\\ 0, & k\neq 1,h\neq 1.\end{cases}$$

Hence the solution of initial value problem (5),(6) has the form

$$u(x,t)=\sum_{k=0}^{\infty}\sum_{h=0}^{\infty}U(k,h)x^k t^h=xt.$$

The same initial problem was solved by Hilal and Elzaki [3] using the Laplace variational iteration method with help of the Laplace transform and a construction of a correction functional in the integral form which leads to uselessly complicated symbolic calculations.

4 Conclusion

In this paper DTM was used to solving linear and nonlinear partial differential equations. The presented method can be easily applied to many problems and is capable of reducing the size of computational work. The main advantage of DTM is to provide the user an analytical approximation to the solution, in many cases, an exact solution, in a rapidly convergent sequence with elegantly computed terms.

Acknowledgement

The work was supported by the project FEKT-S-14-2200 of Brno University of Technology.

References

[1] Biazar, M., Eslami, M.: *A new homotopy perturbation method for solving systems of partial differential equations,* Comput. Math. Appl. vol. 62(1) (2011) 225–234 http://dx.doi.org/10.1016/j.camwa.2011.04.070.

[2] Jang, B.: *Two-point boundary value problems by the extended Adomian decomposition method,* J. Comput. Appl. Math. vol. 219(1) (2008) 253–262.

[3] Hilal, M. A., Elzaki, T. M.: *Solution of Nonlinear Partial Differential Equations by New Laplace Variational Iteration Method,* vol. 2014, Article ID 790714, 5 pp. http://dx.doi.org/10.1155/2014/790714.

[4] Jafari, M. A., Aminataei, A.: *Improved homotopy perturbation method,* International Mathematical Forum, vol. 5, no. 29-32, (2010) 1567–1579.

[5] Jang, M.J., Chen, C. L., Liu, I. C.: Y.C. *Two-dimensional differential transform for partial differential equations,* Appl. Math. Comput. vol. 121 (2001) 261–270.

[6] Mishra, H. K., Nagar, A. K.: *He-Laplace Method for Linear and Nonlinear Partial Differential Equations,* J. Appl. Math. vol. 2012, Article ID 180315, 16 pp. doi:10.1155/2012/180315.

[7] Mohamed, M. A.: *Comparison Differential Transformation Technique with Adomian Decomposition Method for Dispersive Long-wave Equations in (2+1)-Dimensions,* Appl. Appl. Math. vol. 5(2) (2010) 148 –166, http://pvamu.edu/aam.

[8] Wazwaz, A. M.: *The variational iteration method for solving linear and nonlinear systems of PDEs,* Comput. Math. Appl. vol. 54 (2007) 895–902.

[9] Zhou, J. K.: *Differential transformation and its application for electrical circuits,* Huazhong University press, China, (1986)(in Chinese).

Current address

Šmarda Zdeněk
Brno University of Technology
Technická 10, 616 00 Brno
Tel. number: +4205411456099, e-mail: smarda@feec.vutbr.cz

Slovak University of Technology in Bratislava
Faculty of Mechanical Engineering
16th Conference on Applied Mathematics **APLIMAT 2017**
Proceedings

THE REGULARIZATION BY CONSTANT FUNCTIONS OF LAGRANGIANGS LINEAR IN FIRST DERIVATIVES IN FIELD THEORY

SMETANOVÁ Dana (CZ)

Abstract. In this paper we study case of the Lagrangians affine in first derivatives in first order field theory (for example Dirac field Lagrangian is of this type). By the regularization procedure we mean the process how to find appropriate Lepagean equivalent for given Lagrangian (in the sense of the geometric meaning of the regularity). The theory is illustrated on an example in dimension 4. The relations to multisymplectic forms are discussed.

Keywords: Lagrangian, Dirac field Lagrangian, Poincaré–Cartan form, Lepagean equivalent of the Lagrangian, Euler–Lagrange equations, Hamilton equations

Mathematics subject classification: Primary 49S05, 53Z05; Secondary 35R01

1 Introduction

In this paper we study regularity condition in the geometric sense as condition which guarantees an equivalence between sets of Euler–Lagrange and Hamilton equation. Standard Hamiltonian formulation of field theory is found upon the Poincaré–Cartan form. Keeping the requirement on equivalence of the Hamilton and Euler–Lagrange equations as a (geometric) definition of regularity, and considering more general Lepagean equivalents of a Lagrangian then the Poincaré–Cartan equivalent, we obtain a regularity condition, depending not only on a Lagrangian but also on "free" functions of its Lepagean equivalents.

In this way one gets a possibility to "regularize" many Lagrangian systems which are affine, linear and quadratic in first derivatives (singular in the standard sense). Especially, we study the case of the Lagrangians affine in first derivatives.

This theory is applied as regularization by a constant to concrete example in dimension 4. This procedure is possible to use for arbitrary Lagrangian which is same type as Dirac field.

The relations to multisymplectic forms are discussed.

2 Fibred manifolds and Lepagean equivalents

Now we recall the basic tools and notations. Note that in the paper the Einstein summation convention is used.

A surjective submersion $\pi : Y \to X$ is called a *fibred manifold* with a *base* X, $\dim X = n$ and a *total space* Y, $\dim Y = m + n$. For every point $x \in X$, the submanifold $\pi^{-1}(x) \in Y$ is called a *fibre* over x ($\dim \pi^{-1}(x) = m$).

To every fibred chart (V, ψ), $\psi = (x^i, y^\sigma)$, $1 \leq i \leq n$, $1 \leq \sigma \leq m$, on π there exist an *associated fibred chart* on J^1Y, denoted by (V_1, ψ_1), $\psi_1 = (x^i, y^\sigma, y^\sigma_j)$.

We recall Saunders [7] for more details on calculus of fibred manifold and its associated structures.

Definition. (Krupka [2]) An n-form ρ on J^rY is called a *Lepagean equivalent of a Lagrangian* λ (resp. Lepagean n-form) if
1) $h\rho = \lambda$, $\lambda = L\omega_0$,
2) $(n+1)$-form $p_1(d\rho)$ is a $\pi_{r+1,0}$-horizontal form.

For a first order Lagrangian we have all its Lepagean equivalents of order 1 characterized by the following formula

$$\pi_{r+1,r}{}^*\rho = \theta_\lambda + \mu, \tag{1}$$

where θ_λ is the Poincaré–Cartan form of λ,

$$\theta_\lambda = L\omega_0 + \frac{\partial L}{\partial y^\sigma_j}\omega^\sigma \wedge \omega_j,$$

where $\omega_0 = dx^1 \wedge \cdots \wedge dx^n$, $\omega^\sigma = dy^\sigma - y^\sigma_i dx^i$, $\omega_j = i_{\frac{\partial}{\partial x^j}}\omega_0$ and μ is arbitrary at least two contact part.

It is possible that the contact part μ does not depend on the Lagrangian. Note that this part contains the "free" functions (see next paragraph).

A section γ of π is an extremal of λ if and only if

$$J^{2r-1}\gamma^* i_{J^{2r-1}\xi} d\rho = 0 \tag{2}$$

for every π-vertical vector field ξ on Y.

A section δ of the fibered manifold π_{2r-1} is called a *Hamilton extremal* of ρ if

$$\delta^* i_\xi d\rho = 0 \tag{3}$$

for every π_{2r-1}-vertical vector field ξ on $J^{2r-1}Y$.

The equations (2) are called the *Euler–Lagrange equations* and (3) the *Hamilton equations of* ρ, respectively. Notice that while the Euler–Lagrange equations are uniquely determined by λ, Hamilton equations depend upon a choice of μ. Consequently, one gets many different Hamilton theories associated to a given variational problem.

Definition. [4] A Lepagean form ρ on J^1Y is called *regular* if every its Hamilton extremal is holonomic (i.e., $\delta = J^1\gamma$, where γ is an extremal of the Lagrangian).

3 Regularization procedure

In the sequel we shall consider Lepagean forms (1) where μ is 2-*contact*,

$$\rho = L\omega_0 + \frac{\partial L}{\partial y^\sigma_j}\omega^\sigma \wedge \omega_j + g^{ij}_{\sigma\nu}\,\omega^\sigma \wedge \omega^\nu \wedge \omega_{ij}, \tag{4}$$

(summation over all sequences of indices) where $\omega_0 = dx^1 \wedge \cdots \wedge dx^n$, $\omega^\sigma = dy^\sigma - y^\sigma_i dx^i$, $\omega_j = i_{\frac{\partial}{\partial x^j}}\omega_0$, $\omega_{ij} = i_{\frac{\partial}{\partial x^j}}\omega_i$ and where the functions $g^{ij}_{\sigma\nu}$ do not depend on the y^κ_l's and satisfy the conditions

$$g^{ij}_{\sigma\nu} = -g^{ij}_{\nu\sigma}, \quad g^{ij}_{\sigma\nu} = -g^{ji}_{\sigma\nu}, \quad g^{ij}_{\sigma\nu} = g^{ji}_{\nu\sigma}. \tag{5}$$

Note that equations (5) mean that only

$$\binom{m}{2}\cdot\binom{n}{2} = \frac{1}{4}mn(m-1)(n-1)$$

of the $m^2 \cdot n^2$ functions $g^{ij}_{\sigma\nu}$ are independent.

Theorem. (Krupková and Smetanová [4])
Let λ be a first-order Lagrangian, let $\lambda = L\omega_0$ be its expression in a fiber chart (V, ψ), $\psi = (x^i, y^\sigma)$ on Y. Let ρ be a Lepagean equivalent of λ of the form (4), (5). *Assume that the matrix*

$$A^{ij}_{\sigma\nu} = \left(\frac{\partial^2 L}{\partial y^\sigma_i \partial y^\nu_j} - 4g^{ij}_{\sigma\nu}\right) \tag{6}$$

with rows (resp. columns) labelled by the pair (σ, i) (resp. (ν, j)), is regular.

Then ρ is regular, i.e., every Hamilton extremal δ of ρ is of the form $\delta = J^1\gamma$, where γ is an extremal of λ.

The above regularity condition (6) means that Hamilton equations and the Euler–Lagrange equations are equivalent.

By the regularization procedure we understand the process how to find appropriate Lepagean equivalent for given Lagrangian (in the sense of the above condition of the regularity).

In general case the functions $g^{ij}_{\sigma\nu}$ might depend upon the variables x^i, y^σ. An interesting case is regularization by constant functions $g^{ij}_{\sigma\nu}$.

4 Lagrangians affine in the first order variables

In this section we study the Lagrangians affine in the first derivatives, i.e. its expression has the following form $\lambda = (L_0 + L^j_\sigma y^\sigma_j)\omega_0$, where the functions L_0, L^j_σ do not depend on y^σ_j. It is easily seen that for arbitrary indices i, j, σ, ν the above Lagrangians satisfy the conditions $\frac{\partial^2 L}{\partial y^\sigma_i \partial y^\nu_j} = 0$. Hence these Lagrangians are not regular in the standard sense.

Theorem.
Let λ be a first-order Lagrangian affine in the variables y^σ_j, let $\lambda = (L_0 + L^j_\sigma y^\sigma_j)\omega_0$, where the

functions L_0, L^j_σ do not depend on y^σ_j, be its expression in a fiber chart (V,ψ), $\psi = (x^i, y^\sigma)$ on Y. Let ρ be a Lepagean equivalent of λ of the form (4), (5). *Assume that the matrix*

$$A^{ij}_{\sigma\nu} = \left(-4g^{ij}_{\sigma\nu}\right) \tag{7}$$

with rows (resp. columns) labelled by the pair (σ, i) (resp. (ν, j)), is regular.

Then ρ is regular, i.e., every Hamilton extremal δ of ρ is of the form $\delta = J^1\gamma$, where γ is an extremal of λ.

The above theorem is the direct consequence of the theorem (Krupková and Smetanová). It means that the regularization procedure for first order Lagrangian affine in the variables y^σ_j corresponds with problem how to find the coefficients $g^{ij}_{\sigma\nu}$ (i.e., find appropriate regular matrix $(-4g^{ij}_{\sigma\nu})$).

The concrete physical Lagrangian of this type is Dirac field Lagrangian as Lagrangian linear in the variables y^ν_i, with the global coordinates denoted by $(x^\mu, \psi, \bar\psi, \partial_\mu\psi, \partial_\mu\bar\psi)$, $\mu = 1, 2, 3, 4$. Lagrangian takes the form $\lambda = L\omega_0$, where

$$L = \frac{i}{2}\left(\bar\psi\gamma^\mu\partial_\mu\psi + \partial_\mu\bar\psi\gamma^\mu\psi\right) - \bar\psi m\psi$$

and it is regularizable by the above way.

Remark. Relations to the multimplectic theory.

Recall that an $(n+1)$-form Ω on a manifold M is called *multisymplectic* if it is *closed*, and the map $\xi \rightarrow i_\xi\Omega$, mapping vector fields on M to n-forms, is *injective* (Gotay [1]).

A *closed (n+1)-form* α on J^sY will be called *Lepagean* if $p_1\alpha$ is a dynamical form (i.e., $p_1\alpha = E_\sigma\omega^\sigma \wedge \omega_0$, c.f. Krupková [3], Krupková and Smetanová [6]).

We study special type of Lepagean $(n+1)$-form

$$\alpha = d\varrho, \tag{8}$$

where ϱ has following expression (4), (5):

$$\varrho = L\omega_0 + \frac{\partial L}{\partial y^\sigma_j}\omega^\sigma \wedge \omega_j + g^{ij}_{\sigma\nu}\,\omega^\sigma \wedge \omega^\nu \wedge \omega_{ij}, \tag{9}$$

where $\omega_0 = dx^1 \wedge \cdots \wedge dx^n$, $\omega^\sigma = dy^\sigma - y^\sigma_i dx^i$, $\omega_j = i_{\frac{\partial}{\partial x^j}}\omega_0$, $\omega_{ij} = i_{\frac{\partial}{\partial x^j}}\omega_i$, $\lambda = L\omega_0$ is a first order Lagrangian and where the functions $g^{ij}_{\sigma\nu}$ do not depend on the y^κ_l's and satisfy the conditions

$$g^{ij}_{\sigma\nu} = -g^{ij}_{\nu\sigma}, \quad g^{ij}_{\sigma\nu} = -g^{ji}_{\sigma\nu}, \quad g^{ij}_{\sigma\nu} = g^{ji}_{\nu\sigma}. \tag{10}$$

The following Theorem provides sufficient condition to be Lepageagen forms (8), (4) and (5) multisymplectic forms.

Theorem. (Smetanová [9])
Let dim $X \geq 2$. Let α be the Lepagean $(n+1)$-form satisfying the conditions (8), (4) and (5).

Suppose that the form $\eta = g^{ij}_{\sigma\nu}\, dy^\sigma \wedge dy^\nu \wedge \omega_{ij}$ is closed then the form α is multisymplectic on J^1Y.

Proposition. *Let dim $X \geq 2$. Let α be the Lepagean $(n+1)$-form satisfying the conditions (8), (4) and (5). Suppose that the form $\eta = g^{ij}_{\sigma\nu}\, dy^\sigma \wedge dy^\nu \wedge \omega_{ij}$ has a constant coefficients $g^{ij}_{\sigma\nu}$ then the form α is multisymplectic on J^1Y.*

Proof of the above prosition follows from fact that the form $\eta = g^{ij}_{\sigma\nu}\, dy^\sigma \wedge dy^\nu \wedge \omega_{ij}$ with a constant coefficients $g^{ij}_{\sigma\nu}$ is closed (c.f. Theorem Smetanová [9]).

How to find the regular matrix $(-4g^{ij}_{\sigma\nu})$, we can see on concrete example in next section. The different examples of regularization by functions or constant are given in [4], [5], [8].

5 Regularization by constant functions

We study a special case of the affine Lagrangians. In this case we have $X = \mathbb{R}^4$, $Y = \mathbb{R}^2 \times \mathbb{R}^4$, i.e. $J^1Y = \mathbb{R}^2 \times \mathbb{R}^4 \times \mathbb{R}^8$, with the global coordinates on J^1Y, denoted by (V_1, ψ_1), $\psi_1 = (x^i, y^\sigma, y^\sigma_i)$, where $1 \leq i \leq 4$, $1 \leq \sigma \leq 2$.

Let λ be a first-order Lagrangian affine in the variables y^σ_j, let $\lambda = (L_0 + L^j_\sigma y^\sigma_j)\omega_0$, where the functions L_0, L^j_σ do not depend on y^σ_j, be its expression in a fiber chart.

In this case the conditions (5) on the $g^{ij}_{\sigma\nu}$'s mean that we have only 6 independent free functions g^{12}_{12}, g^{13}_{12}, g^{14}_{12}, g^{23}_{12}, g^{24}_{12} and g^{34}_{12}.

Denote $a = 4g^{12}_{12}$, $b = 4g^{13}_{12}$, $u = 4g^{14}_{12}$, $v = 4g^{23}_{12}$, $c = 4g^{24}_{12}$, $d = 4g^{34}_{12}$.

For arbitrary indices i, j, σ, ν the above Lagrangians satisfy the conditions $\frac{\partial^2 L}{\partial y^\sigma_i \partial y^\nu_j} = 0$. The regularization procedure means that we need to find the constants u, v, a, b, c, d such that the matrix (6) is regular, i.e. $\det(-4g^{ij}_{\sigma\nu}) \neq 0$

Let us choose for simplicity $u \neq 0$, $v \neq 0$, $a = 0$, $b = 0$, $c = 0$, $d = 0$ constant functions (in general a, b, c, d, u, v might depend upon the variables x^i and y^σ).

The regularity condition (6) reads

$$\det \begin{pmatrix} 0 & 0 & 0 & 0 & 0 & 0 & 0 & -u \\ 0 & 0 & 0 & 0 & 0 & 0 & -v & 0 \\ 0 & 0 & 0 & 0 & 0 & v & 0 & 0 \\ 0 & 0 & 0 & 0 & u & 0 & 0 & 0 \\ 0 & 0 & 0 & u & 0 & 0 & 0 & 0 \\ 0 & 0 & v & 0 & 0 & 0 & 0 & 0 \\ 0 & -v & 0 & 0 & 0 & 0 & 0 & 0 \\ -u & 0 & 0 & 0 & 0 & 0 & 0 & 0 \end{pmatrix} \neq 0.$$

The Lepagean equivalent (4), (5) of the Lagrangian has the form

$$\begin{aligned} \rho &= (L_0 + L^i_\sigma y^\sigma_i)\, dx^1 \wedge dx^2 \wedge dx^3 \wedge dx^4 + L^i_\sigma\, \omega^\sigma \wedge \omega_i \\ &+ u\, \omega^1 \wedge \omega^2 \wedge \omega_{14} + v\, \omega^1 \wedge \omega^2 \wedge \omega_{23}, \end{aligned}$$

where $\sigma = 1, 2$, $i = 1, 2, 3, 4$.

For our choice of u, v, a, b, c, d the corresponding Hamilton equations are *equivalent* with the Euler–Lagrange equations.

Notice that for every fixed $u \neq 0$ and $v \neq 0$ we have obtained Hamilton equations which are equivalent with the Euler-Lagrange equations.

By analogous way we can obtain the different regularization of the Lagrangians (for example choice $u = 0$, $v = 0$, $a \neq 0$, $b = 0$, $c = 0$, $d \neq 0$). The different choice of the constants satisfying the condition (6) (i.e. choice of the Lepagean equivalent) admits new system of the Hamilton equations. But every of the systems Hamilton equations are equivalent with the original Euler–Lagrange equations.

References

[1] Gotay, M. J. A multisymplectic framework for classical field theory and the calculus of variations, I. covariant Hamiltonian formalism, in: M. Francaviglia, D. D. Holm (Eds.), *Mechanics, Analysis and Geometry: 200 Years After Lagrange*, North-Holland, Amsterdam, 1990, pp. 203–235.

[2] D. Krupka, Some geometric aspects of variational problems in fibered manifolds, *Folia Fac. Sci.* Nat. UJEP Brunensis 14 (1973) 1–65.

[3] Krupková, O., Hamiltonian field theory, *J. Geom. Phys.* 43 (2002), 93–132.

[4] O. Krupková and D. Smetanová, On regularization of variational problems in first-order field theory, In *Proceedings of the 20th Winter School "Geometry and Physics"* (Srní, 2000). Rend. Circ. Mat. Palermo (2) Suppl. No. 66, (2001), 133–140.

[5] O. Krupková and D. Smetanová, Legendre transformation for regularizable Lagrangians in field theory, *Letters in Math. Phys.* 58 (2001) 189–204.

[6] Krupková, O., Smetanová, D., Lepage equivalents of second order Euler-Lagrange forms and the inverse problem of the calculus of variations, *J. Nonlin. Math. Phys.* 16 (2009) 1–16

[7] D. J. Saunders, *The Geometry of Jet Bundles* (London Math. Soc. Lecture Notes Series 142, Cambridge University Press, Cambridge, 1989).

[8] D. Smetanová, On regularization procedure of Lagrangians linear in first derivatives in first-order field theory, In *Proceedings of the 22nd colloquium "Modern Mathematical Methods in Engineering Czech–Polish Colloquium (3mi)"* (Horní Lomná, 2013), Department of Mathematics and Descriptive Geometry (2013), 119–122.

[9] D. Smetanová, On Legagean equivalents and multisymplectic forms, In *15th Conference on Applied Mathematics*, APLIMAT 2016, Bratislava, Slovak University of Technology in Bratislava, 2016, 1004–1009

Current address

Smetanová Dana, RNDr. Ph.D.
Department of Informatics and Natural Sciences,
The Institute of Technology and Bussiness in České Budějovice
Okružní 517/10, 370 01, České Budějovice, Czech Republic
Tel. number: +420 387 842 133, e-mail: smetanova@mail.vstecb.cz

Slovak University of Technology in Bratislava
Faculty of Mechanical Engineering
16th Conference on Applied Mathematics **APLIMAT 2017**
Proceedings

VALIDATION AND CONVERGENCE OF THE ADAPTIVE MONTE CARLO METHOD TOWARD THE LAW OF PROPAGATION OF UNCERTAINTY APPLIED ON SPECIFIC POINT OF CHOSEN SUBSCALE OF ITS-90

SOPKULIAK Peter (SK), PALENČÁR Rudolf (SK), PALENČÁR Jakub (SK), SUROVIAK Emil (SK)

Abstract. The article briefly describes the approach of evaluating calibration using the adaptive method of Monte Carlo and the subsequent validation by the law of uncertainties when applied on the primary realization of the temperature scale, with emphasis on measurement with standard platinum resistance thermometer (SPRT) illustrated by the range (0 ÷ 660) °C of the international temperature scale (ITS-90).

Keywords: Uncertainty, Correlation, Monte Carlo method, International temperature Scale 1990 (ITS-90) Standard Platinum Resistance Thermometer

Mathematics Subject Classification: Primary 65C05

Prologue

One of the most measured quantities, which undoubtedly has a dominant influence in all sectors of the national economy, is the temperature. The measurement of temperature has a significant effects on the quality and efficiency of each production process and ultimately an accuracy of temperature measurement plays more important role in this process. Temperature measurement can be carried out with devices with different levels of accuracy. For listed reasons, as well as taking into account the requirements of the practice is essential to improve temperature measurement at the international level. Many tasks today without a deepening degree of knowledge in science, in technology and in the development of living standards could not be addressed without a fundamental knowledge of the theory of measurement, measuring equipment, physical principle of sensors, their attributes and metrological characteristics.

1 Current status of the issue

The vast majority of published papers submits an approach based on the distribution of uncertainty [2] in not-matrix form, such as [3]. The authors also predict uncorrelated resistance between the defining fixed point (DFP) and generally neglect the impact of correlations. Omitting the correlations between resistance in the DFP does not always correspond to reality and in accordance [4] they can have a significant impact.

The deviation equations model and ratio of the resistance by which the ITS-90 is defined are non-linear, which may cause some doubts with regard to the adequacy of linearization through the Taylor's series expansion in first order terms. Also, there may occur complications with the determination of the sensitivity coefficients. In such cases, Monte Carlo method based on the principle of the propagation of distributions is preferred to use. Articles based on the Monte Carlo method, which follows the recommendations and procedures listed in [5] appear sporadically. The basic principle of both methods is shown on Fig. 1.

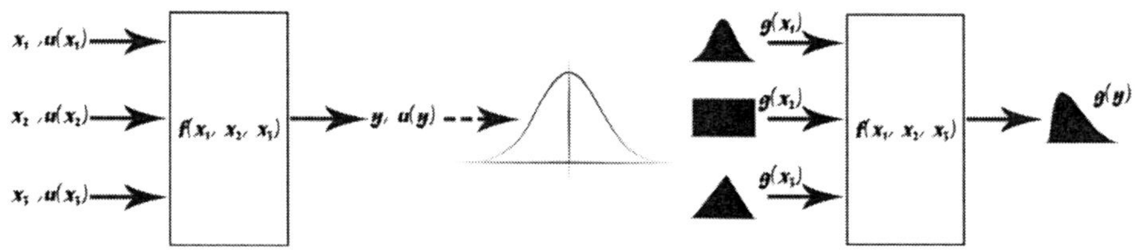

Fig. 1. The law of propagation of uncertainty (right) and the law of propagation of distribution (left)

2 The law of propagation of uncertainty

Whereas the procedure for evaluating by the law of propagation of uncertainty is generally considered to be well executed, its results are used in the validation by the law of propagation of distributions, thus it is possible to compare the results of both methods of calibration. The values used for final validation are from specific calibration of standard platinum resistance thermometer carried by the Slovak Institute of Metrology and evaluated according to the law of propagation of uncertainty. The calibration procedure based on the law of propagation of uncertainty is described e.g. in [6]. The detailed procedure for calculating the uncertainty of each resistor is listed in [7]. Tab. 1 and Tab. 2 shows the values obtained by the calibration procedure based on the law of propagation of uncertainty [6], these values serve also as an input into the evaluation based on the law of propagation of distribution, whereas it is necessary to apply the evaluation on data obtained during the calibration of standard platinum resistance thermometer. Tab. 2. Ratio of resistance in defining fixed point shows the considered correlations and from them resulting covariance between resistances. Tab. 3. Correlation coefficients and covariance
shows a specific value of resistance used in the presented example.

DFP (-)	Resistance R (Ω)	Standard uncertainty of resistance $u(R)$ (Ω)
Sn	46,9397533	$3,85 \cdot 10^{-5}$
Zn	46,7056752	$4,98 \cdot 10^{-5}$
Al	83,7191875	$6,32 \cdot 10^{-5}$
TPW_{Sn}	24,8002001	$1,27 \cdot 10^{-5}$
TPW_{Zn}	24,8001927	$1,27 \cdot 10^{-5}$
TPW_{Al}	24,8001872	$1,27 \cdot 10^{-5}$
TPW^{*}	24,8001933	$1,27 \cdot 10^{-5}$
		*average value from TPW_{SN}, TPW_{ZN}, TPW_{Al}

Tab. 1. Measured values of resistance in defining fixed point

DFP	W_{DFP}	W_{rDFP}
Sn	1,8927167164	1,89279768
Zn	2,5687572661	2,5689173
Al	3,3757482081	3,3760086

Tab. 2. Ratio of resistance in defining fixed point

Correlation coefficients	Value
$r(R_{TPW\,Sn}, R_{TPW\,Zn})$	0,999
$r(R_{TPW\,Sn}, R_{TPW\,Al})$	0,999
$r(R_{TPW\,Zn}, R_{TPW\,Al})$	0,999
$r(R_{TPW\,Sn}, R_{TPW})$	0,999
$r(R_{TPW\,Zn}, R_{TPW})$	0,999
$r(R_{TPW\,Al}, R_{TPW})$	0,999
Other correlations are zero	

Covariance	Value of covariance
$u(R_{TPW\,Sn}, R_{TPW\,Zn})$	$1,6112871 \cdot 10^{-10}$
$u(R_{TPW\,Sn}, R_{TPW\,Al})$	$1,6112871 \cdot 10^{-10}$
$u(R_{TPW\,Zn}, R_{TPW\,Al})$	$1,6112871 \cdot 10^{-10}$
$u(R_{TPW\,Sn}, R_{TPW})$	$1,6112871 \cdot 10^{-10}$
$u(R_{TPW\,Zn}, R_{TPW})$	$1,6112871 \cdot 10^{-10}$
$u(R_{TPW\,Al}, R_{TPW})$	$1,6112871 \cdot 10^{-10}$
Other covariance are zero	

Tab. 3. Correlation coefficients and covariance

R_i (Ω)	t_i (°C)	$u(t_i)$ (°C)
42,02266855	178,96151	$5,8473 \cdot 10^{-4}$

Tab. 4. Evaluation of specific resistance by the law of propagation of uncertainty

3 Application of the Monte Carlo method

When selecting a suitable software environment, it was crucial to create an application which would be easy to use and portable. It was also necessary to consider the efficiency of the calculation of final application and the way of implementation of Marsenne Twister generator. For these reasons the Microsoft Visual Basic .NET programming environment it was chosen, and 32 bit version of operating system has been used due to direct compatibility with newer 64-bit operating systems.

a. Application of Monte Carlo method while calibrating the temperature scale

In order to compare the results of both methods of calibration, it is necessary for the evaluation based on the Monte Carlo method apply on the data obtained by the calibration carried out by the Slovak Institute of Metrology. The values used for evaluating the calibration EOST based on the Monte Carlo method are shown in Tab. 1 and Tab. 2. Calibration software is designed for the ITS-90 sub-range of 0 °C to 660 °C. The temperature T measured by the standard platinum resistance thermometer within the given range will be determined by the relevant inverse function and deviation coefficients a, b, c can be determined from the three known solutions of equation (3.1).

$$W(T_{90}) = \frac{R(T_{90})}{R(273{,}16\ \mathrm{K})} \tag{3.1}$$

b. The calculation of output characteristics

Estimation of the temperature t and its assigned standard uncertainty is a typical representative of the calibration model with one output variable, which process of determination is schematically illustrated on Fig. 3. The functions f_8 up to f_{11} can be represented as follows

$$W = f_8(R_{\mathrm{i}}, R_{\mathrm{TPW}}),\quad \Delta W = f_9(\boldsymbol{a}, W),\quad W_{\mathrm{r}} = f_{10}(\Delta W_{\mathrm{r}}),\quad t = f_{11}(W_{\mathrm{r}}) \tag{3.2}$$

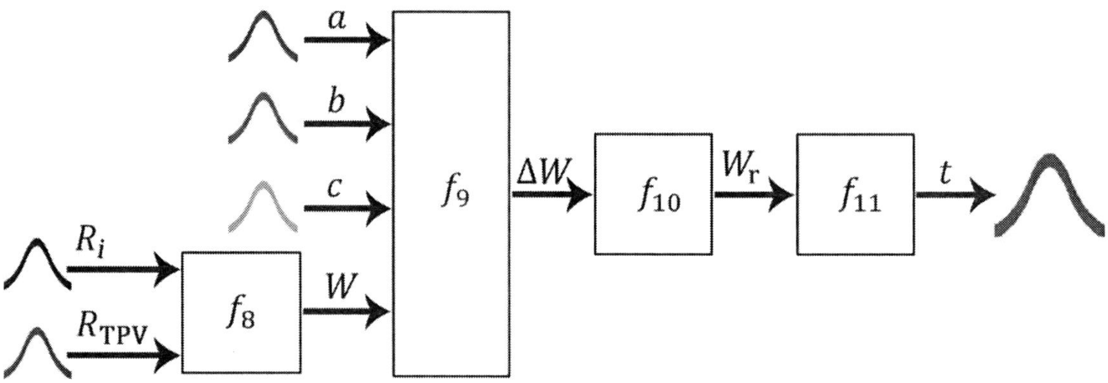

Fig. 2. Model of temperature calculation and its associated standard uncertainty

Inputs to the process are the coefficients a, b, c of the deviation equation in vector format representing the calibration submodel with multiple output variables. Their generated values out of 26 sequences are shown in the histograms below. Furthermore, into the process enters the measured resistance R_i at this specific example it is a scalar vector with a calibrated value of resistance $R \sim 42{,}02266855$ and resistance R_{TPW} which we obtain out of generated values from the normal distribution using the estimate and the corresponding standard deviation (standard uncertainty). Based on the generated input values for resistors and using appropriate relationship we get for our case with $h = 26$ and $M = 10^5$ estimate of the temperature t (°C).

$$t \begin{bmatrix} 178{,}9587285 \\ \vdots \\ 178{,}9638502 \end{bmatrix} = \tag{3.3}$$

Determining the symmetrical reference interval with the specified probability for the estimating output quantity t is obtained from its generated discrete representation (distribution function) by arranging values t at non-decreasing sequence and rules listed in [5]. Using the above we get the following symmetric confidence interval

$$t = [y_{\min} ; y_{\max}] = [178{,}9603671618 ; 178{,}9626585301] \tag{3.4}$$

and following narrowest confidence interval

$$t = [y_{\min} ; y_{\max}] = [178{,}9603671604 ; 178{,}962658518] \tag{3.5}$$

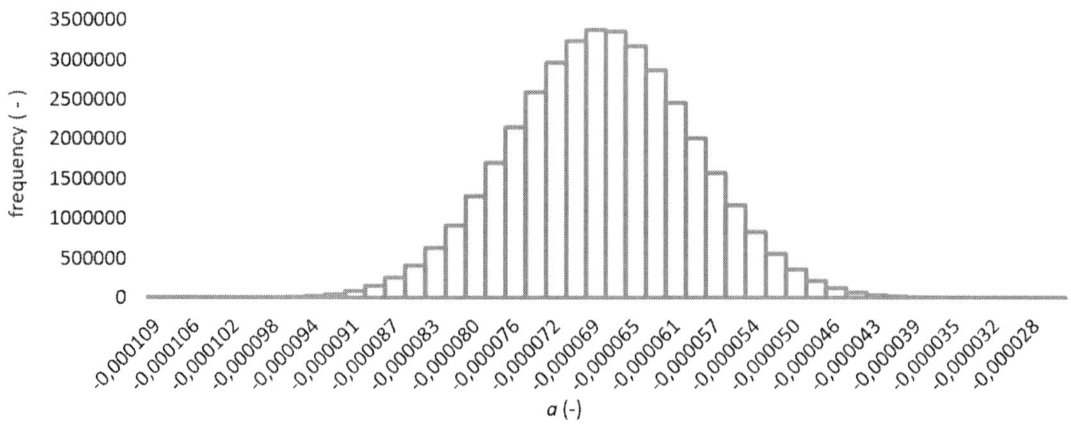

Fig. 3. Histogram of the deviation equation for coefficient a

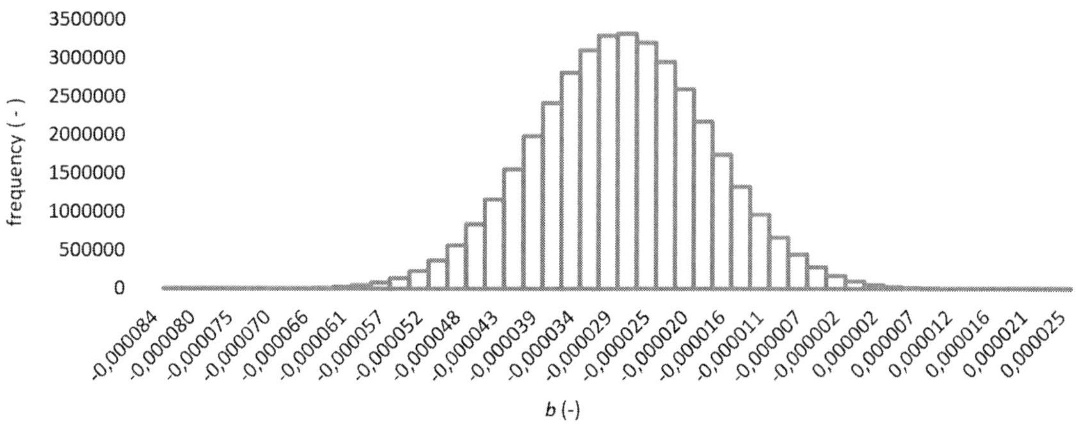

Fig. 4. Histogram of the deviation equation for coefficient b

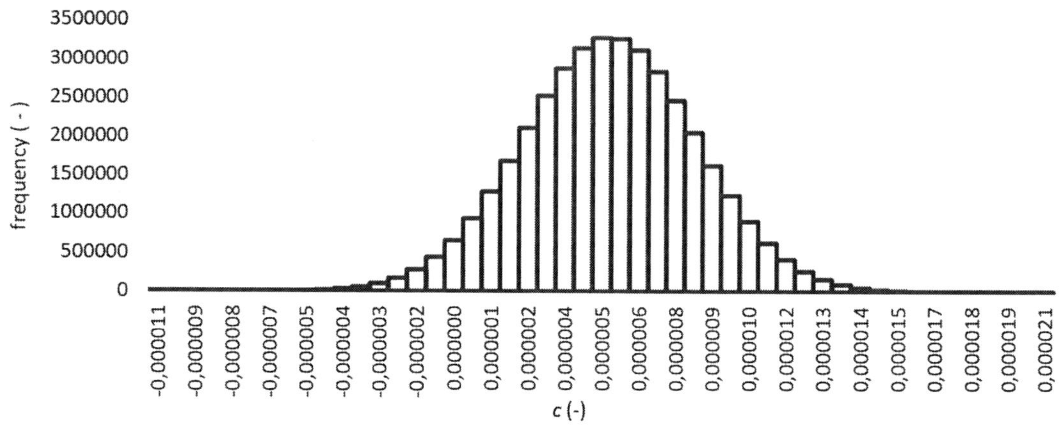

Fig. 5. Histogram of the deviation equation for coefficient c

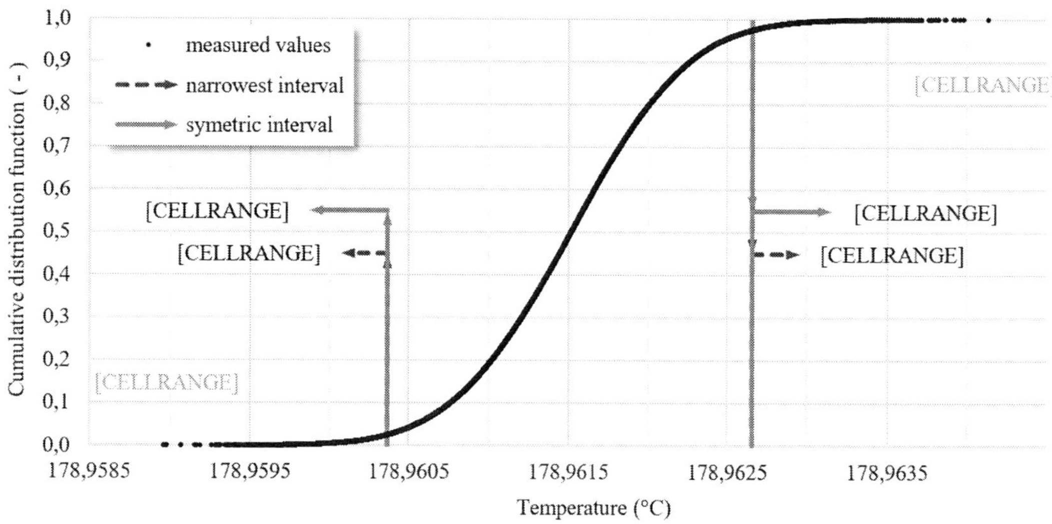

Fig. 6. The distribution function of the output temperature

Statistical characteristics of the resulting estimates are calculated from the partial estimates in each sequence $y^{(h)}$, $u(y^{(h)})$. After carrying out the last sequence h it is possible to calculate the resulting parameters for estimation, where for this case it will

$$\hat{y} = y = \frac{1}{h}\sum_{i=1}^{h} y^{(i)} = 178{,}9615127\ °\mathrm{C} \tag{3.6}$$

$$s_{\hat{y}} = \sqrt{\frac{1}{h(h-1)}\sum_{i=1}^{h}(y^{(i)} - \hat{y})^2} = 3{,}07035 \cdot 10^{-7}\ °\mathrm{C} \tag{3.7}$$

The standard uncertainty

$$\hat{u}(y) = \frac{1}{h}\sum_{i=1}^{h} u(y^{(i)}) = 5{,}848 \times 10^{-4} \tag{3.8}$$

$$s_{\hat{u}(y)} = \sqrt{\frac{1}{h(h-1)}\sum_{i=1}^{h}\left(u(y^{(i)}) - \hat{u}(y)\right)^2} = 2{,}22236 \times 10^{-7}\ °\mathrm{C} \tag{3.9}$$

The lower end point of the narrowest confidence interval

$$\hat{y}_{\min} = \frac{1}{h}\sum_{i=1}^{h} y_{\min}^{(i)} = \frac{1}{26}\left(y_{\min}^{(1)} + y_{\min}^{(2)} + \ldots + y_{\min}^{(26)}\right) = 178{,}960367\ °\mathrm{C} \tag{3.10}$$

$$s_{\hat{y}_{\min}} = \sqrt{\frac{1}{h(h-1)}\sum_{i=1}^{h}\left(y_{\min}^{(i)} - \hat{y}_{\min}\right)^2} = 2{,}46977 \times 10^{-6}\ °\mathrm{C} \tag{3.11}$$

The upper end point of the narrowest confidence interval

$$\hat{y}_{\max} = \frac{1}{h}\sum_{i=1}^{h} y_{\max}^{(i)} = \frac{1}{2}\left(y_{\max}^{(1)} + y_{\max}^{(2)}\right) = 178{,}9626585\ °\mathrm{C} \tag{3.12}$$

$$s_{\hat{y}_{\max}} = \sqrt{\frac{1}{h(h-1)}\sum_{i=1}^{h}\left(y_{\max}^{(i)} - \hat{y}_{\max}\right)^2} = 2{,}23996 \times 10^{-6}\ °\mathrm{C} \tag{3.13}$$

Numerical tolerance δ expressed by the standard uncertainty $\hat{u}(y)$ will

$$\delta = \frac{1}{2} \times 10^{-5} = 0{,}000005 = 5 \times 10^{-6} \tag{3.14}$$

Stabilization criterion determines whether it is necessary to increase the current value of the sequence $h>2$ calculation of Monte Carlo method to the next sequence, in the case if one of the values $2s_{\hat{y}}$, $2s_{\hat{u}(y)}$, $2s_{\hat{y}_{\min}}$ and $2s_{\hat{y}_{\max}}$ becomes greater than the value of δ. After the successful fulfillment of the conditions of the stabilization criteria are final statistical characteristics $y^{(h\times M)}$, $u\left(y^{(h\times M)}\right)$ and $\left[y_{\min}^{(h\times M)}\ ;\ y_{\max}^{(h\times M)}\right]$ determined from all the generated values. In the given example, are the results obtained by the Monte Carlo method following: estimation of the temperature $t = 178{,}9615127\ °\mathrm{C}$, standard uncertainty $u(t) = 5{,}84815 \times 10^{-4}\ °\mathrm{C}$, 95% narrowest interval $[178{,}96036716\ ;\ 178{,}962658518]$ and 95% symmetric interval $[178{,}96036716\ ;\ 178{,}96265853]$.

Statistical Characteristic of output variable t obtained by the adaptive Monte Carlo method is shown in Fig. 7 as a graph of probability distribution, where the model with one output variable t enters the submodel with multiple output variables a, b, c.

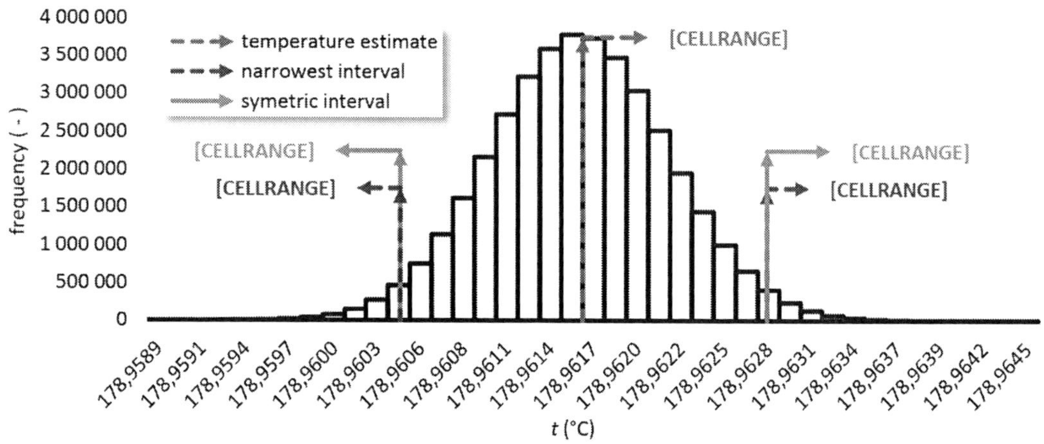

Fig. 7. Probability distribution of output quantity t

c. Validation by the law of propagation of uncertainty and user interface

In cases (as shown in [8] in section 3.2.5), where it is not clear whether the simulation result is correct or not, it is essential that the results of both methods were compared. Valuation is based on comparison of reference intervals obtained by the Monte Carlo method which are compared with the results obtained by the method based on the law of propagation of uncertainty, and it is assessed by calculating whether are the intervals same in a certain numerical tolerance. The numerical tolerance is assessed with regard to endpoints reference intervals, and gives statement of standard uncertainty to the desired valid number of decimal places. For the presented case will

$$d_{\min} = \left| y_{ZŠN} - U_{0,95\,(ZŠN)} - y_{\min(MCM)} \right| = 9{,}995 \times 10^{-7}$$
$$d_{\max} = \left| y_{ZŠN} + U_{0,95\,(ZŠN)} - y_{\max(MCM)} \right| = 2{,}402 \times 10^{-7} \quad (3.15)$$

The calculated absolute differences of the endpoints $d_{\min}$ and $d_{\max}$ are not larger than δ, and therefore is validation by the law of propagation of uncertainty successful. The final comparison of both methods is shown in Tab. 5.

Method	M	t (°C)	$u(y)$ (°C) $\times 10^{-4}$	95% confidence interval	$d_{\min}$; $d_{\max}$ $\times 10^{-7}$ (°C)	Result of validation
ZŠN	-	178,9615122	5,8473	[178,960366 ; 178,9626583]	-	
MCM	2600000	178,9615127	5,84815	[178,960367 ; 178,9626585]	9,995 ; 2,402	✓

Tab. 5. The final results of methods and validation

The user interface of created application for the evaluation and validation of the standard platinum resistance thermometer by the Monte Carlo method is shown in the figure below.

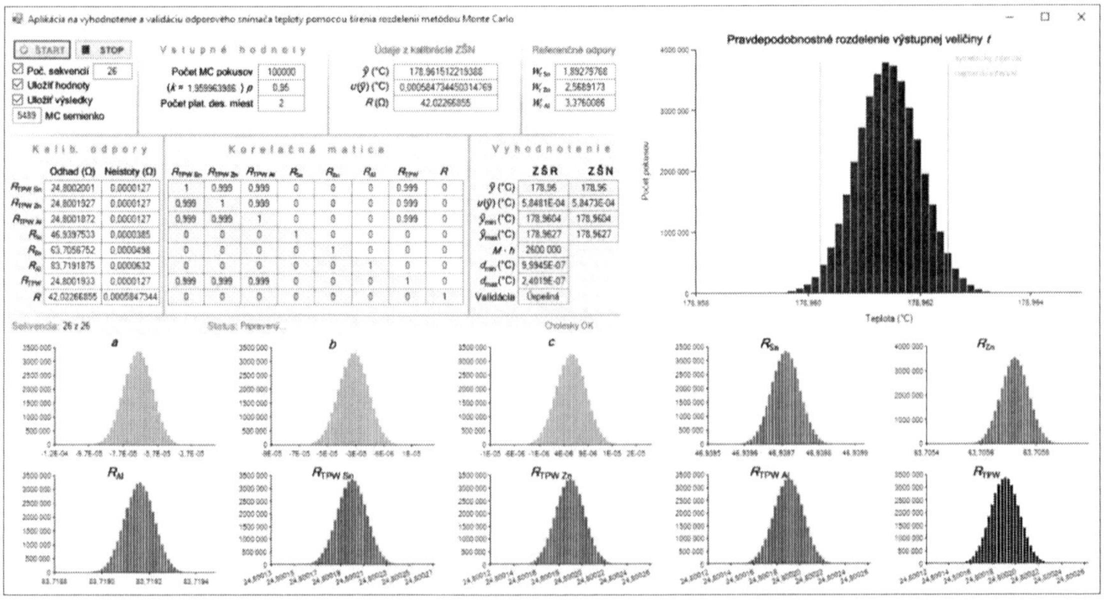

Fig. 8. Graphical user interface at the end of calculation

4 Convergence of stabilizing and validating criteria

The following section shows the progress of the stabilization criteria depending on the particular sequence, and as has been said, each sequence consist from 100 000 pseudorandom generations. As shown in the graph, the values $2s_{\hat{y}}$ (P1) and $2s_{\hat{u}(y)}$ (P2), meet the criteria for stability since the second iteration. Criteria $2s_{\hat{y}_{\min}}$ (P3) and $2s_{\hat{y}_{\max}}$ (P4) are in a particular case dominant and they determine the total number of necessary sequences. Mentioned conditions show signs of stability since 13 sequence, from which it have started to fall and meets the criteria up from 26 iterations. d_{low} and d_{high} criteria used for validation by the law of propagation of uncertainty are met in the thirteenth sequence.

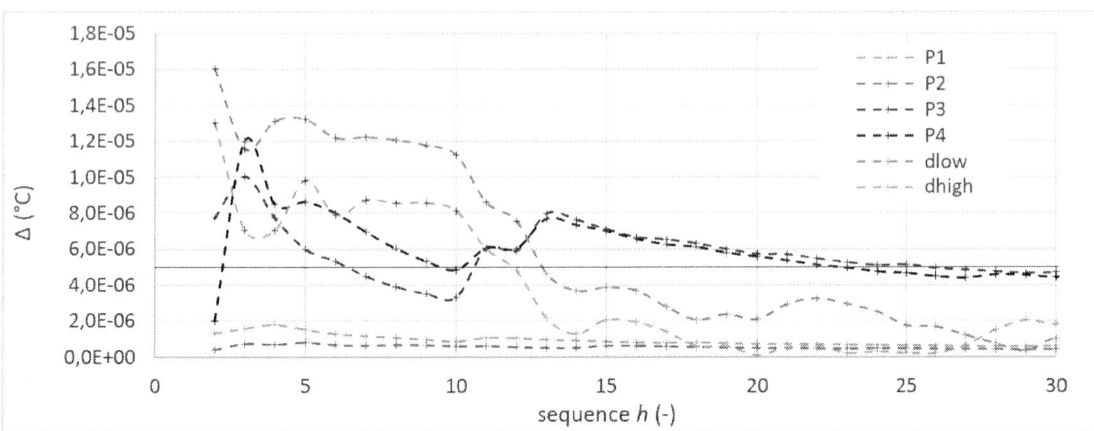

Fig. 9. The progress of criteria up to 30 sequences

The aim of the figure below is to illustrate the progress of stabilization and validation criteria for the number of repetitions several times above the minimum necessary number. In this instance is the number of sequences $h = 200$. With an increasing number of sequences can be seen a gradual shift of the individual parameters where from the mostly stochastic beginning it is gradually passes to an almost systematic reduction. d_{high} criterion used for validation of upper limit by the law of propagation of uncertainty, as the only one slightly increases from the sequence $h = 80$, but is still significantly below the tolerance limit (red line).

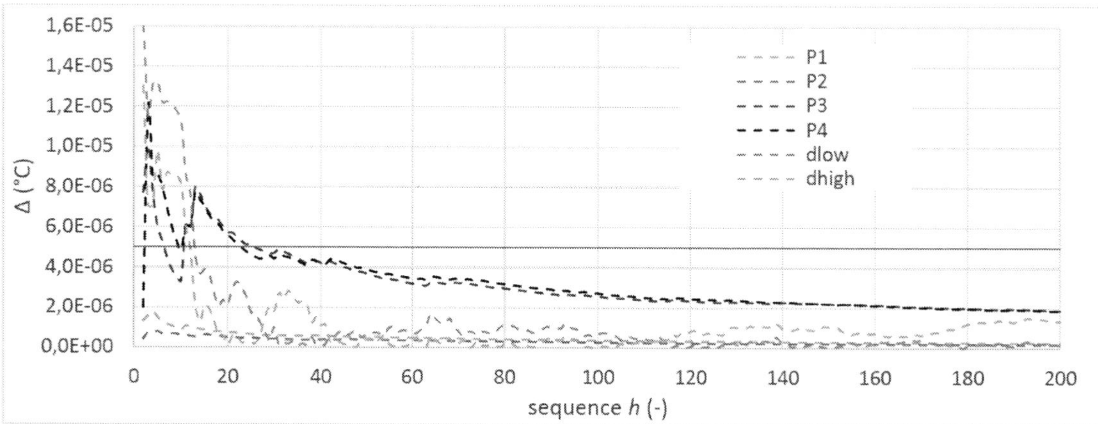

Fig. 10. The progress of criteria up to 200 sequences

Conclusion

The calculation procedure of temperature and its standard uncertainty in the application for calibration of the standard platinum resistance thermometer in the range (0 ÷ 660) °C, carried out by the law of propagation of distributions, has been prepared in accordance with the document [9], [10], [11] whereas in the algorithm strictly integrates general approach for evaluating the measurements listed in annex 1 and 2 GUM [5], [12].

In VB.NET development environment was created for the input values a procedure for generating pseudo-random numbers using the Box-Muller transformation method of eight-dimensional associated probability density function. By the general structure of the program there was a model developed, allowing to generate the values for the correlated and uncorrelated, normally distributed input quantities, by changing variables in the correlation matrix.

This methodology has been appropriately implemented into the created application, while its output is a matrix of relative resistance and coefficients a, b, c of deviation equation of submodel with probability distributions, which subsequently enters into the main model for calculating the estimate of the output quantity t, its assigned standard uncertainty $u(t)$ and the desired confidence interval for the calibrated value of the resistor R_i.

Evaluation of calibration by the Monte Carlo method can be carried out through designed and developed graphical user interface of the application in which it is possible to manage the entire process of calculation the by changing of a input parameters. If desired, it is possible to save all the generated values, intermediate computations and results into the files for the further data analysis.

Algorithms for validation while evaluating calibration of the standard platinum resistance thermometer for a specific sub-range of the ITS-90 are treated in accordance with best practice, while the results of the calibration is statement based on the evaluation of the confidence intervals of the two methods of validating the law of propagation of distributions.

The article also shows the progress of the stabilization and validation criteria in sequential range $h = 200$, which several times exceeds the minimum number of sequences needed for a successful calculation. On a specific example it illustrated the progress and convergence of the individual criteria.

Acknowledgment

Authors would also like to thank the Slovak University of Technology in Bratislava, the grant agency VEGA projects number 1/0604/15, 1/0748/15 and the agency KEGA project number 014STU-4/2015 for their support.

References

[1] PALENČÁR, R. GOLIAŠ, M.: *Stanovenie neistôt pomocou šírenia rozdelení metódou Monte Carlo*. In: Metrológia a skúšobníctvo, p. 2-10, ISSN 1335-2768, 2011.

[2] JCGM 100 - *Evaluation of Measurement Data - Guide to the Expression of Uncertainty in Measurement (GUM)*, GUM 1995 with minor corrections (JCGM, BIPM, Paris,2008).

[3] ĎURIŠ, S., PALENČÁR, R., RANOSTAJ, J.: *Contribution of the SPRT calibration to uncertainty of temperature T90 measured by the calibrated SPRT*. In Measurement Science Review, 8, 5-10, 2008, ISSN 1335-8871, 2008.

[4] PALENČÁR, R. – ĎURIŠ, S. – BROKEŠ. V. – 2015. *Neistoty pri realizácii teplotnej stupnice*. Bratislava: Vydavateľstvo STU, 2015. 159 p. ISBN 978-80-227-4286-3, 2015

[5] JCGM 101 - *Evaluation of measurement data – Supplement 1: Guide to the expression of uncertainty in measurement" – Propagation of distributions using a Monte Carlo method* JCGM, BIPM, Paris 2011.

[6] PALENČÁR, R. DURIŠ, S. BROKEŠ, V.: *Šírenie neistôt pri kalibrácii etalónových odporových snímačov teploty v definičných pevných bodoch : Časť 5 - Softvér*. In: Metrológia a skúšobníctvo, roč. 12, č. 3-4, p. 3-10, 2007.

[7] BIPM. *Uncertainties in the Realization of the SPRT Sub-ranges of the ITS-90*. [online]. Paris, Sèvres: BIPM, 2009. 20 pp

[8] JCGM 104 - *Evaluation of measurement data – An introduction to the "Guide to the expression of uncertainty in measurement" and related documents*, JCGM, BIPM, Paris, 2009

[9] PRESTON, T. H.: *The International Temperature Scale of 1990 (ITS-90)*. Metrologia, 27, 3–10, 1990, ISSN: 0026-1394.

[10] ROSENKRANZ, P.: *Uncertainty Propagation for Platinum Resistance Thermometers Calibrated According to ITS-90*. Int. J. Thermophys, 32,106-119, 2011, ISSN: 0195-928X.

[11] PALENČÁR, R.: *Postup výpočtu kovariancií*. Metrológia a skúšobníctvo, 2, p.22-25, 2000.

[12] JCGM 102 - *Evaluation of Measurement Data, Supplement 2 to the Guide to the Expression of Uncertainty in Measurement (GUM) – Extension to any number of output quantities*, JCGM, BIPM, Paris, 2011.

Current address

Sopkuliak Peter, Ing.
Slovak University of Technology in Bratislava
Námestie slobody 17, 812 31 Bratislava, Slovak Republic
E-mail: xsopkuliak@is.stuba.sk

Palenčár Rudolf, prof. Ing. PhD.
Slovak University of Technology in Bratislava
Námestie slobody 17, 812 31 Bratislava, Slovak Republic
E-mail: rudolf.palencar@stuba.sk

Palenčár Jakub, Ing.
Slovak University of Technology in Bratislava
Námestie slobody 17, 812 31 Bratislava, Slovak Republic
E-mail: jakub.palencar@stuba.sk

Suroviak Emil, Ing.
Slovak University of Technology in Bratislava
Námestie slobody 17, 812 31 Bratislava, Slovak Republic
E-mail: xsuroviak@is.stuba.sk

Slovak University of Technology in Bratislava
Faculty of Mechanical Engineering
16th Conference on Applied Mathematics **APLIMAT 2017**
Proceedings

THE USE OF STATISTICAL METHODS IN VERIFICATION OF A NEW EDUCATIONAL MODEL FOR SAFETY SPECIALISTS AT SELECTED HIGHER SCHOOLS

SOTÁK Vladimír (SK), KURUCOVÁ Zuzana (SK), TIRPÁKOVÁ Anna (SK), MUCHOVÁ Mária (SK), SOTÁK Robert (SK)

Abstract. This paper presents the design of a new educational approach in the area of Occupational Safety and Health (OSH) while taking into consideration specifications of the above mentioned field. Efficiency of the newly proposed approach in the preparation of future safety specialists has also been experimentally verified. The outcomes of our experiment were evaluated by means of non-parametrical statistical methods. The research was done in 2011-2014 and its aim was to propose an optimal model of education in the OSH area through the analysis of the current state in the field.

Keywords: Safety specialist, Education, Mann-Whitney test, Wilcoxon test, Friedman test

Mathematics Subject Classification: 62P30, 90B08, 97B40

1 Introduction

Safety specialist is an occupation which requires qualification in a specifically oriented study field. The need for improvement of technical thinking, ability to analyze, solve problems and apply obtained knowledge in particular practical examples creates a platform for fundamental philosophy in education of safety specialists. From the scope, content and systematization point of view, the protection of health at work plays an important role in organizational, economic, technical, medical, social and educational measures. First-rate quality education will always be a crucial condition for permanent development, implementation of innovations, good research and high quality production. All these attributes represent especially unique abilities of the human brain and its creativity (Beneš, 2014).

Nowadays, most safety specialists are taught and prepared in vocational competences courses offered by entitled organizations. Authorization for these courses is issued by the Ministry of Labor, Social Affairs and Family of the Slovak Republic. Application form and evaluation of the particular educational project are required in the approval process. Preparation of specialists in the field of Occupational safety and health (OSH) at the level of higher education is contemporarily offered at four faculties – Faculty of Materials Science and Technology of the Slovak University of

Technology in Trnava (both bachelor and engineering study programs), Faculty of Engineering of the University of Technology in Košice (engineering study program), Faculty of Mining Industry, Ecology, Management and Geology of the University of Technology in Košice (engineering study program) and Faculty of Education of Constantine The Philosopher University in Nitra (bachelor study program).

According to Act 124/2006 Coll., OSH graduates at a higher school level of education are considered to be specialists equal to graduates of a three week course for safety specialists. All successful graduates obtain the Safety Specialist Attestation and after two years of vocational practice under supervision of an authorized safety specialist, they are allowed to enrol at the Inspectorate of Work for the authorized safety specialist examination. The Slovak legislation takes into consideration neither different length of vocational practice, nor the level of obtained education in the position of safety specialist or authorized safety specialist. The only limiting factor is the Safety Specialist Attestation. This fact frequently leads to obvious qualitative differences in the work of particular safety specialists.

According to personal experience of researchers and tutors, it is important who the tutor of a starting safety specialist is (Adair, 2005). It cannot be presumed that a graduate from a secondary school of textile industry is capable of mastering the principles of production assembly line in full details. Those branches of industry where the principles of OSH are perceived as a part of everyday practice conduct job interviews for particular positions with special emphasis not only on the certificate itself, but also on the obtained education of candidates (Meravý, 2010).
Adequate vocational education can help master their work better, since safety specialists are able to perform in both technical and technological areas of a particular branch. The scope of the course for safety specialists is governed by legislation (Act of the National Council of the Slovak Republic No.124/2006 Coll.). Authorized organizations providing preparation and training of safety specialists are obliged to have an educational project for safety specialists approved by the National Inspectorate of Labor.

It is generally well-known that the occupation of a safety specialist is demanding, since it requires wide orientation in the field and fast and flexible solutions in many different situations. Realizing the overall difficulty of this challenge leads to a conclusion that a three week course is definitely not sufficient. Department of Technology and Information Technologies at the Faculty of Education of the Constantine the Philosopher University in Nitra has chosen a special approach for teaching technically oriented areas since it defines the character and the scope of the curriculum of technical subjects (Bánes, Lukáčová, Sitáš, 2010). The department is situated at the faculty that is primarily focused on education, i. e. the fundamental goal of the department is to educate and prepare technically efficient graduates. From this point of view, it seems suitable to entrench the study program of occupational safety and health at the department. This type of study is closely related to practice since graduates obtain not only a bachelor's degree, but also the Certificate of Safety Specialist (Soták, 2011).

The aim of this paper is to analyze the current state of preparation and training of specialists in the OSH field at the Department of Technology and Information Technologies at the Faculty of Education, Constantine the Philosopher University in Nitra, and based on this analysis to propose an optimal educational model. When verifying the efficiency of the proposed model, valid statistical methods and interpreted obtained results were used.

2 Results

2.1 Research Characteristics and Used Methods

Altogether 40 students of the OSH study field participated in the research which was carried out at the Department of Technology and Information Technologies, Faculty of Education, Constantine the Philosopher University in Nitra in academic years 2011/12, 2012/13, and 2013/14. The aim of the research was to propose a new optimal model of education in the field of OSH based on analysis of current state, while taking into consideration its specifications. School subjects in which this approach was tested were selected through personal experience of a safety specialist directly from practice – Risk Analysis, Total Quality Management and Legislation in the OSH area. Special attention was given to the comparison of current and newly proposed approaches to the process of preparation of future safety specialists at the Department of Technology and Information Technologies. An educational model was produced that is based on the connection between theoretical core and the shell represented by practical examples.

During lectures on the above-mentioned topics, student feedback was collected, especially from students in the external form of study. The so-called daily students were complaining about insipidness in the Law course in which they had to memorize many acts and regulations. When realizing how wide the scope of the subject of Law is, it seems to be impractical to require memorization of so many texts and consequently to be able to use them in practice, since changes in the area of OSH legislation are more than frequent. Any connection between obtained knowledge and real practical demonstrations is absent here. Choosing lecturers from the group of employees at the Inspectorate of Labor was a good solution. Safety specialists must take into consideration also ecological aspects of their proposals and measures - not only for the employer, but for employees as well. The new educational concept is focused especially on the preparation of students for production practice. In the Risk Analysis course, teaching aids (photos and videos) from particular workplaces in different branches of Slovak national industry were used. Students analyzed possible dangers and defects at the workplace that could lead to consequent threat to employees, which helped students learn to deal with appropriate legislative rules and regulations. When processing results of our analyses, the experience from the TQM (Total Quality Management) course was used, especially analytical methods (FMEA, HAZOP, Six sigma, 5WHY, Point method). An important advantage of one of the authors of this paper is his Six sigma-green belt certificate for automotive industry. Thanks to this fact students dealt not only with theoretical materials, but also with those from a real production practice. Our goal also was to teach our students how to work in a team and how to present obtained results.

When proposing this new approach, some objective facts dealing with fundamental didactic phenomena had to be taken into consideration. Didactic principles, goals, educational content, methods, forms and means are the cornerstone of education that cannot be ignored in any serious approach to the educational process. Moreover, it must be clear that the studied case is not covered in so-called classical pedagogy, but in andragogy. Preparation of safety specialists for their occupation is principally no different than preparation of any other specialists. Yet, there are some specifications that need to be respected when creating a new approach. The first specification is related to the educational content that is based on legislative documentation. Laws, regulations, standards, rules, acts and norms are the platform which the educational aims in the cognitive area are built upon. A vast majority of the above-mentioned documents is being changed and innovated too frequently. This phenomenon is caused by several external factors. It is not unusual that some changes occur even in the course of the semester, so the educational content has to be suddenly changed. This requires the teacher's ability to monitor all changes and implement up-to-date knowledge or information into the educational process. It evokes the need for implementation of flexible didactic tools so that teachers are capable of reacting directly to changes with minimal effect on default concept in the educational process.

Students of the OSH study program at the Department of Technology and Information Technologies at Constantine the Philosopher University in Nitra can obtain the certificate of a safety specialist at the end of their bachelor study. Since the scope of knowledge of a safety specialist is strictly defined by legislation, focus was put on those components in education that can be influenced in a way so that the connection between theory and practice can be made. While elaborating the new educational model, the following objectives were kept in mind:

1) To teach our students to think analytically and to present their ideas and conclusions in multicultural environment.
2) To apply computing technology at higher rate with an emphasis on a more frequent use of simulation software.

It was also presumed that by using this new approach, our students could obtain better knowledge and that should be manifested also in the assessment of their work. Efficiency and effectiveness of the new approach in the process of teaching future safety specialists at the Department of Technology and Information Technologies in Nitra were experimentally verified. The whole research was evaluated by means of statistical methods and the obtained outcomes were interpreted.

2.2 Source Data and Basic Numerical Characteristics

At the beginning of the academic year 2011/2012, students of the first year of study were randomly divided into two groups – a control group (CG) and an experimental group (EG). Before the experiment itself started, both groups took a pretest. The pretest tested the students' general knowledge level. The pretest questions were selected from the topics of secondary school curriculum of Physics, History, Chemistry and Civics. This general scope was selected due to heterogeneity of different types of secondary schools which the OSH students had attended. There were graduates of different secondary vocational schools (electrical technician, mechanical engineer, technical and information services graduates) and grouped secondary schools (cooks, waiters) in one class.

The pre-test results of both groups were evaluated in the same way – one point for one correct answer. Since the pretest comprised 20 questions, the maximum obtainable value was 20 points. Results of both groups are shown in Fig. 1.

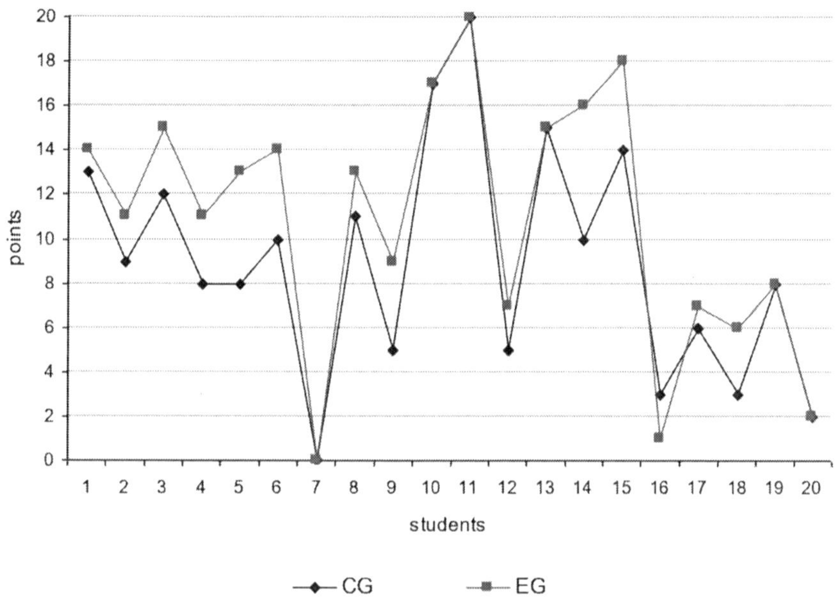

Fig. 1. Results of students (pre-test)

The comparison of results of students from experimental and control groups shows that the average success rate in CG was 45 % and in EG 52 %. These values (number of correct answers) are the realization of two independent random samples defined by ranges $m = 20$ and $n = 20$ out of the two continuous distributions. Our aim was to determine, using statistical methods, which of the differences are statistically significant. Next, null hypothesis H_0: distribution functions of both distributions are equal – was tested against alternative hypothesis H_1: distribution functions of both distributions are different. As the assumption of normal distribution of the observed variables was not confirmed, for testing the null hypothesis the non-parametric Wilcoxon two-sample test (sometimes called Mann-Whitney U test) was used. This test represents the non-parametric analogy to t-test (Markechová et al, 2011). If the calculated value of test criterion Z exceeds the critical level, null hypothesis H_0 is rejected at the level of significance $\alpha = 0.05$ in favor of the alternative hypothesis. Wilcoxon two-sample test was performed by means of STATISTICA software. Having analyzed the input data, the value of test criterion Z and the value of probability p were obtained. The test can also be evaluated by using the value p that refers to the probability of an error that can occur when we reject the tested hypothesis. If the calculated value of probability p is low enough ($p < 0.05$ or $p < 0.01$) the tested hypothesis is rejected (at the level of significance $\alpha = 0.05$ or 0.01).
In our case (calculated in STATISTICA) the following results were obtained: the value of test criterion $Z = -1.082$ and value $p = 0.279$. Since the calculated value of probability p is higher than 0.05, null hypothesis H_0 at the level of significance $\alpha = 0.05$ cannot be rejected. This means that the difference between the two groups with regards to the observed index (pre-test outcomes) is not statistically significant. This means that both groups of students have reached equal level of knowledge in Physics, Chemistry, History and Civics. In chapters of curriculum in particular subjects were taught in a traditional way (CG), and the new model was introduced in EG.

After completing the first year of study and consequently the second one, all the students in CG and EG took a post-test. Both post-tests comprised 20 questions that were evaluated in the same way as

it was in case of the pre-test. Every correct answer was rewarded by one point. Results of both post-tests are shown in Fig. 2 (1st year of study), and in Fig. 3 (2nd year).
The post-test after the 1st year aimed to measure how students managed to process all provided information as well as to measure the extent of their knowledge. It consisted of 20 questions related to the OSH Legislation and TQM. All questions were multiple-choice questions, i. e. students were asked to choose one answer out of three options. The post-test after the 2nd year (Fig. 3) was focused on subjects Risk Analysis, TQM and OSH Legislation. It consisted of 20 questions and there were three pre-defined answers for each question.

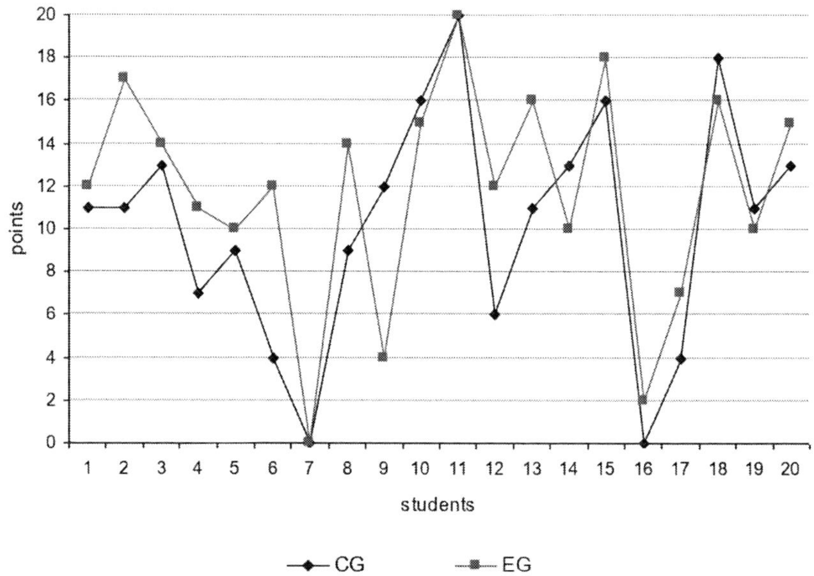

Fig. 2. Results of students after 1st year of study (post-test)

It is obvious Fig. 2 that after the 1st year of study, the average students' success rate in the post-test was 45 % in CG and 59 % in EG.

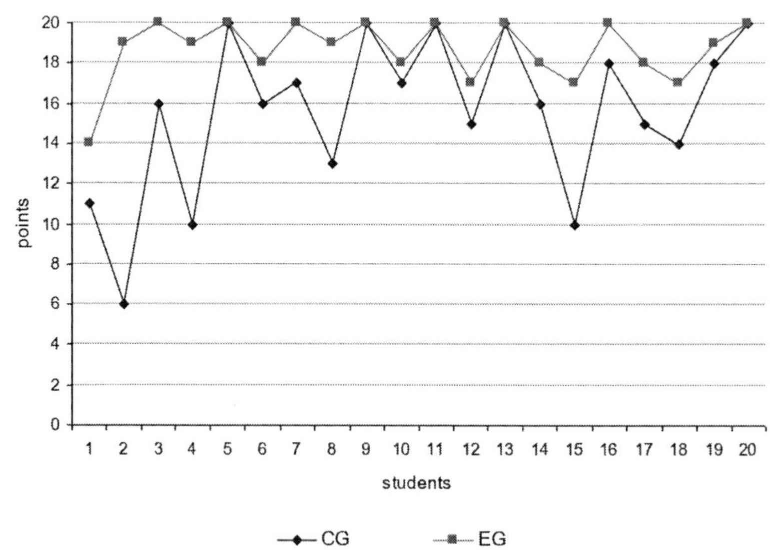

Fig. 3. Results of students after 2nd year of study (post-test)

The figures show that after two years, the average success rate in the post-test is much higher – 78 % in CG and 93 % in EG.
The question was whether these differences in the post-test (post-test results after the 1st and 2nd year of study) outcomes between CG and EG are statistically significant. In order to determine this, the Wilcoxon two-sample test was used again. The STATISTICA software provided us with the values of test criterion *Z* and the values of probability *p* (see Tab. 1).

	Z	*p*
After the 1st year	-1.001	0.317
After the 2nd year	**-2.637**	**0.008**

Tab. 1. Mann-Whitney U Test (Wilcoxon test)

Since the calculated value of probability is $p = 0.317$, the null hypothesis cannot be rejected at the level of significance $\alpha = 0.05$. This means that the differences in post-test results between CG and EG after completing the first year of study are not statistically significant.

On the contrary, the test of the statistical significance of differences in students' post-test results after the 2nd year of study showed the value of probability lower than 0.05 ($p = 0.008$) in Wilcoxon two-sample test. This means that the null hypothesis which states equality of students' results in both control and experimental groups in the post-test is rejected at the level of significance $\alpha =$ 0.05. There are statistically significant differences between students of CG and EG in the post-test results after the 2nd year of study.

Based on the statistical analysis it can be concluded that the new model of preparation of future safety specialists was successful. It is interesting that statistically significant differences between control and experimental groups were confirmed as late as after two years.
Another analysis which was tested in our experiment was the verification of the fact whether students of both CG and EG obtained significantly different results in tests (pre-test and post-test) during three particular years. At first, the differences in the test results of CG during three years were observed.

The goal was to see whether these test result differences are statistically significant. Null hypothesis H_0: Distribution functions of all samples are identical, i. e. test results of three years in each group (CG and EG) are equal – was tested against alternative hypothesis H_1: Distribution functions of all samples are different, i. e. test results of three years in each group are different.
As the assumption of normal distribution of the observed variables was not confirmed, for testing the null hypothesis the non-parametric Friedman test was used. Friedman test is a generalization of the Wilcoxon one-sample test and it is an analogy of the analysis of double sorting variance with one observation in each subclass.

The Friedman test is based on the fact that the observations are ordered in each block separately and the order T_{ij} of the values x_{ij} in the i-th block is determined. Relation $\sum_{i=1}^{n}\sum_{j=1}^{k} T_{ij} = \frac{nk(k+1)}{2}$ can be used as a control calculation. As a testing criterion, the following statistics is used:

$$Q = \frac{12}{nk(k+1)} \cdot \sum_{j=1}^{k}(\sum_{i=1}^{n} T_{ij})^2 - 3n(k+1),$$

which, given the validity of hypothesis H_0 asymptotically χ^2 - distribution with k - 1 degrees of freedom. The tested hypothesis H_0 is rejected at the significance level α if the value of the testing criteria $Q \geq \chi^2_\alpha(k-1)$, where $\chi^2_\alpha(k-1)$ are critical values of χ^2 - distribution with k - 1 degrees of freedom. If the tested hypothesis is rejected in favor of the alternative hypothesis H1 which means that observations in the groups do not provide the same effect, the question remains which files differ significantly from each other. To compare the differences between individual files in the Friedman test, the Neményi method of multiple comparisons is used. In this method, the process is as follows: $\sum_{i=1}^{n} T_{ij} = T_{.j}$ is denoted and $|T_{.j} - T_{.m}|$ for $j < m$ is calculated. If the value $|T_{.j} - T_{.m}|$ is greater than the critical tabulated value for multiple comparisons, the tested hypothesis that observations in the groups do not provide the same effect in the j-th and m-th time phases is rejected on the chosen level of significance α. Critical values for multiple comparisons are given asymptotically by the relationship

$$q_{k,\infty}(\alpha)\sqrt{\frac{1}{12}nk(k+1)},$$

where $q_{k,\infty}(\alpha)$ is a critical value tabulated for $\alpha = 0{,}05$, or $\alpha = 0{,}01$ in the table of critical value T11 and T12 in [8], respectively.

With the use of Friedman test, the value of test criterion Q and value p were obtained. In our case (CG – years), the STATISTICA software provided us with the following results: value of test criterion Q = 11.94521, and value of p = 0.00255. Tab. 2 summarizes the average orders, sums of orders, arithmetic means and standard deviations for values of the observed variable.

Variable	Average order	Sum of orders	Arithmetic mean	Standard deviation
CG – 0th year	1.650	33.000	8.950	5.246
CG – 1st year	1.750	35.000	10.200	5.473
CG – 2nd year	2.600	52.000	15.600	3.966

Tab. 2. Results of Friedman test

Since value of probability p = 0.00255, null hypothesis H_0 is rejected at the level of significance $\alpha = 0.01$. This means that the observed differences are statistically significant. The Friedman test confirmed that the test results of CG during the three years are statistically significantly different.

To see in which years the students obtained different test results, the Nemenyi method of testing contrasts was used. Results of the Nemenyi method are shown in Tab. 3.

Nem. test	0th year	1st year	2nd year
1st year		3	**19 ***
2nd year			**17 ***

Tab. 3. Results of Nemenyi test CG

Tab. 3 shows that based on the results obtained by means of statistical analysis, it is confirmed that statistically significant differences are between the results that students of CG obtained in the test after the 2nd year of study and those results that the same students achieved in the two other years (0th and 1st). The test results after the 0th and the 1st year of study are almost the same, which means that the differences are not statistically significant (see also Fig. 4).

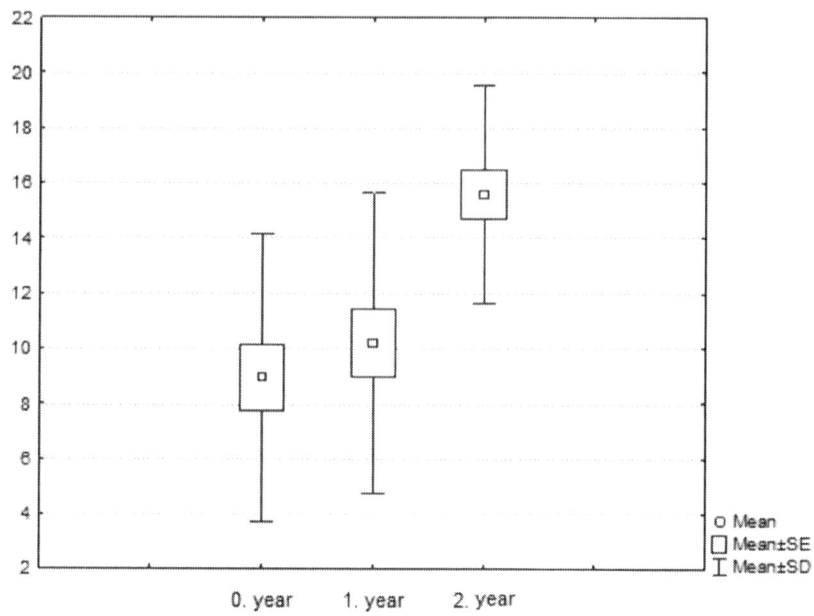

Fig. 4. Visualization of the results CG use of Friedman test

Analogically, differences in the test results of EG students during three years were evaluated. With the use of the Friedman test, the following results were obtained: value of the test criterion $Q = 25.64865$ and value $p = 0.00001$.

Since value $p < 0.001$, null hypothesis H_0 is rejected at the level of significance $\alpha = 0.05$. This means that the observed differences are statistically significant. The Friedman test confirmed that values of the test results in particular years of study are significantly different from the statistical point of view. To compare the differences between particular files in case of the Friedman test, the Nemenyi method was applied. Absolute differences of the sums of orders are shown in Tab. 4.

Nem. test	0th year	1st year	2nd year
1. year		31	**24 ***
2. year			**25 ***

Tab. 4. Results of Nemenyi test EG

Based on the obtained results it was confirmed that statistically significant differences are again between results in the test of students from EG after the second year of study and the results in tests obtained in two previous years (0th and 1st year). Results of the tests of EG students were almost the same after the 0th and 1st year of study – differences are not statistically significant (see Fig. 5).

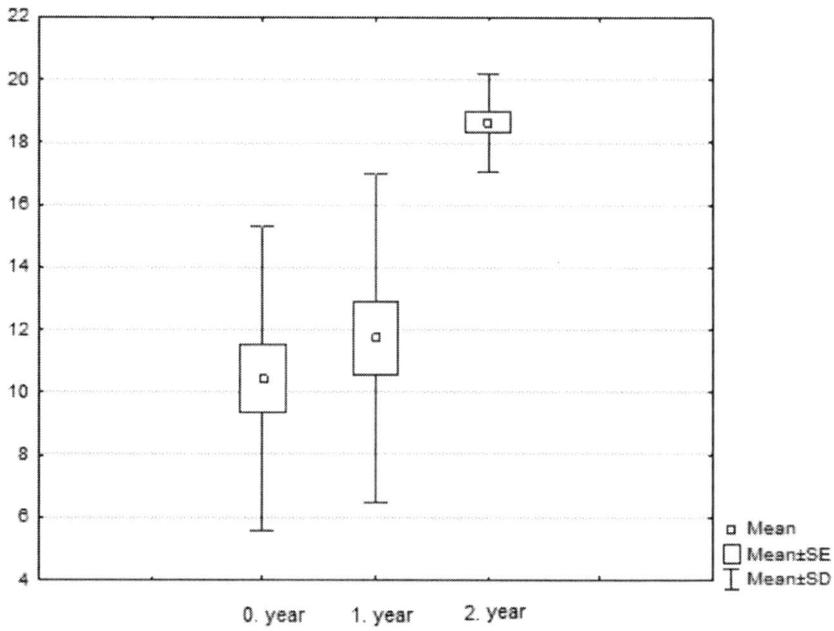

Fig. 5. Visualization of the results EG use of Friedman test

Conclusion

Based on the statistical analysis, it may be said that the new model of preparation of future safety specialists was highly efficient. It is interesting that statistically important differences between control and experimental groups were confirmed as late as after two years of study.

Outcomes obtained by means of the Friedman test confirmed that there are statistically significant differences between students' results in the control group after the 2nd year of study and the results of the same students in two previous years of study. Test results in the 0th and the 1st year of study are approximately the same, which means that the differences are not statistically significant. When using the Friedman test, the same results also for the students in the experimental group were obtained. Both Wilcoxon and Friedman tests confirmed that the proposed educational model for teaching safety specialists at the Department of Technology and Information Technologies at the Faculty of Education of the Constantine the Philosopher University in Nitra is efficient.

The teaching process comprised three school subjects – OSH Legislation, Risk Analysis and TQM. Students in the control group could use traditional study materials and attended traditional lectures. Teaching in the experimental group was performed in a modern way, students were offered to use visual presentations with basic information. The emphasis was put on the presentation of a lecturer and explanation regarding possible implementation of up-to-date information in particular conditions of real practice.

Being taught in the new way, students could better imagine possible ways of implementation of obtained knowledge into practice. In addition, they could get familiar with various types of workplaces through photo documentation. They could also discuss their ideas with the lecturer and consult solutions for concrete issues they faced personally in everyday life.

Recently, the general feature of common pedagogues' efforts is to approach theoretical preparation towards real conditions of practice. Within the Slovakia region, there are some projects oriented toward this trend, also supported by our effort and results. Our research was focused on the design and verification of educational approaches that can meet current requirements. When looking back at our previous effort, it can be said that we managed to bring some new elements and features into the process of preparation of OSH students at our department and designed and applied the so-called project preparation for educational needs. To provide this type of education, CD materials for the area of OSH prepared by the European OSH Agency were used. To make the educational process flexible, on-line teaching materials were also created. To train the risk analysis at higher level, the plan is to create a database of workplaces which can provide students with real practice in particular branches of the national industry.

The experiment confirmed that our newly designed method is effective, which is the real added value in comparison to traditional teaching full of theoretical knowledge. Our approach helps improve analytical thinking and develop team work skills. Besides other positives, verbal skills of the students during presentations significantly improved already during the second year of their study.

In the future, our plan is to develop study materials for OSH students, e. g. a multilingual explanatory dictionary of terms from the area of OSH. Additional research will be carried out on efficiency of the proposed teaching model.

References

[1] ADAIR, J. 2005. *Jak řídit druhé i sám sebe*. Brno: Computer Press, 2005. 172 s. ISBN 80-251-0784-1

[2] BÁNESZ, G. - LUKÁČOVÁ, D. - SITÁŠ, J. 2010. *Technické vzdelávanie v digitálnom prostredí*. Nitra: PF UKF, 2010. 109 s. ISBN 978-80-8094-713-2

[3] BAŠTECKÁ B. 2009. *Psychologická encyklopedie: aplikovaná psychologie*. Praha: Portál, 2009. 520 s. ISBN 978-80-7367-470-0

[4] BENEŠ, M. 2014. *Andragogika 2*. 2. vyd. Praha: Grada Publishing, 2014. 176 s. ISBN 978-80-24748-24-5

[5] BUCKLEY, R. - CAPLE, J. 2004. *Trénink a školení*. Brno: Computer Press, 2004. 304 s. ISBN 80-251-0358-7

[6] ČERNÝ V. 2009. *Jak jednat s různými typy lidí*. Brno: Computer Press, 2009. 264 s. ISBN 978-80-251-2093-4

[7] HOLIČ, Š. 2002. *Rétorika*. Nitra: Enigma, 2002. 86 s. ISBN 80-85471-99-X

[8] MARKECHOVÁ, D. – TIRPÁKOVÁ, A. – STEHLÍKOVÁ, B. 2011. *Základy štatistiky pre pedagógov*. Nitra: FPV UKF v Nitre, 2011. 405 s. ISBN 978-80-8094-899-3

[9] MERAVÝ, J. 2010. *Užitočné informácie pre súčasnú prácu revíznych technikov, elektrotechnikov a projektantov elektrických zariadení*. Trenčín: Ing. Ján Meravý – Lightning, 2010. 161 s. ISBN 978-80-968509-7-6

[10] KOEGEL, T.J. 2009. *Špičková prezentace: Jak zaujmout a přesvědčit posluchače*. Brno: Computer Press, 2009. 192 s. ISBN 978-80-251-2008-8

[11] PRŮCHA, J. 2011. *Multikulturní výchova*. 2. vyd. Praha: Triton, 2011. 167 s. ISBN 978-80-738-7502-2

[12] SOTÁK,V. - KUBALA, A. 2014. *Význam elektrotechniky v odbore BOZP*. Paper supported by KEGA project no. 041UKF-4/2001 entitled „Implementation of modern trends of OSH education in lifelong learning"

Current address

Soták Vladimír, doc. Ing. CSc.
Department of Technology and Information Technologies,
Constantine the Philosopher University in Nitra
Andrej Hlinka 1, 979 74 Nitra, Slovak Republic
Tel. number:+421 37 6408 341, e-mail: vsotak@ukf.sk

Kurucová Zuzana, PaedDr. PhD.
Faculty of Law, Pan-European University
Tomášikova 20, 821 02 Bratislava, Slovak Republic
Tel. number: +421 904 415 154, e-mail: zuzana.kurucova@paneurouni.com

Tirpáková Anna, prof. RNDr. CSc.
Department of Mathematics, Constantine the Philosopher University in Nitra
Andrej Hlinka 1, 979 74 Nitra, Slovak Republic
Tel. number: +421 37 6408 695, e-mail: atirpakova@ukf.sk

Muchová Mária, Ing.
Reydel Automotive Slovakia s.r.o.
Dolné Hony 2, 949 01 Nitra, Slovak Republic
Tel. number: +421 37 690 9226, e-mail: mmuchova@reydel.com

Soták Robert, PaedDr. PhD.
Lomnická 4, 979 74 Nitra, Slovak Republic
Tel. number: +421 903 038 828

Slovak University of Technology in Bratislava
Faculty of Mechanical Engineering
16th Conference on Applied Mathematics **APLIMAT 2017**
Proceedings

BELIEFS OF TEACHER TRAINEES ABOUT THE ISSUE OF SOLVING MATH WORD PROBLEMS

ŠOVČÍKOVÁ Petronela (SK), KLIMENTOVÁ Lucia (SK)

Abstract. In our article we are focused on beliefs of teacher trainees who study Preschool and Elementary Education and Teacher Training for Primary Education at Constantine the Philosopher University in Nitra, Slovakia. By means of an open-ended four-item questionnaire we wanted to examine their beliefs regarding the issue of solving math word problems. In addition, we aimed to help the teacher trainees find effective ways which could improve their teaching skills by encouraging them to ask self-reflective questions.

Keywords: Math word problem, solving, beliefs, teacher trainees

Mathematics Subject Classification: Primary 97C30; Secondary 97C20.

1 Introduction

Recently, teachers' beliefs have been a much discussed research topic in mathematics didactics. The beliefs are understood broadly as the conceptions, views, personal ideologies and values that shape the teaching practice. It is assumed that personal beliefs influence human behaviour and acts, i.e. beliefs act as a teacher's pedagogical predispositions. [1]

Teachers' beliefs play a significant role in their teaching practices. They affect what is taught, how it is taught, and, thus, what is learned in the classroom. Beliefs shape how teachers think and feel about mathematics; they shape mathematics teaching and learning. As teachers' beliefs affect their teaching, it is important to recognize those beliefs. [2]

Pupils first encounter math word problems in the first grade of elementary school. That is what university students who study Preschool and Elementary Education and Teacher Training for Primary Education consider to be crucial regarding math word problems. In addition, the teachers' beliefs influence pupils' beliefs. If the teacher has difficulties with solving and understanding word problems and he/she does not like these tasks, pupils are also likely to have difficulties with solving and understanding math word problems and dislike word problems as well.

2 Theoretical background

Word problem solving is one of the essential components of mathematics problem solving which incorporates real life problems and applications of mathematics. However, many educational researchers report that students express great difficulties in handling a word or story problem.
Word problem is actually a story problem and students normally have to relate between the known and the unknown. [3]

As Gagne (1983) suggests, in the process of mathematical word problem solving students should be able to translate the concrete to the abstract and vice versa. Therefore, a mathematical word problem is usually perceived as a more specific and challenging task than an ordinary mathematics task. [3]

Beliefs about the nature of mathematics influence not only what is taught but also how it is taught. [4]

According to Pehkonen and Törner (1998) an individual's mathematical beliefs are compound of his subjective, experience-based, implicit knowledge on mathematics and its teaching and learning. [5]

In order to bring changes in the teaching of mathematics, it is necessary to investigate what is understood by "good teaching" and, consequently, how one's teaching could be improved. The studies on beliefs help us identify the differences on beliefs, and also the factors that can improve teaching and the development of a favourable learning environment for pupils.

3 Methodology

The research sample consisted of 131 teacher trainees who study Preschool and Elementary Education and Teacher Training for Primary Education at Constantine the Philosopher University in Nitra, Slovakia. The mathematical education which the teacher trainees had received in their secondary school education varied in its content, in relation to the type of the attended secondary school (Graph 1) and the geographical region where the secondary schools are located (Graph 2).

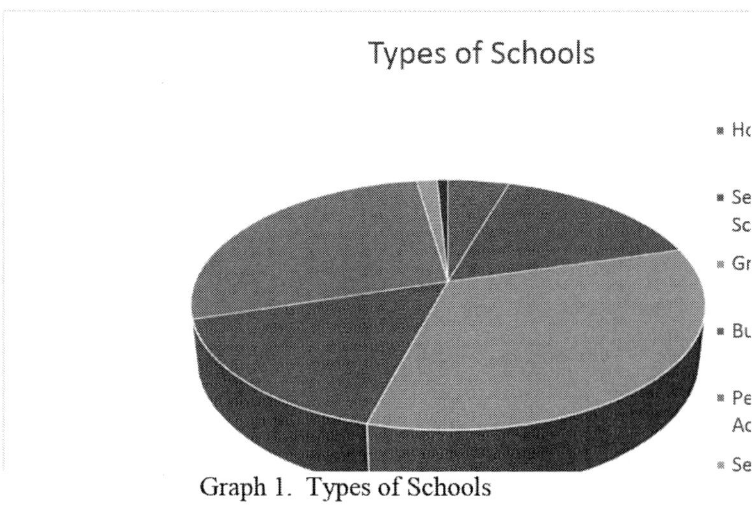

Graph 1. Types of Schools

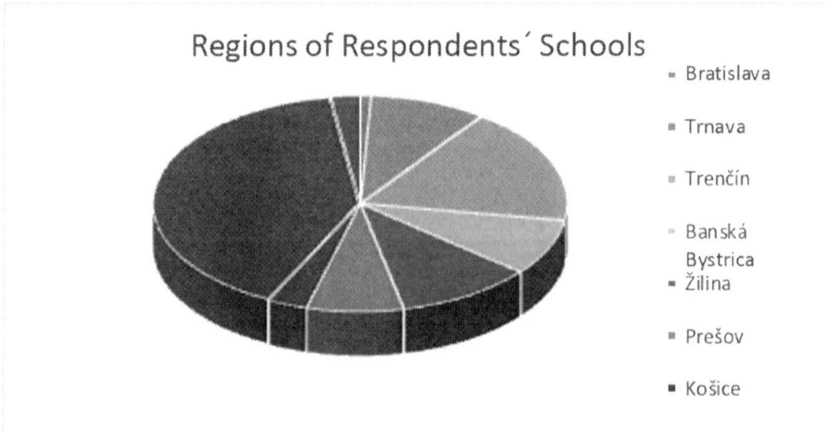

Graph 2. Regions of Respondents´ Schools

By means of an open-ended four-item questionnaire we wanted to identify the teacher trainees' beliefs about the issue of solving math word problems. The questionnaire consisted of the following questions:

When solving math word problems:

- *what is the most important thing for you?*
- *what is the most difficult thing for you?*
- *what causes you most problems?*
- *what would you find helpful?*

4 Findings

As indicated by the teacher trainees' responses, in how they perceive the questions there is a large overlap between the first three questions. The teacher trainees' answers to the three questions were very similar. This suggests that the teacher trainees mostly recognize the most important things (when solving math word problems) as the most difficult and causing them most problems at the same time. Based on their answers, the teacher trainees' beliefs about the most important, difficult and problematic things when solving math word problems can be categorized into the following four groups: (i) the assignment clarity and comprehensibility, (ii) mathematization, (iii) numeric solution, and (iv) miscellaneous.

As far as (i) **the clarity and comprehensibility of an assignment** are concerned, teacher trainees expressed their beliefs in form of answers which were similar to the following:

The most difficult thing is when the texting of the problem is complicated.

The most important thing is to read the assignment of the problem several times.

Many teacher trainees wrote that reading comprehension is a great challenge for them. When they read an assignment, they do not know what is important for the effectiveness of their solving.

Teacher trainees' beliefs regarding the importance and difficulty of (ii) **mathematization** when solving math word problems were recognized in answers similar to the following:

The most complicated thing is to write the word problem in a form using numbers and variables.

I think the most difficult thing is to write an equation, because I don't know how to do it.

I don't know what the principle of problem solving is.

The category of (iii) **numeric solution** refers to numeric mistakes which are often made in solutions of math word problems. Many teacher trainees stated that they have a big problem with numeric solution. They manage to write a correct equation, but they often make numeric mistakes in their calculations and subsequently they do not come at the correct result. Such mistakes, or rather lapses are usually caused by inattention. Covertly, this also shows how over-emphasized the preciseness and correctness of the number result in solving math word problems is in mathematics classes in general; though, figuratively speaking, the art of mathematics lies in the strategies and methods of problem solving. Some teacher trainees also stated that they do not know how to handle variables in equations.

The last broad category of teacher trainees' (iv) **miscellaneous** responses to the first three questions covers the importance and difficulty of attention, concentration, and logical thinking when solving math word problems.

By asking teacher trainees the fourth question we wanted to find out what they consider to be helpful when solving math word problems. We categorized the teacher trainees' beliefs regarding possible helpful factors into the following five categories: (i) practicing, (ii) assistance and tutoring, (iii) group work, (iv) gradation of difficulty level, and (v) exampling and illustration. Teacher trainees expressed their wish for more (i) **practicing** by stating that when solving math word problems they would find helpful *"solving more math word problems"*, *"explanation and more similar math word problems"*, *"solving math word problems and understanding the principles of problem solving"* and the like. Some teacher trainees believe that certain amount of (ii) **assistance and tutoring** would be helpful, as expressed in responses like *"more classes or tutoring with maximum 10 people"*, *"revising how to solve math word problems with somebody who can explain it clearly"*. Teacher trainees also believe that when solving math word problems (iii) **group work** and cooperative learning could be helpful; in their words *"collective solving of math word problems joined with explanation"*, *"if we were allowed to solve math word problems together all the time"*. Another helpful factor stated by the teacher trainees is (iv) **gradation of the difficulty level of math word problems**, i.e. to begin with simple word problems, then continue with slightly harder word problems and finally solve even more complicated problems. This was well expressed in answers like *"to increase the difficulty of word problems and not to begin with complicated problems"* and *"to solve various types of math word problems with different level of difficulty"*. Teacher trainees also believe that when solving math word problems it would be helpful if they were provided with detailed (v) **exampling and illustration** of how to solve various types of word problems, i.e. step-by-step procedures. This was expressed in answers like *"we need to see an example of each type of math word problems for effective learning"* and *"learning to understand a lot of types of problems which are logically sequenced, and the way of solving."* The fifth category of teacher trainees' beliefs about what could be helpful when learning how to solve math word problems directly demonstrates their tendency and even preference to learn by rote rather than become creative mathematical thinkers. That is, however, not really desired in mathematics learning, though, in some cases inevitable.

Based on our experience with teaching mathematics to the teacher trainees included in the research sample we argue that in general they have difficulties with solving math word problems. On the other hand, as indicated by their answers in the questionnaire survey, they are able to identify the nature of the difficulties and see a way to improve their problem solving skills. Their beliefs are, naturally, influenced by their own previous experience as mathematics learners, both in positive and negative manner. As shown by the detailed analysis of their responses in the questionnaire survey, more than 70% of the teacher trainees directly expressed their will to improve their skills in solving math word problems. Furthermore, around 54% of the teacher trainees who expressed their will to improve their problem solving skills want to be active on their own, i.e. autonomous in their learning, whereas the remaining 46% of them prefer to rely on the activity of someone other who could help them, such as teachers, family, friends, and tutoring.

5 Conclusion

Solving math word problems belongs among the most challenging issues in mathematics teaching. Beliefs of teacher trainees are very important because in their future teaching practice they are expected to shape pupils' conceptions, understanding and skills in solving math word problems. In the presented research we were focused on beliefs of 131 teacher trainees about the issue of math word problems. The majority of respondents have difficulties with solving math word problems, and also more or less negative beliefs about teaching to solve math word problems. We assume they have negative beliefs and attitudes to mathematics in general. One of the causes can be their previous experience as mathematics learners at primary and secondary educational level. Like in a chain reaction, teachers' beliefs and attitudes are consequently passed further to their pupils. Therefore, recognition and, if necessary, re-shaping of teacher trainees' beliefs cannot be left on the margin of teacher education.

Acknowledgement

Preliminary results were presented at Autumn Conference for PhD Students supported by European Economic Area grant under project Improving Quality of Higher Education Based on Development of Multilateral Institutional Cooperation (SK 06-IV-01-008).

This paper was created with support of projects APVV-14-0446 - Hodnotenie kompetencií učiteľa and UGA-VII/20/2016 – Analytická geometria kvadratických útvarov v príprave budúcich učiteľov matematiky.

References

[1] LEPIK, M., PIPERE, A. *Baltic – Nordic comparative study on mathematics teachers' beliefs*. Latvia: Acta Pedagogica Vilnensia, 2012, 8 s.

[2] HANNULA, M., OKSANEN, S. *Finnish mathematics teachers' beliefs about their profession expressed through metaphors*. Helsinki: Suomen ainedidaktinen tutkimusseura, 2013, 10 s. ISSN 1799 – 9596.

[3] AHMAD, A., TARMIZI, A.R., NAWAWI, M. *Visual representations in mathematical word problem solving among form four students in Malacca*. USA: Elsevier, 2010.

[4] KISLENKO, K., LEPMANN, L. *Comparing Estonian, Latvian and Finnish teachers' beliefs about good mathematics teaching.* Tartu: Eesti Maaülikool, 2012.

[5] SHAPKOVA, A. *Beliefs on Effective Teaching and Learning of Mathematics (BETLM): Qualitative and Quantitative Perspectives of Latvian Grades 7-9 Mathematics Teachers.* Finland: University of Eastern Finland Joensuu, 2013, 9 s.

Current address

Šovčíková Petronela, Mgr.
Constantine the Philosopher University in Nitra
Tr. A. Hlinku 1, 949 74 Nitra, Slovak Republic
Tel. number: 0908 225 295, e-mail: petronela.sovcikova@ukf.sk

Klimentová Lucia, Mgr.
Constantine the Philosopher University in Nitra
Tr. A. Hlinku 1, 949 74 Nitra, Slovak Republic
Tel. number: 0905 192 183, e-mail: lucia.klimentova@ukf.sk

Slovak University of Technology in Bratislava
Faculty of Mechanical Engineering
16th Conference on Applied Mathematics **APLIMAT 2017**
Proceedings

ABOUT CONSTRUCTION OF TRIANGLE GIVEN BY INSCRIBED CIRCLE, LENGTH OF ONE SIDE AND ONE OTHER PARAMETER

ŠUMNÝ Timotej (SK)

Abstract. We have a geometric problem about construction of a triangle if we know length of its side c, radius of inscribed circle ρ and one other parameter. For solution of this problem we use the method of "a locus of points".

Keywords: locus of points, tangent line, inscribed circle

Mathematics Subject Classification: Primary 97G40

1 Introduction

In the learning process of geometry, we have many types of tasks for triangle construction. But there are only a few tasks which which are typical and those types are solved on the mathematic lessons. By using ICT in the learning process, we can solve unusual geometric tasks. One of the unusual triangle constructions is construction of triangle, if we know parameters: length of side $|\boldsymbol{AB}| = c$, radius of inscribed circle ρ and one other parameter. Typical solution is obtained by using the method of algebraic geometry. It means, that we use formulas for triangle and inscribed circle. These formulas we adjust to one equation. The solving of these equations is useful parameter to solve this task. This method depends on all three parameters, because every task has a special equation. In this paper we discuss this problem by using the method of a locus of points. We find curve determined by parameter values $|\boldsymbol{AB}| = c$, and ρ. This solution is alternative to usual solution to similar geometric task, which are solved by method of analytic geometry. Solution by using special curve like a locus of points is independent on the choice of the third parameter.

1.1 Method of solving geometric task

Methods for solutions of geometric constructions:

1. Method of synthetic geometry:
   a) Locus of points: we use known geometric object to find unknown point. This geometric object can be Euclidian constructible (circle, line..) or approximate Euclidian constructible (ellipse, hyperbole, parabola,..)
   b) Method of plane transformations: In the picture we find an image and a model transformation. In the school geometry we use special types of isometric transformations and similarities on plane.

   Methods of synthetic geometry are difficult to geometric imagination and geometrical manipulating with objects.
2. Method using calculations:
   a) Method algebraic geometry: We use known formulas and expand from this formulas the equations. Solution of this equation is parameter which we can use to solve geometric task. This method is in the principal universal and it can be easy to solve equation but it is difficult to know form of formulas for geometric objects.
   b) Method of analytic geometry: We use a Cartesian right coordinate system and using coordinate we can expend equations and solve it.

Method by using calculus requires intensive knowledge of the geometric object and algebraic manipulation with formulas.

2 Using curve to solve geometric task

One of the methods to solve geometric tasks is method “the locus of points”. In this method we find in the picture of problem such a curve, whose one or more points belong to the solution of the task. In fig. 1 we can see that the triangle sides are tangent to the circle. Problem is that we don’t know where the centre of the circle is. But we know that we can find it on a line m parallel to line $\boldsymbol{AB}$. This parallel line is at the distance of circle radius.

Unknown point $\boldsymbol{C}$ is the intersection of tangent lines from point $\boldsymbol{A}$ and point $\boldsymbol{B}$. The curve, we are looking for, is determined by point $\boldsymbol{C}$. Point $\boldsymbol{C}$ is determined by translated positions of circle along the line m. In fig.1 we can see change of the point C position determined by the tangent to the circle, resulting by change of the positions of circle centre on the line m. Fig. 1 was produced in software GeoGebra 4.2.

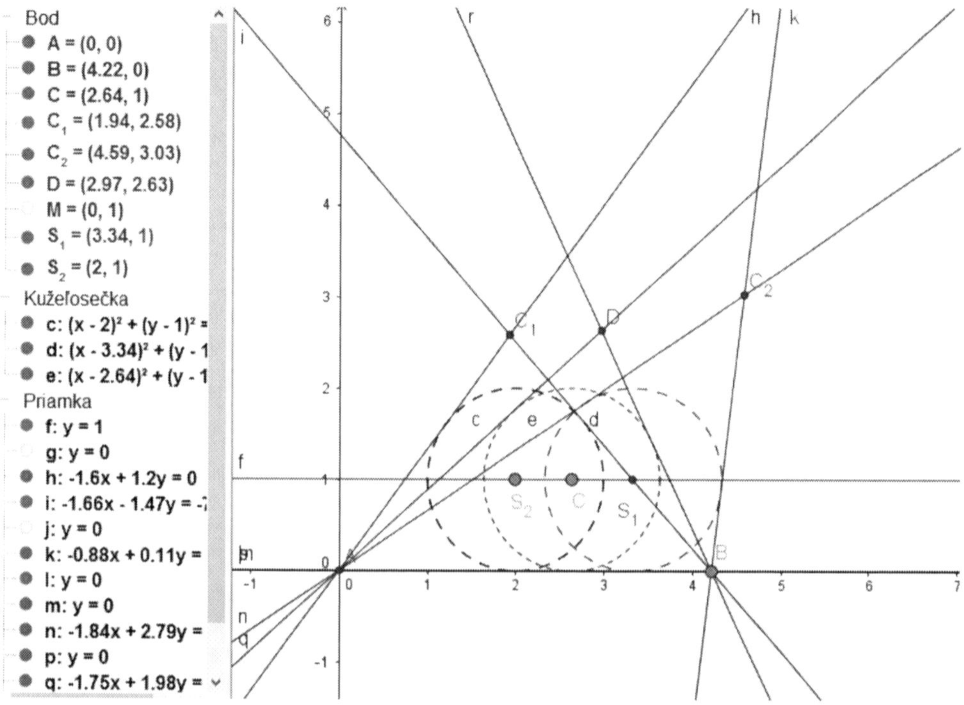

Fig. 1. Three point leading on curve determined c and ρ

2.1 Analytic representation of the curve

Symbols which we use are:
$|\boldsymbol{AB}| = c$, ρ is radius of inscribed circle, m is parameter for x coordinate of the centre of inscribed circle $m \in (0; c)$.

We choose orthogonal Cartesian coordinate system and point $A[0,0]$. Now, we can write formulas. First formula is equation of the circle:

$$(x-m)^2 + (y-\rho)^2 = \rho^2 \tag{1}$$

Tangents are passing through points $\boldsymbol{A[0,0]}$ and $\boldsymbol{B[c,0]}$. Tangent from point $\boldsymbol{A[0,0]}$ has equation:

$$y = qx \tag{2}$$

From (1) and (2) we get:

$$x^2(1+q^2) - 2x(m+q\rho) + m^2 = 0$$

Because our line is tangent to the circle, we use the condition $\boldsymbol{D = 0}$

$$4(m+q\rho)^2 - 4m^2(1+q^2) = 0$$

We get the formula:

$$q = \frac{2m\rho}{m^2 - \rho^2} \qquad m \neq \rho \tag{3}$$

Now for the second tangent line we will use point $B[c, 0]$ and tangent line has equation:

$$y = ax + b \tag{4}$$

From point **B** we get $b = -a.c$
Using formulas (1) and (4) we get:

$$x^2(1 + a^2) - 2x(m + a^2c + a\rho) + 2ac\rho + m^2 + a^2c^2 = 0$$

We are looking for tangent line, therefore we use $D = 0$

$$4(m + a^2c + a\rho)^2 - 4(1 + a^2)(2ac\rho + m^2 + a^2c^2) = 0 \tag{5}$$

From (5) we have determined a:

$$a = \frac{2\rho(c-m)}{c^2 - 2cm + m^2 - \rho^2} \qquad (c - m) \neq \rho\, ; c \neq 2\rho \tag{6}$$

Now, we have everything what we need to define our curve in parametric form $m \in (0; c)$:

$$x = -\frac{ac}{q-a}\ ; y = -qx \qquad q - a > 0$$

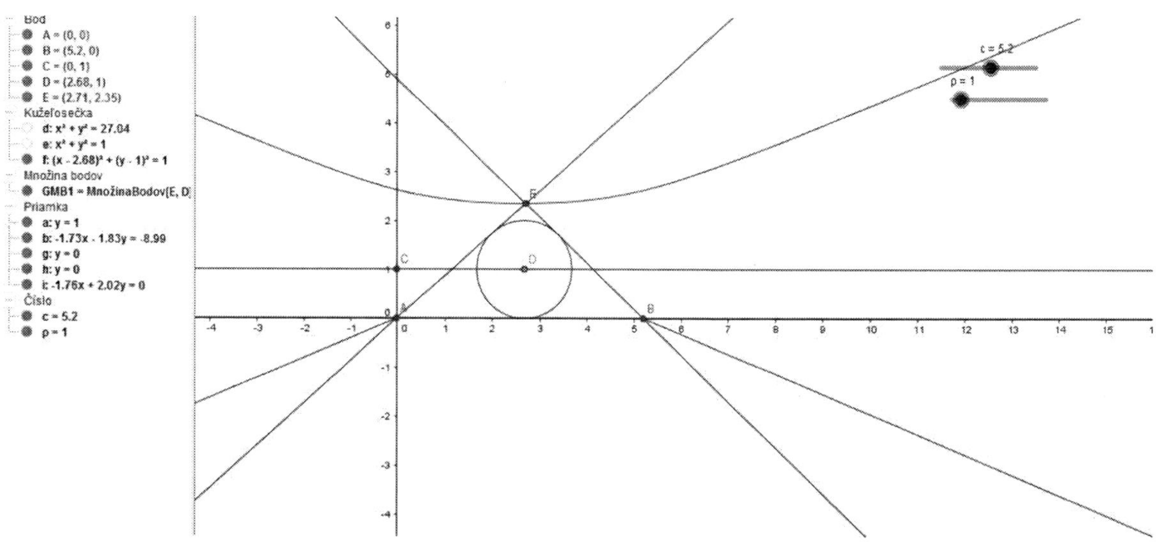

Fig. 2. Locus of point expanded by inscribed circle and length of side

In the fig.2 we have illustrated the curve for $|AB| = 5{,}2$, $\rho = 1$. We constructed this curve in the software GeoGebra 4.2.

Condition for drawing a curve is $y(\frac{c}{2}) > 0$ and this is for $c > 2\rho$.

3 Conclusion

This paper is focused on solving geometry problem by the method of „a locus of points". We found the useful curve to solve geometry task with specific values of parameters, which were the length of the triangle side and the radius of inscribed circle. If we know principia curve, we can approximate this curve by approximation, e.g. by finding points that belong to the locus of points for other parameters. In the other case, we can use software to draw the curve. This curve is useful for task, if we know value of the side length c, radius of inscribed circle ρ and one other parameter (e.g. height to the side c, side a (b) , median to the side (b)). The curve can be used for solutions of task to construct a triangle, if we know parameters: length of side $|\boldsymbol{AB}| = c$, radius of inscribed circle ρ and one other parameter. If we solve the task by using ICT it is better to use orthogonal Cartesian coordinate system. For $\boldsymbol{A}[\mathbf{0},\mathbf{0}]$ and $\boldsymbol{B}[\boldsymbol{c},\mathbf{0}]$ curve was determined in the parametric form, with equations $x = -\frac{ac}{q-a}$; $y = -qx$ for $q = \frac{2m\rho}{m^2-\rho^2}$ and $a = \frac{2\rho(c-m)}{c^2-2cm+m^2-\rho^2}$. Condition for curve existence is $c > 2\rho$. Advantage of this method is in its universality for every similar task.

Acknowledgement

This paper was created with support of project UGA-**VII/21/2016 – Approximate constructions of geometrical examples pointing toward cubic equations.**

References

[1] ŠTALMAŠEK, Ján. *Geometrické konštrukcie*. Slovenské vydavateľstvo technickej literatúry, 1959.

[2] ŠEDIVÝ, O. et al. *Vybrané kapitoly z didaktiky matematiky*. FPV UKF v Nitre, Nitra, Edícia Prírodovedec č. 2001, 78: 125.

[3] DESCARTES, R. Géométria. OIKOYMENH, Praha, 2010, ISBN 978-80-7298-313-1

Current address

Šumný Timotej, Mgr.

Tr. A. Hlinku 1, 949 74 Nitra, Slovak Republic

Tel. number: 0910 205 109, e-mail:timotej.sumny@ukf.sk

Slovak University of Technology in Bratislava
Faculty of Mechanical Engineering
16th Conference on Applied Mathematics **APLIMAT 2017**
Proceedings

A CONTRIBUTION TO SHORTEST PATH ALGORITHMS IN TRANSPORT NETWORKS WITH CAPACITY LIMITATIONS OF EDGES

TEICHMANN Dušan (CZ), DORDA Michal (CZ)

Abstract. The article presents modifications of two well-known pathfinding algorithms – Ford algorithm and Floyd algorithm. The presented modifications enable to find a shortest path which satisfies pre-defined capacity limitations – for example the limitations resulting from known measures and weight of a shipment we want to transport.

Keywords: graph theory, shortest path, Floyd algorithm, Ford algorithm, capacity limitations

Mathematics Subject Classification: Primary 68R10; Secondary 90B20

1 Introduction

The problem of finding paths with optimal parameters in transport networks belongs to elementary optimisation tasks that have been formulated and solved for practical using in transport. If an optimal path is not used when transporting passengers or goods, the transport process is not optimal and evinces potential reserves in its quality (for example from the point of view of its costs or any other criterion of effectivity). In practice, such reserves in effectivity bring lower competitiveness of offered transport services.

Computing methods which have been developed for finding paths can be found in a specialised branch of applied mathematics which is named operations research, more specifically in the discipline called graph theory. In general, pathfinding methods are methods that deal with finding paths with pre-defined criterions of their quality. The criterions can differ in individual examples; the most often used criterions are the path length, the path reliability or the path capacity. A suitable criterion is chosen in accordance with requirements of customers of transport systems or their operators.

2 Optimality criterions of paths in graphs and their importance from the point of view of transport practice

In the following text some chosen quality criterions which are often applied in transport practice are discussed; the criterions are consequently used in the article. In order to define the quality criterions, some formulas provided in the publication [3] are used.

Let us assume a transport network which is represented by an undirected graph $G[V,H]$, where V corresponds to a set of all vertexes of the graph G and H to a set of all edges of the graph G. The vertexes of the graph represent an origin and a destination of the path and vertexes that are important from the point of view of splitting of the network. The edges of the graph correspond to network sections (for example roads). For each edge of the network its weight is known; the weight corresponds to the length of the edge. For the edges some other important capacity limitations (load capacities of bridges or their vertical height clearances) are given. In addition, pieces of information about the shipment regarding capacity limitations of the transport network are known as well.

Let us assume that u is an origin of a path, v is a destination of a path and M is a set of paths leading from the vertex u to the vertex v. Please note that each path from the vertex u to the vertex v is a finite sequence of alternating vertexes and edges that starts in the vertex u and ends in the vertex v and each vertex in the path is visited once.

The first criterion is the path length. It must hold for the length of the path $m(u,v) \in M$ that:

$$d(m(u,v)) = \sum_{h \in m(u,v)} o(h) \tag{1}$$

For the length of the shortest path $m^*(u,v)$ it holds that:

$$d(m^*(u,v)) = \min_{m(u,v) \in M} \left\{ \sum_{h \in m(u,v)} o(h) \right\} \tag{2}$$

The length of the shortest path for which formula (2) holds is called the distance in graph theory.

The second criterion is the path capacity. In transport practice, the capacity of a path represents a different type of limiting parameters. For the capacity of the path $m(u,v) \in M$ we can write:

$$K(m(u,v)) = \min_{h \in m(u,v)} \{r(h)\} \tag{3}$$

It must hold for the capacity of the path $m^*(u,v)$ with the maximal capacity that:

$$K(m^*(u,v)) = \max_{m(u,v) \in M} \left\{ \min_{h \in m(u,v)} \{r(h)\} \right\} \tag{4}$$

Problems that are brought by transport practice evince a high variability. The variability can be demonstrated using two concrete examples.

The first example occurs if several pre-defined optimisation criterions are used for pathfinding; all of the criterions must be taken into account. It is not difficult to show on some practical examples that pathfinding with several optimisation criterions may be contra-productive. Let us explain it using a simple example. Let us consider that we want to take into account two criterions – the path length and the path reliability – for finding a path from an origin to a destination. It is more than obvious that we should try to minimise the length of the path and to maximise its reliability.

Theoretically, if we use such criterions together, the result may be that there is no path which satisfies both criterions. In practice, we can expect in general that the path with the minimum length and the path with the maximum reliability are different. In such cases we must prefer one criterion (the most important for us) over the other criterions. The solving process is divided into two partial phases if we assume two criterions. In the first phase we find a path satisfying the most important criterion; that means we solve a single-criterion problem. Within the second phase we calculate the value of the non-preferred criterion if the first phase has resulted in only single optimal path. If the first phase of the solving process has resulted in several optimal paths (with the same value of the preferred criterion), we choose such path (from the optimal paths we got in the first phase) which has the best value (minimum or maximum) of the non-preferred criterion. If the final path we got after both phases is acceptable for the customers or the operator, the path can be used. If the final path is not acceptable, we must try to find a compromise solution. By the compromise solution we mean such solution for which we do not insist on optimality of the path from the point of view of some criterions. That means it is possible to worsen values of some criterions in order to reach better values of the remaining criterions. Please note that the solving process can be easily generalised also for tasks with more than two criterions.

The second example is from the point of view of its solving easier that the first one; with the second example we can meet if only sole pre-defined criterion is used for optimisation but for pre-defined values of other criterions. For example, in transport we must solve such problem if we want to find a shortest path between an origin and a destination for a transport of a shipment with given weight and measurements. Because the second example is in transport more frequent than the first one the article deals with the second example.

3 State of the Art

As given earlier in the text, theoretical foundations of pathfinding are laid by graph theory, more precisely by algorithms for finding optimal paths in graphs. Such algorithms can be found in many publications, for example in [1], [2] or [3].

From the point of view of the practical application of the pathfinding algorithms, each algorithm consists of two basic phases – a calculation phase and a reconstructive phase. The calculation phase precedes the reconstructive phase. The first (calculation) phase calculates the value of an optimisation criterion. The second (reconstructive) phase finds the path that corresponds to the value of the optimisation criterion we got in the first phase. Principles that are employed in order to calculate the value of the optimisation criterion and to find the optimal path in graphs differ.

Some practical examples have revealed that some algorithms can be applied only for a single type of pathfinding tasks whilst some algorithms are universal. That means they can be successfully used for solving several types of pathfinding tasks. However, it may happen that it is necessary to define for the algorithm how input data should be transformed or how to change or modify some rules and calculation principles of the pathfinding algorithm if we want to use the algorithm for different types of the tasks.

The proposed algorithms that are described in the article are based on algorithms for finding shortest paths in graphs. The pathfinding algorithms that are mentioned most often in different publications (see for example [3]) are Ford algorithm, Dijkstra's algorithm and Floyd algorithm. The algorithms differ one another not only in the way how the length of the shortest path is calculated and how the shortest path is found in the graph but also in tasks for which the corresponding algorithm can be used. Ford algorithm has been primarily created in order to find a shortest path and its length between two given vertexes of a graph. However, after the calculation phase of Ford algorithm distances between the given origin and all other vertexes of the graph are known. Dijkstra's algorithm applies a partially different way of calculation and is primarily intended for finding shortest paths between a given origin and all other vertexes of a graph. And finally, Floyd algorithm is useful if we need to calculate distances between all vertexes of a graph at the same time.

In order to find a path with the maximum capacity an algorithm based on cuts in graphs and shortening of edges (see for example [3]) or Kruskal's algorithm for finding a maximum spanning tree (see for example [1]) can be applied.

In literature, some algorithms for solving combined tasks we are focused on in the article can be found. For example, in publication [1] an algorithm that can be employed for finding a shortest path with the maximum capacity is presented.

4 Problem Formulation

Let us consider a transport network which is represented by an undirected graph $G[V,H]$. For each edge of the network its weight representing its length is known. For the edges some other important capacity limitations (maximum loads of bridges or their vertical height clearances) are given. In addition, pieces of information about the shipment regarding capacity limitations of the transport network are known as well.

Our task is to find a shortest path from the origin to the destination so that no capacity limitation of the transport network is exceeded. In our paper we proceed from Ford algorithm described in publication [3] and Floyd algorithm described in publication [2]. Both algorithms will be modified for finding a shortest path under acceptance of the given weight of the transported shipment.

Let us establish the following notation:

- C is a set of capacity limitations of the edges (it holds $|C| = m$),
- $o(h)$ is the weight of the edge $h \in H$ and represents the length of the connection between the vertexes,
- $c_i(h)$ is the capacity limitation of a type $i \in C$ for the edge $h \in H$ (for example load capacity of a bridge on the edge),

- κ_i is a parameter of the transported shipment for the limitation of the type $i \in C$ (for example weight of the shipment).

Let us assume that the number of the capacity limitations is the same for all the edges. If the capacity limitation of the type $i \in C$ is not applied for the edge $h \in H$ in the real transport network, then $c_i(h) = K_i$, where:

$$K_i = \max_{h \in H}\{c_i(h) + 1\} \tag{5}$$

5 Modification of Ford Algorithm

Step 1
The origin of the path is denoted v_0, the destination of the path is v_n and the remaining vertexes are denoted v_i, where $i = 1,\dots,n-1$. Go to step 2.

Step 2
Evaluate the vertexes of the network according to the following rules:

$$\omega(v_0) = 0,$$
$$\omega(v_i) = \sum_{h \in H} o(h) \text{ for } i = 1,\dots,n,$$

Go to step 3.

Comment:
The notation $\omega(v_i)$ represents a current length of a path which has already been found to the vertex $v_i \in V$. If $\omega(v_i) = \sum_{h \in H} o(h)$, then the length of a path to the vertex $v_i \in V$ is not known.

Step 3
Initialise a set $\overline{V}$ which represents the set of all vertexes for which the length of the path leading from the vertex v_0 (the distance from the vertex v_0) is known. At the beginning of the solution process it holds that $\overline{V} = \{\ \}$. Go to step 4.

Step 4
Initialise sets $\Gamma(v_i)$ for $i = 0,\dots,n$. The set $\Gamma(v_i)$ corresponds to the set of the vertexes that are adjacent to the vertex v_i. At the beginning of the solution process it holds that $\Gamma(v_i) = \{\ \}$ for $i = 0,\dots,n$.
Go to step 5.

Step 5
If the vertex v_n belongs to the set $\overline{V}$, then go to step 10. If not, go to step 6.

Step 6

Find a vertex for which it holds that:

$$\omega(v_r) = \min_{v_i \in V \setminus \overline{V}} \{\omega(v_i)\}$$

Please note that if there are several vertexes with the minimum weight we can choose any of them. Go to step 7.

Step 7
Put the vertexes $v_i \in V \setminus \overline{V}$ with adjacency $p(h) = (v_r, v_i)$ for which it holds that:

$$\kappa_i \leq c_i(h) \text{ for } i \in C$$

into the set $\Gamma(v_r)$. Go to step 8.

Step 8
If for the edge $h \in H$ with adjacency $p(h) = (v_r, v_i)$, where $v_i \in \Gamma(v_r)$, it holds that:

$$\omega(v_r) + o(v_r, v_i) < \omega(v_i)$$

then

$$\omega'(v_i) = \omega(v_r) + o(v_r, v_i)$$
$$\omega(v_i) = \omega'(v_i)$$

If the inequality, which is given above, is not true, then the value $\omega(v_i)$ is not changed. Go to step 9.

Step 9
Put the vertex v_r into the set $\overline{V}$ and go to step 5.

Step 10
For the value of $\omega(v_n)$ it holds:

$$\omega(v_n) = \sum_{h \in m^*(u,v)} o(h)$$

Go to step 11.

Step 11
Initialise a set $\overline{\overline{V}}$ which represents the set of the vertexes which have already been put into the shortest path. In this phase it holds that $\overline{\overline{V}} = \{v_n\}$. Go to step 12.

Step 12
Set $i = 0$, initialise the vertex denoted v_{k_0} and put $v_{k_0} = v_n$. Go to step 13.

Step 13

Has the vertex v_0 been put into the set $\overline{\overline{V}}$? If yes, then go to step 17, if not, go to step 14.

Step 14

Create a set of all the vertexes that are adjacent to the vertex v_{k_i} – $\overline{\Gamma}(v_{k_i})$. The vertex v_j belongs to $\overline{\Gamma}(v_{k_i})$ if $\kappa_p \leq c_p(h)$ for $p \in C$. Go to step 15.

Step 15

Find the vertex $v_{k_{i+1}}$ which belongs to $v_i \in \overline{\Gamma}(v_{k_i}) \setminus \overline{\overline{V}}$ and for which it holds that:

$$\omega(v_{k_{i+1}}) = \omega(v_{k_i}) - o(v_{k_{i+1}}, v_{k_i})$$

Go to step 16.

Step 16

Increase the variable i by 1 – that means $i = i + 1$ – and put the vertex v_{k_i} into the set $\overline{\overline{V}}$. Go to step 13.

Step 17

We have got the following sequence of the edges:

$$v_n = v_{k_0}, v_{k_1}, v_{k_2}, \ldots, v_{k_r} = v_0$$

Go to step 18.

Step 18

Create the shortest path $m^*(u, v)$:

$$u = v_0 = v_{k_r}, (v_{k_r}, v_{k_{r-1}}), v_{k_{r-1}}, \ldots, v_{k_1}, (v_{k_1}, v_{k_0}), v_{k_0} = v_n = v$$

The algorithm is terminated.

6 Modification of Floyd Algorithm

The algorithm finds a sequence of matrices $C^{(0)}, C^{(1)}, \ldots, C^{(n)}$ where n is the number of vertexes.

Step 1

Create an initial matrix $C^{(0)}$ according to the following rules:

$c_{ij}^{(0)} = 0$, if $i = j$;

$c_{ij}^{(0)} = o(v_i, v_j)$, if $i \neq j$, there is an edge leading from v_i to v_j and it holds for the edge that $\kappa_i \leq c_i(h)$ for each limitation of the type $i \in C$;

$c_{ij}^{(0)} = \max_{(v_i, v_j) \in H} \{o(v_i, v_j)\} + 1$ in other cases.
Go to step 2.

Step 2
Initialise $k = 1$
Go to step 3.

Step 3
Calculate elements of a matrix $C^{(k)}$ using the following rule

$$c_{ij}^{(k)} = \min\left\{c_{ij}^{(k-1)}; c_{ik}^{(k-1)} + c_{kj}^{(k-1)}\right\}$$

When calculating the elements of the matrix $C^{(k)}$, the elements for which it holds that $i = j$, $i = k$, $j = k$, $c_{ik}^{(k-1)} = \max_{(v_i, v_j) \in H} \{o(v_i, v_j)\} + 1$ and $c_{kj}^{(k-1)} = \max_{(v_i, v_j) \in H} \{o(v_i, v_j)\} + 1$ do not have to be calculated because the elements do not change in the matrix $C^{(k)}$.
Go to step 4.

Step 4
If $k < n$, then $k = k + 1$ and go back to step 3. If not, the last matrix $C^{(n)}$ is the distance matrix. Go to step 5.

Step 5 (the reconstructive phase is begun by step 5)
Let us define an origin v_p and a destination v_t of a path and go to step 6.

Step 6
Assign a value of $\omega(v_j)$ which corresponds to $c_{pj}^{(n)}$ to each vertex $v_j \in V$ of the graph and go to step 7.

Step 7
Initialise a set $\overline{\overline{V}}$ which represents the set of the vertexes which have already been put into the shortest path. In this phase it holds that $\overline{\overline{V}} = \{v_t\}$. Go to step 8.

Step 8
Set $i = 0$, initialise the vertex denoted v_{k_0} and put $v_{k_0} = v_t$. Go to step 9.

Step 9
Has the vertex $v_{k_i} = v_p$ been put into the set $\overline{\overline{V}}$? If yes, then go to step 13, if not, go to step 10.

Step 10
Create a set of all the vertexes that are adjacent to the vertex v_{k_i} – $\Gamma(v_{k_i})$. The vertex v_j belongs to $\Gamma(v_{k_i})$ if $\kappa_s \le c_s(h)$ for $s \in C$. Go to step 11.

Step 11

Find the vertex $v_{k_{i+1}}$ which belongs to $v_i \in \Gamma(v_{k_i}) \setminus \overline{\overline{V}}$ and for which it holds that:

$$\omega(v_{k_{i+1}}) = \omega(v_{k_i}) - o(v_{k_{i+1}}, v_{k_i})$$

Go to step 12.

Step 12

Increase the variable i by 1 – that means $i = i + 1$ – and put the vertex v_{k_i} into the set $\overline{\overline{V}}$. Go to step 13.

Step 13

We have got the following sequence of the edges:

$v_t = v_{k_0}, v_{k_1}, \ldots, v_{k_r} = v_p$

Go to step 14.

Step 14

Create the shortest path $m^*(u,v)$:

$$u = v_p = v_{k_r}, (v_{k_r}, v_{k_{r-1}}), v_{k_{r-1}}, \ldots, v_{k_1}, (v_{k_1}, v_{k_0}), v_{k_0} = v_t = v$$

The algorithm is terminated.

7 Conclusion

The article is aimed at the modifications of two well-known pathfinding algorithms; our modification can be used for finding a shortest path under acceptance of some additional limitations. The proposed modified algorithms are based on Ford algorithm and Floyd algorithm which are described for example in publications [2] a [3] and as mentioned earlier, the modifications enable to find a shortest path in situations when some additional limitations are given. Such limitations can be for example weight of a transported carriage and related weight limitations of individual edges of the graph. The result that is obtained using one of the presented algorithms is a shortest path for which all the additional limitations are satisfied.

References

[1] PALÚCH, S., PEŠKO, Š. *Kvantitatívne metódy v logistike.* Žilinská univerzita v Žiline, 2006, ISBN 80-8070-636-0, 185 s.

[2] PLESNÍK, J. *Grafové algoritmy.* VEDA, 1983, ISBN 714-023-83, 344 s.

[3] VOLEK, J., LINDA, B. *Teorie grafů – Aplikace v dopravě a veřejné správě.* Univerzita Pardubice, 2012, ISBN 978-80-7395-225-9, 190 s.

Current address

Teichmann Dušan, doc. Ing. Ph.D.
Czech Technical University in Prague / Faculty of Transportation Sciences,
Department of Logistics and Management of Transport
Horská 3, 128 03 Praha 2, Czech Republic
Tel. number:, +420 224 359 171, e-mail: teichdus@fd.cvut.cz

Dorda Michal, Ing. Ph. D.
VSB – Technical University of Ostrava / Faculty of Mechanical Engineering, Institute of Transport
17. listopadu 15, 708 00 Ostrava – Poruba, Czech Republic
Tel. number: +420 597 325 754, e-mail: michal.dorda@vsb.cz

Slovak University of Technology in Bratislava
Faculty of Mechanical Engineering
16th Conference on Applied Mathematics **APLIMAT 2017**
Proceedings

REGIONAL INCOMES STRUCTURE ANALYSIS BASED ON COMPLEX SURVEY DATA

TEREK Milan (SK)

Abstract. The paper deals with the analysis of complex survey data by construction of empirical probability mass function, empirical cumulative distribution function and estimation of medians in domains with aid of sampling weights adjusted regarding to nonresponse. The estimates of median whole gross household incomes for eight Slovak regions on the basis of European Union statistics on income and living conditions in Slovak republic data are calculated and compared.

Keywords: complex survey, sampling weights, estimate of median, empirical probability mass function, empirical cumulative distribution function

Mathematics Subject Classification: 62D05

1 Introduction

Sometimes the data from complex survey are provided for analyses. The survey containing more components such as random sampling, stratification, clustering, ratio estimation and so on is obviously called complex survey. Stratification is used to reduce variances and obtain separate estimates for domains of interest, clustering, often with unequal probabilities, is used to reduce the costs. All these procedures require using of auxiliary information. A survey may be stratified with several stages of clustering and rely on ratio and regression estimation to adjust for other variables. In these cases sampling weights based on auxiliary information and design effects are commonly used to simplify matters and to provide the correct results.

The analysis based on using sampling weights will be described in the paper. The self-weighting and non-self-weighting samples will be studied. The using the sampling weights in construction of empirical probability mass function, empirical cumulative distribution function and in estimation of some population quantities will be described and applied on the data from the survey European Union statistics on income and living conditions 2014 (EU SILC 2014) in Slovak republic[1]. The regional structure of incomes in Slovak republic will be analyzed on the basis of data from

[1] EU SILC is a complex survey.

EU SILC 2014. The estimates of median whole gross household incomes for eight domains[2] – Slovak regions are calculated and compared.

2 Material and Methods

Statistical office of Slovak republic effectuates regular annual survey of mean nominal month wages according to different classifications[3]. Bureau of Labor Statistics in the USA for example regularly publishes the estimates of median and expected hourly wages according to different classifications along with standard errors and some percentiles[4]. Federal statistical office of Switzerland annually publishes median and lower and upper quartile of gross monthly wages according to different classifications[5]. The Consumer Expenditure Survey effectuated regularly in the USA collects information from the Nation's households and families on their buying habits (expenditures), income, and household characteristics. The strength of the survey is that it allows data users to relate the expenditures and income of consumers to the characteristics of those consumers. The survey consists of two components, a quarterly Interview Survey and a weekly Diary Survey, each with its own questionnaire and sample[6]. The surveys EU SILC are annually effectuated in all EU countries including Slovak republic. These surveys concern the households and persons. On the level of household also the different categories of incomes are collected. Similar surveys are regularly realized also in other countries.

The distributions of incomes or wages are obviously skewed and outliers[7] are present. Then, the interpretation power of the mean is very small[8]. Generally in such distributions the mean is not considered as appropriate measure of central tendency. So the mean income is not convenient measure of "typical" income. The median is generally considered as good measure of central tendency because of its stability and robustness toward outliers. Alternatively also some non-traditional measures of location could be interesting as good measures of central tendency for such distributions. The using of the trimmed mean ([7], p. 55), Winsorized mean or M-estimators is recommended[9]. Interesting results provides also traditional measures applied on the reduced data set[10].

2.1 Sampling Weights

Most frequently the means, totals, proportions and ratios are estimated in sample surveys. Sometimes the estimation of median and other quantiles[11] is needed. It is possible with aid of sampling weights. The sampling weights allow to construct also an empirical distribution for the

---

[2] Domain can be defined as subpopulation.
[3] More in details see in [16].
[4] More in details see in [15].
[5] More in details see in [14].
[6] More in details see in [13].
[7] We shall define an outlier in a set of data to be an observation (or subset of observations) which appears to be inconsistent with the remainder of that set of data ([1], p. 7).
[8] More in details see in [4].
[9] More in details see in [10].
[10] More in details see for example in [10].
[11] More in details see for example in [11] or [12].

population[12]. The sampling weights can be calculated using auxiliary information. In general auxiliary information is any non-sample information that can improve the precision of estimates. For example register variables play an important role in many surveys. They are used in creating of appropriate sampling design and/or in the computation of the survey estimates. In both uses they are called auxiliary variables[13].

Suppose we know the size N of finite population U. Symbol x denotes variable under study and also its values, $U = \{1,2,...N\}$ is the set of unit indexes in the population. Symbol S denotes sample from the population – subset containing n units from U. Let's π_i be the probability that unit $i \in U$ will be in random sample. Sampling weights for any sampling design are defined as follows

$$w_i = \frac{1}{\pi_i} \tag{1}$$

Sampling weight of unit i can be interpreted as number of units in the population represented by unit i.

In simple random sample (SRS) each unit in the population has the probability $\pi_i = n/N$ to be in random sample. Sampling weight of each unit in the population is $w_i = 1/\pi_i = N/n$ in this sample design. Each unit in simple random sample represents himself and $N/(n-1)$ of other non-sampled units in the population. In simple random sample

$$\sum_{i \in S} w_i = \sum_{i \in S} \frac{N}{n} = N \tag{2}$$

It is known that in finite population the population mean μ_K of variable x is

$$\mu_K = \frac{1}{N}\sum_{i=1}^{N} x_i \tag{3}$$

and the population total τ is

$$\tau = \sum_{i=1}^{N} x_i = N\mu_K \tag{4}$$

In SRS sample mean $\overline{X}$ is an unbiased estimator of the population mean μ_K. Its value $\overline{x}$ is calculated according to

$$\overline{x} = \frac{1}{n}\sum_{i \in S} x_i \tag{5}$$

To estimate a population total τ, the unbiased estimator $N\overline{X}$ can be used. Its value is calculated according to

[12] In fact it is an empirical distribution of the observation from the population.
[13] More in details see in [8], [9].

$$\hat{\tau} = N\bar{x} \tag{6}$$

With aid of sampling weights population mean and population total can be estimated by

$$\hat{\tau} = \sum_{i \in S} \frac{N}{n} x_i = \sum_{i \in S} w_i x_i \tag{7}$$

$$\hat{\mu}_K = \frac{\hat{\tau}}{N} = \frac{\sum_{i \in S} w_i x_i}{\sum_{i \in S} w_i} \tag{8}$$

The estimators of these and another parameters in stratified sampling, cluster sampling and other sampling designs including its combinations such as for example multistage stratified cluster sampling can be expressed by sampling weights. These weights can be modified in regard to nonresponse and coverage error[14].

2.2 Self-Weighting and Non-Self-Weighting Samples

Sampling weights for all observations units are equal in self-weighting surveys. Self-weighting samples can[15] be considered representative of the population because each observed unit represents the same number of unobserved units in the population. Noting that the term „representative sample“ can have different meanings. In [2], p. 23 – 24, nine different meanings of this term are listed. In this paper this term is used only in one way: a sample is said to be representative with respect to a variable if its relative distribution in the sample is equal to its relative distribution in the population. For example, a sample is representative with respect to the variable gender, if the percentages of females and males in the sample are equal to the percentages of females and males in the population. Sampling weights for all observations units are not equal in non-self-weighting surveys.

2.3 Estimating an Empirical Probability Mass Function, Empirical Cumulative Distribution Function and Some Population Quantities

Suppose the values for the entire population of N units are known. A value of probability mass function (PMF) in x is

$$p(x) = \frac{N_{(x)}}{N} \tag{9}$$

where $N_{(x)}$ is number of units whose value is x. A value of cumulative distribution function (CDF) in x is

$$F(x) = \sum_{y \le x} p(y) \tag{10}$$

[14] More in details, for example in [5].
[15] In the absence of nonsampling errors.

Note that it is probability mass function and cumulative distribution function of observation from the population[16].

Sampling weights allow to construct empirical probability mass function and empirical cumulative distribution function. Empirical probability mass function $\hat{p}(x)$ is defined by the sum of weights for all observations taking on the value x divided by the sum of all the weights:

$$\hat{p}(x) = \frac{\sum_{i \in S; x_i = x} w_i}{\sum_{i \in S} w_i} \tag{11}$$

Empirical cumulative distribution function $\hat{F}(x)$ is

$$\hat{F}(x) = \sum_{y \le x} \hat{p}(y) \tag{12}$$

2.3.1 Estimating of Some Population Quantities

Any population quantity can be estimated from the empirical probability mass function $\hat{p}(x)$ or from empirical cumulative distribution function $\hat{F}(x)$. For example population mean can be estimated by:

$$\hat{\mu}_K = \sum_x x\hat{p}(x) = \frac{\sum_{i \in S} w_i x_i}{\sum_{i \in S} w_i} \tag{13}$$

If cumulative distribution function F were continuous, population median would be defined as value $\tilde{\mu}$ satisfying $F(\tilde{\mu}) = \frac{1}{2}$. In discrete case function $F(x)$ does not attain the value $\frac{1}{2}$. The finite population median is defined to be value $\tilde{\mu}$ satisfying $F(\tilde{\mu}) = \frac{1}{2}$ if such a value exists. Otherwise, a population median is any value in the interval $[\tilde{\mu}_1, \tilde{\mu}_2]$, where $\tilde{\mu}_1$ is the largest value of x in the population with $F(x) < \frac{1}{2}$ and $\tilde{\mu}_2$ is smallest value of x with $F(x) > \frac{1}{2}$. In general, Q_p is $100 \cdot p$ % quantile (percentile) if $F(Q_p) = p$, if such a value exists, otherwise, $Q_p \in [a, b]$ where a is the largest population value of x with $F(x) < p$ and b is the smallest value of x with $F(x) > p$. If $p < \frac{1}{N}$, Q_p is the smallest value of x and if $p > 1 - \frac{1}{N}$, Q_p is the largest value of x.

[16] The „model-free“ or „distribution-free“ approach to sample survey is under consideration (for more details see in [3], p. 8 – 9).

Population quantiles are estimated as follows. Since the empirical cumulative distribution function $\hat{F}$ is a step function, the interpolation is usually needed to find a unique value for the quantile. Let y_1 be the largest value in the sample for which $\hat{F}(y_1) \leq p$ and let y_2 is smallest value in the sample for which $\hat{F}(y_2) \geq p$. Then

$$\hat{Q}_p = y_1 + \frac{p - \hat{F}(y_1)}{\hat{F}(y_2) - \hat{F}(y_1)}(y_2 - y_1) \tag{14}$$

The statistics calculated using weights are much closer to the population quantities. We note that estimators calculated using this method are not necessarily unbiased, however, or numerically stable. For example the estimator $\hat{\sigma}_K^2$ of the population variance is sensitive to roundoff error ([6], p. 293).

3 Analysis of Regional Structure of Incomes on the Basis of EU SILC 2014 data

The analysis of regional structure of incomes was effectuated on the data from the survey EU SILC realized in Slovak republic in 2014. The stratified two-stage survey is used in Slovak republic. A stratification was effected with two stratification variables – region and settlement size. The survey EU SILC 2014 was effectuated on the sample of 6 010 households, 5 490 households and 13 433 individuals 16 and more years old were included to database. Sampling weights were calculated and modified with respect to nonresponse. These weights can be used to inference about the population of Slovak households. Other modified sampling weights involve individuals. In general EU SILC sample data are the data from complex survey.

Data from EU SILC 2014 are concentrated in many sets. Each household has one identification number[17]. The analysis of the whole gross household incomes in eight domains – Slovak regions was realized. The mentioned domains correspond with values of one from stratification variables. Firstly the matching of needed data – sampling weights[18] and whole gross household incomes[19] was effected according to household numbers. Then the matched data were distributed according to regions. Eight sets of data were obtained, one for each region. Each region was analyzed separately. The values of the empirical probability mass function were calculated according to (11) and on the basis of that the values of the empirical cumulative distribution function were calculated by relation (12) for each region. Finally, the estimate of median whole gross household income was calculated according to relation (14) for each region. All calculations were realized in Excel 2013. The obtained results are in Tab. 1.

---

[17] Values of the variables DB030, HB030.
[18] Values of the variable DB090.
[19] Values of the variable HY010.

Region number	Region name	Estimate of median whole gross household income in 2014 (Euros)	Order of region according to median whole gross household income
1	Bratislava	14 491,37	1.
2	Trnava	13 969,12	4.
3	Trenčín	14 368,47	2.
4	Nitra	12 379,67	7.
5	Žilina	14 054,85	3.
6	Banská Bystrica	11 746,41	8.
7	Prešov	13 595,22	5.
8	Košice	13 118,16	6.

Source: own

Tab 1. Regional structure of median whole gross household income in 2014

The estimate of population median whole gross household income for the whole Slovak republic in the year 2014 equals 13 305,83 eur.

4 Conclusions

The obtained ordering of regions according to median whole gross household income is very interesting. The difference between first Bratislava and second Trenčín regions is not very large. The median household incomes of the third Žilina, fourth Trnava and fifth Prešov are also very close. There are bigger differences among last three regions. The median whole gross household income of Banská Bystrica is surprisingly low.

Most large self-weighting samples used in practice are not simple random samples. Mostly, more complex sample designs are used and we work with data from complex survey. The using of special software that is specifically designed for survey data to obtain valid statistics from complex survey data is needed. If instead statistical software that is intended for data fulfilling assumptions of independence and identic distribution is used, the standard errors, confidence intervals and hypotheses testing results produced by the software will be wrong.

Standard statistical methods may be applied to the self-weighting sample to obtain point estimates. A histogram of the sample values displays the approximate frequencies of occurrence in the population, the population mean, median and other population quantities can be estimated by sample mean, median and other sample statistics. When we work with the data from simple random sampling, fulfilling the usual statistical assumptions that observations are independent and identically distributed the calculation of standard errors, confidence intervals and hypotheses testing is possible with aid of the current statistical software.

If the sample is not self-weighting, point estimates of means, totals and other quantities produced by standard statistical software will be also biased. So in this case the using the standard statistical software to point estimation is also inappropriate. It was the case also in mentioned application. The EU SILC sample is non-self-weighting. The point estimation of parameters using sampling weights is needed.

Sometimes the standard statistical methods supposing the independence and identic distribution of observations are applied to the data from complex surveys. In [6], p. 287 – 288 is stated: "When you read the paper or book in which the authors analyze data from the complex survey, see whether they accounted for the data structure in the analysis, or whether they simply ran the raw data through non-survey statistical package procedure and reported the results. If the latter, their inferential results must be viewed with suspicion".

The estimates of median whole gross incomes obtained with aid of sampling weights were also compared with estimates obtained by sample medians[20] in mentioned application. The difference between estimates obtained „with weights" and „without weights" varied from (– 1,82%) to 7,50%. There are not negligible differences. When no weights estimates are used, also ordering of regions according to median income is different: 1. Bratislava, 2. Trenčín, 3. Žilina, 4. Prešov, 5. Košice, 6. Trnava, 7. Nitra, 8. Banská Bystrica.

It is clear that the estimate obtained with aid of finite weights which allow the used sample design, nonresponse and potentially also coverage error better reflects the reality. Note that the steps of the weighting process are quite subjective. Even computing the base weights can be subjective; for example, in many telephone surveys, the number of telephones per household is not obtained and, thus, not used for computing selection probabilities. But there is general agreement in the statistical community that, for estimating descriptive statistics, weights should always be used even though some estimates may not benefit from this practice.

Acknowledgement

The paper was supported by grant from Grant Agency of VEGA no. 1/0092/15 entitled "Modern Approaches to Complex Statistical Surveys Design".

References

[1] BARNETT, V., LEWIS, T.: *Outliers in Statistical Data*. Hoboken: Wiley and Sons, 1994. ISBN 0-471-93094-5.

[2] BETHLEHEM, J.: *Applied Survey Methods. A Statistical Perspective.* Hoboken: Wiley and Sons, 2009, ISBN 978-0-470-37308-8.

[3] COCHRAN, W. G.: *Sampling Techniques*. New York: J. Wiley and Sons, 1977, ISBN 0-471-16240-X.

[4] HALLEY, R. M.: Measures of Central Tendency, Location, and Dispersion in Wage Survey Research. In: *Compensation and Benefits* 2004/36, 39 (2004) p. 39 - 52.

[5] LEVY, P. S., LEMESHOW, S.: *Sampling of Populationas. Methods and Applications*. Fourth Edition. USA: Wiley, 2008, ISBN 978-0-470-04007-2.

[6] LOHR, S. L.: *Sampling: Design and Analysis*. 2nd edition. Boston: Brooks/Cole, 2010, ISBN-10: 0-495-11084-1.

[7] PIEGORSCH, W. W.: *Statistical Data Analysis. Foundations for Data Mining, Informatics, and Knowledge Discovery*. Chichester: Wiley & Sons, 2015, ISBN 978-1-118-61965-0.

[20] Calculated by function Median in Excel (no weights used).

[8] SÄRNDAL, C.-E., SWENSSON, B., WRETMAN, J. (2003): *Model Assisted Survey Sampling*. New York: Springer, 2003, ISBN 0-387-40620-4.

[9] SÄRNDAL, C.-E., LUNDSTRÖM, S. (2005): *Estimation in Surveys with Nonresponse*. USA: Wiley, 2005, ISBN 0-470-01133-5.

[10] TEREK, M.: Odľahlé dáta a charakteristiky polohy v analýzach miezd a príjmov. In: *Revue sociálno-ekonomického rozvoja*, Vol. II, number 1/2016.

[11] TEREK, M.: *Interpretácia štatistiky a dát. Štvrté doplnené vydanie*. Košice: Eqilibria, 2016, ISBN 978-80-8143-177-7.

[12] TOŠENOVSKÝ, J., NOSKIEVIČOVÁ, D.,: *Statistické metody pro zlepšování jakosti*. Ostrava: Montanex, 2000.

[13] Consumer Expenditure Survey. Available at: http://www.bls.gov/cex/faq.htm.

[14] Gross monthly wage. Available at: http://www.bfs.admin.ch/bfs/portal/en/index/themen/03/01/new/nip_detail.html?gnpID=2015-308.

[15] Occupational Employment Statistics. Available at: http://www.bls.gov/oes/current/oes_wa.htm

[16] Priemerná nominálna mesačná mzda zamestnanca v EUR, Sk, indexy (1991 - 2014). Available at: http://www.statistics.sk/pls/elisw/objekt.send?uic=1410&m_sso=2&m_so=15&ic=40.

Current address

Terek Milan, prof. PhD.
University of Economics
Dolnozemská cesta 1, 852 35 Bratislava, Slovak Republic
Tel. number: 421-2-67295713, e-mail: milan.terek@euba.sk

Slovak University of Technology in Bratislava
Faculty of Mechanical Engineering
16th Conference on Applied Mathematics **APLIMAT 2017**
Proceedings

IMPOSSIBLE PICTURES: WHEN ART HELPS SCIENCE EDUCATION

TORRE Matteo (IT)

Abstract. The paper talks about the use of the impossible figures in science education discovering, in particular, the great artistic and communicative value of the artworks of Oscar Reutesvärd, that long before Escher and Penrose, had drawn impossible triangles and stairs through a conscious use of the "Japanese" perspective. In the last part of paper we analyzed the links between art and impossible figures showing how over the centuries their communicative power has been (more or less) deliberately used by artists such as W. Hogarth, or Op Art, movement which tried to involve the observer deceiving his eyes and his brain.

Keywords: Science and Art, impossible figures, Oscar Reutersvärd, R. Penrose, Tribar

Mathematics Subject Classification: Primary 97M80, 97M60; Secondary 81T18

1 Introduction

"*Everybody knows that a thing is inexplicable until some unknowing fool comes along and explains it*". So Einstein said about impossible situations that often need to be addressed when we talk about science at any level, even and especially in teaching. In my opinion, assigning to students, as an alternative to the traditional science education, issues that lead to impossible situations is a very powerful tool to develop the key skills required by OECD-PISA tests and all other forms of assessment of skills to which the students of Italian high school are subjected (INVALSI, *in primis*). The impossible situations have the great benefit of requiring, implicitly, from the student the development of a metacognitive reasoning on the matter, not only a trivial application of the rules, of which it is not able to offer a justification. To better explain this thesis, I propose two mathematical examples.

If you ask to a student of the first year of high school to build a scalene triangle with defined value of sides or angles, he will build it without any problem using a ruler and a goniometer. If instead we propose to the student the realization of a triangle with internal angles that measure 75°, 45° and 65°, he may (or better, should) be able to assert that it is impossible to build this triangle and, if instead he tried to build it accurately using ruler and protractor, you will face an impossible situation which led him to reflect on the nature of the figure and its links to the values of the internal angles. Let's go to the equations. If we asked to a student of the first two years of high school solve the equation $2x = 46$, he will be able, perhaps through the mechanical process of dividing both sides by 2, to easily solve the equation. If you require to solve the equation $0x = 16$, education research [1] tells us that some students proposed as a solution 0, others as solution 16 and only a part (of just under 50%) immediately responds correctly that is an impossible equation. The reasons for these difficulties [2] are to be found in the inability to skip from the "natural" language to symbolic language, i.e. the inability to translate the relationships between the variables in the relationship between symbols and back again. In fact, to solve the second equation you must use a metacognitive reasoning that leads to the translation from the algebraic register to the numerical register (*What number multiplied by 0, results in 16? No one, then the equation is impossible*), or even using only the mechanical rule which, however, add the reasoning that it is impossible to divide by zero and so that the equation is impossible.

From these two examples it is easily seen that when we ask to solve an exercise that will be impossible the student must perform a sort of reasoning on what they have learned, maybe even in a mechanical way, that help them to realize what they really understood. In particular, it avoids the continued use of structures that have been consolidated only repeating mechanical operation that can be used to fix concepts, but that do not help to develop specific skills and interdisciplinary ones. In my opinion this is a very useful way to develop basic skills in high school and that, as we will see later in the paper, can be developed by making use also of artistic and graphic works that will increase the interdisciplinary nature of the teaching proposal.

2 Oscar Reutersvärd: the father of impossible pictures

Going back along the humanity development, we find the theme of the impossible declined in its various sides: for example, in mathematics, it is impossible to square the circle and to rectify the circumference; in the after-war English literature, the Theatre of Absurd, and, in particular, the work of Samuel Beckett *Waiting for Godot*; in the *Divina Commedia*, the theme of the ineffability (XXXIII Canto of Paradise); in Art, the paradoxes and impossible figures.

From this brief introduction we understand that the theme is absolutely fascinating not only for its transversality, but above all for the opportunity to explore each theme through an absolutely fascinating world.

First of all the definition: an impossible object is a type of optical illusion. It consists of a two-dimensional figure which is instantly and subconsciously interpreted by the visual system as representing a projection of a three-dimensional object [3]. In most cases the impossibility becomes apparent after viewing the figure for a few seconds. However, the initial impression of a 3D object remains even after it has been contradicted. There are also more subtle examples of impossible objects where the impossibility does not become apparent spontaneously and it is necessary to consciously examine the geometry of the implied object to determine that it is impossible.

The father of the impossible figures is commonly identified in the Swedish artist Oscar Reutersvärd (1915-2002). Reutersvärd's originality appeared early in his career at the age of 18. In 1934, the school student created a figure, the "Impossible triangle", composed of a series of cubes in perspective, called by Reutersvärd himself "*Opus 1*" (Fig. 1).

Fig. 1. *Opus 1*, O. Reutersvärd, 1934

Reutersvärd's artworks are based on a prospective trick usually called the "Japanese perspective" and which essentially consists in this: an object, or a series of objects, are seen simultaneously in more prospects (at least 2, sometimes 3) under different directions (points of view), but in a way that there is a "weld" between the resulting figures, in a general solution that cannot exist, realistically absurd [4].

Fig. 2. O. Reutersvärd in his studio, 1998

Reutersvärd (Fig. 2) achieved more than 2500 works using the purity of essential figure, all geometric; only the latest works are "contaminated" by the use of watercolors. In fact, he believed that the aesthetic beauty of his creations consisted in the "impossible figures" in themselve, not in the magic that could arise from them, and that instead somehow fascinated and inspired the works of the other great artist who made use of impossible figures: Escher. Reutersvärd contends that he did not need it, confiding only in the pure figure. For example, as well as Escher, also Reutersvärd has transformed some of his "impossible figures" into stairs, but he never felt the need to let it walk by monks or in perpetual descent waters; it is always and only limited to implicitly suggest to those who are looking to visit it with imagination!

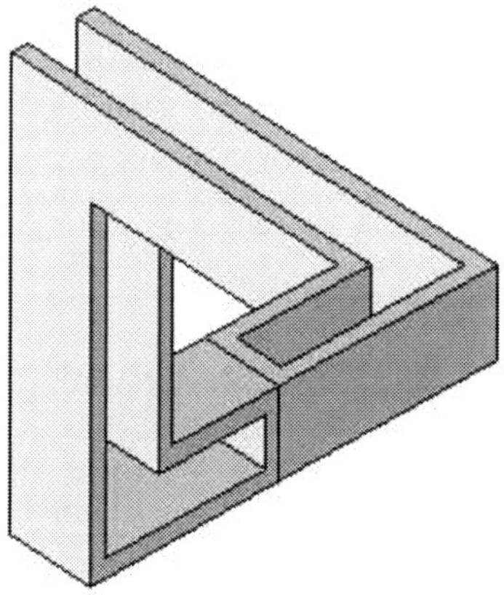

The particular perspective of Reutersvärd's works, with their declared intention to be false, represent a revolt towards the Renaissance scientific perspective and the implicit anthropocentric view [5]. To express his conception of the world Reutersvärd choose this particular perspective probably because in it all parallel lines remain parallel and do not converge visually at any point. In this way we create numerous angles that generate different geometric shapes that join with each other giving rise to an object that does not exist in reality but that our brain has to interpret to be able to extrapolate the realistic side.

Reutersvärd thanks to his impossible figures had important commissions for works in Swedish public buildings as the National Museum of Sweden and the Museum of Modern Art in Stockholm. In 1982 the Swedish Government chose three works to decorate a series of commemorative stamps dedicated to the artist and the same year the artist experimented with new techniques using new media, while remaining true to the theme of optical illusions, creating them using the computer. In 1993 he made the laser an impossible three-dimensional sculpture and in 1995 experienced his creations in virtual reality. In 1996 the Company of Motorways of Helsingborg gave him the construction of a monument to furnish the futuristic highway ring Malmöleden [6]. In the last years of his graphic techniques experimented linked life always linked to new technologies and the improvisation of figures made impossible when listening to Bach Fugues.

If Reutersvärd to recognize the conscious use of the "Japanese Perspective" and the remarkable creativity in using and designing the impossible figures, it is important to remember that in the past the history remembers other artists who made use of impossible figures resorting to forced perspective, such as the painter William Hogarth (1697-1774) who in his famous *False perspective* of 1754 (Fig. 3a) uses fishing line and the lighting of a pipe to make a paradoxical scene that just using the brain can adjust several prospective deceptions that the eye alone cannot fix (fig. 3b). The work of Hogarth in the design and printing are considered important in the development of modern graphic communication languages and his works, also present in many science textbooks, in order to illustrate the deceptions of optical illusions, demonstrate the importance of knowing how to make good use of representation visual tools to communicate correct scientific posts.

Fig. 3a. *False perspective*, W. Hogarth, 1754

Fig. 3b. *False perspective* (visual lines), W. Hogarth, 1754

Fig. 4. *Saint Ignazio glory*, A. Pozzo, Church of Saint Ignazio in Rome, 1685

We have also to remember the spectacular work of the Seventeenth-century Italian artist Andrea Pozzo in the Church of Saint Ignazio in Rome (Fig. 4). When the church was designed, it was planned to have an enormous dome, but the neighboring Dominican monks complained that the resulting structure would cut their library off from light. Commissioning Pozzo to create a painted dome and an adjacent second storey in a flat surface solved the problem. Nevertheless, this clever solution was not without its problems. If you stand away from the correct viewerpoint, the illusio crumbles. Although many people at the time were amazed by the illusion, Pozzo was severely reprimanded by other architects, who criticized the painted architecture's "physical construction". In his defense Pozzo replied that a friend of his would bear "all damage and costs" should the structure ever fall down! [7]

Those mentioned are only some of the most famous examples of artists who have deliberately made use of impossible figures both for communicative reasons or for purely aesthetic reasons. In addition to these, scientists, psychologists and neuroscientists have analyzed the impossible figures and their links with the perception of our brain. We will return in the last paragraph on this topic, recalling other artists also in relation to contemporary art. Now I want to try to analyze the usefulness and potential of impossible figures in teaching science.

3 Impossible pictures: examples and didactic usage

In this first analysis I will speak of the use of some impossible figures, born in the arts or in other areas, to develop key scientific expertise in high school. The impossible figures represent, in fact, a unique meeting point between the mathematical world and the artistic one and a reminiscence of the medieval figurative world, in which proliferated in multiple spatial reading and relief images in which the characters penetrate into each other. The impossibility of the figures drawn byReutersvärd is not always obvious, not everybodyperceived at first sight the incongruity and the paradox of some images, as the human eye tends to see in a two-dimensional description of an object that is still three-dimensional [8]. In Italy, there are already some educational courses based on the analysis as a forecast of the works of Reutersvärd, for children of elementary school, such as the exhibition-workshop designed from the ForMATH research group that works with B. D'Amore, one of the most important world expert of Reutersvärd's artworks [9].

In my opinion the ability to recognize the incongruity is the result of both mathematical and physical skills that the observer must activate with the act of looking. We try to investigate further this idea, thinking of some works of Reutersvärd and other impossible figures.
We start with Reutersvärd "*Opus 1*" of 1934 because it is the first *intentional* impossible object known. Is important to emphasize this intentionality. Impossible objects, as we said previously had already been seen in the arts, not always consciously used by the artists, but here Reutersvärd, bored during a Latin lesson, began to scribble a sort of six-pointed star, and encircled it with 3-dimensional. If we numbers his cubes from 1 to 9 (Fig. 5)

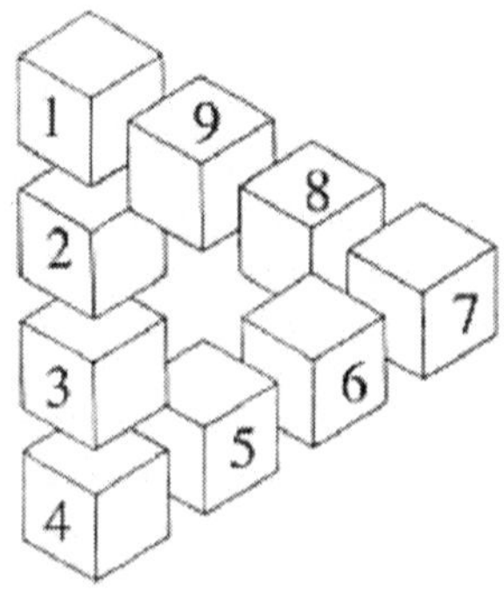

Fig. 5. *Opus 1* (whit numbered cubes), O. Reutersvärd, 1934

and we observed only the "cubes" from 1 to 7 (without the 8th and 9th), the perspective is correct, and has the direction from left to right of the beholder; if you observe the cubes from 4 to 7 to 9 (excluding only the 2nd and 3rd), the outlook is still correct, but has the direction from right to left of the beholder; you can also proceed removing 5 and 6, getting even a proper perspective [6]. What completely changes the question, then, is the attempt to reconcile all these partial versions in one piece, in a single design: you have more points of failure, of inconsistency, which transform the locally correct figure in a globally impossible one. It is this aspect that one sufficiently competent student in geometry, or who knows the essential arguments and not too complex of solid geometry, should detect and that allows him to develop basic soft skills.

Closely linked to this work of Reutersvärd, and very similar, is the "Tribar" or "Triangle of Penrose" (Fig. 6), presented by the the Penroses, father (Lionel Sharples, born in 1898, psychologist) and son (Roger, born in 1931, mathematician, famous for his studies on space-time and black holes), in a paper published in the *British Journal of Psychology* in 1958 [10] and created by the Penroses in a totally independent way from the artworks of Reutersvärd.

Fig. 6. Tribar or *Triangle of Penrose*

The paper, in fact, was inspired by Escher's work, that had captured the interest of the Penrose for the theme, but not Reutersvärd's work, that at the time was unknown by the two. Roger Penrose only discovered Reutersvärd's work in 1984. The paper, published in 1958, was read by the Swedish artist, who was well aware of this impossible object and another figure (known as the the Penrose staircase), was developed by the Penroses independently from Reutersvärd's works.
The tribar appears to be a solid object, made of three straight beams of square cross-section which meet pairwise at right angles at the vertices of the triangle they form. The beams may be broken, forming cubes or cuboids. This combination of properties cannot be realized by any 3-dimensional object in ordinary Euclidean space. Such an object can exist in certain Euclidean 3-manifolds. There

also exist some 3-dimensional solid shapes each of which, when viewed from a certain angle, appear the same as the 2-dimensional depiction of the Penrose triangle (for example Fig. 7, that is a sculpture in Perth, Australia).

Fig. 7. Penrose triangle in Perth, Australia

The tribar also demonstrates 3-fold rotational symmetry: each time the figure is rotated of 120 degrees, it is mapped to itself. This rotation group has three elements, a rotation by 120 degrees, a rotation by 240 degrees, and the identity, a rotation by 360 degrees.
If one third of the tribar figure is mounted on a rotating disk, and illuminated by a stroboscope operating at three times the frequency of rotation of the disc, then a completed image will be created by our eyes and brain. Persistence of vision will allow us to see all three images merged together as one image, thus demonstrating the rotational symmetry of the tribar figure [11]. Can we also observe that the Tribar is generated by rotating of this figure (Fig. 8).

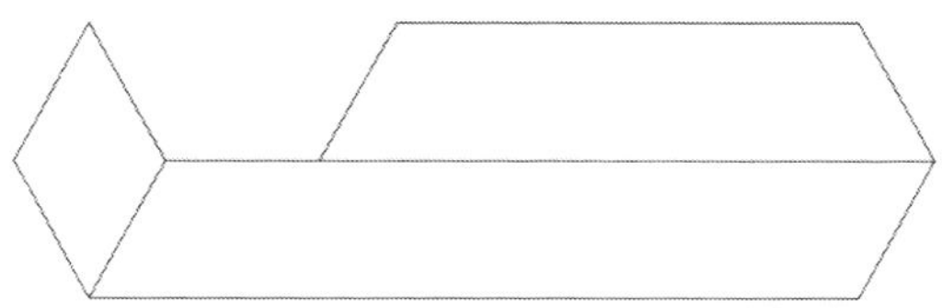

Fig. 8. Polygon from whose full rotation the Tribar is generated

It is possible to construct analogies to the Penrose triangle with other regular polygons to create a "Penrose polygon", but the visual effect is not as striking, and as the sides increase, the object seems merely to be warped or twisted (Fig. 9).

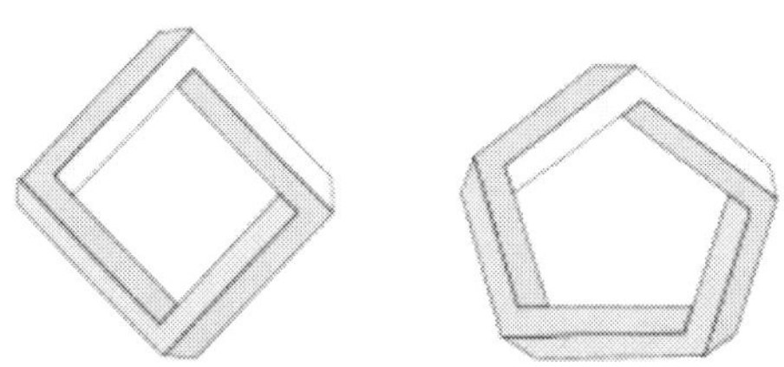

Fig. 9. Polygon from whose full rotation the Tribar is generated

It would be interesting to propose to the students the three-dimensional construction of the Penrose triangle (Fig. 10), which is actually impossible, but that can be created in an almost amazing way through simple prospective escamotages.

As for the building process, it is enough to have paper and scissors, to print and cut out a template available online[1] and then realize the assembly according to the related instructions. The triangle is unreal, but the observation in a certain point can lead our mind to see it, because of a confusion between different levels and perspectives! [12]

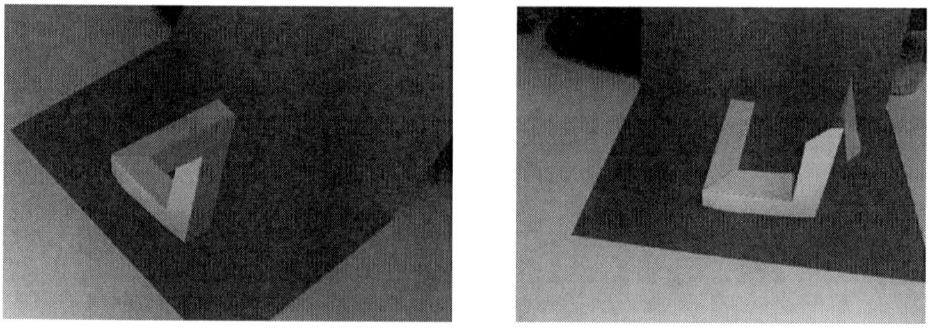

Fig. 10. The construction of a Penrose triangle (two different points of view)

This work can be the starting point for a reflection on astronomy and on the conformation of constellations, which are such only if seen from our point of view, that is the Earth. But since the stars occupy a three-dimensional space (and a very extended one), if we were to go from the Earth, the constellations as we know them would disappear, because the position of the stars would assume a different conformation.

In 1937, Reutersvärd had created his first impossible stairs, much earlier than Escher and Penrose. The staircase had been, in fact, developed with over twenty years in advance by the Swedish artist, but neither Penrose nor Escher knew of his drawings. Inspired, thanks to a radio program, from Mozart's compositional method, described as a "creative automatism" (every creative idea put on paper inspired by a new idea) in 1950 Reutersvärd began drawing a series of impossible objects in the same "unconscious and automatic way" during a trip from Stockholm to Paris [13] (Fig. 11). He did not realize, while drawing, that his figure was a continuous flight of stairs, but the process allowed him to draw ever more complex drawings, step by step.

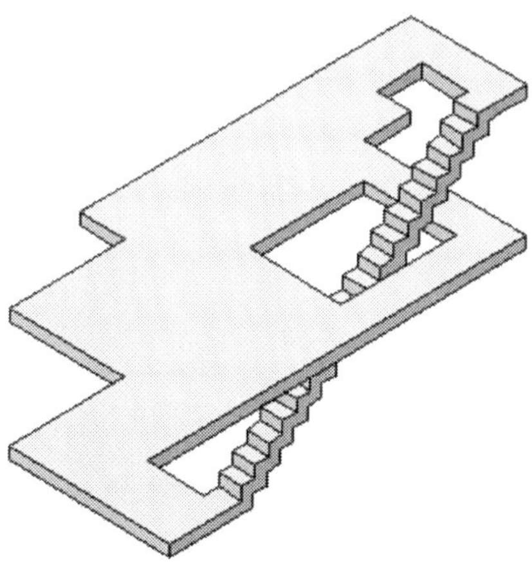

Fig. 11. Impossible staircase by O. Reutersvärd

[1] http://www.coolopticalillusions.com/illusions/coolprinttriangle.pdf

In an indipendent way Penrose draw the Impossible staircase (Fig. 12), a variation on the Tribar, that is a two-dimensional depiction of a staircase in which the stairs make four 90-degrees turns as they ascend or descend, yet forming a continuous loop, so that a person could climb them forever and never get any higher. This is clearly impossible in three dimensions. In their original article, the Penroses noticed that "*each part of the structure is acceptable as representing a flight of steps but the connexions are such that the picture, as a whole, is inconsistent: the steps continually descend in a clockwise direction*". [10]

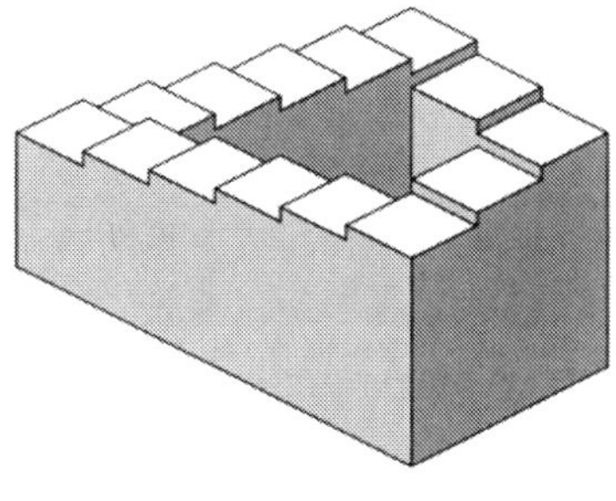

Fig. 12. The Penrose impossible staircase

Our visual system opts for a simple interpretation based on local relationships within the figure, rather than choosing a complex, yet correct, interpretation that takes the entire figure into account. We observe that each stair that is one step clockwise from its neighbor is also one step downward, and so we perceive the staircase as eternally descending. In principle, we could instead perceive the figure correctly as depicting four sets of stairs that are discontinuous, and viewed from a unique perspective, however such a perception never occurs.
Now students, having reasoned on the Tribar and impossible staircases, can analyze more consciously Escher artworks, such as *Waterfall* (Fig. 13) and *Ascent and Descent* (Fig. 14), where respectively Tribar and Penrose staircase appear, recognizing not only a geometric impossibility but also a physical impossibility that defies the laws of gravity.

Fig. 13. *Waterfall*, M.C. Escher, 1961

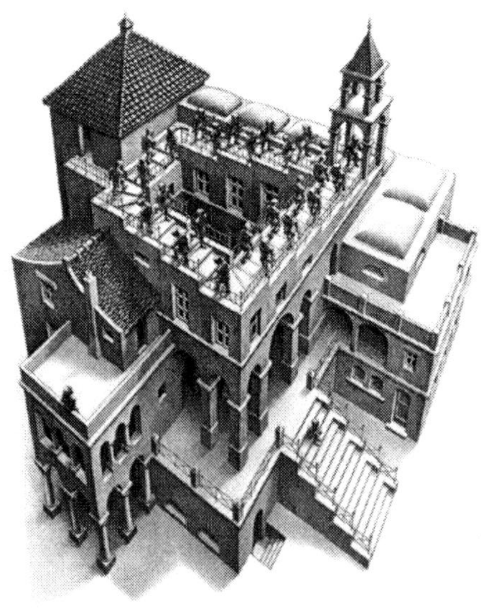

Fig. 14. *Ascent and Descent*, M.C. Escher, 1960

This is another great utility of impossible figures, thanks to which we can develop transversal key skills. In my opinion, supporting the use of impossible figures can promote inclusion and help children with learning difficulties or with space-time recognition deficiencies to understand that a single object can take different perspectives and forms depending on the points of view. This can be a useful starting point for a reflection to convey to them a wider social value: learn to confront and accept different opinions and many ideas that come from social life.

The last impossible figure that we talk about is the Necker cube. The Necker cube is an optical illusion first published as a rhomboid in 1832 by Swiss crystallographer Louis Albert Necker. The Necker cube is an ambiguous line drawing. The effect is interesting because each part of the picture is ambiguous by itself, yet the human visual system picks an interpretation of each part that makes the whole consistent (Fig. 15).

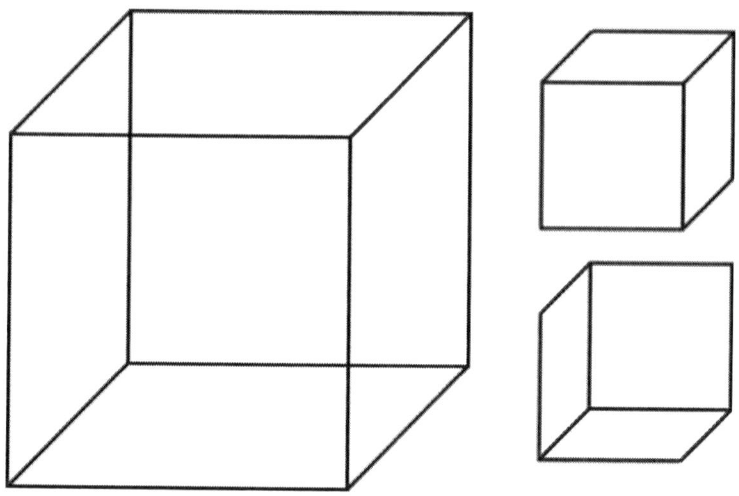

Fig. 15. The Necker cube (left) and his two possible views (right)

The Necker cube is sometimes used to test computer models of the human visual system to see whether they can arrive at consistent interpretations of the image the same way humans do [14]. Humans do not usually see an inconsistent interpretation of the cube. A cube whose edges cross in an inconsistent way is an example of an impossible object, specifically an impossible cube (compare it to Penrose triangle). With the cube on the left, most people see the lower-left face as being in front most of the time. This is possibly because people view objects from above, with the top side visible, far more often than from below, with the bottom visible, so the brain "prefers" the interpretation that the cube is viewed from above. Another reason behind this may be due to the brain's natural preference of viewing things from left to right, therefore seeing the leftmost square as being in front.

There is evidence that, by focusing on different parts of the figure, one can force a more stable perception of the cube. The intersection of the two faces that are parallel to the observer forms a rectangle, and the lines that converge on the square form a "y-junction" at the two diagonally opposite sides. If an observer focuses on the upper "y-junction" the lower left face will appear to be in front. The upper right face will appear to be in front if the eyes focus on the lower junction. Blinking while being on the second perception will probably cause you to switch to the first one. It is possible to cause the switch to occur by focusing on different parts of the cube. If one sees the first interpretation on the right it is possible to cause a switch to the second one by focusing on the base of the cube until the switch occurs to the second interpretation. Similarly, if one is viewing the second interpretation, focusing on the left side of the cube may cause a switch to the first. The Necker cube has shed light on the human visual system. The phenomenon has served as evidence of the human brain being a neural network with two distinct equally possible interchangeable stable states [14].

The Necker cube is therefore another way to develop geometric skills and, at the same time, a simple tool to try to reason with students on the mechanisms related with the vision that our brain controls and imposes on us.

We have described only some impossible figures, because they are the ones we consider most useful in teaching, but many other examples such as Schröder's stairs or the impossible trident can be used to implement my future researches on this argument.

4 Links with contemporary art and future develops

The impossible figures, as we have seen, are an excellent approach to mathematics and are a useful tool to develop geometrical and physical-spatial skills.

Among the artists (contemporary or not) the idea to use mathematics explicitly to build their own works is very common [15] and that, in my opinion, is a further opportunity to virtuously end our first trip to the wonderful world of impossible objects. Besides the various Escher, Hogarth, Reutersvärd, we cannot forget Dali, Arcimboldo, Magritte, but also the artistic movement which has been influenced and who most deliberately have used optical illusions: the Op Art.

The antecedents of Op art, in terms of graphic and color effects, can be traced back to Neo-impressionism, Cubism, Futurism and Dada. *Time Magazine* coined the term *op art* in 1964, in response to Julian Stanczak's show *Optical Paintings* at the Martha Jackson Gallery (Fig. 16), to mean a form of abstract art (specifically non-objective art) that uses optical illusions. Op art perhaps more closely derives from the constructivist practices of the Bauhaus [16].The art form is commonly included in the kinetic art, which explores the illusion dimension. It is an essentially graphic art, based on a strict definition of the operating method. The artists want to get through lines placed in various modular and structural grills, effects that induce a state of perceptive instability. In this way, they stimulate the involvement of the observer [17].

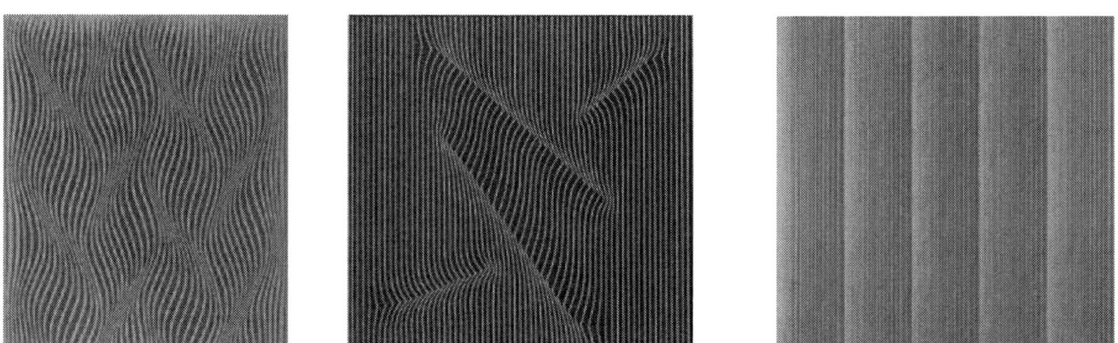

Fig. 16. Three artworks of *Optical Paintings*, J. Stanczak, Martha Jackson Museum, 1964

In particular it is a dynamic visual art that stems from a discordant figure-ground relationship that puts the two planes—foreground and background—in a tense and contradictory juxtaposition. Artists create op art in two primary ways. The first, best known method, is to create effects through pattern and line. Often these paintings are black-and-white, or otherwise grisaille—as in Bridget Riley's famous painting, *Current* (Fig. 17), on the cover of *The Responsive Eye* catalogue (Fig. 18). Here, black and white wavy lines are close to one another on the canvas surface, creating a volatile figure-ground relationship. Starting in 1965 Bridget Riley began to produce color-based op art, however, other artists, such as Julian Stanczak and Richard Anuszkiewicz, were always interested in making color the primary focus of their work. Josef Albers taught these two primary practitioners of the "Color Function" school at Yale in the 1950s. Often, colorist work is dominated by the same concerns of figure-ground movement, but they have the added element of contrasting colors that produce different effects on the eye. For instance, in Anuszkiewicz's "temple" paintings, the juxtaposition of two highly contrasting colors provokes a sense of depth in illusionistic three-dimensional space so that it appears as if the architectural shape is invading the viewer's space.

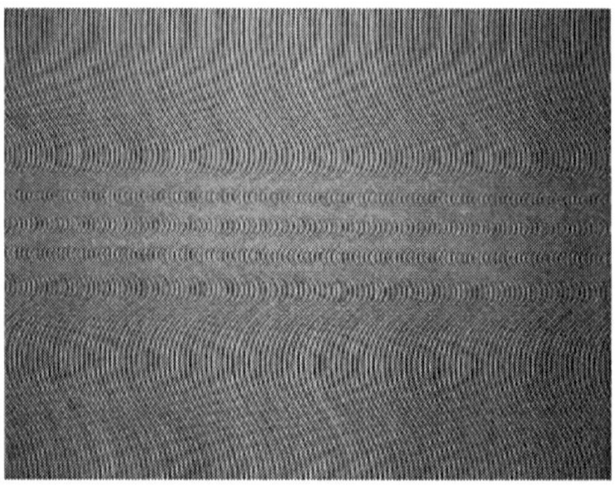

Fig. 17. *Current*, B. Riley, MoMA, 1964

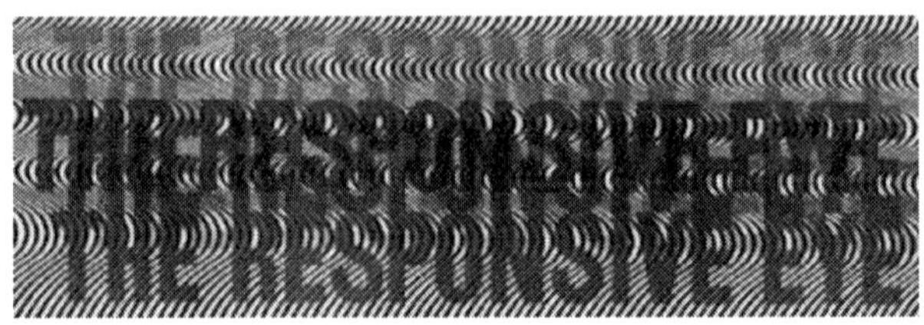

Fig. 18. Cover of *The Responsive Eye* catalogue (particular), 1964

Although has never clearly joined the movement, we can consider the artworks of Reutersvärd as belonging to Op Art. In fact, the creation of a multi-dimensional universe, consisting of plans that communicate with each-other regardless of their unreality, and that lead the viewer to cross them, doing attention to your own inner situations, cognitive and emotional, assigned to the works of artists Reutersvärd and other creators of impossible figures, visual test value and quantitative psychology. Probably M. Duchamp was right when he said that "*the painting should not be only retinal or visual; it should have to deal with the gray matter of your intellect, instead of being purely visual*".

In this regard, it is our firm intention to deepen in a later paper other aspects, related to this last topic, as the staircase of Schröder, the Thiery module, the figure/background Rubin effect. In particular, we want to turn our interest in those habits, those conventions and its automation vision but also those connected with the way that man and mind express themselves and communicate with reality and the external world.

References

[1] ARZARELLO F., BAZZINI L., E CHIAPPINI G., L'Algebra come strumento di pensiero. Analisi teorica e considerazioni didattiche, *Progetto Strategico CNR - TID*, Quaderno n. 6, 1994.

[2] DI CHIARA I., Equazioni di primo e secondo grado, Tesi di laurea in Matematica, Univ. di Palermo, A.A. 2003-2004.

[3] https://en.wikipedia.org/wiki/Impossible_object

[4] D'AMORE B., Oscar Reutersvärd, *Didattica della Matematica nel III millennio. Atti del Convegno Naz.le "Incontri con la matematica"*, n. 14 Castel San Pietro Terme 3-5 novembre 2000, pp. 277-280, Ed. Pitagora, Bologna, 2000.

[5] CALDARELLI M., Cinquant'anni di figure impossibili. *Arte e scienza*, 85, pp. 70-73, 1985.

[6] D'AMORE B., L'opera di Oscar Reutersvärd, *La matematica e la sua didattica*. 3, pp. 240-245, 2002.

[7] SECKEL A., Masters of Deception: Escher, Dali & the Artists of Optical Illusion, Sterling Pub Co Inc, 2007, ISBN 9781402751011.

[8] REUTERSVÄRD O., Impossible coloring book, Perigee Books New York, 1982.

[9] http://www.formath.it/divulgazione-scientifica/le-figure-impossibili-di-oscar-reutersvard/

[10] PENROSE L.S., PENROSE R., Impossible Objects: a special type of visual illusion. *The British Journal of Psychology*, 49, 1958.

[11] http://www.stem2.org/je/tribar.pdf

[12] BARATTI M.T., CARANDINI, A., MAZZUCCO P., TORRE M., Avatar: when a flower becomes mathematics, *Aplimat 2015, XIV International Conference*, Bratislava, 2-4 Feb. 2015.

[13] DEUTSCH D., The paradox of pitch circularity, *Acoustic Today*, June 2010.

[14] https://en.wikipedia.org/wiki/Necker_cube

[15] D'AMORE B., Prospettiva impossibile e arte contemporanea, *Treccani Scuola Dossier*, 2007, pp. 1-7.

[16] https://en.wikipedia.org/wiki/Op_art

[17] TORRE M., From Quantum Physics to Piero Manzoni: the "new" role of the observer, *Aplimat 2015, XIV International Conference*, Bratislava, 2-4 Feb. 2015.

[18] GREGORY R. L., Illusion in Nature and Art, Gerald Duckworth & Company, London, 1973.

[19] BARROW J. D., Impossibility: The Limits of Science and the Science of Limits,

[20] https://en.wikipedia.org/wiki/Oscar_Reutersvärd

[21] http://www.fink.com/papers/impossible.html

[22] http://ilpiccolofriedrich.blogspot.it/2011/09/costruire-figure-impossibili.html

[23] http://im-possible.info/english/articles/trident/trident.html

Current address

Torre Matteo, high school teacher
Liceo Scientifico Statale "B. Cavalieri"
v. Madonna di Campagna 18, Verbania (VB), Italy
E-mail: matteo.torre1984@gmail.com

Slovak University of Technology in Bratislava
Faculty of Mechanical Engineering
16th Conference on Applied Mathematics **APLIMAT 2017**
Proceedings

TREND COMPONENT ANALYSIS OF TIME SERIES

ŤOUPAL Tomáš (CZ), ŠEDIVÁ Blanka (CZ)

Abstract. This paper deals with the problem of trend component estimation particularly for economic time series. The proposed approach could be used in many applications e.g. GDP, unemployment, stock markets etc. There will be presented one method using orthonormal system generated by Gram-Schmidt orthonormalization process from some available linearly independent sequence of time series or functions in an inner product space. Obtained results are then applied to the real data sets of the foreign trade and consumer price index.

Keywords: Time series, trend component, orthonormal system, inner product, foreign trade, consumer price index

Mathematics Subject Classification: Primary 91B84, 91G70; Secondary 62P20

1 Analysis and estimation model of the trend component

The content of this paper is focused on the trend component estimation of time series in specific (macroeconomic) cases. The trend component can be estimated by many different ways and here proposed approach is based on the decomposition of time series into several components (trend, seasonal, cyclical and random component). One of these considered components is required and unknown "trend". The trend is very popular and frequently interpreted word in the world. There is possible to say that this is a very modern term in recent years e.g. in macroeconomics, microeconomics and all subareas of applied economics, statistics, sociology, finance, commerce etc.

Thus, the word "trend" used here can be interpreted as a "suitable fit" of the measured values by some specific curve (smoothed curve). Generally, a trend line reflects the long-term changes in the average behavior of time series such as long-term growth or decline.

1.1 Model

Before the derivation of the trend estimation model itself, there is assumed that the main area of further interest is based on the basic knowledge of macroeconomic theory. This theory assumes the availability of various macroeconomics time series. Some of financial time series can be described by the following (component, index) model

$$X(t) = X_0 \sum_{i=0}^{m} a_i(1 + r_i)^t + \varepsilon_t, \qquad t = 1, 2, \ldots, T, \qquad r_i > -1, \tag{1.1}$$

where by the factor $a_i(1 + r_i)^t$ is denoted mean growth or decline in some component (volume, price, currency, …) of initial value X_0 and by the index r_i is denoted annual increments in the case of $a_i > 0$ or annual losses or stagnation in the case of $a_i \leq 0$. Further, there is used parameter t describing time and parameter m indicates the number of sub-components.

The size of part, that is subject to the influence of the modeled index r_i is given by the value of the parameter $|a_i|$ and the last variable in this model is a time series of the random component ε_t, where the requirements for its properties will be added during the following text.

The previous model given by (1.1) can be rewritten into a new revised form, in which the new convention $X_i = X_0 a_i$ is used.

$$\begin{aligned} X(t) &= X_0 \sum_{i=0}^{m} a_i(1 + r_i)^t + \varepsilon_t \\ &= \sum_{i=1}^{m} X_i(1 + r_i)^t + \varepsilon_t, \qquad t = 1, 2, \ldots, T, \qquad r_i > -1. \end{aligned} \tag{1.2}$$

A commonly used mathematical relationship (description) can be considered as one of several possible models of financial time series and still it can be used to estimation of unknown trend component. In the other worlds, it is an econometrical model for the development of the further research work. For the illustration of considered model there are shown some possible developments (movements) of described time series in the following figures.

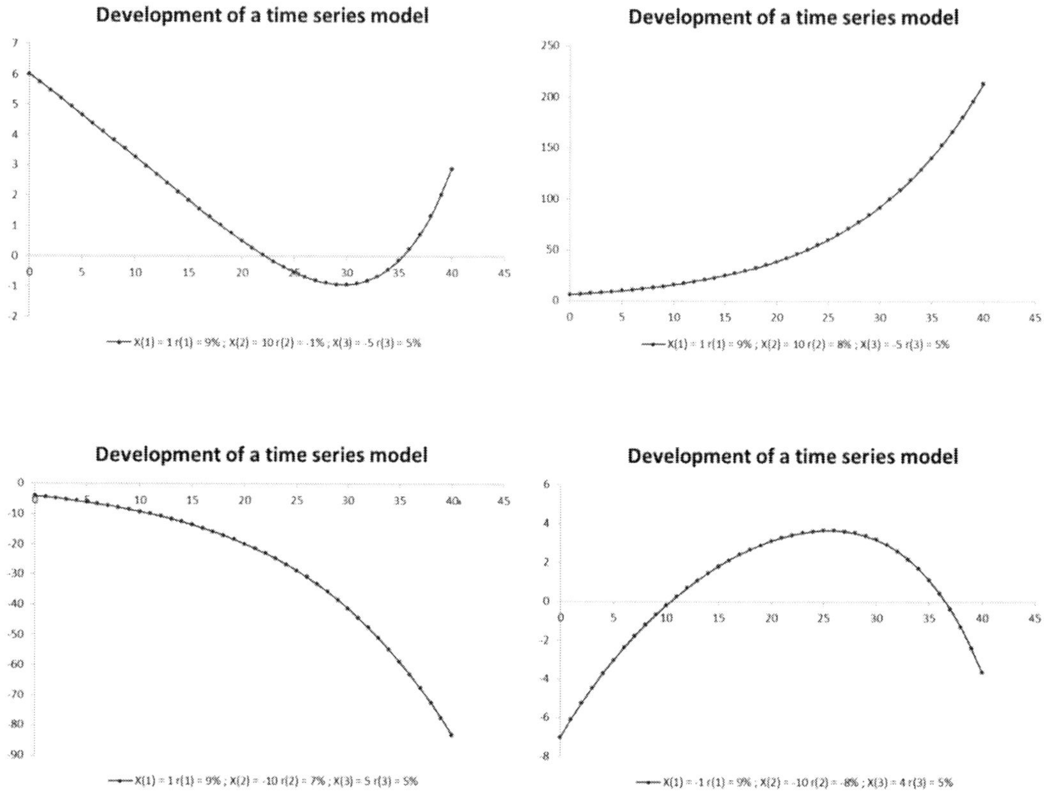

Fig. 1. Possible developments of considered time series

1.2 Formulation of the problem

Generally, there are some sequences of time series (or functions) denoted as $f_i(t)$ where:

1. $i = 1, \dots, m$
2. $t = 1, \dots, T$
3. $m \leq T - 1$
4. $\forall i \in \{0, 1, \dots, m\} \exists t \in \{1, \dots, T\}: f_i(t) \neq 0.$

The principle of these time series models is in their area of application, where the possible solutions of this problem comes. Moreover, there will be assumed the linear independence of this set of some sequences. This is a reason why zero time series or functions are excluded.

Now there is a linearly independent system (or set of elements) $f_i(t)$ for $i = 0, \dots, m$ and $t = 1, \dots, T$, which may be used to generate the required orthonormal system with respect to an inner product. It can be done by the Gram-Schmidt orthonormalization process. We assume that the Gram-Schmidt orthonormalization process methodology is known to the readers including all variables (an inner product, norm, vectors projection, …) and it is also useful to include a more detailed description of the algorithm for computer calculations. A computer algorithm for generating of the required orthonormal base can be interpreted by the following procedure [2]:

$$g_0(t) = \frac{f_0(t)}{\|f_0(t)\|}; \qquad i = 0, \qquad t = 1,..,T,$$
$$h_i(t) = f_i(t) - \sum_{j=0}^{i-1} \langle f_i, g_j \rangle \, g_j(t); \qquad i = 1,..,m, \qquad t = 1,..,T, \tag{1.3}$$
$$g_i(t) = \frac{h_i(t)}{\|h_i\|}.$$

The procedure above and its implementation seems fine in the terms of process, but the problem occurs with the implementation of procedure. This process is mostly "numerically unstable" because the calculated vectors are not completely orthogonal due to errors occurring at an internal display numbers. The loss of orthogonality devalues the whole Gram-Schmidt process. Therefore, there are used some modifications for the numerical stabilization of the whole process [2]:

$$g_0(t) = \frac{f_0(t)}{\|f_0(t)\|}, \qquad i = 0, \qquad t = 1,..,T,$$
$$h_i(t) = f_i(t), \qquad i = 1,..,m, \qquad t = 1,..,T,$$
$$h_i(t) \leftarrow h_i(t) - \langle h_i, g_j \rangle g_j(t), \qquad j = 0, ..., i-1, \tag{1.4}$$
$$g_i(t) = \frac{h_i(t)}{\|h_i\|}.$$

It is a good time now to select one (any) matching sequence to construct an orthonormal system, because this selection is the first step of the method below to estimate the unknown trend components. This suitable system has a large number of possible sequences from which the selection can be made. For the purposes of this paper, here are index lines (sometimes known as interest lines), whose graphical interpretation is shown in the following figure including generated orthonormal system from the possible set of index lines.

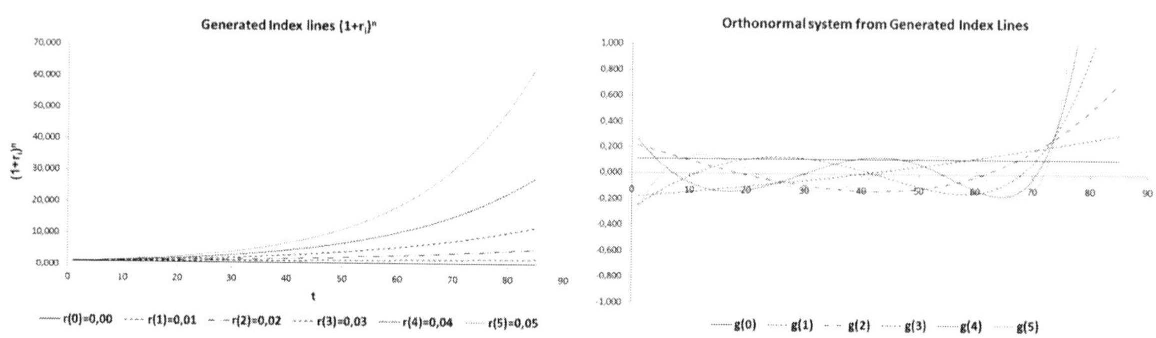

Fig. 2. Generated index lines and orthonormal system

1.3 Least-squares method using orthonormal system

The main principle of the proposed estimate is based on the fact that there will be found an "appropriate" unknown approximation of the trend curve known as $X(t)$ for obtained set of discrete observations for $t = 1, \dots, T$. One standard and well known method, which is often used for obtaining different approximation solutions from overdetermined system (i.e. by this expression is meant a set of equations, where are more equations than unknowns) is called the Least squares method.

Following approximation of unknown trend component from the original time series $X(t)$, which will be denoted as $X_m(t)$, can be given as a linear combination

$$X_m(t) = \sum_{i=0}^{m} a_i g_i(t), \qquad t = 1, \dots, T, \qquad m \leq T - 1, \tag{1.5}$$

so, that the general approach to solving this problem using the least squares method can be described by the following relationship

$$\|X_m(t) - X(t)\|^2 = \sum_{t=1}^{T} (X_m(t) - X(t))^2 \xrightarrow[a_0, \dots, a_m]{} min, \qquad m \leq T - 1, \tag{1.6}$$

where $\|\cdot\|$ indicates the Euclidean norm. The problem with required estimate can be reduced to the fact, where the main objective is to find such an unknown but presumed a_i coefficients which minimize the sum of squared errors between the model to estimate the unknown trend curve and obtained model from the real data set containing T observations.
The object of our interest is therefore k –th estimated coefficient a_k where is used a real vector (calculated) of coefficients X and k –th element of the generated orthonormal system g_k which properties can be found e.g. in [1]. The following estimation of unknown coefficient a_k can be expressed by the relationship

$$a_k = \langle \mathrm{X}, g_k \rangle, \qquad k = 0, \dots, m, \tag{1.7}$$

where the notation $\langle \cdot \rangle$ denotes an inner product and X, g_k are two vectors containing T values. Firstly, you can overwrite the method of least squares as

$$Q(a_0, \dots, a_m) = \|X_m(t) - X(t)\|^2 = \sum_{t=1}^{T} \left(\sum_{i=0}^{m} a_i g_i(t) - X(t) \right)^2, \qquad m \leq T - 1, \tag{1.8}$$

where used function Q denotes the squared Euclidean norm of the difference between the estimated and obtained (assumed) trend curve.

Then there will be used the first partial derivatives to obtain the result for the k –th coefficient. A detailed description of this procedure is in (1.9).

$$
\begin{aligned}
0 = \frac{d}{da_k} Q(a_0, \dots, a_m) &= 2\sum_{t=1}^{T} g_k(t)\left(\sum_{i=0}^{m} a_i g_i(t) - X(t)\right) \\
&= \sum_{i=0}^{m} a_i \sum_{t=1}^{T} g_i(t) g_k(t) - \sum_{t=1}^{T} X(t) g_k(t) \\
&= \sum_{i=0}^{m} a_i \langle g_i, g_k\rangle - \langle \mathrm{X}, g_k\rangle \\
&= a_k - \langle \mathrm{X}, g_k\rangle, \qquad k = 0, \dots, m.
\end{aligned} \tag{1.9}
$$

The second partial derivatives confirm that the Q function reaches its extreme value in the local saddle point $0 = \frac{d}{da_k} Q(a_0, \dots, a_m)$, with the following conditions:

$$
\begin{aligned}
\frac{d}{da_k da_j} Q(a_0, \dots, a_m) &= 1 \qquad \Leftrightarrow \qquad k = j \\
\frac{d}{da_k da_j} Q(a_0, \dots, a_m) &= 0 \qquad \Leftrightarrow \qquad k \neq j.
\end{aligned} \tag{1.10}
$$

To obtain these saddle points is the first derivative set to zero value and the following solution may be found by using the inner product as:

$$
\begin{aligned}
0 &= \frac{d}{da_k} Q(a_0, \dots, a_m) \\
&= a_k - \langle \mathrm{X}, g_k\rangle
\end{aligned} \tag{1.11}
$$

$$
a_k = \langle \mathrm{X}, g_k\rangle, \qquad k = 0, \dots, m.
$$

Further will be assumed, that there is an orthonormal system known as $g_i(t)$ for $i = 0, \dots, m$ and $t = 1, \dots, T$, which is potentially expandable to the value $m = T - 1$ (i.e. this system has the same number of components as the observations of time series). It is also possible to express the estimate of the trend curve $X_m(t)$ by the following formula

$$
X_m(t) = \sum_{i=0}^{m} \langle X, g_i\rangle g_i(t), \qquad t = 1, \dots, T \tag{1.12}
$$

and the difference between the estimated and real trend curve as ε_t^m, $m \leq T - 1$ by the formula

$$
\varepsilon_t^m = X(t) - X_m(t), \qquad t = 1, \dots, T. \tag{1.13}
$$

Then the trend curve from the real data set $X(t)$ can be rewritten with using the preceding notation and expression as

$$X(t) = \sum_{i=0}^{m} \langle X, g_i \rangle g_i(t) + \varepsilon_t^m, \qquad t = 1, \dots, T. \tag{1.14}$$

It is nothing other than an expression of the real observed values using the formula for the trend curves estimation and incurred errors. In this situation, there is also assumed that we have a "specific orthonormal system", which "contains a constant" hereinafter referred to as $g_0(t) = g_0 \in \mathbb{R}_1$ for $t = 1, \dots, T$. Due to these assumptions about extensibility of considered orthonormal system with constant is the main subject of this paper continuing discussion over the specific orthonormal system containing a constant g_0, whose value is obtained as

$$1 = \langle g_0, g_0 \rangle = \sum_{t=1}^{T} g_0^2 = T g_0^2 \Rightarrow g_0 = \frac{1}{\sqrt{T}}, \tag{1.15}$$

and also for each element $g_i(t)$ of the generated orthonormal system for $i = 1, \dots, m$ is

$$0 = \langle g_0, g_i \rangle = \sum_{t=1}^{T} g_i(t) g_0 = g_0 \sum_{t=1}^{T} g_i(t) \Rightarrow \sum_{t=1}^{T} g_i(t) = 0, \qquad i = 1, \dots m. \tag{1.16}$$

In the following discussions will be considered only a specific orthonormal system contains a constant in the finite-dimensional space with a defined inner product, where the equality between the trend curve $X(t)$ from the real obtained data set and the estimated trend curve $X_m(t)$ is expressed as

$$\begin{aligned} \sum_{t=1}^{T} X_m(t) &= \sum_{t=1}^{T} \sum_{i=0}^{m} \langle X, g_i \rangle g_i(t) \\ &= \sum_{i=0}^{m} \langle X, g_i \rangle \sum_{t=1}^{T} g_i(t) \\ &= T g_0 \langle X, g_0 \rangle \\ &= T g_0^2 \sum_{t=1}^{T} X(t) = \sum_{t=1}^{T} X(t). \end{aligned} \tag{1.17}$$

And from this relationship follows the following equality

$$\sum_{t=1}^{T} \varepsilon_t^m = 0. \tag{1.18}$$

If there is used a difference between the estimated and real trend curve ε_t^m and is available any vector from the generated orthonormal system of finite-dimensional space with the inner product $g_k(t)$, $t = 1, \dots, T$, then the sum of their multiples over all obtained observations T is equal to zero, more precisely:

$$
\begin{aligned}
a_k &= \langle X, g_k \rangle = \sum_{t=1}^{T} X(t) g_k(t) \\
&= \sum_{t=1}^{T} \left(\sum_{i=1}^{m} a_i\, g_i(t) + \varepsilon_t^m \right) g_k(t) \\
&= \sum_{i=1}^{m} a_i \sum_{t=1}^{T} g_i(t) g_k(t) + \sum_{t=1}^{T} \varepsilon_t^m g_k(t) \\
&= \sum_{i=1}^{m} a_i \langle g_i, g_k \rangle + \sum_{t=1}^{T} \varepsilon_t^m g_k(t) \\
&= a_k + \sum_{t=1}^{T} \varepsilon_t^m g_k(t) \Rightarrow \sum_{t=1}^{T} \varepsilon_t^m g_k(t) = 0.
\end{aligned}
\tag{1.19}
$$

Thus, the residual (random) component is orthogonal to each component of the generated orthonormal system $g_k(t)$. Therefore

$$
\begin{aligned}
a_k &= \sum_{t=1}^{T} X(t) g_k(t) \\
&= \sum_{t=1}^{T} (X_m(t) + \varepsilon_t^m) g_k(t) \\
&= \sum_{t=1}^{T} X_m(t) g_k(t) + \sum_{t=1}^{T} \varepsilon_t^m g_k(t) \\
&= \sum_{t=1}^{T} X_m(t) g_k(t), \qquad k = 0, \dots, m.
\end{aligned}
\tag{1.20}
$$

It should be noted here, that the total sum of multiples $\varepsilon_t^m g_k(t)$ is equals to zero for $t = 1, \dots, T$ as was demonstrated before and observations from the real data set is replaced by the estimated trend curve and its random component.

2 Real data application

The results of the verification of the proposed model for the trend component estimation and following conclusions are based on two obtained data sets. First one is the set from the balance of payments of the Czech Republic, more specifically from the trade balance and the second one is the set from the Consumer price index. Both data sets contain monthly information.

In the balance of payments of the country is an international movement of goods, services, production factors, receivables and payables with foreign countries. It represents a statistical statement that systematically reflects economic transactions with foreign countries [3]. The balance of payments has many definitions, for example from the technical data sheet of the CNB [2] and the structure itself is not the same in various countries and institutions. The obtained data set is from the statistical reports of the Czech National Bank, which is based on the manuals of the IMF Balance of Payments (Balance of Payments Compilation Guide, 5th Edition, 1993), in which the balance of payments can be divided into the horizontal or vertical structure and for the purpose of this work were used observations from a horizontal structure in millions CZK.

A consumer price index (CPI) measures changes in the price levels of a market basket of consumer goods and services. The obtained data set is again from the statistical reports of the Czech National Bank, where a percentage basis is average value of 2005.

Considered trend curve itself is an approximation of the data set using a continuous curve, that summarizes, where the obtained values were in the past, where these values are now in relation to the past. The application of the proposed model on the real data sets is shown in the figures below where the estimations of unknown trend curves are drawn with dashed lines. The deviations between the observed values and the estimated trend component values, characterized by the dotted line with points, is also shown in the figures.

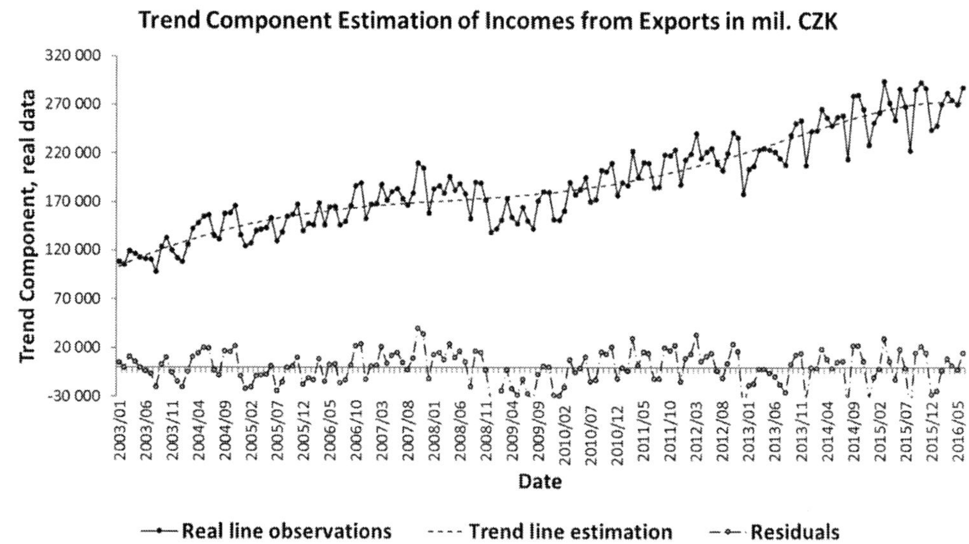

Fig. 3. Trend component estimation of incomes from exports in mil. CZK

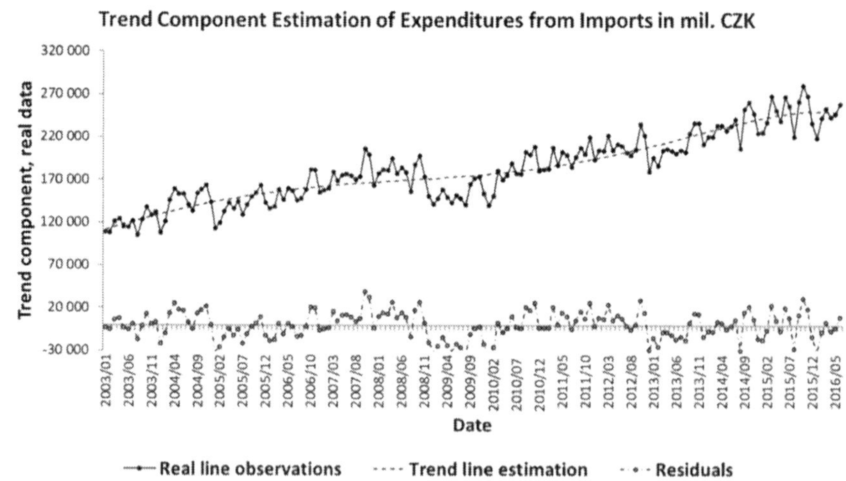

Fig. 4. Trend component estimation of expenditures from imports in mil. CZK

Now we can briefly summarize the obtained results from an economic point of view. The obtained trend curves demonstrate growth potential, which could be caused by the growth of economic activity in the selected country. Thus it is meant the growth of economic activity in the Czech Republic, which is represented by the total amount of expenditure from imports (as well as incomes from exports). Further, there can be described the direction of both trend curves, which has growth potential from due to the general growth in economic activity. The problem between 2008 - 2009 could be done by the "financial crisis".

Here, it is worth noting that from 2003 until the end of 2011 there are no significant decline in import or export volumes and it is evident, that the Czech Republic increases its exports and imports in the aftermath of the first wave of "financial crisis". That method can also be used for short-term forecasts (prediction). Both approximation of unknown trend curves (in the sense of this work) are also useful for their comparison, especially in the long term. This situation is illustrated in the figure below, where the dashed curve indicates a trend curve for the total expenditure from imports and the solid line characterizes the trend curve for the total amount of incomes from exports.

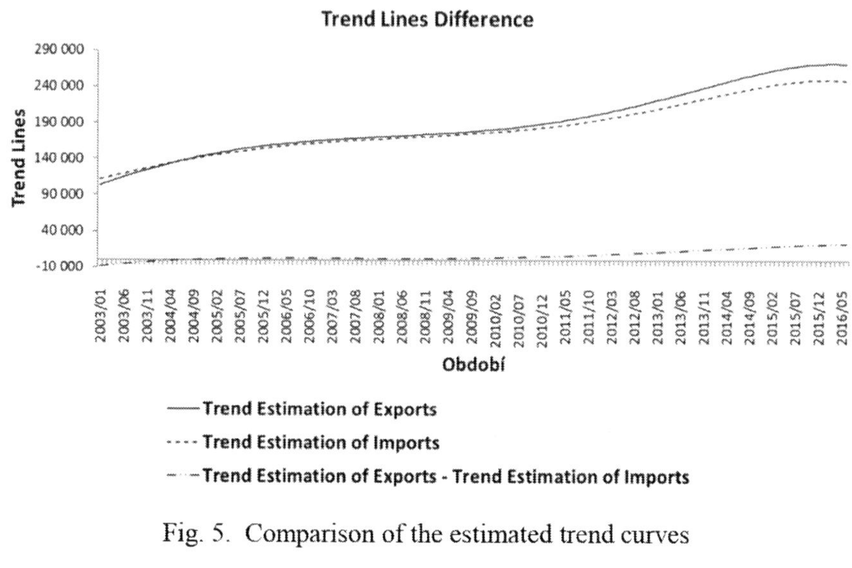

Fig. 5. Comparison of the estimated trend curves

The difference between both trend curves represents a positive or negative value of the trade balance. Generally, there may be said that if the solid line is above the dashed line, then the value of "trend balance" is positive and foreign trade is in a good condition. The Czech Republic increases the total amount of exports more than the total amount of imports in CZK since 2003.

The trend component estimation of the consumer price index (CPI) is shown in the figure below, where the real data set is from 2000 till now. There is necessary to say, that the percentage basis is average 2005 (i.e. average 2005 = 100).

The obtained trend curve demonstrates again growth potential, which could be caused by the growth of economic activity in the selected country as in the case of the trade balance. Of course are these quantities closely linked. It is worth nothing that in the last three years this growth potential is not so important, which could be done by declining agricultural commodity prices and stable food prices, a significant decline in fuel prices, falling prices for electricity and gas and other factors.

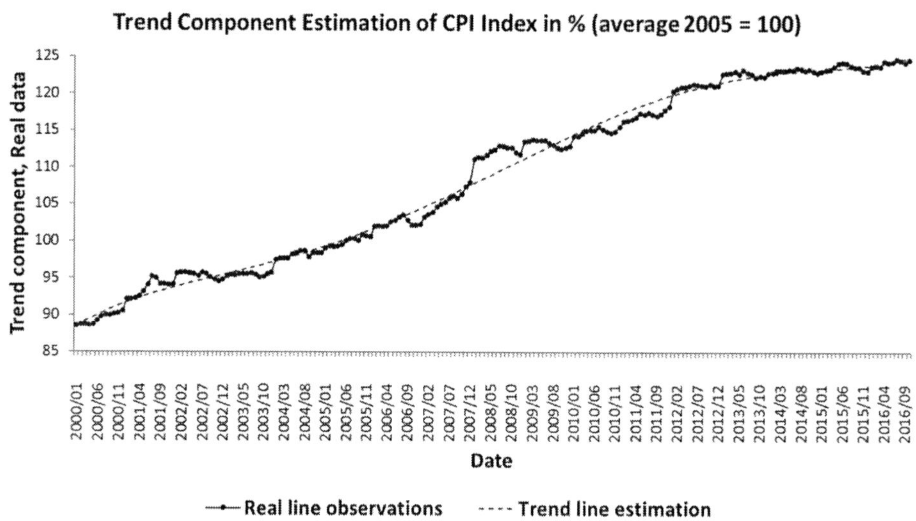

Fig. 6. Trend component estimation of CPI index

The deviations between the observed values and the estimated trend component values in CPI index, characterized by the dotted line with points, is also shown in the last figure.

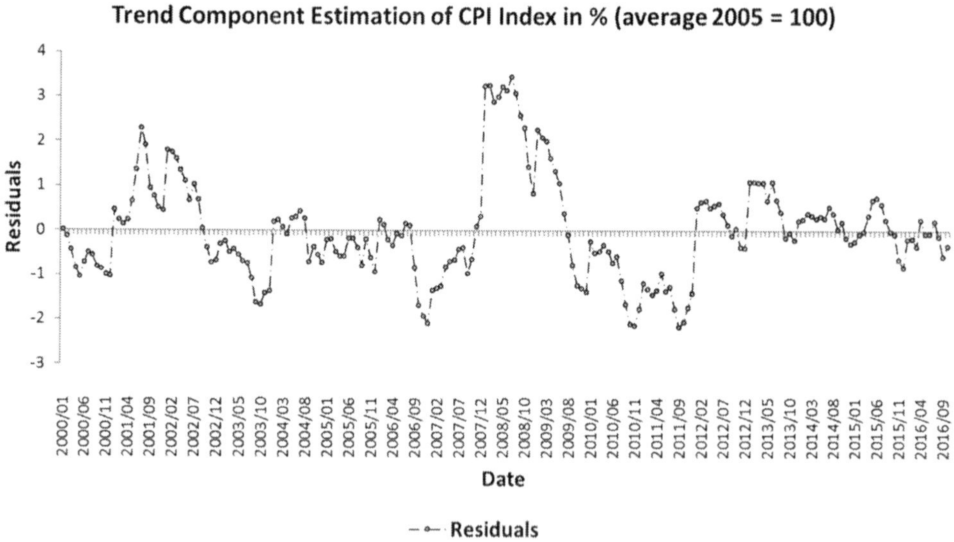

Fig. 7. Residuals from the trend component estimation

References

[1] KUFNER, A. *Geometrie Hilbertova prostoru,* Praha, 1973, SNTL.

[2] TREFETHEN, L.N., BAU, D. *Numerical Linear Algebra,* Philadelphia, 1997 , Society for Industrial and Applied Mathematics.

[3] Česká národní banka (ČNB), 2016. *Platební balance – měsíční.* [online] Available at: <http://www.cnb.cz/cs/statistika/platebni_bilance_stat/platebni_bilance_m/index.html>

Current address

Ťoupal Tomáš, Ing. Ph.D.
Department of Mathematics, Faculty of Applied Sciences, University of West Bohemia
Univerzitni 8, 306 14 Pilsen, Czech Republic
Tel. number: +420 377 63 2674, e-mail: ttoupal@kma.zcu.cz

Šedivá Blanka, RNDr. Ph.D.
Department of Mathematics, Faculty of Applied Sciences, University of West Bohemia
Univerzitni 8, 306 14 Pilsen, Czech Republic
Tel. number: +420 377 63 2618, e-mail: sediva@kma.zcu.cz

Slovak University of Technology in Bratislava
Faculty of Mechanical Engineering
16th Conference on Applied Mathematics **APLIMAT 2017**
Proceedings

NUMERICAL SIMULATIONS FOR ACOUSTIC IMPROVEMENT OF ROMAN THEATRE OF CASSINO

TREMATERRA Amelia (IT), PATERNUOSTO Stefania (IT), LOMBARDI Ilaria (IT)

Abstract. This paper presents the results of acoustic measurements taken inside the Roman theatre of Cassino. The theatre has only been partly rebuilt, with some of the missing stage walls. The theatre is currently used for musical performances. In order to see the effects of installing reflective screens on the stage, the acoustic characteristics have been studied with the aid of a architectural acoustics software, The numerical results show an improvement of the acoustic characteristics especially in the central area of the cavea.

Keywords: room acoustics, historical theatre, simulation, virtual reconstruction, Roman Empire

Mathematics Subject Classification: Primary 68-04, 08A72; Secondary 28-04

1 Introduction

Inside the walls of the ancient Casinum in Cassino, there is a Roman theatre, which was built in the middle of the Augustan period in first century B.C.. Subsequent historical events led it to being abandoned and eventually demolished. Over the centuries, several structures were built inside the theatre. The restoration work was completed in 2000, with only a few parts of the ima cavea remaining of the original theatre. It is a few metres from the road that connects the town centre of Cassino to the Abbey (Fig. 1) and a large square, that covers approximately 3,000 m^2. The cavea stands on the slope of a hill and is oriented to the south-east, with semi-circular steps and orchestra pit. It was originally divided into an ima and summa cavea, with a part of the ima cavea having been rebuilt. In its current state, the theatre has the following dimensions: orchestra pit diameter: 9.2 m; cavea outside diameter: 46 m; stage: 5.0 m x 27.4 m; steps rebuilt in stone and mortar, with a height of 0.40 m and a depth of 0.70 m; slope of cavea 30 °; number of steps along the cavea 19; current capacity is about 3,000 spectators. During the summer, the theatre is used for several events, such as plays and various types of concerts. The acoustics of the theatre cannot be considered satisfactory due to the absence of a vertical wall on the stage. In order to improve the acoustics as well as increase sound diffusion in the cavea, it could be possible to place several reflective screens on the stage [1, 2, 3, 4]. Thus, the aim of this study was to evaluate the acoustic characteristics of the Roman theatre of Cassino in its present state and with help of an architectural acoustics software to simulate the sound field and subsequently evaluate the effects of installing reflective screens on the stage to improve the acoustic characteristics. Current literature reports studies on the acoustics of ancient theatre with the help of virtual reconstructions. There are also studies on the current use of ancient theatres for music performance, concerts and other events. Fig. 2 shows the side view of the theatre in its current state, while Fig. 3 shows the front view.

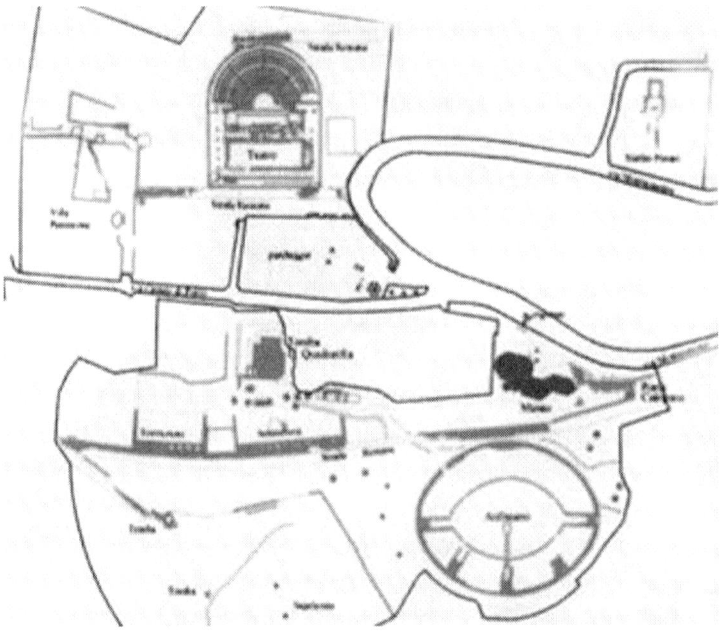

Fig. 1. Theatre location, near the road that connects the centre of Cassino to the Abbey

Fig. 2. Side view in current state

Fig. 3. Front view in current state

2 Acoustic measurements

To objectively evaluate the acoustic behaviour of the Roman theatre, surveys were carried out with a source, placed on the stage at a height of 1.5 m from the floor in the actor position. The sound source was fed with a MLS signal, in accordance to the standard ISO 3382 [5], by impulse response technique. The sound source used for the acoustic measurements consisted of a dodecahedron loudspeaker Peeker Sound JA12 (Peeker Sound Corporation, Reggio Emilia, Italy). MLS signals of order 16 with a length of 5 s were generated by a 01 dB Symphonie system [6, 7]. The response was detected with a ½" GRAS 40 microphone connected to a 01 dB PRE 12 H preamplifier, placed at a height of 1.2 m from the floor [8, 9]. During the acoustic measurements the theatre was completely empty. The receiving points were placed on the I, V, IX, XV and XIX bleacher, in order to obtain information from varying distances from source, along the three radial directions as shown in Fig. 4, which shows the position of the sound source on the stage as well as the five receivers in the cavea set out in three radial directions.

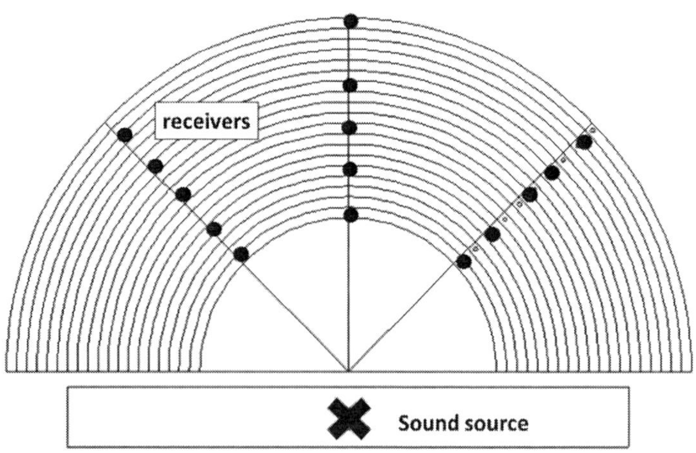

Fig. 4. Position of the sound source (X) on the stage and the five receivers (o) in the cavea, set out in three radial directions

The recorded impulse responses were elaborated with the software, Dirac 4.0, and several acoustic parameters defined in ISO 3382-1. Reverberation time (T_{30}), EDT (Early Decay Time), sound strength (G), clarity (C_{80}), definition (D_{50}), and sound transmission index for speech intelligibility (both RASTI and STI) were analyzed. The typical suggested values of the different monaural acoustic parameters for both speech comprehension and music listening are:

Reverberation time T_{30} should assume values below 1.0 second for a clear perception of speech, while it could assume greater values, around 2.0 seconds for music listening preference.

Clarity C_{80}, expressing balance between early and late arriving energy, should have a higher value if the goal is to separate initial sounds from diffuse ones, making discrete sounds stand apart from each other. In a sound field which is not completely diffuse, clarity C_{80} is uncorrelated to reverberation time, and for the purposes of good listening conditions of music, it is generally reported that C_{80} should be in a range between -2 dB and 2 dB, while it is expected to be above 2 dB if speech perception is a priority.

Definition D_{50} may assume values from 0 to 1.0, but for a good speech comprehension, it is often accepted that D_{50} should have values above 0.5.

Speech Transmission Index (STI) represent a degree of the amplitude modulation in a speech signal, with them both referring to the distortion in speech signals caused by reverberation, echoes, and background noise. The STI Indices can assume values between 0 and 1, being greater than 0.5 for favourable speech conditions.

In open-air theatres, the parameter G assumes a special significance for the assessment of acoustics. Sound strength (G) is representative of the subjective level of sound, and is defined as the gain from sound pressure level, which is produced by the same omnidirectional sound source, with the same power level, in a free field at a distance of 10 m from the sound source. Tab. 1 shows a synthesis of the optimal acoustic parameters values for different listening conditions.

Parameters	EDT, s	T_{30}, s	C_{80}, dB	D_{50}
Optimal values for musical performances	$1.8 < EDT < 2.6$	$1.6 < T_{30} < 2.2$	$-2 < C_{80} < 2$	< 0.5
Optimal values for speech performances	1.0	$0.8 < T_{30} < 1.2$		> 0.5

Tab. 1. Optimal acoustic parameters values for different listening conditions

To measure G "strength" (dB), the sound source was "calibrated" and the emitted sound power level Lw (dB) measured, through the substituents institution technique with an isotropic type B & K 4241sample source. With the same Lw "calibration" set-up, the sound source was measured and fuelled during the measurements in the theatre. G (dB) was calculated with the following formula:

$$G = Lp - Lw + 31\ (dB)$$

Where Lp (dB) is the sound pressure level detected in the cavea along the three directions and on the ten steps, I, III, V, VII, IX, XI, XIII, XV, XVII and XIX (Fig. 5). Lw (dB) is the sound power level of the used source. Fig. 6 shows the values broadband of the strength (G), mediated in the three directions in the cavea, when the sound source is placed on the stage (actor's position). The value of G is low due to the lack of reflective surfaces on the stage. The theatre is an open space, with the stage devoid of any surfaces that may limit possible noise influences coming from outside. Furthermore, the city of Cassino is a junction for rail and road transport infrastructures. There is both a railway line as well as the Rome – Naples motorway a few kilometres from the theatre and the road that leads to the Abbey about 20 metres away. Residual noise LeqA = 47.5 dBA was detected with a sound level metre placed in the square outside the theatre at different points [10]. The frequency spectrum has components that are not negligible at low frequencies, due to the noise generated by transport infrastructures and vehicles passing along the adjacent street. The noise residual level is not high, therefore any influences coming from outside noise can be considered negligible.

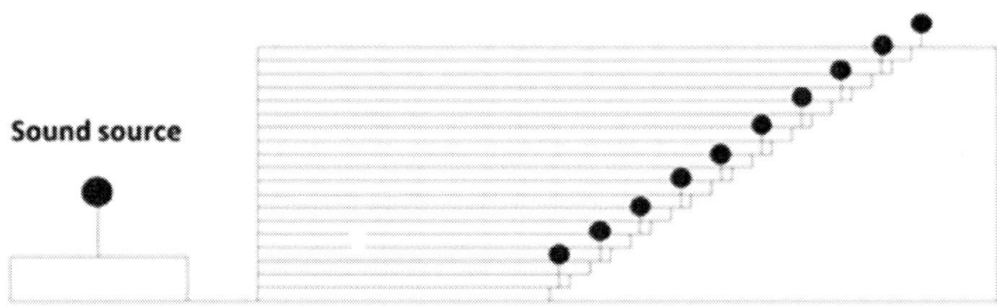

Fig. 5. Sound source on the stage and measurement microphones in the cavea for the G measurement

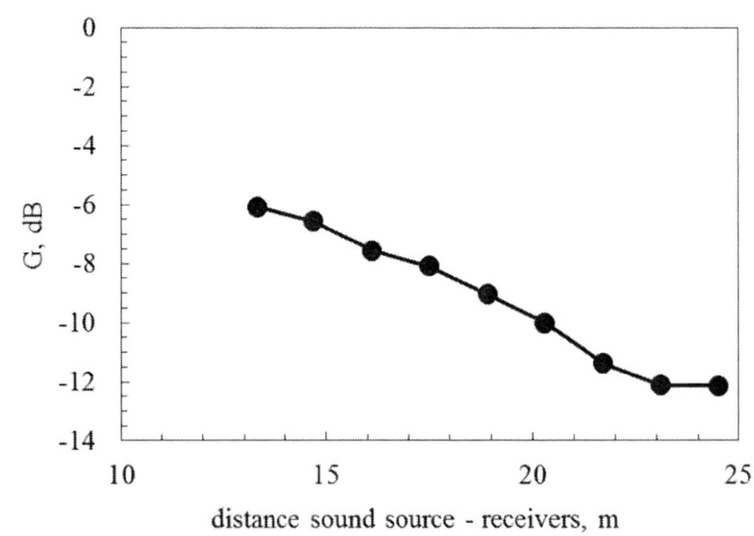

Fig. 6. Average value of G along the cavea, with the sound source on the stage

Fig. 7 shows the average values of the measured acoustic parameters (in octave bands from 125 Hz to 4.0 kHz), with the standard deviation among the results for five receivers, with the sound source on the stage. The acoustic procedure and post processing methodology were similar to those used in other spaces, such as the large theatres of Pompeii [11, 12] and the theatre of Benevento [13] as well as in many other ancient theatres. The measured values confirm that the lack of a roof and back stage wall led to a few sound reflections with short reverberation, as highlighted in values below 0.5 second of T_{30}. Clarity C_{80} and definition D_{50} were higher. The STI (Speech Transmission Index) parameter was also measured, with it equaling STI = 0.65.

An analysis of the measurements shows how the theatre cannot be used for opera or symphonic music, since the stage, with no rear wall, prevents any sound reflections and the listeners only perceive the direct part. To improve acoustics of the theatre, some screens should be installed at rear of the stage, so as to allow for sound reflection, along with a consequent improvement of sound quality. They could also help to isolate the theatre from the outside and avoid any possible negative influences of residual noise during the performances.

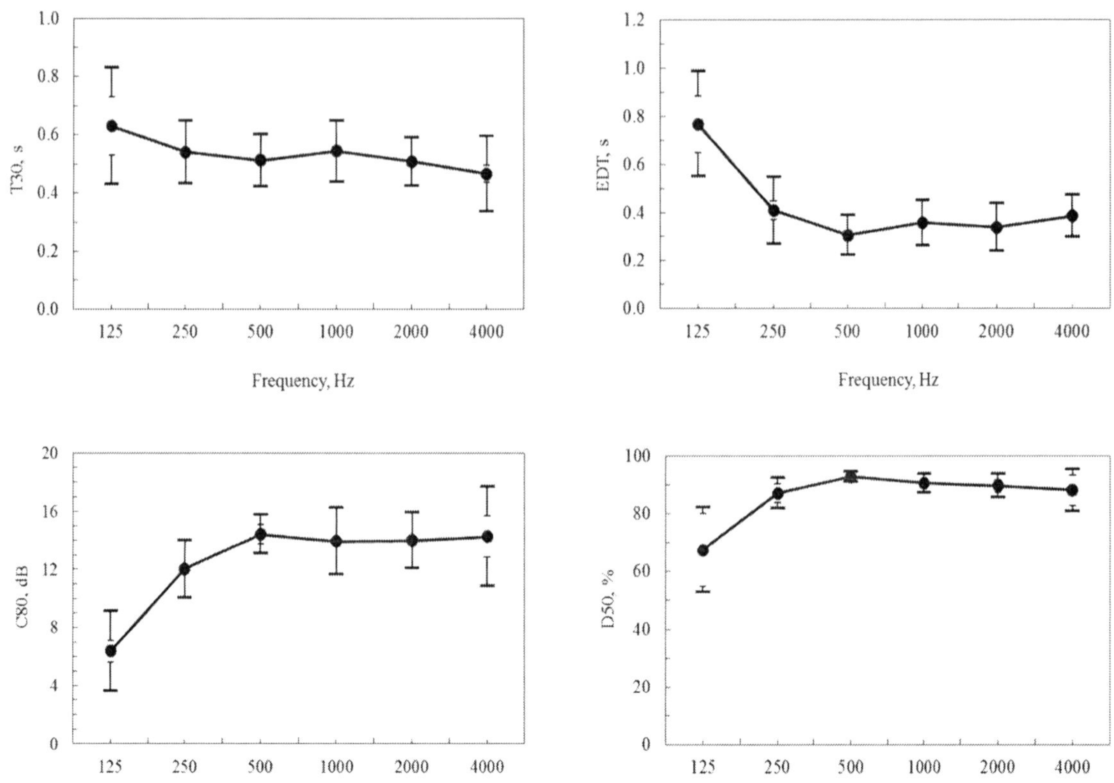

Fig. 7. Average values of the measured acoustic parameters (in octave bands from 125 Hz to 4.0 kHz), with the standard deviation among the results for five receivers, with the sound source on the stage

Fig. 8. Sound source on the stage during the acoustic measurements

In order to assess the acoustic characteristics of the theatre of Cassino in the actual configuration and compare it to other similar theatres, a comparison of the acoustic parameters obtained using virtual models in the theatres studied in the European ERATO project was carried out. In addition to the open-air theatres of Pompeii (a large theatre) and Benevento, the average values of the monaural parameters in the mid frequency bands of 500 Hz and 1.0 kHz are reported in Tab. 2. One of the aims of the ERATO project was to investigate the acoustics of open-air theatres and compare them to smaller, original theatres [14]. This project found that open-air theatres had a very high clarity of sound. Tab. 2 reports the acoustic parameters averaged at mid-frequency bands of 500 Hz and 1.0 kHz calculated through software simulations in some ancient theatres. However, other acoustic parameters were heavily affected by the presence of an audience. Tab. 2 also presents a comparison of acoustics data for ancient theatres with the theatre of Cassino in actual configuration, and shows how this theatre has a shorter reverberation time, so the theatre of Cassino could accommodate theatrical performances.

Theatres	condition	kind of theatre	T_{30} [s]	C_{80} [dB]	STI [-]	diameter [m]
Pompeii (large theatre)	empty	open-air	1.53	4.50	0.62	58.0
Benevento	empty	open-air	1.80	2.00	0.60	40.0
Aspendos	empty	open-air	1.95	1.17	0.53	95.5
Aosta roman	empty	roofed	5.97	-5.28	0.36	60.0
	full		3.49	-2.41	0.43	
Aphrodisias roman	empty	roofed	4.02	-4.21	0.38	45.6
Posillipo roman	empty	roofed	1.10	3.0	0.70	26.0
Cassino	empty	open-air	0.60	12.0	0.65	46.0

Tab. 2. Acoustic parameters averaged at mid-frequency bands of 500 Hz and 1.0 kHz calculated through software simulations in some ancient theatres

3 Virtual model

To virtually investigate the acoustics of the theatre, a 3D software model was created, starting from the geometrical measurements of the building in its current state. Subsequently, this model was integrated with reflective panels on the stage. A sound field generated by the presence of reflective screens on the stage was simulated. A virtual model was simulated using the architectural acoustics software Odeon [15, 16, 17, 18]. This software uses the principles of geometrical acoustics and adopts a hybrid calculation method that combines two classical methods, image source and ray-tracing method. The Odeon program uses a hybrid calculation method, where early reflections are calculated by combining the image- source and a ray tracing method. The late reflections are calculated using a ray tracing process, which generate diffuse secondary sources. In this process rays are sent from the source position detecting image sources up to a certain reflection order and then detecting secondary sources on the surfaces of the room. The secondary sources radiate with a directivity that follow Lamberts Law and an intensity contribution attenuated mainly by the distance and the reflections in the path from source to receiver. The transition order (T.O.) at which the calculation method changes from image-source to ray tracing can be adjusted according to the complexity and shape of the room. If T.O. is set to 0 the calculation method used for all reflections will be the secondary source / ray tracing method. All the other simulation parameters were set as suggested by default in the software [19]. The first post-processing step consisted of comparing the acoustic measurements in the current state with the results obtained from the virtual model. This allowed to adjust the virtual model until obtaining simulated results as close as possible to those measured. The process consisted of minimizing the differences between the calculated and measured values of T_{30} by tuning the absorption coefficients of the room surfaces. The goal was to reduce the difference between the simulated and measured values of T_{30} below 5% [20, 21, 22, 23]. In the virtual model, the receiving points were positioned along three radial directions of the cavea and in the same locations where the microphone was placed during the measurements, while the sound source was only placed on the stage (actor position). In order to assign material properties, some considerations about the materials of the theatre were necessary. This typical of Roman theatres was made of hard materials such as stones. The absence of roofing was simulated in the virtual model by assigning an open roof. This means that the model of the theatre was closed in a box with an absorption coefficient equal to 1.0 assigned to the boundaries, to simulate outgoing propagation of the sound in an open-air space. The seating rows were modelled with a scattering coefficient of s=0.7, while the stage wall and floor were assigned a scattering of s=0.5. Subsequently to the calibration of the model, based on measured data of the acoustic characteristics in the current state, the virtual model was modified to include the reflective panels on the stage. Fig. 9 shows the reflective panels on the stage. The panels have high absorbent coefficients values at low frequencies, while at medium frequencies these values are very low. The measurements of the absorption coefficient were carried out according to the standard EN ISO 10534-2 "Determination of sound absorption coefficient and of the acoustic impedance in impedance tubes - Method of the transfer function" [24, 25, 26]. The measurement of the sound absorption coefficient at normal incidence was carried out by means of an impedance tube. The measuring tube had the following dimensions: inner diameter of 100 mm, length of 560 mm and the distance between the two microphones of 100 mm. For the dimension of the tube, the inner diameter and the distance between the two measuring microphones, the value of the absorption coefficient at normal incidence was valid in the frequency range between 100 Hz - 1.0 kHz. Fig. 10 shows the absorption coefficient values of the reflective sheets in frequency function from 100 Hz to 1.0 kHz [27].

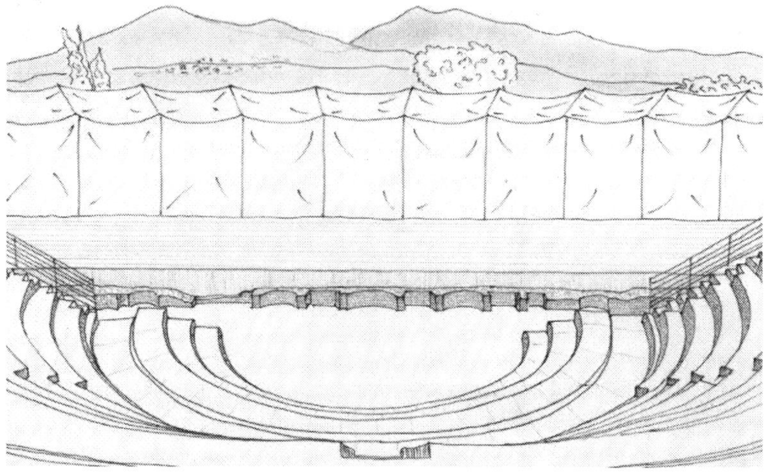

Fig. 9. Reflective panels on the stage

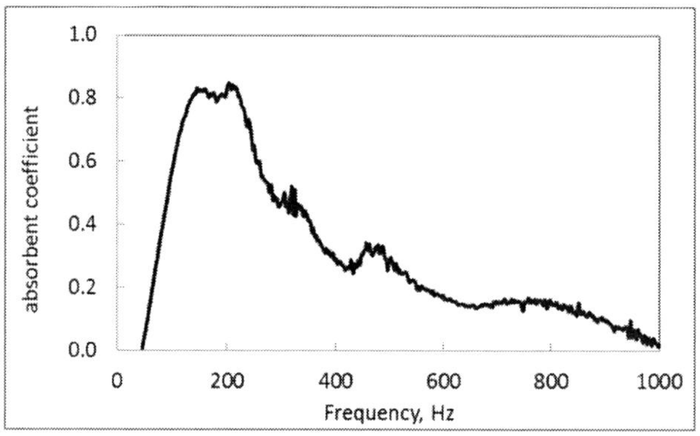

Fig. 10. Reflective panels, absorption coefficient values measured at normal incidence in frequency range of 100 Hz - 1.0 kHz

4 Discussion

The open-air theatre has a dry acoustics with a reverberation time in order of 0.6 s and a high speech intelligibility. The virtual model elaboration of the theatre, using the reflective panels on the stage, has an approximate reverberation time above 1.0 s and a poor speech intelligibility especially for the most distant listening positions. In particular, clarity C_{80} for the roofed version was generally below 4 dB, and D_{50} definition was between 0.6 and 0.7 through all the frequency bands. Fig. 11 shows the T_{30} spatial distribution (A) actual configuration, (B) numerical simulation with reflective panels on the stage, while Fig. 12 shows C_{80} spatial distribution (A) actual configuration, (B) numerical simulation with reflective panels on the stage [28].

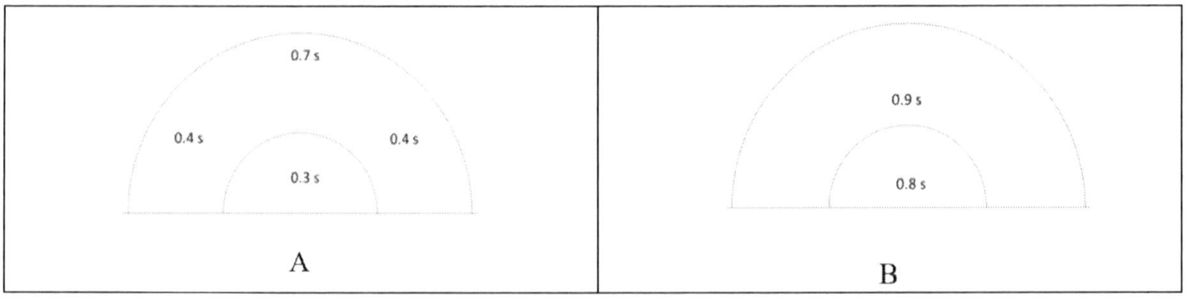

Fig. 11. T_{30} spatial distribution (A) actual configuration, (B) numerical simulation with reflective panels on the stage

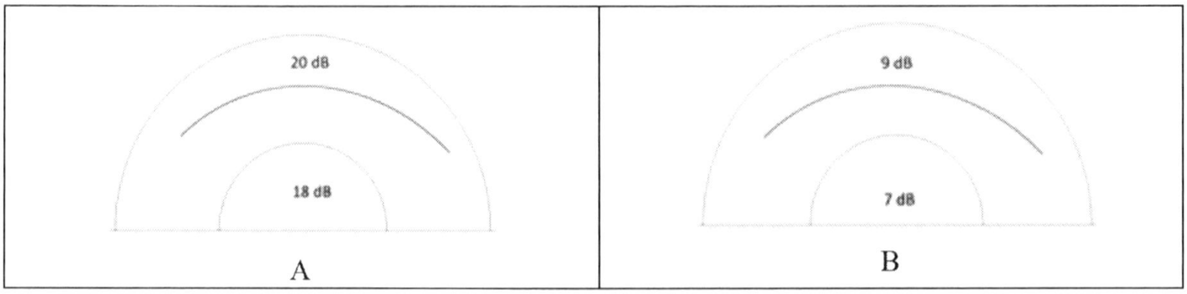

Fig. 12. C_{80} spatial distribution (A) actual configuration,
(B) numerical simulation with reflective panels on the stage

5 Conclusions

This paper has described the acoustic behaviour of the Roman theatre of Cassino, both in its current configuration without stage walls and in the presence of reflective panels on the stage. The results obtained from the elaboration of a virtual model showed that the building with panels is well suited to musical performances. The theatre in its present state is acoustically deaf and is more suitable for spoken performances. The current use of this theatre for musical performances and summer festivals should consider that the acoustics in its present state are highly different from those that the Romans were used to. On the contrary, nowadays, theatres sound like a deaf space where direct sound dominates all the responses and the reflections are generally uncorrelated and too weak to enrich the listeners' experience.

References

[1] VITRUVIO, M.P. *De Architectura.*

[2] SABINE, W.C. *Collected Papers on Acoustics*, Cambridge, Mass, 1923.

[3] MAZZONI, *S. Teatri romani: immagini d'impero e d'attori e altre questioni, Studi di Storia dello spettacolo.* Omaggio a Siro Ferrone, Firenze, 2011, 15-35.

[4] CIANCIO ROSSETTO, P., SARTORIO G.P. *Teatri Greci e Romani: alle origini del linguaggio rappresentato censimento analitico*, 3, Torino, 1994.

[5] ISO 3382-1. Acoustics – Measurement of the reverberation time of rooms with reference to other acoustical parameters

[6] IANNACE, G., TREMATERRA, A. Acoustic measurements and correction of a council room, *Noise and Vibration Worldwide*, 45 (8), 2014, 12-16, doi: 10.1260/0957-4565.45.8.12

[7] IANNACE, G. Acoustic correction of monumental churches with ceramic material: The case of the Cathedral of Benevento (Italy), *Journal of Low Frequency Noise Vibration and Active Control,* 35(3), 2016, 230 – 239, doi: 10.1177/0263092316661028

[8] IANNACE, G., DI GABRIELE, M., SICURELLA, F. Sound Focusing Effects in Horseshoe Plan Theatre, *Acoustics Australia,* 44 (2), 2016, 359-368, doi: 10.1007/s40857-016-0059-2

[9] IANNACE, G., TREMATERRA, A. The rediscovery of Benevento Roman Theatre Acoustics, Journal of Cultural Heritage, 15(6), 2014, 698, doi: http://dx.doi.org/10.1016/j.culher.2013.11.012

[10] IANNACE, G. Effects of noise from wind turbines inside home. *Wind Engineering*, 2016, 40(1), 25-30, doi:10.1177/0309524X15624339

[11] IANNACE, G., TREMATERRA, A., MASULLO, M. The large theatre of Pompeii: Acoustic evolution, *Building Acoustics*, 20(3), 2013, 215-227, doi: 10.1260/1351-010X.20.3.215

[12] BERARDI, U., IANNACE, G., MAFFEI, L. Virtual reconstruction of the historical acoustics of the Odeon of Pompeii, *Journal of Cultural Heritage*, 19, 2016, 555-566, doi: http://dx.doi.org/10.1016/j.culher.2015.12.004

[13] D'AMBROSIO ALFANO, F.R., IANNACE, G., IANNIELLO, C., IANNIELLO, E. "Velaria" in ancient Roman theatres: Can they have an acoustic role?, *Energy and Buildings*, 95, 2015, 98-105, doi: http://dx.doi.org/10.1016/j.enbuild.2015.03.010

[14] RINDEL, J.H. Roman theatres and the revival of their acoustics in the ERATO Project, *Acta Acustica*, 99, 2013, 21-29.

[15] BERARDI, U., IANNACE, G., IANNIELLO, C. Acoustic intervention in a cultural heritage: The chapel of the Royal Palace in Caserta, Italy, *Buildings*, 6(1) 2015, Article number 1. doi 10.3390/buildings6010001

[16] IANNACE, G., BERARDI, U., IANNIELLO, C. Study of a historical Church based on acoustic measurements and computer simulation, *Proceedings 22nd International Congress on Sound and Vibration*, ICSV 2015; Florence; Italy; 12 -16 July 2015; Code 121474

[17] IANNACE, G., TREMATERRA, A., TREMATERRA, P. Acoustic correction using green material in classrooms located in historical buildings. *Acoustics Australia* 41(3), December 2013, 213-218.

[18] BERARDI, U., IANNACE, G., GIORDANO, G., Acoustic characteristics of four subway stations in Naples, Italy. *Proceedings 44th International Congress and Exposition on Noise Control Engineering*, INTER-NOISE 2015; San Francisco; 9 - 12 August 2015; Code 114294

[19] CHRISTENSEN, C.L. ODEON Room Acoustics Software. v.11 manual. www.odeon.dk

[20] ZENG, X., CHRISTENSEN, C.L., RINDEL, J.H. Practical methods to define scattering coefficients in a room acoustics computer model, *Applied Acoustics*, 67, 2006, 771-786.

[21] RINDEL, J.H. Roman theatres and the revival of their acoustics in the ERATO Project, *Acta Acustica*, 99, 2013, 21-29.

[22] BRADLEY, D.T., WANG, L.M. Comparizon of Measured and Computer-Modeled Objective Parameters for an existing Coupled Volume Concert Hall. *Journal of Building Acoustics*, 2007, 14(2), 79-90.

[23] HOWARTH, M.J., LAM, Y.W. An assessment of the accuracy of a hybrid room acoustics model with surface diffusion facility. *Applied acoustics,* 2000, 60(2), 37-43.

[24] EN-ISO-10534-2. Acoustics. Determination of sound absorption coefficient and impedance in impedances tubes. Part 2: Transfer-function method.

[25] TREMATERRA, A., IANNACE, G., NESTI, S., FATARELLA, E., PERUZZI, F. Acoustic properties of nanofibers. *Proceedings of Forum Acusticum.* 7th Forum Acusticum, FA 2014; Poland; 7 -12 September 2014; Code 117181

[26] IANNACE, G. Acoustic properties of nanofibers, Journal Noise & Vibration Worldwide, 2014, 45 (10), 29 – 33. DOI: 10.1260/0957-4565.45.10.29

[27] IANNACE G., IANNIELLO C., IANNIELLO E., Music in an Atrium of a Shopping Center, *Acoustics Australia*, 2015, 43(2), 191-198. DOI 10.1007/s40857-015-0017-4

[28] IANNACE G. The use of historical courtyards for musical performances. Building Acoustics 2016, 23(3-4), 207-222. DOI: https://doi.org/10.1177/1351010X16678219

Current address

Trematerra Amelia, phd
Dipartimento di Architettura e Disegno Industriale, Seconda Università di Napoli
Borgo San Lorenzo, 81031 Aversa, Italy
Tel. number: + 39.3402389882, e-mail: liatrematerra@libero.it

Parternuosto Stefania, student
Dipartimento di Architettura e Disegno Industriale, Seconda Università di Napoli
Borgo San Lorenzo, 81031 Aversa, Italy
E-mail: stefaniapaternuosto@gmail.com

Lombardi Ilaria, student
Dipartimento di Architettura e Disegno Industriale, Seconda Università di Napoli
Borgo San Lorenzo, 81031 Aversa, Italy
E-mail: ilaria.lombardi@studenti.unina2.it

Slovak University of Technology in Bratislava
Faculty of Mechanical Engineering
16th Conference on Applied Mathematics **APLIMAT 2017**
Proceedings

VIRTUAL RECONSTRUCTION AND SOUND FIELD SIMULATION OF THE ODEON OF POSILLIPO

TREMATERRA Amelia (IT), PATERNUOSTO Stefania (IT), LOMBARDI Ilaria (IT)

Abstract. The coast of Campania during the period of Imperial Rome was an ideal place for the residences of Roman patricians due to the healthy climate and beauty of the landscape. In this paper, the virtual reconstruction and the sound field simulation of the Odeon of Posillipo is reported. This analysis is performed using an architectural acoustics software. The sound field simulation gives information that the Odeon was used for speaking.

Keywords: room acoustics, historical theatre, simulation, virtual reconstruction, Roman Empire

Mathematics Subject Classification: Primary 68-04, 08A72; Secondary 28-04.

1 Introduction

The coast of Campania during the period of Imperial Rome was an ideal place for the residences of Roman patricians due to the healthy climate and beauty of the landscape. These villas were often built on particularly eye-catching sites. Villa "Publius Vedius Pollio" was built on the cape of Posillipo with gardens, baths, theatres and a terrace overlooking the sea. Publius Vedius Pollio, was one of the wealthiest men of the late Republic and thanks to the support of Octavian, after the battle of Actium of 31 B.C., he held several important public positions that made him become one of the richest men of the empire. After his death, he bequeathed the Villa to Augustus, who transformed it into an imperial residence. After the fall of Roman Empire, the site was looted and abandoned, with it being rediscovered in the mid-nineteenth century when the first excavations and recovery of artifacts was carried out. Today, the site is a part of the Archaeological Park of Pausilypon. The villa was built at the end of the hill that divides the Gulf of Naples from the Campi Flegrei area, in a difficult to access portion. The site extended over an area of about ten hectares, with a different orographic structure: to the west, there was a promontory overlooking the sea with a sixty metre difference in altitude, while to the east, it gently sloped down towards the beach. The first Greek settlers used to call the hill, Pausilypon, taken from "pausis" which means suspension and "lypon" which means pain, thus a place where griefs end. Vedius Pollio used this to describe his villa, with the term "Posillipo" subsequently being used to describe the coastline. The residence was a seaside villa, without a single nucleus but with more fulcrums arranged on several terraces, according to a typological model already used in Capri and Campi Flegrei. Given its large size and division into fulcrums, the villa was divided into three main parts: maritime, private and public. The maritime part of the complex had a vast coastal with a port, fish ponds, nymphs, artificial caves and triclinium. This part is almost entirely submerged. The private part included a residential nucleus which developed along the eastern side of the hill, gently sloping down towards the sea. It consisted of a number of structures arranged on several terraces connected by porches and cryptoporticos. The public part was used for reception and leisure functions and was in the highest part, to the west of promontory, overlooking the sea. It was composed of several structures and was conceived as a small town for shows. The main structures in this area include the Seiano cave, Necropolis, Temple, Nymphaeum, Theatre and Odeon. Fig. 1 shows the current site of the Archaeological Park of Pausilypon (Theatre and Odeon).

Fig. 1. The current site of the Archaeological Park of Pausilypon

2 The acoustics of the theatres

The study of ancient theatres is not exclusive to neither architectural historians nor archaeologists but also acoustics researchers. The only remaining classical work where the principles of acoustics are expressed is the Vitruvian treatise "De Architectura" [1, 2]. This book was written during the Augustan period in the 1st century B.C.. Book V describes the similarities between the swell of a water mirror and the propagation of sound in the air "...... *the voice spreads in a circular shape, but while the propagation of circles in the water takes place on a flat surface, in the case of a voice, there is a gradual propagation both horizontally and vertically...... Therefore, as for concentric circles that form in water, the same will happen to sound propagation, and it will be sent to the ears of an audience who are in high places, or at the top, with no echo and reverberation, if no obstacles interfere with the spreading of the first sound wave, the second and so on. Therefore, the ancient architects, complying with the natural principles of acoustics on sound emission which tends to rise to the top, realized, according to precise mathematical calculations and to the theory of sound, the bleachers structure of their theatres so that any emitted voice from the scene could arrive clear and pleasing to the ears of the audience*". Moreover, Vitruvius described some useful recommendations for on-site geometry: "*You must take particular care to ensure that site is not "deaf" but is a place where the voice can travel with maximum clarity. This can be achieved if a site is chosen where there are no elements that generate echo*". Additional measures were used in order to improve the acoustics of theatre, such as covering the orchestra with hewn marble slabs so that any discontinuities would generate a greater spread of the sound in all directions; covering the stage with a roof to improve sound reflection towards the auditorium; increasing the size of the stage up to the maximum height of the auditorium so as to make the stage and auditorium one. The stage became a building with columns, stucco and coloured plaster, with the beauty of the scenery having the task of fascinating the spectator, capturing his attention and to transmit with the power of the images the Empire's prominence [3, 4]. The dimensions of the stage were doubled compared to those of Greek theatres. The orchestra was reduced to an exact semicircle; the stand seats were arranged in a straight line, touching all the edges, thus allowing the sound, during propagation, not to be hindered. Vitruvius also described the existence of vessels to be placed under the stand seats to give better sound in the theatre. This fascinating hypothesis, now finds application in acoustic resonators, which are used for the correction of modern theatres. However, the legendary good acoustics of ancient theatres was mainly due to the lay-out of the stand seats, that were aligned and regular in size and acted as scattering surfaces. In the sense that sound perturbation waves advance impacting with the edges of the stand seats, spreading throughout the theatre in all directions, as if the edges of the stand seats were secondary sound sources, significantly improving sound propagation. This complex phenomenon has been only recently understood [5, 6, 7, 8].

3 The Odeon: the origins

The Odeon is a stylish indoor theatre, smaller in size than traditional theatres. It is assumed to have been used for musical, acting, literary and poetic performances. It could hold a few hundred spectators, with it being a more limited audience. The covering made it possible to perform even in adverse weather conditions. The etymology of the word Odeon (indoor theatre) is the place of ode, or singing, even if today it is not known where and how they were executed. The first known Odeons were in Greece, dating from the 5th entury B.C. in Athens. The Odeon of Pericles as well as that of Agrippa were built at the base of the Acropolis, with that of Herodes Atticus being added in the 2nd century A.D, built in honour of his dead wife (it is still possible to admire the beauty of this Odeon today). This type of building was subsequently brought to Rome, where, thanks to the talent of Roman builders sumptuous structures were realized, more stylish and larger than those of the Greeks. Part of secret of this success was due to the ability to realize large and unique bays of covering beams. A secondary problem of the Odeon was lighting and air exchanges, which were carried out with large windows open on the side walls. The Odeon of Pericles in Athens was built around 442. B.C., and according to Pausanias, it was similar to the tent of Xerxes which was a square building with a domed roof and a double row of columns around it. As reported by Vitruvius, the roof was made of the wood from Persian ships defeated at the battle of Salamis. Recent excavations have established that the roof was supported by 90 pillars inside, divided into nine rows of ten, with an orchestra-pit for the choir, a stage for the musicians and the back-wall of the stage with painted decorations. Odeons were built in many cities of the Roman Empire (Syracuse, Catania, Naples, Pompeii, Aosta, Taormina) and frequently used, but only a few authors report their existence. There are only a few descriptions in the treatises of Vitruvius, Tertullian, Cassiodorus. Vitruvius in his treatise "De Architectura" dedicates a great deal of Book V to Greek and Roman theatres, but mentions only once the building built by Pericles in the 5th century B.C., and the existence of a small theatre, at Tralles, when he describes the paintings inside (Book VII).

Today, however, not much is known about the type of covering used, if the roofs were trussed or with suspended platforms and only a few researchers have tried to reconstruct them, with one of them being G.C. Izenour [9, 10].

4 The Odeon of Posillipo

At south of the theatre, with a slight slope, there was a covered building for poetry recitals called the Odeon. It had a central area with a stage, seats and an imperial box (pulvinar) and rich lateral symmetrical rooms. The auditorium had four rows of semi-circular seats, plus two rows of seats in a quarter-circle. Behind the stage, a triplex porch with plastered brick columns was the scenic backdrop that framed and separated two recreational buildings. According to Gunther, the Odeon was created as a hall at the time of Vedius Pollio and, probably under Nero, was later expanded with an imperial box and enriched with frescoes and polychrome marble floors. Without any information to perform acoustic measurements due to the scarcity of existing boundary surfaces the virtual reconstruction of Odeon of Pompeii was used. For the Odeon of Pompeii, the authors carried out acoustic measurements in the current state (building without a roof and with a partially reconstructed stage), with a virtual model of the building before its destruction following the eruption of Mt. Vesuvius in 79 A.D. being realised, starting from the geometrical measurements of the building in its present state [11]. The sound field was simulated in a virtual model by assigning values to the absorption coefficient of the structures in its present state (cavea, scene, walls) and then the acoustic characteristics of the covering were assigned. To virtually reconstruct the acoustics of the Odeon, a model was realized using a 3D design software. The virtual model was then developed with an architectural acoustics software [12, 13, 14, 15]. The software is based on geometric acoustics principles with a tracking technique of sound beams. The maximum dimensions of the building were width 26 m and length 12 m. With the roof, the virtual model had the following geometric characteristics: total surface area 2360 m^2 and volume 4400 m^3. Fig. 2 presents the Odeon in its current state. While Fig. 3 and Fig. 4 show the reconstructed sections of the Odeon. Fig. 5 shows the plant and Fig. 6 shows the interior view of the reconstructed Odeon. Fig. 7 and Fig. 8 show the view of the reconstructed Odeon.

Fig. 2. The Odeon in its current state

Fig. 3. Odeon reconstruction: section

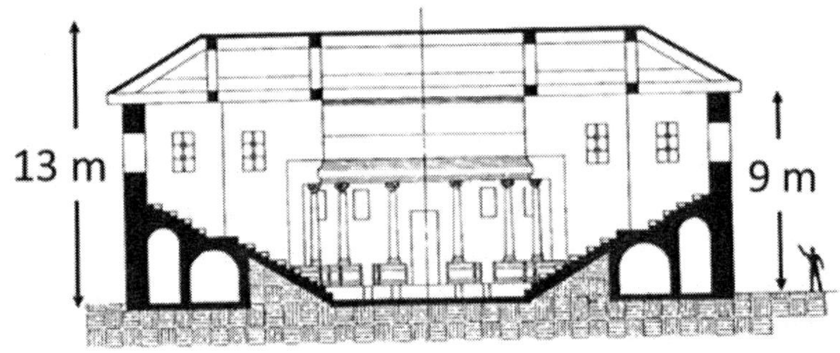

Fig. 4. Odeon reconstruction: section

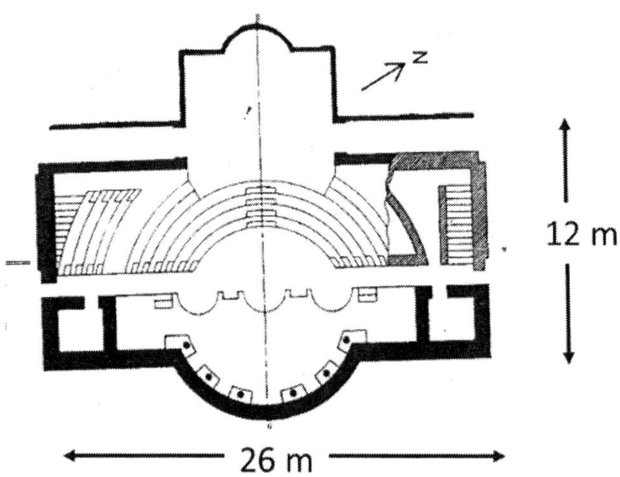

Fig. 5. Odeon reconstruction: plant

5 Sound field simulation

To virtually reconstruct the acoustics of the Odeon, a numerical model was realised with a 3D design software [16, 17], starting from the geometric measurements of the building in its current state. Subsequently, the virtual model was integrated with a covering in analogy to the drawings available in current literature, Izenour G.C. (Fig. 3-5). Then, the virtual model was processed by the architectural acoustics software, Odeon. The architectural acoustics software is based on geometric acoustics principles with a tracking of sound beams technique.

The Odeon program uses a hybrid calculation method, where early reflections are calculated by combining the image- source and a ray tracing method. The late reflections are calculated using a ray tracing process, which generate diffuse secondary sources. In this process rays are sent from the source position detecting image sources up to a certain reflection order and then detecting secondary sources on the surfaces of the room. The secondary sources radiate with a directivity that follow Lamberts Law and an intensity contribution attenuated mainly by the distance and the reflections in the path from source to receiver. The transition order (T.O.) at which the calculation method changes from image-source to ray tracing can be adjusted according to the complexity and shape of the room. If T.O. is set to 0 the calculation method used for all reflections will be the secondary source / ray tracing method.

In the numerical model, the receiving points were positioned in the cavea towards the radial direction, in the same position where the audience would have been seated. The point source was placed on the stage to simulate the presence of actors. The seating rows were modelled with a scattering coefficient of s=0.7, while the stage wall and floor were assigned a scattering of s=0.5 [18, 19, 20]. Fig. 9 and Fig. 10 show the virtual model with sound sources on the stage and receivers in the cavea. The monaural acoustic parameters considered are [21, 22, 23]:

T_{30}, reverberation time, measured in seconds, is the most commonly used descriptor for evaluation of acoustic characteristics of a room. It was defined by W. Clement Sabine in late nineteenth as the time required for, after sound emission interruption from a source, sound pressure level in the room decreases by 60 dB. The value of this parameter is a function of room volume and of total absorption of surfaces and, since the absorption of materials varies with frequency, reverberation time also varies according to the different frequencies considered.

EDT, Early Decay Time, measured in seconds, is the time corresponding to a decrease of 10 dB of the decay curve of the sound pressure level, measured from when the sound is interrupted; this parameter is particularly sensitive to microphone position related to the source and is a particularly significant parameter in comparison between different points of the same environment.

C_{80}, clarity (dB), is a characteristic parameter of goodness of a hall for listening to music; it is a descriptor obtained from the correlation of energy that reaches the listener within the first 80 milliseconds of sound emission (including directed energy and early reflections energy), and the energy that reaches in the following instants.

D_{50}, is a characteristic parameter of goodness of a room for purposes of speech understanding; it is a descriptor that is obtained through the correlation of energy that reaches the point of reception in the first 50 milliseconds of sound emission and entire signal energy, that is until the end of its decay. Experimental results confirm that early reflection energy, adding to that of direct sound, gives a positive contribution to speech understanding.

STI, Speech Transmission Index, represents the degree of amplitude modulation in a speech signal and refers to the distortion in speech signals caused by reverberation, echoes, and background noise.

The indices STI and RASTI can take values between 0 and 1, being greater than 0.5 for favourable speech conditions.

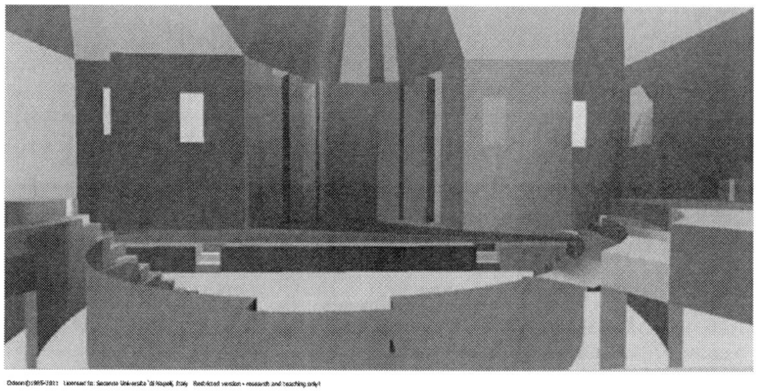

Fig. 6. Odeon reconstruction: interior view

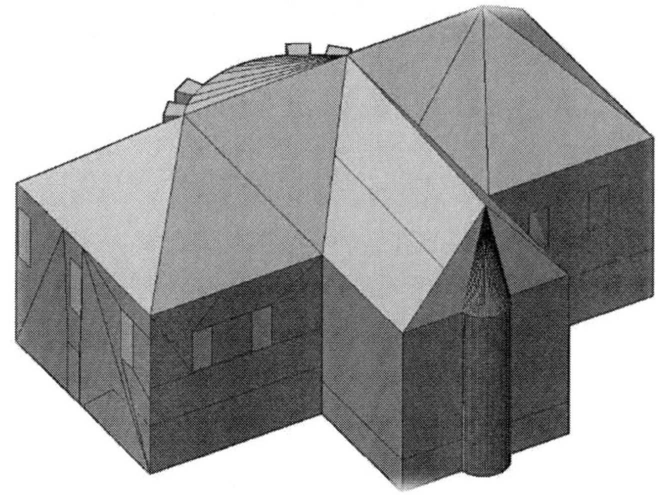

Fig. 7. Odeon reconstruction, external view

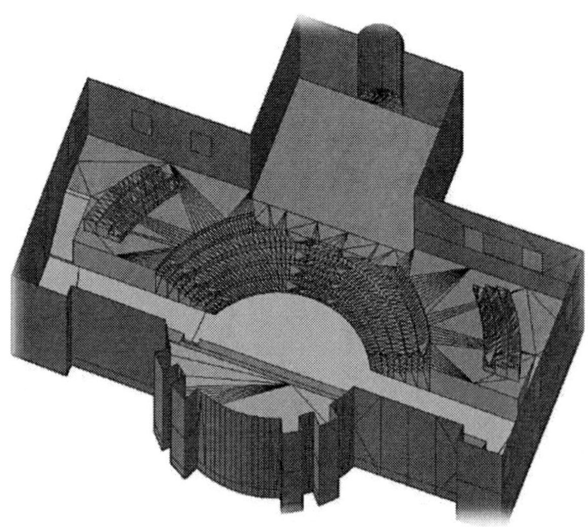

Fig. 8. Odeon reconstruction, internal view

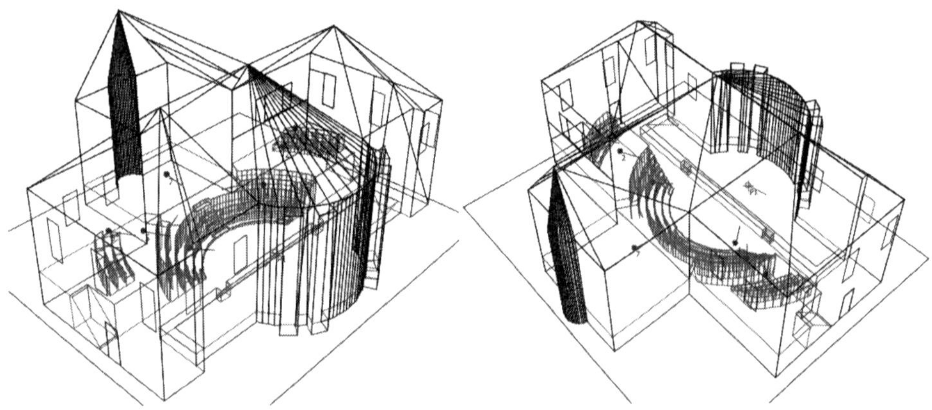

Fig. 9. Virtual model with source on the scene and receivers in the cavea

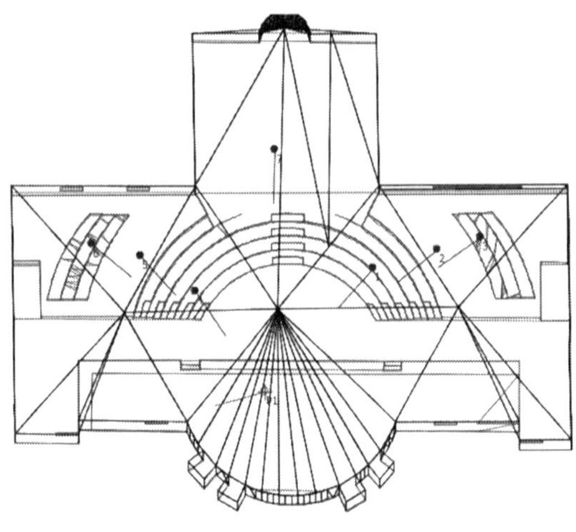

Fig. 10. Virtual model with source on the scene and receivers in the cavea

6 Discussion

Table 1 shows a synthesis of the optimum values of the monaural acoustic parameters both for speech understanding and listening to music. Different parameters are considered, since only reverberation time is not enough to assess the acoustic goodness of a room. For a good speech understanding, definition D_{50} is evaluated (in a room a good condition for speech understanding, there is for D_{50} values > 0.50); for good music listening, parameter C_{80} is evaluated (in a hall C_{80} values in the range: $-2 < C_{80} < 2$ denote a good listening music condition). In the virtual reconstruction, T_{30} (reverberation time) assumes values between 0.8 s and 1.0 s. Voice clarity, measured by the parameter D_{50} (index of speech understanding, minimum value 0, maximum 1) results in the virtual model, with the presence of the roof, $D_{50} = 0.60$, the speech listening condition is good. In contrast, the C_{80} parameter, (clarity, goodness index for listening to music) assumes values greater than 3 dB; the Odeon of Posillipo is poor for listening to music performances.

Descriptors	EDT Seconds, s	T_{30} Seconds, s	C_{80} Decibel, dB	D_{50}
Optimal values for listening to music	$1.8 < EDT < 2.6$	$1.6 < T_{30} < 2.2$	$-2 < C_{80} < 2$	< 0.5
Optimal values for speech listening	1.0	$0.8 < T_{30} < 1.2$		> 0.5

Tab. 1. Optimal acoustic parameters for different listening conditions

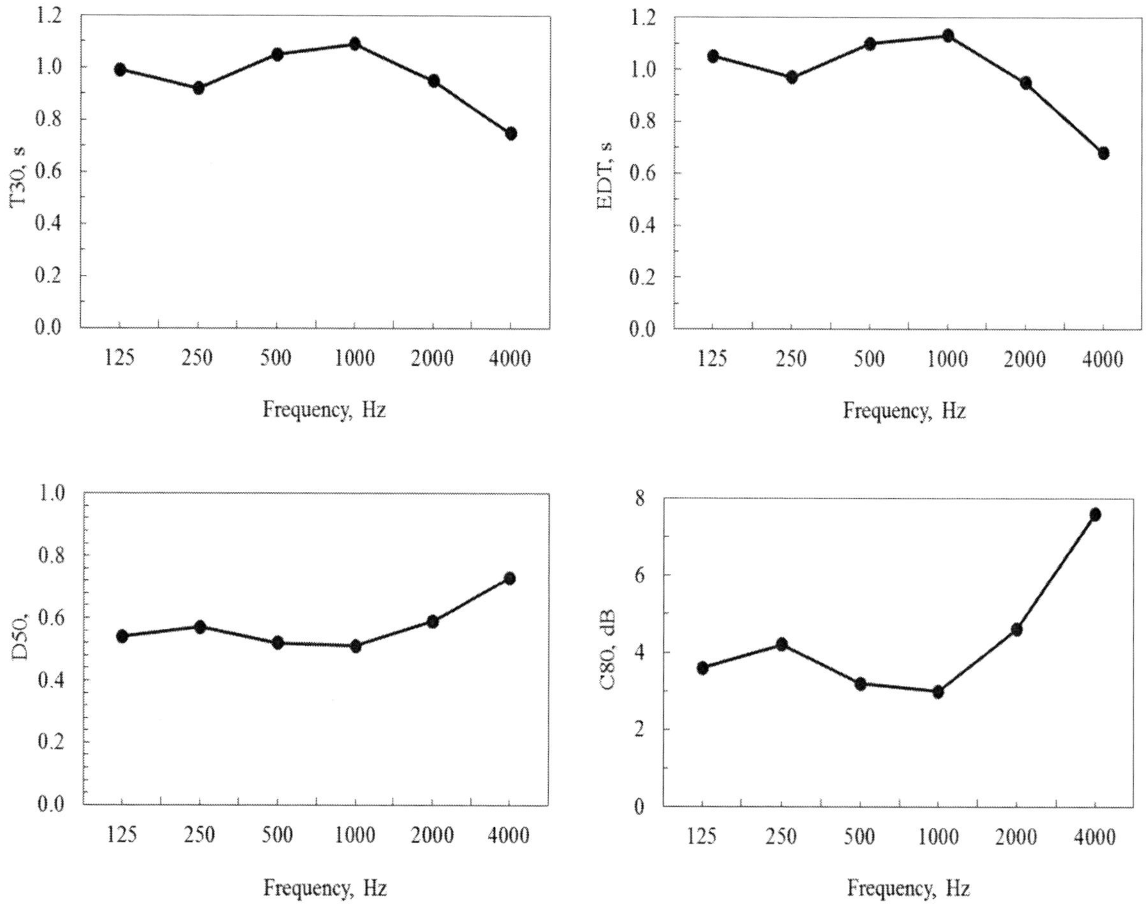

Fig. 11. Monaural acoustic parameters, simulated with the virtual model (STI = 0.7)

In order to assess the acoustic characteristics of the Odeon of Posillipo and compare it to other similar theatres, a comparison of the acoustic parameters obtained using virtual models in the theatres studied in the European ERATO project was carried out. In addition to the open-air theatres of Pompeii (a large theatre) and Benevento, the average values of the monaural parameters in the mid frequency bands of 500 Hz and 1.0 kHz are reported in Tab. 2. One of the aims of the ERATO project was to investigate the acoustics of open-air theatres and compare them to smaller, original theatres [24]. This project found that open-air theatres had a very high clarity of sound. In contrast, the Odeon has a reverberation time similar to a concert hall, a relatively low clarity and a high sound strength. Tab. 2 reports the acoustic parameters averaged at mid-frequency bands of 500 Hz and 1.0 kHz calculated through software simulations in some ancient theatres. However, other acoustic parameters were heavily affected by the presence of an audience. In the Odeon, the reverberation time strongly depends on the ceiling height and size of the windows. Tab. 2 also presents a comparison of acoustics data for ancient theatres with the Odeon of Posillipo and shows how the Odeon of Posillipo had a shorter reverberation time, a stronger direct sounds and a smaller sense of spatiality. The Odeon could accommodate theatrical performances.

Theatres	condition	kind of theatre	T_{30} [s]	C_{80} [dB]	STI [-]	diameter [m]
Pompeii (large theatre)	empty	open-air	1.53	4.50	0.62	58.0
Benevento	empty	open-air	1.80	2.00	0.60	40.0
Aspendos	empty	open-air	1.95	1.17	0.53	95.5
Aosta roman	empty	roofed	5.97	-5.28	0.36	60.0
	full		3.49	-2.41	0.43	
Aphrodisias roman	empty	roofed	4.02	-4.21	0.38	45.6
Posillipo roman	empty	roofed	1.1	3.0	0.70	26.0

Tab. 2. Acoustic parameters averaged at mid-frequency bands of 500 Hz and 1.0 kHz calculated through software simulations in some ancient theatres

7 Conclusions

This paper has described the virtual reconstruction and acoustic simulation of the Odeon of Posillipo, in presence of an original roofing system. The results obtained from the processing of the virtual models, with the aid of an architectural acoustics software have shown how a building with a wooden roof was well suited for speaking. In fact in this Odeon, the Emperor was entertained in the company of his intimated friends. It is possible that the poets Virgil and Horace read from newly created works before their patrons. Furthermore, the room acoustics do not change with the presence of an audience due to the reduced occupancy of the stand seats.

References

[1] VITRUVIO, M.P. *De Architectura.*

[2] SABINE, W.C. *Collected Papers on Acoustics*, Cambridge, Mass, 1923.

[3] MAZZONI, *S. Teatri* romani: immagini d'impero e d'attori e altre questioni, Studi di Storia dello spettacolo. Omaggio a Siro Ferrone, Firenze, 2011, 15-35.

[4] CIANCIO ROSSETTO, P., SARTORIO G.P. *Teatri Greci e Romani: alle origini del linguaggio rappresentato censimento analitico*, 3, Torino, 1994.

[5] DECLERCQ, N.F., DEKEYSER, C.S.A. Acoustic diffraction effects at the Hellenistic amphitheatre of Epidaurus: seat rows responsible for the marvellous acoustics, *Journal Acoust. Soc. Am.* 121(4), 2007, 2011–2022.

[6] LOKKI T., SOUTHERN A., SILTANEN S., SAVIOJA L. Acoustics of Epidaurus – studies with room acoustics modelling methods, *Acta Acustica,* 99, 2013, 40-47

[7] IANNACE, G., MARLETTA, L., SICURELLA, F., IANNIELLO, E. Acoustic measurements in the ear of Dionysius at Syracuse (Italy). 39th International Congress on Noise Control Engineering 2010, INTER-NOISE 2010 ,535.

[8] SHANKLAND, R.S. Acoustics of Greek theatres, *Physics Today* 26, 1973, 30-35.

[9] IZENOUR, G.C. *Theatre Design*, New York, 1977.

[10] IZENOUR, G.C. *Roofed Theatres of Classical Antiquity,* New Haven and London, 1992.

[11] BERARDI, U., IANNACE, G., MAFFEI, L. Virtual reconstruction of the historical acoustics of the Odeon of Pompeii, *Journal of Cultural Heritage*, 19, 2016, 555-566, doi: http://dx.doi.org/10.1016/j.culher.2015.12.004

[12] IANNACE, G., TREMATERRA, A., MASULLO, M. The large theatre of Pompeii: Acoustic evolution, *Building Acoustics*, 20(3), 2013, 215-227, doi: 10.1260/1351-010X.20.3.215

[13] IANNACE, G., TREMATERRA, A. The rediscovery of Benevento Roman Theatre Acoustics, *Journal of Cultural Heritage*, 15(6), 2014, 698, doi: http://dx.doi.org/10.1016/j.culher.2013.11.012

[14] D'AMBROSIO ALFANO, F.R., IANNACE, G., IANNIELLO, C., IANNIELLO, E. "Velaria" in ancient Roman theatres: Can they have an acoustic role?, *Energy and Buildings*, 95, 2015, 98-105, doi: http://dx.doi.org/10.1016/j.enbuild.2015.03.010

[15] IANNACE, G., DI GABRIELE, M., SICURELLA, F. Sound Focusing Effects in Horseshoe Plan Theatre, *Acoustics Australia,* 44 (2), 2016, 359-368, doi: 10.1007/s40857-016-0059-2

[16] IANNACE, G., TREMATERRA, A. Acoustic measurements and correction of a council room, *Noise and Vibration Worldwide*, 45 (8), 2014, 12-16, doi: 10.1260/0957-4565.45.8.12

[17] IANNACE, G. Acoustic correction of monumental churches with ceramic material: The case of the Cathedral of Benevento (Italy), *Journal of Low Frequency Noise Vibration and Active Control,* 35(3), 2016, 230 – 239, doi: 10.1177/0263092316661028

[18] BERARDI, U., IANNACE, G., IANNIELLO, C. Acoustic intervention in a cultural heritage: The chapel of the Royal Palace in Caserta, Italy, *Buildings*, 6(1) 2015, Article number 1. doi 10.3390/buildings6010001

[19] IANNACE, G., BERARDI, U., IANNIELLO, C. Study of a historical Church based on acoustic measurements and computer simulation, *Proceedings 22nd International Congress on Sound and Vibration*, ICSV 2015; Florence; Italy; 12 -16 July 2015; Code 121474

[20] IANNACE, G., TREMATERRA, A., TREMATERRA, P. Acoustic correction using green material in classrooms located in historical buildings. *Acoustics Australia* 41(3), December 2013, 213-218.

[21] BERARDI, U., IANNACE, G., GIORDANO, G., Acoustic characteristics of four subway stations in Naples, Italy. *Proceedings 44th International Congress and Exposition on Noise Control Engineering*, INTER-NOISE 2015; San Francisco; 9 - 12 August 2015; Code 114294

[22] CHRISTENSEN, C.L. ODEON Room Acoustics Software. v.11 manual. www.odeon.dk

[23] ZENG, X., CHRISTENSEN, C.L., RINDEL, J.H. Practical methods to define scattering coefficients in a room acoustics computer model, *Applied Acoustics*, 67, 2006, 771-786.

[24] RINDEL, J.H. Roman theatres and the revival of their acoustics in the ERATO Project, *Acta Acustica*, 99, 2013, 21-29.

Current address:

Trematerra Amelia, phd
Dipartimento di Architettura e Disegno Industriale, Seconda Università di Napoli
Borgo San Lorenzo, 81031 Aversa, Italy
Tel. + 39.3402389882 ; liatrematerra@libero.it

Parternuosto Stefania, student
Dipartimento di Architettura e Disegno Industriale, Seconda Università di Napoli
Borgo San Lorenzo, 81031 Aversa, Italy
stefaniapaternuosto@gmail.com

Lombardi Ilaria, student
Dipartimento di Architettura e Disegno Industriale, Seconda Università di Napoli
Borgo San Lorenzo, 81031 Aversa, Italy
ilaria.lombardi@studenti.unina2.it

Slovak University of Technology in Bratislava
Faculty of Mechanical Engineering
16th Conference on Applied Mathematics **APLIMAT 2017**
Proceedings

RUPTURE FORMATION AND PHASE TRANSITION AS A POSSIBLE CONSEQUENCE OF LIMITING ENVIRONMENTAL CONDITIONS

TURŇA Ľubomír (SK)

Abstract. Any activity in any environment or device often produces waste and changes of the environment. The paper aims to analyze this process on a simple model and determine the conditions of the change of environment characteristics by means of the model parameters.
Results of analyzing the coexistence of biological species in their various relationships, such as cooperation, competition and antinomy, have been known within the mathematical ecology for almost a century. The same mathematical apparatus of evolution equations can be used for monitoring the coexistence of inanimate entities, for example: in particle physics theories, in the behavior of gases and thermodynamics, quantum generator, in the theory of management and so on.
The contribution follows the use of mathematical tools, for example: in the area of information technology, and the analysis of interacting stakeholders and groups in the socio-economic environment.

Keywords: memory of environment, degradation, exploitation, wear, coexistence, non-linear systems, stationary point, (un-)stability

Mathematics Subject Classification: Primary 45A05, 34A34; Secondary 45J05, 34D20

Introduction

The issue of continuous depletion, intolerable devastation and devaluation of the environment is becoming more and more urgent. It is not always only about reducing the environment's values via the said mechanisms, but also about its excessive overload all the way to, or even beyond its ability of natural regeneration, or possibly also about recognition of processes during their manifestation. Such situations often occur in technical practice, when several entities, processes or phenomena coexist, making use of the same facilities, resources, warehouse and storage space and operating time.

The result of this co-owned form depends on whether the coexistence takes a form of a cooperating or competing relationship. With a forced sharing of means of existence, the rigorous view of the problem almost excludes any form of cooperation; perhaps a mutual tolerance and consideration are possible options. However, if follows from the experience, that if strict and bounding criteria are not set, then it is very difficult to rely on stable forms of altruism, especially if the possible lack of resources would lead to fatal consequences. The same co-owned principle is often observed also in the socio-economic environment.

Behavior of dynamic systems of technical and social environments during the use of common resources

The paper aims to follow simpler cases of coexistence in two-component systems in the event of in a limit bounded conditions of shared resources. In the effort to achieve this goal, the methods from the field of nonlinear dynamic systems shall be used.

Problem definition, model proposal and method of solution

The presented simple model deals with the analysis of processes in their mutual interaction, occurring in limited possibilities of their existential environment. The processes can be: exploitation of limited possibilities of the computer, cyber and technical systems in general, or the effort of individuals or collectives to use the social background and its limited resources. For simplicity, the system will consist of two different sets of interests. These sets can be groups of individuals of the collective with the exploitation of the socio-economic background, processes required to carry out a certain activity or utilize bounded storage spaces, and so on.

The gradually drawn amount m of the possibilities, from the limited capacity of the environment M, is proportional to the intensity of activity of both processes, is proportional to the amount of their independent accesses to the environment, or to the amount of realized efforts and interests; it surely depends, in a direct proportionality, on the individual numbers of holders of the given activity m_1, m_2. Instant common consumption will therefore be $m = coef_1.m_1 + coef_2.m_2$, where the coefficients $coef_1, coef_2$ express the intensity of individual approach. For simplicity, at the beginning let each process draw from common initial limited quantity a. Provided that the possibilities or the intensity of execution is proportional to the available resources, then the dynamics of the system can be entered using the following differential system, $a > 0, b \geq 0, c \geq 0$,

$$\dot{m}_1 = (a - b.m_1 - c.m_2).m_1, \quad \dot{m}_2 = (a - b.m_2 - c.m_1).m_2 \ , \tag{1a}$$

$$\dot{m}_1 = a.m_1 - b.m_1^2 - c.m_1.m_2, \quad \dot{m}_2 = a.m_2 - b.m_2^2 - c.m_1.m_2 \ , \tag{1b}$$

in which the behavior of both components is qualitatively similar. The first expression of the equation expresses the subsistence and maintenance of components, the second and the third expression of the equation express the consumption of resources, but at the same time the second expression of the equation also expresses its own oversize and the third expression of the equation expresses the mutual rivalry and antagonism between the components.

The terms $a - b.m_1 - c.m_2$ v.s. $a - b.m_2 - c.m_1$ expresses a contradiction the behavior of the components m_1, m_2 with regard to the possibility of starting resources.

Stationary points of the system (1a), or (1b) – eventually, with their eigenvalues l_1, l_2, depending on the system parameters, may be stable and unstable centers of their surroundings:

$(0,0)$: $l_{1,2}=a>0$ ... permanently/always unstable,

$\left(0,\frac{a}{b}\right)$: $l_1=a.(c-b),\ l_2=-a$ ... it varies, may be stable or unstable,

$\left(\frac{b}{a},0\right)$: $l_1=-a,\ l_2=a.(c-b)$ ... it varies, may be stable or unstable,

$\left(\frac{a}{b+c},\frac{a}{b+c}\right)$: $l_{1,2}=\frac{a.(-b\pm c)}{b+c}$ ... it varies, may be stable or unstable.

Much more interesting than watching the characteristics of stationary points and expressions of the system in their vicinity, is to focus on possible mutually-forced influences of the components, respectively how the processes affect each other. After the introduction of substitutions of $n_1=m_1+m_2$, $n_2=m_1-m_2$ we can obtain

$$\dot{n}_1=a.n_1-b.(n_1^2+n_2^2)/2-c.(n_1^2-n_2^2)/2,\ \ \dot{n}_2=a.n_2-b.n_1.n_2\ . \tag{2}$$

As the consequence of stable coexistence it can be expected $m_1+m_2\approx const>0,\ 0<m_1-m_2<<1$ and then therefore $\dot{n}_1\cong 0,\ n_2<<1$, $n_1^2-2.a.n_1/(b+c)+n_2^2.(b-c)/(b+c)=0$, subsequently $(n_1)_{1,2}=\frac{a\pm\sqrt{a^2-(b^2-c^2).n_2^2}}{2}$. Because $n_1=m_1+m_2\approx const>0,\ 0<n_2=m_1-m_2<<1$ and also $\varepsilon<<1,\ \sqrt{1+\varepsilon}\approx 1+\varepsilon/2$, we can obtain $n_1=2.a/(b+c)+(c-b).n_2^2/a$. After substituting n_1 to the second equation of (2) the result has a structure

$$\dot{n}_2=Q.n_2-R.n_2^3, \tag{3}$$

where Q, R can be expressed through a,b,c: $Q=a.(c-b)/(b+c),\ R=b.(c-b)/a$.

Equation (3) is known in the catastrophic theory and in theory of nonlinear dynamic systems as enabling qualitative changes/leaps. Then combined parameters Q, R are the control parameters and under certain conditions, that rely on the a,b,c, equation (3) describes the formation of Hopf bifurcation, activation of oscillations in the description of Van der Pol oscillator, or the onset of phase transition.

Previous models of coexistence and utilization of shared resources are very similar to the system

$$\dot{m}_1=(a-b.m_1-c.m_2).m_1,\ \ \dot{m}_2=(a-b.m_1-c.m_2).m_2, \tag{4a}$$

$$\dot{m}_1=a.m_1-b.m_1^2-c.m_1.m_2,\ \ \dot{m}_2=a.m_2-b.m_1.m_2-c.m_2^2, \tag{4b}$$

in which instant common consumption will therefore be $m=b.m_1+c.m_2$, where the coefficients b, c express the intensity of individual approach, but in this system the two components are part of the same consumption. For simplicity, at the beginning let each process draw only from common limited quantity a. Provided that the possibilities or the intensity of execution is proportional to the available resources, then the dynamics of the system can be entered using the differential system (4a), (4b), $a>0, b\geq 0, c\geq 0$.

Stationary points (m_{1s}, m_{2s}) of system (4a), (4b) with their eigenvalues λ_1, λ_2, depending on the system parameters, may be stable and unstable centers of their surroundings:

$(0,0)$: $\lambda_{1,2}=a>0$ ... permanently/always unstable,

$\left(\frac{a}{b+c}, \frac{a}{b+c}\right)$: $\lambda_1=-a<0$, $\lambda_2=\frac{a.(c-b)}{b+c}$ ... it varies, may be stable or unstable.

After formal modifications we can come, through a comparison using the same expression in parentheses in system (4a), (4b) to the equation $\frac{\dot{m}_1}{m_1}=(...)=\frac{\dot{m}_2}{m_2}$, and to solving the equation (4a), (4b): $\ln|m_1|=\ln|m_2|+\ln|M|$, $M=const.$:

$$m_1=M.m_2\ . \tag{5a}$$

Taking into account the initial conditions, $t=t_0 \Rightarrow m_1(t_0)=m_{10},\ m_2(t_0)=m_{20} \Rightarrow M=\frac{m_{10}}{m_{20}}$, then

$$m_{10}=M.m_{20}\,. \tag{5b}$$

With regards to (5a), (5b) both equations of system (), we can rewrite them as well

$$\dot{m}_2=\left[a-(b.M+c).m_2\right].m_2\ , \tag{6a}$$

therefore

$$\frac{dm_2}{\left[a-(b.M+c).m_2\right]m_2}=dt \tag{6b}$$

and then the solution of the system (4a), (4b) will be

$$m_2(t)=\frac{a.m_{20}.e^{a.(t-t_0)}}{a+m_{20}.(b.M+c).\left(e^{a.(t-t_0)}-1\right)}\ ,\quad m_1(t)=M.m_2(t)=\frac{m_{10}}{m_{20}}.m_2(t)\,. \tag{7}$$

If $t\to\infty$ then $m_2(t)\to\frac{a.m_{20}}{b.m_{10}+c.m_{20}}$, $m_1(t)\to\frac{a.m_{10}}{b.m_{10}+c.m_{20}}$ and it is the saturation of the function $m_2(t)$ or $m_1(t)$ respectively.

Finally, we can get the information using the first derivations of function [1]

$$\frac{dm_2(t)}{dt}=\frac{m_{20}.a^2.\left(a-(b.M+c).m_{20}\right).e^{a.(t-t_0)}}{\left[a+(b.M+c)m_{20}.\left(e^{a(t-t_0)}-1\right)\right]^2}, \tag{8}$$

determine the conditions of functions monotony $\frac{dm_2(t)}{dt}\begin{pmatrix}<,=,>\\ ?\end{pmatrix}0 \Rightarrow a\begin{pmatrix}<,=,>\\ ?\end{pmatrix}b.m_{10}+c.m_{20}$

and we receive the information using the second derivations of function

$$\frac{d^2m_2(t)}{dt^2}=\frac{m_{20}.a^3\left(a-(b.M+c).m_{20}\right).e^{a.(t-t_0)}.\left(e^{a.(t-t_0)}+a-3(b.M+c)m_{20}\right)}{\left[a+(b.M+c)m_{20}\left(e^{a.(t-t_0)}-1\right)\right]^3} \tag{9}$$

determine the inflection point $t=t_0+\frac{\ln\left(3-\frac{a}{b.m_{10}+c.m_{20}}\right)}{a}$.

Also, in this case, much more interesting than watching the characteristics of stationary points and expressions of the system in their vicinity, how the processes affect each other.
After the introduction of substitutions of $n_1=m_1+m_2$, $n_2=m_1-m_2$ we can obtain

$$\dot{n}_1=-(b+c).n_1^2/2+a.n_1+(c-b).n_1.n_2/2, \quad \dot{n}_2=(c-b).n_2^2/2+a.n_2-(b+c).n_1.n_2/2\ . \tag{10}$$

As the consequence of stable coexistence it can be expected $m_1+m_2\approx const>0,\ 0<m_1-m_2<<1$

and then therefore $\dot{n}_1\cong 0,\ n_2<<1,\ (b+c).n_1^2/2-a.n_1-(c-b).n_1.n_2/2=0$, then subsequently

$n_1.\left[(b+c).n_1/2-a-(c-b).n_2/2\right]=0 \quad \Rightarrow \quad (n_1)_1=0 \vee (n_1)_2=\frac{2.a-(c-b).n_2}{b+c}=\left|\begin{matrix}\text{because}\\ n_2<<1\end{matrix}\right|\cong\frac{2.a}{b+c}$. After

substituting $(n_1)_{1,2}$ to the second equation of (10) the result has a structure:

for $(n_1)_1=0$: $\quad \dot{n}_2=\frac{b-c}{2}n_2^2+a.n_2$,

for $(n_1)_2=\frac{2.a-(b-c).n_2}{b+c}$: $\quad \dot{n}_2=\frac{3.(b-c)}{4}n_2^2-\frac{a}{2}n_2$,

for $(n_1)_2=\frac{2.a-(b-c).n_2}{b+c}=\left|\begin{matrix}\text{because}\\ n_2<<1\end{matrix}\right|\cong\frac{2.a}{b+c}$: $\quad \dot{n}_2=\frac{b-c}{4}n_2^2-\frac{a}{2}n_2$.

---

[1] Expressions $A\ (?>,=,<)\ B$, $A\begin{pmatrix}>,=,<\\ ?\end{pmatrix}B$ means: subject to the fulfillment of comparison A vs B.

As we can see, in the nearby of the conditions $\dot{n}_1 \cong 0,\ n_2 << 1$, where $n_1 = m_1 + m_2,\ n_2 = m_1 - m_2$ and for $m_1 + m_2 \approx const > 0,\ 0 < m_1 - m_2 << 1$ system can "jump" between multiple processes, performs a kind of strange bifurcation.

Conclusions

The paper tried to analyze multiple versions of the simple model, dealing with the residual memory effect occurring during exploitation of the environment. The model works with the possibility of taking into account the accumulation effects of the impacts on the environment. The paper also analyzed, by using multiple models of coexistence, the processes in the event of shared limited resources. Within the proposed model we could show the existence and properties of several stationary points, including the possibility of causing qualitative changes in the behavior of the running process.

References

[1] HAKEN, H.: Dynamics of Synergetics Systems, Springer Verlag, Berlin 1980.

[2] ARNOĽD, V. I., VARČENKO, A. N., GUSEJN-ZADE, S. M. Osobennosti diferenciruemych otobraženij I, II, Nauka, Moskva 1984.

[3] ARNOĽD, V. I.: Teorija katastrof, Nauka, Moskva 1990.

[4] ARNOĽD, V. I.: Osobennosti kaustik i volnovych frontov, Fazic, Moskva 1996.

[5] GILMORE R.: Catastrophe theory for scientists and engineers I, II, JW&Sons, Inst. for Defense analyse, Arlington, Virginia, NY 1981 (Prikladnaja teorija katastrof, Mir, Moskva 1984)

Current address

Turňa Ľubomír, RNDr. CSc.
Department of Applied Informatics, Faculty of Economic Informatics,
University of Economics in Bratislava
Dolnozemská cesta 1, 852 35 Bratislava, Slovak Republic
Tel. number: +421 2 6729 5877, e-mail: stefanluboturna@post.sk

Slovak University of Technology in Bratislava
Faculty of Mechanical Engineering
16th Conference on Applied Mathematics **APLIMAT 2017**
Proceedings

FROM LEONARDO'S YCOCEDRON ABSCISUS VACUUS TO THE FOOTBALL: A GEOMETRIC JOURNEY

UGHI Emanuela (IT)

Abstract. Teaching methods usually require students to look at pictures and drawings related to some mathematical subject, but the manipulation of a concrete model is much more appealing, so that this paper wants to offer a teaching proposal to offer the students the possibility to realize concrete models of a football, so discovering and better mastering its geometric structure.

Keywords: Polyhedron, Leonardo, Ycocedron Abscisus Vacuus, Football, Soccer Ball

Mathematics Subject Classification: 97A99

1 The Ycocedron Abscisus Vacuus

Many know the drawings of polyhedra drawn by Leonardo for "De Divina Proportione" by Luca Pacioli [8]. We are so used to the modern technologic drawing tools that are hardly able to imagine how difficult it was to draw those complex polyhedra at that time.

The history of the cooperation between Leonardo and Pacioli is fascinating: when they met, in Milan, Leonardo was poor in mathematics, not having had a regular education. So he had to learn a lot of mathematics following Pacioli's lessons [3],[4],[11].

Some scholars think that Pacioli made wooden models of the polyhedra, to allow Leonardo to draw them using his "vetro" (glass), a simple tool he describes to help to correctly draw images in perspective. And the result of this work was so perfect that Pacioli indeed was enthousiast, and wrote about "le supreme e leggiadrissime figure de tutti li platonici et Mathematici corpi regulari et dependenti"(supreme and very graceful figures of all platonic and Mathematical regular and dependent solids) drawn by the "ineffabile senistra mano" (ineffable left hand, since Leonardo was left handed) [9].

In particular, the Ycocedron Abscisus corresponds, in modern terminology, to the Truncated Icosahedron. But barely even educated (and sport supporter!) people are conscious that its structure is the same of that of the modern football (soccer ball), and know how many exagons, and pentagons, there are, and how they are arranged.

The first ball having this structure was created in 1962 by the footballer Eigil Nielsen, and rapidly this design became the common design used in football and soccerball. An interesting geometric feature of it is related to the obvious need for the ball to be as close to a sphere as possible; this is usually obtained by inflating a leather polyhedral structure, and to this purpose the truncated icosahedron is a better choice respect to the cuboidal structures used earlier.

Among the celebrations for the 500th anniversary of the appearance of the Divina Proportione, at the end of the last century, several mathematical sculptors made wooden replies of this object, so obtaining wonderful mathematical exhibits: for example, the Ycocedron by George Hart (Wendy Petti Collection) [5], and the one by Nedeljko Adzic (Roma, Sale del Bramante) [2] .

I was very envious for the beauty of their Ycocedrons, and I tried to make one for myself.

But I choose to make it in cardboard; my model of the Ycocedron is so light (just 200 grams) that it is possible to glue the pieces, and to hang it: it has been even done in the Biblioteca Ambrosiana, in Milan, in July 2015, when it was displayed in the same case with the original book by Pacioli [8].

Fig. 1. The model of the Ycocedron Abscisus Vacuus

2 The book "Il poliedro di Leonardo"

While designing the model of the Ycocedron, I also planned to allow anybody to make it on their own; the result of my work has been *Il Poliedro di Leonardo* [10], an activity book that contains all the cardboard pieces (pre-cut and pre-folded) to make joints, and edges, to be assembled to obtain a copy of the Leonardo's polyhedron.

Fig. 2. The book *Il Poliedro di Leonardo*

Fig. 3. Joint and edges around a vertex

This model of the Ycocedron pretends to reply exactly the structure of the wooden model drawn by Leonardo, and requires some time, patience, and manuals skills: there are 60 joints (corresponding to the vertices) and 90 "tubes" (correspondint to the edges) to assemble. There are two kinds of tubes, since some edges are shared by two hexagons, some others by an hexagon and a pentagon. This difference implies a tiny -but important- difference in the angles and in the lenghts of the segments in the net of each tube and of the joint, so that the design of the pieces has been obtained by careful computating all the involved angles and lengths; then the net was drawn using Cabri.

The first attempt, even if geometrically correct, didn't work, and I noticed that, in the sliding pieces, I had to take into account even of the thickness of the paper, so that I had to compute a tiny modification of some pieces, to arrive to the actual design.

Figures 2 and 3 show how the yellow pieces are folded and glued to become joints with 3 branches. They completely disappear when are inserted in the edges (two red and one green). Signs are hidden in the pieces to guide the correct insertion of the branches of the joint in the edges (they almost disappear when the model is complete).

This part of the work is slow, since the pieces have to be glued in several steps, every time waiting for the previous glue to dry.

When all the pieces are ready, the polyhedron can be assembled, carefully copying the structure of the football, perhaps looking at a true football to guide the work.

I can make a copy of the Ycocedron in about 7 hours of work, in several days.

3 Models for children

Perhaps the previous activity is somehow hard or too long for some student. So, I also planned some simpler activities to explore this polyhedron.

The first one has been organized during the European Project MiMa (Mathematics in the Making), aiming to raise primary school children's interest and competence in mathematics.
The MiMa activities occurred (from 2013 to 2015) across the five participant countries (Italy, Hungary, Germany, Portugal and United Kingdom) and involved approximately 40 schools, over 110 teachers with the participation of nearly 1600 students [7].

The MiMa methodology has been planned for children age 8-10, and is based on ten hands-on activities (the MiMa football activity being one of them), in which the children can build some mathematical exhibits, and then show them in a public exhibition, in which they have the responsability to guide the visitors, and to explain the mathematical aspects of their objects.

The MiMa football activity is a collection of cardboard models, first to observe an icosahedron, then an icosahedron equal to the previous one, but having the pieces to be truncated in colour, so that the students can try directly to imagine what shape will remain after the truncation, and, finally, a magnetic puzzle, with 12 pyramids that can be attached to a truncated icosahedron to obtain an icosahedron again.

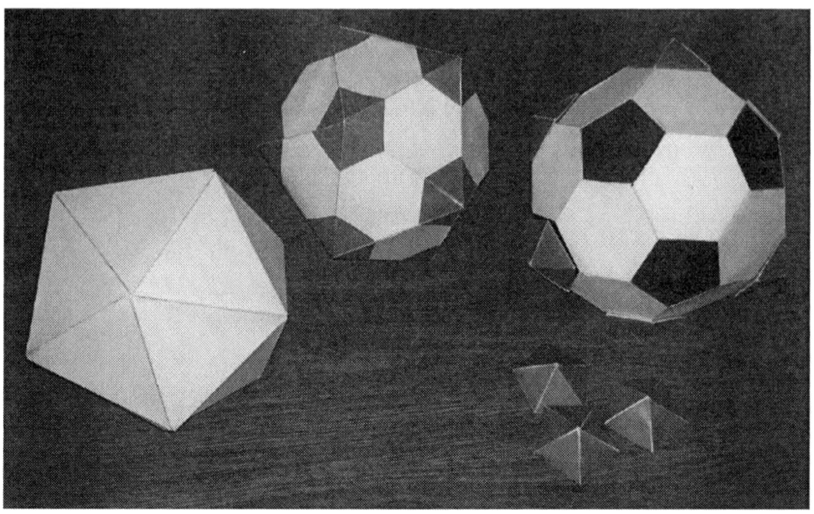

Fig. 4. The MiMa models

To support the teachers' use of the proposal, a rich and complete Toolkit is available in the MiMa website [6], with all the templates to make these objects and a tutorial. To obtain a beautiful result, the templates should be printed (with an usual A4 printer) on heavy paper.

These object are much simpler that the previous Ycocedron, but perhaps they are still a challenge for someone. So, now, here is the last very simple proposal to make a football.

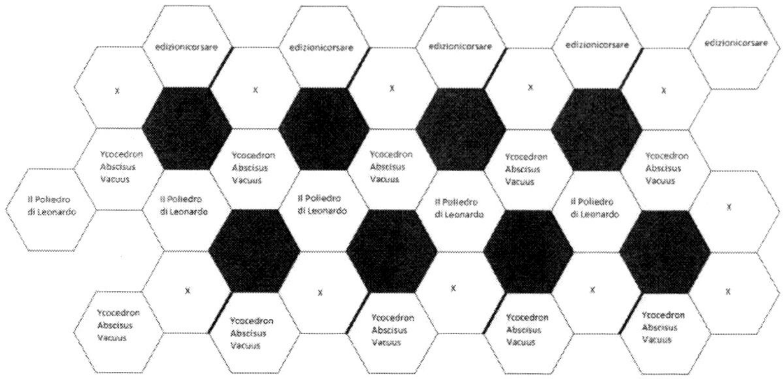

Fig. 5. The net for a holed truncated icosahedron

This template can be downloaded at [1] and printed over a light A4 paper.

Following the directions, all the blue hexagons have to be removed, cutting along the bold segments.

After that, the structure has to be curved, by overlapping some written hexagons over the hexagons signed by "x", and glueing them together (or using a staple or a paper clip).

The result is a very balanced football, in which the pentagons are holes. Its structure is so regular, that it even bounces a bit, when dropped on a table.

4 Conclusion

The first and the last proposal have been shown rarely sometimes during conferences; the second proposal instead has been extensively used and observed with children (during the MiMa project).

After the activity, we asked the involved teachers for feedback: they reported that the climate of the class, during the MiMa activities, was always very serene, and all the children participated, even if perhaps with different levels of consciousness. Some moving moments happened, when children with various problems (dislexia, cognitive retard) were able to "belong to the group" through manual activities.

Moreover, sometimes manual skills vere discovered and appreciated, so giving satisfaction to some child usually poorly performing.

Also, making the puzzle and then playing with it again and again, children really did understand the structure of the polyhedron; sometimes it became clear that a child "understood" the subject, looking at the way he/she was acting, even if he/she was not yet able to fully describe it by words.

Moreover, from the concrete activity, some talented children had the idea to generalize the operation of truncation to different platonic solids, so having an "original" mathematical experience.

The directions to make the object (through a tutorial online) were not completely obvious, so that attention and care were needed to complete the task. I would like to underline that in such a situation a mistake can happen, but there is no need for the teacher to explicitly tell the child that he did something wrong: simply, the object doesn't work the way it should (for example, the polyhedron doesn't close), so that the child can be bothered, but there is no negative emotional charge.

Making an exhibit was just a part of the MiMa proposal: the second (very important) part was the exhibition, in which the children had the chance to explain the mathematical object they made to someone else, often their parents and relatives, so that they were proud to show their work and to share the pleasure for having understood something new.

The teachers monitored this activity, but no pressure nor grades were given: the children spoke just when they felt ready, sometimes after having heard a friend repeating the same explaination several times, so working to learn how to speak correctly about mathematics.

In this way, even with very poor materials, it is possible to offer meaningful mathematical experiences through craft and manipulation, creating a learning environment similar to the Renaissance workshops, where the apprentices learned working with the older artists ; such a situation often allows a very inclusive environment, in which even children usually having low performances can actively cooperate with others, and receive some satisfaction.

References

[1] AMAREMATICA in https://www.dropbox.com/s/5hq5cfn4qurg2gy/Pallone%20-%20AmareMatica.pdf?dl=0 (accessed January 2, 2017).

[2] AZDIC, N. *I Modelli di Leonardo da Vinci,* in http://www.leonardomodelli.com/2012/04/icosaedro-tagliato-vuoto.html (accessed January 2, 2017).

[3] BAGNI, G.T., D'AMORE, B. *Leonardo e la Matematica*, Giunti, Firenze, 2006.

[4] CIOCCI, A. *I poliedri regolari tra arte e geometria - Pacioli e Leonardo tra Euclide e Platone,* Centro Studi "Mario Pancrazi", Quaderno 2, in http://www.centrostudimariopancrazi.it/pdf/quaderno2/pagg%20153-184.pdf (accessed January 1, 2017).

[5] HART, G. W. *Leonardo Project*, in http://www.georgehart.com/sculpture/leonardo-project.html (accessed January 2, 2017).

[6] MIMA in *http://www.mathematicsinthemaking.eu/* (accessed January 2, 2017).

[7] *MiMa - Mathematics in the Making – The Project.* Sheffield Hallam University, Sheffield, 2015, ISBN 978-1-84387-387-7.

[8] PACIOLI, L. *De Divina Proportione.* Anastatic reproduction of the copy of the Biblioteca Ambrosiana in Milan, Silvana Editore, Milano, 2010, ISBN 8836617891.

[9] PACIOLI, L. *De Viribus Quantitatis, Lettera Dedicatoria.* in http://www.uriland.it/matematica/DeViribus/Pagine/ (p.30 – 31) (accessed January 1, 2017).

[10] UGHI, E. *Il Poliedro di Leonardo*. Edizioni Corsare, Perugia, 2013, ISBN 978-88-87938-90-6.

[11] VASARI, G. *Le vite de' più eccellenti pittori, scultori e architettori.* in https://it.wikisource.org/wiki/Le_vite_de%27_pi%C3%B9_eccellenti_pittori,_scultori_e_architettori_(1568) (accessed December 31, 2016).

Current address

Ughi Emanuela, Researcher
Dipartimento di Matematica e Informatica, Università di Perugia
Via Vanvitelli 1, 06123, Perugia, Italy
Tel. number: +390755855008, e-mail: emanuela.ughi@unipg.it

Slovak University of Technology in Bratislava
Faculty of Mechanical Engineering
16th Conference on Applied Mathematics **APLIMAT 2017**
Proceedings

INVESTIGATION OF FRICTIONAL STICK SLIP AND SPRAG SLIP MECHANISMS LEADING TO DISC BRAKE NOISE VIBRATION AND HARSHNESS EFFECTS

ÚRADNÍČEK Juraj (SK), KRAUS Pavel (SK), MUSIL Miloš (SK)

Abstract. Paper describes mechanisms of vibration of disc brake components which can lead into noise effects known as brake Noise Vibration and Harshness (NVH). Stick-slip and Sprag-slip physical effects are mathematically described under simplifying physical assumptions. Self-excited vibration due to stick-slip effect and stability conditions are defined using 1 degree of freedom mechanical system. Response is obtained by numerical solution of ordinary differential equation. Sprag-slip mechanism is demonstrated through multiple degree of freedom elastic system defined by the Finite Element Method (FEM). Kinematic and dynamic behavior is simulated in Multibody Dynamics (MBD) simulation system using Algebraic-differential equations solver.

Keywords: Stick-slip, Sprag-slip, NVH, Matlab, MSC. Adams

Mathematics Subject Classification: Primary 70E18, 74H45, 74H55

1 Introduction

Brakes represent one of the most important safety and performance components of a vehicle. They are important in terms of reliability and braking power, as well as a fluent operation. The unwanted side effect of the braking operation is its occasional squeal and other unwanted harshness effects. Customers understand this strange noise as a functionality defect of the brake system. This results in high warranty payouts and in enormous money losses to carmakers and component producers.

There are several categories of brake noise distinguished according to the frequency range at which they tend to occur or according to the mechanism of generation. In the literature and in many studies we can find different classifications of brake noise, such as judder, hum, groan, squeal, squeak, wire-brush, chatter and moan, among others [1]. NVH is a problem that has been challenging engineers in industry for decades. It is commonly agreed that it occurs due to a friction induced oscillation. There are several mechanisms which can lead into NVH. Most discussed are stick and slip, sprag and slip effects [2], non-conservative follower force [3] and mode coupling [4].

Numerous works have analyzed the phenomenon of brake squeal, ranging from basic studies on mechanisms up to the development of suitable measurement techniques [5]. Nowadays, there is a strong research focus on numerical simulation [6]. Some years ago stability studies in the form of eigenvalue analysis of the linearized system became the state of the art. This kind of linear stability analysis is now used broadly in industry to analyze stability borders and to suggest measures against squeal.

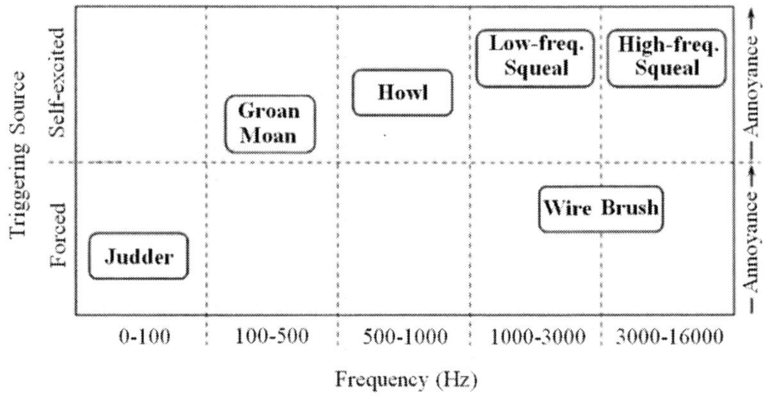

Fig. 1. Brake noise frequency range versus triggering source according to [1]

Nowadays, producers of braking components especially braking pads are strongly bounded by legislation to follow trends of sustainable reduction of environmental burdening and recyclability of materials [7] which makes the process of developing the silent brake system even more challenging.

2 Stick-slip

Stick-slip can be described as surfaces alternating between sticking to each other and sliding over each other, with a corresponding change in the force of friction. Typically, the static friction coefficient between two surfaces is larger than the kinetic friction coefficient. If an applied force is large enough to overcome the static friction, then the reduction of the friction to the kinetic friction can cause a sudden jump in the velocity of the movement. This effect can be demonstrated on simplified representation of the brake pad/disc frictional pair represented by 1 degree of freedom (DOF) mechanical model (Fig. 2), where only pad elasticity in one direction is considered.

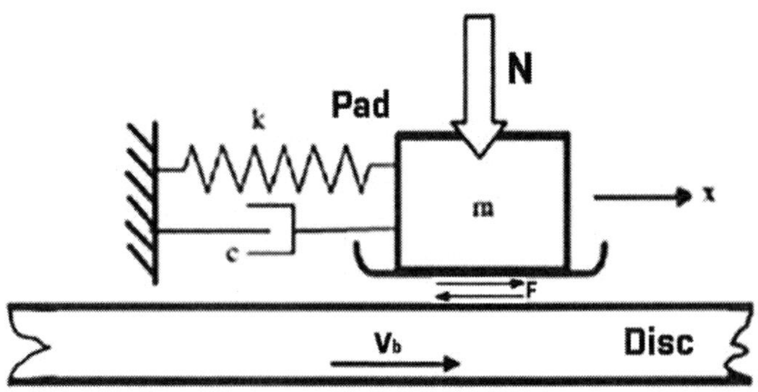

Fig. 2. Single DOF mechanical model of disc brake

Governing differential equation for the model is

$$m\ddot{x}(t) + c\dot{x}(t) + kx(t) = F\,(t, v_{rel}) \tag{1}$$

where,

$$\mathrm{F}(\mathrm{t}, \mathrm{v_{rel}}) = \mu_s \mathrm{N}\left(\mathrm{Sgn}(\mathrm{v_{rel}})\text{-}\sigma \mathrm{v_{rel}}\right) \tag{2}$$

is friction force Fig. 3., μ_s- static coefficient of friction, N - normal force, m - mass, c – damping coefficient, k – stiffness, x - displacement of the mass and its time derivatives representing velocity and acceleration.

The Signum function Sgn (-) operates as

$$\mathrm{Sgn}(v_{rel}) \begin{cases} +1 \text{ for } v_{rel} > 0 \\ -1 \text{ for } v_{rel} < 0 \end{cases} \tag{3}$$

where

$$v_{rel} = (v_b - \dot{x}) \tag{4}$$

is relative velocity acting between moving base (disc) and the mass element (pad), σ is friction coefficient versus relative velocity slope (grade) and v_b disc velocity.

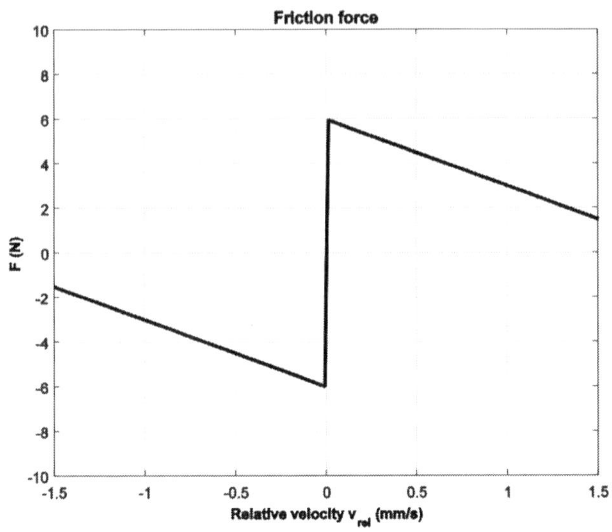

Fig. 3. Friction force for parameters: $\sigma = 0.5$; $\mu_s = 0.6$; $N = 10$ N

Substituting (2), (4) into (1) and rearranging

$$m\ddot{x}(t) + (c - \mu_s N\sigma)\dot{x}(t) + kx(t) = \mu_s N\left(\mathrm{Sgn}\left(v_b - \dot{x}(t)\right) - \sigma v_b\right) \tag{5}$$

The equation (5) shows that the system is nonlinear due to the Sgn(-) function changing the orientation of friction force with the relative velocity. The equation showed also that the friction plays an important role in the damping term $(c - \mu_s N\sigma)$ where damping is along the coefficient of damping c affected also by friction parameters μ_s, σ and normal force N.

The response corresponding to the initial condition was calculated numerically using fifth-order Runge-Kutta integration method included in MATLAB software through ODE45 function.

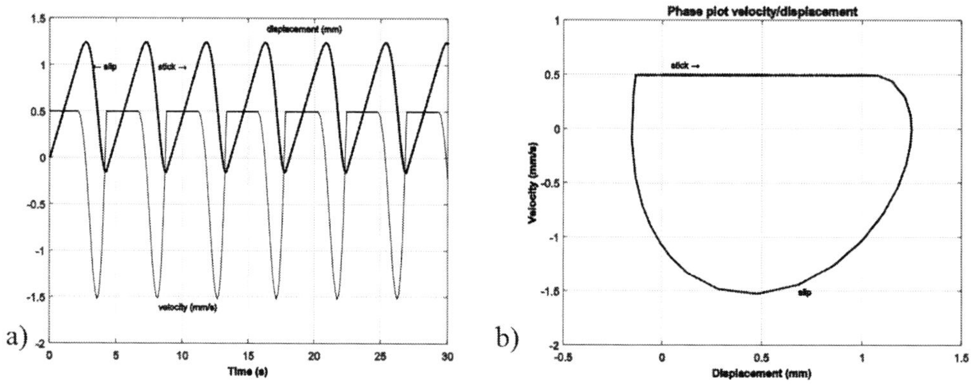

Fig. 4. a) Displacement, velocity of mass, b) phase plot of displacement/velocity of mass for parameters: m = 1 kg; v_b= 0.5 mm/s; σ = 0.5; k = 5 N/mm ; μ_s = 0.6 ; c = 1.5 N/mm.s-1; N = 10 N, and initial conditions x = 0, $dx/dt = v_b$.

When the stick-slip effect occurs allowing $(v_b - \dot{x}) = 0$, system can generate stabile pure self-excitation, Fig. 4. a), which results in limit cycles Fig. 4. b). In this plot the slip mode solutions are represented by continuous arcs of circles corresponding to the solutions of the linear equation. The stick mode solutions are represented by the horizontal lines connected with the arcs, which have constant velocities equal to the base speeds.

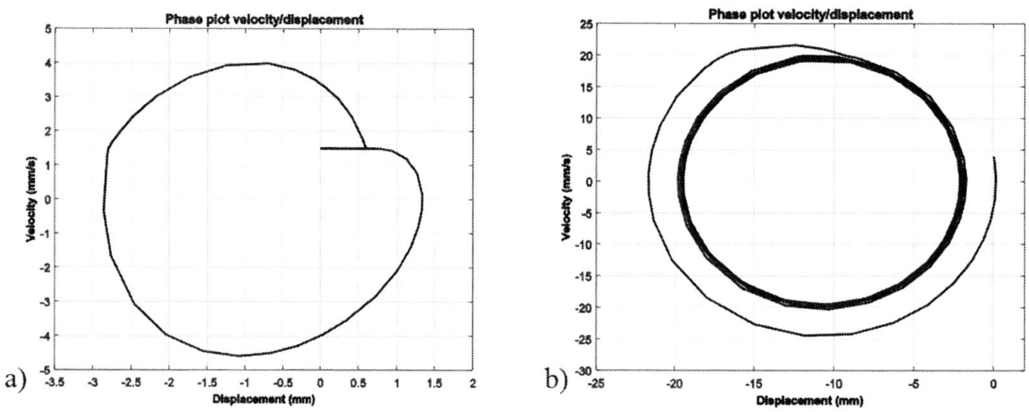

Fig. 5. a) Phase plot for parameters: v_b= 1.5 mm/s; c = 1.5 Ns/mm,
b) Phase plot for parameters: v_b= 20 mm/s; c = 3 Ns/mm.

Increasing the disc speed v_b, the sticking phase is shortening, Fig. 5.a), compared to Fig. 4.b). After critical base speed value, the term $(v_b - \dot{x})$ >0 and stick effect do not occur anymore Fig. 5.b) and equation (5) becomes purely linear

$$m\ddot{x}(t) + (c - \mu_s N\sigma)\dot{x}(t) + kx(t) = \mu_s N(1 - \sigma v_b). \qquad (6)$$

Under these conditions system is in stable self-excited vibration.

When the coefficient of friction curve, Fig. 3, exhibits a large negative slope allowing $\mu_s N\sigma > c$, the system (5) possesses negative damping, which is necessary condition for instability and can result in self-excited vibration instability (Fig. 6).

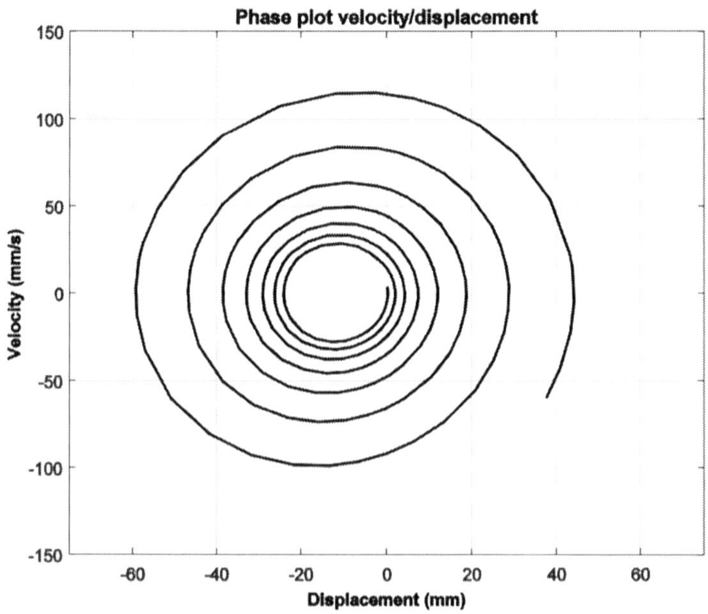

Fig. 6. Phase plot for parameters: v_b= 20 mm/s; c = 2.7 N/mm.s^{-1}

From the numerical analyses of simplified mechanical brake model, it can be seen, that stick-slip effect is source of stabile self-excited vibrations, but when stick effect disappears, due to critical velocity of the disc, system is in purely sliding vibration state and the negative damping can result in self-excited vibration instability. In general, it should be said, that the negative slope of the friction curve is a necessity, but not a sufficient condition for steady sliding instability of the system [8].

Upper mentioned effect accompanied together with the other mechanisms such as mode coupling [4], non-conservative friction force acting as follower force [3] and also kinematic sprag and slip [2] mechanism can lead into brake squeal or other NVH effects.

3 Sprag-slip

The Sprag-slip phenomenon describes the fact that the dynamic instability can occur even with considering constant coefficient of friction. The instability arises from frictional force significant increase above its normal level.

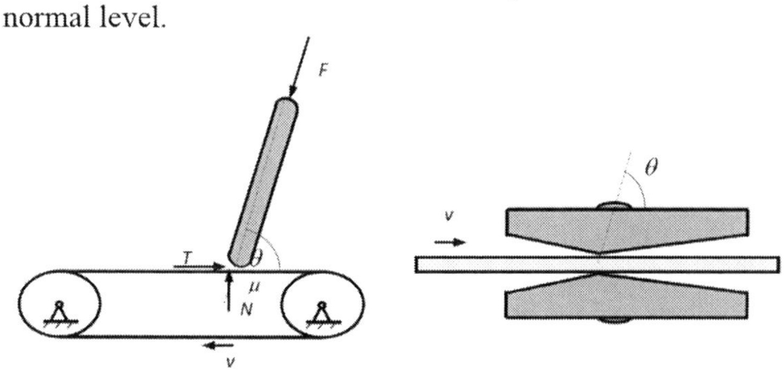

Fig. 7. Sprag-slip phenomenon [9]

For illustration let us consider the system shown in Fig. 7, where an elastic beam is sliding to contact with a moving plate. The friction force increasement arises from locking action of the beam into the moving surface. By assuming $T=\mu N$ and considering the equilibrium of system, the following equations are derived

$$N = \frac{F}{1\text{-}\mu \tan\theta} \tag{7}$$

$$T = \frac{\mu F}{1\text{-}\mu \tan\theta} \tag{8}$$

From these two equations can be seen that, if θ is approaching $\arctan(^1/_\mu)$, then T is approaching infinity and the system becomes unstable.

If considered structure consists of rigid bodies and spragging occurs, the motion can be halted. If flexible bodies are considered with real elastic properties, the flexibility of these components allows to release from the spragging situation by slipping. After the spragging has been relieved, the tight contact situation will rebuild, the contact force again grows, and the next run of spragging-slip takes place [8]. This phenomenon is also known as geometrically-induced or kinematic constraint instability.

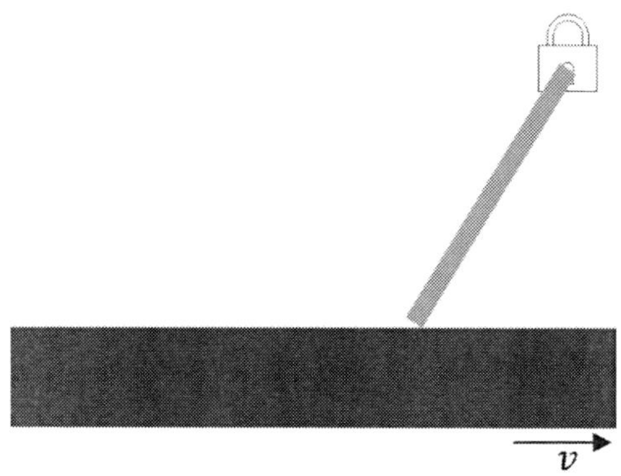

Fig. 8. MBD model

To demonstrate the sprag-slip phenomenon a Multi-body dynamic simulation in MSC. ADAMS is performed. Dynamic model in MSC. ADAMS software is defined using the Lagrange equations of mixed type [10] which leads to a system of differential-algebraic equations. The solution of this system is performed using a numerical integration. For this simulation, the WSTIFF integrator is selected. This solver is stiffly stable, multi-step, variable-order, variable-step backward integrator based on the Gear integration scheme [10,11].

The simulated model contains 2 bodies: a flexible elastic beam and a rigid box, Fig. 8. The flexible beam is defined using modal representation [12], which is obtained from finite element model created in ANSYS. Between the beam and box an Impact contact with Coulomb friction is defined. The characteristic of friction coefficient used in ADAMS can be seen on Fig. 9. The beam is fixed to the ground in the upper end and the box is moving with a constant velocity of v = 15 mm/s.

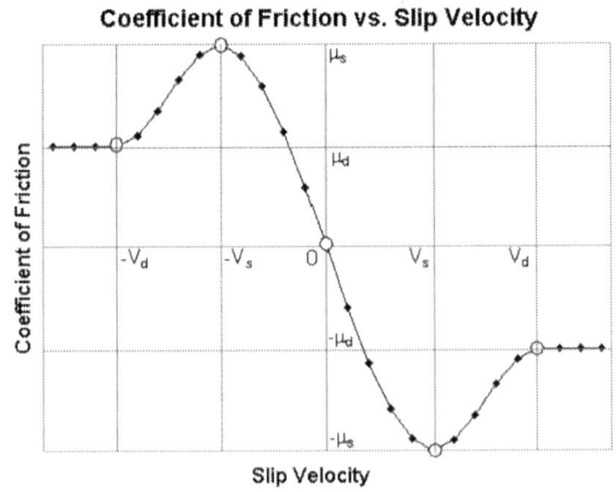

Fig. 9. Coefficient of friction definition in MSC Adams. v_s=0.1 mm/s, v_d=10 mm/s, μ_s=1.8, μ_d=1.6977

The initial angle θ is 60°. The dynamic friction coefficient is defined as 1.6977, which according to (6) and (7) leads to infinite normal and frictional force when angle θ reaches 61°. The results from a 1s simulation can be seen on the Fig. 10 and 11. It can be clearly seen that when the angle θ reaches approximately 61° the beam becomes unstable, Fig. 10, as the normal force jumps to high values, Fig. 11.

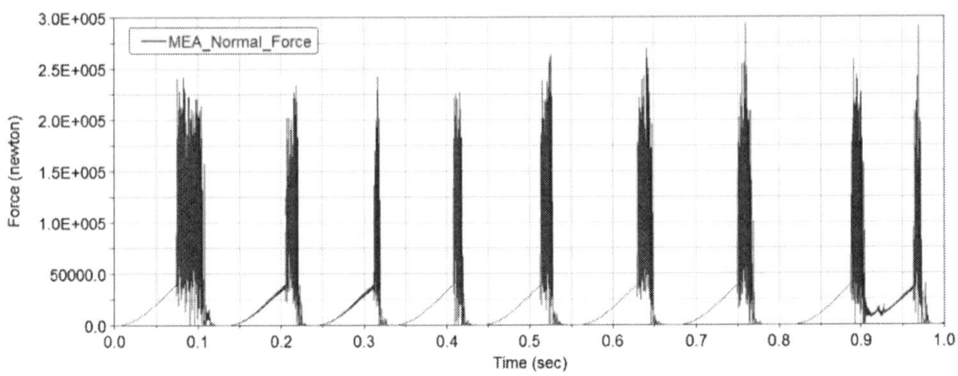

Fig. 10. Normal forces in the beam end tip

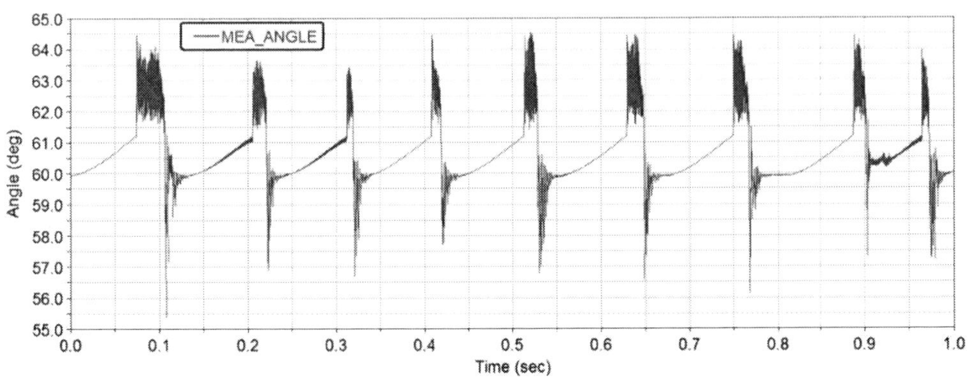

Fig. 11. Angle θ during simulation

According to the results, the multi-degree of freedom model has been destabilized exactly at the same angle θ as the minimum static model predicts. Also, the structure has been able to stabilize itself because the beam has been modeled using a flexible modal representation. This leaded to sprag-slip cycle, confirming that the instability can occur also from geometrical locking.

Conclusions

Stick-slip frictional effect have been described through single degree mechanical system. Nonlinear behavior has been demonstrated using numerical time integration of ordinary differential equation.

From the numerical analyses of simplified mechanical brake model, it can be seen, that stick-slip effect is source of stable self-excited vibrations but when stick effect disappears, due to critical velocity of the disc, system is in purely sliding vibration state and the negative damping can result in self-excited vibration instability.

Sprag-slip frictional effect have been described through multi degree of freedom elastic mechanical system. Nonlinear behavior has been demonstrated using algebraic-differential equations describing physical properties of systems.

From the sprag-slip simulation can be seen, that the instability can be also geometrically induced. Specific geometry state can result in infinite frictional forces which destabilize the structure.

Upper mentioned mechanisms acting in complex disc brake systems can lead into unwanted NVH effects. Understanding to these mechanisms should be the key factor for designing the NVH predicting numerical models.

In the future research, the relation of Stick-slip and Sprag-slip effect to other NVH contributing mechanisms like non-conservative follower force leading to mode lock, should be investigated.

Acknowledgement

The authors would like to thank for financial contribution from the STU Grant scheme for Support of Young Researchers.
This work was supported by the Slovak Research and Development Agency under the contract No. APVV-15-0630.

References

[1] SUCHAL, A. *Influence of Thermal Effect on Dynamics of Disc Brakes and Brake Squeal Propensity.* ,PhD. Thesis, Slovak University of Technology in Bratislava, Slovakia, 2013.

[2] GURAN, A. et al., *Dynamics with friction: modelling, analysis and experiment, part II.*, World scientific publishing Co. Pte. Ltd., ISBN 98-128112-73, 2001.

[3] KRILLOV, O., N., *Nonconservative Stability Problems of Modern Physics.*, De Gruyter Studies in Mathematical Physics 14, ISBN 97-831102703-41, 2013.

[4] ÚRADNÍČEK, J., MUSIL, M., KRAUS, P. Predicting the self-excited vibrations in automotive brake systems. In *Noise and vibration in practice: peer-reviewed scientific proceedings*. 1. vyd. Bratislava: Nakladateľstvo STU, 2016, S. 129-134. ISBN 978-80-227-4563-5.

[5] KINKAID N., M., O'REILLY O. M., PAPADOPULOS, P. Automotive disc brake squeal, In *Journal of Sound and Vibration* 267, 2003, 105–166.

[6] DIHUA G., DONGYING J., A study on disc brake squeal using finite element methods, In *Proceedings of International SAE Congress and Exposition*, 1998.

[7] PECIAR M., FEKETE R.,PECIAR P., *Agglomeration Technologies of Processing Powder Wastes*, Solid State Phenomena, Vol. 244, pp. 121-129, 2016.

[8] CHEN, G., S., *Handbook of friction-vibration interactions*., Elsevier, 2014.

[9] SPURR, R.,T. A theory of brake squeal. *Proc. Inst. Mech. Eng.* 1, pages 33–52, 1961.

[10] RYAN, R., R., *ADAMS – Multibody System Analysis*, Multibody Systems Handbook., Springer Verlag, Berlin Heidelberg New York, 1990, pp. 361 – 402.

[11] MSC.Software Corporation - Adams/Solver Theory: Achieving Robust, Converged Solutions (Course Notes), 2013

[12] MSC.Software Corporation - Flex Body Dynamics and Modal Stress Recovery using Adams (Course Notes), 2012

Current address

Úradníček Juraj, Ing. PhD.
Institute of Manufacturing Systems, Environmental Technology and Quality Management,
Faculty of Mechanical Engineering, Slovak University of Technology in Bratislava
Nám. slobody 17, 812 31 Bratislava, Slovak Republic
Tel. number: +421 57296587, e-mail: juraj.uradnicek@stuba.sk

Kraus Pavel, Ing.
Institute of applied mechanics and mechatronics,
Faculty of Mechanical Engineering, Slovak University of Technology in Bratislava
Nám. slobody 17, 812 31 Bratislava, Slovak Republic
Tel. number: +421 5729 315, e-mail: pavel.kraus@stuba.sk

Musil Miloš, prof. PhD.
Institute of applied mechanics and mechatronics,
Faculty of Mechanical Engineering, Slovak University of Technology in Bratislava
Nám. slobody 17, 812 31 Bratislava, Slovak Republic
Tel. number: +421 5729 389, e-mail: milos.musil@stuba.sk

Slovak University of Technology in Bratislava
Faculty of Mechanical Engineering
16th Conference on Applied Mathematics **APLIMAT 2017**
Proceedings

MIXED PRODUCT IDEALS AND SIMPLICIAL COMPLEXES

UTANO Rosanna (IT)

Abstract. Let I be a squarefree monomial ideal of the polynomial ring $K[x_1, \ldots, x_n]$. We can consider I as the Stanley Reisner ideal of a simplicial complex Δ_I, on the set of vertices $x_1, \ldots, x_n$, whose faces are defined by

$$\Delta_I = \{\{x_{i_1}, \ldots, x_{i_k}\} | i_1 < \cdots < i_k, x_{i_1} \cdots x_{i_k} \notin I\} \qquad [11]).$$

We are interested to special classes of squarefree monomial ideals of the polynomial ring $K[X, Y]$ in two sets of variables $X = \{x_1, \ldots, x_n\}$ and $Y = \{y_1, \ldots, y_m\}$, introduced first in [9]. More precisely, we study two classes of mixed product ideals, $I_r J_r$ and $I_r + J_r$, where I_r (resp. J_r) is the ideal of $K[X, Y]$ generated by all the squarefree monomials of degree r in the variables X (resp. Y) and the simplicial complexes associated to these ideals ([10]).

Keywords: squarefree monomial ideals; simplicial complexes

Mathematics subject classification: 13C15, 13F20

Introduction

A simplicial complex on the vertex set $V = \{x_i, \ldots, x_n\}$ is a collection of subsets of V such that

- $\{x_i\} \in \Delta$ for all $x_i \in V$;
- $F \in \Delta$ and $G \subseteq F$ imply $G \in \Delta$.

An element $F \in \Delta$ is called a face of Δ. If $F \in \Delta$ we define the dimension of F by $\dim F = |F| - 1$ where $|F|$ is the cardinality of the set F, and the dimension of Δ as $\dim\Delta = max\{\dim F : F \in \Delta\}$. Δ is called pure if all maximal faces of Δ, called facets, have the same dimension. If Δ is an arbitrary simplicial complex of dimension $d - 1 \geq 0$ on a vertex set V, if we denote by f_i the number of i-dimensional faces of Δ, then the d-tuple

$(f_0 = |V|, f_1, \ldots, f_{d-1})$ is called the f-vector of Δ.
Let K be a field and $S = K[x_1, \ldots, x_n]$ be the polynomial ring. Let I be a squarefree monomial ideal of S. We associate to any such ideal $I \subseteq S$ the simplicial complex on $V = \{x_1, \ldots, x_n\}$ defined by

$$\Delta_I = \{\{x_{i_1}, \ldots, x_{i_k}\} : x_{i_1} \cdots x_{i_k} \notin I\}.$$

Conversely to a simplicial complex Δ on V it is possible to associate the squarefree monomial ideal of S

$$I_\Delta = (x_{i_1} \cdots x_{i_k} : \{x_{i_1}, \ldots, x_{i_k}\} \notin \Delta).$$

I_Δ is the Stanley Reisner ideal of Δ and $K[\Delta] = S/I_\Delta$ is the Stanley-Reisner ring of Δ. The operations above yeald a 1-1 correspondence between the set of simplicial complexes on $\{x_1, \ldots, x_n\}$ and the set of all squarefree monomial ideals of S. We are interested in special classes of squarefree monomial ideals of the polynomial ring $K[X, Y]$ in two disjoint sets of variables $X = \{x_1, \ldots, x_n\}$ and $Y = \{y_1, \ldots, y_m\}$, introduced first in [9].

Let $R = K[x_1, \ldots, x_n, y_1, \ldots, y_m]$ be a polynomial ring in two disjoint sets of variables over a field K. Given non negative integers k, r, s, t such that $0 \leq k, s \leq n$, $0 \leq r, t \leq m$, $k + r = s + t$, the monomial ideals of R:

$$L = I_k J_r + I_s J_t,$$

where I_k (resp. J_r) is the monomial ideal of R generated by all squarefree monomials of degree k (resp. r) in the variables $x_1, \ldots, x_n$ (resp. $y_1, \ldots, y_m$), are called *ideals of mixed products*. They are squarefree monomial ideals generated in the same degree and they are complete or integrally closed.

If the ideal has the form $I_r J_r$ or $I_r J_{r-1} + I_{r-1} J_r$ then it is the generalized graph ideal of a complete bipartite graph, whose sets of vertices are X and Y. In fact, if G is a complete bipartite graph on the vertex set $\{x_1, \ldots, x_n, y_1, \ldots, y_m\}$, then the generalized graph ideal $I_q(G)$ is an ideal of mixed products of the form

$$I_q(G) = \begin{cases} I_t J_{t-1} + I_{t-1} J_t, & q = 2t - 1, \\ \\ I_t J_t, & \text{q=2t} \end{cases}.$$

We recall that if G is a graph with vertices $\{x_1, \ldots, x_n\}$, then the q-th generalized graph ideal $I_q(G) \subset K[x_1, \ldots, x_n]$ is generated by the squarefree monomials $x_{i_1} \cdots x_{i_q}$ such that x_{i_j} is adjacent to $x_{i_{j+1}}$ for all $1 \leq j \leq q - 1$.

We consider the following classes of monomial ideals arising from the definition of L:
1) $L = I_r + J_r$, with $1 \leq r \leq \min\{m, n\}$,
2) $L = I_r J_r$, with $1 \leq r \leq \min\{m, n\}$.

More precisely, we study two classes of mixed product ideals, $I_r J_r$ and $I_r + J_r$, where I_r (resp. J_r) is the ideal of $K[X, Y]$ generated by all the squarefree monomials of degree r in the variables X (resp. Y).

One process for changing a simplicial complex to get another one is the process of taking subdivisions, based on the join construction. Let Δ_1, Δ_2 be two disjoint simplicial complexes. The join of Δ_1 and Δ_2 is defined to be

$$\Delta_1 \star \Delta_2 := \Delta_1 \cup \Delta_2 \cup \{\sigma \cup \tau | \sigma \in \Delta_1, \tau \in \Delta_2\}.$$

If Δ_1 consists of one (resp. two points), i.e. $\Delta_1 = \{x\}$ (resp. $\Delta_1 = \{\{x,\},\{y\}\}$), then the join $C(\Delta_2) = C(x,\Delta_2) = \{x\} \star \Delta_2$ (resp. $\Sigma(\Delta_2) = \{\{x,\},\{y\}\} \star \Delta_2$ is called the cone (resp. the suspension) of the complex Δ_2. If the ideal has the form $I_r + J_r$, as before, the structure of the associated simplicial complex Δ is not difficult to understand, being the join $\Delta_1 \star \Delta_2$ of two disjoint simplicial complexes consisting of the $(r-1)$-skeletons of two $(r-1)$-simplices on the sets of vertices X and Y respectively, with $I_r = I_{\Delta_1}$ and $J_r = I_{\Delta_2}$. We refer to [1] for basic properties of Stanley-Reisner ideals. In Section N. 1 we study the simplicial complex associated to the squarefree Veronese ideal $I_r \subseteq S = K[x_1,\dots,x_n]$. In Section N. 2 we study the simplicial complexes associated to mixed product ideals ([10]) of the form $I_r + J_r$ and I_rJ_r.

1 Squarefree Veronese ideals

Definition 1.1 An R-module M is called *Cohen-Macaulay* if $M \neq 0$ and $\text{depth}(M) = \dim(M)$ or if $M = 0$.
If R itself is a Cohen-Macaulay module we say that R is a *Cohen-Macaulay ring*. An ideal I of R is Cohen-Macaulay if R/I is a Cohen-Macaulay R-module.

The squarefree Veronese ideal $I_r \subseteq S = K[x_1,\dots,x_n]$ is Cohen-Macaulay. We recall the following

Definition 1.2 Let $I = (X^{\alpha_1},\dots,X^{\alpha_t}) \subseteq S = K[X] = K[x_1,\dots,x_n]$ be a squarefree monomial ideal, with $X^{\alpha_i} = x_1^{\alpha_{i_1}} \dots x_n^{\alpha_{i_n}}$, $\alpha_i = (\alpha_{i_1},\dots,\alpha_{i_n}) \in \{0,1\}^n$, for $i = 1,...,t$. The *Alexander dual* of I is the ideal

$$I^* = \bigcap_{i=1}^{t} \mathfrak{b}_i$$

where $\mathfrak{b}_i = (x_j \mid \alpha_{i_j} = 1)$.

Proposition 1.3 *Let $S = K[x_1,\dots,x_n]$, $I_r \subset S$ with $1 \leq r \leq n$. Then*

$$I_r = I^*_{n-r+1}.$$

Proof. Let $I_r = (x_{i_1}\cdots x_{i_r} \ : \ 1 \leq i_1 < \cdots < i_r \leq n)$. We consider I_r as a Stanley - Reisner ideal of the simplicial complex Δ_{I_r}, that is

$$I_r = I_{\Delta_{I_r}} = (x_F \ : \ F \notin \Delta_{I_r}), \quad \text{where} \quad F = \{x_{i_1},\dots,x_{i_r}\} \quad \text{and}$$

$$x_F = x_{i_1}\cdots x_{i_r}.$$

This implies that the simplicial complex Δ_{I_r} is pure and its facets are all the facets of dimension $r-2$, that are $\sigma = \{x_{i_1},\dots,x_{i_{r-1}}\}$ with $1 \leq i_1 < i_2 < \cdots < i_{r-1} \leq n$.
We want to calculate $\Delta^*_{I_r} = \{\overline{F} \ : \ F \notin \Delta_{I_r}\}$. If we consider a simplex $\sigma \in \Delta_{I_r}$ such that $|\sigma| < q-1$, then $\sigma \cup \{x_{i_r}\} \in \Delta_{I_r}$. Therefore to obtain a minimal non face of Δ_{I_r} we have to consider a facet in Δ_{I_r} and add a vertex not in the facet, that is $F = \{x_{i_1},\dots,x_{i_r}\}$, with $1 \leq i_1 < i_2 < \cdots < i_r \leq n$.

Therefore $\overline{F} = \{i'_1, \dots, i'_{n-r}\}$, with $1 \le i'_1 < \cdots < i'_{n-r} \le n$, are the facets of $\Delta_{I_r} r^*$. The minimal non faces of $\Delta^*_{I_r}$ are $\{i'_1, \dots, i'_{n-r+1}\}$ and its Stanley - Reisner ideal is

$$I_{\Delta^*_{I_r}} = (x_{i'_1} \cdots x_{i'_{n-r+1}} \ : \ 1 \le i'_1 < \cdots < i'_{n-r+1} \le n) = I_{n-r+1}.$$

Remark 1.4 It follows that $I_r = \bigcap (x_{i_1}, \dots, x_{i_{n-rq+1}})$, where the intersection is taken over all subsets with $n - r + 1$ elements $\{i_1, \dots, i_{n-r+1}\} \subseteq \{1, \dots, n\}$.
Hence

$$\dim K[\Delta_{I_r}] = \dim(S/I_r) = \dim S - \mathrm{ht}(I_r) = n - (n - r + 1) = r - 1 = \dim\Delta_{I_r} + 1.$$

Proposition 1.5 *Let $S = K[x_1, \dots, x_n]$ be a polynomial ring. S/I_r is Cohen - Macaulay for every r, that is:*

$$\mathit{dim}(S/I_r) = \mathit{depth}(S/I_r) = r - 1.$$

Proof. If $I \subset K[x_1, \dots, x_r]$ is a squarefree Veronese ideal of degree r, that is $I = I_r$, then I coincides with the Stanley - Reisner ideal $I_\Delta \subset K[x_1, \dots, x_n]$ of the simplicial complex Δ on the vertex set $\{x_1, \dots, x_n\}$ whose facets are the subsets having $(r - 1)$ - elements of $\{x_1, \dots, x_n\}$. Since the simplicial complex Δ is a skeleton of the $r - 1)$ - simplex, it follows that Δ is Cohen - Macaulay (see [5], Example 2.2).
It is

$$f(\Delta_{I_r}) = \left(n, \binom{n}{2}, \dots, \binom{n}{r-1}\right).$$

2 Simplicial complexes associated to $I_r + J_r$ and $I_r J_r$

The ideal $I_r + J_r \subset R = K[x_1, \dots, x_n, y_1, \dots, y_m]$, $r \le \min\{n, m\}$, is generated by all squarefree monomials of degree r in $x_1, \dots, x_n$ and all squarefree monomials of degree r in $y_1, \dots, y_m$. If $\Delta_{I_r+J_r}$, Δ_{I_r} and Δ_{J_r} are the simplicial complexes respectively associated to the squarefree monomial ideals $I_r + J_r \subset R$, $I_r \subseteq K[x_1, \dots, x_n]$, $J_r \subseteq K[y_1, \dots, y_m]$, then $\Delta_{I_r+J_r} = \Delta_{I_r} \star \Delta_{J_r}$and $K[\Delta_{I_r+J_r}] = K[\Delta_{I_r} \star \Delta_{J_r}] \cong K[\Delta_{I_r}] \otimes_K K[\Delta_{J_r}]$.
We want to prove the following:

Proposition 2.1 *Let $\Delta_{I_r+J_r}$, Δ_{I_r} and Δ_{J_r} be the simplicial complexes respectively associated to the squarefree monomial ideals $I_r+J_r \subset R$, $I_r \subseteq K[x_1, \dots, x_n]$, $J_r \subseteq K[y_1, \dots, y_m]$. Then*

$$\mathit{dim}\Delta_{I_r+J_r} = 2r - 3 = \mathit{dim}\Delta_{I_r} + \mathit{dim}\Delta_{J_r} + 1.$$

Proof. The simplicial complex $\Delta_{I_r+J_r}$ consists of all subsets of cardinality $1, 2, \dots, r-1$ of the sets of vertices $X = \{x_1, \dots, x_n\}$ and $Y = \{y_1, \dots, y_m\}$, and all subsets $\{x_{t_1}, \dots, x_{t_i}, y_{\ell_1}, \dots, y_{\ell_j}\}$ of cardinality $i + j$, $2 \le i + j \le 2r - 2$, $1 \le i, j \le r - 1$. It follows that the dimension of $\Delta_{I_r+J_r}$ is the dimension of a facet, that is $2r - 3$. From Remark 1.4, since $\dim\Delta_{I_r} = r - 2$, it follows that

$$\dim\Delta_{I_r+J_r} = 2r - 3 = \dim\Delta_{I_r} + \dim\Delta_{J_r} + 1.$$

The ideal $I_r J_r \subset R = K[x_1, \dots, x_n, y_1, \dots, y_m]$, $r \le \min\{n, m\}$, is generated by all squarefree monomials of degree $2r$ that are products of all monomials of degree r in $x_1, \dots, x_n$ and all squarefree monomials of degree r in $y_1, \dots, y_m$. We want to prove the following:

Proposition 2.2 *Let $\Delta_{I_rJ_r}$, Δ_{I_r} and Δ_{J_r} be the simplicial complex respectively associated to the squarefree monomial ideals $I_rJ_r \subset R$, $I_r \subseteq K[x_1,\dots,x_n]$, $J_r \subseteq K[y_1,\dots,y_m]$. Then*

$$dim\Delta_{I_rJ_r} = max\{m,n\} + r - 2$$

Proof. Let $I_r = (x_1x_2\cdots\cdots x_r,\dots,x_{n-r+1}\cdots\cdots x_n) \subseteq K[x_1,\dots,x_n]$, $J_r = (y_1y_2t\dots ty_r,\dots,y_{m-r+1}t\dots ty_m) \subseteq K[y_1,\dots,y_m]$. Since $I_rJ_r = (x_1x_2\dot{x}_ry_1y_2\dots oty_r,\ \dots,\ x_{n-r+1}t\dots tx_ny_{m-r+1}t\dots y_m) \subseteq R$, then $\Delta_{I_rJ_r}$ contains all subsets of $X = \{x_1,\dots,x_n\}$, all subsets of $Y = \{y_1,\dots,y_m\}$, all subsets of $X \cup \{y_{i_1},\dots,y_{i_{r-1}}\}$, $1 \le i_1 < \dots < i_{r-1} \le m$, all subsets of $Y \cup \{x_{j_1},\dots,x_{j_{r-1}}\}$, $1 \le j_1 < \dots < j_{r-1} \le n$. It follows that facets of $\Delta_{I_rJ_r}$ have dimension $\max\{n+r-2, m+r-2\} = \max\{m,n\} + r - 2$ that is the dimension of $\Delta_{I_rJ_r}$.

Remark 2.3 As a corollary we obtain Proposition 3.1 in [8]

$$\dim(R/I_rJ_r) = \dim\Delta_{I_rJ_r} + 1 = \max\{n,m\} + r - 1 = n + m - (\min\{n,m\} - r + 1).$$

Remark 2.4 The class of mixed product ideals is naturally linked to the subalgebra $K[I_r]$ generated on the field K by the minimal system of generators of the rth-Veronese squarefree ideal I_r of $K[x_1,\dots,x_n]$ and to $K[L]$, where L is a mixed product ideal, in particular $I_1J_1, I_2J_2, I_rJ_r\dots.$ The rth-Veronese squarefree ring has a very important role in many problems of combinatoric algebra and geometry. In particular, in the transportation problems ([10], [3], [4]) the Segre products $K[I_rJ_s]$ of two squarefree Veronese rings are considered from a different point of view. Moreover the initial simplicial complexes associated to the toric ideals of those algebras, studied in [8], are new simplicial complexes. An open problem is to find some possible links with the simplicial complexes studied in this paper.

Bibliography

[1] W. Bruns, J. Herzog, *Cohen-Macaulay Rings*, Cambridge University Press, Rev. Ed. (1997).

[2] G. Failla, *Moves for Segre products of Veronese subrings*, Rendiconti del Circolo Matematico di Palermo, Serie II, Suppl.80 (2008), 161–167.

[3] G. Failla, *On the (1,1) Segre model for business*, Applied Mathematical Sciences, **8**(2014), no. 166, 8329– 8336.

[4] G. Failla, R. Utano, *Connected graphs arising from products of Veronese varieties*, Algebra Colloquium **23**: 2 (2016), 281–292.

[5] J. Herzog, H. Hibi, *Cohen-Macaulay polymatroidal ideals*, Eur. J. Comb. **27**, 513-517 (2006).

[6] C. Ionescu, G. Rinaldo, *Some algebraic invariants related to mixed product ideals*, Arc. Math. **91** (2008), 20–30.

[7] V. Iorfida, *Simplicial complexes present in art and arts facilities*, Aplimat 2015,14th Conference on Applied Mathematics, Proceedings (2015), 406–415

[8] V. Iorfida, G. Restuccia, *Simplicial complexes and new applications*, Applied Mathematical Sciences, Vol. 10 (2016), no. 34, 1653 –1661.

[9] G. Restuccia, R. H. Villarreal, *On the normality of monomial ideals of mixed products*, Comm. Alg., **29** (8), (2001) 3571–3580.

[10] B. Sturmfels, *Groebner bases and convex polytopes*, Univ. Lectures Series, Amer. Math.Soc., **8** (1995).

[11] R. Villarreal, *Monomial Algebras*, Monographs and Textbooks in Pure and Applied Mathematics, Marcel Dekker, Inc. New York, **238** (2001).

Current address

Utano Rosanna
University of Messina
Viale Ferdinando Stagno d'Alcontres, 31, 98166 - Messina, Italy
E-mail: rosanna.utano@unime.it

Slovak University of Technology in Bratislava
Faculty of Mechanical Engineering
16th Conference on Applied Mathematics **APLIMAT 2017**
Proceedings

GAMIFICATION AND GAME-BASED LEARNING – HOW TO IMPLEMENT THEM INTO EDUCATION

VÁCLAVÍKOVÁ Zuzana (CZ)

Abstract. The use of games in the education process is, in our country, usually associated with ICT or digital tools. This paper attempts to clarify the concepts of gamification and game-based learning in connection with the informal, as well as the formal, education process, especially in cases of non-ICT or e-learning. The examples of implementation of gamification and game-based learning in Ostrava´s Science and Technology Center will be introduced. The examples will focus on use in science teaching, but the concept can be successfully used in any subject, and, using the interdisciplinary link, several courses can even be covered in one game.

Keywords: science, education, gamification, game-based learning, LARP game, study module

Mathematics Subject Classification: Primary 97A20, 97A80; Secondary 00A08

1 What is gamification and game-based learning

In context of gamifying the classroom, the words gamification and game-based learning are often used. Gamification, in general, is the idea of adding game elements to a non-game situation. For example, gamification is used in corporate reward programs by adding a scoring system. Unlike gamification, game-based learning relates to the use of games to enhance the learning experience. In our country, these approaches are used in the case of e-learning (existing games as an educational tool). But there are many ways to apply gamification and game-based learning to non-electronic education to enable students to use their knowledge to achieve better results. The obvious benefit of a game is the fact that we can combine many different subjects and activities in one game and the games can be played at school, at home, or indeed anywhere.

2 The Science and Technology Center Ostrava

By formal education we mean education provided by trained teachers in a systematic, targeted way within school. Informal education is the part of one´s development that comes about outside of formal education. It is gleaned in family, peer group, from mass media and the institutions of informal education – in subjects where education is not the main role. Gamification and game-based learning is the best way of connecting these two approaches of education. The Science and Technology Center in Ostrava's Lower Vítkovice Area is an institution of informal education. The science center was opened in 2014 and supporting program has been running since 2013, thanks to Europe Union funds and financial support from regional and Ostrava city council. The goal of the science center is to present science to children, students and everybody in a fun and entertaining way. We want to stimulate interest in these fields by offering opportunities to explore the laws of physics in the world around us. The key to this is interactivity and playfulness. That is why we are trying to make gamification and game-based learning a significant part of the activities that we offer to schools as a supporting program.

3 Games as educational tool – why and how

It is well known that people, especially the children, love playing games. A game provides the opportunity to learn through experience. The best games require and promote problem solving and critical thinking rather than memorizing correct answers, offer opportunities to examine a problem from different perspectives using multiple resources, require collaboration and interdisciplinary perspectives and allow alternative solutions and outcomes instead of a single right answer (see [2]). But above all, as already mentioned, games are the best way of connecting formal and informal education. When we use a game as an educational tool, we have to keep in mind that the main educational element is not the game itself but rather the final analysis like reflection and review after the game (will be specified later). If we use the digital tools for gamification of the classroom or use an existing game for game-based learning without review and reflection, the learning effect is nullified.

3.1 Classification of games

There are many different methods of classifying the games. The Caillois classification depends upon whether, in the games under consideration, the role of competition, chance, simulation or vertigo is dominant. He divides the games into four play forms – agon, alea, mimcry and ilinx (see [1]).

The agon is the kind of game where competitors have an equal chance and the result of the game depends on their endurance, strength, ingenuity (in general) and their knowledge (in case of education) without any help from outside.

In contrast, chance is the crucial element in games of the alea type – the player himself does not influence the result at all. The alea type is not suitable in education, especially for infants, where the important effect is "good ratings for good results of work". But we can use this idea as a part of a game for older students – to show we live in the world where many things depend on luck or on the results of other things that we cannot control. So at some point in a game we can use an element of chance.

Mimicry is typical for education – it is possible to become someone else, often using fantasy, in this game. A player pretends to be someone else; they forget their own personality and takes on the personality of others.

The last category of game, ilinx, is one that employs vertigo. Its essence is the attempt to reduce the perception of stability, induce stupor or trance. Although there are exhibits in science centers that may constitute this type of entertainment (gyroscope, swings), they usually do not carry educational feature.

3.2 Paidia – ludus axis

Each of these forms of game is situated on the axis between "paidia" and "ludus" with respect to the measure of representation of two principles: spontaneous manifestations of the play instinct (paidia), or controlled play according to rules, with the result displaying the superiority either in skill, strength, or good fortune of the winner or winners (ludus). According to Caillois, paidia and ludus are not separate genres but independent forces that form two ends of a continuum on which all games are located. In the case of educational games, a good balance between paidia and ludus is needed.

In fact the majority of games do not fit into one category and by nature combine elements of more than one. From the educational point of view, the agon and mimicry type of game are successfully used. The agon type is represented by the gamification of the classroom (the competition in a scoring mechanism), the typical example of the mimicry type of game is the role playing game or eduLARP game.

4 Role playing game

This kind of game is the most common in the education process across different levels. Depending on depth of fiction in the mimicry game we can talk about the simulation, alternation and characterization (see [4]). Simulation is when the player plays oneself in a fictitious situation or environment, alternation is when the player plays somebody else but without the specification of behavior. Characterization goes much deeper – the player has information about the role they are playing (the characteristics of the person in the narrative are known and precisely defined). The most successful in informal education is a LARP game.

4.1 EduLARP game

The LARP (Live Action Role Play) is a characterization game where the participants represent fictitious parts of a prepared story and they create and perform the story themselves. It is important for the game to be performed in an authentic environment with respect to the plot itself; actors have costumes, scenery and predetermined rules. If the LARP is to have an educational effect – so called eduLARP, it cannot be created universally but always with respect to the age group of the participants. The LARP game is guided by animators who form part of the story, too. The game, itself, is played for several hours. In order to meet the educational requirements, a perfect preparation of both the story itself and material and technical facilities during the LARP is inevitable. LARP is truly a very sophisticated game, and its preparation is not easy. Even in the case of general LARP, it is difficult to create the story, characters, and relationships and so on. In the case eduLARP, the implementation of an educational effect is also necessary. The disadvantage of the LARP is that you can play it with the same group of students only once. So the best thing to do is not create your own game but to cooperate with the science center. At the present time we offer 3 LARPs at the Science and Technology Center in Ostrava (more details about the topics the LARPs are focused on can be seen in [3] or [5]).

The characterization game (not as long or sophisticated as LARP) we use also as a part of study module – in this case the story is much simpler and both the preparation time and game duration are shorter.

4.2 How to work with fiction

When we prepare the mimicry game, we have to pay particular attention to the fiction. The fiction makes the game more attractive and encourages the pupil into action – it takes away the fear of the consequences that could ensue in a real situation. The fiction has to be understandable for the players. The story must not be so complicated as to obscure the point of the game. For example, if the game is targeted on science subject like physics, chemistry or math, the scientific phenomena must behave just like in the real world. It is a good idea to use some real historical facts and/or real places in the story. For example, we use role play in our study module focused on codes and ciphers. One part of the module is a game with the aim of finding treasure (during the game the teams have to communicate using coded messages). We use real facts about the Oak Island in the game (the island is best known for various theories about possible buried treasure or historical artefacts, and the associated exploration). The story in the game, however, is fictional. It is important to explain after the game has been played, which aspects were real and what was fictitious. By using real facts or places we give the motivation for searching for the answers to questions from other areas. The fiction can only be used in areas that are not primary for the learning effect – the unreal parts of story must not be confusing or misleading.

5 Gamification of classroom

In a typical class, a point system is usually implemented (students obtain a point for good work, activity in class, solving tasks, etc.). This system could be very easily be upgraded into a living, breathing classroom game by using role play. It is just a matter of having a good idea. We can create the story and characters with the infrastructure – the several levels with different challenges that have to be done at the level. And let the points shift student or the whole class between the levels. Students do not obtain the points straightaway – they receive indicia and indicia direct them to a task that has to be carried out. This allows room for teamwork and naturally connects the formal classroom learning with informal one. The students must find something outside or bring some answer from somewhere. We can also implement into the game lots of information from other subjects – the indicia could be coded, the question they need to find an answer to could be from different areas, the place with the answer could be situated according to the current topics in history or geography, they can use GPS navigation to find it, and so on. Such method of gamification of classroom has features of interactivity. It gives a space for all students with a strong emphasis on collaboration because the tasks can be targeted to different activities and teamwork is necessary. It produces a real experience and encourages students to apply their knowledge, skills and abilities from other fields, and is playful and entertaining. Moreover, such a game is actually played outside the formal part of education, and does not encroach on the curriculum.

The game can cover a whole study year and the teacher can add new levels on the successful completion of previous ones. Existence of intermediate levels, where players /students reach some special points in the game, is necessary for keeping them motivated. The teacher can prepare some special prizes at these points and, together with the students, make a review, reflection and analysis of the part of the game. There is also room for an element of chance for older students – it is possible to do some random shift in the points scores of the teams based on chance, which would be appropriate to fit the game.

For example, in a game that is aimed at survival of the settlers (teams in class) on the earth polluted and destroyed by wars, natural disasters can strike and people can lose something (roll of the dice, if and what they lose). The other settlers (teams) can help them, but they must also survive. It is important to realize that our actions always affect others – we share common resources for survival. For younger children we can modify "lose something" to "win something more" to give an element of chance.

6 Review analysis as a key education point

When we use games in education (in any form), the final analysis and reflection is the crucial moment for the educational effect. In accordance with the didactic cycle of personal and social education, the reflection/review follows immediately after the action (game) and precedes evaluation. The purpose of the analysis is to help pupils interpret the subjective feelings from the game into an experience applicable in practice. The teacher or animator discusses the progress of the game together with the students – they try to describe important points of the game, explain the knowledge that they used and why they used it, and talk about their feelings about the game, and so on. The teacher does not tell the children what they should do or not do in the game. Review has to be a dialogue about the experience from a game, about the consequences of our reaction, about the balance between collaboration and competition. Thanks to a good review and analysis, the teacher will have sufficient information for the evaluation of game and will be able to upgrade the game for the next time.

7 Conclusion

As has been shown, implementations of game into the conventional techniques of assessing pupils can make the learning process more attractive and entertaining. It can arouse general interest in the search for new information. Many role playing games can also be used as adaptation courses at the beginning of the school year for the classes where students are not familiar with each other. It is also possible to use what institutions of informal education offer or be inspired by existing games. Although currently there are no relevant methods for monitoring the effectiveness of the use of games in the education, it is clear that their implementation in learning can only improve this process.

References

[1] CAILLOIS, R. Hry a lidé. Nakladatelství Studia Ypsilon, Praha, 1998, ISBN 80-902482-2-5

[2] HERRINGTON, J., REEVES, T. C., OLIVER, R. A guide to authentic e-learning, New York: Routledge, 2010.

[3] VÁCLAVÍKOVÁ, Z. Science Education – Formal Versus Informal Education, New Challenges in Education, Verbum, Ružomberok, 2013, pp. 260-277, ISBN 978-80-561-0065-3

[4] VALENTA, J. Metody a techniky dramatické výchovy. Praha: Grada Publishing, a.s., 2008. ISBN 978-80-247-1865-1

[5] URL: http://www.stcostrava.cz/web/guest/hlavni-stranka

Current address

Václavíková Zuzana, RNDr. Ph.D.
University of Ostrava, Faculty of Science, Department of mathematics
30. dubna 22, 702 00 Ostrava, Czech republic
Tel. number: +420 597 09 2143, e-mail: zuzana.vaclavikova@osu.cz

Slovak University of Technology in Bratislava
Faculty of Mechanical Engineering
16th Conference on Applied Mathematics **APLIMAT 2017**
Proceedings

THE APPLICATION OF MATHEMATICAL AND STATISTICAL METHODS TO DETERMINE THE INFLUENCE OF ANODIZING TIME ON THE LAYER THICKNESS

VAGASKÁ Alena (SK), GOMBÁR Miroslav (CZ), KMEC Ján (CZ)

Abstract. The main goal of the present paper is to identify and analyse input and output variables acting during the process of aluminium anodic oxidation based on the application of mathematical and statistical methods and DOE methodology. This contribution builds on our previous experimental work and results, where the set of 26 experimental runs was carried out, data were analysed and prediction model was developed. Unlike our earlier publications, in this paper we focus our attention on the influence of anodizing time on the resulting layer thickness formed at defined current density of 3 Adm^{-2}, as well as on the influence of anodizing time in interaction with other input chemical and physical factors. Another partial aim of this study is to determine the value of the critical time of anodizing process.

Keywords: Design of experiments, aluminium anodic oxidation, prediction model, significant factors, critical value of anodizing time

Mathematics Subject Classification: Primary 62P30; Secondary 62K15

1 Introduction

The identification and analysis of factors (input variables) acting during anodic aluminium oxidation (AAO) process and monitoring of their impact on the resulting AAO layer is the object of our scientific research and experimental work [1], [2], [3], [4], [5]. Despite the large number of scientific publications and industrial studies, e.g. [6], [7], [8], [9], the mechanism of formation and growth layer is still unknown; we do not have some unified theory. So, it is very important to understand how factors and responses relate to each other, and to reveal which factors are influential for which responses [10], [11], [12]. If some processes are very complex and do not exist suitable description or mathematical-physical-chemical models of them, it is necessary to recognize and identify the relationships between considered variables only experimentally.

Unlike most scientific publications in this field of research, where is considered manipulating of only one factor at a time and its impact on the response, in our experimental work we focus on the influence of all relevant factors and their interactions. Design of experiments (DOE) methodology is very useful for this purpose, whereby all such factors are manipulated simultaneously according to rigorously formulated mathematical protocols, so DOE requires fewer experiments than any other approach [13], [14], [15], [16].

This contribution builds on our previous experimental work and results presented in [4], [5]. The main goal of our experimental work is to identify significant factors affecting the thickness of the formed AAO layer and recognize the influence of main effects and their interactions, based on the statistical analysis of experimentally obtained data. It was also important to create soft mathematical models, i.e. semi-empirical models or statistical models [13], [14] based on which it is possible to predict the final thickness of the formed AAO layer for considered factors and their interactions.

In contrast to our previous contributions, in this paper we focus our attention on the influence of anodizing time on the resulting layer thickness formed at defined current density of 3 Adm^{-2}, as well as on the influence of anodizing time in interaction with other input chemical and physical factors. Another partial aim of this study is to determine the value of the critical anodizing time above which will occur dissolving of the formed layer, resulting in deterioration of its quality characteristics.

2 Experimental

Anodic aluminium oxidation in sulphuric acid (GS method) at direct current and at constant current density $J_A = 3\,A \cdot dm^{-2}$ was used for individual test runs of experiment. Individual test runs were performed in the Hull cell of rectangular trapezium shape. The alloy EN AW-1050A H24 with dimensions 100 × 70 × 0.5 [mm] was used for the specimens. The anodizing of aluminum samples was performed based on the DOE methodology corresponding to the central composite design with 26 test runs. Individual test runs were performed on the basis of the design matrix of the experiment created as a combination of individual levels of four investigated factors according to Tab. 1, where experimental conditions are presented. By means of DOE, test runs were performed in random order to eliminate systematic error and to avoid subjective preference of any factor-level.

EXPERIMENT GS							
Method of anodic oxidation:				GS			
Design of experiment:				**Central composite design**			
Number of factors:		4		**Test runs:**		26	
Control factors							
Coded scale	**Natural scale**	**Physical unit**	**Factor levels**				
			-2	-1	0	1	2
x_1	m (H_2SO_4)	[$g.l^{-1}$]	25,00	100,00	175,00	250,00	325,00
x_2	m (Al)	[$g.l^{-1}$]	1,32	3,97	6,62	9,26	11,91
x_3	T	[°C]	2,00	12,00	22,00	32,00	42,00
x_4	t	[min]	10,00	25,00	42,50	60,00	80,00
Constant factors							
material anode/cathode:		AW – 1050A H24		**Voltage:**		12 V	

Tab. 1. Experimental conditions

The influence of input chemical factors x_1, x_2 and physical factors x_3, x_4 on the resulting thickness $\hat{y}$ (denoted as *th* in natural scale) of the formed AAO layer was investigated. It was necessary to obtain function dependence $\hat{y} = f(x_1, x_2, x_3, x_4)$, where x_1 – is amount of sulphuric acid H_2SO_4 in electrolyte, x_2 – concentration of aluminium cations Al in electrolyte, x_3 – the electrolyte temperature, x_4 – anodizing time. DoE standardization (coding) of input factor levels from natural scale to coded scale was performed according to the well-known formula, also mentioned in [5], which presents linear transformation of values of origin variable from interval $< x_{min}, x_{max} >$ to interval $< -2, 2 >$ and provides transformation of factors from original physical unit to dimensionless form. DOE methodology allows us to obtain maximum amount of information of high statistical and numerical correctness in an optimal number of individual test runs. Use scalar products the orthogonality of experiment design was verified, i.e. all columns of design matrix must be perpendicular to each other while nonzero. Due to the orthogonality of the experimental design, we can avoid improper indication of statistical insignificance of factors effects [15]. The orthogonal array design is well suited for estimating main effects.

The thickness of the resulting AAO layer was measured five times at each defined experimental points (the intersections of horizontal and vertical lines with distance of 5 mm) by usage of digital thickness meter MINITEST 4000. The arithmetic average of five measurements, calculated by microprocessor of thickness meter, was taken as an individual measurement. Experimentally obtained data represent an input matrix for further statistical analysis.

3 Results and discussion

Significance of factors affecting the final AAO layer thickness have been indicated based on the performed statistical analysis of experimentally obtained data (EDA, screening analysis, ANOVA, DOE analysis) using software products such as Matlab, Statistica, JMP and QC-Expert [17], [18], [19], [20]. We analysed how factors interact and obtained computational statistical models predicting the thickness of the resulting layer at varied levels of factors. Data analysis was performed with a statistically correct approach including analysis of the basic assumptions and subsequent analysis of the classical regression triplet: data, model, residues. Therefore it can be stated, that results conclusion and interpretation were performed without numerical and statistical incorrectness, what was also confirmed by practical experience in the subject matter field of surface treatment. The fact, that one of the co-authors has been active in the field of surface treatment for many years, is very valuable when interpreting research results. It is well known that statistical techniques are useless unless combined with appropriate subject matter knowledge and experience [12], [21], [22].

The screening analysis showed that at the chosen significance level of $\alpha = 0.05$, the amount of aluminium in electrolyte had no significant effect on the layer thickness created at current density of $J_A = 3\,\mathrm{A \cdot dm^{-2}}$. Therefore, it was possible to develop the statistical prediction model for DOE standardized level values of remaining three factors x_1, x_3, x_4 and their interactions in the form:

$$\hat{y}_{J_A=3A\cdot dm^{-2}} = 13.08 + 4.87x_1 + 1.48x_4 + 1.76x_3 - 1.28x_1x_3 - 2.21x_3^2 + 0.76x_4^3 - 3.46x_4^2x_1 - 1.5x_1x_3x_4 \quad (1)$$

Anodizing time is significant factor affecting the thickness of the formed anodic layer at any used sizes of current densities. This factor influences the response separately and together in interactions with other affecting factors in dependence on size of current density. At current density of $3\,\mathrm{A}\cdot\mathrm{dm}^{-2}$, the anodizing time is a significant factor affecting the thickness of the formed AAO layer at using significance level of $\alpha = 5$ %. Its impact on the variability of the size of layer thickness is 7%. Compared with influence of this factor at current density of $1\,\mathrm{A}\cdot\mathrm{dm}^{-2}$ and $2\,\mathrm{A}\cdot\mathrm{dm}^{-2}$ there is a decrease in the impact (at current density of $1\,\mathrm{A}\cdot\mathrm{dm}^{-2}$ it was 23%).
In the Fig. 1 there are shown graphical representations of the relationship between the thickness of AAO layer formed at current density of $3\,\mathrm{A}\cdot\mathrm{dm}^{-2}$ and anodizing time in dependence on the changing size of the electrolyte temperature at different amount of sulphuric acid in the electrolyte (Fig. 1a) 25 [g.l^{-1}], b) 175 [g.l^{-1}], c) 325 [g.l^{-1}]) in regard to existence of statistically significant interaction of these factors.

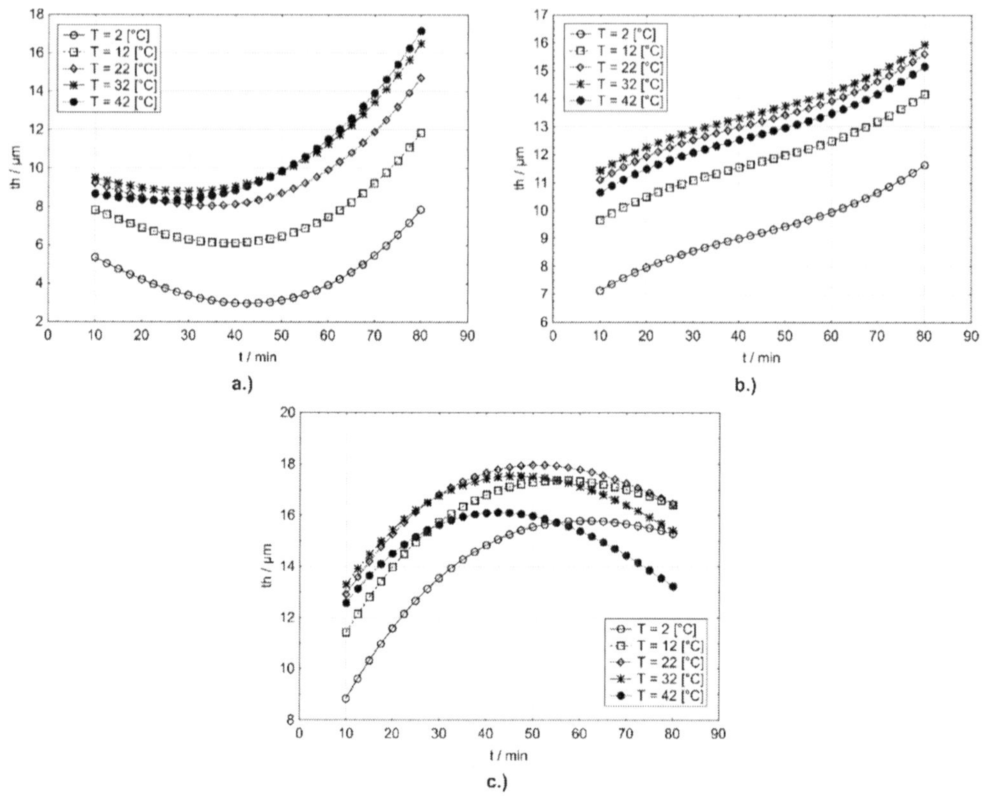

Fig. 1. Interaction effect plot of anodizing time and electrolyte temperature with regards to the layer thickness at different amount of sulphuric acid: a) 25 [g.l^{-1}], b) 175 [g.l^{-1}], c) 325 [g.l^{-1}]

The anodizing time is also statistically significant in interaction with the amount of sulphuric acid in the electrolyte; the impact of this interaction on the size of thickness is 10%. The squared value of anodizing time with sulphuric acid has significant impact on the response; the impact of this interaction is 13%. All of the above interactions are statistically significant not only in terms of statistical analysis of the models, but also in terms of application of anodizing process in practice.

Consider the graph in Fig. 1a) – the AAO layer was formed at minimum amount 25 [g.l^{-1}] of sulphuric acid in electrolyte. We can remark, that the thickness of the formed layer changes with increasing value of the anodizing time in non-linear dependence.

Considering DOE standardization of individual factors and substituting the experimental amount of sulphuric acid 25 [g.l^{-1}] to formula (1), we can express the value of the layer thickness in the form of natural scale as a function of two variables (anodizing time t and electrolyte temperature T):

$$th_{3A\cdot dm^{-2}, m(H_2SO_4)=25gl^{-1}} = f(t,T) = 5.524\cdot 10^{-3}\cdot T^2 + 2.137\cdot 10^{-3} T\cdot t + 0.305\cdot T + 1.774\cdot 10^{-5}\cdot t^3 + 5.65\cdot 10^{-4}\cdot t^2 - 0.149\cdot t + 6.139 \quad (2)$$

Substituting the experimental value of electrolyte temperature of 2 °C to formula (2) we can express the layer thickness as a function of anodizing time:

$$th_{3A\cdot dm^{-2},\ m(H_2SO_4)=25g\cdot l^{-1},\ T=2°C} = f(t) = 1.774\cdot 10^{-5}\cdot t^3 + 5.65\cdot 10^{-4}\cdot t^2 - 0.145\cdot t + 6.726 \quad (3)$$

It is obvious that function (3) $th = f(t)$ is everywhere continuous and differentiable, so by expressing the first derivative of the estimated response function (3) th with respect to time, we can obtain the theoretical speed of layer creation at constant amount of sulphuric acid of 25 [g.l^{-1}] in electrolyte, i.e.:

$$\frac{d}{dt}(th_{3A.dm^{-2}, m(H_2SO_4)=25g.l^{-1}, T=2°C}) = \frac{d}{dt}(1.774\cdot 10^{-5}\cdot t^3 + 5.65\cdot 10^{-4}\cdot t^2 - 0.145\cdot t + 6.726) \quad (4)$$

If we want to identify the intervals on which function $th = f(t)$ is increasing and decreasing according to the anodizing time at defined process conditions, it is necessary to find the critical points of the function (3) and then function's extreme values. The first derivative

$$th'(t) = \frac{d}{dt}(th) = 5.324\cdot 10^{-5}\cdot t^2 + 1.13\cdot 10^{-3}\cdot t - 0.145 \quad (5)$$

is zero at $t_1 = -63.82$ [min] and $t_2 = 42.58$ [min]. But only one of these critical points is suitable for us, because experimental interval of anodizing time is $[10, 80]$ minutes. So, only $t_2 \in [10, 80]$, i.e. the critical anodizing time is $t_k = t_2$ at defined process conditions. This critical point t_k subdivides the domain of th into intervals $[10, t_k)$ and $(t_k, 80]$ on which the first derivative th' is either positive or negative. The sign pattern of th' reveals the behaviour of th. The thickness th has a minimum value $th = 2.96$ [μm] at the point $t_2 = 42.58$ [min], the thickness is decreasing on the left of the minimum value and increasing on its right. We can remark that in the domain of decreasing of the layer thickness, the average rate of change of th with respect to anodizing time is 45% and the average rate of growth of th with respect to anodizing time over the interval $(t_2, 80]$ is 62.3%.

With increasing value of the electrolyte temperature from 2 [°C] to 12 [°C], the critical value t_k of the anodizing time increased to point $t_k = 59.89$ [min], the minimum value of the thickness also increased at the value of $th = 6.10\,[\mu m]$. With the increase in the temperature of the electrolyte, the value of the critical time of anodizing increased by 6%, the minimum value of the thickness of the formed layer increased by 51%. According to Fig. 1, the average rate of growth of th with respect to anodizing time over the interval $(t_k, 80]$ is 48%, the average rate of decay of th over the interval $[10, t_k)$ is 22%. Considering the value of the electrolyte temperature of 22 [°C] the thickness th has a local minimum value $th = 8.05\,[\mu m]$ at the critical point of anodizing time of $t_k = 55.64\,[min]$. With the increase in the temperature of the electrolyte from 12 [°C] up to 22 [°C], the minimum value of the layer thickness increased by 24.2%. The growth rate of th with respect to anodizing time over the interval $(t_k, 80]$ is 45.2%, in the domain of decreasing of the layer thickness (under the critical value of anodizing time $t_k = 55.64\,[min]$), the average rate of change of th is 12.6%. Considering the value of the electrolyte temperature of 32 [°C] the thickness th has a local minimum value $th = 8.79\,[\mu m]$ at the critical point of anodizing time $t_k = 50.93\,[min]$. With increasing electrolyte temperature from 22 [°C] up to 32 [°C], the minimum value of the thickness of the formed layer has increased only by 8%. The average rate of decay of th over the interval $[10, t_k)$ is 7.35% and the average rate of growth of th with respect to anodizing time over the interval $(t_k, 80]$ is 46.6%. As seen in Fig. 1, the function $th = f(t)$ has a local minimum value $th = 8.33\,[\mu m]$ at the critical point of anodizing time of $t_k = 45.66\,[min]$ considering endpoint of the experimental interval of the electrolyte temperature T=42 [°C]. The average rate of growth of th with respect to anodizing time over the interval $(t_k, 80]$ has increased at 51.4%.

Consider the graph in Fig. 1b) – the AAO layer was formed at defined amount 175 [$g.l^{-1}$] of sulphuric acid in electrolyte. As seen in Fig. 1b), the thickness th of the formed layer increases with increasing value of the anodizing time, the function $th = f(t)$ is monotonic on all experimental interval of anodizing time $[10, 80]$ minutes. Considering T=2 [°C], the average rate of growth of th with respect to anodizing time is 38.7%, at T=12 [°C] the average rate of growth of th is 31.7%, at T=32 [°C] it is 28.3% and at T=42 [°C] the average rate of growth of th is 29.7%.

Consider the graph in Fig. 1c) – the AAO layer was formed at defined amount 325 [$g.l^{-1}$] of sulphuric acid in electrolyte. As seen in Fig. 1c), there are two domains on which function $th = f(t)$ is increasing and decreasing according to the anodizing time at defined process conditions. For all considering values of the electrolyte temperature from 2 [°C] to 42 [°C] we can find the critical point of anodizing time t_k (the local maximum value of the thickness th) which subdivides the domain of th into intervals $[10, t_k)$ and $(t_k, 80]$ on which the first derivative th' is either positive or negative. Important characteristics of observed thickness are presented in Tab. 2.

As is shown in Fig. 1c) and in the Tab. 2 the thickness th of the formed layer is increasing on the left of the maximum value and decreasing on its right, i.e. th increases on $[10, t_k)$ and decreases on $(t_k, 80]$. In order to avoid dissolving of the formed layer or deterioration of its quality, the value of the critical anodizing time was determined for various experimental conditions and then verified in technical practice.

Electrolyte temperature [°C]	The rate of growth of *th* on interval [10 min, t_k)	t_k [min] / *th* [μm]	The rate of decay of *th* on interval (t_k, 80 min]
2	43. 94 %	60.62 min / 15.98 μm	3.13 %
12	34.17 %	55.29 min / 17.55 μm	5.43 %
22	28.07 %	50.58 min / 18.12 μm	8.32 %
32	24.18 %	46.33 min / 17.68 μm	12.20 %
42	21.91 %	42.41 min / 16.22 μm	17.84 %

Tab. 2. Maximum values of layer thickness *th* at critical point of anodizing time t_k at defined experimental conditions, i.e. at $J_A = 3\,A\cdot dm^{-2}, m(H_2SO_4) = 25\,g\cdot l^{-1}$ and at various values of electrolyte temperature

Conclusion

This article clarifies the usefulness of application of mathematical and statistical methods when evaluating experimentally obtained data in the process of aluminium anodic oxidation. The usage of DOE methodology allowed us not only identify significant factors and their interactions affecting the response (the thickness of the formed AAO layer), but also to create soft mathematical computational model based on which it is possible to predict the final thickness of the formed AAO layer for considered factors and their interactions.

In contrast to our previous contributions, in this paper we focused our attention on the influence of anodizing time on the resulting layer thickness formed at defined current density of $3\,A\cdot dm^{-2}$, as well as on the influence of anodizing time in interaction with other input chemical and physical factors. The other aim of this study was to determine the critical value of anodizing time and to find out on which intervals of anodizing time the growth of layer thickness with regarding quality occurs and on which it will occur dissolving of the formed layer.

Finally it can be stated that objectives have been achieved. Moreover we know that the results obtained by our experimental work have important benefits for technical practice, because they were practically verified under conditions of real production. And we also realized that statistical techniques are useless unless combined with appropriate subject matter knowledge and experience [11].

Acknowledgement

The paper was supported by grant KEGA 026TUKE-4/2016 of Scientific Grant Agency of the Ministry of Education of Slovak Republic.

References

[1] GOMBÁR, M., VAGASKÁ, A., KMEC, J., MICHAL, P. Microhardness of the coatings created by anodic oxidation of aluminium, *Applied Mechanics and Materials*, vol. 308, 2013, pp. 95-100.

[2] SPIŠÁK, E., GOMBÁR, M., KMEC, J., VAGASKÁ, A., FECHOVÁ, E., MICHAL, P., PITEĽ, J., KUČERKA, D. Effect of the Electrolyte Temperature and the Current Density on a Layer Microhardness Generated by the Anodic Aluminium Oxidation. *Advances in Materials Science and Engineering*. Vol. 2015, art. ID 659846, p 1-9, ISSN 1687-8434

[3] GOMBÁR, M., KMEC, J., BADIDA, M., SOBOTOVÁ, L., VAGASKÁ, A., MICHAL, P. The simulation of the temperature effects on the microhardness of anodic alumina oxide layers. *Metalurgija*. Vol. 53, 2014, pp. 59-62. ISSN 0543-5846

[4] VAGASKÁ, A., GOMBÁR, M., MICHAL, P., KMEC, J. The application of mathematical and statistical methods to determine the critical temperature of the electrolyte during the anodizing process. In 14th *Conference on Applied Mathematics APLIMAT 2015, Proceedings*. Bratislava, 2015, ISBN 978-80-227-4314-3, p. 738 – 747.

[5] VAGASKÁ, A., GOMBÁR, M., MICHAL, P. The application of mathematical and statistical methods in technical practice. In 15th *Conference on Applied Mathematics APLIMAT 2016. Proceedings*. Bratislava, 2016, ISBN 978-80-227-4314-3, p. 1054 – 1063.

[6] LE COZ, F., ARURAULT, L., DATAS, L. Chemical analysis of a single basic cell of porous anodic aluminium oxide templates. *Materials Characterization*, 2010; 61:283-8.

[7] BOCCHETTA, P., SUNSERI, C., CHIAVAROTTI, G., DI QUARTO, F. Microporous alumina membranes electrochemically grown. *Electrochimica Acta*, 2003;48:3175–83.

[8] THOMPSON, G.E. Porous anodic alumina: fabrication, characterization and applications. Elsevier, *Thin Solid Films*, 1997, Vol. 297, Issues 1-2, p. 192–201.

[9] FECHOVÁ, E. Nanocrystalline ferromagnetic powder materials on the basis of the iron. In *Proceedings of 5th International Conference of Mathematics and physics at Technical Universities*. Brno, 2007, ISBN 9788072312740, p. 432-436

[10] ERIKSSON, L., JOHANSSON, E., KETTANEH-WOLD, N., WIKSTRÖM, C. WOLD, S. *Multi-and Megavariate Data Analysis: Part I.* Umetrics Academy, 2006, ISBN -10:91-973730-2-8, 424 p.

[11] ERIKSSON, L., JOHANSSON, E. et all. *Multi-and Megavariate Data Analysis: Part II.* Umetrics Academy, 2006, ISBN -10:91-973730-3-6, 306 p.

[12] BOX, G. E. P., HUNTER, J. S., HUNTER, W. G. *Statistics for Experimenters*. 2008, p. 639, ISBN 978-0-471-71813-0.

[13] ERIKSSON, L., JOHANSSON, E., KETTANEH-WOLD, N., WIKSTRÖM, C. WOLD, S. *Design of Experiments*. Umetrics Academy, 2008, ISBN -10:91-973730-4-4, 459 p.

[14] MELOUN, M. - MILITKÝ, J. *Kompendium statistického zpracování dat*. Praha, ACADEMIA, 2006, ISBN 80-200-1396-2, 982 s.

[15] MORÁVKA, J., MAROŠ, B., MICHALEK, K. Vliv neortogonality plánu experimentu na statistickou korektnost modelu. In *Mezinárodní konference Technical Computing Prague 2008*, 2008, str. 73, http://dsp.vscht.cz/konference_matlab/MATLAB08/

[16] PANDA, A. *Monitorovanie a hodnotenie procesov*. Prešov, FVT TU, 2014, ISBN 978-80-553-1745-8, 162 s.

[17] GABKOVÁ, J., JANIGA, I. Using the software mathematica® in the linear model. In 12th *Conference on Applied Mathematics APLIMAT 2013, Proceedings*. Bratislava, 2013, ISBN 978-802273865-1

[18] FECHOVÁ, E. The support of solving physical problem by the means of mathematical software. In *Proceedings of Conference of Trends in Educations*, Olomouc, 2009, ISBN 9788072203161, p. 412–415.

[19] HREHOVÁ, S. Modelovanie a simulácia v prostredí Matlab. In *Vlastnosti diferenciálnych rovníc a systémov.* Brno: Tribun EU, 2008, p. 53-68, ISBN 9788073996444

[20] HREHOVÁ, S. Using the graphic tool of Matlab for visualization of input data. *Strojárstvo Extra*, 2013, Roč. 17, č. 7-8, s. 84-85, ISSN 1335-2938

[21] PANDA, A., DUPLÁK, J. Comparison of theory and practice in analytical expression of cutting tools durability for potential use at manufacturing of bearings. Applied Mechanics and Materials, 2014, Vol. 616 (2014), p. 300-307. ISSN 1660-9336

[22] VELICHOVÁ, D. Interactive maths with GeoGebra. *International Journal of Emerging Technologies in Learning.*, Vol. 6, Special Issue 2, 2011, p. 31-35, ISSN 18688799

Current address

Vagaská Alena, PaedDr. PhD.
Technical university of Košice, Faculty of Manufacturing Technologies with a seat in Prešov
Bayerova 1, 080 01 Prešov, Slovak Republic
Tel. number: +421 55 602 6430, e-mail: alena.vagaska@tuke.sk

Gombár Miroslav, Ing. PhD.
Department of Mechanical Engineering, Institute of Technology and Business in České Budějovice
Okružní 517/10, 370 01 České Budějovice, Czech Republic
Tel. number: +420 387 842 195, e-mail: gombar.mirek@gmail.com

Kmec Ján, doc. Ing. PhD.
Department of Mechanical Engineering, Institute of Technology and Business in České Budějovice
Okružní 517/10, 370 01 České Budějovice, Czech Republic
Tel. number: +420 387 842 195, e-mail: kmec@mail.vstecb.cz

Slovak University of Technology in Bratislava
Faculty of Mechanical Engineering
16th Conference on Applied Mathematics **APLIMAT 2017**
Proceedings

COMPUTATIONAL PREDICTION AND OPTIMIZATION OF THERMAL CONSUMPTION IN BUILDINGS

VALA Jiří (CZ)

Abstract. Application of advanced materials, structures and technologies in building structures forces new approaches to their design, namely the reliable prediction and optimization of energy requirements to artificial heating and air-conditioning. The paper presents an original robust and effective system computational model of a building, involving its software implementation in MATLAB.

Keywords: Computational modelling, thermal transfer, building structures

Mathematics subject classification: Primary 35K05; Secondary 74A15, 35K10, 65K10.

1 Introduction

Application of advanced materials, structures and technologies in building structures stimulates the development of new approaches to the physical, mathematical and computational analysis of both new and reconstructed buildings. Moreover, European and national directives and technical standards require the evaluation of various thermal certificates, as the European directive [25], forcing the "passive house" properties, taken from [8], for all investment actions starting 2020 and later. Although the first real building actions, connected with [8] in Germany, date back to the last decade of the 20th century, in most other countries, the Czech Republic, they occurred with some delay: Figures 1, 2 and 3 can be seen as a very short history of passive building in Brno city in images. The periodically updated list [24] contains about 500 relevant software tools; however, a physically transparent, intuitive, simple, robust and reliable numerical solver for the evaluation of energy consumption of a building, supporting the optimization of its design, too, is still not available.

In this paper we shall start with the consideration of a building as a thermal system, as presented originally in [18]. The progress in this research area in the last 2 decades can be documented on [17], [12] and [20] and on numerous further references therein. Strong physical and geometrical simplifications are often avoided using some elements of soft computing by [14]: fuzzy sets, rough classifications, genetic algorithms, etc. However, some their assumptions, connecting the thermal analysis with the simplified one-dimensional study of electrical circuits, relying on the finite difference and lumped mass approaches, can be removed or modified, following the weak formulation and utilizing the properties of the generalized Fourier decomposition, as inspired by [11] and sketched in [22].

Fig. 1. Petr Mareček: Family houses – first block, Židlochovice (small town south from Brno, 2006)

Fig. 2. Vize Ateliér (Martin Krč, Rostislav Kubııčcek and Jan Vintr): Family house Vějíř (= Punka), Brno - Bystrc (2009)

In this paper we shall apply the deterministic system approach to the physical, mathematical and computational analysis of the thermal behaviour of a building, including the optimization problem of control of artificial heating, open to various types of generalizations. The

Fig. 3. Projektil architekti (Roman Brychta, Adam Halíř, Ondřej Hofmeister a Petr Lešek): Centre of non-governmental and non-profit organizations in Brno (street Údolní, under Špilberk castle, 2013)

corresponding software implemented in MATLAB (without any requirements to additional toolboxes) is still in development. Thanks to the application of selected climatic data from the international airport Brno - Tuřany, an illustrative example shows the quasi-periodic behaviour of a very simple model object with insufficient heating equipment, to force various types of switching between heating regimes for testing.

Nevertheless, the global study of thermal transfer cannot be reduced to computational techniques, to satisfy technical standards and to save money for energy supply. Although such conceptions as "exergy consumption", "human well-being", "global environmental system" or "sustainable development" are well-known more from election campaigns than from some credible research studies, they have reasonable physical fundamentals, including the complete set of thermodynamic laws, unlike the first one in this paper only: for the case of built-environmental system cf. the review of such theory for closed systems in [19], p. 145, and for open ones in [19], p. 231.

2 Physical and mathematical background

The basic idea of the announced system approach is to decompose a building as a thermal system to particular rooms, walls, ceilings, floors, roof, their constructive and material layers, etc. to certain subsystems and elements, to study them (at least for the following analysis, not obligatory for the software implementation) separately, with the perspective of their coupling with help of (a priori unknown) thermal fluxes. For simplicity, we shall assume the constant temperature inside any chosen room in a fixed time, to avoid complicated calculations referring to computational flow dynamics and to stochastic analysis with the critical lack of relevant data. We shall start with the formulation and discretization of a model direct linear problem on one such element, with references to possible generalizations; namely the nonlinear (power-law) contribution of thermal radiation, as well as the heating control (which needs some optimization strategy) will be discussed.

Let us consider an open set (e. g. a domain) Ω in the 3-dimensional Euclidean space R^3, supplied by the Cartesian coordinate system $x = (x_1, x_2, x_3)$, with its boundary $\partial\Omega$ where the local vector of (formally outward) unit normal $\nu(x) = (\nu_1(x), \nu_2(x), \nu_3(x))$ can be introduced. The usual notation for the Hamilton and Laplace operators $\nabla = (\partial/\partial x_1, \partial/\partial x_2, \partial/\partial x_3)$, $\Delta = \nabla \cdot \nabla$, with the central dot for the scalar product in R^3, will be used. Moreover, let us consider a time interval $I = [0, a]$ with some real positive a; the limit passage $a \to \infty$ is not prohibited, the dot symbol is reserved for partial derivatives with respect to the time $t \in I$. For the brevity of notation, we shall assume an (at least macroscopically) isotropic material (not homogeneous in general) located in Ω, with the hope that the straightforward generalization can be derived by a patient reader.

The conservation principle for a scalar quantity $\theta(x,t)$ on Ω and its flux $\eta(\theta(x,t),x)$, following [3], pp. 5 and 14, and [9], p. 9, reads

$$\rho\dot{\varepsilon}(\theta) + \nabla \cdot \eta(\theta) = \mathcal{F}(\theta) \tag{1}$$

where $\rho(x)$ denotes the material density, $\varepsilon(\theta(x,t),x)$ refers to the enthalpy in the sense of [9], p. 10, and $\mathcal{F}(\theta(x,t),x,t)$ are the sources occurring on $\Omega \times I$; the explicit dependencies on x and t are not emphasized for brevity. Some constitutive equations must be hidden in the evaluation of η and ε from θ. We shall work with the unified boundary condition

$$\nabla\eta(\theta) \cdot \nu = \mathcal{G}(\theta)\,, \tag{2}$$

on $\partial\Omega \times I$, containing the sources $\mathcal{G}(\theta(x,t),x,t)$; some special forms of both $\mathcal{F}$ and $\mathcal{G}$ will be introduced later.

For the thermal performance of buildings, the crucial scalar quantity θ is the (absolute Kelvin) temperature, needed in the conservation principle for (thermal) energy (1). The standard constitutive equations are

$$\varepsilon(\theta) = \mathcal{C}(\theta)\theta\,, \tag{3}$$

$$\eta(\theta) = -\lambda(\theta)\nabla\theta \tag{4}$$

where $\lambda(\theta,x)$ by (4) refers to the thermal conductivity and $\mathcal{C}(\theta,x)$ by (3) to the thermal capacity related to mass unit; frequently also the notation $\kappa(\theta,x) = c(\theta,x)\rho(x)$ is applied for the thermal capacity related to volume unit.

Our definitions of $\mathcal{F}(\theta)$ and $\mathcal{G}(\theta)$ are open to handle relations between different quantities θ, as a priori unknown temperature, velocity components or density, considered as reference variables in conservation principles for energy, momentum and mass. However, some additional generalizations may be needed, as the dependence of ρ on θ. The possible implementation of the complete set of thermodynamic principles (unlike the first one in this paper only) can rely on the proper exergy analysis: for the closed systems cf. [16], p. 145, for the open ones cf. [16], p. 233. Nevertheless, in this paper we shall take into account only one equation of the type (1). Inserting (4) and (3) into (1), we obtain

$$\kappa(\theta)\dot{\theta} - \nabla \cdot (\lambda(\theta)\nabla\theta) = \mathcal{F}(\theta)\,. \tag{5}$$

In particular, for homogeneous materials one can derive $\lambda(\theta) = \mathrm{d}\beta(\theta)/\mathrm{d}\theta$ for a prescribed function $\beta(\theta)$ (independent of x explicitly) where $\eta(\theta) = \nabla\beta(\theta)$, thus (5) simplifies to

$$\varepsilon(\theta) - \Delta\beta(\theta) = \mathcal{F}(\theta)\,. \tag{6}$$

Consequently, using the Kirchhoff and enthalpy transforms by [15], p. 253, then the term $\varepsilon(\theta)$ in (6), unlike (5), can be reduced to certain transformed variable θ, which is accompanied by certain modifications of remaining terms. However, such approach is rather complicated except the case of the constant values of κ and λ (independent of θ) by (3) and (4), generating the so-called thermal diffusivity λ/κ, frequently used in engineering applications.

Our aim is to solve (5), starting with certain initial condition

$$\theta(.,0) = \theta^0 \tag{7}$$

where the knowledge of initial values $\theta^0(x)$ of $\theta(x,t)$ with $t = 0$ for any $x \in \Omega$ is needed. Let us introduce the following notations: (u,v) for the Lebesgue integrals of $u(x)v(x)$ on Ω for appropriate real functions u and v, or for the Lebesgue integrals of $u(x) \cdot v(x)$ on Ω for appropriate real function triples u and v, alternatively, and $\langle u,v\rangle$ for the Hausdorff integrals of $u(x)v(x)$ on $\partial\Omega$ for appropriate real functions u and v. We intend to avoid the deeper analysis in Lebesgue, Sobolev, Bochner, etc. spaces in this paper; however, let us remind that $(.,.)$ and $\langle.,.\rangle$ can be seen as the scalar products in Lebesgue spaces $L^2(\Omega)$, $L^2(\Omega)^3$ and $L^2(\partial\Omega)$ in the sense of [15], p. 11. Using the Green - Ostrogradskiĭ theorem formally, we are able to convert (at least in the sense of distributions) (5) to

$$(\phi, \kappa(\theta)\dot{\theta}) + (\nabla\phi, \lambda(\theta)\nabla\theta) = (\phi, \mathcal{F}(\theta)) + \langle\phi, \mathcal{G}(\theta)\rangle\,. \tag{8}$$

Combining the approach of [15], p. 63 (for thermal convection and radiation), with that of [17] (for solar radiation), let us consider the surface heat sources

$$\mathcal{G}(\theta) = \mathfrak{G}(\theta) + \alpha(\theta)(\theta_\times - \theta) + \sigma(\theta)(\theta_\times^4 - \theta^4)\,, \tag{9}$$

with the first right-hand-side additive term referring to the solar radiation (this needs more detailed explanation, which will be done later), with the second one evaluating the effect of thermal convection and with the third one supplementing the effect of thermal radiation due to the Stefan - Boltzmann law, hiding two additional material characteristics $\alpha(\theta,\theta_\times)$ and $\sigma(\theta,\theta_\times)$; the second one is usually presented in the form $\sigma(\theta,\theta_\times) = \sigma_0 \upsilon(\theta,\theta_\times)$, with certain correction factor $\upsilon(\theta,\theta_\times) \in [0,1]$ and the famous constant $\sigma_0 = 5.67036713.10^{-8}$ W/m^2/K^4 derived by Stefan and Planck, valid for an ideal black body. Consequently from (8) and (10) we have

$$\begin{aligned}(\phi, \kappa(\theta)\dot{\theta}) + (\nabla\phi, \lambda(\theta)\nabla\theta) + \langle\phi, \alpha(\theta,\theta_\times)\theta\rangle + \sigma_0\langle\phi, \upsilon(\theta,\theta_\times)\theta^4\rangle \\ = (\phi, \mathcal{F}(\theta)) + (\phi, \mathfrak{G}(\theta)) + (\phi, \alpha(\theta,\theta_\times)\theta) + \sigma_0\langle\phi, \upsilon(\theta,\theta_\times)\theta_\times^4)\rangle\,,\end{aligned} \tag{10}$$

still with respect to (7).

The volume heat sources $\mathcal{F}(\theta)$ occur (as non-zero) only in rooms and can be expressed as

$$\mathcal{F}(\theta) = \eta(\theta) + \Gamma(\theta,\theta_\times)\kappa^*(\theta,\theta_\times)(\theta - \theta_\times) \tag{11}$$

where the first additive component $\eta(\theta)$ comes from artificial heating (which should be controlled in time carefully, as will be discussed later) and the second one comes from the forced minimal air change by technical standards (limiting the thermal accumulation ability of a building). Here $\Gamma(\theta,\theta_\times)$ denotes the prescribed air change rate and $\kappa^*(\theta,\theta_\times)$ the thermal capacity of air related to volume unit, analogous to $\kappa(\theta)$. The slight modification of (10) due to this decomposition will be not discussed separately: one can see immediately the contribution $(\phi, \Gamma(\theta,\theta_\times)\kappa^*(\theta,\theta_\times)\theta)$ to the left-hand side of (10), whereas all other terms from (11) contribute to the right-hand side.

The linearization of (10) then relies on the choice of κ, λ, $\mathfrak{F}$ and $\mathfrak{G}$ independent of θ, on the choice of κ^*, Γ and α independent of θ and $\theta_\times$ and on the setting $\upsilon = 0$ everywhere. Weakening of such assumptions can be performed thanks to an appropriate iteration process; this will be demonstrated later on the example of replacing the assumption $\upsilon = 0$ (which is too strict for practical calculations, removing the thermal radiation by the Stefan - Boltzmann law at all) by the assumption of independence of υ of θ and $\theta_\times$.

The presence of the surface thermal fluxes $\mathfrak{G}$ in (10) enables us to handle solar radiation, whose evaluation as the sum of beam (direct) and diffusive radiation intensities q_b and q_d comes from the analysis of mutual Sun – Earth positions and from the location and orientation of a building, as explained in [19], [13] and [10].

Let us introduce the surface reflexivity $\mathfrak{R}$, together with certain shading factors $\mathfrak{R}_b$ and $\mathfrak{R}_d$. Let us also consider the geographical latitude ϵ, the azimuth angle γ (depending on the building orientation, related to south), and the angle β between the building surface and the horizontal plane. The evaluation day and year cycles of Sun – Earth positions requires the knowledge of declination $\delta(N) = 0.13027\pi \sin(2\pi(284 + N)/365)$ (presented here only for the standard year length of 365 days for simplicity) and of the day time angle $\omega(\xi) = \pi\xi$ for $\xi = (1 - (t - t_\times))/t_*$; $N \in \{1, \dots, 365\}$ refers to the day in the year, $t_\times$ to the initial day time and t_* (formally) to the standard day length. Further astronomical corrections and generalizations of $\delta(N)$, $\omega(\xi)$, etc., are available, but not presented and implemented here, because of the substantial extension of the following formulae. The same holds for potential dependencies of $\mathfrak{R}$, $\mathfrak{R}_b$ and $\mathfrak{R}_d$ on θ or even on other quantities.

The zenith angle ς comes from the relation

$$\cos\varsigma = \sin\epsilon \sin\delta(N) + \cos\epsilon\cos\delta(N)\cos\omega(\xi)\,, \tag{12}$$

whereas the real angle of incidence of sunrays ϑ needs to calculate

$$\begin{aligned}\cos\vartheta = \cos\varsigma &+ \sin\epsilon(\sin\delta(N)(\cos\beta - 1) + \cos\delta(N)\sin\beta\cos\gamma\cos\omega(\xi))\\ &+ \cos\epsilon(\cos\delta(N)(\cos\beta - 1)\cos\omega(\xi) - \sin\delta(N)\sin\beta\cos\gamma)\\ &+ \cos\delta(N)\sin\beta\sin\gamma\sin\omega(\xi)\,.\end{aligned} \tag{13}$$

Making use of (12) and (13), together with the measurement results for q_b and q_d, we are finally able to evaluate

$$\begin{aligned}\mathfrak{G} = &\mathfrak{R}_b q_b \min(\cos\vartheta, 0)/\min(\cos\varsigma, 0.087) + \mathfrak{R}_d q_d(1 + \cos\vartheta)/2\\ &+ \mathfrak{R}(q_b + q_d)(1 - \cos\vartheta)/2\,.\end{aligned} \tag{14}$$

Consequently (14), referring back to (12) and (13), can be inserted into (10), contributing no additional term dependent on θ.

Up to now, we have considered only one set Ω, although a real building structure consists of a large, but finite number of such sets. However, it is easy to rewrite (10) (involving also (14), which will be not emphasized explicitly in further considerations) respecting (7), too, for more sets with slight modifications: for any fixed set Ω its boundary $\partial\Omega$ is allowed to consist from interfaces to adjacent sets and external environment, consequently $\theta_\times$ come from all such interfaces, creating needed relations to all neigbour sets, etc. Thus we shall not discuss this case separately, assuming such modifications involved in the computational scheme yet. This will be reflected also by the modular structure of the announced MATLAB software.

3 Computational algorithm

To solve (10) with the initial condition (7) analytically is not realistic, except some very simple benchmark problems. The next choice is then to seek for the evolution of θ with help of the multiplicative Fourier decomposition by [1], p. 215, and [6], p. 346, in the form

$$\theta(x,t) = \phi_i(x)\psi_i(t) \tag{15}$$

where ϕ_i with $i \in \{1,\dots,n\}$ for some fixed integer n, open to the limit passage $n \to \infty$, form a sufficiently rich system of functions defined on Ω; the Einstein convention for the sum index i is applied here. The choice of ϕ_i, approximating ϕ in (10), refers frequently to various finite element techniques – cf. [5], p. 299 and 323. Thus we obtain

$$(\phi_j, \kappa\phi_i)\dot{\psi}_i + (\nabla\phi_j, \lambda\nabla\phi_i)\psi_i = (\phi_j, \mathcal{F}) + \langle\phi_j, \mathcal{G}\rangle \tag{16}$$

for any $j \in \{1,\dots,n\}$, which can be rewritten for $\psi = (\psi_1,\dots,\psi_n)^{\mathrm{T}}$, etc., in the obvious matrix form

$$M\dot{\psi}(t) + K\psi(t) = f(t) + \dot{g}(t)\,; \tag{17}$$

f and $\dot{g}$ here are useful to be distinguished in practice due to the type of available input data: g will be used for artificial heating and forced air exchange here, f for all other influences.

In particular, for the zero right-hand side of (17), it is possible to find the general solution of (17) in the form

$$\psi(t) = V\exp(-\Lambda t)c\,, \qquad \dot{\psi}(t) = -V\Lambda\exp(-\Lambda t)c \tag{18}$$

with some constants $c = (c_1,\dots,c_n)^{\mathrm{T}}$; here V denotes the square matrix compound from the eigenvectors and Λ the diagonal square created by the eigennumbers corresponding to the generalized eigenproblem

$$MV\Lambda = KV\,. \tag{19}$$

To verify (19) is easy: inserting (18) into (17) formally and denoting $(0,\dots,0)^T$ by o briefly, we obtain $(-MV\Lambda + KV)\exp(-\Lambda t)c = o$.

Moreover, the multiplication of (19) by V^{T} from the left, the multiplication of the transposed version of (19) by V from the right and their comparison gives $V^{\mathrm{T}}MV\Lambda = \Lambda V^{\mathrm{T}}MV$,, which, together with similar standard manipulations, yields

$$V^{\mathrm{T}}MV = VMV^{\mathrm{T}} = MVV^{\mathrm{T}} = VV^{\mathrm{T}}M = I\,, \qquad K^{-1} = V\Lambda^{-1}V^{\mathrm{T}}\,, \tag{20}$$

etc., with I representing the unit square matrix, which will be useful later.

To find some particular solution of (17) with the non-zero right-hand side, the method of variation of constants from the classical differential calculus can be utilized: instead of (18), replacing c by some $\mathfrak{C}(t)$, we have

$$\psi(t) = V\exp(-\Lambda t)\mathfrak{C}(t)\,, \qquad \dot{\psi}(t) = -V\Lambda\exp(-\Lambda t)\mathfrak{C}(t) + V\exp(-\Lambda t)\dot{\mathfrak{C}}(t)\,, \tag{21}$$

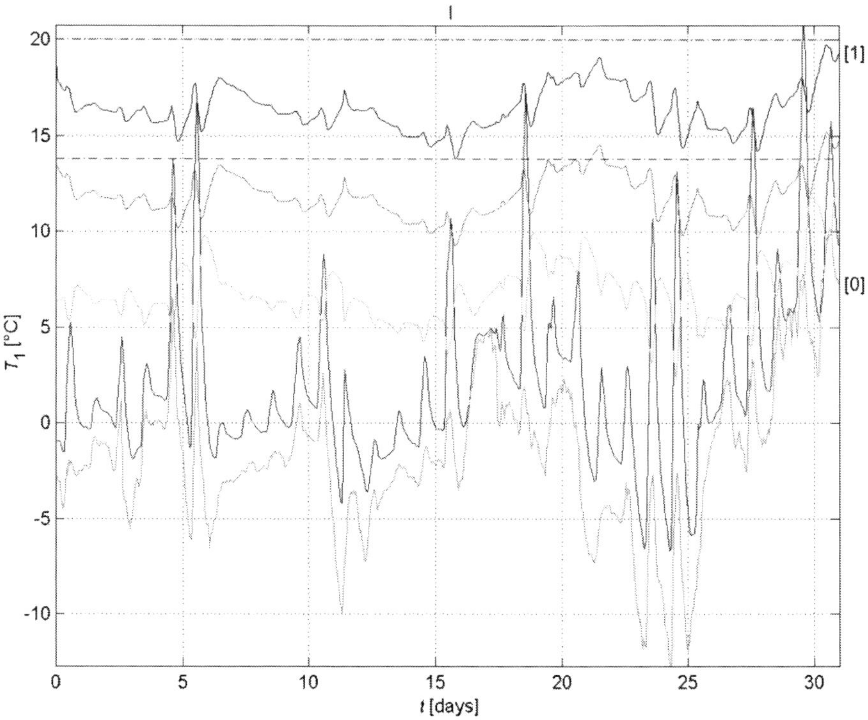

Fig. 4. Temperature in a room and structures in January of the 1st year, artificial heating is active

and inserting (21) into (18), we obtain

$$(-MV\Lambda + KV)\exp(-\Lambda t)\mathfrak{C}(t) + MV\exp(-\Lambda t)\dot{\mathfrak{C}}(t) = f(t) + \dot{g}(t)\,. \tag{22}$$

However, the first left-hand-side additive term vanishes thanks to (19), then the whole equation (22) can be multiplied by V^{T}, which with respect to (20) gives

$$\exp(-\Lambda t)\dot{\mathfrak{C}}(t) = V^{\mathrm{T}}(f(t) + \dot{g}(t))\,. \tag{23}$$

Multiplying (23) by $\exp(\Lambda t)$ from the left, from the direct integration of this equation we receive

$$\mathfrak{C}(t) = \int_0^t \exp(\Lambda\tau)V^{\mathrm{T}}(f(\tau) + \dot{g}(\tau))\,\mathrm{d}\tau\,; \tag{24}$$

$\dot{g}$ here must be understood as the derivative withe respect to τ. Inserting (24) into the first relation (21), we have finally

$$\psi(t) = V\int_0^t \exp(\Lambda(\tau - t))V^{\mathrm{T}}(f(\tau) + \dot{g}(\tau))\,\mathrm{d}\tau\,. \tag{25}$$

The general solution of (17), without any additional assumption, is the sum of right sides of (18) and (25), i. e.

$$\psi(t) = V\exp(-\Lambda t)c + V\int_0^t \exp(\Lambda(\tau - t))V^{\mathrm{T}}(f(\tau) + \dot{g}(\tau))\,\mathrm{d}\tau\,. \tag{26}$$

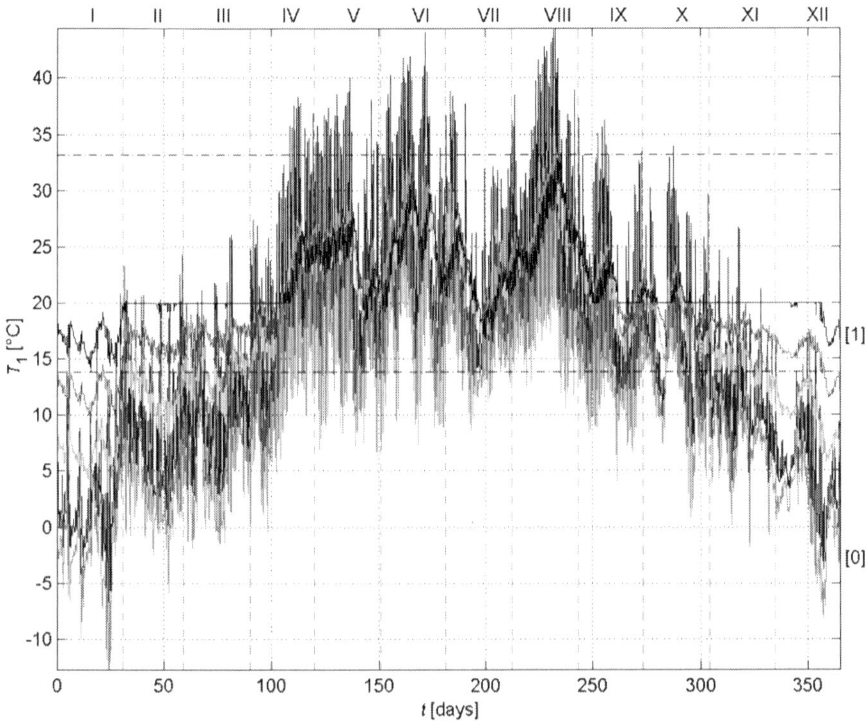

Fig. 5. Temperature in a room and structures during the 3rd year, artificial heating is active

It remains to set c to satisfy (7). Since $\psi^0 = VC$ by (26) at $t = 0$, with help of (20) we obtain $c = V^{\mathrm{T}} M\psi^0$, or alternatively $c = \Lambda^{-1}V^{\mathrm{T}}K\psi^0$,, which, inserted into (26) generates the needed particular solution of (17) satisfying (7)

$$\psi(t) = V\exp(-\Lambda t)V^{\mathrm{T}}\psi^0 + V\int_0^t \exp(\Lambda(\tau - t))V^{\mathrm{T}}(f(\tau) + \dot{g}(\tau))\,\mathrm{d}\tau\,. \qquad (27)$$

Le us notice that the derivation of ψ^0 from θ^0 by (7) may not be trivial: whereas for a hypothetical orthonormal system of ϕ_i with $i \in \{1,\ldots,n\}$ we have just ψ^0 identical with θ^0, in general this can be done only in the least squares (minimum variation) sense, minimizing $(\theta^0 - \phi^{\mathrm{T}}\psi^0, w(\theta^0 - \phi^{\mathrm{T}}\psi^0))$ with some reasonable weight function w defined of Ω (e. g. $w = 1$ everywhere if no better information is available) and an unknown real vector ψ^0 – cf. [4], p. 133.

No better data than $f(t)$ and $\dot{g}(t)$ in rather large time steps h, typically 1 hour, are available. To create a good evaluation algorithm for $\theta(t)$, let us consider both f and g as linear Lagrangian splines on I. Let us understand ψ^k briefly as $\psi(kh)$ with $k \in \{0, 1, \ldots, m\}$, $m = a/h$, and similarly f^k and g^k, too. Starting with $k = 1$, for $0 \le t \le h$ we have $f(t) = f^0(1 - t/h) + f^1 t/h$ and $\dot{g}(t) = (g_1 - g_0)/h$; the exact integration by (27) with $t = h$, after simple formal manipulations (where the software MAPLE, or the MATLAB

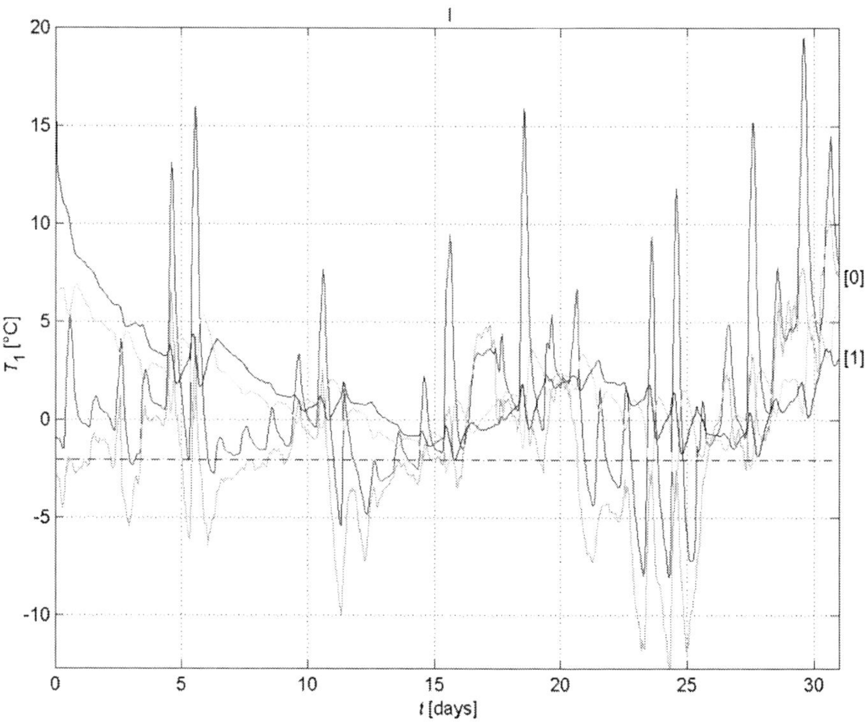

Fig. 6. Temperature in a room and structure in January of the 1st year, artificial heating is inactive

toolbox *symbolic*, can be helpful), results

$$\psi^1 = V\exp(-\Lambda h)V^{\mathrm{T}}M\psi^0 + V(I - \exp(-\Lambda h)) \tag{28}$$
$$\cdot\left(\Lambda^{-1}V^{\mathrm{T}}\frac{g^1 - g^0}{h} - \Lambda^{-2}V^T\frac{f^1 - f^0}{h}\right) + V\Lambda^{-1}V^T f^1 - V\Lambda^{-1}\exp(-\Lambda h)V^T f^0 .$$

The Taylor expansion $I - \exp(-\Lambda h) \approx \Lambda h$, with the error of order h^2, applied to (28), leads to the Euler explicit scheme (unstable for the forced choice of h, due to the available climatic data, frequently), whereas the implicit one can be obtained from the backward integration with the mutually exchanged ψ^0 and ψ^1; for the comparison to the Crank - Nicholson, etc. schemes and the detailed stability and accuracy analysis see [2], pp. 639, 646.

Due to the compatibility with ψ^0 derived from θ^0 by (7), we are allowed to replace ψ^1, f^1, g^1, ψ^0, f^0, g^0 in (28) with ψ^k, f^k and g^k, ψ^{k-1}, f^{k-1} and g^{k-1} for any $k \in \{1, \ldots, m\}$, i. e.

$$\psi^k = V\exp(-\Lambda h)V^{\mathrm{T}}M\psi^{k-1} + V(I - \exp(-\Lambda h)) \tag{29}$$
$$\cdot\left(\Lambda^{-1}V^{\mathrm{T}}\frac{g^k - g^{k-1}}{h} - \Lambda^{-2}V^T\frac{f^k - f^{k-1}}{h}\right) + V\Lambda^{-1}V^T f^k - V\Lambda^{-1}\exp(-\Lambda h)V^T f^{k-1} .$$

This generates the recurrent computational algorithm for the evaluation of ψ by (15) in required time steps everywhere on Ω.

Let us now admit the non-zero values of v on $\partial\Omega$, as promised in the previous section. This forces the following modifications of our approach: now f^k and f^{k-1} must be written as $\widetilde{f}^k +$

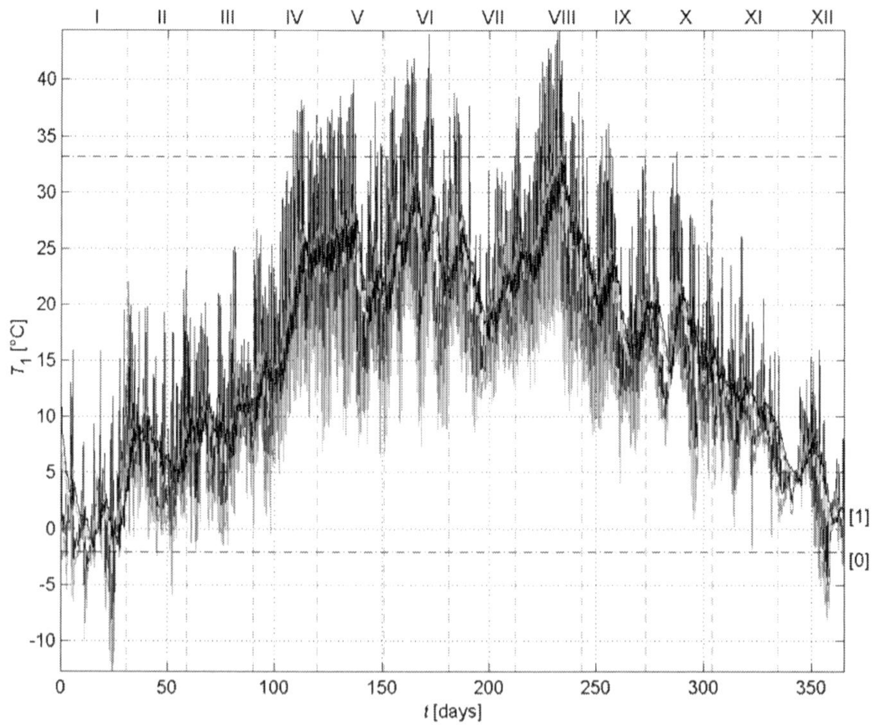

Fig. 7. Temperature in a room and structure during the 3rd year, artificial heating is inactive

$S(\psi^k)^4$ and $\widetilde{f}^{k-1} + S(\psi^{k-1})^4$ where $(.)^4$ means the 4th power of all components of $(.)$ (later also $(.)^3$ the 3rd power) and S is the diagonal square matrix (sparse in practice). The principal difficulty is the poor knowledge of ψ^k a priori (unlike ψ^{k-1}); only some initial estimate ψ^k_* is available, e. g. $\psi^k_* \approx \psi^{k-1}$. For the brevity of notation, let $\widetilde{\psi}^k$ be the result (instead of ψ^k) obtained from (29) with the correct f^{k-1}, but with f^k replaced by $\widetilde{f}^k$. Consequently we have

$$\psi^k = \widetilde{\psi}^k + PS(\psi^k)^4 \tag{30}$$

where $P = V\Lambda^{-}1V^T - V(I - \exp(-\Lambda h))\Lambda^{-2}V^T/h$.

The exact Newton method, as derived in [7], pp. 101, 144, applied to (30), gives

$$(I - 4PS(\psi^k_*)^3)(\psi^k - \psi^k_*) = \widetilde{\psi}^k - \psi^k_* + PS(\psi^1_*)^4 , \tag{31}$$

which can be repeated (with ψ^k_* replaced by ψ^k) in the iteration procedure (if needed, to reach the required accuracy). However, in most application cases the additional term $PS\psi^k$ is not dominating on the right side of (30), consequently the first estimate ψ^k_* on the left side of (31) can be taken as fixed, which slightly disturbs the Newton iteration process, but improves the algorithmic efficiency.

To support the heating control, let us moreover require the conservation of (at least) prescribed temperature level $\overline{\psi}^k$ ($k \in \{1, 2, \ldots, m\}$ again), derived from the prescribed values $\overline{\psi}^k$ in the analogous way as θ^0 from ψ^0. However, the heating power of these equipments

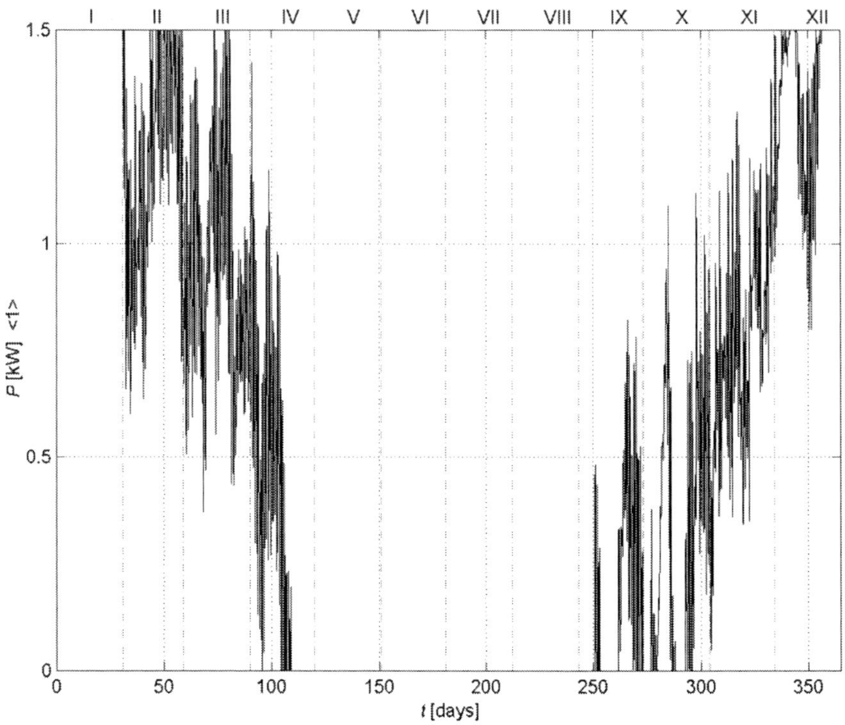

Fig. 8. Actual thermal power of artificial heating during the 3rd year

q^k, valid for $(k-1)h < t \leq kh$, is limited, i. e. $0 \leq q^k \leq q^{\max}$ for some upper bounds $q^{\max}$ for particular rooms (which is a vector of much lesser length than n in practice). Such upper bounds are design parameters of a building, due to the volume, location and assumed exploitation of rooms.

Consequently we have to distinguish between 4 following artificial heating regimes:

1) no heating during the summer season,
2) inactive heating for sufficient temperature outside the summer season, i. e. $\theta^k \geq \overline{\theta}^k$,
3) normal heating: setting of q^k is needed,
4) insufficient heating for $q^k \geq q^{k\max}$ by 3) when the reduction of q^k to $q^{k\max}$ is necessary (and the required temperature is not reached).

Clearly, our principal aim is to evaluate q^k for the regime 3); the modifications for all other regimes are obvious. The switching between all regimes will be demonstrated on the example in the following section.

Let us compose W from the lines of the matrix $V(I - \exp(-\Lambda h))\Lambda^{-1}V^{\mathrm{T}}$ corresponding to the temperature in rooms with artificial heating, and take q as certain additional $\dot{g}$ occurring in (17). Using the least squares technique, we are able to evaluate q^k from the rather simple formula

$$W^T W q^k = W^T(\widehat{\psi}^k - \overline{\psi}^k) \tag{32}$$

where $\widehat{\psi}^k$ corresponds to the hypothetical regime 1), active for $(k-1)h < t \leq kh$ (thus $\widehat{\psi}^k$ can be evaluated by (29) directly). We can see that the system of linear algebraic equations

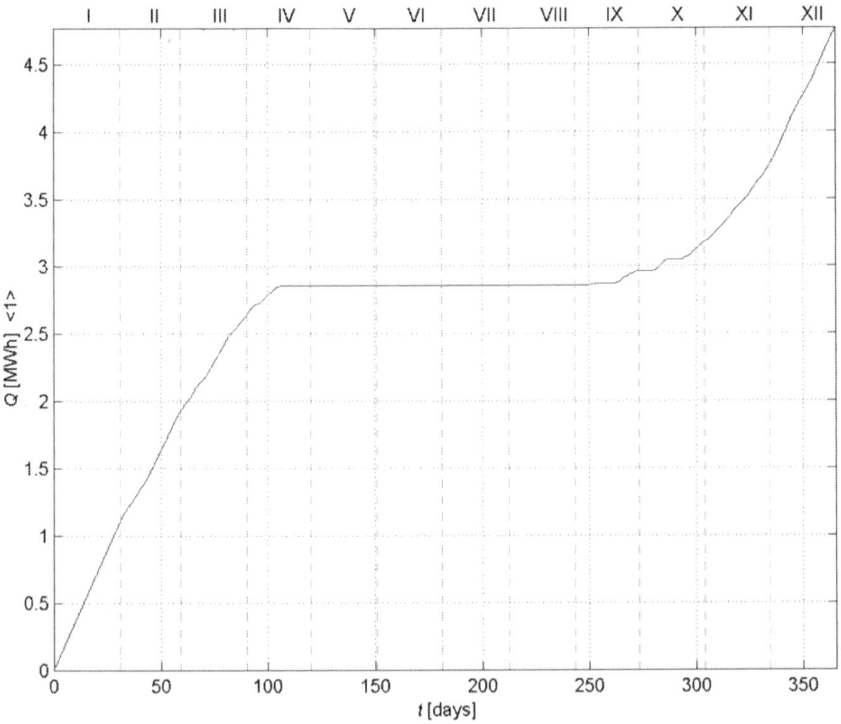

Fig. 9. Total consumption of energy of artificial heating during the 3rd year

(32) is small, thus its solution should be inexpensive; no further inequalities are necessary in the computational algorithm.

4 Illustrative example

The illustrative example presents the thermal analysis of a very simple model building with only one large room, located near Brno. All calculations are based on the one-dimensional model simplification (thus potential "thermal bridges" in building edges may not be handled correctly) and on the application of the finite element method with linear splines as basis functions, with the original software implementation inside the MATLAB environment. However, no additional simplifications like "lumped masses" by ([20]), referring to the analogy to electrical RC-circuits, are needed. The following meteorological data in the time steps $h = 1$ hour, averaged to the periodical form of the so-called reference year, are available: i) the environmental temperature, ii) the total solar radiation intensity, iii) the diffusive component of ii). Such data are sufficient to perform all above sketched evaluations, including the annual energy consumption, crucial for the thermal design of a building. The heating is insufficient, with the aim of proper testing of switching between regimes 1), 2), 3), 4). No summer shading (or air conditioning) is installed, to force the unwanted summer increase of temperature in the room. However, the visible effect of thermal radiation due to the Stefan - Boltzmann law (as the unique contribution disturbing the linearity of the problem except that coming from the inequalities needed in the heating control) is limited to the slight temperature decrease in some nights.

Figure 4 shows the development of temperature in selected locations of constructive and insulation layers of walls in the first month (January) of the first calculation year, starting from the hypothetical initial temperature $\theta^0 = 20$°C everywhere, including the environmental temperature (the lower oscillating curve here) and one indoor temperature, rather close to the required $\overline{\theta} = 20$°C again. The non-negligible variation of indoor temperature corresponds to the strange design of a building, as introduced above. The summer overheating of the room, in some cases even over the prescribed environmental temperature, is caused by the effect of solar radiation, whereas the same radiation in winter, in the direction close to the normal one to most walls and windows, helps to reduce the effect of the insufficient heating.

Figure 5 presents the same computational results as Figure 4, but for the (less synoptic) quasi-periodic temperature development in the reference year (the third one in practice). Figures 6 and 7 are the analogies of Figures 4 and 5 for the case of missing artificial heating, i. e. with the forced regime 1) for the whole year. We can see that, even for good insulated building, such regime is not acceptable for the Central European climatic conditions.

Figure 8, corresponding to Figure 4, records the actual thermal power, whereas Figure 9 shows the total consumption of energy. To present similar figures corresponding to Figure 6, is useless: due to the insufficient heating, the analogy of Figure 8 would show the constant actual thermal power (except the zero time with inactive heating) and Figure 9 the (nearly) linear increase of the total consumption of energy from zero.

5 Conclusion

We have presented a rather simple, robust and effective numerical algorithm and software package for the evaluation of thermal consumption of a building with controlled artificial heating. Its further development should contribute to the optimization of advanced building design. The non-trivial formal verification of existence and uniqueness of above introduced direct and optimization problems, modifying and generalizing selected results from [21], is being prepared for further presentation. Practical validation of numerical results requires time series of relevant data on real energy consumption for a sufficiently simple real building (to suppress external disturbing influences); the first results should be published during this year.

Acknowledgement. This research has been supported by the project LO1408 AdMaS UP (Advanced Materials, Structures and Technologies, Ministry of Education, Youth and Sports of the Czech Republic, National Sustainability Programme I).

References

[1] Barták, J., Herrmann, L., Lovicar, V., and Vejvoda, O.: *Partial Differential Equations of Evolution*. Ellis Horwood, 1991.

[2] Baker, A., and Graves-Morris, D.: *Padé Approximants*. Cambridge University Press, 2010.

[3] Bermúdez de Castro, A.: *Continuum Termodynamics*. Birkhäuser, 2005.

[4] Bochev, P.B., and Gunzburger, M.D.: *Least-Square Finite Element Methods*. Springer, 2009.

[5] Brenner, S. C., and Ridgway Scott, L.: *The Mathematical Theory of Finite Element Methods*. Springer, 2002.

[6] Brigola, R.: *Fourier-Analysis und Distributionen*. Co-Verlag, 2012.

[7] Drábek, P., and Milota, J.: *Methods of Nonlinear Analysis – Application to Differential Equations*. Springer, 2013.

[8] Feist, W.: *Gestaltungsgrundlagen Passivhäuser*. Das Beispiel, Darmstadt, 1999.

[9] Ferziger, J. H., and Periç, M.: *Computational Methods for Fluid Dynamics*. Springer, 2002.

[10] Gliah, O., Kruczek T., and Thibault, J.: The effective sky temperature: An enigmatic concept. *Heat and Mass Transfer* **47** (2011), 1171–1180.

[11] Jarošová, P., and Vala, J.: New approaches to the thermal design of energy saving buildings. *Advanced Materials Research* **1126** (2015), 174–180.

[12] Kämpf J. H., and Robinson, D.: A simplified thermal model to support analysis of urban resource flows. *Energy and Buildings* **39** (2007), 445–453.

[13] Nayak, J. K., and Sukhatme, S. P.: *Principles of Thermal Collection and Storage*. Tata McGraw Hill, 2008.

[14] Pacheco-Vega, A.: *Soft Computing Applications in Thermal Energy Systems*. Springer, 2011.

[15] Roubíček, T.: *Nonlinear Partial Differential Equations with Applications*. Birkhäuser, 2005.

[16] Shukuya, M.: *Exergy – Theory and Applications in the Built Environment*. Springer, 2013.

[17] Škrjanc, I., Župančič, B., Furlan, B., and Kainer, A.: Theoretical and experimental fuzzy modelling of building thermal dynamic response. *Building and Environment* **36** (2001), 1023–1038.

[18] Šťastník, S., and Vala, J.: On the thermal stability in dwelling structures. *Building Research Journal* **52** (2004), 31–55.

[19] Sukhatme, S. P.: *Solar Energy*. Tata McGraw Hill, 1996.

[20] Underwood J. C.: An improved lumped parameter method for building thermal modelling. *Energy and Buildings* **79** (2014), 191–201.

[21] Vala, J.: Computational approaches to some inverse problems from engineering practice. *Programs and Algorithms of Numerical Mathematics* 17 in Dolní Maxov (Czech Republic, 2014), Institute of Mathematics CAS, Prague 2015, 215–230.

[22] Vala, J.: Computational modelling of thermal consumption of buildings with controlled interior temperature. *ICNAAM – 15th International Conference on Numerical Analysis and Applied Mathematics* in Rhodes (Greece, 2016), AIP Conference Proceedings, Melville (USA) 2017, 4 pp., in print.

[23] White, R. E.: *Elements of Matrix Modeling and Computing with MATLAB*. Chapmann & Hall / CRC, 2005.

[24] *Building Energy Software Tools Directory* (US Department of Energy, 2016), available at http://apps1.eere.energy.gov/buildings/tools_directory/.

[25] Directive 2010/31/EU of the European Parliament and of the Council on the Energy Performance of Buildings. *Official Journal of the European Union* **L 153/13**, 2010.

Vala Jiří, prof. Ing. CSc.
Institute of Mathematics and Descriptive Geometry, Faculty of Civil Engineering,
Brno University of Technology
Veveří 331/95, 602 00 Brno, Czech Republic
Tel. number: +420 541 147 602, e-mail: vala.j@fce.vutbr.cz

Slovak University of Technology in Bratislava
Faculty of Mechanical Engineering
16th Conference on Applied Mathematics **APLIMAT 2017**
Proceedings

SOLUTION BEAM ON ELASTIC FOUNDATION WITH STIFFNESS CHANGE IN THE HORIZONTAL DIRECTION

VALKOVIČ Vojtech (SK), JANČO Roland (SK)

Abstract. This article focuses on calculation of deflection, bending moments and transverse forces of the beam deposited on a elastic foundation for changing the stiffness of soil.

Keywords: elastic foundation, beam, probability, FEM, Winkler model, pipe excavation

Mathematics Subject Classification: Primary 60A05, 08A72; Secondary 28E10

1 Introduction

The beam rested on elastic foundation when under the applied external loads, the reaction forces of the foundation are proportional at every point to the deflection of the beam at this point. This assumption was introduced first by Winkler in 1867.

Consider a straight beam supported along its entire length by an elastic medium and subjected to vertical forces acting in the plane of symmetry of the cross section (Fig. 1) [1].

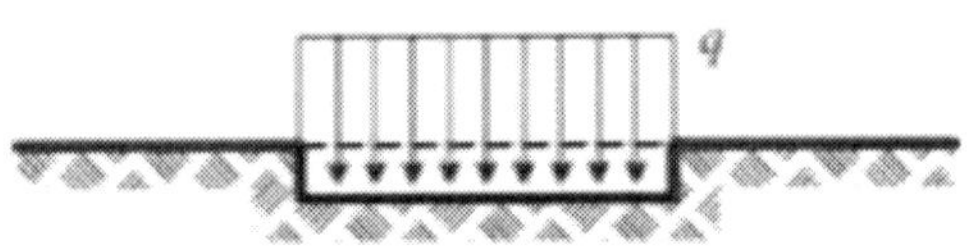

Fig. 1. Deflections of elastic foundations under uniform pressure (Winkler model)

Because of the external loadings the beam will deflect producing continuously distributed reaction forces in the supporting medium. The intensity of these reaction forces at any point is proportional to the deflection of the beam v at this point via the constant k:

$$q_R = kv \tag{1}$$

The reactions act vertically and opposing the deflection of the beam. Hence, where the deflection is acting downward there will be a compression in the supporting medium. Where the deflection happens to be upward in the supporting medium tension will be produced which is not possible.

In spite of all it is assumed that the supporting medium is elastic and is able to take up such tensile forces.

In other words the foundation is made of material which follows Hooke's law. Its elasticity is characterized by the force, which distributed over a unit area, will cause a unit deflection. This force is a constant of the supporting medium called the modulus of the foundation K [Nm^{-3}]. Assume that the beam under consideration has a constant cross section with constant width b [m] which is supported by the foundation. A unit deflection of this beam will cause reaction equal to $K \cdot b$ in the foundation, therefore the intensity of distributed reaction (per unit length of the beam) will be [2]:

$$q_R(x) = Kbv(x) = kv(x) \tag{2}$$

where:

$$k = Kb \tag{3}$$

is the constant of the foundation, known as Winkler's constant, which includes the effect of the width of the beam, and has dimension k [Nm^{-2}].

1.1 Differential equation of equilibrium of a beam on elastic foundation

Consider an infinitely small element enclosed between two vertical cross sections at distance dx apart on the beam into consideration (Fig. 2). Assume that this element was taken from a portion where the beam was acted upon by a distributed loading q(x). The internal forces that arise in section cuts are depicted in Fig. 2.

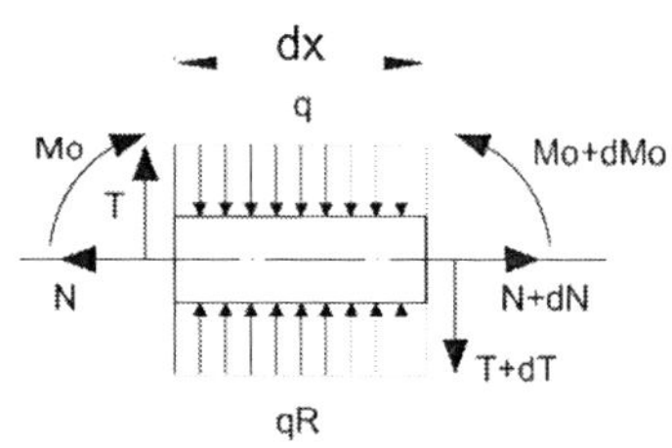

Fig. 2. Differential element of length dx

Considering the equilibrium of the differential element, the sum of the vertical forces gives:

$$T + q_R dx - qdx - (T + dT) = 0 \tag{4}$$

Considering the equilibrium of moments along the left section of the element we get:

$$M_o + mdx + Tdx - (M_o + dM_o) + q_R dx\frac{dx}{2} - qdx\frac{dx}{2} = 0 \tag{5}$$

Using now the well known differential equation of a beam in bending:

$$\frac{d^2v}{dx^2} = -\frac{M_o}{EJ_{ZT}} \tag{6}$$

it can be written:

$$\frac{dT}{dx} = \frac{d^2 M_o}{dx^2} = -EJ_{ZT}\frac{d^4 v}{dx^4} \tag{7}$$

Finally, from the summation of the vertical forces:

$$EJ_{ZT}\frac{d^2 v}{dx^2} + M_o = 0 \tag{8}$$

$$\frac{d^2 M_o}{dx^2} + \frac{k}{EJ_{ZT}} M_o = 0 \tag{9}$$

$$\omega = \sqrt[4]{\frac{k}{EJ_{ZT}}} \tag{10}$$

In the above equation the parameter ω includes the flexural rigidity of the beam as well as the elasticity of the foundation. This factor is called the characteristic of the system with dimension length^{-1}. In that respect 1/ ω is referred to as the so called characteristic length. Therefore, x · ω will be an absolute number [3,5].

2 Beam on elastic foundation

Common examples of practices can be greatly facilitated by a suitable foundation model and the following is possible in some cases to get analytical solutions. Response in the subsoil is then not addressed. Winkler model we are considering a flexible support, the thorium small deformations and beam material apply Hook-law for a replacement for isotropic and homogeneous material. Transverse warming is below the values that are large changes in material beam or subsoil material. Beam placed on a flexible foundation loaded force, with whose values are changing subsoil step change or a gradual change in stiffness (see. Fig. 3) can be calculated by analogy with probabilistic approach and the finite element method [4, 6].

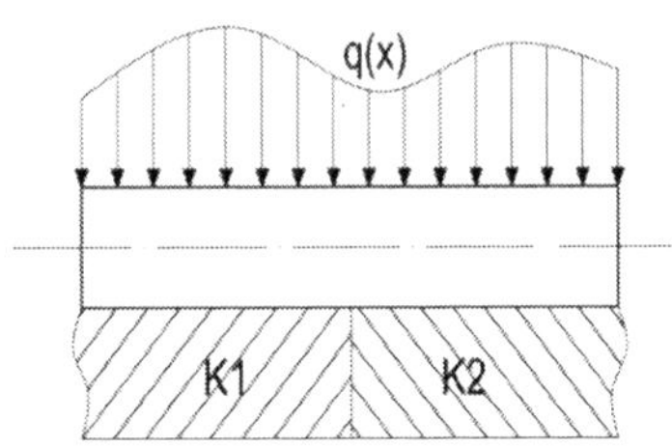

Fig. 3. Beam on two elastic foundation

3 Steps changes in stiffness in the horizontal direction

In large quantities solutions step change in the ground in roles bending beam can be significantly simplified by introducing the concept of equivalent stiffness of the bedrock k_{EKV} [Pa]. This simplification leads to approximate solution. In cases where the stiffness values uncertain, the introduction of equivalent stiffness relatively simple way the whole solution very streamlined. Figure 2 shows the beam.

Continuously distributed responses in each of the subsoil can be replaced with appropriate force F_i. The conditions of force equilibrium in the vertical direction:

$$F_{EKV} = \sum_{i=1}^{n} F_i \tag{11}$$

Assuming the validity of the theory of Winkler elastic foundation and provided that the stiffness k_i are the lengths L_i constant can be for forces F_i derive relations:

$$F_i = \int_{Li} q_{R_i} dx_i = k_i \int_{Li} v_i dx_i \tag{12}$$

where q_{R_i} it is continuously distributed in reaction i-th foundation, v_i is the bending lines in equation i-th foundation.

In a similar fashion can be expressed as the equivalent forces:

$$F_{EKV} = k_{EKV} \int_{L_{EKV}} v_{EKV}\, dx \tag{13}$$

where v_{EKV} equation is equivalent bending lines (approximation of the actual deflection) and L_{EKV} it is equivalent to the length of the beam to which it applies:

$$L_{EKV} = \sum_{i=1}^{n} L_i \tag{14}$$

Substituting 12 and 13 to 11 get a relationship for equivalent stiffness of the subsoil:

$$k_{EKV} = \frac{\sum_{i=1}^{n} \left(k_i \int_{Li} v_i dx_i \right)}{\int_{L_{EKV}} v_{EKV}\, dx} \tag{15}$$

If we consider the beam length L of the free ends of the length stored on the elastic substrate (Fig. 4).

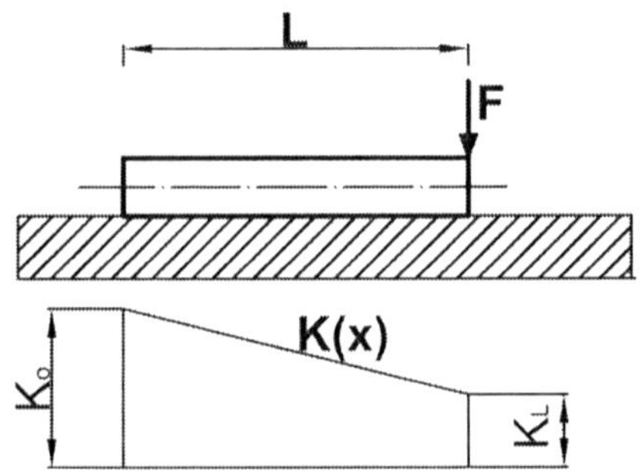

Fig. 4. Changes in stiffness in the horizontal direction on beam

4 Solution by FEM

The beam is on the right loaded with force F (see figure 4). Then we can write that the compression modulus of the subsoil is changed depending on:

$$K(x) = K_0 + \frac{K_L - K_0}{L} x \tag{16}$$

The values for our case are given in Table 1.

L	1 m	**E**	$2\text{x}10^{11}$ Pa
J_{ZT}	$2.16\text{x}10^{-5}$ m^4	**K_0**	$1.4625\text{x}10^{11}$ Nm^{-3}
F	160 000 N	**K_L**	$5\text{x}10^{10}$ Nm^{-3}

Tab. 1. Given values needed to calculate

Example solved with the aid of FEM in which he sends off values for the deflection, bending moment and transverse force.

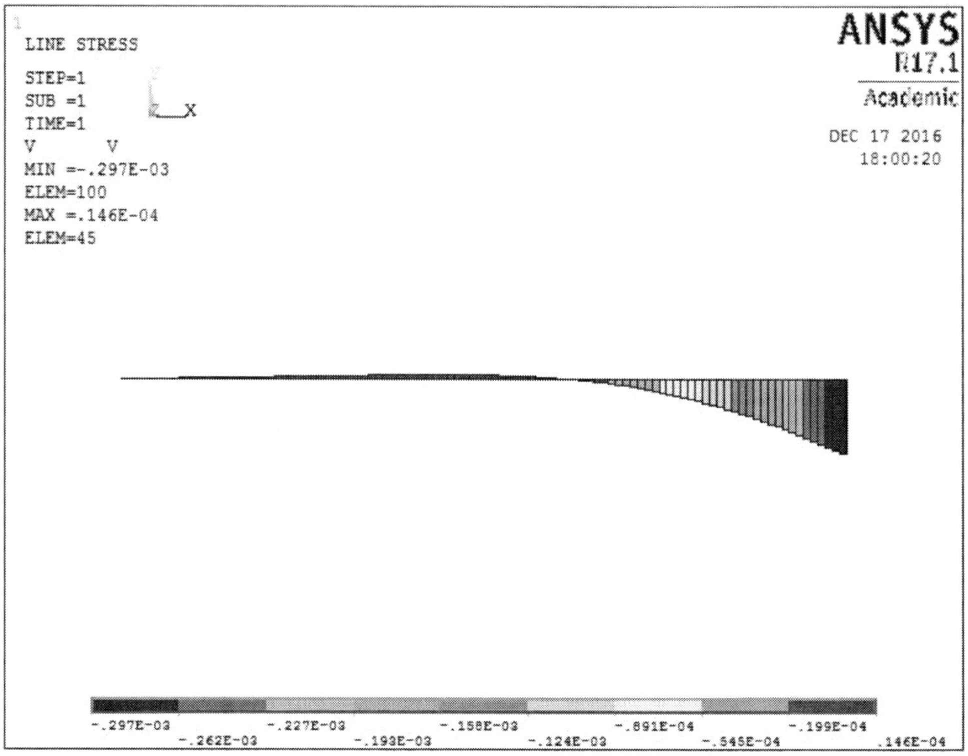

Fig. 5. Deflection of the beam on elastic foundation

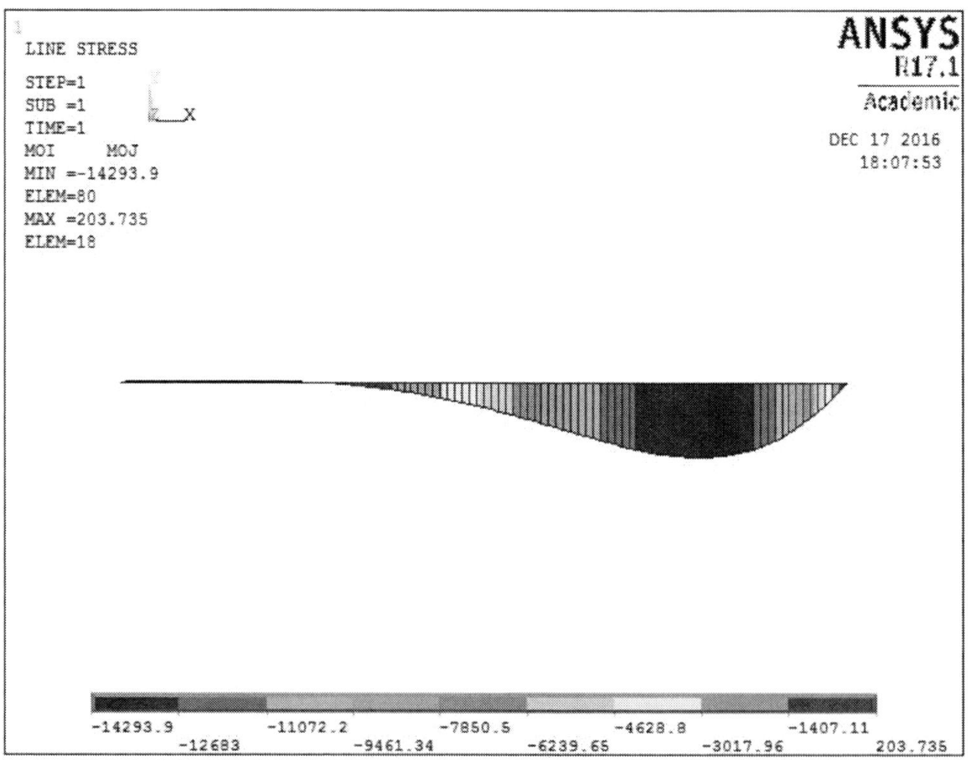

Fig. 6. Bending moment of the beam on elastic foundation

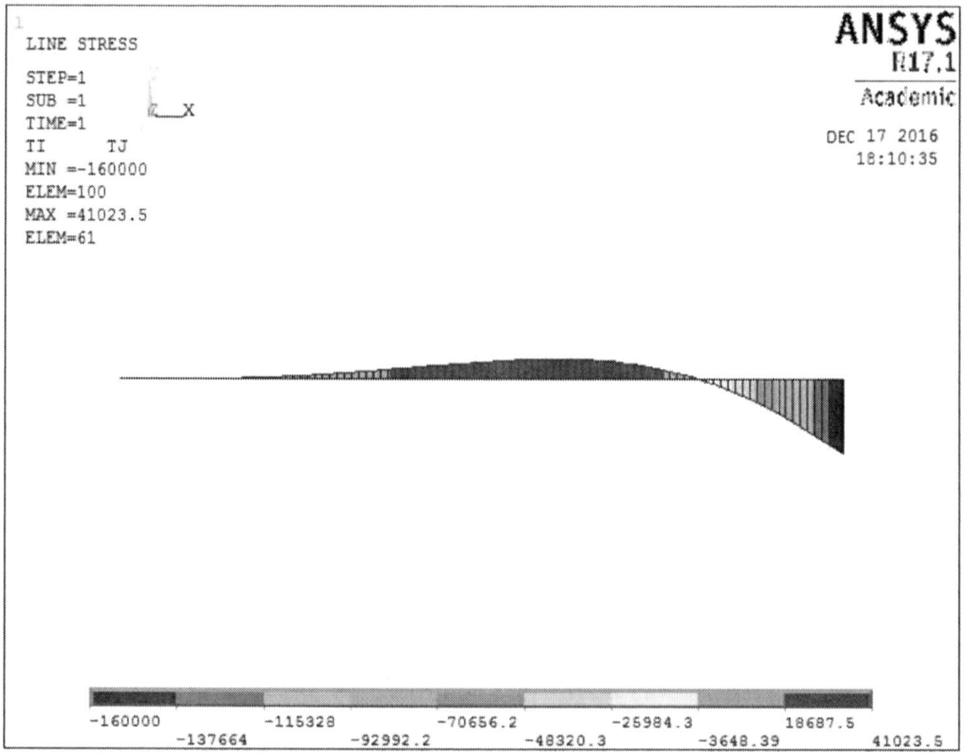

Fig. 7. Transverse force of the beam on elastic foundation

Conclusion

Solving problems beams on an elastic substrate depends on many factors. The simple solution is sufficient Winkler model. For the case of the beam on one side loaded with force F and deposited on a flexible substrate at varying stiffness of the subsoil we were used by FEM calculation while we went see the value figures quoted in the previous chapter. While the maximum deflection was v=2.97x10^{-4} m.

Acknowledgements

The article was supported by the STU Grant scheme for Support of Young Researchers.

References

[1] FRYDRÝŠEK K.: *Beams and Frames on Elastic Foundation 1*, VSB – Technical University of Ostrava, 2006.

[2] FRYDRÝŠEK K, JANČO R.: *Beams and Frames on Elastic Foundation 2*, VSB – Technical University of Ostrava 2008.

[3] ÉLESZTÖS P., BENČA Š.: *Strength rehabilitation of a pipeline system at the critical point by mechanical preloading*, Periodica Polytechnica, Mechanical Engineering Volume 51, Issure 2, 2007, Pages 113-116, DOI: 10.3311/pp.me.2007-2.11, ISSN: 03246051.

[4] FRYDRÝŠEK, K. *Probabilistic Calculations in Mechanics 1*. Department of Mechanics of Materials, Faculty of Mechanical Engineering, VSB – Technical University of Ostrava, ISBN 978-80-2482314-0, Ostrava, Czech Republic, 2010, pp.149.

[5] MUSIL, Miloš. *Passive and active vibration isolation of machines 1*, STU, 2012. 152 s. ISBN 978-80-227-3733-3.

[6] PECIAR, P., ECKERT, M., FEKETE, R. HRNČIAR, V.: *Analysis of Pharmaceutical Excipient MCC Avicel PH102 Using Compaction Equations*. In Journal of Mechanical Engineering - Strojnícky časopis, Vol. 66, No. 1, 2016, pp. 65-82, ISSN 0039-2472.

Current address

Valkovič Vojtech, Ing.
Slovak University of Technology in Bratislava, Faculty of Mechanical Engineering
Nám. slobody 17, 812 31 Bratislava, Slovak Republic
E-mail: vojtech.valkovic@stuba.sk

Jančo Roland, doc. Ing. PhD.
Slovak University of Technology in Bratislava, Faculty of Mechanical Engineering
Nám. slobody 17, 812 31 Bratislava, Slovak Republic
E-mail: roland.janco@stuba.sk

Slovak University of Technology in Bratislava
Faculty of Mechanical Engineering
16th Conference on Applied Mathematics **APLIMAT 2017**
Proceedings

CONSTRUCTIVIST APPROACH TO DOUBLE INTEGRAL VIA GEOGEBRA ENVIRONMENT

VALLO Dusan (SK)

Abstract. The aim of this contribution is to bring some inspiration how to use dynamical geometry software Geogebra in teaching Mathematics and in teachers training education. For this purpose we concerned with some difficult stereometrical problems of computation of volume of intersection of two cylinders. We demostrate numerical computing of double integral via interactive graphics and spreadsheet environment. Some didactical remarks are included in the paper, too.

Keywords: GeoGebra, stereometry, double integral, ICT

Mathematics subject classification: Primary 97G40; Secondary 97D50

1 Introduction

In the theory of mathematical teaching is full accepted an idea that mathematical tasks hold a fundamental role in students' structure of knowledge.The aim of this paper is to bring together a few didactic aspects of the solution of mathematical tasks, specially geometrical tasks related to problems in 3-dimensional Euclidean space. In this direction particularly problematic tasks are positional and metric stereometrical tasks. Student can be successful in their solution under a condition, that one has a geometrical knowledge and needs also an appropriate space and geometric imagination. [1]
A lot of teachers of Mathematics agree in opinion that is important to yield to the students an understandable explanation and a transparent visualization of given problem, else is a high probability that one has to fall down. From this point of view is a great opportunity for using an interactive geometry softwares, which can visualize 3-dimensional space. [2]
There is a lot of dynamical geometry softwares, which can be implemented in educational process for this purpose. The reader can find many way of this implementation in [3], [4], [5], [6].
In our paper we will we demonstrate a potencial of the software Geogebra more closely in solution a specific stereometrical problem of an intersection of two cylinders. A few didactical remarks will be noted, too.

2 Use of Geogebra and computing of volume

In [7] the author present many real-life stereometrical problems, which solution is based on exhaustive method of the evaluation of sums of infinitesimal quantities. Of course, the figures in book are schematic and in many examples the solids are in positions such that is difficult to imagine the space situation for the reader. Presented implementation of GeoGebra in educational process and consecutively in an computing of volume of solids will be helpful in many ways.

Problem 1. *Two cylinders with equal bases' radii r, height v, $0 < r < 2v$, are in a position that their axes perpendicular bisect each other. Calculate the volume of their intersection.*
Solution. We set the Cartesian coordinate system $\langle O, x, y, z\rangle$. The axes of the given cylinders we put congruent with the axes o_x, respectively o_y. The equations of the cylindrical surfaces are in form

$$x^2 + y^2 = r^2 = y^2 + z^2 \tag{1}$$

We identify the intersection of the cylinders by a method of cuts. At first we consider the planar section by the plane O_{xy}. This plane contains the axes of the cylinders and it holds true that the plane O_{xy} intersects one of the cylinders in a rectangle with sides $2r$ and v. By analogy, the intersection of the plane O_{xy} with the second one is a congruent rectangle, too. The axes of the rectangles perpendicular and it implies that their overlapping is a *square* with side $2r$.

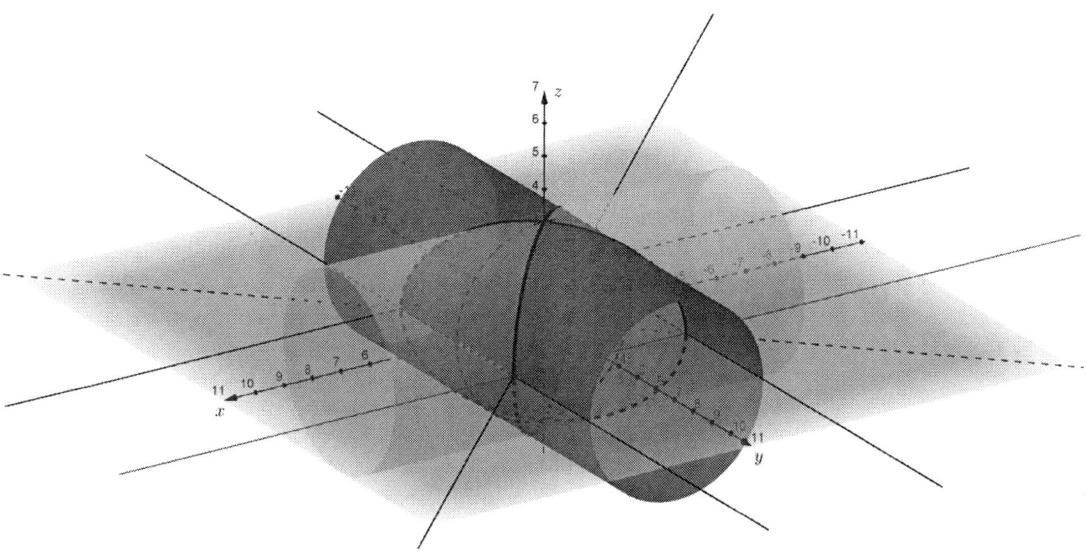

Fig. 1. Intersection of two cylindrical surfaces in Geogebra

A plane $z = m, 0 < m \leq r$ intersects also the cylinders in square. This implies that the volume of common part can be evaluated by method of double integral. We divide the intersection into 16 congruent partitions (see Fig.1) and the elementary domain D one of them is bounded by $0 < x < r, 0 < y < x$. It holds true that

$$V = 16 \iint_D \sqrt{r^2 - x^2}\, dy\, dx = \int_0^r \left(\int_0^x \sqrt{r^2 - x^2}\, dy \right) dx = ... = \frac{16}{3} r^3 \tag{2}$$

Remark. The important point to note here is that the result (2) *does not depend on constant* π. *The solids are rotational and one would naturally expect an occurrence of this constant.*

Using the appropriate software environment a teacher can easy demonstrate the problem more generally. Let us demonstrate some progress.
In GeoGebra one can use the tool *Sliders* in window *Graphics* for input of the different values of the radii r, R of the cylinders. It seem to be interesting to observe this re-new problem from perspective of maths teaching.

Problem 2. *Two cylinders with radii* r, R *and height* v, $0 < r < R < 2v$, *are in a position that their axes perpendicular bisect each other. Calculate the volume of their intersection.*
Solution. We set the Cartesian coordinate system $\langle O, x, y, z \rangle$. From now on we make the assumption the axes of the given cylinders are congruent with the axes o_z, respectively o_x. The equations of the cylindrical surfaces are in form

$$\alpha : \quad x^2 + y^2 = r^2 \tag{3}$$

$$\beta : \quad y^2 + z^2 = R^2 \tag{4}$$

We will show now that the previous method of calculation leads to extremely difficult evaluation of integral.
In Fig. 2 one can see that the intersection curve of the surfaces α, β is not planar and lies on the surface with equation $-x^2 + z^2 = R^2 - r^2$. In the plane O_{xz} the curve is hyperbola.
The domain D of the selected portion of the intersection of the surfaces is bounded by $0 < x < r, 0 < y < \sqrt{r^2 - x^2}$ and it holds true that

$$V_{\frac{1}{8}} = \iint_D \sqrt{R^2 - y^2}\, dy\, dx = \int_0^r \left(\int_0^y \sqrt{R^2 - y^2}\, dy \right) dx, \tag{5}$$

where $y = \sqrt{r^2 - x^2}$.
In [8], p. 454 can be found a formula, which gives a particular result in a form

$$V_{\frac{1}{8}} = \int_0^r \left[\frac{1}{2} \left(y\sqrt{R^2 - y^2} + R^2 \arcsin \frac{y}{R} \right) \right]_0^{\sqrt{r^2 - x^2}} . \tag{6}$$

Roughly speaking to solve the integral (6) is not trivial and a student is brought into of a confounding status. On the other side, this situation gives clever teacher a great opportunity to use the Geogebra to demonstrate the basic idea of the definite integral - approximation of the integral by numerical integration.

The function $z = \sqrt{R^2 - y^2}$ is Riemann-integrable and we will demonstrated the Riemann sums via tools of GeoGebra.

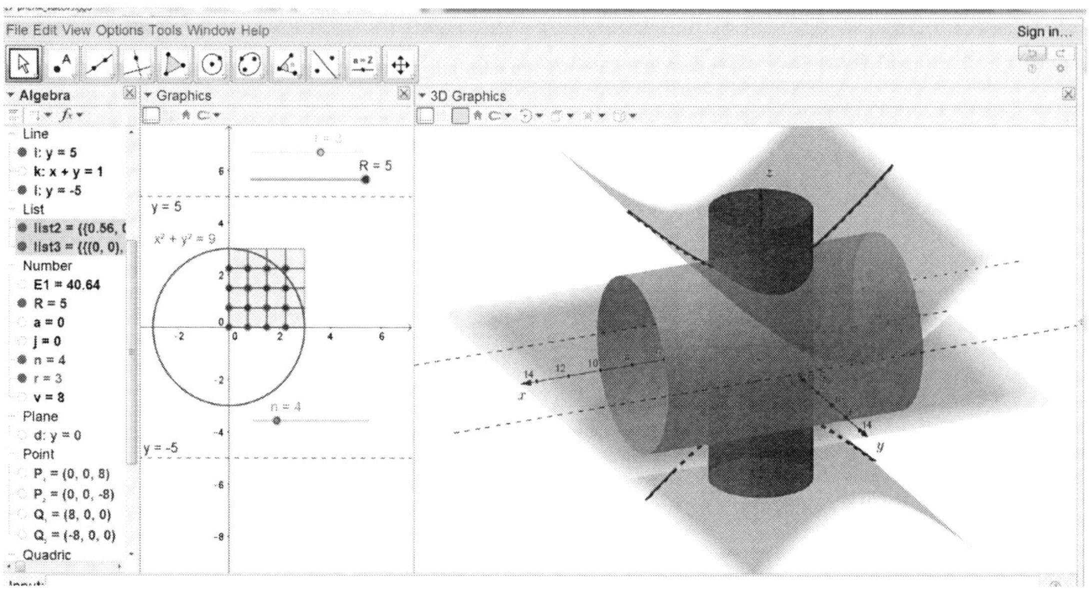

Fig. 2. Complex visualization of the Problem 2 in GeoGebra. The sliders r, R, n provide the interaction of the construction. The *list2* indicates the tesselation of D with the squares and the *list3* maps the points M_i in the bottom left vertex of each smaller square

In *Graphics* window we put sliders $0 < r \leq R < \frac{v}{2}$ and $n \in \{0, 1, 2, \ldots, 20\}$.
Using the tool *Sequence* we tesselate a square with the vertices $O\,[0, 0]$,$A\,[r, 0]$,$B\,[r, r]$, $C\,[0, r]$ into $n \times n$ congruent squares with the side $\frac{r}{n}$. The user can realize this tessellation by command

$$Sequence[Sequence[Polygon[(ir/n, (jr)/n), ((i+1)r/n, jr/n), 4], i, 0, n-1], j, 0, n-1].$$

It is evident that not all vertices ("left bottom vertices") of the sketched squares belong to the domain D.
We use the tool *Sequence* to visualize the points $M_i\,[x_{M_i}, y_{M_i}]$ for $0 \leq i \leq< n^2$ in *Graphic* window.
If $\sqrt{(x_{M_i})^2 + (y_{M_i})^2} \leq r$, then we compute the values $z_i = \sqrt{R^2 - (y_{M_i})^2}$ in *Spreadsheet* window, else we set up $z_i = 0$.

In Fig. 3 reader can see *Graphics* and *Spreadsheet* windows. The input from Graphics window is recorded into the cells *A1:B3*. The rows *6,7* represent sets, which elements can be used as coordinates of the points M_i are in rows *6,7*.
From this implies that the coordinates of the all appointed latticed points M_i are in the row *9* and the column *A*.

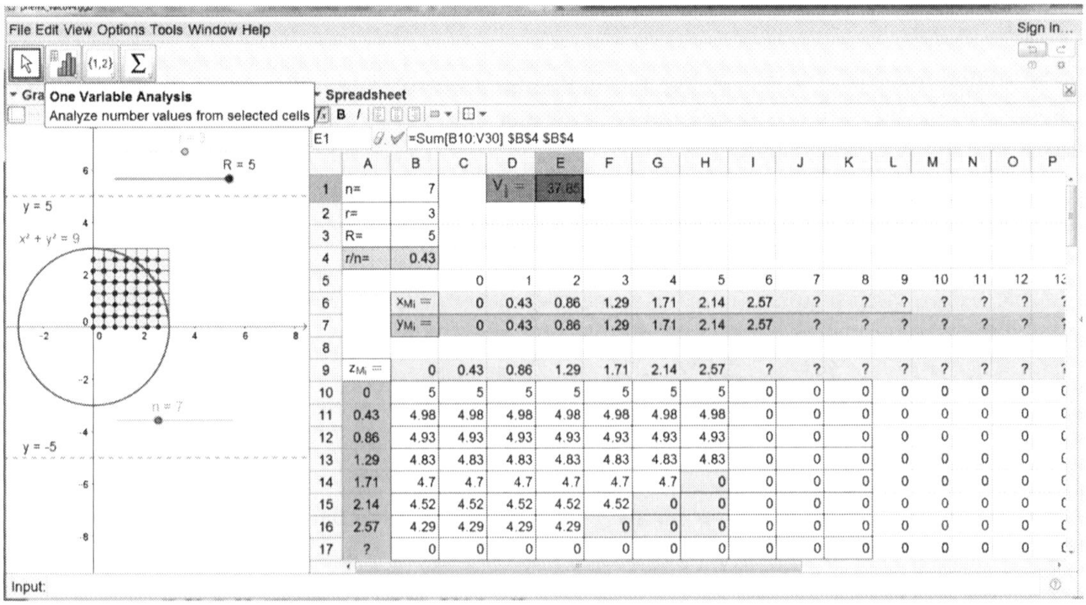

Fig. 3. The value of function $z = \sqrt{R^2 - y^2}$ in points M_i, which do not belong to domain D, is set up equal 0. It is indicated by yellow colour in Spreadsheet

The table in *B10:V30* contains the values of the function $z_i = f(M_i) = \sqrt{R^2 - (M_i)^2}$, if $M_i \in D$, otherwise is set up $f(z_i) = 0$.
According to idea of double integral, the volume of the selected partition of the given solid is computing like a sum of all values multiplied by factor $\left(\frac{r}{n}\right)^2$ and is displayed in the cell *E1*.
In the Tab. 1 the reader can find a few results for $n \in \{1, 2, 5, 10, 15, 20\}$.

r	R	n	$V_{\frac{1}{8}}$
3	5	1	45.00
3	5	2	43.96
3	5	5	41.38
3	5	10	37.82
3	5	15	36.45
3	5	20	35.74

Tab. 1

Remark. The gradual approximation for $n \longrightarrow \infty$ gives the value of the signed volume under the surface β in domain D. The results in Tab. 1 are only rough estimates of the real value for $n \leq 20$. The user / student or teacher can easy redefine n this parameter in Graphic window. The verification of the result can be also check out via [9]. *Online calculator computes the value $V_{\frac{1}{8}} \approx 33.6713$ for $r = 3$, $R = 5$.*

3 Discussion

The spatial imagination and the geometric imagination are important factors of general intelligence of personality of student and both greatly help to student to integrate in society. The forms of imaginations can be developed continually throughout life. The role of teachers of Mathematics is thus indispensable, because a teacher may significantly affect conditions, in which the spatial and geometric imagination should be to cultivated. [10]
In this paper presented solutions of the problems are surprising in many ways. First, the problems are very serious to spatial imagination and it is not easy to imagine the intersection of two cylinders of different radii. We conclude that some visualization via dynamical geometry software is essential in educational process. The linking of several computing environments enables to student to solve the tasks more complex with respect to interdisciplinary relations between Mathematics and Informatics and ones also allow to develop the mathematical competencies.
One could argue that these problems goes beyond the standard mathematics course for teacher's training education, about education of secondary school students not to speak. We note tkat in the theory of didactical constructivism are such problems acceptable criterion for testing an informal student's knowledge. [11]
This kind of tasks may be considered for the students like a motivating challenge and it means also some support of the constructivist approach to maths education, which fully provides ICT.

Acknowledgment

This paper was supported by project KEGA No. 016UKF-4/2016 MINE Slovak Republic titled *The Implementation of Constructivist Approaches to Mathematics Teaching with a Focus on Active Acquisition of Knowledge by Pupils in the Context of Bilingual Education.*

References

[1] Šedivý, O. (2005).: Vyučovanie stereometrie a rozvoj predstavivosti. In: *Rozvíjanie priestorovej a geometrickej predstavivosti.* Nitra: FPV UKF v Nitre, 2005. ISBN: 80-8050-683-1, s. 3-9

[2] Vidermanová, K., Viziová, A., Záhorská, J.: Analýza žiackych chýb pri konštrukcii rezov telies a možnosti ich odstránenia využitím Cabri 3D. In *Acta Mathematica 15*, UKF v Nitre, 2012, ISBN: 978-80-558-0135-3, p. 179-189

[3] Velichová, D.: Interactive maths with Geogebra. In. *International Journal of Emerging Technologies in Learning*, Vol. 6(2011), Special Issue 2, pp. 31-35

[4] Vaníček, J.: *Počítačové kognitivní technologie ve výuce geometrie.* Praha: PF UK v Prahe, 2009. ISBN: 978-80-7290-394-8, s. 212

[5] Gunzel, W. a kol.: *Integrace elektronických prostředí pro počítačem podporovanou výuku matematiky.* České Budějovice: JCU, 2012. ISBN: 978-80-7394-386-8, s. 187

[6] Robová, J.: Programy dynamické geometrie a jejich využití ve výuce stereometrie. In: *Potencial IKT v školskej matematike*, UK v Bratislave, 2009, ISBN: 978-80-223-2754-1, s. 64-71

[7] Natanson, I. P.: *Sčítaní nekonečně malých veličin.* Praha: SNTL v Praze, 1955. s. 72

[8] Bronštejn, I.N., Semenďajev, K.A.:*Príručka matematiky pre inžinierov a pre pracujúcich*, SVTL Bratislava, 1961, s. 748

[9] Double Integral Calculator. URL:http://www.wolframalpha.com/widgets/view.jsp?id=f5f3cbf14f4f5d6d2085bf2d0fb76e8a

[10] Tomková, V.: *Technická neverbálna komunikácia*. PF UKF v Nitre, 2013, ISBN: 978-80-558-0367-8, s. 204

[11] Hejný, M., Kuřina, F.: *Dítě, škola a matematika*. Praha: Portal, 2001, ISBN: 90-7178-581-4, s. 188

Current address

Vallo Dusan, RNDr. PhD.
Department of Mathematics, Faculty of Natural Sciences,
Constantine the Philosopher University in Nitra
Tr. A. Hlinku 1, 949 74 Nitra
Tel. number: +421 37 6408 690, e-mail: dvallo@ukf.sk

Slovak University of Technology in Bratislava
Faculty of Mechanical Engineering
16th Conference on Applied Mathematics **APLIMAT 2017**
Proceedings

NOTES ON PROOFS APPLYING ELEMENTARY METHODS

VARGA Marek (SK)

Abstract. Proofs of mathematical propositions are an essential part of scientific mathematics. In school mathematics, that is mathematics which is covered in school curricula, mathematical proofs are often neglected, mostly due to time demands. In this paper we aim to show that omitting proofs in school and university instruction is not the only way to save time in mathematics lessons. Another option is to provide learners with methods of mathematical reasoning which are as simple as possible.

Keywords: proofs, number series, Euler´s number, arithmetic mean, geometric mean

Mathematics Subject Classification: 97B40

1 Introduction

In recent years certain educational reforms have been carried out in Slovakia which has led to several changes. In relation to mathematics (and natural sciences) education it often seems that the main objective of those processes was to reduce the educational content as much as possible, and spice it with reduced number of lessons. Elementary and secondary school teachers are, thus, forced to turn mathematics into mere rules, theorems and formulas for bare algorithmic use in order to get a solution. Such "abuse of mathematics" is also present in tasks bearing lofty name "real-life related", although they often require only using the formula for n-percent discount calculation, or calculation of a loan instalment, and that does not necessarily demand understanding of the mathematical point.

However, it is believed that mathematics should be one of the school subjects (if not the only one) which contribute to the development of learners' logical thinking. The purpose of mathematics education is to teach learners to use and combine their acquired knowledge in order to solve various tasks. Respecting the principle "from simple to complex", in mathematics lessons learners should first be exposed to exercise-like tasks, and only then to problems which learners have never encountered.

Reasoning and verifying the validity of a formula is frequently hidden in mathematical proofs. While asked for proofs in mathematics lessons, learners are, actually, educated not to automatically believe anything to authorities, and to request explanations and justification of proposed assertions.

Yet, this seems to be – we dare to say – an absolutely undesirable personal quality in modern society, hierarchical occupations, pre-election meetings and the like.

Let us better get back to mathematics education. Due to the scarcity of time it is the mathematical proofs which are usually omitted, even ignored in primary and secondary education. Similarly and consequently, proofs are getting neglected at universities, too. In the following sections we propose another way to prove mathematical theorems, i.e. besides applying traditional methods we show proofs of selected propositions using various elementary methods.

2 Theorems and proofs

Task A. Investigate the convergence of the harmonic series.

Solution 1. In the following we show a solution using integral calculus. It can be easily spotted that for all $x \in \left\langle 0; \frac{1}{n} \right\rangle$ it holds that $1 \geq \frac{1}{1+x}$. Then, however, it also holds (applying the inequalities valid for definite integrals) $\int_0^{\frac{1}{n}} 1\,dx \geq \int_0^{\frac{1}{n}} \frac{1}{1+x}\,dx$, which implies $\frac{1}{n} \geq \ln\left(1+\frac{1}{n}\right)$.

Let us consider series $\sum_{n=1}^{\infty} \ln\left(1+\frac{1}{n}\right)$. For the nth term of its sequence of partials sums σ_n it holds that $\sigma_n = \ln 2 + \ln\frac{3}{2} + \ln\frac{4}{3} + \ldots + \ln\frac{n+1}{n} = \ln(n+1)$. Now, it is apparent that $\lim_{n\to\infty} \sigma_n = \infty$, i.e. $\sum_{n=1}^{\infty} \ln\left(1+\frac{1}{n}\right) = \infty$. This implies that $\sum_{n=1}^{\infty} \frac{1}{n} = \infty$, i.e. the harmonic series diverges.

Solution 2. Here we use bare comparison of fractions. Assume that the harmonic series converges; let its sum be denoted by s. Then, it holds that

$$s = 1 + \frac{1}{2} + \frac{1}{3} + \frac{1}{4} + \frac{1}{5} + \frac{1}{6} + \ldots > \left(\frac{1}{2}+\frac{1}{2}\right) + \left(\frac{1}{4}+\frac{1}{4}\right) + \left(\frac{1}{6}+\frac{1}{6}\right) + \ldots = 1 + \frac{1}{2} + \frac{1}{3} + \ldots = s,$$

which is a contradiction.

Task B. Investigate the convergence of Riemann series.

Solution 1. By Riemann (sometimes also Dirichlet) series we mean number series in form $\sum_{n=1}^{\infty} \frac{1}{n^{\alpha}}$, where $\alpha > 0$. Their convergence can be investigated by means of Cauchy integral criterion. It means that first we calculate improper integral

$$\int_1^{\infty}\frac{1}{x^{\alpha}}dx=\lim_{t\to\infty}\int_1^{t}\frac{1}{x^{\alpha}}dx=\lim_{t\to\infty}\left[\frac{1}{(1-\alpha)x^{\alpha-1}}\right]_1^t=\frac{1}{1-\alpha}\lim_{t\to\infty}\left[\frac{1}{t^{\alpha-1}}-1\right]=\begin{cases}\frac{1}{\alpha-1}; & \alpha>1\\ \infty; & 0<\alpha<1\end{cases}.$$

(For $\alpha=1$ it has been calculated above.) We can see that for $\alpha>1$ the integral in question (and therefore the corresponding series as well) converges, whereas for $0<\alpha<1$ the integral (and the series) diverges.

Solution 2. Once more we apply comparison of fractions. First, let $\alpha>1$. Let us look into sum $\sigma_n=\sum_{m=1}^{n}\frac{1}{m^{\alpha}}$. It holds that $\frac{1}{2^{\alpha}}+\frac{1}{3^{\alpha}}<\frac{2}{2^{\alpha}}=\frac{1}{2^{\alpha-1}}$; $\frac{1}{4^{\alpha}}+\frac{1}{5^{\alpha}}+\frac{1}{6^{\alpha}}+\frac{1}{7^{\alpha}}<\frac{4}{4^{\alpha}}=\frac{1}{4^{\alpha-1}}$; etc.

Therefore, the sum 2^{p-1} of the terms of this series can be bounded as follows:

$$\frac{1}{1^{\alpha-1}}+\frac{1}{2^{\alpha-1}}+\frac{1}{4^{\alpha-1}}+\frac{1}{8^{\alpha-1}}+\ldots+\frac{1}{2^{(p-1)(\alpha-1)}}=\frac{\left(2^{\alpha-1}\right)^{p-1}+\left(2^{\alpha-1}\right)^{p-2}+\ldots+2^{\alpha-1}+1}{2^{(p-1)(\alpha-1)}}=$$

$$=\frac{1}{2^{(p-1)(\alpha-1)}}\cdot\frac{\left(2^{\alpha-1}\right)^{p}-1}{2^{\alpha-1}-1}<\frac{\left(2^{\alpha-1}\right)^{p}}{\left(2^{\alpha-1}\right)^{p-1}\left(2^{\alpha-1}-1\right)}=\frac{1}{1-2^{1-\alpha}}.$$ Sequence $\{\sigma_n\}_{n=1}^{\infty}$ is increasing (since its terms are positive numbers), and it has just been shown that it is also bounded above. It means that the sequence converges; thus, series $\sum_{n=1}^{\infty}\frac{1}{n^{\alpha}}$ has a sum for values $\alpha>1$.

In case $0<\alpha<1$ it is just necessary to realize that $\frac{1}{n^{\alpha}}>\frac{1}{n}$ and recall the divergence of the harmonic series.

Task C. Show that $e=1+\frac{1}{1!}+\frac{1}{2!}+\ldots+\frac{1}{n!}+\mathscr{R}_{n+1}$, where $\mathscr{R}_{n+1}$ represents a calculation error.

Solution 1. For those who are familiar with derivation of exponential function it seems to be wise and succinct to apply Taylor (Maclaurin) expansion of function *f*, i.e.

$$f(x)=f(c)+\frac{f'(c)}{1!}(x-c)+\frac{f''(c)}{2!}(x-c)^2+\ldots+\frac{f^{(n)}(c)}{n!}(x-c)^n+\mathscr{R}_{n+1},$$

where $\mathscr{R}_{n+1}$ is the remainder after the nth term. Naturally, for all $n\in\mathbb{N}$ it holds that $\left[\left(e^x\right)^{(n)}\right]_{x=0}=1$; therefore $e^x=1+\frac{1}{1!}x+\frac{1}{2!}x^2+..+\frac{1}{n!}x^n+\mathscr{R}_{n+1}$. Substituting $x=1$ we obtain $e=1+\frac{1}{1!}+\frac{1}{2!}+\ldots+\frac{1}{n!}+\mathscr{R}_{n+1}$.

Solution 2. As we have already noted, we can hardly make the proof shorter than the previous one; yet, a solution where using the binomial expansion suffices has its special glamour, too. Let $n,k\in\mathbb{N};\ 1<k<n$. From the binomial theorem it follows that

$$\left(1+\frac{1}{n}\right)^n=1+\binom{n}{1}\frac{1}{n}+\binom{n}{2}\frac{1}{n^2}+\ldots+\binom{n}{k}\frac{1}{n^k}+\ldots+\binom{n}{n}\frac{1}{n^n}=$$

$$=1+n\frac{1}{n}+\frac{n(n-1)}{2!}\frac{1}{n^2}+\frac{n(n-1)(n-2)}{3!}\frac{1}{n^3}+\ldots+\frac{n(n-1)\ldots(n-k+1)}{k!}\frac{1}{n^k}+\ldots+\frac{n(n-1)\ldots(n-n+1)}{n!}\frac{1}{n^n}=$$

$$=1+1+\frac{1}{2!}\left(1-\frac{1}{n}\right)+\frac{1}{3!}\left(1-\frac{1}{n}\right)\left(1-\frac{2}{n}\right)+\ldots+\frac{1}{k!}\left(1-\frac{1}{n}\right)\left(1-\frac{2}{n}\right)\ldots\left(1-\frac{k-1}{n}\right)+\ldots+\frac{1}{n!}\left(1-\frac{1}{n}\right)\ldots\left(1-\frac{n-1}{n}\right)=$$

$$=\left[1+1+\frac{1}{2!}\left(1-\frac{1}{n}\right)+\ldots+\frac{1}{k!}\left(1-\frac{1}{n}\right)\left(1-\frac{2}{n}\right)\ldots\left(1-\frac{k-1}{n}\right)\right]+\frac{1}{k!}\left(1-\frac{1}{n}\right)\left(1-\frac{2}{n}\right)\ldots\left(1-\frac{k-1}{n}\right)\times$$

$$\times\left\{\frac{1-\frac{k}{n}}{k+1}+\frac{\left(1-\frac{k}{n}\right)\left(1-\frac{k+1}{n}\right)}{(k+1)(k+2)}+\ldots+\frac{\left(1-\frac{k}{n}\right)\left(1-\frac{k+1}{n}\right)\cdot\ldots\cdot\left(1-\frac{n-1}{n}\right)}{(k+1)(k+2)\cdot\ldots\cdot n}\right\}.$$

The first addend is marked by square brackets; next, we have an estimate of the second addend

$$\frac{1}{k!}\left(1-\frac{1}{n}\right)\left(1-\frac{2}{n}\right)\ldots\left(1-\frac{k-1}{n}\right)\left\{\frac{\left(1-\frac{k}{n}\right)}{k+1}+\frac{\left(1-\frac{k}{n}\right)\left(1-\frac{k+1}{n}\right)}{(k+1)(k+2)}+\ldots+\frac{\left(1-\frac{k}{n}\right)\left(1-\frac{k+1}{n}\right)\cdot\ldots\cdot\left(1-\frac{n-1}{n}\right)}{(k+1)(k+2)\cdot\ldots\cdot n}\right\}<$$

$$<\frac{1}{k!}\left\{\frac{1}{k+1}+\frac{1}{(k+1)^2}+\ldots+\frac{1}{(k+1)^{n-k}}\right\}<\frac{1}{k!}\left\{\frac{1}{k+1}+\frac{1}{(k+1)^2}+\ldots\right\}=\frac{1}{k!}\frac{\frac{1}{k+1}}{1-\frac{1}{k+1}}=\frac{1}{k\cdot k!}.$$

Subsequently we obtain

$$0<\left(1+\frac{1}{n}\right)^n-\left[1+1+\frac{1}{2!}\left(1-\frac{1}{n}\right)+\frac{1}{3!}\left(1-\frac{1}{n}\right)\left(1-\frac{2}{n}\right)+\ldots+\frac{1}{k!}\left(1-\frac{1}{n}\right)\left(1-\frac{2}{n}\right)\ldots\left(1-\frac{k-1}{n}\right)\right]<\frac{1}{k\cdot k!}.$$

If we look at the middle part as a function of n, applying limit laws for $n\to\infty$ we get

$$0<e-\left[1+\frac{1}{1!}+\frac{1}{2!}+\frac{1}{3!}+\ldots+\frac{1}{k!}\right]<\frac{1}{k\cdot k!}.$$

From the latter it follows that sum $1+\frac{1}{1!}+\frac{1}{2!}+\frac{1}{3!}+...+\frac{1}{k!}$ can be used for approximation of e with accuracy $\frac{1}{k \cdot k!}$.

Task D. Prove the AM-GM inequality.

Solution 1. AM-GM inequality denotes the relationship between the arithmetic and geometric mean of n positive numbers, i.e. $AM = \frac{x_1 + x_2 + \dots}{n}$ ... $- \sqrt[n]{x_1 \cdot x_2 \cdot \dots}$ ... M. If we approach this problem in terms of differential calculus of multivariable functions, it suffices to find the extremum of function $w = \sqrt[n]{x_1 \cdot x_2 \cdot \dots}$ at condition $x_1 + x_2 + ... + x_n = a$. The corresponding Lagrange function is $L = \sqrt[n]{x_1 x_2 \dots}$ ... . For $j = 1, 2, ..., n$ it holds that $\frac{\partial L}{\partial x_j} = \frac{1}{n}\frac{x_1 x_2 \dots x_{j-1} x_{j+1} \dots x_n}{(x_1 x_2 \dots} + \lambda = \frac{1}{n}\frac{u}{x_j} + \lambda = 0$; hence $w = -n\lambda x_j$. Therefore, numbers $x_1,\ x_2, ..., x_n$ must be equal, and with respect to the condition we obtain $x_1 = x_2 = \dots$ ... $\frac{a}{n}$. At stationary point $\Lambda\left[\frac{a}{n};\ \frac{a}{n};\ \dots\right]$ function L assumes maximum (i.e. function w assumes constraint maximum), i.e. $\sqrt[n]{x_1 \cdot x_2 \cdot \dots}$ ... $\frac{\cdot x_2 + \dots}{n}$.

Solution 2. This time we approach the given problem applying the method of mathematical induction, more precisely the method of backward induction. First, we show that the given relationship holds for infinitely many natural numbers. The true inequality $\left(\sqrt{a_1} - \sqrt{a_2}\right)^2 \geq 0$ implies that $\frac{a_1 + a_2}{2} \geq \sqrt{a_1 a_2}$, which means that the assertion holds for $n = 2$. Next we consider only even natural numbers, i.e. $n = 2k,\ k \in \mathbb{N}$. Using the just stated and induction assumption we obtain

$$\frac{a_1 + a_2 + ... + a_{2k}}{2k} = \frac{1}{k}\left(\frac{a_1 + a_2}{2} + \frac{a_3 + a_4}{2} + ... + \frac{a_{2k-1} + a_{2k}}{2}\right) \geq$$

$$\geq \frac{\sqrt{a_1 a_2} + \sqrt{a_3 a_4} + ... + \sqrt{a_{2k-1} a_{2k}}}{k} \geq \sqrt[k]{\sqrt{a_1 a_2}\sqrt{a_3 a_4} \dots \sqrt{a_{2k-1} a_{2k}}} = \sqrt[2k]{a_1 a \dots a_{2k}}\ .$$

Assume that AM-GM inequality holds for certain $k \in \mathbb{N}$, i.e. $\frac{x_1 + x_2 + ... + x_k}{k} \geq \sqrt[k]{x_1 x_2 ... x_k}$; we aim to show that the inequality also holds for $n = k-1$. Replace $x_1 = a_1$, $x_2 = a_2$, ..., $x_{k-1} = a_{k-1}$; $x_k = \frac{a_1 + a_2 + ... + a_{k-1}}{k-1}$. Then

$$\frac{a_1 + a_2 + ... + a_{k-1} + \frac{a_1 + a_2 + ... + a_{k-1}}{k-1}}{k} \geq \sqrt[k]{a_1 \cdot a_2 \cdot ... \cdot a_{k-1} \cdot \frac{a_1 + a_2 + ... + a_{k-1}}{k-1}} \Leftrightarrow$$

$$\Leftrightarrow \frac{a_1 + a_2 + ... + a_{k-1}}{k-1} \geq \sqrt[k]{a_1 \cdot a_2 \cdot ... \cdot a_{k-1} \cdot \frac{a_1 + a_2 + ... + a_{k-1}}{k-1}} \Leftrightarrow$$

$$\Leftrightarrow \left(\frac{a_1 + a_2 + ... + a_{k-1}}{k-1}\right)^k \geq a_1 \cdot a_2 \cdot ... \cdot a_{k-1} \cdot \frac{a_1 + a_2 + ... + a_{k-1}}{k-1} \Leftrightarrow$$

$$\Leftrightarrow \left(\frac{a_1 + a_2 + ... + a_{k-1}}{k-1}\right)^{k-1} \geq a_1 \cdot a_2 \cdot ... \cdot a_{k-1} \Leftrightarrow \frac{a_1 + a_2 + ... + a_{k-1}}{k-1} \geq \sqrt[k-1]{a_1 \cdot a_2 \cdot ... \cdot a_{k-1}}.$$

3 Conclusion

Explanations, reasoning and proving belong to daily life, and are also essential in scientific and school mathematics. Unfortunately, at primary and secondary schools mathematical reasoning is often omitted – even in case of the most influential theorems, mostly due to scarcity of time. In university mathematics education mathematical proofs are perceived as the most difficult part of examinations, and that is why it seems more comfortable to "ease off". In this paper we aim to show that it is not a proper decision to omit proofs of important propositions, and it is better to show learners at least proofs where only elementary mathematical methods are applied instead.

References

[1] BUDINSKÝ, B., CARVÁT, J. *Matematika I.* Praha: SNTL/Alfa, 1987. 504 s.

[2] GREBENČA, M. K., NOVOSELOV, S. I. *Kurs matematičeskogo analiza 1.* Moskva: Učpedgiz, 1951. 543 s.

[3] IGNATJEVOVÁ, A. V. *Kurs vyššej matematiky.* Moskva: Vyssaja škola, 1964. 684 s.

[4] PISKUNOV, N. S. *Differencial'noje i integral'noje isčislenija.* Moskva: GITTL, 1957. 844 s.

[5] URL: http://web.science.upjs.sk/jozefdobos/wp-content/uploads/2012/02/2002a.pdf

[6] URL: https://brilliant.org/discussions/thread/proof-of-the-arithmetic-mean-geometric-mean/

Acknowledgement

The author gratefully acknowledges Mgr. Zuzana Naštická for technical translation to English.

Current address

Varga Marek, PaedDr. PhD.
Department of mathematics, FPV UKF
Tr. A. Hlinku 1, 94974 Nitra, Slovak Republic
E-mail: mvarga@ukf.sk

Slovak University of Technology in Bratislava
Faculty of Mechanical Engineering
16th Conference on Applied Mathematics **APLIMAT 2017**
Proceedings

INFINITE SERIES IN CIRCLES

VARGOVÁ Michaela (CZ)

Abstract. In our opinion, one of the obstacles that students face in relation to the concepts of convergence is the confusion and misconception of terms infinite and unbounded. In the paper we show examples of infinite series whose convergence is easily seen from the figures and can be rigorously justified, for instance, by using integral test or direct comparison test. These visual representations may help students overcome some difficulties related to the thorough understanding of this concept.

Keywords: infinite series, convergence, visualisation

Mathematics Subject Classification: Primary 40A05, 97I30

1 Introduction

Infinity, either in its potential or actual form, is undoubtedly one of the central concepts in mathematics. In various mathematical disciplines infinity is often approached from different points of view. In each of the disciplines different attributes and manifestations of infinity are brought to the fore. For instance, infinity can be considered in terms of set cardinality or order, infinite processes or convergence. Understanding the concept of infinity, its particular manifestations or forms, assumes considerable level of cognitive maturity. It is primarily due to this requirement why in education the concept of infinity is a source of many obstacles and difficulties at the acquisition and intimate comprehension of various mathematical infinity-related concepts. That is, apparently, the reason why so many educational researchers focus on the issue of the teaching and learning about the mathematical infinity.

Learners first encounter infinity in “explicit form” (usually as early as during their secondary school studies) in connection with the concept of infinite number sequence and the sum of infinite (geometric) number series. Practical experience of many pedagogues and inferences of various researches (Sierpinska, 1987, Williams, (1991); Monaghan, (2001)) indicate numerous difficulties which learners face (mostly learners at secondary schools and first-year university students, i.e. learners who are on certain level of mental development) in relation to intimate comprehension of the sum of infinite series. As suggested by several studies (Sierpinska, (1987); Williams, (1991); Monaghan, (2001)) and many practicing teachers, and which is also in accordance with our own experience, in terms of learning process the sum of infinite series belongs to difficult and problematic concepts. Many educational researches investigating learners' conceptions of infinity and infinite series (Tall, Vinner (1981); Sierpinska, (1987); Williams, (1991); Monaghan, (2001)) indicate that learners at that age are not able yet to accept that an infinite series can add up to a finite sum.

We suppose that in relation to the task to calculate the sum of an infinite series most difficulties for learners result from the necessity to think of the set of the series terms which "should be counted" as an actually infinite set. Yet, the intuitive learners' conceptions are often bound only to the comprehension of the potentially infinite process embraced in this task.

Another obstacle which learners face in connection with this issue is, in our opinion, mistaking the concepts of infinite and unbounded , which is supported by learners' assertions such as "...but when I add another and another number, it must increase to infinity...". That is to say, the learners' conception corresponds with the belief of Zeno[1] and the ancient Greeks that a sum of infinitely many addends cannot be a finite number. One way how to show such learners (already before providing them with basic information about the sum of infinite series) that their intuitive conceptions are not correct, and, in fact, a sum of infinitely many positive numbers can be bounded and the result is not necessarily "the infinity", is employment of appropriate illustrations and visualisations, sometimes referred to as "proofs without words".

In [1] we tried to highlight the fact that employing graphical representations of the sum of selected infinite geometric series (with both positive and negative quotients) while teaching learners about the sum of infinite series (particularly as a part of propaedeutic of the concept) allows educators to effectively help learners in facing the above mentioned obstacles. Moreover, designing the visual representations of sums of infinite series as dynamical applets enabled us to use also such "proofs without words" in the teaching whose interpretation would be quite plodding if static pictures had been applied instead. When employing "proofs without words" for teaching about infinite series convergence and sums of infinite series we are, naturally, not restricted to use only graphical representations of infinite geometries series sums. Illustrative graphical representations can be also applied in order to "persuade" learners (on certain intuitive level) of the convergence of several other types of infinite series, such as telescoping or Riemann series $\sum_{n=1}^{\infty}\frac{1}{n^{\alpha}}$, $\alpha > 1$. It can be done even before providing learners with the mathematical strategies and tools which serve for rigorous analytical proofs of the convergence of the series.

2 The graphical representations of some convergent infinite series

2.1 Convergence of the infinite series $\frac{1}{1\cdot 2}+\frac{1}{2\cdot 3}+\frac{1}{3\cdot 4}+\ldots+\frac{1}{n(n+1)}+\ldots=\sum_{n=1}^{\infty}\frac{1}{n(n+1)}$

Consider two congruent circles $k_1(S_1, r)$ and $k_2(S_2, r)$ which are externally tangent at point T (Fig.1). Let line t be the common tangent of the two circles, touching circle k_1 at point P_1 and circle

[1] Zeno of Elea (490 – 430 BC), the author of well-known paradoxes, so called Zeno's aporias, of which the most famous are Dichotmy, and Achilles and the tortoise.

k_2 at point P_2. Consider the sequence of such circles $l_n(O_n, r_n)$, $n \in N$, that circle $l_1(O_1, r_1)$ touches line t and is externally tangent to two circles $k_1(S_1, r)$, $k_2(S_2, r)$, and circle $l_2(O_2, r_2)$ is externally tangent to circle $l_1(O_1, r_1)$ and to both circles $k_1(S_1, r)$ and $k_2(S_2, r)$, ..., circle $l_n(O_n, r_n)$ is externally tangent to circle $l_{n-1}(O_{n-1}, r_{n-1})$ and to circles $k_1(S_1, r)$ and $k_2(S_2, r)$. A circle tangent to given three circles can be constructed under any circumstances, which, however, means that the construction of circles $l_n(O_n, r_n)$, $n \in N$, can be continued indefinitely. Let d_n denote the diameters of circles $l_n(O_n, r_n)$, $n \in N$. Then, the diameters $d_1, d_2, ..., d_n, ...$ of circles $l_1, l_2, ..., l_n, ...$ form an infinite sequence, and the sum of the diameters $d_1 + d_2 + ... + d_n + ...$ represents an infinite number series (Fig. 1).

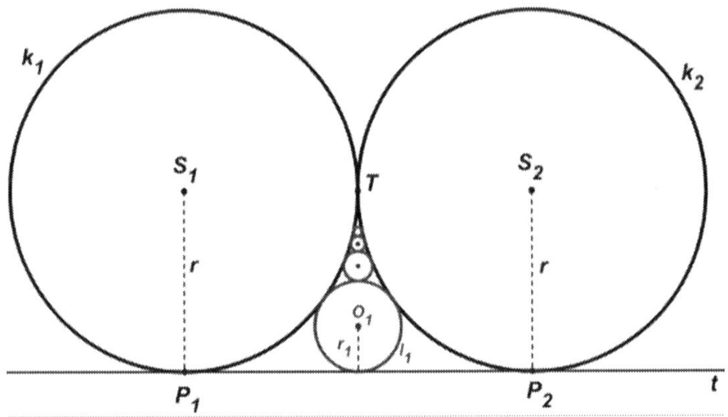

Fig. 1

The centres O_n of circles $l_n(O_n, r_n)$ obviously lie on the common tangent m of circles $k_1(S_1, r)$, $k_2(S_2, r)$ constructed through point T (Fig. 1). Let M be the contact point of circle l_1 and line t. This implies that line segments MT and P_2S_2 are congruent and the nth term of the sequence of partial sums $\{s_n\}_{n=1}^{\infty}$ of series $\sum_{n=1}^{\infty} d_n$, i.e. $s_n = d_1 + d_2 + ... + d_n$, represents the length of segment MO_n. The sequence of partial sums $\{s_n\}_{n=1}^{\infty}$ is increasing (since $d_1 < d_1 + d_2 < d_1 + d_2 + d_3 < ...$) and at the same time for all $n \in N$ it holds that $s_n = d_1 + d_2 + ... + d_n = |MO_n| < P_2S_2 = r$, i.e. sequence $\{s_n\}_{n=1}^{\infty}$ is increasing and above bounded by real number r. As already stated, the construction of circles $l_n(O_n, r_n)$ can be continued indefinitely. Hence the difference $r - s_n = r - (d_1 + d_2 + ... + d_n)$ "can be made" arbitrarily small, i.e. $\lim_{n\to\infty} s_n = \lim_{n\to\infty}(d_1 + d_2 + ... + d_n) = r$, therefore

$$\sum_{n=1}^{\infty} d_n = d_1 + d_2 + ... + d_n + ... = r.$$

We have just shown that the considered infinite series $\sum_{n=1}^{\infty} d_n$ is convergent and its sum is number r.
Now it only remains to determine the form of its nth term d_n; in order to do so, however, it suffices to apply elementary mathematical knowledge.
Let T_n denote the contact point of circles l_n and l_{n+1}. Let V_n be the orthographic projection of the centre O_n of circle $l_n(O_n, r_n)$ on the segment P_2S_2. Let U_n be the orthographic projection of point T_n on the segment P_2S_2 (Fig. 2).

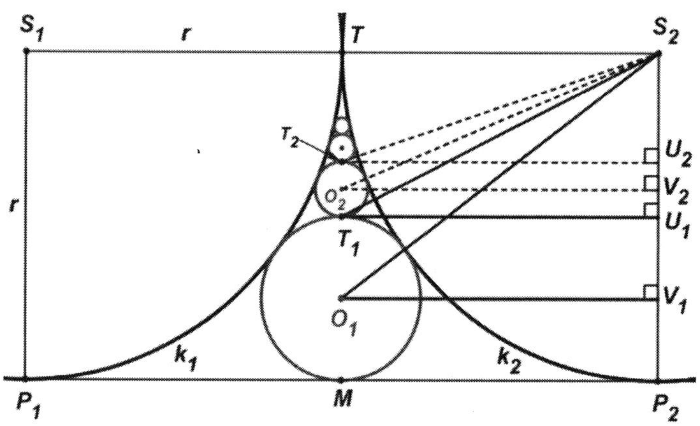

Fig. 2

Applying the Pythagorean theorem in triangle $O_1V_1S_2$ (Fig. 2) we obtain

$$r^2 + (r - r_1)^2 = (r + r_1)^2,$$

which implies that for the radius r_1 of circle $l_1(O_1, r_1)$ it holds $r_1 = \frac{r}{4}$, and $d_1 = \frac{r}{2}$. Then for the length of segment S_2U_1 it holds $|S_2U_1| = r - d_1 = \frac{r}{2}$. Similarly, in triangle $O_2V_2S_2$ it holds

$$r^2 + (|S_2U_1| - r_2)^2 = (r + r_2)^2 \text{ , which is } r^2 + \left(\frac{r}{2} - r_2\right)^2 = (r + r_2)^2.$$

Hence, after rearrangements we obtain $r_2 = \frac{r}{12}$, ergo $d_2 = \frac{r}{6} = \frac{r}{2 \cdot 3}$. Consequently, for the length of segment S_2U_2 it holds

$$|S_2U_2| = \frac{r}{2} - d_2 = \frac{r}{3}.$$

Similarly, in triangle $O_3V_3S_2$ it holds $r^2+\left(\frac{r}{3}-r_3\right)^2=(r+r_3)^2$.

The latter equality implies $r_3=\frac{r}{24}$, ergo $d_3=\frac{r}{12}=\frac{r}{3\cdot 4}$. Therefore, for the length of segment S_2U_3 it holds that $|S_2U_3|=\frac{r}{3}-d_3=\frac{r}{4}$.

Then, apparently, for all $n\in N$ for d_n and the length of segment S_2U_n it holds: $d_n=\frac{r}{n(n+1)}$, $|S_2U_n|=\frac{r}{n+1}$.

This hypothesis can be easily proved by means of mathematical induction.

It has been already shown, that series $\sum_{n=1}^{\infty}d_n$ is convergent and

$$\sum_{n=1}^{\infty}d_n=d_1+d_2+\ldots+d_n+\ldots=r.$$

Hence,

$$\sum_{n=1}^{\infty}d_n=d_1+d_2+\ldots+d_n+\ldots=r\left(\frac{1}{1\cdot 2}+\frac{1}{2\cdot 3}+\frac{1}{3\cdot 4}+\ldots+\frac{1}{n(n+1)}+\ldots\right)=r,$$

which implies that

$$\frac{1}{1\cdot 2}+\frac{1}{2\cdot 3}+\frac{1}{3\cdot 4}+\ldots+\frac{1}{n(n+1)}+\ldots=1.$$

The validity of our result based on geometrical considerations can be further justified. For the nth term of the series $\sum_{n=1}^{\infty}d_n$ it apparently holds that $d_n=\frac{1}{n(n+1)}=\frac{1}{n}-\frac{1}{n+1}$. Therefore, for the nth term of the sequence of partial sums it holds that

$$s_n=a_1+a_2+\ldots+a_n=\frac{1}{1\cdot 2}+\frac{1}{2\cdot 3}+\ldots+\frac{1}{n(n+1)}=\left(1-\frac{1}{2}\right)+\left(\frac{1}{2}-\frac{1}{3}\right)+\ldots+\left(\frac{1}{n}-\frac{1}{n+1}\right)=1-\frac{1}{n+1}.$$

Then $S=\lim_{n\to\infty}s_n=\lim_{n\to\infty}\left(1-\frac{1}{n+1}\right)=1$, which implies

$$\frac{1}{1\cdot 2}+\frac{1}{2\cdot 3}+\frac{1}{3\cdot 4}+\ldots+\frac{1}{n(n+1)}+\ldots=\sum_{n=1}^{\infty}\frac{1}{n(n+1)}=1.$$

Note. The same result can be obtained with the use of Fig. 3, based on the similarity of triangles AD_nP_n, P_nBC, where $n\in N$.

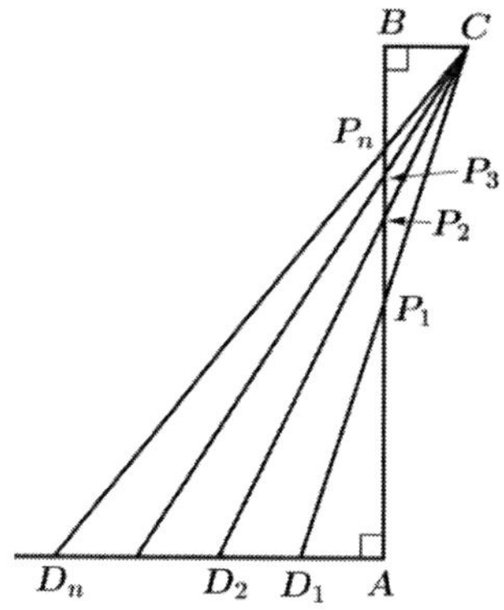

Fig. 3

2.2 Convergence of the infinite series $1+\frac{1}{2^4}+\frac{1}{3^4}+\frac{1}{4^4}+\ldots+\frac{1}{n^4}+\ldots=\sum_{n=1}^{\infty}\frac{1}{n^4}$

Consider two congruent circles $k_1(S_1,r)$ and $k_2(S_2,r)$ which are externally tangent at point T (Fig. 4). Let t be their common tangent line touching circle k_1 at point P_1 and circle k_2 at point P_2. Construct circle $k_3(S_3,r_3)$ which is externally tangent to circles $k_1(S_1,r_1)$, $k_2(S_2,r_2)$ and to line t; next, construct circle $k_4(S_4,r_4)$ externally tangent to circles $k_1(S_1,r_1)$, $k_3(S_3,r_3)$ and to the tangent line t, etc. Since circle $k_3(S_3,r_3)$ is constructed as tangent to circles k_1 and k_2, and circle k_4 as tangent to circles k_3 and k_2, etc.

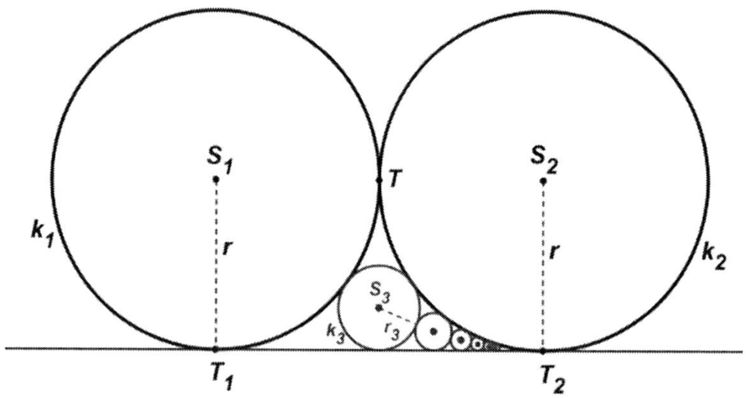

Fig. 4

Apparently, construction of circles $k_n(S_n,r_n)$ can be continued indefinitely, therefore the sum of the areas of circles $k_n(S_n,r_n)$ represents the infinite number series $\pi\sum_{n=3}^{\infty}(r_n)^2$. This series is, however, convergent, since its sum is smaller than number $\frac{r^2}{2}(4-\pi)$, which is the area of the planar figure determined by the line segment T_1T_2 and arcs T_1T, T_2T. Let us now express the lengths of radii r_n of circles $k_n(S_n, r_n)$, where $n \in N, n \geq 3$. Let T_n be the contact points of circles $k_n(S_n, r_n)$ with line t. Let Z_n, where $n \in N$, $n \geq 3$ be the orthogonal projections of the centres S_n, $n \in N$, $n \geq 3$ of circles k_n, $n \in N$, $n \geq 3$ on the line S_2T_2.

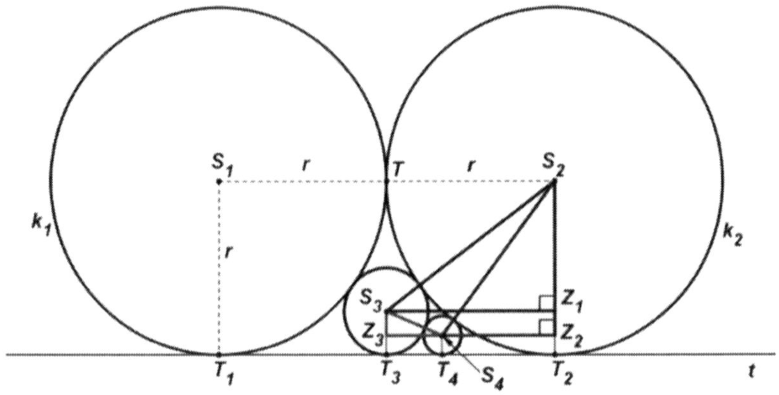

Fig. 5

In triangle $S_2Z_1S_3$ (Fig. 5) based on the Pythagorean theorem it holds that $r^2 + (r - r_3)^2 = (r + r_3)^2$, which implies $r_3 = \frac{r}{4}$. Also, Fig. 5 shows that for the distances between the contact points of circles $k_2(S_2, r)$, $k_3(S_3, r_3)$, $k_4(S_4, r_4)$ it holds $|T_3T_4| + |T_2T_4| = |T_2T_3|$. Applying the Pythagorean theorem in triangles $S_3Z_2S_4$, $S_2Z_3S_4$, $S_2Z_1S_3$ we obtain $|T_3T_4| = 2\sqrt{r_3r_4}$, $|T_2T_4| = 2\sqrt{r_4r}$, ergo $|T_3T_4| + |T_2T_4| = 2\left(\sqrt{r_3r_4} + \sqrt{r_4r}\right) = r$. Hence, after substituting $r_3 = \frac{r}{4}$ for r_4 we obtain $r_4 = \frac{r}{9}$. Consider now circles $k_2(S_2, r), k_4(S_4, r_4)$, $k_5(S_5, r_5)$. Again, it is evident that for the distances of their contact points with the tangent line t it holds $|T_4T_5| + |T_5T_2| = |T_4T_2|$.

Similarly to the previous case, it is easily obtained that $|T_4T_5| = 2\sqrt{r_4r_5}$, $|T_5T_2| = 2\sqrt{rr_5}$, $|T_4T_2| = 2\sqrt{rr_4}$, and then $r_5 = \frac{r}{16}$. Based on the latter we can assume that for all $n \in N$ it holds that $r_n = \frac{r}{(n-1)^2}$. This hypothesis can be easily justified by means of mathematical induction.

The sum of the areas of the infinite sequence of circles $k_n(S_n, r_n)$, where $n \in N, n \geq 3$ is the graphical representation of the infinite series

$$\pi r^2 \left(\frac{1}{2^4} + \frac{1}{3^4} + \frac{1}{4^4} + \ldots \right) = \pi r^2 \sum_{n=2}^{\infty} \frac{1}{n^4}.$$

Based on the above stated, it holds that $\pi r^2\left(\frac{1}{2^4}+\frac{1}{3^4}+\frac{1}{4^4}+...\right)=\pi r^2\sum_{n=2}^{\infty}\frac{1}{n^4}<\frac{r^2}{2}(4-\pi)$, ergo

$$\frac{1}{2^4}+\frac{1}{3^4}+\frac{1}{4^4}+...=\sum_{n=2}^{\infty}\frac{1}{n^4}<\frac{4-\pi}{2\pi}\text{, resp. }1+\frac{1}{2^4}+\frac{1}{3^4}+\frac{1}{4^4}+...=\sum_{n=1}^{\infty}\frac{1}{n^4}<\frac{4-3\pi}{2\pi}.$$

Thus, the series $\sum_{n=2}^{\infty}\frac{1}{n^4}$, and also series $\sum_{n=1}^{\infty}\frac{1}{n^4}$ is convergent.

Note. Series in this form, i.e. $\sum_{n=1}^{\infty}\frac{1}{n^\alpha}$, are referred to as Riemann (sometimes Dirichlet) series. They converge for $\alpha>1$, which means that a convergent series has also been obtained in the given task.

2.3 Convergence of the infinite series $\sum_{n=2}^{\infty}\frac{r}{2+(n-1)^2}$

Let there be a circle $k(A,R)$ and a circle $m\left(M,\frac{R}{2}\right)$ being internally tangent to circle *k* at point *O*. Consider the sequence of circles $k_n(S_n,r_n)$ such that circle k_1 with the centre on the line *AM* is internally tangent to circle *k* and externally tangent to circle *m* and circle k_2 is internally tangent to circle *k* and externally tangent to circles *m*, k_1, ..., and circle k_n is internally tangent to circle *k* and externally tangent to circles *m*, k_{n-1} (Fig. 6*a*). Since circle which touches the given three circles always exists, the construction of circles $k_n(S_n,r_n)$, where $n\in N$, can be continued indefinitely. Is the length of the broken line which connects the centres of circles k_n, $n\in N$, i.e. the length of the broken line $S_1S_2....S_n...$ (Fig. 6*b*) also infinite?

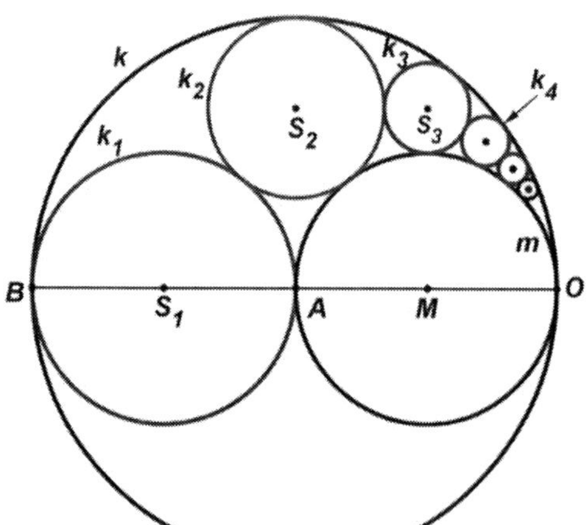

Fig. 6*a*

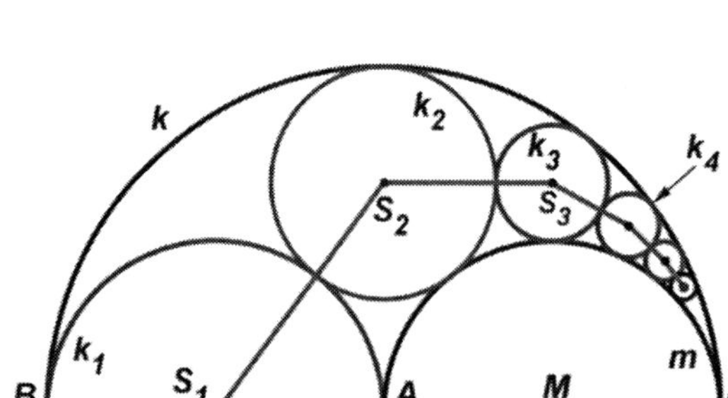

Fig. 6*b*

Let $d(S_1S_2....S_n...)$ denote the length of the broken line $S_2S_3....S_n...$; then it apparently holds that $d(S_2S_3....S_n...) = r_1 + 2r_2 + 2r_3 + ... + 2r_n + ... = r_1 + 2\sum_{n=2}^{\infty} r_n$.

Since the length of the radius r_1 of circle k_1 is obviously finite, in order to answer the above asked question it suffices to investigate if the series $\sum_{n=2}^{\infty} r_n$ converges or diverges. Take notice of one interesting feature of the centres S_n, $n \geq 2$ of circles k_n, and the fact that the sum of their distances from the centres of circles *k*, *m*, i.e. from points *A*, *M*, is constant. Indeed, for all $n \geq 2$ it holds that

$$|AS_n| + |MS_n| = (R - r_n) + \left(\frac{R}{2} + r_n\right) = \frac{3}{2}R.$$

In other words, the centres S_n, $n \geq 2$ of circles k_n lie on ellipse *e* whose foci are points *A*, *M* and the length of its major axis is $\frac{3}{2}R$, i.e. the major vertices of ellipse *e* are points S_1 and *O* (Fig. 7).

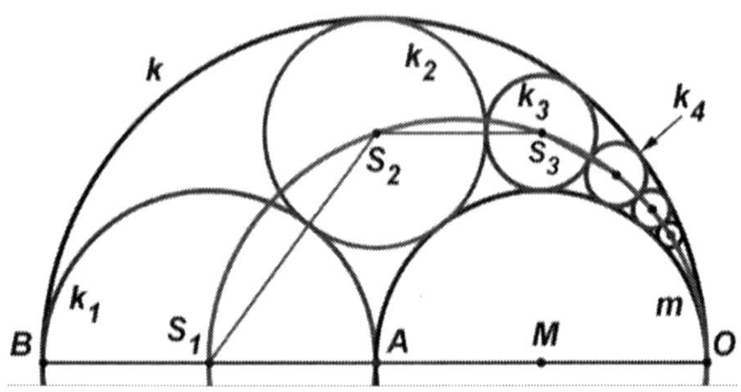

Fig. 7

Evidently, the length of the ellipse arc above each of the segments S_nS_{n+1}, $n \geq 2$, is greater than the length of any of the segments S_nS_{n+1}, $n \geq 2$, i.e. the length $d(S_1S_2....S_n...)$ of the broken line $S_1S_2....S_n...$ is smaller than a half of the circumference of ellipse *e*, which is finite (its approximate length can be determined, for instance, by means of some of the numerical mathematics methods). This, however, means that the sum of series $\sum_{n=2}^{\infty} r_n$ is above bounded by certain real number, in other words, the series $\sum_{n=2}^{\infty} r_n$ is convergent. What is the series the convergence of which we managed to determine based on the figure? Let us now attempt to determine the analytical form of the nth term of the series, i.e. r_n, where $n \geq 2$. In case of constructing the circles $k_n(S_n, r_n)$ for certain number of values $n \in N$ it is evident that the problem would belong to the group of so-called Apollonius' problems which are commonly solved by means of circle inversion.

Consider then the circle inversion $K_i\left[\omega(O,1)\right]$ with centre O and the reference circle ω with radius $r_\omega = 1$. Since circles k, m go through the centre of the circle inversion, point O, their inverses in the circle inversion $K_i\left[\omega(O,1)\right]$ are parallel lines k', m', not going through point O (Fig.8)

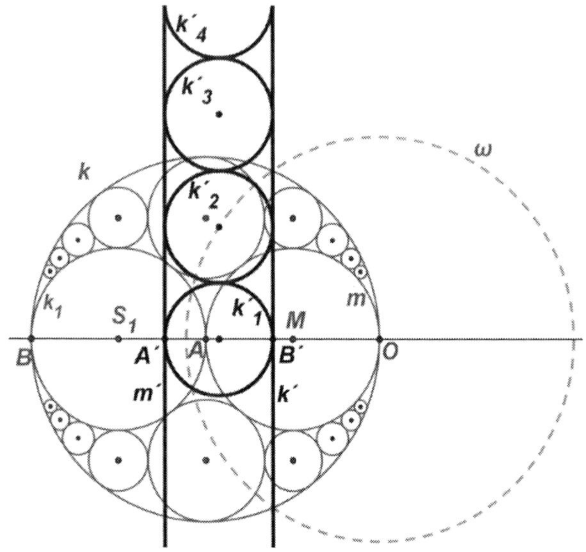

Fig. 8

The construction of circle $k_1(S_1, r_1)$, which is internally tangent to circle k and externally tangent to circle m, is trivial. Since the circle $k_1(S_1, r_1)$ does not go through the centre O of the circle inversion $K_i\left[\omega(O,1)\right]$ and is internally tangent to circle k and externally tangent to circle m, its inverse with respect to $K_i\left[\omega(O,1)\right]$ is a circle $k_1'(S_1', r_1')$, touching the parallel lines k', m' (Fig. 8). Since the circle $k_2(S_2, r_2)$ does not go through the centre O of the circle inversion K_i and is expected to touch circles k, m and k_1, its inverse with respect to K_i, i.e. circle $k_2'(S_2', r_2')$, surely touches the lines k', m' and the circle k_1'. Similarly, circles k_3, k_4, k_5, ... are projected to circles k_3', k_4', k_5', ..., which touch the lines k', m' (Fig. 8). It is evident that all the circles k_1', k_2', k_3', ... have radii of the same length as the circle k_1', which is further denoted by r', i.e. for all $n \in N$ it holds $r_n' = r_1' = r' = \dfrac{|A'B'|}{2}$, where A', B' are the inverses of points A, B with respect to the circle inversion K_i. Fig. 8 also shows that $|A'B'| = |OA'| - |OB'|$. Since points A', B' are the inverses of points A, B with respect to the circle inversion $K_i\left[\omega(O,1)\right]$, it holds that

$$|OA'| \cdot |OA| = 1,\ |OB'| \cdot |OB| = 1,$$

and $|OA| = R$, $|OB| = 2R$, which implies $|OA'| = \dfrac{1}{R}$, $|OB'| = \dfrac{1}{2R}$.

Then, based on Fig. 8 it can be easily seen that

$$2r' = |A'B'| = |OA'| - |OB'| = \frac{1}{R} - \frac{1}{2R} = \frac{1}{2R},$$

which implies $r_n' = r' = \frac{|A'B'|}{2} = \frac{1}{4R}$.

Our objective, however, is to determine the length r_n. Let P, Q be the intersection points of ray OS_n or OS_n' with circle $k_n(S_n, r_n)$, and let P', Q' be the intersection points of the ray with circle $k_n'(S_n', r')$ (Fig. 9).

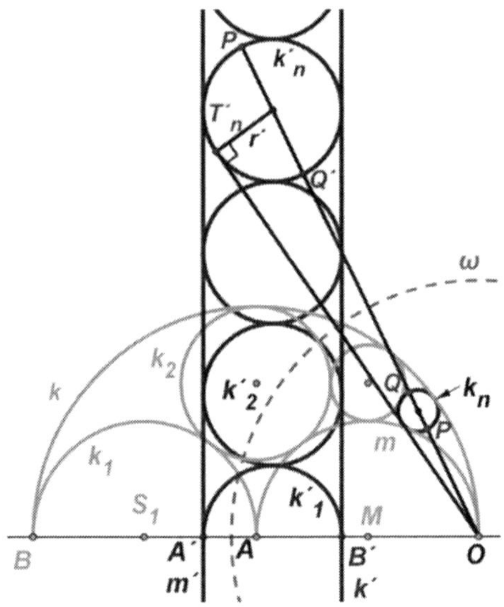

Fig. 9

Then $2r_n = |PQ| = |OQ| - |OP|$, and since points P', Q' are the inverses of points P, Q with respect to the circle inversion $K_i[\omega(O,1)]$, it holds that

$$|OP'| \cdot |OP| = 1,\ |OQ'| \cdot |OQ| = 1, \text{ resp. } |OP| = \frac{1}{|OP'|},\ |OQ| = \frac{1}{|OQ'|}.$$

Construct now the tangent t to the circle k_n', $n \geq 2$, through point O, and let denote their contact point T_n'. Then, based on the Power of a Point Theorem it holds that

$$|OP'| \cdot |OQ'| = |OT_n'|^2.$$

Based on Fig. 9 and applying the previous equalities we obtain

$$2r_n = |QP| = |OQ| - |OP| = \frac{1}{|OQ'|} - \frac{1}{|OP'|} = \frac{|OP'| - |OQ'|}{|OQ'||OP'|} = \frac{2r'}{|OT'_n|^2},$$

Hence for $|OT'_n|^2$ we obtain $|OT'_n|^2 = \frac{r'}{r_n}$ and also $|OT'_n|^2 = \frac{1}{4Rr_n}$.

Take notice of the triangles $OS'_1S'_n$, $OT'_nS'_n$ (Fig. 10).

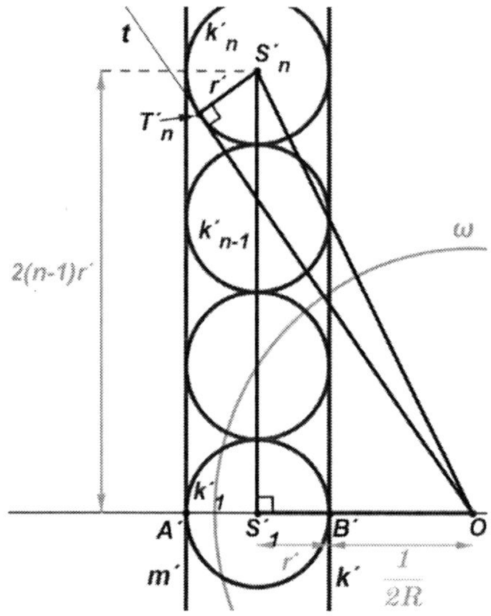

Fig. 10

Since both triangles $OS'_1S'_n$, $OT'_nS'_n$ (Fig. 10) are right-angled, based on the Pythagorean theorem for all $n \in N$ it holds

$$|OS'_1|^2 + |S'_1S'_n|^2 = |OS'_n|^2 \tag{1}$$

$$|OT'_n|^2 + |T'_nS'_n|^2 = |OS'_n|^2 \tag{2}$$

From Fig.10 and applying equality $r' = \frac{1}{4R}$ it is obvious that for the length of segment $S'_1S'_n$, where $n \in N$, it holds

$$|S'_1S'_n| = 2(n-1)r' = 2(n-1)\frac{1}{4R}.$$

Substituting the obtained relationships $r' = \frac{1}{4R}$, $|S'_1S'_n| = 2(n-1)\frac{1}{4R}$, $|OT'_n|^2 = \frac{1}{4Rr_n}$ into equation (1) and (2) we obtain

$$\left(\frac{1}{4R} + \frac{1}{2R}\right)^2 + \left(2(n-1)\frac{1}{4R}\right)^2 = |OS'_n|^2,$$

$$\frac{1}{4Rr_n}+\left(\frac{1}{4R}\right)^2=|OS'_n|^2,$$

hence for r_n we have

$$r_n=\frac{r}{2+(n-1)^2}.$$

To sum up, based on geometrical considerations we managed to show that the length of the broken line $S_1S_3....S_n...$ is finite and can be expressed in the form

$$d(S_1S_3....S_n...)=r_1+2\sum_{n=2}^{\infty}\frac{r}{2+(n-1)^2},$$

which implies that the series $\sum_{n=2}^{\infty}\frac{r}{2+(n-1)^2}$ is convergent.

3 Conclusion

One of the obstacles that learners face in relation to the concepts of convergence and the sum of infinite series is, in our opinion, mistaking the concepts of infinite and unbounded, which is indicated by learners' assesrtions, such as "...but when I add another and another number, it must increase to infinity...". Based on our experience we conclude that incorporating graphical representations of convergent infinite series positively contributes to the learners' comprehension of the given concepts. When incorporating such graphical representations into the educational process, it is not necessarily restricted to only the well-known visualisations of the sum of infinite geometric series. By means of illustrative graphical representations it is possible (on certain intuitive level) to "make learners believe" that also some other types of infinite series are convergent, and this can be done even before learners get familiar with particular convergence tests. In the paper we show examples of infinite series whose convergence is easily seen from the figures and can be rigorously justified, for instance, by calculating the sum of the given series based on the definition, using integral test or direct comparison test.

References

[1] KLEPANCOVÁ, M., SMETANOVÁ, D. Nekonečné rady a ich vizualizácia. *Učitel matematiky*, roč. 23, č. 4, 2015: s. 193-205. ISSN 1210-9037.

[2] Monaghan, J. Young people's ideas of infinity, *Educational Studies in Mathematics*, vol. 48, Issue 2, 2001:pp 239–257, ISSN 1573-0816.

[3] NELSEN, R. B. *Proofs without words*. MAA Service Center, Washington, 1993. ISBN 0-88385-700-6, 152 s.

[4] Sierpinska, A. (1987). Humanities students and epistemological obstacles related to limits. *Educational Studies in Mathematics*, vol. 18, issue 4, 1987, pp: 371-87. ISSN 1573-0816

[5] TALL, D. O., VINNER, S. Concept image and concept definition in mathematics, with particular reference to limits and continuity. *Educational Studies in Mathematics*, vol. 12, no. 2, 1981, pp:151-169. ISSN 1573-0816

[6] VARGA, M. *Úvod do diferenciálneho počtu*. UKF v Nitre, 2015, ISBN 978-80-558-0868-0, 143 s.

[7] Williams, S. R.. Models of limit held by college calculus students. *Journal for Research in Mathematics Education*, vol. 22, issue 3, 1991. pp: 219-236. ISSN 00218251

Current address

Vargová Michaela, Mgr. PhD.
The Institute of Technology and Business in České Budějovice
Okružní 517/10, 370 01 České Budějovice, Czech Republic
Tel. number: +420 387 842 186, e-mail: klepancova@mail.vstecb.cz

Slovak University of Technology in Bratislava
Faculty of Mechanical Engineering
16th Conference on Applied Mathematics **APLIMAT 2017**
Proceedings

MODELLING FREE-FORM CURVES AND SURFACES BY MEANS OF MINKOWSKI OPERATIONS

VELICHOVÁ Daniela (SK)

Abstract. Paper brings information on modelling free-form curves and surfaces by means of Minkowski point set operations performed on two free-form curves. Some special properties of surface patches generated as Minkowski sum and Minkowski product of free-form curve segments are presented. Illustrations of curves and surfaces generated by means of this method are included, Minkowski combinations of two Bézier cubic curve segments, in particular.

Keywords: Minkowski point set operations, Minkowski sum of free-form curves, Minkowski product of free-form curves, intrinsic geometric properties

Mathematics subject classification: Primary 65D17; Secondary 51H30, 68U07.

1 Introduction

In this paper we describe applications of the concept of Minkowski sum and Minkowski product for modelling complex point sets from two free-form curves, and use these concepts for modelling geometric objects that are differential manifolds in the 3-dimensional Euclidean space. Minkowski sum and product of two point sets are ususally defined as point-wise operations based on sum and product of two points in the basic space with fixed reference point O. Both operations are performed on position vectors of these points with respect to the reference point. Minkowski sum of point sets is defined as vector sum and Minkowski product of point sets as exterior (wedge) product of position vectors of all points in the respective sets, details see in [1].

Both forms of Minkowski combinations - summative and multiplicative - will be introduced and analytical representations of resulting 2-parametric family of surface patches, or 2-parametric family of free-form laces in case of patial Minkowski point set operations, will be determined. Thus, intrinsic geometric properties of resulting manifolds can obviously be derived by means of differential characteristics of the two operand manifolds. Details and formulas for calculation of differental properties and intrinsic characteristics of resulting manifolds can be found in [2].

2 Minkowski summative combinations of free-form curves

Let two free-form curve segments in $\mathbf{E}^3$ be represented parametrically by uniform vector maps on the unit interval $I = \langle 0, 1 \rangle \subset \mathbf{R}$

$$\mathbf{k}(u) = \sum_{i=1}^{n} A_i Pl_i(u) = \left(\sum_{i=1}^{n} xa_i Pl_i(u), \sum_{i=1}^{n} ya_i Pl_i(u), \sum_{i=1}^{n} za_i Pl_i(u) \right), u \in I,$$

$$\mathbf{l}(v) = \sum_{i=1}^{n} B_i Pl_i(v) = \left(\sum_{i=1}^{n} xb_i Pl_i(v), \sum_{i=1}^{n} yb_i Pl_i(v), \sum_{i=1}^{n} zb_i Pl_i(v) \right), v \in I,$$

where $A_i = (xa_i, ya_i, za_i), B_i = (xb_i, yb_i, zb_i)$ are vertices of the curve basic determining polygons and Pl_i are interpolation polynomials in the respective form. Minkowski summative combination of the two curves is a two-paranetric family of translation surface patches defined on unit square $I^2 \subset \mathbf{R}^2$ for any $k, l \in \mathbf{R}$ as

$$\mathbf{s}(u,v) = k \sum_{i=1}^{n} A_i Pl_i(u) + l \sum_{i=1}^{n} B_i Pl_i(v) = \sum_{i=1}^{n} (k.A_i Pl_i(u) + l.B_i Pl_i(v)) =$$

$$= (x(u,v), y(u,v), z(u,v))$$

$$x(u,v) = \sum_{i=1}^{n} (k.xa_i Pl_i(u) + l.xb_i Pl_i(v))$$

$$y(u,v) = \sum_{i=1}^{n} (k.ya_i Pl_i(u) + l.yb_i Pl_i(v))$$

$$z(u,v) = \sum_{i=1}^{n} (k.za_i Pl_i(u) + l.zb_i Pl_i(v))$$

Any surface from this family can be generated by translation of k-multiple of one curve segment along the l-multiple of the other one. Differential characteristics of such surface patches can be represented by means of derivatives of the operand curve vector maps and with respect to particular values of parameters $k, l \in \mathbf{R}$.

Illustrations of various forms of surface patches generated as Minkowski summative combinations of two cubic Bézier curve segments are presented in Fig.1, while resulting surface is bicubic Bézier surface patch of degree 3. Partial Minkowski summative combination of the same cubic Bézier curve segments is illustrated in Fig.2, this Bézier curve of order 3 depicted in black colour is located on the bicubic Bézier surface patch from Fig 1.

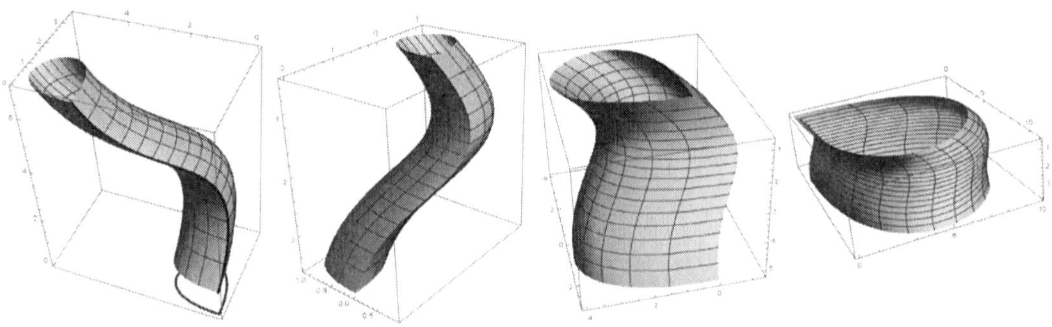

Fig. 1. Minkowski summative combinations of Bézier curve segments

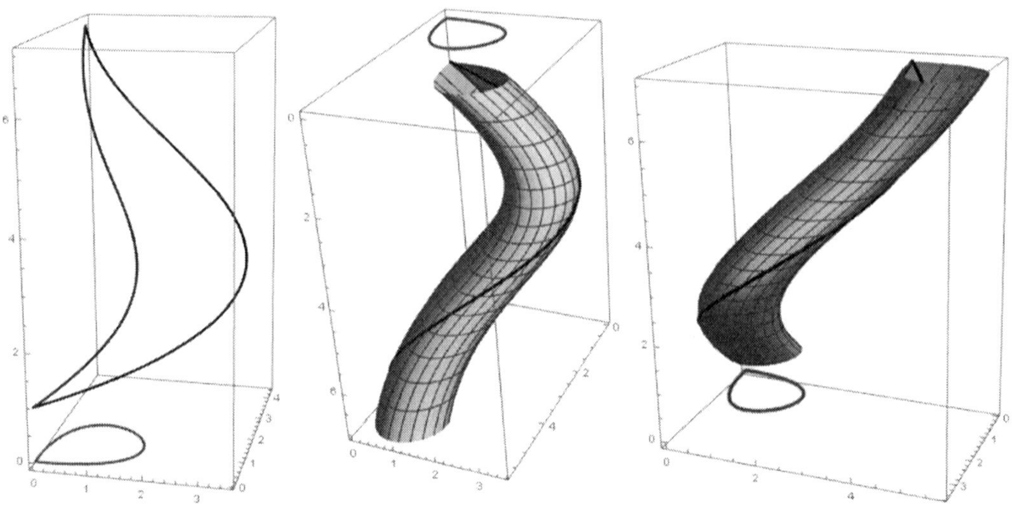

Fig. 2. Minkowski summative combinations of Bézier curve segments

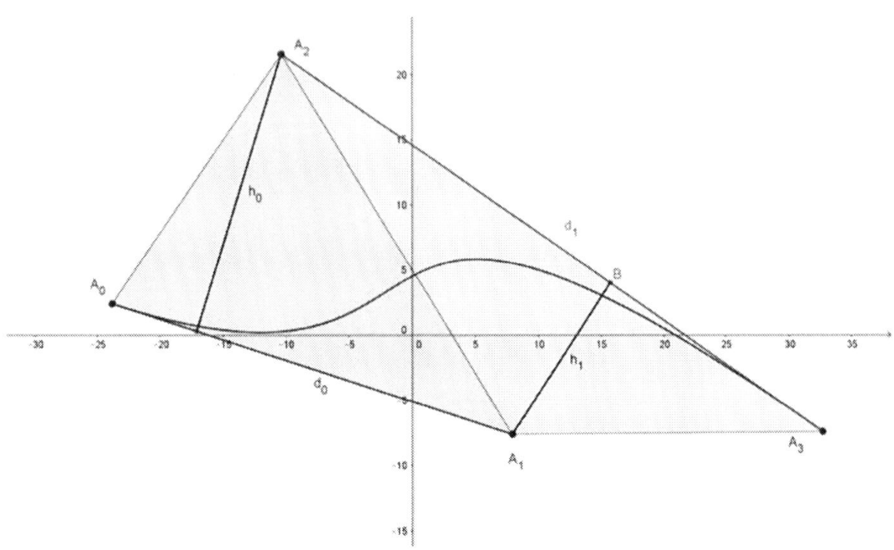

Fig. 3. Triangles determining flection at endpoints of Bézier curve

Formula for calculation of the first curvature at the start point (or end point, respectively) of Bézier curve segment is in the form

$$\kappa(0) = \frac{4area(A_0A_1A_2)}{3\|A_1 - A_0\|^3}, \kappa(1) = \frac{4area(A_1A_2A_3)}{3\|A_3 - A_2\|^3},$$

which can be re-written in terms of triangle $A_0A_1A_2$ or $A_1A_2A_3$ dimensions, length of one side $\|A_0A_1\| = d_0$ or $\|A_2A_3\| = d_1$, and its altitude h_0, or h_1, see in Fig. 3.

$$\kappa(0) = \frac{2h_0}{3d_0^2}, \kappa(1) = \frac{2h_1}{3d_1^2}.$$

Let us suppose that partial Minkowski sum of two Bézier cubic curve segments determined by control points A_0, A_1, A_2, A_3 and B_0, B_1, B_2, B_3 is Bézier cubic segment determined by control points C_0, C_1, C_2, C_3, for which the following holds

$$C_i = A_i + B_i, i = 0, 1, 2, 3.$$

Curvature of this partial Minkowski sum at the start point (and alternatively at the end point) can be then represented by the side d_c, and altitude h_c of triangle $C_0C_1C_2$, which can be re-written in terms of sides and altitudes of the two triangles with original control points (see in Fig. 4.)

$$\kappa(0) = \frac{2h_c}{3d_c^2}$$

$$d_c^2 = d_a^2 + d_b^2 + 2\Delta xa_{01}\Delta xb_{01} + 2\Delta ya_{01}\Delta yb_{01},$$

$$h_c = \frac{1}{4}\sqrt{4c_0^2d_c^2 - (d_c^2 + c_0^2 - c_1^2)^2},$$

$$c_0^2 = a_0^2 + b_0^2 + 2\Delta xa_{02}\Delta xb_{02} + 2\Delta ya_{02}\Delta yb_{02},$$

$$c_1^2 = a_1^2 + b_1^2 + 2\Delta xa_{12}\Delta xb_{12} + 2\Delta ya_{12}\Delta yb_{12},$$

$$\Delta xa_{ij} = xa_j - xa_i, \Delta xb_{ij} = xb_j - xb_i, \Delta ya_{ij} = ya_j - ya_i, \Delta yb_{ij} = yb_j - yb_i.$$

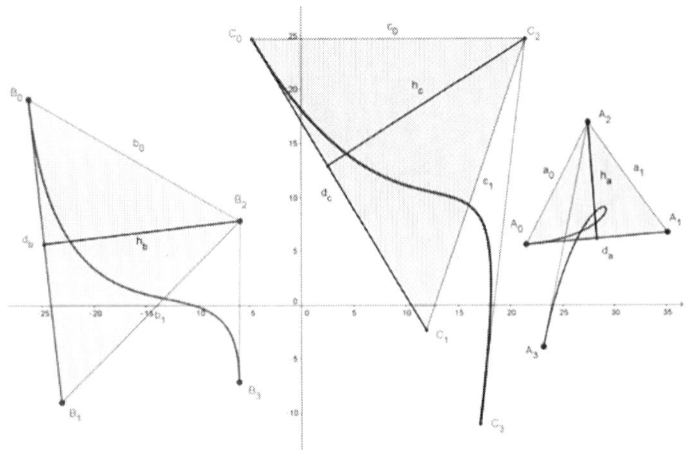

Fig. 4. Triangle determining flection at start point of partial Minkowski sum of two Bézier curves

3 Minkowski multiplicative combinations of free-form curves

Minkowski product of two multiples of free-form curves determined by vector maps $k.\mathbf{k}(u), l.\mathbf{l}(v), u, v \in I \subset \mathbf{R}, k, l \in \mathbf{R}$ is a family of surface patches represented parametrically on unit square $I^2 \subset \mathbf{R}^2$ as

$$\mathbf{p}(u,v) = k.\mathbf{k}(u) \wedge l.\mathbf{l}(v) = k.l \begin{pmatrix} yk(u)zl(v) - zk(u)yl(v) \\ zk(u)xl(v) - xk(u)zl(v) \\ xk(u)yl(v) - yk(u)xl(v) \end{pmatrix}^T, (u,v) \in I^2,$$

where

$$xk(u) = \sum_{i=1}^{n} xa_i Pl_i(u), yk(u) = \sum_{i=1}^{n} ya_i Pl_i(u), zk(u) = \sum_{i=1}^{n} za_i Pl_i(u),$$

$$xl(v) = \sum_{i=1}^{n} xb_i Pl_i(v), yl(v) = \sum_{i=1}^{n} yb_i Pl_i(v), zl(v) = \sum_{i=1}^{n} zb_i Pl_i(v).$$

Surface patches resulting as multiplicative combinations of two free-form curve segments defined by real coefficients $k, l \in \mathbf{R}$ consist of all such points in the space, whose position vectors are $k.l$-multipes of vector products of two position vectors of points on the two curves. In case of cubic curves, the resulting surface patches are of degree six. Partial Minkowski product is generating principle of free-form laces forming a family of homothetic curves of degree six.

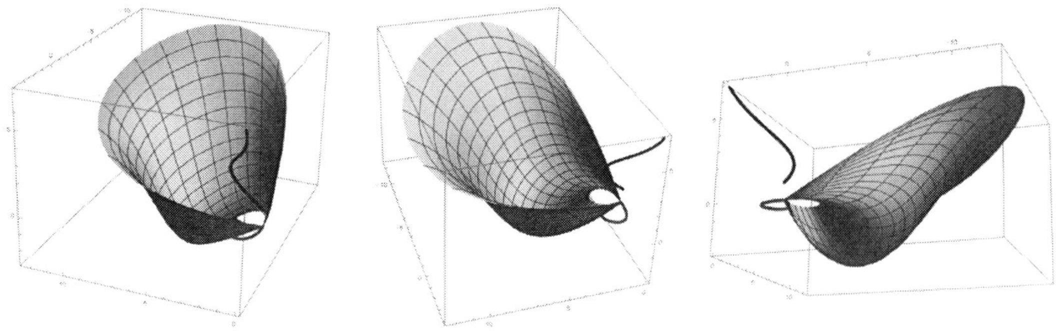

Fig. 5. Minkowski multiplicative combinations of Bézier curve segments

Surface patch generated as Minkowski product of two cubic Bézier curve segments from Fig. 1 is presented in Fig. 5.

Partial Minkowski combination of two cubic free-form curve segments represented by vector functions parameterized for the same parameter on the unit interval $I \in \mathbf{R}$, results in curve of order 6 represented by the following vector map on I

$$\mathbf{s}(u) = \sum_{i=1}^{n} (A_i \wedge B_i) Pl_i(u)$$

$$\mathbf{p}(u) = \begin{pmatrix} yk(u)zl(u) - zk(u)yl(u) \\ zk(u)xl(u) - xk(u)zl(u) \\ xk(u)yl(u) - yk(u)xl(u) \end{pmatrix}^T,$$

which is located on the respective surface patch that is the Minkowski product of differently parameterized original curve segments.

Illustration is in Fig. 6, where Minkowski partial sum of two space Bézier cubic segments is depicted in black colour. This curve is a sextic Bézier curve. Vertices of its basic determining polygon can be obtained as products of the respective vertices of determining basic polygons of operand curves.

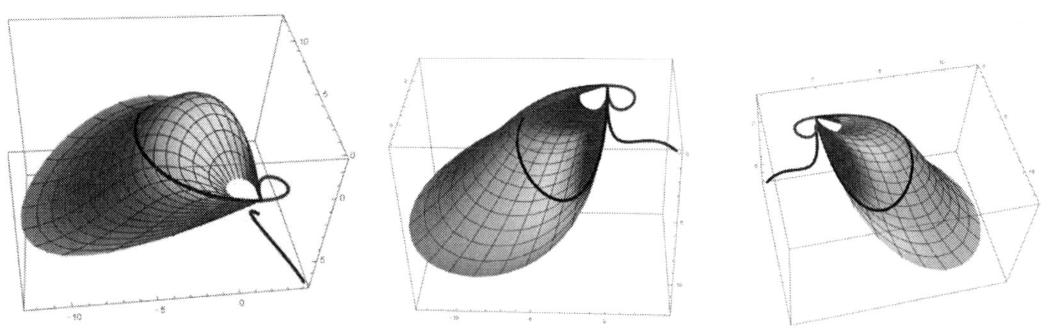

Fig. 6. Minkowski partial multiplicative combinations of Bézier curve segments

4 Conclusions

Minkowski point set combinations serve as powerful means of modelling new forms of curve segments and surface patches, whose specific intrinsic properties can be influenced by form of the two operand curves. In case of free-form curves, resulting manifolds can be recognised as free-form figures determined by specific basic figures related to basic figures of operand manifolds. Thus, the above described new modelling tool is a powerfull deformation algorithm, largly enhancing scope and range of free-form modelling strategies frequently utilised in geometric design and computer graphics applications. It is a tool for interactive design and shape manipulation technique with geometric figures which can be easily implemented and used for practical applications in computer science, in computer aided geometric design and modelling. This natural deformation instrument can be particularly usefull in connection to shape modelling in art and architecture, as described in [4].

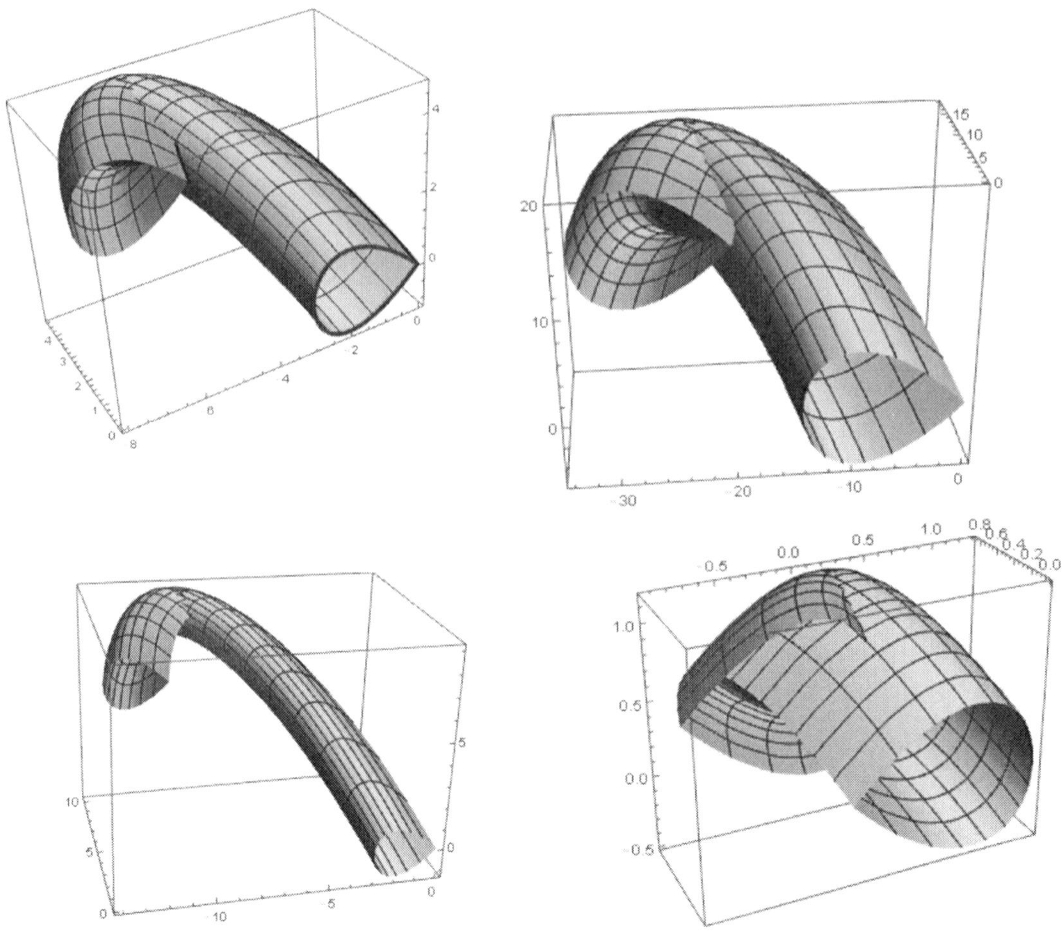

Fig. 7. Minkowski summative combinations of Bézier curve segments

Illustrations of some artistic-like objects generated by the described algorithms are in Fig.7, where forms of Minkowski summative combinations of two Bézeir cubic segments are presented, while in Fig. 8 are presented few forms of lace curves resulting as Minkowski partial summative combinations of the same Bézier cubic segments. Minkowski product and partial product of these cubic curves are depicted in Fig. 9.

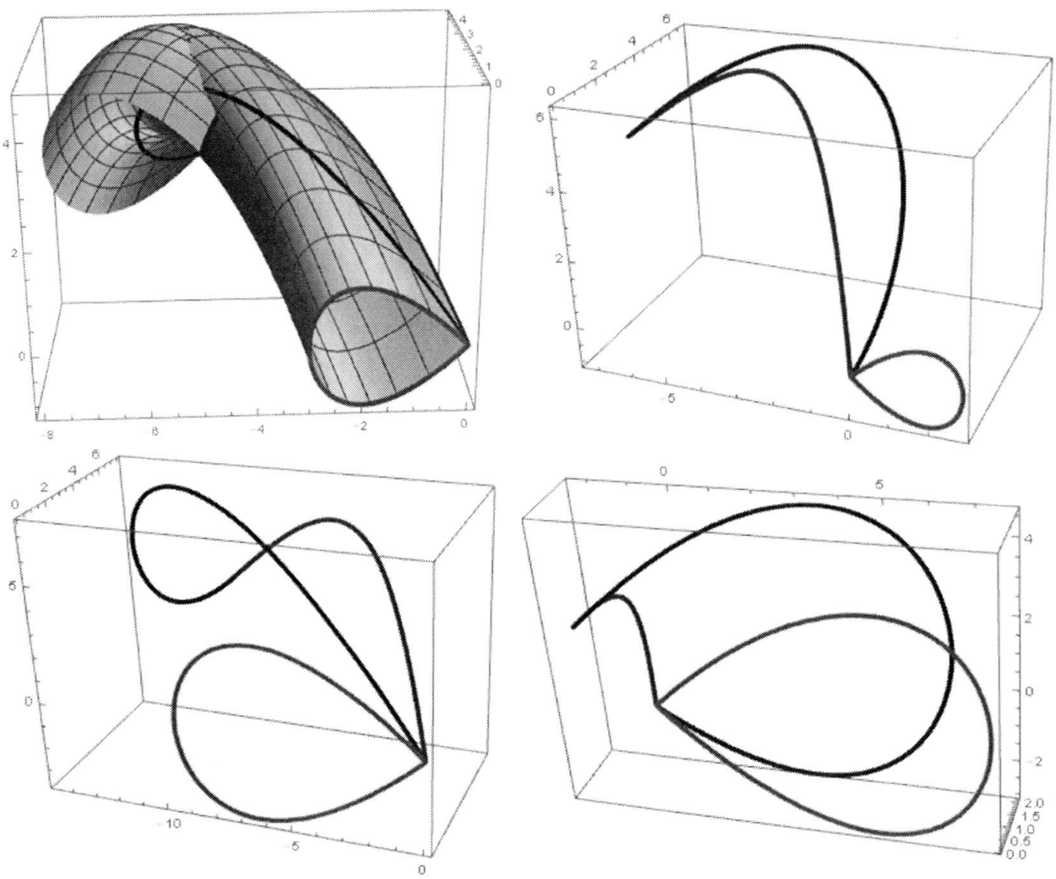

Fig. 8. Minkowski partial summative combinations of cubic Bézier curve segments

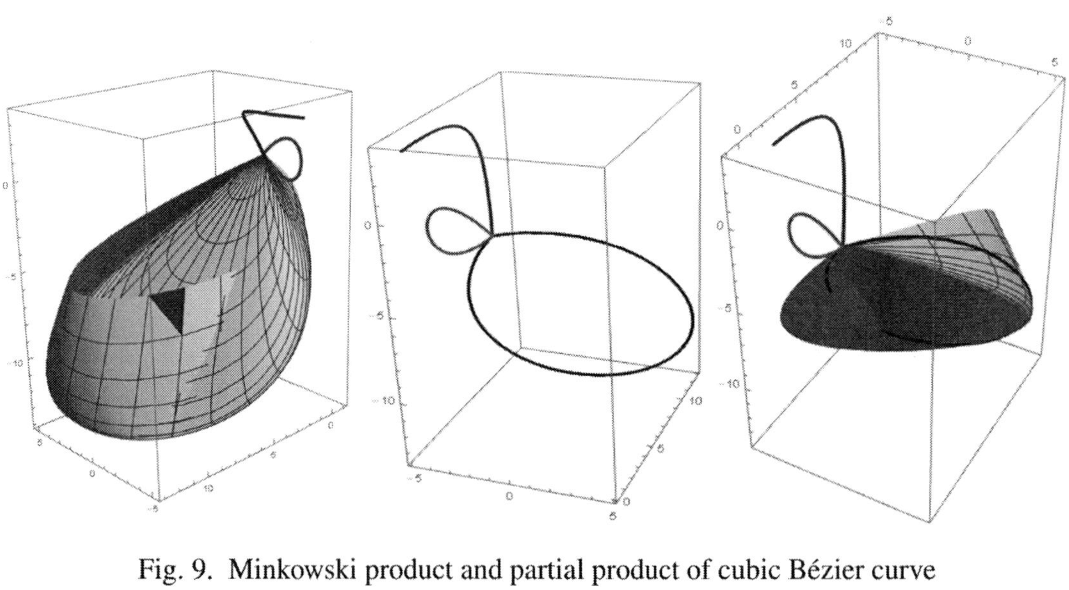

Fig. 9. Minkowski product and partial product of cubic Bézier curve segments

5 Acknowledgement

The author gratefully acknowledges the Scientific Grant Agency APVV of the Slovak Republic for supporting this work under the Grant No. APVV-0161-12.

References

[1] VELICHOVÁ, D. Classification of Manifolds Resulting as Minkowski Operation Products of Basic Geometric Point Sets, in *Journal for Geometry and Graphics*, Volume 19 (2015), No. 1, 13-29.

[2] VELICHOVÁ, D. Minkowski Combinations of Free-Form Curves, in *Aplimat proceedings 2016*, STU Bratislava 2016, Slovakia.

[3] VELICHOVÁ, D. Minkowski Combinations of Curves, in *17th International Conference on Geometry and Graphics ICGG 2016*, Beijing 2016, China, 5 pp.

[4] VELICHOVÁ, D. Minkowski Laces, in *Proceedings of 23rd Conference Geometry-Graphics-Computer*, Lodž 2016, Poland, 53-54.

Current address

Velichová Daniela, doc. RNDr. CSc.
Institute of Mathematics and Physics, Faculty of Mechanical Engineering,
Slovak University of Technology in Bratislava
Nám. slobody 17, 812 31 Bratislava, Slovak Republic
Tel. number: +4212 5729 6136, e-mail: daniela.velichova@stuba.sk

Slovak University of Technology in Bratislava
Faculty of Mechanical Engineering
16th Conference on Applied Mathematics **APLIMAT 2017**
Proceedings

RECONSTRUCTION OF 3D OBJECT DIMENSIONS FROM VIDEORECORDS

VELICHOVÁ Daniela (SK)

Abstract. Paper brings short information about determination of specific characteristic geometric properties of objects from recordings of 3D scenes obtained by means of available camera equipment, which are regarded as relevant image records for the criminalistics explorations. Newly developed tool for 3D reconstruction CamWitt, an interactive tool for the filtration of data recordings and 3D reconstruction is shortly described. Possible generalizations of presented reconstruction methods for development of algorithms of videogrammetry for reconstruction of moving objects from video-records taken by stable camera systems are discussed.

Keywords: photogrammetry, videogrammetry, camera calibration, interior image orientation, reconstruction of 3D objects

Mathematics Subject Classification: Primary 51N05, 51N15; Secondary 68U07, 65D17

1 Introductory Remarks

Paper brings short information about determination of specific characteristic geometric properties of objects from recordings of 3D scenes obtained by means of available camera equipment, which are regarded as relevant image records for the criminalistics explorations. Namely, information about development of a correct algorithm for processing of geometric characteristics is given, and description is presented of reconstruction methods for finding real dimensions and positions of selected objects from views of 3D scenes by means of photogrammetric principles.

Analysis of criminologically relevant images recorded by on-purpose installed cameras with known calibration data is possible using newly developed tool for 3D reconstruction CamWitt. This interactive tool for the filtration of data recordings is shortly described. Automated detection of objects in the recordings and precise determination of dimensional and positional characteristics enabling reliable identification of depicted objects are explained in brief. The presented methods are to be introduced in the criminological praxis, for investigation of acts of crimes, in collecting and analysis of proofs of evidence recorded on video-cameras or two stable cameras, and from any other relevant image recordings for the aims of identification of criminals and committed criminal acts.

Presented theoretical achievements are results of the scientific project granted by Slovak research and development agency, project GEOCRIM number APVV-1061-12, entitled Determination of geometric characteristics of objects on images obtained by visual records relevant for criminology. Main aims of this project is final development of an interactive application for automatic reconstruction of selected objects, i.e. calculations of their real dimensions from two views of these objects in one 3D scene. The two views of important detected points on reconstructed objects have to be selected manually in the two images taken by calibrated cameras statically positioned in the testing space in a given distance.

Prior to programming of this application entitled CamWitt, the reconstruction algorithms based on basic methods of photogrammetry were studied, and implemented for testing necessary calculations in CAS Maple and Mathematica. Methods of epipolar geometry were used for calculations of real dimensions and distances of points using their reconstructed space Cartesian coordinates from coordinates of their views in 2 images from calibrated cameras positioned in a given distance. Then, for testing purposes, a 3D illustration of two central projections of a cube from 2 different centres was developed, presenting possible errors that might occur in calculations due to the not correct detection of two related views of important points in the two images. Illustration of this application is in the following figures, where correct positions in Fig. 1, and one incorrect position WP in Fig. 2 and in Fig. 3 of real points in 3D are presented, as obtained from correctly detected, and not correctly detected views of the point P_3 in one or in both views. In some configurations the two projecting lines form centres S_1 and S_2 do not intersect in a real point, but they form a pair of skew lines in space. Several methods are used to avoid this inaccuracy in calculations.

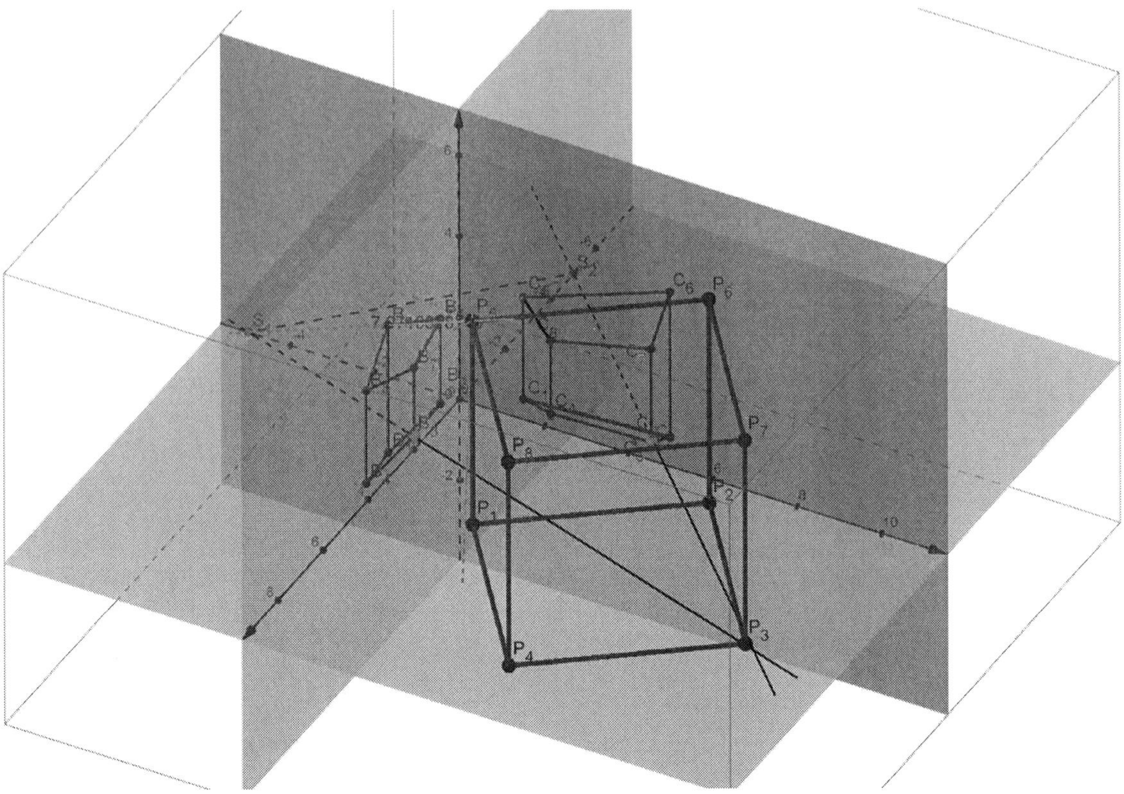

Fig. 1. Two central views of cube from 2 different centres

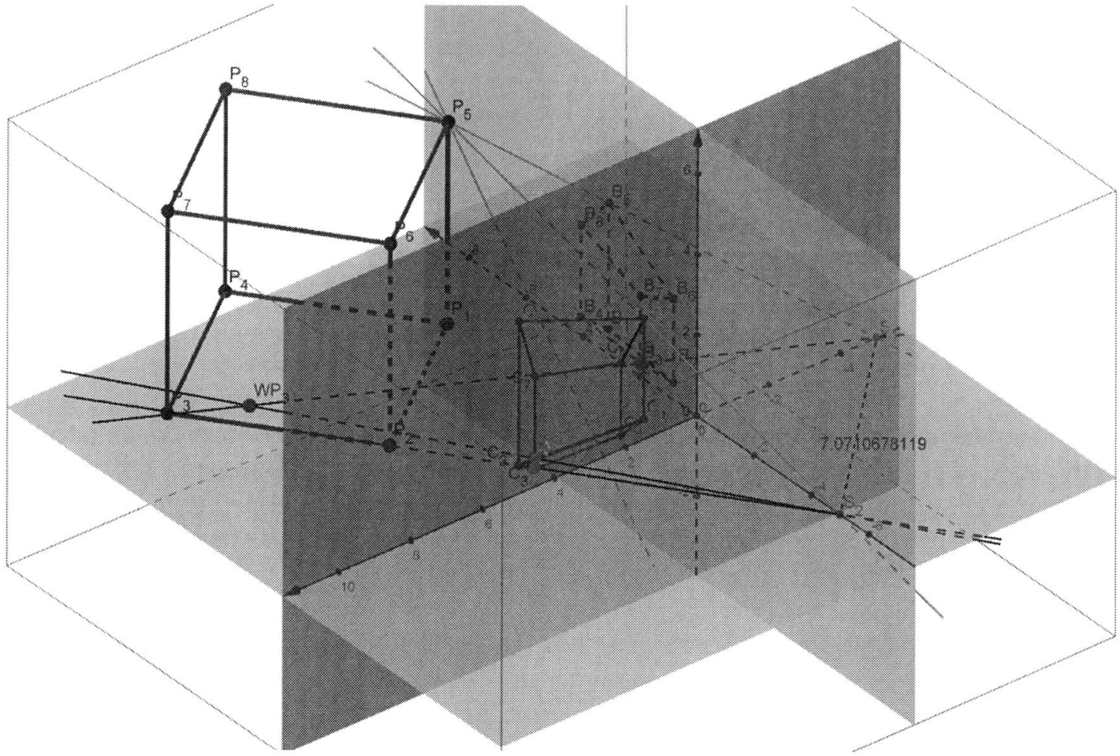

Fig. 2. Illustration of incorrect position of point P_3 detected from incorrect estimation of its view in red image

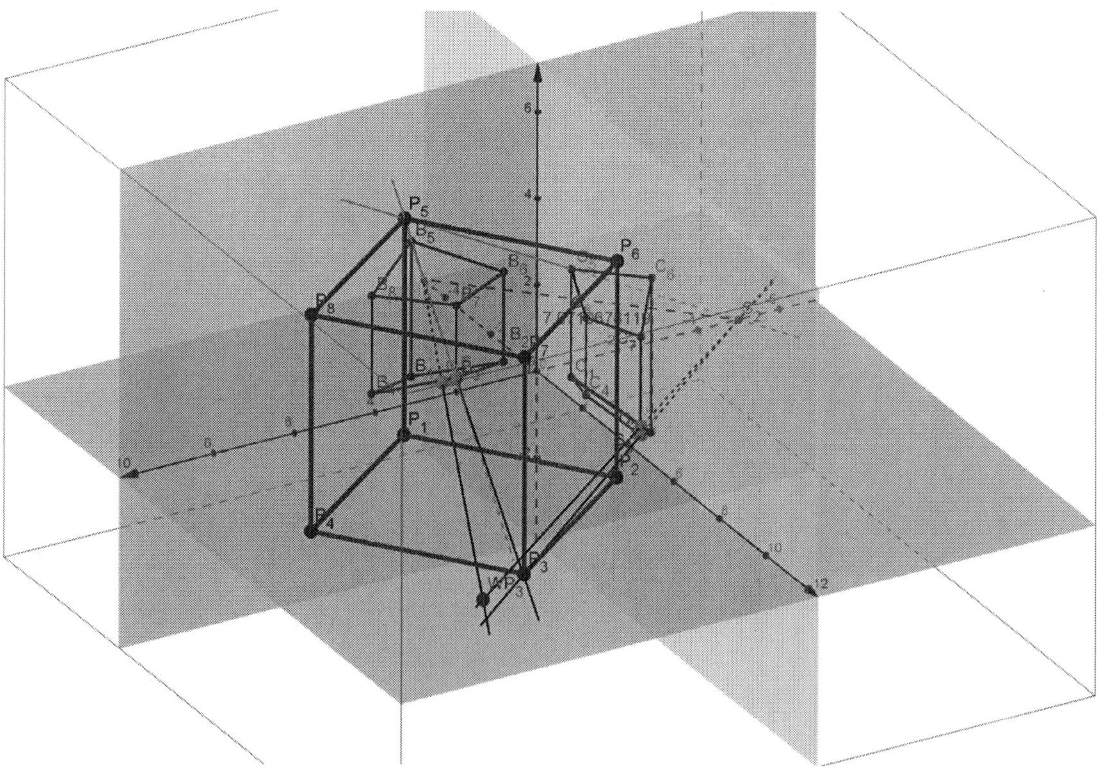

Fig. 3. Illustration of incorrect position of point P_3 detected from incorrect estimations of its view in images

2 Basic Concepts of Photogrammetric Reconstruction

Calculation of real 3D Cartesian coordinates of points detected from a photographic view is realized by means of algorithms based on properties of central or linear perspective, projective transformations represented in matrix form. Each central projection or photographic mapping defines a particular coordinate system in space, called the camera frame. Its origin is placed at the center of projection, the principal ray of the camera is the z coordinate axis, and the principal directions in the photography image plane serve as x and y coordinate axes spanning the vanishing plane of this central projection. Then the central projection in homogeneous coordinates $(x_0 : x_1 : x_2 : x_3) = (1 : x : y : z)$ is expressed as a linear mapping determined by the focal distance d

$$\begin{pmatrix} x_1' \\ x_2' \\ x_3' \end{pmatrix} = \begin{pmatrix} 0 & 0 & 0 & 1 \\ 0 & d & 0 & 0 \\ 0 & 0 & d & 0 \end{pmatrix} \cdot \begin{pmatrix} x_0 \\ x_1 \\ x_2 \\ x_3 \end{pmatrix} \tag{1}$$

Replacing the camera frame by arbitrary world coordinates (x, y, z) and admitting that in the image plane this particular frame is modified by a translation and by scalings, the generalized form of mapping equations for central projections can be represented as

$$\begin{pmatrix} x_1' \\ x_2' \\ x_3' \end{pmatrix} = \begin{pmatrix} 1 & 0 & 0 \\ h_x' & df_x & 0 \\ h_y' & 0 & df_y \end{pmatrix} \cdot \begin{pmatrix} 0 & 0 & 0 & 1 \\ 0 & 1 & 0 & 0 \\ 0 & 0 & 1 & 0 \end{pmatrix} \cdot \begin{pmatrix} 1 & 0 & 0 & 0 \\ z_x & & & \\ z_y & & R & \\ z_z & & & \end{pmatrix} \cdot \begin{pmatrix} x_0 \\ x_1 \\ x_2 \\ x_3 \end{pmatrix} \tag{2}$$

where in the first matrix from the left the intrinsic calibration parameters of the photo are introduced, namely coefficients of the scalings f_x and f_y (which are usually set to value 1) in addition to the focal distance d, and also new coordinates (h_x', h_y') of the image of principal point. Extrinsic calibration parameters are given as entries of the square matric 4x4, where (z_x, z_y, z_z) are world coordinates of the centre of projection and R is a square orthogonal matrix of rank 3.

A photography is called uncalibrated if the exact position of the centre of projection with respect to the photo is unknown. Replacing this exact center by any other point means that the bundle of rays connecting the center with points of the scene is replaced by a collinear transform of this original bundle. Let two central projections be given with different centres and image planes. Thus a general two-views-system consisting of two linear images is defined. The basic geometric property of two-views-system results from the fact that for space points x which are not aligned with the two centres, the two rays of sight from different centres are coplanar and they are located in a plane, which in both linear images appears in an edge view. In this viewing situation the images of the pencil of planes constitute two perspective pencils of lines called epipolar lines. The centres of these pencils of epipolar lines are called epipoles. Each epipole is the image of one centre under the other projection. The projectivity between the two pencils is called epipolar constraint. Detailed description of this property and proofs of the two following theorems can be found in paper [5].

T1: From two uncalibrated images with given projectivity between epipolar lines the depicted object can be reconstructed up to a collinear transformation.

T2: From two calibrated images with given projectivity of epipolar lines the depicted object can be reconstructed up to a similarity.

Therefore, the problem of recovering a 3D scene is reduced to the determination of epipoles. An algorithm for setting up a fundamental matrix for the description of this projective transformation given by suitable number (more than 7) of corresponding images of points in the two photographic images was developed and programmed. Approximated real positions in the space of the depicted points can be then calculated using centres of axes of resulting skew rays through the two centres of projections (points in the same minimal distance from the two skew lines). These calculations were tested in the available computer algebra systems, and finally used as computation package in language C++ in the generated application CamWitt for automatic reconstruction of real dimensions between points detected in two different images. This application enables user to detect manually respective pairs of images of some reference points in the two photo images that are not calibrated, while just the distance between the two centres of projections (foci of the two camera lenses) are given. Automated calculation calibrates the images and returns as result real Cartesian coordinates of the detected points. Real distances of selected pointed pairs of points can be calculated, too. Thus, real dimensions of selected objects in the given images of 3D scenes can be reconstructed up to some precision. Accuracy of these automated calculations depends on the precision of correspondence of point images in the two photos, and on the selection of reference points, while some configurations are restricted (more coplanar points or points located on special quadrics) due to geometric constraints of the underlying algorithm.

3 Basics of Videgrammetry

The above presented ideas can be generalized to development of algorithms of videogrammetry for reconstruction of objects from video-records taken by stable camera systems. Videogrammetry is a new branch of photogrammetry that offers some effective algorithms for direct reconstruction of moving objects from video records. Its methods work in a non-contact mode, can determine and track even very complex point clouds, deliver very precise and reliable results, and can be fully automated. Videogrammetry can solve two major problems: reconstruction of surfaces (body, face, etc.) and determination of trajectories of moving targets. Its methods can inspect permanently provided and stored records, so these can be used for additional measurements and re-measurements, and verification at any time.

The aim of this paper is to present basic principles and algorithms of the fundamentals of videogrammetry, in comparison to classical photogrammetric algorithms used in 3D reconstruction from 2 digital camera images on the principles of epipolar geometry. Usually, the measured object is recorded (imaged) by a certain number of frames taken from different locations. Two frames (stereo model) provide mathematically sufficient information, but it has several serious drawbacks:

a) it does not provide sufficient redundancy in observations for detection and location of important points of the measurements

b) there remain object parts that are completely invisible for the two cameras

c) not suitable geometric configurations of visible object parts may appear leading to inaccuracy of measurements

d) very large objects cannot be covered with only two images at a given accuracy level.

Due to these reasons, in case of a static point cloud, two cameras in different position are sufficient, but for a moving object, up to 4 cameras should be used for simultaneous imaging. The underlying mathematical model, which is going to be introduced in the full paper, is so called bundle method based on principles of perspective projection.

The underlying mathematical model is so called bundle method based on principles of perspective projection. The object points $P_i = (X_i, Y_i, Z_i)$, $i = 1, 2, ..., n$, $n \in N$ in the space are projected by a straight line through several centres of projection $O_j = (X_{0j}, Y_{0j}, Z_{0j})$ onto the points in the image planes with coordinates $P'_{ij} = (x_{ij}, y_{ij}, 0)$. This collinearity condition can be formulated by equations

$$(X_i, Y_i, Z_i)^{\mathrm{T}} = \lambda_{ij} \mathbf{R}_j (x_{ij} - x_{0j}, y_{ij} - y_{0j}, 0 - c_j)^{\mathrm{T}} + (X_{0j}, Y_{0j}, Z_{0j})^{\mathrm{T}}, \quad (3)$$

where the orthographic images of projection centres O_j into the image planes are the principal points $H_j = (x_{0j}, y_{0j}, 0)$ and c_j are the camera related constants, i is the number of projected points, λ_{ij} is the scale factor for image rays ij and $\mathbf{R}_j$ is the rotation matrix between image and object space coordinate systems. In any frame the interior orientation is given by parameters x_{0j}, y_{0j}, c_j and parameters X_{0j}, Y_{0j}, Z_{0j}, and angles of rotations φ_j, ω_j, κ_j from the matrix $\mathbf{R}_j$ represent the exterior orientation. By canceling the scaling factor λ_{ij} and re-arranging the 3-component equation (1) we will receive equations

$$\begin{aligned} x_{ij} &= -c_j f_{ij}^x + x_{oj} = \\ &= -c_j \frac{r_{11j}(X_i - X_{0j}) + r_{21j}(Y_i - Y_{0j}) + r_{31j}Z_i - Z_{0j})}{r_{13j}(X_i - X_{0j}) + r_{23j}(Y_i - Y_{0j}) + r_{33j}(Z_i - Z_{0j})} + x_{0j} \end{aligned}$$

$$(4)$$

$$\begin{aligned} y_{ij} &= -c_j f_{ij}^y + y_{oj} = \\ &= -c_j \frac{r_{12j}(X_i - X_{0j}) + r_{22j}(Y_i - Y_{0j}) + r_{32j}Z_i - Z_{0j})}{r_{13j}(X_i - X_{0j}) + r_{23j}(Y_i - Y_{0j}) + r_{33j}(Z_i - Z_{0j})} + y_{0j} \end{aligned}$$

where $r_{11j}, \ldots, r_{33j}$ are the elements of the matrix $\mathbf{R}_j$.

Depending on the parameters of the equations (4), which may be considered as known or these may be a-priori treated as unknown, while coordinates of the point images x_{ij}, y_{ij} are always regarded as known observed entities, the following problems can be solved:

1) general bundle method - all parameters on the right hand side of (4) are unknown
2) bundle method for metric camera system – interior orientation (x_{0j}, y_{0j}, c_j) is given
3) spatial resection - a) interior orientation and object point coordinates (X_i, Y_i, Z_i) are given, while the exterior orientation has to be determined
   b) only object point coordinates are given, the interior and the exterior orientation has to be determined
4) spatial intersection - both the interior and the exterior orientation are given, and the object point coordinates have to be determined.

The last item – spatial intersection of image rays through the centres of projections – is the main idea of the algorithms used in the reconstruction of real dimensions of objects from two uncalibrated photos taken by stable cameras in the surveilled protected space considered in the CamWitt application.
In videogrammetry, the following problems are distinguished.

A. Point positioning: The aim is to determine 3D coordinates of object points with a specified accuracy. Equations (4) are considered as relating observations x_{ij}, y_{ij} – coordinates of points in the images, to the parameters of the right-hand side as the relation $l = f(x)$. Estimation of x is performed using the Gauss-Markov model of the least square estimation. Linearization and introduction of a true error vector e leads to the form $l - e = \mathbf{A}x$, where $\mathbf{A}$ is $n \times u$ matrix (n is number of observations, u is number of unknown parameters), while in general $n \geq u$ and rank($\mathbf{A}$) = u. The estimation of x is usually attempted as unbiased, minimum variance estimation, performed by means of least squares.
B. Orientation: The aim is to determine parameters of the exterior orientation.

C. Calibration: The aim is to determine the parameters of interior orientation plus additional parameters for the systematic image errors. Many various calibration methods have been developed so far and are used in videogrammetry for different purposes, see e.g. in [4] and [5]. Laboratory calibration is based in using special equipment (goniometers) and various methods for determination of the interior orientation and systematic errors of the used camera, which is a quite expensive and time-consuming procedure. Reference frame calibration uses a reference object (distance bar, or regular polyhedral model) consisting of certain number of precisely measured points, which is imaged. Mathematical model is used, where the required parameters can be derived from the mathematical relations between the reference frame coordinates X, Y, Z and measured pixel coordinates x, y of the image. Point cloud calibration works with a 3D cloud of well signalized points, which need not to be measured beforehand, which is fast and cheap, but a disadvantage is, that to determine all necessary parameters, at least four images from different locations have to be taken. Self-calibration can be considered the most universal and flexible method, based on the expansion of basic equations (4). Introducing corrections Δx_{ij} and Δy_{ij} to image coordinates x_{ij} and y_{ij}, including coordinates x_{0j} and y_{0j} of the principal points and camera constants c_j, new relation system can be derived

$$x_{ij} = c_j f_{ij}^x + x_{0j} + \Delta x_{ij}, \quad y_{ij} = c_j f_{ij}^y + y_{0j} + \Delta y_{ij} \tag{5}$$

The formulation of the added parameter functions is crucial, as it can quite unpleasantly influence the physical distortion of the reconstructed 3D objects.

D. Simultaneous solution: Integration of all previous tasks.
In a general bundle concept all parameters are treated as stochastic variables. This permits to consider a-priori information about these parameters to be included, which covers both situations when parameters are excluded from the considerations or they are treated as free unknowns. The general bundle model may be extended by more equations modeling the observations of geometric constraints in a particular scene, such as parallel straight lines, horizontal and vertical lines, points in specific planes or on quadratic surfaces, see in [3], which stabilize the solution for calibrations and orientation and point

tracking. One of the parameters responsible for the system's overall precision and accuracy is the quality of measurement algorithms for image targets. In applications detecting human movement, the targeting spans from very well defined retrospective targets to the use of natural features. The accuracy of image features measurements varies according to the quality of signal content.

Detailed information about all above concepts can be found in [6] and [7].

4 Conclusions

Basic videogrammetric algorithms and methods can be used in the pletora of applications for practical purposes, when reconstructions of objects, stable or moving in the 3-dimensional scenes, are performed for various purposes. Mathematical analytic relations represented in the paper as algorithms for 3D reconstruction solve basic measurements problems and can help in tracking moving bodies in the recorded space, where calibration can be given due to stability of camera positions. Situation for moving cameras and changing scene can be much more difficult to reconstruct, as calibration is computed on the flow, using self calibration methods, and targeted object must be followed while determined by a fuzzy point cloud with changing set of determining points.

Acknowledgement

The paper was supported by grant from the Slovak Research and Development Agency awarded to the project APVV-1061-12 entitled "GEOCRIM".

References

[1] GRUEN, A. Fundamentals of Videogrammetry. In *Human Movement Science*, 16 (1997), pp.155-187.

[2] GRUEN, A. Processing of Amateur Photographs. In *The Photogrammetric record*, 11 (65), April 1985, pp. 567-579.

[3] HARTLEY, R., ZISSERMAN, A. *Multiple View Geometry in Computer Vision*. Cambridge University Press 2000, 2003, 673 pp.

[4] MA, Y., SOATTO, S., KOŠECKÁ, J., SASTRY, S., S. *An Invitation to 3-D Vision, From Images to Geometric Models*, Sprigler Verlag 2004, 535 pp.

[5] STACHEL, H. Descriptive Geometry Meets Computer Vision. In *Journal for Geometry and Graphics*, Volume 10 (2006), No. 2, pp. 137-153.

[6] VELICHOVÁ, D. Basics of Videogrammetry. In *Proceedings of the 23rd International Symposium on Computer Geometry SCG'2014*, SSGG, Bratislava 2014, SR, pp. 84-88.

[7] VELICHOVÁ, D. Videogrammetry and One of Its Applications. In *Scientific Proceedings of the 13th International Conference on Engineering Graphics BALTGRAF-13*, Vilnius, Lithuania, 2015, eISBN 978-609-457-835-8, pp. 23-28.

Current address

Velichová Daniela, doc. RNDr. CSc.
Institute of Mathematics and Physics, Facuty of Mechanical Engineering,
Slovak University of Technology in Bratislava
Nám. Slobody 17, 812 31 Bratislava, Slovak Republic
Tel. number: +4212 5729 6136, e-mail: daniela.velichova@stuba.sk

Slovak University of Technology in Bratislava
Faculty of Mechanical Engineering
16th Conference on Applied Mathematics **APLIMAT 2017**
Proceedings

REMARKS ON STRONGLY SEPARATELY CONTINUOUS FUNCTIONS $f\colon \ell^p \to \mathbb{R}$

VISNYAI Tomáš (SK)

Abstract. In this paper are investigated the strongly separately continuous functions defined on ℓ^p. We show the sufficient condition when the strongly separately continuous function defined on ℓ^p is continuous and some properties of limit function of sequence of strongly separately continuous functions on ℓ^p.

Keywords: continuity, separate continuity

Mathematics subject classification: Primary 26A15; Secondary 40A30, 54C08

1 Introduction

Separately continuous functions $f\colon \mathbb{R}^m \to \mathbb{R}$ were investigated in several papers (see e.g. [2], [4], [8], [11]). Recall that a function $f\colon \mathbb{R}^m \to \mathbb{R}$ is said to be separately continuous at a point $x_0 = (x_1^0, \dots, x_m^0) \in \mathbb{R}^m$ provided that for each $k = 1, 2, \dots, m$ the function $\varphi_k\colon \mathbb{R} \to \mathbb{R}$ defined by $\varphi_k(t) = f(x_1^0, \dots, x_{k-1}^0, t, x_{k+1}^0, \dots, x_m^0)$ is continuous at x_k^0. It is well known that a function can be separately continuous at x^0 without being continuous at x^0. The standard example illustrating this phenomenon is the function $f\colon \mathbb{R}^2 \to \mathbb{R}$ given by $f(x_1, x_2) = 0$ if $x_1.x_2 \neq 0$, while $f(x_1, x_2) = 1$ if $x_1.x_2 = 0$. This function is separately continuous at $(0,0)$ without being continuous at $(0,0)$. Further example, a function while discontinuous, is not only separately continuous, but its restriction to any straight line is continuous is $f(x,y) = \frac{x.y^2}{x^2+y^4}$ for $(x,y) \neq (0,0)$ and $f(x,y) = 0$ for $(x,y) = (0,0)$.

In the paper [4] the author introduced the notion of strongly separately continuous function $f\colon \mathbb{R}^m \to \mathbb{R}$ at x^0 and obtained the following result: A function $f\colon \mathbb{R}^m \to \mathbb{R}$ is continuous at a point x^0 if and only if it is strongly separately continuous at x^0.

In the paper [14] the authors extended the notions of separately continuous function and strongly separately continuous function to the functions defined on the space ℓ^2 and proved several basic results about such functions. They showed that there is a strongly separately continuous function f defined on ℓ^2, that is discontinuous everywhere on ℓ^2.

In the paper [15] is proved the sufficient condition when the strongly separately continuous function defined on ℓ^2 is continuous.

Finally we show when the property of strong separate continuity is transferred to the limit function of sequence of strongly separately continuous functions on ℓ^p.

2 Definitions and preliminaries

Denote by ℓ^p, $p \geq 1$ the metric space consisting of all sequences $x = (x_j)_{j=1}^{\infty}$ of real numbers such that $\sum_{k=1}^{\infty} |x_k|^p < +\infty$ endowed with the metric d defined by

$$d(x, y) = \left(\sum_{k=1}^{\infty} (x_k - y_k)^p \right)^{1/p}, \qquad p \geq 1$$

for all $x = (x_k) \in \ell^p$, $y = (y_k) \in \ell^p$.

It is known that (ℓ^p, d) is a separable and complete metric space.

If $x^0 \in \ell^p$ and $\delta > 0$, then $B(x^0, \delta) = \{x \in \ell^p : d(x, x^0) < \delta\}$ is an open-ball of x^0 (neighborhood of x^0).

Definition 2.1. a) A function $f\colon \ell^p \to \mathbb{R}$ is said to be *separately continuous at a point* $x^0 = (x_j^0)_{j=1}^{\infty} \in \ell^p$, $p \geq 1$ *with respect to a variable* x_k provided that the function $\varphi_k\colon \mathbb{R} \to \mathbb{R}$ defined by $\varphi_k(t) = f(x_1^0, \ldots, x_{k-1}^0, t, x_{k+1}^0, \ldots)$ is continuous at x_k^0. If f is separately continuous at x^0 with respect to x_k for all $k \in \mathbb{N}$, then f *is said to be separately continuous at* x^0. If f is separately continuous at every point $x^0 \in \ell^p$, $p \geq 1$, then f is said to be *separately continuous on* ℓ^p.

b) A function $f\colon \ell^p \to \mathbb{R}$ is said to be *strongly separately continuous* at a point $x^0 = (x_j^0)_{j=1}^{\infty} \in \ell^p$, $p \geq 1$ *with respect to a variable* x_k provided that for each $\varepsilon > 0$ there exists $\delta > 0$ such that $|f(x) - f(x')| < \varepsilon$ holds for each $x = (x_j)_{j=1}^{\infty} \in B(x^0, \delta)$, and $x' = (x_1, \ldots, x_{k-1}, x_k^0, x_{k+1}, \ldots)$. If f is strongly separately continuous at x^0 with respect to x_k for all $k \in \mathbb{N}$, then $f\colon \ell^p \to \mathbb{R}$, $p \geq 1$ is said to be *strongly separately continuous at* x^0. The function $f\colon \ell^p \to \mathbb{R}$ is said to be *strongly separately continuous on* ℓ^p provided that it is strongly separately continuous at every $x^0 \in \ell^p$, $p \geq 1$.

It is easy to see, that if $f\colon \ell^p \to \mathbb{R}$, $p \geq 1$ is a continuous function at a point $x^0 = (x_k^0) \in \ell^p$, then it is strongly separately continuous function at the point x^0. If $f\colon \ell^p \to \mathbb{R}$, $p \geq 1$ is a strongly separately continuous function at a point $x^0 = (x)_k^0 \in \ell^p$, then it is separately continuous function at the point x^0. The converse of these statements is not true.

Example 1. a) Let $f(x_1, x_2, \ldots) = \frac{x_1 . x_2}{x_1^2 + x_2^2}$ if $x_1 \neq 0$ or $x_2 \neq 0$, and $f(x_1, x_2, \ldots) = 0$ if $x_1 = 0$ and $x_2 = 0$. The function $f\colon \ell^p \to \mathbb{R}$ is everywhere separately continuous on ℓ^p.

b) Let $g(x_1, x_2, \ldots) = 1$ if $\sum_{k=1}^{\infty} |x_k|^p \in \mathbb{Q}$, and $g(x_1, x_2, \ldots) = 0$ if $\sum_{k=1}^{\infty} |x_k|^p \notin \mathbb{Q}$. This function is nowhere separately continuous ($\mathbb{Q}$ being the set of all rational number).

c) Let $f(x_1, x_2, \ldots) = 1$ if all $x_1, x_2 \ldots$ is rational and $f(x_1, x_2, \ldots) = 0$ otherwise. This function is separately continuous at some points of ℓ^p and is not separately continuous at some other points.

Example 2. Let $S = \{x = (x_k) \in l^p : \{k \in \mathbb{N} : x_k \text{ is a rational number}\} \text{ is a finite set}\}$. The sets S and $\ell^p \setminus S$ are dense in ℓ^p.

A subset H of ℓ^p is said to be a *set of type* (P_1) provided the following holds: if $x = (x_k) \in H$, $y = (y_k) \in \ell^p$ and the set $\{k \in \mathbb{N} : x_k \neq y_k\}$ contains at most one element, then $y = (y_k) \in H$.

3 Main results

In paper [4] was proved, that a function $f\colon \mathbb{R}^m \to \mathbb{R}$ is continuous at $x^0 \in \mathbb{R}^m$ if and only if is strongly separately continuous at x^0. In [14] is showed, that similar result is not true for function $f\colon \ell^2 \to \mathbb{R}$.

Now we will investigate this question for a function $f\colon \ell^p \to \mathbb{R}$, $p \geq 1$.

Theorem 3.1. *There exists a function $f\colon \ell^p \to \mathbb{R}$ such that f is strongly separately continuous on ℓ^p and f is discontinuous at every point of ℓ^p.*

Proof. Let $S \subseteq \ell^p$ be a set of type (P_1) such that S and $\ell^p \setminus S$ are dense in ℓ^p (see Example (2)). Let $c \in \mathbb{R}$, $c \neq 0$. Define a function $f\colon \ell^p \to \mathbb{R}$ by $f(x) = c$ for all $x \in S$ and $f(x) = 0$ otherwise. If $x^0 \in \ell^p$, then for every neighbourhood U of x^0 we have $U \cap S \neq \emptyset$; $U \cap (\ell^p \setminus S) \neq \emptyset$, and this yields that f is discontinuous at x^0. On the other hand, let $k \in \mathbb{N}$ and $x^0 = (x_j^0)$, $x = (x_j)$, $x' = (x_j')$ be arbitrary points of ℓ^p such that for all $j \neq k$, $x_j = x_j'$ and $x_k^0 = x_k'$. It is obvious that if $x \in S$, then also $x' \in S$ and if $x \notin S$, then also $x' \notin S$. Hence we always obtain $|f(x) - f(x')| = 0$ so that for each $x^0 \in \ell^p$ and each $k \in \mathbb{N}$ the function f is strongly separately continuous at x^0 with respect to x_k. □

Now, we can formulate a sufficient condition for a strongly separately continuous function to be continuous at the point $x^0 \in \ell^p$, $p \geq 1$.

Theorem 3.2. *Let $f\colon \ell^p \to \mathbb{R}$ be a strongly separately continuous at the point $x^0 = (x_j^0) \in \ell^p$. Let for each $\varepsilon > 0$ there exists $\delta > 0$ and $K \in \mathbb{N}$, such that for all $y = (y_j) \in \ell^p$ for which $\left(\sum_{j=K+1}^{\infty} |x_j^0 - y_j|^p\right)^{1/p} < \delta$ implies*

$$|f(y_1, y_2, \ldots) - f(y_1, \ldots, y_K, x_{K+1}^0, x_{K+2}^0, \ldots)| < \frac{\varepsilon}{2}.$$

Then f is continuous at $x^0 = (x_j^0) \in \ell^p$.

Proof. Let $\varepsilon > 0$. According to the assumption there exists $\delta_0 > 0$ and $K \in \mathbb{N}$ such that for each $y = (y_j) \in \ell^p$ for which $\left(\sum\limits_{j=K+1}^{\infty} |x_j^0 - y_j|^p\right)^{\frac{1}{p}} < \delta_0$ we have

$$|f(y_1, y_2, \ldots) - f(y_1, \ldots, y_K, x_{K+1}^0, x_{K+2}^0, \ldots)| < \frac{\varepsilon}{2}.$$

Since f is strongly separately continuous at $x^0 = \left(x_j^0\right) \in \ell^p$: for all $\varepsilon' > 0$ and for all $k \in \mathbb{N}$ there exists $\delta_k' > 0$ such that for all $x = (x_j) \in \ell^p$ for which $\left(\sum\limits_{j=1}^{\infty}|x_j - x_j^0|^p\right)^{\frac{1}{p}} < \delta_k'$ implies

$$|f(x_1, \dots) - f\left(x_1, \dots, x_{k-1}, x_k^0, x_{k+1}, \dots\right)| < \varepsilon'$$

holds. Let $\varepsilon' = \frac{\varepsilon}{2K}$. Then for each $k \in \{1, 2, \dots, K\}$ there exists $\delta_k > 0$ such that if $x = (x_j) \in B\left(x^0, \delta_k\right)$ then

$$|f(x_1, \dots) - f\left(x_1, \dots, x_{k-1}, x_k^0, x_{k+1}, \dots\right)| < \frac{\varepsilon}{2K}.$$

Let $\delta = \min\{\delta_0, \delta_1, \dots, \delta_k\}$ and $x = (x_j) \in B\left(x^0, \delta\right)$. Then also $(x_1^0, x_2, \dots)$, $(x_1^0, x_2^0, x_3, \dots), \dots, \left(x_1^0, \dots, x_{K-1}^0, x_K, x_{K+1}, \dots\right) \in B\left(x^0, \delta\right)$ which is the subset of $B\left(x^0, \delta_k\right)$, $k = 0, 1, 2, \dots, K$ and

$$\left(\sum_{j=K+1}^{\infty}|x_j - x_j'|^p\right)^{\frac{1}{p}} \le \left(\sum_{j=1}^{\infty}|x_j - x_j^0|^p\right)^{\frac{1}{p}} < \delta \le \delta_0.$$

Then

$$\begin{aligned}
\left|f(x_1, x_2, \dots) - f\left(x_1^0, x_2^0, \dots\right)\right| &\le \left|f(x_1, x_2, \dots) - f\left(x_1^0, x_2, \dots\right)\right| \\
&+ \left|f\left(x_1^0, x_2, x_3, \dots\right) - f\left(x_1^0, x_2^0, x_3, \dots\right)\right| \\
&+ \dots + \\
&+ \left|f\left(x_1^0, \dots, x_{K-1}^0, x_K, x_{K+1}, \dots\right) - f\left(x_1^0, \dots, x_{K-1}^0, x_K^0, x_{K+1}, \dots\right)\right| \\
&+ \left|f\left(x_1^0, \dots, x_{K-1}^0, x_K^0, x_{K+1}, \dots\right) - f\left(x_1^0, \dots, x_K^0, x_{K+1}^0, x_{K+2}^0, \dots\right)\right| \\
&< K \cdot \frac{\varepsilon}{2K} + \frac{\varepsilon}{2} = \varepsilon.
\end{aligned}$$

Hence f is continuous at the point $x^0 = \left(x_j^0\right) \in \ell^p$. □

Corollary 3.3. *Let $f\colon \ell^p \to \mathbb{R}$ be strongly separately continuous on ℓ^p, $p \ge 1$. Let for each $\varepsilon > 0$ there exists $\delta > 0$ and $K \in \mathbb{N}$ such that for all $x, y \in \ell^p$: $\left(\sum_{j=K+1}^{\infty}|x_j - y_j|^p\right)^{1/p} < \delta$ implies*

$$|f(y_1, y_2, \dots) - f(y_1, \dots, y_K, x_{K+1}, x_{K+2}, \dots)| < \varepsilon.$$

Then f is continuous on ℓ^p.

Remark. Theorem 3.1 shows that neither strongly separately continuous nor separately continuous function $f\colon \ell^p \to \mathbb{R}$, $p \ge 1$ belong to the first Baire class. The function f defined in the proof of Theorem 3.1 does not belong to B_1 because the set of all discontinuity points of g is a set of the second Baire category.

Next we prove that if a sequence $(f_n\colon \ell^p \to \mathbb{R})_{n=1}^{\infty}$ converges pointwise to a function $f\colon \ell^p \to \mathbb{R}$ and all f_n are (strongly) separately continuous, then the function f need not be separately continuous.

Theorem 3.4. *There exists a sequence $(f_n\colon \ell^p \to \mathbb{R})_{n=1}^{\infty}$ of strongly separately continuous functions on ℓ^p such that it converges pointwise to a function $f\colon \ell^p \to \mathbb{R}$ which is not separately continuous on ℓ^p.*

Proof. Define for fixed $n \in \mathbb{N}$ the function f_n in the following way. If $x = (x_1, x_2, \dots) \in \ell^p$ then $f_n(x) = f_n(x_1, x_2, \dots) = g_n(x_1)$, where g_n is a continuous function defined on $\mathbb{R}$, where $g_n(\frac{1}{k}) = 1$, $k = 1, 2, \dots, n$, $g_n(x) = 0$ for $x \leq 0$ and $x \geq 2$ and the graph $g_n(x)$ consists of legs of not intercepting isoscele triangles with height equals to 1 at the points $x = \frac{1}{k}$, $k = 1, 2, \dots, n$.

It is ease to see that the $g_n\colon \mathbb{R} \to \mathbb{R}$, $n = 1, 2, \dots$ is continuous on $\mathbb{R}$. Under the definition 2.1 b) every function f_n $(n = 1, 2, \dots)$ is strongly separately continuous at each point $x^0 \in \ell^p$. Moreover f_n is continuous in each point $x^0 = (x_j)^0 \in \ell^p$ because $|f_n(x) - f(x^0)| = |g_n(x_1) - g_n(x_1^0)|$. Considering that the function g_n is continuous on $\mathbb{R}$, f_n is continuous on ℓ^p hence it is strongly separately continuous on ℓ^p. For the limit function $f = \lim_{n\to\infty} f_n$ where $x = (x_j) \in \ell^p$ we have $f(x) = f(x_1, x_2, \dots) = g(x_1)$ where $g(x_1) = 0$ for $x \leq 0$ and $x \geq 2$ and on $(0, 2)$ to the graph belong legs of not intercepting isoscele triangles with height equals to 1 at the points $x = \frac{1}{k}$, $k = 1, 2, \dots$ and otherwise on $(0, 2)$ we have $g(x) = 0$. We see that g is not continuous at the point $x_1 = 0$ from right-hand side. Hence f is not separately continuous at the point $x = (0, 0, 0, \dots)$ for $x_1 = 0$. □

It is natural to ask whether some of various types of convergence of functions which are stronger than the pointwise convergence can guarantee that the limit function of a sequence of (strongly) separately continuous functions on ℓ^p, $p \geq 1$ with respect to this type of convergence is also a (strongly) separately continuous function on ℓ^p, $p \geq 1$. Next we show that there is a weaker type of locally uniform convergence which fulfills this requirement in the case of strongly separately continuous functions on ℓ^p.

Definition 3.5. Let X be a metric space, $(f_n\colon X \to \mathbb{R})_{n=1}^{\infty}$ be a sequence of functions and $x^0 \in X$. A sequence $(f_n)_{n=1}^{\infty}$ is said to converge *weakly locally uniformly* to a function $f\colon X \to \mathbb{R}$ at x^0 if for every $\varepsilon > 0$ there exist $\delta > 0$ and $p \in \mathbb{N}$ such that $|f_n(x) - f(x)| < \varepsilon$ holds for each $n \in \mathbb{N}$ with $n \geq p$ and each $x \in B(x^0, \delta)$.

If a sequence $(f_n)_{n=1}^{\infty}$ converges weakly locally uniformly to a function f at every point $x^0 \in X$, then it is said to converge *weakly locally uniformly* to f on X.

Theorem 3.6. *If a sequence $(f_n\colon \ell^p \to \mathbb{R})_{n=1}^{\infty}$, $p \geq 1$ converges weakly locally uniformly to $f\colon \ell^p \to \mathbb{R}$ at $x^0 \in \ell^p$ and for each $n \in \mathbb{N}$ the function f_n is strongly separately continuous at x^0, then the function f is also strongly separately continuous at x^0.*

Proof. Let $k \in \mathbb{N}$. We will prove that f is strongly separately continuous at x^0 with respect to x_k. Let $\varepsilon > 0$. Since $(f_n)_{n=1}^{\infty}$ converges weakly locally uniformly to f at x^0 there exist an open ball $B(x^0, \delta_1)$ and $m \in \mathbb{N}$ such that $|f_n(x) - f(x)| < \frac{\varepsilon}{3}$ holds for all $n \geq m$ and $x \in B(x^0, \delta_1)$. The function f_m is strongly separately continuous at x^0 with respect to x_k and it follows that there exists $\delta_2 > 0$ such that $|f_m(x) - f_m(x')| < \frac{\varepsilon}{3}$ holds for each $x = (x_j)_{j=1}^{\infty} \in B(x^0, \delta_2)$ and $x' = (x_1, \dots, x_{k-1}, x_k^0, x_{k+1}, \dots)$. Put $\delta = \min\{\delta_1, \delta_2\}$. Then for each $x \in B(x^0, \delta)$ we obtain that $|f_m(x) - f_m(x')| < \frac{\varepsilon}{3}$, $|f_m(x) - f(x)| < \frac{\varepsilon}{3}$ and because $\varrho(x', x^0) \leq \varrho(x^0, x) < \delta$ we have also $|f_m(x') - f(x')| < \frac{\varepsilon}{3}$. Hence, for all $x \in B(x^0, \delta)$ we obtain $|f(x) - f(x')| \leq |f(x) - f_m(x)| + |f_m(x) - f_m(x')| + |f_m(x') - f(x')| < \varepsilon$ and this yields that f is strongly separately continuous at x^0 with respect to x_k. □

Acknowledgement

This paper was supported by Cultural and Educational Grant Agency of the Slovak Republic under the KEGA grant No. 047STU-4/2016.

References

[1] Bruckner,A.M.: Differentiation of Real Functions, *Spinger - Verlag, Berlin - Heidelberg - New York*, 1978

[2] Carrol,F.M.: Separately continuous functions, *Amer, Math. Monthly* **78** (1971), 175

[3] Drahovský,Š. - Šalát,T. - Toma,V.: Points of uniform convergence and oscillation of sequences of functions, *Real Anal. Exchange* **20** (1994-95), 753-767.

[4] Dzagnidze,O.P.: Separately continuous functions in a new sense are continuous *Real Anal. Exchange* **24** (1998-99), 695-702.

[5] Goffman,C.: Reelle Funktionen, *Bibiographiseles Institu, Mannheim - Wien - Zürich*, 1976

[6] Kuratowski,K.: Topologie I, *PWN, Warszawa*, 1958

[7] Legéň,A - Šalát, T.: On some applications of the category method in the theory of sequence spaces (Russian), *Mat.-fyz. čas. SAV* **14**, 217-233.

[8] Mareus,S.: On functions continuous in each variable, *Doklady AV SSSR* **112** (1957), 812-814.

[9] Šalát, T.: On transfinite sequences of B-measurable functions, *Fund. Math.* **LXXVIII** (1973), 157-162.

[10] Sierpínski,W.: Sur les suites transfinies convergentes de fonctions de Baire, *Fund. Math.* **I** (1920), 132-141.

[11] Sierpínski,W.: Sur une propiété de fonctions de deux variables réelles, continues par rapport ã chacune de variables, *Publ. math. Univ. Belgrade* **1**, 125 - 128.

[12] Sikorski,R.: Real Functions I (Polish), *PWN, Warszawa*, 1958

[13] Tolstov,G.P.: On partial derivatives (Russian) *Izv. Akad. Nauk SSSR*, **13** (1949), 425-446.

[14] Činčura,J. - Šalát,T. - Visnyai,T.: On separately continuous functions $f: \ell^2 \to \mathbb{R}$ *Acta Academiae Paedagogicae Agriensis*, Sec. Math. **31** (2004), 11-18.

[15] Visnyai,T.: Strongly separately continuous and separately quasicontinuous functions $f: \ell^2 \to \mathbb{R}$ *Real Analysis Exchange* **38(2)** (2012/2013), 499-510.

[16] Lebesgue,H.: Sur les functions reprezentable analytiquement *J. Math. Pure Appl.* **6** (1905), 139-216.

Current address

Visnyai Tomáš, PaedDr. PhD.
Institute of Information Engineering,
Automation and Mathematics Faculty of Chemical and Food Technology
Slovak University of Technology in Bratislava
Radlinského 9, 812 37 Bratislava
Tel. number: +421 259 325 336, e-mail: tomas.visnyai@stuba.sk

Slovak University of Technology in Bratislava
Faculty of Mechanical Engineering
16th Conference on Applied Mathematics **APLIMAT 2017**
Proceedings

APPLICATION OF INFORMATION TECHNOLOGIES FOR STUDIES OF MATHEMATICS IN RIGA TECHNICAL UNIVERSITY

VOLODKO Inta (LV), CERNAJEVA Sarmite (LV)

Abstract. Utilisation of ICT capabilities in the mathematics studies process facilitates work of teachers and also makes study process more engaging and effective. In order to facilitate the work of teachers and excite students, the Department of Engineering Mathematics created several courses of Mathematics at RTU portal ORTUS, compiled and implemented a series of tests at the ORTUS environment, created Ancillary Course in Elementary Mathematics on the open online platform MOOC and made course of mathematics for pupils of secondary school at RTU ORTUS site.

Keywords: higher education, e-learning, mathematical software in teaching mathematics, students' testing, portal ORTUS

Mathematics Subject Classification: 97D50, 97C70, 97D60

1 Introduction

Nowadays, when realm of the technologies increases the demand for qualified work force with competency in high-tech, latest scientific discoveries, state-of-art technologies and materials and able of developing competitive global business also grows. Therefore, contemporary education system shall be prepared of providing young persons with education that would advance their competitiveness and the level of competency. In context of this, faculty of the mathematics also shall develop contemporary teaching methods based on new technologies that would provide the students with grounds for mathematics knowledge application already while the study process.

2 E-learning at RTU

While here are going on discussions about education quality and its improvement in universities of Latvia, Riga Technical University (RTU) has turned to unified e-learning system ORTUS located at RTU website starting from academic year 2007/2008, which is based on MOODLE (Modular Object-Oriented Dynamic Learning Environment) e-learning software. ORTUS in Latin means *new beginning*. The RTU site ORTUS (www.ortus.lv) facilitates learning process of any student as long as the one desires making learning process more efficient, advanced and expeditious. One can reach the university by means of ORTUS from anywhere at any time. At ORTUS site every student of RTU can:

- obtain the information on syllabus and study results;
- read lecture abstract and other study materials;
- access scientific databases;
- read particularly addressed news and updates on current events;
- use the RTU forum as means of communication with the other university students;
- acquaint oneself with the laws and regulations of the RTU;
- create one's RTU email account.

Each subject has a separate environment where one can both interact with course mates and the teaching staff, as well as find out information about the topics of the subject and the list of teaching staff requirements the students shall commit with for passing the course successfully.

Under the e-learning menu of ORTUS website, there are available following courses developed by Department of Engineering Mathematics - Mathematics I, Mathematics II (for semesters one and two), Discrete Mathematics, Fundamentals of Mathematics. Everyone of department's teaching staff has been working with ORTUS system since academic year 2008/2009. RTU ORTUS e-learning environment renders students with easy access to electronic study materials and tests of mathematics.

Changes of the engineers practical studies are driven by trends of day-to-day work when engineers shall solve intricate combinations of technological and organisational issues, furthermore, the engineering projects will be ever closer constrained with the social and political considerations, new business models. Therefore, changes of practical studies shall be driven considering several important criteria:

- sustainability - this requires a systematic and integrated approach to engineering projects and activities to reduce social and environmental impact of engineering,
- complexity - unexpected phenomenon may appear caused by increasing amounts of information, intricacy of the projects and systems,
- professional cooperation - improve results in all forms of contemporary business activities by learning how to cooperate, not compete,
- risk management - be conscious of unpredictable consequences to health, environmental, safety, and other dimensions that might be induced by complex projects therefore, which requires critical thinking and knowledge in these areas,
- ability of spotting opportunities - sophisticated contemporary projects demand for talented engineers with effective problem-solving customer skills and able satisfy customers' requirements.

Currently, while teaching mathematics at Faculties of the Civil Engineering and the Computer Science and Information Technology, we are also challenged with the matter of exercising the acquired mathematics knowledge in professional activities of specialists. The authors are looking for the answers to the questions in this paper: what is the very thing that should be taught, and how this should be done to make knowledge of mathematics being applicable for professional activity.

While preparing for practical exercises, one shall consider background, intellectual capacity and reasoning faculty of the students. The proportion of contact-hours and self-contained studies shall be adjusted considering complexity and relevance of the theme to the other subjects incorporating mathematics.

3 Application of mathematical software in teaching mathematics

The integrated computer software for automated mathematical calculations MATHCAD has gained vast popularity since the eighties. The software was developed by company Mathcad Soft (US). The calculations within framework of the system are done using familiar mathematical formulas and symbols.

Several versions of the software that were released on a global market within recent years testify the popularity of the program. The software operates at Windows environment and is fairly easy to use due to familiar format of mathematical expressions.

Like other systems, the MATHCAD software has been continuously developed and improved by releasing a wide range of versions. The MATHCAD renders both simple and complex solutions, including. Problems can be solved both numerically and analytically using method of characteristics. The data and results can also be depicted in diagrams.

The MATHCAD has been demonstrated to the RTU students of Faculty of Civil Engineering on mathematics lectures. However, computer cannot resolve the problem the program user does not comprehend and is not able to select the appropriate algorithm. Even though, the algorithm has been formulated, it must be entered in a computable form.

Educators-constructivists enable students examining their own new ideas, evaluate the information, and search for new solutions for various situations. Context of the studies also is important as it shall encourage creative thinking and critical approach [3].

The students of the Faculty of Computer Science and Information Technology had training on programme package MATHEMATICA in parallel to regular mathematics lectures during the period from 2005 to 2014. During the last two academic years, we turned to program package MATLAB. First year students have classes at computer lab for two semesters where they can practice mathematical software aided problem solving close to the problems solved during practical exercises. At the computer lab students learn how to

- set function and define precision of calculated values with the aid of MATHEMATICA or MATLAB;
- perform calculations on matrices and vectors;
- plot graphs of functions defined in Cartesian coordinates, polar coordinates or parametric;
- draw a line or surface in space;
- solve algebraic equation or system of equations;
- calculate limit of the function and apply it for determination of asymptote;
- calculate derivative and apply it for functions' exploration;

- calculate definite, indefinite and multiple integral as well as applying it for practical solutions;
- solve differential equations;
- find the power series expansion for given function and determine interval of the convergence.

Students perfect the acquired knowledge of MATLAB during Numerical Methods course in second study year.

As experience shows, students are learning mathematical software with interest. While learning this, students more easily comprehend algorithm of the problem. Many of those students, who have poor results in mathematics and cannot solve the problem on their own on paper, are doing pretty well on MATLAB or MATHEMATICA environment. Besides, using the mathematical software, students can solve labour-consuming problems in a short time, which would require a lot of time and effort if solved on paper.

Utilisation of ICT capabilities in the mathematics studies process facilitates work of teachers and also makes study process more engaging and effective. This also significantly raises quality of the higher education and preparedness of new professionals for labour market.

4 Assessment of students' knowledge by means of tests

Nowadays, technologies increasingly penetrates not only various areas of our life, but also the field of education. The information technologies are being increasingly used by institutions of higher education. Admittedly, the information technologies enters realm of mathematics as well. There are described 10 main teaching methods and evaluated comparative effectiveness of these methods in literature [2]. Students' testing is mentioned as one of the main methods with high efficiency appreciation.

New ideas and methods usually provoke controversial opinions. Similarly, there are antipodal opinions about the use of tests in tuition, although this merely partly can be considered a new method. Still, here are quite opposite opinions about the tests, their usefulness, and objectivity of the assessment and sometimes the same arguments are used from both sides, i.e., from those who are for and those who are against the knowledge testing.

The main arguments *against* the tests are:

- tests reduce the role of the lecturer;
- the tests do not provide an objective assessment of teachable person, etc.

The main arguments *for*:

- the tests make the education process less labour-intensive;
- the tests enable objective knowledge assessment of students, etc.

The above raises the question: Do students' testing provide objective results and what should be done to obtain these?

Here shall be considered important components of the testing for receiving positive answer to this question. They are:

1) setting objectives of the test;
2) selection of proper question type;
3) exact wording of the question;
4) evaluation and interpretation of the test results.

One must set clear objectives that should be achieved prior to start developing the test. If there are several objectives, their relevance has to be determined, the least important ones should be eliminated, and an equal number of questions should be compiled for the each important objective. The next important point of the test development is selection of the correct question type. When considering the type of question, the compiler should anticipate possible student errors in multiple choice questions. Better option is choosing questions that students have to answer themselves. In addition, the question should be brief and clearly defined, with plain and easy perceivable wording. The question should be made in a manner that does not cause big problems for the student and it could be answered with aid of textbook, by reminiscing facts and algorithms. It is even better, if the questions could be compiled in a manner, which enables verifying intermediate results and tracking errors.

Evaluation and interpretation of results is an essential component of the test. Comprehensive analysis of the results enables drawing conclusions of what aspects should be addressed further, what are the ways for improvement of the tuition methods and quality of the learning process. The mathematical and statistical methods are widely used for summarising and evaluation of the test results. For many of them the computers are effectively used that enables automation and optimisation of the intended tasks.

In order to facilitate the work of teachers and excite students, the faculty of Engineering Mathematics Department of RTU compiled and implemented a series of tests at the ORTUS environment, which replaced most of the term one homework. First year students have to perform 13 tests of seven mathematics chapters during the term one:

1. Linear Algebra:
   1.1. Determinants.
   1.2. Operations on Matrices.
   1.3. Systems of Linear Equations.
2. Vector Algebra:
   2.1. Linear Operations and Scalar Product of Two Vectors.
   2.2. Vector Product of Two Vectors and Triple Scalar Product.
3. Analytical Geometry:
   3.1. Straight Line in Plane.

3.2. Plane and Straight Line in Space.

4. Limits:

   4.1. Calculating Limits 1.

   4.2. Calculating Limits 2.

5. Analysis of Functions:

   5.1. Extreme, the Maximum and Minimum Value of a Function.

   5.2. Points of Inflexion and Asymptotes of Functions.

6. Derivative of Functions of Several Variables:

   6.1. First Order Partial Derivatives of Functions of Several Variables.

   6.2. Higher Order Partial Derivatives of Functions of Two Variables.

   6.3. Extreme of Functions of Two Variables.

7. Complex Numbers.

   7.1. Operations on Complex Numbers of Algebraic Form.

   7.2. Operations on Complex Numbers of Trigonometric and Exponential Form.

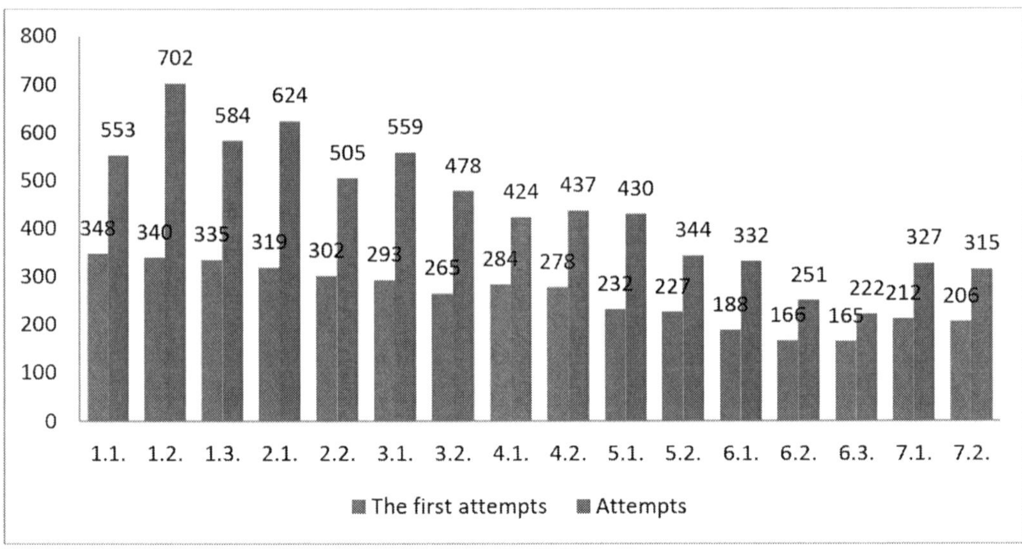

Fig. 1. The number of students who performed the test and the number of attempts

All the above tests are not labour-consuming, each of which contains from two to five problems. Students shall not pick the right answer from multiple-choice question, but shall solve and write the result for every problem in all tests. The test must be executed within two hours as it locks by lapse of two hours. There are allowed three attempts for each test, though the problem changes with each try – the numbers and functions are different, yet the type of the problem remains. The best of achieved results is taken for the final score of the test.

Data of the diagrams in Figures 1 and 2 are derived from results of the first year students of the Faculty of Computer Science and Information Technology (abbreviation FCSIT) of the last academic year (2015/16). There are enrolled 433 students in this course. In Figure 1, the first bar indicates the number of students who have performed the respective tests; the second bar - number of attempts, i.e. how many times the tests have been filled all in all.

Figure 2 shows the number of students who got the highest possible score (percentage of total number of students who performed the tests).

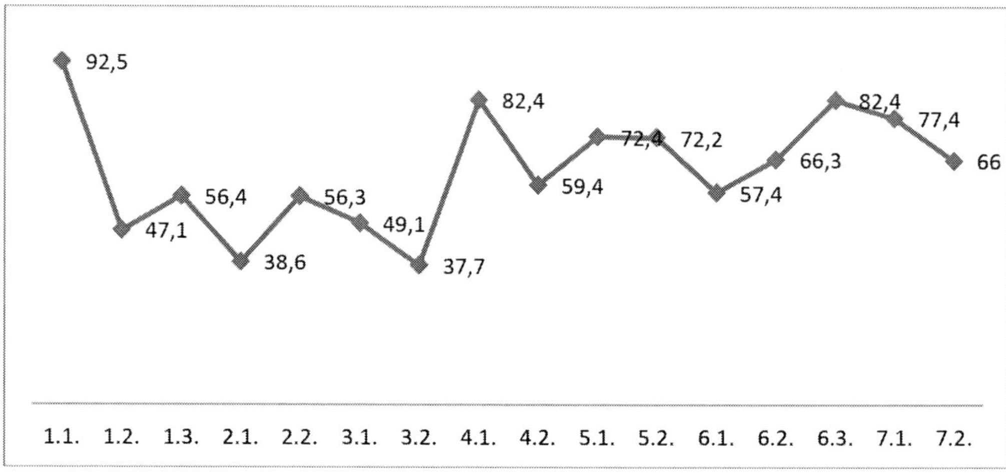

Fig. 2. Percentage of the students who got the highest possible score

By analysing statistics of the tests results, we can conclude:

- about 80% of students (consequently, the most part) perform these tests;
- if a student has not acquired the maximum score at the first attempt, the student performs the test repeatedly;
- from all students who performed the tests, 81% score for the highest rating on average;
- 62% of the students get the highest score already at the first attempt on average;
- if there is a choice, the majority of students opt for performance of the tests at ORTUS, instead of writing homework manually.

5 Ancillary course in Elementary Mathematics on the open online platform mooc.rtu.lv

Open online platform mooc.rtu.lv (MOOC — massive open online course) has been created in autumn 2015 by joint effort of several departments of the Riga Technical University. One of the courses available on this platform is Ancillary Course in Elementary Mathematics. This course enables the students enhancing their knowledge by learning the mathematics in a new, simple and interactive way. The ancillary course comprises six different topics of elementary mathematics curriculum, each of which is divided into three parts: theory, video materials and tests. The user completes the course successfully if one passes all of the tests good (minimum rating – four points). The students shall be very attentive and careful while answering questions of the test as each question can be answered only once. The RTU students' course can be accessed by the ORTUS password. Currently, there is being developed ancillary courses of higher mathematics and physics on the platform.

Ancillary course in elementary mathematics discusses chapters, which the students need to learn for successfully studies of the higher mathematics. The course aims rendering interactive, topical, accessible ancillary educational materials for students and pupils. The main aim of the subject is learning the elementary mathematics, without which successful studies of higher mathematics are unworkable.

There are six different topics of elementary mathematics in the course:

1. Arithmetic operations with fractions (addition, subtraction, multiplication, division).
2. Algebraic expressions, their transformations. Simplification of algebraic expressions. Formulas of abridged multiplication.
3. Linear equations and quadratic equations as well as linear and quadratic inequalities.
4. Domain of a function, range. Values of simple functions. Even and odd function. Ways of representing functions. Graphs of elementary functions.
5. Power properties. Exponential equations and inequalities. Properties of the logarithm. Logarithmic equations and inequalities. Application of properties for solution of equations and inequalities.
6. Trigonometric functions of numeric variables. Trigonometric functions and their characters. Basic trigonometric identities.

As mentioned above, each topic comprises theory, problems and tests. If a student completes the exercise and hands-in the answer, the system shows the right solution with detailed explanation regardless the matter whether the given answer was correct or not. Statistics of the course show that the tests are attracting the greatest interest for the present. Figure 3 shows percentage allocation of the course views by type: three quarters of all views or 75.93% constitute the tests, whereas the theory and problems have got less attention - respectively 12.92% and% 11.15 views.

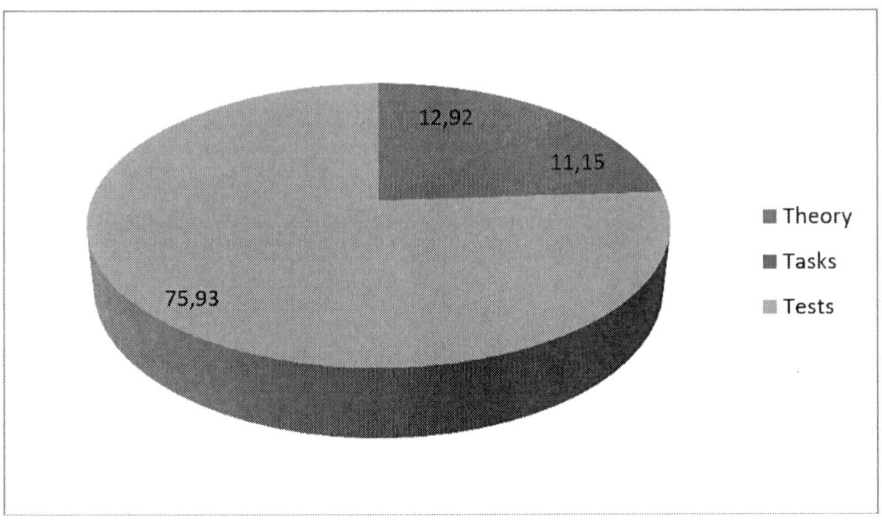

Fig. 3. Allocation of the ancillary course of elementary mathematics views by type (percentage)

As large amounts of learning materials represent the challenge for the students, arrangements of both effective use of visual aids, and communication environment, which inspires students for curiosity and desire for good results, shall be effective. The visual aids facilitate recapitulation process for the students; therefore, these are of great attention in the course. The learning material have been visualised of greatest extent in the MOOC course, it provides possibilities for viewing the material both in general and drill down to the detail. Online courses can be a great tool facilitating professional development where every person can choose a suitable pace. The basics of theory and explanations are provided in parallel to the examples of problem solutions. The material is the aid for the students who are looking for concise information, who are negative about exercising compulsory homework, yet are looking for learning materials suitable for self-contained studies - concise, comprehensible, with pre-set problems. Global statistics of universities show that popularity of free access massive online course or MOOC method growth, though currently, the challenging question is - how to improve students' results, as the global universities, which offer these courses, have concluded that merely ten percent of students have good score.

6 Offer of RTU site ORTUS for secondary school audience

The university have understood however that the student comes from a school desk, thus his interest in exact sciences shall be roused since secondary school classroom. There is alarming fact that almost 60% of students had enrolled for studies of the social sciences and arts, while merely 18% for engineering and natural sciences pursuant to the data provided by the Ministry of Education and Science of Latvia in 2009. This trend could be explained with the secondary school graduates belief that soft sciences are the easiest solution. Though, recruitment experts emphasise that the demand in every respect will grow for specialists with hard sciences background in the future. Therefore, RTU is inclined of providing more opportunities for learning the hard sciences for the site users using pupils' favourite environment - the internet.

Riga Technical University presented new ORTUS site for secondary school audience in 8 January 2010. The purpose of the site is building the peer group of prospective students where pupils could live students' part ahead of time, thus assisting them at their choice of future study program they would enrol after secondary school graduation. The portal will provide support to all young people willing to study regardless of their place of residence, they will be able to learn everything they need to know about students' life at RTU, not visiting the university.

There has been made course of mathematics for pupils of secondary school at RTU ORTUS site, which provides opportunity of additional self-contained studies of mathematics and advances preparation for Central Exam (CE) in mathematics.

Pupils can use following themes of the preliminary course of mathematics:

1. Numbers and operations on them.
2. Algebraic expressions, their identical transformations.
3. Concept of a function. Linear, quadratic, and power functions.
4. Algebraic equations. Systems of equations with two variables.
5. Algebraic inequalities. Systems of inequalities.
6. Exponential functions and their properties. Exponential equations, exponential inequalities.
7. Logarithmic function and their properties. Logarithmic equations. Logarithmic inequalities.

8. Trigonometric functions and their properties. Trigonometric equations. Trigonometric inequalities.
9. Elements of combinatorics. Concept of probability.
10. Vectors in plane and operations on them.
11. Plane geometry.
12. Space geometry.

The large number of subscribers, i.e., 12232 (01.09.2010) for *Preparation Course for Central Exam in* ***Mathematics*** give proof of pupils' interest in materials under the mathematics tab of this site.
The portal provides support to all young people willing to study regardless of their place of residence, they will be able to learn everything they need to know about students' life at RTU, not visiting the university. We care for developing a dialogue with pupils using the ORTUS site, thus facilitating successful beginning of studies at the university and make higher engineering education more accessible to every young person.

It is important to rouse the interest and to prepare a young person for studies already while secondary school, thus building person's confidence of mental faculties and rightness of choice. E-learning site ORTUS is a great support for today's pupils of a secondary school for in-depth learning of hard sciences and for building a link with the RTU students' life.

ORTUS site is an integral part of Quality Management System of the RTU curriculum - ORTUS currently is the most advanced, comprehensive and multifunctional university website in Latvia. It is being visited by more than 10000 users every week. It shall be mentioned that the Information Technology Service of the Riga Technical University has been awarded recognition in nomination *Platīna pele 2009* (The Platinum Mouse 2009) of the Latvian Information and Communication Technology Association (*LIKTA*) for outstanding achievement in building ORTUS site at RTU website.

7 Conclusions

1) The portal ORTUS is an important part of the quality management system of studies of RTU which helps both students and teaching staff in the study process.
2) Using of the computer mathematical systems brings innovation into teaching of mathematics and makes it more attractive. The work with MATLAB, MATHCAD or MATHEMATICA helps students to understand mathematics easier.
3) Assessment of students' knowledge by means of testing enables wide modernization and optimization opportunities of the tuition process, but it cannot be the only evaluation method, though properly compiled tests and proper evaluation system of the results in combination with other assessment methods provide objective results.
4) Author's created and rehearsed material in an open online site platform mooc.rtu.lv can help students and pupils understand separate mathematical questions, give support for acquiring mathematics independently, help prepare for mathematics tests and as a result promote pupils' motivation for an active acquisition of mathematics.

5) It is important to awake interest and prepare a young person for studies in secondary school. The portal ORTUS is a great support for a modern secondary school student to acquire exact sciences in depth and to create links with the life at RTU.

6) Expedience and the necessary amount of the subjects that are included in the study programs of universities are evaluated by the input that each subject renders to achieve the common study goals of the program. To ensure a good quality study process, knowledge, skills and competences that will be provided by acquiring the corresponding subject must be acknowledged, as well as their usefulness in the assessment of the employer. The quality of the students' competences is not dependant only on the amount of knowledge in some subject, but also on the acquired ability and skill to solve problems independently. That is why the goal of teaching mathematics is firstly determined by the necessity for students to be prepared to practically use the acquired subject in life, and that is why study process must be purposefully created.

References

[1] CERNAJEVA, S., VOLODKO, I. Assessment of Students' Knowledge by Means of Tests. In *Rural Environment. Education. Personality. (REEP): Proceedings of the 9th International Scientific Conference*, Latvia, Jelgava, 13-14 May, 2016. Jelgava: Latvia University of Agriculture, 2016, ISBN 978-9984-48-221-7, ISSN 2255-808X, pp.328-333.

[2] DUNLOSKY, J., RAWSON, K.A., MARSH, E.J., NATHAN, M.J., WILLINGHAM, D.T. Improving students' learning with effective learning techniques: Promising directions from cognitive and educational psychology. In *Psychological Science in the Public Interest, Supplement*, 2013, 14 (1), 4-58.

[3] GEIDZS, N.L., BERLINERS, D.C. Pedagogical psychology. *Zvaigzne ABC*., Riga, 1999, 662.p. [In Latvian]

[4] VOLODKO, I., CERNAJEVA, S. Assessment of Students' Knowledge by Means of Tests. In:*Rural Environment. Education. Personality. (REEP): Proceedings of the 9th International Scientific Conference*, Latvia, Jelgava, 13-14 May, 2016. Jelgava: Latvia University of Agriculture, 2016, ISBN 978-9984-48-221-7, ISSN 2255-808X, pp.328-333.

[5] VOLODKO, I., ROSCINA, I. The Use of ORTUS Portal for Teaching Mathematics in Riga Technical University. In: *Некоторые актуальные проблемы современной математики и математического образования: материалы научной конференции "Герценовские чтения - 2012"*, Russia, Санкт Петербург, 16-21 April, 2012. Санкт Петербург: Библиотека Академии наук, 2012, pp.168-173. [In Russian]

[6] VOLODKO, I., ROSCINA, I. Using the Package MATHEMATICA for Teaching Mathematics in Riga Technical University. In: *Труды международной научной конференции „Образование, наука и экономика в вузах и школах. Интеграция в международное образовательное пространство"*, Armenia, Цахкадзор, 24-29 March, 2014. Цахкадзор: 2014, pp.482-485. [In Russian]

Current address

Volodko Inta, professor
Department of Engineering Mathematics, Institute of Applied Mathematics,
Faculty of Computer Science and Information Technology, Riga Technical University
2 Daugavgrivas street, Riga, LV-1007, Latvia
Tel. number: (+371)67089528, e-mail: inta.volodko@rtu.lv

Cernajeva Sarmite, lecturer
Department of Engineering Mathematics, Institute of Applied Mathematics,
Faculty of Computer Science and Information Technology, Riga Technical University
2 Daugavgrivas street, Riga, LV-1007, Latvia
Tel. number: (+371)67089528, e-mail: sarmite.cernajeva@rtu.lv

Slovak University of Technology in Bratislava
Faculty of Mechanical Engineering
16th Conference on Applied Mathematics **APLIMAT 2017**
Proceedings

DEVELOPING MATHEMATICAL MOBILE APPLICATIONS WITHOUT PROGRAMMING SKILLS

VOŠTINÁR Patrik (SK), HANZEL Pavol (SK)

Abstract. In this article we are focusing on developing mathematical mobile applications without programming skills. We will show you how to create mathematical application in environment App Inventor.

Keywords: ICT, App Inventor, mobile application, programming, m-learning

Mathematics subject classification: 97U20, 97G50, 97U50

1 Mobile applications in teaching

We live in times, were mobile devices are part of lifestyle among young people, who can not imagine their existence without it. Most of the users of mobile devices are searching applications in specialized stores with mobile applications. In case of certain application can not be found, users must wait till developers develop that application or pay someone money to develope it. Google and MIT decided to create environment, in which users can develope their own application without previous programming knowledge. This environment is called App Inventor.

App Inventor is based on results of research findings in area of teaching computer systems and on work of the Google company in area of online environment of development. Visual interface is similar to environment of programming language Scratch (by MIT). [1]

In 2015, the MIT App Inventor community consists of nearly 3 million users representing 195 countries. [1]

Creating of mobile applications can represent unique motivational power, by which mobile devices can be used for education. Nowadays in Slovakia are running few researches to include mobile devices in teaching mathematics. One of these researches is running in High School "Gymnáziu Jozefa Gregora Tajovského v Banskej Bystrici" in Banska Bystrica with positive rating from students (Bayerl, Žilková, 2016).

App Inventor can help teachers on each school level to teach easier and students can better understand teaching material, by creating its own application on which they can better practise given study material. With App Inventor teachers can create own applications, which will be used at their classes and motivation for learning will rise among the students. In this article focus will be put on the second option of using App Inventor – display of usage of application developed in this area. Application is named „Matematika" and its main sense is practise of summation, subtraction, multiplication and division of whole numbers. Application is available on Google Play store.

2 APP INVENTOR

App Inventor is free open-source tool, which allows user to create applications for Android platform. By GUI tool user interface is created and by „blocks" codes which are put together like puzzle it is determined how application will work.

Advantage of App Inventor is that all commands are located in same part and in that case they do not have to be memorized, all is needed to search for the command and with simple movement with computer mouse move to the program. All commands must collide, for example if some command code is expecting a number, other code can not be used. In part with the order of commands, application itself knows and recognize when certain code can be used.

Among other advantages of App Inventor environment, belongs support of advanced technologies in mobile devices like GPS sensor, camera, SMS, telephoning, access to contacts, motion sensor – gyroscope, support of control LEGO robots, database, etc.

The App Inventor is online software, we don't need to download it and install it into our computers. All we need is a computer connected to the internet. For login the only thing we need is an account in Google (for example GMAIL account). We can test created application on real devices or in the emulator, which is part of App Inventor.

App Inventor is online tool, meaning to work on it a computer with internet access is needed. If user has created Google account (i.e. GMAIL account), then no registration is needed, all that is needed is to sign in under this account. Applications should be tested on real devices, if real devices are not available it can be tested on emulator, which is part of the App Inventor.

On the web page of App Inventor[1] there are tutorials, which describe programming in this platform.

App Inventor is available on the internet http://appinventor.mit.edu/explore/index-2.html. In order to create application, the user have to click on the orange button "Create apps!" (Fig. 1).

[1] http://appinventor.mit.edu/explore/resources.html

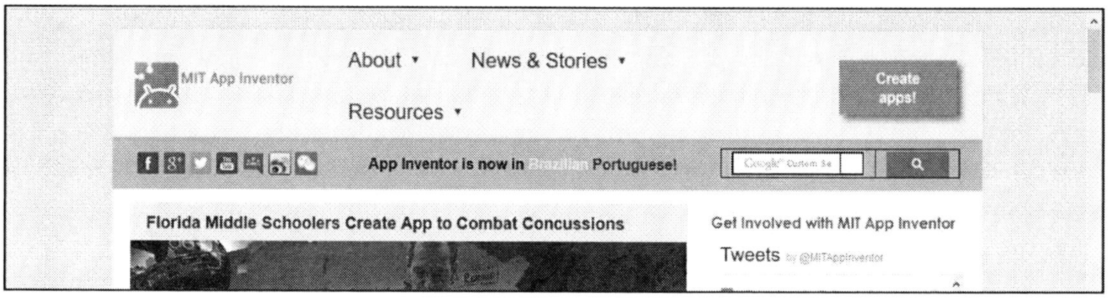

Fig. 1. App Inventor website

After the user presses this button the main screen will appear. On this screen there is a list of created projects (Fig. 2). In case that the user isn't signed in into Google account, login window will appear.

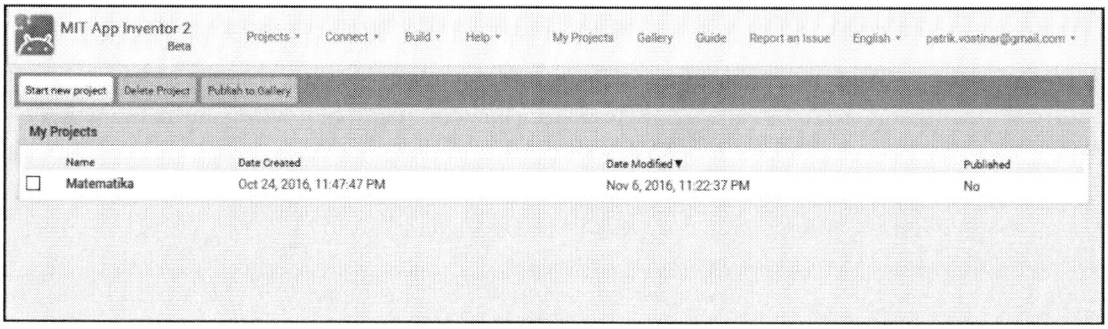

Fig. 2. App Inventor - my projects

To create a new project, we have to click on the button "Start new project". In the dialog box, we have to fill one-word name of our project. After that the tool "Designer" is displayed (Fig. 3).

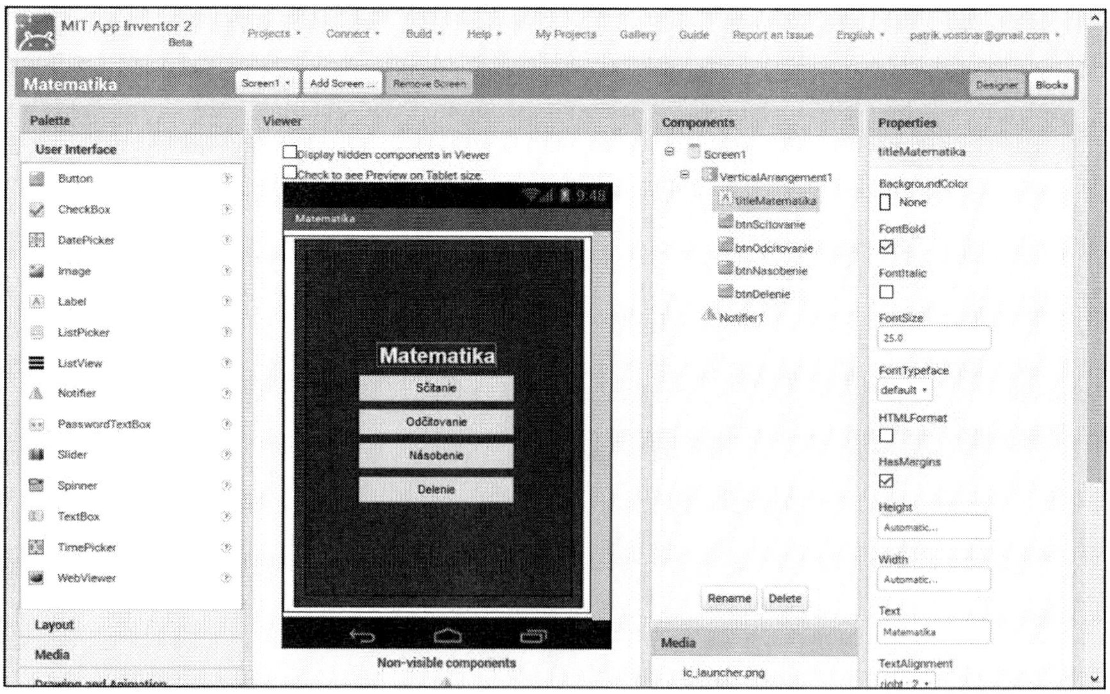

Fig. 3. App Inventor – "Designer" tool

This tool consists of a tools "Palette", "Viewer", a list of all used components "Components", and "Properties".

In the "Palette" there are all available components, which we can use. The "Viewer" is for displaying of how our app will look like. For adding components into our application we have to drag components from the "Palette" to the "Viewer" (drag and drop feature). In the section "Components" there are all the components used in the application arranged hierarchically. In the "Properties", we can change/set the properties for each component (color, text size, etc.).

In addition to "Designer" tool the App Inventor has one more important tool called "Blocks". We can switch between these tools on the bar, which is located under the menu bar (green color). We can see this bar at any part of our project (Fig. 4).

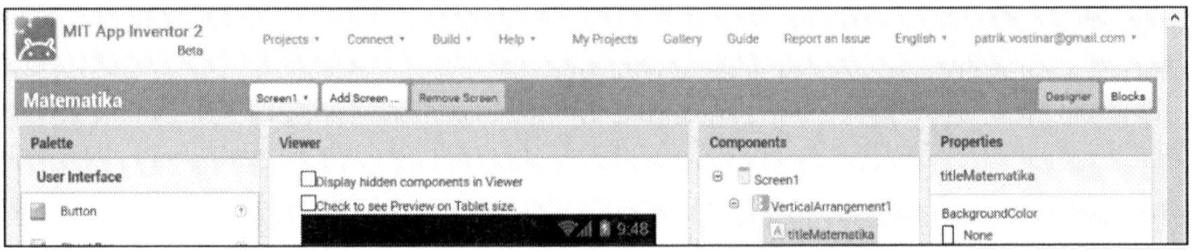

Fig. 4. App Inventor – toolbar with buttons "Designer" and "Blocks"

The tool "Blocks" consists of sections "Blocks" and "Viewer". In the section "Blocks" we can find all the blocks that we may add into our application. In section "Viewer" there are blocks which we are using in our application.

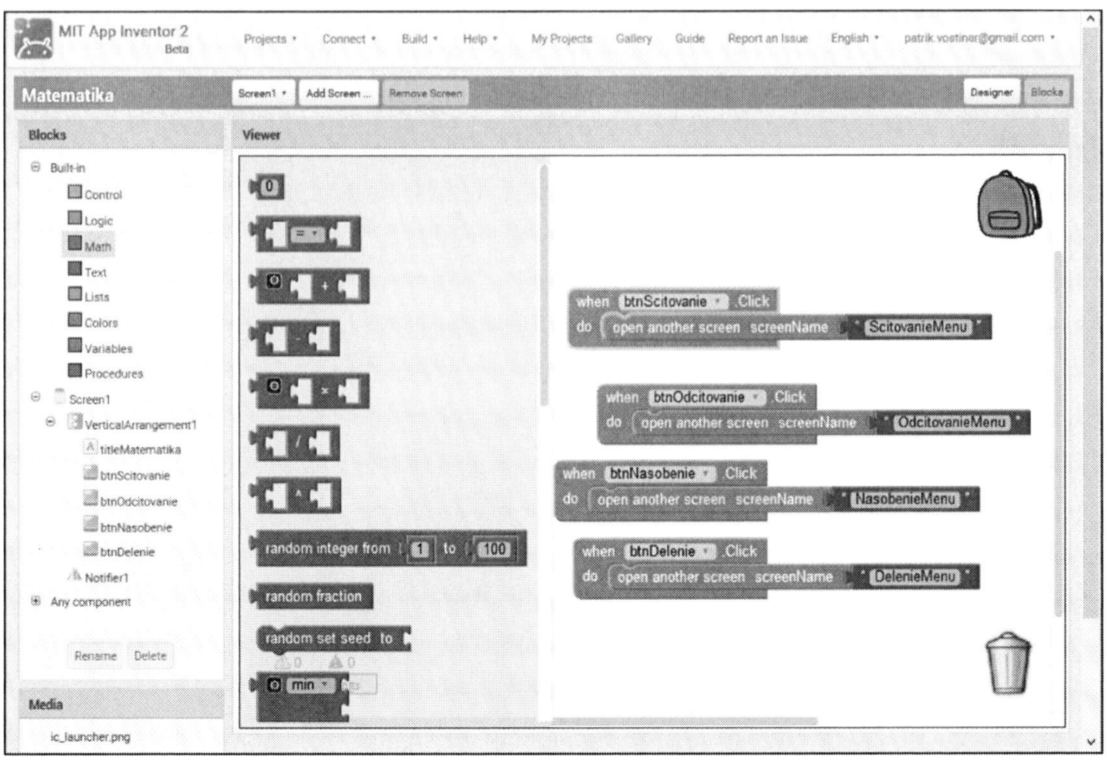

Fig. 5. App Inventor – "Blocks" tool

After our application is created, we can run this application. We have more choices for running the App Inventor application. In the toolbar we can create apk file ("Build →APK"). Another option is to test our app in emulator ("Connect→Emulator"). If we are using an Android device and we have a wireless internet connection, we can start building apps without downloading any software to our computer. We will need to install the App Inventor Companion App [2]on our mobile device ("Connect-->"Al Comparison "). Then we can test our app in real-time on our device.

3 Application „Matematika"

Application "Matematika" was created in the App Inventor environment. The aim was to create simple mobile application, which could be used for practicing addition, subtraction, multiplication and division of integers. Parents can use the application too, when checking their children's homework's. Teachers can use it to increase the motivation of students during mathematics lessons. Another reason for creating this application was to show the possibilities of building applications on this platform. The source code of the application is on the App Inventor webpage:

http://ai2.appinventor.mit.edu/?locale=en#6740536296472576

3.1 Installation of application

Application is free and it is possible to download the application directly onto mobile device by means "Google Play" or on website: (Fig. 6):

https://play.google.com/store/apps/details?id=appinventor.ai_alabama9111.Matematika

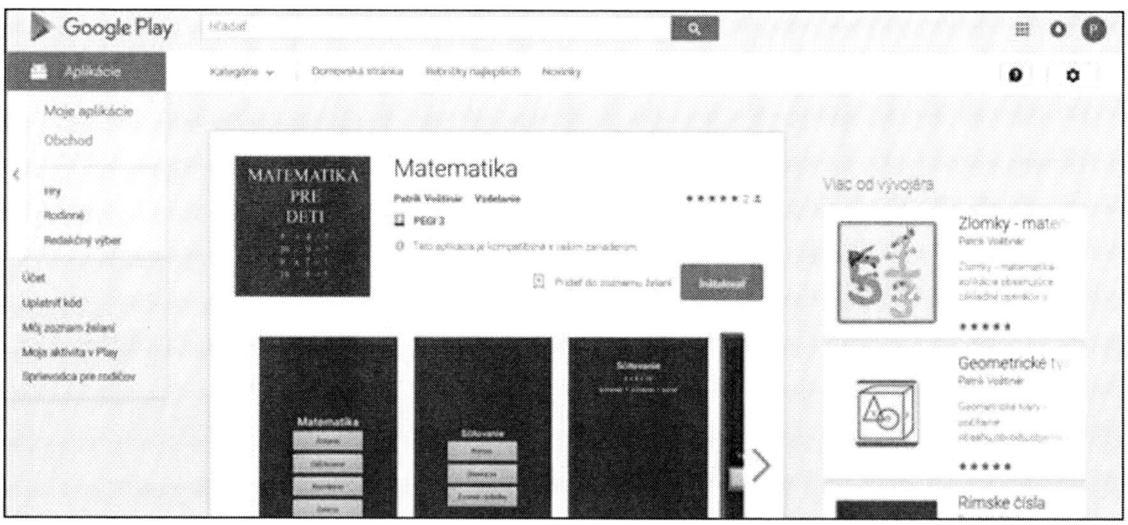

Fig. 6. "Matematika" in Google Play

[2] https://play.google.com/store/apps/details?id=edu.mit.appinventor.aicompanion3

Fig. 7. "Matematika" in mobile Google Play store

Fig. 8. "Matematika" in mobile Google Play store detail

3.2 Structure of the application

After the application starts the main menu is shown (Fig. 9). The user can choose which topic, he /she wants to practice. It is possible to practice addition, subtraction, multiplication and division of integers.

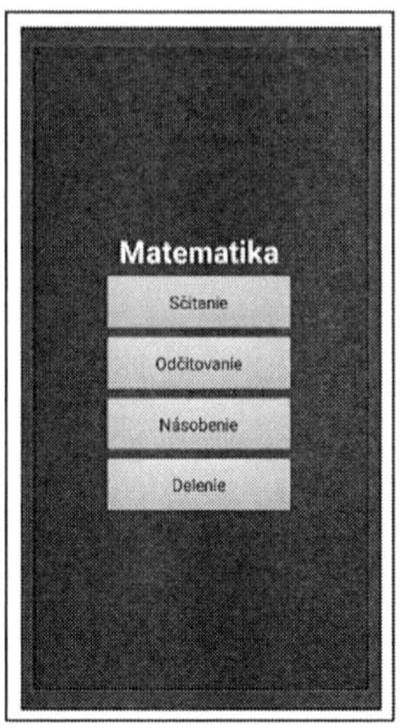

Fig. 9. App - Main menu

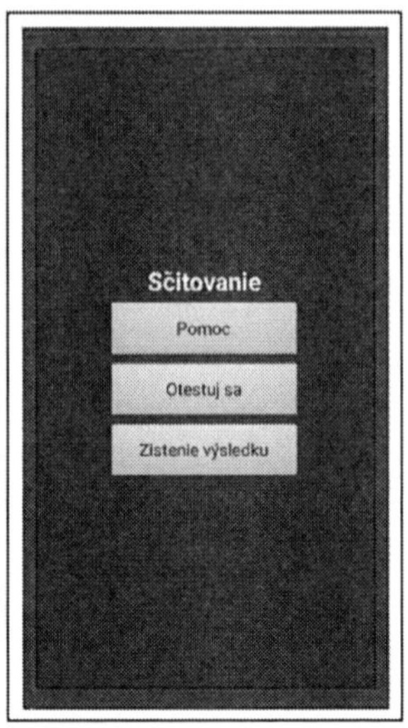

Fig. 10. App - Menu with "Pomoc", "Otestuj sa" and "Zistenie výsledku".

Fig. 11. App - Help of application

After clicking on one of the offered topics, further alternatives are displayed to user: "Pomoc", "Otestuj sa", "Zistenie výsledku" (Fig. 10). If the user chooses the option "Pomoc", on the display there is shown the brief theory connected to the specific topic (Fig. 11). The button "Otestuj sa" offers some exercises to practice the calculation of the chosen topic. Before the task is generated, the user has to select the level of difficulty for solution of the problems. The user may choose from two levels of difficulty - easy and difficult (Fig. 12). The role of the user is to enter the correct answer into the application and then press the button "Potvrdit'". The application checks the given solution and displays text, if the result is correct or incorrect. If the result is correct, the user may press the button "Ďalší príklad" and he could continue with the new task (Fig. 12).

Fig. 12. Pop-up dialog for choosing difficulty

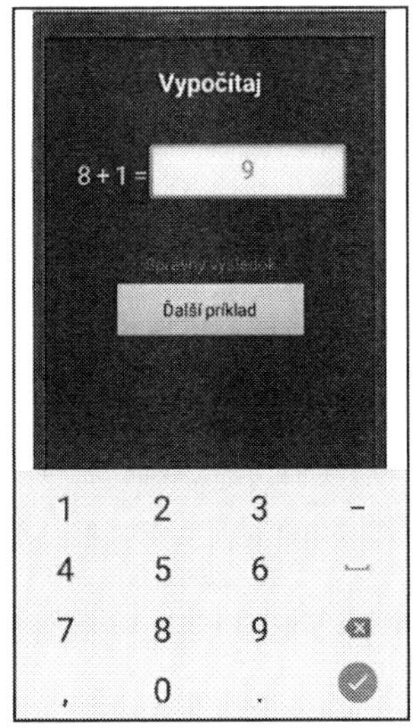

Fig. 13. Random generated task

In the section "Zistenie výsledku", the user can verify the results of arbitrary tasks in the relevant category (Fig. 14). The user enters the task and application solves this task after the button "Zisti výsledok" is pressed (Fig. 15).

Fig. 14. Verification of Result – without answer

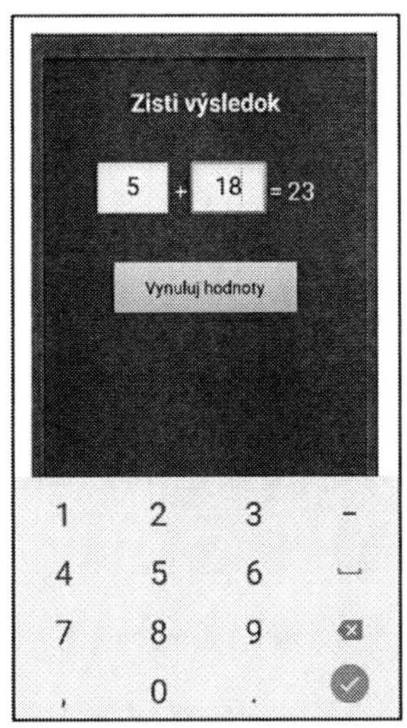

Fig. 15. Verification of Result - with answer

4 Conclusion

In this article we focused on the creating mobile applications without programming skills. The App Inventor can be used for creating applications, which teachers can use during mathematics lessons or they can use it in order to create an application with their students. In this article we have also shown an example of possible mathematical application created in the App Inventor environment.

Acknowledgement

The authors gratefully acknowledge the Scientific Grant Agency KEGA no. 003TTU-4 / 2015: E-learning Courses for Teaching Mathematics at Secondary Schools.

References

[1] WOLBER, D., ABELSON, H. a kol.: 2014. *App Inventor*. 1.vyd. Brno : Computer Press, 2014. 368 s. ISBN 978–80–251–4195–3.

[2] BAYERL, E., ŽILKOVÁ, K.: 2016. Interactive Textbooks in Mathematics Education – What Does It Mean for Students?. In *15th Conference on Applied Mathematics Aplimat 2016.* Bratislava : Nakladateľstvo STU, 2016. s.56-65. ISBN 978–80–227–4531–4.

Current address

VOŠTINÁR Patrik, Mgr.
Department of Mathematics, Faculty of Natural Sciences, Matej Bel University
Tajovského 40, 974 01 Banská Bystrica, Slovak Republic
E-mail: patrik.vostinar@umb.sk

HANZEL Pavol, prof. RNDr. CSc.
Department of Mathematics, Faculty of Natural Sciences, Matej Bel University
Tajovského 40, 974 01 Banská Bystrica, Slovak Republic
E-mail: Pavol.Hanzel@umb.sk

Slovak University of Technology in Bratislava
Faculty of Mechanical Engineering
16th Conference on Applied Mathematics **APLIMAT 2017**
Proceedings

EXAMPLE OF MATHEMATICAL MODELING IN HIGH SCHOOLS

VYSOKÁ Jana (CZ)

Abstract. The aim of this paper is to suggest an expansion of preparation of learning materials designed for university students who have chosen advanced mathematics as an elective, or for high school students who are engaged in mathematics beyond compulsory curriculum. Students will work with a simple example of mathematical modeling in solving simple transport tasks. In this work, we will pay attention to the issue of traffic flow modeling from a macroscopic perspective. First, we derive a mathematical model in the form of partial differential equations and then we will focus on solving this equation using the method of characteristics. Interpretation is presented in a way that should be easily grasped by students.

Keywords: Partial differential equation, method of characteristics, traffic density, flow function

Mathematics Subject Classification: Primary 97-01, Secondary 3501

1 Introduction

Mathematical modeling is a constantly evolving modern discipline, which is used in many areas of human activity. Modeling can be used in various disciplines e.g. technical, biological, economical or social. We meet mathematical models every day together with weather forecasting, the modeling of transport networks, population, inventory management, groundwater flow, the determination of flood areas during floods and civil engineering works etc. Models have become an integral part of a tool for forecasting the development of various processes. Using the appropriate mathematical model has many advantages and helps us to understand easier complex phenomena and processes and contexts, allows the simulation of different possible outcomes. Therefore, modeling simple phenomena using procedures and methods deserves to be included into the curricula of selected schools. The effort of some schools is to enhance the teaching the subjects using of commercial programs for example Matlab or Statistics to demonstrate problems in various subjects such as mathematics, physics, biology or chemistry. From there it is only a small step to the numerical modeling. The graphs in the paper were constructed using a graphical software Derive 6 [6].

2 The density of vehicles and the flow of vehicles

The movement of vehicles on the highway can be described analogously with the movement of fluid – a fluid particle motion. The flow of vehicles on the highway can be (from a certain angle and under given conditions) understood as the flow of fluid particles (e.g. when viewed on the road from the aircraft). Therefore it will be necessary to distinguish the concept of a macroscopic speed as the speed of the flow of vehicles and a microscopic speed as the speed of individual vehicles. The motion of fluid particles is influenced by the movement of surrounding particles but the driving of the driver depends on other factors. The driver adjusts the speed accordingly to rational considerations of the situation in front of him or behind him and perceives other factors which may have impact on the speed. The particles of the fluid may collide in its movement with other particles, the driver of the vehicle tries to avoid the collision with another vehicle, of course. Because vehicles represent units in our model, we will assume that they have the identical length. Further we will assume for simplicity that the vehicle traffic will follow only in one direction. In order to determine the number of vehicles on the section of the highway being measured per a given unit of time, we have to take into account, firstly, the overall density of vehicles and, secondly, their speed. The density of vehicles on the monitored section is a concept that would deserve a more detailed analysis. The density of vehicles could be defined as the ratio of the number of vehicles to the unit length of the road and could be will identified by the symbol ρ:

The density of vehicles = the number of vehicles / the unit length of the road.

For example, if the observed road section is 1000 m long and if 35 vehicles are located in this section, then the density of vehicles is the value $\rho = 0{,}035$. In the case, that measured road segment is blank, then $\rho = 0$. If $\rho = 1$, then we are talking about a traffic jam. We will determine the density of vehicles in a given place x and in a given time t. We will denote it as a function of two real variables in the form $\rho(x,t) = \rho(x(t))$, where the variable x depends on the variable t.

Another important concept of determining the character of road traffic is the flow of vehicles, which can be defined by this way:

The flow of vehicles = the number of vehicles / the time unit.

The flow of vehicles can be defined in many ways, the simplest one is to define the flow of vehicles as the product of the density and the vehicle speed at which the vehicles are moving. We assume always that the speed of the observed group of vehicles is constant. If we introduce a uniform label for the speed of vehicles as v in the text and we use the symbol φ for the flow of vehicles symbol φ, then yields

$$\varphi(\rho) = v\rho. \tag{1}$$

In the event, that the vehicle speed, will be changed as a result of other conditions such as a traffic accident or a speed limitation the speed or changes in the movement of vehicles joined with light signals the flow function need not be in the form of the linear function in the form (1) and may depend on other factors.

2 Characteristic of the road traffic

What means a continuous traffic, a limited traffic or a traffic jam in a practice? We will assume that each vehicle has a length of 4.5 meters and drivers comply to the safe distance (a time lag of 2 seconds). The rule of the distance of 2 seconds means the following situation: Once while driving the vehicle, we will focus on some fixed point in the roadway e.g. a tree or a bollard and from the moment this point misses vehicle in front of you, we deduct 2 seconds. If we miss the same point in less than that, then a controlled vehicle is not in a safe distance from the vehicle ahead. If the vehicle is traveling in the speed of 90 km/h, than it is a distance of approximately 50 meters. In the section of 1000 m long there are a maximum of 18 vehicles in any given time. We can then distinguish the following situations:

1. If the distance between vehicles is 50 m, then on a stretch of road 1000 m long it can move freely from 0 to 18 vehicles. The size of the flow of vehicles will be maximized if the vehicles are not limiting each other during driving. Therefore, the maximum flow of vehicles will be 1,620 vehicles per 1 hour. Whereby, in this case the density is greater the flow is greater. In practice, we are talking about the *continuous traffic*.

2. If a greater number of vehicles i.e. in the range 19 to 100 vehicles will move on the same road segment, then each vehicle is located in the free section of at least 10 m long. In this case, the driver has to keep an attention to the situation before and after the controlled vehicle and adequately respond to any changes. The speed of vehicles will change in certain sections of road and the traffic can be alternately compressed or diluted. The more is traffic dense, the greater the flow of vehicles is less. The traffic is then characterized as *limited*.

3. If the monitored section of road contains more than 100 vehicles, than the driver is forced to stop. This then leads to a local *traffic jam*. If the road contains approximately 200 vehicles, than there is a global jam on the road and all vehicles stand. The flow of vehicles is zero.

We could give illustrative examples of the practice and therefore it will be appropriate to reasonably define the specific form of the function of the flow of vehicles. Suppose that the maximum speed is the same for all vehicles, let it be denoted by the symbol $v_{\max}$. We will omit some reflections on the psychological aspects of driving in a row, for example, if the driver has the vehicle ahead far, then he will accelerate the ride. Based on empirical experience in practice it can be assumed that, if the roadway is empty, therefore the vehicle can move in the maximum speed. Conversely, if the road is full, there is a traffic jam (and the vehicles stop). These properties has for example a function, see [2] in the form

$$v(\rho) = v_{\max}\left(1 - \frac{\rho}{\rho_{\max}}\right). \tag{2}$$

Then we can introduce the flow function in the form

$$\varphi(\rho) = \rho v_{\max}\left(1 - \frac{\rho}{\rho_{\max}}\right), \tag{3}$$

where yields

$$\varphi'(\rho) = v_{\max} - 2\frac{\rho v_{\max}}{\rho_{\max}}.$$

4 Macroscopic model of the vehicle flow

Let us monitor the section of the road on which the vehicle moves in the constant speed in the time interval. The density of vehicles ρ in the given segment and time is specified. The basic consideration will be based on the fact that the change of the number of vehicles in the section being measured by the unit of time corresponds to the difference between the number of cars that enter and leave the given segment. We determine the number of vehicles in the time $t = 0$ and in the time $t = t_0 > 0$, their difference will represent a change of the density of vehicles detected in the time $t = t_0$ and we calculate the number of the vehicles in the time t_0 which leave the measured section in the same time – a change of the flow function measured in the given place. If we denote the change of density in the considered time by the symbol $\Delta\rho_t$ and the change of the flow function in the given section $\Delta\rho_x$, where x is a variable describing the location of the vehicle on the highway, then under the previous considerations is

$$\Delta\rho_t + \Delta\rho_x = 0.$$

This relationship is possible to rewrite under certain assumptions in the form of partial differential equations of the first order in the following form

$$\rho_t + v\rho_x = 0. \tag{4}$$

This equation could be simply verbally explained as follows: The change of the density of vehicles with respect to a certain time interval is identical to the change of the flow function measured in the given section. If the flow function is not given in a linear form, than the equation (4) is changing into the form

$$\rho_t + \varphi_x = 0. \tag{5}$$

Equation (5) can be rewritten under certain assumptions as

$$\rho_t + [\varphi(\rho)]_x = 0. \tag{6}$$

We will be able to determine the specific solution of the given task after we will know the distribution of the density of vehicles in a certain place and time. The most common and most natural choice of conditions is the initial distribution of the density of vehicles, therefore the description of the situation in the beginning; let us denote this function by symbol $\rho(x,0) = \rho_0(x)$. The problem in the form $\rho_t + [\varphi(\rho)]_x = 0$ together with the initial condition $\rho(x,0) = \rho_0(x)$ is called *the initial problem (Cauchy problem)*.

5 The method of characteristics

One of the ways how to describe the solution of the mentioned problem is the way of using the graphical presentation with help of so-called *characteristics*. The equation (4) with the initial condition $\rho(x,0) = \rho_0(x)$ has a simple form, and therefore its solution can be easily guessed. A search function ρ of variables x and $t > 0$ has the form

$$\rho(x,t) = \rho_0(x - vt),$$

This fact can be easily verified by a test. If we are differentiating according to variable x or t, then we are taking the remaining variable as constant and differentiation rules remain unchanged.

The function ρ is constant on the all lines with tangents v in the form $x - vt = c$, where c is a real constant. In other words, the density of the given initial conditions is transmitted in a time on a straight line with the tangent v or the derivative of the function ρ in the direction v is equal to zero. The intersection of this line and the axis t is denoted by the symbol t_0. These lines, on which density is constant, are called *characteristic lines* or *characteristic* corresponding to the given problem. A drawing characteristics is quite simple and we can get a good idea about solving a given task in a predetermined time and place by the using the graphical representation. The importance of characteristics remains unchanged even when we are solving a nonlinear equation (5). Characteristics are defined as a straight line with the tangent $\varphi'(\rho_0)$. The characteristic line with the tangent equal to the value $\varphi'(\rho_0)$ intersects the axis t in the point t_0 and has an equation in the form

$$x = t_0 + \varphi'(\rho_0)t.$$

It is easy to verify by a simply substitution that the given function of the form

$$\rho(x,t) = \rho_0(x - \varphi'(\rho)t)$$

satisfies the equation (5). The value of the solution $\rho(x,t)$ is equal to the corresponding value of the initial conditions $\rho_0(x)$ in the points of this line.

There are some examples of simple situations:

1. Vehicles are going equally in the constant speed, the traffic density is not changed, see Fig. 1.
2. Vehicles are not moving and staying in the row, this is an example illustrating the traffic jam, see Fig. 2.
3. Vehicles are gradually slowing, their density is increasing, and tangents of characteristics are increasing until there is a traffic jam, see Fig. 3.
4. Vehicles are starting to creep, their density is decreasing, tangents of characteristics are decreasing, see Fig. 4.

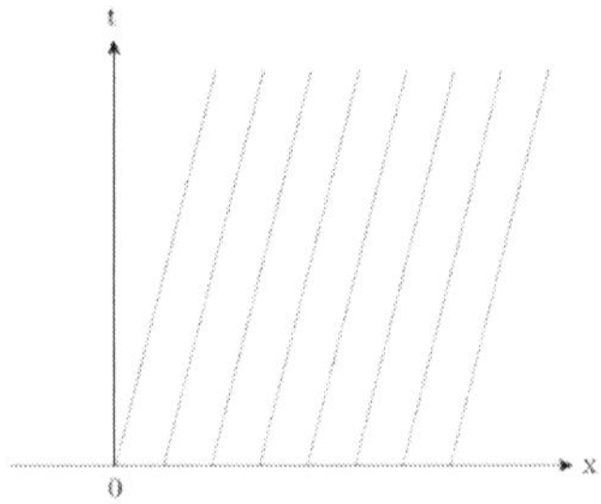

Fig. 1

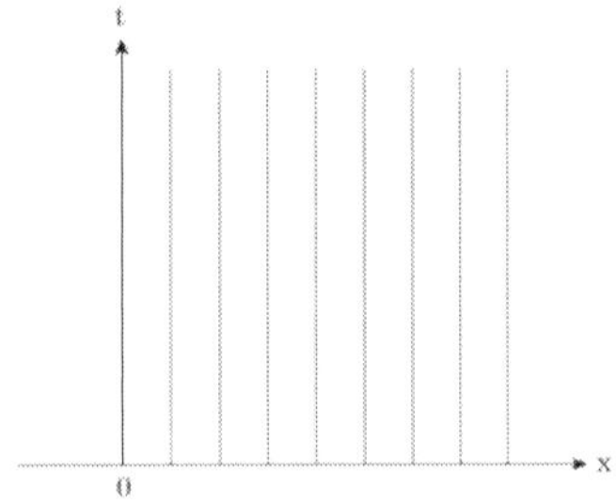

Fig. 2

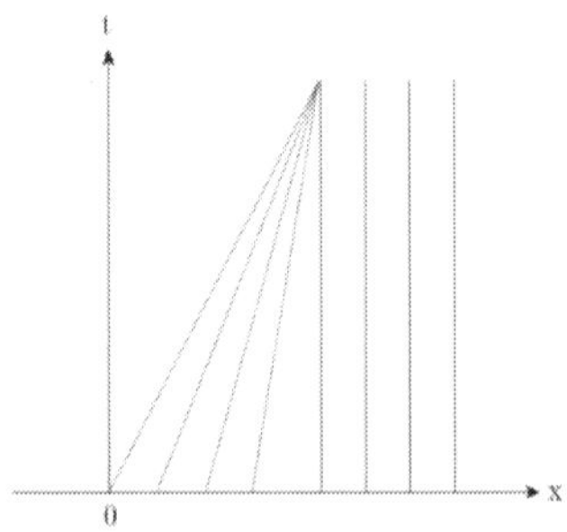

Fig. 3

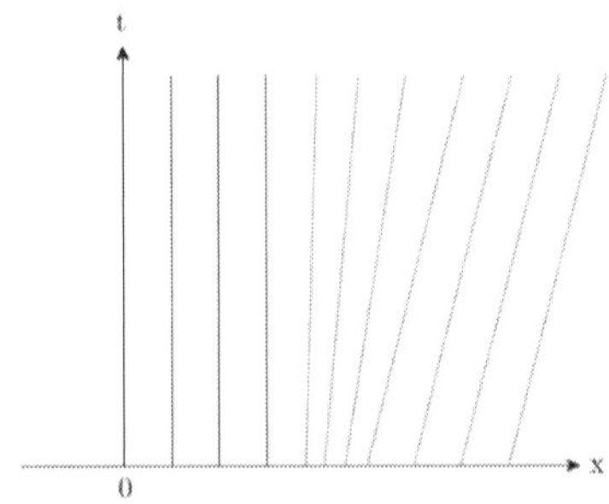

Fig. 4

Combining all these examples it is possible to describe the complicated situation as are the traffic directed by traffic lights or the traffic on sections with reduced speeds. To be able to solve complicated cases of initial conditions, it would be advisable to stop at the cases mentioned in examples 3 and 4. Let us consider the task in the form

$$\rho_t + [v\rho]_x = 0 \tag{7}$$

with the initial condition in the form of the jump $\rho_0(x) = \rho_p$ for $x \geq 0$ and $\rho_0(x) = \rho_l$ for $x < 0$, where ρ_p and ρ_l are given nonnegative real numbers. This problem is called the *Riemann problem*. Then we could distinguish two different situations:

1. $\rho_l < \rho_p$

Characteristics for the value ρ_l are illustrated in the red color and characteristics for the value ρ_p in the blue color, see Fig. 5.

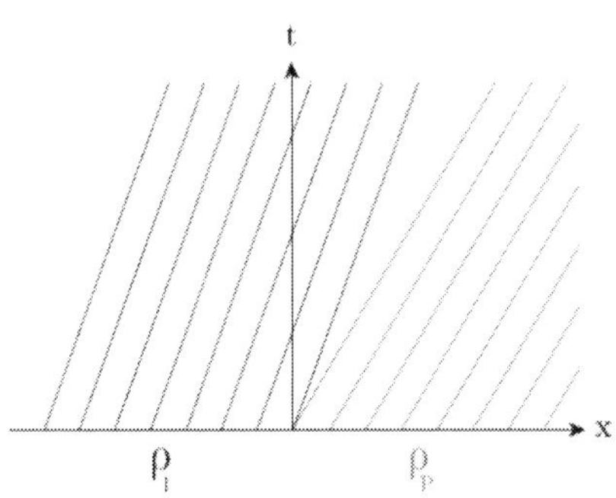

Fig. 5

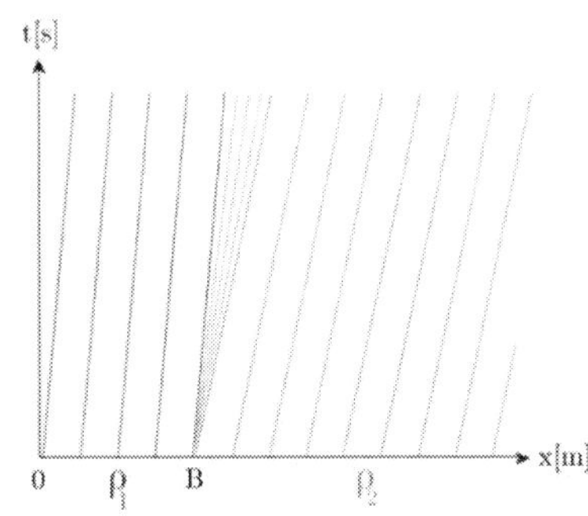

Fig. 6

The Fig. 5 shows that in the period between red and blue areas (areas defined by inequality $\rho_l t < x < \rho_p t$) characteristics are "missing". If we complete an empty part by lines with tangents of values between ρ_l and ρ_p in the green color, than we get continuous information on the transmitted value of the density in the jump, see Fig. 6.

2. $\rho_l > \rho_p$

The Fig. 7 again shows that the characteristics are crossing; it means that we cannot determine a unique value of a density at any time. The direction, in which the information on jump will propagate from the first intersections of the characteristics (so called shock wave), must be only one because the same information is moving in both directions. The tangent of this single characteristic, which represents the propagation of shock wave, see Fig. 8, is given by so called *Rankine-Hugoniot condition* in the form

$$s = \frac{\varphi(\rho_l) - \varphi(\rho_p)}{\rho_l - \rho_p}. \tag{8}$$

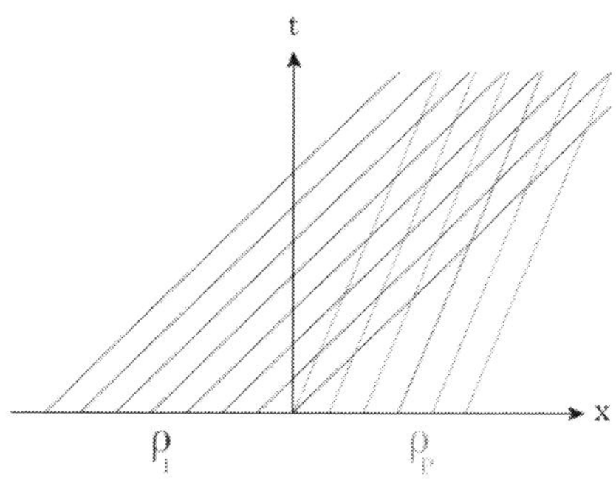

Fig. 7

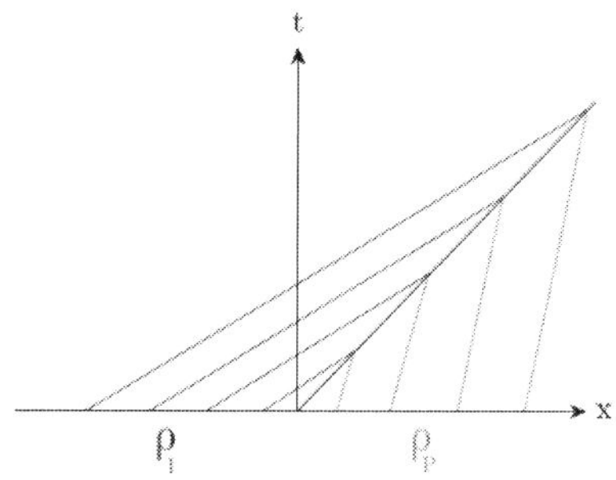

Fig. 8

The derivation of this condition extends beyond the scope of this text but if a student is interested in this topic, we can recommend studying relevant pages in [1].

Exercises:

Draw the following situations by using characteristics. Monitored section is 1 km long, in the first measuring section there is detected the value of the density of vehicles ρ_1 and in the second section, which is measured from the point B, the value of the density of vehicles ρ_2. Sketch a solution and the point B set in a distance of 100 m in the positive part of the axis x .

1. $\rho_1 = 0.021$, $\rho_2 = 0.2$, $v_{max} = 14$ m/s

2. $\rho_1 = 0.2$, $\rho_2 = 0.046$, $v_{max} = 8.3$ m/s

Solution:

1. The density $\rho_1 = 0.021$ represents 21 vehicles per 1 km and vehicles move in the given speed $v_{max} = 14$ m/s, which is 50 km/h. The density $\rho_2 = 0.2$ represents 200 vehicles per 1 km or the traffic jam, vehicles are moving in the zero velocity. Therefore, it is the first described case, where $\rho_l < \rho_p$. It is necessary to calculate the value of the tangents of characteristics corresponding to the given densities ρ_1 and ρ_2 which means to determine the value according to (8). Substituting given values into a following relationship we will get this result:

$$\varphi'(\rho) = v_{max} - \frac{2\rho v_{max}}{\rho_{max}} = 14 - 2\frac{0.021.14}{0.2} \cong 11,$$

Then we can calculate the direction of propagation of the shock wave as follows

$$s = \frac{0 - 0.26}{0.2 - 0.021} \cong -1.47.$$

The resulting graph of characteristics is shown in Fig. 9.

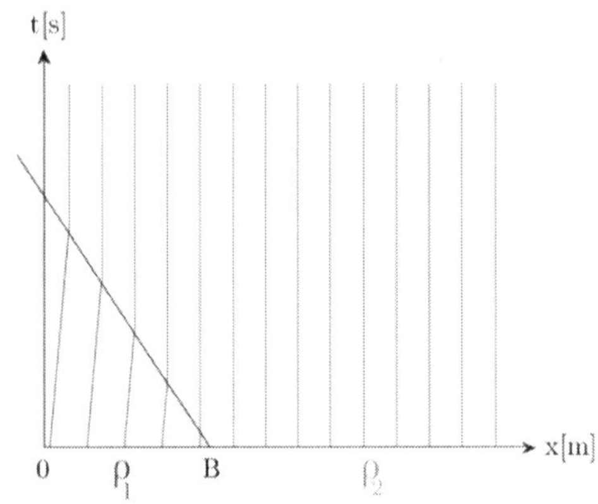

Fig. 9

2. Another example represents a case, where vehicles are in the traffic jam and starting to move taking off at a maximum speed equal to 8.3 m/s = 3 km/h, which is the second described alternative with the condition $\rho_l > \rho_p$. The situation can be illustrated using previous calculations already easily, see Fig. 10.

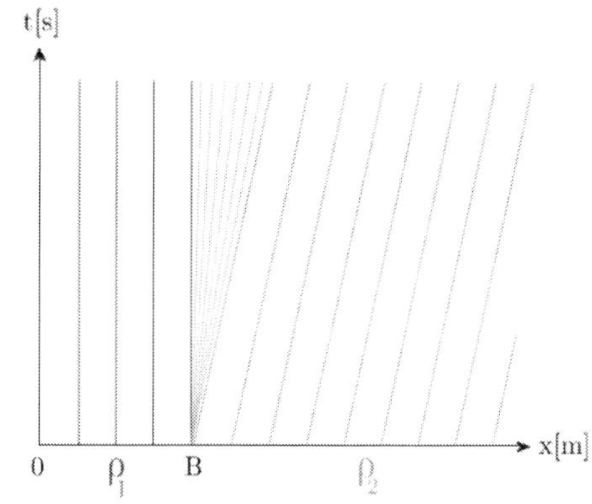

Fig. 10

The initial condition can be defined in a more complex form, where there may be a combination of all the events described above. Monitored section of road with a length of 1000 m is divided into three parts: 0 m - 500 m with a maximum speed of 50 km/ h, vehicles are complying with the safety distance of 2 500 m - 600 m with a maximum density of vehicles simulating a row before traffic lights (maximum speed 0 km/ h) and the section 600 m - 1000 m with a maximum speed of 30 km/ h. The initial condition is specified in the first segment by value ρ_1 = 0.021 vehicles / m (the red color), in the second section of the road ρ_2= 0.2 vehicles/m (the blue color) and in the last part is the initial density value determined by the value ρ_3= 0.046 vehicles/m (the black color), see Fig. 11. In the time $t = 0$ s the traffic light turns to be green and the vehicles start to move. To formulate this problem we have deliberately used the data of the previous examples.

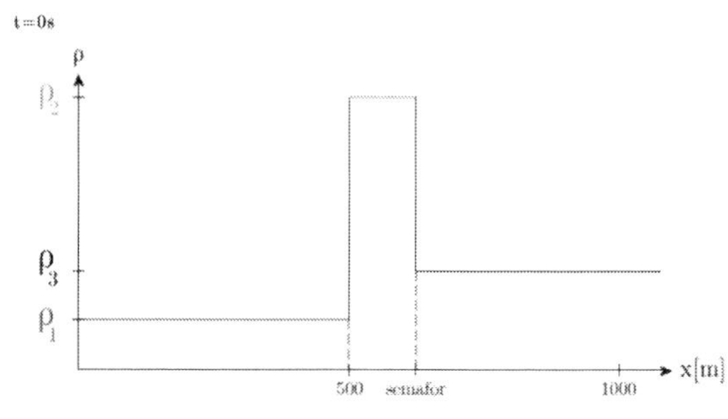

Fig. 11

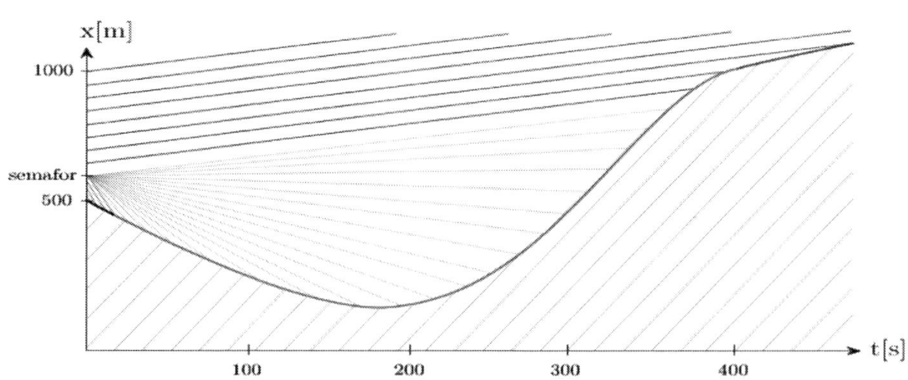

Fig. 12

The complete graph can determine the approximate time and place where there are changes in the flow of the traffic flow. At the point where the red characteristic intersects the green characteristic transferring the value of density approximately equal to 0.2 vehicle/m get in the sum, is the tangent of transition line equals zero (this condition occurs in time approximately 180 s at the point approximately 160 m from the start). In the time about 360 s vehicles are already limited and can freely pass around a traffic light. The map of all characteristics given by the graph in Fig. 12 was necessary to be downloaded in other width, because it would not be appropriately scaled in the text and it was necessary to draw the inverse characteristics to the coordinate system (t,x) (previous graphs were constructed into the coordinate system (x,t)) in order to get well arranged results. Therefore, tangents of characteristics quite precisely do not correspond to the reality, but the essence of the problem is captured.

6 Conclusion

The purpose of this paper was to suggest an expansion of preparation of learning materials designed for students of selected schools. In this work, we focused on solving easy traffic problem by mathematical modeling from a macroscopic perspective. Interpretation was presented in a way that should be easily grasped by students - university students who have chosen advanced mathematics as an elective, or high school students who are engaged in mathematics beyond compulsory curriculum

First, we derived a mathematical model in the form of partial differential equations. In order to apply the equation for an illustrative example of the practice it was be necessary to introduce and understand some important concepts such as the traffic density, the traffic jam, the continuous and reduced traffic or the flow function. Then we focused on solving this equation using the method of characteristics. The theoretical part is complemented by several practical examples describing situations encountered in everyday traffic. The graphs in the paper were constructed using a graphical software Derive 6 [6].

References

[1] DRÁBEK, P., HOLUBOVÁ, G., Elements of Partial Differential Equations, Berlin : Walter de Gruyter, 2007, ISBN: 978-3-11-019124-0

[2] HOKR, M., Transportní procesy. Učební text, Technická univerzita v Liberci, 2005

[3] JELÍNEK, J., VYSOKÁ, J., Modelování dynamiky dopravního proudu, Silnice Železnice, Ostrava - Vítkovice: Konstrukce Media, s.r.o., 2013, roč. 8, č. 2, s. 73-76. ISSN 1801-822X

[4] JELÍNEK, J., VYSOKÁ, J., Porovnání modelů pro výpočet hustoty dopravního toku, Silnice Železnice, Ostrava - Vítkovice: Konstrukce Media, s.r.o., 2013, roč. 8, č. 5, s. 110-112. ISSN 1801-822X

[5] JELÍNEK, J., VYSOKÁ, J., O podchodach v modelirovanii transportnoj dinamiki, Vestnik Astrachanskogo gosudarstvennogo techničeskogo universiteta, Astrachan': Astrachanskij gosudarstvennyj techničeskij universitet, 2014, roč. 2014, č. 1, s. 121-129. ISSN 2072-9502

[6] KUTZLER, B., KOKOL-VOLJC, V., Úvod do Derive 6, OEG, Austria, 2003

[7] Le VEQUE R., Finite Volume Methods for Hyperbolic Problems, Cambridge University Press 2002, pp. 203-213. ISBN 0521009243

[8] Le VEQUE R., Some Traffic Flow Models Illustrating Interesting Hyperbolic Behavior, 2001

Current address

Vysoká Jana, RNDr.
The Institute of Technology and Business in České Budějovice
Okružní 517/10, 370 01 České Budějovice, Czech Republic
Tel number: +420 387 842 131, e-mail: vysoka@mail.vstecb.cz

Slovak University of Technology in Bratislava
Faculty of Mechanical Engineering
16th Conference on Applied Mathematics **APLIMAT 2017**
Proceedings

DEDICATED COMPUTER PROGRAMS VISUALIZING SOME GRAPH THEORY PROBLEMS AS LEARNING ENHANCEMENT

ZAWIŚLAK Stanisław (PL), KOPEĆ Jerzy (PL)

Abstract. In the present paper, some computer programs are roughly described which are used in didactics of the subjects related to graph theory. The programs are dedicated to the following problems: searching - DFS and BFS, bi-criteria TSP, drawing self-complementary graphs as well as checking their isomorphism. The advanatge consists in visualisation of particular problems. Standard packages for graph drawing were also been utilized.

Keywords: enumeration of spanning trees, graph drawing, self-complementary graphs

Mathematics Subject Classification: Primary 68R10, 05C30; Secondary 97K30

1 Graph theory as a taught subject

Graph theory is a branch of mathematics which is commonly considered as a base for all kinds of IT studies. It could be found in almost all curricula of informatics in every type of universities, including universities of technology or economics. It's elements could also be given as part of other subjects like e.g.: discrete mathematics, combinatorics, network theory etc. At the University of Bielsko-Biala (Poland), it is given on bachelor level (subjects: discrete mathematics and algorithmic approach to graph theory) and also on master level (subject: graphs and networks in informatics). Moreover, the application of graph theory is given within the framework of a subject for choice: graphs and mechanics [21, 26] which is an author's specialization. The content of the subject(s) depend(s) on the expected student profile or types of skills which should characterize a graduate student. The teaching process should be atractive and innovative to inspire students to learn and give them professional abilities [6, 10,11]. Usage of different kinds of software makes the learning process more interesting and effective. The goal of the paper is to describe a selected software used for teaching subjects related to graph theory utilized at the University of Bielsko-Biala.

2 Software for visualization of the problems related to graph theory

There is available professional software like e.g. *Pajek, Graphviz* or *Mathematica* (especially graph drawing module) [2, 7, 14, 18, 29]. However, these products are not so widely used as *AutoCAD* or

CATIA in case of mechanical engineers. In references, one can find several proposals of software which had not been internationally distributed and commonly used despite the fact that they were described in well-known international journals [3-6, 18-19]. We describe a series of computer programs written by our students as projects or within the framework of their bachelor theses [23-25]. The standard software has some advantages but in case of IT studies (practically-oriented) writing own programs by students is just developing their programming skills. Moreover, the programs have usually two modules: visualization e.g. graph drawing and procedures which solve a particular problem via implementing a particular algorithm. The relative advantage of own program could be its simplicity due to focusing on one chosen problem. In Fig. 1, two screens are presented for the program dedicated to search algorithms: width, depth and A* - which is not presented in the figures. The second exemplary program (Fig. 2) is dedicated to bi-criteria TSP which is solved here via evolutionary algorithm.

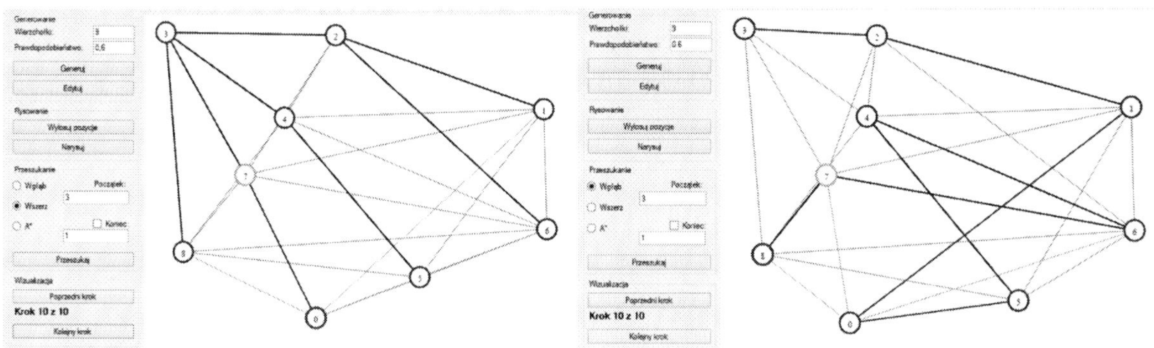

Fig. 1. Program for visualization of searching algorithms: width search and depth first search

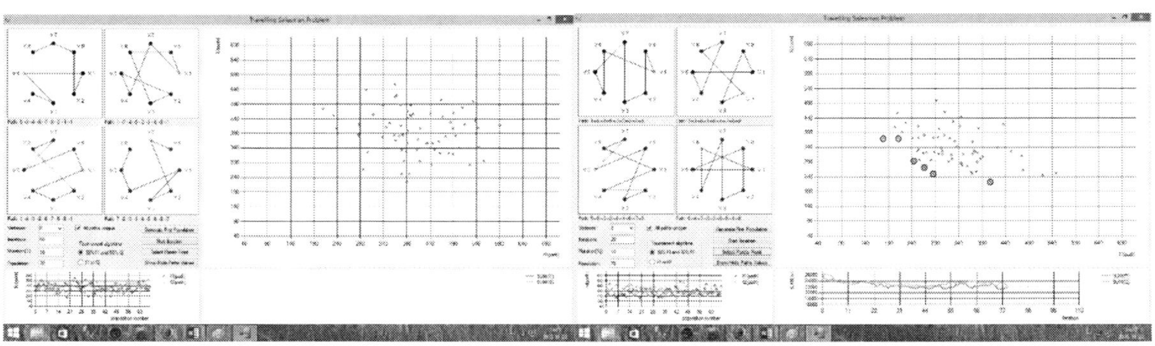

Fig. 2. Computer program visualizing the task of bi-criteria TSP.
The screens are relevant to: initiation of the calculations and results after termination of calculations where points belonging to Pareto front are shown

These types of algorithms tend in proper direction but the solutions are not proved as absolutely perfect. Nevertheless, the algorithm is not too complicated and it could be relatively easy programmed by students [22]. Additionally, the algorithm allows solving bi-criteria problem. The weights of the weighted graph $G(V, E, f1, f2)$ were drawn randomly. The screen is divided in several sub-windows: setting of algorithm parameters e.g. population size ($pop\_size = 70$), probabilities of evolutionary operations, number of iterations until stopping, choice of type of tournament etc. The windows show also the course of total sums of target functions for current population as well as the values of target functions in the $(f1,f2)$–Cartesian co-ordinate system. Moreover, the Pareto front is possible to distinguish and presenting graphically via encircled points.

Four exemplary cycles are given for graphs drawn on a circle, in a right-hand upper corner of the program window. As we can observe, two different ways of presenting vertices have been utilized in the above discussed programs i.e. as circles with vertex number inside (Fig. 1) and as the full dots with vertex number placed nearby (Fig. 2). The screen is divided in the following parts: visualization scene, dialog area as well as charts monitoring of the algorithm rearranged according to the entered number of iterations. In the second case the chart in right hand-lower corner is growing on the screen, self-extending dynamically after each iteration of the performed algorithm which could not be presented in the paper via stable pictures. Advantages and disadvantages of using home-made programs could be discussed. Advantages consist in: development of programming skills be students who are authors. It is possible to use them in didactics for next generations of students. Topics, tasks of programs are changed every academic year so the students could only partly relay on achievements of predecessors. There is a real competition between teams of programmers which improves essentially the teaching process. Further pieces of software will be described in the paper which are dedicated to the problems connected with sc-graphs (i.e. self-complementary graphs) [1,8, 12-13, 15-17, 27] . These create an interesting class of graphs which has unexpected range of applications [9, 20].

3 General remarks on graph drawing

In some theoretical books or lecture notes related to graph theory as well as relevant scientific papers drawings of graphs are sometimes very rare which is amazingly to realize. In case of practically-oriented curricula of studies, on contrary, graphs are presented graphically very frequently. This approach should be used in didactics of subjects connected with graph theory, especially on bachelor level, to visualize the theoretical considerations. This would lead to better understanding of the notions and their properties. Nevertheless, graph drawing itself is a challenging task. The graphs (multi-graphs) can be drawn in different manners, taking into account:

a) vertices: (i) randomly distributed, (ii) placed evenly on a particular (invisible) circle, (iii) in grids' nodes (visible grid is shown in the background by means of thin perpendicular lines) or (iv) in coordinates etc.,

b) edges: (i) drawn as straight lines, (ii) drawn as arcs, (iii) drawn as arbitrary curves, (iv) drawn as straight lines crossing each other under 90° angle etc. In case of weighted graphs – particular weights should be placed near the edges, simultaneously aiming for avoidance of their overlapping.

The methods of graph drawing are discussed yearly in several international conferences especially International Conference named GD (Graph Drawing) - www.graphdrawing.org; e.g.: the 2015 year edition of the conference (http://www.csun.edu/gd2015/) was organized in Los Angeles, USA.

In the present paper, the families of self-complementary graphs for particular n are considered and drawn in different manner – as an exemplary set of graphs, on which several problems could be discussed and visualized. The self-complementary graphs have a property that their complementary graphs are isomorphic to them [1, 27]. They can be characterized by several different features e.g.: symmetry, vertex-transition or sequence of vertices' degrees. Looking through some references, it has been spotted that there are lack of atlases of sc-graphs e.g. for $n = 8$, where $G(V, E)$, $|V| = n$, $|E| = m$ and the reference [1] is rarely cited. Moreover, when the atlases are given, then there is lack of further analysis of features – especially lists of degrees of graph vertices. Therefore additional aims of the paper are: to present an atlas of graphical visualization of a family of sc-graphs and some further analyses e.g. checking of isomorphism.

We would like to show that some graphical representations are sometimes misleading therefore different views of the same graphs could be prepared and compared. We also present checking the isomorphism in particular cases in step by step manner which is usually omitted, however it is a good example for didactics of this abstract and sophisticated notion.

3.1 Graph drawing in case of randomly distributed vertices

The graph vertices can be randomly distributed around the visualization scene (field, area, sub-screen, window). It is a rapid performed option but having many drawbacks. The option of arbitrary displacement of chosen vertex was frequently added in the student programs as peaty simple. Simultaneously we usually add on option of mouse-driven rearrangements. It means that a user can rearranged a given graph to her/his needs or imagination by means of shifting vertices on mouse clicking (click, shift and drop). It is very general manner of presentation to spot the graph at first glimpse, but it is not useful for any other sophisticated purposes and analyses (see Fig. 3a,b).
The problems to solve are e.g. shape and pictogram for a vertex graphical representation: small empty circle, filled circle, rectangle. Position of vertex and edge description, in case of vertex, are placed in the neighborhood of the vertex pictogram or inside it (Fig. 3a,b). The solutions are versatile and they depend on an author of commercial or private (home-made) software performance [22-24].

3.2 Vertices drawn evenly on a circle

The graph vertices can be evenly distributed around a circle or an ellipse which itself is invisible. The advantages are as follows: this approach unifies the visualization, we avoid overlapping

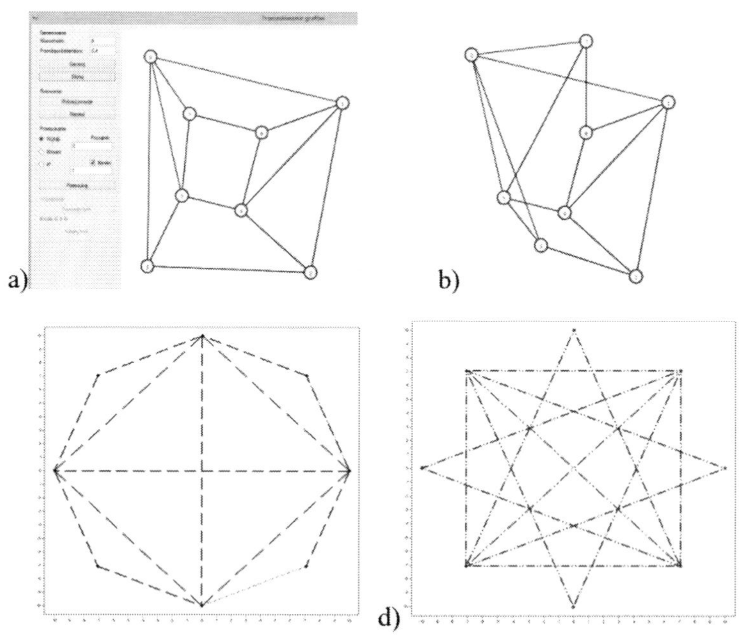

Fig. 3. Graphs drawn in different visualization styles: (a) random vertices positions – 1-st version , (b) random vertices positions – 2-nd version; (c,d) vertices distributed evenly on a circle (Scilab)

of edges and the graph vertices are distributed regularly i.e. shifted by a particular angle $\alpha = 360° / n$ (Fig. 3c,d). Two isomorphic graphs according to different ways of presentations are shown in Fig. 2a,b i.e. via random distribution of vertices using our program. The vertices' numbering starts from 0. The graph is isomorphic to the one presented in Fig. 4a. Whereas two other isomorphic graphs (in Fig. 2c,d) are drawn using "Scilab" – freeware. The graphs are isomorphic but it is almost impossible to recognize this fact by eye inspection.

3.3 Vertices drawn in coordinates and some other approaches

The graph vertices are positioned on the scene according to given values of their (x,y)-coordinates. In general, the graphs can be drawn on plane, sphere, 3D-space etc. We consider here 2D-plane for graph presentation. The coordinates can be prepared by a user of computer program or the software itself can generate special layouts using special algorithms adopting, biomimetic rules from nature as spring action or evolutionary, force or stress relief etc. [24]. The advantages are that the layout is regular, clear, symmetrical and the graph elements are evenly distributed, additionally, planar graphs are shown without crossing edges. Moreover, graphs' layouts can be optimized in such manner that the number of crossing is minimal for non-planar graph etc.

4 Enumeration of sc-graphs for n=8

Family of sc-graphs can have numerous members (Tab. 1) depending on n. We focus our attention on low n=8to present the problem itself. The family of relevant graphs is shown in Figs. 4 - 7. According to several references [1, 28] the cardinality of their set is equal to 10.

n	Vertex degree sequence	Number of graphs	Visualization	Total no of sc-graphs
4	2211	1		1
5	22222	1		2
	33211	1		
8	6644 3311	1	Fig. 4c	10
	5555 2222	2	Fig. 4e, Fig. 4j	
	5544 3322	3	Fig. 4b, Fig. 4g Fig. 4i	
	4444 3333	4	Fig. 4a, Fig. 4d Fig. 4f, Fig. 4h	

Tab. 1. Sequences of vertices degrees for some sc-graphs

One can realize that there is not too easy to check an isomorphism of the graphs representing particular graph.

4.1 Graphical atlas of sc-graphs having 8 vertices

These graphs are arranged in at least two manners: according to the sequences of vertices' degrees (Tab. 1) and according to their view prepared in some special coordinate layouts. The second way was utilized preparing Fig. 4 where the family is divided into three rows: (i) Fig. 4a - is of unique shape in this visualization, (ii) (Fig. 4b) and (Fig. 4c) are similar from a particular point of view (two perpendicular rows of 4 vertices), (iii) Fig. 4d – Fig. 4g are also similar (the same layout of vertices), (iv) Fig. 4h – Fig. 4j are also similar, having a complete rectangular as the outer border line - special version of layout.

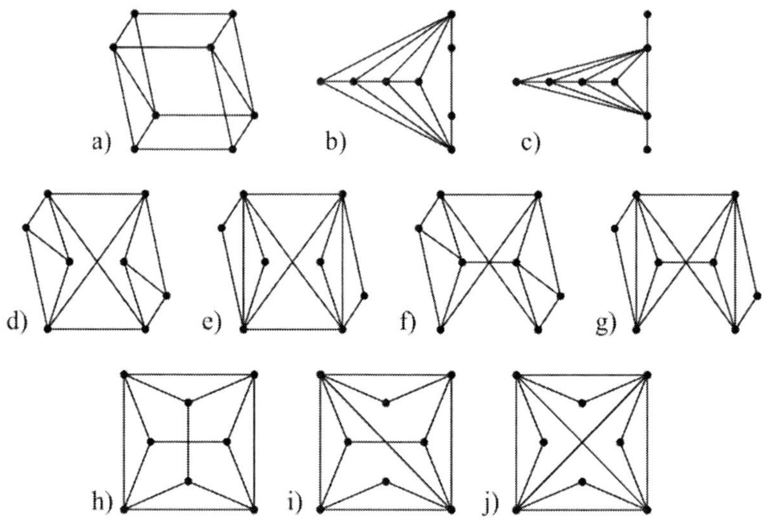

Fig. 4. All sc-graphs having n=8 vertices

The written computer program (Fig. 7) incorporates all graphs given in anonymous source [28]. It was run for every graph from these listed in [1] proving isomorphism of two versions of layouts.

5 Computer programs for drawing sc-graphs

Special computer programs were written to draw graphs and their complementary graphs for $n=8$. The exemplary results are shown in Fig. 5 – in case of student written program. The left hand column in the screen consists of the adjacency matrix of a graph and its (x,y)-coordinates. The window of visualization is organized in the following way: (i) 1-st line - the graph itself and its complementary graphs are presented using particular co-ordinates. The third graph in row is represent both graphs together; (ii) 2-nd line - the pattern is the same but the vertices of graphs are displaced on the circle. The presentation of the graph in special chosen co-ordinates allow for its regular, symmetrical or clear view, whereas adequate versions where vertices are evenly distributed on the circle allows in common representation for easier identification of clique. It is worth to highlight, that in case of the method based on chosen co-ordinates, some edges are overlapping which makes impossible to recognized them separately and in consequence to count them properly. It could be even imagine that the numbers of edges are different but it is a false opinion, because checking their layouts in second rows (on circles) we can spotted that:

- first one has vertices of degree two – 6,7 for source graph and 5,8 for the complementary one;
- second graph: vertices of degree one – 5,8 in source graph and 6,7 in the complementary one.

The second mentioned facts can be only checked studying the visualization done on the circles because in the layout in co-ordinates, the edges overlap one another.

We have the rule that sum of degrees of the particular vertex v from a source graph and from the complementary graph should fulfill the relationship:

$$\deg(v) + \deg(\bar{v}) = n\text{-}1 \qquad (1)$$

For particular data, relevant to graph depicted in Fig. 8 we have: $deg\ (1) = 1$; $\deg(\bar{1}) = 6$; n=8 and formula (1) has the form:

$$\deg(1) + \deg(\bar{1}) = 1 + 6 = 8\text{-}1 = 7 \qquad (2)$$

Chosen graphs from the considered family are presented in Fig. 6 in the way in which they are presented and discussed in [1].

a)

b)

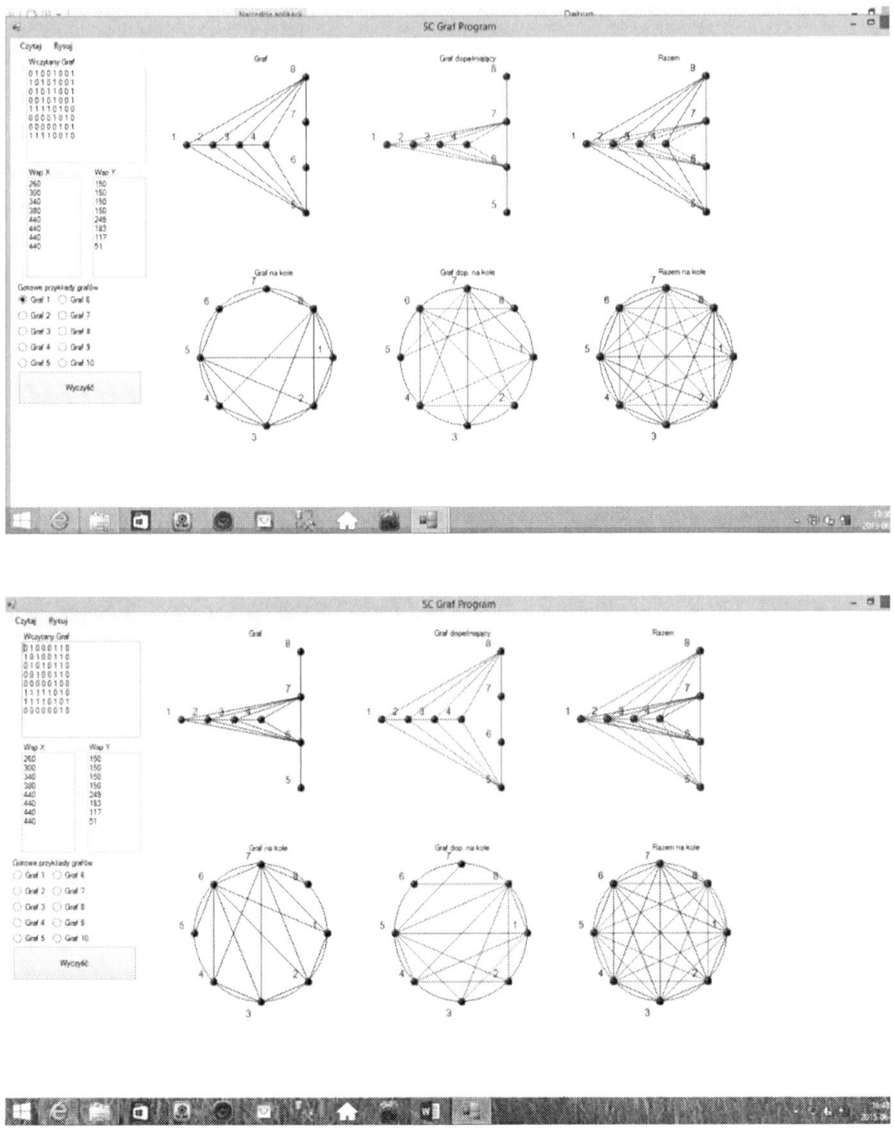

Fig. 5. Graphs plotted using the computer program

The isomorphisms of chosen graphs discussed in the presented paper (shown in different layouts in different figures)are as follows:

- Fig. 6a – Fig. 4c as well as Fig. 5; the graph in Fig. 6a has several edges' crossings, its presentation in Figs. 5 and 7 seems to be more simple;
- Fig. 6b – Fig. 4e; in the layout shown in Fig. 6b the graph has outstanding horizontal form, in case of Fig. 4e it is similar to three other shown in bottom row;
- Fig. 6c – Fig. 4g; conversion of the graph from Fig. 6c into the layout of Fig. 4g consists in lowering vertices 6 and 7 inside the interior of the graph and turning it upside down.

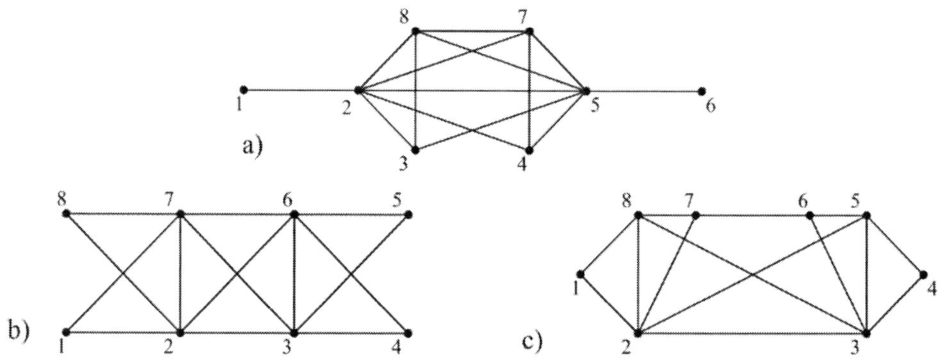

Fig. 6. Selected sc-graphs having n=8 vertices [1]

Summing up, we can state that there is not one classical view of a particular graph - may be only with an exception of Kuratowski's graph where $K_{3,3}$ is usually placed horizontally and K_5 is usually presented with a horizontal lower edge of the pentagon.

6 Isomorphism checking

In this chapter, the exemplary isomorphism of source graph and complementary one will be checked. The analyzed graph is given in Fig. 8.Moreover, the computer program was prepared

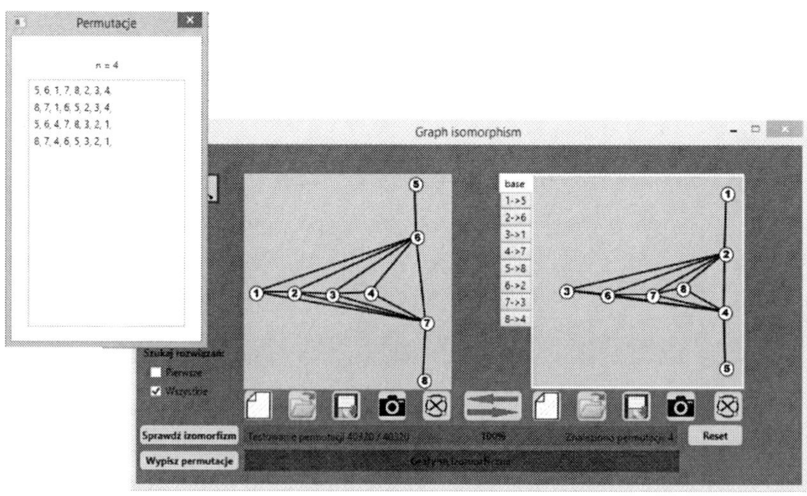

Fig. 7. Print-screen, program for isomorphism checking

aiming for algorithmic checking of isomorphism. In this case, full search algorithm was utilized because it was a student project on a bachelor level. Simultaneously, all the permutations have to be enumerated which was done via three different approaches. It was also a challenging part of the project e.g. the following approaches were utilized: lexicographic, anti-lexicographic and minimal number of transpositions. The advantage of the program is that all possible isomorphism are listed. In Fig. 7, there 4 possible isomorphism associations of respectable vertices (see upper box).

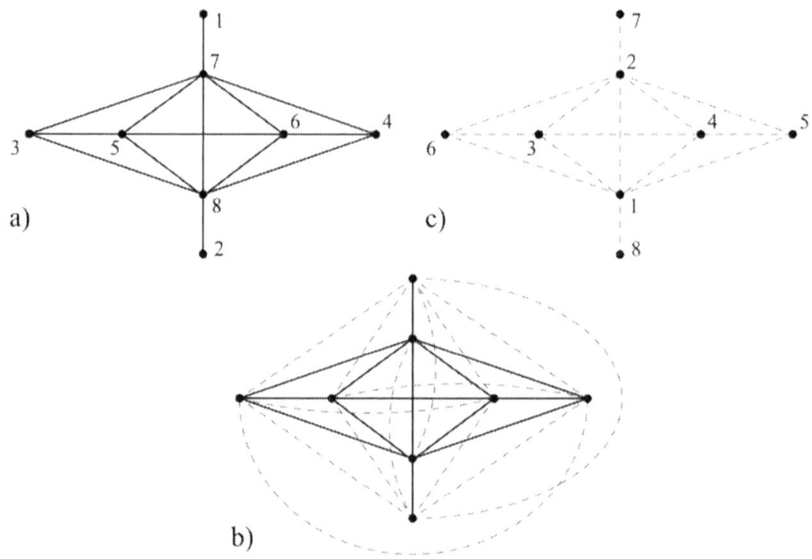

Fig. 8. Analyzed graph: (a)source, (b) both graphs together;(c) rearranged complementary one

Bijection of vertices	Bijection of edges (m=14)	
1↔7	1/ **(1,7)** ↔ (7,2)	8/**(4,8)** ↔ (5,1)
2↔8	2/ **(2,8)** ↔ (8,1)	9/ **(5,6)** ↔ (3,4)
3↔6	3/ **(3,5)** ↔ (6,3)	10/ **(5,7)** ↔ (3,2)
4↔5	4/**(3,7)** ↔ (6,2)	11/ **(5,8)** ↔ (3,1)
5↔3	5/ **(3,8)** ↔ (6,1)	12/ **(6,7)** ↔ (4,2)
6↔4	6/ **(4,6)** ↔ (5,4)	13/ **(6,8)** ↔ (4,1)
7↔2	7/ **(4,7)** ↔ (5,2)	14/ **(7,8)** ↔ (2,1)
8↔1		
List of source graph	List of complementary graph	Rearranged list of complementary graph
(	(	(
(1, 7)	(1,2, 3, 4, 5, 6, 8)	(7, 2)
(2, 8)	(2, 1, 3, 4, 5, 6, 8)	(8, 1)
(3, 5, 7, 8)	(3, 1, 2, 4, 6)	(6,1, 2, 3)
(4, 6, 7, 8)	(4, 1, 2, 3, 5)	(5,1, 2, 4)
(5, 3, 6, 7, 8)	(5,1, 2, 4)	(3, 1, 2, 4, 6)
(6, 4, 5, 7, 8)	(6,1, 2, 3)	(4, 1, 2, 3, 5)
(7, 1, 3 ,4, 5, 6, 7, 8)	(7, 2)	(2, 1, 3, 4, 5, 6, 8)
(8, 2, 3 ,4, 5, 6, 7, 8)	(8, 1)	(1,2, 3, 4, 5, 6, 8)
)	)	)

Tab. 2. Isomorphism details for the graph shown in Fig. 6

Proving the isomorphism (between particular graphs) in algorithmic manner can be done via checking equivalence of particular algebraic structures representing graphs: source and complementary ones based on checking all isomorphism attempts (full search version) i.e. checking all permutations of graph vertices. In case of the discussed program similarity of lists of neighbors was utilized. The lists, given in Tab. 2 are arranged in such a manner that the list consists of sub-lists for particular graph vertices of the graph. In the sub-list, first element is the vertex itself and then there is a sequence of its neighbors in lexicographic order.

The method is based on full search ("brute force" approach) but for greater number of graph vertices is obviously useless. Nevertheless, considered as a tutorial on bachelor level - where focus is given on programming and not on theory itself – such approach could be accepted.

In case of computer programs usually first element of the list is numbered as the '0'-one. In this paper, the adequate permutation was found by a trial and error method, additionally assuming that the vertices of degree 6 should be assigned to these of degree 1 and vice versa, based upon formula (1). The same rule was utilized for vertices of degree 3 and 4. Exemplary assignments are as follows:

$1 \leftrightarrow 7$; deg(1)=1 in source graph, deg(7)=1 in complementary graph,

$2 \leftrightarrow 8$; deg(2)=1 in source graph, deg(8)=1 in complementary graph.

The complete set of data is depicted in Tab. 2. Analogous considerations can be performed for all other sc-graphs. In case of grater n, only algorithmic approach and its computer realization has reasonable meaning where additionally effective algorithm of checking isomorphism is utilized. Exchanging the values in the rearranged list according to the bijection rules, we can obtain the source list. It proves isomorphism of the considered graphs (Fig. 8): source (continuous lines) and complementary one (dashed lines).Both graphs are drawn together in Fig. 8b, but using different approach than utilized in Fig. 5 i.e. aiming for avoidance of overlapping, some edges of the complementary graph are drawn as bows (arcs, lines having particular curvature).

7 Final remarks and conclusions

In the paper, the software dedicated to graph drawing and performance of chosen algorithms is discussed. Special focus is given on the programs written by students within the framework of the subjects related to graph theory. Special family of graphs was especially considered i.e. self-complementary graphs. The exemplary sc-graphs for $n=8$ were listed based on versatile references versions. The following observations have been made:

(i) drawing source sc-graphs in deliberately chosen co-ordinates gives a first glimpse of their layouts and possible similarities or symmetry features;

(ii) drawing source sc-graphs simultaneously with their complementary ones does not prove graphically the analyzed feature i.e. complementarity. Moreover, in case of chosen co-ordinates, it even can give a false view due to the overlapping edges – see e.g. Fig. 5. To avoid this effect, some edges should be drawn as bows (Fig. 8);

(iii) drawing sc-graphs on an invisible circle or ellipse unifies their view however allows for spotting some differences in their layouts. Especially, there are not any overlapping edges, on contrary they are clearly visible. Moreover it allows for graphical checking that the pair of graphs: source and complementary one create a clique (Fig. 5) which graphically confirm the nature of the discussed graphs;

(iv) from the point of view of didactics - the analyzed examples could be a good argument for the statement that versatile graphs views (layouts) should be generated and compared;

(v) exemplary checking of isomorphism of chosen pairs of sc-graphs and their complementary ones has been performed using lists of neighbors as an algebraic structure to represent a graph. It seems to be more effective in proving isomorphism in algorithmic way e.g. in case of matrices exchange of columns and rows in adequate sequence is needed. At the first glimpse, it seems that the problem of self-complementary graphs is purely theoretical one, however to the surprise of the reader we can cited some papers where the sc-graphs were applied in chemistry [20] and mechanics [9]. The wide review of mechanical applications of graphs (in general) can be found in the habilitation thesis [21].

The written computer programs allow for visualization of graphs (especially sc-graphs) and graphical presentations of the graphs in different arrangements i.e.: Cartesian co-ordinate system as well as displacement on an invisible circle. The first method is suitable for identification of graphs, whereas the second version is especially usable in case of presenting both graph simultaneously.

Full list of self-complementary graphs for $n=8$ is given in special arrangements allowing for graphical identification of their properties e.g. one or two axes of symmetry. The discussed approaches could be useful in didactics of subjects related to discrete mathematics and graph theory. The graph theory programs written by students allows them for showing their achieved knowledge and prove managing application of discrete algorithms and versatile visualization of performance of the applied algorithms. The skills of students are also confirmed by participation in domestic programmers competitions.

Acknowledgements

Authors would like to express their gratitude to the students of the University of Bielsko-Biala for a possibility to use programs written within the framework of the subject "Algorithmic approach to graph theory" to prepare Figs. 1,2, 3 (partly), 4,5 and Fig. 7.

References

[1] ALTER, R. Characterization of self-complementary graphs of order 8, *Portugaliae Mathematica*, 1975, 34(3), s. 157-161.

[2] BATAGELJ, V., MRVAR, A. Pajek-program for large network analysis. *Connections*, 1998, 21.2: s. 47-57.

[3] CAPOROSSI, G., HANSEN, P. Variable neighborhood search for extremal graphs: The AutoGraphiX system. *Discrete Mathematics*, 2000, 212.1: s. 29-44.

[4] CARBONNEAUX, Y., LABORDE, J.-M., MADANI, CABRI, R. M. Graph: A tool for research and teaching in graph theory. In: *International Symposium on Graph Drawing*. Springer Berlin Heidelberg, 1995. s. 123-126.

[5] CVETKOVIĆ, D., GUTMAN, I. The computer system GRAPH: A useful tool in chemical graph theory. *Journal of computational chemistry*, 1986, 7.5: s. 640-644.

[6] DAGDILELIS, V., SATRATZEMI, M. DIDAGRAPH: Software for teaching graph theory algorithms. *ACM SIGCSE Bulletin*, 1998, 30.3: s. 64-68.

[7] DE NOOY, W. MRVAR, A., BATAGELJ, V. *Exploratory social network analysis with Pajek.* Cambridge University Press, 2011.

[8] GIBBS, R.A. Self-complementary graphs, *Journal of Combinatorial Theory*, 1974, Series B, 16(2), s. 106-123.

[9] HUSSAIN, M.M., RAHMAN, A., GOMES C. Miniaturization of the half ring monopole antenna using self-complementary method for UWB application, *Antenna Technology: Small Antenna*, IEEE 2014.

[10] MILKOVÁ, E. Multimedia applications and their benefit for teaching and learning at universities. *WSEAS Trans. on Information Science and Applications*, 2008, 5.6: s. 869-879.

[11] MILKOVÁ, E. Multimedia applications: improving student engagement in learning process, *Virtual university VU'09* (Proceedings of 10th international scientific conference, Bratislava, December 10-12, 2009), CD-ROM, E-Academia Slovaca, n.o., Bratislava, 2009.

[12] NAIR, P.S. Construction of self-complementary graphs, *Discrete Mathematics*, 1997, 175, s. 283-287.

[13] PEISERT, W. All self-complementary symmetric graphs. *Journal of Algebra*, 2001, 240(1), s. 209-229.

[14] PEMMARAJU, S., SKIENA, S. S. *Computational Discrete Mathematics: Combinatorics and Graph Theory with Mathematica®*. Cambridge University Press, 2003.

[15] READ, R.C. On the number of self-complementary graphs and digraphs, *Journal of London Math. Soc.*, 1963, 38, s. 99-104.

[16] SUPRUNENKO, D.A. Self-complementary graphs, *Kibernetika*, 1985, 5, pp. 1-6.

[17] TROTIGNON N. On the structure of self-complementary graphs, *Electronic Notes in Discrete Mathematics*, 2005, 22, s. 79-82.

[18] WHITE, D. R., BATAGELJ, V., MRVAR, A. Anthropology analyzing large kinship and marriage networks with ‚Pgraph' and ‚Pajek'. *Social Science Computer Review*, 1999, 17.3: s. 245-274.

[19] WU, M. Teaching graph algorithms using, 2003, 24(4), s. 165-187.

[20] ZAWIŚLAK, S. (supervisor), Perez A., Honkisz J. Graphdrawing, *Proceedings: II Międzynarodowa Konferencja Studentów i Młodych Naukowców*, Bielsko-Biała 2012, s. 141-148.

[21] ZAWIŚLAK, S. (supervisor), Sury J. Graphdrawing by means of forcemethods, Proceedings,*III Międzynarodowa Konferencja Studentów i Młodych* online Java package IAPPGA. *ACM SIGCSE Bulletin*, 2005, 37.4: s. 64-68.

[22] XIAO, L. et all. Novel triple-shape PCU/PPDO interpenetrating polymer networks constructed by self-complementary quadruple hydrogen bonding and covalent bonding, *Polymer Chemistry*, 2014, 5(7), s. 2231-2241.

[23] ZAWIŚLAK, S. *Graph-based methodology as an artificial intelligence aidfor mechanical engineering design*, (Habilitation thesis),WydawnictwoATH, Bielsko-Biała 2010.

[24] ZAWIŚLAK, S., FREJ G. An influence of parameters of the evolutionary algorithm applied for the graph k-partitioning problem, *StudiaInformaticaNaukowców*, Bielsko-Biała 2013, s. 387-398.

[25] ZAWIŚLAK, S. (supervisor), Krysta Ł. Metro map drawing – correctionoptions, Proceedings, *IV Międzynarodowa Konferencja Studentów i Młodych Naukowców*, Bielsko-Biała 2014, s. 167-174.

[26] ZAWIŚLAK, S., RYSIŃSKI, J. (Eds) Graph-based modelling in engineering, Springer, 2017.

[27] ZELINKA, B. Self-complementary vertex-transitive undirected graphs, *Mathematica Slovaca*, 1979, 29(1), s. 91-95.

[28] URL:http://www.columbia.edu/~plm2109/two.pdf - open 18 June 2015.

[29] URL:http://www.graphviz.org/ - open 20 June 2015.

Current address

Zawiślak Stanisław, PhD, D.Sc.
Faculty of Mechanical Engineering and Computer Science,
Department of Informatics and Automation, University of Bielsko-Biala
2 Willowa Street, 43-309 Bielsko-Biała, Poland
Tel. number: +48 8279 352, e-mail: szawislak@ath.bielsko.pl

Kopeć Jerzy, PhD
Faculty of Mechanical Engineering and Computer Science, University of Bielsko-Biala
2 Willowa Street, 43-309 Bielsko-Biała, Poland
Tel. number: +48 8279 360, e-mail: jkopec@ath.bielsko.pl

Slovak University of Technology in Bratislava
Faculty of Mechanical Engineering
16th Conference on Applied Mathematics **APLIMAT 2017**
Proceedings

MATHEMATICS AND ITS APPLICATIONS IN THE ARCHITECTURE AND URBAN DESIGN OF SELECTED PARTS OF MEXICO

ŽDÍMALOVÁ Mária (SK)

Abstract. In this contribution we discuss mathematics and its applications in architecture and urban design. We prepared with students projects devoted to the selected parts of the states of Mexico. The aim was to consider different approaches devoted to the architecture and urbanism.

Keywords: mathematics, grid urban design, measures of compactness, Mexico

Mathematics Subject Classification: Primary 00A69, 93A30

1 Introduction

To find bridge between architecture and mathematics, design, fashion, urbanism and mathematics is very important because of research and teaching purposes. We discuss two topics, connected to the part of Mexico: the urbanistic grids, urbanistic studies.

In Chapter 2 we consider the grid design of the cities. We introduce the concept of the grid in the towns in general, with the short history introduction as well as urbanistic principles and properties connected with Compactness. We show the grid of Oaxaca and Mexico City as well. In Chapter 3 we offer how we use these topics for teaching purposes and creating projects with students of architecture. We show a short selection of their international presentations.

2 Urbanism, mathematics, grids and urbanistic properties applied to Oaxaca and Mexico City

We discuss short brief history of using grid design. The grid plan dates from antiquity and originated in multiple cultures; some of the earliest planned cities were built using grid plans.

The pioneer of urban and city planning is considered according to Aristotle Hippodamus, with the land also divided into three (sacred, public and private). He was marked as the first urban planner if we consider to proper arrangements of cities. He laid out the Piraeus (the port of Athens, for Pericles), and built the city of Rhodes in the form of a theater. In 440 BC he planned the new city of

Thurium (later Thurii), in Magna Graecia, with streets crossing at right angles. His influencis possible to find in many important cities, such as Halicarnassus, Alexandria and Antioch, [31].

Pakistan, India: By 2600 BC, Mohenjo-daro and Harappa, major cities of the Indus Valley Civilization (in what is now Pakistan and Northwestern India), were built with blocks divided by a grid of straight streets, running north-south and east-west. The cities and monasteries of Gandhara (e.g. Sirkap and Taxila), dating from the 1st millennium BC to the 11th century AD, also had grid-based designs. Islamabad, the capital of Pakistan since 1959, was also founded on the grid-plan of the nearby ruined city of Sirkap, see [28].

Giza, Egypt: In Giza and Egypt, a workers' village (2570-2500 BC) housed a rotating labor force and was laid out in blocks of long galleries separated by streets in a formal grid. Many pyramid-cult cities used a common grid planning.

Babylon: Hammurabi (17th century BC) was a king of the Babylonian Empire who made Babylone of the greatest cities in Ancient history. He rebuilt Babylon: the streets of Babylon, [5, 22].

China: The tradition of grid plans is continuous in China from the 15th century BC onward in the Ancient Chinese urban planning of various ancient Chinese states. A capital city should be square on plan. Three gates on each side of the perimeter lead into the nine main streets that crisscross the city and define its grid-pattern.

Mexico and Mesoamerica: We can consider Mexico, where Teotihuacan, near modern-day Mexico City, is the largest ancient grid-plan site in the Americas. The city's grid covered eight square miles. And we can present Tenochtitlan, an advanced Aztecan metropole founded in 1325. Mesoamerica was familiar with grid layout before arrival of Spaniards, [5,22].

Fig. 1. Teotihuacan, Mexico, source: goldenagerpject.org

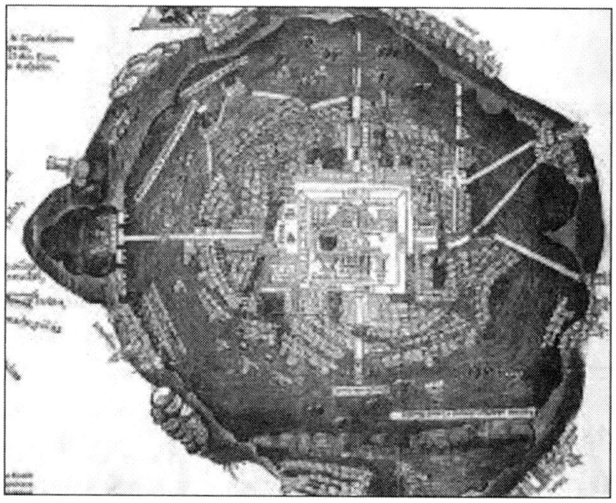

Fig. 2. Tenochtitlan, Mexico, source: sydeth.blogspot.com

Roman Empire: Perhaps the most well-known grid system is that spread through the colonies of the Roman Empire. The archetypal Roman centuriation Roman Grid was introduced to Italy first by the Greeks, with such information transferred by way of trade and conquest, [19].

The ancient Greek grid: Although the idea of the grid was present in Hellenic societal and city planning, it was not pervasive prior to the 5th century BC. However, it slowly gained primacy through the work of Hippodamus of Miletus, who planned and replanned many Greek cities in accordance with this form, [5]. The concept of a grid as the ideal method of town planning had become widely accepted by the time of Alexander the Great. More detailed descrption you can find in [5, 8, 3].

Ancient Roman civilization: "The ancient languages of Europe" [22] founded what is now the city of Marzabotto at the end of the 6th century BC. It was based on Greek Ionic ideas, and it was here that the main east-west and north-south axes of a town (the "decumanus maximus" and "cardo maximus" respectively) could first be seen in Italy [19].

The military expansion of this period facilitated the adoption of the grid form as standard: the Romans established "castra" (forts or camps) first as military centres; some of them developed into administrative hubs. The Roman grid was similar in form to the Greek version of a grid, but allowed for practical considerations, [7, 8, 13].

Japan and Korea: has been politically centralized in the 7 th century AD. Those societies adopted Chinese grid-planning principles in numerous locations. In later periods, some parts of Tokyo were grid-planned, but grid plans are generally rare in Japan, and the Japanese addressing system is accordingly based on increasingly fine subdivisions, rather than grid. Planning with grid continue in Asia through 20 th century. It was influenced under American influence.

Europe: In Europe the idea came from Ancient Roman civilization, which invented land measuring - layout system called „**Centuration**", they were inspired from Ancient Egypt. It became popular in

Renaissance again, and it is very visible in European cities built (or rebuilt) after medieval era, especially in Italy (Florence, Turin, Gramichelle), Spain (Barcelona) and France (Paris).

America: After discovery of Americas and the beginning of European colonial period, colonial empires started to applying their visionary ideas into new cities they built in this excess land. The most geometric, regular, gridded cities lays in the New World (Americas).

A grid is considered the most efficient city layout until today. There are many **advantages** brought by invention of this layout:

- **Regulation** - "Cutting" blocks into regular lot shapes and sizes is making every single building easy to build and rebuild.
- **Consistency**, proporiton, **harmony,**
  - because grid constrain all possibilities into a few, its making it easier to be consistent.
- **Predictability** – planning and expanding city is easier.
- **Navigation** - in the city is much easier.
- **Accessibility** - to one place by routes. When one street is blocked, traffic can easily reroute to the next street.
- **Compactness** - people can live on a smaller area, closer together and closer to other parts of the city.

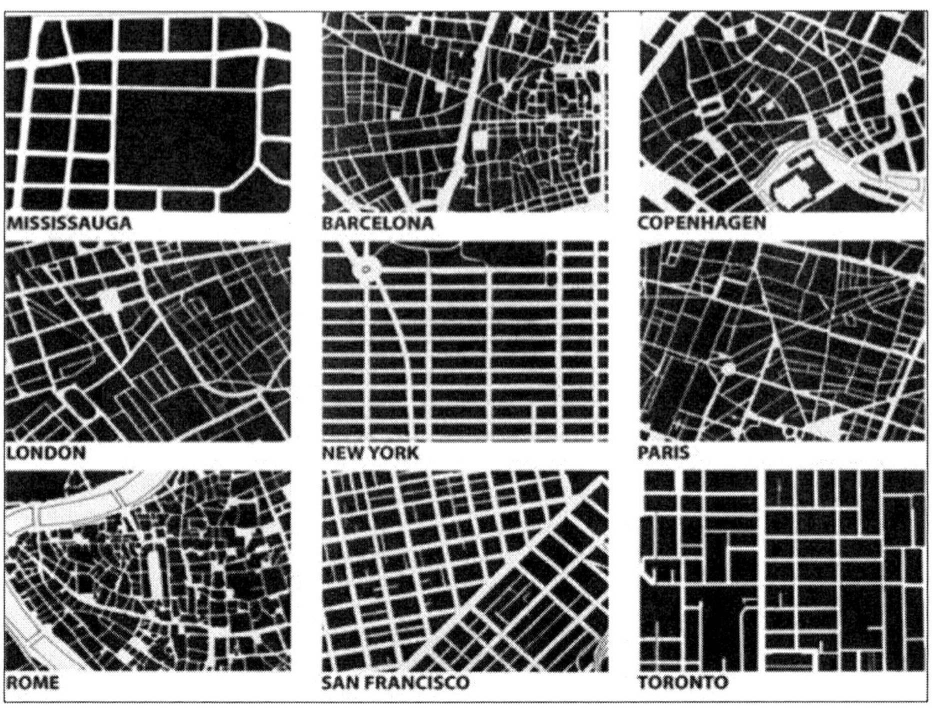

Fig. 3. Example of urbanistic grids of the cities in the world, source: helmofthepublicrealm.com

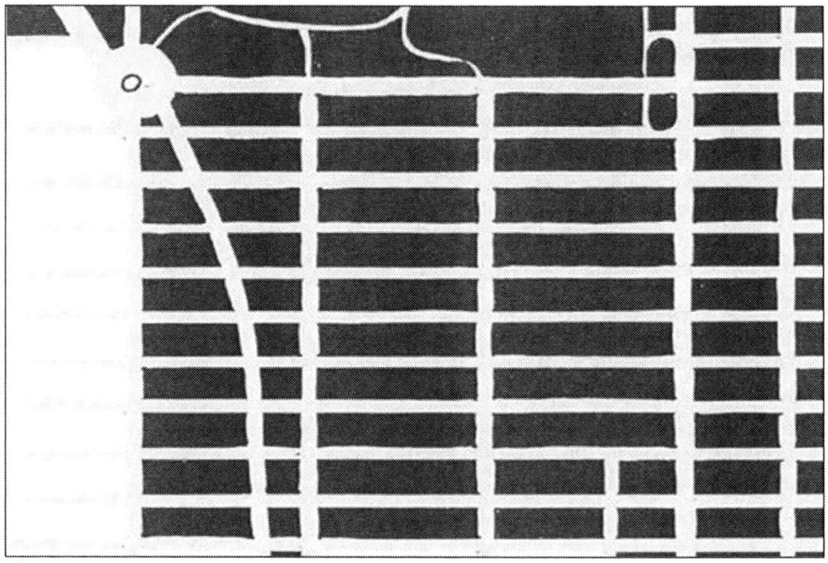

Fig. 4. Example of a rectangle grid, source: helmofthepublicrealm.com

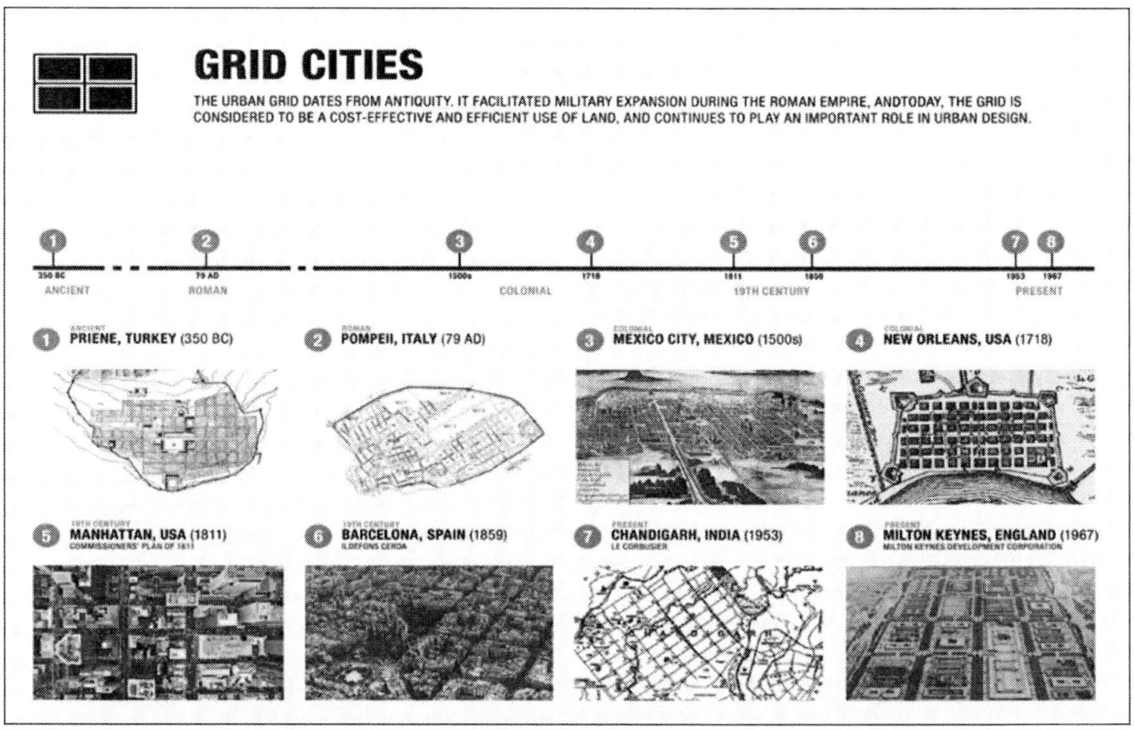

Fig. 5. Example of grid cities, source: helmofthepublicrealm.com

Oaxaca de Juaréz

Oaxaca de Juaréz, [26, 28] is city in United Mexican States, capital city of state Oaxaca. Together with only 10 km distant Monte Albán, the historical centre of Oaxaca is inscribed on the **UNESCO World Heritage property** list. These two distinct, seemingly contrast sites are composing extraordinary two pointed perspective on the history of two worlds – the old one and the new. Monte Albán is one of the earliest cities of mesoamerica. It was important political, economical and ceremonial centre built by **Zapotecs**. It prospered for almost a thousant years but then between AD 500 - 750 was suddenly abandoned. Historical centre of Oaxaca de Juarez is showing **unique urban manners** in "New Spain" from 16th century.

Historical context: (Vanished old gods)

Oaxaca, [26, 28], was founded by Aztecan emperor Ahuitzotl in 1486, relatively shortly before Spanish colonization. This eight emperor in Aztec line is responsible for great expansion of Aztecan empire. Not much is known about the appearance of the city before. Spaniards built their planned city at the expense of what left of the old inhabitants, bringing their God and european **sacred architecture**.

„Oaxaca was one of the southest points of aztecan expansion, also it was quite far : From Oaxaca to Tenochtitlan, capital city of their empire, more than 370 km air distance. Pointing out that Aztecs did not owned a car, nor domesticated horses and actually preferred traveling by foot. The "cornerstone" is a temlpe of GOD", [26, 28]

First spanish expedition arrived in 1521. Amongst them was also a chaplain Juan Diaz. His first mass was said under a large huaje tree. On the same spot today stands Temple of San Juan de Dios, first and the oldest building in the city. The oldest part of temple was built from a native green stone, which is typical for lots of older buildings in Oaxaca. Because of that, city was also nicknamed "la Verde Antequera" or in english "the green Artequera" . City was originally planned to be named Nueva Antequera, after Antequera in Andalúcia, Spain.

Fig. 6. The main city of Oaxaca, Oaxaca, https://en.wikipedia.org/wiki/Oaxaca_City

Location:

Oaxaca is located in Southwestern Mexico. It is the capital and largest city of the Mexican state of the same name. Settlement of the native tribes Zapotecs and Mixtecs in Valley of Oaxaca for thousands of years – in connection with important ancient centres Monte Albán (capital of Aztecs civilization) and Mitla. Aztecs entered the valley later in 1440 and named it "Huaxyacac," a Nahuatl phrase meaning "among the huaje" trees. Arrival of Spanish expedition in 1521 which led to many wars. The Spanish laid out a new town around the existing Zócalo (main square) in 1529 .

Architecture:

Spanish colonial architecture - the layout of the zocalo (Plaza de la Constitución) and surrounding buildings follow the colonial town plan which dictated that the city should have a central plaza with the cathedral and government buildings around the dominant buildings on the city's zócalo were the Catedral de Oaxaca and State Government Palace.

Once the Plaza de la Constitución was established in 1529, Alonso Garcia Bravo used the plaza as point of reference by which to plan the rest of the city, giving it a central location in the new city.

Oaxaca was the first town laid out in the "New Spain" during the 16th century with **square blocks** of approximetly 100 meters per side and planned from a central square. The icons of economic, political and religious powers were built around this central place, giving the city dynamism and contributing to universal urbanism.

City plan and the grid of Oaxaca:

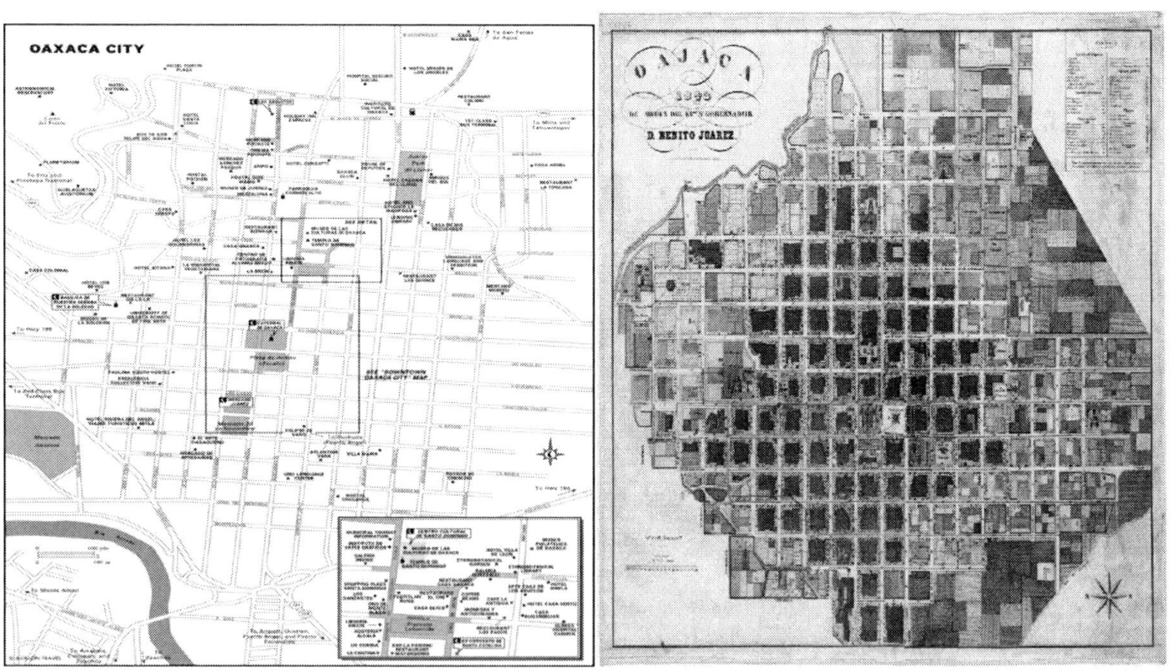

Fig. 7. City plan of Oaxaca, https://en.wikipedia.org/wiki/Oaxaca_City

Mexico City

Basic facts:

- Capital city of Mexico
- Population: City- 8 918 653, Urban- 20 400 000
- Density: 6000/km^2
- Area: 1485/km^2
- Elevation: 2250m
- The city was originally built on an island of Lake Texcoco by the Aztecs in 1325 as Tenochtitlan, which was almost completely destroyed in the 1521 siege of Tenochtitlan, and subsequently redesigned and rebuilt in accordance with the Spanish urban standards.

Geography:

Mexico City, [26, 29] is located in the Valley of Mexico. This valley is located in the Trans-Mexican Volcanic Belt in the high plateaus of south-central Mexico. It has a minimum altitude of 2,200 meters above sea level and is surrounded by mountains and volcanoes that reach elevations of over 5,000 metres. This valley has no natural drainage outlet for the waters that flow from the mountainsides, making the city vulnerable to flooding. "Drainage was engineered through the use of canals and tunnels starting in the 17th century. Mexico City primarily rests on what was Lake Texcoco. Seismic activity is frequent here. Lake Texcoco was drained starting from the 17th century. Although none of the lake waters remain, the city rests on the lake bed's heavily saturated clay. This soft base is collapsing due to the over-extraction of groundwater, called groundwater-related subsidence. Since the beginning of the 20th century the city has sunk as much as nine metres in some areas. This sinking is causing problems with runoff and wastewater management, leading to flooding problems, especially during the rainy season.", see [26, 29].

Fig. 8. Mexico City, source:urbanohio.com

Example of the urbanistic grids of Mexico City and its selected parts:

Fig. 9. Grid of selected part of Mexico City, source:urbanohio.com

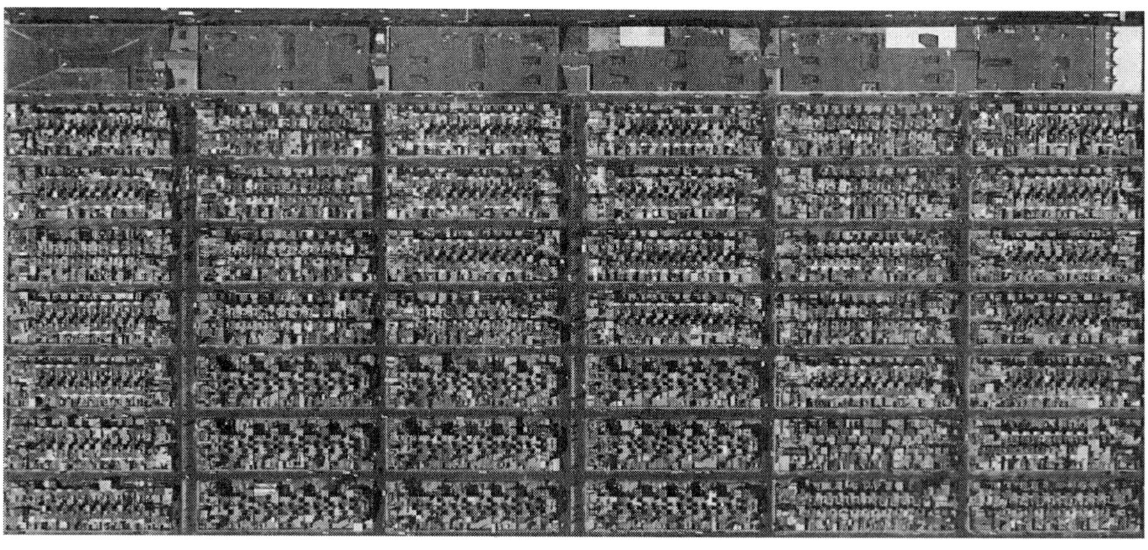

Fig. 10. Grid of selected part of Mexico City, source:urbanohio.com

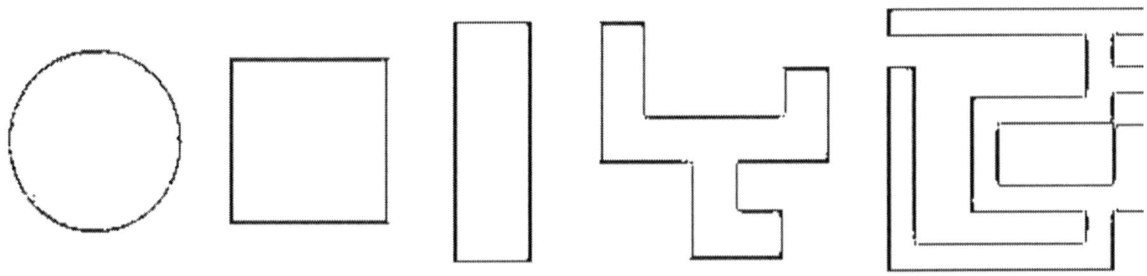

Fig. 11. Compactness Score (Polsby - Popper),
source: https://www.azavea.com/blog/2016/07/11/measuring-district-compactness-postgis/

Compactness

Applying principles of mathematics to urban spaces:

Space is precious. Within our planet, we have limited surface on which we can build. Levels of compactness of 2-dimensional shapes are important while planning a **layout** of the city. Architects also sometimes need 3D shapes to be compact when they want to effectively use the most of the given area. Everyone can easily tell whether or not is some simple 2D shape compact. The most compact 2D shape is definitely a disk. Consider, that circle is just a boundary for a disk. Our world is 3-Dimensional. In nature, the excellent example of compactness is water. The drop of water always wants to form the most compact shape, so its molecules were as close together as possible. In ideal case, with 0 gravity and with no other external aspect affecting water, it would form the perfect ball. Consider that sphere is a boundary for a ball. Ball also contains everything what is inside the sphere. In mathematics, and more specifically in general topology, compactness is a property that generalizes the notion of a subset of Euclidean space being closed and bounded. (Euclidean space itself is not compact since it is not bounded.) An example of a 1-Dimensional compact space is the closed unit interval of real numbers (for example: [0,1]). In two dimensions, closed disks are compact since for any infinite number of points sampled from a disk, some subset of those points must get arbitrarily close either to a point within the disc, or to a point on the boundary. Closed disk contains every point of the whole E2 space (2-Dimensional Euclidean space). A ball contains every point of the whole E3 space. That is the reason why these two shapes are considered as the most compact – we can fit literally the whole space inside. However, some closed and bounded shapes are not as compact as others.

Why squares and cubes work better in real life for now?

„There are few problems with disks and balls in real world. People are not dimensionless points inside Euclidean space. So it would not be really effective if we would just drew a circle and divided it with some roads and call it a city. Also we cannot just build a globe-shaped buildings nor rooms next to each other expecting the outcome to be the most space-saving. Shapes 90^0 have still a high level of compactness but also have the 0% of the negative contact space between two shapes left. There are few problems with disks and balls in real world: People are not dimensionless points inside Euclidean space. Packing rectangular-square rooms together to one floorplan, stacking many rectangular floors on each other, sticking together more building to one rectangular block, and finally, having **90^0** city grid made of many regular block is the easiest solution, " see [33]. That's why the Oaxaca is built inside a square grid see Fig. 8.

Example of counting: "The compactness of an object is often defined as the area divided by the SQUARE of the perimeter. If you multiply this by 4*PI, it will be equal to 1 for the circle and this will be the maximum value of your index. Any other geometric shape has a smaller ratio.

disc : $(4*PI)* PI*R^2 / 4PI^2R^2 = 1$

square : $(4*PI) * C^2 / 16 * C^2 = PI/4$ ~=0.78

Another usefull index could be the lenght/width of the smallest enclosing rectangle. But in this case the square and the circle are alike. As a last recommendation, if you work with perimeter, it is usefull to "smooth" your object before computing the indices, in order to avoid "fractal" effect (especially if your polygons come from raster to polygon conversion)", see [30].

Measures of compactness:

One measure of irregularity surrounding districts is known as compactness. There are a number of ways to measure compactness. When people refer to geographical compactness, they refer to districts of sensible shapes. Odd shapes can indicate gerrymandering. And the idea of single-member district grounded in the representation of territorially organized communities, where each area has its representative, though many dispute this idea of community

We now offer mathematics ways to measure the level of compactness.. A variety of compactness measure have been introduced by lawyers, social scientists, and geographers. Here we offer and highlight some of basic types of measure. A more complete guide may be found by Young [23], Noemi at al. [14], and Altman [1] and Chamber, Miller [6]. Most compactness belong into two broad categories:

1. Dispersion measures.
2. Perimeter – based measures.

Dispersion measures gauge to extent to which the district is scattered over a large area. The simplest dispersion measure is the **length-width test**, which compares the ratio of a district's length to its width. Ratio closer to one are considered more compact. This test had some support on the literature, see Harris [9]. Another type of dispersion measure are given by Reock [14], Boyce and Clark [4], and Kaiser [11]. They are comparison and center of gravidity measures have been adjusted by Hofeller and Grofman [9], Waver and Hess [21]. Dispersion measures have been widely criticized, in part because they consider districts reasonably compact as long as they are concentrated in a well shaped area, see Young [23].

Perimeter measures use the length of the district boundaries to assess compactness. The most common perimeter measure, associated with Schwarzberg [18], involves comparing on the ground that is overly sensitive to small changes in the boundary of a district. While a perfectly square district will receive a score 0,785, a square shape superimposed upon a diagonal grid of city blocks will have a much longer perimeter and a lower score.

We discuss more deeply above mentioned measures and approaches: Reock or Dispersion measure [16], Polsby-Popper measure [15], Area-Convex Hull Score and Schwarzberg Score.

Reock or Dispersion measure: is the ratio of the area of a district to the area of the smallest circle that can enclose the district. The idea behind the measure is that a circle is the most compactness shape, so it compares the district to the smallest circle that can encompass. As the name indicates, it is viewed as a good measure of the dispersion of the area of a district. We can tell that it is an area –based measure that compares each district to a circle, which is considered to be the most compact shape possible. For each district, the test computes the ratio of the area of the district to the area of the minimum enclosing circle for the district. The measure is always between 0 and 1, with 1 being the most compact.

The Polsby – Popper measure: is the area of the district to the area of a circle whose result is equal to the perimeter of the district:

$$4\pi\frac{A}{p^2}$$

This measure tends to give low scores to districts with many crinkles in their boundaries and that are elongated in ways the stretch their boundaries but encompass little territory. In contrast, districts with smooth boundaries that maximize the area enclosed score well. Easily said that the Polsby-Poper measure computes the ratio of the district area to the area of a circle with the same perimeter: 4pArea/(Perimeter^2). The measure is always between 0 and 1, with 1 being the most compact. The Polsby-Popper measure computes one number for each district and the minimum, maximum, mean and standard deviation for the plan. An example see Fig. 11.

The Perimeter measure or test: computes the sum of the perimeters of all districts. The perimeter test computes one number for the whole plan. If somebody compares several plans, the plan with the smallest total perimeter is the most compact.

The Schwarzberg Score: is the ratio of the perimeter if the district to the circumference of a circle whose are is equal to the area of districts. Get the radius of a circle with equal area

$$r = \sqrt{\frac{A}{\pi}}$$

With the radius of a calculated, we can se the following formula to generate the circumference (perimeter): $C = 2\pi r$. We get the Schwarzberg score using the following ratio: $1/\frac{p}{C}$

The Area-Convex Hull Score: is a ratio of the area of the minimum convex, polygon that can enclose the district's geometry.

The basic principle of convexity requires a district to contain the shortest path between every pairs of its points. Circles, square, and triangles are examples of convex shapes, while hooks, stars, and hourglasses are not. The most striking feature of bizarrely shaped districts is that they are extremely non-convex.

The **path based measure** is the probability that district will contain the shortest path between a randomly selected pairs of its points. This measure will always return a number between zero and one, with one being perfectly convex. In ideal case, a measure of compactness should consider the distribution of the population in the district.

Taylor [20] introduced a measure of idetation which compared the number of reflexive (inward-bending) to non-reflexive (outward-bending) angles in the boundary of the district.

At the end we can summarize : Compactness measures:

1. Dispersion measures: Length – Width, Area to Circumscribing Circle, Are to Convex Hull, Boyce – Clark.
2. Other measures: Path-based measure, Schwartzberg measure, Taylor measure.

Lastly, Schneider introduced a measure of convexity using Minkowski addition. For more relation between convex bodies and Minkowski's addition see Schneider [17].

Concrete example 1: According [6] we can show concrete values for some of tests, Fig. 12 :

Dispersion measure	District A	District
Length-Width	0,63	1,00
Area to Convex-Hull	0,57	0,29
Other measures		
Path-Based measure	.84	.42
Schwarzberg	.29	.14
Taylor	.40	.20

Tab. 1. Example of compactness measure

Disctrict A : District B:

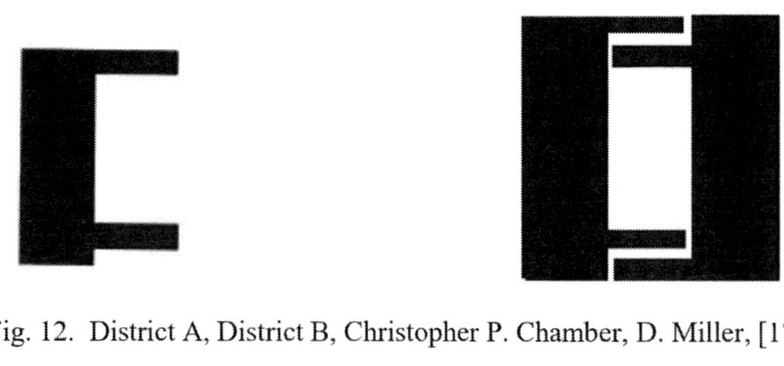

Fig. 12. District A, District B, Christopher P. Chamber, D. Miller, [17]

Concrete example 2, [34]: We can consider different regions, [34] and count the corresponding measures, like is in [34]:

A)

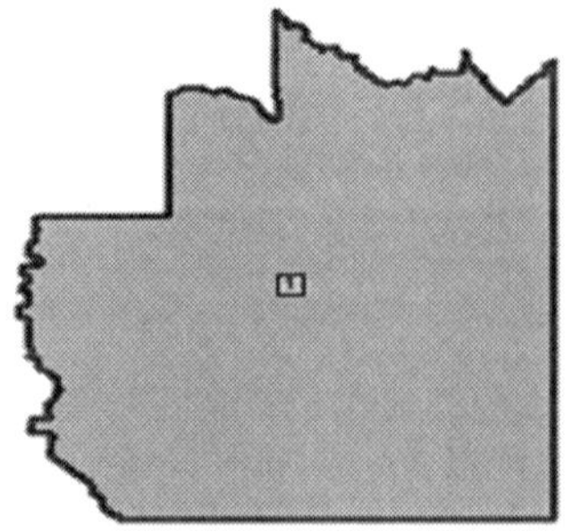

B)

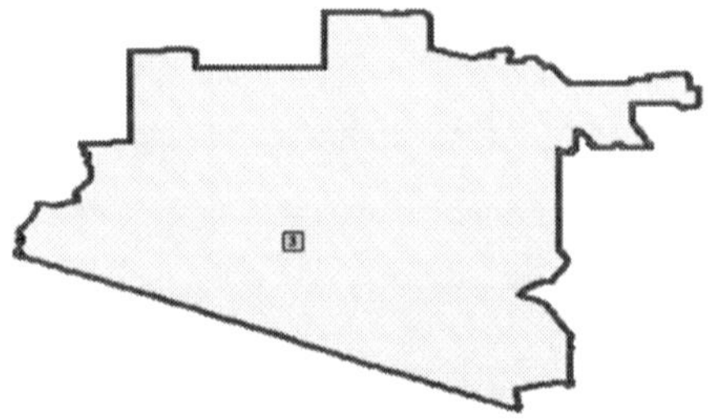

C)

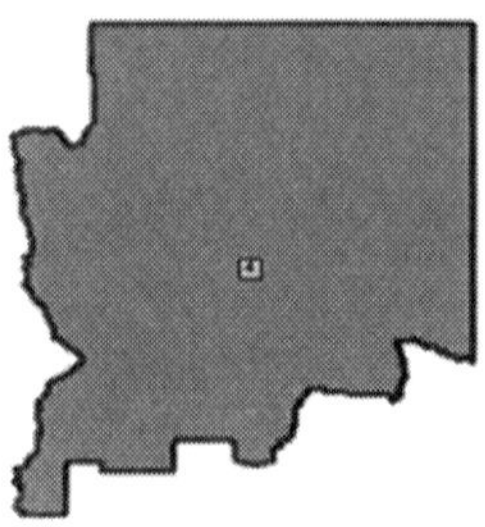

D)

E)

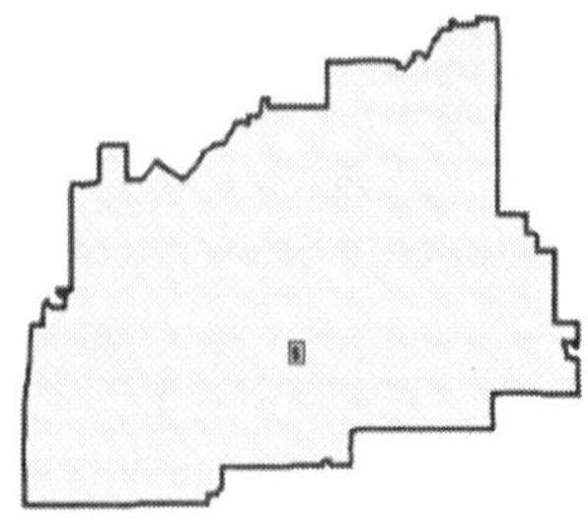

F)

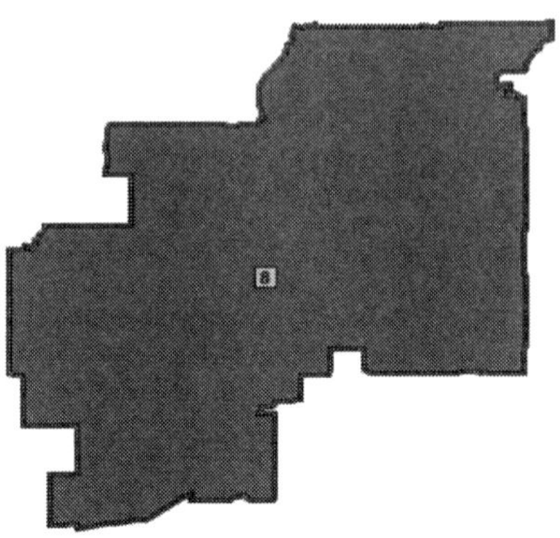

G)

Region	Polsby-Popper measure	Reock Test
A	.44	.64
B	.37	.44
C	.40	.48
D	.28	.43
E	.42	.47
F	.39	.43
G	.59	.60

Tab. 2. Example of regions and Polsby popper measure and Reock Test

The grid A) is the best for the Reock Test. The grid D) and F) is the worst for the Reock Test. For the Polsby-Popper test is the best the grid G) and the worst is the grid D).

5 Teaching experience

It is very necessary to show and motivate students in the first years of studies of architecture. We discussed all before mentioned topics with students during one semester. We used all these topics for students of architecture on the subject mathematics.

We started creating students projects devoted to the mathematics and architecture in 2012. We applied "Project teaching approach", see [2, 10, 11, 24, 25]. For the selected topics and students artworks we created a database of students works at **https:\\archmath.webnode.sk**, and also the parallel webpage at **https://www.facebook.com/Arch-Math-492082727664971/?fref=ts.** In facebook search just: Arch Math.

We started "facebook page" last year, 2016, and we are quite positive. At this moment we have more than 200 likes, we reached 19 countries: Slovakia, Italy, Mexico, India, Belgium, Spain, USA, Croatia, Tunisia, Canada, Peru, Indonesia, Algeria, Czech Republic, Turkey, Argentina, Malaysia, Chile, Serbia.

We present on the webpage students projects from three universities: Slovak University of Technology in Bratislava: Faculty of Civil Engineering, Faculty of Architecture and Academy of Fine Art and Design. This year, winter semester 2016, we have focused and discussed students project devoted to the architecture and urbanism and mathematics of Mexico. We focused on the main 2 areas:

1. **Urbanism of Mexico City-the grids.**
2. **Mathematical-urbanistic measures of compactness.**

Aims: students have to study and collected knowledges devoted to the selected topics. They needed to search informations, as well as mathematics and use it in presentations. Finally they did oral defense and created nice discussions with other students. The main aim is to analyze from mathematical point of view the selected topics.

Mathematical aims: to find and search facts devoted to earthquakes and mathematics and physics behind, the technology of earthquakes systems and principles, urbanism-applications of different types of grids.

6 Conclusion:

We discuss mathematical and urbanistic approach of the selected parts of Mexico. We analyzed mathematical principles of urbanistic grid in general and of Oaxaca and Mexico City. We deal with urbanistic principle like Polsby – Popper measure, Area Convex hull or Schwarzberg Score and others. Finally we used the mentioned topics for teaching purposes. The students prepared projects devoted to these topics. They presented them and finely we published on facebook page. We analyzed as well mathematical and geometrical approach to different designs of buildings.

Acknowledgement

The paper was supported by grant from Grant Agency VEGA 1/0420/15/.

References

[1] M. ALTMAN, *District Principles and Democratic Represantation*, PhD. Thesis, California Institute of Technolgy, 1998, March.

[2] R. BEGHETTO, KAUFMAN, JC, 2013, *Fundamentals of Creativity.* Educational Leadership, Vol. 70, No. 5, pp. 10-15.

[3] M. BELOZERSKAYA LAPITIN, KENNETH: *Ancient Greece: art, architecture and history*, 2004 Los Angels: Getty Publications, pp. 94.

[4] R. R. BOYCE, W. A. V. CLARK, The concept of shape geography, 1964, Geographical Review, 54: pp.561-572.

[5] BURNS, R., DAMASCUS: *A History, Routledge*, 2005., pp.39.

[6] CH. P. CHAMBER, A.D.MILLER, *A Measures of Bizarres,* March 27, 2009, published online

[7] M. GELEMTER, *A history of American arcgitecture buildings in their cultural and technological context*, 2001, pp.15.

[8] HIGGINSF, HANNAH, *The Grid Book*, 2009, p.60

[9] T. HOFFELER, B. GROFMAN, *Comparing the compactness of California congressional ditricts under three different plans:* 1980, 1982, and 1984, 1990. In B. Grofman, editor, Political Gerrymandering and the Court new York: Agathon.

[10] CHARD, SC, *The Project Approach: Six Practical Guides for Teachers. 2009.* These guides areavailable as .pdf files at the following web site: www.projectapproach.org

[11] KATZ LG, CHARD, SC 2000 *Engaging Children's Minds: The Project Approach*, Greenwood. 111

[12] H. F. KAISER, *An objective method for establishimg legislative districts*, 1996, Midwest Jpurnal of Political Scince, 10: pp.200-213.

[13] K. J. KENNETH, Crabgrass frontier, 1985, *The Suburbanization of the United States*, New York, Oxford, University Press, ISBN 0-19-504983-7.

[14] R. G. NIEMI, B. GROFMAN, C. CARLUCCI, T. HOFELLER, *Measuring compactness and the role of a compacness standard in atest for partisan and racial gerrymandering*, 1966, The Journal of politics, 52: 1155-213.

[15] D. D. POLSBY, R.D. POPPER, *The third criterion: Compactness as a procedural safeguard against partisan garymandering*, Yale Law and Policy Rewiev, 9: pp.301-353.

[16] E. C. REOCK, *A note: measuring compactness as a requirement of legislative appotionment*, 1961, Midwest Journal of Political Science, 5: pp.70-74.

[17] R. SCHNEIDER, *Convex Bodies: The Brun – Minkowksi theory*, 1993, Cambridge University Press.

[18] J. E SCHWARZBERG, *Reapportionment, gerrymanders and the ntion of „compactness"*, 1966, Minnesota Law Review, 5-: pp.443-452.

[19] D. STANISLAWSKI, 1946, *The Grid –Patter Town*, Geog. Rev., VI, pp. 105-120., pp,116.

[20] P. J. TAYLOR, *A new shae measure for evaluation electoral district patterns*, 1973, The american Political Science Review, 67: pp.947-950

[21] J. B. WEAVER, S. W. HESS, *A procedure for nonpartisan districting*: Development of computer techniques, 1963, The Yale Law Journal, 73: pp.288-308.

[22] R. WOODARD, *The antient language of Europe*, 2008.

[23] H. P. YOUNG, *Measuring the compactness of legislative distiсts*, 1998, Legislative Studies, Quarterly, 13: pp.105-115.

[24] URL:http://www.communityplaythings.com/resources/articles/2013/the-project-approach-to-teaching-and-learning

[25] URL: http://archmath.webnode.sk

[26] URL:http://www.visitmexico.com/en/mitla-archaeological-zapotec-ceremonial-center

[27] URL: http://wuw.visitslamabad.net/islamabad/files/map-home.asp

[28] URL: http://tanamatales.com/mitla-geometric-patterns-carved-in-stone/

[29] URL: http://www.infracstructuremexico.com/2015/09/17/steady-buildings-earthquakeresistan-design-in-mexico/

[30] URL: http://help.arcgis.com/en/arcgisdesktop/10.0/help/index.html#//00170000003q000000

[31] URL: https://en.wikipedia.org/wiki/Hippodamus_of_Miletus

[32] URL: helmofthepublicrealm.com

[33] URL: https://www.azavea.com/blog/2016/07/11/measuring-district-compactness-postgis/

[34] URL: https://www.google.sk/search?q=polsby%20popper%20score&rct=j

Current address

Ždímalová Mária, Mgr. PhD.
Departmet of Mathematics and Descriptive Geometry, Faculty of Civil Engineering,
Slovak University of Technology in Bratislava
Radlinského 11, 810 05 Bratislava, Slovak Republic
Tel. number: +421 259274419, e-mail: zdimalova@math.sk, maria.zdimalova@stuba.sk

Slovak University of Technology in Bratislava
Faculty of Mechanical Engineering
16th Conference on Applied Mathematics **APLIMAT 2017**
Proceedings

MATHEMATICS, GARDEN AND LANDSCAPE ARCHITECTURE

ŽDÍMALOVÁ Mária (SK), VAVRÍKOVÁ Lucia (SK)

Abstract. In this contribution we discuss students project approach teaching devoted to garden and landscape architecture. We discuss mathematics and its selected topics applied to garden and landscape architecture in the world.

Keywords: mathematics, garden architecture, symmetries, polygons, fractals

Mathematics Subject Classification: Primary 00A69, 93A30

1 Introduction

Garden and landscape architecture is a part of architecture and urban design.

One of the aim of our student project approach teaching was to find a bridge between mathematics garden and landscape architecture. We discuss with students a lot of mathematical topics and we applied them to the garden architecture. We started creating students projects devoted to the mathematics and architecture in 2012. We applied "Project approach teaching", see [1,2,3,4]. For the selected topics and students artworks we created a database of students works at **https:\\archmath.webnode.sk**, and also the parallel webpage at **https://www.facebook.com/Arch-Math-492082727664971/?fref=ts.** In the social network facebook we can search just: Arch Math.

We started "facebook page" last year, in 2016, and we are quite positive. At this moment we have more than 200 likes, we reached 19 countries: Slovakia, Italy, Mexico, India, Belgium, Spain, USA, Croatia, Tunisia, Canada, Peru, Indonesia, Algeria, Czech Republic, Turkey, Argentina, Malaysia, Chile, Serbia.

In this paper we discuss selected topics from mathematics and how they can be applied in garden and landscape architecture.

Students prepared project devoted to mathematics and garden architecture. Students have to prepare projects and do final oral defense. The last part is the open discussion.

We deal and discuss students project devoted to the gardens and landscape architecture. We focused on the following topics connected with landscape and garden architecture. We present students projects from three universities: Slovak University of Technology in Bratislava: Faculty of Civil Engineering, Faculty of Architecture and Academy of Fine Art and Design in Bratislava. We prepared with students projects devoted to different mathematical topics. We created two different pages, where we publish this students artworks.

Aims: students have to study and collected knowledges devoted to the selected topics. They needed to search informations, as well as mathematics and use it in presentations. Finally they did oral defense and created nice discussions with other students. The main aim is to analyze from mathematical, mechanical and physical point of view the selected topics.

Mathematical aims: to study the mentioned topics and its applications in gardens: polygons, tessallations, geometric solids, symmetries, mappings, fractals, surfaces, curves and surfaces of free forms, lights and shadows, geometrical shapes, golden section…

We discuss with students following topics:

1. Fractals and its application in garden architecture and landscape design.

2. Golden section.

3. Symmetries in garden architecture.

4. Surfaces and curves of analytic and "free" form in gardens and landscape architecture.

5. Polygons and geometric solids and its applications to gardens architecture.

1 Fractlas in garden and landscape architecture

The Kaos Garden (Ponte de Lima, Portugal) is an interpretation of the topic belonging to the Fractal Geometry. "The Kaos of the Universe" is a fractal that turns into plants, paths, floors, and globes that offer, through their apparent confusion, the idea of a Cosmic Chaos. Aromatic species, herbs and covering plants are combined with light silver and coloured globes that want to obtain an ideal representation of the stars and the planets of the universe. This leads to a question:

What is a fractal?

It is a natural phenomenon or a mathematical set that exhibits a repeating pattern that is displayed at every scale, see [5,6,7]. If the replication is exactly the same at every scale, it is called a self-similar pattern. An example of this is the Menger Sponge. Fractals can also be nearly the same at different levels. It is also known as expanding symmetry or evolving symmetry. If the replication is exactly the same at every scale, it is called a self-similar pattern. Easily said fractals also include the idea of a detailed pattern that repeats itself. This latter pattern is illustrated in the small magnifications of the Mandelbrot set.

Fractals are different from other geometric figures because of the way in which they scale. Doubling the edge lengths of a polygon multiplies its area by four, which is two (the ratio of the new to the old side length) raised to the power of two (the dimension of the space the polygon resides in). Likewise, if the radius of a sphere is doubled, its volume scales by eight, which is two (the ratio of the new to the old radius) to the power of three (the dimension that the sphere resides in). But if a fractal's one-dimensional lengths are all doubled, the spatial content of the fractal scales by a power that is not necessarily an integer. This power is called the fractal dimension of the fractal, and it usually exceeds the fractal's topological dimension.

As mathematical equations, fractals are usually nowhere differentiable. An infinite fractal curve can be conceived of as winding through space differently from an ordinary line, still being a 1-dimensional line yet having a fractal dimension indicating it also resembles a surface.

The mathematical roots of the idea of fractals have been traced throughout the years as a formal path of published works, starting in the 17th century with notions of recursion, then moving through increasingly rigorous mathematical treatment of the concept to the study of continuous but not differentiable functions in the 19th century by the seminal work of Bernard Bolzano, Bernhard Riemann, and Karl Weierstrass, and on to the coining of the word fractal in the 20th century with a subsequent burgeoning of interest in fractals and computer-based modelling in the 20th century.The term "fractal" was first used by mathematician Benoît Mandelbrot in 1975. Mandelbrot based it on the Latin fractus meaning "broken" or "fractured", and used it to extend the concept of theoretical fractional dimensions to geometric patterns in nature. Fractal patterns have been modeled extensively, albeit within a range of scales rather than infinitely, owing to the practical limits of physical time and space. Models may simulate theoretical fractals or natural phenomena with fractal features. The outputs of the modelling process may be highly artistic renderings, outputs for investigation, or benchmarks for fractal analysis. Some specific applications of fractals to technology are listed elsewhere. Images and other outputs of modelling are normally referred to as being "fractals" even if they do not have strictly fractal characteristics, such as when it is possible to zoom into a region of the fractal image that does not exhibit any fractal properties. Also, these may include calculation or display artifacts which are not characteristics of true fractals.

Modeled fractals may be sounds, digital images, electrochemical patterns, circadian rhythms, etc. Fractal patterns have been reconstructed in physical 3-dimensional space and virtually, often called "in silico" modeling. Models of fractals are generally created using fractal-generating software that implements techniques such as those outlined above. As one illustration, trees, ferns, cells of the nervous system, blood and lung vasculature, and other branching patterns in nature can be modeled on a computer by using recursive algorithms and L-systems techniques. The recursive nature of some patterns is obvious in certain examples—a branch from a tree or a frond from a fern is a miniature replica of the whole: not identical, but similar in nature. Similarly, random fractals have been used to describe/create many highly irregular real-world objects.

Benoit B. Mandelbrot (20 November 1924 – 14 October 2010) was a Polish-born, French and American mathematician with broad interests in the practical sciences, especially regarding what he labeled as "the art of roughness" of physical phenomena and "the uncontrolled element in life."He referred to himself as a "fractalist". He is recognized for his contribution to the field of fractal geometry, which included coining the word "fractal"', as well as developing a theory of "roughness and self-similarity" in nature.

The set is closely related to the idea of Julia sets, which produce similarly complex shapes. Its definition and name are due to Adrien Douady, in tribute to the mathematician Benoit Mandelbrot. Mandelbrot set images are made by sampling complex numbers and determining for each whether the result tends towards infinity when a particular mathematical operation is iterated on it. Treating the real and imaginary parts of each number as image coordinates, pixels are colored according to how rapidly the sequence diverges, if at all.

Fig. 1. Examples of fractals in nature

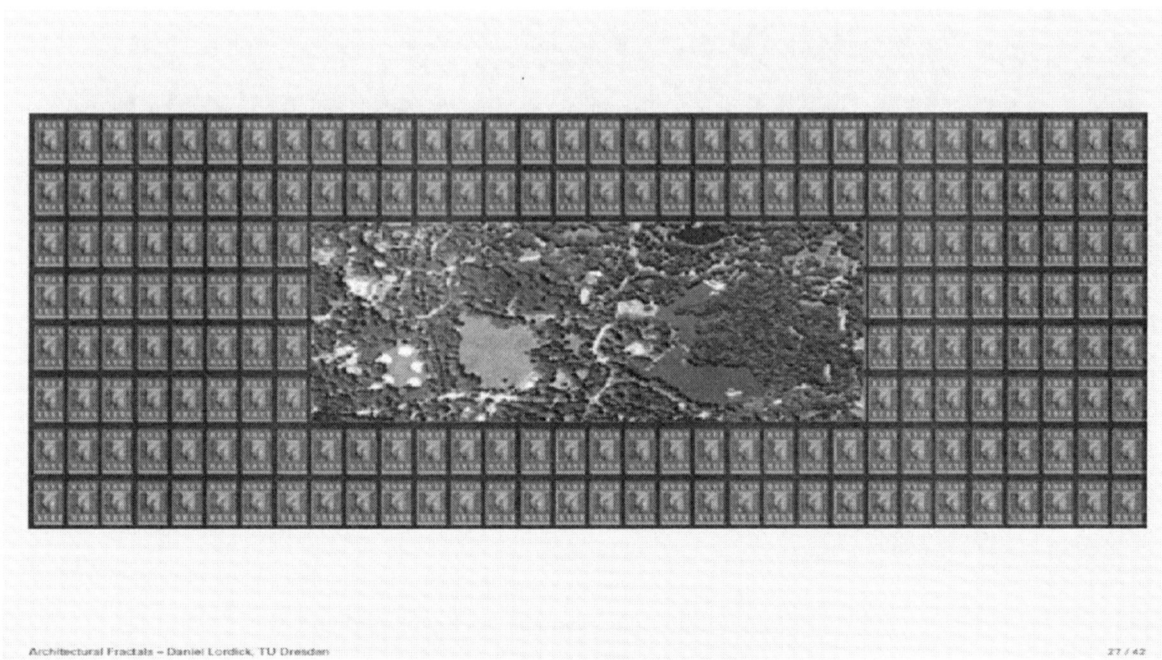

Fig. 2. Daniel Lordick, TU Dresden, Model of Central Park, New York

Fig. 3. Daniel Lordick, Tu Dresden, Architectural Fractals

Fig. 4. Daniel Lordick, Tu Dresden, Architectural Fractals

And finally in architecture:
•Fractals appear in architecture for reasons other than mimicking patterns in Nature.

Fig. 5. Example of fractals in architecture

Fig. 6. Examples of fractals in nature

2 Golden section

An example of using golden section in garden architecture:

Fig. 7. An example pf using golden section in garden architecture

3 Symmetries in landscape and garden architecture

We considered with students different types of geometric symmetries and its applications in gardens and landscape architecture, [8,9,10,11]. We can look at students projects:

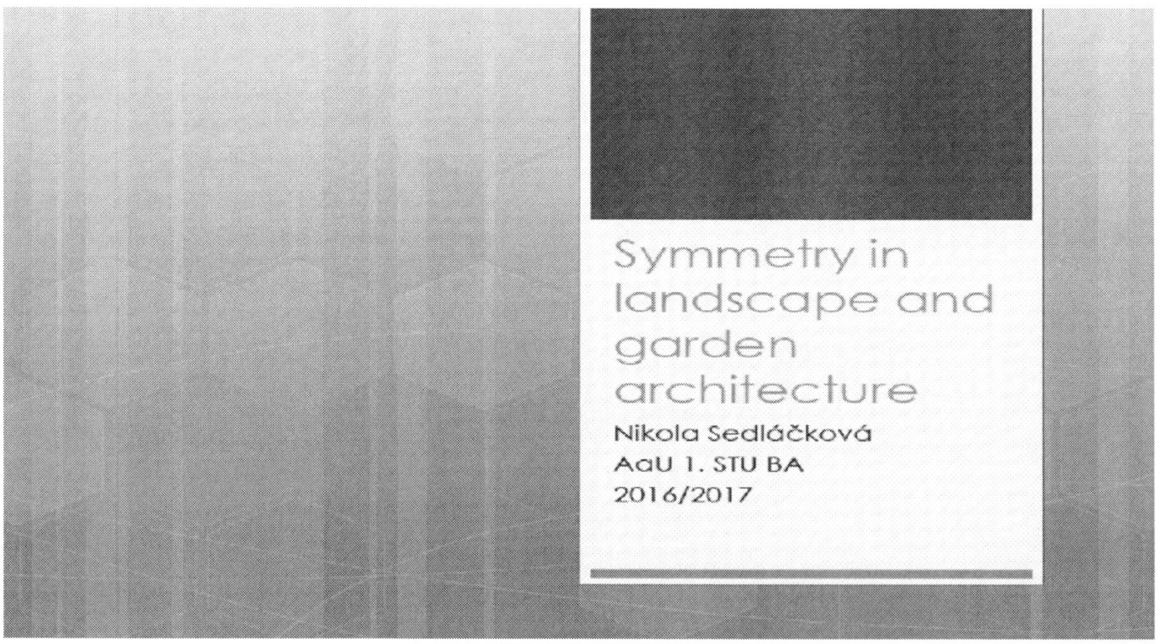

Symmetry

- Definition - an object is invariant to any of various geometric transformations : Reflection(Mirror), Rotation, Translation, Glide reflection
- symmetries preserve distances, angles, sizes, and shapes
- The mathematical study of symmetry is systematized and formalized in a group theory.
- Symmetric patterns occur in nature, and are invented by artists, craftspeople, musicians, choreographers, and mathematicians.

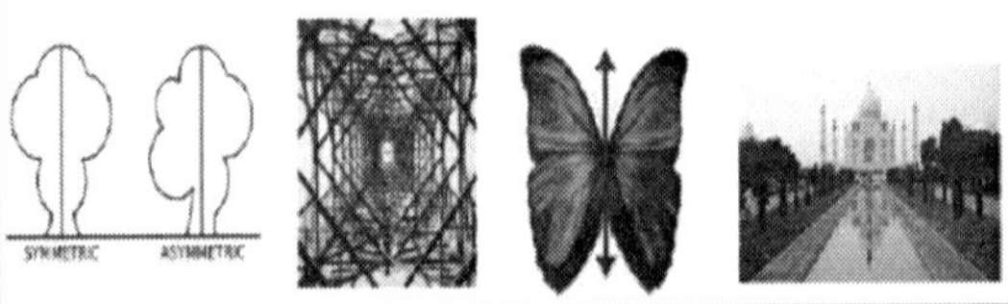

Reflection Symmetry

- = Line Symmetry = Mirror Symmetry
- An object is reflected across a line, which divides the shape into two identical parts.
- The line can be vertical, horizontal or diagonal.

Rotation Symmetry

- The turning of an object or coordinate system by an angle about a fixed point (center).
- An object's degree of rotational symmetry is the number of distinct orientations in which it looks the same.

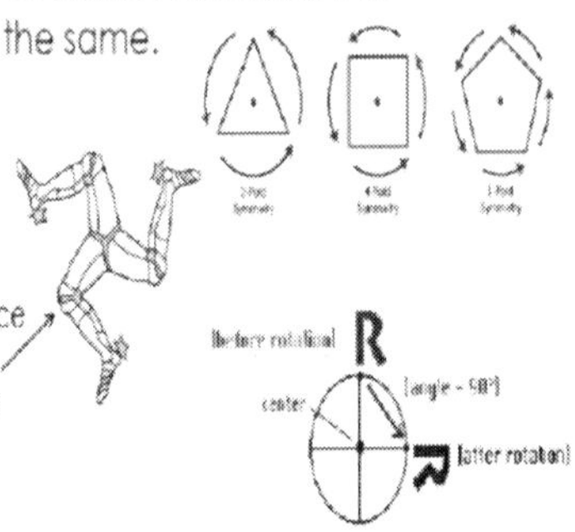

Because its appearance is identical in three distinct orientations, its rotational symmetry is three-fold.

Translation

- Translating an object means moving it without rotating or reflecting it.
- Every translation has a direction and a distance.
- Distances between points, angles, size and shape of the figure remain, just location changes.

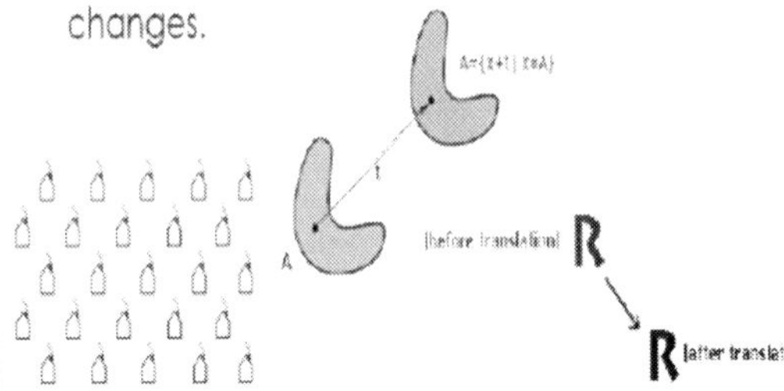

Symmetry in garden architecture

- Symmetry is one of the most common and used features in architecture in general - brings harmony and balance.
- It is the reflection of shared forms, shapes, or angles across a central line or point called the **axis.**
- Style of garden based on symmetry and the principle of imposing order on nature is e.g. **French formal garden** or **English garden.**

Glide reflection

- combines a reflection with a translation along the direction of the mirror line
- the order of these transformations does not matter
- only type of symmetry that involve more than one step

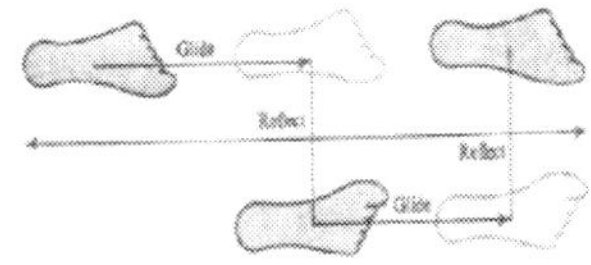
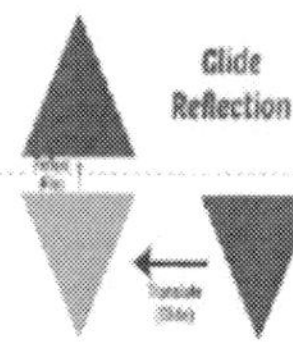

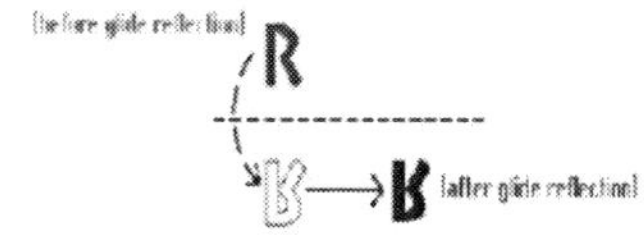

Examples of symmetrical gardens

The Gardens of Versailles, France

- Classic French garden
- one of the most visited public sites in France
- designed by André Le Nôtre, the reign of Louis XIV
- 372 statues, 55 water features, 600 fountains and 35km of canals
- the garden covers 800 hectares which includes 300 hectares of forest and two 'French' gardens, the 'Petit Parc' and the 'Trianon'

Examples of symmetrical gardens

Monet's Garden at Giverny, France

- probably the most famous garden in France - 500 000 people visit every year
- two parts of garden : a) flower garden(Clos Normand) is about one hectare, full of perspectives, symmetries and colours -thousands of flowers - tulips, irises, oriental poppies and peonies

b) water garden – inspiration of Monet's wonderful water lily paintings, full of asymmetries and curves, inspired by the Japanese gardens

The Japanese bridge

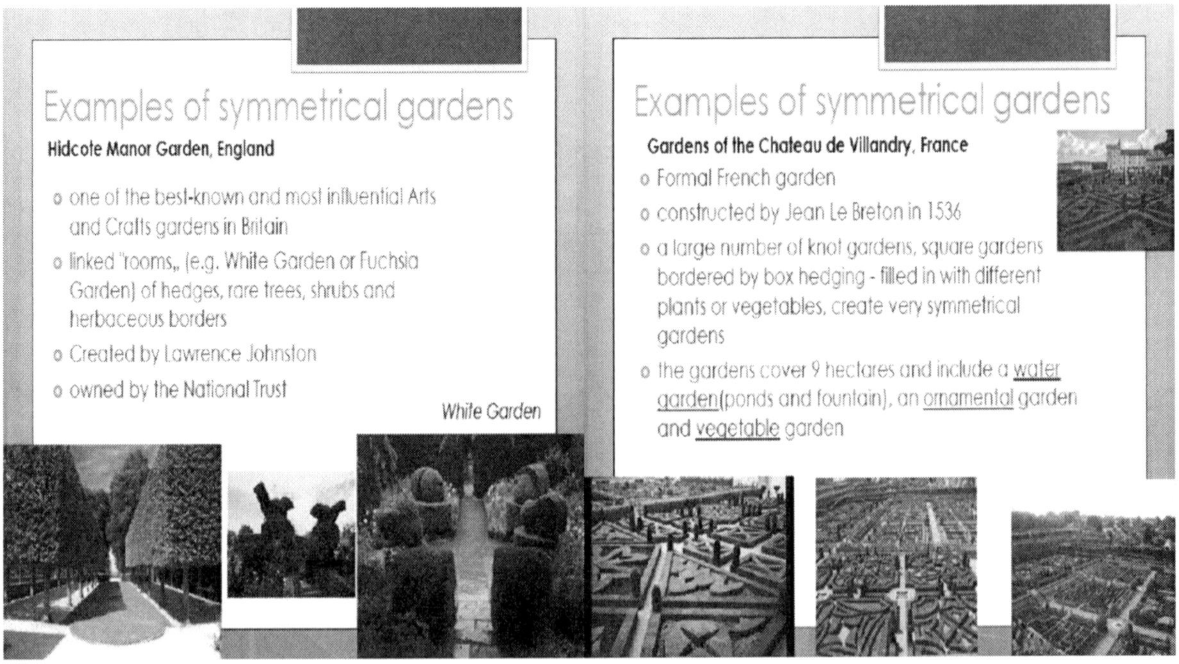

Geometric symmetries are often interpreted in the gardens architecture like:

-point symmetrie: e.g. a water fontaine in the middle of the garden, big central pattern made from plants, a big central geometric shape represented by plants or trees.

-reflection symmetrie: a path in the middle of the garden, with different colours, materials which produce contrast in the garden, path from small rocks, from trees cutted and formed trees in special shapes, a bridge over small river in the garden, path of flowers of the same colours and high. We also can consider garden mazes.

4 Surfaces and curves of implicit expression and also free form in gardens and landscape architecture

We can use curvilinears forms as well as different mathematical curves in the planning of garden and landscape architecture, [10, 11, 12, 13].

A selection from students projects:

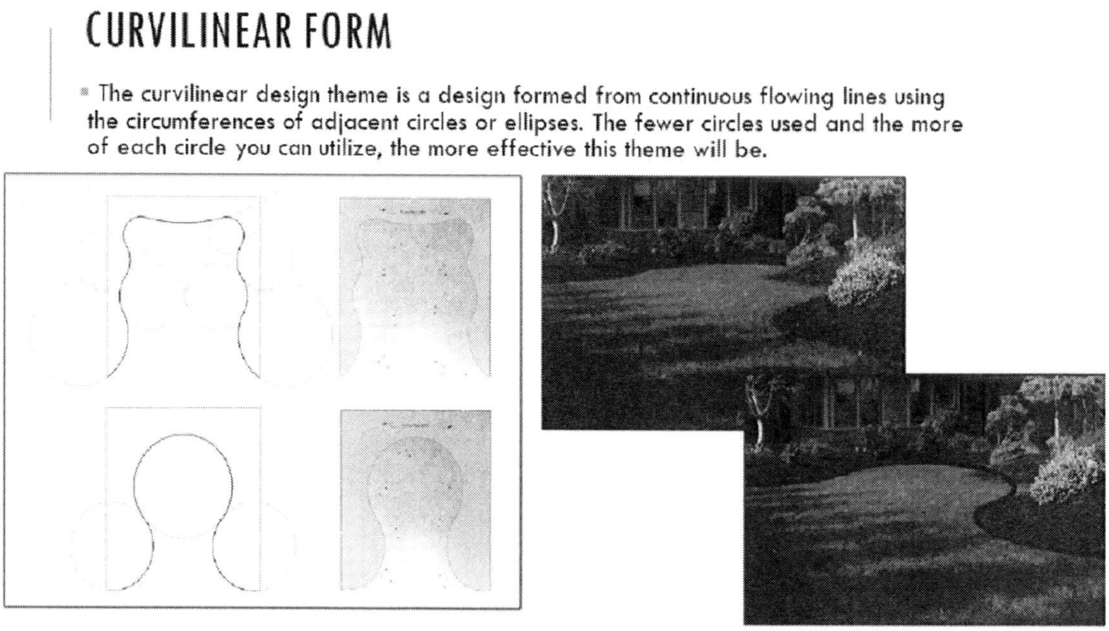

Fig. 8. Example of mathematical curves in garden and landscape architecture

Fig. 9. Example of mathematical curves in garden and landscape architecture

Geometrical curves and surfaces in landscape architecture:

Green walls in buildings:

Fig. 10. Green wall in Emporia

GREEN WALL - EMPORIA

- Emporia is a shopping mall and one of the biggest ones in Scandinavia
- It is situated in the city of Malmö in Sweden
- Emporia opened on 25 October 2012 and the total construction expense was about 2 billion Swedish kronor
- the architect of the Emporia project is Gert Wingårdh of Wingårdh arkitektkontor

Fig. 10. Example of curves in garden architecture

Fig. 10. Example of curves in garden architecture

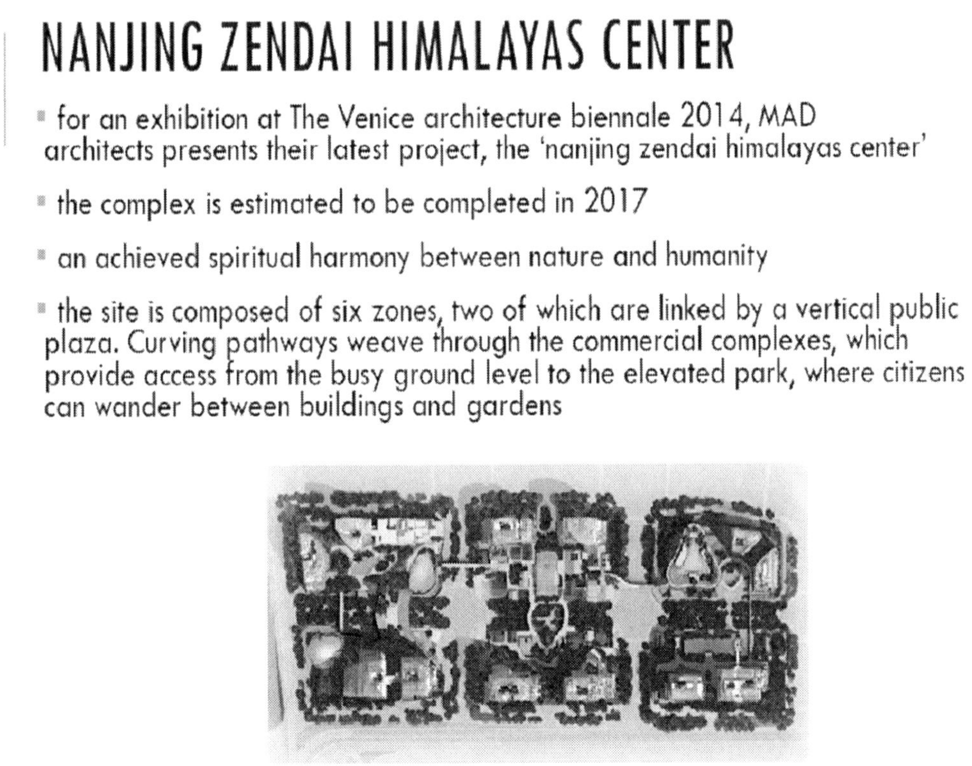

NATIONAL ARBORETUM, CANBERRA

- World Landscape of the Year 2014 from World Architecture Festival
- is a 250-hectare (618 acre) arboretum in Australia
- created after the area was burned out as a result of the Christmas and 2003 Canberra bushfires
- The Himalayan Cedar forest lost about one third of its trees, and the commercial Radiata Pine plantation was burned out, allowing the arboretum to be created
- The winning design by landscape architects Taylor Cullity Lethlean and architects Tonkin Zulaikha Greer proposed 100 forests and 100 gardens focussing on threatened, rare, and symbolic trees from around the world

Fig. 10. Example of mathematical surfaces in landscape architecture

5 Polygons, polyhedrons and geometric solids in landscape and garden architecture

Application of polygons and polyhedral we can find in the following presentation, [14, 15].

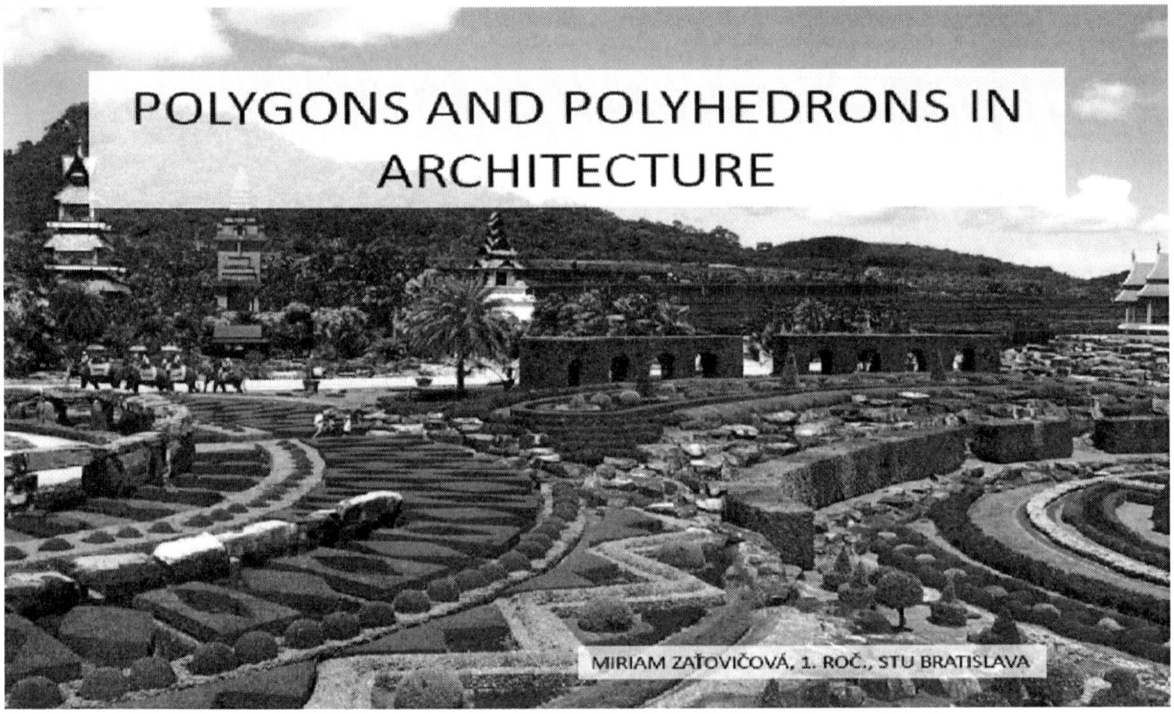

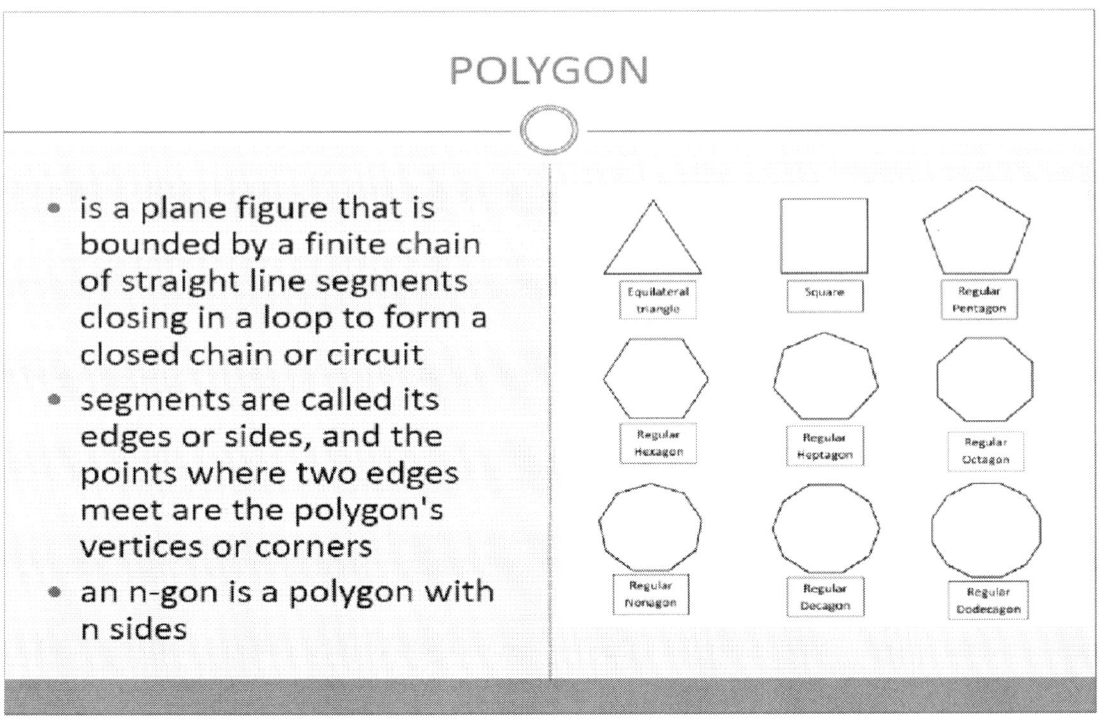

CONVEX POLYGON

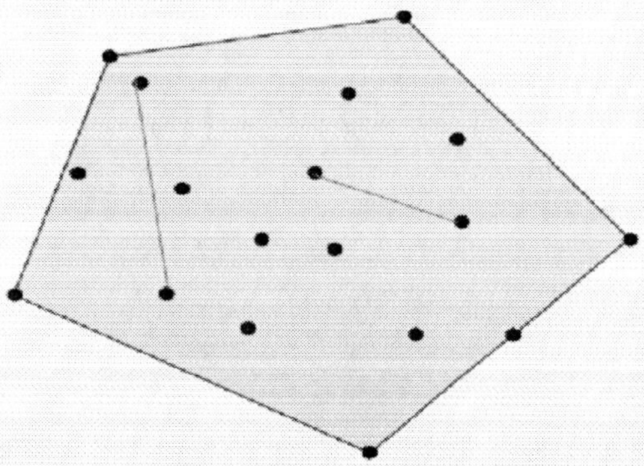

- any line drawn through the polygon (and not tangent to an edge or corner) meets its boundary exactly twice
- as a consequence, all its interior angles are less than 180°
- line drawn between two points inside polygon lies inside the polygon

NON-CONVEX POLYGON

- a line may be found which meets its boundary more than twice

- equivalently, there exists a line segment between two boundary points that passes outside the polygon

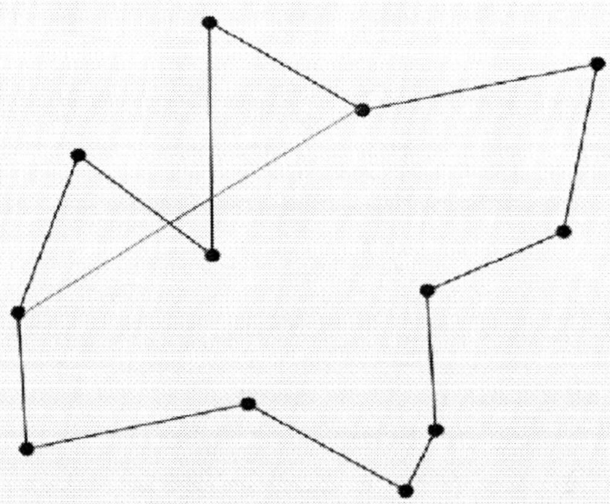

POLYHEDRON

- is a solid in three dimensions with flat polygonal faces
  meeting in pairs along their edges which are straight-line segments, and with the edges meeting in vertex points

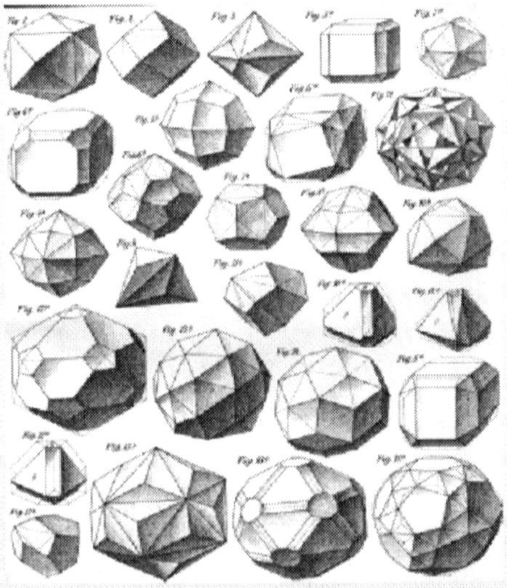

GARDEN OF COSMIC SPECULATION

- is a 30 acre sculpture garden created by landscape architect Charles Jencks at his home near Dumfries, Southwest Scotland
- is not abundant with plants, but sets mathematical formulae and scientific phenomena
- garden is private but usually opens on one day each year through Scotland's Gardens Scheme and raises money for charity

GARDEN OF COSMIC SPECULATION

NONG NOOCH TROPICAL GARDEN

- 500-acre (2.0 km^2) botanical garden and tourist attraction at kilometer 163 on Sukhumvit Road in Chonburi Province, Thailand
- Nongnooch Tansacha purchased the 600-acre (2.4 km^2) plot of land in 1954 with the intentions of developing the land as a fruit plantation
- However, instead decided to plant tropical flowers and plants as a wildlife conservation project

NONG NOCH TROPICAL GARDEN

NONG NOOCH TROPICAL GARDEN

NONG NOOCH TROPICAL GARDEN

NONG NOOCH TROPICAL GARDEN

DUBAI MIRACLE GARDEN

- The garden was launched on valentine's day in 2013
- spread over 72,000 square meters making it the world's largest natural flower garden
- over 45 million flowers planted
- on April 2015, the garden was awarded the Moselle Award for New Garden Experiences of the year
- cost of the project was an estimate of $11 million
- flowers are maintained by re-use of waste water with an average amount of 757,082 litres of water per day

DUBAI MIRACLE GARDEN

DUBAI MIRACLE GARDEN

DUBAI MIRACLE GARDEN

Polygons in islamic gardens:

Created by Zuzana Jurčišinová, AFAD in Bratislava, Slovakia:

Alhambra de Granada

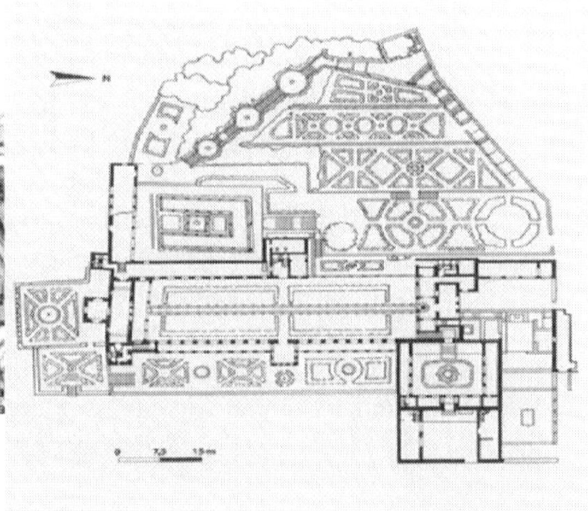

Alhambra did not have a master plan for the total site design, so its overall layout is not orthogonal nor organized. As a result of the site's many construction phases: from the original 9th-century citadel, through the 14th-century Muslim palaces, to the 16th-century palace of Charles V; some buildings are at odd positioning to each other. The terrace or plateau where the Alhambra sits measures about 740 metres (2,430 ft) in length by 205 metres (670 ft) at its greatest width. It extends from west-northwest to east-southeast and covers an area of about 142,000 square metres (1,530,000 sq ft). The Alhambra's most westerly feature is the alcazaba (citadel), a strongly fortified position. The rest of the plateau comprises a number of Moorish palaces, enclosed by a fortified wall, with thirteen towers, some defensive and some providing vistas for the inhábitants. The river Darro passes through a ravine on the north and divides the plateau from the Albaicín district of Granada. Similarly, the Assabica valley, containing the Alhambra Park on the west and south, and, beyond this valley, the almost parallel ridge of Monte Mauror, separate it from the Antequeruela district. Another ravine separates it from the Generalife.

The Circle and its centre is symbol at which all Islamic patterns begin. It emphasizes one god.

The Triangle symbolises human and the principles of harmony.

The Square is the symbol of physical experience and the physical world or materiality.

The Hexagon symbolises heaven.

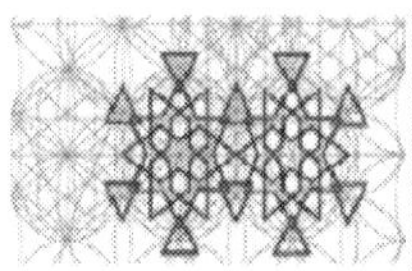

The Star symbolises the spread of Islam.

6 Conclusion

We discussed garden and landscape architecture and its connection to mathematics. We considered with students different topics from mathematics: fractals, polygons, different types of geometric symmetries, curves of free form and prescribed parameters. Students studed mathematical principles of the topics and finally applied them in the garden, landscape architecture and design. As a consequence they prepared projects devoted to the mentioned topics and did an oral defense. The projects are prepared in English. Students considered two streams in projects: complete mathematics discussion and analysing of mathematical facts and its applications in responsible topics in landscape and garden architecture. This brings more positive factors in educational process: Students need to search and study informations, trying to understand mathematical principles, to discover possible applications, to create a project and do an oral defense. As a special bonus is the fact if they are able to apply their own ideas from mathematics in their own artwork. This part belongs to an open discussion. The last part is publishing their projects on international webpage on internet and Facebook page "ARCH MATH". This way their artworks can be presented for other students and abroad countries. In the future we plan an interdisciplinary cooperation (including art, design, dance, music, architecture…) and to deal with different mathematical topics.

Acknowledgement

The paper was supported by grant from Grant Agency VEGA 1/0420/15/.

References

[1] BEGHETTO, RA & KAUFMAN, JC (2013), Fundamentals of Creativity, *Educational Leadership*, Vol. 70, No. 5, pp. 10-15.

[2] CHARD, SC, The Project Approach: Six Practical Guides for Teachers, (2009), These guides areavailable as .pdf files at the following web site: www.projectapproach.org

[3] KATZ LG & CHARD, SC (2000) *Engaging Children's Minds: The Project Approach*, Greenwood. 111.

[4] URL: http://www.communityplaythings.com/resources/articles/2013/the-project-approach-to-teaching-and-learning

[5] URL: http://archmath.webnode.sk

[6] URL: https://en.wikipedia.org/wiki/Fractal

[7] URL: https://en.wikipedia.org/wiki/Mandelbrot_set#/media/File:Mandelbrot_set_image.png

[8] URL: www.study.com

[9] URL: http://mathforum.org/sum95/suzanne/symsusan.htmls

[10] URL: https://www.wikipedia.org/

[11] URL: www.french-gardens.com/gardens

[12] URL: http://www.anglotopia.net/anglophilia/top-12-english-gardens-gardens-in-england-worth-a-visit/

[13] URL: https://www.citylife.sk/miesto/medicka-zahrada

[14] URL: https://en.wikipedia.org/wiki/Polygon
[15] URL: https://en.wikipedia.org/wiki/Polyhedra

Current address

Ždímalová Mária, Mgr. PhD.
Department of Mathematics and Descriptive Geometry, Faculty of Civil Engineering,
Slovak University of technology in Bratislava
Radlinského 11, 810 05 Bratislava, Slovak Republic
Tel. number: +421 259274419, e-mail: zdimalova@math.sk, maria.zdimalova@stuba.sk

Vavríková Lucia, Ing. PhD.
Department of Mathematics and Descriptive Geometry, Faculty of Civil Engineering,
Slovak University of technology in Bratislava
Radlinského 11, 810 05 Bratislava, Slovak Republic
Tel. number: +421 259274411, e-mail: vavrikova@math.sk

Slovak University of Technology in Bratislava
Faculty of Mechanical Engineering
16th Conference on Applied Mathematics **APLIMAT 2017**
Proceedings

AXIOMATIC THEORY OF A RETAIL GRAVITATION FIELD

ZEITHAMER R. Tomáš (CZ), POSPÍŠIL Jiří (CZ)

Abstract. This work presumes that the axioms given below have been met. The retail gravitation field is a vector field that has two components. The first component is a vector field of the intensity of retail gravitation. The intensity of retail gravitation is defined through the scalar potential φ of retail gravitation. In the United States economic investigation has confirmed that the analytic form of scalar potential φ is Newtonian, i.e. the scalar potential is inversely proportional to the square of the distance of the point of observation from the resource point and directly proportional to the population of a large city that attracts retail trade. The second component of the retail gravitation field is the vector field of retail gravitation induction. Induction of retail gravitation is defined through the vector potential $\vec{A}$. The course of economic processes is defined non-relativistically in space and time on the geoid and in the cosmic space near Earth (i.e. within the Hill sphere) assuming that the axiom of unsaturation is the general economic axiom associated with the genetic essence of carbon-based life existing in any part of the universe. Partial differential equations of the retail gravitation field are derived.

Keywords: consilience, geoid, human genome, neuroeconomics, space economics

Mathematics Subject Classification: Primary 35J05, 35Q99; Secondary 35Q61

Introduction

The life of every inhabitant of planet Earth is affected by processes taking place in outer space. The initial boundary of outer space in relation to the Earth has for working purposes been defined based on the Kármán line by the Fédération Aeronautique Internationale (FAI) at a height of 100 km above the surface of the ocean.

Among the many things in outer space is the Sun, a typical class G main-spectrum star.
Basic research of the processes taking place on the star closest to our planet, as well as research of key mechanisms of the effect that solar processes have on terrestrial processes, has gradually led to the rise of a scientific field currently known as "Sun – Earth relations". In addition to basic research on Sun - Earth relations and outer space, there is also applied research of Sun – Earth relations, a field that continues to develop and affect nearly all sectors of national economies. The demands

created by the economic development of countries participating directly or indirectly in research of Sun – Earth relations has led to the birth of a specific economic sector, the space industry.

The strategic position of physics and especially that of theoretical physics in economics was affirmed by the results of research conducted by the Lausanne School of Economics and the Cambridge School of Economics (Holman 2005), (Zeithamer 2012, A), (Zeithamer 2012, B). Representatives of both schools used one of the basic quantitative requirements for determining natural relations in research conducted in any scientific field, namely the requirement that the ratio of input and output signal to noise be greater, ideally much greater, than one. In the latter twentieth and early twenty-first centuries it is also possible to find authors in the Czech School of Economics whose work examined applications of theoretical physics in areas of basic economic research. This group of authors includes, for example, prof. František Drozen (Drozen, 2008), prof. Pavel Hrubý and Ing. Jaromír Kálal (Hrubý & Kálal, 1974).

A classic current example requiring such noise reduction is the detection of gravitational waves (Roehner, 2002), the existence of which was predicted by prof. A. Einstein e.g. in his work from 1916 (Einstein, 1916). Outside the solar system, the theory predicts a number of sources of gravitational waves, which could be detected in the event they reached Earth. The first direct detection of gravitational waves and the first observation of a binary black hole merger were registered on September 14, 2015 at 09:50:45 UTC (Abbott at al., 2016). For the Sun, it has been possible to theoretically determine mechanisms that could be responsible for reliably detectable levels of gravitational radiation (Weinberg, 1972), (Papini & Valluri, 1976), (Křivský & Zeithamer, 1982), (Karmakar & Borah, 2013).

Efforts similar to the detection of gravitational waves can be seen in numerous other multi-disciplinary scientific fields, such as: laser scanning of curved surfaces (Štroner & Pospíšil, 2011) as well as the physical research of economic or neuroeconomic systems in economics of Sun – Earth relations (Neuroeconomics, 2014), (Pérez-Peraza & Libin, 2012).

The consistent and systematic application of the methods of theoretical physics in economics results in a modification of the analytical structure of economics. This was and continues to be one of the reasons we began to look at the mutual interaction of the analytical structures of physics and economics. To enable such analysis, work has begun on a principle of correspondence between economic variables and the physical variables used in one of the most highly developed disciplines of classical physics, classical non-relativistic mechanics (Zeithamer, 2014, A), (Zeithamer, 2014, B) and (Zeithamer, 2014, C).

1 Materials and methods

The application of methods of classical non-relativistic mechanics in microeconomics presented in this work aims to derive a single motion equation for price which describes non-chaotic as well as chaotic fluctuations of price on a market with nearly perfect competition which is defined in references (Zeithamer, 2014, B) and (Zeithamer, 2014, C).

One of the many phenomena which may be used to characterize the United States in the late nineteenth and early twentieth centuries is the rapid change which occurred in retail trade relations in various parts of the country. The common feature of these changes was the flow of retail business from small towns to large cities. However, no general analytical laws were known to describe the rise and distribution of this flow of retail business in space and time. It was also one of the main reasons why in the autumn of 1927 William John Reilly began empirical study of the retail relationships which existed between various cities and towns in a trial locality in the state of Texas. Seven cities were included in this preliminary study, i.e. Fort Worth, Dallas, Houston, Austin, San Antonio, El Paso and Waco (Reilly, 1929). Retail data in these cities was gathered, making it possible to describe and analyze in an empirical way the effect on the retail of each city under various circumstances. A major finding of the preliminary study of retail dynamics was the empirical result that retail trade is attracted from smaller cities and towns to larger cities according to a law with an analytical structure nearly identical to that of Newton´s law of gravitation (Reilly, 1929).

The empirical results of the preliminary study were so persuasive that before it was finished, the study area was expanded from just Texas to the entire country for a three-year period. The nationwide study was carried out from 1927 – 1930 and gradually included 132 cities and towns in 23 states (see references (Reilly, 1929) and (Reilly, 1931)) and it confirmed in an empirical way the analytical relation obtained in the preliminary study.

2 Results and Discussion

Physics is presented with the question of whether it is possible to derive Reilly´s law of retail gravitation from Newton´s law of gravitation. A positive answer to this question was provided in references (Zeithamer, 2015) and (Zeithamer, 2016, A), (Zeithamer, 2016, B).

Here, we work in a coordinate system whose origin is located at the Earth's center of gravity and whose axes are fixed to the geoid. For the description of retail gravitation, this coordinate system approximates with sufficient accuracy an inertial coordinate system (see reference (Votruba, 1977)).

Let A be a city on the geoid with population P_a. Because retail flows are realized in three dimensional space and time, we define the potential of retail gravitation of city A by the relation

$$\varphi(x, y, z) = \varphi(D_a) = \alpha(A, E_A, t)\frac{N_a}{D_a}, \quad (1)$$

where D_a is the distance of the observation place from city A, i.e. $D_a = \sqrt{(x-x_A)^2 + (y-y_A)^2 + (z-z_A)^2}$; $\vec{D}_a = (x-x_A, y-y_A, z-z_A) = (x-x_A)\vec{i} + (y-y_A)\vec{j} + (z-z_A)\vec{k}$ is the position vector of the observation place of retail trade amount attracted in the direction to mass point A (i.e. to city A); $\vec{i}$, $\vec{j}$ and $\vec{k}$ are orthogonal unit vectors. Proportionality constant $\alpha(A, E_A, t)$ is expressed in units $[\alpha] = c.u.\ m^2\ pers.^{-2}$, where A is a mass point with coordinates $x_A(t), y_A(t)$ and $z_A(t)$, i.e. mass point $A = [x_A(t), y_A(t), z_A(t)]$; E_A– is an economic facility of city A; N_a– is the number of long-term economically active adult inhabitants in city A at time t. Vector field $\vec{K}_a$ of the intensity of retail gravitation is determined by the negative gradient of the potential of retail gravitation φ, i.e. $\vec{K}_a = \left(-\frac{\partial \varphi}{\partial x}(D_a), -\frac{\partial \varphi}{\partial y}(D_a), -\frac{\partial \varphi}{\partial z}(D_a) \right)$. This means that the magnitude K_a of the vector of intensity of retail gravitation follows from the last relation, i.e. $K_a = \left|\vec{K}_a\right| = \alpha(A, E_A, t)\frac{N_a}{D_a^2}$.

Then for the intensity vector of retail gravitation we have

$$div\ \vec{K}_a = 0\ , \tag{2}$$

$$rot\ \vec{K}_a = \vec{0}\ , \tag{3}$$

and the potential of retail gravitation is the solution of the Laplace homogeneous partial differential equation

$$\Delta\varphi = 0\ . \tag{4}$$

Any vector field of retail gravitation $\overrightarrow{M_a}$ may be considered the superposition of two fields, one of which is determined by the scalar potential, and the other determined by the vector potential. If we designate the retail gravitation field determined by scalar potential φ as $\vec{K}_a$ and the retail gravitation field determined by vector potential $\vec{A}$ as $\vec{L}_a$, then

$$\overrightarrow{M_a} = \vec{\mathrm{K}}_\mathrm{a} + \vec{\mathrm{L}}_\mathrm{a}\ , \tag{5}$$

and respectively,

$$\overrightarrow{M_a} = -\ grad\ \varphi\ +\ \mathrm{rot}\ \vec{\mathrm{A}}\ .$$

Conclusion

In this work, the construction of basic three-dimensional vector quantities of retail gravitation is derived in an inertial coordinate system, assuming that the field of retail gravitation is a continuous field. These are vector quantities: the vector field of intensity of retail gravitation $\vec{K}_a$ and the vector field of induced retail gravitation $\vec{L}_a$. The construction carried out enables the qualitative and quantitative description of retail gravitation based on classical non-relativistic field theory.

Acknowledgements

The authors are grateful to Mrs. Pavla Jará, Mrs. Věra Hrabovská, Mrs. Hana Košťáková, Ing. Jan Červenka, Bc. Jiří Henzl, Mgr. Klára Kosová, Ing. Olga Martinová, Mgr. Jana Orlová, Mgr. Filip Strych and the National Library of Technology for their great effort and excellent work, which was indispensable in the completion of a large portion of this work. Authors thank for technical assistance to Managing Director Mr. Jiří Břichnáč and to the firm Tomados Ltd. without their contribution would not be possible to finish the article. This paper is dedicated to Mrs. Věra Ruml Zeithamer, Mr. Josef Ruml Zeithamer, Mrs. Anna Ruml, Mr. František Ruml and Ing. Milan Pospíšil.

References

[1] ABBOTT, B. P. at al. Observation of Gravitational Waves from a Binary Hole Merger. *Phys. Rev. Lett.*, vol. 116, 2016: pp. 061102-1 - 061102-16, ISSN 1079-7114 (Online), ISSN 0031-9007 (Print).

[2] DROZEN, F. Modelling of price dynamics and appreciation. *Ekonomický časopis (Journal of Economics)*, vol. 56, 2008: pp. 1033 – 1044, ISSN 0013-3035.

[3] EINSTEIN, A. Näherungweise Integration der Feldgleichungen der Gravitation. *Sitzungsber. K. Preuss. Akad. Wiss.*, vol. 1, 1916: pp. 688 - 696.

[4] GLIMCHER, P. W. and FEHR, E. (Eds.) Neuroeconomics: *Decision Making and the Brain.* 2nd ed. Amsterdam, Netherlands, Elsevier, Academic Press, ISBN 978-0-12-416008-8, s. 577.

[5] HOLMAN, R. at al. *Dějiny ekonomického myšlení.* 3rd ed. Prague, Czech Republic, C. H. Beck, 2005, ISBN 80-7179-380-9.

[6] HRUBÝ, P. and KÁLAL, J. *Metody ekonomického času (Methods of Economic Time)*, Prague. 1st ed. Prague, Czech Republic: Institute of the Czechoslovak Committee for Scientific Management, 1974.

[7] KARMAKAR, P. K. and BORAH, B. Nonlinear Self – Gravitational Solar Plasma Fluctuations with Electron Inertia. *Contributions to Plasma Physics*, vol. 53, 2013: pp. 516 –539, ISSN 1521-3986.

[8] KŘIVSKÝ, L. and ZEITHAMER, T. On the Possibility of Generating Gravitational Radiation by the Sun. *Astrophysics and Space Science*, vol. 85, 1982: pp. 309 – 313, ISSN 0004-640X (Print), ISSN 1572-946X (Online).

[9] PAPINI, G. and VALLURI, S. R. Photoproduction of Gravitational Radiation by Some Astrophysical Objects. *Canadian Journal of Physics*, vol. 54, 1976: pp. 76 – 79, ISSN 0008-4204 (Print), ISSN 1208-6045 (Online).

[10] PÉREZ – PERAZA, J. A., and LIBIN, I. Y. *Highlights in Helioclimatology: Cosmophysical Influences on Climate and Hurricanes.*1st ed. Amsterdam: Elsevier, 2012, ISBN 978-0-12-415977-8.

[11] REILLY, W. J. Methods for the Study of Retail Relationships. *University of Texas Bulletin*, No 2944, Nov, 1929.

[12] REILLY, W. J. *The Law of Retail Gravitation.* 1st ed. New York, Knickerbocker Press, USA, 1931.

[13] ROEHNER, B. M. *Patterns of Speculation: A Study in Observational Econophysics*, 1st ed., Cambridge, UK, Cambridge University Press, 2002, ISBN 978-0-511-17642-5.

[14] ŠTRONER, M. and POSPÍŠIL, J. Systematic Geometrical Errors of Scanning Spherical Surfaces. *Survey Review*, vol. 43, 2011: pp. 731 – 742, ISSN 0039-6265 (Print), ISSN 1752-2706 (Online).

[15] VOTRUBA, V. *Principles of Special Relativity Theory (Základy speciální teorie relativity)*, 2nd ed. Prague, Czech Republic, Academia Publishing House (in czech), 1977.

[16] WEINBERG, S. *Gravitation and Cosmology: Principles and Applications of the General Theory of Relativity.* 1st ed. New York, Wiley, USA, 1972, ISBN 0-471-92567-5.

[17] ZEITHAMER, T. R. (2012 A). Economic Phenomena from the Viewpoint of the Mechanics of Materials. *3rd International Conference on New Horizons in Education (INTE 2012)*, Prague, Czech Republic, *Procedia Social and Behavioral Sciences*, vol. 55, 2012 A: pp. 547 – 553, ISSN 1877-0428.

[18] ZEITHAMER, T. R. Analytical Theory of Monotone Commodity State Development with Inflexion. *3rd International Conference on New Horizons in Education (INTE 2012)*, Prague, Czech Republic, Procedia Social and Behavioral Sciences, vol. 55, 2012 B, pp. 445 – 450, ISSN 1877-0428.

[19] ZEITHAMER, T. R. Methodology of Theoretical Physics in Economics: Examining Price Jerk. *International Educational Technology Conference (IETC, 3–5 September, 2014)*, Chicago, IL,USA, *Procedia Social and Behavioral Sciences*, vol. 176, 2014 A: pp. 933 – 939, ISSN 2146-7366.

[20] ZEITHAMER, T. R. The Correspondence Principle in Theoretical Regional Economics. *Proceedings of the 6th Annual International Scientific Conference: Region in the Development of Society*, October 23rd, 2014, Brno, Czech Republic, 2014 B: pp. 1016 – 1023, ISBN 978-80-7509-139-0.

[21] ZEITHAMER, T. R. Methodology of Theoretical Physics in Economics: Examining Price Jounce and Price Crackle. *The 8th International Scientific Conference INPROFORUM 2014, Proceedings*, České Budějovice, Czech Republic, 2014 C: pp. 143 – 148, ISBN 978-80-7394-484-1 (Print), ISSN 2336-6788 (Online).

[22] ZEITHAMER, T. R. Methodology of Theoretical Physics in Economics: Vector Theory of Retail Gravitation Law. *The 9th International Scientific Conference INPROFORUM 2015, Proceedings*, České Budějovice, Czech Republic, 2015: pp. 199 – 203, ISBN 978-80-7394-536-7 (Print), ISSN 2336-6788 (Online).

[23] ZEITHAMER, T. R. Methodology of Theoretical Physics in Economics: Intensity Vector of Retail Gravitation. *Proceedings of the 15th Conference on Applied Mathematics 2016 (Aplimat 2016)*, Bratislava, Slovak Republic, 2016 A: pp. 1147 – 1153, ISBN 978-15-1082-096-8.

[24] ZEITHAMER, T. R. Theoretical Background for the Globalization of Retail Gravitation. *Proceedings of the 16th International Scientific Conference: Globalization and Its Socio-Economic Consequences*, University of Žilina, The Faculty of Operation and Economics of Transport and Communication, Department of Economics, 5th – 6th October 2016, 2016 B (will be published).

Current address

Zeithamer R. Tomáš, Ing. Ph.D.
University of Economics, Faculty of Informatics and Statistics, Department of Mathematics
Ekonomická 957, 140 00 Prague 4 – Kunratice, Czech Republic
E-mail: zeithamer.tomas@gmail.com

Pospíšil Jiří, prof. Ing. CSc.
Czech Technical University in Prague, Faculty of Civil Engineering,
Department of Special Geodesy
Thákurova 7, 166 29 Prague 6, Czech Republic
E.mail: pospisil@fsv.cvut.cz

Slovak University of Technology in Bratislava
Faculty of Mechanical Engineering
16th Conference on Applied Mathematics **APLIMAT 2017**
Proceedings

THE SOLUTION OF THE HEAT CONDUCTION EQUATION IN 3D ANISOTROPIC ENVIRONMENT AND POSSIBILITIES OF ITS IMPROVEMENT

ZIGUNOVS Maksims (LV), ILTINS Ilmars (LV), RADIN A. Michael (US)

Abstract. This manuscript presents a new method of creating heat conduction solution for the anisotropic space by considering each included heat transfer material's physical characteristics. This manuscript also provides descriptions of new possibilities to make these calculations less time consuming, less time costing and to provide advantages strictly connected to the computer quantity involved in the calculation process.

Keywords: heat conduction, improvement, high performing computing

Mathematics Subject Classification: Primary 34-04, 68-04

1 Introduction

Currently the resource and time effective usage and efficiency need predictions of how the physical processes will be run, what they will influence and how they will influence. For example, to manufacture a car an engineers must consider the possibilities of the car's deformation in possible vehicle accidents that can occur. Therefore, depending on these predictions, car the designers could offer more safety features in cars for their customers.

One of the popularly physical predictions that are used are heat conduction/transfer calculations. In particular, there are several methods applied to make these accurate calculations using software such as COMSOL (computer application) and other ones.

There are more opportunities provided due to the possible processor counts within different type of devices that are now available. As the processor capacity is improving, this means that the current devices can perform more and more parallel tasks as the installed software capacity is improving as well.

In the modern technology the large computing power is available. It is associated with both serial and parallel calculations. The current CPU (central processor unit) processor frequency reaches up to 4 GHz GPU (graphical processor unit) and graphics processor cards have reached up to 1,000 units in a single map, and each processor frequency reaches up to 1GHz. This means that it is

possible to create a computer system that could be designed for solving non-stationary physical phenomena modelling problem of the three-dimensional space.
This paper will describe our developed method of difference scheme realization for parallel calculation of the non-stationary heat conduction in anisotropic 3D space.

2 Methodology of Research

The assumed initial boundary condition for the 3D domain GUΓ is described in [1] as follows:

$$\frac{\partial T}{\partial t}+\frac{v_\varsigma}{R_T}\frac{\partial T}{\partial x_\varsigma}-\frac{1}{R_T}\frac{\partial}{\partial x_\varsigma}\left(\Lambda_{\varsigma x}\frac{\partial T}{\partial x_\varsigma}\right)=\frac{Q_P T}{nR_T},(x_1=x,x_2=y,x_3=z)\in\Gamma,t>0 \quad (1)$$

with the corresponding initial and boundary conditions ($x_1 = x$, $x_2 = y$, $x_3 = z$) ∈ Γ.

The main problem of implementing the difference scheme is connected to the heat conduction tensor. The thermo-dispersion tensor is rewritten in an exponential form with connections between the axes of the system. This approach creates a very sensitive system relative to the input data (sensitivity to initial conditions). This 3D temperature distribution equation (principles are in [6]) is implemented as follows:

$$\frac{\partial T}{\partial t}+V_x\frac{\partial T}{\partial x}+Vy\frac{\partial T}{\partial y}+V_z\frac{\partial T}{\partial z}=\frac{1}{R_T}\frac{\partial}{\partial x}\left(\Lambda_{xx}\frac{\partial T}{\partial x}+\Lambda_{xy}\frac{\partial T}{\partial y}+\Lambda_{xz}\frac{\partial T}{\partial z}\right)+ \quad (2)$$

$$+\frac{1}{R_T}\frac{\partial}{\partial y}\left(\Lambda_{yx}\frac{\partial T}{\partial x}+\Lambda_{yy}\frac{\partial T}{\partial y}+\Lambda_{yz}\frac{\partial T}{\partial z}\right)+\frac{1}{R_T}\frac{\partial}{\partial x}\left(\Lambda_{zx}\frac{\partial T}{\partial x}+\Lambda_{zy}\frac{\partial T}{\partial y}+\Lambda_{zz}\frac{\partial T}{\partial z}\right),$$

where T = temperature, t = time, x, y and z are coordinate directions, Vx, Vy, Vz = heat transferring velocity in x, y, and z directions, Rt – heat retention factor and Λ12 is a heat conduction tensor, where (1) is calculable direction and (2) is a sub direction of direction 1.
We get the following differential scheme for temperature calculations from the time moment "l" to the time moment "l+1":

$$\frac{\overset{l+1}{T_{ijk}}-\overset{l}{T_{ijk}}}{\tau}=\frac{1}{R_T}\frac{\partial}{\partial x}\left(\left(I_x\right)_{i+\frac{1}{2}jk}-\left(I_x\right)_{i-\frac{1}{2}jk}\right)_{x=x^*,y=y^*,z=z^*}+\frac{1}{R_T}\frac{\partial}{\partial y}\left(\left(I_y\right)_{ij+\frac{1}{2}k}-\left(I_y\right)_{ij-\frac{1}{2}k}\right)_{x=x^*,y=y^*,z=z^*}+ \quad (3)$$

$$+\frac{1}{R_T}\frac{\partial}{h^*_z}\left(\left(I_z\right)_{ijk+\frac{1}{2}}-\left(I_z\right)_{ijk-\frac{1}{2}}\right)_{x=x^*,y=y^*,z=z^*}+\frac{QT_{ijk}}{nR_T},$$

where l = current time step, l+1 next time step, i = discrete position of x, j = discrete position of y, k = discrete position of z, Q = heat source produced heat in {i,j,k} [2], T_{ijk} = temperature on the current time step in point {i,j,k}, n = porosity coefficient of material, $h^*_x = (h_{x+1} + h_x) / 2$, where h_x (step in direction of x) = $x_i - x_{i-1}$.

Averaged flux in x, y and z directions are expressed as follows:

$$\left(I_x\right)_{i+\frac{1}{2}jk}=\Lambda_{xx}\left(Exp\left(\rho_{i+\frac{1}{2}jk}\right)T_{i+1jk}-T_{ijk}\right)/\left(\frac{1}{\rho_{i+\frac{1}{2}jk}}\left(Exp\left(\rho_{i+\frac{1}{2}jk}h_x\right)-1\right)\right) \quad (4)$$

$$\left(I_x\right)_{i-\frac{1}{2}jk}=\Lambda_{xx}\left(Exp\left(\rho_{i-\frac{1}{2}jk}\right)T_{ijk}-T_{i-1jk}\right)/\left(\frac{1}{\rho_{i-\frac{1}{2}jk}}\left(Exp\left(\rho_{i-\frac{1}{2}jk}h_x\right)-1\right)\right) \quad (5)$$

$$\left(I_y\right)_{ij+\frac{1}{2}k}=\Lambda_{yy}\left(Exp\left(\rho_{ij+\frac{1}{2}k}\right)T_{ij+1k}-T_{ijk}\right)/\left(\frac{1}{\rho_{ij+\frac{1}{2}k}}\left(Exp\left(\rho_{ij+\frac{1}{2}k}h_y\right)-1\right)\right) \tag{6}$$

$$\left(I_y\right)_{ij-\frac{1}{2}k}=\Lambda_{yy}\left(Exp\left(\rho_{ij-\frac{1}{2}k}\right)T_{ijk}-T_{ij-1k}\right)/\left(\frac{1}{\rho_{ij-\frac{1}{2}k}}\left(Exp\left(\rho_{ij-\frac{1}{2}k}h_y\right)-1\right)\right) \tag{7}$$

$$\left(I_z\right)_{ijk+\frac{1}{2}}=\Lambda_{zz}\left(Exp\left(\rho_{ijk+\frac{1}{2}}\right)T_{ijk+1}-T_{ijk}\right)/\left(\frac{1}{\rho_{ijk+\frac{1}{2}}}\left(Exp\left(\rho_{ijk+\frac{1}{2}}h_z\right)-1\right)\right) \tag{8}$$

$$\left(I_z\right)_{ijk-\frac{1}{2}}=\Lambda_{zz}\left(Exp\left(\rho_{ijk-\frac{1}{2}}\right)T_{ijk}-T_{ijk-1}\right)/\left(\frac{1}{\rho_{ijk-\frac{1}{2}}}\left(Exp\left(\rho_{ijk-\frac{1}{2}}h_z\right)-1\right)\right) \tag{9}$$

$$\rho_{i+\frac{1}{2}jk}=\frac{1}{T_{i+1jk}\Lambda_{xx}}\left(\Lambda_{xy}\frac{T_{ij+1k}-T_{ijk}}{h_y}+\Lambda_{xz}\frac{T_{ijk+1}-T_{ijk}}{h_z}-V_xT_{i+\frac{1}{2}jk}\right) \tag{10}$$

$$\rho_{i-\frac{1}{2}jk}=\frac{1}{T_{i-1jk}\Lambda_{xx}}\left(\Lambda_{xy}\frac{T_{ijk}-T_{ij-1k}}{h_y}+\Lambda_{xz}\frac{T_{ijk}-T_{ijk-1}}{h_z}-V_xT_{i-\frac{1}{2}jk}\right) \tag{11}$$

$$\rho_{ij+\frac{1}{2}k}=\frac{1}{T_{ij+1k}\Lambda_{yy}}\left(\Lambda_{yx}\frac{T_{i+1jk}-T_{ijk}}{h_x}+\Lambda_{yz}\frac{T_{ijk+1}-T_{ijk}}{h_z}-V_yT_{ij+\frac{1}{2}k}\right) \tag{12}$$

$$\rho_{ij-\frac{1}{2}k}=\frac{1}{T_{ij-1k}\Lambda_{yy}}\left(\Lambda_{yx}\frac{T_{ijk}-T_{i-1jk}}{h_x}+\Lambda_{yz}\frac{T_{ijk}-T_{ijk-1}}{h_z}-V_yT_{ij-\frac{1}{2}k}\right) \tag{13}$$

$$\rho_{ijk+\frac{1}{2}}=\frac{1}{T_{ijk+1}\Lambda_{zz}}\left(\Lambda_{zx}\frac{T_{i+1jk}-T_{ijk}}{h_x}+\Lambda_{zy}\frac{T_{ij+1k}-T_{ijk}}{h_y}-V_zT_{ijk+\frac{1}{2}}\right) \tag{14}$$

$$\rho_{ijk-\frac{1}{2}}=\frac{1}{T_{ijk-1}\Lambda_{zz}}\left(\Lambda_{zx}\frac{T_{ijk}-T_{i-1jk}}{h_x}+\Lambda_{zy}\frac{T_{ijk}-T_{ij-1k}}{h_y}-V_zT_{ijk-\frac{1}{2}}\right) \tag{15}$$

Notice that $(I_x)_{i+1/2jk}$ is calculated similar to other parameters' calculations:

$$\left(I_x\right)_{i+\frac{1}{2}jk}\frac{1}{\Lambda_{xx}}\int_{x_i}^{x_{i+1}}Exp\left(\int_{x_0}^{x}\frac{\Lambda_{xy}\frac{\partial T}{\partial y}+\Lambda_{xz}\frac{\partial T}{\partial z}-V_xT}{T\Lambda_{xx}}dx\right)dx=\varphi_{i+1jk}-\varphi_{ijk} \tag{16}$$

$$\left(I_x\right)_{i+\frac{1}{2}jk}=\frac{\varphi_{i+1jk}-\varphi_{ijk}}{\frac{1}{\Lambda_{xx}}\int_{x_i}^{x_{i+1}}Exp\left(\int_{x_0}^{x}\frac{\Lambda_{xy}\frac{\partial T}{\partial y}+\Lambda_{xz}\frac{\partial T}{\partial z}-V_xT}{T\Lambda_{xx}}dx\right)dx} \tag{17}$$

$$
\left(I_x\right)_{i+\frac{1}{2}jk} = \frac{T_{i+1jk} Exp\left(\int_{x_0}^{x_{i-1}} \frac{\Lambda_{xy}\frac{\partial T}{\partial y} + \Lambda_{xz}\frac{\partial T}{\partial z} - V_x T}{T\Lambda_{xx}} dx\right) - T_{ijk}}{\frac{1}{\Lambda_{xx}} \int_{x_i}^{x_{i+1}} Exp\left(\int_{x_0}^{x} \frac{\Lambda_{xy}\frac{\partial T}{\partial y} + \Lambda_{xz}\frac{\partial T}{\partial z} - V_x T}{T\Lambda_{xx}} dx\right) dx} \tag{18}
$$

$$
\left(I_x\right)_{i+\frac{1}{2}jk} = \frac{T_{i+1jk} Exp\left(\int_{x_i}^{x_{i+1}} \frac{\Lambda_{xy}\frac{\partial T}{\partial y} + \Lambda_{xz}\frac{\partial T}{\partial z} - V_x T}{T\Lambda_{xx}} dx\right) - T_{ijk}}{\frac{1}{\Lambda_{xx}} \int_{x_i}^{x_{i+1}} Exp\left(\int_{x_i}^{x} \frac{\Lambda_{xy}\frac{\partial T}{\partial y} + \Lambda_{xz}\frac{\partial T}{\partial z} - V_x T}{T\Lambda_{xx}} dx\right) dx} \tag{19}
$$

$$
\int_{x_i}^{x_{i+1}} \frac{\Lambda_{xy}\frac{\partial T}{\partial y} + \Lambda_{xz}\frac{\partial T}{\partial z} - V_x T}{T\Lambda_{xx}} dx = \int_{x_i}^{x_{i+1}} \frac{\Lambda_{xy}\frac{T_{ij+1k} - T_{ijk}}{h_y} + \Lambda_{xz}\frac{T_{ijk+1} - T_{ijk}}{h_z} - V_x T_{i+\frac{1}{2}jk}}{T_{i+1jk}\Lambda_{xx}} dx = \tag{20}
$$

$$
= \frac{\Lambda_{xy}\frac{T_{ij+1k} - T_{ijk}}{h_y} + \Lambda_{xz}\frac{T_{ijk+1} - T_{ijk}}{h_z} - V_x T_{i+\frac{1}{2}jk}}{T_{i+1jk}\Lambda_{xx}} \left(x_{i+1} - x_i\right)
$$

$$
\int_{x_i}^{x_{i+1}} Exp\left(\int_{x_i}^{x} \frac{\Lambda_{xy}\frac{\partial T}{\partial y} + \Lambda_{xz}\frac{\partial T}{\partial z} - V_x T}{T\Lambda_{xx}} dx\right) dx \approx \tag{21}
$$

$$
\approx \int_{x_i}^{x_{i+1}} Exp\left(\frac{\Lambda_{xy}\frac{T_{ij+1k} - T_{ijk}}{h_y} + \Lambda_{xz}\frac{T_{ijk+1} - T_{ijk}}{h_z} - V_x T_{i+\frac{1}{2}jk}}{T_{i+1jk}\Lambda_{xx}} \left(x_i - x\right)\right) dx =
$$

$$
= \frac{T_{i+1jk}\Lambda_{xx}}{\Lambda_{xy}\frac{T_{ij+1k} - T_{ijk}}{h_y} + \Lambda_{xz}\frac{T_{ijk+1} - T_{ijk}}{h_z} - V_x T_{i+\frac{1}{2}jk}} *
$$

$$
* \left(Exp\left(\frac{\Lambda_{xy}\frac{T_{ij+1k} - T_{ijk}}{h_y} + \Lambda_{xz}\frac{T_{ijk+1} - T_{ijk}}{h_z} - V_x T_{i+\frac{1}{2}jk}}{T_{i+1jk}\Lambda_{xx}} \left(x_{i+1} - x_i\right)\right) - 1 \right)
$$

$$(I_x)_{i+\frac{1}{2}jk} = \Lambda_{xx}\left(Exp\left(\frac{\Lambda_{xy}\frac{T_{ij+1k}-T_{ijk}}{h_y} + \Lambda_{xz}\frac{T_{ijk+1}-T_{ijk}}{h_z} - V_x T_{i+\frac{1}{2}jk}}{T_{i+1jk}\Lambda_{xx}} * (x_{i+1}-x_i) \right) * T_{i+1jk} - T_{ijk} \right) / \tag{22}$$

$$/\left(\frac{1}{\frac{\Lambda_{xy}\frac{T_{ij+1k}-T_{ijk}}{h_y} + \Lambda_{xz}\frac{T_{ijk+1}-T_{ijk}}{h_z} - V_x T_{i+\frac{1}{2}jk}}{T_{i+1jk}\Lambda_{xx}}(x_{i+1}-x_i)} * \right.$$

$$\left. * \left(Exp\left(\frac{\Lambda_{xy}\frac{T_{ij+1k}-T_{ijk}}{h_y} + \Lambda_{xz}\frac{T_{ijk+1}-T_{ijk}}{h_z} - V_x T_{i+\frac{1}{2}jk}}{T_{i+1jk}\Lambda_{xx}}(x_{i+1}-x_i) h_x \right) - 1 \right) \right)$$

According to Λ the definition is (same for x, y and z axes directions):

$$\Lambda_{xx} = \frac{\lambda_{xx}}{n\rho c}, \Lambda_{xy} = \frac{\lambda_{xy}}{n\rho c}, \Lambda_{xz} = \frac{\lambda_{xz}}{n\rho c}. \tag{23}$$

$$\Lambda_{yx} = \frac{\lambda_{yx}}{n\rho c}, \Lambda_{yy} = \frac{\lambda_{yy}}{n\rho c}, \Lambda_{yz} = \frac{\lambda_{yz}}{n\rho c}. \tag{24}$$

$$\Lambda_{zx} = \frac{\lambda_{zx}}{n\rho c}, \Lambda_{zy} = \frac{\lambda_{zy}}{n\rho c}, \Lambda_{zz} = \frac{\lambda_{zz}}{n\rho c}. \tag{25}$$

$$\lambda_{xx} = n\rho c\left(\beta_T V + (\beta_c - \beta_T)\frac{\upsilon_x \upsilon_x}{V_{xx}} \right). \tag{26}$$

$$\lambda_{xy} = n\rho c\left(\beta_T V + (\beta_c - \beta_T)\frac{\upsilon_x \upsilon_y}{V_{xy}} \right). \tag{27}$$

$$\lambda_{xz} = n\rho c\left(\beta_T V + (\beta_c - \beta_T)\frac{\upsilon_x \upsilon_z}{V_{xz}} \right). \tag{28}$$

$$\lambda_{yx} = n\rho c\left(\beta_T V + (\beta_c - \beta_T)\frac{\upsilon_y \upsilon_x}{V_{yx}} \right). \tag{29}$$

$$\lambda_{yy} = n\rho c\left(\beta_T V + (\beta_c - \beta_T)\frac{\upsilon_y \upsilon_y}{V_{yy}} \right). \tag{30}$$

$$\lambda_{yz} = n\rho c\left(\beta_T V + (\beta_c - \beta_T)\frac{\upsilon_y \upsilon_z}{V_{yz}} \right). \tag{31}$$

$$\lambda_{zx} = n\rho c\left(\beta_T V + (\beta_c - \beta_T)\frac{\upsilon_z \upsilon_x}{V_{zx}}\right). \tag{32}$$

$$\lambda_{zy} = n\rho c\left(\beta_T V + (\beta_c - \beta_T)\frac{\upsilon_z \upsilon_y}{V_{zy}}\right). \tag{33}$$

$$\lambda_{zz} = n\rho c\left(\beta_T V + (\beta_c - \beta_T)\frac{\upsilon_z \upsilon_z}{V_{zz}}\right). \tag{34}$$

where n = porosity coefficient of material, q = water density, c = specific heat capacity of water (it is assumed that heat source is located inside water because of assumed chemical reactions in water which produces the heat [3]), β_c,β_T= longitudinal and transverse thermo-dispersivity [1][4], and is V = the absolute velocity, v_x,v_y,v_z = the velocity vectors.

In order to elaborate the parallel calculation method, the difference scheme was decomposed by space and time by applying ADI [5] method.

$$\frac{\overset{l+\frac{1}{3}}{T_{ijk}} - \overset{l}{T_{ijk}}}{\frac{\tau}{3}} = \frac{QT_{ijk}}{nR_T} + \frac{1}{R_T}\frac{\partial}{\partial x}\left(\left(\overset{l+\frac{1}{3}}{I_x}\right)_{i+\frac{1}{2}jk} - \left(\overset{l+\frac{1}{3}}{I_x}\right)_{i-\frac{1}{2}jk}\right)_{x=x^*,y=y^*,z=z^*} + \tag{35}$$

$$+\frac{1}{R_T}\frac{\partial}{\partial y}\left(\left(\overset{l}{I_y}\right)_{ij+\frac{1}{2}k} - \left(\overset{l}{I_y}\right)_{ij-\frac{1}{2}k}\right)_{x=x^*,y=y^*,z=z^*} + \frac{1}{R_T}\frac{\partial}{\partial z}\left(\left(\overset{l}{I_z}\right)_{ijk+\frac{1}{2}} - \left(\overset{l}{I_z}\right)_{ijk-\frac{1}{2}}\right)_{x=x^*,y=y^*,z=z^*}$$

$$\frac{\overset{l+\frac{2}{3}}{T_{ijk}} - \overset{l+\frac{1}{3}}{T_{ijk}}}{\frac{\tau}{3}} = \frac{QT_{ijk}}{nR_T} + \frac{1}{R_T}\frac{\partial}{\partial x}\left(\left(\overset{l+\frac{1}{3}}{I_x}\right)_{i+\frac{1}{2}jk} - \left(\overset{l+\frac{1}{3}}{I_x}\right)_{i-\frac{1}{2}jk}\right)_{x=x^*,y=y^*,z=z^*} + \tag{36}$$

$$+\frac{1}{R_T}\frac{\partial}{\partial y}\left(\left(\overset{l+\frac{2}{3}}{I_y}\right)_{ij+\frac{1}{2}k} - \left(\overset{l+\frac{2}{3}}{I_y}\right)_{ij-\frac{1}{2}k}\right)_{x=x^*,y=y^*,z=z^*} + \frac{1}{R_T}\frac{\partial}{\partial z}\left(\left(\overset{l+\frac{1}{3}}{I_z}\right)_{ijk+\frac{1}{2}} - \left(\overset{l+\frac{1}{3}}{I_z}\right)_{ijk-\frac{1}{2}}\right)_{x=x^*,y=y^*,z=z^*}$$

$$\frac{\overset{l+\frac{3}{3}}{T_{ijk}} - \overset{l+\frac{2}{3}}{T_{ijk}}}{\frac{\tau}{3}} = \frac{QT_{ijk}}{nR_T} + \frac{1}{R_T}\frac{\partial}{\partial x}\left(\left(\overset{l+\frac{2}{3}}{I_x}\right)_{i+\frac{1}{2}jk} - \left(\overset{l+\frac{2}{3}}{I_x}\right)_{i-\frac{1}{2}jk}\right)_{x=x^*,y=y^*,z=z^*} + \tag{37}$$

$$+\frac{1}{R_T}\frac{\partial}{\partial y}\left(\left(\overset{l+\frac{2}{3}}{I_y}\right)_{ij+\frac{1}{2}k} - \left(\overset{l+\frac{2}{3}}{I_y}\right)_{ij-\frac{1}{2}k}\right)_{x=x^*,y=y^*,z=z^*} + \frac{1}{R_T}\frac{\partial}{\partial z}\left(\left(\overset{l+\frac{3}{3}}{I_z}\right)_{ijk+\frac{1}{2}} - \left(\overset{l+\frac{3}{3}}{I_z}\right)_{ijk-\frac{1}{2}}\right)_{x=x^*,y=y^*,z=z^*}$$

The ADI technology is used to compute calculations faster and to make implicit difference scheme, the 5 point stencil per 2 axes direction is used. So the ADI technology provides 3 diagonal matrix usage in the Thomas algorithm. Therefore, it makes very fast and efficient calculations.

The other method, for making faster calculations, it is the calculation time step that changes according to the predictions of how the temperature distribution relative to the previous time step calculation results.

The way that it is done contains several points on each iteration:

1. Make absolute and relative errors between the predicted result and the calculation result;
2. Find out maximal error from 1. point;
3. Calculate next iteration time step by algorithm shown in Fig. 1.;

```
If[(ErrorXYZ ≥ toltol/2), τ1 = τ; τ = τ/τkoef; If[τ < τmin, τ = τmin];];
If[(ErrorXYZ ≤ toltol/2) ∧ (ErrorXYZ ≥ toltol/5), τ1 = τ; τ = τ];
If[(ErrorXYZ ≤ toltol/5), τ1 = τ; τ = τ * τkoef; If[τ > τmax, τ = τmax];];
```

Fig. 1. Algorithm of changing next iteration time step. ErrorXYZ = maximal error between predicted and calculated results, toltol = error threshold, tau1 = temporary variable, tau = time step, taukoef = coefficient of time step changing, taumin = minimal time step, taumax = maximal time step

4. Make calculation prediction about next iteration calculation result.

Another improvement is the normalization of the space and time of the algorithm.
The space normalization starts with finding out the maximal length of axis sizes of the analyzed 3D space. Then it is all the space steps and axis point positions in the previously mentioned formulas are being divided with the maximal length of axis sizes of the analyzed 3D space. The time normalization is being used as a time step and absolute time division with the maximal length of axis sizes of the analyzed 3D space in power of two.

3 Findings/Results

The developed difference scheme was used to produce the temperature distribution calculations for the waste polygon biogas production system. The dimensions of the analyzed 3D area are 90.85 m (horizontal) x 468.02 m (horizontal) x 14 m (vertical). The entire block produces heat during the chemical processes inside the water handling tube system with its reagents. Therefore the heat source dimensions are 30.85 m (horizontal) x 168.17 m (horizontal) x 6 m (vertical). The entire analyzed 3D area has been discretized for 64 point for every axis direction. The boundary conditions are used for the calculation are Robin border conditions for 10C+-0.1% for the bottom surface, 14C+-0.1% for the other surfaces as the average temperature (14C) should be the same as initial conditions. The initial temperature for the air is 14C. The initial time step is 10^{-7} second.

The developed difference scheme became implicit but quite useful in time cost and time efficiency. The example of such calculations in 3D view is illustrated in Fig. 2.

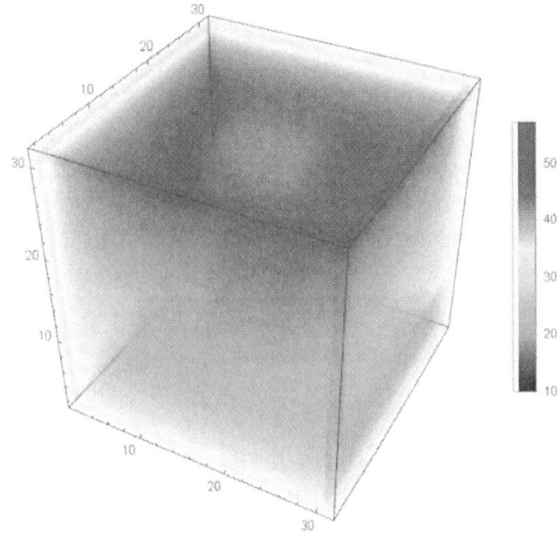

Fig. 2. Implemented software calculation result view in 3D view style. Color gradient represents the temperature inside the 3D space. Central part is the heat source container material

Another calculation result 3D view is shown in Fig. 3.

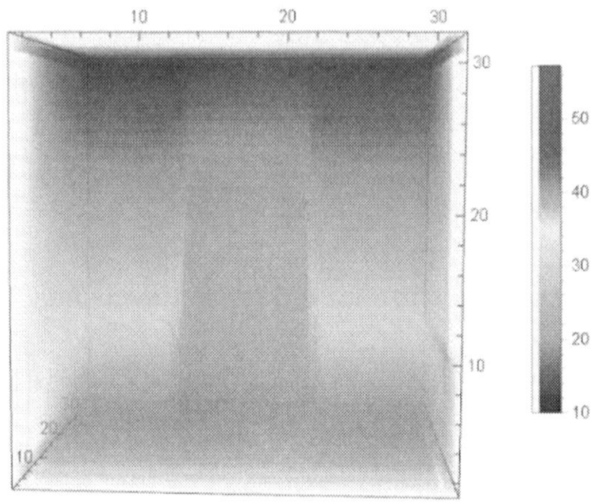

Fig. 3. Calculation result 3D view

A slice of calculation result 3D view is shown in Fig. 4.

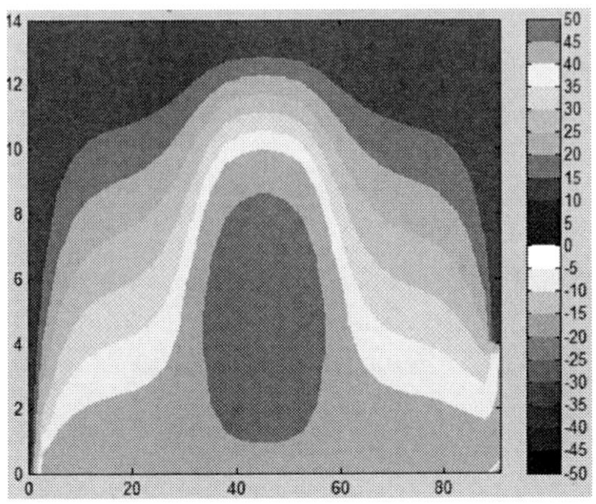

Fig. 4. Calculation result 2D slice

4 Conclusions

A difference scheme for calculating temperature distribution is elaborated to fit the requirements to be monotone, conservative and exponential with the ADI algorithm for the 3D grid. The main advantage of this difference scheme is the reduced calculation time based on the algorithm blocks, which are not connected amongst themselves and therefore reduce the calculation effort. It is quite a strong advantage for the 3D calculations, provided that the solution can use the dynamical time step for each iteration.

References

[1] Rajandrea Sethi, Antonio Di Molfettav. Heat Transport Modeling in an Aquifer Downgradient a Municipal Solid Waste Landfill in Italy. American Journal of Environmental Sciences, 3 106-110, 2007.

[2] Yoshida H., N. Tanaka, H. Hozumi, 1997. Theoretical Study on Heat Transport Phenomena in a Sanitary Landfill. Proceedings Sardinia 97.

[3] Zheng, C., Wang, P., 1998, "MT3DMS: A Modular Three-Dimensional Multispecies Transport Model for Simulation of Advection, Dispersion and Chemical Reactions of Contaminants in Groundwater Systems." University of Alabama.

[4] Zanetti M.C., L. Manna, G. Genon, 1997. Biogas production by means of thermal balances. Proceedings Sardinia 97.

[5] F. MAMPAEY, A numerical technique to increase the stability of the ADI method in solidification simulation, Journal of Computational and Applied Mathematics 28 (1989) 297-308.

[6] Thomas Blomberg, Heat conduction in two and three dimensions, Lund University, Sweden, 1996.

Current address

Zigunovs Maksims, Mg.sc.ing. Ph.D., student in RTU in Automation and Computer Engineering
Riga Technical University, Latvia
Tel. number: +371 28225679, e-mail: maksims.zigunovs@inbox.lv

Iltins Ilmars, Ph.D., Professor in Riga Technical University
Riga Technical University, Latvia
Tel. number: +371 29239517, e-mail: ilmars.iltins@rtu.lv

Radin A. Michael, Ph.D., Associate Professor of Mathematics
Rochester Institute of Technology, United States of America
Tel. number: +1 8562202360, e-mail: michael.radin@rit.edu

Slovak University of Technology in Bratislava
Faculty of Mechanical Engineering
16th Conference on Applied Mathematics **APLIMAT 2017**
Proceedings

DEA METHOD APPLICATION IN SELECTED HOTELS IN THE NITRA REGION

ŽITNÝ Rastislav(SK), GRELL Michal (SK), HYRÁNEK Eduard (SK), MIŠOTA Branislav (SK), JANÁKOVÁ Hana (SK)

Abstract. Watching the performance of tourism facilities there are especially number of beds of facility, number of visitors, number of overnight stays, the average number of nights spent and revenues from accommodation. The paper is focused on selected hotels in Nitra region and important indicators of performance options. The collected data for years 2013 and 2014 are compared with the overall figures for those years in the Slovak Republic and considering the performance of sector in the Nitra region and in the Slovak Republic. The aim of the paper is to evaluate the effectiveness, efficiency and productivity of selected hotel facilities in the Nitra region on the basis of financial and other indicators using data envelopment analysis (DEA) by the size of input's and output's indicators. These indicators can be several species.

Keywords: Data Envelopment Analysis, technical efficiency, indicators of input and output, Nitra region, indicators of tourism performance, input, output factors

Mathematics Subject Classification: Primary 60A05, 08A72; Secondary 28E10

1 Introduction

One of an important indicator of the performance options in addition to its availability are economic indicators such as sales, revenues, assets and liabilities. In terms of regionalization of Slovak republic is possible in the region of Nitra identified five touristic regions of different categories: Nitra has a regional significance (Nitra district, Zlaté Moravce, the greater part of the Levice district and the northern part of the district Nové Zámky), Podunajsko has international meaning (Komarno, southern part of district Nové Zámky), Ipeľ has in the regional category (the eastern district of Levice) Hornonitriansky regional significance (Topoľčany) and the last Dolnopovažie also supra-regional (district Šaľa). A performance of accommodation establishment shows that the Nitra region has a relatively low traffic comparing with other regions as well as the low average number of overnight stays. The paper investigated the proportion of performance in the accommodation services of selected accommodation facilities for 2014 in terms of economic performance and availability of accommodation facilities and their comparison for the whole Nitra governing region and the Slovak Republic in 2014.

2 Model of the the data envelopment analysis and software support

An analysis of the data envelopment analysis (only DEA) represents the tool for modeling of evaluation effectiveness of homogeneous production respectively decision making units (only DMU) [2], [3], whereas under DMU in the next text we can understand the turistical accommodation in Nitra region. This includes any evaluation of the technical efficiency based on the size of inputs and outputs which can be more species and thus models of DEA have the character of multicriterial decision making. Efficiency (only E_k) of individual units in the case of one input and one output is given by relationship of output/input. In case of multiple inputs and more outputs for production is used relative measure of efficiency that is weighted sum of the outputs/inputs (in the next text are referred to as vi and uj uniform weights of inputs and outputs for all the rated entity). However, each unit is oriented differently and therefore is possible to consider also individual weights of inputs and outputs for each unit (in the next text are referred to as vik and ujk individual weights of inputs and outputs for each unit). Weights are derived from the technology that is used and therefore we use the term relative technical efficiency. Hypothetical (virtual) unit is possiblke to define as the weighted average effective units (peer units). DEA models seek individual weight of units so as to maximize their effectiveness and assume that for this problem there is a set of acceptable options, consisting of all permissible combinations of inputs and outputs. The DEA models seek individual weight of units to maximize their effectiveness and assume that for a given problem is a set of acceptable options which consist of all permissible combinations of inputs and outputs. The set of permissible options determines the effective border on which are situated combinations of inputs and outputs of efficient DMU units because it is not supposed that real unit achieves the same or higher outputs with lower inputs [2]. The shape of the efficient border is influenced by the type of revenues of scale which are divided on constatnt (the change in all inputs lead to the proportional change in outputs) and variable (the change in the inputs lead to the less - decreasing returns to scale - or higher - increasing returns of scale - as a proportional change outputs). In the paper we assume that the property is effective when at maximum outputs for given inputs respectively it creates a certain level of output with minimum inputs. Therefore we apply the CCR and BCC models that are oriented on outputs with constatnt or variable returns showing the theoretical background patterns only to the extent that is necessary to solve linear programming applications using already presented software support. CCR model oriented on outputs (CCR-O) supposes that the input is constant and the goal is to maximize the output. It allows to establish the necessary increase outputs to unit U_H be effective. The rate of efficiency is define as the ratio of the weighted sum of inputs and outputs. The model also looking for weights of inputs and outputs to measure scale efficiency was ≥ 1. The unit U_H is effective if $e_H = 1$ and is ineffective if $e_H > 1$.
We formulate primary and dual model.

Primary model:

$$\min e_H = \sum_{i=1}^{m} v_{iH} x_{iH} \quad (1)$$

under conditions

$$\sum_{j=1}^{n} u_{jH} y_{jH} = 1 \quad (2)$$

$$\sum_{i=1}^{m} v_{iH} x_{ik} - \sum_{j=1}^{n} u_{jH} y_{jk} \geq 0, \; k = 1, 2, ..., p \quad (3)$$

$$u_{jH} \geq 0, \; v_{iH} \geq 0. \quad (4)$$

Dual model:
max z_H
under conditions

$$\sum_{k=1}^{p} \lambda_{kH}\, x_{ik} \leq x_{iH},\ i = 1, 2, ..., m \qquad (5)$$

$$z_H y_{jH} - \sum_{k=1}^{p} \lambda_{kH}\, y_{jk} \leq 0,\ j = 1, 2, ..., n \qquad (6)$$

$$\lambda_{kH} \geq 0,\ k = 1, 2, ..., p \qquad (7)$$

For each inefficient unit U_H model finds a virtual unit (as a combination of effective peer units) to which inputs will not exceed the inputs of inefficient operations and its outcomes will equal or exceed the outputs of inefficient units. For the case of CCR-O model is always necessary to calculate with increasing outputs. In the BCC model has an efficient border the convex shape so the virtual drive for inefficient unit U_H will be the convex combinations of sample (peer) units. This means that the previous CCR models would simply expand for condition of convexity $\sum^{p}_{k-1}\lambda_{kH} = 1$. This additional condition in the model is manifested by adding one variable that represents the amount of deviation from constant returns of scale. BCC model is able to oriented on inputs (BCC-I) as well as outputs (BCC-O) analogical with CCR model. In the same way is possible to obtain also target values of inputs and outputs for inefficient units and identify effective unit [2]. The calculations o models are carried out using the software product DEA – *Solver (Learning version 3.0)* and they allow calculations for 50 DMU. The results of calculation are stored in the selected spreadsheet worksheet Excel.

3 Application of DEA method, calculations analysis – Malmquistov index

Table 1 shows the implementation of CCR and BCC-O-O using the input and output indicators in selected 14 accommodation facilities and subsequent analysis of the calculation to the extent described which allows the software support for the years 2013 and 2014.

There is also presenting the measurement of efficiency changes over that period by Malmquistovho output-oriented index. Defining the inputs and outputs is one of the starting steps towards effectiveness analysis and the correlation analysis between values of various inputs and output and combinations there shows Tab. 2.

If is correlation between the input/output very high[1], then the variables can be considered as equivalent to another and can be omitted from the model. When the correlation is very low can be assumed such a variable that is not suitable for our analysis. Based on the principle and taking into account the rule of the sum of inputs and outputs in relation to the number of hotels we chosen these few indicators for analyse: The number of beds, Equity and Profit and Sales and Revenues, while The number of beds is considered the same for both reporting period. The constant returns of scale in CCR models assume that all DMU operates under optimal size. However, in the real economic conditions production units do not work like this and therefore to overcome this problem has been developed BCC model that lets you work with variable returns of scale. Obtained values of efficiency from BCC model are reflecting a pure technical efficiency because a part of this model eliminates inefficiencies caused by the disproportionate size of the

---

[1] It is true that correlations below 0.1 is trivial, from 0.1 to 0.3 is low, from 0.3 to 0.5 middle, over 0.5 is big, and 0.7 to 0.9 is really big and from 0.9 to 1 is almost perfect.

production unit. It follows that the BCC model degrades the efficiency calculated by CCR model on purely technical effectiveness and efficiency of scale. The level of efficiency of scale is possible to evaluate by dividing the value of the measured efficiency of model CCR and model BCC.

DMU	Hotel name	Class (*)	Evaluation (booking)	The type	Level of presentation on the web sites
1	Topky sro Počúvadlo	3	*	P	3 - standard
2	Kaštiel' Mojmírovce, s.r.o.	3	8,0	R	2 - weak
3	Ranč u Bobiho	3	8,6	V	2 - weak
4	Hotel Golden Eagle	4	8,6	M	3 - standard
5	Hotel Banderium	3	9,0	M	3 - standard
6	Hotel St. Florián Štúrovo	3	8,3	P	3 - standard
7	Hotel Thermal Štúrovo	3	8,4	P	3 - standard
8	Park Hotel Tartuf Beladice	3	7,8	R	4 - static, design
9	Hotel River Nitra	4	9,1	M	4 - static, design
10	Hotel Autojas	4	8,2	R	3 - standard
11	Hotel Nitra Novozámocká	3	8,9	V	2 - weak
12	Hotel Vion	3	7,7	V	3 - standard
13	TK Podhájska, Hotel Borinka	3	*	P	3 - standard
14	City Hotel Nitra	3	8,7	M	3 - standard

Explanatory notes: P – the waterfront, R – familiar, V – rural, M - urban

Source: Internet, booking.com, own processes

Tab. 1. The main characteristics of selected accommodation

	(I) Number of beds	(I)Equity	(I)Assets	(O)Profit	(O) Sales and earnings
(I)Number of beds	1	0,544020235	0,46146	0,1261	0,276798029
(I)Equity		1	0,84644	0,5894	0,594180465
(I)Assets			1	0,4218	0,892012554
(O)Profit				1	0,127539004
(O)Sales and earnings					1

Source: own processes

Tab. 2. The correlation between inputs and outputs

In the following tables from 3 to 6 are the results of calculations. In the Tab. 3 and 4 is possible to see that according the model CCR – O in 2013 gained three DMU units 14, 12 and 10 the high level of efficiency. According to model BCC – O in 2013 the high level of efficiency is for six DMU units 14, 13, 12, 10, 4 and 8.

The analysis of the optimal weights has showed that the average values of the optimal weight values of input/output has amounted less than 0.1 and therefore effectiveness was negatively affecting of inputs/outputs. In this point of view at least negatively affect the efficiency input - Number of beds and output - Earnings and Revenue. It is apparent from the projections of each DMU to the effective border of CCR-O model but also BCC-O model.

Model CCR-O		
Rank	DMU	Score
1	14	1
1	12	1
1	10	1
4	8	0,885575
5	13	0,43315
6	9	0,230115
7	4	0,188187
8	5	0,14276
9	7	0,140616
10	11	0,132245
11	6	0,124316
12	2	9,01E-02
13	3	8,33E-02
14	1	6,78E-02

Model BCC-O		
Rank	DMU	Score
1	14	1
1	13	1
1	12	1
1	10	1
1	4	1
1	8	1
7	5	0,999899
8	6	0,99985
9	7	0,448313
10	9	0,249238
11	11	0,139175
12	2	0,138103
13	3	0,11489
14	1	9,42E-02

Source: own processes

Tab. 3 and 4. The order of hotel facilities (DMU) by the score of efficiency (model CCR-O and model BCC-O) in the 2013

Model CCR-O		
Rank	DMU	Score
1	14	1
1	12	1
1	10	1
1	8	1
5	13	0,717615
6	6	0,328552
7	5	0,287349
8	9	0,245063
9	11	0,203997
10	7	0,175735
11	4	0,143827
12	2	0,124952
13	1	9,55E-02
14	3	7,70E-02

Model BCC-O		
Rank	DMU	Score
1	14	1
1	13	1
1	12	1
1	10	1
1	4	1
1	5	1
1	6	1
1	8	1
9	7	0,392166
10	9	0,264205
11	11	0,211591
12	2	0,146668
13	1	0,098409
14	3	7,76E-02

Source: own processes

Tab. 5 amd 6. The order of hotel facilities (DMU) by the score of efficiency (model CCR-O and model BCC-O) in the 2014

The same content calculations are implementing in 2014. Tab. 5 and 6 indicate that in 2014 the CCR-O model achieved a high degree of efficiency the same DMU units 14, 12, 10 and another DMU unit 8.

According to BCC-O model in 2014 reached a high level of efficiency DMU units 14, 13, 12, 10, 4, 8 and the DMU units 5 and 6 too. It is also possible to state of improvement the efficiency between 2013 and 2014 DMU units on one of the CCR-O model and two DMU units of BCC-O model. The optimal values of weights input/output in 2014 show the same trends as in the previous year except the output gain slightly reduce its negative impact on the efficiency of the analysed DMU.

The basic DEA models do not take into account changes in efficiency over the time therefore we have to express this fact used Malmquistov index. As well as the basic DEA models could be formulated in different variants (eg. constant/variable returns of scale, orientation to inputs/outputs, etc.). Total output index are presented in Tab. 7 based on sub-indices (change in relative technical efficiency change and technological advance). According to values of Malmquistovho index is possible to conclude that between the both models in monitored period the accommodation no. 13 (Thermal bath Podhájska, Hotel Borinka) has taken the most effective decisions with the rate of improvement 1.11 respectively 1, 08.

DMU	The outlet o Malmquistov index (constant vr 2013/ variable vr 2014)	Change in technical efficiency 2013/2014	Change in technical progress 2013/2014
1	0.89/0.88	1.41/1.04	0.63/0.84
2	0.95/0.97	1.39/1.06	0.68/0.91
3	0.56/0.55	0.92/0.68	0.61/0.81
4	0.48/0	0.76/1	0.62/0
5	1.01/0	2.01/1	0.5/0
6	1.41/0	2.64/1	0.53/0
7	0.9/0.85	1.25/0.87	0.72/0.98
8	0.58/0.52	1.13/1	0.52/0.52
9	1.07/1.08	1.06/1.08	1.01/1
10	1.06/1.06	1/1	1.06/1.06
11	0.61/0.59	1.54/1.52	0.39/0.39
12	0.9/0.87	1/1	0.9/0.87
13	1.11/1.08	1.66/1	0.67/1.08
14	0.42/0.39	1/1	0.42/0.39

Source: own processes

Tab. 7. Malmquistov index for constant and variable returns of scale for individual houses in the 2013 and 2014

The qualitative change in technological progress in 2014 we can see in options 10 and 13. The change of technical efficiency is noted in options 8 and 13. For example of option 14 can be illustrated a loss of efficiency due to a decline of technological progress despite a high rate of technical efficiency calculated in accordance with both models.

4 Conclusion

The analysis of the data package we found the effectiveness of selected accommodation 14, 13, 12, 10, 4, 5, 6 and 8. Region of Nitra is the site with regional importance the most appropriate way of increasing the performance of accommodation is to intensify the potential of individual regions, with the systematic approach to build a network of attractions and their interdependence and to achieve a transnational level. In this paper is evident that the accommodation efficiency can be achieved by changing the technical efficiency and also technological progress. A possible way seems to be intensifying the involvement of the operators of accommodation facilities in CRS / IRS, GDS and useful presentation on the Web. Tourism in the region of Nitra can also promote tourism website <www.regionnitra.sk>.

Acknowledgement

The paper was made under the grant project VEGA no. 1/0067/15 verification and implementation of modelling business performance in financial decision-making tools.

References

[1] HYRÁNEK, E., GRELL, M., NAGY, L.: *Nové trendy merania výkonnosti podniku pre potreby finančných rozhodnutí*. Bratislava: Vydavateľstvo EKONÓM, 2014. 158s. ISBN 978-80-225-3901-2.

[2] FIALA, P. A KOL.: *Operační výzkum – nové trendy*. Praha: Profesional Publishing, 2010. 239s. ISBN 978-80-7431-036-2.

[3] ŽIŽKA, M.: *Vícekriteriální přístupy k hodnocení výkonnosti podniků*. Vydala Univerzita Tomáše Bati ve Zlíne v edici Qualifying Lecture for Professorship, 2014. 54s. ISBN 978-80-7454-449-1.

[4] *Ubytovacia štatistika CR na Slovensku. Kapacity a výkony UZ CR na Slovensku podľa druhu 2010 – 2014*. Slovenská agentúra pre cestovný ruch. Bratislava, 2015. <http://www.sacr.sk/sacr/statistiky/?no_cache=1&cid=1163&did=9260&sechash=dc9c9e71> [online] (25/08/2015).

[5] MIŠOTA, B.: Charakteristiky open-source redakčných systémov z hľadiska ich využitia v malých a stredných podnikoch = Features an open-source content management systems for their use in small and medium enterprises. *In Journal of knowledge society (časopis znalostní společnosti)*. Roč. 2, č.1 (2014), s.88-100. ISSN 2336-2561.

[6] ORIEŠKA, J., HUDÁČKOVÁ, L. 2013. Svetová hospodárska kríza a slovenské hotelierstvo v rokoch 2007 až 2012. *In Folia Turistica 3*. Banská Bystrica :UMB Ekonomická fakulta, 2013, s. 160-169. ISBN 978-80-8141-040-6.

Current address

Žitný Rastislav, Ing. PhD.
Institute of the Education of Teachers, Faculty of Central European Studies,
Constantine the Philosopher University in Nitra
Dražovská 4, 949 01, Nitra, Slovak Republic
Tel. number: +421 376408 855, e-mail: rzitny@ukf.sk

Grell Michal, Ing. CSc.
Civil association Education–Science-Research
Andrusovova 5, 851 01 Bratislava, Slovak Republic
Tel. number: +421 948048203, e-mail: grell@r15.roburnet.sk

Hyránek Eduard, Doc. Ing. PhD.
Faculty of Business Management, University of Economics in Bratislava
Dolnozemská cesta 1/b, 852 35 Bratislava, Slovak Republic
Tel. number: +421 267295669, e-mail: eduard.hyranek@euba.sk

Janáková Hana, Ing. PhD.
Institute of Management, Slovak University of Technology in Bratislava
Vazovova 5, 812 43 Bratislava, Slovak Republic
Tel. number: +421 918669028, e-mail: hana.janakova@stuba.sk

Mišota Branislav, Ing. PhD.
Institute of Management, Slovak University of Technology in Bratislava
Vazovova 5, 812 43 Bratislava, Slovak Republic
Tel. number: +421 918669133, e-mail: branislav.misota@stuba.sk

EXPERIENCE WORKSHOP

THE EXPERIENCE-CENTERED
MATH/ART MOVEMENT
www.experienceworkshop.org

January 31 – February 2, 2017

Travelling Bridges: Mathematical Art Exhibit at APLIMAT 2017

In Memoriam Mauro Francaviglia

Slovak University of Technology in Bratislava
Faculty of Mechanical Engineering
Institute of Mathematics and Physics
Address: Nám. slobody 17, 812 31 Bratislava, SLOVAKIA

EXPERIENCE WORKSHOP

THE EXPERIENCE-CENTERED MATH/ART MOVEMENT

www.experienceworkshop.org

Travelling Bridges: Mathematical Art Exhibit at APLIMAT 2017 is a memorial event to remember the internationally renowned Italian mathematician Mauro Francaviglia (1953-2013), who has played a key-role in the organization of APLIMAT Conferences together with his wife, Marcella G. Lorenzi. *The exhibit contains artworks, which* **Professor Francaviglia** *has enjoyed as part of* **Bridges Organization's** ***Mathematical Art Exhibit.*** *The exhibition space will also serve as an oasis for intellectual recreation in the conference coffee breaks with several puzzles and mathematical art toys, representing* **Professor Francaviglia** *playful spirit and small bits of* **Experience Workshop**'*s vision of STEAM education.*

STEAM (Science, Technology, Engineering, Arts and Mathematics) integration in learning represents a dynamically developing, but largely unexplored field. We wish to open our growing collection of mathematically inspired artworks, educational tools and other resources for children, families, and teachers to provide a snapshot of the fantastic potentials in collaboration and merging different fields. The collection is based on Bridges Hungary 2010's art exhibition, where artists from Europe, the Middle East, Africa, East Asia, Australia, and North and South America were represented. Artists drew inspiration from the mathematics of fractals, polyhedra, non-Euclidean and four-dimensional geometry, tiling, knot theory, number theory, and more.

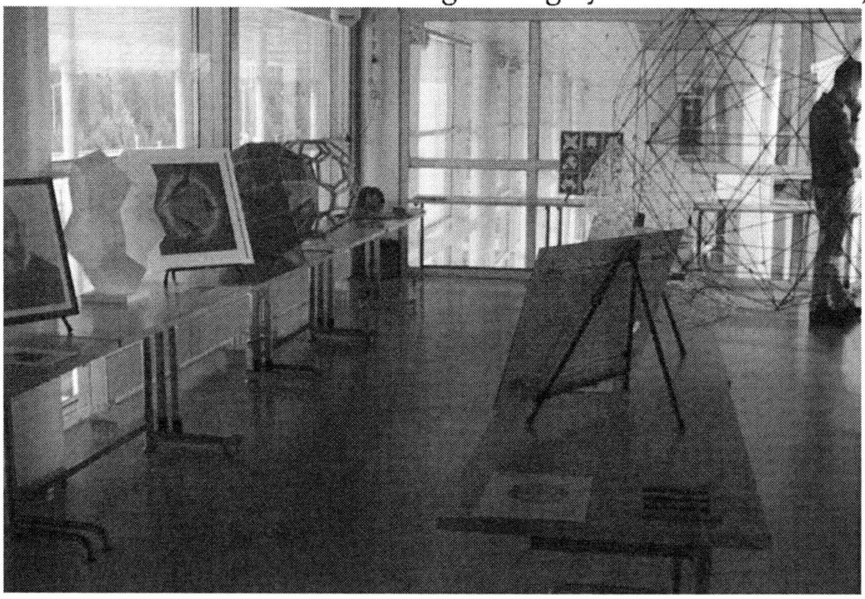

EXPERIENCE WORKSHOP

THE EXPERIENCE-CENTERED MATH/ART MOVEMENT

www.experienceworkshop.org

The best way to learn is to have fun while you learn. **Jomili Cubes** from **László Lukovics** (Slovakia) are great examples of this. There are only 8 different types of modules in the set, but these few simple cubes can be your tool to create a masterpiece. Jomili proves that creating something great requires logic and some patience to capture all that creativity.

We will have a lot of **ITSPHUN pieces** and invite everybody to put them together into something great! Big ideas born from small seeds of creativity. Build your own creations or make Platonic or Archimedean solids.

With **Lux Blox**, brainy builders are now free to make structures that curve, bend, and move. By using **Lux's** futuristic modules, you really can create everything from scientific models in geometry, chemistry, physics and biology to the complex structures of engineering, architecture, contemporary design or a funny robot based on your unlimited fantasy.

EXPERIENCE WORKSHOP

THE EXPERIENCE-CENTERED
MATH/ART MOVEMENT
www.experienceworkshop.org

4Dframe from Korea to build something amazingly complex...

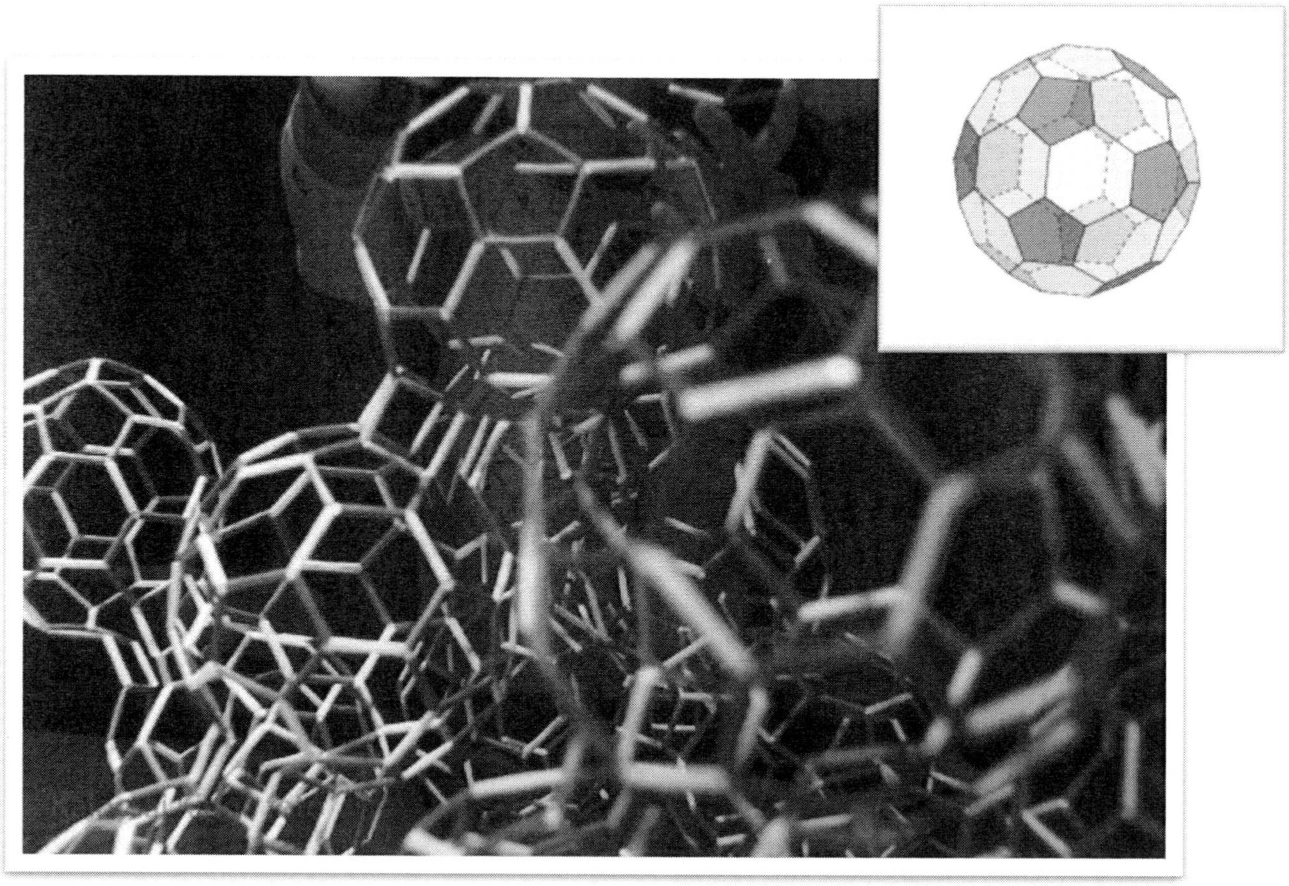

EXPERIENCE WORKSHOP

THE EXPERIENCE-CENTERED
MATH/ART MOVEMENT
www.experienceworkshop.org

>>> *Play, Interact, Cooperate, Discover, and Experience!* The **Experience Workshop – Movement for the Experience-Oriented Education of Mathematics and Arts** has started in 2008. We are most active in Finland and Hungary as a collaborative effort of mathematicians, artists, teachers, parents and children. We have organized various **math-art-education events and interactive exhibitions, workshops, seminars and trainings all over in Europe, Asia and America.** Over the years, more than 30,000 people participated in our events. **Experience Workshop's international network and community** has hundreds of active members like teachers of various subjects, artists, scholars, craftsmen and toymakers from Finland, Hungary and other countries all around the world.

>>> ***Experience Workshop experiments*** with various educational approaches to give opportunity to learn mathematics through the arts, and to make art with mathematics. Our aim is to involve the children, teachers, and families into a vibrant and creative dialogue between the mathematical and artistic way of looking at our world.

>>> ***Experience Workshop's*** main research interests involve, but not limited to STEM and STEAM education; inquiry-based, cooperative, playful and experience-oriented mathematics education; connecting problem-solving processes in science and art education; connecting hands-on activities and digital modeling in mathematics, science, art and design education; science&art connections in learning; phenomenon-based learning and co-teaching; inter-, cross- and transdisciplinary management and trans-curricular leadership in education.

>>> ***Experience Workshop Movement publishes various kind of printed and on-line resources for teachers, parents and students:*** books, apps, science and art albums, teacher resources and scientific articles. Most of our publications and resources are open-access. **Download one of our latest hand-book for teachers and GeoGebra app-collection from www.vismath.ektf.hu/exercisebook**

>>> ***Experience Workshop established International Traveling Exhibitions of Mathematical Art in Hungary and in Finland***, which can travel with **Experience Workshop's** events. The collection includes artworks, scientific modelling tools, math-art puzzles, and other spectacular objects, which have a key-role at **Experience Workshop's** events.

Contact: info@experienceworkshop.org / Website: www.experienceworkshop.org
Facebook: www.facebook.com/experienceworkshop.math.art

EXPERIENCE WORKSHOP

THE EXPERIENCE-CENTERED MATH/ART MOVEMENT

www.experienceworkshop.org

CONTACT

Experience Workshop's Director:
Dr. Kristóf Fenyvesi
(University of Jyväskylä – Bridges Organization – International Symmetry Association)

Phone: +358 40 805 33 24
E-mail: info@experienceworkshop.org; kristof.k.fenyvesi@jyu.fi

AUTHOR INDEX

AUTHOR INDEX

AUTHOR INDEX

Institute of Mathematics and Physics, Faculty of Mechanical Engineering
Slovak University of Technology in Bratislava
Namestie Slobody 17
812 31 Bratislava 1
Slovak Republic

ISBN 978-1-5108-3698-3

9 781510 836983

CURRAN ASSOCIATES INC.
proceedings.com